METHODS OF
TISSUE ENGINEERING

EDITORIAL BOARD

METHODS OF TISSUE ENGINEERING

Edited by

ANTHONY ATALA

Laboratory for Tissue Engineering and Cellular Therapeutics
Childrens's Hospital
Boston, Massachusetts

ROBERT P. LANZA

Advanced Cell Technology
Worcester, Massachusetts

ACADEMIC PRESS

A Harcourt Science and Technology Company

SAN DIEGO   SAN FRANCISCO   NEW YORK   BOSTON   LONDON   SYDNEY   TOKYO

Front Cover Image: Scanning electron micrograph of freeze-fractured poly(ortho ester) scaffold cross section. See figure 1B in Chapter 53.

This book is printed on acid-free paper.

Academic Press
An imprint of Elsevier Science
525 B Street, Suite 1900, San Diego, California 92101-4495, USA
http://www.academicpress.com

Academic Press
84 Theobalds Road, London WC1X 8RR, UK
http://www.academicpress.com

Library of Congress Catalog Card Number: 200188747

International Standard Book Number: 0-12-436636-8

PRINTED IN THE UNITED STATES OF AMERICA
02 03 04 05 06 07 MB 9 8 7 6 5 4 3 2

To Katherine and Christopher

CONTENTS

SECTION II: METHODS FOR CELL DELIVERY VEHICLES

CONTRIBUTORS

Susan Fugett Abu-Absi
Department of Chemical Engineering and Materials Science
University of Minnesota
Minneapolis, Minnesota 55455
Chapter 85

Robert E. Akins
Department of Medical Research
Tissue Engineering and Regenerative Medicine Laboratory
Alfred I. duPont Hospital for Children
Wilmington, Delaware 19899
Chapter 81

Harry R. Allcock
Department of Chemistry
Eberly College of Science
Pennsylvania State University
University Park, Pennsylvania 16802
Chapter 51

Teruo Amagasa
Kanagawa Dental College
Yokosuka, Kanagawa 238-8580
Japan
Chapter 107

Aurelia Amanda
Department of Chemical Engineering
Iowa State University
Ames, Iowa 50011
Chapter 74

Brian C. Anderson
Department of Chemical Engineering
Iowa State University
Ames, Iowa 50011
Chapter 74

Kirk P. Andriano
MacroMed University Research Park
Sandy, Utah 84070
Chapter 53

Dawn R. Applegate
Advanced Tissue Sciences
La Jolla, California 92037
Chapter 104

Takayuki Asahara
Tufts University School of Medicine
St. Elizabeth's Medical Center
Boston, Massachusetts 02135
Chapter 37

Anthony Atala
Laboratory for Tissue Engineering and Cellular Therapeutics
Children's Hospital
Harvard Medical School
Boston, Massachusetts 02115
Chapters 14, 16, 19, 21, 22, 31, 32, 39, 77, 89, 90, 91, 92, 97, 102, 106

Timothy Atalla
Department of Urology
Childrens Hospital
Harvard Medical School
Boston, Massachusetts 02115
Chapter 14

Griet G. Atkins
Poly-Med, Inc.
Anderson, South Carolina 29625
Chapter 49

Kah-Guan Au Eong
The Eye Institute
National Healthcare Group
Department of Ophthalmology
Tan Tock Seng Hospital
Singapore 308433
Chapter 100

François A. Auger
LOEX/Hopital du St. Sacrement
Centre Hospitalier Affilie Universitaire de Quebec
Quebec, Canada G1S 4L8
Chapter 28

Joyce Axelman
Department of Gynecology and Obstetrics
Johns Hopkins University School of Medicine
Baltimore, Maryland 21287
Chapter 33

Stephen F. Badylak
Biomedical Engineering
Purdue University
West Lafayette, Indiana 47907
Chapter 42

Carlos E. Baez
Department of Urology
Children's Hospital
Boston, Massachusetts 02115
Chapter 106

D. Baksh
Institute for Biomaterials and Biomedical Engineering
University of Toronto
Toronto, Ontario M5S 3E3
Canada
Chapter 26

Georg Bartsch
Department of Urology
Children's Hospital
Harvard Medical School
Boston, Massachusetts 02115
Chapter 39

Thierry Berney
Cell Transplant Center
Diabetes Research Institute
University of Miami School of Medicine
Miami, Florida 22136
Chapter 12

Sangeeta N. Bhatia
Bioengineering Department
University of California, San Diego
La Jolla, California 92093
Chapter 8

Lawrence J. Bonassar
Center for Tissue Engineering
University of Massachusetts Medical School
Worcester, Massachusetts 01655
Chapter 93

Kamal H. Bouhadir
Department of Chemical Engineering
University of Michigan
Ann Arbor, Michigan 48109
Chapter 56

Steven T. Boyce
Department of Surgery
University of Cincinnati
Cincinnati, Ohio 45267
Chapter 1

Chad D. Brown
Department of Bioengineering
University of Washington
Seattle, Washington 98195
Chapter 47

Frederick Cahn
Integra Life Sciences Corp.
Plainsboro, New Jersey 08536
Chapter 43

Kimberly A. Carpenter
Poly-Med, Inc.
Anderson, South Carolina 29625
Chapter 50

Kathleen S. Carswell
Reconstructive Technologies
Mountain View, California 94043
Chapter 41

Andrew J. Carter
Smith and Nephew Endoscopy Inc.
Mansfield, Massachusetts 02048
Chapter 95

Elliot L. Chaikof
Departments of Surgery and Engineering
Emory University
School of Chemical Engineering
Georgia Institute of Technology
Atlanta, Georgia 30322
Chapter 70

Thomas Ming Swi Chang
Artificial Cells and Organs Research Centre
Departments of Physiology, Medicine and
  Biomedical Engineering
McGill University
Montreal, Quebec J4S 1K9
Canada
Chapter 86

Christopher S. Chen
Department of Biomedical Engineering
Johns Hopkins School of Medicine
Baltimore, Maryland 21205
Chapter 7

Ho-Yun Chung
Department of Urology
Children's Hospital
Boston, Massachusetts 02115
Chapter 14

Nicolas Christoforou
Department of Gynecology and Obstetrics
Johns Hopkins University School of Medicine
Baltimore, Maryland 21287
Chapter 33

Gislin Dagnelie
Department of Ophthalmology
Johns Hopkins University School of Medicine
Baltimore, Maryland 21205
Chapter 100

Paul D. Dalton
Department of Chemical Engineering and
  Applied Chemistry
Institute of Biomaterials and Biomedical Engineering
University of Toronto
Toronto, Ontario M5S 3E5
Canada
Chapter 63

John E. Davies
Bone Interface Group
Institute for Biomaterials and Biomedical Engineering
Toronto, Ontario M5S 3E3
Canada
Chapter 26

Roger E. De Filippo
Children's Hospital
Harvard Medical School
Boston, Massachusetts 02115
Chapter 19

Robert J. Deans
Osiris Therapeutics, Inc.
Baltimore, Maryland 21231
Chapter 38

Paolo De Coppi
Department of Urology
Children's Hospital
Harvard Medical School
Boston, Massachusetts 02115
Chapter 77

Robert G. Dennis
Institute of Gerontology
Department of Biomedical Engineering
University of Michigan
Ann Arbor, Michigan 48109
Chapters 23, 24

Jonathan H. Dinsmore
Diacrin, Inc.
Charlestown, Massachusetts 02129
Chapter 101

Charles J. Doillon
Oncology and Molecular Endocrinology Research Center
Laval University Medical Center and Laval University
Quebec G1V 4G2
Canada
Chapter 46

DeZheng Dong
Laboratory for Tissue Engineering and Cellular Therapeutics
Children's Hospital
Harvard Medical School
Boston, Massachusetts 02115
Chapter 91

Gary C. du Moulin
Genzyme Biosurgery
Cambridge, Massachusetts 02139
Chapter 5

Elazer R. Edelman
Harvard-Massachusetts Institute of Technology Division of
   Health Sciences and Technology
Massachusetts Institute of Technology
Cambridge, Massachusetts 02139
Chapter 27

Minhua Feng
Department of Chemical Engineering and
   Applied Chemistry
Institute of Biomaterials and Biomedical Engineering
University of Toronto
Toronto, Ontario M5S 3E5
Canada
Chapter 73

Alan M. Flake
The Children's Hospital of Philadelphia
Philadelphia, Pennsylvania 19104
Chapter 38

Lisa Freed
Department of Chemical Engineering
Harvard-Massachusetts Institute of Technology Division of
   Health Sciences and Technology
Cambridge, Massachusetts 02139
Chapter 6

R. Ian Freshney
Department of Medical Oncology
University of Glasgow
Glasgow G61 2HA, Scotland
United Kingdom
Chapter 3

John D. Gearhart
Department of Gynecology and Obstetrics
Johns Hopkins School of Medicine
Baltimore, Maryland 21287
Chapter 33

Frank T. Gentile
Reprogenesis
Cambridge, Massachusetts 02139
Chapter 59

Lucy Germain
LOEX/Pavillon St. Sacrement
Universite Laval
Quebec G1S 4L8
Canada
Chapter 28

William V. Giannobile
Department of Periodontics
The University of Michigan
Ann Arbor, Michigan 48109
Chapter 108

Mattheus F. A. Goosen
Department of Bioresource and Agricultural Engineering
College of Agriculture
Sulton Quaboos University
Muscat
Sultanate of Oman
Chapter 76

David W. Grainger
Department of Chemistry
Colorado State University
Fort Collins, Colorado 80523
Chapter 66

Joel S. Greenberger
Department of Radiation Oncology
University of Pittsburgh School of Medicine
Pittsburgh, Pennsylvania 15213
Chapter 40

May Griffith
University of Ottawa Eye Institute
Ottawa Health Research Institute
Ottawa, Ontario K1H 8L6
Canada
Chapters 9, 83

Kristine E. Groehler
Department of Surgery
University of Minnesota
Minneapolis, Minnesota 55455
Chapter 85

Kristine Guleserian
Department of Surgery
Children's Hospital
Boston, Massachusetts 02115
Chapter 82

Robert Gurny
University of Geneva School of Pharmacy
CH1211 Geneva 4
Switzerland
Chapter 53

Hadar Haddad
Department of Chemical Engineering
Northwestern University
Evanston, Illinois 60208
Chapter 41

Geraldine Hamilton
Department of Drug Delivery and Disposition
UNC School of Pharmacy
Chapel Hill, North Carolina 27599
Chapter 11

Janet Hardin-Young
Organogenesis, Inc.
Canton, Massachusetts 02021
Chapter 105

Daniel A. Harrington
Northwestern University
Evanston, Illinois 60208
Chapter 65

Heather M. Hatch
Department of Pathology, Immunology, and Laboratory
   Medicine
College of Medicine
University of Florida
Gainesville, Florida 32610
Chapter 35

Robert J. Hay
Department of Cell Biology
American Type Culture Collection
Manassas, Virginia 20110
Chapter 4

Kevin E. Healy
Departments of Bioengineering and Materials Science
   and Engineering
University of California, Berkeley
Berkeley, California 94720
Chapters 57, 60

Jorge Heller
Advanced Polymer Systems
Redwood City, California 94063
Chapter 53

Donavon J. Hess
Department of Surgery
University of Minnesota
Minneapolis, Minnesota 55455
Chapter 85

Patrice Hildgen
Department of Pharmacy
Unversity of Montreal
Montreal H3C 3J7
Canada
Chapter 52

Alan S. Hoffman
Department of Bioengineering
University of Washington
Seattle, Washington 98195
Chapter 47

Wei-Shou Hu
Department of Chemical Engineering and Materials Science
University of Minnesota.
Minneapolis, Minnesota 55455
Chapter 85

Johnny Huard
Division of Orthopaedic Surgery
Children's Hospital of Pittsburgh
Pittsburgh, Pennsylvania 15213
Chapter 36

James W. Huckle
Smith and Nephew Group Research Centre
York Science Park
Heslington, York Y010 5DF
United Kingdom
Chapter 95

H. David Humes
Department of Internal Medicine
University of Michigan and VA Medical Center
Ann Arbor, Michigan 48105
Chapter 88

Julia J. Hwang
Northwestern University
Evanston, Illinois 60208
Chapter 65

Laeticia A. Ifeanyi
Department of Gynecology and Obstetrics
Johns Hopkins University School of Medicine
Baltimore, Maryland 21287
Chapter 33

Yoshito Ikada
Department of Medical Engineering
Suzuka University of Medical Science
Uji-shi, Kyoto 611-0011
Japan
Chapter 84

Jeffrey M. Isner
Department of Medicine and Pathology
Tufts University School of Medicine
Department of Vascular Medicine and
    Cardiovascular Research
St. Elizabeth's Medical Center
Boston, Massachusetts 02135
Chapter 37

Noritaka Isogai
Department of Plastic and Reconstructive Surgery
Kinki University Hospital
Osaka 589-8511, Japan
Department of Biochemistry and Molecular Pathology
Northeastern Ohio Universities College of Medicine
Rootstown, Ohio 44272
Chapter 94

Hiroo Iwata
Institute for Frontier Medical Sciences
Kyoto University
Sakyo-ku, Kyoto 606-8507
Japan
Chapter 68

Douglas Jacoby
Diacrin, Inc.
Charlestown, Massachusetts 02129
Chapter 101

Carol Johnston
Laboratory of Reparative Biology and Bioengineering
Department of Plastic Surgery
The University of Texas M. D. Anderson Cancer Center
Houston, Texas 77030
Chapter 10

Heidi Kahler
Department of Gynecology and Obstetrics
Johns Hopkins University School of Medicine
Baltimore, Maryland 21287
Chapter 33

Soverin Karmiol
Department of R&D
BioWhittaker, Inc.
Walkersville, Maryland 21793
Chapter 2

J. M. Karp
Institute for Biomaterials and Biomedical Engineering
University of Toronto
Toronto, Ontario M5S 3E3
Canada
Chapter 26

Adam J. Katz
Department of Plastic Surgery
University of Virginia
Charlottesville, Virginia 22908
Chapter 20

Yusuf Khan
Center for Advanced Biomaterials and Tissue Engineering
School of Biomedical Engineering
Drexel University
Philadelphia, Pennsylvania 19104
Chapter 61

Gilson Khang
Department of Polymer Science and Technology
Chonbuk National University
Chonju 561-756
Korea
Chapter 67

Byung-Soo Kim
Department of Urology
Children's Hospital
Boston, Massachusetts 02115
Chapters 21, 102

Bernd Kinner
Department of Orthopaedic Surgery
Brigham and Women's Hospital
Harvard Medical School
Boston, Massachusetts 02115
Chapters 25, 97

Yukihiko Kinoshita
Kanagawa Dental College
Yokosuka, Kanagawa 238-8580
Japan
Chapter 107

Kelly R. Kirker
Department of Bioengineering
The University of Utah
Salt Lake City, Utah 84112
Chapter 45

Harm-Anton Klok
Max Planck Institute for Polymer Research
D-55128 Mainz
Germany
Chapter 65

Andrea L. Koenig
Department of Chemistry
Colorado State University
Fort Collins, Colorado 80523
Chapter 66

Paul. E. Kosnik
Institute of Gerontology
University of Michigan
Ann Arbor, Michigan 48109
Chapters 23, 24

Hiroshi Kubota
Department of Cell and Molecular Physiology
UNC School of Medicine
Chapel Hill, North Carolina 27599
Chapter 11

Tae-Gyun Kwon
Children's Hospital
Harvard Medical School
Boston, Massachusetts 02115
Chapter 89

Jin-Yao Lai
Laboratory for Tissue Engineering and
    Cellular Therapeutics
Department of Urology
Children's Hospital
Harvard Medical School
Boston, Massachusetts 02115
Chapter 16

William J. Landis
Department of Biochemistry and Molecular Pathology
Northeastern Ohio Universities College of Medicine
Rootstown, Ohio 44272
Chapter 94

Robert Langer
Department of Chemical Engineering
Massachusetts Institute of Technology
Boston, Massachusetts
Chapter 34

Erin B. Larvik
Department of Materials Science and Engineering
Massachusetts Institute of Technology
Boston, Massachusetts
Chapter 34

Cato T. Laurencin
Center for Advanced Biomaterials and Tissue Engineeering
Department of Chemical Engineering
Drexel University
Philadelphia, Pennsylvania 19104
Chapter 61

Rebecca K. Lawson-Smith
Department of Chemical Engineering and
    Applied Chemistry
University of Toronto
Toronto, Ontario M5S 3E5
Canada
Chapter 54

Edward LeCluyse
Department of Drug Delivery and Disposition
UNC School of Pharmacy
Chapel Hill, North Carolina 27599
Chapter 11

Hai Bang Lee
Biomaterials Laboratory
Korea Research Institute of Chemical Technology
Taejon 305-606
Korea
Chapter 67

Il-Woo Lee
Department of Neurosurgery and Brain Center
Daejeon St. Mary's Hospital
Catholic University Medical College

Daejon 301-723
Korea
Chapter 30

Joon Yung Lee
Division of Orthopaedic Surgery
Children's Hospital of Pittsburgh
Pittsburgh, Pennsylvania 15213
Chapter 36

Jin Ho Lee
Department of Polymer Science and Engineering
Hannam University
Daedeog Ku, Taejon 306-791
Korea
Chapter 44

Peter L. Lelkes
School of Biomedical Engineering, Science and
    Health Systems
Drexel University
Philadelphia, Pennsylvania 19104
Chapter 29

Victor S. Lin
Department of Chemical Engineering and Materials Science
Wayne State University
Detroit, Michigan 48202
Chapter 72

Charles Lindberg
Diacrin, Inc.
Charlestown, Massachusetts 02129
Chapter 101

Chia-Yang Liu
Department of Ophthalmology
University of Cincinnati
Cincinnati, Ohio 45221
Chapters 9, 83

Huifei Liu
Department of Cell and Molecular Physiology
UNC School of Medicine
Chapel Hill, North Carolina 27599
Chapter 11

Cynthia Lodestro
Department of Cell and Molecular Physiology
UNC School of Medicine
Chapel Hill, North Carolina 27599
Chapter 11

Helen H. Lu
Center for Advanced Biomaterials and Tissue Engineering
Department of Chemical Engineering
Drexel University
Philadelphia, Pennsylvania 19104
Chapter 61

Tom Luntz
Department of Cell and Molecular Physiology
UNC School of Medicine
Chapel Hill, North Carolina 27599
Chapter 11

Yi Luo
Department of Medicinal Chemistry
The University of Utah
Salt Lake City, Utah 84112
Chapter 45

Peter X. Ma
Department of Biologic and Materials Science
Department of Biomedical Engineering
Macromolecular Science and Engineering Center
University of Michigan
Ann Arbor, Michigan 48109
Chapter 62

Jeff Macdonald
Department of Biomedical Engineering
Department of Cell and Molecular Physiology
UNC School of Medicine
Chapel Hill, North Carolina 27599
Chapter 11

Marcelle Machluf
Children's Hospital
Harvard Medical School
Boston, Massachusetts 02115
Chapter 92

Surya K. Mallapragada
Department of Chemical Engineering
Iowa State University
Ames, Iowa 50011
Chapters 58, 74

Jennifer J. Marler
Department of Surgery
Children's Hospital
Boston, Massachusetts 02115
Chapter 99

David P. Martin
Tepha, Inc.
Cambridge, Massachusetts 02142
Chapter 48

Jerri Martin
Diacrin, Inc.
Charlestown, Massachusetts 02129
Chapter 101

Shojiro Matsuda
Research and Development Department
GUNZE Limited
Ayabe-shi, Kyoto 623-8513
Japan
Chapter 84

Howard W. T. Matthew
Department of Chemical Engineering and Materials Science
Wayne State University
Detroit, Michigan 48202
Chapter 72

John E. Mayer Jr.
Department of Surgery
Children's Hospital
Boston, Massachusetts 02115
Chapter 82

Stephen J. Meraw
Department of Periodontics
The University of Michigan
Ann Arbor, Michigan 48109
Chapter 108

Marc D. Merten
Laboratoire de Pathologie Cellulaire et
    Moléculaire en Nutrition
Vandoeuvre-lès-Nancy 54505
France
Chapter 13

Craig Meyers
Department of Microbiology and Immunology
Pennsylvania State University College of Medicine
Hershey, Pennsylvania 17036
Chapter 18

Antonios G. Mikos
Bioengineering Department
Rice University
Houston, Texas 77251
Chapter 55

Cheryl A. Miller
Department of Chemical Engineering
University of California at Irvine
Irvine, California 92697
Chapter 58

Michael J. Miller
Department of Plastic Surgery
Laboratory of Reparative Biology and Bioengineering
University of Texas M.D. Anderson Cancer Center
Houston, Texas 77030
Chapter 78

R. Damaris Molano
Cell Transplant Center
Diabetes Research Institute
University of Miami School of Medicine
Miami, Florida 33136
Chapter 12

David J. Mooney
Departments of Biomedical Engineering, Chemical
    Engineering, and Biologic and Materials Sciences
University of Michigan
Ann Arbor, Michigan 48109
Chapters 56, 64

Jody Pope Morrison
Diacrin, Inc.
Charlestown, Massachusetts 02129
Chapter 101

Nicholas Moss
Department of Cell and Molecular Physiology
UNC School of Medicine
Chapel Hill, North Carolina 27599
Chapter 11

Gail K. Naughton
Advanced Tissue Sciences
La Jolla, California 92037
Chapter 104

Robert M. Nerem
Institute for Bioengineering and Bioscience
Georgia Institute of Technology
Atlanta, Georgia 30332
Chapter 79

Laura E. Niklason
Departments of Anesthesia and Biomedical Engineering
Duke University
Durham, North Carolina 27708
Chapter 80

Helen M. Nugent
Curis Incorporated
Cambridge, Massachusetts 02139
Chapter 27

Kohei Ogawa
The Laboratory of Tissue Engineering and
    Organ Fabrication
Massachusetts General Hospital
Boston, Massachusetts 02114
Department of Transplantation and Immunology
Kyoto University
Kyoto
Japan
Chapter 87

Gregory M. Organ
Oakland Kaiser Permanente
Oakland, California 94611
Chapter 15

Tadeusz Orlowski
Warsaw Medical School
Tranplantation Institute
Warsaw 00556
Poland
Chapter 71

Anna Orsola
Department of Urology
Children's Hospital
Harvard Medical School
Boston, Massachusetts 02115
Chapter 31

Martin Oudega
The Miami Project to Cure Paralysis
University of Miami School of Medicine
Miami, Florida 33101
Chapter 103

Eleftherios T. Papoutsakis
Department of Chemical Engineering
Northwestern University
Evanston, Illinois 60208
Chapter 41

Nancy L. Parenteau
Organogenesis, Inc.
Canton, Massachusetts 02021
Chapter 105

Sahil A. Parikh
Department of Medicine
Massachusetts General Hospital
Harvard Medical School
Boston, Massachusetts 02115
Chapter 27

Heung Jae Park
Department of Urology
SungKyun Kwan University
School of Medicine
Kangbuk Samsung Hospital
Seoul 110-102
Korea
Chapter 91

Kinam Park
Departments of Pharmaceutics and Biomedical Engineering
Purdue University
West Lafayette, Indiana 47907
Chapter 44

Kook I. Park
Departments of Pediatrics and Pharmacology
Yonsei University College of Medicine
Seoul
Korea
Chapter 34

Charles W. Patrick Jr.
Department of Plastic Surgery
Laboratory of Reparative Biology and Bioengineering
University of Texas M.D. Anderson Cancer Center
Houston, Texas 77030
Chapters 10, 78

Ethan S. Patterson
Department of Gynecology and Obstetrics
Johns Hopkins University School of Medicine
Baltimore, Maryland 21287
Chapter 33

Richard G. Payne
Institute of Biosciences and Bioengineering
Rice University
Houston, Texas 77251
Chapter 55

Donna M. Peehl
Department of Urology
Stanford University School of Medicine
Stanford, California 94305
Chapter 17

Tjorvi Perry
Department of Surgery
Children's Hospital
Boston, Massachusetts 02115
Chapter 82

Martin C. Peters
Department of Biomedical Engineering
University of Michigan
Ann Arbor, Michigan 48109
Chapter 64

Bryon E. Petersen
Department of Pathology
College of Medicine
University of Florida
Gainesville, Florida 32610
Chapter 35

Antonello Pileggi
Cell Transplant Center
Diabetes Research Institute
University of Miami School of Medicine
Miami, Florida 33136
Chapter 12

Mark F. Pittenger
Oriris Therapeutics, Inc.
Baltimore, Maryland 21231
Chapter 38

Hans G. Pohl
Children's Hospital
Harvard School of Medicine
Boston, Massachusetts 02115
Chapter 97

Glenn D. Prestwich
Department of Medicinal Chemistry
The University of Utah
Salt Lake City, Utah 84112
Chapter 45

Judson Ratliff
Diacrin, Inc.
Charlestown, Massachusetts 02129
Chapter 101

Atlantida M. Raya-Rivera
Laboratory for Tissue Engineering
Department of Urology
Children's Hospital
Harvard Medical School
Boston, Massachusetts 02115
Chapter 32

Greg P. Reece
Laboratory of Reparative Biology and Bioengineering
Department of Plastic Surgery
The University of Texas M. D. Anderson Cancer Center
Houston, Texas 77030
Chapter 10

Lola M. Reid
Department of Cell and Molecular Physiology
Program in Molecular Biology and Biotechnology
UNC School of Medicine
Chapel Hill, North Carolina 27599
Chapter 11

Rory P. Remmel
Department of Medicinal Chemistry
University of Minnesota
Minneapolis, Minnesota 55455
Chapter 85

Murielle Rémy-Zolghadri
Laboratoire d'Organogénèse Expérimentale
Hôpital du Saint-Sacrement
Qvebec G1S 4L8
Canada
Chapter 28

Thomas P. Richardson
Departments of Biomedical Engineering and Biologic and
 Materials Sciences
University of Michigan
Ann Arbor, Michigan 48109
Chapter 64

Camillo Ricordi
Cell Transplant Center
Diabetes Research Institute
University of Miami School of Medicine
Miami, Florida 33136
Chapter 12

Geoffrey L. Robb
Department of Plastic Surgery
The University of Texas M.D. Anderson Cancer Center
Houston, Texas 77030
Chapter 78

YiWei Rong
Advanced Cell Technologies and Tissue Engineering
Center for Gastrointestinal and Biliary Disease Biology
UNC School of Medicine
Chapel Hill, North Carolina 27599
Chapter 11

Gregory E. Rutkowski
Department of Chemical Engineering
University of Louisville
Louisville, Kentucky 40292
Chapter 58

Srinivas R. Sadda
Department of Ophthalmology
The Johns Hopkins University School of Medicine
Baltimore, Maryland 21205
Chapter 101

Susan Safley
Departments of Surgery and Engineering
Emory University
School of Chemical Engineering

Georgia Institute of Technology
Atlanta, Georgia 30322
Chapter 70

Jaqueline Sagen
The Miami Project to Cure Paralysis
University of Miami School of Medicine
Miami, Florida 33136
Chapter 103

Gunter Schuch
Children's Hospital
Harvard Medical School
Boston, Massachusetts 02115
Chapter 77

Michael V. Sefton
Department of Chemical Engineering and
    Applied Chemistry
Institute of Biomaterials and Biomedical Engineering
University of Toronto
Toronto, Ontario M5S 3E5
Canada
Chapter 73

Dror Seliktar
ETH Zürich
CH-8092 Zürich
Switzerland
Chapter 79

Robert G. Sellers
Department of Urology
Stanford University School of Medicine
Stanford, California 94305
Chapter 17

Mitchel Seruya
Columbia University College of Physicians and Surgeons
New York, New York 10032
Chapter 80

Shalaby W. Shalaby
Poly-Med, Inc.
Anderson, South Carolina 29625
Chapters 49, 50

Michael J. Shamblott
Department of Gynecology and Obstetrics
Johns Hopkins University School of Medicine
Baltimore, Maryland 21287
Chapter 33

Venkatram Prasad Shastri
Massachusetts Institute of Technology
Cambridge, Massachusetts 02139
Chapter 52

Heather Sheardown
Department of Chemical Engineering
McMaster University
Hamilton, Ontario L8S 4L8
Canada
Chapter 83

Molly S. Shoichet
Department of Chemical Engineering
University of Toronto
Toronto, Ontario M5S3E5
Canada
Chapter 63

Mahmud A. Siddiqi
Department of Gynecology and Obstetrics
Johns Hopkins University School of Medicine
Baltimore, Maryland 21287
Chapter 33

Julie Siegan
Diacrin, Inc.
Charlestown, Massachusetts 02129
Chapter 101

Timothy D. Sielaff
Department of Surgery
University of Minnesota
Minneaplis, Minnesota 55455
Chapter 85

Frank A. Skraly
Metabolix, Inc.
Cambridge, Massachusetts 02142
Chapter 48

Daniel Skuk
Human Genetics Unit
Centre Hospitalier de l'Universite Laval Research Center
Ste-Foy, Quebec G1V 4G2
Canada
Chapter 98

Evan Y. Snyder
Departments of Neurology, Pediatrics, and Neurosurgery
Children's Hospital
Brigham and Women's Hospital
Harvard Medical School
Boston, Massachusetts 02115
Chapter 34

Shay Soker
Laboratory for Cellular Therapeutics and Tissue
    Engineering
Department of Urology
Children's Hospital and Harvard Medical School
Boston, Massachusetts 02115
Chapters 22, 90

Myron Spector
Department of Orthopaedic Surgery
Brigham and Women's Hospital
Harvard Medical School
Boston, Massachusetts 02115
Chapters 25, 96

Jared Sterneckert
Department of Gynecology and Obstetrics
Johns Hopkins University School of Medicine
Baltimore, Maryland 21287
Chapter 33

Ranee A. Stile
Departments of Bioengineering and Materials Science and
    Engineering
Berkeley, California 94720
Chapter 57

Samuel I. Stupp
Northwestern University
Evanston, Illinois 60208
Chapter 65

Fraser Sutherland
Department of Surgery
Children's Hospital
Boston, Massachusetts 02115
Chapter 82

Yang D. Teng
Departments of Neurosurgery and Neurology
The Brigham and Women's Hospital
Children's Hospital
Harvard Medical School
Boston, Massachusetts 02115
Chapter 34

Joe Tien
Department of Biomedical Engineering
Johns Hopkins School of Medicine
Baltimore, Maryland 21205
Chapter 7

Jacques P. Tremblay
Department of Anatomy and Physiology
Laval University and Human Genetics Unit
Chule Research Center
Centre Hospitalier de l'Universite Laval Research Center
Ste. Foy, Quebec G1V 4G2
Canada
Chapter 98

Vickery Trinkaus-Randall
Department of Ophthalmology
Boston University Medical School
Boston, Massachusetts 02118
Chapters 9, 83

Evangelos Tziampazis
Department of Internal Medicine
University Michigan and VA Medical Center
Ann Arbor, Michigan 48105
Chapter 88

Kavid Udompanyanan
Laboratory for Cellular Therapeutics and
    Tissue Engineering
Department of Urology
Children's Hospital
Harvard Medical School
Boson, Massachusetts 02115
Chapter 22

Brian R. Unsworth
Department of Biology
Marquette University
Milwaukee, Wisconsin 53233
Chapter 29

Brian R. Unsworth
Department of Biology
Marquette University
Milwaukee, Wisconsin 43233
Chapter 29

Joseph P. Vacanti
Harvard Medical School
Massachusetts General Hospital
Boston, Massachusetts 02114
Chapter 87

Sarojini Vijayasekaran
Department of Biomaterials Research
Lions Eye Institute
Nedlands, Western Australia 6009
Australia
Chapter 63

Gordana Vunjak-Novakovic
Harvard-Massachusetts Institute of Technology Division of
    Health Sciences and Technology
Cambridge, Massachusetts 02139
Chapter 6

Taylor G. Wang
Department of Mechanical Engineering
School of Engineering
Vanderbilt University
Nashville, Tennessee 37232
Chapter 75

Mitchell A. Watsky
Department of Physiology
University of Tennessee Health Science Center
Memphis, Tennessee 38163
Chapters 9, 83

Colin J. Weber
Departments of Surgery and Engineering
Emory University
School of Chemical Engineering
Georgia Institute of Technology
Atlanta, Georgia 30322
Chapter 70

James D. Weiland
Department of Ophthalmology
Johns Hopkins University School of Medicine
Baltimore, Maryland 21205
Chapter 100

Kyumin Whang
Department of Restorative Dentistry
Division of Biomaterials
The University of Texas Health Science Center
    at San Antonio
San Antonio, Texas 78229
Chapter 60

Simon F. Williams
Tepha, Inc.
Cambridge, Massachusetts 02142
Chapter 48

Stephen Wolfe
Department of Biomedical Engineering
UNC School of Medicine
Chapel Hill, North Carolina 27599
Chapter 11

Kimberly A. Woodhouse
Department of Chemical Engineering and
    Applied Chemistry
University of Toronto
Toronto, Ontario M5S 3E5
Canada
Chapter 54

James R. Wright Jr.
IWK Health Centre Islet Transplantation Laboratory
Dalhousie University
Halifax, Nova Scotia B3H 1V7
Canada
Chapter 69

Xuemei Wu
Laboratory of Reparative Biology and Bioengineering
Department of Plastic Surgey
The University of Texas M. D. Anderson Cancer Center
Houston, Texas 77030
Chapter 10

Arron Xu
Department of Cell and Molecular Physiology
UNC School of Medicine
Chapel Hill, North Carolina 27599
Chapter 11

Hua Yang
IWK Health Centre Islet Transplantation Laboratory
Dalhousie University
Halifax, Nova Scotia B3H 1V7
Canada
Chapter 69

James J. Yoo
Department of Urology
Children's Hospital
Harvard Medical School
Boston, Massachusetts 02115
Chapters 89, 90

Alex Zelikin
Department of Chemical Engineering
Massachusetts Institute of Technology
Cambridge, Massachusetts 02139
Chapter 52

Ruiyun Zhang
Department of Biologic and Materials Science
University of Michigan
Ann Arbor, Michigan 48109
Chapter 62

FOREWORD

The human body is both robust and frail. In good health and in youth, maintenance and repair of organs appear to be effortless. Our bodies thrive on diverse diets and under extremes of climate. The young body remains healthy on either little or extreme exercise. Would that this happy circumstance prevailed. Unfortunately, it does not. Some people are born with defective or malfunctioning tissues or organs. Others sustain life-threatening damage to the body as a consequence of trauma or disease. As we age, we progressively lose the ability to meet physical challenges and to heal injury.

An age-old dream of mankind has been to replace damaged, injured, or worn parts of the body with new, fully functional tissues and organs. Until recently, this has been a fantasy, far from medical reality.

Now change is in the air. The belief is growing that medical science may soon be able to fulfill at least some of these desires. A new era of medicine is dawning, the era of regenerative medicine. Regenerative medicine derives its healing power from information inherent in our genes and cells. We are learning to use such information to rebuild, repair, and restore to normal function tissues that are damaged by disease, injured by trauma, or worn by time.

Methods of Tissue Engineering lays out a clear view of the state of one key element of regenerative medicine, tissue engineering. The volume is not only a description of what can be done today, but also a roadmap for the future. All who wish to embark on the journey into this future will do well to read the contents from cover to cover.

The first section of the book lays out in detail recipes for the care and feeding of a diverse set of human cells. Cells are the indispensable component of new organs and tissues. General directions as well as specific procedures are provided for the cultivation of epithelial, mesenchymal, neuroectodermal, and gonadal cells and the elusive stem cell.

The second section of the book provides a detailed description of methods for creating scaffolds upon which these cells may grow and flourish. To generate new organs or tissues, it is not enough to provide healthy cells. It is necessary to provide forms to help direct their growth. The past several years have seen an explosion in the materials science of tissue engineering. Several natural substances—collagen, alginate, hyaluronic acid, and fibrin—are proving valuable. The world of synthetic polymers is also opening up, including the use of polyanhydrides, polyphosphazines, and poly(amino acids). The new field of hydrogels also offers exciting possibilities. These new materials can be shaped and woven to create intricate structures that mimic natural forms. The ability to generate new materials and structures upon which cells can grow is limited only by the imagination.

The third section summarizes the remarkable progress made to date in creating functional organs. The list of engineered organs is lengthening and may one day be a veritable New Gray's Anatomy of the human body. There are chapters on breast reconstruction, the creation of large as well as small blood vessels, and the growth of cardiac muscle, valves, and arteries. Several chapters are devoted to creating cornea, sections of the gastrointestinal tract, liver, and kidney. Progress has been particularly remarkable for the urogenital tract. Structures have been created for repair of the urethra, penis, testes, and vagina. The work has progressed to rebuilding phalanges of the hand and other small joints. Various methods for regeneration of the nervous system are under investigation, including procedures to regenerate spinal cord, peripheral nerves, and parts of the brain.

Although these descriptions of how to rebuild our bodies from the inside constitute a lengthy tome, it is worth emphasizing that we are still at the beginning of this new and exciting field. The medical community owes Anthony Atala and Robert Lanza a debt a gratitude for assembling such a fine compendium of knowledge. It is on this basis that we will build a happier, healthier tomorrow.

William Haseltine
Human Genome Sciences
Rockville, Maryland

PREFACE

The best advances in medicine are usually those that can accomplish much using simple concepts and methods. The concept behind tissue engineering is attractive for this very reason. It is simple: if there is a tissue deficit, then this can be corrected by giving the patient normal functioning tissue.

The idea of tissue engineering is not quite as new as it seems. The Nobel Laureate Alexis Carrel performed seminal work in the early 1900s that paved the way for today's tissue engineers. Carrel even caught the imagination of the pilot Charles Lindbergh. After his historic first solo flight across the Atlantic Ocean, Lindbergh worked with Carrel at the Rockefeller Institute in New York, with the goal of maintaining viable tissues and organs *in vitro* for subsequent implantation *in vivo*.

Between Carrel's hopes for clinical application and their actualization in the early 1980s, with the use of engineered skin in patients, tissue engineering became a formal interdisciplinary field. Tissue engineering now holds promise for a wide range of congenital and acquired diseases.

Patients are mostly concerned about cures for their diseases, not with the methods behind their treatment modality. Yet, the methods used in the field of regenerative medicine, which involve creating adequately engineered tissues, are important for achieving therapeutic success.

Although there are several textbooks dedicated to various aspects of tissue engineering, no previous work has satisfactorily covered the methods involved in this interdisciplinary field of regenerative medicine. This textbook is a comprehensive reference that combines the tools, experimental protocols, detailed descriptions, and "know-how" for anyone involved in the field of regenerative medicine. It includes comprehensive protocols covering numerous topics, such as stem cells, cell culture, polymer synthesis, encapsulation, bioreactors, gene therapy, drug delivery, therapeutics, and the creation of tissues and organs. This book will serve as a starting point for students, a resource for anyone involved in the life sciences, and a guidebook for the next generation of scientists.

We express our gratitude and appreciation to our many associates and colleagues who as experts in their fields have contributed to this volume, either as members of the Editorial Board or as authors.

Anthony Atala
Robert P. Lanza

SECTION I
METHODS FOR CELL AND TISSUE CULTURE

CHAPTER 1

REGULATORY ISSUES AND STANDARDIZATION

Steven T. Boyce

INTRODUCTION

The emergence of tissue engineering as an academic discipline and global industry has opened unprecedented opportunities for the development of advanced therapies for treatment of congenital or acquired diseases [1,2]. Tissue engineering provides novel combinations of cells, acellular biomaterials, drugs, gene products, or genes that may be designed, specified, fabricated, and delivered either simultaneously or sequentially as therapeutic agents. Although acellular biomaterials or drugs have many predicates, virtually all tissue-engineered therapies have no predicate. Therefore, a general characteristic of tissue engineered therapies at the beginning of the twenty-first century is the absence of standards by which they may be assessed for safety and effectiveness. Despite the lack of standards for assessment, the responsibility of the U.S. Food and Drug Administration (FDA) to protect the public from health risks associated with investigational therapies remains unchanged. FDA's responsibility requires that new therapies approved for use in the United States be safe and effective. Safety criteria for medical devices require that probable benefits to health outweigh probable risks of the therapy, or of the untreated disease. Effectiveness criteria for devices should demonstrate that use of the device for its intended uses, and conditions for use, restore as much as possible of the physiological deficiency caused by the disease in a significant proportion of the target population [3]. These regulatory standards apply to tissue-engineered therapies just as for conventional medical devices, drugs, or biologics. However, the composite nature of most tissue-engineered therapies, together with a lack of predicates, complicates FDA's evaluation of their safety and effectiveness. Nonetheless, FDA has responded very proactively to the tremendous potential for public benefits from tissue engineering by establishing novel approaches to its basic mission of evaluating safety and effectiveness. New initiatives by FDA have pursued consensus development by participation of academics, industry, and government to establish guidance and standards for assessment of composition and performance of engineered tissues. Among the most important of these initiatives is the participation of FDA in the work of the American Society for Testing and Materials (ASTM) to establish standards for tissue-engineered medical products. Development of standards by ASTM promises to facilitate greatly the processes of assessment and introduction of engineered tissues to deliver the greatest variety of medical benefits in the shortest possible time, and at the lowest cost. Participation in the efforts of ASTM to develop standards is voluntary and open to the public. It provides the greatest possibility that comprehensive standards for tissue engineering will be established and will both harmonize with international standards and provide FDA with a broad base of knowledge in the basic and medical sciences. This chapter discusses several issues specific to the regulation of tissue-engineered therapies. However, the discussion does not represent any policy or position of the FDA, nor is it comprehensive in its scope.

SAFETY CONSIDERATIONS

Because the fundamental components of tissues are cells, the majority of tissue-engineered therapies contain cells. In addition, a basic advantage of tissue engineering is conservation or elimination of donor tissue from the recipient. Accomplishment of this advantage depends on proliferation of one or more cellular populations *in vitro*, and/or *in vivo*. Proliferation of cell populations by an engineering approach favors, but does not require, cell culture *in vitro*. Many systems for cell culture contain animal-derived compounds, and other biological or chemical contaminants. In addition, if allogeneic or xenogenic cells are included, transmission of adventitious agents is possible. Both the patient and the laboratory staff must be protected from possible pathogens. To provide adequate assurances of safety in tissue-engineered therapies, FDA has applied standards predominantly from the Center for Biological Evaluation and Research (CBER) [4]. These include, but may not be limited to, media components, tissue acquisition, implant handling and storage, and safety testing of the final product.

MEDIA COMPONENTS

Traditional cell culture media consist of a mixture of essential nutrients (salts, amino acids, vitamins, carbohydrate, fatty acids), buffers, and trace elements [5] that are supplemented with mitogenic factors, which may be animal-derived, synthetic hormones, or recombinant growth factors [6]. Cells of certain types also require coculture with "feeder" cells for propagation [7]. Traditional cell culture seeks to promote selective growth of target cell types with simultaneous suppression of other cell types from the tissue source. But because tissue-engineered therapies may contain multiple cell types [8], multiple media may be used to propagate selective cell cultures, or to incubate constructs that contain multiple cell types. Consequently, the preparation of tissue-engineered materials may involve contact of their components with several media formulations. Each medium has its unique formulation of essential nutrients and supplements that must be considered separately, and in combination with other formulations, to determine patient safety upon direct or indirect exposure. Essential nutrients may carry chemical contaminants not removed during their purification processes. Supplements or feeder cells may contain biological contaminants that infect, or inhibit the growth of, the cells for the final implant. Any contact with animal-derived or human-derived compounds may carry xenogeneic or allogeneic immunogens to the recipient [9]. In general, each component of each medium can be characterized by a Certificate of Analysis that describes its chemical composition and biological tests for pathogens that have been established by FDA [10]. Certificates of Analysis are available from the manufacturers of most commercial media.

Ideally, all media used for fabrication of tissue-engineered implants would have pharmaceutical levels of purity to define their compositions [11]. Unfortunately, this level of biochemical definition has not yet been reached for many cells that are grown *in vitro* for transplantation. Nonetheless, the importance of biochemical purity in cell culture media is recognized and has been studied extensively for more than two decades [6,12–15]. Consequently, there are many examples of biochemically defined media for selective and/or composite culture of human cells [16–20]. Principles established by these examples should be applied by investigators and developers of tissue-engineered implants to reduce or eliminate the use of animal-derived or human-derived compounds in cell culture. Pharmaceutical grade media for tissue-engineered medical products are an achievable goal, and the availability of such products is likely to increase safety to patients. However, media of these kinds may also add substantial cost to the fabrication of tissue-engineered implants.

TISSUE ACQUISITION

The versatility of tissue engineering provides for addition of autologous, allogeneic, or xenogeneic cells to fabricated implants. Safety issues to recipients of autologous cells are generally limited to acquisition of adventitious agents (microbial, viral, prion) in the laboratory. However, incorporation of allogeneic or xenogeneic cells opens the possibility of transmission of pathogens from nonautologous donors. For allogeneic tissues, safety standards have been established for conventional banking of human tissues or tissue extracts.

These standards are similar to blood banking and require the screening of donors for multiple pathogens, among which are human immunodeficiency virus (HIV) and hepatitis types B and C [4,21,22]. For manufacturing procedures of tissue-engineered devices, determination of donor suitability requires the incorporation of pathogen testing into procedures for good manufacturing practices (GMPs) [23]. Similarly to tissue banks, tissue engineers are required to record and track the of history of tissues from the donor through processing to storage and final disposition, whether the tissues are used for implantation or discarded. In the event that a donated tissue must be recalled for detection of a pathogen, it is the responsibility of the distributor of the tissue to notify the physician who transplanted the distributed tissues. To the extent that tissue-engineered implants contain allogeneic components, tissue engineers are tissue bankers and may have the same responsibilities. Requirements for testing also differ if tissues are acquired post mortem, or from living donors. In the latter case, tests for seroconversion are required several months after donation. Until final safety has been determined, tissues or cells are maintained in quarantine, often in a refrigerated or cryopreserved condition. Quarantine of postmortem tissues is also required by tissue banks before release for transplantation. Therefore, tissue engineers who use allogeneic cells may follow safety regulations for tissue banks as a general point of reference. Virtually all the pertinent FDA regulations for safe acquisition and processing of human tissues have been incorporated into the *Standards for Tissue Banking* of the American Association for Tissue Banks [24].

Acquisition of allogeneic tissues also requires proper representation of the intended uses. Both tissue engineering companies and tissue banks distribute tissues or tissue derivatives and may exist as not-for-profit or for-profit businesses. Although safety and effectiveness standards are regulated under federal jurisdiction, tissue donation is regulated under state jurisdiction. Because tissue and organ donation requires consent of the donor or the donor's family, the consent form should make appropriate disclosures of the intended use of the tissues. These disclosures should specify the use for which the acquisition is intended: direct transplantation for therapeutic use, investigative research, or use as a source of materials for fabrication of another product.

Similar standards for safe use of xenogeneic tissues have recently been proposed by FDA [25]. Because the use of xenogeneic tissue holds potentials of enormous supplies of normal tissues from many different species, or tissues from transgenic animals, the development of standards on this subject deserves considerable attention. In addition to safety issues regarding pathogen transmission, immunogenicity and stability of transgenic vectors should be assessed. Furthermore, if genetically modified cells are transplanted, the responses of the recipient to the altered phenotype of the cells require study. Development of a regulatory framework to consider the multitude of possibilities from transplantation of xenogeneic tissues or their derivatives remains a challenge for both FDA and tissue engineers.

IMPLANT FABRICATION AND STORAGE

After acquisition of tissues and their release from quarantine, the process of tissue engineering can begin. Although certain assurances can be provided that allogeneic or xenogeneic tissues are negative for pathogens, autologous tissues are usually not tested for blood-borne pathogens. Consequently, any untested human tissue must be considered to be contaminated with potentially lethal pathogens. Processing of tissues to establish cell cultures, or for direct storage, must provide safety to the laboratory staff. These safety standards for tissue engineering may be adapted from hospital practices for containment of blood-borne pathogens [26]. Generally, these practices provide for protection of staff by gloves, protective clothing, and containment of the tissue or cells. Typically, class II biological safety cabinets [27] are acceptable for containment of cell culture procedures. Additional consideration should be given to containment of nutrient media by use of culture vessels that are open (e.g., petri dishes) or sealed (e.g., roller bottles, filter-capped flasks) to minimize the risks from spills of media that have contacted untested human cells.

SAFETY TESTING OF THE FINAL PRODUCT FOR STERILITY,
ENDOTOXIN, AND MYCOPLASMA

Tissue-engineered implants must also assure safety to patients from biological contaminants acquired during processing. These assurances are provided by testing for sterility,
endotoxins, and mycoplasmas in the final product before release [28–31]. Sterility of an
implant that contains living cells may require sterility tests prior to release, and at the time
of transplantation. Sterility tests may be performed in thioglycollate media [30] or other
acceptable media for detection of microorganisms. Endotoxin determinations may follow
the limulus amoebocyte lysate (LAL) assay [5,28]. Mycoplasmas are a family of intracellular parasites, some of which may be pathogenic to humans. Detection of mycoplasma
may be performed by direct culture in sentry cell culture, by *in situ* hybridization, or by
polymerase chain reaction (PCR) assay [28,32]. Performance of sterility assays is relatively
simple and inexpensive. However, tests for endotoxins and mycoplasmas are more complex and expensive. Consequently, it may not be feasible economically to perform each test
on each preparation of tissue-engineered implant. Alternatively, development of a sampling
algorithm to monitor the preparation of multiple implants may be sufficient. The specific
design of the testing algorithm is often unique to the fabrication protocols for each type of
tissue-engineered implant.

EFFECTIVENESS CONSIDERATIONS

Determination of effectiveness of tissue-engineered devices must identify and measure
specific end points to demonstrate a beneficial change in a disease condition. However, not
all therapies that may be described as tissue engineering require effectiveness data before initiation of marketing. Distinctions between efficacy requirements of biologics and devices are
discussed later in connection with FDA's standards relating to risk and composition. For designs that require efficacy testing, traditional formats of assessment are appropriate. Preclinical testing includes *in vitro* characterization of composition (i.e., anatomy and physiology).
Animal studies determine host responses to implants, kinetic changes to the implant, and
restoration of structure and function of the damaged tissue. However, tissue-engineered devices may require assessments different from those for nonliving implants. Nonliving, nonadsorbable implants (e.g., metallic hip or knee joints, Teflon vascular grafts) are intended to
remain indefinitely and interface with host tissues. Efficacy of implants of these kinds is typically rapid, stable, and durable. By comparison, most tissue-engineered implants contain
an adsorbable component and a cellular component. Although well-characterized standards
exist for the breakdown of acellular degradable polymers [e.g., poly(L-actic acid)] [33], addition of cells to FDA-approved polymers or to polymers not approved by FDA may require
both preclinical and clinical assessments. Tissue-engineered implants that contain cells may
also stimulate synthetic or anabolic processes in transplanted cells or host tissues. Synthetic
processes in tissue-engineered implants may also require assessment before and/or after
transplantation to determine the rate and magnitude of therapeutic effects.

PRECLINICAL ASSESSMENT

During fabrication, and before transplantation, tissue-engineered devices require quality assurance (QA) testing to certify that design specifications for composition are met.
These characteristics may be compared to the anatomy and physiology of the target tissue.
Because tissue engineering is in relatively early stages as a discipline, the anatomy and physiology of the fabricated implant will be a subset of the uninjured tissue in most cases. If
living cells are included, then viability testing may be advisable as a predictive index of the
device's potency to restore anatomy and physiology after grafting. To the extent that devices
from different investigators may seek to repair the same tissue (e.g., bone), QA standards
for all the devices should use the comparative standard of uninjured bone. By this process,
one device may be measured to develop 30% of strength of bone *in vitro*, a second device to
develop 50%, a third device to develop 70%, and so on. Such measures are likely to predict
the magnitude of effectiveness at the time of transplantation, and the rate at which function
may be restored.

Predictions of efficacy usually require animal studies as part of the preclinical assessment. In these studies, end points for tissue-engineered devices can be adapted directly from conventional measures of clinical outcome. By adapting existing end points of outcome in existing animal models, tissue-engineered devices can be compared directly against transplantation of autologous or isogeneic tissues, and against nonliving, nonadsorbable implants. Results of these preclinical studies can generate data to justify initiation of clinical investigations. Without the performance of successful preclinical studies, a reduced probability of successful clinical investigations may be expected.

CLINICAL ASSESSMENT

Virtually all tissue-engineered devices require demonstration of clinical efficacy before marketing approval from FDA can be obtained. Although the investigational therapies with tissue engineering may be very diverse and complex, investigational designs for measurements of efficacy are not. Comparisons of an investigational device to the prevailing standard of treatment as a positive control, or to a placebo treatment as a negative control, are highly likely to generate a valid assessment. However, unlike drugs, placebo devices are more difficult to include in a study design. Therefore, comparison to the prevailing standard of care provides a sound basis for evaluation of tissue-engineered implants [34]. Depending on the intended use and medical indication for which the device is designed, end points for the determination of clinical efficacy can be obtained from conventional diagnosis, treatment, and prognosis of the indication. Also, depending on the disease being treated, the investigational therapy may be compared in a paired-site design within each patient, or randomized among patients in the study population. In general, efficacy can be measured by reduction in clinical symptoms or by improvements in physiologic function. If tissue engineers choose or design end points for clinical efficacy that are outside accepted standards for clinical care, the assessment process may become protracted by a requirement for validation of the novel end point.

In most examples, the process of clinical investigation for tissue-engineered devices is no different from that for nonliving, nonadsorbable devices. Studies are conducted under an Investigational Device Exemption (IDE) and follow the traditional phases (I, II, III) for data collection before submission of a Pre-Marketing Approval (PMA). However, marketing approval for a medical device may also be obtained through the Product Development Protocol (PDP) mechanism as described in the Medical Device Amendments of 1976 [35]. Tissue-engineered therapies that are considered to be biologics may require submission of a Biological License Application (BLA) to FDA, or registration of the manufacturing establishment [36], before marketing can begin.

REGULATORY ACTIVITIES OF FDA

Investigational therapies from tissue engineering have presented unique challenges to the organization of FDA into the Center for Devices and Radiologic Health (CDRH), the Center for Biologics Evaluation and Research (CBER), and the Center for Drug Evaluation and Research (CDER). Perhaps the majority of tissue-engineered implants contain cells that historically were most often reviewed by CBER, and purified polymers that commonly were reviewed by CDRH. Because tissue engineering may also include drug delivery, or drug synthesis by transgenic approaches, certain engineered tissues may also require review by CDER. These new challenges have been addressed very proactively by FDA through dialogue with experts and the public, and the development of intercenter mechanisms to facilitate the review of compound therapies [22,25,37]. These initiatives have resulted in the proposal and establishment of new regulations and policies to determine safety and effectiveness of compound therapies. During the 1990s, considerable efforts by FDA broadened the base of knowledge to which the agency refers and stimulated a consensus approach from which all participants benefit.

TRADITIONAL CLASSIFICATIONS AS DEVICES, BIOLOGICS, OR DRUGS

The Federal Food, Drug and Cosmetic Act of 1938 established authorities for regulation of food, drugs, cosmetics, devices, and animal drugs and required that devices be safe. The Medical Devices Amendments of 1976 and the Safe Medical Devices Act of 1990 define the term "device," provide for classification of all medical devices, establish authorities for advisory panels, prohibit misbranding and adulteration, and require adherence to good manufacturing practices (GMPs) [38]. Exercise of these expanded authorities by FDA sometimes resulted in IDE reviews that required years to complete and placed industrial applicants at competitive disadvantages. Congress responded to these difficulties with the FDA Modernization Act of 1997 [35], which remains current today. This legislation provides time limits for responses from FDA to investigational applications, and for responses by investigators to FDA's questions. Together with the commitment of additional resources from Congress, FDA has greatly decreased both the time required for its reviews of investigative therapies and the backlog of pending reviews.

CBER has also contributed greatly in the past decade to the development of progressive standards for tissue-engineered implants. CBER's initiatives are summarized under the Tissue Action Plan (http://www.fda.gov/cber/tissue/tissue.htm) which addresses regulation and policy development, guidance document development, inspections and compliance, and coordination of scientific and regulatory policies. Among the most far-reaching aspects of the plan is the development of "A Proposed Approach to the Regulation of Cellular and Tissue-Based Products" [21] (Tables 1.1 and 1.2). At present this regulatory framework has not been fully implemented, but it addresses such fundamental issues as transmission of communicable diseases, controls for processing, clinical safety and effectiveness, promotion and labeling, monitoring and education, stem cells, and, demineralized bone. More recently, the subject of xenotransplantation has been addressed by CBER. In December 1999, CBER issued a draft Guidance for Industry [25] that examines the risks of transmission of zoonoses (animal pathogens) by xenotransplantation. The document also identifies both direct risks to recipients of zoonoses and indirect risks of subsequent transmission to "close contacts" (e.g., spouse) of xenotransplant recipients. This draft guidance discusses whether xenotransplant recipients and their close contacts should be prohibited from blood and tissue donation to limit the possibility of disease transmission. These issues represent a very small fraction of the multitude of considerations that are currently under discussion at FDA. Together, CBER and CDRH have addressed the predominant majority of issues concerning tissue-engineered products.

Processes to control the composition of the final product are essential to the assurance of safety for tissue-engineered implants. For this assurance, FDA requires compliance with GMPs [38,39]. These requirements provide for the protection of the recipient and the laboratory staff as already described, and maintenance of comprehensive records of processing procedures. For devices, GMP manufacturing is required when marketing begins, but may be advisable at the phase III (multicenter) level of premarket investigation. Historically, GMP standards did not apply to banked tissues because processing of tissues was not considered to be a manufacturing process. However, good tissue practices (GTPs) are now being developed and will be implemented for acquisition and handling of tissues for transplantation [4,36]. GTPs may apply to handling of tissues from any biological source.

REGULATORY REQUIREMENTS DEPENDENT ON RISK AND COMPOSITION

As discussed earlier, tissue-engineered implants may be viewed primarily as devices or biologics. In most cases, autologous cells combined with polymers, or nonautologous cells with or without polymers, are considered class III (significant risk) devices that require demonstration of effectiveness. However, autologous cells only or an acellular tissue matrix may not require collection of effectiveness data. Living autologous cells that are "manipulated *ex vivo* and intended for structural repair" are not required currently to demonstrate effectiveness [40]. Cells of these kinds may include cultured autologous keratinocytes, chondrocytes, and processed bone marrow. Rather, these autologous cell populations are regulated more similarly to banked tissues, which do not require effectiveness studies because they are considered to be inherently efficacious. Similar low-risk classifications have been applied to acellular matrices derived from human tissue. Consequently, if no effectiveness data are collected, no claims of effectiveness can be made in labeling or advertisement. This

Table 1.1. Regulatory Framework for Cell- and Tissue-Related Products: Relationships among Product Concerns, Product Characteristics, and Regulatory Approaches

Product concern	Product characteristic (product factors)	Industry action required	Regulatory submission
A. Direct transmission of communicable disease (e.g., donor screening and testing)	1. SURGERY (Cells or tissue are removed from and transplanted back into the same person in a single surgical procedure)	1. None	1. None
	2a. AUTOLOGOUS banked/processed/shipped; REPRODUCTIVE from sexually intimate partner	2a. Screening, testing recommended; other GTPs would be required (e.g., record keeping, labeling, product tracking, recalls, notification of communicable disease transmission)	2a, 2b, 2c. No FDA submission. Requirements would be set new final rule for allogeneic tissue-related products under section 361 (finalization of the interim final rule), and in rule making under sections 361 and 351, which would add more products and more specific testing requirements
	2b. ALLOGENEIC, nonviable tissue	2b, 2c. GTPs would be required (e.g., screening, testing, record keeping, labeling, product tracking, recalls, notification of communicable disease transmission)	
	2c. ALLOGENEIC, viable tissue		
B. Control of processing Improper handling or inadequately controlled processing may result in product contamination and consequent communicable disease transmission; or in failure to preserve product integrity and function, and consequent enhanced susceptibility to communicable disease; or in failure to preserve product integrity and function with resulting unsafe or ineffective products.	1. SURGERY (Cells or tissue are removed from and transplanted back into the same person in a single surgical procedure)	1. None	1. None
	2. MINIMALLY MANIPULATED and homologous function and no nontissue components; and and structural, reproductive, or autologous/related-allogeneic metabolic components.	2. GTPs relating to contamination, integrity, and function.	2. No FDA submission regarding processing. Requirements would be set in rule making under section 361.
	3. MORE-THAN-MINIMALLY MANIPULATED or nonhomologous function or nontissue components or unrelated metabolic components.	3. Would have to follow GMPs and have stricter processing controls encompassing clinical safety and effectiveness concerns	3. A marketing application would ordinarily be required to contain a CMC section. If determinations are made that the safety and effectiveness of a product category can be assured by meeting product specifications and processing controls, then applicants would need only to submit a certification that they meet the product specifications and processing controls.
C. Clinical safety (not restricted to communicable disease risks); clinical effectiveness (including use-specific concerns)	1. Product is without any of factors a, b, c, or d.	1. None	1. No FDA submission

(continues)

Table 1.1. (continued)

Product concern	Product characteristic (product factors)	Industry action required	Regulatory submission
Attributes of importance are (a) more than minimal manipulation; (b) nonhomologous use;	2. Product is for local, structural reconstruction or repair and has factors a, b, or c.	2. Would have to gather clinical safety and effectiveness data.	2. Studies would have to be done under IND or IDE; marketing application would have to be submitted [BLA, 510(k) or PMA]; standard for determination of effectiveness would be consistent with that for devices. Standards for manipulated autologous structural cells would be as described in MAS cell policy guidance.
(c) combination with noncell/nontissue components; (d) metabolic use (other than reproductive) except when used autologously or in a close family member).	3. Product is for reproductive or metabolic use with factors a, b, c, or d	3. Would have to gather clinical safety and effectiveness data.	3. Studies would have to be done under IND; marketing application would have to be submitted (BLA); standard for determination of effectiveness would be consistent with that for biologics.
D. Promotion and labeling	All cellular and tissue-based products (excluding cells and tissues that are removed from and transplanted back into the same person in a single surgical procedure).	Clear, accurate, balanced, and nonmisleading labeling	No FDA submission concerning labeling for products regulated only under section 361 (provides claims are limited to those within homologous use). For products regulated under Section 351 or as a device, the usual rules would apply concerning labeling.
E. Baseline knowledge of industry	All cellular and tissue-based products (excluding cells and tissues that are removed from and transplanted back into the same person in a single surgical procedure).	Notification of FDA	Registration and listing under new regulation under 361 or under Section 510 of the FDC Act.

Source: Reprinted from U.S. Food and Drug Administration [21].

Table 1.2. Regulatory Framework for Cell- and Tissue-Related Products: Proposal for Specific Communicable Disease Controls[a]

	Testing[b]								Screening			
	HIV	HCV	HBV	HTLV	CMV	Treponema pallidum	Chlamydia trachomatis	Neisseria gonorrhea	High risk of HIV and hepatitis	CJD screen[d]	TB screen	Quarantine[c]
2a Autologous banked tissue												
Stem cells	R	R	R	R					R			
Other autologous tissue	R	R	R						R			
2b Allogeneic, nonviable tissue	X	X	X			X			X	X	X	R
2c Allogeneic, viable tissue												
Stem cells from family-related donors	X	X	X	X	X	X			X	X	X	
Reproductive tissue from sexually intimate partners	R	R	R	R	R	R	R		R	R	R	

(continues)

11

Table 1.2. *(continued)*

	Testing[b]									Screening		
	HIV	HCV	HBV	HTLV	CMV	*Treponema pallidum*	*Chlamydia trachomatis*	*Neisseria gonorrhea*	High risk of HIV and hepatitis	CID screen[d]	TB screen	Quarantine[c]
Other reproductive tissue (including directed donors)	X	X	X	X	X	X	X	X	X	X	X	X
Other allogeneic viable tissue[e]	X	X	X	X	X	X			X	X	X	R

X, required; R, recommended, for tests; labeling as: tested/negative, or not tested for biohazards will be required. Tissue unsuitable for transplantation may be used for nonclinical research purposes if labeled "Biohazard" or "Untested for Biohazards" and "For Research Use Only."

[a]Notes: Banked tissue for autologous use, from allogeneic family-related donors, from directed reproductive tissue donors, from sexually intimate partners, or in cases where there is a documented urgent medical need from a donor who has a positive risk factor and/or tested positive for an infectious disease agent, will not be required to be destroyed if: (a) the product is labeled "Biohazard" or "Untested for Biohazards," as applicable; (b) autologous tissue is labeled "for autologous use only"; (c) written advance informed consent of the recipient is documented; (d) there is a documented knowledge and authorization of the recipient's physician.

[b]For autologous or allogeneic cord blood donors, a mother's sample may be used for screening and testing.

[c]For allogeneic tissue that can be stored, quarantine for 6 months pending retesting of the donor will be required for all reproductive tissue, excluding sexually intimate partners. For other banked tissue and cells from living donors, quarantine for 6 months pending retest of the donor, or of the mother will be recommended, but where appropriate and feasible, not required.

[d]For dura mater donors, in addition to history for risk factors, a gross and histological examination of brain tissue will be required.

[e]Requirements for HTLV and CMV testing only apply to leukocyte-rich tissue (e.g., stem cells); they will not apply to cornea or skin donors.

Source: Reprinted from U.S. Food and Drug Administration [21].

distinction in regulatory requirements has allowed certain therapies to enter the market-place early, and without effectiveness studies. In selected examples, the clinical performance of these therapies has been very mixed or disappointing. Therefore, the clinician or patient must realize that the availability of a tissue-engineered therapy does not assure that it works. As with other consumer products, it is the responsibility of the end user to take a *caveat emptor* perspective before choosing or prescribing a tissue-engineered therapy.

INTER-CENTER REGULATION AND REVIEW OF TISSUE-ENGINEERED THERAPIES

Most often, tissue-engineered therapies combine cells with polymers, contact xeno-geneic cells, or tissue derivatives during fabrication, and add drugs either before or after transplantation. This level of complexity has stimulated FDA to enlarge and formal-ize its collaborations among CDRH, CBER, and CDER in its review of tissue-engineered therapies. These collaborations are exemplified by formation of the Wound Healing Clin-ical Focus Group [41] and the Guidance for Industry: Chronic Cutaneous Ulcer and Burn Wounds—Developing Products for Treatment [37]. The complexity of tissue-engineered im-plants very often also necessitates review of a therapy by multiple FDA centers. In cases of these types, the primary responsibility for review is held by one center, with consultation from the others as needed. Preliminary discussions with FDA before submission of an appli-cation to study an investigational therapy will help direct the application to the appropriate center.

Historically, the disparate and fragmented policies of FDA brought recognition to the importance of standardization. Beginning in 1997, FDA and the National Institute for Stan-dards and Technology (NIST) initiated a process for development of standards for tissue-engineered medical products (TEMPs). After early meetings, it was decided that the process for standards development should not be directed by the government (NIST and FDA) with comments from industry and academics; rather, it should be a public process with input from government. It was decided by consensus discussion that standards for TEMPs would be developed by the American Society for Testing and Materials (ASTM). This initiative has progressed to the establishment of a separate division of Committee F04 for Medical De-vices of the ASTM [42]. Since 1997, substantial progress has been made in the establishment of standards for TEMPs.

STANDARDIZATION THROUGH THE ASTM

The ASTM introduces itself as follows [43]:

> Established in 1898, ASTM has grown into one of the largest voluntary standards development systems in the world. ASTM is a not-for-profit organization which pro-vides a forum for producers, users, ultimate consumers, and those having a general in-terest (government and academia) to meet on common ground and write standards for materials, products, systems and services. From the work of the 130 standards-writing committees, ASTM publishes more than 10,700 standards each year. These standards and other related technical information are used throughout the world.

Historically, Committee F04 has undertaken responsibility for the development of stan-dards for medical devices. A few examples of the hundreds of subjects under F04 include, but are not limited to, the following: (1) specifications for metals used in surgical implants, (2) test methods for static bending of metallic bone plates, (3) specification of acrylic bone cement, (4) practices for the analysis of retrieved metallic orthopedic implants, (5) criteria for implantable thermoset epoxy plastics, (6) classification of silicone elastomers used in medical applications, and (7) terminology relating to polymeric biomaterials used in med-ical and surgical devices. Based on this broad experience with medical devices, and with the consideration that certain degradable polymers [e.g., poly-(L-lactic acid)] under F04 may be used in tissue-engineered implants, the process for development of standards for TEMPs was placed under the direction of the medical devices committee. Although the ASTM process is voluntary and does not bind either the FDA or any manufacturer to its standards, it is important to consider that FDA frequently refers to ASTM standards in its process of evaluation of investigational therapies. This reference by FDA results from a recognition that the development of standards by ASTM is a consensus process among academics, industry, and government. The importance to FDA of the development of consensus standards for TEMPs is represented by the direct participation of FDA in the ASTM process. It is expected that the investment of effort by all interested participants will result in the greatest number of new therapies becoming available in the shortest time, and at the lowest cost. Therefore, tissue engineers who plan to apply new technologies clinically will benefit from awareness of, and participation in, the ASTM effort for TEMPs.

ORGANIZATIONAL STRUCTURE OF COMMITTEE F04 DIVISION IV FOR TEMPS

Discussions during the early stages of F04 Division IV resulted in the identification of two major areas for standards development: composition and performance. It was also recognized that these standards must be compared to appropriate reference materials for validation. Ten subcommittees (Fig. 1.1) were established to address these complex issues. These subcommittees are organized into three groups: components design, assessment, and, common requirements for TEMPS. Under components design are the subcommittees for tissue-engineered biomaterials, cells, biomolecules, and delivery systems. Under assessment are subcommittees for assessment (i.e., preclinical) and clinical trials. Common requirements for all TEMPs include subcommittees for normal biologic function, tissue characterization, terminology, and microbial safety and adventitious agents. In addition, some 41 task groups under these subcommittees are charged with the development of actual standards in the respective subject areas. During the organizational process, it was observed that three subcommittees (Tissue-Engineered Biomaterials, Normal Biologic Function, and Assessment) were developing standards in task groups for the same tissue or organ system. With recognition that virtually any tissue or organ in the human body may be engineered, the tissue-based task groups were unified to include 10 subject areas: cardiovascular, digestive, endocrine, hematopoietic, integumentary, musculoskeletal, neurosensory, respiratory, reproductive, and urinary. This organizational structure of ASTM Committee F04 Division IV provides a comprehensive framework for development of standards for TEMPs.

PROCESS FOR STANDARDS DEVELOPMENT

Although tissue engineering is considered to be an emerging discipline, the process for the development of ASTM standards is very well established. Subcommittees are responsible for defining their scopes, including the development of standards within their subject area.

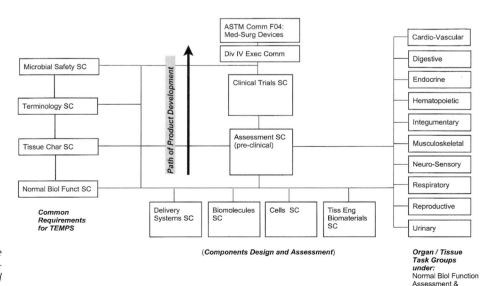

Fig. 1.1. Organizational diagram of the subcommittees (SC) for ASTM Committee F04 Division IV, Tissue-Engineered Medical Products (TEMPs).

Next, task groups are assigned to draft standards relating to specific topics within a subject area. Usually, these specific topics address terminology, materials, and methods for analysis of medical devices, or components of devices that are approved for marketing or under investigation. Drafting of standards for TEMPs may include a classification of risk levels of TEMPs according to an algorithm that assigns an alphanumeric designation to each TEMP based on product components, site of action, therapeutic target, primary modes of action to achieve the therapeutic effect, duration of therapy, and life of the TEMP. Draft standards are submitted from the task group to the subcommittee for review and are subsequently submitted to the Executive Committee of Division IV for balloting by the membership. During the balloting process, there is an opportunity for comments from the membership. All comments from ASTM members are considered, revisions are made by consensus of the membership, and the standard is placed on the ballot for adoption. After the standards developed by the specific task groups have been adopted, the task groups are dissolved until such time as revision of the standard may be required. The subcommittees continue to stand unless the division or committee is reorganized. The details of the ASTM process for standards development are available at the society's Internet website (http://www.astm.org).

RELATIONSHIP WITH FDA

As already noted, the importance of ASTM standards to FDA is emphasized by the federal agency's direct participation in the technical society's organization and process. Appreciation and inclusion by ASTM of standards acceptable to FDA for determination of safety and effectiveness increases the probability that FDA will refer to ASTM standards in evaluation of investigative therapies. However, in the absence of historical standards for tissue engineering, the importance of ASTM standards increases because FDA is not primarily a standards organization. Although FDA is not bound by ASTM standards, the development of these documents by a consensus process may relieve FDA of an independent initiative to develop consensus standards in the community. To the extent that ASTM addresses issues that are also important to FDA, the thoroughness of the ASTM process may satisfy FDA requirements. Participation by FDA in ASTM can help to guide the ASTM process for the greatest regulatory impact. Conversely, awareness by ASTM participants of the challenges that face FDA may influence the direction of standards development. By considering the mutual goals for tissue engineering, and collaborating to achieve those goals, ASTM and FDA are likely to foster a constructive relationship that will result in the delivery of the maximum number of advanced therapies from tissue engineering to provide the greatest medical benefits at the lowest risks to patients.

INTERNATIONAL HARMONIZATION

The present discussion has considered regulatory issues for tissue engineering in the United States. However, tissue engineering is a global discipline that must address the concerns of safety and effectiveness regardless of location. Most, if not all of the issues described thus far (donor suitability, transmission of disease, determination of risk) have also been addressed by the International Standards Organization (ISO) [44]. Therefore, to promote international availability of new therapies, the standards developed in the United States should be consistent with any prevailing ISO standards [29,34,45], and vice versa. However, it should be noted that ISO and U.S. standards may differ depending on continental factors, such as the appearance of bovine spongiform encephalopathy (BSE) in Europe but not in the United States. Nonetheless, awareness and harmonization of international standards will promote tissue engineering as a global discipline and industry.

FUTURE PROSPECTS, ETHICS, AND RESPONSIBILITY

Although some early entrants into therapeutic use of tissue engineering are favorably impressive, engineered tissues generally remain deficient anatomically and physiologically in comparison to their respective healthy tissues. Furthermore, the fabrication of fully functional organs remains an unfulfilled ambition. However, the foundations for true histogenesis and organogenesis have been established for future investigators. The tremendous power of tissue engineering also opens the possibility for deliberate modification of tissues to enhance biological function. Despite this inspirational potential of tissue engineering, enhancement of normal function also raises the chilling specter of human eugenics. Although issues like these are primarily ethical and moral, the U.S. Congress is reviewing legislation that will determine whether U.S. federal funding may support research using fetal tissues or embryonic stem cells for transplantation to humans. If legislation of tissue transplantation is enacted, enforcement of compliance with those laws may become the responsibility of FDA. For such issues, the role of FDA may extend beyond determination of safety and effectiveness. Consequently, it is the responsibility of the proponents of tissue engineering to gain awareness of the regulatory environment in which they work and to participate in the introduction of the benefits that tissue engineering promises. To realize the greatest number of medical benefits from tissue engineering, the discovery of knowledge to fabricate implants *in vitro* must be complemented by participation in the regulatory process, and in the establishment of standards to facilitate the assessment of composition and performance. Acceptance of these broad responsibilities by tissue engineers will greatly facilitate the improvement in quality of human life from tissue engineering.

REFERENCES

1. Morgan, J. R., and Yarmush, M. L. (1999). "Tissue Engineering Methods and Protocols." Humana Press, Totowa, NJ.
2. Skalak, R., and Fox, C. F. (1988). "Tissue Engineering." Alan R. Liss, New York.
3. U.S. Food and Drug Administration (2000). Medical device classification procedures: Determination of safety and effectiveness. 21 CFR 860.7.
4. U.S. Food and Drug Administration (1997). Human tissue intended for transplantation. 21 CFR 1270.
5. Freshney, R. I. (2000). "Culture of Animal Cells: A Manual of Basic Technique," 4th ed. Wiley-Liss, New York.
6. Mather, J. P., and Roberts, P. E. (1998). Serum-free culture. *In* "Introduction to Cell and Tissue Culture: Theory and Technique" (J. P. Mather, and P. E. Roberts, eds.), pp. 129–150. Plenum, New York.
7. Rheinwald, J. G., and Green, H. (1975). Formation of a keratinizing epithelium in culture by a cloned cell line derived from a teratoma. *Cell* (Cambridge, Mass.) **6**, 317–330.
8. Black, A. F., Berthod, F., L'Heureux, N., Germain, L., and Auger, F. A. (1998). In vitro reconstruction of a human capillary-like network in a tissue-engineered skin equivalent. *FASEB J.* **12**, 1331–1340.
9. Meyer, A. A., Manketlow, A., Johnson, M., DeSerres, S., Herzog, M., and Peterson, H. D. (1988). Antibody responses to xenogenic proteins in burned patients receiving cultured keratinocytes grafts. *J. Trauma* **28**, 1054–1059.
10. U.S. Food and Drug Administration (1997). Draft guidance for the submission of chemistry, manufacturing and controls information and establishment description for autologous somatic cell therapy products. *Fed. Regist.* **62**, 1460.
11. Ham, R. G., and McKeehan, W. L. (1979). Media and growth requirements. *Methods Enzymol.* **58**, 44–93.
12. Boyce, S. T., and Ham, R. G. (1983). Calcium-regulated differentiation of normal human epidermal keratinocytes in chemically defined clonal culture and serum-free serial culture. *J. Invest. Dermatol.* **81** (Suppl. 1), 33s–40s.
13. Shipley, G. D., and Pittelkow, M. R. (1988). Growth of normal human melanocytes in a defined medium. *Pigment Cell Res.* **1**, 27–31.
14. Bettger, W. J., and Ham, R. G. (1982). The nutrient requirements of cultured mammalian cells. *Adv. Nutr. Res.* **4**, 249–286.
15. Barnes, D., Sirbasku, D., and Sato, G. (1984). "Methods for the Preparation of Media, Supplements, and Substrata for Serum-free Animal Cultures." Alan R. Liss, New York.
16. Boyce, S. T., and Ham, R. G. (1985). Cultivation, frozen storage, and clonal growth of normal human epidermal keratinocytes in serum-free media. *J. Tissue Cult. Methods* **9**, 83–93.
17. Boyce, S. T., and Williams, M. L. (1993). Lipid supplemented medium induces lamellar bodies and precursors of barrier lipids in cultured analogues of human skin. *J. Invest. Dermatol.* **101**, 180–184.
18. Rosdy, M., and Clauss, L. C. (1990). Terminal epidermal differentiation of human keratinocytes grown in chemically defined medium on inert substrates at the air-liquid interface. *J. Invest. Dermatol.* **95**(4), 409–414.
19. Parenteau, N. L., Nolte, C. M., Bilbo, P., Rosenberg, M., Wilkins, L. M., Johnson, E. W., Watson, S., Mason, V. S., and Bell, E. (1991). Epidermis generated in vitro: Practical consideration and applications. *J. Cell Biochem.* **45**(3), 245–251.
20. Jones, G. E. (1996). "Human Cell Culture Protocols." Humana Press, Totowa, NJ.
21. U.S. Food and Drug Administration (1997). Proposed approach to regulation of cellular and tissue-based products: Availability and public meeting. *Fed. Regist.* **62**, 9721–9722.
22. U.S. Food and Drug Administration (1998). Guidance for industry: Guidance for human somatic cell therapy and gene therapy. *Fed. Regist.* **63**, 36413.
23. U.S. Food and Drug Administration (1999). Suitability determination for donors of human cellular and tissue-based products. *Fed. Regist.* **64**, 52696–52723.
24. AATB Standards Committee (1998). "Standards for Tissue Banking." American Association of Tissue Banks, McLean, VA.
25. U.S. Food and Drug Administration (1999). Guidance for industry: Precautionary measures to reduce the possible risk of transmission of zoonoses by blood and blood products from xenotransplantation product recipients and their contacts. *Fed. Regist.* **64**, 73562–73563.
26. Occupational Safety and Health Administration (1991). Occupational exposure to bloodborne pathogens. 29 CFR, 1910.1030.

27. Benbough, J., and Curran, B. A. (1998). Containment facilities: Design, construction and working practices. *In* "Safety in Cell and Tissue Culture" (G. Stacy, A. Doyle, and P. Hambleton, eds.), pp. 116–134. Kluwer Academic Publishers, Boston, MA.

28. du Moulin, G. C. (1992). Use of the Limulus Amoebocyte Lysate (LAL) assay in living cell therapies. *Biopharmacology* 5, 32–38.

29. U.S. Food and Drug Administration (1999). International conference on harmonisation; guidance on specifications: Test procedures and acceptance criteria for biotechnological/biological products. *Fed. Regist.* 64, 44928–44935.

30. U.S. Food and Drug Administration (1993). Sterility. 21 CFR 610.12.

31. U.S. Food and Drug Administration (1993). Test for mycoplasma. 21CFR 610.30.

32. Doyle, A., and Stacy, G. (1998). Quality control and validation. *In* "Safety in Cell and Tissue Culture" (G. Stacy, A. Doyle, and P. Hambleton, eds.), pp. 102–115. Kluwer Academic Publishers, Boston, MA.

33. American Society for Testing and Materials (1999). Standard specification for virgin poly(L-lactic acid) resin for surgical implants. *In* "1999 Annual Book of ASTM Standards," 13.01, pp. 1440–1442. ASTM, Conshocken, PA.

34. U.S. Food and Drug Administration (1999). International conference on harmonisation; choice of control group in clinical trials. *Fed. Regist.* 64, 51767–51780.

35. U.S. Congress (1997). Food and Drug Administration Modernization Act of 1997. Public Law 105-115.

36. U.S. Food and Drug Administration (1998). Establishment registration and listing for manufacturers of human cellular and tissue-based products. *Fed. Regist.* 63, 26744–26755.

37. U.S. Food and Drug Administration (2000). Guidance for industry: Chronic cutaneous ulcer and burn wounds—developing products for treatment. *Fed. Regist.* 65, 39912.

38. U.S. Food and Drug Administration (1992). Current good manufacturing practices regulations. 21 CFR, 210–211.

39. du Moulin, G. C., Pitkin, Z., Shen, Y. J., Conti, E., Stewart, J. K., Charles, C., and Hamilton, D. (1994). Overview of a quality assurance/quality control compliance program consistent with FDA regulations and policies for somatic cel and gene therapies: A four-year experience. *Cytotechnology* 15, 365–372.

40. U.S. Food and Drug Administration (1996). Guidance on applications for products comprised of living autologous cells manipulated ex vivo and intended for structural repair or reconstruction. *Fed. Regist.* 61, 26523–26524.

41. U.S. Food and Drug Administration (1994). Responses from the Wound Healing Clinical Focus Group at the FDA to the Government Relations Committee of the Wound Healing Society. *Scars and Stripes* 4, 5–12.

42. Picciolo, G. L., Hellman, K. B., and Johnson, P. C. (1998). Meeting report: Tissue engineered medical products standards; the time is ripe. *Tissue Eng.* 4, 5–7.

43. American Society for Testing and Materials (2000). Forward. *In* "1999 Annual Book of ASTM Standards," p. iii. ASTM, Conshohocken, PA.

44. International Standards Organization (2000). "Biological Evaluation of Medical Devices," Parts 1–17. ISO, Geneva, Switzerland.

45. U.S. Food and Drug Administration (1998). International conference on harmonisation; guidance on quality of biotechnological/biological products: Derivation and characterization of cell substrates used for production of biotechnological/biological products; availability. *Fed. Regist.* 63, 50244–50249.

CHAPTER 2

CELL ISOLATION AND SELECTION

Soverin Karmiol

INTRODUCTION

This chapter covers two topics: cell isolation and selection. The division is made for expediency, predominantly. For optimal results it is best for isolation and selection to occur concurrently. Except for the hematopoietic system and certain embryonic tissues, normal cells are intimately associated with extracellular matrices. The various techniques for disrupting the matrices can have an effect not only on the efficiency of digestion but also on the resulting cell population. Also, the cell culture system employed, the basal medium, and the supplements, can not only have an effect on the survival of the cells, by providing the appropriate nutrients and conditions, but also can have a significant selective potential, favoring one cell type over another.

There is a vast literature on the topics of this chapter. The isolation and purification of cells is a complex process. One of the goals of this chapter is to help shed some light on this complexity. For more detailed information on various aspects of cell culture please refer to the texts by Freshney [1], Studzinski [2], and Masters [3].

CELL ISOLATION

It is well established that the same cell type, such as fibroblasts, endothelial cells, or preadipocytes, demonstrates different characteristics in various anatomical sites. It is less well recognized that cells from the same anatomical site may also display different characteristics. The heterogeneity occurs with respect to a particular trait. The trait can be any functional parameter of the cell population, and the phenomenon is dependent on the notion that traits are not necessarily homogeneously distributed within that cell popuation. Fibroblasts demonstrate different phenotypic subpopulations within an anatomical site: see Fries *et al.* [4] and Lekic *et al.* [5] for reviews on murine, human lung, periodontal, and gingival fibroblast heterogeneity. Also, fibroblasts have been observed to possess different responses to the stimulation of collagen release according to whether it is stimulated by tumor or mast cells [6]. Further, separate phenotypes of microvascular endothelial cells from bovine corpus luteum demonstrate heterogeneity with respect to actin, cytokeratin, and vimentin expression and responsiveness to interferon gamma (IFN-γ) [7]. It is reasonable to consider that all cell populations may display heterogeneity with respect to some trait.

The traditional goal in establishing primary cell cultures is to provide sufficiently viable cells in quantities sufficient to ensure that the resulting distribution of the cells reflects the distribution found in the tissue of origin. The use of cell culture as a model of the *in vivo* condition, to a large extent, requires this stipulation. Several factors influence this outcome: anatomical site, age, pathology, normalcy, or degree of ischemia. The type of tissue disaggregation employed will interact with these factors. The diverse nature of tissue architecture requires different techniques for cell dispersal; for example, bone, brain, skeletal muscle, liver, or spleen have different cell–cell and cell–extracellular matrix associations. A consequence of this condition is that no one protocol will serve for all tissue types. However, there are a given number of considerations that encompass most of the issues arising from tissue disaggregation and the establishment of primary culture. The optimization of these considerations for the cell is a fruitful endeavor.

EXPLANTATION

Several strategies are available for isolating cells residing in solid tissues: explantation, enzymatic, mechanical, and chemical disaggregation, perfusion, or a combination of these. One of the earliest techniques in cell isolation and *in vitro* cell propagation was the explantation of tissue fragments. The technique is characterized by reducing the tissue to pieces so small that diffusion of nutrients and gases is not limiting. This is accomplished by trimming and cutting the desired tissue to the appropriate size. The tissue pieces are then introduced into a tissue culture vessel that has been coated with fetal bovine serum or other matrices. The composition of the medium and the nature of the surface coating can have a significant effect on the type of cell ultimately produced, as demonstrated in the isolation of cell types from the glomerulus [8]. The tissue pieces may also be placed under coverslips to support attachment; intimate contact with the tissue culture surface is critical for successful explant culture [9].

Generally, no enzymes are added, though mechanical means can be used to expose the tissue to the tissue culture medium. Primary culture takes longer to establish by the explant method than by the other methods, but the benefits of explantation arise from the potential sensitivity of the desired cell to the enzyme preparation or other manipulations. However, if some exposure to enzymes is permissible, then the tissue can be incubated with suboptimal concentration of enzyme for short periods at room temperature or for longer periods at 4°C. The aim here is not to disaggregate the tissue but to sufficiently compromise the extracellular matrix to allow more efficient migration of the cell of interest.

The choice of using explant primary culture is driven by the concern to maintain a functional membrane surface. Also, it is useful in those circumstances when the available tissue sample is small; enzymatic or mechanical isolation techniques may prove too harsh, and so many cells may be lost that further work becomes impossible. The influence of proteolytic enzymes on the cell surface components is known; for example, surface immoglobulin (Ig) is stripped from B lymphocytes rapidly by pronase and chymotrypsin, less rapidly by trypsin and papain, and not at all by collagenase [10,11]. Also, trypsin releases more radioactively labeled fatty acids from endothelial cell membrane phospholipids than collagenase [12].

Explant primary culture is dependent on the migratory ability of the desired cell type [13]. The disadvantages are due to the potential of a contaminating cell type growing out more rapidly than the cell of interest, and the lack of attachment of the tissue fragments. At the same time, if the cell of interest is the most motile, this technique can be selective.

The loss of phenotypic characteristics of cells due to enzyme digestion is a concern. If the cells are to be used immediately after isolation, an enzyme digest may be counter indicated. However, if the aim is to establish an *in vitro* cell culture population, the initial condition of the cell surface may be less critical. The membrane structures are characterized by undergoing turnover, and for a healthy population of cells it can be assumed that those surface structures will be repaired. However, this situation will be validated for each cell type.

MECHANICAL

Mechanical tissue disruption techniques include vortexing, shaking in a gyrotary shaker, trituration with pipettes of variously sized bore, forcing the tissue through stainless steel or nylon meshes, and teasing tissue apart with fine needles. These procedures generate cell suspensions sooner than enzymatic digestion, but there are limitations. These techniques are not universally effective with all tissues. They are distinguished by the amount of force applied to the tissue, and it is this aspect that can be detrimental. Cells are sensitive to shear, and significant damage may occur. MC3T3-E1 osteoblast-like cells from neonatal mouse calvaria responded to shear by a change in morphology and a gradual decrease in the resting intercellular calcium [14]; endothelial cells respond by growth inhibition due to laminar shear stress, with an increase in p53 expression [15]. However some soft tissues, such as spleen, thymus, lymph nodes, embryonic liver, soft tumors, and possibly brain, can be subjected to mechanical disaggregation without necessarily being accompanied by damage to these cells.

Mechanical techniques have been developed in response to the potential detrimental effects of enzymes during digestion. When appropriate isolation conditions can be established by mechanical means, this is preferable to enzymatic digestion, especially for cellular regenerative therapy; under these conditions the reduction of reagents such as the enzymes would make the sourcing, validation, and other regulatory issues unnecessary. However, mechanical techniques are more effective when coupled to enzyme digestion. The slicing of the tissue with scalpels to reduce the size of the tissue and thereby increase the surface area results in a more efficient isolation.

ENZYME DIGESTION

The type of enzyme employed has an effect on the digestion outcome in terms of efficiency, yield, viability, and toxicity. Isolation of cells from solid tumors using mechanical or enzymatic disaggregation results in different populations: in enzymatically treated tissue, aneuploid subpopulations are under represented in comparison to tissue obtained from mechanical disaggregation [16–18], suggesting that different populations of cells in the tissue demonstrate differing susceptibilities to enzyme preparations. Also, studies comparing rat hepatocyte isolation procedures demonstrated a prolonged, spectrally detectable content of cytochrome P450 [19] and maintenance of the mixed-function oxidase system [20] when EDTA was used instead of collagenase. There are exceptions to these observations: studies have demonstrated that tumor cells can be isolated more effectively by using enzyme digestion than by using mechanical methods [21,22] and that collagenase-isolated pig hepatocytes are more viable than EDTA-isolated hepatocytes [23]. One interpretation of these diverse results is based on the relative sensitivity of cells to enzyme and mechanical disaggregation. The influence of many variables that are difficult to control (e.g., tissue condition) contributes to these varied observations.

Adult rat hepatocytes isolated by trypsin perfusion and collagenase perfusion displayed different properties: collagenase-prepared hepatocytes had twice the attachment efficiency, longer maintenance in culture, higher albumin production, and increased tyrosine aminotransferase activity compared with hepatocytes isolated by trypsin [24]. In the isolation of porcine pancreatic islets by collagenase digestion, high islet yields correlated with low tryptic activities, and poor isolation outcomes were associated with high tryptic activities [25]; the inhibition of trypsin by Pefabloc resulted in reproducibly improved islet isolation [26]. An interesting aspect of cell isolations by exogenous enzyme digestion is the role played by the endogenous secretion of proteolytic enzymes due to the digestion process, and this secretion can be tissue specific [27]. Tissues from different donors respond variably to enzyme digestion [28,29]. Since the pancreas is the site of trypsin synthesis, care needs to be taken in the isolation of pancreatic islet cells to avoid possible overdigestion. Therefore, the presence and activity of endogenous proteolytic enzymes may constitute another aspect of tissue variability, aside from age and pathology, that can affect the outcome of primary culture.

It is recognized that trypsin can damage surface membranes, but it appears that trypsin internalization may also play a role. A study varying the temperature of trypsin during serial passage found that temperatures below 15°C improved viability and multiplication [30]; this is consistent with the influence of temperature on other membrane internalization phenomena. Using lower temperature during the digestion of tissue can be useful if viability is a problem.

Commercial collagenases from *Clostridium histolyticum* are often used in tissue digestion. Varying degress of impurity, efficiency, and toxicity [31,32] characterize the crude preparations. The crude preparations are undefined, containing various collagenases, other proteases, phospholipases, and bacterial by-products like endotoxin. Based on the lack of definition of these collagenase preparations, it is not surprising that there is significant variability between lots from different vendors and also from the same vendor. Attempts to purify crude collagenase preparations have been characterized by reduced tissue digestion efficiency, implying that the other proteases in the crude preparation participated in the digestion. Consequently, the addition of purified collagenase supplemented with other purified proteases can be an effective strategy; care is necessary however to confirm the lack of detrimental effects. The use of defined enzyme systems is becoming more important as the number of cell therapy products increases. Regulatory agencies will require the description of all the components in the reagents used in the preparation of a product; in fulfilling this goal, undefined components, such as the enzyme preparation, will pose a significant challenge.

TRAUMA

Excised tissues suffer from ischemia. Ischemia–reperfusion models are characterized by a burst of reactive oxygen species (ROS). However, damage can occur by the generation of superoxide [33] and lipid peroxidation [34] in the absence of reperfusion. In an apparent paradox, there is sufficient residual O_2 to support the production of superoxide during ischemia; this appears to be the case because the administration of superoxide dismutase (SOD) attenuates the ROS signal [33]. If SOD is added as an antioxidant, catalase needs to be added too, since the product of the dismutation of superoxide is hydrogen peroxide, itself capable of causing damage. Also, free radicals were demonstrated under hypoxic conditions in rat pulmonary artery smooth muscle cells [35].

The administration of antioxidants, such as, vitamin E [36], N-acetyl-L-cysteine or diphenylene iodonium have demonstrated a reduction in ROS levels and subsequent cell death [37]. Also, protease inhibitors such as α_2-macroglobulin and the antioxidant N-acetyl-L-cysteine increased the number of porcine primordial germ cells *in vitro* [38]. Prolonged exposure to ROS leads to either necrosis or apoptosis. Prolonged intense exposure can damage the apoptotic mechanism, leading more directly to necrosis [39]. Growth factors, such as stem cell factor and leukemia inhibitory factor, can inhibit apoptosis and enhance mouse embryo primordial germ cell survival [40].

When adherent cells are detached, they undergo a form of apoptosis termed anoikis [41]. ROS are also implicated as demonstrated by the influence of galectin-3 on apoptosis [42]. The addition of bongkrekic acid, an inhibitor of mitochondrial permeability transition, significantly reduced osteoclasts anoikis, as did the presence of z-VAD-FMK, a caspase inhibitor [43]. In a pancreatic islet cell isolation, preparations that were associated with extracellular matrix remained viable for longer periods in culture and performed better in an insulin secretion assay [44]. Also, fetal pancreatic cells were protected from death by attention to apoptotic and nutritional considerations [45], and N-acetyl-L-cysteine inhibited apoptosis in human germ cells *in vitro* [46]. Some cell types have been difficult to isolate. There are many factors that contribute to the success of a cell isolation and subsequent establishment of a cell culture system. Consideration of cell trauma can benefit the final outcome of a cell separation.

MEASUREMENT OF VIABILITY

The inherent variability of tissue responses to the various manipulations in cell culture requires a means of evaluating the viability of the cells resulting from these manipulations. In optimizing the methods for primary culture, a measure of viability is a convenient endpoint measurement. It is best to evaluate cell viability by methods that address the early onset of trauma. Generally, loss of cell membrane integrity comes late in the trauma process, and consequently there is a tendency to overestimate the viability of the cell [47,48]. For example, the onset and progression of apoptosis occurs without cell membrane disruption, though phospholipid rearrangement does occur; phosphatidylserine, normally found associated with the inner leaflet of the plasma membrane bilayer, is externalized to the outer leaflet during apoptosis [49]. The final disposition of the cell, which may take days to become evident, is to a large extent dependent on detrimental events that initiate apoptosis. On the other hand, damage can be so severe that membrane integrity can be compromised within minutes, and under these conditions assays that evaluate membrane integrity can be useful.

There are various viability assays available to measure membrane integrity. Trypan blue exclusions, the most commonly used method, is based on the property that viable cells exclude this dye and nonviable cells are permeable to it. The assay is implemented by incubating the cell suspension for a short time (<5 min) short periods are chosen because viable cells may suffer under these conditions and begin to take up the dye.

An example of direct measurement involves the uptake of a dye of which only viable cells are capable. Viable cells take up the nonfluorescent dye diacetyl fluorescein. Once inside the cell, the acetyl moieties are released, producing fluorescein, which is fluorescent and to which the cell membrane is impermeable [50,51]. Under the appropriate exitation and emission wavelengths, viable cells fluoresce green; this property can be measured either by visual counting with a fluorescence microscope or by cell sorting. Serum can hydrolyze diacetyl fluorescein, and long-term assays may not be advisable if serum is a component of the medium. Also, the fluorescence of fluorescein is pH dependent, requiring nonbicarbonate buffers. This will generally exclude cell culture media for the viability assay unless they are effectively buffered with nonbicarbonate buffers such HEPES or other organic buffers.

A further improvement in sensitivity is achieved by coupling propidium iodide to diacetyl fluorescein. Propidium iodide is impermeable to viable cells but permeable to nonviable cells. Under these conditions, viable cells fluoresce green and the nonviable cells fluoresce bright red. A comparison with the trypan blue method demonstrated that the diacetyl fluorescein–propidium iodide method is more consistent over prolonged periods of exposure to the dyes [52].

The energy status of the cell, with respect to the adenylate nucleotide concentrations, is an important indicator of viability [53,54]. This relationship is reasonable because the mitochondria are implicated in the onset of apoptosis [55,56]. One parameter of the energy status of a cell is the energy charge [57]; this index is determined by the following ratio: $[ATP] + 0.5[ADP]/[ATP] + [ADP] + [AMP]$. The value for normal cells is approximately 0.9, and a decrease in this value can indicate an early loss of viability. For example, cultured endothelial and P388D1 cells undergoing oxidant stress demonstrated an energy charge decrease from 0.95 to 0.66 in the first minute, while there was no change in trypan blue exclusion until 30 min after the initiation of the stress [58].

SELECTION

Selection is based on unique properties that distinguish one cell from another, such as density, size, unique markers, unique metabolic pathways, and nutritional requirements.

DENSITY AND SIZE

In a centrifugal field the sedimentation rate of particles is dependent on the force applied and the density and viscosity of the liquid. At a given force and viscosity, the sedimentation rate is proportional to the size of the particle and the difference in density between the particle and the density of the liquid. As a consequence, the sedimentation rate approaches zero as the difference between the density of the particle and the density of the liquid approaches zero. Also, the sedimentation rate increases as the centrifugal force increases and decreases as the viscosity of the liquid increases [59].

Because the density and viscosity of the liquid can have a profound effect on the sedimentation rate of particles, various gradient solutions or media have been developed to take advantage of these properties. Iodinated contrast media initially developed for clinical X-ray contrast studies have proven useful in cell isolations. There is a series of these contrast media based on whether the structure is ionic, nonionic, monomeric, or dimeric [60]; the media ranged from metrizoate to metrizamide [61] (the first nonionic version) to Nycodenz, a less toxic medium used in the isolation of neutrophils [62], platelets [63], pancreatic periacinar stellate cells [64], or Kupffer cells [65]. Ficoll, a synthetic polysucrose, has been used extensively in the isolation of a variety of cells, megakaryocytes [66], hepatocytes [67], pancreatic islets cells [68], and neutrophils and lymphoctyes [69]. A combination of sodium diatrizoate (an ionic contrast medium—Hypaque) with Ficoll or sodium metrizoate with Ficoll (Lymphoprep) has been used mainly in the separation of blood cells [70–73]. A relatively new separation medium, Percoll, consisting of a polydisperse silica colloid coated with poly(vinylpyrrolidone) (PVP), has found wide use [74]. A PVP coating renders silica relatively inert, with stability at physiological ionic strength and pH [75]. For good discussions of the of many cell types that have been separated by Percoll from blood and various other organs, see Refs. [75] and [76]. These separation media have found use as a first step in reducing the cellularity of cell mixtures so that other more specific separation techniques can then be employed [77–80].

These gradient media allow the formation of solutions dense enough to band cells according to their buoyant densities without subjecting the cells to high osmotic or ionic stress. The manipulation of the gradient media can separate cells based on density or size. For density-based separations, the gradient medium is made so that the densities encompass the full range of the densities of the cells present in the mixture. This is done to ensure that under the applied force of centrifugation, the cells will reach an equilibrium position, known as the isopycnic position, in which cell density becomes equal to that of the medium. Under this condition size does not play a significant role.

The density of a cell can be influenced by the osmolality of the solution. In a hypertonic medium the cell will shrink and the buoyant density will increase owing to the removal of water, while hypotonic media cause the cells to swell and reduce their buoyant density. The osmolality of the medium needs to be controlled, especially when isopycnic separation is employed. A characteristic of these media is the ability to achieve independently control of density and osmolality; this is one of the features distinguishing these media from solutions such as sucrose.

The separation of cells based on size or velocity requires that the gradient medium be less dense than the cells in the mixture to be separated. Under these gradient conditions, the cells will not experience an isopycnic position, and separation will be determined by size, since large cells move faster than smaller cells. Centrifugation is stopped before the cells reach the bottom of the tube. Isopycnic separations require greater centrifugal speeds than velocity separations. If the cell of interest is sensitive to these higher speeds, care should be taken in the use of isopycnic separations.

Variable densities of these media can be achieved by continuous or discontinuous means. Discontinuous gradients possess abrupt changes or interfaces in density at intervals along the length of the centrifugation tube. A single density change where the separation medium acts as a cushion can perform a cell separation, as in the case of mononuclear leukocytes when a Percoll gradient is used [81]. A series of abrupt density changes also can be used to separate dorsal and ventral dopaminergic neurons from embryonic rat [82] and Y-bearing from X-bearing sperm [83]. Discontinuous density gradients can provide artifacts because cells can aggregate at the interfaces, and there can be a lack of discrimination between cells of different densities depending on the density intervals in the discontinuous gradients [84,85]. Continuous gradients are formed either by gradient makers or in the case of Percoll by self-generating gradients. Percoll will begin to sediment at centrifugal forces greater than 10,000g [86], and since it is a polydisperse colloid, the component particles will sediment at different rates, creating a smooth gradient. Under these circumstances the cell suspension may be mixed with the Percoll and centrifuged at the high speeds necessary to form the gradient; the cells will then be separated based on the gradient formed. Two concerns are present here: the high centrifugation rates necessary to form the gradient may be detrimental to the cells, and it appears that Percoll may be taken up by lysosomes [87,88]. However, Percoll has been used to separate various lysosomal variants of different densities in rat hepatocytes, apparently without influencing the density of the lysosomes [89].

UNIQUE CELL MARKERS

Cells in tissues are heterogeneous with respect to their functions within the tissue. They not only demonstrate unique functions but can be distinguished by unique structures, especially by those that are present on the cell surface. The exceptional selectivity of monoclonal antibodies has spawned various techniques and instruments devoted to exploiting these differences.

Fluorescent-Activated Cell Sorting (FACS)

Fluorescent-activated cell sorting by flow cytometry is one of the most advanced instruments for taking advantage of monoclonal antibody technology. The operation of the instrument itself is based on the ability to generate a stream of single cells that can be selectively diverted by oppositely charged deflecting plates. Initially, the mixture of cells is incubated with the monoclonal antibodies coupled to a fluorescence tag, and cells with the appropriate epitope will bind the tagged antibody. In the instrument, the mixture is made into a stream of cells that is broken into droplets by the application of a vibration transducer. The cells cross a laser beam, and the emission is captured by a photomultiplier and subsequently processed. The cells are also charged during this process so that they can be appropriately deflected by the charged deflecting plates. Because the instrument is able to integrate all this information, a brief charge is applied to the deflecting plates at a set time after the cells of interest have crossed the laser beam, with the result that the tagged cells become deflected and the untagged cells are sent to waste [90]. Generally, separation on a FACS instrument is based on any cell properties that can be measured by light scattering or emission; depending on the complexity of the instrument, sorting can occur on more than one parameter, measuring cell shape [91], cell size, and a variety of traits that are amenable to fluorescence labeling [92].

The identification of cells from the hematopoietic system has utilized flow cytometry to remarkable advantage [93]. These cells are well characterized, and there are many monoclonal antibodies to support work with them. The discrimination with respect to isolation can be quite high, and as a consequence rare cells can be isolated by this method [94].

Immunomagnetic Bead Sorting and Magnetic Activated Cell Sorting (MACS)

Immunomagnetic bead sorting and magnetic activated cell sorting exploit monoclonal antibodies and other selective molecules with magnetism for cell separation. Generally, in this combined technique the cell mixture is incubated with a primary monoclonal antibody for a specific epitope. The magnetic beads are coated with various secondary antibodies, and the cells and beads are incubated together. The cells with the specific marker will be bound to a magnetic particle carrying the secondary antibody. The tube carrying the mixture is placed against a magnet and the beads carrying the cells will be drawn to the magnet. The cells unattached to a magnetic bead can be poured off; this process can be repeated several times to efficiently remove unattached cells. An example of this technique is demonstrated in a study purifying β cells from rat islets of Langerhans [95]. An example of a nonantibody selection utilizing magnetic beads is depicted in a study in which microvascular endothelial cells were purified from within the islets of Langerhans [96]. *Ulex europaeus* agglutinin-1 coated beads were used to bind the microvascular endothelial cells. This isolation demonstrated the unique properties of these endothelial cells, supporting the idea that the anatomical site is important in the phenotypic expression of similar cells. Selection can be positive or negative: positive in that the cell of interest is tagged and held within the magnetic field, or negative where the predominant cell contaminant is tagged and bound and the cell of interest is removed.

The advantage of negative selection is that the cells of interest are not complexed to magnetic beads; large beads may influence subsequent attachment on tissue culture surfaces. If positive selection is necessary, then the removal of the beads from the cells is desirable, and this has been achieved by the use of anti-Fab antiserum [97,98]. The beads may be removed by enzymatic digestion, but that may affect the cell. In another technique, called CELLection, short DNA lengths between the bead and the antibody are detached by DNase treatment [99]. This is the technique using Dynabeads from Dynal.

Another MACS bead design, by Miltenyi Biotec Ltd. can overcome these disadvantages [100], though Dynabeads may have some advantage with regard to the ability to achieve greater cell recovery [101]. The use of the 50-nm-diameter Miltenyi beads requires a separation column. Once on the column, the cells that have the specific marker, and consequently the magnetic bead, will remain in the column as long as the magnetic field is present. The column can be washed, and cells unattached to a magnetic bead will wash off. Once the column has been removed from the magnetic field, the cells attached to a bead can be washed off. A study has used both Dynabeads and a Miltenyi columns system, taking advantage of the relative magnetic fields associated with these respective particles. In this two-stage separation, two cell surface markers are used without removing the first set of magnetic beads. Initially, positive selection occurs with the Miltenyi system and is followed either by a positive or negative selection by mean of Dynabeads [102]: the magnetic strength from the Dynal separator is able to attract the larger Dynabeads but not the smaller Miltenyi beads.

FACS for cell sorting is an expensive instrument, and magnetic sorting has become the method of choice for cell isolation and harvest when monoclonal antibodies are used to separate cells. Both these techniques have been well adapted for work with the hematopoietic cell system rather than for cells from other tissues. The cells of the hematopoietic system are not associated with an extracellular matrix to any great extent and are well adapted to suspension. These two characteristics lend themselves to the conditions necessary to exploit the advantages of these techniques. Nevertheless, cells from solid tissue have been selected by these immunomethods: cultured myogenic cells [103], aortic endothelial cells [104], separation of mammary myoepithelial and luminal cells [105], separation of various sections of the proximal nephron [106], the enrichment of subpopulations of respiratory epithelial cells [107] and selection of kallikrein-containing cells from renal tubular cells [108]. The limitation for the nonhematopoietic cells is that they tend to be stickier and therefore may not flow well on columns; circuitry in FACS exclude clumped cells, thereby improving purity, though at the cost of yield.

Since expression of unique cell markers can be transitory with time in culture, it is important to apply these techniques early in the isolation procedure to increase the likelihood of the presence of the desired marker. Epithelial cells under standard cell culture conditions lose their polarity and unique differentiated functions; the hepatocyte is a notable example. Other cell types also respond to standard cell culture conditions in a similar manner. The need for the cells to remain in suspension for extended periods of time can activate anoikis and reduce cell viability and yield. Further, cells subjected to FACS and immunomagnetic cell separation demonstrate alterations in membrane integrity due to hydrodynamic forces or magnetic fields [109]. Further, beads can detach significant amounts of antigen from the surface of cells with antigen depletion but with the cells still remaining viable and functional [110,111]. Isolation of eosinophils by MACS has demonstrated a reduced response to the chemoattractants LTB4 and PAF, while the initial purification on Percoll did not affect the eosinophils in this manner [112].

One of the distinctions between FACS and MACS is that FACS can be quantitative. Since the degree of fluorescence can be a means of discrimination in FACS, it is possible to isolate cells with varying degrees of fluorescence intensity. In the case of MACS it is an all-or-none affair. However, a combination of MACS and density gradient centrifugation can provide some discrimination with respect to the degree of expression of surface antigen present. A cell with a greater amount of antigen will have more beads attached and thereby, owing to the increased density, those cells will sediment faster in a density gradient [113,114].

The procedures for fluorescence-activated cell and magnetic bead sorting are too useful not to be exploited and offer increased purification when combined [115]. Unmodified protocols taken from the hematopoietic literature may not be fully appropriate for all cells from solid tissues. It may be necessary to develop methods of overcoming the foregoing limitations, but this effort may prove quite useful and in some cases necessary. Stem or progenitor cells have been found in almost all organs of the adult organism. The isolation of these cells will require sensitive isolation techniques. These cells tend to be poorly represented, and the macroisolation techniques that characterize traditional cell culture isolation may not prove sufficient; the existing protocols for hematopoietic cells can be the basis for further development. With consideration of the circumstances mentioned earlier these protocols may be successfully extended to almost any cell sufficiently characterized by unique markers.

Panning

Panning is another immunotechnique, but lectins can be used also to bind certain cell types such as junctional gingival epithelial cells [116] or type II alveolar epithelial cells [117]. Cell culture dishes are conjugated with the appropriate antibody for a specific cell marker. The cell mixture is added to the antibody-coated dishes. The cells displaying the unique marker will attach to the dish. Cells without the marker can be easily removed. This technique is dependent on the tendency of the surface to which the antibodies are attached to have little affinity for the unwanted cells; otherwise, discrimination between the cells would be reduced. Panning can be used in a positive or negative mode: positively to select for the desired cells, and negatively to select and remove the unwanted cells. Enzymatic or mechanical means can be chosen to remove the selected cells. The technique has been used to isolate human trophoblast cells from a mixture of trypsinized placental villi [118], primordial germ cells [119], and fibroblasts from a keratinocyte preparation [120], to effect positive selection of Langerhans cells and keratinocytes [121] and lymphocytes [122].

METABOLIC AND NUTRITIONAL CHARACTERISTICS

The varied metabolic and nutritional requirements of different cells can be exploited for the purposes of selection during the establishment of primary cultures. The conditions are important in that the response of the cells to these manipulations may require defined conditions. For example, selective toxicity may not be apparent in the presence of serum due to the potential detoxification by serum. Also, since serum contains a diverse set of nutrients, the exploitation of different nutrient requirements is undermined in the presence of serum.

METABOLIC

Isozymes are distributed nonhomogeneously among different cell types; nonspecific esterases are a good example of this phenomenon. The differential expression of the esterase isozymes in endothelial cells and pericytes is a basis for discrimination between these two cell types. A concentration of 50 mM L-leucine methyl ester is toxic to endothelial cells but not to pericytes, resulting in the depletion of endothelial cells [123]. Similarly, L-leucyl-L-leucine methyl ester preferentially kills cytotoxic lymphocytes in comparison to helper T cells and B cells. The lysosomal thiol protease dipeptidyl peptidase I converts the, L-leucyl-L-leucine methyl ester to membranolytic products, and the enzyme is present in far higher concentrations in cytotoxic lymphocytes than the other cells, which consequently are depleted [124]. Also, a reduction in granulocytes and monocytes from apheresis-collected cells was achieved by preincubation with L-phenylalanine methyl ester hydrochloride. The CD34+ cells were then purified by means of magnetic bead isolation. The initial purging of the granulocytes and monocytes increased the efficiency of the isolation [125].

Fibroblast contamination is a common problem in tissue culture. One attempt at controlling fibroblast growth has been the exploitation in these cells of the lack of activity of L-amino oxidase, the enzyme that converts D-valine to L-valine. The cells of interest need to be able to adequately execute this conversion. In a medium where D-valine is substituted for L-valine, selectivity has been observed [126–128]. However, there are conflicting reports. Fibroblasts cultured from glioblastoma multiforme and oligodendroglioma were grown in media in which D-valine was substituted for L-valine and compared against a standard medium containing L-valine [129]. All three cell types were inhibited in the D-valine, demonstrating significant morphological changes. Also, though fibroblasts may be inhibited by D-valine, they are not necessarily killed and may grow back once the cells have been replated in medium containing L-valine [130]. Last, there is a report that D-valine does not inhibit human fibroblast growth *in vitro* [131]. Before an attempt is made to inhibit fibroblast overgrowth by D-valine media, however, several issues must be considered. First, the D-valine preparation needs to be low in L-valine contamination. Low amounts of L-valine may not support proliferation but may prolong survival. Also, the optimal concentration of D-valine may not be the original L-valine concentration. D-valine may not be utilized as efficiently as the L-isomer [132]. The relative transport efficiencies of D-valine and the L-isomer may not be the same, and once within the cell, the D-valine needs to be converted to the L-form. The time that is needed for these processes might increase the requirement for the D-form. Ideally, if possible it is best to make a DL-valine-free medium and titrate the appropriate concentration of D-valine required.

The hepatocyte contains an efficient urea cycle. In the absence of L-arginine, any of the urea cycle intermediates can regenerate L-arginine. Traditionally L-ornithine is used to replace L-arginine. The other potential cell contaminants in hepatocyte isolation have no significant urea cycle activity and will not survive in the absence of L-arginine. Apparently, the cross-feeding that may occur when hepatocytes condition the medium for nonproducing arginine cells is very low. The usefulness of L-arginine-free medium is contingent on a functioning urea cycle. Any cell damage affecting the urea cycle directly or indirectly would then affect the viability of the hepatocytes themselves under the arginine-free conditions. If there is suspicion that the hepatocytes may be compromised but still desirable, the use of an L-arginine-free medium may be inadvisable. Generally, this level of selection may not be necessary, since hepatocytes in primary culture can be acquired at high purity by exploiting the large difference in cell size between hepatocytes and the other cells in the liver.

Another cell culture condition that may select for different cell types is the osmolality of the medium. Cells have adaptive responses to osmolality changes but not all to the same extent. Sertoli cells isolated from seminiferous tubules contain contaminating germ cells. Exposure of the cultures to a hypotonic environment for certain times resulted in a selective removal of the germ cells without a noticeable effect on the Sertoli cells [133]. Also, the collecting duct epithelial cells in the medulla of the kidney are subject to a hyperosmotic environment. This characteristic was exploited in the isolation of cells with collecting duct epithelial characteristics; other epithelial cells in the kidney are not so adapted [134]. Further, it was observed that the NaCl concentration could influence the development of collecting duct epithelial cells from neonatal rabbit kidneys [135].

NUTRITIONAL

Even though cells in culture may adapt to the *in vitro* environment, they still maintain significant characteristics of their *in vivo* phenotype as demonstrated in prostate and bronchial epithelial cells [136]. The ability of cells to proliferate and express unique characteristics in cell culture is dependent on growth factors, cytokines, hormones, extracellular matrices, and the nutrient composition of the medium. An understanding of the function of these components is necessary for an understanding of the survival and function of cells *in vitro*.

The elucidation of the nutrient requirements of the cell of interest involves the titration to optimal levels of all the media components [137,138]. To achieve this end, various dose–response growth curves at clonal cell densities are required [139–141]. Clonal conditions are important because high cell densities condition the medium and obscure deficiencies. The process of elucidating the nutrient requirements aids in the ability of the cells to grow at clonal densities [142]. This activity is very time-consuming when traditional methods are used. The exploitation of hardware and software available for high throughput screening can reduce significantly the time needed to complete the titrations.

The development of serum-free media has demonstrated that there are different nutrient requirements for different cell types [143]. Though to a large extent cells require qualitatively the same components in their basal medium, they differ in their quantitative requirements. The various cells of a multicellular organism are surrounded by an extracellular fluid composed of oxygen, carbon dioxide, glucose, lactate, amino acids, and other important nutrients, growth factors, and hormones derived from the plasma and from the cells of that environment [144]. Since different cells have different metabolic programs, these environments are also different. For example, fibroblasts and keratinocytes from the same skin sample will not proliferate in their respective media [145]; differences in the concentrations of calcium and adenine apparently are the most influential in selective growth. Optimal growth of keratinocytes occurs in relatively low calcium concentrations. Calcium concentrations optimal for fibroblasts cause keratinocytes to stratify, and they are no longer capable of serial passaging while the optimal concentration of calcium for keratinocytes is too low to support fibroblast proliferation. Also, mammary epithelial cells and fibroblasts are distinguished in culture by their different responses to substratum conditions and medium components [146].

Calcium has profound effects on the expression of the epithelial cell phenotype. Relatively high calcium concentrations cause the cells to stratify and undergo differentiation as observed in rat esophageal epithelial cells [147,148]. Mouse epidermal keratinocytes grow optimally in 0.03 mM and terminally differentiate in 1.0 mM calcium [149]. Human urothelial cells proliferate in 0.06 mM calcium and differentiate and desquamate at higher concentrations [150]. Serum can also differentiate epithelial cells. Normal human bronchial epithelial cells demonstrate squamous (terminal) differentiation in the presence of serum, whereas lung carcinoma cells do not undergo differentiation in the presence of serum [151]. Also, culture conditions that support the growth of normal luminal epithelial cells do not support the growth of neoplastic cells [152].

In serum-free or chemically defined media, nutrient composition has profound effects on cellular behavior. The requirements for specific nutrients limit the proliferation of cells [153]. The growth factors and other informational molecules, the extracellular matrix as well as for cells that are normally associated with it, interact with the composition of the basal medium to achieve a certain phenotype [154,155]. Embryonic renal collecting duct epithelia responded to various media by demonstrating different differentiation patterns [156]. The MCDB 150 series of media was developed for human keratinocytes [157]. The use of MCDB 153 for the growth of rat salivary gland epithelial cells revealed an inability to subculture extensively in a serum-free environment; however, an increase to optimal levels in the concentration of isoleucine allowed the salivary epithelial cells to be serially subcultured [158]. Human keratinocytes have demonstrated nutrient requirements for inositol [159] and choline [160], and L-serine potentiates the mitogenic effect of epidermal growth factor, serum, insulin, keratinocyte growth factor, and bovine pituitary extract [161]. Also, the presence of ethanolamine has reduced the serum requirement of rat mammary epithelial cells for proliferation [162]. For cell types that respond to serum, serum may support cell performance in culture by providing certain nutrients. The addition of those nutrients to the medium may reduce the requirements for serum.

Under serum-free conditions, nutrient requirements become more evident. In rat liver, high concentrations of pyruvate and lactate enhanced DNA synthesis, with glucose acting synergistically with pyruvate [163]. In the absence of serum, the trace element selenium was found to be essential for clonal growth of fibroblasts and a Chinese hamster cell line [164]. Also, the trace mineral zinc has been demonstrated to have a positive or negative effect on apoptosis in mouse thymocytes depending on the concentration used [165]. Further, differentiation of pig preadipocyte was inhibited in a magnesium-deficient medium [166]. Also, selection of gluconeogenic-competent human proximal tubule cells can be achieved exploiting that metabolic process. The cells in primary were cultured in a hormonally defined serum-free medium without glucose or insulin, and cells not capable of gluconeogenesis would be at a disadvantage [167].

A less thorough approach of medium optimization is possible. Rich media such as M-199 Ham's F-12, or the MCDB series are used individually or are blended. Other combinations are possible. This is a legitimate activity and can produce the desired results. For example, peritoneal mesothelial cells were cultured in serum-free M-199 for the study of thrombin regulation of matrix metalloproteinase activity [168]; immature granulosa cells from rabbits were cultured in M-199 for the study of androstenedione and fibronectin on cytodifferentiation induced by follicle-stimulating hormone [169]. The use of a mixture of Dulbecco's Modified Eagle's Medium with Ham's F-12 along with neurobasal medium in a serum-free environment culturing embryonic rat cholinergic basal forebrain neurons demonstrated that growth factor supplementation was by itself insufficient to compensate for an inadequate nutrient composition [170]. Medium 199 has been combined with MCDB 105 for the culture of human ovarian surface epithelium in a defined environment [171].

CONCLUSION

An attempt has been made to convey the complexity inherent in cell isolation and selection. Tissues are by nature variable, which presents challenges to the establishment of primary cultures. Initial conditions, such as tissue age and tissue handling, are critical because they have a significant effect on the final outcome. The investigator can contribute to the loss of viability and performance by utilizing inappropriate reagents and conditions. Thus the need for care in the selection of reagents and procedures cannot be overemphasized. One of the purposes of this chapter was to present, for difficult isolations, a series of considerations that may prove useful in ultimately achieving success.

REFERENCES

1. Freshney, R. I. (2000). "Culture of Animal Cells: A Manual of Basic Techniques," 4th ed. Wiley, New York.
2. Studzinski, G. P. (1999). "Cell Growth, Differentiation and Senescence: A Practical Approach." Oxford University Press, Oxford, UK.
3. Masters, J. R. W. (2000). "Animal Cell Culture: A Practical Approach." Oxford University Press, Oxford, UK.
4. Fries, K. M., Blieden, T., Looney, R. J., Sempowski, G. D., Silvera, M. R., Willis, R. A., and Phipps, R. P. (1994). Evidence of fibroblast heterogeneity and the role of fibroblast subpopulations in fibrosis. *Clin. Immunol. Immunopathol.* 72, 283–292.
5. Lekic, P. C., Pender, N., and McCulloch, C. A. (1997). Is fibroblast heterogeneity relevant to the health, diseases, and treatments of periodontal tissues? *Crit. Rev. Oral Biol. Med.* 8, 253–268.
6. Dabbous, M. K., Haney, L., Carter, L. M., Paul, A. K., and Reger, J. (1987). Heterogeneity of fibroblast response in host–tumor cell–cell interactions in metastic tumors. *J. Cell. Biochem.* 35, 333–344.
7. Fenyves, A. M., Behrens, J., and Spanel-Borowski, K. (1993). Cultured microvascular endothelial cells (MVEC) differ in cytoskeleton, expression of cadherins and fibronectin matrix. A study under the influence of interferon-gamma. *J. Cell Sci.* 106, 879–890.
8. Oberley, T. D., Yang, A. H., and Gould-Kostka, J. (1986). Selection of kidney cell types in primary glomerular explant outgrowths by in vitro culture conditions. *J. Cell Sci.* 84, 69–92.
9. Holdsworth, S. R., Glasgow, E. F., and Atkins, R. C. (1978). Tissue culture of isolated human glomeruli. *Pathology* 10, 59–67.
10. Russell, S. W., Doe, W. F., Hoskins, R. G., and Cochrane, C. G. (1976). Inflammatory cells in solid murine neoplasms. I. Tumor disaggregation and identification of constitutive inflammatory cells. *Int. J. Cancer* 15, 322–330.
11. Ritson, A., and Bulmer, J. N. (1987). Extraction of leukocytes from human decidua. A comparison of dispersal techniques. *J. Immunol. Methods* 104, 231–236.
12. Kirkpatrick, C. J., Melzner, I., and Goller, T. (1985). Comparative effects of trypsin, collagenase and mechanical harvesting on cell membrane lipids studied in monolayer-cultured endothelial cells and a green monkey kidney cell line. *Biochim. Biophys. Acta* 846, 120–126.
13. Saward, L., and Zahradka, P. (1997). Coronary artery smooth muscle in culture: Migration of heterogeneous cell populations from vessel wall. *Mol. Cell Biochem.* 176, 53–59.
14. Horikawa, A., Okada, K., Sato, K., and Sato, M. (2000). Morphological changes in osteoblastic cells (MC3T3-E1) due to fluid shear stress: Cellular damage by prolonged application of fluid shear stress. *J. Exp. Med.* 191, 127–137.
15. Lin, K., Hsu, P. P., Chen, B. P., Yuan, S., Usami, S., Shyy, J. Y., Li, Y. S., and Chien, S. (2000). Molecular mechanism of endothelial growth arrest by laminar shear stress. *Proc. Natl. Acad. Sci. U.S.A.* 15, 9385–9389.
16. Costa, A., Silvestrini, R., Del Bino, G., and Motta, R. (1987). Implications of disaggregation procedures on biological representation of human solid tumors. *Cell Tissue Kinet.* 20, 171–180.
17. Smeets, A. W., Pauwels, R. P., Beck, H. L., Feitz, W. F., Gaaeraedts, J. P., Debruyne, F. M., Laarakkers, L., Vooijs, G. P., and Ramaaekers, F. C. (1987). Comparison of tissue disaggregation techniques of transitional cell bladder carcinomas for flow cytometry and chromosomal analysis. *Cytometry* 8, 14–19.
18. Konig, J. J., van Dongen, J. W., and Schroder, F. H. (1993). Preferential loss of abnormal prostate carcinoma cells by collagenase treatment. *Cytometry* 14, 805–810.
19. Meredith, M. J. (1998). Rat hepatocytes prepared without collagenase: Prolonged retention of differentiated characteristics in culture. *Cell Biol. Toxicol.* 4, 405–425.
20. Bayad, J., Sabolovic, N., Bagrel, D., Magdalou, J., and Siest, G. (1991). Influence of the isolation method on the stability of differentiated phenotype in cultured rat hepatocytes. *J. Pharmacol. Methods* 25, 85–94.
21. Engelholm, S. A., Spang-Thomsen, M., Brunner, N., Nohr, I., and Vindelov, L. L. (1985). Disaggregation of human solid tumors by combined mechanical and enzymatic methods. *Br. J. Cancer* 51, 93–98.
22. Besch, G. J., Wolberg, W. H., Gilchrist, K. W., Voelkel, J. G., and Gould, M. N. (1983). A comparison of methods for the production of monodispersed cell suspensions from human primary breast carcinomas. *Breast Cancer Res. Treat.* 3, 15–22.
23. Gerlach, J. C., Brombacher, J., Courtney, J. M., and Neuhaus, P. (1993). Nonenzymatic versus enzymatic hepatocyte isolation from pig livers for larger scale investigations of liver cell perfusion systems. *Int. J. Artif. Organs* 16, 677–681.
24. Miyazaki, M., Tsunashima, M., Wahid, S., Miyano, K., and Sato, J. (1984). Comparison of cytologic and biochemical properties between liver cells isolated from adult rats by trypsin perfusion and those isolated by collagenase perfusion. *Res. Exp. Med.* 184, 191–204.
25. Heiser, A., Ulrichs, K., and Muller-Ruchholtz, W. (1994). Isolation of porcine pancreatic islets: Low trypsin activity during the isolation procedure guarantees reproducible high islet yields. *J. Clin. Lab. Anal.* 8, 407–411.

26. Basir, I., van der Burg, M. P., Scheringa, M., Tons, A., and Bouwman, E. (1997). Improved outcome of pig islet isolation by Pefabloc inhibition of trypsin. *Transplant. Proc.* **29,** 1939–1941.

27. Wilhelm, S. M., Eisen, A. Z., Teter, M., Clark, S. D., Kronberger, A., and Goldberg, G. (1986). Human fibroblast collagenase: Glycosylation and tissue-specific levels of enzyme synthesis. *Proc. Natl. Acad. Sci. U.S.A.* **83,** 3756–3760.

28. Heiser, A., Ulrichs, K., and Muller-Ruchholtz, W. (1994). Isolation of porcine pancreatic islets: Low trypsin activity during the isolation procedure guarantees reproducible high islet yields. *J. Clin. Lab. Anal.* **8,** 407–411.

29. White, S. A., Djaballah, H., Hughes, D. P., Roberts, D. L., Contractor, H. H., Pathak, S., and London, N. J. (1999). A preliminary study of the activiation of endogenous pancreatic exocrine enzymes during automated porcine islet isolation. *Cell Transplant.* **8,** 265–276.

30. McKeehan, W. L. (1977). The effect of temperature during trypsin treatment on viability and multiplication potential of single normal human and chicken fibroblasts. *Cell Biol. Int. Rep.* **1,** 335–343.

31. Williams, S. K., McKenney, S., and Jarrell, B. E. (1995). Collagenase lot selection and purification for adipose tissue digestion. *Cell Transplant.* **4,** 281–289.

32. Bowman, C. L., Yohe, L., and Lohr, J. R. (1999). Enzymatic modulation of cell volume in C6 glioma cells. *Glia* **27,** 22–31.

33. Becker, L. B., Vanden Hoek, T. L., Shao, Z.-H., Li, C.-Q., and Schmacker, P. T. (1999). Generation of superoxide in cardiomyocytes during ischemia before perfusion. *Am. J. Physiol.* **277,** H2240–H2246.

34. Lee, Y. B., and Lee, S. M. (2000). Effect of *S*-adenosylmethionine on hepatic injury from sequential cold and warm ischemia. *Arch. Pharm. Res.* **23,** 495–500.

35. Killilea, D. W., Hester, R., Balczon, R., Babal, P., and Gillespie, M. N. (2000). Free radical production in hypoxic pulmonary artery smooth muscle cells. *Am. J. Physiol. (Lung Cell Mol. Physiol.)* **279,** L408–L412.

36. Amann, S., Reinke, C., Valet, G., Moser, U., and Leuenberger, H. (1999). Flow-cytometric investigation of cellular metabolism during oxidative stress and the effect of tocopherol. *Int. J. Vitam. Nutr. Res.* **69,** 356–361.

37. Li, A. E., Ito, H., Rovira, I. I., Kim, K. S., Takeda, K., Yu, Z. Y., Ferrans, V. J., and Finkel, T. (1999). A role for reactive oxygen species in endothelial cell anoikis. *Circ. Res.* **85,** 304–310.

38. Lee, C. K., Weaks, R. L., Johnson, G. A., Bazer, F. W., and Piedrahita, J. A. (2000). Effects of protease inhibitors and antioxidants on in vitro survival of porcine primordial germ cells. *Biol. Reprod.* **63,** 887–897.

39. Chandra, J., Samali, A., and Orrenius, S. (2000). Triggering and modulation of apoptosis by oxidative stress. *Free Radic. Biol. Med.* **29,** 323–333.

40. Pesce, M., Farrace, M. G., Piacentini, M., Dolci, S., and De Felici, M. (1993). Stem cell factor and leukemia inhibitory factor promote primordial germ cells survival by suppressing programmed cell death (apoptosis). *Development (Cambridge, UK)* **118,** 1098–1094.

41. Frisch, S. M., and Ruoslahti, E. (1997). Integrins and anoikis. *Curr. Opin. Cell Biol.* **9,** 701–706.

42. Matarresea, P., Tinari, N., Semeraroa, M. L., Natolib, C., Iacobelli, S., and Maloni, W. (2000). Galectin-3 overexpression protects from cell damage and death by influencing mitochondrial homeostasis. *FEBS Lett.* **473,** 311–315.

43. Sakai, H., Kobayashi, Y., Sakai, E., Shibata, M., and Kato, Y. (2000). Cell adhesion is a prerequisite for osteoclast survival. *Biochem. Biophys. Res. Commun.* **270,** 550–556.

44. Thomas, F. T., Contreras, J. L., Bilbao G., Ricordi, C., Curiel, D., and Thomas, J. M. (1999). Anoikis, extracellular matrix, and apoptosis factors in isolated cell transplantation. *Surgery* **126,** 299–304.

45. Beattie, G. M., Leibowitz, G., Lopez, A. D., Levine, F., and Hayek. A. (2000). Protection from cell death in cultured human fetal pancreatic cells. *Cell Transplant.* **9,** 431–438.

46. Erkkila, K., Hirvonen, V., Wuokko, E., Parvinen, M., and Dunkel, L. (1998). N-acetyl-L-cysteine inhibits apoptosis in human male germ cells in vitro. *J. Clin. Endocrinol. Metab.* **83,** 2523–2531.

47. da Costa, A. O., de Assis, M. C., Marques, E. D., and Plotkowski, M. C. (1999). Comparative analysis of three methods to assess viability of mammalian cells in culture. *Biocell* **23,** 65–72.

48. Mascotti, K., McCullough, J., and Burger, S. R. (2000). HPC viability measurement: Trypan blue versus acridine orange and propidium iodide. *Transfusion (Philadelphia)* **40,** 693–696.

49. Kagan, V. E., Fabisiak, J. P., Shvedova, A. A., Tyurina, Y. Y., Tyurina, V. A., Schor, N. F., and Kawai, K. (2000). Oxidative signaling pathway for the externalization of plasma membrane phosphatidylserine during apoptosis. *FEBS Lett.* **477,** 1–7.

50. Hutz, R. J., DeMayo, F. J., and Dukelow, W. R. (1985). The use of vital dyes to assess embryonic viability in the hamster. *Mesocricetus auratus. Stain Technol.* **60,** 163–167.

51. Zhang, F., Zhang, P., Wang, E., Su, F., Qi, Y., and Wang, Y. (1995). Fluorescein diacetate uptake to determine the viability of human fetal cerebral cortical cells. *Biotechnol. Histochem.* **70,** 185–187.

52. Jones, K. H., and Senft, J. A. (1985). An improved method to determine cell viability by simultaneous staining with fluorescein diacetate–propidium iodide. *J. Histochem. Cytochem.* **33,** 77–79.

53. Redegeld, F. A., Moison, R. M., Koster, A. S., and Noordhoek, J. (1992). Depletion of ATP but not GSH affects viability of rat hepatocytes. *Eur. J. Pharmacol.* **228,** 229–236.

54. Bhatnager, A. (1994). Biochemical mechanism of irreversible cell injury caused by free radical-initiated reactions. *Mol. Cell. Biochem.* **17,** 9–16.

55. Petit, P. X., Lecoeur, H., Zorn, E., Dauguet, C., Mignotte, B., and Gougeon, M. L. (1995). Alterations in mitochondrial structure and function are early events of dexamethasone-induced thymocyte apoptosis. *J. Cell Biol.* **130,** 157–167.

56. Hirsch, T., Marzo, I., and Kroemer, G. (1997). Role of the mitochondrial permeability transition pore in apoptosis. *Biosci. Rep.* **17,** 67–76.

57. Atkinson, D. E. (1968). The energy charge of the adenylate pool as a regulatory parameter. Interaction with feedback modifiers. *Biochemistry* **7,** 4030–4034.

58. Spragg, R. G., Hinshaw, D. B., Hyslop, P. A., Schraufstatter, I. U., and Cochrane, C. E. (1985). Alterations in adenosine triphosphate and energy charge in cultured endothelial and P388D1 cells after oxidant injury. *J. Clin. Invest.* **76**, 1471–1476.

59. Pretlow, T. G., and Pretlow, T. P. (1989). Cell separation by gradient centrifugation methods. *Methods Enzymol.* **171**, 462–482.

60. Stolberg, H. O., and McClennan, B. L. (1991). Ionic versus nonionic contrast use. *Curr. Probl. Diagn. Radiol.* **20**, 47–88.

61. Rickwood, D., and Birnie, G. D. (1975). Metrizamide, a new density-gradient medium. *FEBS Lett.* **50**, 102–110.

62. Ford, T. C., Needle, R., and Rickwood, D. (1987). A rapid method for the preparation of the neutrophil fraction of granulocytes from human blood by centrifugation on isotonic Nycodenz gradients. *Blut* **54**, 337–342.

63. Ford, T. C., Graham, J., and Rickwood, D. (1990). A new, rapid, one-step method for the isolation of platelets from human blood. *Clin. Chim. Acta* **30**, 115–119.

64. Apte, M. V., Haber, P. S., Applegate, T. L., Norton, I. D., McCaughan, G. W., Korsten, M. A., Pirola, R. C., and Wilson, J. S. (1998). Periacinar stellate shaped cells in rat pancreas: Identification, isolation, and culture. *Gut* **43**, 128–133.

65. Heuff, G., Van de Loosdrecht, A. A., Betjes, M. G., Beelen, R. H., and Meijer, S. (1995). Isolation and purification of large quantities of fresh human Kupffer cells, which are cytotoxic against colon carcinoma. *Hepatology* **21**, 740–745.

66. Pretlow, T. G., and Stinson, A. J. (1976). Separation of megakaryocytes from rat bone marrow cells using velocity sedimentation in an ioskinetic gradient of Ficoll in tissue culture medium. *J. Cell. Physiol.* **88**, 317–322.

67. Pretlow, T. G., and Williams, E. E. (1973). Separation of hepatocytes from suspensions of mouse liver cells using programmed gradient sedimentation in gradients of Ficoll in tissue culture medium. *Anal. Biochem.* **55**, 114–122.

68. Lakey, J. R., Cavanagh, T. J., and Zieger, M. A. (1998). A prospective comparison of discontinuous EuroFicoll and EuroDextran gradients for islet purification. *Cell Transplant.* **7**, 479–487.

69. Starkel, P., Sempoux, C., Van Den Berge, N., Stevens, M., De Saeger, C., and Desager, J. P. (1999). CYP 3A proteins are expressed in human neutrophils and lymphocytes but are not induced by rifampicin. *Life Sci.* **64**, 643–653.

70. Rambaldi, A., Borteri, G., Dotti, G., Bellavita, P., Amaru, R., Biondi, A., and Barbui, T. (1998). Innovative two-step negative selection of granulocyte colony-stimulating factor-mobilized circulation progenitor cells: Adequacy for autologous and allogenic transplantation. *Blood* **91**, 2189–2196.

71. Soria, J. C., Gauthier, L. R., Raymond, E., Granotier, C., Morat, L., Armand, J. P., and Boussin, F. D. (1999). Molecular detaection of telomerase-positive circulating epithelial cells in metastatic breast cancer patients. *Clin. Cancer Res.* **5**, 971–975.

72. Ferrero, D., Tarella, C., Cherasco, C., Bondesan, P., Omede, P., Ravaglia, R., Caracciolo, D., Castellino, C., and Pileri, A. (1998). A single step density gradient separation for large scale enrichment of mobilized peripheral blood progenitor cells collected for autotransplantation. *Bone Marrow Transplant.* **21**, 409–413.

73. Sitar, G., Garagna, S., Zuccotti, M., Falacinelli, C., Montanari, L., Alfei, A., Ippoliti, G., Redi, C. A., Moratti, R., Ascari, E., and Forabosco, A. (1999). Fetal erythroblast isolation up to purity from cord blood and their culture in vitro. *Cytometry* **35**, 337–345.

74. Pertoft, H. (2000). Fractionation of cells and subcellular particles with Percoll. *J. Biochem. Biophys. Methods* **44**, 1–30.

75. Pertoft, H., and Laurent, T. (1982). Sedimentation of cells in colloidal silica (Percoll). *In* "Cell Separation; Methods and Selected Applications" (T. G. Pretlow, II, and T. P. Pretlow, eds.), Vol. 1, pp. 115–152. Academic Press, New York.

76. Pharmacia (1995). "Percoll: Methodoly and Applications," 2 ed., Rev. 2. Biotech, Inc., Molecular Biology Reagents Division, Piscataway, NJ.

77. Smits, G., Holzgreve, W., and Hahn, S. (2000). An examination of different Percoll densoty gradients and magnetic activated cell sorting (MACS) for the enrichment of fetal erythroblasts from maternal blood. *Arch. Gynecol. Obstet.* **263**, 160–163.

78. Zavros, Y., Van Antwerp, M., and Merchant, J. L. (2000). Use of flow cytometry to quantify mouse gastric epithelial cell populations. *Dig. Dis. Sci.* **45**, 1192–1199.

79. Lichtenberger, C., Zakeri, S., Baier, K., Willheim, M., Holub, M., and Reinisch, W. (1999). A novel high-purity isolation method for human peripheral neutrophils permitting polymerase chain reaction-based mRNA studies. *J. Immunol. Meth.* **227**, 75–84.

80. Sekizawa, A., Farins, A., Zhen, D. K., Wang, J. Y., Falco, V. M., Elmes, S., and Bianchi, D. W. (1999). Improvement of fetal cell recovery from maternal blood: Suitable density gradient for FACS separation. *Fetal Diagn. Ther.* **14**, 229–233.

81. Ulmer, A. J., and Flad, H.-D. (1979). Discontinuous density gradient separation of human mononuclear leucocytes using Percoll as gradient medium. *J. Immunol. Methods* **30**, 1–10.

82. Silverman, W. F., Alfahel-Kakunda, A., Dori, A., and Barker, J. L. (1999). Separation of dorsal and ventral dopaminergic neurons from embryonic rat mesencephalon by buoyant density fractionation: Disassembling pattern in the ventral midbrain. *J. Neurosci. Methods* **89**, 1–8.

83. Check, M. L., Bollendorf, A., Check, J. H., Hourani, W., Long, R., and McMonagle, K. (2000). Separation of sperm through a 12-layer Percoll column decreases the percentage of sperm staining with aquinacrine. *Arch. Androl.* **44**, 47–50.

84. De Duve, C. (1971). Tissue fractionation, past and present. *J. Cell Biol.* **50**, 20d–55d.

85. Shortman, K. (1972). Physical procedures for the separation of animal cells. *Annu. Rev. Biophys. Bioeng.* **1**, 93–130.

86. Pertoft, H. (1969). Gradient centrifugation in colloidal silica–polysaccharide media. *Biochim. Biophys. Acta* **126**, 594–596.

87. Silverstein, S. C., Steinman, R. M., and Cohen, Z. A. (1977). Endocytosis. *Annu. Rev. Biochem.* **46**, 669–722.

88. Ahlberg, J., Marzella, L., and Glaumann, H. (1982). Uptake and degradation of proteins by isolated rat liver lysosomes. Suggestion of a microautophagic pathway of proteolysis. *Lab. Invest.* **47**(6), 523–532.

89. Niioka, S., Goto, M., Ishibashi, T., and Kadowaki, M. (1998). Identification of autolysosomes directly associated with the proteolysis on the density gradients in isolated rat hepatocytes. *J. Biochem. (Tokyo)* **124**, 1086–1093.

90. Carter, N. P. (1994). Flow sorting. *In* "Flow Cytometry: A Practical Approach" (M. G. Ormerod, ed.), 2nd ed., pp. 55–65. Oxford University Press, Oxford, UK.

91. Geerts, A., Niki, T., Hellemans, K., De Craemer, D., Van Den Berg, K., Lazou, J. M., Stange, G., Van De Winkel, M., and De Bleser, P. (1998). Purification of rat hepatic stellate cells by side scatter-activated cell sorting. *Hepatology* **27**, 590–598.

92. Maftah, A., Huet, O., Gallet, P. F., and Ratinaud, M. H. (1993). Flow cytometry's contribution to the measurement of cell functions. *Biol. Cell* **78**, 85–93.

93. Sutherland, D. R., Anderson, L., Keeney, M., Nayar, R., and Chin-Yee, I. (1996). The ISHAGE guidelines for CD 34+ cell determination by flow cytometry. *J. Hematother.* **5**, 213–226.

94. Orfao, A., and Ruiz-Arguelles, A. (1996). General concepts about cell sorting techniques. *Clin. Biochem.* **29**, 5–9.

95. Hadjivassiliou, V., Green, M. H., and Green, I. C. (2000). Immunomagnetic purification of beta cells from rat islets of Langerhans. *Diabetologia* **43**, 1170–1177.

96. Lou, J., Tripomez, F., Oberholzer, J., Wang, H., Yu, D., Buhler, L., Cretin, N., Mentha, G., Wollheim, C. B., and Morel, P. (1999). Expression of alpha-1 proteinase inhibitor in human islet microvascular endothelial cells. *Diabetes* **48**, 1773–1778.

97. Smeland, E. B., Funderud, S., Kvalheim, G., Gaudernack, G., Rasmussen, A. M., Rusten, L., Wang, M. T., Tindle, R. W., Blomhoff, H. K., and Egeland, T. (1992). Isolation and characterization of human hematopoietic progenitor cells: An effective method for positive selection of CD34+ cells. *Leukemia* **6**, 845–852.

98. Rasmussen, A. M., Smeland, E. B., Erikstein, B. K., Caignault, L., and Funderud, S. (1992). A new method for detachment of Dynabeads from positively selected B lymphocytes. *J. Immunol. Methods* **146**, 195–202.

99. Werther, K., Normark, M., Hansen, B. F., Brunner, N., and Nielsen, H. J. (2000). The use of the CELection kit in the isolation of carcinoma cells from mononuclear cell suspensions. *J. Immunol. Methods* **238**, 133–141.

100. Miltenyi, S., Muller, W., Weichel, W., and Radbruch, A. (1990). High gradient magnetic cell separation with MACS. *Cytometry* **11**, 231–238.

101. Manyonda, I. T., Soltys, A. J., and Hay, F. C. (1992). A critical evaluation of the magnetic cells sorter and its use in the positive and negative sellection of CD45RO+ cells. *J. Immunol. Methods* **149**, 1–10.

102. Partington, K. M., Jenkinson, E. J., and Anderson, G. (1999). A novel method of cell separation based on dual parameter immunomagnetic cell selection. *J. Immunol. Methods* **223**, 195–205.

103. Prattis, S. M., Gebhart, D. H., Dickson, G., Watt, D. J., and Kornegay, J. N. (1993). Magnetic affinity cell sorting (MACS) separation and flow cytometric characterization of neural cell adhesion molecule-positive, cultured myogenic cells from normal and dystrophic dogs. *Exp. Cell Res.* **208**, 453–465.

104. Hoerstrup, S. P., Zund, G., Schoeberlein, A., Ye, Q., Vogt, P. R., and Turina, M. I. (1998). Fluorescence activated cell sorting: A reliable method in tissue engineering of a bioprosthetic heart valve. *Ann. Thorac. Surg.* **66**, 1653–1657.

105. Smalley, M. J., Titley, J., and O'Hare, M. J. (1998). Clonal characterization of mouse mammary luminal epithelial and myoepithelial cells separated by fluorescence-activated cell sorting. *In Vitro Cell. Dev. Biol.—Anim.* **34**, 711–721.

106. Helbert, M. J., Dauwe, S. E., Van der Blest, I., Nouwen, E. J., and de Broe, M. E. (1997). Immunodissection of the human proximal nephron: Flow sorting of S1S2S3, S1S2 and S3 proximal tubular cells. *Kidney Int.* **52**, 414–428.

107. Aitken, M. L., Villalon, M., Verdugo, P., and Nameroff, M. (1991). Enrichment of subpopulations of respiratory epithelial cells using flow cytometry. *Am. J. Respir. Cell Mol. Biol.* **4**, 174–178.

108. Carr, T., Evans, P., Campbell, S., Bass, P., and Albano, J. (1999). Culture of human renal tubular cells: Positive selection of kallikrein-containing cells. *Immunopharmacology* **15**, 161–167.

109. Seidl, J., Knuechel, R., and Kunz-Schughart, L. A. (1999). Evaluation of membrane physiology following fluorescence activation or magnetic cell separation. *Cytometry* **36**, 102–111.

110. Rubbi, C. P., Patel, D., and Rickwood, D. (1993). Evidence of surface antigen detachment during incubation of cells with immunomagnetic beads. *J. Immunol. Methods* **166**, 233–241.

111. Funderud, S., Erikstein, B., Asheim, H. C., Stokke, T., Blomhoff, H. K., Holte, H., and Smeland, E. B. (1990). Functional properties of CD19+ B lymphocytes positively selected from buffy coats by immunomagnetic separation. *Eur. J. Immunol.* **20**, 201–206.

112. Casale, T. B., Erger, R. A., and Rozell, M. D. (1999). Eosinophils isolated by magnetic cell sorting respond poorly to lipid chemoattractants. *Ann. Allergy Asthma Immunol.* **83**, 127–131.

113. Patel, D., and Rickwood, D. (1995). Optimization of conditions for specific binding of antibody-coated beads to cells. *J. Immunol. Methods* **184**, 71–80.

114. Bildirici, L., and Rickwood, D. (2000). Fractionation of differentiating cells using density perturbation. *J. Immunol. Methods* **240**, 93–99.

115. Willems, P., Croockewit, A., Raymakers, R., Holdrinet, R., Van Der Bosch, G., Huys, E., and Mensink, E. (1996). CD34 selection from myeloma peripheral blood cell autographs contain residual tumor cells due to impurity, not to CD34+ myeloma cells. *Br. J. Haematol.* **93**, 613–622.

116. Bampton, J. L., Shirlaw, P. J., Topley, S., Weller, P., and Wilton, J. M. (1991). Human junctional epithelium: Demonstration of a new marker, its growth in vitro and a chracterization by lectin reactivity and keratin expression. *J. Invest. Dermatol.* **96**, 708–717.

117. Castleman, W. L., Northrop, P. J., and McAllister, P. K. (1991). Replication of parainfluenza type-3 virus and bovine respiratory syncytial virus in isolated bovine type-II alveolar epithelial cells. *Am. J. Vet. Res.* **52**, 880–885.

118. Contractor, S. F., and Sooranna, S. R. (1988). Human placental cells in culture: A panning technique using a trophoblast-specific monoclonal antibody for cell separation. *J. Dev. Physiol.* **10**, 47–51.

119. De Felici, M., and Pesce, M. (1995). Immunoaffinity purification of migratory mouse primordial cells. *Exp. Cell Res.* **216**, 277–279.

120. Linge, C., Green, M. R., and Brooks, P. F. (1989). A method for removal of fibroblast from human tissue culture systems. *Exp. Cell Res.* **185**, 519–528.

121. Morhenn, V. B., Wood, G. S., Engleman, E. G., and Oseroff, A. R. (1983). Selective enrichment of human epidermal cell subpopulations using monoclonal antibodies. *J. Invest. Dermatol.* **81**(Suppl. 1), 127s–131s.

122. Wysocki, L. J., and Sato, V. L. (1978). "Panning" for lymphocytes: A method for cell selection. *Proc. Natl. Acad. Sci. U.S.A.* **75**, 2844–2848.

123. Lee, C. S., Patton, W. F., Chung-Welch, N., Chiang, E. T., Spofford, K. H., and Shepro, D. (1999). Selective propagation of retinal pericytes in mixed microvascular cell cultures using L-leucine-methyl ester. *BioTechniques* **25**, 482–488, 490–492, 494.

124. Thiele, D. L., and Lipsky, P. E. (1990). Mechanism of L-leucyl-L-leucine methyl ester-mediated killing of cytotoxic lymphocytes: Dependence on a lysosomal thiol protease, dipeptidyl peptidase I, this is enriched in these cells. *Proc. Natl. Acad. Sci. U.S.A.* **87**, 83–87.

125. Kawano, Y., Takaue, Y., Law, P., Watanabe, T., Abe, T., Okamoto, Y., Makimoto, A., Sato, J., Nakagawa, R., Kajiume, T., Hirao, A., Watanabe, A., and Kuroda, Y. (1998). Clinically applicable bulk isolation of blood CD34+ cells for autografting in children. *Bone Marrow Transplant.* **22**, 1011–1017.

126. Sordillo, L. M., Oliver, S. P., and Akers, R. M. (1988). Culture of bovine mammary epithelial cells in D-valine modified medium: Selective removal of contaminating fibroblasts. *Cell Biol. Int. Rep.* **12**, 355–364.

127. Lazzaro, V. A., Walker, R. J., Duggin, G. G., Phippard, A., Horvath, J. S., and Tiller, D. J. (1992). Inhibition of fibroblast proliferation in L-valine reduced selective media. *Res. Commun. Chem. Pathol. Pharmacol.* **75**, 39–48.

128. Hongpaisan, J. (2000). Inhibition of proliferation of contamination fibroblasts by D-valine in cultures of smooth muscle cells from human myometrium. *Cell Biol. Int.* **24**, 1–7.

129. Coleman, M. T., Hart, R. W., Liss, L., and Yates, A. J. (1979). Changes in growth and morphology of human gliomas and fibroblasts cultured in D-valine medium. *J. Neuropathol. Exp. Neurol.* **38**, 606–613.

130. Frauli, M., and Ludwig, H. (1987). Inhibition of fibroblast proliferation in a culture of human endometrial stromal cells using a medium containing D-valine. *Arch. Gynecol. Obstet.* **241**, 87–96.

131. Masson, E. A., Atkin, S. L., and White, M. C. (1993). D-valine selective medium does not inhibit human fibroblast growth in vitro. *In Vitro Cell. Dev. Biol.—Anim.* **29A**, 912–913.

132. Boebel, K. P., and Baker, D. H. (1982). Comparative utilization of the alpha-keto and D- and L-alpha-hydroxy analogs of leucine, isoleucine and valine by chicks and rats. *J. Nutr.* **112**, 1929–1939.

133. Wagle, J. R., Heindel, J. J., Steinberger, A., and Sanborn, B. M. (1986). Effect of hypotonic treatment on Sertoli cell purity and function in culture. *In Vitro Cell. Dev. Biol.* **22**, 325–331.

134. Isabelle, M. E., Githens, S., Moses, R. L., and Bartell, C. K. (1994). Culture of rat renal medullary tissue in media made hyperosmotic with NaCl and urea. *J. Exp. Zool.* **269**, 308–318.

135. Minuth, W. W., Steiner, P., Strehl. R., Kloth, S., and Tauc, M. (1997). Electrolyte environment modulates differentiation in embryonic renal collecting duct epithelia. *Exp. Nephrol.* **5**, 414–422.

136. Lechner, J. F. (1984). Interdependent regulation of epithelila cell replication by nutrients, hormones, growth factors, and cell density. *Fed. Proc., Fed. Am. Soc. Exp. Biol.* **43**, 116–120.

137. Ham, R. G. (1974). Nutritional requirements of primary cultures. A neglected problem of modern biology. *In Vitro* **10**, 119–129.

138. Zimmerman, A. M., Vierck, J. L., O'Reilly, B. A., and Dodson, M. V. (2000). Formulation of a definded medium to maintain cell health and viability *in vitro. Methods Cell Sci.* **22**, 43–49.

139. Shipley, G. D., and Ham, R. G. (1981). Improved medium and culture conditions for clonal growth with minimal serum protein and for enhanced serum-free survival of Swiss 3T3 cells. *In Vitro* **17**, 656–670.

140. McKeehan, W. L., and Ham, R. G. (1976). Methods for reducing the serum requirement for growth in vitro of nontransformed diploid fibroblasts. *Dev. Biol. Stand.* **37**, 97–98.

141. Ham, R. G. (1974). Unique requirements for clonal growth. *J. Natl. Cancer Inst. (U.S.)* **53**, 1459–1463.

142. Ham, R. G., and McKeehan, W. L. (1979). Media and growth requirements. *Methods Enzymol.* **58**, 44–93.

143. Bettger, W. J., and Ham, R. G. (1982). The nutrient requirements of cultured mammalian cells. *Adv. Nutr. Res.* **4**, 249–286.

144. Lindendbaum, E. S., Tendler, M., and Beach, D. (1995). Serum-free cell culture medium induces acceleration of wound healing in guinea-pigs. *Burns* **21**, 110–115.

145. Peehl, D. M., and Ham, R. G. (1980). Clonal growth of human keratinocytes with small amounts of dialyzed serum. *In Vitro* **16**, 526–540.

146. Wang, S., and Haslam, S. Z. (1994). Serum-free primary cultyure of normal mouse mammary epithelial and stromal cells. *In Vitro Cell. Dev. Biol.—Anim.* **30A**, 859–866.

147. Babcock, M. S., Marino, M. R., Gunning, W. T., and Stoner, G. D. (1983). Clonal growth and serial propagation of rat esophageal epithelial cells. *In Vitro* **19**, 403–415.

148. Jaken, S., and Yuspa, S. H. (1988). Early signals for keratinocyte differentiation: Role of Ca^{2+}–mediated inositol lipid metabolism in normal and neoplastic epidermal cells. *Carcinogenesis (London)* **9**, 1033–1038.

149. Bertolero, F., Kaighn, M. E., Gonda, M. A., and Saffiotti, U. (1984). Mouse epidermal keratinocytes. Clonal proliferation and response to hormones and growth factors in serum-free medium. *Exp. Cell Res.* **155**, 64–80.

150. Kirk, D., Kagawa, S., Vener, G., Narayan, K. S., Ohnuki, Y., and Jones, L. W. (1985). Selective growth of normal adult human urothelial cell in serum-free medium. *In Vitro Cell Dev. Biol.* **21**, 165–171.

151. Lechner, J. F., Haugen, A., McClendon, I. A., and Shamsuddin, A. M. (1984). Induction of squamous differentiation of normal human bronchial epithelial cells by small amounts of serum. *Differentiation (Berlin)* **25**, 229–237.

152. Ethier, S. P., Mahacek, M. L., Gullick, W. J., Frank, T. S., and Weber, B. L. (1993). Differentiatial isolation of normal luminal mammary epithelial cells and breast cancer cells from primary and metastatic sites using selective media. *Cancer Res.* **53**, 627–635.

153. McKeehan, W. L. (1984). Control of normal and transformed cell proliferation by growth factor-nutrient interaction. *Fed. Proc., Fed. Am. Soc. Exp. Biol.* **43**, 113–115.

154. Ham, R. G., and McKeehan, W. L. (1978). Development of improved media and culture conditions for clonal growth of normal diploid cells. *In Vitro* **14**, 11–22.

155. McKeehan, W. L., McKeehan, K. A., Hammond, S. L., and Ham, R. G. (1977). Improved medium for clonal growth of human diploid fibroblasts at low concentrations of serum protein. *In Vitro* **13**, 399–416.

156. Schmacher, K., Strehl, R., Kloth, S., Tauc, M., and Minuth, W. W. (1999). The influence of culture media on embryonic renal collecting duct cell differentiation. *In Vitro Cell. Dev. Biol.—Anim.* **35**, 465–471.

157. Boyce, S. T., and Ham, R. G. (1983). Calcium-regulated differntiation of normal epidermal keratinocytes in chemically defined clonal culture and serum-free culture. *J. Invest. Dermatol.* **81**(Suppl. 1), 33s–40s.

158. Furue, M., Tetsuji, O., Koshika, S., and Asashima, M. (2000). Isoleucine prevents rat salivary gland epithelial cells from apoptosis in serum-free culture. *In Vitro Cell. Dev. Biol.—Anim.* **36**, 287–289.

159. Gordon, P. R., Mawhinney, T. P., and Gilchrest, B. A. (1988). Inositol is a required nutrient for human growth. *J. Cell. Physiol.* **135**, 416–424.

160. Gordon, P. R., Gelman, L. K., and Gilchrist, B. A. (1988). Demonstration of a choline requirement for optimal keratinocyte growth in a defined culture medium. *J. Nutr.* **118**, 1487–1494.

161. Wilkinson, D. I. (1987). L-serine potentiates the mitogenic effects of growth factors on cultured human keratinocytes. *J. Invest. Dermatol.* **88**, 198–201.

162. Ethier, S. P. (1986). Serum-free culture conditions for the growth of rat mammary epithelial cells in primary cultutre. *In Vitro Cell. Dev. Biol.* **22**, 485–490.

163. McGowan, J. A., Russell, W. E., and Bucher, N. L. (1984). Hepatocyte DNA replication: Effects of the nutrients and intermediary metabolites. *Fed. Proc., Fed. Am. Soc. Exp.* **43**, 131–133.

164. McKeehan, W. L., Hamilton, W. G., and Ham, R. G. (1976). Selenium is an essential trace nutrient for growth of WI-38 diploid human fibroblasts. *Proc. Natl. Acad. Sci. U.S.A.* **73**, 2023–2027.

165. Provinciali, M., Di Stefano, G., and Fabris, N. (1995). Dose-dependent opposite effect of zinc on apoptosis in mouse thymocytes. *Int. J. Immunopharmacol.* **17**, 735–744.

166. Tchoukalova, T. D., Grider, A., Mouat, M. F., and Hausman, G. J. (2000). Priming with magnesium-deficient media inhibits preadipocyte differentiation via potential upregulation of tumor necrosis factor-alpha. *Biol. Trace Elem. Res.* **74**, 11–21.

167. Courjault-Gautier, F., Chevalier, J., Abbou, C. C., Chopin, D. K., and Toutain, H. J. (1995). Consecutive use of hormonally defined serum-free media to establish highly differentiated human renal proximal tubule cells in primary culture. *J. Am Soc. Nephrol.* **5**, 1949–1963.

168. Haslinger, B., Mandl-Weber, S., and Sitter, T. (2000). Thrombin suppresses matrix metalloproteinase 2 activity and increases tissue inhibitor of metalloproteinase 1 synthesis un cultured human peritoneal mesothelial cells. *Peritoneal Dial. Int.* **20**, 778–783.

169. Picazo, R. A., Illera, J. C., and Illera, M. (2000). Development of a long-term serum-free culture system for immature granulose cells from diethylstilboestrol-treated prepubertal rabbits: Influence of androstenedione and fibronectin on FSH-induced cytodifferentiation. *J. Reprod. Fertil.* **119**, 279–285.

170. Pongrac, J. L., and Rylett, R. J. (1998). Optimization of serum-free culture conditions for growth of embryonic rat cholinergic basal forebrain neurons. *J. Neurosci. Methods* **84**, 69–76.

171. Elliot, W. M., and Auersperg, N. (1993). Growth of normal human ovarian surface epithelial cells in reduced-serum and serum-free media. *In Vitro Cell. Dev. Biol.* **29A**, 9–18.

CHAPTER 3

MAINTENANCE OF PRIMARY AND EARLY PASSAGE CULTURES

R. Ian Freshney

The first culture to become established after seeding disaggregated cells or primary explants into a culture vessel is known as a primary culture. The primary culture requires frequent observation and medium changes as necessary, depending on the growth rate of the cells. As the culture approaches confluence, it may be subcultured to become a cell line. The transition to cell line usually creates greater homogeneity in the culture, the predominant cell type being determined by the choice of selective medium and substrate. As the cells grow, the culture may need frequent subcultures, requiring careful notes of cell number at the start and finish of each growth cycle and a record of the dilution so that the number of elapsed generations may be calculated.

Propagation naturally favors the proliferative phenotype, and reinduction of differentiation will require an alteration of the culture conditions. Continuous propagation of normal cell lines will also lead ultimately to senescence and cessation of growth, so valuable lines should be stored in liquid nitrogen and only used for a limited number of generations. Cell lines should be characterized to validate the origin and lineage phenotype, confirm the lack of cross-contamination or transformation, and exclude the possibility of microbial contamination.

INITIATION OF CELL LINES

As described in the preceding chapter, tissue disaggregation gives rise to a heterogeneous population of cells, dependent on the origin of the tissue and apparent from the diversity of cell types present in the initial culture. Only a small proportion of the disaggregate will attach, and, of these, only a small proportion will proliferate. As the proliferating component of the culture increases, the ratio of different cell types will change and the total cell mass will increase until the culture vessel's entire growth surface is occupied. At this stage the cellular content will dictate (a) whether to retain the culture and (b) whether it should be subcultured. The culture is called a *primary culture* until the first subculture (also known as *passage* or *transfer*). *Subculture* is the transfer of a culture to a second culture vessel, usually, implying mechanical or enzymatic disaggregation, reseeding, and often division into two or more daughter cultures, depending on the rate of proliferation. After subculture of a primary, the culture becomes a *cell line*, implying propagation and amplification of the culture and the presence of multiple lineages of cells that may or may not be phenotypically distinct. If a specific genotype or phenotype is selected physically (see Chapter 2), or by cloning, the culture becomes a *cell strain*; that is, the culture is homogeneous and expresses characteristic markers [1].

Cells that grow adhering to the substrate are known as *monolayer cultures*, and nonadherent cells are called *suspension cultures*. In general, most normal cells that give rise to cell lines will grow as monolayers. Suspension cultures are often derived from transformed cells but can be derived from normal hematopoietic cells. When monolayer cells cover all the available growth surface, they are said to have become *confluent*, and, if they grow beyond confluence, they are termed *postconfluent*. If postconfluent cultures are maintained under optimal conditions, they may form multilayered cultures (e.g., fibroblasts), or differentiate, or do both. Cultures that are deliberately maintained at high densities to create tissuelike cell densities are called *histotypic cultures*. When histotypic cultures of different lineages are combined, or cells of different lineages are combined and grown to high densities, these are known as *organotypic cultures*.

MAINTENANCE OF THE PRIMARY CULTURE

Observation is critical in the early stages of a primary culture. The presence of adherent, spreading cells indicates that the cells have survived; visual evidence of mitoses and increasing cell number indicates that a component of the culture is increasing; and a clear hyaline cell morphology confirms that the cells are healthy. The medium should be clear and, especially after the first medium change, should show no evidence of particulate material, which would imply microbial contamination. There will always be a certain amount of debris, carried over from the initial disaggregation or generated by apoptosis, but this is usually irregular in shape and should not increase markedly with time. However, visual observation is only the first indication that the culture is free of contamination and it is important to confirm this by appropriate tests (see Chapter 5). Critical morphological analysis will also determine whether the culture is likely to contain the required cells, in as much as most of the major lineages (epithelium, endothelium, fibroblasts, neural cells, etc.) have distinctive morphologies (Fig. 3.1), although full characterization is possible only when replicate cultures are available for immunological, biochemical, or molecular analyses (see Chapter 4).

Two major factors dictate replacement of the medium: (1) the presence of debris from disaggregation and (2) a fall in pH accompanying an increase in cell number. Care should be taken not to replace medium too early, in case some cells are slow to attach; it is best not to replace medium until the third day at the earliest. If an early fall in pH indicates that the cell survival rate is high, and consequently the cell density is high, and it has been confirmed that the culture is not contaminated, then the medium can be changed earlier. In these cases, however, it is preferable to retain the supernatant and seed a second vessel, to prevent the loss of any cells that are slow to attach but may yet be the ones required.

FEEDING PROTOCOL*

OUTLINE

Examine the culture by eye and on an inverted microscope. If indicated by a fall in pH, or if the time since the last medium change exceeds one week, remove the old medium and add fresh medium. Return the culture to the incubator.

MATERIALS

Sterilized

Medium designated for cell type
Pipettes

Not sterilized

70% alcohol
Swabs
Waste beaker or vacuum line

---

*This and subsequent protocols in this chapter are adapted by permission of Wiley-Liss, Inc., a subsidiary of John Wiley & Sons, Inc., from Freshney [2], copyright 2000 by Wiley-Liss, Inc.

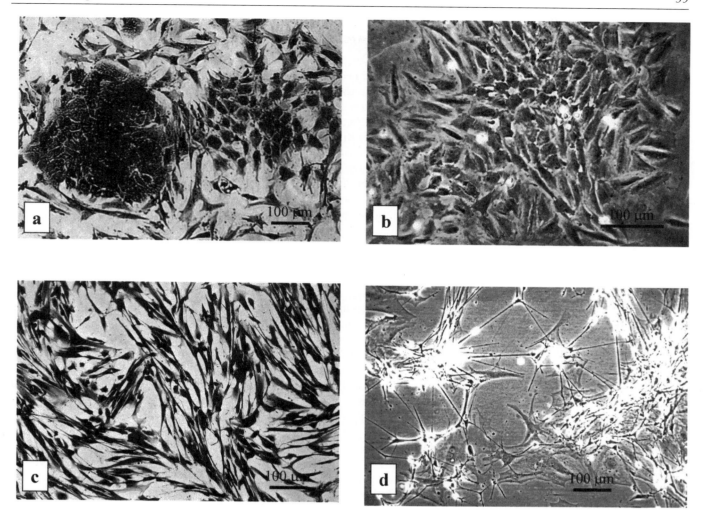

Fig. 3.1. (a) Primary chick embryo kidney. (b) Primary human umbilical vein (HUV) cells. From R. I. Freshney, "Freshney's Culture of Animal Cells: A Multimedia Guide." Copyright 2000 by Wiley-Liss, Inc. Reprinted by permission of Wiley-Liss, Inc., a subsidiary of John Wiley & Sons, Inc. (c) Normal fetal lung fibroblasts. From R. I. Freshney, "Culture of Animal Cells: A Manual of Basic Technique," 4th ed. Copyright 2000 by Wiley-Liss, Inc. Reprinted by permission of Wiley-Liss, Inc., a subsidiary of John Wiley & Sons, Inc.

Nonsterile equipment

   Class II biohazard hood
   Inverted microscope

PROTOCOL

1. Examine culture carefully for signs of contamination or deterioration.
2. Check pH and cell density, and decide whether to replace the medium.
   If feeding is required, proceed as follows.
3. Swab flask with alcohol (check that labeling is retained) and place in hood.
4. Open flask; remove and discard spent medium.
5. Add the same volume ($\sim$0.2–0.5 ml/cm^2) of fresh medium, prewarmed to 37°C.
6. Recap flask and return culture to incubator.

FREQUENCY OF FEEDING

The medium need be changed only if there is a reasonable survival and the cells proliferate. However, even if the cells grow slowly, it is best to feed about once per week, or, if growth is very slow, replace only half the medium once per week.

SELECTION OF MEDIUM

Selective inhibitors (e.g., *cis*-OH-proline [3] or specific growth factors [4]) can inhibit attachment. If they are used, it may be preferable to add them at the first feed. Serum-free medium is to be preferred because it not only allows for definition of the correct selective conditions but also avoids the risk of contamination with adventitious agents (see Chapter 5). The type of medium is usually determined by its selective properties for the cells required (see Chapter 2), but it should be remembered that some cell types may require support from stromal cells (fibroblasts or endothelium) for optimal survival, proliferation and, ultimately, differentiation. Hence it may be advisable to wait until subculture before applying the most stringent selective conditions and then apply different selective conditions to separate flasks to recover different cellular components. Matrix coating of the vessel (e.g., with collagen, laminin, or fibronectin) is often used to promote attachment (see Chapter 2). Reapplying these components, not usually necessary at feeding, must be done at subculture (see heading "Subculture," below).

GAS PHASE

Cultures may be grown with a gas phase of air, with elevated or reduced oxygen, and with CO_2. Most cultures survive satisfactorily with atmospheric concentrations of O_2. A requirement for elevated O_2 is unusual, while reduced O_2 may be of benefit [5]. CO_2 is usually maintained at 5%, but 2% may be preferable for human fibroblasts [6]. If all cultures are to be carried out in sealed vessels, atmospheric CO_2 is usually sufficient, but if unsealed plates are used, exogenous CO_2 must be provided; otherwise the dissolved CO_2, and ultimately the bicarbonate concentration, will be depleted. The bicarbonate concentration of the medium must be adjusted to suit the CO_2 concentration in the gas phase [7]. Elevated CO_2 can be provided from a piped or bottled CO_2 supply, or by placing the flasks in a CO_2 incubator with a slack cap or a gas-permeable cap.

SUBCULTURE

A successful primary culture is the first indication of survival and potential diversity of the cells in the culture. However, for further exploitation it is usually necessary to propagate the culture. Although primary cultures are a valuable indicator of the survival of specific cell types in culture, and can be used to optimize disaggregation, attachment, and initial cell growth, their value is limited by the heterogeneity within the culture and diversity among cultures. There are a number of specific requirements that must be met by cultures for tissue engineering, including confirmation of the homogeneity and lineage fidelity of the propagated cells, the expansion of the culture into sufficient cell mass for reimplantation, and the demonstration of the responsiveness of the cells to the endocrine, paracrine, and matrix signals that enable the expression of the normal differentiated phenotype. The primary culture must be subcultured to allow for expansion, purification, and characterization as well as preservation of the stock in liquid nitrogen.

SUBCULTURE PROTOCOL, MONOLAYER CELLS

OUTLINE

Check status of culture (cell density, freedom from overt signs of contamination, and a healthy appearance). Remove medium, expose cells briefly to trypsin, and remove trypsin. Incubate in residual trypsin, disperse cells in medium, count, dilute, and reseed.

MATERIALS

Sterilized

PBSA: Dulbecco's phosphate-buffered saline without Ca^{2+} and Mg^{2+}
Complete growth medium

Matrix coating medium (MCM): for example, fibronectin, 10 µg/ml; collagen, 30 µg/ml; BSA, 10 µg/ml in LHC basal medium (Biofluids, Inc.) [9]

Trypsin 0.25% in PBSA without Ca^{2+} and Mg^{2+} (PBSA), with or without 1 mM EDTA

Culture flasks, precoated with MCM if required: incubate with MCM, 50 µl/cm², for 2–48 h at 37°C, remove MCM. Use immediately or store at 4°C for 24–48 h.

Pipettes

Not sterilized

Waste beaker or vacuum line

Nonsterile equipment

Class II biohazard hood
Hemocytometer or electronic cell counter
Inverted microscope

PROTOCOL

1. Examine culture carefully for signs of contamination or deterioration and check that the cell density is correct for subculture (just preconfluent, or immediately after reaching confluence).
2. Take to biohazard hood and remove and discard medium.
3. Add PBSA prewash (0.2 ml/cm²) to the side of the flask opposite the cells, to avoid dislodging the cells, rinse the cells, and discard the rinse.
4. Add trypsin (0.1 ml/cm²) to the side of the flask opposite the cells. Then turn the flask over and lay it down, ensuring that the monolayer is completely covered.
5. Wait 15–30 s and withdraw all but a few drops of the trypsin, making sure that the monolayer has not detached. Using trypsin at 4°C helps to prevent this.
6. Incubate for 5–10 min, with the flask lying flat, until the cells round up; when the bottle is tilted, the monolayer should slide down the surface. Do not leave longer than necessary, but do not force the cells to detach before they are ready to do so, or clumping may result.
7. Add medium (0.1–0.2 ml/cm²) and disperse cells by repeated pipetting, taking care not to create a foam. A single-cell suspension is desirable at subculture to ensure an accurate cell count and uniform growth on reseeding.
8. Count cells by hemocytometer (see Chapter 4, Fig. 4.1) or electronic particle counter.
9. Dilute to the appropriate seeding concentration by adding the appropriate volume of cell suspension to a premeasured volume of medium in a culture flask.
10. If the cells are grown in elevated CO_2, gas the flask by blowing the correct gas mixture from a premixed cylinder, or gas blender, through a filtered line into the flask above the medium. Do not bubble gas through the medium. If the normal gas phase is air, as with Eagle's medium with Hanks's salts, this stage may be omitted.
11. Cap the flask(s) and return to the incubator.

SUBCULTURE PROTOCOL, SUSPENSION CELLS

Most normal cells will grow as attached monolayers, but there are some cells (e.g., hematopoietic cells) that may grow in suspension.

OUTLINE

Check cell concentration, viability, and freedom from contamination by eye. Withdraw a sample of cell suspension, count, and seed an aliquot into fresh medium in a new flask at the correct cell concentration to reinitiate a new growth cycle (see heading "Growth Cycle" later in this chapter).

MATERIALS

Sterilized or aseptically prepared

Starter culture
Growth medium [e.g., Eagle's MEM with Spinner salts (S-MEM) or RPMI-1640]
Stirrer flask with magnetic pendulum stirrer (Techne, Bellco)
Pipettes

Nonsterile equipment

Magnetic stirrer
Hemocytometer or cell counter
Class II biohazard hood
Incubator or warm room

PROTOCOL

1. Examine culture carefully for signs of contamination or deterioration. This is more difficult than with monolayer cells, but a poor condition of the cells is indicated by shrinkage and/or granularity. Healthy cells, often found in small clumps in static culture, should look clear and hyaline, with the nucleus visible on phase contrast.
2. Take to biohazard hood.
3. Mix cell suspension and disperse any clumps by pipetting; remove a sample and count.
4. Add appropriate volume of medium to a fresh flask.
5. Add sufficient cells to give a final concentration of 1×10^5 cells/ml for slow-growing cells (36–48 h doubling time) or 2×10^4/ml for rapidly growing cells (12–24 h doubling time).
6. Gas culture if CO_2-buffered; cap and take to incubator.
7. Regular culture flasks should be laid flat as for monolayer culture. Stirrer bottles should be placed on an induction-driven magnetic stirrer and stirred at 60–100 rpm. Deep cultures (>50 mm) should be sparged with CO_2 in air.

GROWTH CYCLE

When a culture is subcultured, the cells enter a *lag period* (Fig. 3.2) after seeding, when they attach and spread out (if monolayer), and reenter cell cycle. Following the lag phase,

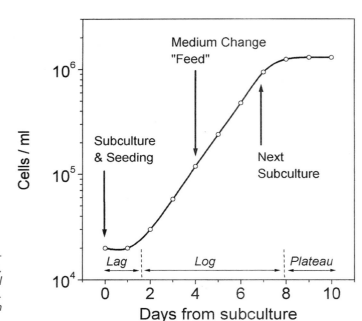

Fig. 3.2. Growth curve and maintenance. Modified with permission of Wiley-Liss, Inc., a subsidiary of John Wiley & Sons, Inc., from R. I. Freshney, "Culture of Animal Cells: A Manual of Basic Technique," 4th ed. Copyright 2000 by Wiley-Liss, Inc. Reprinted by permission of Wiley-Liss, Inc., a subsidiary of John Wiley & Sons, Inc.

the culture enters the *log (or exponential) phase* when the cells double every 18–36 h until they become confluent. Shortly after reaching confluence, the cells enter the *plateau phase*, when proliferation in normal cells ceases or is greatly reduced. Ideally, cultures should be subcultured at the end of the log phase before they enter plateau, since if subcultured from plateau, they will be harder to disaggregate and will spend longer in lag. If the cells grow rapidly or depress the pH, they may be fed half-way through the growth cycle. Cells growing in suspension go through a similar growth cycle on subculture, although the lag phase is often shorter because attachment and spreading are not required, and it is not usual to feed before the next subculture.

SERIAL PROPAGATION

Each subculture can be used to expand the culture. Cells should be frozen after the first and second subcultures (see Chapter 6) as protection against contamination, incubator failure, or other accidental loss, and to preserve stocks for future expansion, if required. A proliferating culture must be diluted at each subculture to reinitiate the growth cycle. The degree of dilution should be determined by counting the cells and diluting them to a concentration compatible with reinitiation of growth after a minimal lag. The seeding concentration should also be chosen to give a culture, ready for the next subculture, after a convenient period, the *subculture interval* (e.g., or 2 weeks).

Since most normal cells will divide in culture for only a fixed number of divisions, usually between 20 and 80, it is important to be aware of the number of divisions, or generations, that the culture has gone through. This is conveniently arranged if the dilution is carried out to a predetermined *split ratio*: that is, if the cells double three times during the subculture interval, then they should be split 1:8 at subculture or passage. This allows calculation of the number of elapsed generations (3 in this case) (Fig. 3.3), and the accumulated number should be written on the flask at each subculture. As this approaches 30 (or whatever number suggests that the cells are approaching senescence, see later), the stock should be replaced from low generation number frozen stock. Note that the passage number in this case is one-third of the generation number; it is only when the cells are split 1:2 at each subculture that passage number and generation number are equal. In general, it is preferable to use generation number because that is what determines life span.

It is important when using split ratios to ensure that the cell concentration remains within the limits proscribed by the growth cycle; that is, the seeding concentration should not fall below that which will give renewed growth with a reasonable lag period, and the cells should not move out of the log phase into plateau before the next subculture. The yield of cells at each subculture should be constant for a given seeding concentration and vessel

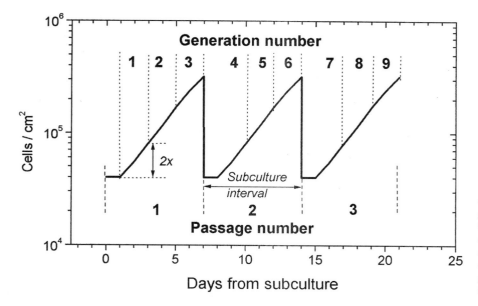

Fig. 3.3. Serial subculture. Modified with permission of Wiley-Liss, Inc., a subsidiary of John Wiley & Sons, Inc., from R. I. Freshney, "Culture of Animal Cells: A Manual of Basic Technique," 4th ed., Fig. 12.2. Copyright 2000 by Wiley-Liss, Inc. Reprinted by permission of Wiley-Liss, Inc., a subsidiary of John Wiley & Sons, Inc.

size. This is best determined by counting the cells at each subculture, but if the large number of cultures being handled precludes this, then representative photographs of each cell line, at each stage of growth, should be kept near the microscope for easy reference.

Although concentration is the parameter used for reseeding at subculture, the limiting parameter for the transition from log phase to plateau is cell density. When cells reach confluence, contact inhibition stops cell migration, and, within one or two further divisions, cell crowding causes a shape change, an alteration in the expression of cell–cell and cell–substrate adhesion molecules and resultant signaling, and the cells withdraw from cycle. Cessation of growth in plateau is, therefore, due to a combination of factors, but principally crowding and exhaustion of growth factors in the medium.

SENESCENCE

Most normal cell lines have a finite life span in culture, typified by normal diploid lung fibroblasts [11] in which the cells stop proliferating at around 50 generations. This is determined by shortening of the telomeres at each cell division [12,13] due to the low telomerase activity in most human somatic cells. Germs cells, and, presumably embryonal stem cells, have high telomerase activity and can therefore restore the telomere ends at each cell division. Tissue stem cells appear to have higher telomerase than committed precursor cells and differentiated cells and may survive longer in culture, although conclusive data on this is not yet available.

IMMORTALIZATION

A number of strategies have been developed to escape senescence in cell lines, including transfection with genes such as the large T gene of SV40 [14,15], Ela early region from adenovirus 5 [16], and E6/E7 from HPV16 [17]. Perhaps one of the most promising to emerge recently is transfection with hTRT telomerase, which appears to overcome senescence without any problems of malignant transformation [18].

TRANSFORMATION

Transformation is due to mutations and deletions in tumor suppressor genes, such as the cell cycle regulatory gene *Rb*, the DNA damage recognition gene *p53*, and the cell–cell adhesion/recognition gene *E-cadherin* [19]. Overexpression, mutation, and/or amplification of one or more oncogenes, such as *ras*, *myc*, or *erb-B2* is also implicated in initiation and progression of tumors [20]. It is characteristic of tumors that the high mutation rate causes the generation of many cells that are unable to survive or differentiate and enter apoptosis, and the cells of the developing tumor, which ultimately kills the patient, may be derived from a very small subclone of the original initiated clone. Since a large number of replicative divisions are involved, it is not surprising to find that at least some members of a tumor population are immortal. While several genes may be involved in the immortalization process, most tumors display elevated telomerase activity.

Many cell types, particularly from rodents that have a high somatic cell telomerase activity, are subject to spontaneous transformation *in vitro*. In some cases, as with the various 3T3 cell lines (NIH 3T3, Balb-c 3T3, Swiss 3T3), this results initially in immortalization but with many of the growth regulatory mechanisms, such as contact inhibition, density limitation of cell proliferation, and anchorage dependence, still functional. Cell lines that have acquired an infinite life span are known as *continuous cell lines*. Under the appropriate culture conditions, chiefly prolonged maintenance at high cell density, these cultures may undergo further transformation, resulting in a loss of growth regulation and acquisition of malignancy (i.e., the ability to form invasive tumors *in vivo*). Cultures derived from tumors of most species are also able to give rise to continuous cell lines; they may be already immortalized and malignantly transformed at the time of explanation, or their genetic instability may result in the acquisition of additional features of transformation.

It would appear that cultures from normal human and avian tissue are remarkably resistant to spontaneous transformation. There are no confirmed records of spontaneous transformation in normal human fibroblasts, and instances of spontaneous epithelial transformation are very rare, leading one to presume that the tissue of origin in these cases may

Table 3.1. Criteria for Transformation

Parameter	Assay	Reference
Anchorage independence	Clonogenicity in agar	[21,22]
Cytogenetic abnormalities	Fluorescence *in situ* hybridization	[23]
Plasminogen activator activity	Chromogenic assay	[24]
	Isoenzyme electrophoresis	[25]
	Solubilization of fibrin clot	[26]
Angiogenesis	Chick chorioallantoic membrane	[27]
	Endothelial cell mitogenesis and migration	[28]
Tumorigenicity	Implantation into nude mice	[29]

Source: From Table 17.1 in R. I. Freshney, "Culture of Animal Cells: A Manual of Basic Technique," 4th ed. Abridged and reprinted by permission of Wiley-Liss, Inc., a subsidiary of John Wiley & Sons, Inc. Copyright 2000 by Wiley-Liss, Inc.

already have had some premalignant changes. For example, HaCaT cells are deficient in both copies of *p53*, suggesting the original donor tissue may have been heterozygous. This implies that given the relatively short culture history of cell lines destined for implantation, *in vitro* transformation is unlikely unless the donor has a genetic or epidemiological predisposition to cancer. The risk of spontaneous *in vitro* transformation is minimal and sustainable for autografts, but scrutiny of the familial genetic history, and genetic screening of tissue or cell samples, may be advisable in the case of allografts.

If immortalization is used to generate a continuous cell line, the potential for malignant transformation is increased, and any attempt to use immortalized cells will require confirmation that they are not malignant when implanted *in vivo*. As a consequence of becoming immortal, cell numbers are less limiting, and material for confirming the absence of malignant transformation is more plentiful.

There are a number of *in vitro* assays that correlate with transformation (Table 3.1) [21–29], such as suspension cloning (anchorage independence), increased plasminogen activator activity, the ability of cell extracts to induce angiogenesis, and tumorigenicity as xenografts in immune-deficient (*Nu/Nu* or SCID) mice. The last two of these, which require in excess of 1×10^7 cells, are of minimal use when a surplus of cells is not available, but the first two may be carried out on relatively small numbers of cells. The most sensitive methods, applicable to small numbers of donor and cultured cells, are screenings for genetic aberrations by PCR or fluorescence *in situ* hybridization.

VALIDATION

There are three main parameters of validation: authentication of the cell line, demonstration of freedom from contamination, and confirmation that the cell line has not transformed.

AUTHENTICATION

A combination of specific assays and good record keeping allows confirmation of origin, provenance, and cell lineage. Donor tissue or cells, or DNA extracted from them, should be stored when any culture is initiated, to allow for comparative DNA profiling at a later stage. Accurate records must be kept regarding the donor and early stages of primary culture (Table 3.2) and later maintenance and subculture. These, together with casual observations and a record of special selective procedures, ultimately form the provenance of the cell line. Cryopreservation details, assays confirming identity (e.g., DNA profiling or histotyp-

Table 3.2. Template for Record of Primary Cultures

Date: . . . .   Time: . . . .   Operator: . . . .

Origin of tissue	Species
	Race or strain
	Age
	Sex
	ID no.
	Tissue
	Site
	Stored tissue/DNA location
	Special attributes
	Clinical contact
Disaggregation	Agent (e.g., trypsin, collagenase)
	Concentration
	Duration
	Diluent
Cell count	Concentration of suspension (C_1)
	Volume (V_1)
	Yield ($Y = C_1 \times V_1$)
	Yield per qram (wet weight) tissue
Seeding	Number of cells seeded
	Type of vessel
	Final concentration
	Volume per flask, dish, or well
Medium	Type
	Batch no.
	Serum type and concentration
	Serum batch no.
	Other additives
	CO_2 concentration

(continues)

Table 3.2. (continued)

Date:	**Time:**	**Operator:** . . . .
Matrix coating	E.g., fibronectin, Matrigel, collagen	
Other conditions		

Source: Modified from Table 13.1 in R. I. Freshney, "Culture of Animal Cells: A Manual of Basic Technique," 4th ed. With permission of Wiley-Liss, Inc., a subsidiary of John Wiley & Sons, Inc. Copyright 2000 by Wiley-Liss, Inc.

ing), lineage (see Chapter 4), and evidence of the absence of transformation, complete the authentication and rule out the possibility of cross-contamination.

CONTAMINATION

Overt contamination with bacteria, yeast, or filamentous fungi is obvious as long as the cultures are maintained in the absence of antibiotics. If the cells are maintained in antibiotics (which is inadvisable because it can lead to the retention of cryptic contaminants), parallel cultures should be kept antibiotic free to allow contaminations to become apparent. Some contaminants, such as mycoplasma and viruses, are not obvious by simple microscopic examination and require specific tests (see Chapter 5). Tests should be performed regularly, and the results should be documented, forming part of the validation process.

TRANSFORMATION

As already discussed, spontaneous transformation is rare and is usually evident from altered morphology and growth pattern (reduction in cell spreading, increased growth rate, piling up and continued growth after confluence, growth in suspension), but it can be confirmed if necessary by cytogenetic analysis [23].

CULTURE CONDITIONS
SUBSTRATES

There is a wide range of tissue culture plastic vessels available for routine culture (e.g., Becton Dickinson, Corning). These may be used as supplied or modified by coating with matrix as in the monolayer subculture protocol given earlier. In general, matrix coating (wet or dry) is used to enhance attachment and cell proliferation, while growth within a gelled matrix like native collagen or Matrigel (Becton Dickinson), gives reduced cell proliferation and enhanced morphogenesis [30]. It is also possible to subculture onto plastic or collagen microcarriers or various microencapsulation media (see Chapters 68–76; for methods and substrates for large-scale production see Chapter 6).

SELECTIVE MEDIA

The correct selective medium should be selected (see later) appropriate to the cell type required. At the first subculture, different media may be used for optimizing growth of different constituents of the population [e.g., MCBD 153 for keratinocytes, MCDB 110 for lung fibroblasts, or MCDB 131 for vascular endothelium [31] (available from BioWhittaker)]. Stromal components may be useful at a later stage for creating organotypic tissue-equivalent cultures (see Chapter 8).

TRANSITION TO CELL LINE

Subculture of the primary culture leads to the establishment of a cell line. This is accompanied by a number of changes in the culture. The *growth fraction*, that is, that proportion of the cell culture which is in cycle, usually increases from 10–50% in the primary

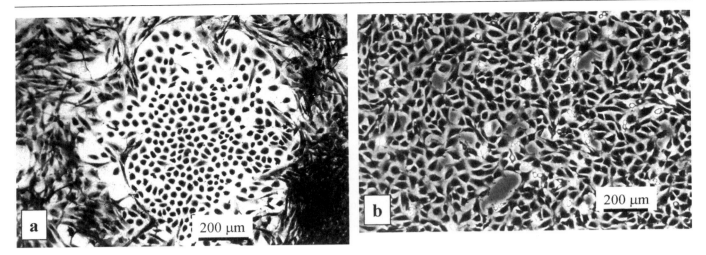

Fig. 3.4. (a) Newborn rat kidney and (b) early plateau phase cells. Reprinted with permission of Wiley-Liss, Inc., a subsidiary of John Wiley & Sons, Inc., from R. I. Freshney, "Culture of Animal Cells: A Manual of Basic Technique," 4th ed., Plates 4 and 7. Copyright 2000 by Wiley-Liss, Inc. Reprinted by permission of Wiley-Liss, Inc., a subsidiary of John Wiley & Sons, Inc.

to 80–90% in an early passage cell line. The heterogeneity of the primary, in terms of both proliferating versus nonproliferating and cell type, is gradually lost in favor of homogeneity, with a uniform proliferative cell type predominating (Fig. 3.4), determined by the selective medium that was used. While the primary may retain some cells that express the differentiated phenotype of the cells *in vivo*, the cell line is made up of predominantly undifferentiated cells, in line with their increased proliferative ability. This leads to the question of the origin of the cells that make up the bulk of the cells in a cell line.

Several stages of development are found in a typical epithelium such as skin. There are slowly cycling stem cells, which are derived originally from the embryonic endodermal or ectodermal germs layers. The stem cells give rise to proliferating precursor cells, which go through a number of generations determined by the required size of the differentiated cell pool. As cells leave the precursor compartment, they withdraw from cycle and become differentiated. It is possible that cell lines are derived from stem cells, proliferating precursor cells, or differentiated cells, if the differentiated cells retain the ability to reenter cycle, like hepatocytes, fibrocytes, or endothelium. If the stem cells retain the longer cell cycle time that they express *in vivo*, and most differentiated cells do not reenter cycle, or may do so with a significant delay, then the most likely phenotype for the cell line is that of the proliferating precursor cell. Relative to the tissue, the cells are dedifferentiated, but this description is probably inaccurate, since the process is more likely one of selection of undifferentiated cells by relative growth rate. Culture of stem cells will require different selection conditions, for example, isolation by surface markers and propagation in the presence of specific growth factors, like LIF [32] or MIP-1α [33], which inhibit differentiation (see Chapters 33–41). Terminally differentiated cells do not generally form cell lines, although if the cell line is made up of genuine precursors, the differentiated phenotype can be recreated.

DIFFERENTIATION

The implication of the foregoing is that cell lines, by and large, are undifferentiated, although they are probably committed to a particular lineage. Their precursor status favors propagation, and constant dilution together with the presence of mitogenic growth factors, such as EGF or PDGF, keeps them in the precursor compartment. Although it may not be necessary for cultures for reimplantation to be fully differentiated, assuming that the *in vivo* environment will stimulate differentiation, it is important to be able to demonstrate that the cells retain the capacity for differentiation. In addition, there are locations (e.g., skin) where to perform its proper role, the culture (i.e., the bulk of it) must be differentiated at the time of transplantation.

INDUCTION OF DIFFERENTIATION

Homotypic cell interaction, mediated by adhesion molecules such as E-cadherin, is required for differentiation in most epithelia, with the formation of characteristic complex intercellular junctions, such as desmosomes, adherens, and tight junctions. They also interconnect via gap junctions, which act to unify the intracellular environment. Hence culture at high cell densities will generally favor differentiation. Heterotypic cell interaction (e.g., between epidermal keratinocytes and dermal fibroblasts) is also required to generate an appropriate basal lamina that separates the two cellular compartments and to which both partners contribute. In addition, diffusible paracrine factors originate in one cell type and bind to receptors in the other. Matrix constituents, particularly the proteoglycans, contribute to this interaction by binding, stabilizing, and translocating growth factors to the correct high-affinity receptor [34].

One function of the basal lamina, in addition to its structural role, is to help in the establishment of epithelial polarity, without which the epithelial cell cannot become fully differentiated. This is simulated in culture by seeding cells either in or on matrix equivalent to that found *in vivo* [35] or by coculturing epithelial cells in organotypic culture with stromal cells in the presence of collagen IV and allowing them to contribute to the basal lamina [36]. While the basal lamina contributes to the geometry of organotypic cultures, the presence of collagen also contributes to a change in cell shape as it retracts [37], and this may be essential for some epithelial cells.

Differentiation can be induced by a number of soluble factors (Table 3.3) including Ca^{2+}, when it is increased from around 0.6 mM to 1.3 mM, cytokines and growth factors such as TGF-β [38], oncostatin M [39], KGF [40], HGF [41], and interferon β, and systemic hormones such as hydrocortisone. Planar polar compounds, such as DMSO [42], HMBA [43], NaBt (Fig. 3.5) and tributyrin [44] are particularly active with selected cell types. Originally described for their activity on Friend murine erythroleukemia cells, they have also been shown to have activity with myeloid cells [45], mammary epithelium [46], colonic epithelium [45], and embryonal stem cells [47].

Finally, the position of the culture, relative to the gas phase, or the constitution of the medium overlying the apical surface of a secretory cell, may be important. Bronchial epithe-

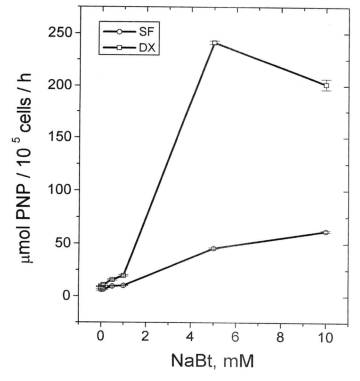

Fig. 3.5. Induction of differentiation by sodium butyrate. A549 cells, derived from a tumor of type II cell or Clara cell origin in the lung, were exposed to sodium butyrate (NaBt) for 3 d in the late log phase of growth and in the absence of serum. Alkaline phosphatase activity (a marker of differentiation in these cells) was determined colorimetrically by the release of p-nitrophenyl phosphate (PNP). The lower graph (open circles) was from cells exposed to varying concentrations of NaBt alone, while the upper graph (open squares) was from cells exposed to NaBt in the presence of 0.25 μm dexamethasone, an analogue of hydrocortisone.

Table 3.3. Differentiation Factors

Inducer[a]	Cell type	Reference
Steroid and related		
Hydrocortisone	Lung alveolar type II cells	[40,48]
	Mammary epithelium	[49]
Retinoids	Bronchial epithelium	[50]
	Endothelium	[51]
	Enterocytes (Caco-2)	[52]
	Embryonal carcinoma	[53]
	Neuroblastoma	[54]
Peptide hormones		
Melanotropin	Melanocytes	[55]
Thyrotropin	Thyroid	[56]
Prolactin, Insulin	Mammary epithelium	[50]
Cytokines		
Nerve growth factor	Neurons	[57]
Epimorphin	Kidney epithelium	[58]
Fibrocyte-pneumocyte factor	Type II pneumocytes	[59]
Interferon-α, β	A549 cells	[49]
Interferon-γ	Neuroblastoma	[60]
CNTF[a]	Type 2 astrocytes	[61]
IL-6, Oncostatin M	A549 cells	[40,49]
KGF	Keratinocytes	[62]
	Prostatic epithelium	[41]
HGF	Kidney (MDCK)	[63]
	Hepatocytes	[64]
TGF-β	Bronchial epithelium	[39]
Vitamins		
Vitamin E	Neuroblastoma	[65]
Vitamin D_3	Monocytes (U937)	[66]
	Osteoblasts	[67]
	Enterocytes (IEC-6)	[68]
Vitamin K	Hepatoma	[69]
	Kidney epithelium	[70]
Minerals		
Ca^{2+}	Keratinocytes	[71]
Planar–polar compounds		
DMSO	Mammary epithelium	[47]
	Hepatocytes	[72]
Sodium butyrate	Type II alveolar cells	(See Fig. 3.5)
	Colon cancer	[73]
HMBA	Erythroleukemia	[74]
Signal transduction modifiers		
Cyclic AMP	Oat cell cancer	[75]
PMA	Bronchial epithelium	[76]
	Mammary epithelium	[77]
	Colon (HT29, Caco-2)	[74,78]

[a]Abbreviations: HMBA, hexamethylene bisacetamide; PMA (TPA), phorbol meristate acetate.

Source: Modified from Tables 16.1 and 16.2 in R. I. Freshney, "Culture of Animal Cells: A Manual of Basic Technique," 4th ed. With permission of Wiley-Liss, Inc., a subsidiary of John Wiley & Sons, Inc. Copyright 2000 by Wiley-Liss, Inc.

lial cells become ciliated [79], and keratinocytes form terminally differentiated cross-linked keratin, only when grown at the gas–liquid interface [36]. Thyroid epithelium requires a liquid supernatant on the apical surface, but this should be a simple salt solution rather that complete medium [57].

ORGANOTYPIC CULTURE

Until the full complement of factors required for inducing differentiation in cells of all types can be defined, it is likely that the only way that complete differentiation can be achieved is by using the correct combination of cell types in organotypic culture (see Chapter 8). This provides the correct geometry, the optimal milieu of matrix and paracrine signaling factors, and the appropriate polarity. There are a number of models currently available, such as filter well inserts (Becton Dickinson, Corning Costar), collagen gels [30], and various synthetic gels (see Section II). Filter well inserts are probably the most reproducible and readily analyzed for *in vitro* studies, although as discussed earlier, other options may have significant advantages for *in vivo* implantation.

REFERENCES

1. Schaeffer, W. I. (1990). Terminology associated with cell, tissue and organ culture, molecular biology and molecular genetics. *In Vitro Cell. Dev. Biol.* **26**, 97–101.
2. Freshney, R. I. (2000). "Culture of Animal Cells: A Manual of Basic Technique," 4th ed., pp. 181–182. Wiley-Liss, New York.
3. Kao, W.-Y., and Prockop, D. I. (1977). Proline analogue removes fibroblasts from cultured mixed cell populations. *Nature (London)* **266**, 63–64.
4. Stanley, M. A., and Greenfield, I. M. (1992). Culture of human cervical epithelial cells. *In* "Culture of Epithelial Cells" (R. I. Freshney, ed.), pp. 135–158. Wiley-Liss, New York.
5. Balin, A. K., Goodman, B. P., Rasmussen, H., and Cristofalo, V. J. (1976). The effect of oxygen tension on the growth and metabolism of WI-38 cells. *J. Cell. Physiol.* **89**, 235–250.
6. Ham, R. G., and McKeehan, W. L. (1978). Development of improved media and culture conditions for clonal growth of normal diploid cells. *In Vitro* **14**, 11–22.
7. Freshney, R. I. (2000). "Culture of Animal Cells: A Manual of Basic Technique," 4th ed., p. 91. Wiley-Liss, New York.
8. Freshney, R. I. (2000). "Culture of Animal Cells: A Manual of Basic Technique," 4th ed., pp. 183–188. Wiley-Liss, New York.
9. Lechner, J. F., and LaVeck, M. A. (1985). A serum free method for culturing normal human bronchial epithelial cells at clonal density. *J. Tissue Cult. Methods* **9**, 43–48.
10. Freshney, R. I. (2000). "Culture of Animal Cells: A Manual of Basic Technique," 4th ed., pp. 189–190. Wiley-Liss, New York.
11. Hayflick, L., and Moorhead, P. S. (1961). The serial cultivation of human diploid cell strains. *Exp. Cell Res.* **25**, 585–621.
12. Holt, S. E., Shay, J. W., and Wright, W. E. (1996). Refining the telomere-telomerase hypothesis of ageing and cancer. *Nat. Biotechnol.* **14**, 836–839.
13. Smith, S., and de Lange, T. (1997). TRF1, a mammalian telomeric protein. *Trends Genet.* **113**, 21–26.
14. Mayne, L. V., Price, T. N. C., Moorwood, K., and Burke, J. F. (1996). Development of immortal human fibroblast cell lines. *In* "Culture of Immortalized Cells" (R. I. Freshney, and M. G. Freshney, eds.), pp. 77–93. Wiley-Liss, New York.
15. Steinberg, M. L. (1996). Immortalization of human epidermal keratinocytes by SV40. *In* "Culture of Immortalized Cells" (R. I. Freshney, and M. G. Freshney, eds.) pp. 95–120. Wiley-Liss, New York.
16. Demers, G. W., Halbert, C. L., and Galloway, D. A. (1994). Elevated wild-type p53 protein levels in human epithelial cell lines immortalised by the human papillomavirus type 16 E7 gene. *Virology* **98**, 169–174.
17. De Silva, R., Whitaker, N. J., Rogan, E. M., and Reddel, R. R. (1994). HPV-16 E6 and E7 genes, like SV40 early region genes, are insufficient for immortalization of human mesothelial and bronchial epithelial cells. *Exp. Cell Res.* **213**, 418–427.
18. Bodnar, A. G., Ouellette, M., Frolkis, M., Holt, S. E., Chiu, C.-P., Morin, G. B., Harley, C. B., Shay, J. W., Lichsteiner, S., and Wright, W. E. (1998). Extension of life-span by introduction of telomerase into normal human cells. *Science* **279**, 349–352.
19. Marshall, C. J. (1991). Tumor suppressor genes. *Cell (Cambridge, Mass.)* **64**, 313–326.
20. Weinberg, R. A., ed. (1989). "Oncogenes and the Molecular Origins of Cancer." Cold Spring Harbor Laboratory Press, Cold Spring Harbor, NY.
21. Macpherson, I., and Montagnier, L. (1964). Agar suspension culture for the selective assay of cells transformed by polyoma virus. *Virology* **23**, 291–294.
22. Freshney, R. I. (2000). "Culture of Animal Cells: A Manual of Basic Technique," 4th ed., pp. 200–202. Wiley-Liss, New York.
23. Ried, T., Liyanage, M., Du Manoir, S., Heselmeyer, K., Auer, G., Macville, M., and Schrock, E. (1997). Tumor cytogenetics revisited: Comparative genomic hybridization and spectral karyotyping. *J. Mol. Med.* **75**, 801–814.
24. Whur, P., Magudia, M., Boston, J., Lockwood, J., and Williams, D. C. (1980). Plasminogen activator in cultured Lewis lung carcinoma cells measured by chromogenic substrate assay. *Br. J. Cancer* **42**, 305–312.

25. Boxman, D. L. A., Quax, P. H. A., Lowick, C. W. G. M., Papapoulos, S. E., Verheijen, J., and Ponec, M. (1995). Differential regulation of plasminogen activation in normal keratinocytes and SCC-4 cells by fibroblasts. *J. Invest. Dermatol.* **104**, 374–378.

26. Unkless, I., Dano, K., Kellerman, G., and Reich, E. (1974). Fibrinolysis associated with oncogenic transformation. Partial purification and characterization of cell factor, a plasminogen activator. *J. Biol. Chem.* **249**, 4295–4305.

27. Danesi, R., Del Bianchi, S., Soldani, P., Campagni, A., La Rocca, R. V., Myers, C. E., Paparelli, A., and Del Tacca, M. (1993). Suramin inhibits bFGF-induced endothelial cell proliferation and angiogenesis in the chick chorioallantoic membrane. *Br. J. Cancer* **68**, 932–938.

28. Jain, R. K., Schlenger, K., Hockel, M., and Yuan, F. (1997). Quantitative angiogenesis assays: progress and problems. *Nat. Med.* **3**, 1203–1208.

29. Pretlow, T. G., Delmoro, C. M., Dilley, G. G., Spadafora, C. G., and Pretlow, T. P. (1991). Transplantation of human prostatic carcinoma into nude mice in Matrigel. *Cancer Res.* **51**, 3814–3817.

30. Berdichevsky, F., Gilbert, C., Shearer, M., and Taylor-Papadimitriou, J. (1992). Collagen-induced rapid morphogenesis of human mammary epithelial cells: The role of the alpha 2 beta 1 integrin. *J. Cell Sci.* **102**, 437–446.

31. Freshney, R. I. (2000). "Culture of Animal Cells: A Manual of Basic Technique," 4th ed., pp. 105–120. Wiley-Liss, New York.

32. Boeuf, H., Hauss, C., Graeve, F. D., Baran, N., and Kedinger, C. (1997). Leukemia inhibitory factor-dependent transcriptional activation in embryonic stem cells. *J. Cell Biol.* **138**, 1207–1217.

33. Graham, G. J., Zhou, L., Weatherbee, J. A., Tsang, M. L.-S., Napolitano, M., Leonard, W. J., and Pragnell, I. B. (1993). Characterization of a receptor for macrophage inflammatory protein 1a and related proteins on human and murine cells. *Cell Growth Differ.* **4**, 137–146.

34. Klagsbrun, M., and Baird, A. (1991). A dual receptor system is required for basic fibroblast growth factor activity. *Cell (Cambridge, Mass.)* **67**, 229–231.

35. Kinsella, J. L., Grant, D. S., Weeks, B. S., and Kleinman, H. K. (1992). Protein kinase C regulates endothelial cell tube formation on basement membrane matrix, Matrigel. *Exp. Cell Res.* **199**, 56–62.

36. Stark, H.-J., Baur, M., Breitkreutz, D., Mirancea, N., and Fusenig, N. E. (1999). Organotypic keratinocyte cocultures in defined medium with regular epidermal morphogenesis and differentiation. *J. Invest. Dermatol.* **112**, 681–691.

37. Sattler, G. A., Michalopoulos, G., Sattler, G. L., and Pitot, H. C. (1978). Ultrastructure of adult rat hepatocytes cultured on floating collagen membranes. *Cancer Res.* **38**, 1539–1549.

38. Masui, T., Wakefield, L. M., Lechner, J. F., LaVeck, M. A., Sporn, M. B., and Harris, C. C. (1986). Type beta transforming growth factor is the primary differentiation-inducing serum factor for normal human bronchial epithelial cells. *Proc. Natl. Acad. Sci. U.S.A.* **83**, 2438–2442.

39. McCormick, C., and Freshney, R. I. (2000). Activity of growth factors in the IL-6 group in the differentiation of human lung adenocarcinoma. *Br. J. Cancer* **82**, 881–890.

40. Thomson, A. A., Foster, B. A., and Cunha, G. R. (1997). Analysis of growth factor and receptor mRNA levels during development of the rat seminal vesicle and prostate. *Development (Cambridge, UK)* **124**, 2431–2439.

41. Ohmichi, H., Koshimizu, U., Matsumoto, K., and Nakamura, T. (1998). Hepatocyte growth factor (HGF) acts as a mesenchyme-derived morphogenic factor during fetal lung development. *Development (Cambridge, UK)* **125**, 1315–1324.

42. Friend, C., Scher, W., Holland, J. G., and Sato, T. (1971). Hemoglobin synthesis in murine virus-induced leukemic cells *in vitro*. 2. Stimulation of erythroid differentiation by dimethyl sulfoxide. *Proc. Natl. Acad. U.S.A.* **68**, 378–382.

43. Marks, P. A., Richon, V. M., Kiyokawa, H., and Rifkind, R. A. (1994). Inducing differentiation of transformed cells with hybrid polar compounds: A cell cycle-dependent process. *Proc. Natl. Acad. Sci. U.S.A.* **91**, 10251–10254.

44. Schröder, C., Eckhert, C., and Maurer, H. R. (1998). Tributyrin induces growth inhibitory and differentiating effects on HT-29 colon cancer cells in vitro. *Int J. Oncol.* **13**, 1335–1340.

45. Tarella, C., Ferrero, D., Gallo, E., Luyca Pagliardi, G., and Ruscetti, F. W. (1982). Induction of differentiation of HL-60 cells by dimethylsulphoxide: Evidence for a stochastic model not linked to the cell division cycle. *Cancer Res.* **42**, 445–449.

46. Rudland, P. S. (1992). Use of peanut lectin and rat mammary stem cell lines to identify a cellular differentiation pathway for the alveolar cell in the rat mammary gland. *J. Cell Physiol.* **153**, 157–168.

47. Wilton, S., and Skerjane, H. (1999). Factors in serum regulate muscle development in P19 cells. *In Vitro Cell. Dev. Biol.—Anim.* **35**, 175–177.

48. McCormick, C., Freshney, R. I., and Speirs, V. (1995). Activity of interferon alpha, interleukin 6 and insulin in the regulation of differentiation in A549 alveolar carcinoma cells. *Br. J. Cancer* **71**, 232–239.

49. Marte, B. M., Meyer, T., Stabel, S., Standke, G. J. R., Jaken, S., Fabbro, D., and Hynes, N. E. (1994). Protein kinase C and mammary cell differentiation: Involvement of protein kinase C alpha in the induction of beta-casein expression. *Cell Growth Differ.* **5**, 239–247.

50. Kaartinen, L., Nettesheim, P., Adler, K. B., and Randell, S. H. (1993). Rat tracheal epithelial cell differentiation *in vitro*. *In Vitro Cell. Dev. Biol.* **29A**, 481–492.

51. Hafny, B. E. L., Bourre, J.-M., and Roux, F. (1996). Synergistic stimulation of gamma-glutamyl transpeptidase and alkaline phosphatase activities by retinoic acid and astroglial factors in immortalised rat brain microvessel endothelial cells. *J. Cell Physiol.* **167**, 451–460.

52. McCormack, S. A., Viar, M. J., Tague, L., and Johnston, L. R. (1996). Altered distribution of the nuclear receptor rar (β) accompanies proliferation and differentiation changes caused by retinoic acid in Caco-2 cells. *In Vitro Cell. Dev. Biol.—Anim.* **32**, 53–61.

53. Mills, K. J., Volberg, T. M., Nervi, C., Grippo, J. F., Dawson, M. I., and Jetten, A. M. (1996). Regulation of retinoid-induced differentiation in embryonal carcinoma PCC4, aza1R cells: Effects of retinoid-receptor selective ligands. *Cell Growth Differ.* 7, 327–337.

54. Ghigo, D., Priotto, C., Migliorino, D., Geromin, D., Franchino, C., Todde, R., Costamagna, C., Pescarmona, G., and Bosia, A. (1998). Retinoic acid-induced differentiation in a human neuroblastoma cell line is associated with an increase in nitric oxide synthesis. *J. Cell. Physiol.* 174, 99–106.

55. Goding, C. R., and Fisher, D. E. (1997). Meeting review—Regulation of melanocyte differentiation and growth. *Cell Growth Differ.* 8, 935–940.

56. Chambard, M., Verner, B., Gabrion, J., Mauchamp, J., Bugeia, J. C., Pelassy, C., and Mercier, B. (1983). Polarization of thyroid cells in culture; evidence for the basolateral localization of the iodide "pump" and of the thyroid-stimulating hormone receptor-adenyl cyclase complex. *J. Cell Biol.* 96, 1172–1177.

57. Levi-Montalcini, R. C. P. (1979). The nerve-growth factor. *Sci. Am.* 240, 68.

58. Hirai, Y., Takebe, K., Takashina, M., Kobayashi, S., and Takeichi, M. (1992). Epimorphin: A mesenchymal protein essential for epithelial morphogenesis. *Cell (Cambridge, Mass.)* 69, 471–481.

59. Post, M., Floros, J., and Smith, B. T. (1984). Inhibition of lung maturation by monoclonal antibodies against fibroblast-pneumocyte factor. *Nature (London)* 308, 284–286.

60. Wuarin, L., Verity, M. A., and Sidell, N. (1991). Effects of interferon-gamma and its interaction with retinoic acid on human neuroblastoma differentiation. *Int. J. Cancer* 48, 136–141.

61. Raff, M. C. (1990). Glial cell diversification in the rat optic nerve. *Science* 243, 1450–1455.

62. Aaronson, S. A., Bottaro, D. P., Miki, T., Ron, D., Finch, P. W., Fleming, T. P., Ahn, J., Taylor, W. G., and Rubin, J. S. (1991). Keratinocyte growth factor. A fibroblast growth factor family member with unusual target cell specificity. *Ann. N.Y. Acad. Sci.* 638, 62–77.

63. Bhargava, M., Joseph, A., Knesel, J., Halaban, R., Li, Y., Pang, S., Golberg, I., Setter, E., Donovan, M. A., Zarnegar, R., Faletto, D., and Rosen, E. M. (1992). Scatter factor and hepatocyte growth factor activities, properties, and mechanism. *Cell Growth Differ.* 3, 11–20.

64. Montesano, R., Soriano, J. V., Pepper, M. S., and Orci, L. (1997). Induction of epithelial branching tubulogenesis in vitro. *J. Cell Physiol.* 173, 152–161.

65. Prasad, K. N., Edwards-Prasad, J., Ramanujam, S., and Sakamoto, A. (1980). Vitamin E increases the growth inhibitory and differentiating effects of tumour therapeutic agents on neuroblastoma and glioma cells in culture. *Proc. Soc. Exp. Biol. Med.* 164, 158–163.

66. Yen, A., Coles, M., and Varvayanis, S. (1993). 1,25-dihydroxy vitamin D3 and 12-O-tetradecanoyl phorbol-13-acetate synergistically induce monocytic cell differentiation: FOS and RB expression. *J. Cell Physiol.* 156, 198–203.

67. Vilamitjana-Amedee, J., Bareile, R., Rouais, F., Caplan, A. I., and Harmand, M. F. (1993). Human bone marrow stromal cells express an osteoblastic phenotype in culture. *In Vitro Cell. Dev. Biol.* 29A, 699–707.

68. Jeng, Y.-J., Watson, C. S., and Thomas, M. L. (1994). Identification of vitamin D-stimulated phosphatase in IEC-6 cells, a rat small intestine crypt cell line. *Exp. Cell Res.* 212, 338–343.

69. Bouzahzah, B., Nishikawa, Y., Simon, D., and Carr, B. I. (1995). Growth control and gene expression in a new hepatocellular carcinoma cell line, Hep40: Inhibitory actions of vitamin K. *J. Cell Physiol.* 165, 459–467.

70. Cancela, M. L., Hu, B., and Price, P. A. (1997). Effect of cell density and growth factors on matrix GLA protein expression by normal rat kidney cells. *J. Cell Physiol.* 171, 125–134.

71. Boyce, S. T., and Ham, R. G. (1983). Calcium-regulated differentiation of normal human epidermal keratinocytes in chemically defined clonal culture and serum-free serial culture. *J. Invest. Dermatol.* 81, 33–40s.

72. Mitaka, T., Norioka, K.-I., and Mochizuki, Y. (1993). Redifferentiation of proliferated rat hepatocytes cultures in L15 medium supplemented with EGF and DMSO. *In Vitro Cell. Dev. Biol.* 29A, 714–722.

73. Velcich, A., Palumbo, L., Jarry, A., Laboisse, C., Racevskis, J., and Augenlicht, L. (1995). Patterns of expression of lineage-specific markers during the in vitro-induced differentiation of HT29 colon carcinoma cells. *Cell Growth Differ.* 6, 749–757.

74. Osborne, H. B., Bakke, A. C., and Yu, J. (1982). Effect of dexamethasone on HMBA-induced Friend cell erythrodifferentiation. *Cancer Res.* 42, 513–518.

75. Tsuji, K., Hayata, Y., Sato, M., *et al.* (1976). Neuronal differentiation of OAT cell carcinoma *in vitro* by dibutyryl cyclic 3′5′-monophosphate. *Cancer Lett.* 1, 311–318.

76. Willey, J. C., Moser, C. E., Jr., Lechner, J. F., and Harris, C. C. (1984). Differential effects of 12-O-tetradecanoylphorbol-13-acetate on cultured normal and neoplastic human bronchial epithelial cells. *Cancer Res.* 44, 5124–5126.

77. Wada, T., Dacy, K. M., Guan, X.-P., and Ip, M. M. (1994). Phorbol 12-myristate 13 acetate stimulates proliferation and ductal morphogenesis and inhibits functional differentiation of normal rat mammary epithelial cells in primary culture. *J. Cell Physiol.* 158, 97–109.

78. Pignata, S., Maggini, L., Zarrilli, R., Rea, A., and Acquaviva, A. M. (1994). The enterocyte-like differentiation of the Caco-2 tumour cell line strongly correlates with responsiveness to cAMP and activation of kinase A pathway. *Cell Growth Differ.* 5, 967–973.

79. De Jong, P. M., Van Sterkenburg, A. J. A., Kempenaar, J. A., Dijkman, J. H., and Ponec, M. (1993). Serial culturing of human bronchial epithelial cells derived from biopsies. *In Vitro Cell. Dev. Biol.* 29A, 379–387.

CHAPTER 4

CELL QUANTITATION AND CHARACTERIZATION

Robert J. Hay

INTRODUCTION

This chapter summarizes and references typical procedures used in the quantitation, identification, and further characterization of cells isolated *in vitro*. Enumeration of cells either suspended in growth medium or adhering to the solid substrate is essential for accurate standardization and analyses within cell systems. The need for full definition not only of the cell lines in question but also of potential contaminants and the cellular environment, is emphasized here and elsewhere in this book. Quality assurance within every laboratory working with cell culture systems is still of major concern, as evidence of lapses is substantial. A wide range of reagents, kits, and services is available to permit key characterizations at multiple genetic and functional levels.

METHODS FOR DETERMINING CELL NUMBER

The choice of methodology and apparatus for quantitating the number of cells in a given preparation depends on the cell population itself, the precision and number of counts required, the need for a simultaneous viability estimate, and the budget. The cell count measures the status of the culture at a given time and is essential when subculturing or assessing the effects of experimental treatments on cells. The cell count can be expressed as number of cells per milliliter of medium or per square centimeter of area of attached surface (anchorage-dependent cells). Most common is the hemocytometer, which is versatile, efficient, accurate, and inexpensive.

HEMOCYTOMETER

The hemocytometer (counting chamber) is a modified glass slide engraved with two counting grids of known areas (Fig. 4.1). A grid contains nine larger squares, each being 1 mm^2 in area.

The hemocytometer is supplied with a glass coverslip of precise thickness. When the coverslip is correctly positioned, the volume of liquid over the grid is a known constant.

Viability counts for the hemocytometer method involve the dye exclusion test, which relies on the ability of living cells to exclude certain stains from crossing the cell membrane. Dead cells are permeable and will take up the stain. The test gives information about the integrity of the cell membrane, but it does not necessarily indicate how the cell is functioning.

1. Preparation

   a. Prepare a solution of the viability stain erythrocin B (C.I. 45430, acid red 51) by dissolving 0.4 g in 100 ml of phosphate-buffered saline (PBS). Alternatively, trypan blue (D.I. 23850, direct blue 14) may be used. Filter the solution to remove particulate debris and store at room temperature.

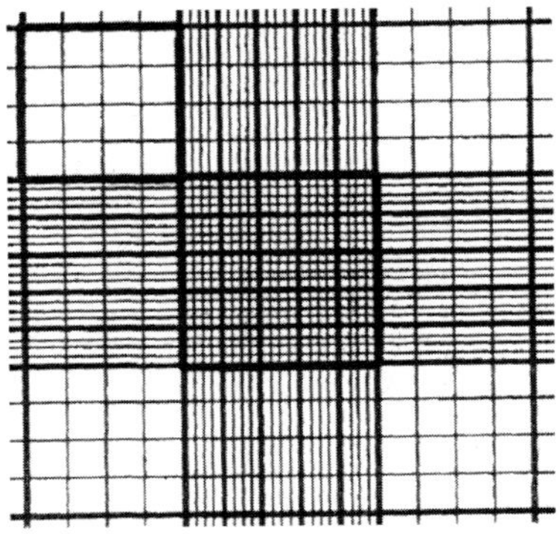

Fig. 4.1. Schematic drawing to illustrate a typical hemocytometer grid. Each of the nine larger squares 1 mm per side. Generally one counts cells in the four larger corner squares (with 16 small squares in each). Since the depth between coverslip and grid surface is 0.1 mm, the count gives cell number in 0.4 mm³.

> b. Clean the surface of the hemocytometer with 70% ethanol, rinse with single-distilled water, and dry thoroughly with lint-free tissue. Care should be taken not to scratch the semisilvered surface.
> c. Clean the coverslip with 70% ethanol, rinse with single-distilled water, and dry thoroughly with lint-free tissue. Wet edges very slightly and place over the grooves and semisilvered counting area. Press down gently and evenly.
> Note: The appearance of interference indicates that the coverslip is properly attached, thereby determining the depth of the counting chamber (0.1 mm).

2. Protocol

> a. Collect a sample from a suspension culture or monolayer culture (cells removed by trypsinization). A single-cell suspension is preferable.
> b. Resuspend sample thoroughly, dilute an aliquot 1:10 or greater with erythrocin B stain, mix well, and collect about 20 µl into the tip of a micropipette.
> c. Immediately transfer just enough of the cell suspension to the edge of the hemocytometer and let the suspension be drawn under the coverslip by capillary action until the fluid runs to the edges of the groove. Fill both chambers.
> Note: Do not overfill or underfill the chamber, as this may lead to changes in its volume due to surface tension.
> d. Transfer the hemocytometer to an upright, light microscope stage.
> e. Using a standard 10× objective, focus on the central area of the grid (see Fig. 4.1).
> f. Move the stage and position the hemocytometer such that the upper left corner square (1 mm²) is in focus. For the upper row, count cells left to right; middle rows, count cells right to left; and for the bottom row, count cells left to right. Count the cells lying within the 1-mm² area, using the subdivisions bound by three parallel lines.
> g. Count cells that lie on the top and left-hand lines of each square but not those on the bottom or right-hand lines. This avoids counting the same cell twice when counting adjacent squares.
> h. Count between 50 and 100 cells per larger (1-mm²) square. Count at least four squares, scoring total viable cells (those excluding erythrocin B) and total cells present.
> i. Count the second chamber by repeating steps 2c–2h.

3. Analysis

> a. Use the average of the two total cell counts from both chambers to calculate N and derive the cell concentration in the sample as follows:
> N_a = average number of cells counted per 1-mm² square (cells per chamber divided by number of squares counted)
> C = unknown concentration of cells = cells/ml
> 10^4 = conversion factor to convert the volume over 1-mm² square to 1 ml

D = dilution factor
$C = N_a \times D \times 10^4$

b. For viable cell number or percent viability, perform the following:
Number of viable cells/ml = $N_v \times D \times 10^4$
N_v = number of unstained cells
D = dilution factor
or

$$\text{Percent (\%) viability} = \frac{\text{number of cells unstained} \times 100}{\text{total number of cells}}$$

4. Precautions and Alternatives

a. For accurate precise quantitation experiments, a total of 500–1000 cells should be counted. The more cells are counted, the more accurate the cell count.
b. If a test sample is less than 100 cells/mm^2 or greater than 1000 cells/mm^2, redilute it. Use only the correct coverslip.
c. Ensure that the coverslip is in the right position.
d. Whenever possible, try to count a minimum of 200 total cells.
e. Resuspend cells just before sampling. This counting method is prone to sampling error due to settling of cells. Count cells soon after trypsinization, as there is a tendency for cells to clump. A single-cell suspension provides more accurate quantitation.
Note: When clumping is a problem, cells may be lysed and nuclei stained at 37°C for one hour in 0.1 M citric acid containing 0.1% crystal violet (C.I.42555; gentian violet). Stained nuclei are counted.
f. Living cells that have been recently trypsinized for either subculture of a primary culture from solid tissue or thawed from a freezing medium containing dimethyl sulfoxide (DMSO) may have leaky membranes, causing them to take up dye.
g. Trypan blue is the most commonly used stain. It has the disadvantage of giving higher background staining if the medium contains substantial serum. It is recommended that cells be suspended in serum-free medium before counting. Erythrocin B has the advantage of clarity, so one is made aware of microbial contamination or turbidity before dilution and use.

Electronic Counters

A number of different devices for rapid electronic counting and sizing of particles in suspension are available, but the basic instrument originally developed by Coulter Electronics in the 1950s is still most commonly used today. Configurations differ according to make, date of manufacture, capabilities of performance, and models, but current instruments consist of a sampling stand, a vacuum control unit, and a pulse-height analyzer. The sampling stand houses the system that detects number and sizes of cells to be counted. A vacuum pump and regulator plus tubing connect this to the vacuum control unit. A manometer within the sampling stand regulates the vacuum applied during an analysis. A beaker with the cells suspended in electrolyte is held on a platform, positioning it to receive the aperture tube, a basic sensing device of the instrument. One electrode is situated within the aperture tube, the other lies externally in the cell suspension. Current passing through the aperture is monitored by the analyzer. In brief, cells suspended in the electrolyte are drawn through the aperture, thus increasing electrical resistance to the flow of current passing through the opening. A pump provides vacuum and the manometer with metering device determines the volume (0.5 ml) of suspension counted. The analyzer with microprocessor defines conditions for particle analysis. Counts and readouts of data are displayed, indicating functions controlled.

The instrument provides a rapid and accurate analysis of the cell suspension virtually free of the operator errors due to fatigue experienced with manual methods. A comparatively high number of cells per sample are counted with each assay, increasing the sensitivity and precision. However, no discrimination is made for viable versus nonviable cells in the population.

1. Preparation

    a. Dissolve the following in one liter of distilled water to provide a suitable PBS electrolyte for counting of cells in suspension: NaC1 (8 g); KCl (0.4 g); $NaH_2PO_4 \cdot H_2O$ (0.2 g); Na_2HPO_4 (1.0 g). One can add formaldehyde (37% w/w) to give a 20% solution in PBS (1.6 parts formaldehyde to 6.4 parts PBS).
    Filter the solution before use using an 0.2-µm membrane filter.

    b. Prepare a uniform cell suspension by standard trypsinization of a flask monolayer culture (if adherent) or by gentle trituration if a suspension culture. Alternatively, cells can be recovered from a dissociated tissue by enzyme treatment.
    Note: It is critical to have a single-celled suspension to achieve an accurate count because clumps of two or more cells will be counted as one cell. Similarly, a suspension with many subcellular particles will yield falsely high counts, since fragments of size above the minimum threshold level will be counted as well.

    c. Follow the manufacturer's directions to calibrate the unit and to set the lower threshold setting, the aperture resistance, and the preset gain. Standard polystyrene microspheres and/or ragweed spores (20-µm diameters) are available from the manufacturers for calibration and repeat standardization purposes.

    d. To make ready for routine use, be sure that the electrolyte reservoir is full and the effluent reservoir is empty. In addition, the aperture must be free of debris and submerged in electrolyte. The external electrode should lie submerged in counting fluid in the sample breaker.

    e. Confirm critical settings, threshold(s), manometer selecting volume to be counted (usually 500 µl), polarity, attenuation, and gain.

    f. After a 5-min instrument warm-up, perform several background counts (should be below 150) in fresh electrolyte and record the average for future use.

2. Protocol

    a. Make an appropriate dilution of the cell suspension to be quantitated using the PBS electrolytes as diluent. The extent of dilution required will be learned by experience, but often a 1:50 will suffice for cells isolated from a T75 monolayer or suspension culture.

    b. It is generally simpler to make the dilution directly in an "accuvet" or 25-ml beaker; then perform the count directly after mixing. Avoid introducing bubbles, which would be counted, producing an erroneously high result.

    c. Place the beaker on the sample platform in the sampling stand and raise the platform to immerse the tip of aperture tube in the cell suspension.

    d. Turn the Reset/Count knob on the sampling stand in the clockwise direction to initiate the count sequence. Vacuum will then be applied to the aperture tube, drawing the cell suspension into the tube.

    e. Watch the aperture monitor on the sampling stand to be certain that the aperture does not become blocked with cell aggregates or debris during operation. This is one major potential source of error.

Analysis

The oscilloscope monitor display on the front panel of the main unit analyzer shows the pulses generated by cells as they pass through the aperture. The threshold(s) will be apparent, and the cell count will be indicated on the count display. Assuming that the unit is set to count cells in 500 µl of suspension and the dilution was 1:50, the count on the display should be multiplied by 100 to give the count per milliliter of the original unknown suspension. Depending on the model of the counter, one may have to subtract the average background before making the calculation. In some models, a coincidence correction may also be required to achieve the optimum accuracy. This factor corrects for situations in which many cells pass through the aperture simultaneously and the count displayed is lower than actual. Consult the instruction manual for particulars.

STAINED ADHERENT CELL CULTURES

Alternative quantitation methods are sometimes necessary—for example, when cell populations are sparse, when one is working with cells in microtiter wells, or when the initial suspension contains multiple cell types not readily recognizable by their morphology after dissociation.

One solution involves fixation and staining of the cell population with subsequent quantitation by eye, using a microscope fitted with a reticle in the eyepiece. If cells are too dense to count directly, photographs may be taken to simplify quantitation. The process is tedious and time-consuming, however, with substantial human error likely, especially over the longer term.

When microtiter plate cell cultures are preferred because of the sheer numbers required, colorimetric assays to quantitate viable cells present are most feasible. The anticancer drug discovery screen sponsored by the National Cancer Institute is an example. With this program, some 20,000 compounds have been tested annually at differing concentrations against a panel of 60 human tumor cell lines [1].

Clonogenic assays can be utilized to quantitate cell proliferation when the cell population is heterogeneous or when culture variables need assessment. This method is sensitive in that a low initial inoculation density per unit volume and surface area is utilized, and cell proliferation over an extended period (2–3 weeks) is usually required. It is not suitable or practical when a larger number of samples is to be analyzed.

MTT Assay

A rapid and convenient method relies on the mitochondrial dehydrogenases in living cells to reduce the yellow, water-soluble tetrazolium compound MTT [3-(4,5-dimethylthiazol-2-yl)-2,5-diphenyl tetrazolium bromide] to an insoluble, dark blue formazan product [2]. With a substantial variety of cell types and configurations, colorimetric data correlate well with numbers of viable cells present.

1. Preparation

   a. Establish test cultures in 96-well microtiter plates using 0.2 ml of the appropriate growth medium and up to 10^4 cells/well, depending on the expected proliferation rate of the population.
   b. Place in a humidified incubator for a period of time (often 1–3 days) sufficient for cells to reach the exponential growth phase. For nonadherent cells, one can omit this step and start with 0.1 ml of a suspension in logarithmic growth.
   c. Prepare dilutions in growth medium of the drugs or factors for testing, attempting to include a range of concentrations to produce effects from maximum to minimum.
   d. Prepare a solution of MTT by dissolving 50 mg/ml in 0.1 M glycine buffer at pH 10.5 Remove particulate matter and sterilize by membrane filtration.

2. Protocol

   a. For adherent cells, remove medium, add 0.2 ml drug/factor dilutions to appropriate well columns of the culture plate; add medium only to control wells. For nonadherent cell types, the drug dilutions (at twice the desired final concentrations) or medium control can be added directly.
   b. Place plates in a humidified incubator for a predetermined exposure period.
   c. Remove medium from all wells and replace with fresh medium without drug or factor. Repeat daily for a sufficient time to allow two to three cell divisions. Plates with nonadherent cells will require centrifugation and careful removal of supernatants for this part of the procedure.
   d. After the final feeding, add 50 µl of MTT solution to each well and place in the dark in a humidified incubator at 37°C. Incubation times of 3–8 h are common.
   e. Remove medium with MTT from all wells. Accomplish this with nonadherent cells after centrifugation.

f. Dissolve the retained formazan by addition of 0.2 ml of DMSO, add 25 μl of glycine buffer, and immediately use an ELISA plate reader to determine the absorbance at 570 nm.

Clone Forming Efficiencies

A second method for quantitating viable cells can be used either to evaluate variables (e.g., sera, defined medium components, substrates) or as an end point to quantitate isolation procedures, such as dissociation, fractionation, and subcultivation. The technique includes measurement of the ability of cells to adhere to the substrate and to respond to added essential ingredients, drugs, growth factors, or other cells by formation of colonies with characteristic morphologies. An example protocol using clone forming efficiencies to evaluate growth promoting properties of serum samples is described.

1. Preparation

   a. Obtain samples from several lots of fetal bovine (or other) serum for testing and have the supplier retain an appropriate volume (10–100 liters) of each for potential order at a later date.
   b. Prepare 0.5- to 1-liter test lots of medium for each serum sample.
   c. Use standard cell lines [e.g., ATCC CCL-75 (WI-38), ATCC CCL-64 (Mv 1 Lu), and ATCC CCL-34 (MDCK)] plus cells from a standard primary tissue dissociation (e.g., lung, see Ref. [3] for methodology). Prepare a series of cultures by inoculating T75 flasks separately with varying-numbers of cells.

2. Protocol

   a. Using as donors for test cells standard stock cultures 2–3 days after the last subculture, resuspend packed cells from the tissue source and from stock cultures separately in the various test media.
   b. Dilute an appropriate aliquot of each cell suspension with a solution of erythrocin B and use a hemacytometer to perform a viable cell count. Add sufficient medium to the initial cell suspensions to yield a convenient cell number per unit volume (e.g., either 10^6 or 10^7 cells/ml).
   c. Make log dilutions by transferring 1.0 ml of each cell suspension at lower dilution to 9.0 ml of test medium (diluent) in a 15-ml plastic tube. Half-log dilutions can be made by transferring 1.0 ml of cell suspension to 2.16 ml of the diluent medium.
   d. Inoculate 1.0 ml of each dilution, beginning with the highest dilution, to each of three T75 flasks containing 8 ml of the test medium and incubate at 37°C.
      Note: Plastic flasks lose CO_2 by diffusion during incubations lasting a week or more. Accordingly, if media with substantial amounts of bicarbonate are being tested, the flasks must be equilibrated and incubated in an air/CO_2 incubator. The inoculation densities can be varied depending on the cell lines and culture configurations. This is usually done by considering the culture surface areas involved.
   e. Renew the fluid on test cultures on the 4th or 5th day and thereafter every 3rd or 4th day. Equilibrate with air/CO_2 if appropriate.
   f. After 12–18 days total (depending on growth rate of cell type selected), remove the fluid and fix the culture with 10% formalin solution or other suitable fixative.
   g. Remove the fixative, rinse the flask interior gently with several changes of tap water, and stain with 1% aqueous toluidine blue or other simple stain for 1–5 min.
   h. Remove the staining solution, rinse out residual fluid with tap water, and examine the flasks under a dissecting microscope.

Analysis

The number of clones per flask and the size and type of colonies that form reflect the quality of serum and medium employed. By retaining representative flasks and records,

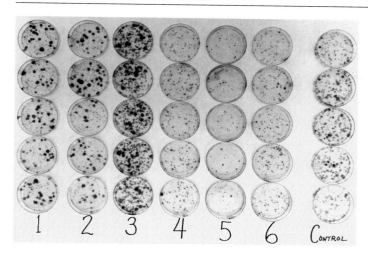

Fig. 4.2. Clone formation in a series of 60-mm petri plates, each with a differing serum being considered for purchase. It is apparent that the medium in group 3 promotes extensive clonal growth. That in the control is less effective but permits an adequate plating efficiency.

investigators can compare results with differing lots of serum obtained over an extended time interval. Thus the degree of uniformity in supply can be quantitated and compared on a year-by-year basis (Fig. 4.2).

1. Count the number of clones developed, keeping in mind that the very small colonies may have developed from secondary growth of single cells sloughed from primary colonies during feeding and handling.

2. Calculate the clone forming efficiency (CFE) as follows:

$$\%\text{CFE} = \frac{\text{clones formed per vessel} \times 100}{\text{number of cells inoculated}}$$

This index will vary extensively with the cell populations, substrates, and media used. Primary clonal cell cultures from a dissociated tissue can yield 1% efficiencies or less. Continuous cell lines often give 10% CFE at least, and some (e.g., ATCC CCL-2.2-HeLa S3) will yield close to 100% CFE under appropriate conditions.

3. The morphology of clones formed after seeding primary tissue cell suspensions can be used as a means to evaluate dissociation and culture conditions of specific component cell populations (see Fig. 4.3 and Hay [3]).

4. Automated devices that will score clones and determine the area of colonies generated are available. An example used in the author's laboratory is shown in Fig. 4.4.

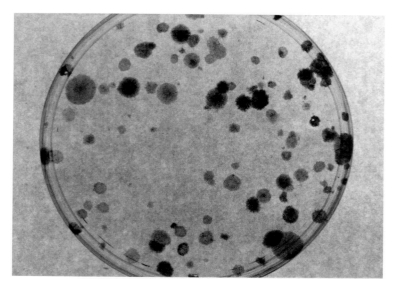

Fig. 4.3. Clonal growth from a suspension of lung cells isolated directly by tissue dissociation. Note the clones of differing morphologies throughout. The method can be utilized to select conditions that favor or inhibit specific cell types of interest.

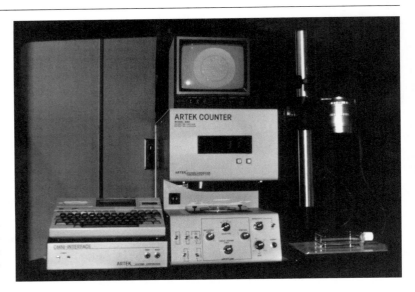

Fig. 4.4. Automated device for quantitating clones present on a plate or flask. Both numbers of clones formed and the average sizes can be determined. The video viewer on the right is used to analyze growth on T75 or larger vessels.

CELL LINE IDENTIFICATION

Thousands of cell lines have been derived from human and other animal tissues. Many are still being exchanged among investigators from different laboratories without sufficient characterization. Wherever cells are grown in culture, there is a risk of inadvertent addition and subsequent overgrowth by cells of another individual or species. Data have been collected over the years by groups offering identification services for cell culture laboratories in the United States and elsewhere. The problem is substantial and has been reviewed extensively [4–7]. Representative data are summarized in Table 4.1.

Techniques in wide use for cell line identification and detection of cell–cell cross-contamination include morphological examination, isoenzymology, cytogenetic analysis after Giemsa banding, immunotyping, and more recently, DNA profiling. These techniques are addressed in this section and the next (Proteins and Cell Line Characterization).

MORPHOLOGICAL EXAMINATION
General Characteristics

The morphological definition of a cell line is only one of its characteristics, and this alone is a notoriously flawed tool for precise identification. Most investigators provide phase contrast micrographs and histological or histoenzymological data on newly described lines; some include electron micrographs, especially where fine structural features are informative. The microscopic gross morphology is usually the first characteristic one records upon establishment of a cell population in culture. Several general texts on cell culture provide extensive photomicrographic features certainly from all major tissue types [8–10].

Protocol for Treating Cells for Evaluation by Electron Microscopy

The presence of desmosomes and keratin filaments is accepted as characteristic of epithelia. Weibel–Palade bodies are specific for endothelial cells of the umbilical vein and other

Table 4.1. Example Summary Data Showing Cell Line Cross-Contamination

Lines received for authentication	Number found to be misidentified	Percent	Ref.
466	75	16	[4]
275	96	35	[5]
252	45	18	[7]

sources. Each of these inclusions can be visualized by electron microscopy (EM). Cells of the islets of Langerhans can also be characterized morphologically by EM demonstration of their specific secretion granules.

A basic technique for preparing cultured cells for such ultrastructural analysis, using the reagents listed in Table 4.2, is outlined next. To prepare a cell monolayer or pellet of cells harvested either from a suspension culture or by dissociation, proceed as follows.

1. Collect a pellet by centrifugation of not more than $1–3 \times 10^6$ cells in a 15-ml polypropylene test tube. Monolayers can be processed directly on plastic culture vessels.

2. Remove culture medium and wash the cells once with stock buffer at 4°C. All steps through step 5 are performed at this temperature.

3. Fix for 1 h in 2% buffered glutaraldehyde. If a pellet is being processed, use an orange stick to dislodge it gently from the test tube sides, thus facilitating diffusion of reagents from all sides of the three-dimensional mass. It is important to retain a mass of cells throughout as opposed to a dispersed population. These larger aggregates will pass through the more viscous mixtures to be used in later steps.

4. Decant off the fixative and rinse three times with buffer for at least 10 min per rinse.

Table 4.2. Reagents Used in Processing Cultured Cells for Electron Microscopy

Reagents[a]	Amount
Stock solutions	
1. Sodium cacodylate buffer, 0.1 M, pH 7.2–7.3	
Na cacodylate	42.8 g
Aqueous phenol red (0.5%)	4 ml
Dissolve in 1 liter of DGDW and adjust pH with 1 M HCl	
2. Glutaraldehyde, 8% aqueous	
Dilute to 4% with DGDW	
3. Osmium tetroxide, 4% aqueous	
4. Uranyl acetate, 0.5% aqueous	
5. Reynolds lead citrate	
Lead nitrate	1.33 g
Trisodium citrate, dehydrate	1.76 g
DGDW	30 ml
Dissolve with stirring for about 30 min. Add 8.0 ml of 1 M NaOH	
(carbonate free) and dilute to 50 ml.	
The solution, which is stable for at least 6 months, should be passed through	
a membrane filter ≤0.22 μm) just before use to avoid small precipitates.	
6. Alcohol series to 100% absolute and propylene oxide	
7. Plastic mixtures	
A. Epon 812	62 ml
Dodecenyl succinic anhydride (DDSA)	100 ml
B. Epon 812	100 ml
Methyl nadic anhydride (MNA)	89 ml
These can be made up in advance and stored at 4°C in well-sealed	
containers for up to 6 months.	
Accelerator (2,4,6-tridimethyl-aminomethyl phenol) (DMP 30) is added	
just before use at 1.5-20%.	
Working solutions	
1. Buffered glutaldehyde (2%): stock buffer and 4% glutaraldehyde stock, 1:1	
2. Buffered osmium tetroxide (1%): osmium and stock buffer, 1:3 4%	
3. Plastic mixture	
Thoroughly mix 7 parts of mixture B with 3 parts of mixture A just before	
use and add the accelerator at 1.5-2.0%.	

[a]Can be obtained from Polysciences, Inc., Warrington, PA.

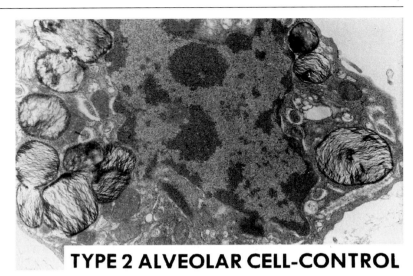

TYPE 2 ALVEOLAR CELL-CONTROL

Fig. 4.5. Electron micrograph showing a type 2 alveolar epithelial cell. The osmiophilic inclusions are characteristic, illustrating the presence of surfactant.

5. Post fix with 1% osmium tetroxide for 1 h.

6. Decant the osmium tetroxide fixative and wash three times with double-glass-distilled water (DGDW) (5 min each).

7. Leave for 16–20 h in 0.5% aqueous uranyl acetate. The low pH of this solution removes glycogen but gives excellent membrane preservation and staining.

8. Pass through an alcohol series (10-30-70-95-100-100%), leaving the sample in each solution for 10 min. For pellet fragments, three subsequent 10-min infiltrations with 1:1 100% alcohol/100% propylene oxide, 100% propylene oxide, and 1:1 propylene oxide/complete plastic mixture are recommended. However, these latter three steps *cannot* be used with monolayers on plastic. In that case, infiltrate directly from 100% alcohol for 30 min with a 1:1 solution of the complete plastic mixture and 100% alcohol.

9. Decant and begin infiltration with 100% complete plastic mixture (7:3 B/A plus 1.5–2.0% DMP-30, see Table 4.2). For pellet fragments this can best be accomplished by placing a fragment no larger than 1 mm in diameter in the upper part of the mixture in a BEEM embedding capsule (Polysciences). The fragment should settle to the bottom (pointed end) of the capsule within a few hours.

10. For a monolayer, the complete plastic mixture is placed in the culture vessel and allowed to infiltrate for 8–18 h. Bubbles can be removed by placing the BEEM capsules or open culture vessels in a vacuum.

11. Polymerize at 65–70°C for 24 h or longer to give a suitable hardness. The B/A mix is rather hard and can be modified depending on the technician's experience and preference.

12. For infiltrated monolayers, use a fine jeweler's saw to cut out the area(s) to be examined and cement sample(s) to a dummy plastic rod. An "instantly" drying form of contact cement (e.g., Superglue) is best. The sample can be oriented for cross-sectioning or to collect sections that are tangential to the monolayer surface (e.g., for desmosomes). For pellet fragments, simply snap off the BEEM capsule, trim as necessary with a single-edged razor blade, and place in the microtome chuck. Use an ultramicrotome for sectioning (see manufacturer's instructions). Silver or gray sections are suitable for examination at magnifications above 15,000. Gold sections can be used at lower magnifications. Collect the sections on a suitable grid, stain for 20–60 s in lead citrate, examine, and photograph as desired. Example photomicrographs are provided as Figs. 4.5 and 4.6.

IMMUNOLOGICAL METHODS AND ANTIGENIC MARKERS

Immunological techniques have been used to confirm the species of cell lines to identify specific, characteristic cell surface and internal markers, and to type the immune products from hybridoma cell lines. Example methodologies are outlined next.

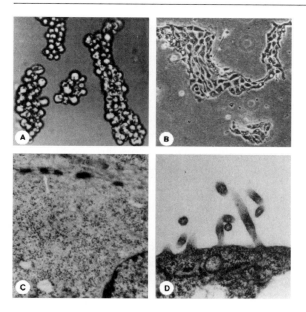

Fig. 4.6. Typical photomicrographs of normal colonic epithelia in vitro. (A) Cryptlike cellular complexes after dissociation and fractionation through Percoll. ×400. (B) Colonies of epithelial cells after 48 h in culture, phase contrast ×125. (C) Transmission electron micrograph (TEM) from section taken tangential to the colony surface showing typical desmosomes (arrow) with electron light and dense bands plus radiating tonofilaments. ×13,000. (D) TEM of a cross section of an epithelial cell showing microvilli with filamentous core. ×33,000.

Fluorescent Antibody Staining to Confirm Species

The indirect fluorescent antibody staining technique is one method for verifying the species of cell lines. The technique involves two general steps. First, species-specific antiserum, produced in rabbits, is used to label test cells plus positive and negative control cell populations. Subsequently, one applies anti-rabbit globulin, produced in goats and coupled to the fluorescent dye fluorescein isothiocyanate (FITC). The second reagent binds to the rabbit antibody that has attached to the target cells and, by virtue of the fluorescence, the antibody–antigen complexes can be visualized.

1. Preparation: To prepare antiserum against cultured cells

    a. Harvest cells of known species by scraping the culture surface with a rubber policeman.
    b. Wash by suspending in PBS with subsequent centrifugation at 200g for 10 min and repeat three times.
    c. Resuspend in PBS such that the viable cell count is about 5×10^5/ml (first week), 10^6/ml (second week), and 10^7/ml (third week).
    d. Inoculate 1 ml into each marginal ear vein of a healthy rabbit twice weekly for 3 weeks, doubling the dose each week.
    e. Administer three or more additional booster injections at 10^7 cells/ml (1 ml/ear) on a weekly basis.
    f. After the third booster injection, perform test bleedings, and collect and serially dilute antisera. Mix an equal volume of cell suspension (10^6/ml) and use the erythrocin B viable staining technique described earlier to evaluate cytotoxicity.
    g. If the titers are satisfactory ≥1:8 collect the blood by cardiac puncture, permit it to clot for 1–2 h at room temperature, and centrifuge at 200g for 15 min.
    h. Remove serum, inactivate complement by heating at 56°C for 30 min, dilute, and distribute in 0.2-ml aliquots for storage at –70°C.

Note: Stulberg [11] provides additional detail.

2. Protocol: To verify species on a test cell line

    a. Harvest by trypsinization if necessary and wash three times by suspending the cells in PBS at pH 7.5 with subsequent centrifugation to form a cell pellet.
    b. Resuspend the washed cells in PBS to give a density of $3–4 \times 10^6$/ml.

c. Mix 0.1 ml of cell suspension and 0.1 of diluted antiserum and place in a humidified incubation chamber at room temperature for 30 min. The appropriate dilution of antisera will have to be determined empirically for each antiserum preparation with positive control cells through an initial preliminary trial.

d. Use three complete changes of PBS to wash samples to remove unabsorbed antiserum and incubate for 30 min in the dark with 0.1 ml of FITC-conjugated, goat anti-rabbit antiserum (obtained commercially).

e. After a final three additional washes with PBS, seal a drop of the final cell suspension under a coverslip. Examine this by fluorescence microscopy at 500× using number 50 barrier and BG12 exciter filters on a Zeiss Universal microscope with an epi-illuminator.

Analysis

Positive reactions are seen as staining of brilliant fluorescence intensity at the cell periphery (Fig. 4.7). Controls consisting of cells of the suspected species, one related species, and a distant species are included with each test. Advantages of this technique over other methods include simplicity and the ability to allow identification of even minor cellular contaminants among populations. Even one contaminating cell among 10,000 can be identified under appropriate conditions and resolution.

Tests for Blood Group and Histocompatibility Antigens

The blood group and human leukocyte antigens (HLA) on the plasma membrane of human lymphoblastic and epithelial cells in culture sometimes provide useful markers for identification. Lack of expression or partial expression of these has been documented in a number of cases. The major histocompatibility system in humans consists of HLA antigens present on the plasma membrane of most nucleated cells. These numerous antigens, which are coded by codominant genes (77 alleles) of five closely linked loci on chromosome 6, provide one of the most polymorphic human group systems known. The antigens are detected routinely by a two-stage, complement-dependent cytotoxicity test, and dye exclusion is used to estimate loss of viability. Standard tests for these can be applied as follows.

1. Preparation for blood group antigen analyses

   a. Obtain a cell suspension by dissociation or harvest and wash twice in PBS by centrifugation with subsequent resuspension as usual.

   b. Perform a cell count and adjust the final cell concentration to $1-2 \times 10^6$/ml.

   c. Place a drop of the suspension on each of the four separate locations on a large microscope slide and add a drop of anti-human A, B, or AB typing antiserum

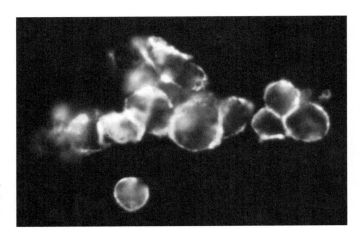

Fig. 4.7. Cells stained with fluorescein-labeled antibody to confirm species identity. The halo of fluorescence around each cell is a typical positive reaction. Absence of label would suggest presence of a cross-contaminant from another species.

to each. Add a drop of PBS to the fourth drop of cell suspension to provide a
negative control.

d. Immediately mix each pool separately with glass rods and observe under low
power for agglutination. The negative control may also show some cell clumping,
but this should be minor compared with the positive test suspension–antiserum
mix.

Analysis for Blood Group Antigens

Experience in the author's laboratory indicates that in many cases the donor's blood
group type is expressed even on lines from malignant tumors. Not infrequently, however,
cells of other human lines from A, B, or AB donors will not react, thus giving a false negative, type O reading. The hypothesis that the presence of false negatives could be due
to removal of antigen during dissociation was been tested by repeating the assay on cells
maintained in suspension culture for 2–18 h. Although this period should be long enough
to allow for replacement synthesis of the surface antigen, inappropriately negative lines
remained negative.

This problem with blood group antigen detection on cultured epithelia should be recognized. The simple test, coupled with others for intraspecies contamination among lines, is
valuable nevertheless in initial screening.

2. Preparation for HLA typing of cultured cells

a. Use a Hamilton dispenser to load wells of plastic histocompatibility typing plates
at 1 μl/well with the desired panel of antisera. Antisera, which are available commercially, are generally obtained from individuals immunized to HLA antigens
by pregnancy or blood transfusions. Monoclonal antibodies can be used alternatively. Typing plates loaded with a spectrum of HLA antisera can also be purchased for routine clinical work. Negative controls should be included with each
cell line and run.

b. Obtain a cell suspension, wash using medium RPMI-1640 without serum, and
adjust the cell concentration such that 1 μl with 10^5 cells can be added to each
well using a single-place Hamilton dispenser.

c. Mix by placing the typing plate against a Yankee pipette shaker and incubate at
20°C for 30 min.

d. Add 5 μl of rabbit complement to each well. Incubate at 20°C for 60 min.

e. Dispense 3 μl of 5% aqueous eosin to each well and wait 2 min for dead cells to
stain.

f. Fill wells with buffered formalin (pH 7.2) and add a cover glass (50×75 mm²)
to flatten the droplets.

g. Observe and use an inverted microscope to record the incidence of staining. The
degree of staining is usually by approximation rather than actual cell count.

Analysis for HLA Types

While analysis for HLA types can be applied successfully for typing some cultured cell
lines, it should be emphasized that modifications will be required in many individual cases.
The major variables are nonspecific antibodies present in the rabbit complement or HLA
antisera.

Rabbit serum is the most satisfactory source of complement for this reaction owing
to the presence of natural antibodies to human cells. The interaction of these with other
cell surface antigens enhances the complement-dependent cytotoxic effect of the anti-HLA
antibody–antigen union. The titers and specificities of these natural antibodies differ even
among pooled rabbit sera. The problem may be overcome by varying the incubation times,
by trying different sources and dilutions of complement, by diluting the rabbit serum with
human serum, or by absorption of the rabbit serum with cultured cells. Each cell line may
have to be examined separately, since the end result depends on the multiple interactions
between antibodies present with the spectrum of antigens on the surface of each cell type.

The presence of given HLA alloantigens on a particular cell line can be confirmed by absorption-inhibiting typing. In this case the ability of the cells to absorb HLA alloantibodies from antisera of known specificity is determined by quantitating the loss of cytotoxic effect after absorption. To accomplish this, the pre- and postabsorption antisera are titrated in parallel against a panel of lymphocytic lines of known HLA profile [12].

KARYOTYPING

Karyology has long been used informatively to monitor for both interspecies and intraspecies contamination among cell lines. Often when cell lines from different species are being examined, the chromosomal constitutions are so dramatically different that even cursory microscopic observations are adequate for verification. In other cases—for example, comparisons among cell lines from closely related primates or among human tumor lines—careful evaluation of banded preparations is required for positive identification.

Described in this section are laboratory procedures for cytogenetic methodologies used to verify cell line identity. There are four basic banding methods: G, Q, C, and R bands. The R band, which shows a mutually reversed pattern to the G band, is not described here. Since both bandings produce equally comprehensive banding information, we have chosen the G-band procedure as routine for chromosome studies.

1. Preparation of stock solutions

   a. *2X SCC*: Mix an equal volume of 0.6 M NaCl and 0.06 M sodium citrate. Sterilize and keep under refrigeration for storage.

   b. *Barium hydroxide*: Prepare a saturated solution by placing an excess amount of BaOH powder in distilled water. Use only the saturated, clear, upper portion of the solution.

   c. *Colcemid*: Add 1 mg of colcemid to 20 ml of double-distilled water. Filter, sterilize, and store frozen in 2-ml vials.

   d. *Fixative*: Just before use, combine 3 parts of chilled absolute methanol and 1 part of glacial acetic acid. Mix well and hold in ice.

   e. *Giemsa stain*: Prepare the stock Giemsa stain by mixing 1 g of Giemsa powder in 66 ml of glycerol. Add 66 ml of methyl alcohol (100%), and incubate overnight to dissolve. Store for 1 month under refrigeration before use.

   f. *Giemsa stain phosphate buffer (G-buffer)*: Dissolve 3.4 g of KH_2PO_4 in 1 liter of double-distilled water; adjust pH to 6.8 with 50% NaOH. Sterilize and keep under refrigeration.

   g. *Hypotonic KCl solution*: Add 0.56 g of KCl to 100 ml of double-distilled water. Sterilize and keep under refrigeration.

   h. *Hypotonic trypsin/versene solution*: To 1 liter of distilled water, add 1 g of dextrose, 0.2 g of disodium EDTA, 0.2 g of KCl, 0.2 g of KH_2PO_4, 2.18 g of $Na_2HPO_4 \cdot 7H_2O$, 2 g of NaCl, 3 ml of phenol red (0.3%), and 0.1 g of trypsin (Difco, 1:250). Adjust the pH to 7.4 with either 1 N HCl or 1 N NaOH, filter-sterilize (0.22 μm), store frozen in aliquots, and thaw before use.

   i. *Trypsin solution*: Dissolve 1 g of trypsin power (Difco 1:250) in 100 ml of sterile double-distilled water and filter-sterilize.

2. Preparation of working stain solutions

   Prepare the Giemsa–trypsin staining solution for G-banding as follows. To 6.5 ml of G-buffer, add 2.5 ml of 100% methanol and mix thoroughly. Then add 0.55 ml of trypsin solution, mix well, add 0.23 ml of stock Giemsa stain, and mix thoroughly again. Prepare this solution just before use. Do not use old or stored complete staining solution.

3. Preparation of chromosome spreads
   The standard method for producing chromosome spreads involves the swelling of arrested mitotic cells by brief exposure to hypotonic saline. These cells are then fixed, applied to slides to optimize spreading, and stained for microscopic observation.

A satisfactory step wise procedure is as follows:

a. To a culture (in a T75 flask) in the exponential growth phase, add colcemid to give a final concentration of 0.05–0.1 µg/ml.

b. Incubate 1–4 h, selecting the length of this period roughly by the cycling time of the cell population under study. Diploid human cells with relatively long doubling times would generally require longer incubation (2–4 h) than would a rapidly proliferating rodent cell line (45–90 min).

c. Gently decant the supernatant and treat the adherent layer with hypotonic trypsin–versene solution to dislodge cells as for a standard subcultivation. Place the cell suspension in a 15-ml centrifuge tube and add serum to make the suspension 10% (v/v) to neutralize further trypsin action on cells. If working with a suspension culture, collect the suspension, spin down to concentrate cells, and resuspend cells in the hypotonic KCl. Then follow steps d–k for anchorage-dependent cells.

d. Collect the suspended cells by centrifugation, discard the supernatant, and resuspend the pellet in the hypotonic KCl solution at room temperature.

e. After 10 min incubation, sediment the cells by centrifugation at 100g for 10 min. Decant most supernatant without disturbing the cell pellet, and leave a small amount of the hypotonic KCl (∼1–2 times volume of cell pellet) for resuspending the cells.

f. Slowly add 3–4 ml of freshly made fixative while agitating the tube manually by gentle pipetting.

g. After 15 min, repeat the centrifugation step, decant the supernatant totally, slowly add 4–5 ml of fresh fixative, mix by agitation, and let stand at room temperature for 15 min.

h. Repeat the centrifugation just described at 100g for 6 min, and remove the supernatant, but leave approximately 10–15 times the volume of cell pellet for resuspending the cells.

i. To prepare slides, add a few drops of the suspension onto a clean, wet, chilled (4°C) slide held at about a 20° angle; blow gently to spread the cell suspension over the slide evenly, and allow to dry completely in air at room temperature. Adjust the room humidity to 40–50%.

j. Use phase contrast optics to examine the preparation for general cell and chromosome quality. The densities can be adjusted by changing the concentration of cells in the suspension.

k. If a conventional karyotype is required, stain with 1% Giemsa stain for 20 min. Rinse in tap water and air-dry.

   Note: The frequency of introduction of artifacts through this method will vary depending on the cell line and the degree of the analyst's experience. Rupturing of cells will occur, for example, and apparent losses or gains in chromosomes will result. However, by counting the chromosomes from 50–100 well-spread metaphases, and analyzing the chromosomes in 15–20 nicely banded metaphases, the cytogeneticist can obtain consistent and reliable evaluations of specific lines.

4. Giemsa staining protocol

A powerful method for cell identification involves karyotype analysis after treatment with trypsin and the Giemsa stain (Giemsa or G-banding). The banding patterns made apparent by this technique are characteristic for each chromosome pair and permit an experienced cytogeneticist to recognize comparatively minor rearrangements found in inversions, deletions, or translocations. Many lines retain multiple marker chromosomes, readily recognizable by this method, which serve to identify the cells specifically and positively.

a. Incubate air-dried slides as described earlier in connection with the preparation of chromosome spreds within 2–7 days after preparation in a dry oven at 60°C for 3–5 h. Then incubate slides in G-buffer for 11 min at 60°C.

Fig. 4.8. *Giesma-banded karyotype of the primate bronchial cell line 4MBr5 (ATCC.CCL–208). A single marker chromosome is located in the majority (60%) of cells examined.*

     b. Drain briefly, flood the slide with about 0.95 ml of the trypsin–Giemsa staining solution, and leave under refrigeration (4–7°C) for 15 min.
     c. Rinse briefly with distilled water and air-dry completely.

5. Analyses
     Examine completely dried slides under bright-field, oil emersion Planapochromat objectives without coverslips. Oil can be placed directly on the slide. However, take care not to scratch the cell surface. Oil must be removed as completely as possible immediately after the use of slides. Generally a few changes in xylene should be satisfactory. A typical banded preparation is shown as Fig. 4.8.
     Construct the standard karyotype by cutting chromosomes from a photomicrograph and arranging them according to arm length, position of centromere, presence of secondary constrictions, and so forth. Consult the *Atlas of Mammalian Chromosomes* [13] for examples of conventionally stained preparations from over 550 species and Chen *et al.* [14] for examples of banded preparations.

DNA PROFILING

     DNA profiling enables one to confirm the authenticity of cell lines and to rule out the possibility of cell line cross-contamination. DNA fingerprinting became possible with the discovery by Jeffreys *et al.* [15] of variable number tandem repeat (VNTR) minisatellite regions of DNA. Traditional use of these loci involved the laborious procedure of Southern blotting. Two strategies have been used for working with VNTR loci. The first strategy employs a single multilocus VNTR probe that can simultaneously illuminate the genotype at many loci. Although highly informative, this strategy is problematic as a routine screening technique because the complex multiple banding patterns are so difficult to interpret [16]. Alternatively, one can use a cocktail of less informative single locus VNTR probes. When the single-locus probe genotypes are combined, a level of discrimination similar to multilocus VNTR fingerprints is possible with the simplicity of interpreting two band patterns. The adaptation of the polymerase chain reaction (PCR) has removed the need to use Southern blotting when one is examining these VNTR loci [17], which means that loci can be typed in hours rather than days.
     Smaller microsatellite loci have been identifed as well. Edwards *et al.* [18] exploited the usefulness of these short tandem repeat (STR) loci in differentiating humans at the DNA level. One significant advantage of STR loci over their minisatellite cousins is the small size of the former, a property that has two advantages. First, it allows multiplex PCR reactions to be developed in which many loci are simultaneously examined in a single reaction. Second, relatively degraded samples have been used for successful PCR amplification of small target loci. We employ a commercially available multiplexed STR system for routinely screening new cell line accessions for authenticity, as well as for validating any subsequent distribution of an authenticated cell line [19].
     For authentication, it is recommended that a traditional liquid extraction method be used to extract DNA from the cell line, as this tends to minimize STR artifacts, which may complicate allele assignment. For validation of subsequent passages of the cell line, more expedient DNA techniques may be used. Comparison with the authentic DNA fingerprint

easily resolves ambiguities. A filter-paper-based DNA extraction method used routinely in the laboratory is also described.

1. Prepare or purchase stock buffers and reagents:

   a. *1X PBS*: NaCl, 135 mM; KCl, 2.5 mM; Na_2HPO_4, 10 mM; KH_2PO_4, 2 mM; pH 7.4

   b. *Low TE buffer*: Tris 10 mM; EDTA 0.1 mM; pH 7.6

   c. Vendor reagents and supplies: UltraClean DNA Bloodspin Kit (MO BIO, cat. no. 12200-250); RNAse A solution (Gentra Systems, cat. no. D-50K6); FTA classic format card (Life Technologies, cat. no. 10786010), which includes FTA purification reagent and TE-4; UltraBARRIER pouch (Life Technologies, cat. no. 64030026); Harris micropunch (Life Technologies, cat. no. 10786069); GenePrint PowerPlex 1.2 System (Promega, cat. no. DC6101), which includes Gold STiR Buffer and PowerPlex 10X Primer Pair Mix; AmpliTaq Gold (Perkin-Elmer, cat. no. N808-0241); Mineral Oil (Perkin Elmer, cat. no. O186-2302); Formamide (AMRESCO, cat. no. 0606-100ML-APP); GS-500 ROX (PE Biosystems, cat. no. 401734); POP-4 (PE Biosystems, cat. no. 402838); 47 cm × 50 μm capillary (PE Biosystems, cat. no. 402839); 10× Genetic Analysis Buffer (PE Biosystems, cat. no. 402824); GS-500 ROX (PE Biosystems, cat. no. 401734)

2. Use of the MOB10 kit to extract a single ampule containing approximately 10^7 cells to prepare genomic DNA from cell lines.

   a. Use the commercially available UltraClean DNA Bloodspin Kit (MO BIO, Solana Beach, CA, cat. no. 12200-250).

   b. Transfer contents of ampule to a 1.5-ml microcentrifuge tube and pellet cells by centrifuging sample at room temperature in a Eppendorf microcentrifuge at 14,000 rpm (20000*g*) for 30 s.

   c. Resuspend pellet in 400 μl of 1× PBS at room temperature.

   d. Add 400 μl of Solution B1 (kit) and 4 μl of RNase A (20 μg/ml), vortex for 15 s, then incubate at room temperature for at least 5 min.

   e. Add 10 μl proteinase K to the sample, vortex for 15 s, then incubate at 65°C for at least 10 min.

   f. Add 400 μl of Solution B2 (kit), vortex for 15 s, and transfer equal portions of the mixture to two spin columns. Then centrifuge sample at room temperature in a microcentrifuge at 11,000 rpm (13,000*g*) for 1 min.

   g. Transfer spin column to a new collection tube, add 500 μl of Solution B3 (kit) to spin column, and centrifuge sample at room temperature in a microcentrifuge at 11,000 rpm (13,000*g*) for 1 min.

   h. Transfer spin column to a new collection tube, add 500 μl of Solution B4 (kit) to spin column, and centrifuge sample at room temperature in a microcentrifuge at 11,000 rpm (13,000*g*) for 1 min.

   i. Transfer spin column to a new collection tube. Then centrifuge sample at room temperature in a microcentrifuge at 14,000 rpm (20,000*g*) for 1 min.

   j. Transfer spin column to a new collection tube, add 200 μl of Solution B5 (kit), and incubate at room temperature for 5 min.

   k. Centrifuge sample at room temperature in a microcentrifuge at 11,000 rpm (13,000*g*) for 1 min.

   l. Quantitate DNA by UV spectrophotometry. Typical yields vary from 5 to 100 μg. Yield will depend on the cell line extracted. Yields tend to be consistent between extractions for any given cell line.

3. Use of FTA paper to prepare genomic DNA from cell lines.

   a. Prepare a suspension consisting of 2.0×10^6 cells/ml. Cells can be washed and resuspended in 1× PBS at the desired concentration. Alternatively, cells at the correct concentration, in ampules retrieved directly from liquid nitrogen storage tanks, have been successfully used with no washing; that is, the cell suspension consisted of growth medium and 5% DMSO.

b. Spot one drop from a 200-µl pipette tip of the cell suspension onto an FTA classic format card (Life Technologies, cat. no. 10786010) and allow card to air-dry thoroughly for at least an hour (this period can be extended to overnight if convenient). The volume of the drop will vary with the cell line and the viscosity of the fluid in the cell suspension. Dried cards should be stored long term at room temperature in a sealed pouch containing desiccant (UltraBARRIER pouch, Life Technologies, cat. no. 64030026).

c. Remove a 2-mm circle from the dried stain using a Harris micropunch (Life Technologies, cat. no. 10786069) and transfer to a 500-µl microcentrifuge tube.

d. Add 200 µl of FTA purification reagent to sample and incubate for 5 min at room temperature.

e. At the end of the incubation, decant or aspirate as much of the purification reagent as possible.

f. Repeat steps d and e two more times.

g. Add 200 µl of TE-4 to each sample and incubate for 5 min at room temperature.

h. At the end of the incubation, decant or aspirate as much of the TE-4 as possible.

i. Repeat steps g and h once more.

j. Allow washed 2-mm circle to air-dry for at least an hour (this period can be extended to overnight if convenient).

k. Sample is now ready for PCR amplification.

PROTOCOL

PCR Amplification and STR Typing of Cell Lines

a. Set up the PCR reaction in a 500-µl microcentrifuge tube. Input amount of DNA should be 1–2 ng or a 2-mm FTA paper punch.

b. Add the following reagents to the sample, with as much sterile, double-distilled water as necessary:

Reagent	Amount (µl)
Gold ST*i*R Buffer	2.5
PowerPlex 10× Primer Pair Mix	2.5
AmpliTaq Gold [5 U/µl]	0.45
Total volume	25.0

c. Mix samples thoroughly by vortexing, then collect samples by a brief microcentrifuge spin. Overlay samples with 25 µl of mineral oil.

d. Incubate samples in a Perkin-Elmer 480 thermal cycler using the following temperature profile:

Cycles	Temperature (°C)	Time (min)
1	95	11.0
1	96	2.0
10	94	1.0
	60	1.0
	70	1.5
22	90	1.0
	60	1.0
	70	1.5
1	60	30.0
1	4	As needed

When FTA paper punches are used as input DNA, the cycle number must be reduced from 22 cycles to 16 cycles to allow collection of the 400–base pair band in the GS-500 internal lane standard.

e. Prepare samples for gel loading by combining the following reagents:

Reagent	Amount (µl)
PCR sample	1.0
GS-500 ROX	1.0
Deionized formamide	24.0
Total volume	25.0

f. Heat denatured samples at 95°C for 2 min, then immediately plunge samples into wet ice and incubate for 10 min. Collect sample volume with a brief microcentrifuge spin.

g. Sample fragments are resolved on a PE Biosystems 310 capillary electrophoresis genetic analyzer through a 47 cm × 50 µm capillary filled with POP-4 in 1× Genetic Analysis Buffer using the following parameters:

Parameter	Value
Injection	15 kV
Injection time	2 s
Temperature	60°C
Run voltage	15 kV
Run time	24 min[a]

[a]The 24-min run time allows collection of the 400–base pair band in the GS-500 internal lane standard.

h. Use PE Biosystems GeneScan 672 v1.0 Data Collection Software to collect the data.
i. Use PE Biosystems GeneScan v3.1 Analysis Software to calculate sample fragment sizes.
j. Use PE Biosystems Genotyper v2.0 Software to make allele assignments.

ANALYSIS

Profiles of eight STR loci (D55818, 135317, 75820, 165539, vWA, TH01, TPOX, and CSF1P0), plus the amalogenin locus are utilized for human cell line identification and gender verification, respectively.

Typical STR profiling results are presented in the electropherograms shown in Figs. 4.9 and 4.10. Figure 4.9 represents the data generated for the analysis of two cell lines derived from the same patient. Figure 4.10 compares the STR profiles generated from two unrelated cell lines.

Although this STR system utilizes fluorescent labels and an automated collection device, it should be noted that various STR formats have been developed that use isotopic, chemiluminescent, or silver staining to detect and type these loci with similar accuracy. A facet of DNA fingerprinting data, that cannot be overemphasized is the need to convert PCR fragment sizes into alleles when developing genotypes for cell lines. This requires the development of an allelic ladder that contains all the commonly observed alleles for each STR locus under examination [20]. This ladder must be run on every gel, as it will serve to normalize the data and minimize errors that inevitably occur in comparisons of fragment size data for different gels. More importantly, this allows different laboratories, which may use different STR formats, to compare their results in an unambiguous manner. Ingredients for the standard ladder are available commercially from PE Biosystems.

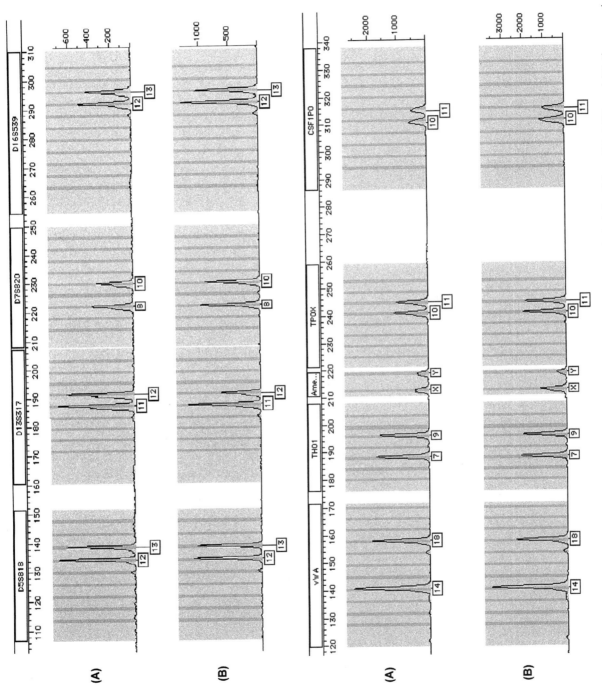

Fig. 4.9. Comparison of identical STR profiles for (A) the normal abdominal aorta endothelial cell line CRL-2473 and (B) the normal femoral artery endothelial cell line CRL-2474 derived from the same male patient. Upper tracings represent the four STR loci D5S818, D13S317, D7S820, and D16S539. The lower tracings represent the four STR loci vWA, TH01, TPOX, and CSF1PO, as well as the amelogenin locus used for gender identification. (Provided through courtesy of A. S. Durkin.)

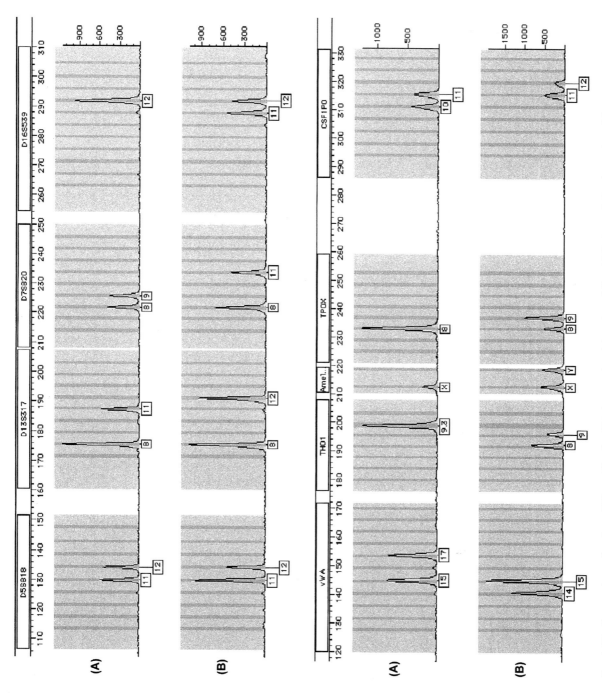

Fig. 4.10. Comparison of unique STR profiles for (A) the unrelated female cell line HTB-14 and (B) the male cell line TIB-161. Upper tracings represent the four STR loci D5S818, D13S317, D7S820, and D16S539. The lower tracings represent the four STR loci vWA, TH01, TPOX, and CSF1PO, as well as the amelogenin locus used for gender identification. (Provided through courtesy of A. S. Durkin.)

PROTEINS AND CELL LINE CHARACTERIZATION
STRUCTURAL
Isoenzymology for Cell Line–Species Verification

The severe problem of cell line cross contamination was documented earlier (Table 4.1). Isozyme analyses performed on homogenates of cell lines from over 25 species have demonstrated utility for species verification. By determining the mobilities of seven different isozyme systems, one can identify the species of origin of cell lines with a high degree of certainty. Required are assays for asparate aminotransferase, glucose 6-phosphate dehydrogenase, lactate dehydrogenase, malate dehydrogenase, mannose 6-phosphate dehydrogenase, nucleoside phosphorylase, and peptidase B. The list can be reduced in some circumstances where the range of cell line species is known to be restricted. The procedures are relatively straightforward, provide consistent results, and do not require expensive equipment.

A kit (AuthentiKit, Innovative Chemistry, Inc.) has been developed for this purpose. Precast 1% agarose gels on a polystyrene film, buffers, enzyme substrates with stabilizers, and appropriate cell line control extracts are available. Upon coupling to power supplies, over 15 different enzyme systems can be evaluated in specifically constructed electrophoretic chambers.

The advantages of the kit include the convenience of ready-made gels and reagents plus the short times required for electrophoretic separations (15–45 min). After drying, the gels can be retained to document cell line characteristics (Fig. 4.11).

Detailed directions for reconstitution of the reagents, enzyme extract preparation, electrophoretic separation, staining and interpretation are provided with the kit [21].

Cytoskeletal Proteins for Tissue Classification

The cytoplasmic intermediate filaments (IF), so called because their apparent diameter is between those of actin and myosin filaments, form an essential part of the cytoskeleton. These diverse, highly elongated, fibrous protein molecules have emerged as extremely useful in cell or tissue classification. Five or more filament subgroups have been described on the basis of polypeptide structure and composition, morphology, and unique immunogenicities. Tissue specificity is readily apparent, reflecting at least in part the degree of exposure to mechanical stress and related function.

Epithelia contain the most varied family of these IF subunits, with over 20 distinct types of IF referred to as the keratins or cytokeratins. They are expressed prominently as specific pairs in keratinized as well as nonkeratinized epithelial tissues.

Vimentin, present in mesodermally derived tissues such as endothelia, fibroblasts, and white blood cells, is the most widely distributed within animal tissues. Desmin, observed mainly in all muscle histotypes, is found throughout the cytoplasm in smooth muscle cells and links adjacent myofibrils together in heart and skeletal striated muscle tissue. Glial filaments are found in glia and in some Schwann cells of peripheral nerves with glial fibrillar acidic proteins.

Of the IF found in nerve cells, the most abundant are the neurofilaments that extend along the axons and provide the main cytoskeletal component. Three types are recognized, NF-L, NF-M, and NF-H, representing low, middle, and high molecular weight neural filament proteins, respectively.

Each of the IF subgroups may be visualized or by electron microscopy or by direct or indirect immunofluorescence staining, one also can demonstrate specific cytoskeletal proteins by immunohistochemical means as indicated next for keratin. Reagents and staining kits are available from a number of commercial sources. The example described here is adapted from a kit (ABC) and materials available from Vector Laboratories in Burlingame, California.

1. Preparation
   Prepare or obtain the items listed in Table 4.3. See Caputo *et al.* [22] for additional source information and detail.

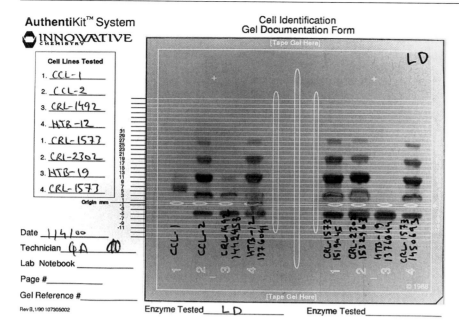

Fig. 4.11. Isoenzyme analysis for lactic dehydrogenase using the Authentikit system. The control and standards CCL1 and CCL2 in the first two left hand lanes are of mouse and human origin, respectively, showing the typical single anodal band for mouse 4 anodal plus 1 cathodal bands for human, respectively. The CRL1492 in the third lane represents a line of rat origin, and the remaining lines are from human sources. The single cathodal band for HTB-19, atypical for human lines generally, usually observed for this human breast cancer line.

2. Protocol

To assay for acidic and basic cytokeratins treat cells from slide chamber cultures, coverslip cultures or cytospin preparations as follows.

a. Rinse briefly in PBS to remove particulate matter and serum components.
b. Fix and permeabilize the cells by exposure for 5 min in methanol-acetone (1:1) at −20°C and allow the preparation to air-dry (2–3 min).
c. Rehydrate in TBS (5 min), and do not allow the specimen to dry out again during the entire staining procedure.
d. If quenching of endogenous peroxidase activity is required, incubate the sections for 30 min in 0.3% H_2O_2 in methanol. Incubation times may be shortened by using higher concentrations of H_2O_2. If endogenous peroxidase activity does not present a problem, this step may be omitted.

Table 4.3. Reagents for Immunoperoxidase Staining to Detect Cytokeratins

Fixative	Absolute methanol–acetone 1:1 Store and utilize at −20°C
PBS	Standard phosphate-buffered saline (Dulbecco's) without divalent cations, pH 7.2
Tris buffer (TBS)	6.075 g of Trizma base dissolved in 1000 ml of double-distilled (DD) water and adjust pH to 7.6
DAB (0.6%)	Diaminobenzidine tetrachloride (0.6 g) dissolved in 100 ml of DD water and stored in 1-ml aliquots at −20°C
BSA	Crystalline-grade bovine serum albumin, 1% in TBS
Blocking serum	Horse serum 2.5% in PBS
H_2O_2	Stock 30% diluted to 0.3% in methanol just before use
Vectastain kit	Kit PK 4002, available from Vector Laboratories, Inc., Burlingame, California
AE1/AE2	Murine monoclonal antikeratins recognize acidic and basic human keratins, respectively. Broad cross-species reactivity exhibited. Available from Boehringer-Mannheim. Use at a 1:250 dilution in PBS

e. Incubate sections for 20 min with diluted normal serum (2.5% horse serum in TBS containing 1% BSA) from the species in which the secondary antibody is made. (If nonspecific staining is not a problem, steps e and f may be omitted.)

f. Blot excess serum from sections.

g. Incubate sections for 30 min with primary antibody diluted in buffer or mouse monoclonal antibodies (AE1 and AE3 at 1:250 in this example). AE1 and AE3 recognize acidic and basic human keratins, respectively, and are broadly specific cross-reactive.

h. Wash slides in TBS for 10 min.

i. Apply a quantity of biotinylated antibody sufficient to cover the smear (included in ABC kit) diluted in TBS–1% BSA (1 drop of antibody to 10 ml of diluent), and incubate at room temperature in a humid environment for 30 min.

j. Wash slides in TBS for 10 min.

k. Apply a quantity of Vectastain ABC reagent sufficient to cover the smear (included in ABC kit) diluted in TBS–1% BSA (2 drops of reagent A to 10 ml of diluent) and then add 2 drops of reagent B; mix immediately and allow to stand for 5 min before use. Incubate in a humid environment for 60 min.

l. Wash slides in Tris for 10 min.

m. Apply a quantity of peroxidase substrate solution sufficient to cover the smear. Make just before use (add 1.0 ml of the 0.6% solution of DAB to 9.0 ml of PBS–1% BSA and then add 3.0 μl of 30% H_2O_2), and incubate slides for 5 min.

n. Wash slides for 5 min in tap water.

o. Counterstain with hematoxylin or other similar stain.

3. Analysis

Positive cells examined under high power light microscopy will exhibit a network of brown keratin filaments distributed throughout the cytoplasm. An example of positively staining human kidney epithelia is shown in Figure 4.12. This technique can be used to demonstrated numerous other cellular structural proteins. One needs only the appropriate murine monoclonal antibody to substitute in step 2.g with application of the other manipulations as presented.

IMMUNOLOGICAL PRODUCTS

Cellular engineering via somatic cell hybridization is now perhaps most extensively established in immunology with production of hybridoma cell lines. These require standard assays for microbial and viral infections plus at least initial verification that the monoclonal antibody of interest is being synthesized and secreted. Subsequently, production of immunoglobulin of the expected isotype provides an expedient compromise assay to indicate functionality.

Ouchterlony Procedure

The classical Ouchterlony technique for characterization of immunoglobulin isotypes is a useful and simple method for application with hybridoma culture supernatants. The precise specificity of the monoclonal antibody produced is clearly the ultimate criterion for identification of a particular hybridoma cell line, but confirmation of isotype is frequently much less time-consuming and less expensive. Satisfactory ELISA and isostrip methods are also available and are preferred for this test when a substantial number of assays are required. Reagents and kits, at least for most murine immunoglobulins, are plentiful. The Ouchterlony procedure is relatively straightforward, requiring inoculation of the test supernatants to wells bored in an agarose layer containing isotyping antisera. Precipitin lines indicate positive matches.

1. Preparation

a. Dissolve agarose to give 0.8% and sodium azide to make 0.025% in PBS by heating in a boiling water bath.

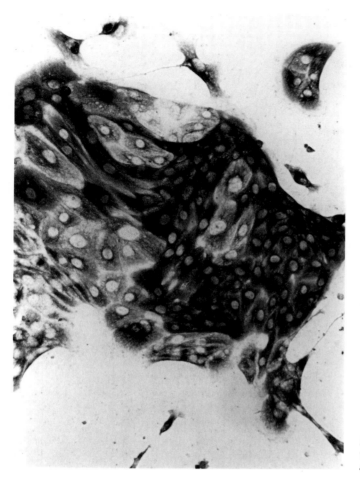

Fig. 4.12. Human fetal kidney cells showing positive staining with AE1/AE3 monoclonals (dark reaction product) for acidic and basic cytokeratins.

b. Add 0.4 ml of 0.5% aqueous trypan blue per 250 ml of solution and mix gently.

c. Pour 2-ml aliquots of the hot solution into 60×15 mm^2 petri plates to form a thin bottom layer. While this is congealing, retain the rest of the agarose solution at 56°C.

d. After the primary layer has congealed, add about 4 ml of the liquid and allow this to solidify as well.

e. Use a tubular well cutter to punch holes in concentric fashion and in the center of the agar layer, and aspirate the plug to form each well.

f. Store plates in the refrigerator until use.

2. Protocol

a. Remove a sample from test hybridoma cultures that are close to maximum in cell density. Collect the supernatants after centrifugation (200g for 10 min).

b. Place 35 μl of the supernatants in the outer wells. If the titer of antibody in the culture fluid is low, it may be necessary to repeat-load the well for that sample up to four times.

c. Place 35 μl of specific anti-mouse immunoglobulin typing antiserum (ICN) in the center well.

d. Maintain the plates at room temperature in a humid environment overnight.

Analysis

Precipitin lines indicating the presence of type-specific antibody–antigen (murine immunoglobulin) complexes should be evident the next morning between wells, with the per-

Table 4.4. Representative Lines[a,b] with Key Structures or Functional Characteristic(s)[c]

Tissue	ATCC No. CRL	Designation	Differentiated function or feature	Comments	Originator/ depositor(s)	Ref.
Adrenal cortex carcinoma	2128	NCI-H295R	Produces androgens	Responsive to angiotensin II	W. E. Rainey	[23]
Bone marrow, stroma, mouse	1972	M2-10B4	Laminin and collagen IV supports human and mouse myelopoiesis	Murine line supports human and mouse myelopoiesis	C. J. Eaves	[24]
Brain cortex	10442	HCN-1A	Tubulin, neurofilament protein, cholecystokinin-8, somatostatin	Derived from cortical tissue of patient undergoing hemispherectomy for intractable seizures, had unilateral megalencephaly	S. Snyder, G. V. Ronnett, et al., Patent No. 5,196,315, Johns Hopkins University	[25]
Breast	2336	HCC1937	BRCA1 mutation, EGP2, 5282C mutant p53+	From a primary ductal carcinoma, lymphoblast line from same donor available as CRL-2337	A. Gazdar	[26]
Colon	2134	LS513	CEA on surface, TGF-β	Multi-drug-resistant (MDR), tumorigenic, wild-type p53	L. Suardet	[27]
Cornea	11515	10.014 pRSV-T (HCE-T)	Positive for AE1/AE3 and 10% AE5	Transfected with SV_{40} plasmid; goes through at least 24 passages; also known as HCE-T	C. R. Kahn	[28]
Fibrosarcoma (murine)	2148	WEHI13VAR	Sensitive to TNF	Useful in bioassay (murine) for TNF	J. A. Armstrong	[29]
DBTRG-05MG	2020	Glioblastoma	Positive for S100/neuron-specific enolase	Negative for PDGF, NCAM, and GFAP	C. A. Kruze	[30]
Kidney	2190	HK-2	Positive for AP/GGP/LAP/cytokeratins	Retain a phenotype of well-differentiated proximal tubule cells	R. A. Zager	[31]
Kidney	2193	As 41	Renin and pro-renin	From mouse transgenic with the renin SV_{40}–T-antigen fusion gene	K. W. Gross	[32]
Liver	2254	AML12	Albumin, human and mouse TFG-α	From a transgenic mouse. Positive for human TGF-α, typical hepatocytes	N. Fausto	[33]
Lung macrophage (rat)	2192	NR8383	Produce TGF-β, IL-1, IL6, esterase+	Phagocytizes zymosan, Fc receptors; macrophage properties	R. J. Helmke	[34]
Lung metastasis	5803	NCI-H1299	Peptide neuromedin B	P53 partial deletion	A. Gazdar and J. D. Minna	[35]
Lung small cancer	5809	NCI-N417	Peptide neuromedin B, EGF receptor+	Neuron-specific enolase and others; from NCI-N231; oncogene profile defined	A. F. Gazdar and J. D. Minna	[35]
Lymphoma (Hodgkins)	2407	NK-92	IL-2 dependent, surface CDs positive	Natural killer cell line surface markers defined	Y. K. Tam and H. G. Klingemann	[36]
Lymphoid tumor (mouse)	2161	SVEC4-10EHR1	Vascular antigen VCAM LDL receptors	SV_{40} transformed, murine, tumorigenic, series of related lines lines	K. A. O'Connell	[37]
Muscle soleus (mouse)	2174	Sol 8	Form myotubes	Serum-deprived cells form myotubes; C_3H mouse	R. Bassel-Duby	[38]
Neuroblastoma (mouse)	2263	N1E-115	Receptors and enzymes for neurotransmitters	From C-1300 clonal isolation; high acetylcholine esterase and tyrosine hydroxylase	E. Richelson	[39]
Pancreas, cystic fibrosis	1918	CFPAC-1	CEA, adenocarcinoma-associated antigen, pancreatic oncofetal antigen, and CA 19-9 antigen	Epithelial morphology, polarization with apical villi, and tight and gap junctions; patient had cystic fibrosis, and CF gene is expressed	R. A. Schoumacher	[40]

(continues)

Table 4.4. *(continued)*

Tissue	ATCC No. CRL	Designation	Differentiated function or feature	Comments	Originator/depositor(s)	Ref.
Pancreas	2119	HPAC	Antigens DU-PAN-2, HMFG1, AUA1, CEA, CA125, CA19-9	Proliferation enhanced by insulin, ILGF-1, EGF, TGF-α-suppressed by glucocorticoids; keratin positive	W. R. Gower	[41]
Pancreas (murine endothelia)	2279	MS1	Endothelial markers, LDL, factor VIII, VEGF receptors	Murine lines, SV$_{40}$ transformed; give rise to benign hemangiomas	J. L. Arbiser	[42]
Pancreas (mouse)	2055	NIT-1	β cell; secretion granules, insulin	From a murine insulinoma; mouse transgenic for SV$_{40}$ large T	E. H. Leiter	[43]
Pancreas (mouse)	2350	Alpha TC1 clone 9	α cell line; glucagon produced	From an adenoma; mouse transgenic for SV$_{40}$ large T	E. H. Leiter	[44]
Pancreas (mouse)	2364	NIT-2	β cell; electron-lucent granules; proinsulin and insulin	From an adenoma; mouse transgenic for SV$_{40}$ large T	E. H. Leiter	[45]
Pituitary, adenoma (rat)	1903	RC-48/C	Luteinizing hormone, growth hormone, ACTH, thyrotropin-β, and prolactin	From a rat anterior pituitary adenoma; gonadotropin; releasing hormone receptors	E. H. Leiter	[46]
Prostate, normal	2221	PZ-HPV-7	Keratins 5 and 8 and HPV genes	HPV18 immortalized human prostate epithelia	D. Peehl	[47]
Prostate endothelium (rat)	2222	YPEN-1	Endothein, MRC, OX-43, acetylated LDL	Adeno-12 immortalized line; form endothelial tubes in Matrigel	K. Yamakazi and K. Pienta	[48]
Retina	2302	ARPE-19	Pigment, CRALBP, and RPE65	Human epithelial line, "spontaneously transformed"; retinal pigment markers and polarity	L. M. Hjelmeland	[49]
Skin keratinocytes	2309	CCD1106 Kertr	Cytokeratins AE1/AE3	Immortalized by exposure to E6/E7 HPV-16; no MHC I or II antigens detected in nonstimulated cells; grow in serum-free medium	L. Vilner	[50]
Thyroid carcinoma	1803	TT	Calcitonin, CEA	Immunoreactive calcitonin present at 7700 pg/10^6 cells in 72 h	S. S. Leong	[51]

[a] The lines are from human source tissue unless otherwise indicated.

[b] Lines included here were acquired by ATCC within the past 10 years.

[c] Visit http://www.atcc.org for additional detail.

tinent hybridoma supernatant and the known anti-immunoglobulin originally present in the center well.

Radial Immunodiffusion for Quantitation

It is often helpful for both hybridoma identification and utility to define immunoglobulin yields quantitatively as well. Once the isotype is known, the amount of immunoglobulin produced by a hybridoma can be determined by means of radial immunodiffusion. The appropriate typing antiserum is mixed with buffered agar. A known concentrate of immunoglobulin from the hybridoma supernatant is then added to an antigen well in the agar layer. As the antigen diffuses into the agar, a ring of antigen–antibody precipitate forms around the well. The diameter of this ring is measured and compared with that of standards and a standard curve to permit quantitation of the unknown placed in the antigen well.

1. Preparation

   a. Dissolve 0.5 g of K_2HPO_4 and 0.6 g of NaCl in 100 ml of double-distilled DD water and adjust to pH 8.0.
   b. Add 1.5 g of Noble agar to 50 ml of buffer and dissolve by heating in a boiling water bath. Store this agar base in the refrigerator in 5- to 10-ml aliquots.
   c. Dilute the typing rabbit anti-mouse antisera 1:10 in the pH 8 buffer and equilibrate in a 56°C water bath. Liquefy agar base in a boiling water bath and equilibrate at 56°C. Mix equal volumes of the two solutions, dispense immediately onto 2×3 cm^2 glass slides, and allow to harden.

2. Protocol
   After preparation of the agar plates proceed to quantitate immunoglobulin as follows.

   a. Collect 25 ml of the cell-free supernatant from the test hybridoma culture, which is at or close to maximum cell density, and cool to 4°C.
   b. Precipitate proteins by addition of 25 ml of a saturated solution of ammonium sulfate at 4°C, and centrifuge at 10,000g for 30 min.
   c. Discard the supernatant and dissolve the precipitate in 1–3 ml of distilled water, noting the final volume and degree of concentration (typically 8- to 16-fold). Store at 60°C until ready for assay.
   d. Use a tubular well cutter to make 12 wells, 3 mm in diameter and 12 mm apart, in the agar on the diffusion plate(s). Aspirate out the plug.
   e. Place 10 µl of antibody concentrate in each test well.
   f. For the standard readings, place 10 µl of serial dilutions of the appropriate, purified mouse immunoglobulin into wells to include concentrations of 1, 0.5, 0.25, and 0.125 mg/ml.
   g. Place the plates in a sealed, humid chamber at room temperature and leave for 16–18 h.

Analysis

Measure the diameter of the precipitation rings, and calculate the immunoglobulin present by referring to radial immunodiffusion diameters for the standards. Many hybridomas produce 25–75 µg of immunoglobulin per milliliter of growth medium.

TISSUE-SPECIFIC ANTIGENS AND FUNCTIONS

A host of different organ-, tissue-, and tumor-specific antigens with related structural and biochemical features are in use for cell line classification and identification. Immunological tests such as those already described for cell surface antigens (HLA) and the keratins, plus hundreds of different monoclonal antibodies, are employed both for the more general classification of histotypes (epithelia, neurons, connective tissue, etc.) and for the specific functional cell lines currently available. A broad range of enzyme assays provides quantification for the metabolites involved to contribute to identification, as do structural features observed at light microscopic and fine structural levels of examination. Representative examples including lines acquired over the past 10 years are summarized in Table 4.4 [23–25] to provide perspective. Note that lines secreting hormones (e.g., androgens, insulin, glucagon, ACTH, etc.), lines producing or responding to growth factors (IL-1, -2, -6; TGF-α and -β; etc.), lines with tissue- or tumor-specific antigens (AE1, AE3, CEA, S100, etc.) cell lines with characteristic structures (muscle, pancreatic α and β granules), and lines with other tissue-specific functions (retinal pigment production, neurotransmitters, endothelial formations, marker enzymes and other proteins, etc.) are all represented.

REFERENCES

1. Boyd, M. R. (1997). The NCI in vitro anticancer drug discovery screen. *In* "Anticancer Drug Development Guide: Preclinical Screening, Clinical Trials, and Approval" (B. Teicher, ed.), pp. 23–42. Humana Press, Totowa, NJ.

2. Mosmann, T. (1983). Rapid colorimetric assay for cellular growth and survival: Application to proliferation and cytotoxicity assays. *J. Immunol. Methods* **65**, 55–63.

3. Hay, R. J. (1979). Identification, separation and culture of mammalian tissue cells. *Biochemistry* **8**, 143–160.

4. Nelson-Rees, W., Daniels, W. W., and Flandermeyer, R. R. (1981). Cross-contamination of cells in culture. *Science* **121**, 446–452.

5. Hukku, B., Halton, D. M., Mally, M., and Peterson, W. D., Jr. (1984). Cell characterization by use of multiple genetic markers. *In* "Eukaryotic Cell Cultures" (R. T. Action, and J. D. Lynn, eds.), pp. 13–31. Plenum, New York.

6. Hay, R. J. (1998). Current methods for authentication of cell lines. *In* "Cell Biology: A Laboratory Handbook" (J. E. Celis, ed.), Vol. 1, pp. 35–42. Academic Press, San Diego, CA.

7. MacLeod, R. A., Dirks, W. G., Matsuo, Y., Kaufmann, M., Milch, H., and Drexler, H. G. (1999). Widespread intraspecies cross-contamination of human tumor cell lines arising at source. *Int. J. Cancer* **83**(4), 555–563.

8. Fogh, J., ed. (1975). "Human Tumor Cells in Vitro." Plenum, New York.

9. Hay, R. J., Park, J.-G., and Gazdar, A., eds. (1994). "Atlas of Human Tumor Cell Lines." Academic Press, San Diego, CA.

10. Freshney, R. I. (2000). "Culture of Animal Cells: A Manual of Basic Technique, 4th ed., Wiley-Liss, New York.

11. Stulberg, C. S. (1973). Extrinsic cell contamination of tissue culture. *In* "Contamination in Tissue Culture" (J. Fogh, ed.), pp. 2–23. Academic Press, New York.

12. Pollack, M. S., Heagney, S. D., Livingston, P. O., and Fogh, J. (1981). HLA—A, B, C, and DR alloantigen to expression on forty-six cultured human tumor cell lines. *J. Natl. Cancer Inst.* **66**, 1003–1012.

13. Hsu, T. C., and Benirschke, K. (1967–1975). "An Atlas of Mammalian Chromosomes," 9 vols. Springer-Verlag, New York.

14. Chen, T. R., Hay, R. J., and Macy, M. L. (1982). Karyotype consistency in human colorectal carcinoma cell lines established in vitro. *Cancer Genet. Cytogenet.* **6**, 93–117.

15. Jeffreys, A. J., Wilson, V., and Thein, S. L. (1985). Hypervariable 'minisatellite' regions in human DNA. *Nature (London)* **314**, 67–73.

16. Gilbert, D. A., Reid, Y. A., Gail, M. H., Pee, D., White, C., Hay, R. J., and O'Brien, S. J. (1990). Application of DNA fingerprints for cell-line individualization. *Am. J. Hum. Genet.* **47**(3), 499–514.

17. Latorra, D., Stern, C. M., and Schanfield, M. S. (1990). Characterization of human AFLP systems apolipoprotein B, phenylalanine hydroxylase, and D1S80. *PCR Methods Appl.* **3**(6), 351–358.

18. Edwards, A., Hammond, H. A., Jin, L., Caskey, C. T., and Chakraborty, R. (1992). Genetic variation at five trimeric and tetrameric tandem repeat loci in four human population groups. *Genomics* **12**, 241–253.

19. Sajantila, A., Puomilahti, S., Johnsson, V., and Ehnholm, C. (1992). Amplification of reproducible allele markers for amplified fragment length polymorphism analysis. *BioTechniques* **12**(1), 16–22.

20. Durkin, A. S., and Reid, Y. A. (1998). Short Tandem Repeat loci utilized in human cell line identification. *ATCC Q. Newsl.* **18**, 1–7.

21. Innovative Chemistry (1988). The Authentikit System. "Handbook for Cell Authentication and Identification," 2nd ed. Innovative Chemistry, Marshfield, MA.

22. Caputo, J. L., Thompson, A., McClintock, P., Reid, Y. A., and Hay, R. (1991). An effective method for establishing human B lymphoblastic cell lines using Epstein-Barr Virus. *J. Tissue Cult. Methods* **13**, 39–44.

23. Rainey, W. E., Bird, I. M., and Mason, J. I. (1994). The NCI-H295 cell line: A pluripotent model for human adrenocortical studies. *Mol. Cell. Endocrinol.* **100**, 45–50.

24. Sutherland, H. J., Eaves, C. J., Lansdorp, P. M., Thacker, J. D., and Hogge, D. E. (1991). Differential regulation of primitive human hematopoietic cells in long-term cultures maintained on genetically engineered murine stromal cells. **Blood 78**, 666–672.

25. Ronnett, G. V., Hester, L. D., Nye, J. S., Connors, K., and Snyder, S. H., (1990). Human cortical neuronal cell line: Establishment from a patient with unilateral megalencephaly. *Science* **248**, 603–605.

26. Tomlinson, G. E., Chen, T. T., Stastny, V. A., Virmani, A. K., Spillman, M. A., Tonk, V., Blum, J. L., Schneider, N. R., Wistuba, I. I., Shay, J. W., Minna, J. D., and Gazdar, A. F. (1998). Characterization of a breast cancer cell line derived from a germ-line BRCA1 mutation. *Cancer Res.* **58**, 3237–3242.

27. Suardet, L., Gaide, A. C., Calmes, J. M., Sordat, B., Givel, J. C., Eliason, J. F., and Odartchenko, N. (1992). Responsiveness of three newly established human colorectal cancer cell lines to transforming growth factors beta 1 and beta 2. *Cancer Res.* **52**, 3705–3712.

28. Kruszewski, F. H., Walker, T. L., and DiPasquale, L. C. (1997). Evaluation of a human corneal epithelial cell line as an in vitro model for assessing ocular irritation. *Fundam. Appl. Toxicol.* **36**, 130–140.

29. Khabar, K. S., Siddiqui, S., and Armstrong, J. A. (1995). WEHI-13VAR: A stable and sensitive variant of WEHI 164 clone 13 fibrosarcoma for tumor necrosis factor bioassay. *Immunol. Lett.* **46**, 107–110.

30. Koochekpour, S., Jeffers, M., Rulong, S., Taylor, G., Klineberg, E., Hudson, E. A., Resau, J. H., and Vande Woude, G. F. (1997). Met and hepatocyte growth factor/scatter factor expression in human gliomas. *Cancer Res.* **57**, 5391–5398.

31. Ryan, M. J., Johnson, G., Kirk, J., Fuerstenberg, S. M., Zager, R. A., and Torok-Storb, B. (1994). HK-2: An immortalized proximal tubule epithelial cell line from normal adult human kidney. *Kidney Int.* **45**, 48–57.

32. Petrovic, N., Black, T. A., Fabian, J. R., Kane, C., Jones, C. A., Loudon, J. A., Abonia, J. P., Sigmund, C. D., and Gross, K. W. (1996). Role of proximal promoter elements in regulation of renin gene transcription. *J. Biol. Chem.* **271**, 22499–22505.

33. Dumenco, L., Oguey, D., Wu, J., Messier, N., and Fausto, N. (1995). Introduction of a murine p53 mutation corresponding to human codon 249 into a murine hepatocyte cell line results in growth advantage, but not in transformation. *Hepatology* **22**, 1279–1288.

34. Henderson, S. A., Lee, P. H., Aeberhardt, E. E., Adams, J. W., Ignarro, L. J., Murphy, W. J., and Sherman, M. P. (1994). Nitric oxide reduces early growth response-1 gene expression in rat lung macrophages treated with interferon-gamma and lipopolysaccharide. *J. Biol. Chem.* **269**, 25239–25242.

35. Mulshine, J., and Johnson, B., eds. (1996). NCI-Navy Medical Oncology Branch Cell Line Supplement. *J. Cell Biochem.* Suppl. **24**, 1–291.

36. Tam, Y. K., Miyagawa, B., Ho, V. C., and Klingemann, H. G. (1999). Immunotherapy of malignant melanoma in a SCID mouse model using the highly cytotoxic natural killer cell line NK-92. *J. Hematother.* **8**, 281–290.

37. O'Connell, K. A., and Edidin, M. (1990). A mouse lymphoid endothelial cell line immortalized by simian virus 40 binds lymphocytes and retains functional characteristics of normal endothelial cells. *J. Immunol.* **144**, 521–525.

38. Daubas, P., Klarsfeld, A., Garner, I., Pinset, C., Cox, R., and Buckingham, M. (1988). Functional activity of the two promoters of the myosin alkali light chain gene in primary muscle cell cultures: Comparison with other muscle gene promoters and other culture systems. *Nucleic Acids Res.* **16**, 1251–1271.

39. Richelson, E. (1990). The use of cultured cells in the study of mood-normalizing drugs. *Pharmacol. Toxicol.*, Suppl. **66**(3), 69–75.

40. Schoumacher, R. A., Ram, J., Iannuzzi, M. C., Bradbury, N. A., Wallace, R. W., Hon, C. T., Kelly, D. R., Schmid, S. M., Gelder, F. B., Rado, T. A., *et al.* (1990). A cystic fibrosis pancreatic adenocarcinoma cell line. *Proc. Natl. Acad. Sci. U.S.A.* **87**, 4012–4016.

41. Gower, W. R., Jr., Risch, R. M., Godellas, C. V., and Fabri, P. J. (1994). HPAC, a new human glucocorticoid-sensitive pancreatic ductal adenocarcinoma cell line. *In Vitro Cell. Dev. Biol.* **30A**, 151–161.

42. Arbiser, J. L., Moses, M. A., Fernandez, C. A., Ghiso, N., Cao, Y., Klauber, N., Frank, D., Brownlee, M., Flynn, E., Parangi, S., Byers, H. R., and Folkman, J. (1997). Oncogenic H-ras stimulates tumor angiogenesis by two distinct pathways. *Proc. Natl. Acad. Sci. U.S.A.* **94**, 861–866.

43. Hamaguchi, K., Gaskins, H. R., and Leiter, E. H. (1991). NIT-1, a pancreatic beta-cell line established from a transgenic NOD/Lt mouse. *Diabetes* **40**, 842–849.

44. Hamaguchi, K., and Leiter, E. H. (1990). Comparison of cytokine effects on mouse pancreatic alpha-cell and beta-cell lines. Viability, secretory function, and MHC antigen expression. *Diabetes* **39**, 415–425.

45. Varlamov, O., Fricker, L. D., Furukawa, H., Steiner, D. F., Langley, S. H., and Leiter, E. H. (1997). Beta-cell lines derived from transgenic Cpe(fat)/Cpe(fat) mice are defective in carboxypeptidase E and proinsulin processing. *Endocrinology (Baltimore)* **138**, 4883–4892.

46. Hurbain-Kosmath, I., Berault, A., Noel, N., Polkowska, J., Bohin, A., Jutisz, M., Leiter, E. H., Beamer, W. G., Bedigian, H. G., Davisson, M. T., *et al.* (1990). Gonadotropes in a novel rat pituitary tumor cell line, RC–4B/C. Establishment and partial characterization of the cell line. *In Vitro Cell. Dev. Biol.* **26**, 431–440.

47. Weijerman, P. C., Konig, J. J., Wong, S. T., Niesters, H. G., and Peehl, D. M. (1994). Lipofection-mediated immortalization of human prostatic epithelial cells of normal and malignant origin using human papillomavirus type 18 DNA. *Cancer Res.* **54**, 5579–5583.

48. Lee, C. W., and Handschumacher, R. E. (1995). Concentrative transport of adenosine in murine splenocytes: Limitation by an ecto-adenosine deaminase. *In Vivo* **9**, 1–5.

49. Dunn, K. C., Aotaki-Keen, A. E., Putkey, F. R., and Hjelmeland, L. M. (1996). ARPE-19, A human retinal pigment epithelial cell line with differentiated properties. *Exp. Eye Res.* **62**, 155–169.

50. Zabrenetzky, L. V., Thompson, A., and Hay, R. J. (1997). The isolation, immortalization and characterization of human fetal keratinocytes. *In Vitro Cell. Dev. Biol.* **33** (Part II), p. 34A.

51. Behr, T. M., Wulst, E., Radetzky, S., Blumenthal, R. D., Dunn, R. M., Gratz, S., Rave-Frank, M., Schmidberger, H., Raue, F., and Becker, W. (1997). Improved treatment of medullary thyroid cancer in a nude mouse model by combined radioimmunochemotherapy: Doxorubicin potentiates the therapeutic efficacy of radiolabeled antibodies in a radioresistant tumor type. *Cancer Res.* **57**, 5309–5319.

CHAPTER 5

ADVENTITIOUS AGENTS AND CHEMICAL TOXICITY

Gary C. du Moulin

Early efforts in the development of tissue-engineered products quickly pointed out the potential for microbial contamination of cells and tissues [1]. Unless effective screening and monitoring procedures were in place, there remained a high likelihood that manufacturing culture systems would become compromised. Moreover, contaminated tissues acquired during procurement could lead to downstream contamination or pose additional infectious risks upon return to the patient of the tissue-engineered product.

DEVELOPMENT OF SCREENING STRATEGIES

The developers of tissue-engineered products and the U.S. Food and Drug Administration (FDA) continue to invest considerable energy in the elucidation of paradigms for the microbiological screening of tissues of autologous, allogeneic, and xenogeneic origin [2]. These algorithms have been published in the scientific literature and in FDA-sponsored documents [3–8]. Omstead *et al.* published a recommendation for a voluntary guidance document that is being further developed by the American Society for Testing and Materials (ASTM) [9]. More recently, the U.S. Pharmacopeia Subcommittee for Cell and Gene Therapy published a draft informational chapter [10,11]. The reader is directed to these important sources of information in developing screening programs for individual product development initiatives.

In addition to the potential for contamination of source tissue, the manufacture of tissue-engineered products may constitute a significant risk to patients if proper controls are not in place. Ensuring microbial integrity during the manufacture of these products, combined with the need to rapidly return short-shelf-lived products to the patient, poses considerable challenges. New microbiological testing methods need to be evaluated and validated to improve sensitivity and minimize time to detection to support the developing tissue engineering industry. This chapter addresses the necessity and challenges for the microbiological and toxicological screening of raw materials, as well as the monitoring of sterility throughout the manufacturing process to ensure the exclusion of microbial contamination.

SOURCES OF ADVENTITIOUS AGENTS

Adventitious agents that can affect the safety of tissue-engineered products can arise from a number of potential sources, including the following:

1. Infected donor tissue arising from autologous, allogeneic, or xenogeneic sources
2. Contamination arising during the process of tissue procurement
3. Contamination occurring during processing and manipulation of the tissue-engineered product
4. Contamination arising from raw materials and components
5. Contamination arising from the manufacturing environment

6. Inadequate container closure systems

7. Improper technique in dispensing the product

Early development of cell therapies and tissue engineering applications in academic settings quickly demonstrated the importance of microbial controls in the manufacture of these products. In 1990 Weisfuse reported an outbreak of hepatitis in cancer patients receiving *ex vivo* lymphokine-activated killer (LAK) cells [12]. Arnow reported an infection in cancer patients infused with LAK cells contaminated with *Aspergillus fumigatus*. Fungal contamination of the cells occurred during processing in a biosafety cabinet contaminated with *Aspergillus fumigatus* spores [13]. Carroll reported in 1993 contamination of pancreatic islet cell preparations in 8 of 26 isolations (31%) [14]. Microorganisms recovered included aerobic, anaerobic, and fungal species.

The FDA recognized the potential risk of these early therapeutic approaches. In their own assessment, a 25% infection rate was observed in some centers [1]. In response, the FDA and, more recently international regulatory agencies, developed communications to provide guidance to developers of tissue-based technologies [15–17].

In developing its approach to the regulation of cellular and tissue-based products, FDA attempted to extrapolate the potential form and function of these products. An infinite number of possibilities relevant to cell source, cell type, and associated biomaterials mandated a tiered approach to regulation [2]. Three objectives emerged:

1. To prevent "unwitting use of contaminated tissues with the potential for transmitting infectious diseases." This was particularly critical if cells or tissues were to be used in an allogeneic setting.

2. To prevent "improper handling or processing that might contaminate or damage tissue." In some cases tissue-engineered products require hundreds of processing steps, and contact with many components, as well as incubation periods ranging from days to months.

3. To ensure clinical safety and effectiveness for these products. Providing unambiguous clinical data demonstrating safety and effectiveness is challenging for makers of cell and tissue-engineered products. Clinical protocols, institutional review board (IRB) approvals, and clinical trial designs that incorporate appropriate controls exhibiting acceptable statistical power and treatment end points require considerable thought and discussion to be relevant and appropriate for this emerging field.

The degree of regulatory oversight would additionally depend upon whether tissues were manipulated extensively, combined with nontissue components, intended to be used for other than their normal function, or used for metabolic purposes.

GENERAL GUIDANCE FOR DEVELOPING SCREENING PROGRAMS

All relevant characteristics of the tissue donor should be noted. At a bare minimum the species, age, sex, and tissue type should be recorded. All tissue-engineered products will require extensive systems of archiving tissue and mechanisms for tracing tissue origin.

ASEPTIC PROCESSING

The quality of cell and tissue harvesting will reflect directly on the ability to successfully process a given tissue. To ensure the acceptability of tissue during manufacture, medical personnel associated with tissue collection should be properly trained or provided with clear directions. Because of the nature of tissues and the inherent levels of bioburden that will be associated with the anatomic area from which a given tissue was harvested, however, acceptable tissue quality cannot always be obtained. Skin, for example, will almost always comprise a bioburden. The number of contaminating microorganisms and the diversity of the species represented can vary considerably depending on the medical condition of the patient and the length of prior hospitalization. Generally speaking, in uncomplicated cases, tissue derived from internal anatomic sites can be harvested aseptically if the site of harvest has been well prepared, has not been damaged, and has been appropriately sanitized.

RAW MATERIALS ASSOCIATED WITH TISSUE COLLECTIONS

Harvested tissue will immediately be transferred into a holding medium, which should maintain sterility and tissue viability until actual processing begins. The media, serum additives, growth factors, and other components should be carefully chosen, screened for adventitious agents, and assessed beforehand for the ability to maintain viability of the tissue.

STORAGE AND SHIPPING

Harvested tissues procured for tissue-engineering processes may have finite and abbreviated shelf lives, measured in hours or days. Plans for transport of tissue to processing or manufacturing sites need to be established and validated before any tissue harvesting program is initiated. Certain tissues will be more sensitive than other tissues to the rigors of time, transport, or temperature. Specific nutritional and holding requirements need to be ascertained through laboratory study to develop optimal conditions for transport.

LABELING

In tissue engineering operations in which the cells and tissues of multiple donors are to be processed, it is critical to develop labeling including numbering systems to ensure segregation between lots. Government regulations require validated systems to ensure proper patient identity.

TRACKING

Record keeping systems for any tissue engineering application should ensure that all tissue could be traced back to its origin. These systems will undergo intense scrutiny to prove that in the event of contamination or infection, prompt and easy accessibility of retained materials will allow source tissue to be traced. These retained samples can then be subjected to additional testing. Mechanisms to track patients are important not only in the case of donor-to-recipient disease transmission, but also to provide "callback" ability if inadvertent contamination of materials or erroneous delivery to the wrong recipient should be discovered.

STORAGE

If there are requirements for temporary or long-term storage of either processed or unprocessed tissues, the conditions for storage should be carefully validated. The quality of cells and tissues can vary greatly depending on their sensitivity and stability under refrigerated or frozen storage conditions. Assays for assessing the conditions of the tissue removed from storage facilities should be developed and validated.

SCREENING OF AUTOLOGOUS TISSUES

Autologous tissue presents the lowest level of risk because the donor is also the recipient. The FDA recommends, but does not require, donor testing for adventitious agents for tissues intended for autologous use [2,4]. Viral screening for autologous banked tissue would include tests for HIV, HTLV-1, HTLV-2, HCV, and HBV. Medical history screening for high risk of HIV and hepatitis is also recommended. In practice however, extensive viral screening for autologous donors prior to processing is problematic. Management of large numbers of tissues within a cell processing facility can be safely conducted if stringent controls are in place to ensure adequate lot segregation.

Autologous tissues that are harvested, transported, or processed along with many other donors will require proper controls as mandated by current good manufacturing practices (GMP) or current good tissue practices (GTP). Robust controls can minimize risk of cross-contamination.

Depending on the source of the autologous tissue, an associated bioburden is possible. Even with autologous tissues collected through seemingly aseptic techniques, bacteria, fungi, or viruses may be carried into the processing area. While antibiotics may suppress or eliminate bacteria or fungi, it should be determined whether viral contamination can persist in the cell culture environment. Campbell reported HIV infection of human cartilage collected postmortem from 9 of 10 patients, who were HIV positive [18].

SCREENING OF ALLOGENEIC TISSUES

Allogeneic cells and tissue may arise from living adult donors, fetal donors, and/or cadaver donors. Besides the documentation of the health status of all donors, viable tissues must undergo viral screening for HIV-1 and HIV-2, HTLV-1 and HTLV-2, HBV, HCV, CMV, and *Treponema pallidum*. Donors at high risk for HIV, hepatitis, Creutzfeld–Jakob disease, and tuberculosis should also be screened. Viral screening employing tests of appropriate sensitivity, must be conducted in approved and certified laboratories.

Mechanisms must be established to ensure that the product can be appropriately quarantined until results of screening are available. This information must be verified before the tissue-engineered product can be released from the manufacturing facility. The FDA has published a draft recommendation for the screening and testing of human tissue intended for transplantation [4]. A tissue donor registry must be established so that new information can be made available to affected recipients.

A number of products have been based on recently established allogeneic cell lines. In these cases, donor testing is conducted following a "points to consider" document describing the characterization of cell lines to produce biologicals [7].

SCREENING OF XENOGENEIC TISSUES

While the prospect of using live cells, tissues, and organs obtained from nonhuman sources is vast, safety concerns related to possible transmission of infectious agents are being carefully addressed. There is an ongoing and active debate on the need to develop screening mechanisms to ensure that the public health is protected. Control of this tissue engineering technology platform encompasses the following framework for control [5,19]:

1. Definition of the composition and function of the xenotransplant team
2. A clinical protocol and consent for all xenotransplant use
3. Animal source screening
4. Surveillance for all xenotransplant recipients
5. A mechanism for hospital infection control
6. A framework for maintaining appropriate records
7. An archive of source animal specimens
8. An archive of patient serum and tissue samples
9. Creation of a national centralized database
10. Creation of a Secretary's Advisory Committee of Xenotransplantation (SACX)

An FDA draft guideline on infectious disease issues in xenotransplantation [5] defines requirements for animal screening, including qualification programs for the donor herd and individual donor animals. Animal screening programs include serological profiling, blood cultures, complete blood count with differential, fecal parasitology, and *in vitro* viral assays performed on peripheral blood lymphocytes. The risk of infection from tissue derived from nonhuman primates was so significant that the FDA has prevented use of these tissues [20]. Cells and tissues derived from xenotransplant sources must show extensive control of origin, relevant genetic traits, animal husbandry expertise, and knowledge of the health status of the herd or colony.

Patients receiving xenotransplants and their close contacts will receive ongoing clinical evaluation. Samples of serum and DNA will be collected and archived at periodic intervals throughout each patient's lifetime.

METHODOLOGIES AVAILABLE FOR DETECTING THE PRESENCE OF BIOLOGICAL AGENTS

The determination of microbial integrity is a basic requirement in the manufacture of tissue-engineered products. The method of screening for biological agents will undergo intense scrutiny by federal regulatory agencies during review of product submissions. Cell or tissue materials arriving as biopsies from donors will need to be screened to rule out infection of the donor tissues or contamination at the biopsy collection site. Sterility tests will become an important quality control component during processing. Critical processing steps will require assurance of sterility. Clinical microbiologic diagnostic protocols, particularly those sanctioned by national and international scientific organizations, should be consulted for specific procedures to detect bacteria and fungi from source materials.

BACTERIAL AND FUNGAL SCREENING

The U.S. Pharmacopeia (USP) sterility test is the "gold standard" for the sterility testing of products regulated by the FDA [21]. Its principle rests upon century-old methods based on the recovery and growth of microorganisms in solid or liquid microbiological growth media. In the USP sterility test, samples are inoculated into two different liquid growth media containing either trypticase soy broth or thioglycollate broth. Samples inoculated into thioglycollate broth are placed into 33°C incubators to encourage growth of anaerobic microorganisms. Samples inoculated into trypticase soy broth are placed into 22°C incubators to culture aerobic microorganisms. The media are examined periodically for the presence of visible turbidity, and procedures to recover and identify the organism(s) are in place. Incubation of samples may last up to 14 days before each assay can be identified as "no growth."

The USP sterility test has distinct disadvantages in the monitoring of sterility for tissue-engineered products. These include extended incubation times, slow microbial growth rates, unintended selectivity, and the inherent variability of microorganisms in their response to culture methods. For many of these products, in-process and final release sterility assays will be completed after the lot of tissue-engineered product has reached its expiration date. FDA has worked with industry and has shown flexibility in identifying acceptable lot release sterility testing points to accommodate these products with short shelf lives. Technologies available that encompass these new methods of detection include growth-based technologies, viability-based technologies, technologies based on cellular components or artifact, and technologies based on nucleic acids. A broad description of these technologies exists in a technical report by the Parenteral Drug Association [22].

VIRUSES

Screening for viruses should be conducted by laboratory organizations carrying appropriate certifications [e.g., those established by the Clinical Laboratory Improvement Act (CLIA)] indicating expertise in performing validated assays with rigorous quality standards, including proficiency testing programs, in place. Laboratories contracted to perform such services should be willing to undergo on-site auditing of procedures, quality systems, and technician training programs.

Viral screening testing for HIV-1 and HIV-2, HTLV-1, HBV, and HCV should be conducted on appropriate samples. Allogeneic donor tissues should not be processed until results of all viral screening have been completed. Automated data management systems that maintain such information should protect the privacy of the donor while clearly providing inclusion or exclusion information regarding the potential donor.

Xenotransplantation viral screening should follow published guidelines [5]. These screening programs should be comprehensive and should identify all appropriate assays that would qualify the tissue for use in humans. The screening program should be tailored to the source of the animal as well as the intended clinical application. Individuals should be chosen to maintain such a program on the basis of their knowledge of veterinary infectious diseases.

Assays should be validated and demonstrate known sensitivity and specificity. Cocultivation assays with appropriate indicator cell lines should be used to facilitate amplification and detection of xenotropic endogenous virus or other xenogenous virus capable of producing infection. Cocultivation should be used in conjunction with observation of cytopathic effects, focus formation, reverse transcriptase assays, or electron microscopy. If the presence of viral agents is suspected, other techniques such as immunoassays for the detection of serologic cross-reactivity, immunofluorescence, Southern blot analysis, polymerase chain reaction, PCR-based reverse transcriptase, or cross-species *in vivo* culturing techniques may be used. Detection of latent viruses may be facilitated by their activation using chemical and irradiation methods.

MYCOPLASMA

Mycoplasma and ureaplasma are the smallest free-living microorganisms [23]. Mycoplasma does not exhibit a cell wall and ranges in size from about 0.2 to 0.3 μm. Mycoplasma can be observed as round or filamentous in cell culture using dark-field or phase contrast microscopy. On solid agar, colonies of mycoplasma can range in diameter from approximately 15 to 300 μm, the larger colonies being distinguished by a typical "fried egg" appearance.

Mycoplasma is ubiquitous and can be isolated from practically all mammals. As such, it has historically been one of the main contaminants of tissue cultures. Mycoplasma tends to be fastidious, requiring preformed nucleic acids supplied by media components. These components may be readily available in the cell culturing materials needed for tissue engineering applications. Mycoplasma could arise from the bovine or other animal-derived culture components, or in biopsy materials from asymptomatic patients.

Routine surveillance for mycoplasma is recommended for all raw materials derived from a human or animal source and is required as a lot release assay for approved biologics. The FDA has published a guideline describing in detail the accepted methods for the cultivation and isolation of mycoplasma [24]. Most quality control laboratories do not have the resources to conduct this assay according to the published guideline, but outside contract laboratories dedicated to mycoplasma isolation are available. Since the guideline-sanctioned assay takes one month of testing to complete, alternative methods have been proposed for the rapid detection of mycoplasma. One such method is based on the use of the polymerase chain reaction [25–28]. With the possibility of increasing sensitivity and specificity, techniques based on enzymatic amplification of DNA target sequences have been attractive [25]. Nissen compared several methods of sample preparation, including primers for genome amplification, to determine the potential for differentiating mycoplasma species [28]. These investigators found that group-specific primers can be useful in the screening of tissue culture for contaminating mycoplasma species.

PRIONS

Another perceived impediment in the use of tissue culture techniques for tissue engineering applications has been the risk of infection by prions causing transmissible spongiform encephalopathy [29]. Transmissible spongiform encephalopathy represents a range of neurodegenerative diseases occurring in man and animals, the most recognized being Creutzfeld–Jakob disease (CJD). Symptoms of CJD include presenile dementia, motor and coordination abnormalities, and death. The unique pathogenesis of these diseases is based on the accumulation of prions in the infected host, with full expression of symptoms becoming apparent approximately 13 years after initial exposure. Infections with CJD have been associated with contaminated depth electrodes, or transplanted tissues including corneas, dura mater, human growth hormone, and gonadotrophin derived from cadaveric pituitaries of infected donors. Since tissue culture production activities are dependent on reagents or components with a bovine origin, the source of this material must be collected from herds that are free of bovine spongiform encephalopathy (BSE).

Despite assurances of BSE-free herds, use of materials of bovine source is anathema in Europe since the well-known recognition of BSE in the United Kingdom [30]. Of approximately 1 million cattle thought to be infected with prions, hundreds of thousands have died of overt BSE in the United Kingdom over the past 10 years. A variant of CJD was subsequently identified in 21 human cases in the United Kingdom and one in France. Extensive epidemiological investigations led to the conclusion that the human disease originated after consumption of infected beef.

The prion is composed of a sialoglycoprotein attached to the cell membrane by way of a glycosylphosphatidylinositol moiety. It has been hypothesized that the normal prion (PrPC) protein may be necessary for latent learning and long-term memory retention [31]. It is the disease induction prion (PrPSc), a variant resulting from a post-translational process, which drastically alters the conformation of the prion. Whereas the normal prion is protease sensitive, monomeric, and detergent soluble, the disease induction prion is protease resistant, polymerized, and detergent insoluble. Brain extracts from cattle infected with BSE have caused the disease in cattle, sheep, mice, pigs, and mink after intracerebral inoculation [32].

The animal industry has engaged in a massive cleansing of potentially infected animals. The routine use in ruminant feeds of meat and bone meal prepared from offal of sheep, cattle, pigs, and chicken was banned in 1988. A number of food products were taken off the shelves, and since 1989 there remains a export ban to the United States and other countries of beef originating in the United Kingdom. There has been no evidence of BSE in the United States. Nevertheless as a preventive measure, FDA in 1997 prohibited the use of mammalian protein in the manufacture of animal feeds.

Since many pharmaceutical and biotechnology-based products use material of bovine origin in manufacturing processes, there has been a significant debate about the potential risk of these materials used as starting materials, ancillary products, or excipients. Twenty-five percent of over-the-counter (OTC) medicines contain bovine gelatin. It is believed unlikely that prion protein could be transmitted through the manufacturing process.

While the developers of tissue-engineered products search for a bovine-serum-free substitute, the use of materials of bovine origin arising from U.S. sources should not be contraindicated if vendor qualification procedures following good manufacturing practices are in place.

TOXINS (ENDOTOXINS)

Regulatory bodies view the absence from biopharmaceutical products of endotoxin as an important indicator of purity [33]. Endotoxin is known to exhibit a number of biological effects on the mammalian cell membrane which can affect secretion and cytokine production or can serve as powerful mitogens [34]. Because of the possibility that wide-ranging biologic effects will be unleashed on cell and tissue culture, raw materials and components used for the manufacture of tissue-engineered products must be assessed for the presence of endotoxin as part of the raw materials qualification process [35].

Biological Effects of Endotoxin in Man

Clinical correlates of endotoxin have been extensively reviewed [36]. For example, uninfected urine has been found to contain less than 20 ng/ml while infected urine ($>1 \times 10^5$ Cfu/ml) has been found to contain 200–300 ng/ml. Cervical or urethral exudates have been found to contain 10 ng/ml. Casey studied the incidence of endotoxemia in pediatric patients undergoing cardiopulmonary bypass [37]. Endotoxemia was detected in 16 of 24 patients. The amount of endotoxin measured in the blood ranged from less than 0.0001 to 0.438 ng/ml of blood. There was no association with perioperative morbidity in this patient population. Garcia reported an LD_{80} of 100 μg/kg *E. coli* endotoxin in the rabbit [38]. This correlated to 70 million endotoxin units (EU) or 7 million ng for a 70-kg adult. Ogawa reported a minimum pyrogenic dose of 1.6×10^{-5} μg/kg in the rabbit [39]. This correlates to 1120 EU or 112 ng for a 70-kg adult. These levels are compared with the endotoxin dose limit for parenteral drugs according to the December 1987 FDA guideline of 350 EU or 35 ng for a 70-kg adult infused over a 1-h period.

Biological Effects of Endotoxin in Mammalian Cell Culture Systems

Endotoxin (lipopolysaccharide, LPS) constitutes the outer membrane component of Gram-negative bacteria. It is an amphipathic structure comprising fatty acid chains, which can interact with the lipid membranes and hydrophobic regions of proteins, and phosphate groups associated with the LPS, which can react with cationic proteins. The effect of endotoxin in tissue culture has been investigated extensively [40–43]. Morphological changes such as surface ruffling and the presence of large vacuoles have been observed in mammalian cell cultures. In higher concentrations endotoxin may be cytotoxic. When cultures are exposed to endotoxin-free media, these changes may be reversible. The presence of endotoxin can affect the secretion of desired cell products such as cytokines or tumor necrosis factor. Endotoxin can also exhibit inhibitory effects. This has been observed in the biosynthesis and release of proteoglycans from cultures of articular cartilage [40]. Endotoxin can also affect adherence, endocytosis, and tumoricidal activity.

Measurement of Endotoxin

Control of endotoxin in the manufacture of tissue-engineered products will become an essential element of any quality control program. One benefit of endotoxin measurement has been the extensive experience regulatory agencies have had since the advent of the *Limulus* amoebocyte lysate assay. Prior to its introduction, the detection of endotoxin could be assessed only through assays of pyrogenicity. Large colonies of rabbits were maintained, and materials to be tested would be implanted. Thermistors placed within the rectums of these caged animals would monitor a temperature rise as indicative of the presence of endotoxin. The historical observation of the combination of lysate prepared from the blood cells of the horseshoe crab (*Limulus polyphemus*) with Gram-negative bacteria resulting in stable clot inaugurated a new era in endotoxin testing. This method, known as the bacterial endotoxins test, is now officially recognized by the U.S. Pharmacopeia as well as by all other international pharmacopoeia [44]. It is an assay that can be validated for a wide range of biopharmaceutical products with published guidance from the FDA [33]. More recently, a chromogenic variation of the assay has resulted in instrumentation allowing full automation of the endotoxin detection process. One other feature of the assay is the ability to conduct the assay prior to the release of products having a short shelf life [45].

Practical Aspects of Endotoxin Testing

Depending on the vendor, media or reagents can vary widely in the amount of endotoxin present. When establishing quality control testing programs, it is important that the development and validation of the LAL assay be given significant attention. The inherent variability in the materials to be tested can render the potential variability in the test method problematic. Alert and action levels for endotoxin in critical reagents and final products should be determined through experience. Development of a robust endotoxin testing program will become a cornerstone of any tissue engineering quality control effort.

FOREIGN PROTEINS

Immunological compatibility for tissue-engineered products should be a design objective [46]. Owing to the incorporation of foreign proteins in the manufacture of tissue-engineered products, however, achieving this level of compatibility may be elusive because the concentration inherent in the tissue-engineered product may be enough to elicit an antibody response. All efforts must be exhausted in locating materials of a recombinant nature before choosing ancillary products composed of proteins. This is particularly true if these ancillary products become part of the final product to be implanted. If materials of a foreign nature, such as enzymes, are used as process components, efforts must be taken to remove these materials as part of downstream processing. Regulatory agencies will require evidence of removal through robust process validation.

MUTAGENICITY/GENOTOXICITY

It is rare to find a mutagenic biomaterial. With the advent of novel biomaterials, however, this possibility cannot be entirely ruled out. ISO 10993 is most often cited as the methodology to be used in determining the ability of a material to induce a genetic abnormality [47]. Three *in vitro* tests have been cited as necessary to assess mutagenicity. These are the Ames salmonella reverse mutation assay, the sister chromatid exchange test, and a chromosomal aberration assay [48].

TOXICOLOGICAL ASSAYS

The evaluation of toxicity is an important aspect of determining the safety of tissue-engineered products. Toxicity needs to be assessed both for the raw materials comprising the product and for the product itself. Before designing a comprehensive general toxicity program, it is prudent to conduct a comprehensive literature search to determine previously identified toxicity profiles of each component. If the resulting search is inconclusive, a series of *in vitro*, *in vivo*, and *in situ* studies should be conducted. Biological Evaluation of Medical Devices (ISO 10993) should be consulted when one is designing a toxicity testing program [47]. Additionally, the U.S. Pharmacopeia addresses toxicity testing [49,50].

In Vitro Studies

In vitro biological reactivity tests are USP methods that provide validated methodologies for determining toxicity *in vitro* [49]. The biological reactivity tests, which include an agar diffusion test, a direct contact test, and an elution test, evaluate the ability of a material to inhibit growth in a standard tissue culture system.

In Vivo Studies

In vivo studies should be conducted for products containing allogeneic and xenogeneic cells. In products containing only autologous cells, only the biomaterial itself needs to be assessed, in addition to the final formulated product. The tests are based on systemic or intracutaneous injections into mice or rabbits [50]. Mice are injected through the tail vein, while rabbits are exposed to the biomaterial intracutaneously. Generally, results are evaluated after a relatively short period of time (3 days). After administration of a single dose, the test animals should exhibit no gross tissue reactions at the injection sites when compared to appropriate controls. Additionally, toxicity should be evaluated for longer periods of time (e.g., 7–14 days).

In Situ Studies

The biomaterials and excipients comprising the final formulated tissue-engineered product should be evaluated through *in situ* studies. All efforts must be made in this trial to determine a dosage regime, site of implantation, and appropriate species that most closely simulates the human application of the product. *In situ* studies are of long duration and should represent at least 10% of the life expectancy of the test animal. At the study's conclusion, detailed histologic or anatomic evaluation of the implantation site should be performed. Examination should include assessments of local tissue reaction and so on.

The Code of Federal Regulations cites the General Safety Test as important in assessing toxicity of the final product [51]. The FDA, however, has determined that because cell-based tissue-engineered products typically involve systems of cultured, live mammalian cells exquisitely sensitive to the *ex vivo* environment, the safety test is not relevant.

MEDIA AND REAGENT PURITY

One of the most challenging areas in the manufacture of tissue-engineered products is the selection of media and reagents necessary for the cultivation of cells and tissue *ex vivo*. These materials are ubiquitous and can range from relatively simple buffered solutions to extraordinarily complex media containing a host of cytokines, growth factors, and monoclonal antibodies. These materials are usually considered to be ancillary products, indicating materials that have a distinct function in manufacture but are not intended to be part of the final product.

Since these qualification programs may not be feasible at early developmental stages, product developers must therefore rely on quality vendors and materials as appropriate for use. Most commercial producers of media and reagents intended for the tissue engineering industry have programs to improve the quality of manufacturing. A number of vendors have already been subjected to FDA audits. Results of these audits can be obtained through the Freedom of Information Act. Developers should prioritize those reagents or media that are the most critical to the process or might have a greater impact upon patient safety. The Certificate of Analysis that accompanies each product should be scrutinized for certain quality parameters such as sterility and endotoxin. These assays should be some of the first developed within the organization intent on developing tissue-engineered products. On-site audits of these vendors is also an important adjunct to raw material qualification.

Lot history files should be constructed for each material so that the specification for each material may be checked against the data supplied on the Certificate of Analysis. These materials will need to be tracked not only by the manufacturer's lot number but also by an internal lot number that correlates to each shipment. This traceability is essential, and the internal lot numbers for reagents and media should be noted in the product records of the tissue-engineered products. The GMPs for materials and components should be consulted when one is developing a program for raw material qualification.

Development of raw materials to support the tissue engineering industry is gaining momentum. Materials chosen for certain applications will need to be validated. It has been noted that certain materials will exhibit significant lot-to-lot variability. This variability must be reconciled if there is to be confidence in the raw material and an ability to control the material during the manufacturing process. Qualification programs should therefore include some ability to measure functionality of the raw material through *in vitro* testing. These assays require their own developmental activities and validation before they can be relied on to assess the functionality of a raw material.

ACKNOWLEDGMENTS

I am indebted to Kathy Grim, Grace Kielpinski, John Duguid, and Russell Herndon, who provided helpful comments and critical review of this manuscript.

REFERENCES

1. U.S. Department of Health and Human Services (1989). "Points to Consider in the Collection, Processing, and Testing of *Ex Vivo* Activated Mononuclear Leukocytes for Administration to Humans." USDHHS, Public Health Service, Food and Drug Administration, Washington, DC.
2. U.S. Department of Health and Human Services (1997). "Proposed Approach to Regulation of Cellular and Tissue Based Products." USDHHS, Public Health Service, Food and Drug Administration, Washington, DC.
3. U.S. Department of Health and Human Services (1996). "Guidance on Applications for Products Comprised of Living Autologous Cells Manipulated *Ex Vivo* and Intended for Structural Repair or Reconstruction." USDHHS, Public Health Service, Food and Drug Administration, Washington, DC.
4. U.S. Department of Health and Human Services (1995). "Draft Document Concerning the Screening and Testing of Donors of Human Tissue Intended for Transplantation." USDHHS, Public Health Service, Food and Drug Administration, Washington, DC.
5. U.S. Department of Health and Human Services (2000). "Draft Public Health Service Guideline on Infectious Disease Issues in Xenotransplantation." USDHHS, Public Health Service, Food and Drug Administration, Washington, DC.
6. U.S. Department of Health and Human Services (1997). "Draft Guidance for the Submission of Chemistry, Manufacturing, and Controls Information and Establishment Description for Autologous Somatic Cell Therapy Products." USDHHS, Public Health Service, Food and Drug Administration, Washington, DC.
7. U.S. Department of Health and Human Services (1993). "Points to Consider in Somatic Cell and Gene Therapy." USDHHS, Public Health Service, Food and Drug Administration, Washington, DC.
8. U.S. Department of Health and Human Services (1993). "Points to Consider in the Characterization of Cell Lines to Produce Biologicals." USDHHS, Public Health Service, Food and Drug Administration, Washington, DC.
9. Omstead, D. R., Baird, L. G., Christenson, L., du Moulin, G. C., Tubo, R., Maxted, D. D., David, J., and Gentile, F. T. (1998). Voluntary guidance for the development of tissue-engineered products. *Tissue Eng.* **4**, 239–266.
10. U.S. Pharmacopeia (2000). Cell and gene therapy products. *Pharmacopeial Forum* **26**, 56–124.
11. Seaver, S. (2000). Announcing USP chapter. Cell and gene therapy products. *Biopharm* **13**, 26–29.
12. Weisfuse, I. B., Graham, D. J., Will, M., Parkinson, D., Snydman, D. R., Atkins, M., Karron, R. A., Feinstone, S., Rayner, A. A., and Fisher, R. I. (1990). An outbreak of hepatitis A among patients treated with interleukin 2 and lymphokine activated killer cells. *J. Infect. Dis.* **161**, 647–652.
13. Arnow, P. M., Houchins, S. G., Richards, J. M., and Chudy, R. (1991). *Aspergillus fumigatus* contamination of Lymphokine Activated Killer Cells infused into cancer patients. *J. Clin. Microbiol.* **29**, 1038–1041.
14. Carroll, P. B., Ricordi, C., Fontes, P., Rilo, H. R., Phipps, J., Tsakis, A. G., Fung, J. J., and Starzl, T. E. (1992). Microbiologic surveillance as part of human islet transplantation. Results of the first 26 patients. *Transplant. Proc.* **24**, 2798–2799.
15. U.S. Department of Health and Human Services (1998). "Guideline for Human Somatic Cell Therapy and Gene Therapy." USDHHS, Public Health Service, Food and Drug Administration, Washington, DC.
16. Committee for Proprietary Medicinal Products (CPMP) (1999). "Points to Consider on Human Somatic Cell Therapy." The European Agency for the Evaluation of Medicinal Products (EMEA).
17. U.S. Department of Health and Human Services (1998). "Establishment Registration and Listing for Manufacturers of Human Cellular and Tissue-based Products: Proposed Rule." USDHHS, Public Health Service, Food and Drug Administration, Washington, DC.
18. Campbell, D. G., and Oakeshott, R. D. (1996). HIV infection of human cartilage. *J. Bone J. Surg. (Br.) Vol.* **78**, 22–25.
19. U.S. Department of Health and Human Services (1999). "Precautionary Measures to Reduce the Possible Risk of Transmission of Zoonoses in Blood and Blood Products from Xenotransplantation Product Recipients or their Contacts." USDHHS, Public Health Service, Food and Drug Administration, Washington, DC.
20. U.S. Department of Health and Human Services (1999). "Public Health Issues Based on the Use of Nonhuman Primate Xenografts in Humans." USDHHS, Public Health Service, Food and Drug Administration, Washington, DC.
21. The United States Pharmacopeia (2000). "Sterility Tests" (USP 24). U.S. Pharmacopeial Convention, Rockville, MD.
22. PDA Task Force on the Evaluatic Validation, and Implementation of New Microbiological Methods (1999). Technical Report No. 33: Evaluation validation, and implementation of new microbiological testing methods. *PDA J. Pharm. Sci. Technol.* **53**, 1–56.

23. Taylor-Robinson, D. (1998). Mycoplasma and ureaplasma. *In* "Manual of Clinical Microbiology" (E. Lennette, A. Balows, W. Hausler, and H. Shadomy, eds.), 5th ed. American Society for Microbiology, Washington, DC.

24. U.S. Department of Health and Human Services (1987). "Recommended Procedures for Detection of Mycoplasmas Contamination in Biological Products Produced in Cell Substrates." USDHHS, Public Health Service, Food and Drug Administration, Washington, DC.

25. Hopert, A., Uphoff, C. C., Wirth, M., Hauser, H., and Drexler, H. G. (1993). Specificity and sensitivity of PCR in comparison with other methods for the detection of mycoplasma contamination in cell lines. *J. Immunol. Methods* **164**, 91–100.

26. Kirchoff, H., and Schmidt, R. (1995). Detection of mycoplasma in cell cultures by mycoplasma PCR ELISA comparison to the culture method. *Biochemica (Boehringer Mannheim)* **1**, 33–35.

27. van Kuppeveld, F. J., Johansson, K. E., Galama, J. M., Kissing, J., Bolske, G., van der Logt, J. T., and Melchens, W. J. (1994). Detection of mycoplasma contamination on cell cultures by a mycoplasma group-specific PCR. *Appl. Environ. Microbiol.* **60**, 149–152.

28. Nissin, E., Pauli, G., and Vollenbroich, D. (1996). Comparison of PCR detection methods for Mycoplasma cultures. *In Vitro Cell. Dev. Biol.—Anim.* **32**, 463–464.

29. Prusiner, S. B. (1997). Prion diseases and the BSE crisis. *Science* **278**, 245–251.

30. Bader, F., Davis, G., Dinowitz, M. (1997). "Assessment of Risk of Bovine Spongiform Encephalopathy," BSE Committee Report. Pharmaceutical Research and Manufacturers of America, Washington, DC.

31. Nishida, N., Katamine, S., Shigematsu, K., Nakatani, A, Sakamoto, N., Hasegawa, S., Nakaoke, R., Atarashi, R., Kataoka, Y., and Miyanoko, T. (1997). Prion protein is necessary for latent learning and long term memory retention. *Cell. Mol. Neurobiol.* **17**, 537–545.

32. Zeidler, M., Stewart, G. E., Barraclough, C. R., Bateman, D. E., Bates, D., Bum, D. J., Colchester, A. C., Durward, W., Fletcher, N. A., Hawkins, S. A., Mackensie, J. M., and Will, R. G. (1997). New variant Creutzfeldt–Jakob disease: Neurological features and diagnostic tests. *Lancet* **350**, 903–907.

33. U.S. Department of Health and Human Services (1987). "Guideline on Validation of the Limulus Amebocyte Lysate Assay Test as an End Product Endotoxin Test for Human and Animal Parenteral Drugs, Biological Products and Medical Devices." USDHHS, Public Health Service, Food and Drug Administration, Washington, DC.

34. du Moulin, G. C., Lynch, S. E., Hedley-Whyte, J., and Broitman, S. A. (1985). Detection of gram-negative bacteremia by *Limulus* amebocyte lysate assay: Evaluation in a rat model of peritonitis. *J. Infect. Dis.* **151**, 148–172.

35. du Moulin, G. C., Price, J. E., Shen, Y., and Osband, M. E. (1992). Use of the *Limulus* amebocyte lysate assay (LAL) in living cell therapies. *Biopharmacology* **5**, 32–54.

36. Hurley, J. C. (1995). Endotoxemia: Methods of detection and clinical correlates. *Clin. Microbiol. Rev.* **8**, 268–292.

37. Casey, W. F., Hauser, G. J., Hannallah, R. S., Midgley, F. M., and Khan, W. N. (1992). Circulating endotoxin and tumor necrosis factor during pediatric cardiac surgery. *Crit. Care Med.* **20**, 1090–1096.

38. Garcia, C., Saladino, R., Thompson, C., Hammer, B., Parsonnet, J., Wainwright, N., Novitsky, T., Fleisher, G. R., and Siber, G. (1994). Effect of a recombinant endotoxin neutralizing protein on endotoxin shock in rabbits. *Crit. Care Med.* **22**, 1211–1218.

39. Ogawa, Y., Murai, T., and Kanoh, S. (1986). Characterization of the pyrogenicity of two different lipopolysaccharides and their lipid A bovine serum albumin complexes. *J. Pharmacobiodyn.* **9**, 722–728.

40. Morales, T. I., Wahl, L. M., and Hascall, V. C. (1984). The effect of bacterial lipopolysacharide on the biosynthesis and release of proteoglycans from calf articular cartilage cultures. *J. Biol. Chem.* **259**, 6720–6729.

41. Edelman, J., Cardozo, C., and Lesser, M. (1989). Lipopolysaccharide stimulates alveolar macrophage adherence *in vivo* and *in vitro*. *Agents Actions* **26**, 287–291.

42. Portoles, M. T., Pagani, R., Ainaga, M. J., DiazLaviada, I., and Municio, A. M. (1989). Lipopolysacharride induced insulin resistance in monolayers of cultured hepatocytes. *Br. J. Exp. Pathol.* **70**, 199–205.

43. Fishel, S., Jackson, P., Webster, J., and Faratian, B. (1988). Endotoxins in culture medium for human *in vitro* fertilization. *Fert. Fertil.* **49**, 108–111.

44. The United States Pharmacopeia (2000). "Bacterial Endotoxins Test" (USP 24). U.S. Pharmacopeial Convention, Rockville, MD.

45. Pitkin, Z., Chhugani, P., Chenette, A., Shen, Y. J., and du Moulin, G. C. (1996). Validation of an automated method of endotoxin testing for use in the end-product testing of *ex vivo* activated T-lymphocytes used in a somatic cell therapy. *Biotechnol. Bioeng.* **50**, 541–547.

46. Olson, M. E., Ceri, H., Morck, D. W., and Lee, C. C. (1995). Biomaterials and the immune system. *In* "Encyclopedic Handbook of Biomaterials and Bioengineering" (D. Altobelli *et al.*, eds.), Vol. I, pp. 493–516. Dekker, New York.

47. "Biological Evaluation of Medical Devices," International Organization of Standardization. (19xx). ISO 10093, Parts 1–17. Geneva, Switzerland.

48. Wallin, R. F., and Upman, P. J. (1995). Evaluating the biological effects of medical devices and materials. *In* "Encyclopedic Handbook of Biomaterials and Bioengineering" (D. Altobelli *et al.*, eds.), Vol. I, pp. 399–432. Dekker, New York.

49. The United States Pharmacopeia (2000). "Biological Reactivity Tests, *In Vitro*" (USP 24). U.S. Pharmacopeial Convention, Rockville, MD.

50. The United States Pharmacopeia (2000). "Biological Reactivity Tests, *In Vivo*" (USP 24). U.S. Pharmacopeial Convention, Rockville, MD.

51. U.S. Department of Health and Human Services (1999). General Safety Test. CFR 21. Part 610.11. USDHHS, Public Health Service, Food and Drug Administration, Washington, DC.

CHAPTER 6

CULTURE ENVIRONMENTS: CELL–POLYMER–BIOREACTOR SYSTEMS

Lisa E. Freed and Gordana Vunjak-Novakovic

INTRODUCTION

One approach to tissue engineering is to provide *in vitro* culture conditions that mimic the biochemical and physical signals that regulate *in vivo* tissue development and maintenance. Our model tissue engineering system has three major components [1]: cells able to express their differentiated phenotype, polymeric scaffolds that provide a three-dimensional (3D) structure for cell attachment and tissue growth, and bioreactor culture vessels that provide an *in vitro* environment in which cell-polymer constructs can develop into functional tissues (Fig. 6.1). The cell–polymer–bioreactor system can be used for controlled *in vitro* studies of tissue development and function, and to generate implants for *in vivo* tissue repair. Specific design requirements of the cell–polymer–bioreactor system depend on the dimensional and functional requirements of the tissue to be engineered. In this chapter, we describe the methods we have developed to engineer two different tissues, cartilage and cardiac muscle. A chronological overview of the tissue engineering studies we have carried out using various cell–polymer–bioreactor systems is provided in Table 6.1 [2–28].

THE CELLS

Ideally, the cells used for tissue engineering should have the capacity to first proliferate and then differentiate *in vitro*, in a manner that can be reproducibly controlled. For cartilage tissue engineering, these criteria can be met by either articular chondrocytes or bone marrow stromal cells (BMSC) [29], and *in vivo* studies have shown that both chondrocyte- and BMSC-based grafts can be used to repair large, full-thickness cartilage defects in rabbit knee joints [26,30–32]. The foregoing criteria have not yet been met for cardiac tissue engineering, since cardiac cells have a very limited proliferative capacity, precursor cell differentiation cannot yet be controlled, and *in vivo* studies in immunocompetent animals have not been published. *In vitro* studies typically use heart cells obtained from embryonic chicks or neonatal rats [10,33,34]. Specific protocols that we have used for cell isolation, and in some cases cell selection and expansion in monolayers, are provided in the subsections that follow.

ARTICULAR CHONDROCYTES

Articular chondrocytes have been obtained from 2- to 4-week-old bovine calves [2] and from 2- to 8-month-old New Zealand white rabbits [3]. Bovine cartilage is harvested from the region between the lateral and medial ridges of the femoropatellar groove extending down to the subchondral bone, whereas lapine cartilage is harvested from the articular surfaces of a variety of joints (e.g., knee, shoulder). Chondrocytes are isolated by enzymatic digestion as described elsewhere [2,35]. In brief, diced tissue is placed in 0.1–0.2% collagenase

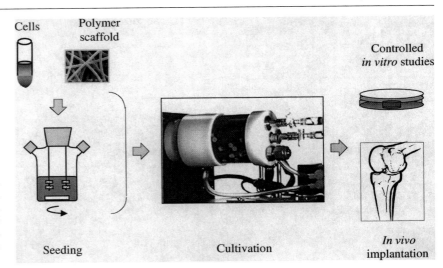

Fig. 6.1. The tissue engineering model system. Cells (obtained from cartilage, bone marrow, or cardiac tissue) are seeded onto three-dimensional polymer scaffolds in spinner flasks and cultured in bioreactors. The resulting engineered tissues can be either used for in vitro research or in vivo tissue repair.

type II (about 10 ml enzyme per gram wet weight, ww) and incubated at 37°C for 15–18 h on an orbital shaker set at 25–30 rpm. Cells are rinsed in phosphate-buffered saline (PBS) without calcium or magnesium and supplemented with 0.02% ethylenediaminetetraacetic acid (EDTA) to minimize aggregation. Cell yields for bovine calf and lapine cartilage range from $3–5 \times 10^7$ cells/g wet tissue. Cells are resuspended in Dulbecco's Modified Eagle's Medium (DMEM) containing 4.5 g/liter glucose and 4 mM glutamine supplemented with 10% fetal bovine serum (FBS), 10 mM N-2-hydroxyethylpiperazine N'-2-ethanesulfonic acid (HEPES), 0.1 mM nonessential amino acids (NEAA), 0.4 mM proline, 50 mg/liter ascorbic acid, 100 U/ml penicillin, 100 mg/liter streptomycin, and 0.5 µg/ml fungizone (optional) [36].

If a sufficient amount of donor cartilage is available, primary chondrocytes can be seeded directly onto polymer scaffolds as described shortly. If only a limited quantity of donor cartilage is available, chondrocytes (bovine and lapine) can be amplified by subculture in the medium just described, supplemented with 5 ng/ml of fibroblast growth factor (FGF-2) [17]. In brief, cells are cultured in monolayers (3×10^5 cells and 10 ml of medium per 100-mm petri dish, or 4×10^3 cells/cm^2) in a humidified 37°C/5% CO$_2$ incubator. Medium is completely replaced every 3–4 days. After 7–10 days, when cell density reaches $4–6 \times 10^4$ cells/cm^2, first-passage (P1) cells are trypsinized (using 0.25% trypsin/1.0 mM EDTA), and replated at 4×10^3 cells/cm^2. After 4–6 more days, when cell density reaches about 1×10^5 cells/cm^2, second-passage (P2) cells are trypsinized and seeded onto polymer scaffolds as described later. During expansion, the chondrocytes undergo about 10 doublings.

BONE MARROW STROMAL CELLS (BMSC)

BMSC have been obtained from 16-day-old embryonic chicks [15] or 2- to 4-week-old bovine calves [24]. Bone marrow is harvested from the tibiae and/or femorae, passed through 16- to 20-gauge needles to make a single cell suspension, and resuspended in DMEM containing 4.5 g/liter glucose and 4 mM glutamine supplemented with 10% FBS, 0.1 mM NEAA, 100 U/ml penicillin, 100 mg/l streptomycin, and 1 ng/ml FGF-2. Nucleated cells are counted and cultured in monolayers (2×10^6 cells and 10 ml of FGF-supplemented medium per 100-mm petri dish, corresponding to 2.5×10^4 cells/cm^2). Medium is completely replaced every 3–4 days, and BMSC are selected based on their ability to adhere to the dish. After 2–3 weeks, when the dishes are approximately 80% confluent, P1 cells are trypsinized and replated at 5×10^3 cells/cm^2. After about 1 more week, when dishes are fully confluent, P2 cells are trypsinized and seeded onto polymer scaffolds as described shortly. During expansion, the BMSC undergo 10–20 doublings.

CARDIAC CELLS

Cardiac cells have been obtained from 1- to 2-day-old neonatal rats [21,22] or 14- to 15-day embryonic chicks [10,22]. Ventricular tissue is harvested, and cells are obtained by enzymatic digestion as already described [37–39]. In brief, diced tissue is incubated in 0.06–0.1% trypsin (about 2 ml of enzyme per heart) at 4°C for 15 h on an orbital shaker set at 25 rpm, then dissociated using 0.1% type II collagenase at 37°C in four or five sequential steps on an orbital shaker set at 50 rpm. The first digestate is discarded, and cells obtained from the subsequent digestions are resuspended in medium. Medium used for neonatal rat cells is DMEM containing 1 g/liter glucose, 110 mg/liter pyruvate, 4 mM glutamine, and 25 mM HEPES, supplemented with 100 U/ml penicillin and 0.5 µg/ml fungizone. Medium used for embryonic chick cells is M199 supplemented with 6% FBS [37]. If desired, the fraction of myocytic cells can be increased to approximately 80% of total by preplating [40]. In brief, cells are preplated in monolayers (approximately $3–5 \times 10^7$ cells and 10 ml medium per 75 cm^2 flask) for two 1-h periods to enrich the fraction of myocytes by allowing attachment of fibroblasts. Cells that remain unattached after the second preplating are used to seed polymer scaffolds. Cell yields range from $1.5–7 \times 10^6$ cells/heart.

THE SCAFFOLDS

An ideal scaffold for tissue engineering should meet the following criteria: reproducible processing into complex, 3D, highly porous structure that permits a spatially uniform cell distribution during cell seeding and minimizes diffusional constraints during *in vitro* cultivation, and controlled degradation at a rate matching that of *in vitro* tissue regeneration, followed by complete elimination of the foreign material to maximize *in vivo* biocompatibility. Of the many classes of biodegradable polymers, we have worked mainly with the poly(α-hydroxyesters) and focused on poly(glycolic acid) (PGA), poly(l-lactic acid) (PLLA), and poly(lactic-*co*-glycolic) acid (PLGA), which are known to be biocompatible and are in widespread clinical use as surgical sutures.

The PGA mesh, with 97% porosity and 13-µm diameter fibers, loses its mechanical integrity over 2 weeks and decreases in mass by 50% over 4 weeks of *in vitro* culture [6] and tested *in vivo* [2,3,26]. In brief, PGA (weight average molecular weight of 69 kDa) is melt-extruded and drawn into an oriented multifilament yarn. The yarn is crimped, cut, carded into a lofty web, needled to form a nonwoven mesh of interlocked fibers, and pressed between heated plates to improve its mechanical stability. The mesh is punched into discs, 5 or 10 mm in diameter, 1, 2, or 5 mm thick, that are sterilized with ethylene oxide (ETO, for 12 h), aerated (for at least 24 h), packaged in trilaminate aluminum foil pouches in a dry box to minimize hydrolytic degradation, and stored at room temperature.

Two surface modifications of the PGA mesh have been studied: hydrolysis (to increase hydrophilicity), and laminin coating (to improve cell attachment and spreading). In brief, surface-hydrolyzed PGA (s-PGA) is made by first incubating PGA discs in 1 N NaOH for 1 min at room temperature. The discs are rinsed with deionized water for 1 min, washed with 200-proof ethyl alcohol and ether for 1 min each, lyophilized, and ETO-sterilized [41]. Surface-hydrolyzed PGA mesh can be coated with laminin by incubating s-PGA discs in 5–7 µg/ml laminin for 4–6 h at 37°C on a orbital shaker at 50–60 rpm, followed by lyophilization and ETO sterilization [23].

PLLA foams with 91% porosity and 100- to 500-µm diameter pores that maintain their size and shape over 2 months of *in vitro* culture were characterized [42] and tested *in vitro* and *in vivo* [2]. In brief, porogen (100- to 500-µm diameter particles of NaCl) is added to a solution of PLLA (number-average molecular weight of 105 kDa) in chloroform at a porogen-to-polymer ratio of 9:1. The vortexed dispersion is cast in 50-mm petri dishes, the solvent is removed by evaporation followed by lyophilization, the porogen is extracted by extensive rinsing in water, and the scaffold is again lyophilized. PLLA foams are cut or punched to the desired dimensions, ETO-sterilized, aerated, and stored in a desiccator at room temperature.

Foams made of an 80:20 blend of PLGA and poly(ethylene glycol) (PLGA-PEG) with 87% porosity and pores 300–500 µm in diameter that maintain their size and shape over

Table 6.1. *Overview of Tissue Engineering Studies We Have Done with the Cell–Polymer–Bioreactor System*

| Cells (expansion; source) | Scaffold | | Construct | | |
	Material	Dimensions	Seeding	Culture	Duration
Articular and costal chondrocytes (primary; bovine and human)	PGA mesh, PLLA sponge	10 mm × (5–10) mm × (2–3) mm 10 mm dia hemisphere	Static dish	Static dish	8 weeks
Articular chondrocyte (expanded; lapine)	PGA mesh	10 mm dia hemisphere 10 mm dia × 2 mm thick	Static dish	Static dish	4 weeks
Articular chondrocyte (primary; bovine)	PGA mesh	10 mm dia × (1–4) mm thick	Static dish	Static dish Static or mixed flask	7 weeks
Articular chondrocyte (primary; bovine)	PGA mesh	10 mm dia × (1–4) mm thick	Static or mixed dish	Static or mixed dish	8 weeks
Articular chondrocyte (primary; bovine)	PGA mesh	10 mm dia × 5 mm thick	Mixed dish	Mixed dish	8 weeks
Articular chondrocyte (primary; bovine)	PGA mesh	5 mm × 5 mm × 2 mm	STLV, Mixed flask	STLV, Mixed flask	1 week
Articular chondrocyte (primary; bovine)	PGA mesh	10 mm dia × 5 mm thick	Mixed dish Mixed flask	Static or mixed dish Static or mixed flask	8 weeks
Articular chondrocyte (primary; bovine)	PGA mesh	10 mm dia × 5 mm thick	Mixed dish	Mixed dish	6 weeks
Articular chondrocyte (bovine) and cardiac myocyte (primary; avian)	PGA mesh	10 mm dia × 2 mm thick	Mixed dish, Mixed flask, STLV, HARV	Mixed dish, Mixed flask, STLV, HARV	3 weeks
Articular chondrocyte (primary; bovine)	PGA mesh	5 mm dia × 2 mm thick	Mixed flask	STLV, RWPV	3 months on earth (STLV) and then 4 months on earth or in space (RWPV)
Articular chondrocyte (primary; bovine)	PGA mesh	5 mm dia × 2 mm thick	Mixed flask	STLV	6 weeks
Articular chondrocyte (primary; bovine)	PGA mesh	5 mm dia × 2 mm thick	Mixed flask	STLV	6 weeks

(continues)

Table 6.1. *(continued)*

Assessment	Results	Reference
H; IH (type II collagen), *in vitro* and *in vivo* GAG and DNA	Feasibility of using human chondrocytes as a cell source to engineer cartilage, PGA mesh better than PLLA sponge	[2]
H (scored), *in vitro* and *in vivo* GAG, DNA, total collagen	Fairly good repair using engineered cartilage as allograft in 3 mm dia × full thickness FPG defect in adult rabbits	[3]
H; Cell number and density, GAG diffusional permeability to glucose	Structure improved by using high initial density of chondrocytes and mixing during cultivation	[4]
H; GAG, DNA, total collagen	Structure improved by mixing during seeding and cultivation	[5]
H; scaffold degradation rate GAG, DNA, total collagen	Properties of PGA mesh characterized	[6]
H; IH (types I and II collagen) GAG, DNA, total collagen	Feasibility of using STLVs to engineer cartilage shown; operating conditions for construct suspension defined	[7]
H; GAG, DNA, total collagen	Structure improved by mixing during seeding and cultivation; hydrodynamic conditions in mixed flasks characterized	[8]
H; GAG, DNA, total collagen diffusional rates of tracers	Diffusional construct permeability decreased during culture	[9]
H; GAG, DNA, total collagen (cartilage constructs)		[10]
H; contractility by video imaging (cardiac constructs)	STLVs better than mixed flasks or dishes for engineering cartilage; Feasibility of cardiac tissue engineering in HARVs demonstrated; cardiac constructs contracted synchronously	
H; TEM; DNA, GAG, total collagen, type II collagen (ELISA); radiolabeled tracer incorporation rates; mechanical properties in static and dynamic compression	Viable engineered cartilage grown in both groups; constructs grown 7 months on earth were larger and stiffer than constructs grown 3 months on earth and then 4 months in space; compressive stiffness of 7 month earth grown cartilage was normal	[11]
H; IH (collagen type II); DNA, GAG, total and type II collagen (ELISA); radiolabeled tracer incorporation rates	Spatial and temporal patterns of chondrogenesis characterized; construct cross sections (6.7 × 5 mm) were continuously cartilaginous; more than 90% of the total collagen was type II	[12]
SEM; DNA, GAG, total collagen; Western blot for collagen types II, IX, and X; pyridinium cross-links	Collagen network in engineered cartilage characterized; network structure and composition comparable to native cartilage; collagen content and cross-linking subnormal	[13]

(continues)

Table 6.1. (continued)

Cells (expansion; source)	Scaffold		Construct		
	Material	Dimensions	Seeding	Culture	Duration
Articular chondrocyte (primary; bovine)	PGA mesh	10 mm in diameter × (2–5) mm thick 5 mm in diameter × 2 mm thick	Mixed flask	Mixed flask	1 day– 6 weeks
BMSC (expanded; avian)	PGA mesh	5 mm dia × 2 mm thick	Mixed flask	Mixed dish	4 weeks
Articular chondrocyte (primary; bovine)	PGA mesh	10 mm dia × 5 mm thick	Mixed flask	Perfused column	7 weeks
Articular chondrocyte (expanded; bovine)	PGA mesh	5 mm dia × 2 mm thick	Mixed flask	Mixed dish	6 weeks
Articular chondrocyte (primary; bovine)	PGA mesh	5 mm dia × 2 mm thick	Mixed flask	Static flask, mixed flask, STLV	6 weeks
Articular chondrocyte (primary; bovine)	PGA mesh	5 mm dia × 2 mm thick	Mixed flask	Static flask, mixed flask, STLV	6 weeks
Articular chondrocyte (primary; bovine)	PGA mesh	5 mm dia × 2 mm thick	Mixed flask	STLV	6 weeks
Cardiac myocyte (primary; rat)	PGA mesh	5 mm dia × 2 mm thick	Mixed flask	Mixed flask	1 week

(continues)

1 month of *in vitro* culture were developed by Shastri *et al.* [43] and tested *in vitro* [24,27]. In brief, porogen (paraffin particles, 300–500 μm) is added to a solution of PLGA-PEG (with weight-average molecular weights of 128 and 10 kDa, respectively) in methylene chloride at a porogen-to-polymer ratio of 4:1. The mixture is compacted in a perforated Teflon mold, the porogen is extracted by using hexane at 50°C, and the solvent is removed by lyophilization. PLGA-PEG foams are cut or punched to the desired dimensions, ETO-sterilized, aerated, and stored in a desiccator at room temperature.

Table 6.1. *(continued)*

Assessment	Results	Reference
Row 5 Row 5 Size distribution of cell aggregates; kinetics of cell seeding; DNA, GAG; H and image analysis (cell distribution)	Seeding in mixed flasks resulted in rapid, high-yield cell attachment; mathematical model describing the kinetics and mechanisms of cell seeding	[14]
H; IH (collagen types II and X, bone sialoprotein and osteopontin); DNA, GAG and total collagen; mineralized area; osteocalcin release; AP activity	Feasibility of using BMSC as a cell source to engineer cartilage and bone	[15]
GAG distribution (MRI); H; GAG	MRI method for quantitiative on line monitoring of GAG distribution in cartilaginous constructs	[16]
H; IH (collagen types II and X; F-actin); DNA, GAG and total collagen; cell doubling times; AP activity;	Chondrocytes expanded 2000-fold in the presence of FGF-2 in monolayers underwent chondrogenesis when cultured on PGA mesh	[17]
H; DNA, GAG, total collagen; electromechanical properties in static and dynamic compression	Construct structure and function could be modulated by flow and mixing; electromechanical parameters correlated with wet weight fractions of GAG, collagen, and water	[18]
H; amounts and distributions of GAG by computer-based image analysis	A method developed for quantitative analysis of GAG distribution in cartilage constructs and explants	[19]
Medium composition (pH, oxygen, glucose, lactate, ammonia, GAG); GAG distributions; amounts of DNA, GAG and collagen; radiolabeled tracer incorporation rates	Construct size and structure improved by medium oxygen tension of 80 mmHg and pH of 7, compared to more anaerobic culture conditions	[20]
Medium composition (oxygen, glucose, lactate, ammonia, LDH); H; IH (sarcomeric tropomyosin); TEM; MTT; DNA and protein; electrophysiological assessment (conduction velocity, maximum capture rate, amplitude, excitation threshold)	Tissue-like region at the construct periphery exhibited relatively homogeneous electrical properties and sustained macroscopic signal propagation on a centimeter scale; tropomyosin, sarcomeres, and intercalated discs demonstrated	[21]

(continues)

All hydrophobic scaffolds must be scrupulously pre-wetted such that all internal air is displaced and cell attachment can occur. PGA mesh can be prewetted by immersion in medium and gentle, repeated pressure with a pair of curved forceps. PLLA and PLGA-PEG foams can be pre-wetted by immersion in 70% ethyl alcohol and then transferred to medium [2]. In addition, to promote the nonspecific adsorption of serum proteins, thereby facilitating cell attachment, all synthetic polymer scaffolds can be incubated for 8–24 h at 37°C in medium containing 10% FBS [6].

Table 6.1. (continued)

Cells (expansion; source)	Scaffold		Construct		
	Material	Dimensions	Seeding	Culture	Duration
Cardiac myocyte (primary; rat)	PGA mesh	5 mm dia × 2 mm thick	mixed dish, mixed flask, HARV	mixed dish, static flask, mixed flask, STLV, HARV	2 weeks
Cardiac myocyte (primary; rat)	PGA mesh, surface-hydrolyzed, laminin-coated	5 mm dia × 2 mm thick	mixed dish, HARV	HARV	1 week
BMSC (selectively expanded; bovine)	PLGA–PEG sponge	5 mm dia × 2 mm thick	Mixed flask	Mixed dish	4 weeks
Articular chondrocytes (primary; bovine)	PGA mesh	5 mm dia × 2 mm thick	Mixed flask	STLV	12 weeks
Articular chondrocytes (expanded; lapine) bone marrow (autologous, lapine)	PGA mesh Collagraft	7 mm × 5 mm × 1 mm 7 mm × 5 mm × 4 mm	Mixed flask	Mixed dish	4 weeks
Articular chondrocyte (primary; bovine) periosteal cells (expanded; bovine)	PGA mesh PLGA–PEG sponge	5 mm dia × 2 mm thick 4.5 mm dia × 1.5 mm thick	Mixed flask	Mixed dish	8 weeks
Articular chondrocyte (bovine)	PGA mesh	5 mm dia × 2 mm thick	Mixed flask	STLV	6 weeks

(continues)

BIOREACTORS

Schematics of the various bioreactor vessels and a summary of their operating parameters and environmental conditions are provided in Fig. 6.2 [44–46].

BIOREACTOR PREPARATION

Spinner flasks (Fig. 6.2a), 12 cm high and 6.5 cm in diameter, are fitted with a stopper and two side arms with caps that are kept loose to permit gas exchange. Flasks are cleaned, autoclave-sterilized, filled with 120 ml of medium, and mixed with a nonsuspended magnetic stir bar (4 cm long, 0.8 cm diameter) at 50–80 rpm.

The slow-turning lateral vessel (STLV) (Fig. 6.2b), which was developed at NASA [47], is configured as two concentric cylinders that have diameters of 5.75 and 2 cm. The annular space is used for tissue culture, and the inner cylinder, which has a hollow core covered with a silicone membrane, serves as a gas exchanger. The high-aspect-ratio vessel (HARV)

Table 6.1. (continued)

Assessment	Results	Reference
Medium composition (oxygen, glucose, lactate, ammonia, LDH); H; IH (sarcomeric α-actin, tropomyosin, muscle desmin, cardiac troponin-T); TEM; MTT; amounts of DNA and protein; contractility (video imaging)	Constructs contained a peripheral tissue-like region; expressing cardiac-specific markers and ultrastructural properties; mixing and high initial cell density improved construct structure; laminar flow conditions enhanced cell viability and metabolism;	[22]
H; IH (sarcomeric tropomyosin); TEM electrophysiological assessment (conduction velocity, maximum capture rate, amplitude, excitation threshold) Western blot for CX-43, CK-MM, MHC	Laminin-coated PGA, low serum medium and laminar flow conditions (HARVs) improved construct structure and electro-physiological properties; constructs had normal conduction velocity; amounts of CX-43, CK-MM, and MHC determined	[23]
H; IH (types I, II, and X collagen, osteonectin, osteopontin); DNA, GAG, total collagen; computer-based image analysis	Mammalian BMSC cultured on PLGA–PEG sponges formed cartilage (in the presence of Dex, Ins, and TGF-β) or bone like tissue (in the presence of Dex, β-GP)	[24]
H; IH (type I collagen); DNA, GAG, total collagen; adhesive strength of construct–explant interface	Integrative potential of engineered cartilage was enhanced by the presence of biosynthetically active cells and trypsin treatment of adjacent cartilage; integration patterns characterized	[25]
H (scored); IH (types II, X collagen); DNA, GAG, collagen; *in vitro* and *in vivo* mechanical properties (indentation testing); computer-based image analysis	Very good repair using composites of engineered cartilage and Collagraft in 5 mm × 5 mm × 7 mm FPG defects in adult rabbits; osteochondral remodeling patterns characterized	[26]
H (semiquantitative) (cells, GAG, mineralization; integration); IH (collagen types II and X; osteocalcin); GAG cell doubling times; AP activity	Structures of cartilage and bone components of tissue composites and integration at the cartilage–bone interface depended on culture duration	[27]
GAG and cell distributions; GAG release into medium; oxygen concentration in medium	Mathematical model of GAG deposition in cartilage constructs	[28]

Abbreviations: AP, alkaline phosphatase; BMSC, bone marrow stromal cells; CK-MM, creatine kinase MM; CX-43, connexin-43; Dex, dexamethasone; ECM, extracellular matrix; FGF-2, fibroblast growth factor; FPG, femoropatellar groove; GAG, glycosaminoglycans; H, histology; IH, immunohistochemistry; Ins, insulin; LDH, lactate dehydrogenase; MHC, sarcomeric myosin heavy chain; MRI, magnetic resonance imaging; MTT, 3-(4,5-dimethylthiazol-2-yl)-2,5-diphenyltetrazolium bromide; PGA, poly(glycolic acid); PLGA, poly(lactic-co-glycolic acid); PEG, poly(ethylene glycol); SEM, scanning electron microscopy; TEM, transmission electron microscopy; TGF-β, transforming growth factor beta; βGP, β-glycerophosphate.

(Fig. 6.2c), which was developed at NASA [48], is configured as a cylinder 1.3 cm high and 10 cm in diameter. The base of the HARV is covered with a silicone membrane that serves as a gas exchanger. The STLV and the HARV are soaked in 70% ethyl alcohol (for ≥12 h), extensively rinsed, dried, and autoclave-sterilized. Bioreactors are then primed with 110 ml of medium to displace all air and mounted on a base that simultaneously rotates the vessel around its central axis at the desired rate (10–45 rpm) and pumps filter-sterilized incubator air over the gas exchange membrane at a rate of approximately 1 liter/min.

The rotating wall perfused vessel (RWPV) (Fig. 6.2d), which is flight hardware developed at NASA, is configured as a 125-ml annular space between two concentric cylinders, with the medium inlet at one end and the medium outlet via a filter on the central cylinder. A flat disc at one end of the central cylinder serves as a viscous pump. The inner and outer cylinders can be rotated at the same or different rates. The RWPV is primed with 125 ml of medium, which is periodically recirculated between the RWPV and an external gas exchanger and periodically exchanged with fresh medium.

Perfused columns (Fig. 6.2e) are custom-made glass tubes (9.5 cm long, 1.7 cm o.d., 1.5 cm i.d.) sealed with silicone stoppers at either end, with the fluid inlet (bottom) and outlet (top) positioned off center, on opposing sides of the column. Each column is filled with 15 ml of medium to displace all air, and medium is continuously recirculated between the column and an external gas exchanger.

Perfused chambers (Fig. 6.2f) are configured as a central tissue culture compartment bounded by a cylindrical membrane, and an annular cell-free space, with an inlet to the central compartment and outlet from the annular space. Each chamber is filled with 3, 10, or 30 ml of medium to displace all air, and medium is continuously recirculated between the chamber and an external gas exchanger.

SEEDING CELLS ONTO SCAFFOLDS

All cell seeding is done in a humidified, 37°C, 5–10% CO_2 incubator. The duration of seeding ranges from 1 h to 3 days, depending on the cell type, the scaffold, and the conditions of seeding.

An ideal method for cell seeding of 3D scaffolds should result in a high yield of attached cells, a high cell density, and spatially uniform distribution of attached cells, to promote rapid, homogeneous tissue development. We have investigated a variety of seeding methods by means of different cell types and concentrations, scaffolds, seeding vessels, mixing conditions, and durations.

Thin (<2 mm) PGA meshes can be seeded with a high concentration of chondrocytes in static petri dishes [2,5]. In brief, 0.1 ml of 4×10^7 cells/ml is adsorbed to a dry PGA mesh (10-mm in diameter and 1.2 mm thick) in a 35-mm petri dish by pipetting the cell droplet onto the dish, adding medium, and gently and repeatedly applying pressure to the mesh with a curved pair of forceps. The dish is cultured statically in a humidified, 37°C/5% CO_2 incubator. An additional 1.5 ml of medium is slowly added after 6 h, and 2.5 ml more medium is added after 24 h.

The use of orbitally mixed petri dishes rather than static dishes allows the seeding of thicker (2–5 mm) PGA mesh with chondrocytes [5,6,8]. In brief, cells are inoculated into 35-mm dishes containing medium and pre-wetted PGA mesh (8×10^6 cells and 6 ml media per mesh (10 mm diameter, 5 mm thick) and placed on an orbital shaker set at 75 rpm.

Rotating vessels (STLVs and HARVs), which can maintain a uniform cell suspension and provide relative velocity between the cells and the scaffolds, can be used to seed chondrocytes and cardiac myocytes on PGA mesh [7,10,22,23]. In brief, scaffolds are prewetted for 12 h and transferred into a medium-primed rotating vessel. The vessel is inoculated with cells (e.g., $5–8 \times 10^6$ cells/scaffold), and the rotation speed is adjusted such that the scaffolds are maintained freely suspended during seeding.

Spinner flasks are at present the preferred system for seeding chondrocytes on PGA mesh, resulting in a relatively uniform spatial distribution of cells within 24 h at a yield of essentially 100% [14]. In brief, pre-wetted scaffolds are threaded onto 4-in.-long 22-gauge needles embedded in the flask stopper and positioned by means of 3-mm-long segments of silicone tubing (two to four needles with one to three scaffolds apiece per flask). Flasks are filled with 120 ml of medium and incubated with stirring at 50–75 rpm. After 8–24 h, the medium is completely replaced and flasks are inoculated with cells ($2–8 \times 10^6$ cells/scaffold).

Spinner flasks can also be used to seed cardiac myocytes on PGA mesh (at a lower yield of 40–60%) [22], and to seed BMSC and periosteal cells on prewetted PLGA-PEG foams [24,27]. A study aimed at improving the yield of cardiac myocyte seeding involved a combination of petri dishes and rotating vessels. In brief, 8×10^6 cells in 0.06 ml were pipetted onto laminin-coated s-PGA mesh in 5-mm dishes and mixed on an *x–y–z* gyrator

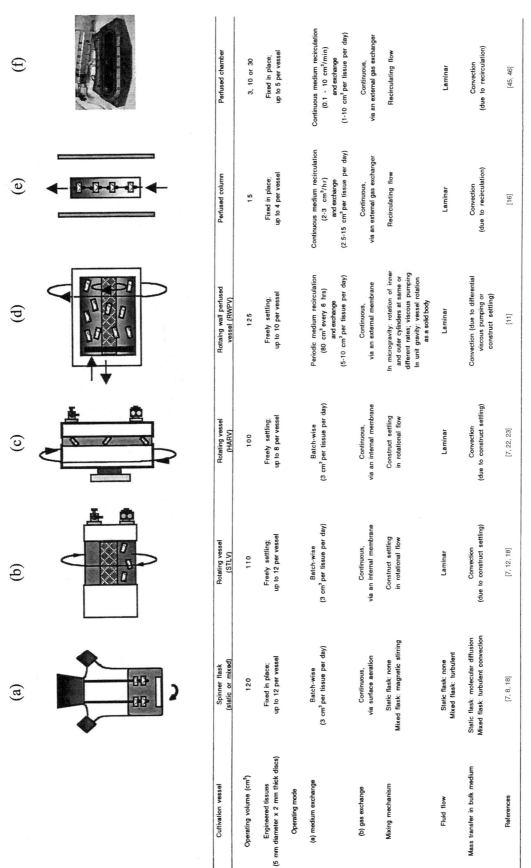

	(a)	(b)	(c)	(d)	(e)	(f)

Cultivation vessel	Spinner flask (static or mixed)	Rotating vessel (STLV)	Rotating vessel (HARV)	Rotating wall perfused vessel (RWPV)	Perfused column	Perfused chamber
Operating volume (cm³)	120	110	100	125	15	3, 10 or 30
Engineered tissues (5 mm diameter x 2 mm thick discs)	Fixed in place; up to 12 per vessel	Freely settling; up to 12 per vessel	Freely settling; up to 8 per vessel	Freely settling; up to 10 per vessel	Fixed in place; up to 4 per vessel	Fixed in place; up to 5 per vessel
Operating mode						
(a) medium exchange	Batch-wise (3 cm³ per tissue per day)	Batch-wise (3 cm³ per tissue per day)	Batch-wise (3 cm³ per tissue per day)	Periodic medium recirculation (80 cm³ every 6 hrs) and exchange (5-10 cm³ per tissue per day)	Continuous medium recirculation (2-3 cm³/hr) and exchange (2.5-15 cm³ per tissue per day)	Continuous medium recirculation (0.1 - 1.0 cm³/min) and exchange (1-10 cm³ per tissue per day)
(b) gas exchange	Continuous, via surface aeration	Continuous, via an internal membrane	Continuous, via an internal membrane	Continuous, via an external membrane	Continuous, via an external gas exchanger	Continuous, via an external gas exchanger
Mixing mechanism	Static flask: none Mixed flask: magnetic stirring	Construct settling in rotational flow	Construct settling in rotational flow	In microgravity: rotation of inner and outer cylinders at same or different rates; viscous pumping In unit gravity: vessel rotation as a solid body	Recirculating flow	Recirculating flow
Fluid flow	Static flask: none Mixed flask: turbulent	Laminar	Laminar	Laminar	Laminar	Laminar
Mass transfer in bulk medium	Static flask: molecular diffusion Mixed flask: turbulent convection	Convection (due to construct settling)	Convection (due to construct settling)	Convection (due to differential viscous pumping or construct settling)	Convection (due to recirculation)	Convection (due to recirculation)
References	[7, 8, 18]	[7, 12, 18]	[7, 22, 23]	[11]	[16]	[45, 46]

Fig. 6.2. Tissue culture bioreactors, their operating conditions, and environmental conditions: (a) spinner flask, (b) slow-turning lateral vessel (STLV), (c) high-aspect-ratio vessel (HARV), (d) rotating wall perfused vessel (RWPV), (e) perfused column, (f) perfused chamber.

at 24 rpm for 1 h; constructs and the remaining (unseeded) cells were then transferred into a 110-ml HARV rotated at 12 rpm [23].

CELL–POLYMER CONSTRUCT CULTIVATION

All cultures are carried out in a humidified, 37°C, 5–10% CO_2 incubator, and medium and constructs are sampled at timed intervals for further analysis.

Petri dishes were used in all of our early studies [2–6], and are still used for engineering skeletal tissues in which the desired constructs are less than 2 mm thick [15,17,26,27] and for studies in which medium contains expensive additives [24]. In dishes, constructs are cultured statically or on an orbital shaker set at 50–75 rpm. The 35-mm dishes typically contain one construct and 5–7 ml of medium per well. Medium is completely replaced every 2–3 days (i.e., 3 ml per construct per day).

Spinner flasks have been used to study the effects of turbulent fluid flow on tissue development [8,18]. In flasks, constructs (4–12/vessel) are cultured fixed in place (threaded on needles, Fig. 6.2a). Each needle holds either three constructs, 5 mm in diameter and 2 mm thick, or two constructs, 10 mm diameter and 5 mm thick; each flask holds two to four needles. Flasks are either stirred at 50–75 rpm or cultured statically. Medium is exchanged batchwise, at a rate of 50% every 2–3 days (i.e., 3 ml per construct per day), to supply fresh nutrients and exogenous regulatory factors and remove waste metabolites while maintaining the presence of cell-secreted regulatory factors.

Rotating bioreactors (STLVs and HARVs) have been used to study the effects of laminar fluid flow on tissue development and are at present the preferred culture system for cartilage tissue engineering if the desired construct thickness exceeds 2 mm [11,12,18] and for cardiac tissue engineering [22,23]. In rotating bioreactors, constructs (8–12/vessel) are cultured without external fixation, in a rotational flow field [7]. Their discoid shape causes the constructs to settle with pitching oscillations, collide with each other, and move laterally along the vessel spin axis. Vessel rotational speed is adjusted such that each construct remains freely suspended. For example, over a 6-week cultivation of engineered cartilage in the STLV, rotation speed is gradually increased from 12 rpm to 45 rpm. Medium is exchanged batchwise (at a rate of 50% every 2–3 days, or 3 ml per construct per day).

The rotating wall perfused vessel (RWPV) can be used to study the effects of microgravity on tissue development. In the RWPV, engineered cartilage (10/vessel) has been cultured for 4 months aboard the *Mir* space station and on earth [11]. In space, the inner and outer cylinders of the bioreactor vessel are differentially rotated to promote convective mixing, whereas on earth, the vessel is rotated as a solid body. Medium is recirculated between the RWPV and an external membrane gas exchanger at a rate of 4 ml/min for 20 min four times a day, and exchanged with fresh medium at a rate of 5–10 ml per construct per day.

Perfused columns can be used in conjunction with magnetic resonance imaging (MRI) to nondestructively monitor the development of engineered cartilage [16]. In perfused columns, constructs (4/vessel) are fixed in place (Fig. 6.2e), and medium is continuously recirculated between the column and an external gas exchanger at a rate of 2–3 ml/h, and periodically replaced at a rate of 2–15 ml per construct per day. At timed intervals, columns are equilibrated with medium containing a MRI contrast agent [gadolinium diethylenetriamine pentacetic acid, $Gd(DTPA)^{2-}$], inserted into an MRI coil, and imaged to map fixed charge density and calculate the distribution of GAG.

A modular bioreactor system containing 8–24 perfused chambers connected in parallel is currently under development for cell and tissue culture in space, with provisions for controlled studies at unit gravity as well as automated control of medium temperature, pH and oxygen tension, automated sampling, and online videomicroscopy [45]. Perfused chambers can be used to culture 3-cm-long, 5-mm-diameter organoids prepared from skeletal myoblasts (C2C12 cells) and collagen gels [49]. Medium is recirculated continuously between the chamber and an external gas exchanger at a rate of 0.2–10 ml/min and periodically exchanged at a rate of 1–10 ml per chamber per day.

ANALYTICAL METHODOLOGIES

Structural assessment of engineered tissues including morphometry, spatial arrangement of cells, and extracellular matrix (ECM) are determined histologically, immunohistochemically by computer-based image analysis [19], and by electron microscopy [11,13,21,22].

Biochemical and molecular analyses of tissue composition include the following: (1) cell number, estimated from the amount of DNA [50], (2) sulfated GAG content, determined by dimethylmethylene blue dye binding [51], (3) total protein content, determined by the BioRad assay (BioRad), (4) total collagen content, determined from hydroxyproline content [52], and cross-linkage [13], (5) type II collagen content, determined by inhibition ELISA [53], and (6) amounts of specific markers for cartilage (type IX collagen) [13] and cardiac muscle (connexin-43, creatine kinase, and sarcomeric myosin heavy chain) [23], as determined by Western blot.

Functional assessment of the cells in engineered tissues include the following: (1) synthesis rates of GAG and collagen, estimated from rates of macromolecular incorporation of radiolabeled tracers [12], (2) metabolic status, assessed from the ratio of glucose consumption and lactate production [20] and from enzymatic conversion of 3-{4,5-dimethylthiazol-2-yl}-2,5-diphenyl tetrazolium bromide (MTT) [21,22], and (3) viability, assessed by medium concentrations of lactate dehydrogenase (LDH) [22] and vital staining (live–dead assay) [23].

Functional evaluation of engineered cartilage includes electromechanical properties (equilibrium modulus, dynamic stiffness, hydraulic permeability, streaming potential), determined by measuring tissue response in static and dynamic radially confined compression [11,18]. Functional assessment of engineered cardiac muscle includes electrophysiological properties (conduction velocity, signal amplitude, excitation threshold, maximum capture rate), which are determined from macroscopic impulse propagation studies done by using an array of extracellular stimulating and recording electrodes [21].

SUMMARY

In this chapter, we discussed the methods we have used to engineer tissues in a cell–polymer–bioreactor system, and some of the underlying principles. Methodological improvements that are likely have a major impact on the field of tissue engineering include the ability to regulate stem cell differentiation (e.g., into cardiac myocytes), the induction of construct vascularization *in vitro*, the development of three-dimensional scaffolds that selectively promote cells to attach and express their differentiated phenotype, the design of bioreactors that can provide physiological levels of mechanical stimuli during construct development, and the extension from animal model systems to human cell sources and scales.

ACKNOWLEDGMENTS

This work was supported by the National Aeronautics and Space Administration (grants NAG9-836 and NCC8-174). The authors thank R. Langer for his helpful advice.

REFERENCES

1. Freed, L. E., and Vunjak-Novakovic, G. (2000). Tissue culture bioreactors. *In* "Principles of Tissue Engineering" (R. P. Lanza, R. Langer, and J. P. Vacanti, eds.), 2nd ed. pp. 143–156. Academic Press, San Diego, CA.

2. Freed, L. E., Marquis, J. C., Nohria, A., Emmanual, J., Mikos, A. G., and Langer, R. (1993). Neocartilage formation in vitro and in vivo using cells cultured on synthetic biodegradable polymers. *J. Biomed. Mater. Res.* 27, 11–23.

3. Freed, L. E., Grande, D. A., Lingbin, Z., Emmanual, J., Marquis, J. C., and Langer, R. (1994). Joint resurfacing using allograft chondrocytes and synthetic biodegradable polymer scaffolds. *J. Biomed. Mater. Res.* 28, 891–899.

4. Freed, L. E., Vunjak-Novakovic, G., Marquis, J. C., and Langer, R. (1994). Kinetics of chondrocyte growth in cell-polymer implants. *Biotechnol. Bioeng.* 43, 597–604.

5. Freed, L. E., Marquis, J. C., Vunjak-Novakovic, G., Emmanual, J., and Langer, R. (1994). Composition of cell-polymer cartilage implants. *Biotechnol. Bioeng.* 43, 605–614.

6. Freed, L. E., Vunjak-Novakovic, G., Biron, R., Eagles, D., Lesnoy, D., Barlow, S., and Langer, R. (1994). Biodegradable polymer scaffolds for tissue engineering. *Bio/Technology* 12, 689–693.

7. Freed, L. E., and Vunjak-Novakovic, G. (1995). Cultivation of cell-polymer constructs in simulated microgravity. *Biotechnol. Bioeng.* 46, 306–313.

8. Vunjak-Novakovic, G., Freed, L. E., Biron, R. J., and Langer, R. (1996). Effects of mixing on tissue engineered cartilage. *AIChE J.* 42, 850–860.

9. Bursac, P. M., Freed, L. E., Biron, R. J., and Vunjak-Novakovic, G. (1996). Mass transfer studies of tissue engineered cartilage. *Tissue Eng.* **2**, 141–150.

10. Freed, L. E., and Vunjak-Novakovic, G. (1997). Microgravity tissue engineering. *In Vitro Cell. Dev. Biol.* **33**, 381–385.

11. Freed, L. E., Langer, R., Martin, I., Pellis, N., and Vunjak-Novakovic, G. (1997). Tissue engineering of cartilage in space. *Proc. Natl. Acad. Sci. U.S.A.* **94**, 13885–13890.

12. Freed, L. E., Hollander, A. P., Martin, I., Barry, J. R., Martin, I., and Vunjak-Novakovic, G. (1998). Chondrogenesis in a cell–polymer–bioreactor system. *Exp. Cell Res.* **240**, 58–65.

13. Riesle, J., Hollander, A. P., Langer, R., Freed, L. E., and Vunjak-Novakovic, G. (1998). Collagen in tissue engineered cartilage: Types, structure and cross-links. *J. Cell. Biochem.* **71**, 313–327.

14. Vunjak-Novakovic, G., Obradovic B., Bursac P., Martin, I., Langer R., and Freed, L. E. (1998). Dynamic seeding of polymer scaffolds for cartilage tissue engineering. *Biotechnol. Prog.* **14**, 193–202.

15. Martin, I., Padera, R. F., Vunjak-Novakovic, G., and Freed, L. E. (1998). In vitro differentiation of chick embryo bone marrow stromal cells into cartilaginous and bone-like tissues *J. Orthop. Res.* **16**, 181–189.

16. Williams, S. N. O., Burstein, D., Freed, L. E., Gray, M. L., Langer, R., and Vunjak-Novakovic, G. (1998). MRI measurements of fixed charge density as a measure of glycosaminoglycan content and distribution in tissue engineered cartilage. *Trans. Orthop. Res. Soc.* **23**, 203.

17. Martin, I., Vunjak-Novakovic, G., Yang, J., Langer, R., and Freed, L. E. (1999). Mammalian chondrocytes expanded in the presence of fibroblast growth factor-2 maintain the ability to differentiate and regenerate three-dimensional cartilaginous tissue. *Exp. Cell Res.* **253**, 681–688.

18. Vunjak-Novakovic, G., Martin, I., Obradovic, B., Treppo, S., Grodzinsky, A. J., Langer, R., and Freed, L. E. (1999). Bioreactor cultivation conditions modulate the composition and mechanical properties of tissue engineered cartilage. *J. Orthop. Res.* **17**, 130–138.

19. Martin, I., Obradovic, B., Freed, L. E., and Vunjak-Novakovic, G. (1999). A method for quantitative analysis of glycosaminoglycan distribution in cultured natural and engineered cartilage. *Ann. Biomed. Eng.* **27**, 656–662.

20. Obradovic, B., Carrier, R. L., Vunjak-Novakovic, G., and Freed, L. E. (1999). Gas Exchange is essential for bioreactor cultivation of tissue engineered cartilage. *Biotechnol. Bioeng.* **63**, 197–205.

21. Bursac, N., Papadaki, M., Cohen, R. J., Schoen, F. J., Eisenberg, S. R., Carrier, R., Vunjak-Novakovic, G., and Freed, L. E. (1999). Cardiac muscle tissue engineering: Towards an in vitro model for electrophysiological studies. *Am. J. Physiol. Heart Circ. Physiol.* **277**, H433–H444.

22. Carrier, R., Papadaki, M., Rupnick, M., Schoen, F. J., Bursac, N., Langer, R., Freed, L. E., and Vunjak-Novakovic, G. (1999). Cardiac tissue engineering: Cell seeding, cultivation parameters and tissue construct characterization. *Biotechnol. Bioeng.* **64**, 580–589.

23. Papadaki, M., Bursac, N., Langer, R., Merok, J., Vunjak-Novakovic, G., and Freed, L. E. (2001). Tissue engineering of functional cardiac muscle: Moleucular, structural, and electrophysiological studies. *Am. J. Physiol. Heart Circ. Physiol.* **280**, H168–H178.

24. Martin, I., Shastri, V., Padera, R. F., Yang, J., MacKay, A. J., Langer, R., Vunjak-Novakovic, G., and Freed, L. E. (2001). Selective differentiation of bone marrow stromal cells cultured on three dimensional polymer foams. *J. Biomed. Mat. Res.* **55**, 229–235.

25. Obradovic, B., Martin, I., Padera, R., Treppo, S., Freed, L., and Vunjak-Novakovic, G. (2001). Integration of engineered cartilage. *J. Orthopaedic. Res.* (in press).

26. Schaefer, D., Martin, I., Jundt, G., Seidel, J., Bergin, J., Grodzinsky, A. J., Vunjak-Novakovic, G., and Freed, L. E. (2000). Tissue engineered composites for the repair of large osteochondral defects. *Trans. Orthop. Res. Soc.* **25**, 619.

27. Schaefer, D., Martin, I., Shastri, V. P., Padera, R. F., Langer, R., Freed, L. E., and Vunjak-Novakovic, G. (2000). In vitro generation of osteochondral composites. *Biomaterials* **21**, 2599–2606.

28. Obradovic, B., Meldon, J. H., Freed, L. E., and Vunjak-Novakovic, G. (2000). Glycosaminoglycan deposition in engineered cartilage: Experiments and mathematical model. *J. Am. Inst. Chem. Engineers* **46**, 1860–1871.

29. Freed, L. E., Martin, I., and Vunjak-Novakovic, G. (1999). Frontiers in tissue engineering: In vitro modulation of chondrogenesis. *Clin. Orthop. Relat. Res.* **367S**, S46–S58.

30. Caplan, A. I., Elyaderani, M., Mochizuki, Y., Wakitani, S., and Goldberg, V. M. (1997). Principles of cartilage repair and regeneration. *Clin. Orthop. Relat. Res.* **342**, 254–269.

31. Kawamura, S., Wakitani, S., Kimura, T., Maeda, A., Caplan, A.I., Shino, K., and Ochi, T. (1998). Articular cartilage repair-rabbit experiments with a collagen gel-biomatrix and chondrocytes cultured in it. *Acta Orthop. Scand.* **69**, 56–62.

32. Schreiber, R. E., Ilten-Kirby, B. M., Dunkelman, N. S., Symons, K. T., Rekettye, L. M., Willoughby, J., and Ratcliffe, A. (1999). Repair of osteochondral defects with allogeneic tissue-engineered cartilage implants. *Clin. Orthop. Relat. Res.* **367S**, S382–S395.

33. Fink, C., Ergun, S., Kralisch, D., Remmers, U., Weil, J., and Eschenhagen, T. (2000). Chronic stretch of engineered heart tissue induces hypertrophy and functional improvement. *FASEB J.* **14**, 669–679.

34. Akins, R., Boyce, R., Madonna, M., Schroedl, N., Gonda, S., McLaughlin, T., and Hartzell, C. (1999). Cardiac organogenesis in vitro: Reestablishment of three-dimensional tissue architecture by dissociated neonatal rat ventricular cells. *Tissue Eng.* **5**, 103–118.

35. Klagsbrun, M. (1979). Large-scale preparation of chondrocytes. *Methods Enzymol.* **58**, 560–564.

36. Sah, R. L.-Y., Kim, Y. J., Doong, J. Y. H., Grodzinsky, A. J., Plaas, A. H., and Sandy, J. D. (1989). Biosynthetic response of cartilage explants to dynamic compression. *J. Orthop. Res.* **7**, 619–636.

37. Barnett, J. V., Taniuchi, M., Yang, M. B., and Galper, J. B. (1993). Co-culture of embryonic chick heart cells and ciliary ganglia induces parasympathetic responsiveness in embryonic chick heart cells. *Biochem. J.* **292**, 395–399.

38. Toraason, M., Luken, M., Breitenstein, M., Krueger, J., and Biagini, R. (1989). Comparative toxicity of allylamine and acrolein in cultured myocytes and fibroblasts from neonatal rat heart. *Toxicology* **56**, 107–113.

39. Springhorn, J. P., and Claycomb, W. C. (1989). Prepoenkephalin mRNA expression in developing rat heart and in cultured ventricular cardiac muscle cells. *Biochem. J.* **258**, 73–78.

40. Maki, T., Gruver, E., Davidoff, A., Izzo, N., Touplin, D., Colucci, W., and Marks, A. (1996). Regulation of calcium channel expression in neonatal myocytes by catecholamines. *J. Clin. Invest.* **97**, 656–663.

41. Gao, J., Niklason, L., and Langer, R. (1998). Surface hydrolysis of poly(glycolic acid) meshes increases the seeding density of vascular smooth muscle cells. *J. Biomed. Mater. Res.* **42**, 417–424.

42. Mikos, A. G., Thorsen, A. J., Czerwonka, L. A., Bao, Y., and Langer, R. (1994). Preparation and characterization of poly (L-lactic acid) foams. *Polymer* **35**, 1068–1077.

43. Shastri, V. P., Martin, I., and Langer, R. (2000). Macroporous polymer foams by hydrocarbon templating. *Proc. Natl. Acad. Sci. U.S.A.* **97**, 1970–1975.

44. Neitzel, G. P., Nerem, R. M., Sambanis, A., Smith, M. K., Wick, T. M., Brown, J. B., Hunter, C., Jovanovic, I. P. Malaviya, P., Saini, S., and Tan, S. (1998). Cell function and tissue growth in bioreactors: Fluid mechanical and chemical environments. *J. Jpn. Soc. Microgr. Appl.* **15**(Suppl. II), 602–607.

45. Vunjak-Novakovic, G., Preda, C., Bordonaro, J., Pellis, N., de Luis, J., and Freed, L. E. (1999). Microgravity studies of cells and tissues: From Mir to ISS. *Am. Inst. Phys. Conf. Proc.* **458**, 442–452.

46. Searby, N. D., de Luis, J., and Vunjak-Novakovic, G. (1998). Design and development of a space station cell culture unit. *J. Aerosp.* **107**, 445–457.

47. Schwarz, R. P., Goodwin, T. J., and Wolf, D. A. (1992). Cell culture for three-dimensional modeling in rotating-wall vessels: An application of simulated microgravity. *J. Tissue Cult. Methods* **14**, 51–58.

48. Prewett, T. L., Goodwin, T. J., and Spaulding, G. F. (1993). Three-dimensional modeling of T-24 Human bladder carcinoma cell line: A new simulated microgravity culture system. *J. Tissue Cult. Methods* **15**, 29–36.

49. Chromiak, J. A., Shansky, J., Perrone, C., and Vandenburgh, H. (1998). Bioreactor perfusion system for the long-term maintenance of tissue-engineered skeletal muscle organoids. *In Vitro Cell. Dev. Biol.—Anim.* **34**, 694–703.

50. Kim, Y. J., Sah, R. L., Doong, J. Y. H., and Grodzinsky, A. J. (1988). Fluorometric assay of DNA in cartilage explants using Hoechst 33258. *Anal. Biochem.* **174**, 168–176.

51. Farndale, R. W., Buttle, D. J., and Barrett, A. J. (1986). Improved quantitation and discrimination of sulphated glycosaminoglycans by the use of dimethylmethylene blue. *Biochim. Biophys. Acta* **883**, 173–177.

52. Woessner, J. F. (1961). The determination of hydroxyproline in tissue and protein samples containing small proportions of this imino acid. *Arch. Biochem. Biophys.* **93**, 440–447.

53. Hollander, A. P., Heathfield, T. F., Webber, C., Iwata, Y., Bourne, R., Rorabeck, C., and Poole, R. A. (1994). Increased damage to type II collagen in osteoarthritic articular cartilage detected by a new immunoassay. *J. Clin. Invest.* **93**, 1722–1732.

CULTURE ENVIRONMENTS: MICROARRAYS

Joe Tien and Christopher S. Chen

INTRODUCTION

Microarrays of cells—cultures in which the spatial and chemical environments of the cells are well defined on the scale of micrometers—possess several advantages over traditional cell cultures, where such local parameters are often undetermined. First, and most important, the controlled microenvironment present in microarrays allows one to characterize the response of cells to specific cell–cell and cell–extracellular matrix binding conditions; these interactions, along with diffusible signals in the culture media, are critical determinants of the tissue-specific cell functions important to tissue engineering applications [1–3]. Second, the spatial arrangement of cells placed in arrays can enhance the automated addressing and screening of cells [4]. Third, the consumption of experimental reagents can be greatly reduced, since fewer cells and smaller amount of media are used in each culture. In this chapter, we describe the reasoning behind the methods we have used to control the adhesive environment of cultured cells, and detail specific protocols for these methods.

Cell–cell and cell–matrix interactions play a central role in the regulation of cellular functions, including cell survival, proliferation, and motility, and tissue-specific activities. While cell–cell interaction is a difficult parameter to control, the adsorption of purified extracellular matrix (ECM) proteins onto culture surfaces has been a mainstay technology in the *in vitro* cultivation of adhesion-dependent cells. Methods to spatially pattern the deposition of ECM ligands on a substrate have provided a local, nondiffusible handle to control cellular behavior. For several systems, such as capillary endothelium [5,6], liver parenchyma [7], keratocytes [8], adipocytes [9], chondrocytes [10], and osteoblasts [11], the interactions between cells and the ECM can regulate the functional phenotype of cultured cells. To elicit a desired cell-specific function, it appears that cells require not only the correct ECM ligands to be present, but also an adhesive environment that induces the cells to adopt a specific cell shape and cytoskeletal architecture [5]. In particular, limiting the degree to which cells spread often results in the emergence of differentiated behaviors. By culturing cells in or on small patches of ECM such that cells cannot spread beyond the edges of those patches, it may be possible to determine how cues from the ECM, coupled with signals from cell spreading, regulate the functional behavior of the cell. We have developed such an approach to study how cell–ECM interactions control the behavior of capillary cells and have used these findings to engineer substrates that cause these cells to proliferate, differentiate, or undergo apoptosis [5,6].

The application of microfabrication to biology has resulted in several methods to produce microarrays of extracellular matrix to which cells can be attached. Most of these methods use photolithography, a light-based technique for patterning surfaces, to define regions on a substrate that cells could attach to, and regions that resist the attachment of cells. The specific chemistries and procedures used to render such patterns vary greatly; one approach is described in depth in Chapter 8 of this volume [12]. In general, these methods suffer from two drawbacks. First, because it is difficult to render a surface completely protein resistant, often an initial pattern of cells breaks down over time: cells migrate in from regions that they adhere to, simultaneously secreting ECM proteins that facilitate the migration of surrounding cells. Second, the need to use specialized lithographic facilities every time in the production of each patterned substrate has limited the adoption of these techniques by biologists.

Our laboratory, in concert with those of G. M. Whitesides (Harvard, Department of Chemistry) and D. E. Ingber (Harvard, Department of Pathology), has developed a technique for patterning cells that relies on non-lithography-based microscale printing of self-assembled monolayers with an elastomeric stamp [12]. This technique is quick (~1 min to pattern a surface, ~1 h to adsorb a pattern of protein, ~1 h to seed a pattern of cells on it), cheap (the chemicals are either readily available or easily synthesized), and convenient (only an initial access to specialized facilities is needed). Alkanethiols spontaneously chemisorb from solution onto gold and silver to form ordered, oriented assemblies called self-assembled monolayers (SAMs); the functional properties of SAMs depend on the end group of the alkanethiol [13]. For instance, a SAM of hexadecanethiol on gold is hydrophobic, whereas a SAM of carboxylic acid-terminated thiol is hydrophilic. Remarkably, a SAM of oligo(ethylene glycol)-terminated ("EG-terminated") thiol resists completely the adsorption of protein [14]. This property motivated our strategy for the patterned attachment of cells: we stamp hexadecanethiol onto a gold-coated surface with a soft, elastomeric stamp whose surface contains a relief of the desired pattern, coat the unstamped regions with a SAM of EG-terminated thiol, and expose the entire surface to ECM protein (usually fibronectin). Because the protein adsorbs onto SAMs of hexadecanethiol but not onto SAMs of EG-terminated thiol, the ECM protein coats only the stamped regions. When seeded on such a substrate, cells can attach and spread only on the protein-coated regions; that is, cells attach and spread only on the stamped regions.

The elastomeric nature of the stamp allows the stamp to make conformal contact with the gold-coated surface across large areas. It also enables us to mold other, hard polymers with the elastomeric stamp as a master. Thus, we can make substrates of rigid plastics that have a topological surface such as an array of wells. By coating the surface with gold and rendering the spaces between wells nonadhesive, we can prepare microarrays of cells in which cells are attached only in the wells.

In this chapter, we describe methods that in various combinations can provide investigators with micrometer-scale control in engineering microculture substrates. We first describe a photolithographic method to produce master templates ("masters") from which stamps ("molds") can be cast. This step, the only one that requires the use of specialized lithographic facilities, needs to be performed only once for each new master. Once a master has been produced, many molds can be cast from it in a typical wet-laboratory environment.

EXPERIMENTAL METHODS
GENERAL MATERIALS AND METHODS

Water is doubly distilled. Ethanol is 200 proof. Glass slides, silicon wafers, or plastic substrates that are coated with a thin layer of gold or silver are prepared by electron beam evaporation of the metals at low pressures ($<10^{-6}$ torr), with samples mounted on a rotating stage.

PRODUCTION OF MASTERS

MATERIALS

1. Silicon wafers (Silicon Sense)
2. Shipley 1813 positive photoresist (Microchem Corp.)
3. Microposit 351 developer (Microchem Corp.)
4. Hexamethyldisilazane (HMDS; Aldrich)

5. UV light source (Karl Suss)
6. Spin-coater (Headway Research)
7. Hot plate, preferably a digital one (Barnstead Thermolyne)

METHODS

Figure 7.1A outlines the process used to create photolithographic masters for casting. This procedure is based on lithography, which uses light passed through a patterned chrome mask to expose a photosensitive resin (the "photoresist") in selected areas. Only the irradiated areas dissolve in the developer.

1. Place a silicon wafer on the spin-coater, with the shiny, smooth side of the wafer facing up.
2. Cover the wafer with ~1 ml of HMDS.
3. Spin for 40 s at 4000 rpm. The wafer should now be hydrophobic.
4. Cover the wafer with ~1 ml of 1813 photoresist.
5. Spin for 40 s at 4000 rpm.
6. Bake for 5 min at 105°C on a hot plate to drive off residual solvent.
7. Remove the wafer from the hot plate. The wafer is now coated with a layer of photoresist ~1.5 μm thick.
8. Place the patterned chrome mask on the wafer, with the chrome-coated side in contact with the photoresist layer.
9. Expose to ~55 mJ/cm^2 of UV irradiation. This is the minimal energy density needed to fully expose ~1.5 μm of 1813 resist. If the power intensity of the UV lamp is unknown, it may be necessary to expose a test wafer for various times and develop it to determine the minimum exposure time needed.
10. Remove the mask and place the exposed wafer in ~50 ml of Microposit 351 developer (diluted 1:5 with water). Exposed photoresist will dissolve in the solution. With gentle agitation of the wafer, development is complete after ~45 s.
11. Remove the wafer from the developer. Rinse thoroughly with water, and dry the wafer in a stream of nitrogen. The wafer should now have a patterned film of photoresist on it.

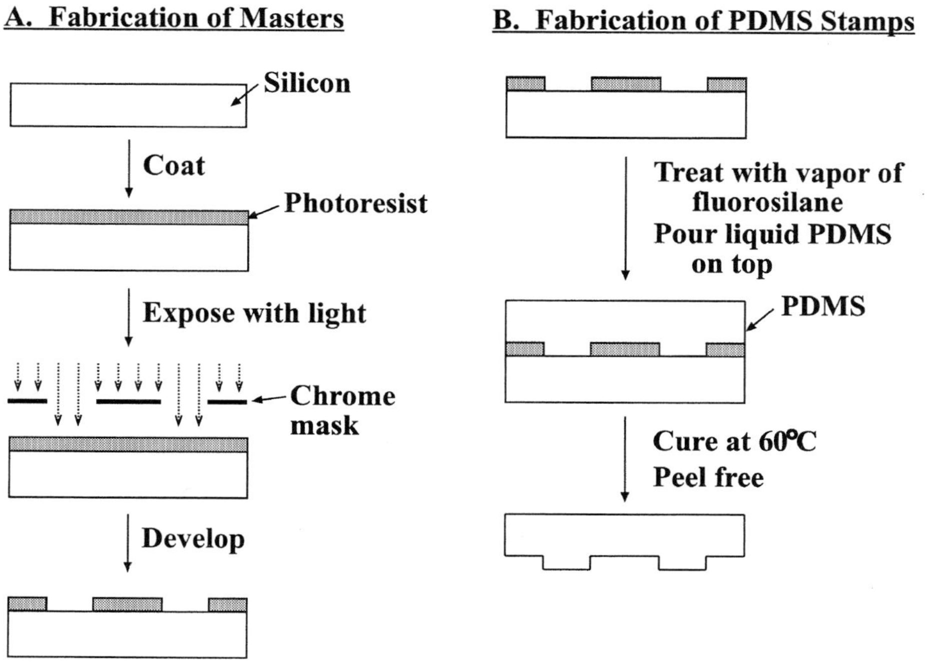

A. Fabrication of Masters

B. Fabrication of PDMS Stamps

Fig. 7.1. Schematic outline of the process used to fabricate lithographic masters and to cast stamps.

PRODUCTION OF MOLDS

MATERIALS

1. Patterned silicon wafer (the "master" from which replicas will be made)
2. Sylgard 184 two-component polydimethylsiloxane (PDMS; Dow Corning)
3. (Tridecafluoro-1,1,2,2,-tetrahydrooctyl)-1-trichlorosilane (United Chemical Technologies)
4. Desiccator
5. Plastic cups, glass slides, petri dishes, and razor blades

METHODS

Figure 7.1B outlines the process for the production of elastomeric molds. Liquid PDMS prepolymer is cast on top of the master, cured, and peeled off to generate a negative replica of the photoresist pattern. Because PDMS will cure and bond to the master, it is necessary to passivate the surface of the wafer beforehand by reaction with the vapor of a fluorosilane.

1. Place a drop or two of fluorosilane on a glass slide.
2. Place the slide and the master in a desiccator, and pump down to ~1 torr overnight.
3. Remove the master from the desiccator. It is now "fluorinated"—that is, coated with a thin, Teflon-like film—and ready to be cast from.
4. Mix ~30 g of PDMS in a plastic cup. The two components of PDMS are combined in a 10:1 w/w ratio (i.e., 30 g of one component and 3 g of the other).
5. Place the fluorinated master, patterned side up, in a petri dish, and pour the liquid PDMS prepolymer on top.
6. Place the petri dish (with master and liquid prepolymer) in a desiccator, and degas for several minutes. Wait until nearly all the bubbles in the PDMS disappear.
7. Remove the petri dish from the desiccator, and place in an oven held at 60°C. Higher temperatures will melt the dish.
8. Wait >1 h. The liquid prepolymer will cure to an elastic solid.
9. Remove the petri dish from the oven. *Gently* remove the PDMS and wafer from the dish. Silicon wafers are brittle. If the silicon wafer was fluorinated well, the PDMS should peel off easily from the wafer. The side of the PDMS that faced the wafer will have a replica of the photoresist pattern.
10. Cut the PDMS into pieces sized for easy handling (~1 cm^2).

MICROCONTACT PRINTING

MATERIALS

1. Patterned PDMS stamp
2. Ethanol
3. Hexadecanethiol (Aldrich), as a 1–10 mM solution in ethanol
4. Glass slides coated with 12 nm of gold (and 1 nm of titanium as adhesion promoter). The thinness of the gold film allows the slide to remain transparent.
5. Q-tips

METHODS

Figure 7.2A outlines the process. Microcontact printing uses the PDMS stamp, inked with hexadecanethiol, to print a SAM directly onto gold.

1. Clean the PDMS stamp by rinsing it with ethanol. Dry in a stream of nitrogen.
2. Dip a Q-Tip into the solution of thiol, and swab a few times across the patterned face of the PDMS.

A. Microcontact Printing

B. Cell Culture

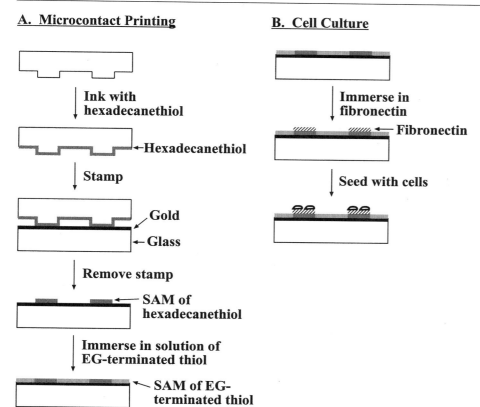

Fig. 7.2. Schematic outline of microcontact printing and cell culture on printed substrates.

3. Dry the PDMS in a stream of nitrogen for 1 min. The PDMS stamp is now coated with a thin film of hexadecanethiol.
4. Clean a gold-coated slide by rinsing it with ethanol and drying with nitrogen.
5. Place the PDMS stamp, patterned and inked side down, onto the gold side of the slide. Apply gentle pressure to the stamp to ensure contact between the stamp and the slide.
6. Remove the stamp after 5–15 s. Stamping for longer times will blur the pattern. Do *not* rinse the stamped slide with ethanol, as the rinse will also blur the pattern. The gold is now coated with a patterned SAM of hexadecanethiol. The stamp can be easily cleaned with ethanol, and may be reused numerous times without any degradation of the stamp.

CELL CULTURE ON PATTERNED SUBSTRATES

MATERIALS

1. Stamped gold-coated slide
2. Ethanol
3. $HS(CH_2)_{11}(OCH_2CH_2)_6OH$ (EG-terminated thiol), as a 1–10 mM solution in ethanol. This thiol is not commercially available. Its synthesis, a three-step procedure, is detailed in Pale-Grosdemange *et al.* [14].
4. Fibronectin, 10 μg/ml in PBS
5. Bovine serum albumin (BSA), 1% in PBS
6. Cells and media. We have patterned the attachment of a variety of cells, such as NIH 3T3 fibroblasts, microvessel endothelial cells [5], macrophages, and epithelial cells (MDCKs). Two formulations of media are needed: normal media and serum-free media supplemented with 1% BSA.

METHODS

Figure 7.2B outlines the process.

1. Cover the stamped gold-coated slide with the solution of EG-terminated thiol for >30 min.
2. Rinse the slide thoroughly with ethanol, and dry with nitrogen. The gold is now coated with patterned SAMs of hexadecanethiol and EG-terminated thiol.
3. Place the patterned slide in a petri dish, and sterilize with ethanol. From now on, all procedures must take place using sterile cell culture protocol.
4. Place ~0.1 ml of fibronectin in a petri dish.
5. Place the slide upside down on the fibronectin.
6. After ~1 h, add ~1 ml of 1% BSA around the slide.
7. Remove the slide after 15 min and rinse several times with PBS. The SAMs of hexadecanethiol on the gold are now coated with a monolayer of fibronectin, whereas the SAMs of EG-terminated thiol remain bare.
8. Trypsinize the cells, wash them to remove trypsin, and suspend them in serum-free media that is supplemented with 1% BSA.
9. Place the protein-coated slide in a petri dish, seed cells at ~1:10 density on top, and incubate.
10. After ~1 h, rinse the slide several times with serum-free media supplemented with 1% BSA. This process removes weakly adherent cells and leaves behind a pattern of cells on the stamped regions of gold.
11. Replace the serum-free media with normal media.

CELL CULTURE ON TOPOLOGICALLY PATTERNED SUBSTRATES

MATERIALS

1. Patterned PDMS stamp
2. Flat, unpatterned PDMS stamp, made by curing PDMS in a petri dish
3. Ethanol
4. Potassium hydroxide
5. Potassium ferricyanide
6. Potassium ferrocyanide
7. Potassium thiosulfate
8. (Tridecafluoro-1,1,2,2,-tetrahydrooctyl)-1-trichlorosilane (United Chemical Technologies)
9. 2-Propanol
10. 1 M nitric acid
11. <100> Silicon wafers coated with 200 nm of silver (and 1 nm of titanium as adhesion promoter)
12. UV-curable polymer (NOA 73 polyurethane, Norland Optical; UVO-114 epoxy, Epotek)
13. Beaker
14. Hot plate
15. Thermometer

METHODS

Occasionally, it is desired to grow cells on contoured substrates. These substrates may consist of two types: those with wells or posts, and those with grooves or pyramidal pits. Substrates of the first type are simply molded against a piece of PDMS. Those of the second type are fabricated by etching silicon anisotropically to generate grooves or pits in a silicon wafer (Fig. 7.3). Replica molding against these masters then generates contoured substrates.

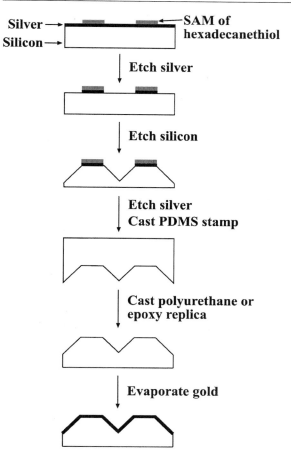

Fig. 7.3. Schematic outline of the fabrication of contoured substrates with grooves or pits.

To fabricate substrates with vertical features (wells or posts):

1. Cover a PDMS stamp with UV-curable prepolymer.
2. Place the stamp upside down on a petri dish.
3. Cure the prepolymer by irradiation with UV light.
4. Remove the stamp. This process leaves a negative replica of the PDMS stamp in the polymer.

To fabricate substrates with slanted features (grooves or pyramidal pits):

1. Microcontact print hexadecanethiol onto the silver-coated silicon wafer.
2. Place the stamped wafer into an aqueous solution of 0.1 M potassium thiosulfate, 0.01 M potassium ferricyanide, and 0.001 M potassium ferrocyanide. This yellowish solution etches bare silver that has not been stamped [15].
3. Etch for ~45 s.
4. Rinse the etched wafer with water. Place the substrate in a silicon etching solution (400 ml of water, 130 ml of 2-propanol, and 90 g of potassium hydroxide, heated to 70–80°C). Bubbles indicate etching.
5. Etch until complete. The solution etches silicon at ~1 μm/min.
6. Remove the remaining silver by placing the wafer in nitric acid. Wash thoroughly with water, and dry in a stream of nitrogen.
7. Place the etched silicon in a desiccator with the fluorosilane, and pump out overnight.
8. Cast a PDMS mold from the fluorinated master. The PDMS mold will have raised grooves or raised pyramids.
9. Mold UV-curable prepolymer with the PDMS mold onto a petri dish, and cure it with exposure to UV light. The molded polymer will have grooves or pits.

To culture cells on these substrates:

1. Evaporate a thin layer of gold (12 nm, with 1 nm of titanium as an adhesion promoter) on the contoured substrates.
2. With a flat PDMS stamp, microcontact-print hexadecanethiol onto the gold-coated substrates.
3. Cover the stamped substrate with the solution of EG-terminated thiol.
4. Coat the stamped surfaces with fibronectin, as before, and seed with cells. The cells will attach on the flat regions of the substrates. To have cells attach in the wells, grooves, or pits in the substrates, microcontact-print with EG-terminated thiol, and coat the unstamped regions with a SAM of hexadecanethiol. It may be necessary to stamp more than once to obtain a complete monolayer of EG-terminated thiol [16].

REFERENCES

1. Folkman, J., and Moscona, A. (1978). Role of cell shape in growth control. *Nature (London)* **273**, 345–349.
2. Adams, J. C., and Watt, F. M. (1993). Regulation of development and differentiation by the extracellular matrix. *Development (Cambridge, UK)* **117**, 1183–1198.
3. Gumbiner, B. M. (1996). Cell adhesion: The molecular basis of tissue architecture and morphogenesis. *Cell (Cambridge, Mass.)* **84**, 345–357.
4. Haggarty, S. J., Mayer, T. U., Miyamoto, D. T., Fathi, R., King, R. W., Mitchison, T. J., and Schreiber, S. L. (2000). Dissecting cellular processes using small molecules: Identification of colchicine-like, taxol-like and other small molecules that perturb mitosis. *Chem. Biol.* **7**, 275–286.
5. Chen, C. S., Mrksich, M., Huang, S., Whitesides, G. M., and Ingber, D. E. (1997). Geometric control of cell life and death. *Science* **276**, 1425–1428.
6. Dike, L. E., Chen, C. S., Mrksich, M., Tien, J., Whitesides, G. M., and Ingber, D. E. (1999). Geometric control of switching between growth, apoptosis, and differentiation during angiogenesis using micropatterned substrates. *In Vitro Cell. Dev. Biol.—Anim.* **35**, 441–448.
7. Hansen, L. K., Mooney, D. J., Vacanti, J. P., and Ingber, D. E. (1994). Integrin binding and cell spreading on extracellular matrix act at different points in the cell cycle to promote hepatocyte growth. *Mol. Biol. Cell* **5**, 967–975.
8. Watt, F. M., Kubler, M. D., Hotchin, N. A., Nicholson, L. J., and Adams, J. C. (1993). Regulation of keratinocyte terminal differentiation by integrin-extracellular matrix interactions. *J. Cell Sci.* **106**, 175–182.
9. Spiegelman, B. M., and Ginty, C. A. (1983). Fibronectin modulation of cell shape and lipogenic gene expression in 3T3-adipocytes. *Cell (Cambridge, Mass.)* **35**, 657–666.
10. West, C. M., Lanza, R., Rosenbloom, J., Lowe, M., and Holtzer, H. (1979). Fibronectin alters the phenotypic properties of cultured chick embryo chondroblasts. *Cell (Cambridge, Mass.)* **17**, 491–501.
11. Vukicevic, S., Luyten, F. P., Kleinman, H. K., and Reddi, A. H. (1990). Differentiation of canalicular cell processes in bone cells by basement membrane matrix components: Regulation by discrete domains of laminin. *Cell (Cambridge, Mass.)* **63**, 437–445.
12. Mrksich, M., Dike, L. E., Tien, J., Ingber, D. E., and Whitesides, G. M. (1997). Using microcontact printing to pattern the attachment of mammalian cells to self-assembled monolayers of alkanethiolates on transparent films of gold and silver. *Exp. Cell Res.* **235**, 305–313.
13. Laibinis, P. E., and Whitesides, G. M. (1992). ω-Terminated alkanethiolate monolayers on surfaces of copper, silver, and gold have similar wettabilities. *J. Am. Chem. Soc.* **114**, 1990–1995.
14. Pale-Grosdemange, C., Simon, E. S., Prime, K. L., and Whitesides, G. M. (1991). Formation of self-assembled monolayers by chemisorption of derivatives of oligo(ethylene glycol) of structure $HS(CH_2)_{11}(OCH_2CH_2)_mOH$ on gold. *J. Am. Chem. Soc.* **113**, 12–20.
15. Xia, Y., Zhao, X.-M., Kim, E., and Whitesides, G. M. (1995). A selective etching solution for use with patterned self-assembled monolayers of alkanethiolates on gold. *Chem. Mater.* **7**, 2332–2337.
16. Mrksich, M., Chen, C. S., Xia, Y., Dike, L. E., Ingber, D. E., and Whitesides, G. M. (1996). Controlling cell attachment on contoured surfaces with self-assembled monolayers of alkanethiolates on gold. *Proc. Natl. Acad. Sci. U.S.A.* **93**, 10775–10778.

CHAPTER 8

CULTURE ENVIRONMENTS: MICROPATTERNED CELL CULTURES AND COCULTURES

Sangeeta N. Bhatia

BACKGROUND

The ability to control cell–surface interactions is of paramount importance in controlling host–biomaterial interactions, influencing cell fate, understanding tissue development, and realizing the potential to tissue-engineer solid organs. The role of tissue organization in many of these applications has been well studied and is ultimately modulated by local receptor-mediated processes that influence cell behavior. The ability to control and study the role of tissue organization with micropatterning tools has provided insight in areas as diverse as angiogenesis, hepatocyte differentiation, calicification of bone-derived cells, stratification of keratinocytes in the epidermis, and neuronal growth cone guidance [1–5]; see Bhatia and Chen [6] for review.

Earlier methods of creating micropatterned cultures that control the cellular microenvironment employed either regional chemical modification of substrates to promote cell adhesion or physical localization of cells on a chemically uniform surface. Examples of chemical modification include photolithographic patterning of glass and subsequent protein immobilization [7], microcontact printing to localize protein [3,8], and photoimmobilization of polymers or adhesive peptides [9]. Physical methods of localization include microfluidic networks to deliver adhesive proteins or live cells directly [10–12]. Similarly, laser-directed cell writing is a new method of physical localization that utilizes a hollow optical fiber coupled with a laser source to direct the placement of individual cells on a target surface [13].

In photolithographic methods, adhesive proteins are typically localized by illumination of coating of photosensitive polymer coatings through a mask. These patterned polymers then act to locally protect the underlying surface from further modification, yielding a pattern of adhesive proteins such as collagen I [14]. We and others have utilized these methods to micropattern cells of a single type and thereby study the role of cell shape, adhesion, and migration. Our laboratory, together with M. Toner (Harvard Medical School, Department of Surgery), has expanded the utility of this tool by using it to localize two distinct cell types on glass—the first by selective adhesion to a patterned adhesive protein (collagen I) and the second by seeding in the presence of adhesive serum proteins, which mediate adhesion of the second cell type to surrounding regions [15]. These micropatterned cocultures enable the examination of key structure–function relationships in tissue engineering. For example, this technique provides a platform to vary cell–cell interactions without altering cell seeding density, even though these variables are typically coupled in conventional cocultures. These homotypic (within a given cell population) or heterotypic (between different cell populations) interactions are important in the function of a wide array of tissues, both developmentally and in the adult.

Another strategy for localizing cells, and one of particular interest in the preservation of micropatterned domains, is creation of localized nonadhesive domains. Poly(ethylene oxide) (PEO) has been widely used as a nonadhesive coating to resist "biofouling" and subsequent cell adhesion resulting from to protein adsorption to surfaces [16]. While still an active area of research, the mechanism by which PEO resists protein adsorption is generally attributed to its hydrophilicity, flexibility, chain mobility, and high steric exclusion volume in water [17]. There have been several reports of the use of microfabrication-derived techniques to localize PEO on artificial substrates to study neurite outgrowth, cell differentiation, and the effect of cell shape on cell fate. These patterning techniques include photoimmobilization of interpenetrated polymer networks [18,19], microcontact printing of PEGylated thiols on gold films [20] (described by Tien and Chen in Chapter 7 of this volume), and use of PEGylated silanes and aldehydes to modify Si-based materials [21]. Typically, these techniques require specific chemistries tailored to the material composition (e.g., gold, glass). Furthermore, some of these PEGylated species are not commercially available. In contrast, one method has been reported that allows coupling of PEO chains to a variety of biocompatible materials solely based on material hydrophobicity. This method reported by Caldwell and coworkers utilizes a commercial triblock polymer (PEO)$_{129}$-(PPO)$_{56}$-(PEO)$_{129}$ (Pluronic F108), which spontaneously adsorbs via the hydrophobic poly(propylene oxide) (PPO) domain to hydrophobic surfaces [22,23]. Here, we describe methods to pattern PEO by combining photolithographic patterning techniques with adsorption of F108. This technique can be utilized to micropattern growth-competent 3T3 murine fibroblasts in 10% serum and retain cell-free domains for at least 2 weeks, as described in detail elsewhere [24].

Thus, photolithographic micropatterning methods can be utilized to customize cell culture environments by localization of both adhesive and nonadhesive moieties for a variety of fundamental studies with an eye toward improving tissue engineering therapies. Applications of these techniques are many—indeed, our lab has focused on using these tools to control and study the role of the microenvironment around hepatocytes *in vitro*. These techniques have allowed us to examine and optimize the role of heterotypic (hepatocyte/fibroblast) cell–cell interactions in hepatocyte function for applications in tissue engineering [25]. Shortly, we will describe methods for microfabricating glass substrates regionally modifying surface chemistry with adhesive and nonadhesive species, micropatterning cell cultures, and micropatterning cocultures. Ultimately, these tools may be useful in many areas, including cell and tissue engineering, tailoring biomaterial implants, and fundamental studies on signaling in cell–cell and cell–matrix interactions.

EXPERIMENTAL METHODS
MICROFABRICATION OF SUBSTRATES

MATERIALS

1. *Borosilicate substrates*: Floating glass wafers 2-in. in diameter and 0.02 in. thick are custom-made (Erie Scientific) to fit a standard 60-mm-diameter (P-60) tissue culture dish.
2. *Chrome masks*: in. × 5 in. chrome masks are fabricated by a high-precision photolithographic process from Corel Draw artwork or produced by commercial printing of emulsion masks on a Linotronics-Hercules 3300 dpi high-resolution line printer.
3. *Cleaning solution (piranha solution)*: 3:1 mixture of 30% H_2SO_4 and H_2O, prepared at the time of the experiment. Extreme caution must be observed during chemical handling because of the corrosive nature of these chemicals. Pyrex containers must be utilized and chemical removal should be performed via aspiration and dilution of acid mixture.
4. *Photoresist*: Theoretically, any photoresist can be utilized that will adhere sufficiently to clean borosilicate, remain intact during subsequent processing, and yet be removed with relative ease after surface modification. We use a positive photoresist, either OCG 825-835 St (Olin-Ciba-Geigy) or Shipley 1813 (Microchem Corp.).

5. Developer: Use developer appropriate for selected photoresist. We use either OCG 934 (Olin-Ciba-Geigy) or MF-319 (Microchem Corp.), diluted with water at a 1:1 ratio.

METHOD

We produced experimental substrates utilizing standard microfabrication techniques at the Integrated Technology Laboratory, University of California at San Diego (http://kesey.ucsd.edu/microtech/). In general, most integrated circuit manufacturing facilities can be used to perform these procedures.

1. Borosilicate wafers are cleaned by placement into wafer carriers. Place in a Pyrex vat, pour piranha solution over wafers, and wait 10 min. Rinse wafers three times in a "dump-to-resistivity tank" and use a "spin-dryer" to nitrogen-dry. If a spin-dryer is not available, manual drying is done by using a N_2 gas stream. This gas stream can either be accessed through a house N_2 gun, or a portable N_2 gas tank connected to tubing.
2. Dehydrate wafers to promote adhesion of photoresist by baking for 60 min at 200° C.
3. Mount wafers on vacuum of a spin-coater chuck and coat with positive photoresist to a uniform layer of approximately 1 μm. In our laboratory this is accomplished as follows: dispense photoresist at 500 rpm for 2 s, spread photoresist at 750 rpm for 6 s, spin at 4000 rpm for 30 s (Fig. 8.1A).
4. Soft-bake for 30 min at 90°C to drive out excess solvent and anneal any stress in the film.
5. To create a latent image of the desired pattern in the resist layer, expose wafers to ultraviolet light in a defined pattern. We expose coated substrates to 365-nm ultraviolet light in a Bottom Side Mask Aligner (Karl Suss) through a 5 in. × 5 in. patterned chrome mask under vacuum-enhanced contact for 3 s at a dose of 10 mW/cm² as depicted in Fig. 8.1B.
6. To produce the final three-dimensional relief image, immerse the exposed photoresist in the appropriate developer. Complete removal of photoresist in exposed areas is critical to subsequent surface modification. Presence of residual photoresist can be assessed by inspection under light interference or fluorescent microscopy

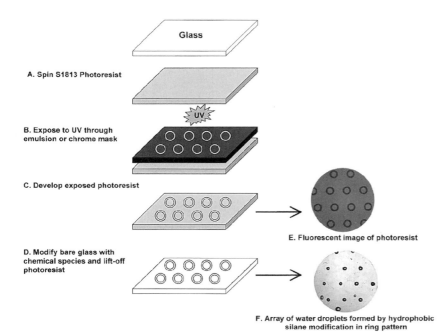

Fig. 8.1. Micropatterning surface chemistry of glass chips. Chips are spin-coated with a light-sensitive polymer known as photoresist. We use a "positive" photoresist that renders the polymer soluble after exposure to UV light through a mask. Masks are made on a computerized drawing program and printed as transparencies. The autofluorescence of photoresist allows easy visualization by conventional fluorescent microscopy as seen in (E). This pattern of photoresist serves to localize subsequent chemical modification of the underlying glass by acting as a protective barrier. Modification of the surface chemistry can then regionally modify surface properties such as hydrophobicity. In (F) the photoresist pattern in (E) has been utilized to localize a hydrophobic, methyl-terminated silane. After removal of the remaining photoresist by a "lift-off" process, regional surface hydrophobicity can be nondestructively visualized by observing the repeating beading pattern of water droplets confined by hydrophobic rings on the surface.

(Fig. 8.1, Step E). We develop by immersion and agitation in a bath of developer for 70 s. Surfaces should then be rinsed three times under running deionized water and cascade-rinsed (if possible) for 2 min (Fig. 8.1C).

7. Postbake patterned discs for 30 min at 120°C to drive off residual solvent and promote film adhesion.

8. Wafers can be stored in closed containers (preferably within a nitrogen box) at room temperature for at least 1 month.

9. Finally, 24 h prior to surface modification, substrates can be exposed to oxygen plasma to dry-etch (remove) a small layer of photoresist. This ensures complete removal of photoresist from exposed borosilicate; however, if substrates are well developed and pattern dimensions are larger than 10 μm, this step may be skipped. We use a parallel plate, Plasma Day Etcher at a base vacuum of 50 mtorr in an O_2 atmosphere and pressure of 100 mtorr at a power of 100 W for 2–4 min, which corresponds to an etch rate of approximately 0.1 μm/min.

SURFACE CHEMISTRY

Local modifications of surface chemistry can be utilized to pattern cells either by patterning adhesive or nonadhesive molecules.

Patterning Molecules to Promote Cell Adhesion

There are a number of techniques for immobilizing adhesive proteins on solid substrates. We describe techniques for covalent coupling of proteins to the surface by a modified technique of Lom *et al.* [7] and Britland *et al.* [26] as well as direct adsorption of proteins from solution.

MATERIALS

1. *3-[2-Aminoethyl)amino] propyl trimethoxysilane (AS)*: AS solution is 2% AS (Aldrich) in water, prepared immediately prior to use. AS should be stored in a nitrogen box or bottle should be gassed with nitrogen prior to closure and storage.

2. *Phosphate-buffered saline (PBS)*: Can be purchased at 10-fold concentration (Biofluids).

3. *Glutaraldehyde*: Stock solution is 25% glutaraldehyde in water (Fisher Chemical). Glutaraldehyde solution is prepared by diluting stock solution in PBS to a final concentration of 2.5% v/v, pH 7.4.

4. *Collagen solution*: Stock collagen I solution is prepared from rat tail tendons to a concentration of approximately 1 mg/ml in 1 mM HCl (described elsewhere [27]). Collagen I solution is prepared by 1:1 dilution of stock solution (final concentration is approximately 500 μg/ml) with water, pH 5.0, prepared immediately prior to use. Alternatively, commercial preparations of collagen (or other desired adhesive proteins) can be substituted for the collagen solution described here.

5. *Acetone*: General laboratory grade (Baxter Diagnostics).

6. *Ethanol*: 70% in distilled water.

METHOD

1. Patterned substrates are handled with wafer tweezers (Fluoroware). Pay special attention to the orientation of the patterned surface of the wafer—transparent substrates can be easily inverted without any obvious differences in appearance. Wafers are rinsed by immersion in distilled, deionized water (DI) in a glass 100-mm petri dish. Repeat.

2. To pattern a protein such as collagen by adsorption, immerse sample in 4 ml of collagen solution in a P-60 petri dish for 1 h at 37°C and skip to step 8.

3. To covalently link a protein to the patterned regions, first immobilize AS by immersion of samples in AS solution for 30 s at room temperature, followed by two rinses in DI water.

4. Dry wafers with a stream of nitrogen gas to avoid drying artifacts.

5. Bake wafers in a closed container for 10 min at 120°C. *Note*: Temperatures greater than 150°C will cause hardening of many photoresists, and removal will be difficulty.

6. Next, soak discs in a covered container of 2.5% v/v solution of glutaraldehyde in PBS (pH 7.4) for 1 h at 25°C, followed by two rinses in fresh PBS. Visually inspect wafers every 15 min to evaluate the integrity of the photoresist—if the glutaraldehyde solution causes peeling of the photoresist, the process will need to be abbreviated.

7. To immobilize collagen, immerse wafers in 4 ml of collagen solution in a P-60 petri dish for 30 min at 37°C.

8. To remove photoresist and expose underlying unmodified glass, float each wafer in acetone in a glass container and sonicate the container in a bath sonicator for 1–15 min. The duration of sonication is empirically determined by observation of the first wafer in each batch—examine wafers for complete removal of photoresist (previously "pink" wafers will appear clear). Treat all wafers in an experimental batch identically to ensure comparability of immobilized protein layers on all substrates. We use 10 ml of acetone in a 100-mm glass petri dish (Fig. 8.1D).

9. Rinse wafers twice by immersion into DI water. As shown by Lom *et al.* [7] and others, modified areas should display differential wetting upon removal of substrate from water, thereby indicating successful patterned surface modification (Fig. 8.1F).

10. Wafers can be stored dry, in a covered container at 4°C for at least 2 weeks. We store wafers on a piece of filter paper (to absorb residual water and prevent sticking) in 60-mm petri dishes. Storage of wafers in solution (i.e., PBS or ethanol) results in transfer of patterned protein to unmodified areas, presumably via desorption from modified areas and adsorption to unmodified areas; therefore, if the immobilized protein will tolerate dehydrated storage conditions, wafers should be stored dry. If immobilized protein requires hydration to maintain its bioactivity, storage time must be empirically determined—in our case, less than 48 h.

Patterning Molecules to Prevent Cell Adhesion

To pattern PEO on a hydrophilic, micropatterned borosilicate wafer, we render the surface hydrophobic by immobilization of a methyl-terminated species followed by adsorption of a triblock polymer, Pluronic F108.

MATERIALS

1. Pluronic F108 (BASF), 1% (w/v) solution in water. Members of this class of polymers have poly(propylene oxide) centers with poly(etheylene oxide) side chains with the following proportions $(PEO)_{129}$-$(PPO)_{56}$-$(PEO)_{129}$ and a molecular weight of 14,600 g/mol. The polypropylene domain adsorbs quasi-irreversibly to hydrophobic surfaces, creating a surface coating of PEO chains; thus surfaces that are hydrophobic can be modified with nonadhesive PEO by adsorption [16].

2. Dimethyltrichlorosilane (Aldrich), 5% (v/v) solution in chlorobenzene (Sigma).

METHOD

1. To render hydrophilic glass hydrophobic, modify patterned borosilicate wafers with methyl-terminated silane by immersion of patterned wafer in a glass petri dish with dimethyltrichlorosilane solution for 5 min, followed by two rinses in chlorobenzene.

2. Strip photoresist by sonication in acetone as described in step 8 under "Patterning Molecules to Promote Cell Adhesion: Method. *Note*: Liquid beading patterns should be observable at this point (Fig. 8.1F).

3. Immerse wafer with methyl-terminated patterned domains in F108 solution for 24 h at room temperature.

4. Rinse twice in water. Store in water or phosphate buffered saline.

MICROPATTERNED CELL CULTURES

In the subsections that follow, we describe methods for culturing and micropatterning primary hepatocytes and an embryonic fibroblast cell line.

Cell Culture

MATERIALS

Fibroblast Culture

1. *Fibroblast feeder layer*: 3T3-J2 mouse fibroblast cell line (provided by H. Green, Department of Physiology and Biophysics, Harvard Medical School [28]). Other cell lines are commercially available.
2. *Fibroblast medium (FM)*: Dulbecco's Modified Eagle's Medium (DMEM, high glucose, L-glutamine, 110 mg/liter sodium pyruvate, Gibco-BRL), bovine calf serum 10% (BCS, Hyclone), penicillin–streptomycin (Boehringer) 100 IU/ml–100 µg/ml.
3. *Trypsin*: 0.1% trypsin (ICN Biomedicals) in PBS.
4. *Versene*: 0.1% EDTA (Boehringer) in PBS.

Hepatocyte Culture

1. Isolation of hepatocytes from rat liver by collagenase perfusion is described in detail elsewhere [29]. Alternatively, isolated, purified hepatocytes may be obtained commercially.
2. Serum-free hepatocyte culture medium (SFM)
   a. Dulbecco's Modified Eagle's Medium with high glucose, without sodium pyruvate, without sodium bicarbonate (Gibco-BRL).
   b. Glucagon (Lilly). Add to final concentration of 7 ng/ml.
   c. Insulin (Squibb). Add to final concentration of 0.5 u/ml.
   d. Epidermal growth factor (Becton Dickinson). Add to a final concentration of 20 ng/ml.
   e. Hydrocortisone (Upjohn). Add to a final concentraion of 7.5 µg/ml.
   f. Penicillin-streptomycin, respectively, 5000 U/ml and 5 mg/ml in 0.9% NaCl (Sigma Chemical). Add 1% v/v to medium.
   g. L-Proline. Add to final concentration of 40 µg/ml (Sigma).
   h. Filter-sterilize medium through a 0.2-µm pore membrane.
3. Hepatocyte culture medium (CM). Add 10% bovine calf serum to SFM.
4. Bovine serum albumin (BSA) solution of 0.05% w/w in water (Sigma). Sterilize by filtration through a 0.45-µm filter and store at 4°C.
5. Sterilized DI water. Prepare two autoclavable containers of 500 ml of DI water. Autoclave to sterilize.
6. Sterilized vessels. Prepare two glass beakers by covering with autoclave wrap. Autoclave.

METHODS

Fibroblast Culture

1. Grow 3T3-J2 cells to preconfluence in 150-cm^2 flasks or other suitable dishes, 10% CO_2, balance moist air.
2. Passage cells by washing with 8 ml of versene and incubating with 2 ml of trypsin solution and 2 ml of versene for ~5 min at 37°C. Resuspend cells in 25 ml of fibroblast media to neutralize trypsin. Approximately 10% of the cells may be inoculated into a fresh tissue culture flask, containing a total of 50 ml of media.
3. A single 150-cm^2 flask will yield 4–8 million cells at confluence.

Hepatocyte Culture

1. Isolate and purify hepatocytes from liver of 2- to 3-month-old adult female Lewis rats weighing 180–200 g and quantify viability by trypan blue exclusion. Routinely, 200–300 million cells can be isolated with viability between 85 and 95%.

2. Assess cell concentration of stock suspension by hemocytometer and record.
3. Store hepatocytes on ice prior to use. In our hands, rat hepatocytes are viable and retain the capacity to attach and spread for at least 6 h on ice.
4. Dilute cell stock suspension in serum-free hepatocyte media (SFM) for wafer patterned with adhesive species (e.g., collagen) or hepatocyte culture media (CM, i.e., with serum) for patterned wafer with PEO species to a final concentration of $1–2 \times 10^6$ hepatocytes/ml.

Micropatterning on Modified Surfaces

METHOD

1. Soak premodified wafer in 70% ethanol for sterilization. Use wafer tweezers to place wafer in P-60 dish with 5 ml of ethanol solution for at least 1 h but not more than 24 h, at room temperature in a sterile hood.
2. To remove residual ethanol from surface of wafer, first pour autoclaved water into sterilized beaker. Flame-sterilize wafer tweezers, allow to cool, and remove wafer from 70% ethanol under sterile conditions. Immerse wafer in water and agitate gently for approximately 10 s, being sure to preserve orientation of the wafer.
3. If using wafers modified with adhesive species, coat wafers with bovine serum albumin to deter nonspecific cell adhesion on unmodified regions. For rat hepatocytes, BSA coating reduces nonspecific cell attachment to glass from 30% to negligible levels [30]. Place sterilized wafers in sterile P-60 dishes, add 4 ml of BSA solution to each dish, and place in incubator for 45 min at 37°C.
4. To remove residual BSA solution, use sterile tweezers to remove wafers from dishes under sterile conditions, and immerse in autoclaved water in sterile beaker, gently agitating for 10 s.
5. Seed hepatocytes by placing 2 ml of hepatocyte solution on each wafer. Agitate solution to disperse cell suspension and place in incubator for 1–1.5 h. Wafers should be periodically agitated (i.e., every 15 min) to promote maximal cell attachment.
6. At this point, selective cell adhesion should render pattern features visible. Typically, to ensure 100% confluence on adhesive areas, hepatocyte seeding is repeated two to three times. Surfaces should be rinsed twice by pipetting and then aspirating 4 ml of media, reseeded with hepatocytes for 1.5 h, and rinsed again. Repeat as necessary.
7. Hepatocytes are then allowed to spread over the remaining modified sites. Rat hepatocytes spread over greater than 10 h [31]; therefore, incubate patterned hepatocytes overnight. Cells (hepatocytes, endothelial cells) on small pattern features will conform to the edges of the pattern (Fig. 8.2) [3,15]. These micropatterned cell cultures can be used in the performance of experimental studies; otherwise proceed with addition of second cell type.

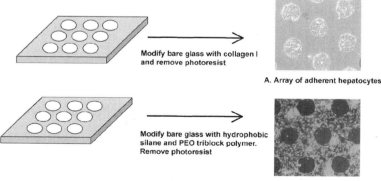

Modify bare glass with collagen I and remove photoresist

A. Array of adherent hepatocytes

Modify bare glass with hydrophobic silane and PEO triblock polymer. Remove photoresist

B. Array of fibroblasts deterred from PEO-modified circles

Fig. 8.2. Micropatterning cell cultures by patterning adhesive ligands or nonadhesive species. Micropatterned surface chemistries can be utilized to pattern mammalian cells in culture. Fig. 8.1 described silane modification and "lift-off." Use of an amine-terminated silane will allow covalent attachment of adhesive proteins (or peptides) to the surface using a glutaraldehyde linker. (A) Primary rat hepatocytes have been patterned on an array of collagen I islands of 500 μm diameter by promoting cell adhesion in these domains. Single-cell resolution is routinely achieved. Alternatively, use of a methyl-terminated silane allows coupling of a commercially available triblock copolymer (Pluronic F108). This allows a simple method to pattern polyethylene oxide (PEO) and thus create cellular micropatterns by preventing cell adhesion.

MICROPATTERNED COCULTURES

Hepatocytes have been cocultured with many secondary cell types ranging from isolated biliary ductal epithilia [32] to cell lines such as murine 3T3 fibroblasts [33]; see Bhatia *et al.* [4] for review. These "cocultures" have been reported to sustain hepatocyte morphology and viability, form functional homotypic gap junctions, and produce elevated levels of various liver-specific markers including albumin synthesis and various detoxification functions. We chose coculture of primary rat hepatocytes with 3T3-J2 fibroblasts to take advantage of their ease of culture and availability, although genetically modified fibroblasts or alternative cell types could also be incorporated into this methodology. Now we describe methods for obtaining micropatterned cocultures.

MATERIALS

1. Hepatocyte culture medium (CM) as earlier.
2. Mitomycin C (Boehringer): 10 μg/ml in fibroblast medium.
3. Fluorescent labels CMFDA (chloromethylfluorescein diacetate, C-2925, Molecular Probes) and CMTMR (chloromethylbenzoylaminotetramethyl rhodamine, C-2927) are utilized to track cells fluorescently.

METHOD

1. To track cells in coculture, fibroblasts and hepatocytes are sometimes fluorescently labeled by incubation in 25 μM dye in media for 45 min, rinsed, and incubated for 30 min prior to plating.
2. To generate a coculture, fibroblasts are trypsinized, resuspended in fibroblast media, and plated in 3 ml of FM per micropatterned culture (Fig. 8.3). Typically 750,000 fibroblasts per dish is sufficient; however, in some cases it is necessary to plate growth-arrested fibroblasts in greater numbers.
3. Growth-arrested fibroblasts are generated by incubating each 150-cm² fibroblast flask with 15 ml of mitomycin C solution for 2 h.
4. Resultant micropatterned cocultures are seen in Fig. 8.3.
5. Continue coculture in CM and proceed with experimental investigation. Preservation of pattern integrity is dependent on both cell types (competence, migration

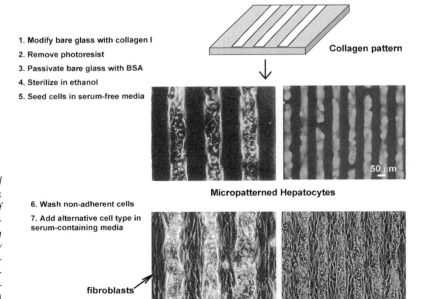

Fig. 8.3. Micropatterning cocultures. Micropatterned surface chemistries, passivating coatings such as bovine serum albumin (BSA), and manipulations of serum content in culture media can be used to pattern two distinct cell types. The spatial configuration of the resulting cocultures can be manipulated by the mask utilized in the photolithography process, allowing control of cell–cell interaction and cell number independently. (A) and (B) Phase contrast micrographs of a micropatterned hepatocyte cultures. (C) and (D) Micropatterned cocultures of hepatocytes and fibroblasts.

rates, etc) and pattern dimensions—for example, our patterns are stable on the order of weeks for collagen-modified areas of larger than a few hundred micrometers.

SPECIAL TOPICS

1. *Microfabrication process*: Facility regulations will largely determine the process utilized for preparation of patterned surfaces. Here, we have detailed the most versatile technique, whereby all biological materials are utilized subsequently in a separate facility. Others have pretreated wafers with organic materials in their own facilities [7]. This approach should also be considered where applicable. Facility specifications with regard to dedicated equipment will also affect problems that may be encountered via contaminants and so on.

   In addition, the overall process will vary with the characteristics of the desired protein. In our case, collagen I retained its bioactivity for hepatocyte attachment and spreading despite treatment with acetone, ethanol, and dehydration. Other proteins may require modifications of this protocol to retain their bioactivity.

2. *Surface modification*: The method described here was intended to be robust. In many instances, users may be able to eliminate certain elements of this process. Exposure to plasma oxygen may be unnecessary if patterns are well developed and not contaminated during storage. Similarly, baking of wafers after AS modification may be unnecessary in some protocols.

3. *Coculture media*: Media must be selected to provide adequate nutrients and buffering, as well as hormone and growth factor stimulation to each individual cell population, to suppress overgrowth of either cell population (here, hydrocortisone). Alternatively, cells must chemically be growth-arrested (here, mitomycin C).

4. *Probing tissue function of micropatterned cocultures*: Cultures of two different cell types often require unique analysis. For example, selective markers of cell function (in our case, albumin or urea synthesis as markers of liver-specific function of hepatocytes), methods to quantitate relative growth of both cell populations (in our case, growth-arresting fibroblasts allowed measurement of total DNA as a vehicle for tracking both cell populations), spatial tracking of both cell populations [in our case, either immunofluorescent staining of cytokeratin, an intermediate filament in hepatocytes as compared to F-actin, a cytoskeletal polymer present in both cell type, or labeling of each cell population with a different long-lived intracellular fluorescent probe such as Molecular Probes (Eugene, OR) Cell Tracker dyes].

REFERENCES

1. Healy, K. E., Thomas, C. H., Rezania, A., Kim, J. E., McKeown, P. J., Lom, B., and Hockberger, P. E. (1996). Kinetics of bone cell organization and mineralization on materials with patterned surface chemistry. *Biomaterials* 17(2), 195–208.
2. Pins, G. D., Toner, M., and Morgan, J. R. (2000). Microfabrication of an analog of the basal lamina: Biocompatible membranes with complex topographies. *FASEB J.* 14(3), 593–602.
3. Chen, C. S., Mrksich, M., Huang, S., Whitesides, G. M., and Ingber, D. E. (1997). Geometric control of cell life and death. *Science* 276, 1425–1428.
4. Bhatia, S. N., Balis, U. J., Yarmush, M. L., and Toner, M. (1999). Effect of cell-cell interactions in preservation of cellular phenotype: Cocultivation of hepatocytes and nonparenchymal cells. *FASEB J.* 13(14), 1883–1900.
5. Hammarback, J. A., McCarthy, J. B., Palm, S. L., Furcht, L. T., and Letourneau, P. C. (1988). Growth cone guidance by substrate-bound laminin pathways is correlated with neuron-to-pathway adhesivity. *Dev. Biol.* 126(1), 29–39.
6. Bhatia, S., and Chen, C. (1999). Tissue engineering at the micro-scale. *Biomed. Microdevices* 2(2), 131–144.
7. Lom, B., Healy, K. E., and Hockberger, P. E. (1993). A versatile technique for patterning biomolecules onto glass coverslips. *J. Neurosci. Methods* 50(3), 385–397.
8. Singhvi, R., Kumar, A., Lopez, G. P., Stephanopoulos, G. N., Wang, D. I., Whitesides, G. M., and Ingber, D. E. (1994). Engineering cell shape and function. *Science* 264, 696–698.
9. Clemence, J. F., Ranieri, J. P., Aebischer, P., and Sigrist, H. (1995). Photoimmobilization of a bioactive laminin fragment and pattern-guided selective neuronal cell attachment. *Bioconjugate Chem.* 6(4), 411–417.
10. Delamarche, E., Bernard, A., Schmid, H., Michel, B., and Biebuyck, H. (1997). Patterned delivery of immunoglobulins to surfaces using microfluidic networks. *Science* 276, 779–781.
11. Folch, A., and Toner, M. (1998). Cellular micropatterns on biocompatible materials. *Biotechnol. Prog.* 14(3), 388–392.
12. Chiu, D. T., Jeon, N. L., Huang, S., Kane, R. S., Wargo, C. J., Choi, I. S., Ingber, D. E., and Whitesides, G. M. (2000). Patterned deposition of cells and proteins onto surfaces by using three-dimensional microfluidic systems. *Proc. Natl. Acad. Sci. U.S.A.* 97(6), 2408–2413.
13. Odde, D. J., and Renn, M. J. (2000). Laser-guided direct writing of living cells. *Biotechnol. Bioeng.* 67(3), 312–318.

14. Britland, S., Perez-Arnaud, E., Clark, P., McGinn, B., Connolly, P., and Moores, G. (1992). Micropatterning proteins and synthetic peptides on solid supports: a novel application for microelectronics fabrication technology. *Biotechnol. Prog.* **8**(2), 155–160.

15. Bhatia, S. N., Yarmush, M. L., and Toner, M. (1997). Controlling cell interactions by micropatterning in co-cultures: Hepatocytes and 3T3 fibroblasts. *J. Biomed. Mater. Res.* **34**(2), 189–199.

16. Li, J., and Caldwell, K. (1996). Plasma protein interactions with Pluronic-treated colloids. *Colloids Surf. B: Biointerfaces* **7**, 9–22.

17. Beebe, D. J., Moore, J. S., Bauer, J. M., Yu, Q., Liu, R. H., Devadoss, C., and Jo, B. H. (2000). Functional hydrogel structures for autonomous flow control inside microfluidic channels. *Nature (London)* **404**, 588, 589–590.

18. Hubbell (1992). Surface physical intepenetrating networks of poly(ethylene terephthalate) and poly(ethylene oxide) with biomedical applications. *Macromolecules* **25**, 226–232.

19. Bearinger, J. P., Castner, D. G., Golledge, S. L., Rezania, A., Hubchak, S., and Healy, K. E. (1997). P(AAm-co-EG) interpenetrating polymer networks grafted to oxide surfaces: Surface characterization, protein adsorption, and cell detachment studies. *Langmuir* **13**(19), 5175–5183.

20. Singhvi, R., Stephanopoulos, G., and Wang, D. I. C. (1994). Effects of substratum morphology on cell physiology—review. *Biotechnol. Bioeng.* **43**(8), 764–771.

21. Saneinejad, S., and Shoichet, M. S. (1998). Patterned glass surfaces direct cell adhesion and process outgrowth of primary neurons of the central nervous system. *J. Biomed. Mater. Res.* **42**(1), 13–19.

22. Neff, J. A., Caldwell, K. D., and Tresco, P. A. (1998). A novel method for surface modification to promote cell attachment to hydrophobic substrates. *J. Biomed. Mater. Res.* **40**(4), 511–519.

23. Neff, J. A., Tresco, P. A., and Caldwell, K. D. (1999). Surface modification for controlled studies of cell-ligand interactions. *Biomaterials* **20**(23-24), 2377–2393.

24. Liu, V., Jastromb, W., and Bhatia, S. (2001). Engineering protein and cell adhesivity using PEO-terminated triblock polymers. *J. Biomed. Mater. Res.* (submitted for publication).

25. Bhatia, S. N., Balis, U. J., Yarmush, M. L., and Toner, M. (1998). Probing heterotypic cell interactions: Hepatocyte function in microfabricated co-cultures. *J. Biomater. Sci. Polym. Ed.* **9**(11), 1137–1160.

26. Britland, S., Clark, P., Connolly, P., and Moores, G. (1992). Micropatterned substratum adhesiveness—A model for morphogenetic cues controlling cell behavior. *Exp. Cell Res.* **198**(1), 124–129.

27. Elsdale, T., and Bard, J. (1972). Collagen substrata for studies on cell behavior. *J. Cell Biol.* **54**(3), 626–637.

28. Rheinwald, J. G., and Green, H. (1975). Serial cultivation of strains of human epidermal keratinocytes: The formation of keratinizing colonies from single cells. *Cell (Cambridge, Mass.)* **6**(3), 331–343.

29. Dunn, J. C., Yarmush, M. L., Koebe, H. G., and Tompkins, R. G. (1989). Hepatocyte function and extracellular matrix geometry: Long-term culture in a sandwich configuration. *FASEB J.* **3**(2), 174–177; erratum: *Ibid.* **3**(7), 1873.

30. Bhatia, S. N., Toner, M., Tompkins, R. G., and Yarmush, M. L. (1994). Selective adhesion of hepatocytes on patterned surfaces. *Ann. N. Y. Acad. Sci.* **745**(1), 187–209.

31. Rotem, A., Toner, M., Bhatia, S., Foy, B. D., Tompkins, R. G., and Yarmush, M. L. (1994). Oxygen is a factor determining in vitro tissue assembly: Effects on attachment and spreading of hepatocytes. *Biotechnol. Bioeng.* **43**(7), 654–660.

32. Guguen-Guillouzo, C., Clement, B., Baffet, G., Beaumont, C., Morel-Chany, E., Glaise, D., and Guillouzo, A. (1983). Maintenance and reversibility of active albumin secretion by adult rat hepatocytes co-cultured with another liver epithelial cell type. *Exp. Cell Res.* **143**(1), 47–54.

33. Donato, M. T., Castell, J. V., and Gomez-Lechon, M. J. (1991). Co-cultures of hepatocytes with epithelial-like cell lines: Expression of drug-biotransformation activities by hepatocytes. *Cell. Biol. Toxicol.* **7**(1), 1–14.

CHAPTER 9

Epithelial Cell Culture: Cornea

May Griffith, Mitchell A. Watsky, Chia-Yang Liu,
and Vickery Trinkaus-Randall

THE CORNEAL EPITHELIUM
Introduction

The corneal epithelium, the outermost cellular layer of the cornea, is exposed to the tear film. In humans, the epithelium is approximately 50 μm thick in the central cornea and slightly thicker in the peripheral areas. Anterior to the epithelium of a healthy human is the tear film, comprising lipid secretions from the meibomian glands of the eyelids and caruncle, and mucous secretions from the goblet cells of the adjacent conjunctiva. The composition of the tear film can be altered by injury. Together with the overlying tear film, the corneal epithelium serves two main functions: to provide a protective barrier between the environment and the inner layers of the cornea, and to act as a smooth refractive surface for the cornea.

The multilayered corneal epithelium can be roughly divided into three cell types. The outermost layer comprises of squamous cells referred to as apical cells. The middle layer consists of polygonal cells called wing cells, and the basal layer consists of a single row of columnar cuboidal cells [1]. The epithelium adheres to a basement membrane that in turn overlies an acellular structure called Bowman's layer. In nonprimate mammals, Bowman's layer is either indistinct or absent.

Taken together, the corneal epithelium is a complex of living cells with unique properties. The objective of this chapter is to provide a synopsis on the state of the art of corneal epithelial culture methodologies, focusing on the human cornea epithelium.

Sources of Corneal Epithelial Cells for Culture

An initial source of epithelial cells is from the central region of the cornea, where the basal cells are proliferative *in vivo*. The corneal epithelium turns over completely every 7–10 days, and the cells are replaced by daughter cells that arise from the basal cells through mitosis. The daughter cells differentiate and migrate upward, finally differentiating into squamous apical surface cells that are sloughed off [2,3].

A second, larger and possibly more accessible source of corneal epithelial stem cells lies at the junction between the cornea and the surrounding conjunctiva. These cells form a thin band at this junction, which is referred to as the corneal limbus.

Development of the Corneal Epithelium

In the developing human embryo, the optic primordium is present during the fourth week of gestation. By the end of the fifth week, the optic cup and lens placode have developed. At this stage, a wave of migrating mesenchymal cells moves into the junction between the cornea and lens to form a *brille*. The brille of cells, together with accumulating extracellular material, forms a wedge between the lens and cornea, separating them [4]. At the start of the fetal period (8 weeks of gestation), the corneal epithelium comprises two layers—a superficial layer of flattened cells and an underlying layer of cuboidal cells [5]. By 26 weeks of gestation, the corneal epithelium has differentiated into a stratified epithelium similar to that of adult corneas, with basal, wing, and apical cells. For a detailed review of corneal development, see Hay [4].

CULTURE OF PRIMARY CORNEAL EPITHELIAL CELLS
TISSUE SOURCES

In many countries, corneas acquired through organ donation programs are stored in eye banks. These corneas are often made available for research after the allowable time for use in transplantation has expired. This time varies from country to country. In the United States, for example, corneas stored in moist chambers are made available for research after 3 days, whereas they are available only after 10–12 days in Canada. In addition, some corneas, including those with low endothelial cell counts but normal epithelial complements or very young corneas, are deemed unsuitable for transplantation. These are immediately made available for research purposes.

Typically, the donor corneas taken for transplantation are stored in a storage medium such as OptiSol (Chiron). However, eye banks that harvest corneas for research as well as transplantation will often store the corneas in medium specified by individual investigators. Whole globes stored in saline-moistened sealed chambers can also be obtained for research purposes from eye banks.

The tissue available for research may or may not have been fully tested for tissue-borne pathogens such as viruses (HIV, hepatitis C), various bacteria, or prions. Hence, caution must be exercised when using donated human corneas.

The probability of starting cell lines and establishing successful cultures from dissociated cells is increased with fresh corneas.

SEVERAL PROTOCOLS FOR ISOLATION OF CELLS FROM DONOR CORNEAS
Mechanical Isolation

MATERIALS: (STERILE)

- *Culture medium 1*: Medium 199 + 10% fetal bovine serum (FBS) + insulin–transferrin–selenium (ITS Premix, Becton-Dickinson: 5 mg/liter insulin; 5 mg/liter transferrin, and 5 μg/liter selenous acid), pH 7.2–7.4
- Optional: OptiSol or DexSol (Chiron)
- *Culture medium 2*: supplemental hormonal epithelial medium (SHEM medium [6]: DMEM/F-12 (Dulbecco's Minimum Essential Medium + Ham's F-12) at 1:1, 15% FBS, 10 ng/ml of epidermal growth factor (EGF), 5 μg/ml insulin, 0.1 μg/ml cholera toxin alpha subunit, 5 mM L-glutamine, 0.5% dimethyl sulfoxide (DMSO), and antibiotics, if needed)
- Phosphate-buffered saline (PBS), Hanks's Balanced Saline Solution (HBSS), or culture medium without serum
- Corneal gill knife (Storz Instruments)

1. Donor human corneas from the eye bank are often stored together with a scleral rim. To remove any contaminating cells, wash each cornea with scleral rim thoroughly in several washes of PBS, HBSS, or medium without serum. Check under an appropriately illuminated dissecting microscope to ensure that the cornea is free from contaminating cells and debris.

2. Place the cornea into a dish precoated with sterile 1% gelatin in water containing culture medium. A 1:1 ratio of Medium 199 to OptiSol has worked best for us. SHEM also works very well. Medium 199 can also be used alone but is suboptimal.

3. Use a corneal gill knife and an appropriately illuminated dissecting microscope to observe the corneal surface. With the epithelial side facing upward, scrape off the epithe-

lium. Make sure that the scraping is confined to the cornea (for limbus, see later: Isolation of Corneal Limbal Cells). Do not apply excessive pressure. Excessive pressure can damage the underlying Bowman membrane, thereby liberating underlying stromal fibroblasts, which are likely to contaminate the epithelial culture.

4. Cut epithelial sheets into small pieces and place into 30-mm² tissue culture dishes. The cells will migrate from the explants onto the dish. Alternatively, the epithelial pieces can be mechanically dissociated by pipetting up and down, or trypsinized to obtain individual cells that can be cultured.

5. Make sure that a good number of cells have settled onto the plastic dish and have spread before replacing the medium (allow up to 2 weeks). If required, feed cultures by adding fresh medium. Floating cells can be transferred into other culture dishes (coated with a thin film of 1% gelatin or fibronectin) and allowed to settle. If contaminating fibroblasts are present, it will be necessary to subculture or clonally select the cells.

Note: If whole globes are received, first decontaminate each globe by soaking in sterile D-PBS containing a large dose of antibiotics [e.g., gentamicin or Ocuflox (ophthalmic antibacterial agent containing 0.3% ofloxacin, Allergan)]. Then excise the cornea together with the surrounding scleral rim. If globes were not enucleated under optimal conditions, rinse well with several changes of sterile D-PBS, followed by a thorough rinse in aqueous 10% Betadine (which is 10% iodine) solution to clean the surface and prevent infection. Wash off iodine with D-PBS.

Another mechanical method for isolating corneal epithelial cells from full-thickness pieces of explanted corneas has been described by Kahn and coworkers [7]. Essentially, pieces of cornea are explanted with epithelium down in contact with the tissue culture substrate to allow epithelial cells to migrate from the explants. We have tried this method very successfully, but with SHEM instead of serum-free keratinocyte growth medium (KGM), as used by Kahn *et al.* [7]. Care is needed to prevent contamination by stromal cells.

Enzymatic Isolation

MATERIALS (STERILE)

- *Culture medium*: DMEM/F-12 (1:1) containing 4% fetal bovine serum (FBS), 1% nonessential amino acid, insulin-transferrin-selenium (ITS Premix, Becton-Dickinson: 5 mg/liter insulin, 5 mg/liter transferrin, and 5 µg/liter selenous acid), 25 µg/ml gentamicin, 100 U/ml penicillin, and 100 µg/ml penicillin–streptomycin, pH 7.2–7.4
- Dispase, 1.2 U/ml (Life Technologies, Calbiochem)
- Alternatively, keratinocyte medium can be used. In this case the defined medium requires no serum but is supplied with EGF and bovine pituitary extract (Life Technologies). The medium should be supplemented with antibiotics, antimycotics, and nonessential amino cids. This avoids problems with lot-to-lot differences in serum.

1. To isolate epithelial cells as intact sheets, remove the corneal endothelium (e.g., with jewelers' forceps). Place the remaining cornea (epithelium and stroma) in MEM containing Dispase II and antibiotics. Dispase is a neutral protease that is used to remove intact epithelial sheets [8]. Incubate the cornea for 1 h at 37°C.

2. Using forceps, carefully tease the epithelium off, it will come off in sheets. Wash carefully in medium to remove contaminating cells and debris.

3. Cut epithelial sheets into small pieces and place into tissue culture dishes, using surface tension to permit the pieces to adhere and spread. Cells will migrate from the explants or can be dissociated to obtain individual cells [9].

ISOLATION OF CORNEAL LIMBAL CELLS

MATERIALS (STERILE)

- *Culture medium*: Medium 199 + 10% fetal bovine serum (FBS) + insulin-transferrin-selenium (ITS Premix, Becton-Dickinson: 5 mg/liter insulin, 5 mg/liter transferrin, and 5 μg/liter selenous acid)
- Optional: OptiSol or DexSol (Chiron)
- Phosphate-buffered saline (PBS), Hanks's Balanced Saline Solution (HBSS), or culture medium without serum
- 8 or 9 mm Trephine (Storz)
- Corneal gill knife (Storz)

1. Obtain donor human corneas that have surrounding scleral rims. Wash thoroughly several times with PBS, HBSS, or medium without serum to remove any contaminating cells.
2. Using a trephine, remove the central cornea, leaving the periphery and the sclera. Alternatively, corneal–scleral rims are often available from corneal surgeons who remove a full-thickness central corneal button for transplantation but leave the rims (including the limbus) intact.
3. Place the cornea into a dish containing culture medium. A 1:1 ratio of Medium 199 to OptiSol or SHEM medium works very well; Medium 199 can also be used alone but is suboptimal.
4. Locate the limbus. The limbal area is the translucent ring (∼1 mm thick) that is at the corneal–conjunctival–scleral junction.
5. Use a corneal gill knife to scrape the cells from the limbus. Make sure that the scraping is confined to the limbus.
6. Limbal cells can grow as explants from the scrapings, or the cells can be dissociated before plating. Make sure that a good number of cells have settled onto the plastic dish and have spread before replacing the medium (allow up to 2 weeks). If required, feed cultures by adding fresh medium. Floating cells can be transferred into other culture dishes (coated with collagen or fibronectin) and allowed to settle.

PROPAGATION OF CELLS IN CULTURE

We have listed four different media options, serum-free and with serum. The serum-free media and SHEM media are more selective for epithelial cells. Medium 199 will allow the cells to propagate at a higher rate, and we have been able to switch media to get the more differentiated phenotype as needed.

MATERIALS

- Culture medium
  Medium with serum such as Complete Medium 199 (as given earlier) or SHEM medium (as before)
  Serum-free medium such as EpiLife with corneal supplement (Sigma or Cascade Biologicals) or keratinocyte serum-free medium and supplement (Life Technologies) with low Ca^{2+} (0.06–0.1 mM)
- 0.05% Trypsin–ethylenediaminetetraacetic acid (trypsin-EDTA)
- Calcium-, magnesium-free PBS

1. Once the cells have settled onto the tissue culture plastic and have spread, replace the growth medium with fresh medium every 2–3 days.
2. Split the cells while they are still in the growth phase (at about 60–70 confluence to be safe). To split the cultures, aspirate the medium, wash the cells with PBS (Ca^{2+}, Mg^{2+} free).
3. Aspirate PBS and add prewarmed PBS containing 0.05 % trypsin–ethylenediaminetetraacetic acid (EDTA). Add 1 ml to 100-mm-diameter tissue culture dishes or 250-ml flasks.

4. Place in 37°C incubator for 1–3 min. Check continually in the microscope for action of trypsin, as cells will round up.

5. Inactivate trypsin with an excess of serum-containing culture medium or soybean trypsin inhibitor. Cells can now be split into aliquots and reseeded. If serum-free medium is used, inhibit the action of the trypsin with soybean trypsin inhibitor so that there is no serum bound to the cells that will influence expression or adherence. At this stage, the cells can be prepared for cryogenic storage.

CRYOGENIC STORAGE OF CELLS

Corneal epithelial cells can be stored for long periods of time (several years) in liquid nitrogen.

1. Follows steps 1–5 under Propagation of Cells in Culture. Once the trypsin has been inactivated, transfer the contents of the flask to a 50-ml centrifuge tube.
2. Centrifuge at $6g$ for 5 min to pellet the cells. A loose pellet of cells will form at the bottom of the centrifuge tube.
3. Carefully aspirate off supernatant without disturbing the cell pellet.
4. Resuspend the cells in freezing medium, which consists of culture medium with 10% FBS containing 10% DMSO. Mix thoroughly.
5. Aliquot at $1–2 \times 10^6$ cells/ml freezing medium (see Chapter 4 or Sigma catalog for cell quantification techniques).
6. Freeze at a rate of –1°C/min, using a controlled-temperature freezing container, such as the Nalgene 5100-0001, which relies on an 2-propanol bath. Place the container in a –80°C freezer overnight to allow the cells to cool.
7. Place cryovials in liquid nitrogen.

RECOVERY OF FROZEN STOCKS

1. Thaw frozen stocks by repeated quick dips and mechanical agitation of each cryovial in a 37°C water bath until only a small bit of ice remains. This will take a couple of minutes.
2. Transfer cells evenly into 2 or more 50-ml tissue culture flasks.
3. Add 5 ml of growth medium (specific for technology of growth being used) to each flask. Use antibiotics if necessary.
4. Incubate at 37°C, 5% CO_2, overnight to allow recovery at a high cell density. The cells can then be subcultured (do not allow cells to become confluent) or used for experiments.

DEVELOPMENT OF CELL LINES WITH EXTENDED LIFE SPANS

For research purposes and for purposes of testing (of drugs and chemicals), the availability of stable, continuous human corneal epithelial cell lines is required. Primary corneal epithelial cell cultures have a finite life span and become senescent after several passages in culture. Current hypotheses for cellular senescence are that the p53 tumor suppressor gene [10], and the retinoblastoma (Rb) gene and p16^{INK4a}, an inhibitor of Rb protein phosphorylation [11], have key roles in limiting the continuous proliferation of cells. Consequently, common immortalization techniques make use of viral genes that disrupt the normal activities of p53 and Rb, and thereby extend the life spans of cells.

According to the telomere hypothesis of cellular senescence, as cells divide, there is progressive shortening of the telomeres within all normal differentiated cells, as the enzyme that facilitates addition of telomeric units, telomerase, is inactive. Senescence occurs when the telomeres have shortened below a predetermined length. By reactivating telomerase, the life spans of cells can therefore be extended [12].

DNA tumor viruses, such as simian virus 40 (SV40), human papilloma viruses (HPV), and adenoviruses, have nucleic acid sequences that can extend the life spans of cells. SV40 early region genes, such as the gene for large T antigen, inactivate p53 and Rb genes [10,11], thereby extending the life spans of cells. HPV E6E7 genes also degrade p53 protein and additionally are able to reestablish telomerase activity in differentiated cells, thereby reversing the telomere shortening that occurs in cells as they count down to senescence [13]. Adenoviral E1A genes code for oncoproteins that cooperate with other oncoproteins to immortalise epithelial cells [14]. Our results to date showed that the HPV16 E6E7 genes were most effective at immortalizing corneal cells without the undesired side effects of transformation seen in SV40-transduced cells [15].

It now seems possible that unwanted changes caused by viral genes may be eliminated by reactivation of genes coding for telomerase [16,17], thereby eliminating much of the rigorous screening process (see later "Selection Protocols for Appropriate Cell Lines") that was required for our existing cell lines. Li *et al.* [18] have reported initial trials with human cornea epithelial cells, using the gene for human telomerase reverse transcriptase (*hTERT*) which directly reactivates telomerase.

PROTOCOL FOR HPV

Producer cells for an amphotropic retrovirus that carries the human papilloma virus (HPV) 16 genes E6 and E7, PA317 LXSN 16E6E7 [19], are available from the American Type Culture Collection (ATCC). The virus also contains a selectable marker, thus stable cell lines can be selected for resistance to the antibiotics geneticin or G418.

MATERIALS (STERILE)

- Laminar flow hood that is certified for biohazard use (e.g., handling of viruses)
- Culture medium for producer cells and transfection: DMEM (4500 mg/liter glucose), 1.5 g/liter sodium bicarbonate, 1 mM sodium pyruvate, 4 mM glutamine, 10% FBS
- Geneticin (G418; Life Technologies or Sigma)
- Polybrene (hexadiamethrine bromide, Sigma)

1. Thaw producer cells and plate densely onto 250-ml flasks, supplementing with culture medium.
2. When the producer cells approach subconfluence, change to fresh medium.
3. Next day, collect culture supernatant. This supernatant contains viral particles.
4. Centrifuge at 3500 rpm (Beckman J6B) to pellet any cells or contaminating debris.
5. Aliquot out the viral supernatant and freeze lots that are not for immediate use at −80°C.
6. To collect supernatant for immediate use, add Polybrene to a concentration of 5–10 μg/ml.
7. Do serial dilutions of viral supernatant/Polybrene in the culture medium.
8. Add 5 ml/25-ml flask of low passage (P1 or P2) primary epithelial cells that have been growing and are at 50–60% confluence. Leave viral supernatants on for 24–48 h.
9. Replace with regular culture medium.
10. Grow cells and passage at least once.
11. Select for stable transfectants using G418. Do a killing curve to determine a concentration that would kill off nonresistant cells in about 7 days. We have found this to be 0.8–1 mg/ml for corneal epithelial cells. After the first medium change as cells begin dying off, the dose can be lowered to about half (400–500 μg/ml).
12. Proceed with other selection criteria to pick appropriate cell lines. It may be necessary to subclone several times to obtain a homogeneous population.

TRANSFECTION WITH OTHER VIRAL CONSTRUCTS

Hybrid viruses containing SV40 and Ad-12-E1A genes have also been used successfully to immortalize corneal epithelial cells [7,20].

SELECTION PROTOCOLS FOR APPROPRIATE CELL LINES
Growth Characteristics and Cell Morphology

Corneal epithelial cell lines should have growth characteristics that are similar to those of the low-passage primary. They should be anchorage dependent and should have doubling times that are close to corresponding primary cells, which is about once approximately every 24 h.

The cell lines should also exhibit contact inhibition; that is, they should not pile up and overgrow. Periodically, remove small aliquots of cells and culture till confluence is reached. Confluent monolayers of corneal epithelial lines should appear similar to monolayers formed by low passage primary cultures by phase microscopy. Both cells have a slightly elongated but cobblestoned morphology.

Biochemical Markers

Cornea epithelial cells express a 64 kD keratin, K3, that is recognized by the AE5 antibody (ICN) [21]. However, trace amounts of K3 keratin are also found in other epithelial cells such as the conjunctiva. A more recently described keratin, K12, has a more restrictive expression to only corneal epithelium and the suprabasal layer of the limbus, as judged by immunostaining and *in situ* hybridization analysis [22]. Therefore, K12 is a better marker than K3 if there is a need to distinguish cornea epithelial cells from surrounding conjunctival epithelial cells. Other markers include expression of epidermal growth factor (EGF) receptors, integrins, and collagenase production.

Soft Agar Growth

Cells from solid tissues, including corneal epithelial cells, will generally grow as adherent monolayers unless they have transformed and have become anchorage independent. This test determines whether immortalized cells are still anchorage dependent: immortalized corneal epithelial cells, like their natural counterparts, should be anchorage dependent for growth and proliferation. Transformed cells, on the other hand, will be able to proliferate in semi solid media such as the agar media described as follows (adapted from Araki-Sasaki *et al.* [20]).

1. Dissolve 0.8% agar in Dulbecco's Minimal Essential Medium (DMEM) containing 20% FBS. Pour 2 ml into the bottom of each well of a 6-well dish (or individual 35-mm dishes) to form an agar bed.
2. Into 0.33% agar, mix 100–1000 cells/ml to form a suspension and pour on top of the agar bed. It is very important to have monodispersed cells and not clumps.
3. Incubate for 3 weeks and then score for colony growth, using a phase contrast microscope. Each colony counted should have more than 4 cells.

For a more definitive assay of tumorigenicity, cells can be injected into the back skin of SCID mice and tumor formation can be observed. Araki-Sasaki *et al.* [20] used a cell suspension of 1×10^4 cells/ml, and injected 1 ml per mouse.

Vero Cell Assay

For cells that were immortalized with viral genes, care needs to be taken to determine whether the lines are shedding viruses. This is accomplished by incubating samples of cell culture supernatant with green monkey kidney cells known as Vero cells. Descriptions of the Vero cell assay can be found in Kahn *et al.* [7] and in Freshney [23].

Karyotyping

Chromosome counting and Giemsa banding techniques will allow a quick check to determine the ploidy of the cell lines and to determine whether there are anomalies in the gross chromosomal morphology of the cell lines. A detailed protocol can be found in Freshney [23].

Electrophysiology

In corneal epithelial cells, ion conductances have been historically examined by means of Ussing chamber methods, along with a limited number of patch–clamp studies. The Ussing chamber studies have determined that the apical layers of the corneal epithelium exhibit a chloride conductance [24], while the basolateral layers exhibit a K+ conductance [25]. Although a host of transport molecules have been described in corneal epithelial cells, the only ion channel characterized to date in epithelial cells from normal corneas is a large conductance K+ channel that dominates the whole-cell conductance in basal epithelial cells. This channel, which is stimulated by swelling, stretching, fenemates, carbachol, and cGMP, is inhibited by barium, quinidine, diltiazem, and Prozac [26–30].

For sorting out immortalized versus transformed cells, characterizing ion currents from nonimmortalized cells provides a set of "physiological" markers for use in screening immortalized cells. Whole-cell currents in cultured human epithelial cells have been characterized by Bockman *et al.* [31]. In Griffith *et al.* [15], immortalized cells were compared with low-passage cultured human corneal cells, as opposed to freshly dissociated cells, because the immortalized cells were maintained in culture and therefore should have currents similar to their nonimmortalized counterparts, also kept in culture. This is an important consideration because rabbit corneal cells placed in culture rapidly change the ion channels they functionally express as early as 24 h after placement in culture [M. A. Watsky, unpublished observations]. Details of the electrophysiological methodology can be found in Chapter 83, Cornea.

THREE-DIMENSIONAL CULTURES
GROWING THREE-DIMENSIONAL CORNEAL EPITHELIUM

MATERIALS (STERILE)

- Collagen-coated tissue culture inserts (Falcon) or Costar that are tissue culture-treated.

Culture Medium

- SHEM (as before) or serum-free medium such as EpiLife with corneal supplement (Sigma or Cascade Biologicals) and keratinocyte serum-free medium and supplement (Life Technologies) with high Ca^{2+} (1.3–1.5 mM)

1. Thaw out corneal epithelial cells the night before and grow as monolayers, or use growing monolayers. Either primary cells or cell lines that have been prescreened for phenotypical similarities to low-passage primary or freshly dissociated cells can be used to reconstruct a three-dimensional corneal epithelium.

2. When ready, trypsinize cells and reseed onto coated culture inserts at about 5×10^5 cells per 30-mm-diameter insert. Supplement with culture medium, both outside the insert and on the inside. For cultures on a hydrated collagen gel bed, refer to Chapter 83: Cornea.

3. Grow until the epithelial cells reach confluence. Remove the medium from within the insert, leaving medium outside the insert. The cells are exposed to air but kept moist and are nourished by medium that has wicked through the collagen-coated membrane to provide an air–liquid interface.

4. The corneal epithelial cells will form a stratified layer in about 2 weeks. Feed by changing the medium 2–3 times per week.

CHARACTERIZING ORGANOTYPIC CORNEAL EPITHELIAL CULTURES
Morphology

The corneal epithelium of native human corneas comprises several layers and cells of different morphologies. The cells of reconstructed human corneal epithelium should have several layers comprising similar cell types. The cells of the outermost surface, the apical surface cells, are squamous (broad, flattened) cells. Each cell is approximately 50 μm in diameter and 4–5 μm thick. Ultrastructurally, depending on the stage of maturation, each apical cells is covered sparsely to thickly with microvilli [32,33] and encircled by tight junctions.

The wing cell layer is 4–5 cells thick at the periphery of the cornea, and thins to 2–3 cells centrally. The wing cells are polygonal, approximately 12–15 μm in diameter, with large ovoid nuclei. Ultrastructurally, these cells contain large numbers of cytoskeletal tonofilaments but few rough endoplasmic reticulum cisternae, mitochondria, or Golgi complexes. The plasma membranes of adjacent cells are elaborately interdigitated. Desmosomes and gap junctions between between basal and apical cells and adjacent wing cells have been described.

The basal cells are low cuboidal cells, approximately 10 μm in width and 15–20 μm in height, with prominent ovoid nuclei. Ultrastructurally, the cytoplasm of basal cells resembles that of wing cells, with interdigitating anterior and lateral plasma membranes. However, these cells have prominent hemidesmosomes associated with keratin fibers on their cytoplasmic aspect [34,35].

Biochemical Markers

As with the monolayer cultures, three-dimensional cultures are also positive for corneal-specific keratin, retain their receptors for EGF, and produce collagenase and matrix molecules. They also express $\alpha_6\beta_4$ integrins that are associated with hemidesmosomes in the basal cells that help to anchor the structure onto its culture substratum [36].

Barrier Function

Each corneal epithelial apical surface cell is also encircled by tight junctions (zona occudens) that act as a semipermeable high-resistance barrier [37,38], preventing the overlying tear fluid and pathogens from entering the stroma and other intraocular tissues. The tight junctions can be visualized by immunohistochemistry using an anti-ZO-1 antibody (Zymed), which is an indicator of tight junction formation [39]. The stratified epithelia should also be able to block sodium fluorescein diffusion through the outer layers through the basal cells.

REFERENCES

1. Trinkaus-Randall, V. (2000). Cornea: Biological responses. *In* "Principles of Tissue Engineering" (R. Lanza, R. Langer, and E. Chick, eds.), 2nd ed., pp. 471–491. Academic Press, San Diego, CA.

2. Hanna, C., and O'Brien, J. E. (1960). Cell production and migration in the epithelium layer of the cornea. *Arch. Ophthalmol. (Chicago)* **64**, 536–539.

3. Hanna, C., Bickness, D. S., and O'Brien, J. (1961). Cell turnover in the adult human eye. *Arch. Ophthalmol. (Chicago)* **65**, 695–698.

4. Hay, E. D. (1980). Development of the vertebrate cornea. *Int. Rev. Cytol.* **63**, 263–322.

5. Watsky, M. A., Olsen, T. W., and Edelhauser. H. W. (1998). "Cornea and Sclera. Duane's Ophthalmology on CD-ROM." Lippincott-Raven, New York.

6. Jumblatt, M. M., and Neufeld, A. H. (1983). Beta-adrenergic and serotonergic responsiveness of rabbit corneal epithelial cells in culture. *Invest. Ophthalmol. Visual Sci.* **24**, 1139–1143.

7. Kahn, C. R., Young, E., Lee, I. H., and Rhim, S. (1993). Human corneal epithelial primary cultures and cell lines with extended lifespan: *In vitro* model for ocular studies. *Invest. Ophthalmol. Visual Sci.* **34**, 3429–3441.

8. Gipson, I. K., and Grill, S. M. (1982). A technique for obtaining sheets of intact rabbit corneal epithelium. *Invest. Ophthalmol. Visual Sci.* **23**, 269–273.

9. Trinkaus-Randall, V., Newton, A. W., and Franzbau, C. (1990). The synthesis and role of integrin in corneal epithelial cells in culture. *Invest. Ophthalmol. Visual Sci.* **31**, 440–447.

10. Wynford-Thomas, D. (1996). p53: Guardian of cellular senescence. *J. Pathol.* **180**, 118–121.

11. Huschtscha, L. I., and Reddel, R. R. (1999). P16^{ink4a} and the control of cellular proliferative life span—commentary. *Carcinogenesis (London)* **20**, 921–926.

12. Colgin, L. M., and Reddel, R. R. (1999). Telomere maintenance mechanisms and cellular immortalization. *Curr. Opin. Genet. Dev.* **9**, 97–103.

13. Rhim, J. S., Trimmer, R., Arnstein, O., and Huebner, R. J. (1981). Neoplastic transformation of chimpanzee cells induced by adenovirus type 12 simian virus 40 hybrid virus (Ad12-SV40). *Proc. Natl. Acad. Sci. U.S.A.* **78**, 13–17.

14. Douglas, J. L., and Quinlan, M. P. (1994). Efficient nuclear localization of the Ad5 E1A 12S protein is necessary for immortalization but not cotransformation of primary epithelial cells. *Cell Growth Differ.* **5**, 475–483.

15. Griffith, M., Osborne, R., Munger, R., Song, Y., Xiong, X., Laycock, N. L. C., Hakim, M., Doillon, C. J., and Watsky, M. A. (1999). A functional human corneal equivalent from cell lines. *Science* **286**, 2169–2172.

16. Bodnar, A. G., Ouelette, M., Frolkis, M., Holt, S. E., Chiu, C.-P., Morin, G. B., Harley, C. B., Shay, J. W., Lichtsteiner, S., and Wright, W. E. (1998). Extension of life-span by introduction of of telomerase into normal human cells. *Science* **279**, 349–352.

17. Jiang, X.-R., Jimenez, G., Chang, E., Frolkis, M., Kusler, B., Sage, M., Beeche, M., Bodnar, A. G., Wahl, G. M., Tlsty, T. D., and Chiu, C.-P. (1999). Telomerase expression in human somatic cells does not induce changes associated with a transformed phenotype. *Nat. Genet.* **21**, 111–114.

18. Li, L., Chang, J. H., and Jester, J. V. (2000). Development of extended life-span human corneal epithelium and stromal keratocytes by transfection with human telomerase reverse transcriptase (hTRT). *Invest. Ophthalmol. Visual Sci.* **41**, S695.

19. Halbert, C. L., Demers, G. W., and Galloway, D. A. (1992). The E6 and E7 genes of human papilloma virus type 6 have weak immortalizing activity in human epithelial cells. *J. Virol.* **66**, 2125–2134.

20. Araki-Sasaki, K., Ohashi, Y., Sasbe, T., Hayashi, K., Watanabe, H., Tano, Y., and Handa, H. (1995). An SV40-immortalized human corneal epithelial cell lines and its characterization. *Invest. Ophthalmol. Visual Sci.* **36**, 614–621.

21. Schermer, A., Galvin, S., and Sun, T.-T. (1986). Differentiation-related expression of a major 64 k corneal keratin *in vivo* and in culture suggests limbal location of corneal epithelial stem cells. *J. Cell Biol.* **103**, 49–62.

22. Liu, C. Y., Zhu, G., Westerhausen-Larson, A., Converse, R., Kao, C. W., Sun, T. T., and Kao, W. W. (1993). Cornea-specific expression of K12 keratin during mouse development. *Curr. Eye Res.* **12**, 963–974.

23. Freshney, R. I. (1994). "Culture of Animal Cells: A Manual of Basic Technique," 3rd ed. Wiley-Liss, New York.

24. Klyce, S. D. (1975). Transport of Na, Cl, and water by the rabbit corneal epithelium at resting potential. *Am. J. Physiol.* **228**, 1446–1452.

25. Rae, J. L., and Farrugia, G. (1992). Whole-cell potassium current in rabbit corneal epithelium activated by fenamates. *J. Membr. Biol.* **129**, 81–97.

26. Farrugia, G., and Rae, J. L. (1992). Regulation of a potassium-selective current in rabbit corneal epithelium by cyclic GMP, carbachol and diltiazem. *J. Membr. Biol.* **129**, 99–107.

27. Farrugia, G., and Rae, J. (1993). Effect of volume changes on a potassium current in rabbit corneal epithelial cells. *Am. J. Physiol. (Cell Physiol.)* **264**, C1238–C1245.

28. Rae, J. L., Rich, A., Zamudio, A. C., and Candia, O. A. (1995). Effect of prozac on whole-cell ionic currents in lens and corneal epithelia. *Am. J. Physiol. (Cell Physiol.)* **38**, C250–C256.

29. Rae, J. L., Dewey, J., and Rae, J. S. (1992). The large-conductance potassium ion channel of rabbit corneal epithelium is blocked by quinidine. *Invest. Ophthalmol. Visual Sci.* **33**, 286–290.

30. Rae, J. L., Dewey, J., Rae, J. S., Nesler, M., and Cooper, K. (1990). Single potassium channels in corneal epithelium. *Invest. Ophthalmol. Visual Sci.* **31**, 1799–1809.

31. Bockman, C., Griffith, C. M., and Watsky, M. A. (1998). Properties of whole-cell ionic currents in cultured human corneal epithelial cells. *Invest. Ophthalmol. Visual Sci.* **39**, 1143–1151.

32. Doughty, M. J. (1990). Morphometric analysis of the surface cells of rabbit corneal epithelium by scanning electron microscopy. *Am. J. Anat.* **189**, 316–328.

33. Nichols, B., Dawson, C. R., and Tongi, B. (1983). Surface features of the conjunctiva and cornea. *Invest. Ophthalmol. Visual Sci.* **24**, 570–576.

34. Kurpakus, M. A., and Jones, J. C. (1991). A novel hemidesmosomal plaque component tissue distribution and incorporation into assembling hemidesmosomes in an *in vitro* model. *Exp. Cell Res.* **194**, 139–146.

35. Owaribe, K., Nishizawa, Y., and Franke, W. V. (1991). Isolation and characterization of hemidesmosomes from bovine corneal epithelial cells. *Exp. Cell Res.* **192**, 622–630.

36. Stepp, M. A., Spurr-Michaud, S., and Tisdale, A. (1990). Alpha 6 beta 4 integrin heterodimer is a component of hemidesmosomes. *Proc. Natl. Acad. Sci. U.S.A.* **87**, 8970–8974.

37. Klyce, S. D., and Crosson, C. E. (1985). Transport processes across the rabbit corneal epithelium: A review. *Curr. Eye Res.* **4**, 323–331.

38. McLaughlin, B. J., Caldwell, R. B., Sasaki, Y., and Wood, T. O. (1985). Freeze fracture quantitative comparison of rabbit corneal epithelial and endothelial membranes. *Curr. Eye Res.* **4**, 951–961.

39. Zieske, J. D., Mason, V. S., Wasson, M. E., Meunier, S. F., Nolte, C. J. M., Fukai, N., Olsen, B. R., and Parenteau, N. L. (1994). Basement membrane assembly and differentiation of cultured corneal cells: Importance of culture environment and endothelial cell interaction. *Exp. Cell Res.* **214**, 621–663.

CHAPTER 10

EPITHELIAL CELL CULTURE: BREAST

Charles W. Patrick Jr., Xuemei Wu, Carol Johnston, and Greg P. Reece

INTRODUCTION

A breast is composed of many various tissue types, including blood vessels, adipose tissue, lymphatic vessels, connective tissue, and mammary glands. This chapter focuses on one tissue type, namely, adipose tissue. With advances in the molecular and cellular biology of adipose tissue, largely from the obesity and diabetes fields, and the maturation of tissue engineering, there has been a resurgence in the potential use of preadipocytes (adipogenic progenitor cells) in clinical strategies. Strategies under current investigation include soft tissue augmentation [1–4] and development of *de novo* breast [1,5] (also see Chapter 78 of this text). Preadipocytes are extremely attractive candidates for tissue engineering. Adipose tissue is uniquely expendable and abundant among most humans. Moreover, preadipocytes can easily be obtained from biopsied or excised fat and from minimally invasive liposuction aspirates. Unlike mature adipocytes, preadipocytes can withstand the mechanical trauma of aspiration and injection, as well as periods of ischemia. In addition, preadipocytes can be expanded into large numbers *ex vivo*, and the biological mechanisms dictating preadipocyte-to-adipocyte conversion are known and can be controlled [6–11].

This chapter describes the harvest, isolation, and *in vitro* culture of rat preadipocytes, as well as polymer seeding and histology of preadipocytes. Epididymal fat pads are utilized as the preadipocyte source, although other adipose tissue stores may be used. Implant integration, histogenesis, and the use of preadipocyte cell lines [12–19] are not discussed. Moreover, the influence of rat age and anatomic site on preadipocyte characteristics is not discussed [20–22]. All techniques can be scaled appropriately for other animal models and human tissue [23–31]. Moreover, the techniques utilize standard laboratory or easily obtainable equipment and supplies.

HARVEST AND ISOLATION OF PREADIPOCYTES

The following steps are required for harvest of epididymal fat pads from rats and isolation of preadipocytes from the fat pads. Steps 1–12 are conducted in an animal necroscopy facility, and steps 13–22 are conducted in a tissue culture facility. The presented procedures are in accordance with the guidelines of the American Association for Accreditation of Laboratory Animal Care (AAALAC) and the National Institutes of Health (NIH). Tables 10.1 and 10.2 list the materials, reagents, and solutions required.

1. Euthanize rat with CO_2.
2. Using electric clippers, shave the mid–lower abdomen.
3. Using Nair, a depilatory cream, completely remove the remaining hair per manufacturer's instructions.
4. Place the rat in supine position and stabilize the limbs (Fig. 10.1).
5. Scrub the harvest site with 70% ethanol.

Table 10.1. Preadipocyte Harvest and Isolation Reagents and Materials

Reagents	
Bovine serum albumin (BSA)	Sigma, St. Louis, MO
Ca^{2+}-, Mg^{2+}-free phosphate buffered saline (PBS)	
Collagenase, type IA	Sigma, St. Louis, MO
Dulbecco's Modified Eagle's Medium (DMEM)	
Fetal bovine serum (FBS)	Sigma, St. Louis, MO
Penicillin–streptomycin–glutamine (P/S/G), 100×	Gibco, Gaithersburg, MD

Materials	
0.22-μm Cellulose acetate syringe filter	Costar, Corning, NJ
250-μm Nylon mesh	Sigma, St. Louis, MO
40-μm Cell strainer	Falcon/Becton Dickinson, Franklin Lakes, NJ
50-ml Conical tube	Falcon/Becton Dickinson, Franklin Lakes, NJ
70% Ethanol wetted cotton balls	
Autoclaved, siliconized beaker	
Cell strainer	Sigma, St. Louis, MO
Electric clipper	
Iris scissors	ASSI, Westbury, NY
Mayo dissecting scissors	ASSI, Westbury, NY
Nair hair remover	
Tissue forceps (two required)	ASSI, Westbury, NY

6. Using dissecting scissors and tissue forceps, aseptically cut through the skin, muscle, and peritoneum along a midline Y-shaped incision, starting from the xiphoid cartilage of the sternum and down and along the inguinal regions of the lower body (Fig. 10.2).
7. Expose the abdominal cavity (Fig. 10.3).
8. Grasp and pull the epididymal adipose tissue (fat pad) with second pair of tissue forceps (Fig. 10.4).
9. Pull the epididymal fat pad until the testis is removed from scrotal sac (Fig. 10.5).
10. Using iris scissors, dissect the fat pad, taking care not to include the internal spermatic artery/vein and caput epididymis (Fig. 10.6).
11. Place harvested fat pad (Fig. 10.7) in a 50-ml conical tube filled with 4°C phosphate-buffered saline (PBS) supplemented with penicillin-streptomycin-glutamino (P/S/G) (Fig. 10.8).

Table 10.2. Preadipocyte Harvest and Isolation Solution Preparation

Complete DMEM (cDMEM) (per 500 ml)	350 ml DMEM 5 ml P/S/G 50 ml FBS
Digestion medium (per 2 fat pads)	4 ml 4°C PBS 3 mg/ml Collagenase 3 mg/ml BSA Sterilize through a 0.22-μm syringe filter
PBS with P/S/G	45 ml 4°C PBS 5 ml P/S/G

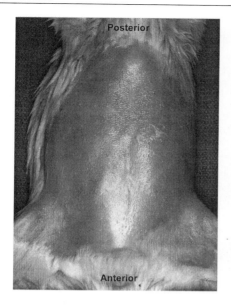

Fig. 10.1. Rat is in supine position, ventral view. Hair of the lower abdomen has been removed via shaving and depilatory cream. Posterior and anterior labels are included to denote rat's head-to-tail orientation.

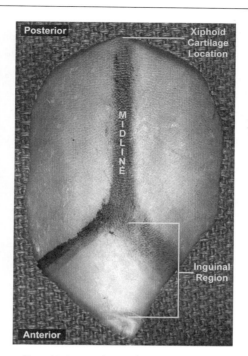

Fig. 10.2. Marking depicting incision to be made through skin, muscle, and peritoneum.

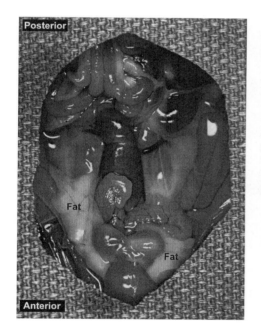

Fig. 10.3. Abdominal viscera in situ, ventral view.

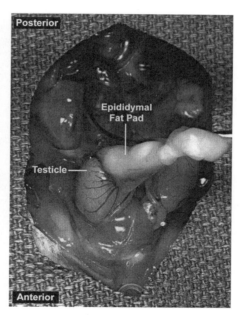

Fig. 10.4. Use of tissue forceps to remove epididymal fat pad.

12. Repeat steps 1–11 for the other epididymal fat pad.
13. Place the fat pads in a tissue culture dish in a biosafety cabinet and aseptically remove the large blood vessels (epididymal branch from internal spermatic artery/vein) and any hard tissue (portion of caput epididymis) with scissors and forceps. This minimizes fibroblast contamination of *ex vivo* cultures.
14. Finely mince the tissue with iris scissors.

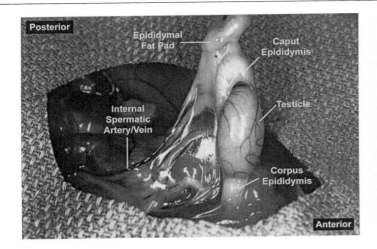

Fig. 10.5. View of epididymal fat pad and surrounding anatomy.

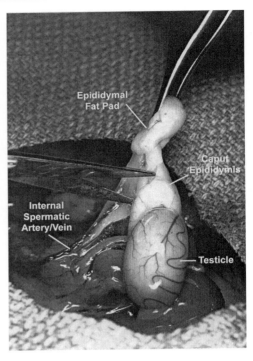

Fig. 10.6. Dissection of epididymal fat pad using iris scissors.

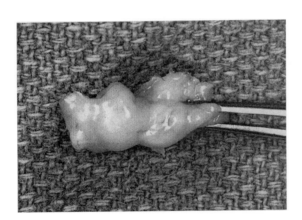

Fig. 10.7. Harvested epididymal fat pad.

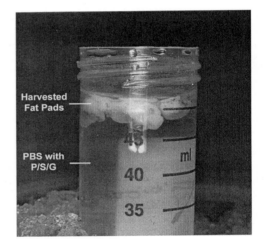

Fig. 10.8. Epididymal fat pads placed in PBS with antibiotics. The 50-ml conical tube is placed in an ice bath.

15. Place the minced tissue into a 50-ml conical tube with digestion medium and then on orbital shaker at a speed of 250 oscillations/min for 20 min at 37°C.
16. Filter the resulting slurry through a 250-μm nylon mesh into a siliconized beaker.
17. Filter the filtrate again through a 40-μm cell strainer into a 50-ml conical tube.
18. Centrifuge the final filtrate at 200g for 10 min at 4°C.
19. Aspirate the supernatant and resuspend the pellet (preadipocytes) in warm complete Dulbecco's Modified Eagle's Medium (cDMEM).
20. Seed the cells into a T75 tissue culture flask. The cell yield is approximately 10^6/ml.
21. Rinse and feed the cells with warm cDMEM after 24 h.
22. Feed the cells every 3 days.

CULTURE OF PREADIPOCYTES

Cells are passed when the cell density is 90%. At this time, the cell number is approximately 10^7/ml. The preadipocytes are passed prior to confluency, since contact inhibition initiates adipocyte differentiation and ceases preadipocyte proliferation. Cells may be frozen, cold-stored, and thawed in accordance with routine cell culture procedures.

Preadipocytes initially possess a fibroblast-like morphology (Fig. 10.9A). Upon reaching confluency, preadipocytes begin to accumulate lipid pools within their cytoplasm (Fig. 10.9B). Lipid droplets continue to grow in volume and finally coalesce to form a unilocular lipid pool within the cell (Fig. 10.9C–F). At this point, a preadipocyte's cytoplasm is 80–90% lipid. Eventually, the lipid pools become buoyant enough to float mature adipocytes to the surface of the culture flask. The amount of lipid loading can be controlled in a dose-dependent manner by varying the amount of fetal bovine serum (FBS) in cDMEM. Ideally, preadipocytes are seeded into polymers prior to differentiation.

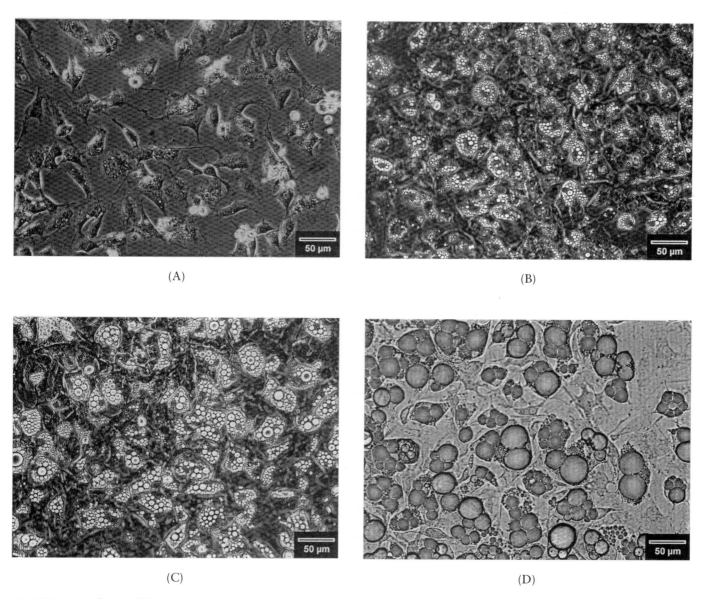

(A)

(B)

(C)

(D)

Fig. 10.9. Growth and differentiation of rat preadipocytes at five points postseeding: (A) 1 day, (B) 4 days, (C) 7 days, (D) 16 days, and (E) 22 days. Images (A)–(C) are phase contrast; images (D) and (E) are bright field. Note accumulation and coalescence of lipid pools as culture time progresses. (F) Image of unilocular lipid pool within a single preadipocyte. Bars denote 50 μm.

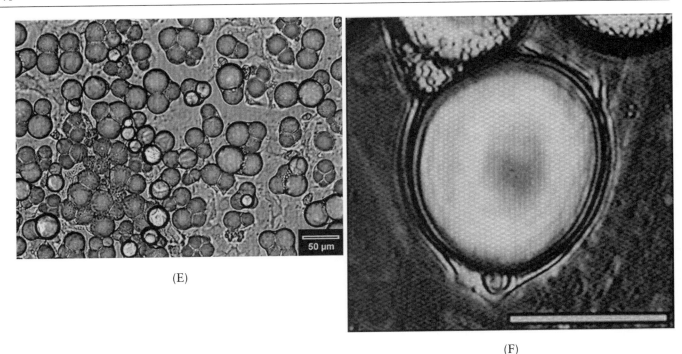

(E)

(F)

Fig. 10.9. (Continued).

POLYMER SEEDING

The following procedure for seeding poly(lactic-*co*-glycolic acid) (PLGA) polymer foams with preadipocytes [5] has proven successful with other biodegradable polymer foams and nonwoven fibers. Prior to seeding, foams are prewetted and sterilized with absolute ethanol for 30 min followed by two sterile saline washes at 20 min/wash and a DMEM wash for 20 min. A 20-µl suspension of preadipocytes (10^5 cells/mL) is injected onto each foam under sterile conditions. Prewetting permits the cell suspension to readily flow throughout the foam. Following 3 h for cell attachment, 24-well culture plates containing one foam/well are filled with 1.5 ml of medium per well. Foams are ready for implantation.

HISTOLOGY

OIL RED O

Adipocyte differentiation *in vitro* is routinely monitored by using oil red O staining for intracellular lipid pools [32] or phase contrast microscopy (lipid appears as phase bright, see Fig. 10.9A–C). The oil red O causes lipid pools to appear red under bright-field microscopy (labeled in Fig. 10.10). Tables 10.3 and 10.4 list the reagents and solutions required. Instructions are as follows.

1. Fix culture with 10% neutral buffer formalin.
2. Rinse with tap water.
3. Stain in oil red O for 10 min.
4. Wash with tap water.
5. Stain for 1 min in acidic Harris hematoxylin.
6. Wash with tap water.
7. Blue in ammonia water.
8. Wash with tap water.
9. Mount with an aqueous mounting medium or acquire pictures immediately. Results: fat, intense red; nuclei, blue.

Table 10.3. Oil Red O Reagents

Acetic acid	Fisher Scientific, Pittsburgh, PA
Ammonia hydroxide	Fisher Scientific, Pittsburgh, PA
Aqueous mounting media	Fisher Scientific, Pittsburgh, PA
Harris hematoxylin	Allegiance, McGaw Park, IL
2-Propanol, 98%	Fisher Scientific, Pittsburgh, PA
Neutral buffered formalin, 10%	Fisher Scientific, Pittsburgh, PA
Oil red O	Fisher Scientific, Pittsburgh, PA
Tap water	

OSMIUM TETROXIDE

An osmium tetroxide (OsO_4) paraffin procedure is used to demonstrate fat within harvested *in vivo* polymer foams [33]. Routine staining outlines only "ghost" cells, since histological processing with organic solvents and alcohols extracts lipid from cells. The OsO_4 chemically combines with fat, blackening it in the process. Fat that combines OsO_4 is insoluble in alcohol and xylene, and the tissue can be processed for paraffin embedding and counterstained. Small fat droplets and individual cells are well demonstrated via this method, whereas gross amounts of fat are not fixed by this diffusion-dependent stain (Fig. 10.11). After staining with OsO_4, foams are processed for paraffin embedding using standard procedures, except that HistoSolve (Shandon Lipshaw), a xylene substitute, is used instead of xylene. Xylene dissolves many biodegradable polymers. Infiltrated foams are cut and oriented in embedding cassettes. Sections 6 μm thick are cut with a microtome (Leica, Wetzlar, Germany), placed on slides, stained with hematoxylin–eosin (H&E), and coverslipped. Sections are analyzed by means of bright-field microscopy (Fig. 10.11). Table 10.5 lists the materials and reagents required. Instructions are as follows.

1. Harvest polymer foams at the appropriate time from rats.
2. Place harvested foams directly in a vial containing 10 ml of 10% formalin and fix overnight at room temperature.
3. After 24 h, remove specimen from formalin and trim. If cross sections are desired, a small piece of the foam to be embedded in cross section is removed at this time.
4. Transfer specimens to individual embedding cassettes that have been appropriately labeled with a pencil.
5. Place all cassettes in a beaker under freely running tap water and wash for 1 h.
6. Wash for 1 h in freely running distilled water.

Table 10.4. Oil Red O Solution Preparation

Acidic Harris hematoxylin	48 ml Harris hematoxylin 2 ml Acetic acid
Ammonia water	3 ml Ammonia hydroxide 1000 ml water
Oil red O stock solution	2.5 g oil red O 500 ml 2-Propanol Mix well
Oil red O working solution	24 ml oil red O stock solution 16 ml Distilled water Mix well and let stand for 10 min. Filter. The filtrate can be used for several hours

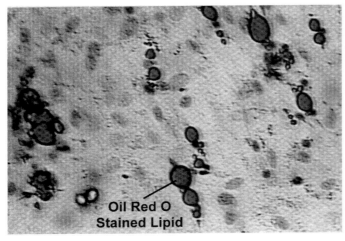

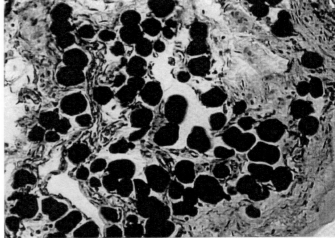

Fig. 10.10. *Preadipocytes in vitro stained with oil red O.*

Fig. 10.11. *OsO₄-stained preadipocytes.*

Table 10.5. OsO₄ Materials and Reagents

Coplin jars	Fisher Scientific, Pittsburgh, PA
Coverslips	Fisher Scientific, Pittsburgh, PA
Distilled water	
Formalin, 10%	Fisher Scientific, Pittsburgh, PA
Tap water	
Osmium tetroxide (OsO₄), 1% (wt/vol), aq	Sigma, St. Louis, MO
Pencil	
Periodic acid, 0.5% (wt/vol), aq	Sigma, St. Louis, MO
Rotating shaker	Fisher Scientific, Pittsburgh, PA
Slides	Fisher Scientific, Pittsburgh, PA
Standard paraffin embedding cassettes	Fisher Scientific, Pittsburgh, PA

7. Transfer cassettes to a Coplin staining jar (4 cassettes per jar) containing 5 ml of 1% aqueous OsO₄ solution.

8. Cap the jars and place on rotating shaker at 60 rpm under a fume hood at room temperature for 1 h.

9. Dispose of the OsO₄ in a hazardous waste container and rinse samples for 30 min in freely running distilled water.

10. Differentiate tissue by placing it in 0.5% aqueous periodic acid for 30 min on rotating shaker at 60 rpm.

11. Wash samples in running tap water for 30 min.

12. At this point samples are ready to be processed for paraffin embedding. Proceed with normal processing procedure using a xylene substitute instead of xylene and counterstain with H&E. Results: fat, black.

REFERENCES

1. Patrick, C. W., Jr., Chauvin, P. B., and Robb, G. L. (1998). Tissue engineered adipose. *In* "Frontiers in Tissue Engineering" (C. W. Patrick Jr., A. G. Mikos, L. V. McIntire, eds.), pp. 369–382. Elsevier Science, Oxford.

2. Katz, A. J., Llull, R., Hedrick, M. H., and Futrell, J. W. (1999). Emerging approaches to the tissue engineering of fat. *Clin. Plast. Surg.* **26**, 587–603.

3. Kral, J. G., and Crandall, D. L. (1999). Development of a human adipocyte synthetic polymer scaffold. *Plast. Reconstr. Surg.* **104**, 1732–1738.

4. Kawaguchi, N., Toriyama, K., Nicodemou-Lena, E., Inou, K., Torii, S., and Kitagawa, Y. (1998). De novo adipogenesis in mice at the site of injection of basement membrane and basic fibroblast growth factor. *Cell Biolog.* **95**, 1062–1066.
5. Patrick, C. W., Jr., Chauvin, P. B., and Reece, G. P. (1999). Preadipocyte seeded PLGA scaffolds for adipose tissue engineering. *Tissue Eng.* **5**, 139–151.
6. MacDougald, O. A., and Lane, M. D. (1995). Transcriptional regulation of gene expression during adipocyte differentiation. *Ann. Rev. Biochem.* **64**, 345–373.
7. Smas, C. M., and SookSul, H. S. (1995). Control of adipocyte differentiation. *Biochem. J.* **309**, 697–710.
8. Butterwith, S. C. (1994). Molecular events in adipocyte development. *Pharm. Ther.* **61**, 399–411.
9. Cornelius, P., MacDougald, O. A., and Lane, M. D. (1994). Regulation of adipocyte development. *Annu. Rev. Nutr.* **14**, 99–129.
10. Brun, R., Kim, J., Hu, E., Altiok, S., and Spiegelman, B. (1996). Adipocyte differentiation: A transcriptional regulatory cascade. *Cell Differ.* **8**, 826–832.
11. Sul, H. S., Smas, C. M., Wang, D., and Chen, L. (1998). Regulation of fat synthesis and adipose differentiation. *In* "Progress in Nucleic Acid Research and Molecular Biology," pp. 317–345. Academic Press, Berkely, CA.
12. Fukai, F., Iso, T., Sekiguchi, K., Miyatake, N., Tsugita, A., and Katayama, T. (1993). An amino-terminal fibronectin fragment stimulates the differentiation of ST-13 preadipocytes. *Biochemistry* **32**, 5746–5751.
13. Spiegelman, B. M., and Ginty, C. A. (1983). Fibronectin modulation of cell shape and lipogenic gene expression in 3T3-adipocytes. *Cell (Cambridge, Mass.)* **35**, 657–666.
14. Pairault, J., and Green, H. (1979). A study of the adipose conversion of suspended 3T3 cells by using glycerophosphate dehydrogenase as differentiation marker. *Proc. Natl. Acad. Sci. U.S.A.* **76**, 5138–5142.
15. Casimir, D. A., Miller, C. W., and Ntambi, J. M. (1996). Preadipocyte differentiation blocked by prostaglandin stimulation of prostanoid FP^2 receptor in murine 3T3-L1 cells. *Differentiation (Berlin)* **60**, 203–210.
16. Green, H., and Kehinde, O. (1979). Formation of normally differentiated subcutaneous fat pads in an established preadipocyte cell line. *J. Cell Physiol.* **101**, 169–171.
17. Green, H., and Kehinde, O. (1975). An established preadipose cell line and its differentiation in culture II. Factors affecting the adipose conversion. *Cell (Cambridge, Mass.)* **5**, 19–27.
18. Green, H., and Meuth, M. (1974). An established pre-adipose cell line and its differentiation in culture. *Cell (Cambridge, Mass.)* **3**, 127–133.
19. Marko, O., Cascieri, M. A., Ayad, N., Strader, C. D., and Candelore, M. R. (1995). Isolation of a preadipocyte cell line from rat bone marrow and differentiation to adipocytes. *Endocrinology (Baltimore)* **136**, 4582–4588.
20. Kirkland, J. L., Hollenberg, C. H., Kindler, S., and Gillon, W. S. (1994). Effects of age and anatomic site on preadipocyte number in rat fat depots. *J. Gerontol.* **49**, B31–B35.
21. Kirkland, J. L., Hollenberg, C. H., and Gillon, W. S. (1996). Effects of fat depot site on differentiation-dependent gene expression in rat preadipocytes. *Int. J. Obes. Relat. Metab. Disord.* **20**(Suppl. 3), S102–S107.
22. Djian, P., Roncari, D. A. K., and Hollenberg, C. H. (1983). Influence of anatomic site and age on the replication and differentiation of rat adipocyte precursors in culture. *J. Clin. Invest.* **72**, 1200–1208.
23. Stanton, L. A., von Venter, M., Litthauer, D., and Oelofsen, W. (1997). Effect of lipoproteins on the differentiation of 3T3-L1 and human preadipocytes in cell culture. *Comp. Biochem. Physiol.* **116B**, 65–73.
24. Teichert-Kuliszewska, K., Hamilton, B. S., Roncari, D. A. K., Kirkland, J. L., Gillon, W. S., Deitel, M., and Hollenberg, C. H. (1996). Increasing vimentin expression associated with differentiation of human and rat adipocytes. *Int. J. Obes.* **20**(Suppl. 3), S108–S113.
25. Barbe, P., Galitzky, J., Gilsezinski, I. D., Riviere, D., Thalamas, C., Senard, J. M., Crampes, F., Lafontan, M., and Berlan, M. (1998). Simulated microgravity increases B-adrenergic lipolysis in human adipose tissue. *Clin. Endocrinol. Metab.* **83**, 619–625.
26. O'Brien, S. N., Mantzke, K. A., Kilgore, M. W., and Price, T. M. (1996). Relationship between adipose stromal-vascular cells and adipocytes in human adipose tissue. *Anal. Quant. Cytol. Histol.* **18**, 137–143.
27. Poznanski, W. J., Waheed, I., and Van, R. (1973). Human fat cell precursors: Morphologic and metabolic differentiation in culture. *Lab. Invest.* **29**, 570–576.
28. Novakofski, J. E. (1987). Primary cell culture of adipose tissue. *In* "Biology of the Adipocyte: Research Approaches" (G. J. Hausman and R. J. Martin, eds.), pp. 160–197. Van Nostrand-Reinhold, New York.
29. Hauner, H., Entenmann, G., Wabitsch, M., Gaillard, D., Aihaud, G., Negrel, R., and Pfeiffer, E. F. (1989). Promoting effect of gluctosteroids on the differentiation of human adipocyte precursor cells cultured in a chemically defined medium. *J. Clin. Invest.* **84**, 1663–1670.
30. Entenmann, G., and Hauner, H. (1996). Relationship between replication and differentiation in cultured human adipocyte precursor cells. *Am. J. Physiol.* **270**, C1011–C1016.
31. Strutt, B., Khalil, W., and Killinger, D. (1996). Growth and differentiation of human adipose stromal cells in culture. *In* "Methods in Molecular Medicine: Human Cell Culture Protocols" (G. E. Jones, ed.), pp. 41–51. Humana Press, Totowa, NJ.
32. Carson, F. L. (1997). *In* "Histotechnology," pp. 152–153. ASCP Press, Chicago.
33. Carson, F. L. (1997). *In* "Histotechnology," pp. 153–154. ASCP Press, Chicago.

CHAPTER 11

LIVER CELL CULTURE AND LINEAGE BIOLOGY

Jeffrey M. Macdonald, Arron Xu, Hiroshi Kubota, Edward LeCluyse,
Geraldine Hamilton, Huifei Liu, YiWei Rong, Nicholas Moss,
Cynthia Lodestro, Tom Luntz, Stephen P. Wolfe,
and Lola M. Reid

INTRODUCTION

Tissue engineering is a fashionable term for a field of studies combining cell culture, stem cell and lineage biology, extracellular matrix chemistry and biology, preparation and use of synthetic scaffoldings, endocrinology and studies on growth factors, and bioreactor design and use. It is hoped that tissue engineering, currently in its infancy, will eventually be able to "engineer" cells into multicellular aggregates *ex vivo* that truly mimic the functional properties of tissue *in vivo*. Although it is premature to claim that one can truly "tissue-engineer" liver, current protocols enable liver cells *ex vivo* to achieve a much closer approximation to "normal" behavior than has been possible heretofore. This chapter presents those state-of-the art protocols for *ex vivo* management of liver as a representative quiescent tissue, since the generic approaches and strategies successful for liver are similar to those found to work for other cell types. Moreover, the exact conditions needed for liver cells from one species are, with rare exceptions, identical, or almost identical, to those needed for liver cells derived from all other species.

Some of the protocols presented (e.g., how to develop a serum-free, hormonally defined medium, or HDM) are largely as given in prior publications [131]; others (e.g., serum-free culture of minimally deviant cell lines or methods for embedding cells in forms of extracellular matrices) have been updated [76,185], and wholly new ones (e.g., conditions for clonogenic expansion of hepatic stem cells and adult cells, biodegradable scaffoldings, and hollow-fiber bioreactors) are presented in protocol form for the first time [76]. Protocols are not given for factors or reagents that are commercially available. Current commercial sources for the reagents are listed in Appendix 2.

GENERAL ISSUES

Tissue engineering of any tissue is done by mimicking the epithelial–mesenchymal relationship and the relevant stem cell and lineage biology.

THE EPITHELIAL–MESENCHYMAL RELATIONSHIP

The epithelial–mesenchymal relationship (Fig. 11.1) constitutes the organizational basis for all metazoan tissues [43,132,136,183]. Epithelia (epidermal, hepatic, pulmonary, intestinal) are wedded to a specific type of mesenchymal cell (stromal, endothelial, or smooth muscle cell), and dynamic interactions between the two are mediated by a set of soluble signals (both autocrine and paracrine), as well as by a set of insoluble signals on the lateral borders of homotypic cells (the lateral extracellular matrix) and on the basal borders (the basal extracellular matrix) between heterotypic cells [10,29,103,137]. Furthermore, endocrine (i.e., systemic) regulation and modulation is often achieved by regulating some

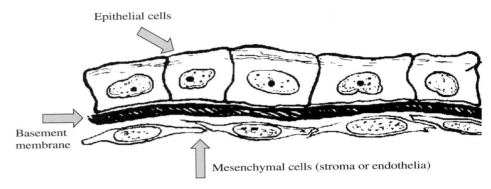

Epithelial cells

Basement
membrane

Mesenchymal cells (stroma or endothelia)

Fig. 11.1. Epithelial–mesenchymal relationship.

aspect of the epithelial–mesenchymal relationship. Normal epithelial cells will not survive for long and will not function properly unless epithelial cells and appropriate mesenchymal cells are cocultured [35,47,136]. Alternatively, the cells may be placed in contact with appropriate extracellular matrix components and cultured in medium containing the soluble signals (either defined components or "conditioned medium") produced by interactions between epithelia and mesenchymal cells.

STEM CELLS AND LINEAGE BIOLOGY

Quiescent tissues (liver, brain), as well as rapidly renewing tissues (skin, intestine), are organized as precursor cell populations (i.e., stem cells and/or committed progenitors) that yield daughter cells undergoing a maturational process ending in apoptotic cells [132,134,156,158,180]. The different maturational stages of cells within a lineage, such as the hepatic lineage (Fig. 11.2), are distinct both phenotypically and in their requirements for *ex vivo* maintenance [11,180]. Therefore, to have cells that behave uniformly under specified culture conditions, it is ideal to purify specific subpopulations of cells at distinct maturational stages. Purification of the cells involves enzymatic dissociation of the tissue followed by fractionation methods that can include immunoselection technologies [157,158]. Most of the culture conditions required for the different maturational stages of liver cells have been defined and involve the use of entirely purified soluble signals and extracellular matrix components [180]. Certain conditions can be used to maintain the cells with reproducible growth properties and others used to put the cells into a state of growth arrest and with expression of particular tissue-specific genes.

Each cellular subpopulation at a specific maturational stage (e.g., stem cells, committed progenitors, diploid adult cells, polyploid adult cells) has a unique phenotype (antigenically, biochemically, morphologically) with its own distinct set of conditions for *ex vivo*

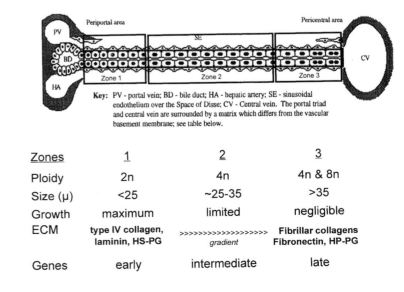

Key: PV - portal vein; BD - bile duct; HA - hepatic artery; SE - sinusoidal endothelium over the Space of Disse; CV - Central vein. The portal triad and central vein are surrounded by a matrix which differs from the vascular basement membrane; see table below.

Zones	1	2	3
Ploidy	2n	4n	4n & 8n
Size (μ)	<25	~25-35	>35
Growth	maximum	limited	negligible
ECM	type IV collagen, laminin, HS-PG	>>>>>>>>>>>>>>>> *gradient*	Fibrillar collagens Fibronectin, HP-PG
Genes	early	intermediate	late

Fig. 11.2. Liver lineage model.

growth versus differentiation. For example, hepatic stem cells must be cocultured with their natural partners, embryonic stroma [76], or ideally stromal feeder cells derived from early embryonic liver [12]. By contrast, the mature hepatocytes will go into growth arrest and express their tissue-specific functions most effectively if cocultured with adult-derived endothelia [47]. If the paracrine signals mediating the epithelial–mesenchymal relationship are known, they can be used to prepare completely defined culture conditions. If they are not known, coculturing of the epithelia and mesenchymal subpopulations is required.

Major Lineage Stages

Stem cells are immature cells that are pluripotent in that they give rise to daughter cells maturing into multiple cell types and have extensive growth potential [46,114,129]. Several broad classes of lineage stages have been defined: totipotent stem cells, determined stem cells, committed progenitors, and adult diploid and polyploid cells.

Totipotent stem cells have the capacity to produce adult cell types derived from all three embryonic germ layers (ectoderm, endoderm, and mesoderm), can enter into the germ line, and have proven ability to self-replicate—that is, to produce daughter cells that are identical to the parent [13,107]. The known and well-characterized totipotent stem cells are found in embryonic tissues and are from preimplantation or peri-implantation embryos [172]. Their derivatives in culture are known as embryonic stem cells (ES cells) and embryonic germ cells (EG cells). The precise conditions for *ex vivo* maintenance of ES cells are described elsewhere [17,98,149].

Determined stem cells, pluripotent cells that give rise to some, but not all, possible adult cell types, have extensive growth potential and questionable ability to self-replicate [129]. The determined stem cells constitute the earliest cells of a given lineage. For example, hepatoblasts are considered to be hepatic stem cells, a type of determined stem cell that gives rise to both biliary cells and hepatocytes [46]. Their derivatives contribute all the parenchymal cells in the adult liver. The determined stem cells for liver have been identified antigenically and purified by multiparametric flow cytometry [76,82]. The biopotency of determined stem cells has been proven rigorously under clonogenic culture conditions in which a single cell is seeded and expanded into a colony of cells with both biliary and hepatocytic fates, [76]. The determined stem cells can expand only when cocultured with appropriate embryonic stroma in culture at low cell density condition.

Committed progenitors are derived from stem cells, have lost pluripotency, and are precursors for a single cell type. The culture conditions for these cells overlaps extensively with that of the determined stem cell, with the caveat that the committed progenitors are able to grow on stromal feeders both from embryonic as well as from neonatal and adult tissues [11,158].

Diploid cells. All fetal and neonatal tissues contain only diploid cells, and these cells are able to undergo complete cell divisions both *in vivo* and *in vitro*. In liver, the number of divisions possible for *adult diploid cells* is unknown but is estimated at 5–8. The clonogenic culture conditions that are successful with hepatic stem cells are permissive for a subpopulation of adult liver cells, hypothesized to be the diploid cells, to grow at extremely low seeding densities and to expand into colonies of cells [76]. However, the propensity of adult liver cells to aggregate (due to their cell adhesion molecules and proteoglycans) means that colony formation by adult cells cannot be defined rigorously as clonogenic. Nevertheless, the conditions do permit analyses of growth of this subpopulation of adult cells at the lowest possible seeding density.

Polyploid cells appear in liver tissue within a few weeks of postnatal life in rodents or within a few years in humans [37,116,143,151]. The majority of *polyploid cells in liver* are tetraploid, although higher levels of ploidy have been observed in apoptotic cells [125, 146]. The percentage of polyploid cells increases with age [2,14,15,158] and after partial hepatectomy [159] and is achieved by cells undergoing nuclear division without cytokinesis. By approximately 6 weeks of age, adult rodents have livers in which the extent of polyploidy is at least 90%, comprising 80% tetraploid cells and 10% octaploid cells. Published data on ploidy in human liver varies greatly, and the interpretations to date are suspect—in some studies owing to the methods of analyzing ploidy, in others owing to the sampling methods used, and in all such work owing to the sparse number of human liver samples analyzed. Prior to the 1970s, some investigators used the assumption that cells with multiple nuclei are polyploid and those with a single nucleus are diploid to define the extent of polyploidy in their analyses of histological sections of tissue. Since it is now well known that most tetraploid cells are mononucleated, the data from these older studies have been discredited. In more recent studies based on valid methods for analyzing ploidy, the authors have used a small piece of the liver or even a needle aspirate [2,4,77,143,174]. Considering the heterogeneous nature of liver cells [48,173,34], such small samples are unlikely to be representative of the tissue as a whole. All studies on human liver are made difficult by the extremely limited supply of reasonable quality samples, which usually are limited in age range as well. Altogether, these difficulties may explain the wide variations in findings on the extent of polyploidy in adult human livers. Some have claimed as few as 10% and others about 30%; the highest numbers reported cite up to 60% of the cells as polyploid. All agree that the percentage of polyploid liver cells increases with age, and that the percentage increases with special significance after 60 years of age [2,37,77,143,164,174]. In summary, polyploidy in human liver and the data correlated with age of donor have yet to be analyzed rigorously.

The properties of the diploid versus polyploid cells are quite distinct. Whereas diploid cells have been found to be capable of colony formation [108–110,168], the tetraploid cells are able to go through only one or two complete cell divisions, represented by the customary findings of studies on routine primary cultures of adult liver cells [10,33]. The few divisions possible include ones in which the cells are thought to go from tetraploid to tetraploid state. Tetraploid cells are thought to be incapable of clonogenic growth, an assumption that has yet to be tested rigorously. Instead, they are ideal for an analysis of highly differentiated functions, some of which are expressed only in the polyploid cells such as certain of the cytochrome P450s [81].

Strategies Influenced by Knowledge of the Lineage Biology

Whereas all forms of progenitors are found in embryonic tissues, most adult tissues contain only determined stem cells and/or committed progenitors [129]. The exceptions are the adult heart (which is thought to have either no progenitors or perhaps only committed progenitors [89,98], resulting in the heart's limited regenerative capacity) and the bone

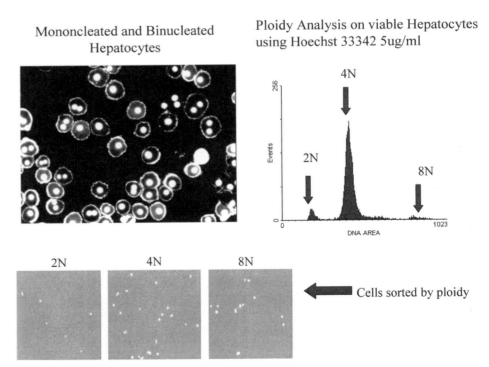

Fig. 11.3. Ploidy and liver cells.

marrow (now being studied as a possible adult reservoir of totipotent stem cells or a novel location for determined stem cells of various types) [82,127,128].

The known properties of the different stages of cells (Table 11.1) help to define their potential in academic, clinical, or industrial programs. Of the types of progenitor studied (e.g., mesenchymal, neuronal, muscle, epidermal), all have been found to be readily cryopreserved [17,25,139] and expanded *ex vivo* [21,30]. However, the embryonic stem cells are especially notable for their ability to survive freezing and to expand without differentiating if maintained under precise culture conditions [17,98,139,149,171]. By contrast, attempts to cryopreserve adult liver cells (predominantly the polyploid cells) have met with limited success, and even that limited success is achieved only by embedding the cells in alginate or a form of extracellular matrix [49,74,88,167,175]. Significant *ex vivo* expansion and the ability to subculture adult liver cells have been observed only with the "small hepatocytes" [108,110,111,168], assumed to be a diploid subpopulation of the liver. Typically, the mature cells undergo one or two rounds of division and then survive for a matter of days in culture, or for a few weeks when supplied with the appropriate extracellular matrix and medium conditions [180].

The ability of totipotent cells and embryonic stem cells to give rise to all, or almost all, possible adult fates makes them appealing as a possible "one serves all" approach for cell therapies and makes them the most exploitable of the classes of stem or progenitor cells [6,126,147]. However, their use in cell transplantation for patients is precluded by their tumorigenic potential [45,106,113,119,126,147]. The tumorigenicity of ES cells when injected at ectopic sites is being investigated extensively, especially by biotechnology companies [6,106,126], in hopes that it can be controlled to enable ES cells to reach their full potential both industrially and clinically.

While determined stem cells are more restricted in their adult cell fates, they have not been found to be tumorigenic, making them the first choice for clinical programs in cell transplantation or for bioartificial organs [6,53,129]. Bone marrow and cord blood transplants, which represent the first forms of progenitor cell therapy, have been performed for years [44,86,105]. More recently, other forms of progenitor cell therapies are being tested in clinical trials. These include mesenchymal progenitor cells [21,60,128], neuronal progenitor cells [1,30,170], and fetal pancreatic islet cell transplants [138,155], and the early data from these trials are very encouraging for the future for progenitor cell therapies as a class.

The problems with determined stem cells include availability of tissues, a particular problem for organs that until now have derived only from brain-dead, beating-heart donors; the need for the development of purification schemes for isolation of the cells; identification of optimal cryopreservation conditions; and defining the *ex vivo* expansion and differentiation conditions.

PURIFICATION OF THE CELLS OF INTEREST
LIVER PERFUSIONS: METHOD FOR PREPARATION OF FRESHLY ISOLATED LIVER CELLS

Liver cells can be isolated readily by using a standard liver perfusion process [9,28]. The method has many variations, but in all its forms it involves the following steps: (1) anesthetizing an animal or obtaining an isolated liver or portion of a liver (human tissue), (2) catheterizing one or more major blood vessels leading into or from the liver (e.g., portal vein and/or the vena cava), (3) perfusing the liver for 10–15 min with a chelation buffer containing EGTA to reduce the calcium concentrations in the liver, (4) perfusing with a second buffer containing collagenase and one or more proteases in a calcium-containing buffer for 10–15 min (rodents) to ~30 min (human), and (5) mechanically dissociating the liver by pressing the digested liver through cheesecloth or raking it with combs and then sequentially filtering the cell suspension through sieves of narrowing mesh size.

Most liver perfusions are done with collagenase preparations that are partially purified. Different companies indicate the degree of purification with a company-specific nomenclature. For example, Sigma indicates the crudest collagenase preparation as type I, and the pure collagenase as type VIII, whereas Boehringer-Mannheim refers to the crudest preparation as type A and the pure collagenase as type D. One must read the company's literature

Table 11.1. Classes and Properties of Cells at Different Maturational Stages

Properties	Totipotent stem cells or ES cells	Determined stem cells	Committed progenitors	Diploid adult cells	Polyploid adult cells
Representatives of the class	Embryonic stem cells (ES cells); cells from pre- or peri-implantation embryos such as morula cells	Hepatic, hemopoietic, neuronal, mesenchymal	Hepatocytic, biliary	Hepatocytes	Hepatocyte
Potency	Pluripotent	Pluripotent	Not pluripotent	Not pluripotent	Not pluripotent
Fate(s)	All possible fates	Restricted fates	Single fate	Single fate	Single fate
Ploidy	Diploid	Diploid	Diploid	Diploid	Polyploid
Growth	Extensive	Extensive	Extensive	5–8 divisions	1–2 divisions
Self-replication?	Yes	Debatable	No	No	No
Survive ischemia?	Yes	Yes	Yes	Yes	No, sensitive to ischemia
Cryopreservation ability	Readily done in suspension	Readily done in suspension	Readily done in suspension	Moderately difficult; requires freezing in adherent or embedded state	Very difficult—only partially successful even when embedded
Tumorigenic?	Yes, if injected at ectopic site and with tumorigenicity reduced but not eliminated by lineage restriction	No	No	No	No
Advantages	Ease in ex vivo expansion and cryopreservation; ability to form all cell types	Safe for cell transplantation; possible to control differentiation to specific cell types *ex vivo* or *in vivo*. Fetal and neonatal livers are richest sources of the hepatic stem cells and committed progenitors; minimal immunogenicity.		Diploid adult cells constitute a major subpopulation of all human livers of all ages; are tolerant of ischemia; can be cryopreserved; have some growth potential; mature to the polyploid cells	Polyploid adult cells express important adult functions (e.g., certain P450s)
Disadvantages	Tumorigenicity precludes use in cell therapies; inability to control reproducibly the differentiation to specific fates	Difficulties in trying to use fetal livers for clinical programs due to legal and ethical considerations; no programs exist yet for neonatal liver procurement		Growth potential of the adult cells may not be adequate to the need of the patient	Found in significant numbers only in adult human liver tissue; very sensitive to ischemia; cannot cryopreserve; extremely sensitive to the variables associated with organ procurement processes

to learn the details of the nomenclature and their implications for the extract or purified factor(s) being sold. Generally, liver perfusions are done with a preparation that is intermediate in purity (e.g., type IV in Sigma's series or type B or C in the Boehringer-Mannheim series), since both collagenase and one or more proteases are required for optimal liver digestion. Moreover, it has been learned only relatively recently that the most effective liver perfusions are achieved with a mixture of purified collagenase and purified elastase at precise ratios [40,122]. One commercial preparation of purified collagenase and elastase, called Liberase (Boehringer-Mannheim), is designed for liver digestion. However, its use has been limited owing to its high cost.

LIVER PERFUSIONS

Prepare stock solutions a day or two ahead of time. Use highly purified, sterile water [prepared, e.g., by using a Milli-Q purification system (Millipore) and then sterilized by autoclaving or by filtration through a Nalgene filtration unit]. The water so prepared will be referred to as MQ water. The stock solutions are prepared as follows.

SUPPLIES AND SETUP

1. Autoclave the following: surgical tools (large forceps, small and large scissors, 2 hemostats) plus two glass funnels, an Erlenmeyer flask (250 ml), 2 sheets of Nytex, thread or suture, and cotton-tipped swabs.
2. *Prepare buffers:* 10× Leffert's, 1 M $CaCl_2$, 0.2 M EGTA, and buffers A, B, and C (see Formulations); culture medium, isotonic Percoll®.
3. Culture plates, petri dishes, 50-ml conical tubes, animal rack, pan to accommodate rack/animal, ice buckets.
4. Fast rat overnight (optional).
5. O_2 (tank), ethanol, collagenase (Sigma type IV), ketamine, bovine serum albumin (BSA), phosphate-buffered saline (PBS), trypan blue.

FORMULATIONS

10× Leffert's buffer (filtered)

250 mM HEPES, 59.57 g
1.15 M NaCl, 69.27 g
50 mM KCl, 3.73 g
10 mM KH_2PO_4, 1.36 g
Combine in as much MQ water as necessary to make 1.0 liter pH to 7.4.

1 M $CaCl_2$ (Autoclave)

73.5 g of $CaCl_2$
MQ H_2O, 500 ml, final vol.

0.2 M EGTA (filtered)

38.04 g of EGTA
MQ H_2O, 500 ml, final vol.

Chelation Buffer (A)

1× Leffert's/0.5 mM EGTA
0.625 ml of 0.2 M EGTA stock in final volume of 250 ml

Collagenase Buffer (B)

1× Leffert's/1 mM $CaCl_2$
0.5 ml of 1 M $CaCl_2$ stock in final volume of 500 ml (digestion buffer)

Within an hour of the time of the perfusion, add the collagenase to the buffer; for rat livers, add 70 mg of collagenase/500 ml; for human livers, add ~500–600 mg per liter of buffer. The exact amount depends on the biological activity of the collagenase sample and must be adjusted also based on empirical findings with the routine samples being digested.

Buffer C

$1\times$ Leffert's/2 mM $CaCl_2$, add 0.5 ml of 1 M $CaCl_2$ stock in final volume of 250 ml. Add 1.5 g BSA at time of use.

Percoll Buffer

After the liver has been digested and just prior to fractionation of the cells, mix the culture medium with 90% isotonic Percoll (Sigma). The ratio of volumes should be 3 parts of culture medium to 1 part of Percoll (e.g., 30 ml of cells in medium + 10 ml of 90% isotonic Percoll). Adjust the ratio of medium to Percoll depending on the extent of fatty deposits in the liver. The more fat, the lower the Percoll ratio to the culture medium.

PLATING MEDIUM AND CULTURE MEDIUM

Use RPMI 1640 (Gibco/BRL) or a 1:1 mixture of Dulbecco's Modified Eagle's Medium and Ham's F-12 (DMEM/F-12, Gibco/BRL) to which is added 20 ng/ml epidermal growth factor (EGF from Collaborative Biomedical Products), 5 µg/ml insulin (Sigma), 10 µg/ml iron-saturated transferrin (Sigma), 4.4×10^{-3} M nicotinamide (Sigma), 0.2% bovine serum albumin (Sigma), 5×10^{-5} M 2-mercaptoethanol (Sigma), 7.6 µequiv free fatty acid mixture (see Appendix 1 for its preparation), 2×10^{-3} M glutamine (Gibco/BRL), 1×10^{-6} M $CuSO_4$, 3×10^{-8} M H_2SeO_3, and antibiotics. For the plating medium add, in addition, 1–2% fetal bovine serum (Hyclone) and keep the cells in it for only 4–6 h until attachment has occurred; this will be referred to as a serum-supplemented medium (SSM) combined with a hormonally defined medium (HDM), or SSM/HDM. After the cells have attached, remove the SSM/HDM, rinse gently, and switch to a serum-free version, which will be called a serum-free, hormonally defined medium (HDM).

PERFUSION SETUP

- 42°C water bath
- Perfusion pump with stopcock three-way valve on two inlet lines (with 5-ml pipettes for bottles)
- Single outlet line (to connect to catheter) that goes through the water bath
- Flow rate at 30 ml/min
- O_2 tank feeding two outlet tubes with screw clamps to control gas flow
- 37°C waterbath
- Low-speed centrifuge
- Good light source for surgery

PREPARE (IN BIOLOGICAL OR LAMINAR FLOW HOOD)

- Turn on water bath (42°C). Rinse pump tubing with ethanol for 5 min, and in distilled water for 5 min.
- Add *collagenase* and *BSA* to *buffers B* and *C*. Aliquot two conical tubes (50 ml) of buffer C on ice.
- Put *buffers A* and *B* in 42°C bath (weight down) and insert lines.
- Fill inlet with *buffer B* up to stopcock, then fill entire line with *buffer A*, eliminating any bubbles.
- Insert O_2 lines into bottles, adjust to a low flow rate (take care that bubbles form above buffer inlets).

- Weigh animal and anesthetize with Ketamine (200 g rat, dose: 0.2 ml).
- Set up rack (to support animal) and pan (to accommodate rack/animal and collect fluids), instruments, two sutures (or pieces of thread), petri dish, and five pieces of tape.

SURGERY

- Restrain limbs of anesthetized animal with tape. Open abdominal skin and cut up centerline. Cut laterally and spread.
- Cut through abdominal muscles up centerline to expose lower end of sternum, but do not open chest cavity. Take care not to nick the liver! Cut laterally and spread sides.
- Use sterile swab to move/sweep intestines out via the right side, exposing vena cava and portal vein.
- Attach hemostat to sternum and flip it upward.
- Use a swab to maneuver liver up against the diaphragm.
- Put a suture around inferior vena cava between renal and hepatic veins, and another around portal vein. Loop each for tying.
- Use a hemostat on the intestine proximal to portal vein to increase tension, then catheterize portal vein.
- Tie off suture, turn on pump, and attach outlet to catheter.
- Watch liver swell and blanch, and cut inferior vena cava below suture.
- Tape down line to catheter for stability.

PERFUSION

- Pump *buffer A* for 2 min to rinse liver well (removal of calcium).
- Push liver back down from diaphragm, then cut through diaphragm.
- Cut open the heart, then tie off suture on inferior vena cava. Open chest cavity for drainage.
- Cut small ligaments that attach liver to diaphragm. Push liver back up against diaphragm to increase flow pressure.
- Pump *buffer A* for about 5 min (150 ml).
- Switch to *buffer B*. Flow until liver is well digested (300–500 ml). Liver should swell and then spread out.
- Finally tissue structure breaks down visibly below the capsule. During this time, free liver from attaching membranes.

CELL ISOLATION

- Remove liver to petri dish and add *buffer C* (room temperature), using about 10–15 ml.
- Comb gently to break capsule, then shake liver with forceps at the junction of the lobes to dislodge cells. Flip liver over and repeat. Ideally, only white connective tissue will remain. Do not comb resistant areas (often at margins of lobes), since this will decrease overall viability. If the liver is not completely reduced to connective tissue, proceed with what shakes out.
- Filter cell suspension into glass funnel with a double layer of Nytex membrane sitting in an Erlenmeyer flask. Rinse petri dish with *buffer C* and add to filtrate. Agitate Nytex membrane to increase flow rate, but do not assist from above.
- Blow O_2 into flask gently; then cover with Parafilm and put into 37°C bath for 5–10 min. Swirl flask gently to resuspend cells, then incubate on ice for 10 min.
- Repeat Nytex filtration into 50-ml conical tubes on ice in hood.
- Divide filtrate between two 50-ml tubes, then spin down at low-speed centrifugation (50*g*) for 1 min.

- Aspirate supernatant (which is cloudy). Add cold *buffer C*, 20 ml/tube, and gently resuspend cells.
- Spin down cells again for 2 min at 60*g* and resuspend as before. Repeat for third spin, again at 2 min at 60*g*. Resuspend in a *combined* volume of 20 ml.

CULTURE

- Count cells in 1:10 diluted trypan blue on a hemacytometer.
- Calculate viability and concentration.
- Calculate for plating; plate cells in SSM/HDM.
- Incubate 1–2 h at 37°C (the lower the viability, the shorter the time).
- Remove media, wash with 1× PBS, then add either SSM or an HDM depending on experimental needs.
- Incubate in CO_2 tissue culture incubator, maintaining cultures with daily media changes.

LIVER PERFUSIONS: SURGICAL PROCEDURES FOR PEDIATRIC OR ADULT HUMAN LIVERS

Pediatric and adult human livers are from brain-dead, beating-heart donors, since the donor organ is destined for organ transplantation, and the liver's exquisite sensitivity to ischemia necessitates that the procurement process occur at the moment of death. Only 1–2% of people who die in the United States have undergone brain death prior to heart arrest. Thus, the number of donor organs per year is very small, on average between 4000 and 5000. Over 95% of these are used successfully for organ transplantation. The remaining 1–5% of the donor organs, or 50–250 livers/year, are rejected for organ transplantation for a variety of reasons, including infections (in such cases the liver goes to investigators studying that type of infection or ischemia), high percentage of fat, or other conditions (in these cases the liver goes to investigators in diverse academic and industrial settings). The rejected livers are shunted to agencies such as the Anatomical Gift Foundation (AGF) or to organizations under contract to the National Institute of Health that handle the distribution process. These livers, ranging in weight from 1500 to 2500 g, are rarely sent as whole livers but rather are partitioned by agency staff members to maximize the number of researchers receiving samples. Samples are shipped to investigators within 10–20 h from the time of removal from the donor or the "clamp time." The samples arrive flushed with transport buffer, most commonly University of Wisconsin solution ("UW" solution or Viaspan from Upjohn), bagged, and on ice. Each researcher receives a piece that is usually about 100–200 g and must be perfused through cut blood vessels exposed on the surface of the sample.

The conditions prior to death and the cold ischemia associated with the transport conditions of the liver or portion of a liver can result in the deterioration of the sample. Thus, the quality of the starting material is extremely variable. For donor organs it is generally accepted that the overall organ integrity and function begins to deteriorate after 18 h post-clamp, and organs will not be used for transplant after this time. In our experience, the quality of the cells prepared from donor organs that have been procured longer than 18–20 h reflects this general phenomenon, and lower yields and viability of the polyploid cell populations are observed compared with fresher organs or tissue. We have observed also that, in general, organs received more than 24 h after clamp time often do not yield cells of adequate quality, nor are the cells able to efficiently attach to culture substrata. However, the time threshold after which a particular organ cannot produce cells of adequate quality is affected by several factors, including age of the donor, proficiency of organ preservation, quality of the tissue perfusion, and disease state of the organ (e.g., extent of cirrhosis and steatosis). For the most part, organs should be a uniform tan or light brown when received; organs that appear "bleached" or dark brown should not be used and generally yield only nonviable or cells that are depleted of cytochrome P450. Medium containing phenol red in contact with liver cells isolated from damaged organs often has a characteristic pink color, especially when mixed with Percoll, which is believed to reflect the depletion of certain macromolecules from damaged cells.

Perfusion of Human Liver Tissue and Liver Cell Isolation

As stated earlier, the quality of the liver cell preparation is a reflection of the quality of the starting tissue. As such, the best sources of tissue for the isolation of liver cells are freshly resected biopsy samples and freshly preserved donor organs (<12 h from clamp time). Of course, these sources of tissue are rarely available to the average academic or industrial investigator. Therefore, strict guidelines must be in place to assure that the quality of the cells is adequate and that proper controls are introduced into every experiment. There are a number of methods described in the literature for the isolation of human liver cells from partial biopsy segments and whole lobes [85,165]. These approaches are essentially modified versions of the original two-step perfusion methods developed by Seglen [150] for the isolation of cells from rat liver.

Protocol for Human Liver Tissue Perfusion

1. Remove liver sample from bag and place in sterile tray.
2. Add some sterile Leffert's buffer to liver to keep it wet during the phase of connecting catheters.
3. The liver surface has been cut during the division of the original liver into samples, revealing many exposed blood vessels. Test for candidate blood vessels that will yield perfusion of a reasonable percentage of the liver by catheterizing, pumping buffer through them, and observing whether the liver tissue swells and where the fluid emerges.
4. Once a blood vessel has been chosen, insert a catheter that fits snugly into the vessel and glue it into place with surgical glue, LocTite Instant Adhesive, medical grade (LocTite Corporation, Rocky Hill, CT). Secure the catheter in place by adding more glue at the cannula–tissue interface.
5. Use glue to seal all other openings on the cut surface. It may be necessary to use a cotton-tipped applicator to seal the larger openings. Either the wooden part only can be used or the cotton tip. The applicator can be reduced to fit in the opening size, and the wooden stick shortened so that no more than a centimeter protrudes. Secure applicators in place by making a glue collar around the edge.
6. Dab dry the cured surface of the liver, and cover the entire cut surface with a thin layer of glue, applied with using a sterile cotton-tipped applicator.
7. Once the glue has dried, place the liver on a weighing boat, connect perfusion tubing to the catheter, and slowly start the perfusion. If no major leaks are observed, carefully place the liver tissue in the perfusion tank containing the chelation buffer. Submerge the liver sample with sufficient buffer to float it.
8. Slowly increase the flow rate for the chelation buffer. Perfuse with chelation buffer for 10–15 min. The flow rate and backpressure will vary with the size of the tissue, and how well it is sealed. (On average, the initial flow rate and backpressure should be between 25 and 50 ml/min and 20 and 40 mmHg, respectively.)
9. Remove chelation buffer and replace with the collagenase buffer. Perfuse with collagenase buffer until liver is digested (~15–30 min). The digestion time will vary depending upon the activity of the collagenase used and the size of the liver. Outflow pressure should not exceed 30–40 mmHg throughout the perfusion procedure. The backpressure will normally decrease as digestion progresses, and the tissue begins to break down. Do not overdigest the liver, for this will lead to overt cell damage and symptoms of oxidative stress.

The quality and integrity of the resulting cell material is dependent on a number of factors, including the time of perfusion (preferably ≤30 min total perfusion time), the flow rate and backpressure values, the quality of the starting material, and the quality of the crude collagenase (nonspecific protease contamination varies from batch to batch).

10. When the perfusion is complete, remove the liver from the tank and place in a sterile bowl or dish of adequate size and then transfer to a sterile hood.
11. Add sufficient plating medium, SSM/HDM, to cover the liver tissue.
12. Remove the dried capsule of glue and catheter from the surface of the tissue. Use tissue forceps and scissors to gently break open the outer Glisson's capsule. With the aid of the tissue forceps, mechanically dissociate the liver by raking the liver with sterile flea

combs (purchased from any veterinary supply) or equivalent instrument, and gently release the liver cells into the medium, leaving behind the connective tissue and any undigested material.

13. The digested material in the culture medium is then filtered through a series of Teflon mesh filters from 1000- to 500- to 100-µm mesh. Use large funnels, and filter into sterile 500- to 600-ml beakers. Scrape off any remaining clumps and aggregates from the filters and transfer to a petri dish. Remove the filter and dispose of it. The clumps and aggregates removed from the filter can be rinsed with Leffert's buffer and frozen for biochemical analyses on the freshly isolated cells, or they can be rinsed and plated for preparing feeder layers (see later) or discarded.

14. Transfer the cells that passed through the filters into centrifuge tubes (the size of the centrifuge tubes will vary according to the amount of material, but the range is 50–200 ml) and wash by low-speed centrifugation (70g for 4 min).

15. The resulting *pellet is enriched for the polyploid liver cells*, which are typically above 25 µm in size. Remove and save the supernatant. Gently resuspend the pellet in a small amount of HDM/SSM. At this point pellets may be pooled if warranted and smaller tubes (50 ml) may be used if the initial centrifugation was performed with larger tubes.

16. The *supernatant* can be kept for the isolation of nonparenchymal cells (e.g., hepatic stellate cells), diploid parenchymal cells, and/or progenitor cells. Spin the supernatant at 200g for 4–5 min to pellet some of the remaining cells. Resuspend the pellet in the SSM/HDM, count, and store appropriately. Spin the supernatant again, but now at 300g. Again resuspend the pellet in the SSM/HDM, count the cells, and store appropriately. Spin the supernatant at 500g for 4–5 min to collect the final remaining cells. Resuspend the pellet in the SSM/HDM. With increasing g force, smaller and smaller cells will pellet and will be increasingly enriched for the diploid subpopulation of cells from the liver. The final spin will include cell debris but will also contain progenitors.

17. Resuspend the pellets (step 15; those enriched for the polyploid cells) in culture medium plus 90% isotonic Percoll; the ratio of volumes should be 3 parts of culture medium to 1 part of Percoll (e.g., 30 ml of cells in medium + 10 ml of 90% isotonic Percoll) (see Reagents section for details on Percoll). [*Note:* If the liver has a high fat content, the Percoll must be altered accordingly by reducing the volume of Percoll used.]

18. Centrifuge the Percoll suspension at 100g for 5 min and then remove the supernatant. [*Note:* This supernatent can be treated similarly to the initial supernatant (see step 16) to collect the diploid subpopulations and other smaller cells.] Care should be taken not to disrupt the top layer of the supernatant, which contains the dead cells, or to contaminate the pellet with these cells. Gently resuspend the pellet in SSM/HDM. Centrifuge for a final time at 70g for 3 min.

19. Resuspend the final cell pellet in 4–5 ml of culture medium per milliliter of packed cell pellet.

20. Place the cell suspension on ice and count the cells in the final suspensions after centrifugation and/or fractionation. Use trypan blue exclusion assay or its equivalent to assess the percentage of viable cells.

PLOIDY ANALYSIS AND FRACTIONATION OF LIVER CELLS
Analysis of Ploidy

Ploidy is a critical variable for many of the functions of cells under consideration. Polyploid cells have less growth potential than do the diploid cells but express critical genes such as *CYP450*; conversely, the diploid cells have the greatest growth potential but express an overlapping set of genes, which nevertheless are somewhat distinct from those expressed by the polyploid cells. Therefore, different experiments may demand cells of distinct ploidy. Ploidy can be analyzed on viable cells and fixed cells, whole cells, or isolated nuclei. The dye used to stain viable cells is Hoechst 33342 (Molecular Probes), and its use has been especially well characterized in lymphocytes. It is membrane permeable and DNA specific. A dye that works well with fixed cells is propidium iodide (Sigma). Analysis of liver cells is more complicated than analysis for lymphocytes, since so many of the cells are polyploid, and many are multinucleated. For example, flow cytometric analyses and sorts cannot distinguish between mononucleated tetraploid cells and binucleated cells in which each nucleus is diploid. Also, analyses of liver cells are dramatically influenced by the quality of cell preparation.

The actual analysis of the cells, whether with Hoechst 33342 dye or propidium iodide, can be by flow cytometer, fluorescence microscopy, confocal microscope, or automated fluorescence image cytometer. The Hoechst dye has a maximal excitation wavelength at 350 nm and an emission wavelength at 461 nm. The propidium iodide has a maximal excitation wavelength at 535 nm and an emission wavelength at 617 nm.

The most accurate measurements of ploidy are made by using flow cytometry and an automated fluorescence image cytometer. Obviously, morphology in combination with ploidy can be determined only by fluorescence microscopy and confocal microscopy, and this is the only way to distinguish binucleated cells from mononucleated cells.

METHOD FOR FIXED CELLS

- Use a single-cell suspension as obtained by liver perfusion.
- Fix cells with a 3:1 mixture of methanol and acetic acid for 5 min.
- Wash three times with PBS, 5 min each time.
- Digest cells with RNase A (50 µg/ml, Sigma) in PBS for 30 min at 37°C.
- Add propidium iodide (Sigma) to a final concentration of 50 µg/ml.
- Analyzed DNA content of cells by flow cytometry (e.g., FACScan).

METHOD FOR ISOLATION OF NUCLEI FROM FRESHLY ISOLATED CELLS

- Freshly isolated cells are washed twice with 5.0 ml of cold saline GM (glucose, 1.1; NaCl, 8.0; KCl, 0.4; $Na_2HPO_4 \cdot 12H_2O$, 0.39; and KH_2PO_4, 0.15 (g/liter) containing 0.5 mM EDTA for chelating free calcium and magnesium ions.
- Cell pellets are then resuspended in 4.0 ml of RSB swelling solution, containing 10 mM NaCl, 1.5 mM $MgCl_2$, 10 mM Tris (pH 7.4), and 0.5 mM EDTA.
- The cell pellets are then vortexed vigorously for 30 s and allowed to stand on ice for 5 min.
- After addition of 0.5 ml of 10% Nonidet P-40 (NP-40) detergent, the samples are vortexed again vigorously for 30 s and put on ice for 30–60 min.
- The samples are then centrifuged at 1000 rpm and the pellets are washed with PBS twice.
- Check with a microscope: the pellets should be purified nuclei, with little cytoplasmic contamination [20].

PLOIDY ANALYSIS ON NUCLEI

- Wash nuclei three times with PBS for 5 min each time.
- Digest nuclei with RNase A (50 µg/ml, Sigma) in PBS for 30 min at 37°C.
- Add propidium iodide (Sigma) to a final concentration of 50 µg/ml.
- Analyze DNA content of the nuclei by flow cytometry (e.g., FACScan analysis).

PLOIDY ANALYSIS ON VIABLE CELLS

- Prepare freshly isolated cells.
- Stain cells with a solution of 5 µg/ml Hoechst 33342 (Molecular Probes) in RPMI-1640 (Gibco) supplemented with fetal bovine serum (2–10%, depending on cells and experimental plans; HyClone), EGTA (0.5 mM), and DNase (0.006%) (Sigma).
- Stain at 37°C for 30 min.
- Put on ice and keep at 4°C until analysis.
- Analyze by flow cytometry or by fluorescence microscopy.

Size

Cells are sieved through filters to enrich for cells of a desired size. For example, filter through a 20-μm filter to obtain the "small hepatocytes," cells averaging 17–18 μm in diameter and known for their ability to form colonies in culture [109,168]. A protocol yielding such adult liver cells able to form colonies and to grow at clonal seeding densities follows.

Fractionation of Adult Liver Cells That Can Be Expanded at Low Seeding Densities

1. Prepare freshly isolated liver cell suspension.
2. Filter the cells through a 30-μm sieve to remove large cells and aggregated cells. Cellular viability should exceed 90% as measured by trypan blue exclusion.
3. If cells that pass through the 30-μm sieve are used to seed the plates, the efficiency of colony formation is approximately 200 colonies per thousand cells seeded or about 20%.
4. If higher efficiencies of colony formation are desired, filter the cells again through sieves (or fractionate with density gradients) and use cells less than 18 μm in diameter. This yields the "small-hepatocyte" fraction.
5. Plate the cells onto STO feeders and into the serum-free HDM used for the hepatic stem cells (see later).
6. Under the conditions specified, colonies of adult cells form at a lower growth rate than those for the hepatic stem cells but are easily visible within a week of culture. By about 20 days of culture, the adult colony-forming cell can generate 120 daughter cells (by contrast, the stem cells in the same time frame will yield a colony with 3000–4000 cells).

Cell Density

- Layer over Ficoll-paque (Pharmacia Biotech) or Percoll (Sigma) and follow manufactures instructions for use. Collect the cells of desired density, rinse, and suspend in the plating medium. The following example describes the isolation of minimally deviant hepatic cell lines.

- *Centrifugal elutriation*: Cells are fractionated very gently by centrifuging them in a special rotor in which the buffer in which the cells are suspended is pumped upward while the centrifuge is spinning. This yields two opposite forces: the centrifugal force in opposition to the direction of the fluid force. When these are carefully calibrated cell fractions of very precise cell densities can be obtained. These procedures as used for liver cells are detailed elsewhere (e.g., [124]).

Antigenic Properties

One can use panning, flow cytometry, columns, and magnetic beads to isolate for immunoselection cells having a desired antigenic profile (e.g., cells expressing connexin 32). Two examples are given: immunoselection to eliminate red blood cells by panning and immunoselection to isolate purified hepatic stem cells.

Immunoselection to Eliminate Specific Cell Types

Elimination of red blood cells is given as an example.

- Coat 150-mm *petri dishes* (not tissue culture dishes) with rabbit anti-rat red blood cell IgG. Let sit overnight at 4°C (100 μl per plate + 10 ml of medium to coat each plate).
- Keep buffer on plate to keep from drying out—BUT remove before adding cells to be panned.
- Add cells to the panning plates at 4°C. After 5 min, gently transfer nonattached cells to a 50-ml sterile tube. Repeat the process two to four times, depending on the extent of contamination by the cells to be eliminated. For example, livers at embryonic days 16–18 are replete with red blood cells. One must pan them four or five times to eliminate the large numbers of red blood cells present. Fewer rounds (once or twice) of panning are required for eliminating red blood cells from adult tissues prepared by perfusion.

• Take a count of remaining cells in solution to establish how many cells you have and to arrive at the correct cell density for plating; check dishes to assess nonspecific losses.

Use of Immunoselection to Purify Hepatic Stem Cells and Committed Progenitors

Methods developed for identification and purification of hepatic progenitors in rodent systems have used multiparametric flow cytometry in combination with multiple fluoroprobe-labeled monoclonal antibodies to purify cells of a defined antigenic profile (Table 11.2). Even when unique antigens are not known, one can obtain significant enrichment for the cells of interest by doing a "negative sort," marking cells not of interest with fluoroprobe-labeled antibodies and then separating the population into those expressing those markers and those not. Secondarily, one can use side scatter, a measure of cytoplasmic complexity caused by mitochondria, ribosomes, and so on. The less mature cells are "agranular" or lower in granularity, whereas the more mature cells are more granular, enabling one to enrich for cell populations of given granularity.

Isolation from Fetal Livers of Hepatoblasts and Hepatic Stem cells (Fig. 11.4)

1. Embryonic livers of a given gestational stage (e.g., E13 fetal livers) are isolated and digested with 800 U/ml collagenase (Sigma) and 20 U/ml thermolysin (Sigma) followed by further digestion with trypsin–EDTA solution (Sigma).
2. Treat the cell suspension with 200 U/ml DNase I (Sigma).
3. The cell suspension should have a cell viability exceeding 90% as determined by trypan blue exclusion.
4. Purification of the stem cells is by multiparametric flow cytometry as given shortly.
5. Cells are stained with monoclonal antibodies (mAbs) and sorted by means of a flow cytometer such as Cytomation's MoFlo or Becton Dickinson FACStarplus.

   a. Block background staining with 20% goat serum and 1% teleostean gelatin (Sigma).
   b. Stain cells with a monoclonal antibody to the class Ia MHC antigen such as anti-RT1A^{a,b,l} B5 (Pharmingen) for the Fisher 344 strain of rats, anti-RT1A OX18 (Pharmingen), anti-rat ICAM-1 1A29 (Pharmingen). (*Note*: The antibodies must

---

Table 11.2. Antigenic Profiles and Phenotype of Colony-Forming Cells Derived from Liver

	Hepatoblasts	Adult liver cells
Antigenic profile	RT1A$^-$, OX18^{low}, ICAM-1$^+$ [profile used for purification of cells by flow cytometric sorting and found on cells forming colonies in culture]	RT1A$^+$, OX18$^+$, ICAM-1$^+$ [profile found on cells forming colonies in culture]
Average cell number in a single colony after 20 days of culture	3000–4000	130
Ploidy of colonies of cells	Diploid	Hypothesized to be diploid^a
Gene expression in culture	α-Fetoprotein, albumin	Albumin^b
Strict mitogens for culture	Insulin, transferrin^c	Insulin, EGF

^{a}The adult cells seeded are a mixture of diploid and polyploid cells; those that emerge in the colonies in culture are likely to be diploid, but the ploidy of the cells that emerge in the colonies is now being tested. If they are diploid, it is unknown whether the diploid cells are selected by the conditions or the tetraploid cells undergo cytokinesis to become diploid.

^{b}More than 90% of the adult liver cells forming colonies in the clonogenic assays expressed albumin and are hepatocytes.

^{c}Adult liver cells are also sensitive to transferrin, but they make it, and so it is an autocrine signal for the adult cells.

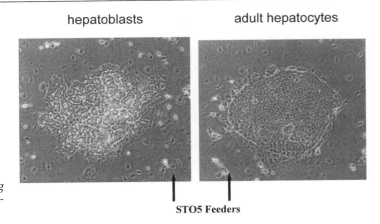

Fig. 11.4. Cultures of clonogenic hepatic stem cells showing morphology of colonies of hepatoblasts versus adult hepatocytes.

be chosen with respect to the species and strain of animal being used as the source of the liver tissue; this is especially true for identifying the class Ia and class Ib MHC antigens; for rats of the Fisher 344 strain, the antibody is against RT1A[a,b,l] B5, and an antibody identifying both class Ia and class Ib antigen in this species and strain is OX18.)

c. Sort for cells that are RT1A[-], OX18[dull], ICAM-1[+], side scatter[high]. (*Note:* The use of high side scatter is relative to that of the major contaminant in fetal livers: hematopoietic cells. In general, the parenchymal cell lineage has a higher level of side scatter or granularity than that of the hematopoietic subpopulations. By contrast, the hepatic stem cells are relatively agranular or have low side scatter in comparison to that in mature parenchymal cells.)

d. The sorted cells should be plated on mouse embryonic stromal feeders (STO cells) treated with mitomycin C and in culture medium supplemented with insulin (5 µg/ml) and transferrin (5 µg/ml). Further details on the culture conditions are given later.

e. The mitomycin C (Sigma cat. no. M-4287) comes in vials, each containing 2 mg of mitomycin C and 48 mg of NaCl. Dissolve 2 mg of mitomycin C in 10 ml of distilled water, aliquot 0.5 ml into 20 cryotubes (100 µg/tube), and store at −20°C or −80°C until use.

FEEDER CELLS

Preparation of Embryonic Stromal Feeders from STO Cells

Original stocks of STO cells can be obtained from the American Tissue Culture Collection (ATCC). Overall the procedure is similar to the original protocol for preparation of STO feeders as described elsewhere [65]. However, several modifications have been found useful for cells that support hepatic stem cells. Briefly, for clonal cultures of hepatoblasts, a subclone, STO5, was isolated from the parent STO cells because of its superior ability to maintain hepatic progenitors *in vitro* [76]. STO5 cells are routinely cultured in DMEM/F-12 supplemented with 7.5% FBS and 1% dimethyl sulfoxide (DMSO) on 10-cm tissue culture dishes. When confluent, the cells are passaged by trypsin–EDTA treatment and diluted 1:3 into fresh plates. Ideally, a confluent 10-cm dish of STO5 cells will yield approximately $2-4 \times 10^6$ cells.

Preparation of Cryopreserved STO5 Feeder Stocks

1. Remove medium from confluent 100-mm plates of STO5 cells.
2. To each 100-mm plate, add 4 ml of medium containing 5 µg/ml mitomycin C. (*Note:* Sigma's basic protocol recommends mitomycin C, 10 µg/ml, for the buffer.)
3. Incubate the plates for between 3 and 4 h.
4. Aspirate the mitomycin C medium from the STO5 cells. Wash each plate twice with 5 ml of Hanks's Balanced Salt Solution (HBSS).
5. Trypsinize the cells and collect the cells in DMEM/F-12 containing 7.5% FBS.

6. Spin down the cells at 1200 rpm for 5 min. Resuspend the cells in DMEM/F-12 plus 7.5% FBS to prepare a cell suspension of 6×10^6 cells/ml.
7. Add an equal volume of cryopreservation buffer (80% FBS + 20% DMSO) dropwise on ice to make a final concentration of 40% FBS, 10% DMSO.
8. Dispense the cells at 1 ml per cryotube (3×10^6 cells/tube).
9. Store the cells at $-80°C$.

PREPARATION OF STO5 FEEDER LAYERS

1. Coat tissue culture plates with a 0.1% (w/v) solution of gelatin (Sigma cat. no. G-1890) in water. Sterilize the gelatin solution by autoclaving. Preheat (at 65°C) the gelatin solution by microwaving before coating the plates.
2. Flood the plates with a few milliliters of the gelatin solution and leave for at least 10 min at room temperature. Aspirate.
3. Take one tube of frozen mitomycin C treated STO5 cells and thaw at 37°C. Transfer the one ml of cell suspension into a conical tube on ice. Add drop wise 9 ml of DMEM/F-12 containing 7.5% FBS.
4. Dispense the suspension into the gelatin-coated plates at a dilution of approximately 4×10^4 cells/cm^2.
5. After cell attachment change the medium to one containing fetal bovine serum. Ideally, the feeder layers should be used within 3–5 days of preparation. However, they last for quite a long time as a monolayer, so they can be used for up to 10 days. Aspirate the medium and rinse with HBSS twice prior to use.

Preparation of Embryonic Liver Stromal Feeders

Embryonic stromal feeders also work for colony formation of rodent hepatic progenitors but have been tested only in high-density cultures (e.g., 50,000–100,000 cells to a matrix-coated transwell and over feeders in 24-well dishes or in 35-mm dishes) [76]. They have not been tested in clonogenic assays and have not been tested at all in human cultures. They are more demanding in their preparation than STO cells but possibly may offer to the cells some advantage that is unique to liver-derived stroma and distinct from that provided by the STO cells.

1. For rodent cultures, use rodent embryonic livers of gestational ages 14–16 days (E14–E16). Earlier embryonic ages also yield active signals but are impractical, since the liver is tiny, hence difficult to remove surgically, and the number of hepatic cells is small. Thus, E14–E16 livers are large enough to yield a number of feeder cultures but still produce the critical paracrine signals. A conditioned medium prepared with the serum-free HDM described shortly works well as long as the stem cells are seeded onto a porous surface (e.g., a transwell) coated with an embryonic matrix substratum (type IV collagen and laminin). Stromal feeders prepared from livers past embryonic day 17 but before E19 work also but tend to yield lower numbers of colonies, and the conditioned medium that results is variably successful. Neonatal livers and adult livers do not work.
2. Isolate 10–20 embryonic livers of the requisite gestational age (E14–E16) and digest the tissue with 800 U/ml collagenase (Sigma) and 20 U/ml thermolysin (Sigma) followed by further digestion with trypsin–EDTA solution (Sigma).
3. Plate the unfractionated cell suspension onto tissue culture dishes (35 or 60 mm) at a seeding density of 10^6 cells/60-mm dish and into a basal medium (e.g., RPMI-1640, Gibco/BRL) to which is added 5% fetal bovine serum (Gibco/BRL), 5 μg/ml insulin (Sigma), 10 μg/ml iron-saturated transferrin (Sigma), 4.4×10^{-3} M nicotinamide (Sigma), 0.2% bovine serum albumin (Sigma), 5×10^{-5} M 2-mercaptoethanol (Sigma), 7.6 μequiv/liter free fatty acid, 2×10^{-3} M glutamine (Gibco/BRL), 3×10^{-8} M H$_2$SeO$_3$, and antibiotics.
4. Culture the cells, with daily medium changes, for approximately one week to permit stromal cells to expand.
5. Disperse the attached cells with trypsin–EDTA or with collagenase, rinse and replate at 200,000 cells/well of a 24-well plate or at 5×10^5 cells/35-mm dish or at 10^6 cells/60-mm dish.

6. Add the plating medium described earlier and culture cells overnight to permit the cells to attach and spread.

7. Progenitor cell populations can be seeded directly onto the feeders if the feeders are treated with mitomycin C (as described earlier) or can be cultured on transwells coated with type IV collagen and laminin and the transwell placed over the feeders.

8. Once the progenitors are added, whether directly to the feeders or via transwells, culture the feeders and the progenitor cells in a *serum-free medium* to which is added 20 ng/ml EGF (Collaborative Biomedical Products), 5 μg/ml insulin (Sigma), 10^{-7} M dexamethasone (Sigma), 10 μg/ml iron-saturated transferrin (Sigma), 4.4×10^{-3} M nicotinamide (Sigma), 0.2% bovine serum albumin (Sigma), 5×10^{-5} M 2-mercaptoethanol (Sigma), 7.6 μequiv/liter free fatty acid, 2×10^{-3} M glutamine (Gibco/BRL), 3×10^{-8} M H_2SeO_3, and antibiotics.

9. Change the medium daily or every other day with freshly prepared medium.

ISOLATION AND ESTABLISHMENT OF MINIMALLY DEVIANT HEPATIC CELL LINES

1. Single-cell suspensions are obtained by incubating fetal livers (e.g., embryonic day 15 to postnatal day 1) with 0.05% trypsin and 0.5 mM EDTA and/or 800 U/ml collagenase (Sigma) at 37°C.

2. The cell suspension should be overlaid on Ficoll-paque (Pharmacia Biotech) for gradient density centrifugation at 450g for 15 min.

3. The cells from the pellet are cultured in a serum-free, homonally defined medium, an HDM, consisting of DMEM/F-12 (Gibco/BRL) to which is added 20 ng/ml EGF (Collaborative Biomedical Products), 5 μg/ml insulin (Sigma), 10^{-7} M dexamethasone (Sigma), 10 μg/ml iron-saturated transferrin (Sigma), 4.4×10^{-3} M nicotinamide (Sigma), 0.2% bovine serum albumin (Sigma), 5×10^{-5} M 2-mercaptoethanol (Sigma), 7.6 μequiv/liter free fatty acid, 2×10^{-3} M glutamine (Gibco/BRL), 1×10^{-6} M $CuSO_4$, 3×10^{-8} M H_2SeO_3 and antibiotics.

4. Plate the cells on tissue culture dishes coated with 10–20 μg/ml collagen type IV (Collaborative Biomedical Products) and 10–20 μg/ml laminin (Collaborative Biomedical Products).

5. After 4 weeks of culture, trypsinize the cells and culture on a feeder layer of mitomycin C treated STO mouse embryonic fibroblast line. Thereafter, the cell lines are maintained on STO feeder cells with HDM. After the establishment of the cell lines, the concentration of EGF should be reduced to 10 ng/ml for all cell cultures.

IDENTIFICATION, PURIFICATION, AND UTILIZATION OF DEFINED SIGNALS: SOLUBLE SIGNALS AND EXTRACELLULAR MATRIX

SOLUBLE SIGNALS: AUTOCRINE, PARACRINE, AND ENDOCRINE FACTORS

Development of Hormonally Defined Media

Cells are regulated in their growth and differentiation by the synergistic effects of soluble signals (autocrine, paracrine, and endocrine) and extracellular matrix. An approach to defining the soluble signals from cell–cell interactions has been to replace the serum supplements in medium with known and purified hormones and growth factors to yield a serum-free, hormonally defined medium, or HDM [5,131,148]. Such media have been developed for many cell types, enabling investigators to have greater control over cells being maintained *ex vivo* [10,30,117,169]. Use of HDM results in selection of the epithelial cell type of interest from primary cultures containing multiple cell types.

Almost all the published HDM are optimized for cell growth. To observe optimal expression of differentiated functions, the HDM must be retailored [131]. Each tissue-specific function requires a discrete set of hormones and growth factors, often at concentrations that differ from those required for cell growth. For example, insulin levels required for growth are typically about 5 μg/ml, whereas those needed for optimal expression of connexins are 100 ng/ml [141,162]. Thus, some of the hormones conducive to growth can markedly inhibit tissue-specific functions. A rule of thumb is to develop an HDM for growth of cells and then use it as a starting point for identifying the conditions needed for differentiation of those cells [32,61,186].

The development of an HDM for a cell type begins by defining the hormonal and growth factor requirements for growth of a minimally deviant neoplastic cell line [32,76]. Since a cell line is already adapted to culture, it can readily be used in clonal growth assays to define growth requirements under serum-free conditions. The HDM for the tumor cells usually contains a subset of the requirements and is, therefore, a starting point in defining the requirements for the normal cellular counterparts to these tumor cells [23,185].

Using the minimally deviant tumor cell line, one determines its clonal growth efficiency (percent of cell colonies that grow at low seeding densities, such as 100–1000 cells/60-mm dish) on tissue culture plastic plates, under serum-supplemented medium (SSM) conditions, or under serum-free medium (SFM) conditions. This will establish the positive and negative control conditions.

Since most differentiated epithelial cells are quite dependent for survival on substrata of extracellular matrix, one should determine the cells' clonal growth efficiency in a serum-supplemented medium, an SSM, and a serum-free medium (SFM), using a simple collagen substratum [32]. Usually, the cells will attach and survive on collagen substrata and in SFM for a week or more. Thus, one can use this condition as a base in which to test the growth effects of soluble signals, one by one. For epithelial cell types, type IV collagen is preferred. However, some epithelial cell types will also accept type I collagen. Since type I collagen substrata are easier and cheaper to prepare, one can try to use them if the cells are growth permissive with type I collagen. The clonal growth efficiency of the minimally deviant tumor cells under these various conditions establishes the critical controls:

1. *Negative control*: a condition (cells on tissue culture plastic and in SFM) in which the cells have minimal survival and growth. This control will determine the extent to which the components in the basal medium are able to facilitate cell survival and growth. For most epithelia, the SFM condition permits no survival at all.
2. *Survival state*: a condition (collagen substratum + SFM) in which the cells can survive but will show limited growth or no growth.
3. *Growth state/positive controls*: a condition (SSM ± collagen) under which the cells will show the highest clonal growth efficiency. It is likely that the minimally deviant tumor cells will grow best on a collagen and in SSM.

In all future experiments, replicate plates of cells are plated under the experimental conditions and under the control conditions, respectively. To test for hormone or growth factor requirements, the cells are plated on collagen substrata and in SSM, and then switched after they have attached (3–4 h) to SFM. The cells are seeded at the lowest density at which they will attach and survive when on collagen and in SFM.

Usually, purified hormones and growth factors are prepared individually and aliquoted as 1000× stocks. The initial concentration tried is based on any information available from *in vivo* or *in vitro* studies on the relevance of the factor to the cell type of interest. The factor is added to cultures that are under survival conditions: SFM and on a collagen substratum. There should be three or four plates for each control or experimental condition, resulting in a total of 12–16 control plates plus three to four times the number of test conditions. Medium changes are done every 3–4 days. The cells are allowed to grow for 1–2 weeks. The number of cell colonies (number to attach and survive) and the diameter of the cell colonies (epithelial cells) or cell density (stroma) are measured. Any hormones or growth factors that increase the number of cell colonies or the colony diameter or density are studied further. Dosage studies are done, again using clonal growth assays, over a 2–3 log range to determine the optimal concentration of the factor in its effects on growth.

Once active factors have been determined, they form the basis for a new control condition: SFM, collagen substratum, plus any biologically active factors at their optimal concentrations. This condition is then used as the base condition to screen again for hormones and growth factors. All the factors are tested again, since a factor found negative in the first screen can prove positive in subsequent screens because of synergies with another soluble signal. Any new growth factor requirements that are identified are again assessed for their optimal concentrations. A cycle consists of the following steps, repeated until no new factors emerge in the screens:

1. Screening for factors on cells under the minimal survival or under the defined "base" conditions
2. Determining the active factors
3. Determining their optimal concentrations
4. Adding them at that concentration to the "base" condition for the next cycle.

Once the active factors have been identified, the optimal concentrations of each one are reassessed in the completed defined medium.

A serum-free HDM has been made for many cell lines with the known repertoire of hormones and growth factors. However, if a specific cell line still grows better in SSM than in the HDM, one must purify from serum or screen tissues for unidentified factors.

One uses the HDM developed for the minimally deviant tumor cells as a starting condition for their normal cellular counterparts. There is one important caveat with respect to normal cells: normal cells will normally not survive at low cell densities in the absence of special conditions such as the use of embryonic stromal feeders, found to support clonogenic assays for both hepatic stem cells and adult liver cells [76]. Thus, if these special conditions are not known, one must develop the HDM by using high density cultures.

General Rules for Developing Defined Medium and Matrix Conditions

1. The primary determinant of cell attachment and survival is a matrix substratum, so the matrix condition should be defined first.
2. The matrix condition required for growth of most cells is a substratum of type IV collagen and laminin; differentiation of most cells calls for a matrix of a fibrillar collagen and fibronectin. Although some cell types will demand additional factors, these conditions are a reasonable starting position for all cell types.
3. If desired, the matrix component can be an extract enriched in extracellular matrix.
4. All cells require transferrin, an iron-carrying protein that is a requirement for the cells' polymerases.
5. All cells require insulin or an insulin-like growth factor (e.g., insulin, IGF-I, IGF-II).
6. All cells are facilitated in culture by reducing agents or antioxidants. An inexpensive one that is potent is selenium. Others that have been used are mercaptoethanol, vitamin E, and ascorbic acid.
7. All cells require a lipid source. For adult liver cells, linoleic acid (bound to bovine serum albumin as a carrier) is sufficient as a primary lipid source, since the adult liver cells have the enzymatic machinery to make all lipid derivatives from this one free fatty acid. However, for most cell types and for all stem cells and other precursor cells, a mixture of free fatty fatty acids along with appropriate carrier proteins (e.g., high-density lipoprotein or HDL) must be given. Appendix 1 gives the protocol for preparing a mixture of free fatty acids that are a strict requirement for survival of hepatic stem cells, other progenitors, and hepatoma cells maintained under serum-free conditions.
8. If cells are on tissue culture plastic, most will attach better and survive longer if a glucocorticoid (e.g., hydrocortisone) is given. The glucocorticoid is not usually required if the cells are on a matrix substratum or directly on a feeder layer.
9. Tumor cells are now known to be transformants of stem cells or other precursor populations [152,153,185], and there is extensive overlap with normal stem cells in phenotypes and in regulation. Defining the requirements for a tumor cell is a stepping-stone toward identifying the conditions for the stem cells. Com-

mon distinctions between normal stem cells and transformed counterparts are as follows:

    a. The level of calcium concentration at which the cells will grow: normal epithelial cells require calcium concentrations below 0.5 mM, whereas tumor cells and stromal cells will grow happily at concentrations two to three times that level.

    b. Normal stem cells strictly require embryonic, and ideally tissue-specific, stromal feeders for survival and growth, whereas tumor cells grow readily without feeders or are tolerant of feeders from diverse sources.

10. Epidermal growth factor (EGF) is a common growth factor requirement for normal epithelial cells. Platelet-derived growth factor (PDGF) is a common requirement that is normal for stromal cells but not for their neoplastic derivatives. In both cases, the autonomy of the cells to a growth factor can be the result of constitutive secretion of the factor or activation of its hormonal pathway intracellularly because of an overexpressed or mutated oncogene [38,87,130,144,176].

11. Clonogenic assays require seeding cells at very low seeding densities. Diploid subpopulations, including tumor cells and stem cells, will do very well at such low seeding densities as long as the proper conditions are used, namely, basal medium, insulin and transferrin, and age- and tissue-specific stromal feeders. Some tumor cells may be qualitatively or quantitatively independent of such feeders.

12. Stem cells do best with the fewest hormones/growth factors possible; many factors drive differentiation. As reported, hepatic stem cells grow clonogenically with only insulin and iron-saturated transferrin and the unknown factor(s) coming from embryonic stromal cells [76].

Influence of Serum-Free, Hormonally Defined Media on Epithelial Cells in Culture

Serum-free, hormonally defined media have been found to select for parenchymal cells even when the cells are on tissue culture plastic [12,32,76,131]. This results, within a few days, in cultures that are predominantly the cell type for which the HDM was developed. However, if the cultures are plated onto tissue culture plastic and in HDM, the life span of the primary cultures is approximately 1 week, at which time, the cells peel off the plates in sheets. Achievement of longer culture life spans under serum-free conditions depends on the use of substrata of matrix components or extracts enriched in extracellular matrix in combination with the serum-free, hormonally defined medium [32].

Tissue-specific gene expression is dramatically improved in cultures under serum-free conditions and especially when serum-free medium is supplemented with only the specific hormones needed to drive expression of a given tissue-specific gene [61,117]. However, mRNA synthesis of tissue-specific genes is not restored by serum-free medium or by any known combination of hormones and growth factors; rather, the improved tissue-specific gene expression in serum-free media or in hormonally defined media is due to post-transcriptional regulatory mechanisms, often an increase in stability of specific mRNA species [61]. Restoration of mRNA synthesis occurs only with serum-free media or with a serum-free, hormonally defined medium containing the specific hormones or factors found to influence a given gene and presented in combination with tissue-specific forms of heparin proteoglycans or their glycosaminoglycan chains, heparins [141,162,186]. Heparins are often bleached in commercial processes to eliminate the brown coloration due to iron deposits; since, however, the bleaching process destroys heparin's biological activity on gene expression, unbleached fractions should be used.

INSOLUBLE SIGNALS: EXTRACELLULAR MATRIX

Extracellular matrix is essential, especially for normal cells to survive and function [10,29,137]. The chemistry and physical features of excellular martrix are known to regulate gene expression and to influence cell morphology [10,58,59,131,132,180]. In the past decade the knowledge of matrix chemistry and biology and of its complexity became better understood [68,91,104,118]. The extracellular matrix is a complex mixture of molecules between and around cells made insoluble by cross-linking. Excellent reviews are available on its chemistry and functions [8,10,68,102,134,137].

It is important to note that there are many types of matrix with distinct chemical composition. Each cell type secretes and is associated with a specific type of matrix. Furthermore, the matrix chemistry changes according to whether the cell is growing or quiescent, and when it is in some pathological condition [23]. Thus, if you want the cell type of interest to mimic its *in vivo* counterpart in a specific physiological or pathological state, you must identify the matrix chemistry associated with the cell in that physiological or pathological state. It is helpful to understand that there is a paradigm governing the making of all types of extracellular matrix, and to be aware of the numerous studies identifying the matrix chemistry in various tissues.

Paradigms in the Construction of Extracellular Matrix

All cells produce an extracellular matrix, and the extracellular matrix in between any given set of cells contains components derived from all the cell types in contact with the matrix. The matrix components present on the lateral borders between homotypic cells, which include cell adhesion molecules (CAMs) [7,70] and proteoglycans [8,68,162], will be referred to as the "lateral matrix." That between the epithelium and a mesenchymal cell partner will be referred to as the "basal matrix," consisting of proteoglycans attached in many ways to the plasma membrane surface and basal adhesion molecules (laminins or fibronectins) that attach the cells to collagen scaffoldings.

Although the matrix chemistry is unique in each tissue, some components are present in all matrices, and the chemistry depends on the maturational stage of the cells [133]. For example, the basal and lateral matrix chemistry in stem cell compartments (e.g., epidermal, liver, or intestine) contains type IV collagen, laminin, hyaluronans, and fetal and tissue-specific forms of heparan sulfate proteoglycans and CAMs [63,92,93,134,163]. The matrix chemistry associated with mature cells is comprised of fibrillar collagens, fibronectins, adult and tissue-specific forms of CAMs, and highly sulfated proteoglycans such as dermatan sulfate proteoglycans or heparin proteoglycans. To date, three distinct developmental patterns have been identified:

• *Liver, intestine, lung*: The cells are in contact with lateral and basal matrices throughout maturation, and there is gradient in the matrix chemistry from the stem cell to the mature cell.

• *Skin and brain*: The cells are in contact with lateral and basal matrices in the early stages of maturation but lose their synthesis of collagens and basal adhesion proteins at a relatively early stage of maturation. Thereafter, the lateral matrix components exhibit a gradient in the chemistry paralleling maturation. Thus, mature keratinocytes and neurons do not synthesize collagens, fibronectins, or laminins, but their level of maturity is correlated with specific forms of CAMs or proteoglycans.

• *Hematopoietic cells*: The cells are in contact with lateral and basal matrices in the early stages of maturation but lose collagen synthesis altogether and make basal adhesion proteins that are missing the cell-binding domains, enabling the cells to float. Qualifications to this pattern are that hematopoietic cells can switch to forms of basal adhesion proteins that do have the cell-binding domains; this switching process is part of specific immunological regulatory mechanisms such as antigen presentation [62,97].

Rules for Cultures on Matrix Substrata

1. Cells attach quickly on an appropriate matrix substratum. Therefore, when cells are being plated on any matrix substrata, they should be added in sufficient volume of medium to allow equal distribution over the plates.

2. Since matrix is a mixture of proteins and carbohydrate, use DNA or RNA stains rather than protein or carbohydrate stains for assessing plating efficiency or clonal growth efficiency of cells on matrix.

3. Cells on complex matrices (Matrigel, biomatrices) do not detach readily. To do growth curves of cells on complex matrices, you may have to scrape the plates, isolate the DNA, and determine DNA content.

4. Antibodies (and many other reagents) will nonspecifically stick to matrices. For antibody staining of cells on matrices, expose the cells to a control antiserum or to blocking reagents first and then to the specific antiserum.

5. Cells can achieve a very high density if cultured on matrix substrata, and especially if proteoglycans or glycosaminoglycans are used. For example, typical saturation densities for normal cells on a 100-mm tissue culture dish are $7–10 \times 10^6$, whereas densities in the presence of appropriate proteoglycans can be greater than 10^8 cells/100-mm dish.

6. In extracting RNA or DNA from cell cultures on collagens, proteoglycans, or complex matrices, *avoid* first detaching the cells enzymatically unless (1) the enzymes are guaranteed to be free of nucleases and (2) detachment does not result in alteration of the RNA levels. Rather, use guanidine or guanidinium solution on the cultures and extract everything on the plate, both cells and matrix. Then use a purification protocol for the RNA or DNA that will eliminate the matrix components (e.g., phenol chloroform extractions and/or centrifuging through cesium chloride gradient). The purification protocol for glycosaminoglycans is so close to that for nucleic acids that the glycosaminoglycans and nucleic acids can be copurified; if this happens, the glycosaminoglycans can readily be eliminated by use of highly purified glycosidases (e.g., heparinase, heparitinase).

Protocols for Preparing Extracellular Matrix Components

Many of the extracellular matrix components can now be purchased commercially, including many of the collagens (type I, type IV), basal adhesion proteins (laminins, fibronectins), and Matrigel. Therefore, the protocols for preparing these are not given here. There are many publications and reviews of protocols that do include them including some given as representative [72,118]. Appendix 2 lists sources for matrix components that are available commercially and are used routinely for cultured cells. The major extracts enriched in extracellular matrix used to date are summarized next.

Purified Matrix Components

Only two collagen types, type I and type IV, and two major basal adhesion proteins, laminin and fibronectin, are used very extensively. All are available commercially (see Appendix 2), so their purification procedures are not presented here. For details on the preparation of these two or any of the other of the many types of collagen, see the reviews of methods by Nimni [118] and Kleinman [72].

Extracts Enriched in Extracellular Matrix

The protocols used in the preparation of tissue extracts enriched in extracellular matrix take advantage of the relative insolubility of extracellular matrix components. The "ECM" and amnionic matrices are insoluble in dilute detergent (or dilute alkali), and the biomatrices are insoluble in salt solutions. Matrigel is prepared by dissolution in strong denaturants and urea, followed conversion back to the insoluble state by dialyzing against PBS or serum-free basal media.

1. *Matrigel*: a urea extract from Engelbreth-Holm-Swerm (ENS) tumors [71], a transplantable mouse embryonal carcinoma that constitutively produces basement membrane components. It has remarkable effects on the differentiation of cells and is available commercially. See the protocol for Matrigel published by Hynda Kleinman and associates [71]. It should be noted that Matrigel is replete with trapped growth factors and nucleic acids. Some of Matrigel's renowned effects are lost if the nucleic acids or these growth factors are eliminated.

2. *ECM (extracellular matrix)*: a simple extract prepared from any type of cell maintained in monolayer culture. The cultures are grown under the conditions desired until the cultures are confluent. The cultures are rinsed with PBS and treated for a few minutes with sterile 1% Triton X-100 prepared in MQ water and then rinsed again with PBS [28]. The

exudate remaining on the dish is enriched in the matrix components synthesized by the cells under the medium conditions. The chemistry of the matrix left behind depends on the cell type used in culture and the conditions under which the cells were grown. The exudate can be used as a substratum for the cells. Newly seeded cells attach readily, grow well, and show improved differentiation, but not to the same extent as on Matrigel or biomatrices.

3. *Matrix from amnions*: amnions extracted with sterile 1% Triton-X-100 and rinsed [90]. The matrix on the side of the epithelial surface is the basal lamina containing type IV collagen and has a shiny appearance. On the other side (the stromal surface) is a matrix enriched for type III collagen that gives a "dull" appearance. You can choose the surface appropriate for your cell type. Studies using this matrix indicate dramatic differences in response according to the side on which a given cell type is plated.

4. *Biomatrix*: a tissue extract prepared by extracting the tissue with NaCl solutions, nucleases, and detergents [135,140]. The extract can be made from any tissue. It is the only matrix that has been shown to result in very stable, long-term ($\geq$6 months), quiescent cultures of normal liver cells. It is partially depleted of proteoglycans and glycosaminoglycans, which are restored by the cells during the first few days of cell culture. However, if you need maximal differentiation within a few days of plating the cells, you must add the proteoglycans or glycosaminoglycans to the medium. Biomatrix can be prepared from any tissue, from any animal, and regardless of whether the tissue is normal or diseased. The following protocols are based on the procedures of Reid and associates [26,140].

ISOLATION OF BIOMATRIX

1. Mince tissue and homogenize in a Waring Blender or with a Polytron homogenizer using 10 vol (to 1 vol of mince) of buffer: 3.4 M NaCl + 0.1 mg/ml soybean trypsin inhibitor + antibiotics (e.g., penicillin–streptomycin) (Buffer D) at concentrations standard for cell culture (all at 4°C). Homogenize thoroughly with pulses of 10–15 s.
2. Centrifuge for 20 min at 10,000 rpm at 4°C in a preparative centrifuge.
3. Save the pellets and resuspend in buffer D. Stir for 1 h at 4°C.
4. Centrifuge for 20 min at 10,000 rpm at 4°C and again save the pellets.
5. Repeat the extractions of the pellets with buffer G until the supernatants after centrifugation are clear and are negative for proteins by Lowry or Bradford assays or by optical density at 280 nm.
6. Rinse pellets twice, each time for 30 min, in serum-free basal medium (e.g., RPMI-1640) containing the soybean trypsin inhibitor at 4°C.
7. Suspend pellets in a serum-free basal medium (must contain calcium and appropriate salts for the nucleases: e.g., RPMI-1640) containing nucleases: suspend each 10 ml volume of pellet in 100 ml of solution containing 1.0 mg of DNase and 5 mg of RNase. Stir at 37°C for 1 h.
8. Centrifuge for 20 min at 10,000 rpm at 4°C and save the pellet.
9. Repeat steps 7 and 8 until the supernatants (after centrifugation) give spectrophotometric readings of less than 0.1 at 260 nm.
10. Suspend pellets in 1% sodium deoxycholate in basal medium for 1 h at room temperature. You must prepare the detergent as a 10× stock in distilled water first—it does not dissolve easily; then, disperse it into any basal medium (e.g., RPMI-1640, DMEM/F-12).
11. Centrifuge samples as before and save the pellets.
12. Rinse the pellets (now considered to be "biomatrix") three times in 30-min rinses with serum-free basal medium (e.g., RPMI-1640).
13. Suspend the pellets in serum-free basal medium (1:5 v/v).
14. Sterilize with 10,000 rad of gamma irradiation.
15. Store at –70°C for long-term storage or at 4°C for samples to be used within a month. (Sterilization is imperative for all samples stored at 4°C.)

TO COAT DISHES, SLIDES, OR ANY SUBSTRATUM

1. Freeze matrix pellets and section with a cryotome. Plate the sectioned material on the culture dish and use a small, sterilized camel hair brush to smear them over the surface.

2. Alternatively, freeze the pellet at liquid nitrogen temperature.

3. Pulverize matrix pellets with a SpexMill Freezer Mill (Metuchen, NJ) into powder, keeping the sample at liquid nitrogen temperature.

4. When the sample is completely pulverized, allow it to come to room temperature. It will become like paint. Use a simple camel hair brush (small artist's brush) to paint the pulverized biomatrix onto dishes, slides, and so on.

5. Sterilize the dishes with 10,000 rad of gamma irradiation. The plates, dishes, and so on, can be stored at 4°C until used. Just prior to use, soak the plates with serum-free basal medium to be used with the cells (e.g., RPMI-1640), all at 37°C in a regular CO_2 incubator.

SYNTHETIC BIODEGRADABLE SCAFFOLDINGS FOR PROGENITOR CELLS

BACKGROUND

A significant portion of the present research in tissue engineering is devoted to the development of novel substrata suitable for *ex vivo* maintenance and expansion of cells, and for *in vivo* transplantation of the cells. In comparison to attachment of adherent cells to a rigid and smooth surface, greater extent of attachment, survival, and function can be achieved by cell adhesion on soft and porous substrata that resemble the macromolecular structure of extracellular matrix [19,180]. In general, degradable scaffolds in various physical forms aim to provide liver cells with substrata suitable for cell attachment, and to maintain cell–matrix as well as cell–cell interactions that are critical for cell signaling and the expression of their differentiated functions [58,112]. Controllable degradation of substratum scaffold is important to the ultimate clearance of the synthetic implant materials within an engineered tissue.

Degradable materials used as substrata in tissue engineering may be divided into natural, synthetic, and hybrid materials. Natural extracellular matrix components (e.g., collagen and laminin) have been used as cell attachment support, as coatings on surfaces of polystyrene, polyurethane, and cross-linked dextran [19], and for encapsulation and formation of collagen sandwich gel systems for *ex vivo* three-dimensional cell culture [54,154]. The advantage of naturally derived polymers is their close compatibility with extracellular matrix environment, thus providing better substrata for *ex vivo* culture of cells. However, they typically lack the necessary mechanical integrity for fabrication into scaffolds or other shaped polymer structures. They also suffer from the limitation of sufficient supply and source variation, which limits their use for tissue engineering. Some naturally derived polymers might also be immunogenic. Synthetic biocompatible and biodegradable polymers are widely used in surgical sutures and constructs for *ex vivo* cell culture, expansion, and cell transplantation [55]. The most widely used synthetic biodegradable polymers, however, are the poly(α-hydroxy acid) family of polymers, including poly(glycolic acid), poly(lactic acid), poly(D,L-lactide-*co*-glycolide) copolymer, and their modified derivatives. Because of their wide range of desirable physical properties, synthetic polymers can be precisely constructed in various formats of defined size, shape, and morphology [101]. Their biodegradation can be controlled by their design and fabrication [39]. However, initial interaction between degradable scaffolds and seeded cells as well as initial host response to tissue engineering devices presents a critical challenge in the design of synthetic polymer [3,57]. Hybrid polymers are generally designed to achieve the desirable physical properties of synthetic polymers and the biocompatibility of natural biopolymers. This may be achieved by chemically coupling bioactive components of natural substrata with synthetic polymers or chemically modifying biopolymers to introduce desired physical properties [55,67].

MICROCARRIERS FOR PROGENITOR CELL POPULATIONS

Commercially available beads for tissue culture are typically nondegradable, rigid, and nonporous. They are designed primarily for limited, short-term *ex vivo* maintenance of the freshly isolated cells or transformed cell lines. Their rigidity and lack of porosity make them unsuitable as substrata for cultures of progenitor cells that die within hours on such substrata [11,76,157,158]. Our group is developing biocompatible and biodegradables porous beads or microcarriers as suitable matrix scaffolds for hepatic progenitors, enabling them to survive, grow, and mature through the entire liver lineage. Porous beads can be readily adapted for *ex vivo* studies in either conventional cell cultures or novel bioreactors of various physical settings and used as carriers of cells and therapeutic agents in surgical implantation. We outline here various protocols used for preparation of porous degradable poly(lactic-*co*-glycolic acid) (PLGA) beads, the conditions of matrix coating of the polymer beads, and culture of hepatoma cell lines and rodent liver cells with the degradable microcarriers.

Materials

- 50% PLGA copolymers with the inherent viscosity of 0.5 and with (cat. no. RG504H) or without terminal carboxyl group modification (cat. no. RG504) (Boehringer Ingelheim Chemical, Winchester, VA)
- RPMI-1640 (Life Technologies, Rockville, MD)
- Fetal bovine serum (FBS) (Collaborative Biomedical Products; Bedford, MA)
- Rat tail type I collagen (Collaborative Biomedical Products)
- Type IV collagen of EHS tumor (Collaborative Biomedical Products)
- *Other reagents*: trypsin, bovine serum albumin (BSA), insulin, streptomycin, penicillin, epidermal growth factor (EGF), transferrin, trypsin, Cytodex-3 collagen beads, mouse monocolonal anti-collagen type I (C2456) antibody from Sigma Chemical (St. Louis, MO)
- Calcein AM, propidium iodide dye (PI dye), Alexa-594 goat anti-mouse IgG (Molecular Probes; Eugene, OR)
- Polyvinyl alcohol, dioxane, methylene chloride, ethanol (Aldrich Chemical (Milwaukee, WI))
- RapidCell beads coated with denatured collagen (ICN Biochemical; Costa Mesa, CA)
- Ultrasonication equipped with a flow-through probe head
- Syringe pump for delivery of polymer solution to the ultrasonic probe head (e.g., Branson Sonifier Cell Disruptor 200, Danbury, CT)
- Cryodewar flask for low temperature extraction of solvent from polymer beads
- Vacuum pump or lyophilizer for evacuation of trace amounts of solvent from polymer beads
- UV source for sterilization of polymer beads

Protocol for Preparation of PLGA Microbeads by Ultrasonic Atomization

PLGA microbeads of diameters under 100 μm can be prepared by an ultrasonic liquid atomization protocol adapted from one reported elsewhere [66]. The protocol described here allows preparation of microcarriers in gram quantities. Microcarriers are formed by dispersing PLGA (or other polymers in suitable solvent) into small droplets of dissolved polymer followed by solidification of the polymer droplets and extraction of the dissolving solvent. The size of the microcarriers formed is influenced by the ultrasonication power delivered to the flow-through ultrasonic probe head, the viscosity of the polymer solution, and the flow rate of polymer delivered through the ultrasonic probe head. In general, smaller droplets, hence smaller polymer beads, may be formed by increasing the sonication power, decreasing polymer viscosity and the flow rate of polymer solution delivered. A variety of second solvents may be used to extract the first solvent from the polymer [66]. In general, the second solvents are nonsolvents for the polymer, but solvents for the first solvent used to dissolve the polymer. The second solvents typically have low melting points, allowing extraction of the first solvent from the polymer beads at low temperature.

20 µm 1000x

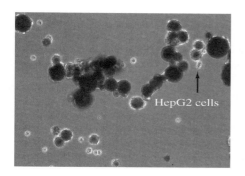

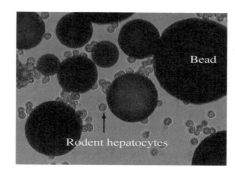

Fig. 11.5. Cells on biodegradable microcarriers. See also color insert.

1. Fully dissolve 1 g of PLGA polymer in 20 ml of dioxane in a clean glass container, and seal the solution to prevent solvent evaporation.

2. Slowly overlay approximately 200 ml of liquid nitrogen over 150 ml of ethanol in a cryodewar flask. Additional liquid nitrogen may be added to ensure that the ethanol is fully frozen and that sufficient liquid nitrogen remains for freezing polymer droplets dispersed by ultrasonic atomization.

3. The polymer solution is then delivered at 4–9 ml/min by a syringe pump to a flow-through ultrasonic microprobe adapted on a Branson Sonifier Cell Disruptor 200. Typically, the power output delivered to the probe is set at about 18 W, with the probe vibration frequency at 20 kHz.

4. Atomized polymer droplets are collected in the liquid nitrogen setup described in step 2. The droplets freeze immediately upon contact with the liquid nitrogen.

5. The first solvent in the solidified polymer beads is then extracted slowly into the slowly melting ethanol at –76°C over 2 days.

6. The beads are then suspended in 100–200 ml of cold hexane at –76°C for an additional 2 days to allow further extraction of the remaining solvent.

7. Trace amounts of polymer solvent are removed under vacuum over 2 days.

8. The dry beads may be sterilized by exposure to UV light over 1 h. The dry beads can be stored at 4°C temporarily or at –20°C for long-term storage.

The foregoing protocol typically yields over 70% polymer mass in preparation. The PLGA beads prepared are amorphous. The microcarriers prepared are heterogeneous in size with the major population typically within the diameter range of 20–40 µm. Figure 11.5 shows scanning electron micrographs (SEM) of a representative population of porous degradable PLGA microbeads (A). Surface pores are highly heterogeneous in size and shape. In contrast to the porous structures of the PLGA beads, commercially available beads are typically nonporous under similar image resolution conditions.

Protocol for Preparation of PLGA Mesobeads by Dispersion of Polymer Emulsion

Larger PLGA mesobeads with approximate diameters of 100–200 µm can be prepared by dispersing water/polymer/water emulsion into small droplets in 0.1% polyvinyl alcohol (PVA) solution, followed by slow extraction of the polymer solvent and solidification of the polymer beads. The dispersion of the polymer emulsion into small droplets is achieved by injecting polymer emulsion into a rapidly stirred 0.1% PVA solution. In general, bead diameter and size depend on the viscosity of emulsion, the rate of polymer emulsion delivered, and the sheer forces used to break the polymer stream in the PVA solution. An increase of emulsion flow coupled with decrease of stirring rate leads to formation of larger beads. The following protocol can be used to prepare gram quantities of the polymer beads.

1. Fully dissolve 1 g of water-soluble PVA. It may be necessary to filter the PVA if small impurities of PVA debris are present.

2. Fully dissolve 1 g of 50% PLGA in 16 ml of dichloromethane in a glass sample container.

3. Add 6 ml of water or desired aqueous buffer to the dissolved PLGA, forming separate water and solvent phases.

4. PLGA emulsion is formed by probe-sonication of the biphasic water–oil polymer solution placed in an ice bath for 1 min. The sonication probe output is typically about 18 W at 20 kHz.

5. The emulsion is injected from a glass syringe at ~16 ml/min through a 21- or 22-gauge needle into the rapidly stirred PVA solution. Continued stirring of the injected emulsion leads to formation of the fine droplets of the polymer emulsion.

6. The suspension of the emulsion is stirred continuously over 24 h to remove the majority of the polymer and to solidify the polymer beads.

7. The solidified beads may be rinsed with water to remove most of the PVA. This is typically followed by evacuation of trace amount of polymers on a vacuum pump or a lyophilizer.

8. The dry beads may be sterilized by exposure to UV light over 1 h. The dry beads are typically stored at 4°C temporarily or at −20° for long-term storage.

Physical Analysis of Degradable Beads

The degradable beads are typically examined for surface morphology by scanning electron microscopy operating at an accelerating voltage (e.g., 5 kV). [*Note*: The use of higher accelerating voltage may lead to polymer surface discharge and distortion of the SEM images. The beads are routinely sputter-coated with gold–palladium (60 and 40%) at ~15 nm thickness prior to acquisition of SEM images.]

The size distribution of degradable beads is determined by using image analysis software (e.g., NIHimage, NIH, Bethesda, MD) to analyze the SEM images. Alternatively, size distribution of the degradable beads was determined by laser diffraction particle size analysis (e.g., Malvern Particle Analyzer from Malvern Instruments, Southborough, MA). Analyses by both methods in general yield highly similar results. In routine preparation only one of the methods may be used, depending on the availability of the instruments. The PLGA beads may be wet-sieve-screened through polypropylene mesh of desired size range.

Total surface specific area is important to the conditions of cell attachment with the beads. The surface area is determined by measuring nitrogen gas adsorption in a gas sorption surface analyzer (e.g., Quantachrome Autosorb Automated Gas Sorption System, Quantachrome Corp., Boynton Beach, FL). Measurement is typically carried with multiple-point Brunauer–Emmett–Teller (BET) measurement to determine the specific surface area of the beads.

The microcarriers prepared by the ultrasonic atomization method yield a specific surface area exceeding 10 m^2/g. Those prepared by dispersion of polymer emulsion yield a specific area exceeding 1 m^2/g. Details of analyses by SEM, laser diffraction, and gas sorption surface analysis are not presented here.

Collagen Coating of the Biodegradable Beads

Coating of collagen type I or IV is achieved mainly by physical adsorption of the collagen in acidic solution to prewetted PLGA beads. Dry PLGA beads were prewetted by suspending into ethanol solution (10–25%) before being rinsed and then fully dispersed in coating solution. The amount of collagen (e.g., type I) used is determined by the specific

area of the degradable beads, generally about 10 μg/cm² is used in the coating of tissue culture surfaces. Collagen dissolved in 10 mM acetic acid is incubated with prewetted beads
suspended in 10 mM acetic acid at 4°C overnight with intermittent mixing. The beads are
then collected in a prechilled polypropylene strainer and quickly suspended into phosphate-
buffered saline (PBS, 10×) at 37°C and incubated for 1 h. The collagen-coated beads are
rinsed and suspended in RPMI-1640 and kept under sterilized conditions at 4°C. No further sterilization of the beads is carried out after collagen coating; all procedures of collagen
coating, except incubation at 4°C, are completed in a sterilized tissue culture hood.

The extent of extracellular matrix coating on the polymer surface is determined qualitatively by the immunochemical analysis of the binding of a monoclonal antibody against the
matrix proteins. For example, to ascertain the coating of type I collagen of bead surface, a
monoclonal antibody against native collagen type I (COL-1, cat. no. C2456, Sigma Chemical, St. Louis, MO) is used. Briefly, excess antibody (1:100 dilution) is incubated with about
10 mg of collagen-coated beads suspended in 1 ml of PBS at 37°C for 1 h. This is followed
by rinsing the beads with 1 ml of PBS twice and incubating in 200 μl of PBS containing
20 μg/ml rabbit anti-mouse Alexa-595 IgG at 37°C in dark for 1 h. Excess of Alexa-595
IgG is then rinsed away with phosphate buffer in the dark. Any collagen on the surface of
the beads can be examined readily by fluorescent images of the immunochemically stained
beads.

MONOLAYER CULTURES OF CELLS
CLASSICAL CELL CULTURE CONDITIONS
A brief description of classical cell culture conditions is presented as background. For
detailed descriptions and discussion of classical cell culture methods, see recently published
books and reviews such as Freshney [28]. All methods of preparing cells for culture start
with the disruption of the tissue and its dispersal into chunks or single-cell suspensions.
In classical cell culture, the dispersed tissue or cells are plated onto an inflexible plastic
substratum that has been exposed to a cationic ionizing gas, which polarizes the polystyrene
in the dishes to reveal a negatively charged layer. Cells attach to the dishes via that negatively
charged surface and subsequently form a more complicated adhesion surface with secreted
forms of extracellular matrix complemented by matrix components from serum. The cells
are suspended in or covered with a liquid medium consisting of a basal medium of salts
and nutrients supplemented with a biological fluid, most commonly serum from animals
routinely slaughtered at commercial slaughterhouses.

The basal media that are commonly used (e.g., RPMI, DME, BME, Waymouth's) were
developed originally for cultures of fibroblasts [28]. Although most of the constituents in
these basal media are requirements for epithelial cells and fibroblasts, some aspects of the
basal media have been redefined for epithelia [28]. Two examples are trace elements and
calcium levels. Highly differentiated epithelia require various trace elements or other factors
that act as cofactors for the enzymes associated with their tissue-specific functions [16–22].
The calcium level in many of the commercially available media is above 1 mM, a concentration that is permissive for the growth of fibroblasts but in recent years has been shown to
inhibit most epithelial cell types [24]. This problem is exacerbated by culturing the cells
in serum-supplemented media, since serum also contributes significantly to the calcium
level. Most normal epithelial cells can grow in a calcium concentration of approximately
0.4 mM [24,145].

Some investigators utilize serum autologous to the cell types to be cultured. More
commonly, however, the serum derives from animals that are routinely slaughtered for
commercial usage, such as cows, horses, sheep, or pigs [28]. Fibroblasts do well in serum-
supplemented media (SSM). By contrast, epithelial cell types such as liver that are cultured
in SSM dedifferentiate rapidly and then die, usually within a week [61]. Tumor cell lines
derived from epithelia are more tolerant of the adverse effects of serum, and will grow and
even express low levels of any tissue-specific functions they can express in SSM; yet, if these
cell lines are switched into serum-free HDM, their differentiated functions are markedly
improved [185].

MONOLAYER CULTURES IN A GROWTH STATE UNDER COMPLETELY DEFINED CONDITIONS

Under completely defined conditions, cells will behave quite reproducibly. Those capable of extensive growth even at clonal seeding densities are the stem cells, committed progenitors, and diploid adult cells, and under growth conditions they will express different genes: "early" genes. By contrast, the mature, polyploid cells will show limited growth, even under optimal growth conditions. They also will require higher seeding densities even for survival, and yet will provide the highest levels of certain of the adult-specific functions such as the cytochrome P450s.

Normal Stem Cells

1. Isolate the stem cells by the protocols given earlier.

2. For epithelial stem cells, you probably will have to use a basal medium that has a low calcium concentration, approximately 0.4 mM, and possibly little or no copper [161,184].

3. Seed the cells at the desired densities onto STO5 feeder cells and into a serum-free, hormonally defined medium consisting of a 1:1 mixture of Dulbecco's Modified Eagle's Medium and Ham's F-12 (DMEM/F-12, Gibco/BRL) to which is added 5 μg/ml insulin (Sigma), 10^{-7} M dexamethasone (Sigma), 10 μg/ml iron-saturated transferrin (Sigma), 4.4×10^{-3} M nicotinamide (Sigma), 0.2% bovine serum albumin (BSA) (Sigma), 5×10^{-5} M 2-mercaptoethanol (Sigma), 7.6 μequiv/liter free fatty acid, 2×10^{-3} M glutamine (Gibco/BRL), 1×10^{-6} M $CuSO_4$, 3×10^{-8} M H_2SeO_3, and antibiotics. *To drive the colonies predominantly or entirely to the hepatocytic fate, add also 20 ng/ml EGF (Collaborative Biomedical Products).*

4. Thereafter, change the medium every day to every 3 days.

5. Even when seeding is done at clonogenic seeding densities, the colonies of cells will be visible easily within a few days and will yield 3000–4000 cells/colony within 20 days. (See Table 11.2 for a summary.)

Normal Mature Cells

1. Isolate the cells by standard protocols for the cell type you are using.

2. Use a basal medium that has a low calcium concentration, approximately 0.4 mM.

3. Plate the cells at densities of $3–4 \times 10^5$/60-mm dish or at least 10^6/100-mm dish. Use dishes coated with type IV collagen, a medium that contains both the hormones and growth factors used in the HDM, and some serum (1–2%) to permit them to attach and to inactivate any enzymes that might have been used in isolating the cells. You can avoid the serum altogether if you use laminin with the collagen and if you use no enzymes in isolating the cells (or have effective ways of inactivating those enzymes without providing a condition that is toxic to cells).

4. Incubate the cells for 4–6 h at 37°C or until they are firmly attached.

5. Rinse the plates, removing debris and floating cells, and feed with the serum-free HDM. Change the medium every day, or at the very least, every other day (prepare the HDM fresh).

6. The cells will grow and will survive up to a month under these conditions.

7. If the cells are stem cells, you can expect to be able to subculture and clone the cells under these conditions. If the cells are adult cells, you can expect several rounds of division and then growth arrest. You may not be able to clone or subculture them at all; at best you will have limited ability to do so. Polyploid cells will not grow under these conditions but may survive for some weeks.

Tumor Cells

1. Tumor cells, especially anaplastic tumor cells, can be plated onto tissue culture plastic or, ideally, on type IV collagen coated dishes or transwells at densities of 10^4 cells/60-mm dish or 10^5 cells/100-mm dish and into HDM/SSM. For lower densities (e.g., clonogenic assays), one must plate them as one would the stem cells (see earlier protocol) and on embryonic stromal feeders.

2. After 4–6 h, the cultures are rinsed and switched into the HDM designed for the tumor cells.

3. The cells will survive and grow under these conditions for 7–10 days.

4. The tumor cells can be subcultured and cloned under the defined conditions as long as soybean trypsin inhibitor (Sigma) is used to inactivate the trypsin used in subculturing and as long as the cells are plated onto collagen or matrix substrata or onto feeders.

5. To date, no HDM with or without an appropriate matrix substratum is able to support a cell line indefinitely as stock culture. To do this requires embryonic stromal feeders in addition to the HDM. Therefore, one or more additional factors from the feeders remain to be defined. If feeders are not being used, then some serum supplementation is required but does not need to be as high as is used customarily. Thus, instead of the usual 10% serum supplementation, one can use 1–2% serum supplementation in combination with the hormones identified as biologically active on the cells.

Table 11.3 gives soluble factors and defined matrix components required for cell growth.

MONOLAYER CULTURES IN A DIFFERENTIATED STATE WITH DEFINED CONDITIONS

Normal Stem Cell

1. Isolate the stem cells by standard protocols for the cell type you are using.

2. For epithelial stem cells, you probably will have to use a basal medium with a calcium concentration above 0.5 mM. Also, copper, critical for aspects of differentiation, must be added.

3. Prepare feeder cells by using STO (ATCC) feeders or stromal feeders prepared from embryonic tissue from which the stem cells are derived. For example, use embryonic liver stroma for hepatic stem cells.

4. Overlay the feeders with a fibrillar collagen or with Matrigel or plate the stem cells onto collagen-coated or Matrigel-coated transwells.

5. Plate cells at desired density (high density of subconfluent to confluent) onto the collagen-embedded feeders or onto the collagen-coated transwells and over feeders.

6. Use a basal medium supplemented with insulin and transferrin and with 1–2% serum for plating.

7. After 4–6 h or when the cells are firmly attached, switch to a serum-free medium supplemented with insulin, transferrin, and other hormones or factors known to drive differentiation and to enhance specific genes (e.g., glucagons for the connexins; EGF, insulin, and glucocorticoids for albumin).

8. Add unbleached, bovine lung derived heparin (Sigma; 20–50 μg/ml) or liver-derived heparin proteoglycans (~10 μg/ml).

9. The timing for full differentiation will vary with the cell type and must be empirically defined but should occur within a few days to a week.

Normal Mature Cells (e.g., polyploid liver cells)

1. Isolate the cells as a single-cell suspension by standard methods.

2. Plate the liver cells at high density, $6–8 \times 10^6$/100-mm dish, into HDM/SSM and onto a flexible and porous scaffolding (e.g., filter, swatch of nylon, transwell, biodegradable microcarrier) coated with fibrillar collagen, ideally type III collagen mixed with fibronectin, or with Matrigel.

3. The cells should attach within 4–6 h. Then rinse the cells with PBS and give them the serum-free HDM retailored to optimize tissue-specific gene expression. Tailor the medium (as described earlier) to suit whichever genes are to be expressed optimally.

4. The medium fed to the cells after 4–6 h (*not* the plating medium) should be supplemented with lung-derived, *unbleached* heparin (Sigma; at 20–50 μg/ml) or ideally the liver-derived (not available commercially) heparin or its corresponding heparin proteoglycan (at ~10 μg/ml).

Table 11.3. Soluble Factors and Defined Matrix Components Required for Growth

Factors	Hepatic stem cells	Committed progenitors	Diploid hepatocytes	Hepatocytes: polyploid; high density	Hepatomas
EGF	Not required	10 ng/ml (hepatocytic but not biliary)	10 ng/ml	10 ng/ml	Variably required Depending on the tumor
Insulin	5 μg/ml	5 μg/ml	5 μg/ml	5 μg/ml	5 μg/ml
Transferrin/Fe	5 μg/ml	5 μg/ml	5 μg/ml	Not required	5 μg/ml
Linoleic acid	Replaced by FFA mixture	Replaced by FFA mixture	Replaced by FFA mixture	5 μg/ml	Replaced by FFA mixture
Free fatty acid (FFA) mixture	7.6 μequiv	7.6 μequiv	7.6 μequiv	Not required	7.6 μequiv
HDL	10 μg/ml	10 μg/ml	10 μg/ml	10 μg/ml (optional)	10 μg/ml
Triiodothyronine	Not required	Not required	1×10^{-9} M (optional)	1×10^{-9} M (optional)	1×10^{-9} M
Prolactin	Not required	Not required	Not required	20 mU/ml	2 mU/ml
Growth hormone	Not required	Not required	Not required	10 μU/ml	10 μU/ml
HS-PG or heparan sulfates (HS)	Unknown	Unknown	Unknown	10 μg/ml of HS-PG or 50 μg/ml of HS (optional)	10 μg/ml of HS-PG or 50 μg/ml of HS (optional)
Trace elements					
Zinc	1×10^{-10} M	1×10^{-10} M	1×10^{-10} M	1×10^{-10} M	1×10^{-10} M
Copper	not essential	not essential	3×10^{-10} M	1×10^{-7} M	1×10^{-7} M
Selenium	3×10^{-10} M	3×10^{-10} M	3×10^{-10} M	3×10^{-10} M	
Matrix conditions for growth	Type IV collagen, laminin (these matrix components are supplied by the feeders if cells are plated directly onto the feeder; if cells are plated onto transwells above the feeders, the matrix components must be provided onto porous, flexible scaffolding)				
Feeders	Embryonic stroma only (e.g., STO cells)	Embryonic stroma (e.g., STO cells); stroma from diverse sources also work	Embryonic stroma (e.g., STO cells); stroma from diverse sources also work	Not required	Embryonic stroma only (e.g., STO cells)

• These culture conditions permit growth at low densities (10^2–10^3 cells/60-mm dish) of hepatoma cell lines (human, rat, and mouse cell lines have been tried) and of stem cells and small hepatocytes (assumed to be diploid hepatocytes). However, polyploid liver cells will undergo only a round or two of division, and so for practical reasons they are plated only at high densities (>10^5 cells/60-mm dish); they will survive and go through a division or two at low seeding densities under the conditions used for clonogenic growth of the diploid cells, but the polyploid cells yield too few cells for routine assays.

• Unless protease inhibitors and/or specific matrix substrata are used, the cells must be seeded in a basal medium supplemented with low levels of serum (1–5%) plus the hormones for a few hours (4–6 h) and then transferred into the serum-free, hormonally defined medium. This is referred to as the plating medium or a mixture of a serum-supplemented medium (SSM) and hormonally defined medium (HDM): a SSM/HDM.

• Glucocorticoid can be hydrocortisone or dexamethasone.

• Linoleic acid must be presented in combination with fatty-acid-free bovine serum albumin (Sigma-Aldrich) at a 1:1 molar ratio. If added alone, it is quite toxic to the cells.

• HS, heparan sulfates; HS-PG, heparan sulfate proteoglycan. These are made by the feeders and so do not need to be supplied exogenously when feeders are used. Their effects are critical for various paracrine signals derived from the feeders. The effects on growth of mature cells are plastic and either negligible or insufficiently studied to be recognized.

• Copper is a trace element found by some to drive differentiation, and its elimination is thought to be helpful in keeping hematopoietic stem cells undifferentiated [4,5]. Given the close relationship between hematopoietic and hepatic lineages, it is likely that copper should be omitted for hepatic stem cells as well.

Table 11.4. Soluble Factors and Matrix Components Required for Differentiation

Factors required for general differentiation and/or for all tissue-specific genes

Factors	Hepatic stem cells	Committed progenitors	Hepatocytes (diploid and polyploid; high density cultures)	Hepatomas
EGF	10 ng/ml for differentiation to hepatocytes; not required for differentiation to biliary epithelia		10 ng/ml	Variably required depending on tumor
Insulin	5 µg/ml	5 µg/ml	5 µg/ml	5 µg/ml
Glucocorticoid	10^{-8} M	10^{-8} M	10^{-8} M	10^{-8} M
Transferrin	5 µg/ml	5 µg/ml	Not required	5 µg/ml
Linoleic acid	Replaced with FFA mixture	Replaced with FFA mixture	5 µg/ml	Replaced with FFA mixture
Free fatty acid (FFA) mixture	7.6 µequiv	7.6 µequiv	Not required	7.6 µequiv
HDL	10 µg/ml	10 µg/ml	10 µg/ml (optional)	10 µg/ml
Heparins (HP)[a] or HP-PG[b]	25–50 µg/ml of unbleached bovine lung derived heparins (or 10 µg/ml of liver-derived heparins) or 10–50 ng/ml of heparin–Proteoglycan			
Trace Elements				
Zinc	1×10^{-10} M	1×10^{-10} M	1×10^{-10} M	1×10^{-10} M
Copper	1×10^{-7} M	1×10^{-7} M	1×10^{-7} M	1×10^{-7} M
Selenium	3×10^{-10} M	3×10^{-10} M	3×10^{-10} M	3×10^{-10} M
Matrix conditions for differentiation	Fibrillar collagen (e.g., type III or type I collagen) plus fibronectin or Matrigel on a porous, flexible scaffolding			
Feeders[c]	Embryonic stroma only (e.g., STO cells); after cells in colonies and at high density, the feeders can be eliminated	Embryonic stroma or other types of stroma; not required after cells are at high density density	Not required	Variably required and when required, it is as for committed progenitors

(continues)

183

Table 11.4. (continued)

Factors	Hepatic stem cells	Committed progenitors	Hepatocytes (diploid and polyploid; high density cultures)	Hepatomas
Factors required for optimal expression of specific genes				
Albumin		EGF (10 ng/ml), insulin (5 μg/ml), glucocorticoids (10^{-8} M)		
IGF-II		Insulin (5 μg/ml), prolactin (20 mU/ml), T_3 (1×10^{-9} M), growth hormone (10 μU/ml)		
TGF-B		EGF (10 ng/ml), insulin (5 μg/ml)		
Connexins		Glucagon at 10 μg/ml; insulin (100 ng/ml), glucocorticoids (10^{-8} M), mixture of free fatty acids plus HDL		

[a]HP = heparins. These are available from multiple sources (see Appendix 2). Be certain to use only unbleached heparins. There is some tissue specificity that is poorly understood. The best HP of those that are commercially available is the bovine lung derived species. Liver-derived heparins are even more potent but are not commercially available.

[b]HP-PG = heparin proteoglycan. It is made in specific regions of the liver [1–3] and has been found to be essential for transcription of tissue-specific genes. The biological activity on transcription is due to the glycosaminoglycan chains, the heparins.

[c]The determined stem cells and the committed progenitors require the feeders for survival and growth but not for the differentiation process. Therefore, it is ideal to plate the stem cells or progenitors on transwells coated with the matrix and then place the transwells over the feeders until the cells become confluent. Then move the transwell to a dish with no feeders and add a serum-free medium with the hormones required for a specific gene and supplement also with the heparin or heparin proteoglycan. The differentiation proceeds fastest in this way.

Table 11.5. Requirements for mRNA Synthesis

Tissue-specific genes
    Each gene is dependent on specific hormones and growth factors presented with tissue-specific heparin proteoglycans
    Cell–cell signals that are lost with enzymatic digestion of tissue into single cell suspensions
    Three-dimensional shape—mRNA synthesis lost or muted in flattened cells
    Serum-free medium, since serum inhibits mRNA synthesis and destabilizes mRNAs of tissue-specific genes
    Near-normal transcription rates achieved with:
        Spheroids, especially in serum-free conditions and supplemented with specific hormones and growth factors
        Monolayers with serum-free medium, specific hormones or growth factors, on porous, flexible substrata coated with
        extracellular matrix (e.g., collagen + fibronectin) and supplements of unbleached heparins or heparin proteoglycans

Common genes
    Each gene is dependent on specific hormones or growth factors and the hormonal effects are independent of proteoglycans
    mRNA synthesis is similar in cell suspensions, cultures, and *in vivo*, and with medium with or without serum
    Very sensitive to shape changes and to changes in proliferation
    Post-transcriptional regulation dominant, with half-life being short *in vivo* in quiescent tissues and long in cultures and
    proliferating tissues

5. Note that the cells will not grow, but will survive, and should be three-dimensional and highly differentiated. The cultures will survive for perhaps 4–6 weeks and will retain most of their differentiated functions. If you are able to add the heparin proteoglycan, you will have near-normal transcription rates for most of the tissue-specific genes. If you use the glycosaminoglycan (GAG), or unbleached heparins, you can observe normal or near-normal transcription rates of some but not all tissue-specific genes. The proteoglycans and GAGs can be tissue specific. So, ideally use the proteoglycan or GAG from the tissue being cultured (e.g., liver-derived heparin proteoglycan for liver cells). With the limited availability of unbleached heparins (and even more so for proteoglycans) you probably will have to screen the available ones and use the one that is most active on your cells. See Table 11.5 for a summary of the requirements for tissue-specific gene expression.

Tumor Cells

1. Prepare stock flasks of tumor cells as single-cell suspensions.

2. Plate the cells at high density (6–8×10^6 cells/100-mm dish) under the same conditions used for the normal cells.

3. The cells will go into growth arrest and, simultaneously, will differentiate to the extent that they are capable of differentiating. Their gene expression capability is a reflection of the lineage stage from which they derive and is always early gene expression, since tumor cells are transformants of stem cells and early progenitors. Therefore, one must study the tumor cells as models of early progenitors and anticipate that the tumor cells can express and regulate only "early" genes, those found expressed and regulated in their normal counterparts, subpopulations of the progenitors. Tumor cells have not been found to be able to mature enough to express late genes such as would be expressed in the polyploid cells. For example, major urinary protein (MUP) is expressed only in the polyploid cells occurring in zone 3 of the liver and has never been found in any tumor cell.

4. The cells will stay in a state of growth arrest for a matter of days (if they are anaplastic tumors) to a few weeks (if they are minimally deviant tumor cells). The well-known ability of tumor cells to secrete enzymes that degrade matrix and/or to overexpress oncogene products (usually products that are the same as a hormone active on the cell or products that activate the hormonal pathways intracellularly) ultimately overrides the growth-arrested/differentiated state caused by the defined matrix and hormonal conditions. The tumor cells will begin to grow in foci, similar to transformation foci, piling up on top of one another. The cells in these foci will be much less differentiated than the growth-arrested cells. If you remove the growing foci of cells and plate them on fresh matrix plates and into fresh medium, they will again go into growth arrest and into a differentiated state.

THREE-DIMENSIONAL SYSTEMS

Three forms of three-dimensional culture systems are described: cells on biodegradable microcarriers kept in suspension, spheroid cultures, and hollow-fiber bioreactors. The first two are ideal for small-scale studies, the usual experiments in most academic labs and in some industrial labs. The bioreactor systems are ideal for industrial-scale analyses or for studies in which a process must be followed noninvasively, using the media stream from the bioreactor, nuclear magnetic resonance (NMR) spectroscopy, or magnetic resonance imaging (MRI).

BIODEGRADABLE MICROCARRIERS
Culture of Cells from a Cell Line on Collagen-Coated PLGA Microcarriers (see Fig. 11.5)

The human hepatoma cell line HepG2 is used here as a model system to demonstrate the use of the porous PLGA beads for culture.

- Prepare a cell suspension from stock flasks of a cell line such as HepG2.
- Prewet PLGA microcarriers, with or without coating with collagen, in the culture media, HDM/SSM, in a 35-mm bacteria culture disk. HepG2 cells are typically seeded with the beads at a seeding density ratio of HepG2 to PLGA microcarriers ranging from 1:1 to 10.
- The attachment and subsequent culture of HepG2 cells with PLGA beads is carried out under stationary conditions in a tissue culture incubator at 37°C, supplied with 5% CO_2.
- The extent of cell attachment is initially examined after overnight incubation.
- The culture is examined for cell density and viability at the time specified after trypsin digestion to dislodge the HepG2 cells from the beads. The medium is changed daily.

Culture of Adult Rodent Liver Cells with Collagen-Coated PLGA Microcarriers

- Adult rodent liver cells are isolated by standard liver perfusion and suspended in HDM/SSM prior to being seeded with the PLGA beads or commercially available beads as controls.
- For seeding and culture of liver cells on collagen-coated degradable microbeads under stationary culture conditions, 10^5–10^6 beads are incubated with 10^6 liver cells in a 35-mm, uncoated bacterial culture dish, first with the HDM/SSM at 37°C, 5% CO_2 overnight.
- The medium is then changed to HDM after 24 h. Cell viability is analyzed typically after the initial 24 h of seeding and on the second day of culture.
- Seeding of cells onto beads or other three-dimensional scaffolds often requires facilitated initial physical contact between cells and the scaffold substrata [27,69]. To increase and facilitate the initial physical contact of the liver cells to the type I collagen coated beads, seeding may be carried out alternatively with constant gentle rocking of the culture vessels. This may be done by placing the tissue culture flask on an orbital shaker in a 37°C incubator. A humidified gas mixture of 15–20% O_2, 5% CO_2, and 75–80% N_2 is supplied continuously to the flask. The tissue culture flasks (T25) are pretreated with 0.1% bovine serum albumin (BSA) to block cell adhesion to the culture flasks.
- Attachment of liver cells to the beads and cell viability is analyzed on the second day of culture. Figure 11.5 shows phase contrast image of rodent liver cells seeded with collagen I coated PLGA mesobeads. Multicell aggregates with or without attachment to the beads may form under the present protocol.

In summary, porous PLGA beads can be readily prepared and used for three-dimensional tissue culture with HepG2 or adult rodent liver cells. In general, HepG2 cells or adult rodent cells do not attach well to uncoated PLGA beads. Although various protocols may be used to coat the polymer beads with matrix proteins, adsorption of collagen under acidic conditions followed by immediate suspension in concentrated neutral buffer at 37°C results in sufficient coating of the collagen on the surface of the polymer. Initial physical contact of cells with collagen coating is important to encourage cell adhesion with the matrix proteins.

SPHEROIDS

The observation by Landry and others [78] that freshly isolated adult rat liver cells cultured under the appropriate conditions can aggregate and form spheroids opened the way for the development of spheroids as a model for the long-term culture of adult liver cells. In spheroid cultures, liver cells are prevented from attaching to the substratum and remain in suspension, forming three-dimensional, multicellular aggregates referred to as spheroids [75,78,84].

Spheroid culture provides a three-dimensional configuration for liver cells allowing the maintenance of an *in vivo*–like cuboidal cell shape and distribution of the cytoskeleton as well as enhanced cell–cell contacts. It has been shown that deposition of extracellular matrix within spheroids takes place, re-creating a more *in vivo*–like microenvironment [78]. This culture system seems to provide more appropriate conditions for the maintenance of a differentiated phenotype *in vitro*. In addition to reestablishment of an *in vivo*–like morphology and ultrastructure, hepatocytes in spheroid culture express many differentiated functions over long-term culture, including albumin secretion [78,179], transferrin secretion, tyrosine amino transferase induction by glucocorticoids [78], ammonia metabolism, urea synthesis, and gluconeogenesis [179]. The main application of the system has been the development of bioartificial livers. The majority of the research on liver spheroids has focused on rat and pig liver cells, with little work on human liver cells.

Factors Affecting Spheroid Formation and Morphology

The spheroid culture model is a technically demanding model, since there are many variables that can affect the outcome, and since the spheroids float, media changes are cumbersome. If all aspects of the culture conditions and methods utilized are not optimal, then the quality of the spheroids formed is compromised, and thereby the results obtained. Some of the most critical factors for spheroid culture are outlined.

Choice of Method for Spheroid Culture

Several different methods have been described in the literature for the generation and culture of spheroids. Typically, hepatocytes are either cultured on bacteriological dishes and placed on a rotary shaker at a constant speed (the shaking method: [84,181]), or they are cultured on a nonadherent substratum in static culture (references for the static method: [75,78]). When large-scale spheroid formation is required, spinner flasks or large Ehrlenmeyer flasks may be susbtituted for the smaller tissue culture vessels. The main factors affected by the method used are the size and shape of the spheroids obtained, and the time required for spheroid formation.

A novel method that combines elements from the shaking and the static methods for spheroid formation has been described [22,50,52]. The method involves the culture of liver cells on six-well tissue culture plates coated with the nonadherent polymer called PHEMA [poly(2-hydroxyethyl methacrylate] and shaking of the cultures on a rotary shaker at a constant speed of 90 rpm. This novel combination serves to prove the efficiency of spheroid formation, reduce the time required to obtain fully formed spheroids, and yield spheroids that are more homogeneous in shape and size. A common concern with the use of spheroid culture is the potential for the formation of necrotic centers. If the spheroid diameter exceeds 250–300 μm, the lack of oxygen and nutrient diffusion will cause cell death at the center of the spheroids. However, the use of this novel method prevents the generation of spheroids that exceed a 250-μm diameter, thereby preventing the formation of necrotic centers, a major improvement to the model.

The combination of PHEMA-coated six-well plates on a rotary shaker is the method of choice in our laboratory, yielding optimal results in combination with the use of SSM/HDM and a seeding density of 5×10^5 viable cells/ml (2 ml/well). Experience in our laboratory with PHEMA coating of plates for spheroid formation has shown that the coating must be smooth and even on the plates, which also must be fully dry prior to use. The PHEMA must be filtered prior to use to remove any undissolved PHEMA or particulate impurities in the solution, which would cause an uneven surface once the ethanol had evaporated from the plates. Uneven coating of the plates due to impurities or poor evaporation of the ethanol would in turn cause poor spheroid formation. The shaking speed will also affect the size of the spheroids formed, increasing speeds resulting in smaller spheroids [115]. However, shear stress forces become an issue at speeds exceeding 110–120 rpm. Earlier studies determined 90 rpm to be optimal for spheroid formation with PHEMA-coated plates [22,50,52].

Liver cells in spheroid culture do not attach and flatten as is observed in conventional monolayer culture; instead, liver cells remained in suspension and retained a rounded cell shape. When the method described here is used, rat liver cells form initial spheroids, irregular in shape and size, after only 24 h in culture. Some unaggregated single cells are also present at this stage, but these are removed from the culture through subsequent medium changes. Spheroids gradually increase in size over the first week of culture, from small aggregates of cells with a diameter of approximately 30–50 μm to fully formed compact spheroids with a diameter of approximately 200 μm. The increase in diameter results from the fusion and restructuring of the initial spheroids to form mature spheroids. Spheroids become more homogeneous in shape and size with time in culture. When fully formed, spheroids appear as tightly packed, dense structures with a distinct smooth outer lining; this outer lining consists of a single layer of flattened liver cells mixed with extracellular matrix.

An interesting observation was made when human liver cells were being cultured as spheroids. The formation of spheroids takes longer with human hepatocytes than with rat hepatocytes, requiring approximately 2–3 extra days to complete spheroid formation. We hypothesize that this is due to the higher level of diploid cells in adult livers; livers from young rodents are greater than 90% polyploid, whereas those from young humans are estimated to be about 50%. Therefore, if spheroid formation is a phenomenon predominantly of the polyploid cells, it may take longer for human liver cell suspensions to form the spheroid, and factors such as heparins that drive the differentiation to the polyploid state may facilitate the kinetics of the process; preliminary evidence with heparins suggests, indeed, that this hypothesis may be true. The hypothesis is supported also by an earlier report that addition of proteoglycans speeded the formation of rat spheroids [75]. An alternative is the use of Eudragit, an artificial polymer that has been shown to enhance spheroid formation [182].

Culture Medium

The composition of the culture medium has been shown to have a major influence on spheroid formation as it does on morphology and function of cultured liver cells [80]. In earlier work carried out to study the effects of different media on spheroid formation and morphology, it was determined that the best ones contained stable forms of insulin, iron-saturated transferrin, and selenium. One that was found especially useful is hepatocyte medium (HM), in which these factors are added to a modified Leibovitz-15 medium that has been specially buffered [50–52]. It is commercially available from Sigma. To obtain optimal spheroid formation and hepatocyte function, HM needs to be supplemented with 2% FBS for at least the first 3 days of culture [52]. Serum has been shown to be essential for the first few days of culture for optimal spheroid formation on an orbital shaker [181]. However, after formation of the spheroids, the serum must be eliminated for optimal expression of tissue-specific genes, especially the P450s that are adversely affected in cultures with serum supplementation (see Table 11.5).

Preliminary studies with rat liver cells indicate that the use of HDM accelerated the rate of spheroid formation. The potential application of this culture medium for improving the efficiency and speed of spheroid formation in human hepatocytes is under investigation.

Hepatocyte Viability and Quality

A major factor affecting spheroid formation is the quality of the liver cells seeded, with preparations having a 90% or greater viability being optimal for spheroid formation.

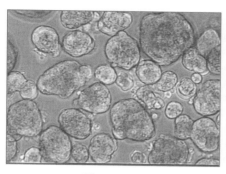
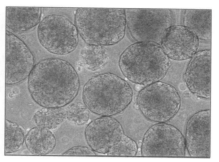

Human   Rat

Fig. 11.6. Three-dimensional spheroidal culture systems.

Dead cells release "sticky"-DNA. Therefore, if there is a high percentage of dead cells and cell debris, the small initial aggregates clump together and yield very large spheroids, often above 300 μm. Spheroids from a hepatocyte preparation with low viability will often be irregular in shape and size, as a result of the incorporation of dead cells and clumps into the spheroids.

After liver cell isolation, the viability of the cells is determined by trypan blue exclusion. Although this is a useful guide to the overall viability of the cells, it does not provide any information on the functional aspects of the cells. So even with a viability of 90% as indicated by trypan blue exclusion, the cells may still not provide good spheroids. The initial stages of spheroid formation are thought to require the expression of functional specific cell surface receptors and adhesions sites [18,31], *de novo* RNA and protein synthesis, as well as active cellular processes such as the reestablishment of junctional complexes [79]. Healthy liver cells in good condition and able to undertake complex cellular processes are required for good spheroid formation. Representative spheroid cultures from rodent and human livers are shown in Fig. 11.6.

Protocol for Establishing Spheroids

SUPPLIES

- Falcon six-well tissue culture plates (Becton Dickinson Labware, Franklin Lakes, NJ)
- PHEMA (Sigma)
- HM (Sigma)
- Fetal bovine serum (HyClone)
- Inova 2000 Orbital Shaker (New Brunswick Scientific, Edison, NJ)
- Sterile transfer pipets (Fisher Scientific, Pittsburgh)

METHOD

1. Six-well tissue culture plates are coated with of 2.5% PHEMA solution in 95% ethanol (2 ml per well). The ethanol is allowed to evaporate, leaving a clear polymer coating in the surface of each well. The plates can be left overnight in a class II safety hood without lids for a more rapid evaporation, or placed in a dry incubator at 37°C (lids on) for 3–4 days. Although slower, the latter method has been found to yield better results.
2. Freshly isolated hepatocytes are resupended in HM supplemented with 2% FBS, 10^{-7} M dexamethasone, and antibiotics ($1\times$ penicillin–streptomycin). These are seeded onto PHEMA-coated plates at a density of 5×10^5 viable cells per milliliter of medium (2 ml of medium per well).
3. Place the plates on an orbital shaker at 90 rpm and in an incubator at 37°C and 5% CO_2.

4. Replace the medium after 24 h of culture and every 2 days thereafter. This is done by tilting the six-well plates at a sharp angle and allowing the spheroids to settle to the bottom of the plate. The medium is then slowly and carefully aspirated off with a sterile plastic transfer pipette. Particular care must be taken on the initial days of culture when the spheroids are small, for loss of spheroids can occur during medium changes. After the first medium change, spheroids often require separation from each other by gentle pipetting with a sterile plastic transfer pipette.

HOLLOW-FIBER BIOREACTOR SYSTEMS

The use of hollow-fiber bioreactors as bioartificial liver devices was first demonstrated by Wolf and Munkelt [177], who cultured hepatocytes in the extracapillary space (ECS) of a commercially available hollow-fiber bioreactor derivative of the original design of Knazek and others [73]. Hollow-fiber bioreactors (HFBRs) are perfusion vessels comprising semi-permeable fibers that are 0.2 mm to as large as 3 mm in outer diameter, with 0.005 μm to 0.7-μm pore sizes and composed of various polymers. There have been various HFBR types described [16,36,96,120,121,142,160,166,178]. Table 11.6 lists the various types of HFBR used in bioartifical liver studies and the experimental parameters and results of these studies. Table 11.7 lists various HFBR cartridges and vendors. There are some generalities in all the experimental designs between these studies, as well as some common factors in the results. Such experimental designs have at least four phases: bioreactor setup and conditioning (pyrogen-free), inoculation, experimental (cell culture analysis), and termination of experiment. A very brief description of each follows. The reader is directed to various references listed in Table 11.6 for details. Either primary hepatocytes from various species or human liver cell lines have been used, and in no case have primary hepatocytes grown [94]. For a complete review of all bioreactor types, see the review by Macdonald and others [94].

Bioreactor Setup and Conditioning

The life support system and the bioreactor proper are often called the bioreactor "loop," and since the inlets and outlets to the tubing meet back at the reservoirs, the entire system forms a loop. All bioreactor loops have at least one reservoir where media are retained and changed, tubing to and from the bioreactor, a pump to circulate the media, and a temperature control system. Bioreactor media can be changed entirely at specific time points, called batch mode, or continuously in a bleed-feed mode. In clinical systems the patient's plasma is fed to the bioreactor effectively in a bleed-feed mode [64,83], but typically all bioartifical liver systems have employed a batch mode (Table 11.5), which ultimately causes pH changes in the media. The pH in nearly all bioartificial liver systems is controlled by buffering with bicarbonate and modulating the percent of carbon dioxide in the gas mixture, although pH controllers have been described [95]. Gas lines are attached to the loop to feed a gas exchange module or the bioreactor directly where integral oxygenation has been engineered into the bioreactor design. We use food grade tubing for media (Pharmed, Cole Parmer, Chicago) and gas-impermeable tubing (Bev-a-line, Cole Parmer) for the gas lines, although gas-permeable tubing such as Silastic (Dow-Corning) may be desirable and is used with the radially scaled-out HFBR by Genespan (Bothell, WA). Typically the various components of the bioreactor loop are autoclaved and then connected in a laminar flow hood or in the incubator that will be used to store the bioreactor for the duration of the experiment. Where bioreactors are autoclavable, the entire loop can be assembled and then autoclaved. The parts should be placed in autoclavable bags and all connections should be tightened with plastic ties to resist contamination and ensure connection integrity. Sterilization can be performed with 70% ethanol in water, gamma irradiation, or ethylene oxide. For example, in HFBRs made from cellulose acetate, typically the bioreactor is sterilized by ethylene oxide and the loop is autoclaved [41], while with HFBRs made from hydrophobic hollow fibers such as polypropylene, the bioreactor is sterilized by ethylene oxide and then, prior to insertion into the bioreactor loop, the membrane is made water permeable by means of 100% dilution of ethanol to 100% water, a process called membrane "wetting." Polyurethane epoxy is the epoxy of choice for construction of bioreactors, and it can be

Table 11.6. Miscellaneous Data on Types of Hollow-Fiber Bioartificial Liver

Hollow-fiber bioreactor type	Ref.	Cell type	Cell density	Total biomass	Culture duration (days)	Vendor website	Company holding patents and/or conducting studies
Conventional							
Hepatocytes in extracapillary space							
ELAD	[166]	Human hepatoma	NC	200 g 2×10^{10} cells (1 g = 10^8 cells)	7.5–10	http://www.vgen.com/	Vitagen Inc.
HepatAssist	[16,142]	Porcine	2.8×10^7 (cells/ml)	5×10^9	0.25	http://www.circebio.com/	Circe Biomedical Inc.
BLSS Sybiol		Porcine Porcine		50 g Continuous addition	7	http://www.upmc.edu/McGowan/ http://www.exten.com/	Excorp Medical Inc. Xenogenics Corp.
Hepatocytes in intracapillary space							
Collagen Type 1 Entrapment (LiverX 2000)	[120,121]	Human cell line, HepG2 and SD rat	$3–5 \times 10^6$ (cells/ml gel)	$5 \times 10^7–10^8$	14	http://www.algenix.com/	Algenix Inc.
Coaxial							
Coaxial	[95]	SD rat	1.5×10^8 (cells/ml)	6.5×10^8	0.25	None	Setec Inc.
Radially scaled out		Porcine	5.6×10^7 (cells/ml)	10^9	35	http://www.genespan.com/	Genespan Inc.
Multicoaxial	[96,178]	Rat				http://www.incara.com/	Incara Pharmaceuticals Inc.
Woven							
Woven	[36]	Porcine	1.6×10^8 (total ECS vol 15 ml)	2.5×10^9 cells	49 (7 weeks)		None
Flatbed hybrid							
Monolayer	[160]	Rat and	$0.5–1.8 \times 10^6$ (cells/cm^2])	Rat, $2.6–5.6 \times 10^7$ Porcine $4.3–9.7 \times 10^7$			None

Table 11.7. Commercially Available Hollow-Fiber Bioreactor Cartridges and Vendors

Hollow-fiber bioreactor type	Type	Vendor website	Vendor or company selling cartridge
Conventional			
Cell-Pharm MicroMouse	a	http://www.unisyntech.com/	Unisyn Technologies Inc.
Cellmax	a	http://www.spectrapor.com/	Spectrum Laboratories Inc.
VirA/Gard	b	http://www.agtech.com/	A/G Technologies Inc.
Romicon	b	http://www.kochmembrane.com/	Koch Membrane Systems Inc.
Plasmaphan	c	http://www.akzo-nobel.com/	Akzo Nobel Fraser AG, Membrana
Sterapore	b	http://www.mrc.co.jp/english/	Mitsubishi Rayon Inc.
Ster-O-Tap	b	http://www.primewater.com/	Prime Water International
Liqui-Cel	b	http://www.celgard.com/	Celgard LLC
Permasep	b	http://www.dupont.com/	Dupont Inc.
Module A5	b	http://www.pcims.com/	PCI Membrane Systems
Hollow fiber cartridges	b	http://www.millipore.com	Millipore Inc.
Microza	b	http://www.pall.com/gelman/	Pall Gelman Laboratory Corporation
Flatbed hybrid			
Technomouse		http://www.integra-biosciences.com/	Integra Biosciences Inc.

[a]Designed and marketed for cell culture, especially for hybridoma cultures for antibody production.
[b]Designed and marketed for filtration of water to clarify, sterilize, or desalinate.
[c]Designed and marketed for filtration of culture media or purification of proteins.

dissolved by 100% ethanol, so membrane wetting should be performed as rapidly as possible. Once the entire loop has been assembled, it is conditioned by flowing typically $1\times$ phosphate buffer first, followed by one to several media changes to remove excess pyrogens or other unwanted materials in the system prior to inoculation. The conditioning period can last a few hours, but typically takes a day or two.

Inoculation

During the inoculation process, it is important to attain a uniform distribution of cells within the hollow-fiber bundle, where mass transfer is sufficient, rather than allowing cells to fall in the space between the fiber bundle and housing. Typical inoculation procedures generate a negative transmembrane pressure from shell side to lumen, either by gravity or by pump in conjunction with rotation of the bioreactor, forcing hepatocytes from the inoculation port to the fiber bundle. Since most cells are at 4°C at the time of inoculation, the temperature change to 37°C should be incremented to 25°C and slowly up to 37°C to avoid a heat shock response. When inoculating with microcarriers such as Cultisphere G collagen (IBN Biomedicals, Aurora, OH) or Cytodex collagen-coated dextran (Pharmacia Biotech, Piscataway, NJ) beads, cell adhesion molecules (CAMs) will create a gelling process during heating that will negatively affect convection. The reader is directed to the various references in Table 11.5 or the websites in Table 11.6 for details of inoculation procedures. Figure 11.7 shows an HFBR from Genespan being inoculated with isolated rat hepatocytes and Cultisphere G microcarriers.

Experimental

The experimental part of all the studies shown in Table 11.6 consists of measuring some common liver function and comparing the results against data from monolayer cul-

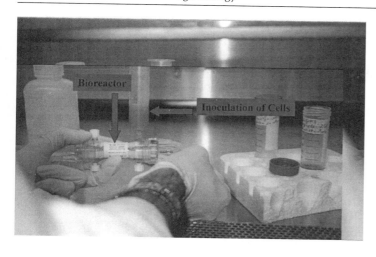

Fig. 11.7. Hollow-fiber bioreactor. See also color insert.

tures and normal human liver functions. There are several common functional features that are necessary for defining success of bioartificial livers. These can be categorized as biotransformatory (detoxification, primary metabolism), synthetic (proteins, storage), and excretory [56,94]. These functions are generally determined by measuring nutrients, xenobiotics, or proteins in media streams, but NMR spectroscopy can noninvasively measure these functions in "real time" within bioreactor cultures [42,95,99,100] or by in-line media stream analysis [123]. Protocols to test the optimal array of liver functions will be difficult to design [56], and the optimum and minimum number of tests for liver function and quality control are still being debated.

Termination of Experiment

The experiment is terminated and cells can be extricated and analyzed for viability and RNA, protein, metabolites, ploidy, and/or surface markers to help determine the degree of differentiation. Extrication is often difficult owing to the extensive extracellular matrix secreted by the culture, so the bioreactor can be fixed with paraformaldehyde or other fixative, stained, and analyzed by microscopy. Once the bioreactor has been extricated from the loop, all the lines should be rinsed with distilled water, then rinsed with 1 N NaOH followed by 1 N HCl to remove adsorbed proteins, and thoroughly rinsed with distilled water. Everything should be appropriately stored or prepared for the next experiment.

APPENDIX 1
PREPARATION OF FREE FATTY ACID (FFA) MIXTURES
Sources of Purified Fatty Acids: See Appendix 2
Preparation of the Stocks

The free fatty acids are prepared by dissolving each of the following individual components in 100% ethanol.

FFA	Comment
Palmitic acid (solid)	1 M stock; soluble in hot alcohol
Palmitoleic acid	1 M stock; readily soluble in alcohol
Oleic acid	1 M stock; readily soluble in alcohol
Linoleic acid	1 M stock; readily soluble in alcohol
Linolenic acid	1 M stock; readily soluble in acohol
Stearic acid (solid)	151 mM stock, soluble in alcohol at 1 g in 21 ml; must be heated

These stocks can be stabilized by bubbling nitrogen through each of them followed by storing at −20°C.

The Free Fatty Acid Mixture Stock Solution

FFA	Concentration (mM)
Palmitic acid	31.0
Palmitoleic acid	2.8
Oleic acid	13.4
Linoleic acid	35.6
Linolenic acid	5.6
Stearic acid	11.6

This yields a combined total of 100 mM free fatty acids. This stock can be stabilized by bubbling through nitrogen followed by storing at –20°C.

Final Solution

Add 76 μl of the free fatty acid mixture stock per liter of culture medium to achieve a final concentration of 7.6 μequiv. The free fatty acids are toxic unless they are presented in the context of purified, fatty-acid-free, endotoxin-free serum albumin (e.g., Pentex type V albumin). Albumin is prepared in the basal medium or PBS to be used and at a typical concentration of 0.1–0.2%.

APPENDIX 2
SOURCES OF REAGENTS: VENDORS

Growth factors, hormones, supplements

Academy Biomedical Co.
Accurate Chemicals
Alfa Aesar
Altech Associates Inc.
Amersham Pharmacia Biotech
Antigenix America Inc.
Biodesign International
BioResource Technology
BIOTREND Chemikalien
Calbiochem
Chemicon International

Clonetics Products
Collaborative Biomedicals
Cortex Biochemicals Inc.
ICN Biomedicals
Novabiochem
Pepro Tech
Per Immune
Sigma-Aldrich
Spectrum Laboratory Products
TCI America
Upstate Biologicals
U.S. Biological

Free fatty acids and lipids

Academy Biomedical Co.
Altech Associates Inc.
Biodesign International
BioResource Technology

Chemicon International
ICN Biomedicals
Per Immune
Sigma-Aldrich

Matrix molecules

Accurate Chemicals
Alexis Corp.
BioChemika
Biosource International
BIOTREND Chemikalien
Calbiochem

CarboMer, Inc.
Chemicon International
Collaborative Biomedicals
EY Laboratories
PolySciences, Inc.
Sigma-Aldrich

Trace elements

Chem Services Inc.
Crescent Chemicals
Gallade Chemical, Inc.
ICN Biomedicals

MV Laboratories, Inc.
Sigma-Aldrich
Spectrum Laboratory Products
Strem Chemicals, Inc.

(continues)

(continued)

Sulfated proteoglycan and glycosaminoglycans

A.G. Scientifics
Alexis Corp.
Alfa Aesar
American Qualex International Inc.
BioChemika
BIOMOL Research Laboratories, Inc.
Calbiochem
CarbMer, Inc.

Chemicon International
Clonetics
ICN Biomedicals
PolySciences, Inc.
Seikagaku USA
Sigma-Aldrich
TCI America
U.S. Biologicals

ACKNOWLEDGMENTS

Funding derives from a National Institutes of Health (NIH) grant, R01-DK52851 (LMR), a National Society for Research into Allergy (NSRA) grant, F32DK097130 (JMM and LMR), a grant from the Johns Hopkins Center for Animal Testing (JMM and LMR), a sponsored research grant (LMR) from Renaissance Cell Technologies (now Incara Cell Technologies) and from Incara Pharmaceuticals, Inc., a U.S. Food and Drug Administration grant, 223-97-3004, from the Center for Drug Evaluation and Research (EL), a grant from DuPont Pharmaceuticals Company (EL), and a grant from Warner-Lambert (EL). Salary support for Geraldine Hamilton derived from funds from Glaxo-Smith-Kline and from Incara Cell Technology; that for Stephen Wolfe is from a Reynolds fellowship through the University of North Carolina graduate school. Jane Bowen, who provided technical assistance in some experiments, was supported by a training grant (NIH 5 T35 DK07386) from the National Institute of Diabetes and Digestive and Kidney Disease.

We thank Lucendia English for technical assistance, Donna Rogers for assistance with the database for the human liver samples and computers, and Dr. J. J. LeMasters, Sherry Franklin, and Dr. YiWei Rong of the ACT Core of the Center for Gastrointestinal and Biliary Disease Biology (CGIBD) for providing some of the preparations of the isolated normal rodent liver cells. Dr. Anthony Hickey provided fruitful discussions. We also thank Dr. B. Bagnell for expert assistance with the SEM studies on the microcarriers, Dr. M. Louey for assistance with the laser diffraction experiments, and Mr. Stacey Long for assistance with surface area measurements.

REFERENCES

1. Alvarez-Buylla, A., and Lois, C. (1995). Neuronal stem cells in the brain of adult vertebrates. *Stem Cells* **13**, 263–272.
2. Anatskaya, O. V., Vinogradov, A. E., and Kudryavtsev, B. N. (1994). Hepatocyte polyploidy and metabolism/life-history traits: Hypotheses testing. *J. Theor. Biol.* **168**, 191–199.
3. Anderson, J. M. (1998). Biocompatibility of tissue-engineered implants. *In* "Frontiers in Tissue Engineering" (C. W. Patrick, Jr., A. G. Mikos, and L. V. McIntire, eds.), pp. 152–165. Pergamon, Oxford and New York.
4. Anti, M., Marra, G., Rapaccini, G. L., Rumi, C., Bussa, S., Fadda, G., Vecchio, F. M., Valenti, A., Percesepe, A., Pompili, M. *et al.* (1994). DNA ploidy pattern in human chronic liver diseases and hepatic nodular lesions. Flow cytometric analysis on echo-guided needle liver biopsy. *Cancer (Philadelphia)* **73**, 281–288.
5. Barnes, D., McKeehan, W. L., and Sato, G. H. (1987). Cellular endocrinology: Integrated physiology in vitro. *In Vitro Cell. Dev. Biol.* **23**, 659–662.
6. Beardsley, T. (1999). Stem cells come of age. *Sci. Am.* **281**, 30–31.
7. Behrens, J. (1993). The role of cell adhesion molecules in cancer invasion and metastasis. *Breast Cancer Res. Treat.* **24**, 175–184.
8. Bernfield, M., Kokenyesi, R., Kato, M., Hinkes, M. T., Spring, J., Gallo, R. L., and Lose, E. J. (1992). Biology of the syndecans: A family of transmembrane heparan sulfate proteoglycans. *Annu. Rev. Cell Biol.* **8**, 365–393.
9. Berry, M. N., and Friend, D. J. (1969). High-yield preparation of isolated rat liver parenchymal cells: A biochemical and fine structure study. *J. Cell Biol.* **43**, 506–520.
10. Brill, S., Holst, P. A., Zvibel, I., Fiorino, A., Sigal, S. H., Somasundaran, U., and Reid, L. M. (1994). Extracellular matrix regulation of growth and gene expression in liver cell lineages and hepatomas. *In* "Liver Biology and Pathobiology" (I. M. Arias, J. L. Boyer, N. Fausto, W. B. Jakoby, D. Schachter, and D. A. Shafritz, eds.), 3rd ed., pp. 869–897. Raven Press, New York.
11. Brill, S., Zvibel, I., and Reid, L. M. (1995). Maturation-dependent changes in the regulation of liver-specific gene expression in embryonal versus adult primary liver cultures. *Differentiation (Berlin)* **59**, 95–102; erratum: *Ibid.*, p. 331.
12. Brill, S., Zvibel, I., and Reid, L. M. (1999). Expansion conditions for early hepatic progenitor cells from embryonal and neonatal rat livers. *Dig. Dis. Sci.* **44**, 364–371.

13. Brinster, R. L. (1974). The effect of cells transferred into the mouse blastocyst on subsequent development. *J. Exp. Med.* **140**, 1049–1056.

14. Brodsky, W. Y., and Uryvaeva, I. V. (1977). Cell polyploidy: Its relation to tissue growth and function. *Int. Rev. Cytol.* **50**, 275–332.

15. Carriere, R. (1967). Polyploid cell reproduction in normal adult rat liver. *Exp. Cell Res.* **46**, 533–540.

16. Chen, S. C., Hewitt, W. R., Watanabe, F. D., Eguchi, S., Kahaku, E., Middleton, Y., Rozga, J., and Demetriou, A. A. (1996). Clinical experience with a porcine hepatocyte-based liver support system. *Int. J. Artif. Organs* **19**, 664–669.

17. Chen, U. (1992). Careful maintenance of undifferentiated mouse embryonic stem cells is necessary for their capacity to differentiate to hematopoietic lineages in vitro. *Curr. Top. Microbiol. Immunol.* **177**, 3–12.

18. Chow, I., and Poo, M. M. (1982). Redistribution of cell surface receptors induced by cell to cell contact. *J. Cell Biol.* **95**, 510–518.

19. Cima, L. G., Vacanti, J. P., Vacanti, C., Ingber, D., Mooney, D., and Langer, R. (1991). Tissue engineering by cell transplantation using degradable polymer substrates. *J. Biomech. Eng.* **113**, 143–151.

20. Crissman, H. A., and Hirons, G. T. (1994). Staining of DNA in live and fixed cells. *Methods Cell Biol.* **41**, 195–209.

21. Deans, R. J., and Moseley, A. B. (2000). Mesenchymal stem cells: Biology and potential uses. *Exp. Hematol.* **28**, 875–884.

22. Dilworth, C., Hamilton, G. A., George, E., and Timbrell, J. A. (2000). The use of liver spheroids as an in vitro model for studying induction of the stress response as a marker of chemical toxicology. *Toxicol. In Vitro* **14**, 169–176.

23. Doerr, R., Zvibel, I., Chiuten, D., D'Olimpio, J., and Reid, L. M. (1989). Clonal growth of tumors on tissue-specific biomatrices and correlation with organ site specificity of metastases. *Cancer Res.* **49**, 384–392.

24. Eckl, P. M., Whitcomb, W. R., Michalopoulos, G., and Jirtle, R. L. (1987). Effects of EGF and calcium on adult parenchymal hepatocyte proliferation. *J. Cell. Physiol.* **132**, 363–366.

25. Ek, S., Ringden, O., Markling, L., Dahlberg, N., Pschera, H., Seiger, A., Sundstrom, E., and Westgren, M. (1993). Effects of cryopreservation on subsets of fetal liver cells. *Bone Marrow Transplant.* **11**, 395–398.

26. Enat, R., Jefferson, D. M., Ruiz-Opazo, N., Gatmaitan, Z., Leinwand, L. A., and Reid, L. M. (1984). Hepatocyte proliferation in vitro: Its dependence on the use of serum-free hormonally defined medium and substrata of extracellular matrix. *Proc. Natl. Acad. Sci. U.S.A.* **81**, 1411–1415.

27. Foy, B., Lee, J., Morgan, J., Toner, M., Tompkins, R. G., and Yarmush, M. L. (1993). Optimization of hepatocyte attachment to microcarriers: importance of oxygen. *Biotechnol. Bioeng.* **42**, 579–588.

28. Freshney, R. I. (2000). "Culture of Animal Cells." Wiley, New York.

29. Furthmayr, H. (1993). Basement membrane collagen: Structure, assembly, and biosynthesis. *In* "Extracellular Matrix: Chemistry, Biology, and Pathobiology with Emphasis on the Liver" (M. Zern and L. M. Reid, eds.), pp. 149–185. Dekker, New York.

30. Gage, F. H. (1994). Neuronal stem cells: Their characterization and utilization. *Neurobiol. Aging* **15**(Suppl. 2), S191.

31. Garrod, D. R., and Nicol, A. (1981). Cell behaviour and molecular mechanisms of cell-cell adhesion. *Biol. Rev. Cambridge Philos. Soc.* **56**, 199–242.

32. Gatmaitan, Z., Jefferson, D. M., Ruiz-Opazo, N., Biempica, L., Arias, I. M., Dudas, G., Leinwand, L. A., and Reid, L. M. (1983). Regulation of growth and differentiation of a rat hepatoma cell line by the synergistic interactions of hormones and collagenous substrata. *J. Cell Biol.* **97**, 1179–1190.

33. Gebhardt, R. (1988). Different proliferative activity in vitro of periportal and perivenous hepatocytes. *Scand. J. Gastroenterol., Suppl.* **151**, 8–18.

34. Gebhardt, R. (1992). Metabolic zonation of the liver: Regulation and implications for liver function. *Pharmacol. Ther.* **53**, 275–354.

35. Gebhardt, R., Wegner, H., and Alber, J. (1996). Perifusion of co-cultured hepatocytes: Optimization of studies on drug metabolism and cytotoxicity in vitro. *Cell Biol. Toxicol.* **12**, 57–68.

36. Gerlach, J. C. (1994). Use of hepatocyte cultures for liver support bioreactors. *Adv. Exp. Med. Biol.* **368**, 165–171.

37. Gerlyng, P., Abyholm, A., Grotmol, T., Erikstein, B., Huitfeldt, H. S., Stokke, T., and Seglen, P. O. (1993). Binucleation and polyploidization patterns in developmental and regenerative rat liver growth. *Cell Proliferation* **26**, 557–565.

38. Gibson, C. W., Lally, E., Herold, R. C., Decker, S., Brinster, R. L., and Sandgren, E. P. (1992). Odontogenic tumors in mice carrying albumin-myc and albumin-rats transgenes. *Calcif. Tissue Int.* **51**, 162–167.

39. Gilding, D. K. (1981). Biodegradable polymers. *In* "Biocompatibility of Clinical Implant Materials" (D. F. Williams, ed.), Vol. 2, pp. 209–232. CRC Press, Boca Raton, FL.

40. Gill, J. F., Chambers, L. L., Baurley, J. L., Ellis, B. B., Cavanaugh, T. J., Fetterhoff, T. J., and Dwulet, F. E. (1995). Safety testing of Liberase, a purified enzyme blend for human islet isolation. *Transplant. Proc.* **27**, 3276–3277.

41. Gillies, R. J., Galons, J. P., McGovern, K. A., Scherer, P. G., Lien, Y. H., Job, C., Ratcliff, R., Chapa, F., Cerdan, S., and Dale, B. E. (1993). Design and application of NMR-compatible bioreactor circuits for extended perfusion of high-density mammalian cell cultures. *NMR Biomed.* **6**, 95–104.

42. Gillies, R. J., Scherer, P. G., Raghunand, N., Okerlund, L. S., Martinez-Zaguilan, R., Hesterberg, L., and Dale, B. E. (1991). Iteration of hybridoma growth and productivity in hollow fiber bioreactors using 31P NMR. *Magn. Reson. Med.* **18**, 181–192.

43. Gittes, G. K., Galante, P. E., Hanahan, D., Rutter, W. J., and Debase, H. T. (1996). Lineage-specific morphogenesis in the developing pancreas: Role of mesenchymal factors. *Development (Cambridge, UK)* **122**, 439–447.

44. Gluckman, E., Broxmeyer, H. A., Auerbach, A. D., Friedman, H. S., Douglas, G. W., Devergie, A., Esperou, H., Thierry, D., Socie, G., Lehn, P. *et al.* (1989). Hematopoietic reconstitution in a patient with Fanconi's anemia by means of umbilical cord blood from an HLA-identical sibling. *N. Engl. J. Med.* **321**, 1174–1178.

45. Greider, C. W. (1998). Telomeres and senescence: The history, the experiment, the future. *Curr. Biol.* **8**, R178–R181.

46. Grisham, J. W., and Thorgeirsson, S. S. (1997). Liver stem cells. *In* "Stem Cells" (C. S. Potter, ed.), pp. 233–282. Academic Press, London.

47. Guillouzo, A. F. M., Langouet, S., Maheo, K., and Rissel, M. (1997). Use of hepatocyte cultures for the hepatotoxic compounds. *J. Hepatol.* **26**, 73–80.

48. Gumucio, J. J., ed. (1989). "Hepatocyte Heterogeneity and Liver Function," Vol. 19. Springer International, Madrid.

49. Guyomard, C., Rialland, L., Fremond, B., Chesne, C., and Guillouzo, A. (1996). Influence of alginate gel entrapment and cryopreservation on survival and xenobiotic metabolism capacity of rat hepatocytes. *Toxicol. Appl. Pharmacol.* **141**, 349–356.

50. Hamilton, G., Westmoreland, C., and George, E. (1996). Liver spheroids as a long-term model for liver toxicity *in vitro*. *Hum. Exp. Toxicol.* **15**, 153.

51. Hamilton, G. A., Barros, S., Jolley, S. L., and LeCluyse, E. L. (1999). Induction of CYP450 enzymes in spheroid cultures of rat and human hepatocytes. *9th North Am. Int. Soc. Stady Xenobiotics Meet.*, Nashville, TN, 1999.

52. Hamilton, G. A., Westmoreland, C., and George, E. (2001). Comparison of different media for the cultures of hepatocytes spheroids and monolayers. *Toxicol. In Vitro* (submitted for publication).

53. Hollands, P. (1997). Comparative stem cell biology. *Int. J. Dev. Biol.* **41**, 245–254.

54. Hu, W. S., Friend, J. R., Wu, F. J., Sielaff, T., Peshwa, M. V., Lazar, A., Nyberg, S. L., Remmel, R. P., and Cerra, F. B. (1997). Development of bioartificial liver employing xenogeneic hepatocytes. *Cytotechnology* **23**, 29–38.

55. Hubbell, J. A. (1995). Biomaterials in tissue engineering. *Bio/Technology* **13**, 565–576.

56. Hughes, R. D., and Williams, R. (1996). Assessment of bioartificial liver support in acute liver failure. *Int. J. Artif. Organs* **19**, 3–6.

57. Ignatius, A. A., and Claes, L. E. (1996). In vitro biocompatibility of bioresorbable polymers: Poly(L,DL-lactide) and poly(L-lactide-*co*-glycolide). *Biomaterials* **17**, 831–839.

58. Ingber, D. E., and Folkman, J. (1989). Mechanochemical switching between growth and differentiation during fibroblast growth factor-stimulated angiogenesis in vitro: Role of extracellular matrix. *J. Cell Biol.* **109**, 317–330.

59. Ingber, D. E., Prusty, D., Sun, Z., Betensky, H., and Wang, N. (1995). Cell shape, cytoskeletal mechanics, and cell cycle control in angiogenesis. *J. Biomech.* **28**, 1471–1484.

60. Jaiswal, R. K., Jaiswal, N., Bruder, S. P., Mbalaviele, G., Marshak, D. R., and Pittenger, M. E. (2000). Adult human mesenchymal stem cell differentiation to the osteogenic or adipogenic lineage is regulated by mitogen-activated protein kinase. *J. Biol. Chem.* **275**, 9645–9652.

61. Jefferson, D. M., Clayton, D. F., Darnell, J. E., Jr., and Reid, L. M. (1984). Posttranscriptional modulation of gene expression in cultured rat hepatocytes. *Mol. Cell. Biol.* **4**, 1929–1934.

62. Jutila, M. A. (1994). Function and regulation of leukocyte homing receptors. *J. Leukocyte Biol.* **55**, 133–140.

63. Kallunki, P., and Tryggvason, K. (1992). Human basement membrane heparan sulfate proteoglycan core protein: A 467-kD protein containing multiple domains resembling elements of the low density lipoprotein receptor, laminin, neural cell adhesion molecules, and epidermal growth factor. *J. Cell Biol.* **116**, 559–571.

64. Kamohara, Y., Rozga, J., and Demetriou, A. A. (1998). Artificial liver: Review and Cedars-Sinai experience. *J. Hepato-Biliary-Pancreatic Surg.* **5**, 273–285.

65. Kennedy, M., Firpo, M., Choi, K., Wall, C., Robertson, S., Kabrun, N., and Keller, G. (1997). A common precursor for primitive erythropoiesis and definitive haematopoiesis. *Nature (London)* **386**, 488–493.

66. Khan, M. A., Healy, M. S., and Bernstein, H. (1992). Low temperature fabrication of protein loaded micropheres. *Proc. Int. Symp. Controlled Relat. Bioact. Mater.* **19**, 518–519.

67. Kim, B. S., Putnam, A. J., Kulik, T. J., and Mooney, D. J. (1998). Optimizing seeding and culture methods to engineer smooth muscle tissue on biodegradable polymer matrices. *Biotechnol. Bioeng.* **57**, 46–54.

68. Kim, C. W., Goldberger, O. A., Gallo, R. L., and Bernfield, M. (1994). Members of the syndecan family of heparan sulfate proteoglycans are expressed in distinct cell-, tissue-, and development-specific patterns. *Mol. Biol. Cell* **5**, 797–805.

69. Kim, S. S., Sundback, C. A., Kaihara, S., Benvenuto, M. S., Kim, B.-S., Mooney, D. J., and Vacanti, J. P. (2000). Dynamic seeding and *in vitro* culture of hepatocytes in a flow perfusion system. *Tissue Eng.* **6**, 39–44.

70. Kim, T. H., Bowen, W. C., Stolz, D. B., Runge, D., Mars, W. M., and Michalopoulos, G. K. (1998). Differential expression and distribution of focal adhesion and cell adhesion molecules in rat hepatocyte differentiation. *Exp. Cell Res.* **244**, 93–104.

71. Kleinman, H. K., McGarvey, M. L., Hassell, J. R., and Martin, G. R. (1983). Formation of a supramolecular complex is involved in the reconstitution of basement membrane components. *Biochemistry* **22**, 4969–4974.

72. Kleinman, H. K., Weeks, B. S., Schnaper, H. W., Kibbey, M. C., Yamamura, K., and Grant, D. S. (1993). The laminins: A family of basement membrane glycoproteins important in cell differentiation and tumor metastases. *Vitam. Horm. (N.Y.)* **47**, 161–186.

73. Knazek, R. A., Guillino, P., Kohler, P. O., and Dedrick, R. L. (1972). Cell culture on artificial capillaries: An approach to tissue growth ex vivo. *Science* **178**, 65–66.

74. Koebe, H. G., Dunn, J. C., Toner, M., Sterling, L. M., Hubel, A., Cravalho, E. G., Yarmush, M. L., and Tompkins, R. G. (1990). A new approach to the cryopreservation of hepatocytes in a sandwich culture configuration. *Cryobiology* **27**, 576–584.

75. Koide, N., Sakaguchi, K., Koide, Y., Asano, K., Kawaguchi, M., Matsushima, H., Takenami, T., Shinji, T., Mori, M., and Tsuji, T. (1990). Formation of multicellular spheroids composed of adult rat hepatocytes in dishes with positively charged surfaces and under other nonadherent environments. *Exp. Cell Res.* **186**, 227–235.

76. Kubota, H., and Reid, L. M. (2000). Clonogenic hepatoblasts, common precursors for hepatocytic and biliary lineages, are lacking classical major histocompatiblity complex class I antigen. *Proc. Natl. Acad. Sci. U.S.A.* **97**, 12132–12137.

77. Kudryavtsev, B. N., Kudryavtseva, M. V., Sakuta, G. A., and Stein, G. I. (1993). Human hepatocyte polyploidization kinetics in the course of life cycle. *Virchows Arch. B* **64**, 387–393.

78. Landry, J., Bernjer, D., Oullet, C., Gayette, R., and Marceau, N. (1985). Spheroidal aggregate culture of rat liver cells: Histotypic reorganization, biomatrix deposition and maintenance of functional activities. *J. Cell Biol.* **101**, 914–923.

79. Landry, J., and Freyer, J. P. (1984). Regulatory mechanisms in spheroidal aggregates of normal and cancerous cells. *Recent Results Cancer Res.* **95**, 50–66.

80. LeCluyse, E. L., Bullock, P. L., and Parkinson, A. (1996). Strategies for restoration and maintenance of normal hepatic structure and function in long-term cultures of rat hepatocytes. *Adv. Drug Deliv. Rev.* **22**, 133–186.

81. LeCluyse, E. L., Madan, A., Hamilton, G., Carroll, K., DeHaan, R., and Parkinson, A. (2000). Expression and regulation of cytochrome P450 enzymes in primary cultures of human hepatocytes. *J. Biochem. Mol. Toxicol.* **14**, 177–185.

82. LeGasse, E., Connors, H., Al-Dhalimym, M., Reitsma, M., Dohse, M., Osborne, L., Wang, X., Finegold, M., Weissman, I. L., and Grompe, M. (2000). Purified hemopoietic stem cells can differentiate into hepatocytes in vivo. *Nat. Med.* **11**, 1229–1234.

83. LePage, E. B., Rozga, J., Rosenthal, P., Watanabe, F., Scott, H. C., Talke, A. M., and Demetriou, A. A. (1994). A bioartificial liver used as a bridge to liver transplantation in a 10-year-old boy. *Am. J. Crit. Care* **3**, 224–227.

84. Li, A. P., Colburn, S. M., and Beck, D. J. (1992). A simplified method for the culturing of primary adult rat and human hepatocytes as multicellular spheroids. *In Vitro Cell. Dev. Biol.* **28A**, 673–677.

85. Li, A. P., Roque, M. A., Beck, D. J., and Kaminski, D. L. (1992). Isolation and culturing of hepatoyctes from human livers. *J. Tissue Cult. Methods* **14**, 139–146.

86. Lian, Z. X., Feng, B., Sugiura, K., Inaba, M., Yu, C. Z., Jin, T. N., Fan, T. X., Cui, Y. Z., Yasumizu, R., Toki, J., Adachi, Y., Hisha, H., and Ikehara, S. (1999). c-*kit*(<low) Pluripotent hemopoietic stem cells form CFU-S on day 16. *Stem Cells* **17**, 39–44.

87. Liebermann, D. A., Hoffman, B., and Steinman, R. A. (1995). Molecular controls of growth arrest and apoptosis: p53-Dependent and independent pathways. *Oncogene* **11**, 199–210.

88. Lin, C., Hou, K. Y., and Zhang, W. X. (1994). Studies of long-term cryopreservation of hepatocytes and their transplantation treating acute hepatic failure in Wistar rats. *Chin. J. Surg.* **32**, 633–635.

89. Lintz, T. J., Parsons, L. M., Hartley, L., Lyons, I., and Harvey, R. P. (1993). Nkx-2.5: A novel murine homeobox gene expressed in early heart progenitor cells and their myogenic descendants. *Development (Cambridge, UK)* **119**, 419–431.

90. Liotta, L. A., Lee, C. W., and Morakis, D. J. (1980). New method for preparing large surfaces of intact human basement membrane for tumor invasion studies. *Cancer Lett.* **11**, 141–152.

91. Lyon, M. (1993). The isolation of membrane proteoglycans. *Methods Mol. Biol.* **19**, 243–251.

92. Lyon, M., Deakin, J. A., and Gallagher, J. T. (1994). Liver heparan sulfate structure. A novel molecular design. *J. Biol. Chem.* **269**, 11208–11215.

93. Lyon, M., and Gallagher, J. T. (1991). Purification and partial characterization of the major cell-associated heparan sulphate proteoglycan of rat liver. *Biochem. J.* **273**, 415–422.

94. Macdonald, J. M., Griffin, J. P., Kubota, H., Griffith, L., Fair, J., and Reid, L. M. (1999). Bioartificial livers. *In* "Cell Encapsulation Technology and Therapeutics" (W. Kuhtreiber, R. P. Lanza, and W. L. Chick, eds.), pp. 252–286. Birkhaeuser, Boston.

95. MacDonald, J. M., Grillo, M., Schmidlin, O., Tajiri, D. T., and James, T. L. (1998). NMR spectroscopy and MRI investigation of a potential bioartificial liver. *NMR Biomed.* **11**, 55–66.

96. Macdonald, J. M., Wolfe, S. P., Chowdhury, I., Kubota, H., and Reid, L. M. (2001). Effect of flow configuration and membrane characteristics on membrane fouling in a novel multi-coaxial hollow fiber bioartificial liver. *Ann. N.Y. Acad. Sci.* (in press).

97. Makgoba, M. W., Sanders, M. E., and Shaw, S. (1989). The CD2-LFA–3 and LFA-1-ICAM pathways: Relevance to T-cell recognition. *Immunol. Today* **10**, 417–422.

98. Maltsev, V. A., Rohwedel, J., Hescheler, J., and Wobus, A. M. (1993). Embryonic stem cells differentiate in vitro into cardiomyocytes representing sinusnodal, atrial and ventricular cell types. *Mech. Dev.* **44**, 41–50.

99. Mancuso, A., Fernandez, E. J., Blanch, H. W., and Clark, D. S. (1990). A nuclear magnetic resonance technique for determining hybridoma cell concentration in hollow fiber bioreactors. *Bio/Technology* **8**, 1282–1285.

100. Mancuso, A., Sharfstein, S. T., Fernandez, E. J., Clark, D. S., and Blanch, H. W. (1998). Effect of extracellular glutamine concentration on primary and secondary metabolism of a murine hybridoma: An in vivo ^{13}C nuclear magnetic resonance study. *Biotechnol. Bioeng.* **57**, 172–186.

101. Marler, J. J., Upton, J., Langer, R., and Vacanti, J. P. (1998). Transplantation of cells in matrices for tissue regeneration. *Adv. Drug Del. Rev.* **33**, 165–182.

102. Martinez-Hernandez, A., and Amenta, P. S. (1993). The hepatic extracellular matrix. I. Components and distribution in normal liver. *Virchows Arch. A* **423**, 1–11.

103. Martinez-Hernandez, A., and Amenta, P. S. (1993). Morphology, localization, and origin of the hepatic extracellular matrix. *In* "Extracellular Matrix: Chemistry, Biology, and Pathobiology with Emphasis on the Liver" (M. Zern and L. M. Reid, eds.). pp. 255–330. Dekker, New York.

104. Martinez-Hernandez, A., and Amenta, P. S. (1995). The extracellular matrix in hepatic regeneration. *FASEB J.* **9**, 1401–1410.

105. Mayani, H., Guilbert, L. J., and Janowska-Wieczorek, A. (1992). Biology of the hemopoietic microenvironment. *Eur. J. Haematol.* **49**, 225–233.

106. Mendiola, M. M., Peters, T., Young, E. W., and Zoloth-Dorfman, L. (1999). Research with human embryonic stem cells: Ethical considerations. *Hastings Cent. Rep.* **29**, 31–36.

107. Mintz, B., and Illmensee, K. (1975). Normal genetically mosaic mice produced from malignant teratocarcinoma cells. *Proc. Natl. Acad. Sci. U.S.A.* **72**, 3585–3589.

108. Mitaka, T., Kojima, T., Mizuguchi, T., and Mochizuki, Y. (1995). Growth and maturation of small hepatocytes isolated from adult rat liver. *Biochem. Biophys. Res. Commun.* **214**, 310–317.

109. Mitaka, T., Mikami, M., Sattler, G. L., Pitot, H. C., and Mochizuki, Y. (1992). Small cell colonies appear in the primary culture of adult rat hepatocytes in the presence of nicotinamide and epidermal growth factor. *Hepatology* **16**, 440–447.

110. Mitaka, T., Mizuguchi, T., Sato, F., Mochizuki, C., and Mochizuki, Y. (1998). Growth and maturation of small hepatocytes. *J. Gastroenterol. Hepatol.* **13**, S70–S77.

111. Mitaka, T., Sato, F., Mizuguchi, T., Yokono, T., and Mochizuki, Y. (1999). Reconstruction of hepatic organoid by rat small hepatocytes and hepatic nonparenchymal cells. *Hepatology* **29**, 111–125.

112. Mooney, D. J., Sano, K., Kaufmann, P. M., Majahod, K., Schloo, B., Vacanti, J. P., and Langer, R. (1997). Long-term engraftment of hepatocytes transplanted on biodegradable polymer sponges. *J. Biomed. Mater. Res.* **37**, 413–420.

113. Morin, G. B. (1995). Is telomerase a universal cancer target? *J. Natl. Cancer Inst.* **87**, 859–861.

114. Morrison, S. J., Shah, N. M., and Anderson, D. J. (1997). Regulatory mechanisms in stem cell biology. *Cell (Cambridge, Mass.)* **88**, 287–298.

115. Moscona, A. (1961). Rotation-mediated histogenic aggregatin of dissociated cells. *Exp. Cell Res.* **22**, 455–475.

116. Mossin, L., Blankson, H., Huitfeldt, H., and Seglen, P. O. (1994). Ploidy-dependent growth and binucleation in cultured rat hepatocytes. *Exp. Cell Res.* **214**, 551–560.

117. Muschel, R., Khoury, G., and Reid, L. M. (1986). Regulation of insulin mRNA abundance and adenylation: Dependence on hormones and matrix substrata. *Mol. Cell. Biol.* **6**, 337–341.

118. Nimni, M. E. (1993). Fibrillar collagens: Their biosynthesis, molecular structure, and mode of assembly. *In* "Extracellular Matrix, Chemistry, Biology, and Pathobiology with Emphasis on the Liver" (M. Zern and L. M. Reid, eds.), pp. 121–148. Dekker, New York.

119. Nozaki, T., Masutani, M., Watanabe, M., Ochiya, T., Hasegawa, F., Nakagama, H., Suzuki, H., and Sugimura, T. (1999). Syncytiotrophoblastic giant cells in teratocarcinoma-like tumors derived from Parp-disrupted mouse embryonic stem cells. *Proc. Natl. Acad. Sci. U.S.A.* **96**, 13345–13350.

120. Nyberg, S. L., Peshwa, M. V., Payne, W. D., Hu, W. S., and Cerra, F. B. (1993). Evolution of the bioartificial liver: The need for randomized clinical trials. *Am. J. Surg.* **166**, 512–521.

121. Nyberg, S. L., Shatford, R. A., Hu, W. S., Payne, W. D., and Cerra, F. B. (1992). Hepatocyte culture systems for artificial liver support: Implications for critical care medicine (bioartificial liver support). *Crit. Care Med.* **20**, 1157–1168.

122. Olack, B. J., Swanson, C. J., Howard, T. K., and Mohanakumar, T. (1999). Improved method for the isolation and purification of human islets of Langerhans using Liberase enzyme blend. *Hum. Immunol.* **60**, 1303–1309.

123. O'Leary, D. J., Hawkes, S. P., and Wade, C. G. (1987). Indirect monitoring of carbon-13 metabolism with NMR: Analysis of perfusate with a closed-loop flow system. *Magn. Reson. Med.* **5**, 572–577.

124. Overturf, K., Al-Dhalimy, M., Finegold, M., and Grompe, M. (1999). The repopulation potential of hepatocyte populations differing in size and prior mitotic expansion. *Am. J. Pathol.* **155**, 2135–2143.

125. Papa, S., Capitani, S., Matteucci, A., Vitale, M., Santi, P., Martelli, A. M., Maraldi, N. M., and Manzoli, F. A. (1987). Flow cytometric analysis of isolated rat liver nuclei during growth. *Cytometry* **8**, 595–601.

126. Pedersen, R. A. (1999). Embryonic stem cells for medicine. *Sci. Am.* **280**, 68–73.

127. Petersen, B. E., Bowen, W. C., Patrene, K. D., Mars, W. M., Sullivan, A. K., Murase, N., Boggs, S. S., Greenberger, J. S., and Goff, J. P. (1999). Bone marrow as a potential source of hepatic oval cells. *Science* **284**, 1168–1170.

128. Pittenger, M. E., Mackay, A. M., Beck, S. C., Jaiswal, R. K., Douglas, R., Mosca, J. D., Moorman, M. A., Simonetti, D. W., Craig, S., and Marshak, D. R. (1999). Multilineage potential of adult human mesenchymal stem cells. *Science* **284**, 143–147.

129. Potten, C. S., ed. (1997). "Stem Cells." Academic Press, London.

130. Rao, U. N., Sonmez-Alpan, E., and Michalopoulos, G. K. (1997). Hepatocyte growth factor and c-MET in benign and malignant peripheral nerve sheath tumors. *Hum. Pathol.* **28**, 1066–1070.

131. Reid, L. M. (1990). Defining hormone and matrix requirements for differentiated epithelia. *In* "Basic Cell Culture Protocols" (J. W. Pollard and J. M. Walker, eds.), Vol. 75, pp. 237–262. Humana Press, Totowa, NJ.

132. Reid, L. M. (1990). Stem cell biology, hormone/matrix synergies and liver differentiation. *Curr. Opin. Cell Biol.* **2**, 121–130.

133. Reid, L. M. (1996). Stem cell-fed maturational lineages and gradients in signals: Relevance to differentiation of epithelia. *Mol. Biol. Rep.* **23**, 21–33.

134. Reid, L. M., Fiorino, A. S., Sigal, S. H., Brill, S., and Holst, P. A. (1992). Extracellular matrix gradients in the space of Disse: Relevance to liver biology. *Hepatology* **15**, 1198–1203.

135. Reid, L. M., Gaitmaitan, Z., Arias, I., Ponce, P., and Rojkind, M. (1980). Long-term cultures of normal rat hepatocytes on liver biomatrix. *Ann. N.Y. Acad. Sci.* **349**, 70–76.

136. Reid, L. M., and Jefferson, D. M. (1984). Culturing hepatocytes and other differentiated cells. *Hepatology* **4**, 548–559.

137. Reid, L. M., and Zern, M. A. (1993). "Extracellular Matrix Chemistry and Biology." Marcel Dekker, NY.

138. Reinholt, F. P., Hultenby, K., Tibell, A., Korsgren, O., and Groth, C. G. (1998). Survival of fetal porcine pancreatic islet tissue transplanted to a diabetic patient. *Xenotransplantation* **5**, 222–225.

139. Resnick, J. L., Bixler, L. S., Cheng, L., and Donovan, P. J. (1992). Long-term proliferation of mouse primordial germ cells in culture. *Nature (London)* **359**, 550–551.

140. Rojkind, M., Gatmaitan, Z., Mackensen, S., Giambrone, M. A., Ponce, P., and Reid, L. M. (1980). Connective tissue biomatrix: Its isolation and utilization for long-term cultures of normal rat hepatocytes. *J. Cell Biol.* **87**, 255–263.

141. Rosenberg, E., Spray, D. C., and Reid, L. M. (1992). Transcriptional and posttranscriptional control of connexin mRNAs in periportal and pericentral rat hepatocytes. *Eur. J. Cell Biol.* **59**, 21–26.

142. Rozga, J., Williams, F., Ro, M. S., Neuzil, D. F., Giorgio, T. D., Backfisch, G., Moscioni, A. D., Hakim, R., and Demetriou, A. A. (1993). Development of a bioartificial liver: Properties and function of a hollow-fiber module inoculated with liver cells. *Hepatology* **17**, 258–265.

143. Saeter, G., Lee, C. Z., Schwarze, P. E., Ous, S., Chen, D. S., Sung, J. L., and Seglen, P. O. (1988). Changes in ploidy distributions in human liver carcinogenesis. *J. Natl. Cancer Inst.* **80**, 1480–1485.

144. Sandgren, E. P., Luetteke, N. C., Qiu, T. H., Palmiter, R. D., Brinster, R. L., and Lee, D. C. (1993). Transforming growth factor alpha dramatically enhances oncogene-induced carcinogenesis in transgenic mouse pancreas and liver. *Mol. Cell. Biol.* **13**, 320–330.

145. Santella, L. (1998). The role of calcium in the cell cycle: Facts and hypotheses. *Biochem. Biophys. Res. Commun.* **244**, 317–324.

146. Sanz, N., Diez-Fernandez, C., Alvarez, A., and Cascales, M. (1997). Age-dependent modifications in rat hepatocyte antioxidant defense systems. *J. Hepatol.* **27**, 525–534.

147. Sarkar, G., and Bolander, M. E. (1995). Telomeres, telomerase, and cancer. *Science* **268**, 1115–1117.

148. Sato, G. H. (1984). Hormonally defined media and long-term marrow culture: General principles. *Kroc Found. Ser.* **18**, 133–137.

149. Schuldiner, M., Yanuka, O., Itskovitz-Eldor, J., Melton, D. A., and Benvenisty, N. (2000). Effects of eight growth factors on the differentiation of cells derived from human embryonic stem cells. *Proc. Natl. Acad. Sci. U.S.A.* **97**, 11307–11312.

150. Seglen, P. O. (1976). Preparation of isolated rat liver cells. *Methods Cell Biol.* **13**, 29–83.

151. Seglen, P. O., Schwarze, P. E., and Saeter, G. (1986). Changes in cellular ploidy and autophagic responsiveness during rat liver carcinogenesis. *Toxicol. Pathol.* **14**, 342–348.

152. Sell, S. (1993). Cellular origin of cancer: Dedifferentiation or stem cell maturation arrest? *Environ. Health Perspect.* **101**, 15–26.

153. Sell, S., and Pierce, G. B. (1994). Maturation arrest of stem cell differentiation is a common pathway for the cellular origin of teratocarcinomas and epithelial cancers. *Lab. Invest.* **70**, 6–22.

154. Shimbara, N., Atawa, R., Takashina, M., Tanaka, K., and Ichihara, A. (1996). Long-term culture of functional hepatocytes on chemically modified collagen gels. *Cytotechnology* **21**, 31–43.

155. Shumakov, V. I., Bliumkin, V. N., Ignatenko, S. N., Skaletskii, N. N., and Kauricheva, N. I. (1983). Transplantation of cultures of human fetal pancreatic islet cells to diabetes mellitus patients. *Klin. Med.* **61**, 46–51.

156. Sigal, S. H., Brill, S., Fiorino, A. S., and Reid, L. M. (1992). The liver as a stem cell and lineage system. *Am. J. Physiol.* **263**, G139–G148.

157. Sigal, S. H., Brill, S., Reid, L. M., Zvibel, I., Gupta, S., Hixson, D., Faris, R., and Holst, P. A. (1994). Characterization and enrichment of fetal rat hepatoblasts by immunoadsorption ("panning") and fluorescence-activated cell sorting. *Hepatology* **19**, 999–1006.

158. Sigal, S. H., Gupta, S., Gebhard, D. F., Jr., Holst, P., Neufeld, D., and Reid, L. M. (1995). Evidence for a terminal differentiation process in the rat liver. *Differentiation (Berlin)* **59**, 35–42.

159. Sigal, S. H., Rajvanshi, P., Gorla, G. R., Sokhi, R. P., Saxena, R., Gebhard, D. R., Jr., Reid, L. M., and Gupta, S. (1999). Partial hepatectomy-induced polyploidy attenuates hepatocyte replication and activates cell aging events. *Am. J. Physiol.* **276**, G1260–G1272.

160. Smith, M. D., Airdrie, I., Cousins, R. B., Ekevall, E., Grant, M. H., and Gaylor, J. D. S. (1997). Development and characterization of a hybrid artificial liver bioreactor with integral membrane oxygenation. *In* "Bioartificial Liver Support Systems" (G. Crepaldi, A. A. Demetriou, and M. Muraca, eds.), pp. 27–35. CIC Edizione Internazionale, Rome.

161. Soderberg, L. S., Barnett, J. B., Baker, M. L., Chang, L. W., Salari, H., and Sorenson, J. R. (1988). Copper(II)2(3,5-diisopropylsalicylate)4 stimulates hemopoiesis in normal and irradiated mice. *Exp. Hematol.* **16**, 577–580.

162. Spray, D. C., Fujita, M., Saez, J. C., Choi, H., Watanabe, T., Hertzberg, E., Rosenberg, L. C., and Reid, L. M. (1987). Proteoglycans and glycosaminoglycans induce gap junction synthesis and function in primary liver cultures. *J. Cell Biol.* **105**, 541–551.

163. Stamatoglou, S. C., and Hughes, R. C. (1994). Cell adhesion molecules in liver function and pattern formation. *FASEB J.* **8**, 420–427.

164. Stein, G. I., and Kudryavtsev, B. N. (1992). A method for investigating hepatocyte polyploidization kinetics during postnatal development in mammals. *J. Theor. Biol.* **156**, 349–363.

165. Strom, S. C., Pisarov, L. A., Dorko, K., Thompson, M. T., Schuetz, J. D., and Schuetz, E. G. (1996). Use of human hepatocytes to study CYP450 gene induction. *Methods Enzymol.* **272**, 388–401.

166. Sussman, N. L., Gislason, G. T., Conlin, C. A., and Kelly, J. H. (1994). The Hepatix extracorporeal liver assist device: Initial clinical experience. *Artif. Organs* **18**, 390–396.

167. Swales, N. J., Luong, C., and Caldwell, J. (1996). Cryopreservation of rat and mouse hepatocytes. I. Comparative viability studies. *Drug Metabol. Dispos.* **24**, 1218–1223.

168. Tateno, C., and Yoshizato, K. (1996). Growth and differentiation in culture of clonogenic hepatocytes that express both phenotypes of hepatocytes and biliary epithelial cells. *Am. J. Pathol.* **149**, 1593–1605.

169. Taub, M., Wang, Y., Szczesny, T. M., and Kleinman, H. K. (1990). Epidermal growth factor or transforming growth factor alpha is required for kidney tubulogenesis in Matrigel cultures in serum-free medium. *Proc. Natl. Acad. Sci. U.S.A.* **87**, 4002–4006.

170. Temple, S., and Alvarez-Buylla, A. (1999). Stem cells in the adult mammalian central nervous system. *Curr. Opin. Neurobiol.* **9**, 135–141.

171. Thomson, J. A., Itskovitz-Eldor, J., Shapiro, S. S., Waknitz, M. A., Swiergiel, J. J., Marshall, V. S., and Jones, J. M. (1998). Embryonic stem cell lines derived from human blastocysts. *Science* **282**, 1145–1146.

172. Thomson, J. A., Itskovitz-Eldor, J., Shapiro, S. S., Waknitz, M. A., Swiergiel, J. J., Marshall, V. S., and Jones, J. M. (1998). Embryonic stem cell lines derived from human blastocysts. *Science* **282**, 1145–1147; erratum: *Ibid.*, p. 1827.

173. Traber, P. G., Chianale, J., and Gumucio, J. J. (1988). Physiologic significance and regulation of hepatocellular heterogeneity. *Gastroenterology* **95**, 1130–1143.

174. Watanabe, T., and Tanaka, Y. (1982). Age-related alterations in the size of human hepatocytes. A study of mononuclear and binucleate cells. *Virchows Arch. B* **39**, 9–20.

175. Watts, P., and Grant, M. H. (1996). Cryopreservation of rat hepatocyte monolayer cultures. *Hum. Exp. Toxicol.* **15**, 30–37.

176. Whitaker, N. J., Bryan, T. M., Bonnefin, P., Chang, A. C., Musgrove, E. A., Braithwaite, A. W., and Reddel, R. R. (1995). Involvement of RB-1, p53, p16INK4 and telomerase in immortalisation of human cells. *Oncogene* **11**, 971–976.

177. Wolf, C., and Munkelt, B. (1975). Bilirubin conjugation by an artificial liver composed of cultured cells and synthetic capillaries. *Trans. Am. Soc. Artif. Intern. Organs* **21**, 16–27.

178. Wolfe, S. P., Hsu, E., Reid, L. M., and Macdonald, J. M. (2001). A novel multi-coaxial tubular bioreactor for adherent cell types. Part 1: Hydrodynamic studies. *Biotechnol. Bioeng.* (in press).

179. Wu, F. J., Friend, J. R., Hsiao, C. C., Zilliox, M. J., Cerra, F. B., and Hu, W. (1996). Efficient assembly of rat hepatocyte spheroids for tissue engineering applications. *Biotechnol. Bioeng.* **50**, 404–415.

180. Xu, A., Luntz, T., Macdonald, J., Kubota, H., Hsu, E., London, R., and Reid, L. M. (2000). Liver stem cells and lineage biology. *In* "Principles of Tissue Engineering" (R. Lanza, R. Langer, and J. Vacanti, eds.), 2nd ed, pp. 559–598. Academic Press, San Diego, CA.

181. Yagi, K., Tsuda, K., Serada, M., Yamada, C., Kondoh, A., and Miura, Y. (1993). Rapid formation of multicellular spheroids of adult rat hepatocytes by rotation culture and their immobilization within calcium alginate. *Artif. Organs* **17**, 929–934.

182. Yamada, K., Kamihira, M., Hamamoto, R., and Iijima, S. (1998). Efficient induction of hepatocyte spheroids in a suspension culture using a water-soluble synthetic polymer as an artificial matrix. *J. Biochem. (Tokyo)* **123**, 1017–1023.

183. Young, H. E., Mancini, M. L., Wright, R. P., Smith, J. C., Back, A. C., Jr., Reagan, C. R., and Lucas, P. A. (1995). Mesenchymal stem cells reside within the connective tissues of many organs. *Dev. Dyn.* **202**, 137–144.

184. Zidar, B. L., Shadduck, R. K., Zeigler, Z., and Winkelstein, A. (1977). Observations on the anemia and neutropenia of human copper deficiency. *Am. J. Hematol.* **3**, 177–185.

185. Zvibel, I., Fiorino, A. S., Brill, S., and Reid, L. M. (1998). Phenotypic characterization of rat hepatoma cell lines and lineage-specific regulation of gene expression by differentiation agents. *Differentiation (Berlin)* **63**, 215–223.

186. Zvibel, I., Halay, E., and Reid, L. M. (1991). Heparin and hormonal regulation of mRNA synthesis and abundance of autocrine growth factors: Relevance to clonal growth of tumors. *Mol. Cell. Biol.* **11**, 108–116.

CHAPTER 12

EPITHELIAL CELL CULTURE: PANCREATIC ISLETS

Thierry Berney, Antonello Pileggi, R. Damaris Molano, and Camillo Ricordi

INTRODUCTION

Islet of Langerhans transplantation is gaining increasing attention and support as a viable strategy to cure type 1 diabetes [1]. An islet transplantation procedure involves a complex sequence of organ procurement, pancreas digestion, purification of the endocrine tissue, and quality control of the islet preparation before transplant can actually take place. Endocrine tissue is dispersed throughout the exocrine pancreas as clusters representing about 1% of the total mass of the gland, so obtaining islet preparations with 50–90% purity in islets implies the need for a 50- to 90-fold enrichment. A mechanically enhanced enzymatic digestion process allows the release of the islets from their surrounding connective and exocrine tissue without destroying their integrity. The digested pancreas is then subjected to a purification process through density gradients, designed to separate islets from contaminating nonendocrine pancreatic cells. Tissue culture or cryopreservation of islets of Langerhans can provide a time lag between isolation and transplantation, allowing the application of immunomodulatory strategies to the graft or to the recipient. Pretransplant preservation of the islet tissue also allows completion of sterility and viability testing on aliquots from the final islet preparation before infusion in the selected recipients.

This chapter reviews the various stages of islet cell manipulation, from the isolation process to tissue culture of purified islet preparations, in the human and in animal models. We also briefly outline the potential of islet culture for *in vitro* differentiation and proliferation.

PRINCIPLES OF ISLET ISOLATION AND CULTURE

ISLET ISOLATION

Several parameters influence the course and success of pancreatic islet isolation. The quality and quantity of islets obtained from a pancreas depend on events occurring before, during, and after pancreas procurement. Prolonged periods of circulatory arrest before organ procurement result in warm ischemic insult with detrimental consequences on islet yield [2,3]. As much care must be taken during pancreas procurement for islet isolation as for whole-organ transplant, and cold ischemia time must be kept as short as possible.

Islet isolation is performed by enzymatic digestion (collagenase) of the pancreas after intraductal distension of the organ, a technique based on the pioneering works of Moskalewski [4] and Lacy [5] in rodents, and Horaguchi in dogs [6]. This method was shown to be efficient in obtaining high islet yields from all mammal species, including the human [7–12]. The distention can be obtained by manual injection of the collagenase with a syringe [13,14] or by connecting the catheter to a peristaltic pump to apply a constant pressure [15,16].

Commercial collagenases are enzyme mixtures, including metalloenzymes that are able to hydrolyze the collagen triple helix. Reproducibility problems in the technique of crude

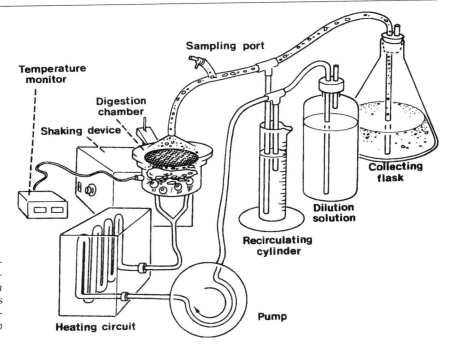

Fig. 12.1. The automated method for islet isolation: schematic representation of the isolation setup. The pancreas is dispersed during a continuous digestion process. The continuous flow allows the islets that are progressively released to be saved from further digestion. From Ricordi and Rastellini [34].

collagenase biological production lead to high lot-to-lot variability in collagenolytic activities [17,18]. The recent availability of highly purified collagenase or new enzyme blends, such as Liberase (Roche-Boehringer-Mannheim, Indianapolis, IN), characterized by reduced lot-to-lot variability, low endotoxin contamination, and better selectivity for exocrine tissue, has allowed improvements in yields and reproducibility of islet isolation, as well as graft function [19–25].

The introduction of the automated method for islet isolation in the late 1980s was a major breakthrough for the feasibility of large-scale islet transplantation in the clinical setting [13]. The procedure is relatively simple and reproducible in all animal species tested-human [13,26], nonhuman primate [27], dog [28], sheep, swine [29,30], rabbit, hamster, rat, mouse [31]. The essential component is a digestion chamber that retains the pancreas during a continuous enzymatic digestion process [13,32,33]. The chamber (250–700 ml for large mammals, 50 ml for rodents) consists of a lower cylinder and an upper conical portion separated by a removable stainless steel screen (400–500 µm for large animal isolation, 230–280 µm for rodent isolation), and contains 1 cm diameter solid glass marbles. The lower portion has one or two inlet ports and an opening for a temperature sensor. The upper portion of the chamber has an outlet at the apex. After intraductal distention, the gland is fully immersed in the chamber with the enzyme solution, the absence of an air–fluid interface reduces the mechanical stress on the pancreatic tissue. The screen retains the gland in the lower portion of the digestive chamber, and the marbles enhance the mechanical disruption during controlled amplitude and frequency shaking. The fluid is circulated through a system of tubing to a heating circuit by means of a peristaltic pump (Fig. 12.1) [34]. Enzymatic and mechanical (shaking and flow) actions allow progressive release of particles from the digesting pancreas, which are continuously removed and preserved from further digestion. The frequency and amplitude of shaking, which must be adapted for each animal species, is reduced to a minimum for the isolation of fragile islets, as in the case of islet isolation from adult porcine pancreata [31].

ISLET PURIFICATION

Islet purification addresses the need to physically separate the islets from the nonendocrine tissue (exocrine fragments, lymph nodes, ductal and vascular tissue) that represents about 99% of the pancreas. The purification method determines the mass and viability of the islets, as well as their purity and immunogenicity [35–37]. Several methods of purifica-

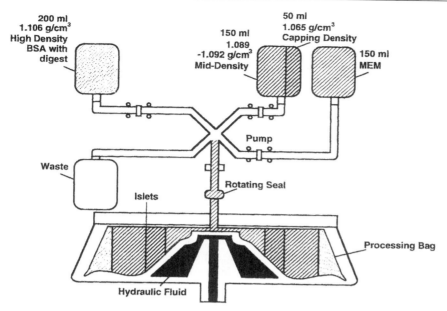

Fig. 12.2. Diagrammatic representation of a discontinuous gradient on the COBE 2991 cell processor. From London et al. [47].

tion have been described [38–45], but separation by density gradient centrifugation is the most widely utilized [5,46,47].

Islets are separated from acinar tissue by isopycnic centrifugation (separation according to differences in density), because the overlapping in size of the islets and exocrine fragments prevents the use of isokinetic centrifugation (separation according to velocity of sedimentation). Isopycnic separation is achieved by centrifuging cells on density gradients long enough for them to reach their gradient of equal density. Acinar cells have densities much higher than other cell types, which renders them suitable for this method [48]. Unfortunately, the swelling and edema of the acinar tissue consecutive to the isolation procedure somewhat lowers its density, leading to some overlapping with that of endocrine tissue, precluding high degrees of purity [47,49,50]. Isokinetic centrifugation after isopycnic separation has been suggested to improve the purity of the islet preparation [37,51,52].

The purification process can be conducted either manually or by means of a semi-automated method, utilizing the COBE 2991 computerized centrifuge system (Fig. 12.2). The COBE 2991 cell processor consists of a centrifuge bowl that bears a doughnut-shaped processing bag in which the pellet sediments to the outside and the lower density layers to the center. The cell processing bag is connected to five input/output tubes through which the tissue to be separated and the gradients are loaded, and the fractions can be collected from top to bottom, under controlled conditions and without deceleration. The introduction of the semiautomated purification method simplified the procedure, and reduced its duration, and optimized islet yields and purity [21,28,53–55]. The pancreatic digest can be loaded on top of the gradient layers, resuspended in preservation solution, or at the bottom, re-suspended in the heaviest density gradient. Advantages of top loading are that the digest is kept in physiological medium for the longest possible time, and centrifugal forces are minimized. Disadvantages include increased cell aggregation, where acinar tissue migrating to the denser gradient layers might drag down islets from the upper interfaces [56,57]. To overcome problems resulting from such migration bottom loading is usually preferred.

Avoiding cellular swelling and edema by using preservation/hyperosmolar solutions before and during the purification process can significantly improve its efficiency [13,58,59]. Discontinuous Ficoll gradients prepared in Euro-Collins solution (Euro-Ficoll) are the most largely utilized for islet separation [13,58,60–65]. Disadvantages of discontinuous gradients are the lower effective cell load [63] and the accumulation of cells at interfaces, which interferes with movement to other levels and may lead to cell aggregation [43,64]. The use of continuous density gradients has been reported to lead to higher yields for large-scale purification with the semiautomated method [25,65].

ASSESSMENT OF THE ISLET PREPARATION

Standardized methods and criteria are essential for the accurately assessment of the isolated islet preparation both quantitatively and qualitatively. This is especially true when one is dealing with islets isolated from humans or large mammals, in which several hundred thousand islets can be recovered. The methods still currently used are the results of a consensus workshop held in 1989 in Minneapolis [32] and are detailed in the subsections that follow.

Islet Number, Mass, and Purity

The determination of the islet mass transplanted into a diabetic recipient is of critical importance. Indeed, it was identified as a major determinant for islet graft function and insulin independence [66]. Islets are not always easy to distinguish from exocrine tissue, lymph nodes, or so-called membrane balls, especially when the degree of purity of the preparation is low. However, they can be easily identified by staining with dithizone (DTZ), a molecule that stains zinc in the insulin granules and gives the islets (Fig. 12.3) a distinctive red color (not shown) [67]. To evaluate the total islet mass available for transplantation, sampled islets are counted according to size, in increments of 50 μm, and their number is converted via an algorithm into an islet equivalent number (IEN), which represents the number of islets in the preparation if all had a diameter of 150 μm. The sampling technique is a critical step in the quantification of the islet preparation, and multiple samples should be collected to minimize counting errors. A proficient observer can roughly estimate the purity of the DTZ-stained preparation, but purity can be better calculated following immunohistochemical stainings of an aliquot from the final preparation. An estimate can also be obtained from the ratio of the calculated total IEN to the measured volume of the preparation pellet.

Islet Viability: Morphological Assessment

Quantitative assays of pancreatic islet cell viability, including measurements of insulin biosythesis, insulin secretion, or respiratory quotient, have been used in studies of basic islet physiology [32]. These are generally cumbersome tests and cannot be performed as rapid assays prior to transplantation. The simultaneous use of fluorometric inclusion and exclusion dyes can quantitate the proportion of viable and damaged cells and can be used as an instant estimate of islet viability. Two dyes extensively studied in this context are acridine orange, a weak base able to penetrate living cells and bind to nucleic acids, causing a green fluorescence, and propidium iodide, an exclusion dye that cannot penetrate living cells but binds to the nucleic acids of dead cells, provoking a bright red fluorescence [68].

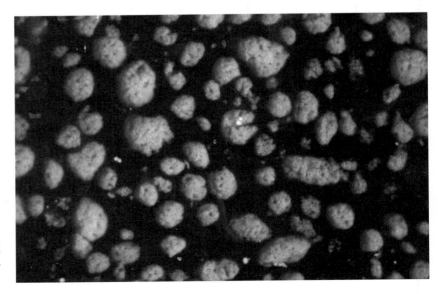

Fig. 12.3. Sample from a human islet isolation procedure. The human islets (lighter) are easily distinguished from the nonendocrine tissue by dithizone staining.

In Vitro Endocrine Function

Islet function *in vitro* is widely assessed by testing glucose-stimulated insulin release. In static incubation methods, islets are incubated sequentially in three periods of 1 h in solutions containing low, high, and low glucose concentrations. Insulin concentration is measured in the supernatants at the end of each incubation period. Islet function is assessed by the ability to both increase insulin release in response to the hyperglycemic challenge and to down-regulate insulin secretion after restoration of nonstimulatory glucose concentrations. Low and high glucose concentrations are typically 1.67 and 16.7 mM, respectively, but these have varied depending on the protocols [69–71]. A preliminary period of tissue culture will stabilize basal insulin release prior to performance of the assay [5,72]. Calculation of the ratio of stimulated over basal insulin release as a stimulation index gives a measure of the secretory capacity. It also allows comparison of function of preparations with different islet sizes, since the quantity of insulin secreted is a function of islet mass [73].

The islet perifusion method [74] provides a dynamic picture of islet endocrine function but is more cumbersome than static incubation. In the former method, islets placed in Millipore chambers are perfused continuously (generally at a flow rate of 1 ml/min) with the same low, high, and low glucose-containing solutions just described, for three consecutive periods of 1 h. Insulin concentration is measured in the effluent at selected intervals to detect both first- and second-phase insulin release.

Data from static incubation and perifusion are useful indicators of the endocrine function of isolated islets, but *in vivo* function cannot always be directly extrapolated from *in vitro* data. For example, in an experiment comparing function of cryopreserved rodent islets after slow or rapid thawing, islet preparation had similar perifusion responses, but totally different *in vivo* function in terms of diabetes reversal after transplantation [74]. Another study investigating *in vitro* glucose sensitivity of human islets has shown that a reduction of glucose concentration in culture media simultaneously decreased the stimulation index, while promoting islet viability [71].

In Vivo Endocrine Function

In rodents, *in vivo* islet function is best assessed in syngeneic models, where confounding variables of allorejection and recurrence of autoimmunity are absent. For human islets, or islets from large animals, a standard method for determination of *in vivo* viability and function is to transplant them under the kidney capsule of nude mice. These mice are immunodeficient owing to congenital thymic aplasia and are unable to reject allo- or xenogeneic tissue. Nude mice can be rendered diabetic by intravenous injection of 185–225 mg/kg streptozotocin a few days prior to transplantation. Viable and functionally competent islets will provide long-term reversal of hyperglycemia following renal subcapsular transplantation in diabetic nude mice [75–77]. In addition to the mere observation of diabetes reversal, intraperitoneal glucose tolerance tests (IPGTT) can be performed by injecting 1 g/kg body weight glucose intraperitoneally into transplanted animals, and measuring the human C-peptide response at various time points (0, 30, 120 min) [77].

A sufficient mass of adult islets can reverse diabetes in nude recipients within 24–72 h of transplantation of fresh islets. Cultured and cryopreserved islets may require 1–2 weeks, whereas 6–10 weeks could be necessary for fetal or porcine neonatal islet-like cell clusters (NICCs) to be functional *in vivo* following transplantation [78,79].

ISLET CULTURE
Islet Tissue Culture

In human clinical islet transplantation, the current favored approach is a short-term (12–48 h), low-temperature (21–24°C) culture prior to the transplant procedure. The purpose of this short-term culture has been principally related to elimination of damaged cells before infusion in the recipient to minimize nonspecific inflammatory reactions at the site of islet transplantation. This short-term period of culture would be, in fact, insufficient to deplete islets from resident passenger leukocytes, as proposed in experimental models [80–83]. Seven-day culture at low temperature alone was required to prolong islet graft survival in models of rat or canine allograft, or even across a xenogeneic human-to-mouse barrier [80–82]. Near-total depletion of rat islets from cells expressing major histocompatibility complex (MHC) class II was demonstrated after a 7-day culture at 24°C [83]. Therefore, pretransplant short-term culture is generally performed to allow islets to recover from the stress endured during the isolation/purification procedure, and to increase the viability and purity of the islet preparation because of the loss of nonviable islets and preferential survival of endocrine over exocrine tissue in culture [84]. In addition, prolonged islet culture might ultimately result in decreased responsiveness to a glucose challenge and in endocrine tissue loss [85,86]. Central necrosis of the cultured islet is a well-known phenomenon [86,87], and apoptotic islet cell death during culture has been demonstrated [88,89]. Factors detrimental to islet survival and function *in vitro* include nitric oxide production [90,91], noxious effects of heterologous proteins carried over by the serum added to the culture media [91], and loss of cellular interactions with components of the normal islet microenvironment within the pancreas [87,89].

Definition of a culture medium optimal and specific for islet culture has met with several difficulties linked to the particularities of the media. In contrast with usual cell cultures, where proliferation rate can be used as an index of medium suitability, islets do not proliferate, and media must be appraised with other parameters, such as long-term viability and functional assays. As a result, islets are still cultured in commercially available media designed for culture of other, simpler cell types, and likely to be suboptimal for the culture of these functionally active miniorgans with a high metabolic rate [92]. Currently, CMRL-1066, supplemented with 10% fetal bovine serum (FBS), antibiotics, and L-glutamine, is the favored culture medium in centers involved in clinical islet transplantation [25,93–95].

The limited study of various tissue culture media in rodents in the late 1970s led to the conclusion that islets should be cultured free-floating, with RPMI 1640 supplemented with 10% FBS as the medium of choice [97]. These culture conditions were extrapolated to human islets [97]. Since then, new studies have compared several commercially available culture media with regard to their ability to maintain islet viability and *in vitro* function in the human and in various animal species. All these studies were conducted with media containing 10% FBS. For human islets, CMRL 1066 was demonstrated to be the medium of choice, as indicated by studies of morphological assessment, microfluorometric membrane integrity assay, and stimulation index in a glucose perifusion assay [98]. The advantage of CMRL 1066 may result from its high contents in nucleotide precursors and ascorbic acid, which are known to increase the insulin secretory capacity. The optimal culture medium might be species dependent, since the optimal insulin responses have been reported with Ham's F-10 or F-12, or TCM 199 media for adult porcine islets [98,99], with Iscove's MEM for rat islets [98], and with CMRL 1066 or TCM 199 for bovine islets [100].

Glucose concentration in the culture medium is at the center of a delicate balance between necessary source of energy and islet toxicity. It should be in the range of 5.6–10 mmol/liter (100–180 mg/dl) [71,101,102]. At lower concentrations, islets may lack fuel to sustain their metabolism, which in turn affects their function and survival [71,101]. At higher concentrations, no increase in the islet energy status is observed, and decreased viability and loss of glucose regulation have been reported [71,102]. In this regard, the D-glucose concentration of CMRL 1066 (100 mg/dl) appears to be optimal.

The role of serum in the medium is another issue. Absence of serum was shown to impair glucose-stimulated insulin release, and to reduce insulin biosynthesis and islet cell survival [96,101,103]. On the other hand, observations that homologous or autologous serum prolonged function and survival of cultured human or porcine islets suggests that heterologous proteins may contribute to the *in vitro* islet loss [104,105]. Additionally, the carryover of xenogeneic proteins of FBS origin with the islets might account for increased antigenicity of the graft. Therefore, there have been several attempts to develop defined serum-free media, supplemented with several components aiming to replace essential nutrients contained in the serum, with improved results in terms of viability and function [86,91,106,107]. Compounds used include bovine or human serum albumin (BSA 1%, HSA 0.1%), hormones or growth factors (insulin-like growth factors I and II, human growth hormone, glucagon), scavenging agents (sodium pyruvate, 6 mmol/liter) and secretagogues such as isobutylmethylxanthine (IBMX, 10–50 μmol/liter) [86,87,91,99,101,106].

Neonatal and fetal porcine islet-like cell clusters need to be cultured in the short term (9 days–3 weeks) before they can aggregate and gain a fully differentiated islet morphology, with a significant proportion of cells differentiating into endocrine cells able to exhibit an appropriate glucose-stimulated insulin secretory response. Optimal *in vitro* maturation is obtained in serum-free medium containing nicotinamide and IBMX [78,79].

Islet Cryopreservation

Islet banking by means of cryopreservation could increase the number of islets available for transplantation at subsequent times. Cryopreserved islets have been used to complement freshly isolated islets to increase the islet transplant mass [108]. Several additional advantages of cryopreservation include selective destruction of exocrine contaminants, reduced immunogenicity by down-regulation of MHC class I expression, and the opportunity to modulate the recipient immune system over an indefinite period of time. However, studies have reported impaired function and reduced *in vivo* viability of islets cryopreserved in DMSO according to established protocols [109]. Additionally, improved clinical results thanks to refined peritransplant management have been recently obtained [110], and supplementation of a fresh islet transplant with cryopreserved islets is no longer considered in the majority of clinical islet transplant centers. The possibility of using cryopreservation to store islets may still represent a useful tool for treatment of the selected recipient by immunomodulatory strategies while the islets are preserved. Islet cryopreservation methods have been extensively described elsewhere [111].

Ex Vivo Proliferation of β Cells

Successful *ex vivo* proliferation of islet β cells would provide an ideal tool for unlimited expansion of the insulin-producing tissue supply. Expansion of differentiated adult β cells has been attempted, and 30,000-fold increases in endocrine cell numbers have been reported. Unfortunately, islet cells tend to lose insulin expression after several passages [112–114].

Another extremely promising approach results from the study of nesidioblastosis, a condition in which neogenesis of endocrine islets occurs by "budding" from the pancreatic ductal epithelium. This phenomenon can be experimentally reproduced in the hamster model by cellophane wrapping of the pancreas [115–117]. Reproduction of human nesidioblastosis *in vitro* has been attempted with partial success [116]. Islet neogenesis-associated protein (INGAP)) is the product of a novel gene recently cloned from hamster pancreata undergoing neogenesis, and present in ilotropin, a protein extract able to induce islet neogenesis from ductal cells [117]. These considerations suggest that pluripotent pancreatic stem cells located in the ductal epithelium are able to differentiate into endocrine cells upon adequate stimulation. A recent study reports *in vitro* differentiation of long-term cultured pancreatic ductal epithelial cells into fully functional islets, able to reverse autoimmune diabetes in NOD mice after syngeneic transplantation [118].

PROTOCOLS
ISLET ISOLATION
Automated Procedure

The automated isolation procedure described herein is the one in use for human islet isolation at our center. It can be applied to most animal species, with minor modifications detailed in Table 12.1.

The digestion chamber is heated to 42°C by recirculating warm Hanks Balanced Salt Solution (HBSS). Pancreas is kept on ice-cold EC and cleaned from fat, lymph nodes, and vessels. The gland is divided in two portions (head, and body–tail) by cutting it at the neck. The pancreatic duct of each portion is cannulated with an 18-G catheter. Prewarmed (30°C) HBSS containing 1.4 mg/ml Liberase HI (Roche-Boehringer-Mannheim, Indianapolis, IN) is used obtain retrograde distension of body and tail and anterograde distension of the head. An appropriate distension of the organ is critical for successful islet isolation. The distended organ is placed in the stainless digestion chamber, which is filled with enzyme solution. The circuit is filled with warm (37°C) enzyme solution circulating at a flow rate of 100 ml/min

Table 12.1. Species-Dependent Parameters for Islet Isolation by Means of the Automated Method

Species	Caliber for pancreas distension (gauge)	Enzyme concentration (mg/ml)	Chamber volume (ml)	Mesh pore size (µm)	Temperature (°C)	Shaking frequency (min)
Human	16–18 Angiocath	Liberase HI, 1.4	500	400–500	36–38	20–50
Monkey	22–24 Angiocath	Liberase HI, 0.47	250	400–500	34–37	Gentle shaking
Dog	20–22 Angiocath	Liberase CI, 0.83	250	400–500	35–37	10–30
Adult pig	20 Angiocath	Liberase PI, 0.75	500	400–500	24–35	5 oscillations every 2 min
Rodent	24 tubing (rats) 27–30 needle (mice)	Liberase RI, 0.17	50	280–300	37	Continuous manual

through a peristaltic pump. The chamber is then placed on the arm of the shaking device set to 20–50 oscillations/min (10-cm excursion). The time of digestion varies, but in general it is carried on for about 16–25 min, maintaining the temperature at approximately 37°C. Samples are collected during the digestion process to verify the progression of the digestion process. When most islets are free from the surrounding acinar tissue and still intact, the digestion is stopped by diluting the enzyme solution in excess volumes of cold (4°C) RPMI supplemented with 10% FBS (flow rate 320 ml/min) and collecting the dispersed pancreatic tissue. The collected digest is washed in RPMI with 10% FBS before proceeding to purification.

Nonautomated Methods
Rodents

Two different enzyme blends can be used for rodent islet isolation. Collagenase type V (Sigma, St. Louis, MO) at a final concentration of 1.5 mg/ml in HBSS, or Liberase RI at a final concentration of 0.17 mg/ml in HBSS supplemented with 25 mM HEPES buffer. The pancreas is distended by injecting the ice cold enzyme solution through the main bile duct, which is clamped at the duodenum. The pancreatic tissue is surgically removed, rinsed in HBSS, immersed in ice-cold enzyme solution, and kept on ice until pancreata from all donors have been collected. Ischemia time should not be longer than one hour. For type V Sigma collagenase isolations, digestion is performed in a 17-min incubation at 37°C, with gentle shaking, after which enzymatic activity is sharply slowed by addition of cold HBSS supplemented with 10% FCS. For Liberase isolations, the digestion is performed in a 30-min static incubation at 37°C, followed by a brief, but more vigorous, manual shaking. Mechanical disruption of the digested pancreatic tissue is achieved by repeated passages through needles of decreasing gauge until complete release of free islets is observed under the microscope, and the tissue is then filtered through a 450-µm screen.

Neonatal Porcine Islet-like Cell Clusters

Isolation of porcine NICCs requires enzymatic digestion without further purification [79]. Donor pancreata are isolated from neonatal piglets (1–3 day old). Total pancreatectomy is performed after complete exsanguination under general anesthesia. The pancreas is carefully dissected from the surrounding tissues and placed in precooled (4°C) HBSS supplemented with 0.25% BSA, 10 mM HEPES, 100 U/ml penicillin, and 0.1 mg/ml streptomycin.

The pancreas is minced into fragments of 1–2 mm^3, and incubated for 16–18 min, at 37°C, in a 2.5-mg/ml collagenase solution (Sigma type V), with gentle shaking. After the digest has been filtered through a 500-µm screen and washed four times in HBSS, it is ready to be cultured.

PURIFICATION

The purification procedure is performed at 4°C. Large-scale islet purification can be obtained with the semiautomated method using the COBE 2991 cell processor. For purification from rodent pancreatic digest, the manual method is preferred.

The Semiautomated Method
Discontinuous Gradients

The pancreatic digest (20 ml of tissue) is resuspended in up to 300 ml of stock Euro-Ficoll (1.121 g/cm^3) and transferred into a 600-ml bag. A standard sterile cell processing set is used, with a peristaltic pump connected to the red COBE line, which will be used to load the gradients. The transfer bag is attached to the green line, and the digest is loaded by gravity into the processing bag.

The centrifuge is spun at 2000 rpm, and the air is vented from the processing bag by releasing the clamp from the green tube and opening the supernatant-out valve (150 ml/min). Then the outputs are clamped and the COBE is stopped. At this point the centrifuge is spun at 1200 rpm, and EuroCollins-Ficoll gradients are loaded from the red line at 90 ml/min as follows: 1.108 (75 ml), 1.096 (75 ml), 1.037 (75 ml), and HBSS (50 ml). After a 3- to 5-min spin, the interfaces are collected through the red line by adjusting the supernatant-out rate to 100 ml/min. Four fractions are collected in separate conical bottles: the first layer is discarded (100–125 ml), fraction 1 contains the highly pure islets located at the 1.037/1.096 interface (about 75 ml), fraction 2 contains fewer islets of lower purity (75 ml), and fraction 3 contains the pellet of acinar tissue. Collected fractions are diluted in excess volume RPMI with 10% FBS and spun twice at 500g for 1 min at 4°C. At this point samples are taken to assess the number of islets and purity of the preparation before culture.

Continuous Gradients

To create continuous gradients, a gradient maker is connected to the red line of the COBE processing bag. The gradient maker consists of two connected 300-ml chambers, with the higher density chamber placed on a magnetic stirrer. The pancreatic digest can be loaded either on top (resuspended in preservation solution) or at the bottom (resuspended in the high-density gradient) of the processing bag. The continuous gradient is loaded at 25–30 ml/min by peristaltic pump. After having loaded 120 ml of the high-density medium or the digest resuspended in the high-density medium, the COBE bowl is spun and the air is released from the bag. The valve between the two chambers is opened to obtain a linear dilution of the heaviest gradient with the lightest to form a continuous density gradient, which is pumped into the processing bag. When the whole gradient has been transferred, UW (80 ml) or HBSS with 2% FBS (100 ml) is loaded on top of it. The excess of air is vented off, and speed is increased (2000–2200 rpm) for 5 min to allow islet and acinar tissue to migrate to their respective densities. Gradients are then unloaded through the red line and collected into one 100- to 150-ml fraction (waste) and 9–11 smaller fractions (30 ml each). Each fraction is washed in RPMI with 10% FBS, and samples are taken for determination of amount and purity.

Manual Method for Rodent Islets

The digest is resuspended in Euro-Ficoll (1.108 mg/cm^3) and 15 ml is loaded into a 50-ml conical tube. To avoid overloading the gradient capacity, a maximum of 2 ml of tissue is resuspended in 15 ml of Euro-Ficoll. Subsequently, 1.096, 1.069, and 1.037 layers of 10 ml each are carefully loaded on top of each other. Islet purification is obtained by centrifugation at 900g for 11 min, followed by two washes in excess volumes of RPMI with 10% FBS at 650g, 4°C. Islets of purity exceeding 90% are routinely obtained. The vast majority of islets will be recovered from the 1.069/1.096 interface.

ISLET ASSESSMENT
Islet Counting and Purity Assessment

The pellet of purified islets is resuspended in 250 ml of CMRL 1066, supplemented with 10% FBS, HEPES (25 mM), L-glutamine (2 mM), sodium pyruvate (1 mM), zinc sulfate (3 mM), ITS + Premix (1%), and ciprofloxacin (20 mg/liter).

Table 12.2. Conversion Factors for the Calculation of the Islet Equivalent Number [a]

Islet diameter range (μm)	IEN conversion factor	Mean islet volume (nl)
50–100	*N*/6	0.29
100–150	*N*/1.5	1.15
150–200	*N* × 1.7	2.98
200–250	*N* × 3.5	6.19
250–300	*N* × 6.3	11.16
300–350	*N* × 10.4	18.29
>350	*N* × 15.8	27.98

[a] *Example*: 100 islets at 50–100 μm + 100 islets at 100–150 μm + 50 islets at 150–200 μm + 20 islets at 200–250 μm + 20 islets at 250–300 μm: IEN = 100/6 + 100/1.5 + (50 × 1.7) + (20 × 3.5) + (20 × 6.3) = 364.4. With a 2500 dilution factor, IEN = 911,000.
The total islet volume is (100 × 0.29) + (100 × 1.15) + (50 × 2.98) + (20 × 6.19) + (20 × 11.16) = 640 nl; with a 2500 dilution factor, this represents a total volume of 1.6 ml. If the volume of the islet pellet was 2 ml, then the purity is 1.6/2 = 80%.
Source: Adapted from Bretzel *et al.* [119].

Triplicate samples of 100 μl are taken from the suspension and transferred to a $10 \times 35 \text{ mm}^2$ counting dish with a 2-mm grid. The sample is covered with phosphate-buffered saline (PBS) and a few drops of a dithizone solution (dithizone 100 mg, DMSO 10 ml, PBS 40 ml), which will stain the islets bright red, is added.

The islets are counted under an inverted light microscope at a 40× magnification. The eyepiece should have a grid calibrated so that 10 divisions of the grid correspond to 250 μm at a 40× magnification.

Islets are counted according to their size and grouped in diameter classes of 50-μm increments. Islets smaller than 50 μm are not counted. The total number of islets is 2500 × the number of islets in the sample. The number of islet equivalents (IEN) of 150 μm diameter is calculated using conversion factors for each group (Table 12.2) [119]. The total islet volume of the preparation can be calculated and its purity can be estimated (Table 12.2).

Islet Viability Assessment with Fluorescent Dyes

A 500-μl sample is taken from the islet preparation and put in small petri dish, whereupon 500 μl of orange acridin dye (0.67 μM) and 500 μl of propidium iodide (75 μM) are added to the islet suspension. Since islet cells die rapidly in this suspension, it must be examined immediately under a fluorescence microscope; a delay will be associated with false positivity. Orange acridin stains living cells, which fluoresce green. Propidium iodide stains dead or nonviable cells, which fluoresce bright orange.

Glucose-Stimulated Insulin Release by Static Incubation

Aliquots of 50 islets are handpicked from the total islet preparation and put in triplicates in 1 ml of culture medium (supplemented CMRL 1066, as earlier) in a non-tissue-culture-treated multiwell plate and incubated at 37°C, in a humidified atmosphere with 5% CO_2. Supernatants are removed and discarded after overnight incubation. Then 500 μl of Krebs–Ringer solution with low glucose (1.67 mM, 30 mg/dl) is added to the islets, and incubation for 1 h proceeds at 37°C, 5% CO_2. Supernatants are removed and discarded.

Next 500 μl of Krebs–Ringer solution with low glucose concentration (1.67 mM, 30 mg/dl) is added to the islets, and incubation for 1 h proceeds at 37°C, 5% CO_2. The supernatants are kept for insulin measurement and stored at −80°C until assayed ("low 1" samples).

Now 500 μl of Krebs–Ringer solution with high glucose concentration (16.7 mM, 300 mg/dl) is added to the islets, and incubation for 1 h proceeds at 37°C, 5% CO_2. The

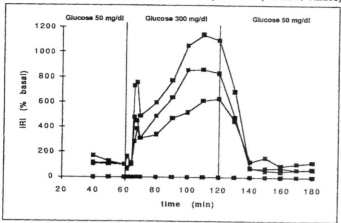

Fig. 12.4. Example of a perifusion test performed on isolated adult human pancreatic islets. From Bretzel et al. [119].

supernatants are kept for insulin measurement and stored at –80°C until assayed ("high" samples).

Finally, 500 µl of Krebs–Ringer solution with low glucose concentration (1.67 mM, 30 mg/dl) is added to the islets, with incubation for 1 h at 37°C, 5% CO_2. The supernatants are kept for insulin measurement and stored at –80°C until assayed ("low 2" samples).

A commercially available ELISA kit (ALPCO, Windham, NH) is used to determine the insulin levels in all three groups of samples.

The stimulation index, calculated by dividing the mean insulin level of the "high" samples by the mean insulin level of the "low 1" samples, provides a measure of the secretory capacity of the islet preparation. Good islet function is further demonstrated by a near return to baseline after the last incubation in low glucose; that is, the insulin level of the "low 2" samples should be approximately equal to that of the "low 1" samples.

Glucose-Stimulated Insulin Release by Islet Perifusion

The principles of the islet perfusion method and the reagents used are the same as in the static incubation method just described, with the major difference that the islets are continuously perifused with the low and high glucose solutions instead of simply incubated.

In brief, aliquots of 50 islets are suspended in Millipore chambers with a 5-µm filter. Chambers are put in an incubator, at 37°C, in a 5% CO_2 atmosphere, and perifused with the Krebs–Ringer solution for 3 consecutive h, during which the glucose concentration is 1.67 mM, 16.7 mM, and 1.67 mM, respectively. The perfusate is delivered by a peristaltic pump, at a rate of 1 ml/min, and the effluent is collected in tubes with a fraction collector.

The effluent is collected every 10 min during the first hour, at 1, 2.5, 5, 7.5, 10, 20, 30, 40, 50 and 60 min during the second hour, and every 10 min during the third hour.

Results of a perifusion assay are best reported as a plot of insulin concentration in collected samples as a function of time (Fig. 12.4), on which basal and stimulated insulin release, and quality of the biphasic response and return to basal secretion are evident. The stimulation index can be calculated as indicated for the static incubation method.

ISLET CULTURE
Culture of Human Islets or Islets from Large Mammals

Islets are cultured in an incubator at 37°C for the first 12 h, and then at 22°C, in a humidified atmosphere with 5% CO_2.

They are cultured free-floating in supplemented CMRL 1066 in non-tissue-culture-treated vented flasks (175 cm^2 culture area).

If islet purity exceeds 90%, a maximum of 30,000 IEN can be placed in each flask, in a total volume of 45 ml.

If islet purity is less than 90%, a lower IEN should be put in each flask to keep the total amount of tissue per flask constant.

The IEN per flask and the number of flasks necessary according to purity can be calculated with the following formulas:

Maximum IEN per flask = 30,000 × % purity

$$\text{Number of flasks} = \frac{\text{IEN}}{30{,}000 \times \% \text{ purity}}.$$

Culture medium and flasks must be changed after 12 h, and every 24 h thereafter.

Culture of Rodent Islets

Islets are cultured in an incubator at 22–37°C, in a humidified atmosphere with 5% CO_2. They are cultured free-floating in supplemented CMRL 1066 in non-tissue-culture-treated vented flasks or petri dishes. A maximum IEN of 500/ml should not be exceeded. Culture medium must be changed after 12 h and every 24 h thereafter.

Culture of Porcine Neonatal Islet Cell Clusters (NICCs)

NICCs are cultured in an incubator at 37°C in a humidified atmosphere with 5% CO_2. They are cultured free-floating in non-tissue culture-treated flasks or dishes in serum-free Ham's F-10 medium, containing 0.5% BSA, 10 mM glucose, 50 μM IBMX, 2 mM L-glutamine, 10 mM nicotinamide, 100 U/ml penicillin, and 100 μg/ml streptomycin. NICCs isolated from one pancreas are distributed into two 175-cm^2 flasks containing 45 ml of medium each. Medium is changed on the first day after isolation and every other day thereafter. NICCs are counted after 9 days in culture, when the cell aggregates have gained a fully differentiated islet morphology.

REFERENCES

1. Hering, B. J., and Ricordi, C. (1999). Results, research priorities, and reasons for optimism: Islet transplantation for patients with type 1 diabetes. *Graft* **2**, 12–27.
2. Lakey, J. R. T., Rajotte, R. V., Warnock, G. L., and Kneteman, N. M. (1995). Pancreas procurement for islet isolation and pancreas storage prior to islet isolation. *In* "Methods in Cell Transplantation" (C. Ricordi, ed.), pp. 421–431. R. G. Landes, Austin, TX.
3. Lakey, J. R. T., Rajotte, R. V., Warnock, G. L., and Kneteman, N. M. (1994). Cold ischemic tolerance of human pancreas: Assessment of islet recover and function. *Transplant. Proc.* **26**, 3416.
4. Moskalewski, S. (1965). Isolation and culture of the islets of Langerhans of the guinea pig. *Gen. Comp. Endocrinol..* **5**, 342.
5. Lacy, P. E., and Kostianovsky M. (1967). Method for the isolation of intact islets of Langerhans from the rat pancreas. *Diabetes* **16**, 35–39.
6. Horaguchi, A., and Merrell, R. C. (1981). Preparation of viable islet cells from dogs by a new method. *Diabetes* **30**, 455–458.
7. Noel, J., Rabinovitch, A., Olson, L., *et al.* (1982). A method for large scale high yield isolation of canine pancreatic islet of Langerhans. *Metabol. Clin. Exp.* **31**, 184–187.
8. Gray, D. W. R., McShane, P., Grant, A., and Morris, P. J. (1984). A method for isolation of islets of Langerhans from the human pancreas. *Diabetes* **33**, 1055–1061.
9. Kneteman, N. M., and Rajotte, R. V. (1986). Isolation and cryopreservation of human pancreatic islets. *Transplant. Proc.* **18**, 182–185.
10. Ricordi, C., Finke, E. H., and Lacy, P. E. (1986). A method for the mass isolation of islets from the adult pig pancreas. *Diabetes* **35**, 649–653.
11. Gotoh, M., Maki, T., Kiyoizumi, T., *et al.* (1985). An improved method for isolation of mouse pancreatic islets. *Transplantation* **40**, 437–438.
12. Sutton, R., Peters, M., McShane, P., *et al.* (1986). Isolation of rat pancreatic islets by ductal injection of collagenase. *Transplantation* **42**, 689–691.
13. Ricordi, C., Lacy, P., Finke, E. H., Olack, B. J., and Scharp, D. W. (1988). An automated method for the isolation of human pancreatic islets. *Diabetes* **37**, 413–420.
14. Ozhato, H., Gotoh, M., Monden, M., *et al.* (1990). Intraductal injection of collagenase solution at the time of harvesting: a possible solution for preservation and collagenase digestion. *Transplant. Proc.* **22**, 782–785.
15. Warnock, G. L., Ellis, D., Rajotte, R. V., Dawidson, I., Baekkeskov, S., and Egebjerg, J. (1988). Studies of the isolation and viability of human islet of Langerhans. *Transplantation* **45**, 957–963.
16. Lakey, J. R. T., Warnock, G. L., Shapiro, A. M., *et al.* (1999). Intraductal collagenase delivery into the human pancreas using syringe loading or controlled perfusion. *Cell Transplant.* **8**, 285–292.
17. Gu, Y. J., Inoue, K., Shinohara, S., *et al.* (1995). Comparison of different collagenases in isolation of adult pig islets. *Cell Transplant.* **4**(Suppl. 1), S49.

18. Klock, G., Kowalski, M. B., Hering, B. J., *et al.* (1996). Fractions from commercial collagenase preparations: Use in enzymic isolation of the islets of Langerhans from porcine pancreas. *Cell Transplant.* **5**, 543–551.

19. Linetsky, E., Bottino, R., Lehmann, R., Alejandro, R., Inverardi, L., and Ricordi, C. (1997). Improved human islet isolation using a new enzyme blend, liberase. *Diabetes* **46**, 1120–1123.

20. Berney, T., Molano, R. D., Cattan, P., *et al.* (2001). Endotoxin-mediated delayed islet graft function is associated with increased intra-islet cytokine production and islet cell apoptosis. *Transplantation* **71**, 125–132.

21. Ulrichs, K., Meyer, T., Klock, G., *et al.* (1998). Monitoring of enzymatic digestions on porcine pancreatic tissue using a simple histological assay. *Transplant. Proc.* **30**, 355.

22. Cavanagh, T. J., Lakey, J. R., Dwulet, F., *et al.* (1998). Improved pig islet yield and post-culture recovery using Liberase PI purified enzyme blend. *Transplant. Proc.* **30**, 367.

23. Brandhorst, H., Brandhorst, D., Hering, B. J., and Bretzel, R. G. (1999). Significant progress in porcine islet mass isolation utilizing liberase HI for enzymatic low-temperature pancreas digestion. *Transplantation* **68**, 355–361.

24. Jahr, H., Pfeiffer, G., Hering, B. J., Federlin, K., and Bretzel, R. G. (1999). Endotoxin-mediated activation of cytokine production in human PBMCs by collagenase and Ficoll. *J. Mol. Med.* **77**, 118–120.

25. Olack, B. J., Swanson, C. J., Howard, T. K., and Mohanakumar, T. (1999). Improved method for the isolation and purification of human islets of Langerhans using Liberase enzyme blend. *Human. Immunol.* **60**, 1303–1309.

26. Ricordi, C., Lacy, P. E., and Scharp, D. W. (1989). Automated islet isolation from human pancreas. *Diabetes* **38**(Suppl. 1), 140–142.

27. Kenyon, N. S., Chatzipetrou, M., Masetti, M., *et al.* (1999). Long-term survival and function of intrahepatic islet allografts in rhesus monkeys treated with humanized anti-CD145. *Proc. Natl. Acad. Sci. U.S.A.* **96**, 8132–8137.

28. Olack, B., Hill, A., Scharp, D., and Lacy, P. (1992). Automated purification of canine islets. *Transplant. Proc.* **24**, 1003–1004.

29. Ricordi, C., Socci, C., Davalli, A. M., *et al.* (1989). Isolation of the elusive pig islet. *Surgery* **107**, 688–694.

30. Ricordi, C., Socci, C., Davalli, A. M., *et al.* (1990). Application of the automated method to islet isolation in swine. *Transplant. Proc.* **22**, 784–785.

31. Ricordi, C., Finke, E. H., Dye, E. S., Socci, C., and Lacy, P. E. (1988). Automated isolation of mouse pancreatic islets. *Transplantation* **46**, 455–457.

32. Ricordi, C., Gray, D. W. R., Hering, B. H., *et al.* (1990). Islet isolation assessment in man and large animals. *Acta Diabetol. Lat.* **27**, 185–195.

33. Warnock, G. L., Kneteman, N. M., Evans, M. G., Dabbs, K. D., and Rajotte, R. V. (1990). Comparison of automated and manual methods for islet isolation. *Can. J. Surg.* **33**, 368–371.

34. Ricordi, C., and Rastellini, C. (1995). Automated method for pancreatic islet separation. *In* "Methods in Cell Transplantation" (C. Ricordi, ed.), pp. 433–438. R. G. Landes, Austin, TX.

35. Gray, D. W. R., Sutton, R., McShane, P., Peters, M., and Morris, P. (1986). Exocrine contamination impairs implantation of pancreatic islets transplanted beneath the kidney capsule. *J. Surg. Res.* **45**, 432–442.

36. Gotoh, M., Maki, T., Satomi, S., Porter, J., and Monaco, A. P. (1986). Immunological characteristics of purified pancreatic islet grafts. *Transplantation* **42**, 387–390.

37. Lau, D., Hering, B. H., El-Ouaghlidi, A., *et al.* (1999). Isokinetic gradient centrifugation prolongs survival of pig islets xenografted into mice. *J. Mol. Med.* **77**, 175–177.

38. Dobroschke, J., Langhoff, G., Seyed Ali, S., *et al.* (1981). Isolation of human islets of Langerhans. *In* "Islet Isolation, Culture, and Cryopreservation" (K. Federlin, and R. G. Bretzel eds.), pp. 32–39. Thieme-Stratton, New York.

39. Lorenz, D., Wolff, H., Lippert, H., *et al.* (1985). Clinical experience in islet transplantation. In "Microsurgical Models in Rats for Transplantation Research" (A. Thiede, E. Deltz, R. Engemann, and H. Hamelmann, eds.), pp. 359–367. Springer-Verlag, Berlin.

40. Kuhn, F., Shultz, H. J., Lorenz, D., *et al.* (1985). Morphological investigation in human islets of Langerhans isolated by the Velcro-technique. *Biomed. Biochim. Acta* **44**, 149–153.

41. Scharp, D., Lacy, P., Ricordi, C., *et al.* (1989). Human islet transplantation in patients with Type 1 diabetes. *Transplant. Proc.* **21**, 2744–2745.

42. Gray, D. W. R., Göhde, W., Carter, N., Heiden, T., and Morris, P. J. (1989). Separation of pancreatic islets by fluorescence-activated sorting. *Diabetes* **38**(Suppl. 1), 270.

43. von Specht, B.U., Finke, M., Eckhardt, A., and Permanetter, W. (1990). Large scale purification of pig islets of Langerhans by fluorescence activated cell sorting. *Horm. Metab. Res. Suppl. Ser.* **25**, 54–57.

44. Winoto-Morbach, S., Ulrichs, K., Leyhausen, G., and Müller-Rucholz, W. (1989). New principle for large scale preparation of purified human pancreas islets. *Diabetes* **38**, 146–149.

45. Samejima, T., Yamaguchi, K., Itawa, H., Morikawa, N., and Ikada, Y. (1998). Gelatin density gradient for isolation of islet of Langerhans. *Cell Transplant.* **7**, 37–45.

46. Ballinger, W. F., and Lacy, P. (1972). Transplantation of intact pancreatic islets in rats. *Surgery* **72**, 175–186.

47. London, N. J. M., Robertson, G. S. M., Chadwick, D. R., James, R. F. L., and Bell, P. R. F. (1995). Adult islet purification. *In* "Methods in Cell Transplantation" (C. Ricordi, ed.), pp. 439–454. R. G. Landes, Austin, TX.

48. Pretlow, T. G., and Pretlow, T. P. (1982). Sedimentation of cells: An overview and discussion of artifacts. *In* "Cell Separation: Methods and Selected Applications" (T. G. Pretlow, and T. P. Pretlow, eds.), pp. 41–60. Academic Press, San Diego, CA.

49. Van den Burg, M. P., Gooszen, H. G., Ploeg, R. J., *et al.* (1990). Pancreatic islet isolation with UW solution: A new concept. *Transplant. Proc.* **22**, 2050–2051.

50. Chadwick, D. R., Robertson, G. M. S., Contractor, H. H., *et al.* (1994). Storage of pancreatic digest before islet purification. The influence of colloids and the sodium to potassium ratio in University of Wisconsin-based preservation solutions. *Transplantation* **58**, 99–104.

51. Hering, B. J., Gramber, D., Ernst, E., Kirchhof, N., Bretzel, R. G., and Federlin, K. (1992). Isokinetic gradients: a new approach to reduce islet graft immunogenicity. *Transplant. Proc.* **25**, 959–960.

52. Gramberg, D., Ernst, E., Liu, X., Hering, B. J., Bretzel, R. G., and Federlin, K. (1994). Isokinetic gradients decrease islet graft immunogenicity. *Transplant. Proc.* **26**, 753.

53. Lake, S. P., Bassett, P. D., Larkins, A., *et al.* (1989). Large-scale purification of human islets utilizing discontinuous albumin gradient on IBM 2991 cell separator. *Diabetes* **38**(Suppl. 1), 143–145.

54. Alejandro, R., Strasser, S., Zucker, P. F., and Mintz, D. H. (1990). Isolation of pancreatic islets from dogs. Semiautomated purification on albumin gradients. *Transplantation* **50**, 207–210.

55. London, N. J. M., Robertson, G. S. M., Chadwich, D. R., *et al.* (1993). Purification of human pancreatic islets by large scale continuous density gradient centrifugation. *Horm. Metab. Res.* **25**, 61–70.

56. Harwood, R. (1974). Cell separation by gradient centrifugation. *Int. Rev. Cytol.* **38**, 369–403.

57. Fritschy, W. M., van Suylichem, P. T. R., Wolters, G. H. J., and van Schilfgaarde, R. (1992). Comparison of top and bottom loading of dextran gradient for rat pancreatic islet purification. *Diabetes Res.* **19**, 91–95.

58. Olack, B., Swanson, C., McLear, M., Longwith, J., Scharp, D., and Lacy, P. E. (1991). Islet purification using Euro-Ficoll gradients. *Transplant. Proc.* **23**, 774–776.

59. Van der Burg, M. P., Guicherit, O. R., Frolich, M., Prins, F. A., Bruijn, J. A., and Gooszen, H. G. (1994). Cell preservation in University of Wisconsin solution during isolation of canine islets of Langerhans. *Cell Transplant.* **3**, 315–324.

60. Lindall, A., Steffes, M., and Sorenson, R. (1969). Immunoassayable insulin content of subcellular fractions of rat islets. *Endocrinology (Baltimore)* **85**, 218–223.

61. London, N. J. M., Toomey, P., Contractor, H., Thirdborough, S. T., James, R. F., and Bell, P. R. (1992). The effect of osmolality on the purity of human islet isolates. *Transplant. Proc.* **24**, 1002.

62. Chadwick, D. R., Robertson, G. M. S., Toomey, P., *et al.* (1993). Pancreatic islet purification using bovine serum albumin: The importance of density gradient temperature and osmolality. *Cell Transplant.* **2**, 355–361.

63. Shortman, K. (1972). Physical procedures for the separation of animal cells. *Annu. Rev. Biophys. Bioeng.* **1**, 93–130.

64. Leif, R. C. (1970). Buoyant density separation of cells. *In* "Automated Cell Identification and Sorting" (G. L. Weid, and G. F. Bahr, eds.), pp. 21–96. Academic Press, New York.

65. Robertson, G. S. M., Chadwick, D. R., Contractor, H., James, R. F. L., and London, N. J. M. (1993). The optimization of large-scale density gradient isolation of human islets. *Acta Diabetol.* **30**, 93–98.

66. Brendel, M. D., Hering, B. J., Schultz, A. O., and Bretzel, R. G., eds. (1999). "International Islet Transplant Registry." Newsl. No. 8. Justus-Liebig-University of Giessen, Giessen.

67. Latif, Z. A., Noel, J., and Alejandro, R. (1988). A simple method of staining fresh and cultured islets. *Transplantation* **45**, 827–830.

68. Bank, H. L. (1988). Rapid assessment of islet cell viability with acridine orange and propidium iodide. *In Vitro* **24**, 266–272.

69. Ashcroft, S. J., Bassett, J. M., and Randle, P. J. (1971). Isolation of human pancreatic islets capable of releasing insulin and metabolizing glucose in vitro. *Lancet* **1**, 888–889.

70. Andersson, A., Borg, H., and Groth, C. G. (1976). Survival of isolated islets of Langerhans maintained in tissue culture. *J. Clin. Invest.* **57**, 1295–1301.

71. Brandhorst, H., Brandhorst, D., Lau, D., Hering, B. J., Federlin, K., and Bretzel, R. G. (1999). Glucose sensitivity of porcine and human islets in vitro. *J. Mol. Med.* **77**, 90–92.

72. Petkov, P., Hahn, H. J., Galabova, R., and Ziegler, M. (1984). Investigations on islets of Langerhans in vitro. VIII. Ultrastructure and insulin secretion of isolated rat islets after different digestion with collagenase. *Acta Histochem.* **51**, 50–60.

73. Morgan, C. R., and Lazarow, A. (1965). Immunoassay of pancreatic and plasma insulin following alloxan injection of rats. *Diabetes* **14**, 669–671.

74. Ricordi, C., Hering, B. J., London, N. J. M., *et al.* (1992). Islet isolation assessment. *In* "Pancreatic Islet Cell Transplantation" (C. Ricordi, ed.), pp. 132–142. R. G. Landes, Austin, TX.

75. Ricordi, C., Scharp, D. W., and Lacy, P. E. (1988). Reversal of diabetes in nude mice after transplantation of fresh and 7 days cultured (24°C) human pancreatic islets. *Transplantation* **45**, 994–996.

76. Ricordi, C., Kneteman, N. M., Scharp, D. W., and Lacy, P. E. (1988). Transplantation of cryopreserved human pancreatic islets into diabetic nude mice. *World J. Surg.* **12**, 861–865.

77. Ricordi, C., Zeng, Y. J., Alejandro, R., *et al.* (1991). In vivo effect of FK506 on human pancreatic islets. *Transplantation* **52**, 519–522.

78. Otonkoski, T., Ustinov, J., Rasilainen, S., Kallio, E., Korsgren, O., and Häyry, P. (1999). Differentiation and maturation of porcine fetal islet cells in vitro and after transplantation. *Transplantation* **68**, 1674–1683.

79. Korbutt, G. S., Elliott, J. F., Ao, Z., Smith, D. K., Warnock, G. L., and Rajotte, R. V. (1996). Large scale isolation, growth, and function of porcine neonatal islet cells. *J. Clin. Invest.* **97**, 2119–2129.

80. Lacy, P. E., Davie, J. M., and Finke, E. H. (1979). Prolongation of islet allograft survival following in vitro culture (24 degrees C) and a single injection of ALS. *Science* **204**, 312–313.

81. Ricordi, C., Lacy, P. E., Sterbenz, K., and Davie, J. M. (1987). Low-temperature culture of human islets or in vivo treatment with L3T4 antibody produces a marked prolongation of islet human-to-mouse xenograft survival. *Proc. Natl. Acad. Sci. U.S.A.* **84**, 8080–8084.

82. Warnock, G. L., Dabbs, K. D., Cattral, M. S., and Rajotte, R. V. (1994). Improved survival of in vitro cultured canine islet allografts. *Transplantation* **57**, 17–22.

83. Lacy, P. E., and Finke, E. H. (1991). Activation of intraislet lymphoid cells causes destruction of islet cells. *Am. J. Pathol.* **138**, 1183–1190.

84. Matas, A. J., Sutherland, D. E., Kretschmer, G., Steffes, M. W., and Najarian, J. S. (1977). Pancreatic tissue culture: Depletion of exocrine enzymes and purification of islets for transplantation. *Transplant. Proc.* **9**, 337–339.

85. Ono, J., Lacy, P. E., Michael, H. E., and Greider, M. H. (1979). Studies of the functional and morphologic status of islets maintained at 24°C for four weeks in vitro. *Am. J. Pathol.* **97**, 489–503.

86. Ling, Z., and Pipeleers, D. G. (1994). Preservation of glucose-responsive islet β-cells during serum-free culture. *Endocrinology (Baltimore)* **134**, 2614–2621.

87. Ilieva, A., Yuan, S., Wang, R. N., Agapitos, D., Hill, D. J., and Rosenberg, L. (1999). Pancreatic islet cell survival following islet isolation: The role of cellular interactions in the pancreas. *Endocrinology (Baltimore)* **161**, 357–364.

88. Paraskevas, S., Duguid, W. P., Maysinger, D., Feldman, L., Agapitos, D., and Rosenberg, L. (1997). Apoptosis occurs in freshly isolated human islets under standard culture conditions. *Transplant. Proc.* **29**, 750–752.

89. Thomas, F. T., Contreras, J. L., Bilbao, G., Ricordi, C., Curiel, D., and Thomas, J. M. (1999). Anoikis, extracellular matrix, and apoptosis factors in isolated cell transplantation. *Surgery* **126**, 299–304.

90. Behboo, R., Carroll, P. B., Trucco, M., and Ricordi, C. (1995). Decreased nitric oxide generation following human islet culture in serum-free media. *Transplant. Proc.* **27**, 3380.

91. Bottino, R., Inverardi, L., Valente, U., and Ricordi, C. (1997). Serum-free medium and pyruvate improve survival and glucose responsiveness of islet β cells in culture. *Transplant. Proc.* **29**, 1978–1979.

92. Clayton, H. A., and London, N. J. (1996). Survival and function of islets during culture. *Cell Transplant.* **5**, 1–12.

93. Oberholzer, J., Triponez, F., Mage, R., *et al.* (2000). Human islet transplantation: lessons from 13 autologous and 13 allogeneic transplantations. *Transplantation* **69**, 1115–1123.

94. Alejandro, R., Lehmann, R., Ricordi, C., *et al.* (1997). Long-term function (6 years) of islet allografts in type 1 diabetes. *Diabetes* **46**, 1983–1989.

95. Socci, C., Falqui, L., Davalli, A. M., *et al.* (1992). Substitution of the endocrine pancreatic function in IDDM patients: The Milan experience. *In* "Pancreatic Islet Cell Transplantation" (C. Ricordi, ed.), pp. 414–422. R. G. Landes, Austin, TX.

96. Andersson, A. (1978). Isolated mouse pancreatic islets in culture: Effects of serum and different culture media on the insulin production of the islets. *Diabetologia* **14**, 397–404.

97. Nielsen, J. H., Brunstedt, J., Andersson, A., and Frimodt-Moller, C. (1979). Preservation of beta cell function in adult human pancreatic islets for several months in vitro. *Diabetologia* **16**, 97–100.

98. Holmes, M. A., Clayton, H. A., Chadwick, D. R., Bell, P. R. F., London, N. J. M., and James, R. F. L. (1995). Functional studies of rat, porcine, and human pancreatic islets cultured in ten commercially available media. *Transplantation* **60**, 854–860.

99. Davalli, A. M., Bertuzzi, F., Socci, C., *et al.* (1993). Paradoxical release of insulin by adult pig islets in vitro. *Transplantation* **56**, 148–154.

100. Coppelli, A., Arvia, C., Giannarelli, R., *et al.* (1996). Long-term survival and function of isolated bovine pancreatic islets maintained in different culture media. *Acta Diabetol.* **33**, 166–168.

101. Ling, Z., Hannaert, J. C., and Pipeleers, D. (1994). Effect of nutrients, hormones and serum on survival of rat islet beta cells in culture. *Diabetologia* **37**, 15–21.

102. Ling, Z., and Pipeleers, D. (1996). Prolonged exposure of human β cells to elevated glucose levels results in sustained cellular activation leading to a loss of glucose regulation. *J. Clin. Invest.* **98**, 2805–2812.

103. Buitrago, A., and Gylfe, E. (1983). Significance of serum for the preservation of insulin secretion during culture. *Med. Biol.* **61**, 133–138.

104. Goldman, H., and Colle, E. (1976). Human pancreatic islets in culture: Effects of supplementing the medium with homologous and heterologous serum. *Science* **192**, 1014–1016.

105. Sakamoto, K., Hatakeyama, E., Kenmochi, T., *et al.* (1998). Improvement of porcine islet culture with porcine serum. *Transplant. Proc.* **30**, 391–392.

106. Clark, S. A., and Chick, W. L. (1990). Islet cell culture in defined serum-free medium. *Endocrinology (Baltimore)* **126**, 1895–1903.

107. Behboo, R., Carroll, P. B., Memarzadeh, S., *et al.* (1994). Improved long-term culture of functional human islets in serum-free medium. *Transplant. Proc.* **26**, 3301.

108. Warnock, G. L., Kneteman, N. M., Ryan, E. A., Rabinovitch, A., and Rajotte, R. V. (1992). Long-term follow-up after transplantation of insulin-producing pancreatic islets into patients with type 1 (insulin-dependent) diabetes mellitus. *Diabetologia* **35**, 89–95.

109. Piemonti, L., Bertuzzi, F., Nano, R., *et al.* (1999). Effects of cryopreservation on in vitro and in vivo long-term function of human islets. *Transplantation* **68**, 655–662.

110. Bretzel, R. G., Brandhorst, D., Brandhorst, H., *et al.* (1999). Improved survival of intraportal pancreatic islet cell allografts in patients with type-1 diabetes mellitus by refined peritransplant management. *J. Mol. Med.* **77**, 140–143.

111. Rajotte, R. V., Lakey, J. R. T., and Warnock, G. L. (1995). Adult islet cryopreservation. *In* "Methods in Cell Transplantation" (C. Ricordi, ed.), pp. 517–524. R. G. Landes, Austin, TX.

112. Rosenberg, L., Metrakos, P., Qi, S. J., and Rajotte, R. (1992). Trophic stimulation of adult human islets in vitro. *Transplant. Proc.* **24**, 3012–3013.

113. Hayek, A., Beattie, G. M., Cirulli, V., Lopez, A. D., Ricordi, C., and Rubin, J. S. (1995). Growth factor/matrix-induced proliferation of human adult beta-cells. *Diabetes* **44**, 1458–1460.

114. Beattie, G. M., Itkin-Ansari, P., Cirulli, V., *et al.* (1999). Sustained proliferation of PDX-1+ cells derived from human islets. *Diabetes* **48**, 1013–1019.

115. Rosenberg, L., Duguid, W. P., Brown, R. A., and Vinik, A. I. (1988). Induction of nesidioblastosis will reverse diabetes in Syrian golden hamster. *Diabetes* **37**, 334–341.

116. Kerr-Conte, J., Pattou, F., Lecomte-Houcke, M., *et al.* (1996). Ductal cyst formation in collagen-embedded adult human islet preparations. A means to the reproduction of nesidioblastosis in vitro. *Diabetes* **45**, 1108–1114.

117. Rafaeloff, R., Pittenger, G. L., Barlow, S. W., *et al.* (1997). Cloning and sequencing of the pancreatic islet neogenesis associated protein (INGAP) gene and its expression in islet neogenesis in hamsters. *J. Clin. Invest.* **99**, 2100–2109.

118. Ramiya, V. K., Maraist, M., Arfors, K. E., Schatz, D. A., Peck, A. B., and Cornelius, J. G. (2000). Reversal of insulin-dependent diabetes using islets generated in vitro from pancreatic stem cells. *Nat. Med.* **6**, 278–282.

119. Bretzel, R. G., Hering, B. J., and Federlin, K. F. (1995). Assessment of adult islet preparations. *In* "Methods in Cell Transplantation" (C. Ricordi, ed.), pp. 455–464. R. G. Landes, Austin, TX.

CHAPTER 13

EPITHELIAL CELL CULTURE: TRACHEAL GLAND CELLS

Marc D. Merten

Human tracheal glands are the main secretory structure of the bronchotracheal tree. To date, tracheal gland physiology is poorly understood. In many bronchopathologies, pharmacological manipulations to control mucus hypersecretion are almost nonexistent. Thus the use of human tracheal gland (HTG) cells in culture is an interesting tool for investigating the mechanisms of regulation of mucus secretion. However, since the first description of the culture technique in 1990, very few data have been available concerning the use of these cultured cells for understanding the mechanisms of bronchial secretion. This chapter describes techniques that optimally facilitate isolation and culture of HTG cells and allow efficient management of the cultured cells in the performance of pharmacological experiments yielding optimal, reliable, and reproducible results to provide better insight into the physiology of bronchial secretion.

INTRODUCTION

Inhaled air arrives clean and sterile at the pulmonary alveolae only after it has encountered the mucociliary clearance. This important system consists of a complex mechanism involving both ciliary beating and mucus secretion, which occur in the bronchotracheal tree. While entrapped by the mucus, particles (including microorganisms) are propelled out by beating. In human airways, mucus originates principally from the submucosal glands [1]. These glands are composed of mucous and serous cells (Fig. 13.1A) and are connected to the bronchotracheal lumen by collecting ducts [2]. Mucous cells secrete mucins, the most widely known macromolecules of bronchial secretion, which have the role of "particle snare." Serous cells are specialized in the secretion of antibacterial proteins such as lactoferrin, lysozyme, and peroxydase and are the primary site in the lung for secretory IgA transcytosis [3]. In addition, gland serous cells secrete an antiprotease—the secretory leukocyte proteinase inhibitor (SLPI)—that also displays antibacterial [4] and antiviral [5] activities. Therefore, tracheal gland serous cells are considered to be major lung defense cells by virtue of their potential antibacterial function.

Tracheal glands are typical acini with a polarized secretion toward the gland lumen. Gland cells contain an abundant rough endoplasmic reticulum and Golgi apparatus as well as numerous secretory granules located at the apex of the cells. Secretion by gland cells is both constitutive and regulated (Fig. 13.1B). Because gland cells possess several identified receptors, they are responsive to many secretagogues. In many bronchopathologies, there is either an overabundance of mucus or a failure in the pulmonary defense system, and the ability to manipulate and/or control tracheal gland secretion pharmacologically would be both worthwhile and pertinent. Poor knowledge of the gland's physiology, however, hinders working this area. A main reason for our relative ignorance of gland physiology is the great difficulty of obtaining biological materials that can be managed to yield reproducible results. Working with biopsy samples or explants in organotypic culture has produced only relatively poor data owing to the extreme complexity of the tissue. Indeed, the richly innervated, vascularized bronchotracheal submucosal tissue also contains smooth

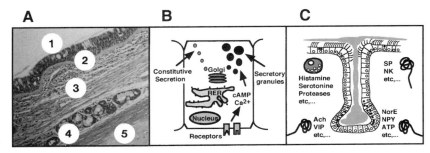

Fig. 13.1. Histophysiology of bronchotracheal glands. (A) Cross section of a human tracheal mucosa showing the mucus layer (1), the surface epithelium lying on a basement membrane (2), the interstitial tissue containing fibroblasts and smooth muscles (3), submucosal seromucous glands (4), and the cartilagenous ring (5). (B) Schematic representation of a tracheal gland cell. As in many secretory cells, tracheal gland cells are polarized, possess a constitutive secretion, and accumulate secretory granules that can be released upon activation of specific receptors. (C) Schematic representation of a human tracheal gland. Both mucous and serous cells contribute to the secretion of mucus, which is then evacuated to the bronchotracheal lumen through collecting ducts. Gland secretion is controlled by the autonomic nervous system, which releases many neurotransmitters and neuromodulators such as acetylcholine (Ach), the vasoactive intestinal peptide (VIP), norepinephrine (NorE), neuropeptide Y (NPY), adenosine triphosphate (ATP), substance P (SP) and neurokinines (NK). It may be provoked by substances derived from mastocytes, as well.

muscle, neuroendocrine cells, and mastocytes. This "rich" environment may expose the glands to various potential physiological secretagogues (Fig. 13.1C). These are neurotransmitters and neuropeptides from the sympathetic, parasympathetic, and sensitive autonomic nervous systems, as well as histamine, serotonin, and proteases from mastocytes, hormones and neuropeptides from neuroendocrine cells, cytokines from immune cells, and so on. This complexity and our relative ignorance of tracheal gland physiology have totally hampered mechanistic investigations using explants. Several teams have thus developed techniques to isolate and culture human tracheal gland (HTG) cells [6,7]. Even though cell isolation and culture conditions have been well described, optimized, and publicized [8], however, few data are available from pharmacological and physiological studies, largely because of the difficulty of performing such experiments in reproducible conditions.

   This chapter describes techniques that facilitate the use of cultured HTG cells for such studies. These techniques have produced high reproducibility, optimal cell responses to pharmacological agonists, rendering this cell culture model a very appropriate and useful tool for studying the mechanisms of regulation of bronchial secretion.

MATERIALS

BRONCHIAL SAMPLE ACQUISITION

   Bronchial samples are obtained from patients undergoing lung cancer surgery or lung transplantation. They can also be obtained from human cadavers if taken less than 12 h from the time of death.

SOLUTIONS AND REAGENTS

   1. *Transporting medium for the surgical specimen*: the medium for collecting explants is composed of RPMI 1640 supplemented with 0.1% Ultroser G (a serum replacement from BioSepra, Villeneuve-la-Garenne, France), 200 µg/ml gentamicin, 5 U/ml amphotericin B (both provided by Sigma), and glucose and sodium pyruvate made up to 10 and 0.33 g/liter, respectively. This medium is transferred in sterile containers in 25-ml aliquots and stored at 4°C for one month.

   2. *Digestion medium*: 110 mM NaCl, 5 mM KCl, 1 mM Na_2HPO_4, 1 mM KH_2PO_4, 1 mM $MgCl_2 \cdot 7H_2O$, 2 mM $CaCl_2 \cdot 2H_2O$, 20 mM N-2-hydroxyethyl-piperazine-N'-2-ethanesulfonic acid (HEPES), 10 mg/ml fraction V human serum albumin, 4 g/liter glucose, and 0.11 g/liter pyruvate, in addition to the enzymes type IA collagenase (200 U/ml), type I-S hyaluronidase (200 U/ml), type I porcine pancreatic elastase (0.1 mg/ml), type II

DNAse (200 U/ml). Storage is at −20°C after sterile filtration in 25-ml aliquots. All reagents are from Sigma.

3. *Trypsinization solution*: 110 mM NaCl, 5 mM KCl, 1 mM Na_2HPO_4, 1 mM KH_2PO_4, 20 mM N-2-hydroxyethyl-piperazine-N'-2-ethanesulfonic acid (HEPES), trypsin (0.05% w/v), 4 g/liter glucose, 0.11 g/liter pyruvate, and 2 mM ethylenediaminetetraacetic acid (EDTA), pH 7.4. Storage is at 4°C after sterile filtration in 25-ml aliquots.

4. *Collagen coating of flasks or multiwell plates*: Collagen, which was found to be the most convenient substrate for HTG cell culture, is prepared from rat tails that have been frozen for at least one week, then defrosted in a 95% ethanol solution for 3 min. Subsequently, strong, sterile pliers are used to break the tails into segments 1 cm long. The isolated tendons are dissected and washed in sterile water. The tendons are gathered and placed into a 1 mM acetic acid solution for 24 h at 4°C (100 ml per tail). After centrifugation to remove the nonsolubilized material, the stock collagen solution is stored at 4°C for no more than 2 months. To achieve functionality of the cells, the tissue culture flasks or multiwell plates must be coated as follows: a 1/100 dilution (from the stock solution) in distilled water is dropped onto the cell surfaces (0.2 ml/cm^2) and left overnight. This collagen solution must be removed just before the cells are seeded, to prevent the collagen film from drying out. Under these conditions about 10 μg/cm^2 of collagen is absorbed.

5. *Culture medium*: DMEM (Dulbecco's Modified Eagles's Medium), Ham's F-12, 50%/50% (v/v) containing 1% Ultroser G, and the following substances made up to the indicated final concentrations: 10 g/liter glucose, 0.33 g/liter pyruvate, 0.2 g/liter leucine, isoleucine and valine, glutamic acid, and cysteine, 3 μM epinephrine [9], and two antibiotics: 100 U/ml penicillin G and 100 μg/ml streptomycin. Glucose, pyruvate, and amino acids are added to the DMEM/F-12 prior to filtration when the media are prepared from powdered mixtures. The Ultroser G and the antibiotics are prepared together in aliquots, to be added just before each medium change. Epinephrine is also prepared independently (stored at −80°C in a 1 mM HCl solution) and added just before use. *Note*: All reagents are available from tissue culture retailers. Ultroser G is provided by BioSepra, in France.

METHODS
ISOLATION OF HUMAN TRACHEAL GLAND CELLS

1. As soon as possible, place the bronchial samples directly from patients in a sterile tube containing a covering volume of transporting medium at 4°C.

2. Before processing, remove samples from the transporting medium, submerge briefly in ethanol, and shake rapidly dry in the tissue culture hood. Spread out the human bronchial tissue on a shallow sterile container and carefully clean the mucous material present on the organ surface with sterile gauze. Soak the surface thoroughly with complete cell culture medium. During the following operations, the surface must be kept continually damp with this solution. Use fine forceps and iris scissors to dissect the mucosa and submucosa, and especially the tissue between the cartilagenous rings. Cut this tissue with the scissors into small strips ($\leq 1 \text{ mm}^3$).

3. Place the strips in a 50-ml sterile tube containing 25 ml of digestion medium and incubate at 37°C for 1 h with gentle shaking.

4. After incubation, centrifuge at *500g* for 10 min. Discard the supernatant.

5. Incubate the pellet with trypsinisation medium for 15 min at 37°C and centrifuge at *50g* for 2 min to separate the undigested tissue and the detached cells. Remove this supernatant and centrifuge it at *500g* for 10 min to retrieve the cells.

6. Subject the undigested tissue to steps 3–5 until the tissue has been completely digested.

7. Count the cells and seed at an initial density of 25,000–45,000 cells/cm^2.

SUBCULTURE OF HUMAN TRACHEAL GLAND CELLS

After isolation, the cells will adhere to the collagen substrate overnight. Their viability will vary according to (1) the length of time between surgery (or death) and collection in the 4°C transporting medium, (2) the age and sex of the donor, and (3) the lifestyle of the donor (e.g., smoker vs nonsmoker). During the first 5–8 days, culture medium must not be changed but supplemented with gentamicin (100 μg/ml) and neomycin (50 μg/ml) and, every 2 days, with amphotericin B (5 μg/ml). Thereafter, medium is changed three times a week. After isolation, microscopic examination indicates the presence of abundant cell debris and different cell types that will grow. Among them, small islets of highly proliferating epithelioid cells can be distinguished [6,8]. When cells reach about 75% confluency, it is necessary to carry out a partial trypsinization to separate HTG cells from all other contaminating cells. This consists of a brief (~3 min) exposure to the trypsinization sulution to allow the harvesting of contaminant cells (fibroblasts, endothelial cells, myocytes, etc.), the tracheal gland cells are much more firmly adherent to the collagen substrate and will not detach until 5–10 min or more of exposure to trypsin. The remaining HTG cell islets will then continue to grow and reach confluency within 6–8 days (see later: Fig. 13.2B). For subcultivation, trypsinize the cells when they reach about 75% confluency but retrieve only one-third to half of the cells. The isolated cells are seeded at 25,000 cells/cm^2 and the original flask can be cultured further—new cells will spread again from the remaining clusters. Cell growth and differentiation at confluency are optimally achieved by using the combination of several factors: type I collagen as a substrate, elevated concentrations of glucose (HTG cells are highly metabolizing cells), the use of Ultroser G as a serum substitute, and addition of epinephrine (which optimizes secretion and polarity). However, no more than four to six passages can be obtained. Reproducibility of results will occur for the only three first passages.

GROWTH CHARACTERISTICS OF HUMAN TRACHEAL GLAND CELLS (FIG. 13.2)

When cells are placed in culture, they undergo a lag phase, during which they attach to the collagen-coated surface. The lag phase can be as long as 24–36 h. Following the lag phase, the cells enter the exponential growth phase, during which the cell number increases logarithmically with time. It should be noted that only a proportion of the cells are proliferating and, as such, contributing to the increase in cell number of the cell population in culture. Cells grow in clusters, and only the ones at the periphery of the clusters will divide and grow. Therefore speed of growth depends on the size and number of clusters in the flasks. Confluent HTG cells consist typically of a single monolayer of cells, and thereafter, the cells enter a stationary phase in which they stop dividing, a phenomenon attributed to contact inhibition. Cells will no longer be able to divide, and thus when a passage is planned, be careful that cell density in the flasks does not reach over 75–90% of confluency. There are two different stages in the stationary phase. In the first stage the cells will progressively develop and acquire their typical differentiated characteristics. A typical feature is

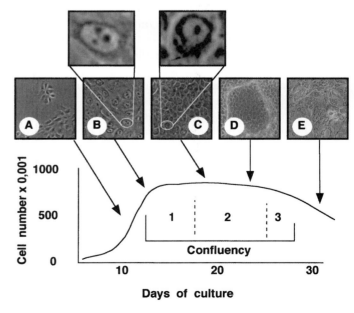

Fig. 13.2. Phase contrast micrographs of cultured human tracheal gland cells at different stages. (A) Cells during the exponential growth phase. After seeding, cells gather in clusters, and cell growth occurs at the periphery of these islets. (B) When cells reach confluency, they undergo phenotypical changes and differentiate to fully secretory cells. These cells are characterized by numerous nucleoles (2–6) in the nucleus. (C) Differentiated cells are characterized by a few nucleoles in the nucleus (1–2) and an accumulation of dark material around the nucleus. (D) At late confluency, numerous domes appear on the cells, as well as a deposit of materials having high refringency. (E) Degeneration of the cells is easily recognizable by flattening and multinucleation of the cells and many signs of apoptosis. Cells can be stimulated and therefore are usable for physiological studies, only during the stage 2 of confluency.

the number (2–6) of nucleoles distinguishable inside the nucleus. In the second stage (which occurs 8–10 days after confluency and lasts for about 2 weeks), the cells will fully express their differentiated characteristics. Only in the second stage is it possible to carry out experiments designed to investigate the action of physiological or pharmacological agents. It is easy to distinguish this stage by several typical characteristics: fewer nucleoles (1–2) inside the nucleus, a dark intracellular material accumulating around the nucleus, an increasing number of domes, and the presence of areas with thick and refringent material deposited onto the cells. Thereafter cells will become senescent and degenerate and will rapidly lose their characteristics as the cell number regularly decreases.

RESULTS

In fundamental as well as in applied research, one of the most interesting aspects of HTG cell culture is its possible use in the study of the physiology, pathology, and pharmacology of bronchial secretion. However, before any investigation can be carried out for this purpose, it is important to check the differentiated functions of HTG cells and to prepare the cells for stimulation.

CRITERIA USED TO CHARACTERIZE CELL PHENOTYPE DIFFERENTIATION OF HTG CELLS

The function of tracheal glands *in vivo* is to produce proteins involved in the antibacterial defense of the airways. These are lactoferrin and lysozyme, or SLPI, which are specific serous secretory markers, and also mucins, which are mucous secretory markers. As cells may adopt a mixed seromucous phenotype *in vitro*, both types of secretory product are secreted concomitantly. Furthermore, according to Finkbeiner *et al.* [10] it is possible to obtain HTG cells of either the serous or the mucous phenotype by varying cell culture conditions. When HTG cells are cultured onto collagen, they differentiate into serous cells, but when they are cultured onto vitronectin, they differentiate into mucous cells. HTG cell culture therefore allows one to obtain both cell types present in the glands. Thus, one principal criterion is the ability of cultures of HTG cells to secrete these proteins at high basal levels. These levels (expressed per minute) are about 300 pg/10^6 cells for SLPI and about 30 pg/10^6 cells for lactoferrin and lysozyme, for instance. Such measurements can easily be carried out by the availability of commercial ELISA kits for all these proteins. To date, no commercial kits are available to measure secretion of mucins.

PREPARATION OF THE CELLS FOR STIMULATION

Several authors have shown that HTG cells are able to respond to many secretagogues by mobilizing second-messenger systems [11,12] or by changes in short-circuit currents [13]. However, they have not demonstrated that these changes are indeed related to the secretory function of the cells, or have they elucidated the extent of any such relationship. It seems therefore of first importance to be able to link an involved second-messenger system to the biological response by the cells. HTG cells are able to respond to secretagogues by stimulating the secretion of defense protein [6,9,14]. However, several necessary and defined conditions must be met before these results can be obtained, and cells must first to be carefully prepared.

1. A period of serum deprivation is necessary to delete from the cells any possible cathecholamine or other agents that may be present in the serum, hence able to interfere with the receptors to be studied. The duration of this period depends on receptor turnover, usually 4–6 h for receptors to neurotransmitters. Thus the cells are exposed to the agent to be tested after they have been rinsed with serum-free medium four times, 1 h.

2. *In vivo*, glands are controlled and actually solicited by the autonomic nervous system, and thus the gland cells are exposed often and briefly to neurotransmitters. If not previously stimulated, cultured cells in standard culture conditions will lose their ability to respond to secretagogues, or will do so only minimally. To obtain optimal responsiveness from the cells, they must be challenged regularly with a cocktail of secretagogues to mimic the *in vivo* situation. The method we use is as follows (see Fig. 13.3): when cells have

Fig. 13.3. Managment of the cells for optimizing responsiveness to secretagogues. As cells reached confluency, they undergo a "training period" where regular stimulations (arrows) are carried out. Eight to 10 days after confluency (corresponding to about 4 "training stimulations"), experiments using the agents or secretagogues to be tested can be carried out.

reached the stationary phase (corresponding to the first stage of confluency), they will be submitted to a so-called training period. Every 2–3 days, cells are first exposed to four times one hour with serum-free medium and then exposed for 30 min to a mixture (in serum-free medium) composed of ATP, carbamoylcholine, and isoproterenol (each at 100 μM concentration). Cells are then rinsed off and new complete culture medium is added. This operation is performed four times. Thereafter, cells will be able to optimally respond to secretagogues during the second stage of confluency.

PERFORMING PHARMACOLOGICAL OR PHYSIOLOGICAL STUDIES ON HTG CELLS IN CULTURE

To study the action of pharmacological or physiological agonists on the secretion of defense proteins, HTG cells are first exposed to four times one hour with serum-free medium and then (in serum-free medium) to the agents to be tested. All the incubations are performed at 37°C. Figure 13.4A shows an example of a 24-well-plate scheme for performing pharmacological or physiological studies on HTG cells in culture. This setup obviously can easily be adapted to 96-well plates. Both a negative control (secretagogue free plus vehicle) and a positive control (100 μM ATP, giving an increase of 65 ± 5% when SLPI secretion is measured) will be compared against the action of the tested agents. For example, Fig. 13.4B presents the kinetics of action of three well-known secretagogues, carbamoylcholine (acting on muscarinic receptors), isoproterenol (acting on β-adrenergic receptors) and phenylephrine (acting on α-adrenergic receptors). In this and in all the experiments mentioned hereafter, SLPI serves as a secretory marker of serous cell secretion. Unstimulated cells constitutively secrete SLPI. Addition of the secretagogues to the cells will lead to exocytosis of the secretory granules and thus to an increase in SLPI concentration in the supernatant. Figure 13.4C integrates the data of Fig. 13.4B to show variation of speed of secretion versus time. Speed is maximal after 10 min exposure, indicating that exocytosis takes place at that time. After 20 min exposure, speed returns to basal levels. Sometimes, according to the agent tested, values at 20 min are significantly less than control (probably indicative of a refilling of the intracellular stores). In all our tests, speed of secretion has been indistinguishable from the control after 30 min exposure. Thus, we advise postponing measurements until the cells have been exposed to the agents to be tested, for at least 30 min. Figure 13.4D expresses the results of Fig. 13.4B as percentages of secretion above control. The significance of the increases in secretion induced by carbamoylcholine, isoproterenol, and phenylephrine will be then assessed by analysis of variance (ANOVA). The difference between the agents or between the concentrations of agents can be isolated by multiple comparison tests such as those of Scheffé, Wilcoxon, or Mann–Whitney.

POSSIBLE APPLICATIONS

One of the major application of the use of HTG cells as a pharmacological tool to investigate tracheal gland secretion is in the design of experiments aimed at better understanding bronchial pathologies. Three main types of study can be carried out for that purpose: the study of agents that act on mucus secretion and as a consequence the mechanisms underlying the physiology of bronchial secretion, and also the pathophysiology and pathopharmacology of HTG cells.

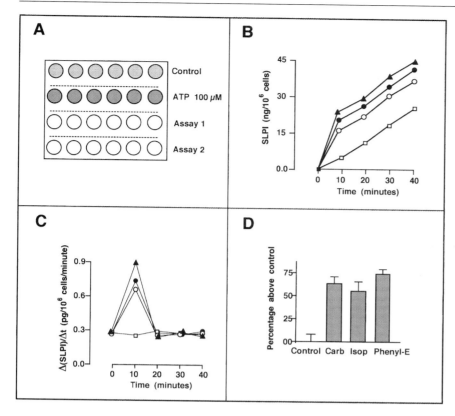

Fig. 13.4. Testing of agents for their abilities to stimulate secretion by HTG cells in culture. (A) When multiwell plates (here a 24-well plate) are used to carry out secretion stimulation assays, both a negative control (vehicle solution) and a positive control (100 μM ATP leading to +65±5% secretion above control) are necessary. (B) Effects of carbamoylcholine (●), isoproterenol (○) and phenylephrine (▲) on SLPI secretion by HTG cells; HTG cells secrete SLPI (□) at a basal and constant level corresponding to constitutive secretion. After exposure of the cells to the secretagogues, SLPI secretion rises. (C) Expression of Δ change in SLPI versus change in time allows variations in speed of secretion to be determined and shows that SLPI exocytosis appears to be maximal 10 min after agent exposure. However, measurements are carried out 30 min after agent exposure, not only to be sure that the entire phenomenon has occurred but also because it is more convenient to make measurements at that time. (D) Results of stimulation of secretion are expressed as a percentage of secretion above the control experiment.

STIMULATION OF MUCUS SECRETION

As described earlier, HTG cells in culture are able to secrete SLPI, which is a protein specific to the serous gland cells, the secretion of which is stimulated by pharmacological agents mimicking the action of neurotransmitters. However, whereas Tournier et al. [6] showed that cultured homogeneous HTG cells are mostly composed of serous cells, Sommerhof et al. [7] observed that by using specific antibodies directed against serous or mucous epitopes all cells could be immunolabeled with both types of antibody. This suggests that cultured HTG cells may carry both the serous and the mucous phenotypes. Furthermore, when cells are incubated for 24 h with ^{3}H[fucose] and ^{35}S[SO$_4$], they produce high molecular weight radiolabeled glycoproteins that are partly proteoglycans (chondroitin and heparan sulfates) and also mucins, as proved by resistance to all hyaluronidase, chondroitinase, heparitinase, and keratanase and sensitivity to β-adrenergic elimination; in addition, the buoyant densities of these hydrolase-resistant radiolabeled macromolecules, as well as the sizes of their glycannic chains, were consistent with those expected from mucins but not from proteoglycans or other glycoproteins [15]. These results suggest that HTG cells are able to produce mucins in vitro. Experiments for testing the effects of pharmacological or physiological agents on mucin secretion by HTG cells are easy to perform. One simply loads the cells for 24 h with ^{3}H[fucose], ^{3}H[glucosamine], or ^{35}S[SO$_4$] before stimulation. These radioprecursors also must be added in the chase (4 × 1 hour rinsing with serum-free medium) period. To avoid any possibility of radiolysis, since labeled glyco- or sulfoconjugates will concentrate in the secretory granules, loading with radioisotopes should not be too elevateted: 10 μCi/ml is an activity that can lead to measurable stimulation of high molecular weight radioconjugates. Figure 13.5 shows an example of data we obtained after loading the cells for 24 h with ^{3}H[fucose] and ^{35}S[SO$_4$]. The effects of increasing concentrations of carbamoylcholine show that cells respond to this agent by a similar secretion of SLPI and sulfoconjugates, while for that agonist, fucoconjugates are less copiously secreted. This is consistent with the observation of heterogeneous composition of the secretory granules [16], which can be differentially secreted according to the stimulus.

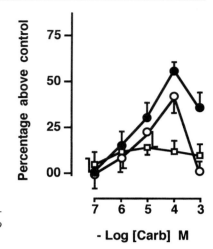

Fig. 13.5. Dose–response curves of the action of carbamoylcholine (Carb) on HTG cell secretion. In a concentration dependent fashion, Carb stimulates secretion of SLPI (●) and also high molecular weight macromolecules labeled with [³H]fucose (□) and [³⁵S]SO₄ (○).

MECHANISMS UNDERLYING PHYSIOLOGY OF BRONCHIAL SECRETION

Nevertheless, the physiology and pharmacology of bronchial secretion remains poorly understood. The complexity of human bronchial tissue makes it difficult to assign the action of an agent to a direct or indirect effect and also to distinguish secretions derived from the glands and from other secretory cells in the tissue (i.e., the goblet cells). Thus, HTG cells in culture appear to be one of the most interesting models for the study of the regulation of bronchial secretion mechanisms. Since the development of techniques for culturing HTG cells, there have been reports of data derived from binding experiments or from the abilities of agents to mobilize second-messenger systems. These reports, which make a strong case, have been few, but their number is increasing. However, because such data can lead to false or misleading interpretations, great caution is necessary. An example of the contribution of the use of the stimulation technique described here is shown in Fig. 13.6. Carbamoylcholine induces an increase in SLPI secretion and was shown to be positively linked to adenylyl cyclase and to the mobilization of intracellular calcium stores [14]. However, neither a forskolin/IBMX mixture (which increases cAMP content of the cells) nor ionomycin (which increases intracellular calcium concentration) can trigger SLPI secretion. This suggests that the cAMP and calcium pathways are not responsible for carbamoylcholine-induced secretion even if they are solicited by the action of the agonist. The forskolin/IBMX mixture is able to potentialize carbamoylcholine-induced secretion in the same magnitude as does the intracellular calcium chelator BAPTA/AM. Conversely, ionomycin significantly reduces magnitude of carbamoylcholine-induced secretion. This suggests that the cAMP pathway should instead be considered to be positive modulator of secretion and the calcium pathway rather a negative modulator of secretion. This also shows that demonstration of action of an agent on the second-messenger system will indicate only the presence of the corre-

Fig. 13.6. Effects of secretagogues and agents on HTG cell secretion. HTG cells respond to Carb by an increase in SLPI secretion. Cells do not respond to a forskolin/IBMX mixture, nor to the calcium ionophore ionomycin, nor to the calcium chelator BAPTA/AM. However, the forskolin/IBMX mixture, ionomycin, and BAPTA/AM modulate Carb-induced SLPI secretion. This suggests that the cAMP and the intracellular calcium pathways that are mobilized by Carb are not responsible for SLPI secretion.

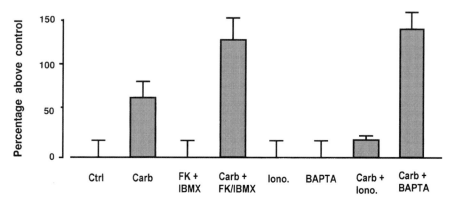

sponding receptor but give no information about its potential effect. In that respect, very few data about the action of secretagogues on defense protein secretion by HTG cells in culture are available.

Cultured HTG cells were shown to be responsive to cholinergic and adrenergic agonists [6,9], and to the purinergic agents ATP and UTP [14]. They respond to the physiological neurotransmitters acetylcholine and norepinephrine by an increase or a decrease in secretion, respectively [17], evidencing both a positive and negative control of secretion by HTG cells. Thus, HTG cells appear to be particularly responsive to agents present in the airways, the action of which, at cellular and molecular level, is only, beginning to be documented.

PATHOPHYSIOLOGY AND PATHOPHARMACOLOGY OF HTG CELLS

Asthma, acute and chronic bronchitis, and cystic fibrosis are the most widespread of the pulmonary pathologies in which mucus hypersecretion is observed. In these diseases, if overabundance of mucus is a determinant feature of morbidity, there is currently no real treatment able to control glandular secretion. Therefore HTG cells in culture may represent a pertinent tool to identify, to understand, and to decipher mechanisms of bronchial secretion linked to these pathologies as well as to analyze the effects of putative agents aimed at decreasing mucus secretion. In spite of the great potential interest in studying these cells for this purpose, very few data are available to date. Previous data showed a failure in the secretory mechanism by HTG cells in culture when cells were treated by bacterial LPS [18] or in cells derived from cystic fibrosis patients [17]. HTG cells were shown to be responsive to histamine [12], but no further development has been done to assess possible implications of HTG cell responsiveness to histamine in asthma. HTG cells were also shown to actively and produce cytokine [19] at high levels, suggesting that they participate strongly in the cytokinic network of the bronchotracheal tree. Although partial and sparse, the data offer a clear indication that HTG cells in culture are responsive to stimuli linked to pathologies. Thus their cell culture technique can be a useful tool for unraveling mechanisms leading to mucus hypersecretion observed in these pathologies.

CONCLUSION

Fortified by precise, up-to-date descriptions of culture techniques, a growing number of teams are isolating and culturing HTG cells. However, pertinent and physiologically significant use of these cultured cells depends on the acquisition of fully differentiating functions that can be attained in very specific conditions. These conditions were determined by analyzing the growth supplement requirements, the substrate requirements, and the culture methods. Our first objective was to establish conditions that allowed HTG cells to grow and especially to differentiate in culture. We observed that differentiation depends on parameters of two types: (1) cell culture conditions: the collagen substrate, an elevated concentration of glucose, and supplementation with epinephrine are important to HTG cell growth and differentiation; and (2) managment of cells after confluency: cells respond 8–10 days after confluency has been reached, and responses can be best measured after a "chase" period and recurrent secretion stimulations.

As a result, when cells are isolated and cultured from a 10-cm^2 surgically resected sample, for example, more than 100 multiwell plates can be obtained. This relatively important quantity of cells is sufficient to realize many fundamental or applied studies regarding the action of physiologically or pharmacologically active substances on receptors linked to secretion, especially for experiments designed to investigate the involvement of intracellular pathways in the effector–secretion coupling.

The mechanisms of secretion in the complex human tracheobronchial mucosa remain poorly explored. Cultures of HTG cells seem to provide an obvious interesting experimental model, which indeed has proved to be useful in certain investigations regarding pathologies. Current culture procedures render the HTG cell culture one of the best models available for the study of the cellular and molecular mechanism of human bronchial secretion.

Acknowledgments

The author's work on this subject was supported by grants from the Association Française de Lutte contre la Mucoviscidose. The author would like to thank Catherine Figarella for advice and fruitful discussions.

References

1. Read, L. (1960). Measurement of the bronchial mucous gland layer: A diagnostic yardstick in chronic bronchitis. *Thorax* **15**, 132–141.

2. Meyrick, B., Sturgess, J. M., and Read, L. (1969). A reconstruction of the duct system and secretory tubules of the human bronchial submucosal gland. *Thorax* **69**, 729–736.

3. Basbaum, C. B., Jany, B., and Finkbeiner, W. E. (1990). The serous cell. *Annu. Rev. Physiol.* **52**, 97–113.

4. Hiemstra, P. S., Maassen, R. J., Stolk, J., Heinzel-Wieland, R., Steffens, G. J., and Dijkman, J. H. (1996). Antibacterial activity of antileukoprotease. *Infect. Immun.* **64**, 4520–4524.

5. McNeely, T. B., Dealy, M., Dripps, D. J., Orenstein, J. M., Eisenberg, S. P., and Wahl, S. M. (1995). Secretory leukocyte protease inhibitor: A human saliva protein exhibiting antihuman immunodeficiency virus 1 activity in vitro. *Clin. Invest.* **96**, 456–464.

6. Tournier, J. M., Merten, M., Meckler, Y., Hinnrasky, J., Fuchey, C., and Puchelle, E. (1990). Culture and characterization of human tracheal gland cells. *Am. Rev. Respir. Dis.* **141**, 1280–1288.

7. Sommerhof, C. P., and Finkbeiner, W. E. (1990). Human tracheobronchial submucosal gland cells in culture. *Am. J. Respir. Cell Mol. Biol.* **2**, 41–50.

8. Merten, M. (1996). Human tracheal gland cells in primary culture. *In* "Methods in Molecular Medecine" (G. E. Jones, ed.), pp. 201–216. Humana Press, Totowa, NJ.

9. Merten, M. D., Tournier, J. M., Meckler, Y., and Figarella, C. (1993). Epinephrine promotes growth and differentiation of human tracheal gland cells in culture. *Am. J. Respir. Cell Mol. Biol.* **9**, 172–178.

10. Finkbeiner, W. E., Shen, B. Q., Mrsny, R. J., and Widdicombe, J. H. (1993). Induction of mucous phenotype in cultures of glands from human airways leads to loss of CFTR and chloride secretion. *Pediatr. Pulmonol.* **9**, 187.

11. Jacquot, J., Merten, M. D., Millot, J. M., Sébille, S., Ménager, M., Figarella, C., and Manfait, M. (1995). Asynchronous dynamic changes of intracellular free Ca^{++} and possible exocytosis in human tracheal gland cells induced by neutrophil elastase. *Biochem. Biophys. Res. Commun.* **212**, 307–316.

12. Jacquot, J., Spilmont, C., Maizières, M., Millot, J. M., Sébille, S., Merten, M., Kammouni, W., and Manfait, M. (1996). Intracellular free Ca^{2+} response to histamine is decreased in cystic human tracheal gland cells but not with human neutrophil elastase. *FEBS Lett.* **386**, 123–127.

13. Yamaya, M., Finkbeiner, W. E., and Widdicombe, J. H. (1991). Ion transport by cultures of human tracheobronchial glands. *Am. J. Physiol.* **261**, L485–L490.

14. Merten, M. D., Breittmayer, J. P., Figarella, C., and Frelin, C. (1993). ATP and UTP increase secretion of the bronchial inhibitor by human tracheal gland cells in culture. *Am. J. Physiol.* **265**, L479–L484.

15. Merten, M. D., Tournier, J. M., Meckler, Y., and Figarella, C. (1992). Secretory proteins and glycoconjugates synthesized by human tracheal gland cells in culture. *Am. J. Respir. Cell Mol. Biol.* **7**, 598–605.

16. Hinnrasky, J., Chevillard, M., and Puchelle, E. (1990). Immunocytochemical demonstration of quantitative differences in the distribution of lysozyme in human airway secretory granule phenotypes. *Biol. Cell* **68**, 239–243.

17. Merten, M. D., and Figarella, C. (1993). Constitutive hypersecretion and insensitivity to neurotransmitters by cystic fibrosis tracheal gland cells. *Am. J. Physiol.* **264**, L93–L99.

18. Kammouni, W., Figarella, C., and Merten, M. (1997). CF-like alteration by *Pseudomonas aeruginosa* lipopolysaccharide of protein secretion by human tracheal gland cells. *Biochem. Biophys. Res. Commun.* **241**, 305–311.

19. Kammouni, W., Figarella, C., Marchand, S., and Merten, M. (1997). Altered cytokine production by cystic fibrosis tracheal gland cells. *Infect. Immun.* **65**, 5176–5183.

CHAPTER 14

EPITHELIAL CELL CULTURE: KIDNEY

Ho-Yun Chung, Timothy Atalla, and Anthony Atala

INTRODUCTION

Various treatment modalities for patients with renal failure are currently being used clinically, depending on the individual's existing kidney function. The treatments include dietary and medical measures, dialysis, and transplantation. Patients with progressive renal failure eventually must receive a kidney transplant improve or restore renal function. However, renal transplantation is severely limited by donor kidney availability. Each year, patients with renal failure are placed on waiting lists for organ transplantation, and a good portion of affected individuals die during this period [1].

Efforts to maintain, improve, and restore renal function, have been pursued over the years, however, these attempts have not been fruitful [2–4]. Cell-based transplantation has been proposed as an alternate modality to meet the challenges posed by the small donor pool. Tissue engineering approaches using cells have been suggested as a method of enhancing kidney tissue function [3–8]. Augmentation of either isolated or total renal function with kidney cell expansion is a feasible solution. Investigative studies are being conducted continually, using various cell transplantation approaches [6–8].

The most important component in the achievement of functioning kidney tissue is renal cells. Although numerous studies have shown various renal cell culture methods, kidney cell expansion has not been formally approached for the purpose of engineering kidney tissue [9–20]. The challenge associated with renal cell culture and the associated expansion system is due to the unique structural and cellular heterogeneity present within the kidney. Each structure is composed of different cell types with specific cellular function. The kidney is a complex organ with multiple functions. These include waste excretion, body homeostasis, electrolyte balance, solute transport, and hormone and erythropoietin production. To bioengineer functioning kidney tissue, different approaches may be attempted. Partial renal function could be achieved by obtaining specific cell populations for the production of deficient factors. These could be obtained by growing and expanding particular cell populations and transplanting them back in the *in vivo* environment.

The kidneys are critical to body homeostasis because of their excretory, regulatory, and endocrinological functions. The excretory function is initiated by filtration of blood at the glomerulus, and the regulatory function is provided by the tubular segments. Glomerular epithelial cells in culture have shown to synthesize type IV collagen, glycosaminoglycans, and prostaglandins [12,21–23]. Glomerular mesangial cells in culture are able to synthesize and release tissue-type plasminogen activator (t-PA) and plasminogen activator inhibitor 1 (PAI-1) [24]. Constitutive expression of renin, pre-proendothelin 1 transcripts, endothelin 1 (ET-1), cyclooxygenase, and lipoxygenase have been detected in cultured mesangial cells [25–29]. Primary proximal tubule cells express a single saturable Na^+-dependent glucose carrier with a high affinity for the nonmetabolizable glucose analogue [30,31]. Therefore, these cell populations could be selected, grown, and expanded in the effort to augment specific renal function.

CELL CULTURE METHODS FOR GLOMERULAR AND MESANGIAL CELLS

Two techniques are commonly used to grow glomerular cells: direct explant and enzyme dissociation systems (Protocol 1, shown shortly) [9,10]. Isolated glomerular cells adhere to culture plates initially within several hours (enzymatic dissociation method) or several days (explant method). Cell proliferation and migration should be observed thereafter. Glomerular epithelial cells are the predominant cell type in outgrowth initially. The early outgrowth from glomeruli also contains other cell types, such as endothelial cells, macrophages, and contractile mesangial cells. However, mesangial cell populations eventually dominate the culture within a few weeks (Protocol 2).

Several methods of glomerular cell purification have been used. One technique employs a cell cloning method [3,4]. Investigators have successfully isolated and cultivated glomerular epithelial and mesangial cells from explants [32]. Other methods, such as enzymatic digestion of glomeruli, delayed preparation of glomeruli, and use of selective cytotoxic agents, have been developed.

PROTOCOL 1. GLOMERULAR CELL CULTURE

Growth Medium

> RPMI 1640 + 3T3 cell conditioned medium (1:1)
> 20% Fetal calf serum (FCS)
> Insulin (200 mU/ml)
> Penicillin (100 U/ml), streptomycin (100 µg/ml)

Digestion solution

> 0.1% Collagenase (190 U/mg)
> 0.2% Trypsin (12,700 BAEE U/mg)
> 0.01% Deoxyribonuclease (1115 Kienitz U/mg)

Method

1. Kidneys are perfused with Hanks's Balanced Salt Solution (HBSS).
2. The renal capsule is removed and the cortical tissue is mechanically separated from the medulla.
3. The isolated cortical tissue is minced and successively passed through 250-, 150- and 75-µm sieves.
4. Renal tissue fragments, which consist of glomerular structures, are thoroughly rinsed with HBSS and transferred to a conical tube.
5. After several washes, the glomerular tissue structures are placed in HBSS, containing digestion solution and incubated at 37°C for 20 min.
6. Subsequently, the glomerular suspension is pipetted vigorously to dissociate cell aggregates and rinsed several times in HBSS.
7. The glomerular cells are plated and incubated in a 37°C CO_2 incubator.

PROTOCOL 2. GLOMERULAR MESANGIAL CELL CULTURE

Growth Medium

> RPMI 1640
> 20% Fetal calf serum (FCS)
> Penicillin (100 U/ml), streptomycin (100 µg/ml)

Method

1. Follow the glomerular cell culture procedure described in Protocol 1.
2. After 2–3 weeks, mesangial cells appear in culture and predominate over epithelial cells.
3. Subculture of cells with trypsin further eliminates glomerular epithelial cells, and mesangial cell population is obtained.

CELL CULTURE METHODS FOR TUBULAR CELLS

Renal tubular cells are heterogeneous. Several cell culture methods have been tried with various success rates. One simple method is the use of the tissue explant technique. One study reports that growing minced explants of human renal cortex on collagen matrix and serum-free media results in the outgrowth of epithelial monolayers with proximal tubular properties [29]. Simple explant techniques have been largely superseded by the application of more carefully controlled fractionation methods using density sedimentation or gradients, such as Ficoll or Percoll [14]. Another method would be to use a differential sieving and magnetic removal of glomeruli preloaded with iron oxide, as shown shortly in Protocol 3 [33]. Yet another method involves initiation of culture from well-identified microdissected tubule segments. Cultures of microdissected proximal tubules, cortical and medullary thick ascending limbs, early distal tubules, and cortical and medullary collecting ducts have been successfully achieved and characterized [15,16,34–38]. Immunodissection techniques using cell-membrane-specific antibodies have been described in the literature (see Protocol 3) [12,17,39–41]. This method allows the isolation of homogeneous cell populations and production of large cell quantities. However, the yield is usually variable, depending on the density of membrane expression of the antigen and actual affinity of the antibodies used.

PROTOCOL 3. PROXIMAL TUBULAR CELL CULTURE

Growth Medium

> DMEM + Ham's F-12 (1:1)
> 15 mM HEPES
> 20 mM Sodium bicarbonate
> Penicillin (100 U/ml), streptomycin (100 µg/ml)
> Insulin (5 µg/ml)
> Transferrin (5 µg/ml)
> Hydrocortisone (5×10^{-8} M)

Other Materials

> 0.5% Iron oxide solution
> Collagenase (10 mg/ml)
> Soybean trypsin inhibitor (10 mg/ml)
> 253- and 85-µm sieves

Method

1. Kidneys are perfused initially with PBS (4°C) and then with 0.5% iron oxide solution in PBS.
2. The renal capsule is removed and the cortical tissue is dissected. Subsequently, cortical tissue is gently fragmented with a homogenizer.
3. The fragmented cortical tissue is successively passed through 253- and 85-µm sieves.
4. Tissue consisting of tubules and glomeruli are transferred to a conical tube containing 40 ml of medium.
5. A sterile magnetic stir bar is placed in the tube to remove glomerular cells to which iron oxide is bound.
6. Soybean trypsin inhibitor and collagenase are added to the tubular suspension and incubated for 2 min at 23°C.
7. The tubular suspension is centrifuged for 5 min at 500 rpm, followed by two wash cycles with the medium.
8. The tubular cells are plated and incubated in a 37°C CO_2 incubator.

PROTOCOL 4. CORTICAL COLLECTING TUBULE CELL CULTURE

Growth Medium

> DMEM

10% fetal bovine serum (decomplemented)
Penicillin (100 U/ml), streptomycin (100 µg/ml)
2 mM Glutamine
Bovine serum albumin

Method

1. Mouse IgG$_3$ is purified from *rct*-30 culture medium and each culture plate is treated with 5 ml of PBS containing 200 µg of purified IgG$_3$ (*rct*-30) and incubated overnight at 4°C.
2. Kidneys are obtained and the capsules are removed. Cortical tissue is mechanically dissected and minced.
3. The tissue is transferred to a conical tube containing 24 ml of 0.1% collagenase in Krebs buffer and incubated at 37°C for 40 min under a water-saturated 7% CO_2 condition.
4. The cell suspension is passed through a sieving mesh to remove undigested tissue. The resulting suspension is centrifuged for 7 min at 200g.
5. Aliquots (1 ml) of the renal cortical cell mixture ($1-3 \times 10^8$ cells) are placed to cover each of the culture plates coated with IgG$_3$ (*rct*-30), and the samples are incubated for 1–3 min at 24°C.
6. The plates are rinsed three to five times with PBS.
7. The remaining cells bound to the plate are grown and expanded.

REFERENCES

1. OPTN/SRAR 1988-1994. UNOS; DOT/HRSA/DHHS (1995). "Annual Report of the U.S. Scientific Registry of Transplant Recipients and the Organ Procurement and Transplantation Network." VSDHHS, Washington, DC.
2. Kramer, P., Wigger, W., Rieger, J., *et al.* (1977). A new and simple method for treatment of overhydrated patients resistent to diuretics. *Klin. Wochenschr.* **55**, 1121.
3. Amiel, G. E., and Atala, A. (1999). Current and future modalities for functional renal replacement. *Urol. Clin. North Am.* **26**, 235–246.
4. Atala, A., Schlussel, R. N., and Retik, A. B. (1995). Renal cell growth in vivo after attachment to biodegradable polymer scaffolds. *J. Urol.* **153**, 4.
5. Humes, D. H., Buffington, D. A., MacKay, S. M., Funke, A. J., and Weitzel, W. F. (1999). Replacement of renal function in uremic animals with a tissue engineered kidney. *Nat. Biotechnol.* **17**, 451–455.
6. Fung, L. C. T., Elenius, K., Freeman, M., Donovan, M. J., and Atala, A. (1996). Reconstitution of poor EGFr-poor renal epithelial cells into tubular structures on biodegradable polymer scaffold. *Pediatrics* **98**(Suppl), S631.
7. Yoo, J. J., Ashkar, S., and Atala, A. (1996). Creation of functional kidney structures with excretion of kidney-like fluid in vivo. *Pediatrics* **98S**, 605.
8. Amiel, G. E., Yoo, J. J., and Atala, A. (2000). Renal tissue engineering using a collagen-based kidney matrix. *Tissue Eng. Suppl.*
9. Kreisberg, J. I., Hoover, R. L., and Karnovsky, M. J. (1978). Isolation and characterization of rat glomerular epithelial cells in vitro. *Kidney Int.* **14**, 21.
10. Burlington, H., and Cronkite, E. P. (1973). Characteristics of cell cultures derived from renal glomeruli. *Proc. Soc. Exp. Biol. Med.* **142**, 143.
11. Kreisberg, J. I., and Karnovsky, M. J. (1983). Glomerular cells in culture. *Kidney Int.* **23**, 439.
12. Foidart, J. B., Dechenne, C. A., and Mahieu, P. R. (1981). Tissue culture of normal rat glomeruli: Characterization of collagenous and noncollagenous basement membrane antigens on the epithelial and mesangial cells. *Diagn. Histopathol.* **4**, 71.
13. Detrisac, C. J., Sens, M. A., Garvin, A. J., *et al.* (1984). Tissue culture of human kidney epithelial cells of proximal tubule origin. *Kidney Int.* **25**, 383.
14. Kreisberg, J. I., Pitts, A. M., and Pretlow, T. G. (1977). Separation of proximal tubule cells from suspensions of rat kidney cells in density gradients of Ficoll in tissue culture medium. *Am. J. Pathol.* **86**, 591.
15. Chuman, L., Fine, L. G., Cohen, A. H., *et al.* (1982). Continuous growth of proximal tubular kidney epithelial cells in hormone-supplemented serum-free medium. *J. Cell Biol.* **94**, 506.
16. Currie, M. G., Cole, B. R., DeSchryver-Kecskemeti, K., *et al.* (1983). Cell culture of renal epithelium derived from rabbit microdissected cortical collecting tubules. *Am. J. Physiol.* **244**, F724.
17. Spielman, W. S., Sonnenburg, W. K., Allen, M. L., *et al.* (1986). Immunodissection and culture of rabbit cortical collecting tubule cells. *Am. J. Physiol.* **251**, F348.
18. Smith, W. L., and Garcia-Perez, A. (1985). Immunodissection: Use of monoclonal antibodies to isolate specific types of renal cells. *Am. J. Physiol.* **248**, F1.
19. Stanton, R. C., Mendrick, D. L., Rennke, H. G., *et al.* (1986). Use of monoclonal antibodies to culture rat proximal tubule cells. *Am. J. Physiol.* **251**, C780.
20. Allen, M. L., Nakao, A., Sonnenburg, W. K., *et al.* (1988). Immuno-dissection of cortical and medullary thick ascending limb cells from rabbit kidney. *Am. J. Physiol.* **255**, F704.

21. Foidart, J. B., Pirard, Y. S., Winand, R. J., *et al.* (1979). Tissue culture of normal rat glomeruli: Glycosaminoglycan biosynthesis by homogeneous epithelial and mesangial cell populations. *Renal. Physiol.* **3**, 169.

22. Kreisberg, J. I., Karnovsky, M. J., and Levine, L. (1982). Prostaglandin production by homogeneous cultures of rat glomerular epithelial and mesangial cells. *Kidney Int.* **22**, 355.

23. Nitta, K., Simonson, M. S., and Dunn, M. J. (1991). The regulation and role of prostaglandin biosynthesis in cultured bovine glomerular endothelial cells. *J. Am. Soc. Nephrol.* **2**, 156.

24. Lacave, R., Rondeau, E., Ochi, S., *et al.* (1989). Characterization of plasminogen activator and its inhibitor in human mesangial cells. *Kidney Int.* **35**, 806.

25. Sakamoto, H., Sasaki, S., Hirata, Y., *et al.* (1990). Production of endothelin-1 by rat cultured mesangial cells. *Biochem. Biophys. Res. Commun.* **169**, 462.

26. Sakamoto, H., Sasaki, S., Nakamura, Y. *et al.* (1992). Regulation of endothelin-1 production in cultured rat mesangial cells. *Kidney Int.* **41**, 350.

27. Zoja, C., Orisio, S., Perico, N., *et al.* (1991). Constitutive expression of endothelin gene in cultured human mesangial cells and its modulation by transforming growth factor_, thrombin, and a thromboxane A_2, analogue. *Lab. Invest.* **64**, 16.

28. Chansel, D., Dussaule, J. C., Ardaillou, N., *et al.* (1987). Identification and regulation of renin in human cultured mesangial cells. *Am. J. Physiol.* **252**, F32.

29. Kurtz, A., Jelkmann, W., Pfeilschifter, J., *et al.* (1985). Role of prostaglandins in hypoxia-stimulated erythropoietin production. *Am. J. Physiol.* **249**, C3.

30. Sakhrani, L. M., Badie-Dezfooly, B., Trizna, W., *et al.* (1984). Transport and metabolism of glucose by renal proximal tubular cells in primary culture. *Am. J. Physiol.* **246**, F757.

31. Suzuki, T., Fujikura, K., and Takata, K. (1996). Na^+-dependent glucose transporter SGLT1 is localized in the apical plasma membrane upon completion of tight junction formation in MDCK cells. *Histochem. Cell Biol.* **106**, 529.

32. Foidart, J. B., Dechenne, C. A., Mahieu, P., *et al.* (1979). Tissue culture of normal rat glomeruli: Isolation and characterization of two homogeneous cell lines. *Invest. Cell Pathol.* **2**, 15.

33. Taub, M. (1997). Primary kidney cells. *Methods Mol. Biol.* **75**, 153.

34. Wilson, P. D., Dillingham, M. A., Breckon, R., *et al.* (1985). Defined human renal tubular epithelia in culture: Growth, characterization, and hormonal response. *Am. J. Physiol.* **248**, F436.

35. Burg, M., Green, N., Sohraby, S., *et al.* (1982). Differentiated function in cultured epithelia derived from thick ascending limbs. *Am. J. Physiol.* **242**, C229.

36. Valentich, J. D., and Stokols, M. F. (1986). An established cell line from mouse kidney medullary thick ascending limbs: I. Cell culture techniques, morphology, and antigenic expression. *Am. J. Physiol.* **251**, C299.

37. Wilson, P. D., Anderson, R. J., Breckon, R. D., *et al.* (1987). Retention of differentiated characteristics by cultures of defined rabbit kidney epithelia. *J. Cell. Physiol.* **130**, 245.

38. Merot, J., Bidet, M., Gachot, B., *et al.* (1989). Electrical properties of rabbit early distal convoluted tubule in primary culture. *Am. J. Physiol.* **257**, F288.

39. Dewitt, D. L., Rollins, T. E., Day, J. S., *et al.* (1981). Orientation of the active site and antigenic determinants of prostaglandin endoperoxide synthase in the endoplasmic reticulum. *J. Biol. Chem.* **256**, 10375.

40. Garcia-Perez, A., and Smith, W. L. (1983). Use of monoclonal antibodies to isolate cortical collecting tubule cells: AVP induces PGE release. *Am. J. Physiol.* **244**, C211.

41. Grenier, F. C., Rollins, T. E., and Smith, W. L. (1981). Kinin-induced prostaglandin synthesis by renal papillary collecting tubule cells in culture. *Am. J. Physiol.* **241**, F94.

CHAPTER 15

EPITHELIAL CELL CULTURE: GASTROINTESTINAL TRACT

Gregory M. Organ

BACKGROUND

Short bowel syndrome (SBS) refers to a condition of malabsorption and malnutrition following major resection of small intestine which may also include some or all of the large intestine. The primary clinical signs and symptoms of SBS are intractable diarrhea, steatorrhea, weight loss, dehydration, malnutrition, and vitamin and electrolyte deficiencies, as well as macrocytic anemia [1].

Although adaptation of the remaining intestine occurs, it is often not sufficient to maintain metabolism and nutrition on an oral diet [2,3]. The survival of patients with SBS, particularly infants and children, has improved primarily because of advances in and usage of total parenteral nutrition (TPN) both during hospitalization and at home [4].

Numerous innovative surgical techniques for increasing intestinal surface area and improving absorption have been developed [5–9]. None of the currently available surgical alternatives for treatment of SBS is sufficiently safe and effective for routine and widespread use. Transplantation of small intestine seemingly offers the most reasonable alternative for patients with near-fatal loss of intestine and therefore no hope for adaptation, and for those with intolerance to TPN. Despite recent improvements in immunosuppression and preservation of harvested small intestine, the major clinical obstacles to intestinal transplantation continue to be allograft rejection, graft-versus-host disease, and cytomegalovirus infection [10,11]. The development of flymphoproliferative complications is also a major concern. The adverse effects to growth and development of chronic corticosteroid use in children are a significant limitation. Thus, mortality and morbidity remain high with intestinal transplantation. Furthermore, when these immunological problems are resolved, the application of this therapy will still be limited by donor scarcity, the other significant problem facing the entire field of whole-organ transplantation.

Transplanting selected cell populations is a method of creating new tissue substitutes for the replacement and repair of lost or damaged tissue [12]. The emerging field of tissue engineering involves a multidisciplinary effort, merging the fields of cell and molecular biology, materials science, and surgical reconstruction to engineer new tissue [13].

A variety of strategies have been adopted by numerous investigators to engineer virtually every mammalian tissue. Our laboratory has utilized highly porous synthetic polymer devices that serve as a surface for these adhesion-dependent cells and as a delivery device to implant large numbers of cells to a given anatomic location. The design parameters of the device allow for the diffusion of gases and nutrients as well as the elimination of cellular waste products in the host implantation site. The scaffold functions as a template for the three-dimensional reorganization of the transplanted cells as they form new tissue [14,15]. The morphogenesis of the implanted cell–polymer construct involves the ingrowth of fibrovascular elements from the host implantation site into the graft. As the bioerodable scaffold disappears, the implanted cells express a differentiated phenotype, resulting in a permanently engrafted new tissue. A successfully engrafted and functioning new three-dimensional tissue is a true chimera of implanted parenchymal elements from the donor and mesenchymal elements (endothelial cells, fibroblasts, lymphatic, and neural cells) from the recipient. The term "chimeric neomorphogenesis" has been coined to describe this approach [14].

The epithelial cells of the small intestine, referred to as intestinal epithelial cells (IEC) or enterocytes, are a rapidly renewing cell population characterized by continuous growth and differentiation *in vivo* [16–19]. Intense mitotic activity is limited to the lower two-thirds of the crypts and is balanced by continuous loss of differentiated cells at the villus tip. This pattern of programmed cell death has been termed apoptosis [20]. IEC have a rapid cell turnover, with a mean cell life span of 2–3 days in most animals. Approximately 75% of IEC are differentiated, mature absorptive columnar villus cells, and 25% consist of their precursors, the rapidly proliferating undifferentiated cuboidal crypt cells. Differentiation of these mitotically active crypt cells is accompanied by dramatic changes in enzyme and transport activities, along with changes in morphology, including the appearance of a well-organized brush border at the luminal surface and a more columnar cell shape. The four main differentiated cell types of the intestinal epithelium—villus columnar, mucous goblet, enteroendocrine, and Paneth cells—originate from pluripotent stem cells located in the crypts of Lieberkuhn [18,19]. Excellent reviews of IEC isolation, culture, and metabolic characterization are available [21–24]. Enterocytes have been isolated by mechanical [25,26], enzymatic [22,27,37], and chelation methods [28–30] from a variety of different experimental animals and man [31]. Most of these techniques yield large numbers of cells, and some provide for reasonable separation between the villus and crypt cells.

Several broad conclusions about enterocyte isolation can be made from the earlier work of a number of different authors [22,32–36]. No single technique of enterocyte isolation is equally effective for all animal species. The rat has served as the primary model system for studying enterocytes, although valuable contributions have come from studies in guinea pigs [37,38], chickens [22], mice [35,36], hamsters [39,40], and rabbits [26,27]. Columnar cells situated near the villus tip tend to be morphologically more senescent and easily dislodged. Cuboidal crypt cells are not removed without distension of the intestinal wall and some mild mechanical dissociation. Purely mechanical methods of isolation lead to excessive cell damage, with release of proteolytic enzymes resulting in further cell damage, loss of glycolytic activity and consistently poorer viability. Techniques utilizing a variety of different enzymes and/or chelators result in improved viability and yield a greater proportion of the morphologically younger cuboidal cells found in the crypts of Lieberkuhn.

In our initial approach to tissue engineering neointestinal mucosa, we used suspensions of individual or small clusters of IEC [41]. Weiser's chelating method for isolating IEC in a villus–crypt cell gradient was modified to harvest IEC on a large scale [42,43]. A syngeneic Lewis rat model was developed for selective isolation of highly proliferative, crypt cell enriched fractions of IEC. IEC were seeded onto nonwoven fibers of poly(glycolic acid) (PGA) to form constructs. IEC-PGA constructs were rolled around a silastic stent and implanted into various vascular beds. Histological analysis confirmed that the grafts contain cells that have the cytological characteristics of IEC and form a stratified epithelium and have positive cytokeratin immunohistochemical staining features of epithelial cells; moreover, labeling of implanted IEC with fluorescent vital dyes confirms the origin of the engrafted cells, and the engrafted cells maintain remarkable proliferative capacity up to 28 days after implantation. At any of the observed time points out to 60 days, there was no evidence of neomucosa that had remodeled with typical small intestinal villus–crypt morphology. The histological appearance of the transplanted IEC was reminiscent of fetal intestinal development, arrested at the stratified epithelial stage.

Advances in developmental biology have highlighted the importance of epithelial–mesenchymal interactions, particularly in intestinal organogenesis [44,45]. Evans and Potten described a dramatically improved method of utilizing neutral protease and collagenase to isolate IEC [46]. This method yields intact crypts with their pericryptal fibroblast, thus maintaining the epithelial–mesenchymal interaction. The resulting crypt cell aggregates

(IECA) have been shown to have significantly improved viability, greater *in vitro* proliferation, and greater expression of differentiated function.

Employing this method of isolating IECA has led to several reports of successful engraftment, organotypic differentiation, and function of tissue-engineered neointestinal mucosa [47–52]. Perhaps, the most critical factor in this multistep process is the method of IEC isolation.

LABORATORY PROTOCOLS

PREPARATION

Nonfasted syngeneic male Lewis rats (Charles River, Wilmington, MA) served as epithelial cell donors and recipients. Animals should be housed and maintained in accordance with guidelines for the care of laboratory animals published by the National Institutes of Health. Donors and recipients should be allowed water and standard rat chow *ad libitum* and maintained in 12-h light and dark cycles before and after operation.

Poly(glycolic acid) (PGA) nonwoven fiber scaffold (15-μm fiber diameter, average pore diameter 250 μm, Davis & Geck, Danbury, CT) is supplied as multiple layered sheets and must be maintained in a desiccated environment except when the scaffold is being prepared for construct formation. The stock sheets of PGA should be handled gently with sterile forceps, scissors, and technique in a laminar flow tissue culture hood. Individual layers can be easily separated and cut to the desired dimensions for construct formation and placed into the appropriate-sized tissue culture dish and packaged for gas sterilization with ethylene oxide.

Implantation of cell–polymer constructs is optimally performed with optical magnification. Recipient animals should be placed in a cage maintained in a warm environment (30–35°C) in the immediate postoperative period to minimize stress.

SMALL-BOWEL HARVEST

The donors are four to eight neonatal Lewis rats weighing 6–26 g [48]. Under Methoxyflurane (Pittman-Moore Inc., Mundelein, IL) inhalation anesthesia, Cefazolin (Eli-Lilly, Indianapolis, IN) 1 g/kg is administered intramuscularly. The abdomen is shaved and prepped with 70% 2-propyl alcohol, followed by Povidone iodine, and draped in a sterile fashion. The abdominal viscera are removed through a midline laparotomy. The stomach and distal ileum are intubated with the male end of a tapered polypropylene tubing connector (0.125 in. i.d., Cole-Parmer, Chicago, cat. no. L-06539-37) and secured with a single ligature of 3-0 silk applied proximal to the connector's flange. The tubing of a primed 150-ml Buretrol (Baxter Health Care Corp., cat. no. 2C7505S, Deerfield, IL) is connected to the tubing connector–intestine complex via a three-way stopcock with male Luer Lok (Cole-Parmer, cat. no. L06464 Chicago). Peroperative lavage is performed with 30 ml of Balanced Salt Solution (BSS) (Alcon, Deerfield, IL) at room temperature to remove fecal material and mucous debris. The Buretrol is elevated 2.5 ft above the donor and allowed to perfuse the intestine by gravity drainage. Sterile applicators are used in a rolling, prograde fashion during the lavage phase to gently massage residual intestinal contents out of the lumen. The intestine is lavaged with 10 ml of Dulbecco's Modified Eagle's Medium (DMEM) (JRH Biosciences, Lenexa, KS) at 0–4°C to institute a state of cellular hypometabolism. With the intestine mildly dilated, the stopcocks are closed, and the bowel is transected proximal to the stomach and distal to the ileal stopcocks. The mesenteric attachments are sharply divided, and the bowel is measured and placed in a glass petri dish on ice in DMEM. The donor is sacrificed by overdosage with Methoxyflurane. All solutions contain Fungizone (0.25 μg/ml) (Biowhittaker, Walkersville, MD), and penicillin (100 μg/ml)–streptomycin (100 μg/ml) (JRH Biosciences, Lenexa, KS).

INTESTINAL EPITHELIAL CELL AGGREGATE ISOLATION

After harvesting of the donor bowel, the small intestine is slit longitudinally and sharply diced into sections 2–3 mm^2, transferred to a T25 flask (Corning Glass Works, Horseheads, NY), and washed eight times in 25-ml changes of Hank's Balanced Salt Solution (HBSS) (Life Tecnologies, Grand Island, NY) at 0–4°C with vigorous shaking on an orbital shaking

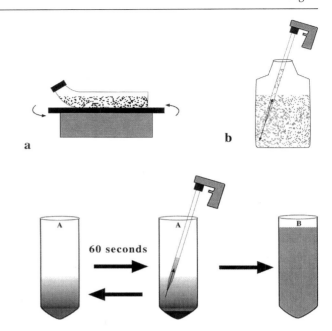

Fig. 15.1. Collagenase/dispase method for IECA isolation (Evans et al., 1992). (a) 1 mm³ diced segments of small intestine are placed into a T25 flask with 20 ml of ES on orbital shaking platform (80 cycles/min × 25 min at 25°C). (b) Pipette suspension vigorously for 30 s with 10 ml pipette and transfer to 60 ml conical tube A. (c) Allow suspension to sediment under gravity for 60 s. Carefully remove the supernatant and transfer to conical tube B at 0–4°C. Resuspend sediment in tube A with 20 ml ES and repeat step c twice. Add 10 ml of DMEM-S to conical tube B, mix and centrifuge at 8g for 2 min at 0–4°C. Carefully aspirate supernatant and discard. Resuspend pellet in tube B with 20 ml of DMEM-S and repeat centrifugation and supernatant aspiration cycles 4 times to eliminate isolated cells and residual debris. Resuspend pellet (100–200 μl) in DMEM (3–5 ml) to form final suspension for construct formation and assays.

platform at 80 cycles/min. The contents are transferred to a sterile petri dish and sharply diced into pieces 1 mm³ and transferred into a T25 flask for digestion with enzyme solution (ES): 0.1 mg/ml Dispase, neutal protease type I (Boehringer-Mannheim) and 300 U/ml collagenase type XI (Sigma) in HBSS, pH 7.4. All solutions contain penicillin (100 U/ml)–streptomycin (100 μg/ml) (Sigma) and fungizone (0.25 μg/ml) (Biowhittaker). The diced segments of small intestine are placed in a T25 flask with 20 ml of ES on an orbital shaking platform for 80 cycles/min for 25 min at 25°C (Fig. 15.1a). The suspension is pipetted vigorously for 30 s with a 10-ml pipette and transfered to 60-ml conical tube A (Fig. 15.1b). The suspension is allowed to sediment under gravity for 60 s (Fig. 15.1c), and the supernate is carefully removed and transfered to conical tube B at 0–4°C. Then the sediment is resuspended in tube A with 20 ml of ES and the gravity sedimentation is repeated twice. Next 10 ml of DMEM-S [Dulbecco's Modified Eagle's Medium, 1× high glucose, with L-glutamine (Life Technologies) + 2.5% fetal calf serum (lot no. 1007757, Life Technologies) + 2% sorbitol] is added to conical tube B, mixed, and centrifuged at 8g for 2 min at 0–4°C. After careful aspiration, the supernatant is discarded. The pellet is resuspended in tube B with 20 ml of DMEM-S, and the centrifugation–supernatant aspiration cycle is repeated four times. To form the final suspension for assays and construct formation, the pellet. (100–200 μl) is suspended in 3–5 ml of DMEM.

ASSAYS

The final suspension of isolated IECA is counted in a standard hemocytometer (Hausser Scientific, Horsham, PA). The dimensions of IECA, which are estimated to contain 600–800 cells in a single crypt aggregate, preclude performing hemocytometric counts in the standard fashion (i.e., applying the suspension to the trough of the hemocytometer and allowing the chamber to fill by capillary action). We have modified the procedure by applying 12 μl of the final suspension to the surface of the hemocytometer, placing the coverslip, and then counting four quadrants in the standard fashion. Hemocytometer counts are performed in triplicate.

The final IECA suspension is assayed for cell protein, measured in triplicate, as a separate measure of quantitative cellular mass. Viability is determined by mitochondrial reduction of a tetrazolium salt (MTT) to formazan, measured spectrophotometrically (OD 570 nm) in triplicate [53].

To quantify seeding efficiency, a cohort of IECA-PGA constructs (n = 2–4) are seeded with 500 μl of the final suspension in a separate 24-well plate and incubated for 60 min in a 5% CO_2 incubator at 37°C. Constructs are removed from their original seeding well and

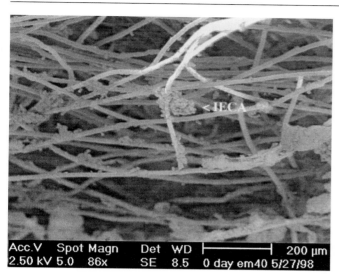

Fig. 15.2. Scanning electon micrograph of freshly seeded IECA-PGA construct. IECA is firmly adherent to PGA fiber. SEM courtesy of DJ Mooney, University of Michigan, Department of Chemical Engineering.

placed into a secondary well. MTT activity is measured in the original seeding wells, and grafting efficiency is determined by the percentage of MTT activity of the residual IECA in the original seeding well divided by the MTT activity of the final suspension. IECA cell mass delivered is derived by multiplying the MTT-generated grafting efficiency by the number of IECA seeded.

Donor neonatal intestine yields 779 ± 224 IECA/cm. When there are eight donors, the final suspension typically contains $10,645 \pm 2979$ IECA/ml with a cell protein concentration of 2.18 ± 0.08 mg/ml. Relative viability of IECA of 25.5 ± 1.5 OD/mg cell protein is to be expected. Seeding efficiency is $83\% \pm 1.1$, resulting in delivery of $4,418 \pm 1,236$ IECA/cm^2. (All data in this paragraphed expressed as standard error of the mean.)

IECA-PGA CONSTRUCT FORMATION

To begin the process, 500 µl of the final IECA suspension is gently seeded onto 1-cm^2 patches ($n = 4$–8) of nonwoven polymer fibers (15-µm diameter, pore size 150–200 µm, 4 ply at 80–100 µm thickness, Davis & Geck) of poly(glycolic acid) (PGA) in a 24-well plate. The seeding technique for optimal distribution and attachment efficiency calls for applying the final IECA suspension a single drop at a time, in a smooth, circular motion, starting at the center of the PGA and proceeding toward the edges (Fig. 15.2). An additional 500 µl of DMEM is gently added to each well. IECA-PGA constructs, grouped in cohorts of four constructs per 24-well plate, are placed in a 5% CO$_2$ incubator at 37°C for 1–2 h prior to implantation.

CONSTRUCT IMPLANTATION

Four constructs are implanted as a single cohort. Upon removal from the incubator, the IECA-PGA constructs are placed on ice. The constructs are gently rolled around a sterile Silastic stent (o.d. 2.2 mm) (Dow Corning, Midland, MI) and placed on the eviscerated omentum of a male Lewis rat (50–225 g). The omentum is atraumatically folded around the construct and secured with 7-0 Prolene (Ethicon, Somerville, NJ) sutures placed in a nonischemic fashion, parallel to the gastroepiploic arcade. The omentum is returned to the peritoneal cavity, 1–2 ml of sterile normal saline is added to the abdominal cavity for postoperative fluid resuscitation, and the incision is closed in two layers with 7-0 Maxon polyglyconate monofilament suture, T-30 taper needle (Davis & Geck).

DISCUSSION

The development of tissue-engineered neointestinal mucosa has proceeded in a systematic, stepwise fashion to its current state. Critical empiric observation and progressively more quantitative assessment of the end point of our work combined with applying advances made in the field of epithelial biology have dramatically moved this endeavor closer to clinical application.

The future of tissue engineering neointestinal mucosa must address important obstacles. Developing methods of IECA isolation from mature small intestinal epithelium is a rate-limiting obstacle for the whole field of intestinal epithelial biology. Because of the significant immunological issues involved in transplanting intestine, creating methods for isolating IECA from an SBS patient's own intestine takes on additional importance. In addition, the techniques of culturing and clonally expanding autologous IECA in a sustained fashion will be a significant advance. The appropriate strategies of reaching the critical mass needed for successful engraftment are evolving. Because the intestinal epithelium is the most rapidly proliferating tissue in the human body, this affords a unique opportunity for the field of tissue engineering and epithelial biology. Assessing the role of growth factors and extracellular matrix molecules, particularly *in vitro*, will require development of more stringent quantitative measures for these large, multicellular aggregates. Clinical demonstration of absorptive function of tissue-engineered neomucosa will also play an important role in moving this approach toward clinical applicability. The eventual attainment of a neuromuscular coat for peristalsis, as required for a functional, tissue-engineered neointestine, will have a profound impact.

Although many challenges lie ahead for the entire field of tissue engineering, tremendous strides have been made over the past decade. The foundation for therapeutic implementation of tissue-engineered organs and tissues is being established today. It is hoped and expected that someday, equipping patients with tissue-engineered organs and tissues may be as routine as coronary bypasses are today [54].

ACKNOWLEDGMENTS

The guidance of Marvin McMillan, M.D., the technical assistance of Harold Robinson, the histological evaluation of Betsy Schloo, M.D., of the Deborah Heart and Lung Center in Brown Mills, NJ, and the graphics assistance of Jay Alexander are gratefully appreciated. This work was funded by the Department of Surgical Research, Columbia Michael Reese Hospital and Medical Center, Chicago, a LISS II award from the Living Institute for Surgical Studies, University of Illinois at Chicago College of Medicine, Chicago, and Kaiser Foundation Research Institute Grant DCBI 11-97299 (OAK).

REFERENCES

1. Tilson, M. D. (1980). Pathophysiology and treatment of short bowel syndrome. *Surg. Clin. North Am.* **60**(5), 1273–1284.
2. Williamson, R. C. N. (1978). Intestinal adaptation. Part I. Structural, functional and cytokinetic changes. *N. Engl. J. Med.* **298**(25), 1393–1402.
3. Williamson, R. C. N. (1978). Intestinal adaptation. Part II. Mechanisms of control. *N. Engl. J. Med.* **298**(26), 1444–1450.
4. Grosfeld, J. L., Rescorla, F. J., and West, K. W. (1986). Short bowel syndrome in infancy and childhood: Analysis of survival in 60 patients. *Am. J. Surg.* **151**, 41–46.
5. Diego, M. D., Miguel, E., Lucen, C. M., *et al.* (1982). Short gut syndrome: A new surgical technique and ultrastructural study of the liver and pancreas. *Arch. Surg. (Chicago)* **117**, 789–795.
6. Pokorny, W. J., and Fowler, C. L. (1991). Isoperistaltic intestinal lengthening for short bowel syndrome. *Surg. Gynecol. Obstet.* **172**, 39–43.
7. Ricotta, J., Zuidema, F. D., Gadacz, R. J., and Sadri, D. (1981). Construction of an ileocecal valve and its role in massive resection of the small intestine. *Surg. Gynecol. Obstet.* **152**, 310–314.
8. Thompson, J. S., Vanderhoot, J. A., Antonson, D. L., *et al.* (1984). Comparison of techniques for growing small bowel neomucosa. *J. Surg. Res.* **36**(4), 401–406.
9. Weber, T. R., Vane, D. W., and Grosfeld, J. L. (1982). Tapering enteroplasty in infants with bowel atresia and short gut. *Arch. Surg. (Chicago)* **117**, 684–688.
10. Reyes, J., Bueno, J., Kocoshis, S., *et al.* (1998). Current status of intestinal transplantation in children. *J. Pediatr. Surg.* **33**(2), 243–254.
11. Schraut, W. H., Leck, C. W., and Tsujinaku, Y. (1986). Intestinal preservation of small bowel grafts by vascular washout and cold storage. *In* "Small Bowel Transplantation: Experimental and Clinical Fundamentals" (E. Deltz, A. Thiele, and A. Hammelman, eds.), pp. 65–73. Springer-Verlag, Heidelberg.
12. Russell, P. S. (1985). Selective cell transplantation. *Ann. Surg.* **201**, 255–262.
13. Langer, R., and Vacanti, J. P. (1993). Tissue engineering. *Science* **260**, 920–926.

14. Vacanti, J. P., Morse, M. A., Saltzman, W. M., Domb, A. J., Perez-Atayde, A., and Langer, R. (1988). Selective cell transplantation using bioabsorbable artificial polymers as matrices. *J. Pediatr. Surg.* **23**(1), 3–9.

15. Vacanti, J. P. (1988). Beyond Transplantation, Third annual Jason Mixter lecture. *Arch. Surg. (Chicago)* **123**, 545–549.

16. Carnie, A. B., Lamerton, L. F., and Steel, G. G. (1965). Cell proliferation studies in the intestinal epithelium of the rat. I. Determination of kinetic parameters. *Exp. Cell Res.* **39**, 528–538.

17. Carnie, A. B., Lamerton, L. F., and Steel, G. G. (1965). Cell proliferation studies in the intestinal epithelium of the rat. II. Theoretical aspects. *Exp. Cell Res.* **39**, 539–553.

18. Cheng, H., and Leblond, C. P. (1974). Origin, differentiation and removal of the four main epithelial cell types in the mouse small intestine. I. Columnar cell. *Am. J. Anat.* **141**, 461–479.

19. Cheng, H., and Leblond, C. P. (1974). Origin, differentiation and removal of the four main epithelial cell types in the mouse small intestine. V. Unitarian theory of the origin of the four epithelial cell types. *Am. J. Anat.* **141**, 537–562.

20. Potten, C. S., Wilson, J. W., and Booth, C. (1997). Regulation and significance of apoptosis in the stem cells of the gastrointestinal epithelium. *Stem Cells* **15**, 82–93.

21. Hulsmann, W. C., Van den Berg, J. W. O., and DeJonge, H. R. (1981). Isolation of intestinal mucosa cells. *Methods Enzymol.* **67**, 665–673.

22. Kimmich, G. A. (1975). Preparation and characterization of isolated intestinal epithelial cells and their use in studying intestinal transport. *Methods Membr. Biol.* **151**(5), 51–115.

23. Pinkus, L. M. (1981). Separation and use of enterocytes. *Methods Enzymol.* **77**, 154–162.

24. Quaroni, A., and May, R. J. (1980). Establishment and characterization of intestinal epithelial cell cultures. *Methods Cell Biol.* **21B**, 403–427.

25. Harrison, D. D., and Webster, H. L. (1969). The preparation of isolated intestinal crypt cells. *Exp. Cell Res.* **55**, 257–260.

26. Schulz, S. G., Fuisz, R. E., and Curran P. F. (1966). Amino acid and sugar transport in rabbit ileum. *J. Gen Physiol.* **49**, 849.

27. Santos, M., Nguen, B. T., and Thompson, J. S. (1992). Factors affecting in vitro growth of harvested enterocytes. *Cell Transplant.* **1**, 299–306.

28. Stern, B. L. (1966). Some biochemical properties of suspensions of intestinal epithelial cells. *Gastroenterology* **51**, 855–867.

29. Weiser, M. M. (1973). Intestinal epithelial cell surface membrane glycoprotein synthesis. I. An indicator of cellular differentiation. *J. Biol. Chem.* **248**(7), 2536–2541.

30. Weiser, M. M. (1973). Intestinal epithelial cell surface membrane glycoprotein synthesis. II. Glycosyltransferases and endogenous acceptors of the undifferentiated cell surface membrane. *J. Biol. Chem.* **248**(7), 2542–2548.

31. Moyer, M. P. (1983). Culture of human gastrointestinal epithelial cells. *Proc. Soc. Exp. Biol. Med.* **174**, 12–15.

32. Eade, O. E., Andre-Ukena, S., and Beeken, W. L. (1981). Comparative viabilities of rat intestinal epithelial cells prepared by mechanical, enzymatic and chelating methods. *Digestion* **21**, 25–32.

33. Van Corven, E., De Jong, M., and Van Os, C. (1986). Enterocyte isolation procedure specifically effects ATP-dependent calcium-transport in small intestinal plasma membranes. *Cell Calcium.* **7**, 89–99.

34. Watford, M., Lund, P., and Krebs, H. (1979). Isolation and metabolic characteristics of rat and chicken enterocytes. *Biochem. J.* **178**, 589–596.

35. Whitt, D. D., and Savage, D. C. (1988). Stability of enterocytes and certain enzymatic activities in suspensions of cells from the villous tip to the crypt of Lieberkuhn of the mouse small intestine. *Appl. Environ. Microbiol.* **54**(10), 2398–2404.

36. Whitt, D. D., and Savage, D. C. (1988). Influence of indigenous microbiota on activities of alkaline phosphatase, phosphodiesterase I, and thymidine kinase in mouse enterocytes. *Appl. Environ. Microbiol.* **54**(10), 2405–2410.

37. Evans, E. M., Wrigglesworth, J. M., Burdett, K., and Pover, W. F. R. (1971). Studies on epithelial cells isolated from guinea pig small intestine. *J. Cell Biol.* **51**, 452–464.

38. Porteous, J. W., Furneaux, H. M., Pearson, C. K., Lake, C. M., and Morrison, A. (1979). Poly(adenosine diphosphate ribose) synthetase activity in nuclei of dividing and non-dividing but differentiated intestinal epithelial cells. *Biochem. J.* **180**, 455–463.

39. Gaginella, T. S., Haddad, A. C., Go, V. W., and Phillips, S. F. (1977). Cytotoxicity of ricinoleic acid (castor oil) snd other intestinal secretagogues on isolated intestinal epithelial cells. *J. Pharmacol. Exp. Ther.* **201**(1), 259–266.

40. Gore, J., and Hoinard, C. (1989). Na$^+$/K$^+$ exchange in isolated hamster enterocytes, Its major role in intracellular pH regulation. *Gastroenterology* **97**, 882–887.

41. Organ, G. M., and Vacanti, J. P. (1997). Tissue engineering neointestine. *In* "Principles of Tissue Engineering" (R. P. Lanza, R. Langer, and W. L. Chick, eds.), pp. 441–462. Academic Press, San Diego, CA.

42. Organ, G. M., Mooney, D. J., Hansen, L. K., Schloo, B., and Vacanti, J. P. (1993). Design and transplantation of enterocyte–polymer constructs: A small animal model for neointestinal replacement in short bowel syndrome. *Surg. Forum* **44**, 432–436.

43. Organ, G. M., and Vacanti, J. P. (1995). Enterocyte transplantation for tissue engineering of neointestinal mucosa. *In* "Methods of Cell Transplantation" (C. Ricordi, ed.). R.G. Landes Press, Austin, TX.

44. Kedinger, M., Simmon-Assmann, P., Lacroix, B., Marxer, A., Hauri, H. P., and Haffen, K. (1986). Fetal gut mesenchyme induces differentiation of cultured intestinal endodermal and crypt cells. *Dev. Biol.* **113**, 474–483.

45. Kedinger, M., Simmon-Assmann, P., Alexandre, E., and Haffen, K. (1987). Importance of a fibroblastic support for in vitro differentiation of intestinal endodermal cells and for their response to glucocorticoids. *Cell Differ.* **20**, 171–182.

46. Evans, G. S., Flint, N., Somers, A. S., Eyden, B., and Potten, C. S. (1992). The development of a method for the preparation of rat intestinal epithelial cell primary cultures. *J. Cell Sci.* **101**, 219–231.

47. Choi, R. S., Riegler, M., Pothoulakis, C., *et al.* (1998). Studies of brush border enzymes, basement membrane components and electrophysiology of tissue-engineered neointestine. *J. Pediatr. Surg.* **33**(7), 991–997.

48. Organ, G. M., Kahn, A. F., Robinson, H. T., *et al.* (1998). Organotypic differentiation of tissue engineered neointestinal mucosa: The role of mesenchymal–endodermal interaction. *Surg. Forum* **49**, 583–585.

49. Organ, G. M., Robinson, H. T., *et al.* (1998). Utilization of crypt cell-enriched aggregates to tissue-engineer intestinal neomucosa. *Tissue Eng.* **4**(4), 482.

50. Kaihara, S., Kim, S. S., Benvenuto, M., *et al.* (1999). Successful anastamosis between tissue-engineered intestine and native small bowel. *Transplantation* **67**(2), 241–245.

51. Kaihara, S., Kim, S. S., Kim, B. S., *et al.* (1999). Long-term follow-up of tissue-engineered intestine after anastamosis to native small bowel. *Transplantation* **69**(9), 1927–1932.

52. Kim, S. S., Kaihara, S., Benvenuto, M. S., *et al.* (1999). Regenerative signals for intestinal epithelial organoid units transplanted on biodegradable polymer scaffolds for tissue engineering of small intestine. *Transplantation* **67**(2), 227–233.

53. Plumb, J. A., Milroy, R., and Kaye, S. B. (1989). Effects of the pH dependence of 3-(4,5-dimethylthiazol-2-yl)–2,5-diphenyl-tetrazolium bromide–formazan absorption on chemosensitivity determined by a novel tetrazolium-based assay. *Cancer Res.* **49**(16), 4435–4440.

54. Langer, R. S., and Vacanti, J. P. (1999). Tissue engineering: The challenges ahead. *Sci. Am.* **280**(4), 86–89.

CHAPTER 16

EPITHELIAL CELL CULTURE: UROTHELIUM

Jin-Yao Lai and Anthony Atala

INTRODUCTION

The urothelium forms a barrier that provides resistance between the urine and the underlying connective tissue. It is composed of multiple cell layers: a basal layer that attaches to the connective tissue substratum, an intermediate cell layer that is one to two layers thick, and a superficial cell layer composed of highly differentiated umbrella cells that line the luminal surface of the bladder [1].

A few decades ago, it was believed that urothelial cells had a natural senescence that was hard to overcome. Normal urothelial cells could be grown in the laboratory setting, but only for a limited period of time (a few weeks) and without any demonstrated expansion. The culture of urothelial cells had been hampered by inadequate media, which facilitated fibroblast overgrowth. These difficulties were initially overcome by the use of 3T3 feeder layers. More recently, the development of serum-free media has permitted the clonal growth of urothelial cells [2]. The optimization of nutrients, particularly the reduction in calcium concentration and the omission of serum, favors proliferation rather than differentiation, extending the life of the culture and permitting serial propagation [2,3].

Three methods for urothelial cell isolation have been reported.

1. Mechanical techniques, including microdissection to strip off the urothelium or directly scraping off the cells with a scalpel. These methods were reported to be successful in large mammals such as humen, pigs, and dogs [4–11]. However, the technique is difficult in small animals such as rats or mice [12].

2. Harvesting of urothelial cells from urine [13]. This method, which is associated with cells contaminated from the prostate, kidney, or vagina, is not suitable for large-scale cell culture.

3. Enzymatic urothelial cell retrieval by dispase, trypsin, or collagenase [14–17]. Such methods are most suitable for smaller animals (e.g., rats or mice) in which mechanical manipulation is difficult.

MATERIALS
Urothelial Cell Growth Medium

1. Serum-free keratinocyte growth medium (Keratinocyte SFM, Gibco, Grand Island, NY) is supplied as a basal medium to which the following aliquots (also supplied) are added: 5 ng/ml human recombinant epidermal growth factor (EGF) and 50 µg/ml of bovine pituitary extract.

2. Alternatively, the keratinocyte growth medium BulletKit (Clonetics) can be used. It is supplied as a basal medium to which the following aliquots are added: 0.1 ng/ml human recombinant EGF, 5 µg/ml insulin, 0.5 µg/ml hydrocortisone, 0.15 mM calcium, 2 ml of bovine pituitary extract, 50 µg/ml gentamicin, and 50 ng/ml amphotericin B.

Enzymes for Urothelial Cell Preparation

1. Dispase (protease type IX from *Bacillus polymyxa*): Dissolve 250 mg of dispase in 100 ml of serum-free Dulbecco's Modified Eagle's Medium (DMEM). Filter-sterilize and store in 10-ml aliquots at –20°C.

2. Trypsin

    a. *Routine subculture of cells*: Trypsin (Difco; 1:250) 0.05% solution containing 0.02% ethylenediaminetetraacetic acid (EDTA) is made up in $NaHCO_3$-buffered calcium- and magnesium-free Hanks's Balanced Salt Solution (HBSS).

    b. *Urothelial cell suspensions*: 0.25% trypsin solution containing 0.02% EDTA is made up in $NaHCO_3$-buffered calcium- and magnesium-free HBSS.

3. *Collagenase*: 0.1% collagenase (Worthington type IV) is made up in a phosphate-buffered saline solution.

PREPARATION OF UROTHELIAL CELLS

All the procedures should be carried out in a class II laminar flow hood using aseptic techniques.

MECHANICAL TECHNIQUE-MICRODISSECTION FOR PRIMARY EXPLANT [12]

1. The tissue specimen must be placed in a sterile container with a small quantity of phosphate-buffered saline or growth medium containing penicillin and streptomycin. The tissue should be processed as soon as possible.

2. The bladder is transferred to a sterile 90-mm petri dish and washed with phosphate-buffered saline or growth medium containing penicillin and streptomycin to remove red blood cells.

3. The mucosa of the bladder is dissected from the underlying submucosa under a dissecting microscope.

4. The mucosa is cut into 1-mm pieces. With the mucosal side down, these pieces are placed on a 90-mm petri dish, allowing 5–10 min for the tissue fragments to attach to the petri dish.

5. Add the urothelial cell growth medium slowly. The outgrowth of cells can usually be observed after 48 h of incubation.

MECHANICAL TECHNIQUE-SCRAPING [5,7]

1. Repeat steps 1 and 2 of the microdissection technique.

2. Transferr the bladder specimen to a petri dish containing urothelial cell growth medium. Usually for a specimen sized 1×1–2×2 cm^2, 3.5 ml of medium is adequate. The urothelial cells are detached by gentle scraping.

3. Transferr the cell suspension to six wells of a 24-well cell culture vessel (0.5 ml in each well) by means of a 5-ml pipette or a 5-ml syringe with a No. 23 needle.

4. Add another 0.5 ml of fresh growth medium to every well and incubate for 24–48 h, then change medium.

ENZYME RETRIEVAL METHOD

1. For large animals, the urothelial cells can be obtained by a combination of the microdissection and enzymatic retrieval methods [4,15,17].

    a. Wash the specimen as described for the microdissection technique.

    b. Strip the mucosa and excessive fat from the muscle by microdissection.

    c. Transferr the mucosa to a 10×10 cm^2 dish containing an 8×8 cm^2 plastic rack with 10 sharp metal pins along each edge. With the mucosa side up, the tissue is stretched and incubated overnight at 4°C in sterile urothelial growth medium containing 2.5 mg/ml dispase.

    d. Evacuate the medium–dispase solution and scarpe the epithelial cells from the underlying connective tissue with two flexible cell scrapers.

    e. Transferr the cells to a culture dish, resuspended them in 20 ml of 0.25% trypsin–1 mM EDTA, and incubate 15–30 min at 37°C.

    f. Bring the suspension up to 50 ml with growth medium containing 5% fetal bovine serum in a sterile conical tube and spin down at 1000 rpm for 5 min to pelletize the cells and remove the trypsin.

    g. Carefully remove the supernatant and wash the cells in growth medium.

    h. Resuspend the cells in a final concentration of $6–8 \times 10^5$ cells/ml and are plate them.

2. For small animal such as rats or mice [14,16]

    a. Irrigate the bladders with phosphate-buffered saline or antibiotic-containing growth medium.

    b. Evert the bladders, with the luminal side facing outward.

    c. Close the bladder neck surgically to make "bladder balls."

    d. Immerse the bladders in 2–4 ml of 1% collagenase (type IV) and place them in a shaking water bath at 37°C for 1 h.

    e. Remove the bladders and centrifuge the suspension at 1000 rpm for 4 min.

    f. Wash the pellets with HBSS.

    g. Resuspend in growth medium and seed the cells at a concentration of 5×10^6 cells/ml.

SEEDING OF UROTHELIAL CELLS

Urothelial cells can be seeded onto tissue culture dishes, onto 3T3 feeder layers, or onto extracellular matrices. In our experience, the extracellular matrix is less successful than 3T3 feeder layers.

1. Tissue culture dish

    a. Diluted the suspension of cells in an appropriate volume and seed onto the tissue culture dish at a density of $6–8 \times 10^5$ cells/cm^2.

    b. After the cells have attached (1–2 days), change the medium every 2–3 days. Cultures are incubated in 37°C.

    c. The cells should approach confluence within 10 days and should be subcultured before they reach confluence, Cultures held at confluence may differentiate.

2. 3T3 feeder layers

    a. Grow 3T3 cells in DMEM with 10% fetal bovine serum.

    b. For preparation of the feeder layer, lethally irradiat confluent cultures of 3T3 cells for 35 min with a dose of 4500 rads from a cobalt source.

    c. After irradiation, trypsinize the 3T3 cells and replate them at a 1–10 dilution. Within 2 h, inoculate the urothelial cells on the irradiated 3T3 cells.

    d. Seed the urothelial cells at a concentration of 2.5×10^4 cells/cm^2 in urothelial growth medium.

    e. Clones of epithelial cells may not be apparent among the 3T3 feeder cells for several days, but with increasing incubation, small islands of 20–50 cells should be easily visible.

    f. The culture should approach confluence in 10 days. The urothelial cells may be subcultured at seeding densities of $1.25–2.5 \times 10^4$ cells/cm^2 onto the new irradiated 3T3 cells.

3. Extracellular matrix (ECM)

    a. ECM may be made from 3T3 cells or human fibroblasts. Once confluent, the cultured cells are lysed with 1% Triton-X-100 in 0.0125 M ammonium hydroxide for 5–10 min and washed in balanced salt solution twice before use.

    b. Urothelial cells may be seeded onto the matrix at a density no less than 5×10^4 cells/cm^2.

    c. Change medium after the cells have become attached (1–2 days).

SUBCULTURE OF UROTHELIAL CELLS

1. Incubate nearly confluent dishes of urothelial cells in phosphate-balanced salt solution with 0.05% EDTA for at least 5 min.
2. Remove the phosphate-balanced salt solution with 0.05% EDTA.
3. Add 0.05% trypsin–EDTA and incubate for 2–3 min.
4. Cell lifting can be observed under the microscope.
5. Trypsinization is stopped by adding soybean trypsin inhibitor as soon as cell lifting is complete.
6. Transfer the cells to a 50-ml conical tube and centrifuge at 1000 rpm for 4 min.
7. Remove the supernatant. Resuspend the pellet in an appropriate density and replateit to the dish.

REFERENCES

1. Hicks, R. M. (1975). The mammalian urinary bladder: An accommodating organ. *Biol. Rev. (Cambridge Philos. Soc.)* **50**(2), 215–246.
2. Tobin, M. S., Freeman, M. R., and Atala, A. (1994). Maturational response of normal human urothelial cells in culture is dependent on extracellular matrix and serum additives. *Surg. Forum* **45**, 786.
3. Barlow, Y., and Pye, R. J. (1997). Keratinocyte culture. *Methods Mol. Biol.* **75**, 117–119.
4. Rahman, Z., Reedy, E. A., and Heatfield, B. M. (1987). Isolation and primary culture of urothelial cells from normal human bladder. *Urol. Res.* **15**(6), 315–320.
5. Cilento, B. G., Freeman, M. R., Schneck, F. X., *et al.* (1994). Phenotypic and cytogenetic characterization of human bladder urothelia expanded in vitro. *J. Urol.* **152**(2, pt. 2), 665–670.
6. Southgate, J., Hutton, K. A., Thomas, D. F., *et al.* (1994). Normal human urothelial cells in vitro: Proliferation and induction of stratification. *Lab. Invest.* **71**(4), 583–594.
7. Wolf, J. S., Soble, J. J., Ratliff, T. L., *et al.* (1996). Ureteral cell cultures. I. Characterization and cellular interactions. *J. Urol.* **156**(3), 1198–1203.
8. Guhe, C., and Follman, W. (1994). Growth and characterization of porcine urinary bladder epithelial cells in vitro. *Am. J. Physiol.* **266**(2, pt. 2); F298–F308.
9. Fauza, D. O., Fishman, S. J., Mehegan, K., *et al.* (1998). Videofetoscopically assisted fetal tissue engineering: Bladder augmentation. *J. Pediatr. Surg.* **33**(1), 7–12.
10. Yoo, J. J., Meng, J., Oberpenning, F., *et al.* (1998). Bladder augmentation using allogenic bladder submucosa seeded with cells. *Urology* **51**(2), 221–225.
11. Rosenbloom, J., Abrams, W. R., Rosenbloom, J., *et al.* (1997). Expression of microfibrillar proteins by bovine bladder. *Urology* **49**(2), 287–292.
12. Johnson, M. D., Bryan, G. T., Reznikoff, C. A. (1985). Serial cultivation of normal rat bladder epithelial cells in vitro. *J. Urol.* **133**(6), 1076–1081.
13. Linder, D. (1976). Culture of cells from the urine and bladder washing of adults. *Somatic Cell Genet.* **2**(3), 281–283.
14. Chlapowski, F. J., and Haynes, L. (1979). The growth and differentiation of transitional epithelium in vitro. *J. Cell Biol.* **83**(3), 605–614.
15. Truschel, S. T., Ruiz, W. G., Shulman, T. *et al.* (1999). Primary uroepithelial cultures: A model system to analyze umbrella cell barrier function. *J. Biol. Chem.* **274**(21), 15020–15029.
16. van der Kwast, T. H., van Rooy, H., and Mulder, A. H. (1989). Establishment and characterization of long-term primary mouse urothelial cell cultures. *Urol. Res.* **17**(5), 289–293.
17. Bonar, R. A., Reich, C. F., and Sharief, Y. (1977). Canine urinary bladder epithelial cells: Preparation for cell cultures by enzyme dispersion. *Urol. Res.* **5**(2), 87–94.

CHAPTER 17

EPITHELIAL CELL CULTURE: PROSTATE

Donna M. Peehl and Robert G. Sellers

P rostatic cell cultures provide a valuable tool with which to investigate the biology of the normal prostate and changes that take place in benign and malignant diseases. From early work with explant cultures in relatively uncontrolled conditions, techniques have advanced to include monolayer culture of prostatic epithelial as well as stromal cells in serum-free media and in sophisticated three-dimensional histocultures. Characterization of these cultures and validation as representative models of prostatic structure and function improve as information is gained from microarray analyses of gene expression in tissues and cells. Transformed cell lines created by the introduction of immortalizing or oncogenic genes and spontaneously immortalized, tumorigenic cell lines established from primary cancers or metastatic tissues now span the range of prostatic carcinogenesis. New xenograft and transgenic models of prostate cancer and derived cell lines permit *in vitro/in vivo* comparative studies.

INTRODUCTION

Initial attempts to culture prostatic cells, which date back to the late 1970s, utilized primarily explant culture techniques [1–5]. During the 1980s, investigators focused on developing monolayer cultures from disaggregated cell suspensions and formulating optimized serum-free culture media [6–8]. The 1990s introduced additional complexity into tissue culture design, with emphasis on coculture of epithelial and mesenchymal cells [9] and three-dimensional growth in reconstituted basement membranes or gels [10–15]. Cell culture models to simulate interactions between prostatic cancer cells and bone, the primary site of metastasis of prostate cancer, were initiated [16]. Taking advantage of space age technology, scientists even grew prostatic organoids under microgravity-simulated conditions in efforts to promote normal histogenesis and differentiation [17,18]. Currently, many investigators hope to take advantage of the identification of new markers of cell lineage and differentiation and malignant potential that are being derived from gene microarray analyses to better characterize prostatic cell culture models. Efforts to develop additional established cell lines and xenograft or transgenic animal models that reflect the spectrum of early and advanced stages of prostate cancer progression are also being emphasized.

PROSTATIC EPITHELIAL CELL CULTURES
PRIMARY CULTURES

A variety of more or less similar methods are available to establish primary cultures of human prostatic epithelial cells derived from normal, benign prostatic hyperplasia (BPH), or malignant tissues. Small tissue samples are obtained from radical or suprapubic prostatectomy specimens, from transurethral resections of the prostate, or from biopsy material. Specimens are processed in such a way that histological assessment of the tissue of origin can be obtained from adjacent fixed or frozen sections. Tissues are typically disaggregated with enzyme cocktails, then single cells or small clumps of cells are inoculated into dishes

Table 17.1. Primary Culture of Human Prostatic Epithelial Cells

1. Collect tissue in tube containing 5 ml of cold, sterile HBS (HEPES-buffered saline: see Table 17.6).
2. Store tissue at 4°C for up to several hours if unable to process immediately.
3. Transfer prostate tissue from tube into sterile dish (one piece per 60-mm dish).
4. Rinse tissue twice with HBS (5 ml/rinse).
5. Leave tissue in final HBS rinse. Use sterile scissors and tweezers to mince into small pieces (~1 mm^3).
6. With a broken-tip, stuffed pipette, transfer pieces to a sterile 15-ml centrifuge tube. Pellet pieces (full-speed, clinical centrifuge, 5 min).
7. Resuspend pieces in "complete PFMR-4A" (Table 17.3) containing 40 U/ml type 1 collagenase (Sigma-Aldrich, St. Louis, MO; cat. no. C-0130).
8. Rock at 37°C overnight.
9. Pipette digested tissue vigorously with a 5-ml pipette to break up clumps.
10. Pellet tissue (full-speed, clinical centrifuge, 5 min).
11. Rinse twice with HBS, repeating steps 9 and 10.
12. Resuspend pellet in 5 ml of complete PFMR-4A and transfer to one 60-mm, collagen-coated dish (Table 17.2). Incubate at 37°C in humidified incubator containing 95% air and 5% CO$_2$.

containing specialized media which have been optimized for the growth of prostatic epithelial cells. The methods used by our laboratory [19] are shown in Tables 17.1–17.8.

The protocol that we use for establishing primary cultures of human prostatic epithelial cells is given in Table 17.1. Typically, we dissect a small piece of tissue (~10 mm^3) from a radical prostatectomy specimen. The cut edges of the specimen are inked to mark the area, then the prostate is fixed, cut, and blocked. Histological assessment of tissue sections adjacent to the area removed for culture reveals whether the tissue of origin was normal, BPH, or malignant. The tissue for culture is minced, digested overnight with collagenase, and inoculated into collagen-coated dishes (Table 17.2) containing complete PFMR-4A (Tables 17.3–17.5). After approximately one week of incubation, at which time a million or more cells have grown from the original tissue specimen, the cells are aliquoted and frozen at 10,000 cells per ampule to maintain the stock (Tables 17.6–17.8). As needed, ampules are thawed and cells are serially passaged in complete MCDB 105 (Table 17.8).

Variations on Primary Culture of Prostatic Epithelial Cells

We have described modifications of our standard methods for the culture of prostatic epithelial cells derived from material acquired by way of needle biopsies instead of from radical prostatectomy specimens [20]. Bologna *et al.* [21] have described a method of culturing prostatic epithelial cells from prostatic fluid obtained by massage. Other formulations for media for establishment and propagation of human prostatic epithelial cells have been

Table 17.2. Collagen-Coated Dishes

Add 1 part of cold collagen type I (Vitrogen) (Cohesion Technology, Palo Alto, CA) to 3 parts of sterile 0.013 N HCl; mix.
Using sterile plastic pipettes, distribute 0.2-ml aliquots into 35-mm dishes, 0.3-ml aliquots into 60-mm dishes, or 0.5-ml aliquots into 100-mm dishes.
Tilt and shake dishes until collagen solution is evenly distributed over bottom of each dish.
Stack in cupboards.
Allow to dry at least 3 days before use.

Table 17.3. Complete PFMR-4A: Mix together sterile components; store up to 2 weeks at 4°C

Component[a]	Concentration of stock	Amount to add	Final concentration
PFMR-4A		500 ml	
CT	100 μg/ml	50 μl	10 ng/ml
EGF	100 μg/ml	50 μl	10 ng/ml
BPE	14 mg/ml	1.4 ml	40 μg/ml
PEA	0.1 M	500 μl	0.1 mM
HC	10 mg/ml	50 μl	1 μg/ml
SE	3×10^{-4} M	50 μl	3×10^{-8} M
GENT	40 mg/ml	1.25 ml	100 μg/ml
RA	0.1 μg/ml	50 μl	0.01 ng/ml
IN	4 mg/ml	500 μl	4 μg/ml
VIT E	2.3×10^{-2} M	50 μl	2.3×10^{-6} M

[a]See Table 17.4 for preparation and storage of PFMR-4A; for preparation and storage of all other components (abbreviations as in Table 17.5), see Table 17.5.

described [22–34]. Epithelial cells have also been grown from the prostates of rodents by a variety of methods [6,35–38].

Characteristics of Primary Cultures of Prostatic Epithelial Cells

Two distinctive populations of cells comprise the epithelia of normal or BPH tissues. These are the basal cells, believed to contain the stem cells of the epithelium, and the differentiated luminal secretory cells [39]. These cells are distinguishable by a variety of markers, which have been applied to cultures of prostatic epithelial cells. Labeling patterns derived from immunocytochemical evaluation of cell cultures with these markers have not yielded definitive answers regarding the status of these cells. We have reported that all cells in our primary cultures of epithelial cells express keratin 5, a marker of the basal phenotype, but that a subpopulation also simultaneously expresses keratins 8 and 18, markers of the secretory, differentiated phenotype [40]. Concurrent expression of both basal and secretory markers is seen in regenerating epithelial cells of the prostate [41], suggesting that the phenotype of cultured cells is that of a replicative, regenerating epithelium. New techniques to catalog gene and protein expression of purified subpopulations of cells from tissues will undoubtedly reveal a plethora of novel markers for application to cell cultures [42].

Table 17.4. PFMR-4A

	STOCK 1A, 50×		
Component	MW (Da)	Concentration in stock	Amount in stock (g/2 liters)
L-Arginine · HCl	210.7	1×10^{-1} M	42.14
Choline chloride	139.6	5×10^{-3} M	1.40
L-Histidine · HCl · H$_2$O	209.6	1×10^{-2} M	4.19
L-Isoleucine	131.2	3×10^{-2} M	7.87
L-Leucine	131.2	1×10^{-1} M	26.24
L-Lysine · HCl · H$_2$O	182.7	2×10^{-2} M	7.31
L-Methionine	149.2	3×10^{-2} M	8.95

Gentle warming while stirring will help components dissolve (but not too hot!). After thawing, must warm at 37°C to dissolve precipitate. Storage: –20°C for 1 year in 400-ml aliquots.

(continues)

Table 17.4. (continued)

| STOCK 1B, 50× | | | |
Component	MW (Da)	Concentration in stock	Amount in stock (g/2 liters)
L-Phenylalanine	165.2	3×10^{-3} M	0.99
L-Serine	105.1	1×10^{-2} M	2.10
L-Threonine	119.1	1×10^{-1} M	23.82
L-Tryptophan	204.2	1×10^{-3} M	0.41
L-Valine	117.2	1×10^{-2} M	2.34

After thawing, warm at 37°C to dissolve precipitate. Storage: −20°C for 1 year, dark, in 400-ml aliquots.

| STOCK 1C, 50× | | | |
Component	MW (Da)	Concentration in stock	Amount in stock (g/2 liters)
L-Tyrosine	181.2	3×10^{-3} M	1.09

Dissolve solid in 100 ml of 4 N NaOH, dilute to 2 liters with H_2O. Storage: −20°C for 1 year in 400-ml aliquots.

| STOCK 2, 100× | | | |
Component	MW (Da)	Concentration in stock	Amount in stock (g/liter)
Biotin	244.3	3×10^{-5} M	0.007
Ca · pantothenate	238.3	1×10^{-4} M	0.024
Niacinamide	122.1	3×10^{-5} M	0.004
Pyridoxine · HCl	205.7	3×10^{-5} M	0.006
Thiamine · HCl	337.5	1×10^{-4} M	0.034
KCl	74.6	3.8×10^{-1} M	28.348

Storage: −20°C for 1 year in 200-ml aliquots.

| STOCK 3, 100× | | | |
Component	MW (Da)	Concentration in stock	Amount in stock (g/liter)
$Na_2HPO_4 \cdot 7H_2O$	268.1	8.1×10^{-2} M	21.72
Folic acid	441.4	3×10^{-4} M	0.13

Completely dissolve $Na_2HPO_4 \cdot 7H_2O$ before adding folic acid. Be sure that folic acid completely dissolves. Storage: −20°C for 1 year, dark, in 200-ml aliquots.

(continues)

EPITHELIAL CELL LINES

Immortal cell lines are not readily derived from human prostatic tissues. For many years, most investigators relied on three cell lines to use as experimental models of prostate cancer. These were PC-3, LNCaP, and DU 145, all derived from metastases, for review, see Peehl [43]. Other less commonly used cell lines include TSU-Pr1, JCA, and ALVA-31, for review, see Bosland *et al.* [44].

Variations of Cell Lines

Immortal and/or tumorigenic prostatic cell lines representing the spectrum of prostate cancer progression have been developed from primary cultures by the introduction of oncogenes or exposure to chemicals or radiation, for reviews, see [45–47]. New lines of serially transplantable prostatic xenografts have been developed as well—for review, see

Table 17.4. (continued)

Component	MW (Da)	Concentration in stock	Amount in stock (g/500 ml)
	STOCK 4, 100×		
$FeSO_4 \cdot 7H_2O$	278.0	3×10^{-4} M	0.04
$MgCl_2 \cdot 6H_2O$	203.3	5.2×10^{-2} M	5.29
$MgSO_4 \cdot 7H_2O$	246.5	1.6×10^{-2} M	1.97
$CaCl_2 \cdot 2H_2O$	147.0	9.2×10^{-2} M	6.76

Add 5 drops of concentrated HCl per 500 ml of stock 4 to prevent precipitate of ferric hydroxides. Storage: room temperature for 1 year in bottles tightly sealed with caps wrapped in Parafilm, sterile, in 50-ml aliquots.

Component	MW (Da)	Concentration in stock	Amount in stock (g/100 ml)
	STOCK 5, 1000×		
Phenol red · Na salt	376.4	5.9×10^{-3} M	0.22

Storage: room temperature for 1 year, sterile, in 20-ml aliquots.

Component	MW (Da)	Concentration in stock	Amount in stock (g/liter)
	STOCK 6B, 100×		
Sodium pyruvate	110.0	2×10^{-1} M	22.0

Storage: −20°C for 1 year in 200-ml aliquots.

Component	MW (Da)	Concentration in stock	Amount in stock (g/liter)
	STOCK 6C, 100×		
Riboflavin	376.4	1×10^{-5} M	0.004

Riboflavin is difficult to dissolve and requires extensive stirring. Cover with foil while stirring. Storage: −20°C for 1 year, dark, in 200-ml aliquots.

Component	MW (Da)	Concentration in stock	Amount in stock (g/200 ml)
	STOCK 7, 100×		
L-Cystine	240.3	1.5×10^{-2} M	0.72

Dissolve in ~180 ml of H_2O; while stirring, add concentrated HCl dropwise until solubilized. Bring to final volume of 200-ml with H_2O. Make fresh.

(continues)

Stearns *et al.* [48] and several immortal cell lines have been derived from these xenografts [49].

Characteristics of Cell Lines

Many of the established prostatic cell lines express few features that distinguish them from generic undifferentiated epithelial cells. However, some have unique features that make them suitable for particular lines of investigation. The LNCaP cell line, for example, has long been used as a model of androgen regulation because it expresses androgen receptors (albeit mutated) and prostate-specific antigen (PSA) [50]. Derivatives of this cell line that preferentially migrate to bone provide a much-needed model of metastasis [51]. The ability of some of the prostatic cancer cell lines to express neuroendocrine markers in response to appropriate stimuli reflects the occurrence of this phenotype in tissues [52].

Table 17.4. (continued)

STOCK 8, 100×			
Component	**MW (Da)**	**Concentration in stock**	**Amount in stock (g/liter)**
L-Asparagine	150.1	2×10^{-2} M	3.00
L-Proline	115.1	6×10^{-2} M	6.91
Putrescine · 2HCl	161.1	2×10^{-4} M	0.032
Vitamin B_{12}	1355.4	1×10^{-4} M	0.136

Storage: −20°C for 1 year, dark, in 200-ml aliquots.

STOCK 9, 100×			
Component	**MW (Da)**	**Concentration in stock**	**Amount in stock (g/liter)**
L-Aspartate	133.1	2×10^{-2} M	2.66
L-Glutamate	147.1	2×10^{-2} M	2.94
L-Alanine	89.09	2×10^{-2} M	1.78
Glycine	75.1	2×10^{-2} M	1.50

Aspartate and glutamate are added to 900 ml of H_2O containing 1 ml of stock 5 (phenol red). While stirring, 4 N NaOH is added dropwise to maintain neutrality (pinkish orange) as the aspartate and glutamate dissolve. Alanine and glycine are then added and dissolved. Bring to final volume of 1 liter with H_2O. Storage: −20°C for 1 year in 200-ml aliquots.

STOCK 10, 100×			
Component	**MW (Da)**	**Concentration in stock**	**Amount in stock (g/liter)**
Hypoxanthine	136.1	3×10^{-3} M	0.408
6,8-Thioctic acid	206.3	1×10^{-4} M	0.021
Myo-inositol	180.2	1×10^{-1} M	18.02
Thymidine	242.2	3×10^{-3} M	0.73
$CuSO_4 \cdot 5H_2O$	249.7	1×10^{-6} M	0.00025

Hypoxanthine is dissolved in 100 ml of boiling H_2O before adding to stock. Thioctic acid is dissolved in a few drops of 1 N NaOH, diluted with 10 ml of H_2O, then added to stock. $CuSO_4 \cdot 5H_2O$ is prepared by dissolving 0.025 g in 1 liter of H_2O; 10 ml of this is added to the stock. Storage: −20°C for 1 year, dark, in 200-ml aliquots.

STOCK 11, 1000×			
Component	**MW (Da)**	**Concentration in stock**	**Amount in stock (g/100 ml)**
$ZnSO_4 \cdot 7H_2O$	287.5	5×10^{-4} M	0.014

Storage: room temperature for 1 year, sterile, in 10-ml aliquots.

(continues)

Reviews describing the properties of most of the available cell lines can be found in Peehl, Webber *et al.* [43,44,46,47,53].

PROSTATIC STROMAL CELL CULTURES
PRIMARY CULTURES

Experimental models for studying the biology of prostatic epithelial cells would not be complete without the inclusion of prostatic stromal cells. Mesenchymal–epithelial interactions are key elements in prostatic development, differentiation, hormonal response,

Table 17.4. (continued)

Final preparation of PFMR-4A (20 liters)

Add to ~14 liters of H$_2$O in carboy while stirring

Stock	Concentration	Amount (ml)
1A	50×	400
1B	50×	400
1C	50×	400
2	100×	200
3	100×	200
5	1000×	20
6B	100×	200
6C	100×	200
7	100×	200
8	100×	200
9	100×	200
10	100×	200

Add the following dry ingredients

Component	MW (Da)	Final concentration	Amount (g/20 liter)
Glucose	180.2	7×10^{-3} M	25.23
NaCl	58.45	1×10^{-1} M	116.9
KH$_2$PO$_4$	136.1	4.3×10^{-4} M	1.17
L-Glutamine	146.1	2×10^{-2} M	58.44
HEPES	238.3	3×10^{-2} M	142.98

Bring pH to 7.4 with 4 N NaOH, then add

Component	MW (Da)	Final concentration	Amount (g/20 liter)
NaHCO$_3$	84.0	1.4×10^{-2} M	23.52

Bring to a final volume of 19,780 ml with H$_2$O. Mix well. Filter sterilize through 0.22-μm filter. Storage: −20°C for 1 year, sterile, in 494.5-ml aliquots. To finish preparation, thaw, then add 5 ml of sterile Stock 4 and 0.5 ml of sterile Stock 11 to each aliquot of 494.5 ml. Store at 4°C for up to 2 months.

and carcinogenesis [54,55]. To initiate stromal cell cultures, prostatic tissues are typically obtained, characterized, and processed by means of methods similar to those used to establish epithelial cell cultures. However, different substrata and media are used to promote the preferential growth of stromal cells instead of epithelial cells. The method used by our laboratory to establish primary cultures of human prostatic stromal cells [56] is shown in Table 17.9. Tissues are procured and processed as for epithelial cell culture. However, the digested tissue fragments are inoculated into medium MCDB 105 with 10% fetal bovine serum (FBS) instead of into complete PFMR-4A, and dishes are not coated with collagen. Epithelial cells also attach, but they degenerate over time and are lost after one or two passages.

Variations on Culture of Prostatic Stromal Cells

Methods describing the culture of human prostatic stromal cells have been published by several investigators [23,25,57–61]. Stromal cells are also cultured from rodent prostates by techniques slightly different from those used for human tissues [38]. Preliminary efforts to develop optimal serum-free culture systems for prostatic stromal cells have been described [27,38,56,62].

Table 17.5. Media Supplements

Cholera toxin (CT)

Add 1 ml of sterile H_2O to sterile vial containing 1 mg of cholera toxin (List Biological Laboratories, Campbell, CA). Swirl to dissolve. Transfer 1 ml of this 1 mg/ml solution to a sterile tube containing 9 ml of sterile H_2O; mix. Transfer ~1-ml aliquots of this 100 µg/ml solution to sterile tubes; store at 4°C. Do not freeze.

Epidermal growth factor (EGF)

Add 1 ml of sterile H_2O to sterile vial containing 100 µg of epidermal growth factor (Collaborative Biomedical Products, Bedford, MA). Swirl to dissolve. Transfer ~100-µl aliquots of this 100 µg/ml solution to sterile tubes; store at –20°C until needed, then thaw and store at 4°C for 2 months. Do not refreeze.

Bovine pituitary extact (BPE)

BPE is obtained from Hammond/Cell Tech (Alameda, CA) and is stored at –70°C. To prepare for use in cell culture, thaw a bottle at 37°C and swirl to mix. Centrifuge at 13,218g for 10 min to pellet insoluble material. Pass supernatant through a Whatman no. 1 filter, followed by passage through 0.8- and 0.45-µm filters. Finally, sterile filter through 0.22-µm filters (may take several). The extract concentration is 14 mg/ml. Store 2-ml aliquots in sterile tubes at –20°C until needed, then thaw and store at 4°C for 2 months. Do not refreeze.

Phosphoethanolamine (PEA)

Prepare a 0.1 M solution by dissolving 0.7 g of *o*-phosphorylethanolamine (Sigma-Aldrich, St. Louis, MO; cat. no. P-0503) in ~45 ml of H_2O; bring to final volume of 50 ml. Filter sterilize, then store 5-ml aliquots in sterile tubes at –20°C. May be frozen and thawed.

Hydrocortisone (HC)

Prepare a 10 mg/ml solution by dissolving 100 mg of hydrocortisone (Sigma-Aldrich, St. Louis, MO; cat. no. H-0888) in 10 ml of 100% ethanol. Distribute 1-ml aliquots into sterile tubes. Store tightly sealed at room temperature.

Selenium (SE)

Prepare a 3×10^{-2} M solution by dissolving 0.039 g of selenous acid (H_2SeO_3) (J.T. Baker, Phillipsburg, NJ; cat. no. 0310-4) in 10 ml of H_2O. Filter-sterilize, then transfer 0.1 ml of this solution to 9.9 ml of sterile H_2O; mix. Distribute 1-ml aliquots of this sterile 3×10^{-4} M solution into sterile tubes; store at room temperature.

Gentamicin (GENT)

Prepare a 40 mg/ml solution by dissolving 4 g of gentamicin sulfate (Gemini Bio-Products, Woodland, CA) in 100 ml of H_2O. Filter-sterilize. Distribute 5-ml aliquots into sterile tubes; store at room temperature.

Retinoic acid (RA)

Prepare a 1 mg/ml (3×10^{-3} M) solution by dissolving 1 mg of all-*trans* retinoic acid (Sigma-Aldrich, St. Louis, MO; cat. no. R-2625) in 1 ml of DMSO. This solution is considered to be sterile. Store at –20°C, dark, in 25-µl aliquots in sterile amber tubes. May be frozen and thawed twice; then discard. Transfer 20 µl of a 1 mg/ml stock solution of all-*trans* retinoic acid to a tube containing 1.98 ml of DMSO. Mix well. Transfer 100 µl of this 10 µg/ml solution to a tube containing 9.9 ml of DMSO. Mix well. This solution is considered sterile and is 0.1 µg/ml (3×10^{-7} M). Store at –20°C, dark, in 250-µl aliquots in sterile amber tubes. Thaw fresh each day and discard any leftover.

(continues)

Characterization of Prostatic Stromal Cells

The prostate is somewhat unique in that a large proportion of its stroma is composed of smooth muscle [63]. The science of prostatic stromal cell culture has matured from simply growing fibroblast-like cells that were incompletely characterized to careful consideration and documentation of smooth muscle differentiation [64]. A number of laboratories have described culture conditions that promote the development of the smooth muscle phenotype in prostatic stromal cells *in vitro*. We reported that transforming growth factor β (TGF-β) promotes the expression of smooth muscle markers in cultured human prostatic stromal cells [65]. Other factors, such as basic fibroblast growth factor, epidermal growth factor,

Table 17.5.   (continued)

Insulin (IN)

Add 25 ml of sterile H_2O to sterile vial containing 100 mg of insulin (Sigma-Aldrich, St. Louis, MO; cat. no. I-2767). Swirl to dissolve. Distribute ~2-ml aliquots into sterile tubes. Store at −20°C. After thawing an aliquot, store at 4°C for up to 2 months (do not refreeze).

Vitamin E (VIT E)

Transfer 100 μl of 100% α-tocopherol (Sigma-Aldrich, St. Louis, MO; cat. no. T-3634) to a sterile tube containing 9.9 ml of DMSO. Tocopherol is very sticky, so try to get all of it rinsed off the tip into the DMSO. Vortex to solubilize. This solution is considered sterile and is 2.3×10^{-2} M. Store at −20°C in the dark, in 1-ml aliquots in sterile amber tubes. May be frozen and thawed.

and platelet-derived growth factor, block the induction of the smooth muscle phenotype by TGF-β [66]. Other investigators have reported that differentiation of cultured prostatic stromal cells is modulated by adrenergic receptor antagonists [67,68], estradiol, dexamethasone, and androgen [69]. It will be interesting to identify the molecular pathways that regulate smooth muscle differentiation in prostatic stromal cells and to study that information in comparison to pathways delineated in smooth muscle cells from other tissues.

STROMAL CELL LINES

Immortal lines of prostatic stromal cells are few in number, but several have been derived from the rat ventral prostate [70] and the human prostate [71,72].

Table 17.6.   Culture Solutions

Component	HEPES-buffered saline (HBS)	
	Concentration in solution (mM)	Amount in solution (g/20 liters)
HEPES	30	238.3
Glucose	4	14.42
KCl	3	4.48
NaCl	130	151.94
$Na_2HPO_4 \cdot 7H_2O$	1	5.36
Phenol red · Na salt	0.0033	0.02

Dissolve HEPES in ~19 liters of H_2O. Bring pH to 7.6 with 4 N NaOH. Add rest of components, dissolve, and bring to final volume of 20 liters with H_2O. Storage: −20°C for 1 year, sterile, in 500-ml aliquots. Store thawed HBS at 4°C for several months.

0.2% Trypsin/0.2% EDTA (T/E)

Dissolve 0.4 g of trypsin (Sigma-Aldrich, St. Louis, MO; cat. no. T-8003) and then 0.4 g of EDTA · Na_4 salt in 180 ml of HBS.
Bring to final volume of 200 ml with HBS.
Storage: −20°C, sterile, in 5- or 10-ml aliquots. Thaw just before use. Do not freeze/thaw more than a few times.

0.1% Trypsin inhibitor (TI)

Dissolve 1 g of trypsin inhibitor (Sigma-Aldrich, St. Louis, MO; cat. no. T-9003) in 900 ml of HBS.
Bring to final volume of 1000 ml with HBS.
Storage: 4°C, sterile, in 50-ml aliquots.

Table 17.7. Freezing and Thawing Prostatic Epithelial Cells

To freeze

1. Rinse cells with HBS (see Table 17.6).
2. Add T/E (see Table 17.6) (0.25 ml/60-mm dish, 0.5–1 ml/100-mm dish).
3. Incubate at 37°C until cells detach (1–5 min).
4. Resuspend cells with TI (see Table 17.6). Use ~5 ml to resuspend the cells and transfer to a centrifuge tube, then another ~5 ml to rinse the dish again.
5. Pellet cells (full-speed, clinical centrifuge, 5 min).
6. Resuspend cell pellet in complete PFMR-4A (see Table 17.3) plus 10% DMSO and 10% FBS at 10^4 cells/ml. Transfer 1-ml aliquots of cell suspensions in freezing medium to each cryotube. Seal tightly.
    When freezing multiple ampules of cells (>20), keep the cells resuspended in freezing medium on ice while aliquotings. (At every aliquot of 10 or so, ampules should be transferred to 4°C.)
7. Transfer ampules to 4°C. Do not put in styrofoam trays or similar carriers; be sure that air can freely circulate around the ampules so that they will cool quickly. Store at 4°C for exactly 3 h.
8. Transfer ampules to –70°C, overnight (again, be sure that air can freely circulate around the ampules).
9. Transfer ampules to liquid N_2 freezer.

To thaw

1. Remove ampules from liquid N_2 freezer.
2. Place ampules in 37°C water bath.
3. Agitate ampules to thaw quickly.
4. After making sure that each ampule is tightly sealed, squirt area around lip with 95% EtOH. Use a Kimwipe to blot excess EtOH from depression in cryotube cap, but do not touch lip area.
5. Let dry thoroughly.
6. Transfer contents of each ampule to a centrifuge tube containing 10 ml of HBS. Pellet cells (full-speed, clinical centrifuge, 5 min).
7. Resuspend cell pellet in complete MCDB 105 (see Table 17.8) and transfer to a collagen-coated dish (see Table 17.2). Incubate at 37°C in humidified incubator with 95% air/ 5% CO_2.
8. Feed cells every 3–4 days with complete MCDB 105 and passage when ~50% confluent.

NOVEL APPLICATIONS OF PROSTATIC CELL CULTURES

One of the goals of researchers working with prostatic cell cultures is to create experimental models that accurately reflect the proliferation, differentiation, apoptotic, and transformation processes that occurred in the tissues of origin. The prostate is a complex organ, so this goal has been difficult to achieve. In particular, attention to several essential elements is believed to be key to achieving this goal. One element is the elaborate three-dimensional architecture and glandular structure of the prostate. The other element is the stromal–epithelial interactions that are intrinsic to the three-dimensional architecture but also to growth regulation and tissue homeostasis. These elements have not yet been perfectly attained in culture, but recent work has brought these goals closer to reality.

STROMAL–EPITHELIAL INTERACTIONS

Paracrine-acting factors secreted by epithelial or stromal cells are undoubtedly important mediators of stromal–epithelial interactions in the prostate. Numerous factors with paracrine-acting properties have been identified from studies of prostatic epithelial or stromal cell cultures. These factors include nerve growth factor [73], interleukin 6 [74] and TGF-β [64], to name a few. Potentiation of tumor formation by prostatic cancer cell lines

Table 17.8. Complete MCDB 105[a]

Component	Concentration of stock	Amount to add	Final concentration
MCDB 105		500 ml	
CT	100 μg/ml	50 μl	10 ng/ml
EGF	100 μg/ml	50 μl	10 ng/ml
BPE	14 mg/ml	357 μl	10 μg/ml
PEA	0.1 M	500 μl	0.1 mM
HC	10 mg/ml	50 μl	1 μg/ml
SE	3×10^{-4} M	50 μl	3×10^{-8} M
GENT	40 mg/ml	1.25 ml	100 μg/ml
RA	0.1 μg/ml	50 μl	0.01 ng/ml
IN	4 mg/ml	500 μl	4 μg/ml
VIT E	2.3×10^{-2} M	50 μl	2.3×10^{-6} M

[a] Mix together sterile components. Store up to 2 weeks at 4°C. MCDB 105 was purchased from Sigma-Aldrich, St. Louis, MO (cat. no. M-6395) and prepared according to the manufacturer's instructions. For preparation and storage of all other components, see Table 17.5.

in response to stromal cells has been frequently observed, but it was not until the late 1990s that the active factors mediating this process began to be identified [75,76]. A noteworthy finding was the observation that stromal cells derived from carcinomas uniquely promote tumor formation of immortalized but nontumorigenic prostatic epithelial cells [77]. Reviews focusing on the roles of peptide factors in the regulation of growth and differentiation in the prostate can be found in Culig *et al.* [78], Djakiew [79], and Lee *et al.* [80].

THREE-DIMENSIONAL CULTURES

A review by Hoffman [81] discusses the attributes of creating cell cultures in two or three dimensions. For example, sensitivity of prostatic cancer cells to drugs differs in three-dimensional versus monolayer culture [82–85], but whether the results obtained from three-dimensional cultures more accurately reflect anticipated responses in cancer patients has yet to be substantiated. Hormone sensitivity and differentiation into secretory epithelia are other topics being investigated in three-dimensional cultures of prostatic cells and tissues [86–89].

TRANSGENICS

The development of transgenic mice with oncogenic genes targeted to the prostate has provided new models of prostatic carcinogenesis. Genes that have been targeted to the prostate include SV-40 large T antigen, *ras*, polyoma virus middle T antigen, *int-2*, *wap*, *bcl-2*, and keratinocyte growth factor [90,91]. Cell lines have been derived from prostatic tissues of some of these animals, including lines from SV-40 or polyoma virus transgenics [92,93] and from p53-deficient mice [94].

Table 17.9. Primary Culture of Prostatic Stromal Cells

1. Follow steps 1–11 of Primary Culture of Prostatic Epithelial Cells protocol (Table 17.1).
2. Resuspend pellet in 5 ml of MCDB 105 with 10% FBS and 100 μg/ml gentamicin. Transfer to one 60-mm dish (not collagen-coated). Incubate.
3. Feed every 3–4 days after cells attach until semiconfluent.

PROSTATE RECONSTITUTION MODELS

Another valuable tool for prostate research has been the mouse reconstitution model of cancer developed by Thompson and colleagues [95]. Stromal or epithelial cells are dissociated from the urogenital sinus of embryonic mice and maintained *in vitro* for a short period while new genes are introduced via retrovirus. Various combinations of mesenchyme and epithelium are reconstituted in a collagen matrix and grafted under the renal capsule of host animals. Over time, the implants are recovered and assessed. Cell lines have been isolated from these tissues [96].

CONCLUSIONS

Epithelial and stromal cell cultures are valuable tools for the study of benign and malignant diseases of the prostate. Technological developments in other fields are expanding the range of experimental possibilities and facilitating novel applications. For example, video imaging of contractions of cultured prostatic smooth muscle cells provides an *in vitro* model for testing drugs that might relieve urinary obstruction caused by BPH [97]. PC-3 cells have been stably transfected with genes for green fluorescent protein (GFP) or luciferase to track metastatic pathways [98,99]. Implants of human bone tissues into severe combined immunodeficient (SCID) mice, combined with prostatic cancer xenografts, are being tested as an improved approach to study bone–cancer interactions [100]. Synergy of developments in diverse fields promises to yield novel bioengineered cell culture models of prostatic biology and disease.

REFERENCES

1. Franks, L. M. (1980). Primary cultures of human prostate. *Methods Cell Biol.* **21B**, 153–169.
2. Heatfield, B. M., Sanefuji, H., and Trump, B. F. (1980). Long-term explant culture of normal human prostate. *Methods Cell Biol.* **21B**, 171–194.
3. Lechner, J. F., Babcock, M. S., Marnell, M., Narayan, K. S., and Kaighn, M. E. (1980). Normal human prostate epithelial cell cultures. *Methods Cell Biol.* **21B**, 195–225.
4. Malinin, T. I., Claflin, A. J., Block, N. L., and Brown, A. L. (1980). Establishment of primary cell cultures from normal and neoplastic human prostate gland tissue. *Prog. Clin. Biol. Res.* **37**, 161–80.
5. Merchant, D. J., Clarke, S. M., Ives, K., and Harris, S. (1983). Primary explant culture: An in vitro model of the human prostate. *Prostate* **4**, 523–542.
6. McKeehan, W. L., Adams, P. S., and Fast, D. (1987). Different hormonal requirements for androgen-independent growth of normal and tumor epithelial cells from rat prostate. *In Vitro Cell. Dev. Biol.* **23**, 147–152.
7. Peehl, D. M. (1985). Serial culture of adult human prostatic epithelial cells. *J. Tissue Cult. Methods* **9**, 53–60.
8. Chaproniere, D. M., and McKeehan, W. L. (1986). Serial culture of single adult human prostatic epithelial cells in serum-free medium containing low calcium and a new growth factor from bovine brain. *Cancer Res.* **46**, 819–824.
9. Bayne, C. W., Donnelly, F., Chapman, K., Bollina, P., Buck, C., and Habib, F. (1998). A novel coculture model for benign prostatic hyperplasia expressing both isoforms of 5 alpha-reductase. *J. Clin. Endocrinol. Metab.* **83**, 206–213; erratum: *Ibid.*, p. 910.
10. Fong, C. J., Sherwood, E. R., Sutkowski, D. M., Abu-Jawdeh, G. M., Yokoo, H., Bauer, K. D., Kozlowski, J. M., and Lee, C. (1991). Reconstituted basement membrane promotes morphological and functional differentiation of primary human prostatic epithelial cells. *Prostate* **19**, 221–235.
11. Freeman, M. R., Bagli, D. J., Lamb, C. C., Guthrie, P. D., Uchida, T., Slavin, R. E., and Chung, L. W. (1994). Culture of a prostatic cell line in basement membrane gels results in an enhancement of malignant properties and constitutive alterations in gene expression. *J. Cell. Physiol.* **158**, 325–336.
12. Ma, Y. L., Fujiyama, C., Masaki, Z., and Sugihara, H. (1997). Reconstruction of prostatic acinus-like structure from ventral and dorsolateral prostatic epithelial cells of the rat in three-dimensional collagen gel matrix culture. *J. Urol.* **157**, 1025–1031.
13. Olbina, G., Miljkovic, D., Hoffman, R. M., and Geller, J. (1998). New sensitive discovery histoculture model for growth-inhibition studies in prostate cancer and BPH. *Prostate* **37**, 126–129.
14. Turner, T., Bern, H. A., Young, P., and Cunha, G.R. (1990). Serum-free culture of enriched mouse anterior and ventral prostatic epithelial cells in collagen gel. *In Vitro Cell. Dev. Biol.* **26**, 722–730.
15. Webber, M. M., Bello, D., Kleinman, H. K., and Hoffman, M. P. (1997). Acinar differentiation by non-malignant immortalized human prostatic epithelial cells and its loss by malignant cells. *Carcinogenesis (London)* **18**, 1225–1231.
16. Sourla, A., Doillon, C., and Koutsilieris, M. (1996). Three-dimensional type I collagen gel system containing MG-63 osteoblasts-like cells as a model for studying local bone reaction caused by metastatic cancer cells. *Anticancer Res.* **16**, 2773–2780.
17. Zhau, H. E., Goodwin, T. J., Chang, S. M., Baker, T. L., and Chung, L. W. (1997). Establishment of a three-dimensional human prostate organoid coculture under microgravity-simulated conditions: Evaluation of androgen-induced growth and PSA expression. *In Vitro Cell. Dev. Biol.—Anim.* **33**, 375–380.
18. Margolis, L., Hatfill, S., Chuaqui, R., Vocke, C., Emmert-Buck, M., Linehan, W. M., and Duray, P. H. (1999). Long term organ culture of human prostate tissue in a NASA-designed rotating wall bioreactor. *J. Urol.* **161**, 290–297.

19. Peehl, D. M. (1992). Culture of human prostatic epithelial cells. *In* "Culture of Epithelial Cells" (R. I. Freshney, ed.), pp. 159–180. Wiley-Liss, New York.

20. Peehl, D. M., Wong, S. T., Terris, M. K., and Stamey, T. A. (1991). Culture of prostatic epithelial cells from ultrasound-guided needle biopsies. *Prostate* **19**, 141–147.

21. Bologna, M., Vicentini, C., Corrao, G., Festuccia, C., Muzi, P., Tubaro, A., Biordi, L., and Miano, L. (1993). Early diagnosis of prostatic carcinoma may be achieved through in vitro culture of tumor cells harvested by prostatic massage. *Eur. Urol.* **24**, 148–155.

22. Chopra, D. P., Grignon, D. J., Joiakim, A., Mathieu, P. A., Mohamed, A., Sakr, W. A., Powell, I. J., and Sarkar, F. H. (1996). Differential growth factor responses of epithelial cell cultures derived from normal human prostate, benign prostatic hyperplasia, and primary prostate carcinoma. *J. Cell. Physiol.* **169**, 269–280.

23. Cronauer, M. V., Eder, I. E., Hittmair, A., Sierek, G., Hobisch, A., Culig, Z., Thurnher, M., Bartsch, G., and Klocker, H. (1997). A reliable system for the culture of human prostatic cells. *In Vitro Cell. Dev. Biol.—Anim.* **33**, 742–744.

24. Cussenot, O., Berthon, P., Cochand-Priollet, B., Maitland, N. J., and Le Duc, A. (1994). Immunocytochemical comparison of cultured normal epithelial prostatic cells with prostatic tissue sections. *Exp. Cell Res.* **214**, 83–92.

25. Delos, S., Carsol, J. L., Ghazarossian, E., Raynaud, J. P., and Martin, P. M. (1995). Testosterone metabolism in primary cultures of human prostate epithelial cells and fibroblasts. *J. Steroid Biochem. Mol. Biol.* **55**, 375–383.

26. Gilad, E., Laudon, M., Matzkin, H., Pick, E., Sofer, M., Braf, Z., and Zisapel, N. (1996). Functional melatonin receptors in human prostate epithelial cells. *Endocrinology (Baltimore)* **137**, 1412–1417.

27. Krill, D., Shuman, M., Thompson, M. T., Becich, M. J., and Strom, S. C. (1997). A simple method for the isolation and culture of epithelial and stromal cells from benign and neoplastic prostates. *Urology* **49**, 981–988.

28. McKeehan, W. L., Adams, P. S., and Rosser, M. P. (1984). Direct mitogenic effects of insulin, epidermal growth factor, glucocorticoid, cholera toxin, unknown pituitary factors and possibly prolactin, but not androgen, on normal rat prostate epithelial cells in serum-free, primary cell culture. *Cancer Res.* **44**, 1998–2010.

29. Mitchen, J., Oberley, T., and Wilding, G. (1997). Extended culturing of androgen-responsive human primary epithelial prostate cell isolates by continuous treatment with interstitial collagenase. *Prostate* **30**, 7–19.

30. Pantel, K., Dickmanns, A., Zippelius, A., Klein, C., Shi, J., Hoechtlen-Vollmar, W., Schlimok, G., Weckermann, D., Oberneder, R., Fanning, E., et al. (1995). Establishment of micrometastatic carcinoma cell lines: A novel source of tumor cell vaccines. *JNCI, J. Natl. Cancer Inst.* **87**, 1162–1168.

31. Pretlow, T. G., Ogrinc, G. S., Amini, S. B., Delmoro, C. M., Molkentin, K. F., Willson, J. K., and Pretlow, T. P. (1993). A better defined medium for human prostate cancer cells. *In Vitro Cell. Dev. Biol.—Anim.* **29A**, 528–530.

32. Robinson, E. J., Neal, D. E., and Collins, A. T. (1998). Basal cells are progenitors of luminal cells in primary cultures of differentiating human prostatic epithelium. *Prostate* **37**, 149–160.

33. Varani, J., Dame, M. K., Wojno, K., Schuger, L., and Johnson, K. J. (1999). Characteristics of nonmalignant and malignant human prostate in organ culture. *Lab. Invest.* **79**, 723–731.

34. Zwergel, T., Kakirman, H., Schorr, H., Wullich, B., and Unteregger, G. (1998). A new serial transfer explant cell culture system for human prostatic cancer tissues preventing selection toward diploid cells. *Cancer Genet. Cytogenet.* **101**, 16–23.

35. Danielpour, D., Kadomatsu, K., Anzano, M. A., Smith, J. M., and Sporn, M.B. (1994). Development and characterization of nontumorigenic and tumorigenic epithelial cell lines from rat dorsal-lateral prostate. *Cancer Res.* **54**, 3413–3421.

36. Lipschutz, J. H., Foster, B. A., and Cunha, G. R. (1997). Differentiation of rat neonatal ventral prostates grown in a serum-free organ culture system. *Prostate* **32**, 35–42.

37. Ravindranath, N., and Dym, M. (1999). Isolation of rat ventral prostate basal and luminal epithelial cells by the STAPUT technique. *Prostate* **41**, 173–180.

38. Taketa, S., Nishi, N., Takasuga, H., Okutani, T., Takenaka, I., and Wada, F. (1990). Differences in growth requirements between epithelial and stromal cells derived from rat ventral prostate in serum-free primary culture. *Prostate* **17**, 207–218.

39. Bonkhoff, H., Stein, U., and Remberger, K. (1994). Multidirectional differentiation in the normal, hyperplastic, and neoplastic human prostate: Simultaneous demonstration of cell-specific epithelial markers. *Hum. Pathol.* **25**, 42–46.

40. Peehl, D. M., Leung, G. K., and Wong, S. T. (1994). Keratin expression: A measure of phenotypic modulation of human prostatic epithelial cells by growth inhibitory factors. *Cell Tissue Res.* **277**, 11–18.

41. Verhagen, A. P., Aalders, T. W., Ramaekers, F. C., Debruyne, F. M., and Schalken, J. A. (1988). Differential expression of keratins in the basal and luminal compartments of rat prostatic epithelium during degeneration and regeneration. *Prostate* **13**, 25–38.

42. Liu, A. Y., True, L. D., LaTray, L., Nelson, P. S., Ellis, W. J., Vessella, R. L., Lange, P. H., Hood, L., and van den Engh, G. (1997). Cell-cell interaction in prostate gene regulation and cytodifferentiation. *Proc. Natl. Acad. Sci. U.S.A.* **94**, 10705–10710.

43. Peehl, D. M. (1994). The male reproductive system: Prostatic cell lines. *In* "Atlas of Human Tumor Cell Lines" (R. J. Hay, J.-G. Park, and A. Gazdar, eds.), pp. 387–411. Academic Press, San Diego, CA.

44. Bosland, M. C., Chung, L. W. K., Greenberg, N. M., Ho, S., Isaacs, J. T., Lane, K., Peehl, D. M., Thompson, T. C., van Steenbrugge, G. J., and van Weerden, W. M. (1996). Recent advances in the development of animal and cell culture models for prostatic cancer research. *Urol. Oncol.* **2**, 99–128.

45. Rhim, J. S., Peehl, D. M., Webber, M. M., Jay, G., and Dritschilo, A. (1996). In vitro multistep human prostate epithelial cell models for studying prostate carcinogenesis. *Radiat. Oncol. Invest.* **3**, 326–329.

46. Webber, M. M., Bello, D., and Quader, S. (1996). Immortalized and tumorigenic adult human prostatic epithelial cell lines: Characteristics and applications. Part I. Cell markers and immortalized nontumorigenic cell lines. *Prostate* **29**, 386–394.

47. Webber, M. M., Bello, D., and Quader, S. (1997). Immortalized and tumorigenic adult human prostatic epithelial cell lines: Characteristics and applications. Part 2. Tumorigenic cell lines. *Prostate* **30**, 58–64.

48. Stearns, M. E., Ware, J. L., Agus, D. B., Chang, C. J., Fidler, I. J., Fife, R. S., Goode, R., Holmes, E., Kinch, M. S., Peehl, D. M., Pretlow, T. G., 2nd, and Thalmann, G. N. (1998). Workgroup 2: Human xenograft models of prostate cancer. *Prostate* **36**, 56–58.

49. Sramkoski, R. M., Pretlow, T. G., 2nd, Giaconia, J. M., Pretlow, T. P., Schwartz, S., Sy, M. S., Marengo, S. R., Rhim, J. S., Zhang, D., and Jacobberger, J. W. (1999). A new human prostate carcinoma cell line, 22Rv1. *In Vitro Cell. Dev. Biol.—Anim.* **35**, 403–409.

50. Horoszewicz, J. S., Leong, S. S., Chu, T. M., Wajsman, Z. L., Friedman, M., Papsidero, L., Kim, U., Chai, L. S., Kakati, S., Arya, S. K., and Sandberg, A. A. (1980). The LNCaP cell line—a new model for studies on human prostatic carcinoma. *Prog. Clin. Biol. Res.* **37**, 115–132.

51. Wu, T. T., Sikes, R. A., Cui, Q., Thalmann, G. N., Kao, C., Murphy, C. F., Yang, H., Zhau, H. E., Balian, G., and Chung, L.W. (1998). Establishing human prostate cancer cell xenografts in bone: Induction of osteoblastic reaction by prostate-specific antigen-producing tumors in athymic and SCID/bg mice using LNCaP and lineage-derived metastatic sublines. *Int. J. Cancer* **77**, 887–894.

52. di Sant'Agnese, P. A., and Cockett, A. T. (1996). Neuroendocrine differentiation in prostatic malignancy. *Cancer (Philadelphia)* **78**, 357–361.

53. Webber, M. M., Bello, D., and Quader, S. (1997). Immortalized and tumorigenic adult human prostatic epithelial cell lines: Characteristics and applications. Part 3. Oncogenes, suppressor genes, and applications. *Prostate* **30**, 136–142.

54. Hayward, S. W., Cunha, G. R., and Dahiya, R. (1996). Normal development and carcinogenesis of the prostate. A unifying hypothesis. *Ann. N.Y. Acad. Sci.* **784**, 50–62.

55. Hayward, S. W., Rosen, M. A., and Cunha, G.R. (1997). Stromal-epithelial interactions in the normal and neoplastic prostate. *Br. J. Urol.* **79**, 18–26.

56. Peehl, D. M., Sellers, R. G., and Wong, S. T. (1998). Defined medium for normal adult human prostatic stromal cells. *In Vitro Cell. Dev. Biol.—Anim.* **34**, 555–560.

57. Berthaut, I., Portois, M. C., Cussenot, O., and Mowszowicz, I. (1996). Human prostatic cells in culture: Different testosterone metabolic profile in epithelial cells and fibroblasts from normal or hyperplastic prostates. *J. Steroid Biochem. Mol. Biol.* **58**, 235–242.

58. Collins, A. T., Robinson, E. J., and Neal, D. E. (1996). Benign prostatic stromal cells are regulated by basic fibroblast growth factor and transforming growth factor-beta 1. *J. Endocrinol.* **151**, 315–322.

59. Gleave, M. E., Hsieh, J. T., von Eschenbach, A. C., and Chung, L. W. (1992). Prostate and bone fibroblasts induce human prostate cancer growth in vivo: Implications for bidirectional tumor-stromal cell interaction in prostate carcinoma growth and metastasis. *J. Urol.* **147**, 1151–1159.

60. Planz, B., Kirley, S. D., Wang, Q., Tabatabaei, S., Aretz, H. T., and McDougal, W. S. (1999). Characterization of a stromal cell model of the human benign and malignant prostate from explant culture. *J. Urol.* **161**, 1329–1336.

61. Story, M. T., Hopp, K. A., and Meier, D. A. (1996). Regulation of basic fibroblast growth factor expression by transforming growth factor beta in cultured human prostate stromal cells. *Prostate* **28**, 219–226.

62. Kassen, A., Sutkowski, D. M., Ahn, H., Sensibar, J. A., Kozlowski, J. M., and Lee, C. (1996). Stromal cells of the human prostate: Initial isolation and characterization. *Prostate* **28**, 89–97.

63. Shapiro, E., Hartanto, V., and Lepor, H. (1992). Anti-desmin vs. anti-actin for quantifying the area density of prostate smooth muscle. *Prostate* **20**, 259–267.

64. Kooistra, A., Elissen, N. M., Konig, J. J., Vermey, M., van der Kwast, T. H., Romijn, J. C., and Schroder, F. H. (1995). Immunocytochemical characterization of explant cultures of human prostatic stromal cells. *Prostate* **27**, 42–49.

65. Peehl, D.M., and Sellers, R.G. (1997). Induction of smooth muscle cell phenotype in cultured human prostatic stromal cells. *Exp. Cell. Res.* **232**, 208–215.

66. Peehl, D. M., and Sellers, R. G. (1998). Basic FGF, EGF, and PDGF modify TGFbeta-induction of smooth muscle cell phenotype in human prostatic stromal cells. *Prostate* **35**, 125–134.

67. Boesch, S. T., Corvin, S., Zhang, J., Rogatsch, H., Bartsch, G., and Klocker, H. (1999). Modulation of the differentiation status of cultured prostatic smooth muscle cells by an alpha1-adrenergic receptor antagonist. *Prostate* **39**, 226–233.

68. Smith, P., Rhodes, N. P., Ke, Y., and Foster, C. S. (1999). Influence of the alpha1-adrenergic antagonist, doxazosin, on noradrenaline-induced modulation of cytoskeletal proteins in cultured hyperplastic prostatic stromal cells. *Prostate* **38**, 216–227.

69. Zhang, J., Hess, M. W., Thurnher, M., Hobisch, A., Radmayr, C., Cronauer, M. V., Hittmair, A., Culig, Z., Bartsch, G., and Klocker, H. (1997). Human prostatic smooth muscle cells in culture: Estradiol enhances expression of smooth muscle cell-specific markers. *Prostate* **30**, 117–129.

70. Gerdes, M. J., Dang, T. D., Lu, B., Larsen, M., McBride, L., and Rowley, D. R. (1996). Androgen-regulated proliferation and gene transcription in a prostate smooth muscle cell line (PS-1). *Endocrinology (Baltimore)* **137**, 864–872.

71. Webber, M. M., Trakul, N., Thraves, P. S., Bello-DeOcampo, D., Chu, W. W., Storto, P. D., Huard, T. K., Rhim, J. S., and Williams, D. E. (1999). A human prostatic stromal myofibroblast cell line WPMY-1: A model for stromal-epithelial interactions in prostatic neoplasia. *Carcinogenesis (London)* **20**, 1185–1192.

72. Roberson, K. M., Edwards, D. W., Chang, G. C., and Robertson, C. N. (1995). Isolation and characterization of a novel human prostatic stromal cell culture: DuK50. *In Vitro Cell. Dev. Biol.—Anim.* **31**, 840–845.

73. Dalal, R., and Djakiew, D. (1997). Molecular characterization of neurotrophin expression and the corresponding tropomyosin receptor kinases (trks) in epithelial and stromal cells of the human prostate. *Mol. Cell. Endocrinol.* **134**, 15–22.

74. Degeorges, A., Tatoud, R., Fauvel-Lafeve, F., Podgorniak, M. P., Millot, G., de Cremoux, P., and Calvo, F. (1996). Stromal cells from human benign prostate hyperplasia produce a growth-inhibitory factor for LNCaP prostate cancer cells, identified as interleukin-6. *Int. J. Cancer* **68**, 207–214.

75. Tam, N. N. C., Phil, M., Wang, Y. Z., and Wong, Y. C. (1997). The influence of mesenchyme of neonatal seminal vesicle and embryonic urogenital sinus on the morphologic and functional cytodifferentiation of Dunning prostatic adenocarcinoma: Roles of growth factors and proto-oncogenes. *Urol. Oncol.* **3**, 85–93.

76. Olumi, A. F., Dazin, P., and Tlsty, T. D. (1998). A novel coculture technique demonstrates that normal human prostatic fibroblasts contribute to tumor formation of LNCaP cells by retarding cell death. *Cancer Res.* **58**, 4525–4530.

77. Olumi, A. F., Grossfeld, G. D., Hayward, S. W., Carroll, P. R., Tlsty, T. D., and Cunha, G. R. (1999). Carcinoma-associated fibroblasts direct tumor progression of initiated human prostatic epithelium. *Cancer Res.* **59**, 5002–5011.

78. Culig, Z., Hobisch, A., Cronauer, M. V., Radmayr, C., Hittmair, A., Zhang, J., Thurnher, M., Bartsch, G., and Klocker, H. (1996). Regulation of prostatic growth and function by peptide growth factors. *Prostate* **28**, 392–405.

79. Djakiew, D. (2000). Dysregulated expression of growth factors and their receptors in the development of prostate cancer. *Prostate* **42**, 150–160.

80. Lee, C., Kozlowski, J. M., and Grayhack, J. T. (1997). Intrinsic and extrinsic factors controlling benign prostatic growth. *Prostate* **31**, 131–138.

81. Hoffman, R. M. (1993). To do tissue culture in two or three dimensions? That is the question. *Stem Cells* **11**, 105–111.

82. Vescio, R. A., Redfern, C. H., Nelson, T. J., Ugoretz, S., Stern, P. H., and Hoffman, R. M. (1987). In vivo-like drug responses of human tumors growing in three-dimensional gel-supported primary culture. *Proc. Natl. Acad. Sci. U.S.A.* **84**, 5029–5033.

83. Perrapato, S. D., Slocum, H. K., Huben, R. P., Ghosh, R., and Rustum, Y. (1990). Assessment of human genitourinary tumors and chemosensitivity testing in 3-dimensional collagen gel culture. *J. Urol.* **143**, 1041–1045.

84. O'Connor, K. C. (1999). Three-dimensional cultures of prostatic cells: Tissue models for the development of novel anti-cancer therapies. *Pharm. Res.* **16**, 486–493.

85. Hedlund, T. E., Duke, R. C., and Miller, G. J. (1999). Three-dimensional spheroid cultures of human prostate cancer cell lines. *Prostate* **41**, 154–165.

86. Nevalainen, M. T., Harkonen, P. L., Valve, E. M., Ping, W., Nurmi, M., and Martikainen, P. M. (1993). Hormone regulation of human prostate in organ culture. *Cancer Res.* **53**, 5199–5207.

87. Lopes, E. S., Foster, B. A., Donjacour, A. A., and Cunha, G. R. (1996). Initiation of secretory activity of rat prostatic epithelium in organ culture. *Endocrinology (Baltimore)* **137**, 4225–4234.

88. Geller, J., Sionit, L. R., Connors, K., and Hoffman, R. M. (1992). Measurement of androgen sensitivity in the human prostate in in vitro three-dimensional histoculture. *Prostate* **21**, 269–278.

89. Geller, J., Sionit, L., Connors, K., Youngkin, T., and Hoffman, R. M. (1993). Expression of prostate-specific antigen in human prostate specimens in in vitro three-dimensional histoculture. *In Vitro Cell. Dev. Biol.—Anim.* **29A**, 523–524.

90. Sharma, P., and Schreiber-Agus, N. (1999). Mouse models of prostate cancer. *Oncogene* **18**, 5349–5355.

91. Green, J. E., Greenberg, N. M., Ashendel, C. L., Barrett, J. C., Boone, C., Getzenberg, R. H., Henkin, J., Matusik, R., Janus, T. J. and Scher, H. I. (1998). Workgroup 3: Transgenic and reconstitution models of prostate cancer. *Prostate* **36**, 59–63.

92. Foster, B. A., Gingrich, J. R., Kwon, E. D., Madias, C., and Greenberg, N. M. (1997). Characterization of prostatic epithelial cell lines derived from transgenic adenocarcinoma of the mouse prostate (TRAMP) model. *Cancer Res.* **57**, 3325–3330.

93. Jorcyk, C. L., Liu, M. L., Shibata, M. A., Maroulakou, I. G., Komschlies, K. L., McPhaul, M. J., Resau, J. H., and Green, J. E. (1998). Development and characterization of a mouse prostate adenocarcinoma cell line: Ductal formation determined by extracellular matrix. *Prostate* **34**, 10–22.

94. Hanazono, M., Nakagawa, E., Aizawa, S., and Tomooka, Y. (1998). Establishment of prostatic cell line "Pro9ad" from a p53-deficient mouse. *Prostate* **36**, 102–109.

95. Thompson, T. C., Timme, T. L., Park, S. H., Ren, C., Baley, P. A., Eastham, J. A., Sehgal, I., Yang, G., and Kadmon, D. (1995). Tissue and cell–cell interactions in prostate cancer progression. *Cancer (Philadelphia)* **75**(Suppl.), 1885–1891.

96. Baley, P. A., Yoshida, K., Qian, W., Sehgal, I., and Thompson, T. C. (1995). Progression to androgen insensitivity in a novel in vitro mouse model for prostate cancer. *J. Steroid Biochem. Mol. Biol.* **52**, 403–413.

97. Corvin, S., Bosch, S. T., Eder, I., Thurnher, M., Bartsch, G., and Klocker, H. (1998). Videoimaging of prostatic stromal-cell contraction: An in vitro model for studying drug effects. *Prostate* **37**, 209–214.

98. Yang, M., Jiang, P., Sun, F. X., Hasegawa, S., Baranov, E., Chishima, T., Shimada, H., Moossa, A. R., and Hoffman, R. M. (1999). A fluorescent orthotopic bone metastasis model of human prostate cancer. *Cancer Res.* **59**, 781–786.

99. Rubio, N., Villacampa, M. M., and Blanco, J. (1998). Traffic to lymph nodes of PC-3 prostate tumor cells in nude mice visualized using the luciferase gene as a tumor cell marker. *Lab. Invest.* **78**, 1315–1325.

100. Nemeth, J. A., Harb, J. F., Barroso, U., Jr., He, Z., Grignon, D. J., and Cher, M. L. (1999). Severe combined immunodeficient-hu model of human prostate cancer metastasis to human bone. *Cancer Res.* **59**, 1987–1993.

CHAPTER 18

EPITHELIAL CELL CULTURE: THREE-DIMENSIONAL CERVICAL SYSTEM

Craig Meyers

Studies concerning the development, maintenance, and transition from a normal to a disease state of human uterine cervical tissue have used monolayer cell culture systems and/or cells from abnormal tissues. We have developed an *in vitro* organotypic (raft) culture system capable of reproducing three-dimensional cervical epithelial tissue architecture. These *in vitro* grown cervical tissues resemble their *in vivo* counterparts morphologically and have been used as a model system for the study of cervical carcinogenesis. A mixture of type I collagen and fibroblasts is used to make a dermal equivalent.

Cervical keratinocytes are placed on top of the dermal equivalent. Then the dermal equivalent with cervical keratinocytes on top is lifted onto a wire grid. The keratinocytes are now at the air–liquid interface and never come in contact with the culture media. The cultures are fed by diffusion through the dermal equivalent in a process similar to what occurs *in vivo*. Under these conditions cervical keratinocytes will stratify and differentiate over a period of 7–10 days. The tissue that develops morphologically and biochemically resembles that of the cervix *in vivo*. We describe in detail how to prepare these three-dimensional cervical epithelial organotypic (raft) culture tissues.

INTRODUCTION

The purpose of this chapter is to describe a technique for growing three-dimensional cervical epithelium *in vitro*. The morphology of the cervix is described only briefly, to provide the reader with fundamental understanding. Those who are interested in a more extensive description of the morphology should read a general text; for a good example, see Ferenczy and Wright [1]. Based on gross anatomy, the cervix is divided into two general sections, the vaginal portion (exocervix or ectocervix) and the cervical canal (endocervix). The ectocervix has a convex elliptical surface delineated by the anterior and posterior vagina fornices. The endocervix connects the external os with the isthmus of the uterus. Histologically the cervix is lined by squamous and columnar epithelium. The ectocervix is lined with nonkeratinized squamous epithelium, which is morphologically similar to the vaginal epithelium. The nonkeratinized squamous epithelium is divided into a basal cell layer, a stratum spinosum, and the superficial zone. The basal layer is composed of two cell types, the true basal cell and the parabasal cell. Ectocervical basal cells behave like stem cells, whereas the parabasal cells constitute the actively replicating cells of the tissue [2]. The stratum spinosum is similar to the spinosum of keratinized squamous epithelium. The superficial zone contains the most differentiated cells. These cells are flattened and have an enlarged glycogenated cytoplasmic area. The endocervix consists of a single layer of columnar cells that secrete mucin. Separating the ectocervix and the endocervix is the transformation zone, characterized histologically by the presence of metaplastic epithelium. Essentially all cervical neoplasia initiates at this transitional epithelium as measured by the extension and limits of cancer precursors that correspond with the distribution of the transformation zone.

Numerous investigators have developed an assortment of techniques to culture epithelial cells in tissue culture. The most accurate technique to mimic the *in vivo* biology of epithelial tissues *in vitro* has been the organotypic (raft) culture system. In this system epithelial cells are placed on top of a dermal equivalent and lifted to an air–liquid interface for growth and differentiation. Placing human skin explants onto the reticular aspect of split-thickness sections of pig skin [3] was one of the first successful methods for epithelial organotypic cultures. After 14 days of growth and differentiation of the epithelium at the air–liquid interface, characteristic basal, spinous, granular, and squamous layers were observed. The functionality of epithelium grown in this manner was demonstrated by successfully grafting it onto patients with third-degree burns. De-epidermized human skin flaps were used in place of pig skin with equivalent results [4–6].

Dermal equivalents consisting of collagen matrices greatly simplified the use of organotypic cultures. When liver epithelial cells were placed on a collagen matrix and raised to the air–liquid interface, they underwent appropriate differentiation. These liver epithelial organotypic cultures were maintained at the air–liquid interface by letting them float on the media surface [7]. It was this floating that gives the system the commonly used name of "raft" culture. The liver epithelium grown in this manner not only was morphologically accurate but also demonstrated functional features comparable to those present in the *in vivo* state. The introduction of silicon chambers [8] and stainless steel grids [9] to hold rafts has greatly improved the ease of working with the cultures. The accuracy of the differentiated phenotype *in vitro* was greatly improved by the introduction of fibroblasts into the collagen matrix [10–12]. Raft culture technology has been and is used for studies involving numerous types of epithelial tissue for diverse studies. Today most laboratories use a form of a technique described by Asselineau and Pruniéras [13].

Here we describe an *in vitro* organotypic (raft) culture system for growing three-dimensional cervical tissue. This system facilitates the study of the normal and diseased growth and differentiation of cervical epithelium, including cervical intraepithelial neoplasia and carcinoma. This raft culture system also allows for investigations of infectious agents known to target the cervix. Our laboratory has used this system to investigate the productive life cycles of human papillomavirus [14–21] and herpes simplex virus [22].

KERATINOCYTE GROWTH MEDIUM

Materials

1. Nalgene 20-liter carboy (Fisher Scientific, Springfield, NJ, cat. no. 02-963-2B).
2. Trypsin–EDTA (Life Technologies, Gaithersburg, MD, cat. no. 25300-047).
3. Distilled, deionized water.
4. Powdered Dulbecco's Modified Eagle's Medium, 4500 mg/liter D-glucose and L-glutamine, no sodium pyruvate or sodium bicarbonate, 5-liter packages (Life Technologies, cat. no. 12100-103).
5. Powdered F-12 Nutrient Mixture (Ham), with L-glutamine, no sodium bicarbonate, 5-liter packages (Life Technologies, cat. no. 21700-109).
6. Sodium bicarbonate (Life Technologies, cat. no. 11810-025).
7. Adenine (Sigma Chemical Company, St. Louis, MO, cat. no. A-2786).
8. Concentrated HCl (Fisher Scientific, cat. no. A144-212).
9. Insulin (Sigma Chemical Company, cat. no. I-6634).
10. Transferrin (Sigma Chemical Company, cat. no. T-1147).
11. Phosphate-buffered saline (PBS): NaCl (J.T. Baker Inc., Phillipsburg, NJ, cat. no. 3624-05), KH_2PO_4 (Fisher Scientific, cat. no. P285-500), Na_2HPO_4 (Fisher Scientific, cat. no. S374-1), KCl (Fisher Scientific, cat. no. P217-500).
12. 3,3′,5-Triiodo-L-thyronine (T3) (Sigma Chemical Company, cat. no. T-6397).
13. Penicillin–streptomycin, (Life Technologies, cat. no. 15140-023).
14. Hydrocortisone (Sigma Chemical Company, cat. no. H-0888).

15. HEPES, free acid, ultrol grade (Calbiochem-Novabiochem Corporation, La Jolla, CA, cat. no. 391338).
16. Cholera enterotoxin (ICN Pharmaceuticals Inc., Costa Mesa, CA, cat. no. 190329).
17. Fetal bovine serum (HyClone Laboratories, Inc., Logan, UT, cat. no. A-1115).
18. KGM (Clonetics, San Diego, CA, cat. no. CC-3107).
19. Epidermal cell growth factor (Collaborative Biomedical Products, Bedford, MA, cat. no. 40001).

Procedure

In our laboratory we have found that the most efficient and reproducible way to grow normal cervical keratinocytes is to use the keratinocyte growth medium, E medium [23] with 5 ng/ml epidermal cell growth factor (EGF), and a mitomycin C treated J2 3T3 feeder cell layer. We have found that frozen stocks thaw best into KGM medium, with a switch after 2 days to E medium with a mitomycin C treated J2 3T3 feeder cell layer. This procedure allows for the best growth characteristics (Fig. 18.1). We have found that by using these techniques, we can passage normal human cervical epithelial cells two to three times.

To the 20-liter Nalgene carboy, add 16.25 liters of distilled and deionized water. Add three 5-liter packages of powdered Dulbecco's Modified Eagle's Medium, with 4500 mg/liter D-glucose and L-glutamine, no sodium pyruvate or sodium bicarbonate. Then add one 5-liter package of powdered F-12 Nutrient Mixture (Ham), with

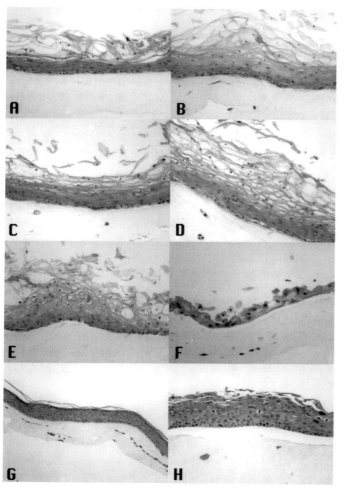

Fig. 18.1. Hematoxylin and eosin staining pattern of organotypic (raft) culture tissues. (A)–(E) Ectocervical keratinocytes used were thawed into KGM medium and transferred to E medium with mitomycin C treated J2 3T3 feeder cells. Cultures were initiated on normal ectocervical keratinocyte rafts by using (A) 0.5×10^6 keratinocytes, (B) 0.75×10^6 keratinocytes, (C) 1.0×10^6 keratinocytes, (D) 1.5×10^6 keratinocytes, and (E) using 1.75×10^6 keratinocytes. (F) Normal ectocervical keratinocyte rafts, using the same ectocervical keratinocytes used for (A)–(E) except that these 1×10^6 keratinocytes were thawed and continuously grown in KGM medium until used to initiate the raft culture. (G) Normal foreskin keratinocyte rafts, using 1×10^6 keratinocytes to initiate the cultures. (H) Human papillomavirus type 18 productively infected ectocervical keratinocyte rafts [15]. See also color insert.

L-glutamine, no sodium bicarbonate. Rinse the excess powder out of the packages. When the powder is completely dissolved, do the following:

1. Add 61.375 g of tissue culture grade sodium bicarbonate and allow to dissolve completely.
2. Add 20 ml of 1.8×10^{-1} M adenine, prepared earlier by dissolving 486 mg of adenine in 15 ml of sterile distilled water. Then by adding approximately 10 drops of concentrated HCl until the adenine is dissolved, bring the volume to 20 ml. Stock solutions can be stored at $-20°C$.
3. Add 20 ml of 5 mg/ml insulin, prepared earlier by dissolving 100 mg of insulin in 20 ml of 0.1 N HCl. Stock solutions can be stored at $-20°C$.
4. Add 20 ml of 5 mg/ml transferrin, prepared earlier by dissolving 100 mg of transferrin in 20 ml of sterile PBS. To make PBS, add to 700 ml of distilled, deionized water 8 g of NaCl, 0.2 g of KH_2PO_4, 1.15 g of Na_2HPO_4, and 0.2 g of KCl, dissolve completely and bring final volume to 1 liter. Stock solutions can be stored at $-20°C$.
5. Add 20 ml of 2×10^{-8} M T3, prepared earlier by dissolving 13.6 mg of T3 in 100 ml of 0.02 N NaOH, then adding 0.1 ml of this solution to 9.9 ml of sterile PBS, and further diluting 1.0 ml of this solution in 99 ml of sterile PBS. Stock solutions can be stored at $-20°C$.
6. Add 200 ml of penicillin–streptomycin solution (10,000 U/ml and 10,000 μg/ml, respectively).
7. Add 20 ml of 0.4 mg/ml hydrocortisone, prepared earlier by dissolving 25 mg of hydrocortisone in 5 ml of 100% ethanol, diluting 4.8 ml of this solution in 55.2 ml of 1 M HEPES buffer, pH 7. (The excess can be divided into aliquots and stored at $-20°C$.)
8. Add 20 ml of 10^{-7} M cholera enterotoxin, prepared earlier by adding 1 mg of cholera enterotoxin to 100 ml of sterile distilled water. Store stock solution at $4°C$ in the dark.

Finally, add 6.25 ml of concentrated HCl. Mix solution and check the pH; add more HCl if necessary to bring to approximately pH 7.1. Mix well and filter-sterilize 950 ml into sterile 1-liter bottles, using a low protein binding 0.2 μM filter. Store at $4°C$ in the dark. Before using, add 50 ml of fetal bovine serum (5%) to each bottle. Fetal bovine serum is variable in essential components necessary for proper growth and differentiation of the raft culture tissues. Serum lots should therefore be tested and the lot showing favorable results purchased in bulk.

KGM medium purchased from Clonetics is prepared according to the manufacturer's instructions.

J2 3T3 MAINTENANCE MEDIUM

Materials

1. Liquid Dulbecco's Modified Eagle's Medium (DMEM), 4500 mg/liter D-glucose and L-glutamine, no sodium pyruvate (Irvine Scientific, Santa Ana, CA, cat. no. 9031).
2. Newborn calf serum, heat inactivated (Life Technologies, cat. no. 26010-025).
3. Gentamicin (Life Technologies, cat. no. 15710-015).

Procedure

To one 500-ml bottle of DMEM, add 50 ml of heat-inactivated newborn calf serum and 1.25 ml of gentamicin (10 mg/ml).

FEEDER CELL LAYER

Materials

1. 100-mm dishes (Fisher Scientific, cat. no. 08-772E).
2. Mitomycin C (Boehringer-Mannheim, Indianapolis, IN, cat. no. 107-409).

Procedure

Using standard cell culture techniques [24], grow J2 3T3 cells in plastic tissue culture dishes, using the J2 3T3 maintenance medium. Do not allow cells to grow to confluence or for more than 20 passages. Continuous rodent fibroblast cell lines have a high tendency to spontaneously transform and lose their ability to induce differentiation of human keratinocytes in organotypic culture. When J2 3T3 cells reach near confluence, remove media, wash once with 1–2 ml of trypsin–EDTA, then add 2 ml of trypsin–EDTA. Incubate at 37°C until cells lift off the plastic. Plates usually require tapping to help cells detach. Split J2 3T3 cells 1:3 to 1:10 according to need. Plates of J2 3T3 cells to be treated with mitomycin C are allowed to just reach confluence.

A mitomycin C stock solution is made by adding 2 mg of mitomycin C to 5 ml of sterile PBS (final concentration, 0.4 mg/ml). The stock solution is filter-sterilized, wrapped in foil, and stored at 4°C. The solution will remain good for 2–3 weeks until a heavy precipitate forms. To a confluent plate of J2 3T3 cells, add the 0.4 mg/ml stock solution of mitomycin C to a final concentration of 8 µg/ml. Incubate the mitomycin C treated plate of J2 3T3 cells at 37°C for at least 2 h, but no more than 4 h. Then wash the J2 3T3 monolayer three times with 3–5 ml of sterile PBS. Trypsinize the mitomycin-treated J2 3T3 cells as already described, but after treatment and trypsinization, use E medium plus 5 ng/ml EGF to plate them. Split mitomycin C treated J2 3T3 cells 1:2 to 1:5 as needed.

ORGANOTYPIC CULTURES

Materials

1. Sodium hydroxide (Fisher Scientific, cat. no. S318-500).
2. Rat tail type I collagen (Collaborative Biomedical Products, cat. no. 40236).
3. Dichromate acid cleaning solution (Thomas Scientific, Swedesboro, NJ, cat. no. 2903-E10).
4. 50-ml polypropylene conical tubes (Fisher Scientific, cat. no. 14-959-49A).
5. Six-well cluster dishes (Costar Corporation, Cambridge, MA, cat. no. 3506).
6. 3 1/16′-in. 40-mesh 010 SS wire cloth circles (Williams & Mettle Co., Houston, TX, cat. no. 20368).
7. Long forceps (Fisher Scientific, cat. no. 10-316A).
8. One liter beakers (Fisher Scientific, cat. no. 02-540P).
9. Lab spoon (Fisher Scientific, cat. no. 14-375-10).

Procedures: Dermal Equivalent (collagen matrix) and Lifting the Collagen Matrix/ Keratinocytes to the Air–Liquid Interface

A 10× reconstitution buffer is made by adding 2.2 g of sodium bicarbonate and 4.77 g of HEPES to 75 ml of 0.062 N NaOH. Dissolve completely, then bring the final volume to 100 ml with 0.062 N NaOH. Filter-sterilize, aliquot, and store at –20°C. Make 10× DMEM by dissolving a 5-liter package of powdered DMEM without sodium bicarbonate into 500 ml of sterile distilled, deionized water. Although the powder will not completely dissolve, warming the solution in a waterbath will help. Filter-sterilize, aliquot, and store at –20°C. As the precipitate will persist, always take care to have the solution as homogeneous a suspension as possible before using it. The precipitate will dissolve when all components of the dermal equivalent are mixed together. Keep the rat tail type I collagen between 0 and 4°C until dermal equivalent preparation has been completed. At temperatures above 4°C the rat tail type I collagen will begin to solidify. The collagen should also never be frozen. Calculate the number of raft tissue cultures you will need to grow. Each raft culture requires a total of 2.5 ml of collagen–fibroblast mixture. We have found 50-ml conical tubes easy to manage in an ice bucket while working in a laminar flow hood. An appropriate number of tubes is decided by the total volume of collagen–fibroblast mixture required. For ease in manipulation, we recommend preparing 30–45 ml of collagen–fibroblast mixture volume per tube.

The collagen–fibroblast mixture is made by combining 1/10 of the final volume of $10\times$ reconstitution buffer, 1/10 of the final volume of $10\times$ DMEM, 8/10 of the final volume of rat tail type I collagen, 2.5×10^5 J2 3T3 cells per milliliter of collagen mixture, and 1.2 μl of 10 N NaOH per milliliter of the final volume. Be sure all components are chilled before combining. Always make 2.5 ml more volume (i.e., calculate for one extra raft) than is needed to make up for adherence of the viscous collagen in pipettes. Prepare the required number of J2 3T3 cells by standard trypsinization protocols as described earlier. Note that for organotypic cultures, the J2 3T3 cells are not treated with mitomycin C.

Determine cell number, and then pellet the trypsinized J2 3T3 cells by centrifugation and aspirate off the trypsin solution. Resuspend the cell pellet in the predetermined amount of $10\times$ reconstitution buffer and $10\times$ DMEM. The collagen–fibroblast mixture must be made on ice.

Aliquot the $10\times$ reconstitution buffer/$10\times$ DMEM/J2 3T3 cell mixture into the 50-ml conical tubes on ice. Add the appropriate amount of rat tail type I collagen (8/10 the final volume) to each tube. Add the proper amount of 10 N NaOH to each tube of the mixture of $10\times$ reconstitution buffer, $10\times$ DMEM/J2 3T3 cells, and collagen. Rat tail collagen is very viscous and will adhere to the pipette; therefore, using the same pipette to dispense collagen to all the tubes will help prevent excessive loss of collagen left in the pipette. Screw lids tightly onto tubes, and gently invert the tubes repeatedly until the solution is fully mixed. Care should be taken not to introduce many bubbles to the solution.

Using a 10-ml pipette, aliquot 2.5 ml of the collagen–fibroblast mixture into wells of six-well cluster dishes. Again, it is best to use one pipette and try to prevent bubbles as much as possible. Some bubbles will always occur and a few will not interfere with the cultures. Place the six-well cluster dishes containing the collagen–fibroblast mixture in a 37°C incubator until they solidify, which should occur in approximately 1 h. Commercially available rat tail type I collagen can vary significantly in its concentration. We recommend using a collagen lot with a concentration of approximately 4 mg/ml. The collagen–fibroblast matrix can remain in the incubator overnight before the next step in the protocol is begun.

After solidification, carefully add 2–3 ml of E medium (no EGF) to the top of each collagen matrix. Excessive force when adding medium can sometimes damage the collagen matrix. The matrix can be left in the incubator for up to a week, with medium changed every 2 days. Special care is also required in the aspiration of medium. To avoid aspirating the collagen matrix with the medium, tilt the six-well cluster dish until you can see the top of the collagen and take care not to touch the collagen matrix when aspirating. Removal of all the medium is not necessary.

Numerous methods and procedures have been described for initiating and continuing normal cervical epithelial cell lines [25–36]. The purpose of this chapter is not to review this topic, but to describe in detail how to grow three-dimensional cervical epithelial tissue cultures. We have tried several methods to initiate and continue normal cervical epithelial cell lines, including purchasing them commercially. The most important factors we have identified in our laboratory are the time the cells spend in monolayer culture (shorter is better) and the type of medium used.

Place your cervical keratinocytes on top of the collagen matrix. The total E medium (no EGF) volume should be 3–4 ml. Incubate for 1 day at 37°C to allow the cells to attach. We have found that raft tissue growth is affected by the number of cells placed on top of the collagen matrix and by the time the cells remain submerged on the collagen matrix. For best results, use more than 1×10^6 keratinocytes per collagen matrix and keep the cells submerged on the collagen matrix (Fig. 18.1).

To lift the collagen matrix/keratinocytes to the air–liquid interface, the 3 1/16′-in. 40-mesh 010 SS wire cloth circles are first bent at the edge, at three places equidistant from each other. The bends are to form legs that will raise the mesh approximately 2 mm from the bottom of the tissue culture dish. The wire cloth circles are then soaked in dichromate acid cleaning solution for 1–2 h. Next, they are rinsed for 24–48 h in a beaker with gently running tap water and for an additional 2 h with gently running distilled water. After rinsing, the wire cloth circles are placed into a 1-liter beaker, covered with foil, and autoclaved, followed by a 30-min dry cycle. Then sterile long forceps are used to place the sterile wire cloth circles into 10-cm tissue culture plates. Next one removes medium from the collagen matrix topped by keratinocytes. A sterile lab spoon is used to carefully separate the collagen matrix from the wall of the six-well cluster dishes. The six-well cluster dish is tilted, and the sterile lab spoon is slipped under the lower side of the collagen matrix until it can be lifted. The collagen matrix is then placed onto the wire cloth circles in the 100-mm tissue culture dish. The lab spoon is quickly moved between the collagen matrix and the wire cloth circles, smoothing and spreading the matrix out as much as possible. The matrix will quickly attach to the wire cloth circles. One to four matrices can be placed onto a single 3 1/16′-in. 40-mesh 010 SS wire cloth circle. From this point on we will refer to the collagen matrix epithelial cells on the wire cloth circle as raft cultures.

During the first feeding of the raft cultures, some medium may be left on top of the raft. However, after the first overnight incubation at 37°C the top of the raft will dry. At the air–liquid interface, all feeding of the organotypic tissue is done by diffusion through the dermal equivalent. The diffusion of medium components creates a gradient mimicking the *in vivo* situation, where nutrients diffuse through the dermis to the epithelium. Therefore, it is very important not to get medium on top of the raft; this ends the gradient, disturbing cell orientation and proper differentiation.

When feeding the raft culture, the surface tension of the wire cloth circle will help the medium to completely contact the bottom where the rafts touches the wire cloth circle. It is usually helpful to gently tap the plate or wire cloth circles to help the medium spread evenly underneath. Air pockets under the wire cloth circles should be avoided. If they occur, remove the medium and feed again.

Cultures should be refed every other day. When removing medium by vacuum suction, do not touch the collagen matrix itself, or it will be rapidly sucked off the wire cloth circle. Add new medium as described earlier.

Normal human cervical epithelial raft culture tissues take approximately 7–10 days to fully stratify and differentiate (Fig. 18.1). Using the protocols as described here should provide reliable and reproducible cervical epithelial tissues, faithfully mimicking the morphology of the *in vivo* counterpart. We have provided a demonstration of ectocervical epithelium grown in the raft culture (Fig. 18.1). Endocervical epithelium grows equally well in the system. Cervical keratinocytes can be used directly from the initiating culture. The keratinocyte lines can also be frozen and thawed before use, and equally acceptable results can be obtained if a couple of precautions are followed. First as already mentioned, it is necessary to thaw and plate frozen normal keratinocyte lines initially into KGM medium (Clonetics) and switch to E medium with a mitomycin C treated J2 3T3 feeder layer approximately 48 h later (Fig. 18.1, cf. A–E with F). Second, each batch of frozen keratinocytes should be tested to determine how long (i.e., after how many passages) the keratinocytes will remain viable in monolayer culture after thawing.

Raft cervical epithelial culture tissues can be harvested and used for histology, immunohistochemistry, electron microscopy, biochemical, and molecular biological studies. Sufficient amounts of nucleic acids or proteins can be isolated from a single raft culture tissue for several analyses. We have described techniques that reproducibly work for our laboratory. These techniques can be adapted to individual needs.

Some possible areas of research using the cervical epithelial raft culture system are as follows:

1. Investigation of the molecular mechanisms of terminal differentiation.
2. Study of human hormonal influences on cervical epithelial growth and differentiation.
3. Study of cervical tumorigenesis (Fig. 18.1H).
4. Substitution of human fibroblasts for murine fibroblasts to determine species specificity.
5. Substitution of fibroblasts from different anatomical sites to determine cross-talk specificity between epithelial cells and fibroblasts.
6. Use of human fibroblast isolated from cervical tumors of different types and from different stages of tumorigenesis. These materials can be used to study how the

underlying fibroblast from different types and stages of tumorigenesis change and affect tissue differentiation.

7. Infectious agents of the cervix can be studied by using raft cultures (Fig. 18.1H) [37,38].

8. The raft culture system can be used for toxicological investigations.

9. The raft culture system can be adapted for the growth and differentiation of epithelium from any part of the body (Fig. 18.1G).

REFERENCES

1. Ferenczy, A., and Wright, T. C. (1994). Anatomy and histology of the cervix. *In* "Blaustein's Pathology of the Female Genital Tract" (R. J. Kurman, ed.), pp. 185–201. Springer-Verlag, New York.

2. Averette, H. E., Weinstein, G. D., and Frost, P. (1970). Autoradiographic analysis of cell proliferation kinetics in human genital tissues. I. Normal cervix and vagina. *Am. J. Obstet. Gynecol.* **108**, 8–17.

3. Freeman, A. E., Igel, H. J., Herrman, B. J., and Kleinfeld, K. L. (1976). Growth and characterization of human skin epithelial cell cultures. *In Vitro* **12**, 352–362.

4. Régnier, M., Pautrat, G., Pauly, G., and Pruniéras, M. (1984). Natural substrates for the reconstruction of skin *in vitro*. *Br. J. Dermatol.* **111**(Suppl. 27), 223–224.

5. Régnier, M., Pruniéras, M., and Woodley, D. (1981). Growth and differentiation of adult human epidermal cells on dermal substrates. *Front. Matrix Biol.* **9**, 4–35.

6. Woodley, D., Saurat, J. T., Pruniéras, M., and Régnier, M. (1982). Pemphoid, pemphigus and Pr antigens in adult human keratinocytes grown on nonviable substrates. *J. Invest. Dermatol.* **79**, 23–29.

7. Michalopoulos, G., and Pitot, H. C. (1975). Primary culture of parenchymal liver cells on collagen membranes. *Exp. Cell Res.* **94**, 70–78.

8. Fusenig, N. E., Amer, S. M., Boukamp, P., and Worst, P. K. M. (1978). Characteristics of chemically transformed mouse epidermal cells *in vitro* and *in vivo*. *Bull. Cancer* **65**, 271–280.

9. Lillie, J. H., MacCallum, D. K., and Jepsen, A. (1980). Fine structure of subcultivated stratified squamous epithelium grown on collagen rafts. *Exp. Cell Res.* **125**, 153–165.

10. Bell, E., Merrill, C., and Solomon, D. (1979). Characteristics of a tissue-equivalent formed by fibroblasts cast in a collagen gel. *J. Cell Biol.* **83**, 398a.

11. Bell, E., Sher, S., Hull, B., Merrill, C., Rosen, S., Chamson, A., Asselineau, D., Dubertret, L., Coulomb, B., Lapiere, C., Nusgens, B., and Neveux, Y. (1983). The reconstitution of living skin. *J. Invest. Dermatol.* **81**, 2s–10s.

12. Chamson, A., Finley, J., Hull, B., and Bell, E. (1982). Differentiation and morphogenesis of keratinocytes grown of contracted collagen lattices. *J. Cell Biol.* **95**, 59a.

13. Asselineau, D., and Pruniéras, M. (1984). Reconstitution of 'simplified' skin: Control of fabrication. *Br. J. Dermatol.* **111**(Suppl. 27), 219–222.

14. Mayer, T. J., and Meyers, C. (1998). Temporal and spatial expression of the E5a protein during the differentiation-dependent life cycle of human papillomavirus type 31b. *Virology* **248**, 208–217.

15. Meyers, C., Mayer, T. J., and Ozbun, M. A. (1997). Synthesis of infectious human papillomavirus type 18 in differentiating epithelium transfected with viral DNA. *J. Virol.* **71**, 7381–7386.

16. Meyers, C., Frattini, M. G., Hudson, J. B., and Laimins, L. A. (1992). Biosynthesis of human papillomavirus from a continuous cell line upon epithelial differentiation. *Science* **257**, 971–973.

17. Ozbun, M. A., and Meyers, C. (1996). Transforming growth factor β1 induces differentiation in human papillomavirus-positive keratinocytes. *J. Virol.* **70**, 5437–5446.

18. Ozbun, M. A., and Meyers, C. (1997). Characterization of late gene transcripts expressed during vegetative replication of human papillomavirus type 31b. *J. Virol.* **71**, 5161–5172.

19. Ozbun, M. A., and Meyers, C. (1998). Temporal usage of multiple promoters during the life cycle of human papillomavirus type 31b. *J. Virol.* **72**, 2715–2722.

20. Ozbun, M. A., and Meyers, C. (1998). Human papillomavirus type 31b E1 and E2 transcript expression correlates with vegetative viral genome amplification. *Virology* **248**, 218–230.

21. Ozbun, M. A., and Meyers, C. (1999). Two novel promoters in the upstream regulatory region of human papillomavirus type 31b are negatively regulated by epithelial differentiation. *J. Virol.* **73**, 3505–3510.

22. Visalli, R. J., Courtney, R. J., and Meyers, C. (1997). Infection and replication of herpes simplex virus type 1 in an organotypic epithelial culture system. *Virology* **230**, 236–243.

23. Wu, Y.-J., Parker, M., Binder, N. E., Beckett, M. A., Sinard, J. H., Griffiths, C. T., and Rheinwald, J. G. (1982). The mesothelial keratins: A new family of cytoskeletal proteins identified in cultured mesothelial cells and nonkeratinizing epithelia. *Cell (Cambridge, Mass.)* **31**, 693–703.

24. Freshney, R. I. (1994). "Culture of Animal Cells: A Manual of Basic Technique." Alan R. Liss, New York.

25. Alitalo, K., Halila, H., Vesterinen, E., and Vaheri, A. (1982). Endo- and ectocervical human uterine epithelial cells distinguished by fibronectin production and keratinization in culture. *Cancer Res.* **42**, 1142–1146.

26. Delvenne, P., Al-Saleh, W., Gilles, C., Thiry, A., and Boniver, J. (1995). Inhibition of growth of normal and human papillomavirus-transformed keratinocytes in monolayer and organotypic cultures by interferon-γ and tumor necrosis factor-α. *Am. J. Pathol.* **146**, 589–598.

27. Gilles, C., Piette, J., Rombouts, S., Laurent, C., and Foidart, J.-M. (1993). Immortalization of human cervical keratinocytes by human papillomavirus type 33. *Int. J. Cancer* **53**, 872–879.

28. Gorodeski, G. I., Eckert, R. L., Utian, W. H., and Rorke, E. A. (1990). Maintenance of *in vivo*-like keratin expression, sex steroid responsiveness, and estrogen receptor expression in cultured human ectocervical epithelial cells. *Endocrinology (Baltimore)* **126**, 399–406.

29. Masup, T. (1995). Establishment of an outgrowth culture system to study growth regulation of normal human epithelium. *In Vitro Cell. Dev. Biol.—Anim.* **31**, 440–446.

30. Rheinwald, J. G., and Green, H. (1975). Serial cultivation of strains of human epidermal keratinocytes: The formation of keratinizing colonies from single cells. *Cell (Cambridge, Mass.)* **6**, 331–344.

31. Stanley, M. A., and Parkinson, E. K. (1979). Growth requirements of human cervical epithelial cells in culture. *Int. J. Cancer* **24**, 407–414.

32. Turyk, M. E., Golub, T. R., Wood, N. B., Hawkins, J. L., and Wilbanks, G. D. (1989). Growth and characterization of epithelial of epithelial cells from normal human uterine ectocervix and endocervix. *In Vitro Cell Dev. Biol.* **25**, 544–556.

33. Vesterinen, E., Leinikki, P., and Saksela, E. (1975). Cytopathogenicity of cytomegalovirus to human ecto- and endocervical epithelial cells *in vitro*. *Acta Cytol.* **19**, 473–481.

34. Vesterinen, E., Nedrud, J. G., Collier, A. M., Walton, L. A., and Pagano, J. S. (1980). Explantation and subculture of epithelial cells from human uterine ectocervix. *Cancer Res.* **40**, 512–518.

35. Wilbanks, G. D., Leipus, E., and Tsurumoto, D. (1980). Tissue culture of the human uterine cervix. *Methods Cell Biol.* **21B**, 29–50.

36. Woodworth, C. D., Bowden, P. E., Doniger, J., Pirisi, Barnes, W., Lancaster, W. D., and DiPaolo, J. A. (1988). Characterization of normal human exocervical epithelial cells immortalized *in vitro* by papillomavirus types 16 and 18 DNA. *Cancer Res.* **48**, 4620–4628.

37. Meyers, C., Frattini, M., and Laimins, L. A. (1998). Tissue culture techniques for the study of human papilloma viruses in stratified epithelia. *In* "Cell Biology: A Laboratory Handbook" (J. E. Celis, ed.), Vol. 1, pp. 513–520. Academic Press, San Diego, CA.

38. Meyers, C. (1997). Organotypic (raft) epithelial tissue culture system for the differentiation-dependent replication of papillomavirus. *Methods Cell Sci.* **18**, 201–210.

CHAPTER 19

EPITHELIAL CELL CULTURE: VAGINAL CELL RECONSTRUCTION

Roger E. De Filippo and Anthony Atala

INTRODUCTION

A variety of pathological disorders exist, affecting the external genitalia and mandating extensive surgical intervention [1]. Male genital reconstruction affords the most currently reported long term clinical success with tissue engineering applications and substantiated suitability for urethral reconstruction [2,3]. Certainly other disorders like cloacal malformations and exstrophy can result in severe genital ambiguity for both males and females. However, there is a paucity of information regarding the reconstruction of female external genitalia and vaginal reconstruction employing the techniques described with tissue engineering.

This chapter outlines the techniques employed for expanding vaginal epithelial and smooth muscle cells *in vitro* with adjunctive histochemical measures employed for cellular characterization. These preliminary steps set the stage for larger three-dimensional constructs, which can eventually be applied toward larger animal models.

METHODS OF CELL CULTURE

EXPERIMENTAL GOALS

To preserve the integrity of cellular therapeutic regimes manufactured in the laboratory, certain principles must be considered. The investigator must begin with a reliable tissue source for cellular expansion that is easily accessible for biopsy. Both the level of technical difficulty in accessing these tissues and the tissue yield are important considerations for the investigator. If the separation of different cell layers is required, then the degree to which a particular specimen can be manipulated for microdissection must also be considered. When a substantial cellular colony can be established and maintained with the proper medium, the investigator must demonstrate that cellular integrity is preserved both *in vitro* and *in vivo* through a variety of histochemical and molecular techniques, which may include immunohistochemistry, immunolabeling, Western blot analyses, and analysis by means of reverse transcriptase polymerase chain reaction (RT-PCR).

CULTURE MEDIUM

A variety of commercially manufactured culture media are available for epithelial and smooth muscle cell growth. We prefer Dulbecco's Modified Eagle's Medium supplemented with 10% fetal bovine serum (DMEM/FBS) for smooth muscle cells. The vaginal epithelial cells (keratinocytes) are cultured in serum-free medium specifically for keritinocytes, supplemented with bovine pituitary extract and epidermal growth factor (K-SFM).

PROCEDURES FOR CELL HARVEST

The availability of an abundance of easily retrievable tissue sources is imperative for the success of any experimental design involving animal models and tissue engineering. In our experience, the New Zealand white rabbit has served as a excellent source of vaginal tissue that can be harvested through a simple, midline, transabdominal approach allowing for good exposure during the harvest of tissue. The rabbit's vagina has ample size and girth and allows for excellent tissue yield during each procedure. The harvested specimen is transported in sterile culture medium to the laboratory, where the process of separating the individual tissue layers begins.

TISSUE PROCESSING AND CELL CULTURE

Several wash cycles with phosphate-buffered saline (PBS) containing ethylenediamenetetracetic acid (EDTA) are performed. The specimen is finally placed into a clean reservoir of culture medium until the process of microdissection begins.

Cell Isolation and Culture

Epithelial and smooth muscle cells are grown separately, and isolation of the individual cell types involves one of two processes that consist of either an explant method or enzymatic digestion [4].

Explant Method

The explant method begins with careful microdissection with sterile instruments under loop magnification, separating the epithelial and seromuscular layers. The investigator may find that detubularizing the vagina into a flat segment facilitates the dissection. With experience, this exercise can be performed without difficulty. Small portions of the tissue are individually placed onto culture dishes, where they dry and adhere to the surface. The pieces of tissue are incubated with the appropriate medium at 37°C in air and 5% CO_2 undisturbed until a sufficient colony of progenitor cells develops from the tissue islets, which usually takes approximately 5–7 days. The explants can be removed by gentle suction and the cells maintained with scheduled replacement of the medium every 24–48 h.

Enzymatic Digestion

The fastidious nature of epithelial cells sometimes makes growth to large quantities difficult. Therefore, a method of enzymatic digestion has also been applied for the processing and culture of vaginal epithelial cells with good success in achieving ample colony sizes. Powder forms of collagenase type IV and dispase, a neutral protease, are combined and suspended with approximately 25 ml of K-SFM. This collagenase–medium solution is then filtered into a 50-ml tube to ensure sterility. The vaginal specimen is cut into several large pieces, immersed into the enzymatic solution, and vigorously shaken for 25–30 min at 37°C. With gentle pipette suction, the cell–fluid suspension is transferred to another 50-ml tube and centrifuged at low revolutions for 5 min. Finally, the supernatant is removed and the cell pellet resuspended in medium and distributed into culture dishes.

CELL EXPANSION

Passage of the cells is performed by first removing the culture medium and washing the cells with PBS-EDTA. The cells are incubated with a trypsin–EDTA solution, prepared from stock ingredients, and monitored under the microscope until cell separation is observed. Gentle pipette suction removes the cell–trypsin solution into a 50-ml Falcon tube with serum-containing medium to inactivate the trypsin. The cells are centrifuged at low revolutions for 5 min. The cell pellet is resuspended to a predetermined volume with fresh medium and partitioned equally among several more culture dishes for expansion.

CELL CHARACTERIZATION

HISTOLOGY AND IMMUNOHISTOCHEMISTRY

Cells are transferred and cultured onto chamber slides, fixed with 4% buffered formaldehyde, and processed. The cells are exposed to antigen-specific primary antibodies applied to the surface of the cell. Broadly reacting anti-cytokeratin and anti–smooth muscle α-actin antibodies are routinely employed. Negative controls are treated with plain serum instead of primary antibody. Positive controls will consist of antigen-exposed cells. After washing with phosphate-buffered saline, the chamber slides are incubated with a biotinylated secondary antibody and washed again. A peroxidase reagent is added and, upon substrate addition, the sites of antibody deposition will be visualized as a brown precipitate. Counterstaining is performed with Gill's hematoxylin.

MOLECULAR ANALYSIS

Western blot analysis is one method widely used in our laboratroy for cell characterization at a molecular level. Monoclonal antibodies α-actin, myosin, and cytokeratins AE1/AE3 were used in our experiments to compare protein expression with cells and controls cultured *in vitro* to confirm the maintenance of epithelial and smooth muscle cell lines. The cells are homogenized in cold lysis buffer and the soluble protein supernatant collected. A BioRad DC protein assay kit is used for quantification of the protein samples. Equal concentrations of protein are loaded and separated on SDS-PAGE gel and probed overnight at 4°C with the primary antibody. Peroxide-conjugated anti-mouse secondary antibody is complexed and detected with an enhanced chemiluminescent system. Polymerase chain reactions can also be concomitantly performed for additional qualification of the cell types.

CONCLUSION

Techniques described by tissue engineering can now be used to create many possibilities in reconstructive materials in the laboratory. Successfully accomplishing cell growth and expansion will ultimately lead to larger animal studies for autologous tissue replacement and ultimately to clinical trials that will prove to be beneficial for patients who require reconstructive procedures.

REFERENCES

1. Hendren, H. W. (1998). Cloacal malformations. *In* "Campell's Urology" (P. C. Walsh, A. B. Retik, E. D. Vaughan, and A. J. Wein, eds.), 7th ed., pp. 1991–2001. Saunders, Philadelphia.
2. Atala, A., Guzman, L., and Retik, A. B. (1999). A novel inert collagen matrix for hypospadias repair. *J. Urol.* **162**(3, pt. 2), 1148–1151.
3. Chen, F., Yoo, J. J., and Atala, A. (2000). Experimental and clinical experience using tissue regeneration for urethral reconstruction. *World J. Urol.* **18**(1), 67–70.
4. Williams, D. W., and Wynford-Thomas, D. (1989). Human thyroid epithelial cells. *Methods Mol. Biol.* **5**, 139–149.

CHAPTER 20

MESENCHYMAL CELL CULTURE: ADIPOSE TISSUE

Adam J. Katz

The eventual development of tissue-engineered soft tissue equivalents will impact the treatment of numerous reconstructive challenges. Many emerging tissue engineering strategies necessitate the use of cells. Cells from autologous tissue sources are ideal for many reasons. Fat tissue, in particular, is unique because of its unmatched accessibility and expendability, and liposuction is an excellent method with which to *harvest* autologous adipose cells. Liposuction is appealing to patients because it is minimally invasive and potentially improves physical appearance. It is equally appealing to the physician/scientist because it can easily yield large volumes of tissue, making research and clinical endeavors practical. Now, a growing body of evidence substantiates that liposuctioned tissue contains adipogenic stem cells that can be isolated, culture-expanded, and developmentally manipulated. Although there is still much to learn, a broad foundation for soft tissue engineering now exists.

INTRODUCTION
CLINICAL NEED

Emerging tissue engineering strategies represent a promising and innovative solution to many clinical challenges. The eventual development of tissue-engineered fat equivalents for reconstructive and augmentation purposes will be most welcomed by nearly every surgical discipline and will prove to be especially useful for plastic and reconstructive surgeons. The clinical applications for which tissue-engineered fat will be particularly useful are vast and varied and can be loosely categorized into reconstructive, cosmetic, corrective, and orthotic indications. *Reconstructive* challenges potentially benefiting from a soft tissue equivalent include congenital malformations, complex traumatic wounds, pressure sores, and oncologic-related defects (Fig. 20.1). From a *cosmetic* perspective, the potential use of engineered soft tissue implants is just as boundless. *Correctional* uses for engineered fat tissue equivalents might include urinary incontinence or vocal cord insufficiency in which a stable, long-lasting "bulking agent" is required. Finally, engineered soft tissue constructs might prove useful for various *orthotic*-related applications in which normal soft tissues have atrophied.

CURRENT PRACTICE AND FUTURE DIRECTIONS

Standard approaches to soft tissue reconstruction currently include the use of various surgical flaps. All are associated to some degree with operative risk, technical difficulty, costly operative time, hospital stay, and donor site morbidity. Alternative approaches to soft tissue reconstruction and augmentation have traditionally included alloplastic and allogeneic products such as Teflon paste, silicone implants, and bovine collagen. More recent options include autologous injectable collagen and dermal allograft scaffolds. Each of these methods, however, is associated with certain drawbacks. Alloplastic materials are subject to potential migration and associated with foreign body reactions, allergic reactions, and

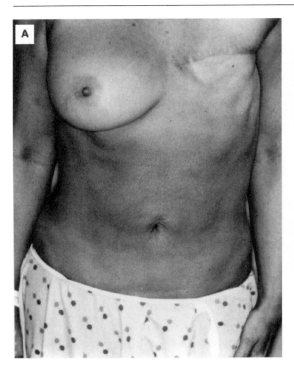
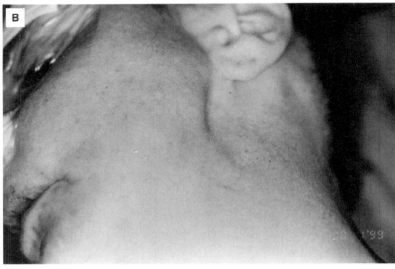

Fig. 20.1. Clinical examples of soft tissue defects. A mastectomy defect (A) and a postsurgical contour defect of the neck (B) offer good examples of patients who would benefit from engineered autologous soft tissue equivalents.

extrusion. Allogeneic materials also include risks of allergic reactions and infection transmission and ultimately fail to integrate into the recipient site for any extended time period. Current autologous products also have questionable long-term efficacy profiles.

Autologous fat transfer represents a logical approach to soft tissue reconstruction and enhancement. Successful free fat transplantation, however, remains an elusive clinical endeavor because it is usually accompanied by a significant volume loss. In short, issues of *predictability* and *reproducibility* remain obstacles that prevent more widespread clinical use. Emerging tissue engineering strategies represent an innovative approach to these important issues. Tissue engineering will soon yield techniques that enable the (*re*)generation of various tissues using cell-based implants fabricated *ex vivo*, as well as strategies that exploit the innate developmental plasticity of a given tissue *in situ*.

TISSUE ENGINEERING OF ADIPOSE TISSUE

Compared with tissues such as skin, bone, and cartilage, relatively little effort has been directed toward the tissue engineering of fat. The most logical means to generate or (*re*)generate adipose tissue for reconstructive or cosmetic purposes is to mimic to the best extent possible the events of natural development. There is no doubt that such achievements will require a thorough understanding of adipose tissue on a cellular and molecular level. Whether adipose tissue is engineered *ex vivo* (e.g., as autologous cell-based implants) or induced *in situ*, a common theme remains the controlled developmental manipulation of adipo-derived cells in a way that strives to *recapitulate normal adipose development*.

Adipose tissue is historically perceived as a static collection of lipid-engorged cells that can enlarge but not proliferate; in fact, it is a surprisingly dynamic tissue composed of mature adipocytes as well as stromal–vascular fraction (SVF) cells such as fibroblasts, pericytes, endothelial cells, and adipogenic progenitor cells, or *preadipocytes*. It was once thought that the trauma inherent in the suction process rendered tissue obtained by liposuction nonviable. It has subsequently been shown by our lab and others that this is not true [1,2]. Our studies confirm the viability of significant populations of mature adipocytes and preadipocytes in liposuction-harvested human adipose tissue.

ISOLATION, CULTURE, AND DEVELOPMENTAL MANIPULATION OF LIPO-DERIVED CELLS

As mentioned, there are many cell types found in adipose tissue. This chapter focuses on the cells proposed to be most integral to the development of engineered soft tissue equivalents: adipogenic precursor cells, or preadipocytes, and mature adipocytes.

PREADIPOCYTES

Cell Sources and Tissue Procurement

Most work relating to the cellular and molecular mechanisms of preadipocyte growth and differentiation is and has been done using cell lines or primary cells derived from animals. Only a small amount of work has used primary *human cells* derived from subcutaneous depots. Cell lines often used include 3T3-L1, 3T3-F422A, and Ob17. In general, cell lines represent an earlier stage of development than preadipocytes that are isolated from primary tissues. Cell lines have the advantage of being clonally derived and are therefore a defined, homogeneous population of cells presumably at the same stage of differentiation. On the other hand, cultures derived from primary tissues contain cells at varying stages of development. While this is not as scientifically controlled as cell lines, it probably represents a more accurate reflection of the normal *in vivo* condition [3].

A common animal source of preadipocytes is the epididymal fat pad of rats. Despite the widespread use of this source, however, one must wonder what similarities, if any, rat epididymal preadipocytes have to preadipocytes derived from human subcutaneous fat. This matter is worth consideration for several reasons. First, subcutaneous fat is the most likely source of preadipocytes for tissue engineering strategies for fat. Second, most engineered constructs will ultimately be used for indications that involve placement in subcutaneous sites. And, as will be discussed further, there are definite depot-specific differences in the growth and differentiation capacity of fat.

It is intuitive that one's own (autologous) tissues are the ideal source of cells for tissue engineering. Using cells from another person or animal will require immunosuppressive medications similar to those necessary for organ transplantation and will carry all the expense and risks associated with such drugs. Critical shortages of most human tissues and organs severely limit the practice of many current medical therapies. In similar fashion, most autologous tissues cannot be obtained without invasive procedures and/or are not expendable in amounts necessary to make future tissue engineering strategies practical. Fat tissue, however, is the exception. In fact, fat is an *ideal source* of autologous cells for tissue engineering strategies because it is abundant, expendable, and readily obtainable for research and clinical uses. Human adipose tissue can be obtained in two main ways: excision or liposuction. From the perspective of an end-stage clinical therapy, however, liposuction represents a far more appealing and practical method of tissue harvest to both patient and physician, given its minimally invasive nature and the potential to improve physical appearance. And though liposuction was originally intended for the *removal* of unwanted fat tissue (and is still predominately utilized as such), it is also an excellent way to *harvest* autologous tissue for tissue engineering research and emerging technologies.

Processing Liposuctioned Tissue: Novel Issues and Novel Approaches

The isolation and culture of specific cells from liposuctioned fat tissue first requires that the "raw" effluent be washed and separated (i.e., dissociated) into its cellular components. Liposuctioned tissue presents two unique challenges: volume and viscosity. Methods currently used for tissue dissociation are tedious, time-consuming, labor intensive, subject to contamination, and generally limited to the laboratory setting. To overcome the technical challenges associated with the dissociation of liposuctioned fat, our lab has designed a novel device for the expedient filtration, cleansing, and dissociation of large volumes (100–1500 ml) of human liposuctioned tissue. The ultimate purpose of this device is to make the production-scale isolation of cells from raw liposuction effluent practical and efficient. A motorized prototype of this device has been fabricated, complete with a heat source for optimal enzymatic dissociation (Fig. 20.2). The device design allows for the direct transfer

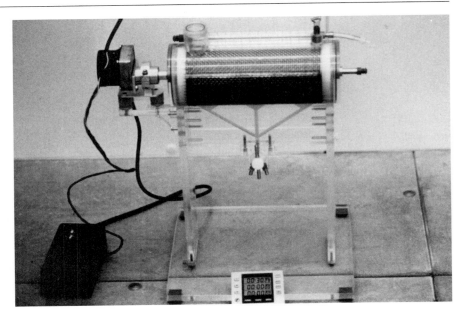

Fig. 20.2. Prototype of a novel device for processing large volumes of liposuctioned adipose tissue. Harvested tissue can be transferred directly to a steel mesh cylinder that resides within a watertight plastic housing unit. The cylinder revolves on its long axis to assist with straining of the tissue sample. Waste fluids egress through pores in the steel mesh and drain inferiorly through a stopcock valve.

of harvested tissue from collection vessel to device without exposure to air. As a single, self-sufficient unit, it enables the efficient processing (filtration, cleansing, and dissociation) of raw liposuctioned tissue in an enclosed, sterile environment.

Adipodissociation: Methods

Once in the lab, raw liposuction effluent should be strained and thoroughly rinsed under sterile conditions to clean and separate tissue fragments from blood, serum, tumescent solution, oil, and so on. This tissue refinement step serves to optimize enzymatic activity during dissociation. We use phosphate-buffered saline (PBS) for this "tissue-refinement" step (Fig. 20.3). Once a refined tissue sample has been obtained, it is ready for dissociation. There are many published methods of enzymatic and/or mechanical dissociation. The main objective, of course, is to find a method that results in quick and effective tissue dissociation while at the same time optimizes cell viability. Our lab has modified enzymatic methods described by Hauner *et al.* [4]. Lipo-derived tissue that has been refined is incubated in a collagenase–PBS solution with a final collagenase concentration of 300 U/ml. The refined tissue is incubated in this solution at 37°C with constant agitation for 30–45 min (Fig. 20.3). The volume of collagenase solution to refined tissue volume is at least 1:1. After adequate dissociation, medium with 10% fetal bovine serum is used to dilute the digestate. The entire digestate volume is then filtered through 250-μm mesh (i.e., the steel mesh cylinder described in Figs. 20.2 and 20.3) and allowed to layer by gravity for 5 min. Two main layers subsequently develop: an upper layer of buoyant mature adipocytes and free oil, and a lower layer of buffer and collagenase solution that contains preadipocytes (stromal–vascular cells) (Fig. 20.4). The two layers are easily isolated by pipette aspiration and further processed for culture and manipulation.

Isolation of Stromal–Vascular Cells

After gravity sedimentation, the resulting lower layer (containing the stromal–vascular cell fraction) is pipetted into centrifuge tubes (250 ml) and centrifuged at 400*g* for 10 min (Fig. 20.5). The supernatant is discarded, and the pelleted stromal–vascular cell fraction is resuspended in an erythrocyte lysing buffer for 10 min at room temperature with constant shaking. The cells are pelleted again at 400*g* for 10 min and resuspended in plating medium [a 1:1 mixture of Dulbecco's Modified Eagle's Medium (DMEM) and Ham's F-12 nutrient mixture with 10% fetal bovine serum and antibiotic/antimycotic supplements]. The cell suspension is then filtered through 100-μm nylon mesh (Fig. 20.5), washed two more times, and counted. Counting, using standard techniques, is accomplished with a hemacytometer and trypan blue exclusion.

Fig. 20.3. Washing and dissociation of liposuctioned adipose tissue. Tissue pieces retained within the mesh cylinder are rinsed thoroughly with buffer (A) and then incubated with collagenase solution (B). The cylinder is rotated constantly during collagenase digestion (C) to help enhance the tissue dissociation process. As dissociation proceeds, cells are able to exit the cylinder's mesh pores, while undigested tissue fragments are retained within the cylinder. After 30–45 min of incubation, the resulting cell suspension (digestate) is collected via the inferior outlet valve.

Plating and Culture of Stromal–Vascular Cells

Isolated stromal cells are plated onto culture flasks at concentrations ranging from 20,000 to 100,000 cells/cm^2 and grown in DMEM/Ham's F-12 (1:1) with 10% fetal bovine serum and antibiotic/antimycotic supplements. To expedite cell proliferation, bovine fibroblast growth factor may be added (10 ng/ml) to the medium. To passage, confluent cultures are trypsin-released, washed, counted, and replated into new culture flasks. In all cases, cells are given fresh medium at least three times per week and incubated at 37°C with 5% CO_2.

Differentiation of Lipo-Derived Preadipocytes

For adipogenic induction, cells that reach confluence are rinsed twice with PBS and then induced with serum-free adipogenic medium containing 1:1 DMEM/Ham's F-12, 15 mM NaHCO$_3$, 15 mM HEPES, 33 μM biotin, 17 μM pantothenate, 10 μg/ml transferrin, 100 nM hydrocortisone, 66 nM insulin, 0.2 nM T3, 100 U/ml penicillin, and 100 μg/ml streptomycin. Isobutylmethylxanthine (IBMX) is added to the cultures at a concentration of 0.25 mM for the first 4–5 days of adipogenic induction. The adipogenic media should be changed every 2–3 days. Over the next 1–3 weeks, the accumulation of intracellular lipid (i.e., differentiation) is confirmed by microscopic analysis and staining with oil red O.

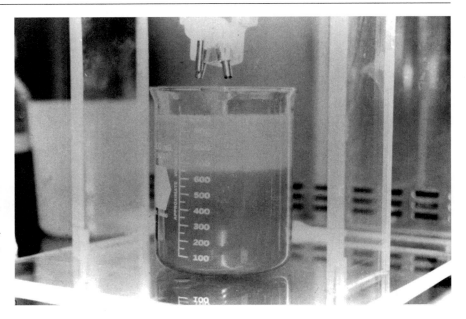

Fig. 20.4. Gravity layering of the tissue digestate. After collagenase digestion, the cell suspension is collected and allowed to layer by gravity for 5 min. Two layers subsequently develop: an upper layer of buoyant mature adipocytes, and a lower layer consisting of stromal–vascular cells suspended in collagenase solution–media. Each layer can be separated by aspiration and processed further for cells of interest.

MATURE ADIPOCYTES

The cellular biology of mature adipocytes has been technically difficult to study in the past because these cells are buoyant in culture medium and do not attach easily to culture surfaces. With the introduction of "ceiling culture" methods by Sugihara et al., mature fat cells were shown to exist in culture for long periods of time and to have the potential to become fibroblast-like cells [5,6]. This process has been termed dedifferentiation by some, and merely delipidation by others. Regardless of what term is applied, lipid-filled adipocytes have the capacity to revert to a lipid-free, fibroblast-like morphology under certain environmental conditions. After these changes have occurred, the cells are able to proliferate to confluency and be passaged. Even more, when exposed to adipogenic growth conditions, such dedifferentiated adipocytes are able to reaccumulate intracytoplasmic lipid droplets (i.e., to redifferentiate).

The objective of the ceiling culture technique is to obtain a purified population of adherent mature adipocytes. Our initial experience suggested that even with the ceiling technique, "stray" stromal cells trapped within the viscous, oily phase of the layered digestate attached to the flask surface. These "contaminating" stromal cells prevent the purification and accurate evaluation of mature adipocyte dedifferentiation. As such, we modified the ceiling culture technique by instituting serial ceiling passages (Fig. 20.6).

Isolation, Culture, and Dedifferentiation of Mature Adipocytes

Liposuction-harvested tissue is refined and dissociated as described earlier. After layering of the digestate and isolation of the mature adipocyte fraction, the cells are rinsed with 10% FBS media and centrifuged at 180g for 10 min. This sequence is repeated as necessary to obtain a clean, purified population of mature adipocytes. The purified adipocytes are then serially filtered through 250- and 200-μm nylon mesh to achieve a uniform cell suspension. Adipocytes are then counted in a counting chamber, Hoescht stain 33258 (fluorescent dye) is used to stain nuclei by adding Hoescht (1 mg/ml in DMSO) to a sample of the cell suspension until a final concentration of 4 μg/ml has been achieved (Fig. 20.7). This, of course, requires use of a fluorescent microscope. We use Hoescht stain because it is nearly impossible to distinguish adipocytes proper from free oil droplets, even at high power magnification. In our experience, trypan blue does not work as well as Hoescht stain.

For culture and dedifferentiation, refined mature adipocytes are plated by means a modification of the ceiling culture technique reported by Sugihara et al. [5,6]. More specifically, mature adipocytes are incubated at a concentration of 50,000 to 100,000 cells per 25-cm^2 culture flask. The ceiling culture technique requires that the flask be *completely filled* with

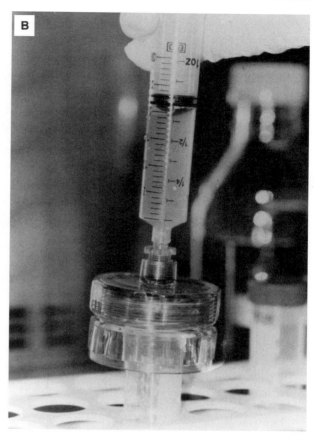

Fig. 20.5. Purification of stromal–vascular cells. Stromal cells are concentrated by centrifuging the lower layer of gravity-layered digestate. A cell pellet forms (A), which is resuspended in erythrocyte lysing buffer and then filtered through nylon mesh (B). The cells are then washed one or two more times, assessed for viability and numbers, and finally plated.

medium. Initially, the flask is turned upside down so that the buoyant mature adipocytes will float to the upper surface of the flask (which is actually the flask bottom in an inverted position) and adhere (Fig. 20.6). The cultures are maintained at 37°C and 5% CO_2 in a plating medium consisting of DMEM and Ham's F-12 (1:1), supplemented with 10% fetal bovine serum (pH 7.4) and antibiotic/antimycotic supplements.

After 48 h in the first ceiling culture, all media (containing nonadherent cells) in a given culture flask are transferred into a sterile 50-ml conical centrifuge tube. Cells that have adhered in the first flask are fed new media and evaluated under the microscope. This usually reveals the presence of many fibroblast-like cells, which, at this early time point, are considered to be "contaminating" stromal cells rather than dedifferentiated adipocytes. The non-adherent cells that have been transferred to a conical tube are then serially filtered through 200- and 150-μm nylon mesh to remove aggregates of dead cells and then centrifuged at 180g for 10 min. Again, the floating mature adipocyte layer is aspirated by pipette, counted, and plated into a second ceiling culture flask. After another 48 h, the second ceiling culture flask is turned right-side up to permit observation the cells that have attached to the flask. If contamination from fibroblast-like cells is noted, the washing and passage process is repeated (passage to third ceiling culture, and so on) until a pure population of adherent lipid-filled adipocytes is obtained.

When absence of contamination from fibroblast-like cells 48 h after a given plating (e.g., first, second, or third passage) has been confirmed, adipocytes are left undisturbed in ceiling culture for 10 days to allow the cells to attach *firmly* to the surface of the culture flask. After 10 days in ceiling culture, the flask is turned to normal position and the media (5 ml) changed every 2 days. Over 2–3 weeks in this culture setting, the mature adipocytes

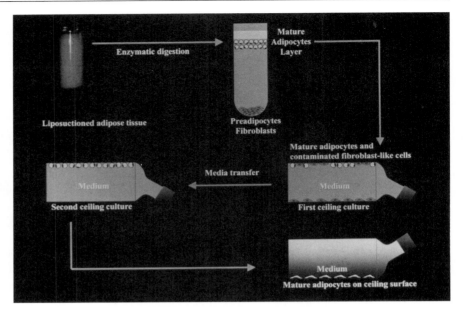

Fig. 20.6. Schematic of serial ceiling culture technique. Mature adipocytes are washed and counted as described in the text. They are then plated in ceiling culture, which entails filling culture flasks completely with medium and turning them upside down to allow buoyant mature adipocytes to attach to flask bottoms. This procedure is repeated every 48 h until a pure population of adherent adipocytes has been obtained.

will dedifferentiate, or delipidate into fibroblast-like cells reminiscent of cells from the stromal–vascular fraction (Fig. 20.8). These dedifferentiated adipocytes can then be grown to confluence and passaged using similar methods described earlier for preadipocytes. In addition, they can be redifferentiated into mature adipocytes using methods also described earlier.

PRESENT AND FUTURE

As mentioned before, a thorough understanding of, and ability to control preadipocyte differentiation will be integral to developing techniques for the regeneration of adipose tissue. It is hypothesized that cell-based constructs will survive implantation well enough to establish an equilibrium of cell turnover that permits optimal host integration and long-term volume (i.e., contour) maintenance. Preadipocytes—which exist in the stromal–vascular fraction of adipose tissue—are the logical cell candidate on which to base the fabrication of cell-based, tissue-engineered constructs. Given their smaller size and lack of intracytoplasmic lipid, preadipocytes seem to be able to tolerate the mechanical trauma and ischemia

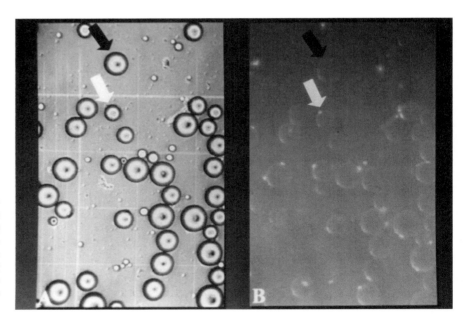

Fig. 20.7. Fluorescent staining of mature adipocytes. Counting of mature adipocytes is facilitated by staining nuclei with Hoescht stain and evaluating them with fluorescent microscopy (B). When standard microscopic techniques (A) are used, it is extremely difficult to distinguish free oil droplets (black arrows) from mature adipocytes (white arrows).

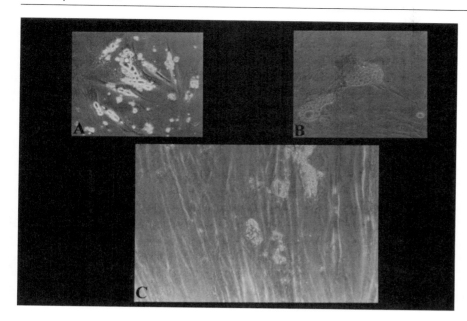

Fig. 20.8. Dedifferentiation of mature adipocytes. After several weeks in culture, mature adipocytes gradually dedifferentiate (delipidate). Initially, the cells become spindle shaped with multiloculated intracytoplasmic lipid droplets (A, B). Ultimately, they become fibroblast-like cells devoid of lipid (C) that closely resemble primary cultures of preadipocytes (stromal vascular cells).

associated with harvest and implantation better than more fragile mature adipocytes [7]. In addition, under appropriate conditions preadipocytes are able to proliferate and differentiate into mature, lipid-synthesizing and lipid-storing cells both *in vitro* and *in vivo* [8–12]. Factors that have been shown to have a positive effect on *in vitro* preadipocyte proliferation include the insulin-like growth factor (IGF) family, the fibroblast growth factors (FGF), transforming growth factor β (TGF-β), epidermal growth factor (EGF), platelet-derived growth factor (PDGF), and 17β-estradiol. Though contradictory results exist, several studies have interestingly shown that preadipocytes from massively obese persons proliferate to a greater extent than do preadipocytes from lean individuals [13,14].

Most studies demonstrate that preadipocyte proliferation and preadipocyte differentiation are inversely related. In other words, preadipocytes must undergo growth arrest (not necessarily contact inhibition) to undergo differentiation. Following the report of a *serum-free* adipogenic medium in the literature, subsequent studies revealed that the *in vitro* differentiation of human preadipocytes is enhanced by the combined presence of glucocorticoids *and* insulin [4,15,16]. In addition, elevated levels of cAMP have been shown to enhance adipogenesis. More recently, various transcription factors implicated in the control of adipogenesis and the ligands that activate them have been reported [17–19]. Factors that inhibit adipogenic induction or stimulate delipidation of differentiated cells include tumor necrosis factor α, EGF, TGF-β, and endothelin-1 [20–24].

The development of a reliable culture system to study the dedifferentiation and redifferentiation of human adipocytes will likely prove useful for delineating the mechanisms that influence and control adipogenic equilibrium. The ultimate means to long-term, predictable volume maintenance of future engineered implants may prove to depend just as much on *preventing dedifferentiation* as on inducing neovascularization and differentiation. Not surprisingly, the *in vivo* control of adipogenesis will prove to be far more complex than *in vitro* studies suggest because *all fat is NOT the same*. There are now documented differences in the growth and differentiation of fat cells derived from different adipose tissue sites [25–28]. For example, preadipocytes from different sites of the same individual respond quite differently to a specific adipogenic stimulus *on a molecular level* [29]. Further work along these lines may elucidate "optimal" harvest sites for adipose-related tissue engineering goals, if any such sites exist. Moreover, the same research efforts may provide significant insights into the pathogenesis and treatment of obesity.

In short, the engineering of adipose tissue/soft tissue implants is at an early stage. A solid foundation pertaining to the harvest, isolation, and culture of primary human adipo-derived cells now exists. Rapid advances in developmental biology continue to provide insights into the mechanisms of adipose growth and differentiation. The eventual development of soft tissue equivalents is sure to improve the quality of life of many individuals.

REFERENCES

1. Moore, J. H., Jr., Kolaczynski, J. W., Morales, L. M., *et al.* (1995). Viability of fat obtained by syringe suction lipectomy: Effects of local anesthesia with lidocaine. *Aesthetic Plast. Surg.* **19**(4), 335–339.
2. Novaes, F., dos Reis, N., and Baroudi, R. (1998). Counting method of live fat cells used in lipoinjection procedures. *Aesthetic Plast. Surg.* **22**(1), 12–15.
3. Butterwith, S. C. (1994). Molecular events in adipocyte development. *Pharmacol. Ther.* **61**(3), 399–411.
4. Hauner, H., Schmid, P., and Pfeiffer, E. F. (1987). Glucocorticoids and insulin promote the differentiation of human adipocyte precursor cells into fat cells. *J. Clin. Endocrinol. Metab.* **64**(4), 832–835.
5. Sugihara, H., Yonemitsu, N., Miyabara, S., *et al.* (1986). Primary cultures of unilocular fat cells: Characteristics of growth in vitro and changes in differentiation properties. *Differentiation (Berlin)* **31**(1), 42–49.
6. Sugihara, H., Yonemitsu, N., Miyabara, S., *et al.* (1987). Proliferation of unilocular fat cells in the primary culture. *J. Lipid Res.* **28**(9), 1038–1045.
7. Patrick, C. W., Jr., Chauvin, P. B., and Robb, G. L. (1998). Tissue engineered adipose tissue. *In* "Frontiers in Tissue Engineering" (C. W. Patrick, Jr., A. G. Mikos, and L. V. McIntire, eds.), pp. 369–382. Elsevier Science, New York.
8. Green, H., and Kehinde, O. (1979). Formation of normally differentiated subcutaneous fat pads by an established preadipose cell line. *J. Cell. Physiol.* **101**, 169–172.
9. Poznanski, W. J., Waheed, I., and Van, R. (1973). Human fat cell precursors. Morphologic and metabolic differentiation in culture. *Lab. Invest.* **29**(5), 570–576.
10. Van, R. L., Bayliss, C. E., and Roncari, D. A. (1976). Cytological and enzymological characterization of adult human adipocyte precursors in culture. *J. Clin. Invest.* **58**(3), 699–704.
11. Van, R. L., and Roncari, D. A. (1978). Complete differentiation of adipocyte precursors. A culture system for studying the cellular nature of adipose tissue. *Cell Tissue Res.* **195**(2), 317–329.
12. Van, R. L., and Roncari, D. A. (1982). Complete differentiation in vivo of implanted cultured adipocyte precursors from adult rats. *Cell Tissue Res.* **225**(3), 557–566.
13. Ng, C. W., Poznanski, W. J., Borowiecki, M., *et al.* (1971). Differences in growth in vitro of adipose cells from normal and obese patients. *Nature (London)* **231**, 445.
14. Roncari, D. A., Lau, D. C., and Kindler, S. (1981). Exaggerated replication in culture of adipocyte precursors from massively obese persons. *Metab. Clin. Exp.* **30**(5), 425–427.
15. Deslex, S., Negrel, R., and Ailhaud, G. (1987). Development of a chemically defined serum-free medium for differentiation of rat adipose precursor cells. *Exp. Cell Res.* **168**(1), 15–30.
16. Hauner, H., Entenmann, G., Wabitsch, M., *et al.* (1989). Promoting effect of glucocorticoids on the differentiation of human adipocyte precursor cells cultured in a chemically defined medium. *J. Clin. Invest.* **84**(5), 1663–1670.
17. Spiegelman, B. M. (1997). Peroxisome proliferator-activated receptor gamma: A key regulator of adipogenesis and systemic insulin sensitivity. *Eur. J. Med. Res.* **2**(11), 457–464.
18. Spiegelman, B. M. (1998). PPAR-gamma: Adipogenic regulator and thiazolidinedione receptor. *Diabetes* **47**(4), 507–514.
19. Tontonoz, P., Kim, J. B., Graves, R. A., *et al.* (1993). ADD1: A novel helix-loop-helix transcription factor associated with adipocyte determination and differentiation. *Mol. Cell. Biol.* **13**(8), 4753–4759.
20. Hauner, H., Petruschke, T., and Gries, F. A. (1994). Endothelin-1 inhibits the adipose differentiation of cultured human adipocyte precursor cells. *Metab. Clin. Exp.* **43**(2), 227–232.
21. Hauner, H., Rohrig, K., and Petruschke, T. (1995). Effects of epidermal growth factor (EGF), platelet-derived growth factor (PDGF) and fibroblast growth factor (FGF) on human adipocyte development and function. *Eur. J. Clin. Invest.* **25**(2), 90–96.
22. Petruschke, T., and Hauner, H. (1993). Tumor necrosis factor-alpha prevents the differentiation of human adipocyte precursor cells and causes delipidation of newly developed fat cells. *J. Clin. Endocrinol. Metab.* **76**(3), 742–747.
23. Petruschke, T., Rohrig, K., and Hauner, H. (1994). Transforming growth factor beta (TGF-beta) inhibits the differentiation of human adipocyte precursor cells in primary culture. *Int. J. Obes. Relat. Metab. Disord.* **18**(8), 532–536.
24. van de Venter, M., Litthauer, D., and Oelofsen, W. (1994). Catecholamine stimulated lipolysis in differentiated human preadipocytes in a serum-free, defined medium. *J. Cell. Biochem.* **54**(1), 1–10.
25. Fried, S. K., and Kral, J. G. (1987). Sex differences in regional distribution of fat cell size and lipoprotein lipase activity in morbidly obese patients. *Int. J. Obes.* **11**(2), 129–140.
26. Fried, S. K., Russell, C. D., Grauso, N. L., *et al.* (1993). Lipoprotein lipase regulation by insulin and glucocorticoid in subcutaneous and omental adipose tissues of obese women and men. *J. Clin. Invest.* **92**(5), 2191–2198.
27. Hauner, H., Ditschuneit, H. H., Pal, S. B., *et al.* (1988). Fat distribution, endocrine and metabolic profile in obese women with and without hirsutism. *Metab. Clin. Exp.* **37**(3), 281–286.
28. Hauner, H., and Entenmann, G. (1991). Regional variation of adipose differentiation in cultured stromal–vascular cells from the abdominal and femoral adipose tissue of obese women. *Int. J. Obes.* **15**(2), 121–126.
29. Adams, M., Montague, C. T., Prins, J. B., *et al.* (1997). Activators of peroxisome proliferator-activated receptor gamma have depot-specific effects on human preadipocyte differentiation. *J. Clin. Invest.* **100**(12), 3149–3153.

CHAPTER 21

MESENCHYMAL CELL CULTURE: SMOOTH MUSCLE

Byung-Soo Kim and Anthony Atala

INTRODUCTION

Smooth muscle (SM) is a functionally critical component of a variety of cardiovascular, gastrointestinal, and urinary tissues (e.g., blood vessel, intestine, and bladder). The functions of the SM element in these tissues include providing appropriate mechanical properties to withstand fluid pressures (e.g., blood pressure and urodynamic pressure), maintaining tissue structures as a fluid conduit (e.g., blood vessel) or reservoir (e.g., bladder), and contracting to regulate blood pressures or transfer fluids. Thus, any attempt to engineer these tissues must include the development of functional SM. In this section, we describe the methods of smooth muscle cell (SMC) isolation, culture, and characterization.

ISOLATION

SMCs can be isolated using either enzyme digestion methods or explant methods. The enzyme digestion method takes a much shorter period to obtain a certain amount of cells, and produces a larger cell number than the explant method. On the other hand, large amounts of cells are lost during the enzyme digestion procedures, and these the explant method is preferable if only small amounts of tissue specimens are available.

MATERIALS

All the materials utilized for the cell isolation procedure must be sterile. Surgical instruments should be washed and autoclaved, and buffer solutions and culture medium should be filtered.

1. Equipment

   a. Scissors
   b. No. 11 scalpel blade
   c. Glass slide
   d. Forceps

2. Buffer solution and culture medium
   a. Hank's Balanced Salt Solution (HBSS)
   b. Culture medium: Medium 199 supplemented with 20% (v/v) fetal bovine serum (FBS), 2mM L-glutamine, 100 U/ml penicillin, and 0.1 mg/ml streptomycin.
3. Enzyme solutions used in enzyme digestion method

   a. Enzymatic digestion buffer for aortic SMCs [0.125 mg/ml elastase (Sigma Chemical Co., St. Louis, MO), 1.0 mg/ml collagenase (CLS type I, 204 U/mg, Worthington Biochemical Corp., Freehold, NJ), 0.250 mg/ml soybean trypsin inhibitor (type 1-S, Sigma), and 2.0 mg/ml crystallized bovine serum albumin (BSA) in Medium 199]

b. Enzymatic digestion buffer for bladder or intestinal SMCs [0.35 mg/ml papain (crude type II, 2.8 U/mg, Sigma), 0.35 mg/ml collagenase (CLS type I, 138 U/mg, Worthington), 0.250 mg/ml soybean trypsin inhibitor, and 2.0 mg/ml BSA in Medium 199]
c. Spinner flask (100 ml, Bellco Glass Inc., Vineland, NJ)
d. 100-μm Nitex filter (Tetko, Inc., Briarcliff Manor, NY)

METHODS

Collected tissue specimens must be kept at 4°C until cell isolation process begins (preferably within 12 h). The cell isolation procedures must be performed in a laminar flow hood using aseptic techniques.

Isolation of SM Tissue

SM tissue must be isolated from the organ of interest prior to the cell isolation procedures. Once the SM tissue has been isolated, SMCs can be isolated from the SM tissue by either the enzyme digestion method or explant method.

1. Isolation from vascular tissue [1,2]

   a. Wash the vessel with HBSS twice.
   b. Transfer the specimen to petri dishes containing HBSS.
   c. Cut and open the specimen longitudinally with scissors and place the luminal side upward.
   d. Remove the endothelium and intima by scraping off the cell layer with a sterile no. 11 scalpel blade.
   e. Strip SMC-containing media from the adventitia or remove adventitia by blunt dissection with scissors.
   f. Wash the specimen with serum free (SF)-HBSS twice.

2. Isolation from urinary tissue [3–6]

   a. Cut a 1×1 cm^2 area from the tissue.
   b. Wash the specimen with HBSS twice.
   c. Remove fat and connective tissue by dissection with scissors.
   d. Cut SM bundle from the serosal side of the specimen.

3. Isolation from intestine [7]

   a. Cut 5–10 cm of intestine.
   b. Wash the specimen with HBSS extensively several times.
   c. Cut the specimen longitudinally.
   d. Remove fat and connective tissue by dissection with scissors.
   e. Scrape the luminal side with the edge of a sterile glass slide.
   f. Wash the specimen with HBSS twice.

Enzyme Digestion Method

1. Mince SM tissue into multiple small pieces with microsurgical scissors until the diameter of the pieces is 1–2 mm.
2. Place the small tissue pieces into a spinner flask (100 ml) containing an enzymatic digestion buffer.
3. Incubate the tissue suspension in the spinner flask at 37°C for 90 min for vascular SM or 30 min for intestinal or urinary SM.
4. Filter the suspension through a 100-μm Nitex filter.
5. Centrifuge the filtrate at 200*g* for 5 min.
6. Resuspend the pellet in culture medium, place the cells in tissue culture plates, and culture in a humidified incubator (37°C, 5% CO_2).

Explant Method

1. Mince SM tissue into multiple small pieces with microsurgical scissors until the diameter of the pieces is 1 mm.
2. Distribute the small tissue pieces evenly on the bottom of 25-cm^2 tissue culture flasks at a density of 2–4 pieces/cm^2.
3. Add a minimum volume of culture medium (about 1–2 ml) to the flask, just enough to keep the tissues moist and small enough to prevent tissue floating.
4. Transfer flasks gently into a humidified incubator (37°C, 5% CO_2) and leave for 2 h. During this time, the tissue pieces should adhere to the culture flasks.
5. Once the tissue pieces have attached, gently add 15 ml of culture medium.
6. Change medium every 3 days. Cells will migrate out of the tissue pieces on day 4–5.
7. When the culture is approximately 70% confluent, remove the tissue pieces with a Pasteur pipette and aspirator.

CULTURE

MAINTENANCE

The SMCs are cultured in tissue culture flasks in a humidified 5% CO_2 atmosphere with the medium changed every 3 days. SMCs were maintained in growth media containing 20% (v/v) FBS until the first passage. Subsequent cultures were supplemented with 10% (v/v) FBS.

PASSAGE

The primary cell culture should be confluent within 10 days. Passage the culture within 10 days.

1. Remove culture medium from the flask.
2. Wash the flask with a small amount of phosphate-buffered saline/ethylenediamine-tetraacetic acid (PBS-EDTA, 0.53 mM EDTA). This procedure will remove any trace amount of serum and trypsin inhibitor left in the flask after removal of the medium.
3. Add enough trypsin–EDTA (0.5% trypsin, 0.53 mM EDTA in HBSS) to cover the bottom of the flask (e.g., 1.5 ml for a 25-cm^2 flask).
4. Incubate at 37°C for 3–5 min.
5. Agitate the flask vigorously to detach the cells.
6. Examine the cells under a microscope to ensure that all the cells have detached. If some cells are still attached, use a cell scraper to detach the cells.
7. Add an equal volume of culture medium containing serum to the flask to stop the action of trypsin.
8. Mix the solution in the flask gently by agitation.
9. Collect the cell suspension in a centrifuge tube and centrifuge at 200g for 5 min.
10. Remove the supernatant and add 5–10 ml of fresh medium, and resuspend the cell pellet in the medium.
11. Plate the cells in 3–5 flasks at densities higher than 1500 cells/cm^2 (cell proliferation may cease at lower densities). Add medium.
12. Culture the passaged cells in a incubator and change medium every 3 days.

FREEZING AND THAWING CELLS

SMCs can be frozen and stored in liquid nitrogen for several years. The frozen cells can be thawed whenever needed and cultured again. To improve cell viability, cell freezing must be performed slowly (decrease by 1°C per minute) and the frozen cells must be than rapidly (within 3 min).

1. Freezing cells

   a. Trypsinize the cultured cells, collect cell suspension in a centrifuge tube, and centrifuge.
   b. Remove the supernatant and add appropriate amount (1 ml per 1–10 × 10^6 cells) of freezing medium [50% (v/v) basal medium, 40 (v/v) % FBS, and 10% (v/v) DMSO (dimethyl sulfoxide)]. The freezing medium must be ice cold.

    c. Resuspend the cell pellet in the freezing medium.

    d. Transfer 1-ml aliquots to cryogenic tubes (Nunc, Roskilde, Denmark).

    e. Put the cryogenic tubes in a cell-freezing box (Nalgene or Sigma) filled with iso-propyl alcohol.

    f. Leave the cryogenic tubes in the box at −80°C overnight and store the cryogenic tubes in liquid nitrogen.

2. Thawing cells

    a. Put cryogenic tubes in a 37°C water bath and gently agitate to thaw the cells rapidly.

    b. Transfer thawed cell suspension to a centrifuge tube and add approximately 10-fold volume of warm medium.

    c. Centrifuge at 200*g* for 5 min.

    d. Remove the supernatant and add fresh medium.

    e. Resuspend the cell pellet in the medium.

    f. Check the cell viability by staining with trypan blue and examining under a microscope.

    g. Plate the cells in flasks and culture.

CHARACTERIZATION

CYTOSKELETAL MARKERS

SMCs express specific cytoskeletal proteins, smooth muscle α-actin, and smooth muscle myosin [8–11]. The detailed protocols for immunohistochemistry to stain these proteins are described in Chapter 4 in this volume.

GROWTH CHARACTERISTICS

Cell Growth Kinetics

Cell growth kinetics can be examined by culturing cells and counting cells at multiple time points.

1. Seed cells in six-well plates at a density of 1500 cells/cm^2. This lower plating density ensure that cells do not reach confluence and cease to proliferate for at least 96 h, during which time the cell growth experiment is performed.

2. Trypinize cells in at least three wells and use a hemocytometer or Coulter counter to count cells (for detailed protocol, see Chapter 4 in this volume).

3. Plot the result as cell number (log scale) versus time. The slope in the graph is the specific cell growth rate. The doubling time can be calculated by log2 divided by the slope.

DNA Synthesis

Relative rate of DNA synthesis can be determined by measuring [^{3}H]thymidine incorporation into DNA.

1. Tissue culture medium containing 3 μCi/ml of [^{3}H]thymidine is added to each well of cultured cells and incubated for times ranging from 2 to 10 h in a 5% CO_2 incubator at 37°C.

2. Cold (nonradioactive) thymidine is added to yield 1 mM, and cells are incubated for 1 h.

3. Glutaraldehyde is added to yield 1%, and the cells are left at room temperature for 15 min. This step and the following washes are done in a fume hood to prevent exposure to glutaraldehyde fumes.

4. Wash twice with PBS. All washes in this and following steps must be properly disposed as radioactive waste.

5. Wash with cold methanol twice for 10 min each time.

6. Allow to air-dry in hood.

7. Add a solution of cold 15% trichloroacetic acid (TCA) in PBS, and place on ice for 15 min.

   8. Remove TCA solution.

   9. Repeat steps 7 and 8.

  10. Add 0.3 N NaOH to each dish, and place on a shaker for 30 min.

  11. Vigorously wash plate with NaOH solution and transfer solution to 40-ml scintillation vial.

  12. Wash the plate with 0.3 N NaOH solution and add this to the scintillation vial.

  13. Add concentrated acetic acid to neutralize the base (e.g., 33 μl acid for 1.5 ml base). This step must performed in the hood to prevent exposure to acetic fumes.

  14. Vortex to mix thoroughly.

  15. Add 10 ml of scintillation fluid and mix with a vortex mixer.

  16. Count in a scintillation counter.

Protein Synthesis

   The synthesis rates of collagen, one of the major proteins synthesized by SMCs, can be determined by quantification of [^{3}H]proline incorporation into collagen [12].

   1. Culture medium containing 3 μCi/ml of tritiated L-2,3,4,5-proline and sodium ascorbate (50 mg/ml) is added to culture, and incubated for 24 h in a 5% CO_2 incubator at 37°C. A laboratory coat, eye protection, and disposable gloves must be worn when handling [^{3}H]proline.

   2. After the incubation, place the culture on ice bath and chill at 0°C.

   3. The medium is collected in 15-ml centrifuge tubes and can be stored at –20°C.

   4. Aliquot 1.2 ml of the medium fraction to a microcentrifuge tube.

   5. Add 300 μl of cold 50% TCA to the microcentrifuge tube containing the medium fraction, and place the tube at 4°C for more than 15 min to precipitate proteins.

   6. Using liquid nitrogen and a 37°C water bath, repeat freezing and thawing the culture flask containing the cell fraction three times.

   7. Add 1.5 ml of cold 10% TCA to the flask, and scrape cells with a cell scraper. TCA aqueous solution is very corrosive. Gloves must be worn.

   8. Collect the cell suspension into another microcentrifuge tube and place the tube at 4°C for more than 15 min to precipitate proteins.

   9. Centrifuge both tubes at 3000 rpm for 15 min.

  10. Discard supernatant (radioactive waste).

  11. Use a vortex mixer to resuspend pellets in 1.5 ml of cold 10% TCA.

  12. Repeat steps 9–10.

  13. Add 200 μl of 0.2 N NaOH to both tubes to dissolve the pellet, and vortex well to resuspend pellet.

  14. Put both tubes in water bath at 37°C for 20–30 min.

  15. Add 300 μl of a solution of 1 M Tris HCl (pH 7.6) and 0.002% phenol red to both tubes and mix well.

  16. Adjust pH of the solution at 6.8–7.4 (red color) with HCl or NaOH.

  17. Prepare 50 mM N-ethylmaleimide (NEM), 100 mM benzamidine, and 50 mM $CaCl_2$ in H_2O. This solution should be fresh.

  18. Add 100 μl of this solution to both tubes and mix well.

  19. Add collagenase solution (collagenase form III, 10 U/ml, Advance Biofactures Corp., in 0.05 M Tris, 0.01 M $CaCl_2$, pH 7.2) to both tubes, mix well, and incubate at 37°C for 2 h. Shake occasionally during the incubation.

  20. Add very cold solution of 20% TCA and 0.5% tannic acid.

  21. Vortex and incubate at 4°C for 15 min.

  22. Centrifuge at 3000 rpm for 30 min.

  23. Collect each supernatant to a scintillation vial, add 5 ml of scintillation cocktail, mix it well, and measured the rate of disintegration per minute (collagen synthesis rate) with a liquid scintillation counter.

REFERENCES

1. Kolpakov, V., Rekhter, M. D., Gordon, D., Wang, W. H., and Kulik, T. J. (1995). Effect of mechanical forces on growth and matrix protein synthesis in the in vitro pulmonary artery. *Circ. Res.* **77**, 823–831.
2. Kim, B. S., Putnam, A. J., Kulik, T. J., and Mooney, D. J. (1998). Optimizing seeding and culture methods to engineer smooth muscle tissue on biodegradable polymer matrices. *Biotechnol. Bioeng.* **57**, 46–54.
3. Oberpenning, F., Meng, J., Yoo, J. J., and Atala, A. (1999). De novo reconstitution of a functional mammalian urinary bladder by tissue engineering. *Nat. Biotechnol.* **17**, 149–155.
4. Yoo, J. J., Meng, J., Oberpenning, F., and Atala, A. (1998). Bladder augmentation using allogenic bladder submucosa seeded with cells. *Urology* **51**, 221–225.
5. Schot, R., van Asseltm, E., and van Mastrigtm, R. (1993). A method for isolating smooth muscle cells from pig urinary bladder with low concentrations of collagenase and papain: The relation between calcium concentration and isolated cell length. *Urol. Res.* **21**, 49–53.
6. Glerum, J. J., van Mastrigt, R., Romijn, J. C., and Griffiths, D. J. (1987). Isolation and individual electrical stimulation of single smooth-muscle cells from the urinary bladder of the pig. *J. Muscle Res. Cell Motil.* **8**, 125–134.
7. Maruyama, I., Yoshida, C., Kobayashi, M., Oyamada, H., and Momose, K. (1987). Preparation of single smooth muscle cells from guinea pig taenia coli by combinations of purified collagenase and papain. *J. Pharmacol. Methods* **18**, 151–161.
8. Rovner, A. S., Murphy, R. A., and Owens, G. K. (1986). Expression of smooth muscle and nonmuscle myosin heavy chains in cultured vascular smooth muscle cells. *J. Biol. Chem.* **261**, 14740–14745.
9. Owens, G. K., Loeb, A., Gordon, D., and Thompson, M. M. (1986). Expression of smooth muscle-specific alpha-isoactin in cultured vascular smooth muscle cells: Relationship between growth and cytodifferentiation. *J. Cell Biol.* **102**, 343–352.
10. Baskin, L. S., Howard, P. S., Duckett, J. W., Snyder, H. M., and Macarak, E. J. (1993). Bladder smooth muscle cells in culture: I. Identification and characterization. *J. Urol.* **149**, 190–197.
11. Eddinger, T. J., and Murphy, R. A. (1991). Developmental changes in actin and myosin heavy chain isoform expression in smooth muscle. *Arch. Biochem. Biophys.* **284**, 232–237.
12. Rothman, A., Kulik, T. J., Taubman, M. B., Berk, B. C., Smith, C. W. J., and Nadal-Ginard, B. (1992). Development and characterization of a cloned rat pumonary arterial smooth muscle cell line that maintains differentiated properties through multiple subcultures. *Circulation* **86**, 1977–1986.

CHAPTER 22

MESENCHYMAL CELL CULTURE: CARDIAC-DERIVED MUSCLE CELLS

Shay Soker, Kavid Udompanyanan, and Anthony Atala

INTRODUCTION

Cultured cardiac myocytes offer a homogeneous population of single cells that can be manipulated *in vitro*. The cells can be analyzed in a controlled environment without interference of adjacent, nonmyocyte cells such as fibroblasts. Cardiac myocyte cultures serve as the standard model system to investigate basic functions of the adult myocardium, such as ion channels, contractility, and regulation of cardiac gene expression, under normal and pathological conditions [1,2]. Cardiac myocytes are terminally differentiated, and they stop dividing shortly after birth [3]. Because of the low proliferative activity of cardiomyocytes and the lack of cardiac myocyte cell lines, primary cultures of cardiomyocytes have been used frequently. Primary cardiomyocyte cultures exhibit normal behavior, such as spontaneous beating, but they require special culture conditions and do not remain in good condition for a long time period [4]. The chapter describes methods for the isolation of cardiac myocytes, possible cell sources, culture conditions, and available options for characterization.

SOURCES FOR CARDIAC MYOCYTES

There are two aspects of the possible source for cell retrieval, the species and the age. For research applications, most studies are carried out in small-animal models such as chickens, mice, rats, cats, guinea pigs, and rabbits [5,6]. Rats and rabbits are the most common species used for cardiac models. These relatively small animals can be treated with different diets or pharmacological agents to study their effects on various cardiac functions. Cardiomyocytes can be isolated at different time points during the study, and cell function can be analyzed *in vitro*. Recent advances in mouse genetics led to the generation of genetically altered mice. Cardiac myocytes may be isolated from mutant mice and grown *in vitro* to determine the role of specific genes in cardiomyocyte function [7]. The results may be correlated later with the *in vivo* phenotype of the mutant. The size of the animal's heart should be taken into consideration, since cardiomyocytes do not proliferate rapidly and tend to lose some of their properties over time. The initial culture should consist of a large number of viable cardiomyocytes.

The age of the animal or the patient at the time of tissue retrieval also plays a role in the rate of cell proliferation. Proliferating cardiomyocytes can be retrieved from fetuses or neonatal rats before the cells undergo terminal differentiation [3,8]. Studies have looked at the effect of the donor's age on the viability and culture characteristics of cardiomyocytes [9]. Adult-differentiated cardiomyocytes did not divide initially but later dedifferentiated and began to proliferate. Cardiomyocytes from embryos or young donors dedifferentiated more rapidly and grew faster *in vitro*. However, the amount of cardiomyocytes retrieved from embryos or young donors is smaller than that from adult donors.

An interesting source of cardiomyocytes consists of progenitor cells that can be isolated from the bone marrow or from an embryonic stem cell culture. Marrow stromal cells that attached to the dish were treated with 5-azacytidine to generate spontaneously beating clones. These cells expressed cardiac muscle markers and had a cardiomyocyte-like action potential [10]. Murine embryonic stem cells were differentiated spontaneously *in vitro* into contracting cardiac myocytes. The cells expressed cardiac genes that are characteristic of mouse cardiac differentiating cells *in vivo* [11].

SPECIFIC CULTURE CONDITIONS FOR CARDIOMYOCYTES
BASAL MEDIA AND SERA
Different basal media have been used for culturing cardiomyocytes such as Minimum Essential Medium, Dulbecco's Modified Eagle's Medium, Basal Eagle's Medium, and Medium 199 (MEM, DMEM, BME, and M-199) [4,5,12–14]. Cardiomyocytes can be maintained in the presence of 5–20% serum, and the medium is replaced every 2–5 days. However, the addition of serum to the medium is controversial [15]. Lack of serum in the medium prevents the growth of "contaminants" such as fibroblasts [16]; on the other hand, serum enhances attachment, survival, and growth of cardiomyocytes [4]. Preventing the growth of rapidly proliferating contaminants can be achieved by using antimitotic agents such as cytosine-β-D-arabinofuranoside [4].

CULTURE DISH PREPARATION
Precoating the dishes with serum or laminin (1–12 μg/ml) significantly enhanced cardiac myocyte attachment and preserved cell structure [4]. Higher plating density may have a beneficial effect on cell survival and preservation of differentiated phenotype [17]. Because cardiomyocytes do not attach rapidly to the surface of the culture dish, quickly adhering fibroblasts can be purified out of the culture by transferring all the nonadherent cells to a fresh dish about 2 h after seeding [13].

SPECIAL ADDITIVES
Serum-free media supplemented with hormones and additives such as insulin, hydrocortisone, and transferrin can support cardiomyocyte growth similar to serum-containing media [16,18]. Low levels of calcium in the media enhance cardiomyocyte proliferation and repress the expression of sarcomeric proteins, probably owing to dedifferentiation of the cells [19]. To enhance cardiomyocyte proliferation *in vitro* it is possible to use growth factors that are known to induce cell proliferation. Platelet-derived growth factor (PDGF) binds to specific receptors on the surface of chick cardiomyocytes and significantly induces their proliferation [20]. The most active form is the BB homodimer of PDGF. PDGF has no effect on calcium mobilization in the cells. Other supplements such as insulin, metabolic substrates, and extracellular matrix proteins have been shown to promote cardiomyocyte growth in culture, especially under serum-free conditions [4,16,18].

CELL CONFLUENCE AND PASSAGES
Cardiomyocyte cultures should be kept at relatively high cell concentrations because their confluence directly influences cell–cell contact. It was shown that cell contact can promote spontaneous beating of cultured cardiomyocytes [17]. Addition of β-adrenergic agonist such as isoproterenol-enhanced cell hypertrophy, and cell–cell contact resulted in increasing spontaneous beating [17]. Once confluent, the culture is passaged by means of trypsin digestion as normally used for other cells in culture. During time in culture and over passages, cardiomyocytes do not maintain their original phenotype, and they undergo dedifferentiation. Yet, cardiomyocytes may be cultured for more than 20 passages and over 6 months, while the cells still express cardiac-specific contractile proteins such as cardiac myosin heavy chain [9].

METHODS FOR ISOLATION OF CARDIAC MYOCYTES

Three main techniques are used to isolate cardiac cells [5,21–23]: the explant method, perfusion of the heart tissue, via the aorta, with proteolytic enzyme-containing buffers, and mechanical shearing of the cardiac tissue.

THE EXPLANT METHOD

The first method used for isolation of cardiomyocytes was the explant technique, in which pieces of cardiac muscle tissue were kept in culture until the myocytes migrated out and were cultured separately [5].

1. After sacrifice, the entire heart is isolated and washed extensively with 70% ethanol. The apical portion is dissected and cut in a staggered fashion and placed in buffer containing 90 mM NaCl, 30 mM KCl, 5.6 mM glucose, 42 mM sucrose, 2 mM $NaHCO_3$ and 2 mM HEPES.

2. A surgical blade is used to cut the heart tissue into small pieces (about 1 mm^2), and the pieces are transferred evenly to an empty dish (6 or 10 cm).

3. A small volume of medium is added, and the explants are left to adhere to the surface of the dish for 2–5 min before medium is added carefully.

4. Medium can be changed every 3 days.

5. Cardiomyocytes will migrate out of the tissue, and the tissue pieces can be removed when they cover approximately 70% of the dish.

6. When reaching confluence, the cells are trypsinized and transferred into fresh dishes.

ENZYMATIC DIGESTION OF AN INTACT HEART

1. The intact heart is placed in Tyrode solution (137 mM NaCl, 5.4 mM KCl, 0.5 mM $MgCl_2$, 1.8 mM $CaCl_2$, 11.6 mM HEPES, and 5 mM glucose at pH 7.4), and the aorta is cannulated. The heart is then placed on a Langendorff perfusion apparatus with constant flow (diagrams in Refs. [5,22]).

2. The heart is perfused for about 5 min with calcium-free Tyrode solution followed by a 5-min perfusion with calcium-free Tyrode solution containing 0.6 mg/ml collagenase II and optionally with 0.12 mg/ml protease XIV.

3. The last step is perfusion with a low calcium solution (137 mM NaCl, 0.18 mM $CaCl_2$, 1.2 mM KH_2PO_4, 1.2 mM $MgSO_4$, 6 mM HEPES, and 10 mM glucose at pH 7.2–7.4).

4. After the enzymatic digestion has been completed, the heart is placed in a petri dish containing low-calcium solution, and single cells can be dispersed by agitation and manual trituration.

5. Suspended cells can be filtered and washed with culture medium and plated in tissue culture dishes.

MECHANICAL DISPERSION

Single-cell cultures can be obtained by mincing the cardiac muscle tissue and plating the mixture. Cardiomyocytes are obtained from small samples of fetal and young hearts by mincing followed by enzymatic digestion [24].

1. The heart tissue is dissected into small (1 mm^2) pieces in balanced NaCl solution with HEPES buffer.

2. The small tissue pieces are placed in a manual homogenizer, which produces gentle sheering effects to disperse the tissue.

3. Optionally, enzymatic digestion with collagenase II or trypsin may be performed to efficiently separate the cardiomyocytes.

4. The mixture is centrifuged to concentrate the cells and to discard the digested extracellular matrix. The cell pellet is resuspended in the culture medium and seeded in tissue culture dishes. Addition of EDTA to the plating medium, as a chelating agent, facilitates the dispersion of the cells.

CHARACTERIZATION OF CARDIOMYOCYTES IN CULTURE

Different culture conditions may affect different parameters of cardiomyocyte functionality. Thus, assessing various parameters during cell culture is essential for reliable evaluation of cardiomyocyte function under specific *in vitro* conditions.

MICROSCOPIC EXAMINATION

Cell shape and size are good indicators for cardiomyocyte phenotype, and they can easily be monitored microscopically during the culture period. Initially, cardiac myocytes have a rodlike shape with rectangular "stepped" ends; cross-striations are observed clearly [2,5,12]. Over time in culture, the ends of the cells tend to become more rounded. After about 2 weeks in culture, most of the cardiomyocytes will spread, probably because of changes in the cytoskeleton that allow adoption to the two-dimensional surface. Precoating the culture dishes with adhesion molecules and the presence of serum in the culture media promote cell spreading and prevent cells from becoming slim and elongated [4]. Spontaneous cell beating is another microscopic parameter used to evaluate cardiomyocytes [17]. Information on the intracellular structure can be obtained by electron microscopy [17].

ELECTROPHYSIOLOGICAL PROPERTIES

The most important parameters, which can be tested, are the electrophysiological properties [2,5,25,26]. The common tested parameters are resting and active potentials of the cells and transmembrane resistance.

Electrophysiological Properties

The cells are placed in Iscove's Modified Dulbecco's Medium containing 1.5 mM CaCl, 4.23 mM KCl, and 25 mM HEPES at pH 7.4 and room temperature. Action potentials are recorded by conventional microelectrodes. Glass microelectrodes are filled with 3M KCl (dc resistance of 15, 30 MΩ), and membrane potential can be measured using a patch–clamp unit.

Intracellular Ca^{2+}

Cardiomyocytes are loaded with Fura 2-AM [7] and are excited at 340 and 380 nm. The resulting fluorescence is captured and recorded [7]. Fluorescence emission is collected over 3-ms intervals. The ration of fluorescence at 340 to 380 nm is calculated as a function of intracellular Ca^{2+}.

Cell Shortening

Cardiomyocyte shortening is measured after induction of electric contraction at 0.5 Hz [7]. Cell images are recorded and then measured.

GROWTH PROPERTIES

Other parameters that reflect on cultured cardiomyocyte functionality are DNA and protein synthesis [9,14,17,19] and metabolic aspects such as synthesis of fatty acids, respiration, and energy reserves [26].

DNA Synthesis

DNA synthesis is measured by incubation of the cells in the presence of [^{3}H]thymidine (1 mCi/ml) for 12 h followed by measurements of cell-associated radioactivity.

Protein Synthesis

Protein synthesis can be measured by seeding the cells in a multiwell dish and by preparation of a whole-cell lysate from each well at various time points. Total protein content in the lysates is measured by commercially available kits.

Fatty Acids Content

For measurements of fatty acids, cardiomyocyte cell pellets are homogenized in methanol, and the phospholipid fraction is separated and analyzed by gas chromatography.

EXPRESSION OF CARDIAC-SPECIFIC PROTEINS

The expression of cardiac-specific proteins [9,12,16,18,27] can be measured by means of the Western blot analysis. Cardiomyocyte whole-cell lysate is separated on 4% SDS-PAGE and gels are blotted onto membranes. The membranes can be probed with antibodies against myosin heavy chain. Alternatively, cardiomyocytes grown on laminin-coated coverslips can be fixed with ice-cold methanol and permeabilized with 1:1 methanol–acetone. The cells are immunostained with anti-myosin antibodies and visualized by means of fluorescent secondary antibodies [12].

REFERENCES

1. Jacobson, S. L., and Piper, H. M. (1986). Cell cultures of adult cardiomyocytes as models of the myocardium. *J. Mol. Cell. Cardiol.* **18**, 661–678.

2. New, R. B., Zellner, J. L., Hebbar, L., Mukherjee, R., Sampson, A. C., Hendrick, J. W., Handy, J. R., Crawford, F. A., Jr., and Spinale, F. G. (1998). Isolated left ventricular myocyte contractility in patients undergoing cardiac operations. *J. Thorac. Cardiovasc. Surg.* **116**, 495–502.

3. Tam, S. K., Gu, W., Mahdavi, V., and Nadal-Ginard, B. (1995). Cardiac myocyte terminal differentiation. Potential for cardiac. *Ann. N. Y. Acad. Sci.* **752**, 72–79.

4. Piper, H. M., Jacobson, S. L., and Schwartz, P. (1988). Determinants of cardiomyocyte development in long-term primary culture. *J. Mol. Cell. Cardiol.* **20**, 825–835.

5. Mitcheson, J. S., Hancox, J. C., and Levi, A. J. (1998). Cultured adult cardiac myocytes: future applications, culture methods, morphological and electrophysiological properties. *Cardiovasc. Res.* **39**, 280–300.

6. Mitra, R., and Morad, M. (1985). A uniform enzymatic method for dissociation of myocytes from hearts and stomachs of vertebrates. *Am. J. Physiol.* **249**, H1056–H1060.

7. Wolska, B. M., and Solaro, R. J. (1996). Method for isolation of adult mouse cardiac myocytes for studies of contraction and microfluorimetry. *Am. J. Physiol.* **271**, H1250–H1255.

8. Soonpaa, M. H., Koh, G. Y., Klug, M. G., and Field, L. J. (1994). Formation of nascent intercalated disks between grafted fetal. *Science* **264**, 98–101.

9. Li, R. K., Mickle, D. A., Weisel, R. D., Carson, S., Omar, S. A., Tumiati, L. C., Wilson, G. J., and Williams, W. G. (1996). Human pediatric and adult ventricular cardiomyocytes in culture: Assessment of phenotypic changes with passaging. *Cardiovasc. Res.* **32**, 362–373.

10. Makino, S., Fukuda, K., Miyoshi, S., Konishi, F., Kodama, H., Pan, J., Sano, M., Takahashi, T., Hori, S., Abe, H., *et al.* (1999). Cardiomyocytes can be generated from marrow stromal cells in vitro. *J. Clin. Invest.* **103**, 697–705.

11. Metzger, J. M., Lin, W. I., and Samuelson, L. C. (1996). Vital staining of cardiac myocytes during embryonic stem cell cardiogenesis in vitro. *Circ. Res.* **78**, 547–552.

12. Bugaisky, L. B., and Zak, R. (1989). Differentiation of adult rat cardiac myocytes in cell culture. *Circ. Res.* **64**, 493–500.

13. Carrier, R. L., Papadaki, M., Rupnick, M., Schoen, F. J., Bursac, N., Langer, R., Freed, L. E., and Vunjak-Novakovic, G. (1999). Cardiac tissue engineering: Cell seeding, cultivation parameters, and tissue construct characterization. *Biotechnol. Bioeng.* **64**, 580–589.

14. Clark, W. A., Rudnick, S. J., Simpson, D. G., LaPres, J. J., and Decker, R. S. (1993). Cultured adult cardiac myocytes maintain protein synthetic capacity of intact adult hearts. *Am. J. Physiol.* **264**, H573–H582.

15. Mather, J. P. (1998). Making informed choices: Medium, serum, and serum-free medium. How to choose the appropriate medium and culture system for the model you with to create. *Methods Cell Biol.* **57**, 19–30.

16. Kessler-Icekson, G., Sperling, O., Rotem, C., and Wasserman, L. (1984). Cardiomyocytes cultured in serum-free medium. Growth and creatine kinase. *Exp. Cell Res.* **155**, 113–120.

17. Clark, W. A., Decker, M. L., Behnke-Barclay, M., Janes, D. M., and Decker, R. S. (1998). Cell contact as an independent factor modulating cardiac myocyte hypertrophy and survival in long-term primary culture. *J. Mol. Cell. Cardiol.* **30**, 139–155.

18. Freerksen, D. L., Schroedl, N. A., and Hartzell, C. R. (1984). Control of enzyme activity levels by serum and hydrocortisone in neonatal rat heart cells cultured in serum-free medium. *J. Cell. Physiol.* **120**, 126–134.

19. Harayama, H., Koide, M., Obata, K., Iio, A., Iida, M., Matsuda, N., Akins, R. E., Yokota, M., Tuan, R. S., and Saito, H. (1998). Influence of calcium on proliferation and phenotype alteration of cardiomyocyte *in vitro. J. Cell. Physiol.* **177**, 289–298.

20. Shimizu, T., Kinugawa, K., Yao, A., Sugishita, Y., Sugishita, K., Harada, K., Matsui, H., Kohmoto, O., Serizawa, T., and Takahashi, T. (1999). Platelet-derived growth factor induces cellular growth in cultured chick ventricular myocytes. *Cardiovasc. Res.* **41**, 641–653.

21. Weishaar, R. E., and Simpson, R. U. (1986). Isolation, stages of differentiation, and long term maintenance of adult rat myocytes. *Cell. Biol. Int. Rep.* **10**, 745–753.

22. Tytgat, J. (1994). How to isolate cardiac myocytes. *Cardiovasc. Res.* **28**, 280–283.

23. Dow, J. W., Harding, N. G., and Powell, T. (1981). Isolated cardiac myocytes. I. Preparation of adult myocytes and their homology with the intact tissue. *Cardiovasc. Res.* **15**, 483–514.

24. Li, R. K., Mickle, D. A., Weisel, R. D., Zhang, J., and Mohabeer, M. K. (1996). In vivo survival and function of transplanted rat cardiomyocytes. *Circ. Res.* **78**, 283–288.

25. Eschenhagen, T., Fink, C., Remmers, U., Scholz, H., Wattchow, J., Weil, J., Zimmermann, W., Dohmen, H. H., Schafer, H., Bishopric, N., *et al.* (1997). Three-dimensional reconstitution of embryonic cardiomyocytes in a collagen matrix: A new heart muscle model system. *FASEB J.* **11**, 683–694.

26. Dow, J. W., Harding, N. G., and Powell, T. (1981). Isolated cardiac myocytes. II. Functional aspects of mature cells. *Cardiovasc. Res.* **15**, 549–579.

27. Akins, R. E., Boyce, R. A., Madonna, M. L., Schroedl, N. A., Gonda, S. R., McLaughlin, T. A., and Hartzell, C. R. (1999). Cardiac organogenesis in vitro: Reestablishment of three-dimensional tissue architecture by dissociated neonatal rat ventricular cells. *Tissue Eng.* **5**, 103–118.

CHAPTER 23

MESENCHYMAL CELL CULTURE: FUNCTIONAL MAMMALIAN SKELETAL MUSCLE CONSTRUCTS

Paul E. Kosnik and Robert G. Dennis

Described is a method for engineering skeletal muscle constructs *in vitro*, using mammalian cells. Skeletal muscle is engineered *de novo* from primary cells of neonatal, adult, or aged mammals, or from myogenic cell lines. These constructs are termed *myooids* because they are musclelike in morphology, excitability, and contractility. Myooids are cylindrical, with a diameter that is controlled by the tissue engineer and ranges from 0.1 to 1.0 mm. Myooid length is determined by the placement of the artificial tendons, typically from 1 to 2 cm apart. The myotubes and fibroblasts of the myooid reside within an extracellular matrix generated by the cells themselves. Myooids are excitable, and contract to generate directed force and power. Myooid excitability and contractility are measured by attaching one of the artificial tendons to a force transducer and electrically exciting the myooid. We present methods for performing such evaluations in Chapter 24. Once formed, myooids remain viable in culture for several months, the life span depending upon several factors under the control of the tissue engineer.

PREPARATION OF SOLUTIONS AND MEDIA

Unless otherwise indicated, solutions and media should be prepared and stored before isolation and culture of muscle cells. The volumes of each medium described are appropriate for the engineering of approximately two dozen myooids and their maintenance in culture for about 2 months.

CULTURE MEDIA

Growth medium (GM)
    400 ml Ham's F-12 nutrient mixture (GibcoBRL, 11765-054)
    100 ml fetal bovine serum (FBS, GibcoBRL, 10437-036)
    100 units/ml penicillin G (Sigma, P-3414)
    Prepare aseptically and store at 4°C until use.
Differentiation medium (DM)
    465 ml Dulbecco's Modified Eagle's Medium (DMEM, GibcoBRL, 11995-065)
    35 ml horse serum (HS, GibcoBRL, 16050-114)
    100 units/ml penicillin G
    Prepare aseptically and store at 4°C until use.
Culture medium (CM), for use with C2C12 and $10T_{1/2}$ cells
    465 ml DMEM
    35 ml FBS
    100 units/ml penicillin G
    Prepare aseptically and store at 4°C until use.

Sodium azide solution (stock)
    500 ml isotonic saline
    0.05% NaN$_3$ (m/v)
    Store at room temperature until use.
Preincubation medium (PIM)
    40 ml DM
    5 ml sodium azide solution
    Filter-sterilize with 0.22-μm filter; prepare immediately before use.

TISSUE HARVESTING AND DISSOCIATION MEDIA

Mincing solution, for storing neonatal muscle during harvesting [1]
    8 mM NaH$_2$PO$_4$
    22.6 mM NaHCO$_3$
    116 mM NaCl
    5.3 mM KCl
    5.6 mM glucose
    Mix in deionized water, bring solution to pH 7.4
    Filter-sterilize with 0.22-μm filter and store at 4°C until use.
Dispase and collagenase (D&C) solution for dissociation of tissues from adult animals
    4 U/ml dispase (activity 0.85 U/mg, Gibco BRL, 17105-041)
    100 U/ml type IV collagenase (activity 206 U/mg, Worthington, CLS 4)
    Mix in DMEM and filter-sterilize with 0.22-μm filter immediately before use.
Dispase and collagenase solution for dissociation of neonatal (ND&C) tissues
    4 U/ml dispase (0.85 U/mg, GibcoBRL, 17105-041)
    100 U/ml type IV collagenase (Worthington, CLS 4, activity 206 U/mg)
    Mix in mincing solution and filter-sterilize with 0.22-μm filter immediately prior to use.

CULTURE PLATE SUBSTRATE COATING SOLUTIONS

Substrate coating solution (SCS) (sufficient for 25 plates, 35 mm in diameter at 1 μg laminin/cm^2)
    75 ml Dulbecco's Phosphate Buffered Saline (DPBS) (GibcoBRL, 14190-136, pH 7.2)
    250 μg natural mouse laminin (GibcoBRL, 23017-015)
    Mix aseptically, do not filter. Prepare immediately before use.
Artificial tendon coating solution (ATCS)
    1.0 ml DPBS
    50 μg natural mouse laminin
    Mix aseptically in a 1.5-mm capped cryovial immediately before use.

PREPARATION OF CULTURE DISHES

The general method of tissue engineering myooids from primary cells is illustrated in Fig. 23.1. Each myooid is engineered in a standard 35-mm culture dish, permitting functional evaluation of individual myooids and limiting contamination to single plates. Prepare each culture dish (Falcon, Becton Dickinson Labware, Franklin Lakes, NJ, 1008) by pouring into it 1.5 ml Sylgard (type 184 silicone elastomer, Dow Chemical Corporation, Midland, MI).* The Sylgard substrate allows artificial tendons to be pinned in place, and cells do not normally adhere to the hydrophobic Sylgard. This allows control of cell adhesion by coating the substrate with cellular adhesion molecules at the desired density. If culture plates of

---

*We recommend the use of only type 184 Sylgard, because type 184 is manufactured with platinum catalyst, whereas other types of Sylgard use potentially toxic catalysts.

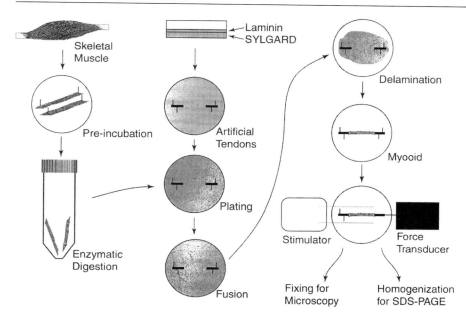

Fig. 23.1. Summary of method for isolating cells from skeletal muscle tissue and engineering the cells into a new skeletal muscle.

different diameters are used, adjust the Sylgard volume to result in a substrate thickness of 1.5–2.0 mm. Thinner substrates do not allow adequate seating of minutien pins to hold the artificial tendons in place, and thicker substrates preclude the use of higher-powered objectives to view myooids on inverted microscopes. After pouring the Sylgard, cap and leave the culture dishes on a level surface for 1 week to cure, then store for at least 2 weeks before use to reduce cell toxicity [2].

One week before the primary tissues are harvested, or about 10 days before the cell plating, coat the Sylgard substrate with the adhesion molecules to promote cell adhesion. For primary cultures from rodents, use 0.5- to 1.5-µg natural mouse laminin per square centimeter. Laminin densities below this range generally result in inadequate cell adhesion and proliferation for the formation of myooids, and densities greater than this inhibit detachment of the cells from the substrate during myooid formation. Coat the Sylgard substrate with laminin by addition of 3 ml of substrate-coating solution (SCS) per 35-mm-diameter culture dish. To whet the entire hydrophobic substrate with the SCS, swirl the blunt end of a Pasteur pipette around the periphery of each plate to form a uniform meniscus. The laminin is deposited on the Sylgard as the SCS evaporates. To accelerate SCS evaporation, leave the dishes in a biological safety hood (Forma Scientific, Marietta, OH, model 1184) with the blower running. Remove the remaining salt crystals by adding 3 ml of DPBS to each culture dish, allowing the crystals to dissolve, and aspirating the fluid. The laminin does not persist beyond about 2 weeks, even when stored at 4°C, so the laminin coating should be prepared one day before placement of the artificial tendons and applied within several days of use of the culture plates. Laminin-coated plates should not be stored before use.

PREPARATION AND PLACEMENT OF ARTIFICIAL TENDONS

We have found laminin-coated silk suture to be an easily prepared and effective material to use as artificial tendons for mammalian myooids [2]. Prepare artificial tendons one day after laminin-coating the culture plates by cutting 6-mm-long segments of size 0 braided silk suture (Ethicon, Somerville, NJ, metric size 3.5). Hold each segment at the center with a pair of no. 5 forceps (Fine Science Tools, Foster City, CA) and gently fray one of the ends by hand. Then immerse the segment in the anchor coating solution, making certain to whet the frayed end of the artificial tendon. Place each pair of artificial tendons in the center of a laminin-coated 35-mm culture dish, separated by 12 mm at their closest point, with the frayed ends facing each other. Pin each artificial tendon in place with two 0.10-mm-diameter minutien pins (Fine Science Tools, cat. no. 26002-10) to prevent rotation. Allow the anchors to dry in a biological safety cabinet with the blower running. Add 3 ml of GM to each culture dish, cover the plates, and arrange the dishes in an array in a biological safety cabinet such that each dish can be exposed to radiation from the UV lamp. Expose the dishes to the UV lamp (wavelength 253.7 nm) for 60–90 min to sterilize the substrate and artificial tendons.

After ultraviolet sterilization, place the culture dishes in an incubator at 5% CO_2 and 37°C for 5–8 days of presoaking in GM before plating of cells. The substrate and artificial tendons are presoaked for three reasons: (1) the serum in the GM contains soluble adhesion molecules that allow for improved cell adhesion to the substrate; (2) the presoaking process allows for the removal of plates contaminated during the preparation of the artificial tendons before experimental cells are introduced; and (3) the presoaking process removes toxins from the silk suture that otherwise result in a ring of inhibited cell growth 2–5 mm (diameter) around each anchor [2]. Complete integration of the cells with the silk fibers of the artificial tendons is essential for myooid formation, and presoaking is a vital step in preparing the artificial tendons for cellular adhesion. With inadequate presoaking, the cells do not attach to the anchors and the success rate for myooid formation is less than 5%. Longer presoaking times result in a reduced success rate, presumably because the laminin coating on the Sylgard substrate becomes degraded. Storage of the laminin-coated dishes at 4°C for 2 weeks before presoaking also results in reduced cell adhesion to the Sylgard substrate.

ISOLATION OF SATELLITE CELLS AND FIBROBLASTS
PRIMARY CELLS FROM YOUNG, ADULT, OR AGED MICE AND RATS (2–37 MONTHS OLD)

Harvest satellite cells and fibroblasts from the soleus muscles of young, adult, or old rats and mice by surgical removal of the desired whole muscles. The soleus muscle is ideal because of the relatively greater number of satellite cells associated with oxidative than with glycolytic fibers [3]. Remove the muscles, cut away fat and fascia, and place the muscles in a 60-mm-diameter culture dish with sterile phosphate-buffered saline (PBS). All skeletal muscles from adult rats or mice that we have tried work well with this protocol, including plantaris, gastrocnemius, anterior tibialis brevis, extensor digitorum longus, soleus, and biceps femoris.

Dry each whole muscle with a paper towel and weigh to determine the tissue mass before enzymatic dissociation. Then with a single-edged razor blade, slice the whole muscles longitudinally, from tendon to tendon, into strips of 20–30 mg. Use the razor blade to bisect the muscle longitudinally on the lid of a 100-mm culture dish, and run the closed tips of a pair of no. 5 forceps along the length of the muscle to release it from the razor blade. Use 0.10-mm-diameter minutien pins to pin the muscle strips at slack length in 35-mm culture dishes with Sylgard, without laminin coating, two strips per dish. Add 3 ml of PIM to each dish. Cover the dishes and expose to ultraviolet light for about an hour; then place them in a water-saturated incubator at 37°C and 5% CO_2 for 40–50 h of preincubation. Preincubation before enzymatic digestion of muscle results in more dense cell cultures and more rapid formation of confluent monolayers of cells, presumably because of the activation of satellite cells and the increased yield of myoblasts.

After preincubation, inspect each plate with preincubated muscle strips and discard any that are infected. Enzymatically digest the muscle strips to release satellite cells and fibroblasts. Digest four muscle strips in each 50-ml conical vial (Falcon, cat. no. 352098). Place 20–30 ml of freshly prepared and filter-sterilized dissociation (D&C) solution in each vial. Under aseptic conditions, remove each muscle strip from the Sylgard-coated plate, manipulating the strip by the minutien pins with a pair of no. 5 forceps. Sterilize a pair of no. 5 forceps in 70% EtOH, tap the forceps to remove any excess EtOH, and remove one of the minutien pins from one of the muscle strips. Pull the remaining pin out with the muscle strip still attached, and slide the muscle off the pin and into the conical vial with the dissociating solution. Observe sterile technique and dip the forceps in 70% EtOH between each use.

Incubate the strips in a shaker bath at 37°C for about 3–4 h. Accelerate digestion of the muscle strips by occasionally shaking each vial by hand. When the tissue is fully digested, centrifuge it at 1000g for 15 min and discard the supernatant. Reconstitute the cells by adding 2 ml of GM for every myooid formed. Use 10–13 mg of tissue (mass before digestion) for each myooid. For example, a rat soleus of 150 mg would be cut into four to six strips, dissociated, and reconstituted with 24 ml of GM to ultimately produce 12 myooids.

PRIMARY CELLS FROM NEONATAL RATS (1–2 DAYS OLD)

To engineer myooids from neonatal rat muscles, harvest myoblasts and fibroblasts from the fore and hind limbs. Order pups to be delivered in litters of 11–14 within one day of birth (Charles River, Wilmington, MA). Upon arrival, place the neonates in 15-cm glass petri dishes, six or seven in each dish. Cap the dishes and place on ice for 60–90 min as a general anesthesia. In a biological safety cabinet, hold the anesthetized pups by the tail and completely immerse in 70% EtOH to minimize possibility of contamination of muscles. Cut off the paws with scissors, and remove the skin by first cutting along the back and around the tail. Then pull the skin off the neonate with two pairs of forceps, being careful not to rupture the abdominal wall when pulling the skin away from the umbilicus. Discard the skin and paws and remove the muscles from the fore and hind limbs with two pairs of forceps. Graefe forceps work well, one serrated (Fine Science Tools, cat. no. 1050-10) pair and one with teeth (Fine Science Tools, cat. no. 1053-10). Place the muscle tissue in 8 ml of mincing solution in a 60-mm-diameter culture dish. When all the muscle has been removed from the litter, add 12 ml of ND&C to the mincing solution and dissociate the tissue by magnetic stirring at room temperature for 1.5 h. Remove the magnetic stir bar and centrifuge the slurry for 15 min at 1000g. Resuspend the pellet in 20 ml of 0.1% (w/v) collagenase type IV (Worthington) in Ham's F-12 nutrient medium (GibcoBRL) and place it in a shaker bath at 37°C for 1.5 h. Then centrifuge the solution for 15 min at 1000g and resuspend the pellet in GM. Count the cells with a hemacytometer, and adjust the solution to 5×10^6 cells/ml. Plate 2 ml of the cell suspension onto each 35-mm culture dish with artificial tendons in place, and place the culture dishes in a water-saturated incubator with 5% CO_2 at 37°C.

PLATING AND CULTURE OF PRIMARY CELLS

Replace the spent medium with GM every 48 h until the cells become confluent, typically 3–8 days. Replace the spent medium with DM at the next regular medium change after the cells reach confluence, and replace medium three times per week thereafter. Shortly after switching to DM, myocytes fuse to form multinucleated myotubes that begin to contract spontaneously.

CULTURES OF C2C12 AND 10T$_{1/2}$ CELLS

Myooids can be engineered from myoblasts of the C2C12 cell line (American Type Culture Collection, CRL-1772, Manassas, VA) when cocultured with 10T$_{1/2}$ fibroblasts (American Type Culture Collection, CCL-226, Manassas, VA) [4]. Although myooids can be formed from ratios of C2C12 to 10T$_{1/2}$ cells of 30–80%, there are no differences in the excitability or contractility of the myooids formed from different initial ratios of cell types [4]. To engineer myooids from cell lines, combine C2C12 myoblasts with equal numbers of 10T$_{1/2}$ fibroblasts in CM and plate the cells at 1×10^5 cells/cm^2 in 35-mm culture dishes with Sylgard, laminin, and artificial tendons. Replace spent media with CM every 48–72 h. C2C12–10T$_{1/2}$ myooids frequently do not detach completely from the Sylgard substrate. To aid detachment, release the myooid from the substrate by running the closed tips of a pair of sterile no. 5 forceps between the myooid and the substrate. Allow the myooids to remodel for at least 72 h before functional evaluation.

MYOOID FORMATION

When the cultured cells reach confluence, the myoblasts fuse into multinucleated myotubes and the myotubes begin contracting spontaneously. The spontaneous contractions lead to delamination of the monolayer, which detaches from the substrate while remaining attached to the artificial tendons (Fig. 23.2). This process begins peripherally and proceeds radially inward. If the cells do not detach from the substrate within about 30 days of cell

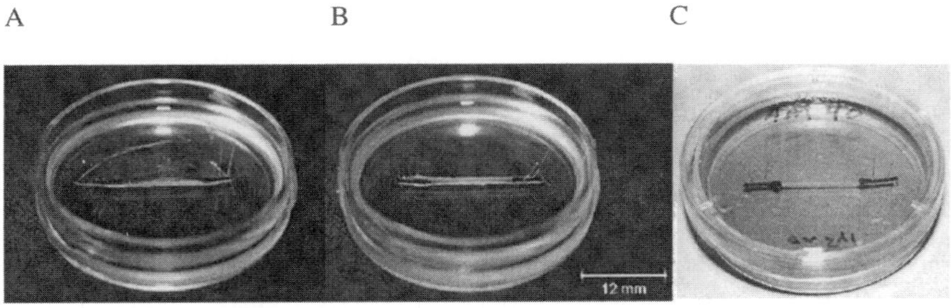

A   B   C

Fig. 23.2. Myooid formation occurs by peripheral delamination of the sheet of myotubes and fibroblasts (A). The cells detach everywhere from the substrate but remain attached to the artificial tendons (B), while remodeling into a cylindrical skeletal muscle construct, the myooid (C).

plating, either there are not enough cells or there is too much laminin on the Sylgard substrate for the release of the monolayer. In this case, plate at a higher cell density or decrease the initial laminin density before introduction of the cells. Occasionally the cells detach from the substrate in small patches, rather than as a coherent sheet, and myooids do not form. This problem is usually the consequence of poor laminin coating or inadequate presoaking of the culture plates and can be eliminated by using only fresh culture dishes that have been properly laminin-coated and presoaked.

Once the sheet of myotubes and fibroblasts has delaminated from the substrate, the sheet rolls up between the artificial tendons and lifts off the substrate, being suspended above it by tension generated by the cells within the construct (Fig. 23.2). The cells rapidly remodel in response to the new mechanical environment, and within 72 h the majority of the remaining cells are aligned along the long axis of the construct, now termed the myooid (Fig. 23.3). The construct continues to develop and has many of the characteristics of skeletal muscle, such as organized sarcomeres (Fig. 23.4). Myooid diameter at the time of myooid formation is a result of the cell source, cell number, and laminin density [2]. Higher laminin densities delay myooid formation but increase the initial diameter of the myooids. Greater cell densities decrease the time in culture before myooid formation and increase initial myooid diameter. However, very high cell concentrations (>10^6 cells/cm^2) can result in cells that delaminate too soon after cell plating, before the cells have formed a coherent sheet, thus preventing myooid formation.

After myooid formation, myooid diameter decreases with time. Diameter decreases more rapidly for myooids with larger initial diameters [2]. Regardless of initial diameter, myooid diameter decreases most rapidly initially, then slows as it reaches about 200 μm. Decreases in diameter continue until the myooid breaks, usually when the diameter is less than 100 μm. Myooid life span, defined as the time from myooid formation until myooid breaking, depends on myooid cell source, plating density, and feeding frequency. Myooids with large passive baseline tensions tend to have the shortest life spans [4]. Myooids from

Fig. 23.3. Myooids are cylindrical in cross section, and are composed of myotubes and fibroblasts that are principally aligned with the long axis of the myooid.

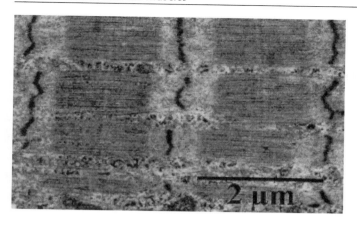

Fig. 23.4. An electron micrograph of a longitudinal section of a myooid reveals the organization of sarcomeres, the contractile units of skeletal muscle.

neonatal rats have the highest baseline tensions, followed by adult and then aged rat myooids. For a given cell source, myooid life span is greatest for cultures of high initial cell plating density and high laminin concentrations on the substrate, since these two variables produce myooids with large initial diameters [2]. By decreasing the frequency of media changes to once per week, one can also increase myooid life span.

ACKNOWLEDGMENTS

The authors thank Dr. William Kuzon for his suggestions and scientific input on the development of the myooid and Dr. Krystyna Pasyk for light and electron microscopy. Dr. John Faulkner provided significant scientific, financial, and infrastructural support and resources without which this method would not have been developed. This work was supported by the University of Michigan Center for Biomedical Engineering Research, the Contractility Core of the Nathan Shock Center the Basic Biology of Aging, and AG-06157.

REFERENCES

1. Cognard, C., Constantin, B., Rivet-Bastide, M., Imbert, N., Besse, C., and Raymond, G. (1993). Appearance and evolution of calcium currents and contraction during the early post-fusional stages of rat skeletal muscle cells developing in primary culture. *Development (Cambridge, UK)* **117**, 1153–1161.
2. Dennis, R. G., and Kosnik, P. E. (2000). Excitability and isometric contractile properties of mammalian skeletal muscle constructs engineered *in vitro. In Vitro Cell. Dev. Biol. Anim.* **36**(5), 327–335.
3. Schultz, E., and McCormick, K. M. (1994). Skeletal muscle satellite cells. *Rev. Physiol. Biochem. Pharmacol.* **123**, 213–257.
4. Dennis, R. G., Kosnik, P. E., Gilbert, M. E., and Faulkner, J. A. (2000). Excitability and contractility of skeletal muscle engineered from primary cultures and cell lines. *Am. J. Physiol. Cell.* **280**, C288–C295.

CHAPTER 24

MESENCHYMAL CELL CULTURE: INSTRUMENTATION AND METHODS FOR EVALUATING ENGINEERED MUSCLE

Robert G. Dennis and Paul E. Kosnik

Described is a method and instrumentation for evaluating the excitability and isometric contractility of engineered skeletal muscle constructs *in vitro*. The constructs, termed *myooids*, are engineered with synthetic tendons at each end and are muscle like in morphology, excitability, and contractile function (see Chapter 23). After formation, myooids are affixed to a force transducer and electrically stimulated between parallel electrodes for functional evaluation. Minor modifications will allow the method to be applied to any *in vitro* engineered musculoskeletal construct.

INSTRUMENTATION

The measurement of the excitability and contractility of myooids requires standard laboratory equipment as well as custom instrumentation.

MICROSCOPES

Use an inverted microscope to view the engineered constructs during development before functional evaluation, since the condensation on the inside of the culture plate lid and the long focal distances required prevent the viewing of the cells with a conventional microscope. Equip the inverted microscope with an eyepiece reticle to measure myooid diameter. Use a microscope length standard to calibrate the eyepiece reticle in micrometers per division. Use a stereo dissecting microscope with a long focal length for myooid evaluations. A focal length of at least 10 cm permits enough working distance for the investigator to attach the myooids to the force transducer.

FORCE TRANSDUCER

The maximum range of the force transducer should be greater than the force generated by myooids (~2000 μN), with a resolution of about 1 μN. The mechanical compliance of the transducer load element should be less than 100 μm for full-scale loads, since a larger compliance will change the contractile properties of the myooid because of the force–length relationship of skeletal muscle tissues. Force transducer bandwidth should be at least dc to 150 Hz to allow for rapid changes in myooid force generation as well as measurements of baseline forces generated by myooids. The load element of the force transducer should project over the edge of the culture dish to attach to one of the artificial tendons of the myooid (Fig. 24.1). An appropriate force transducer may not be commercially available, but a custom-built force transducer that meets the requirements for this application has been described in detail [1,2].

Amplify the signal from the force transducer such that the full-range output of the force transducer approximately fills the range of the analog input channel of the data acquisition system. Calibrate the force transducer in units of force (μN/V) rather than in units of mass (mg/V). Typically, force transducers are calibrated by hanging weights of known mass from the load element. The force transducer is rotated 90° to measure a vertical force vector during this calibration, yet horizontal force vectors are measured during myooid functional evaluation. The high sensitivity of the force transducers required for functional evaluation of myooids will not permit vertical calibration because the weight of the force transducer load element is often greater than the forces generated by the myooids or exceeds the maximum range of the force transducer. Therefore, calibrate the force transducer in position (with a horizontal force vector), using a microforce generator such as that described by Minns [3] or as modified by Kosnik [2]. Mount the force transducer such that its position can be adjusted in all three axes with a micromanipulator (Newport, Irvine, CA, model MT-XYZ). Construct the force transducer/micromanipulator assembly such that it can be moved freely below the stereo dissecting microscope during myooid functional evaluations (Fig. 24.1).

DATA ACQUISITION SYSTEM AND SOFTWARE

The data acquisition system includes a stimulus pulse generator (described shortly) and at least one analog input channel to record the force trace. All analog signals are converted to discrete digital values by the data acquisition system (Fig. 24.2). A minimum resolution of 12 bits is required. For excitability and contractility measures, a sampling rate of 200 samples/s is adequate. Two analog input channels are required to simultaneously record the stimulus pulses and force traces. Coordination of the onset of data acquisition and electrical stimulation must also be provided, such that analog data acquisition begins about a second before the initiation of electrical stimulation. Standard MIO (multiple input/output) data acquisition boards, such as the AT-MIO-16E-1, available from National Instruments (Austin, TX), are adequate for this application.

ELECTRICAL STIMULATOR

The evaluation of engineered skeletal muscle requires precise control of electrical stimulus parameters. Generate stimuli digitally, using the same software that is used for the measurement of force traces. In general, the excitability of engineered tissues is low, so a power amplifier that can amplify both current and voltage is required (Fig. 24.2).

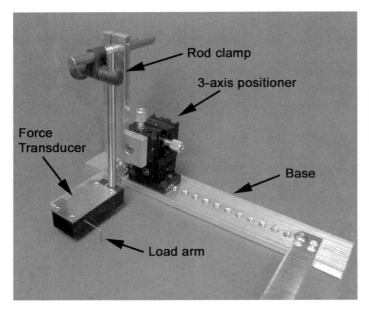

Fig. 24.1. Force transducer configuration (oblique view).

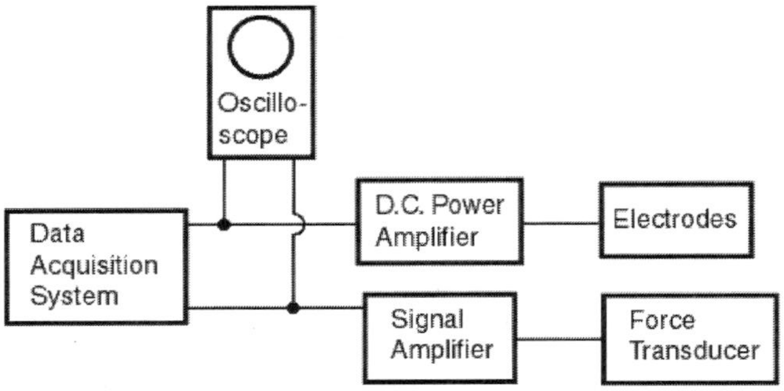

Fig. 24.2. Schematic diagram of data acquisition system.

Stimulus Pulses

Myooid twitches require a single square-wave pulse of known amplitude (voltage and current) and duration (pulse width). While the amplifier will substantially increase the pulse amplitude, pulse width remains unchanged after amplification. Pulse widths of 20–6000 μs are required, with a resolution of 5 μs. For pulse trains, the stimulation frequency should be adjustable over the range of 2–150 Hz, for train durations of 200 ms to 4 s.

Power Amplifier

After amplification, the pulse voltage should be adjustable over the range of 0.0–100 V, with a resolution of at least 0.1 V. At the onset of a stimulus pulse, the impedance (resistance) of the culture medium is very low, so the current can be extremely high. Confirm the stimulation amplitude by measuring it as close to the electrodes as possible with a digital oscilloscope and a 10× probe. The long pulse widths used for myooids have a significant dc component, so it is necessary to use a dc-coupled power amplifier. The Crown model DC300A-2 (Crown International Inc., Elkhart IN), or equivalent is appropriate for this application.

Electrodes

The myooids are electrically stimulated by application of a transverse electrical field between parallel electrodes. To ensure a uniform electrical field across the myooid, employ parallel wire or plate electrodes separated by 5–7 mm. Minimize the surface area of the plates to reduce the total current flux through the culture medium. Use electrodes about as long as the length of the contractile tissue, and place the electrodes equidistant on either side of the myooid. Suitable electrode materials include platinum or stainless steel (stainless steel alloys 302, 303, and 304 are preferable). Construct the electrodes such that the solder joint between the electrodes and the electrode wires is not immersed in the culture medium. Coat the solder joint with silicone adhesive if necessary. Because of the high voltages and currents used for stimulation, NEVER handle the electrodes when the power amplifier is ON. Mount the electrodes on a movable nonconductive platform to allow easy placement and removal from the culture medium. A platform of Delrin, 0.5 in. (~1.3 mm) thick and 2.4 in. × 2.4 in. (~6 mm × 6 mm) square, with a means for clamping the electrode holders in place, is shown in Fig. 24.3.

TEMPERATURE CONTROLLER

During functional evaluations, maintain the temperature of the myooid at 37°C on a heated aluminum plate. A 0.5-in.-thick aluminum plate, approximately 14 in. wide and 8–10 in. deep, with adhesive rubber feet and a heating pad affixed to the bottom, is adequate. Maintain temperature control manually in an open-loop configuration, or with

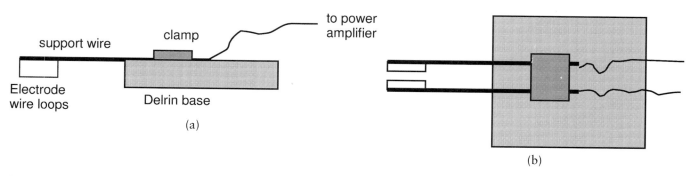

Fig. 24.3. (a) Side view and (b) top view of the electrode configuration. The plastic (Delrin) base allows the electrodes to be placed or removed easily. The support wires are constructed from heavy-gauge solid copper wire, with insulation left intact where the wire is exposed. The electrode wire loops are constructed from 36-gauge solid platinum wire, or alternatively may be constructed from stainless steel shim stock, if parallel plate electrodes are preferred over wire loop electrodes. The support wires are clamped to the plastic block by means of an insulated clamp and screw. The height of the base is set to allow the support wires to just clear the top of the open culture dishes in which the myooids will be functionally evaluated.

temperature feedback in a closed-loop configuration. In an open-loop system, adjust the heating pad setting before evaluation of myooids so that the temperature of the culture media is maintained at $37 \pm 0.5°C$, and confirm the temperature regularly. If a closed-loop temperature control system is employed, place the temperature sensor either directly in the medium or as close to the culture dish as possible. Electrically insulate the temperature sensor if it is placed in the medium and exposed to electrical fields from the electrodes.

ADDITIONAL OPTIONAL INSTRUMENTATION
Pulse Expander

A pulse expander will allow electrical stimulation pulses of very short duration to be detected, even with low sampling rates. For example, if the stimulus pulse width t is 200 µs, the minimum sampling frequency F to ensure that each pulse is detected is $F = 1/t = 5000$ Hz. Detection of each stimulus pulse is important, for example, in the measurement of the latent period (see next section). Because the pulse width and other stimulation parameters are known, during myooid evaluation it is necessary to detect only the time of the onset of stimulation. This is particularly true if LabVIEW data acquisition software is used, because this software is based on data-driven data flow and the exact execution time of parallel subroutines is generally not well defined. A "pulse expander" circuit (Fig. 24.4) allows the stimulation pulse onset time to be detected when short-duration pulse widths and slow sampling rates are being used. This circuit does not change the duration of the stimulus pulse delivered to the myooid, rather, it allows the data acquisition system to detect stimulation pulses at low sampling rates by widening each pulse before detection (Fig. 24.4). If the component values specified in Fig. 24.4 are used, pulse widths narrower than 100 µs can be reliably detected at sampling rates as low as 200 samples/s.

MEASUREMENT OF EXCITABILITY AND ISOMETRIC CONTRACTILITY

Allow at least 3 days after myooid formation before functional evaluation to permit cellular remodeling. Define the age of each myooid either from the time of cell plating or from the time of myooid formation, depending upon the experiment. In either case, record the time to myooid formation from cell plating for each myooid so the measurements to be converted from one system to the other as necessary. Carry out myooid evaluations in the following order.

PREHEAT THE ALUMINUM PLATFORM AND ADDITIONAL MEDIA

Preheat the aluminum platform to a stable temperature as described earlier. Preheat sufficient culture media to add 2 ml to each culture dish of each myooid to be evaluated.

MEASURE MYOOID DIAMETER

Use the eyepiece reticle on the inverted microscope to measure the diameter of each myooid. Because the myooids are suspended in tension and there are no significant lateral

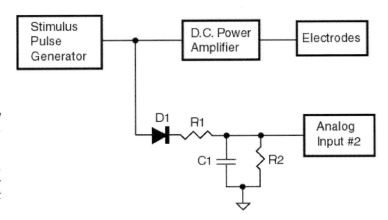

Fig. 24.4. Pulse expander circuit. Values for the individual components may vary with the data acquisition system employed. Using the National Instruments board, (cat. no. AT-MIO-16E-1) the components are as follows: R1, 1 kΩ, 1/8 W; R2, 100 kΩ, 1/8 W; D1, 1N914; C1, 0.1-µF ceramic capacitor. The stimulus pulse generator can be either a separate, stand-alone device or it can be an analog output channel from the data acquisition system.

forces on the myooid from surrounding tissue, the myooids are generally circular in cross section. Assume that the myooids are cylindrical and calculate the estimated cross-sectional area (CSA) from the smallest diameter measured along the length of the myooid.

TURN ON AND TEST THE STIMULATOR/DATA ACQUISITION SYSTEM

Before attaching the myooid to the force transducer, turn on and verify the function of the data acquisition system. Place the electrodes in a culture dish filled with physiologic saline, and apply a pulse train of 1.2-ms pulses at 40 V and 40 Hz. The rapid formation of bubbles at the electrodes verifies the function of the pulse generator and amplifier. The actual stimulus parameters being delivered to the myooid should be regularly verified by using an oscilloscope to directly measure the voltage across the electrodes.

Record the force transducer offset before each use. If the data acquisition system does not allow the reduction of the force transducer offset to zero, subtract the offset from all data files before analysis and storage. If the force transducer offset is stable between uses, regularly record the offset value to confirm that the force transducer has not been damaged. If the offset is normally stable, but changes suddenly, recalibrate the force transducer to be sure that it has not been damaged and that the sensitivity or linearity of the transducer has not changed.

AFFIX THE FORCE TRANSDUCER TO ONE OF THE SUTURE ANCHORS

To attach myooids to the force transducer, remove one of the minutien pins from one of the artificial tendons and seal the other minutien pin with canning wax to the load element of the force transducer. Seal the paraffin with a soldering iron at 150°C (300°F). Avoid excessive heating of the myooid by adding 2 ml of differentiation medium (see Chapter 23) and restricting contact time of the soldering iron with the joint to less than 2 s. Once the myooids have been attached to the force transducer, raise the force transducer using the three-axis micromanipulator to pull the minutien pin from the Sylgard substrate. The myooid should remain immersed in the culture medium while the minutien pin can move unobstructed above the substrate.

ADJUST THE MYOOID LENGTH

The plateau of the length–tension curve for myooids generally occurs at a length about 5–10% less than the length at which the myooids were held during culture [4]. For this reason, shorten myooids by 5% (0.6 mm) from the length at which they were cultured before measuring contractile properties.

PLACE ELECTRODES IN BATH

Position the electrodes so that the myooid is parallel to and directly between the electrode wires. Record the electrode spacing.

MEASURE BASELINE FORCE P_b

Measure isometric force with respect to the baseline force P_b. Myooid P_b typically is of the same magnitude as the maximum isometric tension generated by the myooid. It correlates with the life span of myooids in culture and the cross-sectional area of the myooid that is composed of fibroblasts [2]. Myooids often spontaneously contract, as shown in Fig. 24.5, in which spontaneous contractions precede the electrically elicited twitch. The presence of signal noise and spontaneous contractions complicate the determination of P_b, as shown in Fig. 24.6. When spontaneous contractions are present, it is not appropriate to determine P_b by simply averaging the force signal prior to the application of the stimulus twitch pulse. Two methods of measuring P_b may be employed in this case. If spontaneous contractions are noted, they can be attenuated by the application of a single supramaximal twitch pulse. If this does not prove effective, or if spontaneous contractions are detected only after the experiment has been completed, determine P_b by calculating the average baseline value between spontaneous contractions.

For example, force data from a myooid engineered from C2C12 cells show spontaneous contractions before the twitch (Fig. 24.6). The spontaneous twitches S occur at about 3 Hz

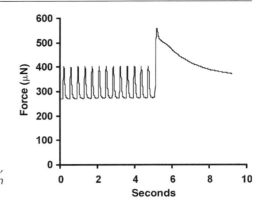

Fig. 24.5. Spontaneous contractions before the first twitch stimulus pulse. Often, spontaneous contractions are attenuated or disappear entirely after the application of a single stimulus pulse to elicit a twitch.

with an amplitude of about 125 μN, nearly half the amplitude of the peak twitch force *T*. As an example, two methods are employed to determine P_b. In method A, all force data points from time 0.0–0.85 s are averaged. For method B, only the force values between the second spontaneous twitch and the elicited twitch (time 0.7–0.85 s) are averaged. Method B gives the correct P_b because the forces generated by the spontaneous twitches are excluded from the calculation. Using the incorrect method, A, P_b is calculated to be 303 μN. The correct method, B, results in a P_b of 275 μN. Since, as explained shortly, the peak twitch tension P_t is calculated by subtracting P_b from the peak value on the force trace, the difference in this example results in an error in the calculation of P_t of approximately 10% when the incorrect method A is used.

APPLY A SINGLE SUPRAMAXIMAL STIMULATION PULSE AND RECORD PEAK TWITCH FORCE P_t

Elicit myooid twitches by applying a single electrical pulse (Fig. 24.7). Carry out preliminary experiments to determine the appropriate stimulus pulse parameters to elicit a maximal twitch for each experimental apparatus. Using parallel wire electrodes 5.0–7.0 mm apart, we have determined that a single pulse of 70-V amplitude and 4-ms pulse width always elicits a maximal twitch from a myooid. Stimulate the myooid in the plateau region of the force–voltage relationship, a practice known as "supramaximal stimulation." Supramaximal stimulation is stimulation greater than the minimum necessary to elicit a maximal contraction, but not high enough to damage the muscle. Verify supramaximal stimulation experimentally by confirming that a small reduction in either stimulation parameter (amplitude or pulse width) does not result in a significant reduction in twitch force, and that repeated stimulation at the same level results in identical twitch forces, indicating that injury to the tissue is not occurring. It should be possible to repeat the twitch dozens or hundreds

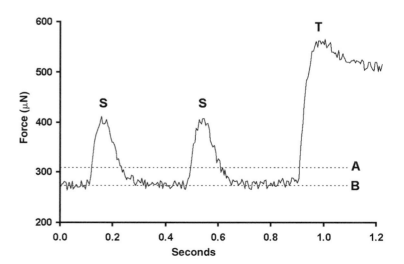

Fig. 24.6. Determination of baseline force P_b.

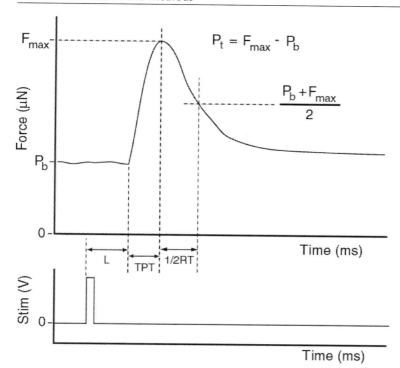

Fig. 24.7. Isometric and dynamic twitch parameters: L, latent period, P_t, peak twitch force, P_b, baseline force, TPT, time to peak twitch tension, ½RT, half-relaxation time, F_{max}, maximum force of the twitch, which includes the baseline force.

of times, allowing a brief rest period of 30–60 s between twitches, and obtain values for P_t that do not change with time. Calculate P_t by subtracting P_b from the maximum force value on the twitch trace (Fig. 24.7).

MEASURE THE EXCITABILITY: RHEOBASE (R_{50}) AND CHRONAXIE (C_{50})

Myooids are cultured aneurally and do not exhibit excitability typical of adult muscle. As a consequence, electrical stimulation pulses required to elicit maximal twitch and tetanus are generally an order of magnitude greater than those for normally innervated whole muscles evaluated *in vitro*. Excitability measurements are important because experimental interventions such as crude nerve extracts, chronic electrical stimulation, genetic modifications, or nerve–muscle coculture could influence excitability.

Excitability is classically defined as the stimulus pulse amplitude and duration required to depolarize the membrane of a single cell (Fig. 24.8). Rheobase is the minimum stimulus *amplitude* required to depolarize a cell membrane, when a long stimulus duration is used. Chronaxie is the minimum stimulus *duration* required to elicit the same response when the stimulus amplitude is set to twice the rheobase. High values of rheobase and chronaxie denote low tissue excitability. Rheobase measurements depend upon the conditions of the experimental setup, including electrode configuration, so direct rheobase comparisons are valid only for tissues tested in the same experimental setup. We have modified the classical definitions of rheobase and chronaxie to describe the bulk properties of an engineered tissue. Because the excitability of engineered tissues involves the depolarization of many cell membranes, these measurements describe the average excitability of the tissue (Fig. 24.8).

To distinguish the bulk tissue excitability from the classical measurements of rheobase and chronaxie, we denote the bulk tissue rheobase with the symbol R_{50}, and the bulk tissue chronaxie with the symbol C_{50} (Fig. 24.8); both these values are determined when the stimulus is adequate to elicit a twitch of 50% of P_t. To measure the rheobase and chronaxie of myooids, rheobase (R_{50}) is defined as the voltage of a 4-ms-wide pulse that elicits a twitch force of 50% of P_t. To normalize the rheobase, divide the stimulation voltage by the electrode separation (in millimeters) and report R_{50} as electric field strength in units of volts per millimeter. To determine the chronaxie (C_{50}), adjust the stimulation voltage to twice R_{50}, and reduce the pulse width until the elicited twitch is once again 50% of P_t (Fig. 24.8).

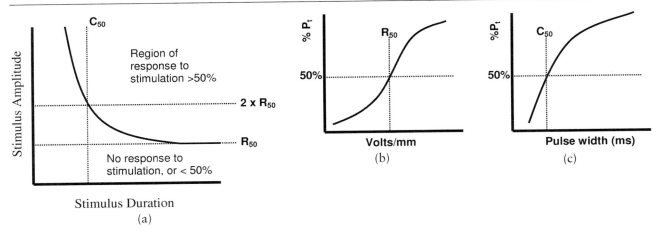

Fig. 24.8. *Measurement of excitability of myooids. The bulk tissue excitability is quantified in terms of rheobase (R_{50}) and chronaxie (C_{50}). (a) Classical definition of chronaxie and rheobase, modified to describe the bulk tissue excitability of myooids. The curve shows combinations of stimulus pulse amplitude and duration for which the elicited twitch amplitude is 50% of P_t. Above the curve, the response is exceeds 50% P_t, below the curve it is under 50% P_t. (b) Determination of myooid R_{50} by adjustment of the stimulus pulse amplitude to achieve 50% P_t, while pulse duration is maintained at 4 ms. (c) Determination of C_{50} with the stimulation amplitude set to $2 \times R_{50}$.*

MEASURE PEAK ISOMETRIC FORCE (P_0)

Determine peak isometric force P_0 by stimulating the myooid at 40 V and 40 Hz with 1.20-ms pulses for a train duration of 2 s. Calculate P_0 by subtracting P_b from the maximum force during stimulation. Calculate specific isometric peak force (sP_0) by dividing P_0 by the myooid CSA. Alternatively, the sP_0 may be calculated from the CSA of myotubes, as determined by an immunohistochemical technique. Because total CSA measurements include void spaces, extracellular material, and nonmuscle cells, the use of total CSA for the calculation of sP_0 yields a conservative value.

ADDITIONAL MEASUREMENTS
Length–Tension Curves

Generate length–force curves by applying stimulus trains of 40 Hz at 40 V and 1.2-ms pulse width for 2 s to achieve maximum tetanus. Adjust the length of the myooid between 50% reduction and 40% increase of length from the initial myooid length of 12 mm. For length changes of the myooid, calculate based upon the contractile region of the myooid only, excluding the length of the artificial tendons. Allow 30 s between contractions for the myooid to recover resting baseline force.

Force–Frequency Curves

Use a pulse width of 1.2 ms, an amplitude of 40 volts, a train duration of 2 s, and frequencies in the range of 3–100 Hz to collect data. Allow 30 s between stimulus trains for the myooid to recover.

HISTOLOGY

Since myooids are small and fragile, they are difficult to section either frozen or embedded in paraffin. However, epoxy resin (EPON) embedding has proven effective for myooid sectioning for both light and electron microscopy. Pin myooids at L_c and fix for 4 h at 4°C in Karnovsky's solution (0.1 M sodium cacodylate buffer with 3% formaldehyde and 3% glutaraldehyde at pH 7.4). Then rinse three times (30 min, 30 min, and 4 h) with cacodylate buffer (pH 7.4) with 7.5% sucrose. Postfix myooids in 1% osmium tetroxide for 2 h at room temperature, dehydrate in graded concentration of ethanol and propylene oxide, and embed in EPON (Eponate 12 resin, Ted Pella Inc., Redding, CA) for microscopy. ATPase staining used for the identification of myosin isoform expression in muscle does not yield satisfactory results with myooids because of the presence of developmental myosin isoforms [5].

DATA ANALYSIS AND INTERPRETATION
SYSTEM NOISE AND FILTERING OF DATA

Raw data files (unfiltered) should be permanently retained for all twitch and tetanic contractions. The best method of noise reduction and data storage is to eliminate noise before amplification. The most common types of noise in muscle mechanics experiments are 60-Hz line noise and "salt and pepper" noise. Identify and eliminate sources of 60-Hz noise, either by shielding or by moving the sources away from the data acquisition system. Do not filter 60-Hz noise from muscle force trace signals, since myooid force traces contain frequencies that are near 60 Hz, and filtering out the noise will significantly alter the signal. Filtering of data is often required to remove "salt and pepper" noise, which normally occurs as narrow spikes superimposed over the signal. Linear filters and rolling average schemes often do not yield satisfactory results when large-amplitude, narrow-duration noise spikes are present. In this case, use a simple nonlinear digital filter: a median filter of rank 2 or 3, which is available as a standard filter in LabVIEW data acquisition software. Either implement the filter directly in the data acquisition program or apply the filter to the data post hoc.

DYNAMIC TWITCH PARAMETERS (LATENT PERIOD, *TPT*, *dP/dt*, ½*RT*)

Save to disk the raw data for the twitch and peak isometric force for each myooid for later analysis. Calculate dynamic twitch parameters, time to peak twitch force (*TPT*) and half-relaxation time (½*RT*), from the raw data traces (Fig. 24.7). The latent period is the time delay between the onset of stimulation and the onset of force generation. Determine *TPT* by measuring the time from the end of the latent period until P_t is achieved, and determine ½*RT* by measuring the time from P_t until the twitch force decayed to ½P_t (Fig. 24.7). Determine the rate of twitch force generation (*dP/dt*) by measuring the slope between 20 and 80% of P_t. To measure the latent period, or the dynamic twitch parameters accurately, use a sampling rate of 1000 samples/s.

NORMALIZATION OF DATA

For comparisons between myooids of different geometry or mass, it is necessary to normalize the excitability and contractility data. As a general rule, length changes should be normalized by dividing by the initial myooid length (contractile section only, excluding the artificial tendons). Forces should be normalized by the CSA, and the method for determining the CSA should be the same for all myooids to be compared. The dynamic twitch parameters *TPT* and ½*RT* should be normalized by dividing by P_t, but note that the rate of force development, *dP/dt*, does not require normalization. Other measurements should be normalized appropriately before comparisons are made.

ACKNOWLEDGMENTS

The authors thank Dr. William Kuzon for his suggestions and scientific input on the development of the myooid and Dr. Krystyna Pasyk for light and electron microscopy. Dr. John Faulkner provided significant scientific, financial, and infrastructural support and resources without which this method would not have been developed. This work was supported by the University of Michigan Center for Biomedical Engineering Research, the Contractility Core of the Nathan Shock Center the Basic Biology of Aging, and AG-06157.

REFERENCES

1. Dennis, R. G. (1996). Measurement of pulse propagation in single permeabilized muscle fibers by optical diffraction. Doctoral thesis, University of Michigan, Ann Arbor.
2. Kosnik, P. E. (2000). Contractile properties of engineered skeletal muscle. Doctoral thesis, University of Michigan, Ann Arbor.
3. Minns, H. G. (1971). A voltage-controlled force generator for calibrating sensitive transducers. *J. Appl. Physiol.* **30**, 895–896.
4. Dennis, R. G., and Kosnik, P. E. (2000). Excitability and isometric contractile properties of mammalian skeletal muscle constructs engineered in vitro. *In Vitro Cell. Dev. Biol.—Anim.* **36**(5), 327–335.
5. Guth, L., and Samaha, F. J. (1972). Erroneous interpretations which may result from application of the "myofibrillar ATPase" histochemical procedure to developing muscle. *Exp. Neurol.* **34**, 465–475.

CHAPTER 25

MESENCHYMAL CELL CULTURE: CARTILAGE

Bernd Kinner and Myron Spector

INTRODUCTION

Cartilage tissue engineering is an emerging scientific field with important clinical end points. There are, however, many forms of cartilage in the body that vary greatly with respect to the phenotypic characteristics of the cells; the chemical composition, structure, and mechanical properties of the extracellular matrix; the degree of vascularity and innervation; and function. This wide variability challenges both approaches to form the specific cartilage types *in vitro* and approaches designed to facilitate regeneration *in vivo*. Culture of cartilage cells is important in providing a tool for the investigation of their behavior under defined conditions and for the expansion of cells for cell-based therapies.

Since chondrocytes were first isolated and maintained in cell culture in the late 1960s [1,2], there has been remarkable progress in the technology for chondrocyte cell expansion, and in methods for evaluating their proliferation and biosynthetic activity. Our understanding of the influence of various environmental factors on the phenotypic expression of this cell type has also advanced. Cell culture methods implemented today have been validated for the clinical use of isolated cells for a novel cartilage repair procedure [Genzyme]. In some respects chondrocyte isolation and maintenance of primary cell cultures of certain types of cartilage (viz., articular cartilage) are relatively easy because only a single cell type is involved. Contamination with a different cell type with different growth characteristics, which could jeopardize the results, therefore is less likely with cartilage than with other tissues.

There are three types of cartilage: hyaline cartilage (e.g., tracheal and articular), fibrocartilage (e.g., meniscus and intervertebral disc), and elastic cartilage (e.g., ear). The spherical or rounded shape of the cells that reside in lacunae is the common feature of the cartilages. The cartilage types vary greatly, however, with respect to the composition and structure of the extracellular matrix (Table 25.1).

Chondrocytes comprise the single cellular component of adult hyaline cartilage and are considered to be terminally differentiated, thus being highly specialized. Their main function is to maintain the cartilage matrix, synthesizing types II, IX, and XI collagen; the large aggregating proteoglycan aggrecan; the smaller proteoglycans biglycan and decorin; and several specific and nonspecific matrix proteins that are expressed at defined stages during growth and development. The cells that are relatively sparse account for less than 5% of the tissue volume. Three zones can be distinguished in articular cartilage, comprising different cell morphology, distribution, and density. Studies have also indicated phenotypic differences among the cells in these zones. The cells in the superficial layer synthesize a phospholipid that has been identified as lubricin [3–5]. Moreover, compared with cells in the deep region of cartilage, a greater percentage of cells in the top layer have been found to express the gene for a muscle actin isoform, smooth muscle α-actin [6,7]. One of the challenges encountered in the monolayer culture expansion of articular chondrocytes is

Table 25.1. Anatomy and Morphology of Cartilaginous Tissues

	Articular cartilage	Meniscal fibrocartilage	Intervertebral disc
Cell type	Chondrocytes	Fibrochondrocytes	Fibrochondrocytes and chondrocytes
Cell density	1.4–1.7×10^4 cells/mm^3	Fusiform superficial cells Ovoid cells in deeper zones	5.8×10^4 cells/mm^3 cartilage end plate 15×10^4, annulus 9×10^4 and nucleus pulp, 4×10^4 cells/mm^3
Collagen type	Collagen II (V, VI, IX, X, XI) 65–80% w/w	Collagen I (55–65% dry wt) Collagen II, II, V, VI (5–10% dry wt)	Nucleus: collagen II (collagen VI, IX, XI) Annulus: collagen I, II, III, V, IX
Fiber orientation	Superficial: parallel Deep: orthogonal	Circumferential	Collagen I/II: radial in opposing concentrations
Proteoglycans	Aggrecan, 4–7% w/w		Large aggregating PG, but monomers are smaller than in hyaline cartilage
Glycosaminoglycans[a]	CS, KS	C-6-S (40%) C-4-S (10–20%) DS (20–30%) KS (15%)	C-6-S,KS
Intrinsic repair capacity	Low	Low	Low
Synthetic activity	Low	Low	Low
Mitotic activity	Low	Low	Low

[a]C-6-S, Chondroitin 6-sulfate; C-4-S, chondroitin 4-sulfate; DS, dermatan sulfate; KS, keratan sulfate.

that the cells lose their specific phenotype, adopting a fibroblastic morphology, and they cease to synthesize articular cartilage–specific products, type II collagen and aggrecan [8]; instead the cells produce type I collagen and, under certain conditions, type X. This process is commonly referred to as dedifferentiation. If these cells are to be used for subsequent cell-based therapies it is important to show that they are able to regain their chondrocytic phenotype [9,10].

Additional issues need to be dealt with in the isolation and maintenance of cells from fibrocartilage (meniscus and intervertebral disc) where the phenotype of the cell is less characterized and contamination and overgrowth of stromal fibroblasts and vascular-derived cells is possible. The majority of meniscal cells can be assigned to two relatively distinct populations based on their morphological characteristics. Cells in the more superficial layer of the meniscus are oval or fusiform, therefore having a more fibroblastic appearance, whereas those in the deeper zones of the tissue are rounded or polygonal. There has been much debate over the classification of these meniscal cells as fibroblasts or chondrocytes. They are sometimes referred to as fibrochondrocytes. Although they might look like chondrocytes that make up hyaline cartilage, these cells predominantly produce a type I collagen.

The intervertebral disc also comprises two distinct cell types: chondrocytic cells predominantly found in the nucleus pulposus and fibroblastic cells found in the annulus fibrosus. The percentage of these two distinct populations varies with age, as cell number decreases. Associated with the cell type, proteoglycan content is highest in the nucleus and progressively decreases from the most central region of the disc to the peripheral regions. Paralleling proteoglycan content, the amount of type II collagen also decreases from the central to peripheral zones of the disc.

These unique and complex cellular and biochemical features, which are crucial for the mechanical behavior of the various cartilage structures, are of prominent importance for the *in vitro* engineering of these tissues. Most research efforts have been made in the field of articular cartilage repair because of its accessibility and homogeneous tissue characteristics, and because of its clinical relevance.

CELL ISOLATION AND CULTURE PROCEDURES
TISSUE DIGESTION FOR ISOLATION OF CELLS
Articular Chondrocytes

This chapter describes the isolation and *in vitro* culture of animal and human articular chondrocytes, as well as fibrochondrocytes from menisci and intervertebral discs. Isolation of chondrocytes is essentially a two-step procedure: digestion of the tissue and isolation the cells from the tissue digest. In the past most authors performed sequential digestion of cartilage using selected enzymes, including collagenase, hyaluronidase, and trypsin [11]. More recently collagenase alone has been found to be adequate to sufficiently digest these tissues.

Cartilage is harvested under aseptic conditions and transferred to the tissue culture lab in a transport medium comprising either cold phosphate-buffered saline (PBS) supplemented with 1% antibiotic–antimycotic solution or Dulbecco's Minimum Essential Medium (DMEM) with 1% of antibiotic–antimycotic. In a biosafety cabinet (BSC 10,000 for human tissue) or laminar flow hood, the cartilage is dissected free from connective tissue or bone and minced into small pieces ($\sim$1–3 mm^3). If mincing is done in a culture dish, the presence of 1 ml of 0.1% collagenase has proven useful to prevent cartilage pieces from sticking to the dish. The tissue is rinsed several times using cold PBS-AB and transferred to a digestion vessel after weighing. For tissue samples weighing $\leq$300 mg we use a 50-ml tube as the digestion vessel and for a larger amount of tissue we use a 225-ml spinner flask.

Our digestion medium comprises DMEM and Ham's F-12 nutrient medium (F-12), with 1% of an antibiotic–antimycotic solution. The literature suggests a wide range of collagenase concentration (e.g., 0.1–0.3%), depending on the collagenase used. In our own work we have used 400 U/mg tissue. Suggestions of the use of fetal bovine serum (FBS) within the digest vary as well from 0 to 10%. Importantly, FBS is less effective in terminating the collagenase enzymatic reaction than for stopping the action of trypsin. The digestion procedure for cartilage samples often proceeds under these conditions for 24 h. However, a wide range of digestion times (4–24 h) is favored in literature. The period required for digestion of cartilage specimens (especially human) may vary significantly. Prolonged digestion, however, is generally tolerated. Remaining tissue may even be subjected to further digestion using fresh collagenase (0.25%) and trypsin (0.1%) for 2–3 h.

Upon discontinuing the digestion, the cell suspension may be poured through a nylon mesh (0.7-μm mesh size) to remove any matrix debris or undigested particles,[*] followed by centrifugation at 500g or 1500 rpm for 10 min. To remove any residual collagenase, the cell pellet is washed three times with PBS. After washing, the cells are counted in a hemacytometer and a trypan blue exclusion test is used to assess their viability. A uniform cell population with viability greater than 70% is the goal. Plating to tissue culture vessels is performed using $1-3 \times 10^4$ cells/cm^2. Freshly isolated cells usually need 3–4 days to adhere. Incubation takes place at 37°C and 5% CO_2.

The yield of cells per milligram of fresh tissue varies considerably, depending on species, age, and health of the tissue. For adult human subjects the cell yield might be low, especially in a clinical setting where only 150–350 mg of cartilage is available. The number of cells obtained from the digestion procedure in such a case has been reported to be approximately 1500 cells/mg [12].

Fibrochondrocytes from the Meniscus and Intervertebral Disc

The harvest and dissection procedures are essentially the same as for articular cartilage, outlined earlier. Owing to the anatomy of the meniscus and intervertebral disc, however, it is much more difficult to obtain a homogeneous cell population. Debate continues over whether these tissues comprise different cell types (e.g., fibroblasts and chondrocytes) or are different phenotypes of a single cell type, the chondrocyte. As is the case for articular cartilage, different digestion protocols have been proposed. Our own procedure employs a two-step digestion: digestion with trypsin–EDTA for 2 h on a shaker at 37°C followed by collagenase (0.2%) treatment for 3–4 h.

SUBCULTURING OF CELLS FOR THEIR EXPANSION

Standard tissue culture techniques are used for the subculturing of chondrocytes. Prolonged expansion of cells in monolayer may be necessary owing the slow replication rate of the cells, especially adult human cells. Growth kinetics in literature for standardized procedures like that associated with the commercially available Genzyme Carticel Product are reported as 0.3 doubling per day, but very different numbers may be achieved depending on patient age and cartilage quality [13]. Thus considerable time may be needed to obtain a sufficient amount of cells.

CULTURE CONDITIONS

There are many ways to influence the behavior of cells in culture, depending on whether cell expansion or phenotypic stability is the primary focus. For tissue engineering purposes, it is possible that the focus may change through the process. Selection of media has an important impact on cell growth and phenotype *in vitro*. After cell isolation, the expansion of cells is generally the first goal, and media that promote cell growth are employed [14]. Once a sufficient number of cells have been obtained, cells might be implanted into the defect or seeded into a scaffold and implanted into a defect after an additional period in culture. At this point the focus of culturing may have to promote the biosynthetic activity of the cells to produce tissue-specific matrix molecules. The media may have to be changed to enhance this step.

Polypeptide growth factors play a major role in the regulation of cell behavior, including that of chondrocytes (Table 25.2). Among the most influential of these factors identified for articular chondrocytes are insulin-like growth factor I (IGF-I), basic fibroblast growth factor (FGF-2), and transforming growth factor β (TGF-β). Selected growth factors—individually

---

[*]The use of a cell strainer is optional. Especially after prolonged digestion the suspension essentially comprises single cells, and one might want to avoid loss of chondrocytes in a straining procedure especially if biopsy samples are small (<300 mg).

Table 25.2. Cytokines Evaluated for Their Effects on Chondrocyte Growth and Matrix Production

TGF-β	Differentiation	↑Proteoglycan synthesis ↓Collagen synthesis	[16–18]
FGF-α	Mitogenic, differentiation	↑Matrix synthesis (synergistic with insulin)	[19–21]
IGF-1	Mitogenic, differentiation	↑Matrix synthesis	[22,23]
PDGF-bb	Mitogenic		[24,25]
BMP-7 (rhOP-1)	Differentiation	↑Proteoglycan synthesis	[26,27]
RhBMP-2	Differentiation	↑Collagen synthesis	[28,29]

or in combination—may be added to the growth medium. Chondrocyte responsiveness to growth factor has been shown to change during development; and after skeletal maturity, there is a profound decline in the levels of DNA synthesis and cell replication in response to the known chondrocyte growth factors [15]. (See Refs. [16–29] in Table 25.2.)

To promote matrix production, culture media can be supplemented with ascorbic acid (25 μg/ml), which has been found to increase the amount of matrix production during short-term culture. However, it was also found to inhibit the transcription of cartilage-specific matrix genes in long-term culture [30]. TGF-β was found to promote differentiation of chondrocytes and thus support matrix synthesis [31–34]. In addition to being a potent mitogen for articular chondrocytes, FGF-2 supports their differentiated state in a three-dimensional culture system [19] and augments articular cartilage repair *in vivo*. Other experiments have shown synergetic effects of TGF-β and FGF-2 [35]. However, some growth factors may elicit seemingly opposite effects under different experimental conditions. They interact to modulate their respective actions, creating effector cascades and feedback loops of intercellular and intracellular events that control articular chondrocyte functions [36].

Because the actions of growth factors are not yet completely understood or sometimes even contradictory, it is difficult to recommend a single growth factor—or a growth factor cocktail—to promote cartilage repair either during *in vitro* or *in vivo* articular cartilage tissue engineering.

Fibrochondrocytes have also been shown to be sensitive to variations in the composition of culture media. When meniscal fibrochondrocytes were cultured in DMEM, cells had a polygonal shape. Upon culture in Ham's F-12, however, more fusiform-shaped cells were found [37]. A difference in responsiveness to a mixture of platelet-derived growth factor (PDGF) and antibiotic has also been reported. Cells isolated from the central portion of the meniscus did not respond to stimulation *in vitro*, whereas cells of the periphery did [38]. This finding, which has been explained in terms of the difference in the signal transduction pathways for various cytokine receptors, additionally supports the notion of distinct fibrochondrocyte populations within the meniscus [39]. Moreover, the data available suggest that there is a difference between the isolated cells of the medial and lateral meniscus. In rabbit and human menisci, the population doubling time of cells derived from the medial meniscus was much shorter than that for cells from the lateral meniscus [37].

Special culture conditions (e.g., culturing cells in a "micromass") have been used to demonstrate maintenance of the chondrocyte phenotype during the process of *in vitro* culture. Various hydrogels (including agarose, collagen gel, and alginate gel), culture systems (e.g., a rotating wall "bioreactor"), and scaffolds have also been implemented to facilitate the maintenance of the chondrocyte phenotype in culture. All these devices provide a three-dimensional framework that provides the conditions necessary for the cells to maintain their characteristic rounded shape and continued expression of chondrocyte-specific genes. Specific features of these systems are discussed shortly.

The use of serum (viz., fetal bovine serum) continues to be a principal issue in the culture of chondrocytes and fibrochondrocytes. There are two important points in this respect. First, serum introduces into the experiment another variable that is difficult to control, especially because bovine serum can differ considerably from lot to lot. It has been proposed that every new lot to be tested separately before use. These tests are difficult, however, and routine testing is often not possible. A second concern is that there is always the possibility of an immunogenic response and the transmission of infectious agents (e.g., prions) when FBS is used in a patient.

In light of the potential problems associated with the use of serum, alternative culture media have undergone investigation. One of the more widely examined medium supplements is an insulin–transferrin (partially iron-saturated)–sodium selenite (ITS) formulation. Insulin is a polypeptide hormone that promotes the uptake of glucose and amino acids and may owe its observed mitogenic effect to this property. Transferrin is an iron transport protein. Iron is an essential trace element, but, it can be toxic in the free form; thus to nourish cells in culture, it is supplied bound to transferrin in serum. Selenium is an essential trace element normally provided by serum. These factors enhance cell proliferation and decrease the serum requirement. The cell proliferation resulting from the use of only 2–4% serum supplemented with ITS is reported to be similar to medium containing 10% serum [40–43].

Micromass and Pellet Culture

Micromass cultures have traditionally been used to investigate embryonic limb bud development and cartilage differentiation [44–50]. These culture conditions have been found to promote chondrocytic differentiation of cells isolated from chicken embryos. Only recently has this culture method attracted interest for tissue engineering, and it is now being used to induce hypoxic conditions in the process of transdifferentation of mesenchymal stem cells and dermal fibroblasts into chondrocyte-like cells [51].

After isolation, cells are resuspended at a high concentration (containing 2×10^7 cells/ml) spotted in 10-μl drops onto multiwell plates (viz., 6-well plates). Cells are allowed to adhere for 90 min at 37°C before the culture dish is flooded with standard growth medium.* The high density of the seeded cells promotes a prolonged maintenance of the differentiated state of the cells.

Alternatively, cells can be cultured in pellet form [52–54]. In one procedure 250,000 cells are gently spun down in a cervical 15-ml tube. The pellet that in obtained is subsequently cultured to form a free floating micromass after several hours. The main disadvantages— which may have prevented the widespread use of this method—are uncertain cell survival (due to insufficient nutrient supply to the cells in the center) and cell loss (due to incomplete cell attachment). This culture method, despite questions about its reliability, may be of value for certain specific questions dealing with chondroid differentiation.

Agarose

Two percent (w/v) agarose (low temperature gelling agarose) is prepared, autoclaved, and plated into a culture dish or well plate producing a thin layer (0.5 ml on a 3.5-cm dish, 1.5 ml on a 6-cm dish, or 4.5 ml on a 10-cm dish). In the same manner, a 4% Seaplaque agarose is prepared, autoclaved and cooled. After cooling, a 1:1 mixture of cell solution and gel is prepared ($1–4 \times 10^6$ cells/ml gel). The use of warmed pipettes will prevent the gel from solidifying inside the pipettes. The cultures are left at room temperature until the agarose is gelled, and the cells are allowed to settle (20–30 min at 37°C) before medium is added.

---

*The use of tissue culture polystyrene (TCPS) is recommended (especially when chamberslides are used), because it is difficult to establish a droplike micromass culture on glass slides, where surface tension is much lower. Because glass has much better wetting properties than TCPS, cells will not be contained within the defined are of the droplet.

Agarose-encased cells have also been prepared and stored frozen at −20°C. For many evaluation procedures extraction buffers have been directly added to the gel. To remove the agarose and debris, the samples are centrifuged at high speed at 4°C. As a rule, the low melting point of agarose make it impossible to embed these gels in paraffin. As an alternative for histological procedures, glycol methacrylate embedding has been employed successfully.

The advantages of agarose cultures include the ease with which they can be prepared and their low costs. However, for histological examinations they can be rather troublesome, and recovering cells may be difficult. Another disadvantage is the low replication rate of cells grown in agarose gels or on agarose-coated dishes.

Collagen Gel

As alternative to agarose gels, collagen gels have been widely used to examine chondrocytes in attachment-free culture systems [55–57]. To prepare the collagen gel, a 0.3% collagen solution (purified type I collagen) in 0.001 M acetic acid is used after UV sterilization. Then 9 volumes of this collagen solution are added dropwise with gentle shaking at 0°C to 10 volumes of 2× concentrated stock solution of the basic growth medium. One volume of a chondrocyte suspension (6×10^6 cells/ml) is added to the prepared medium–collagen mixture. The final solution is plated to a multiwell plate and allowed to gel by incubation at 37°C. The final concentration will be 0.135%. Complete medium is added after gelling has occurred. Cells may be released from their collagen gel matrix by soaking in a 0.3% collagenase solution for 45 min at 37°C. Histological analysis can be carried out with paraffin and plastic embedding. Standard preparation/isolation techniques for proteoglycans and collagens can be used.

Under these conditions chondrocytes retain their typical rounded morphology and accumulate metachromatic matrix. Glycosaminoglycan synthesis is maintained over several weeks and the major collagen type produced is type II collagen [56].

Alginate Gel

Cells ($1–4 \times 10^6$ cells/ml) are resuspended in sterile 0.15 M NaCl containing low-viscosity alginate gel (1%) and slowly expressed dropwise through a 22-gauge needle into a 102 mM $CaCl_2$ solution. After polymerization has proceeded at room temperature for 10 min, the $CaCl_2$ is discarded and the pelleted beads are washed three times in 0.15 M NaCl and once in complete culture medium. Beads are incubated in culture dishes or well plates at 70–80 beads/ml.

The alginate gel can be solubilized by adding 3 volumes of a 55 mM Na citrate in 0.15 M NaCl solution and incubating this at 37°C for 10 min. The isolated chondrocytes can be recovered with their pericellular matrix by centrifuging at 500g for 10 min, followed by trypsin–EDTA treatment for 5 min to recover a dispersed cell suspension.

Dynamic Culture Conditions

The ability of cells to maintain their differentiated phenotype and synthetize extracellular matrix, which eventually will build up the new tissue in question, requires exposure of these cells to an appropriate environment. This includes the availability and distribution of suitable nutrients as well as the exposure of the cells and their extracellular matrix to appropriate signalling stimuli (growth factors, mechanical stimuli).

Several investigators have shown the importance of fluid flow in the *in vitro* synthesis of cartilaginous tissues [58,59]. The constant availability of fresh media, the mechanical action of shear stress on the cells, and the ability to actively transport nutrients through an increasingly dense extracellular matrix are some of the reasons for using perfusion systems as a basic component in the *in vitro* engineering of cartilaginous tissue. However, optimal conditions such as flow rate or the design of the bioreactor (perfusion system vs rotating vessel [60–66]) have yet to be defined.

Scaffolds

It has been shown that three-dimensional cultures support cartilage-specific gene expression [51]. This is not only true for alginate [67], agarose [10], and collagen gels [56], but also for synthetic scaffolding devices. The matrix used to support the chondrocytes for implantation needs to fulfill certain mechanical requirements. Since must gels do not meet these specifications, porous scaffolds have become the focus of investigation; both synthetic polymers and modified natural polymers are being used.

Numerous materials have been proposed, and are being developed as scaffolds for use in tissue engineering and cartilage repair (see also Section II). What is the rationale for scaffold selection? Selecting, or even more designing and constructing, an ideal matrix for cartilage tissue engineering requires careful consideration of numerous factors [68]. The following questions must be answered: Which material should be used? What should the degradation rate be? What is the optimal pore size?

The following general properties are crucial matrix requirements for tissue engineering in general [69,70]:

1. Cell adhesion to the matrix.
2. Neither the polymer nor degradation products can be toxic or antigenic.
3. Reproducible three-dimensional structure.
4. Porosity exceeding 90%. Maximized surface area for cell–matrix interactions, allowing for cell migration and the deposition of extracellular matrix. Supporting diffusion of nutrients and gas exchange during *in vitro* culture.
5. Complete degradation after completing the supportive function.
6. Steady state between matrix production of the cells and degradation of the matrix.

Moreover, a matrix for cartilage tissue engineering must meet certain specific requirements.

1. The matrix should support cartilage-specific matrix production (collagen type II and aggrecan). Our studies showed a considerable difference between the scaffolds, even if only collagen type, pore size, or method of cross-linking is changed [71,72]
2. The matrix should provide enough mechanical support for allowing early mobilization of the treated joint [73,74].
3. Cell migration sufficient to achieve a good bonding to the adjacent host tissue should be allowed.

Poly(α-hydroxy esters)—poly(L-lactic acid) (PLA), poly(glycolic acid) (PGA), and their copolymers—have a long history of use as synthetic biodegradable materials and have been used as surgical suture material, plates, and screws for fracture fixation devices. They have been approved by the U.S. Food and Drug Administration (FDA) for certain clinical uses and are therefore very attractive as scaffolds for cartilage tissue engineering. Degradation rate can be controlled during the fabrication process (weeks–years), and degradation products are generally nontoxic when released in small amounts at a slow rate. However, at high concentrations the degradation products, lactic and glycolic acid, can cause a decrease in local pH resulting in tissue damage [70,75].

Porous type I collagen sponge-like matrices have been researched extensively for regeneration of tissues in skin, bone, knee meniscus, articular cartilage, esophagus, and muscle. Copolymers with glycosaminoglycans have been applied for skin, bone, and cartilage regeneration. More recently type II collagen, with or without copolymers, has also been investigated [71,72]. Stiffness and degradation rate can be controlled by cross-linking. Cell morphology is dependent on collagen type and pore diameter. It was shown that a greater percentage of canine chondrocytes retained their spherical shape in a type II collagen matrix, compared with cells in a type I collagen matrix [72]. Moreover chondrocytes in the type II matrix synthesized more glycosaminoglycans. Pore size is especially critical to obtain uniform cell distribution throughout the matrix. The higher the porosity, the more cells are able to penetrate the matrix and finally to attach to the collagen fibers.

Grande *et al.* [68] could show a marked variability with respect to how chondrocytes responded to culture on various materials (PGA, collagen gel, porous collagen). Bioabsorbable polymers such as PGA enhanced proteoglycan synthesis, whereas collagen matrices stimulated synthesis of collagen [68].

SEEDING METHODS

Seeding requirements for scaffolds to make implants for potential clinical use include a high yield of cells from the biopsy, to maximize the utilization of donor cells, a high proliferation rate, to minimize the time in suspension for anchorage-dependent and shear-sensitive cells, and high cell density and spatially uniform distribution of attached cells, for rapid and uniform tissue regeneration.

Static Cell Seeding

Most investigators prefer static seeding of the scaffolds using a pipette although not meeting all of the aforementioned requirements. The cell suspension is delivered to the scaffold at a desired concentration (7–15×10^4 cells/mm^3 matrix) [60,62,78,79] using aliquots of 25–50 μm of cell suspension. After 10 min the scaffold is flipped over onto the other side and another aliquot of 25–50 μm is administered. Alternatively cells can be injected into the matrix with a very fine needle. To increase cell distribution, so-called vacuum seeding can be used. The scaffold is placed into a tube and the cells are administered by a pipette. After that, a vacuum is created by introducing a needle and syringe and repeatedly aspirating the air of the seeding vessel to force cells into the matrix [76]. Seeding chambers have been developed to get more uniform seeding and higher seeding efficiency and to prevent cells from attaching to the culture dish [77]. After seeding the constructs are incubated for about 2 h to allow cells to attach. After 2 h another aliquot of growth medium is added carefully, to avoid washing out of cells not yet attached. After another 24 h the complete medium is added.

We prefer multiwell plates (6 or 12 well), which we precoat, with 2% agarose to prevent cells form attaching or migrating to the culture dish.

Capillary forces and gravity force cells into the matrix depending on whether dry or prewetted matrices are used. The capillary forces of prewetted matrices are smaller than those of dry matrices. However, we found that using dry matrices can lead to collapse of pores due to the exceedingly high capillary forces, at least when a highly porous collagen–glycosaminoglycan copolymer is used.

Seeding efficiency does not generally exceed 50–70% using static seeding conditions and cell distribution is not uniform throughout the scaffold. A higher cell density is always found at the periphery of the scaffold.

Dynamic Cell Seeding

To obtain a more uniform cell distribution and a higher efficiency of cells attaching to the scaffold, the cell suspension or the scaffolds may be agitated, to allow additional forces (fluid flow) to contribute to the distribution of the cells. Seeding efficiency is very dependent on seeding method, cell source, and cell density.

Cells are resuspended at the desired concentration and incubated with the scaffold on a nutator/shaker. This practice provides a multidirectional flow within the cell suspension. The time needed for this seeding method varies between 10 min and 4 h.

Many dynamic seeding techniques use a spinner flask. Scaffolds are fixed to a pin (8 cm long, 22-gauge needle) inserted into the stopper of the flask. The cell suspension is then exposed to a unidirectional turbulent flow created by a nonsuspended magnetic stir bar set at 50–80 rpm [70]. In one protocol, 12 scaffolds per 120-ml spinner flask are incubated with a cell suspension containing 60×10^6 cells/120 ml. Seeding time extends over 1–3 days. The seeding efficiency reported for this particular method approaches 100% [65].

Burg *et al.* [80] compared the two different seeding environments, static and dynamic, and three proliferation environments, static, dynamic, and bioreactor, for a total of six possible methods. The six seeding and proliferation combinations were analyzed following a 1-week total culture time. It was determined that for this specific system, dynamic seeding followed by a dynamic proliferation phase was the least promising method and dynamic seeding followed by a bioreactor proliferation phase was the most promising [80].

Similar obervations were made by Kim *et al.* [81]. The density of cells adherent to the matrices was a function of cell concentration in the seeding solution. Under all conditions, there were both a larger number (approximately 1 order of magnitude) and a more uniform distribution of cells adherent to the matrices with dynamic than with static seeding methods. In comparison to the static method, the dynamic seeding methods, also resulted in new tissues that had a higher cellularity, and more uniform cell distribution.

None of the seeding techniques has been validated so far, and in any event seeding efficiency is also dependent upon the matrix used. Thus these procedures have yet to be accepted for clinical use. However, the few reports focusing on seeding technique indicate that a dynamic seeding techniques seem to provide a higher cell yield, more uniform distribution of the cells, and higher efficiency of the seeding process in terms of the cells used.

PHENOTYPIC CHANGES: GENE EXPRESSION

Freshly isolated human articular chondrocytes express cartilage-specific type II collagen and continue to do so for several days to weeks in monolayer culture. More reliable and less unstable with respect to chondrocytic gene expression are chondrocytes isolated from young animals or of embryonic origin. During prolonged culture and serial subculture, however, these cells begin to express type I and III collagens [10]. This change in gene expression can be altered by the culture conditions as mentioned earlier (e.g., plating at low density leads to a accelerated loss of chondrocyte-specific gene expression). If cells are placed in suspension culture in spinner flasks or agarose-coated dishes, or seeded in high density micromass cultures or on different matrices, these changes in gene expression are at least partially reversible.

However, other studies indicate that the process of redifferentiation may not be complete [72]. Understanding the mechanisms that control chondrocyte differentiation and chondrogenesis will provide important insights into the regulation of cartilage formation during development and growth, and will also aid in development of rational strategies for the repair of cartilage defects.

Chondrocyte Cell Lines

For specific questions, different chondrocyte cell lines (Table 25.3), which retain at least some chondrocyte-specific properties, can be used [80–98]. Most retain the ability to synthesize large aggregating proteoglycans. However, type II collagen synthesis is more easily lost. These cells have a high proliferative capacity that can be dissociated from the ability to express chondrocyte-specific phenotype. In summary, these cell lines can provide useful tools for answering certain questions regarding tissue engineering. A detailed discussion, however, is beyond the scope of this chapter.

Contraction

One of the major concerns during *in vitro* and *in vivo* tissue engineering is the contraction of scaffolds; especially, collagen gel matrices, were shown to contract during *in vitro* culture. Moreover, the poor integration of the engineered tissue to the host cartilage might be expected to be related to contraction of the implant. Contraction of porous scaffolds can be overcome by producing a stiffer matrix, namely, by altering the fiber size or the cross-linking of the collagen network [99]. A connective tissue cell that displays muscle-like, contractile activity, the "myofibroblast," was first described 30 years ago [100]. The contraction of fibroblasts in certain connective tissues (viz., dermis) has subsequently been found to play a critical role in wound healing [101] and has been identified as the cause of pathological contractures (e.g., Dupuytren contracture) [102]. This contractile behavior has been found to be enabled by the expression of a specific muscle actin, α-smooth muscle actin (SMA). The recent literature contains descriptions of several SMA-expressing musculoskeletal connective tissue cells including articular chondrocytes [6,103] meniscus cells [104], and intervertebral disc cells [105]. Moreover, these SMA-containing cells have demonstrated the capacity to contract a collagen–glycosaminoglycan analogue of extracellular matrix *in vitro* [7,99,104,105].

Table 25.3. Cell Lines Available for Specific Questions

Origin	Cell line
Viral oncogenes	Avian myelocytomatosis virus strain MC29 [82] recombinant retrovirus (NIH/J-2) carrying the *myc* and *raf* oncogenes [83]
Rat chondrosarcoma cell line-human chondrosarcoma	Type II collagen stable [84]; HCS-2/8 [85,86] HCS-SA [85] Proteoglycan stable 105KC [87] Type II, X procollagen, and aggrecan OUMS-27 [88], CS-OKB [89]
Spontaneous, rat calvaria	CFK2 [90] RCJ 3.1 [91] R-tTA-24 [92]
Human, SV-40-containing vectors	MCT [93] SVRAC [94–96] TC6 [97] C-20/A4 [98]

Others have reported articular chondrocyte-mediated contraction of a poly-(glycolic acid) scaffold [106] and fibrin gel [107] *in vitro*. The roles of SMA-expressing cells in the formation, remodeling, healing, and pathology of musculoskeletal tissues have yet to be determined. It has been proposed that the contractile forces generated by SMA-incorporated cytoplasmic apparatus may facilitate the cell's manipulation of its extracellular matrix to impart tissue-specific architecture. These contractile forces may be disadvantageous when the cells are seeded or when they migrate into scaffolds employed for tissue engineering in that they may be responsible for distorting the matrix and collapsing the pores.

REFERENCES

1. Green, W. T., Jr. (1971). Behavior of articular chondrocytes in cell culture. *Clin. Orthop. Relat. Res.* 75, 248–260.
2. Manning, W. K., and Bonner, W. M., Jr. (1967). Isolation and culture of chondrocytes from human adult articular cartilage. *Arthritis Rheum.* 10, 235–239.
3. Jay, G. D., Britt D. E., and Cha, C. J. (2000). Lubricin is a product of megakaryocyte stimulating factor gene expression by human synovial fibroblasts. *J. Rheumatol.* 27, 594–600.
4. Flannery, C. R., Hughes, C. E., Schumacher, B. L., Tudor, D., Aydelotte, M. B., Kuettner K. E., and Caterson, B. (1999). Articular cartilage superficial zone protein (SZP) is homologous to megakaryocyte stimulating factor precursor and Is a multifunctional proteoglycan with potential growth-promoting, cytoprotective, and lubricating properties in cartilage metabolism. *Biochem. Biophys. Res. Commun.* 254, 535–541.
5. Hills, B. A. (2000). Boundary lubrication in vivo. *Proc.—Inst. Mech. Eng. H* 214, 83–94.
6. Kim, A. C., and Spector, M. (2001). Distribution of chondrocytes containing alpha–smooth muscle actin in human articular cartilage. *J. Orthop. Res.* 18, 749–755.
7. Kinner, B., and Spector, M. (2001). Smooth muscle actin expression by human articular chondrocytes and their contraction of a collagen-glycosaminoglycan matrix in vitro. *J. Orthop. Res.* 19, 233–241.
8. Holzer, J., Abbott, J., Lash, J., and Holtzer, A. (1960). The loss of phenotypic traits by differentiated cells in vitro. I. Dedifferentation of cartilage cells. *Proc. Natl. Acad. Sci. U.S.A.* 46, 1533–1542.
9. Norby, D. P., Malemud, C. J., and Sokoloff, L. (1977). Differences in the collagen types synthesized by lapine articular chondrocytes in spinner and monolayer culture. *Arthritis Rheum.* 20, 709–716.
10. Benya, P. D., and Shaffer, J. D. (1982). Dedifferentiated chondrocytes reexpress the differentiated collagen phenotype when cultured in agarose gels. *Cell (Cambridge, Mass.)* 30, 215–224.
11. Srivastava, V. M., Malemud, C. J., Hough, A. J., Bland, J. H., and Sokoloff, L. (1974). Preliminary experience with cell culture of human articular chondrocytes. *Arthritis Rheum.* 17, 165–169.
12. Peterson, L., Minas, T., Brittberg, M., Nilsson, A., Sjögren-Jansson, E., and Lindahl, A. (2000). Two- to 9-year outcome after autologous chondrocyte transplantation of the knee. *Clin. Orthop. Relat. Res.* 374, 212–234.
13. Mayhew, T. A, Williams, G. R., Senica, M. A., Kuniholm, G., and Du Moulin, G. C. (1998). Validation of a quality assurance program for autologous cultured chondrocyte implantation. *Tissue Eng.* 4, 325–334.
14. Saadeh, P. B., Brent, B., Mehrara, B. J., Steinbrech, D. S., Ting, V., Gittes, G. K., and Longaker, M. T. (1999). Human cartilage engineering: Chondrocyte extraction, proliferation, and characterization for construct development. *Ann. Plast. Surg.* 42, 509–513.

15. Guerne, P. A., Blanco, F., Kaelin, A., Desgeorges, A., and Lotz, M. (1995). Growth factor responsiveness of human articular chondrocytes in aging and development. *Arthritis Rheum.* **38**, 960–968.

16. Dounchis, J. S., Goomer, R. S., Harwood, F. L., Khatod, M., Coutts, R. D., and Amiel, D. (1997). Chondrogenic phenotype of perichondrium-derived chondroprogenitor cells is influenced by transforming growth factor-beta 1. *J. Orthop. Res.* **15**, 803–807.

17. Tajima, K., Yamakawa, M., Katagiri, T., and Sasaki, H. (1997). Immunohistochemical detection of bone morphogenetic protein-2 and transforming growth factor beta-1 in tracheopathia osteochondroplastica. *Virchows Arch.* **431**, 359–363.

18. Johnstone, B., Hering, T. M., Caplan, A. I., Goldberg, V. M., and Yoo, J. U. (1998). In vitro chondrogenesis of bone marrow-derived mesenchymal progenitor cells. *Exp. Cell Res.* **238**, 265–272.

19. Martin, I., Vunjak-Novakovic, G., Yang, J., Langer, R., and Freed, L. E. (1999). Mammalian chondrocytes expanded in the presence of fibroblast growth factor 2 maintain the ability to differentiate and regenerate three-dimensional cartilaginous tissue. *Exp. Cell Res.* **253**, 681–688.

20. Dunham, B. P., and Koch, R. J. (1998). Basic fibroblast growth factor and insulinlike growth factor I support the growth of human septal chondrocytes in a serum-free environment. *Arch. Otolaryngol. Head Neck Surg.* **124**, 1325–1330.

21. Fujisato, T., Sajiki, T., Liu, Q., and Ikada, Y. (1996). Effect of basic fibroblast growth factor on cartilage regeneration in chondrocyte-seeded collagen sponge scaffold. *Biomaterials* **17**, 155–162.

22. Loeser, R. F., and Shanker, G. (2000). Autocrine stimulation by insulin-like growth factor 1 and insulin-like growth factor 2 mediates chondrocyte survival in vitro. *Arthritis Rheum.* **43**, 1552–1559.

23. Chopra, R., and Anastassiades, T. (1998). Specificity and synergism of polypeptide growth factors in stimulating the synthesis of proteoglycans and a novel high molecular weight anionic glycoprotein by articular chondrocyte cultures. *J. Rheumatol.* **25**, 1578–1584.

24. Kieswetter, K., Schwartz, Z., Alderete, M., Dean, D. D., and Boyan, B. D. (1997). Platelet derived growth factor stimulates chondrocyte proliferation but prevents endochondral maturation. *Endocrine* **6**, 257–264.

25. Lohmann, C. H., Schwartz, Z., Niederauer, G. G., Carnes, D. L., Jr., Dean, D. D., and Boyan, B. D. (2000). Pretreatment with platelet derived growth factor-BB modulates the ability of costochondral resting zone chondrocytes incorporated into PLA/PGA scaffolds to form new cartilage in vivo. *Biomaterials* **21**, 49–61.

26. Klein-Nulend, J., Louwerse, R. T., Heyligers, I. C., Wuisman, P. I., Semeins, C. M., Goei, S. W., and Burger. E. H. (1998). Osteogenic protein (OP-1, BMP-7) stimulates cartilage differentiation of human and goat perichondrium tissue in vitro. *J. Biomed. Mater. Res.* **40**, 614–620.

27. Huch, K., Wilbrink, B., Flechtenmacher, J., Koepp, H. E., Aydelotte, M. B., Sampath, T. K., Kuettner, K. E., Mollenhauer, J., and Thonar, E. J. (1997). Effects of recombinant human osteogenic protein 1 on the production of proteoglycan, prostaglandin E2, and interleukin-1 receptor antagonist by human articular chondrocytes cultured in the presence of interleukin-1beta. *Arthritis Rheum.* **40**, 2157–2161.

28. Sellers, R. S., Peluso, D., and Morris, E. A. (1997). The effect of recombinant human bone morphogenetic protein-2 (rhBMP-2) on the healing of full-thickness defects of articular cartilage. *J. Bone J. Surg., Am. Vol.* **79A**, 1452–1463.

29. Sellers, R. S., Zhang, R., Glasson, S. S., Kim, H. D., Peluso, D., D'Augusta, D. A., Beckwith, K., and Morris, E. A. (2000). Repair of articular cartilage defects one year after treatment with recombinant human bone morphogenetic protein-2 (rhBMP-2). *J. Bone J. Surg., (Am. Vol.)* **82A**, 151–160.

30. Sandell, L. J., and Daniel, J. C. (1988). Effects of ascorbic acide on collagen mnRNA levels in short term chondrocyte cultures. *Connect. Tissue Res.* **17**, 11–22.

31. van Osch, G. J., van der Veen, S. W., Buma, P., and Verwoerd-Verhoef, H. L. (1998). Effect of transforming growth factor-beta on proteoglycan synthesis by chondrocytes in relation to differentiation stage and the presence of pericellular matrix. *Matrix Biol* **17**, 413–424.

32. Redini, F., Galera, P., Mauviel, A., Loyau, G., and Pujol, J. P. (1988). Transforming growth factor beta stimulates collagen and glycosaminoglycan biosynthesis in cultured rabbit articular chondrocytes. *FEBS Lett.* **234**, 172–176.

33. Livne, E. (1994). In vitro response of articular cartilage from mature mice to human transforming growth factor beta. *Acta Anat.* **149**, 185–194.

34. Bujia, J., Pitzke, P., Kastenbauer, E., Wilmes, E., and Hammer, C. (1996). Effect of growth factors on matrix synthesis by human nasal chondrocytes cultured in monolayer and in agar. *Eur. Arch. Otorhinolaryngol.* **253**, 336–340.

35. Mattey, D. L., Dawes, P. T., Nixon, N. B., and Slater, H. (1997). Transforming growth factor beta 1 and interleukin 4 induced alpha smooth muscle actin expression and myofibroblast-like differentiation in human synovial fibroblasts in vitro: Modulation by basic fibroblast growth factor. *Ann. Rheum. Dis.* **56**, 426–431.

36. Trippel, S. B. (1995). Growth factor actions on articular cartilage. *J. Rheumatol., Suppl.* **43**, 129–132.

37. Webber, R. J., Harris, M. G., and Hough, A. J., Jr. (1985). Cell culture of rabbit meniscal fibrochondrocytes: Proliferative and synthetic response to growth factors and ascorbate. *J. Orthop. Res.* **3**, 36–42.

38. Spindler, K. P., Mayes, C. E., Miller, R. R., Imro, A. K., and Davidson, J. M. (1995). Regional mitogenic response of the meniscus to platelet-derived growth factor (PDGF-AB). *J. Orthop. Res.* **13**, 201–207.

39. Webber, R. J. (1990). In vitro culture of meniscal tissue. *Clin. Orthop. Relat. Res.* **252**, 114–120.

40. Guilbert, L. J., and Iscove, N. N. (1976). Partial replacement of serum by selenite, transferrin, albumin and lecithin in haemopoietic cell cultures. *Nature (London)* **263**, 594–595.

41. McKeehan, W. L., and Ham, R. G. (1976). Methods for reducing the serum requirement for growth in vitro of non-transformed diploid fibroblasts. *Dev. Biol. Stand.* **37**, 97–98.

42. McKeehan, W. L., and Ham, R. G. (1976). Stimulation of clonal growth of normal fibroblasts with substrata coated with basic polymers. *J. Cell Biol.* **71**, 727–734.

43. McKeehan, W. L., Hamilton, W. G., and Ham, R. G. (1976). Selenium is an essential trace nutrient for growth of WI-38 diploid human fibroblasts. *Proc. Natl. Acad. Sci. U.S.A.* **73**, 2023–2027.

44. Sawyer, L. M., and Goetinck, P. F. (1981). Chondrogenesis in the mutant nanomelia. Changes in the fine structure and proteoglycan synthesis in high density limb bud cell cultures. *J. Exp. Zool.* **216**, 121–131.

45. Gay, S. W., and Kosher, R. A. (1984). Uniform cartilage differentiation in micromass cultures prepared from a relatively homogeneous population of chondrogenic progenitor cells of the chick limb bud: effect of prostaglandins. *J. Exp. Zool.* **232**, 317–326.

46. Kosher, R. A., Gay, S. W., Kamanitz, J. R., Kulyk, W. M., Rodgers, B. J., Sai, S., Tanaka, T., and Tanzer, M. L. (1986). Cartilage proteoglycan core protein gene expression during limb cartilage differentiation. *Dev. Biol.* **118**, 112–117.

47. Kulyk, W. M., Rodgers, B. J., Greer, K., and Kosher, R. A. (1989). Promotion of embryonic chick limb cartilage differentiation by transforming growth factor-beta. *Dev. Biol.* **135**, 424–430.

48. Kulyk, W. M., and Reichert, C. (1992). Staurosporine, a protein kinase inhibitor, stimulates cartilage differentiation by embryonic facial mesenchyme. *J. Craniofacial Genet. Dev. Biol.* **12**, 90–97.

49. DeLise, A. M., Stringa, E., Woodward, W. A., Mello, M. A., and Tuan, R. S. (2000). Embryonic limb mesenchyme micromass culture as an in vitro model for chondrogenesis and cartilage maturation. *Methods Mol. Biol.* **137**, 359–375.

50. Kameda, T., Koike, C., Saitoh, K., Kuroiwa, A., and Iba, H. (2000). Analysis of cartilage maturation using micromass cultures of primary chondrocytes. *Dev. Growth Differ.* **42**, 229–236.

51. Nicoll, S. B., Wedrychowska, A., Smith, N., and Bhatnager, R. S. (1998). A new approach to cartilage tissue engineering using human dermal fibroblasts seeded on three-dimmensional polymer scaffolds. *Mater. Res. Symp. Proc. Soc.* **530**.

52. Solursh, M. (1991). Formation of cartilage tissue in vitro. *J. Cell. Biochem.* **45**, 258–260.

53. Chen, Q., Johnson, D. M., Haudenschild, D. R., and Goetinck, P. F. (1995). Progression and recapitulation of the chondrocyte differentiation program: Cartilage matrix protein is a marker for cartilage maturation. *Dev. Biol.* **172**, 293–306.

54. Lennon, D. P., Haynesworth, S. E., Arm, D. M., Baber, M. A., and Caplan, A. I. (2000). Dilution of human mesenchymal stem cells with dermal fibroblasts and the effects on in vitro and in vivo osteochondrogenesis. *Dev. Dyn.* **219**, 50–62.

55. Yasui, N., Osawa, S., Ochi, T., Nakashima, H., and Ono, K. (1982). Primary culture of chondrocytes embedded in collagen gels. *Exp. Cell Biol.* **50**, 92–100.

56. Kimura, T., Yasui, N., Ohsawa, S., and Ono, K. (1984). Chondrocytes embedded in collagen gels maintain cartilage phenotype during long-term cultures. *Clin. Orthop. Relat. Res.* **186**, 231–239.

57. Kawamura, S., Wakitani, S., Kimura, T., Maeda, A., Caplan, A. I., Shino, K., and Ochi, T. (1998). Articular cartilage repair. Rabbit experiments with a collagen gel-biomatrix and chondrocytes cultured in it. *Acta Orthop. Scand.* **69**, 56–62.

58. Rotter, N., Aigner, J., Naumann, A., Planck, H., Hammer, C., Burmester, G., and Sittinger, M. (1998). Cartilage reconstruction in head and neck surgery: Comparison of resorbable polymer scaffolds for tissue engineering of human septal cartilage. *J. Biomed. Mater. Res.* **42**, 347–356.

59. Sittinger, M., Schultz, O., Keyszer, G., Minuth, W. W., and Burmester, G. R. (1997). Artificial tissues in perfusion culture. *Int. J. Artif. Organs* **20**, 57–62.

60. Freed, L. E., Vunjak-Novakovic, G., and Langer, R. (1993). Cultivation of cell–polymer cartilage implants in bioreactors. *J. Cell. Biochem.* **51**, 257–264.

61. Freed, L. E., Langer, R., Martin, I., Pellis, N. R., and Vunjak-Novakovic, G. (1997). Tissue engineering of cartilage in space. *Proc. Natl. Acad. Sci. U.S.A.* **94**, 13885–13890.

62. Freed, L. E., Hollander, A. P., Martin, I., Barry, J. R., Langer, R., and Vunjak-Novakovic, G. (1998). Chondrogenesis in a cell-polymer-bioreactor-system. *Exp. Cell Res.* **10**, 58–65.

63. Obradovic, B., Carrier, R. L., Vunjak-Novakovic, G., and Freed, L. E. (1999). Gas exchange is essential for bioreactor cultivation of tissue engineered cartilage. *Biotechnol. Bioeng.* **63**, 197–205.

64. Potter, K., Butler, J. J., Horton, W. E., and Spencer, R. G. (2000). Response of engineered cartilage tissue to biochemical agents as studied by proton magnetic resonance microscopy. *Arthritis Rheum.* **43** 1580–1590.

65. Vunjak-Novakovic, G., Martin, I., Obradovic, B., Treppo, S., Grodzinsky, A. J., Langer, R., and Freed, L. E. (1999). Bioreactor cultivation conditions modulate the composition and mechanical properties of tissue-engineered cartilage. *J. Orthop. Res.* **17**, 130–138.

66. Wu, F., Dunkelman, N., Peterson, A., Davisson, T., De La Torre, R., and Jain, D. (1999). Bioreactor development for tissue-engineered cartilage. *Ann. N.Y. Acad. Sci.* **875**, 405–411.

67. Bonaventure, J., Kadhom, N., Cohen-Solal, L., Ng, K. H., Bourguignon, J., Lasselin, C., and Freisinger, P. (1994). Reexpression of cartilage-specific genes by dedifferentiated human articular chondrocytes cultured in alginate beads. *Exp. Cell Res.* **212**, 97–104.

68. Grande, D. A., Halberstadt, C., Naughton, G., Schwartz, R., and Manji, R. (1997). Evaluation of matrix scaffolds for tissue engeneering of articular cartilage grafts. *J. Biomed. Mater. Res.* **34**, 211–220.

69. Freed, L. E., Grande, D. A., Lingbin, Z., Emmanuel, J., Marquis, J. C., and Langer, R. (1994). Joint resurfacing using allograft chondrocytes and synthetic biodegradable polymer scaffolds. *J. Biomed. Mater. Res.* **21**, 891–899.

70. Freed, L. E., Vunjak-Novakovic, G., Biron, R. J., Eagels, D. B., Lesnoy, D. C., Barlow, S. K., and Langer, R. (1994). Biodegradable polymer scaffolds for tissue engeneering. *Bio/Technol.* **12**, 689–693.

71. Nehrer, S., Breinan, H. A., Ramappa, A., Hsu, H.-P., Minas, T., Shortkroff, S., Sledge, C. B., Yannas, C. B., and Spector, M. (1998). Chondrocyte-seeded collagen matices implanted in a chondral defect in a canine model. *Biomaterials* **19**, 2313–2328.

72. Nehrer, S., Breinan, H. A., Ramappa, A., Young, R. G., Shortkroff, S., Louie, L. B., Sledge, C. B., Yannas, C. B., and Spector, M. (1997). Matrix collagen type and pore size influence behavior of seeded canine chondrocytes. *Biomaterials* **18**, 769–776.

73. Delaney, J. P., O'Driscoll, S. W., and Salter, R. B. (1989). Neochondrogenesis in free intraarticular periosteal autografts in an immobilized and paralyzed limb. An experimental investigation in the rabbit. *Clin. Orthop. Relat. Res.* **248**, 278–282.

74. O'Driscoll, S. W., and Salter, R. B. (1986). The repair of major osteochondral defects in joint surfaces by neochondrogenesis with autogenous osteoperiosteal grafts stimulated by continuous passive motion. An experimental investigation in the rabbit. *Clin. Orthop. Relat. Res.* **208**, 131–140.

75. Athanasiou, K. A., Agrawal, C. M., Barber, F. A., and Burkhart, S. S. (1998). Orthopaedic applications for PLA-PGA biodegradable polymers. *Arthroscopy* **14**, 726–737.

76. Ponticiello, M. S., Schinagl, R. M., Kadiyala, S., and Barry, F. P. (2000). Gelatin-based resorbable sponge as a carrier matrix for human mesenchymal stem cells in cartilage regeneration therapy. *J. Biomed. Mater. Res.* **52**, 246–255.

77. Mizuno, S., and Glowacki, J. (1996). Chondroinduction of human dermal fibroblasts by demineralized bone in three-dimensional culture. *Exp. Cell Res.* **227**, 89–97.

78. Freed, L. E., Marquis, J. C., Nohria, A., Emmanual, J., Mikos, A. G., and Langer, R. (1993). Neocartilage formation in vitro and in vivo using cells cultured on synthetic biodegradable polymers. *J. Biomed. Mater. Res.* **27**, 11–23.

79. Freed, L. E., Martin, I., and Vunjak-Novakovic, G. (1999). Frontiers in tissue engineering. In vitro modulation of chondrogenesis. *Clin. Orthop. Relat. Res., Suppl.* **367**, S46–S58.

80. Burg, K. J., Holder, W. D., Jr., Culberson, C. R., Beiler, R. J., Greene, K. G., Loebsack, A. B., Roland, W. D., Eiselt, P., Mooney, D. J., and Halberstadt, C. R. (2000). Comparative study of seeding methods for three-dimensional polymeric scaffolds. *J. Biomed. Mater. Res.* **51**, 642–649.

81. Kim, B. S., Putnam, A. J., Kulik, T. J., and Mooney, D. J. (1998). Optimizing seeding and culture methods to engineer smooth muscle tissue on biodegradable polymer matrices. *Biotechnol. Bioeng.* **57**, 46–54.

82. Gionti, E., Pontarelli, G., and Cancedda, R. (1985). Avian myelocytomatosis virus immortalizes differentiated quail chondrocytes. *Proc. Natl. Acad. Sci. U.S.A.* **82**(9), 2756–2760.

83. Horton, W. E., Jr., *et al.* (1988). An estabished rat cell line expressing chondrocyte properties. *Exp. Cell Res.* **178**(2), 457–468.

84. Mukhopadhyay, K., *et al.* (1995). Use of a new rat chondrosarcoma cell line to delineate a 119-base pair chondrocyte-specific enhancer element and to define active promoter segments in the mouse pro-alpha I(II) collagen gene. *J. Biol. Chem.* **270**(46), 27711–27719.

85. Takigawa, M., *et al.* (1991). Establishment from a human chondrosarcoma of a new immortal cell line with high tumorigenicity in vivo, which is able to form proteoglycan-rich cartilage-like nodules and to respond to insulin in vitro. *Int. J. Cancer.* **48**(5), 717–725.

86. Nishida, T., *et al.* (1998). Demonstration of receptors specific for connective tissue growth factor on a human chondrocytic cell line (HCS-2/8). *Biochem. Biophys. Res. Commun.* **247**(3) 905–909.

87. Block, J. A., *et al.* (1991). Synthesis of chondrocytic keratan suphate-containing proteoglycans by human chondrosarcoma cells in long-term cell culture. *J. Bone J. Surg. Am. Vol.* **73**(5), 647–658; erratum: *Ibid.* **73A**(8), 1274.

88. Kunisada, T., *et al.* (1998). A new human chondrosarcoma cell line (OUMS-27) that maintains chondrocytic differentiation. *Int. J. Cancer* **77**(6), 854–859.

89. Chano, T., *et al.* (1998). Characterization of a newly established human chondrosarcoma cell line, CS-OKB. *Virchows Arch.* **432**(6), 529–534.

90. Bernier, S. M. and Goltzman, D. (1993). Regulation of expression of the chondrocytic phenotype in a skeletal cell line (CFK2) in vitro. *J. Bone Miner. Res.* **8**(4), 475–484.

91. Grigoriadis, A. E., Heersche, J. N., and Aubin, J. E. (1988). Differentiation of muscle, fat, cartilage, and bone from progenitor cells present in a bone-derived clonal cell population: Effect of dexamethasone. *J. Cell Biol.* **106**(6), 2139–2151.

92. Bergwitz, C., *et al.* (2000). A versatile chondrogenic rat calvaria cell line R-tTA-24 that permits tetracycline-regulated gene expression. *Histochem. Cell Biol.* **113**(2), 145–150.

93. Terkeltaub, R. A., *et al.* (1998). Bone morphogenetic proteins and bFGF exert opposing regulatory effects on PTHrP expression and inorganic pyrophosphate elaboration in immortalized murine endochondral hypertrophic chondrocytes (MCT cells). *J. Bone Miner. Res.* **13**(6), 931–941.

94. Thenet, S., *et al.* (1992). SV40-immortalization of rabbit articular chondrocytes: Alteration of differentiated functions. *J. Cell Physiol.* **150**(1), 158–167.

95. Goldring, M. B. *et al.* (1994). Interleukin-1 beta-modulated gene expression in immortalized human chondrocytes. *J. Clin. Invest.* **94**(6), 2307–2316.

96. Lefebvre, V., Garofalo, S., and de Crombrugghe, B. (1995). Type X collagen gene expression in mouse chondrocytes immortalized by a temperature-sensitive simian virus 40 large tumor antigen. *J. Cell Biol.* **128**(1-2), 239–245.

97. Mataga, N., *et al.* (1996). Establishment of a novel chondrocyte-like cell line derived from transgenic mice harboring the temperature-sensitive simian virus 40 large T-antigen gene. *J. Bone Miner. Res.* **11**(11), 1646–1654.

98. Moulton, P. J., *et al.* (1997). Detection of protein and mRNA of various components of the NADPH oxidase complex in an immortalized human chondrocyte line. *Br. J. Rheumatol.* **36**(5), 522–529.

99. Lee, C. R., Breinan, H. A., Nehrer, S., and Spector, M. (2000). Articular cartilage chondrocytes in type I and type II collagen-GAG matrices exhibit contractile behavior in vitro. *Tissue Eng.* **6**, 555–565.

100. Majno, G., Gabbiani, G., Hirschel, B. J., Ryan, G. B., and Statkov, P. R. (1971). Contraction of granulation tissue in vitro: Similarity to smooth muscle. *Science* **173**, 548–550.

101. Gabbiani, G. (1992). The biology of the myofibroblast. *Kidney Int.* **41**, 530–532.

102. Janmey, P. A., and Chaponnier, C. (1995). Medical aspects of the actin cytoskeleton. *Curr. Opin. Cell Biol.* **7**, 111–117.

103. Wang, Q., Breinan, H. A., Hsu, H. P., and Spector, M. (2000). Healing of defects in canine articular cartilage: distribution of nonvascular alpha-smooth muscle actin-containing cells. *Wound Repair Regeneration* **8**, 145–158.

104. Mueller, S. M., Schneider, T. O., Shortkroff, S., Breinan, H. A., and Spector, M. (1999). Alpha-smooth muscle actin and contractile behavior of bovine meniscus cells seeded in type I and type II collagen-GAG matrices. *J. Biomed. Mater. Res.* **45**, 157–166.

105. Schneider, T. O., Mueller, S. M., Shortkroff, S., and Spector, M. (1999). Expression of alpha-smooth muscle actin in canine intervertebral disc cells in situ and in collagen-glycosaminoglycan matrices in vitro. *J. Orthop. Res.* **17**, 192–199.

106. Martin, I., Jakob, M., Schaefer, D., Heberer, M., Vunjak-Novakovic, G., and Freed, L. E. (2000). Cartilaginous tissues generated in vitro from serially passaged calf, bovine and adult human articular chondrocytes. *Proc. 46th Annu. Meit. Orthop. Res. Soc.*, Orlando, FL, p. 618.

107. Hunter, C. J., Shieh, A. C., Nerem, R. M., and Levenston, M. E. (2000). Effects of collagens I and II and chitosan on chondrocyte behavior in fibrin gel cultures. *Proc. 6th World Biomater. Congr.*, Kamuela, HI.

CHAPTER 26

MESENCHYMAL CELL CULTURE: BONE

John E. Davies, J. M. Karp, and D. Baksh

INTRODUCTION

Bone engineering, or iatrogenic bone growth, has been practiced for centuries in the repair of fractures and, in modern medicine, for at least several decades by the implantation of dental and orthopedic devices and bone-substitute materials. Bone tissue engineering should therefore be considered to be a more recent, and novel, approach to this long-established field, where the driving force is the requirement to restore, or augment, bone stock.

Bone serves a multitude of functions in the body, including locomotor support, calcium homeostasis, and the provision of a sequestered environment for hematopoiesis, and can thus be considered to be both an organ and a tissue. Weinmann and Sicher [1] simply, but eloquently, provided the essential distinction between these classifications in the phrase "bone is a tissue, bones are organs." Bone tissue engineering has recently been extensively reviewed [2–5], and although it is quite conceivable that tissue engineering will, in the future, be employed to regenerate both bone and bones, we shall restrict ourselves herein to engineering bone tissue.

The matrix of bone is uniquely elaborated by populations of osteogenic cells derived from mesenchymal, or stromal, stem cells that differentiate into fully secretorily active mature osteoblasts. With the advent of growth factor purification, protein engineering and, even more recently the emergence of gene therapy, these essentially biochemical and molecular biological approaches to bone tissue engineering are now feasible [6–10]. Such biological approaches to bone tissue engineering do not always require cell culture or the seeding and growth of osteogenic cells on three-dimensional scaffolds. However, in this chapter, we focus on scaffold-driven bone tissue engineering strategies, which are dependent upon an *in vitro* stage of autologous cell culture, to the exclusion of those involving growth factor and gene therapy. These scaffolds, ideally, should biodegrade and may be of either inorganic or polymer composition (or composites of these two), many of which have been recently reviewed in detail elsewhere [11,12].

Several animal studies have shown that it is possible to culture osteogenic cells on three-dimensional scaffolds and achieve bone formation. But, such studies have raised several issues that demand careful attention when one is designing a bone tissue engineering strategy. For example, some calcium phosphate scaffolds demonstrate very limited ability to degrade *in vivo* [13], while others degrade too rapidly in the presence of explanted marrow [14]. Both inorganic [15] and organic [16] scaffolds have been employed whose geometries preclude invasion by bone cells and tissue. Nevertheless, some inorganic and organic biodegradable scaffolds have been shown to permit tissue invasion [17] and to result in bony union of critical-sized defects and restoration of function in ruminants [14] and lagomorphs [18], following seeding with autologous marrow-derived cells.

Human osteochondral repair has also been reported in popular press accounts of human cells seeded in an inorganic scaffold of very limited degradation [19]; human cells seeded in similar scaffolds have been shown to elaborate histologically identifiable tissue following implantation in the nude mouse [20,21]. However, reliance on immunologically incompetent animals for such studies should be treated with caution because anecdotal evidence is emerging that suggests that some cells, which demonstrate an ability to produce bone in a nude mouse, do not produce bone in the autologous species.

The purpose of this chapter is to discuss methods of facilitating the elaboration of bone tissue in culture by osteogenic cells, distinguishing between bone and other forms of biologically driven mineralization, and describing the culture conditions conducive to *in vitro* bone matrix elaboration in three-dimensional tissue engineering scaffolds.

THE OSTEOGENIC CELL SOURCE

Like other matrix-producing connective tissue cells, bone cells are derived from so-called mesenchymal stem cells (MSC). While there is considerable debate concerning the definition of stem cells [22,23], which is not the focus of this chapter, the crucial issue is that the proliferative capacity of these undifferentiated cells provides the opportunity to create large autologous cell populations by expansion, *in vitro*, of a small number of cells harvested from the host. By entering discrete differentiation pathways, MSCs can differentiate into a number of connective tissue phenotypes that include bone, cartilage, tendon/ligament, and muscle [24–28]; thus the culture microenvironment is crucial in dictating the cell differentiation pathway. It is salutary to point out that although a multitude of cell lines are also employed to study specific aspects of connective tissue (including bone) cell behavior, we believe that realistic tissue engineering strategies, currently, can be based only on primary cells, and thus cell lines and tumor-derived cells are not addressed herein.

Essentially, two strategies can be employed to achieve MSC-mediated tissue repair. First, MSC populations are expanded, in culture, in an assumed undifferentiated state. Following introduction into the host, and in the presence of local environmental cues *in vivo*, the MSCs will then differentiate into osteogenic cells, which will be responsible for the tissue regeneration process. Second, culture-expanded MSCs are directed *ex vivo* into the osteogenic lineage prior to implantation, to accelerate the regeneration of bone. In fact, it has been shown that the cells responsible for bone nodule formation arise from single colonies [29,30]. Thus both the expansion phase of each strategy and reliance on the proliferative capacity of these cells are of crucial importance to methods in tissue engineering: it has been estimated that less than 1% of the stromal layer of rat and mouse marrow cells is osteogenic [31] and, in humans, the available stem cell pool in the total marrow cell population decreases from 1:10,000 in newborns to $1:2 \times 10^6$ at the age of 80 [32]. To avoid the necessity of marrow biopsy to harvest MSC, attempts have been made to derive osteogenic cells from peripheral blood progenitor cells but, to date, these have been unsuccessful [33].

The common challenge in the two approaches to MSC-mediated tissue repair is to demonstrate the capacity of the expanded cell population to produce bone tissue. In the first, matrix synthesis would occur only *in vivo*; but for human applications, proof of principle has to be demonstrated either *in vitro* or by the employment of an immunologically incompetent animal model. Similarly, in the second strategy one has to demonstrate effective differentiation along the chosen pathway, in this case the osteogenic pathway. Here we are, again, faced with two alternatives. While bone is a connective tissue—that is, an extracellular matrix made by, and containing, the cells responsible for its elaboration—it can be argued that differentiation along the osteogenic pathway *in vitro* can be reliably demonstrated only by the elaboration of bone matrix in culture. Alternatively, since differentiation is marked by the intracellular expression of various matrix precursor molecules, it can be argued that demonstration of intracellular markers of osteogenic differentiation, of which alkaline phosphatase, osteocalcin, and collagen are by far the most commonly employed, is sufficient to demonstrate the efficacy of this approach. Both these arguments are valid. However, some authors rely on intracellular markers of osteogenic differentiation to the exclusion of any evidence for matrix elaboration. This is particularly troublesome in view of the evidence that cells expressing markers of osteogenic differentiation can transdifferentiate into other connective tissue lineages [34,35]. Thus, considerable effort is directed toward demonstrating the capacity of MSCs to differentiate into specific connective tissue cell types and elaborate matrix *in vitro* [36].

BONE MARROW CULTURE

Bone cultures originally made use of organ cultures or cells enzymatically extracted from bone or osteoid, while marrow was employed as a source of nonadherent hematopoietic cells. Indeed, the adherence of marrow stroma to the culture flask was initially used only to provide a necessary microenvironment for the growth of the hematopoietic stem cells and their differentiated phenotypes [37–39]. More recently, this adherence of the marrow stroma was harnessed as a means of isolating stromal from other marrow cells [24,25, 27,40].

Friedenstein [41] was the first to show, by means of diffusion chamber experiments, that bone marrow contains cells capable of proliferation and bone formation. In fact, it is generally accepted that there are at least two populations of osteoprogenitor cells. The first is capable of constitutive differentiation *in vitro*, while the other uncommitted but inducible osteoprogenitor population undergoes osteoblastic differentiation only in the presence of added differentiation stimuli [42].

Maniatopoulos *et al.* [43] used dexamethasone-stimulated bone marrow cultures to morphologically and biochemically characterize bone elaborated *in vitro*. In fact, such cultures can be employed to study the complete cascade of bone formation events from cement line formation to the mineralization of osteoid [44,45], and also to mimic bone remodeling by monitoring osteoclastic resorption of the formed extracellular matrix [44]. The use of stromal bone marrow cultures to obtain bone cell cultures started in 1988 with the rat protocol developed by Maniatopoulos *et al.*, and current bone marrow culture protocols are still based on this original protocol [19]. We focus herein on this widely accepted bone marrow culture protocol and the modifications that are currently used in our laboratory.

CELL CULTURE MEDIUM

The cell culture medium from the original protocol by Maniatopoulos *et al.* [43] is based on the alpha modification of Minimal Essential Medium (α-MEM), supplemented with antibiotics, antifungal agents, and 15% heat-inactivated fetal bovine serum (FBS). The addition of FBS renders the medium nondefined and can impose considerable variability in culture characteristics depending on the batches or lot numbers provided by various suppliers [46]. Therefore, the practice of obtaining a sufficiently large stock of FBS to avoid interlot variation and enhance culture reproducibility is highly recommended. It is worth noting that the majority of commercial suppliers of serum will provide small samples for batch testing and reserve batches from which the customer can chose the most appropriate. Before purchasing a serum batch, we routinely batch-test at least six sera by conducting comparative bone nodule assays.

Proliferation Media

As mentioned earlier, the challenge of cell-seeded, scaffold-based, tissue engineering strategies is to expediently supply large numbers of appropriate progenitors to the repair and/or regeneration site. All approaches to date involve culturing MSCs from whole bone marrow. The initial cell numbers adhering to the culture substrate are low, [47,48], and these are passaged to achieve expansion [24,25,27,49,50]. Some authors have used what is referred to in the literature as Complete Medium [27,36,51,52], which comprises low-glucose Dulbecco's MEM (DMEM-LG) with 10% fetal bovine serum; while others have used α-MEM as a source of essential nutrients and amino acids [21,40]. Serum lots, in these cases, are selected based on their ability to facilitate cell proliferation but minimize differentiation. Others have studied the effects of adding various growth factors either alone or in combination to Complete Medium, including epidermal growth factor (EFG), platelet-derived growth factor (PDGF), and transforming growth factor β1 (TGF-β1), which have been shown to be potent mitogens for proliferation [21,53,54]. Morphologically, the adherent stromal cell population in expansion cultures appears to be fibroblastic. However, cultures derived from various murine strains also comprise up to 80% of other cell types including monocytes, macrophages, and endothelial cells, which are contaminants and persist for several weeks [40]. Consequently, care should be exercised before proposing such heterogeneous cell populations for modeling tissue engineering therapeutic procedures. Primary human bone marrow cultures also contain CD14- and CD45-positive cells (markers for macrophages and lymphohematopoietic cells, respectively), but such cell types comprise only approximately 5% of the total population. By second passage these round cells have been reduced in frequency (1–2% of total cells) owing to the lack of adherence after passaging [52].

Standard Medium for Osteogenic Cultures

The standard culture medium comprises α-MEM 85% (v/v), FBS 15% (v/v) and the following antibiotic–fungizone solution: penicillin G (167 U/ml); gentamicin (50 μg/ml), and amphotericin B (0.3 μg/ml). Usually the antibiotic–fungizone solution is prepared as a 10× stock solution in α-MEM, filtered at a pore size of 0.1 μm and stored at 4°C for up to 2 weeks. A useful variation is the addition of the fluorescent antibiotic tetracycline (9 μg/ml), which labels the elaborating bone matrix in culture. It is necessary to add this only at the last culture refeed and, on examination under ultraviolet light, the fluorescent signal may be used as a quantitative measure of bone growth during the culture period [55,56].

Medium Supplements

Additional supplements are added to the standard culture medium to facilitate bone formation *in vitro*. In the original protocol these supplements are dexamethasone (Dex) (10^{-8} M), ascorbic acid (AA) (50 μg/ml), and β-glycerophosphate (β-GP) (10 mM) [35]. However, variations in the concentrations of these additions have emerged and will be discussed separately.

Dexamethasone

Dexamethasone is a synthetic glucocorticoid and stimulates osteogenic progenitor cell differentiation. This concentration-dependent [57,58] stimulation (range from 10^{-9} M to 10^{-7} M) is required for the formation of bone tissue *in vitro* by marrow stromal populations [59]. *In vivo*, however, dexamethasone results in decreased osteogenesis. This discrepancy has been explained by a stimulatory effect on uncommitted osteoprogenitors [42] that is accompanied by an inhibitory effect on more undifferentiated osteogenic cells, leading to an initial burst of osteogenesis *in vivo* but a subsequent depletion of the osteogenic population and bone loss [60]. Dex has been employed in many mammalian cultures including mouse [61], rabbit [18], ferret [62], and human [63].

β-Glycerophosphate (β-GP)

β-GP is added to the culture medium as a source of organic phosphate to facilitate the matrix mineralization process. While the inorganic phosphate concentration of α-MEM is in the physiological range, the organic phosphate concentration is below physiological [64]. *In vivo*, various organic phosphates act as substrates for the cell membrane associated enzyme, alkaline phosphatase (AP). With the action of AP, inorganic phosphate is cleaved from β-GP as a cell-mediated, rather than nonspecific, process. While the use of β-GP has been contraindicated in chondrocyte cultures, our experience is that this organic phosphate source is more reliable than elevating the inorganic phosphate in osteogenic cultures. However, cells also release alkaline phosphatase into the medium and thus may convert organic to inorganic phosphate to cause a nonspecific precipitation of mineral, or what may be called ectopic or dystrophic mineralization [65].

Clearly, the higher the level of alkaline phosphatase expression, as is found in preosteoblasts, the more likelihood there is of dystrophic mineralization occurring in the culture. What is not so universally understood is that the presence of dystrophic mineralization can impede the further differentiation of these cells and compromise the formation of a bony extracellular matrix in such cultures [63]. Therefore, in murine cultures we routinely reduce to 5 mM the 10 mM β-GP, as employed by Maniatopoulos *et al.* [43], to avoid dystrophic

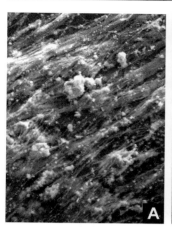 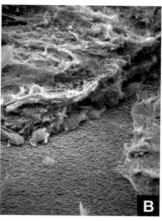

Fig. 26.1. Scanning electron micrographs of (A) dystrophic mineralization, and (B) bone nodule formation in human bone marrow cultures. Dystrophic mineralization in this case was the result of the early addition of β-glycerophosphate to the culture (see text) and has resulted in a cell-mediated precipitation of calcium phosphate mineral (white clumps), which are lying on, and in between, extended cell processes, and collagen fibers. This cell-mediated "biomineralization" has been described by some [25] as intramembranous ossification. In the more structured matrix (B), the culture substrate is covered with the mineralized globules of the cement line and, above, a thick layer of collagenous and mineralized extracellular bone matrix. Reproduced from Parker et al. [63] with permission. The complete sequence of bone matrix elaboration events is described in Davies [45]. Field widths for (A) and (B), 65 μm.

mineralization. We have found human cultures to be far more sensitive to β-GP addition, and we employ 3.5 mM which is added only after the appearance of cell multilayering. Without this precaution, human cultures will display considerable dystrophic mineralization. Some authors have stated that human cells do not form bone nodules in culture, as seen in other species [25], describing a "biomineralization" [25,36] that, they suggest, models intramembraneous ossification [25]. While such mineralization is inevitably biologically mediated, as a result of the presence of high levels of AP, this matrix does not resemble bone matrix. However, morphologically identifiable bone matrix can be elaborated in the right culture conditions as illustrated by comparing Figs. 26.1A and B.

Ascorbic Acid

Ascorbic acid (AA) is a required cofactor for the hydroxylation of proline. Hydroxyproline is an essential amino acid in the stabilization of the collagen triple helix, without which stable collagen fibers cannot be assembled. Collagen is the predominant bone protein, and thus the importance of AA for bone cultures is evident. Most studies that use bone marrow cultures report the presence of 50 μg/ml AA, as in the original protocol of Maniatopoulos et al. [43]. However, most commercial α-MEM formulations already contain about 50 μg/ml AA. Since AA is relatively unstable, about 9% a day is lost in storage 5°C [66], and the actual AA concentration is therefore dependent on the time that the α-MEM is stored before use. When the concentration of AA must be known, it is better to prepare ascorbic acid deficient α-MEM and add the required amount of fresh AA. The half-life of ascorbic acid at room temperature is 15.5 h [66], and during culture at 37°C, most of the AA degrades on the first day. This is a problem when cells are refed less frequently, as in human, or bioreactor (dynamic flow), cultures; and thus long-acting ascorbic acid (ascorbic acid 2-phosphate) should be employed [63].

CELL ISOLATION

To illustrate the variations that may occur in cell isolation procedures, we provide outlines of the protocols we use for rat and human marrow, respectively. We obtain rat bone marrow, cells according to the method described earlier [43–45]. Briefly, femora from young adult (115–125 g; 40–43 days) Wistar rats are excised aseptically and transferred into antibiotic–fungizone solution. After 10 min in this solution the femora are transferred into fresh solution. This washing procedure is repeated three times. Subsequently, the femora are transferred into α-MEM to wash away the antibiotic–fungizone solution. Epiphyses are then removed from the femora aseptically with scissors, and the marrow from each diaphysis is flushed out with fully supplemented standard culture medium, using a syringe with a 20-gauge needle. Marrow is collected in a sterile 50-ml tube and transferred to a culture flask in fully supplemented standard culture medium. For mouse bone marrow cultures, we use animals 6–8 weeks old, and the cells are isolated as in the rat. We generally pool the marrow from several animals (usually three) to reduce the effects of interanimal variation.

For human bone marrow cultures [63], from either trabecular bone biopsies or marrow aspirations, the harvest is flushed several times with phosphate-buffered saline (PBS) to generate a cell suspension, which in turn is sieved and subsequently centrifuged at 500g for 10 min at room temperature. Cells are suspended in PBS, fractionated on a Ficoll density gradient, and centrifuged at 400g for 45 min. Cells isolated from the gradient interface are seeded at 2×10^5 cells/cm^2, or 1×10^6 cells/cm^3, and cultured in α-MEM containing 15% fetal bovine serum, 10% antibiotic solution (100 mg/ml penicillin G, 50 mg/ml gentamicin sulfate, 0.3 mg/ml amphotericin B), 50 mg/ml L-ascorbic acid, and 10^{-8} M dexamethasone.

CELL SEEDING ON TWO-DIMENSIONAL SUBSTRATES

Depending on the purpose of the end-point application, whether it be expanding the adherent cell population to produce a high cell concentration for tissue engineering applications, or passaging the osteogenic cell culture several times to attain a more homogeneous cell population for studying bone formation, a two-dimensional tissue-culture-treated surface will generally be required.

The cell suspension is transferred into either T25 (with total volume of 5 ml) or T75 (with total volume of 15 ml) flasks and maintained in an incubator with humidified atmosphere consisting of 95% air and 5% CO_2 at 37°C, and 100% relative humidity. The medium is changed after 24 h to remove primarily nonadherent hematopoietic cells, and subsequently three times a week for rat and murine cultures and twice a week for human cultures.

On day 5 (rat) or, approximately, day 16 (human) the cells are subcultured. For passaging, the cells are enzymatically released using 0.01% trypsin. These trypsinized cells are then plated onto tissue culture dishes at approximately 10^4 cells per square centimeters. For this step, a cell strainer (grid size 70 μm) can be used to avoid cell clumps and achieve a more uniform culture.

Design Considerations for Three-Dimensional Scaffolds

The tissue engineering strategy of the type discussed in this chapter involves the development of suitable scaffolds for MSC seeding and later, *in vivo*, new tissue support. The appropriate carrier should allow a three-dimensional distribution of cells *in vitro*, thereby accelerating bone healing *in vivo*. The natural open-pore geometry of trabecular bone macrostructure provides a good starting point for the design of scaffolds for bone tissue engineering [67,68], which can be modeled with either inorganic or organic synthetic scaffolds (Fig. 26.2).

Scaffold Macroporosity

Scaffold macroporosity has been known to be a critical factor in cell migration and bone matrix elaboration *in vitro*. We showed that matrices with a nominal pore size of 200 μm result in occlusion of pores by cells [69]. These observations led others to employ similar scaffolds of 500 μm nominal pore size that permitted three-dimensional tissue growth *in vitro* [70], although some authors have reported that pore sizes ranging from 200 to 400 μm encourage migration, attachment, and proliferation of osteoblast-like cells [71]. Thus, as differentiating osteogenic cells migrate over a surface *in vitro*, they are affected not only by the surface roughness and chemistry [72,73] but also by the scaffold macroporosity.

This is most clearly demonstrated by comparing the cell colonization of a series of scaffolds of different macroporosity but identical chemistry [17] (Fig. 26.3). The complete occlusion of the surface pores poses a serious problem for bone tissue engineering applications, since pore occlusion prevents cellular penetration and matrix elaboration within the pores themselves. Pore bridging and occlusion occur with both inorganic [17,69,70] and organic [16,74] scaffold materials, but we have also shown that these effects can be avoided by increasing the nominal macropore size [17,18,75,76]. Not only does macroporosity influence *in vivo* performance but also, the rate and distribution of tissue ingrowth through the scaffold will be affected by the size and number of interconnecting channels [77].

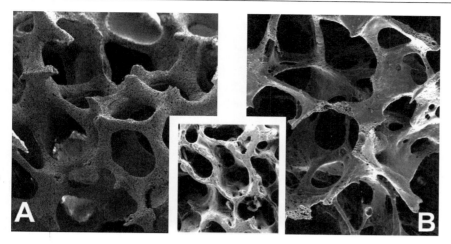

Fig. 26.2. Scanning electron micrographs of both inorganic and organic biodegradable scaffolds that can be processed to mimic the trabecular structure of natural human bone tissue (inset), which displays a highly interconnected macroporosity. (A) Calcium polyphosphate scaffold, provided by Prof. S. Y. Kim of the School of Materials Engineering, Yeungnam University, Kyongasan, Korea. (B) Poly(lactic-co-glycolic acid) (PLGA) scaffold provided by BoneTec Corp., Toronto, Canada. Field width each image, 2.58 mm.

Interconnected Macroporosity

It is important to distinguish between the phenomenon of pore occlusion, which is solely dependent upon macroporosity, and that of bone ingrowth, which is a function of both macroporosity and macropore interconnectivity. This is illustrated in the work of Kadiyala *et al.* [15], who employed a macroporous biphasic calcium phosphate scaffold (similar to that illustrated in the inset to Fig. 26.3) with culture-expanded MSC to regenerate a critical-sized bone defect in rat femora. Their results revealed that MSC-loaded materials showed significantly more bone formation at 4 weeks and even more bone formation by 8 weeks compared with cell-free implants. However, their histology also revealed that bone grew preferentially along the implant–host interface but not throughout the bulk material. Ohgushi *et al.* [13,78,79] reported comparable results with the same material, indicating that the biphasic calcium phosphate ceramic combined with expanded bone marrow cells showed healing potential for the treatment of bone defects. Nevertheless, the lack of interconnecting macroporosity in this biphasic material resulted in no bone formation in the bulk scaffold. Similarly, Ishaug-Riley *et al.* [16], who used poly(lactic-*co*-glycolic acid) foams of 300–500 μm pore size, made by a solvent-casting particulate leaching technique, found that bone cells and tissue were only found within 300 μm of the outer edge of the scaffold. Such results clearly illustrate that fully interconnected pores of the appropriate size are required to facilitate cell migration throughout the macrostructure.

Cell Seeding on Three-Dimensional Scaffolds

To seed cells on three-dimensional porous substrates, we subculture at 80% confluency and use a 10-ml syringe fitted with a 20-gauge needle to dispense the cells. This gauge size

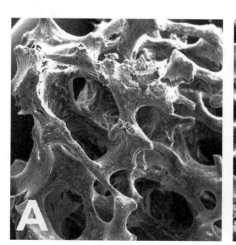
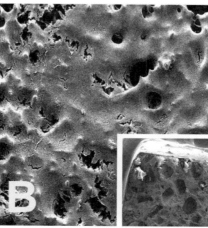

Fig. 26.3. Scanning electron micrographs of rat bone marrow cells grown on calcium polyphosphate scaffolds of different nominal pore size. (A) The cell population is completely covering all scaffold surfaces throughout the sample. (B) The surface of a scaffold with a nominal pore size of 150 μm is almost completely covered with a thin cell sheet, which is bridging the majority of pore openings. Inset shows the freeze-fractured specimen of a 400 μm-pore-size biphasic calcium phosphate material with no interconnecting macroporosity; again, the cell sheet is limited to the outer boundaries of the sample. Field widths (A) and (B), 1.47 mm; inset, 2.25 mm. Reproduced from Baksh and Davies [76] with permission.

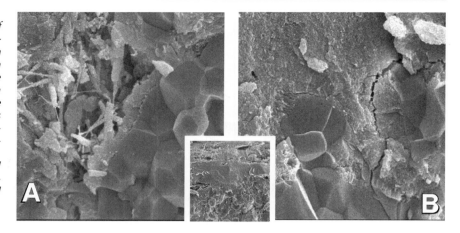

Fig. 26.4. Scanning electron micrographs of details of bone growth on the type of calcium polyphosphate scaffold illustrated in Fig. 26.2A. Inset shows a low magnification of the bone scaffold interface with osteocyte lacunae visible in the bone above. (A) High magnification of the bone–scaffold interface showing abundant mineralized collagen fibers and a mineralized cement line interfacial matrix. (B) Interdigitation of the cement line matrix with micropores of the scaffold surface, indicative of bone bonding. The geometrical bodies are the individual grains of the calcium polyphosphate scaffold material. Field widths (A) and (B), 6.5 μm; inset, 43 μm. From Baksh [81].

is chosen because the diameter allows passage of whole cells averaging 10–15 μm diameter through the tip without rupturing cell membranes. Constant pressure is applied to the syringe to the point that the cells are dispensed dropwise onto the surface. Employing this technique ensures that the cells contact the three-dimensional substrate prior to contacting the well-plate surface. To optimize cell adhesion to the scaffold, rather than the culture vessel, we use bacteriological grade (BG) polystyrene plates, which are hydrophobic and thus do not promote cell attachment. Anchorage-dependent cells, such as osteogenic cells, require relatively hydrophilic surfaces to spread and migrate [80]. We have generally found 1 h a sufficient time to allow for seeding [81], following which the scaffold can be cultured in either static or dynamic flow conditions. In either format it is possible to demonstrate the morphological hallmarks of *in vitro* bone formation throughout the structures of both inorganic and organic three-dimensional scaffolds. Examples are provided in Figs. 26.4 and 26.5.

Increasing Yield with Dynamic Culture Conditions

It is difficult to sustain long-term experiments without contamination and promoting the differentiation of the desired cell population in conventional static culture conditions [82]. Consequently, dynamic systems that permit continuous movement of the medium throughout the substrate during the culture period are preferred. Although many methods can be employed to achieve such dynamic conditions [83–86], we have chosen the simple expedient of a modified culture rotisserie [81]. The latter is used to provide continuous distribution of culture media throughout a porous structure and to eliminate contact with the walls of the container. Once the cells have been seeded and allowed to migrate within

Fig. 26.5. Scanning electron micrographs of rat bone matrix being elaborated in the body of a PLGA scaffold as illustrated in Fig. 26.2B. Inset shows that cell seeding and bone matrix growth have not occluded the interconnecting macroporosity of this material (cf. Fig. 26.2 inset), while (A) clearly shows the globules of cement line matrix deposited on the scaffold surface (polymer bulk in foreground), which is overlaid with an abundant collagenous matrix containing cells. (B) Higher magnification of an area of the scaffold; the overlying collagenous matrix and cells have been removed to illustrate the adaptation of the cement line mineralized matrix to the convoluted morphology of this polymer scaffold. Field widths, (A), 60 μm; (B), 25.5 μm, inset, 2.58 mm.

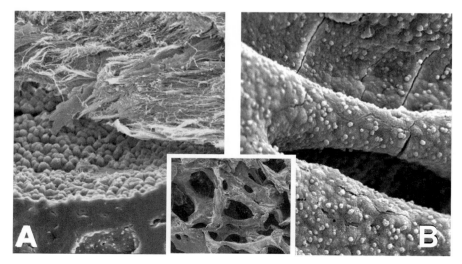

the interior of the scaffold, the culture microenvironment plays a significant role in maintaining the cell population at the interstices. A dynamic culture system provides continuous exchange of nutrients and metabolic products from the scaffold microenvironment to the surrounding fluid environment, and sustains a more viable and metabolically active cell population [76,81,86].

CONCLUDING REMARKS

We have focused on bone marrow culture conditions that facilitate phenotypic expression of osteogenic cells on both two- and three-dimensional substrates across several species. Successful culture of osteogenic cells depends on not only the cells themselves but, particularly for tissue engineering, on the geometry of the scaffold. Of course, the ultimate test of the suitability of a specific culture regime is the growth of new bone *in vivo*, the restoration of bone stock, and recovery of function. Increasing reports of the use of both inorganic and organic scaffolds, combined with either primary or passaged marrow cells, indicate that at least some of the requirements of bone tissue engineering using scaffold-driven strategies are being met. It can reasonably be assumed that further advances in such cell-seeded, scaffold-based protocols, perhaps augmented with growth factor or gene therapies, will soon render human bone tissue engineering a clinical reality.

ACKNOWLEDGMENTS

The authors are grateful to Morris Hosseini for his insights into the culture conditions described, Elaine Parker for her work with the human bone cell populations, and Robert Chernecky for assistance with the scanning electron microscopy. This, like our ongoing work, is financially supported by the Natural Sciences and Engineering Research Council (NSERC) of Canada, the Ontario Research and Development Challenge Fund (ORDCF), BoneTec Corp., Toronto, Canada, and an NSERC graduate scholarship to one of us (DB).

REFERENCES

1. Weinmann, J. P., and Sicher, H. (1955). "Bone and Bones," 2nd ed. Mosby, St. Louis, MO.
2. Niklason, L. E. (2000). Engineering of bone grafts. *Nat. Biotechnol.* **18**, 929–930.
3. Vacanti, C. A., and Bonassar, L. J. (1999). An overview of tissue engineered bone. *Clin. Orthop. Relat. Res.* **367**(Suppl.), S375–S381.
4. Bruder, S. P., and Fox, B. S. (1999). Tissue engineering of bone. Cell based strategies. *Clin. Orthop. Relat. Res.* **367**(Suppl.), S68–S83.
5. Fleming, J. E., Jr., Cornell, C. N., and Muschler, G. F. (2000). Bone cells and matrices in orthopedic tissue engineering. *Orthop. Clin. North Am.* **31**, 357–374.
6. Service, R. F. (2000). Tissue engineers build new bone. *Science* **289**, 1498–1500.
7. Reddi, A. H. (2000). Morphogenesis and tissue engineering of bone and cartilage: Inductive signals, stem cells, and biomimetic biomaterials. *Tissue Eng.* **6**, 351–359.
8. Goldstein, S. A., Patil, P. V., and Moalli, M. R. (1999). Perspectives on tissue engineering of bone. *Clin. Orthop. Relat. Res.* **367**(Suppl.), S419–S423.
9. Boden, S. D. (1999). Bioactive factors for bone tissue engineering. *Clin. Orthop. Relat. Res.* **367**(Suppl.), S84–S94.
10. Breitbart, A. S., Grande, D. A., Mason, J. M., Barcia, M., James, T., and Grant, R. T. (1999). Gene-enhanced tissue engineering: Applications for bone healing using cultured periosteal cells transduced retrovirally with the BMP-7 gene. *Ann. Plast. Surg.* **42**, 488–495.
11. Kyriacos, A. A., and Agrawal, C. M., eds. (2000). Orthopedic polymeric biomaterials: Applications of biodegradables. *Biomaterials* (*Spec. Issue*), Vol. 21.
12. Davies, J. E., ed. (2000). "Bone Engineering." em squared Inc., Toronto.
13. Ohgushi, H., Okumura, M., Masuhara, K., Goldberg, V. M., Davy, D. T., and Caplan, A. I. (1992). Calcium phosphate block ceramic with bone marrow cells in a rat long bone defect. *In* "CRC Handbook of Bioactive Ceramics" (T. Yamamuro, L. L. Hench, and J. Wilson, eds.), vol. 2, pp. 235–238. CRC Press, Boca Raton, FL.
14. Petite, H., Viateau, V., Bensaid, W., Meunier, A., de Pollak, C., Bourguignon, M., Oudina, K., Sedel, L., and Guillemin, G. (2000). Tissue-engineered bone regeneration. *Nat. Biotechnol.* **18**, 959–963.
15. Kadiyala, S., Jaiswal, N., and Bruder, S. P. (1997). Culture-expanded, bone marrow-derived mesenchymal stem cells can regenerate a critical-sized segmental bone defect. *Tissue Eng.* **3**, 173–185.
16. Ishaug-Riley, S. L., Crane, G. M., Gurlek, A., Miller, M. J., Yasko, A. W., Yaszemski, M. J., and Mikos, A. G. (1997). Ectopic bone formation by marrow stromal osteoblast transplantation using poly(DL-lactic-co-glycolic acid) foams implanted into the rat mesentery. *J. Biomed. Mater. Res.* **36**, 1–8.
17. Baksh, D., Davies, J. E., and Kim, S. (1998). Three dimensional matrices of calcium polyphosphates support bone growth in vitro and in vivo. *J. Mater. Sci. Mater. Med.* **9**, 743–748.
18. Holy, C. E., Fialkov, J. A., Shoichet, M. S., and Davies, J. E. (2000). In vivo models for bone tissue-engineering constructs. *In* "Bone Engineering" (J. E. Davies, ed.), Chapter 46, pp. 496–504. em squared Inc., Toronto.

19. Charles, H. (1998). The body shop. *In* "Life Magazine" (R. Friedman, ed.), pp. 50–57. Time Inc., New York.

20. Yoshikawa, T., Ohgushi, H., Uemura, T., Ueda, Y., Nakajima, H., Enomoto, Y., Ichijima, K., Takakura, Y., and Tateishi, T. (2000). Bone regeneration with cultured human bone grafts. *In* "Bioceramics. Proceedings of the 13th International Symposium on Ceramics in Medicine" (S. Giannini and A. Moroni, eds.), Vol. 13. World Scientific, New York.

21. Mendes, S. C., de Bruijn, J. D., Bakker, K., van Apeldoorn, A. A., Platenburg, P. P., Tibbe, G. J. M., and van Blitterswijk, C. A. (2000). Human bone marrow stromal cells for bone-tissue engineering: In vitro and in vivo characterization. *In* "Bone Engineering" (J. E. Davies, ed.), Chapter 47, pp. 505–515. Em Squared Inc., Toronto.

22. Potten, C. S., and Loeffler, M. (1990). Stem cells: Attributes, cycles, spirals, pitfalls and uncertainties. Lessons for and from the crypt. *Development (Cambridge, UK)* **110**, 1001–1020.

23. Morrison, S. J., Shah, N. M., and Anderson, D. J. (1997). Regulatory mechanisms in stem cell biology. *Cell. (Cambridge, Mass.)* **88**, 287–298.

24. Caplan, A. I., and Bruder, S. P. (1997). Cell and molecular engineering of bone regeneration. *In* "Principles of Tissue Engineering" (R. P. Lanza, R. Langer, and W. L. Chick, eds.), pp. 579–584. Elsevier, Amsterdam.

25. Jaiswal, N., Haynesworth, S. E., Caplan, A. I., and Bruder, S. P. (1997). Osteogenic differentiation of purified culture-expanded human mesenchymal stem cells in vitro. *J. Cell. Biochem.* **64**, 295–312.

26. Malaval, L., Liu, F., Roche, P., and Aubin, J. E. (1999). Kinetics of osteoprogenitor proliferation and osteoblast differentiation in vitro. *J. Cell. Biochem.* **74**, 616–627.

27. Colter, D. C., Class, R., DiGirolamo, C. M., and Prockop, D. J. (2000). Rapid expansion of recycling stem cells in cultures of plastic-adherent cells from human bone marrow. *Proc. Natl. Acad. Sci. U.S.A.* **97**, 3213–3218.

28. Kale, S., Biermann, S., Edwards, C., Tarnowski, C., Morris, M., and Long, M. W. (2000). Three-dimensional cellular development is essential for ex vivo formation of human bone. *Nat. Biotechnol.* **18**, 954–958.

29. Owen, M. E. (1998). The marrow stromal cell system. *In* "Marrow Stromal Stem Cells in Culture" (J. N. Beresford and M. E. Owen, eds.), pp. 88–110. Cambridge University Press, Cambridge, UK.

30. Prockop, D. J. (1997). Marrow stromal cells as stem cells for nonhematopoietic tissues. *Science* **276**, 71–74.

31. Aubin, J. E. (1999). Osteoprogenitor cell frequency in rat bone marrow stromal cell populations: Role for heterotypic cell-cell interactions in osteoblast differentiation. *J. Cell. Biochem.* **72**, 396–410.

32. Caplan, A. I. (1991). Mesenchymal stem cells. *J. Orthop. Res.* **9**, 641–650.

33. Lazarus, H. M., Haynesworth, S. E., Gerson, S. L., and Caplan, A. I. (1997). Human bone marrow-derived mesenchymal (stromal) progenitor cells (MPCs) cannot be recovered from peripheral blood progenitor cell collections. *J. Hematother.* **6**, 447–455.

34. Nuttall, M. E., Patton, A. J., Olivera, D. L., Nadeau, D. P., and Gowen, M. (1998). Human trabecular bone cells are able to express both osteoblastic and adipocytic phenotype: Implications for osteogenic disorders. *J. Bone Miner. Res.* **13**, 371–382.

35. Aubin, J. E. (2000). Osteogenic cell differentiation. *In* "Bone Engineering" (J. E. Davies, ed.), Chapter 3, pp. 19–30. em squared Inc., Toronto.

36. Pittenger, M. F., Mackay, A. M., Beck, S. C., Jaiswal, R. K., Douglas, R., Mosca, J. D., Moorman, M. A., Simonetti, D. W., Craig, S., and Marshak, D. R. (1999). Multilineage potential of adult human mesenchymal stem cells. *Science* **284**, 141–147.

37. Dexter, T. M., Allen, T. D., Lajtha, L. G., Schofield, R., and Lord, B. I. (1973). Stimulation of differentiation and proliferation of haemopoietic cells in vitro. *J. Cell. Physiol.* **82**, 461–473.

38. Dexter, T. M., and Lajtha, L. G. (1974). Proliferation of haemopoietic stem cells in vitro. *Br. J. Haematol.* **38**, 525–530.

39. Dexter, T. M., Allen, T. D., and Lajtha, L. G. (1977). Conditions controlling the proliferation of haemopoietic stem cells in vitro. *J. Cell. Physiol.* **91**, 335–344.

40. Phinney, D. G., Kopen, G., Isaacsson, R. L., and Prockop, D. J. (1999). Plastic adherent stromal cells from the bone marrow of commonly used strains of inbred mice: Variations in yield, growth, and differentiation. *J. Cell. Biochem.* **72**, 570–585.

41. Friedenstein, A. J. (1990). Osteogenic stem cells in the bone marrow. *In* "Bone and Mineral Research" (J. N. M. Heersche and J. A. Kanis, eds.), pp. 243–270. Elsevier Science Publishers (Biomedical Division), Amsterdam.

42. Turksen, K., and Aubin, J. E. (1991). Positive and negative immunoselection for enrichment of two classes of osteoprogenitor cells. *J. Cell Biol.* **114**, 373–384.

43. Maniatopoulos, C., Sodek, J., and Melcher, A. H. (1988). Bone formation in vitro by stromal cells obtained from bone marrow of young adult rats. *Cell Tissue Res.* **254**, 317–330.

44. Davies, J. E., Chernecky, R., Lowenberg, B., and Shiga, A. (1991). Deposition and resorption of calcified matrix in vitro by rat marrow cells. *Cells Mater.* **1**, 3–15.

45. Davies, J. E. (1996). In vitro modeling of the bone/implant interface. *Anat. Rec.* **245**, 426–445.

46. Tsuji, T., Hughes, F. J., McCulloch, C. A., and Melcher, A. H. (1990). Effects of donor age on osteogenic cells of rat bone marrow in vitro. *Mech. Ageing Dev.* **51**, 121–132.

47. Friedenstein, A. J., Latzinkik, N. V., Gorskaya, Y. F., Luria, E. A., and Moskvina, I. L. (1992). Bone marrow stromal colony formation requires stimulation by haemopoietic cells. *Bone Miner.* **18**, 99–213.

48. Castro-Malaspina, H., Gay, R. E., Resnick, G., Kappor, N., Meyers, P., Chiarieri, D., McKenzie, S., Broxmeyer, H. E., and Moore, M. A. S. (1980). Characterization of human bone marrow fibroblast colony-forming cells (CFU-F) and their progeny. *Blood* **56**, 289–301.

49. Ohgushi, H., and Caplan, A. (1999). Stem cell technology and bioceramics: From cell to gene engineering. *J. Biomed. Mater. Res. (Appl Biomater.)* **48**, 913–927.

50. Bruder, S. P., Jaiswal, N., and Haynesworth, S. E. (1997). Growth kinetics, self-renewal, and the osteogenic potential of purified human mesenchymal stem cells during extensive subcultivation and following cyropreservation. *J. Cell. Biochem.* **64**, 278–294.

51. Haynesworth, S. E., Goshima, J., Goldberg, V. M., and Caplan, A. I. (1992). Characterization of cells with osteogenic potential from human marrow. *Bone* **13**, 81–88.

52. Lazarus, H. M., Haynesworth, S. E., Gerson, S. L., Rosenthal, N. S., and Caplan, A. I. (1995). Ex-vivo expansion and subsequent infusion of human bone-marrow-derived stromal progenitor cells (mesenchymal progenitor cells)— implications for therapeutic use. *Bone Marrow Transplant.* **16**, 557–564.

53. Martin, I., Muraglia, A., Campanile, G., Cancedda, R., and Quarto, R. (1997). Fibroblast growth factor-2 supports ex vivo expansion and maintenance of osteogenic precursors from human bone marrow. *Endocrinology (Baltimore)* **138**, 4456–4462.

54. Kimura, A., Katoh, O., and Kuramoto, A. (1988). Effects of platelet derived growth factor, epidermal growth factor and transforming growth factor-B on the growth of human marrow fibroblasts. *Br. J. Haematol.* **69**, 9–12.

55. Todescan, R., Lowenberg, B. F., Hosseini, M. M., and Davies, J. E. (1996). Tetracycline fluorescence: A new method to quantitative bone produced in vitro. *Proc. 5th World Biomater. Congr., 1996*, Abstr. No. 721.

56. Davies, J. E., Krasnoshtein, F., and Todescan, R. (2001). Quantification of bone mass, and osteoclast resorption, in vitro using thin calcium phosphate ceramic films. *J. Appl. Biomater.* (in press).

57. Canalis, E. (1985). Effect of growth factors on bone cell replication and differentiation. *Clin. Orthop. Relat. Res.* **193**, 246–263.

58. Maniatopoulos, C. (1988). Development and characterization of an in vitro system permitting osteogenesis by stromal cells isolated from bone marrow of young adult rat. Ph.D. Thesis, University of Toronto, Faculty of Dentistry, Toronto.

59. Hosseini, M. M., Peel, S. A. F., and Davies, J. E. (1996). Collagen fibres are not required for initial matrix mineralization by bone cells. *Cells Mater.* **10**, 357–365.

60. Baron, J., Huang, Z., Oerter, K. E., Bacher, J. D., Cutler, G. B. (1992). Dexamethasone acts locally to inhibit longitudinal bone growth in rabbits. *Am. J. Physiol.* **263**, E489–E492.

61. Qu, Q., Perala-Heape, M., Kapanen, A., Dahllund, J., Salo, J., Vaananen, H. K., and Harkonen, P. (1998). Estrogen enhances differentiation of osteoblasts in mouse bone marrow culture. *Bone* **22**, 201–209.

62. Graziano, V. M. A. (1998). The ferret: A potential in vitro small animal model for the study of osteogenesis and osteoclasis. M.A.Sc. Thesis, University of Toronto, Faculty of Dentistry, Toronto.

63. Parker, E., Shiga, A., and Davies, J. E. (2000). Growing human bone in vitro. *In* "Bone Engineering" (J. E. Davies, ed.), Chapter 6, pp. 63–77. em squared Inc., Toronto.

64. Tenenbaum, H. C. (1981). Role of organic phosphate in mineralization of bone in vitro. *J. Dent. Res.* **60**, 1586–1590.

65. Tenenbaum, H. C., McCulloch, C. A., Fair, C., and Birek, C. (1989). The regulatory effect of phosphates on bone metabolism in vitro. *Cell Tissue Res.* **257**, 555–563.

66. Feng, J., Melcher, A. H., Brunette, D. M., and Moe, H. K. (1977). Determination of L-ascorbic acid levels in culture medium: Concentrations in commercial media and maintenance of levels under conditions of organ culture. *In Vitro* **13**, 91–99.

67. Panda, R., Teung, P., Danforth, S. C., and Safari, A. (1998). Fabrication of calcium phosphate based ceramics with controlled porosity for bone scaffolds. *In* "Bioceramics. Proceedings of the 11th International Symposium on Ceramics in Medicine" (R. Z. LeGeros and J. P. LeGeros, eds.), Vol. 11, pp. 727–730. World Scientific, New York.

68. Holy, C. E., Shoichet, M. S., Campbell, A. A., Song, L., and Davies, J. E. (1998). Preparation of a novel calcium phosphate-coated scaffold for bone tissue engineering. *In* "Bioceramics. Proceedings of the 11th International Symposium" (R. Z. LeGeros and J. P. LeGeros, eds.), Vol. 11, pp. 509–512. World Scientific, New York.

69. Rout, P. G. J., Tarrant, S. F., Frame, J. W., and Davies, J. E. (1988). Interactions between primary bone cell cultures and biomaterials. PART 3: A comparison of dense and macroporous hydroxyapatite. *In* "Biomaterials and Clinical Applications" (A. Pizzoferrato, P. G. Ravaglioli, and A. J. C. Lee, eds.), pp. 591–596. Elsevier, Amsterdam.

70. Yoshikawa, T., Ohgushi, H., and Tamai, S. (1996). Intermediate bone forming capability of prefabriacated osteogenic hydroxyapatite. *J. Biomed. Mater. Res.* **32**, 481–492.

71. Boyan, B. D., Hummert, T. W., Dean, D. D., and Schwartz, Z. (1996). Role of material surfaces in regulating bone and cartilage cell response. *Biomaterials* **17**, 137–146.

72. Petite, H., Kacem, K., and Triffitt, J. T. (1996). Adhesion, growth and differentiation of human bone marrow stromal cells on non-porous calcium carbonate and plastic substrata: Effects of dexamethasone and 1,25 dihydroxyvitamin D3. *J. Mater. Sci. Mater. Med.* **7**, 665–671.

73. Davies, J. E, and Baldan, N. (1997). Scanning electron microscopy of the bone-bioactive implant interface. *J. Biomed. Mater. Res.* **36**, 429–440.

74. Holy, C. E, Shoichet, M. S., and Davies, J. E. (1997). Bone marrow cell colonization of, and extracellular matrix expression on, biodegradable polymers. *Cells Mater.* **7**, 223–234.

75. Holy, C. E., Davies, J. E., and Shoichet, M. S., (1997). Bone tissue engineering on biodegradable polymers: Preparation of a novel poly(lactide-*co*-glycolide) foam. *In* "Biomaterials, Carriers for Drug Delivery, and Scaffolds for Tissue Engineering" (N. A. Peppas, D. J. Mooney, A. G. Mikos, and L. Brannon-Peppas, eds.), pp. 272–274. AIChE Press, New York.

76. Baksh, D., and Davies, J. E. (2000). Design strategies for 3-dimensional in vitro bone growth in tissue-engineering scaffolds. *In* "Bone Engineering" (J. E. Davies, ed.), Chapter 45, pp. 488–495. em squared Inc., Toronto.

77. LeGeros, R. Z., and LeGeros, J. P. (1995). Calcium phosphate biomaterials: Preparation, properties, and biodegradation. *In* "Encyclopedic Handbook of Biomaterials & Bioengineering Part A: Materials" (D. L. Wise, D. J. Trantolo, D. E. Altobelli, M. J. Yaszemski, J. D. Gresser, and E. R. Schwartz, eds.), Vol. 2, pp. 1429–1463. Dekker, New York.

78. Ohgushi, H., Okumara, M., Yoshikawa, T., Inouem, K., Senpuku, N., and Tamai, S. (1992). Bone formation process in porous calcium carbonate and hydroxyapatite. *J. Biomed. Mater. Res.* **26**, 885–895.

79. Ohgushi, H., Okumura, M., Masuhara, K., Goldberg, V. M., Davy, D. T., and Caplan, A. I. (1992). Osteogenic potential of bone marrow sustained by porous calcium phosphate ceramics. *In* "CRC Handbook of Bioactive Ceramics" (T. Yamamuro, L. L. Hench, and J. Wilson, eds.), Vol. 2, pp. 229–233. CRC Press, Boca Raton, FL.

80. Zygourakis, K. (1996). Quantification and regulation of cell migration. *Tissue Eng.* **2**, 1–16.

81. Baksh, D. (1999). Comparison of 3-dimensional calcium phosphate scaffolds for candidate bone tissue constructs. M.A.Sc. Thesis, University of Toronto, Department of Chemical Engineering and Applied Chemistry and the Institute of Biomaterials and Biomedical Engineering, Toronto.

82. Sittinger, M., Buija, J., Hammer, C., Minuth, W. W., and Burmester, G. R. (1996). Tissue engineering and autologous transplant formation: Practical approaches with resorbable biomaterials and new cell culture techniques. *Biomaterials* **17**, 237–242.

83. Freed, L. E., Vunjak-Novakovic, G., and Langer, R. (1993). Cultivation of cell-polymer cartilage implants in bioreactors. *J. Cell. Biochem.* **51**, 237–242.

84. Sittinger, M., Schultz, O., Keyszer, G., Minuth, W. W., and Burmester, G. R. (1997). Artificial tissues in perfusion culture. *Int. J. Artif. Organs* **20**, 57–62.

85. Burg, K. J., Holder, W. D. Jr., Culberson, C. R., Beiler, R. J., Greene, K. G., Loebsack A. B., Roland, W. D., Eiselt, P., Mooney, D. J., and Halberstadt, C. R. (2000). Comparative study of seeding methods for three-dimensional polymeric scaffolds. *J. Biomed. Mater. Res.* **51**, 642–649.

86. Mueller, S. M., and Glowacki, J. (2000). Construction and regulation of 3-dimensional bone tissue in vitro. *In* "Bone Engineering" (J. E. Davies, ed.), Chapter 44, pp. 473–487. em squared Inc., Toronto.

CHAPTER 27

MESENCHYMAL CELL CULTURE: ENDOTHELIAL CELL TISSUE ENGINEERING

Helen M. Nugent, Sahil A. Parikh, and Elazer R. Edelman

INTRODUCTION

The confluent monolayer of endothelial cells that line blood vessels provides a thromboresistant lining and plays a crucial role in controlling hemostasis, vascular tone, inflammation, and smooth muscle cell proliferation. Compromised endothelial function is a characteristic of vascular diseases and has been implicated in contributing to the accelerated arteriopathies that follow vascular interventions. Almost 1.5 million interventional procedures are performed each year for attempted relief of the complications of cardiovascular diseases. While bypass grafts, angioplasty, and stenting provide initial clinical success; early thrombosis and late restenosis can actually heighten disease with accelerated arteriopathies characterized by the rapid proliferation of smooth muscle cells and their accumulation within the tunica intima [1–5].

The normal blood vessel is a complex structure of extraordinary sophistication that is comprised of three concentric tunics [6]. The tunica intima, located at the blood vessel wall–lumen interface, is lined by a single layer of endothelial cells supported by a basement membrane below which resides a sparse layer of vascular smooth muscle cells [6]. As a continuous monolayer, the endothelium provides structural boundaries to the circulating blood in the lumen and the vascular wall, and serves as a selectively permeable thromboresistant surface. The endothelium also secretes a myriad of powerful biochemical mediators that regulate blood cell trafficking, vasomotor tone, cell growth, and vascular remodeling [7]. The smooth muscle cells in the intima principally serve as responsive elements that set vascular tone [8]. The bulk of the smooth muscle cells reside in the middle of the blood vessel wall, the tunica media, in lamellar units bound by elastic bands or lamina. The contraction and relaxation of these units allow the artery to constrict or dilate, regulating blood flow [6]. The outer layer, the tunica adventitia, is composed of a loose fibrous network of fibroblasts. The vessels that nourish the blood vessel wall, the vasa vasorum, and the nerves that supply neural control, the vasa nervosum, are also found in the adventitia [6].

Native atherosclerosis and the accelerated arteriopathies that follow therapy share a common initial pathological event, loss of endothelial cell integrity and function [9,10]. Damage to the endothelium removes endothelial compounds, and a chain of events is set into motion that leads to the proliferation and migration of smooth muscle cells into the blood vessel lumen, resulting in obstructive arterial lesions [11]. These lesions have been termed *restenosis*, and in 30–50% of cases the condition is of such significance that tissue and organ integrity are impaired. Exogenous analogues of endothelial vasoregulatory compounds have been characterized, isolated, and their *in vivo* and *in vitro* effects extensively examined; yet no pharmaceutical approach has successfully limited the proliferative complications of vascular interventions [12]. In most cases, disappointing clinical results have followed promising tissue culture and preclinical animal experimentation [12]. This list includes even compounds mechanistically expected to yield the greatest effect, including heparin and related antithrombotic compounds [13–15], the angiotensin-converting-enzyme inhibitors [16,17], the calcium channel blockers [18–22], steroids, and other anti-inflammatory agents [23]. In major part this reflects the complexity of human disease. The number of cell types involved, the complex mechanisms, and the broad period of time over which these events occur make it unlikely that administration of an isolated agent at one point in time will control all vascular reparative events [24]. Thus, tissue- and cell-based therapies are increasingly viewed as a promising alternative [24]. A blood vessel, under normal conditions, is able to heal itself by complex mechanisms of cell signaling, migration, and proliferation. Tissue engineering seems to be most promising in this regard. Replacing key cellular components of the blood vessel (i.e., endothelial cells) may restore this reparative potential in cases where individual vascular cell–derived agents have failed to generate a therapeutic effect [24,25].

ENDOTHELIAL CELL SEEDING OF VASCULAR GRAFTS

One important clinical intervention in advanced arterial disease states is the replacement of stenotic or occluded arteries with bypass grafts. In large-diameter vessels, this can be accomplished through the use of autologous veins or arteries or by using prosthetic grafts composed of materials such as Dacron or expanded poly(tetrafluoroethylene). The high flow rates of large-diameter grafts provide long-term patency rates of 85–95% with minimal pharmacological therapy [26]. Bypass of smaller diameter, low-flow vessels (i.e., coronary arteries) usually requires the use of autologous vessels, since the performance of prosthetic vascular grafts is poor. Smaller diameter grafts (<5 mm) have met with early thrombotic complications and late myointimal hyperplasia, often leading to total graft occlusion [27]. In fact, fewer than 50% of small-diameter femoro popliteal grafts remain patent 5 years postimplantation [28]. The lack of satisfactory long-term patency of small-diameter grafts has been attributed to the inherent thrombogenicity of their luminal interface.

One of the first artificial vascular grafts was developed in 1952 out of the fabric Vinyon N [29]. In the following years, researchers have focused on the production of an ideal synthetic vascular graft. Many of the complications that threaten long-term graft viability have been investigated, and it has become clear that many complications are related to the artificial graft's inability to combat one or more of the phases of the vascular response to injury. An ideal artificial vascular graft would be one that resists thrombosis, inflammation, and neointimal proliferation. It is believed that such a graft would be able to maintain long-term patency *in vivo*. Clearly, the lack of viable endothelial cells on the luminal surface of polymeric grafts is a significant cause of synthetic graft thrombogenicity and intimal proliferation within the graft. Herring and coworkers were the first to report the successful isolation of endothelial cells and their subsequent transplantation onto synthetic vascular grafts using a method they termed endothelial cell seeding [30].

Suspensions of isolated canine venous endothelial cells were used to precoat 6-mm Dacron grafts prior to infrarenal implantation in a canine model [31]. At 4 weeks postimplantation, they observed 76% patency rates for seeded grafts versus 22% for unseeded grafts [31,32]. Explanted grafts possessed an intact endothelial cell lining supported by smooth muscle cells along with penetrating vasa vasorum. Work by other groups confirmed that seeding Dacron grafts with endothelial cells prior to implantation into animals resulted in the recovery of endothelial cells lining the lumen of grafts after explantation [33–37]. Since these studies, numerous labs have focused on the preparation of other polymer surfaces to enhance endothelial cell attachment.

ENDOTHELIALIZATION OF POLYMER SURFACES

The limited usefulness of artificial vascular grafts has motivated the development of an optimal material for endothelial cell growth. It is known that endothelial cells, in the presence of serum, will adhere and spread on moderately wettable polymers [38]. For example, in the series of hydroxyethyl methacrylate–methyl methacrylate (HEMA-MMA) copolymers, the highest number of adhering endothelial cells and the most extensive spreading were found on the moderately wettable copolymer 25 HEMA/75 MMA [39]. Fewer cells adhered onto more hydrophilic, polyHEMA (PHEMA), or hydrophobic copolymers, polyMMA. This is in agreement with the low and reversible protein adsorption usually observed on extreme hydrophilic or hydrophobic surfaces.

It is possible, by altering the surface of PHEMA, for vascular endothelial cells to attach and populate a PHEMA surface [40]. This was achieved by brief treatment of PHEMA with concentrated sulfuric acid. Hydrolytic etching can expand the use of PHEMA as a biomaterial without modifying the overall physicochemical properties of the material. It was found that other polymer surfaces could also be optimized for endothelial cell attachment and growth. Poly(ethyleneterephthalate) (PET) surfaces were treated by plasma processes with oxygen and ammonia and also in the presence of a gas mixture to verify the effects on human umbilical vein endothelial cells (HUVEC) [41]. The results showed that the plasma treatments improved PET biocompatibility to HUVEC. The role of substrate surface properties in the extent of endothelialization of polymer surfaces has also been examined. Absolom and coworkers found for a wide range of polymer surfaces that both porcine and bovine endothelialization was directly related to polymer surface tension and that cell morphology and extracellular matrix production were influenced by substrate surface properties [42]. Other graft materials that have been investigated for endothelial cell attachment and growth include perfluorosulfonic acid (Nafion) [43,44], expanded poly(tetrafluoroethylene) (ePTFE) [45], denucleated ePTFE [45], poly(ether urethane urea) (PEUU) [46], and modified PEUU [46].

Because a desired outcome of seeding vascular grafts with endothelial cells is to decrease graft thrombogenicity, effects of polymer materials on endothelial cell fibrinolytic function have been the focus of several groups. One study revealed that HUVEC cultured on PTFE and polyurethane released into the conditioned media more tissue-type plasminogen activator as well as plasminogen activator inhibitor 1, two major components of the fibrinolytic system, than did the same cells grown on tissue culture polystyrene [47]. The same result was obtained by Storck and coworkers using poly(vinyl chloride) plastic as a support material [48]. Cenni *et al.* investigated both cytokine production and adhesive protein expression by HUVEC after contact with PET and Dacron [49,50]. These molecules, as mediators of adhesion, could play a role in the response of tissues to an artificial implant. They observed a significant increase in expression of endothelial leukocyte adhesion molecule 1 (ELAM-1) and an insignificant increase in cytokine production after contact with PET. Similarly, after contact with knitted Dacron, they observed a significant increase in ELAM-1 expression, while there was a significant decrease in platelet endothelial cell adhesion molecule 1 (PECAM-1) expression and no significant variations of either intercellular adhesion molecule 1 (ICAM-1) or vascular cell adhesion molecule 1 (VCAM-1) expression. These findings emphasize the importance of the effects of the polymer matrix on endothelial cell function. Differences in patency and intimal proliferation after implantation of endothelialized grafts may be due to differences in endothelial cell function depending on the type of graft and polymer chosen.

COVALENTLY ATTACHED PEPTIDES TO PROMOTE ENDOTHELIAL CELL ADHESION

The failure of vascular grafts is usually from platelet-initiated thrombosis and ingrowth of smooth muscle cells and fibroblasts [51]. Endothelial cell attachment to vascular grafts is desired to provide a bioactive and biological blood–graft interface, while attachment of other cell types, such as platelets and smooth muscle cells, would be detrimental to graft survival. An approach to obtain endothelial-cell-specific adhesion is to covalently graft synthetic peptides, which are based on the receptor binding domains of cell adhesion proteins, onto the material surface. Endothelial cells are known to have several receptors that are involved in adhesion to the extracellular matrix [52–55]. Integrins, which are synthesized and expressed on the surface of endothelial cells, recognize the Arg-Gly-Asp (RGD) sequence.

Lin and coworkers grafted RGD-containing peptides on a polyurethane copolymer [56]. *In vitro* endothelial cell adhesion experiments showed that without the presence of serum in culture medium, RGD-grafted polyurethanes dramatically enhanced cell attachment and spreading. Holland and coworkers obtained similar results with the RGD sequence as endothelial cell adhesion, spreading, and growth performance significantly increased for cells plated on RGD attached to a starch-coated polystyrene surface compared with a fibronectin-coated surface [57]. Hubbell and coworkers also immobilized a synthetic peptide containing the sequence Arg-Glu-Asp-Val (REDV) on the otherwise nonadhesive glycophase glass and PET that had been surface-modified with pol(yethylene glycol) (PEG) [58,59]. When the REDV sequence was immobilized on cell nonadhesive substrates, endothelial cells attached and spread, but fibroblasts, vascular smooth muscle cells, and platelets did not. They also found that the endothelial monolayers on REDV-grafted substrates were nonthrombogenic. Two more recent novel studies have further exploited the preferential adherence of endothelial cells to peptide adhesion motifs [60,61]. For example, Zhang *et al.* engineered a biological surface using a combination of self-assembling oligopeptide monolayers and microcontact printing [60]. The adhesion motif, RADS, was used in combination with hexaethylene glycol thiolate, which resists nonspecific adsorption of proteins and cells. The use of microcontact printing resulted in patterns consisting of areas either supporting or inhibiting cell adhesion. The patterns were capable of aligning cells in a well-defined manner, which led to specific endothelial cell arrays and pattern formation. This approach to the selective adhesion of endothelial cells, coupled with nonadhesion of deleterious cell types, may provide a solution to the failures of vascular grafts.

SEEDING GRAFTS WITH GENETICALLY ENGINEERED CELLS

The performance and versatility of prosthetic vascular grafts composed of synthetic materials may be improved by lining the luminal surface with endothelial cells genetically modified to promote repopulation, prevent thrombosis, or secrete therapeutic proteins. Wilson and coworkers explored the possibility of using the vascular endothelial cell as a target for gene replacement therapy [62]. Recombinant retroviruses were used to transduce the *lacZ* gene into endothelial cells harvested from mongrel dogs. Prosthetic vascular grafts seeded with the genetically modified cells were implanted as carotid interposition grafts. Analysis of the grafts 5 weeks after implantation revealed genetically modified endothelial cells lining the luminal surface of the graft. Because a significant limitation to utilizing genetically modified endothelial cells to seed artificial grafts before implantation is poor cell adhesion to the graft lumen, Falk and coworkers incorporated changes to improve cell adherence in a canine carotid interposition graft model [63]. The lumina of control grafts were coated with fibronectin and were seeded with transduced endothelial cells immediately after G418 selection. These grafts were incubated for 2 days before implantation. In contrast, experimental grafts had fibronectin forcefully squeezed through the interstices before endothelial cell seeding. The endothelial cells were allowed to recover for 5 days after G418 selection, and the grafts were incubated for 4 days before implantation. Grafts were explanted after 30 days and evaluated for patency, thrombus, and the presence of transduced endothelial cells. No significant differences in patency rates were seen between groups at this time point. However, both the thrombus-free surface area and area of endothelialization by transduced cells were improved for the experimental grafts.

SOURCE OF ENDOTHELIAL CELLS AND SEEDING TECHNIQUES

Initial attempts at seeding vascular grafts resulted in widely variable seeding efficiencies. Initially endothelial cells were mechanically or enzymatically harvested from autologous venous structures and mixed with whole blood or tissue culture media prior to incubation with the synthetic graft material [31,33,64]. In the 1970s, the ability to harvest and grow pure cultures of endothelial cells was reported [65–69]. Methods were described for the culture of human umbilical vein cells [65,66], bovine aortic cells [67], porcine aortic cells [68], and pulmonary artery cells [69]. Regardless of the vascular bed or species from which the cells are harvested, one technique has prevailed as the method of choice by most researchers. Cells are enzymatically dislodged from the blood vessel wall by collagenase digestion. Primary endothelial cell cultures are generally maintained in standard tissue culture media, supplemented with fetal bovine serum, and incubated at 5% CO_2 and 37°C.

Various methods for determining the purity of endothelial cell cultures are available in the literature. They consist of the presence of factor VIII related antigen, Weibel–Palade bodies, uptake of acetylated low-density lipoprotein, and the absence of smooth muscle cell α-actin [70]. Techniques for endothelial cell seeding have been also been reported and are discussed shortly. The primary determinants of seeding success appear to include the source of the endothelial cell, the incubation time prior to implantation or exposure to fluid flow, and the substrate on which the cells are seeded.

One must balance considerations of immune rejection, vascular bed differences, and cell function in determining the optimal vessel from which to obtain cells for graft seeding. Autografts are tolerated better than allografts and xenografts; however, a characteristic of vascular disease is endothelial dysfunction. Patients with systemic disease that predisposes them to obstructive vascular disease may have dysfunctional endothelial cells throughout different vascular beds, precluding the use of the patient's own cells to seed bypass segments. Moreover, the biochemical functions of venous versus arterial and microvascular versus macrovascular endothelial cells are likely so dissimilar that placement of endothelial cells from one bed into another may result in suboptimal cell performance. Xenogeneic cells, porcine endothelial cells implanted in a canine thoracoabdominal aortic bypass model, have been shown to inhibit graft thrombosis to the same extent as synthetic vascular grafts seeded with autologous canine endothelial cells [71]. These experiments raised the possibility of using xenogeneic or allogeneic cells for endothelial cell seeding of vascular grafts. The successful use of both xenogeneic and allogeneic endothelial cells in animal models of vascular injury will be further discussed in a later section.

The use of autologous endothelial cells could pose another set of problems. A vascular graft seeded with autologous cells would need to be created in time for surgery, and often there is little time between the diagnosis of vascular disease and surgery. To provide a graft lined with endothelial cells, there must be adequate time for cell harvest, isolation, identification, selective growth, graft preparation, seeding, incubation, characterization, and implantation. Not surprisingly, the incubation time is critical to seeding efficiency and ultimately graft performance. Data available in the literature suggest that longer incubation times are better for complete endothelialization. For example, one study demonstrated that endothelial cells adhered to grafts with 20-min incubations, but adherence was far better if the time was extended 4.5-fold [72]. Similar reports have confirmed that incubation times of greater than 60 min and up to 8 h result in higher endothelial cell retention rates after implantation and/or exposure to flow conditions [73]. Therefore, it appears that longer incubations for cell seeding result in enhanced cell retention upon the graft. However, the solution may not be this simple, for the incubation times will also vary depending on the graft material. For example, a study by Sugawara et al. has shown that while an incubation time of 8 h is sufficient for maximal endothelial cell retention on collagen-coated knitted Dacron grafts, incubation lasting up to 24 h is necessary to achieve similar cell adhesion on fibronectin-coated ePTFE grafts [74].

Although it may be preferred to harvest cells, seed the lumen, and then implant the graft during the same procedure, this allows no time for error, cell characterization, or long incubation times, should time be necessary for any of these reasons. Some investigators advocate using completely endothelialized grafts for implantation by employing a multistep procedure that separates cell harvest and seeding from graft implantation after an appropriate incubation of the seeded grafts [75]. Several groups have demonstrated the feasibility of preincubating seeded vascular grafts to generate a confluent, endothelial cell monolayer prior to graft implantation [76–82]. The time required to achieve a confluent endothelial lining after in vitro graft culture ranges from 3 days to 2 weeks. In both animals and humans, these grafts perform as well as or better than single-procedure harvest/seeding grafts. While preincubating grafts requires multiple procedures for cell harvest and subsequent graft implantation, the practice may offer the added benefit of having a stabilized confluent endothelial lining in place prior to exposure to blood flow. The use of either xenogeneic or allogeneic cells for the formulation of preincubated grafts would further improve the utility of these grafts, since it could greatly reduce or eliminate the time needed between diagnosis and surgery for patients whose own cells had to be isolated.

Along with duration of cell seeding prior to exposure to fluid flow, the method of cellular application has proven to be an important factor. Initially, investigators relied upon gravity to seed graft material, resulting in irregular deposition of endothelial cells along the lumen of the vascular graft [31]. Rotating the graft material about cell-rich seeding suspensions [83–85], application of an external vacuum [86], or use of electrostatic seeding techniques [87,88] all seem to improve seeding efficiency, adhesion strength, and circumferential uniformity. In 1997 Bowlin and Rittgers reported a novel electrostatic endothelial cell seeding device for small-diameter (<6 mm) vascular grafts [87]. The unique feature of this technique is the induction of a temporary positively charged surface on the graft luminal surface. The induced charge then dissipates after removal of the graft from the device. After endothelial cell adhesion to the graft's lumen, the surface is able to revert back to its nonthrombogenic, negatively charged surface.

GRAFT SURFACE: PRECOATING WITH SPECIFIC COMPOUNDS

In addition to coating grafts with proteins intended to increase endothelial cell adhesion, several investigators have attempted to enhance endothelial cell proliferation and inhibit smooth muscle cell proliferation by impregnating grafts with endothelial cell growth factors or other compounds. Greisler *et al.* showed that precoating ePTFE with a fibrin glue containing fibroblast growth factor 1 (FGF-1) prior to implantation into the infrarenal position of rabbits increased capillary ingrowth and endothelial proliferation [89,90]. Since FGF-1 also stimulates smooth muscle cell proliferation, this group undertook additional studies; they found that adding an optimal ratio of FGF-1 to heparin along with the fibrin glue limited smooth muscle cell migration and proliferation [91,92]. They also found that this pretreatment enhanced endothelial cell adherence under flow conditions [93] and resulted in decreased platelet deposition along the length of the graft [94,95]. Studies of grafts coated with fibrin glue and vascular endothelial growth factor (VEGF) along with heparin have shown stimulation of endothelial cell proliferation and suppression of vascular smooth muscle cell proliferation *in vitro* [96]. In another study, it was found that endothelial cells were able to grow to confluency on surfaces coated with an albumin–heparin conjugate. Furthermore, endothelial cells adhering and proliferating on albumin–heparin coatings significantly decreased the number of platelets that adhere to the surface [97]. It is hoped that such formulated endothelial cell grafts will accelerate reendothelialization of seeded grafts, while limiting other undesirable cellular responses.

CLINICAL EXPERIENCE WITH ENDOTHELIALIZED VASCULAR GRAFTS

Although many trials in animals have indicated higher patency rates for endothelial cell seeded grafts in comparison to unseeded grafts, early clinical trials of seeded grafts met with mixed results. Direct comparison of these clinical trials is difficult because graft position, size, and endothelial seeding technique vary. However, some important trends have emerged. In 1987 Herring *et al.* reported that of 17 patients undergoing femoropopliteal bypass with autologous endothelialized ePTFE grafts, those receiving seeded grafts exhibited a 2.5-fold higher graft patency rate than those receiving unseeded control grafts at 1 year postimplantation [98]. This study indicated the clinical feasibility of using seeded grafts in humans. However, at nearly the same time, Zilla *et al.* reported conflicting results in patients undergoing distal femoropopliteal bypass with autologously endothelialized grafts [99]. In these patients, objective serum markers and platelet survival studies revealed incomplete endothelialization of the grafts, raising the question of whether endothelial cell seeding would prove clinically successful [99]. Despite the early conflicting data, further trials were initiated to demonstrate higher patency rates for endothelial cell seeded grafts in comparison to unseeded grafts. Indeed, other groups demonstrated decreased platelet adherence in endothelial cell seeded grafts 6–12 months after lower limb arterial reconstruction [100,101].

Long-term clinical trials with endothelial cell seeded vascular grafts have also met with similar success. Leseche *et al.* reported high patency rates for seeded grafts versus control grafts with the use of a two-stage seeding procedure of autologous venous endothelial cells in above-knee femoropopliteal bypass grafts [102]. Primary patency of seeded grafts was

95% at 3 months, 89% at 10 and 48 months, and 67% at 67 and 76 months, respectively, demonstrating for the first time long-term patency of endothelial cell seeded grafts in humans. The next long-term trial was reported serially in 1994 [103], 1997 [104], and 1999 [105]. Over the entire follow-up period of the trial, autologous venous endothelial cell seeded femoropopliteal bypass grafts demonstrated higher patency rates than unseeded grafts. Further research in endothelial cell seeding of vascular grafts will likely remain a top priority in vascular medicine.

INTRAVASCULAR STENT ENDOTHELIALIZATION

The intravascular stent is one of the most successful devices designed both to eliminate coronary stenoses and to prevent their reoccurrence. However, intravascular stents are limited by thrombosis [106,107] and in-stent restenosis [108]. It was thought that perhaps expeditious stent endothelialization could decrease stent thrombosis and restenosis. In 1997 Van Belle and coworkers described a strategy to decrease stent thrombosis by increasing the rate of stent endothelialization [109]. This strategy relied on the local delivery of VEGF, an endothelial mitogen, into the vascular wall. Van Belle *et al.* deployed stents in the iliac arteries of rabbits coupled with the delivery of a single dose of recombinant human VEGF to the vessel wall via balloon catheter. Stent endothelialization and thrombus deposition was evaluated at 4 and 7 days. It was found that VEGF infusion increased stent endothelialization at days 4 and 7 and decreased thrombus deposition on the stents at day 7.

Another therapeutic approach to stent-related thrombosis and restenosis is to coat the stents with endothelial cells before implantation. Stents seeded with genetically engineered endothelial cells provide a unique solution to these problems in that the endothelial cells can be modified to produce large amounts of antithrombotic or antiproliferative compounds. Dichek and coworkers inserted the gene for either bacterial β-galactosidase or human tissue type plasminogen activator into cultured sheep endothelial cells [110,111]. The endothelial cells were seeded onto stents and grown until the stent was covered. When the stents were expanded and exposed to pulsatile flow *in vitro*, substantial cell retention was observed on the lateral stent surface. These cells were also judged viable and healthy after trypsin removal by their ability to proliferate to confluence with the same kinetics as control cells.

PERIVASCULAR ENDOTHELIAL CELL IMPLANTATION

It is possible that methods that focus on seeding endothelial cells at the luminal interface address only that aspect of the endothelium as a nonthrombogenic barrier. This is difficult to achieve in practice and may not be necessary for secretion of biochemical compounds that regulate vascular homeostasis. Moreover, it is not clear that endothelial control of vascular injury and smooth muscle cell proliferation requires that the endothelial cells reside at the their original position. It has recently become evident that the biochemical regulation imposed by the endothelium is equal to or of greater importance than its barrier function and may be active even if these cells are at a distance from the lumen. We reasoned that we could implant many more cells, and that they may survive for a longer period of time, in the perivascular space and still achieve growth control. We have shown that when porcine or bovine aortic endothelial cells (PAE and BAE, respectively) are cultured on denatured collagen matrices, the engrafted endothelial cells display normal growth kinetics and remain viable in large numbers within the matrices of the porous collagen material [112–114]. More importantly, they retain their biological activity and the biochemical markers of normal endothelial cells. For example, the production of heparan sulfate proteoglycan, uptake of acetylated low-density lipoprotein, and inhibition of FGF-induced smooth muscle cell mitogenesis by engrafted cells were indistinguishable from endothelial cells grown in standard tissue culture dishes.

When these endothelial cell scaffolds were implanted around balloon-denuded rat carotid arteries, there was an 88% reduction in intimal hyperplasia [112]. The endothelial cell implants inhibited intimal proliferation to a greater extent than heparin, a potent inhibitor of smooth muscle cell proliferation *in vitro*. None of the controls had an effect. Only the engrafted endothelial cells inhibited growth—not the matrix alone nor the control grafts, with nonendothelial, Chinese hamster ovary (CHO) cells. The control exerted by the

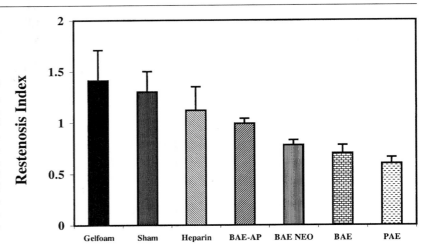

Fig. 27.1. Effects of perivascular endothelial cell implants on experimental restenosis one month after balloon injury of porcine carotid arteries. Normal bovine (BAE) and porcine (PAE) cells reduced experimental restenosis by 46 and 54%, respectively, in comparison to control Gelfoam and balloon injury alone (Sham). Perivascular-released heparin had no significant effect. BAE transfected with vector alone (BAE-NEO) significantly reduced intimal thickening by 42%, however BAE with reduced perlecan levels (BAE-AP) produced only an insignificant reduction of 26%. Modified from Nugent et al. [113,114].

endothelial cell implants in this animal model appeared to result solely from the biochemical effects of the engrafted cells rather than from a mechanical barrier function. Furthermore, the combined secretion of all endothelial compounds, at physiological doses, appeared to be superior to the administration of an isolated endothelial-derived product.

Further studies in our laboratory showed that this technology remained effective in a more complex animal model where the administration of isolated compounds have failed [113]. Both bovine and porcine aortic endothelial cell implants inhibited intimal hyperplasia (Fig. 27.1) and eliminated occlusive thrombosis (Fig. 27.2) in a porcine carotid artery overstretch injury model. Heparin release devices that had been effective in the rat did not produce a similar reduction in this model. Similar levels of inhibition of intimal hyperplasia by xenogeneic and allogeneic cells were obtained despite an elevation of serum antibodies to the xenogeneic cells and an increase in cellular infiltration into the xenografts. These data suggest that the delivery of isolated endothelial-derived compounds such as heparin may be sufficient for the inhibition of intimal hyperplasia in less complex animal models, but is insufficient for the treatment of vascular injury in more sophisticated models. Perhaps only the cell-based delivery of all endothelial compounds in physiological doses and subject to physiological control can regulate complex diseases like experimental and clinical restenosis.

COUPLING OF PERIVASCULAR ENDOTHELIAL CELL TRANSPLANTATION WITH MOLECULAR MODIFICATION TECHNOLOGY

While tissue engineering allows isolated endothelial cells to be engrafted within three-dimensional polymer matrices with retention of growth kinetics and biochemical function,

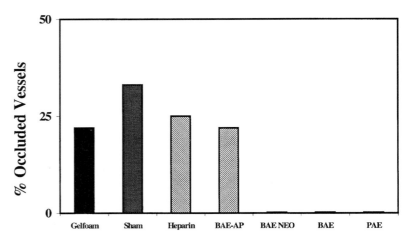

Fig. 27.2. Effects of perivascular endothelial cell implants on occlusive thrombosis one month after balloon injury of porcine carotid arteries. Normal bovine (BAE) and porcine (PAE) cells eliminated thrombosis, compared with control Gelfoam and balloon injury alone (Sham). Perivascular-released heparin had no significant effect. BAE transfected with vector alone (BAE-NEO) also eliminated thrombosis, however BAE with reduced perlecan levels (BAE-AP) had no effect. Modified from Nugent et al. [113,114].

it also enables the implantation of cells that have been genetically modified to produce more or less of specific compounds. Implants bearing modified endothelial cells might then be used to dissect the various aspects of the vascular response to injury. In a recent study in our lab, we generated stably transfected clones of bovine endothelial cells to express high levels of an antisense vector targeting domain III of perlecan, an especially important regulator of vascular homeostasis that is derived from endothelial cell [114]. Transfected cells produced significantly less perlecan *in vitro* and showed a reduced ability to inhibit FGF-2 binding to and mitogenic activity in cultured smooth muscle cells. When three-dimensional polymeric matrices containing these transfected endothelial cells were implanted adjacent to injured porcine carotid arteries, an interesting divergence of biological effect of the modified implants was observed. Endothelial cell implants with reduced perlecan expression (BAE-AP), were less effective than implants containing endothelial cells transfected with vector only (BAE-NEO) at inhibiting experimental restenosis (Fig. 27.1) yet showed a complete loss of the antithrombotic effect observed with the nontransfected cellular implants (Fig. 27.2). It is likely that endothelial control over intimal thickening results from a combination of perlecan and other secreted cell-based products such as nitric oxide, endothelins, prostaglandins, and a myriad of growth factors, cytokines, and vasoreactive agents [9,10]. By removing an important factor (perlecan) that directly interacts in these processes, the implanted cells displayed a decreased ability to reduce intimal thickening. These observations may explain why single endothelial-derived products, aimed at one of the cellular events thought to be involved in either thrombosis or smooth muscle cell proliferation, do not lead to full control of restenosis. Cell-based therapies offer new tools for added insight into vascular repair, and when coupled with molecular modification technology, are unlimited in their usefulness to further dissect the complex biology of vascular injury.

CONCLUSION

A wide range of techniques now exists for the delivery of endothelial cells to the site of blood vessel injury. The transition from experimental models of vascular injury to a clinical reduction in intimal hyperplasia may be enhanced by further elucidation of the biological mechanisms that determine growth after severe or prolonged vascular injury. The site and method of endothelial cell delivery as well as the type and number of cells delivered will dictate its effect on thrombosis and intimal hyperplasia after single acute or prolonged severe injury. The latter model may more resemble clinical coronary arterial instrumentation, characterized by deep intimal and medial disruption. The continued investigation into experimentally effective antiproliferative cell therapies will determine the degree of success of these techniques at limiting human restenosis. The innovative application of endothelial cells as a physiologically responsive drug delivery platform for cardiovascular therapy necessitates successful integration among the fields of polymer chemistry, cell biology, vascular biology, cardiovascular surgery, and clinical cardiology.

REFERENCES

1. Ip, J. H., Fuster, V., Badimon, L., Badimon, J., Taubman, M. B., and Chesebro, J. H. (1990). Syndromes of accelerated atherosclerosis: Role of vascular injury and smooth muscle proliferation. *J. Am. Cell. Cardiol.* **15**, 1667–1687.
2. McBride, W., Lange, R. A., and Hillis, L. D. (1988). Restenosis after successful coronary angioplasty: Pathophysiology and prevention. *N. Engl. J. Med.* **318**, 1734–1738.
3. Liu, M. W., Roubin, G. S., and Spencer, B. K. (1989). Restenosis after coronary angioplasty. Potential biologic determinants and role of intimal hyperplasia. *Circulation* **79**, 1374–1387.
4. Fanelli, C. R., and Aronoff, R. (1990). Restenosis following coronary angioplasty. *Am. Heart J.* **119**, 357–366.
5. O'Keefe, J. H., Rutherford, B. D., McConahay, D. R., *et al.* (1990). Multivessel coronary angioplasty from 1980 to 1989: Procedural results and long-term outcome. *J. Am. Coll. Cardiol.* **16**, 1097–1102.
6. Burkitt, H. G., Young, B., and Heath, J. W. (1993). "Wheater's Functional Histology: A Text and Colour Atlas," 3rd ed. Churchill-Livingstone, New York.
7. Luscher, T. F., and Barton, M. (1997). Biology of the endothelium. *Clin. Cardiol.* **20**(Suppl. 2), II-3-II-10.
8. Schwartz, S. M. (1997). Smooth muscle migration in atherosclerosis and restenosis. *J. Clin. Invest.* **100**(Suppl.), S87–S89.
9. Davies, M. G., and Hagen, P. O. (1993). Vascular endothelium: A new horizon. *Ann. Surg.* **218**, 593–609.
10. Rubanyi, G. M. (1993). The role of endothelium in cardiovascular homeostasis and diseases. *J. Cardiovasc. Pharmacol.* **22**, S1–S14.

11. Edelman, E. R., Nugent, M. A., Smith, L. T., and Karnovsky, M. J. (1992). Basic fibroblast growth factor enhances the coupling of intimal hyperplasia and proliferation of vasa vasorum in injured rat arteries. *J. Clin. Invest.* **89,** 465–473.

12. Casterella, P. J., and Teirstein, P. S. (1999). Prevention of coronary restenosis. *Cardiol. Rev.* **7,** 219–231.

13. Ellis, S. G., Roubin, G. S., Wilentz, J., Douglas, O. S., and King, S. B. (1989). Effect of 18- to 24-hour heparin administration for prevention of restenosis after uncomplicated coronary angioplasty. *Am. Heart J.* **117,** 777–782.

14. Lehmann, K. G., Doria, R. J., Feuer, J. M., *et al.* (1991). Paradoxical increase in restenosis rate with chronic heparin use: Final results of a randomized trial. *J. Am. Coll. Cardiol.* **17,** 181A (abstr.).

15. Leung, W.-H., Kaplan, A. V., Grant, G. W., *et al.* (1990). Local delivery of antithrombin agent reduces platelet deposition at the site of balloon angioplasty. *Circulation* **82,** III-428 (abstr.).

16. Powell, J. S., Clozel, J.-P., Muller, R. K. M., *et al.* (1989). Inhibitions of angiotensin-converting enzyme prevent myointimal proliferation after vascular injury. *Science* **245,** 187–189.

17. MERCATOR (1992). Does the new angiotensin converting enzyme inhibitor cilazapril prevent restenosis after percutaneous transluminal coronary angioplasty? *Circulation* **86,** 100–110.

18. Hoberg, E., Schwartz, F., Schomig, A., *et al.* (1990). Prevention of restenosis by verapamil. The verapamil angioplasty study (VAS). *Circulation* **82,** III-428.

19. Whitworth, H. B., Roubin, G. S., Hollman, J., *et al.* (1986). Effect of nifedipine in recurrent stenosis after percutaneous coronary angioplasty. *J. Am. Coll. Cardiol.* **8,** 1271–1276.

20. Handley, D. A., van Valen, R. G., Melden, M. K., and Saunders, R. N. (1986). Suppression of rat carotid lesion development by the calcium channal blocker PN-200-110. *Am. J. Pathol.* **124,** 88–93.

21. Henry, P. D., and Bentley, K. I. (1981). Suppression of atherogenesis in cholesterol-fed rabbits treated with nifedipine. *J. Clin. Invest.* **68,** 1366–1369.

22. El-Sanadiki, M. N., Cross, K. S., Murray, J. J., *et al.* (1990). Reduction of intimal hyperplasia and enhanced reactivity of experimental vein bypass grafts with verapamil treatment. *Ann. Surg.* **212,** 87–96.

23. Pepine, C. J., Hirshfeld, J. W., MacDonald, R. G., *et al.* (1990). A controlled trial of corticosteroids to prevent restenosis after coronary angioplasty. *Circulation* **81,** 1753–1761.

24. Edelman, E. R. (1999). Vascular tissue engineering: Designer arteries. *Circ. Res.* **85,** 1115–1117.

25. Zarge, J. I., Huang, P., and Greisler, G. P. (1997). Blood vessels. *In* "Tissue Engineering" (R. Lanza, R. Langer, and W. Chick, eds.), pp. 349–364. R. G. Landes, New York.

26. Antiplatelet Trialists' Collaboration (1994). Collaborative overview of randomised trials of antiplatelet therapy. II. Maintenance of vascular graft or arterial patency by antiplatelet therapy. Antiplatelet Trialists' Collaboration. *Br. Med. J.* **308,** 159–168.

27. Bos, G. W., Poot, A. A., Geugeling, T., van Aken, W. G., and Feijen, J. (1998). Small-diameter vascular graft prostheses: Current status. *Arch. Phys. Biochem.* **106,** 100–115.

28. Pevec, W. C., Darling, R. C., L'Italien, G. J., and Abbott, W. M. (1992). Femoropopliteal reconstruction with knitted, nonvelour Dacron versus expanded polytetrafluoroethylene. *J. Vasc. Surg.* **16,** 60–65.

29. Voorhees, A., Jaretski, A., and Blakemore, A. H. (1952). The use of tubes constructed from Vinyon 'N' cloth in bridging arterial defects. *Ann. Surg.* **135,** 332–336.

30. Herring, M., Gardner, A., and Glover, J. (1978). A single staged technique for seeding vascular grafts with autologous endothelium. *Surgery* **84,** 498–504.

31. Herring, M., Gardner, A., and Glover, J. (1979). Seeding endothelium onto canine arterial prostheses. The effects of graft design. *Arch. Surg. (Chicago)* **114,** 679–682.

32. Herring, M. B., Dilley, R., Jersild, R. A., Jr., Boxer, L., Gardner, A., and Glover, J. (1979). Seeding arterial prostheses with vascular endothelium. The nature of the lining. *Ann. Surg.* **190,** 84–90.

33. Graham, L. M., Vinter, D. W., Ford, J. W., Kahn, R. H., Burkel, W. E., and Stanley, J. C. (1979). Cultured autogenous endothelial cell seeding of prosthetic vascular grafts. *Surg. Forum* **30,** 204–206.

34. Graham, L. M., Vinter, D. W., Ford, J. W., Kahn, R. H., Burkel, W. E., and Stanley, J. C. (1980). Endothelial cell seeding of prosthetic vascular grafts: Early experimental studies with cultured autologous canine endothelium. *Arch. Surg. (Chicago)* **115,** 929–933.

35. Graham, L. M., Burkel, W. E., Ford, J. W., Vinter, D. W., Kahn, R. H., and Stanley, J. C. (1982). Expanded polytetrafluoroethylene vascular prostheses seeded with enzymatically derived and cultured canine endothelial cells. *Surgery* **91,** 550–559.

36. Schmidt, S. P., Hunter, T. J., Sharp, W. V., Malindzak, G. S., and Evancho, M. M. (1984). Endothelial cell-seeded four-millimeter Dacron vascular grafts: Effects of blood flow manipulation through the grafts. *J. Vasc. Surg.* **1,** 434–441.

37. Schmidt, S. P., Hunter, T. J., Hirko, M., *et al.* (1985). Small-diameter vascular prostheses: Two designs of PTFE and endothelial cell-seeded and nonseeded Dacron. *J. Vasc. Surg.* **2,** 292–297.

38. van Wachem, P. B., Beugeling, T., Feijen, J., *et al.* (1985). Interaction of cultured human endothelial cells with polymeric surfaces of diffrent wettabilities. *Biomaterials* **6,** 403–408.

39. van Wachem, P. B., Hogt, A. H., Beugeling, T., *et al.* (1987). Adhesion of cultured human endothelial cells onto methacrylate polymers with varying surface wettability and charge. *Biomaterials* **8,** 323–328.

40. McAuslan, B. R., and Johnson, G. (1987). Cell responses to biomaterials. I: Adhesion and growth of vascular endothelial cells on poly(hydroxyethyl methacrylate) following surface modification by hydrolytic etching. *J. Biomed. Mater. Res.* **21,** 921–935.

41. Ramires, P. A., Mirenghi, L., Romano, A. R., Palumbo, F., and Nicolardi, G. (2000). Plasma-treated PET surfaces improve the biocompatibility of human endothelial cells. *J. Biomed. Mater. Res.* **51,** 535–539.

42. Absolom, D. R., Hawthorn, L. A., and Chang, G. (1988). Endothelialization of polymer surfaces. *J. Biomed. Mater. Res.* **22,** 271–285.

43. McAuslan, B. R., Johnson, G., Hannan, G. N., Norris, W. D., and Exner, T. (1988). Cell responses to biomaterials. II: Endothelial cell adhesion and growth on perfluorosulfonic acid. *J. Biomed. Mater. Res.* **22**, 963–976.

44. Steele, J. G., Johnson, G., Norris, W. D., and Underwood, P. A. (1991). Adhesion and growth of cultured human endothelial cells on perfluorosulphonate: Role of vitronectin and fibronectin in cell attachment. *Biomaterials* **12**, 531–539.

45. Wigod, M. D., and Klitzman, B. (1993). Quantification of *in vitro* endothelial cell adhesion to vascular graft material. *J. Biomed. Mater. Res.* **27**, 1057–1062.

46. Brunstedt, M. R., Ziats, N. P., Rose-Caprara, V., *et al.* (1993). Attachment and proliferation of bovine aortic endothelial cells onto additive modified poly(ether urethane ureas). *J. Biomed. Mater. Res.* **27**, 483–492.

47. Zhang, J. C., Wojta, J., and Binder, B. R. (1995). Growth and fibrinolytic parameters of human umbilical vein endothelial cells seeded onto cardiovascular grafts. *J. Thorac. Cardiovasc. Surg.* **109**, 1059–1065.

48. Storck, J., Razek, H., and Zimmerman, E. R. (1996). Effect of polyvinyl chloride plastic on the growth and physiology of human umbilical vein endothelial cells. *Biomaterials* **17**, 1791–1794.

49. Cenni, E., Granchi, D., Ciapetti, G., *et al.* (1997). Expression of adhesion molecules on endothelial cells after contact with knitted Dacron. *Biomaterials* **18**, 489–494.

50. Cenni, E., Granchi, D., Ciapetti, G., *et al.* (1997). Cytokine production and adhesive protein expression by endothelial cells after contact with polyethylene terephthalate. *Biomaterials* **17**, 2071–2076.

51. Yeager, A., and Callow, A. D. (1988). New graft materials and current approaches to an acceptable small diameter vascular graft. *Trans. Am. Soc. Artif. Intern. Organs* **34**, 88–94.

52. Massia, S. P., and Hubbell, J. A. (1991). Human endothelial cell interactions with surface-coupled adhesion peptides on a nonadhesive glass substrate and two polymeric biomaterials. *J. Biomed. Mater. Res.* **25**, 223–242.

53. Cheresh, D. A. (1987). Human endothelial cells synthesize and express Arg-Gly-Asp directed adhesion receptor involved in attachment to fibrinogen and von Willebrand factor. *Proc. Natl. Acad. Sci. U.S.A.* **84**, 6471–6475.

54. Dejana, E., Languino, L. R., Colella, S., *et al.* (1988). The localization of platelet gbIIb-IIIa-related protein in endothelial cell adhesion structures. *Blood* **71**, 566–572.

55. Charo, I. F., Bekeart, L. S., and Philips, D. R. (1987). Platelet glycoprotein IIb-IIIa-like proteins mediate endothelial cell attachment to adhesive proteins and the extracellular matrix. *J. Biol. Chem.* **262**, 9935–9938.

56. Lin, H.-B., Sun, W., Mosher, D. F., *et al.* (1994). Synthesis, surface and cell-adhesion properties of polyurethanes containing covalently grafted RGD-peptides. *J. Biomed. Mater. Res.* **28**, 329–342.

57. Holland, J., Hersh, L., Bryhan, M., Onyiriuka, E., and Ziegler, L. (1996). Culture of human vascular endothelial cells on an RGD-containing synthetic peptide attached to a starch-coated polystyrene surface: Comparison with fibronectin-coated tissue culture grade polystyrene. *Biomaterials* **17**, 2147–2156.

58. Hubbell, J. A., Massia, S. P., Desai, N. P., and Drumheller, P. D. (1991). Endothelial cell-selective materials for tissue engineering in the vascular graft via a new receptor. *Biotechnology* **9**, 568–572.

59. Massia, S. P., and Hubbell, J. A. (1992). Vascular endothelial cell adhesion and spreading promoted by the peptide REDV of the IIICS region of plasma fibronectin is mediated by integrin alpha 4 beta 1. *J. Biol. Chem.* **267**, 14019–14026.

60. Zhang, S., Yan, L., Altman, M., *et al.* (1999). Biological surface engineering: A simple system for cell pattern formation. *Biomaterials* **20**, 1213–1220.

61. Thomas, C. H., Lhoest, J. B., Castner, D. G., McFarland, C. D., and Healy, K. E. (1999). Surfaces designed to control the projected area and shape of individual cells. *J. Biomech. Eng.* **121**, 40–48.

62. Wilson, J. M., Birinyi, L. K., Salomon, R. N., *et al.* (1989). Implantation of vascular grafts lined with genetically modified endothelial cells. *Science* **244**, 1344–1346.

63. Falk, J., Townsend, L. E., Vogel, L. M., *et al.* (1998). Improved adherence of genetically modified endothelial cells to small-diameter expanded polytetrafluoroethylene grafts in a canine model. *J. Vasc. Surg.* **27**, 902–908.

64. Belden, T. A., Schmidt, S. P., Falkow, L. J., and Sharp, W. V. (1982). Endothelial cell seeding of small-diameter vascular grafts. *Trans. Am. Soc. Artif. Intern. Organs* **28**, 173–177.

65. Gimbrone, M. A., Cotran, R. S., and Folkman, J. (1974). Human vascular endothelial cells in culture: growth and DNA synthesis. *J. Cell Biol.* **60**, 673–684.

66. Jaffe, E. A., Nachman, R. L., Becker, C. G., and Mimick, C. R. (1973). Culture of human endothelial cells derived from umbilical veins. *J. Clin. Invest.* **52**, 2745–2756.

67. Booyse, F. M., Sedlak, B. J., and Rafelson, M. E. (1975). Culture of arterial endothelial cells: Characterization and growth of bovine aortic endothelial cells. *Thromb. Diath. Haemorrh.* **34**, 825–839.

68. Slater, D. N., and Sloan, J. M. (1975). The porcine endothelial cell in culture. *Atherosclerosis* **21**, 259–272.

69. Ryan, U. S., Clements, E., Habilston, D., and Ryan, J. W. (1978). Isolation and culture of pulmonary artery endothelial cells. *Tissue Cell* **10**, 535–554.

70. Rosenthal, A. M., and Gotlieb, A. I. (1990). Macrovascular endothelial cells from porcine aorta. *In* "Cell Culture Techniques in Heart and Vessel Research" (H. M. Piper, ed.), pp. 117–129. Springer-Verlag, Berlin.

71. Muhl, E., Gatermann, S., Iven, H., Dendorfer, A., and Bruch, H. P. (1996). Local application of vancomycin for prophylaxis of graft infection: Release of vancomycin from antibiotic-bonded Dacron grafts, toxicity in endothelial cell culture, and efficacy against graft infection in an animal model. *Ann. Vasc. Surg.* **10**, 244–253.

72. Kent, K. C., Oshima, A., and Whittemore, A. D. (1992). Optimal seeding conditions for human endothelial cells. *Ann. Vasc. Surg.* **6**, 258–264.

73. Sugawara, Y., Miyata, T., Sato, O., Kimura, H., Namba, T., and Makuuchi, M. (1997). Rapid postincubation endothelial retention by Dacron grafts. *J. Surg. Res.* **67**, 132–136.

74. Sugawara, Y., Miyata, T., Sato, O., Kimura, H., Namba, T., and Makuuchi, M. (1997). Rapid postincubation endothelial retention by Dacron grafts. *J. Surg. Res.* **67**, 132–136.

75. Bordenave, L., Remy-Zolghadri, M., Fernandez, P., Bareille, R., and Midy, D. (1999). Clinical performance of vascular grafts lined with endothelial cells. *Endothelium* 6, 267–275.

76. Prendiville, E. J., Coleman, J. E., Callow, A. D., *et al.* (1991). Increased in-vitro incubation time of endothelial cells on fibronectin-treated ePTFE increases cell retention in blood flow. *Eur. J. Vasc. Surg.* 5, 311–319.

77. Zilla, P., Deutsch, M., Meinhart, J., *et al.* (1994). Clinical in vitro endothelialization of femoropopliteal bypass grafts: An actuarial follow-up over three years. *J. Vasc. Surg.* 19, 540–548.

78. Kadletz, M., Magometschnigg, H., Minar, E., *et al.* (1992). Implantation of in vitro endothelialized polytetrafluoroethylene grafts in human beings. A preliminary report. *J. Thorac. Cardiovasc. Surg.* 104, 736–742.

79. Leseche, G., Ohan, J., Bouttier, S., Palombi, T., Bertrand, P., and Andreassian, B. (1995). Above-knee femoropopliteal bypass grafting using endothelial cell seeded PTFE grafts: Five-year clinical experience. *Ann. Vasc. Surg.* 9(Suppl.), S15–S23.

80. Li, J. M., Menconi, M. J., Wheeler, H. B., Rohrer, M. J., Anderson, F. A., Jr., and Cutler, B. S. (1993). Experimental femoral vein reconstruction with expanded polytetrafluoroethylene grafts seeded with endothelial cells. *Cardiovasc. Surg.* 1, 362–368.

81. Magometschnigg, H., Kadletz, M., Vodrazka, M., *et al.* (1992). Prospective clinical study with in vitro endothelial cell lining of expanded polytetrafluoroethylene grafts in crural repeat reconstruction. *J. Vasc. Surg.* 15, 527–535.

82. Kadletz, M., Moser, R., Preiss, P., Deutsch, M., Zilla, P., and Fasol, R. (1987). In vitro lining of fibronectin coated PTFE grafts with cryopreserved saphenous vein endothelial cells. *Thorac. Cardiovasc. Surg.* 35, 143–147.

83. Lindblad, B., Wright, S. W., Sell, R. L., Burkel, W. E., Graham, L. M., and Stanley, J. C. (1987). Alternative techniques of seeding cultured endothelial cells to ePTFE grafts of different diameters, porosities, and surfaces. *J. Biomed. Mater. Res.* 21, 1013–1022.

84. Gerlach, J., Kreusel, K. M., Schauwecker, H. H., and Bucherl, E. S. (1989). Endothelial cell seeding on PTFE vascular prostheses using a standardized seeding technique. *Int. J. Artif. Organs* 12, 270–275.

85. Mazzucotelli, J. P., Roudiere, J. L., Bernex, F., Bertrand, P., Leandri, J., and Loisance, D. (1993). A new device for endothelial cell seeding of a small-caliber vascular prosthesis. *Artif. Organs* 17, 787–790.

86. van Wachem, P. B., Stronck, J. W., Koers-Zuideveld, R., Dijk, F., and Wildevuur, C. R. (1990). Vacuum cell seeding: A new method for the fast application of an evenly distributed cell layer on porous vascular grafts. *Biomaterials* 11, 602–606.

87. Bowlin, G. L., and Rittgers, S. E. (1997). Electrostatic endothelial cell seeding technique for small-diameter (6 mm). vascular prostheses: Feasibility testing. *Cell Transplant.* 6, 623–629.

88. Bowlin, G. L., and Rittgers, S. E. (1997). Electrostatic endothelial cell transplantation within small-diameter (6 mm). vascular prostheses: A prototype apparatus and procedure. *Cell Transplant.* 6, 631–637.

89. Greisler, H. P., Cziperle, D. J., Kim, D. U., *et al.* (1992). Enhanced endothelialization of expanded polytetrafluoroethylene grafts by fibroblast growth factor type 1 pretreatment. *Surgery* 112, 244–255.

90. Gray, J. L., Kang, S. S., Zenni, G. C., *et al.* (1994). FGF-1 affixation stimulates ePTFE endothelialization without intimal hyperplasia. *J. Surg. Res.* 57, 596–612.

91. Kang, S. S., Gosselin, C., Ren, D., and Greisler, H. P. (1995). Selective stimulation of endothelial cell proliferation with inhibition of smooth muscle cell proliferation by fibroblast growth factor-1 plus heparin delivered from fibrin glue suspensions. *Surgery* 118, 280–287.

92. Greisler, H. P., Gosselin, C., Ren, D., Kang, S. S., and Kim, D. U. (1996). Biointeractive polymers and tissue engineered blood vessels. *Biomaterials* 17, 329–336.

93. Gosselin, C., Vorp, D. A., Warty, V., *et al.* (1996). ePTFE coating with fibrin glue, FGF-1, and heparin: Effect on retention of seeded endothelial cells. *J. Surg. Res.* 60, 327–332.

94. Zarge, J. I., Huang, P., Husak, V., *et al.* (1997). Fibrin glue containing fibroblast growth factor type 1 and heparin with autologous endothelial cells reduces intimal hyperplasia in a canine carotid artery balloon injury model. *J. Vasc. Surg.* 25, 840–849.

95. Zarge, J. I., Gosselin, C., Huang, P., Vorp, D. A., Severyn, D. A., and Greisler, H. P. (1997). Platelet deposition on ePTFE grafts coated with fibrin glue with or without FGF-1 and heparin. *J. Surg. Res.* 67 4–8; errata: *Ibid.* 74(2), 197 (1998); 75(1), 90 (1998).

96. Weatherford, D. A., Sackman, J. E., Reddick, T. T., Freeman, M. B., Stevens, S. L., and Goldman, M. H. (1996). Vascular endothelial growth factor and heparin in a biologic glue promotes human aortic endothelial cell proliferation with aortic smooth muscle cell inhibition. *Surgery* 120, 433–439.

97. Bos, G. W., Schaarenborg, N. M., Poot, A. A., *et al.* (1999). Endotheliallization of crosslinked albumin-heparin gels. *Thromb. Haemostasis.* 82, 1757–1763.

98. Herring, M., Baughman, S., and Glover, J. (1985). Endothelium develops on seeded human arterial prosthesis: A brief clinical note. *J. Vasc. Surg.* 2, 727–730.

99. Zilla, P., Deutsch, M., Meinhart, J., *et al.* (1994). Clinical in vitro endothelialization of femoropopliteal bypass grafts: An actuarial follow-up over three years. *J. Vasc. Surg.* 19, 540–548.

100. Ortenwall, P., Wadenvik, H., Kutti, J., and Risberg, B. (1990). Endothelial cell seeding reduces thrombogenicity of Dacron grafts in humans. *J. Vasc. Surg.* 11, 403–410.

101. Ortenwall, P., Wadenvik, H., and Risberg, B. (1989). Reduced platelet deposition on seeded versus unseeded segments of expanded polytetrafluoroethylene grafts: Clinical observations after a 6-month follow-up. *J. Vasc. Surg.* 10, 374–380.

102. Leseche, G., Ohan, J., Bouttier, S., Palombi, T., Bertrand, P., and Andreassian, B. (1995). Above-knee femoropopliteal bypass grafting using endothelial cell seeded PTFE grafts: Five-year clinical experience. *Ann. Vasc. Surg.* 9(Suppl.), S15–S23.

103. Zilla, P., Deutsch, M., Meinhart, J., *et al.* (1994). Clinical in vitro endothelialization of femoropopliteal bypass grafts: An actuarial follow-up over three years. *J. Vasc. Surg.* **19**, 540–548.

104. Meinhart, J., Deutsch, M., and Zilla, P. (1997). Eight years of clinical endothelial cell transplantation. Closing the gap between prosthetic grafts and vein grafts. *Am. Soc. Artif. Intern. Organs. J.* **43**, M515–M521.

105. Deutsch, M., Meinhart, J., Fischlein, T., Preiss, P., and Zilla, P. (1999). Clinical autologous in vitro endothelialization of infrainguinal ePTFE grafts in 100 patients: A 9-year experience. *Surgery* **126**, 847–855.

106. Serruys, P. W., de Jaegere, P., Kiemeneij, F., *et al.* (1994). A comparison of balloon-expandable-stent implantation with balloon angioplasty in patients with coronary artery disease. Benestent Study Group. *N. Engl. J. Med.* **331**, 489–495.

107. George, B. S., Voorhees, W. D., Roubin, G. S., *et al.* (1993). Multicenter investigation of coronary stenting to treat acute or threatened closure after percutaneous transluminal coronary angioplasty: Clinical and angiographic outcomes. *J. Am. Coll. Cardiol.* **22**, 135–143.

108. Serruys, P. W., Emanuelsson, H., van der Giessen, W., *et al.* (1996). Heparin-coated Palmaz-Schatz stents in human coronary arteries. Early outcome of the Benestent-II Pilot Study. *Circulation* **93**, 412–422.

109. Van Belle, E., Tio, F. O., Couffinhal, T., Maillard, L., Passeri, J., and Isner, J. M. (1997). Stent endothelialization. Time course, impact of local catheter delivery, feasibility of recombinant protein administration, and response to cytokine expedition. *Circulation* **95**, 438–448.

110. Dichek, D. A., Neville, R. F., Zwiebel, J. A., *et al.* (1989). Seeding of intravascular stents with genetically engineered endothelial cells. *Circulation* **80**, 1347–1353.

111. Flugelman, M. Y., Virmani, R., Leon, M. B., Bowman, R. L., and Dichek, D. A. (1992). Genetically engineered endothelial cells remain adherent and viable after stent deployment and exposure to flow in vitro. *Circ. Res.* **70**, 348–354.

112. Nathan, A., Nugent, M. A., and Edelman, E. R. (1995). Tissue engineered perivascular endothelial cell implants regulate vascular injury. *Proc. Natl. Acad. Sci. U.S.A.* **92**, 8130–8134.

113. Nugent, H. M., Rogers, C., and Edelman, E. R. (1999). Endothelial implants inhibit intimal hyperplasia after porcine angioplasty. *Circ. Res.* **84**, 384–391.

114. Nugent, M. A., Nugent, H. M., Iozzo, R. V., Sanchack, K., and Edelman, E. R. (2000). Perlecan is required to inhibit thrombosis after deep vascular injury and contributes to endothelial cell-mediated inhibition of intimal hyperplasia. *Proc. Natl. Acad. Sci. U.S.A.* **97**, 6722–6727.

CHAPTER 28

MESENCHYMAL CELL CULTURE: BLOOD VESSELS

Lucie Germain, Murielle Rémy-Zolghadri, and François A. Auger

T he replacement of small-diameter vascular prostheses by nonautologous veins is the subject of intensive research. Attempts to produce synthetic, hybrid artificial or biological conduits are numerous and dependent on progress in technology and comprehension of the mechanisms implicated in their reconstruction. The production of a completely biological vascular prosthesis is described in this chapter. Exclusively composed of human cells embedded in their own extracellular matrix, this graft could be produced autologously and offers advantages related to its biological composition. The steps to produce such a living structure are detailed, and some important areas, such as the purity and the phenotype of cells, and the application of mechanical strain to the construct, are pointed out as essential factors for the success of our tissue-engineered blood vessel (TEBV). The main characteristics obtained with this prosthesis are presented, as well as directions future research could take to improve this human living prosthesis.

INTRODUCTION

The replacement of small-diameter blood vessels (≤ 6 mm) is a clinical situation frequently encountered by vascular and cardiothoracic surgeons [1]. Indeed, with the increased life expectancy, and nutritional and sedentary conditions of life in our industrialized countries, vascular pathologies have also rapidly escalated in number: coronary reconstruction and limb salvage have become common surgical procedures.

Thus, in most cases, the use of the autologous saphenous vein, which is considered to be excellent autologous graft material in small-diameter reconstructions [2–4], is the preferred solution [1]. However, this vein is not available in one-third of patients, and the use of alternative prostheses becomes necessary. Artificial materials are then proposed and are currently represented by commercially available polyester prostheses (Dacron) and extended poly(tetrafluoroethylene) (ePTFE) ones [1,4,5]. Unfortunately, these biomaterials are not hemocompatible enough to allow a long-term patency rate (>5 years) and to be implanted without the administration of an adequate anticoagulotherapy [6]. Despite numerous efforts to increase their biological properties [1], these biomaterials must be radically improved for their routine long-term use to be successful in small-diameter blood vessel reconstructions [6].

Accordingly, the introduction of autologous cells such as endothelial cells onto the biomaterial to cover the lumen and play a biological role seems to be a promising alternative [7–10]. However, the attempt to produce a completely biological substitute may be a more attractive solution with significant advantages [11]. During the last decades, progress in the field of tissue engineering has allowed investigation in this area [11–15]. Our laboratory was involved in such a project and, on the basis of previously published blood vessel models composed of living cells embedded in an exogenous collagen matrix [16–23], we used the self-assembly approach to develop the first completely biological tissue-engineered blood vessel (TEBV) constituted of living human cells in the absence of any synthetic or exogenous material [24].

The TEBV produced has a morphological and histological structure closely comparable to that of a natural blood vessel (Fig. 28.1). It is composed of an endothelium lying on a subendothelial layer of collagen fibers, media, and adventitia, respectively, composed of smooth muscle cells (SMCs) and fibroblasts surrounded by their own extracellular matrix, as in a native vessel (Fig. 28.1B). In the absence of culture support, this TEBV does not collapse (Fig. 28.1A), shows an impressive burst strength resistance (2500 mmHg) (Fig. 28.2), and can be easily handled, as was proven by testing in femoral interposition prostheses in dogs [24]. In light of its composition, this TEBV has the distinct advantage of being completely biological and thus, presents the possibility of being produced autologously with a wide range of inner diameters ($\geq$3 mm), is fully biocompatible, and is able to renew itself over time if damaged.

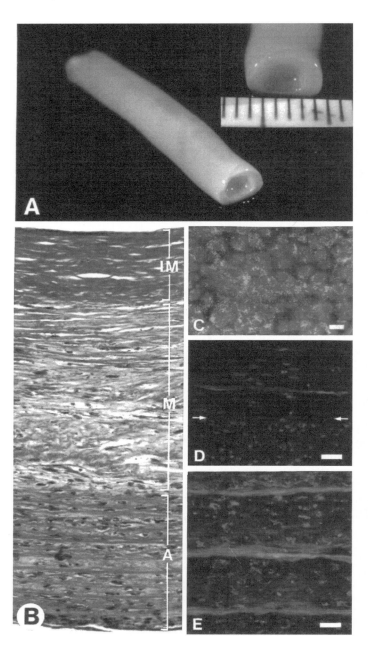

Fig. 28.1. Organization of the TEBV. (A) Macroscopic view of a mature TEBV (9 weeks of adventitial maturation). The vessel is self-supporting when removed from the culture medium (open lumen, 3 mm). Note that the various layers now form a continuous vascular wall (inset, graduation, 1 mm). (B) Paraffin cross section of the vascular wall stained with Masson's trichrome shows collagen in blue-green and cells in dark purple. Aside from an oversized internal elastic lamina (IM = 125 μm), the histology is similar to that of a muscular artery with a large media (M = 320 μm) and a surrounding adventitia (A = 325 μm). (C) En face view of the endothelium seeded on the IM. Cytoplasmic green fluorescence reveals cell viability, metabolic activity, and degree of confluence. Red fluorescence (Dil-ac-LDL uptake) confirms cell viability and the endothelial nature of the cells. Blue fluorescence shows the characteristic of von Willebrand factor expression in ECs (orange = red + green; pink = red + green + blue). Scale bar, 25 μm. (D) Frozen cross section of the media–adventitia junction (arrows) stained for desmin (nuclei are stained blue). Scale bar, 50 μm. (E) Frozen cross section of the adventitia double stained for elastin (green) and vimentin (red). Scale bar, 50 μm. From L'Heureux et al. [24]. Courtesy of FASEB J.

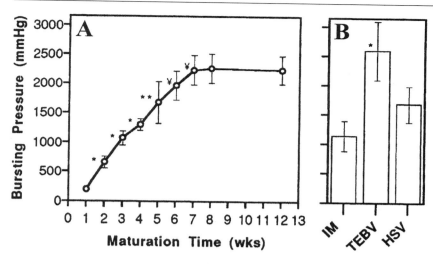

Fig. 28.2. Mechanical strength of the TEBV. (A) Burst strength of tissue-engineered adventitia as a function of maturation time: *significantly different from preceding point (p < 0.001; **p < 0.005, ¥ p < 0.05); Student's t test, n = 8–13. (B) Burst strength of rehydrated IM alone (n = 6), matured TEBV (9 weeks of adventitial maturation, n = 5), and human saphenous veins (HSV, n = 2); *significantly different from HSV (p < 0.05 ANOVA). From L'Heureux et al. [24]. Courtesy of FASEB J.

Consequently, the production of these autologous TEBV conduits, even though their preparation is quite long (4 months), could be the ideal solution for most clinical situations and could possibly eliminate reinterventions and anticoagulotherapies associated with biomaterial prosthesis implantations.

STEPS FOR THE TEBV PRODUCTION

The elaboration of the TEBV requires several steps, which must be appropriately coordinated to obtain the complete structure illustrated in Fig. 28.3. The technology used in our laboratory could be qualified as a "self-assembly" method [12] because we prepare *in vitro* sheets of cells (SMCs or fibroblasts) surrounded by their own *de novo* produced extracellular matrix. These sheets could be easily handled and wrapped around a tubular support to obtain a cylindrical structure (Fig. 28.3). The combination of this method with the appropriate cell culture conditions led to the fusion of the sheets and the restructuring of the tissue layers by the cells [24].

The first to be wrapped around the support is a fibroblastic sheet, which is dehydrated to keep only the matrix structure: this is the IM (inner membrane) (Figs. 28.1B and 28.3). Then a sheet of SMC is wrapped around the IM to produce a media-like tissue (M) (Figs. 28.1B and 28.3), and finally a living fibroblastic sheet is added for the reconstruction of the adventitia (A) (Figs. 28.1B and 28.3). The ultimate step is the introduction of endothelial cells into the lumen after the removal of the support, and their adherence to the IM to obtain a confluent endothelium (Figs. 28.1B and 28.3).

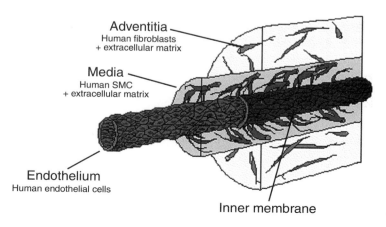

Adventitia
Human fibroblasts
+ extracellular matrix

Media
Human SMC
+ extracellular matrix

Endothelium
Human endothelial cells

Inner membrane

Fig. 28.3. Layers of the TEBV. The TEBV is produced by rolling the inner membrane (IM) around a tubular support, then rolling the media layer, and the adventitia. Finally, after removal of the support, endothelial cells are introduced into the lumen to produce the endothelial layer (in red).

MATERIALS AND METHODS
CELL HARVESTING

The procedures described are routinely used in our laboratory. The compositions of the chemicals cited in all the subsections are as given.

- *DME-HAM*: a combination of the Dulbecco–Vogt modification of Eagle's Medium (DME) with Ham's F-12 in a 3:1 proportion (Gibco BRL, Burlington, Ontario, Canada)
- *Endothelial cell culture medium*: M-199 (Sigma) supplemented with 20% Fetal Clone II serum (FCS) (HyClone, POI NorsScience, Avrora, Ontario, Canada), 2 mM L-glutamine, 50 U/ml heparin, 25 µg/ml endothelial cell growth factor supplement (Sigma, Inc. St. Louic, MO), 100 U/ml penicillin G, and 25 µg/ml of gentamicin
- *SMC and fibroblast medium*: DME-HAM supplemented with 10% fetal calf serum (HyClone), 100 U/ml of penicillin G, and 25 µg/ml gentamicin
- *Calcium-free HEPES buffer*: 10 mM HEPES, 119 mM NaCl, 6.7 mM KCl, and 11 mM glucose, at pH 7.35
- *PBS 10×*: 1.4 M NaCl, 27 mM KCl, 80 mM Na_2HPO_4, 14 mM KH_2PO_4; the solution is diluted to 1×, pH 7.4, for use
- *Trypsin–EDTA* solution: 0.05% trypsin 1-300 (ICN Biochemical, Montreal, Quebec, Canada) containing 0.01% EDTA (Becker Chemical Co., Phillipsburg, NJ) in PBS. Storage at –20°C (several months) and 4°C (a few days)
- *Freezing medium*: 60% DME-HAM, 30% FCS, 10% dimethyl sulfoxide (DMSO).

Endothelial Cells

Human endothelial cells (EC) are harvested by the enzymatic digestion of the vein of the umbilical cord (10–30 cm long). The use of thermolysin [25] instead of collagenase [26] allows the production of pure cultures of EC with a 100% success rate. The harvesting proceeds as follows.

- Collect umbilical cords in DME-HAM, 10% Fetal Clone II Serum (FCS), 100 UI/ml penicillin G (Sigma), 25 µg/ml gentamicin sulfate (Schering Inc., Pointe-Claire, Quebec, Canada), and 0.5 µg/ml amphotericin B. Proceed within 12 h after delivery.
- Cannulate the cord at each extremity, rinse with calcium-free HEPES buffer, then fill with thermolysin (Sigma) (250 µg/ml in HEPES buffer with 5 mM $CaCl_2$) for 30 min in a warm bath (37°C).
- Collect the effluent and the subsequent wash (10–20 ml of HEPES buffer).
- Pellet the cells (10 min at 500*g*), resuspend in culture medium, and plate on gelatin-coated flasks (one 25-cm² flask previously coated with 0,2% gelatin per umbilical cord). The procedure for the gelatin coating of culture flasks is described shortly.

In general, 3–5 days is necessary to obtain cells at 90% confluence.

Smooth Muscle Cells

Human smooth muscle cells are harvested from umbilical cords using the explant method described by Ross [27]. Briefly, the protocol is the following:

- After harvesting EC (enzymatic treatment), open the vein longitudinally and fix on a dissection board with the luminal face on top.
- Dissect and cut squares of the media. Place five to eight explants in each 25-cm² culture flask.
- Add culture medium to the explants (squares of tissue) in such a manner that the plate surface is just re-covered by a thin film of liquid.
- Place the cells to be cultured in an incubator at 37°C in a 8% CO_2 air atmosphere, 100% humidity, during 4 days before the first medium replacement.
- Then, change the culture medium three times a week.
- When cells become numerous, increase the volume of culture medium to 5 ml.

• When the cells begin to detach from the explants, passage them as described in the section entitled Passaging.

Fibroblasts

Human fibroblasts are harvested from human skin biopsies according to the following procedure:

• Harvest surgically removed skin in DME-HAM, 10% FCS, 100 UI/ml penicillin G, 25 µg/ml gentamicin sulfate, and 0.5 µg/ml amphotericin B.
• Wash the sample abundantly in PBS until the supernatant becomes clear.
• Cut the skin in 5-mm-wide rectangular pieces and place them in a solution of thermolysin (Sigma) (500 µg/ml PBS) overnight at 4°C.
• Separate the epidermis from the dermis with forceps. Cut the dermis into smaller pieces and place in a collagenase H (Boehringer-Mannheim, Montreal, Quebec, Canada) solution (0.07% in culture medium) for 3 h at 37°C with agitation.
• At the end of the incubation period, collect the supernatant and centrifuge at 500g for 10 min. Resuspend the cells for counting.
• Plate the cells in 75-cm^2 flasks at a density of 10^6/flask.
• Change the culture medium three times a week.

Cells reached 90% confluence approximately 5 days after plating.

CELL CULTURE AND AMPLIFICATION

For the primary culture step and the amplification, the basic media compositions are those classically mentioned in the literature [28–30]. Depending on the cell type, culture media must be supplemented with internally tested and validated sera, adequate additives, and growth factors. The compositions of culture media used in our laboratory were listed earlier. The culture medium is changed three times a week, using 5 ml for 25-cm^2 and 15 ml for 75-cm^2 culture flasks. Cultures are kept at 37°C in an incubator with an 8% CO_2 air atmosphere, at 100% humidity.

GELATINATION OF CULTURE FLASKS

The coating of culture flasks with gelatin is necessary for EC amplification. A 1% gelatin solution, prepared in water and filtered with a 0.22-µm sterile filter, can be stored at 4°C for several months. For use, the sterile solution is diluted in sterile water to a final concentration of 0.2%, then put in the flasks (10 ml/75-cm^2 flask, 3 ml/25-cm^2 flask, and 3 ml/9-cm^2 petri dish) and left overnight at 4°C. The solution is then carefully removed and the flasks are air-dried during 24 h. The coated flasks can be stored at 4°C.

PASSAGING

Continued cell growth may be maintained by transferring the cells into new flasks at a lower density, after their detachment by trypsin (passage). Depending on the cell type, normal cells are kept for 5–10 passages. During this period, the cell phenotype may be maintained and the major characteristics preserved. It is important to appropriately monitor each culture during the passages, as described in connection with the disccussion, later, of cell purity and identification. Cells are passaged when they reach 90% confluence and are systematically plated at 10,000 cells/cm^2. The procedure is the following:

• Aspirate the medium, wash rapidly once with the trypsin–EDTA solution.
• Add trypsin–EDTA (just enough to cover the surface of the flask), incubate at 37°C until cell detachment (3–10 min), monitor with a phase contrast microscope.
• Add an equal volume of culture medium to stop the action of the trypsin.
• Collect, count, and pellet (10 min at 500g) the cells.
• Resuspend the cells at the desired concentration. Replate in new culture flasks for amplification.

CRYOPRESERVATION

Cells may be stored frozen in liquid nitrogen (1 or 2×10^6 cells/ml) for several years. With an appropriate technique, cell viability after thawing is characteristically 95%.

Cell Freezing

The freezing protocol is as follows:

- Perform the first four steps of cell passaging, then proceed on crushed ice for the following steps.
- Add the freezing medium to the pellet to obtain the desired cell concentration.
- Aliquot 1 ml/cryovial (Nalgene, vWR Canlab, Ville Mont Royal, Quebec, Canada) and rapidly transfer to −80°C in Styrofoam box for 24 h before transferring the vials to liquid nitrogen.
- Keep the cells in liquid nitrogen until their use.

Cells are thawed by means of the following steps:

Cell Thawing

- Remove the vial from liquid nitrogen and thaw in warm water (37°C).
- As soon as the medium is liquid, add 10 ml of cold culture medium. Estimate the cell viability by trypan blue exclusion.
- Centrifuge 10 min at 500*g* to pellet the cells.
- Resuspend in culture medium at the desired concentration and plate at 10,000 cells per square centimeter.

CELL IDENTIFICATION AND PURITY

To characterize the lineage of the cultured cells and their purity, coverslips are prepared for immunocytochemistry with specific markers at each amplification step: assessment of the von Willebrand factor intracytoplasmic inclusions for the EC and of smooth muscle α-actin for the SMC (Fig. 28.4). Concerning the fibroblasts, to date, none of the commercially available markers we have tried have shown convincing specific staining. Therefore, fibroblasts are identified by their lack of both smooth muscle α-actin and von Willebrand factor staining. To assess the purity of the cultures, it is very important to process the cultures for negative markers (e.g., smooth muscle α-actin staining of endothelial cells), since it is much easier to detect a single positive SMC in otherwise negative endothelial cell cultures than a negative SMC in endothelial cells that are otherwise positive for von Willebrand factor (Fig. 28.4). The protocol for coverslip preparation is the following:

- Put 10 coverslips (16 × 16 mm) in a 60-mm-diameter petri dish (gelatin-coated for EC).
- Prepare a cell suspension containing 4000 cells/50 μl of the corresponding culture medium (depending on cell type).
- Put a drop of 50 μl on each coverslip.
- Let the cells adhere during a minimum of 2 h in incubators at 37°C in an 8% CO_2 air atmosphere, at 100% humidity.
- Then add 5 ml of the corresponding medium and culture until 85% confluence is reached (2–4 days, depending on the cell type). Fix the cells according to the following steps:
  - Wash 3 times with 5 ml of DME-HAM alone.
  - Add 5 ml of fixative solution (depending on the antibody used for the characterization): 90% acetone: 10% water solution, 10 min at −20°C in the case of von Willebrand factor staining and 100% methanol, 10 min at −20°C in the case of α-actin staining.
  - Wash extensively with PBS.
  - Then, stock at 4°C in PBS until use. For long-term conservation add 0.1% sodium azide to the PBS.

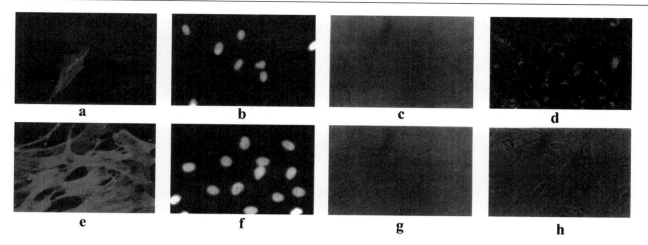

Fig. 28.4. Characterization of human cell types in culture. (a)–(d) Umbilical vein endothelial cells. (e)–(g) Smooth muscle cells. (h) Fibroblasts. (a) and (e) α-Actin immunofluorescent staining; (b) and (f) Hoechst staining of nuclei; (c), (g) and (h) phase contrast microscopy; (d) von Willebrand factor immunofluorescent staining. (a)–(c) and (e)–(g) are photographs of the same fields. Original magnification ×40. From Germain et al. [11]. Courtesy of Med. Biol. Eng. Comput.

Coverslips can be used for classical immunofluorescent staining of the cells with markers, and these cultures can be carefully analyzed for cell identification and purity determination.

CELL SHEET PRODUCTION AND PROCESSING
The self-assembly approach is based on the ability of fibroblasts and SMC to produce and organize an extracellular matrix when cultured in media containing ascorbic acid. The procedures for the production of SMC and fibroblast sheets are similar except that the length of the culture time varies: SMC sheets are obtained in 2–3 weeks (depending on the cell strain), whereas fibroblasts are cultured for 4 weeks.

The following steps explain these procedures.

Preparation of Cell Sheets
- Detach the cells (SMC or fibroblasts) at 90% confluence from the flask according to the protocol described earlier under Passaging.
- Replate 750,000 cells in each 75-cm^2 culture flask.
- Add culture medium supplemented with 50 μg/ml extemporaneously prepared ascorbate (Sigma).
- Change the culture medium three times a week for the appropriate culture period.

Cell Sheet Wrapping
- Open the culture flasks and using forceps, carefully detach the sheet from the culture flask.
- Wrap around a tubular support [preferentially of extended poly(tetrafluoroethylene) and chosen according to the desired inner diameter].
- Culture in 50-ml tubes in the following culture medium: DME-HAM, 10% FCS, 100 UI/ml penicillin G, 25 μg/ml gentamicin sulfate, and 50 μg/ml extemporaneously prepared ascorbate.
- Change the medium three times a week during the appropriate maturation period.

Depending on the reconstructed vessel layer (IM, media, or adventitia), the time period necessary for the complete maturation of each layer is different, as indicated in the following subsections.

IM Maturation and Processing

The IM layer is composed of a collagen matrix free of living cells. For its production, the wrapped sheet of fibroblasts is maintained in culture during 30 days before its dehydration by air-drying. The dehydrated IM is rinsed in distilled water and preserved for several months at −20°C until use.

Media Maturation

The media wrapped around the IM are kept in culture during one week before the rolling of the adventitial fibroblast sheet.

Adventitia Maturation

The adventitial layer (i.e., the sheet of fibroblasts) is wrapped around the media–IM construct and kept in culture during an optimal time, which was determined to be 7 weeks after rolling [24]. This period is necessary for the acquisition of an optimal strength for resistance to bursting by the structure. Indeed, as illustrated in Fig. 28.2, the resistance of the adventitia increases over time to reach a maximum value after 7 weeks in culture.

ENDOTHELIALIZATION OF THE TEBV

The last step before the implantation of the TEBV is endothelialization. For this purpose, the constructed tissue is removed from the support and filled with a suspension of freshly trypsinated EC as described under Passaging. The construction is rotated during 3 h to allow the attachment of the endothelial cells to the IM. Then the TEBV is maintained in culture conditions allowing the nutrition of the internal EC by a circulatory loop of feeding medium and of the fibroblasts and SMC by an bath of culture medium.

Seven additional days must be allowed to complete the maturation of the reconstructed TEBV before its implantation.

DISCUSSION

Very few solutions are available for vascular pathologies that call for the replacement of a segment of an altered or obstructed vessel and none are completely and definitively satisfactory. The current best method of reconstruction is the use of an autologous saphenous vein, when available [3]. This tissue encounters some morphological adaptations when transplanted in limb arteries and exhibits patency rates of $81 \pm 6\%$ 5 years after implantation [3]. When it is impossible to use this vein, alternative prostheses made of artificial materials (Dacron or ePTFE) are proposed by surgeons, depending on the patient's expectancy of life and health [5]. The use of another source of prostheses (i.e., veins or arteries) harvested from human umbilical cords [31], from human sources [32,33], or from animal ones [34,35] was transiently tested but is no longer in routine clinical use because of the numerous disadvantages involved. Thus, new strategies to improve graft patency for long-term implantation periods must be developed, and in this domain, the production of an autologous biological prosthesis could provide the best solution. This challenge passed from dream to reality in 1998 when we published the first biological tissue-engineered blood vessel produced *in vitro* [24]. Composed exclusively of human cells, its construction is based on several criteria, the most important being the property of the cells to produce and organize their own extracellular matrix *in vitro* in the presence of appropriate cell culture nutrients and with adequate cell culture three-dimensional supports and periods of maturation. This autoassembly approach leads to the structuralization of well-organized tissues [12].

In this respect, the use of appropriate sera in culture media is essential not only for cell nutrition but also for the maintenance of the cells' characteristics. The latter must be carefully monitored during culture. Indeed, we demonstrated in 1997 that for human dermal fibroblasts, the choice of serum can lead to differentiation or to significant changes in the cell phenotype [36]. Thus, the choice of serum is of crucial importance for maintaining cell characteristics and properties [36,37]. The addition of ascorbic acid to the culture medium allows cells to produce an abundant extracellular matrix, as shown in two-dimensional culture conditions as well as three-dimensional ones [24,38,39]. The application of an adequate mechanical strain to cells (i.e., rolling around a tubular support) allows the reorientation of the cells and their extracellular matrix as indicated by works published earlier [17,40–43]. The surrounding matrix is composed of collagen fibers [24], laminin, fibronectin, and various glycosaminoglycans (unpublished data) in the case of SMC and fibroblast (Fig. 28.5) cultures and of elastin for the adventitial layer (Fig. 28.1E) [24]. This is the first time that a vascular prosthesis prepared *in vitro* has displayed such characteristics; none of the other biological vascular prostheses have presented similar patterns of differentiation and function [21,22]. The absence of elastin in the model produced by Weinberg and Bell, who used comparable cell types of bovine origin [22], might be due to the shorter period of maturation of the construction *in vitro* (4 weeks) and to the level of proteases. Concerning Niklason's model [21], we can tentatively attribute this failure to the absence of fibroblast cells in the model and/or to the different culture medium. The shorter maturation period *in vitro* for their substitutes is another factor. The prostheses implanted by Hirai and Matsuda show "fiber-like elastin" after a 12-week implantation period in the dog [18]. Moreover, the interesting model developed by Niklason *et al.* [21] was produced under pulsatile conditions, which are known to favor cell organization and maturation; but despite this characteristic, elastic fibers were lacking. The TEBV we produced (in the absence of pulsatile conditions) displayed a well-organized matrix as shown by transmission electron microscopy analyses (Fig. 28.5) and presented an impressive resistance to bursting pressure (Fig. 28.2). Thus, we believe that the application of pulsatile conditions will improve our present model. This additional culture parameter is under investigation in our laboratory.

The introduction of endothelial cells into the lumen of the TEBV led to the formation of a complete endothelium after a 7-day maturation period, as described in Fig. 28.6. This endothelium exhibited an elegant cobblestone morphology, and the cells were successfully assessed for the production of intracytoplasmic Weibel–Palade bodies (Fig. 28.1C). The functionality of this layer was measured by the incorporation of Dil-ac-LDL (Fig. 28.1C) and by the inhibition of platelet adhesion (Fig. 28.6). The importance of the biological functions of this endothelium is evident and has a crucial importance for the success of small-diameter vascular conduits [9]. Nevertheless, the point to be stressed is the necessity of obtaining a complete endothelium that exhibits an appropriate phenotype (i.e., antithrombotic) at the time of implantation. Indeed, this monolayer has displayed a capacity to shift from an antithrombogenic state to a thrombogenic one [44], depending on environmental stimuli (circulatory ones as well as prosthesis materials) [45]. Moreover, even if *in vivo* endothelialization of synthetic prostheses occurs in animal models, this endothelialization is a phenomenon rarely observed in human beings [46]. The endothelium is also able to shift promptly from an antiadhesive surface to a proadhesive one, being responsible for the extravasation of abundant leukocytes and the initiation of a rapid inflammation process. This characteristic is dependent on the expression of specific adhesive molecules by the membrane of activated endothelial cells and varies with the environmental conditions applied to the endothelium [47]. Thus, the endothelial layer controls early as well as late events after

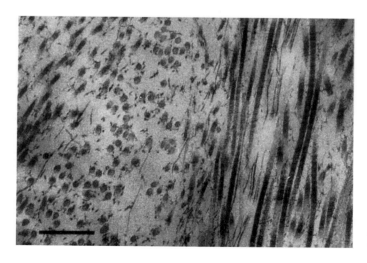

Fig. 28.5. *Extracellular matrix ultrastructure. Transmission electron micrograph of the adventitial matrix. Uranyl acetate and lead citrate stain. Scale bar, 500 nm. From L'Heureux et al. [24]. Courtesy of FASEB J.*

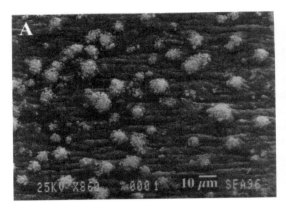
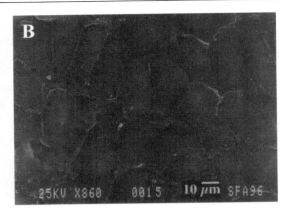

Fig. 28.6. *Endothelium view in scanning electron microscopy analysis. Scanning electron micrographs of unendothelialized IM (A) promoted platelet adhesion and activation whereas endothelialized IM (B) almost completely inhibited the process. From L'Heureux et al. [24]. Courtesy of FASEB J.*

graft implantation. Work is in progress at our institution to further assess the phenotypic state of this tissue.

The properties of the human TEBV were promising enough to test its implantation in a dog femorofemoral interposition. Because of the xenogeneic situation, human EC were not added to the TEBV, and the prostheses were implanted for only a week before explantation: an encouraging 50% patency rate was obtained despite these unfavorable clinical conditions [24]. An autologous situation could be more realistic, since it allows the addition of an endothelial cell monolayer, which represents a good parameter for success. But the patent grafts showed a well-preserved structure without early tearing or dilatation.

In conclusion, the human TEBV produced in our laboratory displays promising characteristics and its production, even if lengthy, could be largely compensated for by its intrinsic properties, which are, until now, not shared with other models produced *in vitro*.

CONCLUSION

Since the first clinical attempt to implant a small-diameter vascular graft, many advances have been made to produce acceptable vascular prostheses. Progressive tissue engineering methods lead to a novel approach based on an autologous self-assembly approach [12]. The model presented in this chapter offers interesting advantages in comparison to earlier ones, and this new autologous tissue self-assembly technology, if combined with the exciting potential of gene therapy [48], could pave the way to future clinical improvements for the treatment of vascular pathologies.

ACKNOWLEDGMENTS

The authors are grateful to Mrs. Cindy Hayward de Mendoza for skillful revision of the manuscript, and to all present and past members of the LOEX for their contributions to the work presented in this review. This work was supported by the Medical Research Council of Canada, the Heart and Stroke Foundation, and the Fonds de la Recherche en Santé du Québec. LG is recipient of scolarships from the Fonds de la Recherche en Santé du Québec and the Medical Research Council of Canada. MR-Z is the recipient of scholarships from the Association en Recherche sur la Cancer, the Ministère de l'Education du Québec, and the Institut National en Santé et Recherches Médicales.

REFERENCES

1. Zdrahala, R. J. (1996). Small caliber vascular grafts. Part I: State of the art. *J. Biomater. Appl.* **10**, 309–329.
2. Veith, F. J., Gupta, S. K., Ascer, E., White-Flores, S., Samson, R. H., Scher, L. A., Towne, J. B., Bernhard, V. M., Bonier, P., Flinn, W. R., Astelford, P., Yao, J. S. T., and Bergan, J. J. (1986). Six-year prospective multicenter randomized comparison of autologous saphenous vein and expanded polytetrafluoroethylene grafts in infrainguinal arterial reconstructions. *J. Vasc. Surg.* **3**, 104–114.
3. Archie, J. P., Jr. (1994). Femoropopliteal bypass with either adequate ipsilateral reversed saphenous vein or obligatory polytetrafluoroethylene. *Ann. Vasc. Surg.* **8**, 475–484.

4. Abbott, W. M., and Vignati, J. J. (1995). Prosthetic grafts: When are they a reasonable alternative? *Semin. Vasc. Surg.* **8**, 236–245.

5. Abbott, W. M. (1997). Prosthetic above-knee femoral-popliteal bypass: Indications and choice of graft. *Semin. Vasc. Surg.* **10**, 3–7.

6. Greenwald, S. E., and Berry, C. L. (2000). Improving vascular grafts: The importance of mechanical and haemodynamic properties. *J. Pathol.* **190**, 292–299.

7. Deutsch, M., Meinhart, J., Fischlein, T., Preiss, P., and Zilla, P. (1999). Clinical autologous in vitro endothelialization of infrainguinal ePTFE grafts in 100 patients: 9-year experience. *Surgery* **126**, 847–855.

8. Meinhart, J., Deutsch, M., and Zilla, P. (1997). Eight years of clinical endothelial cell transplantation closing the gap between prosthetic grafts and vein grafts. *Am. Soc. Artif. Intern. Organs J.* **43**, M515–M521.

9. Zilla, P., von Oppell, U., and Deutsch, M. (1993). The endothelium: A key for the future. *J. Cardiovasc. Surg.* **8**, 32–60.

10. Zilla, P. (1999). Endothelial cell transplantation. *Semin. Vasc. Surg.* **12**, 52–63.

11. Germain, L., Rémy-Zolghadri, M., and Auger, F. A. (2000). Tissue engineering of vascular system: From capillaries to larger blood vessels. *Med. Biol. Eng. Comput.* **38**, 232–240.

12. Auger, F. A., Rémy-Zolghadri, M., Grenier, G., and Germain, L. (2000). The self-assembly approach for organ reconstruction by tissue engineering. *E-biomed* **1**, 75–86.

13. Jankowski, R. J., and Wagner, W. R. (1999). Directions in cardiovascular tissue engineering. *Clin. Plast. Surg.* **26**, 605–616.

14. Langer R., and Vacanti, J. P. (1993). Tissue engineering. *Science* **260**, 920–926.

15. Niklason, L. E., and Langer, R. S. (1997). Advances in tissue engineering of blood vessels and other tissues. *Transplant. Immunol.* **5**, 303–306.

16. Ziegler, T., and Nerem, R. M. (1994). Tissue engineering a blood vessel: Regulation of vascular biology by mechanical stresses. *J. Cell. Biochem.* **56**, 204–209.

17. Hirai, J., and Matsuda, T. (1995). Self-organized, tubular hybrid vascular tissue composed of vascular cells and collagen for low-pressure-loaded venous system. *Cell Transplant.* **4**, 597–608.

18. Hirai, J., and Matsuda, T. (1996). Venous reconstruction using hybrid vascular tissue composed of vascular cells and collagen: Tissue regeneration process. *Cell Transplant.* **5**, 93–105.

19. L'Heureux, N., Germain, L., Labbé, R., and Auger, F. A. (1993). *In vitro* construction of human vessel from cultured vascular cells: A morphologic study. *J. Vasc. Surg.* **17**, 499–509.

20. Matsuda, T., Akutsu, T., Kira, K., and Matsumoto, H. (1989). Development of hybrid compliant graft: Rapid preparative method for reconstruction of a vascular wall. *Trans. Am. Soc. Artif. Intern. Organs* **35**, 553–555.

21. Niklason, L. E., Gao, J., Abbott, W. M., Hirschi, K. K., Houser, S., Marini, R., and Langer, R. (1999). Functional arteries grown in vitro. *Science* **284**, 489–493.

22. Weinberg, C. B., and Bell, E. (1986). A blood vessel model constructed from collagen and cultured vascular cells. *Science* **231**, 397–400.

23. Ziegler, T., Alexander, R. W., and Nerem, R. M. (1995). An endothelial cell-smooth muscle cell co-culture model for use in the investigation of flow effects on vascular biology. *Ann. Biomed. Eng.* **23**, 216–225.

24. L'Heureux, N., Pâquet, S., Labbé, R., Germain, L., and Auger, F. A. (1998). A completely biological tissue-engineered human blood vessel. *FASEB J.* **12**, 47–56.

25. Germain, L., Rouabhia, M., Guignard, R., Carrier, L., Bouvard, V., and Auger, F. A. (1993). Improvement of human keratinocyte isolation and culture using thermolysin. *Burns* **19**, 99–104.

26. Jaffe, E., Nachman, R. L., Becker, C. G., and Minick, C. R. (1973). Culture of human endothelial cells derived from umbilical vein. Identification by morphologic and immunologic criteria. *J. Clin. Invest.* **52**, 2745–2756.

27. Ross, R. (1971). The smooth muscle cell. II. Growth of smooth muscle in culture and formation of elastic fibers. *J. Cell Biol.* **50**, 172–186.

28. Campbell, J. H., and Campbell, G. R. (1993). Culture techniques and their applications to studies of vascular smooth muscle. *Clin. Sci.* **85**, 501–513.

29. Jaffe, E. A. (1980). Culture of human endothelial cells. *Transplant. Proc.* **12**, 49–53.

30. Pinney, E., Liu, K., Sheeman, B., and Mansbridge, J. (2000). Human three-dimensional fibroblast cultures express angiogenic activity. *J. Cell. Physiol.* **183**, 74–82.

31. Dardik, H. (1989). Modified human umbilical vein allograft. *In* "Vascular Surgery" (R. Brutherford, ed.), pp. 474–480. Saunders, Philadelphia.

32. Tice, D., and Zerbino, V. (1972). Clinical experience with preserved human allografts for vascular reconstruction. *Surgery* **72**, 260–267.

33. Vermassen, F., Degrieck, N., De Kock, L., Eeckhout, K., Van Landuyt, K., and Derom, F. (1994). The use of homologous veins in arterial reconstruction. An experimental study. *Acta Chir. Belg.* **94**, 277–283.

34. Dale, W., and Lewis, M. (1976). Further experiences with bovine arterial grafts. *Surgery* **80**, 711–721.

35. Rosenberg, N., Thompson, J., Keshishian J., and Vander Werf, B. (1976). The modified bovine arterial graft. *Arch. Surg. (Chicago)* **111**, 222–226.

36. Moulin, V., Auger, F. A., O'Connor-McCourt, M., and Germain, L. (1997). Fetal and postnatal sera differentially modulate human dermal fibroblast phenotypic and functional features in vitro. *J. Cell. Physiol.* **171**, 1–10.

37. Grenier, G., Remy-Zolghadri, M., Auger, F. A., and Germain, L. (1999). Use of a three-dimensional model for the study of phenotypic modulation of smooth muscle cells. *39th Annu. Mat. Am. Soc. Cell Biol.* Washington, DC, 1999.

38. Berthod, F., Germain, L., Guignard, R., Lethias, C., Garrone, R., Damour, O., Van Der Rest, M., and Auger, F. A. (1997). Differential expression of collagen XII and XIV in human skin and in reconstructed skin. *J. Invest. Dermatol.* **108**, 737–742.

39. Peterkofsky, B. (1972). The effect of ascorbic acid on collagen polypeptide synthesis and proline hydroxylation during the growth of cultured fibroblasts. *Arch. Biochem. Biophys.* **152**, 318–328.

40. Kanda, K., Matsuda, T., and Oka, T. (1992). Two-dimensional orientational response of smooth muscle cells to cyclic stretching. *Am. Soc. Artif. Intern. Organs J.* **38**, M382–M385.

41. Kanda, K., and Matsuda, T. (1993). Behavior of arterial wall cells cultured on periodically stretched substrates. *Cell Transplant.* **2**, 475–484.

42. Kanda, K., Matsuda, T., and Oka, T. (1993). Mechanical stress induced cellular orientation and phenotypic modulation of 3-D cultured smooth muscle cells. *Am. Soc. Artif. Intern. Organs J.* **39**, M686–M690.

43. Lopez Valle, C. A., Auger, F. A., Rompre, P., Bouvard, V., and Germain, L. (1992). Peripheral anchorage of dermal equivalents. *Br. J. Dermatol.* **127**, 365–371.

44. Cines, D. B., Pollak, E. S., Buck, C. A., Loscalzo, J., Zimmerman, G. A., McEver, R. P., Pober, J. S., Wick, T. M., Konkle, B. A., Schwartz, B. S., Barnathan, E. S., McCrae, K. R., Hug, B. A., Schmidt, A.-M., and Stern, D. M. (1998). Endothelial cells in physiology and in the physiopathology of vascular disorders. *Blood* **91**, 3527–3561.

45. Rémy, M., Bordenave, L., Bareille, R., Gorodkov, A., Rouais, F. and Baquey, C. (1994). Endothelilal cell compatibility testing of various prosthetic surfaces. *J. Mater. Sci.: Mater. Med.* **5**, 808–812.

46. Wu, M. H.-D., Shi, Q., Wechezak, A. R., Clowes, A. W., Gordon, I. L., and Sauvage, L. R. (1995). Definitive proof of endothelialization of a Dacron arterial prosthesis in a human being. *J. Vasc. Surg.* **21**, 862–867.

47. Remy, M., Valli, N., Brethes, D., Montis, K., Dobrova, N. B., Novikova, S. P., Gorodkov, A. J., and Bordenave, L. (1999). *In vitro* and *in situ* intercellular adhesion molecule-1 (ICAM-1) expression by endothelial cells lining a polyester fabric. *Biomaterials* **20**, 241–251.

48. Clowes, A. W. (1996). Improving the interface between biomaterials and the blood: The gene therapy approach. *Circulation* **93**, 1319–1320.

CHAPTER 29

NEUROECTODERMAL CELL CULTURE: ENDOCRINE CELLS

Peter I. Lelkes and Brian R. Unsworth

INTRODUCTION

Our long-term goal is to engineer functional neuroendocrine tissue constructs, which may become clinically useful as implants in the treatment of debilitating neurodegenerative diseases, such as Parkinson's disease, or for alleviating chronic pain. Parkinson's disease is characterized by the gradual loss of dopaminergic as well as other catecholaminergic neurons, mainly in the basal ganglia, but also in the substantia nigra and the locus ceruleus. In the United States alone, this debilitating progressive disease afflicts more than 500,000 patients, men and women alike. Up to 50,000 new cases are diagnosed each year, and there is no cure in sight. The conventional, conservative treatment involves replacing the missing neurotransmitter(s) by systemic administration of levodopa, or L-dopa. However, the debilitating side effects and the recurrence of the symptoms eventually outweigh any benefits of prolonged treatment with these drugs.

For more than two decades, tissue transplantation of autologous adult (and also of fetal) adrenal tissue has been attempted as a possible approach to treat Parkinson's disease. The rationale for this highly experimental procedure was to transplant tissues containing cells that endogenously synthesize and secrete the missing neurotransmitters (i.e., catecholamines, such as dopamine). While showing some (transient) measure of subjective improvement in a few patients, the objective success of this experimental treatment remains to be proven. For instance, of concern is the minimal survival/functionality of the catecholaminergic tissue fragments, which can be obtained in only small numbers from the patients' own adrenal glands. For a review see Kordower et al. [1].

In parallel and following essentially the same rationale, Perlow et al. [2] suggested transplanting not the tissue fragments but rather the isolated parenchymal cells, namely, chromaffin cells that synthesize the missing catecholamines. Neuroendocrine cells of the sympathoadrenal lineage also contain abundant opiate peptides, such as endorphins. Thus, in addition to their potential usefulness in treating neurodegenerative diseases, transplantation of these cells might be indicated for attempting to alleviate chronic pain [3,4].

Increasing numbers of laboratories are seeking a potential cure for neurodegenerative diseases by injecting embryonic or adult neuronal stem or progenitor cells. Recent promising results confirm that such cells indeed home into the target regions and locally differentiate into neurons [5,6]. A caveat common to all strategies for transplanting individual cells is the notion that while some (stem) cells might home in on a certain target region, the number of cells that actually survive locally is too minute to warrant the conclusion that they may be functionally relevant. Moreover, to date there is no conclusive proof for the functional integration of individual (transplanted) neurons.

An alternate strategy to transplanting individual (stem) cells is to generate *in vitro* functional neuroendocrine tissue constructs, which might stand a better chance of surviving *in vivo* and being functionally integrated locally. The focus of this chapter is to describe some of the relevant methodologies used in our laboratory for engineering functional, differentiated neuroendocrine tissue constructs. Specifically we will focus on employing a transformed, bipotential cell line of neuronal/neuroendocrine origin and a unique tissue culture venue: rotating wall vessel (RWV) bioreactors. Using this combination of cells and biotechnology, we have successfully generated differentiated, macroscopic neuroendocrine organoids.

CELLS

NEUROENDOCRINE CELLS OF THE SYMPATHOADRENAL LINEAGE

By definition, neuroendocrine cells are endocrine cells that synthesize neurohormones and secrete them into the bloodstream. Chromaffin cells of the adrenal medulla are among the best characterized neuroendocrine cells and have in the past been intensely studied in terms of, for example, the mechanisms of neurosecretion, many of which they share with bona fide neurons [7,8]. Developmentally, chromaffin cells are catecholaminergic secretory cells, which originate from neural crest derived cells of the sympathoadrenal (SA) lineage (Fig. 29.1). The term "sympathoadrenal" refers to the bipotentiality of the committed SA precursors, which migrate out of the neural crest. Depending on local environmental cues, these SA precursor cells differentiate into either catecholaminergic sympathetic neurons or chromaffin cells of the adrenal medulla. The latter process is an example of "cellular neoteny" [9]. In this chapter we will mainly focus on cultures of chromaffin and PC12 cells derived from the adrenal medulla.

The major "task" of adrenal medullary chromaffin cells is to synthesize and secrete catecholamines, a particular group of biogenic amines derived from tyrosine, which in addition to being neurohormones serve as neurotransmitters. Figure 29.2 presents in a simplified scheme the cascade of catecholamine synthesis: several specific enzymes mediate the sequential conversion of tyrosine to dopa, dopamine, norepinephrine, and finally epinephrine.

In most species, from avian to humans, two phenotypes of adrenal medullary chromaffin cells coexist in the same organ. These two distinct populations, initially identified based on distinct morphological characteristics, contain either norepinephrine (NE cells) or epinephrine (E cells). NE- and E-cells differ in a number of specific markers, including the expression of a functional form of the enzyme phenylethanolamine-*N*-methyltransferase

Differentiation of Cells of the Sympathoadrenal Lineage

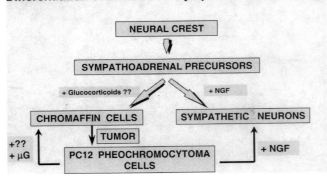

Adrenal Medullary Neuroendocrine Chromaffin Cells

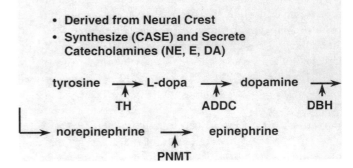

Fig. 29.1. *Schematic representation of the differentiation pathways for cells of the sympathoadrenal lineage. The involvement in the neuronal differentiation of neurotrophins, such as nerve growth factor (NGF) is well established; the role of glucocorticoids in the neuroendocrine/differentiation is less certain. Indicated in this scheme is also the possible role of microgravity (µG, RWV) in the differentiation of PC12 cells toward the neuroendocrine phenotype.*

Fig. 29.2. *Schematic representation of the cascade of catecholamine synthesis in chromaffin cells, CASE, catecholamine-synthesizing enzymes; NE, norepinephrine; E, epinephrine; DA, dopamine; TH, tyrosine hydroxylase; ADDC, aromatic dopa decarboxylase; DBH, dopamine-β-hydroxylase; PNMT, phenylethanolamine-N-methyl transferase.*

(PNMT), the terminal enzyme in the cascade of catecholamine-synthesizing enzymes, which converts NE to E. It is believed that chromaffin cells expressing NE and E represent two distinct, terminally differentiated forms of the SA lineage. Until very recently chromaffinergic differentiation of SA precursors was strictly correlated with the presence/action of glucocorticoids [10]. However, studies using knock out mice indicate that chromaffinergic differentiation can also occur in the absence of functional glucocorticoid receptors [11]. This finding supports our prior notion that cell–cell contacts, homotypic or heterotypic, might be important in the initial differentiation of SA precursors toward the neuroendocrine/chromaffin cell phenotype [12].

Reliable methods for isolating large numbers of bovine adrenal medullary chromaffin cells were initially developed in the late 1970s [13,14]. These methods involve mechanical mincing and repeated enzymatic digestion of the adrenal medulla with collagenase and/or a mixture of proteolytic enzymes (collagenase/dispase and trypsin) followed by purification of the ensuing mixed cell population over a density gradient (Renographin or Ficoll-Hypaque). The yield of these initial preparations was about $1–2 \times 10^6$ cells from three to four glands, with a purity exceeding 80% chromaffin cells. Since these cells were intended for acute experiments only, without long-term culturing, this relatively small contamination with other, not further characterized cells was of no initial concern. However, since adult chromaffin cells are terminally differentiated and do not proliferate, contamination of chromaffin cell preparations with other cells became a problem as soon as several laboratories attempted to establish long-term chromaffin cell cultures. Without further purification steps [15,16] or the addition of cytostatic agents such as mytomycin or cytosine arabinoside (Ara C), to inhibit the proliferation of the contaminant cells, primary cultures are quickly overgrown by fibroblasts and endothelial cells.

Subsequent modifications of the basic methodology [17–19] have significantly increased the purity to over 99% and the yield to (theoretically) more than 10^7 purified chromaffin cells per bovine adrenal medulla.

Another important development, the capability to separate out NE- and E-cells, was achieved by a slight modification of some of the prior purification methods [20]. The use of Percoll gradients allows for fractionation of the purified chromaffin cells into highly enriched populations of cells that synthesize/secrete epinephrine and norepinephrine. This distinction is important, since epinephrine-secreting cells seem to be further along the neuroendocrine pathway and thus further away from their neuronal origin. Studies indicate that PNMT is differentially regulated in the two chromaffin cell subpopulations [21].

Rat Chromaffin Cells

The most detailed methodology for isolating and culturing chromaffin cells from adult, neonatal, and embryonic rat adrenal glands is found in an excellent chapter by Unsicker *et al.* [22]. By exactly following these protocols, we have been able to successfully isolate and culture rat chromaffin cells from embryonic, postnatal, and adult animals. Isolation of purified medullae from adult animals, and even more from neonates and embryos, is rather tedious; the cell yield is small. In our hands we usually do not obtain more than 20,000 chromaffin cells per adult rat medulla. Thus, isolating large numbers of neonatal or adult rat chromaffin cells for tissue engineering purposes is impractical and tantamount to mass slaughter. The advantage of working with primary cell cultures is that there are very well-defined markers of cell-specific differentiation available, thus providing a wide array of molecular probes. Furthermore, cells are relatively easily available from all stages of development from embryonic day 16.5 through adult. However, since, with the exception of the early embryonic stages, chromaffin cells are terminally differentiated, the use of primary rat chromaffin cells for tissue engineering purposes will be of limited value until small bioreactors (1-ml volume) become available.

PC12 Pheochromocytoma Cells

Based on the foregoing discussion, it is obvious that the most useful system for regenerating neuroendocrine tissues is one that combines phenotypic plasticity, abundance, and ease of isolation/maintenance with the availability of molecular analytical tools. PC12 pheochromocytoma cells offer such a system. This clonal cell line was derived from a spontaneous tumor, pheochromocytoma, of the adrenal medulla of New England Deaconess rats [23]. At the end of 2000 there were more than 5500 peer-reviewed papers listed on Medline in which these cells were used for a various purposes, such as models for neuronal differentiation, neuroendocrine secretion, neurotoxicology, signal transduction and apoptosis [24,25]. Many of the results initially obtained in PC12 cells have been faithfully validated in nontransformed cells of neuronal/neuroendocrine origin.

A remarkable trait of PC12 cells is their bipotentiality, a feature that renders these cells particularly suitable for generating both neuronal and neuroendocrine tissue constructs. Our own work has focused on the latter, less explored aspect of PC12 cell differentiation. In generating neuroendocrine tissue equivalents, we have used a methodology that comprises culturing the cells in a unique venue, the rotating wall vessel (RWV) bioreactors, discussed in detail later. The reduced shear stress environment simulates microgravity, favors colocalization of cells, and their assembly into "organoids" or tissue equivalents. Furthermore, this novel culture environment appears to specifically up-regulate the neuroendocrine pathway of differentiation [26–29].

Over the years PC12 cells have undergone numerous phenotypic variations, in part because of differences in how different laboratories culture these cells, including use of different basal media (e.g., RPMI vs DMEM) and level of supplementation with sera. Moreover, special subclones (e.g., PC12-h, PC18, etc.) have been produced that differ from the parental PC12 cells in some basic properties, such as expression of distinct receptor subtypes, and responsiveness to nerve growth factor (NGF). It is clear that great caution is necessary in comparing results obtained with "generic" PC12 cells from various laboratories. In the future, it might be prudent to standardize the culture methods for what should be termed PC12 cells. Alternatively, modern molecular biological techniques should be employed to clearly identify the genotypic basis for the phenotypic differences among the confounding multitude of today's "PC12 cells."

"Our" PC12 cells were originally obtained from the late Gordon Guroff of the National Institutes of Health (NIH). Stocks of the cells are routinely frozen at a density of 2×10^6 cells/ml in normal culture medium (see later) supplemented with 10% dimethyl sulfoxide as cryoprotectant and stored in a cryogenic freezer. Conventional two-dimensional culture of the cells in tissue culture flasks or dishes follows standard procedure [12]. Briefly, upon thawing, the cells are added to prewarmed cell culture medium and centrifuged (600 rpm for 5 min). The supernatant containing the cryoprotectant is removed, and the cells are resuspended in 25 ml of culture medium and placed into a T150 cell culture flask. The choice of the cell culture flask seems to be important; in our hands, T150 flasks (Nunclon) provide the best venue for routine culture of PC12 cells. The cells are routinely cultured in a medium comprising 85% Dulbecco's Modified Eagle's Medium (DMEM), high glucose, (4.5 g/liter), 7.5% fetal calf serum, and 7.5% horse serum. The medium also contains 2 mM L-glutamine and a mixture of antibiotics (100 IU/ml penicillin and 100 µg/ml streptomycin) to prevent bacterial contamination. Occasionally the use of amphotericin B (fungizone at 2 µg/ml) is indicated.

In conventional two-dimensinal cultures, the cells are "fed" every 2–3 days: that is, 90% of the old medium is removed and replaced with the same volume of new medium. Regular feeding is a must for maintaining the cells in their undifferentiated state in conventional two-dimensional culture. Otherwise, especially at higher cell densities, neurotrophic factors synthesized/secreted by the cells and accumulating in the medium will initiate cellular differentiation along the neuronal pathway. This can clearly be observed by the increasing number of cells that extend neurites. In the continued (>10 days) presence of neurotrophic growth factors, such as NGF, the cells develop an elaborate neuronal network and become *de facto* indistinguishable from sympathetic neurons. Figure 29.3 shows micrographs of PC12 cells in conventional two-dimensionsional tissue culture, maintained either undifferentiated (A) or exposed for 14 days to 50 ng/ml NGF (B).

Once the cells have reached confluence (usually in a week, ~10–15×10^6 cells in a large T150 flask), the cultures are "split" at a ratio of roughly 1:8. It has to be pointed out that the adhesion of PC12 to the substratum is rather loose and that *undifferentiated* cells

PC12 Cells in Conventional 2-D Culture

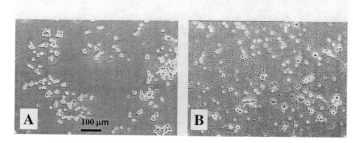

Fig. 29.3. Phase contrast micrographs of PC12 cells in a conventional two-dimensional cell culture, (A) control and Panel (B) +50 ng/ml NGF (details in text).

remain mainly rounded or trapezoidal (see Fig. 29.3) and will not spread like conventional anchorage-dependent cells. Therefore the appearance of a "confluent monolayer" of PC12 cells is quite different from that of an epithelial cell monolayer. Propagation/splitting of PC12 cells is accomplished by mechanically dislodging the cells, that is, by literally banging with the palm of the hand against the side walls of the culture flask. Subsequently, 3 ml of the cell suspension is added to a new flask containing 22 ml of fresh culture medium. The remainder of the cells either is used for experiments, frozen for future use, or discarded. For each experiment, the number of cells to be used is determined in a hemocytometer. To obtain suspensions of single cells, we filter the cells through nylon filters with a pore size of 32 μm. As a rule, we tend to avoid using trypsin on these cells to prevent phenotypic alterations, as evidenced, for example, by the enhanced appearance of flattened, anchorage-dependent subclones.

OTHER TRANSFORMED CELLS OF THE SA LINEAGE

Two other permanent cell lines of the sympathoadrenal lineage might warrant further consideration for tissue engineering purposes. Approximately 10 years ago, a v-*myc*-transformed continuous cell line (MAH cells) was generated from early (day-14.5) embryonic rat adrenal medullary chromaffin precursor cells [30]. These cells can differentiate toward the sympathetic neuronal phenotype in response to neurotrophic factors and coexpress several neurotransmitter enzyme genes. However, the culture of these cells requires a rather complex and costly medium. Future studies might lead to a simplified medium in which the cells can be expanded. More recently, Eaton *et al.* [31] produced a continuous chromaffin cell line by transforming primary cultures of both rat and bovine chromaffin cells. Interestingly, both the MAH cells and the new continuous chromaffin cell lines seem to resemble PC12 cells, in that their initial assessment suggests that individual cells might differentiate toward the NE-type chromaffin cells—that is, that they do not express *functional* PNMT. It remains to be tested whether either of these cell lines, when expanded and cultured in three-dimensions in RWV bioreactors, will, like the PC12 cells, differentiate toward an epinephrinergic chromaffin phenotype.

THREE-DIMENSIONAL CELL CULTURE IN ROTATING WALL VESSEL BIOREACTORS

PRINCIPLE OF RWV BIOREACTORS

Approximately 10 years ago engineers and scientists at NASA introduced rotating wall vessel (RWV) bioreactors as a novel modality for maintaining cells during space flight. The dynamic environment in these rotatory bioreactors combines a number of beneficial factors, such as low shear ($\sim$0.5 dyn/cm^2), effective mass transfer, and randomization of the gravitational vector ("simulated" microgravity of $\sim$10^{-2}g). This unique cell culture environment has been found to facilitate the generation of macroscopic, functional three-dimensional tissue constructs [32–34]. The principle of RWV bioreactors is shown schematically in Fig. 29.4. In brief, RWV bioreactors are closed-culture vessels with zero headspace (i.e., completely filled with cell culture fluid), which rotate as a whole body around a horizontal axis. For generating three-dimensional tissue constructs, single-cell suspensions are inocu-

Principle of RWV Bioreactors

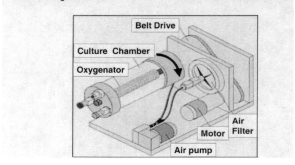

Fig. 29.4. Principal components of an STLV-type RWV bioreactor.

RWV Bioreactor (STLV)

Fig. 29.5. STLV-type RWV bioreactor in action: arrow indicates macroscopic organoids suspended in a rotating STLV.

lated at high density in the presence or absence of suitable cell culture beads or other scaffolds. A crucial step is the proper (empirical) choice of the initial rotational speed, which maintains the suspended cells/particles quasi-stationary with the rotating fluid in a state of "continuous free fall" [35,36]. This minimal exposure of the cells to shear stress and spatial colocation of the particles/cells leads to the rapid formation of cell aggregates and facilitates their subsequent differentiation into macroscopic tissue like assemblies (Fig. 29.5). As these "organoids" grow in size, the rotational speed is adjusted to compensate for their increased settling rates (see later).

RWV bioreactors come in various configurations. Slow-turning lateral vessels (STLVs, Fig. 29.6A), intended primarily for microcarrier cell culture and explant tissue cultures, are longitudinal cylinders with a center core membrane oxygenator. High aspect ratio vessels (HARVs, Fig. 29.6B), used for cells that require a higher oxygen tension, are discoid vessels that incorporate a flat membrane oxygenator in the back wall; they have been successfully used for both suspension and anchorage-dependent cell types. To date, several dozen cell types have been cultured in both types of RWV bioreactor, which are commercially available as Rotary Cell Culture Systems. Up-to-date information on these bioreactors, including a recent bibliography, can be obtained from the manufacturer's website (http://www.synthecon.com).

Results obtained in RWV bioreactors must be compared with cells cultured under "control" conditions. At this point, there is little consensus on what constitutes the most appropriate control environment. Traditionally, three-dimensional cultures are generated by seeding cells onto three-dimensional scaffolds or by maintaining cells in suspension/aggregate cultures, under either dynamic or static conditions. In the latter approach, cells are inoculated into either conventional, small spinner flasks (50–250 ml in volume, e.g., from Bellco), or maintained in inert, gas-permeable Teflon bags (7–50 ml, e.g., from American Fluoroseal) (Fig. 29.7). The major disadvantage of stirred spinner flasks, as of most other conventional "dynamic" cell culture venues, is the need for excess physical forces (e.g., turbulent flow, high shear stress) to maintain cells in suspension and provide adequate mass transfer and oxygenation. High levels of turbulent fluid shear forces have been shown to be detrimental to long-term assembly and differentiation of three-dimensional cultures [37]. By contrast, the minimal mass transfer under static conditions (e.g., in culture bags) leads to the rapid establishment of necrotic cores in cell aggregates exceeding approximately 1 mm in diameter. In conventional tissue culture, the two-dimensional environment and/or an inappropriate substrate (plastic) tend to alter gene expression, prevent tissue-specific differentiation, and lead to rapid dedifferentiation.

One possible solution for an appropriate control, especially for the STLV, is to turn the base by 90° and thus rotate around a vertical rather than a horizontal axis. Under these conditions, the low shear stress component is not altered, while efficient mass transfer

Different Configurations of RWV Bioreactors

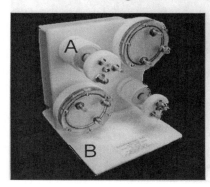

- **A: STLV**
  - Slow Turning Lateral Vessel

- **B: HARV**
  - High Aspect Ratio Vessel

mechanically stirred spinner flask

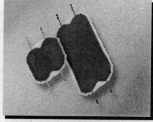

static cell culture bags

Fig. 29.7. "Control" tissue culture venues. For details, see text.

Fig. 29.6. Reusable RWV bioreactors mounted on a four-station rotator base: (A) slow turning lateral vessel (STLV) and (B) high aspect ratio vessel (HARV). Courtesy of Synthecon, Inc.

is maintained. However the unique aspect of randomization of the gravitational vector is eliminated.

PRACTICAL CONSIDERATIONS FOR GENERATING NEUROENDOCRINE ORGANOIDS IN RWV BIOREACTORS

The general aspects of the care and maintenence of RWV bioreactors have been described, and basic instructions are available from the manufacturer. In this section we will focus on a few of the special "little hints and tricks" that facilitate successful implementation of RVW biotechnology for tissue engineering of neuroendocrine organoids.

Type of RWV and Medium

In the past, we have used both the STLV and the HARV types of RWV bioreactor for studying the generation of macroscopic three-dimensional neuroendocrine constructs and the mechanisms of PC12 cell differentiation in this environment. We specifically use STLVs in conjunction with cell culture (microcarrier) beads for long-term cultures. HARVs (especially the disposable 10-ml version) serve as the venue of choice in the absence of beads for short-term studies aimed at analyzing cellular/molecular processes involved in the initial steps of cell–cell interactions/aggregation. For long-term cultures, the high initial cell densities of the inoculum ($2–5 \times 10^5$ cells/ml) that are needed for the rapid formation of cell aggregates require special attention to the medium. We routinely monitor cellular metabolism by measuring pH, pO_2, pCO_2, and glucose levels, refeeding if necessary several times a day (see below) mainly based on the rate of glucose consumption. Our medium contains 4.5 g/liter glucose, which, if left unattended, rapidly (within <24 h) is exhausted, especially at the highest cell densities ($>2 \times 10^6$ cells/ml). Therefore, whenever cells are refed, the new medium has to be supplemented with additional glucose (added from a 10% stock solution). Also, since the enhanced metabolism in RWV, especially at high densities, tends to render the medium acidic, it is prudent to reinforce the buffering capacity of the medium by adding 25 mM HEPES. It cannot be stressed enough that proper maintenance of the physiological conditions in the medium, while time-consuming and labor intensive, is one of the most essential ingredients for ensuring the reproducibility and success of the experiments.

Assembly of the RWV Bioreactors

New, reusable RWV bioreactors, must be detoxified following the manufacturer's procedures. Subsequently, these reusable bioreactors can be sterilized by autoclaving. However, parts that tend to wear out over time (e.g., the oxygenator membrane of the STLV and the stainless steel ports) should be replaced on a regular basis. Also, excess tightening of these ports should be avoided, since this quickly destroys the Teflon lids.

The assembly of the STLVs is more complex than that of the HARVs. Of crucial importance for both systems is the complete removal of all air bubbles. Residual air bubbles negate the proper functioning of the system as a low-shear/microgravity-simulating environment. Filling an STLV is harder because the geometry of this type of RWV bioreactor allows for air bubbles to hide away, underneath the Teflon lid and around the O-ring. In general, the autoclavable parts of the STLV are assembled per the manufacturer's instructions, all screws are hand-tightened, and two barrels of disposable 10-ml syringes are attached to the open syringe ports. The vessel is tilted to about 45° and filled with the cell suspension through the inlet port. Additional cell culture medium is added until the fluid completely fills the apparatus: that is, until the fluid starts to rise in the syringe attached to the outlet port. The inlet port is then closed, theoretically providing for zero headspace, and pushing excess fluid into the syringe barrels. However, in practice, the irregular shape of the inlet port always allows some air pockets to be trapped under both the Teflon lid and the silicon O-rings. These bubbles, dislodged by repeatedly (gently) banging the assembled STLV on the table, are maneuvered underneath one of the syringe ports and finally removed by reciprocal movement of the medium in and out of both syringe ports.

The assembly and filling of HARVs, especially the disposable ones, is less complex. First, the barrel of a disposable syringe (with Luer-Lock fittings) is attached aseptically to each of the inlet/outlet ports. Depending on the vessel size, we use 5-ml syringes for the 10-ml vessels and 10- or 30-ml syringes for the 55-ml vessels. The vessel, with the open, large-bore "maintenance" port excentrically located between lower (inlet) and upper (outlet) syringe ports, is tilted at an angle of 45°, and the medium, containing the cells, is gradually added by means of a disposable 10-ml pipette. The maintenance port is capped as soon as it is filled with liquid. This procedure helps to avoid trapping an air bubble underneath the maintenance port. Additional culture medium is then added through the (lower) inlet port until fluid passes into the upper syringe, while taking care that the lower syringe barrel never runs empty. Then both Luer-Locks are closed—first the inlet, then the outlet. If carried out properly, this procedure practically ensures that the HARV will be filled without any air bubbles being introduced. Excess medium is removed by pipette from the upper and lower barrels. The barrels are removed, and the syringe ports sterilized by wiping with a cloth soaked in 70% ethanol.

To Bead or Not to Bead

Ideally, tissue constructs that can be assembled without having to resort to using (nonbiodegradable) microcarrier beads might be preferable for subsequent implantation. Unfortunately, in the absence of "vascular channels," the size of such "cells-only organoids" tends to be limited, presumably owing to limitations of oxygen diffusion/nutrient supply through a densely packed tissuelike assembly. By contrast, by using microcarrier beads, we obtained viable large "tissues" (up to 1 cm in diameter), which enwrap numerous beads without signs of necrosis or other kinds of deterioration (e.g., enhanced apoptosis) in the center of these constructs. Since PC12 cells do not readily adhere to cell culture beads of any kind (even collagen-coated ones, such as Cytodex-2), we let the cells initially attach for 24 h under static conditions to porous, collagen-based microcarrier beads, such as Cultisphere, precoated with either collagen type I (10 µg/ml) or poly(L-ornithine) (30 µg/ml), prior to initiating the RWV cultures. A major advantage of Cultisphere is the ease with which these collagen-based beads can be processed for routine histology, immunohistochemistry (including frozen sections), and electron microscopy.

Initiation of an Experiment

The physical principle of culturing cells in RWV bioreactors dictates that the rotational speed be adjusted to counterbalance the tendency of the particles to settle. Hence, the rotational speed must be empirically chosen to keep the inoculum in constant suspension, without adding centrifugal forces. The speed of rotation, which is crucial for the experimental outcome, is determined by the specific density of the inoculum. Single cells obviously sediment much more slowly than large cell assemblies, or cells growing on tissue culture beads. As the assemblies grow in size, the rotational speed must be adjusted to maintain the cell assemblies in continuous free fall or suspension. In our experience, individual cells, cultured without beads, are kept in suspension at 8 rpm. As the cell aggregates (cell organoids) grow in size, the number of revolutions is accelerated to approximately 20–22 rpm. Cells on culture beads, such as Cytodex-3, are initially cultured at a speed of 15–18 rpm, while cells growing on less dense (porous) microcarrier beads, such as Cultisphere are initiated at an initial speed of about 12 rpm. As the organoids grow in size, and more beads/cells aggregate, the rotational speed is increased to as much as 32–35 rpm. Regrettably, setting the initial rotational velocity as well as the subsequent adjustment as the organoids grow in size, is more an empirical art than science. In the absence of a fundamental study, which would yield objective parameters for setting the rotational velocities, procedural consistency is the only way to ensure experimental reproducibility.

Humidity Control

Maintenance of 100% humidity in the incubator is critical. As the air from the incubator is forced through the oxygenation membrane into an RWV, less than saturated humidity in the incubator is prone to lead to a slow loss of the fluid and gradual accumulation of air bubbles in the vessels. Air bubbles, particularly encountered in HARVs, which have a large-surface-area oxygenation membrane, induce turbulence, which in turn distorts the trajectories of the particles suspended in the bioreactors and can thwart tissuelike assembly and differentiation.

Feeding

Nutrients depleted from the medium must be replenished in a timely manner to maintain homeostasis in the cultures, especially at high densities. Feeding frequency is a function of the cell density, the metabolic activity of the cells, and also of the medium composition. As discussed, we use the rate of consumption of glucose and the drop in pH as our prime markers for metabolic activity. Moreover, in a closed system, such as the STLV, the levels of pO_2 and pCO_2 must be carefully monitored. By contrast, in the more open system of the HARV, with its large-surface-area oxygenator membrane, the levels of pO_2 and pCO_2 in the medium essentially reflect that of the ambient environment in the incubator. In our experience, in rapidly growing cells/organoids, up to two-thirds of the medium needs to be replenished daily. In high-density cultures, that is, cultures containing more than 2×10^6 cells/ml, the medium must be replenished more frequently, up to three times a day for fast growing and/or metabolically active cells.

Traditionally, in preparation for feeding the cultures, an RWV is dismounted from the base and placed into a laminar flow hood. The cells/tissue constructs are allowed to settle at the bottom of the vessels. The medium is initially sampled aseptically via one of the ports, using a 1-ml (tuberculin) syringe, for determination of the vital signs (pH, pO_2, glucose, etc.). Two larger syringes are then attached to the ports, one of them filled with new medium, while the other serves as the reservoir for the waste. Great care must be exerted not to reintroduce air bubbles while refeeding. It is best to first open the syringe port containing the new medium and then gently aspirate the spent medium via the second syringe. This way, up to half the old medium can be removed while refeeding of the culture occurs concomitantly. To avoid introducing air bubbles, the "inlet" port is closed first, leaving a small amount of the fresh medium in the syringe. As a caveat, the time and effort needed to remove extra air bubbles is far in excess of the little time it takes to ensure that no air bubbles are introduced at the time of feeding.

Preliminary results, which show that intracellular signaling pathways are rapidly activated each time the rotation of RWV cultures is initiated or stopped, leave us to question the validity/usefulness of the traditional way of feeding the cells. Rather, we have begun to sample/feed the culture "on the fly" (i.e., all the operations described earlier on vessels as they are rotating).

Reusable vs Disposable HARVs

The manufacturer's claims to the contrary not withstanding, disposable HARVs can be reused several times, following washing and sterilization. This is especially useful and cost-

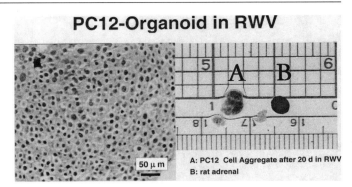

Fig. 29.8. PC12 organoid cultured for 20 days in STLV-type RWV bioreactor. Left: Histological H&E section; bar 50 μm. Right: Comparison of the size of a PC12 organoid (A) to that of an intact (adult) rat adrenal (B).

effective when one is carrying out large numbers of short-term (<24 h) experiments. In the absence of facilities to sterilize the vessels by gamma irradiation or ethylene oxide, the most straightforward way to sterilize the "disposable" vessels (which cannot be autoclaved) is by immersing them (and the port caps) in 0.25 M sodium hydroxide. After 15 min, the vessels are removed and generously (at least six times rinsed with pyrogen-free water). Finally, the vessels are filled with distilled water containing concentrated penicillin–streptomycin–fungizone (5× solution), soaked for one hour, and rinsed twice. HARVs and port caps are placed into a hood under UV illumination (20 min each for top and bottom sides) and air-dried overnight in the hood. Prior to use, the vessels are rinsed once with medium, to assure that the pH remains neutral—that is, that all the NaOH has been removed. As a caveat, the disposable vessels are very sensitive to the application of excess pressure during either filling or cleaning of the system through the syringe ports. Pushing or pulling the plungers too hard when adding or removing fluid will irreversibly damage the vessels, both new and reused, and cause leakage, thus ruining the experiment. By carefully heeding these caveats, we have successfully reused "disposable" vessels more than a dozen times.

ANALYSIS OF NEUROENDOCRINE ORGANOIDS ASSEMBLED IN RWV BIOREACTORS

By following the foregoing procedures, we have been able to generate large macroscopic tissue assemblies in RWV bioreactors. The dimensions of these organoids approach, or in some cases even surpass that of the parental organs (Fig. 29.8). Histological hematoxylin-and-eosin (H&E) sections suggest a tissue like assembly/organization of the cells without a necrotic core in the center of the constructs, which contrasts with the findings for conventional spheroid cultures exceeding 1–2 mm in diameter. A detailed discussion of our past and ongoing studies, which focus on characterizing the cell/molecular biology of PC12 cell assembly into differentiated, functional neuroendocrine organoids, is beyond the scope of this chapter. Briefly, we are using complementary immunohistochemical, biochemical, and molecular biological techniques to analyze the basic morphology and the signal transduction pathways involved in the expression of histiotypic differentiation markers in our "organoids" [26,27]. More recently, we have begun to use cDNA arrays to evaluate the patterns of differential gene expression. Our aim is to establish the genotypic base for the phenotypic changes (altered cellular signaling pathways, expression of intercellular junctions, synthesis and secretion of extracellular matrix, etc.), which characterize the transition from mere aggregates in suspension culture to the tissuelike constructs in RWV bioreactors (Fig. 29.9).

METHODOLOGICAL ISSUES FOR FUTURE RESEARCH

As shown by many others and us, the RWV environment is ideal for assembling individual cells into "organoids" and differentiating these assemblies into tissuelike constructs. Obviously, "real" tissues and organs comprise a number of distinct cell types. Indeed, early on, the capability to coculture dissimilar cells was one of the major goals of RWV bioreactor-based biotechnology. However, it appears that the simple approach of concomitantly throwing together dissimilar cell types will have to be refined, conceptually as well

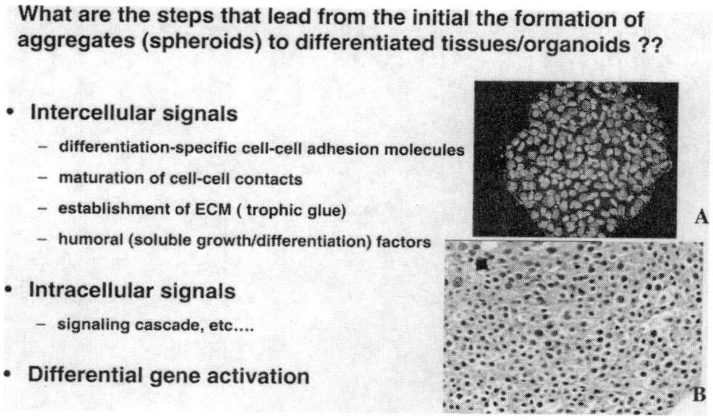

What are the steps that lead from the initial the formation of aggregates (spheroids) to differentiated tissues/organoids ??

- **Intercellular signals**
  - differentiation-specific cell-cell adhesion molecules
  - maturation of cell-cell contacts
  - establishment of ECM (trophic glue)
  - humoral (soluble growth/differentiation) factors

- **Intracellular signals**
  - signaling cascade, etc....

- **Differential gene activation**

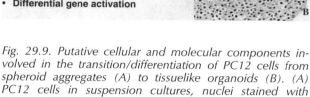

FUTURE DIRECTIONS

- **RWV: Low Shear and/or Simulated Microgravity?**
  - **Extension of this Study to True Microgravity (Space Flight)**

- **Genomics of RWV/Simulated vs. True Microgravity**

- **Vascularization from Within**
  - **Scaffoldig/Tissue Engineering**

- **Encapsulation of Neuroendocrine Organoids and Implantation**
  - **Treatment of Neurodegenerative Disease and Alleviation of Chronic Pain**

Fig. 29.9. Putative cellular and molecular components involved in the transition/differentiation of PC12 cells from spheroid aggregates (A) to tissuelike organoids (B). (A) PC12 cells in suspension cultures, nuclei stained with Hoechst (H H33258 Bisbenzimide) DNA stain. (B) H&E staining of PC12 organoid. Original magnifications, 200×.

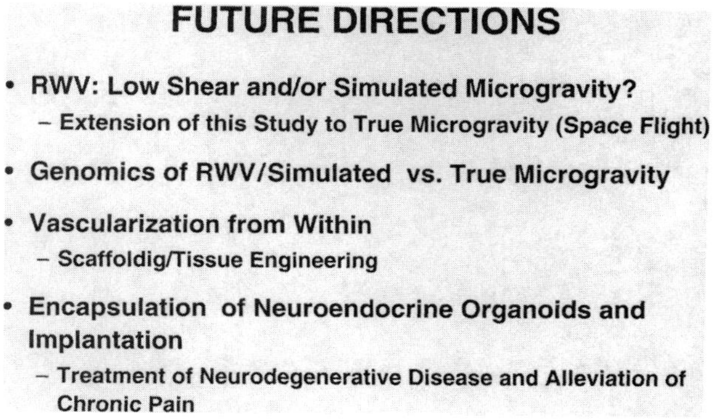

Fig. 29.10. Future methodological developments for neuroendocrine organoids generated in RWV bioreactors. For details, see text.

as methodologically. Among the major issues that will have to be addressed in detail is the sequential seeding of dissimilar cell types onto (biodegradable) scaffolds with the capability of creating constructs in a vectorial/hierarchical fashion. Paracrine interactions between dissimilar cells are a prerequisite for proper differentiation and functioning in many organs—for example, between chromaffin and cortical cells in the adrenal. Secondly, and probably intimately related to the first issue, is the vascularization of the ensuing constructs/organs. Generally speaking, no sizable "piece of meat" is viable and functional unless it is vascularized through and through. Specifically, in the case of a neuroendocrine organ, the direct interaction of the parenchymal cells with, and capability to secrete their products into, the local microvasculature is a fundamental prerequisite for its proper function (Fig. 29.10).

Finally, methodologies must be developed to ascertain that tissue constructs generated in RWV bioreactors remain viable, differentiated, and functional after removal from the culture venue. Also, upon implantation, the vasculature in these organoids should be capable of anastomosing with that of the host. An entirely different ball game is the generation of neuroendocrine tissue constructs that will be immunologically silent, and accepted by the host without the need for continual immunosuppression.

ACKNOWLEDGMENTS

We are grateful for the continued support of our work by grants from the National Aeronautics and Space Administration (NASA NAG 9-172, NCC8-173, and NAG2-1436). We also would like to acknowledge the significant contributions of our past and current collaborators and students.

REFERENCES

1. Kordower, J. H., Goetz, C. G., Freeman, T. B., and Olanow, C. W. (1997). Dopaminergic transplants in patients with Parkinson's disease: Neuroanatomical correlates of clinical recovery. *Exp. Neurol.* **144**, 41–46.
2. Perlow, M. J., Kumakura, K., and Guidotti, A. (1980). Prolonged survival of bovine adrenal chromaffin cells in rat cerebral ventricles. *Proc. Natl. Acad. Sci. U.S.A.* **77**, 5278–5281.
3. Michalewicz, P., Laurito, C. E., Pappas, G. D., Lu, Y., and Yeomans, D. C. (1999). Purification of adrenal chromaffin cells increases antinociceptive efficacy of xenotransplants without immunosuppression. *Cell Transplant.* **8**, 103–109.
4. Hains, B. C., Chastain, K. M., Everhart, A. W., McAdoo, D. J., and Hulsebosch, C. E. (2000). Transplants of adrenal medullary chromaffin cells reduce forelimb and hindlimb allodynia in a rodent model of chronic central pain after spinal cord hemisection injury. *Exp. Neurol.* **164**, 426–437.
5. Schwarz, E. J., Alexander, G. M., Prockop, D. J., and Azizi, S. A. (1999). Multipotential marrow stromal cells transduced to produce L-DOPA: Engraftment in a rat model of Parkinson disease. *Hum. Gene Ther.* **10**, 2539–2549.
6. Björklund, A., and Lindvall, O. (2000). Cell replacement therapies for central nervous system disorders. *Nat. Neurosci.* **3**, 537–544.
7. Rosenheck, K., and Lelkes, P. I., eds. (1987). "Stimulus-Secretion Coupling in Chromaffin Cells," 2 vols. CRC Press, Boca Raton, FL.

8. Aunis, D. (1998). Exocytosis in chromaffin cells of the adrenal medulla. *Int. Rev. Cytol.* **181**, 213–320.

9. Anderson, D. J. (1993). Molecular control of cell fate in the neural crest: The sympathoadrenal lineage. *Annu. Rev. Neurosci.* **16**, 129–158.

10. Anderson, D. J., and Axel, R. (1986). A bipotential neuroendocrine precursor whose choice of cell fate is determined by NGF and glucocorticoids. *Cell (Cambridge, Mass.)* **47**, 107910–107990.

11. Finotto, S., Krieglstein, K., Schober, A., Deimling, F., Lindner, K., Brühl, B., Beier, K., Metz, J., Garcia-Arraras, J. E., Roig-Lopez, J. L., Monaghan, P., Schmid, W., Cole, T. J., Kellendonk, C., Tronche, F., Schütz, G., and Unsicker, K. (1999). Analysis of mice carrying targeted mutations of the glucocorticoid receptor gene argues against an essential role of glucocorticoid signalling generating adrenal chromaffin cells. *Development (Cambridge, UK)* **126**, 2935–2944.

12. Mizrachi, Y., Naranjo, J. R., Pollard, H. B., and Lelkes, P. I. (1990). PC12 Cells differentiate into chromaffin cell like phenotype in co-culture with adrenal medullary endothelial cells. *Proc. Natl. Acad. Sci. U.S.A.* **87**, 6161–6185.

13. Brooks, J. C. (1977). The isolated bovine adrenomedullary chromaffin cell: A model of neuronal excitation-secretion. *Endocrinology (Baltimore)* **101**, 1369–1378.

14. Schneider A. S., Herz, R., and Rosenheck, K. (1977). Stimulus-secretion coupling in chromaffin cells isolated from bovine adrenal medulla. *Proc. Natl. Acad. Sci. U.S.A.* **74**, 5036–5040.

15. Role, L. W., and Perlman, R. L. (1980). Purification of adrenal medullary chromaffin cells by density gradient centrifugation. *J. Neurosci. Methods* **2**, 253–265.

16. Unsicker, K., and Müller, T. H. (1981). Purification of bovine adrenal chromaffin cells by differential plating. *J. Neurosci. Methods* **4**, 227–241.

17. Waymire, J. C., Bennett, W. F., Boehme, R., Hankins, L., Gilmer-Waymire, K., and Haycock, J. W. (1983). Bovine adrenal chromaffin cells: High-yield purification and viability in suspension culture. *J. Neurosci. Methods* **7**, 329–335.

18. Livett, B. G., Boksa, P., Dean, D. M., Mizobe, F., and Lindenbaum, M. H. (1983). Use of isolated chromaffin cells to study basic release mechanisms. *J. Auton. Nerv. Syst.* **7**, 59–86.

19. Artalejo, C. R., García, A. G., and Aunis, D. (1987). Chromaffin cell calcium channel kinetics measured isotopically through fast calcium, strontium, and barium fluxes. *J. Biol. Chem.* **262**, 915–926.

20. Moro, M. A., López, M. G., Gandía, L., Michelena, P., and García, A. G. (1990). Separation and culture of living adrenaline- and noradrenaline-containing cells from bovine adrenal medullae. *Anal. Biochem.* **185**, 243–248.

21. Cahill, A. L., Eertmoed, A. L., Mangoura, D., and Perlman, R. L. (1996). Differential regulation of phenylethanolamine-N-methyltransferase expression in two subpopulations of bovine chromaffin cells. *J. Neurochem.* **67**, 1217–1224.

22. Unsicker, K., Reichert-Preibsch, H., and Hofmann, H. D. (1989). Chromaffin cell cultures and their use for assaying effects of nerve growth factor and other growth factors. *In* "Nerve Growth Factors" (R. A. Rush, ed.), pp. 149–167. Wiley, Chichester, UK.

23. Greene, L. A., and Tischler, A. S. (1976). Establishment of a noradrenergic clonal line of rat adrenal pheochromocytoma cells, which respond to nerve growth factor. *Proc. Natl. Acad. Sci. U.S.A.* **73**, 2424–2428.

24. Fujita, K., Lazarovici, P., and Guroff, G. (1989). Regulation of the differentiation of PC12 pheochromocytoma cells. *Environ. Health Perspect.* **80**, 127–142.

25. Shafer, T. J., and Atchison, W. D. (1991). Transmitter, ion channel and receptor properties of pheochromocytoma (PC12) cells: A model for neurotoxicological studies. *Neurotoxicology* **12**, 473–492.

26. Lelkes, P. I., Galvan, D. L., Hayman, G. T., Goodwin, T. J., Chatman, D. Y., Cherian, S., Garcia, R. M. G., and Unsworth, B. R. (1998). Simulated microgravity conditions enhance differentiation of cultured PC12 cells towards the neuroendocrine phenotype. *In Vitro Cell. Dev. Biol. Anim.* **34**, 316–324.

27. Lelkes, P. I., and Unsworth, B. R. (1998). Cellular signaling mechanisms involved in the 3-dimensional assembly and differentiation of PC12 pheochromocytoma cells under simulated microgravity in NASA Rotating Wall Vessel Bioreactors. *In* "Advances in Heat and Mass Transfer in Biotechnology" (S. Clegg, ed.), pp. 35–41. American Society of Mechanical Engineers, New York.

28. Unsworth, B. R., and Lelkes, P. I. (1998). Growing tissues in microgravity. *Nat. Med.* **4**, 901–907.

29. Unsworth, B. R., and Lelkes, P. I. (1998). The use of rotating wall bioreactors for the assembly of differentiated tissue-like organoids. *In* "Advancements in Tissue Engineering" (L. Savage, ed.), pp. 113–132. International Business Communications, Southborough, MA.

30. Birren, S. J., and Anderson, D. J. (1990). A v-*myc*-immortalized sympathoadrenal progenitor cell line in which neuronal differentiation is initiated by FGF but not NGF. *Neuron* **4**, 189–201.

31. Eaton, M. J., Frydel, B. R., Lopez, T. L., Nie, X. T., Huang, J., McKillop, J., and Sagen, J. (2000). Generation and initial characterization of conditionally immortalized chromaffin cells. *J. Cell. Biochem.* **79**, 38–57.

32. Schwarz, R. P., Goodwin, T. J., and Wolf, D. A. (1992). Cell culture for three-dimensional modeling in rotating-wall vessels: An application of simulated microgravity. *J. Tissue Cult. Methods* **14**, 51–58.

33. Goodwin, T. J., Prewett, T. L., Wolf, D. A., and Spaulding, G. F. (1993). Reduced shear stress: A major component in the ability of mammalian tissues to form three-dimensional assemblies in simulated microgravity. *J. Cell. Biochem.* **51**, 301–311.

34. Goodwin, T. J., Schroeder, W. F., Wolf, D. A., and Moyer, M. P. (1993). Rotating-wall vessel coculture of small intestine as a prelude to tissue modeling: Aspects of simulated microgravity. *Proc. Soc. Exp. Biol. Med.* **202**, 181–192.

35. Wolf, D. A., and Schwarz, R. P. (1991). Analysis of gravity-induced particle motion and fluid perfusion flow in the NASA-designed rotating zero-head-space tissue culture vessel. *NASA Techn. Pap.* **3143**, 1–12.

36. Tsao, Y. D., Goodwin, T. J., Wolf, D. A., and Spaulding, G. F. (1992). Responses of gravity level variations on the NASA/JSC bioreactor system. *Physiologist* **35**, S49–S50.

37. Hua, J., Erickson, L. E., Yiin, T. Y., and Glasgow, L. A. (1993). A review of the effects of shear and interfacial phenomena on cell viability. *Crit. Rev. Biotechnol.* **13**, 305–328.

CHAPTER 30

NEUROECTODERMAL CELL CULTURE: GLIA AND NEURONS

Il-Woo Lee

INTRODUCTION: NERVE REGENERATION AND NEURAL TISSUE ENGINEERING

Tissue engineering of the nervous system has two major goals. One is application in the treatment of diseases of the nervous system, including central and peripheral nerve injury, Parkinson disease, Huntington chorea, chronic intractable pain, hearing impairment, and retinal damages. The other is supply of innervation to the tissue-engineered organ substitutes that need neural control for optimal function, such as urinary bladder, intestine, muscle, and tendon [1]. Nevertheless there are problems to be solved to reach the goals. For example, the peripheral nervous system, owing to its complex and numerous long-distance interconnections, must heal by true regeneration, since healing by scar prevents the reestablishment of electrical connectivity. In certain situations, underlying disease, such as diabetes or drug-induced neuropathy, or extensive damage, such as large nerve defects or avulsion injuries, leads to negligible recovery [2]. In contrast, central nervous system (CNS) tissues, which include the brain and spinal cord, show limited healing under almost all circumstances. A major impediment to healing is the inability of adult neurons to proliferate *in vivo* and to be cultivated *in vitro*. Fortunately, developments in neuroscience since the 1990s have showed us possibilities of CNS regeneration [3,4], and these findings provide a strong basis for hope in the field [5].

Neural tissue engineering may provide a means of reducing health care costs and improving the quality of life by accelerating healing, providing alternatives to current treatments, and enabling repair where it is not currently feasible. Neural tissue engineering also entails efforts to replace neuroactive molecules in the CNS, to re-create neural circuitry *in vitro*, and to design functional neural interfaces. For this purpose, neural tissue engineering seeks to provide new strategies to optimize and enhance regeneration through the use of three-dimensional scaffolds, the delivery of growth-promoting molecules, and the use of neuronal support cells or genetically engineered cells [6]. Among these, although cellular transplantation is a major component of strategies, culturing techniques of various neuronal and gliar cell are also very essential. This chapter concentrates mainly on isolation and culturing methods of neuronal or glial cells, which are used for tissue engineering in the nervous system. Because all neuronal and glial cells could not possibly be included, several cell types that are widely used in the field of neural tissue engineering are emphasized: PC12 cell lines, Schwann cells, olfactory ensheathing cells, and hippocampal and spinal cord neurons.

PC12 CELL LINE

The PC12 cell line was cloned in 1976 from a transplantable rat pheochromocytoma. The original tumor appeared in an X-irradiated male rat of the New England Deaconess

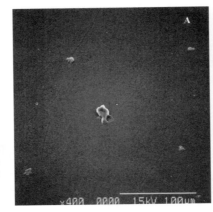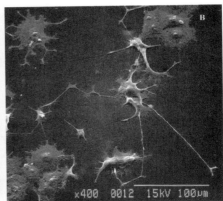

Fig. 30.1. Scanning electron microscopic findings of PC12 cells cultured without (A) or with (B) NGF for 4 days. Note the generated neurites from PC12 cells with NGF treatment. Original magnification ×400.

Hospital strain. In general, PC12 cells exhibit many of the phenotypical properties associated with pheochromocytoma cells and their nonneoplastic counterparts, adrenal chromaffin cells. For instance, they synthesize, store, release (in response to typical secretagogues and depolarizing agents), and take up considerable levels of catecholamines. Thus, they have served as a major model for elucidating the molecular mechanisms underlying each of these events. PC12 cells respond to nerve growth factor (NGF) (Fig. 30.1) and undergo a dramatic change in phenotype wherein they require many of the properties characteristic of sympathetic neurons. Among the salient responses to NGF are the cessation of proliferation, the generation of long neurites, the appearance of electrical excitability, hypertrophy, and a number of changes in composition associated with acquisition of a neuronal phenotype. These features have promoted a large number of studies dedicated, in particular, to uncovering the mechanism of action of NGF and, in general, to unraveling the molecular mechanisms by which neuroblasts differentiate into mature, postmitotic neurons. The last 20 years have seen an ever-increasing use of PC12 cells to approach a number of fundamental problems related to neuronal cell differentiation and function. This widespread utility of the line is attributable in part to its relative stability, homogeneity, high degree of differentiation and of differentiative capacity, robust response to NGF and dramatic change in phenotype brought about by this factor, fidelity to many of the features of normal neuroblasts and neurons, potential for genetic manipulation, and the accrual of a large number of studies regarding its characterization.

For those considering the PC12 cell line for their studies, it is appropriate at this point to consider some of the limitations of this system. First, PC12 cells are tumor derived and, therefore, by definition must differ from nontransformed cells in their behavior. In this respect, it is interesting that the defects responsible for their transformed phenotype have yet to be revealed. Second, despite their attractiveness and utility for studying neuronal differentiation, PC12 cells cannot be taken as an exact model of developing neurons. In certain respects, they are less than perfect counterparts even for chromaffin cells and sympathetic neurons. For example, there is no evidence that they are capable of receiving or forming synaptic connections or that they form dendrites. Thus, in light of these realities, it is paramount to consider that the line should be used only as a starting point for investigations and that findings made with it ultimately must be verified with primary tissue [7].

METHODS
Culture Medium for Cell Growth

1. 85% RPM1 1640 medium (available from Gibco/BRL, Gaithersburg, MD, and many other suppliers in either powdered or liquid form), 10% horse serum (heat inactivated at 56°C for 30 min), 5% fetal bovine serum, 25 units/ml penicillin, and 25 µg/ml streptomycin. Pretesting of horse serum batches is prudent, because not all are satisfactory for optimal cell growth and maintenance. At present, consistently good results have been obtained with "donor" horse serum purchased from IRH Biosciences (Lenexa, KS).

2. If necessary, PC12 cells can be adapted to grow in Dulbecco's Modified Eagle's Medium (DMEM) supplemented as just outlined. The adaptation is more successful if carried out in stages, with intermediate mixtures of the two media. Also, although the cells may be maintained for up to several weeks with medium containing horse serum alone, it has been our experience that continued propagation of the line requires supplementation with fetal bovine serum.

3. PC12 cells will tolerate fungizone at 2.5 µg/ml. In cases of bacterial infections that escape the penicillin and streptomycin present in the medium, treatment with 50 µg/ml of gentamicin generally has proved to be effective.

Culture Substrate

PC12 cells adhere poorly to tissue culture plastic. Therefore, it is recommended strongly that stock cultures be maintained on an adhesive coating, to which all the cells attach well. The most useful and cost-effective coating for plastic culture dishes used for PC12 cell culture has proved to be rat tail collagen. It may be either purchased from commercial sources (e.g., Boehringer-Mannheim, cat. no. 1, 179, 179, or Sigma, cat. no. C 7661) or prepared in the laboratory.

PC12 cells also attach well to substrates coated with polylysine or polyornithine. Tissue culture dishes can be coated by treatment for 4 h with a 1-mg/ml solution of filter-sterilized poly(L-ornithine hydrobromide) (MW 30,000–70,000 Da) or poly(L-lysine hydrobromide) (MW 150,000–300,000 Da) prepared in 150 mM sodium borate buffer, pH 8.4.

General Maintenance of Cultures

1. The doubling time of PC12 cells is quite long: 2.5–4 days. Cultures are maintained with the complete medium described earlier.

2. Optimal proliferation is obtained by feeding the cultures every 2–3 days. Only about two-thirds of the medium is exchanged, ensure to that a portion of "conditioned" medium remains to facilitate growth. Care should be taken to avoid damaging or tearing the substrate when fresh medium is added.

3. Cultures are maintained at 36–37°C in a water-saturated atmosphere containing 7.5% CO_2. Growth also appears to be adequate in atmospheres containing 5 or 10% CO_2.

4. When the cultures reach confluency, they are subcultured at a ratio of 1:3–1:4. Under these conditions, subculturing is carried out every 7–10 days. For subculturing, the cells are detached from the substrate by forceful aspiration of the medium through a Pasteur or tissue culture pipette. Repeated trituration of the cells before replating decreases the formation of clumps.

Preparation and Storage of Frozen PC12 Cells

1. The cells are detached from culture dishes, harvested by centrifugation, and suspended in complete medium supplemented with 10% dimethyl sulfoxide (DMSO).

2. The cell suspension then is aliquoted into plastic cryovials and is chilled on ice. The cells are slow-frozen to at least –60°C at a rate of approximately –1°C/min.

3. Alternatively and more economically, the chilled cell-containing vials (containing approximately 1 ml of cell suspension) may be placed overnight in a freezer maintained at –70 to –80°C.

SCHWANN CELL CULTURE

The Schwann cell stimulates axonal regeneration in the central nervous system [8–10] and probably has a key role in the cellular transplantation methods in neural tissue engineering. Looking toward the therapeutic use of Schwann cell populations in nerve repair, a likely application would involve autotransplantation [11]. In such a paradigm, a small biopsy of nerve from a patient could yield a sufficient number of Schwann cells, after purification and proliferation *in vitro*, to provide sufficient graft material to repair peripheral or central nerve damage not conducive to direct nerve grafting. Such paradigms require the use of adult-derived Schwann cell populations. Until recently, little emphasis had been placed on the acquisition of Schwann cells from adult animals or from human donors. Peripheral nerves retrieved from adult rats or human organ donors can be processed to obtain relatively pure populations of Schwann cells, despite the high fibroblast and extracellular matrix contents of these nerves. The Schwann cell isolation method was described first by Morryssey *et al.* [12], and we are continuing to test modifications of this technique.

METHODS
Preparing Media
1. *Fibroblast inhibition medium*: 10% fetal calf serum (FCS)-DMEM + 10 μl/ml Ara-C stock
2. *Schwann cell culture medium* (10% FCS-DMEM-GGF-FSK): 10% 17C5-DMEM + purified recombinant human glial growth factor 2, 2.3 μg + 5 mM forskolin, 15 μl/15 ml
3. *Freezing medium*: 40 ml of 10% FCS-DMEM + 10 ml of DMSO + 50 ml of FCS

Preparing PDL Coating Tissue Culture Flask
1. Prepare fresh PDL [poly(D-lysine) (80 μg/ml)]. Thaw 0.5 nil aliquot of 4 mg/ml PDL and add to 30 ml of sterile Milli Q H$_2$O. Vortex.
2. Precoat 175 tissue culture flasks by adding 30 ml of PDL (80 μg/ml) and rotating until it can be seen that entire bottom of flask has been coated. Allow to stand for 20 min. Remove PDL by aspiration with a sterile Pasteur pipette and wash twice with 5 ml of sterile Milli Q H$_2$O. Flasks are ready for immediate use or can be air-dried under a hood and stored for later use. To ensure drying, leave opened flasks in the hood overnight.

Isolation of Primary Rat Schwann Cells
From Adult

Peripheral nerves retrieved from adult rats or human organ donors can be processed to obtain relatively pure populations of Schwann cells, despite the high fibroblast and extracellular matrix content of these nerves. To dissect sciatic nerves of adult Sprague-Dawley rats, the animals are ether-anesthetized, and their hindlimbs are shaved and cleaned with Betadine and alcohol. In a laminar flow hood, the skin over the legs is incised and retracted. Different sets of sterile scissors are used sequentially to cut the muscles over the femur until the nerve is visible and to expose the nerve from the knee to the spinal column, and to cut the sciatic nerve free, place it in L-15 with 50 U/ml of penicillin and 0.05 mg/ml of streptomycin, and cut it into pieces some 2–3 cm long.

From Pups

Obtain two litters of 3-day-old Sprague-Dawley rat pups. Inject 0.15 ml of somlethal intraperitoneally into each rat pup, three animals at a time. Dissect one animal at a time, in order of injection. Inject three more animals when only one dead animal remains. This eliminates waiting for the animals to die. After the animal has died (about 3 min) transfer entire body to beaker of 70% EtOH. Place beaker with animal in dissecting hood and remove animal from beaker with forceps. Hold animal firmly with one hand and remove skin by grasping firmly and tearing from back of neck. Place animal dorsal side up on inside of lid of a sterile 100-mm dish, which is on the platform of a binocular microscope. Soak dissecting instruments in absolute EtOH before use. Shake off excess ethanol and allow to air-dry in dissecting hood. Repeat as necessary. Using the binocular microscope as a dissecting aid, aseptically dissect sciatic nerves by making an incision through the dorsal leg muscle from the bottom of the spinal cord to the first leg joint with two fine forceps. When the sciatic nerve is exposed, grasp it with forceps and cut it at the extreme ends with scissors.

Procedure for Nerves from Adults and from Pups
1. Place the nerves in a sterile 35-mm tissue culture dish containing 5 ml of L-15 PS. Remove nerves from all animals and combine in the L-15 PS. Remove connective tissue from nerves with fine forceps and discard.

2. Transfer nerves to a conical polypropylene tube containing 2 ml of dissociation medium. Make certain the inside of the Pasteur pipette and the sides of the tube are wetted with medium beforehand to prevent the nerves from sticking.

3. Incubate the nerves in a 37°C water bath for 15 min. Shake the tube gently every few minutes. Spin at 1000 rpm for 5 min.

4. Decant off the dissociation medium and replace with 2 ml of fresh dissociation medium.

5. Repeat step 3.

6. Decant off the dissociation medium and replace with 2 ml of 10% FCS-DMEM. Transfer to a sterile 35-mm petri dish.

7. Using a 10-ml syringe with a 21-gauge needle, triturate nerves by aspirating and expelling liquid through the syringe, being careful to avoid bubbles.

8. Use a 23-gauge needle to repeat step 7.

9. Sterilize 2 in.^{2} of nylon mesh by immersing in a petri dish containing EtOH for 5 min. Remove and place on a sterile, dry surface and allow to air-dry in the laminar flow hood.

10. With sterile forceps, place the sterile 20-μm nylon mesh over the opening of a sterile 15-ml conical tube. Use a pipette to slowly filter the cell suspension through the mesh into the tube.

11. Rinse the petri dish that had contained the nerves with 4 ml of 10% FCS-DMEM, Add this solution to the nerves through the nylon mesh. Centrifuge for 5 min at 1000 rpm (150g).

12. Decant the media from the tube and replace with 1 ml of 10% FCS-DMEM. Dissociate the cell pellet by repeated pipetting with a 1-ml pipette.

13. Add 10 μl of cell suspension to 90 μl of trypan blue in an Eppendorf tube, and vortex. Cells that exclude this dye are viable.

14. Count cells in a hemacytometer by loading a 10-μl cell suspension onto each chamber of the hemacytometer along the grooves on the edge of the chamber. This is best done with an Eppendorf pipettor. Count unstained (live) cells in 2–4 sections measuring 1 mm^2, divide the total number of cells counted by the number of 1-mm^2 grids counted, and multiply this number by the dilution factor(10) times 10^4 to arrive at the number of cells per milliliter.

15. Since the cells were resuspended in 1 ml, the number obtained in step 14 is also the total number of cells. Add [(total number of cells/0.33 $\times$ 10^6) – 1] ml to the cell suspension (for 5 million cells per 15 ml). Add 15 ml of cell suspension per T75 flask precoated with PDL (5 million cells per flask). Label and place in 37°C incubator regulated with 10% CO$_2$.

Inhibition of Fibroblast Proliferation

1. At least 24 h after the isolation of primary rat Schwann cells has been completed, remove all 10% FCS-DMEM from flasks.

2. Carefully wash the cells by twice adding and removing 8 ml of prewarmed 10% FCS-DMEM with a pipette.

3. Refeed culture flasks by adding 15 ml of prewarmed fibroblast inhibition medium per flask. Place in 37°C incubator regulated with 10% CO$_2$.

4. Two to three days later, remove all fibroblast inhibition medium from the flasks.

5. Wash the cells as described in step 2. Refeed the culture flasks by adding 15 ml of 10% FCS-DMEM-GGF-FSK to each flask. Place in 37°C incubator regulated with 10% CO$_2$.

Schwann Cell Purification

1. When the cells have grown to confluence, remove all 10% FCS-DMEM-GGF-FSK from the flasks. Wash the cells by adding 8 ml of phosphate-buffered saline (CMF [calcium- and magnesium-free]-PBS) to the flasks. Aspirate off all PBS with a sterile Pasteur pipette and add 4 ml of prewarmed 0.25% trypsin–EDTA solution (calcium free) to each flask.

Monitor microscopically until cells begin to detach from the flasks. Sheets of cells can be seen to move across the flasks.

2. Add 4 ml of 10% FCS-DMEM per flask to inhibit enzyme activity, pipette cells up and down to disperse them, and transfer the cell suspension to one 15-ml conical tube per flask. Rinse flasks with another 4 ml of medium to improve recovery, and return the suspension to the centrifuge tube containing the initial harvest. Centrifuge at 1000 rpm for 5 min.

3. Decant off all but 100 μl of supernatant and add a 500-μl aliquot of anti-Thy-1.1 antibody (1:1000) to the cell pellet. Resuspend the cell pellet by carefully pipetting liquid until no cell clumps appear. If multiple flasks have been trypsinized, cells can be combined in one 50-ml conical tube at this point, maintaining the ratio of 500 μl of anti-Thy 1.1 per pellet.

4. Incubate for 10 min in a 37°C water bath, gently mixing tube contents every few minutes.

5. Remove the cell suspension from the water bath. Add to the cell suspension 250 μl of freshly thawed rabbit complement per cell pellet, place back in the 37°C water bath, and incubate for 30 min, gently mixing tube contents every 10 min.

6. After 30 min, remove the cell suspension from the 37°C water bath and add 10 ml of 10% FCS-DMEM. Centrifuge the suspension at 1000 rpm for 5 min.

7. Discard the supernatant and resuspend the cells in 1 ml of 10% FCS-DMEM per cell pellet by carefully pipetting liquid repeatedly until no cell clumps appear.

8. Count cells in a hemacytometer as described earlier to determine the number of cells per milliliter.

9. Multiply the number of cells per milliliter by the number of milliliters in which the cells were resuspended to calculate the total number of cells in the tube. The volume of 10% FCS-DMEM-GGF-FSK to add in milliliters is [(total number of cells)/2×10^6] cells per 15 ml. Transfer cell suspension to PDL-coated T75 flasks, 15-ml. Place in 37°C incubator regulated with 10% CO_2.

10. Monitor periodically and observe the cultures for fibroblast levels and viability of the Schwann cells. Microscopic examination should reveal a population of attached bipolar Schwann cells that may or may not be phase bright (Fig. 30.2). Fibroblasts can be distinguished from Schwann cells by their flattened appearance. If the fibroblast levels are greater than 5%, the Schwann cell purification procedure should be repeated.

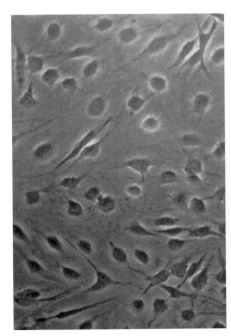

Fig. 30.2. Phase contrast photomicrograph of pure Schwann cells cultured from sciatic nerve of an adult Sprague-Dawley rat. Original magnification ×400.

Freezing Cells

1. When the flasks become confluent, trypsinize cells as described in 1 and 2 of the preceding procedure for Schwann cell purification.

2. Count cells from one flask as described in 7 and 8 of the Schwann cell purification procedure. Since the cells were resuspended in 1 ml, the number obtained is also the total number of cells. The volume of freezing medium to add in milliliters per tube is [(total number of cells)/(2×10^6)]. Spin at 1000 rpm for 5 min.

3. Remove all the medium from all tubes, leaving just enough medium to keep the pellet wet.

4. Resuspend the cells in one tube with the volume of freezing medium, which will give 2×10^6 cells/ml calculated in step 2.

5. Quickly aliquot the cell suspension into cryovials, 1 ml/tube.

6. Immediately place the cryovials in a styrofoam box (cover with aluminum foil if no lids are available, to avoid spilling contents of tubes if the box is dropped) and freeze cells by quickly placing in –70°C freezer overnight.

7. Repeat steps 4, 5 and 6 for each tube, resuspending the cells in a tube only after the cryovials from the preceding tube have been placed in the freezer.

8. Transfer frozen cells to the liquid nitrogen storage tank and enter their location in the tank into the storage tank log book.

OLFACTORY ENSHEATHING CELLS

The olfactory bulb (OB) is a structure of the central nervous system in which axonal growth occurs throughout the lifetime of the organism. A major difference between the OB and the remaining CNS is the presence of ensheathing glia in the first two layers of the OB. Ensheathing glia display properties that might be involved in the process of regeneration, and they appear to be responsible for the permissibility of the adult OB to axonal growth. In fact, transplants of ensheathing glia can be used as promoters of axonal regeneration within the adult CNS. The axonal growth-promoting properties of ensheathing glia make the study of this cell type interesting for understanding the mechanisms underlying axonal regeneration. This cell type does not exhibit the phenotypic features of any glial population described thus far [13–15]. Cultures of ensheathing glia can be obtained not only from embryonic and early postnatal animals but also from adult olfactory bulb. In 2000 adult human olfactory ensheathing cells were isolated and transplanted to the demyelinated rat spinal cord, where they elicited remyelination [16]. Table 30.1 reviews the work on the

Table 30.1. Comparison of Immunocytochemical Properties of Multipolar Glia Cultured from Olfactory Nerve and Glomerular Layers (ONGL) Known Glial Populations

	ONGL glia	Astrocytes	Schwann cells
Vimentin	+	+	+
GFAP	+	+	+
EDI	–	–	–
Rip	–	–	–
Ran2	–	+	–
A2B5	–	+	–
HNK1	–	+	–
MAP2	–	+	Not determined
Laminin	+	+	+
Fibronectin	+	+	–
L1/G4	+	+	+
NGFR	+	–	+
MBP	+	–	+

Source: Ramon-Cueto and Nieto-Sampedro [13].

culture of olfactory ensheathing cells from the adult rat and gives the characteristics that differentiate ensheathing glia from other peripheral and central glial populations.

METHODS
Culture

Adult male (2.5 months) Wistar rats were killed by decapitation, and the olfactory bulb was dissected in Hanks' Balanced Salt Solution (HBSS).

1. After the careful removal of pia mater, the olfactory nerve and glomerular layers (ONGL) were dissected together, away from the rest of the bulb. The two fractions (ONGL and remainder of the OB) were treated identically, but cultured separately from each other, as described shortly.

2. The tissue was washed twice with Ca^{2+}-, Mg^{2+}-free HBSS, diced in small fragments, and incubated with trypsin (Gibco BRL, 0.1% w/v) at 37°C for 15 min. Trypsinization was stopped by adding a 1:1 mixture of Dulbecco's Modified Eagle's Medium (DMEM) and Hams F-12 medium (D/F-12, mixture, Sigma, St. Louis, MO), supplemented with 10% fetal bovine serum (FBS: Seromed lot no. 5S02; Biochrom KG, Berlin); the tissue was centrifuged at 800 rpm for 3 min, resuspended in a 1:1 mixture of DMEM and Ham's F-12 plus 10% FBS (D/F-10S) and this last step repeated.

3. After suspension in 1 ml of culture medium, single-cell dissociation was achieved by 15–20 passes through a fire-polished siliconized Pasteur pipette.

4. The cells from ONGL and those from the rest of the bulb were plated separately in two flasks (Falcon, 25 cm^2) pretreated for 1 h at room temperature with poly(L-lysine) (Sigma; average MW 25,000; 50 μg/ml, 15 mM sodium borate buffer, pH 8.4).

5. Viable cells were seeded at a density of 6.5×10^6 per flask (cells from two bulbs per flask) and maintained in D/F-10S, supplemented with 2 mM glutamine, 10 U/ml penicillin. 10 μg/ml streptomycin, and 50 μg/ml gentamicin. The flasks were incubated at 37°C, 5% CO_2, and the medium was changed every 2 days.

6. Eight days after plating, the cells were detached from the flask with trypsin (0.25%, w/v) for 3 min at 37°C and plated at a density of 10,000 cells per well in eight chamber glass culture slides (LabTek, Miles Scientific, Naperville, IL) pretreated with poly(L-lysine) (50 μg/ml).

7. The cells were grown for 2 days in the LabTek chambers and then processed for immunocytochemistry.

CHARACTERIZATION

The following primary antibodies were used.

1. Mouse anti-vimentin, mouse anti-GFAP (glial fibrillary acidic protein) and mouse anti-microtubule associated protein 2 (anti-MAP2) were monoclonal IgGs from Boehringer-Mannheim (Indianapolis, IN) and were used at dilutions of 1:500 (the first two) and 1:2000 (the latter).

2. Mouse anti-ED1 was a monoclonal IgG from Serotee (Oxford, UK) used at a dilution of 1:2500.

3. Anti-Laminin and anti-fibronectin were rabbit polyclonal monospecific antibodies from F-Y Labs Inc. (San Mateo, CA) and Gibco/BRL (Basel, Switzerland), respectively; they were used at a dilution of 1:500.

4. Mouse anti–nerve growth factor receptor (anti-NGFR, IgG 192) was a monoclonal IgG, affinity-purified from ascites fluid on a protein A-Sepharose column and used at 5 μg/ml.

5. Rabbit (polyclonal) and mouse (monoclonal ascites fluid) anti-L1/G4 were both used at 1:200 dilution.

6. The supernatant of mouse hybridoma Rip was used at a 1:10 dilution.

7. Guinea pig polyclonal anti–myelin basic protein (MBP) was used at a 1:500 dilution.

8. Ran 2 cells were obtained from the American Type Culture Collection and the supernatants containing the antibodies A2B5 and IINK-1 were used at a dilution of 1:2.

9. Rhodamine-conjugated goat anti-rabbit or anti-mouse, fluorescein-conjugated goat anti-mouse IgG and IgM (Boehringer-Mannheim) and biotinylated anti–guinea pig (Vector, Burlingame, CA) were used as secondary antibodies at a dilution of 1:200.

10. Fluorescein-conjugated streptavidin (Amersham, UK) was used at a 1:100 dilution.

METHODS

1. Cells were immunostained for GFAP, vimentin, laminin, fibronectin, MAP2, and ED1 in the culture flasks after 8 days *in vitro* (DIV), at the stage prior to trypsinization and secondary culture plating.

2. After fixation in 4% paraformaldehyde (10 min, room temperature), the cells were washed with phosphate-buffered saline (PBS), pH 7.3, and incubated in PBS containing 0.1% Triton-1% normal goat serum (15 min, 4°C) followed by the same mixture containing the primary antibodies (24 h, 4°C). When staining for fibronectin and laminin, the use of Triton was avoided. After repeated washing with PBS, the cells were incubated with secondary antibodies (1 h, 25°C).

3. Secondary cultures in LabTek slides were immunostained 2 days after plating with antibodies to GFAP, vimentin, L1 polyclonal, L1 monoclonal, Ran2, ED1, Rip, and MAP2, following essentially the same protocol just described.

4. Incubations with anti-NGFR, anti-laminin, anti-fibronectin, A2135, HNK-1, and anti-MBP were performed without Triton, since immunostaining was better when the integrity of the membrane was preserved.

5. The cells were treated with the primary antibody in culture medium (37°C, 5% CO_2, 1 h), washed with culture medium, and then incubated with secondary antibody under the same conditions, followed by fixation (4% paraformaldehyde, 10 min) and washing with PBS. All the slides were mounted with glycerol–PBS (1:1) and examined in a fluorescence microscope. The proportion of immunoreactive cells was determined by counting 16 fields in each of three wells from three independent cultures at 200× magnification (expressed as percentage ± standard deviation) [13].

PRIMARY DISSOCIATED CELL CULTURES OF RAT HIPPOCAMPAL NEURONS

The hippocampus is a structure that has fascinated neuroscientists for generations because of its remarkable capacity for synaptic function and as a source of seizure activity. Moreover, the hippocampus is a source of a relatively homogeneous population of neurons with well-characterized properties typical of central nervous system neurons in general, and this is why the culture of hippocampal neurons is included in this chapter. Pyramidal neurons, the principal cell type in the hippocampus, are estimated to account for 85–90% of the total neuronal population. Hippocampal pyramidal cells exhibit other features of interest for cell biological studies. For example, they have a characteristic, well-defined shape, and they make direct connection with one another and with the population of endogenous interneurons [7,17,18].

Hippocampal cultures typically are prepared from 18- to 19-day fetal rats or from fetal mice at a comparable stage of development. At this stage, the generation of pyramidal neurons, which begins in the rat at about E15, is essentially complete, but the generation of dentate granule cells, which largely occurs postnatally, has scarcely begun. The tissue still is easy to dissociate, the meninges are removed readily, and the number of glial cells still is relatively modest [7,18].

METHODS: ASTROGLIAL CELL CULTURES

One or two week before, perform the following steps.

1. Sacrifice rat pups, remove the brains, transfer them to a dish containing HBSS, and dissect out the cerebral hemispheres.

2. Carefully strip away the meninges, mince the tissue into small pieces, then transfer it to a 15-ml tube in a final volume of 12 ml of HBSS.

3. Add 1.5 ml each of trypsin (Gibco 1:250) and 1% DNase (Sigma DN-25), incubate at 37°C for 15 min.

4. Aspirate off the supernatant, add 5 ml of HBSS, invert the tube several times, and let stand for 5 min. Repeat this step twice more to allow residual trypsin to diffuse from the tissue. Bring the final volume to about 5 ml with HBSS.

5. Dissociate the cells by vigorously pipetting up and down with a narrow-bore Pasteur pipette (usually about 15–20 times).

6. Use a hemacytometer to determine the cell density.

7. Dilute the cell suspension to about 100,000 cells/ml, and plate into 60-ml dishes at 4 ml per dish.

8. Glial cultures are fed twice a week. The cultures generally reach confluence within 10–14 days after plating and are used for about 2 weeks after they become confluent.

9. One day before the preparation of hippocampal cultures is scheduled, select glial dishes to be used for coculturing. Remove the medium from these dishes, add 6 ml of neuronal maintenance medium, and return them to the incubator.

Day 1
1. Place coverslips in porcelain staining racks and rinse in water (2 times, 10 min each).
2. Clean in concentrated HNO_3 for 18–36 h.

Day 2
3. Rinse twice for 1 h, then twice for 30 min in MilliQ water.
4. Tap off excess H_2O and dry in oven (1 h at 60°C).
5. Place racks in beakers, cover with foil, and sterilize with dry heat (225°C for 6 h).

Day 3: All Steps Must Be Performed in Laminar Flow Hood
6. Melt the sterilized paraffin in hot water bath (100°C).
7. Place coverslips in a 60-mm microbiological (not tissue culture) petri dish, and apply 3 small drops of sterile, melted paraffin (about 0.5 mm high) near the outer edge of each coverslip. This practice keeps the coverslips from resting directly on the glial cells during coculturing. Sterilize with UV irradiation for 30 min.
8. Dissolve poly(L-lysin) (MW 30,000–70,000, Sigma P2636) (1 mg/ml) in 0.1 M borate buffer (pH 8.5) and filter-sterilize.
9. Cover each coverslip with poly(L-lysin) solution (about 6 drops) and leave overnight (about 18 h).

Day 4
10. Rinse coverslips with sterile water (two washes, 2 h each).
11. Remove final rinse, add 4 ml of plating medium, and store the dishes in an incubator until ready. Hippocampal neurons will be plated into these dishes. The plating media consist it Eagle's Minimal Essential Medium (MEM) with Earle's salts 500 ml + extra glucose (600 mg/liter) 0.3 g + horse serum (FC 10%) 50 ml + 100× glutamine 5 ml.
12. The day before preparation of hippocampal cultures is scheduled, select the dishes containing a confluent monolayer of glial cells to be used for coculturing.
13. Remove the medium from the dishes, add 6 ml of maintenance medium, and return to incubator until ready. The maintenance medium consists of Eagle's MEM with Earle's salts 500 ml + 100× N_2 supplement, 5 ml + Ovalbumin (FC 0.1%, Sigma A5503) 0.5 g + 100× glutamine, 5 ml + extra glucose (600 mg/liter) 0.3 g + 1 mM sodium pyrate.

Day 5
14. Euthanize a pregnant Sprague-Dawley rat and wipe the abdomen with 70% ethanol before making any incisions. Remove the uterus, and place the fetuses in a sterile petri dish.
15. Decapitate the fetuses, dissect out the brains, and place them in HEPES-buffered, calcium- and magnesium-free balanced salt solution (prepared from 10× Hanks's BSS, not

1× Hanks's BSS). Keep brains in the following BSS at all times to prevent the tissue from drying out.

HEPES-buffered calcium- and magnesium-free BSS: 10× HBSS, 100 ml, + HEPES (FC 15 mM), 3.57 g, + distilled water up to 1000 ml; adjust the pH to 7.4 with NaOH.

17. Remove the hippocampi under a dissecting microscope at 10–15× magnification. Strip away the meninges; then collect the dissected hippocampi in a small petri dish in BSS.

18. Use a Pasteur pipette to transfer all the hippocampi from one litter to a 15-ml centrifuge tube. Bring the total volume to 4.5 ml with BSS, then add 0.5 ml of 2.5% trypsin (trypsin 1:250, Gibco cat. no. 610-5095). Incubate at 37°C for 15 min. The hippocampus will settle to the bottom of the tube.

19. Gently pipette off the trypsin solution, add 5 ml of BSS, shake the tube to mix, and let stand for 5 min. Repeat this step twice more to allow residual trypsin to diffuse from the tissue. Bring the final volume to about 5 ml with BSS (proportionally less if fewer than 10 fetuses have been used).

20. Dissociate the cells by vigorously pipetting up and down, first in a normal Pasteur pipette, then in a pipette whose tip has been fire-polished to about half the normal diameter (to a little less than 1 mm). Continue pipetting until no chunks of tissue remain (usually about 15–20 times). To prevent foaming, push the BSS out against the side of the centrifuge tube, rather than into the liquid at the bottom.

21. Determine the density of cells in a hemacytometer. The yield should be about 500,000 cells per hippocampus. Also determine the fraction of viable, trypan blue excluding cells. The trypan blue stock solution consists of 0.4% trypan blue (Sigma T8154), mixed 1:1 with a few drops of the cell suspension and allowed to stand for 4 min before the fraction of dye-excluding cells is counted. If everything has gone well, 85–95% of the cells should exclude trypan blue.

22. Add the desired number of viable cells to each of the dishes containing the poly(L-lysine)-treated coverslips in MEM with 10% horse serum. Our standard plating density is 150,000 cells per 60-mm petri dish (1000 cells/cm²). Much lower and higher plating densities (5000–800,000) can be used. Plating must be done rapidly to prevent the pH of the medium from becoming basic. Remove a few dishes from the incubator at a time, add the appropriate volume of cells to each with a micropipette, and swirl the dishes.

23. After 3–4 h, transfer the coverslips, with neurons attached, to dishes containing glial cells in maintenance medium. Turn the coverslips over so the neurons are facing down, toward the glia.

24. To reduce glial proliferation, add Ara-C (5 mM) after 2–3 days.

25. Once a week, remove about 2 ml of medium and replace it with fresh maintenance medium. It is important not to change the media completely; the neurons depend on a "conditioning." Therefore, replace about one-third of the medium (2 ml) at each feeding.

SPINAL CORD NEURONAL CULTURE FROM RAT EMBRYO

Dissociated cell cultures of embryonic spinal cord provide useful systems for studying various aspect of spinal motor neurons, including nerve growth and turning, synaptogenesis, and transmitter secretion. For CNS regeneration, some researchers have found that placing transplants of fetal spinal cord tissue placed in the lesion site results in an even greater extent of anatomical reorganization of host CNS pathways. The transplant-mediated anatomical plasticity results in a greater functional recovery than after lesion alone [7,19–21]. The protocol for dissection and culture methods of fetal spinal cord neurons from rat is briefly described as follows.

METHODS
1 Day Before

1. Coat the plates with poly(L-lysine).
2. Dissection

    a. Sacrifice a pregnant Sprague-Dawley rat with dry ice.

    b. Perform hysterectomy and take out four to five embryos (embryonic day 15).

    c. Dissect the embryos out the sacs and rinse them briefly in 70% ethanol. Place them in Hanks's solution without sodium bicarbonate.

    d. Clean the ventral surface of the embryo of any remaining viscera, such as the aorta.

    e. Make a ventral cut, in a rostral (from where the head was removed) to caudal direction, through the vertebral bodies, to expose the cord throughout its length. Alternatively, the vertebral pedicles can be cut on each side, and the strip of vertebral bodies, removed *en bloc*.

    f. Using the medulla as a "handle," pull straight up (toward the microscope) on the cord, freeing it from the vertebral canal. Try to minimize side-to-side movement of the cord to ensure that the dorsal root ganglia will remain attached to the cord. Rinse the cord at least once with L-15 medium to free it of debris and blood.

    g. Place dissected spinal cord in a new plate with 1 or 2 ml of Hanks's solution over ice (Fig. 30.3).

    h. Transfer the washed cord to the second 35-mm petri dish filled with L-15 medium. It is very important to do the dissection as fast as possible (<1 h) because if the work is prolonged, the survival of cells will decrease dramatically. It also very important to dissect the dorsal root ganglia out of the spinal cord.

3. Plating (under the hood).

    a. Use two sterile no. 10 scalpels to chop the tissue into very small pieces. Remove the solution with a Pasteur pipette.

    b. Add 5 ml of trypsin for 8 min (12 mg/5 ml Hanks's solution at room temperature).

    c. Add 5 ml of trypsin inhibitor (12 mg/5 ml Hanks's solution at room temperature).

    d. Centrifuge at 1000 rpm for 10 min at 4°C.

    e. Remove the supernatant with a 10-ml pipette and rinse the pellet in 5 ml of Hanks's solution. Make sure to resuspend the cells.

    f. Centrifuge at 1000 rpm for 10 min at 4°C.

    g. Repeat steps e and f twice.

    h. Remove the supernatant and resuspend the pellet in 3–4 ml of Hanks's solution.

    i. Triturate carefully with a Pasteur pipette 10 times.

    j. Repeat step i with a preheated Pasteur pipette to make the tip narrower. Repeat this one more time with a smaller tip opening. Triturate 10 times per size for 30 triturations in total.

    k. Centrifuge at 1000 rpm for 30 s at 4°C.

    l. Without disturbing the pellet, take 10 μl of the cell suspension and place in an Eppendorf tube with 90 μl of trypan blue.

    m. Take 10 μl of the cells suspension in trypan blue and place in the hemacytometer.

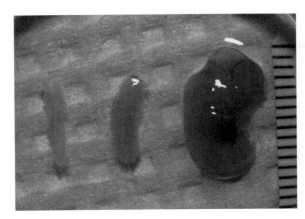

Fig. 30.3. Stages of embryonic spinal cord dissection (right to left): day-15 Sprague-Dawley rat embryo; the isolated vertebral column; the spinal cord without dorsal root ganglia.

n. Plate the cells in a 48-well plate (10,000 cells per well) or a 24-well plate (20,000 cells per well). Use medium E for final plating, buffered with 2.2 g/liter sodium bicarbonate before the cell suspensial is prepared.

o. Place the plate in the CO_2 incubator at 37°C and wait for 2 days until the cells are attached on the plate. Remember that the cells should be plated in less than 3 h after the rat has been sacrificed.

ACKNOWLEDGMENT

This work was supported by the Korea Science and Engineering Foundation (KOSEF) through the Bio-Medicinal Resources Research Center at PaiChai University and Roche Foundation in the program year of 2000.

REFERENCES

1. Yoo, J. J., and Lee, I. W. (1998). Tissue engineering: in nervous system. *In* "Tissue Engineering: Concepts and Application," First Edition, pp. 243–266. Korea Press, Seoul.

2. Bjorklund, A. (1994). "Functional Neural Transplantation," pp. 157–195. Raven Press, New York.

3. Cheng, H., Cao, Y., and Olson, L. (1996). Spinal cord repair in adult paraplegic rats: Partial restoration of hind limb function. *Science* **273**, 510–513.

4. Li, Y., Field, P. M., and Raisman, G. (1997). Repair of adult rat corticospinal tract by transplants of olfactory ensheathing cells. *Science* **277**, 2000–2002.

5. Young, W. (1996). Spinal cord regeneration. *Science* **273**, 451.

6. Valentini, R. F., and Aebischer, P. (1997). Strategies for the engineering of peripheral nervous tissue regeneration. *In* "Principles of Tissue Engineering" (R. Lanza, R. Langer, and W. Chick, eds.), pp. 671–684. R.G. Landes, Austin, TX.

7. Banker, G., and Goslin, K. (1998). "Culturing Nerve Cells," 2nd ed., p. 15. A Bradford Book, MIT Press, Cambridge, MA.

8. Guénard, V., Kleitman, N., Morrisey, T. K., Bunge, R. P., and Aebischer, P. (1992). Syngeneic Schwann cells derived from adult nerves seeded in semipermeable guidance channels enhanced peripheral nerve regeneration. *J. Neurosci.* **2**, 3310–3320.

9. Richardson, P. M., McGuiness, U. M., and Aguayo, A. J. (1980). Axon from CNS neurones regenerate into PNS grafts. *Nature (London)* 264–265.

10. Aebischer, P., Salessiotis, A. N., and Winn, S. R. (1989). Basic fibroblast growth factor released from synthetic guidance channels facilitates peripheral nerve regeneration across long nerve gap. *J. Neurosci. Res.* **23**, 282–289.

11. Vrbova, G., Clowry, G., Nogradi, A., and Sieradzan, K. (1994). *In* "Transplantation of Neural Tissue in the Spinal Cord," First Edition, p. 663. R.G. Landes, Austin, TX.

12. Morryssey, T. K., Kleitman, N., and Bunge, R. P. (1991). Isolation and functional charaterization of Schwann cells derived from adult peripheral nerve. *J. Neurosci.* **11**(8), 2432–2442.

13. Ramon-Cueto, A., and Nieto-Sampedro, M. (1992). Gliar cell from adult rat olfactory bulb: Immunocytochemical properties of pure cultures of ensheathing cells. *Neuroscience* **47**, 213–220.

14. Ramon-Cueto, A., and Valverde, F. (1995). Olfactory bulb ensheathing glia; a unique cell type with axonal growth promoting properties. *Glia* **14**, 163–173.

15. Devon, R., and Doucette, R. (1992). Olfactory ensheathing cells myelinate dorsal root ganglion neurites. *Brain Res.* **589**, 175–179.

16. Kato, K., Honmou, O., Uede, T., Hashi, K., and Kocsis, J. D. (2000). Transplantation of human olfactory ensheathing cells elicits remyelination of demyelinated rat spinal cord. *Glia* **30**, 209–218.

17. Fedoroff, S., and Richardson, A. (1997). "Protocols for Neural Cell Cultures," 2nd ed. Humana Press, Totowa, NJ.

18. Swanson, L. W., Kohler, C., and Bjorkland, A. (1987). *In* "Limbic Region: The Septohippocampal System," pp. 125–277. Elsevier, New York.

19. Iwashita, Y., Kawaguchi, S., and Murata, M. (1994). Restoration of function by replacement of spinal cord segments in the rat. *Nature (London)* **367**, 167–170.

20. Freed, W. J. (1996–1997). Cell transplantation; Brain. *In* "Yearbook of Cell and Tissue Transplantation" (R. P. Ranza and W. L. Chick, eds.), pp. 163–173. Kluwer Academic Publishers, Dordrecht, The Netherlands.

21. Bregman, B. S., *et al.* (1993). Recovery of function after spinal cord injury: Mechanisms underlying transplant-mediated recovery of function differ after spinal cord injury in new born and adult rats. *Exp. Neurol.* **123**, 3–16.

CHAPTER 31

GONAD CELL CULTURE: TESTIS

Anna Orsola and Anthony Atala

The three main cellular components of the testis are Leydig cells, Sertoli cells, and germ cells. These cells are responsible for the two basic functions of the testis: steroidogenesis and spermatogenesis, which are under endocrine control by the gonadotropins. Even though luteinizing hormone (LH) acts on the Leydig cells and follicle-stimulating hormone (FSH) acts on Sertoli cells, these two testicular functions are not independent, and there is a complex interaction between them [1]. The Leydig cells drive spermatogenesis via the secretion of testosterone, which acts on the Sertoli cells to create an environment that will enable normal progression of germ cells through stage VII of the spermatogenic cycle. Both testosterone secretion and the expression of Leydig cells are under paracrine regulation by factors diffusing from the seminiferous tubule and from the Sertoli cells.

The high correlation between the functions of the testicular cells has implications when these cells are studied in an isolated manner and when the *in vitro* findings the correlated with the *in vivo* situation [2]. Acute steroidogenic factors, growth factors, cytokines, inhibins, and activins secreted by the Sertoli cells seem to be involved in the regulation of Leydig-cell-differentiated functions, both *in vitro* and *in vivo* in animal and human cells [3]. For example, the steroidogenic activity of Leydig cells cultured alone declines with time and becomes almost undetectable after 6–8 days in culture, whereas in coculture with Sertoli cells, testosterone secretion remains elevated and at a plateau for up to a month [4]. As a consequence, there is a strong tendency with *in vitro* systems to use cocultures of at least two of these types of cell.

HARVESTING OF LEYDIG, SERTOLI, AND GERM CELLS
ANIMAL CELL ISOLATION

Cell harvesting of testicular tissue is based mainly on enzymatic dissociation and separation methods. Most of the protocols use a Percoll gradient centrifugation method that separates cell fractions based on size. In the normal adult testis, Leydig cells comprise only a small percentage of the total weight. Total Leydig cell volume quantitatively determined in rat, buffalo, and six common mammals including man is fairly constant, making up 9–16% of the testicular volume [5]. These cells, however, are responsible for the secretion of 95% of the total testosterone.

Leydig Cells

Our method of rat Leydig cell isolation is a slight variation of classical descriptions [6,7]. Sixty-day-old male Sprague-Dawley rats stimulated with human chorionic gonadotropin (hCG) are used as a source of Leydig cells. Injection of a single 100-IU dose of hCH per rat increases Leydig cell size significantly with a maximum cross-sectional area after 24 h [8]. Immediately after castration, the testes are placed in ice-cold M-199 medium supplemented with bovine serum albumin (4 g/liter) and buffered with sodium bicarbonate (2.2 g/liter). Under sterile conditions, the testes are decapsulated after repeated rinsing in the medium just described. The decapsulated testes are incubated for 30 min with medium containing type IV collagenase (Sigma) at a concentration of 0.05% in a shaking water bath. A low concentration of soybean trypsin inhibitor can be added in this step. After the tubes have been inverted several times and sediment once, the supernatant contains the interstitial tissue and peritubular cells (Leydig cells), whereas the pellet contains the seminiferous tubules with both the Sertoli and the spermatic cells. Thus, both Leydig and Sertoli cells can be purified separately or from the same tissue.

To retrieve the Leydig cells, the supernatant is sequentially filtered through sterile 140- and 73-μm steel meshes. After the filtrate has been centrifuged and washed, the resulting cell pellet that contains the crude Leydig cell fraction is layered at a concentration of 100×10^6 cells over each 35 ml of a discontinuous Percoll gradient that is centrifuged for 20 min at 800g at 4°C. An additional tube containing only marker beads of a known density (Pharmacia Biotech) should also be centrifuged as a reference to distinguish four visible bands; band I has the lowest density and band IV the highest [9]. Band III encompases a 3-ml range of the gradient and averages 70% Leydig cells, which will stain darkly for 3β-hydroxysteroid dehydrogenase (3β-HSD) activity. A 1-ml fraction of this band will contain 90–95% pure Leydig cells and represents 1.5–2% of the initial number of interstitial cells isolated. This band corresponds to a Percoll concentration of 38–52% (v/v) or in gradient fractions to a density of 1.054–1.096 g/ml [10]. The band should be harvested and the remaining Percoll removed from the cells by washing with medium and centrifuging. Further treatment with a second continuous Percoll might yield a more purified Leydig cell suspension [11]. It has also been suggested that precursor cells in band II can be converted to Leydig cells in culture by hCG [12].

Sertoli Cells

The method used to isolate Sertoli cells depends mainly on the need to also retrieve Leydig cells or spermatogonia from the same original tissue. A modification of the method of testicular digestion described by Mather and Sato [13] and Verhoeven and Cailleau [6] allows isolation of both Leydig and Sertoli cells. After the initial enzymatic digestion, the pellet that contains the tubules is washed twice with phosphate-buffered saline (PBS) and allowed to sediment at gravity. The tubules are minced with a scalpel and submitted to a second enzymatic digestion with either 0.1% trypsin or collagenase and soybean trypsin inhibitor, at the same concentrations. The cells are collected by centrifugation and washed several times. The resulting Sertoli cell aggregates are resuspended in the desired volume. After this step, if the experiment requires coculture with Leydig cells, the cells obtained and purified from the initial supernatant are combined with the Sertoli cell fractions.

With slight modifications, this protocol applies to testes not only from rats and mice but also from different animals such as pigs [14–16] and hamster [17]. Variations of this method include vascular perfusion of collagenase via the testicular artery [18] or use of centrifugal elutriation before the Percoll step [19]. Other density gradient media used, such as metrizamide, yield similar cell populations [20]. Alternatively, a method that does not use Percoll has been utilized with younger rats [21]. After the enzymatic digestion, the seminiferous tubules are allowed to settle under gravity, and the supernatant containing the interstitial cells is removed and collected. The tubules are resuspended in medium and allowed to resettle, and the new supernatant is collected. The pooled supernatants are then centrifuged, yielding a pellet that contains the Leydig cells ready to plate.

Spermatogonia

Spermatogonia can be isolated in rats and mice of different ages. Briefly, the different types of spermatogonium can be isolated from single cell suspensions by using a combination of enzymatic digestion and separation of seminiferous tubule cells by centrifugal elutriation, followed by density gradient centrifugation in Percoll. In this manner, populations of type A spermatogonia, type B spermatogonia, spermatocytes in different stages, and Sertoli cells can be obtained with purities of 51–89% [22,23]. Sertoli cells possess a wide range of densities and sizes and might overlap with germ cells. Consequently, a second Percoll can be used to obtain a more purified cell population. The total time necessary to isolate germ cells should not exceed 5–6 h, and it should be minimized to prevent significant cell death. With larger testes and thus older rats, the handling time is reduced, increasing total cell number and cell viability, but this also contributes to more contamination by other cell types. Thus, the purity and the cell yield depend on several parameters that will affect the isolation of premeiotic and meiotic cells, such as animal age, duration of the harvesting procedure, viability of the cells, and accurate identification. To successfully separate the leptotene and zigotene primary spermatocytes, a fluorescence-activated cell sorting system based on flow cytometry has been developed [24]. Similarly, an immunomagnetic cell sorting procedure has been shown to be effective in isolating large quantities of viable spermatogonia [25]. Germ cells survive only briefly in culture, but spermatogonial stem cells have been successfully isolated and maintained in culture for approximately 4 months [26].

In vitro organ cultures have also been used successfully. When germ cell differentiation is the aim of the study, small testicular fragments (>100 fragments/testis) can be cultured by using a mesh grid on a dish. This practice allows for future examination of the tissue with standard histological procedures [27,28].

HUMAN MALE GONADAL CELL ISOLATION

The isolation of human male gonadal cells is performed by means of methods similar to those just described. Aging reduces testicular cell number and function. Therefore, if the tissue is from older individuals, it would be advisable to first examine the tissue histologically, since there seems to be a good correlation between normal histological features and good *in vitro* cell yield [4]. Briefly, the protocol includes digestion of the tissue with a mixture of collagenase, deoxyribonuclease, and soybean trypsin inhibitor in $NaHCO_3$ (15 nM), Ham's F-12 medium, and Dulbecco's Modified Eagle's Medium. Similarly to animal cells, the digested tissue is filtered (mesh 400) and centrifuged. The resulting pellet is resuspended in fresh medium and submitted twice to unit gravity sedimentation to separate the seminiferous tubule fragments and interstitial cells. The tubule fragments are filtered (mesh 170), sedimented, and plated as a Sertoli cell enriched preparation; the final washing steps decrease the persistence of germ cells in the tubular fragments. The interstitial cells remaining in suspension after the first sedimentation contain the Leydig cell fraction. The Leydig cells are purified after the pellet has been applied to a Percoll discontinuous gradient as described earlier. As in animal Leydig cells [10,29,30] there seems to be heterogeneity in human Leydig cell populations according to the buoyant density of the cells [31]. Normally, the cell band between 34 and 60% Percoll ($1.05 \leq$ buoyant cell density < 1.08) contains the majority of Leydig cells [32].

CELL CHARACTERIZATION

The trypan blue exclusion method should be used for cell viability and counting before plating. Characterization to confirm cell origin should be the next step. Leydig cells are identified by their microscopic appearance, which consists of a bright yellow halo under phase contrast microscopy. They contain numerous lipid droplets and have round nuclei and prominent nucleoli. PAS/hematoxilin staining shows bright red PAS-positive granules in an extensive cytoplasm and the nuclear heterochromatin distributed around the periphery of the nuclei. Increased testosterone production as a response to hCG, LH, and forskolin, in a dose-dependent manner is also a characteristic feature of these cells [33].

As with other steroidogenic cells, Leydig cells selectively express 3β-hydroxysteroid dehydrogenase (3β-HSD). This enzyme allows the specific characterization of these cells by direct staining methods or by immunostaining [34]. Direct staining methods rely on the transformation of a substrate (dehydroepiandrosterone) by the enzyme, whereas the immunostaining methods use an antibody directed against 3β-HSD. Briefly, 1 ml of cell suspension is incubated with 1 ml of β-nicotinamide adenine dinucleotide (NAD, 8 mg/ml), 1 ml of nitroblue tetrazolium (NTB, 2 mg/ml), and 0.1 ml of dehydroepiandrosterone (DEA, 1 mg/ml). As a negative control, the DEA should be omitted. The suspension should be incubated for 60 min in a shaking water bath at 37°C, 95% O_2 and 5% CO_2. Cells should be counted on a hemocytometer and the number of stained and total cells should be determined, indicating the percentage of Leydig cells. A similar staining method is that developed by Browning *et al.* [9] from a modification of Steinberger *et al.* [35].

Sertoli cells have a typical epithelial appearance with rounded nuclei and wide, clear, spread-out cytoplasm. Human Sertoli cells have typical granulations present under light microscopy [36]. They also have intracellular lipid droplets that can be identified by means of Nile red. When stained with PAS/hematoxilin they exhibit a characteristic red cytoplasm [23]. Immunohistochemical detection of transferrin, a typical product of these cells, can be used as an immunological criteria to test cell purity [37].

Germ cells can be fixed with Bouin's solution and stained with PAS/hematoxilin. The presence of mitosis, nuclear size, and cytoplasmic characteristics can be used as identification criteria [23]. Type A spermatogonia contain large, finely grained nuclei, weak staining chromatin with one or two large pale nucleoli, and a thin cytoplasm. Type B spermatogonia are small, with a large, dark, central heterochromatic mass. Preleptotene spermatocytes are even smaller and have a dense nucleus with a woolly chromatin. Germ cells can also be fixed in Carnoy's solution and stained with Giemsa for nuclear criteria [23]. Under these conditions, Sertoli cells can be differentiated because of the uniform size and oval shape of the nuclei, together with faint chromatin and small nucleoli. Type A spermatogonia are round, 1.5–3 times larger than Sertoli cell nuclei, and have either a diffuse or a reticulated nuclei. Type B spermatogonia nuclei are slightly smaller and have a very dark chromatin. Preleptotene and leptotene spermatocytes are much smaller and very dark. Zygotene spermatocytes are similar in size to Sertoli cell nuclei, and pachytene spermatocytes have a sex vesicle. Type A spermatogonia, type B spermatogonia, and differing size classes of primary spermatocytes should be separable from other germ cells types with minimal overlap. Type A spermatogonia, both undifferentiated and differentiated, express the c-*kit* receptor that is not present in the somatic cells and can be detected immunohistochemically.

PLATING AND CULTURE CONDITIONS

Culture of gonadal cells from multiple animals (rat, hamster, pig, etc.) and from humans is difficult because of the loss of specific functions under subculture and prolonged culture conditions. Normal adult Leydig cells are nonproliferating and cease to grow after the first trypsinization for cell culturing [38]. Even though some groups have been able to develop cell lines from testis, the identity of these cells has not been absolutely confirmed [38]. As in many other cell types, different media and supplements can be used to culture these cells, and the one chosen depends mainly on the purpose of the experiment. If the aim is to maintain Leydig cells in culture for a long time, it is important to avoid fibroblast overgrowth, whereas if the main interest is steroidogenic production, then hormonal supplementation is important. In general, hormone-supplemented, serum-free, or low-serum medium is used to reduce fibroblast overgrowth [13,39]. The medium can either be a 1:1 mixture of Dulbecco's Modification of Eagle's Medium (DMEM) and Ham's F-12, Eagle's Minimum Essential Medium (MEM), or M-199. Using DMEM and F-12, Mather and Sato showed that after 2 weeks in culture, the medium could be changed to medium containing 5% fetal bovine serum without overgrowth of fibroblasts. In that study, viable cells were obtained for periods of 1 month and 3 weeks for Sertoli and Leydig cells, respectively [13]. An alternative medium that we have successfully used in our laboratory is M-199 with a low concentration (1%) of fetal bovine serum. Sodium bicarbonate, HEPES buffer, and antibiotics should be added. The medium can be supplemented with hormones such as insulin, transferrin, epidermal growth factor (EGF), and FSH for both cell types, plus growth hormones, Somatomedin C, and retinoic acid for Sertoli cells.

Cells should be plated at a known concentration after measurement of cell viability by means of the trypan blue exclusion method. Even though several plate sizes and plating densities can be used, most of the literature uses 24-well plates at a wide range of cell densities, from 50,000 cells/well to 10^6 cells/well [3,19]. The shape, attachment, proliferation, and expression of gene products by the Leydig cells can be modified by altering the composition of the extracellular matrix [40]. In monocultures of germ cells, contaminant cells can be eliminated by utilizing the property of many nongerm cells of attaching to plastic surfaces [22,41]. In coculture with Sertoli cells, plating densities are usually 1:1 [3]. A four-fold increase in total measured steroid output has been observed under these conditions [3]. To enhance cell attachment without using serum, extracellular basement-membrane-coated dishes (obtained from primary culture of corneal endothelial cells) can be used [4]. Other alternatives are dishes coated with collagen, type I fibronectin, and laminin. Both cell populations can be mixed, or a two-chamber culture system without direct cell–cell interaction can be used [3,42]. Two-compartment systems have been developed to simulate *in vitro* the blood–testis barrier that allows germ cell differentiation [43]. Similarly, primary cultures of Sertoli cells maintained in conventional cultures do not retain many of the structural and functional properties of their *in vivo* counterparts. Sertoli cell phenotype is better maintained by incorporating certain environmental parameters, intrinsic to the testis, such as high cell density, a unique extracellular matrix, a semipermeable support between the basal plasma membrane of the cells and the interstitium, chemically distinct microenvironments at the apical and basal surfaces of the cells, and cell-to-cell interactions among Sertoli cells and other testicular cell types [44]. Germ cells are frequently plated together with Sertoli cells, to which they provide nutritional support. Sophisticated cocultivation chambers, with either a nylon mesh or a Micropore filter, have been utilized to study the culture patterns of rat Sertoli, peritubular and spermatogenic cells. This system also allows the study of the polarized secretion of Sertoli cell-specific proteins [45].

Stimulation can be done by pretreating the cocultures with LH, FSH, and hCG at variable concentrations. Recently, glial cell line–derived neurotropic factor (GDNF) has been shown to have a stimulatory effect on the proliferation of rat Sertoli cells and to regulate cell fate of stem cells for spermatogenesis [46,47]. If testosterone production is the goal, the cells can be boosted with hCG at ranges of 10 mIU/ml medium [10], with 0.1–100 ng/ml LH [37], with DHT [48], or with FSH [49]. The medium should then be sampled from 4 h after to up to 24 h following stimulation for hormonal level measurements. The sampled medium should be immediately stored at –20°C until the hormonal level is measured by using a radioimmunoassay [50]. At low cell concentrations, Leydig cells from gradient fractions of lower density (1.054–1.064 g/ml) seem to produce slightly less testosterone in response to hCG stimulation than Leydig cells from more dense fractions (1.070–1.096 g/ml) [10].

REFERENCES

1. Sharpe, R. M., Maddocks, S., and Kerr, J. B. (1990). Cell–cell interactions in the control of spermatogenesis as studied using Leydig cell destruction and testosterone replacement. *Am. J. Anat.* **188**, 3–120.
2. Saez, J. M. (1994). Leydig cells: Endocrine, paracrine, and autocrine regulation. *Endocr. Rev.* **15**, 574–626.
3. Verhoeven, G., and Cailleau, J. (1990). Influence of coculture with Sertoli cells on steroidogenesis in immature rat Leydig cells. *Mol. Cell. Endocrinol.* **71**, 239–251.
4. Lejeune, H., Sanchez, P., and Saez, J. M. (1998). Enhancement of long-term testosterone secretion and steroidogenic enzyme expression in human Leydig cells by co-culture with human Sertoli cell-enriched preparations. *Int. J. Androl.* **21**, 129–140.
5. Kothari, L. K., Patni, M. K., and Jain, M. L. (1978). The total Leydig cell volume of the testis in some common mammals. *Andrologia* **10**, 218–222.
6. Verhoeven, G., and Cailleau, J. (1988). Testicular peritubular cells secrete a protein under androgen control that inhibits induction of aromatase activity in Sertoli cells. *Endocrinology (Baltimore)* **123**, 2100–2110.
7. Schumacher, M., Schafer, G., Holstein, A. F., and Hilz, H. (1978). Rapid isolation of mouse Leydig cells by centrifugation in Percoll density gradients with complete retention of morphological and biochemical integrity. *FEBS Lett.* **91**, 333–338.
8. Hodgson, Y. M., and de Kretser, D. M. (1984). Acute responses of Leydig cells to hCG: Evidence for early hypertrophy of Leydig cells. *Mol. Cell. Endocrinol.* **35**, 75–82.
9. Browning, J. Y., D'Agata, R., and Grotjan, H. E., Jr. (1981). Isolation of purified rat Leydig cells using continuous Percoll gradients. *Endocrinology (Baltimore)* **109**, 667–669.
10. Hedger, M. P., and Eddy, E. M. (1987). The heterogeneity of isolated adult rat Leydig cells separated on Percoll density gradients: An immunological, cytochemical, and functional analysis. *Endocrinology (Baltimore)* **121**, 1824–1838.
11. Browne, E. S., Sohal, G. S., and Bhalla, V. K. (1990). Characterization of functional Leydig cells after purification on a continuous gradient of percoll. *J. Androl.* **11**, 379–389.
12. Murono, E. P., and Washburn, A. L. (1990). Evidence that Leydig precursors localize in immature band two cells isolated on Percoll gradients. *J. Steroid Biochem. Mol. Biol.* **37**, 675–680.
13. Mather, J. P., and Sato, G. H. (1979). The use of hormone-supplemented serum-free media in primary cultures. *Exp. Cell Res.* **124**, 215–221.

14. Sordoillet, C., Savona, C., Chauvin, M. A., de Peretti, E., Feige, J. J., Morera, A. M., and Benahmed, M. (1992). Basic fibroblast growth factor enhances testosterone secretion in cultured porcine Leydig cells: Site(s) of action. *Mol. Cell. Endocrinol.* **89**, 163–171.

15. Kukucka, M. A., and Misra, H. P. (1994). Isolation and culture of highly enriched populations of Leydig cells from guinea-pig (*Cavia porcellus*) testes. *Andrologia* **26**, 217–224.

16. Lejeune, H., Chuzel, F., Sanchez, P., Durand, P., Mather, J. P., and Saez, J. M. (1997). Stimulating effect of both human recombinant inhibin A and activin A on immature porcine Leydig cell functions in vitro. *Endocrinology (Baltimore)* **138**, 4783–4791.

17. Niedziela, M., and Lerchl, A. (1999). Isolation method of Leydig cells from mature male Djungarian hamsters (*Phodopus sungorus*) and their steroidogenic activity in vitro. *Andrologia* **31**, 157–161.

18. Klinefelter, G. R., Hall, P. F., and Ewing, L. L. (1987). Effect of luteinizing hormone deprivation in situ on steroidogenesis of rat Leydig cells purified by a multistep procedure. *Biol. Reprod.* **36**, 769–783.

19. Sun, X. R., Hedger, M. P., and Risbridger, G. P. (1993). The effect of testicular macrophages and interleukin-1 on testosterone production by purified adult rat Leydig cells cultured under in vitro maintenance conditions. *Endocrinology (Baltimore)* **132**, 186–192.

20. Georgiou, M., and Payne, A. H. (1987). Functional and physical characteristics of rat Leydig cell populations isolated by metrizamide and Percoll gradient centrifugation. *Biol. Reprod.* **37**, 335–341.

21. Khan, S., Teerds, K., and Dorrington, J. (1992). Growth factor requirements for DNA synthesis by Leydig cells from the immature rat. *Biol. Reprod.* **46**, 335–341.

22. Rivarola, M. A., Sanchez, P., and Saez, J. M. (1985). Stimulation of ribonucleic acid and deoxyribonucleic acid synthesis in spermatogenic cells by their coculture with Sertoli cells. *Endocrinology (Baltimore)* **117**, 1796–1802.

23. Bucci, L. R., Brock, W. A., Johnson, T. S., and Meistrich, M. L. (1986). Isolation and biochemical studies of enriched populations of spermatogonia and early primary spermatocytes from rat testes. *Biol. Reprod.* **34**, 195–206.

24. Mays-Hoopes, L. L., Bolen, J., Riggs, A. D., and Singer-Sam, J. (1995). Preparation of spermatogonia, spermatocytes, and round spermatids for analysis of gene expression using fluorescence-activated cell sorting. *Biol. Reprod.* **53**, 1003–1011.

25. von Schonfeldt, V., Krishnamurthy, H., Foppiani, L., and Schlatt, S. (1999). Magnetic cell sorting is a fast and effective method of enriching viable spermatogonia from Djungarian hamster, mouse, and marmoset monkey testes. *Biol. Reprod.* **61**, 582–589.

26. Nagano, M., Avarbock, M. R., Leonida, E. B., Brinster, C. J., and Brinster, R. L. (1998). Culture of mouse spermatogonial stem cells. *Tissue Cell.* **30**, 389–397.

27. Haneji, T., Maekawa, M., and Nishimune, Y. (1984). Vitamin A and follicle-stimulating hormone synergistically induce differentiation of type A spermatogonia in adult mouse cryptorchid testes in vitro. *Endocrinology (Baltimore)* **114**, 801–805.

28. Tajima, Y., Watanabe, D., Koshimizu, U., Matsuzawa, T., and Nishimune, Y. (1995). Insulin-like growth factor-I and transforming growth factor-alpha stimulate differentiation of type A spermatogonia in organ culture of adult mouse cryptorchid testes. *Int. J. Androl.* **18**, 8–12.

29. Bhalla, V. K., Rajan, V. P., Burgett, A. C., and Sohal, G. S. (1987). Interstitial cell heterogeneity in rat testes. I. Purification of collagenase-dispersed Leydig cells by unit gravity sedimentation and demonstration of binding sites for gonadotropin in light cells versus enhanced steroidogenesis in heavier cells. *J. Biol. Chem.* **262**, 5313–5321.

30. Dirami, G., Poulter, L. W., and Cooke, B. A. (1991). Separation and characterization of Leydig cells and macrophages from rat testes. *J. Endocrinol.* **130**, 357–365.

31. Simpson, B. J., Wu, F. C., and Sharpe, R. M. (1987). Isolation of human Leydig cells which are highly responsive to human chorionic gonadotropin. *J. Clin. Endocrinol. Metab.* **65**, 415–422.

32. Lejeune, H., Skalli, M., Sanchez, P., Avallet, O., and Saez, J. M. (1993). Enhancement of testosterone secretion by normal adult human Leydig cells by co-culture with enriched preparations of normal adult human Sertoli cells. *Int. J. Androl.* **16**, 27–34.

33. Khanum, A., and Dufau, M. L. (1990). A cAMP independent inhibitory action of high doses of forskolin in rat Leydig cells. *J. Steroid Biochem. Mol. Biol.* **37**, 669–674.

34. Mendelson, C., Dufau, M., and Catt, K. (1975). Gonadotropin binding and stimulation of cyclic adenosine $3':5'$-monophosphate and testosterone production in isolated Leydig cells. *J. Biol. Chem.* **250**, 8818–8823.

35. Steinberger, E., Steinberger, A., and Vilar, O. (1966). Cytochemical study of delta-5-3-beta-hydroxysteroid dehydrogenase in testicular cells grown in vitro. *Endocrinology (Baltimore)* **79**, 406–410.

36. Lipshultz, L. I., Murthy, L., and Tindall, D. J. (1982). Characterization of human Sertoli cells in vitro. *J. Clin. Endocrinol. Metab.* **55**, 228–237.

37. Rouiller-Fabre, V., Carmona, S., Merhi, R. A., Cate, R., Habert, R., and Vigier, B. (1998). Effect of anti-Mullerian hormone on Sertoli and Leydig cell functions in fetal and immature rats. *Endocrinology (Baltimore)* **139**, 1213–1220.

38. Nagpal, M. L., W. D., Calkins, J. H., and Lin, T. (1994). Transformation and immortalization of Leydig cells from Sprague-Dawley rat by an early genetic region of simian virus 40 DNA. *Cell Tissue Res.* **3**, 459–465.

39. Zhuang, L. Z., Phillips, D. M., Gunsalus, G. L., Bardin, C. W., and Mather, J. P. (1983). Effects of gossypol on rat Sertoli and Leydig cells in primary culture and established cell lines. *J. Androl.* **4**, 336–344.

40. Vernon, R. B., Lane, T. F., Angello, J. C., and Sage, H. (1991). Adhesion, shape, proliferation, and gene expression of mouse Leydig cells are influenced by extracellular matrix in vitro. *Biol. Reprod.* **44**, 157–170.

41. Morena, A. R., Boitani, C., Pesce, M., De Felici, M., and Stefanini, M. (1996). Isolation of highly purified type A spermatogonia from prepubertal rat testis. *J. Androl.* **17**, 708–717.

42. Wu, N., and Murono, E. P. (1994). A Sertoli cell-secreted paracrine factor(s) stimulates proliferation and inhibits steroidogenesis of rat Leydig cells. *Mol. Cell. Endocrinol.* **106**, 99–109.

43. Steinberger, A., and Klinefelter, G. (1993). Sensitivity of Sertoli and Leydig cells to xenobiotics in in vitro models. *Reprod. Toxicol.* 7, 23–37.

44. Hadley, M. A., Byers, S. W., Suarez-Quian, C. A., Djakiew, D., and Dym, M. (1988). In vitro models of differentiated Sertoli cell structure and function. *In Vitro Cell. Dev. Biol.* 24, 550-557.

45. Ueda, H., Tres, L. L., and Kierszenbaum, A. L. (1988). Culture patterns and sorting of rat Sertoli cell secretory proteins. *J. Cell Sci.* 89, 175–188.

46. Hu, J., Shima, H., and Nakagawa, H. (1999). Glial cell line-derived neurotropic factor stimulates sertoli cell proliferation in the early postnatal period of rat testis development. *Endocrinology (Baltimore)* 140, 3416–3421.

47. Meng, X., Lindahl, M., Hyvonen, M. E., Parvinen, M., de Rooij, D. G., Hess, M. W., Raatikainen-Ahokas, A., Sainio, K., Rauvala, H., Lakso, M., Pichel, J. G., Westphal, H., Saarma, M., and Sariola, H. (2000). Regulation of cell fate decision of undifferentiated spermatogonia by GDNF. *Science* 287, 1489–1493.

48. Hardy, M. P., Kelce, W. R., Klinefelter, G. R., and Ewing, L. L. (1990). Differentiation of Leydig cell precursors in vitro: A role for androgen. *Endocrinology (Baltimore)* 127, 488–490.

49. Benahmed, M., Tabone, E., Grenot, C., Sanchez, P., Chauvin, M. A., and Morera, A. M. (1986). Paracrine control of Leydig cell activity by FSH dependent proteins from Sertoli cells: An in vitro study. *J. Steroid Biochem.* 24, 311–315.

50. Kerr, J. B., Robertson, D. M., and de Kretser, D. M. (1985). Morphological and functional characterization of interstitial cells from mouse testes fractionated on Percoll density gradients. *Endocrinology (Baltimore)* 116, 1030–1043.

CHAPTER 32

GONAD CELL CULTURE: OVARIAN CELLS

Atlantida M. Raya-Rivera and Anthony Atala

INTRODUCTION

Under the regulation of pituitary gonadotropins, the ovaries produce the hormones that stabilize the female phenotype, regulate sexual function, and serve as ova reservoirs. The ovaries undergo marked changes from the time of birth to the postmenopausal period. The human ovary at birth contains approximately 750,000 oocytes, 400,000 at the end of puberty, and just a few thousand at the time of menopause [1–3].

The follicle, the functional unit of the ovary, is composed of the oocyte surrounded by follicular epithelium. In the mature ovary each month, a group of primary follicles are recruited, and by day 6–8 of the menstrual cycle, one follicle becomes mature or dominant and will be able to ovulate and luteinize [1]. The factors that initiate the growth of resting primordial follicles remain elusive [4]. Once the primordial follicles start to develop, the oocytes commence growing and the granulosa cells begin to divide. The cell-to-cell interactions between the thecal and granulosa cells are essential for follicular development. Recruitment of thecal cells from the ovarian stromal stem cells involves cellular proliferation as well as the induction of thecal-cell-specific functional markers.

Granulosa cells have been shown to produce *c-kit* ligand stem cell factor (KL). KL directly stimulates ovarian stromal cell proliferation and is thought to promote early follicular development by inducing the proliferation and organization of stromal stem cells (thecal cell organizer) around small follicles [5].

The growth and maturation of the oocytes and the proliferation and differentiation of the surrounding somatic cells are closely related. Oocyte growth also seems be stimulated by the *c-kit* ligand produced by the granulosa cells [5]. The oocytes induce the expression of proteins within the zona pellucida (ZP) [4], which may give rise to two different types of granulosa cell, which differentiate into the adjacent cumulus cells and are closely associated with the oocyte, and the second part of granulosa cells expressing FSH receptors, which are closely associated with the thecal cells. It has also been postulated that granulosa cells arise from a population of stem cells within the ovary [6,7].

Gonadotropins induce the functional differentiation of the thecal and granulosa cells and are essential for the maintenance and development of the growing follicles. At the early antral stage, FSH binds exclusively to the FSH receptors in the granulosa cells, whereas LH binds its cognate receptors to the thecal cells [8]. It is presumed that changes in thecal cell function occur that increase the ability of the thecal cells to synthesize aromatizable androgens from cholesterol [9].

In the granulosa cells, the LH receptors are induced on the preovulatory stage of differentiation by the actions of FSH. The intrafollicular androgens and estradiol act via their respective receptors (androgen and estrogen receptors) to enhance the response of the granulosa cells to FSH. This enhanced responsiveness leads to the FSH-induced expression of numerous genes (LH receptors, RIIb, aromatase, and inhibin α and β subunits in the granulosa cells) and a subsequent increase in steroidogenesis [8].

The recruited follicles not destined to ovulate undergo degeneration. The LH peak occurs at the middle of the menstrual cycle. Ovulation of the dominant follicle occurs 16–23 h after the LH peak and 24–38 h after the onset of the LH surge [1]. The most likely mechanism by which gonadotropin (hCG) induces ovulation is an increase in the synthesis of maturation-inducing steroids (MIS) [10].

Luteinization is a process by which follicular granulosa and thecal cells become non-mitotic and establish a specific, stable luteal cell phenotype [4]. The biochemical changes associated with luteinization include the marked and sustained induction of P450scc [11], the transient expression of the progesterone receptor [4,12], and a subsequent increase in progesterone production. The LH surge also leads to a change in the expression of specific forms of the prolactin receptor [13]. These changes may allow the newly formed corpus luteum to respond more effectively or by a different mechanism to the pituitary prolactin [4].

METHODS FOR GRANULOSA CELL CULTURE
GRANULOSA CELL MEDIUM PREPARATION

The medium of choice for culturing granulosa cells is McCoy's 5A (Gibco BRL) suplemented with L-glutamine, 2.2 g/liter sodium bicarbonate, 100 IU/ml penicillin, 100 mg/ml streptomycin, 2% fetal bovine serum, and 50–100 μl/liter insulin.

GRANULOSA CELL ISOLATION
Mechanical Isolation

Based on the method of Knecht *et al.* [14], mechanical isolation can be accomplish as follows.

1. Under anesthetic conditions, the ovaries are harvested.
2. Random punctures are made on the follicular surfaces with a fine 27-gauge needle.
3. The follicles are gently squeezed with a blunt probe to express the cells into McCoy's 5A medium in a 10-cm plate.
4. The cells are collected by filtering the medium through a 430-μm-mesh stainless steel grid and centrifuged for 5 min at 1000 rpm.
5. The supernatant is discarded and the pellet is resuspended. The cells are counted and plated.

A cell viability of 30–35% is obtained, as assessed by trypan blue exclusion. The mixture contains less than 5% of other cell types.

Enzymatic Isolation

Our laboratory has successfully used the following method of enzymatic isolation, which was modified from Tetsuka and Hillier [15].

1. The ovaries are harvested.
2. Random punctures are made on the surface of the ovaries with a fine 27-gauge needle, and the ovaries are carefully teased into small pieces.
3. The cells are incubated in Medium 199 (Gibco BRL) containing 2000 U/ml collagenase type I (Sigma), 80 U/ml deoxyribonuclease type III (Sigma), and 1% bovine serum albumin (BSA) (Sigma) at 37°C for 10–30 min.
4. The follicles are separated by size into large (>400 mm), medium (200–400 mm), and small (<200 mm) groups by filtering them sequentially through three nylon meshes (600, 300, 212 μm). The follicles are washed with medium and centrifuged to remove the collagenase. If needed, the follicles can be plated at this time. The medium obtained from the last filtration step contains the granulosa cells.
5. The granulosa cells are centrifuged for 5 min at 1000 rpm. The cells are counted and plated (1×10^6) into a 10-cm plate containing McCoy's 5A medium supplemented with L-glutamine.
6. Four days after seeding, the cells are lifted by using 0.05% trypsin–EDTA (Sigma) at a pH of 7.4, and passaged.

The medium is changed every 4 days. The viability obtained is 60–80%, as assessed by trypan blue exclusion, and contains less than 5% of other cell types is present.

Isolation by Chelation

Based on the Campbell method [16], isolation by chelation proceeds as follows:

1. The ovaries are punctured with a 27-gauge needle.
2. The ovaries are incubated for 15 min in McCoy's 5A medium containing 0.2% BSA, 6.8 mM ethylen glycol tetraacetic acid (EGTA) (Sigma), at a pH of 7.4, at 37°C, and in 5% CO_2–95% air.
3. Centrifuge for 5 min at 1000 rpm. The supernatant is discarded and the tissue is resuspended in McCoy's 5A medium containing 0.5 M sucrose (Sigma), 1.8 mM EGTA and 0.2% BSA, at pH 7.4, and incubated for 10 min at 37°C in 5% CO_2–95% air.
4. The solution is diluted with 3 volumes of media without sucrose and centrifuged for 5 min at 1000 rpm. The medium is removed and resuspended with McCoy's 5A by gentle vortexing.
5. The ovaries are carefully expressed with a blunt spatula to release the granulosa cells into a 10-cm plate with media.
6. The medium is collected from the plate and centrifuged at 1000 rpm for 5 min.
7. The supernatant is aspirated, and resuspended with 3 ml of McCoy's 5A medium. The viable cells are counted by using trypan blue and plated at a density of 1–2 × 10^5 cells/well on 24-well plates. The viability obtained is usually 80%, with 95% purity.

METHODS FOR CULTURING FOLLICLES

FOLLICLE CULTURE MEDIA

Ovarian follicles are commonly grown in Dulbecco's Modified Eagle's Medium (DMEM, Gibco BRL), or Minimal Essential Medium alpha (α-MEM; Gibco BRL), which includes precursors of DNA and is suitable for rapidly dividing cell types. McCoy's 5A medium supplemented with L-glutamine can also be used. The medium is usually supplemented with 5% serum or, in serum-free conditions, supplemented with insulin, transferrin, and selenium [17,18].

FOLLICLE ISOLATION

Mechanical Isolation

In large animals, mechanical dissection is preferable to avoid damage to the cells [17].

1. The ova are cut into 0.5-mm fragments and placed in a glass container with α-MEM and 1 mg/ml BSA.
2. Each piece is examined with a dissecting microscope and dissected by using a 21-gauge needle attached to a 1-ml syringe.

Enzymatic Isolation

Rapid enzymatic digestion is generally used for small animals. Slow enzymatic digestion is usually used for animals with larger and more fibrous ovaries [19].

1. The ovaries are carefully teased into small pieces (measuring approximately 20 × 10 × 3 mm³) and incubated in 1× Medium 199 (or McCoy's 5A + L-glutamine) containing 2472 U of collagenase type I, 180 U of pancreatic deoxyribonuclease type I per milliliter of mixture (5 ml/ova), and 1% BSA (all from Sigma) at 37°C for 1 h.
2. An equal volume of Medium 199 (or McCoy's 5A + L-glutamine) with 5% BSA is added to the mixture and gently swirled. Place the vial in a refrigerator for 36 h.
3. Approximately every 12–14 h the ovarian mixture is agitated to assist the digestion process.
4. The follicles are separated by size into large (>400 mm), medium (200–400 mm), and small (<200 mm) groups by sequential filtration through three nylon meshes. The filters are washed with medium to retrieve the follicles. Medium is added and the suspension is

centrifuged to remove the collagenase. The pellet is resuspended with medium, and the follicles are plated.

Discontinuous Percoll Gradient Isolation

This method has been used in pig ovaries to isolate primordial follicles [20]. After the follicles have been isolated enzymatically, the following procedure can be followed.

1. Before starting the procedure, Percoll solution concentrations are prepared. Percoll gradient concentrations of 80, 60, 40, and 20% are prepared and vortexed in sterile tubes.
2. In an Oakridge tube, 2 ml of each Percoll concentration is added slowly, starting with 80% and ending with 20%. The mixture is refrigerated at 4°C for 2 h.
3. The enzymatically dissociated contents of one ovary are diluted in medium, are loaded on the 8-ml discontinuous Percoll gradient concentration tube, and are centrifuged at 2100g for 15 min; 1-ml fractions are collected.
4. Most of the primordial follicles are recovered from the fourth sample. The fourth fractions from more than one ovary may be pooled. Phosphate-buffered saline (PBS) is added and the mixture is centrifuged at 3000 rpm for 5 min.

Flow Cytometry

For large-scale recovery of approximately 30,000 follicles or more, flow cytometry is recommended [20].

1. The fourth fraction previously obtained from the Percoll gradient protocol is isolated in 2 ml of PBS.
2. The fraction is filtered through a 35-μm nylon mesh for cell sorting.
3. The sample is sorted for 30–40 min in a FACS 420 cell sorter, equipped with a 70-mm flow nozzle and a 488-nm argon laser line at 400 mW with forward light scattering. Two sorting windows are used: a left sort to collect somatic cells and a right sort to recover purified primordial follicles.

NONSPHERICAL FOLLICLE CULTURE

The following simplified method of culture may have several advantages over the spherical culture systems, perhaps allowing improved oxygenation, nutrition, and access of hormonal support [18].

Follicle Plating

The follicles are diluted with medium containing fetal bovine serum, 5 g/ml. The serum makes the follicles become nonspherical within 24–48 h.

Growth

After the follicles have been plated, FSH is added during the first 2 weeks to stimulate growth and produce antrumlike structures [17].

Ovulation

To induce ovulation, hCG, 10,000 U/ml, is used. For oocyte maturation, maturation-inducing steroid is added [10,21]. EGF is also another efficient stimulus [22]. The follicles respond to these substances by releasing the oocyte with surrounding cumulus cells into the medium. The oocytes can be fertilized and developed at least to the hatched blastocyst stage [23].

MULTIPLE FOLLICLE CULTURE

Multiple follicle culture consists of placing the ovarian fragments or whole ovary in culture dishes containing McCoy's 5A medium enriched with 5% fetal bovine serum, or with growth factors such as insulin, tranferrin, and selenium. This method has been used to study hormone responsiveness and ovulation [10]. Other utilities include growth of primordial follicles from ovarian fragments [24]. Degeneration of tissues owing to inadequate oxygenation is a problem in larger tissue fragments [18].

REFERENCES

1. Carr, B. R., and Bradshaw, K. D. (1998). Disorders of the ovary and female reproductive tract. *In* "Harrison's Principles of Internal Medicine" (A. S. Fauci, ed.), Chapter 337, pp. 2097–2115. McGraw-Hill, New York.
2. Hsueh, A. J. W., Billing, H., and Tsafriri, A. (1994). Ovarian follicle atresia: A hormonally controlled apoptotic process. *Endocr. Rev.* **15**, 707–724.
3. De Pol, A., Marzona, L., Vaccina, F., Negro, R., Sena, P., and Forabosco, A. (1998). Apoptosis in different stages of human oogenesis. *Anticancer Res.* **18**, 3457–3462.
4. Richards, J. (1994). Hormonal control of gene expression in the ovary. *Endocr. Rev.* **15**, 725–751.
5. Parrot, J. A., and Skinner, M. K. (2000). Kit Ligand actions on ovarian stromal cells: Effects on thecal cell recruitment and steroid production. *Mol. Reprod. Dev.* **55**, 55–64.
6. Rodgers, R. J., Lavranos, T. C., van Wezel, I. L., and Irving-Rodgers, H. F. (1999). Development of the ovarian follicular epithelium. *Mol. Cell. Endocrinol.* **151**, 171–179.
7. Van Deerlin, P. G., Cekleniak, N., Coutifaris, C., Boyd, J., and Strauss, J. F., III. (1997). Evidence for the oligoclonal origin of the granulosa cell population of the mature human follicle. *J. Clin. Endocrinol. Metab.* **82**, 3019–3024.
8. Richard, J. S. (1980). Maturation of ovarian follicles: Actions and interactions of pituitary and ovarian hormones on follicular cell differentiation. *Physiol. Rev.* **60**, 51–89.
9. Bogovich, K., Richards, J. S., Reichert, L. E., Jr. (1981). Obligatory role of LH in the initiation of preovulatory follicular growth in the pregnant rat: Effects of hCG and GSH on LH receptors and steroidogenesis in theca, granulosa and luteal cells. *Endocrinology (Baltimore)* **109**, 860–867.
10. Pinter, J., and Thomas, P. (1999). Induction of ovulation of mature oocytes by the maturation-inducing steroid 17,20b,21-trihidroxy-4 pregnen-3-one in the spotted Seatrout. *Gen. Comp. Endocrinol.* **115**, 200–209.
11. Oonk, R. B., Krasnow, J. S., Beattie, W. G., and Richards, J. S. (1989). Cyclic AMP- dependent and -independent regulation of cholesterol sidechain cleavage cytochrome P450 gene expression in human adipose tissue. *J. Biol. Chem.* **264**, 21934–21942.
12. Park, O.-K., and Mayo, K. E. (1991). Trasient expression of progesterovbne receptor messenger RNA in ovarian granulosa cells after the preovulatory LH surge. *Mol. Endocrinol.* **5**, 967–978.
13. Clarke, D. L., and Linzer, D. I. H. (1993). Changes in PRL receptor expression during pregnancy in the mouse ovary. *Endocrinology (Baltimore)* **133**, 224–232.
14. Knecht, M., Katz, M. S., and Catt, K. J. (1981). Gonadotropin-releasing hormone inhibits cyclic nucleotide accumulation in cultured rat granulosa cells. *J. Biol. Chem.* **256**, 34–36.
15. Tetsuka, M., and Hillier, S. G. (1996). Androgen receptor gene expression in rat granulosa cells: The role of follicle-stimulating hormone and steroid hormones. *Endocrinology (Baltimore)* **137**, 4392–4397.
16. Campbell, K. L. (1979). Ovarian granulosa cells isolated with EGTA and hypertonic sucrose: Cellular integrity and function. *Biol. Rep.* **21**, 773–786.
17. Abir, R., Stephen, F., Mobberley, M. A., Moore, P. A., Margara, R. A., and Winston, R. M. L. (1997). Mechanical isolation and in vitro growth of preantral and small antral human follicles. *Fertil. Steril.* **68**, 682–688.
18. Hartshorne, G. M. (1997). In vitro culture of ovarian follicles. *Rev. Reprod.* **2**, 94–104.
19. Roy, S. K., and Treacy, B. J. (1993). Isolation and long-term culture of human preantral follicles. *Fertil. Steril.* **59**, 783–790.
20. Greenwald, G. S., and Moor, R. M. (1989). Isolation and preliminary characterization of pig primordial follicles. *J. Reprod. Fertil.* **87**, 561–571.
21. Calogero, A. E., Burrello, N., Negri-Cesi, P., Loredana, P., Palumbo, M. A., Cianci, A., Sanfilippo, S., and D'Agata, R. (1996). Effects of corticotropin-releasing hormone on ovarian estrogen production in vitro. *Endocrinology (Baltimore)* **137**, 4161–4166.
22. Spears, N., Boland, N. I., Murray, A. A., and Gosden, R. G. (1996). The establishment of follicular dominance in co-cultured mouse ovarian follicles in co-culture mouse ovarian follicles. *J. Reprod. Fertil.* **106**, 1–6.
23. Cortvrindt, R., Smita, J., and Van Steirteghem, A. (1996). In vitro maturation, fertilization and embryo development of immature oocytes from early preantral follicles from prepuberal mice in a simplified culture system. *Hum. Reprod.* **11**, 2656–2666.
24. Yu, N., and Roy, S. K. (1999). Development of primordial and prenatal follicles from undifferentiated somatic cells and oocytes in the hamster prenatal ovary in vitro: Effect of insulin. *Biol. Reprod.* **61**, 1558–1567.

CHAPTER 33

STEM CELL CULTURE: PLURIPOTENT STEM CELLS

Michael J. Shamblott, Joyce Axelman, Jared Sterneckert, Nicolas Christoforou, Ethan S. Patterson, Mahmud A. Siddiqi, Heidi Kahler, Laeticia A. Ifeanyi, and John D. Gearhart

H uman pluripotent stem cells (hPSCs) have been derived from the inner cell mass, of blastocysts [embryonic stem (ES) cells], and primordial germ cells of the developing gonadal ridge [embryonic germ (EG) cells]. Like their mouse counterparts, hPSCs can be maintained in culture in an undifferentiated state, and upon differentiation generate a wide variety of cell types. Embryoid body (EB) formation, an important step in the process of *in vitro* differentiation of these stem cells, has been used to derive neurons and glia, vascular endothelium, hematopoietic cells, cardiomyocytes, and glucose-responsive insulin-producing cells from mouse PSCs (mPSCs). EBs generated from human EG cell cultures have also been found to contain a wide variety of cell types, including neural cells, vascular endothelium, muscle cells, and endodermal derivatives. Methods for the growth and differentiation of human and mouse pluripotent stem cells are presented here.

INTRODUCTION

Mouse pluripotent stem cells (mPSCs) have been derived from two embryonic sources. ES cells are derived from the inner cell mass (ICM) of preimplantation embryos [1,2], while EG cells are derived from primordial germ cells (PGCs) that during normal development migrate to and colonize the gonad, eventually forming eggs and sperm [3,4]. Both mouse ES and EG cells demonstrate germ line transmission in experimentally produced chimeric mice [5,6]. Mouse ES and EG cells share many characteristics, such as high levels of alkaline phosphatase (AP) activity and the presence of specific cell surface antigens. Other important shared characteristics include growth as multicellular colonies, normal and stable karyotypes, the ability to be continuously passaged, and the capability to differentiate into cells derived from all three embryonic germ layers. Pluripotent stem cell lines that share most of these characteristics have also been reported for chicken [7], mink [8], hamster [9], pig [10,11], rhesus monkey [12], and common marmoset [13]. Human pluripotent stem cells derived from PGCs [14] and ICM [15] were described in 1998.

Many different mouse ES and EG cell lines exist, and many can be obtained commercially or through collaboration. Some common ES cell lines are ES-D3, developed by Doetschman (American Type Culture Collection CRL-1934); J1, developed by Jaenisch and coworkers [16]; and R1, developed by Nagy and coworkers [17]. EG cell lines are less frequently used, but the EG1 cell line developed by Stewart and coworkers [5] is an excellent choice. Although each of these cell lines has been used in *in vitro* differentiation and transgenic mouse experiments, one usually finds strong personal preferences among experienced users, especially in the generation of transgenic animals. Although mouse ES and EG cells appear to be morphologically similar and to behave similarly *in vitro* and *in vivo* differentiation paradigms, there are some indications that they differ in DNA methylation pattern and imprinting [6].

When mPSCs differentiate *in vitro*, they form complex three-dimensional cell aggregates termed embryoid bodies (EBs). Some early developmental processes are recapitulated within the environment of an EB, resulting in a haphazard collection of precursors and more fully differentiated cells from a wide variety of lineages. Through this intermediate step, mPSCs can generate cells of the hematopoietic lineage [18,19], cardiomyocytes [20,21], neurons [22], glial precursors [23], skeletal muscle [24], vascular endothelial cells [25], visceral endoderm [26,27], and glucose-responsive insulin-producing cells [28].

When human EG cells differentiate, they also form EBs comprised of endodermal, ectodermal, and mesodermal derivatives [14]. Methods have been developed to grow the constituent cells of human EBs, known as embryoid body-derived (EBD) cells. These cells are capable of robust long-term proliferation in culture. Mixed-cell EBD cultures and clonally isolated EBD cell lines simultaneously express a wide array of mRNA and protein markers that are normally associated with distinct developmental lineages. The proliferation and expression characteristics of these cells suggest they may be useful in the study of human cell differentiation and as a resource for cellular transplantation therapies [29].

mPSC METHODS

There are many different methods for growing mPSCs. The decision of which methods to use is often based on anecdote or habit rather than on any proven superiority, and the variety is often a source of confusion to beginning practitioners seeking the "best" method. It is often helpful to consider the source and ultimate use of mPSC when deciding on a method or methods. It is advisable to carefully follow the protocols supplied by the distributor of the cells when you begin your work. This includes the passage ratio and timing, choice of feeder layers, and choice of growth medium components. Each of these considerations will be discussed in detail.

The ultimate use of the cells should also dictate their handling. If the cells are to be used for a particular *in vitro* differentiation paradigm, the choice of methods can be based on those published with the differentiation protocol, but the exact nature of the methods is often unreported or of little importance. In this case it is best to pick a standard protocol, such as those presented here, and stick with it. This is in contrast to the use of cells for the generation of transgenic mice. This difficult and expensive procedure places heavy demands on the cells and on the practitioner to ensure the desired result: cells that can contribute to the germ line. Indeed anecdotal evidence suggests that choice of ES cell and methodology play a greater role in the success or failure of transgenic mouse experiments.

Most mPSC lines require or grow best on a feeder layer of cells that are mitotically inactive but metabolically active. The exact nature of the contribution of the feeder layer is largely unknown. However, secretion of leukemia inhibitory factor (LIF) and other cytokines, the presentation of transmembrane growth factors, and the production of supportive extracellular matrix components have all been attributed to feeder layers.

The two most common feeder layer cell types are primary cultures of mouse embryonic fibroblasts (MEFs) and the immortalized STO mouse fibroblast line. Both have benefits and drawbacks. MEFs are primary cell cultures, so a ready supply of mice must be available, and the task of preparing them is time consuming. These problems are alleviated by the purchase of MEFs from companies like StemCell Technologies (Vancouver, B.C., Canada).

PROCEDURE FOR PREPARING PRIMARY MEFs

Any strain of mice can be used to prepare embryonic fibroblasts. For neomycin-resistant fibroblasts, any strain of mice carrying a *neo*-resistance cassette can be used as long as the transgene or mutation is not embryonic lethal.

1. Set up timed matings by placing one to three females into a cage containing a fertile male. Check for a vaginal plug the following morning and designate the day of observed plug as day 0.5 of gestation.

2. On day 12–13 of gestation females should be visibly pregnant and the embryos palpable. Pregnant females are killed by cervical dislocation or CO_2 inhalation.

3. Open the skin, then wash the body wall with 70% EtOH. Open the body wall, remove the uterus, and place it in a dish with phosphate-buffered saline (PBS).

4. Remove embryos from the uterus and separate them from the placenta and surrounding membranes.

5. Transfer the embryos to a fresh dish of PBS to rinse off blood.

6. Use forceps to open the abdomen and remove the internal organs and head of each embryo. Rinse the embryos in another dish of PBS.

7. Prepare 24 well plates by adding 1 ml of calcium- and magnesium-free PBS to each well. Wash the embryos four times in individual wells.

8. Add 1 ml of trypsin (0.05%–0.53 mM EDTA) to 15-ml tubes; then transfer embryos into individual tubes. Shake vigorously to break up embryos. Incubate at 4°C for 1 h to overnight, and incubate at 37°C for 10 min.

9. After incubation, add 10 ml of fresh media [Dulbecco's Modified Eagle's Medium (DMEM), 10% fetal bovine serum (FBS), 0.1 mM nonessential amino acids (NEAA), 100 U/ml penicillin, 100 µg/ml streptomycin], and pipette to mix. Then let the debris settle.

10. Transfer the supernatant into 100 mm tissue culture dishes and incubate for approximately 3 days. If necessary, check for neomycin resistance by placing a couple of drops of cells in a 6-well dish with 300 µg/ml G418 added to the medium.

11. Freeze cells in a controlled-rate (1°C/min) freezing vessel (Nalgene, 5100-0001) to −80°C, then store at −80°C or in liquid nitrogen.

USING MEFs AS A FEEDER LAYER FOR mPSC

Once primary MEFs have been prepared, they can be split approximately 1:25 before they become senescent. It is most convenient to fully expand MEFs, mitotically inactivate them by γ irradiation, then freeze the cells in single- or two-use aliquots. Although mitomycin C (10 µg/ml in growth medium for 2–3 h) can be used to mitotically inactivate MEFs, there is a substantial risk of carryover into the stem cell culture. For this reason, γ irradiation is preferred unless an irradiator is unavailable. Most hospitals, oncology centers, and large research facilities operate radioactive source irradiators and will allow the use of this equipment for a fee. Smaller cathode ray tube source irradiators, such as the Faxitron model 43855D, can be purchased for $25,000–$30,000.

1. Thaw one vial of primary mouse embryonic fibroblasts in a 100-mm dish.
2. Change the medium the following day.
3. When confluency has been attained, subculture 1:5 (five 100-mm dishes), and when confluent, again subculture 1:5.
4. When cells are 75% confluent, irradiate with 3000 rad γ radiation or an amount of X-ray energy that will stop cell division while allowing the cells to survive for approximately 1 week.
5. Let cells recover for several hours to overnight and either use or freeze cells (1 vial per dish) in a controlled-rate (1°C/min) freezing vessel to −80°C before storing at −80°C or in liquid nitrogen.

STO fibroblasts are a rapidly growing continuous cell line created by spontaneous transformation of SIM mouse fibroblasts [30]. Their properties of thioguanine and ouabain resistance are not commonly used when these cells are employed as feeder layers, but do contribute to their name (SIM–Thioguanine–Ouabain). STO cells grow well in culture, so supply is not limiting. However, if STO cells are grown too densely (>90% confluence) fast-growing variant subpopulations can become dominant. These variant STOs are not supportive as feeder layers and can become contact independent and destroy the stem cell culture. To guard against this, plating densities should be kept low, and multiple frozen stocks with carefully recorded passage information should be kept. Frozen stocks can then be thawed and used without further expansion. Unlike MEFs, STO cells should not be irradiated prior to cryopreservation.

Using STO as a Feeder Layer for mPSC

STO cells can be purchased from the American Type Culture Collection (ATCC no. CRL 1503).

1. Expand STO cells and freeze cells in a controlled-rate (1°C/min) freezing vessel to –80°C before storing at –80°C or in liquid nitrogen (1 ampule/100-mm tissue culture dish).

2. Thaw STO cells into 8–10 100-mm tissue culture dishes.

3. When cells are approximately 70% confluent, irradiate at 5000 rad or an amount of X-ray energy that will stop cell division while allowing the cells to survive for approximately 1 week.

4. Let cells recover for several hours, trypsinize, and plate onto gelatinized plates or dishes (0.1% gelatin for 15–30 min). Optimal plating density will need to be determined empirically, however 5×10^4 STO per 96-well plate and 7×10^6 STO per 100-mm dish are generally acceptable.

mPSC Growth

It is important to be meticulous when culturing mPSCs in an undifferentiated state. As discussed earlier, a feeder layer of mitotically inactivated MEFs or STO cells should be plated in mPSC growth medium at least 2 h prior to the addition of mPSCs. Great care should be taken to fully disaggregate mPSCs during routine passage, and cells should not be allowed to become too dense. Medium should be replaced approximately every 1–2 days when it begins to turn yellow. The quality of reagents used to make mPSC growth and differentiation media can have a very large impact on resultant differentiation. This is most noticeable in the choice of fetal bovine serum. Companies like Life Technologies (Gibco/BRL) prescreen batches of serum for mouse ES cell toxicity, but it is still important to screen new batches of serum for toxicity and ability to support mPSC growth and differentiation. Many batches of serum from HyClone have also been found to be acceptable.

FIBROBLAST MEDIA FOR FEEDER LAYERS

Dulbecco's Modified Eagle Medium (DMEM)
10% inactivated fetal calf serum (FCS). Incubate at 56°C for 30 min to inactivate
0.1 mM nonessential amino acids (NEAA)
100 U/ml penicillin
100 µg/ml streptomycin

mPSC MEDIA

Dulbecco's Modified Eagle Medium (DMEM)
15% inactivated fetal calf serum (ES-qualified and/or batch-tested)
1 mM sodium pyruvate
0.1 mM nonessential amino acids
100 U/ml penicillin
100 µg/ml streptomycin
2 mM L-glutamine
0.1 mM β-mercaptoethanol or 0.46 mM monothioglycerol (MTG, Sigma)
1000 U/ml LIF or LIF-containing conditioned media

PLATING A FEEDER LAYER

1. Add 5 ml of autoclaved 0.1% gelatin to a 10-cm tissue culture plate and place at room temperature for 15–30 min. Remove liquid by aspiration.
2. Quickly thaw a frozen vial of irradiated MEF or STO cells in a sterile 37°C water bath.
3. Dilute by adding 1 ml of thawed cells to 10 ml of fibroblast media.
4. Spin at 1000 rpm for 5 min.
5. Remove supernatant.
6. Resuspend cells in 10 ml of fibroblast medium.
7. Add suspension to gelatinized plate and place in incubator for 2 h to overnight at 37°C with 5% CO_2, 90% humidity.

PLATING mPSC FROM FROZEN STOCKS

1. Quickly (<5 min) thaw a vial of frozen mPSCs in a sterile 37°C water bath.
2. Dilute by adding 1 ml of thawed mPSCs to 10 ml of fibroblast medium.
3. Spin at 1000 rpm for 5 min, remove supernatant, resuspend cells in 10 ml of mPSC medium.
4. Remove fibroblast medium from feeder layer.
5. Add resuspended cells to feeder layer and place in incubator at 37°C with 5% CO_2, 90% humidity.
6. Colonies will appear in about 24 h. Initially they appear as small refractile colonies or single cells under phase contrast microscopy. No large clumps of cells should be present. Within 1–3 days, mPSCs will generate large numbers of multicellular colonies. Cultures should have their media changed daily, especially when the media becomes yellow (acidic) in <24 h.

PASSAGING mPSC

1. When a 10-cm plate is approximately 80% confluent (~3 days after plating), mPSC colonies should be trypsinized and replated so that they remain pluripotent.
2. Remove mPSC media by aspiration.
3. Wash the plates twice with 5 ml of sterile calcium- and magnesium-free 1× PBS.
4. Add 1 ml of trypsin (0.05% trypsin, 0.53 mM EDTA) to the 10-cm plate, incubate at 37°C for 5 min, pump up and down repeatedly and gently until no large aggregates are present, then place suspension in a 15-ml tube.
5. Spin at 1000 rpm for 5 min.
6. Remove supernatant.
7. Resuspend cells in 10 ml of mPSC medium.
8. Add 1 ml of the cell suspension to each of 10 plates with feeder layers and 9 ml of mPSC medium. It is very important to gently but thoroughly mix the cells and medium or the cells will be unevenly atributed. The best results are obtained by tipping each dish and slowly pipetting the mixture two or three times.
9. Place in incubator at 37°C with 5% CO_2, 90% humidity.

FREEZING mPSC

1. Follow procedure for passing mPSC up through number 6.
2. Resuspend cells in mPSC freezing media (50% fetal calf serum, 10% DMSO, 40% DMEM) quickly but thoroughly by pumping repeatedly. You can freeze cells at 10^6 cells per ml or more: simply resuspend cells from 1 plate in 1 ml.
3. Quickly transfer cells to a controlled-rate (1°C/min) freezing vessel and transfer to a −80°C freezer.
4. For long-term (>1 yr) storage, ES cells should be transfered to liquid nitrogen storage.

ELIMINATION OF FEEDER LAYERS AND THE USE OF FEEDER-INDEPENDENT ES CELLS

The presence of a feeder layer in mPSC culture is beneficial, but a second cell type can be a complicating factor in expression studies. Three strategies to reduce or eliminate this problem are feeder subtraction, growth of cells on acellular matrices, and the use of feeder-layer-independent ES cells.

Methods of reducing the contaminating feeder cells should be employed when their presence interferes with downstream cell manipulation or expression analysis. Feeder layer cells can be a particular problem during attempts to differentiate mPSCs in solution. Analysis of differentiation by means of sensitive methods such as reverse transcriptase polymerase chain reaction (RTPCR), immunocytochemistry, or fluorescence-activated cell sorting (FACS) can be confounded by the presence of cells not derived from stem cells. In this regard, it is important to note that no feeder subtraction is 100% efficient, and thus a "feeder-only" negative control should be performed. Feeder layer substraction is based on the principle that feeder cells settle out of suspension and reattach to tissue culture dishes more rapidly than mPSCs. The usual result of this subtraction is a suspension from which 80–95% of feeder cells have been removed, along with the loss of about 20–50% of the ES cells. The remaining mPSCs are plated on tissue culture plasticware coated with gelatin. Many ES cell lines can be kept from differentiating on gelatin-coated plastic for short periods (days); however, they will undergo morphological changes suggestive of differentiation and should not be continuously passaged without careful thought to the possible repercussions. The requirement for LIF is increased in the absence of feeder layers, so it is suggested that a study be performed to determine the minimal effective dose of LIF.

FEEDER LAYER SUBTRACTION

1. Prepare a sufficient number of gelatin-coated tissue culture dishes to receive the mPSCs after subtraction. Place 5–10 ml of autoclave sterilized 0.1% gelatin into a 10-cm tissue culture plate and incubate at room temperature for 30–60 min. Remove gelatin by aspiration and allow the remaining liquid to evaporate.
2. Trypsinize ES cells on feeders as usual. It is important that the cells not be in clumps.
3. Pellet cells, resuspend in 20–30 ml of mPSC growth medium. Two or three 10-cm plates of cells may be resuspended in this volume.
4. Transfer the resuspended cells to an uncoated 10-cm tissue culture dish. Return the dish to the incubator for 30–45 min.
5. Collect the supernatant from the dish and transfer it to a fresh uncoated 10-cm tissue culture dish. Return to the incubator for 30–45 min.
6. Collect the supernatant. Cells can be transferred to gelatin-coated dishes for culturing, or used in other procedures.

Gelatin-coated plasticware is one example of a feeder-cell-free growth surface for mPSCs. In most cases, this surface is suboptimal for the maintenance of pluripotency. The extracellular matrix deposited by MEFs or STO cells also provides surfaces on which one can try to grow mPSCs. In this procedure, feeder cells are grown until confluent, then rinsed in PBS, lysed with 0.5% NP-40 (monitored under phase contrast microscopy for rapid cell lysis that leaves nuclei intact). The matrix is then rinsed several times with PBS and used as a plating substrate. As with plating on a gelatin substrate, mPSCs require high levels of exogenous LIF to remain undifferentiated on this surface.

Feeder layer contamination can be avoided entirely by the use of mPSC lines that have been derived in the absence of, or weaned from, feeder cells. There are many such lines available by collaboration. The most common feeder-independent ES line is CCE, described by Robertson [31], Keller [19], and their colleagues. These cells can be obtained from Stem-Cell Technologies. The CCE line has been used for *in vitro* differentiation, but germ line transmission in transgenic mice has not been established. CCE cells are grown on gelatin-coated tissue culture plates in standard mPSC media. CCEs require significantly more LIF than feeder-dependent ES cells (in our hands they require double the amount of purified LIF to remain fully undifferentiated).

In culture, CCE cells bear little resemblance to most feeder-layer-dependent ES cells. Isolated CCE cells appear cuboidal, with a distinct plasma membrane, cytoplasm, nucleus, and prominent nucleoli when viewed by Hoffman Modulation Contrast microscopy. Individual cells within small clusters will remain easily distinguishable, with sharp cell borders. Proliferating CCEs do not form typical three-dimensional, tightly packed mPSC cell colonies, but instead grow in rapidly expanding flat carpets of cells. As CCEs become confluent, individual cells become less distinct, and the surface of the colonies takes on a granular appearance. Fully confluent CCEs appear as an essentially contiguous, granular sheet of cells covering 50–80% of the surface area of the culture dish. CCEs may be passaged with routine techniques and can be split up to 1:40 when cells cover 50–70% of the dish or when the growth media becomes acidic (yellow) after overnight growth. If passage is infrequent, cell differentiation will result. This can be recognized by the appearance of large, well-defined fibroblast-like cells, often at the edges of expanding sheets of cells.

ALTERNATE SOURCES OF LEUKEMIA INHIBITORY FACTOR

Leukemia inhibitory factor (LIF) is a necessary supplement for maintaining the pluripotency of ES and EG cells in culture media. Commercially supplied purified LIF can become very expensive in large-scale experiments. A less expensive alternative is the use of cell lines that have been stably transfected with a LIF expression plasmid. Our laboratory has transfected African green monkey kidney cells (CV-1) with a vector containing the cytomegalovirus promoter driving human LIF expression. Growth medium conditioned by several days of culture of these CV-1 cells is then filtered and frozen in aliquots at $-80°C$ for storage. Aliquots are assayed for the ability to maintain undifferentiated ES or EG cells. A second alternative to commercial LIF is production in a bacterial expression system. These alternatives make sense only if large-scale mPSC growth is anticipated.

Harvesting and Assay of LIF-Containing Conditioned Media

1. Culture transfected CV-1 cells in DMEM, 10% fetal calf serum, 0.1 mM nonessential amino acids, 100 U/ml penicillin, and 100 μg/ml streptomycin. Cells are grown until approximately 70–90% confluent, whereupon the medium is changed. Afterward, the cells are grown without changing the medium for 2 days. This medium is then harvested, filtered through a 0.22-μm filter, and frozen at $-80°C$ in 1-ml aliquots.

2. To assay CV-1-conditioned media, grow ES cells on a feeder layer as described earlier. Two 30-min feeder subtractions are performed and the remaining cells are resuspended at a density of 2×10^5 cells/ml of ES medium without LIF. Plate 50 μl of the cells (10^4 cells) into each well of a gelatinized 24-well plate containing 1 ml of the appropriate media supplemented with increasing amounts of conditioned media or 1000 U/ml LIF (positive control). The media should be changed every day for a week. The cell morphology of the samples cultured with CV-1 conditioned media is compared against the LIF-supplemented media. About 3 ml of CV-1-conditioned media per 500 ml of ES media is generally sufficient.

MOUSE ES CELL EMBRYOID BODY FORMATION: ON PLATE, HANGING DROPLET, AND SUSPENSION CULTURE

Undifferentiated mPSCs can give rise to many different cell types. An important, but not essential, step in this process is the formation of embryoid bodies (EBs). This can be accomplished by the removal of leukemia inhibitory factor (LIF) from the medium in which mPSCs are cultured. This is done to mPSCs growing on a plate, in hanging droplets, or in suspension. Typically, mPSCs growing on a feeder layer are feeder-subtracted and plated on gelatinized plates for 1–2 days in the presence of LIF. The cells are dissociated with 1 ml of 0.25% trypsin–EDTA solution (Gibco/BRL) at 37°C for 5 min. To perform on-plate differentiation, simply remove LIF from the growth medium. This results in a heterogeneous collection of EBs in 2–3 days. Often, a more controlled EB formation process is required.

Undifferentiated mPSCs (on gelatin after feeder subtraction) are trypsinized, and then the trypsin is deactivated by adding 9 ml of mPSC differentiation medium. The cell concentration is measured by means of a hemocytometer. The cells are diluted to a final concentration of 5000 cells/ml.

For the hanging droplet method, 500-1000 cells are applied to the lid of a tissue culture dish in a volume of 20 μl. A standard 100-mm dish can take about 30–40 droplets. A multichannel pipettor is very useful for this process. The bottom of the dish is filled with about 5 ml of PBS to keep a humid environment in the dish. The dishes with the hanging droplets stay in the incubator (37°C, 5% CO_2, 90% humidity) for 48 h. The hanging droplets are then removed by pipette and placed in 10 ml of ES differentiation medium in a bacteriological (not tissue culture) petri dish (the surface of the petri dish does not allow the embryoid bodies to adhere). About 100 hanging droplets can be resuspended in 10 ml of medium. The forming embryoid bodies should stay in suspension (in the incubator) for another 2–5 days, depending on how dense and how large they are. Fresh medium (3–5 ml) should be added every 2 days to ensure that the cells have adequate nutrition. At the end of the embryoid body formation the medium containing the embryoid bodies is gently transferred to a standard 10-ml tissue culture dish to allow the EBs to adhere to the dish surface. Generally the embryoid bodies adhere within the first 24 h. At that point the medium should be replaced with fresh stem cell differentiation medium containing any factors that would allow the differentiation into a desired cell type.

The second method allows the retrieval of a much larger number of embryoid bodies. After the undifferentiated stem cells have been diluted to a final concentration of 5000 cells/ml, 10 ml of that medium is placed a standard 10-mm bacteriological petri dish in the absence of LIF. The petri dish then is placed into an incubator under standard conditions. The embryoid bodies are ready for plating on a standard tissue culture dish within 4–7 days.

The choice of method, the exact timing, and the choice of factor to be used depend largely on the desired outcome. In most cases, the hanging drop method generates the most uniform EBs. A review of differentiation paradigms is beyond the scope of this chapter, but a literature search will provide many detailed examples. We have found that it is best to initially follow a single detailed protocol before experimentation. This includes the mPSC line, the timing of EB formation, and the source of all reagents, especially serum and growth factors. Various mPSC cultures can be expected to differ in the timing and outcome of differentiation.

ES CELL DIFFERENTIATION MEDIA

IMDM (Iscove's modified MEM) or DMEM

15% inactivated FCS (batch-tested)

1 mM sodium pyruvate

0.1 mM NEAA

100 U/ml penicillin, 100 μg/ml streptomycin

2 mM L-glutamine

0.1 mM β-mercaptoethanol [0.46 mM monothioglycerol (MTG, Sigma) can be substituted]

Any other factors that may enrich for a particular cell type [e.g., retinoic acid, dimethyl sulfoxide (DMSO), and cytokines]

HUMAN EG CELL GROWTH, EMBRYOID BODY FORMATION, AND IMMUNOHISTOCHEMICAL ANALYSIS

Like their mouse counterparts, human EG cells are derived from primordial germ cells plated on a feeder layer and grown in the presence of LIF, basic fibroblast growth factor, and forskolin [14]. Unlike mouse EGs, the culture of human EGs is currently very difficult. Multicellular human EG colonies are relatively insensitive to trypsin or other disaggregating compounds, probably because of the extraordinarily tight intracellular adhesion that can be observed using electron microscopy (MJS, unpublished results). Inefficient disaggregation of human EGs (as well as mPSCs), inevitably leads to differentiation through the process of embryoid body formation. The impact of inefficient disaggregation and a low plating efficiency results in human EG cultures that cannot be cryopreserved by using standard techniques. For this reason, the culture of human EGs remains a difficult and specialized procedure.

During the routine culture of human EG cells, approximately 10% of the colonies spontaneously differentiate in the presence of LIF to form EBs. The EBs can be harvested with forceps or pipette without damaging the EG culture. The EBs can then be disaggregated by incubation in Collagenase/Dispase (Roche Biochemicals) and plated into a variety of growth and differentiation environments to generate a wide variety of mixed cell populations termed embryoid body derived (EBD) cells [29].

Unlike human EG cells, EBD cells are remarkably easy to grow in culture. They grow as a monolayer without a requirement for feeder cells, are very sensitive to trypsin, and have few soluble factor requirements. EBD cultures are karyotypically normal and many have undergone more than 70 population doublings before entering senescence. One outcome of their growth characteristics is that they can routinely generate clonal lines derived from single cells. EBD cells are also highly amenable to routine genetic manipulation by means of lipofection and electroporation, as well as retroviral and lentiviral infection.

Individual EBD cultures differ widely in their expression pattern, but most EBD cell cultures and individual clonal lines express a developmentally broad array of gene products, including markers associated with cell types that during normal development derive from endoderm, ectoderm, and mesoderm [29]. Multilineage gene expression by EBD cells may suggest that they are relatively undifferentiated progenitor/precursor cells and may be amenable to a variety of *in vitro* and *in vivo* differentiation paradigms. In this respect, it is important to realize that EBD cells are not similar to existing stem cell cultures, so significant alterations to differentiation paradigms and novel approaches will likely be required before EBD cells can be used in tissue engineering or therapeutic studies.

CONCLUSION

Because of their developmental plasticity, predictable differentiation, and superior growth characteristics, mouse pluripotent stem cells are a vital tool for the investigation of developmental processes and for initial work into the establishment of human cellular therapies. Mouse ES cells with various genetic manipulations have proved to be valuable tools for the elucidation of gene function during differentiation.

Human pluripotent stem cells, such as ES, and EG, and EBD cells, are a relatively new resource with tremendous but untested potential. It is likely that findings based on mPSCs will be valuable in the differentiation of human cells; however, existing human pluripotent cells are not equivalent to their mouse counterparts, and significant challenges lie ahead. For therapies based on human pluripotent stem cells to succeed, the fields of tissue engineering, transplantation surgery, interventional radiology, and stem cell biology will need to unite in a meaningful way. Engineered matrices and cell delivery systems must be developed in tandem acknowledgment of, the specialized needs of source cells. Surgical cell/tissue delivery techniques and immunomodulation methods must be developed, along with methods to monitor the distribution of cells in the patient. Once these challenges have been met human stem cells may be of therapeutic value in the treatment of diseases such as amyotrophic lateral sclerosis (ALS), spinal cord injury, stroke, multiple sclerosis (MS), Parkinson's disease, heart and muscle diseases, metabolic diseases of the liver, and diabetes.

REFERENCES

1. Evans, M. J., and Kaufman, M. H. (1981). Establishment in culture of pluripotential cells from mouse embryos. *Nature (London)* **292**, 154–156.
2. Martin, G. R. (1981). Isolation of a pluripotent cell line from early mouse embryos cultured in media conditioned by teratocarcinoma stem cells. *Proc. Natl. Acad. Sci. U.S.A.* **78**, 7634–7638.
3. Matsui, Y., Toksoz, D., Nishikawa, S., Nishikawa, S., Williams, D., Zsebo, K., and Hogan, B. L. (1991). Effect of steel factor and leukaemia inhibitory factor on murine primordial germ cells in culture. *Nature (London)* **353**, 750–752.

4. Resnick, J. L., Bixler, L. S., Cheng, L., and Donovan, P. J. (1992). Long-term proliferation of mouse primordial germ cells in culture. *Nature (London)* **359**, 550–551.

5. Stewart, C., Gadi, I., and Bhatt, H. (1994). Stem cells from primordial germ cells can reenter the germ line. *Dev. Biol.* **161**, 626–628.

6. Labosky, P., Barlow, D., and Hogan, B. (1994). Mouse embryonic germ (EG) cell lines: Transmission through the germline and differences in the methylation imprint of insulin-like growth factor 2 receptor (IGF2r) gene compared with embryonic stem (ES) cell lines. *Development (Cambridge, UK)* **120**, 3197–3204.

7. Pain, B., Clark, M. E., Shen, M., Nakazawa, H., Sakurai, M., Samarut, J., and Etches, R. J. (1996). Long-term in vitro culture and characterisation of avian embryonic stem cells with multiple morphogenetic potentialities. *Development (Cambridge, UK)* **122**(8), 2339–2348.

8. Sukoyan, M. A., Vatolin, S. Y., Golubitsa, A. N., Zhelezova, A. I., Semenova, L. A., and Serov, O. L. (1993). Embryonic stem cells derived from morulae, inner cell mass, and blastocysts of mink: Comparisons of their pluripotencies. *Mol. Reprod. Dev.* **36**(2), 148–158.

9. Doetschman, T., Williams, P., and Maeda, N. (1988). Establishment of hamster blastocyst-derived embryonic stem (ES) cells. *Dev. Biol.* **127**(1), 224–227.

10. Wheeler, M. B. (1994). Development and validation of swine embryonic stem cells: A review. *Reprod. Fertil. Dev.* **6**(5), 563–568.

11. Shim, H., Gutierrez-Adan, A., Chen, L., BonDurant, R., Behboodi, E., and Anderson, G. (1997). Isolation of pluripotent stem cells from cultured porcine primordial germ cells. *Biol. Reprod.* **57**, 1089–1095.

12. Thomson, J. A., Kalishman, J., Golos, T. G., Durning, M., Harris, C. P., Becker, R. A., and Hearn, J. P. (1995). Isolation of a primate embryonic stem cell line. *Proc. Natl. Acad. Sci. U.S.A.* **92**(17), 7844–7848.

13. Thomson, J. A., Kalishman, J., Golos, T. G., Durning, M., Harris, C. P., and Hearn, J. P. (1996). Pluripotent cell lines derived from common marmoset (*Callithrix jacchus*) blastocysts. *Biol. Reprod.* **55**(2), 254–259.

14. Shamblott, M. J., Axelman, J., Wang, S., Bugg, E. M., Littlefield, J. W., Donovan, P. J., Blumenthal, P. D., Huggins, G. R., and Gearhart, J. D. (1998). Derivation of pluripotent stem cells from cultured human primordial germ cells. *Proc. Natl. Acad. Sci. U.S.A.* **95**(23), 13726–13731.

15. Thomson, J. A., Itskovitz-Eldor, J., Shapiro, S. S., Waknitz, M. A., Swiergiel, J. J., Marshall, V. S., and Jones, J. M. (1998). Embryonic stem cell lines derived from human blastocysts. *Science* **282**, 1145–1147.

16. Li, E., Bestor, T. H., and Jaenisch, R. (1992). Targeted mutation of the DNA methyltransferase gene results in embryonic lethality. *Cell (Cambridge, Mass.)* **69**(6), 915–926.

17. Nagy, A., Rossant, J., Nagy, R., Abramow-Newerly, W., and Roder, J. C. (1993). Derivation of completely cell culture-derived mice from early-passage embryonic stem cells. *Proc. Natl. Acad. Sci. U.S.A.* **90**(18), 8424–8428.

18. Wiles, M. V., and Keller, G. (1991). Multiple hematopoietic lineages develop from embryonic stem (ES) cells in culture. *Development (Cambridge, UK)* **111**(2), 259–267.

19. Keller, G., Kennedy, M., Papayannopoulou, T., and Wiles, M. V. (1993). Hematopoietic commitment during embryonic stem cell differentiation in culture. *Mol. Cell. Biol.* **13**, 473–486.

20. Klug, M., Soonpaa, M., and Field, L. (1995). DNA synthesis and multinucleation in embryonic stem cell-derived cardiomyocytes. *Am. J. Physiol.* **269**, H1913–H1921.

21. Rohwedel, J., Sehlmeyer, U., Shan, J., Meister, A., and Wobus, A. (1996). Primordial germ cell-derived mouse embryonic germ (EG) cells *in vitro* resemble undifferentiated stem cells with respect to differentiation capacity and cell cycle distribution. *Cell Biol. Int.* **20**, 579–587.

22. Bain, G., Kitchens, D., Yao, M., Huettner, J. E., and Gottlieb, D. I. (1995). Embryonic stem cells express neuronal properties in vitro. *Dev. Biol.* **168**(2), 342–357.

23. Brustle, O., Jones, K. N., Learish, R. D., Karram, K., Choudhary, K., Wiestler, O. D., Duncan, I. D., and McKay, R. D. (1999). Embryonic stem cell-derived glial precursors: A source of myelinating transplants. *Science* **285**, 754–756.

24. Rohwedel, J., Maltsev, V., Bober, E., Arnold, H.-H., Hescheler, J., and Wobus, A. (1994). Muscle cell differentiation of embryonic stem cells reflects myogensis *in vivo*: Developmentally regulated expression of myogenic determination genes and functional expression of ionic currents. *Dev. Biol.* **164**, 87–101.

25. Wang, R., Clark, R., and Bautch, V. (1992). Embryonic stem cell-derived cystic embryoid bodies form vascular channels: An *in vitro* model of blood vessel development. *Development (Cambridge, UK)* **114**, 303–316.

26. Abe, K., Niwa, H., Iwase, K., Takiguchi, M., Mori, M., Abe, S.-I., Abe, K., and Yamura, K.-I. (1996). Endoderm-specific gene expression in embryonic stem cells differentiated to embryoid bodies. *Exp. Cell Res.* **229**, 27–34.

27. Doetschman, T. C., Eistetter, H., Katz, M., Schmidt, W., and Kemler, R. (1985). The *in vitro* development of blastocyst-derived embryonic stem cell lines: Formation of visceral yolk sac, blood islands and myocardium. *J. Embryol. Exp. Morph.* **87**, 27–45.

28. Soria, B., Roche, E., Berna, G., Leon-Quinto, T., Reig, J. A., and Martin, F. (2000). Insulin-secreting cells derived from embryonic stem cells normalize glycemia in streptozotocin-induced diabetic mice. *Diabetes* **49**(2), 157–162.

29. Shamblott, M., Axelman, J., Littlefield, J., Blumenthal, P., Huggins, G., Cui, Y., Cheng, L., and Gearhart, J. (2001). Human embryonic germ cell derivatives express a broad range of developmentally distinct markers and proliferate extensively in vitro. *Proc. Natl. Acad. Sci. U.S.A.* **98**(1), 113–118.

30. Martin, G. R., and Evans, M. J. (1975). Differentiation of clonal lines of teratocarcinoma cells: Formation of embryoid bodies in vitro. *Proc. Natl. Acad. Sci. U.S.A.* **72**, 1441–1444.

31. Robertson, E., Bradley, A., Kuehn, M., and Evans, M. (1986). Germ-line transmission of genes introduced into cultured pluripotential cells by retroviral vector. *Nature (London)* **323**, 445–448.

CHAPTER 34

STEM CELL CULTURE: NEURAL STEM CELLS

Yang D. Teng, Kook I. Park, Erin B. Larvik, Robert Langer, and Evan Y. Snyder

INTRODUCTION

Neural stem cells (NSCs) are primordial and uncommitted cells that have been believed to give rise to the vast array of more specialized cells of the central nervous system [1–7]. They are operationally defined by their abilities (1) to differentiate into cells of all neural lineages (i.e., neurons of multiple subtypes, oligodendroglia, and astroglia) in multiple regional and developmental contexts (i.e., to be multipotent), (2) to self-renew (i.e., to give rise also to new NSCs with similar potential), and (3) to migrate and populate developing and/or degenerating CNS regions. An unambiguous demonstration of monoclonal derivation of progeny is essential to the definition: that is, a single cell must possess these attributes. In the past two decades, it has been recognized that cells abstracted from the CNS from a variety of structures and at different developmental stages, including adulthood, possess stemlike properties [1,2,6]. Such NSCs can be propagated in culture and reimplanted into both rodent and higher mammalian (including primate) brains, where they can reintegrate appropriately and steadily express foreign genes [8]. Some studies, including ours, provide hope that the use of NSCs may circumvent some limitations of available graft material and gene transfer vehicles and make feasible a variety of therapeutic strategies [9–11]. This significant advancement has led neurobiologists to speculate about how such phenomena might be harnessed both for therapeutic advantage and for better understanding of developmental mechanisms [1,2].

The ready availability of unlimited quantities of NSCs derived from the human brain holds great interest for basic and applied neuroscience, including therapeutic cell replacement and gene transfer following transplantation. We summarize here the combination of epigenetic and genetic procedures for perpetuating rodent (in particular, murine) and human neural stem cell lines. We also include our routine procedures for *in vitro* maintenance of NSCs of both kinds as well as methods for their preparation for surgical transplantation. Finally we discussed the critical roles that NSCs may play for tissue engineering.

MURINE NSCs

Cells from the CNS with stemlike properties have been successfully isolated from the embryonic, neonatal, and adult murine CNS. They can be propagated *in vitro* by many effective and safe means that rely on both epigenetic and genetic strategies. The epigenetic approach includes mitogens such as epidermal growth factor (EGF) [12,13] or basic fibroblast growth factor (bFGF) [14,15]. The genetic method consists of gene transfer with propagating genes such as v-*myc* [16] or large T-antigen [T-Ag] [17]. Importantly, maintaining murine NSCs (mNSCs) in a proliferative state in culture does not subvert the cells' ability to respond to normal developmental cues *in vivo* following transplantation. Upon entering the *in vivo* environment, mNSCs withdraw from the cell cycle, interact with host cells, and differentiate to express cellular markers for neuronal and glial lineages [18,19]. A subpopulation will remain as quiescent, undifferentiated cells intermixed seamlessly among more differentiated host and donor cells. These extremely plastic cells migrate and differentiate in a temporally and regionally appropriate manner, particularly following implantation into germinal zones. In addition, they can express foreign reporter and therapeutic genes *in vivo*, and they are capable of neural cell replacement [17,20,21]. They participate in normal development along the rodent neuaxis, largely independent of the initial region from which they were isolated, attesting to their exceptional plasticity. For example, multipotent, clonal mNSC lines generated from neonatal murine cerebellum can integrate not only back into the developing cerebellum but throughout the entire neuroaxis of the immature and adult central and peripheral nervous system in a nontumorigenic, cytoarchitecturally appropriate manner. Differentiation fate appears to be determined by site-specific microenvironmental factors: cells from the same clonal line differentiated into neurons or glia based on their site of engraftment. Transplanted NSCs that differentiated into ultrastructurally identifiable neurons appeared, for example, to participate in synaptogenesis, suggesting functional as well as anatomic integration. Similar phenomena have been reported for NSCs obtained from rodent hippocampus [22], spinal cord [23], and cortex [1]. Therefore, NSCs, as modeled by these various mNSC lines, may have the potential for repair of, or transport of genes into, the central nervous system. Furthermore, such plasticity at the cellular level provides one mechanism for a developmental strategy whereby multipotent progenitors migrate, with commitment to cell type not occurring until after interaction with their microenvironment. While mouse and rat generally seem to behave similarly under most circumstances, for as-yet unclear reasons the experience of a number of labs has shown mouse NSCs to be even easier than those of rat to propagate and maintain for prolonged periods. In this chapter, because most of our lab's rodent experience has been with mouse, mNSCs will be our focus. However, most concepts can be extrapolated to rat.

MAINTENANCE AND PROPAGATION OF MURINE NSCS *IN VITRO*

The most widely used mNSC line in our lab and that of our collaborators is C17.2. C17.2 cells, a prototypical NSC clone, are originally derived from neonatal mouse cerebellum as described elsewhere [24]. Briefly, derivation of NSC lines can be most readily accomplished by dissecting the primary structure of interest (e.g., cerebellum, olfactory bulb, cortex, hippocampus, spinal cord) from immature (fetal or newborn) mice and incubating the tissue in 0.5% trypsin in phosphate-buffered saline (PBS). (If the tissue is abstracted from adult animals, dissociation will probably require collagenase as well as longer incubation times.) Tissue is triturated and washed twice before cultures are plated on uncoated tissue culture dishes in serum-containing medium. Serum at this stage both inactivates the trypsin and ensures the health of the dissociated cells. At this stage, cells can then be propagated in serum-free medium with mitogens—EGF (10–20 ng/ml) plus bFGF (10–20 ng/ml) plus leukocyte inhibitory factor (LIF) (10 ng/ml)—or they can be transduced with a gene that enhances propagation by, it is believed, working downstream of these mitogens. (For the growth factor selection process, see later section on human NSCs.) If one elects to use the genetic approach to propagate NSC clones, then the primary culture just described is infected 24–48 h after plating, by incubation with the avian *myc* or viral *myc* [v-*myc*] vector PK-VM-2; the parent retrovirus vector in which v-*myc* DNA was inserted was pneoMLV [25]. Cells are then cultured in Dulbecco's Modified Eagles Medium (DMEM) + 10% fetal calf serum (FCS) + 5% horse serum (HS) + 2 mM glutamine for 3–7 days, until cultures appear to have undergone at least two doublings. Cultures are then trypsinized and seeded at 3–10% confluence in DMEM + 10% FCS + 0.3 mg/ml G418. Neomycin-resistant colonies are typically observed within 1–2 weeks. Chosen colonies are then isolated by brief exposure to trypsin inside plastic cloning cylinders after about 2 weeks. Colonies were then replated and expanded on uncoated 24-well CoStar plates. At confluence, these cultures were further passaged and expanded first to 35-mm and then to 60-mm and ultimately to 100-mm uncoated Corning tissue culture dishes. Cells infected with control vectors without propagating genes did not survive beyond this passage in serum-containing medium. Early passages of the expanded colonies were frozen in Nunc cryostat tubes at 10^6 cells/vial in DMEM + 30% FCS + 12% dimethyl sulfoxide (DMSO). (This method of freezing also is effective for most murine NSC lines.)

NSC lines such as C17.2 can then be transfected with additional genes, such as *lacZ*, the gene encoding *E. coli* β-galactosidase (βGal). This procedure is done by plating a recent 1:10 split of the cell line of interest onto 60-mm tissue culture plates. Twenty-four to forty-eight hours after plating, the cells are incubated with a *lacZ*-encoding retroviral vector (e.g., BAG) plus 8 µg/ml Polybrene for 1–3 h. Cells are then cultured in fresh feeding medium (DMEM + 10% FCS + 5% HS + 20 mM glutamate) for approximately 3 days until they appear to have undergone at least two doublings. The cultures are then trypsinized and seeded at low density (50–5000 cells on a 100-mm tissue culture dish). If your newly inserted transgene contains a selection marker distinct from your first selection marker (e.g., a non-neomycin-based selection marker such as hygromycin or pyromycin), then you can select for the new infectants as just described. After about 3 days, well-separated colonies are isolated by brief exposure to trypsin inside plastic cloning cylinders, as already described. If your new transgene does not have a new selection marker or also employs *neo*, identification of infectants is still possible. Select colonies as described and proceed to plate them in 24-well plates. At confluence, these cultures can be passaged to and expanded in tissue culture dishes of increasingly larger surface area. A representative dish from each clone can then be stained histochemically for your new gene of interest (e.g., for βGal expression via Xgal histochemistry [17], directly in the culture dish). The percentage of Xgal+ blue cells is then assessed microscopically. The clones with the highest percentage of blue cells are then used for future studies. The ability to detect the *lacZ* gene product by Xgal histochemistry *in vivo* and/or by anti-βGal immunohistochemistry makes such *lacZ*-expressing cells ideal for transplantation studies.

Preferred murine cerebellar NSC lines were grown in DMEM + 10% FCS + 5% HS + 2 mM glutamine on tissue culture dishes coated with poly(L-lysine) (PLL). They are either fed weekly with equal parts of conditioned medium from confluent cultures and fresh medium, or split (1:10) weekly into fresh medium. Lines should not routinely be split at dilution greater than 1:10, although this is clearly possible. For important transplant studies, a given thaw of cells should not have been passaged for longer than 4–8 weeks. It is better thaw out an earlier vial. Therefore, it is advisable to have a stockpile of hundreds of early freezes. Subclones can be abstracted by seeding 50–5000 cells on 10-cm dishes in the usual medium, or in 1:3 conditioned medium from confluent culres. Well-separated colonies can again be isolated using cloning cylinders. One should be aware that for reasons still unclear, changes in brands of tissue culture plastic, type of substrate coat, and type, concentration, and lot of serum can change the phenotypes displayed by NSC lines.

PREPARATION OF MURINE NSCs *IN VITRO* FOR TRANSPLANTATION

For preparation of cells for transplantation, a subconfluent (90%) 10-cm culture dish of adherent mNSCs (or cells of the line of interest) is washed three times with phosphate-buffered saline (PBS). If cells are allowed to become too confluent for more than 48 h prior to transplantation, they begin to elaborate an extracellular matrix that causes them to become clumpy and results in poor engraftment and, often, autonomous clusters of cells. In addition, the cells begin to exit the cell cycle and differentiate, which also predisposes the culture to poor engraftment. The more immature and proliferative the cells, the better the engraftment. Of course, it is reassuring to allow sister plates to proceed (in parallel assessments *in vitro*) to confluence and differentiation to ensure that the cells still do possess the ability to differentiate into mature neural phenotypes. It is better to use many dishes of 90% confluent cells than fewer dishes with more cells that have become heavily confluent.

Cells that have been prepared appropriately for transplantation (we usually do a split 48 h before the intended grafting procedure) are then well trypsinized, gently but thoroughly triturated into well-dissociated single-cell suspensions, and resuspended (in feeding medium plus 0.05% w/v trypan blue, to aid with localization of the injected material) to give a high cellular concentration ($3–6 \times 10^4$ cells/µl), which is nevertheless free flowing, hence will not obstruct the needle used for implantation. Although we employ finely drawn glass micropipettes, one can employ a Hamilton syringe. It is important to ensure that the cells neither settle out of suspension (which can occur within as little as 30 s) nor clump. Hence, we maintain the cells on ice and constantly triturate them gently. One should be careful not to implant vehicle from which the cells have settled out, under the false impression that one has implanted cells. Interpretation of such data, as one might imagine, will be entirely erroneous.

For NSCs engineered via retrovirus to express a transgene (e.g., for βGal-transduced cell lines), the "helper virus–free" status needs to be demonstrated prior to transplantation. Otherwise, host-derived cells may be falsely interpreted as being donor-derived. The test can be down by confirmation of the failure of supernatants from the cell lines to produce neomycin-resistant colonies or Xgal+ colonies in naive cells (e.g., 3T3 fibroblasts).

While the most efficient method for detecting cells *in vivo* is via the presence of a transgene, the risk of down-regulation of a transgene exists, for reasons that remain unclear. Such down-regulation, which ususally occurs unpredictably, presents a problem so considerable that multiple simulataneous methods of donor cell detection *in vivo* should also be put in place prior to an important transplantation study. For example, FISH (fluorescence *in situ* hybridization) against the *lacZ* gene itself can be effective. Using NSCs derived from a male animal and implanting them in a female animal allows one to do FISH against the Y chromosome. Most easily, one can preincubate the NSCs in bromodeoxyuridine (BrdU) 48 h prior to transplantation and subsequently identify the cells *in vivo* through the use of an anti-BrdU antibody that will indicate donor-derived cells via their immunopositive nucleus [26]. Most NSCs undergo very few *in vivo* cell divisions following transplantation, usually not enough to "dilute" the BrdU marker.

HUMAN NSCs

Emulating their rodent counterparts, stable clones of NSCs have been isolated from the human fetal telencephalon [26–34]. These self-renewing clones give rise *in vitro* to all fundamental neural lineages. Following transplantation into germinal zones of the brain of the newborn mouse, they participate in aspects of normal development, including migration along established migratory pathways to disseminated CNS regions, differentiation into multiple developmentally and regionally appropriate cell types, and nondisruptive interspersion with host progenitors and their progeny. Readily genetically engineered *ex vivo*, human NSCs (hNSCs) are capable of expressing foreign transgenes *in vivo* in these disseminated locations [26–34]. The secretory products from these NSCs can cross-correct a prototypical genetic metabolic defect in neurons and glia *in vitro*, further supporting their gene therapy potential. Finally, human NSCs can replace specific deficient neuronal populations [26].

For research and potential clinical applications, hNSCs have been studied and found feasible for cryopreservation. In addition, these cells can be propagated by both epigenetic and genetic means that are comparably safe and effective. The observations encourage investigations of NSC transplantation for a range of disorders. Thus we have tested various culture conditions and genes for those that optimally allow for the continuous, rapid expansion and passaging of human neural stem cells. Among them, v-*myc* (the p110 gag-myc fusion protein derived from the avian retroviral genome) seems to be the most effective gene; we have also identified a strict requirement for the presence of mitogens (FGF-2 and EGF) in the growth medium, in effect constituting a conditional perpetuality or immortalization (LIF has been useful for blunting potential senescence). In general, our monoclonal, nestin-positive, human neural stem cell lines perpetuated in this way divide every 40 h and stop dividing upon mitogen removal, undergoing spontaneous morphological differentiation and up-regulating markers of the three fundamental lineages in the CNS (neurons, astrocytes, and oligodendrocytes). Therefore hNSC lines retain basic features of epigenetically expanded human neural stem cells. Clonal analysis confirmed the stability, multipotency, and self-renewability of the cell lines. Finally, hNSC lines can be transfected and transduced by using a variety of procedures and genes encoding proteins for marking purposes (e.g., *lacZ*) and of therapeutic interest [e.g., brain-derived neurotrophic factor (BDNF)].

The isolation, propagation, characterization, cloning, and transplantation of NSCs from the human CNS followed a tack established by prior experience with the successful murine NSC clone C17.2 (propagated following transduction of a constitutively downregulated v-*myc* [24,35]) and with growth-factor-expanded murine NSC clones [36,37].

Isolation, Selection, Maintenance, and Propagation of Human NSCs *in Vitro*

A suspension of primary dissociated neural cells (5×10^5 cells/ml), prepared from the telencephalon (particularly the periventricular region) of an early second trimester human fetus (e.g., 13 or 15 weeks), as detailed elsewhere [26,27], was plated on uncoated tissue culture dishes (Corning) first in serum-containing medium as described earlier for mouse for about 24–48 h and then in the following "growth medium": Dulbecco's Modified Eagle's Medium (DMEM) + Ham's F-12 medium (1:1) supplemented with N-2 medium (Gibco), to which was added bFGF (10–20 µg/ml) + heparin (8 µg/ml) and EGF (10–20 µg/ml) and LIF (10 ng/ml). Cultures were then put through the following "growth factor" selection process. Cells were transferred to bFGF-containing, serum-free medium alone for 2–3 weeks; then they were cultured in EGF-containing, serum-free medium alone for another 2–3 weeks; subsequently they were returned to bFGF-containing, serum-free medium alone for 2–3 weeks. Finally they were maintained in serum-free medium containing bFGF plus LIF. At each stage of selection, large numbers of cells died or failed to survive passaging. What were left following the selection process were passageable, immature, proliferative cells that were responsive to both FGF and EGF (i.e., coexpressed both receptors)—qualities essential, in our view, for operationally defining an NSC. Medium was changed every 5–7 days. Cell aggregates were dissociated in trypsin–EDTA (0.05%) when they exceeded 10–20 cell diameters in size and replated in growth medium at 5×10^5 cells/ml. Some dissociated NSCs were plated on slides coated with poly(L-lysine) (PLL) (Nunc) in DMEM + 10% FBS and processed weekly for immunocytochemistry (ICC). In most cases, differentiation occurred spontaneously. For astrocytic maturation, clones were cocultured with primary dissociated embryonic CD-1 mouse brain [38]. Polyclonal populations of the hNSCs were then separated into single clonal lines, either by serial dilution alone (i.e., one cell per well) or by first infecting the cells with a retrovirus (allowing there to be a molecular marker of clonality—i.e., the proviral integration site) and then performing serial dilution. The transgenes transduced via retrovirus were either nontransforming propagation enhancement genes, such as v-*myc*, or purely reporter genes such as *lacZ*, or both.

For retrovirus-mediated gene transfer, two xenotropic, replication-incompetent retroviral vectors were used to infect hNSCs. A vector encoding *lacZ* was similar to BAG [24] except for bearing a PG13 xenotropic envelope. An amphotropic vector encoding v-*myc* was generated by means of the ecotropic vector described for generating murine NSC clone C17.2 [39] to infect the GP+envAM12 amphotropic packaging line [40]. No helper virus was produced. Infection of bFGF-maintained human neural cells with either vector (titer: 4×10^5 CFU) followed similar procedures detailed elsewhere [24,39].

For cloning of hNSCs, cells were dissociated as already described, diluted to 1 cell/15 µl and plated at 15 µl/well of a Terasaki or 96-well dish. Wells with single-cells were noted immediately. Single-cell clones were expanded and maintained in bFGF-containing growth medium. Single cells grow best when conditioned medium from dense hNSC cultures comprises at least 20–50% of the growth medium. Monoclonality was confirmed by identifying in all progeny a single and identical genomic insertion site on Southern analysis for either the *lacZ*- or the v-*myc*-encoding provirus as detailed earlier [39]. The v-*myc* probe was generated by nick translation labeling with ^{32}P-labeled deoxycytidine 5'-triphosphate (dCTP); a probe to the *neo* sequence of the *lacZ*-encoding vector was generated by PCR utilizing [^{32}P]dCTP.

Cryopreservation of hNSCs is done by resuspending post-trypsinized human cells in a freezing solution composed of 10% DMSO, 50% FBS, and 40% bFGF-containing growth medium. Afterward the temperature of cells is brought down slowly, first to 4°C for an hour, then to –80°C for 24 h, and then to –140°C.

Preparation of Human NSCs *in Vitro* for Transplantation

Cultured cells grow as a combination of adherent cells and floating clusters within T25 flasks. For efficient engraftment, all cells must be collected and must be well dispersed into a suspension of individual cells. To accomplish this, all medium and cells (including those adherent which are mechanically dislodged) are transferred into a 15-ml centrifuge tube, and centrifuged for 3 min at 1000 rpm. Following removal of the supernatant, 0.7 ml of trypsin–EDTA (0.05% trypsin, 0.53 mM EDTA) is added to the centrifuge tube, and the cells are again triturated briefly before a 5-min incubation at 37°C to facilitate dissociation of cells from each other. Further triturating is required to break up pellets and reach true single-cell suspension status. (Trypsin can also be added to the original flask to retrieve cells that may have still been adherent to the flask.) Trypsinization is then terminated by adding 0.7 ml of trypsin inhibitor (0.25 mg/ml in PBS) into the tube (or the flask), and triturating the mixture thoroughly. After a 3-min centrifugation at 1000 rpm, and removal of supernatant, cells are washed at least twice by resuspending them in 10 ml of PBS. Procedures for additional labeling of cells with trackers such as DiI or Hoechst can be done at this stage according to protocols suggested by the manufacturers. Then cells can be resuspended with small volume of PBS with a sufficient amount of trypan blue (~0.05% w/v) to permit localization of the injected suspension. An ideal injection concentration of cells (i.e., $5–10 \times 10^4$ cells/μl) can be achieved by cell counting and final volume adjustment. The remaining transplantation procedures are similar to those employed for murine NSCs.

NEURAL STEM CELLS AND TISSUE ENGINEERING

As mentioned earlier, the transplantation of NSCs into areas of injury can be useful for cell replacement and/or for delivery of therapeutic genes. Some of the most impressive examples of this, in pilot studies, were observed in rat models of traumatic spinal cord injury (SCI), and mouse models of hypoxic–ischemic (HI) cerebral injury [41,42]. Although NSCs appear to have the capacity to repopulate hemisectioned spinal cord or HI-injured brain, their ability to re-form connections is often limited by the vast amount of parenchyma loss. Since the core of the injury changes rapidly to a cystic cavity, even the most capable NSC may need intrinsic organization and a template to guide restructuring. Furthermore, large volumes of cells will not survive if located more than a few hundred micrometers from the nearest capillary [42]. Hence, *we hypothesized that three-dimensional, highly porous "scaffolds" composed of polyglycolic acid (PGA), if cotransplanted with NSCs into the infarction cavity, might facilitate re-formation of structural and funtional circuits.* As detailed in other chapters of this book, PGA is a synthetic biodegradable polymer used widely in clinical medicine. Highly hydrophilic, PGA loses its mechanical strength rapidly over 2–4 weeks in the body. The scaffold might initially provide a matrix to guide cellular organization and growth, allow diffusion of nutrients to the transplanted cells, become vascularized, and then disappear, obviating concerns over long-term biocompatibility.

To test this hypothesis for spinal cord repair, a multicomponent, degradable, synthetic PGA scaffold of specified architecture and seeded with NSCs was designed to guide regeneration, direct cell replacement, impede glial scar formation, and mitigate secondary injury. The implantation of the scaffold seeded with NSCs in an adult rat hemisection model of spinal cord injury led to robust long-term improvement in function relative to the lesion control group. At 70 days postinjury, the scaffold-with-cells group exhibited coordinated weight-bearing stepping compared with movement of two to three hindlimb joints in the lesion control group. Tract tracing demonstrated corticospinal tract fibers passing through the injury epicenter to the caudal side of the cord. Histological and immunocytochemical analysis including GAP-43 immunostaining further suggested that functional recovery was the product of a significant reduction in scar formation, reduced secondary tissue loss resulting in an increase in preserved tissue, and possible regeneration of damaged tissue [41]. In addition, the NSC-PGA polymer combination was transplanted into the cerebral infarction cavity via a glass micropipette 4–7 days following induction of an experimental HI insult [42]. The NSCs were observed to completely impregnate the PGA matrix, and the NSC-PGA unit seemed to refill the infarction cavity, even becoming vascularized by the host. The NSCs seeded on polymers displayed robust engraftment, foreign gene expression, and differentiation into neurons and glia within the region of HI injury. Neuronal tracing studies with DiI and biotinylated dextrose amine (BDA) showed that the long-distance neuronal circuitry between donor-derived and host neurons in both cerebral hemispheres may have been re-formed through the corpus callosum in some instances [42].

These preliminary findings suggest that NSC-PGA complexes may facilitate even further the differentiation of host and donor neurons, and enhance the ingrowth/outgrowth of such cells to help promote reformation of structural–functional spinal cord and cortical tissue. Together the studies suggest that neural stem cells may play a role in tissue engineering for the central nervous system. Indeed, they further reinforce the idea that, for CNS repair, NSCs may serve as "glue" that holds many repair strategies together: cell replacement, gene therapy, and biomaterial tissue engineering.

ACKNOWLEDGMENTS

The work was supported by Project ALS and by grants 80877 and 80787 from the U.S. National Institutes of Health. KIP was supported by grant 981-0713-097-2 from the Basic Research Program and BDRC of the Korean Science and Engineering Foundation, and HMP-98-N-1-0003 of the Ministry of Health & Welfare, Republic of Korea.

REFERENCES

1. McKay, R. D. G. (1997). Stem cells in the central nervous system. *Science* **276**, 66–71.
2. Gage, F. H., and Christen, Y., eds. (1997). "Isolation, Characterization, and Utilization of CNS Stem Cells. Research and Perspectives in Neuroscience." Springer-Verlag, Berlin.
3. Morrison, S. J., Shah, N. M., and Anderson, D. J. (1997). Regulatory mechanisms in stem cell biology. *Cell (Cambridge, Mass.)* **88**, 287–298.
4. Stemple, D. L., and Mahanthappa, N. K. (1997). Neural stem cells are blasting off. *Neuron* **18**, 1–4.
5. Edwards, B. E., Gearhart J. D., and Wallach E. E. (2000). The human pluripotent stem cell: Impact on medicine and society. *Fertil. Steril.* **74**, 1–7.
6. Vescovi, A. L., and Snyder, E. Y. (1999). Establishment and properties of neural stem cell clones: Plasticity in vitro and in vivo. *Brain Pathol.* **9**, 569–598.
7. Alvarez-Buylla, A., and Lois, C. (1995). Neuronal stem cells in the brain of adult vertebrates. *Stem Cells* **13**, 263–272.
8. Martinez-Serrano, A., and Snyder, E. Y. (1998). Neural stem cell lines for CNS repair. *In* "CNS Regeneration: Basic Science and Clinical Applications" (M. Tuszynski and J. Kordower, eds.), pp. 203–250. Academic Press, San Diego, CA.
9. Snyder, E. Y., and Vescovi, A. L. (2000). The possibilities/perplexities of stem cells. *Nat. Biotechnol.* **18**, 827–828.
10. Snyder, E. Y., and Macklis, J. D. (1996). Multipotent neural progenitor or stem-like cells may be uniquely suited for therapy for some neurodegenerative conditions. *Clin. Neurosci.* **3**, 310–316.
11. Park, K. I., Liu, S., Flax, J. D., Nissim, S., Stieg, P. E., and Snyder, E. Y. (1999). Transplantation of neural progenitor and stem cells: Developmental insights may suggest new therapies for spinal cord and other CNS dysfunction. *J. Neurotrauma* **16**, 675–687.
12. Sulston, J., Schieenberg, E., White, J., and Thompson, N. (1983). The embryonic cell lineage of the nematode *Caenorhabditis elegans. Dev. Biol.* **100**, 64–119.
13. Qian, X., Davis, A. A., Goderie, S. K., and Temple, S. (1997). FGF2 concentration regulated the generation of neurons and glia from multipotent cortical stem cells. *Neuron* **18**, 81–93.
14. Cepko, C. L. (1988). Retrovirus vectors and their applications in neurobiology. *Neuron* **1**, 345–353.
15. Cepko, C. L. (1989). Immortalization of neural cells via retrovirus-mediated oncogene transduction. *Annu. Rev. Neurosci.* **12**, 47–65.
16. Walsh, C., and Cepko, C. L. (1988). Clonally related cortical cells show several migration patterns. *Science* **241**, 1342–1345.
17. Price, J., Turner, D., and Cepko, C. L. (1987). Lineage analysis in the vertebrate nervous system by retrovirus-mediated gene transfer. *Proc. Natl. Acad. Sci. U.S.A.* **84**, 156–160.
18. Holt, C. E., Bertsch, T. W., Ellis, H. M., and Harris, W. A. (1988). Cellular determination in the *Xenopus* retina is independent of lineage and birth date. *Neuron* **1**, 15–26.
19. Kimmel, C. B., and Warga, R. M. (1987). Cell lineages generating axial muscle in the zebrafish embryo. *Nature (London)* **327**, 234–237.
20. Cepko, C. L., Turner, D., Price, J., Ryder, E., Snyder, E. Y. (1987). Retrovirus-mediated gene transfer and expression in the nervous system. *In* "Gene Transfer Vectors for Mammalian Cells" (J. H. Miller and M. P. Calos, eds.), pp. 15–18. Cold Spring Harbor Lab., Cold Spring Harbor, NY.
21. Turner, D., and Cepko, C. L. (1987). A common progenitor for neurons and glia persists in rat retina late in development. *Nature (London)* **328**, 131–136.
22. Kempermann, G., and Gage, F. H. (2000). Neurogenesis in the adult hippocampus. *Novartis Found. Symp.* **231**, 220–235.
23. Shihabuddin, L. S., Horner, P. J., Ray, J., and Gage, F. H. (2000). Adult spinal cord stem cells generate neurons after transplantation in the adult dentate gyrus. *J. Neurosci.* **20**, 8727–8735.
24. Snyder, E. Y., Deitcher, D. L., Walsh, C., Arnold-Aldea, S., Hartwieg, E. A., and Cepko, C. L. (1992). Multipotent neural cell lines can engraft and participate in development of mouse cerebellum. *Cell (Cambridge, Mass.)* **68**, 33–51.
25. Kaplan, P. L., Simon, S., Cartwright, C. A., and Eckhart, W. (1987) cDNA cloning with a retrovirus expression vector: Generation of a pp60c-src cDNA clone. *J. Virol.* **61**, 1731–1734.

26. Flax, J. D., Aurora, S., Yang, C., Simonin, C., Wills, A. M., Billinghurst, L. L., Jendoubi, M., Sidman, R. L., Wolfe, J. H., Kim, S. U., and Snyder, E. Y. (1998). Engraftable human neural stem cells respond to developmental cues, replace neurons, and express foreign genes. *Nat. Biotechnol.* **16,** 1033–1039.

27. Vescovi, A. L., Parati, E. A., Gritti, A., Poulin, P., Ferrario, M., Wanke, E., Frolichsthal-Schoeller, P., Cova, L., Arcellana-Panlilio, M., Colombo, A., and Galli, R. (1999). Isolation and cloning of multipotential stem cells from the embryonic human CNS and establishment of transplantable human neural stem cell lines by epigenetic stimulation. *Exp. Neurol.* **156,** 71–83.

28. Vescovi A. L., and Snyder E. Y. (1999). Establishment and properties of neural stem cell clones: Plasticity in vitro and in vivo. *Brain Pathol.* **9,** 569–598.

29. Fricker, R. A., Carpenter, M. K., Winkler, C., Greco, C., Gates, M. A., and Bjorklund, A. (1999). Site-specific migration and neuronal differentiation of human neural progenitor cells after transplantation in the adult rat brain. *J. Neurosci.* **19,** 5990–6005.

30. Carpenter, M. K., Cui, X., Hu, Z. Y., Jackson, J., Sherman, S., Seiger, A., and Wahlberg, L. U. (1999). In vitro expansion of a multipotent population of human neural progenitor cells. *Exp. Neurol.* **158,** 265–278.

31. Svendsen, C. N., Caldwell, M. A., Shen, J., ter Borg, M. G., Rosser, A. E., Tyers, P., Karmiol, S., and Dunnett, S. B. (1997). Long-term survival of human central nervous system progenitor cells transplanted into a rat model of Parkinson's disease. *Exp. Neurol.* **148,** 135–146.

32. Eriksson, P. S., Perfilieva, E., Bjork-Eriksson, T., Alborn, A. M., Nordborg, C., Peterson, D. A., and Gage, F. H. (1998). Neurogenesis in the adult human hippocampus. *Nat. Med.* **4,** 1313–1317.

33. Roy, N. S., Wang, S., Jiang, L., Kang, J., Benraiss, A., Harrison-Restelli, C., Fraser, R. A., Couldwell, W. T., Kawaguchi, A., Okano, H., Nedergaard, M., and Goldman, S. A. (2000). In vitro neurogenesis by progenitor cells isolated from the adult human hippocampus. *Nat. Med.* **6,** 271–277.

34. Brustle, O., Choudhary, K., Karram, K., Huttner, A., Murray, K., Dubois-Dalcq, M., and McKay, R. D. (1998). Chimeric brains generated by intraventricular transplantation of fetal human brain cells into embryonic rats. *Nat. Biotechnol.* **16,** 1040–1044.

35. Snyder, E. Y., Taylor, R. M., and Wolfe, J. H. (1995). Neural progenitor cell engraftment corrects lysosomal storage throughout the MPS VII mouse brain. *Nature (London)* **374,** 367–370.

36. McKay, R. D. G. (2000). Stem cells–hype and hope. *Nature (London)* **406,** 361–364.

37. Weiss, S., Reynolds, B. A., Vescovi, A. L., Morshead, C., Craig, C., and van der Kooy, D. (1996). Is there a neural stem cell in the mammalian forebrain? *Trends Neurosci.* **19,** 387–393.

38. Moretto, G., Xu, R. Y., Walker, D. G., and Kim, S. U. (1994). Co-expression of mRNA for neurotrophic factors in human neurons and glial cells in culture. *J. Neuropathol. Exp. Neurol.* **53,** 78–85.

39. Ryder, E. F., Snyder, E. Y., and Cepko, C. L. (1990). Establishment and characterization of multipotent neural cell lines using retrovirus vector mediated oncogene transfer. *J. Neurobiol.* **21,** 356–375.

40. Markowitz, D., Goff, S., and Bank, A. (1988). Construction and use of a safe and efficient amphotropic packaging cell line. *Virology* **167,** 400–406.

41. Teng, Y. D., Lavik, E., Qu, X., Park, K. I., Ourednik, J., Langer, R., and Snyder, E. Y. (2000). Transplantation of neural stem cells seeded in biodegradable polymer scaffold ameliorates long-term functional deficits resulting from spinal cord hemisection in adult rats. *Program & Abstr. Am. Soc. Neural Transplant. Repair* **7,** 61.

42. Park, K. I., Lavik, E., Teng, Y. D., and Snyder, E. Y. (2000). Transplantation of neural stem cells (NSCs) seeded in biodegradable polyglycolic acid (PGA) scaffolds into hypoxic–ischemic (HI) brain injury. *Soc. Neurosci. Abstr.* **26,** 868.

CHAPTER 35

STEM CELL CULTURE: LIVER STEM CELLS

Bryon E. Petersen and Heather M. Hatch

Hepatic oval "stem" cells are a small subpopulation of cells found in the liver when hepatocyte proliferation is impeded and followed by some type of hepatic injury. The hepatic stem cells can be stimulated to proliferate using a 2-acetylaminofluorase/hepatic injury (i.e., CCl4, partial hepatectomy) protocol. These cells are believed to be bipotential: able to differentiate into hepatocytes and bile ductular cells. Isolation of a highly pure population of this cell population has been considered to be difficult. Thy-1 is a cell surface marker used to identify hematopoietic stem cells in conjunction with CD34 and lineage-specific markers. Thy-1 antigen is not normally expressed in adult liver but is expressed in fetal liver, most probably on the hematopoietic cells. It has been shown that hepatic oval cells express high levels of Thy-1 as well as, α-fetoprotein (AFP), γ-glutamyl transpeptidase (GGT), cytokeratin 19 (CK-19), OC.2, and OV-6, all known markers for oval cell identification. When the Thy-1 antibody is used as a new marker for the detection of oval cells, a highly pure population can be obtained. We have used flow cytometry to establish a method to isolate a 95–97% pure Thy-1$^+$ oval cell population. Utilizing cell-sorting techniques in combination with the Thy-1 antibody will facilitate both *in vivo* and *in vitro* studies of hepatic oval cells.

THE HEPATIC OVAL CELL COMPARTMENT

There is a strong interest in identifying hepatic stem cells with respect to their origin, mechanism of activation/function, and final lineage destination. Oval cells are the primary candidates. Adequate data have been gathered showing that oval cells exist in the regenerating liver, but their place of origin and their role in liver development, regeneration, and carcinogenesis remains enigmatic. Oval cells increase in number when hepatocyte proliferation is suppressed. 2-acetylaminofluorene (2-AAF) given prior to and during hepatic injury, as by two-third partial hepatectomy (PHx), results in suppression of hepatocyte proliferation and expansion of the oval cell compartment [1,2]. When metabolized by hepatocytes, 2-AAF blocks the cyclin D1 pathway in the hepatocyte cell cycle. Oval cells then arise in the periportal region of the liver, from cells not yet identified. Morphologically, oval cells are small ($\sim$10 µm), with a large nucleus-to-cytoplasm ratio and an oval nucleus (hence the name) [3]. They have similarities to bile ductular cells in their distinct isoenzyme profiles, expressing certain keratin markers (e.g., CK-19), and γ-glutamyl transpeptidase (GGT), yet may also express high levels of α-fetoprotein (AFP). Monoclonal antibodies, such as OV6, OC.2, and BD1, also aid in their identification and characterization [4–7].

Traditionally, oval cells have been assumed to have a liver origin. Recently, however, we found in the rat that bone marrow stem/progenitor cells are capable of becoming hepatic oval cells able to engraft into the recipient liver, then differentiate into mature hepatocytes [8]. In these initial studies and confirmatory murine studies [9], the frequency of engraftment was less that 1%. The low rate may be attributable either to a low seeding efficiency or to the very low numbers of oval cell progenitors existing in adult rat bone marrow. Alternatively, the time points tested (days 9 and 13 post–hepatic injury) may have been too early in the repair process. Considering that very low numbers of murine hematopoietic stem cells (as low as 1 cell) can reconstitute all hematopoietic pathways in a transplanted recipient mouse, the relatively low efficiency of donor engraftment observed in our earlier studies could be accounted for by any of these three possibilities. Higher rates of bone marron-to liver restitution have been reported in damaged human liver months after transplant [10].

Returning to the rat model, Evarts *et al.* showed that the oval cell compartment is activated extensively in the 2-AAF/PHx model, a variant of the Solt–Farber protocol [11,12]. In further studies, the same investigators showed that proliferation of oval cells and their differentiation into hepatocytes during the early stages of carcinogenesis are associated with an activation of stellate cells. It has been suggested that perisinusoidal stellate cells may regulate the developmental fate of the progenitor cells, either directly by secreting growth factors such as hepatocyte growth factor (HGF), transforming growth factors α (TGF-α), and β (TGF-β), or indirectly via effects of extracellular matrix (ECM) components induced by urokinase up-regulation [13,14]. Progenitor cell proliferation and differentiation may also be regulated by autocrine production of TGF-α, acidic fibroblast growth factor (aFGF), and IGF-II (insulin-like growth factor II), since it has been shown that oval cells can produce all these factors [15]. Hence, hepatic injury–induced changes in cytokines and growth factors appear to modulate *in situ* oval cell proliferation/differentiation within the liver.

Lemire and Fausto showed that a very small number of cells localized in the canals of Hering of adult normal rat liver expresses the fetal form of AFP mRNA, suggesting that this may be the compartment in which the oval cells arise [16]. Marceau *et al.* suggested that bile ductular cells expressing the fetal form of AFP in the adult liver are unlikely to be the putative stem cells because they lack other markers, and in contrast to hepatoblasts, they do not react with the monoclonal antibody BPC5 [17]. However, we showed that when the bile ductular epithelium is destroyed by exposure to methylene dianaline (DAPM), 24 h prior to hepatic damage (2-AAF/hepatic injury, PHx or CCl$_4$), oval cell proliferation is inhibited [18]. Our study was the first to show a direct association between an intact bile ductular epithelium and oval cell activation, but it does not prove that oval cells arise from bile ductular cells; DAPM could have a direct or indirect toxic effect upon the oval cells. What this study does show is that a portion of oval cells arises in the periportal regions, presumably in the canals of Hering [19]. All studies suggest that oval cells lie somewhere in the hepatic architecture and have a close association with bile duct epithelium, but the exact location is still unknown.

HEPATIC REGENERATION

Compensatory hyperplasia of the liver, most often referred to as liver regeneration, takes place after the occurrence of mild or severe injury, such as that resulting from a surgical partial hepatectomy or the widespread injury caused by hepatotoxic agents like carbon tetrachloride (CCl$_4$) or acetaminophen. Under normal conditions, hepatocytes exhibit a minimal replicative activity: only 1 in every 20,000 has been observed to undergo mitotic division [20]. Twelve to 15 h after surgical removal of two-thirds of the liver mass (partial hepatectomy, PHx), hepatocytes move from the G$_0$ resting phase of the cell cycle into G$_1$, the first replication phase of the cell cycle [21,22]. Hepatocytes located around the portal triad (lobular zone 1) are the first to undergo DNA synthesis [23–27], which gradually progresses to include the hepatocytes located around the central vein (zone 3). A large peak of DNA synthesis is observed at about 24 h post-PHx and a second but smaller peak at 48 h. The smaller peak reflects DNA synthesis occurring in nonparenchymal cells (NPC) and zone 3 hepatocytes. Unlike hepatocytes, NPCs exhibit a more diffuse DNA synthesis throughout the liver lobule [28]. The original liver mass is restored within 10 days [29].

Several animal models of chemical hepatotoxicity have been developed to identify the mechanisms regulating the hepatic replication response following injury. Among the most extensively investigated agents are carbon tetrachloride (CCl$_4$), which causes necrosis of the centrilobular regions of the liver, and allyl alcohol (AA), which causes periportal necrosis [30–34]. After administration of a single dose, CCl$_4$ readily induces peroxidation of the endoplasmic reticulum membrane lipids, followed by mitochondrial dysfunction, inhibition of protein synthesis, and fat accumulation within the first 12 h. Necrosis of centrilobular hepatocytes and inflammatory cell influx becomes apparent after 24–48 h [35–39]. DNA synthesis begins at about 24 h peaks at 48 h and returns to normal levels by 96 h. On a molecular level in terms of growth factor regulation, CCl_4 differs only slightly from that after PHx [40]. The major difference seen between CCl_4 and PHx is the inflammatory response induced by CCl_4, hence the cytokines released by the inflammatory cells and possibly factors released by dying hepatocytes.

MATERIALS

Carbon tetrachloride (CCl_4): 99% pure HPLC grade) and 2-acetylaminofluorene (2-AAF) can be purchased from Aldrich Chemical Co. (St. Louis, MO). Crystals of 2-AAF are incorporated into time-released pellets (70 mg/pellet over a 28-day release, 25 mg/day) by Innovative Research Inc. (Sarasota, FL). Thy 1.1 can be purchased from PharMingen Inc. (San Diego, CA). Male Fischer 344 rats (125–150 g) can be obtained from Fredericks Laboratories (Frederick, MD). Microscope Superfrost Plus slides, buffered Formalin-Fresh, and dextran sulfate are obtained from Fisher Scientific (Pittsburgh, PA).

METHODS
2-AAF Exposure

Studies of liver regeneration following hepatic injury (PHx, CCl_4, etc.) have provided insight regarding the molecular mechanisms by which the liver heals itself while simultaneously maintaining homeostasis for the body. Upon understanding the regenerative process under normal conditions with respect to signaling pathways and cell-to-cell interactions, we can begin to investigate the role hepatic stem cells plays in this regeneration process.

To activate the oval cell compartment, hepatocyte proliferation must be suppressed and followed by severe hepatic injury. 2-AAF has been shown to suppress hepatocyte proliferation. 2-AAF pellets were chosen over powder based upon ease and time involved in administration. The timeline of events for oval cell proliferation is represented in Fig. 35.1. Briefly, the animals are anesthetized under Metofane inhalation. The abdomen is shaved and sterilized with an ethanol swab, and a small incision is made in the lower quadrant of the abdomen. A 2-AAF pellet is inserted subcutaneously. When the pellet has been inserted, the incision is closed. It usually takes three sutures to close the wound. The wound is then sterilized again with an ethanol swab and the animal is placed in a sterile recovery cage. This procedure takes approximately 3–5 min from start to finish. The animals recover quickly from the anesthetizing agent and regain normal daily functions. The animals are monitored for bleeding and coloration of eyes and ears for about 2 h postsurgery and then placed back to their normal caging. Using 2-AAF pellets alleviates undue stress to the animals brought

Timeline for Oval Cell Activation

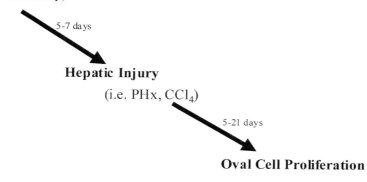

2-AAF exposure
(Inserted subcutaneously)

5-7 days

Hepatic Injury
(i.e. PHx, CCl_4)

5-21 days

Oval Cell Proliferation

Fig. 35.1. Timeline of events for initiation of oval cell proliferation. The presence of 2-AAF is necessary to inhibit hepatocyte proliferation and to allow extended proliferation of oval cells. The diagram represents the different stages of oval cell proliferation.

on by multiple 2-AAF exposures through oral administration (i.e., gavage), and it reduces the risk of human exposure, which often exists when a chemical is handled in powder form.

Compound Delivery

An LD_{50} dose of carbon tetrachloride as determined by Lewis was used [41]. A single dose of 1.9 ml/kg (1500 mg/kg) of body weight in a 1:1 vol/vol dilution in corn oil is administered by intraperitoneal (IP) injection.

Partial Hepatecomy

One of the 2-AAF/hepatic injury models includes using a partial hepatecomy procedure. For the PHx procedure, rats are hepatectomized under general anesthesia according to the methods described by Higgins and Anderson [42]. The procedure for the mouse is the same except the individual lobes are removed separately to prevent bile duct obstruction.

Briefly, animals are anesthetized by Metofane inhalation. The abdomen is shaved and sterilized using a Betadine-ethanol antiseptic scrub. A small incision is made along the mid-line of the abdomen (~1 in. in length), just below the xyphoid process. Following this, the xyphoid process is located and removed. With the right and left lobes of the liver non exposed from the abdominal cavity, the lobes are ligated with 2-0 black braided silk suture material and excised. Any excessive bleeding is cauterized with a cauterizing pen. The animal is then closed, first by suturing the abdominal muscles together (about 5–6 sutures, 3-0 silk) then by suturing the outside epidermis (about 5–6 staples). The wound is scrubbed again with Betadine–ethanol antiseptic to clean the area around the sutures. The animals are placed in a sterile area and monitored following the operation for any postoperative stress, then placed back into their normal caging for the remainder of the experiments. Again, this is a very simple operation. The whole procedure take only minutes, and all animals are expected to recover and regain normal daily activities.

PROTOCOL FOR TWO-STEP RAT PERFUSION

Refer to Seglan [43].

Materials:

Peristaltic pump
96400-16 Masterflex tubing (Fisher or VWR)
1-ml glass sterile pipette
1.25-in. 20-gauge winged polyethylene catheter (JELCO, Johnson & Johnson)
Nylon mesh-(Tetko Inc. 3-100/35 Nitex, 006 3-0100/35)
Solution A: 250 ml of Strom and Michalopoulos (S&M) solution
Solution B: 250 ml of S&M solution supplemented with $10\times$ $CaCl_2$ $(H_2O)_2$ to a final concentration of $1\times$ and collagenase H or type I. Collagenase can be obtain either from Boehringer–Mannheim or Sigma. Use type I or type H.
Nembutal (Abbotts) or methoxyflurane (Metofane, Pittman-Moore)

Also required are dissecting and iris scissors, forceps, hemostats, 2-0 suture silk, 95% EtOH, 250-ml beaker covered with nylon mesh and 100-ml beaker (both autoclaved).

S&M solution

500 mg KCl
8.3 g NaCl
2.4 g HEPES
190 mg NaOH
Makes one liter. Adjust pH to 7.4 and filter through 0.2-μm filter.

$10\times$ CaCl$_2$

$(H_2O)_2$ solution: 6.36 g/liter.

Method

1. *Warm circulating water bath to 40°C.* The optimal temperature for collagenase is 37°C, but the solution will cool slightly as it passes through the tubing. To combat this, make sure the tubing is very long. This will allow for the excess to be placed in the water, with only a short segment of the tubing exposed to room temperature.

2. *Warm both bottles of solution A and B* (solution B will not have the collagenase added to it yet). Solution A will be used to remove extracellular calcium necessary to maintain intercellular adhesion. Solution B with the $CaCl_2$ and collagenase (add the collagenase 10 min prior to perfusion procedure) needs the calcium as a cofactor for function. Use lead donuts to keep the bottles in place while in the water bath.

3. *Thread rubber tubing through the peristaltic pump.* On one end insert the 1-ml glass pipette, on the other end take a tuberculin syringe, and cut off the winged end, and place the cut end into the tubing. This is where the catheter will attach. Place the 1-ml pipette into solution A, place the excess tubing in the water bath, and allow the solution to flow through the tubing and circulate back into the bottle.

4. *Turn pump on to recirculate the solution.* The initial perfusion rate should be about 15–20 ml/min.

5. *Anesthetize the rat with Nembutal* (100 μl/100 g). Nembutal is metabolized by the liver, and Metofane is eliminated through the lungs. Depth of anesthesia is checked by pinching the back feet and watching for abdominal muscle contractions. The amount of Nembutal can vary depending on how fast you want the animal to go under. Be careful, though, not to give the rat too much, which will kill it. You need the heart to continue to beat until you actually sever the hepatic vein and clamp off the superior vena cava.

6. *Secure the rat to the perfusion table by taping its extremities.* The legs should be positioned to ensure that the animal's body is stretched and tight.

7. *Rinse the abdomen and lower thorax of the rat with 95% EtOH.* Be careful not to get EtOH on the tape because it will dissolve the adhesive on the tape. Perform a midline incision through the skin and peritoneum; make a subsequent cut perpendicular to the laparotomy incision on both sides of the animal at the level 5 mm below the subcostal margin. DO'NOT ENTER THE CHEST CAVITY. A pneumothorax is rapidly fatal for the animal.

8. *Retract the wound edges* with either hemostats or bulldog clamps and position them out of the way.

9. *Displace abdominal viscera toward the lower right quadrant and expose the inferior vena cava (IVC).* Cotton swabs or 2 × 2 sponges can be used to retract the mesentery for greater exposure of the IVC. When the IVC is exposed, take a pair of curved forceps and work them under the IVC and through the other side. Once through, open the forceps slightly, grasp the 2-0 silk, and pull through. Once through, take the ends and begin to tie a knot. DO NOT TIGHTEN COMPLETELY. Leave a small loop. This is where you will insert the catheter and secure to the IVC.

10. *Turn off pump and free any bubbles that may have collected in the tubing.* Turn pump on and release the bubbles from the tubing, making sure that the fluid is at the end of the syringe. Turn off pump and place the syringe next to the animal.

11. *Cannulate the IVC, using a 20-gauge catheter*, making sure that the catheter will be going through the opening of the 2-0 silk knot. Once the catheter has been placed inside the vessel, secure the catheter in place by closing the 2-0 silk knot. Secure the syringe to the catheter. Being careful not to pierce the vessel wall when inserting the catheter. Also, take care that the knot to hold the catheter is not too tight, which would restrict the flow of the perfusate.

12. *Turn the pump on*, and watch to make sure the catheter has been positioned properly and that displacement will not occur. At this time cut the hepatic portal vein with the dissecting scissors. Make sure that the vein is completely severed and that the fluid can flow unrestricted out of the body. This will create an outflow tract. When the portal vein is cut, open the chest cavity and expose the superior vena cava (SVC). Taking a pair of curved

hemostats, clamp off the SVC and place the hemostats aside as gently as possible, to prevent disturbing the catheter. Refer to Fig. 35.2.

13. *With a moistened cotton swab, gently massage the lobes of the liver* to assist in clearing of the blood within the sinusoids. This should be done until the lobes look completely blanched and the effluent is clear. A blanched liver will have a pale brown color.

14. *Allow solution A to run for about 10–15 min.* At this time add the collagenase to solution B with the $CaCl_2$ already added. Mix the collagenase completely by inverting the bottle several times. The amount of collagenase will vary depending upon its activity. Usually it is about 160–180 mg into the 250 ml. Solution B will now have a light brown coloration.

15. *Shut off pump, remove the 1-ml glass pipette, and put it in the solution B bottle.* Then resume pump flow and decrease the flow rate to about 12 ml/min. The perfusion should take about 22 min. This time will vary, so at about 15 min after the collagenase has started to enter the liver, watch the liver for changes. Toward the end of the perfusion, the liver will begin to show signs of trabeculation and fragmentation. This is an indication of breaking down of the extracellular matrix. (The liver looks somewhat netlike.)

16. *Prior to harvesting the liver, place about 50-ml of ice-cold S&M into a 100-ml beaker.* This is where you will place the excised liver. Keep the top of the beaker covered at all times to limit any contamination.

17. *Use curved Mayo scissors* to remove the lobes and place into the ice-cold S&M. If the perfusion worked, the lobes will be very loose and hard to pick up.

18. *This part should take place in a laminar flow hood.* Mince/tease apart the tissue so that you break the Glisson capsule. Grasp the tissue and shake/agitate it until the loosened cells are dispersed. The S&M will have light brown coloration—the better the perfusion, the darker the color. If the perfusion was successful, the only thing remaining will be the undigested matrix–vessel portion.

19. *Carefully decant the supernatant through a sterile nylon mesh that covers a 250-ml beaker.* The mesh is placed over the beaker, held in place by aluminum foil secured by a rubber band, and autoclaved. Open this only in the laminar flow hood. Pass the supernatant through the mesh, and wash the mesh with an adequate volume of S&M to collect all the cells. Aliquot this new supernatant into 50-ml conical tubes.

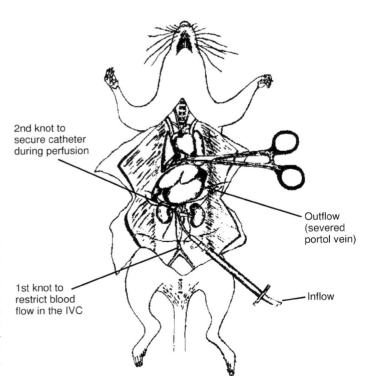

Fig. 35.2. Cartoon drawing of rat perfusion. First, tighten the first knot to restrict the blood flow in the IVC. Then cannulated the IVC with a 20-gauge catheter, making sure that the catheter will be going through the opening of the 2-0 silk knot. When the catheter is in place inside the vessel, secure the catheter into place by closing the 2-0 silk knot. Secure the syringe to the catheter (inflow). Also, take care in that the knot to hold, to the catheter is not too tight, to avoid restricting the flow of the perfusate. Cut the hepatic portal vein to allow perfusate to flow freely out of the body (outflow). Bulldog clamps or hemostats are used to clamp off the SVC to restrict the flow of the perfusate out of the hepatic portal vein.

2nd knot to secure catheter during perfusion

Outflow (severed portol vein)

1st knot to restrict blood flow in the IVC

Inflow

20. *Wash cells in cold S&M solution with centrifugation at 50g for 5 min, three times* taking care not to disturb the pellet of cells at the bottom of the tube. Save the supernatant from these tubes. The oval cells are in the nonparenchymal cell (NPC) fraction and are in this supernatant. Keep the collected supernatants on ice until all three washes of the hepatocytes have been performed. Once all the supernatants have been collected from the three washes, centrifuge them at 400g for 5 min. Aspirate the supernatant from these tube (no need to save it), leaving the cell pellets at the bottom of the tube. Resuspend the cells in 50 ml of S&M for one last wash. Centrifuge for 5 min at 400g, aspirate the supernatant off, and resuspend the cells in Iscove media, and place on ice until ready for immunotagging with Thy-1 antibody.

Check cell viability by trypan blue exclusion: 100% viability will yield about 70 million hepatocytes per milliliter of pellet. About 10 ml of pellet can be obtained for hepatocytes, and the NPC fraction will contain about $2–3 \times 10^8$ cells, of which the oval cells will be 10–20% of the total population of NPCs.

Flow-Cytometry
Cell Sorting

Oval cell isolation is performed by means of flow cytometry. Briefly, the nonparenchymal cell fraction has been determined to contain the hepatic oval cell population as described by Yaswen *et al.* [44]. Immunohistochemistry was performed on the parenchymal and nonparenchymal fractions to be sure that the cells of interest were in the NPC fraction. In support of Yaswen's findings, we also found that the NPC fraction contained the highest percentage of oval cells [45]. A portion ($\sim 60–80 \times 10^6$ of the total 200×10^6 cells) of the NPC fraction is further purified by using flow cytometry. Fluorescein isothiocyanate (FITC)-conjugated anti-rat Thy 1.1 (1 mg/million cells) is used to label the target cells. The cell fraction was incubated with the antibody for 20 min at 4°C, rinsed twice in $1\times$ phosphate-buffered soline–1% fetal bovine serum for 5 min each time, and stored in the dark on ice until sorting. A FITC mouse IgG_1 is used as an isotype control. A Becton Dickinson FACStar flow cytometer is used to sort the cell into two populations: Thy 1.1 positive (Thy 1.1$^+$) and Thy 1.1 negative (Thy 1.1$^-$). The animals in which the liver is perfused will have one of the smaller lobes surgically removed prior to the cell isolation procedure. This tissue can be used as an internal control for light microscopy as well as for in situ hybridization techniques. The excised liver tissue is divided in half and fixed in either 10% buffered formalin or placed in optimal cutting temperature (OCT) compound, frozen in cold 2-methylbutane (Fisher Scientific) and stored at –80°C.

Cell Cycle Analysis

When cell cycle analysis is required, it can be performed in the following manner. First 1×10^6 Thy-1.1 positive cells are fixed in ice-cold 70% ethanol at 4°C overnight. Following centrifugation (5 min at 3000 rpm), the ethanol is removed and 1 ml of propidium iodide staining (PI) solution, consisting of 50 μg/ml PI, 100 U/ml RNase A in Ca^{2+}- and Mg^{2+}-free PBS + glucose is added and incubated for 30 min at room temperature. Then flow cytometric cell cycle analysis can be performed.

APPLICATIONS OF ISOLATED ADULT HEPATIC OVAL CELLS

Today, stem cell research lies in the center of considerable debate, both in the public and scientific arenas. In the middle of this controversy lies the issues of using embryonic and fetal stem cell for biomedical research. Although embryonic and fetal stem cells are getting the most attention, it is the adult stem cell that may prove to be the most useful. There have been a number of methods used to isolate oval cells. Pack *et al.* [46] modified the methods of Yaswen *et al.* [44] and reported that by using centrifugal elutriation, they could obtain $15–70 \times 10^6$ oval cells per rat liver (80–90% purity), with the number of cells obtained dependent on the length of time the rats were fed a choline-deficient/ethionine (CDE) diet. A major flaw in their method is that as much as 20% of the isolated cells are nonoval cells, leading to concerns with respect to purity since it is not certain how the other cell population may influence the oval cells in culture. The use of flow cytometry to isolate oval cells has allowed us to obtain a 95–97% pure population, thereby reducing the confounding effects of other cells in culture. The yields of oval cells we obtained by flow cytometry are similar to those reported by Pack, although the time required to sort the cells is longer. We sort only a portion of the NPC fraction, and the yield and high purity of Thy-1$^+$ oval cells is adequate for immunohistochemical characterization as well as for establishing cultures of the isolated oval cells. When larger numbers of Thy-1$^+$ cells are needed (e.g., transplantation), more of the NPC fraction could be sorted.

This technique of cell sorting may also prove be useful in future studies dealing with *in vitro* and bioengineering issues. For instance, there now can be a more detailed investigation of these cells in culture to determine the signals involved in the pathways for both proliferation and differentiation. This unique population could possibly be used in gene/cell therapies, such as α_1 antitrypsin deficiency and Crigler–Najjar syndrome, to restore normal function to the patient's otherwise normal liver, if the oval cell can be transfected with reasonable efficiency. With the ever-expanding world of bioengineering, this stem cell population may be very useful in providing the basic cell type needed in tissue engineering aspects as they apply to artificial liver devices.

Each year there are approximately 14,000 patients waiting for liver transplants for various reasons, with only about 4000 organs available for transplantation. Every year more than 25 million Americans will be afflicted with liver diseases, with more than 43,000 deaths per year [47]. There are few effective treatments for most life-threatening liver diseases, except for liver transplantation. It may be possible to sort out Thy-1$^+$-positive cells from a patient with a failing liver and expand them in culture to numbers high enough to be transplanted back into the patient's own liver, thereby giving such patients time and critical cell numbers to regenerate their own livers. The use of a patient's own cells could eliminate rejection issues and years of immunosuppressive medication, in addition reduce the number of patients waiting for liver transplants.

REFERENCES

1. Thorgeirsson, S. S., Evarts, R. P., Bisgaard, H. C., Fujio, K., and Hu, Z. (1999). Hepatic stem cell compartment: Activation and lineage commitment. *Proc. Soc. Exp. Biol. Med.* **204**, 253–260.
2. Evarts, R. P., Nakatsukasa, H., Marsden, E. R., Hsia, C.-C., Dunsford, H. A., and Thorgeirsson, S. S. (1990). Cellular and molecular changes in the early stages of chemical hepatocarcinogenesis in the rat. *Cancer Res.* **50**, 3439–3444.
3. Farber, E. (1956). Similarities in the sequence of early histologic changes induced in the liver of the rat by ethionine, 2-acetylaminofluorene, and 3′-methyl-4-dimethylaminoazbenzene. *Cancer Res.* **16**, 142–149.
4. Yang, L., Faris, R. A., and Hixson, D. C. (1993). Phenotypic heterogeneity with clonogenic ductual cell populations isolated from normal adult rat liver. *Proc. Soc. Exp. Biol. Med.* **204**, 280–288.
5. Faris, R. A., Monfils, B. A., Dunsford, H. A., and Hixson, D. C. (1991). Antigenic relationship between oval cells and subpopulation of hepatic foci, nodules, and carcinomas inducted by the "resistant hepatocyte" model system. *Cancer Res.* **51**, 1308–1317.
6. Hixson, D. C., Faris, R. A., and Thompson, N. L. (1990). An antigenic portrait of the liver during carcinogenesis. *Pathobiology* **58**, 65–73.
7. Shiojiri, N., Lemire, J. M., and Fausto, N. (1991). Cell lineage and oval cell progenitors in rat liver development. *Cancer Res.* **51** 2611–2620.
8. Petersen, B. E., Bowen, W. C., Patrene, K. D., Mars, W. M., Sullivan, A. K., Murase, N., Boggs, S. S., Greenberger, J. S., and Goff, J. P. (1999). Bone marrow as a potential source of hepatic oval cells. *Science* **284**, 1168–1170.
9. Theise, N. D., Nimmakayalu, M., Gardner, R., Illei, P. B., Morgan, G., Teperman, L., Henegariu, O., and Krause, D. S. (2000). Liver from bone marrow in humans. *Hepatology* **32**(1), 11–16.
10. Theise, N. D., Badve, S., Saxena, R., Henegariu, O., Sell, S., Crawford, J. M., and Krause, D. S. (2000). Derivation of hepatocytes from bone marrow cells in mice after radiation-induced myeloablation. *Hepatology* **31**(1), 235–240.
11. Evarts, R. P., Hu, Z., Fujio, K., Marsden, E. R., and Thorgeirsson, S. S. (1993). Activation of hepatic stem cell compartment in the rat: Role of transforming growth factor-β, hepatocyte growth factor, and acidic fibroblast growth factor in early proliferation. *Cell Growth Differ.* **4**, 555–561.
12. Evarts, R. P., Nagy, P., Nakatsukasa, H., Marsden, E., and Thorgeirsson, S. S. (1989). *In vivo* differentiation of rat oval cells into hepatocytes. *Cancer Res.* **49**, 1541–1553.
13. Fausto, N. (1991). Protooncogenes and growth factors associated with normal growth and abnormal liver growth. *Dig. Dis. Sci.* **36**, 653.
14. Bisgaard, H. C., Snatoni-Rugiu, E., Nagy, P., and Thorgeirsson, S. S. (1998). Modulation of the plasminogen activatior/plasmin system in rat liver regenerating by recruitment of oval cells. *Lab. Invest.* **78**(3), 237–246.

15. Fausto, N. (1994). Liver stem cells. *In* "The Liver Biology and Pathobiology" (I. M. Arias, J. L. Boyer, N. Fausto, W. B. Jakoby, D. A. Schachter, and D. A. Shafritz, eds.), 3rd ed., pp. 1501–1518. Raven Press, New York.

16. Lemire, J. M., and Fausto, N. (1991). Multiple alpha-fetoprotein RNAs in adult rat liver: Cell type-specific expression and differential regulation. *Cancer Res.* **51**, 4656–4664.

17. Marceau, N., Blouin, M.-J., Noel, M., Torok, N., and Loranger, A. (1992). The role of bipotential progenitor cells in liver ontogenesis and neoplasia. *In* "The Role of Cell Types in Hepatocarcinogenesis" (A. E. Sirica, ed.), pp. 121–149. CRC Press, Boca Raton, FL.

18. Petersen, B. E., Zajac, V. F., and Michalopoulos, G. K. (1997). Exposure to methylene dianaline inhibits oval cell proliferation. *Am. J. Pathol.* **151**(4), 905–909.

19. Saxena, R., Theise, N. D., and Crawford, J. M. (1999). Microanatomy of the human liver—Exploring the hidden interfaces. *Hepatology* **30**, 1339–1346.

20. Steer, C. J. (1995). Liver regeneration: Serial review. *FASEB J.* **9**, 1396–1400.

21. Bucher, N. L. R. (1963). Regeneration of mammalian liver. *Int. Rev. Cytol.* **15**, 245–300.

22. Michalopoulos, G. K. (1990). Liver regeneration: Molecular mechanisms of growth control. *FASEB J.* **4**, 176–187.

23. Fausto, N., and Webber, E. M. (1994). Liver regeneration. *In* "The Liver Biology and Pathobiology" (I. M. Arias, J. L. Boyer, N. Fausto, W. B. Jakoby, D. A. Schachter, and D. A. Shafritz, eds.), pp. 1059–1086, Raven Press, New York.

24. Rabes, H. M. (1976). Kinetics of hepatocellular proliferation after partial resection of the liver. *Prog. Liver. Dis.* **5**, 83–99.

25. Petersen, B. E. (1996). Hepatic stem cells and growth factor regulation in liver regeneration following toxic injury. Ph.D. Dissertation, Graduate School of Public Health, Department of Environmental and Occupational Health and Toxicology, University of Pittsburgh.

26. Grisham, J. W. (1962). A morphologic study of deoxyribonucleic acid synthesis and cell proliferation in regenerating rat liver; autoradiography with thymidine-H^3. *Cancer Res.* **22**, 842–849.

27. Diehl, A. M., and Rai, R. (1996). Review: Regulation of liver regeneration by pro-inflammatory cytokines. *J. Gastroenterol. Hepatol.* **11**(5), 466–470.

28. Bucher, N. L. R., and Swaffield, M. N. (1964). The rate of incorporation of labeled thymidine into deoxyribonucleic acid of regenerating rat liver in relationship to the amount of liver excised. *Cancer Res.* **24**, 1611–1625.

29. Michalopoulos, G. K., and DeFrances, M. C. (1997). Liver regeneration. *Science* **27**, 60–66.

30. Edwards, M. J., Keller, B. J., Kauffman, F. C., and Thurman, R. G. (1993). The involvement of Kupffer cells in carbon tetrachloride toxicity. *Toxicol. Appl. Pharmacol.* **119**, 275–279.

31. Reid, W. D. (1972). Mechanism of allyl alcohol-induced hepatic necrosis. *Experientia* **28**, 1058–1070.

32. Badr, M. Z., Belinsky, S. A., Kauffman, F. C., and Thurman, R. G. (1986). Mechanism of hepatotoxicity to periportal regions of the liver lobule due to allyl alcohol: Role of oxygen and lipid peroxidation. *J. Pharmacol. Exp. Therap.* **238**, 1138–1142.

33. Leduc, E. H., and Wilson, J. W. (1958). Injury to liver cells in carbon tetrachloride poisoning. *AMA Arch. Path.* **65**, 147–156.

34. Edwards, M. J., Keller, B. J., Kauffman, F. C., and Thurman, R. G. (1993). The involvement of kupffer cells in carbon tetrachloride toxicity. *Toxicol. Appl. Pharmacol.* **119**, 275–279.

35. Mehendale, H. M. (1989). Mechanism of the lethal interaction of chlordecone and CCl_4 at non-toxic doses. *Toxicol. Lett.* **49**, 215–222.

36. Myren, J., Bang, S., Linnestad, P., Stave, R., Hanssen, L. E., Dolva, L. O., Serck-Hanssen, A., Arnesen, K., Stromme, J., Beraki, K., and Vagene, S. (1989). Liver cell necrosis and regeneration following injections of carbon tetrachloride. *Acta Pathol. Microbiol. Immunol. Scand.* **97**, 334–345.

37. Rechnagel, R. O., and Glende, E. A. (1973). Carbon tetrachloride hepatotoxicity: An example of lethal cleavage. *CRC Crit. Rev. Toxicol.* **2**, 263–275.

38. Kim, H. J., Odend'hal, S., and Bruckner, J. V. (1990). Effect of oral dosing vehicles on the acute hepatotoxicity of carbon tetrachloride in rats. *Toxicol. Appl. Pharmacol.* **102**, 34–44.

39. Leevy, C. M., Hollister, R. M., Schmid, R., MacDonald, R. A., and Davidson, C. S. (1959). Liver regeneration in experimental carbon tetrachloride intoxication. *Proc. Soc. Exp. Biol. Med.* **102**, 672–681.

40. Lindross, P. M., Zarnegar, R., and Michalopoulos, G. K. (1991). Hepatocyte growth factor (Hepatopoietin A) rapidly increases in plasma before DNA synthesis and liver regeneration stimulated by partial hepatectomy and carbon tetrachloride administration. *Hepatology* **13**, 743–752.

41. Lewis, R. J., Sr., ed. (1992). SAX Dangerous Properties of Industrial Material, 8th ed., Vol. III. Van Nostrand-Reinhold, New York.

42. Higgins, G. M., and Anderson, R. M. (1931). Experimental pathology of the liver: I. Restoration of the liver of the white rat following partial surgical removal. *Arch. Pathol.* (*Chicago*) **12**, 186–202.

43. Seglan, P. O. (1976). Preparation of isolated rat liver cells. *Methods Cell Biol.* **13**, 29–83.

44. Yaswen, P., Hayner, N. T., and Fausto, N. (1984). Isolation of oval cells by centrifugal elutriation and comparison with other cell types purified from normal and preneoplastic livers. *Cancer Res.* **44**, 324–331.

45. Petersen, B. E., Goff, J. P., Greenberger, J. C., and Michalopoulos, G. K. (1998). Hepatic oval cells express the hematopoietic stem cell marker Thy-1 in the rat. *Hepatology* **27**(2), 433–445.

46. Pack, R., Heck, R., Dienes, H. P., Oesch, F., and Steinber P. (1993). Isolation, biochemical characterization, long-term culture, and phenotype modulation of oval cells from carcinogen-fed rats. *Exp. Cell Res.* **204**, 198–209.

47. American Liver Foundation Annual Report: Taken from the ALF Fact Sheet published on their Website: http:/gi.ucsf.edu/alf/info/factsheet.html

CHAPTER 36

STEM CELL CULTURE: MUSCLE STEM CELLS

Joon Yung Lee and Johnny Huard

INTRODUCTION

Skeletal muscle development is a complex pathway of gene regulation and expression that allows immature myoblasts to fuse and form mature myofibers. Temporally distinct lineages of myogenic cells can be identified at embryonic, fetal, perinatal, and even adult muscle tissue. Recent investigations have delivered strong evidence that myogenic cells isolated from skeletal muscle are heterogenic, and a small proportion of these cells display stem cell characteristics. Muscle stem cells are defined as self-renewing cells that produce myogenic progeny, arise early in development, and persist throughout life. From a practical standpoint, it has been important to isolate these cells because transplantation of stem cells can be an effective therapy for many diseases, including muscular dystrophies and nonmuscle diseases [1,2]. The stem cells seem to survive transplantation much better than other muscle cells, including early myoblasts [3–5]. These stem cells have been shown to be pluripotent and are characteristically distinguishable from myoblasts. Traditionally, it has been difficult to identify muscle stem cells because there have been no definitive markers of these cells. However, recent studies have led to several ways of enriching and identifying the stem cells from other myogenic cells. Because behavioral and morphological characteristics of stem cells are similar to those of myoblasts, the two distinct cell-types have often been confused in the literature as well as in experiments.

In this chapter, we review general characteristics of myoblasts and muscle stem cells, discuss methods currently available to enrich and identify muscle stem cells, and briefly review the possible applications of these cells in disease states.

MYOBLASTS

Based on the recent discovery that muscle-derived cells are very heterogeneous, it is likely that the nomenclature of myogenic cells may change over the next few years. The current definition of myoblasts is muscle-derived cells that can fuse with each other to form multinucleated myofibers. However, in-depth molecular characterizations have led to the identification of distinct subpopulations of "myoblasts" that do not necessarily fit into the traditional definition. Myoblasts exist as mononucleated cells, and they appear to have the capacity to proliferate in early stages. However, at later stages they are often observed as nonmitotic and withdrawn from the cell cycle. When isolated as a group, they appear to be in heterogenic stages of differentiation. They have distinct patterns of gene expression, leading to maturation and formation of myofibers.

Myoblasts in general do not express skeletal muscle myosins, but they do express desmin and muscle regulatory factors [6–8]. These regulatory factors may be one of the following: *Myf-5*, *MyoD*, myogenin, and *MRF4*. Typically, *Myf-5* and *MyoD* are expressed early, and myogenin and *MRF4* are expressed in later stages of maturation [9–11]. This sequence of gene expression defines the life cycle of myoblasts. Those that lack expression of myogenin and *MRF4* are referred to as early myoblasts, and those that express myogenin/*MRF4* are late myoblasts. Expression of myogenin seems to coincide with myoblast withdrawal from the cell cycle and its commitment to terminal differentiation.

Myoblast subpopulations exist as experimentally distinct types. Clonal analysis was [12a] first used to identify the diverse nature of myoblasts. Individual mononucleated cells from muscle were isolated and the types of progeny were determined. Some cells were identified to form multinucleate myotubes and others remained mononucleated [12b]. Additional subcloning experiments have shown that there are multiple types of chicken-derived myoblasts that are committed to form either slow or fast muscle types, independent of innervations [13,14]. This commitment to a distinct lineage is stable through multiple generations and is also stable *in vivo* [15]. Therefore, the formation of fast or slow muscle types seems dependent on distinct types of myoblasts [16–18].

Mouse myoblast subpopulations are probably the best defined. There have been four major groups named according to location and time of appearance [19,20]. They are named somatic, embryonic, fetal, and satellite cells. Among these subgroups of myoblasts, satellite cells have recently been a focus of interest. Because of their unique phenotype both *in vivo* and *in vitro*, a detailed discussion of these cells is warranted.

SATELLITE CELLS

Satellite cells have acquired the particular interest of many scientists because they are persistent in adults and seem to be most resilient when transplanted into animals. They are named satellite cells because of their location: they are primarily found within the basal lamina adjacent to muscle fibers. There are distinct subtypes of satellite cells that form different types of myofibers [21,22]. Those that lie adjacent to slow fibers express slow myosin, and those adjacent to fast fibers do not express slow myosin [21]. The majority of satellite cells remain committed to the type-specific lineage and do not interconvert to other phenotypes [22].

When cultured in a monolayer system, subpopulations of satellite cells remain mononucleated and desmin positive, dividing to produce fusing and nonfusing myoblasts [23]. When transplanted into a host muscle, satellite cells can either fuse with host myofibers or fuse together into myotubes, which will consequently differentiate into muscle fibers [4,5,24–45]. However, the vast majority of transplanted satellite cells die rapidly following the injection [4,5,31,36,42], a result that may be partly related to inflammatory reactions [43]. Indeed, an improvement of cell survival was achieved by blocking inflammation, but a dramatic loss of injected cells was still observed [5,43].

It has been recently observed that a small minority of satellite cells entirely survives postimplantation [4,5]. Although the mechanism by which these specific muscle-derived cells display a high cell survival is unclear, their ability to rapidly fuse with various types of host myofiber may help to improve their survivability in skeletal muscle [5]. In fact, it was observed in 2000 that the fusion of myoblasts with host myofibers posttransplantation is muscle fiber dependent [44,45]. These results suggest that myoblasts fuse with myofibers expressing the same type of myosin heavy chain (MyHC); consequently, matching host muscle and donor myoblasts for MyHCs improves myoblast transfer therapy [44,45]. Alternatively, the specific muscle-derived cells, which can fuse with myofibers of both types, can also be used to improve the efficiency of myoblast transplantation [44,45].

It has been hypothesized that the same subpopulation of satellite cells has the capacity to differentiate into other mesenchymal lineages such as bone [46–49]. Whether this small subpopulation of satellite cells represents muscle stem cells remains to be seen. However, it is clear that this subpopulation of satellite cells is phenotypically different from myoblasts and satellite cells and should be targeted for identification, purification, and characterization of muscle stem cells.

MUSCLE STEM CELLS

In general, stem cells are defined as immature cells that are capable of self-renewal as well as production of progeny that can undergo terminal differentiation. Muscle stem cells are those that do not express markers of mature muscle tissue, but are capable of producing new muscle stem cells and fusing progeny that form myoblasts and myofibers [8,50]. Myoblasts, prior to expression of myogenin, initially appear to fit the definition of a muscle stem cell. However, these early myoblasts do not have sufficient proliferative capacity to produce large numbers of progeny [8,51]. Therefore, we must look at muscle cells even more immature than early myoblasts to identify a true population of muscle stem cells. Recent functional and gene expression data seem to indicate that muscle stem cells are distinct from early myoblasts.

Earlier subcloning experiments attempted to identify cells that remain mononucleate and do not differentiate into myofibers. These studies indicated that cells that are capable of forming colonies were minorities in comparison to the total number of mononucleate cells [8,23,52]. The data indicated that the number of muscle stem cells in a given muscle tissue is very small in comparison to cells that are already committed to becoming differentiated myofibers [23]. In this case, a simple functional assay for muscle stem cells would be to identify muscle-derived cells that are capable of forming colonies. These colonies should be able to give rise to a heterogeneous population of progeny cells at all stages of myogenesis. By definition, then, a pure population of muscle stem cells is difficult to obtain because these cells would rapidly differentiate to form a mix of cells at various levels of differentiation.

As discussed earlier, multiple recent reports have suggested that a subpopulation of satellite cells may be muscle stem cells. Transplantation experiments with satellite cells [4,5] suggest that new myofibers formed in the host are formed by a minority of satellite cells. The majority of the transplanted satellite cells die shortly after implantation. One explanation of the data is that only a minority of satellite cells are true muscle stem cells that can proliferate and produce large quantities of progeny [4,5,49].

Gene expression of satellite cells has been extensively studied [9–11,53,54]. The satellite cells do not express any muscle regulatory factors unless they are activated by injury or in a culture system [7,11,53]. Satellite cells reside in a quiescent state within the basal lamina adjacent to a myofiber. When muscle is injured, however, the satellite cells are activated and they begin to proliferate, expressing a series of muscle regulatory factors that allow terminal differentiation [8,54]. Therefore, the focus has been to identify genes that are expressed prior to the activation of satellite cells in the hope of isolating a specific marker for muscle stem cells.

There are several proteins identified in unactivated satellite cells. These include myocyte nuclear factor alpha (MNF-α), c-met, and Bcl-2 [11,55–57]. MNF-α is a nuclear protein in quiescent satellite cells [55]. With activation, however, it is expressed in conjunction with muscle regulatory factors and is believed to be involved in maintaining the myogenic capacity of satellite cells [55]. The cell–cell signaling receptor c-met is believed to be required for myogenic migration [58]. Bcl-2 has been identified in approximately 1–4% of mononucleate cells in muscle colonies [56]. This expression seems to be limited to mononucleate cells, and the protein is not coexpressed with late markers of myogenesis [56]. Bcl-2, as inhibitor of apoptosis, is expressed in many different tissues. The data seems to suggest that Bcl-2 is expressed in progenitor myoblasts, but it would be difficult to use Bcl-2 as a marker for muscle stem cells because of its ubiquity. Expression of c-met and MNF-α, however, may be useful in identifying early myogenic progenitor cells.

Other recently emerging markers include Sca-1, KDR, and CD34 [59,60]. A 1999 *in vivo* study demonstrated that in human hematopoietic cells, the CD34$^+$KDR$^+$ population had the highest pluripotent characteristics [60]. Flk-1, a mouse homologue of KDR, was identified in a specific subpopulation of satellite cells that may have stem cell–like characteristics [49]. Sca-1 was also identified in repopulation studies in mice that indicated that Sca-1$^+$ mesenchymal stem cells are able to regenerate dystrophin-expressing myogenic cells in mdx mice [59]. Therefore, Sca-1 and KDR seem to be consistently able at least in mesenchymal and hematopoietic stem cells, to identify early progenitors. Whether this will also be true in muscle progenitors remains to be explored. CD34 has traditionally been used to identify hematopoietic progenitor cells. It has also been used in mesenchymal stem cells to identify pluripotent populations. However, studies suggest that expression of CD34 may be reversible and too unreliable for this purpose [61]. CD34 is probably a marker of activated stem cells, but it is not necessarily expressed in all stem cells. Although the reversibility of CD34 expression is unknown in muscle-derived stem cells, the use of CD34 as a marker of muscle-derived stem cells should at least be used with caution.

In summary, the marker studies and functional assays indicate that myogenic cells are heterogenic. Furthermore, a minority of satellite cells seems to harbor defining characteristics of a muscle stem cell. For this reason, much recent research has focused on isolation and purification of satellite cells, and further defining the subpopulation of satellite cells with stem cell–like properties.

METHODS OF PURIFICATION OF SATELLITE CELLS

Numerous isolation methods of muscle cells have been published. However, in our laboratory, we have used a serial technique of enzymatic digestion, mechanical separation, and "preplating" to fractionate myogenic cells from other muscle-derived cells. We initially used desmin as a marker for enrichment of myogenic cells. Each fraction then was characterized by using multiple different markers, including M-cadherin, for satellite cells, and Sca-1/Flk-1/Bcl-2 for muscle stem cells. This technique was initially described by Rando and Blau [62]. We have tried the "preplating" technique on human, mouse, and rabbit muscle cells. This technique does not seem to give consistent results in human muscle cells. The contents of human muscle tissue seem to be variable according to the age of the patient from whom it was isolated (unpublished results). Also, each fraction seems to be much more heterogenic than mouse and rabbit cells. However, for mice and rabbits the following technique has consistently yielded good fractions of desmin-positive cells. In mouse cells, this technique also yields fractions enriched in satellite cells and muscle stem cells.

ISOLATION OF MUSCLE-DERIVED CELLS

The muscles from a muscle biopsy (approximately 1 g from hind limb for mouse, thigh for rabbit) were dissected free of bone and minced into slurry. This slurry was then digested by serial 1-h incubation at 37°C in 0.2% type XI collagenase, dispase (grade II, 240 U), and 0.1% trypsin. The cell suspension was passed through 18-, 20-, and 22-gauge needles and centrifuged at 3000 rpm for 5 min. Cells were then suspended in the growth medium (DMEM supplemented with 10% fetal bovine serum, 10% horse serum, 0.5% chick embryo extract, and 2% penicillin–streptomycin).

PREPLATING FOR ISOLATION OF DIFFERENT CELL POPULATIONS

The isolated mixture of cells is plated in collagen-coated flasks [5,62]. After approximately an hour, the supernatant is withdrawn from the flask and replated in a fresh collagen-coated flask (Fig. 36.1). The cells that adhered rapidly within this one-hour incubation were mostly fibroblasts [5,44,45,62]. The serial replating of the supernatant was repeated when 30–40% of the cells had adhered to each flask. Each preplating stem enriches the desmin-positive cell fraction (Fig. 36.2). By the fifth to sixth serial plating, the culture is enriched with small, round cells [4,44,45,49]. These cells are usually 80–95% desmin positive, indicating myogenic origin [5,44,45,49].

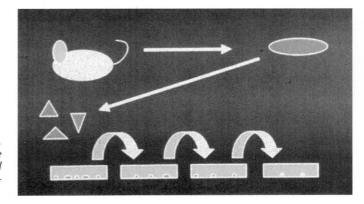

Fig. 36.1. Schematic diagram of the preplating technique. A muscle biopsy is performed from an animal and the tissue is enzymatically digested to yield a mixture of cells. The serial plating at specific time points allows fractionation of the muscle-derived cells.

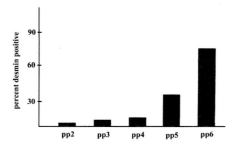

Fig. 36.2. Preplating technique concentrates desmin-positive myogenic cells as the number of plating increases.

All muscle-derived cells are grown in a 20% serum concentration media (10% fetal bovine serum, 10% horse serum). The high concentration of the serum keeps cells from fusing and differentiating into myofibers [5]. We have noted that when the serum concentration drops below 5%, the mouse myoblasts readily fuse to form myofibers (unpublished data).

Most importantly, the late preplate population should be kept at less than 60% confluency. Higher confluence will allow these cells easier cell–cell contact and trigger them to terminally differentiate [5,44,45].

FURTHER PURIFICATION OF MUSCLE STEM CELLS BY CLONAL ISOLATION

Clonal isolation of muscle stem cells can be accomplished from the late preplate cells. Either a limited dilution method or direct visualization of cells can be used to isolate a clonal population. It is extremely difficult to maintain a homogeneous population of cells. These stem cells have enormous propensity to differentiate, and most often a heterogeneous population of myogenic cells at all different stages of differentiation will result. Therefore, if clonal isolation is done, the cells should be characterized at the earliest possible passages. We have used limited dilution method for isolation of colonies of muscle-derived cells from the sixth serial preplate with good results [49]. We always analyze the cells at either first or second passage.

CHARACTERIZE PREPLATE FRACTIONS USING VARIOUS MARKERS

As shown in Table 36.1, late preplate fractions express multiple myogenic markers, including desmin+, myoD+/−, and myogenin−/+. In addition, analysis by means of the reverse transcriptase polymer chain reaction showed that the late preplate cells were positive for c-*met* and *MNF*, two genes that are expressed at an early stage of myogenesis [8]. A smaller percentage of late preplate cells express m-cadherin (−/+), a satellite cell-specific marker [63],

Table 36.1. Immunochemical and RT-PCR Analysis of Late Preplate Cells and Fibroblasts*a*

	Late Preplates		Fibroblasts	
	Immuno	**RT PCR**	**Immuno**	**RT PCR**
Decmin	+	nd	−	nd
CD34	+	+	−	−
Bcl-2	+	nd	−	nd
FLK-1	+	nd	−	nd
Sca-1	+	nd	−	nd
M-Cadherin	−/+	nd	−	nd
Myogenin	−/+	+	−	−
C-met	nd	+	nd	−
MNF	nd	+	nd	−
MyoD	+/−	+	nd	−

*a*Key: −, <2%; +, >95%; +/−, 40–80%; −/+, 5-30%; nd, not determined.

but a higher percentage of cells express Bcl-2, a marker limited to cells in the early stages of myogenesis [57], and CD34, a marker identified in human hematopoietic progenitor cells as well as stromal cell precursors in bone marrow [64–67]. The late preplate cells also tend to be highly positive for the expression of Flk-1, a mouse homologue of the human KDR gene, which was identified in 1999 as a marker of hematopoietic cells with stem cell–like characteristics [60]. Similarly, these cells were also found positive for Sca-1, a marker present in subpopulations of both skeletal muscle and hematopoietic cells with stem cell–like characteristics [59]. Finally, these cells are also CD45 and c-*kit* negative: see Table 36.1 Lee *et al.* [49].

The salient features of these results are as follows. First, it is a very effective method of fractionating myogenic cells from other contaminating cells such as fibroblasts. This is indicated by a higher percentage of desmin-positive cells with higher preplating. Second, it may also be an effective method of fractionating muscle stem cells from satellite cells. As the number of preplating increases, the M-cadherin-positive satellite cells are left behind, and the concentration of Sca-1+/Flk-1+/CD34+ cells increase. This may indicate an effective method of enriching muscle stem cell populations. Third, as the preplating number increases, the total number of cells isolated dramatically decreases. This behavior would support earlier evidence that the number of muscle stem cells in a given muscle tissue is very small.

CELL PURIFICATION BY FLOW CYTOMETRY ANALYSIS

Flow cytometrics is another possible way to sort and purify these cells. Until recently, specific cell surface markers were not available for this purpose. With recent advancements in identifying myoblast progenitors earlier in lineage than satellite cells, it may now be possible to sort these cells by fluorescence-activated techniques (FACS). An obvious target of cell surface markers to be used consists of CD34, Bcl-2, Sca-1, and KDR or Flk-1. Use of such methods has been reported in hematopoietic cells to isolate pluripotent stem cells from bone marrow [59,60]. One report indicated that a complete restoration of the hematopoietic compartment was possible in sublethally irradiated mice when they were intravenously injected with Sca-1-enriched marrow cells [59]. More surprisingly, a partial restoration of dystrophin expression was possible in *mdx* mice when the similar population of FACS sorted cells was used [59]. This report not only highlighted the possibility of treating muscular disorders by using systemic delivery of mesenchymal stem cells, but it also hinted at a possible relationship between muscle-derived stem cells and pluripotent mesenchymal stem cells that are derived from bone marrow.

ARE MUSCLE-DERIVED STEM CELLS PLURIPOTENT?

For many years now, it has been known that nonmyogenic cells can become muscle cells under certain external stimulations [68]. For example, dermal fibroblasts were one of the first nonmyogenic cells shown to convert to the myogenic phenotype. They were initially shown to fuse with myotubes *in vitro* to form a mosaic system [69]. It was subsequently shown that dermal fibroblasts could form dystrophin-expressing myotubes and myofibers in *mdx* mice [70]. Others have subsequently reproduced the data both *in vitro* and *in vivo* [71]. Other examples of nonmyogenic cells differentiating into muscle include bone marrow stromal cells [59,72,73], neural epithelial cells, and chicken pregastrulation epiblast cells [74]. This data indicates that possible early progenitors exist in multiple tissues, and these early progenitors may be able to interconvert to different mesenchymal tissues. This would indicate that muscle stem cells also have the capacity to differentiate into tissues other than skeletal muscle.

Marshall Urist [75] introduced the concept of the existence of osteoprogenitors in muscle tissue when he implanted demineralized bone matrix into skeletal muscle and formed bone. He subsequently isolated compounds now known as bone morphogenetic proteins (BMPs), which were responsible for the osteoinductive effect. However, he did not delineate which muscle-derived cells were responsible for this ectopic bone formation. Katagiri *et al.* [76] made an important link when they reported that a myogenic cell line, C2C12, can become osteogenic upon treatment with BMP-2. Subsequently, C2C12 was shown to become adipogenic when treated with long chain fatty acids or thiazolidine-diones [77]. Thus, a subpopulation of myogenic cells seems to be capable of differentiating into multiple lineages. Recently we have identified that a subpopulation of mouse satellite cells fractionated by the preplating technique have osteogenic potential [46–48]. These late preplates were shown to express high levels of alkaline phosphatase, an osteogenic marker, in response to BMP-2 stimulation [46]. A clonal analysis from the mouse late preplate cells has shown that a pure population of osteoprogenitors can be obtained from muscle [49]. A majority of these cells seem to already express myogenic markers, such as desmin, but they readily lose these markers and begin expressing osteogenic markers such as osteocalcin and alkaline phosphatase in response to BMP-2 [46,49]. Similarly, human muscle cells also seem to harbor osteoprogenitor cells that can express high levels of alkaline phosphatase in response to BMP-2 [47,48]. Unfortunately, no further data are available in human cells. Another recent report used a clonal population of muscle-derived stem cells to heal a non-healing bone defect with the help of BMP-2 [49]. No muscle cell has been shown to differentiate into nonmuscle lineages without an external growth factor stimulus. However, this data indicates that at least a subpopulation of muscle stem cells have the potential for multilineage differentiation. Thus, muscle tissue holds a promising potential for use in tissue engineering. Skeletal muscle is an enormous reservoir for cells, and if these pluripotent stem cells can be efficiently harvested, it would have a tremendous impact in the clinical realm.

TISSUE ENGINEERING APPROACHES BASED ON MUSCLE-DERIVED STEM CELLS

INHERITED DISEASES

The most obvious diseases for use of muscle stem cells are the muscular dystrophies. Muscular dystrophies are characterized by progressive muscle wasting and weakness. Duchenne muscular dystrophy (DMD), which is X-linked, is one of the most common and severe inherited myopathies. It is characterized by a lack of dystrophin expression in the sarcolemma of muscle fibers [78–81]. Dystrophin functions to maintain the muscle membrane integrity. Absence of this protein in DMD patients results in damage to the membrane during muscle contraction, leading to muscle necrosis [82–84]. Currently, there is no treatment for this disease. The goal of therapy for DMP is to restore dystrophin to diseased muscles, restructuring the integrity of the membranes to allow normal function and consequently alleviate the characteristic muscle weakness.

Various approaches have been explored to deliver dystrophin into skeletal muscle. These methods include myoblast transplantation and gene delivery based on nonviral and viral vectors.

Myoblast Transplantation

Myoblast transplantation uses normal myoblasts as vehicles for delivery of dystrophin [24,25,85–92]. This method, however, has been hindered by immune response, poor dispersion of injected myoblasts, and rapid loss of the injected cells [5,26,27,32–35, 38,40,41,43].

In this regard, purified muscle stem cells would be perfect candidates for transplantation. As stated earlier, subpopulations of satellite cells seem to be able to survive much better than other myoblasts when implanted into mice [4,5]. If these cells represent stem cells, and if they can be harvested in an efficient manner, they may dramatically improve dystrophin delivery to diseased muscles. In addition, a recent report that hematopoietic stem cells can partially reconstitute dystrophin in *mdx* mice is very encouraging [59]. These hematopoietic stem cells were delivered systemically by a tail vein injection in mice. For the first time, this makes possible treatment of "uninjectable" muscles, such as the diaphragm. We have also reported that the injection of a clonal population of muscle-derived stem cells is capable of being systemically delivered throughout the bloodstream and partially restoring dystrophin in *mdx* muscle [49]. Whether muscle-derived stem cells can also be injected systemically to treat such difficult-to-reach locations remains to be seen.

Gene Therapy

Virus-mediated gene delivery may be a promising method for delivery of dystrophin. However, current vectors are limited by various technical problems. Direct injection of retroviral vectors does not efficiently transduce muscle fibers, since muscle fibers do not actively divide [93]. A low-level retroviral transduction occurs in immature and adult regenerating muscles owing to regenerating muscles, possibly through myoblasts [93,94]. Adenoviral vectors have been extensively studied as gene delivery vehicles to skeletal muscle [95–100]. However, the adenovirus displays a reduced efficiency of transduction in mature myofibers that is due to inefficient penetration of vectors into basal lamina [101,102]. Also, first-generation adenoviral vectors induce immune responses, which leads to the rejection of transduced cells [7,103–107]. Herpes simplex virus has also been investigated. This vector also is limited by differential transducibility of mature and immature fibers, cytotoxicity, and immunological problems [108,109].

One way to circumvent these problems is to devise an *ex vivo* method of gene transfer [36,46–48,110–112]. *Ex vivo* methods involve isolation of desired cells from a patient, expansion of these cells in a laboratory, transduction of the cells *in vitro*, and reimplantation into a desired site. This method has been successfully used to deliver dystrophin in dystrophic muscle [110] as well as to deliver BMP-2 via adenoviral vector and consequently heal a segmental bone defect [112]. In humans, this method was used in 1999 in a clinical trial of arthritis using the interleukin 1 receptor antagonist protein [113]. The advantages of this method are severalfold. First, you can increase the transduction efficiency, thus increasing the amount of desired protein expression. Also, you have the option of using a selection marker (such as neomycin) to select the cells after transduction, to obtain even higher protein expression. In addition, you have the option of choosing the type of cells to transduce. This is an added advantage of the *ex vivo* method that allows an enormous flexibility. For example, muscle stem cells would be an ideal choice for *ex vivo* gene therapy for muscular dystrophies. Since they have the capacity to survive implantation longer than any other myogenic cells, if they are genetically engineered with anti-inflammatory substances, their survival may be even more prolonged. In this regard, myoblasts engineered to express interleukin 1 receptor antagonist protein have been shown to be capable of longer survival [5].

REPAIR OF INJURED MUSCLES

Muscle contusion, strain, ischemia, and laceration are common recreational and traumatic injuries. It has been shown that at the time of injury, satellite cells are released from their native positions and are activated. They begin to differentiate into myotubes and myofibers to facilitate muscle healing. Although very efficient, this process can be incomplete, mostly as a result of fibrous tissue formation. This development of scar formation is often a limitation to complete recovery [114,115].

There are numerous animal models of muscle injuries [116–123]. Our muscle laceration model involves incision of 75% of the width and 50% of the thickness of the gastrocnemius muscle in mice [124,125]. Our contusion model is also in mice, and it involves dropping a 16-g iron ball from a height of 100 cm onto the gastrocnemius muscle [126]. These are highly reproducible models with distinct reparative processes identifiable by histology. We have observed muscle myofiber regeneration at 7–10 days of injury, but a dramatic decrease of reparative activity at 14 days. This decline continues until 35 days [124–126]. Appearance of fibrosis coincides with the decline of muscle regeneration and seems to hinder the healing process.

Both muscle stem cell transplantation and *ex vivo* gene therapy are excellent candidates for treatment of muscle injuries. The injured muscle releases many growth factors that may be involved in modulating the sequence of muscle healing. These proteins act to stimulate satellite cells to release, proliferate, and terminally differentiate [90,115,127,128]. Delivery of growth factors by *ex vivo* methods is a logical step in augmenting muscle healing. Several growth factors that can enhance myoblast proliferation have been identified. Basic fibroblast growth factor (bFGF), insulin-like growth factor 1 (IGF-1), and nerve growth factor (NGF) were shown to significantly enhance myoblast proliferation. Acidic fibroblast growth factor (aFGF), bFGF, IGF-1, and NGF were shown to facilitate myoblast differentiation into myotubes. The data seems to indicate that bFGF, IGF-1, and NGF are logical candidates for *ex vivo* delivery [124,126].

Muscle stem cells, again, are perfect vehicles for *ex vivo* gene delivery in muscle injuries. The isolated muscle stem cells are transduced *in vitro*, with the desired gene carrying the vector and reinjected into the host. These stem cells then fuse to form postmitotic myotubes and myofibers and begin producing growth factors. These stem cells should also be able to proliferate and produce progeny that can replenish the satellite cell population. Therefore, these stem cells should repopulate the lost myogenic cells as well as deliver growth factors capable of enhancing the healing process. This technique has been tried with adenoviral [36], retroviral [129], and herpes simplex viral vectors [111]. Myoblast transplantation was shown to improve muscle regeneration in muscle experimentally injured with myonecrotic agents [35]. The nature of muscle injuries is that they are usually localized, allowing direct injection of these cells into the site of injury. If this technique can be perfected, the transplantation of genetically engineered muscle stem cells holds great promise for treatment of muscle injuries.

Bone Healing

When Urist [75] initially described and isolated [130] bone morphogenetic protein, he invoked the idea of healing bone by molecular biological means. Currently, methods of treating segmental bone defects and nonunions with the help of BMPs are fervently being investigated. Among these methods is the use of gene therapy to deliver BMPs. Retrovirus, adenovirus, and adeno-associated virus have been proposed and used in experimental settings to deliver BMP *in vitro* and *in vivo* [131]. As described earlier, direct use of these vectors has many disadvantages. For example, retroviral vectors require actively dividing cells for integration of the foreign gene, and they incorporate randomly into the host DNA, possibly transforming proto-oncogenes to oncogenes [131]. The first-generation adenovirus seems to induce host immune response that may decrease the efficiency of infection [107].

Ex vivo gene delivery should be a method of choice for augmentation of bone healing. Any population of cells that is easily isolated from a patient can potentially be used as a delivery vehicle for BMP. Recently, identification of stem cells in both bone marrow and skeletal muscle has raised the question of whether these cells can be used as a vehicle for BMP delivery. As described in earlier sections, these stem cells have been shown to differentiate into osteogenic cells upon stimulation with BMP [46–48,76]. Thus, these populations of cells can both act as carriers of BMP and participate in new bone formation by differentiating into osteogenic cells. Primary muscle-derived cells are easy to obtain via muscle biopsy and are simple to culture/cryopreserve in a laboratory setting [46–48]. Muscle biopsy is an easy procedure to do, especially in trauma settings, where nonhealing and segmental defects most often occur. Therefore, primary muscle-derived cells may be ideal vehicles for gene delivery to bone by *ex vivo* methods.

Recently, bone marrow stromal cells were used with *ex vivo* gene therapy to heal segmental bone defects in syngeneic rats [112]. In our laboratory, we have examined whether primary muscle-derived cells (a heterogenic population) as well as purified muscle stem cells can augment bone healing. Using a critical-sized defect model in mice, we were able to show that a heterogenic population of primary muscle cells transduced with adenovirus to produce BMP-2 can significantly augment bone healing. We observed that a small subpopulation of these cells was able to differentiate into osteocytes *in vivo* [49]. When a purified population of muscle stem cells was used, a significantly higher percentage of cells differentiated into osteocytes *in vivo* [49]. Therefore, it seems likely that muscle-derived stem cells can effectively deliver BMP-2 to diseased bone as well as differentiate into osteogenic cells to enhance bone healing.

Other Orthopedic Applications

Intra-articular joint diseases are causes of morbidity in a large percentage of the current population. As the population becomes older but remains active, joint injuries including arthritis, ligamental tears, osteochondral injuries, and meniscal tears will remain a significant load to health care costs. Currently, management of these injuries is either conservative (rest and immobilization in hopes of healing) or surgical (replacement or repair of the damaged joint). The conservative treatments are usually unsatisfactory. Surgical managements include reconstruction of ligaments, total joint replacements, and cartilage debridements, all of which are highly invasive. Tissue engineering applied to these orthopedic injuries offers a less invasive and more physiological repair process.

Direct and *ex vivo* gene delivery methods for arthritis have been investigated [113]. The *ex vivo* approaches have involved synovial cells [132], but a decline of gene expression was noted 5–6 weeks after transduction. This short-term expression may be due to various other factors, including fast turnover rates of synoviocytes. Muscle-derived stem cells may offer theoretical advantages to synovial cells. Stem cells have the capacity to form postmitotic myotubes and myofibers and may last longer than synovial cells. This may translate to longer expression of the protein of interest [133].

Ex vivo gene delivery to an intra-articular surface with muscle-derived cells has been attempted. Adenovirus carrying the marker gene β-galactosidase was used to transduce and implant autologous muscle cells onto intra-articular tissues in the rabbit knee. Tissues found to be expressing β-galactosidase included synovial lining, meniscal surface, and cruciate ligament. In contrast to this result, *ex vivo* gene transfer with synovial cells only resulted in expression of the reporter gene in synovial cells. This may be due to the propensity of myoblasts to fuse with many different cells. This may be an added advantage of using immature muscle cells for gene therapy to the joint, since longer expression may result.

Anterior cruciate ligament (ACL) is the second most frequently injured knee ligament. Because of its low metabolic turnover, however, the ACL does not heal. Complete tears of the ACL often require surgical intervention using auto- or allograft materials. Recent research has been directed at augmentation of substituted ligament with *ex vivo* gene delivery. *In vivo* data suggests that platelet-derived growth factor (PDGF), transforming growth factor β (TGF-β), and epidermal growth factor (EGF) promote ligament healing. Direct injection of these growth factors is unlikely to help because the high local concentration will quickly decline. *Ex vivo* gene delivery of these growth factors using muscle-derived stem cells, for the reasons discussed earlier, may also be feasible here. We have, in fact, demonstrated the feasibility of delivering marker gene in the ACL with muscle-based *ex vivo* gene transfer [134]. In some aspects, the ultimate goal should be to do away with surgical reconstruction all together, allowing primary healing to occur at the ends of the acutely torn ACL. For this to occur, sustained delivery of growth factors will definitely be crucial, since primary reanastomosis of the torn ACL has never been successful.

The meniscal injuries, which are primarily injuries of a young age group, ultimately lead to early arthritis if not treated properly. Also, the low vascularity and low turnover of the meniscus result in the failure of certain injuries to this structure to heal without surgical intervention. Because the meniscus normally functions as a load transmitter to the knee, compromise of this cartilage will ultimately lead to formation of an arthritic joint. Historically, the treatment includes menisectomy, meniscal repair, or meniscal transplant. It has been seen that growth factors such as PDGF are capable of enhancing meniscal healing [135]. Since sustained release of the protein will be necessary for the successful healing, the delivery vehicle for the growth factors will be crucial for this structure. Investigations of ways of using muscle stem cells to deliver marker genes and growth factors to rabbit meniscus are under way, and preliminary results show the feasibility of the myoblast-mediated *ex vivo* gene transfer in the meniscus [136]. Whether muscle stem cells will be capable of allowing the meniscus to form appropriate collagen to reestablish its structure without formation of fibrosis remains to be seen.

FUTURE DIRECTIONS

We have only scratched the surface of understanding the phenotype of muscle stem cells. Most of the research has been concentrated on their use and behavior in the myogenic lineage. Muscle stem cells' potential to differentiate into other mesenchymal lineages has only recently attracted the interest of scientists. We do not yet know whether these cells are capable of differentiating into hematopoietic, chondrogenic, adipogenic, or fibrocytic lineage. The functional assays for these cells will need to be refined to show that these cells are truly pluripotent.

Much more research is needed to isolate and purify muscle stem cells in an efficient manner. Current understanding of expression markers of these cells does not allow us to effectively purify these cells in quantities needed for clinical use. Use of cell sorters to identify and further purify these cells must be investigated. However, further characterization of these cells with the expression patterns of their surface markers will be necessary if we are to easily identify the stem cell population.

In the long run, molecular and biochemical understanding of stem cells, not just from muscle but from all different tissues in the body, will greatly expand the treatment options for currently incurable diseases. It will also improve the healing of various tissues of the musculoskeletal system.

Acknowledgments

The authors thank Marcelle Pellerin and Ryan Pruchnic for their technical assistance. This work was supported in part by grants to JH from the National Institutes of Health (1 P60 AR44811-01, 1PO1 AR45925-01), the Pittsburgh Tissue Engineering Initiative (PTEI), and the William F. and Jean W. Donaldson Chair at Children's Hospital of Pittsburgh.

References

1. Miller, J. B., and Boyce, F. M. (1995). Gene therapy by and for muscle cells. *Trends Genet.* **11**, 163–165.
2. Blau, H. M., and Springer, M. L. (1995). Muscle-mediated gene therapy. *N. Engl. J. Med.* **333**, 1554–1556.
3. Partridge, T. A., and Davies, K. E. (1995). Myoblast-based gene therapies. *Br. Med. Bull.* **51**, 123–137.
4. Beauchamp, J. R., Pagel, C. N., and Partridge, T. A. (1997). A dual-marker system for quantitative studies of myoblast transplantation in the mouse. *Transplantation* **63**, 1794–1797.
5. Qu, Z., Balkir, L., van Deutekom, J. C., Robbins, P. D., Pruchnic, R., and Huard, J. (1998). Development of approaches to improve cell survival in myoblast transfer therapy. *J. Cell Biol.* **14**, 1257–1267.
6. George-Weinstein, M., Foster, R. R., Gerhart, J. V., and Kaufman, S. J. (1993). *In vitro* and *in vivo* expression of alpha 7 integrin and desmin define the primary and secondary myogenic lineages. *Dev. Biol.* **156**, 209–229.
7. Smith, T. A. G., Mehaffey, M. G., Kayda, D. B., Saunders, J. M., Yei, S., Trapnell, B. C., McClell, A., and Kaleko, M. (1993). Adenovirus mediated expression of therapeutic plasma levels of human factor IX in mice. *Nat. Genet.* **5**, 397–402.
8. Miller, J. B., Schaefer, L., and Dominov, J. A. (1999). Seeking muscle stem cells. *Curr. Top. Dev. Biol.* **43**, 191–219.
9. Smith, T. H., Block, N. E., Rhodes, S. J., Konieczny, S. F., and Miller, J. B. (1993). A unique pattern of expression of the four muscle regulatory factors distinguishes somatic from embryonic, fetal, and newborn mouse myogenic cells. *Development (Cambridge, UK)* **117**, 1125–1133.
10. Smith, T. H., Kachinsky, A. M., and Miller, J. B. (1994). Somite subdomains, muscle cell origins, and the four muscle regulatory factor proteins. *J. Cell Biol.* **127**, 95–105.
11. Cornelison, D. D., and Wold, B. J. (1997). Single-cell analysis of regulatory gene expression in quiescent and activated mouse skeletal muscle satellite cells. *Dev. Biol.* **191**, 270–283.
12. (a) Konigsberg, I. R. (1963). Clonal analysis of myogenesis. *Science* **140**, 1273–1284.
    (b) Law, P. K., Goodwin, T. G., and Wang, M. G. (1988). Normal myoblast injections provide genetic treatment for murine dystrophy. *Muscle Nerve* **11**, 525–533.
13. Miller, J. B., and Stockdale, F. E. (1987). What muscle cells know that nerves don't tell them. *Trends Neurosci.* **10**, 325–329.
14. Feldman, J. L., and Stockdale, F. E. (1991). Skeletal muscle satellite cell diversity: Satellite cells form fibers of different types in cell culture. *Dev. Biol.* **143**, 320–334.
15. DiMario, J. X., Fernyak, S. E., and Stockdale, F. E. (1993). Myoblasts transferred to the limbs of embryos are committed to specific fiber fates. *Nature (London)* **362**, 165–167.
16. Butler, J., Cosmos, E., and Brierly, J. (1982). Differentiation of muscle fiber types in aneurogenic brachial muscle of the chick embryo. *J. Exp. Zool.* **224**, 65–80.
17. Crow, M. T., and Stockdale, F. E. (1986). Myosin expression and specialization of the earliest muscle fibers of the developing avian limb. *Dev. Biol.* **113**, 238–254.
18. Condon, K., Siberstein, L., Blau, H., and Thompson, W. J. (1990). Differentiation of fiber types in aneural musculature of the prenatal rat hindlimb. *Dev. Biol.* **138**, 275–295.
19. Cossu, G., and Molinaro, M. (1987). Cell heterogeneity in the myogenic lineage. *Curr. Top. Dev. Biol.* **23**, 185–208.
20. Miller, J. B. (1992). Myoblast diversity in skeletal myogenesis: How much and to what end? *Cell (Cambridge, Mass.)* **69**, 1–3.
21. Rosenblatt, J. D., Parry, D. J., and Partridge, T. A. (1996). Phenotype of adult mouse muscle myoblasts reflects their fiber type of origin. *Differentiation (Berlin)* **60**, 39–45.

22. Barjot, C., Rouanet, P., Vigneron, P., Janmot, C., d'Albis, A., and Bacou, F. (1998). Transformation of slow or fast twitch rabbit muscles after cross-reinnervation of low frequency stimulation does not alter the *in vitro* properties of their satellite cells. *J. Muscle Res. Cell Motil.* **19**, 25–32.

23. Baroffio, A., Hamann, M., Bernheim, L., Bochatonpiallat, M. L., Gabbiani, G., and Bader, C. R. (1996). Identification of self-renewing myoblasts in the progeny of single human muscle satellite cells. *Differentiation (Berlin)* **60**, 47–57.

24. Partridge, T. A. (1991). Myoblast transfer: A possible therapy for inherited myopathies. *Muscle Nerve* **14**, 197–212.

25. Karpati, G., Pouliot, Y., Zubrzycka-Gaarn, E., Carpenter, S., Ray, P. N., Worton, R. G., and Holland, P. (1989). Dystrophin is expressed in *mdx* skeletal muscle fibers after normal myoblast implantation. *Am. J. Pathol.* **135**, 27–32.

26. Karpati, G., Holland, P., and Worton, R. G. (1992). Myoblast transfer in DMD: Problems and interpretation of efficiency. *Muscle Nerve* **15**, 1209–1210.

27. Tremblay, J. P., Malouin, F., Huard, J., Bouchard, J. P., Satoh, A., and Richards, C. L. (1993). Results of a blind clinical study of myoblast transplantations without immunosuppressive treatment in young boys with Duchenne Muscular Dystrophy. *Cell Transplant.* **2**, 99–112.

28. Morgan, J. E., Watt, D. J., Slopper, J. C., and Partridge, T. A. (1988). Partial correction of an inherited defect of skeletal muscle by graft of normal muscle precursor cells. *J. Neurol. Sci.* **86**, 137–147.

29. Morgan, J. E., Hoffman, E. P., and Partridge, T. A. (1990). Normal myogenic cells from newborn mice restore normal histology to degenerating muscle of the mdx mouse. *J. Cell Biol.* **111**, 2437–2449.

30. Morgan, J. E., Pagel, C. N., Sherat, T., and Partridge, T. A. (1993). Long-term persistence and migration of myogenic cells injected into pre-irradiated muscle of mdx mice. *J. Neurol. Sci.* **115**, 191–200.

31. Fan, Y., Maley, M., Beilharz, M., and Grounds, M. (1996). Rapid death of injected myoblasts in myoblast transfer therapy. *Muscle Nerve* **19**, 853–860.

32. Huard, J., Bouchard, J. P., Roy, R., Malouin, F., Dansereau, G., Labrecque, C., Albert, N., Richards, C. L., Lemieux, B., and Tremblay, J. P. (1992). Human myoblast transplantation: Preliminary results of 4 cases. *Muscle Nerve* **15**, 550–560.

33. Huard, J., Roy, R., Bouchard, J. P., Malouin, F., Richards, C. L., and Tremblay, J. P. (1992). Human myoblast transplantation between immunohistocompatible donors and recipients produces immune reactions. *Transplant. Proc.* **24**, 3049–3051.

34. Huard, J., Guerette, B., Verreault, S., Tremblay, G., Roy, R., Lille, S., and Tremblay, J. P. (1994). Human myoblast transplantation in immunodeficient and immunosuppressed mice: Evidence of rejection. *Muscle Nerve* **17**, 224–234.

35. Huard, J., Verreault, A., Roy, R., Tremblay, M., and Tremblay, J. P. (1994). High efficiency of muscle regeneration following human myoblast clone transplantation in SCID mice. *J. Clin. Invest.* **93**, 586–599.

36. Huard, J., Acsadi, G., Massie, B., and Karpati, G. (1994). Gene transfer to skeletal muscles by isogenic myoblasts. *Hum. Gene Ther.* **5**, 949–958.

37. Mendell, J. R., Kissel, J. T., Amato, A. A., King, W., Signore, L., Prior, T. W., Sahenk, Z., Benson, S., McAndrew, P. E., and Rice, R. (1995). Myoblast transfer in the treatment of Duchenne's muscular dystrophy. *N. Engl. J. Med.* **333**, 832–838.

38. Gussoni, E., Pavlath, P. K., Lancot, A. M., Sharma, K., Miller, R. G., Steinman, L., and Blau, R. M. (1992). Normal dystrophin transcripts detected in DMD patients after myoblast transplantation. *Nature (London)* **356**, 435–438.

39. Gussoni, E., Blau, H. M., and Kunkel, L. M. (1997). The fate of individual myoblasts after transplantation into muscles of DMD patients. *Nat. Med.* **3**, 970–977.

40. Kinoshita, I., Vilquin, J. T., Guerette, B., Asselin, I., Roy, R., and Tremblay, J. P. (1994). Very efficient myoblast allotransplantation in mice under FK506 immunosuppression. *Muscle Nerve* **17**, 1407–1415.

41. Vilquin, J. T., Wagner, E., Kinoshita, I., Roy, R., and Tremblay, J. P. (1995). Successful histocompatible myoblast transplantation in dystrophin-deficient mdx dystrophin. *J. Cell Biol.* **131**, 975–988.

42. Beauchamps, J. R., Morgan, J. E., Pagel, C. N., and Partridge, T. A. (1994). Quantitative studies of the efficacy of myoblast transplantation. *Muscle Nerve (Suppl.)* **261**, 18.

43. Guerette, B., Asselin, I., Skuk, D., Entman, M., and Tremblay, J. P. (1997). Control of inflammatory damage by anti-LFA-1: Increase success of myoblast transplantation. *Cell Transplant.* **6**(2), 101–107.

44. Qu, Z., and Huard, J. (2000). Matching host muscle and donor myoblasts for myosin heavy chain improves myoblast transfer therapy. *Gene Ther.* **7**, 428–437.

45. Qu, Z., and Huard, J. (2000). The influence of muscle fiber type in myoblast-mediated gene transfer to skeletal muscle. *Cell Transplant.* **9**, 503–517.

46. Bosch, P., Musgrave, D. S., Lee, J. Y., Cummins, J., Shuler, F., Ghivizzani, S. C., Evans, C., and Robbins, P. D. (2001). Osteoprogenitor cells within skeletal muscle. *J. Orthop. Res.* **18**, 933–944.

47. Musgrave, D. S., Bosch, P., Ghivizzani, S., Robbins, P. D., Evans, C. H., and Huard, J. (1999). Adenovirus-mediated direct gene therapy with bone morphogenetic protein-2 produces bone. *Bone* **24**, 541–547.

48. Musgrave, D. S., Bosch, P., Lee, J. Y., Pelinkovic, D., Ghivizzani, S., Whalen, J., Niyibizi, C., and Huard, J. (2000). *Ex vivo* gene therapy to produce bone using different cell types. *CORR* (accepted for publication).

49. Lee, J. Y., Qu-Petersen, Z., Cao, B., Kimura, S., Jankowski, R., Cummins, J., Usas, A., Gates, C., Robbins, P., Wernig, A., and Huard, J. (2000). Clonal isolation of muscle derived cells capable of enhancing muscle regeneration and bone healing. *J. Cell Biol.* **150**(5), 1085–1099.

50. Slack, J. M. W. (1991). "From Egg to Embryo," 2nd ed. Cambridge University Press, Cambridge, UK.

51. Nicolas, J. F., Mathis, L., and Bonnerot, C. (1996). Evidence in the mouse for self-renewing stem cells in the formation of a segmented longitudinal structure, the myotome. *Development (Cambridge, UK)* **122**, 2933–2946.

52. Rutz, R., and Hauschka, S. D. (1982). Clonal analysis of vertebrate myogenesis. VII. Heritability of muscle colony type through sequential subclonal passages *in vitro*. *Dev. Biol.* **91**, 399–411.

53. Maley, M. A., Fan, Y., Beilharz, M. W., and Grounds, M. D. (1994). Intrinsic differences in MyoD and myogenin expression between primary cultures of SJL/J and BALB/C skeletal muscle. *Exp. Cell Res.* **211**, 99–107.

54. Koishi, K., Zhang, M., McLennan, I. S., and Harris, A. J. (1995). MyoD protein accumulates in satellite cells and is neurally regulated in regenerating myotubes and skeletal muscle fibers. *Dev. Dyn.* **202**, 244–254.

55. Garry, D. J., Yang, Q., Bassel-Duby, R., and Williams, R. S. (1997). Persistent expression of MNF identifies myogenic stem cells in postnatal muscles. *Dev. Biol.* **188**, 280–294.

56. Miller, J. B., Dunn, J. J., and Dominov, J. A. (1997). Myogenic stem cells reside in a small sub-set of muscle cells that express Bcl-2 and are resistant to apoptosis. *Dev. Biol.* **186**, 252 (abstr.).

57. Dominov, J. A., Dunn, J. J., and Miller, J. B. (1998). Bcl-2 expression identifies an early stage of myogenesis and promotes clonal expansion of muscle cells. *J. Cell Biol.* **142**, 537–544.

58. Bladt, F., Riethmacher, D., Isenmann, S., Aguzzi, A., and Birchmeier, C. (1995). Essential role for the c-met receptor in the migration of myogenic precursor cells into the limb bud. *Nature (London)* **376**, 768–771.

59. Gussoni, E., Soneoka, Y., Strickland, C. D., Buzney, E. A., Khan, M. K., Flint, A. F., Kunkel, L. M., and Mulligan, R. C. (1999). Dystrophin expression in the mdx mouse restored by stem cell transplantation. *Nature (London)* **401**, 390–394.

60. Ziegler, B. L., Valtieri, M., Porada, G. A., De Maria, R., Muller, R., Masella, B., Gabbianelli, M., Casella, I., Pelosi, E., Bock, T., Zanjani, E. D., and Peschle, C. (1999). KDR receptor: A key marker defining hematopoietic stem cells. *Science* **285**, 1553–1558.

61. Goodell, M. A. (1999). CD34+ or CD34-: Does it really matter? *Blood* **94**, 2545–2547.

62. Rando, T. A., and Blau, H. M. (1994). Primary mouse myoblast purification, characterization, and transplantation for cell-mediated gene therapy. *J. Cell Biol.* **125**, 1275–1287.

63. Irintchev, A., Zeschnigk, M., Starzinski-Powitz, A., and Wernig, A. (1994). Expression pattern of m-cadherin in normal, denervated, and regenerating mouse muscle. *Dev. Dyn.* **199**, 326–337.

64. Andrews, R. G., Singer, J. W., and Bernstein, I. D. (1986). Monoclonal antibody 12-8 recognizes a 115-kd molecule present on both unipotent and multipotent hematopoietic colony-forming cells and their precursors. *Blood* **67**, 842–845.

65. Civin, C. I., Strauss, L. C., Brovall, C., Fackler, M. J., Schwartz, J. F., and Shaper, J. H. (1984). Antigenic analysis of hematopoiesis. III. A hematopoietic progenitor cell surface antigen defined by a monoclonal antibody raised against KG-1a cells. *J. Immunol.* **133**, 157–165.

66. Fina, L., Molgaard, H. V., Robertson, D., Bradley, N. J., Monaghan, P., Delia, D., Sutherland, D. R., Baker, M. A., and Greaves, M. F. (1990). Expression of the CD34 gene in vascular endothelial cells. *Blood* **75**, 2417–2426.

67. Simmons, P. J., and Torok-Storb, B. (1991). CD34 expression by stromal precursors in normal human adult bone marrow. *Blood* **78**, 2848–2853.

68. Cossu, G. (1997). Unorthodox myogenesis: Possible developmental significance and implications for tissue histogenesis and regeneration. *Histol. Histopathol.* **12**, 755–760.

69. Chaudhari, N., Delay, R., and Beam, K. G. (1989). Restoration of normal function in genetically defective myotube by spontaneous fusion with fibroblasts. *Nature (London)* **341**, 445–447.

70. Gibson, A. J., Karasinski, J., Relvas, J., Moss, J., Sheratt, T. G., Strong, P. N., and Watt, D. J. (1995). Dermal fibroblasts convert to a myogenic lineage in mouse muscle. *J. Cell Sci.* **108**, 207–214.

71. Salvatori, G., Lattanzi, L., Coletta, M., Aguanno, S., Vivarelli, E., Kelly, R., Ferrari, G., Harris, A., Mavilio, F., Molinary, M., and Cossu, G. (1995). Myogenic conversion of mammalian fibroblasts induced by differentiating muscle cells. *J. Cell Sci.* **108**, 2733–2739.

72. Ferrari, G., Cusella-De Engelis, G., Coletta, M., Paolucci, E., Stornaiuolo, A., Cossu, G., and Mavilio, F. (1998). Muscle regeneration by bone marrow-derived myogenic progenitors. *Science* **279**, 1528–1530.

73. Pereira, R. F., Halford, K. W., O'Hara, M. D., Leeper, D. B., Sokolov, B. P., Pollard, M. D., Bagasra, O., and Prockop, D. J. (1995). Cultured adherent cells from marrow can serve as long-lasting precursor cells for bone, cartilage, and lung in irradiated mice. *Proc. Natl. Acad. Sci. U.S.A.* **92**, 4857–4861.

74. George-Weinstein, M., Gerhart, J., Reed, R., Flynn, J., Callihan, B., Mattiacci, M., Miehle, C., Foti, G., Lash, J. W., and Weintraub, H. (1996). Skeletal myogenesis: The preferred pathway of chick embryo cells *in vitro*. *Dev. Biol.* **173**, 279–291.

75. Urist, M. R. (1965). Bone: Formation by autoinduction. *Science* **150**, 893–899.

76. Katagiri, T., Yamaguchi, A., Komaki, M., Abe, E., Takahashi, N., Ikeda, T., Rosen, V., Wozney, J. M., Fujisawa-Sehara, A., and Suda, T. (1995). Bone morphogenetic protein-2 converts the differentiation pathway of C2C12 myoblasts into the osteoblast lineage. *J. Cell Biol.* **127**, 1755–1766.

77. Grimaldi, P. A., Teboul, L., Inadera, H., Gaillard, D., and Amri, E. Z. (1997). Transdifferentiation of myoblasts to adipoblasts: Triggering effects of fatty acids and thiazolidinediones. *Prostaglandins, Leukotrienes Essent. Fatty Acids* **57**, 71–75.

78. Hoffman, E. P., Brown, J., and Kunket, L. M. (1987). Dystrophin: The protein product of the Duchenne Muscular Dystrophy locus. *Cell (Cambridge, Mass.)* **51**, 919–928.

79. Arahata, K., Ishiura, S., Ishiguro, T., Tsukahara, T., Suhara, Y., Eguchi, C., Ishihara, T., Nonaka, I., Ozawa, E., and Sugita, H. (1988). Immunostaining of skeletal and cardiac muscle surface membrane with antibody against Duchenne Muscular Dystrophy peptide. *Nature (London)* **333**, 861–863.

80. Sugita, H., Arahata, K., Ishiguro, T., Sohara, Y., Tsukahara, T., Ishiura, J., Eguchi, C., Nonaka, I., and Ozawa, E. (1988). Negative immunostaining of Duchenne Muscular Dystrophy (DMD) and mdx muscle surface membrane with antibody against synthetic peptide fragment predicted from DMD cDNA. *Proc. Jpn. Acad.* **64**, 37–39.

81. Zubryzcka-Gaarn, E. E., Bulman, D. E., Karpati, G., Burghes, A. H. M., Belfall, B., Klamut, H. J., Talbot, J., Hodges, R. S., Ray, P. N., and Worton, R. G. (1988). The Duchenne Muscular Dystrophy gene is localized in the sarcolemma of human skeletal muscle. *Nature (London)* **333**, 466–469.

82. Bonilla, E. C. E., Samitt, A. F., Miranda, A. P., Hays, G., Salviati, S., Dimauro, S., Kunkel, L. L., Hoffman, E. P., and Rowland, L. P. (1988). Duchenne Muscular Dystrophy: Deficiency of dystrophin at the muscle cell surface. *Cell (Cambridge, Mass.)* **54**, 447–452.

83. Watkins, S. C., Hoffman, E. P., Slayte, H. S., and Kunkel, L. M. (1988). Immunoelectron microscopic localization of dystrophin in myofibers. *Nature (London)* **333**, 863–866.

84. Menke, A., and Jokush, H. (1991). Decreased osmotic stability of dystrophin less muscle cells from the mdx mice. *Nature (London)* **349**, 69–71.

85. Watt, D. J., Lambert, K., Morgan, J. E., Partridge, T. A., and Sloper, J. C. (1982). Incorporation of donor muscle precursor cells into an area of muscle regeneration in the host mouse. *J. Neurol. Sci.* **57**, 31–331.

86. Watt, D. J., Morgan, J. E., and Partridge, T. A. (1984). Use of mononuclear precursor cells to insert allogenic genes into growing mouse muscles. *Muscle Nerve* **7**, 741–750.

87. Morgan, J. E., Watt, D. J., Slopper, J. C., and Partridge, T. A. (1988). Partial correction of an inherited defect of skeletal muscle by graft of normal muscle precursor cells. *J. Neurol. Sci.* **86**, 137–147.

88. Morgan, J. E., Hoffman, E. P., and Partridge, T. A. (1990). Normal myogenic cells from newborn mice restore normal histology to degenerating muscle of the mdx mouse. *J. Cell Biol.* **111**, 2437–2449.

89. Morgan, J. E., Pagel, C. N., Sherrat, T., and Partridge, T. A. (1993). Long-term persistence and migration of myogenic cells injected into pre-irradiated muscles of mdx mice. *J. Neurol. Sci.* **115**, 191–200.

90. Allamedine, H. S., Dehaupas, M., and Fardeau, M. (1989). Regeneration of skeletal muscle fiber from autologous satellite cells multiplied *in vitro. Muscle Nerve* **12**, 544–555.

91. Allamedine, H. S., and Fardeau, M. (1990). Muscle reconstruction by satellite cell graft. *J. Neurol. Sci.* **99**, 126.

92. Partridge, T. A., Morgan, J. E., Coulton, G. R., Hoffman, E. P., and Kunkel, L. M. (1989). Conversion of mdx myofibers from dystrophin negative to positive by injection of normal myoblasts. *Nature (London)* **337**, 176–179.

93. Dunckley, M. G., Davies, K. E., Walsh, F. S., Morris, G. E., and Dickson, G. (1992). Retroviral-mediated transfer of dystrophin minigene into mdx mouse myoblasts *in vitro. FEBS Lett.* **296**, 128–134.

94. Dunckley, M. G., Wells, D. J., Walsh, F. S., and Dickson, G. (1993). Direct retroviral-mediated transfer of a dystrophin minigene into mdx mouse in muscle *in vivo. Hum. Mol. Genet.* **2**, 717–723.

95. Quantin, B., Perricaudet, L. D., Tajbakhsh, S., and Mandell, J. L. (1992). Adenovirus as an expression vector in muscle cells *in vivo. Proc. Natl. Acad. Sci. U.S.A.* **89**, 2581–2581.

96. Ragot, T., Vincent, M., Chafey, P., Vigne, E., Gilgenkrantz, H., Couton, B., Cartaud, J., Briand, P., Kaplan, J. C., Perricaudet, M., and Kahn, A. (1993). Efficient adenovirus mediated gene transfer of a human mini-dystrophin gene to skeletal muscle of mdx mice. *Nature (London)* **361**, 647–650.

97. Vincent, M., Ragot, T., Gilgenkrantz, H., Couton, D., Chafey, P., Grégoire, A., Briand, P., Kaplan, J. C., Kahn, A., and Perricaudet, M. (1993). Long-term correction of mouse dystrophic degeneration by adenovirus-mediated transfer of a mini-dystrophin gene. *Nat. Genet.* **5**, 130–134.

98. Ascadi, G., Jani, A., Huard, J., Blaschuk, K., Massie, B., Holland, P., Lochmuller, H., and Karpati, G. (1994). Cultured human myoblasts and myotubes show markedly different transducibility by replication-defective adenovirus recombinants. *Gene Ther.* **1**, 338–340.

99. Ascadi, G., Jani, A., Massie, B., Simoneau, M., Holland, P., Blaschuk, K., and Karpati, G. (1994). A differential efficiency of adenovirus-mediated *in vivo* gene transfer into skeletal muscle cells of different maturity. *Hum. Mol. Genet.* **3**, 579–584.

100. Huard, J., Lochmuller, H., Ascadi, G., Jani, A., Holland, P., Guérin, C., Massie, B., and Karpati, G. (1995). Differential short-term transduction efficiency of adult versus newborn mouse tissues by adenoviral recombinants. *Exp. Mol. Pathol.* **62**, 131–143.

101. van Deutekom, J. C. T., Floyd, S. S., Booth, D. K., Oligino, T., Krisky, D., Marconi, P., Glorioso, J., and Huard, J. (1998). Implications of maturation for viral gne delivery to skeletal muscle. *Neuromuscular Disord.* **8**, 135–148.

102. van Deutekom, J. C. T., Hoffman, E. P., and Huard, J. (1998). Muscle maturation: Implications for gene therapy. *Mol. Med. Today* **4**, 214–220.

103. Engelhardt, J. F., Litzky, L., and Wilson, J. M. (1994). Prolonged transgene expression in cotton rat lung with recombinant adenovirus defective in E2A. *Hum. Gene Ther.* **5**, 1217–1229.

104. Engelhardt, J. F., Ye, X., Doranz, B., and Wilson, J. M. (1994). Ablation of E2A in recombinant adenoviruses improves transgene persistence and decreases inflammatory responses in mouse liver. *Proc. Natl. Acad. Sci. U.S.A.* **91**, 6196–6200.

105. Yang, Y., Nunes, F. A., Berencsi, K., Gonczol, E., Engelhardt, J. F., and Wilson, J. M. (1994). Inactivation of E2A in recombinant adenoviruses improves the prospect for gene therapy in cystic fibrosis. *Nat. Genet.* **7**, 362–369.

106. Yang, Y., Nunes, F. A., Berencsi, K., Furth, E. E., Gonczol, E., and Wilson, J. M. (1994). Cellular immunity to viral antigens limits E1-delited adenoviruses for gene therapy. *Proc. Natl. Acad. Sci. U.S.A.* **91**, 4401–4407.

107. Vilquin, J. T., Guerette, B., Kinoshita, I., Roy, B., Goulet, M., Gravel, C., Roy, R., and Tremblay, J. P. (1995). FK506 immunosuppression to control the immune reactions triggered by first-generation adenovirus-mediated gene transfer. *Hum. Gene Ther.* **6**, 1391–1400.

108. Huard, J., Feero, W. G., Watkins, S. C., Hoffman, E. P., Rosenblatt, D. J., and Glorioso, J. C. (1996). The basal lamina is a physical barrier to HSV mediated gene delivery to mature muscle fibers. *J. Virol.* **70**, 8117–8123.

109. Huard, J., Akkaraju, G., Watkins, S. C., Cavalcoli, M. P., and Glorioso, J. C. (1997). LacZ gene transfer to skeletal muscle using a replication defective Herpes Simplex virus type 1 mutant vector. *Hum. Gene Ther.* **8**, 439–452.

110. Floyd, S. S., Clemens, P. R., Ontell, M. R., Kochanek, S., Day, C. S., Yang, J., Hauschka, S. D., Balkir, L., Morgan, J., Moreland, M. S., Feero, G. W., Epperly, M., and Huard, J. (1998). Ex vivo gene transfer using adenovirus-mediated full-length dystrophin delivery to dystrophic muscles. *Gene Ther.* 5, 19–30.

111. Booth, D. K., Floyd, S. S., Day, C. S., Glorioso, J. C., Koveske, I., and Huard, J. (1997). Myoblast mediated *ex vivo* gene transfer to mature muscle. *J. Tissue Eng.* 3, 125–133.

112. Lieberman, J. R., Daluiski, A., Stevenson, S., Wu, L., McAllister, P., Lee, Y. P., Kabo, J. M., Finerman, G. A. M., Berk, A. J., and Witte, O. N. (1999). The effect of regional gene therapy with bone morphogenetic protein-2-producing bone-marrow cells on the repair of segmental femoral defects in rats. *J. Bone Jt. Surg.* 81, 905–917.

113. Evans, C. H., and Robbins, P. D. (1999). Gene therapy of arthritis. *Intern. Med.* 38, 233–239.

114. Hurme, T., Kalima, H., Lehto, H., and Jarvinen, M. (1991). Healing of skeletal muscle injury: An ultrastructual and immunohistochemical study. *Med. Sci. Sports Exercise* 23, 808–810.

115. Hurme, T., and Kalimo, H. (1992). Activation of myoblast precursor cells after muscle injury. *Med. Sci. Sports Exercise* 24, 197–205.

116. Jarvinen, M., and Sorvari, T. (1975). Healing of a crush injury in rat striated muscle. *Acta Pathol. Microbiol. Scand. Sect. A* 83, 259–265.

117. Carlson, B. M., and Faulner, J. A. (1983). The regeneration of skeletal muscle fibers following injury: A review. *Med. Sci. Sports Exercise* 15, 187–196.

118. Garrett, W. E., Saeber, A. V., Boswick, J., Urbaniak, J. R., and Goldner, L. (1984). Recovery of skeletal muscle after laceration and repair. *J. Hand Surg. [Am.]* 9A, 683–692.

119. Garrett, W. E. (1990). Muscle strain injuries: Clinical and basic aspects. *Med. Sci. Sports Exercise* 22, 436–443.

120. Nikolaou, P. K., MacDonald, B. L., Glisson, R. R., Seaber, A. V., and Garrett, W. E. (1987). Biomechanical and histological evaluation of muscle after controlled strain injury. *Am. J. Sports Med.* 15, 9–14.

121. Taylor, D. D., Dalton, J. D., Seaber, A. V., and Garrett, W. E. (1993). Experimental muscle strain injury, early functional and structural deficits and the increased risk for reinjury. *Am. J. Sports Med.* 21, 190–194.

122. Crisco, J. J., Jolk, P., Heinen, G. T., Connell, M. D., and Panjabi, M. M. (1994). A muscle contusion injury model, biomechanics, physiology, and histology. *Am. J. Sports Med.* 22, 702–710.

123. Hughes, C., Hasselman, C. T., Best, T. M., Martinez, S., and Garrett, W. E. (1995). Incomplete, intrasubstance strain injuries of the rectus femoris muscle. *Am. J. Sports Med.* 23, 500–506.

124. Menetrey, J., Kasemkijwattana, C., Day, C. S., Bosch, P., Vogt, M., Fu, F. H., Moreland, M. S., and Huard, J. (2000). Growth factors improve muscle healing in vivo. *J. Bone Jt. Surg. (Br.)* 82(1), 131–137.

125. Menetrey, J., Kasemkijwattana, C., Fu, F. H., Moreland, M. S., and Huard, J. (1998). Suturing versus immobilization of a muscle laceration: A morphological and functional study. *Am. J. Sports Med.* 27(2), 222–229.

126. Kasemkijwattana, C., Menetrey, J., Day, C., Bosch, P., Buranapanitkit, B., Moreland, M. S., Fu, F. H., Watkins, S. C., and Huard, J. (1998). Biologic intervention in muscle healing and regeneration. *Sports Med. Arthro Rev.* 6, 95–102.

127. Schultz, E. (1989). Satellite cells behavior during skeletal muscle growth and regeneration. *Med. Sci. Sports Exercise* 21, 181–186.

128. Schultz, E., Jaryszak, D. L., and Valiere, C. R. (1985). Response of satellite cells to focal skeletal muscle injury. *Muscle Nerve* 8, 217–222.

129. Salvatori, G., Ferrari, G., Messogiorno, A., Servidel, S., Colette, M., Tonalli, P., Giarassi, R., Cosso, G., and Mavillo, F. (1993). Retroviral vector-mediated gene transfer into human primary myogenic cells leads to expression in muscle fibers *in vivo*. *Hum. Gene Ther.* 4, 713–723.

130. Urist, M. R., Huo, Y. K., Brownell, A. G., Hohl, W. M., Buyske, J., Lietze, A., Tempst, P., Hunkapiller, M., and DeLange, R. J. (1984). Purification of bovine bone morphogenetic protein by hydroxyapatite chromatography. *Proc. Natl. Acad. Sci. U.S.A.* 81, 371–375.

131. Robbins, P. D., Tahara, H., and Ghivizzani, S. C. (1998). Viral vectors for gene therapy. *Trends Biotechnol.* 16, 35–40.

132. Bandara, G., Mueller, G. M., Balea-Lauri, J., *et al.* (1993). Intraarticular expression of biologically active interleukin-1 receptor antagonist protein by *ex vivo* gene transfer. *Proc. Natl. Acad. Sci. U.S.A.* 90, 10764–10768.

133. Day, C. S., Kasemkijwatana, C., Menetrey, J., Floyd, S. S., Booth, D., Moreland, M. S., Fu, F. H., and Huard, J. (1997). Myoblast-mediated gene transfer to the joint. *J. Orthop. Res.* 15, 894–903.

134. Menetrey J., Kasemkijwattana, C., Day, C. S., Bosch, P., Fu, F. H., Moreland, M. S., and Huard, J. (1999). Direct-, fibroblast- and myoblast-mediated gene transfer to the anterior cruciate ligament. *Tissue Eng.* 5, 435–442.

135. Spindler, K. P., Mayes, C. E., Miller, R. R., Imro, A. K., and Davidson, J. M. (1995). Regional mitogenic response to the meniscus to platelet-derived growth factor (PDGF-AB). *J. Orthop. Res.* 13, 201–207.

136. Kasemkijwattana, C., Menetrey, J., Goto, H., Niyibizi, C., Fu, F., and Huard, J. (2000). The use of growth factors, gene therapy and tissue engineering to improve meniscal healing. *Mater. Sci. Eng.* 13, 19–28.

CHAPTER 37

STEM CELL CULTURE: ENDOTHELIAL PROGENITOR CELLS FOR REGENERATION

Takayuki Asahara and Jeffrey M. Isner

STEM AND PROGENITOR CELLS FOR REGENERATION

In the past decade, researchers have defined committed stem or progenitor cells from various tissues, including bone marrow, peripheral blood, brain, liver, and reproductive organs, in both adult animals and humans. While most cells in adult organs are composed of differentiated cells, which express a variety of specific phenotypic genes adapted to each organ's environment, quiescent stem or progenitor cells are maintained locally or in the systemic circulation and are activated by environmental stimuli for physiological and pathological tissue regeneration.

Tissue replacement in the body takes place by two mechanisms. One is the replacement of differentiated cells by newly generated populations derived from residual cycling stem cells. Blood cells are a typical example for this kind of regeneration. Whole hematopoietic lineage cells are derived from a few self-renewal stem cells by regulated differentiation under the influence of appropriate cytokines and/or growth factors. The second mechanism is the self-repair of differentiated functioning cells, preserving their proliferative activity. Hepatocytes, endothelial cells, smooth muscle cells, keratinocytes, and fibroblasts are considered to possess this ability. Following physiological stimulation or injury, factors secreted from surrounding tissues stimulate cell replication and replacement. However these fully differentiated cells are still limited by senescence and by their inability to incorporate into remote target sites.

Thus, quiescent stem or progenitor cells in most adult tissues are mobilized in response to environmental stimuli when an emergent regenerative process is required, while during a minor event, neighboring differentiated cells are relied upon.

IDENTIFICATION OF ENDOTHELIAL PROGENITOR CELLS

Available evidence suggests that hematopoietic stem cells (HSCs) and endothelial progenitor cells (EPCs) [1,2] are derived from a common precursor (hemangioblast) [3–5]. Growth and fusion of multiple blood islands in the yolk sac of the embryo ultimately give rise to the yolk sac capillary network [6]; after the onset of blood circulation, this network differentiates into an arteriovenous vascular system [1]. The integral relationship between the elements that circulate in the vascular system—the blood cells—and the cells that are principally responsible for the vessels themselves—endothelial cells (ECs)—is implied by the composition of the embryonic blood islands. The cells destined to generate hematopoietic cells are situated in the center of the blood island and are termed hematopoietic stem cells. EPCs, or angioblasts, are located at the periphery of the blood islands. In addition to this spatial association, HSCs and EPCs share certain antigenic determinants, including Flk-1,

Tie-2, and CD34. These progenitor cells have consequently been considered to derive from a common precursor, putatively termed a hemangioblast [3–5].

The identification of putative HSCs in peripheral blood and BM and the demonstration of sustained hematopoietic reconstitution with these HSCs transplants have constituted inferential evidence for HSC in adult tissues [7–10]. In 1997 the related descendants—endothelial progenitor cells—were isolated along with HSCs in hematopoietic organs. Flk-1 and a second antigen, CD34, shared by embryonic EPCs and HSCs, were used to detect putative EPCs from the mononuclear cell fraction of the peripheral blood [11]. *In vitro* these cells differentiated into endothelial lineage cells and, in animal models of ischemia, heterologous, homologous, and autologous EPCs were shown to incorporate into sites of active neovascularization.

KINETICS OF ENDOTHELIAL PROGENITOR CELLS

In 1999 bone marrow (BM) transplantation (T) experiments have demonstrated the incorporation of BM-derived EPCs into foci of physiological and pathological neovascularization [12]. Wild-type mice were lethally irradiated and transplanted with BM harvested from transgenic mice in which constitutive *LacZ* expression is regulated by an EC-specific promoter, Flk-1 or Tie-2. The tissues in growing tumors, healing wound ischemic skeletal, and cardiac muscles, and cornea micropocket surgery have shown localization of Flk-1 or Tie-2 expressing endothelial lineage cells derived from BM in blood vessels and stroma around vasculatures. Similar incorporation was observed in physiological neovascularization in uterus endometrial formation following induced ovulation as well as estrogen administration [12].

Earlier investigators have shown that wound trauma causes mobilization of hematopoietic cells, including pluripotent stem or progenitor cells in spleen, BM, and peripheral blood. Consistent with the EPC/HSC common ancestry, data from our laboratory show that mobilization of BM-derived EPCs constitutes a natural response to tissue ischemia [13]. The former murine BMT model presented direct evidence of enhanced BM-derived EPC incorporation into the foci of corneal neovascularization following the development of hindlimb ischemia. Light microscopic examination of corneas excised 6 days after micropocket injury and concurrent surgery to establish hindlimb ischemia demonstrated a statistically significant increase in cells expressing β-galactosidase in the corneas of mice with and without an ischemic limb [13]. This finding indicates that circulating EPCs are mobilized endogenously in response to tissue ischemia, after which they may be incorporated into neovascular foci to promote tissue repair.

Having demonstrated the potential for endogenous mobilization of BM-derived EPCs, we considered that iatrogenic expansion and mobilization of this putative EC precursor population might represent an effective means of augmenting the resident population of ECs that is competent to respond to administered angiogenic cytokines. Such a program might thereby address the issue of endothelial dysfunction or depletion that may compromise strategies of therapeutic neovascularization in older, diabetic, and/or hypercholesterolemic animals and patients. Granulocyte–macrophase colony-stimulating factor (GM-CSF) which stimulates hematopoietic progenitor cells and myeloid lineage cells, as well as nonnematopoietic cells including BM stromal cells and ECs, has been shown to exert a potent stimulatory effect on EPC kinetics. Such cytokine-induced EPC mobilization could enhance the neovascularization of severely ischemic tissues as well as *de novo* corneal vascularization.

The mechanisms whereby these EPCs are mobilized to the peripheral circulation are in the stages early of definition. Among other growth factors, vascular endothelial growth factor (VEGF) is the most critical for vasculogenesis and angiogenesis [14–16]. Data published in 1999 indicate that VEGF is an important factor for the kinetics of EPC as well [17]. In response to recombinant VEGF protein, when administered systemically, EPCs were mobilized from the bone marrow into the peripheral circulation, resulting in an augmentation of neovascularization as shown in the mouse cornea micropocket assay.

CIRCULATING ENDOTHELIAL CELL VERSUS ENDOTHELIAL PROGENITOR CELLS

While several endothelial specific genes are generally considered to distinguish EPCs from HSCs, there exists no epitope whose expression is restricted exclusively to either fully differentiated ECs or EPCs. There are at least three lines of evidence, however, which suggest that EPCs constitute the preponderance of such circulating BM-derived endothelial lineage cells. First, earlier work has shown that freshly isolated CD34+ cells display a paucity of EC-specific markers, in contrast to plated cells cultured for 7 days [11]. Second, recent work from our own laboratory has shown that in contrast to EPCs, heterologously transplanted differentiated ECs rarely incorporate into foci of neovascularization. Third, earlier work suggests that the number of differentiated ECs circulating in peripheral blood identified using P1H12 antibody [18], ranges between 2 and 3 per milliliter, whereas the population of circulating EPCs in normal individuals based on work from our own laboratory is in the range of $0.5–1 \times 10^3$ per milliliter of blood.

THERAPEUTIC VASCULOGENESIS INDUCED BY ENDOTHELIAL PROGENITORS

The regenerative potential of stem cells is under intense investigation. *In vitro*, stem and progenitor cells possess the capability of self-renewal and differentiation into organ-specific cell types. *In vivo*, transplantation of these cells may reconstitute organ systems, as shown in animal models of diseases [11,19–23]. In contrast, differentiated cells do not exhibit such characteristics. Human EPCs have been isolated from the peripheral blood of adult individuals, expanded *in vitro*, and committed into an endothelial lineage in culture [11]. The transplantation of these human EPCs has been shown to facilitate successful salvage of limb vasculature and perfusion in athymic nude mice with severe hindlimb ischemia, while differentiated ECs (human microvascular ECs) failed to accomplish limb-saving neovascularization [20].

These experimental findings call into question certain fundamental concepts regarding blood vessel growth and development in adult organisms. Postnatal neovascularization had been considered to be synonymous with proliferation and migration of preexisting, fully differentiated ECs resident within parent vessels (i.e., angiogenesis) [24]. The finding that circulating EPCs may home to sites of neovascularization and differentiate into ECs *in situ* is consistent with "vasculogenesis" [1], a critical paradigm for establishment of the primordial vascular network in the embryo. While the proportional contributions of angiogenesis and vasculogenesis to postnatal neovascularization remain to be clarified, our findings together with the latest reports from other investigators [25,26] suggest that growth and development of new blood vessels in the adult are not restricted to angiogenesis but encompass both embryonic mechanisms. As a corollary, augmented or retarded neovascularization—whether endogenous or iatrogenic—likely includes enhancement or impairment of vasculogenesis.

Moreover, the observation that circulating EPCs home to foci of neovascularization suggests potential utility as autologous vectors for gene therapy. For treatment of regional ischemia, neovascularization could be amplified by transfection of EPCs to achieve highly localized constitutive expression of angiogenic cytokines and/or provisional matrix proteins. For antineoplastic therapies, EPCs could be transfected with or coupled to antitumor drugs or angiogenesis inhibitors.

Future studies will clarify the mechanisms and circumstances that may be responsible for modulating the contribution of vasculogenesis to postnatal neovascularization. Specifically in this regard, it is intriguing to consider the possibility that certain angiogenic growth factors that are acknowledged to promote both angiogenesis and vasculogenesis in the embryo, but have been assumed to promote neovascularization exclusively by angiogenesis in the adult, may in fact promote migration, proliferation, and mobilization of EPCs from BM. The possibility that modulation of vasculogenesis can be used therapeutically to augment as well as inhibit neovascularization deserves further investigation.

Endothelial Progenitor Cell Culture

EPCs can be enriched by antiger-positive selection from mononuclear cells of peripheral blood (PB) or BM cells. CD34 [11,25], KDR/Flk-1 [11,25], AC133 [27], and others have been reported as antigens for EPC selection. The number of positive cells, however, is very limited, and single-population culture is not always successful. For that reason, we employ whole mononuclear cell culture to augment differentiation and proliferation of EPC in endothelial specific culture conditions.

Total human PB mononuclear cells (hPBMCs) were isolated from blood of volunteers by density gradient centrifugation with Histopaque 1077 (Sigma, St. Louis, MO) [11]. Cells were plated on culture dishes coated with human fibronectin (Sigma) and maintained in EC basal medium 2 (EBM-2) (Clonetics, San Diego, CA). The medium was supplemented with 5% fetal bovine serum (FBS), human VEGF-1, human fibroblast growth factor 2 (FGF-2), human epidermal growth factor (EGF), insulin-like growth factor 1 (IGF-1), and ascorbic acid. After 3 or 4 days in culture, nonadherent cells were removed by washing with phosphate-buffered saline (PBS), new media applied, and the culture maintained through days 7–10. The result was cells with a spindle shape, EC-like morphology [11].

These differentiating cells could be shown to endocytose acetylated low-density lipoprotein (acLDL) and to bind *Ulex europaeus* agglutinin 1 (UEA-1), consistent with endothelial lineage cells. Analysis by fluorescence-activated cell sorter (FACS) after I week in culture disclosed that the vast majority of the cells assumed not only morphologic but also qualitative properties of ECs. Among $88.5 \pm 4.6\%$ antigenically defined cells, more man 70% of adherent cells expressed EC-specific antigens, including VEGF receptor 2 [VEGFR-2 (KDR), $77 \pm 4.1\%$], vascular endothelium (VE)–cadherin ($78.1 \pm 8.2\%$), and CD31 ($72 \pm 8.8\%$). Positive immunostaining for CD14 was identified in $90.7 \pm 1.8\%$ of the cells. The cell surface adhesion molecule CD34 and the integrin $\alpha v \beta 3$ were expressed by 28.6 ± 9.5 and $11.7 \pm 2.2\%$ of the cells, respectively. Further characterization excluded significant contamination by hematopoietic lineage cells such as T lymphocytes (CD3, $2.6 \pm 1.8\%$) or macrophages.

Ex vivo expansion of EPCs was derived from an hPBMC population of healthy adult subjects. Incubation for 7–10 days with endothelial mitogens, including VEGF, bFGF, IGF, and EGF, resulted in 80- to 90-fold expansion of cells expressing the EC-specific antigens KDR, CD31, and VE–cadherin. Earlier analyses of embryonic neovascularization suggest that coexpression of flk-1 (KDR), and VE-cadherin denote the point of divergence of ECs from hematopoietic lineages [28]. Moreover, a combination of monoclonal antibodies prepared against flk-1/KDR, VE-cadherin, CD31, Tie-1, and Tie-2 have been interpreted to define most intermediate stages during differentiation of embryonic stem cell–derived ECs [28–31]. The capacity to take up acLDL as well as UEA-1 further characterize ECs [32].

Cultured EPCs, as opposed to freshly isolated CD34-antigen-positive (CD34$^+$) EPCs [11], are now being used in experiments for three reasons. First, the number of EPCs obtained by *ex vivo* expansion (3.5×10^4 from 1 ml of whole blood) exceed 5 the number of CD34$^+$ cells that can be freshly isolated (0.5×10^4/ml blood). Second, purity and quality of EPCs in a cultured population are superior to those of freshly isolated CD34$^+$ cells; since CD34$^+$ was originally described as the prototypical antigen expressed by both HSCs and endothelial lineage cells, hematopoietic cells may contaminate freshly isolated CD34$^+$ cells. Indeed, pilot studies demonstrated that the extent of neovascularization achieved following transplantation of freshly isolated CD34$^+$ cells was inferior to that of culture-expanded EPCs. Third, for therapeutic strategies designed to employ transplanted cells that constitutively express pro- or anti-angiogenic factors, gene transfer of EPCs is facilitated by the use of culture-committed versus less differentiated CD34$^+$ EPCs [T. Asahara, unpublished data].

Under the described conditions, contamination by other cell lines, including lymphocytes, macrophages, and dendritic cells, was minimized as indicated by being limited to absent expression of CD3, CD19, CD68, CD83, and CD86. Incubation of similar mononuclear cell cultures with other cytokines such as GM-CSF or tumor necrosis factor α (TNF-α) has been reported to favor isolation of dendritic cells [33,34]; in contrast, VEGF appears to inhibit dendritic cell maturation from CD34$^+$ precursors [35,36]. Because cytokine composition of the culture media may influence *in vitro* mononuclear cell differentiation [W. Kalka-Moll, unpublished data], the cytokine mixture employed here for EPCs, containing VEGF, basic FGF, IGF, and EGF, appears to preferentially promote endothelial lineage differentiation.

Isolation of circulating mononuclear cells for *ex vivo* EPC expansion was carried out by using peripheral blood from human donors. Isolation of circulating EPCs from human subjects thus appears realistic for harvesting EPCs for therapeutic neovascularization in future clinical applications. The feasibility of retrieving cells from peripheral blood has been established [7]. Augmented mobilization of bone marrow derived EPCs may be achieved by using several cytokines, including GM-CSF [13], similar to the approach utilized in preparation for stem cell transplants [37,38]. The potential value of this approach is that it supplies substrate (i.e., a source population of robust ECs) that may complement current strategies of ligand administration for patients in whom depleted and/or dysfunctional ECs preclude an optimal response to cytokine supplements.

REFERENCES

1. Risau, W., Sariola, H., Zerwes, H.-G., Sasse, J., Ekblom, P., Kemler, R., and Doetschman, T. (1988). Vasculogenesis and angiogenesis in embryonic stem cell-derived embryoid bodies. *Development (Cambridge, UK)* **102**, 471–478.

2. Pardanaud, L., Altman, C., Kitos, P., and Dieterien-Lievre, F. (1989). Relationship between vasculogenesis, angiogenesis and haemopoiesis during avian ontogeny. *Development (Cambridge, UK)* **105**, 473–485.

3. Flamme, I., and Risau, W. (1992). Induction of vasculogenesis and hematopoiesis in vitro. *Development (Cambridge, UK)* **116**, 435–439.

4. His, W. (1900). Leoithoblast und angioblast der Wirbelthiere. *Abh. K. Ges. Wiss. Math. Phys.* **22**, 171–328.

5. Weiss, M., and Orkin, S. H. (1996). In vitro differentiation of murine embryonic stem cells: new approaches to old problems. *J. Clin. Invest.* **97**, 591–595.

6. Risau, W., and Flamme, I. (1995). Vasculogenesis. *Annu. Rev. Cell Dev. Biol.* **11**, 73–91.

7. Brugger, W., Heimfeld, S., Berenson, R. J., Mertelsmann, R., and Kanz, L. (1995). Reconstitution of hematopoiesis after high-dose chemotherapy by autogous progenitor cells generated ex vivo. *N. Engl. J. Med.* **333**, 283–287.

8. Kessinger, A., and Armitage, J. O. (1991). The evolving role of autologous peripheral stem cell transplantation following high-dose therapy for malignancies. *Blood* **77**, 211–213.

9. Sheridan, W. P., Begley, C. G., and Juttener, C. (1992). Effect of peripheral-blood progenitor cells mobilised by filgrastim (G-CSF) on platelet recovery after high-dose chemotherapy. *Lancet* **339**, 640–644.

10. Shpall, E. J., Jones, R. B., and Bearman, S. I. (1994). Transplantation of enriched CD34-positive autologous marrow into breast cancer patients following high-dose chemotherapy. *J. Clin. Oncol.* **12**, 28–36.

11. Asahara, T., Murohara, T., Sullivan, A., Silver, M., van der Zee, R., Li, T., Witzenbichler, B., Schatteman, G., and Isner, J. M. (1997). Isolation of putative progenitor endothelial cells for angiogenesis. *Science* **275**, 965–967.

12. Asahara, T., Masuda, H., Takahashi, T., Kalka, C., Pastore, C., Silver, M., Kearney, M., Magner, M., and Isner, J. M. (1999). Bone marrow origin of endothelial progenitor cells responsible for postnatal vasculogenesis in physiological and pathological neovascularization. *Circ. Res.* **85**, 221–228.

13. Takahashi, T., Kalka, C., Masuda, H., Chen, D., Silver, M., Kearney, M., Magner, M., Isner, J. M., and Asahara, T. (1999). Ischemia- and cytokine-induced mobilization of bone marrow-derived endothelial progenitor cells for neovascularization. *Nat. Med.* **5**, 434–438.

14. Carmeliet, P., Ferreira, V., Breier, G., Pollefeyt, S., Kieckens, L., Gertsenstein, M., Fahrig, M., Vandenhoeck, A., Kendraprasad, H., Eberhardt, C., Declercq, C., Pawling, J., Moons, L., Collen, D., Risau, W., and Nagy, A. (1996). Abnormal blood vessel development and lethality in embryos lacking a single VEGF allele. *Nature (London)* **380**, 435–439.

15. Ferrara, N., Carver-Moore, K., Chen, H., Dowd, M., Lu, L., O'Shea, K. S., Powell-Braxton, L., Hilan, K. J., and Moore, M. W. (1996). Heterozygous embryonic lethality induced by targeted inactivation of the VEGF gene. *Nature (London)* **380**, 439–442.

16. Shalaby, F., Rossant, J., Yamaguchi, T. P., Gertsenstein, M., Wu, X.-F., Breitman, M. L., and Schuh, A. C. (1995). Failure of blood-island formation and vasculogenesis in Flk-1 deficient mice . *Nature (London)* **376**, 62–66.

17. Asahara, T., Takahashi, T., Masuda, H., Kalka, C., Chen, D., Iwaguro, H., Inai, Y., Silver, M., and Isner, J. M. (1999). VEGF contlibutes to postnatal neovascularization by mobilizing bone marrow-derived endothelial progenitor cells. *EMBO J.* **18**, 3964–3972.

18. Solovey, A., Lin, Y., Brown, P., Choong, S., Wayner, E., and Hebbel, R. P. (1997). Circulating activated endothelial cells in sickle cell anemia. *N. Engl. J. Med.* **337**, 1582–1590.

19. Evans, J. T., Kelly, P. P., O'Neill, E., and Garcia, J. V. (1999). Human cord blood CD34+CD38-cell transduction via lenlivirus-based gene transfer vectors. *Hum. Gene Ther* **10**, 1479–1489.

20. Kalka, C., Masuda, H., Takahashi, T., Kalka-Moll, W. M., Silver, M., Kearney, M., Li, T., Isner, J. M., and Asahara, T. (2000). Transplantation of ex vivo expanded endothelial progenitor cells for therapeutic neovascularization. *Proc. Natl. Acad. Sci. U.S.A.* **97**, 3422–3427.

21. Flax, J. D., Aurora, S., Yang, C., Simonin, C., Wills, A. M., Billinghurst, L. L., Jendoubi, M., Sidman, R. L., Wolfe, J. H., Kim, S. U., and Snyder, E. Y. (1998). Engraftable human neural stem cells respond to developmental cues, replace neurons, and express foreign genes. *Nat. Biotechnol.* **16**, 1033–1039.

22. Lindvall, O., Brundin, P., Widner, H., Rehncrona, S., Gustavii, B., Frackowiak, R., Leenders, K. L., Sawle, G., Roth-well, J. C., and Marsden, C. D. (1990). Grafts of fetal dopamine neurons survive and improve motor function in Parkinson's disease. *Science* **247**, 574–577.

23. Anklesaria, P., Kase, K., Glowacki, J., Holland, C. A., Sakakeeny, M. A., Wright, J. A., FitzGerald, T. J., Lee, C. Y., and Greenberger, J. S. (1987). Engraftment of a clonal bone marrow stromal cell line in vivo stimulates hematopoietic recovery from total body irradiation. *Proc. Natl. Acad. Sci. U.S.A.* **84**, 7681–7685.

24. Folkman, J. (1971). Tumor angiogenesis: Therapeutic implications. *N. Engl. J. Med.* **285**, 1182–1186.

25. Shi, Q., Rafii, S., Wu, M. H.-D., Wijelath, E. S., Yu, C., Ishida, A., Fujita, Y., Kothari, S., Mohle, R., Sauvage, L. R., Moore, M. A. S., Storb, R. F., and Hammond, W. P. (1998). Evidence for circulating bone marrow-derived endothelial cells. *Blood* **92**, 362–367.

26. Hatzopoulos, A. K., Folkman, J., Vasile, E., Eiselen, G. K., and Rosenberg, R. D. (1998). Isolation and characterization of endothelial progenitor cells from mouse embyros. *Development (Cambridge, UK)* **125**, 1457–1468.

27. Peichev, M., Naiyer, A. J., Pereira, D., Ahu, Z., Lane, W. J., Williams, M., Oz, M. C., Hicklin, D. J., Witte, L., and Moore, M. A. S. (2000). Expression of VEGFR-2 and AC133 by circulating human CD34+ cells identifies a population of functional endothelial precursors. *Blood* **95**, 952–958.

28. Nishikawa, S., Hirashima, M., Matsuyoshi, N., and Kodama, H. (1998). Progressive lineage analysis by cell sorting and culture identifies Flk1+VEcadherin+cells at a diverging point of endothelial and hematopoietic lineages. *Development (Cambridge, UK)* **125**, 1747–1757.

29. Vittet, D., Prandini, M. H., Berthier, R., Schweitzer, A., Martin-Sisteron, H., Uzan, G., and Dejana, E. (1996). Embry-onic stem cells differentiate in vitro to endothelial cells through successive maturation steps. *Blood* **88**, 3424–3431.

30. Yamaguchi, T. P., Dumont, D. J., Conlon, R. A., Breitman, M. L., and Rossant, J. (1993). *flk*-1, and *flt*-related receptor tyrosine kinase is an early marker for endothelial cell precursors. *Development (Cambridge, UK)* **118**, 489–498.

31. Choi, K., Kennedy, M., Kazarov, A., Papadimitriou, J. C., and Keller, G. (1998). A common precursor for hematopoietic and endothelial cells. *Development (Cambridge, UK)* **125**, 725–732.

32. Rafii, S., Shapiro, F., Rimarachin, J., Nachman, R. L., Ferris, B., Weksler, B., Moore, M. A. S., and Asch, A. S. (1994). Isolation and characterization of human bone marrow microvascular endothelial cells: Hematopoietic progenitor cell adhesion. *Blood* **84**, 10–19.

33. Reid, C., Stackpoole, A., Meager, A., and Tikerpae, J. (1992). Interactions of tumor necrosis factor with granulocyte-macrophage colony-stimulating factor and other cytokines in the regulation of dendritic cell growth in vitro from early bipotent CD34+ progenitors in human bone marrow. *J. Immunol.* **149**, 2681–2688.

34. Caux, C., Dezutter-Dambuyant, C., Schmitt, D., and Banchereau, J. (1992). GM-CSF and TNF-alpha cooperate in the generation of dentritic Langerhans cells. *Nature (London)* **360**, 258–261.

35. Gabrilovich, D., Ishida, T., Oyama, T., Ran, S., Kravtsov, V., Nadaf, S., and Carbone, D. P. (1998). Vascular endothe-lial growth factor inhibits the development of dentritic cells and dramatically affects the differentiation of multiple hematopoietic lineages in vivo. *Blood* **92**, 4150–4166.

36. Gabrilovich, K. I., Chen, H. L., Girgis, K. R., Cunningham, H. T., Meny, G. M., Nadaf, S., Kavanaugh, D., and Carbone, D. P. (1996). Production of vascular endothelial growth factor by human tumors inhibits the functional maturation of dendritic cells. *Nat. Med.* **10**, 1096–1103.

37. Siena, S., Bregni, M., Brando, B., Ravagnani, F., Bonadonna, G., and Gianni, A. M. (1989). Circulation of CD34+ hematopoietic stem cells in the peripheral blood of high-dose cyclophosphamide-treated patients: Enhancement by intravenous recombinant human granulocyte-macrophage colony-stimulating factor. *Blood* **74**, 1905–1914.

38. Duhrsen, U., Villeval, J. L., Boyd, J., Knnourakis, G. K., Morstyn, G., and Metcalf, D. (1988). Effect of recombinant granulocyte colony-stimulating factor on hematopoietic progenitor cells in cancer patients. *Blood* **72**, 2074–2082.

CHAPTER 38

Stem Cell Culture: Mesenchymal Stem Cells from Bone Marrow

Mark F. Pittenger, Alan M. Flake, and Robert J. Deans

SOURCES OF MULTIPOTENTIAL ADULT CELLS

Several tissues in the adult body contain progenitor cells that can proliferate and then differentiate to provide organ-specific cell types. Examples include the proliferative keratinocytes found in skin, hepatocytes responding to liver damage, intestinal crypt cells that replenish the absorptive epithelium cells, and osteoblasts actively forming new bone and becoming osteocytes. These progenitor cells appear to have their differentiation limited to a defined lineage. On the other hand, stem cells exist in the adult that can form many differentiated cell types under the appropriate conditions. The hematopoietic stem cell (HSC) found in bone marrow is the most studied example, providing eosinophils, erythrocytes, megakaryocytes, osteoclasts, and B and T cells. These differentiated progeny of the HSC provide a variety of essential functions in the body. The bone marrow also contains mesenchymal stem cells (MSCs), cells capable of differentiating into several connective tissue cell types including osteocytes, chondrocytes, adipocytes, tenocytes, myocytes, and bone marrow stromal cells.

Experiments implanting whole bone marrow suggested that marrow contained cells capable of forming new bone at ectopic sites [1,2]. To understand the nature of this ectopic bone, the *in vitro* cultivation of cells from bone marrow was first explored by Alexander Friedenstein and colleagues, who cultivated cells from marrow expressed from long bones of the guinea pig [3,4]. They showed that the culture expanded, fibroblastic cells could become osteocytes when implanted into recipient animals. Later research led by Maureen Owen demonstrated that similar cells could be isolated from the rabbit [5,6]. Work from several labs showed that bone marrow derived cells could form bone and cartilage and additional tissues, and immortalized mouse and rat cell lines with multilineage potential were also isolated [7–10]. Several reviews have examined the characterization and experimental uses of nonhuman, bone marrow derived MSCs [10–13].

The first isolation, *in vitro* cultivation, and characterization of human mesenchymal stem cells (hMSCs) was accomplished by Steve Haynesworth and Arnold Caplan, using marrow taken from the posterior iliac crest of volunteer donors [14]. They furthered the use of the ceramic implant as a carrier for hMSCs in implantation experiments that demonstrated both bone and cartilage derived from the *in vitro* expanded cells (Fig. 38.1). They also were able to generate monoclonal antibodies that proved useful for characterizing the proliferating multipotential cells [15]. The development of *in vitro* assays for differentiation of hMSCs to osteocytes [16–18], chondrocytes [19–21], and adipocytes [22] has allowed for greater understanding of the requirements for hMSC commitment to these distinct lineages.

It was demonstrated in 1999 and 2000 that individual, clonal hMSCs can be expanded from a single cell to provide a homogeneous population that will differentiate to the os-

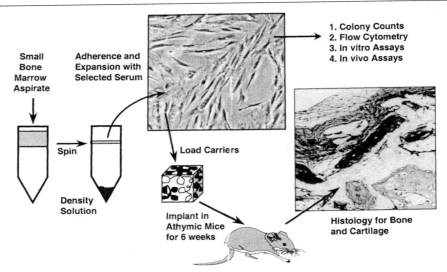

Fig. 38.1. The bone marrow-derived hMSCs are isolated based on their ability to attach and grow in medium containing fetal bovine serum from selected lots. The hMSCs expand as fibroblast-like cells and their ability to differentiate to specific cell lineages is tested in vivo and in vitro.

teocyte, chondrocyte, and adipocyte lineages in specific, defined *in vitro* conditions, providing unequivocal evidence of their stem cell nature [22,23]. Engraftment experiments with hMSCs injected into the mouse heart demonstrated that the cells engraft in the heart and undergo differentiation to a striated myocytic cell type [24]. Evidence that hMSCs should prove safe to administer to patients was shown by the reinfusion of autologous hMSCs into volunteers, with no deleterious effects after more than 5 years [25]. Promising data on the therapeutic effects of hMSCs in aiding engraftment following bone marrow transplantation was published in 2000 [26]. Additionally, children with the debilitating disease osteogenesis imperfecta have been infused with bone marrow that contains hMSCs, and the early results suggested a promising therapeutic effect [27]. Over the last 5–10 years, much research has gone into understanding these important stem cells found in all adults, and this body of work includes steps that are important for the development of therapeutic uses of hMSC [28–31]. Evidence in mouse reported in 1999 suggests that adult stem cells may be found that are capable of differentiating to additional lineages, beyond those suspected by their tissue of origin [32,33]. This is likely true for the human MSC from adult bone marrow, and these cells would appear to be the ideal candidate cell for development of therapeutic tissue regeneration and tissue engineering.

ISOLATION AND INITIAL PLATING OF HUMAN MSCS

The marrow aspirate is most easily taken by a qualified medical professional from the posterior iliac crest with the donor face down on the examination table. Only about 20 ml of marrow are taken per side. A 50-ml syringe containing 6000 units of heparin is used with a strong, rapid pull on the plunger and a turn of the needle to assure that the bevel is in the marrow space. Immediate mixing of heparin with the aspirate to prevent any clot formation will ensure good stem cell recovery. Samples can be stored or shipped at room temperature for 24 h with excellent cell recovery.

We routinely culture the mesenchymal stem cells from bone marrow in Dulbecco's Modified Eagle's Medium (DMEM, containing 1 g/liter of glucose) with 10% fetal bovine serum (FBS) from selected lots. The process of selecting serum is important to ensure the propagation of MSCs, and poor serum lots will not produce cells capable of multilineage differentiation. A description of the selection process has been published and involves *in vivo* implantation of carriers containing cells grown in each of the serum lots [34]. We recommend screening lots of FBS for each species of interest. One report suggests that adding fibroblast growth factor 2 (FGF-2) to FBS obviates the need for serum selection [35], but we have not found this to be optimal. Until a serum-free medium utilizing defined growth factors and components is developed, the requirement for serum selection will continue.

A detailed procedure for hMSC isolation from adult donors has been published [36]. Briefly, the sample is washed, and the cells transferred to a 50-ml conical tube containing a density gradient solution of 1.073, and subjected to centrifugation at 3000g for 30 min at room temperature. The majority of the erythrocytes will be in the pellet. The nucleated cells will be found at the interface and in the upper layer. To confirm this, it is useful to culture cells from different positions in the gradient and observe their phenotype. To remove the density medium, the upper layer is carefully withdrawn and diluted with two volumes of phosphate-buffered saline (PBS) and centrifuged again. The cells are resuspended in medium, counted using a hemocytometer, and transferred to a culture flask at a density of about 160,000 nucleated cells/cm^2. The mesenchymal stem cells will attach and begin to divide, and the first small colonies will be noticeable in 4–6 days. The medium is changed two or three times a week, and any contaminating red cells are rinsed away also with the medium changes. The colonies will expand and the cells should be passaged before the colonies begin to merge, usually at about 14 days. The attached cells are easily subcultured using 0.05% trypsin in 0.53 mM ethylenediamine tetraacetic acid (EDTA).

hMSC EXPANSION AND CHARACTERIZATION

Once established, the hMSCs grow rapidly as attached, well-spread fibroblastic cells, becoming contact inhibited as they approach confluence. The cells should be subcultured 1:3 when they become 80–90% confluent. By convention, the passage number is increased whenever the cells are trypsinized and replated; that is, the primary or P_0 culture becomes the P_1 culture as the cells are passaged into the new flasks or assay dishes. Typically, we generate 100–200 million cells from a 25-ml marrow aspirate by the end of first passage (P_1). The hMSCs can be expanded manyfold, but it has not been possible under current *in vitro* conditions to expand them indefinitely while maintaining their multilineage potential. We typically utilize the hMSCs in passages 1–4 and see a slow decrease in *in vitro* differentiation potential beyond passage 7. However, hMSCs at passage 12 maintain a normal karyotype and telomerase activity [22]. The hMSCs do eventually decrease their *in vitro* proliferative potential, becoming enlarged and flattened and showing signs of cell senescence.

As with most cell populations, the hMSC population is most easily characterized by the expressed surface proteins detectable by flow cytometry. We have utilized more than 70 antibodies to analyze the cell surface molecules on hMSCs [22,36]. No unique marker has been found that positively identifies the hMSCs, and they express surface molecules often associated with distinctive differentiated cell types. The surface molecules include integrins, growth factor and cytokine receptors, and self-antigens such as major histocompatibility complex. The hMSCs are characterized by the absence, as well as presence, of particular surface molecules. We have found the hMSCs to have a combination of surface molecules, but this combination was not found on other cell types we analyzed. For example, while the cells express endoglin, they do not appear to express other endothelial markers such as von Willebrand factor or E-selectin. The *in vitro* cultured hMSCs are negative for the hematopoietic markers CD14 and CD34. It should be recognized that the expanded hMSC population that has undergone *in vitro* expansion may differ in its cell surface molecule expression from the hMSC resident in undisturbed marrow. Nevertheless, flow cytometry offers a convenient and powerful tool for characterizing the hMSCs prior to further analysis.

DIFFERENTIATION ASSAYS *IN VITRO*

Several *in vitro* differentiation assays have been developed which provide a rigorous test of the multilineage potential of the hMSCs. As the assays afford temporal lineage progression, the cells can be assayed at intervals to analyze the degree of differentiation. They provide a method of testing the cells' osteogenic, adipogenic, and chondrogenic capabilities (Fig. 38.2). To date, we have not found another cell type capable of a similar degree of differentiation.

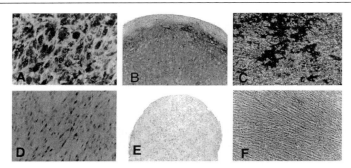

Fig. 38.2. The in vitro assays provide a rigorous test of the culture-expanded hMSCs ability to differentiate to the (A) adipogenic lineage, shown by staining of the lipid vacuoles with oil red O, (B) chondrogenic lineage, shown by morphology and immunostaining with an antibody to the articular collagen type II, (C) osteogenic lineage, shown by staining for alkaline phosphatase and mineral deposition. Primary fibroblasts cultured under identical conditions do not show evidence of lineage differentiation (D–E).

Osteogenic differentiation is initiated in subconfluent monolayers of hMSCs by changing the medium to 50 μM ascorbic acid 2-phosphate, 10 mM β-glycerol phosphate, and 100 nM dexamethasone in DMEM with 10% FBS [18]. This medium is changed twice weekly, and differentiation is first noticed as the cells form aggregates or "nodules" at about one week. The nodules are found to be alkaline phosphatase positive, and they progress to form a highly mineralized bone matrix. In this monolayer format, individual cells do not form osteocytes fully surrounded by matrix and are therefore classified as osteoblasts.

Adipogenic differentiation also occurs in the presence of serum-containing medium, but unlike the osteogenic assay, the hMSCs must be postconfluent before they can differentiate [22]. The adipogenic induction medium contains 0.5 mM methyl isobutylxanthine, 1 μM dexamethasone, 10 μg/ml insulin, and 100–200 μM indomethicin in DMEM with 10% FBS. The hMSCs are treated with this medium for 3 days, and then the medium is changed to DMEM with 10% FBS and insulin alone for 24 h. The cells are treated a second time with the adipogenic induction medium for 3 days, followed by a day in the insulin-containing medium. The induction medium is added a third time, followed by the change to the insulin-containing medium as a maintenance medium. Small refractile lipid vacuoles will be apparent in a portion of the hMSCs at the end of the first induction, and more cells commit to the adipogenic lineage with each treatment. Following the third treatment, the adipogenic cells will continue to accumulate lipid, and the vacuoles will enlarge over time and coalesce. We have maintained these adipogenic cells in culture for at least 3 months, and they have the characteristic appearance of white fat cells containing one or two large lipid vacuoles. The adipogenic differentiation also occurs with the hMSCs grown on a three-dimensional matrix material.

Chondrogenic differentiation of hMSCs is very inefficient when the cells are in a monolayer format. Instead, the hMSCs are cultured as a micromass pellet of cells, created by gentle centrifugation [19–21]. The medium composition is very important and contains no FBS. It consists of DMEM supplemented with 10 ng/ml TGF-β3, 100 nM dexamethasone, 50 μg/ml ascorbic acid 2-phosphate, 100 μg/ml sodium pyruvate, 40 μg/ml proline, and a commercial preparation of insulin, transferrin, selenium, and linoleic acid. To initiate the chondrogenic cultures, the hMSCs are harvested with trypsin–EDTA and resuspended in the chondrogenic medium. Aliquots of approximately 2.5×10^5 hMSCs are centrifuged at 600g for 5 min in polypropylene conical tubes. The cells do not adhere to the tubes, and over 24 h they form a ball of cells about 1 mm in diameter. The medium is changed three times per week, and the cell pellets are gently dislodged to keep them free floating. Little change is evident during the first week. During the second and third weeks of culture the pellets become chondrogenic and enlarge as a result of the synthesis of abundant extracellular matrix material. Upon sectioning and histochemical analysis of the differentiating pellets, one observes abundant expression of proteoglycans including aggrecan, and type II collagen. The chondrogenic differentiation of hMSCs can also be performed in a three-

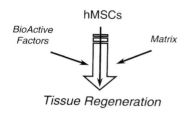

Fig. 38.3. *Mesenchymal stem cells provide a basis for tissue regeneration. When combined with knowledge of growth factors and cytokines that promote differentiation, and suitable scaffold or matrix material for delivery to a particular tissue site, they may be suitable for many applications.*

dimensional matrix, such as a collagen sponge, a format that may be more amenable to mechanical testing and tissue engineering objectives [37].

The *in vitro* differentiation assays provide for controlled differentiation of hMSCs to these lineages. We have evaluated the differentiating cultures by multiple means, including reverse transcriptase polymerase chain reaction analysis, and the cells appear to differentiate only to the desired lineage [22]. We do not detect unpredictable or unwanted cell types among the differentiating hMSCs. This differs from all reports investigating the differentiation of embryonic stem cells (ES cells) or embryonic germ cells, which form multiple and unpredictable cell types as they differentiate.

In Vivo DIFFERENTIATION OF hMSCS IN ADULT TISSUES

The MSCs have been implanted into a variety of adult tissues to assess their differentiation and therapeutic potential. Similar to the *in vitro* experimental data, implanting MSCs into tissue sites results in differentiated cells of the appropriate phenotype. The most extensive experiments have been performed in orthopedic applications using MSCs from several species. A large segmental gap in the femur that has been filled with a biocompatible carrier that fails to heal even after an extended period can be effectively regenerated by the addition of MSCs to the carrier. This has been demonstrated in rat and canine models, and also with human MSCs implanted into the immunocompromised athymic rat [38–40]. Rabbit MSCs have been utilized to regenerate the articular surface in an experimental model of joint disease [41,42]. The MSCs isolated from rabbit bone marrow were seeded onto a collagen matrix and demonstrated to form new tendon in the severed Achilles tendon of experimental animals. The regenerated tendons had excellent mechanical properties [43]. One of the major challenges for cell-based therapeutics is the development of suitable biomatrices that serve as an initial cell support but are readily resorbed by the body (Fig. 38.3). The MSCs are compatible with, and propagate on, most biomaterials in a variety of formats.

The differentiation potential of stem cell populations is often dependent on the assay system applied. The ability to induce specific differentiation patterns *in vitro* does not always correspond to the more "physiological" *in vivo* potential of a given stem cell population. In addition, the effects of manipulation of stem cells for tissue engineering or cellular transplantation applications cannot be assessed outside the complexity of *in vivo* systems. For instance, the limitations of homing, transendothelial migration, and engraftment of systemically administered cells, or the response to physiological stimuli such as tissue injury or degeneration, cannot be replicated *in vitro*. For these reasons, it is important to assess cellular differentiation potential in living systems. The *in vivo* investigation of human stem cells obviously requires the use of surrogate xenogeneic systems. The primary obstacles to this approach are the host immune response against xenogeneic antigens present on cells, and issues of species specificity of the microenvironment that impact donor cell viability and function. The immune response can be avoided by the use of sufficiently immunodeficient animal models, most notably the nude mouse, the SCID mouse, or the NOD/SCID mouse [44–46]. Since, however, the mouse is relatively short-lived, long-term assessment of human cell engraftment is difficult, even under the ideal circumstances. This, in combination with other significant limitations, prompted our interest in the fetal sheep model as an *in vivo* assay system for human MSCs.

THE FETAL SHEEP MODEL AS AN ASSAY SYSTEM FOR
HUMAN MESENCHYMAL STEM CELLS

An alternative environment permissive for human cells is the preimmune sheep fetus [47–49]. Via the normal process of self-antigen recognition, the early gestational fetal recipient processes foreign allo- or xenoantigen as "self" with deletion of donor reactive lymphocyte clones [50,51]. In addition to immunological tolerance, there are other advantages for the fetal sheep model. No irradiation or other conditioning regimen is used, so there is no perturbation of the normal receptive environment. The transplantation of cells during development may offer a maximal opportunity for distribution of stem cells into many niches, as the developmental process of hematogenous distribution and migration of stromal elements into developing hematopoietic environments occurs. The sheep lives for many years, allowing true long-term assessment of transplanted cell populations. Finally, because human and sheep DNA and proteins are widely disparate with respect to sequence homology, human-specific markers can be utilized for the unequivocal detection and characterization of human cells by a variety of methodologies.

We have utilized the sheep model for the analysis of engraftment and ultimate cell fate of human MSCs [52]. In this model, human MSCs can be transplanted into fetal lambs by intraperitoneal injection under direct visualization or ultrasound guidance as early as 45 days gestation (term = 145 days). In our studies, we have transplanted hMSCs at either 65 or 85 days gestation. These time points were chosen to assess the requirement for immunological tolerance for engraftment of hMSC. Beyond 75 days of gestation, fetal lambs develop the capacity to reject allogeneic skin grafts [53] and demonstrate allogeneic or xenogeneic hematopoietic engraftment failure [54], which appears to be immune system mediated. To assess distribution of engraftment, and to screen for human cell engraftment in tissues prior to immunohistochemical assessment, we utilized PCR for human-specific β-2 microglobulin DNA sequences. We assessed differentiation of human cells in various tissues by different techniques: (1) characteristic morphology with anti-human β-2 microglobulin staining, (2) immunohistochemical double staining for anti-human β-2 microglobulin and a second differentiation specific marker; (3) *in situ* hybridization for human *ALU* DNA sequences combined with markers specific for muscle differentiation. A further option is the positive staining of tissue sections with human-specific differentiation markers, tested to not cross-react with sheep tissues. Use of these techniques allowed us to confirm site-specific differentiation of hMSCs to a number of cell types, with persistence of human cells for up to 13 months after transplantation. Transplanted human cells were documented to undergo site-specific differentiation into chondrocytes, adipocytes, skeletal myocytes, cardiomyocytes, bone marrow stromal cells, and thymic stroma. Surprisingly, long-term engraftment was observed even when hMSCs were transplanted at 85 days, after the expected development of immune competence. Thus, the hMSCs maintain their multipotential capacity after transplantation and appear to have immunological characteristics that allow persistence in a xenogeneic environment.

Beyond the fate mapping application outlined earlier, future studies in the sheep model will assess the response of engrafted hMSCs to physiological stimuli such as tissue injury, regeneration, and inflammation. The capacity for specific environments to direct differentiation of hMSCs will be tested by direct injection into sites such as the CNS, liver, or skeletal muscle. In addition, strategies for manipulation of the *in vivo* behavior of hMSCs can be tested for efficacy in the sheep model. Such strategies might include transgene-encoded expression of specific homing receptors, chemokines, or growth factors to target hMSCs to specific tissues, facilitate migration within tissues, regulate proliferation of hMSCs, or to direct specific pathways of differentiation. Such studies will be critical for the development of the full therapeutic potential of hMSCs.

The fetal sheep studies just described provide additional evidence for the safety of the hMSC in clinical applications. Another important model for assessing *in vivo* safety of MSCs is the nonhuman primate. Intravenous infusion of baboon MSCs (bMSCs) at doses up to 20 million cells/kg have been shown to pose no short- or long-term toxicity, and no ectopic tissue formation [55]. The infused bMSCs initially showed a broad tissue distribution, which at later times showed detectable engraftment primarily in bone marrow, with secondary retention in spleen, lung, and liver. Bone marrow biopsies taken as late as 1.5 years postinfusion were still positive for infused, gene-marked bMSCs, and necropsy results from over 70 tissues failed to demonstrate any abnormal pathology.

These *in vivo* studies in the baboon and sheep demonstrated that infused adult MSCs adapt to the tissue environment in which they were placed, with corresponding lineage differentiation. This microenvironmental response likely reflects the stem cell plasticity of the MSCs, another example of the crossover of adult stem cells in transplant models to unprecedented lineages. Other examples of such lineage crossover or transdifferentiation include the reconstitution of the hematopoietic compartment by muscle stem cells isolated from a muscle biopsy by their dye exclusion properties [32]. An additional advantage of the xenogeneic transplant models is that separation or reisolation of implanted MSCs may be facilitated by standard cell enrichment techniques. The stem cell properties of the HSC have primarily been defined through sequential adoptive transfer experiments, in which full hematopoietic reconstituting activity can be transferred serially from animal to animal by reisolation of the infused stem cells. Similar rigorous analysis of the adult mesenchymal stem cell can be tested in a xenogeneic engraftment model, whether utilizing the sheep fetus or an adult transplant model. By employing genetic marking, it is possible to prove cell lineage relationships and demonstrate maintenance of the multipotential hMSC in diverse tissues.

FUTURE CONSIDERATIONS

The ability to replicate mesenchymal stem cell differentiation *in vitro* is a critical achievement toward defining the molecular events regulating stem cell biology. In an example of the dissection of a biochemical pathway regulating mesenchymal lineage decisions reported by Jaiswal *et al.*, the osteogenic versus adipogenic fate was shown to be regulated through the activity of the MAP kinase ERK-2 [56]. Adaptation of differentiation assays for high-throughput screening strategies will facilitate the identification of morphogenic regulatory compounds. Reproducible assays for mesenchymal lineages are vital to gene profiling and gene discovery efforts to characterize mesengenic processes at the molecular level. Development of chemically defined media supporting stem cell expansion and differentiation will facilitate the study of many aspects of mesenchymal stem cell biology. A chemically defined medium is also crucial for biochemical approaches to isolate molecules produced by stem cells, such as for *ex vivo* support of hematopoiesis [57] or molecules regulating immune recognition [58]. The potential to use hMSCs allogeneically [58] would mean that cells from one donor may be used in many recipients, providing a tremendous improvement in quality control and cost savings that would truly facilitate the therapeutic development of hMSC technologies for tissue regeneration. While much remains to be learned concerning the *in situ* differentiation responses of the hMSC, it is clear that they provide the ideal cellular basis for tissue engineering for many tissues.

REFERENCES

1. Urist, M. R., and McLean, F. C. (1952). Osteogenic potence and new bone formation by induction in transplants to the anterior chamber of the eye. *J. Bone Jt. Surg. (Am. Vol.)* **34**, 443–470.
2. Tavassoli, M., and Crosby, W. H. (1968). Transplantation of marrow to extramedullary sites. *Science* **161**, 548–556.
3. Friedenstein, A. J., Piatetzky-Shapiro, I. I., and Petrakova, K. V. (1966). Osteogenesis in transplants of bone marrow cells. *J. Embryol. Exp. Morphol.* **16**, 381–390.
4. Friedenstein, A. J., Chailakhjan, R. K., and Lalykina, K. S. (1970). The development of fibroblast colonies in monolayer cultures of guinea pig bone marrow and spleen cells. *Cell Tissue Kinet.* **3**, 393–403.
5. Ashton, B. A., Allen, T. D., Howlett, C. R., Eaglesom, C. C., Hattori, A., and Owen, M. (1980). Formation of bone and cartilage by marrow stromal cells in diffusion chambers *in vivo*. *Clin. Ortho. Relat. Res.* **151**, 294–307.
6. Owen, M. E., Cave, J., and Joyner, C. J. (1987). Clonal analysis in vitro of osteogenic differentiation of marrow CFU-F. *J. Cell Sci.* **87**, 731–738.
7. Gregoriadis, A. E., Heersche, J. N. M., and Aubin, J. E. (1988). Differentiation of muscle, fat, cartilage, and bone from progenitor cells present in a bone derived clonal cell population: Effect of dexamethasone. *J. Cell Biol.* **106**, 2139–2151.
8. Poliard, A., Nifuji, A., Lamblin, D., Plee, E., Forest, C., and Kellerman, O. (1995). Controlled conversion of an immortalized mesodermal progenitor cell towards the osteogenic, chondrogenic, or adipogenic pathways. *J. Cell Biol.* **130**, 1461–1472.
9. Dennis, J. E., Merriam, A., Awadallah, A., Yoo, J. U., Johnstone, B., and Caplan, A. I. (1999). A quadripotent mesenchymal progenitor cell isolated from the marrow of mice. *J. Bone Miner. Res.* **14**, 700–709.

10. Pereira, R. F., Halford, K. W., O'Hara, M. D., Leeper, D. B., Sokolov, B. P., Pollard, M. D., Bagasra, O., and Prockop, D. J. (1995). Cultured adherent cells from marrow can serve as long lasting precursor cells for bone, cartilage, and lung in irradiated mice. *Proc. Natl. Acad. Sci. U.S.A.* **92**, 4857–4861.

11. Owen, M. E., and Friedenstein, A. J. (1988). Stromal stem cells: Marrow-derived osteogenic precursors. *Ciba Found. Symp.* **136**, 42–60.

12. Caplan, A. I. (1991). Mesenchymal stem cells. *J. Orthop. Res.* **9**, 641–650.

13. Aubin, J. E., Bellows, C. G., Turksen, K., Liu, F., and Heersche, J. N. M. (1992). Analysis of the osteoblast lineage and regulation of differentiation. *In* "Chemistry and Biology of Mineralized Tissues" (H. Slavkin and P. Price, eds.), pp. 267–275. Elsevier, Amsterdam.

14. Haynesworth, S. E., Goshima, J., Goldberg, V. M., and Caplan, A. I. (1992). Characterization of cells with osteogenic potential from human bone marrow. *Bone* **13**, 81–88.

15. Haynesworth, S. E., Baber, M., and Caplan, A. I. (1992). Cell surface antigens on human marrow-derived mesenchymal cells are detected by monoclonal antibodies. *Bone* **13**, 69–80.

16. Bruder, S. P., Jaiswal, N., and Haynesworth, S. E. (1997). Growth kinetics, self-renewal, and osteogenic potential of purified human mesenchymal stem cells during extensive subcultivation and following cryopreservation. *J. Cell. Biochem.* **64**, 278–294.

17. Cheng, S.-L., Yang, J. W., Rifas, L., Zhang, S.-F., and Avioli, L. V. (1994). Differentiation of human bone marrow osteogenic stromal cells in vitro: Induction of the osteoblast phenotpye by dexamethasone. *Endocrinology (Baltimore)* **134**, 277–286.

18. Jaiswal, N., Haynesworth, S. E., Caplan, A. I., and Bruder, S. P. (1997). Osteogenic differentiation of purified, culture expanded human mesenchymal stem cells in vitro. *J. Cell. Biochem.* **64**, 295–312.

19. Barry, F. P., Johnstone, B., Pittenger, M. F., Mackay, A. M., and Murphy, J. M. (1997). Modulation of the chondrogenic potential of human bone marrow-derived mesenchymal stem cells by TGFβ_1 and TGFβ_3. *Transact. Orthop. Res. Soc.* **22**, 228.

20. Mackay, A. M., Beck, S. C., Murphy, J. M., Barry, F. P., Chichester, C. O., and Pittenger, M. F. (1998). Chondrogenic differentiation of cultured human mesenchymal stem cells from marrow. *Tissue Eng.* **4**, 415–428.

21. Yoo, J. U., Barthel, T. S., Nishimura, K., Solchaga, L., Caplan, A. I., Goldberg, V. M., and Johnstone, B. (1998). The chondrogenic potential of human bone marrow derived mesenchymal progenitor cells. *J. Bone Jt. Surg. (Am. Vol.)* **80A**, 1745–1757.

22. Pittenger, M. F., Mackay, A. M., Beck, S. C., Jaiswal, R. K., Douglas, R., Mosca, J. D., Moorman, M. A., Simonetti, D. W., Craig, S., and Marshak, D. R. (1999). Multilineage potential of adult human mesenchymal stem cells. *Science* **284**, 143–147.

23. Muraglia, A., Cancedda, R., and Quarto, R. (2000). Clonal mesenchymal progenitors from human bone marrow differentiate in an hierarchical model. *J. Cell Sci.* **13**, 1161–1166.

24. Toma, C., Pittenger, M. F., Byrne, B. J., and Kessler, P. D. (2001). Human mesenchymal stem cells differentiate into a cardiomyocyte-like phenotype in the adult murine heart. In revision.

25. Lazarus, H. M., Haynesworth, S. E., Gerson, S. L., Rosenthal, N. S., and Caplan, A. I. (1995). Ex vivo expansion and subseqeunt infusion of human bone marrow derived stromal progenitor cells (mesenchymal progenitor cells): Implications for therapeutic use. *Bone Marrow Transplant.* **15**, 935–942.

26. Koc, O. N., Gerson, S. L., Cooper, B. W., Dyhouse, S. M., Haynesworth, S. E., Caplan, A. I., and Lazarus, H. M. (2000). Rapid hematopoietic recovery after coinfusion of autologous blood stem cells and culture expanded marrow mesenchymal stem cells in advanced breast cancer patients receiving high dose chemotherapy. *J. Clin. Oncol.* **18**, 307–316.

27. Horwitz, E. M., Prockop, D. J., Fitzpatrick, L. A., Koo, W. W. K., Gordon, P. L., Neel, M., Sussman, M., Orchard, P., Marx, J. C., Pyeritz, R. E., and Brenner, M. K. (1999). Transplantability and therapeutic effects of bone marrow-derived mesenchymal cells in children with osteogenesis imperfecta. *Nat. Med.* **5**, 309–313.

28. Caplan, A. I., Reuben, D., and Haynesworth, S. E. (1998). Cell-based tissue engineering therapies: Influence of whole body physiology. *Adv. Drug Delivery Rev.* **3**, 3–14.

29. Deans, R. J., and Moseley, A. B. (2000). Mesenchymal stem cells: Biology and potential clinical uses. *Exp. Hematol.* **28**, 875–884.

30. Pittenger, M. P., and Marshak, D. R. (2001). Mesenchymal stem cells of human adult bone marrow. *In* "Stem Cells" (D. R. Marshak, R. L. Gardner, and A. Gottlieb, eds.), pp. 288–294. Cold Spring Harbor Laboratory Press, Cold Spring Harbor, NY.

31. Pittenger, M. F., and Mackay, A. M. (2001). Multipotential human mesenchymal stem cells. *Graft* **3** (in press).

32. Jackson, K. A., Tiejuan, M., and Goddell, M. A. (1999). Hematopoietic potential of stem cells isolated from murine skeletal muscle. *Proc. Natl. Acad. Sci. U.S.A.* **96**, 14482–14486.

33. Bjornson, C. R. R., Rietze, R. L., Reynolds, B. A., Magli, M. C., and Vescovi, A. L. (1999). Turning brain into blood: A hematopoietic fate adopted by adult neural stem cells *in vivo*. *Science* **283**, 534–537.

34. Lennon, D. P., Haynesworth, S. E., Bruder, S. P., Jaiswal, N., and Caplan, A. I. (1996). Development of a serum screen for mesenchymal progenitor cells from bone marrow. *In Vitro Cell. Dev. Biol.* **32**, 602–611.

35. Martin, I., Muraglia, A., Campanile, G., Cancedda, R., and Quarto, R. (1997). Fibroblast growth factor-2 supports ex vivo expansion and maintenance of osteogenic precursors from human bone marrow. *Endocrinology (Baltimore)* **138**, 4456–4462.

36. Pittenger, M. F., Mbalaviele, G., Mosca, J. D., Black, M., and Marshak, D. R. (2000). Adult mesenchymal stem cells. *In* "Primary Mesenchymal Cells" (M. R. Koller, B. O. Palsson, and J. R. W. Masters, eds.), pp. 189–207. Kluwer Academic Publishers, Dordrecht, The Netherlands.

37. Ponticello, M. S., Schinagl, R. M., Kadiyala, S., and Barry, F. P. (2000). Gelatin based resorbable sponge as a carrier matrix for human mesenchymal stem cells in cartilage regeneration therapy. *J. Biomed. Mater. Res.* **52**, 246–255.

38. Kadiyala, S., Jaiswal, N., and Bruder, S. P. (1997). Culture-expanded, bone marrow-derived mesenchymal stem cells can regenerate a critical-sized segmental bone defect. *Tissue Eng.* **3**, 173–185.

39. Bruder, S. P., Krauss, K. H., Goldberg, V. M., and Kadiyala, S. (1998). The effects of implants loaded with autologous mesenchymal stem cells on the healing of canine segmental bone defects. *J. Bone Jt. Surg. (Am. Vol.)* **80A**, 985–996.

40. Bruder, S. P., Kurth, A. A., Shea, M., Hayes, W. C., Jaiswal, N., and Kadiyala, S. (1998). Bone regeneration by implantation of purified, culture-expanded human mesenchymal stem cells. *J. Orthop. Res.* **16**, 155–162.

41. Wakitani, S., Goto, T., Pineda, S. J., Young, R. G., Mansour, J. M., Caplan, A. I., and Goldberg, V. M. (1994). Mesenchymal cell-based repair of large full-thickness defects of articular cartilage. *J. Bone Jt. Surg. (Am. Vol.)* **76A**, 579–592.

42. Grande, D. A., Southerland, R., and Manji, R. (1995). Repair of articular defects using mesenchymal stem cells. *Tissue Eng.* **1**, 345–352.

43. Young, R. G., Butler, D. L., Weber, W., Caplan, A. I., Gordon, S. L., and Fink, D. J. (1997). Use of mesenchymal stem cells in a collagen matrix for Achilles tendon repair. *J. Orthop. Res.* **16**, 406–413.

44. Dick, J. E. (1994). Future prospects for animal models created by transplanting human haematopoietic cells into immune-deficient mice. *Res. Immun.* **145**, 380–384.

45. Dorrell, C., Gan, O. I., Pereira, D. S., Hawley, R. G., and Dick, J. E. (2000). Expansion of human cord blood CD34(+)CD38(−) cells in ex vivo culture during retroviral transduction without a corresponding increase in SCID repopulating cell (SRC) frequency: Dissociation of SRC phenotype and function. *Blood* **95**, 102–110.

46. Larochelle, A., Vormoor, J., Hanenberg, H., Wang, J. C., Bhatia, M., Lapidot, T., Moritz, T., Murdoch, B., Xiao, X. L., Kato, I., Williams, D. A., and Dick, J. E. (1996). Identification of primitive human hematopoietic cells capable of repopulating NOD/SCID mouse bone marrow: Implications for gene therapy. *Nat. Med.* **2**, 1329–1337.

47. Flake, A. W., Harrison, M. R., Adzick, N. S., and Zanjani, E. D. (1986). Transplantation of fetal hematopoietic stem cells in utero: The creation of hematopoietic chimeras. *Science* **233**, 776–778.

48. Zanjani, E. D., Pallavicini, M. G., Flake, A. W., Ascensao, J. L., Langlois, R. G., Reitsma, M., MacKintosh, F. R., Stutes, D., Harrison, M. R., and Tavassoli, M. (1992). Engraftment and long-term expression of human fetal hemopoietic stem cells in sheep following transplantation in utero. *J. Clin. Invest.* **89**, 1178–1188.

49. Zanjani, E. D., Flake, A. W., Rice, H., Hedrick, M., and Tavassoli, M. (1994). Long-term repopulating ability of xenogeneic transplanted human fetal liver hematopoietic stem cells in sheep. *J. Clin. Invest.* **93**, 1051–1055.

50. Goodnow, C. (1996). Balancing immunity and tolerance: Deleting and tuning lymphocyte repertoires. *Proc. Natl. Acad. Sci. U.S.A.* **93**, 2264–2271.

51. Sprent, J. (1995). Central tolerance of T cells. *Int. Rev. Immunol.* **13**, 95–105.

52. Liechty, T. C., M. Shaaban, A. R., A. F., Moseley, A. B., Deans, R., Marshak, D. R., and Flake, A. W. (2000). Human mesenchymal stem cells engraft and demonstrate site-specific differentiation after in utero transplantation in sheep. *Nat. Med.* **6**, 1282–1286.

53. Silverstein, A. M., Prendergast, R. A., and Kraner, K. L. (1964). Fetal response to antigenic stimulus. IV. Rejection of skin homografts by the fetal lamb. *J. Exp. Med.* **119**, 955–964.

54. Zanjani, E., Almeida-Porada, G., Ascensao, J., MacKintosh, F., and Flake, A. W. (1997). Transplantation of hematopoietic stem cells in utero. *Stem Cells* **15**, 79–93.

55. Devine, A., Bartholomew, A., Nelson, M., Mahmud, N., Moran, S., Hardy, W., Sturgeon, C., Hewett, T., Chung, T., Sher, D., Weissman, S., Mosca, J., Deans, R., Moseley, A., and Hoffman, R. (2001). Mesenchymal stem cells home to and engraft in the bone marrow of non-human primates. *Exp. Hematol.* **29**, 244–255.

56. Jaiswal, R. K., Jaiswal, N., Bruder, S. P., Mbalaviele, G., Marshak, D., and Pittenger, M. F. (2000). Adult human mesenchymal stem cell differentiation to the osteogenic or adipogenic lineage is regulated by mitogen-activated protein kinase. *J. Biol. Chem.* **275**, 9645–9652.

57. Majumdar, M. K., Thiede, M. A., Mosca, J. D., Moorman, M., and Gerson, S. L. (1998). Phenotypic and functional comparison of cultures of marrow-derived mesenchymal stem cells and stromal cells. *J. Cell. Physiol.* **176**, 57–66.

58. McIntosh, K., and Bartholomew, A. (2001). Stromal cell modulation of the immune system. *Graft* **3**, 324–328.

CHAPTER 39

STEM CELL CULTURE: CHONDROGENIC STEM CELLS

Georg Bartsch and Anthony Atala

D amaged cartilage may not regenerate into normal functional tissue. As a result, many strategies have been proposed for the regeneration of normal cartilage using stem cells and tissue engineering [1].

This chapter describes three approaches that are being used for the isolation of chondrogenic stem cells:

1. Embryonic stem cells
2. Mesenchymal stem cells from bone marrow
3. Mesenchymal stem cells from connective tissue

CHONDROGENIC STEM CELLS FROM EMBRYONIC SOURCES

PRIMARY CHONDROGENIC STEM CELLS FROM CHICK EMBRYOS [2,3]

1. Wing buds of 18/19 to 24/25 stage [3a] chicken embryos are dissected at the distal wing bud, with the excised region extending 0.3–0.4 mm from the distal apex of the limb to the proximal cutting edge.

2. After the ectoderm has been stripped, mesenchymal tissue is treated with dispase (0.8 U/ml).

3. Either only the isolated distal subridge or the whole wing bud mesoderm is incubated for 20 min in 0.25% trypsin and then dissociated into a mesenchymal cell suspension.

4. The cells are plated between 0.5 and 2×10^5 cells in 10 μl of differentiation medium on a plastic tissue culture plate. As differentiation medium, Ham's F-12 containing 10% fetal calf serum (FCS), 2 mM glutamine, and antibiotics is used. Ethanol in final concentrations ranging from 0.1 to 4% is added to the medium.

5. After attachment for 1–2 h at 37°C, the pellet is flooded with 4 ml of differentiation medium.

6. Before analyses, cultures are incubated for 1–3 days at 37°C in a 5% CO_2/95% air atmosphere.

MODIFIED PROTOCOL FOR STEM CELLS FROM CHICK EMBRYOS [3A]

1. Wing buds of 18/19 to 24/25 stage [3a] chicken embryos are dissected at the distal wing bud, with the excised region extending 0.3–0.4 mm from the distal apex of the limb to the proximal cutting edge.

2. A modified protocol can be done without the digestion process, with the mesenchymal tissue cut into 1-mm^3 pieces that are explanted into fibronectin-coated petri dishes. For fibronectin coating, the 35-mm petri dishes are incubated overnight at 37°C with 1 ml of sterile water containing 20 μg of human plasma fibronectin.

3. After 10–20 min of adherence, the tissue pieces are supplied with 1–2 ml of Ham's F-12 medium supplemented with 1% fetal bovine serum (FBS) and CR-ITS Premix (5 μg/ml each of insulin, transferrin, and sodium selenite).

4. Cultures are maintained at $37°C$ in a 5% CO_2/95% air atmosphere.

CHONDROGENIC EMBRYONIC STEM CELL LINES
ATDC5 Cells [4–6]

1. The ATDC5 cell line is derived from a mouse embryonal carcinoma from the Riken Gene Bank (Ibaraki, Japan).

2. ATDC5 cells are cultured in a 1:1 mixture of Dulbecco's Modified Eagle's (DMEM) and Ham's F-12 medium containing 5% FBS, 10 μg/ml human transferrin, and 3×10^{-8} M sodium selenite at $37°C$ in a humidified atmosphere of 5% CO_2 in air.

3. For chondrogenic induction, 10 μg/ml bovine insulin is added to the medium. Inoculum density of the cells is as follows: 2×10^4 cells/well in a 24-well plate, 4×10^4 cells/well in a 12-well plate, or 6×10^4 cells/well in a 6-well plate for 21 days.

The Mesoblastic Immortalized Clone C1, Derived from Mouse Teratocarcinoma [7–9]

1. C1 cells are routinely grown in DMEM supplemented with 10% FBS on tissue culture dishes.

2. For chondrogenic induction, C1 cells are seeded at 3×10^5 cells per 10 ml of DMEM supplemented with 10% FCS on bacterial untreated plastic dishes.

3. After 10 days the medium is changed to differentiate the formed aggregates in the plastic dishes. The differentiation medium consists of DMEM supplemented with 1% fetal calf serum (FCS) and 10^{-6} M dexamethasone. Differentiation is performed for different time periods up to 50 days.

Embryonic Stem Cell Line D3 [10]

1. D3 cells are cultured on a feeder layer of mitomycin C inactivated mouse embryonic fibroblasts.

2. For cultivation, DMEM, supplemented with 15% FCS, 2 mM L-glutamine, 5×10^{-5} M β-mercaptoethanol, and nonessential amino acids is used.

3. For differentiation, aliquots of 20 μl of differentiation medium (with 20% FCS instead of 15%) containing 200, 500, or 800 cells are cultivated in "hanging drops" for 2 days.

4. The cells are cultured in suspension on bacterial petri dishes for 3 days.

5. To enhance chondrogenesis, bone morphogenic protein 2 (BMP-2) in concentrations between 2 and 10 ng/ml, or BMP-10 in concentrations of 2 ng/ml, or transforming growth factor $\beta1$ (TGF-$\beta1$) in a concentration of 2 ng/ml, is added to the differentiation medium.

6. The formed embryoid bodies are cultured up to 30 days.

MESENCHYMAL STEM CELLS FROM BONE MARROW
RECOVERY OF MESENCHYMAL STEM CELLS FROM BONE MARROW

The recovery of bone marrow depends on the species. In humans, 20–30 ml of bone marrow aspirate is collected from the iliac crest into a syringe containing 6000 U of heparin to prevent clotting [11–13]. In rodents and chicken bone marrow is collected by flushing femurs and tibias with culture medium [14–16]. In rabbits, the iliac crest or tibia is punctured, and 7–8 ml of marrow is aspirated into a syringe with 3000 U of heparin [17,18]. Regardless of the species, several methods have been described for stem cell isolation. Three methods are described here.

Using Culture Plates for Cell Isolation [17]

1. The marrow is washed once with Dulbecco's phosphate-buffered saline (DPBS).

2. Cells are spun down at 900g for 5 min.

3. Cells are resuspended in DMEM with 10% FBS.

4. Cells are seeded at a density of 20×10^6 cells/100-mm dish and grown for 14 days in a 5% CO_2 atmosphere at $37°C$. After this time period, the cells can be differentiated.

Using Cloning Rings for Cell Isolation [16]

1. The marrow is washed once with DPBS.

2. Adherent subpopulations are cultured in permissive conditions at 33°C with 50 ng/ml gamma-interferon added to the base medium (base medium consists of BGJ$_b$, Gibco, containing 10% of a selected FBS), to promote growth in a 5% CO_2 atmosphere at 37°C.

3. Colonies are isolated by digestion with trypsin–EDTA (0.05% trypsin and 0.53 mM EDTA) for 5 min at 37°C while enclosed within stainless steel cloning rings.

4. Mesenchymal cells are expanded in permissive conditions and recloned by limiting dilution.

Using the Percoll Gradients for Cell Isolation [11–13,19]

1. The marrow is washed once with DPBS.

2. The cells are loaded onto 25 ml of Percoll at a density of 1.073 g/ml in a 50-ml conical tube.

3. The cells are spun down in the Percoll at 1100g for 30 min at 20°C.

4. The cells are collected from the interface and washed with two volumes of DPBS.

5. The cells are spun down at 900g for 5 min and resuspended in DMEM low glucose (1.0 g/liter glucose) with 10% FBS from selected lots.

6. The cells are plated at a density of 2×10^5 cells/cm^2 in a 5% CO_2 atmosphere at 37°C.

DIFFERENTIATION OF BONE MARROW-DERIVED MESENCHYMAL STEM CELLS INTO A CHONDROGENIC LINEAGE [11–13,16,17,19]

1. Cells are trypsinized and counted, and $2–2, 5 \times 10^5$ cell aliquots are spun down at 500g in 15-ml polypropylene conical tubes.

2. The FBS-containing medium is replaced by a defined differentiation medium. This medium consists of DMEM high glucose (4.5 g/liter glucose) supplemented with ITS (6.25 μg/ml insulin, 6.25 μg/ml transferin, 6.25 μg/ml selenous acid), 5.35 μg/ml linoleic acid, and 1.25 μg/ml bovine serum albumin. Sodium pyruvate (1 mM or 100 μg/ml), 100 nM dexamethasone, and 37.5–50 μg/ml ascorbate 2-phosphate are also added. TGF-β1 or TGF-β3 is added to the medium in a concentration of 0.5–10 ng/ml.

3. After the FBS-containing medium has been replaced with 0.5 ml of the differentiation medium, cell pellets are spun down at 500g for 5 min and cultured in a 5% CO_2 atmosphere at 37°C for 3 weeks.

MESENCHYMAL STEM CELLS FROM CONNECTIVE TISSUE

PROGENITOR CELL ISOLATION FROM PERIOSTEUM

Several protocols are reported in literature. Three protocols are described here.

Cell Isolation from Calf Periosteum [20]

1. Skin and the underlying muscular–fibrous connective tissue are removed from the metacarpal region of 18-month-old calves up to the periosteum.

2. The exposed periosteum is removed aseptically and placed into Earle's medium with penicillin and streptomycin at 100 μg/ml.

3. The washed periosteum is cut into 1×1 cm^2 pieces and placed with the osteogenic side downward onto 133-mm standard petri dishes.

4. Cells are fed with a modification of DMEM with 75 mg/liter of glutamine and 10% fetal calf serum. Medium is changed every 3 days.

5. Periosteal osteoprogenitor cells are cultured for 4 weeks until confluency.

Cell Isolation from Chick Periosteum [21,22]

1. Tibial periostea of 1-week-old white Leghorn chicks are exposed by dissecting muscles and fasciae on the the anteromedial surface.

2. Periostea are digested with collagenase–trypsin.

3. Liberated cells are plated at a density of 1.0×10^6 cells/100 mm plastic tissue culture dish in BGJb medium (Gibco) supplemented with 10% FCS. Only viable cells, determined with trypan blue (0.4%), are seeded.

4. Cells are incubated for 10–12 days at 37°C in 95% humified air plus 5% CO_2, with medium changed every 3 days.

Chondrogenic Differentiation from Periosteum-Derived Cells [20,22,23]

1. Cells are lifted off the culture plate with trypsin–EDTA (0.25% trypsin, 1 mM EDTA) for 5–10 min.

2. Onto petri dishes containing a gel–medium fluid, 5×10^5 cells/cm^2 are transferred.

3. The gel contains a thin film of high-T_m agarose and a low-T_m agarose for cell suspension. High-T_m agarose (1% in water) is autoclaved for 15 min and used when above 90°C. Dishes are rapidly coated, and excess agarose is removed by aspiration. After agarose has solidified at 22°C, low-T_m agarose is used for cell suspension. Low-T_m agarose (2%) is autoclaved for 15 min, equilibrated to 37°C, and mixed with an equal volume of DMEM at 37°C to give a final concentration of 1% agarose in DMEM.

4. In DMEM, resuspended cells are mixed with an equal volume of low T_m agarose at 37°C to reach the concentration of 0.5% agarose. The following volumes of gel–medium fluid are placed for each dish: 0.5 ml for 35-mm dishes, 1.0 ml for 60-mm dishes, and 2.5 ml for 100-mm dishes. Dishes are held at 37°C in a CO_2 incubator for 5 min before gelation at 4°C for 15 min.

5. Cells are fed with growth medium and cultured in an environment of 7.5% CO_2 and 92.5% air.

6. Culture medium used for chondrogenic differentiation in gel culture is DMEM containing 4.8 g/liter of glucose, supplemented with 10% FCS, 50 µg/mlm of ascorbate, 1 mM cysteine, 1 mM pyruvate, 100 U/ml of penicillin, and 100 µg/ml of streptomycin.

7. Cells are assayed after 2 weeks in culture.

CHONDROGENIC STEM CELL ISOLATION FROM MUSCLE CELLS [18,24]

1. Newborn rabbit leg skeletal muscle is harvested from New Zealand white rabbits under sterile conditions.

2. Specimens are placed into minimal essential medium with Earle's salts (EMEM) supplemented with $3 \times$ antibiotic–antimycotic solution for at least 10 min.

3. The muscle is finely minced with scissors.

4. The media and tissue are centrifuged at 150g for 10 min, the supernatant is discarded, and the tissue is transferred to a sterile bottle containing a magnetic stir bar.

5. The tissue is digested with collagenease–dispase solution consisting of 250 U/ml CLSI (Worthington, collagenase, Freehold, NJ) and 33 U/ml dispase (Collaborative Research, Cambridge, MA), in a ratio of 1:4:15 (v/v/v) of tissue to collagenase/dispase to EMEM.

6. Tissue is digested for approximately 45 min.

7. The digested tissue is centrifuged at 300g for 20 min.

8. The supernatant is discarded.

9. The cell pellet is resuspended in EMEM supplememted with 10% horse serum with penicillin–streptomycin antibiotics.

10. The suspension is filtered through a 20-µm Nitex filter, and an aliquot of the cells is counted on the hemocytometer.

11. One each 100-mm gelatin coated dish, 10^5 cells are plated.

12. Cells are grown for 7–10 days until confluency, with medium changes every 3 days.

13. Cells are detached from the dish with 0.025% trypsin in a solution of 3:1 DPBS without Ca^{2+}, Mg^{2+}, and DPBS-EDTA.

14. Trypsin is neutralized with horse serum and the suspension is centrifuged at 150g for 20 min.

15. The supernatant is discarded, the cell pellet is resuspended in EMEM with 10% horse serum, and cells are filtered through a 20-μm Nitex filter.

16. 2×10^6 cells/ml are frozen down in a freezing medium containing EMEM with 10% horse serum and 7.5% dimethyl sulfoxide (DMSO). Cells are frozen slowly in a freezing chamber to $-80°C$.

17. After at least 16 h, cells are thawed and plated at 10^5 cells per 100 mm gelatin-coated culture dish and grown to confluency. This culture consists of mesenchymal stem cells.

18. For chondrogenic differentiation, the mesenchymal stem cells are treated with medium supplemented with dexamethasone at concentrations ranging from 10^{-10} M to 10^{-6} M for up to 5 weeks.

REFERENCES

1. Johnstone, B., and Yoo, J. U. (1999). Autologous mesenchymal progenitor cells in articular cartilage repair. *Clin. Orthop. Relat. Res.* 367(Suppl.), S156–S162.
2. Kulyk, W. M., and Hoffman, L. M. (1996). Ethanol exposure stimulates cartilage differentiation by embryonic limb mesenchyme cells. *Exp. Cell Res.* 223(2), 290–300.
3. Kosher, R. A., and Rodgers, B. J. (1987). Separation of the myogenic and chondrogenic progenitor cells of undifferentiated limb mesenchyme. *Dev. Biol.* 121(2), 376–388.
3a. Hamburger, V., and Hamilton, H. L. (1951). A series of normal stages in development of the chick embryo. *J. Morphol.* 88, 49–92.
4. Fujishige, K., *et al.* (1999). Alteration of cGMP metabolism during chondrogenic differentiation of chondroprogenitor-like EC cells, ATDC5. *Biochim. Biophys. Acta* 1452(3), 219–227.
5. Shukunami, C., *et al.* (1997). Cellular hypertrophy and calcification of embryonal carcinoma-derived chondrogenic cell line ATDC5 in vitro. *J. Bone Miner. Res.* 12(8), 1174–1188.
6. Shukunami, C., *et al.* (1996). Chondrogenic differentiation of clonal mouse embryonic cell line ATDC5 in vitro: Differentiation-dependent gene expression of parathyroid hormone (PTH)/PTH-related peptide receptor. *J. Cell Biol.* 133(2), 457–468.
7. Poliard, A., *et al.* (1995). Controlled conversion of an immortalized mesodermal progenitor cell towards osteogenic, chondrogenic, or adipogenic pathways. *J. Cell Biol.* 130(6), 1461–1472.
8. Kellermann, O., *et al.* (1990). An immortalized osteogenic cell line derived from mouse teratocarcinoma is able to mineralize in vivo and in vitro. *J. Cell Biol.* 110(1), 123–132.
9. Poliard, A., *et al.* (1999). Lineage-dependent collagen expression and assembly during osteogenic or chondrogenic differentiation of a mesoblastic cell line. *Exp. Cell Res.* 253(2), 385–395.
10. Kramer, J., *et al.* (2000). Embryonic stem cell-derived chondrogenic differentiation in vitro: Activation by BMP-2 and BMP-4. *Mech. Dev.* 92(2), 193–205.
11. Mackay, A. M., *et al.* (1998). Chondrogenic differentiation of cultured human mesenchymal stem cells from marrow. *Tissue Eng.* 4(4), 415–428.
12. Lennon, D. P., *et al.* (2000). Dilution of human mesenchymal stem cells with dermal fibroblasts and the effects on in vitro and in vivo osteochondrogenesis. *Dev. Dyn.* 219(1), 50–62.
13. Pittenger, M. F., *et al.* (1999). Multilineage potential of adult human mesenchymal stem cells. *Science* 284, 143–147.
14. Leboy, P. S., *et al.* (1991). Dexamethasone induction of osteoblast mRNAs in rat marrow stromal cell cultures. *J. Cell. Physiol.* 146(3), 370–378.
15. Bruder, S. P., *et al.* (1990). Osteochondral differentiation and the emergence of stage-specific osteogenic cell-surface molecules by bone marrow cells in diffusion chambers. *Bone Miner.* 11(2), 141–151.
16. Dennis, J. E., *et al.* (1999). A quadripotential mesenchymal progenitor cell isolated from the marrow of an adult mouse. *J. Bone Miner. Res.* 14(5), 700–709.
17. Johnstone, B., *et al.* (1998). In vitro chondrogenesis of bone marrow-derived mesenchymal progenitor cells. *Exp. Cell Res.* 238(1), 265–272.
18. Grande, D. A., *et al.* (1995). Repair of articular cartilage defects using mesenchymal stem cells. *Tissue Eng.* 1(4), 345–353.
19. Yoo, J. U., *et al.* (1998). The chondrogenic potential of human bone-marrow-derived mesenchymal progenitor cells. *J. Bone Jt. Surg. (Am. Vol.)* 80(12), 1745–1757.
20. Bahrami, S., *et al.* (2000). Periosteally derived osteoblast-like cells differentiate into chondrocytes in suspension culture in agarose. *Anat. Rec.* 259(2), 124–130.
21. Nakahara, H., *et al.* (1990). Bone and cartilage formation in diffusion chambers by subcultured cells derived from the periosteum. *Bone* 11(3), 181–188.
22. Nakata, K., *et al.* (1992). Collagen gene expression during chondrogenesis from chick periosteum-derived cells. *FEBS Lett.* 299(3), 278–282.
23. Benya, P. D., and Shaffer, J. D. (1982). Dedifferentiated chondrocytes reexpress the differentiated collagen phenotype when cultured in agarose gels. *Cell (Cambridge, Mass.)* 30(1), 215–224.
24. Young, H. E., *et al.* (1995). Mesenchymal stem cells reside within the connective tissues of many organs. *Dev. Dyn.* 202(2), 137–144.

CHAPTER 40

STEM CELL CULTURE: HEMATOPOIETIC STEM CELLS

Joel S. Greenberger

Hematopoietic stem cells represent one of the first multiplelineage, multipotential cell populations to be studied in animal models *in vitro* in the setting of human transplantation. Each area of study of hematopoietic stem cells requires specific methodologies, and research advances in each area have led to new sets of problems.

IDENTIFICATION OF HEMATOPOIETIC STEM CELLS BY ANIMAL MODEL SYSTEMS

Limiting dilution techniques have been used to demonstrate that infusion of bone marrow cells into lethally irradiated mice results in a dose-dependent increase in the number of spleen colonies detected on the surface of spleens at 9–10 days [1]. Cells-forming detectable colonies at these earlier time points were later shown to be positive, committed progenitor cells rather than totipotential (lymphoid/myeloid reconstituting) hematopoietic stem cells [2]. Competitive repopulation assay provides a true *in vivo* quantitative method for totipotential hematopoietic stem cells. This assay [3], compared with one in which irradiated animals received hematopoietic cells from the bone marrow of two different donor mouse strains, was antigenically distinct and had separate cell surface markers. The "competitive" ability to repopulate the hematopoietic system of recipient animals was used to quantitate the frequency of totipotential cells from each donor.

Separation techniques to identify subsets of mouse hematopoietic progenitor cells by surface phenotype led to the isolation of Sca-1$^+$lin$^-$ subsets either intensely bright surface antigen or for the Thy-1 (anti-CDw90) (Pharmingen, San Diego, CA) surface antigen or Thy-1^{low} [4]. Exclusion of the mitochondrial-activated dye rhodamine-123 also correlated with the quiescence date of totipotential reconstituting hematopoietic stem cells [5]. Sophisticated fluorescence-activated cell sorters with the capacity to sort for multiple lineages with as many as seven different subset sorting capabilities led to the isolation of more refined subsets of hematopoietic stem cells from mouse bone marrow or spleen. Such techniques, and the ability to carry out large numbers of experiments with mice, led to the observation that a single cell could explain totipotential hematopoietic repopulation [6].

In vivo transplantation techniques are carried out by means of intravenous injection of sorted subsets of hematopoietic cells. There is also the availability of genetically inbred mouse strains for study. However, a problem persists in that very large numbers of cells are required to get sufficient cell populations for in *vivo* reconstitution. Furthermore, the expense of waiting for *in vivo* reconstitution and analyzing donor origin of multiple hematopoietic lineages in the reconstituted host limits the speed of experimentation with respect to the problem of expanding hematopoietic stem cells. When the best available techniques for *in vitro* expansion of stem cells were need, a 1.2-fold to 1.4-fold increase in reconstituting cells was recently reported in 1997 [7]. The severe combined immunodeficient/human (SCID/HU) and nonobese diabetic SCID (NOD/SCID) immune-deficient mouse strains [8] have provided a valuable vehicle with which to study engraftment by human hematopoietic stem cell subsets. However, the problem with inadequate cell numbers is exacerbated in the human xenotransplant model. While as little as one mouse stem cell can reconstitute all pathways in the recipient mouse, one estimate is that a thousand or more cells are required to reconstitute an SCID/SCID mouse if one is using human donor cells [9].

IN VITRO ASSAYS FOR GROWTH OF HEMATOPOIETIC STEM CELLS

In vitro assays for colony-forming multilineage progenitor cells have been available since the early 1980s [10]. Assays for single cells of mouse or human origin with the capacity to produce multiple-lineage colonies containing lymphohematopoietic cells have been reported [7]. The capacity of circulating hematopoietic stem cells to home to the hematopoietic microenvironment and differentiate into multilineages has been utilized to develop the colony-forming unit Dexter, colony-forming unitcobblestone (CFU-D and CFU-C) , and long-term culture initiating cell (LTCIC) assays [11–15]. These assays rely upon the capacity of single cells to attach to an underlying stromal cell layer of either whole bone marrow or a permanent murine cell line [16]. Attachment and proliferation of the hematopoietic cell on the adherent layer leads to the formation of cobblestone islands. Cobblestone island formation was first recognized in murine long-term bone marrow cultures (LTBMCs) as flattened areas associated with long-term *in vitro* hematopoiesis. LTBMCs [17] rely upon the availability of 17-hydroxycorticosteroid in culture medium with maintenance of cobblestone islands. Attempts to reconstitute Dexter-type cultures with single populations of cells that are sorted prior to explant have not reproduced successful long-term hematopoiesis (for periods exceeding one year) in certain mouse strains in which cultures are initiated by explant of the entire contents of the femur and tibia of an adult mouse [18].

The LTCIC assay [19] provides a system with which to quantitate primitive hematopoietic progenitor cells, since cell suspensions containing large numbers of LTCICs have been shown to be associated with reconstitution capacity of experimental animals or humans *in vivo*. However, LTCICs do not run an accurate measure of totipotential hematopoietic stem cells. While the LTCIC population of human umbilical cord blood, peripheral-blood-mobilized stem cells, or bone marrow does contain cells with totipotential repopulating ability, cobblestone islands and LTCIC biologic behavior *in vitro* can be produced by hematopoietic progenitor cell lines dependent on interleukin 3 (IL-3) [20]. Expansion of LTCIC numbers from assorted populations of hematopoietic cells has been correlated with increased reconstitution capacity, so that there is clear overlap between these populations. The LTCIC assay is also useful in detecting the success with which pools of hematopoietic stem cells or single stem cells can be expanded *in vitro*.

Single-cell cloning experiments represent the most effective technique by which to quantitate the multilineage lymphohematopoietic capacity of cells thought to represent stem cells. Since there remains continued controversy over the surface phenotype of the totipotential reconstituting stem cell, whether it be CD34$^+$ for human Sca-1/Y6$^+$ or CD34$^-$ for mouse Sca-1/Y6$^-$, or contained with other subsets of phenotypically defined cells, it is clear that only through single-cell cloning studies can one determine the accuracy of the surface phenotype as a reflection of true biological behavior. For this reason, the term "stem cell candidate" [21] has been used by scientists studying hematopoietic stem cells as defined by particular patterns of cell surface phenotype [22].

While single-cell cloning assays provide new evidence that some cells have multilineage differentiation capacity, there is an associated problem of cell compatibility with the *in vitro* culture conditions chosen. For example, cocultivation of single cells with AFT024 (a murine fetal liver line, that supports culture of transplantable murine stem cells) fetal liver cells [23] in the presence of stem cell factor (SCF), FLT-3 ligand (FLT-3L), and IL-7 has been shown to allow amplification over 4 weeks of several thousand cells from a single hematopoietic cell.

Phenotyping of these thousands of cells has shown natural killer (NK) cell, B-lymphocyte, T-lymphocyte, myeloid, megakaryocyte, and erythroid differentiation capacity. Such evidence strongly indicates that the particular cell producing numerous progeny in so many different lineages was in fact a stem cell. However, single-cell cloning assays may miss other

Table 40.1. *Assays for Detection of Hematopoietic Stem Cells*

In vivo assays
  Competitive repopulation assay [10]
  NOD/SCID, SCID/HU repopulation assay [5]

In vitro assays
  Long-term bone marrow culture (LTBMC) longevity [13]
  Long-term culture initiating cell (LTCIC) [12,19]
  Murine long-term culture longevity [18]

Single-colony assay techniques
  Single-cell doublet and microdissection with expansion [23]
  Single-cell expansion assay [10]

cells that do not adapt to a particular *in vitro* culture condition but would function *in vivo*. Recent data have shown heterogeneity between CD34$^+$lin$^-$ subsets from different sources. For example, three different human umbilical cord blood samples yielded different frequencies of CD34$^+$lin$^-$ cells; and of these final sorted numbers, the frequency with which single cells formed multilineage differentiated progeny in the coculture system with fetal liver cells varied (Tables 40.1, 40.2).

Assays studying micromanipulated and separated cell doublets appear to be a valuable method by which to show that daughter cells of the first cell division are hematopoietic stem cells. Initial studies using fluorochrome dyes for CD34 and lineage markers showed an increased capacity of thrombopoietin (TPO) and FLT-3L to induce a conservative division of single CD34$^+$lin$^-$ stem cell candidates [22]. Experiments isolating doublets of the first division and coculture in the expansion culture system on fetal liver stroma also demonstrated a significant capacity of these cells to form multilineage daughters (Table 40.1). The problems associated with adaptability of cells to culture media, the fragility of cells following sorting, and the suboptimal probability of cell transfer are borne out in such experiments. Table 40.2 shows that 666 single cells from one experiment with cells plated at 1 cell per well formed 110 doublets. A very small proportion of 220 micromanipulated and separated daughters were transferable and could be identified as having survived the transfer, a very small subset of which produced two multilineage daughter cells. Table 40.2 classifies seven basic assay types and provides literature references.

HUMAN HEMATOPOIETIC STEM CELL TRANSPLANTATION

The capacity to freshly transfer human umbilical cord blood, bone marrow, or granulocyte-colony-stimulating factor (G-CSF) immobilized peripheral blood stem cells to a human transplant recipient and result in totipotential donor-origin hematopoiesis provides strong evidence that such cell populations contained hematopoietic stem cells [24]. Early controversy on this point focused on the possibility that the transplantation of large numbers of cells actually transfused committed progenitor cells but not totipotential stem cells. During the past 20 years, there has been strong evidence that true stem cells are in fact transplanted during human transplantation. This evidence comes from the detection of chromosome markers associated with one clonal line in multiple lineages of differentiated cells, the observation that those stem cell sources contained *in vitro* high numbers of LTCIC or large colony-forming cells, which are more effective at reconstituting hematopoiesis, and the observation that clonal succession of engrafted stem cells led to varying patterns of hematopoiesis at different times after transplant. These donor cellular sources have been successfully cryopreserved, thawed, and reinfused with successful multilineage hematopoietic engraftment.

Problems in attempting to expand stem cell numbers from these human cellular sources continue. Batch-type bioreactors [25], profusion devices [26,27], and tissue culture on stro-

Table 40.2. Single-Cell Expansion of Single Cord Blood CD34+lin− Cells Cultured in Thrombopoietin (TPO) Plus FLT-3 Ligand (FLT-3L) at the Time of Doubling, Microdissection of Doublet, Separation, and Expansion

Frequency

666 Single cells plated in TPO plus FLT-3L   16.5% of single cells plated
110 Doublets detected
220 Cells transferred to AFT024 culture
146 Cells viably transferred (confirmed)
8 Doublets reveal ≥ daughter myeloid/lymphoid (stem cell)   7.3% of doublets, 1.2% of single cells
3 Doublets reveal both daughters are myeloid/lymphoid (stem cell)   2.7% of doublets, 0.5% of single cells
(lines 3, 4, and 7, column 1, daughter A)

Daughter A

Well 1: total neutrophil NK cells	Well 2: total T cells myeloid	Well 3: total B cells	Well 4 total NK cells	Grand total of cells
CD15/CD56	CD203/CD13+33	CD19	CD16	
1. 280/0/500	7/0/460	0/23	0/305	1548
2. 29/5/352	0/140/350	0/300	6/300	1302
3. 733/0/1170[a]	0/160/1339	0/1000	8/1000	4509
4. 77/0/95[a]	0/24/27	0/105	0/200	427
5. 106/0/115	0/0/72	0/48	0/39	274
6. 98/20/100	22/49/82	14/123	8/95	400
7. 558/0/600[a]	10/40/500	0/750	0/550	2400
8. 482/2/499	7/43/616	0/722	7/672	2459

Daughter B

Well 1: total neutrophils NK cells	Well 2: total T cells myeloid	Well 3: total B cells	Well 4 total NK cells	Grand total of cells
CD15/CD56	CD203/CD13+33	CD19	CD16	
0	0	0	0	0
0	0	0	0	0
5/96/174	0/0/250	0/200	0/180	724
79/0/100	0/22/45	0/50	0/28	223
0	0	0	0	0
0	0	0	0	0
533/6/746	5/33/460	0/220	0/220	3801
0	0	0	0	0

[a]Both daughters are myeloid/lymphoid (stem cell).

mal cell lines or irradiated feeder layers led to the production of increased cell numbers, but no detectable increase in the number of reconstituted hematopoietic stem cells. The number of cells required for a successful transplantation has been shown to increase with the body weight of the recipient. Human umbilical cord blood transplants were more successful in children and infants than in adults despite the observation that the amplification potential of single cord blood cells *in vitro* appeared to exceed that of adult bone marrow or G-CSF peripheral blood cells. Homing of intravenously injected donor stem cell sources in the marrow microenvironment has been shown to be significantly affected by cell surface expression of the vascular cell adhesion molecule 1 (VCAM-1) and availability of the VCAM-1 ligand on bone marrow stromal cell surfaces [28]. Homing of cells to the microenvironmental niches of the hematopoietic stem cell in the bone marrow microenvironment has not been conclusively linked to a particular stromal cell type. Since controversy persists with respect to the phenotype or phenotypes of the totipotential reconstituting stem cell, similar controversy continues regarding the phenotype or phenotypes of the bone marrow stromal cell (mesenchymal stem cell) with a capacity for supporting the homing and return to quiescence of totipotential hematopoietic stem cells.

QUIESCENCE

The importance of quiescence in successful bone marrow transplantation experimental models and in humans has been continually emphasized. Transfusion of hematopoietic stem cell candidates, if not accompanied by return of a subset of these cells to quiescence, would result in continued growth differentiation and completion of the hematopoietic stem cell compartment. Clonal succession of hematopoietic stem cells has been shown *in vivo* and *in vitro* [29]. Clonal succession is assumed to require that a subset of hematopoietic stem cells either maintain or assume a quiescent state following homing into the marrow microenvironment. Heterogeneity of the cycling status of hematopoietic stem cell candidates has been identified in umbilical cord blood, G-CSF-mobilized peripheral blood, or bone marrow. Trafficking of cells from one site to another is known to occur during ontogeny, with migration of hematopoietic stem cells from the yolk sac to the metanephric, splanchnic region of the developing embryo to the fetal liver, spleen, and bone marrow. Trafficking of hematopoietic stem cells prior to birth is documented by the high frequency of detection of totipotential hematopoietic stem cells in human umbilical cord blood at the time of birth and the rapid disappearance of these cells from the circulation after birth. Since stem cells can be mobilized from the marrow by G-CSF infusion (a result of combination chemotherapy exposure of the adult human), there appear to be complex signals, both embryological and stress-induced, that govern trafficking. Trafficking cells may remain quiescent if visible surface markers associated with cycling cells do not prevent return to quiescence.

These data provide invaluable information to direct future experimental research efforts in growing hematopoietic stem cells (i.e., expanding totipotential hematopoietic cells to increase numbers of cells with no detectable differentiation).

CURRENT RESEARCH EFFORTS AND FUTURE DIRECTIONS

One approach toward developing the technology for expanding hematopoietic stem cells has relied on the success of single-cell culture. There is a need to control observation of cells in real time, with detection of cell doubling within minutes of cell fission, and a need to limit further cell division after a first doubling. Several hypotheses went into the design of the combinatorial cell culture system that facilitates new techniques of experimentation in the control of quiescence, cell doubling, and manipulation of the microenvironment [30].

A schematic of the combinatorial cell culture system is shown in Fig. 40.1. This device was designed and constructed to test several hypotheses:

1. Stimulation of a first division of a true hematopoietic stem cell in the hematopoietic microenvironment is followed by restriction of a second division. The correlate to this hypothesis is that unrestricted further divisions after a first conserved division will result

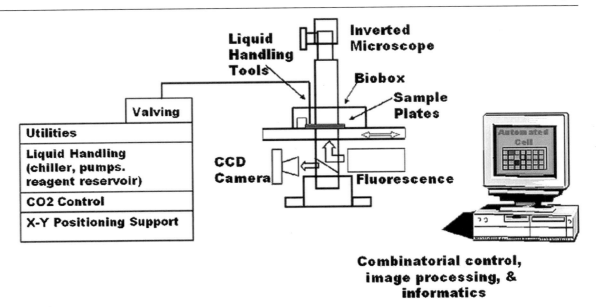

Fig. 40.1. Schematic of the combinatorial cell culture system.

in negative feedback from daughter or granddaughter cells, moving all cells out of the true stem cell compartment.

2. Return to quiescence after the first division of a hematopoietic stem cell is required to "reset" transcriptional machinery for further division without differentiation.

3. Conservative cell division of hematopoietic stem cells is possible without contact with cells in the hematopoietic microenvironment.

4. Different concentrations and combinations of growth factors/cytokines are required for stimulation of division compared to maintenance in quiescence.

To test these hypotheses, it was obvious that the parameter of time had to be controlled. It was also obvious that multiple single cells had to be cultured in restricted environments separated from other cells. The concept of negative feedback and restriction of the stem cell phenotype of a true stem cell by associated nonstem cells required single-cell culture.

The combinatorial cell culture system combines a miniature biobox controlling temperature, carbon dioxide, oxygen, and humidity with a movable motorized stage that is computer controlled. Each of hundreds of tissue culture wells containing a single cell is moved under a cooled charge-coupled device (CCD) camera for capture of the image and storage of data with imagining processing software. Software designed to return each cell to the crosshairs in the imaging system required a cell tracking program to identify how far each cell has moved from its prior position. Search mode software design and automated tracking were linked to pattern recognition software to determine whether cell shape change had led to cell division. Intervention at the time of early cell division, and pre- or postfission, has been designed to facilitate removal of growth factors and insertion of quiescence factors. The computer software recognition of cell division and pre- or postfission states allows strategies for automated intervention such as replacement of quiescence factors. A miniaturized robot and a liquid handling system required to remove growth factors during the insertion of quiescent factors from tissue culture wells containing small volumes (10–40 μl).

Sterility, humidity control, and maintenance of single cells within a controlled microenvironment within each tissue culture well, free of cellular debris and other particles that could be confused by the software as another cell, remained major hurdles to overcome.

The system has been designed and has now been used in automated experiments tracking stem cell candidate movement in various combinations of growth factors and on differ-

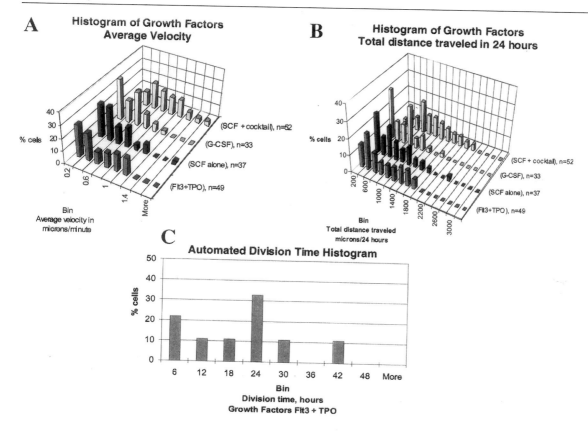

Fig. 40.2.

ent cell surface matrices. Figure 40.1 shows a schematic of the combinatorial cell culture system. Figure 40.2 shows the data from representative experiments as follows: (A) histogram of growth factors, average velocity in microns per minute, (B) histogram of growth factors, total distance traveled in 24 hours, and (C) automated division time histogram for growth factors, FLT-3 and TPO (hours).

Automated cell phenotyping to determine whether daughter cells are conserved with the CD34$^+$lin$^-$ phenotype has been carried out by removal of growth factor and replacement of liquid with fluorescein isothiocyanate (FITC)-conjugated anti-CD34 (HPCA-2) (Becton Dickinson, San Jose, CA), and phycoerythrin (PE) conjugated antilineage markers (Beckton Dickinson) for phenotype assessment of the daughter cells.

Automation of these experiments with hundreds of single CD34$^+$lin$^-$ human hematopoietic stem cell candidates is in progress to attempt to answer each of the foregoing hypotheses.

New strategies are now possible for expansion of hematopoietic stem cells, including *in vivo* expansion in xenotransplant or transgenic species, amplification in batch-type bioreactors with the appropriate combination of stromal cells, and extracellular matrix proteins and growth factors. Other strategies calling for profusion bioreactors in circulatory systems, include the use of microgravity bioreactors.

Amplification of true totipotential reconstituting hematopoietic stem cells remains the "holy grail" of experimental hematology.

The techniques of tissue engineering, as applied to the growth of hematopoietic stem cells, will provide assistance to many tissue engineering areas of different cell types, and successful use of these techniques will provide the directive for experimental hematologists to expand true hematopoietic stem cells.

REFERENCES

1. Till, J. E., and McCulloch, E. A. (1961). A direct measurement of the radiation sensitivity of normal mouse bone marrow. *Radiat. Res.* **14**, 213–220.

2. Voura, E. B., Billia, F., Iscove, N. N., and Hawley, R. G. (1997). Expression mapping of adhesion receptor genes during differentiation of individual hematopoietic precursors. *Exp. Hematol.* **25**, 1172–1179.

3. Trevisan, M., Yan, X.-Q., and Iscove, N. N. (1996). Cycle initiation and colony formation in culture by murine marrow cells with long-term reconstituting potential *in vivo*. *Blood* **88**, 4149–4158.

4. Randall, T. D., and Weissman, I. L. (1998). Characterization of a population of cells in the bone marrow that phenotypically mimics hematopoietic stem cells: Resting stem cells or mystery population? *Stem. Cells* **16**, 38–48.

5. Bhatia, M., Bonnet, D., Murdoch, B., Gan, O. I., and Dick, J. E. (1998). A newly discovered class of human hematopoietic cells with SCID-repopulating activity. *Nat. Med.* **4**, 1038–1043.

6. Osatake, O., Kenichi, H., Hamada, H., and Nakauchi, H. (1996). Long-term lymphohematopoietic reconstitution by a single CD34-low/negative hematopoietic stem cell. *Science* **273**, 242–245.

7. Miller, C. L., and Eaves, C. J. (1997). Expansion *in vitro* of adult murine hematopoietic stem cells with transplantable lymphomyeloid reconstituting ability. *Proc. Natl. Acad. Sci. U.S.A.* **94**, 13648–13653.

8. Peled, A., Petit, I., Kollet, O., Magid, M., Ponomaryov, T., Byk, T., Nagler, A., Ben-Hur, H., Many, A., Shultz, L., Lider, O., Alon, R., Zipori, D., and Lapidot, T. (1999). Dependence of human stem cell engraftment and repopulation of NOD/SCID mice on CXCR4. *Science* **283**, 845–848.

9. Van Zant, G., de Haan, G., and Rich, I. N. (1997). Alternatives to stem cell renewal from a developmental viewpoint. *Exp. Hematol.* **25**, 187–192.

10. Smith, L. G., Weissman, I. L., and Heimfeld, S. (1991). Clonal analysis of hematopoietic stem cell differentiation *in vivo*. *Proc. Natl. Acad. Sci. U.S.A.* **88**, 2788–2792.

11. Ploemacher, R., and Brons, N. (1988). Cells with marrow and spleen repopulating ability and forming spleen colonies on day 16, 12 and 8 are sequentially ordered on the basis of increasing rhodamine-123 retention. *J. Cell. Physiol.* **136**, 531–536.

12. Neben, S., Anklesaria, P., Greenberger, J. S., and Mauch, P. (1993). Quantitation of murine hematopoietic stem cells *in vitro* by limiting dilution analysis of cobblestone area formation on a clonal stromal cell line. *Exp. Hematol.* **21**, 438–444.

13. Bennaceur-Griscelli, A., Tourino, C., Izac, B., Vainchenker, W., and Coulombel, L. (1999). Murine stromal cells counteract the loss of long-term culture initiating cell potential induced by cytokines in CD34 + CD38low/negative human bone marrow cells. *Blood* **94**, 529–538.

14. Zandstra, P. W., Conneally, E., Petzer, A. L., Piret, J. M., and Eaves, C. J. (1997). Cytokine manipulation of primitive human hematopoietic cell self-renewal. *Proc. Natl. Acad. Sci. U.S.A.* **94**, 4698–4730.

15. Moore, K. A., Pytowski, B., Witte, L., Hicklin, D., and Lemischka, I. R. (1997). Hematopoietic activity of a stromal cell transmembrane protein containing epidermal growth factor-like repeat motifs. *Proc. Natl. Acad. Sci. U.S.A.* **94**, 4011–4016.

16. Moore, K. A., Ema, H., and Lemischka, I. R. (1997). *In vitro* maintenance of highly purified, transplantable hematopoietic stem cells. *Blood* **89**, 4337–4347.

17. Dexter, T. M., Allen, T. D., Lajtha, L. G., Schofield, R., and Lord, B. I. (1973). Stimulation of differentiation and proliferation of hematopoietic cells *in vitro*. *J. Cell. Physiol.* **82**, 461–474.

18. Sakakeeny, M.A., and Greenberger, J. S. (1982). Granulopoiesis longevity in continuous bone marrow cultures and factor dependent cell line generation: Significant variation among 28 inbred mouse strains and outbred stocks. *JNCI, J. Natl. Cancer. Inst.* **68**, 305–317.

19. Roy, V., and Verfaille, C. M. (1999). Expression and function of cell adhesion molecules on fetal liver, cord blood and bone marrow hematopoietic progenitors: Implications for anatomical localization and developmental stage specific regulation of hematopoiesis. *Exp. Hematol.* **27**, 302–312.

20. Greenberger, J. S., Epperly, M. W., Fisher, A. M., Cote, G. J., Goldring, S. M., and Glowacki, J. (1995). Hematopoietic progenitor cells expressing the transgene for human calcitonin receptor bind in a "juxtacrine" pair with bone marrow stromal cells expressing the transgene for human calcitonin. *Exp. Hematol.* **23**, 750 (abstr.).

21. Yagi, M., Ritchie, K. A., Sitnicka, E., Storey, C., Roth, G. J., and Bartelmez, S. (1999). Sustained *ex vivo* expansion of hematopoietic stem cells mediated by thrombopoietin. *Proc. Natl. Acad. Sci. U.S.A.* **96**, 8120–8131.

22. Goff, J. P., Shields, D. S., and Greenberger, J. S. (1998). The influence of cytokines on the growth of kinetics and immunophenotype of daughter cells resulting from the first division of single CD34+Thy-1+lin-cells. *Blood* **92**, 4098–4107.

23. Miller, J. S., McCullar, V., Punzel, M., Lemischka, I. R., and Moore, K. A. (1999). Single adult human CD34+lin-/CD38-progenitors give rise to natural killer cells, B-lineage cells, dendritic cells, and myeloid cells. *Blood* **93**, 96–106.

24. Fruehauf, S., Haas, R., Conradt, C., Murea, S., Witt, B., Mohle, R., and Hunstein, W. (1995). Peripheral blood progenitor cell (PBPC) counts during steady-state hemopoiesis allow to estimate the yield of mobilized PBPC after filgrastim (R-metHuG-CSF)-supported cytotoxic chemotherapy. *Blood* **85**, 2619–2626.

25. Koller, M. R., Manchel, I., Brott, D. A., and Palsson, B. O. (1996). Donor-to-donor variability in the expansion potential of human bone marrow cells is reduced by accessory cells but not by soluble growth factors. *Exp. Hematol.* **24**, 1484–1493.

26. Szilvassay, S. J., Bass, M. J., Van Zant, G., and Grimes, B. (1999). Organ-selective homing defines engraftment kinetics of murine hematopoietic stem cells and is compromised by *ex vivo* expansion. *Blood* **93**, 1557–1566.

27. Spangrude, G. J., Brooks, D. M., and Tumas, D. B. (1995). Long-term repopulation of irradiated mice with limiting numbers of purified hematopoietic stem cells: *In vivo* expansion of stem cell phenotype but not function. *Blood* **85**, 1006–1016.

28. Imai, K., Kobayashi, M., Wang, J., Ohiro, Y., Hamada, J.-I., Cho, Y., Imamura, M., Musashi, M., Kondo, T., Kosokawa, M., and Asaka, M. (1999). Selective transendothelial migration of hematopoietic progenitor cells: A role in homing of progenitor cells. *Blood* **93**, 149–156.

29. Mauch, P., Greenberger, J. S., Botnick, L. E., Hannon, E. C., and Hellman, S. (1980). Evidence for structured variation in self-renewal capacity within long-term bone marrow cultures. *Proc. Natl. Acad. Sci. U.S.A.* **77**, 2927–2930.

30. Greenberger, J. S., Goff, J. P., Bush, J., Bahnson, A., Koebler, D., Athanassiou, H., Domach, M., and Houck, R. K. (1999). Expansion of hematopoietic stem cells *in vitro* as a model system for human tissue engineering. *Clin. Plastic. Surg.* **26**, 569–578.

CHAPTER 41

STEM CELL CULTURE: LYMPHOID CELLS

Hadar Haddad, Kathleen S. Carswell, and Eleftherios T. Papoutsakis

U tilization of T lymphocytes for cellular immunotherapy requires the *ex vivo* generation of large numbers of cells (10^9–10^{11}). Here we present important culture methods that dramatically influence cell expansion and biological activity, including those popularly used for the study of LAK, NK, TIL, and antigen-specific cytotoxic T lymphocytes. Several systems have been used to culture T cells; this chapter discusses properties of tissue culture bags, hollow-fiber reactors, and agitated vessels. Optimal ranges for the amounts of various media additives are outlined, as are the common cytokines supplemented to these solutions. Routinely used activation methods such as lectins, antibodies, and antigen-presenting cells are discussed. Relevant surface-antigen markers are also briefly noted. Detailed, step-by-step protocols for isolation of mononuclear cells from peripheral blood, activation of T cells with monoclonal antibodies or phytohemagglutinin, generation of B-cell lines, and the chromium release assay are included. This chapter will aid those requiring fundamental laboratory techniques in the immunotherapy research setting.

INTRODUCTION

Cellular immunotherapy, in which a patient is treated with large doses of *ex vivo* expanded immune cells, with or without genetic modifications, to eradicate aberrant, malignant, or virally infected cells, offers potentially attractive treatment for diseases such as HIV or iatrogenic viral [cytomegalovirus (CMV) or Epstein–Barr virus (EBV)] infections. The efficacy of these therapies is dose dependent, for patient survival increases proportionately with the dose of effector cells [1]. As a result, treatment regimens require large numbers of cells, typically 10^9–10^{11} cells per patient, making *ex vivo* expansion of the cells necessary. In expanding T cells for immunotherapy, it is important not only to generate large numbers of cells, but also to ensure that the cells are biologically active upon transfusion back to the patient.

CELL TYPES AND SOURCES

Immune cells which have been used for cellular immunotherapy treatments include lymphokine-activated killer cells (LAK), tumor-infiltrating lymphocytes (TIL), lymph node lymphocytes (LNL), purified populations of natural killer cells (NK), and isolated populations of cytotoxic T lymphocytes (CTL) specifically directed toward the target cells. These cells are obtained from a variety of sources throughout the body (Table 41.1).

LYMPHOKINE-ACTIVATED KILLER CELLS (LAK)

LAK cells are not a homogeneous cell type, but rather a group of cells with a characteristic activity. Upon incubation in high doses of interleukin 2 (IL-2), peripheral blood lymphocytes acquire cytolytic activity toward a variety of tumor targets otherwise resistant to unstimulated lymphocytes [2]. This nonspecific cytotoxicity, which is not restricted to molecules of the major histocompatibility complex (MHC), is attributed to natural killer

Table 41.1. Cell Types Used in Adoptive Immunotherapy and Their Sources

Cell type	Specificity	Source
Lymphokine-activated killer	Nonspecific (non–MHC restricted)	Any lymphocyte source; peripheral blood (PBL) is the most common
Natural killer	Nonspecific (non–MHC restricted)	Peripheral blood
Tumor-infiltrating lymphocyte (TIL)	Variable: depending on the immunogenicity of the tumor	Primary solid tumor
Lymph node lymphocyte (LNL)	Variable	Tumor-draining lymph nodes
Antigen-specific cytotoxic T lymphocyte (CTL)	Highly specific	Allogeneic: PBL Autologous: PBL or TIL

(NK) cells [3]. However, lymphocytes cultured from other lymphoid tissues such as the thymus or lymph nodes may derive LAK activity from $CD3^+$ T cells, a subset that is also capable of non-MHC-restricted cytotoxicity. One drawback of LAK cell therapy is that systemic administration of high doses of IL-2 is required for the cells to remain active after infusion. IL-2 infusions have caused a number of side effects including capillary leak syndrome, cardiac arrhythmias, anemia, thrombocytopenia, and gastrointestinal distress [4].

NATURAL KILLER (NK) CELLS

The mechanism by which NK cells recognize targets has yet to be completely identified; however, it is known that MHC class I molecules expressed on the target cells play an important role in this process [5,6]. NK cells possess receptors for MHC class I that inhibit NK-cell lytic activity upon binding. Thus, target cells expressing reduced levels of MHC class I do not engage these receptors and lysis of the target cell occurs [7,8].

TUMOR-INFILTRATING LYMPHOCYTES (TIL)

Lymphocytes isolated from solid tumors were first used by Rosenberg *et al.* [9]. The TIL were expected to be more concentrated in cancer-specific T cells than peripheral blood derived LAK cells. In early trials, TIL were cultured in the presence of IL-2, resulting in populations in which approximately 30% of the expanded cells exhibited highly specific, MHC-restricted cytotoxicity [10]. Adoptive transfer of TIL, unlike LAK therapy, was shown to be effective without systemic administration of IL-2, although low doses of IL-2 were shown to enhance the antitumor effects [11]. This lower IL-2 requirement substantially reduces the side effects of immunotherapy.

LYMPH NODE LYMPHOCYTES (LNL)

Tumor-draining lymph node lymphocytes (LNL) are critical frontline cells in the immune response. Additional interest in these cells arises because the yield of immunologically active cells can be several orders of magnitude greater than that from tumor sources [12]. LNL have been used in recent clinical immunotherapy protocols in which patients were "immunized" with irradiated tumor cells and LNL were collected from lymph nodes draining the site of immunization. The LNL were expanded *ex vivo* and reinfused. These protocols were used to treat melanoma, advanced renal cell cancer, and malignant gliomas [13,14].

ANTIGEN-SPECIFIC CYTOTOXIC T LYMPHOCYTES (CTL)

A disadvantage of using TIL or LAK cells for immunotherapy is that large expansion of irrelevant effector cells may occur in the presence of IL-2. Attention has therefore focused on developing methods to isolate and culture cytotoxic T lymphocytes (CTL) specifically directed against the desired target. Limiting dilution culture methods involving stimulation with antigen-presenting cells (APC), as well as anti-CD3 and anti-CD28 monoclonal antibodies, have been developed to isolate individual CD8+, cytomegalovirus-specific CTL [15,16]. Clinical trials have been conducted using allogeneic, antigen-specific CTLs against CMV and HIV following bone marrow transplants [17,18]. Immunodominant peptides from several tumor antigens have also been used to expand antigen-specific CTL from both peripheral blood and TIL sources [19–22].

CULTURE MEDIUM, CYTOKINES, AND ISOLATION OF PERIPHERAL BLOOD MONONUCLEAR CELLS

CULTURE MEDIUM

Media of several different types have been used to culture human T lymphocytes. RPMI-1640, usually supplemented with a variety of additives, is the most common serum-containing medium. IMDM (Iscove's Modified Dulbecco's Medium) has also been used. These media are typically supplemented with 5–10% serum (fetal calf, bovine, or human AB) or plasma. In addition to serum or plasma, L-glutamine is almost always included (1–3 mM) as a carbon source. Antibiotics such as penicillin (50–200 U/ml), streptomycin (50–100 U/ml), and/or gentamicin (10–20 μg/ml) are usually included to prevent contamination, although media used in clinical samples typically exclude these additives. Other supplements sometimes added include HEPES (2–25 mM), sodium bicarbonate, sodium pyruvate (1 mM), nonessential amino acids (0.1–1 mM), and 2-mercaptoethanol (25–50 μM).

For clinical applications it is preferable not to use serum, especially nonhuman serum, because it introduces a multitude of unknowns into the system, including potential sources of known and unknown viral pathogens. Serum-free media have been developed for culturing lymphocytes and are being used extensively. AIM V (Life Technologies, Gaithersburg, MD) and X-VIVO (BioWhittaker, Walkersville, MD) are commonly used for human T lymphocytes. As with serum-containing media, these media are sometimes supplemented with L-glutamine, antibiotics, and other additives already discussed.

CYTOKINES

Interleukin 2, originally called T-cell growth factor (TCGF), is a standard cytokine included in almost all *ex vivo* T-cell expansion protocols. IL-2 binds to the IL-2 receptor (IL-2R) and triggers progression through the cell cycle. Naïve and resting T cells express a low-affinity IL-2R composed of only the β and γ chains. Once activated, the cells begin synthesizing the α chain (CD25), which associates with the β and γ chains, generating a high-affinity IL-2R [23,24]. While IL-2 is a common supplement to any medium used for T-cell expansion, high doses of IL-2 induce non-MHC-restricted cytolytic activity associated with LAK and NK cells [25].

Other cytokines have been reported to have beneficial effects on T-cell cultures and are included in some protocols. In combination with IL-2, IL-12 has been shown to enhance cytotoxic activity in comparison to cultures containing only IL-2 or IL-12 [26]. Addition of IL-1β increases production of interferon gamma (IFN-γ). Although IL-1β does not augment T-cell expansion when combined with IL-2, it has been shown to lead to preferential outgrowth of CD3+CD8+ cells. T cells with this phenotype are usually associated with cytotoxic activity, and indeed these cultures were also shown to specifically kill autologous tumor cells [27]. IL-4 has also been shown to synergize with IL-2 in mitogen-activated cells, resulting in increased proliferation and IL-2 production in comparison to either cytokine alone [28,29].

ISOLATION OF PERIPHERAL BLOOD MONONUCLEAR CELLS (PBMC)

When peripheral blood is used as a cell source, the mononuclear cell fraction, which includes the lymphocytes, must be isolated. This can be achieved by density gradient centrifugation. Not only does this procedure remove red blood cells (RBC), but it also eliminates nonviable cells. Several solutions are available, such as Ficoll-Paque and Percoll, which take advantage of differences in the density of mononuclear cells and other elements in the blood. Ficoll-Paque is an aqueous solution with a density of 1.077 g/ml. Mononuclear cells accumulate on top of the Ficoll-Paque layer because they have a lower density. RBC, granulocytes, and nonviable cells, which have a higher density, pellet at the bottom of the tube.

PROCEDURE: DENSITY GRADIENT SEPARATION
All reagents should be warmed to 37°C.

1. In a 50-ml conical tube, combine 12.5 ml of medium and 12.5 ml of whole blood.
2. Gently layer 14 ml of Ficoll-Paque underneath the medium/blood layer by placing the tip of the pipette at the bottom of the tube.
3. Centrifuge for 20 min at 450g with no brake.
4. Carefully pipette the white interfacial layer.
5. Wash the cells once with phosphate-buffered saline (PBS).
6. Resuspend in medium and determine cell concentration.

CULTURE SYSTEMS
Many of the early animal model experiments and clinical trials made use of well plates and tissue culture flasks (T-flasks) [30,31]. However, owing to the large volumes required to obtain the number of cells needed for successful therapies, smaller culturing systems have quickly become prohibitive. Larger static vessels such as gas-permeable tissue culture bags have gained wide acceptance because of their ease of manipulation [32]. Hollow-fiber cartridge bioreactors allow for very high cell densities and therefore require less physical space [33,34]. The use of agitated suspension culture systems (bioreactors) for the successful expansion of T cells was reported in 2000 [35], although T lymphocytes were grown earlier in spinner flasks for metabolic studies [36,37]. In traditional static culture systems (well plates and T-flasks), LAK and TIL have been cultured at densities up to 2×10^6 cells/ml. Therefore, to produce the desired 10^{11} cells, the final culture volume required is 50–150 liters. Cell culture operations of this magnitude require large amounts of space and labor to incubate the growing cells and handle them during feeding and harvesting procedures [34].

TISSUE CULTURE BAGS
Use of gas-permeable tissue culture bags, typically 1.5 or 3 liters in size, has become the most common method of culturing T cells for use in clinical immunotherapy protocols [38,39]. Maximum cell densities similar to those seen in T-flasks (2×10^6 cells/ml) can be generated. The larger volume reduces the number of vessels that need to be manipulated to as few as 33 bags to achieve the same 100-liter final volume required per treatment. These bags are fitted with Luer-Lok connections to facilitate media transfers using standard sterile tubing kits, a measure that may lessen the chance for contamination during culture [32].

HOLLOW-FIBER REACTORS
Hollow-fiber reactors have been used to expand TIL from metastatic melanoma [34] and ovarian carcinoma [33]. Between 3.5×10^7 and 10^9 TIL from melanoma patients were inoculated into the extrafiber space of the reactors and subsequently expanded 124- to 1170-fold to yield $1.5–5.4 \times 10^{10}$ TIL over a 14- to 32-day period [34]. Medium containing IL-2 was perfused through the lumen of the hollow fibers, and the average use was 4.3 liters per 10^{10} TIL harvested. This system resulted in an estimated 80% reduction in technical labor time and in incubator space requirements for TIL expansion, in comparison to culture in bags. A disadvantage of the hollow-fiber bioreactor systems is that cell proliferation and other characteristics cannot be accessed by direct sampling. Analysis of the status of the cultures was made based on glucose consumption rates. Another difficulty is that large nutrient, metabolite, pO_2 and pH gradients occur in these reactors [40]. We have recently

shown [41,42] that pO_2 and pH have profound effects on the activation and proliferation of T cells, and these two key culture parameters must be effectively controlled to achieve large numbers of quality T cells.

AGITATED VESSELS

Suspension culture in agitated vessels, such as spinner flasks and stirred-tank bioreactors, has not been widely applied to LAK, TIL, or CTL expansion. Agitated culture systems provide several advantages over static culture. They allow for a homogeneous culture environment without the presence of nutrient or waste product gradients. Controlled agitated bioreactors also permit maintenance of culture parameters such as pH and dissolved oxygen.

Natural killer cells have been cultured in 250-ml spinner flasks and in 750-ml stirred bioreactors [43]. Swartz *et al.* [44] reported that lymphocytes derived from TIL or LAK cultures grow poorly or not at all in traditional spinner flask–type suspension cultures, even when inoculated at relatively high densities. In 2000 we showed [35] that primary human T cells can be effectively cultured in 100-ml spinner flasks and 2-liter bioreactors, and we have investigated the effects of shear forces produced by agitation and sparging-based aeration on the expansion of T cells. Primary T cells can be grown at agitation rates of up to 120 rpm in spinner flasks and 180 rpm in bioreactors with no immediate detrimental effects on proliferation. Exposure to agitation and sparging did however cause marked downregulation of the IL-2 receptor.

METHODS OF ACTIVATION/STIMULATION

In vivo activation occurs in the lymph nodes when a T cell encounters an antigen-presenting cell (APC) with the appropriate peptide–MHC conjugate. Two interactions occur that signal the T cell to begin proliferating and differentiating into an effector cell. The first signal is provided by binding of the T-cell receptor (TCR) to the peptide–MHC of the APC. Signal transduction in the T cell occurs via the closely associated CD3 complex. The second signal, termed the co-stimulatory signal, is provided by interaction of CD28 on the T cell with B7 on the APC [23,45,46]. This signal synergizes with the TCR-CD3 signal to enhance T-cell proliferation and cytokine production.

T cells that have never been activated are called "naïve" and require both signals for activation. After activation, the cells return to a resting state in which they are again quiescent and do not display effector functions such as cytokine secretion or cytolytic activity. However, resting cells are distinct from naïve cells in that they require only the first signal for subsequent reactivation.

In vitro activation attempts to mimic one or both of these signals. Several methods are routinely used to induce T-cell proliferation and differentiation, including the use of mitogenic lectins, monoclonal antibodies, and antigen-presenting cells.

LECTINS

Lectins are a class of mitogens that bind to specific glycoproteins and induce cellular division. Lectins capable of stimulating proliferation of lymphocytes, both T and B cells, are available. Phytohemagglutinin (PHA), a lectin isolated from red kidney beans, is often used to activate human T cells. The mechanism by which PHA induces T-cell proliferation is not well understood, but it is thought to trigger the same growth response mechanisms as recognition of antigen [46,47]. Other lectins that can be used to activate T cells include concanavalin A (con A) and pokeweed mitogen (PMW), although PMW stimulates both T and B cells. Activation with lectins is a simple, inexpensive method for generating large numbers of T cells.

PHA and other lectins do not distinguish between antigenic specificities and are therefore called polyclonal activators. Because T cells with a particular antigenic specificity comprise a minute fraction of the entire T-cell population ($\leq 1\%$), activation with PHA will not enhance the fraction of T cells with the antigenic specificity of interest. However, if the starting population contains a high fraction of antigen-specific T cells, as would be present in cell populations isolated from a solid tumor, a tumor-draining lymph node, or an individual suffering from a specific infection (e.g., HIV), then the resulting population will also contain a large fraction of the desired specific T cells [48].

ANTIBODIES

The use of monoclonal antibodies (mAb) directed against the CD3 complex provides a more physiological means of activating T cells. Binding of anti-CD3 mAb is thought to mimic recognition of antigen by the TCR and trigger signal transduction [49]. Unfortunately, repeated cycles of activation with anti-CD3 mAb have been shown to result in a decline of cytolytic activity [50]. As with PHA activation, antibody activation is polyclonal and does not affect naïve T cells, which require a co-stimulatory signal for activation [51].

To improve activation, anti-CD28 mAb has been used in combination with anti-CD3 mAb, providing the co-stimulatory signal. Use of both antibodies has been shown to be more effective at inducing proliferation [52], a result of activation of naïve T cells and enhanced activation of the remainder of the T-cell population [53]. Furthermore, a more rapid T-cell response is generated, usually identifiable within 2 days of activation, versus 3–5 days for PHA or anti-CD3 mAb alone. Anti-CD28 mAb has also been used to augment PHA activation with similar results [54].

When work begins with a purified population of T cells (i.e., after sorting or panning), activation with soluble antibodies may not be effective. Immobilization of antibodies to tissue culture surfaces has been shown to enhance proliferation [55], thought to be a reflection of the need for TCR cross-linking [56]. This is supported by the observation that the use of immobilized rather than soluble antibodies removes the need for accessory cells and reduces the required concentration when a purified population of T cells is used [57]. Soluble antibodies were also found to interfere with antigen-specific responses, and extensive washing was required to remove residual antibody and realize optimal responses [15].

PROCEDURE: ACTIVATION WITH PHA OR ANTIBODIES

Other lectins or combinations of antibodies can be substituted. PHA is stored in a lyophilized powder and needs to be reconstituted in culture medium. The reconstituted PHA should be stored at 4°C for no more than a week.

1. Isolate PBMC by density gradient centrifugation as described earlier.
2. Resuspend at 1×10^6 cells/ml in culture medium and place in a T-flask.
3. Add 100 U/ml IL-2.
4. Add PHA (1–5 μg/ml), anti-CD3 (range 20 ng/ml to 1 μg/ml), or anti-CD3 and anti-CD28 mAb (in a 1:1 ratio), and mix well.
5. Place culture in a humidified incubator at 37°C, 5% CO_2.
6. After 3–5 days evaluate the culture by microscopy and flow cytometry. The cells will have increased in size, and the culture will contain greater than 90% CD3-positive cells with up-regulated CD25 expression (see Fig. 41.1).

ANTIGEN-PRESENTING CELLS

Coculture of T cells with primed APC most closely mimics the physiological mode of activation. Dendritic cells (DC) and B cells can activate both naïve and memory T cells, more importantly, they provide antigen-specific activation, allowing generation of a large population of T cells with the same antigenic specificity [58,59]. Owing to difficulties in isolating and culturing DC, B cells were until recently the primary APC used for *in vitro* activation. B-cell lines can be generated *in vitro* by immortalization with the Epstein–Barr virus (EBV) [60,61]. These lymphoblastoid cell lines (EBV-LCL) are then used to activate antigen-specific T cells that recognize EBV antigens [62]. This technique has been utilized in immunotherapies to treat EBV infection and EBV-induced lymphoma in immunosuppressed patients [63,64]. Additionally, these cell lines can be transfected with plasmids and induced to express other antigens [48,65]. Thus EBV-LCL can be used to activate EBV-specific T cells or T cells with other specificities.

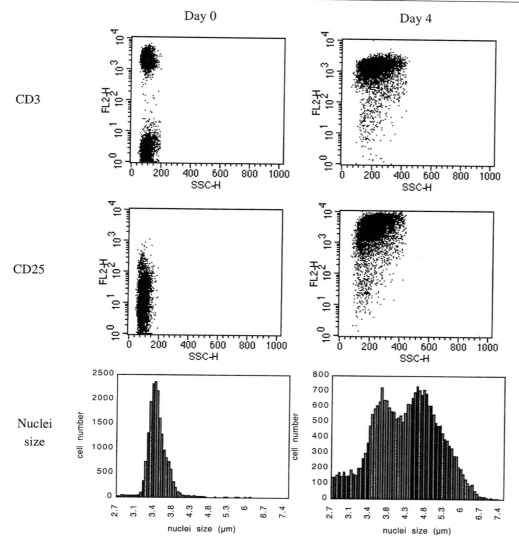

Fig. 41.1. Changes in CD3 and CD25 (α-chain of IL-2R) expression and nuclei size after activation. PBMC were activated with 100 ng/ml of anti-CD3 and 100 ng/ml anti-CD28 monoclonal antibodies in AIM V medium containing 100 U/ml IL-2 on day 0. The culture was maintained in a humidified incubator with 5% CO_2. Flow cytometry was used to measure CD3 and CD25 expression. Nuclei size was measured with a Coulter Counter Multisizer. On day 0 the culture contains 56% CD3-positive cells; and of the CD3-positive population, 36% are positive for CD25. Four days after activation, the cells are 99% CD3 positive, CD25 expression has been up-regulated such that the cells are 99% positive, and nuclei size has increased.

Recent developments in DC isolation and culture techniques have made working with DC more straightforward and less labor intensive. Two subsets of DC are present in circulating blood, CD11c$^+$ and CD11c$^-$, and these cells can be isolated from fresh or cultured blood. When one is working with fresh blood, the isolated DC population is 50% CD11c$^+$ and 50% CD11c$^-$. If cultured blood is used, the resulting population is 80–90% CD11c$^+$ and 10–20% CD11c$^-$, because the CD11c$^-$ subset does not survive in culture without the addition of specific cytokines. Therefore, if the CD11c$^-$ subset is desired, one must start with fresh blood [66].

Dendritic cells can also be generated from proliferating CD34$^+$ precursors (0.06% in peripheral blood) or from nonproliferating CD14$^+$ monocytes (5–10% in PBMC). Because of the low frequency of CD34$^+$ cells in adult blood, utilizing CD34$^+$ cells as the starting population for generation of DC requires increasing the number of peripheral blood CD34$^+$ cells by stem cell mobilization via administration of a hematopoietic growth factor, such as granulocyte colony-stimulating factor (G-CSF), and/or a chemotherapeutic agent. No pretreatment is necessary when CD14$^+$ cells are used as the starting population. Protocols starting with CD34$^+$ cells require granulocyte–macrophage CSF (GM-CSF) and tumor necrosis factor α (TNF-α), while protocols for CD14$^+$ cells call for GM-CSF and IL-4 [66]. Detailed protocols for generation of myeloid DC from monocytes, cord blood CD34$^+$ cells, bone marrow CD34$^+$ cells, and peripheral blood CD34$^+$ cells are available in *Dendritic Cells: Biology and Clinical Applications* [66].

Although surface markers identifying DC will vary depending on origin, means of purification, and state of activation, mature DC can be characterized by specific functions and features. For example, mature DC have the phenotype of a lineage-negative leukocyte (e.g., CD3⁻, CD19⁻, CD20⁻, CD16⁻, CD56⁻), express high levels of MHC class I and class II molecules and high levels of adhesion and co-stimulatory molecules, such as CD86 and CD40. Functionally, they are potent stimulators of naïve helper and cytotoxic T cells [67].

DC must be primed for antigen presentation to be effective activators of antigen-specific T cells. This is accomplished by simple incubation with an antigen-specific peptide (1–5 µg/ml) and β_2-microglobulin (1–5 µg/ml) [58,59] or with soluble antigen [68,69], depending on the subpopulation of T cells, CD8⁺ or CD4⁺, respectively, that one is interested in activating. Once primed, the DC can be cocultured with PBMC for T-cell activation.

SUPPLEMENTAL PROCEDURE: GENERATION OF EBV-LCL

Generating EBV-LCL is relatively simple, although close and constant attention must be given to the developing cell lines, which are fragile in the early stages of immortalization. Maintaining the cultures requires both feeding by retention (pelleting the cells, removing some supernatant, and adding fresh medium) and by dilution (adding fresh medium to the culture) and will depend on the concentration of cells in the culture and how rapidly the cells are growing. Cultures should be kept highly confluent, especially in the first 4 weeks, as cultures that are diluted too rapidly may not survive. The entire process takes approximately 4–6 weeks.

EBV-containing medium can be prepared by culturing B95-8 (ATCC, American Type Culture Collection, CRL-1612) cells for 3 days (the medium should be yellowish), centrifuging the culture to remove the cells, and filtering the supernatant through a 0.45-µm filter. Aliquots of the supernatant can be stored in liquid nitrogen until ready for use. If the supernatant is not potent enough for immortalization, the B95-8 cells can be activated with 3 mM butyric acid or 20 ng/ml tissue plasminogen activator (TPA).

PBMC are isolated as described earlier and used as the starting cell population in this protocol. Most cells will die off early in the culture, since specific cytokines necessary for their growth are not included. B lymphocytes not infected or immortalized will also die off. Cyclosporin A (CsA) is added to prevent the T cells present from being activated by the infected B cells and outgrowing the initially small number of immortalized B cells.

1. Pellet 10^7 PBMC in a 15-ml Falcon tube.
2. Decant the supernatant and resuspend in 0.5 ml of RPMI. Add 4 ml of B95-8 supernatant.
3. Incubate for 1 h in a 37°C water bath.
4. Add CsA (final concentration 1 µg/ml).
5. Add 50 µl to each well of a 96-well plate. Sterile water should be added to the perimeter wells to reduce evaporation. (*Note*: Flat-bottomed plates allow cell clumps to be visualized more easily than round-bottomed plates.)
6. Incubate overnight in a humidified incubator.
7. Combine 5 ml of medium and CsA (1 µg/ml) in a polypropylene tube and add 50 µl to each well (feeding by dilution).
8. After 1 week, feed the cells by retention: centrifuge the plate for 5 min at 150*g*, remove half the volume by pipetting (being careful not to disturb the pellet), add 50 µl of medium containing 1 µg/ml CsA, and mix gently.

Feed the cells once a week for the first 3–4 weeks and increase the feeding frequency as the cells begin to grow more rapidly. Look for cell clumps 1–2 weeks after infection. Continue to add CsA once a week for 5–6 weeks. As the cells become more concentrated and the clumps become larger, begin feeding by dilution. Alternate between feeding by dilution and retention, slowly increasing the culture volume. Not all wells will produce cell lines. Transfer cells to increasingly larger well plates as they become more concentrated and grow more rapidly. Established cell lines will require feeding approximately every other

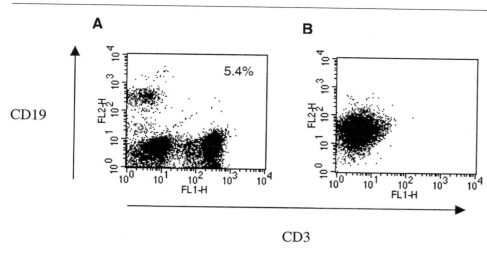

A

5.4%

CD19

FL2-H
FL1-H

B

FL2-H
FL1-H

CD3

Fig. 41.2. Generation of EBV-LCL. (A) On day 0 the PBMC culture contains 5.4% CD19-positive (B cells) and 38% CD3-positive (T cells). The culture was infected with B95-8 supernatant, treated periodically with 1 μg/ml of cyclosporin A, and maintained by weekly replacement of media. (B) After 12 weeks, flow cytometry shows that all cells are CD19+CD3−.

day. Stain cells with anti-CD3 and anti-CD19 mAb and analyze by flow cytometry. Cell lines should be negative for CD3 and positive for CD19 (see Fig. 41.2).

SURFACE ANTIGEN EXPRESSION

Analysis of the repertoire of surface markers expressed on cells has become an indispensable tool in cell culture. Flow cytometry is a fast and simple technique that measures expression on the single-cell level. This method can be used to identify the types of cells present in a culture, their fraction in the culture, their size, and which receptors they express. Apoptosis and intracellular cytokine expression can also be measured. Several markers can be assayed at the same time, allowing intricate questions to be addressed.

Numerous books are available that discuss the theory and specific protocols for a variety of applications. *Practical Flow Cytometry* [70], *Flow Cytometry: First Principles* [71], and *Introduction to Flow Cytometry* [72] all provide good background and information on staining procedures and data analysis. Here we will only discuss some receptors important in the study of T cell cultures.

CD3: CD3 is a complex of five polypeptides that associate into dimers ($\gamma\varepsilon$, $\delta\varepsilon$, and $\zeta\zeta$ or $\zeta\eta$). Association of the CD3 complex with the TCR is required for TCR expression, thus all T cells express this marker. The CD3 complex is responsible for signal transduction upon TCR triggering [73].

CD4/CD8: These coreceptors help stabilize binding of the TCR to the peptide–MHC complex [74]. They also augment signal transduction through the CD3 complex and enhance activation of the T cell upon target cell binding [75,76]. In general, coreceptor expression defines the effector function a T cell acquires upon activation and maturation. CD4+ T cells recognize antigenic peptides associated with MHC class II molecules and generally function as helper T cells. Antigenic peptides associated with MHC class I molecules are recognized by CD8+ T cells. Once activated, these cells differentiate into effector cells called cytotoxic T lymphocytes (CTL), which function to eliminate pathogen-infected cells or altered self cells. CD4 and CD8 bind to conserved regions of the MHC class II and MHC class I, respectively.

CD25: CD25 is the α chain of the IL-2 receptor. Binding of IL-2 to its receptor leads to increased proliferation and gain of effector functions. Upon activation, expression of CD25 is up-regulated with peak levels between days 2 and 8. If no further stimulation is provided, CD25 expression decreases back to levels associated with resting T cells [47,77,78].

CD28: CD28 is a co-stimulatory molecule that binds to the B7 surface antigen on APC, triggering a signal in the T cell that synergizes with the TCR/CD3 signal [79]. This leads to proliferation and differentiation of effector functions. Signaling through CD28 is required for activation of naïve T cells [53,80].

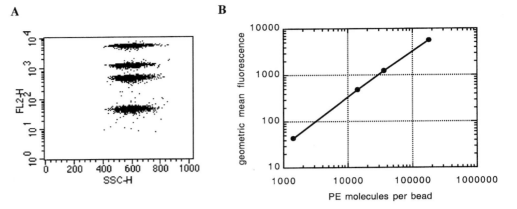

Fig. 41.3. The QuantiBRITE bead system. (A) Flow cytometry plot of QuantiBRITE beads with four levels of PE molecules per bead: 1400, 14,000, 36,600 and 182,000. (B) A standard curve is then plotted which relates geometric mean fluorescence with the number of PE molecules per bead.

CD62L: Also called L-selectin or lymph node homing receptor, CD62L is an adhesion molecule involved in T-cell circulation and homing to the lymph nodes. It binds CD34 and GlyCAM on high endothelial venules (HEV) and mediates rolling, the initial step in extravasation. When activated, T cells down-regulate CD62L expression, allowing exit from the lymph nodes [81].

CD45RO/RA: Isoforms of CD45 identify naïve (CD45RA^{high}CD45RO^{low}) and memory (CD45RA^{low}CD45RO^{high}) T-cell populations.

QUANTITATIVE ANALYSIS: QUANTIBRITE BEADS

Specialized beads have been developed to relate mean fluorescence intensity, detected by using flow cytometry, to the number of receptors present per cell. The QuantiBRITE system (Becton Dickinson, San Jose, CA) consists of beads with four known levels of phycoerythrin (PE) molecules. The fluorescence intensity of each of the beads is used to construct a standard curve relating fluorescence intensity and the number of PE molecules (see Fig. 41.3), which in turn is proportional to the number of receptors per cell. To determine the actual number of receptors per cell, the stoichiometry of antibody-to-antigen binding must be known. Even without this information, however, the QuantiBRITE system provides relative numbers and allows for comparison of samples analyzed on different days [82,83].

EFFECTOR FUNCTION

There are two major effector functions of interest for T lymphocytes. Cells carrying the CD8 antigen typically develop into cytolytic, or cytotoxic, T cells whose function is to recognize and kill pathogen-infected or malignant cells. Cells carrying the CD4 antigen typically differentiate into helper T cells that secrete cytokines to recruit other immune cells such as macrophages, B cells, and cytotoxic T cells. The profile of cytokines secreted by the helper T cells determines the type of immune response generated.

CYTOLYTIC ACTIVITY

Several assays are available for measurement of cytolytic activity, including fluorescence-based and radioactive assays. In all cases target cells are labeled and cocultured with T cells for 4–5 h, after which target cell death is assessed. Varying effector-to-target cell (E:T) ratios are used to ensure that the observed effect is not due to an element that negatively influences the reaction. Results are presented as percent cytolytic activity versus E:T ratio (see Fig. 41.4).

The most common method used to measure cytolytic activity is the chromium release assay. The isotope ^{51}Cr is beta active, with a prominent gamma line at 0.32 MeV, and has a half-life of 28.7 days. Precautions should be taken to shield any radioactive materials with lead, and general radiation safety techniques should be employed. Target cells are labeled with 50–200 μCi of sodium chromate, Na$_2$[^{51}Cr]O$_4$ by simple incubation for 1–2 h in a humidified incubator. The Cr diffuses freely into the cells. Interactions with cytoplasmic proteins limit diffusion of Cr out of the cell [23,46]. Residual Cr is removed by thoroughly

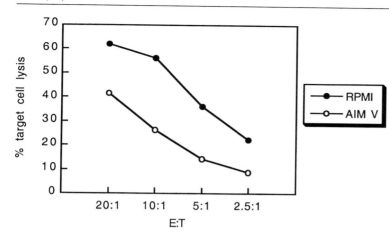

Fig. 41.4. Cytolytic activity measured by the standard Cr release assay. PBMC were activated with 5 μg/ml of PHA and cultured in RPMI or AIM V medium supplemented with 100 U/ml IL-2. Cultures were maintained in a humidified 5% CO_2 incubator. Allogeneic EBV-LCL were used as target cells and labeled with 100 μCi of sodium chromate for 2 h. Target cells (T) and T cells (E) were combined at several E:T ratios in a round-bottomed 96-well plate and placed in a humidified incubator for 4 h. The plate was centrifuged for 10 min, and 100 μl of supernatant from each well was counted with a liquid scintillation counter.

washing the target cells with media. Effector cells should also be washed prior to the assay, especially when soluble antibodies were used for activation, as disccussed earlier. A fixed number of target cells (1000–5000) is placed in each well of a round-bottomed 96-well plate, and various concentrations of effector cells are added. After the target and effector cells have been combined, the plate should be centrifuged briefly to aggregate the cells, allowing for better cell interactions.

As a control, medium is added instead of effector cells to several wells to measure the amount of Cr that naturally leaks out of the target cells during the assay. This spontaneous release (SR) should be less than 20%. Lysing the cells with a mild detergent such as Triton X-100, Saponin, or Tween-20 provides a measure of the total amount of Cr retained by the target cells, the so-called total release (TR). All conditions and controls should be performed in triplicate.

After 4–5 h in a humidified incubator, the plate should be centrifuged and the supernatants transferred to scintillation vials and counted in a liquid scintillation counter. Percent target cell lysis is calculated by

$$\frac{\text{Experimental release} - \text{spontaneous release}}{\text{Total release} - \text{spontaneous release}} \times 100.$$

Modifications to the basic assay include addition of monoclonal antibodies (e.g., anti-MHC class I or anti-MHC class II) to block the activity of a particular subset of cells or specific cytokines to augment the reaction. Such variations can provide information on which cells are responsible for the cytolytic activity or which mechanism dominates (Fas/FasL, granzymes) in the system.

Fluorometric assays have also been described [84]. Calcein AM is an acetoxymethyl ester of calcein that passively diffuses across the cell membrane. Inside the cell, calcein AM is converted by esterases to the polar fluorochrome calcein, which is retained by the cells. Analogous to the Cr release assay, lysis by effector cells releases the calcein into the supernatant, which is then measured. Spontaneous and total release must also be measured, and cytolytic activity is calculated by an analogous equation.

Assays based on flow cytometry are also available for measurement of cytolytic activity. Target cells can be labeled with a fluorochrome (e.g., PKH-1) or dye [e.g., $DiO_{18}(3)$] that is stably incorporated in the cell membrane [85–87] and fluoresces at 520–550 nm (FL1 detector). It is important that the label not be released from the membrane during the time frame of the assay, as this could lead to effector cell labeling. Target and effector cells are then mixed and incubated for 4–5 h at different E:T ratios, as with the Cr release assay. Propidium iodide (PI, fluoresces at 575–675 nm, FL2 detector) is used to label nonviable cells. Live target cells will be FL1 positive and FL2 negative, while dead target cells will be positive for both FL1 and FL2 (Fig. 41.5). As a control, target cells alone are stained with PI. An advantage of this type of assay is that the actual E:T ratios can be calculated.

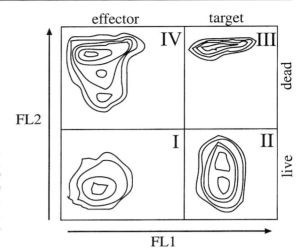

Fig. 41.5. *Representative plot of a cytolytic activity assay based on flow cytometry. A fluorescent label, detected in the FL1 channel, is used to label the target cells. The target cells and T cells are combined at various ratios and incubated for approximately 4 h. PI, detected in the FL2 channel, is added to label nonviable cells just prior to flow cytometry analysis. The quadrants identify the cells as follows: quadrant I, live effector cells; quadrant II: live target cells; quadrant III, dead target cells; and quadrant IV, dead effector cells.*

Target Cell Types

Many cell types can be used as target cells. Frequently used cell lines include K562 (ATCC CCL-243), a human erythromyeloid leukemia cell line that is sensitive to natural killer cell activity; Daudi (ATCC CCL-213), a B-cell line sensitive to LAK (lymphokine-activated killer cell) activity; P815 (TIB-64), a murine mastocytoma cell line sensitive to CTL (cytotoxic T cell) activity [88]; tumor cells, and B-lymphoblastoid cell lines (B-LCL). B-LCL can be generated by immortalization of human B cells with the Epstein–Barr virus as discussed earlier. These cell lines are simple to generate and maintain and are an easily accessible source of autologous target cells. The choice of target cell will depend on the system being tested.

PROCEDURE: CR RELEASE ASSAY

Effector-to-target ratios may need to be adjusted for each system and can range widely. The number of target cells per well will also vary depending on availability of target and effector cells and the desired E:T ratios. As few as 1000 target cells can be used. RPMI is the medium used in this example.

1. *Prepare target cells*: Pellet 500,000 cells in a microcentrifuge tube. Decant the supernatant and resuspend the pellet in 100 μl of 1 mCi/ml sodium chromate (100 μCi). Place the tube in a shielded container in a humidified incubator for 2 h. Wash the cells four times with medium and resuspend in 10 ml of RPMI (50,000 cells/ml).

2. *Prepare effector cells*: Wash 5×10^6 cells three times with medium and resuspend in 1 ml of RPMI. Perform a series of 1:2 dilutions generating solutions (≥ 300 μl each) with the following concentrations: 5×10^6, 2.5×10^6, 1.25×10^5, and 6.25×10^4 cells/ml.

3. In each well of a round-bottomed 96-well plate, combine 100 μl of target cells and 100 μl of effector cells, three wells per condition.

4. In three wells combine 100 μl of target cells and 100 μl of medium (SR control). For the TR control, combine 100 μl of target cells and 100 μl of 1% v/v Triton X-100.

5. Centrifuge the plate briefly and place in a shielded container in a humidified incubator for 4 h.

6. Centrifuge the plate for 10 min at 150*g*.

7. Transfer 100 μl of supernatant from each well into a scintillation vial containing 5 ml of scintillation fluid. Use a liquid scintillation counter to measure the experimental release and use the foregoing equation to calculate the percentage of target cell lysis.

CYTOKINE PRODUCTION

Sandwich enzyme-linked immunosorbent assays (ELISA) are the most commonly used method of measuring cytokine production from CD4+ T cells. Detailed protocols of assays of these types can be found in a variety of sources [23,46,89]. A wide range of ELISA kits are commercially available, often with precoated plates. Bioassays are also available for measurement of some cytokines.

ACKNOWLEDGMENTS

We thank Mary Kirklin and Dr. James Perkins at the Evanston Hospital (Evanston, IL) for providing blood samples. Interleukin 2 was a generous gift from Chiron Corporation. HH was supported by the Whitaker Foundation. KSC was partially supported by predoctoral biotechnology training grant (GM08449) from the National Institutes of Health.

REFERENCES

1. Cheever, M., and Chen, W. (1997). Therapy with cultured T cells: Principles revisited. *Immunol. Rev.* **157**, 177–194.
2. Grimm, E. A., Robb, R. J., Roth, J. A., Neckers, L. M., Lachman, L. B., Wilson, D. J., and Rosenberg, S. A. (1983). Lymphokine-activated killer cell phenomenon. III. Evidence that IL-2 is sufficient for direct activation of peripheral blood lymphocytes into lymphokine-activated killer cells. *J. Exp. Med.* **158**, 1356–1361.
3. Phillips, J. H., and Lanier, L. L. (1986). Dissection of the lymphokine-activated killer phenomenon. Relative contribution of peripheral blood natural killer cells and T lymphocytes to cytolysis. *J. Exp. Med.* **164**, 814–825.
4. Chang, A. E., and Shu, S. (1992). Immunotherapy with sensitized lymphocytes. *Cancer Invest.* **10**, 357–369.
5. Karlhofer, F. M., Ribaudo, R. K., and Yokoyama, W. M. (1992). MHC class I alloantigen specificity of Ly-49+ IL-2-activated natural killer cells. *Nature (London)* **358**, 66–70.
6. Storkus, W. J., Salter, R. D., Alexander, J., Ward, F. E., Ruiz, R. E., Cresswell, P., and Dawson, J. R. (1991). Class I-induced resistance to natural killing: identification of nonpermissive residues in HLA-A2. *Proc. Natl. Acad. Sci. U.S.A.* **88**, 5989–5992.
7. Long, E. O., and Wagtmann, N. (1997). Natural killer cell receptors. *Curr. Biol.* **7**, 615–618.
8. Timonen, T., and Helander, T. S. (1997). Natural killer cell–target cell interactions. *Curr. Opin. Cell Biol.* **9**, 659–666.
9. Rosenberg, S. A., Packard, B. S., Aebersold, P. M., Solomon, D., Topalian, S. L., Toy, S. T., Simon, P., Lotze, M. T., Yang, J. C., Seipp, C. A., *et al.* (1988). Use of tumor-infiltrating lymphocytes and interleukin-2 in the immunotherapy of patients with metastatic melanoma. A preliminary report. *N. Engl. J. Med.* **319**, 1676–1680.
10. Topalian, S. L., Solomon, D., and Rosenberg, S. A. (1989). Tumor-specific cytolysis by lymphocytes infiltrating human melanomas. *J. Immunol.* **142**, 4520–4526.
11. Topalian, S. L., Muul, L. M., Solomon, D., and Rosenberg, S. A. (1987). Expansion of human tumor infiltrating lymphocytes for use in immunotherapy trials. *J. Immunol. Methods* **102**, 127–141.
12. Triozzi, P. L. (1987). Identification and activation of tumor-reactive cells for adoptive immunotherapy. *J. Immunol. Methods* **102**, 127–141.
13. Chang, A. E., Aruga, A., Cameron, M. J., Sondak, V. K., Normolle, D. P., Fox, B. A., and Shu, S. (1997). Adoptive immunotherapy with vaccine-primed lymph node cells secondarily activated with anti-CD3 and interleukin-2. *J. Clin. Oncol.* **15**, 796–807.
14. Plautz, G. E., Barnett, G. H., Miller, D. W., Cohen, B. H., Prayson, R. A., Krauss, J. C., Luciano, M., Kangisser, D. B., and Shu, S. (1998). Systemic T cell adoptive immunotherapy of malignant gliomas. *J. Neurosurg.* **89**, 42–51.
15. Riddell, S. R., and Greenberg, P. D. (1990). The use of anti-CD3 and anti-CD28 monoclonal antibodies to clone and expand human antigen-specific T cells. *J. Immunol. Methods* **128**, 189–201.
16. Riddell, S. R., Rabin, M., Geballe, A. P., Britt, W. J., and Greenberg, P. D. (1991). Class I MHC-restricted cytotoxic T lymphocyte recognition of cells infected with human cytomegalovirus does not require endogenous viral gene expression. *J. Immunol.* **146**, 4414–4420.
17. Greenberg, P. D., Reusser, P., Goodrich, J. M., and Riddell, S. R. (1991). Development of a treatment regimen for human cytomegalovirus (CMV) infection in bone marrow transplantation recipients by adoptive transfer of donor-derived CMV-specific T cell clones expanded in vitro. *Ann. N.Y. Acad. Sci.* **636**, 184–195.
18. Riddell, S. R., Greenberg, P. D., Overell, R. W., Loughran, T. P., Gilbert, M. J., Lupton, S. D., Agosti, J., Scheeler, S., Coombs, R. W., and Corey, L. (1992). Phase I study of cellular adoptive immunotherapy using genetically modified CD8+ HIV-specific T cells for HIV seropositive patients undergoing allogeneic bone marrow transplant. The Fred Hutchinson Cancer Research Center and the University of Washington School of Medicine, Department of Medicine, Division of Oncology. *Hum. Gene Ther.* **3**, 319–338.
19. Labarriere, N., Pandolfino, M. C., Raingeard, D., Le Guiner, S., Diez, E., Le Drean, E., Dreno, B., and Jotereau, F. (1998). Frequency and relative fraction of tumor antigen-specific T cells among lymphocytes from melanoma-invaded lymph nodes. *Int. J. Cancer* **78**, 209–215.
20. Rivoltini, L., Kawakami, Y., Sakaguchi, K., Southwood, S., Sette, A., Robbins, P. F., Marincola, F. M., Salgaller, M. L., Yannelli, J. R., Appella, E., *et al.* (1995). Induction of tumor-reactive CTL from peripheral blood and tumor-infiltrating lymphocytes of melanoma patients by in vitro stimulation with an immunodominant peptide of the human melanoma antigen MART-1. *J. Immunol.* **154**, 2257–2265.
21. Salgaller, M. L., Weber, J. S., Koenig, S., Yannelli, J. R., and Rosenberg, S. A. (1994). Generation of specific anti-melanoma reactivity by stimulation of human tumor-infiltrating lymphocytes with MAGE-1 synthetic peptide. *Cancer Immunol. Immunother.* **39**, 105–116.
22. Stevens, E. J., Jacknin, L., Robbins, P. F., Kawakami, Y., el Gamil, M., Rosenberg, S. A., and Yannelli, J. R. (1995). Generation of tumor-specific CTLs from melanoma patients by using peripheral blood stimulated with allogeneic melanoma tumor cell lines. Fine specificity and MART-1 melanoma antigen recognition. *J. Immunol.* **154**, 762–771.
23. Kuby, J. (1997). "Immunology." Freeman, New York.

24. Taniguchi, T., and Minami, Y. (1993). The IL-2/IL-2 receptor system: A current overview. *Cell* (*Cambridge, Mass.*) **73**, 5–8.

25. Grimm, E. A., and Owen-Schaub, L. (1991). The IL-2 mediated amplification of cellular cytotoxicity. *J. Cell. Biochem.* **45**, 335–339.

26. Cesano, A., Visonneau, S., Clark S. C., and Santoli, D. (1993). Cellular and molecular mechanisms of activation of MHC nonrestricted cytotoxic cells by IL-12. *J. Immunol.* **151**, 2943–2957.

27. Baxevanis, C. N., Dedoussis, G. V., Gritzapis, A. D., Stathopoulos, G. P., and Papamichail, M. (1994). Interleukin 1 beta synergises with interleukin 2 in the outgrowth of autologous tumour-reactive CD8+ effectors. *Br. J. Cancer* **70**, 625–630.

28. Mitchell, L. C., Davis, L. S., and Lipsky, P. E. (1989). Promotion of human T lymphocyte proliferation by IL-4. *J. Immunol.* **142**, 1548–1557.

29. Spits, H., Yssel, H., Takebe, Y., Arai, N., Yokota, T., Lee, F., Arai, K., Banchereau, J., and de Vries, J. E. (1987). Recombinant interleukin 4 promotes the growth of human T-cells. *J. Immunol.* **139**, 1142–1147.

30. Lafreniere, R., and Rosenberg, S. A. (1985). Successful immunotherapy of murine experimental hepatic metastases with lymphokine-activated killer cells and recombinant interleukin 2. *Cancer Res.* **45**, 3735–3741.

31. Rosenberg, S. A., Spiess, P., and Lafreniere, R. (1986). A new approach to the adoptive immunotherapy of cancer with tumor-infiltrating lymphocytes. *Science* **233**, 1318–1321.

32. Yannelli, J. R., Thurman, G. B., Dickerson, S. G., Mrowca, A., Sharp, E., and Oldham, R. K. (1987). An improved method for the generation of human lymphokine activated killer cells. *J. Immunol. Methods* **100**, 137–145.

33. Freedman, R. S., Tomasovic, B., Templin, S., Atkinson, E. N., Kudelka, A., Edwards, C. L., and Platsoucas, C. D. (1994). Large-scale expansion in interleukin-2 of tumor-infiltrating lymphocytes from patients with ovarian carcinoma for adoptive immunotherapy. *J. Immunol. Methods* **167**, 145–160.

34. Knazek, R. A., Wu, Y. W., Aebersold, P. M., and Rosenberg, S. A. (1990). Culture of human tumor infiltrating lymphocytes in hollow fiber bioreactors. *J. Immunol. Methods* **127**, 29–37.

35. Carswell, K., and Papoutsakis, E. (2000). Culture of human T cells in stirred bioreactors for cellular immunotherapy applications: Shear, proliferation, and the IL-2 receptor. *Biotechnol. Bioeng.* **68**, 328–338.

36. Sand, T., Condie, R., and Rosenberg, A. (1977). Metabolic crowding effect in suspension of cultured lymphocytes. *Blood* **50**, 337–346.

37. Truskey, G., Nicolakis, D., DiMasi, D., Haberman, A., and Swartz, R. (1990). Kinetic studies and unstructured models of lymphocyte metabolism in fed-batch culture. *Biotechnol. Bioeng.* **36**, 797–807.

38. Levine, B. L., Cotte, J., Small, C. C., Carroll, R. G., Riley, J. L., Bernstein, W. B., Van Epps, D. E., Hardwick, R. A., and June, C. H. (1998). Large-scale production of CD4+ T cells from HIV-1-infected donors after CD3/CD28 costimulation. *J. Hematother.* **7**, 437–448.

39. Robinet, E., Certoux, J. M., Ferrand, C., Maples, P., Hardwick, A., Cahn, J. Y., Reynolds, C. W., Jacob, W., Herve, P., and Tiberghien, P. (1998). A closed culture system for the ex vivo transduction and expansion of human T lymphocytes. *J. Hematother.* **7**, 205–215.

40. Piret, J., Devens, D., and Cooney, C. (1991). Nutrient and metabolite gradients in mammalian cell hollow fiber bioreactors. *Can. J. Chem. Eng.* **69**, 421–428.

41. Carswell, K., Weiss, J., and Papoutsakis, E. (1999). Low oxygen tension enhances the stimulation and proliferation of human T lymphocytes in the presence of interleukin-2. *Cytotherapy* **2**, 25–37.

42. Carswell, K., and Papoutsakis, E. (2000). Extacellular pH affects the proliferation of cultured human T cells and their expression of the interleukin-2 receptor. *J. Immunother.* **23**, 669–674.

43. Pierson, B. A., Europa, A. F., Hu, W. S., and Miller, J. S. (1996). Production of human natural killer cells for adoptive immunotherapy using a computer-controlled stirred-tank bioreactor. *J. Hematother.* **5**, 475–483.

44. Swartz, R., Haberman, A., DiMasi, D., Jacobson, B., Langes A., Grise, M., Nicolakis, D., and Truskey, G. (1988). Activation and expansion of cells for adoptive immunotherapy. *In* "Tissue Engineering" (R. Skalak and C. Fox, eds.), pp. 299–312. Alan R. Liss, New York.

45. Linsley, P. S., and Ledbetter, J. A. (1993). The role of the CD28 receptor during T cell responses to antigen. *Annu. Rev. Immunol.* **11**, 191–212.

46. Janeway, C., and Travers, P. (1996). "Immunobiology: The Immune System in Health and Disease." Current Biology Ltd., London.

47. Hviid, L., Felsing, A., and Theander, T. (1993). Kinetics of human T-cell expression of LFA-1, IL-2 receptor, and ICAM-1 following antigenic stimulation in vitro. *J. Clin. Lab. Immunol.* **40**, 163–171.

48. Trimble, L., Perales, M., Knazek, R., and Lieberman, J. (1996). Serum enhances the ex vivo generation of HIV-specific cytotoxic T cells. *Biotechnol. Bioeng.* **50**, 521–528.

49. Garbrecht, F., Russo, C., and Weksler, M. (1988). Long-term growth of human T cell lines and clones on anti-CD3 antibody-treated tissue culture plates. *J. Immunol. Methods* **107**, 137–142.

50. Crossland, K., Lee, V., Chen, W., Riddell, S., Greenberg, P., and Cheever, M. (1991). T cells from tumor-immune mice nonspecifically expanded in vitro with anti-CD3 plus IL-2 retain specific function in vitro and can eradicate disseminated leukemia in vivo. *J. Immunol.* **146**, 4414–4420.

51. Londei, M., Grubeck-Loebenstein, B., de Berardinis, P., Greenall, C., and Feldmann, M. (1988). Efficient propagation and cloning of human T cells in the absence of antigen by means of OKT3, interleukin 2, and antigen-presenting cells. *Scand. J. Immunol.* **27**, 35–46.

52. Flens, M., Mulder, W., Bril, H., von Blomberg van de Flier, M., Scheper, R., and van Lier, R. (1993). Efficient expansion of tumor-infiltrating lymphocytes from solid tumors by stimulation with combined CD3 and CD28 monoclonal antibodies. *Cancer Immunol. Immunother.* **37**, 323–328.

53. June, C., Ledbetter, J., Linsley, P., and Thompson, C. (1990). Role of the CD28 receptor in T-cell activation. *Immunol. Today* **11**, 211–216.

54. Gmünder, H., and Lesslauer, W. (1984). A 45-kDa human T-cell membrane glycoprotein functions in the regulation of cell proliferative responses. *Eur. J. Biochem.* **142**, 153–160.

55. Van Lier, R., Brouwer, M., Rebel, V., Van Noesel, C., and Aarden, L. (1989). Immobilized anti-CD3 monoclonal antibodies induce accessory cell-independent lymphokine production, proliferation and helper actvity in human T lymphocytes. *Immunology* **68**, 45–50.

56. Lamers, C., van de Griend, R., Braakman, E., Ronteltap, C., Bénard, J., Stoter, G., Gratama, J., and Bolhuis, R. (1992). Optimization of culture conditions for activation and large-scale expansion of human T lymphocytes for bispecific antibody-directed cellular immunotherapy. *Int. J. Cancer* **51**, 973–979.

57. Geppert, T., and Lipsky, P. (1987). Accessory cell independent proliferation of human T4 cells stimulated by immobilized monoclonal antibodies to CD3. *J. Immunol.* **138**, 1660–1666.

58. Schultze, J., Michalak, S., Seamon, M., Dranoff, G., Jung, K., Daley, J., and Delgado, J. (1997). CD40-activated human B cells: An alternative source of highly efficient antigen presenting cells to generate autologous antigen-specific T cells for adoptive immunotherapy. *J. Clin. Invest.* **100**, 2757–2765.

59. Peshwa, M., Page, L., Qian, L., Yang, D., and van Schooten, W. (1996). Generation and ex vivo expansion of HTLV-1 CD8+ cytotoxic T-lymphocytes for adoptive immunotherapy. *Biotechnol. Bioeng.* **50**, 529–540.

60. Murray, R., Young, L., Calender, A., Gregory, C., Rowe, M., Lenoir, G., and Rickinson, A. (1988). Different patterns of Epstein-Barr virus gene expression and of cytotoxic T-cell recognition in B-cell lines infected with transforming (B95.8) or nontransforming (P3HR1) virus strains. *J. Virol.* **62**, 894–901.

61. Bejarano, M., Masucci, M., Klein, G., and Klein, E. (1988). T-cell-mediated inhibition of EBV-induced B-cell transformation: Recognition of virus particles. *Int J. Cancer* **42**, 359–364.

62. Misko, I., Sculley, T., Schmidt, C., Moss, D., Soszynski, T., and Burman, K. (1991). Composite response of naive T cells to stimulation with the autologous lymphoblastoid cell line is mediated by CD4 cytotoxic T-cell clones and includes an Epstein-Barr virus-specific component. *Cell. Immunol.* **132**, 295–307.

63. Heslop, H., Ng, C., Li, C., Smith, C., Loftin, S., Krance, R., Brenner, M., and Rooney, C. (1996). Long-term restoration of immunity against Epstein-Barr virus infection by adoptive transfer of gene-modified virus-specific T lymphocytes. *Nat. Med.* **2**, 551–555.

64. Rooney, C., Smith, C., Ng, C., Loftin S., Sixbey, J., Gan, Y., Srivastava, D., Bowman, L., Krance, R., Brenner, M., and Heslop, H. (1998). Infusion of cytotoxic T cells for the prevention and treatment of Epstein-Barr virus-induced lymphoma in allogeneic transplant recipients. *Blood* **92**, 1549–1555.

65. Nazaruk, R., Rochford, R., Hobbs, M., and Cannon, M. (1998). Functional diversity of the CD8+ T-cell response to Epstein-Barr virus (EBV): Implications for the pathogenesis of EBV-associated lymphoproliferative disorders. *Blood* **91**, 3875–3883.

66. Schuler, G., Lutz, M., Bender, A., Thurner B., Roder, C., Young, J., and Romani, N. (1999). A guide to the isolation and propagation of dendritic cells. *In* "Dendritic Cells: Biology and Clinical Applications" (M. Lotze, and A. Thomson, eds.), pp. 515–533. Academic Press, San Diego, CA.

67. Clark, G., and Hart, D. (1999). Phenotypic characterization of dendritic cells. *In* "Dendritic Cells: Biology and Clinical Applications" (M. Lotze, and A. Thomson, eds.), pp. 555–577. Academic Press, San Diego, CA.

68. Chen, B., Shi, J., Smith, J., Choi, D., Geiger, J., and Mulé, J. (1998). The role of tumor necrosis factor α in modulating the quantity of peripheral blood-derived, cytokine-driven human dendritic cells and its role in enhancing the quality of dendritic cell function in presenting soluble antigens to CD4+ T cells in vitro. *Blood* **91**, 4652–4661.

69. Sallusto, F., and Lanzavecchia, A. (1994). Efficient presentation of soluble antigen by cultured human dendritic cells is maintained by granulocyte/macrophage colony-stimulating factor plus interleukin 4 and downregulated by tumor necrosis factor α. *J. Exp. Med.* **179**, 1109–1118.

70. Shapiro, H. (1995). "Practical Flow Cytometry." Wiley-Liss, New York.

71. Givan, A. (1992). "Flow Cytometry: First Principles." Wiley-Liss, New York.

72. Watson, J. (1991). "Introduction to Flow Cytometry." Cambridge University Press, Cambridge, UK.

73. Gupta, S. (1989). Mechanisms of transmembrane signalling in human T cell activation. *Mol. Cell. Biochem.* **91**, 45–50.

74. Bierer, B. E., and Burakoff, S. J. (1988). T cell adhesion molecules. *FASEB J.* **2**, 2584–2590.

75. Chalupny, N. J., Ledbetter, J. A., and Kavathas, P. (1991). Association of CD8 with p56lck is required for early T cell signalling events. *EMBO J.* **10**, 1201–1207.

76. Eichmann, K., Ehrfeld, A., Falk, I., Goebel, H., Kupsch, J., Reimann, A., Zgaga-Griesz, A., Saizawa, K., M., Yachelini, P., and Tomonari, K. (1991). Affinity enhancement and transmembrane signaling are associated with distinct epitopes on the CD8 alpha beta heterodimer. *J. Immunol.* **147**, 2075–2081.

77. Biselli, R., Matricardi, P., D'Amelio, R., and Fattorossi, A. (1992). Multiparametric flow cytometric analysis of the kinetics of surface molecule expression after polyclonal activation of human peripheral blood T lymphocytes. *Scand. J. Immunol.* **35**, 439–447.

78. Caruso, A., Licenziati, S., Corulli, M., Canaris, A., Francesco, M. D., Fiorentini, S., Peroni, L., Fallacar, F., Dima, F., Balsari, A., and Turano, A. (1997). Flow cytometric analysis of activation markers on stimulated T cells and their correlation with cell proliferation. *Cytometry* **27**, 71–76.

79. Bluestone, J. A. (1995). New perspectives of CD28-B7-mediated T cell costimulation. *Immunity* **2**, 555–559.

80. Turka, L. A., Ledbetter, J. A., Lee, K., June, C. H., and Thompson, C. B. (1990). CD28 is an inducible T cell surface antigen that transduces a proliferative signal in CD3+ mature thymocytes. *J. Immunol.* **144**, 1646–1653.

81. Chao, C., Jensen, R., and Dailey, M. (1997). Mechanisms of L-selectin regulation by activated T cells. *J. Immunol.* **159**, 1686–1694.

82. Lenkei, R., Gratama, J., Rothe, G., Schmitz, G., D'hautcourt, J., Arekrans, A., Mandy, F., and Marti, G. (1998). Performance of calibration standards for antigen quantitation with flow cytometry. *Cytometry* **33**, 188–196.

83. Serke, S., von Lessen, A., and Huhn, D. (1998). Quantitative fluorescence flow cytometry: A comparison of the three techniques for direct and indirect immunofluorescence. *Cytometry* **33**, 179–187.

84. Lichtenfels, R., Biddison, W., Schulz, H., Vogt, A., and Martin, R. (1994). CARE-LASS (calcein-release-assay), an improved fluorescence-based test system to measure cytotoxic T lymphocyte activity. *J. Immunol. Methods* **172**, 227–239.

85. Kroesen, B., Mesander, G., ter Haar, J. G., Thé, T. H., and de Leij, L. (1992). Direct visualisation and quantification of cellular cytotoxicity using two colour flourescence. *J. Immunol. Methods* **156**, 47–54.

86. Mattis, A., Bernhardt, G., Lipp, M., and Forster, R. (1997). Analyzing cytotoxic T lymphocyte activity: A simple and reliable flow cytometry-based assay. *J. Immunol. Methods* **204**, 135–142.

87. Slezak S., and Horan, P. (1989). Cell-mediated cytotoxicity. A highly sensitive and informative flow cytometric assay. *J. Immunol. Methods* **117**, 205–214.

88. De Jong, R., Brouwer, M., Rebel V., van Seventer, G., Miedema F., and van Lier, R. (1990). Generation of alloreactive cytolytic T lymphocytes by immobilized anti-CD3 monoclonal antibodies. Analysis of requirements for human cytolytic T-lymphocyte differentiation. *Immunology* **70**, 357–364.

89. Coligan, J., Kruisbeek, A., Margulies, D., Shevach E., and Strober, W. (1991). "Current Protocols in Immunology." Wiley, New York.

SECTION II

METHODS FOR CELL DELIVERY VEHICLES

CHAPTER 42

MODIFICATION OF NATURAL POLYMERS: COLLAGEN

Stephen F. Badylak

INTRODUCTION

Most tissue engineering approaches to the restoration and repair of damaged tissues require a scaffold material upon which cells can attach, proliferate, and differentiate, hopefully into a functionally and structurally appropriate tissue for the body location into which it is placed. A variety of scaffold materials are available, each with different physical properties and each associated with a specific and unique host response when implanted in mammalian hosts. All scaffold materials are subjected to certain processing steps before use in therapeutic devices. These processing procedures can moderately or severely affect both the native material properties and the subsequent clinical utility of scaffolds intended for various tissue engineering applications.

Scaffold materials can be either synthetic or naturally occurring. Synthetic materials such as poly(L)-(lactic acid) and poly(glycolic acid) have received considerable attention for tissue engineering applications and have shown promise in preclinical animal studies and some early human clinical trials. Synthetic materials have predictable and reproducible mechanical and physical properties (e.g., tensile strength and pore size) and can be manufactured with great precision. However, synthetic materials tend to elicit a foreign material type of response in the host: specifically, a mixed polymorphonuclear and mononuclear inflammatory cell infiltrate and associated fibrous connective tissue deposition. This type of response is not always compatible with optimal constructive tissue remodeling. Therefore, naturally occurring materials such as purified collagen and extracellular matrix, which theoretically should not cause as great a "foreign" material response, have been investigated as alternatives to synthetic scaffolds for their utility as tissue engineering scaffolds.

The most commonly used naturally occurring scaffold material has been the structural protein collagen. Collagen is a naturally occurring, highly conserved protein that is ubiquitous among mammalian species and accounts for approximately 30% of all body proteins [1]. Therefore, inherent commonalities exist in amino acid sequence and epitope structure across species lines. Bovine and porcine type I collagen provide a readily available source of scaffold material for numerous applications and have been shown to be very compatible with human systems.

Collagen can be extracted from tissues such as tendons and ligaments, solubilized, and then reconstituted into fibrils that can be fashioned into a variety of shapes and sizes that mimic body structures such as heart valves, blood vessels, and skin [2]. The reconstituted protein is usually stabilized by various methods of cross-linking and must be sterilized prior to surgical use.

Collagen is generally treated as a "self" tissue by recipients into which it is placed and is subjected to the fundamental biological processes of tissue degradation and integration into adjacent host tissues when left in its native ultrastructure. Certain treatment methods, however, significantly alter the mechanical and physical properties of collagen-based materials and may negatively affect the natural physiological processes of cell attachment and proliferation and tissue remodeling. For example, exposure to cross-linking agents such as glutaraldehyde and subsequent sterilization with gamma irradiation can render a collagen-based material biologically inert and incite a host foreign material response [3].

Collagen is a natural substrate for cellular attachment, growth, and differentiation in its native state. In addition to its desirable structural properties, collagen has functional properties. Certain sequences of the collagen fibrils are chemotactic and promote cellular proliferation and differentiation. For the above-mentioned reasons and others, collagen has become a favorite substrate for tissue engineering.

Collagen provides considerable strength in its natural polymeric state. The necessary and required mechanical and physical properties of tissue engineered products for use in cardiovascular, orthopedic, and other body systems often depend upon chemical manipulation of collagen-based scaffolds. The source of collagen, either purified from animal sources or as an integral component of a more complex extracellular matrix, and its treatment prior to use are important variables in the design of tissue-engineered devices. Several methods of cross-linking and sterilization can be utilized to alter (i.e., usually decrease) the rate of *in vivo* degradation or to change the mechanical properties of collagen. These methods include glutaraldehyde treatment, carbodiimide treatment, dye-mediated photooxidation, exposure to polyepoxy compounds, and glycerol treatment. Commonly used methods of terminal sterilization include gamma or electron beam irradiation or ethylene oxide treatment.

Naturally occurring biomaterials derived from extracellular matrix (ECM), which include collagen in addition to other naturally occurring structural and functional proteins, have also been used as scaffolds for tissue engineering applications [4]. The secreted ECMs of cell-enhanced polymer scaffolds such as Apligraf and Dermagraft are examples of such applications. Small-intestinal submucosa (SIS) is a naturally occurring ECM scaffold that is utilized without a cellular component (e.g., SurgiSIS). These materials have been spared treatment with protein cross-linking agents and instead are treated by methods that decellularize, disinfect, and terminally sterilize the material for a variety of medical uses. Some naturally occurring matrix materials such as Perigard (bovine pericardium) and Alloderm (decellularized human dermis) are subjected to a protein cross-linking step as part of the material processing procedure.

ECM-based materials such as those just described are subject to *in vivo* degradation just as purified collagen biomaterial scaffolds are subject to degradation. The degree to which these materials retain their natural mechanical properties and the degree to which they are subject to degradation are affected by the amount of cross-linking that occurs. The methods used to treat collagen and other naturally occurring matrix materials and the effect of these methods upon the host response to these tissue engineering scaffolds is the subject of this chapter.

GLUTARALDEHYDE TREATMENT OF COLLAGEN-BASED MATERIALS

Treatment of collagen-based materials with glutaraldehyde, a five-carbon aldehyde, is a long-established and historically favoral method of stabilization and sterilization. Glutaraldehyde induces cross-links in the lysyl amino acid residues of adjacent collagen monomers [1]. This structural change reduces immunogenicity by neutralizing antigenic epitopes, reduces the rate of *in vivo* degradation, and can, with appropriate procedures, sterilize this naturally occurring collagen-based biomaterials. Methods for glutaraldehyde treatment vary but commonly use concentrations ranging from 0.200 to 0.625%. The induced protein cross-links cause significant changes in the mechanical properties of the native materials such as reduced stress relaxation [5,6] and increased extensibility under applied stress [7,8]. Although these properties may have certain advantages for specific applications, the protein cross-linking is associated with consequences that are often considered less desirable. For example, glutaraldehyde-treated tissues are prone to calcification and subject to fibrous encapsulation following implantation. *In vivo* degradation of the collagen scaffold is virtually eliminated following glutaraldehyde treatment; thus, these biomaterials tend to persist for the life of the host.

The functionality of glutaraldehyde-treated devices is often limited by the degree of calcification that subsequently occurs (e.g., in porcine heart valves) [9,10]. A variety of treatments have been used to prevent calcification of glutaraldehyde-treated tissues, none with any measurable degree of success. These methods include the use of chondroitin sulfate [1], protamine [11], diphosphonates [1,12], glutamic acid [13,14], various amino acids [15], extraction of lipids [16,17], α-amino oleic acid [18], and trivalent metal ions [19,20]. Calcification remains one of the banes of glutaraldehyde treatment for collagen-based biomaterials.

An additional detrimental side effect of glutaraldehyde treatment has been the tendency of such treated tissues to release into the surrounding environment cytotoxic by-products such as monomeric glutaraldehyde, hemiacetals, and alcohol condensation products [21]. Persistent, low-grade, local tissue inflammation is common as a result of these by-products [22,23], and cell growth (e.g., endothelial cells) on these biomaterials is markedly diminished [24]. Therefore, the intended remodeling of glutaraldehyde-treated scaffolds, especially larger three-dimensional scaffolds, is very limited.

The summated result of glutaraldehyde treatment is a collagen-based material that is sterile and antigenically neutral, persistent *in vivo* rather than degradable, cytotoxic rather than friendly to cell growth, causing encapsulation rather than integration of host tissues, and having mechanical properties that differ markedly from those of natural collagen. Although some of these new properties may be desirable for specific applications, the negative effects appear to outweigh the positive effects. Therefore, glutaraldehyde treatment has lost favor among the various methods used to treat naturally occurring biomaterials as tissue engineering scaffolds.

A Sample Method of Glutaraldehyde Treatment

One method of glutaraldehyde treatment proceeds in four steps as follows.

1. Fresh tissue or purified collagen is harvested and rinsed in 50 mM HEPES, pH 7.4, prior to use. The material is immersed at room temperature in a solution containing 0.6% glutaraldehyde buffered with 50 mM HEPES at pH 7.4 for 24 h.
2. The fixed tissue/collagen is then transferred to a solution of 0.2% glutaraldehyde at 4°C for 7 days to ensure that cross-linking is complete [25].
3. The cross-linkage is stabilized by reducing the fixed tissue/collagen with 10 mM sodium borohydride at 4°C for a minimum of 16 h [26].
4. The tissue/collagen is rinsed three times in 50 mM HEPES, pH 7.4, and stored in isotonic saline at 4°C for 24 h prior to implantation [25].

Ethanol has been shown to reduce the degree of calcification associated with glutaraldehyde fixation [25]. The tissue/collagen therefore can be treated with HEPES-buffered ethanol (80% ethanol in 50 mM HEPES pH 7.4) at room temperature for 24 h following treatment with sodium borohydride [27,28].

Measuring Percent of Cross-Linking by Glutaraldehyde

Since lysine is the amino acid in collagen most susceptible to fixation, and glutamic acid is most inert and remains essentially unreacted, the degree of cross-linking can be estimated by using the following formula [29]:

$$\text{Degree of cross-linking (\%)} = \left[1 - \frac{(\text{Lys/Glu})_{\text{fixed}}}{(\text{Lys/Glu})_{\text{fresh}}}\right]100$$

where Lys = lysine concentration by amino acid analysis, Glu = glutamic acid concentration by amino acid analysis, fixed = cross-linked tissue, and fresh = non-cross-linked tissue.

STABILIZATION OF COLLAGEN-BASED BIOMATERIALS BY CARBODIIMIDE TREATMENT

Protein cross-links within collagen-based materials can be induced by methods other than glutaraldehyde treatment [30,31]. Carbodiimides such as cyanamide (a water-soluble compound) and 1-ethyl-3(-3-dimethyl aminopropyl) carbodiimide hydrochloride have been used with moderate success. The cytotoxic by-products found with glutaraldehyde appear to be avoided with the carbodiimides, while mechanical properties that more closely match those of natural tissue can be maintained to a greater degree than is found following glutaraldehyde treatment. Carbodiimide-treated biomaterials, like glutaraldehyde-treated materials, tend to be encapsulated by the host following implantation, as opposed to being incorporated into surrounding tissues. Either outcome may be desirable depending upon the application, but host cell infiltration and neomatrix deposition are generally considered to be more favorable for tissue engineering applications.

Studies have shown that carbodiimide treatments result in less cross-linking of collagen-based materials than is found with the use of glutaraldehyde or the epoxy compound poly-(glycidyl ether) [32]. Treatment of collagen-based materials with carbodiimides results in a lower shrinkage temperature than treatment with more stringent cross-linking agents such as glutaraldehyde. Although cyanamide has been classified with formaldehyde as a fixative, it is a monofunctional molecule and is not by definition a cross-linking agent. Rather, monofunctional molecules "mask" certain amino acid residues and are referred to as masking agents. Cyanamide may not be as useful as a cross-linking agent for fresh tissues as it is for purified collagen.

A SAMPLE METHOD OF CARBODIIMIDE TREATMENT

A variety of carbodiimides have been used to fix tissues according to the following procedure; these include such compounds as cyanamide and 1-ethyl-3(-3 dimethyl aminopropyl) carbodiimide hydrochloride (EDC).

1. Fresh tissue is harvested and thoroughly rinsed in PBS, pH 7.4, prior to use.
2. The tissue is immersed at room temperature in a 0.2 M acetate-buffered solution of 1% cyanamide (or 1.7% EDC) at pH 5.5 for 72 h.
3. The tissue is then thoroughly rinsed for 24 h in sterile water and equilibrated in 1% glycine for 24 hours prior to implantation [26,32,33].

COLLAGEN STABILIZATION BY DYE-MEDIATED PHOTOOXIDATION

Unlike reagents that depend upon the incorporation of cross-linking moieties into the protein matrix, the process of photooxidation does not depend or rely upon such additions. The principle of photooxidation is based upon the visible-light-induced catalysis of proteins that have been sensitized by prior exposure to a photosensitizing dye such as rose bengal or methylene blue [34–38]. Specific amino acids such as histidine, methionine, tyrosine, and tryptophan within naturally occurring proteins can become oxidized following such treatment and thus provide for mechanical stability and reduced calcification in comparison to that seen with glutaraldehyde. A more cell-friendly substrate for growth of host cells appears to result from this method of protein cross-linking, as opposed to cross-linking in the glutaraldehyde or carbodiimide. Collagen-based biomaterials treated with dye-mediated photooxidation are either nonbiodegradable *in vivo* or very slowly degraded. Host tissue ingrowth is slow but does eventually occur.

Dye-mediated photooxidation has been used for several decades to oxidize specific amino acids such as tryptophan, histidine, tyrosine, and threonine in proteins. Specific amino acids can be oxidized by the irradiation of visible light in the presence of a suitable photosensitizer. More recently, photooxidation has been used to stabilize intact collagen-based tissues [39,40].

Photooxidation has not received widespread acceptance for several reasons. The degree of cross-linking cannot be tightly controlled and, thus, less predictable results are obtained in comparison to cross-linking with glutaraldehyde. In addition, the site-specific application of the catalyst can be problematic when required for tissues in an *in situ* setting. When using this method *in vitro*, the methodologies can be implemented more accurately and precisely than *in situ*.

Photooxidation of a protein in the presence of a photocatalyst and sufficient oxygen, under controlled conditions of pH and temperature, can result in a collagen-based substance that is stable to enzymatic degradation without stiffening of the matrix [41]. This process results in increased stability of proteinaceous materials that may be more useful as a scaffold biomaterial for soft tissue applications than more labile non-cross-linked collagen or ECM-based scaffolds. Cytotoxic by-products like those found with glutaraldehyde and formaldehyde are avoided, and calcification is not problematic.

A Sample Method of Dye-Mediated Photooxidation

The following method of dye-mediated photooxidation has been used.

1. Purified collagen or naturally occurring ECM is first soaked or exposed to an aqueous buffer of high osmolality.
2. The material is then incubated in an aqueous solution that includes specific photooxidative catalysts that promote the formation of inter- and intramolecular bonds by oxidizing the material. For example, rose bengal [37] and methylene blue in an aqueous acetic acid solution at low temperature (0–5°C) are typical catalysts.
3. Finally, the soaked material is exposed to light that includes a range of wavelengths that are selectively absorbed by the catalyst (e.g., 200–230 μm for methylene blue). Irradiation under such conditions is affected by temperature, pH, and oxygen concentration such that cross-linking of the material occurs to varying degrees.

To photooxidize intact tissue (e.g., blood vessel), the tissue is immersed for 44 h in chilled PBS (pH 7.4) containing a dye catalyst while being exposed to a broad-wavelength light source. The dye is composed of 0.1% (w/w) each of sodium iodide, potassium iodide, and elemental iodine in a solution of PBS containing 2% ethanol, buffered to pH 6.5 [42,43].

COLLAGEN STABILIZATION BY POLYEPOXY COMPOUNDS

Polyepoxy compounds (PC) have been utilized as protein cross-linking agents [44–47] for purified collagen and for intact tissues. Mammalian tissues such as arteries, veins, and heart valves have been cross-linked with PC and have been shown to have biomechanical properties that more closely match the native tissue than tissue that has been cross-linked with glutaraldehyde [48–50]. Denacol is a proprietary mixture of glycerol poly(glycidyl ethers) containing approximately half bifunctional and half trifunctional oxy groups. The type and degree of fixation that occur are dependent upon temperature, concentration, and pH. As might be expected, the degree of fixation increases as a function of increased temperature, increased concentration of PC, and higher pH values.

A wide variety of polyepoxy compounds are available to induce chemical cross-linking of collagen-based materials, and many different PC have been investigated for their ability to fix tissues. Among these, glycerol poly(glycidyl ether) [51,52], poly(ethylene glycol diglycidyl ether) (PGDE), sulfonated poly(ethylene oxide) (PEO-SO$_3$) [26], Epon812 [51,52], Ex-810 (Denacol) [53,54], and Ex-313 [29,53,54] are most commonly described.

The epoxy groups react with amino groups of purified collagen or mammalian tissues and result in the reduction of free amino groups. When monofunctional PC are used, "masking" of protein epitopes occurs. Bifunctional and trifunctional PC result in true cross-linking of the amino groups.

Polyepoxy stabilization of collagen may be more desirable than glutaraldehyde stabilization when retention of naturally occurring mechanical properties is important. Polyepoxy compounds appear to cause less calcification *in vivo* than glutaraldehyde-treated materials [55]. The irreversible cross-linking that occurs as a result of PC treatment results in a collagen-based material that is essentially nondegradable *in vivo*. These materials tend to be encapsulated by mammalian host tissues and little, if any, integration of host tissues into PC-treated collagen materials can be expected to occur.

A SAMPLE METHOD OF POLYEPOXY TREATMENT

1. The epoxy compounds are slowly dissolved in $Na_2CO_3/NaHCO_3$ (pH 9.5–10.0) buffer containing 30% ethanol to make a 4–5% solution.

2. Freshly harvested tissues are thoroughly rinsed in PBS, pH 7.4, prior to being immersed in the solution for 4 days at room temperature [51,52].

3. Fixed tissues are rinsed in PBS prior to use.

COLLAGEN STABILIZATION BY DEHYDRATION AND THERMAL METHODS

Theoretically, many more amino acid residues are available within collagen-based materials for forming cross-links by dehydration methods and by thermal methods compared with sites available when glutaraldehyde is used. Severe dehydration and thermal (i.e., increased-temperature) methods involve both the amino and carboxyl residues on the collagen fibrils, which may allow these two different methods to act synergistically.

Intuitively, a greater degree of cross-linking within collagen-based materials should make these materials less susceptible to degradation and to host tissue ingrowth. Although the rate at which degradation and tissue ingrowth occur is usually decreased following dehydration or thermal denaturation of collagen-based tissues, these phenomena do indeed occur. The ability to retain water is decreased in tissues treated with thermal cross-linking or severe dehydration. In one study, vascular grafts treated with thermal cross-linking were implanted in the abdominal aorta location of dogs. After 3 weeks, these cross-linked grafts were almost totally resorbed, with no evidence for encapsulation or foreign body reaction [56].

A separate study showed that tissue ingrowth into collagen-based materials that have been cross-linked by dehydrothermal methods occurs to a much greater extent than that which follows glutaraldehyde treatment [33]. Therefore, for tissue engineering applications, where tissue and cellular ingrowth is desirable, methods such as dehydrothermal treatment as opposed to glutaraldehyde or polyepoxy compounds can be employed to produce scaffold stabilization. Neither dehydration nor thermal cross-linking methods are widely used for collagen-based products currently in clinical use. The reasons for this may be the structural changes that occur at the molecular level when these methods are employed. For example, loss of water molecules causes changes in collagen quaternary structure and increased density of the solid fraction of tissue composition. In addition, these methods provide little if any type of sterilization, and all such treated materials need to be subjected to additional treatments to render them sterile and suitable for clinical use.

A combination of dehydration and thermal treatment (i.e., dehydrothermal treatment) increases the denaturation temperature for collagen by removing water [57]. Removal of water from collagen results in the formation of interchain cross-links that are the result of condensation reactions either by esterification or amide formation. The practical consequences of these changes include the failure of rehydration of thermally treated and/or severely dehydrated collagen-based materials to return the materials to 100% of pretreatment hydration levels. Whether this lack of complete rehydration affects phenomena such as tissue ingrowth, cell adherence, and other matrix cell interactions remains to be determined.

A SAMPLE METHOD FOR DEHYDROTHERMAL CROSS-LINKING

1. The purified collagen or collagen matrix is placed in a vacuum oven at 22°C and subjected to a vacuum of less than 100 mtorr for 1 h [33].

2. The temperature is then increased to 110°C.

3. Specimens can be left in the oven at 50°C for 1–3 days.

PREPARATION OF COLLAGEN-BASED MATRICES WITHOUT CROSS-LINKING

Several sources of naturally occurring extracellular matrix, including the urinary bladder and the small intestine, have been used as scaffolds for tissue replacement and remodeling [58–63]. These materials have retained their native three-dimensional ultrastructure and, when used as scaffolds for tissue repair, have been associated with constructive remodeling in numerous mammalian systems [64–68]. The small-intestinal submucosa ECM is acellular, and the preparation of the matrix for clinical use avoids all attempts at intentional cross-linking. Published methods for preparation of this scaffold material include treatment with 0.1% peracetic acid followed by terminal sterilization methods that include gamma irradiation, electron beam irradiation, and ethylene oxide treatment.

The primary difference between the treatment of these collagen-based scaffolds and the treatment of the scaffold materials described earlier is that no method of intentionally stabilizing the matrix is made. These scaffolds are subject to rapid degradation following implantation, infiltration by host cells, and replacement by host tissues. Without deliberate attempts to stabilize the matrix, mechanical properties of collagen-based ECMs such as SIS are very similar, if not identical, to those of the naturally occurring ECM. However, no masking of epitopes occurs, and antigenic structure remains unchanged.

Unless these ECMs are harvested from one host and immediately implanted into a recipient as a transplant tissue [61,62], some method of sterilizing the material is required. Irradiation of collagen-based matrices has proven to be effective without adversely affecting the scaffold remodeling properties. The effects of gamma irradiation and electon beam irradiation upon structural proteins in collagen-based materials has not been determined with certainty. Alterations in collagen fibrillar structure such as the formation of loops and helical twisting have been reported [69]. Variability in fibrillar diameter and increases in the degree and range of collagen fiber crimping have also been reported [70]. It is generally accepted that both gamma and electron-beam irradiation cause a decrease in tensile strength of collagen-based materials, likely due to breaks or scission of the collagen chains. It appears unlikely that free radical agents cause structural protein changes. Rather, the structural changes appear to be dependent upon the energy of the photon- or electron beam that is used, the radiation geometry, and the optical absorbance of the extracellular medium [71,72].

In summary, collagen-based materials that are not intentionally stabilized by chemical agents but require terminal sterilization through irradiation are affected somewhat by a loss of mechanical strength. It does not appear that the functionality of various proteins within complex ECMs is significantly or adversely affected by the energy requirements for sterilization.

A Sample Method of Non-Cross-Linked Collagen-Based ECM Sterilization

1. Harvested tissue ECM is rinsed in water or phosphate-buffered saline.
2. Rinsed material is subjected to 0.1% peracetic acid for 2 h.
3. The material is subjected to a series of rinses in buffered and unbuffered saline to bring pH to 7.0.
4. Peracetic acid treated specimen is then subjected to terminal sterilization with gamma or electron-beam irradiation (1.5 or 2.5 mrad, respectively).

SUMMARY

A variety of methods have been used to process and sterilize collagen-based scaffolds for medical use. Most methods employ mechanisms that cross-link protein moieties, thus altering the natural mechanical and physical properties of the collagen matrix. The most significant effects are found in the degree to which the collagen can be degraded vs encapsulated following implantation in mammalian hosts and the degree to which the scaffolds support and promote host cell attachment, proliferation, and differentiation. Each method has certain advantages and disadvantages. It is likely that all methods will continue to be selectively utilized for collagen-based biomaterials that are intended as scaffolds for tissue engineering applications.

ACKNOWLEDGMENT

The author thanks Jason Hodde for his assistance in the preparation of this book chapter.

REFERENCES

1. Nimni, M. E., Cheung, D., Strates, B., Kodama, M., and Sheikh, K. (1987). Chemically modified collagen: A natural biomaterial for tissue replacement. *J. Biomed. Mater. Res.* **21**, 741–771.
2. Berthiaume, F., and Yarmush, M. L. (1995). Tissue engineering. *In* "The Biomedical Engineering Handbook" (J. D. Bronzino, ed.), pp. 1556–1566. CRC Press, Boca Raton, FL.
3. Sato, K. (1983). Radiation sterilization of medial products. *Radioisotopes* **32**, 431.
4. Weinberg, C. B., and Bell, E. (1986). A blood vessel model constructed from collagen and cultured vascular cells. *Science* **231**, 397–400.
5. Talman, E. A., and Boughner, D. R. (1995). Glutaraldehyde fixation alters the internal shear properties of porcine aortic heart valve tissue. *Ann. Thorac. Surg.* **60**, S369–S373.
6. Mayne, A. S., Christie, G. W., Smaill, B. H., Hunter, P. J., and Barratt-Boyes, B. G. (1989). An assessment of the mechanical properties of leaflets from four second-generation porcine bioprostheses with biaxial testing techniques. *J. Thorac. Cardiovasc. Surg.* **98**, 170–180.
7. Trowbridge, E. A., and Crofts, C. E. (1986). The standardization of gauge length: Its influence on the relative extensibility of natural and chemically modified pericardium. *J. Biomech.* **19**, 1023–1033.
8. Lee, J. M., Haberer, S. A., and Boughner, D. R. (1989). The bovine pericardial xenograft: I. Effect of fixation in aldehydes without constraint on the tensile viscoelastic properties of bovine pericardium. *J. Biomed. Mater. Res.* **23**, 457–475.
9. Trowbridge, E. A., Lawford, P. V., Crofts, C. E., and Roberts, K. M. (1988). Pericardial heterografts: Why do these valves fail? *J. Thorac. Cardiovasc. Surg.* **95**, 577–585.
10. Golomb, G., Schoen, F. J., Smith, M. S., Linden, J., Dixon, M., and Levy, R. J. (1987). The role of glutaraldehyde-induced cross-links in calcification of bovine pericardium used in cardiac valve bioprostheses. *Am. J. Pathol.* **127**, 122–130.
11. Golomb, G., and Ezra, V. (1991). Prevention of bioprosthetic heart valve tissue calcification by charge modification: Effects of protamine binding by formaldehyde. *J. Biomed. Mater. Res.* **25**, 85–98.
12. Levy, R. J., Wolfrum, J., Schoen, F. J., Hawley, M. A., and Lund, S. A. (1985). Inhibition of calcification of bioprosthetic heart valves by local controlled-release diphosphonate. *Science* **228**, 190–192.
13. Grabenwoger, M., Grimm, M., Eybl, E., Leukauf, C., Muller, M. M., Plenck, H., Jr., and Bock, P. (1992). Decreased tissue reaction to bioprosthetic heart valve material after glutamic acid treatment. A morphological study. *J. Biomed. Mater. Res.* **26**, 1231–1240.
14. Grimm, M., Grabenwoger, M., Eybl, E., Moritz, A., Bock, P., Muller, M. M., and Wolner, E. (1992). Improved biocompatibility of bioprosthetic heart valves by L-glutamic acid treatment. *J. Cardiovasc. Surg.* **7**, 58–64.
15. Moritz, A., Grimm, M., Eybl, E., Grabenwoger, M., Grabenwoger, F., Bock, P., and Wolner, E. (1991). Improved spontaneous endothelialization by post-fixation treatment of bovine pericardium. *Eur. J. Cardiothorac. Surg.* **5**, 155–160.
16. Rossi, M. A., Braile, D. M., Teixeira, M. D. R., Souza, D. R. S., and Peres, L. C. (1990). Lipid extraction attenuates the calcific degeneration of bovine pericardium used in cardiac valve bioprosthesis. *J. Exp. Pathol.* **71**, 187–196.
17. Jorge-Herrero, E., Fernandez, P., Guiterrez, M., and Castillo-Olivares, J. L. (1991). Study of the calcification of bovine pericardium: Analysis of the implication of lipids and proteoglycans. *Biomaterials* **12**, 683–689.
18. Girardot, J. M., Gott, J. P., Guyton, R. A., and Schoen, F. J. (1991). A novel technology inhibits calcification of bioprosthetic heart valves: A sheep pilot study. *Trans. Soc. Biomater.* **14**, 60.
19. Webb, C. L., Glowers, W. E., Boyd, J., Rosenthal, E. L., Shoen, F. J., and Levy, R. J. (1990). Al+++ binding studies and metallic cation effects on bioprosthetic heart valve calcification in the rat subdermal model. *Trans. Am. Soc. Artif. Intern. Organs* **36**, 56–59.
20. Levy, R. J., Schoen, F. J., Flowers, W. B., and Staelin, S. T. (1991). Initiation of mineralization in bioprosthetic heart valves: Studies of alkaline phosphatase activity and its inhibition by $AlCl_3$ or $FeCl_3$ preincubations. *J. Biomed. Mater. Res.* **25**, 905–935.
21. Zilla, P., Fullard, L., Trescony, P., Meinhart, J., Bezuidenhout, D., Gorlitzer, M., Humon, P., and von Oppell, U. (1997). Glutaraldehyde detoxification of aortic wall tissue: A promising perspective for emerging bioprosthetic valve concepts. *J. Heart Valve Dis.* **6**, 510–520.
22. Hoffman, D., Gong, G., Liao, K., Macaluso, F., Nikolic, S. D., and Frater, R. W. M. (1992). Spontaneous host endothelial growth on bioprostheses: Influence of fixation. *Circulation* **86**, 75–79.
23. Gendler, E., Gendler, S., and Nimni, M. E. (1984). Toxic reactions evoked by glutaraldehyde-fixed pericardium and cardiac valve tissue bioprosthesis. *J. Biomed. Mater. Res.* **18**, 727–736.
24. Leukauf, C., Szeles, C., Salaymeh, L., Grimm, M., Grabenwoger, M., Losert, U., Moritz, A., and Wolner, E. (1993). *In vitro* and *in vivo* endothelialization of glutaraldehyde-treated bovine pericardium. *J. Heart Valve Dis.* **2**, 230–235.
25. Vyavahare, N. R., Hirsch, D., Lerner, E., Baskin, J. Z., Zand, R., Schoen, F. J., and Levy, R. J. (1998). Prevention of calcification of glutaraldehyde-cross-linked porcine aortic cusps by ethanol preincubation: Mechanistic studies of protein structure and water-biomaterial relationships. *J. Biomed. Mater. Res.* **15**, 577–585.

26. Park, K. D., Yun, J. Y., Han, D. K., Yeong, S. Y., Kim, Y. H., Choi, K. S., Kim, H. M., Kim, H. J., and Kim, K. T. (1994). Chemical modification of implantable biologic tissue for anti-calcification. *Am. Soc. Artif. Intern. Organs J.* 40, M377–382.

27. Vyavahare, N., Hirsch, D., Lerner, E., Baskin, J. Z., Schoen, F. J., Bianco, R., Kruth, H. S., Zand, R., and Levy, R. J. (1997). Prevention of bioprosthetic heart valve calcification by ethanol preincubation. Efficacy and mechanisms. *Circulation* 95, 479–488.

28. Lee, C. H., Vyavahare, N., Zand, R., Kruth, H., Schoen, F. J., Bianco, R., and Levy, R. J. (1998). Inhibition of aortic wall calcification in bioprosthetic heart valves by ethanol pretreatment: Biochemical and biophysical mechanisms. *J. Biomed. Mater. Res.* 42, 30–37.

29. Tu, R., Quijano, R. C., Lu, C. L., Shen, S., Wang, E., Hata, C., and Lin, D. (1993). A preliminary study of the fixation mechanism of collagen reaction with a polyepoxy fixative. *Int. J. Artif. Organs* 16, 537–544.

30. Girardot, J. M., and Girardot, M. N. (1996). Amide cross-linking: An alternative to glutaraldehyde fixation. *J. Heart Valve Dis.* 5, 518–525.

31. Hopwood, D. (1985). Cell and tissue fixation, 1972-1982. *Histochem. J.* 17, 389–442.

32. Pereira, C. A., Lee, M. M., and Haberer, S. A. (1990). Effect of alternative cross-linking methods on the low strain rate viscoelastic properties of bovine pericardial bioprosthetic material. *J. Biomed. Mater. Res.* 24, 345–361.

33. Weadock, K., Olson, R. M., and Silver, F. H. (1983). Evaluation of collagen cross-linking techniques. *Biomater. Med. Devices Artif. Organs* 11, 293–318.

34. Gurnani, S., Arifuddin, M., and Augusti, K. T. (1966). Effect of visible light on amino acids. I. Tryptophan. *Photochem. Photobiol.* 5, 495–505.

35. Gurnani, S., and Arifuddin, M. (1966). Effect of visible light on amino acids. II. Histidine. *Photochem. Photobiol.* 5, 341–345.

36. Weil, L., Seibles, T. S., and Herskovits, T. T. (1965). Photo-oxidation of bovine insulin sensitized by methylene blue. *Arch. Biochem. Biophys.* 111, 308–320.

37. Jori, G., Galiazzo, G., Marzotto, A., and Scoffone, E. (1968). Selective and reversible photo oxidation of the methionyl residues in lysozyme. *J. Biol. Chem.* 243, 4272–4278.

38. Kenkare, U. W., and Richards, F. M. (1966). The histidyl residues in ribonuclease S. Photo-oxidation in solution and in single crystals; the iodination of histidine-12. *J. Biol. Chem.* 241, 3197–3206.

39. Moore, M. A., Bohachevsky, I. K., Cheung, D. T., Boyan, B. D., Chen, W. M., Bickers, R. R., and McIlroy, B. K. (1994). Stabilization of pericardial tissue by dye-mediated photooxidation. *J. Biomed. Mater. Res.* 28, 611–618.

40. Moore, M. A., Chen, W. M., Phillips, R. E., Bohachevsky, I. K., and McIlroy, B. K. (1996). Shrinkage temperature versus protein extraction as a measure of stabilization of photo-oxidized tissue. *J. Biomed. Mater. Res.* 32, 209–214.

41. Mechanic, G. L. (1992). Process for cross-linking collagenous material and resulting product. U.S. Pat. 5,147,514.

42. Moore, M. A., and Phillips, R. E. (1997). Biocompatibility and immunologic properties of pericardial tissue stabilized by dye-mediated photo oxidation. *J. Heart Valve Dis.* 6, 307–315.

43. Moore, M. A. (1997). Pericardial tissue stabilized by dye-mediated photo oxidation: A review article. *J. Heart Valve Dis.* 6, 521–526.

44. Imamura, E., Sawatani, O., Koyanagi, H., Noishiki, Y., and Miyata, T. (1989). Epoxy compounds as a new cross-linking agent for porcine aortic leaflets: Subcutaneous implant studies in rats. *J. Cardiovasc. Surg.* 4, 50–57.

45. Norijiri, C., Noishiki, Y., and Koyanagi, H. (1987). Aorta-coronary bypass grafting with heparinized vascular grafts in dogs. *J. Thorac. Cardiovasc. Surg.* 93, 867–877.

46. Masuoka, M., and Nakamura, M. (1985). Effect of catalysts on epoxy tannage. *Leather Chem.* 30, 223–232.

47. Noishiki, Y., Kodaira, K., Furuse, M., and Miyata, T. (1989). Method of preparing antithrombogenic medical materials. U.S. Pat. 4,806,595.

48. Tu, R., McIntyre, J., Hata, C., Lu, C. L., and Quijano, R. C. (1991). Dynamic internal compliance of a vascular prosthesis. *Trans. Am. Soc. Artif. Intern. Organs* 37, M470–472.

49. Lee, J. M., Thyagarajan, K., Pereira, C., McIntyre, J., and Tu, R. (1991). Cross-linking of a prototype bovine artery xenograft: Comparison of the effects of glutaraldehyde and four poly(glycidyl ether) compounds. *Artif. Organs* 15, 303.

50. McIntyre, J., Tu, R., Wang, E., Hata, C., DeJonge, M., and Quijano, R. C. (1991). Biomechanical properties of a new biomaterial—A polyepoxy tanned bovine artery. *Trans. Soc. Biomater.* 14, 119.

51. Xi, T., Lui, F., and Xi, B. (1992). Effect of pretreatment with epoxy compounds on the mechanical properties of bovine pericardial bioprosthetic materials. *J. Biomater. Appl.* 7, 61–75.

52. Xi, T., Ma, J., Tian, W., Lei, X., Long, S., and Xi, B. (1992). Prevention of tissue calcification on bioprosthetic heart valve by using epoxy compounds: A study of calcification tests *in vitro* and *in vivo*. *J. Biomed. Mater. Res.* 26, 1241–1251.

53. Sung, H. W., Shen, S. H., Tu, R., Lin, D., Hata, C., Noishili, Y., Tomizawa, Y., and Quijano, R. C. (1993). Comparison of the cross-linking characteristics of porcine heart valves fixed with glutaraldehyde or epoxy compounds. *Am. Soc. Artif. Intern. Organs J.* 39, M532–536.

54. Sung, H. W., Witzel, T. H., Hata, C., Tu, R., Shen, S. H., Lin, D., Noishiki, Y., Tomizawa, Y., and Quijano, R. C. (1993). Development and evaluation of a pliable biological valved conduit. Part II: Functional and hemodynamic evaluation. *Int. J. Artif. Organs* 16, 199–204.

55. Imamura, E., Sawatani, O., Koyanagi, H., Noishiki, Y., and Miyata, T. (1988). Anticalcification effects of epoxy compounds as cross-linking agent. *Jpn. J. Artif. Organs* 17, 1101–1103.

56. Ma, X. H., Noishiki, Y., Yamane Y., Iwai, Y., Marato, D., and Matsumoto, A. (1996). Thermal cross-linking for biologically degradable materials. Preliminary report. *Am. Soc. Artif. Intern. Organs J.* 42, M866–871.

57. Yannas, I. V., and Tobolsky, A.V. (1967). Cross-linking of gelatine by dehydration. *Nature (London)* 215, 509–510.

58. Badylak, S. F., Lantz, G. C., Coffey, A. C., and Geddes, L. A. (1989). Small intestinal submucosa as a large diameter vascular graft in the dog. *J. Surg. Res.* **47,** 74–80.

59. Badylak, S. F., Kropp, B., McPherson, T., Liang, H., and Snyder, P. W. (1998). Small intestinal submucosa: A rapidly resorbed bioscaffold for augmentation cystoplasty in a dog model. *Tissue Eng.* **4,** 379–388.

60. Cobb, M. A., Badylak, S. F., Janas, W., and Boop, F. A. (1996). Histology after dural grafting with small intestinal submucosa. *Surg. Neurol.* **46,** 389–394.

61. Atala, A., Guzman, L., and Retik, A. B. (1999). A novel inert collagen matrix for hypospadias repair. *J. Urol.* **162,** 1148–1151.

62. Chen, F., Yoo, J. J., and Atala, A. (1999). Acellular collagen matrix as a possible "off the shelf" biomaterial for urethral repair. *Urology* **54,** 407–410.

63. Dahms, S. E., Piechota, H. J., Dahiya, R., Lue, T. F., and Tanagho, E. A. (1998). Composition and biomechanical properties of the bladder acellular matrix graft: Comparative analysis in rat, pig and human. *Br. J. Urol.* **82,** 411–419.

64. Badylak, S., Arnoczky, S., Plouhar, P., Haut, R., Mendenhall, V., Clarke, R., and Horvath, C. (1999). Naturally occurring extracellular matrix as a scaffold for musculoskeletal repair. *Clin. Orthop.* **367S,** S333–S343.

65. Kropp, B. P., Sawyer, B. D., Shannon, H. E., Rippy, M. K., Badylak, S. F., Adams, M. C., Keating, M. A., Rink, R. C., and Thor, K. B. (1996). Characterization of small intestinal submucosa-regenerated canine detrusor: Assessment of innervation, in vitro compliance and contractility. *J. Urol.* **156,** 599–607.

66. Kropp, B. P., Ludlow, J. K., Spicer, D., Rippy, M. K., Badylak, S. F., Adams, M. C., Keating, M. A., Rink, R. C., Birhle, R., and Thor, K. B. (1998). Rabbit urethral regeneration using small intestinal submucosa onlay grafts. *Urology* **52,** 138–142.

67. Yoo, J. J., Meng, J., Oberpenning, F., and Atala, A. (1998). Bladder augmentation using allogenic bladder submucosa seeded with cells. *Urology* **51,** 221–225.

68. Huynh, T., Abraham, G., Murray, J., Brockbank, K., Hagen, P. O., and Sullivan, S. (1999) Remodeling of an acellular collagen graft into a physiologically responsive neovessel. *Nat. Biotechnol.* **17,** 1083–1086.

69. Tzaphlidou, M., Kounadi, E., Leontiou, I., Matthopoulos, D. P., and Glaros, D. (1997). Influence of low doses of γ-irradiation on mouse skin collagen fibrils. *Int. J. Radiat. Biol.* **71,** 109–115.

70. Belkoff, S. M., and Haut, R. C. (1992). Microstructurally based model analysis of gamma-irradiated tendon allografts. *J. Orthop. Res.* **10,** 461–464.

71. Fujisato, T., Tomihata, K., Tabata, Y., Iwamoto, Y., Burczak, K., and Ikada, Y. (1999). Cross-linking of amniotic membranes. *J. Biomater. Sci. Polym. Ed.* **10,** 1171–1181.

72. Redpath, J. L. (1986). UV-type damage associated with ionizing radiation: A review. *Int. J. Radiat. Biol. Relat. Stud. Phys. Chem. Med.* **50,** 191–203.

CHAPTER 43

MODIFICATION OF NATURAL POLYMERS: COLLAGEN–GLYCOSAMINOGLYCAN COPOLYMERS

Frederick Cahn

INTRODUCTION

Collagen–glycosaminoglycan (collagen–GAG) copolymer is the key functional component of artificial skin grafts for skin replacement (dermis and epidermis). The artificial skin graft was engineered to meet specific design criteria for a satisfactory clinical substitute for tissue autograft. Thus, the processes and assays for the collagen–GAG copolymer are keyed to the design objectives of the artificial skin graft.

COLLAGEN–GAG DESIGN PRINCIPLES FOR ARTIFICIAL SKIN GRAFTS

THE CLINICAL NEED FOR AN ARTIFICIAL SKIN GRAFT

An intact epidermis provides a barrier to microbial invasion and is required for normal physiological homeostasis. Epidermis provides essential physiological functions, such as preventing loss of moisture or fluid, protection from ultraviolet light, and support of thermoregulation. The dermis, because of the strength and elasticity of its collagen- and elastin-rich extracellular matrix, provides the essential mechanical functions of skin. The dermis is also a vascularized tissue that provides nutrition to both the dermis and epidermis, and its transport of immune system components is an essential part of the barrier function of the skin. In the case of a full-thickness injury to skin, both skin layers are lost and these essential functions must be restored [1].

Modern surgical treatment of full-thickness injury to the skin begins with a thorough excision of all necrotic or diseased tissue, resulting in a surgically created full-thickness wound with a surface of viable vascularized tissue [2,3]. This open wound must be closed to halt the local and systematic consequences of the open wound and to prevent the further loss of tissue.

The clinical requirements for the closure of a surgically excised full-thickness wound can be divided into two stages [4]. Stage 1 addresses the requirements for patient survival that immediately follow the excisional surgery; stage 2 addresses the long-range need for a permanent closure of the wound that avoids disfiguring scars and crippling contractures.

Both stages of closure may be achieved by a "split-thickness" skin autograft. That is, a sheet of the patient's own skin, including its epidermis and a portion of its dermis that usually can be harvested from a healthy donor site. A skin autograft provides a solution to each stage of wound closure: In stage 1, it creates an immediate physiological closure of the wound that reduces local symptoms of inflammation, fluid loss, and wound contraction, and (for large wounds) reduces fever and hypermetabolic systemic responses. In stage 2, it regenerates a permanent and functional skin by reestablishing vascular connections with the wound bed and by allowing normal tissue remodeling to regenerate the dermis.

The typical result of the application of a skin graft is a satisfactory and permanent repair of the dermal and epidermal components of skin, without significant scar or contraction of the wound. However, these advantages are achieved at the expense of another wound at the donor site. Frequently, because sufficient healthy donor skin is not available, surgical compromises result in a delay in definitive treatment and/or less satisfactory results, such as scar tissue formation.

DESIGN INPUTS FOR AN ARTIFICIAL SKIN GRAFT

Yannas, Burke, *et al.* have described the design principles of an artificial skin graft in publications and patents [4–6]. The basic objective of an artificial skin graft is to enable the surgeon to close a surgically excised wound with a graft material that emulates the functions of the skin autograft, but without its chief disadvantage, the donor site wound.

Thus, the design of the artificial skin graft must address both stages of wound closure to achieve (1) immediate physiological closure of the wound and (2) a permanent vascularization of the graft and regeneration of the dermal and epidermal skin layers, without introducing fibrosis, scar, or contracture. These design principles are summarized in Table 43.1.

STRUCTURE AND FUNCTION OF AN ARTIFICIAL SKIN GRAFT

Yannas and Burke found a solution to these design requirements in a bilayer device that they called "artificial skin." Artificial skin consists of a porous collagen–GAG dermal regeneration layer and a silicone temporary epidermal-substitute layer that is firmly bound to it. The clinical utility of this artificial skin graft for treating surgically excised wounds was demonstrated in clinical trials on burn patients [7,8]. Artificial skin has since been approved by regulatory authorities in the United States and the European Union and is available as a commercial product.

The bilayer artificial skin graft provides an immediate physiological closure of the wound that inhibits inflammation and minimizes the formation of granulation tissue, wound contraction, fluid loss, and the systemic effects of an open wound [7,9]. The porous collagen–GAG layer achieves the stage 1 functional requirements for biocompatibility and low inflammation, and contractile cells are not observed in either stage of healing [10]. The highly hydrophilic biomaterial and the porous design make the artificial skin adherent to the wound bed [4]. The proprietary purified medical grade collagen and glycosaminoglycan components that are used in commercial manufacture of the artificial skin graft are free of endotoxin and immunogenic impurities, as well. The silicone layer contributes the stage 1 requirements for low moisture transmission and impermeability to microorganisms, as well the appropriate tear strength and handling characteristics. Because the silicone layer is not degradable under physiological conditions, these properties persist until a surgeon removes the layer in a stage 2 procedure. Both materials are very flexible and conformable to the wound surface.

The stage 1 initial wound closure is followed by vascularization of the dermal layer and regeneration of a permanent dermal tissue; the original material from the dermal regeneration layer is degraded and remodeled over a period of 1 or 2 months. Upon adequate vascularization of the dermal layer (typically about 2–3 weeks), and availability of donor epithelial tissue, the stage 2 surgical procedure is performed in which the temporary silicone layer is removed, and a meshed layer of ultrathin epidermal autograft is placed over the neodermis. Cells from this epidermal autograft migrate and grow to form an intact, normal, epidermis, thereby completing the stage 2 closure of the wound and regenerating a functional, autologous dermis ("neodermis") and epidermis. Analysis of biopsy samples from patients in a clinical trial showed no evidence of scar formation [10].

Table 43.1. Design Inputs for Artificial Skin Grafts

Clinical requirements	Design requirements	Design details
Stage 1: Immediate physiological wound closure	**Dermal functions**	Highly biocompatible
	Reduce wound inflammation	Noninflammatory
	Inhibit wound contraction	Avoid expression of contractile cell phenotypes
	Prevent systemic physiological responses	Non antigenic Endotoxin free
	Epidermal functions	Impermeable to microorganisms
	Barrier to fluid and moisture loss	Low moisture flux rate
	Barrier to infection	Biocompatible
	Maintain these properties until stage 2	
	Other Requirements	Tear strength
	A wound bed of viable tissue	Surface energy that allows wetting and peel strength
	Artificial skin adherent to wound	Low bending rigidity
	Handling and suturing	Elastic Conformable to wound bed
Stage 2: Permanent wound closure with intact epidermis and a regenerated dermis (without fibrosis or scar)	**Dermal functions** Regenerate vascularized nonfibrotic tissue that supports epidermis ("neodermis")	Porous dermal layer High pore volume Appropriate mean pore size Appropriate thickness Invasion of normal fibroblasts and capillaries
	Strength and elasticity of neodermis must be similar to normal skin and not scar	Synthesized extracellular matrix fibers must be similar in structure to normal dermis and must not scar
	Capable of supporting viable epidermis	
	Permanent	Metabolic disposal of exogenous biomaterials
	Minimal contracture	No toxic metabolites
	No persistent biomaterials that could limit future skin function or growth	
	Epidermal functions	The polymeric graft material is removed after the vascularized neodermis is formed
	Reconstitute confluent and permanent epidermis	
	Maintain control of infection	An epidermal autograft seeds the neodermis with autologous cells
	Maintain control of fluid loss	Epidermal cells grow to confluence and differentiate to regenerate intact epidermis

DESIGN OF A COLLAGEN–GAG COPOLYMER FOR DERMAL REGENERATION

The following design considerations and experimental results led Burke and Yannas to choose a collagen–GAG copolymer for the dermal regeneration layer that meets the design critera listed in Table 43.1 [5,11].

1. Collagen and glycosaminoglycan are principal components of normal extracellular matrix, so it was expected that these materials would be inherently biocompatible, weakly immunogenic, degradable, and remodelable by normal physiological mechanisms, resulting in nontoxic metabolites.

2. There were existing cross-linked collagen sponges that had received regulatory approval as implantable medical devices (for other applications).

3. The biodegradation rate of collagen membranes can be adjusted over a wide range by controlling the cross-link density; however, heavily cross-linked collagen is a stiff, brittle material. Animal implantation studies showed inflammatory responses when collagen sponges were cross-linked sufficiently to have a residence time of several weeks.

4. Coprecipitation of chrondroitin-6 sulfate and certain other glycosaminoglycans with collagen yields sponges that are more resistant to collagenase degradation at equivalent cross-linking densities.

5. Cross-linking of the coprecipitate enables the GAG to be partially retained at neutral pH and physiological ionic strength.

6. This methodology can be used to synthesize sponges that can be cross-linked sufficiently to have adequate strength and to be degraded at optimal rates. Animal implantation studies demonstrated that appropriately cross-linked collagen–GAG sponges exhibited minimal inflammatory and encapsulation responses and appropriate *in vivo* degradation rates and residence time. In addition, nonfibrotic cellular ingrowth and more normal connective tissue fibers were observed, suggesting the possibility of tissue regeneration.

7. Collagen–GAG sponges had significantly higher moduli of elasticity and higher energy to fracture than collagen controls. Thus addition of chrondroitin 6-sulfate can be used not only for controlling biodegradation rate but also as a means of controlling mechanical behavior.

8. Sponges made from collagen–GAG coprecipitates (prepared as described shortly) exhibited a more open pore structure than a collagen control. The porosity of biomaterials is known to affect the response of tissue in contact with them.

9. Chrondroitin 6-sulfate has wider commercial availability than other glycosaminoglycans having similar properties.

QUANTITATIVE VARIABLES OF THE DERMAL REGENERATION LAYER

Two quantitative characteristics of the collagen–GAG dermal layer have been shown to be critical for its biological activity: pore size and cross-linking density. These parameters affect the performance of artificial skin in inhibiting wound contraction of full-thickness dermal excisions in the guinea pig. This assay has shown an optimal value for the average pore size around 100 µm. Cross-linking density must be above a threshold value for the material to show biological activity in this assay [14].

OTHER APPLICATIONS OF COLLAGEN–GAG CHEMISTRY

Devices fabricated from collagen–GAG have also been demonstrated to promote peripheral nerve regeneration in animal models [12,13].

COLLAGEN–GAG PROCESSES FOR ARTIFICIAL SKIN GRAFTS

The method of making the artificial skin graft is summarized here, along with some comments on its rationale. The operations described in the subsections that follow are carried out under aseptic conditions.

PREPARATION OF A COLLAGEN DISPERSION

A suspension of 0.25% w/v of fibrous collagen (e.g., from bovine hide or tendon) is dispersed by means of a suitable homogenizer in 0.05 M acetic acid at about pH 3.2 and a temperature below 20°C. Since this pH is below the isoelectric point of collagen, a viscous suspension or gel is formed as the collagen molecules swell [14]. Electron micrographs of fibers from this gel show that the characteristic collagen banding at 64 nm, a feature of the quaternary structure of collagen fibers, is lost. However, the collagen is not denatured, as demonstrated by infrared spectroscopic measurement of the triple helical tertiary structure of the collagen molecules [5,14].

ADDITION OF THE CHONDROITIN 6-SULFATE

A 0.1% w/v solution of chondroitin 6-sulfate is added slowly into the homogenizer to a final concentration of about 8 wt% (e.g., the volume of c6s added is 20% of the volume of collagen suspension). A precipitate is formed by ionic bonding between the cationic collagen and anionic chondroitin-6-sulfate. The precipitation reaction can be observed by a decrease in viscosity of the suspension [14].

INCREASE OF DENSITY BY CENTRIFUGATION

The density of the mixture is increased by centrifugation. An aliquot of supernatant equal in volume to half the original volume is removed. The target density of the suspension is about 0.5% w/v, dry weight.

LYOPHILIZATION

The gel is poured into trays and leveled. The trays are placed on chilled shelves of a lyophilizer. Freezing of the suspension leads to a phase separation, in which crystals of ice form one phase and compressed, hydrated collagen fibers become another. The result is a frozen, porous sponge. The cooling rate of the suspension determines pore size and shape; average pore size is one of the critical quantitative parameters affecting the biological activity of artificial skin [14]. Lyophilization of the frozen sponge produces a dry sponge.

DEHYDROTHERMAL CROSS-LINKING

The pore structure of the collagen sponge would quickly collapse upon rehydration [15]. An initial cross-linking of the dry sponge to preserve its porosity as well as to prevent subsequent elution of the chondroitin 6-sulfate is accomplished by extreme drying at 105°C at below 100 mtorr for about 1–5 days. Covalent links can be formed in gelatin (and by inference in collagen) under these conditions [16]. The result is a sponge that does not collapse when rehydrated [15]. Denaturation of collagen does not take place at this high temperature because the collagen is dry [17].

COATING

To form the bilayer artificial skin device, the sponge is now coated with a medical grade of silicone adhesive, and the adhesive is allowed to cure [14].

REHYDRATION

The sponge is then rehydrated in 0.05 M acetic acid. This acidic pH is the same as that used to form the collagen–GAG precipitate, so the ionic bonds between the collagen and the glycosaminoglycan will be maintained.

GLUTARALDEHYDE CROSS-LINKING

Further cross-linking of the collagen component of the device is accomplished by soaking in a solution of 0.25% w/v glutaraldehyde in 0.05 M acetic acid for 24 h [14]. The reaction of glutaraldehyde with collagen is slow at low pH. However, subsequent analysis demonstrates that cross-links are formed under these conditions. The time, concentration, and temperature parameters of glutaraldehyde cross-linking determine cross-link density, which controls the degradation rate of the collagen when exposed to collagenase, as well as the *in vivo* residence time of the collagen–GAG material.

WASHING

The device is washed in multiple washes of water to remove residual glutaraldehyde and acetic acid. The concentration of glutaraldehyde in the final wash should be below 5 ppm [14].

STORAGE

The device is now ready for use. It is not terminally sterilized. It is stored in 70% 2-propanol, as a preservative.

PREPARATION FOR USE AND USE

The device is prepared for use by soaking in isotonic saline. Since the device serves as a graft rather than as a wound dressing, it is cut to fit the wound shape and sutured or stapled in place on the excised would bed with the collagen–GAG sponge in contact with the wound bed [7]. The device is left in contact with the wound bed or about 2–3 weeks. After a vascularized dermal layer has formed, the silicone layer is easily removed, and the ultrathin meshed epidermal autograft is applied. Final closure of the wound with a confluent epithelium occurs in about another week [7].

COLLAGEN–GAG ASSAYS FOR ARTIFICIAL SKIN GRAFTS

PORE SIZE DISTRIBUTION

Pore size is measured by scanning electron microscopy of a cut edge of the sponge. Average pore size can be measured by stereology [18].

CROSS-LINKING DENSITY

A physical–chemical measure of cross-linking density, expressed as average molecular weight between cross-links, can be determined by mechanical stress–strain measurements [6]. The molecular weight between cross-links (M_c) is inversely proportional to the number of cross-links per unit volume. Values of M_c can be determined by measuring the stress–strain behavior of thermally denatured collagen–GAG composites [19].

COLLAGEN DENATURATION BY INFRARED SPECTROSCOPY

The infrared spectroscopic analysis of collagen for helical content is described by Yannas *et al.* [6]. This test is used as a measure of collagen denaturation due to the processing steps. A thin film, approximately 1- to 20 μm thick, is made by drying a collagen suspension thoroughly on a smooth poly(tetrafluoroethylene) surface. The mid-infrared transmission spectrum is measured, and the ratio of the absorbance band at 1235 cm^{-1} ("helical marker band") to the band at 1450 cm^{-1} is calculated. (An absorption band in the far infrared at 345 cm^{-1} can also be used as a helical marker.)

COLLAGENASE DEGRADATION RATE

As stated earlier, a key variable that determines the biological effectiveness of artificial skin is the *in vivo* rate of degradation. A correlation exists between *in vivo* degradation rate and *in vitro* degradation rate of collagen–GAG sponges by bacterial collagenase [20].

Sensitivity of collagen–GAG to bacterial collagenase is measured by assay of the generation of free amino groups by the enzyme. A summary of the method follows [20]. About 1–2 mg of cross-linked collagen sponge is finely ground, then incubated for 5 h at 37°C in 0.08% w/v of collagenase from *Clostridium histolyticum* at pH 7.4. Undigested particles are filtered out, and ninhydrin and hydrindantin added at pH 5 and incubated in a boiling water bath for 20 min. The color is extracted with 50% (w/w) propanol. After 15 min the absorbance at 600 nm is measured, and standardized with a solution of 0.002 M L-leucine. The absorbances are compared with a sample of un-cross-linked collagen treated by the foregoing procedure.

Collagenase sensitivity can also be measured mechanochemically by the change in force necessary to hold a sample at fixed extension when exposed to *C. histolyticum* collagenase. The method is as follows [11]. Collagen films, in the form of tape, were extended to a strain of $4.0 \pm 0.5\%$ in the presence of a solution of collagenase (40 U/ml), and the force induced on the tape was recorded as a function of time. The force was found to be representable by a single negative exponential of the time, hence a plot of the logarithmic force versus time yields a straight line. The slope of the straight line yields the rate of enzymatic degradation of the collagen by the collagenase.

IN VIVO WOUND HEALING ASSAYS

The stage 1 wound closure and dermal regeneration capabilities of artificial skin prototypes are predicted by means of an *in vivo* assay [4,7]. Guinea pig studies are carried out as described by Yannas *et al.* [21]. A more detailed protocol follows.

Guinea Pig Surgery and Application of Artificial Skin

Female Hartley guinea pigs, 1–2 months of age, weighing 400–500 g each, are randomly assigned to groups containing an appropriate number of animals per test article and housed in large cages, four to a cage. After surgery, they are housed in individual cages. Food and water are given *ad libitum*. Food is commercial guinea pig formula, which is withdrawn the night before surgery.

Guinea pigs are shaved and the residual hair removed with a commercial depilatory (Nair) either the morning of a study or the day before. The hair is removed from the entire back and halfway down the sides. Tetracycline is given subcutaneously at a dose rate of 0.1 mg/kg. The guinea pigs are anesthetized with halothane at a concentration of 2.5%. After its back has been prepped with Betadine, each animal is laid on a sterile surgical towel and is ready for surgery.

The recipient site is marked with Mercurochrome to the size of the graft, 1.5×3.0 cm^2, and then prepped with 70% 2-propanol and draped. The graft is placed on the mid-portion of the back, slightly to the left of the spine, with the length of the graft going with the length of the animal.

An incision using a no. 10 surgical blade is made around the perimeter of the 1.5×3.0 cm^2 area down to the panniculus carnosus. One corner of the area is picked up with forceps. Keeping tension on the corner, the surgical blade is used to excise the area down to the panniculus carnosus. After excision, the site is covered with a sterile dressing sponge to stop any bleeding. The artificial skin is placed in the recipient site and sutured in place using 5-0 Ethicon suture. Ten sutures are placed in each graft. Neosporin ointment is used along the wound edge to decrease the chance of any infection. The graft is covered with a sterile sponge and two wraps of Elastoplast. The guinea pigs are placed in a warm environment to recover from anesthesia.

To compare the effect of the artificial skin graft to control wounds, the same surgical procedure is performed except the artificial skin graft was not sutured into the wound. The open wound is bandaged as with the grafted sites and allowed to heal as an open wound.

The grafts and wounds are examined at periodic intervals and rebandaged if necessary until the animal is sacrificed. At periodic intervals following surgery, the animals are sacrificed and the graft site is removed. The tissue specimen is placed in 10% formalin, processed, and stained with hematoxylin–eosin for histological evaluation.

Contraction

At about 1-week intervals, the bandages can be removed and the wound areas estimated by photography or tracings. The wound areas are plotted against time since grafting. In this assay, successful artificial skin prototypes increased the time that the wound area is reduced by half from 8 days (for an ungrafted control) to greater than 15 days.

Histology

For histopathological analysis, the wound sites are excised from each animal, including 2–5 mm of normal surrounding tissue, and fixed in 10% phosphate-buffered formalin for histological processing. (Since the silicone membrane is routinely sheared off the collagen–GAG matrix and lost during paraffin block sectioning, whether it has naturally separated from the graft or was lost during tissue processing is not readily observed.)

Foreign Body Giant Cells (FBGC)

In observations of FBGC, there is an important distinction between those with identifiable foreign matter (starch granules, cotton, hair, suture, etc.) and substances associated with the collagen–GAG matrix in the absence of recognizable foreign matter. FBGC associated with foreign particles in the interface between matrix and host tissue bed generally usually indicate foreign matter that unavoidably contaminates the wound during the surgical procedure. When found within the matrix, it may indicate contamination of the matrix during its manufacture.

FBGC associated with the matrix fibers in the absence of recognizable foreign matter are also occasionally observed. Large numbers of them may indicate an insufficiently biocompatible matrix.

Neutrophilic Infiltration

Typically, few neutrophils are seen in the matrix during healing. Heavy infiltrations or other signs of infection may exclude a sample from further analysis.

Other Cellular Responses

At about 4 days, the first observed event is the migration of sparse numbers of mononuclear cells resembling monocytes and macrophages to the deep recesses of the collagen–GAG lattice.

By day 7, the lower collagen–GAG matrix should show ingrowth of buds and tufts of mesenchymal cells with oval to spindle configurations. Small endothelial cell lined vascular spaces may be found at the base of these tufts. By day 10, typical fibroblasts are recognized. From day 10 on, the mesenchymal tissue proliferates and differentiates into moderately vascular fibroblastic connective tissue with birefringent collagen fibers until lattice spaces are filled by about day 22.

Also by day 7, if the silicone membrane has separated from the collagen–GAG matrix, new connective tissue that originates from the dermis at the wound edge spreads laterally and centripetally and grows both over and into the collagen–GAG matrix below the silicone membrane. Epidermal proliferation trails behind this connective tissue proliferation. If the silicone membrane remains attached to the matrix, epidermal proliferation extends between the membrane and the surface of the matrix. At day 22, this connective tissue growing in from the wound edge becomes confluent with the connective tissue growing in upwardly from the bottom. Final epidermal migration and coverage usually occurs by 48 days.

Matrix Resorption

Matrix resorption and connective tissue proliferation appear to occur simultaneously. At day 4, mononuclear cells are seen lined up along matrix fibers singly, in clusters, and occasionally with the appearance of syncytial, multinucleate cells. After day 7, significant numbers of definitive multinucleate giant cells are present in intimate surface contact with matrix fibers. Loss of matrix substance with fiber fragmentation is evident at day 14. The population density of multinucleate cells appears to peak and plateau between 22 and 30 days and complete disappearance by day 70, when matrix resorption in all animals should be complete.

Inflammation and Immune Reactions

Typically, few neutrophils are seen in the matrix during healing. Heavy infiltrations or other signs of infection should exclude a sample from further analysis. However, the 10- to 35-day period is characterized by an advancing growth of epidermis and connective tissue over the collagen–GAG matrix from the wound edge, below the silicone. Eventually there will be a complete sloughing of the silicone membrane to leave an open wound. This sloughing may be accompanied by significant localized acute inflammation at the surfaces of the wound that are still devoid of epidermal covering. After 40 days, when wounds will be covered with epidermis, neutrophilic infiltration should be essentially nonexistent. Thus, when the mechanical silicone protective covering becomes weakened and lost, there is a local risk of acute inflammation. Otherwise, there should be no acute inflammatory response to the collagen–GAG matrix material alone.

The presence of lymphocytes in the collagen–GAG matrix should be consistent over the healing period. The infiltrations range from trace to mild and are always diffuse; they are not in aggregates and never in lymphoid follicles. Plasma cells are not seen. Eosinophils are usually not seen in the collagen–GAG matrix before day 20. After 20 days, eosinophils may be present in trace to moderate numbers in the matrix as well as in normal skin adjacent to the collagen–GAG graft, in the normal skin of control animals, and in the healed open wound scars. The presence of eosinophils thus does not appear to indicate an allergic response.

Comparison of Artificial Skin Graft and Open Wound Healing in Guinea Pigs

Control, ungrafted wounds will typically show prominent granulation tissue and marked to severe inflammation, especially at the open wound surfaces; severe inflammation is characteristic of healing of open wounds. Under this protocol (which does not include a stage 2 procedure), guinea pig wounds ultimately contract to a thin scar regardless of whether an artificial skin graft is used. The final healed wounds in the control animals and grafted animals are similar after wound contraction is complete.

REFERENCES

1. Williams, W. G., and Phillips, L. G. (1996). Pathophysiology of the burn wound. *In* "Total Burn Care" (D. N. Herndon, ed.), pp. 63–69. W. B. Saunders, Philadelphia, PA.
2. Schulz, J. T., 3rd, Tompkins, R. G., and Burke, J. F. (2000). Artificial skin. *Annu. Rev. Med.* **51**, 231–244.
3. Cahn, F. (2000). Technologies and characteristics of tissue-engineered skin substitutes. *Enbiomed.* **1**, 145–155.
4. Yannas, I. V., and Burke, J. F. (1980). Design of an artificial skin. I. Basic design principles. *J. Biomed. Mater. Res.* **14**, 65–81.
5. Yannas, I. V., Gordon, P., Burke, J. F., and Huang, C. (1977). Multilayer membrane useful synthetic skin. U.S. Pat. 4,060,081.
6. Yannas, I. V., Burke, J. F., Gordon, P. L., *et al.* (1980). Design of an artificial skin. II. Control of chemical composition. *J. Biomed. Mater. Res.* **14**, 107–131.
7. Burke, J. F., Yannas, I. V., Quinby, W. C., Jr., Bondoc, C. C., and Jung, W. K. (1981). Successful use of a physiologically acceptable artificial skin in the treatment of extensive burn injury. *Ann. Surg.* **194**, 413–428.
8. Heimbach, D., Luterman, A., Burke, J., Cram, A., Herndon, D., Hunt, J., Jordan, M., McManus, W., Solem, L., Warden, G., and Zawacki, B. (1988). Artificial dermis for major burns. A multi-center randomized clinical trial. *Ann. Surg.* **208**, 313–320.
9. King, P. (2000). Artificial skin reduces nutritional requirements in a severely burned child. *Burns* **26**, 501–503.
10. Stern, R., McPherson, M., and Longaker, M. T. (1990). Histologic study of artificial skin used in the treatment of full-thickness thermal injury. *J. Burn Care Rehabil.* **11**, 7–13.
11. Yannas, I. V., Gordon, P. L., Huang, C., Silver, F. H., and Burke, J. F. (1981). Crosslinked collagen-mucopolysaccharide composite materials. U.S. Pat. 4,280,954.
12. Chamberlain, L. J., Yannas, I. V., Hsu, H. P., Strichartz, G., and Spector, M. (1998). Collagen-GAG substrate enhances the quality of nerve regeneration through collagen tubes up to level of autograft. *Exp. Neurol.* **154**, 315–329.
13. Chamberlain, L. J., Yannas, I. V., Hsu, H. P., Strichartz, G. R., and Spector, M. (2000). Near-terminus axonal structure and function following rat sciatic nerve regeneration through a collagen-GAG matrix in a ten-millimeter gap. *J. Neurosci. Res.* **60**, 666–677.
14. Yannas, I. V. (1992). Tissue regeneration by use of collagen-glycosaminoglycan copolymers. *Clin. Mater.* **9**, 179–187.
15. Yannas, I. V., Burke, J. F., and Stasikelis, P. J. (1985). Method for preserving porosity in porous materials. U.S. Pat. 4,522,753.
16. Yannas, I. V., and Tobolsky, A. V. (1967). Crosslinking of gelatin by dehydration. *Nature (London)* **215**, 509–510.
17. Yannas, I. V. (1972). Collagen and gelatin in the solid state. *Rev. Macromol. Chem.* **C7**, 49B–104B.
18. Dagalakis, N., Flink, J., Stasikelis, P., Burke, J. F., and Yannas, I. V. (1980). Design of an artificial skin. Part III. Control of pore structure. *J. Biomed. Res.* **14**, 511–528.
19. Treloar, L. R. G. (1958). "The Physics of Rubber Elasticity," 2nd ed. Clarendon Press, London.
20. Yannas, I. V., Burke, J. F., Huang, C., and Gordon, P. L. (1975). Correlation of in vivo collagen degradation rate with in vitro measurements. *J. Biomed. Mater. Res.* **9**, 623–628.
21. Yannas, I. V., Burke, J. F., Orgill, D. P., and Skrabut, E. M. (1982). Wound tissue can utilize a polymeric template to synthesize a functional extension of skin. *Science* **21**, 174–176.

CHAPTER 44

MODIFICATION OF NATURAL POLYMERS: ALBUMIN

Jin Ho Lee and Kinam Park

INTRODUCTION

Albumin, the most abundant protein constituent in blood, is synthesized in the liver to serve as a source of amino acids in cases of malnutrition. The general structure and binding properties of albumin are well characterized [1–5]. Human serum albumin (HSA) constitutes 60% of the total protein mass in plasma: albumin concentration in plasma is about 40 mg/ml, while the total protein concentration in plasma is about 70 mg/ml. It is a carbohydrate-free, single-chain protein. The fatty-acid-free human albumin has a molecular weight of 67,000 Da (585 amino acid residues including one tryptophan). The shape of this protein is a prolate ellipsoid with a size of about $150 \times 40 \times 40$ $Å^3$, within which 17 disulfide bridges act to stabilize the conformation. It contains a comparatively large number of polar and charged residues (about 200 residues), and thus it is highly soluble in water and negatively charged at pH 7.4 (the isoelectric point is 4.7–5.5, depending on the amount of bound fatty acid). The nitrogen content in the fatty-acid-free HSA is 15.7–15.8 % w/w. HSA does not form dimers or higher order oligomers in its native state but may do so during isolation, concentration, and storage.

Albumin has many important functions in the body. It is a transport protein; that is, it binds and transports low molecular weight hydrophobic compounds in the blood stream. Among the substances bound by albumin are bilirubin, urobilin, fatty acids, bile acid salts, hormones, and many extraneous substances, such as penicillin, sulfonamides, and mercury. A single albumin molecule, for example, can bind 25–50 molecules of bilirubin at the same time (association constant of $10^{7.7}$) [4,6,7]. Because of its relatively high concentration and relatively small size, albumin is responsible for almost 80% of the colloid osmotic pressure of the plasma.

ALBUMIN IMMOBILIZATION

Adsorption of albumin to biomaterial surfaces is known to significantly reduce surface-induced platelet adhesion and activation [8–11]. Both the total number of adherent platelets and the extent of platelet activation are reduced on the albumin-adsorbed surfaces. Platelet membrane does not have receptors for albumin molecules, and this may be one of the reasons for poor adhesion of platelets to the albumin-adsorbed surfaces [12]. Coating of biomaterial surfaces with albumin for improved blood compatibility was first described in the late 1960s [13]. At the time, albumin coating was considered to be an approach for preparing thromboresistant surfaces. Human and bovine serum albumin have been most frequently used in surface modification of biomaterials. Albumin immobilization on biomaterial surfaces can be classified into physical adsorption and covalent grafting categories.

Physical Adsorption

The most common and simple approach for albumin immobilization is physical adsorption onto biomaterials [14]. Because albumin is adsorbed tightly onto hydrophobic surfaces but it has poor affinity for hydrophilic surfaces [15–17], it is easy modify hydrophobic surfaces with albumin (by simple exposure to the albumin solution). The thermodynamics of albumin adsorption has been treated theoretically and experimentally by various investigators aiming to understand the mechanism of protein adsorption on different surfaces [18]. Various factors, such as surface charge [19,20], surface free energy and hydrophilicity [15–17,21–23], chemical structure [24], and surface morphology [25], are known to affect the adsorption behavior of albumin. The exact contribution of each factor and the interplay among the factors, however, are only poorly understood. It is noted that albumin that has been physically adsorbed can be exchanged by other blood proteins that have higher binding affinity to the surface.

Several groups have attempted to engineer surfaces with enhanced albumin affinity. One such attempt was based on the ability of albumin to bind free fatty acids with high affinity. The albumin affinity to the surface increased by grafting of C_{16} or C_{18} hydrocarbon chains to the biomaterial surfaces [26–30]. Those alkyl-chain-modified surfaces were reported to have improved thromboresistance in an acute canine *ex vivo* experiment [31]. Another approach was based on the observation that albumin has very strong binding sites for bilirubin and a number of triazine dyes that have structural similarities to bilirubin [32]. When poly(ether urethane) was incorporated into high molecular weight dextran modified by cibacron blue, the derivatized poly(ether urethane) surface showed increased albumin affinity [33]. Although these approaches appear to be feasible and useful, the adsorbed albumin is still not covalently grafted on the surfaces, and thus the approaches are effective only for short periods of blood contact.

Covalent Grafting

For long-term beneficial effect of the immobilized albumin, it is desirable to graft to biomaterial surfaces through covalent bonding. Covalent grafting of albumin usually requires chemically active groups, such as amine, hydroxyl, or carboxyl groups, on the biomaterial surface [14]. These groups can be introduced to the surface by plasma surface modification [34–36] or by chemical modifications of the polymer surface [37,38]. For example, albumin was linked to the amino groups on silicone surfaces previously activated with amino silanes through glutaraldehyde cross-linking [37]. Albumin was also attached to BrCN-activated hydrogels, which were grafted onto the silicone surface [38].

For covalent grafting of albumin to surfaces without chemically active functional groups, albumin can be modified with functional groups that react with chemically inert surfaces [14]. Albumin, after modification with azide groups, can be grafted covalently to various surfaces by ultraviolet (UV) irradiation [39,40]. The photografting of albumin to chemically inert polymer surfaces exploits the generation of nitrene from azide by UV irradiation. In one study, 4-azido-2-nitrophenyl albumin (ANP-albumin) was prepared by reacting albumin with 4-fluoro-3-nitrophenyl azide. Photolysis of the phenyl azide group of ANP-albumin by UV light yielded highly reactive triplet nitrene, which extracts hydrogen atoms and bonds covalently with the surface. It was observed that the concentration of ANP-albumin at 0.05 mg/ml in the adsorption solution was enough to prevent platelet adhesion completely on glass treated with dimethyldichlorosilane (DDS) [40].

The azide group can also be activated by temperature increase. ANP-albumin was grafted onto polypropylene by thermal activation of the azide group to overcome a limitation of the UV irradiation method [41]. Since UV light does not penetrate most materials used in the fabrication of biomaterials, albumin grafting onto the surface of biomaterials inside an assembled device is not possible with UV activation. Since the azide groups are activated to yield nitrene at 100°C, biomaterial devices that are stable at that temperature can be grafted with albumin by exposing them to 100°C. The minimum concentration of ANP-albumin in the adsorption solution required for the surface passivation was 5 mg/ml, which was orders of magnitude higher than that required by the photoactivation approach. Even though the conformation of albumin is expected to be altered by introduction of phenyl azides and heat treatment, it was observed that the grafted albumin was effective in the prevention of platelet adhesion. This is understandable, since prevention of platelet adhesion was based on blocking of the surface rather than by presenting certain structures to the platelets arriving at the surface.

Gamma irradiation was also used for covalent grafting of albumin to the biomaterial surfaces. Albumin was first functionalized by reacting with glycidyl acrylate to introduce double bonds [42]. The functionalized albumin was adsorbed to the DDS–glass surface and then exposed to gamma irradiation in aqueous solution. In grafting of the functionalized albumin onto DDS–glass, two important variables determining the grafting efficiency were the concentration of functionalized albumin in adsorption solution and the gamma irradiation time. As the concentration of functionalized albumin increased from 1 mg/ml to 30 mg/ml, platelet activation decreased progressively. Similar results were obtained when the functionalized albumin was grafted by gamma irradiation on polymers such as polypropylene, polycarbonate, or poly(vinyl chloride) [43]. Albumin grafting on the fully assembled blood oxygenators by using this approach brought about 70% reduction in platelet adhesion and thrombus formation compared with the control oxygenators.

CROSS-LINKING

Albumin can be readily cross-linked by using water-soluble cross-linking agents such as glutaraldehyde. Arterial grafts coated or impregnated with cross-linked albumin were shown to prevent thrombus formation, to reduce potential embolization, and to favor the endothelialization process. Many investigators have prepared impervious vascular grafts by treating porous grafts (mainly knitted Dacron arterial grafts) with cross-linked albumin [44]. The grafts coated or impregnated with cross-linked albumin are blood-tight during implantation. This eliminates the need for preclotting of the grafts. The knitted Dacron vascular grafts, which were made impervious by coating with glutaraldehyde-cross-linked albumin, were implanted in the canine thoracic aorta for periods ranging from 4 h to 6 months [45,46]. The luminal surface of the albumin-coated grafts did not show major early thrombotic deposits, while the control preclotted graft developed a continuous, thick thrombotic matrix during early implantation times. It was also reported that albumin coating resulted in a faster healing. It was suggested that the enhanced healing capacity was due to the coating's improved adhesion to the Dacron substrate after drying, blood compatibility, and cytocompatibility. It is also likely that gradual degradation and dissolution of albumin creates increasingly large pores in the initially impervious grafts, allowing tissue ingrowth.

One of the disadvantages of using cross-linked, albumin-coated prostheses was that the albumin layer on the surface became fragile. Thus, albuminated grafts were stored in ethanol solution [47]. Ethanol, however, is cytotoxic and extensive rinsing is required before implantation. The cross-linked albumin coating and ethanol preservation also resulted in a somewhat slower rate of healing. To avoid using ethanol, Dacron grafts were coated with cross-linked albumin and then dried by critical point drying [48]. The grafts were rehydrated at the time of implantation by soaking in heparinized saline. It was reported that the healing rate was faster with a critical point dried prosthesis than with one stored in ethanol, but slightly slower than that with the preclotted control. Instead of critical point drying, a freeze drying method can be used for the albumin-coated grafts [49]. The albuminated grafts were marginally stiffer than the preclotted control. Comparison of the pathology of the albuminated and preclotted expanded grafts after implantation in the thoracic aorta of dogs suggested significant differences in short-term healing sequence due to the presence of albumin. The healing characteristics and surface thrombogenicity of the two systems after one month, however, were similar. They also found that the cross-linked albumin coating provides some reinforcement of the fabric and improves the dimensional stability of the grafts.

The healing and stability of albuminated grafts in dogs after sterilization by either ethylene oxide (EtO) or gamma radiation was evaluated [50–52]. Pathological analysis of the explanted grafts after implantation in the thoracic aorta of dogs indicated that gamma radiation was preferable to over EtO because the former resulted in faster rates of healing.

The albumin coating reportedly delayed the thrombotic response and fibrinolytic activity, and the extent of healing of the radiation-sterilized graft was equivalent to that achieved by preclotted prostheses in the medium and long terms.

The rate of *in vivo* degradation of cross-linked albumin coated on knitted Dacron vascular prostheses was investigated by using radiolabeled protein [53,54]. Albumin was cross-linked with either glutaraldehyde or carbodiimide. It was found that the rate of biodegradation was dependent on the site of implantation. Degradation of the cross-linked albumin occurred more rapidly in the peritoneal cavity, where less than 20% of the albumin remained after 4 weeks, whereas approximately 30% of the albumin remained in the thoracic aorta. They also found that glutaraldehyde-cross-linked albumin induced an inflammatory response and a delay in healing, while carbodiimide-cross-linked albumin caused only a mild tissue reaction. *In vivo* animal studies indicated that the albumin-coated graft showed excellent biocompatibility in terms of blood T-cell behavior and acid phosphatase activity at the implant site [55]. The resorption rate of cross-linked albumin was such that only traces were still present after 1 month of implantation, and these were no longer observable after 2 months of implantation.

ALBUMIN MODIFICATION WITH HEPARIN

Heparin is a mucopolysaccharide with highly negative charge due to the presence of sulfate groups. It is mainly used as an anticoagulant. Heparinized polymeric materials, first described by Gott *et al.* [56], are generally accepted as relatively thromboresistant. Excellent reviews are available on the preparation and properties of heparinized materials [57–60]. Albumin has been modified with heparin to improve blood–tissue compatibility and/or to enhance the hydrophilicity of albumin.

ALBUMIN–HEPARIN CONJUGATES

Albumin–heparin (Alb-Hep) conjugate was prepared by using the coupling agent 1-ethyl-3-(3-dimethylaminopropyl)carbodiimide (EDC) to couple albumin and heparin at pH 5.1–5.2 [61]. EDC was used instead of glutaraldehyde because carbodiimide was shown to be nontoxic [54,62–64]. Alb-Hep conjugate was then separated from unreacted albumin and heparin by using DEAE–cellulose and cibacron blue Sepharose chromatography, respectively. Alb-Hep conjugates have been frequently used as surface coating material for improved blood–tissue compatibility. Like other proteins, Alb-Hep conjugates can be immobilized to the surface by physical adsorption or covalent grafting [61]. Biomaterials preadsorbed with Alb-Hep conjugates showed 70% reduction in platelet adhesion, significant prolongation of blood clotting time, and prolonged recalcification time when compared with the control surfaces [63,64].

Alb-Hep conjugates are thought to combine the individual beneficial effects of heparin and albumin. The ability of heparin of Alb-Hep conjugates to bind antithrombin III (AT III) was maintained even after the Alb-Hep conjugates were immobilized onto polystyrene surfaces by physical adsorption [65] or covalently coupling [66]. A large amount of AT III was bound to the polystyrene surfaces preadsorbed with Alb-Hep conjugates via the heparin moiety of the conjugates [65]. It is, of course, possible that AT III adsorbs to certain portions of the surface that were not covered with Alb-Hep conjugates [66].

Surfaces coated with Alb-Hep conjugates were also used as substrates for seeded endothelial cells (ECs) [67–69]. Human umbilical vein ECs were grown to confluency on Alb-Hep conjugate that was covalently immobilized on plasma-treated polystyrene [67]. The ECs cultured on this surface released von Willebrand factor and prostacyclin in the amounts comparable to those by ECs grown on fibronectin-coated tissue culture polystyrene surface. When basic fibroblast growth factor (bFGF) was bound to the surface-immobilized Alb-Hep conjugates, as shown in Fig. 44.1, a confluent monolayer of ECs was formed more rapidly upon seeding of the cells [69].

CROSS-LINKED ALBUMIN–HEPARIN GELS

Heparinized albumin gels were prepared as potential sealants of prosthetic vascular grafts [70]. Albumin was first cross-linked by using EDC and *N*-hydroxysuccinimide (NHS).

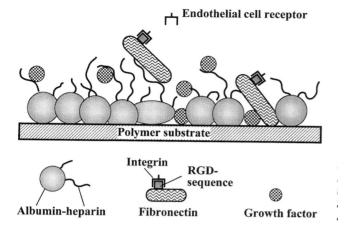

Endothelial cell receptor

Integrin   RGD-sequence

Albumin-heparin   Fibronectin   Growth factor

Fig. 44.1. Schematic representation of adhesion of endothelial cells to the polymer substrate coated with albumin–heparin conjugates. Integrin on endothelial cell membranes binds to cell adhesive proteins, such as fibronectin and growth factor, which are known to interact with heparin (RED, arginine-glycinet, aspartic acid). From Bos et al. [69].

Then, heparin solution, including EDC and NHS, was pipetted onto an air-dried, cross-linked albumin gel to prepare heparin-immobilized albumin gel. Cross-linked albumin gels and heparinized albumin gels were studied with respect to their *in vitro* stability, binding of bFGF, and cellular interactions. Gelation of albumin occurred within 5 min after addition of EDC and NHS, but only part of the added heparin was immobilized in the cross-linked albumin gel. Human umbilical vein ECs adhered and subsequently spread on albumin gels as well as heparinized albumin gels, but proliferation was observed only in the heparinized albumin gel. This result may be related to the observation that binding of bFGF was higher to heparinized albumin gel than to nonheparinized albumin gel. Growth of the ECs occurred only on heparinized albumin gel loaded with bFGF, not on bFGF-loaded albumin gel. Seeding of ECs on heparinized albumin gel significantly reduced the number of platelets adhering to substrates and, as expected, no spreading of platelets was observed on substrates seeded with ECs.

ALBUMIN–HEPARIN MICROSPHERES

Albumin has been widely used in preparation of microspheres for drug delivery and other biomedical applications [71,72]. Alb-Hep conjugates were also used in making microspheres for drug delivery applications [73,74]. One of the rationales in using Alb-Hep conjugates was that the heparin immobilized in the microspheres may increase hydrophilicity of the microspheres and enhance blood and tissue compatibility. Alb-Hep conjugates, prepared by using EDC [61] as described earlier; were made into microspheres by emulsifying the aqueous solution of the conjugates followed by stabilization (i.e., cross-linking) with glutaraldehyde [75]. The sizes of the prepared microspheres were in the range of 5–35 μm, depending on the emulsification conditions.

Alb-Hep conjugate microspheres were used as carriers for adriamycin (ADR), a cytostatic agent [76,77]. Heparin and ADR are known to form ionic complexes [78], and this property allowed loading of ADR up to 33% w/w, depending on the heparin content of the conjugate. ADR release from the microspheres was based on ion exchange, and thus an *in vitro* ADR release study showed dependency on the ionic strength of the release medium: 90% of the drug was released within 45 min in saline solution, whereas only 30% was released in nonionic medium (distilled water) [77].

ALBUMIN MODIFICATION WITH PEG
ALBUMIN-CROSS-LINKED PEG HYDROGELS

Poly(ethylene glycol) (PEG), which is also called poly(ethylene oxide) (PEO) when the molecular weight is higher than about 10,000 Da, is neither ionized nor very polar, but its water solubility is extremely high, at least up to temperatures slightly below 100°C [79]. The hydrophilicity of PEG seems to be related to hydration of the ether oxygens by a good structural fit between water molecules and the polymer that is apparently unique to the PEG structure [80–82]. PEG has been one of the most widely used polymers in the surface modification of biomaterials [83–86].

Recently, development of hybrid biomaterials based on synthetic and biological polymers has attracted a significant attention. These hybrid biomaterials, referred to as "bioartificial polymeric materials," have been synthesized in an effort to merge the good mechanical properties of the synthetic polymers with the good biocompatibility of biological polymers [87]. Albumin-cross-linked PEG hydrogels are classified as one of bioartificial polymeric materials. The use of PEG in the synthesis of this family of hydrogels was expected to lower the immunogenicity of this material [88] and reduce the protein adsorption or cell adhesion [89]. Albumin-cross-linked PEG hydrogels were prepared by polymerization of bifunctionalized PEG with bovine serum albumin (BSA) [87,90,91]. The schematic procedure of the preparation of PEG-BSA hydrogel is shown in Fig. 44.2 [90]. PEG was activated by *p*-nitrophenyl chloroformate in acetonitrile at 60°C to obtain PEG dinitrophenylcarbonate. The polymerization between the activated PEG (molecular weights of 2000–35,000 Da) and BSA was carried out at ambient temperature in an aqueous buffered solution at pH 8.5–9.4. Urethane bonds were formed between the activated OH groups at the ends of the PEG and the NH$_2$ groups of the BSA. A three-dimensional network was then obtained, with pore size and reticulation controlled by the length of the PEG chain. This hydrogel had good stability in various conditions of pH, temperature, and ionic strength, and excellent equilibrium water content (>95%) in a physiological solution. As the molecular weight of PEG increased, the cross-link density decreased but the swelling rate increased [90]. The hydrogels were highly deformable, with good elastic behavior [91]. This hydrogel also showed

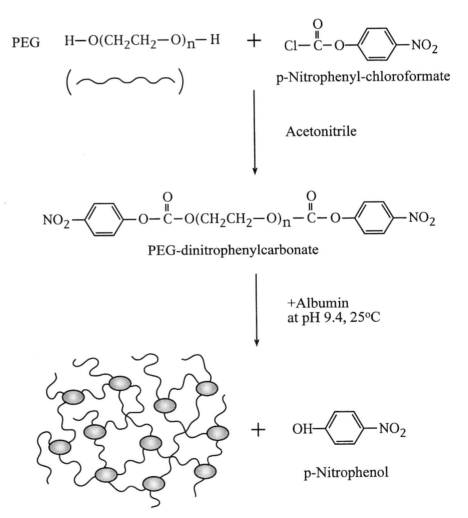

Fig. 44.2. Synthesis of a hydrogel based on PEG and albumin cross-linker. From D'Urso and Fortier [90].

Three-dimensional network of PEG & albumin

good biocompatibility when it was implanted subcutaneously in mice: after 1 month of implantation, the fibrous capsule formed around the implant was thin and the inflammatory tissue was limited [92].

PEG-BSA hydrogels were used to immobilize various enzymes, such as acid phosphatase, L-asparaginase, alginase, apyrase, and glutaminase [92–96]. Enzymes were mixed in the monomer solution to permit their immobilization during synthesis of the hydrogel in a one-step procedure [92]. Both stability and activity of the immobilized enzymes were significantly improved. Acid phosphatase immobilized into PEG-BSA hydrogel had a half-life greater than 200 h under operational conditions at 37°C compared with 72 min observed for the native enzyme under the same conditions [93]. L-Asparaginase immobilized into the hydrogel retained more than 90% of its initial activity after 50 days of incubation at 37°C, compared with a half-life of 2 days for the native enzyme [94]. The L-asparaginase-immobilized hydrogels implanted into the peritoneal cavity of rats were able to deplete the plasma level of asparagines for 7 days [95]. The *in vivo* evaluations showed that a large amount (>75%) of the initial enzyme activity was still available 1 month after implantation, but a fibrous capsule had formed around the implant. Histological analysis showed a good biocompatibility of the hydrogel even 6 months after implantation.

PEG-BSA hydrogels were also used for controlled delivery of drugs that are used in wound dressing [97,98]. The hydrogels allowed release of hydrophilic as well as hydrophobic substances, and even of small protein molecules. Studies with seven different drugs (acetaminophen, cortisone, hydrocortisone, methylene blue, prednisone, tetracycline, and theophylline) and one protein (lysozyme) showed that the drug release from the hydrogel matrix was a Fickian diffusion-controlled process. The diffusion was faster if the hydrogels were made with higher molecular weight PEG. This is because high molecular weight PEG resulted in more porous hydrogels.

ALBUMIN NANOSPHERES SURFACE-MODIFIED WITH PEG DERIVATIVES

HSA nanospheres with a size of around 100 nm were prepared using a pH-coacervation method [99]. Particles of this size, if they have not been sequestered previously by the reticuloendothelial system (RES), are known to have a good chance to escaping from the vascular system to target sites outside the circulation. For delaying the uptake by the RES of nanospheres, as well as microspheres, the spheres are usually covered with PEG. The surface-grafted PEG molecules exert steric repulsion against other objects that may come in contact with the spheres. This steric repulsion results in longer circulation in the blood than the control spheres without the grafted PEG molecules.

HSA nanospheres (10 nm in diameter) were also modified with PEG-containing copolymers, such as those of poly(thioetheramido acid) (PTAA) and PEG (Fig. 44.3A) and poly(amidoamine) (PAA) and PEG (Fig. 44.3B), which provided an effective steric barrier on the nanospheres [100]. The PTAA and PAA provide an easy route for attachment of PEG chains and carry acidic and basic groups, respectively. The nanospheres were produced by using a pH–coacervation method and cross-linked with glutaraldehyde. The surface-modified nanospheres showed reduced plasma protein adsorption on the particle surface compared with the control particles. It was suggested that the surface-modified albumin nanospheres could avoid uptake by the RES and could be used as drug carriers targeting areas outside the vascular circulation.

ALBUMIN MODIFICATION WITH VINYL POLYMERS

HSA molecules modified with vinyl groups were used as a cross-linking agent in vinyl polymerization. Since the molecules modified by vinyl group were still able to be degraded by enzymes, hydrogels made of the albumin cross-linker were also degraded by enzymes. To make the (vinyl–albumin) (i.e., the albumin modified by vinyl groups), albumin was first alkylated with glycidyl acrylate (Fig. 44.4, top). The water-soluble vinyl monomers, such as acrylic acid, acrylamide, or 1-vinyl-2-pyrrolidinone, were then polymerized in the presence of the alkylated albumin as a cross-linking agent (Fig. 44.4, bottom) [101–107]. The degree of chemical cross-linking and the kinetics of the hydrogel degradation were controlled by

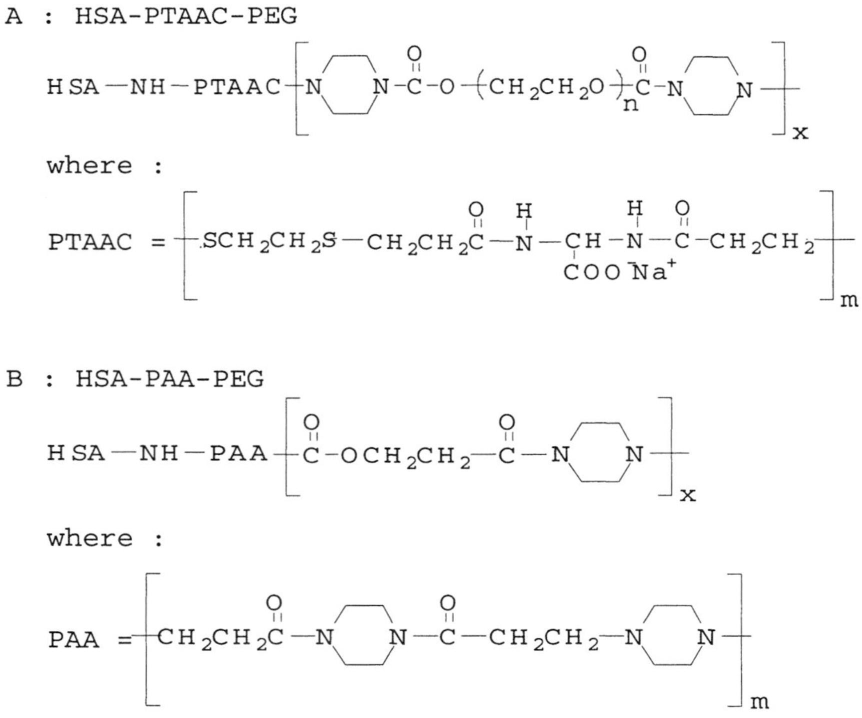

A : HSA-PTAAC-PEG

$$HSA-NH-PTAAC \left[N \underset{}{\bigcirc} N-\overset{O}{\overset{\|}{C}}-O-(CH_2CH_2O)_n \overset{O}{\overset{\|}{C}}-N \underset{}{\bigcirc} N \right]_x$$

where :

$$PTAAC = \left[SCH_2CH_2S-CH_2CH_2\overset{O}{\overset{\|}{C}}-\overset{H}{\overset{|}{N}}-\overset{|}{\underset{COO^-Na^+}{CH}}-\overset{H}{\overset{|}{N}}-\overset{O}{\overset{\|}{C}}-CH_2CH_2 \right]_m$$

B : HSA-PAA-PEG

$$HSA-NH-PAA \left[\overset{O}{\overset{\|}{C}}-OCH_2CH_2-\overset{O}{\overset{\|}{C}}-N \underset{}{\bigcirc} N \right]_x$$

where :

$$PAA = \left[CH_2CH_2\overset{O}{\overset{\|}{C}}-N \underset{}{\bigcirc} N-\overset{O}{\overset{\|}{C}}-CH_2CH_2-N \underset{}{\bigcirc} N \right]_m$$

Fig. 44.3. Structures of human serum albumin–PEG (HSA-PEG) conjugates. (A) HSA–poly(thioetheramido acid)–PEG (HSA-PTAAC-PEG). (B) HSA–poly(amido-amine)–PEG (HSA-PAA-PEG). From Lin et al. [100].

varying the degree of albumin functionality and the concentration of albumin [101,102]. Albumin-cross-linked networks were shown to be susceptible to proteolytic degradation by various enzymes, such as pepsin, trypsin, and chymotrypsin [101,102]. The mode of degradation (i.e., surface degradation or bulk degradation) was dependent on the degree of albumin modification. When the degree of albumin modification was low (e.g., <30% of the available amine groups for modification), surface degradation was dominant [106]. These hydrogels were shown to remain in the stomach of dogs for up to 60 h [105]. Such a long gastric retention, natually, led to much improved bioavailability of a loaded drug, flavin mononucleotide [108]. The absorption of flavin mononucleotide from the hydrogel dosage form was extended more than 50 h, whereas the absorption from control dosage form was limited to several hours owing to restricted absorption from the upper small intestine.

Albumin-cross-linked hydrogels were also made into semi-interpenetrating networks by using another polymer [109]. An albumin-cross-linked poly(vinylpyrrolidone) (PVP) hydrogel was exposed to the second monomer solution of acryloxyethytrimethylammonium chloride (AETAC). AETAC molecules diffused into the PVP gel, and a semi-interpenetrating network was formed upon polymerization by gamma irradiation. This type of approach is

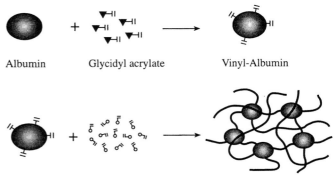

Albumin   Glycidyl acrylate   Vinyl-Albumin

Vinyl-Albumin   Vinyl monomer   Albumin-crosslinked hydrogel

Fig. 44.4. Introduction of vinyl groups to albumin by reaction with glycidyl acrylate (top), and formation of albumin-cross-linked hydrogels by using vinyl–albumin and vinyl monomers (bottom).

useful for combining two different monomers with different properties into one hydrogel system.

OTHER ALBUMIN MODIFICATIONS

Albumin has been chemically modified to improve transendothelial transport across endothelia of various types, including those of the blood–brain barrier. Such modifications include glycosylation [110,111], cationization [112–114], complexation with gold [115, 116], and insulin [117]. Albumin can easily adsorb to colloidal gold particles to stabilize them. The albumin-adsorbed colloidal gold particles (albumin–gold conjugates) have been used as a macromolecular tracer in ultrastructural studies of vascular permeability [115,116]. Albumin–gold conjugates are also used as a control marker in examination of specific interaction between a certain protein and its receptors on cell membranes [118–120]. It was assumed that complexing albumin molecules with insulin would enhance albumin adsorption to brain microvessels and thus transport across them [117]. Such modified albumin molecules can be used as carriers for the delivery of pharmacologically active substances into the brain parenchyma for enzyme replacement therapy and to treat some tumors and infections of the central nervous system [113,121].

SUMMARY

Albumin has been modified with a number of agents for various applications. Albumin can be cross-linked to form a gel or solid materials, or it can be used as a component of more complex materials. Albumin is one of the most biocompatible molecules, and thus materials made of albumin are thought to be highly biocompatible. The biodegradable nature of albumin without producing any undesirable molecules makes it even more attractive in applications of controlled drug delivery and tissue engineering. For tissue engineering applications, albumin can be used in conjunction with other molecules having supplementary properties. Albumin can be mixed with other biodegradable synthetic polymers or with naturally occurring extracellular matrices to make scaffolds. The versatility of albumin that enables it to be formulated into various physical forms and sizes makes it a highly useful molecule for various biomedical applications.

REFERENCES

1. Rosener, V. M., Oratz, M., and Rothschild, M. A. (1977). "Albumin Structure, Function, and Use." Pergamon, New York.
2. Peters, T. (1970). Serum albumin. *Adv. Clin. Chem.* **13**, 37–111.
3. Peters, T. (1975). Serum albumin. *In* "The Plasma Proteins: Structure, Function, and Genetic Control" (F. W. Putnam, ed.), pp. 133–181. Academic Press, New York.
4. Peters, T. (1985). Serum albumin. *Adv. Protein Chem.* **37**, 161–245.
5. Andrade, J. D., and Hlady, V. (1987). Plasma protein adsorption: The big twelve. *Ann. N.Y. Acad. Sci.* **516**, 158–172.
6. Kragh-Hansen, U. (1981). Molecular aspects of ligand binding to serum albumin. *Pharmacol. Rev.* **33**, 17–53.
7. Weiss, C. (1983). Functions of blood. *In* "Human Physiology" (R. F. Schmidt and G. Thews, eds.), pp. 331–357. Springer-Verlag, Berlin.
8. Kim, S. W., and Feijen, J. (1985). Surface modification of polymers for improved blood compatibility. *CRC Crit. Rev. Biocompat.* **1**, 229–260.
9. Park, K., Mosher, D. F., and Cooper, S. L. (1986). Acute surface-induced thrombosis in the canine ex vivo model: Importance of protein composition of the initial monolayer and platelet activation. *J. Biomed. Mater. Res.* **20**, 589–612.
10. Absolom, D. R., Zingg, W., and Neumann, A. W. (1987). Protein adsorption to polymer particles: Role of surface properties. *J. Biomed. Mater. Res.* **21**, 161–171.
11. Amiji, M., Park, H., and Park, K. (1992). Study on the prevention of surface-induced platelet activation by albumin coating. *J. Biomater. Sci., Polym. Ed.* **3**, 375–388.
12. Brynda, E., Houska, M., Pokorna, Z., Cepalova, N. A., Moiseev, Y. V., and Kalal, J. (1978). Irreversible adsorption of human serum albumin onto polyethylene film. *J. Bioeng.* **2**, 411–418.
13. Chang, T. M. S. (1969). Removal of endogeneous and xenogeneous toxins by a microencapsulated absorbent. *Can. J. Physiol. Pharmacol.* **47**, 1042–1045.
14. Amiji, M. M., Kamath, K. R., and Park, K. (1995). Albumin-modified biomaterial surfaces for reduced thrombogenicity. *In* "Encyclopedic Handbook of Biomaterials and Bioengineering. Part B. Applications" (D. L. Wise, D. J. Trantolo, D. E. Altobelli, M. J. Yaszemski, J. D. Gresser, and E. R. Schwartz, eds.), pp. 1057–1070. Dekker, New York.
15. Brynda, E., Cepalova, M. A., and Stol, M. (1984). Equilibrium adsorption of human serum albumin and human fibrinogen on hydrophobic and hydrophilic surfaces. *J. Biomed. Mater. Res.* **18**, 685–693.

16. Lee, S. H., and Ruckenstein, E. (1988). Adsorption of proteins onto polymeric surfaces with different hydrophilicities: A case study with bovine serum albumin. *J. Colloid Interface Sci.* **125**, 365–379.

17. Uyen, H. M. W., Schakenraad, J. M., Sjollema, J., Noordmans, J., Jongebloed, W. L., Stokroos, I., and Busscher, H. J. (1990). Amount and surface structure of albumin adsorbed to solid substrata with different wettabilities in a parallel plate flow cell. *J. Biomed. Mater. Res.* **24**, 1599–1614.

18. Norde, W. (1992). Energy and entropy of protein adsorption. *J. Dispersion Sci. Technol.* **13**, 363–377.

19. Nossell, H. L., Wilneer, G. D., and LeRoy, E. C. (1969). Importance of polar groups for initiating blood coagulation and aggregating platelets. *Nature (London)* **221**, 75–76.

20. Sawyer, P. N. (1984–1985). Electrode-biologic tissue interactions at interfaces: A review. *Biomater. Med. Devices Artif. Organs* **12**, 161–196.

21. Golander, C. G., Lin, Y. S., Hlady, V., and Andrade, J. D. (1990). Wetting and plasma-protein adsorption studies using surfaces with a hydrophobicity gradient. *Colloids Surf.* **49**, 289–302.

22. Walker, D. S., Garrison, M. D., and Reichert, W. M. (1993). Protein adsorption to HEMA/EMA copolymers studied by integrated optical techniques. *J. Colloid Interface Sci.* **157**, 41–49.

23. Warkentin, P., Walivaara, B., Lundstrom, I., and Tengvall, P. (1994). Differential surface binding of albumin, immunoglobulin G and fibrinogen. *Biomaterials* **15**, 786–795.

24. Lyman, D. J. (1991). Bulk and surface effects on blood compatibility. *J. Bioact. Compat. Polym.* **6**, 283–295.

25. Wilkins, E., and Radford, W. (1990). Biomaterials for implanted closed loop insulin delivery system: A review. *Biosens. Bioelectron.* **5**, 167–213.

26. Munro, M. S., Quattrone, A. J., Ellsworth, S. R., Kulkarni, P., and Eberhart, R. C. (1981). Alkyl substituted polymers with enhanced albumin affinity. *Trans. Am. Soc. Artif. Intern. Organs* **27**, 499–503.

27. Frautshi, J. R., Munro, M. S., Lloyd, D. R., and Eberhart, R. C. (1983). Alkyl derivatized cellulose acetate membranes with enhanced albumin affinity. *Trans. Am. Soc. Artif. Intern. Organs* **24**, 242–244.

28. Eberhart, R. C., Munro, M. S., Frautschi, J. R., Lubin, M., Clubb, F. J., Miller, C. W., and Sevastianov, V. I. (1987). Influence of endogenous albumin binding on blood material interactions. *Ann. N.Y. Acad. Sci.* **516**, 78–95.

29. Pitt, W. G., and Cooper, S. L. (1988). Albumin adsorption on alkyl chain derivatized polyurethanes: I. The effect of C-18 alkylation. *J. Biomed. Mater. Res.* **22**, 359–382.

30. Tsai, C. C., Huo, H. H., Kulkarni, P., and Eberhart, R. C. (1990). Biocompatible coatings with high albumin affinity. *Trans. Am. Soc. Artif. Intern. Organs* **36**, M307–M310.

31. Grasel, T. G., Pierce, J. A., and Cooper, S. L. (1987). Effects of alkyl grafting on surface properties and blood compatibility of polyurethane block copolymers. *J. Biomed. Mater. Res.* **21**, 815–842.

32. Leatherbarrow, R. J., and Dean, P. D. G. (1980). Studies on the mechanism of binding of serum albumin to immobilized Blue F3GA. *Biochem. J.* **189**, 29–34.

33. Keogh, J. R., Velander, F. F., and Eaton, J. W. (1992). Albumin-binding surfaces for implantable devices. *J. Biomed. Mater. Res.* **26**, 441–456.

34. Sipehia, R., and Chawla, A. S. (1982). Albuminated polymer surfaces for biomedical application. *Biomater. Med. Devices Artif. Organs* **10**, 229–246.

35. Sipehia, R., Chawla, A. S., and Chang, T. M. S. (1986). Enhanced albumin binding to polypropylene beads via anhydrous ammonia gaseous plasma. *Biomaterials* **7**, 471–473.

36. Kiaei, D., Horbett, T. A., and Hoffman, A. S. (1991). Albumin retention by glow discharge deposited polymers. *Artif. Organs* **15**, 302.

37. Guidoin, R. G., Awad, J., Brassard, A., Domurado, D., Lawny, F., Wetzer, J., Barbotin, J. N., Calvot, C., and Broun, G. (1976). Blood compatibility of silicone rubber chemistry coated with cross-linked albumin. *Biomater. Med. Devices Artif. Organs* **4**, 205–224.

38. Hoffman, A. S., Schmer, G., Harris, C., and Kraft, W. G. (1972). Covalent bonding of biomolecules to radiation grafted hydrogels on inert polymer surfaces. *Trans. Am. Soc. Artif. Intern. Organs* **18**, 10–17.

39. Matsuda, T., and Inoue, K. (1990). Novel photoreactive surfacemodification technology for fabricated devices. *Trans. Am. Soc. Artif. Intern. Organs* **36**, M161–M164.

40. Tseng, Y. C., Kim, J., and Park, K. (1993). Photografting of albumin onto dimethyldichlorosilane-coated glass. *J. Biomater. Appl.* **7**, 233–249.

41. Tseng, Y. C., Mullins, W. M., and Park, K. (1993). Albumin grafting on to polypropylene by thermal activation. *Biomaterials* **14**, 392–400.

42. Kamath, K. R., Park, H., Shim, H. S., and Park, K. (1994). Albumin grafting on dimethyldichlorosilane-coated glass by gamma-irradiation, *Colloids Surf. B: Biointerfaces* **2**, 471–479.

43. Kamath, K. R., and Park, K. (1994). Surfacemodification of polymeric biomaterials by albumin grafting using gamma-irradiation, *J. Appl. Biomater.* **5**, 163–173.

44. Lee, J. H., Khang, G., and Lee, H. B. (2000). Blood-leak proof vascular grafts. *In* "Biomaterials Engineering and Devices: Human Applications" (D. L. Wise, D. J. Trantola, K. U. Lewandrowski, J. D. Gresser, M. V. Cattaneo, and M. J. Yaszemski, eds.), pp. 161–179. Humana Press, Totowa, NJ.

45. Domurado, D., Guidoin, R., Marois, M., Martin, L., Gosselin, C., and Awad, J. (1978). Albuminated Dacron prostheses as improved blood vessel substitutes. *J. Bioeng.* **2**, 79–91.

46. Guidoin, R., Synder, R., Martin, L., Botzko, K., Marois, M., Awad, J., King, M., Domurado, D., Bedros, M., and Gosselin, C. (1984). Albumin coating of a knitted polyester arterial prostheses: An alternative to preclotting. *Ann. Thorac. Surg.* **37**, 457–465.

47. Guidoin, R., Martin, L., Marois, M., Gosselin, C., King, M., Gunasekera, K., Domurado, D., Sigot-Luizard, M. F., Sigot, M., and Blais, P. (1984). Polyester prostheses as substitutes in the thoracic aorta of dogs. II. Evaluation of albuminated polyester grafts stored in ethanol. *J. Biomed. Mater. Res.* **18**, 1059–1072.

48. Guidoin, R. G., King, M. W., Awad, J., Martin, L., Domurado, D., Marois, M., Sigot-Luizard, M. F., Gosselin, C., Gunasekera, K., and Gagnon, D. (1983). Albumin-coated and critical-point dried polyester prostheses as substitutes in the thoracic aorta of dogs. *Trans. Am. Soc. Artif. Intern. Organs* **29**, 290–295.

49. Roy, J., King, M., Synder, R., Guidoin, R., Martin, L., Botzko, K., Marois, M., Awad, J., and Gosselin, C. (1985). Polyester (Dacron) arterial prostheses treated with cross-linked and freeze-dried albumin. *Am. Soc. Artif. Intern. Organs J.* **8**, 166–173.

50. Guidoin, R., Synder, R., King, M., Martin, L., Botzko, K., Awad, J., Marois, M., and Gosselin, C. (1985). A compound arterial prostheses: The importance of the sterilization procedure on the healing and stability of albuminated polyester grafts. *Biomaterials* **6**, 122–128.

51. Benslimane, S., Guidoin, G., Marceau, D., King, M., Merhi, Y., Rao, T. J., Martin, L., Lafreniere-Gagnon, D., and Gosselin, C. (1987). Albumin-coated polyester arterial prostheses: Is xenogenic albumin safe? *Biomater. Artif. Cells Artif. Organs* **15**, 453–481.

52. Baquey, C., Sigot-Luizard, M. F., Friede, J., Prud'hom, R.E., and Guidoin, R. G. (1987). Radiosterilization of albuminated polyester prostheses. *Biomaterials* **8**, 185–189.

53. Benslimane, S., Guidoin, R., Roy, P. E., Friede, J., Hebert, J., Domurado, D., and Sigit-Luizard, M. F. (1986). Degradability of cross-linked albumin as an arterial polyester prosthesis coating in in vitro and in vivo rat studies. *Biomaterials* **7**, 268–272.

54. Benslimane, S., Guidoin, R., Mourad, W., Hebert, J., King, M.W., and Sigot-Luizard, M. F. (1988). Polyester arterial grafts impregnated with cross-linked albumin: The rate of degradation of the coating in vivo. *Eur. Surg. Res.* **20**, 12–17.

55. Marois, Y., Chakfe, N., Guidoin, R., Duhame, R. C., Roy, R., Marois, M., King, M. W., and Douville, Y. (1996). An albumin-coated polyester arterial graft: In vivo assessment of biocompatibility and healing characteristics. *Biomaterials* **17**, 3–14.

56. Gott, V. L., Whiiffen, J. D., and Datton, R. C. (1963). Heparin bonding on colloidal graphite surfaces. *Science* **142**, 1297.

57. Leininger, R. I. (1975). Polymeric materials that don't clot blood. *Chem. Tech. (Leipzig)* **5**, 172–176.

58. Ehrlich, J. (1975). Long term thromboresistance of heparinized materials. *Polym. Eng. Sci.* **15**, 281–285.

59. Wilson, J. E. (1981). Heparinized polymers as thromboresistant biomaterials. *Polym.-Plast. Technol. Eng.* **16**, 119–220.

60. Bjork, I., and Lindahl, U. (1982). Mechanism of the anticoagulant action of heparin. *Mol. Cell. Biochem.* **48**, 161–182.

61. Hennink, W. E., Feijen, J., Ebert, C. D., and Kim, S. W. (1983). Covalently bound conjugates of albumin and heparin: Synthesis, fractionaltion and characterization. *Thromb. Res.* **29**, 1–13.

62. Merhi, Y., Roy, R., Guidoin, R., Hebert, J., Mourad, W., and Benslimane, S. (1989). Cellular reactions to polyester arterial prostheses impregnated with cross-linked albumin: In vivo studies in mice. *Biomaterials* **10**, 56–58.

63. Hennink, W. E., Dost, L., Feijen, J., and Kim, S. W. (1983). Interaction of albumin-heparin conjugate preadsorbed surfaces with blood. *Trans. Am. Soc. Artif. Intern. Organs* **29**, 200–205.

64. Hennink, W. E., Kim, S. W., and Feijen, J. (1984). Inhibition of surface induced coagulation by preadsorption of albumin-heparin conjugates, *J. Biomed. Mater. Res.* **18**, 911–926.

65. Hennink, W. E., Ebert, C. D., Kim, S. W., Breemhaar, W., Bantjes, A., and Feijen, J. (1984). Interaction of antithrombin III with preadsorbed albumin-heparin conjugates. *Biomaterials* **5**, 264–268.

66. van Delden, C. J., Engbers, G. H. M., and Feijen, J. (1995). Interaction of antithrombin III with surface-immobilized albumin-heparin conjugates. *J. Biomed. Mater. Res.* **29**, 1317–1329.

67. Bos, G. W., Scharenborg, N. M., Poot, A. A., Engbers, G. H. M., Terlingen, J. G. A., Beugeling, T., van Aken, W. G., and Feijen, J. (1998). Adherence and proliferation of endothelial cells on surface-immobilized albumin-heparin conjugate. *Tissue Eng.* **4**, 267–279.

68. Bos, G. W., Scharenborg, N. M., Poot, A. A., Engbers, G. H. M., Beugeling, T., van Aken, W. G., and Feijen, J. (1999). Blood compatibility of surfaces with immobilized albumin-heparin conjugate and effect of endothelial cell seeding on platelet adhesion. *J. Biomed. Mater. Res.* **47**, 279–291.

69. Bos, G. W., Scharenborg, N. M., Poot, A. A., Engbers, G. H. M., Beugeling, T., van Aken, W. G., and Feijen, J. (1999). Proliferation of endothelial cells on surface-immobilized albumin-heparin conjugate loaded with basic fibroblast growth factor. *J. Biomed. Mater. Res.* **44**, 330–340.

70. Bos, G. W., Scharenborg, N. M., Poot, A. A., Engbers, G. H. M., Beugeling, T., van Aken, W. G., and Feijen, J. (1999). Endothelialization of cross-linked albumin-heparin gels. *Thromb. Haemostasis* **82**, 1757–1763.

71. Bogdansky, S. (1990). Natural polymers as drug delivery systems. *In* "Biodegradable Polymers as Drug Delivery Systems" (M. Casin and R. Langer, eds.), pp. 240–246. Dekker, New York.

72. Arshady, R. (1990). Albumin microspheres and microcapsules: Morphology of manufacturing techniques. *J. Controlled Release* **14**, 111–131.

73. Kwon, G. S., Bae, Y. H., Kim, S. W., Cremers, H., and Feijen, J. (1991). Preparation and characterization of microspheres of albumin-heparin conjugates. *J. Colloid Interface Sci.* **143**, 501–512.

74. Cremers, H. F. M., Kwon, G., Bae, Y. H., Kim, S. W., Verrijk, R., Noteborn, H. P. J. M., and Feijen, J. (1994). Preparation and characterization of albumin-heparin microspheres. *Biomaterials* **15**, 38–48.

75. Burger, J. J., Tomlinson, E., Mulder, E. M. A., and McVie, J. G. (1985). Albumin microsphers for intra-arterial tumour targeting. I. Phamaceutical aspects. *Int. J. Pharm.* **23**, 333–344.

76. Cremers, H. F. M., Feijen, J., Kwon, G., Bae, Y. H., Kim, S. W., Noteborn, H. P. J. M., and McVie, J. G. (1990). Albumin-heparin microspheres as carriers for cytostatic agents. *J. Controlled Release* **11**, 167–179.

77. Cremers, H. F. M., Verrijk, R., Noteborn, H. P. J. M., Kwon, G., Bae, Y. H., Kim, S. W., and Feijen, J. (1994). Adriamycin loading and release characteristics of albumin-heparin conjugate microspheres. *J. Controlled Release* **29**, 143–155.

78. Menozzi, M., and Arcamone, F. (1978). Binding of adriamycin to sulphated mucopolysaccharides. *Biochem. Biophys. Res. Commun.* **80**, 313–318.

79. Bailey, F. E., and Koleske, J. Y. (1976). "Poly(Ethylene Oxide)." Academic Press, New York.

80. Bailey, F. E., and Callard, R. W. (1959). Thermodynamic parameters of poly(ethylene oxide) in aqueous solution. *J. App. Polym. Sci.* **1**, 373–374.

81. Blandamer, M. J., Fox, M. F., Powell, E., and Stafford, J. W. (1969). A viscometric study of poly(ethylene oxide) in *t*-butyl alcohol/water mixture. *Makromol. Chem.* **124**, 222–231.

82. Kjellander, R., and Florin, E. (1981). Water structure and changes in thermal stability of the system poly(ethylene oxide)-water. *J. Chem. Soc. Faraday Trans. I* **77**, 2053–2077.

83. Amiji, M., and Park, K. (1993). Surface modification of polymeric biomaterials with poly(ethylene oxide), albumin, and heparin for reduced thrombogenicity. *J. Biomater. Sci. Polym. Ed.* **4**, 217–234.

84. Lee, J. H., Lee, H. B., and Andrade, J. D. (1995). Blood compatibility of polyethylene oxide surfaces. *Prog. Polym. Sci.* **20**, 1995.

85. Harris, J. M. (1992). "Poly(Ethylene Glycol) Chemistry: Biotechnical and Biomedical Applications." Plenum, New York.

86. Harris, J. M., and Zalipsky, S. (1997). "Poly(Ethylene Glycol) Chemistry and Biological Applications." American Chemical Society, Washington, DC.

87. D'Urso, E. M., and Fortier, G. (1994). New hydrogel based on polyethylene glycol cross-linked with bovine serum albumin. *Biotechnol. Tech.* **8**, 71–76.

88. Katre, N. V. (1993). The conjugation of proteins with polyethylene glycol and other polymers. Altering properties of proteins to enhance their therapeutic potential. *Adv. Drug Delivery Rev.* **10**, 91–114.

89. Brooks, D. E., van Alstine, J. M., Sharp, K. A., and Stocks, S. J. (1992). PEG-derivatized ligands with hydrophobic and immunological specificity: Applications in cell separation. *In* "Poly(Ethylene Glycol) Chemistry: Biotechnical and Biomedical Applications," (J. M. Harris, ed.), pp. 57–71. Plenum, New York.

90. D'Urso, E. M., and Fortier, G. (1994). New bioartificial polymeric material: Poly(ethylene glycol) cross-linked with albumin. I. Synthesis and swelling properties. *J. Bioact. Compat. Polym.* **9**, 367–387.

91. Gayet, J. C., He, P., and Fortier, G. (1998). Bioartificial polymeric material: Poly(ethylene glycol) cross-linked with albumin. II. Mechanical and thermal properties. *J. Bioact. Compat. Polym.* **13**, 179–197.

92. D'Urso, E. M., Jean-Francois, J., Doillon, C. J., and Fortier, G. (1995). Poly(ethylene glycol)-serum albumin hydrogel as matrix for enzyme immobilization: Biomedical applications. *Artif. Cells, Blood Substitutes, Immobilization Biotechnol.* **23**, 587–595.

93. D'Urso, E. M., and Fortier, G. (1996). Albumin-poly(ethylene glycol) hydrogel as matrix for enzyme immobilization: Biochemical characterization of cross-linked acid phosphates. *Enzyme Microb. Technol.* **18**, 482–488.

94. Jean-Francois, J., and Fortier, G. (1996). Immobilization of L-asparaginase into a biocompatible poly(ethylene glycol)-albumin hydrogel: I: Preparation and in vitro characterization. *Biotechnol. Appl. Biochem.* **23**, 221–226.

95. Jean-Francois, J., D'Urso, E. M., and Fortier, G. (1997). Immobilization of L-asparaginase into a biocompatible poly(ethylene glycol)-albumin hydrogel: Evaluation of performance in vivo. *Biotechnol. Appl. Biochem.* **26**, 203–212.

96. Belgoudi, J., and Fortier, G. (1999). Poly(ethylene glycol)-bovine serum albumin hydrogel as a matrix for enzyme immobilization. In vitro biochemical characterization. *J. Bioact. Compat. Polym.* **14**, 31–53.

97. Gayet, J. C., and Fortier, G. (1995). Drug release from new bioartificial hydrogel. *Artif. Cells, Blood Substitutes, Immobilization Biotechnol.* **23**, 1995.

98. Gayet, J. C., and Fortier, G. (1996). High water content BSA-PEG hydrogel for controlled release device: Evaluation of the drug release properties. *J. Controlled Release* **38**, 177–184.

99. Lin, W., Coombes, A. G. A., Davies, M. C., Davis, S. S., and Illum, L. (1993). Preparation of sub 100 nm human serum albumin nanospheres using a pH-coacervation method. *J. Drug Target.* **1**, 237–243.

100. Lin, W., Garnett, M. C., Davies, M. C., Bignotti, F., Ferruti, P., Davis, S. S., and Illum, L. (1997). Preparation of surface-modified albumin nanosphere. *Biomaterials* **18**, 559–565.

101. Park, K. (1988). Enzyme-digestible swelling hydrogels as platforms for long-term oral drug delivery: Synthesis and characterization, *Biomaterials* **9**, 435–441.

102. Shalaby, W. S. W., and Park, K. (1990). Biochemical and mechanical characterization of enzyme-digestible hydrogels. *Pharm. Res.* **7**, 816–823.

103. Shalaby, W. S. W., Peck, G., and Park, K. (1991). Release of dextromethorphan hydrobromide from freeze-dried enzyme-degradable hydrogels. *J. Controlled Release* **16**, 355–364.

104. Shalaby, W. S. W., Blevins, W. E., and Park, K. (1991). Enzyme-degradable hydrogels: Properties associated with albumin-cross-linked polyvinylpyrrolidone hydrogels. *In* "Water-soluble Polymers, Synthesis, Solution Properties, and Application." (S. W. Shalaby, C. L. McCormick, and G. B. Butler, eds.), pp. 484–492. Americaln Chemical Society, Washington, DC.

105. Shalaby, W. S. W., Blevins, W. E., and Park, K. (1991). Gastric retention of enzyme-digestible hydrogels in the canine stomach under fasted and fed conditions: A preliminary analysis using new analytical techniques. *In* "Polymeric Drugs and Drug Delivery Systems." (R. Dunn and R. M. Ottenbrite, eds.), pp. 237–248. American Chemical Society, Washington, DC.

106. Shalaby, W. S. W., Chen, M., and Park, K. (1992). A mechanistic assessment of enzyme-induced degradation of albumin-cross-linked hydrogels. *J. Bioac. Compat. Polym.* **7**, 257–274.

107. Shalaby, W. S. W., Blevins, W. E., and Park, K. (1992). The use of ultrasound imaging and fluoroscopic imaging to study gastric retention of enzyme-digestible hydrogels. *Biomaterials* **13**, 289–296.

108. Shalaby, W. S. W., Blevins, W. E., and Park, K. (1992). In vitro and in vivo studies of enzyme-digestible hydrogels for oral drug delivery. *J. Controlled Release* **19**, 131–144.

109. Shalaby, W. S. W., Jackson, R., Blevins, W. E., and Park, K. (1993). Synthesis of enzyme-digestible, interpenetrating hydrogel networks by gamma-irradiation. *J. Bioact. Compat. Polym.* **8**, 3–23.

110. Williams, S. K., Devenny, J. J., and Bitensky, M. W. (1981). Micropinocytotic ingestion of nonenzymatically glucosylated proteins by capillary endothelium. *Microvasc. Res.* **28**, 311–321.

111. Predescu, D., Simionescu, M., Simionescu, N., and Palade, G. E. (1988). Binding and transcytosis of glycoalbumin by the microvascular endothelium of the murine myocardium: Evidence that glycoalbumin behaves as a bifunctional ligand. *J. Cell Biol.* **107**, 1729–1738.

112. Smith, K. R., and Borchardt, R. T. (1989). Permeability and mechanism of albumin, cationized albumin, and glycosylated albumin transcellular transport across monolayers of cultured bovine brain capillary endothelial cells. *Pharm. Res.* **6**, 466–473.

113. Pardridge, W. M., Triguero, D., Buciak, J., and Yang, J. (1990). Evaluation of cationized rat albumin as a potential blood-brain barrier drug transport vector. *J. Pharmacol. Exp. Ther.* **255**, 893–899.

114. Shimon-Hophy, M., Wadhwani, K. C., Chandrasekaran, K., Larson, D., Smith, Q. R., and Rapoport, S. I. (1991). Regional blood-brain barrier transport of cationized bovine serum albumin in awake rats. *Am. J. Physiol.* **261**, R478–R483.

115. Handley, D. A., and Chien, S. (1987). Colloidal gold labelling studies related to vascular and endothelial function, hemostasis and receptor-mediated processing of plasma macromolecules. *Eur. J. Cell Biol.* **43**, 163–174.

116. Villashi, S. (1989). Preparation and application of albumin-gold complex. *In* "Colloidal Gold." (M. A. Hayat, ed.), pp. 163–174. Academic Press, Orlando, FL.

117. Vorbrodt, A. W., Dobrogowska, D. H., and Lossinsky, A. S. (1994). Ultrastructural study on the interaction of insulin-albumin-gold complex with mouse brain microvascular endothelial cells. *J. Neurocytol.* **23**, 201–208.

118. Park, K., Albrecht, R. M., Simmons, S. R., and Cooper, S. L. (1986). A new approach to study the adsorbed protein layer on biomaterials: Immunogold staining techniques. *J. Colloid Interface Sci.* **111**, 197–212.

119. Park, K., and Park, H. (1989). Application of video-enhanced interference reflection microscopy to the study of platelet-surface interactions. *Scanning Microsc. Suppl.* **3**, 137–146.

120. Park, K. (1989). Factors affecting efficiency of colloidal gold staining: pH-dependent stability of protein-gold, conjugates. *Scanning Microsc.* **3**, 15–25.

121. Pardridge, W. M., Kumagai, A. K., and Eisenberg, J. B. (1987). Chimeric peptides as a vehicle for peptide pharmaceutical delivery through the blood–brain barrier. *Biochem. Biophys. Res. Commun.* **146**, 307–313.

CHAPTER 45

MODIFICATION OF NATURAL POLYMERS: HYALURONIC ACID

Yi Luo, Kelly R. Kirker, and Glenn D. Prestwich

Hyaluronic acid (HA) is an abundant nonsulfated glycosaminoglycan component of synovial fluid and the extracellular matrix. A combination of unique physicochemical properties and biological functions suggest HA as an attractive building block for new biocompatible and biointeractive materials with applications in drug delivery, tissue engineering, and viscosupplementation. This chapter highlights the chemical modification of HA and medical applications of these HA-based biomaterials. Important new products have already reached the marketplace, the approval and introduction of an increasing number of medical devices and new drugs using HA-derived biomaterials can be anticipated in the present decade.

INTRODUCTION

Hyaluronic acid (hyaluronan, HA) is a naturally occurring polysaccharide consisting of 200–10,000 repeating disaccharide units. It is composed of (β-1,4)-linked D-glucuronic acid and (β-1,3)-N-acetyl-D-glucosamine (Fig. 45.1) and has a molecular weight ranging from 1×10^5 to 5×10^6 Da. HA is an abundant glycosaminoglycan (GAG) found in the extracellular matrix (ECM) of all higher animals, and it is the only nonsulfated GAG. HA forms highly viscous aqueous solutions and takes on an expanded random coil structure as a result of strong hydrogen bonding. The coiled structure of HA can trap approximately 1000 times its weight in water. These characteristics give it unique physicochemical properties as well as distinctive biological functions [1]. HA has been implicated in the water homeostasis of tissues, in the regulation of permeability of other substances by steric exclusion phenomena, and in the lubrication of joints [2]. HA plays an important role in the structure and organization of the ECM, including the maintenance of extracellular space, the transport of ion solutes and nutrients, and the preservation of tissue hydration. HA concentrations increase whenever rapid tissue proliferation regeneration and repair occur.

HA also binds specifically to proteins in the ECM, on cell surface receptors, and within the cell cytosol. These protein–ligand interactions are important in stabilizing the cartilage ECM [2,3], regulating cell adhesion and motility [4,5], mediating proliferation and differentiation [6], and in the action of growth factors [7]. HA has also been implicated in morphogenesis and embryonic development [8], in cancer [9–15], in modulating inflammation [13], in stimulating angiogenesis and healing [16], and as a protective coating [17]. In most situations, HA signaling occurs via cell surface receptors, such as CD44 and RHAMM, but cytosolic and nuclear proteins are also implicated in signaling.

The actions of HA appear to be related to molecular size. For example, high molecular weight HA inhibits angiogenesis, while low molecular weight HA stimulates angiogenesis [18]. At a molecular level, HA can act as a scavenger molecule for oxygen-derived toxic free radical species [18]. This unique suite of properties makes HA an attractive building

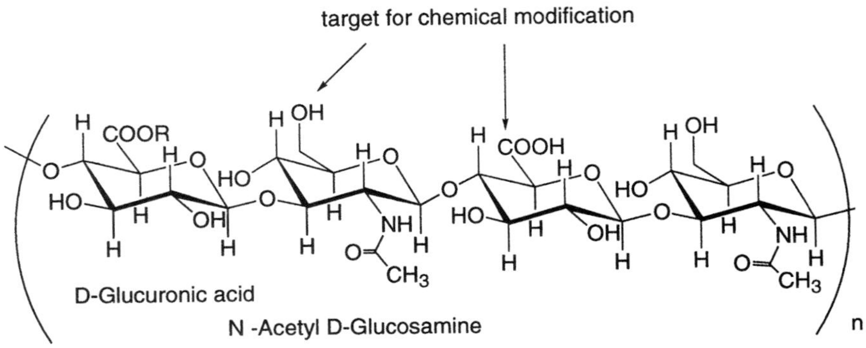

Fig. 45.1. Tetrasaccharide fragment of HA showing the disaccharide repeat units (R = H).

block for new biocompatible and biointeractive materials that have medical applications in drug delivery, tissue engineering, and viscosupplementation [19]. This chapter highlights the chemistry and medical applications of these HA-based biomaterials.

Approved uses for highly purified HA greatly increased in the 1990s in surgery, drug delivery, and cosmetics [20]. The unique rheological properties of high molecular weight HA ($>1 \times 10^6$ Da) are exploited in clinical treatment of osteoarthritis [21,22] and in ophthalmic surgery. HA has shown beneficial properties in corneal transplantation, intraocular lens implantation, and the treatment of cataracts, vitroretinal diseases, and tympanic membrane perforations. In addition, HA solutions have been investigated as space-filling substances in laryngeal augmentation, such as vocal fold injection, and are in clinical use for urinary incontinence in women and in pediatric urology for the treatment of vesicourethral reflux [23]. HA injections also facilitate nerve growth [24], and topical HA can improve wound healing. High levels of high molecular weight HA or of an HA–sucrose formulation can enhance the healing of wounds in collagen-containing tissues, including skin, bone, and mucosa, with limited scarring [16,25]. In addition, HA is used as an adjuvant for ophthalmic drug delivery [26] and can enhance the absorption of drugs and proteins via mucosal tissues [27–29]. The rapid skin permeability and epidermal retention of HA prolongs the pharmacokinetic half-life of topically delivered drugs and growth factors [30].

However, the poor biomechanical properties of this soluble, natural polymer currently preclude many direct applications in medicine. Thus, to obtain more mechanically and chemically robust materials, it has become necessary to chemically modify natural HA. The resulting HA derivatives have physicochemical properties that are significantly different from those of the native polymer but maintain the biocompatibility and biodegradability, as well as the potential pharmacological and therapeutic properties, of HA itself. This chapter explores the chemical modification of HA (several architectures are shown in Fig. 45.2) and the potential uses of these biomaterials in tissue engineering and drug delivery.

CHEMICAL MODIFICATIONS
MODIFICATION OF THE HA CARBOXYL GROUP
Modification of HA by Esterification

Fidia Advanced Biopolymers (Abano Terme, Padua, Italy) manufactures a number of HA-derivatized materials collectively called HYAFF (R = alkyl group, Fig. 45.1). The HYAFF materials [31] are prepared by alkylation of the tetrabutylammonium salt of HA with an alkyl halide in dimethyl formamide (DMF) solution. At higher percentages of esterification, solubility in water is reduced. HYAFF materials can be extruded to produce membranes and fibers, lyophilized to obtain sponges, or treated by spray-drying, extraction, and evaporation to produce microspheres. These polymers show good mechanical strength when dry, but the hydrated materials are less robust [32].

The biocompatibility and biodegradability of the HYAFF materials have been studied extensively. The degree of esterification influences the size of hydrophobic patches, which produces a polymer chain network that is more rigid and stable, and less susceptible to enzymatic degradation [33,34]. Drug release from HYAFF-based devices was examined for entrapped or covalently attached molecules. For example, release of the steroids hydrocor-

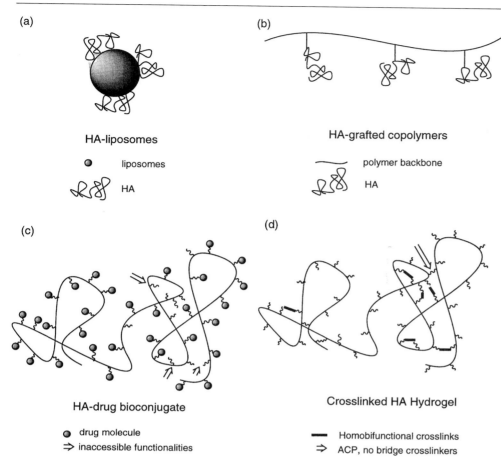

(a)

HA-liposomes

- liposomes

HA

(b)

HA-grafted copolymers

—— polymer backbone

HA

(c)

HA-drug bioconjugate

- drug molecule
⇒ inaccessible functionalities

(d)

Crosslinked HA Hydrogel

— Homobifunctional crosslinks
⇒ ACP, no bridge crosslinkers

Fig. 45.2. Schematic structure of HA derivatives: (a) HA–liposome composite, (b) HA-grafted co-polymer, (c) HA–drug bioconjugate, and (d) cross-linked HA.

tisone and α-methylpredisolone from microspheres fabricated from different HYAFF materials was examined with the drug either dispersed or bound to the polymer. While the hydrocortisone diffused out of the microspheres in 10 min, the release rate of the covalently bound drug was found to show zero-order release, requiring more than 100 h, consistent with ester bond hydrolysis [31,35]. Preclinical *in vivo* evaluations in a rabbit animal model demonstrated that an α-methylpredisolone carried by a HYAFF 11 formulation could maintain its anti-inflammatory and chondroprotective properties [36] and increase the residence time of the drug in the tear fluid [37,38]. Microspheres and thin films of HYAFF 11 have been found to be the most suitable physical forms for drug delivery systems.

Studies of macromolecule diffusion and release suggested a possible use of HYAFF 11 microspheres for peptide delivery. Furthermore, the mucoadhesive properties of HYAFF materials make realistic drug delivery via nasal, oral, and vaginal mucosal routes—for examples intranasal delivery of insulin in sheep [39] and vaginal delivery of calcitonin in rats [40,41] and flu vaccine in rodents [42]. Microspheres could also deliver recombinant human granulocyte–macrophage colony-stimulating factor (huGM-CSF) [43].

HA plays fundamental roles during embryonic development [44–46] and wound healing [47,48], suggesting a use for HA derivatives to create a suitable environment for the growth of cells derived from organ biopsy samples. In particular, stem elements or adults cells can be stimulated to divide and differentiate in this embryo-like environment [49,50]. HYAFF products also have water and vapor transport properties similar to those of commercial products for wound coverage, such as Bioprocess, Biobrane II [a composite of poly(dimethylsiloxane) on nylon fabric], and OpSite (polyurethane dressing with an adhesive layer of vinyl ether applied to one side).

The cultivation of fibroblasts in ECM-like structures is receiving widespread attention not only because of the need to provide a dermal substitute to epithelial sheets, but also because these living constructs may be applied in other pathologies such as ligament/tendon repair, vascular prosthesis, and dermal augmentation. As expected, cultured human fibroblasts actively proliferated in 30 days until the degradation of HYAFF 11 nonwoven mesh tests-commenced, and morphological tests on paraffin-embedded specimens demonstrated that cells migrated through the nonwoven mesh and populated both sides of the biomaterial. Furthermore, immunohistochemical analysis revealed the presence of collagen I, III, and fibronectin fibers, which were evident after only 15 days of culture in the HYAFF 11 scaffold, and allowed preparation of a living dermal equivalent [51]. *In vitro* studies of Laserskin membranes, which are produced with regular laser-made microperforations from HYAFF 11, showed that keratinocytes grew on these membranes and expressed a proliferative phenotype within the microperforations. The epithelial cells cultured on the membranes were able to differentiate, and after 15 days of standard culture, the investigators observed the formation of several distinct layers—starting from a basal proliferative through an upper keratinized zone [52–54]. A three-dimensional (3D) artificial skin [55] was developed for the treatment of second- and third-degree burns, chronic wounds such as larger phlebopathic ulcers [52,56], and diabetic foot ulcers [57]. In clinical trials on extensive burns of humans, the underlying wound bed of this 3D scaffold-treated area showed evidence of new collagen synthesis and organization into dermislike structure. The wounds were completely healed 7 weeks after burn excision, with no unusual scarring [58].

In vitro culture of chondrocytes on a HYAFF 11 nonwoven mesh was investigated in tissue engineering procedures for cartilage reconstruction. One month after subcutaneous implantation of the scaffold seeded with human chondrocytes in an athymic nude mouse model, the development of tissue similar to hyaline cartilage was observed [59]. Moreover, pluripotent mesenchymal stem cells, which are capable of giving rise to differentiated elements of several kinds (myoblasts, chondroblasts, osteoblasts, adipocytes, fibroblasts), could be cultured within a HYAFF scaffold. The mesenchymal cells within the scaffold could be induced toward a chondrocyte phenotype with the addition of basic fibroblast growth factor (bFGF) [60–62]. Similarly, culture-expanded bone marrow derived mesenchymal progenitor cells differentiate into chondrocytes or osteoblasts implanted subcutaneously *in vivo* in combination with a HYAFF 11 sponge as well as when porous calcium phosphate ceramic is used as the delivery vehicle. When coated with fibronectin, HYAFF 11 sponge bound 130% more cells than coated ceramics, which indicates the HA-based delivery vehicles are superior to porous calcium ceramic. Additionally, the HA-based vehicles have the advantage of degradation/resorption characteristics that allow complete replacement of the implant with newly formed tissue [63].

Modification of HA with Hydrazides, Amines, and Carbodiimides

The chemical modification of HA by means of its carboxylic functions, hydrazides, and carbodiimide compounds has been studied in great detail [64,65]. The modification of HA with hydrazides is performed in water at pH 4.75, with continuous addition of aqueous HCl to maintain the mild acidic pH. Although the carbodiimide reaction fails to give efficient coupling to amines, the high levels of *O*- to *N*-acyl migration gave *N*-acylureas with potentially important medical uses, as discussed later [66–68]. Alternatively, carbodiimide-mediated coupling to dihydrazide compounds such as adipic dihydrazide (ADH) provided an efficient and mild route for the introduction of multiple pendant hydrazide groups for further derivatization with drugs [69], for making biochemical probes [70], or for cross-linking [67,71]. For example, HA–Taxol bioconjugates were synthesized by the conjugation of HA-ADH and ester-activated Taxol. The conjugate showed selective binding and uptake by human breast cancer cells, human ovarian cancer cells, and human colon tumor cells. Increased cytotoxicity required cellular uptake of the conjugate followed by hydrolytic release of the active free Taxol [72,73]. Targeting can be accomplished by receptor-mediated uptake of the HA–drug bioconjugate that releases the drug only inside the target cell, thereby enhancing selectivity for cancerous cells [72].

As indicated earlier, attempted carbodiimide-mediated coupling of primary amines to HA, to generate polymeric amides, resulted in negligible coupling became of the protonation of the required nucleophilic nitrogen at the reaction pH. In 1999 a modified method was developed for the coupling of polyfunctional amines to HA through mediation of carbodiimide–active ester [74]. HA ($M_w > 1$ MDa) was reacted with 30-fold excess of the amine component at pH 6.8 in the presence of a soluble carbodiimide and 1-hydroxybenzotriazole (HOBt) with DMSO/H_2O (1:1) as solvent.

Many HA–drug bioconjugates have been prepared by using the method of carbodiimide activation. HA–mitomycin C and HA–epirubicin conjugates were synthesized and found to enhance the selective delivery of the parent antitumor drugs into regional lymph nodes and cancerous tissues through HA receptor CD44 [75]. Mitomycin C was coupled to HA (1200 kDa) through an amide bond in a DMF–water cosolvent with a water-soluble carbodiimide (EDCI) as the coupling agent. The HA–mitomycin C conjugate exhibited potent antimetastatic effects against Lewis lung carcinoma xenograft at an extremely low dose (0.01 mg/kg), whereas free mitomycin C had no effect. Similarly, adriamycin was reported to couple to HA through an amide bond linkage [76]. Epirubicin was coupled to HA after the synthesis of acetylated HA to increase its solubility in organic solvents. Similarly, daunomycin, 5-fluorouracil, and cytosine arabinoside have been coupled to HA. The same carbodiimide technology was applied to prepare HA–superoxide dismutase (HA-SOD) conjugate, which was synthesized by a water-soluble carbodiimide coupling of HA and bovine Cu/Zn-SOD or bacterial Mn-SOD at pH 4.8 and 4°C for 20 h. Some 20% of the amino groups on SOD were conjugated to HA, and the HA-SOD conjugate exhibited a more pronounced anti-inflammatory effect than SOD or HA alone [77].

MODIFICATION OF THE HA HYDROXYL GROUPS
Modification of HA by Sulfation

HA is an ideal hydrophilic coating for a variety of medical devices, including catheters, guide wires, and sensors. Various *in vitro* and *in vivo* tests have been conducted over the years by Biocoat Inc. (Fort Washington, PA) to demonstrate biocompatibility of the HA coating Hydak. However, to obtain the optimum blood-compatible material for use in the body, approaches were devised for sulfating the OH group present in the HA molecule.

The sulfation of HA (M_w 150–200 kDa) with sulfur trioxide pyridine complex in DMF produced different degrees of sulfation, HyalS$_x$, x = 1–4 [78]. The sulfated hyaluronic acid HyalS$_{3.5}$ was then immobilized onto plasma-processed polyethylene (PE) by using a diamine poly(ethylene glycol) derivative and a water-soluble carbodiimide. The thrombin time test and platelet adhesion behavior indicated that this procedure was promising for the preparation of blood-compatible, antithrombotic PE surface [79,80]. In addition, the anticoagulant polysaccharide was photoimmobilized on a poly(ethylene terephthalate) (PET) film [81]. Sulfated HA (HyalS$_x$) was conjugated with azidoaniline in the presence of a water-soluble carbodiimide to prepare azidophenylamino-derivatized HyalS$_x$. The derivative was dried on a PET surface before UV irradiation. A micropatterning profile of the anticoagulant polysaccharides was achieved with photolithography, and platelet adhesion was reduced on the sulfated hyaluronic acid areas. Surfaces coated with sulfated HA exhibited a marked reduction of cellular attachment, fouling, and bacterial growth compared with uncoated surfaces [82]. The coating was stable to degradation by chondroitinase and hyaluronidase and acquired new antithrombogenic properties [83].

Other Chemistries

Several other modifications have been accomplished with the hydroxyl groups in the HA molecule. Butyric acid, which is known to induce cell differentiation and to inhibit the growth of a variety of human tumors, was conjugated to HA [84] via the reaction between butyric anhydride and HA (M_w 85 kDa) *sym*-collidinium salt, in the presence of pyridine or dimethylaminopyridine, with DMF as solvent, at room temperature for 24 h. The results of cell culture on a human breast cancer cell line (MCF-7) suggested that the conjugation of butyrate residue to HA led to an increased half-life of the active component. In addition, HA butyrate offered a novel drug delivery system targeted specifically at tumor cells.

The anthracycline antibiotics adriamycin and daunomycin were coupled to HA via cyanogen bromide (CNBr) activation [76]. This reaction scheme is purported to lead to the attachment of the drug via a urethane bond to one of the hydroxylic functions of the polysaccharide, but no spectroscopic verification was provided.

A wider utilization of HA biomaterials can be envisaged if two obstacles can be overcome: enzymatic degradation and poor mechanical properties. Approaches to solving the first problem are based on the formation of chemically modified HA sols and hydrogels. Approaches to the second problem include the preparation of HA composite materials by grafting of other polymers or by formation of interpenetrating networks.

MODIFICATION OF HA HYDROGEL FORMATION

HA can be readily degraded by tissue hyaluronidases and free radicals, which commonly occur during inflammation. A small number of breaks in the molecular structure can result in a profound decrease in its molecular size, resulting in drastic alterations in viscosity. Currently, several cross-linking chemistries are being investigated to improve the stability of HA toward biodegradation.

Auto-Cross-Linked Polymer

Auto-cross-linked polymer (ACP, Fidia) consists of inter- and intramolecularly esterified HA (200 kDa) in which both the carboxyl groups and hydroxyl groups belong to HA molecules. ACP is a white powder and, upon hydration with water, gives rise to a transparent gel [85]. This novel biomaterial has shown promising results for several applications. For example, it can be used as a barrier between organs to reduce postoperative adhesions after abdominal and gynecological surgery [86–88]. In addition, the biocompatible, biodegradable, porous ACP behaves as a scaffolding for reparative cells to regenerate and to integrate when it is placed in natural reparative tissue defects, such as osteochondral defects in a rabbit knee model [89]. Subcutaneous implantation of ACP porous sponges seeded with chondrocytes or osteoblasts were found to regenerate bone and cartilage [63,90].

HA Cross-Linked with Diepoxide Cross-Linkers

Laurent and colleagues [91] cross-linked high concentration HA (average molecular mass 1.5 MDa, 50–175 mg/ml) in an alkaline environment with 1,2,3,4-diepoxybutane and sodium borohydride (50°C, 2 h). Using similar chemistry, HA (average molecular mass 870 kDa) at high concentration was allowed to react with either ethylene glycol diglycidyl ether [92] or polyglycerol poly(glycidyl ether) [93] in 0.1 N NaOH at 60°C with ethanol as a cosolvent. The gels obtained had a high water content (95%) even though high concentrations of cross-linker had been used. This HA gel was specifically degraded *in vitro* and *in vivo* by hydroxyl radicals, and the inflammatory responsive degradation proceeded via surface erosion [94,95]. This HA gel has been investigated for use as inflammation (stimulus)-responsive degradable matrices and as an electrically responsive and pulsatile release system [92] for implantable drug delivery. Hydrogels prepared by reaction of HA with 1,4-butanediol diglycidyl ether in the presence of 0.5% NaOH gave a porous material that was activated with periodate [96] and then modified with an 18 amino acid peptide containing the cell attachment domain sequence Arg-Gly-Asp (RGD). The presence of this peptide markedly enhanced cell attachment to the HA gels. Following cell attachment [MG63 cells (human osteosarcoma)], cells actively proliferated and colonized the pores of the matrix. This material may prove useful for the maintenance of cells as a scaffold for enhancing tissue repair and could have clinical implications for wound healing therapies [97].

Photo-Cross-Linking of HA

A methacrylate anhydride modified HA (14% modification) was synthesized by the reaction between HA and excess methacrylate anhydride; this derivative was photo-cross-linked to form a stable hydrogel by using ethyl eosin 0.5% w/v in 1-vinyl-2-pyrrolidone and triethanolamine as an initiator under argon ion laser with wavelength of 514 nm. The use of *in situ* photopolymerization of an HA derivative resulting in the formation of a cohesive gel enveloping the injured tissue may provide isolation from surrounding organs and thus prevent the formation of adhesions [98]. A preliminary cell encapsulation study was also successfully performed with islet of Langerhans.

HA Cross-Linked by Glutaraldehyde

HA strands fabricated from the cation-exchanged sodium hyaluronate (M_w 1.6 MDa) were cross-linked in glutaraldehyde aqueous solution [99]. Afterward, the strand surfaces were remodeled by the absorption of poly(D-lysine) and poly(L-lysine). The polypeptide-resurfaced HA strands have good biocompatibility and positive advantages for cellular adhesion.

HA Cross-Linked by Trivalent Iron

Intergel (FeHA, formerly Lubricoat) is a formulation of HA electrostatically cross-linked with trivalent iron. This gel is intended for postsurgical applications to prevent adhesions [100].

HA Cross-Linked with Carbodiimide Chemistry

Incert, under development by Anika Therapeutics, Inc. (Woburn, MA), is a bioresorbable sponge derived from cross-linked HA with a biscarbodiimide coupling agent in the presence of 2-propanol–H_2O as cosolvent [19]. It adheres to tissues without the need for sutures and retains its efficacy even in the presence of blood. Recently, it was found to be effective at preventing postoperative adhesions in a rabbit cecal abrasion study. In addition, Ossigel, which incorporates bFGF, was examined for acceleration of fracture healing.

Building on the hydrazide modification already described, hydrogels have been prepared using bishydrazide, trishydrazide, or polyvalent hydrazide compounds as cross-linkers [71]. HA (average molecular weight 1.5–2.0 MDa) would react with the hydrazide cross-linkers in the presence of a water-soluble carbodiimide at pH 3.5–4.7. Depending on the reaction conditions and the molar ratios of the reagents involved, gels with physico-chemical properties ranging from soft-pourable gels to more mechanically rigid and brittle gels could be obtained. Applications of this chemistry are in development by Clear Solutions Biotech, Inc. (Stony Brook, NY).

HA-ADH can be cross-linked using commercially available homobifunctional cross-linkers [67]. HA-ADH was dissolved in 0.1 M $NaHCO_3$ to give a concentration of 15 mg/ml, and the cross-linker was added in solid form. Gelation was observed within 30–90 s. The gels appeared clear and colorless after washing with water.

Recently, an *in situ* polymerization technique was developed by cross-linking HA-ADH with a macromolecular cross-linker, PEG–dialdehyde, under physiological conditions [101]. Biocompatible and biodegradable HA hydrogel films with well-defined mechanical strength were obtained after the evaporation of solvent. Drug release from these HA hydrogel films was investigated, and preliminary data with the HA films showed accelerated reepithelialization during wound healing.

A low water content HA hydrogel film [102] was made by cross-linking HA (1.6 MDa) film with a water-soluble carbodiimide as a coupling agent in H_2O containing a water-miscible nonsolvent of HA (e.g., ethanol or acetone). The highest degree of cross-linking that gave a low water content hydrogel was achieved in 80% ethanol. This film, having 60% water content, remained stable for 2 weeks after immersion in buffered solution. The cross-linking of HA films with a water-soluble carbodiimide in the presence of L-lysine methyl ester prolonged the *in vivo* degradation of HA film. The higher resistance to hydrolytic degradation might be attributable to amide bonds, in contrast to those cross-linked through ester bonds.

Hylans

Hylan is a hydrogel formed by cross-linking HA containing residual protein with formaldehyde (soluble gel) or divinyl sulfone [103]. In an alkaline environment at 20°C (10–15 min), the divinyl sulfone reagent is believed to react with hydroxyl groups of HA, forming cross-links. Similar cross-linking was obtained by using dimethylolurea, ethylene oxide, and polyisocyanate reagents. Soluble hylan is a high weight-average molecular weight form (8–23 MDa) of HA (5–6 MDa) that exhibits enhanced rheological properties compared to HA alone. Hylan gel is an insoluble form that has greater elasticity (at all frequencies) and viscosity (at low shear rates) than soluble hylan and retains the high biocompatibility of the native macromolecule (nonimmunogenic, noninflammatory, and nontoxic). The hylans are produced by Biomatrix, now owned by Genzyme (Cambridge, MA).

Hylan exhibits biocompatibility identical to that of native HA and has been investigated in a number of medical applications. In the treatment of degenerative joint disease and rheumatoid arthritis, hylan was found to protect cartilage and prevent further chondrocyte injury. However, the effect was found to be reversible and viscosity dependent [104]. The hylan product Synvisc was developed specifically as a device for viscosupplementation therapy in osteroarthritis of the knee. It can increase the viscoelastic properties of the synovial fluid and the intercellular matrix of the synovial tissue and capsule [105,106].

Results from the clinical trials of hylan B gel slurry injections (Hylaform gel) suggested that it provides a safe, effective alternative for soft tissue augmentation [107–109]. Additionally, the injection of viscoelastic hylan B gel gave a durable augmentation of the soft tissues in the vocal fold in a rabbit model [110]. The gel can remain in place for 12 months, allowing the ingrowth of newly formed connective tissue after only a month. The new soft tissue contained collagen, HA, fibroblasts, and a few new vessels, without causing inflammation and adverse reactions. Finally, hylan has been used in plastic surgery for intradermal implantations and cosmetic injection [111]. Similarly, Restylane (Q-Med Inc., Sweden), a partially cross-linked HA derivative, has been available for soft tissue augmentation in Europe for more than 4 years [109]. However, neither Restylane nor hylan gel is currently approved by the FDA.

Hylagel is an engineered hylan gel being investigated as an adjuvant to prevent post-surgical adhesions. Finally, hylans can be used as drug carriers by the incorporation of therapeutic agents. For example, the *in vitro* biological activity of cytokine α-interferon was enhanced by approximately 40% as a result of its covalent attachment to hylan matrix [112].

Multiple-Component Cross-Linking of HA

Another way to prepare hyaluronan-based networks is based on the Passerini reaction and the Ugi reaction [113]. In the Passerini reaction, an aqueous solution of hyaluronan is mixed with aqueous glutaraldehyde (or other water-soluble dialdehyde) and added to a known amount of cyclohexylisocyanide. In the Ugi four-component reactions, the same mixture contains a diamine, as well. The degree of cross-linking is controlled by the amount of dialdehyde and diamine. Passerini reactions are kinetically faster and give good yields of a single condensation product. The ensuing hydrogels are transparent and mechanically stable; they swell in aqueous salt solutions depending on ion strength, and exhibit values of the compression modulus markedly dependent on the degree of cross-linking.

COMPOSITES

Synthetic polymers have been prepared as biomedical devices, and different physicochemical and mechanical properties are required for a given application. Synthetic polymers often have optimized mechanical properties but suffer from insufficient biocompatibility. In contrast, biocompatible biopolymers often have suboptimal mechanical properties. Blending synthetic polymers with biological macromolecules can yield composite materials that feature the desired properties of the individual polymers. Specifically, HA has been blended with other materials to produce novel biomaterials with the desired physicochemical, mechanical, and biocompatible properties. Two different synthetic polymers, poly(vinyl alcohol) (PVA) and poly(acrylic acid) (PAA), were blended with either collagen or HA [114]. HA-PAA sponges were prepared by dissolving both polymers in water at different ratios, lyophilizing, and cross-linking them by thermal treatment at 130°C under vacuum for 24 h. HA-PVA hydrogels were prepared by dissolving both polymers in water at different ratios and subjecting these mixtures to eight cycles of freeze–thawing.

HA–Liposome Composites

HA has been incorporated into liposomes. For example, HA was combined with cyclosporin A (CsA) and encapsulated within phospholipid liposomes [115]. HA (M_w 10 and 1000 kDa) powder was added into and hydrated with CsA–liposome composites to provide a final solution concentration of 2.5 wt% HA. Developed as a topically administered pharmaceutical agent, the solution was found to effectively treat skin disorders while minimizing systemic circulation. These compositions can also be administered orally, parenterally, and intrarectally.

Bioadhesive liposomes, in which HA is the surface-anchored bioadhesive ligand on a liposome surface, were prepared by the preactivation of HA with a carbodiimide and then added to a suspension of multilamellar liposomes consisting of phosphatidylcholine, phosphatidylethanolamine, and cholesterol [116]. In principle, an HA-coated liposome functionally resembles the PEG-coated "stealth" liposomes. These were investigated for their ability to act as site-adherent and sustained-release carriers of epidermal growth factor for the topical therapy of wounds and burns [117].

HA–Gelatin Composites

A porous matrix composed of gelatin and HA was prepared by dipping a gelatin–HA water-soluble sponge into 90% w/v acetone–water mixture with a small amount of a carbodiimide (EDCI) as cross-linking agent [118]. This sponge-type biomaterial was constructed for either wound dressings or scaffolds for tissue engineering. The sponge, impregnated with silver sulfadiazine, was found to facilitate the epidermal healing rate in a rat model.

HA–Alginate Composites

The association of alginate with another polysaccharide has been proposed to combine alginate's gel-forming effects with the properties of the partner macromolecule. Alginate–HA gels were prepared through the diffusion of calcium into alginate–HA mixtures. The resulting gels with a weight ratio of alginate to HA up to 0.50, had satisfactory mechanical properties. This composite matrix might be suitable as a biopolymeric carrier, or for articular surgery applications, because of its stability in synovial fluid [119,120].

HA–Collagen Composites

Composite materials consisting of HA and collagen have been prepared by complexing both components into a coagulate followed by cross-linking, with either glyoxal or starch dialdehyde as the cross-linking agent [121]. Atelocollagen was suspended in 0.5 M acetic acid, and the addition of an HA solution resulted in formation of coagulated material. The fabricated membrane was then chemically cross-linked with starch dialdehyde and glyoxal by immersing the membrane into a solution of this cross-linker. This membrane was quite resistant to collagenase and permitted fibroblast growth. Other composite materials consisting of HA and collagen were prepared by cross-linking the dried HA–collagen coagulates with polyethylene oxide and hexamethylene diisocyanate [122].

In addition, a hydroxyapatite–collagen–HA composite material was prepared by adding hydroxyapatite particles to an HA solution followed by blending with an aqueous dispersion of collagen fibers [123]. The final material, consisting of 90% hydroxyapatite, 9.2% collagen, and 0.8% w/w HA, is biocompatible and mechanically robust; it can be used as a bone defect filler. In 1999 a porous collagen–HA matrix was prepared by cross-linking collagen with HA–aldehyde made from the periodate oxidation of HA [124]. The presence of HA within the collagen matrix supported new bone formation, and this collagen–HA matrix has potential uses for the delivery of growth factors or as an implantable-cell-seeded matrix.

High concentrations of HA (1 mg/ml) in cell culture medium not only inhibit fibroblast contraction of a floating collagen fibrillar matrix (CFM) but also stimulate fibroblast migration on the CFM [52]. HA incorporation into an artificial skin material (collagen–gelatin sponge) accelerates the ingrowth of granulation tissue, thus providing a more suitable graft bed to support a skin graft [125]. Further investigation of HA covalently bound to a collagen matrix found significantly reduced collagen contraction in fibroblast cultures. Such results warrant consideration of HA as an alternative biomaterial for building a dermal substitute and tissue engineering scaffold [126].

HA–Carboxymethylcellulose Composites

A bioabsorbable membrane, Seprafilm, has been developed as a physical barrier for prevention of postsurgical adhesions. Genzyme scientists prepared this material by blending two anionic polymers, HA and carboxymethylcellulose (CMC) [127], followed by carbodiimide-mediated modification. An anionic–cationic cross-linked network produces a rather fragile biomaterial. Seprafilm was approved by the FDA in 1996 for use in patients undergoing abdominal or pelvic laparotomy, to reduce the incidence, extent, and severity of postoperative adhesion [128]. The use of Seprafilm is limited to accessible areas that can be fully covered. Since for many adhesions the location is unpredictable or inaccessible, the general clinical utility of Seprafilm is compromised. In addition, Seprafilm is reported to suffer from handling difficulties, hampering its acceptance by surgeons. The same technology has been expanded to reduce the development of adhesions to synthetic nonabsorbable mesh, such as hernia repair products. Sepramesh is a composite of polypropylene mesh and a foam form of the Seprafilm membrane. This composite reduces the development of adhesions to the surface of the hernia repair mesh without compromising the long-term abdominal wall incorporation into the mesh [129]. Similar technology has also been used in presurgical coating, which would theoretically limit tissue trauma and prevent desiccation during surgery. However, a Genzyme-conducted clinical trial of Sepracoat, an HA-CMC solution applied in presurgical manipulation, failed to meet efficacy end points.

HA-Grafted Copolymers

A variety of HA-grafted copolymers have been prepared in recent years. In one example, several polyampholyte comb-type copolymers consisting of poly(L-lysine) (PLL) main chains, a DNA binding site, and an HA side chain with cell-specific ligands were prepared to target sinusoidal endothelial cells of liver [130]. The reducing end of HA and ε-amino groups of PLL were covalently coupled by reductive amination, using sodium cyanoborohydride to obtain the resulting comb-type copolymers (PLL-*graft*-HA). The PLL backbone selectively formed the polyion complex with DNA even in the presence of the HA side chain. In addition, the PLL-*graft*-HA-DNA complex may form a multiphase structure in which the hydrophobic PLL-DNA complex was surrounded by a hydrated shell of free HA. Complex formation with free HA chains was considered to be essential for directing the complex to target cells.

A PEG–grafted-HA copolymer was prepared by coupling HA with methoxy–PEG–hydrazide in the presence of EDCI as a coupling agent at acidic pH [131]. The copolymer is expected to be used for the delivery of water-soluble peptides. For example, when loaded with insulin, the loaded insulin appeared to partition into PEG moieties and intermolecular interactions, preventing the conformational changes of insulin. This appears to be a reasonable method to prevent drug leakage and to achieve degradation-controlled insulin release. Moreover, such a heterogeneously structured polymeric solution may be useful as an injectable therapeutic formulation for ophthalmic or arthritis treatment with a suitable drug.

CONCLUSION

HA is an important starting material for preparation of new biocompatible and biodegradable polymers that have applications in drug delivery, tissue engineering, and viscosupplementation. HA derivatives containing a variety of versatile functional groups can be produced with varying chemical structures and morphology. In this way, the rate of degradation, the degree of hydration, the cellular responses, and the overall tissue response can be manipulated. Approval and introduction of an increasing number of medical devices and new drugs using HA-derived biomaterials can be anticipated in the next decade.

Acknowledgments

We thank the University of Utah, Clear Solutions Biotech, Inc. (Stony Brook, NY), the U.S. Department of Defense, and the Center for Biopolymers at Interfaces (University of Utah) for financial support of research developed in our laboratories.

REFERENCES

1. Laurent, T. C., Laurent, U. B. G., and Fraser, J. R. E. (1995). Functions of hyaluronan. *Ann. Rheum. Dis.* **54**, 429–432.
2. Fraser, J. R. E., Laurent, T. C., and Laurent, U. B. G. (1997). Hyaluronan: Its nature, distribution, functions and turnover. *J. Intern. Med.* **242**, 27–33.
3. Dowthwaite, G. P., Edwards, J. C. W., and Pitsillides, A. A. (1998). An essential role for the interaction between hyaluronan and hyaluronan binding proteins during joint development. *J. Histochem. Cytochem.* **46**, 641–651.
4. Hardwick, C., Hoare, K., Owens, R., Hohn, H. P., Hook, M., Moore, D., Cripps, V., Austen, L., Nance, D. M., and Turley, E. A. (1992). Molecular cloning of a novel hyaluronan receptor that mediates tumor cell motility. *J. Cell Biol.* **117**, 1343–1350.
5. Collis, L., Hall, C., Lange, L., Ziebell, M. R., Prestwich, G. D., and Turley, E. A. (1998). Rapid hyaluronan uptake is associated with enhanced motility: Implications for an intracellular mode of action. *FEBS Lett.* **440**, 444–449.
6. Entwistle, J., Hall, C. L., and Turley, E. A. (1996). Receptors: Regulators of signalling to the cytoskeleton. *J. Cell. Biochem.* **61**, 569–577.
7. Cheung, W. F., Cruz, T. F., and Turley, E. A. (1999). Receptor for hyaluronan-mediated motility (RHAMM), a hyaladherin that regulates cell responses to growth factors. *Biochem. Soc. Trans.* **27**, 135–142.
8. Toole, B. P. (1997). Hyaluronan in morphogenesis. *J. Intern. Med.* **242**, 35–40.
9. Gotoda, T., Matsumura, Y., Kondo, H., Saitoh, D., Shimada, Y., Kosuge, T., Kanai, Y., and Kakizoe, T. (1998). Expression of CD44 variants and its association with survival in pancreatic cancer. *Jpn. J. Cancer Res.* **89**, 1033–1040.
10. Herrlich, P., Sleeman, J., Wainwright, D., Konig, H., Sharman, L., Hilberg, F., and Ponta, H. (1998). How tumor cells make use of CD44. *Cell Adhes. Commun.* **6**, 141–147.
11. Baumgartner, G., Boltzman, L., and Hamilton, G. (1998). The impact of extracellular matrix on chemoresistance of solid tumors—experimental and clinical results of hyaluronidase as additive to cytostatic chemotherapy—Symposium on novel aspects in chemoresistance, Vienna, 20th March 1998—Editorial. *Cancer Lett.* **131**, 1–2.
12. Baumgartner, G., Gomar-Hoss, C., Sakr, L., Ulsperger, E., and Wogritsch, C. (1998). The impact of extracellular matrix on the chemoresistance of solid tumors—experimental and clinical results of hyaluronidase as additive to cytostatic chemotherapy. *Cancer Lett.* **131**, 85–99.
13. Gerdin, B., and Hallgren, R. (1997). Dynamic role of hyaluronan (HYA) in connective tissue activation and inflammation. *J. Intern. Med.* **242**, 49–55.
14. Delpech, B., Girard, N., Bertrand, P., Courel, M. N., Chauzy, C., and Delpech, A. (1997). Hyaluronan: Fundamental principles and applications in cancer. *J. Intern. Med.* **242**, 41–48.
15. Hall, C. L., Yang, B. H., Yang, X. W., Zhang, S. W., Turley, M., Samuel, S., Lange, L. A., Wang, C., Curpen, G. D., Savani, R. C., Greenberg, A. H., and Turley, E. A. (1995). Overexpression of the hyaluronan receptor RHAMM is transforming and is also required for H-ras transformation. *Cell (Cambridge, Mass.)* **82**, 19–28.
16. Boyce, D. E., Thomas, J. H., Moore, K., and Harding, K. (1997). Hyaluronic acid induces tumour necrosis factor-α production by human macrophages *in vitro*. *Br. J. Plast. Surg.* **50**, 362–368.
17. Partsch, G., Schwarzer, Ch., Neumuller, J., Dunky, A., Petera, P., Broll, H., Ittner, G., and Jantsch, S. (1989). Modulation of the migration and chemotaxis of PMN cells by hyaluronic acid. *Z. Rheum.* **48**, 123–128.
18. Chen, W. Y. J., and Abatangelo, G. (1999). Functions of hyaluronan in wound repair. *Wound Repair Regeneration* **7**, 79–89.
19. Vercruysse, K. P., and Prestwich, G. D. (1998). Hyaluronate derivatives in drag delivery. *Crit. Rev. Ther. Drug Carrier Syst.* **15**, 513–555.
20. Sutherland, I. W. (1998). Novel established applications of microbial polysaccharides. *Trends Biotechnol.* **16**, 41–46.
21. Pasquali, R. I., Guerra, D., Taparelli, F., Georgountoz, A., and Frizziero, L. (1999). *In* "New Frontiers in Medical Sciences: Redefining Hyaluronan" (G. Abatangelo and P. Weigel, eds.), p. N28. Abbazia di Praglia, Padua, Italy.
22. Punzi, L., Pianon, M., Schiavon, F., and Todesco, S. (1999). Mechanism of the action of hyaluronan in arthropathies. *In* "New Frontiers in Medical Sciences: Redefining Hyaluronan" (G. Abatangelo and P. Weigel, eds.), p. N32. Abbazia di Praglia, Padua, Italy.
23. Hallen, L., Johansson, C., Dahlqvist, A., and Laurent, C. (1999). The potential use of hyaluronan based compounds in laryngeal augmentative surgery. *In* "New Frontiers in Medical Sciences: Redefining Hyaluronan" (G. Abatangelo and P. Weigel, eds.), p. N22. Abbazia di Praglia, Padua, Italy.
24. Seckel, B. R., Jones, D., Hekimian, K. J., Wang, K. K., Chakalis, D. P., and Costas, P. D. (1995). Hyaluronic acid through a new injectable nerve guide delivery system enhances peripheral nerve regeneration in the rat. *J. Neurosci. Res.* **40**, 318–324.
25. Jorgensen, T., Moss, J., Nicolajsen, H. V., and Nielsen, L. S. (1998). PCT Int. Appl. WO 9822114A1.
26. Saettone, M. F., Monti, D., Torracca, M. T., and Chetoni, P. (1994). Mucoadhesive ophthalmic vehicles: Evaluation of polymeric low-viscosity formulations. *J. Ocul. Pharmacol.* **10**, 83–92.
27. Morimoto, K., Yamaguchi, H., Iwakura, Y., Morisaka, K., Ohashi, Y., and Nakai, Y. (1991). Effects of viscous hyaluronate-sodium solutions on the nasal absorption of vasopressin and an analogue. *Pharm. Res.* **8**, 471–474.
28. Miller, J. A., Ferguson, R. L., Powers, D. L., Burns, J. W., and Shalaby, S. W. (1997). Efficacy of hyaluronic acid/nonsteroidal anti-inflammatory drug systems in preventing postsurgical tendon adhesions. *J. Biomed. Mater. Res. (Appl. Biomater.)* **38**, 25–33.
29. Gowland, G., Moore, A. R., Willis, D., and Willoughby, D. A. (1996). Marked enhanced efficacy of cyclosporin when combined with hyaluronic acid. Evidence from two T cell-mediated models. *Clin. Drug Invest.* **11**, 245–250.
30. Brown, T. J., Alcorn, D., and Fraser, J. R. E. (1999). Absorption of hyaluronan applied to the surface of intact skin. *J. Invest. Dermatol.* **113**, 740–746.

31. Benedetti, L. M., Topp, E. M., and Stella, V. J. (1990). Microspheres of hyaluronic acid esters—fabrication methods and *in vitro* hydrocortisone release. *J. Controlled Release* **13**, 33–41.

32. Iannace, S., Ambrosio, L., Nicolais, L., Rastrelli, A., and Pastorello, A. (1992). Thermomechanical properties of hyaluronic acid-derived products. *J. Mater. Sci.: Mater. Med.* **3**, 59–64.

33. Campoccia, D., Hunt, J. A., Doherty, P. J., Zhong, S. P., Oregan, M., Benedetti, L., and Williams, D. F. (1996). Quantitative assessment of the tissue response to films of hyaluronan derivatives. *Biomaterials* **17**, 963–975.

34. Benedetti, L., Cortivo, R., Berti, T., Berti, A., Pea, F., Mazzo, M., Moras, M., and Abatangelo, G. (1993). Biocompatibility and biodegradation of different hyaluronan derivatives (HYAFF) implanted in rats. *Biomaterials* **14**, 1154–1160.

35. Benedetti, L. M., Joshi, H. N., Goei, L., Hunt, J. A., Callegaro, L., Stella, V. J., and Topp, E. M. (1991). Dosage forms from polymeric prodrugs: Hydrocortisone esters of hyaluronic acid. *New Polym. Mater.* **3**, 41–48.

36. Drobnik, J. (1991). Hyaluronan in drug delivery. *Adv. Drug Delivery Res.* **7**, 295–308.

37. Hume, L. R., Lee, H. K., Benedetti, L. M., Sanzgiri, Y. D., Topp, E. M., and Stella, V. J. (1994). Ocular sustained delivery of prednisolone using hyaluronic acid benzyl ester films. *Int. J. Pharm.* **111**, 295–298.

38. Kyyronen, K., Hume, L., Benedetti, L. M., Urtti, A., Topp, E., and Stella, V. (1992). Methyl-prednisolone esters of hyaluronic acid in opthalmic drug delivery. *Int. J. Pharm.* **80**, 161–169.

39. Ilium, L., Farraj, N. F., Fisher, A. N., Gill, I., Miglietta, M., and Benedetti, L. M. (1994). Hyaluronic acid ester microspheres as a nasal delivery system for insulin. *J. Controlled Release* **29**, 133–141.

40. Richardson, J. L., Ramires, P. A., Miglietta, M. R., Rochira, M., Bacelle, L., Callegaro, L., and Benedetti, L. M. (1995). Novel vaginal delivery systems for calcitonin. 1. Evaluation of HYAFF calcitonin microspheres in rats. *Int. J. Pharm.* **115**, 9–15.

41. Bonucci, E., Ballanti, P., Ramires, P. A., Richardson, J. L., and Benedetti, L. (1995). Prevention of ovariectomy osteopenia to rats after vaginal administration of HYAFF 11 microspheres containing calcitonin. *Calcif. Tissue Int.* **56**, 274–279.

42. Singh, M., Briones, M., and O'Hagan, D. (1999). Hyaluronic acid biopolymers for mucosal delivery of vaccines. *In* "New Frontiers in Medical Sciences: Redefining Hyaluronan" (G. Abatangelo and P. Weigel, eds.), p. N34. Abbazia di Praglia, Padua, Italy.

43. Nightlinger, N. S., Benedetti, L., Soranzo, C., Pettit, D. K., Pankey, S. C., and Gombotz, W. R. (1995). *Proc. Int. Symp. Controlled Release Bioact. Mater.*, Seattle, WA, pp. 738–739.

44. Rooney, P., and Kumar, S. (1993). Inverse relationship between hyaluronan and collagens in development and angiogenesis. *Differentiation (Berlin)* **54**, 1–9.

45. Wheatley, S. C., Isacke, C. M., and Crossley, P. H. (1993). Restricted expression of the hyaluronan receptor, CD44, during postimplantation mouse embryogenesis suggests key roles in tissue formation and patterning. *Development (Cambridge, UK)* **119**, 295–306.

46. Peterson, P. E., Pow, C. S. T., Wilson, D. B., and Hendrickx, A. G. (1993). Distribution of extracellular matrix components during early embryonic development in the macaque. *Acta Anat.* **146**, 3–13.

47. Knudson, C. B., and Knudson, W. (1993). Hyaluronan-binding proteins in development, tissue homeostasis, and disease. *FASEB J.* **7**, 1233–1241.

48. Siebert, J. W., Burd, A. R., MaCarthy, J. G., Weinzweig, J., and Ehrlich, H. P. (1990). Fetal wound healing: A biochemical study of scarless healing. *Plast. Reconstr. Surg.* **85**, 495–502.

49. Cortivo, R., De Galateo, A., Castellani, I., Brun, P., Giro, M. G., and Abatangelo, G. (1990). Hyaluronic acid promotes chick embryo fibroblast and chondroblast expression. *Cell Biol. Int. Rep.* **14**, 111–122.

50. Shepard, S., Becker, H., and Hartmann, J. X. (1996). Using hyaluronic acid to create a fetal-like environment *in vitro*. *Ann. Plast. Surg.* **36**, 65–69.

51. Denizot, F., and Lang, R. (1986). Rapid colorimetric assay for cell growth and survival. *J. Immunol. Methods* **89**, 271–277.

52. Pianigiani, E., Andreassi, A., Taddeucci, P., Alessandrini, C., Fimiani, M., and Andreassi, L. (1999). A new model for studying differentiation and growth of epidermal cultures on hyaluronan-based carrier. *Biomaterials* **20**, 1689–1694.

53. Burn, P., Cortivo, R., Zavan, B., Vecchiato, N., and Abatangelo, G. (1999). In vitro reconstructed tissues on hyaluronan-based temporary scaffolding. *J. Mater. Sci., Mater. Med.* **10**, 683–688.

54. Andreassi, L., Casini, L., Trabucchi, E., Diamantini, S., Rastrelli, A., and Donati, L. (1991). Human keratinocytes cultured on membranes composed of benzyl ester of hyaluronic acid suitable for grafting. *Wounds* **3**, 116–126.

55. Soranzo, C., Abaatangelo, G., and Callegaro, L. (1996). Artificial skin containing as support biocompatible materials based on hyaluronic acid derivatives. PCT Int. Appl. WO 9633750.

56. Zacchi, V., Soranzo, C., Cortivo, R., Radice, M., Brun, P., and Abatangelo, G. (1998). *In vitro* engineering of human skin-like tissue. *J. Biomed. Mater. Res.* **40**, 187–194.

57. Caravaggi, C., Faglia, E., L.D., P., De Giglio, R., Cavaiani, P., Mantero, M., Gino, M., Quarantiello, A., Sommariva, E., and Pritelli, C. (1999). Tissue engineering in the treatment of diabetic foot ulcers. *In* "New Frontiers in Medical Sciences: Redefining Hyaluronan" (G. Abatangelo and P. Weigel, eds.), p. N9. Abbazia di Praglia, Padua, Italy.

58. Harris, P. A., Francesco, F., Barisono, D., Leigh, I. M., and Navsaria, H. A. (1999). Use of hyaluronic acid and cultured autologous keratinocytes and fibroblasts in extensive burns. *Lancet* **353**, 35–36.

59. Aigner, J., Tegeler, J. A., Hutzler, P., Campoccia, D., Pavesio, A., Hammer, C., Kastenbauer, E., and Naumann, A. (1998). Cartilage tissue engineering with novel nonwoven structured biomaterial based on hyaluronic acid benzyl ester. *J. Biomed. Mater. Res.* **42**, 172–181.

60. Radice, M., Burn, P., Cortivo, R., Scapinelli, R., Battaliard, C., and Abatangelo, G. (2000). Hyaluronan-based biopolymers as delivery vehicles for bone-marrow-derived mesenchymal progenitors. *J. Biomed. Mater. Res.* **50**, 101–109.

61. Wakitani, S., Goto, T., Pineda, S. J., Young, R. G., Mansour, J. M., Caplan, A. I., and Goldeberg, V. M. (1994). Mesenchymal qell-based repair of large, full-thickness defects of articular cartilage. *J. Bone Jt. Surg. (Am. Vol.)* **76**-A(4), 579–592.

62. Butnariu-Ephrat, M., Robinson, D., Mendes, D. G., Halperin, N., and Nevo, Z. (1996). Resurfacing of goat articular cartilage by chondrocytes derived from bone marrow. *Clin. Orthop. Relat. Res.* **330**, 234–243.

63. Solchaga, L. A., Goldberg, V. M., and Caplan, A. I. (1999). Augmentation of the repair of osteochondral defects by autologous bone marrow in a hyaluronic acid-based delivery. In "New Frontiers in Medical Sciences: Redefining Hyaluronan" (G. Abatangelo and P. Weigel, eds.), p. N55. Abbazia di Praglia, Padua, Italy.

64. Prestwich, G. D., Marecak, D. M., Marecek, J. F., Vercruysse, K. P., and Ziebell, M. R. (1997). Controlled chemical modification of hyaluronic acid: Synthesis, applications and biodegradation of hydrazide derivatives. *J. Controlled Release* **53**, 99.

65. Kuo, J.-W., Swann, D. A., and Prestwich, G. D. (1994). Water-insoluble derivatives of hyaluronic acid and their methods of preparation and use. U.S. Pat. 5,356,883.

66. Kuo, J.-W., Swann, D. A., and Prestwich, G. D. (1991). Chemical modification of hyaluronic acid by carbodiimides. *Bioconjugate Chem.* **2**, 232–241.

67. Pouyani, T., Harbison, G. S., and Prestwich, G. D. (1994). Novel hydrogels of hyaluronic acid: Synthesis, surface morphology, and solid-state NMR. *J. Am. Chem. Soc.* **116**, 7515–7522.

68. Pouyani, T., Kuo, J.-W., Harbison, G. S., and Prestwich, G. D. (1992). Solid-state NMR of N-acylureas derived from the reaction of hyaluronic acid with isotopically-labeled carbodiimides. *J. Am. Chem. Soc.* **114**, 5972–5976.

69. Pouyani, T., and Prestwich, G. D. (1994). Functionalized derivatives of hyaluronic acid oligosaccharides: Drug carriers and novel biomaterials. *Bioconjugate Chem.* **5**, 339–347.

70. Pouyani, T., and Prestwich, G. D. (1994). Biotinylated hyaluronic acid: A new tool for probing hyaluronate-receptor interactions. *Bioconjugate Chem.* **5**, 370–372.

71. Vercruysse, K. P., Marecak, D. M., Marecek, J. F., and Prestwich, G. D. (1997). Synthesis and *in vitro* degradation of new polyvalent hydrazide cross-linked hydrogels of hyaluronic acid. *Bioconjugate Chem.* **8**, 686–694.

72. Luo, Y., and Prestwich, G. D. (1999). Synthesis and selective cytotoxicity of a hyaluronic acid-antitumor bioconjugate. *Bioconjugate Chem.* **10**, 755–763.

73. Luo, Y., Ziebell, M. R., and Prestwich, G. D. (2000). A hyaluronic acid-Taxol antitumor bioconjugate targeted to cancer cells. *Biomacromolecules* **1**, 208–218.

74. Bulpitt, P., and Aeschlimann, D. (1999). New strategy for chemical modification of hyaluronic acid: Preparation of functionalized derivatives and their use in the formation of novel biocompatible hydrogels. *J. Biomed. Mater. Res.* **47**, 152–169.

75. Akima, K., Ito, H., Iwata, Y., Matsuo, K., Watari, N., Yanagi, M., Hagi, H., Oshima, K., Yagita, A., Atomi, Y., and Tatekawa, I. (1996). Evaluation of antitumor activities of hyaluronate binding antitumor drugs: Synthesis, characterization and antitumor activity. *J. Drug Target.* **4**, 1.

76. Cera, C., Terbojevich, M., Cosani, A., and Palumbo, M. (1988). Anthracycline antibiotics supported on water-soluble polysaccharides: Synthesis and physicochemical chracterization. *Int. J. Biol. Macromol.* **10**, 66–74.

77. Sakurai, K., Miyazaki, K., Kodera, Y., Nishimura, H., Shingu, M., and Inada, Y. (1997). Anti-inflammatory activity of superoxide dismutase conjugated with sodium hyaluronate. *Glycoconjugate J.* **14**, 723–728.

78. Magnani, A., Albanese, A., Lamponi, S., and Barbucci, R. (1996). Blood-interaction performance of different sulphated hyaluronic acids. *Thromb. Res.* **81**, 383–395.

79. Favia, P., Palumbo, F., and D'Agostino, R. (1997). Grafting of functional groups onto polyethylene by means of RE glow discharges as first step to the immobilization of biomolecules. *Polym. Prepr.* **38**, 1039–1040.

80. Favia, P., Palumbo, F., D'Agostino, R., Lamponi, S., Magnani, A., and Barbucci, R. (1998). Immobilization of heparin and highly-sulfated hyaluronic acid onto plasma-treated polyethylene. *Plasma Polym.* **3**, 77–96.

81. Chen, G. P., Ito, Y., Imanishi, Y., Magnani, A., Lamponi, S., and Barbucci, R. (1997). Photoimmobilization of sulfated hyaluronic acid for antithrombogenicity. *Bioconjugate Chem.* **8**, 730–734.

82. Hoekstra, D. (1999). Hyaluronan-modified surfaces for medical devices. *Med. Devices Diagn. Ind.*, 48–58.

83. Barbucci, R. (1999). Hyaluronan derivatives: Chemical modifications and biochemical applications. In "New Frontiers in Medical Sciences: Redefining Hyaluronan" (G. Abatangelo and P. Weigel, eds.), p. N4. Abbazia di Praglia, Padua, Italy.

84. Coradini, D., Pellizzaro, C., Miglierini, G., Daidone, M. G., and Perbellini, A. (1999). Hyaluronic acid as drug delivery for sodium butyrate: Improvement of the anti-proliferative activity on a breast-cancer cell line. *Int. J. Cancer* **81**, 411–416.

85. Mensitieri, M., Ambrosio, L., and Nicolais, L. (1996). Viscoelastic properties modulation of a novel autocrosslinked hyaluronic acid polymer. *J. Mater. Sci.: Mater. Med.* **7**, 695–698.

86. De Iaco, P. A., Stefanetti, M., Pressato, D., Piana, S., Dona, M., Pavesio, A., and Bovicelli, L. (1998). A novel hyaluronan-based gel in laparoscopic adhesion prevention: Preclinical evaluation in an animal model. *Fertil. Steril.* **69**, 318–323.

87. De Iaco, P. (1999). In "New Frontiers in Medical Sciences: Redefining Hyaluronan," p. N12. Abbazia di Praglia, Padua, Italy.

88. Lise, M. (1999). Prevention of adhesions in abdominal surgery. In "New Frontiers in Medical Sciences: Redefining Hyaluronan" (G. Abatangelo and P. Weigel, eds.), p. N24. Abbazia di Praglia, Padua, Italy.

89. Caplan, A. I., Solchaga, A. I., and Goldberg, V. M. (1999). Hyaluronic acid-based polymers in the treatment of osteochondral defects. In "New Frontiers in Medical Sciences: Redefining Hyaluronan" (G. Abatangelo and P. Weigel, eds.), p. N8. Abbazia di Praglia, Padua, Italy.

90. Solchaga, L. A., Dennis, J. E., Goldberg, V. M., and Caplan, A. I. (1999). Hyaluronic acid-based polymers as cell carriers for tissue-engineered repair of bone and cartilage. *J. Orthop. Res.* **17**, 205–213.

91. Laurent, T. C., Hellsing, K., and Gelotte, B. (1964). Cross-linked gels of hyaluronic acid. *Acta Chem. Scand.* **18**, 274–275.

92. Tomer, R., Dimitrijevic, D., and Florence, A. T. (1995). Electrically controlled release of macromolecules from cross-linked hyaluronic acid hydrogels. *J. Controlled Release* **33**, 405–413.

93. Yui, N., Okano, T., and Sakurai, Y. (1993). Photo-responsive degradation of heterogeneous hydrogels comprising crosslinked hyaluronic acid and lipid microspheres for temporal drug delivery. *J. Controlled Release* **26**, 141–145.

94. Yui, N., Okano, T., and Sakurai, Y. (1992). Inflammation responsive degradation of crosslinked hyaluronic acid gels. *J. Controlled Release* **22**, 105–116.

95. Yui, N., Nihira, J., Okano, T., and Sakurai, Y. (1993). Regulated release of drug microspheres from inflammation responsive degradable matrices of crosslinked hyaluronic acid. *J. Controlled Release* **25**, 133–143.

96. Glass, J. R., Dickerson, K. T., Stecker, K., and Polarek, J. W. (1996). Characterization of a hyaluronic acid-Arg-Gly-Asp peptide cell attachment matrix. *Biomaterials* **17**, 1101–1108.

97. Cooper, M. L., Hansbrough, J. F., and Polarek, J. W. (1996). The effect of an arginine-glycine-aspartic acid peptide and hyaluronate synthetic matrix on epithelialization of meshed skin graft interstices. *J. Burn Care Rehabil.* **17**, 108–116.

98. Burns, J. M., Skinner, K., Colt, J., Sheidlin, A., Bronson, R., Yaacobi, Y., and Goldberg, E. P. (1995). Prevention of tissue injury and postsurgical adhesions by precoating tissues with hyaluronic acid solutions. *J. Surg. Res.* **59**, 644–652.

99. Hu, M., Sabelman, E. E., Lai, S., Timek, E. K., Zhang, F., Hentz, V. R., and Lineaweaver, W. C. (1999). Polypeptide resurfacing method improves fibroblast's adhesion to hyaluronan strands. *J. Biomed. Mater. Res.* **47**, 79–84.

100. diZerega, G. S. (1999). *In* "New Frontiers in Medical Sciences: Redefining Hyaluronan" (G. Abatangelo and P. Weigel, eds.), p. N13. Abbazia di Praglia, Padua, Italy.

101. Luo, Y., Kirker, K. P., and Prestwich, G. D. (2000). Cross-linked hyaluronic acid hydrogel films: New biomaterials for drug delivery. *J. Controlled Rel.* **69**, 169–184.

102. Tomihata, K., and Ikada, Y. (1997). Crosslinking of hyaluronic acid with water-soluble carbodiimide. *J. Biomed. Mater. Res.* **37**, 243–251.

103. Larsen, N. E., and Balazs, E. A. (1991). Drug delivery systems using hyaluronan and its derivatives. *Adv. Drug Delivery Rev.* **7**, 279–293.

104. Larsen, N. E., Lombard, K. M., Parent, E. G., and Balazs, E. A. (1992). Effects of hylan on cartilage and chondrocyte cultures. *J. Orthop. Res.* **10**, 23–32.

105. Pozo, M. A., Balazs, E. A., and Belmonte, C. (1997). Reduction of sensory responses to passive movements of inflamed knee joints by hylan, a hyaluronan derivative. *Exp. Brain Res.* **116**, 3–9.

106. Adams, M. E. (1993). An analysis of clinical studies of the use of crosslinked hyaluronan, hylan, in the treatment of osteoarthritis. *J. Rheumatol.* **20**, 16–18.

107. Larsen, N. E., Pollak, C. T., Reiner, K., Leshchiner, E., and Balazs, E. A. (1993). Hylan gel biomaterial: Dermal and immunologic compatibility. *J. Biomed. Mater. Res.* **27**, 1129–1134.

108. Larsen, N. E., Pollak, C. T., Reiner, K., Leschiner, E., and Balazs, E. A. (1994). Hylan gel for soft tissue augmentation. *In* "Biotechnology and Bioactive Polymers" (C. Gebelein and C. Carrahar, eds.), pp. 25–33. Plenum, New York.

109. Krauss, M. C. (1999). Recent advances in soft tissue augmentation. *Semin. Cutaneous Med. Surg.* **18**, 119–128.

110. Hallen, L., Johansson, C., and Laurent, C. (1999). Cross-linked hyaluronan (hylan B gel): A new injectable remedy for treatment of vocal folds insufficiency-an animal study. *Acta OtoLaryngol.* **119**, 107–111.

111. Piaquadio, D., Jarcho, M., and Glotz, R. (1997). Evaluation of hylan B gels as a soft tissue augmentation implant material. *J. Am. Acad. Dermatol.* **36**, 544–549.

112. Larsen, N. E., Parent, E., and Balazs, E. (1999). *In* "New Frontiers in Medical Sciences: Redefining Hyaluronan," p. N21. Abbazia di Praglia, Padua, Italy.

113. Crescenzi, V., Tomasi, M., and Francescangeli, A. (1999). New routes to hyaluronan-based networks and supramolecular assemblies. *In* "New Frontiers in Medical Sciences: Redefining Hyaluronan" (G. Abatangelo and P. Weigel, eds.), p. N11. Abbazia di Praglia, Padua, Italy.

114. Cascone, M. G., Sim, B., and Downes, S. (1995). Blends of synthetic and natural polymers as drug delivery systems for growth hormone. *Biomaterials* **16**, 569–574.

115. Marriott, C., Martin, G. P., and Brown, M. (1998). Hyaluronic acid drug delivery system. PCT Int. Appl. WO 9812024.

116. Yerushalmi, N., and Margalit, R. (1998). Hyaluronic acid-modified bioadhesive liposomes as local drug depots: Effects of cellular and fluid dynamics on liposome retention at target sites. *Arch. Biochem. Biophys.* **349**, 21–26.

117. Yerushalmi, N., Arad, A., and Margalit, R. (1994). Molecular and cellular studies of hyaluronic acid-modified liposomes as bioadhesive carriers for topical drug delivery in wound healing. *Arch. Biochem. Biophys.* **313**, 267–273.

118. Chio, Y. S., Hong, S. R., Lee, Y. M., Song, K. W., Park, M. H., and Nam, Y. S. (1999). Studies on gelatin-containing artificial skin: II. Preparation and characterization of cross-linked gelatin-hyaluronate sponge. *J. Biomed. Mater. Res.* **48**, 631–639.

119. Oerther, S., Gall, H. L., Payan, E., Lapicque, F., Presle, N., Hubert, P., Dexheimer, J., Netter, P., and Lapicque, F. (1999). Hyaluronate-alginate gel as a novel biomaterial: Mechanical properties and formation mechanism. *Biotechnol. Bioeng.* **63**, 206–215.

120. Oerther, S., Payan, E., Lapicque, F., Presle, N., Hubert, P., Muller, S., Netter, P., and Lapicque, F. (1999). Hyaluronate-alginate combination for the preparation of new biomaterials: Investigation of the behavior in aqueous solutions. *Biochim. Biophys. Acta* **1426**, 185–194.

121. Rehakova, M., Bakos, D., Vizarova, K., Soldan, M., and Jurickova, M. (1996). Properties of collagen and hyaluronic acid composite materials and their modification by chemical crosslinking. *J. Biomed. Mater. Res.* **30**, 369–372.

122. Soldan, M., and Bakos, D. (1997). Complex matrix atelocollagen-hyaluronic acid. *In* "Advances in Medical Physics, Biophysics and Biomaterials," pp. 58–61. Stara Lesna, Slovak Republic.

123. Bakos, D., Soldan, M., and Vanis, M. (1997). Hydroxyapatite-collagen-hyaluronic acid composite as bone defects filler. *In* "Advances in Medical Physics, Biophysics and Biomaterials," pp. 54–56. Stara Lesna, Slovak Republic.

124. Liu, L., Thompson, A. Y., Heidaran, M. A., Poser, J. W., and Spiro, R. C. (1999). An osteoconductive collagen/hyaluronate matrix for bone regeneration. *Biomaterials* **20**, 1097–1108.

125. Murashit, T., Nakayama, Y., Hirano, T., and Ohashi, S. (1996). Acceleration of granulation tissue ingrowth by hyaluyronic acid in artificial skin. *Br. J. Plast. Surg.* **49**, 58–63.

126. Huang-Lee, L. L. H., and Nimni, M. E. (1994). Crosslinked CNBr-activated hyaluronan-collagen matrices: Effects on fibroblast contraction. *Matrix Biol.* **14**, 147–157.

127. Burns, J. W., Burgess, L., Skinner, K., Rose, R., Colt, M. J., and Diamond, M. P. (1996). A hyaluronate based gel for the prevention of postsurgical adhesions: Evaluation in two animal species. *Fertil. Steril.* **66**, 814–821.

128. Bowers, D., Raybon, B., and Wheeless, C. R. (1999). Hyaluronic acid-carboxymethyl-cellulose film and perianastomotic adheisons in previously irradiated rats. *Am. J. Obstet. Gynecol.* **181**, 1335–1338.

129. Hooker, G. D., Taylor, B. M., and Driman, D. K. (1999). Prevention of adhesion formation with use of sodium hyaluronate-based bioresorbable membrane in a rat model of ventral hernia repair with polypropylene mesh—A randomized, controlled study. *Surgery* **125**, 211–216.

130. Asayama, S., Nogawa, M., Takei, Y., Akaike, T., and Maruyama, A. (1998). Synthesis of novel polyampholyte comb-type copolymers consisting of a poly(L-lysine) backbone and hyaluronic acid side chains for a DNA carrier. *Bioconjugate Chem.* **9**, 476–481.

131. Moriyama, K., Ooya, T., and Yui, N. (1999). Hyaluronic acid grafted poly(ethylene glycol) as a novel peptide formulation. *J. Controlled Release* **59**, 77–86.

CHAPTER 46

MODIFICATION OF NATURAL POLYMERS: FIBRINOGEN–FIBRIN

Charles J. Doillon

Fibrin is a natural and biodegradable polymer that supports cells and assists cell differentiation in biological processes. Fibrinogen, the monomer, is purified from plasma and subsequent precipitation procedures. Polymerization occurs in the presence of thrombin and $CaCl_2$. Fibrin matrices resulting from plasma and precipitate are designed as glue, adhesive, and sealant. Furthermore, three-dimensional systems made of cell-seeded fibrin (plasma, precipitate plasma, and protein in crystalline form) can assist tissue engineering and help us to arrive at a better understanding of cell behavior, since the resulting matrices mimic *in vivo* the microgeometry in wound healing and other pathologies. Fibrin can be shaped into gels, foams, sheets, and beads and can be treated by cross-linking methods. Moreover, fibrin is used as a growth factor delivery system.

INTRODUCTION

Fibrinogen–fibrin, an important extracellular matrix component, is involved in hemostasis, wound healing, and cancer. Fibrinogen has a molecular weight between 330,000 and 340,000 and is composed of a dimer of three peptide chains of three different types, α, β, and γ, connected through disulfide bridges [1]. Upon blood coagulation activation, fibrinogen is polymerized into a fibrin network in which blood cells and proteins are physiologically trapped. Fibrinogen, a plasma protein (2–5 mg/ml in blood) is clotted by thrombin (activated prothrombin). Thrombin, a proteolytic enzyme, releases 2 molecules each of fibrinopeptides A and B, which are split off from the resulting fibrin monomer (α and β). These fibrin monomers aggregate mainly because of hydrogen bonding and thus produce the resulting fibrin matrix. Fibrin becomes insoluble by the action of thrombin and plasma factor XIIIa (or fibrin-stabilizing factor, a plasma transglutaminase on fibrinogen) in the presence of calcium ions. Thrombin further activates factor XIII to the transaminase factor XIIIa, cross-linking in the γ-γ dimer. Afterward, fibrin degradation occurs by activation of plasminogen into plasmin.

Fibrin is formed at sites of tissue injury and provides a temporary matrix to support the initial response of endothelial, epithelial, and mesenchymal cells needed for tissue repair. Fibrin plays a bioactive role through specific receptor-mediated interactions with cells. On the other hand, fibrin polymer represents a natural, biodegradable, and biocompatible matrix. There are many methods reported for fibrin preparation (see Table 46.1). The purity of fibrinogen differs from one method to another, and it ranges from blood plasma to the crystalline protein. The result of production or reconstitution of fibrin matrices has been termed "fibrin gel" when a purified fibrinogen source was used, "fibrin clot" with plasma, and "fibrin glue" with cryoprecipitate plasma. However, other terms such as "fibrin sealant" and "fibrin-adhesive tissue" are used as well, to indicate their clinical application or their

Table 46.1. Fibrinogen Purifications and Fibrin Preparations

Fibrinogen (Fbg) source	Anticoagulant	Purification step	Concentration	Sterilization	Thrombin	CaCl$_2$ (mM)	Antifibrinolytic	Ref.
Autologous blood	129 mM Tri-Na citrate (1:9 to blood)	Ethanol precipitation (0°C)	43 mg/ml (33 mg/ml final)		150 NIHU/ml 1:0.3 to Fbg	100	None	[9]
Blood	Citrate	Freezing >1 h, then slow thaw (overnight/4°C)					None	[51]
American Red Cross blood		Freezing cryoprecipitate	120 mg/ml	Tri-(n-butyl) phosphate Tween 80 octoxynol			None	[16]
Blood	10% Na citrate	Ammonium sulfate precipitation Cryoprecipitate			1000 NIHU/ml	40	ε-amino caproic acid (3.33 mg/ml)	[35]
Single-donor cryoprecipitate					Bovine 500–2000 U/ml	100	None	[52]
Human blood pool	10% Na citrate	100% ethanol, then frozen −18°C/20 min			5000 NIHU/ml	40	ε-amino caproic acid (20 mg/ml)	[53]
Freeze-dried blood			95–120 mg/ml fibrinogen		Human 500 NIHU/ml	40	Aprotinin 3000 KIU/ml	[3] Biocol, biotransfusion
Human blood, freeze-dried			70–115 mg/ml fibrinogen	Two steps: vapor, heat	Bovine 500 UI/ml	40	Aprotinin 3000 KIU/ml	Tissucol/Tisseel (Immuno)
Human blood	Na citrate	Cool (4°C) 10% ethanol precipitation		Tri-(n-butyl) phosphate, Tween 80	Bovine 500 NIHU/ml	55	Aprotinin 10,000 KIU/ml	[15]
Human blood pool	Citrate phosphate dextrose	Freeze (−80°C/24 h) then thaw (4°C/12 h)	31 mg/ml			40	Aprotinin 3000 KIU/ml	[7]

production method. For example, fibrin tissue adhesive corresponds to products originally made from plasma proteins to form a fibrin clot.

FIBRIN PREPARATION
COLLECTION OF BLOOD

Autologous or pooled donor blood (Red Cross) is usually collected in 10% sodium citrate (1:9 of a 129 mM trisodium citrate solution to blood) and centrifuged at 3200–4000 rpm for 10 min. Supernatant (plasma) is aseptically separated from blood cells. A Plasma-Saver (Haemometics Co.) can be used to facilitate blood and plasma collection [2,3]. The plasma fraction is aseptically transferred into siliconized tubes and used to extract fibrinogen, mostly by precipitation. Heterologous cryoprecipitate is used in most cases, but autologous cryoprecipitate, fresh plasma, fresh-frozen plasma, or platelet-rich plasma are other alternatives [4]. The American Red Cross (Rockville, MD) produces concentrated human fibrinogen complex (120 mg/ml) from screened, fresh-frozen pooled human plasma. A high concentration of fibrinogen is generally needed for binding adhesive and sealant properties, while low concentrated fibrinogen is more appropriate to improve wound healing [5].

PRECIPITATION PROCEDURES
Cryoprecipitation

Fresh cryoprecipitate for a fibrinogen source in the preparation of a fibrin matrix can be produced routinely. One unit of a patient's blood produces 6–10 ml of autologous fibrin matrix. However, the use of stored cryoprecipitated plasma appears to be superior in efficiency and cost-effectiveness. Autologous methods of obtaining fibrinogen eliminate the risk of blood-borne disease transmission but are associated with more inconvenient donation procedures [6].

The cryoprecipitation method consists of freezing plasma at –20 or –80°C for at least 1 h up to 24 h. Afterward, the container is placed at 4°C to allow slow thawing overnight. Once the content has thawed, it is centrifuged for 10 min at 4000 rpm. The serum fraction is decanted, leaving the cryoprecipitate. The tube is gently shaken to suspend the cryoprecipitate, which is used to produce the fibrin matrix [7].

Combination of Cold and Ethanol Precipitations

Fibrinogen is precipitated from blood plasma by using ethanol (10%) at low temperature (–2°C). After centrifugation, the precipitate can be resolubilized in buffer containing NaCl (0.06 M), ε-amino caproic acid (0.02 M), arginine (0.02 M), and glycine (0.06 M) at pH 7.5. The solution is then diafiltered against the same buffer, sterile-filtered, and concentrated to obtain a protein concentration of 120 mg/liter [8].

Ethanol precipitation (1.4 M) of fibrinogen from plasma can be produced at 0°C. The precipitate is separated by centrifugation (1500g for 20 min) to obtain an average of 28 mg/ml fibrinogen. The supernatant is decanted and the precipitate is dissolved by incubation at 37°C [9].

The combination of ethanol and freezing at –18 to –20°C is a method that has been shown to improve significantly the production of fibrinogen (fibrin tissue adhesive) in comparison to the precipitation by ammonium sulfate [10]. It has been reported that this combination is more efficient, resulting in 3–4 ml for 100 ml of a patient's blood obtained in 60 min [11].

Ammonium Sulfate Precipitation

Proteins are precipitated with saturated ammonium sulfate solution and centrifuged at 3200 rpm for 3 min; the supernatant is discarded and the precipitate is collected. Ammonium precipitation yields more fibrinogen (2 ml for 72 ml of blood obtained in 45 min), but the bonding power of the resulting fibrin is likely to be weaker than that of cryoprecipitate [10,11].

Poly(ethylene glycol) Precipitation

Precipitation of plasma with poly(ethylene glycol) (30% w/v PEG in distilled water) can be another alternative. Comparison studies demonstrated a better recovery of fibrinogen with ammonium sulfate precipitation and with higher bond strengths than those with PEG [12].

THROMBIN ISSUE

Fibrinogen solution is combined with bovine or human thrombin to produce fibrin. Human recombinant thrombin is preferred. Concentrated human thrombin (300-unit vial, American Red Cross) is produced from screened fresh-frozen plasma. The manufacturing process for thrombin also includes viral inactivation by solvent–detergent treatment discussed shortly.

Thrombin is reconstituted with sterile water and serially diluted in 40 mM $CaCl_2$ solution (55–100 mM has been reported) to an average final concentration of 15 U/ml. Thrombin can be replaced by Batroxobin (from snake venom, Pentapharm Ltd, Switzerland) to polymerize fibrinogen, but it must be removed from the mixture with fibrinogen prior to use.

FACTOR XIII ISSUE

Factor XIII is copurified with fibrinogen in most products. However, fibrinogen can be enriched with exogenous factor XIII at 0.15 unit (1.5 μg) per milligram of fibrinogen. This procedure makes the fibrin more stable, but the long-term effect is limited.

ANTIFIBRINOLYTIC AGENTS

Proteolytic inhibitors such as the following aprotinin [3000 kallikrein inhibitor units (KIU)/ml: Trasylol)], tranexamic acid (30 mg/ml), and ε-amino caproic acid (25 mg/ml) can be introduced during gel making. This implement remains controversial. However, aprotinin has greater antifibrinolytic effect when produced as an inhibitor of plasmin. A synergistic effect of tranexamic acid and aprotinin has been reported [13].

APPLICATION DEVICES

Thrombin solution (including $CaCl_2$) and fibrinogen solution are prepared separately, in syringes, for example [14]. Some companies have developed devices to deliver or apply the different solutions. A double-syringe delivery system assures mixing of the components, and a dual-needle design prevents clogging with by permitting convenient one-handed application (Duoflo dispenser kit by Immuno, Inc., Vienna, Austria; an autologous fibrin sealant delivery system by Micromedics, St Paul, MN). A spray set can be installed on these devices to facilitate fibrin application for endoscopic surgery.

STERILIZATION

Sterilization can be achieved by gamma radiation at 2.5 Mrad.

Others use solvent–detergents to attempt to inactivate virus. A mixture of tri(*n*-butyl) phosphate and polysorbate (Tween 80) has been reported [15]. This treatment has been applied to frozen plasma derived from 10% ethanol precipitation. After centrifugation, the fibrinogen-rich precipitate is recovered, solubilized in Tris–citrate–lysine buffer, and concentrated. The concentrate is treated with a mixture of 0.3% tri(*n*-butyl) phosphate and 1% Tween 80. A second precipitation step is performed under the same conditions to further eliminate solvent–detergents. The precipitate is then solubilized in a Tris–NaCl buffer, concentrated, filter-sterilized and freeze-dried [15]. Octoxynol can also be introduced as a detergent [16].

Another procedure consists of filtering the cryoprecipitate and heating the solution at 60.5°C for 10 h, the proteins being protected during this procedure by addition of glucose and sorbitol [8]. The stabilizers are removed after pasteurization by dilution and ethanol precipitation of the fibrinogen (10% at –2°C). After centrifugation, the precipitate is resolubilized in a buffer containing NaCl (0.06 M), ε-amino caproic acid (0.02 M), arginine (0.02 M), and glycine (0.06 M) at pH 7.5. The solution is then diafiltered against the same buffer, filter-sterilized and concentrated to obtain a protein concentration of 120 mg/liter.

Aprotinin is then added to the final concentration at 250 KIU/ml [8]. One advantage of heat treatment is that antigenicity can be modified while keeping the hemostatic performance.

STORAGE

Frozen fibrinogen cryoprecipitate can be stored for up to 5 years at $-30°C$. It can then be thawed for immediate use or stored for up to 5 days in a standard blood bank refrigerator. Fibrin glue such as Tisseel, from Immuno, Ltd (Vienna, Austria, and Baxter, CA) can be stored for 12 months between 2 and 8°C.

PURIFIED FIBRINOGEN: APPLICATION TO THREE-DIMENSIONAL CELL CULTURE SYSTEMS

CELL-EMBEDDED FIBRIN MATRIX

Human fibrinogen (from a blood bank) or bovine fibrinogen (e.g., type I-S, >75% protein clottable from Sigma, St Louis, MO) can be used. Fibrinogen (3.0 mg/ml) solution is mixed to thrombin (50 U/ml) at a ratio of 1:0.02 (v/v) in Hanks's Balanced Salt Solution (HBSS). Fibrin gels are obtained within a minute in wells of different sizes (e.g., 6- and 24-well plates) at 37°C.

Cells can be introduced in four manners:

- Polymerization of fibrinogen simultaneously with cell seeding [17].
- Seeding cells on top of the gel [18] at various cell densities (e.g., 5×10^4 cells/cm^2).
- Making a hole in the middle of the polymerized fibrin matrix by vacuum-sucking with a Pasteur pipette down to the bottom of the wells [19].
- Seeding cells primarily on the bottom of wells and secondarily after cell confluence is reached. A fibrin gel (containing cells) is then overlaid [20].

One application is the design of a three-dimensional angiogenesis system in which endothelial cells can form capillaries within the malleable fibrin matrix. The system consists of a confluent cell monolayer on the bottom of culture wells (24 gelatin-coated multiwell plates) and endothelial cells within the gel. For the latter, fibrinogen dissolved in HBSS is mixed with endothelial cells (5×10^4 per milliliter) and the solution is distributed in wells. Thrombin is added in each well (1.75 mg/ml) at a ratio of 1:0.03 v/v and mixed with the fibrinogen solution by gentle agitation [20].

VARIATIONS AND ALTERNATIVES TO PREPARATION

Coculture System

To mimic endothelial cell–cell interactions, cell–extracellular matrix interactions, wound healing, tumor invasion, or tumor-associated angiogenesis, a variety of three-dimensional culture systems can be developed (i.e., tissue engineering). Among them, a system combines a fibroblast-embedded collagen gel sandwiched in an endothelial-cell-embedded fibrin gel. The procedure consists of preparing separately collagen gels and fibrin gels. Collagen gels are prepared by rapidly raising the pH and ionic strength of a collagen dispersion (3 mg/ml of type I collagen extracted from rat tail tendon) and mixing fibroblasts at 10^4 cells/ml of collagen solution prior to gelation. Next 500 μl of cell-containing collagen solution is distributed in 24 multiwell plates. After gelation at 37°C for 30 min, culture medium is added and gels are released from the plastic with of induce gel contraction; stabilization of the culture to obtain gels smaller than that made of fibrin, occurs in 3 to 5 days. After gelation of a fibrin solution at 37°C (15 min), a second layer of the same solution is laid down onto the preformed fibrin gel [20].

Fibrin gel is prepared as described earlier. Half the volume of the fibrinogen solution (containing cells) is laid down into the well and polymerized to form a first layer. Contracted collagen gel is laid down on the first layer of endothelial-cell-embedded fibrin. A second layer of endothelial-cell-embedded fibrin is overlaid onto the collagen like a sandwich. The reconstituted matrix system is then covered with medium.

Combination of Other Extracellular Matrix Compounds

Glycosaminoglycans, such as hyaluronic acid and heparin, and glycoproteins, such as fibronectin, can be combined with the fibrinogen solution, prior to polymerization with thrombin [19,21]. The concentrations of glycosaminoglycans and glycoproteins vary from 1 to 1000 µg/ml (w/v to fibrinogen solution), while the final amount of fibrinogen is kept as the control fibrin matrix. It has been reported that fibrin interacts with glycosaminoglycans, particularly hyaluronic acid and chondroitin sulfate, altering the rate of fibrin polymer formation and the apparent size of fibrin fibrils, whereas heparan sulfate had no effect [22].

Soluble collagen (from rat tail tendon, or bovine collagen) can be combined to the fibrinogen solution as described for glycosaminoglycans. Ratio of 1:0.02 to 1:0.25 (w/w) results in stable fibrin matrix. Collagen dispersion is prepared as described for collagen gel and mixed within the fibrinogen solution prior to polymerization. Moreover, collagen can be replaced by gelatin at different ratios.

MODIFICATIONS IN PROCEDURES AND APPLICATIONS
FIBRIN AS A GROWTH FACTOR DELIVERY SYSTEM

Fibrin has properties that allow it to slowly release different growth factors to a healing wound. Fibrin is a porous structure in which exchanges can easily occur. Fibroblast growth factor 2 (FGF-2 or basic FGF-bFGF) binds specifically and saturably to fibrinogen and fibrin with high affinity, and this property may have implications for the localization of its effect at sites of tissue injury [23]. Therefore, fibrin can be used as carrier for FGF. For example, FGF-2 has been attached to heparin that was cross-linked to fibrin matrices during coagulation by factor XIII and used successfully in nerve guide tubes [24]. FGF-1 and FGF-2 incorporated into a fibrin film (air-dried fibrin) or fibrin gel induced endothelial cell and fibroblast proliferation *in vitro* and rapid healing in animals [21,25,26]. Among different extracellular matrix analogues, fibrin sustained the biological activity of FGF, particularly that associated with heparin incorporated in fibrin [21,27]. The method consists of diluting the growth factor at the desired concentration (e.g., FGF at 100 µg/ml to fibrinogen solution; FGF-2 at 10 ng/ml) in HBSS, which is mixed with the solutions of fibrinogen prior to polymerization with thrombin and CaCl$_2$. Similar experiments were conducted with FGF-1 (acidic FGF: aFGF) at 8 µg/ml [28] and 11 ng/ml combined with heparin [29]. For example, the combination of fibrin and FGF-1 has been reported to facilitate endothelialization of vascular grafts [30]. Demineralized bone powder comtaining bone morphogenetic proteins can also be introduced in a fibrin matrix to enhance bone healing [5].

CROSS-LINKING PROCEDURES

To reinforce a fibrin matrix, different cross-linking procedures have been proposed. These include methylation with diazomethane [31], use of formaldehyde (1% formaldehyde solution at room temperature for 20–60 min [32]) or glutaraldehyde (0.5% glutaraldehyde aqueous solution obtained after the fibrin film has been equilibrated in Tris buffer [33]), and gamma irradiation at 10 Mrad.

Polymerization of fibrinogen by laser is another alternative. Here the fibrinogen is dyed with indocyanine green and the diode laser energy wavelength is 808 nm [34,35]. This method can be applied to reinforce sutured closures and to improve healing of anastomoses.

BIOMATERIALS
Fibrin Glue: Fibrin Adhesive and Sealant

Among the fibrin glues, Tisseel, mentioned earlier, is a product that has been extensively investigated for various applications. Tisseel consists of freeze-dried powder in a kit together with a solution of human fibrinogen (120 mg/ml), thrombin (500 IU/ml), CaCl$_2$(40 mM/liter), and aprotinin (3000 KIU/ml). The substances in the kit are used to prepare two components: the sealer and the thrombin solution. After preheating at 37°C, the components are reconstituted with their solutions and drawn up in separate syringes. To obtain the sealer solution, protein concentrate is dissolved in the stock solution of fibrinolysis inhibitor (aprotinin) or a dilution of it, where applicable. Dried thrombin is dissolved in CaCl$_2$ solution to yield the thrombin solution. The two components are mixed either immediately before application to the recipient surface or in *in situ* with a special dual syringe (Duploject). The adhesive is formed when the fibrin and thrombin compounds come in contact. Other fibrin glues are based on the same pattern as that of Tisseel [36,37].

Fibrin Foam

Fibrin foam can be obtained either by whipping a dilute fibrinogen solution, and freeze-drying, or by clotting the dissolved fibrinogen simultaneously with the addition of hydrogen peroxide. Foams can be used as filling materials for defects, in which they induce rapid healing and epithelialization.

Fibrin Sheets

Fibrin sheets can be prepared either by air-dried fibrin gel on a support (glass) or from freeze-dried fibrinogen dissolved in citrate buffer. For the latter procedure, the fibrinogen dispersion is then mixed with thrombin and injected onto a support membrane. After incubation for 30 min, the fibrin gel dispersion is pressed between two glass plates for 30 min, with the liquid being removed by pressure. The fibrin sheet is then dried and stored in a desiccator [38]. Very slowly increasing the temperature leads to shrinkage of the length of fibrin film [39]. Bioplast was developed for temporary tissue replacement in which fibrin is produced by adding $CaCl_2$, then heated (120–140°C) under pressure (300–600 kg/cm^2). Glycerol is used as an additive for molding (plasticizer) under high temperature and pressure, while formaldehyde reinforces the clot. Adsorption can vary from 2 weeks to 6 months [32].

Fibrin Beads

Fibrin has been prepared in a form of beads for drug delivery system [40]. The method consists of producing fibrin that is emulsified into the oil phase, forming droplets (oleic acid is the surfactant). After curing for 1 h, glutaraldehyde solution (0.5% v/v) is added to minimize coagulation. The recovery of fibrin beads is simply done by decanting the oil phase and washing the residual with diethyl ether once and then with a mixture of 2-propanol and *n*-hexane (1:3) containing 0.2% w/v Tween 80 [40].

Combination with Other Polymers

To enhance hemostasis, the fibrin sealant can be used in combination with cellulose, collagen, or gelatin sponge. The sponge is usually impregnated with fibrinogen solution upon implantation. Others combine thrombin and eventually factor XIII to gelatin or collagen, and after freeze-drying, they produce a sponge which is then impregnated with a fibrinogen solution upon implantation [41].

On the other hand, un-cross-linked fibrin and factor XIII cross-linked fibrin can be used to coat synthetic polymers [42]. More recently, an adsorbable fibrin adhesive bandage (called "AFAB", developed by the U.S. Army) could be put in place at the time of injury or surgery [43].

OTHER APPLICATIONS

Fibrin has been used as a tissue adhesive (e.g., for skin grafting) and as a hemostatic compound in burn treatment and wound repair. Associated with a collagen-based hemostatic agent, fibrin can control hemorrhage in liver and spleen injuries [44]. Fibrin also seals leaks of air (e.g., pulmonary surgery [45]) or fluid (e.g., ventricular septal defect [46], cerebrospinal fluid leakage [47]), secures anastomoses (colonic, tracheal), and prevents tendon adhesion [48]. Fibrin can be a carrier for local delivery of antibiotics [49], corticoids [38], and immobilize enzymes [50].

CONCLUSION

Further development for significant hemostatic, strengthening, and adhesive properties of the fibrin matrix is still needed to be addressed in tissue engineering and biomaterial point of view. It has been suggested that the concentrations of both fibrinogen and thrombin are key factors in the final product effects on the rate of wound healing. The strength of fibrin sealant is proportional to fibrinogen concentration, and sealant polymerization time is related to thrombin concentration. High concentrations of fibrinogen may inhibit wound healing, but this statement is subjected to controversy. Nevertheless, the need for a variety of formulas would have its own clinical requirements for hemostasis, strength, adhesiveness, and/or healing.

References

1. Doolittle, R. F., Everse, S. J., and Spraggon, G. (1996). Human fibrinogen: Anticipating a 3-dimensional structure. *FASEB J.* **10**, 1464–1470.

2. Tawes, R. L., Jr., Sydorak, G. R., and DuVall, T. B. (1994). Autologous fibrin glue: The last step in operative hemostasis. *Am. J. Surg.* **168**, 120–112.

3. Sirieix, D., Chemla, E., Castier, Y., Massonnet-Castel, S., Fabiani, J. N., and Baron, J. F. (1998). Comparative study of different biological glues in an experimental model of surgical bleeding in anesthetized rats: Platelet-rich and -poor plasma-based glue with and without aprotinin versus commercial fibrinogen-based glue. *Ann. Vasc. Surg.* **12**, 311–316.

4. Martinowitz, U., and Schulman, S. (1995). Fibrin sealant in surgery of patients with a hemorrhagic diathesis. *Thromb. Haemostasis* **74**, 486–492.

5. Lasa, C., Jr., Hollinger, J., Drohan, W., and MacPhee, M. (1995). Delivery of demineralized bone powder by fibrin sealant. *Plast. Reconstr. Surg.* **96**, 1409–1417.

6. Spotnitz, W. D. (1995). Fibrin sealant in the United States: Clinical use at the University of Virginia. *Thromb. Haemostasis* **74**, 482–485.

7. Spotnitz, W. D., Falstrom, J. K., and Rodeheaver, G. T. (1997). The role of sutures and fibrin sealant in wound healing. *Surg. Clin. North Am.* **77**, 651–660.

8. Chabbat, J., Tellier, M., Porte, P., and Steinbuch, M. (1994). Properties of a new fibrin glue stable in liquid state. *Thromb. Res.* **76**, 525–526.

9. Kjaergard, H. K., and Weis-Fogh, U. S. (1994). Important factors influencing the strength of autologous fibrin glue, the fibrin concentration and reaction time. *Eur. Surg. Res.* **26**, 273–278.

10. Siedentop, K. H., Park, J. J., and Sanchez, B. (1995). An autologous fibrin tissue adhesive with greater bonding power. *Arch. Otolaryngol. Head Neck Surg.* **121**, 769–772.

11. Park, M. S., and Cha, C. I. (1993). Biochemical aspects of autologous fibrin glue derived from, ammonium sulfate precipitation. *Laryngoscope* **103**, 193–196.

12. Silver, F. H., Wang, M.-C., and Pins, G. D. (1995). Preparation and use of fibrin glue in surgery. *Biomaterials* **16**, 891–903.

13. Pipan, C. M., Glasheenm, W. P., Matthew, T. L., Gonias, S. L., Hwang, L. J., Jane, J. A., and Spotnitz, W. D. (1992). Effects of antifibrinolytic agents on the life span of fibrin sealant. *J. Surg. Res.* **53**, 402–407.

14. Hartman, A. R., Galanakis, D. K., Honig, M. P., Seifert, F. C., and Anagnostopoulos, C. E. (1992). Autologous whole plasma fibrin gel: Intraoperative procurement. *Arch. Surg. (Chicago)* **127**, 357–359.

15. Burnouf-Radosevich, M., Burnouf, T., and Huart, J. J. (1990). Biochemical and physical properties of a solvent-detergent-treated fibrin glue. *Vox Sang.* **58**, 77–78.

16. Fricke, W. A., and Lambe, M. A. (1993). Viral safety of clotting factor concentrates. *Semin. Thromb. Hemostasis* **19**, 54–56.

17. Gillery, P., Bellon, G., Coustry, F., and Borel, J.-P. (1989). Cultures of fibroblasts in fibrin lattices: Models for the study of metabolic activities of the cells in physiological conditions. *J. Cell. Physiol.* **140**, 483–490.

18. Pepper, M. S., Vassalli, J. D., Orci, L., and Montesano, R. (1993). Biphasic effect of transforming growth factor-beta 1 on in vitro angiogenesis. *Exp. Cell Res.* **204**, 356–363.

19. Fournier, N., and Doillon, C. J. (1994). In vitro effects of extracellular matrix and growth factors on endothelial cell migration and vessel formation. *Cells Mater.* **4**, 399–408.

20. Janvier, R., Sourla, A., Koutsilieris, M., and Doillon, C. J. (1997). Stromal fibroblasts are required for PC-3 human prostate cancer cells to produce capillary-like formation of endothelial cells in a three-dimensional co-culture system. *Anticancer Res.* **17**, 1551–1557.

21. DeBlois, C., Côté, M.-F., and Doillon, C. J. (1994). Heparin-FGF-fibrin complex: In vitro and in vivo applications to collagen-based materials. *Biomaterials* **15**, 665–672.

22. LeBoeuf, R. D., Gregg, R. R., Weigel, P. H., and Fuller, G. M. (1987). Effects of hyaluronic acid and other glycosaminoglycans on fibrin, polymer formation. *Biochemistry* **26**, 6052–6057.

23. Sahni, A., Odrljin, T., and Francis, C. W. (1998). Binding of basic fibroblast growth factor to fibrinogen and fibrin. *J. Biol. Chem.* **273**, 7554–7559.

24. Sakiyama-Elbert, S. E., and Hubbell, J. A. (2000). Development of fibrin derivatives for controlled release of heparin-binding growth factors. *J. Controlled Release* **65**, 389–402.

25. Sirois, E., Côté, M. F., and Doillon, C. J. (1993). Growth factors and biological support for endothelial cell lining: In vitro study. *Int. J. Artif. Organs* **16**, 609–619.

26. Doillon, C. J., DeBlois, C., Côté, M.-F., and Fournier, N. (1994). Bioactive collagen sponge as connective tissue substitute. *Mater. Sci. Eng.* **C2**, 43–49.

27. Roy, F., DeBlois, C., and Doillon, C. J. (1993). Extracellular matrix analogs as carrier for growth factors: In vitro fibroblast behavior. *J. Biomed. Mater. Res.* **27**, 389–397.

28. Pandit, A. S., Wilson, D. J., and Feldman, D. S. (2000). Fibrin scaffold as an effective vehicle for the delivery of acidic fibroblast growth factor (FGF-1). *J. Biomater. Appl.* **14**, 229–242.

29. Zarge, J. I., Huang, P., Husak, V., Kim, D. U., Haudenschild, C. C., Nord, R. M., and Greisler, H. P. (1997). Fibrin glue containing fibroblast growth factor type 1 and heparin with autologous endothelial cells reduces intimal hyperplasia in a canine carotid artery balloon injury model. *J. Vasc. Surg.* **25**, 840–848.

30. Greisler, H. P., Cziperle, D. J., Kim, D. U., Garfield, J. D., Petsikas, D., Murchan, P. M., Applegren, E. O., Drohan, W., and Burgess, W. H. (1992). Enhanced endothelialization of expanded polytetrafluoroethylene, grafts by fibroblast growth factor type 1 pretreatment. *Surgery* **112**, 244–250.

31. Osbahr, A. J. (1980). The polymerization of fibrinogen under the influence of diazomethane modification. *Biomaterials* **1**, 100–102.

32. Wood, C. B., Capperauld, I., and Blumgart, L. H. (1976). Bioplast fibrin buttons for liver biopsy and partial hepatic resection. *Ann. R. Coll. Surg. Engl.* **58**, 401–404.

33. Ho, H.-O., and Chen, C.-Y. (1993). Diffusion characteristics of fibrin films. *Int. J. Pharm.* **90**, 95–100.

34. Ashton, R. C., Jr., Libutti, S. K., Oz, M. C., Lontz, J. F., Lemole, G. M., Jr., and Lemole, G. M. (1992). The effects of laser-assisted fibrinogen bonding on suture, material. *J. Surg. Res.* **53**, 39–44.

35. Gleich, L. L., Wang, Z., Pankratov, M. M., Aretz, H. T., and Shapshay, S. M. (1995). Tracheal anastomosis with the diode laser and fibrin tissue adhesive: An in vitro and in vivo investigation. *Laryngoscope* **105**, 494–496.

36. Dunn, C. J., and Goa, K. L. (1999). Fibrin sealant: A review of its use in surgery and endoscopy. *Drugs* **58**, 863–866.

37. Chan, M. W., Schwaitzberg, S. D., Demcheva, M., Vournakis, J., Finkielsztein, S., and Connolly, R. J. (2000). Comparison of poly-N-acetyl glucosamine (P-GlcNAc) with absorbable collagen (Actifoam), and fibrin sealant (Bolheal) for achieving hemostasis in a swine model of splenic hemorrhage. *J. Trauma* **48**, 454–457.

38. Senderoff, R. I., and Sokoloski, T. D. (1993). A practical approach to evaluate the effect of the aqueous stagnant layer on drug release from polymeric devices. *Int. J. Pharm.* **91**, 235–240.

39. Nakatani, A. I., and Ferry, J. D. (1988). Thermally induced conformational changes in fibrin film. *Thromb. Res.* **52**, 361–366.

40. Ho, H.-O., Hsiao, C.-C., Chen, C.-Y., Sokoloski, T. D., and Sheu, M.-T. (1994). Fibrin-based drug delivery systems. II. The preparation, and characterization of microbeads. *Drug Dev. Ind. Pharm.* **20**, 535–540.

41. Sugitachi, A., and Sakamoto, I. (1985). A new approach to wound healing. *In* "Biomaterials in Artificial Organs" (J. P. Paul, J. M. Courtney, J. D. S. Gaylor, and T. Gilchrist, eds.), pp. 318–326. VCH Publishers, Weinheim.

42. Bense, C. A., and Woodhouse, K. A. (1999). Plasmin degradation of fibrin coatings on synthetic polymer substrates. *J. Biomed. Mater. Res.* **46**, 305–314.

43. Cornum, R., Bell, J., Gresham, V., Brinkley, W., Beall, D., and MacPhee, M. (1999). Intraoperative use of the absorbable fibrin adhesive bandage: Long term effects. *J. Urol.* **162**, 1817–1820.

44. Schelling, G., Block, T., Gokel, M., Blanke, E., Hammer, C., and Brendel, W. (1988). Application of a fibrinogen-thrombin-collagen-based hemostyptic, agent in experimental injuries of liver and spleen. *J. Trauma* **28**, 472–473.

45. Bergsland, J., Kalmbach, T., Balu, D., Feldman, M. J., Caruana, J. A., and Gage, A. A. (1986). Fibrin seal—An alternative to suture repair in experimental, pulmonary surgery. *J. Surg. Res.* **40**, 340–345.

46. Seguin, J. R., Frapier, J.-M., Colson, P., and Cahptal, P. A. (1992). Fibrin sealant for early repair of acquired ventricular septal defect. *J. Thorac. Cardiovasc.* **104**, 748–754.

47. Sierra, D. H. (1993). Fibrin sealant adhesive systems: A review of their chemistry, material properties and clinical applications. *J. Biomater. Appl.* **7**, 309–312.

48. Frykman, E., Jacobsson, S., and Widenfalk, B. (1993). Fibrin sealant in prevention of flexor tendon adhesions: An experimental study in the rabbit. *J. Hand Surg. [Am.]* **18A**, 68–70.

49. Greco, F., de Palma, L., Spagnolo, N., Rossi, A., Specchia, N., and Gigante, A. (1991). Fibrin-antibiotic mixtures: an in vitro study assessing the possibility of using a biologic carrier for local drug delivery. *J. Biomed. Mater. Res.* **25**, 39–51.

50. Dillon, J. G., Wade, C. W., and Daly, W. H. (1976). Enzyme immobilization on fibrin. *Biotechnol. Bioeng.* **18**, 133–139.

51. Sierra, D. H., Nissen, A. J., and Welch, J. (1990). The use of fibrin glue in intracranial procedures: Preliminary results. *Laryngoscope* **100**, 360–363.

52. Stechison, M. T. (1992). Rapid polymerizing fibrin glue from autologous or single-donor blood: Preparation and indications. *J. Neurosurg.* **6**, 626–628.

53. Park, J. J., Cintron, J. R., Siedentop, K. H., Orsay, C. P., Pearl, R. K., Nelson, R. L., and Abcarian, H. (1999). Technical manual for manufacturing autologous fibrin tissue adhesive. *Dis. Colon Rectum* **42**, 1334–1338.

CHAPTER 47

MODIFICATION OF NATURAL POLYMERS: CHITOSAN

Chad D. Brown and Allan S. Hoffman

Chitosan is a natural polymer that has been investigated for tissue engineering applications because of its biodegradability, biocompatibility, availability of reactive groups, and structural similarities to glycosaminoglycans. Chitosan can be easily fabricated into porous scaffolds, hydrogels, fibers, and microspheres. The use of chitosan as a tissue engineering scaffold has been investigated for a variety of tissues, including bone, liver, neural tissue, vascular grafts, cartilage, and skin. Finally, chitosan has been utilized for its ability to deliver proteins, such as growth factors, in a controlled and stable manner, which may prove beneficial in many tissue engineering strategies for the promotion of tissue growth and angiogenesis.

BACKGROUND

Chitosan is a natural linear copolymer of $\beta(1 \rightarrow 4)$-linked glucosamine and N-acetyl-D-glucosamine and is obtained from the base-catalyzed partial N-deacetylation of chitin (Fig. 47.1) [1,2]. Chitin is the second most abundant natural polymer next to cellulose and is the major constituent of the shells of shrimp and crab [3]. Chitosan can be produced with a wide range of molecular weights (inherent viscosity) and average degrees of deacetylation. Chitosan is insoluble in neutral aqueous solution; however, chitosan becomes protonated and dissolves in dilute acidic solutions, such as glutamic acid, hydrochloric acid, lactic acid, acetic acid, formic acid, and butyric acid [4,5]. Thus, tissue engineering scaffolds and protein delivery devices of chitosan can be fabricated under milder conditions than those of poly(lactide-co-glycolide), which require organic solvents such as methylene chloride. Owing to its positive charge, chitosan has been utilized to deliver negatively-charged compounds such as acidic proteins, glycosaminoglycans, and DNA [6–16]. The recent increased use of chitosan in biomedical applications has been motivated by its biodegradability and apparent biocompatibility [17]. Chitosan is degraded by lysozyme to the natural by-products of glucosamine and N-acetyl glucosamine, which are incorporated into glycoproteins or excreted as carbon dioxide [17–20]. The cytotoxicity of chitosan is dose dependent and has been shown to decrease with lower molecular weights and lower degrees of deacetylation [21]. The toxicity of chitosan also varies depending on the type of counterion present [22]. Further, chitosan has been used as a component in wound healing formulations and is an ingredient in the 3M product Tegasorb [5]. Chitosan has been shown to accelerate early phase healing of open skin wounds by increasing the rate of infiltration of polymorphonuclear cells and the production of collagen by fibroblasts [23]. Another attractive feature of chitosan is the availability of reactive hydroxyl and amine groups for the conjugation of cell ligands, drugs, or polymer grafts [24–26]. More recently, chitosan has been investigated as a tissue engineering scaffold because of its structural resemblance to

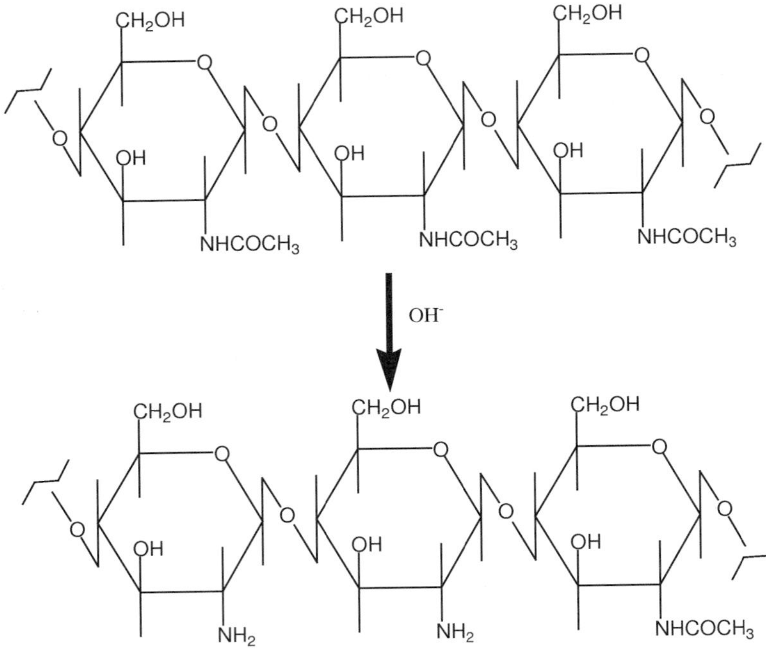

Fig. 47.1. Base-catalyzed N-deacetylation of chitin to
produce chitosan: β(1 → 4)-linked glucosamine-co-
N-acetyl-ᴅ-glucosamine.

glycosaminoglycans, an important structural element of the extracellular matrix of many tissues [27–29].

PREPARATION OF PARTIALLY DEACETYLATED CHITOSAN

Crude chitin can be obtained from the shells of shrimp and crab after proteolysis with sodium hydroxide and removal of calcium carbonate with hydrochloric acid. There are two distinct methods by which chitin can be N-deacetylated to produce chitosan: the alkali fusion procedure and the aqueous sodium hydroxide procedure [1,2]. The alkali fusion procedure involves the fusion of chitin with potassium hydroxide pellets in a nickel crucible under a stream of nitrogen. The melt is stirred for 30 min at 180°C. The hot melt is then poured slowly into ethanol. The precipitate is collected and washed with ethanol and water until the washings are neutral. The chitosan is purified by dissolution in 5% acetic acid followed by precipitation in dilute alkali. This step is repeated three times to obtain crude chitosan. The crude chitosan can be converted to the hydrochloride salt by dissolution in 0.1 N hydrochloric acid at 50°C, followed by the addition of concentrated hydrochloric acid to precipitate the chitosan. After cooling to room temperature, the chitosan salt can be redissolved by boiling in water. After cooling, concentrated hydrochloric acid is added to reprecipitate the chitosan. The mixture is subsequently heated until the solid dissolves. The solution is allowed to cool and the chitosan salt is filtered out. The product is washed with ethanol and ether before being dried under vacuum. This procedure results in chitosan that has an average deacetylation of approximately 95%. Further deacetylation can be achieved by repeating the purification step.

The aqueous sodium hydroxide procedure involves treating the chitin with a 40% sodium hydroxide solution for 6 h at 115°C in the absence of air. The mixture is cooled, filtered, and washed with water until the washings are neutral. This crude chitosan is purified by dissolution in 10% acetic acid. This solution is centrifuged, and the supernatant is treated dropwise with 40% sodium hydroxide to precipitate the chitosan. When the solution reaches pH 7, the solution is centrifuged and the chitosan is repeatedly washed with water, ethanol, and ether before being dried under vacuum. Chitosan produced in this manner will have an average deacetylation of 85%.

To achieve varying degrees of deacetylation, chitosan can be N-acetylated with acetic anhydride [30–32]. Chitosan is dissolved in 1–2% acetic acid and diluted with either methanol or ethanol. Acetic anhydride is added at 0.1–1.0 mole of acetic anhydride per mole of free amine. The mixture is allowed to stand overnight at room temperature. The degree of acetylation of the resulting viscous solution or gel can be determined by means of nuclear magnetic resonance (NMR) spectroscopy.

CHITOSAN SCAFFOLD PREPARATION

Freeze-drying techniques have been used to prepare porous tissue engineering scaffolds from both chitosan and chitin [32–34]. Madihally *et al.* have investigated the effect of freezing conditions on pore size and morphology [33]. Chitosan solutions were prepared in 0.2 M acetic acid at chitosan concentrations between 1 and 3%. Scaffolds were prepared by adding 3–5 ml of solution to flat glass tubes, which were subsequently frozen at –20, –78, or –196°C and lyophilized. These scaffolds were then rehydrated in either dilute sodium hydroxide or ethanol to stabilize the scaffold by removing the acetate counterions. It was found by scanning electron microscopy (SEM) that the pore diameter could be controlled between 40 and 250 μm by varying the freezing temperature, with lower temperatures providing smaller pore sizes (Fig. 47.2). In another study, Oungbho and Müller examined the effect of glutaraldehyde cross-linking on chitosan scaffold morphology [32]. Briefly, 2.5% by weight chitosan was dissolved in 1% acetic acid. Glutaraldehyde was added at 1.33% the

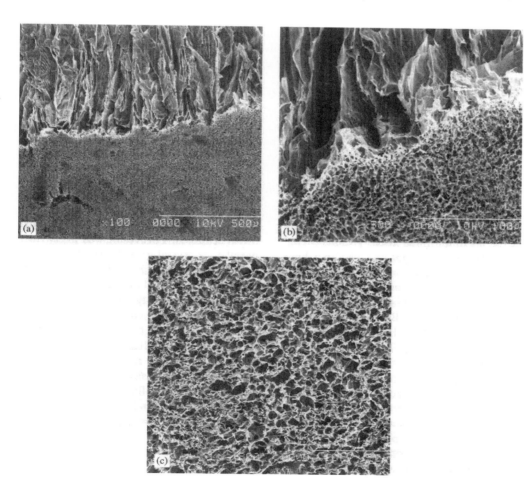

Fig. 47.2. SEM images of chitosan scaffolds prepared by lyophilization of a 2% by weight chitosan solution in a cylindrical glass tube. (a) Low-magnification view of fracture zone showing surface and interior pore structure. (b) Higher-magnification of fracture zone. (c) High-magnification of pore structure at glass-contacting surface. Reprinted from Biomaterials, vol. 20, S. V. Madihally and H. W. T. Matthews, "Porous chitosan scaffolds for tissue engineering," pp. 1133–1142. Copyright 1999, with permission from Elsevier Science.

concentration of chitosan weight and allowed to react at room temperature for 24 h before lyophilization to produce a porous scaffold. It was determined that glutaraldehyde-cross-linked chitosan sponges were less porous than control chitosan sponges. Finally, Chow and Khor have examined the fabrication of porous chitin matrices [34]. Chitin was dissolved in a 5% lithium chloride–dimethyl acetamide solution at a concentration of 0.5% by weight. Approximately 100 ml of this chitin solution was loaded with 0.3–3.0 g of calcium carbonate, which forms a stable suspension. These solutions were allowed to evaporate for 2 days to produce chitin–calcium carbonate gels. These gels were then treated with 1 N HCl at room temperature for 2 h to generate carbon dioxide formation within the gel. The resulting gels were then washed in deionized water to remove residual solvents and calcium by-products, and dried in air. This technique resulted in chitin matrices with pore sizes between 100 and 1000 μm.

Most strategies to produce chitosan scaffolds involve a rinsing step in ethanol or sodium hydroxide to stabilize the scaffold. This is an undesired step when scaffolds have been loaded with protein for *in vivo* release. Recently, we have developed chitosan–glycerol films where glycerol acts as a hydrogen bonding agent with chitosan and aids in the removal of acetic acid during the drying of the film, thus stabilizing the film without a rinsing step [35]. Briefly, a 10-mg/ml solution of 84% deacetylated chitosan in 1% acetic acid and a 10-mg/ml solution of glycerol in distilled water were prepared. These solutions were mixed at a 1:1 ratio, and 5 ml was added to a 10-ml Teflon beaker. The solutions were allowed to evaporate at room temperature and form thin uniform films. The films were further dried under vacuum to remove residual solvent. Films containing glycerol demonstrated less swelling and were mechanically more stable than films prepared without glycerol.

CHITOSAN FORMULATIONS FOR TISSUE ENGINEERING APPLICATIONS

This section presents examples of a variety of formulations containing chitosan that have been studied for the tissue engineering of various tissues. A summary of chitosan modifications and tissues investigated is presented in Table 47.1.

BONE

The use of chitosan for the tissue engineering of bone has been widely investigated [36–43]. This is in part due to the apparent osteoconductive properties of chitosan. In fact, mesenchymal stem cells cultured in the presence of chitosan have demonstrated an increase in differentiation to osteoblasts compared to cells cultured in the absence of chitosan [39]. Further, it is speculated that chitosan may enhance osteoconduction *in vivo* by entrapping growth factors at the wound site [40]. Tamura and Yoshihara have investigated the use of carboxymethyl–chitin/hydroxyapatite composites as bone repair scaffolds [37,38]. Carboxymethyl–chitin is a water-soluble derivative (pH 7.4) of chitin prepared by the treatment of chitin with monochloroacetic acid. This modification was achieved by suspending chitin powder in 40–50% sodium hydroxide with sodium dodecyl sulfate. This suspension

Table 47.1. *Summary of Chitosan Modifications Utilized in Tissue Engineering Applications*

Modification	Tissue	Ref.
Carboxymethylated	Bone	[37,38,41]
Tripolyphosphate cross-linked	Bone	[36]
Glutaraldehyde cross-linked	Liver	[44,45]
Sugar conjugation	Liver, neural	[45,48]
Protein blending	Liver, neural, skin	[10,29,46,47,55–57]
Glycosaminoglycans	Cartilage, skin, vascular	[11–13,50,55–57]
Hydroxypropylated	vascular	[51,52]

was frozen at −20°C for 12 h. The frozen alkali chitin was thawed in 2-propanol, and monochloroacetic acid powder was added until the solution was neutralized. The resulting carboxymethyl–chitin was purified by extraction into water, and salts were removed by dialysis. The purified carboxymethyl–chitin obtained was approximately 70% carboxymethylated. Carboxymethyl–chitin/hydroxyapatite composites were prepared by mixing an aqueous solution of 3% carboxymethyl–chitin with 20% by weight hydroxyapatite. The mixture was frozen in liquid nitrogen and freeze-dried. The resulting composite scaffold contained pores with diameters less than 100 μm. These scaffolds, along with scaffolds prepared from carboxymethyl–chitin and hydroxyapatite alone were inserted into 4-mm holes drilled into the femur of rabbits. After 12 weeks postsurgery, bone defects treated with carboxymethyl–chitin/hydroxyapatite composites demonstrated a significant increase in bone thickness and extension ratio.

Muzzarelli *et al.*, who used dicarboxymethyl chitosan to electrostatically complex bone morphogenetic protein 7 (BMP-7), have investigated its effects in an osteoporotic model [41]. Scaffolds were prepared by dissolving dicarboxymethyl chitosan and BMP-7 in water (pH 6). This solution was rotoevaporated at 50°C for 20 h to produce a gel. Femoral defects in rats were filled with dicarboxymethyl chitosan–BMP-7 complexes or with dicarboxymethyl chitosan, or were left open as a control. Defects treated with dicarboxymethyl chitosan–BMP-7 complexes demonstrated improved bone regeneration with bone tissue that was well organized and mineralized.

Finally, Park *et al.* have investigated the use of a chitosan sponge loaded with platelet-derived growth factor (PDGF) for its ability to regenerate periodontal bone [36]. In this study, chitosan was cross-linked by using tripolyphosphate and freeze-dried to obtain a chitosan sponge with 100-μm pores. The cross-linked sponges were submersed overnight in a solution containing PDGF and freeze-dried. The resulting sponges were loaded with 100–400 ng of PDGF. These sponges were subsequently seeded with rat osteoblastic cells. Sponges containing PDGF exhibited greater cellular attachment and proliferation than sponges without PDGF or polystyrene controls. PDGF released from the chitosan sponge with a large initial burst followed by a lower continuous release for 6 days. Sponges containing PDGF were implanted into calvarial defects in rats and monitored for their bone regenerative properties. Defects treated with sponges without PDGF demonstrated formation of inner and outer bone tables at four weeks; however, complete reunion of the bone was not obtained. Defects treated with sponges loaded with PDGF demonstrated an increase in new bone formation and rapidly calcified. It was also observed that the sponge material had begun to degrade and was being replaced with newly formed bone.

LIVER

Several investigators have modified chitosan for use in a tissue-engineered liver [29,44–46]. Kawase *et al.* cultured rat hepatocytes on glutaraldehyde-cross-linked chitosan gels [44]. Briefly, 500 mg of 100% deacetylated chitosan was dissolved in 20 ml of a 1% acetic acid–50% methanol solution. This solution was cross-linked with 2 ml of 5% glutaraldehyde. Approximately 3 ml of this solution was added to a 6-well plate and allowed to gel. The gels were then washed with 50 mM PBS–50% methanol. Rat hepatocytes cultured *in vitro* on these gels retained their spherical shape and had higher urea synthesis activity than control cells plated on a collagen surface. Furthermore, hepatocytes cultured on glutaraldehyde-cross-linked chitosan produced less lactate dehydrogenase, a marker of cell damage. This group of investigators then examined the effect of modifying chitosan with various sugar moieties [45]. This was accomplished by preparing a chitosan solution in an acetic acid–methanol solvent as described earlier and reacting the free amine groups on chitosan with an excess of aldehyde groups of the sugar to form a Schiff base. The Schiff base was then reduced overnight with sodium borohydride in the dark at room temperature. Finally, the sugar-modified chitosan was glutaraldehyde-cross-linked as already described. It was found that chitosan modified with fructose, lactose, and maltose demonstrated increased cell attachment over control chitosan surfaces. These surfaces also maintained high urea synthesis and lower rates of lactate dehydrogenase secretion. Further, it was shown that collagen films that were glutaraldehyde-cross-linked to thin chitosan–fructose films demonstrated cell behavior similar to that of the chitosan–fructose films and were superior to films of collagen alone.

A second group of researchers examined the effect of blending 89% deacetylated chitosan with collagen, albumin, and gelatin to improve the biocompatibility and surface roughness of the chitosan [29,46]. Chitosans of various molecular weights were prepared as a 2% by weight solution in 2% acetic acid and blended with a 2% by weight protein solution at a 7:3 ratio of chitosan to protein. Polystyrene petri dishes were coated with the blended chitosan–protein solution and heated at 50°C to form uniform films. The films were then treated with 0.5 M sodium hydroxide for 30 min. The films were washed in deionized water and ethanol before UV sterilization. It was found that rat hepatocyte attachment was greatest on chitosan–collagen films, while fetal porcine hepatocytes preferred chitosan–albumin films. Urea and total protein synthesis was highest on scaffolds containing chitosan of a medium molecular weight (750,000 Da). These researchers then used chitosan–albumin microspheres, (as described later) to deliver endothelial cell growth factor (ECGF) to subcutaneous sites of a rat. The release of ECGF induced neovascularization of the subcutaneous sites.

Chitosan–albumin scaffolds containing fetal porcine hepatocytes were then transplanted to the neovascularized subcutaneous sites. These cell-seeded scaffolds formed liver-like tissue structures after 2 weeks *in vivo* and contained 30–40% the mass of the transplanted cells. Scaffolds transplanted to subcutaneous sites that were not pretreated with chitosan–albumin–ECGF microspheres died within 1 day of transplantation.

NEURAL TISSUE

The ability of chitosan to aid in the development of engineered neural tissue has been studied by several investigators [47–49]. Pappas *et al.* studied the effect of chitosan blended with collagen, albumin, or gelatin as scaffolding material for bovine chromaffin cell attachment [47]. First 2% by weight of 93% deacetylated chitosan was dissolved in 2% acetic acid. This solution was blended with 2% collagen, albumin, or gelatin at a ratio of chitosan to protein of 7:3. The blended solutions were then extruded through a 23-gauge needle into a 0.5 M sodium hydroxide–methanol solution under high flow rate. The resulting porous fibers were allowed to gel completely for 20 min before being washed with distilled water and ethanol. The fibers were UV-sterilized and stored wet until use. Bovine chromaffin cells were found to attach to the collagen-blended chitosan scaffolds more readily than the other protein blends. Further, after 14 days *in vivo*, the bovine chromaffin cells were found to be not only adhered to the chitosan–collagen scaffold but also well integrated into the scaffold material.

Another group has examined the effect of covalently linking chitosan to agarose to examine the effect on neurite extension into the hydrogel scaffold [48]. In this study 85% deacetylated chitosan with a molecular weight of 130 kDa was linked to agarose using 1,1′-carbonyldiimidazole (CDI). A 4-ml gel of 1.5% agarose was dehydrated in a series of acetone washing steps. The agarose gel was then activated with 30 mg/ml CDI in dry acetone for 1 h. The gel was subsequently rehydrated in 100 mM sodium bicarbonate buffer (pH 8.5). Next, an aqueous solution of 0.435 μM chitosan was added to the activated agarose and allowed to couple overnight with gentle agitation. The solution was gelled at 4°C, washed with deionized water, and lyophilized before being reconstituted as a 1% solution in phosphate-buffered saline (PBS). Sections of embryonic dorsal root ganglia were cultured within the gels and examined for neurite extension. It was found that the neurite extensions in the chitosan-coupled agarose gels were 41% longer than in the agarose controls.

VASCULAR GRAFTS

Vascular grafts are another area in which chitosan has found utility as a tissue engineering scaffold. The ability of chitosan to support tissue growth for a replacement vascular graft as well as to modify existing grafts has been investigated [50–52]. One group has investigated the use of composite chitosan and glycosaminoglycan membranes to support endothelial cell proliferation [50]. Porous chitosan–glycosaminoglycan membranes were formed by lyophilization with pore diameters, between 40 and 500 μm. These membranes were shown to support endothelial proliferation while inhibiting smooth muscle cell growth.

Another group has investigated the use of a commercially available (Nippon Soda Co., Tokyo), water-soluble form of chitosan, hydroxypropyl chitosan [51,52]. This derivative of chitosan contains hydroxypropyl groups covalently linked through either the chitosan hydroxyl or amine group. Hydroxypropyl chitosan was used to load the pores of a 3-mm Gore-Tex poly(tetrafluoroethylene) graft with basic fibroblast growth factor (bFGF). The release of bFGF from the graft was hypothesized to increase the endothelialization of the graft. Briefly, 4% hydroxypropyl chitosan was dissolved in a 180-μg/ml solution of bFGF and loaded into 5-cm-long graft sections under sufficient pressure to cause the solution to leak from the pores. The grafts were then freeze-dried, and the loading of bFGF was determined to be 2.71 μg/graft. These grafts were then implanted between the artery and vein of dogs. After 6 weeks, the grafts were harvested and examined for reendothelialization. It was found that grafts containing the hydroxypropyl chitosan–bFGF mixture were significantly more endothelialized than control grafts. In vivo release studies in rabbits demonstrated that 40% of the bFGF was released from the graft after 24 h when hydroxypropyl chitosan was used as the carrier. Grafts containing bFGF without a chitosan carrier released all the bFGF within 24 h.

CARTILAGE

Sechriest et al. have utilized a scaffold comprised of ionically cross-linked chondroitin sulfate A and chitosan to support chondrogenesis [11]. A 1% solution of 85% deacetylated chitosan was dissolved in 2% acetic acid. Then 500 μl of this solution was added to a 24-well polystyrene tissue culture plate and heated for 6 h. Next, 500 μl of a 10% chondroitin sulfate A solution in distilled water was added to the chitosan residue and allowed to ionically cross-link for 18 h. The composite membranes were then washed with PBS prior to cell seeding with bovine articular chondrocytes. Chondrocytes seeded on chitosan–chondroitin sulfate-A matrices exhibited discrete points of attachment and maintained a round morphology suggesting receptor-mediated attachment. Chondrocytes on control polystyrene surfaces exhibited a fibroblastic morphology. Further, chondrocytes on chitosan–chondroitin sulfate A surfaces produced higher amounts of collagen II and maintained limited mitosis, both characteristics of differentiated chondrocytes.

Hanson and Grzanna have also investigated the use of chitosan to aid in the repair of damaged cartilage [53,54]. Chitosan scaffolds were prepared by freezing solutions of 1–1.5% by weight chitosan in 1 M acetic acid. The scaffolds were subsequently placed in 1 M sodium hydroxide and washed with phosphate buffer. The resulting gels were then refrozen and lyophilized before seeding with chondrocytes. Scaffolds made from solutions with lower concentrations of chitosan demonstrated better cellular permeation and attachment. Further, the chondrocytes appeared to proliferate in the scaffold and synthesized extracellular matrix and proteoglycans. Additional scaffolds were prepared by mixing 5–40% by weight arabinogalactans with a 2% chitosan solution in acetic acid. These solutions were neutralized with sodium hydroxide and subsequently washed and lyophilized. Chondrocytes adhered and proliferated on these scaffolds, forming cell clusters after 2 weeks in culture. The chitosan scaffolds containing chondrocytes immunostained postively for collagen II and keratan sulfate. Further, the chondrocytes synthesized proteoglycans similar to aggrecan, a marker for hyaline cartilage.

SKIN

Chitosan has been used as a structural element in the fabrication of a tissue-engineered skin replacement [55–57]. These scaffolds were prepared by dissolving type I and III bovine collagen, chondroitin sulfate, and 95% deacetylated chitosan in 0.1% acetic acid. The solutions were mixed, and 1 ml was added to a 12-well plate. The mixture was frozen overnight at −70°C followed by lyophilization. The resulting scaffolds were then treated with 70% ethanol and rinsed with Dulbecco's Modified Eagle's Medium. These scaffolds were coseeded with human fibroblasts and human endothelial cells and cultured for 10 days. These cellularized scaffolds were then seeded with human keratinocytes and cultured for an additional 21 days. It was found that scaffolds coseeded with fibroblasts and endothelial cells developed new extracellular matrix originating from the fibroblasts. Further, the endothelial cells formed capillary-like structures within this extracellular matrix. These tissues also demonstrated differentiated layers typical of native skin including a basal layer, stratum spinosum, stratum granulosum, stratum corneum, and a multistratified epidermis. This construct has proven valuable as an *in vitro* angiogenic model.

Another group has used chitosan–heparin membranes to accelerate the reepithelialization of open skin wounds [12,13]. The membranes were formed by dissolving the hydrochloric acid salt of 84% deacetylated chitosan in distilled water at a 1% concentration by weight. Twenty milliliters of this solution was added to a petri dish, and the solvent was evaporated at 70°C for 16 h. The membrane was then neutralized with 1 M sodium hydroxide for 3 h and washed with distilled water. After the membrane had been dried at 70°C for 2 h, 30 ml of a 1% heparin solution in 0.2 M phosphate buffer (pH = 6.4) were added and allowed to incubate overnight. The membrane was then rinsed with distilled water and sterilized by means of ethylene oxide gas. Wounds created *in vitro* from mastectomy specimens were covered with chitosan–heparin membranes, chitosan membranes, or 2% fetal calf serum. It was found that 9 out of 10 wounds treated with chitosan–heparin membranes completely healed, versus only 3 wounds treated with chitosan alone and zero wounds treated with only 2% fetal calf serum. Further, wounds treated with chitosan–heparin membranes were completely reepithelialized. Treatment of *in vivo* wounds of human subjects undergoing skin grafting procedures demonstrated complete reepithialization of wounds treated with chitosan–heparin membranes 12 days postsurgery versus 15 days for wounds left untreated. It is hypothesized that chitosan–heparin complexes aid in reepithialization by binding growth factors in the wound site and providing stabilization and protection from degradation. This technique may prove helpful in the treatment of hard-to-heal wounds.

STRATEGIES TO PROMOTE ANGIOGENESIS USING CHITOSAN

The integration of vasculature into implanted engineered tissues is essential in the ultimate viability of the tissue. Chitosan formulations have been used to deliver angiogenic growth factors and have demonstrated the stimulation of angiogenesis *in vivo*. Chitosan–albumin microspheres have been fabricated to deliver endothelial cell growth factor (ECGF), an angiogenic factor [10]. The microspheres were formed by preparing a 2% by weight solution of 89% deacetylated chitosan in 2% acetic acid. This solution was blended with a 2% by weight solution of bovine albumin at a 7:3 ratio (chitosan to albumin). The chitosan–albumin blend was delivered dropwise through a 23-gauge needle into a 1:1 solution of 0.5 M sodium hydroxide–methanol solution and allowed to gel completely. The resulting microspheres ranged from 400 to 600 μm in diameter. The microspheres were subsequently loaded with 6 mg of ECGF per gram of microspheres by incubating them overnight in an ECGF solution. *In vitro* release studies demonstrated that after an initial burst releasing 25% of the loaded protein, approximately 5% of the loaded ECGF was released per day. Microspheres loaded with ECGF were injected into the groin fascia of rats. These formulations led to extensive neovascularization after 7 days of implantation, while rats treated with injections of ECGF without a chitosan carrier demonstrated no significant vascularization.

Another group found that the incorporation of nerve growth factor into chitosan channels designed for nerve regeneration resulted in increased angiogenesis but had no effect on nerve regeneration [58]. Chitosan channels were formed by dipping 22-gauge stainless steel needles into a 36-mg/ml solution of 81% deacetylated chitosan in 5% acetic acid followed by drying in air. This procedure was repeated eight times. Next, the chitosan-coated needle was heated to prevent the swelling and dissolution of the chitosan. After 2 h of heating at 100°C, the chitosan channel was pulled off the needle. The channels were placed between the distal and proximal ends of severed sciatic nerves of mice. It was found that there was no improvement in nerve regeneration; however, chitosan channels containing nerve growth factor did lead to an increase in the number and size of blood vessels in comparison to chitosan channels without nerve growth factor.

SUMMARY

Chitosan is a natural biodegradable and biocompatible polymer that can easily be fabricated into various tissue engineering scaffolds. The versatility of chitosan to be modified and combined with other polymers has allowed its use in developing a wide range of tissues such as bone, liver, neural tissue, vascular grafts, cartilage, and skin. Chitosan's ability to stabilize and deliver proteins has allowed the incorporation of growth factors into many of these strategies and has led to methods of promoting angiogenesis in tissue-engineered structures.

REFERENCES

1. Horton, D., and Lineback, D. R. (1965). N-Deacetylation: Chitosan from chitin, *In* "Methods in Carbohydrate Chemistry," (R. L. Whistler, J. N. BeMiller, and M. L. Wolfrom, eds.), pp. 403–406. Academic Press, New York and London.

2. Horowitz, S. T., Roseman, S., and Blumenthal, H. J. (1957). The preparation of glucosamine oligosaccharides. I. Separation. *J. Am. Chem. Soc.* **79**, 5046–5049.

3. Sandford, P. A., and Steinnes, A. (1991). Biomedical applications of high-purity chitosan. *In*: "Water-Soluble Polymers: Synthesis, Solution Properties, and Applications" (S. W. Shalaby, C. L. McCormick, and G. B. Butler, eds.), pp. 430–445. American Chemical Society, Washington, DC.

4. Demarger-Andre, S., and Domard, A. (1994). Chitosan carboxylic acid salts in solution and in the solid state. *Carbohydr. Polym.* **23**, 211–219.

5. Illum, L. (1998). Chitosan and its use as a pharmaceutical excipient. *Pharm. Res.* **15**, 1326–1331.

6. Jameela, S. R., Misra, A., and Jayakrishnan, A. (1994). Cross-linked chitosan microspheres as carriers for prolonged delivery of macromolecular drugs. *J. Biomater. Sci. Polymer Ed.* **6**, 621–632.

7. Calvo, P., *et al.* (1997). Novel hydrophilic chitosan-polyethylene oxide nanoparticles as protein carriers. *J. Appl. Polym. Sci.* **63**, 125–132.

8. Lueben, H. L., *et al.* (1997). Mucoadhesive polymers in peroral peptide drug delivery. IV. Polycarbophil and chitosan are potent enhancers of peptide transport across intestinal mucosae *in vitro*. *J. Controlled Release* **45**, 15–23.

9. Illum, L., Farraj, N. F., and Davis, S. S., Chitosan as a novel nasal delivery system for peptide drugs. *Pharm. Res.* **11**, 1186–1189.

10. Elcin, Y. M., Dixit, V., and Gitnick, G. (1996). Controlled release of endothelial cell growth factor from chitosan-albumin microspheres for localized angiogenesis: *In vitro* and *in vivo* studies. *Artif. Cells, Blood Substitutes, Immobilization Biotechnol.* **24**, 257–271.

11. Sechriest, V. F., *et al.* (2000). GAG-agumented polysaccharide hydrogel: A novel biocompatible and biodegradable material to support chondrogenesis. *J. Biomed. Mater. Res.* **49**, 534–541.

12. Kratz, G., *et al.* (1997). Heparin-chitosan complexes stimulate wound healing in human skin. *Scand. J. Plast. Reconstr. Hand Surg.* **31**, 119–123.

13. Kratz, G., *et al.* (1998). Immobilized Heparin accelerates the healing of human wounds *in vivo*. *Scand. J. Plast. Reconstr. Hand Surg.* **32**, 381–385.

14. MacLaughlin, F. C., *et al.* (1998). Chitosan and depolymerized chitosan oligomers as condensing carriers for *in vivo* plasmid delivery. *J. Controlled Release* **56**, 259–272.

15. Richardson, S. C. W., Kolbe, H. J. V., and Duncan, R. (1999). Potential of low molecular mass chitosan as a DNA delivery system: Biocompatibility, body distribution and ability to complex and protect DNA. *Int. J. Pharm.* **178**, 231–243.

16. Roy, K., *et al.* (1999). Oral gene delivery with chitosan-DNA nanoparticles generates immunologic protection in a murine model of peanut allergy. *Nat. Med.* **5**, 387–391.

17. Tomihata, K., and Ikada, Y. (1997). *In vitro* and *in vivo* degradation of films of chitin and its deacetylated derivatives. *Biomaterials* **18**, 567–575.

18. Onishi, H., and Machida, Y. (1999). Biodegradation and distribution of water-soluble chitosan in mice. *Biomaterials* **20**, 175–182.

19. Pangburn, S. H., Trescony, P. V., and Heller, J. (1982). Lysozyme degradation of partially deacetylated chitin, its films and hydrogels. *Biomaterials* **3**, 105–108.

20. Hirano, S., Tsuchida, H., and Nagao, N. (1989). N-acetylation in chitosan and the rate of its enzymic hydrolysis. *Biomaterials* **10**, 574–576.

21. Schipper, N. G. M., *et al.* (1997). Chitosans as absorption enhancers for poorly absorbable drugs 2: Mechanism of absorption enhancement. *Pharm. Res.* **14**, 923–929.

22. Carreno-Gomez, B., and Duncan, R. (1997). Evaluation of the biological properties of soluble chitosan and chitosan microspheres. *Int. J. Pharm.* **148**, 213–240.

23. Ueno, H., *et al.* (1999). Accelerating effects of chitosan for healing at early phase of experimental open wound in dogs. *Biomaterials* **20**, 1407–1414.

24. Nishiyama, Y., *et al.* (1999). Regioselective conjugation of chitosan with laminin-related peptide, Tyr-Ile-Gly-Ser-Arg, and evaluation of its inhibitory effect on experimental cancer metastasis. *Chem. Pharm. Bull.* **47**, 451–453.

25. Ouchi, T., *et al.* (1989). Synthesis and antitumor activity of chitosan carrying 5-fluorouracils. *Makromol. Chem.* **190**, 1817–1825.

26. Shantha, K. L., Bala, U., and Rao, K. P. (1995). Tailor-made chitosans for drug delivery. *Eur. Polym. J.* **31**, 377–382.

27. Muzzarelli, R., *et al.* (1988). Biological activity of chitosan: Ultrastructural study. *Biomaterials* **9**, 247–252.

28. Chandy, T., and Sharma, C. P. (1990). Chitosan—as a biomaterial. *Biomater., Artif. Cells, Artif. Organs* **18**, 1–24.

29. Elcin, Y. M., Dixit, V., and Gitnick, G. (1998). Hepatocyte attachment on biodegradable modified chitosan membranes: *In vitro* evaluation for the development of liver organoids. *Artif. Organs* **22**, 837–846.

30. Hirano, S., Tsuneyasu, S., and Kondo, Y. (1981). Heterogeneous distribution of amino groups in partially N-acetylated derivatives of chitosan. *Agric. Biol. Chem.* **45**, 1335–1339.

31. Hirano, S., and Yamaguchi, R. (1976). N-acetylchitosan gel: A polyhydrate of chitin. *Biopolymers* **15**, 1685–1691.

32. Oungbho, K., and Müller, B. W. (1997). Chitosan sponges as sustained release drug carriers. *Int. J. Pharm.* **156**, 229–237.

33. Madihally, S. V., and Matthew, H. W. T. (1999). Porous chitosan scaffolds for tissue engineering. *Biomaterials* **20**, 1133–1142.

34. Chow, K. S., and Khor, E. (2000). Novel fabriation of open-pore chitin matrixes. *Biomacromolecules* **1**, 61–67.

35. Brown, C. D., *et al.* (2001). Release of PEGylated granulocyte-macrophage colony-stimulating factor from chitosan/glycerol films. *J. Controlled Release* **72**, 35–46.

36. Park, Y. J., *et al.* (2000). Platelet derived growth factor releasing chitosan sponge for periodontal bone regeneration. *Biomaterials* **21**, 153–159.

37. Tamura, H., *et al.* (2000). Carboxymethyl-chitin and hydroxyapatite composite for bone repairing. *Polym. Prepr.* **41**, 1032–1033.

38. Yoshihara, Y., *et al.* (1997). Study of carboxymethyl-chitin and hydroxyapatite composite for bone repairing. *Adv. Chitin Sci.* **2**, 682–687.

39. Klokkevold, P. R., *et al.* (1996). Osteogenesis enhanced by chitosan (poly-n-acetyl glucosaminoglycan) *in vitro*. *Periodontology* **67**, 1170–1175.

40. Muzzarelli, R. A. A., *et al.* (1993). Osteoconductive properties of methyl-pyrrolidinone chitosan in an animal model. *Biomaterials* **14** (1993), 925–929.

41. Muzzarelli, R. A. A. *et al.* (1997). Osteoinduction by chitosan-complexed BMP: Morpho-structural responses in an osteoporotic model. *J. Bioact. Compat. Polym.* **12**, 321–329.

42. Wan, A. C. A., Khor, E., and Hastings, G. W. (1998). Preparation of a chitin-apatite composite by *in situ* precipitation onto porous chitin scaffolds. *J. Biomed. Mater. Res.* **41**, 541–548.

43. Ito, M., *et al.* (1999). Effect of hydroxyapatite content on physical properties and connective tissue reactions to a chitosan-hydroxyapatite composite membrane. *J. Biomed. Mater. Res.* **45**, 204–208.

44. Kawase, M., *et al.* (1997). Application of glutaraldehyde-crosslinked chitosan as a scaffold for hepatocyte attachment. *Biol. Pharm. Bull.* **20**, 708–710.

45. Yagi, K., *et al.* (1997). Effectiveness of fructose-modified chitosan as a scaffold for hepatocyte attachment. *Biol. Pharm. Bull.* **20**, 1290–1294.

46. Elcin, Y. M., *et al.* (1999). Xenotransplantation of fetal porcine hepatocytes in rats using a tissue engineering approach. *Artif. Organs* **23**, 146–152.

47. Elcin, A. E., Elcin, Y. M., and Pappas, G. D. (1998). Neural tissue engineering: Adrenal chromaffin cell attachment and viability on chitosan scaffolds. *Neurol. Res.* **20**, 648–654.

48. Dillon, G. P., *et al.* (1998). The influence of physical structure and charge on neurite extension in a 3D hydrogel scaffold. *J. Biomater. Sci. Polymer Ed.* **9**, 1049–1069.

49. Desheng, W., and Dinglin, Z. (1997). Tissue engineering study on repairment of injured nerve gap in rat. *Shengwu Yixue Gongchengxue Zazhi* **14**, 108–110.

50. Madihally, S. V., *et al.* (1996). Polysaccharide-based scaffolds for a cell-seeded vascular graft. *211th Am. Chem. Soc. Nat. Meet.* New Orleans, LA.

51. Yamamura, K., Nabeshima, T., and Sakurai, T. (1999). Use of hydroxypropylchitosan acetate as a carrier for growth factor release. *In* "Tissue Engineering of Prosthetic Vascular Grafts" (P. Zilla and H. P. Greisler, eds.), pp. 599–604. R. G. Landes, Austin, TX.

52. Yamamura, K., *et al.* (1995). Sustained release of basic fibroblast growth factor from the synthetic vascular prosthesis using hydroxypropylchitosan acetate. *J. Biomed. Mater. Res.* **29**, 203–206.

53. Hanson, J. C., *et al.* (2000). Chondrocyte seeded chitosan scaffolds for cartilage repair. *6th World Biomater. Congr.*, Hawaii.

54. Grzanna, M., *et al.* (2000). Arabinogalactan-chitosan polymers as chondrocyte scaffolds for repair of cartilage defects. *6th World Biomater. Congr.*, Hawaii.

55. Black, A. F., *et al.* (1998). *In vitro* reconstruction of a human capillary-like network in a tissue-engineered skin equivalent. *FASEB J.* **12**, 1331–1340.

56. Black, A. F., *et al.* (1999). A novel approach for studying angiogenesis: A human skin equivalent with capillary-like network. *Cell Biol. Toxicol.* **15**, 81–90.

57. Duplan-Perrat, F., *et al.* (2000). Keratinocytes influence the maturation and organization of the elastin network in a skin equivalent. *J. Invest. Dermatol.* **114**, 365–370.

58. Santos, P. M., *et al.* (1991). Nerve growth factor: Increased angiogenesis without improved nerve regeneration. *Otolaryngol. Head Neck Surg.* **105**, 12–25.

CHAPTER 48

POLYMERS BIOSYNTHESIZED BY MICROORGANISMS: POLYHYDROXYALKANOATES

David P. Martin, Simon F. Williams, and Frank A. Skraly

INTRODUCTION

Polyhydroxyalkanoates (PHAs) are a class of naturally occurring polyesters that are produced by fermentation [1]. For certain tissue engineering applications, they are rapidly becoming good alternatives to synthetic absorbable polymers, such as polylactides (PLA) and polyglycolides (PGA), since it is possible to tailor the properties of PHAs to particular tissue engineering applications. The PHA polymers have an extraordinarily wide range of mechanical properties and produce a less acidic environment when they are absorbed *in vivo*—which is more conducive to cell growth. It is also possible to tailor the absorption rates of these polymers, balancing tissue generation with scaffold absorption. Most of the PHA polymers are thermoplastics and can be easily fabricated into different types of tissue engineering scaffold by using conventional techniques including melt molding and solution casting.

Polyhydroxyalkanoates have the general structure shown in Fig. 48.1. Generally speaking, the polymers contain optically active monomers and are isotactic. Although the first PHA polymer was discovered as early as 1925, no significant commercial production was established until the late 1970s when ICI (subsequently Zeneca) developed a process using a microorganism called *Ralstonia eutropha* (formerly *Alcaligenes eutrophus*). At that time, interest in the PHA polymers was primarily as environmentally friendly biodegradable replacements for oil-derived plastic materials. Two PHA polymers, poly(3-hydroxybutyrate) (PHB) and poly(3-hydroxybutyrate-*co*-3-hydroxyvalerate) (PHBV), were produced at scale for these applications. Production ultimately ended, however, because the polymers, produced from wild-type microorganisms, remained expensive as commodity materials. Nevertheless, evaluation for use in medical applications has continued, and these polymers have been used as scaffolds in several tissue engineering and guided tissue regeneration studies.

In the mid-1980s, several research groups identified the genes responsible for the production of PHA polymers, and over the next decade they succeeded in developing transgenic fermentation methods for making these polymers. This technology is currently being used by Metabolix, Inc. (Cambridge, MA) to develop more economical PHA production systems for commodity and specialty uses of the polymers [2]. A second company, Tepha, Inc. (Cambridge, MA), is applying the technology in the field of tissue engineering [3].

The development of transgenic methods to make PHA polymers has three significant advantages when applied to the development of tissue-engineered products. First, it provides an extensive polymer design space that allows PHA polymers to be tailored to the needs of each tissue engineering application. In fact, using this approach it is possible to produce PHA polymers with properties similar to a range of real tissues as shown in Fig. 48.2. Second, the transgenic methods allow the selection for PHA production of well-recognized

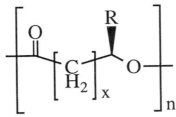

Fig. 48.1. General chemical formula for polyhydroxyalkanoates. The group R is typically an alkyl group containing 1–10 carbon atoms, but it can also be hydrogen. The molecular weight of a PHA is typically between 100,000 and 1,000,000, so n is an appropriate number based on the monomer type(s). The number of methylene carbons, x, is typically 1, but can be as many as 4. The group R should not be confused with the designation (R) to indicate chemical configuration, although PHAs are typically of the (R) configuration. PHA examples given in the text:

Polymer		R group	X
Poly(3-hydroxybutyrate)	*PHB*	*Methyl*	*1*
Poly(3-hydroxybutyrate-co-3-hydroxyvalerate)	*PHBV*	*Methyl and ethyl*	*1*
Poly(4-hydroxybutyrate)	*P4HB*	*Hydrogen*	*2*
Poly(3-hydroxyoctanoate-co-hydroxyhexanoate)	*PHOH*	*Pentyl and propyl*	*1*

fermentation hosts that may be more suitable for use in medical material than uncharacterized wild-type organisms. For example, the transgenic approach allows PHA polymers to be produced in *Escherichia coli* K12, the workhorse of the biopharmaceutical industry. Third, PHA polymers with ultrahigh molecular weights can be produced with the transgenic technology. These high molecular weight polymers can offer significantly enhanced material properties, such as improved tensile strength and elongation to break, and they are particularly valuable in the fabrication of highly porous tissue engineering scaffolds. Taken together, these advantages derived from the transgenic approach have significantly increased interest in the use of PHA polymers in tissue engineering.

Over 100 different hydroxy acid monomers have been incorporated into PHA polymers under controlled fermentation conditions [4]. Although not all these monomers are suitable for the preparation of tissue engineering scaffolds, the variety of monomers that can be brought together provides a class of materials with an extensive range of properties from weak to strong, stiff to flexible, and rigid to elastic. Examples of the breadth of properties are given for several PHA polymers in Table 48.1 [5,6] in comparison to polyglycolide and polylactide. Whereas it will be noted that the latter materials are relatively rigid, inelastic polymers, the PHA family encompasses materials that not only have properties similar to those of polyglycolide and polylactide but also are much more flexible and elastic polymers, such as poly(4-hydroxybutyrate) (P4HB), poly(3-hydroxybutyrate-*co*-4-hydroxybutyrate) (PHB-4HB), and poly(3-hydroxyoctanoate-*co*-3-hydroxyhexanoate) (PHOH). For example, PHB has an elongation to break of 4%, whereas P4HB has an elongation to break of about 1000%. Moreover, by combining the PHB and P4HB monomer constituents into a copolymer, namely, poly(3-hydroxybutyrate-*co*-4-hydroxybutyrate) it is possible to prepare

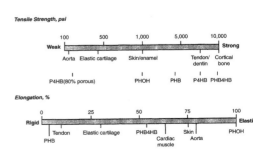

Fig. 48.2. PHA design space. Tailoring PHA properties to specific tissue types.

Table 48.1. Thermal and Mechanical Properties of Selected Bioabsorbable Polymers

Polymer[a]	T_m(°C)	T_g(°C)	Tensile strength (psi)	Modulus (psi)	Elongation (%)	Ref.
PHB	175	0	5000	110,000	4	Vendor data[b]
PHBV	145	2	3000	160,000	17	[5]
PHOH	50	−35	1000	~1,000	~400	Vendor data[b]
P4HB	60	−51	7500	9,400	~1000	Vendor data[b]
PHB-4HB			3700		444	[6]
PGA	210	35				[5]
PDLLA	Amorphous	~51	4600	300,000	5	[5]
PLLA	~170	~57	6000	300,000	4	[5]

[a]PHBV, poly(3-hydroxybutyrate-*co*-3-hydroxyvalerate) containing 11% 3-hydroxyvalerate; PHOH, poly(3-hydroxyoctanoate-*co*-3-hydroxyhexanoate) containing 10% 3-hydroxyhexanoate and a trace of 3-hydroxydecanoate; PHB-4HB, poly(3-hydroxybutyrate-*co*-4-hydroxybutyrate) containing 16% 4-hydroxybutyrate.
[b]Metabolix, Inc., and Tepha. Inc.

PHA polymers with elongation values ranging between 4 and 1000%. These more flexible and elastomeric polymers are beginning to be used fairly extensively in the development of tissue-engineered cardiovascular and orthopedic devices.

Besides tailoring mechanical and thermal properties, it is also possible to produce PHA polymers with different rates of absorption. Subcutaneous implants of PHB and PHBV, for example, are slowly absorbed *in vivo*. Typically this process takes well over one year [7]. In contrast, P4HB is more rapidly absorbed, and the rate appears to be dependent on the configuration of the implant. Thin coating films used in the construction of a tissue-engineered heart valve, for example, were absorbed completely in 6–8 weeks. Larger porous subcutaneous implants are typically absorbed in 10–40 weeks depending upon the level of porosity. Importantly, P4HB is relatively stable in tissue culture media, allowing prolonged cell support *in vitro*.

A number of reports have described the use of PHA polymers in tissue engineering. Two PHA polymers, PHOH and P4HB, have been used successfully in the development of tissue-engineered heart valves [8,9,15–17]. The former has also been evaluated in a scaffold for large arterial vessels [10]. Foams derived from P4HB and seeded with cells currently show significant promise as cardiovascular patches and have been used with remarkable results to augment arteries [11]. There have also been several studies in which PHA polymers have served as guided tissue regeneration templates to repair pericardial membranes after open heart surgery [12] and also to repair atrial septal defects [13]. Given the flexibility to tailor PHA properties to specific applications, it is highly likely that more extensive use of these polymers will occur as more materials become available.

In the listings that follow, weight-average molecular weight M_w was determined by gel permeation chromatography.

MATERIALS

Poly(3-hydroxybutyrate) (PHB)

Poly(3-hydroxybutyrate), derived from *R. eutropha,* with a melting temperature of 172°C (Aldrich, Milwaukee, WI)

Poly(3-hydroxybutyrate), medical grade, derived from *E. coli*, M_w ~1 million g/mol; melting temperature of 175°C (Tepha, Inc., Cambridge, MA)

Poly(3-hydroxybutyrate-co-3-hydroxyvalerate)(PHBV)

Three different compositions of PHBV, containing 5, 8, and 12 wt % derived from *R. eutropha* (Aldrich, Milwaukee, WI)

Poly(4-hydroxybutyrate) (P4HB)

Poly(4-hydroxybutyrate), medical grade, derived from *E. coli*; M_w ~500,000 g/mol; melting temperature of 60°C (Tepha, Inc., Cambridge, MA)

Poly(3-hydroxybutyrate-co-4-hydroxybutyrate) (PHB4HB)

A range of compositions of PHB4HB, medical grade, derived from *E. coli*; M_w ~500,000 g/mol; melting temperature 40–160°C (Tepha, Inc., Cambridge, MA)

Poly(3-hydroxyoctanoate-co-3-hydroxyhexanoate) (PHOH)

Poly(3-hydroxyoctanoate-*co*-3-hydroxyhexanoate) derived from *P. putida*; M_w ~100,000 g/mol; melting temperature 50–55°C (Tepha, Inc., Cambridge, MA)

PURIFICATION

Some commercial samples of PHA polymers contain contaminants, such as cellular debris and endotoxin, that should be removed prior to use in any tissue engineering applications. (Removal of these contaminants is actually best achieved during the initial preparation rather than as a subsequent step.) The following subsections describe procedures that have been used to further purify non medical grades of PHA polymers.

REMOVAL OF CELLULAR DEBRIS

Cellular debris can generally be removed from PHA polymers by dissolving the polymer in hot chloroform, filtering to remove impurities, and precipitating the polymer with methanol (5–10 volumes). The precipitated polymer may be washed with methanol and dried under vacuum to constant weight.

TREATMENT OF ENDOTOXIN

Endotoxin contamination from the production process or from the microbial host can be inactivated or removed by using oxidizing agents, such as hydrogen peroxide and organic peroxides. Generally, the best results are obtained by dissolving the polymer in an organic solvent and treating the solution with a peroxide [3]. An alternative treatment that uses sodium hydroxide has also been described to remove endotoxin from PHB, but this procedure was performed during the initial extraction process [14].

CHARACTERIZATION OF POLYHYDROXYALKANOATES
COMPOSITIONAL ANALYSIS OF POLYHYDROXYALKANOATES

The composition of most PHA polymers may be readily determined by gas chromatographic (GC) analysis of the esters derived by transesterifying the polymers with an alcohol. The best results are usually obtained by preparing butyl esters of the PHA hydroxy acid monomers and comparing these products against standards. A typical procedure for a PHB polymer is as follows.

1. A sample of PHB polymer is subject to butanolysis at 110°C for about 3 h in 2–3 ml of a mixture containing (by volume) 90% 1-butanol and 10% concentrated hydrochloric acid, with 2 mg/ml benzoic acid added as an internal standard.
2. The water-soluble components of the resulting mixture are extracted by addition of 3 ml of water and vigorous shaking.
3. After settling or mild centrifugation, 1 μl of the upper organic phase is injected onto a GC column at a split ratio of 1:50 and an overall flow rate of 2 ml/min. Analysis is performed on an SPB-1 fused-silica capillary GC column (Supelco, Bellefonte, PA; 30 m; 0.32 mm i.d.; 0.25-μm film) or similar column with the following temperature profile: 80°C, 2 min; 10°C/min to 250°C; 250°C, 2 min. The standard used to test for the presence of 3-hydroxybutyrate units in the polymer is sodium 3-hydroxybutyrate or a highly purified sample of PHB, which form a butyl ester derivative upon butanolysis. The standard should

always be run alongside the unknowns so that the conditions of the butanolysis (time, temperature) do not need to be controlled very precisely.

DETERMINATION OF PHA MOLECULAR MASS

Molecular masses of PHA polymers may be determined by gel permeation chromatography as follows:

1. Isolated polymers are dissolved in chloroform (HPLC grade) at approximately 1 mg/ml. Chloroform is the preferred solvent for GPC analysis of PHAs.
2. Samples (50 µl) can be analyzed by chromatography on a Waters (Milford, MA) Styragel HT6E or similar column at a flow rate of 1 ml of chloroform per minute at room temperature, using a refractive index detector.
3. Molecular mass is determined relative to polystyrene standards of narrow polydispersity (Polysciences Inc., Warrington, PA). It should be noted that the hydrodynamic volume of a PHA in chloroform solution may differ significantly from that of polystyrene of similar molecular mass. Thus, this type of measurement provides only a relative measure of weight-average molecular weight, based on the polystyrene standard samples. Laser light scattering and other absolute methods of molecular mass determination may provide additional molecular mass information.

THERMAL ANALYSIS OF PHA POLYMERS

A differential scanning calorimeter such as the Perkin-Elmer Pyris 1 may be used to measure the thermal transitions of PHA polymers during heating and cooling cycles. The analysis proceeds as follows.

1. Samples of approximately 4–8 mg are encapsulated in aluminum pans.
2. A typical profile for the determination of melting point (T_m) and glass transition temperature (T_g) of P4HB, for instance, would be as follows: hold at 25°C for 2 min; heat to 195°C at 10C°/min; hold at 195°C for 2 min; cool to –80°C at 300 C°/min; hold at –80°C for 2 min; heat to 195°C at 10 C°/min. The T_g is determined during the second heating cycle and is noted as the midpoint for the change in heat capacity of the polymer.
3. If the polymer is slow to crystallize from the melt, a melting peak may not be observed in the second heating cycle. In any case, the heating cycle used for the determination of T_m will affect the determination of T_m, since the T_m will depend on the thermal history of the sample.

PREPARATION OF POROUS PHA MATRICES

Porosity can be introduced into PHA polymers by particulate leaching and by phase separation strategies. A combination of these two techniques is particularly desirable to generate interconnected highly porous matrices. Fiber-based approaches for the manufacture of porous PHA matrices are also very attractive but beyond the scope of this chapter.

PARTICULATE LEACHING

Particulate leaching is done by dispersing a particulate material, typically a water-soluble salt such as sodium chloride, throughout the PHA. The particulate–polymer composite can be melted and molded into the desired form, the polymer allowed to crystallize, and the particulate leached out in water, leaving a porous PHA matrix. A typical procedure for the creation of a porous P4HB tissue engineering scaffold by particulate leaching is as follows.

1. P4HB is dissolved in dry acetone (5% wt/vol) to yield a viscous solution. Gentle warming to 55°C may be necessary to fully dissolve the polymer.
2. Sodium chloride particles are sieved to the appropriate size and mixed uniformly throughout the polymer solution to produce a thick paste. The acetone solvent is allowed to evaporate. The ratio of polymer to salt can be varied to produce the desired porosity;

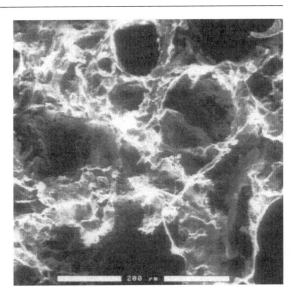

Fig. 48.3. ESEM photomicrograph of a porous P4HB foam produced by salt leaching. Size bar is 200 μm.

beyond a 1:10 weight ratio of polymer to salt, however, the mixture becomes difficult to melt-mold.

3. After complete evaporation of the solvent, the polymer–salt mixture can be pressed into the desired shape by using a heated mold (125°C). The optimal temperature will depend on the type of PHA and the salt/polymer ratio.

4. After the polymer has been allowed to crystallize, the salt is leached out overnight, using several changes of distilled water. The leaching process is monitored by testing for chloride in the water using silver nitrate solution. Complete removal of the salt can be confirmed by elemental analysis of the porous polymer. Figure 48.3 shows an environmental scanning electron microscope (ESEM) image of a P4HB matrix produced by salt leaching.

PHASE SEPARATION

Phase separation is done by precipitating the polymer out of solution, typically by cooling a polymer solution or by the introduction of a nonsolvent. If the polymer solidifies rapidly during this process, and subsequently crystallizes, a porous network of polymer can be produced. While particulate leaching allows greater control over the size and distribution of the pores, it is generally limited to lower porosity by the difficulty in molding composites of relatively high particulate content. Phase separation allows for production of more highly porous materials, but the pores are generally small (5–25 μm), and the process must be optimized to control the pore size and distribution. A porous P4HB matrix may be produced according to the following method.

1. P4HB is dissolved in anhydrous dioxane (5% wt/vol.). Preferably, the P4HB has M_w at least 500,000 (by GPC) to ensure the formation of a viscous solution and a strong foam. Gentle warming to 60°C may be necessary to fully dissolve the polymer.

2. The polymer solution is poured into a mold that allows rapid heat transfer. The fixture is rapidly cooled to 5°C or lower, to freeze the dioxane and precipitate the polymer. The composite is kept frozen near 0°C for 30–60 min to allow the P4HB to crystallize before solvent removal. The composite must be kept cold prior to dioxane removal to prevent melting of the solvent with loss of foam structure.

3. The dioxane is removed by freeze-drying or by leaching into cold methanol, water, or other nonsolvent for the polymer. Figure 48.4 shows an ESEM image of a P4HB matrix produced by thermally induced phase separation.

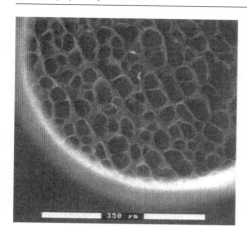

Fig. 48.4. ESEM photomicrograph of porous P4HB foam produced by thermally induced phase separation of a dioxane solution of P4HB. Size bar is 350 μm.

PHA MATRICES PRODUCED BY COMBINING PARTICULATE LEACHING AND PHASE SEPARATION

Generally, a combination of particulate leaching and phase separation results in more highly porous PHA matrices.

1. P4HB (M_w > 500,000 by GPC) is dissolved in anhydrous dioxane (5–35% wt/vol) to produce a viscous solution. Gentle warming to 50°C may be necessary to fully dissolve the polymer. Salt particles of the chosen particle size are mixed with the polymer solution in the desired ratio.

2. The polymer solution–salt mixture is transferred into a mold that allows rapid heat transfer. The fixture is rapidly cooled below 5°C (typically –20°C) to freeze the dioxane and precipitate the polymer. The composite is kept frozen near 0°C for 30–60 min to allow the polymer to crystallize before solvent removal. The composite must be kept cold to prevent melting of the dioxane and loss of the pore structure prior to removal of the solvent.

3. The dioxane is removed by freeze-drying or by leaching into cold methanol, water or other nonsolvent for the polymer. Figure 48.5 shows an ESEM image of a P4HB foam produced by a combination of salt leaching and thermally induced phase separation.

STERILIZATION

Most PHAs can be sterilized in cold ethylene oxide gas. Care must be taken because, depending on the material, elevated temperatures during the sterilization cycle can distort the PHA device, especially highly porous foams. Also, sufficient time must be allowed for proper degassing of the sterilized PHA. Care must be exercised when gamma irradiation is used for sterilization because the radiation treatment can affect the molecular mass of the

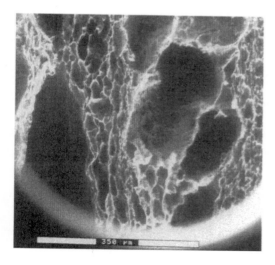

Fig. 48.5. ESEM photomicrograph of porous P4HB foam produced by combination of salt leaching and thermally induced phase separation of a dioxane solution of P4HB. Size bar is 350μm.

polymer and potentially cause cross-linking reactions. Treatment in 70% ethanol may also be useful as a sterilization treatment.

TISSUE ENGINEERING APPLICATIONS

Several papers describe the use of PHAs in animal experiments for cardiovascular tissue engineering applications [8–11,15–17]. The polymeric scaffolds were produced as porous foams—for example, by salt leaching or phase separation—or used in combination with a nonwoven mesh of PGA. Cold ethylene oxide gas was used to sterilize the constructs. To enhance cell attachment and improve the wettability of the polymer scaffold, the constructs were prewet in a protein solution such as collagen, laminin, or culture medium prior to seeding with cells. For seeding, vascular cells (smooth muscle, fibroblast, and endothelial cells) were harvested from ovine carotid arteries, cultured, and expanded *in vitro*. Current experiments are aimed at identifying alternate cell sources and optimizing the scaffold. The PHA scaffolds were seeded with at least one million cells per square centimeter of construct. Both cell seeding with mixed cell cultures or sequential seeding with isolated cell types were successful. After cell seeding, the constructs were cultured for several days to a few weeks prior to implantation into the donor sheep or testing *in vitro* under pulsatile flow. The progress of tissue formation and polymer degradation was followed by harvesting the tissue-engineered constructs at scheduled time points.

CONSTRUCTION OF A POROUS P4HB HEART VALVE SCAFFOLD

A salt-leached foam of P4HB has been assembled into a scaffold for a tissue-engineered heart valve according to the following procedure [9].

1. P4HB was combined with salt particles (80–240 μm) and pressed into a film between two sheets of Mylar. The final porosity was about 80%, although a combination of salt leaching and phase separation could be used for higher porosity and to improve the flexibility of the material.
2. The porous foam was cut into several strips. One strip was used to create an outer conduit. A second strip was placed within the conduit, folded into thirds, and trimmed to replicate the three-leaflet architecture of a native heart valve.
3. The conduit was sealed and the layers laminated together by applying heat with a small handheld, temperature-controlled heating iron (75°C).
4. The assembled valve was sterilized with cold ethylene oxide gas.

CONSTRUCTION OF A POROUS P4HB FOAM AS A SCAFFOLD FOR AUGMENTATION OF THE PULMONARY ARTERY

A P4HB foam was been prepared by a combination of phase separation and salt leaching and used as a patch for augmenting the pulmonary artery in an ovine model according to the following procedure [11].

1. A solution of P4HB (3% in dioxane, M_w 750,000) was combined with salt particles (180–240 μm) in a ratio of 2:1 (wt/wt).
2. The salt–P4HB–solvent mixture was cast as a 1.6-mm-thick sheet between sheets of aluminum foil.
3. The material was frozen at –26°C. The foil was removed, the sample was transferred in the cold, and the dioxane was removed by freeze-drying.
4. After removal of the solvent, the salt was leached out into distilled water at 37°C.
5. The foam was used to replace a portion of the pulmonary artery. Cell-seeded and unseeded foams were used.

CONSTRUCTION OF A P4HB-PGA COMPOSITE HEART VALVE SCAFFOLD

A composite material of P4HB and PGA has been assembled into a scaffold for a tissue-engineered heart valve according to the following procedure [16].

1. A nonwoven mesh of PGA (1 mm thick. 63 mg/ml, ~30 μm fiber diameter) was dipped into a solution of P4HB (1% in tetrahydrofuran). Excess polymer solution was removed and the remainder was allowed to air-dry. The P4HB formed a weblike coating between the PGA fibers.

2. The mesh was cut into several strips. One strip was used to create an outer conduit. A second strip was placed within the conduit, folded into thirds, and trimmed to replicate the three-leaflet architecture of a native heart valve.

3. The conduit was sealed and the layers laminated together by applying heat with a small handheld, temperature-controlled heating iron (75°C).

4. The assembled valve was sterilized with cold ethylene oxide gas.

5. The scaffold was seeded with ovine vascular cells and cultured *in vitro* prior to implantation into the donor sheep.

ACKNOWLEDGMENT

The authors thank our collaborators at Children's Hospital, Boston and Massachusetts General Hospital for their valuable contributions to the tissue engineering component of this work.

REFERENCES

1. Madison, L. L., and Huisman, G. W. (1999). Metabolic engineering of Poly(3-Hydroxyalkanoates): From DNA to plastic. *Microbiol. Mol. Biol. Rev.* **63**, 21–53.
2. Williams, S. F., and Peoples, O. P. (1996). Biodegradable plastics from plants. *CHEMTECH* **26**, 38–44.
3. Williams, S. F., Martin, D. P., Horowitz, D. M., and Peoples, O. P. (1999). PHA applications: Addressing the price performance issue. I. Tissue engineering. *Int. J. Biol. Macromol.* **25**, 111–121.
4. Steinbüchel, A., and Valentin, H. E. (1995). Diversity of bacterial polyhydroxyalkanoic acids. *FEMS Microbiol. Lett.* **128**, 219–228.
5. Endelberg, I., and Kohn, J. (1991). Physico-mechanical properties of degradable polymes used in medical applications: a comparative study. *Biomaterials* **12**, 292–303.
6. Doi, Y., Segawa, A., and Kunioka, M. (1990). Biosynthesis and characterization of poly(3-hydroxybutyrate-*co*-4-hydroxybutyrate) in *Alcaligenes eutrophus*. *Int. J. Biol. Macromol.* **12**, 106–111.
7. Yahia, L. H., Chaput, C. J., Selmani, A., and Rivard, C.-H. (1995). In vitro and in vivo biodegradation of three microbial poly(hydroxybutyrate-*co*-hydroxyvalerate) polymers. *Mater. Clin. Appl.* **12**, 519–526.
8. Stock, U. A., Nagashima, M., Khalil, P. N., Nollert, G. D., Herden, T., Sperling, J. S., Moran, A., Lien, J., Martin, D. P., Schoen, F. J., Vacanti, J. P., and Mayer, J. E., Jr. (2000). Tissue-engineered valved conduits in the pulmonary circulation. *J. Thorac. Cardiovasc. Surg.* **119**, 732–740.
9. Sodian, R., Hoerstrup, S. P., Sperling, J. S., Martin, D. P., Daebritz, S., Mayer, J.E., Jr., and Vacanti, J. P. (2000). Evaluation of biodegradable, three-dimensional matrices for tissue engineering of heart valves. *Am. Soc. Artif. Intern. Organs J.* **46**, 107–110.
10. Shum-Tim, D., Stock, U., Hrkach, J., Shmoka, T., Lien, J., Moses, M. A., Stamp, A., Taylor, G., Moran, A. M., Landis, W., Langer, R., Vacanti, J. P., and Mayer, J. E., Jr. (1999). Tissue engineering of autologous aorta using a new biodegradable polymer. *Ann. Thorac. Surg.* **68**, 2298–2304.
11. Stock, U. A., Sakamoto, T., Hatsuoka, S., Martin, D. P., Nagashima, M., Moran, A. M., Moses, M. A., Khalil, P. N., Schoen, F. J., Vacanti, J. P., and Mayer, J. E., Jr. (2000). Patch augmentation of the pulmonary artery using bioabsorbable polymers and autologous cell seeding. *J. Thorac. Cardiovasc. Surg.* **119**, 732–740.
12. Duvernoy, O., Malm, T., Ramstrom, J., and Bowald, S. (1995). A biodegradable patch used as a pericardial substitute after cardiac surgery: 6- and 24-month evaluation. *J. Thorac. Cardiovasc. Surg.* **43**, 271–274.
13. Malm, T., Bowald, S., Karacagil, S., Bylock, A., and Busch, C. (1992). A new biodegradable patch for closure of atrial septal defect. *Scand. J. Thorac. Cardiovasc. Surg.* **26**, 9–14.
14. Lee, S. Y., Choi, J.-L., Han, K., and Song, J. Y. (1999). Removal of endotoxin during purification of poly(3-hydroxybutyrate) from Gram-negative bacteria. *Appl. Environ. Microbiol.* **65**, 2762–2764.
15. Sodian, R., Sperling, J. S., Martin, D. P., Stock, U., Mayer, J. E., Jr., and Vacanti, J. P. (1999). Tissue engineering of a trileaflet heart valve- Early in vitro experiences with a combined polymer. *Tissue Eng.* **5**, 489–494.
16. Hoerstrup, S. P., Sodian, R., Sperling, J. S., Martin, D. P., Schoen, F. J., Vacanti, J. P., and Mayer, J. E., Jr. (2000). Trileaflet heart valve grown in vitro. *Circulation* **102**(Suppl. III), III44–III49.
17. Sodian, R., Hoerstrup, S. P., Sperling, J. S., Daebritz, S., Martin, D. P., Moran, A. M., Kim, B. S., Schoen, F. J., Vacanti, J. P., and Mayer, J. E., Jr. (2000). Early in vivo experience with tissue-engineered trileaflet heart valves. *Circulation* **102**(19 Suppl. 3), III22–III29.

CHAPTER 49

SYNTHESIS OF SYNTHETIC POLYMERS: ALIPHATIC CARBONATE-BASED POLYMERS

G. G. Atkins and S. W. Shalaby

After a brief discussion of the relevance of aliphatic carbonate to the design of absorbable macromolecular chains, the syntheses of specific polymeric systems comprising carbonate moieties for use as biomaterials are documented.

INTRODUCTION

As a class of polymers, polycarbonates are similar to polyesters in several respects. The aromatic versions of polyesters and polycarbonates are biostable and are not discussed in this chapter. Meanwhile, the chemistry of aliphatic carbonate, hence their physical properties, limit their use as an independent class of polymers compared with aliphatic polyesters. With the exception of a few examples of the step-growth type of polycarbonate [1], the high entropy associated with the carbonate-based aliphatic chains compromises, considerably, their tendency to crystallize into structurally useful materials. Therefore, most aliphatic polycarbonates are low-melting crystalline solids, amorphous materials with a low glass transition temperature, or materials that exist as viscous liquids at room temperature. Additionally, the chemical reactivity and specifically hydrolytic stability is much higher than that of those associated with polyester chains based on 2-hydroxy acids (e.g., glycolic or lactic acid). Based on their physicochemical properties, aliphatic carbonates have been used almost exclusively to modulate the properties of traditional absorbable polyesters [2–4]. And, most of this chapter deals with copolymers comprising trimethylene carbonate.

APPROACHES TO THE SYNTHESIS OF BIOMEDICALLY SIGNIFICANT CARBONATE-BASED ABSORBABLE MACROMOLECULES AND RATIONALE

With the exception of the polycarbonates made from *trans*-1,4-cyclohexane dimethanol, or poly(CHDM carbonate) [1], which are crystalline, practically nonabsorbable materials, polymers comprising trimethylene carbonate (TMC) are most relevant to biomedical applications. Meanwhile, the chemical reactivity and the six-membered ring TMC make it well suited for polymerization or copolymerization with lactones, such as glycolide and lactide under typical ring-opening conditions when a hydroxylic initiator (e.g., 1-decanol) and an organometallic catalyst (e.g., stannous octoate) are used at practical temperatures to achieve high conversion and favorable polymer–monomer equilibrium. In concert with the foregoing, note that carbonates are most useful as components of copolymeric ester chains: (1) low molecular weight poly(CHDM carbonate) has been reacted with cyclic lactones, such as *p*-dioxanone, to produce absorbable polymeric chains [1]; (2) low molecular weight poly(trimethylene carbonate) has been reacted with one or two lactones, such as lactide or glycolide, to produce crystalline absorbable copolymers; and (3) low molecular weight copolymers of TMC and other monomers, such as glycolide or *p*-dioxanone, have been end-grafted with one of the latter monomers to produce absorbable crystalline materials. Since most of the studies on systems based on absorbable aliphatic carbonate have been conducted in industrial laboratories, typical examples of their preparations are derived from the patent literature.

PREPARATION METHODS OF ALIPHATIC CARBONATE-BASED SYSTEMS DERIVED FROM STEP-GROWTH PREPOLYMERS [1]

Preparation methods are exemplified by the synthesis and characterization of poly (hexamethylene carbonate)–*p*-dioxanone copolymer. Such synthesis was conducted in two steps. In the first step, a step-growth polycarbonate polymer was prepared and then end-grafted in the second step by *p*-dioxanone.

PREPOLYMER PREPARATION

A flame-dried, 250-ml, single-neck flask was charged with 36.6 g of 1,6-hexanediol and 69.2 g of diphenyl carbonate. The flask was fitted with a mechanical stirrer, an adapter, and a receiver. The reaction flask was held under vacuum at room temperature for about 18 h. The reaction was then conducted under nitrogen at 180°C for 1.5 h, at 200°C for 1 h, and at 220°C for 3 h; then the flask was cooled to room temperature. Under high vacuum (0.1 mmHg) the reaction flask was gradually heated to 200°C and maintained there for about 18 h. The resulting polyhexamethylene carbonate was isolated, ground, and dried under vacuum (0.1 mmHg). The polymer had an I.V. of 0.38 dl/g.

POLYMER PREPARATION

A flame-dried, 100-ml, single-neck flask was charged with 5 g of the polyhexamethylene carbonate just described and dried at 60°C under 0.1 mmHg vacuum for 24 h. Under nitrogen, 20 g (0.196 mol) of *p*-dioxanone and 0.02 ml of stannous octoate (0.33 M in toluene) were added to the flask, which was then fitted with a flame-dried mechanical stirrer and an adapter. The reaction flask was held under high vacuum (0.1 mmHg) at room temperature for about 24 h. The reaction was then conducted in an oil bath at 110°C for 8 h under nitrogen. The copolymer was isolated, ground, and dried at 80°C under high vacuum to remove any unreacted monomer. The copolymer had an I.V. of 0.38 dl/g.

PREPARATIVE METHODS OF THE ALIPHATIC TMC-BASED COPOLYMERS

Preparation of aliphatic TMC-based copolymers is exemplified by the synthesis of amorphous and crystallizable copolymers made from trimethylene carbonate and one or more lactones.

SYNTHESIS OF AMORPHOUS COPOLYMERS MADE FROM TRIMETHYLENE CARBONATE AND DL-LACTIDE [5]

Into a glass tube containing 2.5 g of trimethylene carbonate were added successively 2.5 g of DL-lactide and 0.25 ml of a solution of tin octoate [tin(II)-2-ethylhexanoate] in toluene (136.5 mg in 20 ml). The test tube was evacuated several times to eliminate the solvent. The test tube was sealed by fusion under a nitrogen atmosphere, suspended in an oil bath maintained at a temperature of 190°C, and kept there for 2.5 h. After being cooled to ambient temperature, the glass test tube was broken open and any glass adhering to the polymer was removed. To remove unreacted monomer, the polymer was reprecipitated from chloroform–petroleum ether and dried under vacuum.

Investigation by NMR spectroscopy (250 MHz ^{1}H-NMR, CDCl$_3$) gave a level of reaction of 96% for DL-lactide and 97% for trimethylene carbonate. The polymer was found to consist of 51% by weight of trimethylene carbonate units. The crude product had an inherent viscosity of 1.05 dl/g (measured in a 0.1% solution in chloroform at 25°C). The purified polymer had a glass transition temperature of +9°C (differential scanning calorimetry, heating rate 5°C/min.).

BLOCK/SEGMENTED ABSORBABLE COPOLYMERS

The family of block/segmented adsorbable copolymers can be represented by those made from TMC and glycolide or lactide under different reaction conditions.

Block Copolymers of Glycolide and Trimethylene Carbonate [6]

Two copolymers of glycolide and trimethylene carbonate (GT-1 and GT-2) were prepared by means of the sequential addition method. Both copolymers were made with 100% trimethylene carbonate (TMC) in the first stage and 100% glycolide (G) in subsequent stages. In overall composition the copolymers were similar. The difference between the two copolymers was that one was polymerized in two stages whereas the other was a three-stage copolymer.

Preparation of Copolymer GT-1

A kettle was charged with 65.10 g of TMC, 4.09 mg of $SnCl_2 \cdot 2H_2O$, and 7.8 μl of diethylene glycol. The mixture was heated to 165°C for 15 min and then the temperature was increased to 180°C over a period of 15 min. Following this, 134.9 g of glycolide was added and the temperature was increased to 210°C over a period of 30 min. The temperature was maintained at 210°C for 1.5 h.

Preparation of Copolymer GT-2

A kettle was charged with 65.10 g of TMC, 4.09 mg of $SnCl_2 \cdot 2H_2O$, and 7.8 μl of diethylene glycol. The mixture was heated to 165°C for 15 min and then the temperature was increased to 180°C over a period of 15 min. Following this, 20.2 g of glycolide was added and the temperature was increased to 195°C over a period of 20 min. After holding at 195°C for 10 min, 114.7 g of glycolide was added. The temperature was increased to 215°C over a period of 15 min and maintained at 210°C for 45 min.

Block Copolymers of TMC and L-Lactide [7]

The preparations of two typical polymers (TL-1 and TL-2) are discussed next.

Preparation of Copolymer TL-1

Freshly distilled trimethylene carbonate (12.95 g, 127 mmol) was melted together with dried, recrystallized L-lactide (2.03 g, 14.1 mmol); then the mixture was syringed into a 15-ml polymerization tube. The catalyst (73 μl of 3.0×10^{-2} M stannous octoate in toluene) was added, and the tube was degassed by freezing, pumping, and thawing twice. After sealing under vacuum, the tube was immersed in an oil bath at 160°C for 60 h. The tube was cracked, and 10 g of the crude polymer was dissolved in chloroform (250 ml) and precipitated into 2-propanol. The dried polymer, 8.6 g, had a reduced viscosity of 1.53 dl/g (0.1% solution in dioxane).

Preparation of Copolymer TL-2

A freshly distilled sample of trimethylene carbonate (12.95 g), 2.03 g of recrystallized L-lactide, and 7.5 μl of 1.0 M stannous octoate in toluene was placed inside a 160°C oil bath for 16 h. The ampule was cracked, and 12.9 g was yielded after being twice reprecipitated from a tetrahydrofuran (THF) solution. The weight-average molecular weight was 87,000, and the number-average molecular weight was 13,760, as determined by gel permeation chromatography (GPC) in THF. The GPC system was calibrated with polystyrene standards.

Two- and Three-Step Synthesis of TMC Block Copolymers [8]

TMC block copolymers were prepared under conditions that differ from those described earlier. The polymers in this section are exemplified by copolymers TMC-G1 and TMC-G2, which are discussed next.

Copolymers of TMC and Glycolide (TMC-G1 and TMC-G2)

TMC-G1 was prepared by drying glycolide (2214 g, 19.1 mol) in a reactor at $24 \pm 2°C$ overnight. Then, 778.6 g (7.63 mol) of trimethylene carbonate and 1.0 g of stannous octoate, both of which had been dried overnight at $24 \pm 2°C$ under vacuum, were added to the reactor. The temperature within the reactor was brought to 150°C with stirring, and polymerization was allowed to proceed for 2 h. Then, 1026.7 g (8.85 mol) of glycolide and 2973.3 g (20.63 mol) of lactide, dried for 24 h under vacuum at $24 \pm 2°C$, were added to the reactor and polymerization was allowed to continue at 150°C for another 20 h and 15 min. The temperature of the polymerization was raised to 190°C for 2 h and 45 min, and then the polymer was extruded.

The resulting copolymer contained 20% by weight of blocks of a random copolymer of glycolide and trimethylene carbonate, having 20 mol% glycolide and 80 mol% trimethylene carbonate, and 80% by weight of blocks of a random glycolide–lactide copolymer having 30 mol% glycolide and 70 mol% lactide.

TMC-G2 was prepared by drying glycolide (166.03 g, 1.43 mol), 580.00 g (5.73 mol) of trimethylene carbonate, 3.21 g of diethylene glycol, and 1.0 g of stannous octoate overnight at 24°C under vacuum in a reactor. The temperature within the reactor was brought to 160°C with stirring and polymerization was allowed to proceed for 10 h and 30 min. When polymerization was complete, 348.83 g (3 mol) of glycolide and 3901.15 g (27.07 mol) of lactide, which had been dried for 24 h at $24 \pm 2°C$, were added to the reactor with no stirring, and the temperature in the reactor was raised to 170°C. Stirring was restarted, and polymerization was allowed to continue at 170°C for another 19 h.

The resulting copolymer contained 15% by weight of blocks of a random copolymer of glycolide and trimethylene carbonate having 20 mol% glycolide and 80 mol% trimethylene carbonate, and 85% by weight of blocks of a random glycolide–lactide copolymer having 10 mol% glycolide and 90 mol% lactide.

Copolymer of TMC and Lactide (TMC-LG)

TMC-LG was prepared by drying lactide (260.08 g, 1.809 mol), 739.2 g (7.25 mol) of trimethylene carbonate, 3.21 g of diethylene glycol, and 1.0 g of stannous octoate in a reactor overnight at $24 \pm 2°C$ under vacuum. The temperature within the reactor was brought to 160°C and polymerization was allowed to proceed with stirring for 7 h. When polymerization was complete, 670.6 g (5.78 mol) of glycolide and 3329.4 g (23.1 mol) of lactide, which had been dried for 24 h $24 \pm 2°C$, were added to the reactor with no stirring, and the temperature in the reactor was raised to 170°C. Stirring was restarted and polymerization was allowed to continue at 170°C for another 22.75 h. The polymer was extruded and post-treated to remove unreacted monomers.

The resulting copolymer contained 20% by weight of blocks of a random copolymer of lactide and trimethylene carbonate having 20 mol% lactide and 80 mol% trimethylene carbonate, and 80% by weight of blocks of a random glycolide–lactide copolymer having 20 mol% glycolide and 80 mol% lactide.

BLOCK COPOLYMERS WITH POLY(ETHYLENE GLYCOL) AND TRIMETHYLENE CARBONATE [9]

A 250-ml flask was charged with PEG-14000 (50 g, 0.0036 mol). The flask was placed in a vacuum oven and the PEG was dried overnight under vacuum at 70°C, with P_2O_5 as a drying agent. The flask was then placed in a glove bag under nitrogen. Glycolide (25.0 g, 0.21 mol) and trimethylene carbonate (25.0 g, 0.24 mol) were charged to the flask, and the contents were melted and mixed under nitrogen. The monomer mixture was then quickly transferred into a stirred reactor, which had been heated under a nitrogen flow to 165°C. Stannous octoate (0.16 ml, 4.9×10^{-4} mol) was quickly charged to the reactor with the use of a syringe. The polymer melt was stirred at 40 rpm for approximately 3 h at 165°C. This time period corresponded to a maximum in the melt viscosity. The polymer was discharged from the reactor and allowed to cool to room temperature. A portion of the crude polymer (42.8 g) was dissolved in CH_2Cl_2 (250 ml) and reprecipitated dropwise into rapidly stirred absolute ethanol (3000 ml). After filtration and drying to constant weight, the reprecipitation yield was determined to be 96%.

The inherent viscosity of the polymer (0.5 g/dl, in CHCl$_3$ at 30°C) was 0.38 dl/g. The composition was analyzed by ^{1}H-NMR spectroscopy and was found to be 34:41:25 wt% Gly/PEO/TMC. The T_g of the polymer was 11°C, the T_m was 59°C.

COPOLYMERS OF SUBSTITUTED TRIMETHYLENE CARBONATE [10]

The four systems discussed in the concluding sections of this chapter exemplify copolymers of substituted trimethylene carbonate.

Preparation of ABA Block Copolymer of 5,5-Dimethyl-1,3-dioxan-2-one (DMTMC) and Caprolactone (CL): B = 1:1 DMTMC:CL, A = DMTMC, A:B = 80:20

An oven-dried, silanized glass 150-ml resin flask was equipped with a mechanical stirrer (with a Teflon paddle), an argon inlet, a serum cap on one port, and a glass stopper on the remaining port. To the flask were added freshly dried and purified DMTMC (4.15 g, 31.9 mmol), caprolactone (3.64 g, 31.9 mmol), and 2,2-dimethylpropanediol (7.5 mg, 0.072 mmol). The flask was evacuated and filled with argon several times, then immersed in an oil bath at 160°C. Stirring was initiated, and after 5 min the catalyst solution, 25 µl of a 1.0 M solution of stannous octoate in toluene, was added. Noticeable thickening occurred in about 20 min; after 1.5 h, the oil bath was lowered and the flask was evacuated briefly to remove some of the unreacted monomer that had condensed on the upper part of the flask. Heating was resumed, and DMTMC (6.64 g, 51 mmol) was added, followed by an additional 25 µl of the catalyst solution. After an additional 10 min, more DMTMC (26.57 g, 204.2 mmol) was added and the mixture was stirred with continued heating at 160°C. In about 30 min, the mixture became too thick to be stirred, but heating was continued for an additional hour, when the polymerization was terminated.

The viscous polymer was scooped out of the flask (37.8 g recovery), dissolved in dioxane (300 ml), and precipitated in a blender into water (1200 ml). The polymer was then washed twice in the blender with water, filtered, and dried in vacuum at 45°C overnight. The yield was 32.9 g (80%). Weight-average molecular weight, as determined by GPC (THF), was found to be 121,000.

Preparation of ABA Block Copolymer of DMTMC and Caprolactone (CL): B = 1:1 DMTMC:CL, A = DMTMC, A:B = 70:30

A polymerization similar to that just described was carried out in a 1-liter resin flask. The first (B-block) stage employed DMTMC (39.04 g 300 mmol), caprolactone (34.24 g 300 mmol), 2,2-dimethylpropanediol (30 mg 0.29 mmol), and 150 µl of 1 M stannous octoate in toluene. After heating for 2 h at 160°C, the reactor was evacuated briefly and all the remaining DMTMC (182.2 g, 1400 mmol) was added at once. An additional 150 µl of catalyst solution was added, and stirring continued for 2.5 h until the polymer became too viscous to stir. Heating at 160°C was continued overnight. Then the polymer was removed from the flask, dissolved in 2.5 liters of dioxane, and precipitated in batches into water (~10 liters). After washing and drying as before, the polymer weighed 226 g (89%).

The weight-average molecular weight was determined by GPC (THF) to be 110,000; caprolactone content was determined by ^{1}H NMR analysis to be 17% (theory, 15%).

Preparation of ABA Block Copolymer of DMTMC and Trimethylene Carbonate (TMC): B = TMC, A = DMTMC, A:B = 80:20

A polymer was prepared as in the first example except that the initial charge consisted of TMC (6.12 g, 60 mmol) and 2,2-dimethylpropanediol (10.3 mg, 0.1 mmol). The flask was immersed in an oil bath at 160°C, then after 5 min, 25 µl of 1 M stannous octoate was added. In 30 min the TMC had polymerized to a viscous material; this was sampled, and then DMTMC, (31.12 g, 240 mmol) was added all at once. The poly(TMC) prepolymer gradually dissolved in the DMTMC, and the mixture became homogeneous and eventually very viscous. After a total time of 3.5 h, the reaction was stopped and worked up as described in the first example.

The yield was 27.7 g (74%). Trimethylene carbonate content was determined by ^{1}H NMR spectroscopy to be 23.6% (theory, 20%). The weight-average molecular weight by GPC (THF) of the prepolymer was found to be 37,000, final polymer had a weight-average molecular weight of 105,000. Differential scanning calorimetry of the final polymer showed a glass transition temperature of 0°C and a melting temperature of 71°C.

Block Copolymerization of DMTMC and TMC in Xylene Solution

In an oven-dried 100-ml resin flask were combined DMTMC (7.81 g, 60 mmol), TMC (6.13 g, 60 mmol), and dimethylpropanediol (3 mg). The flask was evacuated to 0.1 mmHg for 10 min, then filled with dry argon. Xylene (15 ml), dried by distilling from sodium metal, was added by syringe to the flask, which was then immersed in an oil bath at 150°C. After stirring for 5 min, tin octoate (25 ml of a 1.0 M solution in toluene) was added. The solution became very viscous over a 2-h period; a sample was taken for GPC and NMR evaluation. Additional DMTMC (10.41 g, 80 mmol) was added to the flask, and the mixture was stirred at 150°C for an additional 3.5 h. The polymer was removed and was dissolved in 350 ml of dioxane, precipitated into methanol (1100 ml), washed with additional methanol, and dried.

The first sample taken for GPC evaluation showed a weight-average molecular weight of 142,000. The GPC sample was also precipitated into methanol, and the polymer was washed with methanol and dried. NMR analysis of the precipitated sample showed a TMC content of 51% and a DMTMC content of 49%. From the carbonyl carbon region of the 100-MHz carbon spectrum, it was determined that the carbonate groups of the polymer consisted of 27% DMTMC-DMTMC linkages, 28% TMC-TMC linkages, and 45% DMTMC-TMC linkages.

The final polymer yielded 9.0 g (78%). The weight-average molecular weight was determined by GPC (THF) to be 168,00. ^{1}H NMR analysis showed the TMC content to be 32% (theory, 30%). The carbonyl region of the spectrum showed 48% DMTMC-DMTMC linkages, 36% DMTMC-TMC linkages, and 16% TMC-TMC linkages; calculated values assuming that only DMTMC-DMTMC linkages are formed in the second stage were as follows: 50% DMTMC-DMTMC, 31% DMTMC-TMC, and 19% TMC-TMC.

REFERENCES

1. Bezwada, R. S., Hunter, A. W., and Shalaby, S. W. (1991). Bioabsorbable copolymers of polyalkylene carbonate/*p*-dioxanone. U.S. Pat. (to Ethicon, Inc.) 5,037,950.
2. Shalaby, S. W. (1985). Fibrous materials for biomedical applications. *In* "High Technology Fibers: Part A" (M. Lewin and J. Preston, eds.), Chapter 3. Dekker, New York.
3. Shalaby, S. W. (1988). Bioabsorbable polymers. *In* "Encyclopedia of Pharmaceutical Technology" (J. C. Boylan and J. Swarbrick, eds.), Vol. 1, p. 465. Dekker, New York.
4. Shalaby, S. W., and Johnson, R. A. (1994). Synthetic absorbable polyesters. *In* "Biomedical Polymers: Designed-to-Degrade Systems" (S. W. Shalaby, ed.), pp. 1–34. Hanser Publishers, New York.
5. Buchholz, B. (1997). Copolymers of trimethylenecarbonate and optionally inactive lactides. U.S. Pat. (to Boehringer Ingelheim KG) 5,610,266.
6. Jarrett, P. K., Rosati, L., and Casey, D. J. (1990). Segmented absorbable copolymer. U.S. Pat. (to American Cyanamid Company) 5,252,701.
7. Tang, R., Mares, F., Boyle, W. J., Chiu, T., and Patel, K. (1990). Homopolymers and copolymers having recurring carbonate units. U.S. Pat. (to Allied-Signal, Inc.) 5,145,945.
8. Muth, R. R., Totakura, N., and Liu, C. (1994). Absorbable block copolymers and surgical articles made therefrom. U.S. Pat. (to United States Surgical Corporation) 5,322,925.
9. Casey, D. J., Jarrett, P. K., and Rosati, L. (1986). Diblock and triblock copolymers. U.S. Pat. (to American Cyanamid Company) 4,716,203.
10. Mares, F., Boyle, W. J., Tang, R., Patel, K. M., Kotliar, A. M., and Chui, T. (1996). Polycarbonate-based block copolymers and devices. U.S. Pat. (to United States Surgical Corporation) 5,531,998.

CHAPTER 50

SYNTHESIS OF SYNTHETIC POLYMERS: DIOXANONE- AND DIOXEPANONE-BASED POLYMERS

K. A. Carpenter and S. W. Shalaby

The chapter briefly discusses the design and general approaches to the synthesis of absorbable chain molecules comprising sequences derived from p-dioxanone (PD) and its higher homologue 1,5-dioxanpan-2-one (DOP). Then approaches to the synthesis of specific polymeric systems for use as biomaterials are provided.

INTRODUCTION

p-Dioxanone (PD) and 1,5-dioxepane-2-one (DOP) are key examples of a family of cyclic ether esters that have been used for the synthesis of absorbable polymers with more flexible chains than most of their polyester counterparts [1–3]. More specifically, the six-membered ring ether–lactone (PD) can be polymerized to produce polymeric materials with more flexible chains, lower glass transition temperature T_g, and higher compliance compared with those derived from the six-membered ring lactones, lactide, or glycolide. Similarly, while poly-ε-caprolactone is a crystalline material that melts at 65–72°C, depending on its molecular weight, its ether analogue, POD (or 5-oxa-ε-caprolactone), produces liquid polymers at comparable degrees of polymerization [4–6].

APPROACHES TO THE SYNTHESIS OF BIOMEDICALLY SIGNIFICANT PD- AND POD-BASED ABSORBABLE MACROMOLECULES AND RATIONALE

Both PD and POD can be easily polymerized under typical ring-opening polymerization conditions. These usually entail the use of a hydroxylic initiator (e.g., 1-decanol) and an organometallic catalyst, such as stannous octoate. The polymerization temperature varies between about 90 and 225°C, depending on the type of monomer used, the sought molecular weight, and percentage of monomer conversion.

POLY-P-DIOXANONE (PPD) AND COPOLYMERS

PPD is a crystalline material that melts at about 115°C, undergoes glass transition below room temperature, and absorbs in 6–8 months. Accordingly, PPD has been used as an absorbable monofilament suture and other ligating devices. From the synthesis perspective, PPD is thermodynamically unstable at moderate to elevated temperatures. An unacceptable monomer–polymer equilibrium can be encountered during melt polymerization and processing in the melt. Provisions are made under these circumstances to avoid excessive formation of monomer, which is difficult to remove and can, in turn, compromise the end-use performance of PPD devices. Meanwhile, a great deal of effort has been directed to modify PPD using several copolymerization strategies toward accelerating its absorption rate, increasing its thermal stability, increasing its stability against gamma radiation to allow its radiation-sterilization, and lowering its modulus to produce more compliant oriented devices. Since most of the studies addressing these modifications were conducted by industrial investigators, experimental details on the synthesis of PPD and its copolymers are practically limited to examples disclosed in the patent literature.

POLY-1,5-DIOXEPANE-2-ONE (PDOP) AND COPOLYMERS

PDOP, at practical degrees of polymerization, is a viscous liquid with a slower absorption profile than that of PPD. To date, this has limited the pharmaceutical and biomedical application of the homopolymer (PDOP). Meanwhile, DOP and PDOP exhibit certain desirable features that are to be underscored. For example, PDOP is made by the ring-opening polymerization of a seven-membered ring monomer to produce a high molecular weight material with only a trace amount of monomer at a broad range of reaction temperatures, moreover, PDOP is a liquid that can be formulated with several types of pharmaceutical excipients, and DOP can be easily copolymerized with the key monomers, which are commonly used in preparing most of the absorbable materials presently used both clinically and in absorbable polymers research. Almost all available studies on this system have been directed toward the use of DOP as a comonomer to modulate the properties of other polymers and include those dealing with copolymerization with ε-caprolactone (to provide low-melting, highly crystalline materials due to oxygen–methylene isomorphic replacement), glycolide (to produce absorbable materials with higher compliance than polyglycolide), and PD (to yield compliant materials). Since the DOP-based polymers are relatively new, most of the experimental studies on these systems have been originated in industry and are covered, for the most part, in the patent literature.

PREPARATIVE METHODS OF PD- AND DOP-BASED POLYMERS

The methods of preparing PD- and DOP-based polymers are based on experimental schemes disclosed in the pertinent patent literature.

PD-BASED SYSTEMS

Five examples were derived from recent patents and represent typical schemes for the synthesis of PD-based polymers, which may be used to construct absorbable scaffolds for tissue engineering.

Crystalline Copolymers of *p*-Dioxanone and Lactide [7]

The procedure described here deals with synthesis of a copolymer by reacting L-lactide with a mixture of a preformed poly-*p*-dioxanone and PD to yield a composition based on a 90:10 combination of PD and L-lactide.

A thoroughly dried, mechanically stirred 1.5-gal stainless steel reactor was charged with 1800 g (17.632 mol) of *p*-dioxanone, 3.9 ml of 1-dodecanol, and 1.92 ml of stannous octoate (0.33 M solution in toluene). The contents of the reactor were held under high vacuum at room temperature for about 16 h. The reactor was purged with nitrogen. The reaction mixture was heated to 110°C and maintained there for 5.25 h. A sample of the polymer was removed [inherent viscosity (IV), 0.54 dl/g; unreacted monomer content, 25.5%] and 200 g (1.3877 mol) of L-lactide was added. The temperature was raised to and maintained at about 125°C for about 2 h. The polymer was isolated, ground, and dried for 48 h at 80°C and 0.1 mmHg to remove any unreacted monomer. A weight loss of 22.3% was observed. The resulting polymer had a melting temperature of about 102°C by hot stage microscopy, an inherent viscosity of 2.15 dl/g, a crystallinity content of about 33% by X-ray diffraction, and a PDO/PL molar ratio of 93.3/6.7 as determined by NMR spectroscopy.

Preparation of a Crystalline Copolyester of *p*-Dioxanone and Polyester of 1,3-Propanediol and 1,4-Phenylene-bis-oxyacetic Acid [8]

To a dry, 250-ml, single-neck, round-bottomed flask was added 9.0 g of poly(trimethylene1,4-phenylene-bis-oxyacetate) having an inherent viscosity in hexafluorisopropyl alcohol of 0.63 dl/g. The latter polymer was prepared as described by Koelmel and coworkers [8] by the condensation of trimethylene glycol and dimethyl-1,4-phenylene-bis-oxyacetate. The contents of the flask were dried by exposure to a high vacuum (<0.05 mmHg) for several hours at room temperature, followed by heating at 90°C for 12 h under high vacuum. The following materials were then added to the dried contents of the flask, under a dry nitrogen atmosphere: 81.0 g of *p*-dioxanone (0.794 mol) and 0.0160 g of stannous octoate (3.960 mmol). At this point, a flame-dried vacuum-tight stainless steel paddle stirrer and an adapter with a hose connection were attached to the charged reaction flask, and the pressure in the reaction assembly was reduced to a low level for several hours. The reaction flask was then vented with nitrogen and placed in an oil bath. Under a dry nitrogen atmosphere, the reaction mixture was heated, with initially rapid mechanical stirring, according to the following temperature/time sequence: 80°C for 1.0 h, 90°C for 27.0 h, and 80°C for 88.0 h. Stirring was slowed and eventually stopped when the viscosity of the polymerizing mass became so great that further stirring was virtually prevented.

The resulting copolyester was isolated after chilling in liquid nitrogen and was then ground. After exposure to vacuum at room temperature for 16 h, the ground copolyester was heated at 80°C and a pressure of 0.05 mmHg for 16 h to remove unreacted *p*-dioxanone from the desired copolyester product; an 11.2% weight loss was observed. The resulting copolyester product had an inherent viscosity of 1.74 dl/g measured at 25°C at a concentration of 0.1 g/dl in hexafluoroisopropyl alcohol; its melting point (determined by thermal microscopy) was 107–111°C. The resulting polymer was reported to have an improved stability over the homopolymer of PD when sterilized by gamma radiation.

Preparation of a Crystalline Copolyester of *p*-Dioxanone and Polyester of 1,3-Propanediol and Malonic Acid [9]

Under a nitrogen atmosphere at 60°C, 8.4 g of poly(trimethylene malonate) (prepared as described by Koelmel and Shalaby) [9] was poured into a dry 250-ml round-bottomed flask. The contents of the flask were then thoroughly dried by exposure to vacuum (<0.05 mmHg) for several hours at room temperature, followed by heating at 110°C for 18 h under vacuum. Under a nitrogen atmosphere at room temperature, 47.6 g (0.466 mol) of *p*-dioxanone (PDO) and 0.00628 g of stannous octoate (1.55×10^{-5} mol) were next added to the dried contents of the flask. (The stannous octoate was added as a solution in toluene.) A flame-dried, vacuum-tight, stainless steel paddle stirrer and glass adapter with a gas port were attached to the charged reaction vessel, and the pressure in the reaction assembly was subsequently reduced to a low level for several hours. The reaction vessel was then purged with nitrogen, closed off, and placed in an oil bath. Under nitrogen at one-atmosphere pressure, the reaction mixture was heated, with initially rapid mechanical stirring to facilitate dissolution of the polyester in the monomer according to the following temperature–time sequence: 75°C for 0.5 h and 100°C for 8.5 h. The stirring rate was slowed as the viscosity of the polymerizing mass increased. The resulting copolyester was isolated as described above, ground, and subsequently dried at room temperature for 16 h by exposure to vacuum. The ground product was then heated to 80°C and a pressure of 0.05 mmHg for 52 h to remove unreacted *p*-dioxanone from the copolyester product; a 22.8% weight loss was observed. The resulting copolyester product had an inherent viscosity of 1.97 dl/g, measured at 25°C, and a concentration of 0.1 g/dl in hexafluoroisopropenol (HFIP), a melting point of 99-104°C. (determined by thermal microscopy), and a crystallinity of 38% (determined by X-ray diffraction).

Preparation of a Crystalline Copolymer of *p*-Dioxanone and Poly(alkylene oxides) [10]

In the copolymerization of PD and poly(alkylene oxide), a block copolymer of ethylene and propylene oxide was end-grafted with *p*-dioxanone at a 3.9–96.1 weight ratio to increase the compliance of the PD homopolymer. For this, a flame-dried, 250-ml, round-bottomed, single-neck flask is charged with 4.18 g of Pluronic TM F-68 (M_w 8350) block copolymer of propylene oxide and ethylene oxide. The reaction flask is held under high vacuum at 80°C for about 18 h. After cooling to room temperature, the reaction flask is charged with 102.1 g (1.0 mol, 96.1 wt%) of PDO and 0.101 ml of stannous octoate (0.33 M solution in toluene). The contents of the reaction flask are held under high vacuum at room temperature for about 16 h. The flask is fitted with a flame-dried mechanical stirrer and an adapter. The reactor is purged with nitrogen three times before being vented with nitrogen. The temperature of the reaction mixture is heated to 110°C and maintained there for about an hour, lowered to 90°C, maintained at this temperature for 24 h, and then lowered to 80°C and maintained for 3 days. The copolymer is isolated and dried for 16 h at 60°C, for 16 h at 70°C, and for 32 h at 80°C under high vacuum (0.1 mmHg) to remove any unreacted monomer (~16%). The copolymer has an inherent viscosity of 3.43 dl/g in HFIP, and a melting point range of 110–120°C.

Crystalline, High Compliance 91.1:8.9 Copolymer of *p*-Dioxanone and ε-Caprolactone [11]

The strategy for producing a high-compliance, highly crystalline copolymer without compromising the melting temperature of the parent copolymer was carried out with a pre-formed polydioxanone copolymer end-grafted with ε-caprolactone to produce a diblock system. For this, a flame-dried, 250-ml, round-bottomed single-neck flask is charged with 10 g (8.9 wt%) of monohydroxy-terminated polycaprolactone prepolymer with a weight-average molecular weight of 10,000 as determined by gel permeation chromatography (GPC) supplied by Scientific Polymer Products, Inc. The reaction flask is held under high vacuum at 80°C for about 64 h. After cooling to room temperature, the reaction flask is charged with 102.1 g (1.0 mol, 91.1 wt%) of 1,4-dioxan-2-one and 0.101 ml of stannous octoate (0.33 M solution in toluene). The contents of the reaction flask are held under high vacuum at room temperature for about 16 h. The flask is fitted with a flame-dried mechanical stirrer and an adapter. The reactor is purged with nitrogen three times before being vented with nitrogen. The reaction mixture is heated to 90°C, and maintained at this temperature for about 75 min. The temperature of the reaction mixture is raised to 110°C and maintained at this temperature for about 4 h, lowered to 90°C, maintained at this temperature for 24 h, and then lowered to 80°C and maintained at 80°C for 3 days. The copolymer is isolated and dried for 8 h at 60°C and for 8 h at 70°C under high vacuum (0.1 mmHg) to remove any unreacted monomer (~ 15%). The copolymer has an inherent viscosity of 2.18 dl/g in HFIP at 25°C and a melting point range by hot-stage microscopy between 108 and 112°C. The copolymer is melt-spun, drawn, and annealed to prepare oriented, dimensionally stable filaments by means of conventional extrusion techniques. The mechanical and biological properties of these filaments are reported in Table 50.1 [12], which also includes these properties for a *p*-dioxanone homopolymer for comparison.

The data from Table 50.1 show that a filament prepared from a block copolymer of ε-caprolactone and *p*-dioxanone has equivalent straight and knot tensile strength relative to such properties for filaments prepared from a *p*-dioxanone homopolymer; but it has significantly enhanced flexibility, as demonstrated by the reduction in Young's modulus relative to the filaments prepared from the *p*-dioxanone homopolymer. Additionally, complete absorption *in vivo* occurred within 210 days, which is comparable to the rate of absorption for a *p*-dioxanone homopolymer.

DOP-Based Systems

At practical molecular weights, the homopolymer of DOP is a viscous liquid [5,13] and may find use as a slow-absorbing matrix for three-dimensional cell cultures. Meanwhile, the three examples of DOP copolymers described next may be useful as scaffolds for tissue engineering.

Preparation of 1,5-Dioxepan-2-one [6]

To a one-liter, three-necked flask equipped with a dropping funnel and condenser was added 248 g (4.0 mol) of ethylene glycol. Four grams (0.074 mol) of sodium methoxide

Table 50.1. Mechanical and Biological Properties of PCL/PDO Diblock Copolymer[a]

Fiber properties	Annealed (6 h at 80°C, 5% relaxation)
Diameter	7.96 mils
Tensile strength	80,023 psi
Knot tensile strength	49,759 psi
Elongation	40%
Young's modulus	199.836 psi
In vitro BSR[b]	
4 days	84% BSR
7 days	7% BSR
In vivo absorption	
after implantation[c]	100
5	89
91	83
119	36
154	0
210	

[a]PDS violet monofilament polydioxanone suture.
[b]Breaking strength retention (BSR) *in vitro* is the percentage of original straight tensile strength remaining after the indicated number of days in phosphate buffer; pH 7.27 at 50°C.
[c]Median percentage of original cross-sectional area remaining after intramuscular implantation in rats for the indicated number of days, determined according to the procedures described in Bezwada *et al.* [12].

was stirred in the glycol, and upon dissolution, 344 g (4.0 mol) of methyl acrylate was added dropwise. The mixture was heated at 85°C for 17 h, cooled to room temperature, then stirred overnight with 150 g of water-presoaked Amberlite IR-120 (sulfonated ion-exchange resin). The Amberlite was filtered, and the filtrate was treated with 10 ml of tetra isopropyl orthotitanate catalyst and distilled under reduced pressure (0.05 torr). The initial fraction, boiling point ambient to 80°C, was mostly water, catalyst, and compound II. The liquid that distilled over at 80–85°C was chilled in dry ice and thus became very viscous. Upon trituration with anhydrous diethyl ether, white crystals of (1,5-dioxepan-2-one: DXO) formed which were quickly filtered. The crude material was rapidly recrystallized from anhydrous ether to give colorless crystals, melting at 35°C, which were stored in a dry container in a freezer.

Preparation of 75:25 Glycolide–DOP Copolymer [4]

A glycolide–DXO copolymer of initial composition 75:25, mol/mol, was prepared by charging 11.25 g (96.98 mmol) of glycolide, 3.75 g (32.33 mmol) of DXO, 0.0153 g (0.1293 mmol) of 1,6-hexanediol (initiator), and 0.065 ml of 0.033 M stannous octoate

Table 50.2. Properties of 75:15 G/DOP Copolymer

IV (HFIP)[a]	1.34 dl/g
Conversion	
T_m (hot stage microscopy)	200–220°C
T_m (DSC)	215–225°C
Crystallinity (x-ray)	32%
Final composition (proton NMR)	75.8% glycolate units, 24.8% DOP units

[a]Inherent viscosity, tested at 25°C and at a concentration of 0.1 g/d in hexafluoroisopropyl alcohol [4].

Table 50.3. Comparative Polymer and Fiber Properties of Polycaprolactone and Isomorphic CL/DOP Copolymers[a]

Properties	ε-Caprolactone– 1,5-Dioxepan-2-one			
	100/0	95/5	90/10	85/15
Inherent Viscosity	1.80	1.82	1.57	1.66
T_m, °C, DSC, reheating	70	66	54	56
Fibers				
Diameter, mils	8.6	7.6	7.0	6.8
Tensile strength, psi	44,800	57,300	54,500	44,100
Knot strength, psi	37,900	35,300	36,300	44,100
Elongation, %	124	235	173	120
Young's modulus, psi	95,000	105,000	106,000	118,000
Inherent viscosity in $CHCl_3$	1.90	1.81	1.77	1.66
Crystallinity, %	40	42	43	42

[a]Copolymers of Lactide and/or glycolide with 1,5-dioxepan-2-one [4].

solution in toluene (2.1×10^{-6} mol), to a 50-ml ampule. The ampule was sealed under partial vacuum and then held at 190°C for 18 h. Upon cooling to room temperature, an off-white, tough copolymer was obtained. The latter was ground and devolatilized at 110°C for 16 h under vacuum to give a copolymer with the properties described in Table 50.2.

Isomorphic Copolymers of ε-Caprolactone and 1,5-Dioxepan-2-one [13]

The first member of this copolymer family was made as an 85:15 ε-caprolactone–DOP system. Purified ε-caprolactone (19.4 g, 0.170 mol), 1,5-dioxepan-2-one (3.48 g, 0.030 mol), and a catalytic amount of stannous octoate (0.06 ml of 0.33 M toluene solution, 0.020 mmol) were heated in a sealed ampule, equipped for magnetic stirring, for 22 h at 170°C. The polymer was isolated and melt-spun into continuous filaments. The properties of the polymer and drawn filaments are presented in Table 50.3 [4].

REFERENCES

1. Shalaby, S. W. (1985). Fibrous materials for biomedical applications. In "High Technology Fibers: Part A" (M. Lewin and J. Preston, eds.), Chapter 3. Dekker, New York.
2. Shalaby, S. W. (1988). Bioabsorbable polymers. In "Encyclopedia of Pharmaceutical Technology" (J. C. Boylan and J. Swarbrick, eds.), Vol. 1, p. 465. Dekker, New York.
3. Shalaby, S. W., and Johnson, R. A. (1994). Synthetic absorbable polyesters. In "Biomedical Polymers: Designed-to-Degrade Systems" (S. W. Shalaby, ed.), pp. 1–34. Hanser Publishers, New York.
4. Kafrawy, A., Mattei, F. V., and Shalaby, S. W. (1984). Copolymers of lactide and/or glycolide with 1,5-dioxepan-2-one). U.S. Pat. (to Ethicon, Inc.) 4,470,416.
5. Kafrawy, A., and Shalaby, S. W. (1986). Copolymers of 1,5-dioxepan-2-one and glycolide as absorbable monofilaments. *J. Bioact. Biocompat. Polym.* **1**, 431.
6. Kafrawy, A., and Shalaby, S. W. (1987). A new method for the synthesis of 1,5-dioxepan-2-one. *J. Polym. Sci., Polym. Chem. Ed.* **25**, 2629.
7. Bezwada, R. S., Shalaby, S. W., Newman, H., Jr., and Kafrawy, A. (1987). Crystalline copolymers of P-dioxanone and lactide and surgical devices made therefrom. U.S. Pat. (to Ethicon, Inc.) 4,643,191.
8. Koelmel, D. F., Jamiolkowski, D. D., Shalaby, S. W., and Bezwada, R. S. (1985). Poly(p-dioxanone) polymers having improved radiation resistance. U.S. Pat. (to Ethicon, Inc.) 4,546,152.
9. Koelmel, D., and Shalaby, S. W. (1985). Absorbable crystalline alkylene malonate copolyesters and surgical devices therefrom. U.S. Pat. (to Ethicon, Inc.) 4,559,945.
10. Bezwada, R. S., and Shalaby, S. W. (1991). Crystalline copolymers of p-dioxanone and poly(alkylene oxides). U.S. Pat. (to Ethicon, Inc.) 5,019,094.
11. Bezwada, R. S., Shalaby, S. W., and Erneta, M. (1991). Crystalline copolymers of p-dioxanone and ε-caprolactone. U.S. Pat. (to Ethicon, Inc.) 5,047,048.
12. Bezwada, R. S., Shalaby, S. W., Kafrawy, A., and Newman, H. (1987). Crystalline p-dioxanone-glycolide copolymers and surgical devices made therefrom. U.S. Pat. (to Ethicon, Inc.) 4,653,497.
13. Shalaby, S. W. (1980). Isomorphic Copolymers of ε-caprolactone and 1,5-dioxepan-2-one. U.S. Pat. (to Ethicon, Inc.) 4,190,720.

CHAPTER 51

SYNTHESIS OF SYNTHETIC POLYMERS: POLYPHOSPHAZENES

Harry R. Allcock

The polyphosphazene class of polymers offers numerous advantages over traditional materials for tissue engineering developments. In particular, facile changes in organic side group structure allow properties such as hydrolytic sensitivity, mechanical properties, and surface characteristics to be tuned over a wide range. Moreover, biological compatibility of the hydrolysis products can be ensured via the choice of side groups and by the ability of the polymer backbone to break down to the pH-buffered combination of phosphate and ammonia. Examples are given of the use of this system in solid bioerodible matrices, hydrogels, microspheres, and controlled-surface materials.

BACKGROUND

The field of tissue engineering has traditionally been constrained by the limited number of polymers available as platforms for cell adhesion and proliferation, and for bioerosion. A newly evolving polymer system based on the polyphosphazene class of materials offers many opportunities for advances in this field.

Polyphosphazenes constitute a very broad and diverse class of polymers with the molecular structure shown in **I**. The side groups, R, are organic units such as alkoxy, aryloxy, amino, alkyl, or aryl units, which control many of the polymer properties. The backbone is a flexible chain, often 15,000 or more repeating units in length, which has the biological advantage that in the presence of specific side groups, it can be hydrolyzed to the pH-buffered combination of phosphate and ammonia.

$$\left[-N = \overset{\overset{\textstyle R}{|}}{\underset{\underset{\textstyle R}{|}}{P}} - \right]_n$$

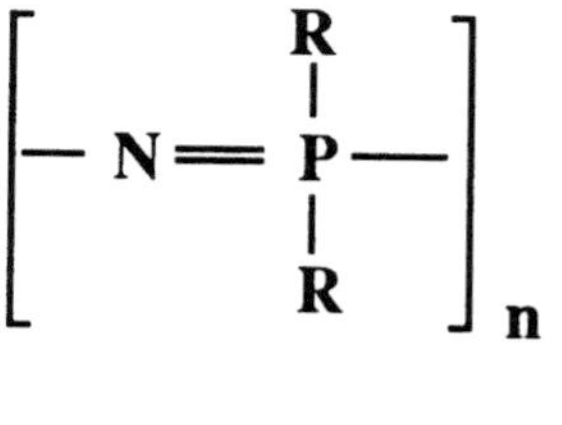

(I)

Various methods are available for the synthesis of these polymers, but the most widely used route is summarized in Fig. 51.1 [1–4]. In this process, the ring-opening polymerization of a phosphazene cyclic trimer (**II**) yields a reactive macromolecular intermediate (**III**), which is used for chlorine replacement reactions to yield organic derivatives of the types shown in **IV–VI**. Because literally hundreds of different organic reagents can be used in this process, and two or more different types of side group can be introduced into the same polymer,

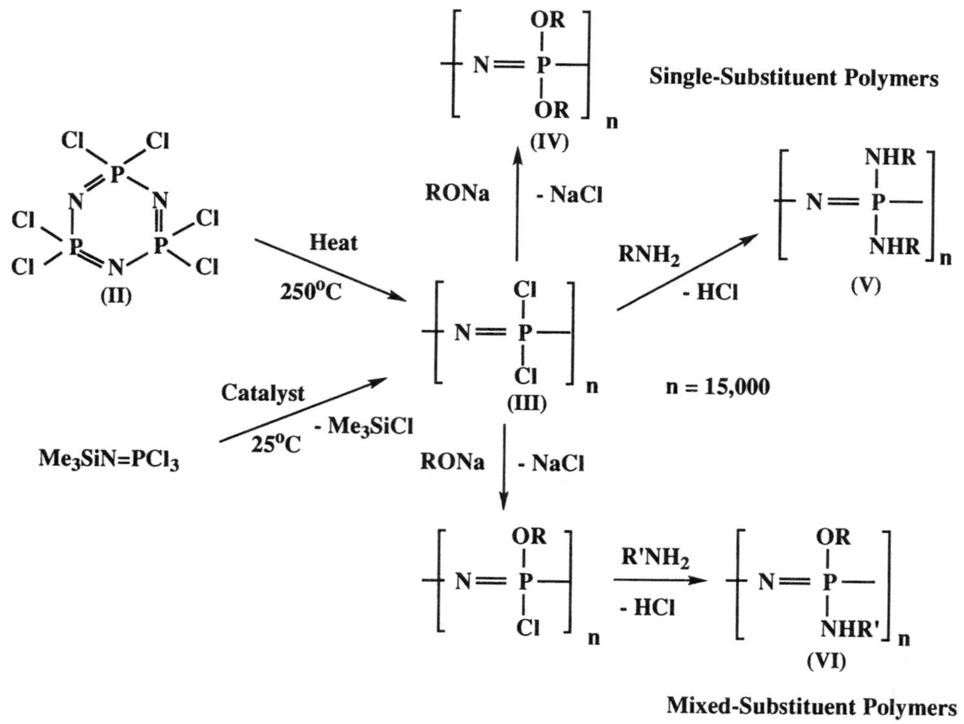

Fig. 51.1. Macromolecular substitution route to polyphosphazenes.

the opportunities for structural variations and property tuning in this system are almost unique [4]. More than 700 different polyphosphazenes have been reported in the literature.

REASONS FOR USE IN TISSUE ENGINEERING

Six features of these polymers make them especially useful for tissue engineering applications, and these are summarized as follows.

1. The polymer properties can be controlled and optimized for tissue engineering by the introduction of different organic side groups. For example, materials with hydrophobic surfaces can be produced by the use of certain side groups, and water-soluble polymers can be formed by the introduction of other side units.

2. Different property combinations are possible by the introduction of two or more different types of side group, and by varying the ratios of them in each macromolecule. For example, variations in side group ratios can change a polymer from biostable to bioerodible, or alter the rate of bioerosion.

3. Some side groups, such as amino acid esters, glucosyl, glyceryl, lactate, glycolate, or imidazolyl, sensitize the polymer to hydrolytic breakdown to nontoxic small molecules such as the free side group, phosphate, and ammonia.

4. Hydrogels are readily accessible from water-soluble polyphosphazenes, including hydrogels that expand or contract in response to changes in temperature or pH. The hydrogels may be hydrolytically stable or bioerodible.

5. Specific polyphosphazenes are excellent materials for the preparation of microspheres using mild, ionic cross-linking techniques.

6. Polyphosphazenes can undergo surface chemistry to optimize cell adhesion and proliferation and to tune the surface to be more hydrophobic or hydrophilic, or to have biofunctionality.

DESIGN PRINCIPLES

CONTROL OF PROPERTIES THROUGH SIDE GROUP CHANGES

The characteristics of a polyphosphazene are determined by a combination of the properties generated by the backbone and those of the side groups. The backbone is an essential

Table 51.1. Side Groups (R) of Interest for Tissue Engineering:

$$\left[-N = \underset{\underset{R}{|}}{\overset{\overset{R}{|}}{P}} - \right]_n$$

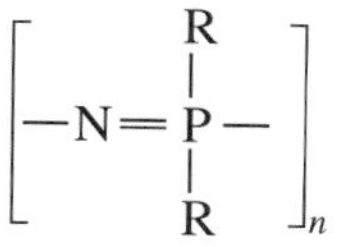

For hydrophobic character

$-OCH_2CF_3$ $-O(CH_2)_{3-11}CH_3$ $-O-\langle\text{phenyl}\rangle$ $-O-\langle\text{phenyl}\rangle-CH_3$

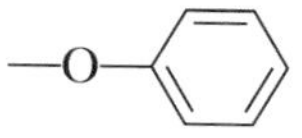

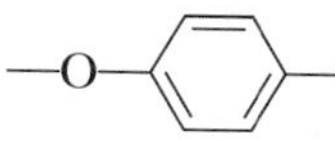

For hydrophilic character or water solubility

$-NHCH_3$ $-O(CH_2)_xNH_2$ $-O(CH_2)_xOH$

$-O(CH_2CH_2O)_{1-16}CH_3$ $-(OCH_2)_xNH_2$ $-(OCH_2)_xOH$

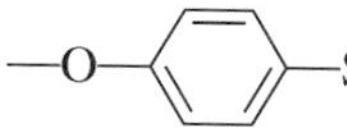

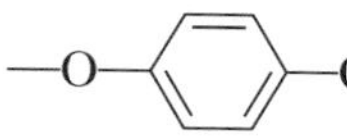

$-O-\langle\text{phenyl}\rangle-SO_3H$ $-O-\langle\text{phenyl}\rangle-COONa$

For bioerosion

$-OC_2H_5$ $-N(R)CH_2COOC_2H_5$ $-OCH_2COOC_2H_5$

$-OCH(CH_3)COOC_2H_5$ $-OCH_2CH(OH)CH_2OH$ $-NH(CH_2)_3OH$

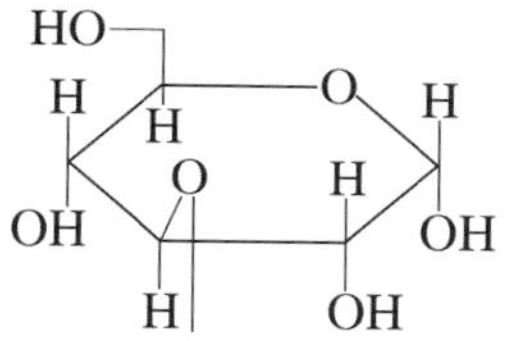

platform that is the crucial component for allowing macromolecular substitution to take place and bioerosion to benign products to occur, and for providing materials with flexibility and radiation stability. But the side groups control solubility in different solvents, biological response, cross-linking, surface character, and ease of bioerosion.

Hydrophobic side groups such as aryloxy, C_3-C_{12}-alkoxy, or fluoroalkoxy groups give polymers that are insoluble in water but soluble in organic solvents. Others, such as alkyl ether units like $-OCH_2CH_2OCH_2CH_2OCH_3$, or side groups with $-OH$, $-NH_2$, or SO_3H units, favor solubility in water. Some, like amino acid esters, $NH(R)COOR'$, may be hydrophilic or hydrophobic, depending on R and R'. Polymer flexibility or rigidity depends on the size and polarity of the side units. Very bulky side group give rise to rigid or crystalline polymers. Flexible side groups often yield amorphous elastomers. More than 250 types of different side group have been linked to the polyphosphazene chain. Table 51.1 lists a small number of those that have proved to be particularly interesting for tissue engineering applications [5–12].

MIXED-SUBSTITUENT POLYMERS

Several different methods are available to synthesize polyphosphazenes that have two or more different types of side group on the same macromolecule. One approach is to allow two different reagents to replace the chlorine atoms in **III**, either simultaneously or sequentially. Another method is to use a "living" cationic polymerization process to prepare block copolymers [13,14]. Block copolymers that have phosphazene and poly(ethylene oxide) or phosphazene and poly(dimethylsiloxane) blocks were developed in the late 1990s [15,16].

Random distribution of two or more different types of side group along the polyphosphazene backbone usually generates amorphous, elastomeric materials. Block copolymers may be phase-separated solids or amphiphilic materials.

The strength of the mixed-substitution approach is that the polymer materials' and biological properties can be tuned to a fine degree by changing the ratios of the two types of side group. Thus, nearly all properties, from surface character to bioerodibility, can be controlled through the side group ratios. A few examples of mixed-substituent species are given later.

SYNTHESIS ISSUES

Several restrictions must be kept in mind when one is designing a polymer for biological applications. First, if the final polymer is to bear a functional group on the side chains, such as $-OH$, or NH_2, or $COOH$, these groups must be introduced *after* the macromolecular substitution step. Otherwise, the reagent, with two functional groups, would cross-link the chains before halogen replacement was complete. Thus, functional units must be protected during primary substitution, and deprotected at a later stage. For example, ester groups are used to protect carboxylic acid groups. Other protection–deprotection processes involve the use of groups protected by (THP) [17] or (BOC) groups [18,19].

Second, if precise main chain lengths and narrow molecular weight distributions are needed, the polymers must be either fractionated after synthesis or prepared by another process. The living cationic polymerization of phosphoranimines [13,14] gives $(NPCl_2)_n$ with precise control of the chain length, as well as access to block copolymers.

SOLID BIOERODIBLE POLYMERS

Most polyphosphazenes are stable to water at physiological pH. However, a few different side groups sensitize the polymers to hydrolytic breakdown. In most of these cases, the hydrolysis products are benign small molecules such as phosphate, low concentrations of ammonia, and the released organic side group. Table 51.1 shows a number of side groups that promote hydrolytic breakdown of polyphosphazenes. Their sensitizing influence depends on pH and other factors, but the imidazolyl group generates the fastest hydrolysis rates at near-neutral pH. Equally important, the actual rate of hydrolytic breakdown will depend on which cosubstituent groups are present and on the ratios of the two.

The amino acid ester derivatives have been studied in greatest detail by the groups of Allcock [5,20,21], Schacht *et al.* [22–24], and Grolleman *et al.* [25–27] but mainly from the viewpoint of controlled drug release. One thing is clear, that the polymer hydrolysis products are phosphate, ammonia, amino acid, and the alcohol from the ester group, and that this combination of products is compatible with the growth of mammalian cells. Imidazole-bearing polymers have also been examined for the release of drugs and proteins [28], especially when hydrolysis-retarding cosubstituent groups, such as 4-methylphenoxy units, are also present.

For example, Laurencin *et al.* [29–31] examined a polymer with 20% imidazolyl and 80% 4-methylphenoxy side groups as a medium for bone regeneration. Osteoblast cells adhere to and grow well on this polymer during the first 3 days. However, further increases in the imidazole percentage in the polymer reduce both cell adhesion and growth. By contrast, films of a polyphosphazene with 50% ethyl glycinate and 50% 4-methylphenoxy side groups promoted the adhesion, growth, and spreading of osteoblast cells [32–34]. A porous, three-dimensional construct of the same polymer, with 20-μm interconnected channels, promoted osteoblast proliferation for at least 21 days.

A different approach to bone regeneration has been explored by Brown, Allcock, and their coworkers [35,36]. In this system, solid, porous composite matrices were prepared from poly[bis(carboxyphenoxy)phosphazene], $[NP(OC_6H_4COOH)_2]_n$, (PCPP) [37], and hydroxyapatite produced at body temperature by the interaction of calcium phosphates.

The calcium ions present in the system cross-link the polymer chains and bind the hydroxyapatite into a strong composite. The introduction of hydrolysis-sensitizing cosubstituent groups into the polymer is expected to increase the rate of hydrolytic breakdown. Although osteoblast colonization experiments have not yet been conducted on this material, its physical properties and the opportunities for property tuning appear promising for bone regeneration studies.

Veronese *et al.* [38] used films of a polyphosphazene with ethyl alanato side groups for the controlled release of drugs to assist the healing of bone and deep periodontal tissue. Bone regrowth was promoted without any inflammatory response in both rat and rabbit models.

Langone *et al.* [39] and Aldini and coworkers [40] have studied the use of amino acid esters for nerve regeneration *in vivo*. Tubular nerve guides, prepared from two different polyphosphazenes, were sutured to the ends of rat nerves. First, poly[bis(ethylalanato)phosphazene] showed a slow and gradual absorption of the conduit without any evidence of local or general toxicity. However, the rate of nerve regeneration in this conduit was not sufficiently different from that of autologous grafts. It was proposed that this polymer is a good candidate for the controlled release of growth factors for nerve regeneration.

By contrast, a polyphosphazene with 70% ethyl alanato and 30% imidazolyl side groups promoted the regeneration of rat sciatic nerve cable and fiber bundles. Clear advantages were seen for the use of the polyphosphazene tubular nerve guides over the more frequently used (nonerodible) silicone tubes.

Overall, it appears that polyphosphazenes with amino acid ester side groups have considerable potential as solid, bioerodible matrices. Related polymers with sugar, glyceryl, or other hydrolysis-sensitizing side units have similar potential but have not yet been studied beyond the mechanistic chemistry stage.

HYDROGELS

A hydrogel is a cross-linked, hitherto water-soluble polymer. Numerous different polyphosphazenes are known that bear alkyl ether side chains and are soluble in, and stable to, aqueous media at pH 7 [41]. Examples are shown in Table 51.2. These polymers can be cross-linked readily by means of gamma radiation [42] or by photosensitized ultraviolet techniques [43] and, when cross-linked, they absorb water to form hydrogels. Bioerosion capability can be built into the system in the form of a few hydrolysis-sensitizing side groups per chain of the types discussed earlier.

The value of polyphosphazene hydrogels in tissue engineering stems from several factors. First, they show a lower critical solution temperature behavior (LCST) in which they absorb water below the LCST and collapse with expulsion of water above that temperature [44]. The LCST depends on the structure of the alkyl ether side chains [45]. As shown in Table 51.2, at least one of these LCSTs is close to human body temperature. Second, the presence of acidic cosubstituent groups, such as OC_6H_4COOH units shown in structure **VII**, makes the gels responsive to pH changes [46].

(VII)

As the pH is lowered, gel collapse occurs. At higher pH values, the gels expand. Calcium or other multivalent ions also cause contraction of the gels (Fig. 51.2). Sodium or potassium ions will reverse this change. Third, enzymes or other proteins can be trapped within the

Table 51.2. Lower Critical Solution Temperatures of Polyphosphazenes[a]:

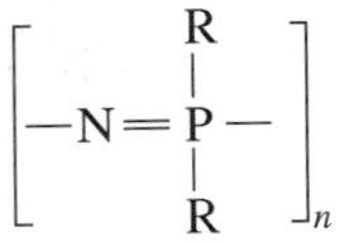

$$\left[-N = \underset{\underset{R}{|}}{\overset{\overset{R}{|}}{P}} - \right]_n$$

R	LCST (°C)
$-OCH_2CH_2CH_2OCH_3$	30
$-OCH_2CH_2OCH_2CH_2OCH_3$	65
$-OCH_2CH_2OCH_2CH_2OC_2H_5$	38
$-OCH_2CH_2OCH_2CH_2OC_4H_9$	51
$-OCH_2CH_2OCH_2CH_2NH_2$	None
$-OCH_2\overset{\overset{CH_2OCH_3}{\|}}{C}HOCH_3$	44
$-OCH_2\overset{\overset{CH_2OCH_2CH_2OCH_3}{\|}}{C}HOCH_2CH_2OCH_3$	38
$-OCH_2\overset{\overset{CH_2OCH_2CH_2OCH_2CH_2OCH_3}{\|}}{C}HOCH_2CH_2OCH_2CH_2OCH_3$	50
$-OCH_2\overset{\overset{CH_2OCH_2CH_2OCH_2CH_2OCH_2CH_2OCH_3}{\|}}{C}HOCH_2CH_2OCH_2CH_2OCH_2CH_2OCH_3$	61

[a]Hydrogels derived from these polymers collapse when heated above the LCST.

gel structure during the cross-linking process [47]. The enzymes retain most of their their activity in spite of the radiation exposure. However, expansion or contraction of the gel in response to external factors can either shut down the biological activity or, in principle, open the gel sufficently to permit escape of the trapped species. Finally, gels of this type have been shown to have antibacterial activity [48], a property that may reflect the ability of the alkyl ether side chains to disrupt bacterial cell membranes.

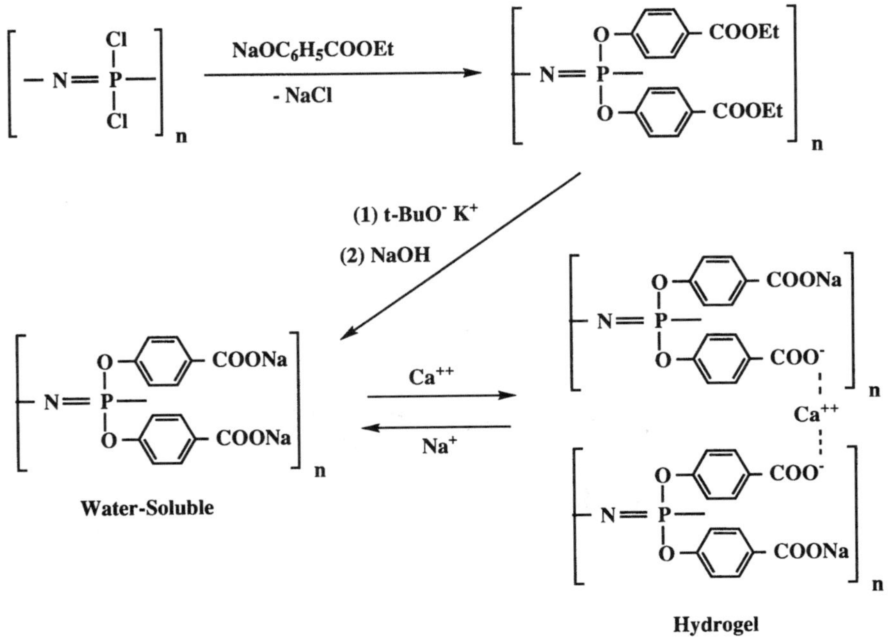

Fig. 51.2. Ionic cross-linking of a polyphosphazene electyrolyte.

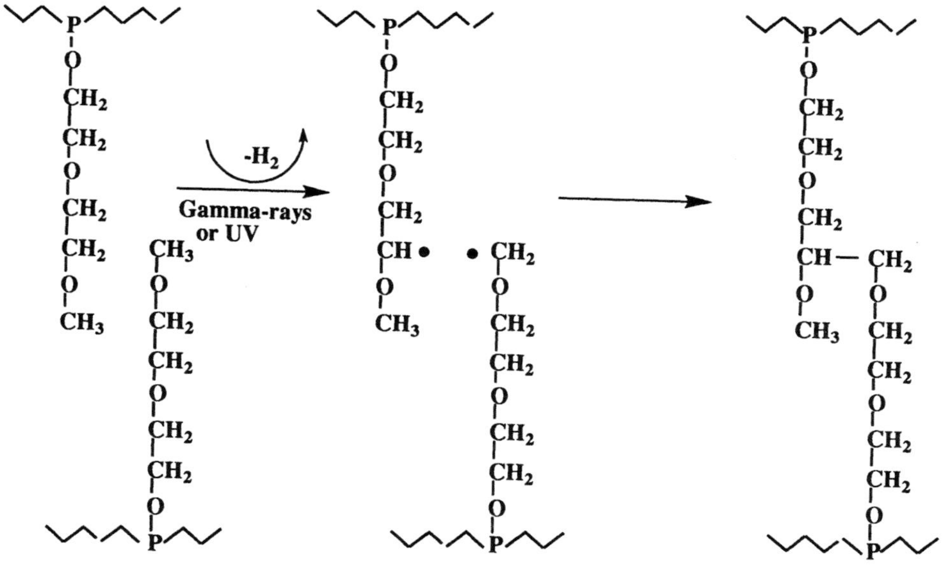

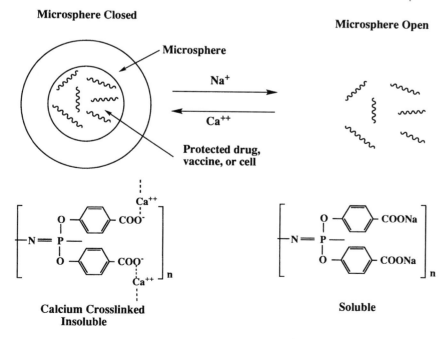

Fig. 51.3. Radiation cross-linking of [NP(OCH₂CH₂OCH₂CH₂OCH₃)₂]n to convert a water-soluble polymer to a hydrogel.

Fig. 51.3. Radiation cross-linking of $[NP(OCH_2CH_2OCH_2CH_2OCH_3)_2]n$ to convert a water-soluble polymer to a hydrogel.

From a tissue engineering viewpoint, hydrogels of this type could be useful as frameworks for the proliferation of mammalian cells and/or for the controlled release of growth factors. In addition, because the cross-linking process is radiation controlled, compartments within a gel may be constructed by three-dimensional focused radiolysis. This could enable different tissue lines to be cultivated in different parts of the device.

MICROSPHERES

An important aspect of tissue engineering is the isolation of mammalian cells within microspheres, either as a prelude to organ reconstruction or to provide protection against antibodies. The use of microspheres for the controlled release of growth factors or antimicrobial agents is another area of interest. Traditionally, microsphere encapsulants have been produced from naturally occurring macromolecules, such as alginates, which can be cross-linked ionically by divalent cations such as Ca^{2+}. However, the variable chain lengths and compositions of these polymers are complicating factors in their use.

Fig. 51.4. Formation of microspheres for cell encapsulation, or drug or vaccine delivery.

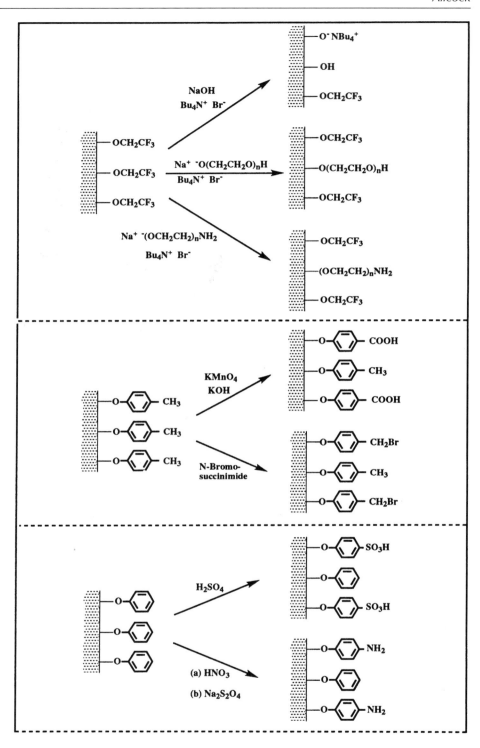

Fig. 51.5. Surface reactions carried out on various polyphosphazenes.

An alternative approach is to use a water-soluble synthetic polymer with a reproducible molecular weight and molecular weight distribution that bears carboxylic acid units in the side group structure. These provide sites for ionic cross-linking under mild aqueous conditions. The polyphosphazene system allows such species to be synthesized via macromolecular substitution and protection–deprotection methods (Fig. 51.4). The polymer shown in Fig. 51.4 is insoluble in water in its free acid form, soluble in base as its salt, and insoluble in acid. The process was originally developed by Cohen, Langer and coworkers [49] as a basis for cell, protein, or drug microencapsulation systems of the type schematized in

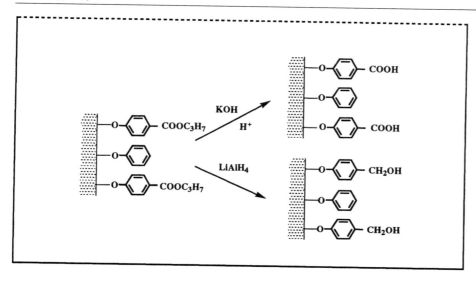

Fig. 51.5. (Continued).

Fig. 51.4 [50,51]. Bioerodibility can be designed into the system using one of the hydrolytic sensitizing systems shown in Table 51.1. Moreover, the properties can be tuned further by the use of water-solubilizing ethyleneoxy cosubstituents (Table 51.2).

Although not yet investigated beyond their use as microspheres for hybridoma mammalian cells [49] and protein [50,51] or vaccine microencapsulation [52], this system has clear advantages for the assembly of three-dimensional constructs of mammalian cells, and for the segregation of different cell lines into different regions of a device.

SURFACES OPTIMIZED FOR TISSUE ENGINEERING

Cell proliferation on a polymer surface depends on the nature of that surface. Factors such as hydrophobicity or hydrophilicity and the types of polymer side group that populate the interface all influence cell adhesion and proliferation. These factors are especially crucial for blood compatibility and endothelial cell overgrowth of cardiovascular devices.

Very few polymers combine appropriate bulk properties for tissue engineering with desirable surface properties. An answer to this problem is to choose a polymer for its bulk properties and then modify the surface regions selectively by chemical transformations.

This has been accomplished for a number of polyphosphazenes, as shown in Fig. 51.5. Here, the side groups that are positioned at the interface are exposed to chemical reagents, while their counterparts in the bulk material are not. Surface reactions, such as sulfonation [53], nitration/reduction [54], hydrolysis [55], hydrogenation, and oxidation reactions [56] have been carried out with different polyphosphazenes, both to change hydrophilicity and biocompatibility and to provide functional groups for binding oligopeptides, enzymes, or drug molecules [57,58]. In addition, MEEP [58] hydrogels have been grafted to the surface of a number of other polymers to create a possible matrix for cell immobilization. Preliminary experiments have shown that different surface structures on the same polyphosphazene have a stong influence on mammalian cell adhesion and spreading.

CONCLUSIONS

It will be clear that the polyphosphazene system has the potential to be useful in several different areas of tissue engineering. At this point the chemistry is at a far more advanced stage than the biological developments. Nevertheless, enough preliminary biological work has been carried out to illustrate some of the possibilities, and a wider application for these materials can be foreseen.

REFERENCES

1. Allcock, H. R., and Kugel, R. L. (1965). Synthesis of high polymeric alkoxy- and aryloxyphosphonitriles. *J. Am. Chem. Soc.* 87, 4216–4217.
2. Allcock, H. R., Kugel, R. L., and Valan, K. J. (1966). High molecular weight poly(alkoxy- and aryloxyphosphazenes). *Inorg. Chem.* 5, 1709–1715.

3. Allcock, H. R., and Kugel, R. L. (1966). High molecular weight poly(diaminophosphazenes). *Inorg. Chem.* 5, 1716–1718.

4. Mark, J. E., Allcock, H. R., and West, R. (1992). "Inorganic Polymers," pp. 61–140. Prentice Hall, Englewood Cliffs.

5. Allcock, H. R., *et al.* (1977). Phosphazene compounds. Synthesis of poly[(amino acid alkyl ester)phosphazenes]. *Macromolecules* 10(4), 824–30.

6. Allcock, H. R., Pucher, S. R., and Scopelianos, A. G. (1994). Poly[(amino acid ester) phosphazenes] as substrates for the controlled release of small molecules. *Biomaterials* 15(8), 563–569.

7. Allcock, H. R., Pucher, S. R., and Scopelianos, A. G. (1994). Synthesis of poly(organophosphazenes) with glycolic acid ester and lactic acid ester side groups: Prototypes for new bioerodible polymers. *Macromolecules* 27(1), 1–4.

8. Allcock, H. R., and Kwon, S. (1988). Glyceryl polyphosphazenes: Synthesis, properties, and hydrolysis. *Macromolecules* 21(7), 1980–1985.

9. Crommen, J. H. L., Vandorpe, J., and Schacht, E. H. (1992). Synthesis and properties of water-soluble polyphosphazenes with aminoalcoholic functions. *Makromol. Chem.* 193(6), 1261–1271.

10. Allcock, H. R., and Scopelianos, A. G. (1983). Synthesis of sugar-substituted cyclic and polymeric phosphazenes and their oxidation, reduction, and acetylation reactions. *Macromolecules* 16(5), 715–719.

11. Allcock, H. R., and Pucher, S. R. (1991). Polyphosphazenes with glucosyl and methylamino, trifluoroethoxy, phenoxy, or (methoxyethoxy)ethoxy side groups. *Macromolecules* 24(1), 23–34.

12. Allcock, H. R., Fuller, T. J., and Matsumura, K. (1982). Hydrolysis pathways for aminophosphazenes. *Inorg. Chem.* 21(2), 515–521.

13. Allcock, H. R., *et al.* (1996). Living cationic polymerization of phosphoranimines as an ambient temperature route to polyphosphazenes with controlled molecular weights. *Macromolecules* 29(24), 7740–7747.

14. Allcock, H. R., *et al.* (1997). Polyphosphazene block copolymers via the controlled cationic, ambient temperature polymerization of phosphoranimines. *Macromolecules* 30, 2213–2215.

15. Nelson, J. M., *et al.* (1998). Synthesis of the first organic polymer/ polyphosphazene block copolymers: Ambient temperature synthesis of triblock poly(phosphazene -ethylene oxide) copolymers. *Macromolecules* 31(3), 947–949.

16. Prange, R., and Allcock, H. R. (1999). Telechelic syntheses of the first phosphazene siloxane block copolymers. *Macromolecules* 32(19), 6390–6392.

17. Allcock, H. R., and Kim, C. (1991). Photochromic polyphosphazenes with spiropyran units. *Macromolecules* 24(10), 2846–2851.

18. Allcock, H. R., and Chang, J. Y. (1991). Poly(organophosphazenes) with oligopeptides as side groups: Prospective biomaterials. *Macromolecules* 24(5), 993–999.

19. Allcock, H. R., Pucher, S. R., Turner, M. L., and Fitzpatrick, R. J. (1992). Poly(organophosphazenes) with poly(alkyl ester) side groups. *Macromolecules* 25, 5573–5577.

20. Allcock, H. R., Pucher, S. R., and Scopelianos, A. G. (1994). Poly[(amino acid ester) phosphazenes]: Synthesis, crystallinity, and hydrolytic sensitivity in solution and the solid state. *Macromolecules* 27(5), 1071–1075.

21. Allcock, H. R., Pucher, S. R., and Scopelianos, A. G. (1994). Poly(amino acid ester) phosphazenes as substrates for the controlled release of small molecules. *Biomaterials* 15, 563–569.

22. Crommen, J. H. L., Schacht, E. H., and Mense, E. H. G. (1992). Biodegradable polymers. II. Degradation characteristics of hydrolysis-sensitive poly[(organo) phosphazenes]. *Biomaterials* 13(9), 601–611.

23. Crommen, J. H. L., Schacht, E. H., and Mense, E. H. G. (1992). Biodegradable polymers. I. Synthesis of hydrolysis-sensitive poly[(organo) phosphazenes]. *Biomaterials* 13(8), 511–520.

24. Schacht, E., *et al.* (1996). Biodegradable polyphosphazenes for biomedical applications. *Adv. Biomater. Biomed. Eng. Drug Delivery Syst.*, pp. 81–85.

25. Grolleman, C. W. J., *et al.* (1986). Studies on a bioerodible drug carrier system based on polyphosphazene. Part I. Synthesis. *J. Controlled Release* 4, 119–131.

26. Grolleman, C. W. J., *et al.* (1986). Studies on a bioerodible drug carrier system based on a polyphosphazene. Part II. Experiments in vitro. *J. Controlled Release* 4(2), 119–131.

27. Grolleman, C. W. J., *et al.* (1986). Studies on a bioerodible drug carrier system based on a polyphosphazene. Part III. Experiments in vivo. *J. Controlled Release* 4(2), 133–142.

28. Laurencin, C. T., Koh, H. J., Neenan, T. X., Allcock, H. R., and Langer, R. (1987). Controlled release using a new bioerodible polyphosphazene matrix system. *J. Biomed. Mater. Res.* 21(10), 1231–1246.

29. Laurencin, C. T., *et al.* (1993). Use of polyphosphazenes for skeletal tissue regeneration. *J. Biomed. Mater. Res.* 27(7), 963–973.

30. Ibim, S. E. M., *et al.* (1994). Colchicine delivery systems to joints based on degradable poly(organophosphazenes): In vitro studies. *Proc. Int. Symp. Controlled Release Bioact. Mater.*, pp. 1022–1078.

31. Laurencin, C. T., *et al.* (1988). Sustained-release pharmaceuticals containing a bioerodible phosphazene polymer matrixes. *Pct Int. Appl.* 29.

32. Laurencin, C. T., *et al.* (1996). A highly porous 3-dimensional polyphosphazene polymer matrix for skeletal tissue regeneration. *J. Biomed. Mater. Res.* 30(2) 133–138.

33. Laurencin, C. T., *et al.* (1997). Biodegradable polyphosphazene/poly(lactide-co-glycolide) blends. *Proc. Int. Symp. Controlled Release Bioact. Mater.*, pp. 1022–1078.

34. Ibim, S. M., *et al.* (1998). In vitro release of colchicine using polyphosphazenes: The development of delivery systems for musculoskeletal use. *Pharm. Dev. Technol.* 3(1), 55–62.

35. Reed, C. S., *et al.* (1996). Thermal stability and compressive strength of calcium deficient hydroxyapatite-poly[bis(carboxylatophenoxy)phosphazene] composites. *Chem. Mater.* 8(2), 440–447.

36. Tenhuisen, K. S., *et al.* (1996). Low temperature synthesis of a self-assembling composite: Hydroxyapatite-poly[bis(sodium carboxylatophenoxy)phosphazene]. *J. Mater. Sci.: Mater. Med.* 7(11), 673–682.

37. Allcock, H. R., and Kwon, S. (1989). An ionically cross-linkable polyphosphazene: Poly[bis(carboxylatophenoxy) phosphazene] and its hydrogels and membranes. *Macromolecules* **22**(1), 75–79.

38. Veronese, F. M., *et al.* (1998). Polyphosphazene membranes and microspheres in periodontal diseases and implant surgery. *Biomaterials* **20**(1), 91–98.

39. Langone, F., *et al.* (1995). Peripheral nerve repair using a poly(organophosphazene) tubular prosthesis. *Biomaterials* **16**(5), 347–353.

40. Aldini, N. N., *et al.* (1997). Peripheral nerve reconstruction with bioabsorbable polyphosphazene conduits. *J. Bioact. Compat. Polym.* **12**, 3–13.

41. Allcock, H. R., *et al.* (1986). Polyphosphazenes with etheric side groups: Prospective biomedical and solid electrolyte polymers. *Macromolecules* **19**, 1508–1512.

42. Allcock, H. R., *et al.* (1988). Hydrophilic polyphosphazenes as hydrogels: Radiation cross-linking and hydrogel characteristics of poly[bis(methoxyethoxyethoxy) phosphazene]. *Biomaterials* **9**(6), 509–513.

43. Nelson, C. J., Coggio, W. D., and Allcock, H. R. (1991). Ultraviolet radiation-induced cross-linking of poly[bis(2-(2-methoxyethoxy)ethoxy)phosphazene]. *Chem. Mater.* **3**(5), 786–787.

44. Allcock, H. R., and Dudley, G. K. (1996). Lower critical solubility temperature study of alkyl ether based polyphosphazenes. *Macromolecules* **29**(4), 1313–1319.

45. Allcock, H. R., *et al.* (1992). Poly(organophosphazenes) with poly(alkyl ether) side groups: A study of their water solubility and the swelling characteristics of their hydrogels. *Macromolecules* **25**(21), 5573–5577.

46. Allcock, H. R., and Ambrosio, A. M. A. (1996). Synthesis and characterization of pH-sensitive poly(organophosphazene) hydrogels. *Biomaterials* **17**(23) 2295–2302.

47. Allcock, H. R., Pucher, S. R., and Visscher, K. B. (1994). Activity of urea amidohydrolase immobilized within poly[di(methoxyethoxyethoxy) phosphazene] hydrogels. *Biomaterials* **15**(7), 502–506.

48. Allcock, H. R., *et al.* (1992). Antibacterial activity and mutagenicity studies of water-soluble phosphazene high polymers. *Biomaterials* **13**(12), 857–862.

49. Cohen, S., *et al.* (1990). Ionically cross-linkable polyphosphazene: A novel polymer for microencapsulation. *J. Am. Chem. Soc.* **112**(21), 7832–7833.

50. Cohen, S., Allcock, H. R., and Langer, R. (1993) Cell and enzyme immobilization in ionotropic synthetic hydrogels. *6th Meet., Recent Adv. Pharm. Ind. Biotechnol., Minutes Int. Pharm. Technol. Symp.*, 36–48.

51. Cohen, S., *et al.* (1993). Design of synthetic polymeric structures for cell transplantation and tissue engineering. *Clin. Mater.* **13**(1–4), 3–10.

52. Payne, L. G., *et al.* (1998). Poly [di(carboxylatophenoxy) phosphazene] (PCPP) is a potent immunoadjuvant for an influenza vaccine. *Vaccine* **16**(1), 92–98.

53. Allcock, H. R., Fitzpatrick, R. J., and Salvati, L. (1991). Sulfonation of (aryloxy)- and (arylamino)phosphazenes: Small-molecule compounds, polymers, and surfaces. *Chem. Mater.* **3**(6), 1120–1132.

54. Allcock, H. R., and Kwon, S. (1986). Covalent linkage of proteins to surface-modified poly(organophosphazenes): Immobilization of glucose-6-phosphate dehydrogenase and trypsin. *Macromolecules* **19**(6), 1502–1508.

55. Allcock, H. R., *et al.* (1996). Controlled formation of carboxylic acid groups at polyphosphazene surfaces: Oxidative and hydrolytic routes. *Chem. Mater.* **8**(12), 2730–2738.

56. Allcock, H. R., Fitzpatrick, R. J., and Salvati, L. (1992). Oxidation of poly[bis(4-methylphenoxy) phosphazene] surfaces and chemistry of the surface carboxylic acid groups. *Chem. Mater.* **4**(4), 769–775.

57. Allcock, H. R., Fitzpatrick, R. J., and Salvati, L. (1991). Functionalization of the surface of poly[bis(trifluoroethoxy) phosphazene] by reactions with alkoxide nucleophiles. *Chem. Mater.* **3**(3), 450–454.

58. Allcock, H. R., Fitzpatrick, R. J., and Visscher, K. (1992). Thin-layer grafts of poly[bis((methoxyethoxy)ethoxy)phosphazene] on organic polymer surfaces. *Chem. Mater.* **4**(4), 775–780.

CHAPTER 52

SYNTHESIS OF SYNTHETIC POLYMERS: POLY(ANHYDRIDES)

Venkatram Prasad Shastri, Alex Zelikin, and Patrice Hildgen

Synthetic polymers have found wide-ranging applications in medicine. Degradable polymers are of particular interest in development of bioerodible implants, scaffolds for tissue engineering, and controlled drug delivery systems. Poly(anhydrides), which belong to the class of poly(esters), as a consequence of their unique chemistry have found broad applications in medicine. These applications include orthopedic devices and drug delivery systems. By varying the monomer composition in the polymer, polymers that possess degradative lifetimes varying from a few days to few months can be obtained. The chemical lability of the anhydride bond allows for near-zero-order release of drugs from polyanhydride devices. By changing the chemical nature of the anhydride end groups in this monomer, polyanhydrides that can be cured upon exposure to radiation can be synthesized. These polymers can be cured to complex shapes. Poly(anhydrides) have been shown to possess excellent soft and hard tissue compatibility and have been successfully developed into both biomaterials for tissue repair and drug delivery systems. A poly(anhydride)-based device for the local delivery of chemotherapeutic agent is approved by the FDA for the treatment of brain cancer. This chapter describes the synthesis of poly(anhydrides) that have been widely explored in medicine.

INTRODUCTION

In keeping with the scope of this book, this chapter is focused on general synthetic procedures and methodologies. Details with respect to the synthesis or use of polymers of specific compositions are referred appropriately. Material properties and biomedical applications are discussed in general terms with appropriate reference to original research articles.

Poly(anhydrides) belong to the general glass of synthetic biodegradable polymers (Scheme 52.1). Poly(anhydrides), were first synthesized in the early 1900s as an alternative to poly(esters) in textile applications [1]. However, their hydrolytic instability was an undesirable attribute. In the early1980s, Langer recognized that this very hydrolytic instability that rendered these polymers unsuitable for textile applications made them excellent materials for medical applications, such as drug delivery and orthopedic prosthesis, where device degradability is desirable [2].

The degradation in poly(anhydrides) occurs via the water-labile anhydride linkages in the backbone. The anhydride linkage, which is obtained by the condensation of carboxylic acids, imparts certain unique properties to these polymers. When appropriate monomers are incorporated in the polymer, the degradation of a device derived from poly(anhydrides) occurs primarily at the surface, similar to a dissolving bar of soap [3]. This degradation pattern is different from bulk erosion, which is characteristics of the poly(α-hydroxy acids),

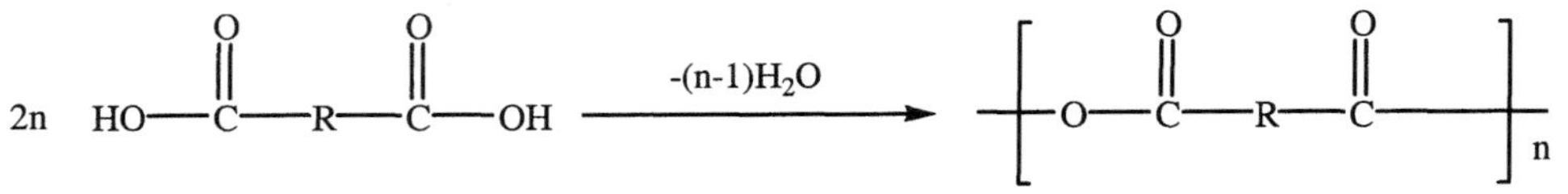

$$2n \quad HO-\overset{\overset{O}{\|}}{C}-R-\overset{\overset{O}{\|}}{C}-OH \quad \xrightarrow{-(n-1)H_2O} \quad \left[O-\overset{\overset{O}{\|}}{C}-R-\overset{\overset{O}{\|}}{C}\right]_n$$

Scheme 52.1.

wherein degradation occurs throughout the polymer device (Fig. 52.1). This phenomenon of "surface erosion" is a consequence of the balance between the high lability of the anhydride linkage, the lipophilicity of the individual monomer units, and the solubility of the monomers at physiological pH. Owing to the relatively hydrophobic nature of the dicarboxylic acid monomers, the uptake of water is restricted to surface of the polymer device and as a result, the degradation of the device becomes solely a function of oligomer and monomer solubility. As a consequence, device degradation occurs by the erosion of insoluble oligomers followed by the dissolution of monomer units [3]. This has enormous benefits in the development of orthopedic implants and devices for drug delivery.

BIOMEDICAL APPLICATIONS OF POLY(ANHYDRIDES)

Fracture fixation is typically augmented with metallic pins, rods, and screws. Two major drawbacks to using metallic implants are the need for secondary implant retrieval surgery and the occurrence of bone atrophy due to stress shielding. It has been proposed that a resorbable implant would eliminate these drawbacks and improve fracture healing and new bone formation [4]. For a polymer-based implant to be clinically viable, however, it must be mechanically strong enough to bear the necessary stresses while exhibiting good biocompatibility and resorption characteristics. The bulk eroding characteristics of poly(α-hydroxy acids) [poly(L-lactic acid) (PLA) and poly(glycolic acid) (PGA)] can cause devices to fail prematurely due to rapid loss in device mass and integrity [5]. This can lead to a localized buildup of acidic degradation products, which has been implicated in a strong inflammatory response observed to these implants [6–8]. The surface-eroding characteristics of poly(anhydrides) offers three key advantages over PLA and PGA: (1) retention of mechanical properties over the degradative lifetime of the device, owing to the maintenance of mass to volume ratio, (2) minimal changes in local acidity, owing to lower solubility and concentration of degradation products, and (3) potentially enhanced integration of the device with surrounding bony tissue. In studies carried out in rodents it has been shown that poly(anhydrides), including photocurable poly(anhydrides), possess excellent soft-tissue compatibility [9,10]. Furthermore, biocompatibility studies in rabbits have shown that in osteocompatibility, photocurable poly(anhydrides) are comparable to PLA and that the former implants show enhanced integration with surrounding bone in comparison to PLA controls [10].

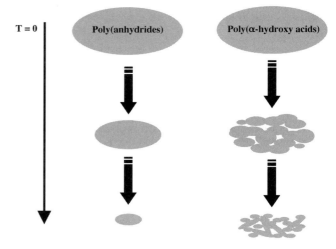

Fig. 52.1. Schematic representation of erosion behavior in poly(anhydride) and poly(α-hydroxy acid) devices.

From a drug delivery standpoint, the "surface erosion" of poly(anhydride)-based devices allows for a near-zero-order release kinetics and improved protection of the drug, since penetration of water into the device is minimal [9,11,12]. This is extremely advantageous for the delivery of molecules such as peptide drugs, which possess a very short half-life in aqueous environments [11,13]. In 1997 the U.S. Food and Drug Administration approved the Gliadel implant, a poly(anhydride) wafer for the delivery of *N*,*N*-bis(2-chloroethyl)-*N*-nitrosourea (BCNU), for site-specific chemotherapy in the treatment of glioblastoma multiforme, a lethal form of brain cancer [14,15].

SYNTHESIS OF POLY(ANHYDRIDES)

Poly(anhydrides) can be classified in general into two categories: linear saturated poly(anhydrides) and linear unsaturated poly(anhydrides). The latter class of poly(anhydrides) can be cross-linked (cured) by chemical means or by exposure to radiation to yield degradable networks [12,16,17]. The linear poly(anhydrides) can be further classified into polymers that possess a purely anhydride backbone and an anhydride–imide backbone [18–20]. The latter class of linear polymers can be derived from naturally occurring monomers such as amino acids [20]. Poly(anhydride-*co*-imides) possessing glycine and alanine derivatives have been synthesized and have been shown to possess good mechanical properties and good hard- and soft-tissue biocompatibility in rodents [20,21].

Poly(anhydrides) and poly(anhydride-*co*-imides) are typically synthesized by condensation polymerization [1,22]. This approach involves first the activation of the free carboxylic acid followed by coupling, with the concurrent elimination of small-molecule by-products. The following three approaches have been extensively studied (Scheme 52.2):

1. Condensation of the acid chloride with the free acid (dehydrochlorination) [23], in the presence of an acid acceptor (reaction 2.1 in Scheme 52.2)
2. Coupling of the free acid by means of phosgene or diphosgene [24] in the presence of an acid acceptor (reaction 2.2). This approach is a variant of the first, the difference being that the acid chloride is generated *in situ* with concurrent elimination of carbon dioxide and hydrochloride salt as the by-products.
3. Melt–condensation polymerization of the mixed anhydrides (acetate esters) (reaction 2.3) [25,26].

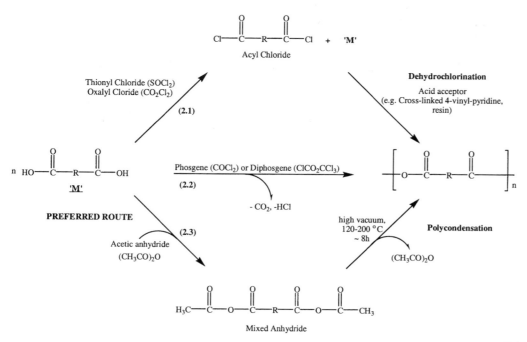

Scheme 52.2.

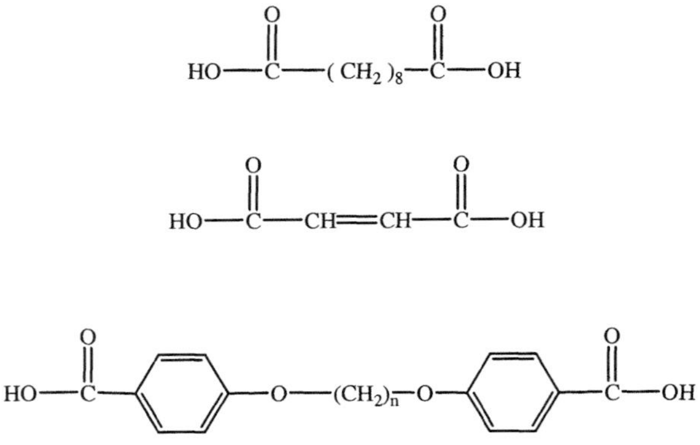

Sebacic Acid [SA]

Fumaric Acid [FA]

n=3: 1,3-bis (p-carboxyphenoxy) propane [CPP]
n=6: 1,6-bis-(p-carboxyphenoxy) hexane [CPH]

Scheme 52.3.

Among the three approaches, the melt–condensation polymerization is the preferred route. Polymers possessing weight-average molecular weights (M_w) in excess of 200 kDa have been synthesized by using coordination catalysts [26]. Some of the common monomers used in the synthesis of poly(anhydrides) for biomedical applications are shown in Scheme 52.3.

SYNTHESIS OF α, ω-BIS(*p*-CARBOXYPHENOXY)ALKANES (CPP AND CPH) (SCHEME 52.4)

The synthesis of CPP was first described by Conix in 1958 [22,25]. A typical synthesis of CPP and CPH is as follows. A 2-liter, three-neck, round-bottomed flask is equipped with a mechanical stirrer in the center neck and a water-cooled condenser and dropping funnel in the remaining necks. Next 2.2 mol of sodium hydroxide (100 g) is slowly dissolved in 500 ml of distilled water, cooled, and then transferred to the round-bottomed flask. Then 1.05 mol of *p*-hydroxybenzoic acid (PHBA, 145 g) is added slowly to the sodium hydroxide solution with constant mechanical stirring. Upon complete dissolution of PHBA, 0.5 mol of the α,ω-dibromoalkane (for CPP: 1,3-dibromopropane, 102 g; for CPH: 1,6-dibromohexane, 123 g) is transferred into the dropping funnel and added dropwise over a 2-h period. Throughout the addition period the reaction mixture is stirred vigorously and heated to a reflux temperature. Periodically water is added to the reaction mixture to lower the viscosity and enable efficient mixing. After 3 h a white thick paste is obtained. To this 0.55 mol of sodium hydroxide (22 g), dissolved in 200 ml of water, is added slowly. It is

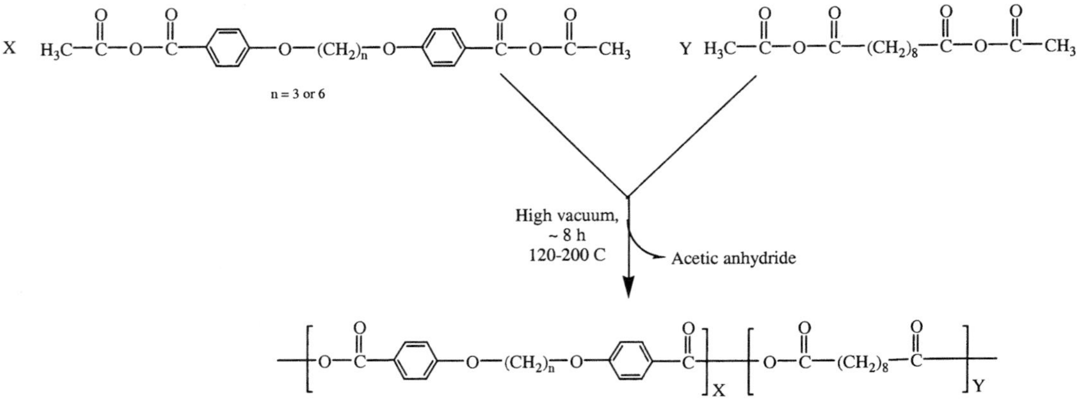

Scheme 52.4.

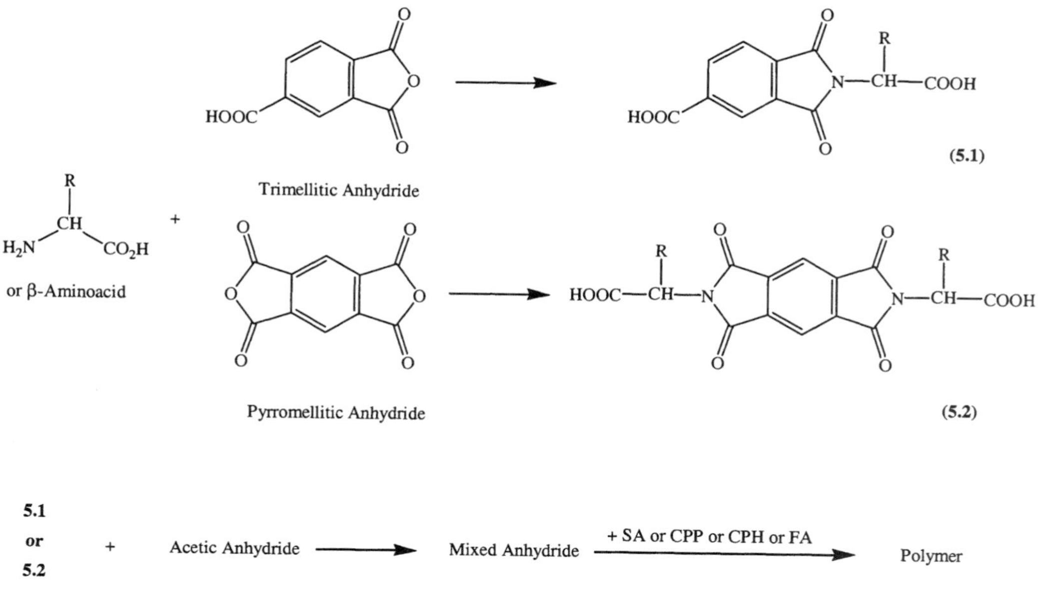

Scheme 52.5.

important to carry out this addition slowly to avoid an exotherm. The reaction mixture is then cooled to room temperature and left standing overnight to yield the disodium salt of CPP or CPH as a white powder. The white powder is isolated by filtration and then washed with 2 × 200 ml of methanol. The wet precipitate is then dispersed in 2 liters of water and heated to 60°C. To this white slurry, concentrated sulfuric acid is added dropwise to acidify the solution (pH 2). Adequate precautions should be taken to prevent accidental exposure to sulfuric acid. The white precipitate (CPP or CPH) is then filtered and washed with 200 ml of distilled water and 2 × 200 ml of acetone and then dried in a vacuum oven at 80°C.

SYNTHESIS OF IMIDE-CONTAINING MONOMERS (SCHEME 52.5) [19,20]

In addition to monomers having a purely hydrocarbon backbone such as sebacic acid (SA), CPP, and CPH, monomers possessing imide linkages in the backbone have been be synthesized. These monomers are obtained by the condensation of trimellitic and pyromellitic anhydride with amino acids such as alanine, glycine, and β- and α,ω-amino acids such as β-alanine. A typical procedure for the synthesis of trimellityl and pyromellityl imido derivative is as follows. A 1-liter round-bottomed flask equipped with a water-cooled condenser is charged with 1 mol of the trimellitic anhydride (182 g) or pyromellitic anhydride (218 g) and 400 ml of N,N-dimethylformamide (DMF). To this mixture, 1.05 mol of the amino acid (2.1 mol in the case of pyromellitic anhydride) is added and the reaction mixture is refluxed for 3 h. The crude product is then isolated either by precipitating the reaction mixture in cold distilled water or by filtration of the solid obtained upon cooling the reaction mixture to room temperature. The solid so collected is then washed with diethyl ether to remove any unreacted starting material. The pure product is then recrystallized from water.

SYNTHESIS OF THE MIXED ANHYDRIDE OF THE DIACID MONOMER (REACTION 2.3, SCHEME 52.2) [26]

In a typical procedure, 50 g of the diacid monomer and 500 ml of fresh acetic anhydride are transferred to a single-neck, round-bottomed flask and then heated under a stream of dry nitrogen until all or almost all of the diacid is dissolved. It is important to ensure that the acetic anhydride is free of acetic acid, for the presence of this acid can significantly lower the yield of the reaction. Hence reagent grade or freshly distilled acetic anhydride should be used. The temperature of the reaction mixture is maintained below 100°C to minimize the oligomerization of the activated diacid monomer. At the end of the reaction, the excess acetic anhydride and acetic acid is removed under reduced pressure and the reaction mixture is concentrated to about 100 ml and chilled overnight in the refrigerator. The pure mixed anhydride is isolated by filtration as white needle-shaped crystals and then washed with dry ether, followed by drying under vacuum.

SYNTHESIS OF POLY(ANHYDRIDES) BY MELT–POLYCONDENSATION (REACTION 2.3, SCHEME 52.2) [26]

Mixed anhydrides of the desired diacid monomers in the desired molar or weight ratio are fed into a glass tube or a round-bottomed flask that has been flame-dried to ensure low moisture and equipped with an inert gas inlet, a vacuum outlet, and a magnetic stir bar. The reaction mixture is flushed with dry argon or nitrogen and heated to a temperature between 150 and 180°C, depending of the monomer composition in the feed, at which time a melt is obtained. This melt is flushed with inert gas to remove any dissolved oxygen, and the reaction vessel is then evacuated to a vacuum of 10^{-4} torr. The by-product of the condensation reaction, acetic anhydride, is collected in a chilled trap. Periodically, the reaction vessel is filled with nitrogen and reevacuated to ensure efficient removal of acetic anhydride. The duration of the reaction ranges from 1.5 to 8 h, depending on the scale of the polymerization and temperature. In the case of monomers such as fumaric acid, derivatives of amino acid based imide derivatives of trimellitic anhydride (reaction 5.1, Scheme 52.5), and pyromellitic anhydride (reaction 5.2), a lower temperature with a longer reaction time is desirable. The procedure described earlier is optimal for the synthesis of polymers up to 50-g scale. Synthesis on a larger scale should be carried out in either a high vacuum glass or steel reactor vessel. After the completion of the polymerization, the reaction flask (vessel) is cooled to room temperature and the polymer is dissolved in a chlorinated hydrocarbon solvent such as methylene chloride or chloroform. The typical ratio of polymer to solvent is 1:5. The objective is to get a dilute solution of the polymer with low viscosity. The pale yellow to light brown polymer solution so obtained is then added dropwise to hexane or petroleum ether with vigorous stirring to precipitate the polymer as a white to off-white and in some cases pale cream-colored powder. The ratio of the polymer solution to the hexane or petroleum ether is typically 1:10. This powder is collected by vacuum filtration, dissolved in chloroform, and then further purified by precipitating it in diethyl ether. The polymer is then collected and flushed with dry argon and dried in a vacuum oven at room temperature and then stored at –20°C under vacuum until further use.

PHOTOCURABLE POLY(ANHYDRIDES) [17]

As mentioned earlier, the utility of poly(anhydride) as a biomaterial for orthopedic applications such as bone cement, bone pins, and screws is dictated by its handling characteristics and mechanical properties. Now a new class of poly(anhydrides) that undergoes photocuring has been developed [17]. These novel polymers possess photopolymerizable methacrylate moieties at their chain ends and can be cured chemically or upon exposure to radiation into high-strength, degradable networks. The rapid curing characteristics allow for fabrication of these polymers into complex shaped devices such as bone screws. These devices have been shown to undergo surface erosion with retention of mechanical properties under physiological conditions. The rapid curing also allows for *in situ* formation of these networks such as in a bony site [17]. These polymers have also been shown to have excellent soft-tissue and osteocompatibility [10].

SYNTHESIS OF METHACRYLATED SEBACIC ACID, CPP, AND CPH (SCHEME 52.6)

A single-neck, round-bottomed flask is heated under a stream of argon by using a heat gun and then dried to room temperature to ensure dry conditions. This round-bottomed flask is then charged with X moles of SA, CPP, or CPH, $3-4X$ moles of methacrylic anhydride, and 200 ppm of hydroquinone inhibitor. The reaction mixture is warmed to 70°C and stirred until all the monomer has dissolved. The solution is then diluted with methylene

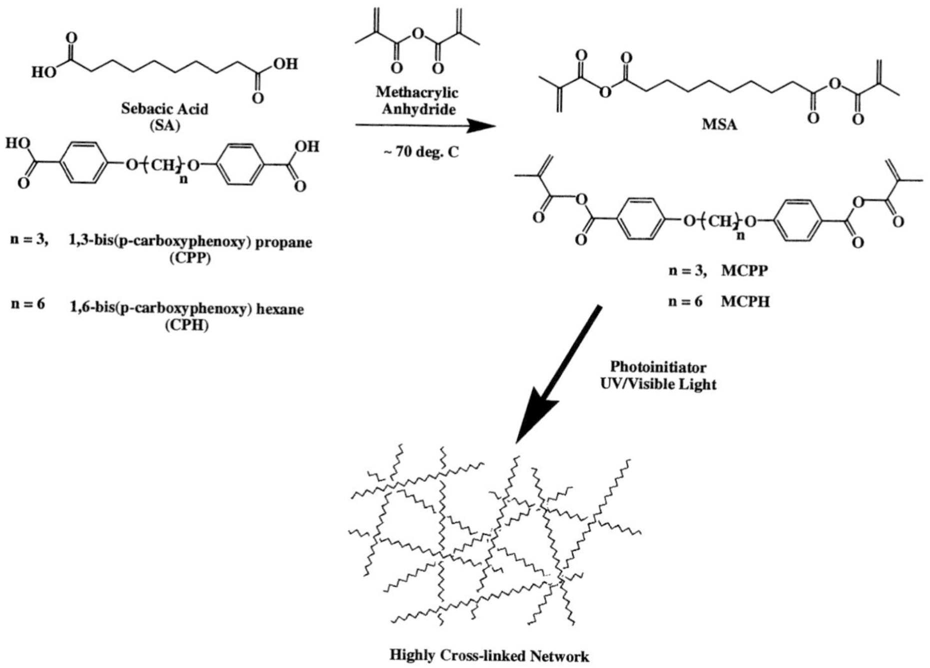

Scheme 52.6.

chloride (5:1 v/v), filtered, and precipitated into a chilled solution of petroleum ether. The ratio of the reaction mixture to petroleum ether is typically 1:10 by volume. The product is isolated as a white waxy solid (methacrylated SA, MSA) or a white waxy powder (MCPP and MCPH). To ensure complete removal of unreacted methacrylic anhydride and methacrylic acid by-product, the product is washed two more times with petroleum ether before being dried under vacuum at room temperature. The final product is flushed with argon and then stored at –20°C. Proper precautions, such as gloves, face shield, and lab coat, must be used at all times when handling methacrylic anhydride and their derivatives, which are high in dermal toxicity.

PHOTOCURING OF METHACRYLATED MONOMERS [10,17]

Curing of the methacrylated monomers can be initiated by both chemical and photochemical means. The photochemical approach is preferred because it allows enhanced spatial and temporal control over the curing. The extent, depth, and area of curing can be varied by varying the intensity of the light source, the spot size of the beam, and the duration of exposure. One could potentially imprint intricate patterns in three-dimensional space by using a combination of computer-aided design and machining techniques (CAD/CAM) and principles of photolithography. This type of enhanced control has immediate applications in drug delivery and tissue engineering. An example in the context of drug delivery would be controlling the regimen of drug release from a device. One could achieve this by controlling the area and spacing between cured and semicured regions in a device. A similar concept may be applied to the design of scaffold for tissue engineering when the goal is to obtain certain temporal changes in scaffold dimensions, structure, degradation, and mechanical properties that complement tissue growth and differentiation.

A typical photopolymerization involves mixing the monomers of interest in the desired ratio with other additives and a single photoinitiator or a combination of photoinitiators and then placement of this mixture in a mold and subsequent exposure to a light source to initiate the curing. The typical concentration of the initiator is 0.1–1% w/w with respect to the reactive monomer concentration. The light source and the propensity of the monomer to undergo photopolymerization (cure speed) dictate the choice of the photoinitiator. For a UV light source, the typical photoinitiators are derivatives of benzophenone in combination with amine accelerators. For deep blue and visible light source such as blue light used in dentistry, a visible initiator such as camphorquinone is optimal.

MATERIAL CHARACTERISTICS

Poly(anhydrides) possessing M_w in excess of 200 kDa can be obtained by addition of inorganic catalyst during the melt polymerization [26]. By varying the ratios of monomers in the feed, polymers with degradative lifetimes ranging from a few days to over several months can be obtained [27]. One such well-studied system consists of the copolymers of SA and CPP. These polymers have been used for the delivery of BCNU and local anesthetics from microsphere formulations [15,28,29]. In general, the solubility of the polymer and its crystallinity decrease with the increased incorporation of more rigid and hydrophobic monomers units such as CPP and CPH [12]. Polymers derived from CPH possess greater solubility in common solvents such as methylene chloride and chloroform than their CPP counterparts. Furthermore, the degradation of copolymers derived from SA and CPP or CPH is increased with increasing SA content in the copolymer [27]. In general, an increase in the content of higher molecular weight monomer in a copolymer increases the degradative lifetime. This is because an increase in the hydrophobicity (lipophilicity) of the copolymer has led to diminished water uptake.

REFERENCES

1. Bucher, J., and Slade, W. (1909). The anhydrides of isophthalic and terephthalic acids. *J. Am. Chem. Soc.* **32**, 1319.
2. Rosen, H., Chang, J., Wnek, G., Linhardt, R., and Langer, R. (1983). Biodegradable poly(anhydrides) for controlled drug delivery. *Biomaterials* **4**, 131.
3. Tamada, J. A., and Langer, R. (1993). Erosion kinetics of hydrolytically degradable polymers. *Proc. Natl. Acad. Sci. U.S.A.* **90**, 552–556.
4. Rokkanen, P. U. (1991). Adsorbable materials in orthopedic surgery. *Ann. Med.* **23**, 109–115.
5. Pistner, H., Stallforth, H., Gutwald, R., Muhling, J., Reuther, J., and Michel, C. (1994). Poly(L-lactide): A long term degradation study in vivo. Part II. Physico-mechanical behavior of implants. *Biomaterials* **15**, 439–450.
6. Bergsma, E., Rozema, F., Bos, R., and de Bruijn, W. (1993). Foreign body reaction to resorbable poly(L-lactide) bone plates and screws used for the fixation of unstable zygomatic fractures. *J. Oral Maxillofacial Surg.* **51**, 666–670.
7. Bergsman, J., de Bruijn, W., Rozema, F., Boss, R., and Boering, G. (1995). Late degradation tissue response to poly(L-lactide) bone plates and screws. *Biomaterials* **16**, 25–31.
8. Martin, C., Winet, H., and Bao, J. (1996). Acidity near eroding polylactide-polyglycolide in vitro and in vivo in rabbit tibial bone chambers. *Biomaterials* **17**, 2373–2380.
9. Leong, K., D'Amore, P., and Langer, R. (1986). Bioerodible poly(anhydrides) as drug carrier matrices. II. Biocompatibility and chemical reactivity. *J. Biomed. Mater. Res.* **20**, 51.
10. Shastri, V., Marini, P., Padera, R., Kirchain, S., Tarcha, P., and Langer, R. (1999). Osteocompatibility of photopolymerizable anhydride networks. *Mater. Res. Soc. Symp. Proc.* **530**, 93–98.
11. Langer, R. (1998). Drug delivery and targeting. *Nature* (*London*) **392**(Suppl.), 5–10.
12. Domb, A., and Langer, R. (1988). Polyanhydrides: Stability and novel composition. *Makromol. Chem., Macromol. Symp.* **19**, 189–200.
13. Langer, R. (1990). New methods of drug delivery. *Science* **249**, 1527–1533.
14. Brem, H., Kader, A., Epstein, J., Tamargo, R., Domb, A., Langer, R., and Leong, K. (1989). Biocompatibility of a biodegradable, controlled-release polymer in the rabbit brain. *Sel. Cancer Ther.* **5**, 55–65.
15. Brem, H., Piantadosi, S., Burger, P., Walker, M., Selker, R., Vick, N., Black, K., Sisti, M., Brem, S., and Mohr, G., *et al.* (1995). Placebo-controlled trial of safety and efficacy of intraoperative controled delivery by degradable polymers of chemotherapy for recurrent gliomas. The polymer-brain tumor treatment group. *Lancet* **345**, 1008–1012.
16. Domb, A., Mathiowitz, E., Ron, E., Giannos, S., and Langer, R. (1991). Polyanhydrides. IV. Unsaturated and crosslinked polyanhydrides. *J. Polym. Sci., Part A* **29**, 571–579.
17. Anseth, K., Shastri, V., and Langer, R. (1999). Photopolymerizable degradable polyanhydrides with osteocompatibility. *Nat. Biotechnol.* **17**, 156–159.
18. de Abajo, J., Babe, S., and Fontan, J. (1971). Copolyanhydride-Imides. *Angew. Makromol. Chem.* **19**, 121–134.
19. Ganzalez, J., de Abajo, J., Babe, S., and Fontan, J. (1976). Polyanhydride-Imides. *Angew. Makromol. Chem.* **55**, 85–96.
20. Staubli, A., Ron, E., and Langer, R. (1990). Hydrolytically degradable amino acid containing polymers. *J. Am. Chem. Soc.* **112**, 4419–4424.
21. Uhrich, K., Gupta, A., Thomas, T., Laurencin, C., and Langer, R. (1995). Synthesis and characterization of degradable poly(anhydride-co-imide). *Macromolecules* **28**, 2184–2193.
22. Conix, A. (1958). Aromatic polyanhydrides: A new class of high melting fiber-forming polymers. *J. Polym. Sci.* **29**, 343.
23. Leong, K., Simonte, V., and Langer, R. (1987). Synthesis of poly(anhydrides): Melt-polycondensation, dehydrochlorination, and dehydrative coupling. *Macromolecules* **20**, 705.
24. Domb, A., Ron, E., and Langer, R. (1988). Poly(anhydrides). 2. One-step polymerization using phosgene or diphosgene as coupling agents. *Macromolecules* **21**, 1925–1929.

25. Conix, A. (1966). Poly(1,3-bis(*p*-carboxyphenoxy)-propane anhydride). *Makromol. Synth.* **2**, 95–99.
26. Domb, A., and Langer, R. (1987). Synthesis of high molecular weight poly(anhydrides). *J. Polym. Sci.* **23**, 3375.
27. Domb, A., Elmalak, O., Shastri, V., Ta-Shma, Z., Masters, D., Ringel, I., Teomim, D., and Langer, R. (1997). Polyanhydrides. *In* "Handbook of Biodegradable Polymers," pp. 135–159. Harwood Academic Publishers, New York.
28. Masters, D., Berde, C., Dutta, S., Turek, S., and Langer, R. (1993). Sustained local anesthetic release from bioerodible polymer matrices: A potential method for prolonged regional anesthesia. *Pharm. Res.* **10**, 1527–1532.
29. Masters, D., Berde, C., Dutta, S., Griggs, C., Hu, D., Kupsky, W., and Langer, R. (1993). Prolonged regional nerve blockade by controlled release of local anesthetic from biodegradable polymer matrix. *Anesthesiology* **79**, 340–346.

CHAPTER 53

Synthesis of Synthetic Polymers: Poly(ortho esters)

Kirk P. Andriano, Robert Gurny, and Jorge Heller

INTRODUCTION

Tissue engineering involves the development of functional tissue with the ability to replace missing or damaged tissue. This may be achieved either by transplanting cells seeded into a porous material having open pores or by relying on ingrowth of cells into such a material, which in both cases develops into normal tissue. The latter process is known as tissue induction. An open-pore structure is important for tissue ingrowth into the interior of the scaffold and for the exchange of nutrients and waste products between its interior and exterior. This process can be enhanced by the release of soluble signaling factors such as growth factors from the scaffold that attracts cells to the area of tissue repair. In either case, the porous material acts as a scaffold, which provides cells with a temporary three-dimensional platform for cell adhesion, proliferation, and differentiation [1].

The synthetic absorbable polymers most often utilized for three-dimensional porous scaffolds in tissue engineering are the poly(α-hydroxy acids) [1,2]. These are the homopolymers of poly(L-lactide) (PLLA) and poly(glycolide) (PGA), as well as poly(DL-lactide-co-glycolide) (PLGA) copolymers. However, these polymers were originally developed as bioerodible sutures, and they undergo a bulk erosion process accompanied by abrupt release of lactic and glycolic acids with possible toxic effects.

For tissue engineering applications, as well as drug delivery applications, surface erosion, as opposed to bulk erosion, is much preferred for two important reasons. First, surface erosion leads to a gradual mass loss, which not only minimizes possible toxic effects of hydrolysis products, such as reduced osteogenesis, but the device also retains structural integrity for a major portion of the erosion process, a property that is clearly of interest for bone ingrowth and fracture fixation applications. Second, when a surface-eroding polymer is used to fabricate device, the pores will gradually increase in size as the erosion proceeds, which may be a significant advantage for bone ingrowth applications.

Poly(ortho esters) are a family of hydrophobic polymers that undergo a heterogeneous hydrolysis process that is predominantly, but not exclusively, confined to the polymer–water interface [3]. A new family of poly(ortho esters) was described in the late 1990s and details of the hydrolysis process were elucidated [4–6]. In this family, short segments of poly(glycolic acid), or poly(lactic acid) are incorporated into the polymer backbone, and when these segments hydrolyze, lactic or glycolic acid is slowly released. Because ortho ester linkages are susceptible to acid-catalyzed hydrolysis, varying the concentration of the α-hydroxy acid segments in the polymer backbone can control the erosion rate of the polymer.

In this chapter we briefly review synthesis, hydrolysis, and drug release studies of poly(ortho esters) and then describe use of this polymer in bone ingrowth applications.

EXPERIMENTAL
MATERIALS

3,9-diethylidene 2,4,8,10-tetraoxaspiro[5.5]undecane was prepared as described elsewhere [7]. Diol glycolides were prepared also, in accordance with establish protocols [4]. All other materials were purchased from Aldrich and used as received.

POLYMER SYNTHESIS

In a dry box, 8.64 g (60 mmol) of *trans*-cyclohexanedimethanol, 4.13 g (35 mmol) of 1,6-hexanediol, and 0.80 g (5 mmol) of 1,6-hexanediol glycolide were dissolved in 40 g of tetrahydrofuran (THF). Then, 21.44 g (101 mmol) of 3,9-diethylidene-2,4,8,10-tetraoxaspiro[5.5]undecane was weighed into a round-bottomed flask and added to the diol solution with the aid of 20 g of THF, in several portions. The flask was sealed and removed from the dry box and rapidly connected to a condenser and a nitrogen inlet, whereupon a few drops of *p*-toluenesulfonic acid solution (10 mg/ml) were added.

After the reaction exotherm had subsided, the solution was slowly dropped into 3 liters of stirred methanol containing about 1000 ppm of triethylamine. After isolation by filtration and drying in a vacuum oven at 40°C for 24 h, 33.3 g (95% yield) of polymer was obtained.

PREPARATION OF SCAFFOLDS

PLGA scaffolds were prepared from a commercially available copolymer of 50:50 poly(DL-lactic-*co*-glycolic acid) purchased from Medisorb Technologies International (Cincinnati, OH) Poly(ortho ester) (POE) scaffolds were prepared from a polymer prepared as just described. Polymer characteristics are shown in Table 53.1.

Milled polymer (PLGA or POE) with particle diameters less than 300 µm was thoroughly dry-mixed with 65 wt% sieved NaCl salt particles (Aldrich Chemical Co., Tokyo, Japan). Polymers were milled in air at room temperature to a fine powder and sieved through a 300-µm mesh. Salt particle diameters after sieving ranged from 106 to 355 µm. For the POE-NaCl dry-mix, 10 wt% micronized NaHCO$_3$ (99.5%, Nacali Tesque, Inc., Kyoto, Japan) was added to the dry mix to protect the acid-sensitive POE polymer from degradation during scaffold processing. Choice of salt particle diameters was based on the work of Robinson *et al.* [8], which suggest that pore sizes of this range will support bone ingrowth in porous absorbable polymer scaffolds.

Six hundred milligrams of PLGA-NaCl or 500 mg of POE-NaCl-NaHCO$_3$ dry mix was poured into a circular–cylindrical Teflon mold (6 mm in diameter and 40 mm in length) and hot-compression-molded. For PLGA composites, the press temperature was 80°C and the compressive load was 350 g. For the POE composites, the press temperature was 100°C and the compressive load was 500 g. Hot-compression-molding was accomplished by placing the Teflon mold with dry mix into a 100-mL graduated glass cylinder and adding 100- or 50-g steel weights to achieve the desired compressive load. Finally, the glass cylinder–Teflon

Table 53.1. Polymer Characteristics

Polymer property	Poly(DL-lactic-*co*-glycolic acid)	Poly(ortho ester)
Molecular weight Da[a]	95,000	47,000
Polydispersity	1.94	1.79
Inherent viscosity (ml/g)[b]	43.4	25.6
Glass transition temperature, °C[c]	40–45	73

[a] Determined using gel permeation chromatography with polystyrene standards.
[b] Determined using 0.5% polymer in chloroform (w/w) at 25°C.
[c] Determined using differential scanning calorimetry against indium standards at a heating rate of 10°C/min.

mold–steel weight assembly was placed in a preheated electric oven for 1 h at the desired temperature. The glass cylinder was convenient for keeping the load centered and was easy to handle.

The polymer–salt composites were removed from their molds after cooling to ambient temperature, with compressive loads maintained during cooling. Four composites samples were placed in 2 liters of Milli-Q-filtered deionized-distilled water (0.2-μm filter) for 48 h at room temperature, and the water was changed every 12 h. NaCl and NaHCO$_3$ were leached out, leaving porous scaffolds with geometry identical to the shape of the mold. Scaffolds were air-dried for a minimum of 12 h and then vacuum-dried at 0.05 torr and room temperature for at least 12 h. The resulting scaffolds, prepared with a 65:35 ratio of salt to polymer, had porosities ranging from 65 to 70%. The procedure described by Thomson *et al.* was used to determine porosity [9]. For scaffolds with large pore sizes such as those used in this study, mercury porosimetry underestimates porosity [10]. A more accurate determination of porosity (ε) is possible by measuring the weight, height, and diameter of each sample. From these measurements the apparent density of the scaffold, ρ^*, was calculated and the porosity determined:

$$\varepsilon = 1 - \frac{\rho^*}{\rho}$$

where ρ is the density of the solid polymers.

The experimentally determined density of 50:50 PLGA was 1.35 g/ml and that for the POE was 1.00 g/ml. This porosity range is similar to that of trabecular bone and, as suggested by Robinson, the minimum porosity needed for bone ingrowth into porous degradable polymer scaffolds [8].

Scanning electron micrographs of the freeze-fractured cross-sectional areas of PLGA and POE scaffold disks show an open, interconnected pore structure (Fig. 53.1).

GELATIN COATING

Scaffolds were coated with cross-linked acidic gelatin to provide uniform surface properties for cell attachment. Cross-linked acidic gelatin was chosen to coat scaffolds because it is a known carrier for delivery of growth factors such as basic fibroblast growth factor (bFGF) [11], which is a powerful angiogenic factor that could be used for vascular formation within the scaffold interior. No immobilization of growth factors has yet been carried out.

A custom-made Plexiglas jig and a razor blade, were used to cut cylindrical scaffold discs, 6 mm in diameter and 3 mm thick, from the original scaffold. Scaffold discs were prewetted by filtering 5 ml of deionized-distilled water through the specimen. This was followed by filtration with 10 ml of room temperature 1% acidic gelatin solution. The

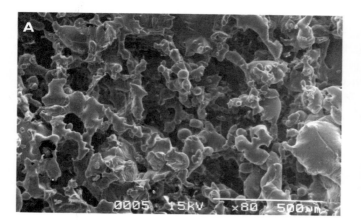 

Fig. 53.1. Scanning electron micrographs of freeze-fractured scaffold cross sections: (A) 50:50 PLGA scaffold and (B) POE scaffold. Original magnification 80×.

solution was prepared weight-to-weigh with acidic gelatin from alkaline-processed bovine bone (Nitta Gelatin Co., Osaka, Japan, M_w 99,000 Da, isoelectric point 4.9) and Milli-Q-filtered deionized-distilled water. Finally, the scaffold discs were immersed in excess 1% acidic gelatin solution at 37°C for 30 min. The gelatin coating was cross-linked by immersion of the scaffold discs into excess 5 mM glutaraldehyde–phosphate-buffered saline (PBS), for 12 h at 4°C. The cross-linking solution was prepared with 25% (w/v) glutaraldehyde solution (Wako Pure Chemical Industries, Osaka, Japan) and commercially available PBS (Dulbecco's PBS 0.1 M, pH 7.4, Nissue Pharmaceutical Ltd., Tokyo). The cross-linking reaction was neutralized by incubating treated scaffolds in excess 50 mM glycine (Nacalai Tesque, Inc., Kyoto, Japan) solution for 60 min at 37°C, followed by soaking in excess deionized-distilled water for 30 min (3×) and dried (air-drying at room temperature for 12 h followed by vacuum-drying at 0.05 torr and room temperature for a minimum of 12 h). The water content of the gelatin microspheres cross-linked by this method was reported to be 90% [12]. It was anticipated that when hydrated, the cross-linked gelatin coating on scaffold discs would have a similar water content, since the cross-linking conditions were identical.

RESULTS

POLYMER SYNTHESIS, HYDROLYSIS, AND PROPERTIES

A generic representation of polymer synthesis is shown in Scheme 53.1. A detailed characterization of the polymer has been published [5].

The hydrolysis of a polymer prepared using 1,6-hexanediol monoglycolide is shown in Scheme 53.2. A detailed study of the hydrolysis process has been published [6].

The major advantage of this polymer is that its properties and erosion rates can be independently varied by controlling the nature of the R group in the diol and the latent acid diol, and by varying the relative proportion of these two diols. The effect of latent acid concentration in the polymer on erosion rates is shown in Fig. 53.2 [13].

The data, which show a good correlation between latent acid concentration and erosion rates, also show that a substantial induction period is observed for very low latent acid concentrations. This induction period has been discussed [6] and is a consequence of the highly hydrophobic nature of the polymer: no hydrolysis can take place until sufficient water has penetrated the polymer matrix. The induction period can be decreased by increasing polymer hydrophilicity and the amount of latent acid diol, and by using lower molecular weight materials. While not desirable in many applications, an induction can be advantageously used to delay polymer erosion or the release of an active agent incorporated into the polymer.

DRUG RELEASE

The polymer has been extensively investigated as a bioerodible drug delivery platform in a variety of applications, and the release of certain compounds that can affect the behavior of cells is clearly of interest in tissue engineering. Of particular interest is the ability to release macromolecules (e.g., bone morphogenic proteins in a controlled manner).

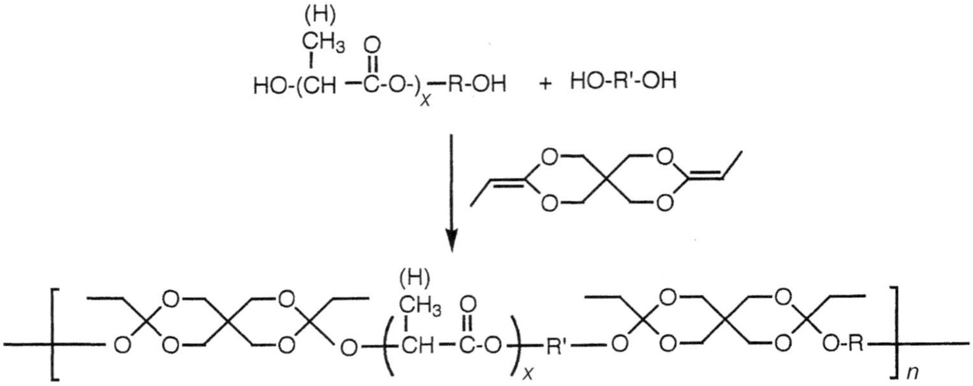

Scheme 53.1.

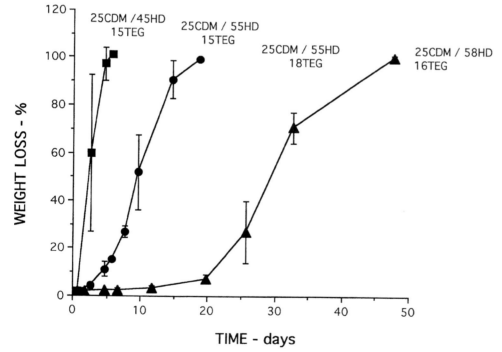

Scheme 53.2.

Poly(ortho esters) have been investigated as a delivery systems for peptides and proteins with the objective of delivering these compounds by well-defined kinetics without an initial burst and with full retention of activity. Since it is known that many proteins lose activity when exposed to an organic solvent–water interface, a solventless method of preparing such delivery systems was used.

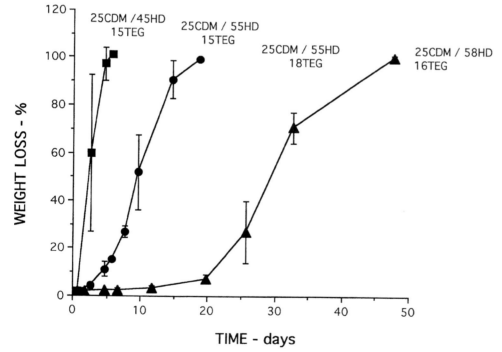

Fig. 53.2. Effect of triethyleneglycol monoglycolide (TEG mono-GL) concentration on erosion rates for poly(ortho esters) prepared from 3,9-diethylidene-2,4,8,10-tetraoxaspiro[5.5]undecane and mixtures of diols as shown: (■), 15 mol% TEG mono-GL; (●), 5 mol% TEG mono-GL; (▲), 1 mol% TEG mono-GL; 0.05 M phosphate buffer, pH 7.4, 37°C. CDM is trans-cyclohexanedimethanol, HD is 1,6-hexanediol, TEG is triethylene glycol. Reprinted with permission from Ng et al. [13].

Because mechanical properties of poly(ortho esters) can be manipulated within very wide limits by using suitable diol pairs, polymers that can be extruded at moderate temperatures ranging between 50 and 70°C can be readily prepared, and such temperatures are low enough to permit many dry proteins to survive the extrusion process with full retention of their activity.

Figure 53.3 shows release of the model protein fluorescein isothiocyanate–bovine serum albumin (FITC-BSA) from a polymer extruded at 70°C, as well as weight loss of the extruded rods. These results are highly encouraging in that excellent linear kinetics with concomitant weight loss has been achieved with only a negligible initial burst. The induction period has been discussed [6] and is the result of the highly hydrophobic nature of the polymer, which makes water penetration difficult. While in many drug delivery applications an induction period is not desirable in some applications, such as delivery of vaccines, a delay prior to vaccine release is of interest. A delay prior to the onset of degradation is also of interest in fracture fixation applications, where the device must retain high initial strength and then gradually weaken as bone healing tales place.

The method has been validated by extruding a rod containing 15 wt% recombinant human growth hormone (rhGH) at 70°C. When the protein was extracted from the rods, it was found to contain 90.5% native protein, which compares very favorably to 95.2% native protein in the rhGH prior to extrusion.

DISCUSSION

One of the key design characteristics of absorbable polymer scaffolds in tissue engineering of bone is the coordination of controlled polymer mass loss with new tissue formation [14]. This appears difficult to achieve with the bulk hydrolyzing poly(α-hydroxy acid) polymers that undergo a sudden mass loss only after a prolonged period of reactive hydrolysis [15–19]. Absorbable scaffolds based on these polymer systems may limit new tissue formation because of the initial lack of change in pore size and porosity. For some weight-bearing applications, a precipitous mass loss could lead to massive structural collapse of the scaffold before bone formation is mature enough to mechanically handle normal *in vivo* stresses. Surface-hydrolyzing polymer systems that undergo controllable mass loss at the onset of hydrolysis may allow for an increase in new tissue formation and a gradual transfer of load to maturing bone in weight-bearing applications.

In addition, in a rabbit tibial bone chamber model, PLGA copolymers have been shown to significantly reduce osteogenesis of healing bone during polymer erosion [20,21]. Controlled polymer mass loss from surface-hydrolyzing absorbable polymers could possibly minimize the negative effect that bulk hydrolysis appears to have on osteogenesis.

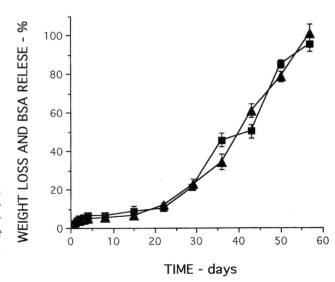

Fig. 53.3. Release of FITC-BSA from poly(ortho ester) rods prepared by extruding a mixture of protein and polymer at 70°C, 0.05 M phosphate buffer, pH 7.4, 37°C. Polymer prepared from 3,9-diethylidene-2,4,8,10-tetraoxaspiro[5.5]undecane and 1,4-pentanediol/1,6-hexanediol monoglycolide (95:5). Rods, 1 × 10 mm²; FITC-BSA loading, 15 wt%. (▲), Weight loss; (■), FITC-BSA release.

Table 53.2. *Bone Mineral Density[a]: n = 3, mean ±SD*

Time (weeks)	PLGA–gelatin (mg/cm^2)	POE–gelatin (mg/cm^2)
6	43.1 ± 2.5	57.2 ± 1.1
12	38.8 ± 5.0	53.0 ± 8.3

[a] Determined by dual-energy x-ray absorptiometry.

Figure 53.4 shows percent mass loss and percent change in inherent viscosity for 50:50 PLGA and POE scaffold discs after exposure to phosphate-buffered saline, pH 7.4, at 37°C for various time periods. These two polymer scaffolds have shown very different *in vitro* degradation kinetics. The rapid decrease in viscosity (indicative of molecular weight change) for PLGA samples before any real significant mass loss has been well documented for bulk hydrolyzing polymers [22]. The linear response in mass loss for POE (r^2 = 0.9800) with little initial decrease in viscosity suggests that the rate of hydrolysis at the surface is greater than in the interior. This early mass loss also suggests that pore volume may be increasing initially at a greater rate in the POE scaffolds than for PLGA scaffolds.

Table 53.2 summarizes results of bone mineral density measurements in noncritical calvarial defects of rabbits at 6 and 12 weeks after implantation of gelatin-coated PLGA and POE scaffolds discs. A greater amount of average new bone formation was found in the POE scaffolds than in the PLGA scaffolds at both 6 weeks (24.7% more bone) and 12 weeks (26.8% more bone). Scaffold porosity was similar to that of human cancellous bone (65–70%), and pore size range (106–300 μm) was similar to that for poly(DL-lactide) scaffolds reported to support bone ingrowth by osteoconduction into porous absorbable polymer scaffolds [8]. The report of Robinson *et al.* was the first manuscript to evaluate the effects of porosity and pore size on bone ingrowth into synthetic absorbable polymer scaffolds. These results, like ours, showed extremely limited bony ingrowth by osteoconduction. As already mentioned, *in vitro* results suggested that in POE scaffolds pore volume increases with increasing exposure time, suggesting in turn that improved coordination of polymer mass loss to new bone formation appeared to be achieved and may be the reason for the small observed difference. However, further more detailed studies are needed to confirm or reject this hypothesis.

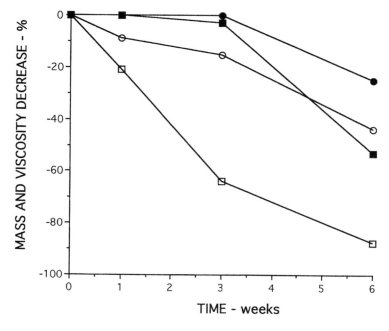

Fig. 53.4. *Percent mass loss (n = 3, mean ± SD) and percent change in inherent viscosity (data are the pool of three samples) of scaffolds exposed to phosphate-buffered saline at pH 7.4 and 37°C: (○), PLGA percent mass loss; (□), PLGA percent change in inherent viscosity; (●), POE percent mass loss; (■), POE percent change in inherent viscosity.*

Figure 53.5 shows photographs of 6- and 12-week histological sections of PLGA/gelatin and POE/gelatin calvarial/implant cross sections. At 6 weeks, the PLGA defect was filled with an inner fibrous and an outer fibroadipose layer separated by a thin central mass of loose fibrovascular tissue containing large numbers of multinucleated foreign body giant cells, lymphocytes, monocytes, and plasma cells surrounding multiple clear pore spaces lined by multinucleated foreign body giant cells. There were three moderate-sized bone trabeculae present. At 6 weeks, the POE defect was filled with an inner layer of fibrous tissue and an outer layer of fibroadipose tissue with a multiple pore spaces between. The latter was composed of thick fibrous trabeculae separating large pore spaces containing small amounts of protein fibrils and granules. There were prominent multinucleated foreign body giant cells lining the pore spaces and moderate numbers of lymphocytes and monocytes scattered through the fibrous septa. There were seven small bone trabeculae at the ends of the defect. At 12 weeks, the PLGA defect was filled with inner and outer fibrous layers fused together around small amounts of central fibroadipose tissue containing a few small foreign body granulomas with pore spaces. There were three moderate sized bone trabeculae present. At 12 weeks, the POE defect was filled with inner and outer fibrous layers separated by large multiple void spaces with multiple broad fibroadipose septa, lined by multinucleated foreign body giant cells. There were minimal lymphocytes and other types of chronic inflammatory cells. There were 15 small bone trabeculae present at one edge.

The local tissue responses to PLGA and POE polymer scaffolds were similar at 6 weeks regardless of the mode of polymer degradation. The tissue response was characterized by a mild inflammatory reaction, with chronic inflammatory cells (lymphocytes, monocytes, and plasma cells) and foreign body giant cells present. The only real difference observed was the morphological appearance of the scaffold cross sections (Figure 53.1). Bulk-hydrolyzing PLGA scaffolds had partially collapsed, while surface-hydrolyzing scaffolds had retained their original three-dimensional structural shape.

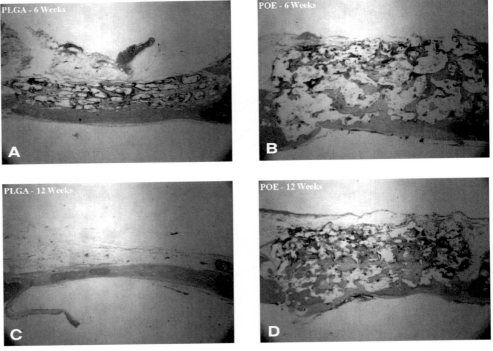

Fig. 53.5. Photographs of 6- and 12-week histological sections of PLGA–gelatin and POE–gelatin–calvarial implant cross sections: (A) PLGA at 6 weeks, (B) POE at 6 weeks, (C) PLGA at 12 weeks, and (D) POE at 12 weeks.

CONCLUSIONS

Preliminary *in vitro* and *in vivo* results suggest that improved coordination of polymer mass loss to new bone formation may be achieved better in scaffolds constructed with POE than with PLGA, The histological responses of these two polymer systems were similar. The persistence of structural integrity for POE scaffolds after 6 and 12 weeks, in comparison to 50:50 PLGA, suggests that POE scaffolds may have some advantage over PLGA scaffolds in resisting normal *in vivo* stresses for tissue engineering of bone.

REFERENCES

1. Langer, R., and Vacanti, J. (1993). Tissue engineering. *Science* **260**, 920–926.
2. Langer, R., Vacanti, J. P., Vacanti, C. A., Atala, A., Freed, L. E., and Vunjak-Novakovic, G. (1995). Tissue engineering: Biomedical applications. *Tissue Eng.* **1**(2), 151–161.
3. Heller, J. (1993). Poly(ortho esters). *Adv. Polym. Sci.* **41**, 92–107.
4. Ng, S. Y., Taylor, M. S., and Heller, J. (1997). Synthesis and erosion studies of self-catalyzed poly(ortho esters). *Macromolecules* **30**, 770–772.
5. Sintzel, M. B., Heller, J., Ng, S. Y., Taylor, M. S., Tabatabay, C., and Gurny, R. (1998). Synthesis and characterization of self-catalyzed poly(ortho ester). *Biomaterials* **19**, 791–800.
6. Schwach-Abbellaoui, K., Heller, J., and Gurny, R. (1999). Hydrolysis and erosion studies of autocatalyzed poly(ortho esters) containing lactoyl-lactyl acid dimers. *Macromolecules* **32**, 301–307.
7. Ng, S. Y., Penhale, D. W. H., and Heller, J. (1992). Poly(ortho esters) by the addition of diols to a diketene acetal. *Macromol. Synth.* **11**, 23–26.
8. Robinson, B., Hollinger, J. O., Szachowicz, E., and Brekke, J. (1995). Calvarial bone repair with porous D,L-polylactide. *Otolaryngol. Head Neck Surg.* **112**(6), 707–713.
9. Thomson, R. C., Yaszemaski, M. J., Powers, J. M., and Mikos, A. G. (1995). Fabrication of biodegradable polymer scaffolds to engineer trabecular bone. *J. Biomater. Sci., Polym. Ed.* **7**(1), 23–38.
10. Mikos, A. G., Thorsen, A. J., Czerwonka, L. A., Bao, Y., Langer, R., Winslow, D. N., and Visconti, J. P. (1994). Preparation and characterization of poly(L-lactic acid) foams. *Polymer* **35**, 1068.
11. Muniruzzaman, M. D., Tabata, Y., and Ikada, Y. (1998). Complexation of basic fibroblast growth factor with gelatin. *J. Biomater. Sci. Polymer Edn* **9**(5), 459–473.
12. Tabata, Y., Hijikata, S., and Ikada, I. (1994). Enhanced vascularization and tissue granulation by basic fibroblast growth factor impregnated in gelatin hydrogels. *J. Controlled Release* **31**, 189–199.
13. Ng, S. Y., Lopez, E., Zherebin, Y., Barr, J., Schacht, E., and Heller, J. (2000). Development of a poly(ortho ester) prototype with a latent acid in the polymer backbone for 5-fluorouracil delivery. *J. Controlled Release* **65**, 367–374.
14. Hollinger, J. O., Jamiolkowski, D. D., and Shalaby, S. W. (1995). Bone repair and a unique class of biodegradable polymers: The poly(α-esters). *In* "Biomedical Applications of Synthetic Biodegradable Polymers" (J. O. Hollingery ed.), pp. 197-222. CRC Press, Boca Raton, FL.
15. Li, S., Garreau, J., and Vert, M. (1990). Structure-property relationships in the case of the degradation of massive aliphatic poly(alpha-hydroxy-acids) in aqueous media. Part 1: Poly(DL-lactic acid). *J. Mater. Sci.: Mater. Med.* **1**, 123–130.
16. Li, S., Garreau, J., and Vert, M. (1990). Structure-property relationships in the case of the degradation of massive aliphatic poly(alpha-hydroxy-acids) in aqueous media. Part 2: Degradation of lactide-glycolide copolymers: PLA37.5GA25 and PLA75GA. *J. Mater. Sci.: Mater. Med.* **1**, 131–139.
17. Li, S., Garreau, J., and Vert, M. (1990). Structure-property relationships in the case of the degradation of massive aliphatic poly(alpha-hydroxy-acids) in aqueous media. Part 3: Influence of morphology on Poly(L-lactic acid). *J. Mater. Sci.: Mater. Med.* **1**, 198–206.
18. Therin, M., Christel, P., Li, S., Garreau, H., and Vert, M. (1992). In vivo degradation of massive poly(alpha-hydroxy acids): Validation of in vitro. *Biomaterials* **13**, 594–600.
19. Andriano, K. P., Pohjonen, T., and Tormala, P. (1994). Processing and characterization of absorbable polylactides for use in surgical implants. *J. Appl. Biomater.* **5**, 133–140.
20. Winet, H., and Bao, J. Y. (1997). Comparative bone healing near eroding polylactide-polyglycolide implants of differing crystallinity in rabbit tibial bone chamber. *J. Biomater. Sci., Polym. Ed.* **8**(7), 517–532.
21. Winet, H., and Bao, J. Y. (1998). Fibroblast growth factor-2 alters the effect of eroding polylactide-polyglycolide on osteogenesis in the bone chamber. *J. Biomed. Mater. Res.* **40**(4), 567–576.
22. Anderson, J. M. (1995). Perspectives on the *in vivo* responses to biodegradable polymers, *In* "Biomedical Applications of Synthetic Biodegradable Polymers" (J. O. Holinger ed.), pp. 223–233. CRC Press, Boca Raton, FL.

CHAPTER 54

SYNTHESIS OF SYNTHETIC POLYMERS: POLY(AMINO ACIDS)

Rebecca K. Lawson-Smith and Kimberly A. Woodhouse

P oly(amino acid) design and synthesis are becoming invaluable tools in the field of tissue engineering. Since advantageous biocompatibility properties usually are found in materials that incorporate poly(amino acids), scientists have been very interested in synthesizing these amino acids in the laboratory. This interest has developed into a major field of synthetic chemistry known as poly(amino acid) synthesis. The four major objectives in this field are as follows.

1. To verify the structure of naturally occurring peptides as determined by degradation techniques.
2. To study the relationship between structure and activity of biologically active proteins and peptides and establish their molecular mechanisms.
3. To synthesize peptides that are of medical importance, such as hormones and vaccines.
4. To develop new peptide-based immunogens.

The methods described in this chapter for the synthesis of poly(amino acids) are applicable to most polymeric peptides.

A poly(amino acid) is any polymer comprising monomer amino acids, covering the broad spectrum from oligomers to proteins and includes biologically active proteins. For the purposes of this chapter, synthetic polymeric peptides are subdivided into size categories. These size categories generally correspond to the most convenient methods for their synthesis. Short oligomers (<6 residues) are referred to as peptides, while the slightly longer versions can be called oligopeptides. Because of their size, the synthesis of these oligomers is less complex than that of longer chain polymeric peptides. Since peptides and oligopeptides can usually be purchased from a chemical supplier, their synthesis is not discussed in this chapter.

The next largest group is the group of polypeptides consisting of poly(amino acids) with fairly simple primary sequences longer than oligopeptides and generally, although not always, of lower molecular weight than proteins. Typically, these polypeptides are formed by chemical methods. Interestingly, the methodology for synthesizing these polymers has remained relatively unchanged since its conception, roughly 40 years ago. Solid phase and solution phase syntheses are still the two most common methodologies used for the production of most polypeptides. Over the last several years, some slight changes have been made in techniques for the blocking of the amino acid side chain groups, and developments in automation and purification have improved the efficiency and yield of the process. As well, newer methodologies have been developed to deal with a special class of peptides, cyclic peptides. In the final group, there are high molecular weight polypeptides with complicated secondary and tertiary structures. These poly(amino acids) are generally referred to as proteins if they are produced within an organism; biological synthesis is the most effective way of synthesizing them. As the field of recombinant biology grows, new and improved methods for recombinant protein synthesis are being developed. Some of those methods are highlighted in this chapter, with a focus on protein synthesis for tissue engineering.

Although some methods are better suited for producing certain sizes of poly(amino acids), as technologies are improved, their applications begin to overlap. For example, chemical methods can be used to synthesize some small proteins. These synthetic versions, termed "synthetic proteins," can be called "biologically active proteins" if they are produced synthetically but still retain the complexity that gives them their biological function. Some synthetic proteins are produced by using chemical techniques with additional steps for folding and the formation of disulfide bonds. Similarly, as recombinant techniques are refined and the study of protein design grows, these techniques can be used to produce engineered proteins, ones that are not normally produced by organisms. What follows in this chapter, then, is a description of new and traditional methods for chemical and biological synthesis of poly(amino acids).

CHEMICAL SYNTHESIS OF POLY(AMINO ACIDS)
BLOCKING OF THE AMINO ACID SIDE GROUPS

Amino acids are defined by their structure, which consists of a central carbon atom (called the α-carbon) surrounded by four groups, three of which are reactive. As illustrated in Fig. 54.1, these groups are an unreactive hydrogen atom, an amine group at the α-terminus, a carboxyl group at the ω-terminus, and a side chain group. It is the side chain group that distinguishes one amino acid from the other. To control the steps in any chemical synthesis of peptides and to prevent branching of the polymer, it is necessary to block the side chain and effectively shield it from the reaction. This task is accomplished by covalently attaching a large, nonreactive or barely reactive chemical group to the side chain. Common side chain blocking groups consist of methyl groups or cyclic hydrocarbons. Table 54.1 lists some readily available amino acids with their common blocking groups.

Approximately 35 commercial companies world wide manufacture and sell custom-made peptides and a variety of premade peptides through catalogs [1]. Generally, these companies offer a large variety of blocked amino acid derivatives that are suitable for most syntheses of poly(amino acids). Individuals requiring unusual derivatives can speak to companies about custom-made amino acids, but it is more than likely that the onus will be on the individual to produce his or her own.

SOLID PHASE SYNTHESIS

Solid phase synthesis is the one-dimensional growth of a polypeptide from a solid support by the controlled sequential addition of monomers [2]. This method allows each amino acid to be selected according to the desired sequence of the final peptide. Since the addition of each amino acid can be confirmed before the next is attached, solid phase synthesis allows ultimate control in the design of poly(amino acids) and polypeptides. Unfortunately, the length of peptide that can be produced is limited by the procedure. Early machines could produce only polypeptides with fewer than 40 residues [3]. Advances in technology now allow synthesis of molecules containing up to 130 residues [2]. Today, most solid phase synthesis methodologies are best suited for producing small to medium-length polypeptides whose primary structure is nonrepetitive yet crucial for its function. There are an endless

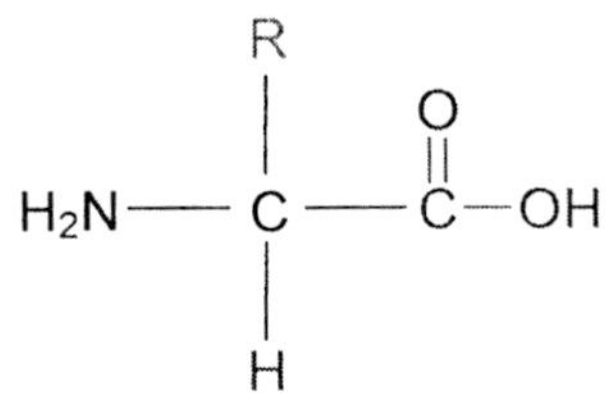

Fig. 54.1. An amino acid with its reactive groups. Each amino acid is distinguished by its side chain group. The carboxyl terminus and amine terminus of two amino acids can react to form a peptide bond.

Table 54.1. A Selection of Amino Acids and Their Most Common Side Chain Blocking Groups[a]

Amino Acid	Blocking Groups	Amino Acid	Blocking Groups
Arginine	Boc, Tos, Mts, MBS	Histadine	im-benzyl
Asparagine	Tmob, Trt, *b*-benzyl		Boc, Trt, Bom, Tos
	OtBu, *o*-Bzl, *o-c*-Hex	Hydroxyproline	*o*-acetyl
	α-b-n-2hydroxyethyl	Lysine	*ε*-CBZ, *ε*-Boc
	α-b-n-3hydroxypropyl		*ε*-Fmoc, Mtt
Cysteine	*s*-benzyl, *s*-CBZ	Methionine	O
	Acm, Trt, 4-MeBzl	Ornithine	*d*-CBZ
Glutamine	Tmob, Trt	Serine	*o*-Acetyl, *o*-CBZ
	OtBu, *g*-benzyl, *g*-ethyl		Bzl, *o-t*-Butyl
	o-Bzl, *o-c*-Hex	Threonine	*t*-Butyl, Bzl
	g-n-hexyl, *g*-methyl	Tryptophan	CHO, Mts
	n-3-hydroxypropyl	Tyrosine	*o*-Acetyl, *o*-CBZ,
			TBu, Br-Z, Bzl

[a] Amino acids with relatively unreactive side chain groups, like alanine, seldom require blocking.

variety of applications of these molecules, among which are raising anti-peptide antibodies, studying enzyme substrates and the binding properties of viral proteins, and identifying and locating gene products [4].

The traditional condensation polymerization reactions employed for peptide synthesis were basically ineffective before Dr. Merrifield [5] began designing the process that is still used today. The purification that was necessary after the addition of each amino acid greatly decreased the process yield. The yield was decreased so much that to produce such molecules was unreasonably time-consuming and expensive. Merrifield's innovation improved yield by eliminating the multiple purification steps and automating the process to alleviate the hours of manual labor. The result, as shown in Fig. 54.2, was a cyclic process that first anchored an amino acid—actually the terminal amino acid for the proposed peptide—to a solid resin via its carboxyl group. The resin is completely insoluble in the various solvents; hence all the intermediate peptide products are also rendered insoluble. They can, therefore, be purified by dissolving the unwanted by-products and reagents and washing them away. The second amino acid, with all but one of its reactive groups protected, is then added to the first through the steps of activation, coupling, and deprotection. The incoming amino acid is "activated" by converting its terminal carboxylic acid to an ester. The ester and amine terminus of the anchored amino acid are "coupled," forming a peptide linkage. Next, the protecting group on the amino end is removed, "deprotecting" it, and the whole process

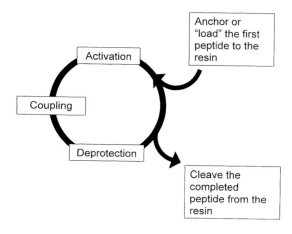

Fig. 54.2. The cyclic process of solid phase peptide synthesis. After the primary amino acid has been anchored to a resin, the cycle of deprotection, activation, and coupling begins. An amine end-protecting group is removed from the attached residue, to which the incoming activated amino acid is coupled. In this way, amino acids are sequentially added until a poly(amino acid) of highly controlled sequence has been produced. Once the full polypeptide has been synthesized, it is chemically removed from the resin.

is repeated with succeeding amino acids until the entire polypeptide has been synthesized. A more detailed description of this process is depicted in Fig. 54.3. Upon completion, the peptide is chemically cleaved from the resin and collected. The covalent attachment of the initial amino acid to the resin drastically improves the yield, and the repetitive nature of the process readily allows for automation.

One major concern in the synthesis of peptides is racemization, the change in rotation of the amino acid along the chain backbone [6]. Various studies involving peptide synthesis have also investigated the structure of the polypeptide to determine whether racemization had occurred. Racemization has been known to occur, but not to any great extent, and it is reasonably easy to control. The easiest way to measure racemization is to compare the optical rotation of the synthesized peptide with that of a compound known to have the desired optical rotation [2]. Unfortunately, racemization analysis calls for a sample of peptide taken from the resin, which makes this test difficult to do during automated synthesis.

Besides testing for racemization, other tests are performed on growing polypeptides to ensure that the series of reactions are successful [7–9]. If even one residue in the sequence is not successfully joined to the polymer, a waste product is created. Without online testing, the undesired poly(amino acids) would have to be filtered out after all the reactions were completed, resulting in extremely low yield. With online testing, the synthesis is monitored, either spectrophotometrically or chemically, throughout the entire process, and steps are repeated until sufficient efficiency is achieved.

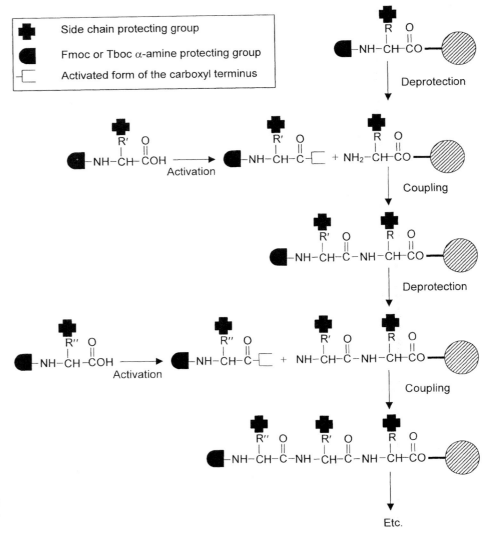

Fig. 54.3. A detailed description of the Merrifield solid phase peptide synthesis method.

Resins

Merrifield selected a styrene–divinyl benzene copolymer as the resin for his initial auto-mated protocol [10]. Selection of resin is still a critical design criterion. Ideally, the resin will be nonreactive during coupling but can be functionalized for the covalent attachment of the terminal residue. The resin itself, as well as the resin–amino acid anchor, should be resistant to most solvents yet still allow chemical cleavage once the synthesis has been completed. For Merrifield, this meant that the aromatic rings of the polymer were partially chloromethy-lated with chloromethyl methyl ether, with anhydrous stannic chloride as a catalyst [2,10]. The resulting chlorinated group reacted readily with the carboxylic group forming an acti-vated ester. The simple protocol in Protocol 1 describes a method for activating a styrene resin and covalently attaching the terminal amino acid.

PROTOCOL 1. ANCHORING THE PRIMARY AMINO ACID TO A WANG RESIN[1]

EQUIPMENT AND REAGENTS

- Round bottomed flask and drying tube
- 250-ml flasks
- RotoMix mechanical shaker (or equivalent)
- Vacuum chamber
- Wang resin (Advanced ChemTech)
- HOBt
- Diisopropylcarbodiimide (DIC)
- Chloroform
- Dimethylformamide (DMF)
- Carboxyllic acid
- Primary amino acid (DMAP)
- Acetic anhydride
- Pyridine
- Dichloromethane (DCM)
- Methanol (MeOH)

METHOD

1. Prepare a solution of 9:1 (v/v) chloroform/DMF.
2. In a round-bottomed flask, suspend the resin in approximately 15 ml of the chloroform–DMF per gram of resin.
3. In a separate flask, dissolve 1.5–2.5 equiv (relative to the resin) of the carboxylic acid in a minimum amount of DMF. Add the same equivalency of HOBt.
4. Stir the mixture until the HOBt dissolves, adding enough DMF to bring it into solution if necessary. Add this solution to the resin.
5. In a separate flask, dissolve 0.1 equiv (relative to the resin) of DMAP in a minimum amount of DMF.
6. Add 1.0 equiv (relative to the amino acid) of DIC to the resin mixture, then add the DMAP solution. Equip the flask with a drying tube.
7. Agitate with a mechanical shaker 2–3 h at room temperature.
8. Add 2 equiv (relative to the resin) of acetic anhydride and pyridine to the reaction flask. Mix 30 min at room temperature to end-cap any unreacted hydroxyl groups on the resin.
9. Remove a small sample of the resin and wash it with DCM. Use the Kaiser test to test for free amino groups. If the test is positive for amino groups, add 1 equiv of acetic anhydride and pyridine to the reaction flask and mix for 30 min.

---

[1]Copyright Advanced Chem Tech, 1999. Used with permission. See www.peptide.com/support/resinprotocols.htm.

10. Filter the resin in a fine-porosity sintered glass funnel. Wash three times with DMF. Wash three times with DCM. Wash three times with MeOH.

11. Dry the resin *in vacuo* to a constant weight.

Most companies that supply equipment for automated synthesizers will provide resin columns that are already activated. The columns contain the plastic solid support that either bears reactive groups for attachment of an amino acid or is preloaded. (Preloaded columns already have the first residue anchored to the support). When one is using columns that contain only reactive groups, the first amino acid residue must be attached via its carboxyl group, a method detailed in Protocol 1. It is important to realize that the type of linker provided by the resin will determine the type of anchoring bond, hence the kind of peptide that results from the synthesis after cleavage. For example, resins with a peptide acid linker (PAC) result in a peptide with a ω-carboxylic acid terminus, while peptide amine linkers (PAL) result in amine-terminated peptides.

Other resins besides poly(styrene) have been investigated for solid state peptide synthesis (SSPS) and are being used, although they are not nearly as common because the poly(styrene) variety is so effective. The newest of these resins are poly(ethylene glycol-*graft*-polystyrene) (PEG-PS) [11] and cross-linked ethoxylate acrylate resin (CLEAR) [12], which are increasing in popularity because of their favorable mechanical and swelling properties. If unusual chemistry is being used for SSPS, an alternative resin may be more appropriate. The protocol for activating poly(styrene) could be used as a guideline, should a researcher require a solid support that is not provided by suppliers. State-of-the-art technology for resins is currently focused on the shape of the support, not the polymer chemistry. One company has been developing a system that uses individual wafers instead of beads, a technique that is reported to result in higher purity products [1] and to allow the simultaneous synthesis of many peptide chains.

Fmoc and Tboc Chemistry

After the first amino acid has been anchored to the resin, the synthesis cycles through a sequence of steps, the first of which is deprotection. As discussed in the earlier section on blocking chemistry for the side chain groups, there are three reactive groups comprising an amino acid. The overall scheme for solid phase synthesis relies on the fact that all but one of the reactive groups is blocked, as was highlighted in Fig. 54.3. While the blocking of the side groups is relevant for all chemical poly(amino acid) synthesis, blocking of the other groups is crucial for SSPS. 9-Fluorenylmethyoxycarbonyl (Fmoc) and *tert*-butyloxycarbonyl (Tboc) groups are used to block the amine terminus of the residue involved in coupling [13,14]. They are depicted in Fig. 54.4.

All amino acids used in SSPS begin with the Fmoc or Tboc groups attached. These groups are removed during deprotection. The method for removing these protecting groups is a delicate matter and varies depending on which of the groups was used. Either way, it is essential that none of the other bonds, particularly any of the side chain blocking groups or the amino acid–resin anchoring bond, be disturbed during the cleavage of these groups. To remove Fmoc groups from a growing chain, the resin and its attached peptide are covered in a 20% piperidine solution in dimethylformamide (DMF) for 30 min [15].

Activation and Coupling

Once the amine end of the most recently added amino acid has been exposed, it is coupled to an activated form of the next residue. The design of any coupling method should

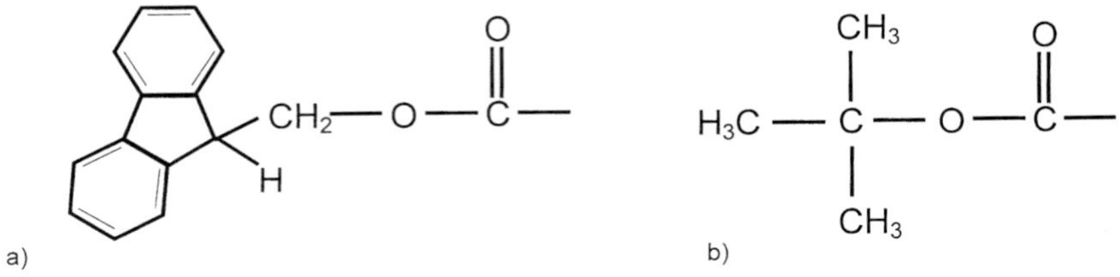

Fig. 54.4. Chemistry of the (a) Fmoc and (b) Tboc α-amine blocking groups.

maximize reaction efficiency while minimizing side product formation and racemization. There are four major coupling methods used in solid phase peptide synthesis. Each of these methods is associated with an activating or activated compound for which it is named: active esters, preformed symmetrical anhydrides, acid halides, and *in situ* reagents [16]. The active ester, symmetrical anhydride, and acid halide methods all involve activating the incoming amino acid prior to its addition to the column, while the *in situ* methods activate the amino acid when it is in the column. Fmoc synthesis favors methods that employ active ester derivatives; they offer advantages like high reactivity of isolated intermediates and simplicity of released by-products [16]. Pentafluorophenyl (OPfp) ester derivatives of amino acids are the ones most commonly used in the active ester methods. These (OPfp) ester varieties perform much better than their *p*-nitrophenyl and N-hydroxysuccinimide (ONSu) predecessors, which were associated with side product formation [17].

Because of their versatility, *in situ* methods are also very popular. In these methods, an amino acid is added to the column with a coupling agent that converts the amino acid to a highly reactive form. Many coupling agents have been investigated for use in solid phase peptide synthesis. Of them, the carbodiimides are the most widely used. They work by converting amino acids to symmetrical anhydrides that convert to esters in the presence of additives. Dicyclohexyl carbodiimide (DCC), which has by-products that are soluble in tetrafluoroacetic acid (TFA) and diisopropylcarbodiimde (DIPCDI), which has by-products that are soluble in DMF, are best used during Tboc and Fmoc syntheses, respectively [18]. The early carbodiimide methods are slowly being replaced with methods that use pyrol and aminium salt coupling agents [19]. The pyrols and aminium salts convert an amino acid to an ester in the presence of a base. (Benzotriazol-1-yloxy)tris(dimethylamino)phosphonium hexafluorophosphate (BOP), (benzotirazol-1-yloxy)tris(pyrrolidino)phosphonium hexafluorophosphate (PyBOP), N-[(1H-benzotriazol-1-yl)(dimethylamino) methylene]-N-methyl-methanaminium hexafluorophospate N-oxide (HBTU) and its tetrafluoroborate salt TBTU, and finally N-[(dimethylamino)-1H-1,2,3-triazolo[4,5-*b*]pyridin-1-yl- methylene]-N-methylmethanaminium hexafluorophospate N-oxide (HATU), are the best examples of the newest coupling agents. BOP, PyBOP, HBTU, HATU, and TBTU have shown to be very effective coupling reagents, acting with little side product formation and no significant racemization [19,20]. In 2000 Li and Xu published a report of a novel coupling agent they have developed. Use of their reagent [(1H-benzotriazol-1-yloxy)-N,N-dimethylmethaniminium hexachloroantimonate (BOMI)] results in less racemization and shorter reaction times than are reported with other *in situ* coupling agents [20]. It is also worth noting that coupling agents have been used with preformed ester derivatives of amino acids in an attempt to improve efficiency.

The typical way of maximizing the speed and efficiency of the coupling reaction is to add a catalyst. The catalysts 1-hydroxy-7-azabenzotriazole (HOAt) and 1-hydroxybenzotriazole (HOBt) are used with preformed esters and during *in situ* methods. HOAt has been described as being superior to HOBt as an additive for solid phase synthesis [21]. Together, HATU and HOAt enhance coupling yields, shorten coupling times, and reduce racemization better than most other combinations. Another effective additive for solid phase synthesis coupling reactions is diisopropylethylamine (DIEA). Although it is not a catalyst, DIEA is a base that activates aminium salts and pyrols. The best methods for activation and coupling in solid phase synthesis rely on carefully combining catalysts, additives, and activating agents, rather than any one compound on its own.

Monitoring the Reactions

In solid phase peptide synthesis, to ensure that the desired peptide structure is achieved, it is important for all the reactions to go to completion before the next stage is started. There are two precautions that can be taken to aid in achieving the desired structure. The first is to apply Le Châtelier's principle and perform the synthesis such that there is always a large excess of reactants [2]. The second precaution is to monitor the reactions as they proceed. After deprotection and after coupling are the best times to sample the synthesis for reaction completion. Automatic synthesizers generally do this with spectrophotometric techniques, such as nuclear magnetic resonance (NMR) or infrared spectroscopy (IR), although near-IR

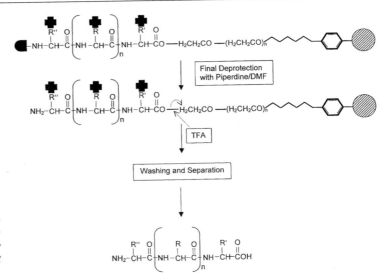

Fig. 54.5. The reaction chemistry in removal of a completed peptide from the resin. The strong base used with Fmoc synthesis is shown. Cleavage of the peptide from the resin also results in deprotection of the side chain blocking groups.

spectroscopy is the preferred method [8,9]. The Kaiser test, using the ninhydrin colormetric technique, can also be used, wherein the addition of ninhydrin and the application of heat turn a solution of exposed amine groups blue. Merrifield used the Kaiser test, but it is difficult to perform during the synthesis process [22].

Table 54.2. The Deprotection Step Reagents Compatible with Various Side-Group-protected Amino Acids[a]

Amino acid	Blocking group	Fmoc TFA/TMSBr	HF	Tboc TFMSOTf	TMSOTf
Arg	Boc	×			
	Tos		×		
	Mts	×		×	×
	MBS				×
Asn	Tmob, Trt	×			
Asp	O-*t*-Bu	×			
	Obzl, OcHex		×	×	×
Cys	Acm	×	×	×	×
	Trt	×			
	MBzl		×	×	
Gln	Tmob, Trt	×			
Glu	OcHex		×		×
	Obzl		×	×	
	O-*t*-BU	×			
His	Bom		×	×	×
	Boc, Trt	×			
	Dnp, Tos		×	×	
	Trt	×			
Lys	Cl-Z		×	×	×
	Boc, Fmoc, Mtt	×			
Met	O			×	×
Ser	Bzl		×	×	×
	t-Bu	×			
Thr	*t*-Bu	×			
	Bzl		×	×	×
Trp	Mts				×
	CHO		×	×	×
Tyr	*t*-Bu	×			
	Bzl, Cl-Bzl				×
	Br-Z		×	×	×

[a]An × indicates that the reagent is compatible with the given amino acid.

Removal of the Completed Peptide from the Resin

The final stage in peptide synthesis is to cleave the completed poly(amino acid) from its resin and remove the side chain protecting groups. Often, both these steps can be completed at once with one cleaving reagent, as shown in Fig. 54.5. The choice of cleaving reagent is determined by the type of α-amine blocking group that was used during the rest of the synthesis [11]. Table 54.2 is a comprehensive list of side-chain-protected amino acids that are compatible with particular solvents commonly used for Fmoc or Tboc cleavage. The solvents indicated are able to successfully cleave substrate from the resin and at the same time remove the side chain blocking groups of the amino acids listed. If Tboc chemistry is used to synthesize the peptide, the final product is cleaved by using a strong acid like hydrogen fluoride (HF), which can be purchased in liquid form [23]. The resin is distilled in HF for about an hour while being held at a temperature of 0–5°C. If Fmoc chemistry is used to synthesize the peptide, a milder acid can be used. The typical choice is tetrafluoroacetic acid (TFA) [11]. A sample protocol using TFA for Fmoc substrates is included in Protocol 2.

PROTOCOL 2. USING TFA TO CLEAVE A SUBSTRATE FROM A WANG RESIN[2]

EQUIPMENT AND REAGENTS

- The Wang resin with polypeptide substrate attached
- Round-bottomed flask
- Other flask large enough to contain the resin
- Fine sintered glass filter
- Refrigeration unit
- Vacuum chamber
- Dimethylformamide (DMF)
- Dichloromethane (DCM)
- 20% piperidine (v/v) in DMF
- Trifluoroacetic acid (TFA)
- Cold ether

METHOD

1. If the last N-terminal Fmoc group is to be removed, place the resin in a round-bottomed flask and add 20% (v/v) piperidine in DMF until the resin is just covered. Let the mixture stand for 30 min.
2. Transfer the resin to a sintered glass funnel with fine porosity and apply vacuum. Wash with DMF three times. Wash with DCM three times to remove the DMF.
3. If no protecting groups are to be removed, the cleavage cocktail can consist of a 1:1 mixture of TFA/DCM. Cover the resin with this solution and let stand for a minimum of 20 min.
4. If side chain protecting groups are going to be removed, add the appropriate scavengers to the cocktail, according to their chemistry.
5. Filter the resin in the glass funnel and wash with a small portion of the cleavage cocktail.
6. Combine the filtrates and add 8–10 times the volume of cold ether. If necessary, keep the mixture at 4°C overnight to precipitate the product.
7. Use the glass funnel to filter the precipitate and wash it with cold ether.

During the removal of the side chain protecting groups, carbocations are released that have the potential to interfere with the exposed side chain. Various additives may be required to "scavenge" for these ions, avoiding damage to any side chain groups. Thiol and phenol compounds and water are typical scavengers that can be added [24].

---

[2]Copyright Advanced Chem Tech, 1999. Used with permission. See www.peptide.com/support/resinprotocols.

Automatic Synthesizers

As mentioned numerous times throughout this section, the typical method for the solid phase synthesis of poly(amino acids) is automatic synthesis (see, e.g., Protocol 3).

PROTOCOL 3.   Using an Automatic Synthesizer for Solid Phase Peptide Synthesis[3]

EUIPMENT AND REAGENTS

- Synergy automatic synthesizer from PEBiosystems (or the equivalent)
- Reagent bottles for the synthesizer
- Glass 250-ml flasks or beakers
- Glass fine-sintered funnel
- Peptides required for the synthesis of the polymer
- Preloaded Merrifield resin (Applied Biosystems)
- Anhydrous HOBt
- HBTU
- Double-distilled water
- Dimethylformamide (DMF)
- 10% ammonium bicarbonate in water
- 30% acetic acid in water
- Dimethyl sulfoxide (DMSO)
- 6 M urea in water or 6 M urea in 20% acetic acid

METHOD

1. Dissolve the peptide in water to desired concentration. If the peptide is not water soluble, use the following guidelines:
   a. Mix the peptide with a very small amount of the appropriate diluent, just to dissolve the peptide. Then dilute to the desired concentration with water.
   b. For acidic peptides use a base, like 10% ammonium bicarbonate; for basic peptides use an acid like 30% acetic acid, for very hydrophobic peptides use DMSO, for peptides that tend to aggregate use 6 M urea or 6 M urea with 20% acetic acid.
2. Prepare 0.5 M HOBt in DMF by dissolving 13.5 g of anhydrous HOBt in 200 ml of DMF, in a 250-ml flask.
3. Prepare 0.45 M HBTU-HOBt solution by adding the solution prepared in step 2 to 37.9 g of HBTU contained in a flask. Stir for about 15 min with a magnetic stirring bar until the HBTU is dissolved. Filter the solution through a fine-pore-size sintered glass funnel. Pour the filtered solution into an appropriate bottle for attachment to a peptide synthesizer. This solution is stable at room temperature for at least 6 weeks.
4. Add the resin and the appropriate amino acids. Some systems require the use of premeasured amino acid columns. Otherwise, use the dissolved peptides in the appropriate bottle for attachment to the synthesizer.
5. Press Start.

There are at least four companies that produce automatic synthesizers; their models cover a wide range of features and prices. Recognizing that there are new models being developed as this section is written and that advances in technology will support improvements in the synthesizer market, it is difficult to give an "up to the minute" report on automatic synthesizers. However, Table 54.3 provides a reasonably thorough list of peptide synthesis devices, their basic features, and the contact information for the manufacturers who develop them.

---

[3] Adapted from www.pebio.com/pa/343905/343905.html.

SOLUTION PHASE SYNTHESIS

The earliest method for polymerizing amino acids was a solution phase method. It is probably the simplest methodology for synthesis, as well as the most easily understood, since it follows all the fundamental thermodynamic and kinetic principles of other common polymerization techniques. Both step-growth polymerizations in the form of a series of condensation reactions and chain-growth polymerization by initiation and propagation have been performed with amino acids [25]. Figure 54.6 illustrates the two forms of producing poly(amino acids) in the solution phase. Just as in typical polymer reactions, solution phase synthesis is ideal for making poly(amino acids) with high degree of polymerization. For example, Sigma produces poly(lysine) in a range of molecular weights from 5000 to 500,000. However, the ability to produce these large poly(amino acids) comes at the cost of little control over the sequence. For this reason, most poly(amino acids) produced by solution phase synthesis are made from only one or two amino acids. Inability to control the specific sequence or even to control the synthesis enough give a uniform molecular length limits the possible applications that solution phase polymers can achieve. These kinds of polymers are used more for their structural properties and nontoxicity in materials engineering than for the highly specific roles that are fulfilled by solid phase synthesized polymers [26]. Typically, high molecular weight poly(amino acids) that can be produced only by solution phase

Table 54.3. Some Automatic Solid-Phase Peptide Synthesizers, Their Manufacturers, and Features

Manufacturer	Model	Features	Contact information
Advanced ChemTech	Model 90	Table top, dual synthesis, 2-h. cycle time	www.peptide.com
	Model 348	Table top, 48-peptide synthesis, mg output	
	Model 396	Table top, 96-peptide synthesis	
	Model 400	Floor model, kg output	
Applied Biosystems	431A	Fmoc and Tboc capable; resin sampler	www.pebio.com
	433A	Fmoc, Tboc, or Fastmoc capable; feedback	
	631	Fastmoc capable; online monitoring	
Labortec AG	SP4000-LAB	Semiautomatic, table top, 0.5–10g resin	www.bachem.com
	SP4000-PRO	Semiautomatic, table top, 10–100 g resin	
	SP4000-PPS	Semiautomatic, floor model, 6–8 kg resin	
PerSeptive Biosystems	Pioneer	Fmoc continuous flow, 32-peptide synthesis; highly automated, 16-min cycle time	www.pbio.com

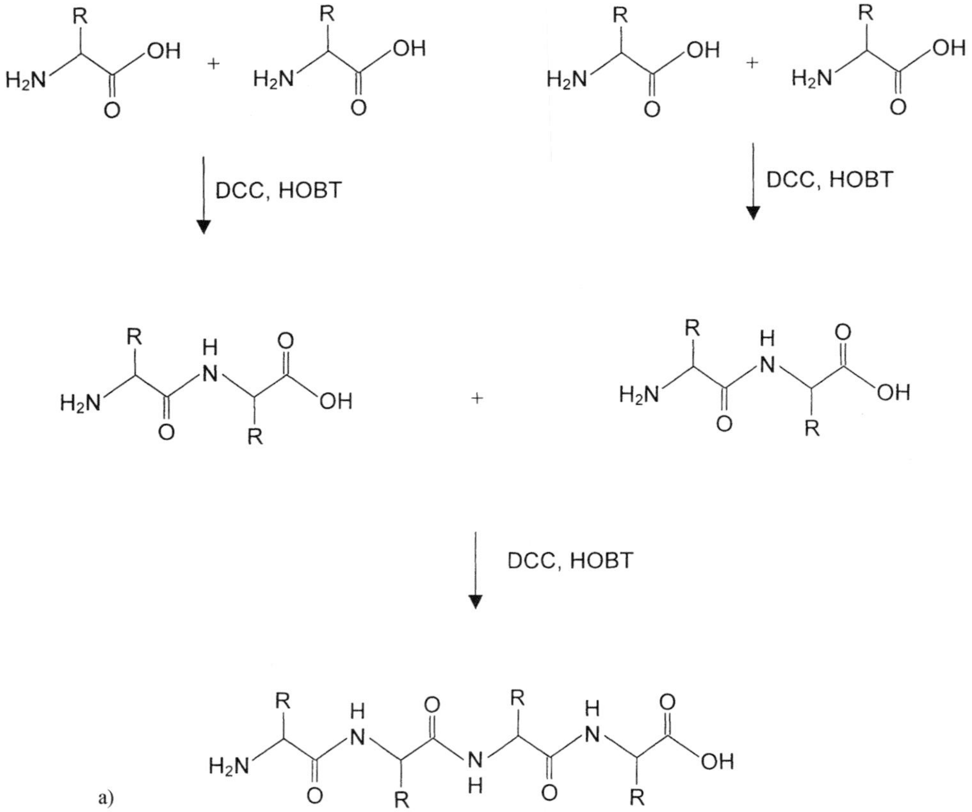

Fig. 54.6. Step-growth versus chain-growth polymerization in solution phase synthesis. Step-growth polymerization (a) is used to synthesize sequential linear polypeptides, while chain-growth polymerization (b) is used for the synthesis of other copolymers.

methods are being used as simple models in structure–property studies of natural polypeptides and proteins. They can also be used in microencapsulation, drug delivery devices, and biotechnology as substrates in the isolation of plasma membranes or as enzyme inhibitors [27–29].

Choosing a protocol for solution phase synthesis is dependent on the desired composition of the final polymer. Most poly(amino acids) are produced by radical polymerization, but this method is suitable only for homopolymers and heteropolymers of random or block composition. A method initiated in the mid-1950s for radical polymerization of amino acids is the N-carboxyanhydride (NCA) method [30]. The NCA method is very reliable and is suitable for producing most homopolymers and random heteropolymers. The design of this method is not attributed to any one individual but was the result of efforts from a group of people. Sela [31], Blout [32], and Katchalski-Katzir [30] performed the earliest syntheses using the NCA method, which is depicted in Fig. 54.6. Initially, they could consistently produce only homopolymers, but random heteropolymers can now be synthesized. With recent investigations into living free radical polymerization, block copolymers are also being produced, but with some difficulty [31]. Implementing a carbodiimide linkage between the α-amine of one homopolymer with the ω-carboxyl group of another homopolymer can also produce block copolymers [32].

The first method for synthesizing poly(amino acids) employed a condensation reaction [25]. As mentioned in the section on solid phase synthesis, this method was relatively ineffective because of the elaborate purification steps that followed the actual synthesis. Usually the yield was so poor that synthesizing polypeptides in this way would have been impractical. Fortunately yield has improved as the technology for separation and purification has improved, since step-growth condensation polymerization is still the only method available for producing high molecular weight poly(amino acids) that have some structural

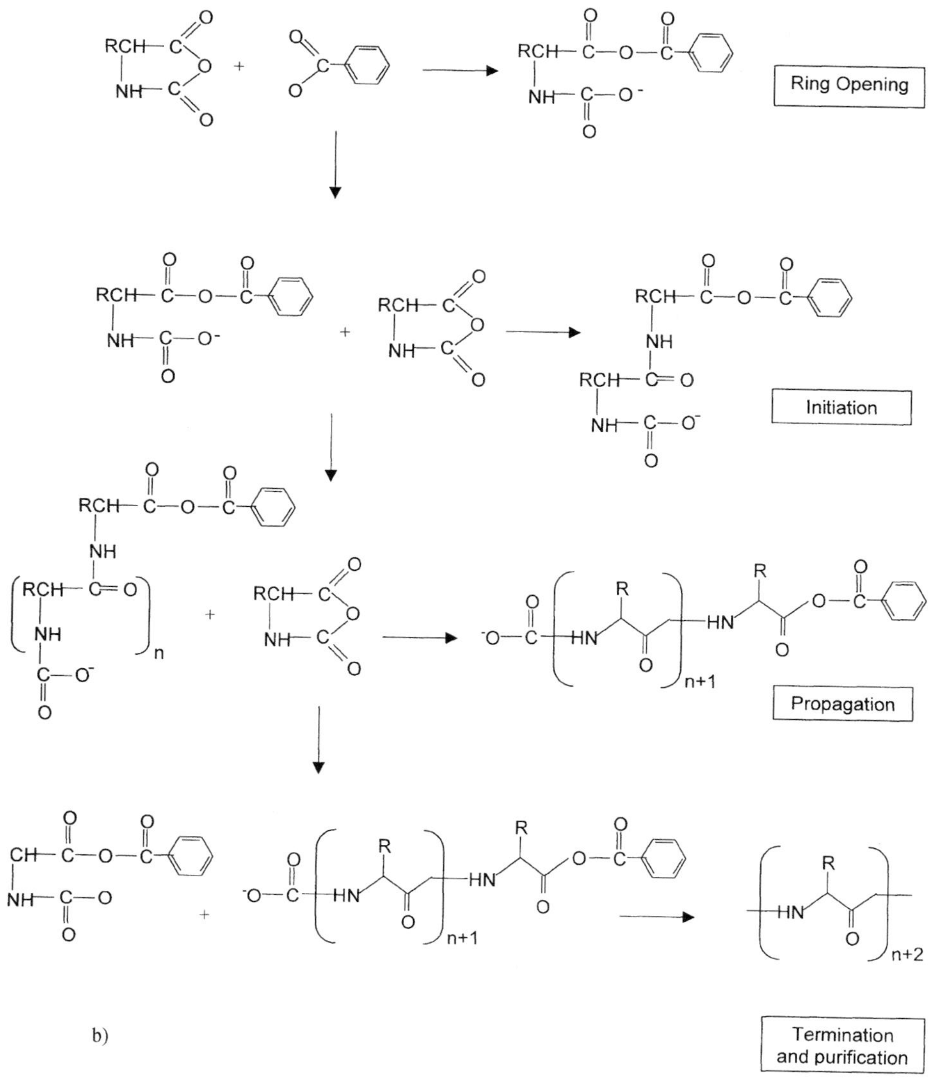

Fig. 54.6. (continued).

definition. New techniques hope to further improve reaction yields [33]. The perfect examples are heteropolymers of the form $-(ABC)_n-$, where A, B, and C are different amino acids and their order in the repeat unit is important. Methods for producing these various homopolymer and heteropolymer poly(amino acids) by solution phase synthesis are discussed next.

Homopolymers

Homopolymeric poly(amino acid) synthesis has changed very little since its inception many years ago. It is the simplest form of polymerizing amino acids, with only minor variations in reaction times or solvents between methods. The initiation can be performed with primary amines, but since such polymerizations are amine initiated, there are concerns associated with premature termination and racemization as well as side reactions [34]. Most of these problems occur when the reaction occurs at high temperatures and over prolonged activation times [34]. The trend in initiation is toward strong bases, which can effectively initiate the polymerization of N-carboxyamino acids (NCAs) with very little premature termination and racemization [34,35]. Side reactions may still be a problem and should be given some consideration before this method is selected. The amino acids used for NCA solution phase synthesis have protecting groups on their side chains to prevent the occurence of unwanted reactions. A list of possible blocking groups was given earlier. An example of a simple protocol for using the NCA solution phase methodology to (synthesize homopoly(amino acids) given in Protocol 4.

PROTOCOL 4.   SOLUTION PHASE SYNTHESIS OF A HOMOPOLYMER, POLY(L-GLUTAMATE) [32]

EQUIPMENT AND REAGENTS

- *g*-Benzyl-L-glutamate-*N*-carboxyanhydride
- Freshly distilled dry dioxane
- 0.5 N NaOH, stored at −30°C
- 0.5 M HCl
- 95% Ethanol
- Anhydrous ether
- Flasks, stopper
- Filter
- Vacuum chamber

METHOD

1. Weigh 2.0 g of glutamate-NCA into an Erlenmeyer flask. Add 50 ml of dry dioxane and stir to mix.
2. Add enough NaOH at −30°C to give an anhydride to initiator ratio of 100.
3. Stopper the flask and allow it to stand at room temperature for 2 h. The mixture should become viscous.
4. Isolate the polymer by pouring the solution into about 7 times its volume of 95% ethanol (roughly 1.5 liters). Add enough HCl to neutralize the initiator.
5. Filter and wash with anhydrous ether.
6. Dry in a vacuum and store at 2°C.

Block Copolymers

The simplest method for producing block copolymers of poly(amino acids) is linking two preformed homopolymers through a condensation reaction [32]. Although the field of living free radical polymeriztion (LFRP) presents some exciting options for block copolymers of amino acids, the methods used are not well established [31]. Sela and Berger attempted a version of LFRP in 1955, wherein polypeptides containing free amino groups as initiators continued the polymerization of the peptide chain with other amino acid NCAs, but side reactions and premature termination became a problem [31].

Heteropolymers

As mentioned earlier, two distinguishable types of heteropolymers can be formed from amino acids: the random heteropolymers and the fixed-order heteropolymers. The differences in chemistry require different synthetic processes so, each is considered separately in this discussion.

Truly random copolymers can only be obtained from NCAs under very specific conditions: namely, when the affinity for adding one amino acid to the end of a growing chain is equal to the affinity of adding another, irrespective of which amino acid is on the growing end [36]. To complicate matters, since the conformation of a growing peptide chain can change as each residue is added, more than just the terminal amino acid may be responsible for determining the affinity of adding the next amino acid. For the case of poly(amino acids), truly random copolymers can never be guaranteed, and so the definition is broadened to include any heteropolymer in which the order of amino acids is not completely determined. In many cases, the order can be statistically determined, but even then it is never entirely controllable. However, the composition of these statistical poly(amino acids) is controllable. In any other polymerization of more than one distinct monomer, the composition of the polymer is determined by the composition of the feed and the respective reactivity ratios of the monomers. This rule also applies in the polymerization of poly(amino acids). Therefore, NCA methods for synthesizing homopolymers can be implemented to synthesize random heteropolymers of amino acids [37], an example of which is outlined in Protocol 5.

PROTOCOL 5.   SOLUTION PHASE SYNTHESIS OF STATISTICAL AMINO ACID COPOLYMERS [37]

EQUIPMENT AND REAGENTS

- *N*-Carboxyanhydrides of Nε-carbobenzoxy-L-lysine
- *N*-Carboxyanhydrides of co–amino acids
- Hydrogen bromide (HBr)
- Glacial acetic acid (HAc)
- Dry dioxane
- Triethylamine (TEA)
- 0.5 M sodium acetate in water
- Flasks
- Vacuum chamber
- Distilled, deionized water (ddH$_2$O)

METHOD

1. Prepare a solution of 33% w/v of HBr in the acetic acid.
2. Prepare a solution of 2–3% amino acid anhydride in dry dioxane such that the relative concentrations of the peptides in the feed solution are the same as the desired relative concentrations for the polymer.
3. Add TEA in an anhydride-to-TEA ratio of 100. TEA will initialize the polymerization.
4. Allow the reaction to proceed under argon at room temperature for 14 days.
5. After the polymerization is complete, remove the Boc blocking groups by addition of the HBr-HAc solution prepared in step 1.
6. Dialyze the product against 0.5 M sodium acetate, then dialyze again against ddH$_2$O.
7. Dry in vacuum.

Here, the composition of the final copolymer is controlled by the composition of the feed. Reactivity ratios of the various amino acid derivatives that are used for solution phase synthesis of polymers are not available. They can be determined experimentally by performing amino acid analysis on a polymer composed of a known mixture of amino acid monomers.

The approach to synthesizing heteropoly(amino acids) of a fixed sequence is entirely different. Typically, condensation polymerizations are used for poly(amino acids) of the general form $-(ABC)_n-$, where A, B, and C are different amino acids and their order in the repeat unit is important. It follows that this step-growth method can also be applied to poly(amino acids) with any number of residues in the repeat unit, as long as their order within the unit is fixed. The general procedure for condensation polymerization first synthesizes a precursor molecule, either by solid phase methods or by condensation and coupling mechanisms. This precursor molecule has the defined structure of the desired repeat unit. One terminus of these short oligopeptides is then activated to readily react with the functional group at the other terminus. The easiest way of doing this is by activating the carboxyl terminus and employing the carbodiimide method as was discussed in the SPPS section of this chapter [38]. The oligopeptides then couple, linking end-to-end to form the final polymer. Improvements on this method involve the use of preactivated amino acids and oligopeptides [39]. Succinimide esters are the most common activating groups.

When combined with TFA in DMSO, they form the required amide bond with few side products. A sample of this method is outlined in Protocol 6 as an example of synthesizing heteropoly(amino acids) with a defined sequence in the solution phase.

PROTOCOL 6.   SOLUTION PHASE SYNTHESIS OF A SEQUENTIAL POLY(AMINO ACID) COPOLYMER: (ALA-PRO-GLY)$_n$ AS THE EXAMPLE [39]

EQUIPMENT AND REAGENTS

- Flasks
- Vacuum chamber
- Magnetic stirrer and stir bar
- Ice bath
- Hot plate or hot water bath
- Cylinder of P_2O_5 gas
- Boc-Ala-OH
- HCl-H-Pro-OSu
- HBr-H-Gly-OSu
- Distilled deionized water (ddH$_2$O)
- Sodium carbonate (NaHCO$_3$)
- Chloroform
- *N*-Methylmorpholine (MMP)
- Isobutyl chloroformate
- Trifluoroacetic acid (TFA)
- Dimethyl sulfoxide (DMSO)
- Triethylamine (TEA)
- Diethyl ether
- Methylene chloride

METHODS

1. Preparation of Boc-Ala-Pro-OH
   a. Prepare a 10% w/v solution of sodium bicarbonate in ddH$_2$O.
   b. Weigh 5 g of Boc-Ala-OH into a clean dry flask and add 50 ml of the solution from step a.
   c. Recrystallize from ethyl acetate.
2. Preparation of Boc-Ala-Pro-Gly-OSu
   a. Weigh 0.5 g of Boc-Ala-Pro-OH into a flask and dissolve in 20 ml of chloroform. Stir and keep chilled at −15°C.
   b. Add 0.20 ml of MMP and 0.25 ml of isobutyl chloroformate while continuing to stir, and chill for 30 min. Add 0.45 g of HBr-H-Gly-OSu.
   c. Prepare a solution of 0.2 ml MMP in 4 ml of chloroform. Add this solution dropwise to the reaction vessel.
   d. Remove cooling bath. Allow reaction to proceed at room temperature for 18 h.
   e. Wash with ddH$_2$O three times and dry in vacuum.
3. Polymerization
   a. React the Boc-Ala-Pro-Gly-OSu peptide with TFA. The result will be a solid salt.
   b. Dissolve 4.16 mmol of salt in 3 ml of DMSO with stirring. Add 1.15 ml of TEA. Continue to stir at room temperature for 18 h.
   c. Add another 1.15 ml of TEA and continue stirring at room temperature for 4 days, and then at 50°C for 1 day.
   d. Add diethyl ether and centrifuge to separate. Wash many times with diethyl ether. Wash with methylene chloride.
   e. Dry in vacuum under P_2O_5.

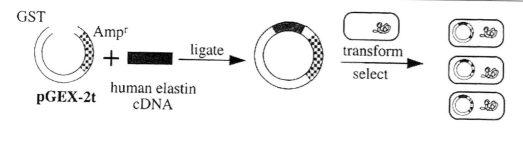

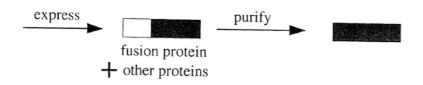

Fig. 54.7. An overview of recombinant protein synthesis [C. Bellingham, personal communication].

BIOLOGICAL SYNTHESIS OF POLY(AMINO ACIDS)

RECOMBINANT METHODS

There are many obvious advantages to using biological systems for the synthesis of poly(amino acids), particularly if these molecules are large proteins of complicated structure. Typically, biological materials are inexpensive and produce homogeneous molecules of precisely the same molecular weight and structure. Post-translational modifications are often performed as part of the synthesis [40]. The major drawback to this procedure is yield. For many reasons, very large volumes of broth and cells often are needed to produce even the smallest amount of product. When the expressed protein is large and is easily recognizable as "non-native" the cells are likely to supress further expression or may commit suicide. Even when the cells can express large quantities of the desired protein, the process of separating it from the other expressed proteins, described next, is very difficult, expensive, and laborious [41].

A cDNA copy of mRNA has been made and the mRNA is removed by an alkali treatment that destroys RNA but does not affect DNA. A duplex DNA is made from the cDNA strand in a process depicted in the Fig. 54.7. The now double-stranded DNA can be inserted into a plasmid, which is a piece of circular DNA that exists in bacteria and is capable of self-replicating as well as transferring genes between bacteria. To insert the DNA, a restriction enzyme is used to open the plasmid circle. The duplex DNA is mixed with the opened plasmid in the presence of ligating enzymes, which join the ends of the DNA to the opened ends of the plasmid. This plasmid is then introduced to a host cell, usually a bacterial cell like *E. coli*, which takes the plasmid into the intracellular cavity. As the bacteria multiply, the plasmid is transferred to each daughter cell, creating an entire population of cells able to synthesize the desired protein. If the plasmid also contains an anitbiotic-resistant gene, any bacteria containing the plasmid or a copy of the plasmid can perform its living functions in the presence of antibiotic. Thus, cells containing the engineered plasmid can be separated from those without by the addition of antibiotic and subsequent removal of dead cells. The remaining cells are grown and multiplied in growth media, and at an optimum cell density they are chemically induced to begin protein synthesis. Upon completion, the cell membranes are lysed and the dead cells are removed from the protein mixture. Once the desired protein has been separated from the other proteins in the batch, the attached fusion protein is removed in purification steps.

FURTHER READING

For more information or details on the synthesis methods discussed in this chapter, consult these excellent texts:

1. Watson, J., D., Gilman, M., Witkowski, J., and Zoller, M. (1996). "Recombinant DNA," 2nd ed. Scientific American Books, New York.

2. Chan, W. C., and White, P. D., eds. (2000). "Fmoc Solid Phase Peptide Synthesis: A Practical Approach," Vol. 222. Oxford University Press, New York.
3. Fields, G. B., ed. (1991). "Methods in Enzymology: Solid Phase Peptide Synthesis," Vol. 289. Academic Press, San Diego, CA.

REFERENCES

1. Young-Kreegar, K. (1996). Immunological applications top list of peptide-synthesis services. *Scientist* 10(13), 18–19.
2. Merrifield, R. B. (1965). Automated synthesis of peptides: solid phase synthesis, a simple and rapid synthetic method, has now been automated. *Science* 150, 178–184.
3. Bayer, E., Eckstein, H., Hagele, K., Konig, W., Bruning, W., Hagenmaier, H., and Parr, W. (1970). Failure sequences in the solid-phase synthesis of polypeptides. *J. Am. Chem. Soc.* 92, 1735–1738.
4. Plaue, S., Muller, S., Briand, J. P., and Van Regenmortel, M. H. (1990). Recent advances in solid-phase peptide synthesis and preparation of antibodies to synthetic peptides. *Biologicals* 18, 147–157.
5. Merrifield, R. B. (1963). Solid phase peptide synthesis. I. The synthesis of a tetrapeptide. *J. Am. Chem. Soc.* 85, 2149–2154.
6. Riester, D., Wiesmuller, K. H., Stoll, D., and Kuhn, R. (1996). Racemization of amino acids in solid-phase peptide synthesis investigated by capillary electrophoresis. *Anal. Chem.* 68, 2361–2365.
7. Hodges, R. S., and Merrifield, R. B. (1975). Monitoring of solid phase peptide synthesis by an automated spectrophotometric Picrate Method. *Anal. Biochem.* 65, 241–272.
8. Yan, B. (1998). Monitoring the progress and the yield of solid-phase organic reactions directly on resin supports. *Acc. Chem. Res.* 31, 621–630.
9. Fischer, M., and Tran, C. (1999). Investigation of solid-phase peptide synthesis by the near-infrared multispectral imaging technique: A detection method for combinatorial chemistry. *Anal. Chem.* 71, 2255–2261.
10. Merrifield, R. B. (1968). The automatic synthesis of proteins. *Sci. Amer.* 218, 56–74.
11. Kates, S. A., Cammish, L. E., and Albercicio, F. (1999) Continuous-flow solid phase peptide synthesis using polystyrene resins. *J. Pept. Res.* 53, 682–683.
12. Sucholeiki, I. (1999). New developments in solid phase synthesis supports. *Mol. Diversity* 4, 25–30.
13. Wong, T. W., and Merrifield, R. B. (1980). Solid-phase synthesis of thymosin α1 using tert-butyloxycarbonylaminoacyl-4-(oxymethyl)phenylacetamidomethyl-resin. *Biochemistry* 19, 3233–3238.
14. Fields, C. G., Lloyd, D. H., Macdonald, R. L., Otteson, K. M., and Noble, R. L. (1991). HBTU activation for automated Fmoc solid phase peptide synthesis. *Pept. Res.* 4(2), 95–101.
15. White, P. D., and Chan, W. C. (2000). Basic principles. *In* "Fmoc Solid Phase Peptide Synthesis: A Practical Approach" (W. C. Chan and P. D. White, eds.), Vol. 222, pp. 61–68. Oxford University Press, New York.
16. Albericio, F., and Carpino, L. A. (1997). Coupling reagents and activation. *In* "Methods in Enzymology" (G. B. Fields, ed.), Vol. 289, pp. 29–44. Academic Press, San Diego, CA.
17. Otvos, L., Dietzschold, B., and Kisfaludy, L. (1987). Solid phase peptide synthesis using tert-butyloxycarboylamino acid pentafluorophenyl esters. *Int. J. Pept. Protein Res.* 30, 511–514.
18. Izdebski, J., Pachulska, M., and Alicija, O. (1994). N-cyclohexyl-N′–isopropylcarbodiimide: a hybrid that combines the structural features of DCC and DIC. *Int. J. Pept. Protein Res.* 44, 414–419.
19. Alewood, P., Alewood, D., Miranda, L., Love, S., Meutermans, W., and Wilson, D. (1997). Rapid in situ neutralization protocols for Boc and Fmoc solid-phase chemistries. *In* "Methods in Enzymology" (G. B. Fields, ed.), Vol. 289, pp. 29–44. Academic Press, San Diego, CA.
20. Li, P., and Xu, J. C. (2000). (1H-Benzotriazol-1-yloxy)-N,N-dimethylmethaniminium hexacloroantimonate (BOMI), a novel coupling reagent for solution and solid phase peptide synthesis. *J. Pept. Res.* 55, 110–119.
21. Carpino, L. A. (1993). 1-Hydroxy-7-azabenzotriazole. An efficient peptide coupling additive. *J. Am. Chem. Soc.* 115, 4397–4398.
22. Sarin, V. K., Kent, S. B., Tam, J. P., and Merrifield, R. B. (1981). Quantitative monitoring of solid phase peptide synthesis by the ninhydrin assay. *Anal. Biochem.* 117, 147–157.
23. Stewart, J. M., (1991). Cleavage methods following Boc-based solid phase peptide synthesis. *In* "Methods in Enzymology" (G. B. Fields, ed.), Vol. 289, pp. 29–44. Academic Press, San Diego, CA.
24. Wellings, D. A., and Atherton, E. (1997). Standard Fmoc protocols. *In* "Methods in Enzymology" (G. B. Fields, ed.), Vol. 289, pp. 44-67. Academic Press, San Diego, CA.
25. Katchalski-Katzir, E. (1974). Poly(amino acids): Achievements and Prospects. *In* "Peptides, Polypeptides, and Proteins" (E. R. Blout, F. A. Bovey, M. Goodman, and N. Lotan, eds.), pp. 1–13. Wiley, Toronto.
26. Dolnik, V., and Novotny, M. (1993). Separation of amino acid homopolymers by capillary gel electrophoresis. *Anal. Chem.* 65, 563–567.
27. Nishikawa, M., Takemura, S., Takakura, Y., and Hashida, M. (1998). Targeted delivery of plasmid DNA to hepatocytes in vivo: Optimization of the pharmacokinetics of plasmid DNA/galactosylated poly(L-lysine) complexes by controlling their physicochemical properties. *J. Pharmacol. Exp. Ther.* 287, 408–415.
28. Halle, J. P., Bourassa, S., Leblond, F. A., Chevalier, S., Beaudry, M., Chapdelaine, A., Cousineau, S., Saintonge, J., and Yale, J. F. (1993). Protection of islets of Langerhans from antibodies by microencapsulation with alginate-poly-L-lysine membranes. *Transplantation* 55, 350–354.
29. Choi, Y. H., Liu, F., Park, J. S., and Kim, S. W. (1998). Lactose-poly(ethylene glycol)-grafted poly-L-lysine as haepatoma cell-targeted gene carrier. *Bioconjugate Chem.* 9, 708–718.
30. Katchalski-Katzir, E. (1997). Synthesis, structure and function of poly-a-amino acids—the simplest of protein models. *Cell. Mol. Life Sci.* 53, 780–789.

31. Sela, M., and Berger, A. (1955). The terminal groups of poly- α-amino acids. *J. Am. Chem. Soc.* 77, 1893–1898.

32. Blout, E. R., and DesRoches, M. E. (1959). The preparation of high molecular weight polypeptides. *J. Amer. Chem. Soc.* 81, 370–372.

33. Sakakibara, S. (1999). Chemical synthesis of proteins in solution. *Biopolymers* 51, 279–296.

34. Blout, E. R. (1962). Synthesis and chemical properties of polypeptides. *In* "Polyamino acids, Polypeptides, and Proteins" (M. A. Stahman, ed), pp. 3–12. University of Wisconsin Press, Madison.

35. Fasman, G. D., Idelson, M., and Blout, E. R. (1961). The synthesis and conformation of high molecular weight poly-ε-carbobenzyloxy-L-lysine and Poly-L-lysine HCl. *J. Amer. Chem. Soc.* 83, 709–712.

36. Penczek, S. (2000). Cationic ring opening-polymerization: Major mechanistic phenomena. *J. Polym. Sci. Polym. Chem.* 38, 1919–1933

37. Arfmann, H-A., Labitzke, R., Lawaczeck, R., and Wagner, K. G. (1974). Aromatic amino acid-lysine copolymers. Conformation and specificity of nucleotide interaction. *Biochimie* 56, 53–60.

38. Kimura, T., Takai, M., Masui, Y., Morikawa, T., and Kakakibara, S. (1981). Strategy for the synthesis of large peptides: An application to the total synthesis of human parathyroid hormone. *Biopolymers* 20, 1823–1832.

39. Lorenzi, G. P., Doyle, B. B., and Blout, E. R. (1971). Synthesis of polypeptides and oligopeptides with the repeating sequence L-Alanyl-Lprolylglycine. *Biochem. J.* 10, 3046–3051.

40. Darnell, J., Lodish, H., and Baltimore, D. (1990). "Molecular Cell Biology," 2nd ed. Scientific American Books, New York.

41. Watson, J. D., Gilman, M., Witkowski, J., and Zoller, M. (1996). "Recombinant DNA," 2nd ed. Scientific American Books, New York.

42. Bellingham, C. (1999). Personal communication.

CHAPTER 55

SYNTHESIS OF SYNTHETIC POLYMERS: POLY(PROPYLENE FUMARATE)

Richard G. Payne and Antonios G. Mikos

BACKGROUND

Poly(propylene fumarate) (PPF) is an unsaturated linear polyester (Fig. 55.1). PPF has been suggested for use as a scaffold for guided tissue regeneration, often as part of an injectable bone replacement composite [1–4]. PPF has been used for controlled release of antibiotics [5] and as a substrate for osteoblast culture [6]. In addition, PPF has been modified with poly(ethylene glycol) to form a cross-linkable copolymer [7–9].

Degradation of the ester bonds via hydrolysis yields propylene glycol and fumaric acid [10], which are biocompatible and readily removed from the body. The double bond along the backbone of the polymer permits cross-linking *in situ*, which causes a moldable composite to harden within 10–15 min [1]. Mechanical properties and degradation time of the composite may be controlled by varying the PPF molecular weight [1,11]. Therefore, preservation of the double bonds and control of molecular weight during PPF synthesis are critical issues.

Several schemes have been used to synthesize PPF. Gresser *et al.* reported making PPF via a two-step process whereby equimolar quantities of diethyl fumarate and propylene glycol were combined in an acid-catalyzed reaction, first at 250°C at 760 mmHg and then at 220°C at 1 mmHg [12]. This reaction can be difficult to control if loss of the volatile propylene glycol leads to stoichiometric imbalance.

Domb *et al.* have produced PPF via direct condensation of complementary trimers [13]. Bis(2-hydroxypropyl fumarate) and bis(hydrogen maleate) trimers were formed, then reacted together, producing PPF polymer. Although the reactants in the condensation reaction are nonvolatile, the reaction is still sensitive to unbalanced stoichiometry.

Kharas *et al.* have analyzed several reactions (diethyl fumarate with a variety of diols and catalysts) [14]. It was determined that the method with the highest yield was the reaction of diethyl fumarate with 10% molar excess propylene glycol in the presence of an acid catalyst, zinc chloride ($ZnCl_2$). This reaction requires a maximum temperature of 200°C.

Peter *et al.* [15] used a modification of a method developed by Yaszemski *et al.* [16] to synthesize PPF in a two-step reaction. In the first step fumaryl chloride was reacted with a large excess of propylene glycol in the presence of a proton scavenger to produce the bis(2-hydroxypropyl fumarate) trimer. In the second reaction, this trimer was self-transesterified at 160°C, producing PPF with the side product of propylene glycol. The advantage of this system is that since the only reactant in the second step is the trimer, there are no stoichiometric concerns. An intermediate purification step was included to remove the proton scavenger before the transesterification [15]. NMR results have indicated partial loss of the double bonds due to HCl addition [15]. This synthesis method virtually ensures that

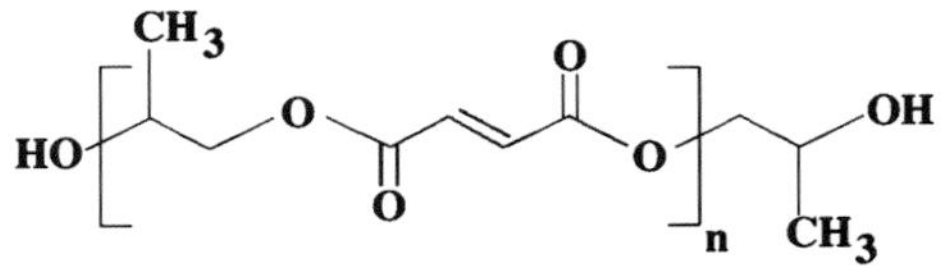

Fig. 55.1. Poly(propylene fumarate).

the resulting polymer is hydroxy-terminated, which presents advantages for performing additional chemistry, such as the covalent linking of peptides to PPF [17]. This method is described in the protocol section.

PROTOCOL

MATERIALS

Fumaryl chloride (FuCl) (Acros, Pittsburgh, PA)
Propylene glycol (PG) (Acros, Pittsburgh, PA)
Potassium carbonate (Fisher, Pittsburgh, PA)
Methylene chloride (Fisher, Pittsburgh, PA)
Hydroquinone (Hq) (Fisher, Pittsburgh, PA)

REACTIONS
FIRST PHASE

The FuCl is purified by distillation 1 day prior to use. The FuCl and PG are reacted in a 1:3 molar ratio. The PG is mixed with double its volume of methylene chloride and the potassium carbonate in a 2:1 molar ratio (PG to potassium carbonate) in the bottom of a three-neck reaction flask with overhead mechanical stirring. The fumaryl chloride is mixed with an equal volume of methylene chloride and placed in an addition funnel in one of the reaction vessel necks. Purified nitrogen is purged through the top of the addition funnel and passes out through a tube attached to an adapter in the third neck of the reaction vessel. This outlet tube allows the nitrogen and any gases produced in the reaction to bubble through a water trap.

At the start of the reaction, the vessel is placed in an ice bath, and the FuCl–methylene chloride solution is added dropwise to the stirred PG–methylene chloride–potassium carbonate suspension. The reaction is highly exothermic, producing HCl. Care must be taken to control the rate of addition of the FuCl to prevent overheating of the reaction mixture. Potassium carbonate is used to scavenge the protons given off in the reaction.

After all the FuCl–methylene chloride solution has been added, the ice bath is removed and the reaction mixture is allowed to stir for an additional 3 h. The product of this reaction is bis(2-hydroxypropyl fumarate) (BHPF).

INTERMEDIATE PURIFICATION STEP

An equal volume of water is shaken with the reaction mixture and separated to extract the potassium carbonate. The residual BHPF is then reextracted from the water phase by shaking it with methylene chloride. The organic phases are combined and shaken with a brine solution. Following separation, the organic phase is then dried by mixing it with sodium sulfate. The suspension is filtered to remove the sodium sulfate and the resulting solution is added dropwise to cold ethyl ether to precipitate the BHPF. After precipitation, a Rotavapor is used to evaporate residual solvent.

SECOND PHASE

Hydroquinone is added in a molar ratio of 0.002:1, Hq to BHPF. Hydroquinone is used to inhibit undesired reactions, such as thermal cross-linking. The BHPF is placed in a round-bottomed, three-neck reaction vessel and agitated via mechanical stirrer. A cold water condenser with thermometer is attached to one of the reaction vessel necks. At its outlet, a round-bottom flask is placed on ice to collect the condensate (propylene glycol).

The system is slowly purged with purified nitrogen through the third neck of the reaction vessel. A bubbler at the outlet of the system monitors nitrogen flow.

At the start of the reaction, the reaction vessel is submerged in an oil bath. The temperature is raised to 100°C, the nitrogen flow is stopped, and vacuum is slowly applied. In the course of 45 min, the pressure is reduced to approximately 0.1 mmHg. At this point, the unreacted PG is distilled and condensed. When full vacuum is achieved, the temperature is raised to 160°C over 1 h. The temperature is maintained at 160°C for a minimum of 5 h. Propylene glycol is produced during the reaction and is removed and condensed. The longer the reaction proceeds, the higher the molecular weight of the resulting polymer. When the desired molecular weight is achieved, the system is filled with nitrogen and cooled for purification.

FINAL PURIFICATION

MATERIALS

Methylene chloride (Fisher, Pittsburgh, PA)
Ethyl ether (Fisher, Pittsburgh, PA)

METHOD

The PPF is dissolved in an equal volume of methylene chloride. The PPF–methylene chloride solution may be added dropwise to cold ethyl ether. This measure, which removes the hydroquinone as the PPF precipitates, may also result in a significant loss of product. The hydroquinone may be left in the product to inhibit side reactions and increase shelf life, in which case this step may be ignored. The remaining solvent is removed via Rotavapor, followed by vacuum drying.

CONCLUSIONS

The unsaturated nature of poly(propylene fumarate) allows for it to be cross-linked *in situ*. This property gives PPF utility in a variety of applications, such as injectable composite for guided bone regeneration. Certain mechanical and degradative characteristics of this material, which make it useful for tissue engineering, are determined by properties that can be controlled during polymer synthesis.

REFERENCES

1. Peter, S. J., Kim, P., Yasko, A. W., Yaszemski, M. J., and Mikos, A. G. (1999). Crosslinking characteristics of an injectable poly(propylene fumarate)/β-tricalcium phosphate paste and mechanical properties of the crosslinked composite for use as a biodegradable bone cement. *J. Biomed. Mater. Res.* **44**, 314–321.
2. Yaszemski, M. J., Payne, R. G., Hayes, W. C., Langer, R., Aufdemorte, T. B., and Mikos, A. G. (1995). The ingrowth of new bone tissue and initial mechanical properties of a degrading polymeric composite scaffold. *Tissue Eng.* **1**, 41–52.
3. Lewandrowski, K.-U., Cattaneo, M. V., Gresser, J. D., Wise, D. L., White, R. L., Bonassar, L., and Trantolo, D. J. (1999). Effect of a poly(propylene fumarate) foaming cement on the healing of bone defects. *Tissue Eng.* **5**, 305–316.
4. Domb, A. J., Manor, N., and Elmalak, O. (1996). Biodegradable bone cement compositions based on acrylate and epoxide terminated poly(propylene fumarate) oligomers and calcium salt compositions. *Biomaterials* **17**, 411–417.
5. Gerhart, T. N., Roux, R. C., Horowitz, G., Miller, R. L., Hanff, P., and Hayes, W. C. (1988). Antibiotic release from an experimental biodegradable bone cement. *J. Orthop. Res.* **6**, 585–592.
6. Peter, S. J., Lu, L., Kim, D. J., Stamatas, G. N., Miller, M. J., Yaszemski, M. J., and Mikos, A. G. (2000). Effects of transforming growth factor β1 released from biodegradable polymer microparticles on marrow stromal osteoblasts cultured on poly(propylene fumarate) substrates. *J. Biomed. Mater. Res.* **50**, 452–462.
7. Suggs, L. J., Payne, R. G., Yaszemski, M. J., Alemany, L. B., and Mikos, A. G. (1997). Synthesis and characterization of a block copolymer consisting of poly(propylene fumarate) and poly(ethylene glycol). *Macromolecules* **30**, 4318–4323.
8. Suggs, L. J., Krishnan, R. S., Garcia, C. A., Peter, S. J., Anderson, J. M., and Mikos, A. G. (1998). *In vitro* and *in vivo* degradation of poly(propylene fumarate-*co*-ethylene glycol) hydrogels. *J. Biomed. Mater. Res.* **42**, 312–320.
9. Suggs, L. J., and Mikos, A. G. (1999). Development of poly(propylene fumarate-*co*-ethylene glycol) as an injectable carrier for endothelial cells. *Cell Transplant.* **8**, 345–350.
10. He, S., Timmer, M. D., Yaszemski, M. J., Yasko, A. W., Engel, P. S., and Mikos, A. G. (2001). Synthesis of biodegradable poly(propylene fumarate) networks with poly(propylene fumarate)–diacrylate macromers as crosslinking agents and characterization of their degradation products. *Polymer* **42**, 1251–1260.
11. Peter, S. J., Miller, S. T., Zhu, G., Yasko, A. W., and Mikos, A. G. (1998). *In vivo* degradation of a poly(propylene fumarate)/β-tricalcium phosphate injectable composite scaffold. *J. Biomed. Mater. Res.* **41**, 1–7.

12. Gresser, J. D., Hsu, S.-H., Nagaoka, H., Lyons, C. M., Nieratko, D. P., Wise, D. L., Barabino, G. A., and Trantolo, D. J. (1995). Analysis of a vinyl pyrrolidone/poly(propylene fumarate) resorbable bone cement. *J. Biomed. Mater. Res.* **29**, 1241–1247.
13. Domb, A. J., Laurencin, C. T., Israeli, O., Gerhart, T. N., and Langer, R. (1990). The formation of propylene fumarate oligomers for use in bioerodible bone cement composites. *J. Polym. Sci., Part A: Polym. Chem.* **A28**, 973–985.
14. Kharas, G. B., Kamenetsky, M., Simantirakis, J., Beinlich, K. C., Rizzo, A.-M. T., Caywood, G. A., and Watson, K. (1997). Synthesis and characterization of fumarate-based polyesters for use in bioresorbable bone cement composites. *J. Appl. Polym. Sci.* **66**, 1123–1137.
15. Peter, S. J., Suggs, L. J., Yaszemski, M. J., Engel, P. S., and Mikos, A. G. (1999). Synthesis of poly(propylene fumarate) by acylation of propylene glycol in the presence of a proton scavenger. *J. Biomater. Sci. Polymer Edn* **10**, 363–373.
16. Yaszemski, M. J., Payne, R. G., Hayes, W. C., Langer, R., and Mikos, A. G. (1996). *In vitro* degradation of a poly(propylene fumarate)-based composite material. *Biomaterials* **17**, 2127–2130.
17. Jo, S., Engel, P. S., and Mikos, A. G. (2000). Synthesis of poly(ethylene glycol)-tethered poly(propylene-*co*-fumarate) and its modification with GRGD peptide. *Polymer* **41**, 7595–7604.

CHAPTER 56

SYNTHESIS OF HYDROGELS: ALGINATE HYDROGELS

Kamal H. Bouhadir and David J. Mooney

A lginate hydrogels are widely utilized in biomedical applications such as dental impression materials and wound dressings. However, alginate hydrogels have a limited range of mechanical properties, lack cellular interactions, and dissolve in an uncontrollable manner following the loss of calcium ions. A variety of covalent cross-linking techniques have been utilized to increase the range of mechanical properties of these hydrogels. Moreover, cell adhesion ligands have been successfully incorporated into alginate hydrogels to promote cellular proliferation and differentiation. Ionic and covalent modification of alginate hydrogels resulted in degradable gels with controllable degradation over a wide range of time periods. These modified alginate hydrogels show promise in tissue engineering applications.

INTRODUCTION

There is a tremendous need for tissues and organs to replace tissue and organs lost to disease, trauma, or congenital defects. This has led to the development of the field of tissue engineering, where natural tissue is created by transplanting cells on a biomaterial matrix [1]. This field has been defined as a "combination of the principles and methods of the life sciences with those of engineering to elucidate fundamental understanding of structure–function relationships in normal and diseased tissues, to develop materials and methods to repair damaged or diseased tissues, and to create entire tissue replacement" [2]. By utilizing a combination of synthetic polymer matrices and cells, one can create three-dimensional cell–cell interactions and tissue formation [3]. These matrices provide the three-dimensional space for cells to proliferate, excrete their extracellular matrices (ECM), and organize to form natural tissue. Therefore, it is desirable to utilize a biodegradable polymer as the cell transplantation matrix. A critical need in tissue engineering applications is the development of biocompatible and biodegradable synthetic extracellular matrices.

ALGINATE HYDROGELS

Hydrogels, intensively investigated in a variety of biomedical applications including tissue engineering, have been formed from a wide selection of both synthetic and natural polymers. One such polymer is alginate, a natural polysaccharide extracted from seaweed. Hydrogels formed from alginate are considered to be biocompatible [4,5] and have been utilized as dental impression material, wound dressings [6,7], and more recently as cell immobilization matrices [8–10]. Alginate is a linear polysaccharide composed of two uronic acids, α-D-mannuronic acid (M) and β-L-guluronic acid (G) [11]. Hence, alginate is composed of block polymers of sodium poly(L-guluronate), sodium poly(D-mannuronate), and alternating sequences of both sugars. An attractive feature of aqueous alginate solutions is their gelling properties in the presence of divalent cations (e.g., Mg^{2+}, Ca^{2+}, Sr^{2+}, Ba^{2+}) [12,13]. It is well known that divalent cations ionically cross-link the carboxylate groups of the polyguluronate block polymer in sodium alginate.

Divalent cations are introduced into alginate hydrogels following two methodologies, external and internal gelation [14]. External gelation is widely used for micro- and macrobead formation, where aqueous alginate solutions are added dropwise into aqueous calcium chloride solutions. The surface of the alginate drop gels immediately after contact with the calcium ions in solution, forming a gelatinous membrane around the drop. Afterward, calcium ions diffuse further into the membrane and cross-link additional sites. The depth of penetration of divalent cations depends on the initial alginate concentration, the divalent cation type, and concentration. Internal gelation of alginate hydrogels is accomplished by physically dispersing solid calcium salt particles in aqueous alginate solutions. Typically, these salts have a very low solubility in aqueous solutions, such as the sulfates and carbonates. Calcium ions from the solid particles dissolve slowly and cross-link alginate. The driving force for the complete dissolution of the solid particles is the consumption of the soluble calcium ions by alginate. The mechanical properties of alginate hydrogels can be varied in both external and internal gelation approaches by choosing the right divalent ion type and concentration [15]. However, the range of mechanical properties achieved is narrow.

ALGINATE IN TISSUE ENGINEERING

The gentle gelling conditions for the formation of alginate hydrogels made them attractive for tissue engineering applications. Hydrogels of alginates mixed with cells and drugs could potentially be transplanted endoscopically into patients, thus providing a minimally invasive method for cell transplantation and drug delivery applications. For example, cartilage regeneration has been demonstrated with alginate hydrogels [16]. Alginate solutions that were mixed with chondrocytes, gelled with aqueous calcium chloride solutions, and transplanted in athymic mice, formed cartilage after 3 months. The newly formed cartilage in these studies resembled native cartilage structurally and biochemically [17]. The feasibility of using injectable chondrocyte–alginate gel suspensions has also been demonstrated in the treatment of vesicoureteral reflux in pigs [18,19]. Moreover, alginate hydrogels have been shown to protect and deliver vascular endothelial growth factors (VEGF) over extended times *in vitro*. The released VEGF was functional and more potent than an equivalent mass of VEGF added directly to the cell culture medium [20,21]. Alginate has also been used as a mean to protect VEGF incorporated in other polymers such as poly(lactic-*co*-glycolic acid) [22,23].

LIMITATIONS OF ALGINATE HYDROGELS

The use of alginate hydrogels in tissue engineering applications is limited, however, owing to their mechanical properties, lack of cellular interactions, and uncontrollable degradation. It is desirable to form hydrogels with a wide range of mechanical properties enabling one to engineer a variety of tissues (e.g., muscle, cartilage, and bone). In addition, mammalian cells do not specifically interact with alginate hydrogels because the receptors required for binding are absent. Alginate hydrogels promote minimal protein adsorption owing to their hydrophilic nature, and as a result, most cells do not adhere to alginate hydrogels. Furthermore, ionically cross-linked alginate hydrogels dissolve in an uncontrollable manner following the loss of divalent cations into the surrounding medium. After the dissolution of the gels, high and low molecular weight alginates are released.

MODIFICATIONS OF ALGINATE HYDROGELS
IONIC COMPLEXES

Several approaches have been utilized to improve the mechanical stability of alginate hydrogels. The most common approach is the use of cationic polymers [e.g., poly(L-lysine), poly(ethyleneimine), poly(allyl amine), and poly(vinyl amine)] to coat the external surface of alginate hydrogels [24,25]. The ammonium groups in these polymers form an ionic complex with the carboxylate groups in alginate. This polyanion–polycation complex membrane stabilizes, strengthens, and enhances the long-term stability of alginate hydrogels [26,27]. However, these forms of alginate are typically used for immunoisolated cell transplantation, not to create tissues integrated with the host tissue.

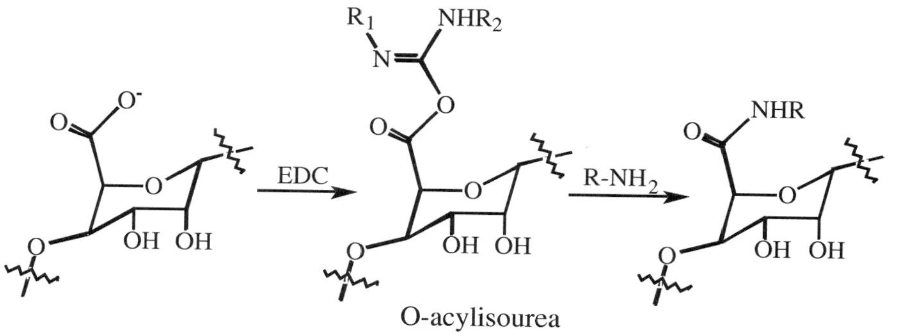

Fig. 56.1. The mechanism of EDC activation of alginate.

COVALENT CROSS-LINKING

Covalent cross-linking has also been utilized to enhance the mechanical properties of alginate hydrogels. In one approach, epichlorohydrin was used to cross-link the hydroxyl groups in alginate [28,29]. However, this molecule is highly toxic and is not suitable for tissue engineering applications. A water-soluble carbodiimide, 1-ethyl-3-(dimethyl aminopropyl) carbodiimide (EDC), has been utilized instead to activate the carboxylate groups of alginate hydrogels followed by a bifunctional cross-linker to covalently cross-link these groups [30,31]. A variety of cross-linkers has been used, including the methyl ester of L-lysine, adipic dihydrazide, and poly(ethylene glycol)–diamines (1000 and 3400 Da) (Fig. 56.1) [30,31]. Alternatively, one can oxidize alginate to generate aldehydes, which may readily be used as cross-linking sites. The mechanical properties of hydrogels formed with both approaches can be controlled by the type of cross-linker used and the cross-link density [32]. In the following sections, we will discuss the protocols for cross-linking with both approaches.

Carbodiimide Chemistry

An aqueous alginate solution (2% w/w) was prepared in 0.1 M 4-morpholinoethanesulfonic acid (MES) buffer and 0.5 M sodium chloride, and the pH was adjusted to 6. At pH 6, the EDC activation of the carboxylate groups is optimal [30]. The viscous solution was filtered through a 0.45-μm filter to get rid of the aggregates. An aliquot of aqueous alginate (15 g) was then transferred to a conical tube (50 ml) and N-hydroxysuccinimde (NHS, 87.5 mg, 0.76 mmol) was added with stirring followed by EDC (291.4 mg, 1.52 mmol). EDC reacts with the carboxylate group to form an N-acylisourea intermediate that is very reactive with a short half-life. NHS reacts with the N-acylisourea to form the N-hydroxysuccinimde ester that is more stable than the N-acylisourea intermediate. The mixture was agitated for 30 s and the cross-linker was added (methyl ester L-lysine, PEG–diamine, or adipic dihydrazide), and the mixture was shaken vigorously for 30 s and cast between two glass plates separated by a 2-mm spacer. The N-hydroxysuccinimide ester reacts with the amine groups of the cross-linkers to form covalent amide bonds. The gels were allowed to set for 24 h and cut into disks (12.7 mm in diameter). The gels were then placed in double-distilled water to remove unreacted materials and by-products.

Carbodiimide chemistry is more suitable for synthesizing preformed matrices than *in situ* gel formation in the presence of cells. Upon completion of the cross-linking of the hydrogels, these matrices are thoroughly purified, to eliminate all unreacted EDC, NHS, and other contaminants. Different molds can be used to perform gels into the desired size and shape. The gels are then frozen, lyophilized, and stored until needed. Cells could be seeded on these scaffolds and transplanted afterward.

Oxidation of Alginate

Another approach to improving the stability of alginate hydrogels is to create additional active sites for covalent cross-linking. Alginate has been oxidized with sodium periodate to generate aldehyde groups (Fig. 56.2) [33]. Poly(ethyleneimine) was used to cross-link oxidized alginate via the formation of Schiff bases between the amine groups in poly(ethyleneimine) and the aldehyde groups in oxidized alginate. However, Schiff bases

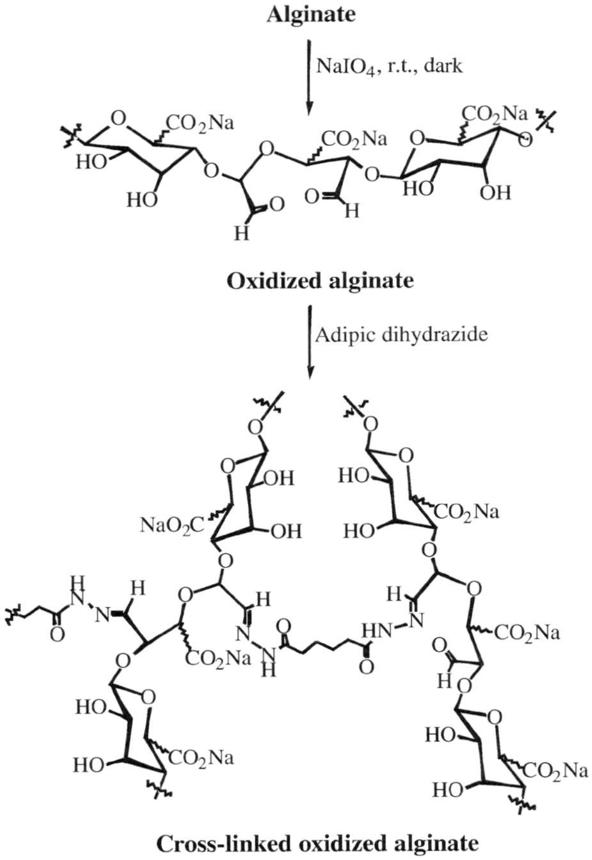

Alginate

Fig. 56.2. *Synthesis and cross-linking oxidized alginate.*

are relatively unstable in aqueous solutions, and it is advantageous to cross-link oxidized alginates with a functional group that is more reactive and more stable. One such group is the hydrazide group that reacts with aldehyde groups to form hydrazone bonds (Fig. 56.2). Therefore, a suitable bifunctional cross-linker is the water-soluble adipic dihydrazide molecule.

To prepare oxidized alginate, aqueous solutions of sodium alginate were oxidized in the dark with sodium periodate at room temperature following a modified procedure reported previously [34]. In short, a 1-liter Erlenmeyer flask was wrapped with aluminum foil and charged with sodium alginate (8.0 g). Double-distilled water (800 ml) was added, and the mixture was stirred until the solid dissolved. An aqueous solution of sodium periodate (0.25 M, 162 ml) was added, and the reaction was stirred for 24 h at room temperature. Ethylene glycol (2.3 ml) was then added to the reaction mixture to quench any unreacted periodate. The reaction was stirred for 0.5 h at ambient temperature and the solution was filtered and exhaustively dialyzed [Spectra/Pro membrane with a molecular weight cutoff (MWCO) of 3500 Da] against double-distilled water for 3 days. The water was changed at least three times a day. The solutions were then concentrated to around 100 ml, and freeze-dried under reduced pressure to yield a white product (6.9 g, 86%). Results from IR spectroscopy (KBr pellet) gave the following wavenumbers of maximum absorption peaks: 3336, 2942, 1730, 1622, 1406, 1321, 1159, 1117, and 1026 cm^{-1}.

The amount of sodium periodate used in these reactions was varied to form alginates with different degrees of oxidation. The experimental degree of oxidation was determined by measuring the percentage of sodium periodate that was consumed in each reaction. This process generated pendant aldehyde groups that were used as cross-link sites. Oxidized alginate of different aldehyde contents has been cross-linked with adipic dihydrazide to form hydrogels. Fourier transform infrared (FTIR) spectroscopic analysis of the dried hydrogels indicated the disappearance of the peak at 1730 cm^{-1}. An additional band was detected at

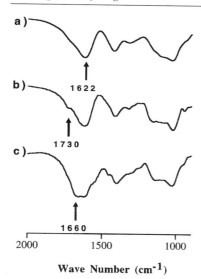

Fig. 56.3. FTIR spectra of (a) sodium alginate, (b) oxidized alginate (100% equiv of perio-date), and (c) cross-linked oxidized alginate; arrows point at the carbonyl-stretching vibra-tional bands of the carboxylate (1622), aldehyde (1730), and hydrazide (1660) groups.

1660 cm^{-1} corresponds to the stretching vibration of the carbonyl in the hydrazide group (Fig. 56.3).

Hydrogels were formed following the mixing of aqueous solutions of oxidized alginate and adipic dihydrazide. One can control the mechanical properties of these hydrogels by varying the cross-linking densities (ionic and covalent). The compressive modulus of cross-linked oxidized alginate hydrogels was measured to determine and compare the physical strength of hydrogels formed at different conditions. A wide range of mechanical proper-ties could be achieved with these materials by changing the concentration of the polymer, covalent cross-linker, and ionic cross-linker [34].

CELL INTERACTIONS
COUPLING GRGDY TO ALGINATE

Mammalian cells do not have the necessary receptors to recognize and bind to alginate hydrogels. Adhesion is necessary, however both for long-term viability of cells and to con-trol cell growth and function on polymers [35]. To improve the cellular interactions with alginate, one can couple cell adhesion ligands to the alginate backbone. A water-soluble carbodiimide, EDC, was used to activate the carboxylate groups in alginate, as mentioned earlier. The amine terminal of the pentapeptide, GRGDY, was then reacted with the ac-tivated carboxylate groups to form amide bonds (Fig. 56.4). This procedure was used to couple the ligands to alginate in aqueous solutions as well as the surface of preformed al-ginate discs. Trace ^{125}I[GRGDY] was added to the pentapeptide solution to quantify the percentage of peptide that is coupled to alginate. Alginate hydrogels with coupled cell ad-hesion ligands have promoted cell adhesion and proliferation. Mouse skeletal myoblasts seeded on the surface of these hydrogels adhered, proliferated, and fused into multinucle-ated myofibrils. Furthermore, myoblasts expressed heavy-chain myosin, a differentiation marker for skeletal muscle [36].

COUPLING GRGDY TO OXIDIZED ALGINATE

Cell adhesion ligands have also been successfully incorporated into oxidized alginate following two different approaches: carbodiimide and reductive amination chemistry. In the first approach, EDC was used to activate the carboxylate groups that were allowed to react with the amine terminal of GRGDY (Gly-Arg-Gly-Asp-Tyr) to form amide bonds. The second approach was to allow the amine group of the peptide to react with the pendant aldehyde group of oxidized alginate to form an imine bond (Fig. 56.4). The imine bond was then selectively reduced with sodium cyanoborohydride to form an amine bond while maintaining the reactivity of the unreacted aldehyde groups for further cross-linking with adipic dihydrazide [34].

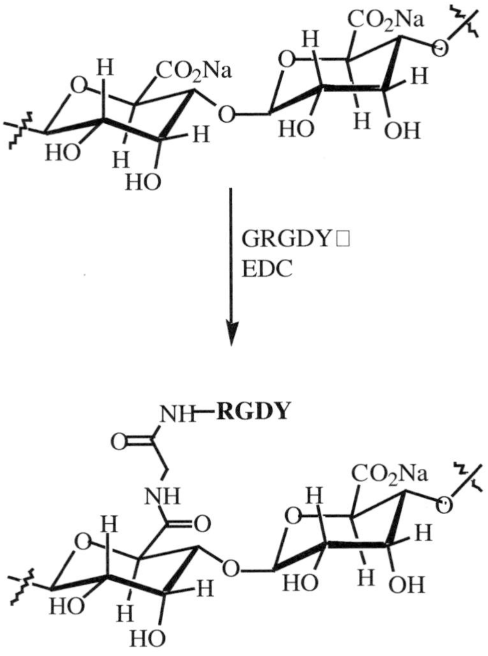

Fig. 56.4. Coupling GRGDY peptide to alginate backbone.

Carbodiimide Chemistry Protocol

An aqueous solution of oxidized alginate (1.8 g, 60 ml) in MES buffer (pH 6.5) containing 0.5 M NaCl was mixed with sulfo-N-hydroxysuccinimide (sulfo-NHS, 99 mg, 0.45 mmol). An aqueous solution of GRGDY (90 μl, 10 mg/ml) was added, and the reaction was stirred for 10 min. A solution of 100 μl of freshly prepared of EDC (145.7 mg, 0.76 mmol) in MES buffer (1.0 ml, pH 6.5) was added, and the reaction was stirred for 12 h at ambient temperature. The solution was then exhaustively dialyzed (MWCO 1000 Da) against 1 liter of double-distilled water for 3 days, frozen, and lyophilized. The degree of peptide incorporation was found to be 58%.

Reductive Amination Protocol

Oxidized alginate (1.8 g) was dissolved in aqueous sodium bicarbonate buffer (60 ml, pH 9.6). An aqueous solution of GRGDY (90 μl, 10 mg/ml) was added, followed by the careful addition of sodium cyanoborohydride (NaCNBH$_3$, 150 mg, 2.4 mmol). The reaction was stirred for 24 h at ambient temperature. The solution was then exhaustively dialyzed (MWCO 1000 Da) against 1 liter of double-distilled water for 3 days, frozen, and lyophilized to yield a white powder (68% peptide incorporation).

DEGRADATION OF MODIFIED ALGINATE HYDROGELS

It is critical that biomaterials in tissue engineering applications be able to degrade over time and release low molecular weight oligomers that can be excreted from the body. However, commercially available alginate has a high molecular weight and mammalian cells do not appear to secrete enzymes that hydrolyze the β-glycoside bonds in alginate [37]. As a result, it is desirable to break down alginate into low molecular weight polymers before hydrogel formation for certain applications. Cross-linking these polymers with degradable cross-linkers yields degradable hydrogels suitable for tissue engineering applications. Different approaches has been utilized to break down alginates including gamma irradiation, acid hydrolysis, and more recently periodate oxidation.

GAMMA IRRADIATION

The radiolytic effect of aqueous alginate solutions has been studied in food irradiation processes [38]. Alginate chains in solution are broken down into low molecular weight fragments when exposed to gamma irradiation from a cobalt source [39,40]. Moreover, alginate powder has also been broken down upon exposure to gamma irradiation (5 Mrad, 2.83 h) [41]. Low molecular weight alginate (M_n 9370 Da, polydispersity index 2.5) has been cross-linked with calcium ions to form porous alginate beads with interconnected pores [40]. The porous beads were transplanted subcutaneously on the back of Lewis rats. These beads maintained their porous structure and allowed cell invasion *in vivo*.

ACID HYDROLYSIS

Sodium alginate has also been hydrolyzed under acidic conditions to yield the low molecular weight oligomers, sodium polyguluronate (6 kDa) [34]. In short, sodium alginate (75 g) was dissolved in 3.5 liters of double-distilled water followed by the addition of aqueous hydrochloric acid (390 ml, 3 M). The reaction was refluxed for 5 h to hydrolyze the β-glycoside bonds connecting the guluronic and mannuronic acids in the alternating blocks. The poly(guluronic acid) and poly(mannuronic acid) blocks remained intact. The solution was cooled to room temperature and centrifuged. The supernatant liquid containing the guluronic and mannuronic acid sugars was decanted. The solid containing poly(mannuronic acid) and poly(guluronic acid) oligomers was collected, combined, and suspended in 5 liters of double-distilled water. Sodium chloride (29.25 g) was added to the solution followed by aqueous sodium hydroxide (50 ml, 4 M) to form the water-soluble oligomers sodium polyguluronate and polymannuronate. The pH was then adjusted to 2.2 with concentrated hydrochloric acid (20 ml, 12 M). At this pH, poly(guluronic acid) is precipitated, whereas sodium polymannuronate is soluble. The suspension was centrifuged and washed with 500 ml of double-distilled water. The solid was suspended in 1 liter of water and 3 g of sodium chloride was added to the suspension. The pH of the solution was adjusted to 7.5 with aqueous sodium hydroxide (4 M) to form sodium polyguluronate. The solution was mixed with activated carbon (20 g) and filtered, and the product was precipitated with the addition of 2 liters of 95% ethanol, centrifuged, and freeze-dried under reduced pressure to yield a white solid (38.0 g): M_w 6200 Da, M_w/M_n 1.14. Spectroscopic results were as follows:

IR (KBr pellet): wavenumbers of 3430.5, 2913.2, 1616.4, 1412.5, 1320.5, 1126, 1092, and 1028 cm^{-1}

^{1}H NMR (360 MHz, D$_2$O): chemical shifts (ppm) δ 3.90 (broad, singlet), 4.07 (br, s), 4.45 (br, s), and 5.03 (br, s)

^{13}C NMR (360 MHz, D$_2$O): chemical shifts (ppm) δ 67.64, 69.87, 71.73, 82.94, 103.66, and 178.29.

SYNTHESIS OF POLY(ALDEHYDE GULURONATE) (PAG)

Sodium polyguluronate was oxidized with sodium periodate to generate pendant aldehyde groups necessary for covalent cross-linking with adipic dihydrazide [34]. Sodium polyguluronate (40.0 g) was dissolved, in 400 ml of double-distilled water in an Erlenmeyer flask wrapped with aluminum foil. A solution of sodium periodate (0.5 M, 400 ml) was added, and the reaction was stirred for 19 h at room temperature. Ethylene glycol (40 ml) was then added to quench the periodate ions that did not react. The reaction was stirred for 10 min at room temperature, filtered, and dialyzed (MWCO 1000 Da) against double-distilled water for 3 days. The solution was then concentrated under reduced pressure to 100 ml and freeze-dried under reduced pressure to yield a white product (24.8 g, 62%): M_w 6 kDa, M_w/M_n 1.4. IR (spectroscopic results) (KBr pellet) gave wavenumbers of 3336.5, 2942.4, 1724.4, 1616.4, 1405.8, 1321.1, 1159.4, 1117.2, and 1025.8 cm^{-1}.

Cross-linked PAG hydrogels has been formed by mixing aqueous solutions of PAG and adipic dihydrazide. The hydrazone bond, formed between the hydrazide groups and the aldehyde groups in PAG, is susceptible to hydrolytic cleavage in aqueous solutions. This results in the complete degradation and dissolution of the hydrogel in the surrounding medium. After the dissolution of the hydrogel, low molecular weight oligomers are released. These oligomers are expected to clear from the body. In addition, the degradation rate of these hydrogels can be controlled by varying the concentration of the polymer and the

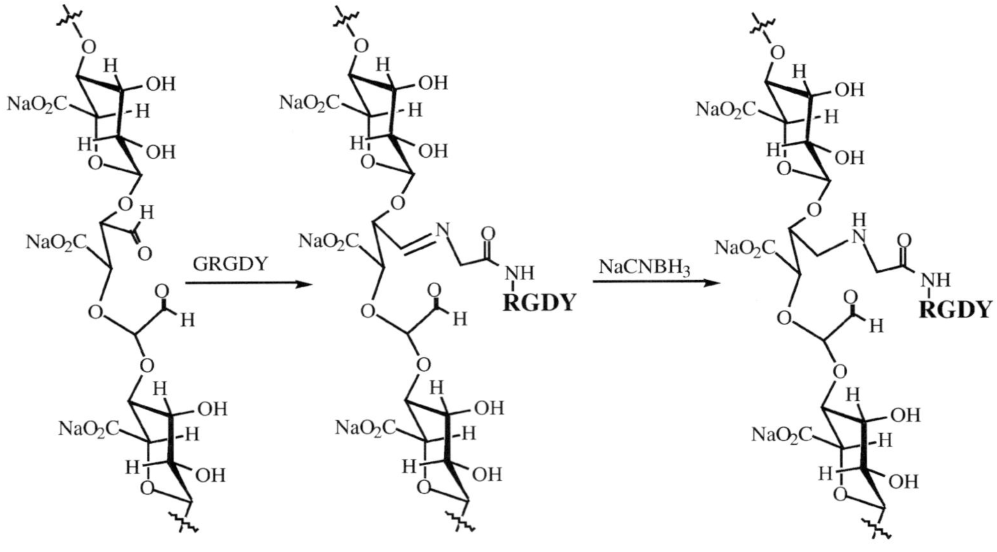

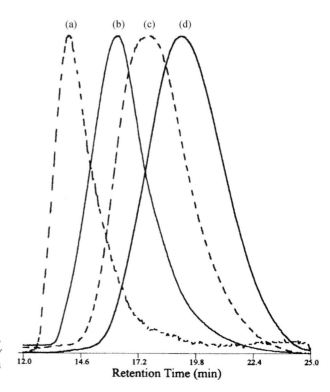

Fig. 56.5. Reductive amination of pendant aldehyde groups.

cross-linker (ionic and covalent) [34,42]. One can form hydrogels that degrade from days to months with this approach (Fig. 56.7).

PERIODATE OXIDATION

In another approach to breaking down alginate chains, the periodate oxidation is used to generate aldehyde groups. Polysaccharides depolymerize under the conditions of the periodate oxidation to yield low molecular weight polymers [43]. Alginate depolymerizes in a similar manner depending on the initial concentration of sodium periodate (Fig. 56.6). Therefore, one can directly oxidize the high molecular weight (300 kDa) alginate to yield oxidized alginate with lower molecular weight, and one can control the molecular weights of oxidized alginate by changing the concentration of periodate used in these reactions.

Fig. 56.6. Representative chromatograms of (a) sodium alginate, (b) 25%, (c) 50%, and (d) 100% oxidized alginate as detected by the differential refractive index detector. The solvent is 0.1 M NaNO₃ (0.05% NaN₃) at a flow rate of 0.7 ml/min [44].

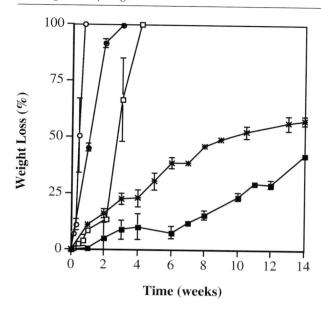

Fig. 56.7. Percentage weight loss of cross-linked oxidized alginate hydrogels as a function of time. Hydrogels were formed at (■) 150 mM adipic dihydrazide and 50 kDa oxidized alginate, (●) 100 mM adipic dihydrazide and 50 kDa oxidized alginate, (○) 100 mM adipic dihydrazide and 6 kDa oxidized alginate, (□) 150 mM adipic dihydrazide and 6 kDa oxidized alginate, (*) 150 mM adipic dihydrazide, 6 kDa oxidized alginate, and 40 mM calcium chloride. All hydrogels were prepared with 10% w/w oxidized alginates (100% equiv periodate) in double-distilled water [42,44].

As the concentration of sodium periodate was increased, the average molecular weight of oxidized alginate decreased to reach 50 kDa with 100 equiv of periodate (Fig. 56.6). Low molecular weight oxidized alginate was cross-linked with adipic dihydrazide to form biodegradable hydrogels [44]. These hydrogels degrade in weeks to months, depending on the concentrations of the ionic and covalent cross-links (Fig. 56.7).

CONCLUSIONS

Alginate can be modified covalently to make it more suitable for tissue engineering applications. Covalently cross-linked alginate hydrogels have a wide range of mechanical properties that are desirable in engineering a variety of tissues. Cell adhesion ligands have been coupled to alginate and have promoted cellular proliferation and differentiation. These hydrogels degrade in a controllable manner over a wide range of time periods, from weeks to months. During the degradation of the gels, low molecular weight chains are released and are expected to clear from the body. Modified alginate hydrogels are useful for drug delivery and tissue engineering applications.

REFERENCES

1. Langer, R., and Vacanti, J. P. (1993). *Tissue Eng. Sci.* **260**, 920–926.
2. Nerem, R. M., and Sambanis, A. (1995). Tissue engineering: From biology to biological substitutes. *Tissue Eng.* **1**, 3–13.
3. Bouhadir, K. H., and Mooney, D. J. (1998). In vitro and in vivo models for the reconstruction of intracellular signaling. *Ann. N.Y. Acad. Sci.* **842**, 188–194.
4. Klock, G., Pfeffermann, A., Ryser, C., Grohn, P., Kuttler, B., Hahn, H. J., and Zimmermann, U. (1997). Biocompatibility of mannuronic acid-rich alginates. *Biomaterials* **18**, 707–713.
5. Schmidt, R. J., Chung, L. Y., Andrews A. M., Spyratou, O., and Turner, T. D. (1992). Biocompatibility of wound management products: A study of the effects of various polysaccharides on murine L929 fibroblasts proliferation and macrophage respiratory burst. *J. Pharm. Pharmacol.* **45**, 508–513.
6. Schmidt, R. J., Chung, L. Y., Andrews, A. M., Spyratou, O., and Turner, T. D. (1993). Biocompatibility of wound management products: A study of the effects of various polysaccharides on Murine l929 fibroblast proliferation and macrophage respiratory burst. *J. Pharm. Pharmacol.* **45**, 508–513.
7. Doyle, J. W., Roth, T. P., Smith, R. M., Li, Y., Dunn, and R. M. (1996). Effect of calcium alginate on cellular wound healing process modeled in vitro. *J. Biomed. Mater. Res.* **32**, 561–568.
8. Chang, P. L., Hortelano, G., Tse, M., and Awrey, D. E. (1994). Growth of recombinant fibroblasts in alginate microcapsules. *Biotechnol. Bioeng.* **43**, 925–933.
9. Wang, T., Lacik, I., Brissova, M., Anilkumar, A. V., Prokop, A., Hunkler, D., Green, R., Shahrokhi, K., and Powers, A. C. (1997). An encapsulation system for the immunoisolation of pancreatic islets. *Nat. Biotechnol.* **15**, 358–362.
10. Cook, W. (1986). Alginate dental impression materials: Chemistry, structure, and properties. *J. Biomed. Mater. Res.* **20**, 1–24.

11. Sutherland, I. W. (1991). Alginates. *In* "Biomaterials: Novel Materials from Biological Sources" (D. Byron ed.), pp. 309–331. Stockton Press, New York.

12. Wang, Z., Zhang, Q., Konno, M., and Saito, S. (1993). Sol–gel transition of alginate solution by the addition of various divalent cations: ^{13}C-NMR spectroscopic study. *Biopolymers* **33**, 703–711.

13. Honghe, Z. (1997). Interaction mechanism in sol–gel transition of alginate solutions by addition of divalent cations. *Carbohydr. Res.* **302**, 97–101.

14. Quong, D., Neufeld, R. J., Skjak-Braek, G., and Poncelet, D. (1998). External versus internal source of calcium during the gelation of alginate beads for DNA encapsulation. *Biotechnol. Bioeng.* **57**, 438–446.

15. Moe, S., Skja-Braek, G., Smidsrod, O., and Ichijo, H. (1994). Calcium alginate gel fibers—influence of alginate source and gel structure on fiber strength. *J. Appl. Polym. Sci.* **51**, 1771–1775.

16. Paige, K. T., Cima, L. G., Yaremchuk, M. J., Scloo, B. L., Vacanti, J. V., and Vacanti, C. A. (1995). De novo cartilage generation using calcium alginate–chondrocyte constructs. *Plast. Reconstr. Surg.* **97**, 168–178.

17. Paige, K. T., Cima, L. G., Yaremchuk, M. J., and Vacanti, J. V. (1995). Injectable Cartilage. *Plast. Reconstr. Surg.* **96**, 1390–1398.

18. Atala, A., Cima, L. G., Kim, W., Paige, K. T., Vacanti, J. V., Retik, A. B., and Vacanti, C. A. (1993). Injectable alginate seeded with chondrocytes as a potential treatment for vesicoureteral reflux. *J. Urol.* **150**, 745–747.

19. Atala, A., Kim, W., Paige, K. T., Vacanti, C. A., and Retik, A. B. (1994). Endoscopic treatment of vesicoureteral reflux with a chondrocyte–alginate suspension. *J. Urol.* **152**, 641–643.

20. Peters, M. C., Isenberg, B. C., Rowely, J. A., and Mooney, D. J. (1998). Release from alginate enhances the biological activity of vascular endothelial factor. *J. Biomater. Sci., Polym. Ed.* **9**, 1267–1278.

21. Kawada, A., Hiura, N., Tajima, S., and Takahara, H. (1999). Alginate oligosaccharides stimulate VEGF-mediated growth and migration of human endothelial cells. *Arch. Dermatol. Res.* **291**, 542–547.

22. Sheridan, M. H., Shea, L. D., Peters, M. C., and Mooney, D. J. (2000). Bioadsorbable polymer scaffolds for tissue engineering capable of sustained growth factor delivery. *J. Controled Release* **64**, 91–102.

23. Murphy, W. L., Peters, M. C., Kohn, D. H., and Mooney, D. J. (2001). Sustained release of Vascular Endothelial Growth Factor from mineralized poly(lactide-*co*-glycolide) scaffolds for tissue engineering. *Biomaterials* **21**, 2521–2527.

24. Clayton, H. A., James, R. F. L., and London, N. J. M. (1993). Islet microencapsulation: A review. *Acta Diabetol.* **30**, 181–189.

25. Shoichet, M. S., Li, R. H., White, M. L., and Winn, S. R. (1996). Stability of hydrogels used in cell encapsulation: An in vivo comparison of alginate and agarose. *Biotechnol. Bioeng.* **50**, 374–381.

26. Thu, B., Bruheim, P., Espevik, T., Smidsrod, O., SoonShiong, P., and Skjak-Braek, G. (1996). Alginate polycation microcapsules. I. Interaction between alginate and polycation. *Biomaterials* **17**, 1069–1079.

27. Wang, F. F., Wu, C. R., and Wang, Y. J. (1992). Preparation and application of poly(vinylamine)/alginate microcapsules to culturing of a mouse erythroleukemia cell line. *Biotechnol. Bioeng.* **40**, 1115–1118.

28. Fundueanu, G., Nastruzzi, C., Carpov, A., Desbrieres, J., and Rinaudo, M. (1999). Physicochemical characterization of Ca–alginate microparticles produced with different methods. *Biomaterials* **20**, 1427–1435.

29. Moe, S. T., Skjak-Braek, G., Elgsaeter, A., and Smidsrod, O. (1993). Swelling of covalently crosslinked alginate gels: Influence of ionic solutes and nonpolar solvents. *Macromolecules* **26**, 3589–3597.

30. Rowley, J., Bouhadir, K. H., Petrovsky, D., Wang, S. W., and Mooney, D. J. (2001). Synthesis and characterization of covalently crosslinked alginates. Submitted for publication.

31. Eiselt, P., Lee, K. Y., and Mooney, D. J. (1999). Rigidity of two-component hydrogels prepared from alginate and poly(ethylene glycol)-diamines. *Macromolecules* **32**, 5561–5566.

32. Lee, K. Y., Rowely, J. A., Eiselt, P., Moy, E. M., Bouhadir, K. H., and Mooney, D. J. (2000). Controlling mechanical and swelling properties of alginate hydrogels independently by cross-linker type and cross-link density. *Macromolecules* **33**, 4291–4294.

33. Birnbaum, S., Pendleton, R., Larsson, P., and Mosbach, K. (1981). Covalent stabilization of alginate gels for the entrapment of living whole cells. *Biotechnol. Lett.* **3**, 393–400.

34. Bouhadir, K. H., Hausman, D. S., and Mooney, D. J. (1999). Synthesis of cross-linked poly(aldehyde guluronate) hydrogels. *Polymer* **40**, 3575–3584.

35. Price, L. S. (1997). Morphological control of cell growth and viability. *BioEssays* **19**, 941–943.

36. Rowley, J. A., Madlambayan, G., and Mooney, D. J. (1999). Alginate hydrogels as synthetic extracellular matrix materials. *Biomaterials* **20**, 45–53.

37. Al-Shamkhani, A., and Duncan, R. (1995). Radioiodination of alginate via covalently-bound tyrosinamide allows for monitoring of its fate in vivo. *J. Bioact. Compat. Polym.* **10**, 4–13.

38. Delincee, H. (1989). Radiolytic effects in food. *Proc. Int. Workshop Food Irradiat.*, pp. 160–179.

39. King, K. (1994). Changes in the functional properties and molecular weight of sodium alginate following γ-irradiation. *Food Hydrocolloids* **8**, 83–96.

40. Purwanto, Z. I., Broeck, L. A. M., Schols, H. A., Pilnik, W., and Voragen, A. G. J. (1998). Degradation of low molecular weight fragments of pectin and alginates by gamma-irradiation. *Acta Aliment.* **27**, 29–42.

41. Eiselt, P., Yeh, J., Latvala, R. K., Shae, L. D., and Mooney, D. J. (2000). Porous carriers for biomedical applications based on alginate hydrogels. *Biomaterials* **21**, 1921–1927.

42. Lee, K. Y., Bouhadir, K. H., and Mooney, D. J. (2000). Degradation behavior of covalently cross-linked poly(aldehyde guluronate) hydrogels. *Macromolecules* **33**, 97–101.

43. Painter, T. J. (1988). Control of depolymerisation during the preparation of reduced dialdehyde cellulose. *Carbohyr. Res.* **179**, 259–268.

44. Bouhadir, K. H., Alsberg, E., and Mooney, D. J. (2001). Hydrogels for combination delivery of antineoplastic agents. *Biomaterials* **22**, 2625–2630.

CHAPTER 57

SYNTHESIS OF HYDROGELS: ENVIRONMENTALLY SENSITIVE HYDROGELS BASED ON N-ISOPROPYLACRYLAMIDE

Ranee A. Stile and Kevin E. Healy

INTRODUCTION

Tissues and organs, such as cartilage [1–3], bone [4], blood vessels [5], peripheral nerves [6], the liver [7], skin [8], and the pancreas [9], have been studied extensively in many tissue engineering initiatives. A common approach to the repair of damaged or diseased tissues and organs employs three-dimensional (3D) polymer matrices that act as temporary analogues for the extracellular matrix (ECM) and foster regeneration both *in vitro* and *in vivo*. Isolated cells are seeded into the polymer matrices, or scaffolds, and the 3D structures guide the cells' organization and development into tissues and organs [10, 11]. Ideally, the polymer scaffolds impart qualities that mimic the *in vivo* environment while providing a milieu that allows the cells to proliferate, differentiate, maintain their natural phenotype, and ultimately function as a tissue or an organ. Natural polymers, such as alginate [3], hyaluronic acid [6], and type I collagen [2,8], and synthetic polymers, such as poly(glycolic acid) [1,5], poly(lactic acid) [7], and copolymers of poly(glycolic acid) and poly(lactic acid) [4], have been used to synthesize scaffolds for tissue and organ repair. The mechanical properties of the scaffold dictate whether the matrix must be implanted surgically. More pliable scaffolds may be implanted by less invasive means (e.g., arthroscopically). In some instances, these more flexible or "injectable" matrices may be better suited for treating irregularly shaped defects, since it may be difficult to form rigid scaffolds into aberrant configurations.

ISSUES WITH CURRENT INJECTABLE SCAFFOLDS

From a clinical perspective, injectable scaffolds are extremely desirable because the implantation procedure is less invasive. Injectable scaffolds have been fabricated using such materials as calcium alginate [3] and Pluronics [12] [i.e., block copolymers of poly(ethylene oxide) and poly(propylene oxide)]. Calcium alginate is immunogenic [13] while Pluronics, which form associative networks as a function of temperature and polymer concentration, lack structural integrity. In addition, Pluronics tend to be unstable due to surface dissolution which results from the dissipation of packed micelles by dilution [14]. Other injectable materials, such as hyaluronic acid (HA), tend to be too soft and too difficult to handle [15]. Scaffolds that are too soft may lack adequate mechanical integrity to support tissue and organ formation. Finally, injectable systems that polymerize *in situ* increase the possibility that potentially toxic residual monomers, cross-linkers, and reaction catalysts will be present in the body following the formation of the matrix.

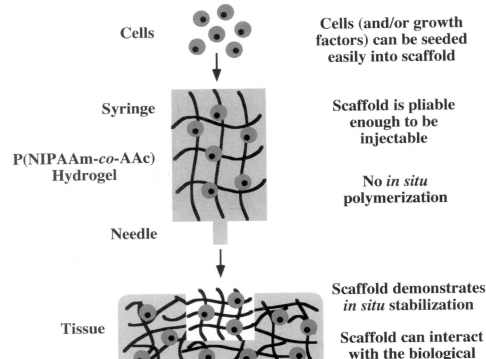

Cells

Cells (and/or growth factors) can be seeded easily into scaffold

Syringe

Scaffold is pliable enough to be injectable

P(NIPAAm-*co*-AAc) Hydrogel

No *in situ* polymerization

Needle

Tissue

Scaffold demonstrates *in situ* stabilization

Scaffold can interact with the biological environment

Fig. 57.1. Schematic representation of an ideal injectable polymer scaffold.

AN IDEAL INJECTABLE POLYMER SCAFFOLD

An "ideal" injectable polymer scaffold for tissue regeneration should be (1) synthesized prior to implantation, eliminating the potential adverse effects of *in situ* polymerization; (2) be pliable enough at room temperature (RT; i.e., ≈22°C) to allow for the incorporation of cells and macromolecules (e.g., growth factors); (3) be amenable to minimally invasive implantation procedures; (4) demonstrate *in situ* stabilization (i.e., an increase in rigidity) upon placement into the body, improving the mechanical integrity of the matrix to better support tissue or organ formation; (5) be minimally toxic; and (6) be capable of interacting with the host tissue and biological environment on a molecular level. A schematic representation of this ideal polymer scaffold is presented in Fig. 57.1.

EXPLOITING PHASE SEPARATION TO DEVELOP INJECTABLE POLYMER SCAFFOLDS

One strategy for developing an injectable polymer scaffold is to employ the principles of lower critical phase separation. Polymer mixtures which demonstrate this behavior phase-separate as the temperature is increased above the lower critical solution temperature (LCST) [16,17]. This type of phase separation is generally regarded as a phenomenon governed by the balance of hydrophilic and hydrophobic moieties on the polymer chain [18] and driven by a negative entropy of mixing [18–20]. In addition, the temperature dependence of certain molecular interactions, such as hydrogen bonding and hydrophobic effects, contributes to this type of phase separation [21]. At the LCST, the hydrogen bonding between the polymer and water becomes unfavorable compared with polymer–polymer and water–water interactions [20], and an abrupt transition occurs as the hydrated hydrophilic macromolecule quickly dehydrates and changes to a more hydrophobic structure [20,22].

Poly(*N*-isopropylacrylamide) [P(NIPAAm)] phase separates from water at an LCST of approximately 32°C [19]. Figure 57.2 gives a schematic representation of the P(NIPAAm)–water phase diagram and a depiction of the phase behavior of P(NIPAAm) chains and cross-linked hydrogels. P(NIPAAm) chains are soluble in water below the LCST, but precipitate at the LCST, where the mixture turns cloudy because the hydrophobic groups in the polymer chain form insoluble aggregates [23]. This behavior is reversible, and the P(NIPAAm) chains

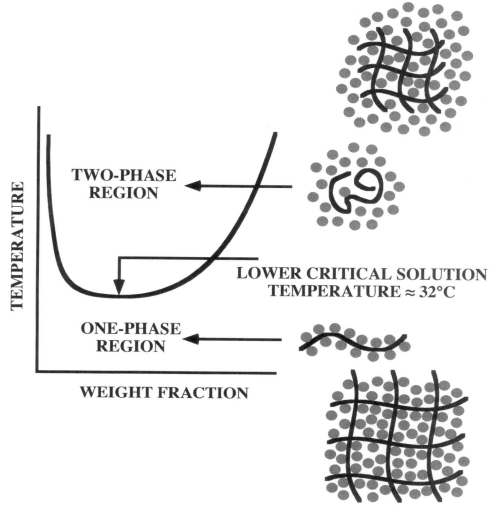

Fig. 57.2. Schematic representation of the P(NIPAAm)–water phase diagram with depictions of the phase behavior of P(NIPAAm) chains and cross-linked hydrogels. Adapted from Heskins and Guillet [19], Taylor and Cerankowski [18], and Hoffman [24].

will dissolve again in water as the temperature is lowered below the LCST, although the kinetics of redissolution are often slower than the precipitation process [24]. P(NIPAAm) hydrogels, formed by polymerizing the hydrophilic NIPAAm monomer with small amounts of cross-linking agents, experience a volume phase transition at the LCST and collapse substantially as the temperature is increased above the LCST [25,26]. During the volume phase transition, the hydrogel expels a large amount of its pore water and generally becomes stiff and opaque [24,25]. This behavior is reversible, and the P(NIPAAm) hydrogel will reswell in water as the temperature is decreased below the LCST, but at a rate slower than that of the initial deswelling process [24]. Okano and coworkers [27–29] have studied the deswelling kinetics of homopolymer and hydrophobically modified P(NIPAAm) hydrogels. These hydrogels tend to shrink rapidly at first and then deswell slowly, taking more than a month to reach equilibrium deswelling. A "dense skin layer," seen on the surface of the hydrogels above the LCST, is believed to hinder the release of hydrogel water and limit the shrinkage rate [29].

One of the most common ways of manipulating the phase behavior of P(NIPAAm) chains and hydrogels is by the addition of more hydrophilic or more hydrophobic monomers. Copolymerization of NIPAAm with a more hydrophilic monomer typically increases the LCST of P(NIPAAm) copolymers [9,18] and copolymer hydrogels [25,30,31], while incorporation of a more hydrophobic monomer tends to have the opposite effect [21,30]. Additionally, more hydrophilic monomers decrease both the extent of aggregation experienced by P(NIPAAm) copolymer chains [9] and the extent of temperature-sensitive volume change demonstrated by P(NIPAAm) copolymer hydrogels [28,31,32].

P(NIPAAm) homopolymer and copolymer chains and cross-linked hydrogels have been studied for use in a number of diverse applications including solute recovery [33], solute delivery [25,32], bioseparations [34], catalytic reaction control [35], and chromatography [36]. P(NIPAAm)-based systems have also been examined in combination with cells. For example, Bae and coworkers [9,37,38] used P(NIPAAm) copolymers in the development of a biohybrid artificial pancreas, Shimizu *et al.* [39] employed P(NIPAAm) copolymers in cell microencapsulation applications, and Okano and coworkers [40,41] investigated the 3D manipulation of cell sheets cultured on P(NIPAAm) surfaces. The evolution of most of these applications was based on the unique phase behavior of P(NIPAAm) in aqueous media.

P(NIPAAm)-BASED HYDROGELS AS INJECTABLE POLYMER SCAFFOLDS

Injectable P(NIPAAm)-based hydrogels have been developed for tissue engineering applications [31]. The hydrogels are synthesized by simultaneously polymerizing and cross-linking NIPAAm and acrylic acid (AAc). These loosely cross-linked hydrogels are extremely pliable and fluidlike at 22°C and can be injected through a small-diameter aperture (2 mm) without demonstrating appreciable macroscopic fracture following injection (Fig. 57.3a). Since the LCST of the P(NIPAAm-co-AAc) hydrogels is ~34°C [31], these networks exhibit a phase transition when heated from 22°C to body temperature (i.e., 37°C), during which the rigidity and, consequently, the mechanical integrity of the hydrogel significantly increase (Fig. 57.3b). Furthermore, the covalent cross-links are crucial because they give the matrix even greater structural integrity and physical stability [42].

The AAc comonomer is used in the P(NIPAAm-co-AAc) hydrogel for two important reasons. AAc is more hydrophilic than NIPAAm and acts to decrease the temperature-induced collapse of the P(NIPAAm-co-AAc) hydrogel [28,31,32]. As discussed earlier, P(NIPAAm) homopolymer hydrogels collapse significantly and release a large fraction of pore water at the LCST. This extensive collapse would be an undesirable characteristic of a scaffold for tissue repair, since after being injected into a defect inside the body, the collapsed scaffold would fill only a small fraction of the defect volume; thus, the potential for successful tissue regeneration would be substantially decreased. In addition, extensive collapse of the matrix could force cells out of the scaffold and/or damage the cells. When a critical ratio of NIPAAm to AAc is used, the P(NIPAAm-co-AAc) hydrogels demonstrate a minimal volume change during the phase transition and retain their pore water, in comparison to P(NIPAAm) hydrogels [31]. The second reason to include AAc in the P(NIPAAm-co-AAc) hydrogels is that the −COOH groups can be functionalized to incorporate biological recog-

INJECTABILITY   **CHANGE IN RIGIDITY**

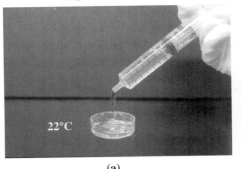
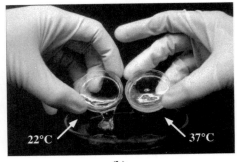

(a)   (b)

Fig. 57.3. (a) A P(NIPAAm-co-AAc) hydrogel being injected through a 2-mm diameter aperture. (b) The change in rigidity demonstrated by the P(NIPAAm-co-AAc) hydrogel upon heating from 22°C to 37°C.

nition capabilities, allowing the hydrogel to interact with the biological environment at a molecular level.

Most of the elements listed for the ideal polymer scaffold are present in the P(NIPAAm-*co*-AAc) hydrogels. The matrices are injectable and offer the benefit of *in situ* stabilization without the potential adverse effects of *in situ* polymerization. Furthermore, the AAc groups can be modified with biologically relevant macromolecules, enabling the hydrogel to interact with cells and/or tissues. The sections that follow detail the synthesis and characterization of these unique P(NIPAAm-*co*-AAc) hydrogels, as well as the behavior of cells within the matrices.

SYNTHESIS OF P(NIPAAm-*co*-AAc) HYDROGELS
MATERIALS

NIPAAm, N, N'-methylenebisacrylamide (BIS), ammonium peroxydisulfate (AP), N, N, N', N'-tetramethylethylenediamine (TEMED), and AAc were obtained from Polysciences, Inc. (Warrington, PA). Deuterium oxide (D_2O) was purchased from Aldrich (Milwaukee, WI). Phosphate-buffered saline (PBS; without magnesium chloride, without calcium chloride) was obtained from GIbco BRL (Grand Island, NY). All materials were used as received.

METHODS

The method used to synthesize the P(NIPAAm-*co*-AAc) hydrogels was reported in 1999 [31]. A schematic representation of the hydrogel synthesis is presented in Fig. 57.4. Dry nitrogen gas was bubbled for 15 min through a mixture of 2.443 g (21.6 mmol) of NIPAAm,* 0.057 g (0.792 mmol) of AAc, 0.005 g (0.0325 mmol) of BIS, and 50 ml of PBS (pH 7.2 ± 0.1) in a two-neck flask to remove dissolved oxygen. Following the nitrogen gas purge, 0.020 g (0.0876 mmol) of AP and 200 μl (1.3 mmol) of TEMED were added as the initiator and accelerator, respectively. The mixture was stirred vigorously for 15 s and allowed to polymerize at 22°C for 19 h under ambient fluorescent lighting in a 250-ml glass beaker covered with a glass plate. Following the polymerization, the P(NIPAAm-*co*-AAc) hydrogel was washed three times, 15–20 min each, in excess ultrapure water (UPW) to remove unreacted compounds. These hydrogels contained 96.5% mol/mol total monomer NIPAAm, 3.5% mol/mol total monomer AAc, and 0.14% mol/mol (total monomer + crosslinker) BIS.

CHARACTERIZATION OF P(NIPAAm-*co*-AAc) HYDROGELS

To fully characterize the P(NIPAAm-*co*-AAc) hydrogels, it is crucial to examine the matrices from both a materials science and a tissue engineering perspective. It is not enough to study the material properties from a pure materials science standpoint; one must also have a thorough understanding of the tissue engineering application, and appropriate tests should be developed to investigate the material properties as they pertain to the function of the hydrogel in that application. While some analyses may not be considered to comprise "traditional" materials characterization tests, these studies give invaluable insight into how the hydrogels would behave in a specific tissue engineering application.

NUCLEAR MAGNETIC RESONANCE (NMR) SPECTROSCOPIC STUDIES

P(NIPAAm-*co*-AAc) hydrogel samples are dried in a vacuum desiccator for 9 days to remove the hydrogel water. Once dried, the samples are immersed in excess D_2O and allowed to swell for 4 days. This process should be repeated a second time to ensure sufficient replacement of the hydrogel water with D_2O and adequate conversion of the −NH and −OH groups to −ND and −OD groups, respectively [43,44]. The D_2O-swollen hydrogel samples are analyzed by means of solid state ^1H magnetic angle spinning (MAS)-NMR spectroscopy.

---

*The amounts of NIPAAm and AAc reported in 1999 [31] were slightly different because a different lot of AAc was used. When the amounts of NIPAAm and AAc given in this chapter are used, the results are not statistically different than those reported earlier [31].

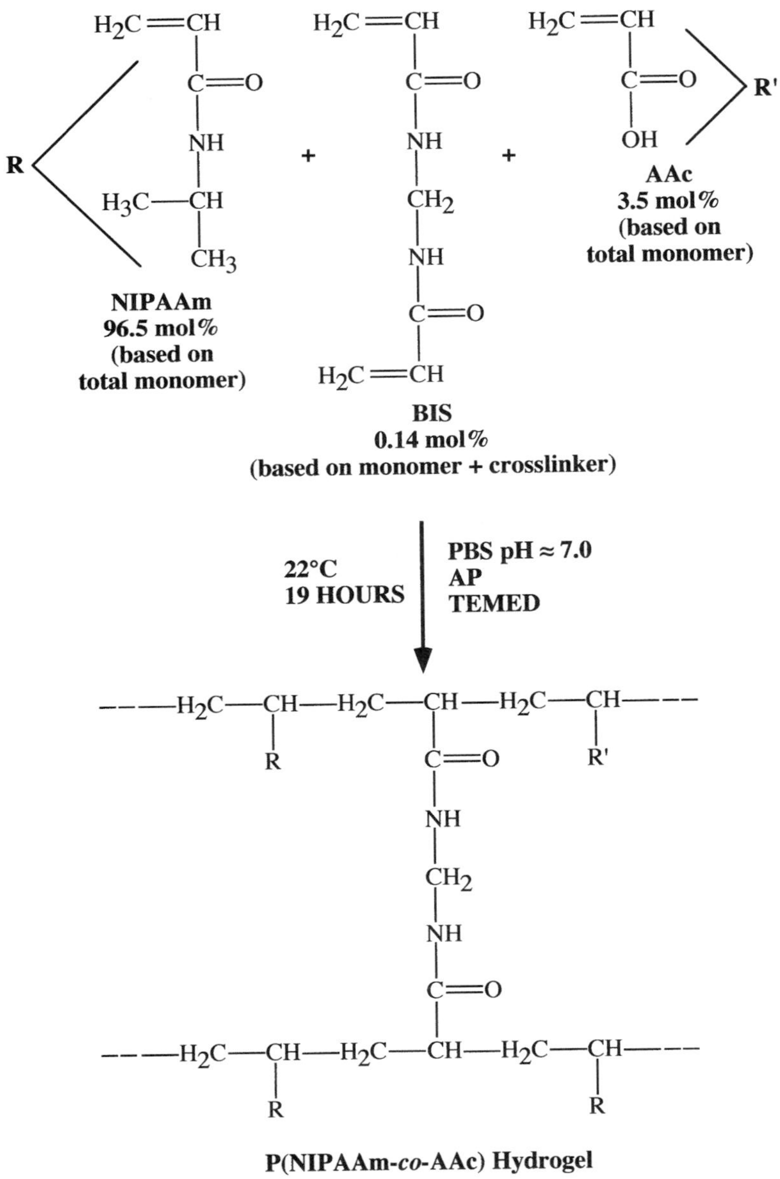

Fig. 57.4. Schematic representation of the P(NIPAAm-co-AAc) hydrogel synthesis.

P(NIPAAm-*co*-AAc) Hydrogel

MAS-NMR spectroscopy is used because this technique tends to increase spectral resolution [45]. In MAS-NMR spectroscopy, the sample is rotated rapidly (~2–4 kHz) about an axis inclined at a 54.7° angle relative to the magnetic field [46]. At this "magic angle," the chemical shift dispersion is averaged to its isotropic value. In the absence of MAS, chemical shift anisotropy is observed because the shielding of the nucleus is generally not symmetrical in three dimensions.

The solid state ^{1}H-MAS-NMR spectrum of a D_2O-swollen P(NIPAAm-*co*-AAc) hydrogel is shown in Fig. 57.5 [31]. These NMR data were consistent with both the chemical structure of the hydrogel and other published reports [44]. Peak 1 at 0.921 ppm represents the $-CH_3$ protons in NIPAAm, peak 2 at 1.350 ppm represents the $-CH_2$ protons in the polymer backbone, peak 3 at 1.781 ppm represents the $-CH$ proton in the polymer backbone, and peak 6 at 3.658 ppm represents the 2-propyl $-CH$ group in NIPAAm. The theoretical intensities of peaks 1, 2, 3, and 6 should be in a ratio of roughly 6:2:1:1. The measured intensities were in a ratio of 6.2:1.9:1.1:1. Peak 4 at 2.4 ppm and peak 5 at 2.8 ppm in the P(NIPAAm-*co*-AAc) hydrogel spectrum are not assigned. These small

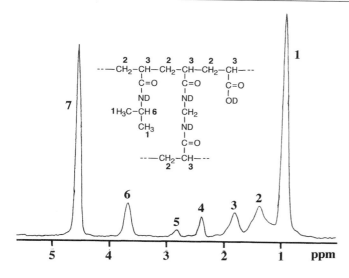

Fig. 57.5. Solid state ^1H MAS NMR spectrum of the P(NIPAAm-co-AAc) hydrogel. The peak numbers correspond to the hydrogen marked in the structural schematic. Reprinted from Stile et al. [31] with permission.

peaks were not present in the NIPAAm monomer, AAc monomer, or BIS spectra; however, they were seen in the P(NIPAAm-*co*-AAc) spectrum and may represent isomers and/or end groups of polymer chains formed by different termination steps (e.g., disproportionation vs coupling). Peak 7 at 4.547 ppm represents residual protons in the D_2O used to replace the hydrogel water. Importantly, there were no peaks representing NIPAAm vinyl protons, suggesting that the P(NIPAAm-*co*-AAc) hydrogel was washed thoroughly enough to remove residual monomer.

LCST DETERMINATION

A UV–visible spectrophotometer with a water-regulated single-cell holder is used to determine the LCST of the hydrogel samples. The transmittance of visible light (wavelength 500 nm, path length 1 cm) through the hydrogel is recorded as the hydrogel temperature is varied with a refrigerated bath. A K-type thermocouple attached to a digital thermometer is used to measure the hydrogel temperature. The heating rate should be 0.1–0.25°C/min. At the start of each experiment, the spectrophotometer should be calibrated with UPW. Once a plot of transmittance versus temperature has been obtained, the LCST is judged to be the initial break point of the curve [47].

LCST data are shown in Fig. 57.6 with P(NIPAAm) hydrogels shown for comparison [31]. The P(NIPAAm) hydrogel phase transition is very sharp, spanning approximately 0.4°C, while the P(NIPAAm-*co*-AAc) hydrogel phase transition is much broader and extends roughly 5°C. The P(NIPAAm) hydrogel LCST was determined to be 31.2 ± 0.8°C and the P(NIPAAm-*co*-AAc) hydrogel LCST was found to be 34.4 ± 0.5°C ($p < 0.01$). The LCST of the P(NIPAAm) hydrogel agreed with published reports [20–22,25,33]. The

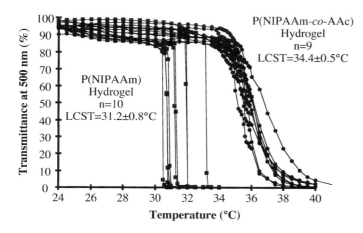

Fig. 57.6. Transmittance versus temperature data for P(NIPAAm) (squares) and P(NIPAAm-co-AAc) (circles) hydrogels. Each line represents a single experiment with one hydrogel sample. Reprinted from Stile et al. [31] with permission.

extremely sharp transition observed was consistent with results published earlier and has been attributed to the hydrophobic/hydrophilic balance of the side groups on the polymer chain, which leads to rapid dehydration of the polymer as the temperature is increased above the LCST [20,22].

The increased LCST of the P(NIPAAm-co-AAc) hydrogel also agreed with earlier studies [25,28]. In general, the addition of more hydrophilic monomers to a P(NIPAAm) hydrogel increases the LCST because the monomer hinders the dehydration of the polymer chains and acts to expand the collapsed structure. Specifically, hydrophilic AAc tends to increase the LCST of NIPAAm-AAc copolymers because the ionized $-COO^-$ groups (at a pH of 7.2) are sufficiently soluble to counteract the aggregation of the hydrophobic temperature-sensitive elements [23]. Also, the repulsion of the $-COO^-$ groups may impede the collapse induced by the NIPAAm components, increasing the LCST. In addition to an elevated LCST, the P(NIPAAm-co-AAc) hydrogel exhibited a broader transition than the P(NIPAAm) hydrogel, indicating decreased swelling thermosensitivity (i.e., the degree of swelling change with external temperature change [28]). Copolymer hydrogels composed of NIPAAm and a more hydrophilic monomer have demonstrated decreased swelling thermosensitivity [28,32] because the monomer prevents the formation of a compact shrunken structure. It is important to note that the LCST of the P(NIPAAm-co-AAc) hydrogel can be tailored by changing the amount of AAc included in the polymerization formulation.

WATER CONTENT DETERMINATION OF HYDROGELS AT 22°C AND 37°C

A freeze-drying technique is used to assess the water content (vol%) of the hydrogels. For 22°C water content studies, a syringe is used to transfer 5.0-ml hydrogel samples into preweighed cylindrical copper molds (internal diameter 26 mm), and the sample weight is determined. The samples are quenched in liquid nitrogen ($\sim$ −196°C) and freeze-dried at a pressure of 30 mtorr. During the freeze-drying process, the samples warm from −106°C to 22°C over a period of 24 h. The freeze-dried hydrogels are weighed upon removal from the freeze-dryer, and an estimate of the water content is calculated based on the weight difference of the samples before and after freeze-drying. Using the weight difference and the density of water at 22°C, the volume percent of water in the hydrogel at 22°C can be determined.

For 37°C water content studies, a syringe is used to transfer 5.0-ml hydrogel samples into preweighed glass vials (internal diameter 23 mm; volume 25 ml) containing 3 ml of PBS. PBS is added to the glass vials to simulate the environment the hydrogels would be exposed to inside the body. The glass vials are placed in an incubator at a temperature of 37°C and a humidified atmosphere of 5% CO_2(95% air; 21% O_2) for 3 days. The weight of the heated hydrogel samples is determined, and the volume of each heated sample is estimated by the displacement of 37°C UPW. The time course of the test is short enough to neglect substantial swelling of the hydrogels in water. In addition, P(NIPAAm)-based hydrogels typically expel water at 37°C, so significant swelling at 37°C would not be anticipated. The glass vials containing the P(NIPAAm-co-AAc) samples are quenched in liquid nitrogen ($\sim$ −196°C). The samples are freeze-dried for 24 h and weighed at the completion of the freeze-drying process. An estimate of the water content in the heated P(NIPAAm-co-AAc) hydrogels is calculated based on the weight difference of the samples before and after freeze-drying. Then the weight difference and the density of water at 37°C can be used to determine the volume percent of water in the hydrogel at 37°C.

The water content data are presented in Table 57.1 [31]. At 22°C, the P(NIPAAm) and P(NIPAAm-co-AAc) hydrogels contained more than 90% water. When heated to 37°C, the P(NIPAAm-co-AAc) hydrogel contained significantly more water than the P(NIPAAm) hydrogel (93.3 ± 5.4% vs 43.7 ± 7.7%; $p < 0.01$). The low water content of the P(NIPAAm) hydrogel was not unexpected, since these hydrogels collapse substantially and expel a large amount of pore water when heated to temperatures above the LCST [24–26].

Table 57.1. Water Content at 22 and 37°C

Hydrogel	Water content (%)	
	At 22°C	At 37°C
P(NIPAAm)	91.5 ± 0.8^{a}	43.7 ± 7.7^{a}
P(NIPAAm-*co*-AAc)	92.6 ± 0.7^{a}	93.3 ± 5.4^{b}

[a]$n = 8.$
[b]$n = 12.$

CHANGE IN VOLUME ESTIMATION BETWEEN 22°C AND 37°C

To measure the temperature-dependent volume change of the P(NIPAAm-*co*-AAc) hydrogels, 5.0-ml samples are immersed in 3 ml of PBS and heated in an incubator from 22°C to 37°C for 6 days under a humidified atmosphere of 5% CO_2. The volume of the heated hydrogel samples is determined via 37°C water displacement. The change in volume between 22°C and 37°C for the P(NIPAAm-*co*-AAc) hydrogels is computed by subtracting the 22°C volume from the 37°C volume and dividing by the 22°C volume.

Changes in hydrogel volume data are presented in Table 57.2 [31]. The addition of AAc to the P(NIPAAm) hydrogel significantly decreased the extent of collapse exhibited by the P(NIPAAm-*co*-AAc) matrix in comparison to the P(NIPAAm) hydrogel ($p < 0.01$). The P(NIPAAm) hydrogel collapsed over 90% of its volume when heated from 22°C to 37°C. Such extensive collapse is commonly observed with P(NIPAAm) hydrogels [24–26]. Conversely, the P(NIPAAm-*co*-AAc) hydrogel demonstrates minimal volume change when heated from 22°C to 37°C, owing to the collapse resistance provided by the $-COO^-$ groups (e.g., $-COO^-$ repulsion).

The volume change data presented in Table 57.2 were obtained without preswelling the samples in PBS before heating. If the P(NIPAAm-*co*-AAc) hydrogels are swollen in PBS for 3 and 6 days before heating, the volume change increases significantly ($p < 0.01$) [31], and the hydrogel collapses more, owing to a decrease in the $-COO^-$ collapse resistance in the presence of ions.

RHEOLOGY OF THE P(NIPAAM-CO-AAC) HYDROGEL AT 22°C AND 37°C

Mechanical properties of the P(NIPAAm-*co*-AAc) hydrogels are obtained on a rheometer using a parallel plate configuration in oscillatory mode. Sample volumes of about 0.5 ml are collected with a 3-ml syringe. Prior to sample loading, a plastic collar is placed around the lower plate, creating a reservoir. Once the sample has been loaded and the upper fixture lowered, silicone fluid is placed in the reservoir to prevent sample dehydration during data collection. Each sample is tested first at 22°C and then at 37°C. The temperature of the lower plate is maintained with a recirculating water bath. Linear viscoelastic data are collected over a frequency range of 0.001–10 Hz. Preliminary experiments have demonstrated that the rheological properties of P(NIPAAm-*co*-AAc) hydrogels are independent of the applied strain and are unaffected by the silicone fluid.

Table 57.2. Change in Hydrogel Volume When Heated from 22°C to 37°C

Hydrogel	Volume change (%)[a]
P(NIPAAm)	-92.0 ± 0.0
P(NIPAAm-*co*-AAc)	$+5.3 \pm 6.1$

[a]$n = 3.$

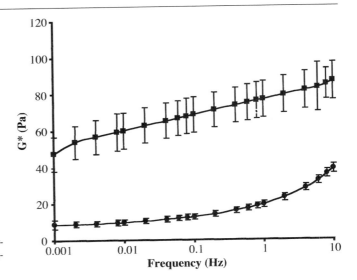

Fig. 57.7. *Complex modulus (G*) of the P(NIPAAm-co-AAc) hydrogel at 22°C (circles) and 37°C (squares) as a function of frequency. Frequency is on a log scale.*

Plots of the complex modulus (G^*) as a function of frequency and temperature are shown in Fig. 57.7 [31]. At 22°C, the P(NIPAAm-*co*-AAc) hydrogel rheology was characteristic of an extremely soft, cross-linked solid. As the frequency decreased, the modulus approached a constant value in the range of 1–10 Pa. When the temperature was increased, the gel became significantly more rigid and solidlike. Over the range of frequencies tested, G^* increased significantly when the temperature increased from 22°C to 37°C ($p < 0.01$). The rheology data for the P(NIPAAm-*co*-AAc) hydrogels at 37°C were similar to data reported for collagen gels [48]. The water in the P(NIPAAm-*co*-AAc) hydrogel did not appear to contribute substantially to the measured viscoelastic properties, since the magnitude of the loss modulus (G'') at 22 and 37°C over the range of frequencies tested was at least one order of magnitude higher than would be predicted from viscous dissipation of water alone. The variance in the data was due primarily to inconsistent covering of the parallel plates with the hydrogel samples.

From a tissue engineering perspective, the increase in stiffness demonstrated by the P(NIPAAm-*co*-AAc) hydrogels represents *in situ* stabilization. At 22°C, the P(NIPAAm-*co*-AAc) hydrogel is extremely pliable and injectable through a small-diameter aperture. As the temperature is increased above the LCST to 37°C, the hydrogel rigidity increases, and the matrix, in essence, stabilizes into a more substantial structure that may better support tissue growth.

P(NIPAAm-*co*-AAc) HYDROGELS FOR ARTICULAR CARTILAGE REGENERATION

The original focus for the development of these P(NIPAAm-*co*-AAc) hydrogels was to aid in the repair of articular cartilage defects. Articular cartilage is a complex tissue that exhibits unique compositional, structural, and material characteristics but possesses limited capacity for regeneration. Articular cartilage covers the surface of diarthrodial joints [49] and functions to absorb energy and distribute loads uniformly to subchondral bone [50,51]. In addition, its nearly frictionless surface allows unrestricted joint motion [52,53]. Since a healthy articular surface plays such a crucial role in normal joint function, the degeneration of articular cartilage is a significant clinical issue. Each year, millions of people in the United States are affected by deficiencies in joint function resulting from damaged or diseased articular cartilage [54]. Furthermore, the consequences of these articular cartilage defects are extremely costly to the health care industry [55]. Traditional methods used to treat articular cartilage defects suffer from numerous limitations and generally fail to produce functionally equivalent repair tissue. The significance of articular cartilage defects has prompted the development of alternative methods to restore normal joint function.

The idea of transplanting cells to generate new hyaline cartilage developed after the first successful isolation of chondrocytes [56]. Later, the concept was expanded to include trans-

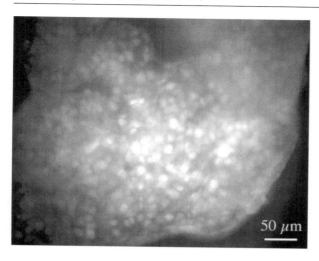

Fig. 57.8. In situ fluorescent viability study of bovine articular chondrocytes in the P(NIPAAm-co-AAc) hydrogel after 28 days of in vitro culture. The hydrogels were not swollen in PBS prior to cell loading. The chondrocytes were stained with fluorescein diacetate. Reprinted from Stile et al. [31] with permission.

planting cells seeded in an artificial matrix [57]. Since then, natural and synthetic materials acting as artificial templates to foster articular cartilage regeneration have been studied incessantly [1,3,58–65]. While this area of research has shown promise, many of the polymer scaffolds suffer from deficiencies discussed earlier, such as the invasiveness of the implantation procedure, the use of in situ polymerization, and the inability of the matrix to interact with the native tissue.

Injectable P(NIPAAm-co-AAc) hydrogels have been seeded with bovine articular chondrocytes and cultured in vitro, and in situ fluorescent viability studies have been performed [31]. The chondrocytes were viable for at least 28 days of in vitro culture (Fig. 57.8). In addition, the chondrocytes maintained a round shape, a typical characteristic of differentiated articular chondrocytes [66]. Clumps of cells, as well as individual cells, were observed in the samples. Tissue, consisting of individual cells surrounded by an ECM, was formed in the P(NIPAAm-co-AAc) hydrogels (Fig. 57.9) [31]. The structure of this tissue resembled the histoarchitecture of native articular cartilage and was comparable to that of tissue formed in other cartilage regeneration studies [1,3]. The formation of tissue suggests that the isolated chondrocytes synthesized and secreted ECM components after being seeded into the hydrogels, a further indication that the hydrogels sustained cell viability. The cells in the tissue exhibited a round morphology and were located within lacunae; moreover, the ECM of the tissue formed stained positive for Alcian blue (pH 1.0), indicating the presence of sulfated polysaccharides [67]. Alcian blue staining is generally accepted as a marker for the differentiated articular chondrocyte phenotype [1,3].

Large cell clusters were visible in the P(NIPAAm-co-AAc) hydrogels. Since these matrices are generally not considered to be degradable, it seems appropriate to ask, "Where are

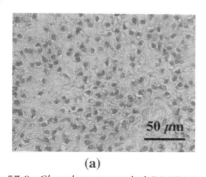

(a)   (b)

Fig. 57.9. Chondrocyte-seeded P(NIPAAm-co-AAc) hydrogel stained with (a) hematoxylin and eosin and (b) Alcian blue. The sections were obtained after 36 days of in vitro culture. The hydrogels were not swollen in PBS prior to cell loading. Reprinted from Stile et al. [31] with permission.

the hydrogels?" Large pores may conceivably be formed in the hydrogels above the LCST as a result of the aggregation of the NIPAAm components [68]. In addition, the groups of cells may actually deform the matrices on the length scale of the cell and either push or fracture the hydrogels to create larger spaces.

These cell studies demonstrated that the P(NIPAAm-*co*-AAc) hydrogels sustained bovine articular chondrocyte viability *in vitro* and were not cytotoxic. Preliminary *in vivo* biocompatibility studies have been performed with the P(NIPAAm-*co*-AAc) hydrogels in a *nu/nu* nude (thymus-deficient) mouse model. The mice did not suffer any adverse effects from the hydrogels over a 7-month period, and there was minimal evidence of fibrous tissue encapsulation of the matrices. Although these preliminary studies are encouraging, more extensive studies are needed to better ascertain the biocompatibility and immunogenicity of the P(NIPAAm-*co*-AAc) hydrogels.

Degradability is an important characteristic of scaffolds for tissue engineering applications, as the risk of infection or other complications (e.g., impedance of new tissue function) can arise when a material remains in the body permanently. P(NIPAAm) hydrogels have been synthesized by means of hydrolytically cleavable cross-linkers composed of ethylene glycol, L-lactide, and ε-caprolactone. These P(NIPAAm) hydrogels degraded into a low-viscosity liquid after 13 days at 37°C. Thus, it is possible to incorporate biodegradation into the P(NIPAAm)-based hydrogels.

P(NIPAAm-*co*-AAc) HYDROGELS FOR BONE REGENERATION

While the results of the bovine articular chondrocyte studies were encouraging, the P(NIPAAm-*co*-AAc) hydrogel did not interact with the cells at a molecular level. Therefore, the hydrogel could not act as an artificial ECM through exploitation of the natural associations between cells and the native ECM. To induce interactions between the P(NIPAAm-*co*-AAc) hydrogel and the biological system, the matrix was functionalized with biologically active synthetic peptides [69].

The amino acid sequence -arginine-glycine-aspartic acid- (-RGD-), a ubiquitous cell-binding domain found in many ECM proteins (e.g., fibronectin and vitronectin) and recognized by cell surface receptors called integrins [70–72], has been extensively studied [73–80]. The -RGD- peptide has been covalently grafted to two-dimensional (2D) substrates [73–79] or within 3D networks [80]. Other cell-binding domains have been investigated as well, including heparin-binding domains, such as -Phe-His-Arg-Arg-Ile-Lys-Ala- (-FHRRIKA-) [81–83]. Studies have shown that a more extensive cell response (e.g., cell attachment, spreading, formation of discrete focal contacts, and organized cytoskeletal assembly) was obtained when both the -RGD- and heparin-binding domains of fibronectin were provided [81,84].

To avoid the technical limitations involved with investigating ligand–receptor interactions within 3D matrices, we have focused on initial studies of mammalian cell–material interactions on 2D substrates. In our earlier work, we examined the behavior of bone-forming cells on 2D biomimetic surfaces containing both the integrin-binding (i.e., -RGD-) and heparin-binding (i.e., -FHRRIKA-) domains of bone sialoprotein (BSP) [73,82,83]. Knowledge gained from these 2D experiments has provided the foundation for examining cell behavior within 3D networks. We describe an approach to the study of the biomolecular associations between bone-forming cells and P(NIPAAm-*co*-AAc) hydrogels modified with -RGD- and -FHRRIKA- peptides. It is important to note that this approach can be generalized to study any biomolecular interactions within the P(NIPAAm-*co*-AAc) hydrogels.

Synthetic peptides containing the -RGD- and -FHRRIKA- sequences have been covalently grafted to the AAc moieties in the P(NIPAAm-*co*-AAc) hydrogel (Fig. 57.10) [69]. Chemical modification of the hydrogel was confirmed via solid state ^{1}H MAS NMR spectroscopy, LCST studies, and volume change studies [69], as described in the preceding sections. These characterization studies revealed one limitation with the functionalization process, however. Owing to the extensive swelling in buffer required for modification, the peptide-grafted P(NIPAAm-*co*-AAc) hydrogels collapsed significantly when heated to

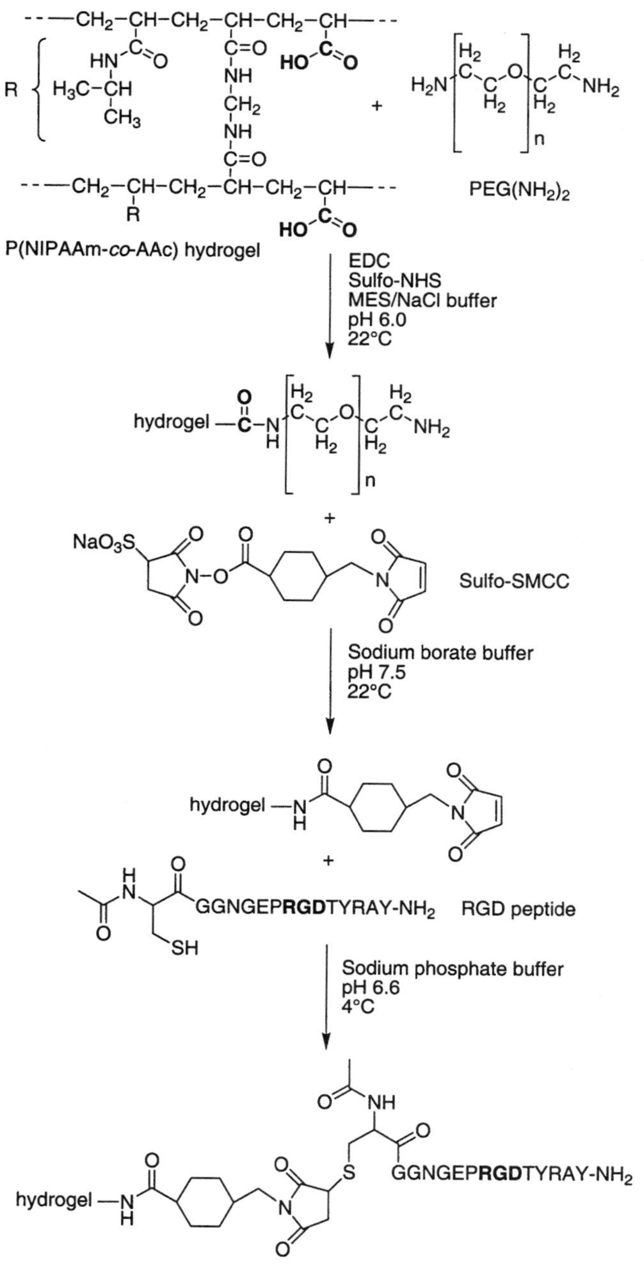

Fig. 57.10. Schematic representation of the peptide-functionalization of P(NIPAAm-co-AAc) hydrogels. Reprinted from Stile et al. [69] with permission.

37°C [31,69]. This property is undesirable from a clinical perspective, but the conjugation method can be modified to circumvent this issue.

Rat calvarial osteoblasts (RCO) were seeded into the peptide-modified P(NIPAAm-co-AAc) hydrogels and cultured in vitro. After 1 day of in vitro culture, RCO could be seen spreading within the peptide-modified hydrogels [69]. Spread cells were not seen in the buffer-control P(NIPAAm-co-AAc) hydrogels after 1 day of in vitro culture. With increased time in culture, a considerable amount of cell proliferation was qualitatively observed in the peptide-modified P(NIPAAm-co-AAc) hydrogels, and this proliferation was qualitatively greater than that observed in control P(NIPAAm-co-AAc) hydrogels (Fig. 57.11). Quantitative analyses made with CyQuant (Molecular Probes, Eugene, OR) supported the observation of cell proliferation within these hydrogels (Fig. 57.12). In situ fluorescent studies confirmed the viability of the RCO within the peptide-modified P(NIPAAm-co-AAc) hydrogels for at least 21 days of in vitro culture (Fig. 57.13).

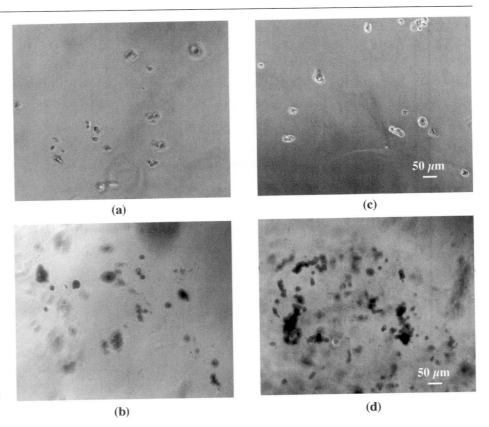

Fig. 57.11. In situ phase contrast image of RCO within the control P(NIPAAm-co-AAc) hydrogel after (a) 1 day and (b) 8 days of in vitro culture, and within peptide-modified P(NIPAAm-co-AAc) hydrogels after (c) 1 day and (d) 8 days of in vitro culture. Reprinted from Stile et al. [69] with permission.

SUMMARY

P(NIPAAm-*co*-AAc) hydrogels have been developed for tissue engineering applications. These hydrogels approach the "ideal" polymer scaffold because they are injectable through a small-diameter aperture at 22°C; they are not synthesized via *in situ* polymerization; they demonstrate *in situ* stabilization at 37°C, yielding more rigid structures; they are pliable enough at 22°C for the incorporation and relatively uniform distribution of cells and other macromolecules; and they can be functionalized with bioactive elements, making them ca-

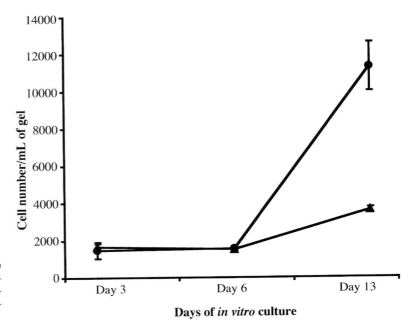

Fig. 57.12. Quantitative analysis of RCO proliferation within control P(NIPAAm-co-AAc) hydrogels (triangles) and peptide-modified P(NIPAAm-co-AAc) hydrogels (circles). Reprinted from Stile et al. [69] with permission.

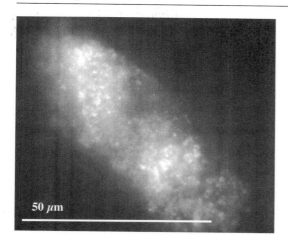

Fig. 57.13. In situ fluorescent viability study of RCO, stained with calcein AM, within the peptide-modified P(NIPAAm-co-AAc) hydrogels after 7 days of in vitro culture.

pable of interacting with the host tissue and biological environment. Furthermore, these hydrogels support cell viability, proliferation, and phenotypic expression *in vitro*. Thus, these P(NIPAAm-*co*-AAc) hydrogels have the potential to be used as scaffolds for the regeneration of tissues and organs.

REFERENCES

1. Freed, L. E., Grande, D. A., Lingbin, Z., Emmanual, J., Marquis, J. C., and Langer, R. (1994). Joint resurfacing using allograft chondrocytes and synthetic biodegradable polymer scaffolds. *J. Biomed. Mater. Res.* **28**, 891–899.
2. Kawamura, S., Wakitani, S., Kimura, T., Maeda, A., Caplan, A. I., Shino, K., and Ochi, T. (1998). Articular cartilage repair. Rabbit experiments with a collagen gel–biomatrix and chondrocytes cultured in it. *Acta Orthop. Scand.* **69**, 56–62.
3. Paige, K. T., Cima, L. G., Yaremchuk, M. J., Vacanti, J. P., and Vacanti, C. A. (1995). Injectable cartilage. *Plast. Reconstr. Surg.* **96**, 1390–1398.
4. Whang, K., Tsai, D. C., Nam, E. K., Aitken, M., Sprague, S. M., Patel, P. K., and Healy, K. E. (1998). Ectopic bone formation via rhBMP-2 delivery from porous bioabsorbable polymer scaffolds. *J. Biomed. Mater. Res.* **42**, 491–499.
5. Shinoka, T., Shum-Tim, D., Ma, P. X., Tanel, R. E., Isogai, N., Langer, R., Vacanti, J. P., and Mayer, J. E., Jr. (1998). Creation of viable pulmonary artery autografts through tissue engineering. *J. Thorac. Cardiovasc. Surg.* **115**, 536–545.
6. Seckel, B. R., Jones, D., Hekimian, K. J., Wang, K.-K., Chakalis, D. P., and Costas, P. D. (1995). Hyaluronic acid through a new injectable nerve guide delivery system enhances peripheral nerve regeneration in the rat. *J. Neurosci. Res.* **40**, 318–324.
7. Cusick, R. A., Lee, H., Sano, K., Pollok, J. M., Utsunomiya, H., Ma, P. X., Langer, R., and Vacanti, J. P. (1997). The effect of donor and recipient age on engraftment of tissue-engineered liver. *J. Pediatr. Surg.* **32**, 357–360.
8. Ellis, D. L., and Yannas, I. V. (1996). Recent advances in tissue synthesis *in vivo* by use of collagen–glycosaminoglycan copolymers. *Biomaterials* **17**, 291–299.
9. Vernon, B., Gutowska, A., Kim, S. W., and Bae, Y. H. (1996). Thermally reversible polymer gels for biohybrid artificial pancreas. *Macromol. Symp.* **109**, 155–167.
10. Cima, L. G., Vacanti, J. P., Vacanti, C., Ingber, D., Mooney, D., and Langer, R. (1991). Tissue engineering by cell transplantation using degradable polymer substrates. *J. Biomech. Eng.* **113**, 143–151.
11. Hubbell, J. A., and Langer, R. (1995). Tissue engineering. *Chem. Eng. News*, March 13, pp. 42–53.
12. Cao, Y., Rodriguez, A., Vacanti, M., Ibarra, C., Arevalo, C., and Vacanti, C. A. (1998). Comparative study of the use of poly(glycolic acid), calcium alginate and Pluronics in the enginering of autologous porcine cartilage. *J. Biomater. Sci. Polymer Edn* **9**, 475–487.
13. Sims, C. D., Butler, P. E. M., Casanova, R., Lee, B. T., Randolph, M. A., Lee, W. P. A., Vacanti, C. A., and Yaremchuk, M. J. (1996). Injectable cartilage using polyethylene oxide polymer substrates. *Plast. Reconstr. Surg.* **98**, 843–850.
14. Jeong, B., Bae, Y. H., and Kim, S. W. (1999). Thermoreversible gelation of PEG-PLGA-PEG triblock copolymer aqueous solutions. *Macromolecules* **32**, 7064–7069.
15. Hunziker, E. B., Nixon, A. J., and Johnstone, B. (1997). Cell-based articular cartilage repair: Research considerations. *Proc. 43rd Ann. Meet. Orthop. Res. Soc.*, San Francisco.
16. Casassa, E. F. (1977). Phase equilibrium in polymer solutions. *In* "Fractionation of Synthetic Polymers: Principles and Practices" (L. H. Tung, ed.), pp. 3–55. Dekker, New York.
17. Kamide, K. (1990). "Thermodynamics of Polymer Solutions: Phase Equilibria and Critical Phenomena." Elsevier, New York.
18. Taylor, L. D., and Cerankowski, L. D. (1975). Preparation of films exhibiting a balanced temperature dependence to permeation by aqueous solutions—a study of lower consolute behavior. *J. Polym. Sci., Polym. Chem. Ed.* **13**, 2551–2570.
19. Heskins, M., and Guillet, J. E. (1968). Solution properties of poly(*N*-isopropylacrylamide). *J. Macromol. Sci., Chem. Ed.* **A2**, 1441–1455.

20. Schild, H. G. (1992). Poly(*N*-isopropylacrylamide): Experiment, theory, and application. *Prog. Polym. Sci.* **17**, 163–249.

21. Bae, Y. H., Okano, T., and Kim, S. W. (1991). "On–off" thermocontrol of solute transport. I. Temperature dependence of swelling of *N*-isopropylacrylamide networks modified with hydrophobic components in water. *Pharm. Res.* **8**, 531–537.

22. Bae, Y. H., Okano, T., and Kim, S. W. (1990). Temperature dependence of swelling of cross-linked poly(*N*, *N'*-alkyl substituted acrylamide) in water. *J. Polym. Sci., Polym. Phys. Ed.* **28**, 923–936.

23. Chen, G., and Hoffman, A. S. (1995). Graft copolymers that exhibit temperature-induced phase transitions over a wide range of pH. *Nature* (*London*) **373**, 49–52.

24. Hoffman, A. S. (1991). Environmentally sensitive polymers and hydrogels: "Smart" biomaterials. *MRS Bull.* September, 42–46.

25. Hoffman, A. S., Afrassiabi, A., and Dong, L. C. (1986). Thermally reversible hydrogels: II. Delivery and selective removal of substances from aqueous solutions. *J. Controlled Release* **4**, 213–222.

26. Hirokawa, Y., and Tanaka, T. (1984). Volume phase transition in a nonionic gel. *J. Chem. Phys.* **81**, 6379–6380.

27. Bae, Y. H., Okano, T., and Kim, S. W. (1991). "On–off" thermocontrol of solute transport. II. Solute release from thermosensitive hydrogels. *Pharm. Res.* **8**, 624–628.

28. Yoshida, R., Sakai, K., Okano, T., and Sakurai, Y. (1994). Modulating the phase transition temperature and thermosensitivity in *N*-isopropylacrylamide copolymer gels. *J. Biomater. Sci., Polym. Ed.* **6**, 585–598.

29. Yoshida, R., Uchida, K., Kaneko, Y., Sakai, K., Kikuchi, A., Sakurai, Y., and Okano, T. (1995). Comb-type grafted hydrogels with rapid de-swelling response to temperature changes. *Nature* (*London*) **374**, 240–242.

30. Yoshida, R., Sakai, K., Okano, T., Sakurai, Y., Bae, Y. H., and Kim, S. W. (1991). Surface-modulated skin layers of thermal responsive hydrogels as on–off switches. I. Drug release. *J. Biomater. Sci. Polymer Edn* **3**, 155.

31. Stile, R. A., Burghardt, W. R., and Healy, K. E. (1999). Synthesis and characterization of injectable poly(*N*-isopropylacrylamide)-based hydrogels that support tissue formation *in vitro*. *Macromolecules* **32**, 7370–7379.

32. Vakkalanka, S. K., Brazel, C. S., and Peppas, N. A. (1996). Temperature- and pH-sensitive terpolymers for modulated delivery of streptokinase. *J. Biomater. Sci. Polymer Edn* **8**, 119–129.

33. Freitas, R. F. S., and Cussler, E. L. (1987). Temperature sensitive gels as extraction solvents. *Chem. Eng. Sci.* **42**, 97–103.

34. Monji, N., and Hoffman, A. S. (1987). A novel immunoassay system and bioseparation process based on thermal phase separating polymers. *Appl. Biochem. Biotechnol.* **14**, 107–120.

35. Park, T. G., and Hoffman, A. S. (1991). Immobilization of *Arthrobacter simplex* in thermally reversible hydrogels: Effect of gel hydrophobicity on steroid conversion. *Biotechnol. Prog.* **7**, 383–390.

36. Lakhiari, H., Okano, T., Nurdin, N., Luthi, C., Descouts, P., Muller, D., and Jozefonvicz, J. (1998). Temperature-responsive size-exclusion chromatography using poly(*N*-isopropylacrylamide) grafted silica. *Biochim. Biophys. Acta* **1379**, 303–313.

37. Bae, Y. H., Vernon, B., Han, C. K., and Kim, S. W. (1998). Extracellular matrix for a rechargeable cell delivery system. *J. Controlled Release* **53**, 249–258.

38. Vernon, B., Kim, S. W., and Bae, Y. H. (2000). Thermoreversible copolymer gels for extracellular matrix. *J. Biomed. Mater. Res.* **51**, 69–79.

39. Shimizu, S., Yamazaki, M., Kubota, S., Ozasa, T., Moriya, H., Kobayashi, K., Mikami, M., Mori, Y., and Yamaguchi, S. (1996). *In vitro* studies on a new method for islet microencapsulation using thermoreversible gelation polymer, *N*-isopropylacrylamide-based copolymer. *Artif. Organs* **20**, 1232–1237.

40. Okano, T., Yamada, N., Sakai, H., and Sakurai, Y. (1993). A novel recovery system for cultured cells using plasma-treated polystyrene dishes grafted with poly(*N*-isopropylacrylamide). *J. Biomed. Mater. Res.* **27**, 1243–1251.

41. Yamato, M., Kushida, A., Konno, C., Kikuchi, A., Sakurai, Y., and Okano, T. (1999). 2D and 3D manipulations of cell sheets using temperature-responsive culture surfaces for reconstruction of tissue architectures. *Spring Meet. Mater. Res. Soc.* p. 475, San Francisco.

42. Peppas, N. A. (1987). "Hydrogels in Medicine and Pharmacy." CRC Press, Boca Raton, FL.

43. Loudon, G. M. (1988). "Organic Chemistry," 2nd ed. Benjamin/Cummins, Menlo Park, CA.

44. Tokuhiro, T., Takayuki, A., Mamada, A., and Tanaka, T. (1991). NMR study of poly(*N*-isopropylacrylamide) gels near phase transition. *Macromolecules* **24**, 2936–2943.

45. Badiger, M. V., Rajamohanan, P. R., Kulkarni, M. G., Ganapathy, S., and Mashelkar, R. V. (1991). Proton MAS-NMR: A new tool to study thermoreversible transition in hydrogels. *Macromolecules* **24**, 106–111.

46. Jelinski, L. W. (1984). Modern NMR spectroscopy. *Chem. Eng. News* **62**, 26–47.

47. Schild, H. G., and Tirrell, D. A. (1990). Microcalorimetric detection of lower critical solution temperatures in aqueous polymer solutions. *J. Phys. Chem.* **94**, 4352–4356.

48. Tan, J., and Saltzman, M. (1999). Influence of synthetic polymers on neutrophil migration in three-dimensional collagen gels. *J. Biomed. Mater. Res.* **46**, 465–474.

49. Mow, V. C., Fithian, D. C., and Kelly, M. A. (1990). Fundamentals of articular cartilage and meniscus biomechanics. *In* "Articular Cartilage and Knee Joint Function: Basic Science and Arthroscopy" (J. W. Ewing, ed.), pp. 1–18. Raven Press, New York.

50. Mow, V. C., Kuei, S. C., Lai, W. M., and Armstrong, G. C. (1980). Biphasic creep and stress relaxation of articular cartilage in compression: Theory and experiment. *J. Biomech. Eng.* **102**, 73–84.

51. Mow, V. C., and Lai, W. M. (1980). Recent developments in synovial joint biomechanics. *SIAM Rev.* **22**, 275–317.

52. Mow, V. C., and Mak, A. F. (1986). Lubrication of diarthrodial joints. *In* "Handbook of Bioengineering" (R. Shalak and S. Chien, eds.), pp. 1–5, 34. McGraw-Hill, New York.

53. Dowson, D. (1981). Basic tribology. *In* "Introduction to the Biomechanics of Joints and Joint Motion" (D. Dowson and V. Wright, ed.), pp. 49–60. Institute of Mechanical Engineering, London.

54. Data obtained from the American Academy of Orthopedic Surgeons and the National Center for Health Statistics (1997).

55. Yelin, E., and Callahan, L. F. (1995). The economic cost and social and psychological impact of musculoskeletal conditions. *Arthritis Rheum.* **38**, 1351–1362.

56. Smith, A. (1965). Survival of frozen chondrocytes isolated from cartilage of adult mammals. *Nature (London)* **205**, 782–784.

57. Green, W. T. J. (1977). Articular cartilage repair: Behavior of rabbit chondrocytes during tissue culture and subsequent allografting. *Clin. Orthop. Relat. Res.* **124**, 237–250.

58. Freed, L. E., Marquis, J. C., Nohria, A., Emmanual, J., Mikos, A. G., and Langer, R. (1993). Neocartilage formation *in vitro* and *in vivo* using cells cultured on synthetic biodegradable polymers. *J. Biomed. Mater. Res.* **27**, 11–23.

59. Wakitani, S., Kimura, T., Hirooka, A., Ochi, T., Yoneda, M., Yasui, H., Owaki, H., and Ono, K. (1989). Repair of rabbit articular surfaces with allograft chondrocytes embedded in collagen gel. *J. Bone J. Surg.* **71B**, 74–80.

60. Robinson, D., Halperin, N., and Nevo, Z. (1990). Regenerating hyaline cartilage in articular defects of old chickens using implants of embryonal chick chondrocytes embedded in a new natural delivery substance. *Calcif. Tissue Int.* **46**, 246–253.

61. Hendrickson, D. A., Nixon, A. J., Grande, D. A., Todhunter, R. J., Minor, R. M., Erb, H., and Lust, G. (1994). Chondrocyte–fibrin matrix transplants for resurfacing extensive articular cartilage defects. *J. Orthop. Res.* **12**, 485–497.

62. Chu, C. R., Coutts, R. D., Yoshioka, M., Harwood, F. L., Monosov, A. Z., and Amiel, D. (1995). Articular cartilage repair using allogenic perichondrocyte-seeded biodegradable porous polylactic acid (PLA): A tissue-engineering study. *J. Biomed. Mater. Res.* **29**, 1147–1154.

63. Messner, K. (1993). Hydroxylapatite supported Dacron plugs for repair of isolated full-thickness osteochondral defects of the rabbit femoral condyle: Mechanical and histological evaluations from 6–48 weeks. *J. Biomed. Mater. Res.* **27**, 1527–1532.

64. Robinson, D., Efrat, M., Mendes, D. G., Halperin, N., and Nevo, Z. (1993). Implants composed of carbon-fiber mesh and bone-marrow-derived, chondrocyte-enriched cultures for joint surface reconstruction. *Bull. Hosp. J. Dis.* **53**, 75–82.

65. Hanlon, J., Tubo, R., Estridge, T., Binette, F., and McPherson, J. (1997). Collagen–polyethylene glycol (PEG) cross-linked gels: A biocompatible, injectable matrix for delivery of chondrocytes to articular cartilage defects. *Proc. 43rd Ann. Meet. Orthop. Res. Soc.* p. 539, San Francisco, CA.

66. Glowacki, J., Trepman, E., and Folkman, F. (1983). Cell shape and phenotypic expression in chondrocytes. *Proc. Soc. Exp. Biol. Med.* **172**, 93–98.

67. Lev, R., and Spicer, S. S. (1964). Specific staining of sulphate groups with alcian blue at low pH. *J. Histochem. Cytochem.* **12**, 309.

68. Tanaka, T., Ishiwata, S., and Ishimoto, C. (1977). Critical behavior of density fluctuations in gels. *Phys. Rev. Lett.* **38**, 771–774.

69. Stile, R. A., and Healy, K. E. (2001). Thermo-responsive peptide-modified hydrogels for tissue regeneration. *Biomacromolecules* **2**, 185–194.

70. Pierschbacher, M. D., and Ruoslahti, E. (1984). Cell attachment activity of fibronectin can be duplicated by small synthetic fragments of the molecule. *Nature (London)* **309**, 30–33.

71. Ruoslahti, E., and Pierschbacher, M. D. (1986). Arg-Gly-Asp: A versitile cell recognition signal. *Cell (Cambridge, Mass.)* **44**, 517–518.

72. Albelda, S. M., and Buck, C. A. (1990). Integrins and other cell adhesion molecules. *FASEB J.* **4**, 2868–2880.

73. Rezania, A., and Healy, K. E. (1999). Integrin subunits responsible for adhesion of human osteoblast-like cells to biomimetic peptide surfaces. *J. Orthop. Res.* **17**, 615–623.

74. Rezania, A., Thomas, C. H., Branger, A. B., Waters, C. M., and Healy, K. E. (1997). The detachment strength and morphology of bone cells contacting materials modified with a peptide sequence found within bone sialoprotein. *J. Biomed. Mater. Res.* **37**, 9–19.

75. Bearinger, J. P., Castner, D. G., and Healy, K. E. (1998). Biomolecular modification of p(AAm-*co*-EG/AA) IPNs supports osteoblast adhesion and phenotypic expression. *J. Biomater. Sci. Polymer Edn* **9**, 629–652.

76. Massia, S. P., and Hubbell, J. A. (1990). Covalent surface immobilization of Arg-Gly-Asp- and Tyr-Ile-Gly-Ser-Arg-containing peptides to obtain well-defined cell-adhesive substrates. *Anal. Biochem.* **187**, 292–301.

77. Drumheller, P. D., and Hubbell, J. A. (1994). Polymer networks with grafted cell adhesion peptides for highly biospecific cell adhesive substrates. *Anal. Biochem.* **222**, 380–388.

78. Drumheller, P. D., Elbert, D. L., and Hubbell, J. A. (1994). Multifunctional poly(ethylene glycol) semi-interpenetrating polymer networks as highly selective adhesive substrates for bioadhesive peptide grafting. *Biotechnol. Bioeng.* **43**, 772–780.

79. Rowley, J. A., Madlambayan, G., and Mooney, D. J. (1999). Alginate hydrogels as synthetic extracellular matrix materials. *Biomaterials* **20**, 45–53.

80. Moghaddam, M. J., and Matsuda, T. (1993). Molecular design of three-dimensional artificial extracellular matrix: Photosensitive polymers containing cell adhesive peptide. *J. Polym. Sci., Polym. Chem. Ed.* **31**, 1589–1597.

81. Dalton, B. A., McFarland, C. D., Underwood, P. A., and Steele, J. G. (1995). Role of the heparin-binding domain of fibronectin in attachment and spreading of human bone-derived cells. *J. Cell Sci.* **108**, 2083–2092.

82. Rezania, A., and Healy, K. E. (1999). Biomimetic peptide surfaces that regulate adhesion, spreading, cytoskeletal organization, and mineralization of matrix deposited by osteoblast-like cells. *Biotechnol. Prog.* **15**, 19–32.

83. Rezania, A., Johnson, R., Lefkow, A. R., and Healy, K. E. (1999). Bioactivation of metal oxide surfaces: I. Surface characterization and cell response. *Langmuir* **15**, 6931–6939.

84. Woods, A., Couchman, J. R., Johansson, S., and Hook, M. (1986). Adhesion and cytoskeletal organization of fibroblasts in response to fibronectin fragments. *EMBO J.* **5**, 665–670.

CHAPTER 58

PROCESSING OF POLYMER SCAFFOLDS: SOLVENT CASTING

Gregory E. Rutkowski, Cheryl A. Miller, and Surya K. Mallapragada

INTRODUCTION

Solvent casting is a simple method for fabricating constructs for tissue engineering. The polymer is dissolved in a suitable solvent and poured into a mold. The solvent is then removed, leaving the polymer set in the desired shape. This method is limited in the shapes that can be obtained. Typically flat sheets and tubes are the only shapes that can be formed, but flat sheets can be stacked and shaped for more complex configurations (see Chapter 59, Membrane Lamination). The films can be made porous by the leaching of particles, such as salt crystals, embedded within the polymer. Also, the solvent–polymer mixture can be placed into a nonsolvent for the polymer that is also miscible in the first solvent. This method, also known as phase inversion or separation (see Chapter 62, Phase Separation) is used to form asymmetric porous membranes.

The main advantage of solvent casting is the ease of fabrication without the need of specialized equipment. Also, because the fabrication occurs at room temperature, the degradation rate of the solvent-cast biodegradable polymer is lower than that of compression-molded films (Fig. 58.1). The primary disadvantage of solvent casting is the possible retention of toxic solvent within the polymer. This can be overcome by allowing the polymers to fully dry and then using a vacuum to further remove the solvent. Also, the use of solvents can denature proteins and other molecules incorporated into the polymer, though this can also occur in films fabricated by other techniques such as compression molding.

CHOICE OF SOLVENT

Factors to consider when choosing a solvent are the solvent's power, evaporation rate, viscosity, solvent retention, and toxicological properties. Solvents are classified by their intermolecular forces (Table 58.1) [1]. These are the factors that influence how the solvent interacts with the polymer, and they can affect the solvent power and retention. The solubility parameter is useful for determining the strength of the solvent. Solvents will more readily dissolve polymers with like solubility parameters. The solubility parameters for common solvents are shown in Table 58.2 [1,2]. The solubility parameter for polymers can be determined experimentally by using various solvents or calculated based on group contributions [3]. Other methods have also been developed that incorporate dispersion forces, polarity, and hydrogen bonding into estimations of the solubility parameter [4,5].

The evaporation rate is dependent on the solvent boiling point. Solvents with low boiling points will evaporate at a faster rate. The drying of the solvent-cast film occurs in two stages. In the first stage, the solvent evaporates from the film at a rate comparable to that of the pure solvent alone. Eventually, the film will gel and the evaporation will be limited by the diffusion of the solvent through the polymer. During this stage other factors such as

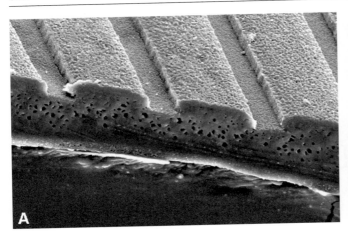

Fig. 58.1. Comparison of (A) solvent-cast and (B) compression-molded films kept in media at 37°C for 2 weeks [G. E. Rutkowski, C. A. Miller, and S. K. Mallapragada, unpublished results].

polymer type, physical state (glassy or rubbery), and steric hindrance of the solvent are more important that the volatility. More solvent is retained in glassy polymers than in rubbery ones. The shape of the solvent molecule relative to the polymer determines the diffusion of the solvent through the matrix. There will always be a small amount of solvent within the film due to equilibrium with the polymer. The amount of solvent retained within the polymer film is dependent on the interaction of the solvent with the polymer.

Different environmental conditions can affect the amount of solvent retained in the film. Evaporation of a solvent at certain conditions is determined by its vapor pressure. Humidity in the air can decrease the rate of vaporization. Also, placing the film under vacuum will further evaporate the solvent. Heating the film will vaporize more solvent but can also start to degrade the polymer.

The viscosity of a fluid is a measure of its resistance to flow. It is dependent on the temperature, intermolecular forces, and the molecular weight of the solvent. For a polymer–solvent solution, the effective viscosity is also dependent on the solvent–polymer interaction as well as the polymer structure and molecular weight. Although mathematical models, such as the Mark–Houwink–Sakurada equation, can be used for predicting solution viscosity [6], experimental work is still necessary to determine the parameters used in such models. Viscosity, which determines how well the solution flows during the casting procedure, is important for solvent-cast films. It also ensures uniformity in the thickness of the film. A less viscous solution would be necessary for the production of thin films.

The final factor of importance in choosing a solvent is the toxicity of the solvent. Because of their volatile nature, organic solvents are very flammable, and precautions should

Table 58.1. Solvent Power

Solvent type	Dispersion forces	Polar forces	Hydrogen bonding	Examples
Oxygenated solvent, alcohols	Moderate	Moderate/high	Donors	Methanol, ethanol
Oxygenated solvent, nonalcohol	Moderate/high	High	Strong acceptors	Diethyl ether, ethyl acetate
Aliphatic hydrocarbon	Low	None	None	Hexane
Aromatic hydrocarbon	High	Low	Weak acceptors	Benzene, toluene
Chlorinated solvents	Moderate/high	High	Strong acceptors	Chloroform, methyl chloride

Source: B. P. Whim and P. G. Johnson, eds. "Directory of Solvents," Table 1.1. Chapman & Hall, London, 1996. Reprinted with kind permission from Kluwer Academic Publishers.

Table 58.2. Evaporation Rate and Retention of Various Solvents [1,2]

Solvent	Relative evaporation rate (diethyl ether = 1.0)	Vapor pressure (kPa at 20°C)	Solubility parameter $(J/cm^3)^{1/2}$	Viscosity (cP at 20°C)
Methanol	6.3	13.1	29.7	0.59
Acetone	2.0	24.7	20.5	0.33
Ethyl acetate	2.9	10.0	18.6	0.50
Methylene chloride	1.7	47	20.2	0.43
Dimethylformamide	100	0.5	12.1	0.8
Tetrahydrofuran	2.1	17.3	19.2	0.54
Chloroform	2.5	20.0	19.4	0.57
Dimethyl acetamide	172	0.06	22.1	0.9
Diethyl ether	1.0	58.8	15.6	0.24

be made when they are used in the production of solvent-cast films. Since an equilibrium amount of solvent will be retained in the polymer film, the solvent chosen should be retained only at levels that are not toxic to the cells. Toxicity studies should be done on the solvent-cast films as well as on pure polymer to ensure that no additional cell death is due to the solvent.

With the wide variety of polymers available for tissue engineering and the large number of solvents, choosing an appropriate combination can be difficult. Table 58.3 lists several polymers used in tissue engineering with their commonly used solvents [1–13]. In general, most polymers are soluble in some type of organic solvent. Poly(ortho esters), though, can be used to create devices without the use of solvents. The solvent finally selected should be one that can be removed from the polymer film to reduce toxicity while still ensuring proper casting capabilities.

SOLVENT CASTING

In the fabrication of solvent-cast films, the polymer is usually dissolved in a solvent to a concentration of 10–30%. The concentration depends on the viscosity of the final solution and the demands for the solvent-cast film. A low concentration will ensure uniform film thickness but will make thinner films. Repeated casting and drying can be done to obtain thicker films. Low concentrations also require longer drying times to ensure adequate evaporation of the solvent from the film.

Table 58.3. Common Polymers Used in Tissue Engineering

Polymer	Example	Solvent	Ref.
Biodegradable polymers			
Poly(α-hydroxy esters)	Poly 3-hydroxybutyrate	Methylene chloride	[7]
Poly(ε-caprolactone)	Polycaprolactone	Acetone	[8]
Poly(ortho esters)	Various	No + applicable	[9]
Poly(anhydrides)	Poly(L-lactide)	Chloroform	[8]
Poly(phosphazenes)	Poly(phosphazene)	Methylene chloride	[10]
Poly(amino acids)	Poly(L-lysine)	Water	[11]
Nonbiodegradable polymers			
Poly(ether glycols)	Poly(ethylene glycol)	Water and most organics	[12]
Poly(urethane)	Poly(oxytetramethylene glycol)	Dimethylformamide at 102°C	[13]

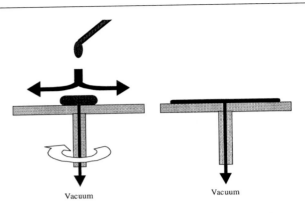

Fig. 58.3. Formation of ultrathin films by spin casting.

Fig. 58.2. Image of microfabricated surface [G. E. Rutkowski, C. A. Miller, and S. K. Mallapragada, unpublished results].

The polymer solution can be poured onto various surfaces to obtain the necessary shapes. Glass coverslips and petri dishes are common items used in the formation of flat thin films. Nonstick surfaces such as Teflon can also be used to aid in the removal of the film from the mold. Other surfaces may require the use of release agents that may later affect the growth of cells seeded on the substrate. Besides making flat films, solvent casting can be used to form tubular conduits. A rod is dipped into the polymer solution and then dried. Glass mandrels are typically used for rods [8], though rods of poly(vinyl alcohol) offer the advantage of removal of the water by dissolution without the possibility of deformation of the tube. Reactive ion etching can be used to fabricate micropatterned dies from silicon wafers. Such dies can be used to make films with micropatterned surfaces that can physically guide cell growth (Fig. 58.2).

A spin coater can be used to form ultrathin films (<5 μm) (Fig. 58.3). The spin coater consists of a mount that is rotated from 1–5000 rpm. A vacuum can be drawn through the mount to ensure the mold stays firmly in place. The rotational speed and acceleration and the spinning time can all be controlled. Once the mold is set in place and is set spinning, the polymer solution is dropped onto the spinning surface and centrifugal forces cause the solution to spread into very thin films of uniform thickness. The rotation speed, solution viscosity, and density are all factors that determine the thickness of the film.

Solvent casting can also exploit the characteristics of the polymer itself to create unique films. Films made from block copolymers can provide features not found with homopolymers. For example, films made with poly(g-benzyl L-glutamate)–co-poly(ethylene) (PBLG-PEO) dissolved in chloroform can affect the adhesion and proliferation of cells on the surface [14]. PBLG is a hydrophobic polymer, while PEO is hydrophilic. Solvent-cast films of PBLG-PEO formed microphase-separated structures that influenced cell adhesion depending on the amount of PEO. Langmuir–Blodgett films can be made by placing the polymer solution on a layer of water. In this type of film, the block copolymer aligns itself in a monolayer such that the PEO end is facing toward the water and the PBLG is away. The films can then be transferred from the water surface to a more suitable substrate. Such films allow cell adhesion on one side and prevent it on the other.

After the solution has been poured, the film must be allowed to dry. Initially, the solvent evaporates from the solution at a rate dictated by the volatility of the pure solvent. After the film gels, evaporation is limited by the diffusion of the solvent through the polymer matrix. The drying time will vary depending on the thickness of the film and the volatility of the solvent. Additional removal of the solvent can be obtained by subjecting the film to a vacuum. In general the film should be allowed to dry at normal pressure until it becomes firm before the vacuum is applied. The vacuum should be gradually lowered to prevent the flash vaporization of residual solvent within the polymer matrix. This could lead to

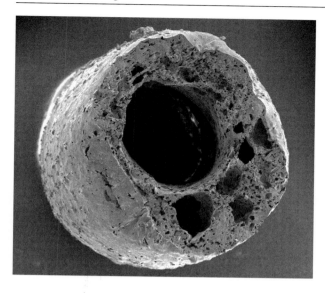

Fig. 58.4. Void formation in solvent-cast film [G. E. Rutkowski, C. A. Miller, and S. K. Mallapragada, unpublished results].

deformations in the film to the large voids (Fig. 58.4). In general, the film should be allowed to dry for 6–8 h and then placed under vacuum for 24 h to ensure adequate removal of solvent.

Once the dried film has been removed from the mold, it is ready for use. Additional processing may be necessary to obtain the final form. The film may be cut into different shapes or even rolled to form tubular conduits. Also, the surface may be modified by various factors to promote cell adhesion or cell proliferation.

SURFACE ANALYSIS

Solvent-cast films for tissue engineering may incorporate various factors onto the surface to aid in cell adhesion or to promote cellular growth. Various analytical methods have been developed to characterize the surface of these films. Initially, light and scanning electron microscopy can be used to characterize the surface topology by ordinary visual inspection. More advanced techniques can be used to determine detailed topography, as well as to assess to bulk and surface chemistry of the films.

Atomic force microscopy (AFM) can be used to evaluate the surface topography of the film nondestructively. In AFM, laser light is reflected off cantilever arm that moves in response to changes in the level of the film surface. The resolution is approximately 0.1 nm. The scanning of the cantilever over the surface may cause deformation of soft objects if the tip is allowed to interact with weakly held molecules on the surface.

Chemical information on the molecular structure of the bulk film material can be obtained from attenuated total reflection Fourier transform infrared (ATR-FTIR) spectroscopy based on the vibrational bands of bonded units. Surface composition can be evaluated by using electron spectroscopy for chemical analysis (ESCA), or X-ray photoelectron spectroscopy (XPS) and secondary ion mass spectrometry (SIMS). ESCA can analyze surface depth down to 100 Å, while SIMS only reaches 10–15 Å. ESCA uses incident X-rays to liberate core-level electrons that are detected by an energy spectrum analyzer. This method can be used to study polymer surface dynamics as well as protein adsorption kinetics [15]. SIMS can be used to complement the information gained from ESCA. Time-of flight SIMS combined with chemical spectra can be used to determine chemical species, even adsorbed protein, on the film surface [16].

Contact angle measurements can be made to determine the wettability of the film surface. When a drop of liquid is in contact with the film surface and is in equilibrium with the surrounding vapor, the angle between the vapor–liquid interface and the liquid–solid interface can be used to determine the surface energy. The surface energy gives an indication of the polarity of functional groups in the top 2–10 Å. The hydrophobicity or hydrophilicity of the film surface is important in determining cellular interactions with the film surface.

These techniques provide valuable tools to evaluate solvent-cast films for use as support structures for tissues. Care should be taken when interpreting results, since several methods are done in environments very dissimilar from the biological cases. For example, ESCA and SIMS are done under ultrahigh vacuum, which can affect solvent retained within a film. Though there are a wide variety of techniques available for surface characterization, each method should be evaluated in detail to determine appropriateness for analyzing solvent-cast films.

SUMMARY

Solvent casting is a simple and inexpensive method for creating substrates for tissue engineering applications [17]. Numerous solvents are available for the dissolution of polymers. Care must be taken in the choice of solvent to ensure polymer stability while eliminating any toxic effects on the seeded cells. This method allows for the formation of simple shapes, such as flat sheets and tubes, that can be combined with lamination, particulate leaching, and spin casting to create more complex structures. Also, since casting is done at room temperature, heat-induced degradation of the polymer does not occur, as with compression molding and extrusion. Once films have been fabricated, several analytical methods, such as ESCA, SIMS, and contact angle measurements, can be used to characterize the polymer structure as well as materials adsorbed onto the surface of the substrate.

REFERENCES

1. Whim, B. P., and Johnson, P. G., eds. (1996). "Directory of Solvents," 1 ed. Chapman & Hall, London.
2. Perry, R. H., and Green, D. W., eds. (1984). "Perry's Chemical Engineering Handbook," 6 ed. McGraw-Hill, New York.
3. Burrell, H. (1955). Solubility parameters. *Interchem. Rev.* **14**, 3–16, 32–46.
4. Hansen, C. M., and Skaarup, K. (1967). Three-dimensional solubility parameter—Key to paint component affinities. III. Independent calculation of the parameter components. *J. Paint Technol.* **39**, 511–514.
5. Nelson, R. C., *et al.* (1970). Solution theory and the computer—Effective tools for the coatings chemist. *J. Paint Technol.* **42**, 644–652.
6. Rudin, A. (1982). "The Elements of Polymer Science and Engineering." Academic Press, San Diego, CA.
7. Brandl, H., *et al.* (1995). Biodegradation of cyclic and substituted linear oligomers of poly(3-hydroxybutyrate). *Can. J. Microbiol.* **41**(Suppl. 1), 180–186.
8. Hoppen, H. J., *et al.* (1990). Two-ply biodegradable nerve guide: Basic aspects of design, construction, and biological performance. *Biomaterials* **11**(May), 286–290.
9. Heller, J. (1990). Development of poly(ortho esters): A historical overview. *Biomaterials* **11**(9), 659–665.
10. Veronese, F. M., *et al.* (1999). Polyphosphazene membranes and microspheres in periodontal diseases and implant surgery. *Biomaterials* **20**, 91–98.
11. Hwang, J. J., *et al.* (1999). Organoapatite growth on an orthopedic alloy surface. *J. Biomed. Mater. Res.* **47**, 504–515.
12. Ash, M., and Ash, I. (1990). "Polymers and Plastics. What Every Chemical Technologist Wants to Know..." Vol. VI. Chem. Publ. Co., New York.
13. Schierholz, J. M., *et al.* (1997). Controlled release of antibotics from biomedical polyurethanes: Morphological and structural features. *Biomaterials* **18**(12), 839–844.
14. Cho, C. S., Kotaka, T., and Akaike, T. (1993). Cell adhesion onto block copolymer Langmuir-Blodgett films. *J. Biomed. Mater. Res.* **27**(2), 199–206.
15. Andrade, J. D. (1985). "Surface and Interfacial Aspects of Biomedical Polymers," Vols. 1 and 2. Plenum, New York.
16. Mantus, D. S., *et al.* (1993). Static secondary ion mass spectroscopy of adsorbed proteins. *Anal. Chem.* **65**, 1431–1437.
17. Miller, C. A., Shanks, H., Witt, A., Rutkowski, G. E., Miller, and Mallapragada, S. K. (2001). Oriented Schwann cell growth on micropatterned biodegradable polymer substrates. *Biomaterials* **22**, 1263–1269.

CHAPTER 59

PROCESSING OF POLYMER SCAFFOLDS: MEMBRANE LAMINATION

Frank T. Gentile

Membrane lamination for encapsulated cell therapy is being investigated for the delivery of a wide range of products stemming from small molecules (e.g., dopamine, enkephalins) to very large gene products (e.g., growth factors, immunoglobulins). A large number of disease states have been studied in small and large animal models and in human subjects. Disease targets include liver failure, type I diabetes, chronic pain, Alzhemeir's, amyotrophic lateral sclerosia, Huntington's chorea, and Parkinson's disease. All the devices involve some level of immunoprotection of xenogeneic or allogeneic cells. The first one is nearing regulatory approval in the United States as a liver-assist device (i.e., a bridge to liver transplant). Key points in the use of materials to create these devices are as follows: (1) biocompatible membrane, device components, and matrix materials; (2) strong implants; and (3) materials that can be made into membranes with the proper transport properties. This chapter focuses on methods of affecting or quantifying membrane strength and transport properties. Both these properties will be critical to any commercial implantation system. Membrane transport properties will be critical to all systems, whether implanted or extracorporeal-based devices. Methods of membrane manufacture can yield membranes having two to four times the strength of traditional phase inversion membranes. With new tools to measure small and large molecular weight transport, membranes with the desired combination of immunoisolation and product delivery rates can be selected for many applications, potentially affecting a wide variety of disease treatments.

INTRODUCTION

An important area for biomaterials is the delivery of active substances to specific sites *in vivo*. Traditionally this area has been dominated by degradable and nondegradable polymer capsules containing one or more drugs. The substances are mixed with a polymer matrix during fabrication and then released over time through the material or as the material degrades. Proper control of the release kinetics is important here. One example is the zero-order release kinetics achieved in ethylene–vinyl acetate copolymer (EVAc) rods used to release chemotherapeutic agents in the brain [1]. Over the last two decades, researchers have attempted to deliver substances from hybrid bioartificial delivery vehicles comprising a membrane laminate over a cellular component encapsulated within the membrane (i.e., membrane-encapsulated cell therapy).

The goal of encapsulated cell therapy research is to develop implants containing living xenogeneic or allogeneic cells to treat serious and disabling human conditions. The enabling concept is straightforward: cells or small clusters of tissue are surrounded by a selective membrane laminate that allows unhindered passage of oxygen and other required metabolites, releases bioactive cell secretions, but limits the transport of the larger cytotoxic agents of the body's immune defense system.

Target applications for encapsulated cell therapy include chronic pain, Parkinson's disease, and type I diabetes, as well as other disabilities caused by loss of secretory cell function that cannot be adequately treated by current organ transplantation or drug therapies. In addition, conditions potentially capable of responding to local sustained delivery of growth factors and other biological response modifiers have been studied with this approach. Cross-species immunoisolated cell therapy has been validated in small and large animal models of chronic pain [2], Parkinson's disease [3], type I diabetes [4–6], and acute liver failure (extracorporeally) [7], and has been studied by several groups in animal models of Huntington's chorea [8], hemophilia [9], Alzheimer's disease [10,11], amyotrophic lateral sclerosic [12], and epilepsy. Of these, acute liver failure appears to be the first that will be approved for commercial use in humans [7].

Encapsulation of tissues has generally taken two forms: microencapsulation and macroporous membrane lamination (intravascular and extravascular). In microencapsulation, one or several cells are encapsulated in many spherical dispersions (100–300 μm in diameter); in macroencapsulation, large numbers of cells or cell clusters are transplanted in one or several relatively large capsules (for hollow fibers, typical dimensions are 0.5–6 mm in diameter, with a total length of 0.5–10 cm). Advantages of the latter approach include better mechanical and chemical stability and ease of retrievability if warranted or desired.

STRUCTURE–PROPERTY RELATIONSHIPS FOR POLYMERS USED IN MEMBRANE LAMINATION

It is important that the membrane material and selective membrane properties stay within an appropriate range over time. Membrane degradation involves the changing of both the physical and transport properties over time as a result of interaction with the *in vivo* environment. The main properties that can be changed are pore sizes (and the overall pore size distribution) and the diffusive mass transfer coefficients.

A wide variety of membrane materials can be used for artificial organs of these types. One of the more extensively used materials for this application is poly(acrylonitrile-*co*-vinyl chloride) [P(AN-VC)]. P(AN-VC) is a statistical copolymer made from acrylonitrile and vinyl chloride monomers. It has been used in many cell–membrane lamination studies. Other candidate materials include poly(acrylonitrile) (PAN), poly(sulfone) (PS), poly(ether sulfone) (PES), poly(vinylidine difluoride) (PVDF), poly(amides) (PA), poly(carbonate) (PC), poly(ether imides) (PEI), poly(propylene) (PP), and poly(ethylene) (PE). In addition to these materials made from mostly commodity polymers, membranes have been prepared from composite structures—for example, membranes made from poly(acrylonitrile-*co*-vinyl chloride)–poly(ethylene oxide) [13].

The physical makeup of membranes used in membrane lamination is determined by the metabolic requirements of the cells to be encapsulated, the size of the therapeutic substance(s) to be released, the degree of required *immunoprotection*, and adequate tissue biocompatibility. The most critical transport properties of membranes are determined by the metabolic requirements of the encapsulated cell. The membrane must have sufficient passage of nutrients for encapsulated cells to remain viable and functional. While maintaining cell function, the membrane must also have pores that are large enough to allow the therapeutic agent free passage to the target site. If the encapsulated cells require immunoprotection, the membranes must reject the entrance of immunological elements into the capsule. In addition, membrane morphology has a strong impact on the biocompatibility at the host–membrane interface. The transport properties and external morphologies can be manipulated using techniques described later.

Semipermeable membranes have been used extensively in studies of macroencapsulated insulin secreting cells for the treatment of type I diabetes. Semipermeable phase inversion membranes fabricated using a P(AN-VC) polymer have been used to encapsulate insulin secreting cells [14–16]. The technique employed by these groups involves placing the islets of Langerhans, which contain insulin-secreting β cells, in a sealed, hollow-fiber membrane that allows the insulin to be secreted via a glucose stimulus while cell viability is maintained by the diffusing into the implant of oxygen and nutrients.

Similar approaches to the encapsulation of insulin-secreting cells have been under-taken using microporous membranes fabricated with poly(urethane) (PU) [17] and poly(2-hydroxyethyl methacrylate) (PHEMA) [18]. The PU membrane was formed by dissolution of an entrapped pore former, whereas the PHEMA membrane was formed by a cross-linking agent.

Membrane lamination over islets of Langerhans has also been studied in devices that are intravascular, whether implanted or extracorporeal. Although such systems are still dif-fusion based, a connective component may also be present. The design of these devices places the cells around the membrane, and blood is passed through the membrane lumen. Studies using the P(AN-VC) copolymer have been used in multiple fiber cartridges primarily by Chick *et al.* [4]. Other membrane materials such as semipermeable PS membranes have been evaluated by Sun *et al.* [19,20]. Segawa *et al.* [21] used multiple fiber devices made with poly(vinyl alcohol) (PVA) to macroencapsulate insulin-secreting cells.

Sullivan *et al.* [22] used a similar approach with a single large-bore membrane in an implantable artificial pancreas through which blood is passed. Single-fiber devices made of PA semipermeable membranes have also been used by Catapano *et al.* [23] in this system.

Other endocrine cells besides islets of Langerhans have been studied in the projected treatment of hypoparathyroidism [24] by means of immunoprotective P(AN-VC) mem-branes. Christenson *et al.* [25] evaluated immunodeficiencies of the thymus with a similar membrane.

Microporous PS fiber bundles have been used as extracorporeal devices when seeded with hepatocytes for use as a liver-assist device [7]. Heparinized blood was pumped through the cartridge to increase the removal of harmful metabolites. This device may soon be ap-proved by the Food and Drug Administration for marketing as a bridge to liver transplan-tation. Macrocapsules have also been used for the treatment of growth deficiencies [26] and to relieve chronic pain [2].

Membrane lamination of cells for the treatment of neurodegenerative diseases such as Parkinson's in rodents and nonhuman primates has been studied by Aebischer *et al.* [27,28], who used P(AN-VC) as the membrane material. Encapsulation and transplantation of cells that secrete nerve growth factor (NGF) were investigated Hoffman *et al.* [10] and Winn *et al.* [11], who demonstrated survival of basal forebrain cholinergic neurons in the fimbria–fornix-lesioned rat. Learning and memory disabilities in Alzheimer's disease are associated with the loss of these neurons.

MEMBRANE LAMINATION PROCESSING
PHASE INVERSION

The majority of thermoplastic ultrafiltration (UF) and microfiltration (MF) membranes used to encapsulate cells are manufactured from homogeneous polymer solutions by phase inversion. Ultrafiltration membranes have pore sizes ranging from 5 nm to 0.1 μm, while microfiltration (or microporous) membranes have pores ranging from 0.5 to 3 μm. Phase inversion is a very versatile technique, allowing for the formation of membranes with a wide variety of nominal molecular weight cutoffs, permeabilities, and morphologies. The morphology and membrane properties depend on thermodynamic parameters and kinetics of the process. The polymer is dissolved in an appropriate solvent. The solution is then cast as a flat sheet or extruded as a hollow fiber. As part of the casting or extrusion procedure, the polymer solution is precipitated by a phase transition, which can be brought about by a change in temperature or solution composition. This process involves the transfer of a single-phase liquid polymer solution into a two-phase system consisting of a polymer-rich phase that forms the membrane structure and a second liquid polymer-poor phase that forms the membrane pores. Any polymer can be used, as long as it will form a homogeneous solution that under certain temperatures and compositions will separate into two phases. Thermodynamic and kinetic parameters such as chemical potential of the components and the free energy of mixing of the components determine the manner in which the phase separation takes place [29]. The process can be described by a polymer–solvent–nonsolvent ternary phase diagram.

THERMAL GELATION PHASE INVERSION

The thermally induced phase inversion process utilizes a polymer dissolved at an elevated temperature in a latent solvent (one that shows a lower solvency for a particular polymer at lower temperatures), which because of the loss of solvent power by heat removal will produce a solution that forms a gel when cooled [30]. The nonvolatile latent solvents must then be extracted from the gel, using another liquid that is a solvent for the latent solvent and a nonsolvent for the polymer. The thermal gelation process is capable of yielding asymmetric and isotropic microporous and ultrafiltration structures.

DIFFUSION-INDUCED PRECIPITATION

Diffusion-based precipitation requires solvent removal that results in the insolubility of the polymer. In one method, the solvent in which the polymer is dissolved is removed by evaporation, contact with a nonsolvent vapor, or total immersion in a nonsolvent bath. The evaporation of a volatile solvent as the membrane is cast creates a dense homogeneous structure. The vapor or immersion techniques rely on the diffusion of the nonsolvent into the solution precipitating the polymer due to the decreased solubility.

POST-TREATMENTS OF DENSE FILMS

Certain types of MF membrane are prepared by either mechanical stretching or chemical etching of dense films. For example, poly(tetrafluoroethylene) membranes are prepared by subjecting the film to a tensile stress. Poly(carbonate) membranes are prepared by a track etching process [31].

The membranes that have been evaluated for macroencapsulation have generally been UF or MF types. A UF membrane is defined as retaining species in the 300–300,000 molecular weight range, depending on the membrane [32]. Most xenograft cell transplantations require a UF membrane, while allografts or immunoprivileged cells may be successfully encapsulated in MF grade membranes that retain species in the range of 0.1–10 μm and solely inhibit host cell–transplanted cell contact.

MEMBRANE PROPERTIES

STRENGTH

The resilience of a medical device in the *in vivo* environment ultimately limits its success. The design of the final device configuration will determine the extent of strength required from the membrane. Note that membrane strength is in general inversely proportional to the diffusive transport in a homologous series. The membrane must also exhibit some degree of flexibility to remain intact during implantation and retrieval. If some other device component is used as a strength-bearing member, the choice of membrane structure, dimensions, composition, and materials may be limited to those that optimize transport properties.

If membrane strength is limiting for the overall device strength, then the membrane must be manufactured with certain considerations in mind. For example, membrane dimensions, composition, and structure may have to be altered to increase the strength. Choosing a material with which to cast the membrane, that is inherently stronger (i.e., more ordered), or has a higher molecular weight, should increase the overall mechanical properties. UF or MF membranes can be fabricated with macrovoids within the wall or as an open cell foam where the microvoids are interconnected. By incorporating techniques that increase this isoreticulated structure within the membrane wall, the tensile strength can be increased. The strength can also be improved by increasing the cross-sectional area of the membrane by thickening the walls. Decreasing the overall membrane porosity will also serve to increase the overall membrane strength. Examples of both macrovoid containing and isoreticulated structures are shown in Figs. 59.1 and 59.2, respectively.

The outer morphology of the membranes can be altered during fabrication or by a post-treatment to improve the reaction required for a successful implant. Using various phase inversion techniques, the outer surface of the membrane can range from a rejecting skin to a structure that is large enough to allow cells to enter into the wall itself (~10 μm in diameter). The combination of proper membrane transport and outer morphologies may

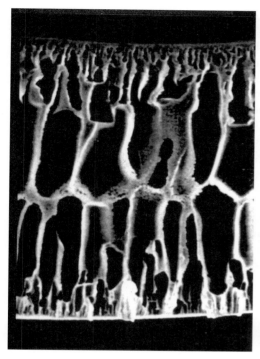

Fig. 59.1. Membrane containing macrovoids.

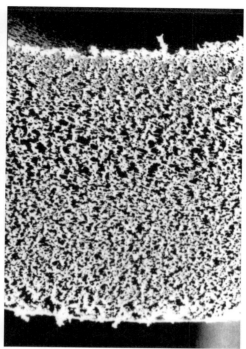

Fig. 59.2. Membrane showing isoreticulated structures.

also be achieved using composite membranes. Boggs *et al.* [33] have used such membranes for the treatment of type I diabetes.

TRANSPORT CHARACTERISTICS

Many different transport measurements including hydraulic permeability, solute rejection, and diffusive coefficient, can be used to determine, phenomenological transport characteristics of an encapsulation membrane. Hydraulic permeability and solute rejection relate to convective processes in which bulk fluid motion is driven by a transmembrane pressure difference. These convective resistances for water and solute flow are captured in hydraulic permeability and solute rejection measurements, respectively. Diffusion is the process by which molecules are driven by concentration gradients and move from a region of high concentration to low concentration via Brownian-like motion. Measurements of the convective and diffusive properties of an encapsulation membrane will indicate the capacity to maintain immunoisolation.

Convective Techniques

The hydraulic permeability (HP) of a membrane is determined from the convective water flux at a set transmembrane pressure. Hydraulic permeability is normalized with respect to the exposed surface area and the transmembrane pressure, resulting in the units of flux/area/pressure. Contemporary hemodialysis membranes range in hydraulic permeability from 2–6 ml/[h m² mmHg] for low-flux membranes to 10–200 ml/[h m² mmHg] for high flux membranes [34]. This performance parameter, which is proportional to the percentage of surface pores that are continuous through the membrane wall, averages the pore size, pore size distribution, and tortuosity into a single parameter. Experimental measurements can be compared to theoretical fluxes calculated from the Hagen–Pouseille equation [35]:

$$Q_f = \frac{\Delta p n \pi r^4}{8 l \eta \tau} \qquad (1)$$

where Q_f = volumetric water flux (m^3/s], n = number of pores (dimensionless), η = viscosity (kg/[m·s]) , r = pore radius (m), τ = tortuosity (dimensionless), l = membrane thickness (m), and Δp = pressure difference across the membrane (N/m^2).

Skin layer thickness for asymmetric ultrafiltration of poly(phenylene oxide) (PPO) membranes was measured by using a solution of colloidal gold particles, and the size distribution of open pores was determined by permporometry. Assuming a tortuosity of 1, Eq. (1) was then used to calculate water fluxes for the PPO membranes, which were in good agreement with the measured values. This comparison of measured and theoretical pure water fluxes indicated that the membrane morphological parameters found with permporometry and the colloidal gold particle method were relevant for the membrane transport properties [35].

In a second study, pore radii were calculated for Nucleopore membranes based on water flux measurements [36]. These calculations were used to estimate the hydrodynamic thickness of adsorbed dextran (negligible), poly(ethylene oxide) (PEO) (6–7 nm), and poly(vinylpyrrolidone) (PVP) (3–6 nm) used as tracers in diffusion experiments.

Membrane pore size can also be assessed by determining the ultrafiltrate flow rate. The ultrafiltrate flux J_f, is defined as the volumetric flow Q_f per unit membrane area A according to [37]:

$$J_f = \frac{Q_f}{A} \tag{2}$$

The rate of solute removal, M, is given by

$$M = Q_f c_f \tag{3}$$

where c_f is the ultrafiltrate solute concentration. The ultrafiltrate solute concentration is related to the bulk concentration in the retentate, c_{wb}, by the observed sieving coefficient, S, according to

$$S = \frac{c_f}{c_{wb}} \tag{4}$$

The solute rejection coefficient R often appears in the literature, and is defined as

$$R = 1 - S \tag{5}$$

Equations (2)–(5) can be combined to define the solute flux J_s in terms of the ultrafiltrate flux and the observed sieving coefficient as

$$J_s = J_f c_{wb}(1 - R) \tag{6}$$

Therefore, a solute flux through a membrane can be calculated given the ultrafiltrate flux, the bulk solute concentration, and the membrane rejection coefficient for the solute.

A range of solute sizes is used to generate a rejection coefficient profile representative of the membrane. This profile can be characterized by the nominal molecular weight cutoff (nMWCO), which is defined as the solute M_w that is reduced in concentration by a log order upon convective transport through the membrane (or 90% rejection). The position of a rejection profile (nMWCO) represents an average pore size, while the slope of the curve represents the pore size distribution.

Concentration polarization caused by boundary layer formation during the rejection measurement can have a tremendous effect on the measured rejection curves. To obtain rejection profiles that represent the properties of the membrane instead of hydrodynamic effects, operating flow conditions must be controlled based on the fiber geometry. This involves setting luminal flow rates to establish a set wall shear rate and setting the permeate flux to minimize concentration polarization [38].

Solutes commonly used to generate rejection curves include globular proteins [bovine serum albumin (BSA), MW = 67 kDa; IgG; ovalbumin, MW 59 kDa; and myoglobin, MW 16 kDa] and polysaccharides such as dextran and Ficoll. These solutions can be run singly or as a mixture of different size tracers. Protein detection systems include UV spectrophotometry for single-component solutions and size exclusion chromatography coupled with

UV spectrophotometry for protein mixtures. For enhanced sensitivity, enzymes such as lactate dehydrogenase or pyruvate kinase can be used with their respective enzymatic assay, or a fluorescein-tagged protein coupled with fluorescence detection.

Polydisperse dextran solutions (2,000–2,000,000 g/mol) are commonly used to generate membrane rejection curves. Size exclusion chromatography with refractive index detection is used to analyze reservoir and filtrate concentrations as a function of molecular weight, and fluorescein-tagged dextrans with fluorescence detection can be used for enhanced sensitivity. Membrane-fouling issues inherent in protein rejection curves are minimized by using dextran solutions, with their low binding capacity to many polymeric membrane structures [39]. The effect of protein adsorption on pore shutdown for poly(sulfone) and poly(ether sulfone) membranes has been demonstrated [40,41]. Rein *et al.* [42] showed a decrease in nMWCO from 40 kDa to 14 kDa upon exposure to a BSA solution, which is consistent with a decrease in pore size with protein adsorption. A comparison between the protein and dextran rejection curves based on solute size shows good agreement for the BSA-treated dextran curve.

Diffusive Techniques

The convective MWCO characterization test provides information on the membrane sieving properties and is dominated by pressure-driven convective transport processes. Molecular transport in most immunoisolation devices is governed by diffusive processes, which are driven by concentration gradients instead of pressure gradients. The diffusive flux can be described according to Fick's law [43]:

$$F = -\frac{D_{\text{eff}}\Delta c}{\Delta x} \tag{7}$$

where F = diffusive flux per unit area of section (g/[cm^2·s]), D_{eff} = effective diffusion coefficient (cm^2/s), and $\Delta c/\Delta x$ = solute concentration gradient across the membrane thickness (g/cm^4). Equation (7) can be simplified to the expression

$$F = k_{\text{m}}\Delta c \tag{8}$$

where membrane thickness, diffusivity, and partition coefficient are incorporated into the overall membrane mass transfer coefficient k_{m} with the units of centimeters per second.

Equation (7) states that the diffusive flux is inversely proportional to the membrane thickness, but the selectivity is independent of the thickness [34]. This has led to the development of extremely thin membranes that meet the separation requirements at an acceptable transmembrane diffusive flux. (Note that the same is true for membranes used in convective applications.) Commonly used asymmetric membranes incorporate a thin separating layer in series with a support substrate. The support substrate provides minimal transport resistance and a majority of the mechanical strength. Experimental determination of the device diffusive properties in both low and high range molecular weights is necessary to understand the encapsulated cellular environment.

Molecular transport across an immunoisolation device is influenced by the steric repulsion supplied by the pores of the membrane as well as the bulk porosity. Diffusive properties of smaller molecular weight species are governed by the overall membrane porosity, while larger molecular weight diffusive properties are governed by the membrane pore size. Techniques have been developed to measure the diffusive mass transfer for high-flux, low molecular weight species (180–1000 g/mol) as well as low-flux, high molecular weight species (60,000–150,000 g/mol) for P(AN-VC) hollow-fiber membranes [42].

The experimental apparatus for small molecular weight diffusion measurements involves a dialysis setup in which a tracer solution flows around the outside of the fiber, while a sampling solution flows down the fiber lumen. These flow rates are adjusted to minimize boundary layer formation on both the inside and outside of the fiber. Diffusion coefficients are calculated from the concentration difference between lumen and bath at a set lumen flow rate. This test was developed to measure the diffusion coefficients through these relatively higher water flux membranes for glucose (MW 186), vitamin B$_{12}$ (MW 1.3 kDa), and cytochrome C (MW 13.4 kDa).

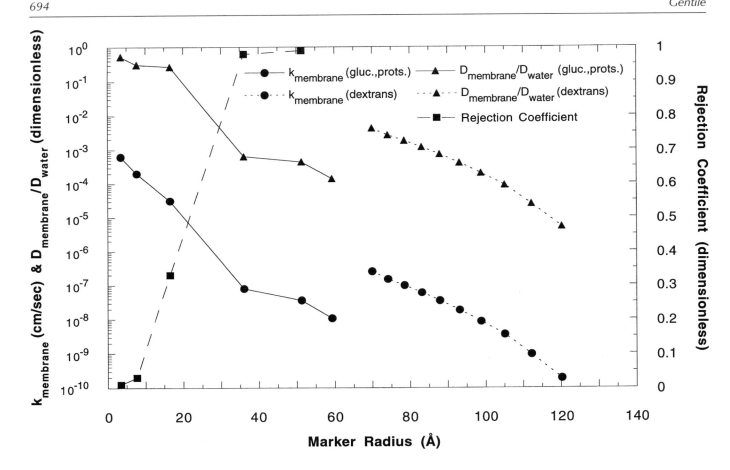

Fig. 59.3. Comparison of the diffusion of dextran and water across a P(AN-VC) membrane as a function of dextran molecular weight.

For the large molecular species, membrane resistances are far greater than the boundary layer resistances, and thus the experimental design does not involve flow [44]. The measurement can be made with the tracer diffusing across the membrane in either direction, both directions resulting in the same value for the diffusion coefficient. This test has been developed to measure the protein species myoglobin, BSA, and IgG, and dextran species from 50,000 to 300,000 MW. To enhance the sensitivity of the measurement, fluorescently tagged species were used. Figure 59.3 is a graph for the diffusion of dextran across a P(AN-VC) membrane with respect to that through water as a function of dextran molecular weight.

SUMMARY

Membrane lamination for encapsulated cell therapy is being investigated for the delivery of a wide range of products stemming from small molecules (e.g., dopamine, enkephalins) to very large gene products (e.g., growth factors, immunoglobulins). The key points in the use of materials to create these devices are as follows: (1) biocompatible membrane, device components and matrix materials; (2) strong implants; and (3) materials that can be made into membranes with the proper transport properties. This chapter has attempted to focus on methods of affecting or quantifing membrane strength and transport properties. Both these properties will be critical to any commercial implantation system. Membrane transport properties will be critical to all systems, whether implanted or extracorporeal-based devices. Methods of membrane manufacture can yield membranes having two to four times the strength of traditional phase inversion membranes. With new tools to measure small and large molecular weight transport, membranes with the desired combination of immunoisolation and product delivery rates can be selected for many applications potentially affecting a wide variety of disease treatments.

REFERENCES

1. Wahlberg, L. U., Almqvist, P. M., Glantz, M. J., and Boethius, J. (1996). Polymeric controlled-release amsacrine chemotherapy in an experimental glioma model. *Acta Neurochir.* **138**(11), 1323–1329.

2. Sagen, J., Wang, H., Tresco, P. A., and Aebischer, P. (1993). Transplants of immunologically isolated xenogeneic chromaffin cells provide a long-term source of pain-reducing neuroactive substances. *J. Neurosci.* **13**(3), 2415–2423.

3. Aebischer, P., Goddard, M., Signore, A. P., and Timpson, R. L. (1994). Functional recovery in hemiparkinsonian primates transplanted with polymer encapsulated PC12 cells. *Exp. Neurol.* **126**(2), 151–158.

4. Chick, W. L., Perna, J. J., Lauris, V., Law, D., Galletti, P. M., Panol, G., Whittemore, A. D., Like, A. A., Colton, C. K., and Lysaght, M. J. (1977). Artificial pancreas using living beta cells: Effects of glucose homeostasis in diabetic rats. *Science* **197**, 780–782.

5. Tze, W. J., Tai, J., Wong, F. C., and Davis, H. R. (1980). Studies with implantable artificial capillary units containing rat islets on diabetic dogs. *Diabetologia* **19**, 541–545.

6. Sharp, D. W., Mason, N. S., and Sparks, R. E. (1984). Islet immuno-isolation: The use of artificial organs to prevent tissue rejection. *World J. Surg.* **8**, 221–229.

7. McClaughlin, B. E., Tosone, C. M., Custer, L. M., and Mullon, C. (1999). Overview of extracorporeal liver support systems and clinical results. *Ann. N.Y. Acad. Sci.* **18**, 310–325.

8. Emerich, D. F., Hammang, J. P., Baetge, E. E., and Winn S. R. (1994). Implantation of polymer-encapsulated human nerve growth factor-secreting fibroblasts attenuates the behavior and neuropathological consequences of quinolinic acid into rodent striatum. *Exp. Neurol.* **131**(1), 141–150.

9. Brauker, J. (1992). Bioarchitecture of polymers determines soft tissue response to implants *ACS Polym. Sci. Eng., Biomater. 21st Cent. Meet.* Palm Springs, CA.

10. Hoffman, D., Breakfield, X. O., Short, M. P., and Aebischer, P. (1993). Transplantation of a polymer-encapsulated cell line genetically engineered to release NGF. *Exp. Neurol.* **122**, 100–106.

11. Winn, S., Hammang, J., Emerich, D. F., Lee, A., Palmiter, R., and Baetge, E. E. (1994). Polymer-encapsulated cells genetically modified to secrete human nerve growth factor to promote the survival of axotomized septal cholinergic neurons. *Proc. Natl. Acad. Sci. U.S.A.* **91**, 2324–2328.

12. Aebischer, P., Schluep, M., Deglon, N., Joseph, J. M., Hirt, L., Heyd, B., Goddard, M., Hammang, J. P., Zurn, A. D., Kato, A. C., Regli, F., and Baetge, E. E. (1996). Intrathecal delivery of CNTF using encapsulated genetically modified xenogeneic cells in amyotrophic lateral sclerosis patients. *Nat. Med.* **2**(6), 696–699.

13. Shoichet, M. S., Winn, S. R., Athavale, S., Harris, J. M., and Gentile, F. T. (1994). Poly(ethylene oxide)-grafted thermoplastic membrane for use as cellular hybrid bioartificial organs in the central nervous system. *Biotechnol. Bioeng.* **43**, 563–572.

14. Scharp, D. W., Swanson, C. J., Olack, B. J., Latta, P. P., Hegre, O. D., Doherty, E. J., Gentile, F. T., Flavin, K. S., Ansara, M. F., and Lacy, P. E. (1994). Viability of encapsulated human islets following subcutaneous implantation in Type I and Type II diabetic patients and normal controls. *Diabetes* **43**, 1167–1170.

15. Hegre, O. D., Lacy, P. E., Dionne, K. E., Gerasimidi-Vazeou, A., Gentile, F. T., Aebischer, P., Laurance, M., Fiore, D., Gardiner, A., Hazlett, T., and Purzycki, M. (1992). Macroencapsulation of islets of Langerhans. *Diabetes. Nutr Metab.* **5**, 159–162.

16. Altman, J. J., Manoux, A. A., Callard, P., McMillan, P., Solomon, B., Rosen, J., and Galletti, P. (1985). Long term plasma glucose normalization in experimental diabetic rats with macroencapsulated implants of benign human insulinomas. *Diabetes* **35**, 625–633.

17. Zondervan, G. L., Hoppen, H., Pennings, A. J., Fritschy, W., Wolters, A., and van Schilfgaarde, R. (1992). Design of a polyurethane membrane for the encapsulation of islets of Langerhans. *Biomaterials* **13**, 136–141.

18. Ronel, S. H., and D'Andrea, M. J. (1983). Macroporous hydrogel membranes for a hybrid artificial pancreas. I. Synthesis and chamber fabrication. *J. Biomater. Res.* **17**, 855–864.

19. Sun, A., Parasious, W., Healy, G. M., Vacek, I., and Macmorine, H. (1977). The use of capillary units containing cultured islets of Langerhans. *Diabetes* **26**, 1136–1139.

20. Sun, A., Parisius, W., Macmorine, H., Sefton, M., and Stone, R. (1980). An artificial endocrine pancreas containing cultured islets of Langerhans. *Artif. Organs* **4**, 275–278.

21. Segawa, M., Kakano, H., Nakagawa, K., Kanahiro, K., Nakajima, Y., and Shiratori, T. (1987). Effect of hybrid artificial pancreas on glucose regulation in diabetic dogs. *Transplant. Proc.* **19**, 985–988.

22. Sullivan, S., Maki, T., Borland, K., Mahoney, M. D., Solomon, B. A., Muller, T. E., Monoco, A. P., and Chick, W. L. (1991). Biohybrid artificial pancreas: long-term implantation studies in diabetic, pancreatectomized dogs. *Science* **252**, 718–721.

23. Catapano, G., Iorio, G., Drioli, E., Lombardi, C. P., Crucitti, F., Doglietto, G. B., and Bellantone, M. (1990). Theoretical and experimental analysis of a hybrid bioartificial membrane pancreas: A distributed parameter model taking into account starling fluxes. *J. Membr. Sci.* **52**, 351–378.

24. Aebischer, P., Russell, P. C., Christenson, L., Panol, G., Monchik, J. M., and Galletti, P. M. (1986). A bioartificial parathyroid. *Am. Soc. Artif. Intern. Organs* **32**, 134–137.

25. Christenson, L., Aebischer, P., and Galletti, P. M. (1988). Encapsulated thymic epithelial cells as a potential treatment for immunodeficiences. *ASAIO J.* **34**, 681–685.

26. Hymer, W. C., Wilbur, D. L., Page, R., Hibbard, E., Kelsey, R. C., and Hatfield, J. M. (1981). Pituitary hollow fiber units *in vivo* and *in vitro*. *Neuroendocrinology* **32**, 339–344.

27. Aebischer, P., Winn, S. R., and Galletti, P. M. (1988). Transplantation of neural tissue in polymer capsules. *Brain Res.* **448**, 364–368.

28. Aebischer, P. (1994) Cell therapy in the CNS: First clinical results and future promise. *American Society for Artificial Internal Organs Natl. Meet.*, San Francisco.

29. Strathmann, H. (1985). Production of microporous media by phase inversion processes. *In* "Material Science of Synthetic Membranes," pp. 65–195. American Chemical Society, Washington, DC.

30. Kesting, R. E. (1985). Phase inversion membranes. *In* "Material Science of Synthetic Membranes," pp. 1–65. American Chemical Society, Washington, DC.

31. Mulder, M. (1991). "Basic Principles of Membrane Technology." Kluwer Academic Publishers, London.

32. Lonsdale, H. K. (1982). The growth of membrane technology. *In* "Elsevier Journal of Membrane Science," pp. 75–95. Scientific Publishing, New York.

33. Boggs, D., Khare, A., McLarty, D., Pauley, R., and Sternberg, S. M. (1994). Membrane for immunoisolation. *Proc. 6th Ann. Meet. North Amer. Membr. Soc.*, Breckenridge, CO.

34. Lysaght, M. J., and Baurnmeister, U. (1993). *In* "Kirk-Othmer Encyclopedia of Chemical Technology," 4th ed., Vol. 8. Wiley, New York.

35. Cuperus, F. P., and Smolders, C. A. (1991). Characterization of UF membranes: Membrane characterization and characterization techniques. *Adv. Colloid Interface Sci.* **34**, 651–666.

36. Davidson, M. G., and Deen, W. M. (1987). Hydrodynamic theory for hindered transport of flexible maromolecules in porous membranes. *J. Membr. Sci.* **49**, 1021–1033.

37. Ofstrun, N. J., Colton, C. K., and Lysaght, M. J. (1986). Determination of fluid and solute removal rates during hemofiltration. *In* "Hemofiltration," pp. 68–91. Springer-Verlag, Berlin.

38. Tkacik, G., and Michaels, S. (1991). A rejection profile test for ultrafiltration membranes & devices. *Bio/Technology* **9**(10), 941–946.

39. Leypoldt, J. K., Figon, R. P., and Henderson, L. W. (1983). Dextran seiving coefficients of hemofiltration membranes. *Am. Soc. Artif. Intern. Organs Trans.* **29**, 678–690.

40. Mochizuki, S., and Zydney, A. L. (1992). Effect of protein adsorption on the transport characteristics of asymmetric ultrafiltration membranes. *Biotechnol. Prog.* **8**, 553–561.

41. Robertson, B., and Zydney, A. L. (1987). Polarization and adsorption effects on sieving in membrane protein filtration. *Am. Soc. Artif. Intern. Organs Trans.* **33**(3), 118–122.

42. Rein, D. H., Chandonait, S., Cain, B. M., and Dionne, K. E. (1994). Characterization of protein fouled membranes for immunoisolation. *American Society for Artificial Internal Organs Natl. Meet.*, San Francisco.

43. Crank, J. (1989). "Mathematics of Diffusion," 2nd ed. Oxford University Press, Oxford, UK.

44. Dionne, K. E., Cain, B. M., Li, R. H., Doherty, E. J., Rein, D. H., Lysaght, M. L., and Gentile, F. T. (1996). Transport characterization of membranes for immunoisolation. *Biomaterials* **17**(3), 257–266.

CHAPTER 60

Processing of Polymer Scaffolds: Freeze-Drying

Kyumin Whang and Kevin E. Healy

INTRODUCTION

Polymer scaffolds have been used as surrogates for the native extracellular matrix (ECM) for regeneration of bone, cartilage, liver, skin, and other tissues [1–9]. Polylactide (PL), polyglycolide (PG), and their copolymers (PLG) are attractive materials for ECM analogues because they degrade by random hydrolysis when implanted, and their degradation products are ultimately expelled from the body as carbon dioxide and water [10]. An ideal scaffold should have sufficient porosity for diffusion of nutrients and clearance of wastes, as well as adequate mechanical stability to support and transfer loads; in addition, the surface chemistry of the material must promote cell adhesion and intracellular signaling so that the cells express their normal phenotype [11]. For rapid cell growth, the scaffold also must have optimal microarchitecture (e.g., pore size, shape, and specific surface area). The effect of implant pore size on tissue regeneration is emphasized by experiments demonstrating optimum pore sizes of 45 μm for fibroblast ingrowth [12], between 20 and 125 μm for regeneration of adult mammalian skin [9], and 30–350 μm, depending on the mechanism, for regeneration of bone [12–14]. Thus, a major goal in fabricating scaffolds for tissue regeneration is to accurately control pore size and porosity.

Macroporous, biodegradable polymer scaffolds have been prepared by numerous techniques including solvent casting/particle leaching [7,15,16], phase separation [12], solvent evaporation [17], and fiber bonding to form a polymer mesh [15,18]. These methods are discussed in detail in other chapters. In this chapter, we address a novel processing technique for the fabrication of highly porous PLG scaffolds that have the added benefit of being amenable to the incorporation of protein-based growth and differentiation factors at the time of processing. More specifically, this involves fabricating scaffolds with porosity greater than 90% and the ability to control pore sizes ranging between 20 and 200 μm. The processing method consists of creating an emulsion by homogenization of a polymer solvent solution and water, rapidly cooling the emulsion to lock in the liquid state structure, and removing the solvent and water by freeze-drying [13,19].

MATERIALS

Poly(DL-lactide/glycolide) (PLG), 85:15 molar ratio; Birmingham Polymers, Birmingham, AL

Methylene chloride (MC), Baxter Diagnostics, McGraw Park, IL

Ultrapure water: ASTM grade I, 18 MΩ·cm

$CaCl_2$, NaCl, and bovine serum albumin (BSA), Sigma-Aldrich, St. Louis, MO

Poly(ethylene glycol) (PEG), Fluka Chemical Corp., Ronkonkoma, NY

EQUIPMENT

Handheld homogenizer, Omni Int., Waterbury, CT

Mercury intrusion porosimeter (MIP), Autoscan 33 Porosimeter, Quantachrome Corp., Syosset, NY

A modified freeze-dryer is needed because commercial freeze-dryers are not capable of freeze-drying MC or protecting the vacuum pump from MC vapors. This system consisted of a −110°C condenser (VirTis) connected to a liquid nitrogen trap (Vir-Tis, Gardiner, NY), connected to a chemically resistant vacuum pump (RV12, BOC Edwards Vacuum Products, Wilmington, MA) capable of pulling a vacuum in this system to about 20 mtorr.

FABRICATION PROCESS

A schematic diagram of the fabrication process is shown in Fig. 60.1. Two immiscible solutions are first prepared, an organic phase and a water phase. The organic phase is prepared by dissolving PLG with a specific inherent viscosity (η_{inh}) in MC such that it will have the desired weight-per-volume percent of total emulsion. Hydrophobic bioactive factors and surfactants can also be dissolved in this phase for incorporation and delivery and for control of scaffold microarchitecture. The water phase is prepared from ultrapure water with or without various dissolved additives such as hydrophilic bioactive factors for incorporation and delivery, salts such as $CaCl_2$ or NaCl, and/or surfactants to help control microarchitecture. The organic and water phases are added together in a glass test tube at 40% v/v water, forming two immiscible layers. From preliminary studies, this volume percent of water produced the thickest and most stable emulsion suitable for fabricating scaffolds. Other combinations have led to either melting or phase inversion (i.e., formation of microspheres). These immiscible layers are homogenized with a handheld homogenizer set at various speeds, poured into a suitable mold (e.g., copper or glass), and quenched by quickly placing the mold onto a copper block maintained near liquid nitrogen temperature (∼−196°C). The samples are then freeze-dried in a custom-built freeze-dryer at a pressure of 20 mtorr and a temperature beginning at −110°C. After allowing the temperature inside

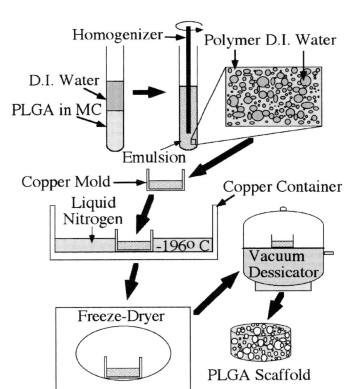

*Fig. 60.1. Schematic of the emulsion freeze-drying process. Two immiscible solutions (an organic phase and a water phase) are first prepared and added together in a glass test tube at 40% v/v water to form two immiscible layers. These immiscible layers are homogenized, poured into a suitable mold (e.g., copper or glass), quenched in liquid nitrogen, and freeze-dried in a custom built freeze-dryer at a pressure of 20 mtorr and a temperature beginning at −110°C and slowly warming to room temperature over 12 h. Reprinted from Whang et al. [13], Polymer **36**, pp. 837–842, Copyright 1995, with permission from Elsevier Science.*

the emulsion to equilibrate to −110°C for 1 h, the condenser is turned off and the condenser and emulsion allowed to slow warm to room temperature over 12 h. Finished samples are placed in a vacuum desiccator at room temperature for storage and further removal of any residual solvent [13,20–22]. To minimize protein loss during the fabrication process, an interpenetrating network of poly(acrylamide-*co*-ethylene glycol) was grafted onto the homogenizer and glassware to eliminate protein adsorption [23].

CHARACTERIZATION OF SCAFFOLDS

The microarchitecture of the scaffolds can be analyzed qualitatively by means of low-voltage (1.3-kV) scanning electron microscopy (SEM) and quantitatively by means of mercury intrusion porosimetry (MIP) to determine pore size distributions, specific pore area, median pore diameter, and porosity as described elsewhere [20–22]. In general, all emulsion freeze-dried scaffolds made using the range of weight percent of PLG described shortly and 40% v/v water had porosities exceeding 90% and specific surface areas of approximately 10 m^2/g. The scaffolds have very highly interconnected pores with large pore size distributions. The largest pores were greater than 200 μm. However, specific median pore sizes were evident as processing conditions were varied. The scaffolds also had a nonporous skin on one side, the side of the emulsion exposed to atmosphere after being poured into the mold. Scaffolds that were thick, homogeneous, physically manageable, and without many voids (pores >1 mm) were considered to be good scaffolds [24].

PROCESSING VARIABLES AFFECTING SCAFFOLD MICROARCHITECTURE

Other than the volume-per-volume percent of water, the effects of all other variables on scaffold microarchitecture and physical quality were investigated. Because this process has so many variables, evaluating their effects by using a full factorial experimental design would require an almost endless number of experiments. For example, in a process with 4 variables, each with 3 possible levels (settings) the minimum number of experiments to evaluate the variables affecting scaffold microarchitecture is $3^4 = 81$. We used Taguchi experimental analysis to minimize the number of experiments requiring statistically designed orthogonal arrays and to determine the most important variables to control [24,25]. Variables that were investigated include PLG concentration, PLG molecular weight, homogenizing speed, and additional emulsion stabilizers and surfactants. Table 60.1 summarizes the effects of these variables on scaffold microarchitecture and physical quality.

PLG CONCENTRATION

The effect of polymer weight fraction (5 and 10% w/v polymer) on scaffold microarchitecture and physical quality was evaluated. Polymer concentrations greater than 10% w/v were not suitable for scaffold fabrication because the high viscosity of the organic phase prevented adequate homogenization. Samples with 5% w/v PLG demonstrated substantial

Table 60.1. Effects of Processing Variables on Scaffold Median Pore Size and Physical Quality[a]

Processing variable	Median pore size	Physical quality
PLG % w/v (5 vs 10%)	10% w/v	10% w/v
PLG η_{inh}	↑↑↑↑↑	↑↑↑↑
Homogenizing speed	↑↑	↑↑
[CaCl$_2$]	↓↓↓	↑↑
[NaCl]	↓↓↓↓	↑↑
PEG M_w	↓	↓
[BSA]	↓↓↓↓	↑↑↑

[a]↑, processing variable increases scaffold properties, ↓, processing variable decreases scaffold properties.

melting that distorted the microarchitecture, while samples with 10% w/v PLG had better physical quality. Hence, 5% w/v PLG was disregarded as a viable processing variable [24]. We have also tried an intermediate value of 7.5% w/v with success.

PLG MOLECULAR WEIGHT

The effect of PLG η_{inh} of 0.57 and 0.88 dl/g were evaluated. An increase in PLG η_{inh} increased median pore size and physical characteristics. Median pore size is directly affected by the viscosity of the polymer solution, with samples of higher η_{inh} being more viscous [24]. This observation may be due to two factors: according to Stokes' law, the increase in η_{inh} decreases the rate of creaming by increasing the density of the polymer solution and the viscosity of the emulsion [24,26], and according to the Stokes–Einstein expressions for the diffusion coefficient, the increase in the viscosity of the polymer solution decreases the rate of flocculation [24,27].

HOMOGENIZING SPEED

Scaffolds made with homogenizing speeds of 5000 and 17,500 rpm revealed that the higher homogenizing speed resulted in a larger median pore size and in better physical properties [24]. This is counterintuitive, since in theory an increase in shear rate should decrease the water phase sphere size, hence pore size. However, the increase in the shear rate induces air bubbles in the emulsion and thus leads to scaffolds with increased pore size. Although pore size was increased, this method of processing is variable and difficult to control. At much lower shear rates, the introduction of air bubbles is not a factor.

EMULSION STABILIZERS AND SURFACTANTS

The effects of emulsion stabilizers and surfactants and their concentrations on scaffold microarchitecture and physical quality were investigated. Emulsion stabilizers and their concentrations tested were $CaCl_2$ (0, 0.01, 0.05, 0.1 M) and NaCl (0, 0.01, 0.1, 1 M). These electrolytes were added to the water phase before the homogenization step. From Derjaguin–Landau–Verwey–Overbeck (DLVO) theory, it is known that the presence and concentration of electrolytes like $CaCl_2$ and NaCl can alter the stability of oil-in-water emulsions by affecting the Debye length, hence the electrical repulsive force between particles [13,28,29]. Although DLVO theory was derived for colloid particles in a polar medium (i.e., oil-in-water emulsion), the opposite of the emulsion used here, this theory was employed in this work in conjunction with the concept of increasing long-range electrical and repulsive forces to increase emulsion stability. Various electrolyte concentrations were chosen to allow close packing of the particles to increase ultimate porosity and at the same time not allow coalescence (i.e., breakdown of the emulsion) [29,30].

The increase in $CaCl_2$ or NaCl concentration helped stabilize the emulsion allowing the formation of thick scaffolds with smaller median pore sizes [24]. Since the electrolytes are in the water (discontinuous) phase, the $CaCl_2$ or NaCl charges the water particle surface, increasing long-range electrostatic repulsive forces through the nonpolar MC solvent. Thus, the increase in electrolyte concentration appears to increase the electrostatic repulsive forces, as well as the Debye length that helps the stability of smaller particles in the emulsion by preventing aggregation and ultimately coalescence [27].

The two surfactants studied, PEG and BSA, were added to the water phase before the homogenization step [24]. PEG and BSA were added to the emulsion to act as surfactant and steric stabilizers and to prevent coalescence of dispersed particles [26,27,31]. Note that other surfactants such as Pluronics, Span, and Tween can also be used, and depending on the surfactant, they can be added to the water or the organic phase. The PEG concentration was fixed at 0.1% w/v because phase inversion occurred at other concentrations, and the effect of PEG weight-average molecular weight (M_w 20,000 or 35,000 Da) was investigated instead [24]. The PEG concentration was chosen to be in the range of critical micelle concentrations (cmc), which is in the range of 10^{-5}–10^{-4} M [26]. The cmc is defined as the concentration in which micelle formation becomes critical, that is, the concentration at which the surfactant efficacy becomes significant [26]. Different molecular weights can also

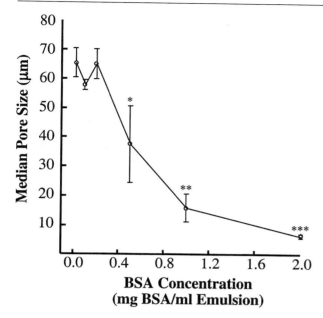

Fig. 60.2. Effect of BSA concentration on median pore size. Scaffold fabrication conditions were as follows: η_{inh}, 0.53; PLG, 10% w/v; homogenizing speed, 17,500 rpm; concentration of $CaCl_2$, 0.1 M; and PEG M_w, 35,000 Da. At concentrations less than 0.2 mg BSA/ml emulsion, scaffold median pore sizes were between 55 and 70 μm. Between concentrations of 0.2 and 1 mg BSA/ml emulsion, BSA acted to reduce median pore size with increasing concentrations. At concentrations greater than 1 mg BSA/ml emulsion, a practical minimum median pore size limit of about 6.5 ± 0.8 μm was reached. * = p < 0.01 different median pore size vs all groups. ** = p < 0.01 smaller median pore size vs all groups except 2 mg BSA/ml emulsion. *** = p < 0.01 smaller median pore size vs all groups except 1 mg BSA/ml emulsion. Reprinted from Whang et al. [22], Biomaterials 21, with permission from Elsevier Science.

be used; care is necessary, however, since the weight-average molecular weight M_w affects the cmc [26].

The increase in PEG M_w decreased pore size. Though this was the least important variable in comparison to other factors, it was still significant in controlling median pore size [24]. Note that the smaller the M_w, the larger the number of PEG molecules, since the same % w/v was used. The increase of M_w decreases the cmc, and together with the higher PEG concentration, cmc is more readily achieved or surpassed [26]. Thus, a lower cmc is favored for increased emulsion stability via steric stabilization, since micelles are more easily formed, allowing smaller particles to be formed.

For BSA, the effect of concentration (emulsion of 0.02, 0.01, 0.2, 0.5, 1, or 2 mg/ml) was assessed [22]. Scaffolds loaded with 0.01 mg BSA/ml emulsion were also made without any $CaCl_2$ or PEG to see if their effects were additive. The BSA concentration did not affect median pore size significantly until the concentration exceeded 0.2 mg BSA/ml emulsion (Fig. 60.2). At concentrations less than 0.2 mg BSA/ml emulsion, median pore sizes were between 55 and 70 μm. Between concentrations of 0.2 and 1 mg BSA/ml emulsion, BSA acted as an effective surfactant and colloid protector to decrease dispersed (water) phase particle size (i.e., median pore size); thus, the higher the concentration, the smaller the median pore size. At concentrations greater than 1 mg BSA/ml emulsion, the surfactant effect seemed to reach a practical limit, and a minimum median pore size of about 6.5 ± 0.8 μm was reached. All scaffolds had very high porosity and specific surface area (>90% and ~10 m²/g, respectively) [22].

The control of median pore size via regulation of protein concentration is due to the ability of proteins to affect the stability of emulsions. Proteins accumulate and adsorb at interfaces of emulsions, allowing them to act as surfactants and colloid protectors [32–34]. The emulsion interface used in this work contains a hydrophobic phase (MC/PLG), and dehydration of hydrophobic layers have been shown to lower the Gibbs free energy of the interface favoring adsorption of the surrounding protein to the hydrophobic surface [34]. Albumin, with its high conformational adaptability to changing environmental conditions, tends to adsorb to most surfaces even if dehydration and/or other electrostatic interactions are unfavorable. This capacity is further aided by the large net negative charge on albumin that reduces the structural stability of this molecule in solution, and increases the likelihood of spontaneous adsorption [34]. This preferential adsorption of protein (BSA) at the interface affects emulsion stability and consequently the median pore size of the scaffold. However, when the $CaCl_2$ and PEG were removed, the surfactant effect of the BSA (0.01 mg/ml emulsion) was more pronounced and the median pore size reduced significantly

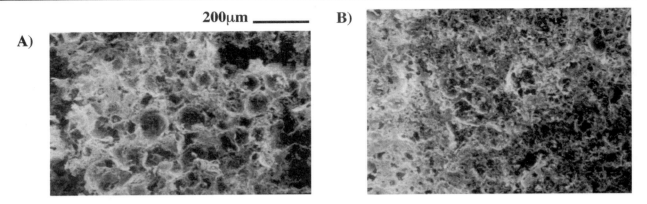

Fig. 60.3. Low-voltage scanning electron photomicrograph of uncoated BSA loaded scaffolds with median pore sizes of (A) 58 ± 2 μm and (B) 33 ± 8 μm. Scaffold fabrication conditions for (A) were as follows: η_{inh}, 0.53; PLG, 10% w/v; homogenizing speed, 17,500 rpm; concentration of $CaCl_2$, 0.1 M; PEG M_w, 35,000 Da; concentration of BSA, 0.01 mg/ml emulsion. Scaffold in (B) did not have any $CaCl_2$ or PEG. Note that these scaffolds do not look as porous because contrast was lost as a result of using uncoated samples. Reprinted from Whang et al. [22], Biomaterials **21**, with permission from Elsevier Science.

from 58 ± 2 μm to 33 ± 8 μm (Fig. 60.3). The MIP analyses of pore size distributions are shown in Fig. 60.4. Again large pore size distributions are shown, but with clear distinction in median pore size between groups 1 and 2. Thus, the effects of $CaCl_2$ and PEG were not additive to those of BSA.

PREFERRED METHOD OF SCAFFOLD FABRICATION

Based on the foregoing information as summarized in Table 60.1, a preferred set of conditions for making a scaffold with controlled pore size and good physical quality would be a as follows: PLG concentration, 10% w/v; PLG η_{inh} = 0.88 dl/g, [$CaCl_2$] = 0.01 M; [NaCl] = 0; homogenizing speed, 17,500 rpm; and PEG M_w = 25,000 Da. These parame-

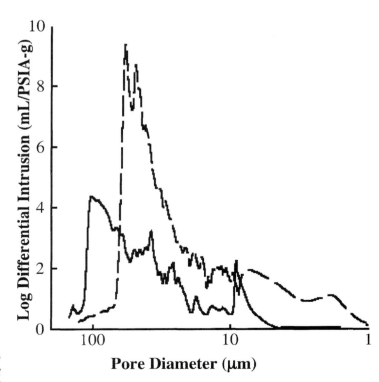

Fig. 60.4. MIP pore size distribution of scaffolds shown in Fig. 60.3. Note that there is a wide distribution of M_w among the samples, but a clear difference in median pore sizes is evident. Reprinted from Whang et al. [22], Biomaterials **21**, with permission from Elsevier Science.

—— BSA loaded scaffolds with median pore size of 58±2 μm
— - BSA loaded scaffolds with median pore size of 33±8 μm

Table 60.2. Processing Conditions and Median Pore Sizes of Scaffolds Having Large and Small Pore Sizes[a]

Scaffold	PLG η_{inh} (dR/g)	Homogenezer speed (rpm)	Concentration (M)		PEG M_w (Da)	BSA concentration (mg/ml)	Pore size (μm)
			CaCl$_2$	NaCl			
Large	0.88	17,500	0.01	—	35,000	—	57 ± 5
Small	0.53	17,500	0.1	—	35,000	2	6.5 ± 1
Small	0.81	3,500	—	1	35,000	—	4.6 ± 1

[a]All scaffolds had good physical quality and >90% porosity.

ters yielded scaffolds with good physical quality and median pore size of 57 ± 5 μm [24]. Scaffolds with the smallest median pore size were fabricated with additions of high concentrations of either BSA (6.5 ± 1 μm at 2 mg/ml emulsion) or NaCl (4.6 ± 1 μm at 1 M). The processing conditions and median pore sizes are summarized in Table 60.2.

REFERENCES

1. Cooper, M. L., Hansbrough, J. F., Spielvogel, R. L., Cohen, R., Bartel, R. L., and Naughton, G. (1991). *In vivo* optimization of a living dermal substitute employing cultured human fibroblasts on a biodegradable polyglycolic acid or polyglactin mesh. *Biomaterials* **12**, 243–248.

2. Freed, L. E., Marquis, J. C., Nohria, A., Emmanual, J., Mikos, A. G., and Langer, R. (1993). Neocartilage formation *in vitro* and *in vivo* using cells cultured on synthetic biodegradable polymers. *J. Biomed. Mater. Res.* **27**, 11–23.

3. Haberstadt, C., Anderson, P., Bartel, R., Cohen, R., and Naughton, G. (1992). Physiological cultured skin substitutes for wound healing. *Mater. Res. Soc. Symp. Proc.* **252**, 323–330.

4. Lucas, P. A., Laurencin, C., Syftestad, G. T., Domb, A., Goldberg, V. M., Caplan, A. I., and Langer, R. (1990). Ectopic induction of cartilage and bone by water-soluble proteins from bovine bone using a polyanhydride delivery vehicle. *J. Biomed. Mater. Res.* **24**, 901–911.

5. Mikos, A. G., Wald, H. L., Sarakinos, G., Leite, S. M., and Langer R. (1992). Biodegradable cell transplantation devices for tissue regeneration. *Mater. Res. Soc. Symp. Proc.* **252**, 353–358.

6. Mikos, A. G., Bao, Y., Cima, L. G., Ingber, D. E., Vacanti, J. P., and Langer, R. (1993). Preparation of poly(glycolic acid) bonded fiber structures for cell attachment and transplantation. *J. Biomed. Mater. Res.* **27**, 183–189.

7. Mooney, D. J., Cima, L., Langer, R., Johnson, L., Hansen, L. K., Ingber, D. E., and Vacanti, J. P. (1992). Principles of tissue engineering and reconstruction using polymer–cell constructs. *Mater. Res. Soc. Symp. Proc.* **252**, 345–352.

8. Vacanti, C. A., Cima, L. G., Ratkowski, D., Upton, J., and Vacanti, J. P. (1992). Tissue engineering growth of new cartilage in the shape of a human ear using synthetic polymers seeded with chondrocytes. *Mater. Res. Soc. Symp. Proc.* **252**, 367–374.

9. Yannas, I. V., Lee, E., Orgill, D. P., Skrabut, E. M., and Murphy, G. F. (1989). Synthesis and characterization of a model extracellular matrix that induces partial regeneration of adult mammalian skin. *Proc. Natl. Acad. Sci. U.S.A.* **86**, 933–937.

10. Cutright, D. E., Perez, B., Beasley, J. D., Larson, W. J., and Posey, W. R. (1974). Degradation rates of polymers and copolymers of polylactic and polyglycolic acids. *Oral Surg. Oral Med. Oral Pathol.* **37**, 142–152.

11. Healy, K. E., Tsai, D., and Kim, J. E. (1992). Osteogenic cell attachment to degradable polymers. *Mater. Res. Soc. Symp. Proc.* **252**, 109–114.

12. Klawitter, J. J., and Hulbert, S. F. (1971). Application of porous ceramics for the attachment of load bearing internal orthopedic applications. *J. Biomed. Mater. Res. Symp.* **2**, 161–229.

13. Whang, K., Thomas, C. K., Nuber, G., and Healy, K. E. (1995). A novel method to fabricate bioabsorbable scaffolds. *Polymer* **36**, 837–842.

14. Robinson, B. P., Hollinger, J. O., Szachowicz, E. H., and Brekke, J. (1995). Calvarial bone repair with porous D,L-polylactide. *Otolaryngol. Head Neck Surg.* **112**, 707–713.

15. Brauker, J. H., Carr-Brendel, V. E., Martinson, L. A., Crudele, J., Johnston, W. D., and Johnson, R. C. (1995). Neovascularization of synthetic membranes directed by membrane microarchitecture. *J. Biomed. Mater. Res.* **29**, 1517–1524.

16. Holy, C. E., Shoichet, M. S., and Davies, J. E. (2000). Engineering three-dimensional bone tissue *in vitro* using biodegradable scaffolds: Investigating initial cell-seeding density and culture period. *J. Biomed. Mater. Res.* **51**, 376–382.

17. Mikos, A. G., Sarakinos, G., Leite, S. M., Vacanti, J. P., and Langer, R. (1993). Laminated three-dimensional biodegradable foams for use in tissue engineering. *Biomaterials* **14**, 323–330.

18. Wang, N., Butler, J. P., and Ingber, D. E. (1993). Mechanotransduction across the cell surface and through the cytoskeleton. *Science* **260**, 1124–1127.

19. Healy, K. E., Whang, K., and Thomas, C. H. (1998). Method of fabricating emulsion freeze-dried scaffold bodies and resulting products. U.S. Pat. 5,723,508.

20. Whang, K., Tsai, D. C., Nam, E. K., Aitken, M., Sprague, S. M., Patel, P. K., and Healy, K. E. (1998). Ectopic bone formation via rhBMP-2 delivery from porous bioabsorbable polymer scaffolds. *J. Biomed. Mater. Res.* **42**, 491–499.

21. Whang, K., Healy, K. E., Elenz, D. R., Nam, E. K., Tsai, D. C., Thomas, C. H., Nuber, G. W., Glorieux, F. H., Travers, R., and Sprague, S. M. (1999). Engineering bone regeneration with bioabsorbable scaffolds with novel microarchitecture. *Tissue Eng.* 5, 35–51.

22. Whang, K., Goldstick, T. K., and Healy, K. E. (2000). A biodegradable polymer scaffold for delivery of osteotropic factors. *Biomaterials* 21, 2549–2551.

23. Bearinger, J. P., Castner, D. G., Golledge, S. L., Hubchak, S., and Healy, K. E. (1997). P(AAm-*co*-EG) interpenetrating polymer networks grafted to oxide surfaces: Surface characterization, protein adsorption, and cell detachment studies. *Langmuir* 13, 5175–5183.

24. Whang, K. (1998). A novel bioabsorbable scaffold useful for controlled drug release and tissue regeneration. Dissertation, Northwestern University, Evanston, IL.

25. Phadke, M. S. (1989). "Quality Engineering Using Robust Design." Prentice-Hall, Englewood Cliffs, NJ.

26. Schramm, L. L. (1992). Petroleum emulsions: Basic principles. *In* "Emulsions, Fundamental Applications in the Petroleum Industry" (L. L. Schramm, ed.), pp. 1–49. American Chemical Society, Washington, DC.

27. Everett, D. H. (1989). "Basic Principles of Colloid Science," pp. 30–53. Royal Society of Chemistry, London.

28. Everett, D. H. (1989). "Basic Principles of Colloid Science," pp. 115–143. Royal Society of Chemistry, London.

29. Garbassi, F., Morra, M., and Occhiello, E. (1994). "Polymer Surfaces: From Physics to Technology," pp. 3–48. Wiley, West Sussex, England.

30. Isaacs, E., and Chow, R. (1992). Practical aspects of emulsion stability. *In* "Emulsions, Fundamental Applications in the Petroleum Industry" (L. L. Schramm, ed.), pp. 51–77. American Chemical Society, Washington, DC.

31. Napper, D. H. (1977). Steric stabilization. *J. Colloid Interface Sci.* 58, 390–407.

32. Verrecchia, T., Huve, P., Bazile, D., Veillard, M., Spenlehauer, G., and Couvreur, P. (1993). Adsorption/desorption of human serum albumin at the surface of poly(lactic acid) nanoparticles prepared by a solvent evaporation process. *J. Biomed. Mater. Res.* 27, 1019–1028.

33. MacRitchie, F. (1987). Consequences of protein adsorption at fluid interfaces. *In* "Proteins at Interfaces: Physicochemical and Biochemical Studies" (J. L. Brash and T. A. Horbett, eds.), pp. 176–178. American Chemical Society, Washington, DC.

34. Norde, W., and Lyklema, J. (1991). Why proteins prefer interfaces? *J. Biomater. Sci. Polymer. Edn* 2, 183–202.

CHAPTER 61

PROCESSING OF POLYMER SCAFFOLDS: POLYMER–CERAMIC COMPOSITE FORMS

Cato T. Laurencin, Helen H. Lu, and Yusuf Khan

As an alternative or supplement to autografts and allografts, polymers and ceramics have been used separately and in combination as hard-tissue replacements for dental and orthopedic applications, as each material has intrinsic properties making it appropriate for certain applications. Several biodegradable polymers are the subjects of research projects and in clinical use for musculoskeletal applications. Poly(ortho esters), poly(anhydrides), poly(phosphazenes), and poly(amino acids) have all been examined as bone replacements owing to their unique degradation and mechanical properties. The poly(α-hydroxy acid) family of degradable polymers, including poly(lactic acid) (PLA), poly(glycolic acid) (PGA), and its copolymer poly(lactic-*co*-glycolic acid) (PLAGA), have been used extensively as fixation plates, screws, and pins, as well as drug delivery devices and tissue engineering scaffolds. There are several different ceramics that have been used alone and in conjunction with polymers for orthopedic applications including tricalcium phosphate, tetracalcium phosphate, hydroxyapatite, and composites based on bioactive material. These ceramics have been combined with several degradable and nondegradable polymers to improve the polymers' strength, attachment to bone, porosity, and ability to encourage bony ingrowth. Of particular interest here is the combination of PLAGA and hydroxyapatite into one multifunctional composite form, applicable to tissue engineering. With this in mind, three different methods are described for creating a porous, interconnected composite scaffold of PLAGA and hydroxyapatite: a polymer–ceramic film formed by solvent casting, polymer–ceramic constructs synthesized by a solvent–aggregation method, and polymer–ceramic constructs synthesized by using the gel–microsphere method.

INTRODUCTION

Tissue engineering can be defined as the application of biological, chemical, and engineering principles toward the repair, restoration, or regeneration of living tissues using biomaterials, cells, and factors, alone or in combination [1]. Both polymers and ceramics possess distinctly different intrinsic properties; individually, each material has been used extensively as a biomaterial in the regeneration of living tissues, and these applications are well documented in the literature [2–28]. For example, ceramics have been utilized in the repair of hard tissue involving dental and musculoskeletal applications. Polymers have been used extensively throughout the body as both temporary and permanent replacements for arteries, bone, and joints, and plastic reconstruction, just to name a few.

A single material type does not always provide the necessary mechanical and/or chemical properties desired for a particular application. In these instances, composite materials that combine the advantages of both materials may be most appropriate. Of particular interest here is the combination of polymers and ceramics into one multifunctional composite form, applicable in tissue engineering. Polymer–ceramics are designed primarily as a synthetic alternative to biological grafts such as autografts and allografts. This chapter briefly discusses the application of these two types of material, alone or in combination, for dental and musculoskeletal tissue regeneration.

CONVENTIONAL BONE GRAFTS

AUTOGRAFTS

The current gold standard in bone grafts is the autograft, tissue usually taken from the iliac crest of the patient and transferred to the affected site. Structurally, the autograft possesses both osteoconductive and osteoinductive capabilities, providing a framework into which new bone tissue and vasculature can grow, while stimulating the regeneration of new bone by promoting the differentiation of mesenchymal cells into bone-forming osteoblasts. There is, however, a limited supply of autograft tissue, and morbidity is often associated with the donor site; moreover, the surgery necessary to procure the graft causes added pain and other complications.

ALLOGRAFTS

The primary clinical alternative to the autograft is the allograft, tissue donated by another individual or procured from cadaver. This eliminates the problems of donor-site morbidity and supply limitations. However, there are increased risks of disease transfer and implant rejection. Although it also provides an osteoconductive structure for bone ingrowth, the necessary graft sterilization process reduces its osteoinductive potential and often results in a structure with compromised mechanical properties [2].

There is currently a sizable market for bone graft substitutes. In the United States alone, about 6.2 million fractures occur every year [3], of which 500,000 cases require some form of bone grafting [4]. Approximately 350,000 of these fractures are treated with autografts, and the rest by allografts [2]. Moreover, the average cost of a bone grafting procedure is estimated to be $5000 [4], making total health care costs per year as high as $2.5 billion. Given the limitations associated with biological grafts, engineers and clinicians are working together to develop bone graft substitutes.

RATIONALE FOR COMPOSITE BIOMATERIALS

Biological bone is composed of both organic and inorganic phases. Of the inorganic phase, approximately 70% is calcium phosphate [4], which is mostly in the form of a semi-crystalline mineral called hydroxyapatite and is largely responsible for the mechanical integrity of bone [5]. Several types of ceramic have been found to be biocompatible and to promote bone growth and implant stability. Examples are coralline, calcium sulfate, calcium phosphate, 45S5 bioactive glass, and hydroxyapatite. These materials have been used in powder, paste, particulate, and block forms, and they can be combined with polymers via dry mixing or compression molding or formed on the surface of polymers through chemical modifications (see Table 61.1). However, the ceramic material is often nondegradable and, owing to its brittle nature and poor tensile properties, is often mechanically incompatible with natural bone.

Several classes of polymers are currently considered as candidates for bone repair. Poly(methyl methacrylate) (PMMA) has been used as bone cement since the early 1960s and is widely used today, alone and sometimes in combination with other materials such as titanium fibers [6] and ceramics [7]. Gamma-irradiated polyethylene is used extensively in joint replacement implants because of its high wear resistance and biocompatibility [8]. Several biodegradable polymers are currently the subject of research projects and in clinical use for musculoskeletal applications. Poly(ortho esters), poly(anhydrides), poly(phosphazenes), and poly(amino acids) have all been examined as bone replacements owing to their unique degradation and mechanical properties. The poly(α-hydroxy acid) family of degradable polymers, including poly(lactic acid) (PLA), poly(glycolic acid), and its copolymer poly(lactic-*co*-glycolic acid), have been used extensively as fixation plates,

Table 61.1. Ceramics Used in Hard-Tissue Applications

Ceramic	Description	Applications	Properties
Calcium phosphate	Ceramic used alone or with polymers	Bone void filler	Osteoconductive, degrades moderately slowly
Hydroxyapatite	Crystalline calcium phosphate, major ceramic component of bone	Osteoconductive and mechanically strong additive to polymer bone cements and bone replacements	Osteoconductive, degrades very slowly
Coralline	Sea coral chemically converted to HA	Bone void filler	Osteoconductive; interconnected pore structure similar to trabecular bone
Calcium sulfate	Gypsum or plaster of paris	Bone void filler used extensively for dental applications	Degrades in 30–60 days with considerable loss in mechanical properties
Bioactive glass	Glass containing Na, P, and Ca	Chemically bonds to bone	Osteoconductive and osteoinductive, but degrades very slowly
Bone like apatite	Mineral formed on polymer surfaces by simulated body fluid	Facilitates bone bonding to surface	Adheres to polymers to form mineralized layer

screws, and pins [9], as well as drug delivery devices and tissue engineering scaffolds. PLA, PGA, and PLAGA have the added advantage of being approved by the U.S. Food and Drug Administration (FDA) [9] and are easily manipulated into porous structures, with mechanical properties approaching trabecular bone.

To capitalize on their advantages and minimize their shortcomings, ceramic materials have been combined with various degradable and nondegradable polymers to form composite biomaterials for osseous repair. Several current research and commercially available products are reviewed.

CALCIUM PHOSPHATE BASED COMPOSITES

Calcium phosphate was first used as a bone defect filler for craniofacial and dental applications in the early 1970s [5]. Since then, it has been processed into different forms for various orthopedic applications. Calcium phosphates such as tricalcium phosphate, tetracalcium phosphate, and hydroxyapatite differ in their biocompatibility, crystallinity, and degradability. Tricalcium phosphate (TCP) has been utilized in conjunction with other ceramics and bone cements [5], as well as several polymers to improve both its mechanical and osteoconductive properties. Now TCP is used as a bone replacement and delivery vehicle for human recombinant bone morphogenetic protein 2 (BMP-2). Enhanced bone growth was seen in TCP combined with BMP-2 in comparison to TCP alone [10]. Kikuchi *et al.* [11] combined TCP with *co*-poly(L-lactide) (CPLA) to form a polymer–ceramic composite that possesses TCP's osteoconductive potential and mechanical strength, along with the degradability of the CPLA. The Young's modulus of the composite was found to be twice that of CPLA alone, while there was no loss in bending strength with the addition of the TCP to the CPLA. The biocompatibility of the composite was assessed and determined to be similar to that of control wells containing cells alone [11].

Calcium phosphates have also been combined with bone cements to form resorbable composites with adequate mechanical properties. Beruto *et al.* [7] added TCP powder to poly(methyl methacrylate) (PMMA) bone cement to form resorbable bone cement with improved interconnectivity and mechanical properties to better support new bone ingrowth.

The addition of TCP to PMMA increased the porosity of the bone cement but lowered the compressive strength to a point similar to that found with other porous ceramic implants.

HYDROXYAPATITE

Hydroxyapatite has been used extensively alone and in combination with polymers to form composite forms of a bone substitute. Sea coral, which is calcium carbonate, has been used as a bone graft substitute [12] because of its interconnectivity and biomimetic nature. This material, which is first processed to convert the calcium carbonate into hydroxyapatite [4], is distributed commercially by Interpore Cross International, Inc. as ProOsteon 500 and 500R. The mechanical properties of the intact coral, however, proved insufficient to sustain physiological loads commonly beared by cortical bone [4,12].

In general, hydroxyapatite (HA) is used in particulate form, the size of which is customized according to the application. HA particles of 15 μm in diameter have been combined with PMMA bone cement to promote bone ingrowth and attachment to prosthetic devices being secured by the cement [13]. Morita *et al.* found that the addition of HA did not decrease the tensile, compressive, or bending strength of the cement, despite poor bonding between the cement and the HA particles. Bone cement with HA particles, however, measured an increased tensile bonding strength to bone in comparison to cement without HA particles, implying that the HA bonded to the bone directly. A study by Vallo *et al.* [14] examined the effect of the amount of HA particles added to PMMA-based bone cement on the cement's density, pore size, and mechanical properties. It was found that bone cement density decreased while porosity increased with increasing HA content, and an ultimate decrease in yield stress was measured after an initial increase. The long-term decrease in yield stress with increasing HA content was attributed to poor bonding between HA and bone cement, leading to larger void spaces within the cement and ultimately larger regions of insufficient bonding between cement and HA. However, small amounts of HA added to bone cement did increase both its yield stress and its fracture toughness.

Our laboratory has explored extensively the possibility of utilizing a polymer/composite scaffold for bone tissue regeneration and repair. Porous scaffolds containing poly(lactide-*co*-glycolide) and HA crystals were formed and seeded with osteoblasts harvested from rat calvaria. After only 24 h, osteoblasts were seen to adhere to the exterior surface of the scaffold and also to migrate into the pore structure [15]. In a separate study, the influence of HA content on mechanical and degradation properties was reported [16]. Addition of HA to the scaffold increased the compressive modulus by over 400%, while decreasing mass loss and therefore scaffold degradation over a 6-week period. Finally, the ability of the PLAGA-HA scaffold to support osteoblast proliferation and differentiation as well as mineral formation over 21 days was examined [17]. It was found that cells proliferated up to 21 days and formed a mineralized layer on the PLAGA-HA scaffold, indicating cellular differentiation. The scaffold used in these studies combined the degradability of the PLAGA with the mechanical support of HA to form a tissue engineering replacement for bone defects.

BIOACTIVE MATERIAL BASED COMPOSITES

Bioactive materials are biocompatible materials with the added ability of enhancing bone formation and bonding to surrounding bone tissue. Bioactive glasses, in particular 45S5 bioactive glass, have been widely studied as a potential biomaterial for use in musculoskeletal tissue engineering [18,19]. The ability of 45S5 bioactive glass to bond to bone was first reported by Hench *et al.* [20] in the early 1970s. Through interfacial and cell-mediated reactions, bioactive glass develops a calcium-deficient, carbonated calcium phosphate surface layer that allows it to chemically bond to surrounding bone. The ability of an implant to form a chemical interface with surrounding tissue is critical in eliminating implant loosening, one of the major causes of synthetic implant failure. In addition to its ability to bond to bone, bioactive glass has been found to support osteoblast adhesion, growth, and differentiation. Moreover, bioactive materials have been shown to induce the differentiation of mesenchymal cells into osteoblasts [21,22].

Despite their osteointegrative, osteoconductive, and osteoinductive nature, bioactive glasses alone have limited application in load-bearing situations owing to mechanical mis-

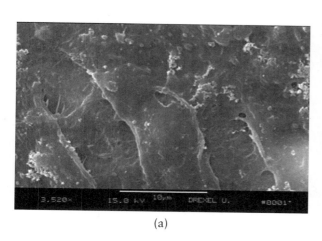

(a)   (b)

Fig. 61.1. (a) The solvent-casted composite surface promoted the formation of mineralized nodules by human osteoblast-like cells after only 1 week of culture. SEM micrograph, 15 keV, original magnification, 3520×. (b) Saos-2 cultured on porous PLAGA-BG composite for 3 weeks. Note the extensive coverage of the spheres by cell growth and the formation of surface Ca-P nodules. 20 keV, original magnification, 81×.

match with surrounding bone. However, these materials can be used in combination with polymers to form composite materials having bone repair potential. Fujita *et al.* [23] added wollastonite glass–ceramic to PMMA to formulate a bioactive bone cement. In comparison to PMMA alone, the bioactive bone cement showed a higher bonding strength between the bone cement and the femurs of dogs, with increased bone resorption over PMMA alone. Marcolongo *et al.* [24] blended polysulfone with bioactive glass fibers to form composite rods, and the rods were placed in rabbit femurs to examine the formation of a mechanical bond between bone tissue and the composite implant. After 6 weeks, interfacial strength was examined and found to be more than twice that of polysulfone alone.

Polymer–ceramic composites can also be fabricated by inducing direct deposition of the bioactive calcium phosphate layer on polymer surfaces. In this approach, the polymer is immersed in simulated body fluid (SBF) with an ion concentration simulating interstitial fluid. Upon incubating the material in the SBF, an apatite layer gradually forms on the surface. Calcium phosphate layers have been successfully formed on titanium [25,26], organic polymers [27], ceramic–polymer composite fibers [24], ethylene-vinyl alcohol [28], and poly(lactide-*co*-glycolide) [29]. Murphy *et al.* [29] reported on the formation of an apatite layer on the surfaces of a porous, three-dimensional scaffold and found an increase in compressive modulus over polymer alone from 50 kPa to 300 kPa after 16 days of incubation.

Our laboratory has also developed composites of bioactive glass and biodegradable polymers, and these materials formed calcium phosphate layers *in vitro*. Since bioactive glass exhibits high compressive strength, it can act as a mechanical reinforcer for materials with lesser mechanical strength. We fabricated composite films (solvent casting) of bioactive glass and 50:50 poly(lactide-*co*-glycolide) copolymer, microsphere-based, three-dimensional, porous constructs. These composites were able to form calcium phosphate layers on their surfaces, and they supported rapid and abundant growth of human osteoblasts and osteoblast-like cells when cultured *in vitro* (Fig. 61.1). Moreover, these cells formed a mineralized matrix on these composite materials. Future work will examine the *in vivo* healing and bone bonding potential of these glass–polymer composite materials.

PROTOCOL A: POLYMER–CERAMIC FILM FORMED BY SOLVENT CASTING

INTRODUCTION

Polymer–ceramic thin films can be made by first melting the polymer in an organic solvent, then adding the desired ceramic in particulate form. The solvent is then allowed to evaporate and a polymer-ceramic film of desired thickness can be produced.

MATERIALS

- Polymer [e.g., poly(lactide), poly(glycolide), poly(lactide-*co*-glycolide)]
- Organic solvent (e.g., methylene chloride)
- Ceramic granules (e.g., hydroxyapatite, bioactive glass; size, 1–500 μm)
- Teflon dish
- Erlenmeyer flask or beaker
- Stir bar, stirrer

METHODS

1. Weigh out polymer according to desired (polymer weight/solvent volume) ratio.
2. Add organic solvent to the Erlenmeyer flask or beaker and cover.
3. Stir the polymer–solvent mixture until the polymer has dissolved.
4. Add ceramic granules to the polymer–solvent mixture.
5. Mix until the granules have been distributed homogeneously in the mixture.
6. Under a chemical hood, slowly pour the polymer–solvent solution into the Teflon dish.
7. Allow the solvent to slowly evaporate.

Useful Notes

- By varying the ratio of polymer weight to solvent volume, thin films of different thicknesses can be produced.
- The rate of evaporation is a key parameter in producing homogeneous thin films. Porous films can be formed by accelerating solvent evaporation. To decrease evaporation rate, the Teflon dish can be put in a (–20°C) freezer and allowed to evaporate overnight.
- Teflon paper or spray can be used in place of a Teflon dish.

PROTOCOL B: POLYMER–CERAMIC CONSTRUCTS SYNTHESIZED BY SOLVENT–AGGREGATION METHOD

INTRODUCTION

Polymer–ceramic constructs based on microspheres can be fabricated by the solvent–aggregation method where the microspheres were first formed from traditional water oil/water emulsions. Solvent-aggregated polymer–ceramic matrices can then be fabricated by mixing solvent, salt particles, ceramic granules, and the prehardened microspheres. A three-dimensional structure of controllable porosity is formed based on this method combined with salt leaching and microsphere packing.

FABRICATION OF MICROSPHERES

Materials

- Polymer [e.g., poly(lactide), poly(glycolide), poly(lactide-*co*-glycolide)]
- Organic solvent (e.g., methylene chloride)
- Borosilicate glass vials (volume capacity, 10–30 ml)
- Poly(vinyl alcohol) (PVA) solution, 1 wt%
- Vortex machine
- Mechanical stirrer (rpm range, 100–1000)
- 1000-ml beaker
- Vacuum line or equivalent
- Lyophilizer
- Desiccators

Methods

1. Weigh out polymer according to desired ratio of polymer weight to solvent volume.
2. Add polymer and organic solvent to the glass vial and cover.
3. Vortex the polymer–solvent mixture until the polymer has dissolved.
4. Add the polymer solution dropwise to the stirring 1% PVA solution.
5. Stir the 1% PVA–polymer solution at 300 rpm for at least 4 h to allow evaporation of the solvent.
6. Collect the microspheres via vacuum filtration.
7. Wash with deionized water and allow to air-dry for at least 2 h at room temperature.
8. Lyophilize the microspheres for another 24 h to remove any remaining solvent.
9. Store the microspheres in a desiccator prior to use.

Useful Notes

- The volume of 1% PVA is dependent on the volume of polymer–solvent mixture. For example, with 20 ml of polymer–solvent mixture, about 600 ml of 1% PVA solution should be used.
- The final microsphere size distribution is a function of stirring speed in 1% PVA.
- Microspheres will harden as stirring time increases.
- Microspheres can be sieved into selected sizes by using mechanical sifting systems.

SOLVENT–AGGREGATION METHOD

Materials

- Polymer microspheres of selected diameter (e.g., 2.0 g)
- Organic solvent (e.g., methylene chloride, 2.0 ml)
- NaCl particles of selected size (e.g., 2.0 g)
- Ceramic granules of desired size (e.g., hydroxyapatite, bioactive glass, 2.0 g)
- Teflon or stainless steel mold
- Carver press
- Deionized water

Methods

1. Measure out desired amounts of polymeric microspheres, NaCl, and ceramic particles.
2. Dry-mix the microspheres, NaCl particles, and ceramic granules according to selected ratio of weight to weight to weight.
3. While stirring the mixture, slowly add a small amount of solvent.
4. Place the mixture in the Teflon mold.
5. Use the Carver press to apply compressive load to the mold.
6. Lyophilize the resultant constructs for 24 h.
7. Immerse the constructs in 300 ml of 37°C deionized water for at least 24 h leach out the NaCl particles.
8. Lyophilize the constructs for 24 h.

Useful Notes

- A very small amount of solvent is needed in this method. It is used to dissolve the outer surface of the microsphere and to allow microspheres to bond and ceramic particles to aggregate on the surface.

- The Teflon mold is should have the dimensions of the desired structure (cylinder, etc.).
- The size of the NaCl crystals controls both porosity and pore size of the three-dimensional construct.
- Compression of the construct is not required, although it has been found to increase mechanical properties of the structure.

PROTOCOL C: POLYMER-CERAMIC CONSTRUCTS SYNTHESIZED BY GEL MICROSPHERE METHOD

INTRODUCTION

The gel microsphere method is similar to solvent aggregation, since it is also based on the microsphere method. However, in this case, during the microsphere formation process, the spheres were isolated before complete solvent removal. In this gel state, they aggregated to form a three-dimensional structure. Void-forming NaCl and ceramic particles can then be added to create a porous structure (Fig. 61.2).

MATERIALS

- Polymer [e.g., poly(lactide), poly(glycolide), poly(lactide-*co*-glycolide)]
- Organic solvent (e.g., methylene chloride, 2.0 ml)

1 hour stirring time

1:15 hour stirring time

1:30 hour stirring time

2 hour stirring time

Fig. 61.2. The effect of stirring time on the gel microsphere matrix with 50 wt% hydroxyapatite particles. With an increase in stirring time, the microspheres begin to form from the gel. 20 keV, original magnification, 100×. Reprinted from Biomaterials (in press), Borden, M. D., Attawia, M., Khan, Y., and Laurencin, C. T., "Tissue Engineered Microsphere-Based Matrices for Bone Repair," Fig. 2. Copyright 2001, with permission from Elsevier Science.

- NaCl particles of selected size (e.g., 2.0 g)
- Ceramic granules of desired size (e.g., hydroxyapatite, bioactive glass, 2.0 g)
- Construct mold
- Centrifuge tubes
- Deionized water (e.g., 300 ml)

METHODS

1. Follow steps 1–4 in Protocol B.
2. Stir the 1% PVA/polymer solution for 1–2 h to allow formation of gel microspheres.
3. Remove most of the PVA and transfer the gel microspheres + PVA into a centrifuge tube.
4. Remove the remaining PVA after the microspheres have settled to the bottom.
5. Weigh out desired amounts of NaCl and ceramic particles.
6. Mix and stir together the gel microspheres, NaCl particles, and ceramic granules according to selected weight:weight:weight ratio.
7. Place the mixture in the Teflon mold and air-dry for 24 h.
8. Remove the constructs from the mold and immerse them in 37°C deionized water for at least 24 h to leach out the NaCl particles.
9. Lyophilize the constructs for 24 h.

REFERENCES

1. Laurencin, C. T., Ambrosio, A. A., Borden, M. D., and Cooper, J. A. (1999). Tissue engineering: Orthopaedic applications. *In* "Annual Review of Biomedical Engineering" (M. L. Yarmush, ed.), p. 19. Annual Reviews, Palo Alto, CA.
2. Stevenson, S. (1999). Biology of bone grafts. *Orthop. Clin. North Am.* **30**(4), 543–552.
3. Bostrom, M. P. G., Saleh, K. J., and Einhorn, T. A. (1999). Osteoinductive growth factors in preclinical fracture and long bone defect models. *Orthop. Clin. North Am.* **30**(4), 647–658.
4. Shors, E. C. (1999). Coralline bone graft substitutes. *Orthop. Clin. North Am.* **30**(4), 599–613.
5. Tay, B. K. B., Patel, V. V., and Bradford, D. S. (1999). Calcium sulfate- and calcium phosphate-based bone substitutes. *Orthop. Clin. North Am.* **30**(4), 615–623.
6. Topoleski, L. D., Ducheyne, P., and Cuckler, J. M. (1998). Flow intrusion characteristics and fracture properties of titanium-fibre-reinforced bone cement. *Biomaterials* **19**(17), 1569–1577.
7. Beruto, D. T., Mezzasalma, S. A., Capurro, M., Botter, R., and Cirillo, P. (2000). Use of α–tricalcium phosphate (TCP) as powders and as an aqueous dispersion to modify processing, microstructure, and mechanical properties of polymethylmethacrylate (PMMA) bone cements and to produce bone-substitute compounds. *J. Biomed. Mater. Res.* **49**, 498–505.
8. Kurtz, S. M., Muratoglu, O. K., Evans, M., and Edidin, A. A. (1999). Advances in the processing, sterilization, and crosslinking of ultra-high molecular weight polyethylene for total joint arthroplasty. *Biomaterials* **20**(18), 1659–1688.
9. Hollinger, J. O., Brekke, J., Gruskin, E., and Lee, D. (1996). Role of bone substitutes. *Clin. Orthop. Relat. Res.* **324**, 55–65.
10. Laffargue, P. H., Heldebrand, H. F., Rtaimate, R., Frayssinet, P., Amoureux, J. P., and Marchandise, X. (1999). Evaluation of human recombinant bone morphogenetic protein-2-loaded tricalcium phosphate implants in rabbits' bone defects. *Bone* **25**(2S), 55S–58S.
11. Kikuchi, M., Tanaka, J., Koyama, Y., and Takakuda, K. (1999). Cell culture test of TCP/CPLA composite. *J. Biomed. Mater. Res., Appl. Biomater.* **48**, 108–110.
12. Cornell, C. N. (1999). Osteoconductive materials and their role as substitutes for autogenous bone grafts. *Orthop. Clin. North Am.* **30**(4), 591–598.
13. Morita, S., Furuya, K., Ishihara, K., and Nakabayashi, N. (1998). Performance of adhesive bone cement containing hydroxyapatite particles. *Biomaterials* **19**, 1601–1606.
14. Vallo, C. I., Montemartini, P. E., Fanovich, M. A., Porto Lopez, J. M., and Cuadrado, T. R. (1999). Polymethylmethacrylate-based bone cement modified with hydroxyapatite. *J. Biomed. Mater. Res., Appl. Biomater.* **48**, 150–158.
15. Attawia, M. A., Herbert, K. M., and Laurencin, C. T. (1995). Osteoblast-like cell adherence and migration through 3-dimensional porous polymer matrices. *Biochem. Biophys. Res. Commun.* **213**(2), 639–644.
16. Devin, J. E., Attawia, M. A., and Laurencin, C. T. (1996). Three-dimensional degradable porous polymer–ceramic matrices for use in bone repair. *J. Biomater. Sci., Polym. Ed.* **7**(8), 661–669.
17. Laurencin, C. T., Attawia, M. A., Elgendy, H. E., and Herbert, K. M. (1996). Tissue engineered bone-regeneration using degradable polymer: The formation of mineralized matrices. *Bone* **19**(1) 93S–99S.
18. Hench, L. L. (1991). Bioceramics: From concept to clinic. *J. Am. Ceram. Soc.* **74**(7), 1487–1510.
19. Ducheyne, P., Bianco, P., Radin, S., and Schepers, E. (1992). Bioactive materials: Mechanisms and bioengineering considerations. *In* "Bone Bonding" (P. Ducheyne, T. Kokubo, and C. A. van Blitterswijk, eds.), pp. 1–12. Reed Healthcare Communications.

20. Hench, L. L., Splinter, R. J., Allen, W. C., and Greenlee, T. K. (1971). Bonding mechanisms at the interface of ceramic prosthetic materials. *J. Biomed. Mater. Res.* **2**(part 1), 117–141.

21. Schepers, E., de Clercq, M., Ducheyne, P., and Kempeneers, R. (1991). Bioactive glass particulate material as a filler for bone lesions. *J. Oral Rehab.* **18**(5), 439–452.

22. Gatti, A. M., Valdre, G., and Andersson, O. H. (1994). Analysis of the in vivo reactions of a bioactive glass in soft and hard tissue. *Biomaterials* **15**, 208–212.

23. Fujita, H., Ido, K., Matsuda, Y., Iida, H., Oka, M., Kitamura, Y., and Nakamura, T. (2000). Evaluation of bioactive bone cement in canine total hip arthroplasty. *J. Biomed. Mater. Res.* **49**, 273–288.

24. Marcolongo, M., Ducheyne, P., Garino, J., and Schepers, E. (1998). Bioactive glass fiber/polymeric composites bond to bone tissue. *J. Biomed. Mater. Res.* **39**, 161–170.

25. Li, P., and Ducheyne, P. (1998). Quasi-biological apatite film induced by titanium in a simulated body fluid. *J. Biomed. Mater. Res.* **41**(3), 341–348.

26. Yan, W. Q., Nakamura, T., Kawanabe, K., Nishigochi, S., Oka, M., and Kokubo, T. (1997). Apatite layer-coated titanium for use as bone bonding implants. *Biomaterials* **18**(17), 1185–1190.

27. Miyaji, F., Kim, H. M., Handa, S., Kokubo, T., and Nakamura, T. (1999). Bonelike apatite coating on organic polymers: Novel nucleation process using sodium silicate solution. *Biomaterials* **20**(10), 913–919.

28. Oyane, A., Minoda, M., Miyamoto, T., Takahashi, R., Nakanishi, K., Kim, H. M., Kokubo, T., and Nakamura, T. (1999). Apatite formation on ethylene–vinyl alcohol copolymer modified with silanol groups. *J. Biomed. Mater. Res.* **47**(3), 367–373.

29. Murphy, W. L., Kohn, D. H., and Mooney, D. J. (2000). Growth of continuous bonelike mineral within porous poly(lactide-co-glycolide) scaffolds in vitro. *J. Biomed. Mater. Res.* **50**, 50–58.

CHAPTER 62

PROCESSING OF POLYMER SCAFFOLDS: PHASE SEPARATION

Ruiyun Zhang and Peter X. Ma

This chapter covers new techniques for preparing synthetic biodegradable polymer scaffolds from polymer solutions by phase separation. Several protocols for fabricating highly porous scaffolds related to different phase separation processes are included. The crystallization of the solvent in the polymer solution induces solid–liquid phase separation. The foam obtained by this solid–liquid phase separation process has a highly anisotropic tubular morphology with an internal ladderlike structure. The foam with isotropic continuous pore network is prepared by thermally induced liquid–liquid phase separation. A synthetic fibrous matrix with a fiber diameter on the nanometer scale is prepared by a thermally induced gelation process. Nanofibrous matrices with macroporous architecture are prepared by combining a porogen leaching technique and a thermally induced gelation process. Composite foams of biodegradable polymers and bone mineral–like apatites are prepared by means of a solid–liquid phase separation process and a biomimetic process.

INTRODUCTION

Tissue engineering offers a new promising approach to the creation of biolological alternatives for implants and prostheses [1]. In this approach, a highly porous scaffold is necessary to accommodate cells and to guite their growth and tissue regeneration in three dimensions. Synthetic biodegradable polymers such as poly(L-lactic acid) (PLLA), poly(glycolic acid) (PGA), and poly(D,L-lactic acid-co-glycolic acid) (PLGA) have been widely used as scaffolds for cell transplantation and tissue engineering. Several methods have been reported for the preparation of highly porous scaffolds from these synthetic biodegradable polymers. Particulate leaching is a well-documented technique for fabricating porous foams for tissue engineering [2,3]. Textile technologies are also widely used to fabricate biodegradable woven and nonwoven fabrics for tissue engineering [4,5].

Emulsion freeze-drying [6], gas foaming [7], and three-dimensional printing [8] have also been exploited to fabricate scaffolds. Now a new technique for preraring highly porous biodegradable polymer scaffolds—thermally induced phase separation of polymer solutions and subsequent removal—has generated considerable interest [9–14].

The controlled phase separation process has been used for years in the preparation of porous polymer membranes. Phase separation of the polymer solution can be induced in several ways, including non-solvent-induced phase separation, chemically induced phase separation, and thermally induced phase separation (TIPS) [15]. In the TIPS process, a relative new technique for preparing porous membranes [16], the temperature of a polymer solution is decreased to induce phase separation, that is, to form one phase having a high polymer concentration (polymer-rich phase) and a second phase that is low in polymer concentration (polymer-lean phase). After the solvent has been removed by extraction, evaporation, or sublimation, the polymer in the polymer-rich phase solidifies into the skeleton,

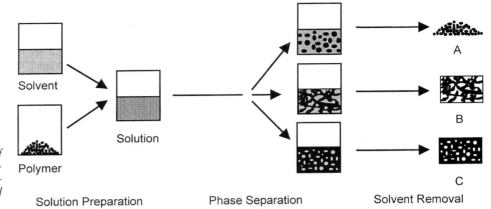

Fig. 62.1. Schematic illustration of scaffold fabrication with phase separation processes: (A) powder, (B) scaffolds with continuous network, and (C) foam with closed pores.

and the spaces originally occupied by the solvent in the polymer-lean phase become pores of the polymer foam. The pore morphology of the porous membrane varies depending on the polymer, solvent, concentration of the polymer solution, and the phase separation temperature. The membranes obtained from this process usually have pore diameters of several micrometers, which is usually not suitable as a scaffold for tissue engineering. A scaffold must have a pore size big enough for cell seeding, a surface area high enough for cell adhesion, and diffusivity sufficient for nutrients and metabolites to permeate. In this chapter we will focus on the development of the thermally induced phase separation techniques for the fabrication of scaffolds with controlled pore morphologies and properties for cell transplantation and tissue engineering applications (Fig. 62.1). The protocols developed in our laboratory will be used as examples to address these issues.

MATERIALS
POLYMERS
Poly(L-lactic acid) (PLLA), poly(D,L-lactic acid-*co*-glycolic acid) (85/15) (PLGA85), and poly(D,L-lactic acid-*co*-glycolic acid) (50:50) (PLGA50) with inherent viscosity of approximately 1.6, 1.4, and 0.5, are from Boehringer Ingelheim (Ingelheim, Germany)

Poly(D,L-lactic acid-*co*-glycolic acid) (75:25) (PLGA75) with inherent viscosity of 0.5–0.65 is from Medisorb Technologies International (Cincinnati, OH)

Poly(DL-lactide) (PDLLA) with a molecular weight of 103,000 is from Sigma Chemical Co. (St. Louis, MO)

SOLVENTS
Dioxane, tetrahydrofuran (THF), *N,N*-dimethylformamide (DMF), pyridine, methanol, and benzene are from Aldrich Chemical Co. (Milwaukee, WI)

Hydroxyapatite [$Ca_{10}(PO_4)_6(OH)_2$] (HAP), and all salts for preparation of simulated body fluid (SBF) are from Aldrich.

SOLID–LIQUID PHASE SEPARATION
Thermally induced phase separation, based on the thermodynamic and kinetic behavior of the polymer solution with the variation of the solution temperature, is a complicated process [16,17]. The solvent crystallization can induce phase separation of the polymer solution when the temperature decreases. We define this phase separation process as a solid–liquid phase separation process. In this case, the crystallization temperature (freezing point) of the solvent is higher than the liquid–liquid phase separation temperature of the polymer solution. When the temperature of the solution decreases, the solvent crystallizes and the polymer is expelled from the solvent crystallization front; that is, the polymer solution undergoes a solid–liquid phase separation. The morphology of the solvent crystals changes with the solvent used, the polymer concentration, the crystallization temperature, and the temperature gradient applied to the polymer solution. Foams with a variety of pore morphologies as negative replicas of the solvent crystals can be obtained.

PREPARATION OF POLYMER MATRICES

Foams of PLLA and PLGA are prepared with solid–liquid phase separation by decreasing the temperature of the polymer solutions to induce solvent crystallization and the subsequent sublimation of the solvent [11,12]. Typically, the foam is prepared with the following steps. First the chosen polymer is dissolved in a solvent (typically dioxane), under magnetical stirring, at 50°C for 2 h. Then 2 ml of the polymer–dioxane solution is added to a Teflon vial (5 ml, prewarmed to 50°C). The vial containing the solution is quickly transferred to a refrigerator or freezer at a preset temperature to solidify the solvent and induce solid–liquid phase separation. The solidified mixture is kept cold for 2 h and then transferred into a freeze-drying vessel at a temperature between −5°C and −10°C in an ice–salt bath. The samples are then freeze-dried at 0.5 mmHg for at least a week to ensure the complete removal of the solvent.

The foams prepared from solid–liquid phase separation of the polymer solution have highly anisotropic tubular morphology with an internal ladderlike structure (Fig. 62.2) [11]. The channels are parallel to the direction of solidification (heat transfer direction), and have repeating partitions with nearly uniform spacing perpendicular to the solidification direction. The diameter of the channels and the spacing between repeating partitions in the channel change with the cooling rate, the polymer concentration, and the solvent used. For a selected polymer–solvent system, the average pore size increases with increasing quenching temperature.

The pore structure from this process is attributed to the crystallization of the solvent. When the temperature of the polymer solution is lower than the freezing point (crystallization temperature) of the solvent (~12°C), the solvent crystallizes and the polymer phase is expelled from the crystallization front as "impurities." A continuous polymer-rich phase is formed by aggregation of polymer expelled from every single solvent crystal. After solvent crystals have been sublimated, a foam is formed with pore morphology as the fingerprint of the solvent crystals. The temperature gradient along the solidification direction (from sample surface to sample center) may have led to the highly anisotropic pore structure.

PREPARATION OF POLYMER–HAP COMPOSITE MATRICES

Composite foams of polymer and hydroxyapatite can be prepared with the solid–liquid phase separation of a polymer–HAP–solvent mixture and subsequent sublimation of the solvent [11]. The polymer–HAP–solvent mixture is obtained by adding HAP powder to the prepared polymer solution. The mixture is stirred at room temperature overnight to obtain a homogeneous mixture. The mixture is quenched to induce phase separation, and the solvent is then sublimated under vacuum with the same process as described for polymer matrices. In this procedure, 2 ml of PLLA–HAP–dioxane mixture is used instead of polymer solution in a Teflon vial to obtain the composite foam. The final composition of the PLLA–

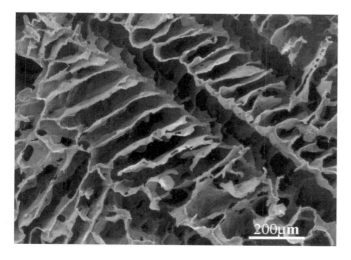

Fig. 62.2. SEM micrograph of a PLLA scaffold prepared with a thermally induced solid–liquid phase separation process.
From Zhang and Ma [11], p. 449. © 1999 by John Wiley & Sons.
Reprinted by permission of John Wiley & Sons.

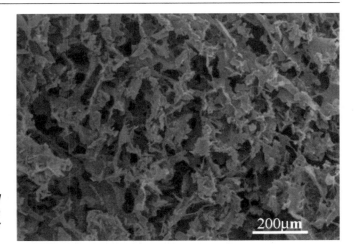

Fig. 62.3. SEM micrograph of a PLLA-HAP composite scaffold prepared with a thermally induced solid–liquid phase separation process. From Zhang and Ma [11], p. 449. © 1999 by John Wiley & Sons. Reprinted by permission of John Wiley & Sons.

HAP composite foam is determined by the concentration of the polymer solution and HAP content in the mixture.

The solid–liquid phase separation technique yields a cocontinuous structure of interconnected irregular pores in the polymer–HAP foam that is very different from that in pure polymer foam (Fig. 62.3) [11]. The irregular pores range in size from several micrometer to about 300 µm. The walls of the pores are composed of both polymer and HAP. The HAP particles added in the polymer solution perturb the crystallization of the solvent, change the solvent crystallization front by impeding the crystal growth, and result in the formation of more irregular solvent crystals. On the other hand, both HAP particles and the polymer are expelled from the crystallization front and form a phase that is rich in polymer–HAP. The HAP particles are randomly distributed in the polymer matrix. After the sublimation of the solvent, this phase that is rich in polymer–HAP forms a continuous skeleton of the polymer–HAP foam, and the spaces originally taken by the solvent crystals become pores of the foam. As a result of irregular solvent crystal growth, the irregular pores become the dominant morphology of the polymer–HAP composite foam.

With the solid–liquid phase separation procedure, the microstructure of the composite foam can be controlled by varying the polymer concentration, the HAP content, the quenching temperature, the polymer, and the solvent utilized. The pore size increases with increasing quenching temperature. Both pore size and porosity increase with decreasing polymer concentration and HAP content. The mechanical properties of the composite foams are significantly improved over those of pure polymer foams [11]. Composites containing HAP, a material with good osteoconductive properties, are promising scaffolds for bone tissue engineering [18].

PREPARATION OF POLYMER–APATITE COMPOSITE MATRICES BY A BIOMIMETIC PROCESS

A biomimetic process is applied to form apatite in polymer foams obtained from solid–liquid phase separation. The polymer foams are prepared by solid–liquid phase separation as described in connection with polymer matrices. Simulated body fluid (SBF) is prepared by dissolving reagent grade chemicals (NaCl, $NaHCO_3$, KCl, $K_2HPO_4 \cdot 3H_2O$, $MgCl_2 \cdot 6H_2O$, $CaCl_2$, and Na_2SO_4) in deionized water. The inorganic ion concentrations are 1.5 times those of human blood plasma. The fluid is buffered at a pH value of 7.4 at 37°C with tris(hydroxymethyl) aminomethane [$(CH_2OH)_3CNH_2$] and hydrochloric acid (HCl). Five rectangular polymer foam specimens with dimensions of 12 mm × 8 mm × 6 mm are immersed into 100 ml of SBF in a glass bottle maintained at 37°C. The SBF is renewed every other day. After incubation for various periods of time, the specimens are removed from the fluid, immersed overnight in 100 ml of deionized water to remove the soluble inorganic ions, and dried at room temperature [12].

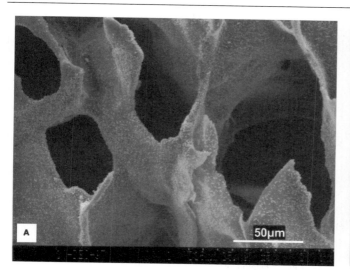

Fig. 62.4. SEM micrographs of a PLLA–apatite scaffold prepared by incubating a PLLA foam in a simulated body fluid (SBF) for 30 days.

The composite foam obtained have a large number of apatite microparticles growing on the surfaces of pore walls (Fig. 62.4). The apatite particles are similar to the apatite in natural bone in composition and structure, as demonstrated with scanning electron microscopy (SEM), energy-dispersive spectroscopy (EDS), X-ray diffraction (XRD), and Fourier transform IR (FTIR) analyses [12]. The particle number and size are controlled by several factors, such as incubation time, ionic concentration of the SBF, polymer surface area, and surface modification. The average particle diameter, density (number of particles per unit surface area), and total apatite mass increase with incubation time. The mechanical properties of this novel composite matrix are significantly improved over the pure polymer matrix, and they increase with the incubation time [12]. Because the formed apatite particles on the pore walls mimic the bone mineral, an improved osteoconductivity is expected.

LIQUID–LIQUID PHASE SEPARATION

When the crystallization temperature of the solvent is much lower than the phase separation temperature of an amorphous polymer solution, the thermally induced liquid–liquid phase separation takes place by decreasing temperature of the polymer solution with an upper critical solution temperature. Figure 62.5 shows a typical schematic equilibrium phase diagram of the amorphous polymer solution system. The spinodal curve divides the liquid–liquid phase separation region into a thermodynamically metastable region (between the binodal and the spinodal) and a thermodynamically unstable region (area enclosed by the spinodal). At high temperatures the solution is homogeneous. When a homogeneous polymer solution is quenched to a temperature–composition point below the binodal decomposition temperature, there is a force driving the system to separate into a polymer-rich phase and a polymer-lean phase by one of two phase separation mechanisms: nucleation and growth, or spinodal decomposition. In a polymer solution of very low concentration, when the temperature–composition point is located in the metastable region, the possible structure formation consists of droplets of the polymer-rich phase dispersed in the matrix of the polymer-lean phase. In this case, the powderlike polymer solid is obtained after the solvent has been removed (Fig. 62.1). When the composition–temperature point is located in the unstable region, a bicontinuous structure, in which both the polymer-rich phase and polymer-lean phases are completely interconnected, will form as a result of the spinodal decomposition. The foams with a continuous pore network structure are obtained from this phase-separated system after removal of the solvent (Fig. 62.1). When the composition–temperature point is located in the metastable region of a solution with very high polymer concentration, droplets of the polymer-lean phase are formed in the matrix of the polymer-rich phase. Foam with a closed-pore structure is obtained (Fig. 62.1).

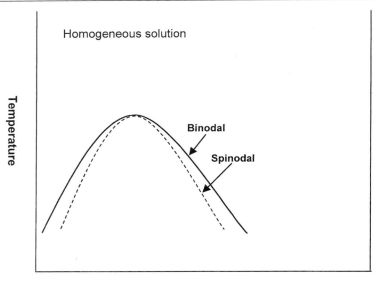

Fig. 62.5. Schematic equilibrium temperature–composition phase diagram for a polymer solution system.

In a solution of a semicrystalline polymer, the phase separation is more complicated because of the crystallization of the polymer. If the temperature of the polymer solution is low enough, the solution will experience driving forces both for liquid–liquid phase separation and for polymer crystallization. In this situation, the relative kinetics of liquid–liquid phase separation and polymer crystallization will determine the final structure of the system. Generally, liquid–liquid phase separation is faster than polymer crystallization. The solution undergoes liquid–liquid phase separation first and then polymer crystallization takes place in the polymer-rich phase, and the morphology of the phase-separated solution depends mainly on the liquid–liquid phase separation. In some cases, the polymer solution gels during the cooling process [19]. Gelation is the process in which the whole polymer solution solidifies into a gel, a network of physically cross-linked polymer chains with solvent trapped in the network. In the semicrystalline polymer solution, the interlocking of the small crystal agglomerates may play a key role for the formation of the gel. The highly porous structure is obtained by removing the solvent from the gel.

When the polymer crystallization temperature is higher than the spinodal temperature, the crystallization of the polymer in the solution takes place before the spinodal phase separation during the cooling process. If the polymer solution is held long enough at a temperature higher than spinodal temperature, the solution will experience a crystallization-induced phase separation, another type of solid-liquid phase separation. The polymer-rich phase forms by nucleation and growth of the polymer crystals. The polymer platelets formed are suspended in the matrix of the polymer-lean phase or precipitated from the solution. When the polymer concentration is high enough, gelation can also take place in this phase separation process, whereupon it can be used to fabricate polymer foams.

PREPARATION OF CONTINUOUS NETWORK POLYMER MATRICES

Thermally induced liquid–liquid phase separation of the polymer solution can be realized by choosing a solvent with a freezing point lower than the phase separation temperature of the polymer solution. For poly(α-hydroxyl acids), a mixture of dioxane and water is used to attain liquid–liquid phase separation [13,14]. A clear polymer solution (previously warmed to 60°C) is quenched to a preset temperature to induce liquid–liquid phase separation. The frozen solution is then transferred into a freeze-drying vessel at a temperature between –5°C and –10°C in an ice–salt bath. After freeze-drying at 0.5 mmHg for a week, the solvent is completely removed, leaving a porous foam.

Depending on where in the temperature–composition phase diagram of the polymer solution the phase separation occurs, either a powdery material or a foam with an isotropic continuous pore network (Fig. 62.6) can be obtained by means of liquid–liquid phase sep-

Fig. 62.6. SEM micrograph of a PLGA scaffold prepared with a thermally induced liquid–liquid phase separation. From Ma and Zhang [13], p. 69. © 1999 by John Wiley & Sons. Reprinted by permission of John Wiley & Sons.

aration. The final structure of the polymer matrix from a liquid–liquid phase separation is dependent on the concentration of the polymer solution, the quenching temperature, and the molecular weight of the polymer. Generally, a powderlike structure is attained from a polymer solution with a very low concentration, while foam with an isotropic pore structure is obtained from a polymer solution that is relatively higher in concentration. At the same concentration, the polymer with higher molecular weight is favorable to the formation of a uniform isotropic pore structure. In a certain polymer concentration range, different quenching temperatures result in foams with different pore structures. The lower quenching temperature (higher cooling rate) generally results in an interconnected pore network with a uniform pore size. A higher quenching temperature (lower cooling rate) results in larger pores with a wider pore size distribution.

The pore size of the foam obtained from liquid–liquid phase separation usually ranges from several micrometers to several tens of micrometers. The coarsening of the separated phase during the late stage of phase separation is another factor that affects system phase morphology [20]. When the sample is held at the phase separation temperature, the droplets tend to increase in the average size, with a corresponding decrease in the number of droplets per unit volume. The driving force for such an increase in droplet average size is the tendency of the system to minimize its interfacial free energy by minimizing the interfacial area between the polymer-rich and polymer-lean phases. This phenomenon can be used to control the pore size of foams prepared from liquid–liquid phase separation, especially for porous structures of polymer scaffolding for tissue engineering, in which a proper pore size is necessary for cell seeding.

PREPARATION OF POLYMER MATRICES WITH NANOFIBROUS NETWORK

To prepare highly porous PLLA nanofibrous matrices from thermally induced gelation of PLLA solution, a suitable solvent system must be chosen. Typically, the processing steps are as follows.

1. A 2-ml PLLA solution (previously warmed to 50°C) is added in a Teflon vial. The vial containing PLLA solution is immediately placed in a refrigerator or freezer at a chosen temperature to gel. Several solvent systems (e.g., THF, DMF, pyridine, THF–methanol, dioxane–methanol, dioxane–H_2O, dioxane–acetone) have been used to obtain polymer gels [13]. The gelation time depends on the temperature, the solvent, and the PLLA concentration of the solution [13]. After gelation, the gel is kept at the gelling temperature for another 2 h before the next step.

2. The vial containing the gel is immersed in distilled water for solvent exchange. The water is changed three times a day for 2 days.

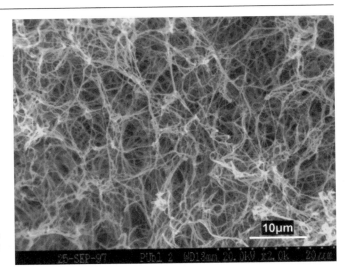

Fig. 62.7. SEM micrograph of a PLLA matrix with nanofibrous network prepared with a thermally induced gelation process (gelled at −20°C).

3. The gel is removed from the water, blotted with a piece of filter paper, and placed in a freezer at −20°C for 2 h.

4. The frozen gel is placed in a freeze-drying vessel at about −5 to −10°C in an ice–salt bath, then freeze-dried under vacuum (<0.5 mmHg) for a week.

With the gelation procedure, a three-dimensional nanofibrous network is obtained from PLLA solution (Fig. 62.7). The fibrous network has fiber diameters ranging from 50 to 500 nm. The fiber network formation depends on the gelling temperature and the solvent of the PLLA solution. In general, at a low gelation temperature, the nanofibrous structure is formed. The gelation temperature for fibrous structure also depends on the solvent used [13]. For a PLLA-THF solution, the gelation temperature for nanofibrous structure formation must be lower than about 15°C. The porosity of the nanofibrous matrix decreases with increasing polymer concentration, while the mechanical properties (Young's modulus and tensile strength) increase with the polymer concentration.

The gelation procedure offers several advantages in the fabrication of fibrous matrices. There are almost no equipment requirements, unlike the case of fabricating nonwoven fabrics with textile technology. The procedure is also much simpler. The diameter of the fiber, which is on the nanometer scale similar to natural extracellullar matrices, is very difficult if not impossible to achieve with textile technology. The process can be used to directly fabricate a scaffold into the anatomical shape of a body part with a mold. The surface-to-volume ratio is much higher than those of fibrous nonwoven fabrics fabricated with the textile technology or foams fabricated with other techniques. A higher surface area is believed to enhance the cell attachment. For many cell types, cell migration, growth, and differentiated function are all dependent on the cell attachment. Therefore, the nanofibrous synthetic extracellullar matrices might provide a better environment for cell attachment, proliferation, and function.

PREPARATION OF POLYMER MATRICES WITH PLATELETLIKE STRUCTURE

Highly porous PLLA matrices having a plateletlike structure are prepared from PLLA solution by thermally induced gelation at relatively high temperatures. The solvents used in preparation of nanofibrous matrices can also, be used to fabricate matrices with a plateletlike structure. The procedure for preparing foams with plateletlike structure is the same as described in connection with nanofibrous networks. However, the gelation temperature should be higher and dependent on the solvent used. For a PLLA-THF solution, foams with plateletlike structure form at a gelation temperature higher than 20°C (Fig. 62.8). The nucleation and growth of the PLLA crystals at a temperature higher than that of the spin-

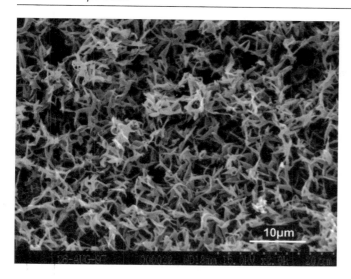

Fig. 62.8. SEM micrograph of a PLLA scaffold with plateletlike structure prepared with a thermally induced gelation process (gelled at 23°C).

odal phase separation temperature may be the cause of the formation of the plateletlike structure.

PREPARATION OF NANOFIBROUS MATRICES WITH MACROPORE STRUCTURE

A three-dimensional fibrous matrix prepared from the gelation of a PLLA solution has an average pore size (spacing between fibers) of several micrometers. However, scaffolds for tissue engineering must have pores of at least several tens of micrometers, to accommodate cell seeding. To obtain scaffolds with both nanofibrous structure and pore size large enough for cell seeding, a combination of gelation and porogen leaching technique is used.

Sugar particles are used as a porogen to obtain scaffolds with both nanofibrous structure and macropore architecture. The prewarmed polymer solution is dripped slowly onto sugar particles in a mold and then cooled to a preset temperature to gel. The gel–sugar composite is then immersed in distilled water to simultaneously extract the solvent and leach the sugar from the composite. The gel is removed from the water after 4 days, blotted with a piece of filter paper, and placed in a freezer at –20°C for 2 h. The frozen gel is freeze-dried at –5 to –10°C in an ice–salt bath under vacuum (<0.5 mmHg) for a week.

By combining the gelation and porogen leaching techniques, a three-dimensional nanofibrous matrix with macropore architecture is prepared from a PLLA–sugar composite (Fig. 62.9). Unlike the pore wall of foams prepared with the traditional salt-leaching tech-

Fig. 62.9. SEM micrograph of a PLLA scaffold prepared by combining a sugar-leaching technique and a thermally induced gelation process.

nique, the pore wall of the new matrices is a nanofibrous network. As a scaffold for tissue engineering, the fibrous microstructure may enhance the attachment of the cells, and the macropore architecture may give a pore structure that is desirable for cell spatial distribution.

REFERENCES

1. Langer, R. S., and Vacanti, J. P. (1999). Tissue engineering: The challenges ahead. *Sci. Am.* **280**, 86–89.
2. Mikos, A. G., Thorsen, A. J., Czerwonka, L. A., Bao, Y., Langer, R., Winslow, D. N., and Vacanti, J. P. (1994). Preparation and characterization of poly(L-lactic acid) foams. *Polymer* **35**, 1068–1077.
3. Ma, P. X., and Langers, R. A. (1999). Fabrication of biodegradable polymer foams for cell transplantation and tissue engineering. *In* "Tissue Engineering Methods and Protocols" (J. Morgan and M. Yarmush, eds.), pp. 47–56. Humana Press, Totowa, NJ.
4. Ma, P. X., and Langers, R. (1995). Degradation, structure and properties of fibrous nonwoven poly(glycolic acid) scaffolds for tissue engineering. *In* "Polymers in Medicine and Pharmacy" (A. G. Mikos *et al.*, eds.), pp. 99–104. MRS, Pittsburgh, PA.
5. Ma, P. X., Shin'oka, T., Zhou, T., Shum-Tim, D., Lien, J., Vacanti, J. P., Mayer, J., and Langer, R. (1997). Biodegradable woven/nonwoven composite scaffolds for pulmonary artery engineering in a juvenile lamb model. *Trans. Soc. Biomater.* **20**, 295.
6. Whang, K., Thomas, C. H., and Healy, K. E. (1995). A novel method to fabricate bioabsorbable scaffolds. *Polymer* **36**, 837–842.
7. Mooney, D. J., Baldwin, D. F., Suh, N. P., Vacanti, J. P., and Langer, R. (1996). Novel approach to fabricate porous sponges of poly(D,L-lactic-*co*-glycolic acid) without the use of organic solvents. *Biomaterials* **17**, 1417–1422.
8. Park, A., Wu, B., and Griffith, L. G. (1998). Integration of surface modification and 3D fabrication techniques to prepare patterned poly(L-lactide) substrates allowing regionally selective cell adhesion. *J. Biomater. Sci. Polymer Edn* **9**, 89–110.
9. Lo, H., Ponticiello, M. S., and Leong, K. W. (1995). Fabrication of controlled release biodegradable foams by phase separation. *Tissue Eng.* **1**, 15–28.
10. Schugens, C., Maquet, V., Grandfils, C., Jerome, R., and Teyssie, P. (1996). Polylactide macroporous biodegradable implants for cell transplantation. II. Preparation of polylactide foams by liquid–liquid phase separation. *J. Biomed. Mater. Res.* **30**, 449–461.
11. Zhang, R., and Ma, P. X. (1999). Poly(alpha-hydroxy acids)/hydroxyapatite porous composites for bone tissue engineering: 1. Preparation and morphology. *J. Biomed. Mater. Res.* **44**, 446–455.
12. Zhang, R., and Ma, P. X. (1999). Porous poly(L-lactic acid)/apatite composites created by biomimetic process. *J. Biomed. Mater. Res.* **45**, 285–293.
13. Ma, P. X., and Zhang, R. (1999). Synthetic nano-scale fibrous extracellular matrix. *J. Biomed. Mater. Res.* **46**, 60–72.
14. Nam, Y. S., and Park, T. G. (1999). Porous biodegradable polymeric scaffolds prepared by thermally induced phase separation. *J. Biomed. Mater. Res.* **47**, 8–17.
15. Young, A. T. (1986). Microcellular foams via phase-separation. *J. Vac. Sci. Technol.* **4**, 1128–1133.
16. vande Witte, P., Dijkstra, P. J., vanden Berg, J. W. A., and Feijen, J. (1996). Phase separation processes in polymer solutions in relation to membrane formation. *J. Membr. Sci.* **117**, 1–31.
17. Bansil, R., and Liao, G. D. (1997). Kinetics of spinodal decomposition in homopolymer solutions and gels. *Trends Polym. Sci.* **5**, 146–154.
18. Ma, P. X., Zhang, R., Xiao, G., and Franceschi, R. (2001). Engineering new bone tissue in vitro on highly porous poly(alpha-hydroxyl acids)/hydroxyapatite composite scaffolds. *J. Biomed. Mater. Res.* **54**, 284–293.
19. Guenet, J.-M. (1992). "Thermoreversible Gelation of Polymers and Biopolymers." Academic Press, London and San Diego, CA.
20. Aubert, J. H. (1990). Structural coarsening of demixed polymer-solutions. *Macromolecules* **23**, 1446–1452.

CHAPTER 63

PROCESSING OF POLYMER SCAFFOLDS: POLYMERIZATION

Paul D. Dalton, Sarojini Vijayasekaran, and Molly S. Shoichet

INTRODUCTION

Scaffolds formed as a consequence of polymerization have been used for tissue engineering and offer advantages over other scaffolding techniques because of the simplicity of the process. While several polymers can be polymerized *in situ* [1–3], few result in a cell-invasive scaffold that is also porous. For example, poly(ethylene glycol)-*multi*-acrylate and poly(2-hydroxyethyl methacrylate) (PHEMA) can be cross-linked or polymerized *in situ*; however, the resulting structure is a gel and not a scaffold with interconnecting pores large enough for penetration by cells. By manipulating the polymerization conditions, cell-invasive scaffolds can be formed from PHEMA and poly[N-(2-hydroxypropyl) methacrylamide] (PHPMA). Briefly, a monomer mixture is polymerized within a mold in the presence of a solvent in which the monomer is soluble and the polymer insoluble. The solubility transition during polymerization results in a two-phase, bicontinuous structure of polymer and solvent (Fig. 63.1). The scaffold is therefore created as a consequence of polymerization, and leaching of porogens is unnecessary to generate interconnecting pores. For cell-invasive scaffolds of PHEMA, often referred to as PHEMA sponges, the excess solvent is typically water.

PHEMA sponges were first prepared in the 1960s, with their first clinical uses being breast augmentation [4,5] and nasal cartilage replacement [6]. The PHEMA sponges withstood autoclaving, were easy to form into different shapes, and with their "spongy" physical properties, felt like soft tissue to the surgeon. Some investigators hailed PHEMA sponges as a soft-tissue replacement with a multitude of potential applications [5,7]. Long-term *in vivo* calcification, however, hindered the development of PHEMA sponges [8–10] until the early 1990s, when Chirila used PHEMA sponges for an artificial cornea project [11]. The PHEMA sponge has since been developed as a porous skirt to an artificial cornea [12–18] that has had some early clinical success [19]. PHEMA scaffolds are cell invasive, thereby serving as the anchor between corneal tissue and the transparent, noninvasive central core. PHEMA sponges have also been incorporated into an orbital implant, with muscle penetrating into the PHEMA sponge component [20,21].

PHEMA sponges have demonstrated cell invasion both *in vitro* and *in vivo*, as is mandatory for tissue engineering applications. The scaffold forms as a consequence of polymerization, allowing manipulation of the synthetic construct during synthesis. This chapter describes the methodology of manufacturing reproducible PHEMA sponges in the research laboratory.

REAGENTS

The monomer and initiating systems for PHEMA sponges are inexpensive and are off-the-shelf chemicals. The chemicals listed in Tables 63.1 and 63.2 can be purchased from

Fig. 63.1. A PHEMA scaffold that is easily pene-trated by cells.

Aldrich (Milwaukee, WI). Water is usually distilled and deionized using Millipore Milli-RO 10 Plus and Milli-Q UF Plus (Bedford, MA) at 18-MΩ resistance.

METHODS

The key steps in the synthesis of PHEMA sponges for biomedical purposes are as follows:

> Mold preparation
> Formulation preparation
> Dispensing/injecting formulation into molds
> Polymerization and formation of scaffold
> Soxhlet extraction of scaffold
> Sterilization
> Cellular penetration of scaffolds
> Histological processing

MOLD PREPARATION

The molds should be ready before the polymerization formulation is prepared. Molds should be either disposable (i.e., glass or polypropylene) or thoroughly cleanable with alcohol (i.e., Teflon). Since PHEMA adheres to polystyrene, steel, and Tygon tubing, it is difficult to maintain the quality of the surfaces when these materials are used as the mold.

FORMULATION PREPARATION

The monomer mixture is straightforward to prepare: simply add all the liquids listed in Table 63.1, *except for the accelerating agent* (N,N,N',N'-tetramethylethylenediamine,

---

Table 63.1. Typical Formulation Used to Prepare a Cell-Invasive Scaffold

---

1 g (20 wt%) HEMA (monomer)
4 g (80 wt%) H_2O (solvent/nonsolvent)
0.25 wt% (2.4 μl)[a,b] EDMA (cross-linking agent)
0.5 wt% (46 μl)[b] APS 10% aqueous solution, made new each half-day (initiator)
0.5 wt% (6.4 μl)[b] TEMED (accelerating agent)

---

[a]Initiator and cross-linking and accelerating agents are calculated as weight percent of the monomer.
[b]Do not add if HEMA has been spiked with EDMA.

Table 63.2. Useful Information for Possible Components of the Monomer Mixture

Name	Chemical structure	Density (g/ml)	Molecular weight (Da)	CAS no.
2-Hydroxyethyl methacrylate (HEMA)		1.08	130.1	868-77-9
Methyl methacrylate (MMA)		0.94	100.1	80-62-6
Methacrylic acid (MA)		1.02	86.1	79-41-4
Poly(ethylene glycol) methacrylate (PEGMA)		1.105 ($n = 6, 7$)	Variable	
Ethylene dimethacrylate (EDMA)		1.05	198.2	97-90-5
N,N,N',N'-Tetramethylethylene diamine (TEMED/TMDA)		0.78	116.1	110-18-9
Hydroquinone monomethyl ether (MEHQ)		1.55	124.1	150-76-5
Ammonium persulfate (APS)		1.98	228.2	7727-54-0
Sodium metabisulfite (SMBS)		1.48	190.1	7681-57-4

TEMED) to a glass vial. Place the vial in an ultrasonic bath for 3 min, taking care that the temperature of the bath remains constant. After the accelerating agent has been added, gently agitate, or roll, the mixture for 30 s. Avoid rigorous movements that will introduce oxygen into the mixture and alter the polymerization kinetics. The formulation can now be injected into the mold.

DISPENSING/INJECTING FORMULATION INTO MOLDS

The formulation can be withdrawn into a disposable syringe with an 18-gauge needle and then dispensed into the appropriate mold, having the desired shape and dimensions of the final product. It is important when withdrawing/dispensing the formulation to avoid introducing bubbles. The formulation should be injected before phase separation (i.e., whitening) occurs.

POLYMERIZATION AND FORMATION OF SCAFFOLD

Since temperature can influence polymerization kinetics, it is imperative to maintain a constant temperature within the molds during polymerization. Time of polymerization is approximately 2–10 h at room temperature, depending on the exact composition of the formulation. Formulations with lower initiator and/or cross-linking agent generally require more time for the scaffold to form. Molds are commonly left overnight, or placed in a 50°C oven for 3 h after the initial 2–10 h to ensure high conversion and minimal residual monomer. Phase separation and gelation are two physical changes that can be observed during polymerization [22]. Phase separation, or whitening, of the formulation can be followed spectroscopically at 550 nm after 5–60 min (depending on the reaction kinetics). It is best to use disposable polystyrene cuvettes for light measurements because the polymer adheres to quartz surfaces. After phase separation, the mixture will remain in the liquid form until the phase-separated particles gel, which can take an additional 30 s to 1 h, depending on kinetics. It is important to note that the scaffold is a result of phase separation, then gelation; if the reverse occurs, a gel (not a scaffold) will result. There is a phase change at ~75% water, when gelation precedes phase separation; these gels are not cell invasive. It is best to retain at least 80% water in the formulation to preserve the cell invasiveness of the scaffold [12]. The pore size ranges from 2 to 80 μm, depending on the formulation.

SOXHLET EXTRACTION OF SCAFFOLD

A study by Cifková *et al.* [23] revealed that the water extracted from PHEMA gels and injected intradermally in rats is mildly irritating. It is important, therefore, to remove all cytotoxic moieties from PHEMA sponges before use with tissue. This is especially important if acrylamide, a potent neurotoxin, is a comonomer. Soxhlet extraction is the most convenient and effective technique for removing residual monomer and any unreacted initiator. Small samples may need to be placed into a thimble, or an embedding cassette, to prevent them from being dragged into the outlet flow of the sample container. If modified PHEMA sponges are temperature sensitive or contain degradable particles, low-temperature extractions can be performed with an additional cooling apparatus attached to the sample container or with the Soxhlet system. When working under vacuum, use a capillary-bleed inlet adapter on a two-neck, round-bottom flask to generate a stream of air bubbles to prevent bumping. Extraction of PHEMA sponges is typically performed overnight. The minimum number of washes necessary to remove cytotoxic components is unknown, but may be as low as one or two 40-ml extractions. Do not use any grease between the joints—clean well with chloroform or acetone if grease is present.

STERILIZATION

PHEMA sponges can be autoclaved, preferably immersed in water inside a container with the lid partially unscrewed to prevent pressure buildup. PHEMA sponges exhibit minimal degradation with exposure to such temperatures. Gamma irradiation (25 kGy) is also possible; however, either cross-linking or chain scission may result in a slight volume change [24].

CELLULAR PENETRATION OF SCAFFOLDS

Standard cell seeding techniques can be used to implant sterile PHEMA sponges in the desired tissue site or to seed them with cells for *in vitro* culture [25–27]. To enhance cellular penetration of PHEMA sponges, the less porous "skin" that surrounds the molded samples can be removed [12], either with a lathe (frozen samples) or with a razor blade (semifrozen scaffolds −78°C, 10 min). Immersion in liquid nitrogen should be avoided, as sponges tend to shatter; but this technique may be useful for scanning electron microscopy (SEM) imaging. (Soxhlet extraction reduces cracking.) Interestingly, PHEMA sponges manufactured in an ultrasonic bath have no skin [28]; however the heat generated in the sonicator may affect polymerization kinetics.

HISTOLOGICAL PROCESSING

General morphology and ultrastructural changes in explanted or cultured PHEMA sponges can be observed by optical and transmission electron microscopy (TEM) on tissue processed in epoxy resin. The excised implants should be stored in 2.5% glutaraldehyde for 24–48 h and postfixed in 1% osmium tetroxide. Samples are dehydrated with graded ethanol solutions and then immersed in propylene oxide. Samples are then infiltrated with epoxy resin for 72 h, with fresh changes of resins daily, followed by embedding. Semithin sections (2 μm) for light microscopy and ultrathin sections (0.1 μm) for TEM are cut on an ultramicrotome with a diamond knife. Stains for light microscopic examination include hematoxylin and eosin, alizarin red, Masson's trichrome, and toluidine blue. Except for toluidine blue staining, sections must be deplasticized by immersion in sodium ethoxide for 2 min and hydrated through graded ethanol solutions.

Inflammatory cell populations in explanted PHEMA sponges can be identified by immunocytochemical localization of macrophages using antibodies to macrophages on frozen and paraffin sections of tissue fixed in 2% paraformaldehyde [29]. Neutrophils can be detected by using enzyme histochemistry on frozen sections fixed in citrate–acetone–formaldehyde, incubated in naphthol AS-D chloroacetate at pH 6.3 for 15 min at 37°C in the dark, and counterstained with hematoxylin. Smooth muscle α-actin expression has been used as a chemical marker for myofibroblast transformation and can be detected by immunocytochemical localization methods in cell culture [30].

Extracellular matrix production in PHEMA explants can be detected by using autoradiography. Collagen produced by cells, and its type, can be detected and identified using L-[2,3-³H] proline (precursor) and immunocytochemical localization of collagen types I–VI with antibodies to human and bovine collagen [31]. This demonstrates that PHEMA sponges allow both cell invasion and deposition of extracellular matrix, forming a stable three-dimensional tissue scaffold.

Cell density and viability can be determined by using a modification of the fluorescent staining technique of Poole *et al.* [32]. This highly sensitive and reliable technique uses two types of fluorescent dyes: 5-chloromethylfluorescein diacetate (CMFDA) and ethidium homodimer 1 to detect live and dead cells, respectively. Samples of the explanted tissue at different end points are incubated with both CMFDA and ethidium homodimer 1 at 37°C for 24 h. Thereafter the specimens are fixed briefly in 2% paraformaldehyde and frozen in optimum cutting temperature (OCT) embedding medium. Frozen sections can be collected on slides coated with aminopropyl triethoxysilane and viewed on a fluorescent microscope, thereby enabling a semiquantitative analysis of dead and live cells. CMFDA penetrates into live cells and produces an intense green color by reacting with active intracellular organelles. Conversely, ethidium homodimer 1 enters dead cells with disrupted membranes, binds to nucleic acids, and produces a bright red fluorescence, thus permitting the analyst to distinguish dead from live cells.

TIPS AND KNOW-HOW

The ratio of monomer to water is the single most important factor in controlling the resultant morphology [13]. Other variables that can change the final morphology include temperature during polymerization, amount of dissolved oxygen in monomer mixture and the concentration and type of chemicals in the monomer mixture. It is important to fix all these parameters when investigating the effect of changing one.

HEMA should be a clear liquid: if any yellowing is observed, the monomer needs to be discarded. HEMA is commonly distilled under vacuum before use, but once the inhibitor, hydroquinone monomethyl ether (MEHQ), is removed, its storage life is short. If high-purity HEMA with less than 50 ppm MEHQ is purchased, and the bottle is replaced regularly, the HEMA can simply be used as received. It is best to choose a particular source of HEMA, maintain a small supply from that distributor, and order a new bottle every 6 months.

There is usually some residual ethylene dimethacrylate (EDMA) in purchased HEMA; for sensitive or degradation studies, EDMA can be removed by extraction in hexane [15]. For general use, EDMA does not need to be removed from HEMA, and for consistency between batches it is recommended to spike the HEMA with a known concentration of EDMA. This prevents variations in the reaction kinetics and overcomes errors associated with delivering small volumes of EDMA to the monomer mixture. The cross-linking agent plays a complex role in scaffold formation by affecting reaction kinetics, the consumption of HEMA, and thus the kinetics of phase separation vs gelation. Unlike PHEMA gels, there is not a strong correlation between the cross-linking agent concentration and mechanical strength of PHEMA scaffolds [20]. A typical EDMA concentration used for scaffolds is 0.1–0.25 wt%.

A fresh solution of ammonium persulfate (APS) should be prepared for each formulation, and the easiest technique for this repetitive process is to accurately weigh out $0.1(\times)$g and top up the vial with water to $1(\times)$g. Do not tare the weighing balance between the addition of APS and water. PHEMA sponges manufactured with low APS concentrations are generally weak and tacky—if a slow reaction is desired, use lower accelerating agent concentrations rather than initiator. In general, keep the APS concentration above 0.3 wt% of the monomer concentration. Higher concentrations may be necessary for lower monomer concentrations; for example, 0.5 wt% is recommended for a 10:90 mixture of HEMA and water. TEMED is used as received and dispensed into the monomer mixture from a micropipette. TEMED is a much more powerful electron donor than other accelerating agents, such as sodium metabisulfite (SMBS) and ascorbic acid, and consequently, the reaction rate is much faster. Settling or stratification of the morphology has been observed when SMBS is the accelerating agent [14].

Nitrogen or helium gas can be bubbled through the monomer mixture for a constant time (~15 min) to remove dissolved oxygen, or the monomer can be used as provided. Do not compare PHEMA sponges that have been bubbled with an inert gas to ones that have not. Removal of dissolved oxygen will affect the rate of reaction and, consequently, the morphology. Using the monomer as provided is acceptable, however consistency is necessary in each preparation.

CONCLUSIONS AND FUTURE DIRECTIONS

Cell-invasive PHEMA scaffolds manufactured by the polymerization of HEMA in excess water are convenient to make, autoclavable, and inexpensive. Moreover, the chemistry outlined in this chapter allows PHEMA scaffolds to be prepared reproducibly. PHEMA scaffolds will not erode in water, degradation products will not released, and the cell–surface interface with the scaffold will be stable. Surface modification, such as the addition of peptides or tethering of bioactive molecules, is possible with this synthetic scaffold, taking advantage of the hydroxyl functionality [33–35].

PHEMA sponges can be manufactured with very low (~0.05 wt%) quantities of crosslinking agent and any EDMA impurities can be removed. If PHEMA scaffolds were modified to degrade, then experiments to assess the swelling of the implant and the cytotoxicity of degradation products would be necessary [36].

Scaffold formation by means of polymerization has advantages over other approaches for scaffold generation. The scaffold is formed during polymerization from polymer-rich and polymer-poor regions, so the shape and morphology can potentially be manipulated by external stimuli, such as light or centrifugal forces [37]. Such control over scaffold formation can result in a multitude of morphologies and geometries that meet the desired tissue-engineered application.

ACKNOWLEDGMENTS

The authors thank Traian Chirila for reviewing this manuscript and Shaily Sanghavi for her contributions. We are grateful for funding from the Whitaker Foundation to MSS.

REFERENCES

1. Stile, R. A., Burghardt, W. R., and Healy, K. E. (1999). Synthesis and characterization of injectable poly(N-isopropylacrylamide)-based hydrogels that support tissue formation in vitro. *Macromolecules* **32**, 7370–7379.
2. Quinn, C. P., Pathak, C. P., Heller, A., and Hubbell, J. A. (1995). Photo-crosslinked-copolymers of 2-hydroxyethyl methacrylate, poly(ethylene glycol) tetra-acrylate and ethylene dimethacrylate for improving biocompatibility of biosensors. *Biomaterials* **16**, 389–396.
3. Elwell, M. J., Ryan, A. J., Gruenbauer, H. J. M., and Van Lieshout, H. C. (1996). In-situ studies of structure development during the reactive processing of model flexible polyurethane foam systems using FT-IR spectroscopy, synchrotron SAXS, and rheology. *Macromolecules* **29**, 2960–2968.
4. Kliment, K., Štol, M., Fahoun, K., and Stockar, B. (1968). Use of spongy Hydron in plastic surgery. *J. Biomed. Mater. Res.* **2**, 237–243.
5. Simpson, B. J. (1969). Hydron: A hydrophilic polymer. *Biomed. Eng.* **4**, 65–68.
6. Voldrich, Z., Tomanek, Z., Vacik, J., and Kopecek, J. (1975). Long term experience with poly(glycol monomethacrylate) gel in plastic operations of the nose. *J. Biomed. Mater. Res.* **9**, 675–685.
7. Calnan, J. S., Pflug, J. J., Chhabra, A. S., and Raghupati, N. (1971). Clinical and experimental studies of polyhydroxyethylmethacrylate gel ("Hydron") for reconstructive surgery. *Br. J. Plast. Surg.* **24**, 113–124.
8. Sprincl, L., Kopecek, J., and Lim, D. (1973). Effect of the structure of poly(glycol monomethacrylate) gel on the calcification of implants. *Calcif. Tissue Res.* **13**, 63–72.
9. Šprincl, L., and Novák, M. (1981). The initial stage of calcification in porous hydrophilic polymers. *J. Biomed. Mater. Res.* **15**, 437–440.
10. Cerveny, J., and Sprincl, L. (1981). The calcification of poly(glycol methacrylate) gels in experimental and clinical practice. *Polym. Med.* **11**, 71–78.
11. Crawford, G. J., Constable, I. J., Chirila, T. V., Vijayasekaran, S., and Thompson, D. E. (1993). Tissue interaction with hydrogel sponges implanted in the rabbit cornea. *Cornea* **12**, 348–357.

12. Chirila, T. V., Constable, I. J., Crawford, G. J., Vijayasekaran, S., Thompson, D. E., Chen, Y.-C., Fletcher, W. A., and Griffin, B. J. (1993). Poly(2-hydroxyethyl methacrylate) sponges as implant materials: *In vivo* and *in vitro* evaluation of cellular invasion. *Biomaterials* **14**, 26–38.

13. Chirila, T. V., Chen, Y.-C., Griffin, B. J., and Constable, I. J. (1993). Hydrophilic sponges based on 2-hydroxyethyl methacrylate. 1. Effect of monomer mixture composition on the pore-size. *Polym. Int.* **32**, 221–232.

14. Clayton, A. B., Chirila, T. V., and Dalton, P. D. (1997). Hydrophilic sponges based on 2-hydroxyethyl methacrylate. III. Effect of incorporating a hydrophilic crosslinking agent on the equilibrium water content and pore structure. *Polym. Int.* **42**, 45–56.

15. Clayton, A. B., Chirila, T. V., and Lou, X. (1997). Hydrophilic sponges based on 2-hydroxyethyl methacrylate. V. Effect of crosslinking agent reactivity on mechanical properties. *Polym. Int.* **44**, 201–207.

16. Crawford, G. J., Chirila, T. V., Vijayasekaran, S., Dalton, P. D., and Constable, I. J. (1996). Preliminary evaluation of a hydrogel core-and-skirt keratoprosthesis in the rabbit cornea. *J. Refract. Surg.* **12**, 525–529.

17. Chirila, T. V. (1997). Artificial cornea with a porous polymeric skirt. *Trends Polym. Sci.* **5**, 346–348.

18. Lou, X., Dalton, P. D., and Chirila, T. V. (2000). Hydrophilic sponges based on 2-hydroxyethyl methacrylate. Part VII: Modulation of sponge characteristics by changes in reactivity and hydrophilicity of crosslinking agents. *J. Mater. Sci.: Mater. Med.* **11**, 319–325.

19. Hicks, C. R., Crawford, G., Chirila, T., Wiffen, S., Vijayasekaran, S., Lou, X., Fitton, J., Maley, M., Clayton, A., Dalton, P., Platten, S., Ziegelaar, B., Hong, Y., Russo, A., and Constable, I. (2000). Development and clinical assessment of an artificial cornea. *Prog. Retinal Eye Res.* **19**, 149–170.

20. Hicks, C. R., Clayton, A. B., Vijayasekaran, S., Crawford, G. J., Chirila, T. V., and Constable, I. J. (1999). Development of a poly(2-hydroxyethyl methacrylate) orbital implant allowing direct muscle attachment and tissue ingrowth. *Ophthalmol. Plast. Reconstr. Surg.* **15**, 326–332.

21. Hicks, C. R., Morris, I. T., Vijayasekaran, S., Fallon, M. J., McAllister, J., Clayton, A. B., Crawford, G. J., Chirila, T. V., and Constable, I. J. (1999). Correlation of histological findings with gadolinium enhanced MRI scans during healing of a PHEMA orbital implant in rabbits. *Br. J. Ophthalmol.* **83**, 616–621.

22. Chirila, T. V., Higgins, B., and Dalton, P. D. (1998). The effect of synthesis conditions on the properties of poly(2-hydroxyethyl methacrylate) sponges. *Cell. Polym.* **17**, 141–162.

23. Cifková, I., Brynda, E., Mandys, V., and Štol, M. (1988). Irritation effects of residual products derived from p(HEMA) gels. II. Compounds extracted from hydrogels. *Biomaterials* **9**, 372–375.

24. Chou, K. F., Han, C. C., and Lee, S. (2000). Water transport in 2-hydroxyethyl methacrylate copolymer irradiated by γ rays in air and related phenomena. *J. Polym. Sci., Part B: Polym. Phys.* **38**, 659–671.

25. Plant, G. W., Chirila, T. V., and Harvey, A. R. (1998). Implantation of collagen IV/poly(2-hydroxyethyl methacrylate) hydrogels containing Schwann cells into the lesioned rat optic tract. *Cell Transplant.* **7**, 381–391.

26. Ziegelaar, B. W., Fitton, J. H., Clayton, A. B., Platten, S. T., Maley, M. A. L., and Chirila, T. V. (1999). The modulation of corneal keratocyte and epithelial cell responses to poly(2-hydroxyethyl methacrylate) hydrogel surfaces: Phosphorylation decreases macrophage collagenase production in vitro. *Biomaterials* **20**, 1979–1988.

27. Holy C. E., Shoichet, M. S., and Davies, J. E. (2000). Engineering 3-D bone tissue in vitro using biodegradable scaffolds: Investigating initial cell-seeding density and culture period. *J. Biomed. Mater. Res.* **51**, 376–382.

28. Dziubla, T., and Lowman, A. (2000). Evaluation and characterization of PHEMA sponges as implantable drug delivery systems. *Trans. 6th World Biomater. Congr.*, Vol. I, p. 160.

29. Vijayasekaran, S., Fitton, J. H., Hicks, C. R., Chirila, T. V., Crawford, G. J., and Constable, I. J. (1998). Cell viability and inflammatory response in hydrogel sponges implanted in the rabbit cornea. *Biomaterials* **19**, 2255–2267.

30. Desmouliere, A., Rubbia-Brant, L., Grau, G., and Gabbiani, G. (1992). Heparin induces α-smooth muscle actin in cultured fibroblasts and in granulation tissue myofibroblasts. *Lab. Invest.* **67**, 716–726.

31. Chirila, T. V., Thompson-Wallis, D. E., Crawford, G. J., and Constable, I. J. (1996). Production of neocollagen in hydrogel sponges implanted in the rabbit cornea. *Graefe's Arch. Clin. Exp. Ophthalmol.* **234**, 193–198.

32. Poole, C. A., Brookes, N. H., Gilbert, R. T., Beaumont, B. W., Crowther, A., Scott, L., and Merrilees, M. J. (1996). Detection of viable and non-viable cells in connective tissue explants using fixable fluoroprobes 5-chloromethylfluorescein diacetate and ethidium homodimer-1. *Connect. Tissue Res.* **33**, 233–241.

33. McAuslan, B. R., and Johnson, G. (1987). Cell responses to biomaterials I: Adhesion and growth of vascular endotheila cells on poly(hydroxyethyl methacrylate) following surface modification by hydrolytic etching. *J. Biomed. Mater. Res.* **21**, 921–935.

34. Bergethon, P. R., Trinkaus-Randall, V., and Franzblau, C. (1989). Modified hydroxyethylmethacrylate hydrogels as a modelling tool for the study of cell–substratum interactions. *J. Cell Sci.* **92**, 111–121.

35. Hern, D. L., and Hubbell, J. A. (1998). Incorporation of adhesion peptides into nonadhesive hydrogels useful for tissue resurfacing. *J. Biomed. Mater. Res.* **39**, 266–276.

36. Bruining, M. J., Blaauwgeers, H. G. T., Kuijer, R., Pels, E., Nuijts, R., and Koole, L. H. (2000). Biodegradable three-dimensional networks of poly(dimethylaminoethyl methacrylate). Synthesis, characterization and in vitro studies of structural degradation and cytotoxicity. *Biomaterials* **21**, 595–604.

37. Dalton, P. D., and Shoichet, M. S. (2001). Creating porous tubes by centrifugal forces for soft tissue applications. *Biomaterials* **22**, 2661–2669.

CHAPTER 64

PROCESSING OF POLYMER SCAFFOLDS: GAS FOAM PROCESSING

Thomas P. Richardson, Martin C. Peters, and David J. Mooney

Tissue engineering holds great promise as a method of providing fully functional organs to counter the growing problem of donor organ shortage. Current approaches to form these tissues *in vitro* typically utilize hybrid devices composed of biodegradable polymer scaffolds and cells from the tissues. Numerous approaches have been developed to form and process polymers for use in tissue engineering, and each distinct process possesses unique features and utility to form scaffolds for tissue engineering applications. Significant advances continue to be made regarding these techniques, and one in particular, gas foaming, carries great promise. Gas foaming is marked by the ability to form highly porous polymeric scaffold foams without the use of high temperatures or organic solvents. By eliminating the requirement of high temperatures and organic solvents, large bioactive molecules, including growth factors, can be incorporated into the polymer while maintaining their biological activity. These polymer scaffolds can serve as sustained delivery vehicles for proteins necessary to induce cellular response (e.g., migration and proliferation) and as substrata for cell adhesion, both important considerations for the development of *in vitro* tissues. Work in our laboratory has focused on utilizing this technique to process copolymers of lactic and glycolic acids, and to encapsulate proteins and plasmid DNA encoding proteins known to alter cellular behavior to aid in tissue engineering. This chapter discusses the theory and procedure of gas foaming, with particular attention to practical aspects of foam sponge processing.

INTRODUCTION

Tissue engineering aims to provide functional organs or partial tissue replacements for patients with organ failure, serious injury, or disease [1]. Researchers have focused on both *in vivo* and *in vitro* approaches for providing functional tissue replacements by utilizing numerous polymer preparations combined with seeding of cells found in the particular tissue (e.g., chondrocytes, osteoblasts, etc.) [1]. In addition, significant progress has been achieved with preparations that induce the recipient's own tissues to respond to the device and provide a tissue that is nearly the functional equivalent of the damaged or absent one.

Current strategies are aimed at developing biodegradable scaffolds into which either cells are seeded directly or tissue-inductive factors (e.g., growth factors) are encapsulated; combinations of both strategies are considered, as well [2,3]. This approach assumes that the polymer provides necessary structural features enabling cell invasion, proliferation, extracellular matrix deposition, and cellular organization, eventually leading to a fully functional organized tissue. Another assumption is that these cellular processes occur at or near the rate of polymer degradation. In recent years, numerous polymeric structures (e.g., films, tubes, sponges, etc.) have been developed by various processes and tested for the ability to provide varying mechanical strengths, porosities, degradation rates, and release of bioactive molecules [4–8]. This work has demonstrated that various approaches to the fabrication of polymers have distinct utilities for achieving the end goal of functional tissue replacements. These approaches are distinguished by ease of processing, differential polymer porosities, variable surface-to-volume ratios, and compatibilities with bioactive molecules. Common techniques that have been developed for tissue engineering applications are briefly reviewed in the next section and the gas foaming technique is then described in the following sections. This chapter is focused on processing poly(lactic-*co*-glycolic acid) (PLGA), a very common tissue engineering scaffold material.

PROCESSING OF POLYMERS FOR USE IN TISSUE ENGINEERING

Polymers of various types have been used in tissue engineering applications, but this chapter focuses on the processing of PLGA. Polymers consisting of lactic and glycolic acids have been widely used for tissue engineering and have been utilized as biodegradable sutures for over 25 years, indicating appropriate biocompatibility [9]. Several approaches to the fabrication of PLGA scaffolds have been developed, with characteristic advantages and disadvantages, though all except gas foaming require processing of PLGA in its liquid state. These processing techniques are the subject of an excellent review [10] and thus are discussed only briefly here. There are four widely used techniques: solvent casting, phase separation, melt molding, and gas foaming, which are covered in detail in the following sections. Each of these processing techniques possesses specific utility for tissue engineering applications.

The remainder of the chapter is focused on an alternative technique that does not require either organic solvents or high temperatures. This process, gas foaming, is described in detail, including the theory of formation, the fabrication parameters and composition, the applications of gas-foamed sponges, a standard protocol for preparation, and a brief description of characterization techniques.

Solvent casting [4,11] involves the dissolution of PLGA in an organic solvent (e.g., methylene chloride), mixing with a water-soluble porogen, usually a salt, and casting the solution into a predefined three-dimensional mold. The solvent is subsequently allowed to evaporate, and the resulting scaffold is then leached to remove the salt, leaving pores within the scaffold of sizes predefined by the particle dimensions of the salt. The advantages of this processing technique are as follows: the controlled porosity, dictated by the mass of the salt crystals included in the mixture; the pore size, controlled by the dimensions of the salt crystal; the varying degree of crystallinity, as determined by the polymer composition and intrinsic viscosity; and the ratio of surface area to volume, as determined by the proportion of salt to polymer. The major disadvantage is the use of organic solvents to cast the polymer, which may decrease the activity of bioinductive molecules (e.g., proteins) and may remain in small quantities after fabrication.

An alternative technique of forming scaffolds utilizes controlled phase separation [12]. This approach takes advantage of the separation characteristics of polymer dissolved in organic solvent (e.g., molten naphthalene). When PLGA and solvent are mixed and homogenized, the bioactive factor is introduced into the mixture. By decreasing the temperature of the mixture, a liquid–liquid phase separation is induced, and subsequently, each phase solidifies. The bioactive molecule is trapped in the polymer and the organic fraction is removed by sublimation. The advantages of this technique are that bioactivity of small molecules (e.g., peptides) may be retained and that larger scaffolds can be produced. While heat is not required, larger molecules subject to denaturation (e.g., growth factors, enzymes) are exposed to organic solvents. Thus, phase separation processing may result in loss of activity of these molecules. Further, residue from the organic solvent could remain, and this could compromise the biocompatibility of the scaffolds.

The process of melt molding utilizes high temperatures to melt PLGA for mixing with a porogen, and the mixture is subsequently molded into the desired shape. The porogen, often gelatin microspheres in this procedure, is subsequently leached, leaving behind a porous scaffold [13]. This method has many of the same advantages as solvent casting, including the ability to control pore size, and the range of degrees of porosity and crystallinity. An additional advantage is that organic solvents are not required, thus eliminating the possibility of residual organic solvents in the scaffolds. The requirement for high temperatures to melt the PLGA may alter the tertiary and quaternary structures of biomolecules required for biological activity.

GAS FOAMING

Gas foaming is a processing technique that requires neither high temperatures nor organic solvents to process PLGA into highly porous scaffolds. This enables the incorporation of bioactive molecules (e.g., proteins, plasmid DNA) with minimal loss of activity. The process allows the formation a continuous PLGA scaffold from an originally discontinuous mixture of PLGA particles (Fig. 64.1). PLGA particles and a porogen, typically sodium chloride, are equilibrated with high-pressure CO_2, and the CO_2 pressure is then rapidly decreased. The drop in pressure causes nucleation of the bubbles from thermodynamically unstable CO_2 [14], expansion of these bubbles due to diffusion of increasing amounts of gas into the bubbles, and a resulting expansion of the individual polymer particles around the porogen particles. The discontinuous PLGA particles fuse to form a continuous, open-pore material. The porogen can then be leached, leaving a highly porous, open-pore PLGA scaffold (Fig. 64.2). Gas foaming encompasses many of the same parameters as the other processing techniques, and thus has many of the same advantages: the control of pore size and porosity, and the ability to form large scaffolds. In addition, this approach has additional advantages. Owing to the high porosity of the scaffold, high densities of cells can be introduced into the device, relative to low porosity PLGA preparations. Further, the high porosity may minimize difficulties due to limited diffusion of nutrients and wastes into and from the nascent tissue, thereby increasing cell viability.

Gas foaming possesses an advantage from a manufacturing perspective in that it is amenable to continuous processing and thus is not limited to batch processing [15]. Perhaps the single greatest advantage of this technique is that PLGA scaffolds can be formed in the absence of organic solvents and high temperatures. This may be critical, as proteins can be denatured by contact with organic solvents, and at the interface of organic solvents with the aqueous buffer containing the protein. Plasmid DNA and large proteins, including growth factors and enzymes, may potentially be incorporated into the scaffolds with minimal degradation [2,16]. Further, since the ultimate goal of preparing scaffolds is to introduce them into the body as either tissue replacements or inductive devices to regenerate tissue, the lack of organic residues in gas-foamed scaffolds is a significant advantage.

THEORY

A critical aspect of this process is the nucleation and growth of gas pores in the scaffolds following the creation of thermodynamic instability. There are three general types of nucleation in a system such as this, as described by classical nucleation theory: homogeneous, heterogeneous, and mixed [17]. Briefly, homogeneous nucleation occurs when CO_2 molecules, originally dissolved in the polymer at high pressure, combine in response to the thermodynamic driving force for phase change to produce small gas bubbles homoge-

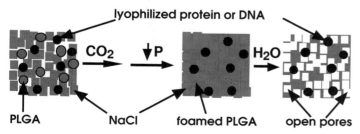

Fig. 64.1. Schematic of gas foam processing. PLGA (hashed circles) is mixed with protein or DNA (solid circles) and a porogen (NaCl, squares) and compression-molded. The scaffold is equilibrated with high-pressure CO_2 gas for 24 h and the pressure is rapidly reduced. CO_2 pore nucleation and growth lead to expansion of discontinuous PLGA particles to form a continuous matrix around the porogen particles (hashed squares). Leaching of the porogen from the matrix by exposure to water leaves a highly porous, interconnected scaffold.

Fig. 64.2. Gas foam processing produces highly porous PLGA scaffolds. (A) Low-power and (B) high-power scanning electron micrographs of gas-foamed PLGA scaffolds (85:15 PLGA) demonstrate the highly porous, interconnected structure of gas-foamed PLGA.

neously throughout the polymer. The free energy changes controlling this process are related to the pressure of the gas, and the surface energy of the polymer–bubble interface. When the pressure is released, CO_2 bubbles initially nucleate and reach a critical size. The free energy of the bubbles is reduced by further expansion of the bubbles due to CO_2 diffusion from the surrounding polymer into the bubble. The process of heterogeneous nucleation is driven by gas bubble nucleation at any number of interfaces. These include gas interfaces with the polymer, a particle incorporated into the sponge (e.g., NaCl, sucrose), or both. These interfaces provide thermodynamically favorable sites for bubble formation, and this process can potentially be used to control the pore structure within the scaffolds by regulating where pores form. Mixed nucleation is a combination of homogeneous and heterogeneous nucleation. Mixed nucleation is hypothesized to be the mechanism controlling pore formation in gas foam processing, due to the multiple components in the molded disk (PLGA, NaCl).

FABRICATION AND COMPOSITION PARAMETERS

There are several parameters controlling the fabrication of highly porous PLGA scaffolds. Included among these are the type of gas used and the porogen characteristics and amounts. PLGA gas-foamed scaffolds are fabricated with CO_2 gas. Other gases (e.g., He, N_2) have been investigated for the ability to form porous scaffolds, though limited success has been achieved [7]. The high porosity achieved when CO_2 is used does not depend significantly on the release rate of gas pressure. Increasing the release rate demonstrates no effect on scaffold porosities, as very rapid release resulted in a decrease in porosity only to 93%. Success with CO_2, relative to other gases, has been achieved presumably because of its increased solubility in the polymer, resulting from interactions of CO_2 with carbonyl groups of the PLGA [7,18].

Porosity can also be modified by the porogen included in the disk by changing the size and amount of NaCl crystals, the most commonly used porogen. Typically, NaCl is sieved to a relatively narrow range of dimensions (e.g., 250 µm < d < 450 µm), and by altering the sieve dimensions, pore size can be altered concomitantly. Further, by the weight percent of NaCl incorporated into the sponge during processing, higher porosities can be achieved [e.g., >98%, M. C. Peters and D. J. Mooney, unpublished observations].

Other parameters for fabrication, including the pressure applied to mold the PLGA-NaCl composite and the time required for NaCl leaching (efficiency of leaching), have not been as extensively studied and are not considered to be primary control parameters affecting PLGA scaffold mechanical properties, erosion kinetics, and porosity. Leaching, however, directly affects the release rate of incorporated proteins and is discussed in the preparation protocol.

USES FOR GAS-FOAMED SCAFFOLDS

Since the first description of the gas foaming process to form PLGA sponges [14], these composites have been employed with considerable success as delivery devices, targeting proteins and plasmid DNA [7,16,19], and for cell transplantation [19].

Protein delivery is one application for gas-foamed sponges. Proteins (e.g., growth factors) can be incorporated into the disk during the fabrication step and delivered in a controlled manner. A typical release profile from the disk is shown in Fig. 64.3. This profile demonstrates a controlled, sustained release of vascular endothelial growth factor (VEGF), important for blood vessel development. Importantly, the VEGF released from gas-foamed scaffolds retains high bioactivity, both *in vitro* and *in vivo*. This suggests that other large, complex proteins might be readily delivered using gas-foamed sponges. Recent work has extended these studies to include modifying the PLGA by mineralization [20], providing a calcium phosphate rich PLGA scaffold containing VEGF for use in bone tissue engineering.

In addition to proteins, plasmid DNA can be successfully delivered [16]. Recently, a plasmid encoding platelet-derived growth factor was released from PLGA gas-foamed disks and studied both *in vitro* and *in vivo*. Plasmid DNA incorporated in the scaffold resulted in high transfection efficiencies, indicating that cells in the scaffolds take up active DNA and modify their gene expression profiles. This study suggests a wide-ranging utility for plasmid DNA delivery to aid in the development of large tissue engineering, by promoting the growth and differentiation of cells in the scaffold, and by providing angiogenic factors necessary to sustain the development of large tissues.

Gas-foamed sponges have also been investigated for use as cell transplantation devices [19,21]. Cell transplantation may provide functionally equivalent tissue constructs for use as partial organ replacements, or for tissue regenerative templates. However, cell transplantation approaches are limited by transport of nutrients into and wastes out of the interior of the scaffold [17,22]; thus vascularization is required for both survival of the transplanted cells and development of larger tissues requiring mature vascularization for maintenance of function. Efficient vascularization might be increased by combining cell transplantation with angiogenic molecule delivery from the scaffolds. Work using different processing techniques to form tissue constructs of cartilage [23,24], adipose [25], and bone [26] indicates that PLGA possesses great potential to form tissues of many different types.

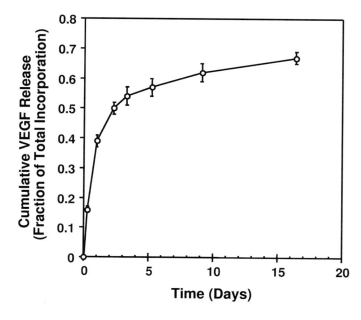

Fig. 64.3. Controlled release of vascular endothelial growth factor from gas-foamed PLGA scaffolds. VEGF (3 μg) and [^{125}I-VEGF] (0.5 μCi) were incorporated in the PLGA matrix (85:15 PLGA) prior to compression molding and gas foaming. The scaffold was leached with double distilled water for 24 h and placed in phosphate-buffered saline. Released VEGF was quantified by removing aliquots of the release medium and counting radioactivity using a gamma counter. Data represent the average of quadruplicate determinations and error bars represent standard error.

PREPARATION PROTOCOL

This section describes a typical protocol for the preparation of a PLGA scaffold by gas foam processing. This process is highlighted by formation of the initial PLGA-NaCl composite disk, gas foaming of the disk with CO_2, and leaching to form the porous scaffold. The protocol yields a 3-mm-thick sponge of 13-mm radius.

DISK FORMATION

A total of 800 mg of material is required to form a 13 mm × 3 mm disk. Typically, 760 mg of NaCl and 40 mg of PLGA are used. Alternatively, sucrose or other porogens can be incorporated as the porogen. This ratio of salt (or sucrose) to polymer results in a highly porous scaffold with sufficient mechanical strength (compression modulus > 75 kPa). If the PLGA content is significantly reduced, sponges exhibit inconsistent porosity and mechanical strength. Sieved NaCl (250 μm < d < 425 μm) is mixed thoroughly with ground and sieved PLGA (85:15, 106 μm < d < 250 μm). This mixture is compression-molded in a sponge die for 1 min at 1500 psi, using a Carver press to form a disk. If protein is to be included in the scaffold, then a volume of the protein dissolved in buffer can be added to the PLGA in the siliconized glass tube, frozen with liquid N_2, and lyophilized prior to mixing with NaCl.

GAS FOAMING

Upon completion of the compression molding of the PLGA-NaCl (±protein) into a disk, the composite is placed in the pressure vessel (Fig. 64.1). The pressure of CO_2 is increased to at least 800 psi for equilibration with the PLGA disk. Equilibration time should be at least 16 h [7] to ensure an adequate dissolution of CO_2 prior to pressure release. After incubation with CO_2, the pressure is released over 2–3 min. The rate at which the gas is released does not significantly alter the integrity of the sponges; however, very rapid release rates might slightly affect sponge porosity [7].

LEACHING

The porogen is leached from the disk following gas foaming. This process results in open, interconnected pores devoid of the porogen. Sponges formed using 760 mg of NaCl per disk (13 mm × 3 mm) are leached with 25 ml of double-distilled water [7,21]. An important consideration for analytical usage is consistency of leach times: within a defined PLGA-NaCl composite, NaCl leaching will occur with predictable rates, and protein release occurs concomitantly. Therefore, scaffolds should only be leached for 24 h, as NaCl leaching is complete by this time [7,21]. Further leaching will affect PLGA disk degradation and erosion, and, more importantly, protein release. To quantify accurately protein incorporation and release kinetics and to predict the concentration of protein at any point in the assay, these initial parameters should be maintained within a given experimental protocol. After leaching has finished, the disks may be characterized (e.g., with respect to porosity, protein release, etc.).

CHARACTERIZATION

Numerous factors affect the processing and resultant integrity of gas-foamed PLGA disks. After fabrication, an initial characterization may be performed to assess disk integrity, mechanical strength, porosity, the release of factors, and degradation profiles. Scanning electron microscopy (SEM) enables an analysis of the gross morphology of the disk and provides a visual record of the pores formed [7,21]. The range of scale available by SEM enables visualization of both large fields of view and detailed microstructures in the disk. To assess mechanical properties (e.g., compression modulus), the disks are subjected to mechanical testing to measure the deformation of the disk in response to applied stress [21]. Porosimetry is performed to measure of the void volume of the three-dimensional construct. The porosity is a combination of both open and closed pores, which can be identified via SEM and quantified by measuring the intrusion volume of mercury [27]. Alternatively, atomic force microscopy can be used to gain more detailed information of pore structure [28]. Degradation and erosion of the PLGA can be measured as a function of loss of disk mass or a loss of mechanical integrity with time. As protein or DNA delivery vehicles, PLGA can be assayed for release of these factors by measuring the amount of factor present in a release buffer (e.g., phosphate-buffered saline, cell culture medium) as a function of time [7].

SUMMARY

Gas foam processing of polymers into tissue engineering matrices has unique advantages, including the ability to form scaffolds containing bioactive factors. Eliminating the requirement of organic solvents and high temperatures for fabrication enables the delivery of large, complex molecules such as growth factors, enzymes, and plasmid DNA with retention of bioactivity to promote inductive tissue engineering applications.

REFERENCES

1. Langer, R., and Vacanti, J. P. (1993). Tissue engineering. *Science* **260**, 920–926.
2. Peters, M. C., and Mooney, D. J. (2000). Growth factor delivery from tissue engineering matrices: Inducing angiogenesis to enhance transplanted cell engraftment. *ACS Symp. Ser.* **752**.
3. Mooney, D. J., Sano, K., Kaufmann, M., Majahod, K., Schloo, B., Vacanti, J. P., and Langer, R. (1997). Long-term engraftment of hepatocytoes transplanted on biodegradable polymer sponges. *J. Biomed. Mater. Res.* **37**, 413–420.
4. Mikos, A. G., Thorsen, A. J., Czerwonka, L. A., Bao, Y., and Langer, R. (1994). Preparation and characterization of poly (L-lactic acid) foams. *Polymer* **35**, 1068–1077.
5. Mikos, A. G., Bao, Y., Cima, L. G., Ingber, D. E., Vacanti, J. P., and Langer, R. (1993). Preparation of poly(glycolic acid) bonded fiber structures for cell attachment and transplantation. *J. Biomed. Mater. Res.* **27**(2), 183–189.
6. Mooney, D. J., Mazzoni, C. L., Organ, G. M., Puelacher, W. C., Vacanti, J. P., and Langer, R. (1994). *Mater. Res. Soc. Symp. Proc.* **331**, 47.
7. Sheridan, M. H., Shea, L. D., Peters, M. C., and Mooney, D. J. (2000). Bioabsorbable polymer scaffolds for tissue engineering capable of sustained growth factor delivery. *J. Controlled Release* **64**(1–3), 91–102.
8. Mooney, D. J., Breuer, M. D., McNamara, K., Vacanti, J. P., and Langer, R. (1995). Fabricating tubular devices from polymers of lactic and glycolic acid for tissue engineering. *Tissue Eng.* **1**(2), 107–118.
9. Frazza, E. J., and Schmitt, E. E. (1971). A new absorbable suture. *J. Biomed. Mater. Res.* **5**(2), 43–58.
10. Lu, L., and Mikos, A. G. (1996). The importance of new processing techniques in tissue engineering. *Mater. Res. Soc. Bull.*, pp. 28–32.
11. Ishaug, S. L., Crane, G. M., Miller, M. J., Yasko, A. W., Yaszemski, M. J., and Mikos, A. G. (1997). Bone formation by three-dimensional stromal osteoblast culture in biodegradable polymer scaffolds. *J. Biomed. Mater. Res.* **36**(1), 17–28.
12. Lo, H., Ponticiello, M. S., and Leong, K. W. (1995). Fabrication of controlled release biodegradable foams by phase separation. *Tissue Eng.* **1**(1), 15.
13. Thomson, R. C., Yaszemski, J. M., Powers, J. M., and Mikos, A. G. (1995). Fabrication of biodegradable polymer scaffolds to engineer trabecular bone. *J. Biomater. Sci. Polymer Edn* **7**, 23.
14. Mooney, D. J., Baldwin, D. F., Suh, N. P., Vacanti, J. P., and Langer, R. (1996). Novel approach to fabricate porous sponges of poly(D,L-lactic-*co*-glycolic acid) without the use of organic solvents. *Biomaterials* **17**, 1417–1422.
15. Park, C. B., Baldwin, D. F., and Suh, N. P. (1995). Effect of the pressure drop rate on cell nucleation in continuous processing of microcellular polymers. *Polym. Eng. Sci.* **35**(5), 432–440.
16. Shea, L. D., Smiley, E., Bonadio, J., and Mooney, D. J. (1999). DNA delivery from polymer matrices for tissue engineering. *Nat. Biotechnol.* **17**(6), 551–554.
17. Colton, J. S., and Suh, N. P. (1987). The nucleation of microcellular thermoplastic foam with additives. Part I. Theoretical considerations. *Polym. Eng. Sci.* **27**(7), 485–492.
18. Kazarian, S. G., Vincent, M. F., Bright, F. V., Liotta, C. L., and Eckert, C. A. (1996). Specific intermolecular interaction of carbon dioxide with polymers. *J. Am. Chem. Soc.* **118**, 1729–1736.
19. Peters, M. C., Polverini, P. J., and Mooney, D. J. (2001). Submitted.
20. Murphy, W. L., Kohn, D. H., and Mooney, D. J. (2000). Growth of continuous bonelike mineral within porous poly(lactide-*co*-glycolide) scaffolds in vitro. *J. Biomed. Mater. Res.* **50**(1), 50–58.
21. Harris, L. D., Kim, B. S., and Mooney, D. J. (1998). Open pore biodegradable matrices formed with gas foaming. *J. Biomed. Mater. Res.* **42**(3), 396–402.
22. Mooney, D. J., Kaufmann, P. M., Sano, K., Schwendeman, S. P., Majahod, K., Schloo, B., Vacanti, J. P., and Langer, R. (1996). Localized delivery of epidermal growth factor improves the survival of transplanted hepatocytes. *Biotechnol. Bioeng.* **50**, 422–429.
23. Spain, T. L., Agrawal, C. M., and Athanasiou, K. A. (1998). New technique to extend the useful life of a biodegradable cartilage implant. *Tissue Eng.* **4**(4), 343–352.
24. Lohmann, C. H., Schwartz, Z., Niederauer, G. G., Carnes, D. L., Dean, D. D., and Boyan, B. D. (2000). Pretreatment with platelet derived growth factor-BB modulates the ability of costochondral resting zone chondrocytes incorporated into PLA/PGA scaffolds to form new cartilage in vivo. *Biomaterials* **21**(1), 49–61.
25. Patrick, C. W., Chauvin, P. B., Hobley, J., and Reece, G. P. (1999). Preadipocyte seeded PLGA scaffolds for adipose tissue engineering. *Tissue Eng.* **5**(2), 139–151.
26. Boyan, B. D., Lohmann, C. H., Somers, A., Niederauer, G. G., Wozney, J. M., Dean, D. D., Carnes, D. L., and Schwartz, Z. (1999). Potential of porous poly-D,L-lactide-*co*-glycolide particles as a carrier for recombinant human bone morphogenetic protein-2 during osteoinduction in vivo. *J. Biomed. Mater. Res.* **46**(1), 51–59.

27. Goldstein, A. S., Zhu, G., Morris, G. E., Meszlenyi, R. K., and Mikos, A. G. (1999). Effect of osteoblastic culture conditions on the structure of poly(DL-lactic-*co*-glycolic acid) foam scaffolds. *Tissue Eng.* 5(5), 421–434.

28. Ferdous, A. J., Stembridge, N. Y., and Singh, M. (1998). Role of monensin PLGA polymer nanoparticles and liposomes as potentiator of ricin A immunotoxins in vitro. *J. Controlled Release* 50(1–3), 71–78.

CHAPTER 65

CELL–SYNTHETIC SURFACE INTERACTIONS: SELF-ASSEMBLING BIOMATERIALS

Julia J. Hwang, Daniel A. Harrington, Harm-Anton Klok, and Samuel I. Stupp

INTRODUCTION

One of the strategies of modern tissue engineering is the use of isolated cells that then grow within three-dimensional matrices to replace the function and structure of damaged or diseased tissues [1–3]. Since cells that comprise tissue are substrate dependent, they can often grow and function only when they are attached to an appropriate surface [4]. Some of the most frequently used synthetic matrix materials for the repair and regeneration of tissue are degradable, biocompatible polymers such as poly(L-lactic acid), poly(glycolic acid), and their copolymers. Variable mechanical properties, degradation rates, and surface properties can be achieved by controlling composition, chain length, and processing of the materials [5,6]. The ability to control these properties is an important requirement for tissue engineering, for it typically dictates later effects on cell adhesion, growth, and function.

Surface chemistry has been shown to influence the adhesion of a variety of cultured cells (e.g., fibroblasts and endothelial cells) [7]. Studies have shown that surface chemistry influences the composition, quantity, and conformation of adsorbed proteins on material surfaces. The relative adsorption of serum fibronectin and vitronectin in response to surface chemistry is believed to be responsible for initial cell adhesion to substrates [8]. Silanized glass and self-assembled monolayers of alkanethiols have been used as model surfaces for studying the interactions among material surface properties, protein adsorption, and cellular responses [9]. Unfortunately, these methods are not applicable for modifying the surfaces of tissue engineering scaffolds.

Self-assembling biomaterials that form layered or lamellar structures would allow one to carefully control surface properties and ensure that the same chemistries are repeatedly exposed over time as a result of the substrate's degradation [10,11]. The molecules of the model system studied here contain cholesteryl moieties and short chains of oligo(L-lactic acid) covalently bound through an ester bond. Cholesterol was selected because of its thermodynamic affinity for cell membranes and its ability to change the cell membrane's transport and mechanical properties [12,13]. In fact, cholesterol, a significant component in the membranes of all eukaryotic cells, may be able to couple universally to cells in ways that are different from well-known ligand–receptor interactions. Cholesterol was also selected because of its well-known mesogenic nature, allowing its derivatives to self-order into liquid crystalline substances. The oligomeric chains of L-lactic acid are biodegradable and could potentially interact with common tissue engineering matrices of poly(L-lactic acid) or its copolymers by cocrystallization or secondary bonds. In this way, the self-assembling structures could be used to modify the internal surfaces of three-dimensional scaffolds fabricated from lactic acid and glycolic acid homopolymers or copolymers.

In this chapter we report on the synthesis of a series of self-assembling biomaterials with molecular features designed to interact with cells and scaffolds for tissue regeneration. The self-assembling biomaterials can be synthesized in good yields with low polydispersities in the range of 1.05–1.15 [11]. These molecules form layered structures that can be described as smectic phases and can also order into single-crystal stacks with an orthorhombic unit cell. When cast on glass substrates, the layered structures were found to promote fibroblast adhesion and spreading, thus behaving as bioactive substrates [10]. The method for synthesizing cholesteryl-(L-lactic acid) derivatives is versatile and is applicable for any molecule with acid functionality. Examples of functionalized derivatives would include molecules containing covalently attached anti-inflammatory drugs such as indomethacin or chromophores such as rhodamine or pyrene for bioimaging. We also report the procedure for using these self-assembling biomaterials to modify the internal surfaces of poly(L-lactic acid) scaffolds, which are processed via the salt-leaching method [14]. This coating method should be of general use for all self-assembling biomaterials.

MATERIALS

Synthesis of Cholesteryl–(L-lactic acid)$_{\bar{n}}$

All listed items were of commercial quality, obtained from Aldrich Chemical Co. (Milwaukee, WI) unless otherwise noted.

1. Oven-dried glass round-bottomed flask, rubber septa, and stir bar
2. Cholesterol, recrystallized from ethanol
3. L-Lactide, recrystallized from ethyl acetate
4. Triethylamine (TEA), stored over KOH
5. Dichloromethane (CH_2Cl_2), freshly distilled from P_2O_5
6. Toluene freshly distilled from sodium–benzophenone
7. N, N'-Dimethylformamide (DMF) stored over molecular sieves (4 Å)
8. $Al(Et)_3$ and tin(II)(2-ethylhexanoate)$_2$ [$Sn(Oct)_2$]
9. Methanol (MeOH)
10. Preheated oil bath at 80°C and 150°C
11. 0.25-mm Glass plates (Merck), precoated with silica gel F254
12. Merck silica gel 60 (0.040–0.063 mm, 230–400 mesh, 60 Å)

Synthesis of Cholesteryl–(L-lactic acid)$_{\bar{n}}$ Derivatives

1. Cholesteryl-(L-lactic acid)$_{\bar{n}}$
2. (Dimethylamino)pyridinium-4-toluenesulfonate (DPTS) prepared according to a literature procedure [15]
3. Diisopropylcarbodiimide (DIPC)
4. 1-(3-Dimethylaminopropyl)-3-ethylcarbodiimide HCl (EDC) and 4-dimethylaminopyridine (DMAP)
5. MeOH, CH_2Cl_2, EtAc, DMF, brine solution, and $MgSO_4$
6. Biofunctional molecules: indomethacin, rhodamine B, 1-pyrenebutyric acid, and cholesterylchloroformate
7. 4-(*tert*-Butoxy)benzoic acid, 4-aminobenzoic acid, and benzyl-4-hydroxybenzoate
8. Pd/C, Celite, TEA, and $KHSO_4$ solution

Preparation of Glass Surfaces

1. Water purified by passage through a Milli-Q purification system (resistivity of 18.2 MΩ·cm)
2. Concentrated sulfuric acid and concentrated hydrochloric acid
3. Solution of concentrated ammonium hydroxide and 30% hydrogen peroxide (4:1 by volume) prepared at the time of the experiment

4. Glass coverslips (22 mm in diameter) to fit 12-well tissue culture plates (Costar) or 2.7-cm diameter glass petri dishes with optically flat bottoms (custom made)
5. Nitrogen gas filtered with 0.22-μm filters (Millipore)
6. Solvent-compatible 0.22-μm filters (AllTech, Deerfield, IL)

Scaffold Production

1. Poly(L-lactic acid), MW ~50,000 (Polysciences)
2. Chloroform, sodium chloride (Fisher Scientific, Pittsburgh, PA)
3. Pyrex petri dishes with lids in 50- and 150-mm sizes (Fisher Scientific)
4. Standard testing sieves in 125- to 500-μm size (Fisher Scientific)
5. Cylindrical steel punches (General Tools)
6. Tex-Wipe sterile cloths (Tex-Wipe)

Scaffold Coating

1. Dip-coat method

    a. 40-μm cell strainers (Falcon)
    b. 50-mm glass petri dishes (Fisher Scientific, Pittsburgh, PA)
    c. Timer
    d. Tetrahydrofuran (Fisher Scientific)

2. Gravity-feed method

    a. 40-μm nylon mesh (Falcon)
    b. *Scaffold holder*: 24/40 glass stir rod bearing (Ace Glass)
    c. *Feed tube*: 14/20 reflux condenser with a Vigrieux fitting (custom), which helps disrupt the solution flow through the tube and promotes a more uniform coverage of the scaffold surface.
    d. *Solution reservoir*: 25-ml separatory funnel with 14/20 fitting
    e. Tetrahydrofuran (Fisher Scientific)

Fibroblast Cell Culture

1. 3T6 mouse fibroblast cell line (CCL-96, American Type Culture Collection, Rockville, MD)
2. *Fibroblast medium*: Dulbecco's Modified Eagle's Medium (DMEM, high-glucose, L-glutamine, 110 mg/l sodium pyruvate, Gibco-BRL, Grand Island, NY), fetal bovine serum 10% (HyClone, Logan, UT), penicillin–streptomycin (Gibco-BRL) 100 U/ml–100 μg/ml.
3. Trypsin (0.25%) and trypan blue (Gibco-BRL)
4. Hemocytometer

METHODS

SYNTHESIS OF CHOLESTERYL–(L-LACTIC ACID)$_{\overline{n}}$

All reactions were performed under a nitrogen atmosphere. Depending on the desired length of the oligo(L-lactic acid) segment, the polymerizations were performed in toluene solution or in bulk.

Solution Polymerization

As a typical example, the synthesis of an oligomer with a targeted length of 20 lactic acid residues is described. As expected for a "living" polymerization, the molecular weights can be controlled by the controlling the molar ratio of L-lactide to initiator, and well-defined oligomers with polydispersities (M_w/M_n) around 1.1 were obtained. This method is particularly suited for the synthesis of cholesteryl-(L-lactic acid)$_{\overline{n}}$ oligomers with *n* exceeding 20, which can be easily isolated by precipitation of the reaction mixture in excess methanol.

Owing to the increased solubility of the lower molecular weight oligomers in methanol, the isolated yields dropped rapidly when this methodology was applied for the preparation of cholesteryl-(L-lactic acid)$_{\bar{n}}$ oligomers with *n* under 20.

1. Initiate polymerization by forming the aluminum alkoxide that is generated *in situ* upon addition of 1 equiv Al(Et)$_3$ to 3 equiv cholesterol. Specifically, add 1.9 M toluene solution of Al(Et)$_3$ [3 ml, 5.7 mmol Al(Et)$_3$] to a mixture of L-lactide (24.65 g, 17.1 mmol) and cholesterol (6.61 g, 17.1 mmol) in toluene (120 ml).
2. Stir this mixture for 15 min at room temperature and then place it in a preheated oil bath at 80°C.
3. After 5 h, quench the polymerization with MeOH (5 ml).
4. Precipitate the warm reaction mixture in MeOH (1500 ml).
5. Filter the solids and vacuum-dry at room temperature.
6. Pass the crude product over a short column (SiO$_2$, CH$_2$Cl$_2$-MeOH 100:5 v/v). Yield: 16.8 g (54%).

Bulk Polymerization

As a typical example, the synthesis of an oligomer with a targeted length of 10 lactic acid units is described. Lower molecular weight cholesteryl-(L-lactic acid)$_{\bar{n}}$ oligomers (*n* < 20) were synthesized in bulk at 150°C, with tin(II)(2-ethylhexanoate)$_2$ [Sn(Oct)$_2$] as a catalyst. In the absence of an alcohol as a coinitiator, this procedure does not allow accurate control of molecular weight. However, when an alcohol coinitiator is used in combination with a catalytic amount of Sn(Oct)$_2$, oligomers with defined molecular weights (corresponding to the molar ratio of L-lactide to initiator) can be prepared. Performing the polymerization in bulk eases the workup and strongly improves the isolated yields of these short chain oligomers. Although the Sn(Oct)$_2$-catalyzed bulk polymerization of L-lactide is not strictly a "living" process, the molecular weights can be adjusted by varying the ratio of L-lactide to cholesterol, and after workup oligomers with fairly narrow polydispersities (~1.2) are obtained.

1. Place a mixture of L-lactide (10.19 g, 71 mmol) and cholesterol (5.44 g, 14 mmol) in a preheated oil bath at 150°C and stir until contents are molten.
2. Add a solution of Sn(Oct)$_2$ in toluene [1 ml, 0.37 g Sn(Oct)$_2$/ml].
3. Stir the reaction mixture at 150°C for 5 h.
4. Cool the reaction mixture to room temperature.
5. Triturate the residual solid with MeOH (100 ml).
6. Filter the solids and vacuum-dry at room temperature. Yield: 7.82 g (50%).

SYNTHESIS OF CHOLESTERYL-(L-LACTIC ACID)$_{\bar{n}}$ DERIVATIVES

Derivatives of cholesteryl-(L-lactic acid)$_{\bar{n}}$ can be prepared by direct esterification of the secondary alcohol terminus of the oligomer with the carboxylic acid of the functional unit. These preparations are illustrated in Fig. 65.1. In the case of cholesterol, indomethacin, and pyrene, these were first coupled to a linker unit to form aromatic acids for facile coupling to the oligomer by means of the DIPC/DPTS methodology [15].

Cholesteryl-(L-lactic acid)$_{\bar{n}}$-4-(Indomethacin)benzoate

As a typical example, the derivatization of a cholesteryl-(L-lactic acid)$_{\overline{34}}$ oligomer is described.

1. DIPC (0.11 ml, 0.70 mmol) was added dropwise to a solution of cholesteryl-(L-lactic acid)$_{\overline{34}}$ (1.87 g, 0.66 mmol), 4-(indomethacin)-benzoic acid (0.38 g, 0.80 mmol), and DPTS (0.20 g, 0.68 mmol) in CH$_2$Cl$_2$ (10 ml).

    a. Synthesis of 4-(indomethacin)-benzoic acid

        i. Add Pd-C (0.47 g, 50 wt% H$_2$O) to a solution of benzyl-4-(indomethacin)-benzoate (1.21 g, 2.13 mmol) in mixture of EtAc (80 ml) and EtOH (40 ml).

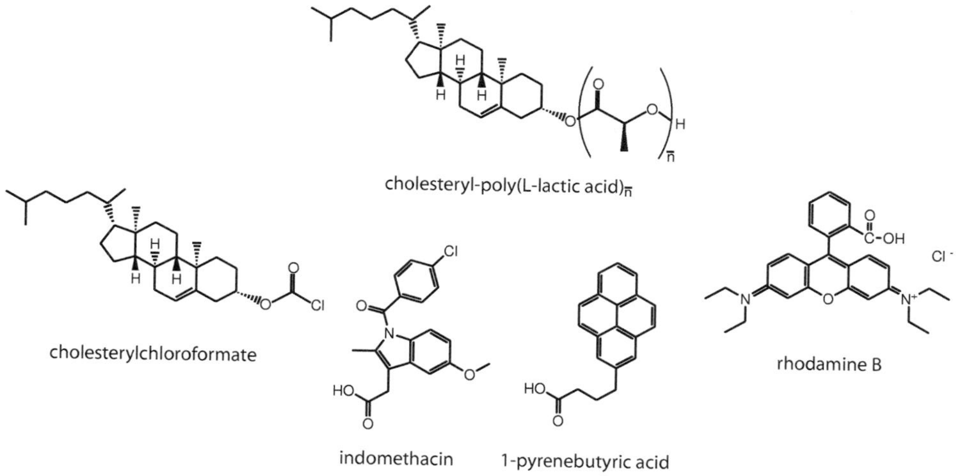

Fig. 65.1. *Cholesteryl–poly(L-lactic acid)$_{\overline{n}}$ and additional terminal biofunctional units.*

   ii. Stir overnight at room temperature under a hydrogen atmosphere.
  iii. Filter the reaction mixture over a short plug of Celite.
  iv. Evaporate to dryness.

 b. Synthesis of benzyl-4-(indomethacin)-benzoate

     i. Add DIPC (1.0 ml, 6.40 mmol) to a solution of indomethacin (1.0 g, 2.80 mmol), DPTS (1.24 g, 4.21 mmol), and benzyl-4-hydroxybenzoate (0.77 g, 3.37 mmol) in DMF (30 ml).
    ii. Stir the solution overnight at room temperature.
   iii. Pour the reaction mixture into water (250 ml).
   iv. Extract the aqueous mixture with CH$_2$Cl$_2$ (3×).
    v. Wash the CH$_2$Cl$_2$ extracts with water (1×) and brine (1×).
   vi. Separate the organic phase and dry over MgSO$_4$.
  vii. Filter and evaporate to dryness.
 viii. Triturate the oily residue with MeOH.
   ix. Filter the solids and vacuum-dry. Yield: 1.24 g (79%).

2. Stir overnight at room temperature.
3. Evaporate the reaction mixture to dryness.
4. Redissolve the residue in a minimal amount of CH$_2$Cl$_2$ and precipitate in MeOH.
5. Solids were filtered and vacuum-dried. Yield: 1.90 g (88%).

Cholesteryl-(L-lactic acid)$_{\overline{n}}$–4-(Cholesterylcarbamate)benzoate

As a typical example, the derivatization of a cholesterol-(L-lactic acid)$_{\overline{34}}$ oligomer is described.

1. Add DIPC (20 µl, 0.13 mmol) dropwise to a solution of cholesteryl-(L-lactic acid)$_{\overline{34}}$ (0.215 g, 0.076 mmol), benzoic acid 4-cholesterylcarbamate (0.051 g, 0.093 mmol), and DPTS (0.023 g, 0.078 mmol) in CH$_2$Cl$_2$ (25 ml).
Synthesis of benzoic acid 4-cholesterylcarbamate:

 a. Add a solution of cholesterylchloroformate (1.0 g, 2.23 mmol) in CH$_2$Cl$_2$ (10 ml) to a mixture of 4-aminobenzoic acid (0.306 g, 2.23 mmol) and triethylamine (0.31 ml, 2.23 mmol) in CH$_2$Cl$_2$ (20 ml).
 b. Add TEA (0.31 ml, 2.23 mmol) dropwise to the reaction mixture and stir overnight at room temperature.
 c. Wash the reaction mixture with an aqueous KHSO$_4$ solution (1×), water (1×), and brine (1×).
 d. Separate the organic phase, dry over MgSO$_4$, filter, and evaporate to dryness.

     e. Triturate the residue with MeOH.
     f. Filter solids and vacuum-dry. Yield: 0.92 g (74%).

2. Stir overnight at room temperature.
3. Evaporate the reaction mixture to dryness.
4. Redissolve the residue in a minimal amount of CH_2Cl_2 and precipitate in MeOH.
5. Filter solid and vacuum-dry. Yield: 0.21 g (83%).

Cholesteryl-(L-lactic acid)$_{\bar{n}}$ –Indomethacin

As a typical example, the derivatization of a cholesteryl-(L-lactic acid)$_{\overline{25}}$ oligomer with the anti-inflammatory drug indomethacin is described.

1. Add EDC (0.096 g, 0.50 mmol) to an ice-cooled solution of cholesteryl-(L-lactic acid)$_{\overline{25}}$ (0.507 g, 0.23 mmol), indomethacin (0.162 g, 0.45 mmol), and DMAP (0.038 g, 0.31 mmol) in CH_2Cl_2 (60 ml).
2. Stir overnight in the ice bath.
3. Evaporate the reaction mixture to dryness.
4. Redissolve the residue in CH_2Cl_2 (5 ml) and precipitate in MeOH (100 ml).
5. Filter solid and vacuum-dry. Yield: 0.49 g (84%).

Cholesteryl-(L-lactic acid)$_{\bar{n}}$ –Rhodamine B

As a typical example, the derivatization of a cholesteryl-(L-lactic acid)$_{\overline{25}}$ oligomer with rhodamine B is described.

1. Add EDC (0.096 g, 0.50 mmol) to an ice-cooled solution of cholesteryl-(L-lactic acid)$_{\overline{25}}$ (0.52 g, 0.24 mmol), rhodamine B (0.21 g, 0.44 mmol), and DMAP (0.035 g, 0.29 mmol) in CH_2Cl_2 (60 ml).
2. Stir overnight in ice bath.
3. Evaporate the reaction mixture to dryness.
4. Redissolve the residue in CH_2Cl_2 (5 ml) and precipitate in MeOH (100 ml).
5. Filter solid and vacuum-dry.
6. Purify by column chromatography (SiO_2, CH_2Cl_2-MeOH 100:10 v/v). Yield i: 0.050 g.
7. Evaporate the MeOH phase and purify by column chromatography (SiO_2, CH_2Cl_2-MeOH 100/10 v/v) to provide a second batch of the product. Yield ii: 0.060 g.

Cholesteryl-(L-lactic acid)$_{\bar{n}}$ –4-(1-Pyrenecarboxylate)benzoate

As a typical example, the end functionalization of a cholesteryl-(L-lactic acid)$_{\overline{25}}$ –4-(hydroxy)benzoate oligomer with pyrene is described.

1. Add DIPC (50 μL, 0.32 mmol) dropwise to a mixture of cholesteryl-(L-lactic acid)$_{\overline{25}}$ –4-(hydroxy)benzoate (0.103 g, 0.045 mmol), 1-pyrenebutyric acid (0.025 g, 0.10 mmol), and DPTS (0.047 g, 0.16 mmol) in CH_2Cl_2 (15 ml).
Synthesis of cholesteryl-(L-lactic acid)$_{\overline{25}}$ –4-(hydroxy)benzoate:

     a. Add DIPC (0.70 ml, 4.47 mmol) dropwise to a mixture of cholesteryl-(L-lactic acid)$_{\overline{25}}$ (1.0 g, 0.53 mmol), 4-(*tert*-butoxy)-benzoic acid (0.5 g, 2.63 mmol), and DPTS (0.88 g, 2.99 mmol) in CH_2Cl_2 (60 ml).
     b. Stir the reaction mixture at room temperature overnight.
     c. Evaporate the reaction mixture to dryness.
     d. Redissolve the residue in CH_2Cl_2 (20 ml) and precipitate in MeOH (400 ml).
     e. Filter the solids and vacuum-dry. Yield i: 0.90 g.
     f. Redissolve the residue in CH_2Cl_2 (80 ml) and cool in an ice bath.
     g. Add 15 ml of TFA dropwise and stir on ice for one hour. Continue stirring at room temperature for another hour.
     h. Evaporate reaction mixture to dryness.

    i. Redissolve the residue in CH_2Cl_2 (10 ml) and precipitate in MeOH (200 ml).
    j. Filter the solids and vacuum-dry. Yield ii: 0.74 g.

2. Stir overnight at room temperature.
3. Evaporate the reaction mixture to dryness.
4. Redissolve the residue in CH_2Cl_2 (5 ml) and precipitate in MeOH (100 ml).
5. Filter solids and vacuum-dry.
6. Purify the crude product by column chromatography (SiO_2, CH_2Cl_2-MeOH 100:2.5 v/v). Yield: 0.075 g (66%).

Cleaning of Glass Slides

1. Immerse coverslips (placed in a custom-made glass coverslip rack) or petri dishes in fuming hot sulfuric acid bath for 10 min.
2. Rinse coverslips thoroughly with purified water.
3. Place the coverslips in warm (~60°C) mixture of ammonium hydroxide and hydrogen peroxide mixture (4:1 by volume) for 10 min.
4. Rinse coverslips thoroughly with purified water.
5. Place coverslips in HCl solution at room temperature.
6. Rinse coverslips thoroughly with purified water.
7. Dry coverslips with filtered nitrogen gas (0.22-μm filter).
8. Autoclave.

Preparation of Films for Cell Studies

The following methods produce smooth films of the bioactive molecules described earlier, with thicknesses on the order of 1000–1500 Å. Use aseptic conditions at room temperature.

1. Dissolve the materials in CH_2Cl_2 to make a 2 wt% solution and sterilize by passage through a solvent-compatible Millipore-type (0.22-μm) filter membrane.
2. Solution-cast

    a. Pipette 2 ml of the polymer solution into each 2.7-cm-diameter petri dish.
    b. Be sure that the solution spreads over the entire dish surface.
    c. Cover each dish with a larger glass petri dish and slowly dry in a chemical hood to promote formation of a smooth, even surface.

3. Spin-coat

    a. Mount a clean coverslip on the vacuum chuck of a spin coater.
    b. Pipette enough solution to cover the entire surface of the substrate (approximately 0.5 ml for a coverslip).
    c. Spin film at 2000 rpm for 1 min.

4. Anneal the films at an appropriate temperature and time under vacuum to promote formation of an ordered liquid crystalline phase.*

Scaffold Production

The method described here will produce a 50-mm disk of porous poly(L-lactic acid) (PLLA), approximately 1 mm thick, from which approximately 12–20 individual scaffolds can be obtained, depending on size. This method is adapted from Mikos et al. [14]. Clean, aseptic tools and materials should be used to produce optimal scaffolds for later use in cell culture.

---

*The transition temperatures for the self-assembling coating should be predetermined by differential scanning calorimetry and variable-temperature optical microscopy under crossed polarized lenses. Annealing should occur in a temperature regime that promotes ordering in the coating (e.g., a liquid–crystalline phase regime) but safely below its isotropization temperature.

1. In a glass vial, dissolve 0.5 g of PLLA in 4 ml of chloroform for 60 min. The solution may be sonicated for 5 min to aid in dissolution.

2. Sieve the salt to a desired size range. Place 4 g of salt into the 50-mm petri dish and tap lightly to evenly distribute in the dish.[*]

3. Add the entire polymer solution to the salt, and swirl gently to promote even distribution. Cover the petri dish with its own lid, and a second larger lid, to slow down solvent evaporation. Leave this to dry on a level surface in a chemical hood for 24 h, followed by 24 h under vacuum.

4. Use a circular steel punch to punch the polymer–salt composites into small disks (e.g., 10-mm diameter).

5. Place individual scaffolds from one petri dish in 75 ml of ultrapure deionized water for 48 h, changing water every 6 h. A clean Teflon stir bar may be used in the leaching container to provide gentle agitation (40 rpm). Dry the leached porous foams under vacuum for 24 h.

6. To remove the thin polymer film from the bottom of each scaffold, moisten a 30-mm Tex-Wipe strip in chloroform and lay it across a petri dish. Lightly touch each scaffold to the cloth strip for 3 s and quickly remove. Dry all samples under vacuum for 8 h.[†]

SCAFFOLD COATING

Two convenient coating methods are described. The dip-coat method is easy to set up and gentle on fragile scaffolds. The gravity-feed method ensures better coating of all internal surfaces but can cause crumbling or cracking in brittle samples that are low in mechanical integrity. At high porosities, the gravity-feed method is better suited for small-pore scaffolds (<250 μm), while the dip-coat method is appropriate for more fragile, large-pore scaffolds (>250 μm). An example gravity-feed setup is shown in an exploded view in Fig. 65.2. In regular use, the upper and lower pieces fit together, trapping the scaffold snugly within the glassware. This promotes fluid flow through the entire cross section of the scaffold disk, rather than simply diverting along its edges.

Sterilize scaffolds under UV for 30 min prior to coating. All coating equipment should be autoclaved (or presterilized), and all chemicals should be filtered prior to use. Polymer concentrations and coating times may be adjusted to modify coating thickness.

1. Dip-coat method

   a. In a glass vial, dissolve 1 wt% (w/v) of the coating molecule in 10 ml of tetrahydrofuran. Pour the polymer solution into a small petri dish until almost full.

   b. Place a scaffold in a nylon cell strainer and dip the entire assembly into the solution (3 s). Remove the strainer and allow the scaffold to dry briefly (10 s) before dipping again. Repeat three times for each scaffold and dry on a glass dish under vacuum for 24 h.

   c. Anneal samples under vacuum to promote ordering of the self-assembling surface layers. See note on transition temperatures, page 747.

2. Gravity-feed method

   a. Prepare a 1 wt% polymer solution in tetrahydrofuran in the same manner used for the dip-coat method. Each scaffold will require 5 ml of solution.

   b. Place a single scaffold between two sheets of nylon mesh. Position the layered

---

[*]The ratio of salt to polymer will define the porosity of the scaffolds. Adjust these to match your needs; the values given here will produce scaffolds with interconnected pores and high porosity.

[†]The casting method results in a thin polymer film that effectively seals one face of the scaffold and interferes later in coating the internal surfaces. In removing this, the dissolution time should be kept to 3 s or less to prevent dissolution of more than just this bottom skin.

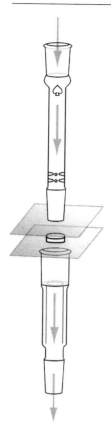

Fig. 65.2. Schematic view of gravity-feed coating apparatus. Scaffold disk fits snugly within the glass column to promote uniform coating by the dilute polymer solution. Diagram used with permission from Cambridge Soft Corp. and Ace Glass Inc.

sample in the scaffold holder and insert the feed tube so that it surrounds the scaffold as shown in Fig. 65.2.[‡]

c. Load the solution reservoir with the polymer solution and allow the solution to flow down through the scaffold. Flow time will vary with pore size and porosity, typically between 3 and 30 s.[*]

d. Carefully remove feed tube and scaffold. Dry the scaffold on a glass dish in air, then under vacuum for 24 h.

e. Anneal samples under vacuum to promote ordering of the self-assembling surface layers.

Fibroblast Cell Culture on Films of Self-Assembling Biomaterials

1. Grow 3T6 cells to preconfluence in T75 flasks at 37°C in a humidified 10% CO_2 atmosphere.
2. Passage cells by washing with PBS and incubating with 2 ml of trypsin.
3. Collect cells by centrifugation and resuspend in growth media.
4. Seed fibroblasts onto substrates at 10,000 cells/cm^2.
5. Change media every 2 days.

[‡]This setup may vary based on scaffold size; the equipment shown in Fig. 65.2 is designed for scaffolds approximately 10 mm in diameter. The important features are a close fit of the upper glassware around the scaffold (to ensure that the entire cross section is coated), and the ability the lower glassware to support the scaffold without interfering with fluid flow. In our example, the upper opening of the 24/40 bearing is small enough to prevent the scaffold disk from falling through, but large enough to allow for good fluid flow. The nylon mesh simply adds stability to the scaffold, and still allows the 14/20 joint of the upper glassware to fit securely around the sample.

[*]Both the dip-coat and the gravity feed methods assume interconnected pores, which typically requires 90% or greater porosity.

REFERENCES

1. Kim, S. S., and Vacanti, J. P. (1999). The current status of tissue engineering as potential therapy. *Semin. Pediatr. Surg.* 8, 119–123.
2. Naughton, G. K., and Mansbridge, J. N. (1999). Human-based tissue-engineered implants for plastic and reconstructive surgery. *Clin. Plast. Surg.* 26, 579–586.
3. Freed, L. E., Martin, I., and Vunjak-Novakovic, G. (1999). Frontiers in tissue engineering. *Clin. Orthop. Relat. Res.* 367S, S46–S58.
4. Boyan, B. D., Hummert, T. W., Dean, D. D., and Schwartz, Z. (1996). Role of material surfaces in regulating bone and cartilage cell response. *Biomaterials* 17, 137–146.
5. Kim, B.-S., and Mooney, D. J. (1997). Engineering smooth muscle tissue with a predefined structure. *J. Biomed. Mater. Res.* 41, 322–332.
6. Wake, M. C., Gupta, P. K., and Mikos, A. G. (1996). Fabrication of pliable biodegradable polymer foams to engineer soft tissues. *Cell Transplant.* 5, 465–473.
7. McClary, K. B., Ugarova, T., and Grainger, D. W. (2000). Modulating fibroblast adhesion, spreading, and proliferation using self-assembled monolayer films of alkylthiolates on gold. *J. Biomed. Mater. Res.* 50, 429–439.
8. Tziampazis, E., Kohn, J., and Moghe, P. V. (2000). PEG-variant biomaterials as selectively adhesive protein templates: Model surfaces for controlled cell adhesion and migration. *Biomaterials* 21, 511–520.
9. Chen, C. S., Mrksich, M., Huang, S., Whitesides, G. M., and Ingber, D. E. (1998). Micropatterned surfaces for control of cell shape, position, and function. *Biotechnol. Prog.* 14, 356–363.
10. Hwang, J. J., Iyer, S. N., Li, L.-S., and Stupp, S. I. (2001). Submitted for publication.
11. Klok, H.-A., Hwang, J. J., Stupp, S. I. (2001). Submitted for publication.
12. Yeagle, P. L. (1985). Cholesterol and the cell membrane. *Biochim. Biophys. Acta* 822, 267–287.
13. Yeagle, P. L. (1991). Modulation of membrane function by cholesterol. *Biochimie* 73, 1303–1310.
14. Mikos, A. G., Thorsen, A. J., Czerwonka, L. A., Bao, Y., Langer, R., Winslow, D. N., and Vacanti, J. P. (1994). Preparation and characterization of poly(L-lactic acid) foams. *Polymer* 35, 1068–1077.
15. Moore, J. S., and Stupp, S. I. (1990). Room temperature polyesterification. *Macromolecules* 23, 65–70.

CHAPTER 66

CELL–SYNTHETIC SURFACE INTERACTIONS: TARGETED CELL ADHESION

Andrea L. Koenig and David W. Grainger

This chapter describes methods used to attach and stabilize various attachment-dependent mammalian cell types on biomaterials surfaces. Such research is increasingly important to encourage cell recruitment and healing at the tissue–biomaterials interface. Rapid cellular and tissue integration is required in long-term implants, drug delivery devices, and artificial tissue scaffolds used for functional organ regeneration to prevent problems with inflammatory and immune cascades from resulting in chronic foreign body response and device failure. The most widely studied methods for specifically targeting cells to a biomaterial surface exploit interactions of cell–surface integrin receptors with adsorbed or covalently attached extracellular matrix molecules (e.g., fibronectin) or with RGD, the well-known cell-binding peptide motif. This chapter reviews the many ways of attaching cells to surfaces through cell receptors; discusses the use of some non-integrin-binding molecules for cell adhesion; surveys solid surfaces, polymers, and gel matrices used as cell-adhesive biomaterials; briefly describes experimental protocols utilized to monitor attachment-dependent cell growth, proliferation, and intracellular signaling events on surfaces; and compiles an extensive reference collection targeting published primary literature describing current technology, research strategies, and scientists endeavoring to make progress in this field.

CELLS AT BIOMATERIAL SURFACES

Despite continuous development of new biomaterials for longer term deployment in diverse medical and biotechnology applications, including biosensors [1,2], drug delivery devices [3–6], orthopedic and dental implants, catheters, gene therapy [7,8], pacemakers, stents, artificial organs, bioassays, tissue engineering [8–12], and prosthetics [13], few materials exhibit requisite capability to sustain continued cell attachment and integration of tissues *in vivo*. Nevertheless, device integration with tissue is a necessity to avoid chronic inflammation, foreign body reactions, and infection in an implant healing scenario.

Advances in tissue scaffolds, functional organ and tissue regeneration, and spontaneously integrating biomaterials rely upon the proven ability to promote, maintain, and organize the adhesion of required biomolecular components and specific cells on and within biomaterials indefinitely upon implantation. While event sequences and physiological cues that link simple initial cell–substrate adhesion and proliferation to longer term three-dimensional tissue neogenesis are far from understood, the adhesion of cells to substrates is a critical event for all attachment-dependent cell types. Expression of cell phenotype and differentiation also depends on this initial event. Also hinging on successful cell–surface recognition are important events secondary to attachment, including cell-based processing and organization of extracellular matrix (ECM) material, intracellular signals, intercellular communication, and coculture or functional integration with other cell types necessary to form tissue architectures and histologies.

Hence, while biomaterial governance of cell adhesion may not completely determine long-term stability and success of tissue-engineered biomaterials, cell–surface interactions are very important and perhaps essential to control in early stages of biomaterial–tissue interfacing. Interfacial chemistry and surface topology are critically important determinants of interactions between biomaterials and physiological fluids, cells, and tissues. Because of ubiquitous nonspecific protein adsorption to the often primitive or unnatural chemistries utilized in most clinically applied biomaterials (e.g., metals and alloys, synthetic polymers, ceramics) [14–17], foreign body reactions, both acute and chronic, make immediate cell attachment and tissue integration problematic. Competing with cell–surface adhesion and proliferation to promote healing and integration, the classic implant foreign body response is a consequence of adverse surface recognition events by immunomodulating cells that react in complex ways at protein-coated biomaterial surfaces to stimulate inflammatory reactions [18]. There are also neutrophil, monocyte, and macrophage attachment-mediated reactions resulting from yet poorly understood surface recognition and signaling phenomena. Thus, control of cell–surface adhesion is important to natural tissue regeneration and integration as well as to inhibition of foreign body response, chronic inflammation, and eventual device rejection.

This chapter seeks to present current information and methods used by researchers in this field to directly and chemically attach specific cell-adhesive ligands to attract certain cell types to biomaterial surfaces. The ultimate objective of this strategy is the stable attachment, proliferation, and migration of cells on synthetic biomaterials with the intent of promoting tissue integration, cell phenotypic preservation, and natural cell differentiation.

Because synthetic or adapted natural chemistries found in implanted biomaterials are often suboptimal promoters of tissue in growth, cell attachment, or integration, host–implant interfaces are often chemically modified to improve host response. Typically, surface modification strategies (coatings, bioimmobilization, deposition, films, chemical grafting, and derivatization) are used to impart desirable chemistry and thereby facilitate specific interfacial interactions between biomaterials and biological milieu. Various techniques, including self-assembled monolayer organic films [19–25], lithography [26,27], polymer coatings and thin-film deposition [28], grafting [29–31], and plasma modification [32–34], have been described and are divided along two general strategies: passive biorepulsion and active cell integration. The passive coating approach uses modification to hinder attachment or interaction of all biological components, including proteins, peptides, bacteria and cells (Fig. 66.1). Typically, lubricious, highly hydrated hydrogel coatings are used [e.g., polyacrylates, poly(ethylene glycol) (PEG), poly(vinyl alcohol) (PVA), dextrans, heparin]. By contrast, the active coating approach uses surface chemical modification to recruit specific

Direct Surface Adsorption:

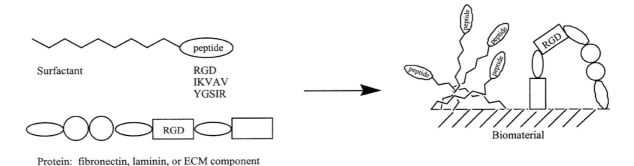

Fig. 66.1. Modified biomaterial surfaces that adherent to cells by direct adsorption of ECM peptides or ECM proteins to create an adhesive surface.

Surface Chemical Immobilization (Homogeneous or Micropatterned):

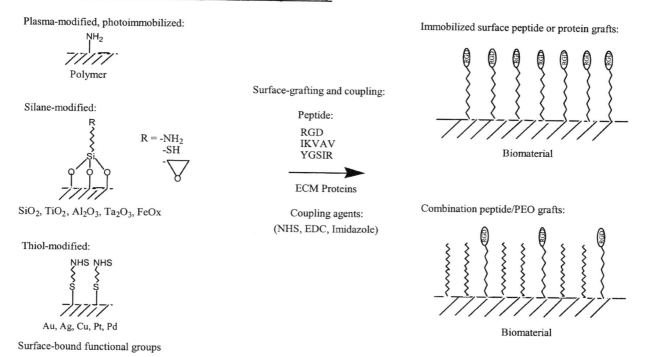

Fig. 66.2. Modified biomaterial surfaces that use various surface coupling strategies adherent to cells by covalent attachment of ECM peptides (see reference [91]).

matrix proteins and desired cell types to implant surfaces, promoting a stable environment locally at the implant–host interface that is conducive to cell attachment, proliferation, and tissue integration (Fig. 66.2). These modified biomaterials surfaces are capable of either (1) selectively adsorbing intact, functional, soluble ECM proteins that in turn can interact with receptors from a targeted cell type or (2) presenting ligand motifs that interact with cell surface receptors directly. Frequently, early interactions between soluble biomacromolecules (proteins, peptides, growth factors) determine the fate of cell and tissue events occurring much later at the implant surface. Hence, most active cell integration strategies attempt to control these early events by providing specific surface chemistries known to bind soluble cell-adhesive ECM proteins that, in turn, attract cells to attach to biomaterials via receptor-mediated events. In fact, a primary advantage of the tissue-engineered device over generic biomaterials is the ability to control cell growth, proliferation, and differentiation on a bio-compatible surface with the intent of generating functional tissue over a longer term. There-fore, one ultimate technology goal is to create new materials with biologically interactive surfaces to prompt fabrication of bioengineered organs and to regenerate tissues [9–11].

Many biomaterials demonstrate at least limited abilities to bind cells under controlled conditions in *in vitro* assays, where cell culture medium, ECM, and other serum proteins and attachment factors can be biased or manipulated to adsorb in desirable ways, facilitating cell adhesion. Decades of use of tissue culture polystyrene have proven that rather unsophisticated surface modification of hydrophobic polystyrene (intrinsically unsupport-ive of cell attachment) by means of oxidizing chemistry changes amounts, compositions, and conformations of trace serum proteins adsorbed to these oxidized surfaces, facilitat-ing successful cell culture after surface treatment [33–37,42]. However, in more practical applications, physiological conditions that mandate uncontrolled competition of soluble trace ECM with other more abundant serum proteins directly with biomaterial surfaces (multicomponent serum milieu) determine cell–surface attachment. Competitive protein ad-sorption conditions provide a more relevant model to predict and facilitate desired *in vivo* cellular events with biomaterials, such as cell adhesion, spreading, proliferation, differentiation, signaling cascades, migration, and eventually successful coculture and ultimately angiogenesis [38–43].

Production of surfaces that successfully support cell adhesion has historically been done empirically: certain surface chemistries are known to support cell attachment under controlled conditions [44]. Research continues to try to explain, for its predictive design value, why more wettable, polar surfaces support greater cell attachment from competitive multicomponent ECM or full serum media [14,22,35,45,46], and why surface chemistry–cell influences are negligible if only ECM proteins (e.g., collagen or fibronectin) are present at sufficient concentrations to adsorb [14,35–37,45,47]. In addition to attachment, surface chemistry is important to events signaled downstream of initial cell attachment behavior. These latter events are just as critical to the tissue-engineered device, since cell attachment alone is necessary but not sufficient for sustained cell–surface integration and proliferation. Cytoskeletal construction and remodeling yielding cell flattening, phenotypic shape changes and spreading, membrane–skeletal associations including focal contacts and receptor clustering, and subsequent "outside-in" signaling activities responsible for gene regulation, phenotypic expression, and proliferation are important events [48] that must be preserved over the longer term for surface-resident cells in the implant scenario.

EXTRACELLULAR MATRIX AND CELL ADHESION MOTIFS

Various soluble serum components (proteins, peptides) are well known to facilitate cell attachment, spreading, and proliferation [35–37,45,47], and to exhibit surface chemistry dependence in their ability to do so [35,49–52]. Continued discovery and modern cell and molecular biology studies of endogenous ECM components have prompted elucidation of conserved protein primary sequence motifs deemed essential for cell surface receptor-mediated attachment to ECM proteins. Minimalist strategies have ascertained small, common peptide sequences in ECM proteins that are responsible for mediating cell attachment to ECM-adsorbed biomaterials. Two general strategies that attempt to exploit this molecular-level information on cell–protein adhesive interactions use ECM and peptide surrogates to modify biomaterials. Amino acid sequences within several ECM proteins have been synthetically duplicated as small peptides and appropriated as either soluble or surface ligands to enhance cell interactions with synthetic biomaterials surfaces [11,12,53–56]. A survey of the extensive literature regarding peptide and ECM surface modifiers indicates that all peptides shown to be effective promoters of cell adhesion derive from a small set of conserved short sequences found in ECM proteins. Several surface modification routes are used to make these bioactive motifs available to cell receptors on biomaterial surfaces.

Probably the most studied class of cell-adhesive molecules comprises ECM-soluble serum proteins, including fibronectin [56–58], vitronectin [59], collagen families [60–63], thrombospondin [64], osteopontin [65], fibrinogen [66], von Willebrand factor [67], and laminin [68–71]. Of these, classes of fibronectins ubiquitously found in virtually all physiological fluids as well as on cell surfaces have been most studied for cell adhesion [47]. Within the primary amino acid sequences of several of these ECM proteins are conserved peptide motifs—either arginine-glycine-aspartic acid-serine (RGDS), tyrosine-isoleucine-glycine-serine-arginine (YIGSR), or isoleucine-lysine-valine-alanine-valine (IKVAV) domains—known to support cellular adhesion mediated through the integrin family of cell surface receptors [72]. Receptor recognition, binding, and subsequent signaling events characteristic of full ECM protein–cell integration are effectively conveyed by these shorter ECM sequences immobilized on surfaces. These primary amino acid sequences important to cell–surface attachment have been extensively studied as integrin-stimulating peptides [11,12,32, 53–56,68,73]. By far the most extensively studied peptide adhesion ligand is the arginine-glycine-aspartic acid or RGD sequence [72] found in the tenth type III repeat unit of the ECM glycoprotein fibronectin. This fibronectin RGD sequence resides in the central part of this protein [57]. Recent evidence indicates that RGD does not act alone in adsorbed fibronectin, but is thought to work in tandem with flanking sequences in this protein to promote cell receptor recognition and full adhesion events [74–76].

Typically, fibronectin adsorbed to surfaces increases cell attachment [77,78]. The intrinsic binding affinity of fibronectin for cell receptors seems to be surface or conformationally modulated: cell receptor binding affinity for soluble fibronectin is substantially less than that found on surface-adsorbed fibronectin [50–52,58,77–80]. The synthetic RGD sequence alone duplicates many of fibronectin's cell adhesion properties, but perhaps not all. This is remarkable, considering fibronectin's substantial, multidomain size of roughly 450 kDa, and recent suggestions that other protein domains are involved in complete cell binding as well [74,75].

Significantly, the RGD adhesion sequence is common to collagen type I, thrombospondin, fibrinogen, fibronectins, vitronectin, osteopontin, bone sialoprotein I, and likely other proteins yet undiscovered. Soluble and tethered forms of RGD peptides can be utilized as cell adhesion inhibitors and promoters, respectively. Commercial sources of these soluble peptides are numerous [53,81]. Because these molecules are readily available, less than 100 times as massive as their ECM parental "sources," and effective adhesion promoters when immobilized, they are now widely applied in various biomedical and biotechnology device settings.

Second, while RGD immobilization (details later) has certainly received most attention as a cell-targeting moiety, new classes of biomaterials have been designed with the ability to adsorb specific adhesive protein classes [22,29], peptides [53,82–85], carbohydrates [45, 86–89], phospholipids [90], and nucleic acids [91] to surfaces as cell-specific ligands.

RECEPTOR-MEDIATED CELL–SURFACE ADHESION

Mechanisms of cell–surface adhesion are extensively studied and documented. Both receptors and their target ligands have been reviewed [11,12,53–55,92,93]. Focus in this chapter is limited to aspects of enhancing cell–receptor binding to ligand-immobilized surfaces that promote further subsequent development of effective attachment, spreading, and signaling responses.

Cell integrin receptors that recognize natural ECM adhesive motifs in both proteins and truncated peptides are widely expressed on most mammalian cell types as heterodimeric transmembrane proteins consisting of α and β transmembrane subunits [94,95]. Families of related cell integrins that recognize the RGD sequence as an adhesive ligand, either alone or within complete ECM proteins, have been identified as $\alpha_5\beta_1$, $\alpha_V\beta_1$, $\alpha_M\beta_2$, $\alpha_{IIb}\beta_3$, $\alpha_V\beta_3$, $\alpha_V\beta_6$, $\alpha_i\beta_j$ [96–98], although from the 16 known α subunits and 8 known β subunits, many other pairing permutations are possible. Integrin-mediated binding of cells to ECM proteins adsorbed onto surfaces results in events significant to biomaterials and tissue engineering [93], including the adherence of the cells to these surfaces and activation of intracellular signal transduction cascades [40,41] resulting in gene expression [99], and the recruitment and reorganization of cytoskeletal components responsible for cell phenotypic expression [100]. Thus, integrin-mediated cell attachment to biomaterials requires their natural recognition of domains within ECM proteins or relevant immobilized adhesive peptide sequences such as RGD. This receptor-mediated approach to integrate biomaterial with tissue-derived cells *in vivo* requires that the surface-resident integrin peptide ligands remain surface accessible, conformationally and chemically stable, and immobilized at a sufficient density to support receptor–ligand binding. Several related issues are identified:

1. Physiological fluids contain many other surface active components including proteins, proteoglycans, proteases, fatty acids, lipids and platelets that might bury, alter, obscure, or remove these ligands from biomaterials.

2. Cell integrin receptor expression in culture cells is often different from patterns of receptor expression seen *in vivo*, despite demonstrated capabilities of both types of cell context to facilitate integrin–ECM binding. Additionally, temporal sequences of integrin expression may change over time with cell differentiation or phenotypic alteration.

3. Short peptide sequences recognized by many α,β-integrins retain only primary sequence information; that is, they are too small for secondary, tertiary, or quaternary structural transitions. Hence, these short peptides are stable to thermal-, solvent-, or surface-induced denaturation. Longer synthetic, biomimetic peptide sequences, particularly conju-

gates with flanking spacers, other peptide units, or longer biopolymer constructs, may be more susceptible to structural perturbations or enzyme alteration that reduce integrin recognition at a surface.

4. The soluble RGDS sequence is known to block the ability of platelets to participate in platelet aggregation [15,18] and to inhibit fibroblast adhesion to fibronectin precoated surfaces [101].

CELL–PEPTIDE AND PROTEOGLYCAN BINDING

Although several proteins that contain the RGDS peptide sequence serve as ligands for integrins, other peptide sequences also bind integrins and can be immobilized on biomaterial surfaces and utilized for cell binding and attachment. Examples of various peptide sequences derived from ECM molecules such as vitronectin, laminin, collagen, and fibrinogen have also been used to mediate cell attachment [44]. Furthermore, more unique peptides may be immobilized to surfaces to target a specific cell type for adhesion. For example, the peptide REDV has been used to selectively adhere endothelial cells to biomaterials and not fibroblasts, smooth muscle cells, or platelets [102]. The peptide sequence DGEA, found in type I collagen [47], was shown to block the binding of vascular smooth muscle cells [103] but not the binding of fibrosarcoma and osteosarcoma cell lines to type I collagen [104]. Growth factor immobilization represents a more recent "whole-molecule" immobilization approach to serum-free cell culture. Insulin [105–107] and epidermal growth factor, either homogeneously [108–110] or patterned [111] have shown the ability to target cell adhesion without other ECM proteins.

Furthermore, other researchers have used peptides known to bind to biological molecules other than integrins that are also found on surfaces of many cell types. Glycosaminoglycans (GAGs), such as heparan sulfates, are found on most mammalian cells and are recognized as ligands for many known proteins, such as antithrombin III, basic fibroblast growth factor, and fibronectin. By creating a peptide analogous to the heparin-binding amino acid motif, Dee and coworkers [112] were able to bind rat calvarial osteoblasts to glass coverslips via the ligand KRSR; however, the RDGS sequence did not support adhesion of this same cell type, indicating integrin-independent cellular binding. In 1999 Sakiyama *et al.* [113] used four different heparin-binding domains of antithrombin III to bind to GAGs found on dorsal embryonic chick neurons. The results indicate that these peptides in fibrin gels could be used as therapeutic agents for regeneration of peripheral nerves by serving as nerve guide tubes.

Peptides and proteins need not be the only biological molecules to tether and support cellular growth on surfaces. It is well known that mammalian cells have a dense forest of carbohydrates on their surfaces, often acting as ligands for lectins found on different mammalian cell types. This interaction between carbohydrates and their lectin receptors is important for cellular migration and for cells to adhere to one another. In one study [114], clusters of galactose tethered to PEO surfaces supported the binding of primary hepatocytes through the asialoglycoprotein receptor and promoted cellular spreading. Thus, there is a great potential to exploit additional, ubiquitous cell–surface biological molecules to attach specific cell types to biomaterials.

ADHESION MOLECULE IMMOBILIZATION

Known exploited targeted cell–surface adhesion methods reported to date include (1) directly adsorbing ECM proteins or related peptides, such as RGDS, to engineered surfaces optimized to retain active forms of these proteins and peptides (Fig. 66.1), (2) preparation of materials surfaces used to covalently link minimal binding fragments of ECM proteins and/or peptides to surfaces (Fig. 66.2), and (3) the use of known cell-specific ligands to target various cell and tissue types to surfaces. It is impossible within the scope of this chapter to describe every experimental protocol and technique currently used to target specific cell types to biomaterials. Hence, a few examples are described along with the current primary literature reference compendium that encompasses many of the methods currently used in this rapidly evolving field of study and provides more precise experimental details.

To successfully integrate mammalian cells with materials to regenerate functional tissues, biomaterial surfaces must foster an environment in which the cell type of interest actively proliferates and behaves in a manner similar to physiological conditions (phenotypic preservation). This also includes the maintenance of the cell phenotype within the proliferating context without either malignant transformation or regressive adaptation where cells may lose their differentiated phenotype, contain atypical nuclei, or proliferate at high rates at specific focal areas [115]. Design of biomaterials with a controlled hierarchy of structures and chemistries that optimize regeneration of specific cell types is imperative for continued cell viability and maintenance of function *in vivo*—properties essential for tissue-engineered devices. This approach of designing synthetic implant surfaces that actively stimulate cellular growth and proliferation has been recently termed "biomimetic engineering of materials" [92]. Fabrication of biomaterials that improves the spontaneous induction of tissue integration requires a sophisticated hybridization of chemistry, biochemistry, cell biology, and materials engineering. Because bacteria and immune surveillance cells (e.g., macrophages) also have receptors for ECM proteins, performance parameters for these materials include encouragement of rapid cell integration without either adverse immune responses [116–119] or bacterial colonization [120,121]. Each of the bulk material physical characteristics beneath such engineered surfaces (in the form of solid polymer, metal, or ceramic monoliths, gels, sheets, fibers, textiles, composites, or particles) can be considered separately for surface modification in specific applications. However, each bulk materials chemistry presents associated challenges for surface derivatization to permit effective and controlled immobilization of cell adhesion ligands.

Several common bioimmobilization options used to modify surfaces with recognizable ligands [122,123] are summarized in Figs. 66.1 and 66.2. Materials texture, topology, and porosity also play a large role in mediating cell attachment behavior independent of chemistry [124–127]. However, effects of combinations of ligand bioimmobilization and surface topology on cell attachment are not well investigated.

Many cell types have been cultured *in vitro* on biomaterials surfaces, including skin [128], bone [112,129–132], nerve [3–5,69,113,133–137], cornea [42], endothelial cells [102,138], spinal cord [139], hematopoeitic cells [118,119,140,141], platelets [142, 143], stem cells [144–146], hepatocytes [147–150], sensory hair cells [151], islets of Langerhans (pancreatic) cells [143,152], uterine cells [153], kidney [12,154], and fibroblasts [19, 20,22,155] to study aspects of cell–material and cell–cell communication as well as cell culture and organogenesis. Now we turn to a discussion of a few examples of some of these cell–surface studies in the context of targeted cell adhesion.

HEMATOPOEITIC CELLS

All implanted materials activate immunosurveillance cells, including monocytes, neutrophils, and macrophages, producing both acute and chronic immune responses leading to inflammation and local adverse responses. Macrophages have the integrin receptor Mac-1, which recognizes adsorbed proteins on surfaces [18]. New biomaterials that minimize or control immune cell reactions using implant chemistry, drug (e.g., cytokine) release, or novel receptor-based ligands to cue desirable immune and/or inflammatory responses are a promising approach. Kao and coworkers [117] reported that human macrophages were able to attach to poly(ethylene glycol)-based gel networks grafted with the RGDS cell-binding domain. Although foreign body giant cell (FBGC) formation characteristic of foreign body response was observed, depending on the immobilized orientation of the displayed RGD, interleukin 4 (IL-4) cytokine stimulation of FBGC formation at the implant was controlled both by use of a delivery system that limited IL-4 participation and by injecting a monoclonal antibody against IL-4.

PERIPHERAL AND CENTRAL NERVOUS SYSTEM (CNS)

Nerve regeneration is a compelling need in both central and peripheral nervous system trauma and pathologies. The use of biomaterials for guided regeneration is often required for directional nerve growth; but nerve cells often require special interfacial properties to use synthetic nerve guides. Furthermore, while peripheral nerves are more conducive to regrowth, neurons from the CNS cannot regenerate spontaneously following injury, except for in the presence of peripheral nerve cell basal lamina or on a surface that mimics this basal lamina [156]. In one example, fluoropolymers, commonly used *in vivo* in a variety of implants and known to be inert in the CNS and generally nonadhesive [49] are grafted with cell-adhesive peptides YIGSR, IKVAV, or RGDS. These grafted surfaces supported the *in vitro* growth of primary hippocampal neurons, demonstrating that peptide-modified surfaces were capable of mimicking the laminin–hippocampal nerve cell interactions [157]. Another set of CNS regeneration studies found that cortical astrocytes from rats were able to grow on increasingly hydrophobic materials and could form a cellular matrix as long as cell-membrane-associated proteins were capable of adsorbing to the surfaces at sufficient density and/or conformation [133].

DERMAL AND ENDOTHELIAL CELLS

Perhaps the cell–materials integration technology closest to commercial markets seeks to provide synthetic or biohybrid materials as scaffolds for dermal wound repair. Collagen is a major component of the ECM and is used extensively in various forms for biomaterials because of its known abilities to support cell attachment, proliferation, differentiation, and migration. Collagen comprises a family of at least 15 distinct proteins with distinct functions. Each collagen type has specific histological distribution and function. Collagen type I contains both RGD and DGEA sequences known to interact with integrins. A number of cell types have been shown to attach and move efficiently through various collagen matrices. However, this attachment and migration is also dependent on additional cell-adhesive molecules, including cell surface ligands for integrins [79,96,97], schematically shown in Fig. 66.3. Immobilized collagen peptoid collagen mimetic materials comprising Gly-Pro-Nleu helical repeats show cell surface binding activity. This binding is

Gels and Soft Materials:

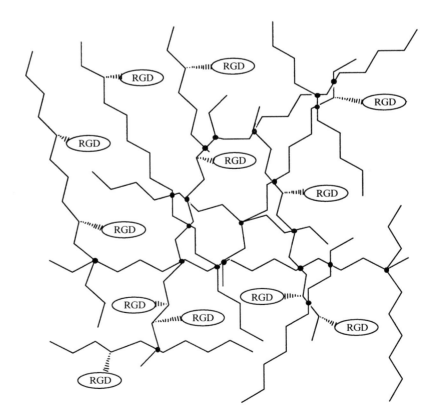

Fig. 66.3. Example of RGD peptides cross-linked to a gel matrix. Gel matrix can be a number of different materials: PEG, elastin, collagen, HA, dextrans, cellulose, and alginates.

inhibited by peptide KDGEA, known to interact with the $\alpha_2\beta_1$ integrin [158]. In addition to direct collagen modification with RGD ligands [60], another study covalently modified collagen–glycosaminoglycan matrices with an RGD-containing peptide to enhance cell interaction with implants comprising these materials [89]. Both these biopolymer matrix components are important members, forming mature fibrillar networks of ECM molecules found in basement membrane and native ECM. The three cell types tested—fibroblasts, keratinocytes, and endothelial cells, all important for wound healing—demonstrated increased attachment to these peptide-modified matrices over GAG–collagen materials alone. Thus, improved cell–material interactions in collagen are mediated through cell-based integrin recognition of the RGD sequence in the matrix.

Skeletal

Bone structural implants of many types have long been used for space-filling and structural replacements in both the medical and dental fields. Titanium alloys have often been used as orthopedic implant materials because of their strength and biointerfacial properties with bone cells *in vivo* [159]. Titanium rods surface-coated by means of gold–thiol chemistry with the peptide sequence RGDC and transplanted into rat femurs [82,83] exhibited a significant increase in bone formation observed around the peptide-modified titanium implants versus unmodified implants. Alkylsilanes terminated with RGD peptides have been used to modify titanium surfaces, using its native oxide for coupling to enhance cell adhesion *in vitro* [82,160]. Silicon oxide has been modified with analogous RGD–silanes as well, and shown to promote spatial control of cell attachment [161]. Thus, the presence of the RGDC sequence may prove to enhance long-term metal–osseointegration by bone cell integrin mediated binding to peptide ligands on implant surfaces.

Cell Adhesion Surfaces for Various Cell Types

Many other examples of RGD and other ligand-immobilized surfaces for cell adhesion have been reported. Predictably, cell adhesion is often demonstrated on surfaces that are otherwise nonadhesive in serum, most notably hydrogels (Fig. 66.3), polylactides, and fluorinated or hydrophobic polymers. Specifically, numerous natural and synthetic hydrogels have been grafted with RGD peptides [42,84,85,102,131,139,162–165]. Hydrophobic polymers that have been modified with RGD include fluoropolymers [69,118,166], polylactides [167], Dacron [168], and poly(propylene-*co*-fumarate) [169]. Other RGD-modified polymers include various biomedical polyurethanes [71,170,171].

Additionally, other RGD-presenting biomaterials constructs include RGD-grafted lipids as cell-adhesive biomembrane mimics [172,173], triple-helix collagen-type lipopeptides [61, 62,79], and countless covalently attached or derivatized organic thin-film designs for surface modification. Specific RGD-immobilized peptide thin films include functionalized silanes for active oxide surfaces [24,82,161,174], functionalized thiol-modified metals [82,83,175], and *in situ* surface-grafted polymer surfaces [28,176,177].

Cell Attachment in Serum versus Serum-free Conditions

Cell attachment, spreading, and anchoring are critical to the survival and growth of most nonhematopoietic cell types. Attachment under normal physiological conditions is a complex dynamic of cell–surface, chemomechanical, and biochemical stimuli that rely on receptor specificity for both soluble and substrate-resident cues in the presence of multiple nonspecific proteins. In tissue culture, cells are typically grown in specialized nutrient media supplemented with 10% complete serum comprising over 200 soluble proteins and peptides. Confluent cultured monolayers of attached cells most frequently employ commercial surfaces such as tissue culture grade polystyrene (TCPS: i.e., polystyrene treated by oxygenated gas plasma to create a more hydrophilic oxidized polymer surface). This disposable surface, well known as the commercial cell culture standard, provides a surface chemistry that adsorbs sufficient amounts of trace ECM proteins from serum-supplemented media to promote cell attachment [35–37,50,51,88]. Most attachment-dependent cell types will affix to and proliferate on serum-protein-adsorbed TCPS surfaces. While this is a relatively simple and general technique to attach cells to surfaces, it is also a nonspecifically mediated attachment, since serum media contain trace amounts of many distinct proteinaceous attachment factors such as fibrinogen, collagen, vitronectin, fibronectin, as well as a higher abundance of other soluble proteins, including albumin and immunoglobulins, that block cell adhesion on surfaces.

Hence, substrates successful in promoting cell attachment in a physiological environment are those that have surface chemistry facilitating adsorption of sufficient amounts of ECM proteins in the presence of competing non-cell-adhesive proteins (e.g., albumin) [14]. Because elucidation of surface chemistries that provide this ECM selectivity has resulted from empirical, often trial-and-error, materials analysis, current design of improved surface chemistries to promote cell attachment and proliferation with ECM proteins is limited to more primitive concepts of hydrophilic/hydrophobic balance [19,38,45,46,50–58,88]. Biomaterials that intrinsically lack desired cell attachment properties [e.g., bacteriological grade polystyrene, polylactides, Dacron, poly(tetrafluoroethylene) (PTFE), silicone elastomers] are often surface-modified with polar groups to modify serum protein deposition patterns to enhance ECM uptake and improve cell–surface recognition. Plasma-modified implant devices, coatings, and chemical grafting are often used with this simple objective, without much molecular understanding of what synthetic surfaces are actually doing in physiological media to provide this enhancement. Biased uptake of desirable ECM adhesive components onto suboptimal surfaces to confer improved cell attachment can also be achieved by precoating these substrates with ECM components directly. Excess fibronectin, laminin, and collagen monocomponent preadsorption to substrates has been shown to improve attached cell density in a number of biomaterials contexts, *more or less independent of surface chemistry* [35–37,157]. Commercial cell adhesion products (e.g., Pronectin, Biocoat, and Matrigel) are rich in soluble ECM-like fragments or mimicking components or polymers that promote ECM or cell attachment to surfaces (Fig. 66.1) for exactly this purpose.

Hence, direct adsorption of soluble ECM components can be successfully used to facilitate and guide cell attachment. Patterning of ECM surface deposition has been used to pattern cell attachment, directional migration, and proliferation spatially on surfaces [45,125,127,178,179], based on the principles of surface chemistry to spatially select either cell-adhesive [31,180] or nonadhesive protein adsorption. Alternatives include more directed two-dimensional or three-dimensional [126] surface spatial control, using chemically lithographed patterns of nonadsorbing chemistry [e.g., dextran or poly(ethylene glycol) grafted chains or gels] combined with immobilized polar chemistries known to promote ECM protein deposition, or the use of directly immobilized patterns of ECM peptides or other cell receptor ligands [44,179,181,182]. Nonetheless, passively facilitated cell attachment via serum protein surface adsorption is problematic under physiological conditions in proteinaceous milieu in which smaller non-cell-adhesive proteins (e.g., albumin, 66 kDa, 40 mg/ml in serum) are naturally biased to outcompete large ECM proteins (e.g., fibronectin, 450 kDa, 300 µg/ml in serum) for surface sites. Current knowledge of surface determinants that can selectively bind desired soluble proteins from serum to surfaces at sufficient densities and viable receptor-recognized conformations is limited. Furthermore, serum-based culture is time- and temperature-sensitive, expensive, potentially a carrier of viruses, prions, and pathogens and, therefore, not always desirable as a media supplement to promote cell attachment. Direct immobilization of requisite, minimized cell attachment factors had advantages.

Thus, for more specific, tailored interactions between cells and surface-adsorbed proteins, single-serum ECM proteins (e.g., collagen or fibronectin) or adhesive peptide sequences such as RGD are immobilized to achieve more controlled experimental cell culture conditions. In the case of RGD peptides, cell attachment to surfaces can be maximized [183]. Other methods specifically design a protein "capture surface" to bind cells to surfaces, even without integrin receptor involvement. For example, cells can be captured by using monoclonal antibodies absorbed to surfaces directed against cell membrane proteins [70] or by using the extensively studied biotin–(strept)avidin chemistry by adsorbing avidin to surfaces and attaching biotinylated endothelial cells to these surfaces [138,184]. This latter technique takes advantage of the high-affinity interaction (K_d 10^{-15} M) between avidin and biotin, forming a noncovalent bond that is unaffected by most extremes of pH, organic solvents, and other denaturing reagents [185]. Direct chemical or biochemical cell immobilization to surfaces bypasses "normal" cell receptor–surface interactions to achieve adhesion. Yet, important integrin-mediated signal transduction events necessary for phenotypic expression, proliferation, and other possibly necessary downline events are bypassed as well. Such approaches require further study to evaluate their value to longer term cell biomaterials devices and tissue engineering.

EXPERIMENTAL PROTOCOLS FOR ACCESSING CELL–SURFACE INTERACTIONS

To target cells to specific biomaterial surfaces, it is necessary to provide prospective new, improved materials that not only fail to elicit adverse immune responses, but do facilitate the selective adsorption or covalent attachment of biological molecules, including proteins, peptides, oligosaccharides, or lipids that support cell adhesion and stimulate cell proliferation and growth. Most mammalian cell types use integrins to recognize these ligands, and attachment of either ECM proteins or these minimal cell-binding peptide sequences facilitates attachment of most cell types to biomaterial surfaces. Although the most convenient route to achieve this is by surface adsorption, the problems with this method (loss of protein three-dimensional conformation, lack of recognition or availability of cell-binding motif, and insufficient ECM density available for integrin recognition and binding) often preclude effective cell–surface attachment. By covalently linking ECM-like peptides to biomaterials, presentation of the minimal amino acid sequence known to support cell adhesion is more directly controlled both in spatial patterns and density. Furthermore, peptides can be produced synthetically, and they are generally more stable to pH, heat, storage, and desiccation fluctuations in biotechnology applications than high molecular weight serum or ECM proteins [44].

Many methods have been reported that covalently link RGD, IKVAV, YIGSR, and other less-studied motifs described herein to various biomaterials, and the design of these surfaces may contain patterns that regulate cell growth at specific regions on these surfaces [125–127,181]. Some examples of peptides covalently linked to materials include glass [174], quartz [177], PVA [176], polyurethanes [71,170,171], other synthetic polymers [157,167,178,186], polystyrene and polyterephthalate films [91,187], methacrylate [181], PEG [164,178], acrylic acid [164], dextran hydrogels [181,182], silicon [126,127], pregrafted polymers [61], elastin [188], collagen [60], alginate hydrogels [131,172], and agarose [132].

Other methods use materials of these types, but then simply adsorb proteins or peptides to the material, or else allow cells to directly adhere to the material. A few examples (and methods referenced therein) are (1) solid surfaces (titanium, titanium dioxide, titanium carbide [159], gold, stainless steel [189], and calcium phosphate), (2) polymer matrices (fluorine-containing polymers [49,69,118,157], biodegradable polylactide copolymers [135,162], ethylene–vinyl acetate copolymer [189,190], polystyrene [91], polypropylene, polyurethane [71], and dextran–methacrylate copolymers [181,182]), (3) organic thin films (polymer coatings, alkylsilanes [22,30,46,112,126,129,130,134,141,174,179], alkanethiols terminated with various functional groups [19–21,23–25], and poly(ethylene oxide) (PEO) [121]), and (4) three-dimensional hydrogels (PVA, carboxymethylcellulose, alginates, poly(acrylates) [2,144,152,191] and poly(acrylamides), collagen [7,89,143], PEG-grafted scaffolds [121,135,140,162,163,191], fibrin [113], hyaluronic acid (HA)-modifed materials [86–88,153]).

Protein adsorption to the surfaces of these materials with the intent of selectively promoting cell–surface adhesion might proceed by using one of the following strategies to simply immerse these clean substrates into proteinaceous milieu.

1. Preadsorption with monocomponent ECM (fibronectin, collagen I) solutions followed by albumin masking or cell adhesion assays. Because ECM monoadsorption impacts cell adhesion independent of surface chemistry by imparting sufficient ECM density and conformation, patterning ECM deposition provides spatial control of cell adhesion [50,51].

Alternatively, patterns of a nonadsorptive surface chemistry (e.g., PEG), can be used with monocomponent ECM solutions to pattern cell adhesion (all or nothing deposition) [21–27, 161,175,178–180]. Protein-resistant surface chemistry fails to accumulate sufficient ECM to support cell attachment in designated patterned regions, while other areas bind ECM proteins from monocomponent solutions.

2. Competitive selection of protein deposition from multicomponent (e.g., serum) solutions [14]. Surface chemistry (such as TCPS) with known ability to preserve fibronectin (or ECM protein) surface density, conformation, and integrin recognition when adsorbed from serum dilutions can be exposed to such milieus for several hours to promote cell attachment (standard cell culture conditions). Patterned surfaces using distinct regions of ECM-selective surface chemistry with other chemistry known to promote adsorption of non-cell-adhesive proteins (e.g., albumin) naturally permits cell patterning by exploiting differential serum protein binding (on alkylated of fluorinated surfaces) [17,19,20,22,27,29,34–37,49,92,118, 133,134].

A number of biochemical and molecular biological protocols can then be applied to study the immobilized peptide (or biological molecule of choice) on the biomaterial, and then the attachment, proliferation, and growth of cells that these modified surfaces support. A brief outline of some more common methods is provided:

Enzyme-Linked Immunosorbent Assays (ELISAs)

Commonly throughout biological research, ELISAs utilize the specificity of antibodies to a particular antigen of interest. Reagents are diverse and sourced by many commercial vendors in numerous configurations and detection schemes. Also, a large number of experiments can be performed at a very small scale. (A typical ELISA plate contains 96 wells, which corresponds to 96 potential experiments, and each well has a volume of 200–300 μl, depending on the particular ELISA plate.) Quantification of proteins/peptides, lipids, oligosaccharides, and so forth adsorbed or covalently linked to surfaces can be conducted by ELISA experiments using antibodies directed against these molecules, followed by the addition of an enzyme-tagged secondary antibody, which produces a colorimetric product when substrate is added during the development step. An ELISA plate-reader determines optical absorbance from each well and thus quantifies the relative amount of material initially bound to the plate. Because like receptors, antibodies can recognize antigen conformation and probe from the solution phase, ELISAs can produce data on "recognizable" fractions of adsorbed proteins that more direct radiolabeling experiments cannot discern.

Cellular Analysis and Count

Cell–surface attachment can be observed directly by means of phase contrast microscopy and cell number quantified by removal of cells from the surfaces of cell culture dishes or flasks using trypsin, a sterile cell scraper, or low temperature [192], followed by counting the cells with a hemocytometer or an instrument, such as a Coulter Counter particle and cell size analyzer, or else a flow cytometer, a commercial laser-based cell counter. Fluorescence markers, either endocytosed by cells or attached to cell-specific antibodies, are readily available and can be used to enhance sensitivity and distinguish cell viability [193].

Mitogenic Assay

Cell viability can be monitored by a number of specific methods that correlate to the particular cell type of interest. A mitogenic assay is one way of measuring cell proliferation upon cellular stimulation by a mitogenic factor, such as a growth factor, an antigen, or a synthetic molecule. Cell activation is determined by the incorporation of a radiolabeled nucleotide into cellular DNA. In this assay, metabolic uptake of [³H]thymidine is a simple procedure, requiring only the addition of the labeled base to the cellular growth medium, followed by an incubation period at 37°C for 48–72 h, after which the amount of radioactivity is determined by using a scintillation counter. Actively proliferating cells incorporate more [³H]thymidine into their DNA than cells that are proliferating slowly or not at all.

IMMUNOFLUORESCENCE TAGGING

Cytoskeletal reorganization within a cell that has attached to a surface is one of the events that occurs when cells spread on a surface after attachment [48]. Immunofluorescence microscopy utilizes fluorescently tagged antibodies to mark specific, accessible intracellular cytoskeletal components or organelles, thus demonstrating where a particular protein of interest is located within the cell. Formation of focal adhesions, for example, can be studied by immunofluorescence by staining components such as paxillin, actinin, or vinculin with probe-tagged antibodies to these cytoskeletal proteins [19,20].

POLYACRYLAMIDE GEL ELECTROPHORESIS (PAGE), WESTERN BLOTS, AND NORTHERN BLOTS

Sodium dodecyl sulfate (SDS)-PAGE [194] and immunoblotting techniques (Western blots), can be used to study the production of various proteins, protein modifications (such as phosphorylation), or protein transcripts (mRNA anaylsis by Northern blots) [195,196] that are normally produced by cells in physiological environments. Cells can be removed from the culture surface and lysed, and proteins can be extracted, which then can be run on PAGE gels for size analysis and also for identification by means of antibodies specific to individual proteins. For Northern blots, mRNA is isolated from the cytosol, and a labeled antisense probe can be used to hydrogen bond with the mRNA sequence of interest captured onto a membrane. Commercially popularized polymerase chain reactions (PCR) may also be utilized to amplify the amount of mRNA isolated to enhance sensitivity to small cell numbers.

ADDITIONAL ASSAYS

Kinetic assays can be used to study long-term versus short-term cellular growth [30]. Aberrant cells grown on nonoptimal surfaces can be monitored by identification of expressed abnormal phenotypic markers, such as proteinslike growth factors or transcription factors, or cytosolic secondary messengers expressed during malignant transformation or apoptosis. Of course, many more methods available but not mentioned here are used regularly or produced by biotechnology companies in the form of commercial protein, peptide, cell, DNA, isolation/detection/analysis kits. Many of these protocols, described in the methods sections of the literature listed in the References, are available commercially in various useful forms as kits (e.g., Molecular Probes, Pierce, Pharmacia Biotech, and Clonetics all have catalogs that contain useful product and assay information) or can be found in biochemical manuals and textbooks [197] or queried via the Internet.

FUTURE DEVELOPMENTS

Bioimmobilization strategies have traditionally focused on coupling chemistries that have changed only incrementally over decades. Hence, little focus on such chemistry was attempted here. Rather, the bioactive candidates selected for surface coupling have evolved as cell drug and molecular targets have been discovered. Hence, it is likely that future developments will follow tradition: incremental improvements in surface chemistry coupling methods will likely attract less attention than the new ligand targets they couple. Additionally, molecular biology and genomics methods will facilitate accelerated discovery, iterative minimization, and optimization and application of new ligands that target specific receptor-adhesive signaling functions. Challenges exist in surface density control of immobilized ligands [198], and in mixed ligand copresentation of multiple targets that may synergize or uniquely modulate cell responses via aggregation in suspension or upon receptor recognition [199].

Adhesion implies contact between two opposing surfaces. Focus to date has been directed nearly entirely on one receptor (integrin) and a few binding ligands, primarily RGD. A newer approach involves modification of the cell surface to modulate adhesion. The simplest concept might be pharmacological or genetic regulation of cell surface receptor expression, exploiting the dynamics of cell receptor densities with artificial soluble stimuli. A more attractive strategy extends the exciting and pioneering work of Bertozzi and coworkers [200], who directly and specifically modify the cell glycocalix chemically to provide functional groups for a number of possible cell surface modifications. This approach could readily include direct coupling of cells to surfaces, addition of artificial "receptor-like" molecules, or use of ECM ligands that facilitate cell–cell surface communications. Asymmetric (i.e., cell apical, or basal surface) control of varied ligands and receptors that promote either selective cell–surface or cell–cell communication could form a basis for controlling three-dimensional cell constructs for tissue architectures.

Last, fibronectin, elastin, collagen, and other extracellular proteins are often first naturally excreted in soluble form by the cell and then enzymatically processed into insoluble fibrillar matrices *in situ* on surfaces postadsorption, often with evolving hierarchical and anisotropic fibrillar structure. Peptide or saccharide immobilization fails to produce this processing or evolving matrix structure. Because endogenous ECM fibrillar structure is often found coaligned with the attached, spread cell's cytoskeleton in natural matrix postprocessing [48], such alignment is not possible in grafted ligand systems. Methods of synthetically duplicating molecular "ropes" of aligned ECM-type ligands, or ligand patterns of linear resolution commensurate with cell receptor spacings, might facilitate cytoskeleton construction and selective control of different cell signaling/proliferative responses.

ABBREVIATIONS

EDC:	Ethyl-3-(3-dimethylaminopropyl)carbodiimide hydrochloride
DGEA:	Aspartic acid-glycine-glutamic acid-alanine
FBGC:	Foreign body giant cell
IKVAV:	Isoleucine-lysine-valine-alanine-valine
KRSR:	Lysine-arginine-serine-arginine
NHS:	N-Hydroxysuccinimide
REDV:	Arginine-glutamic acid-aspartic acid-valine
RGD:	Arginine-glycine-aspartic acid
RGDC:	Arginine-glycine-aspartic acid-cysteine
RGDS:	Arginine-glycine-aspartic acid-serine
SDS-PAGE:	Sodium dodecyl sulfate-polyacrylamide gel electrophoresis
TCPS:	Tissue culture polystyrene
YIGSR:	Tyrosine-isoleucine-glycine-serine-arginine

REFERENCES

1. Thomas, C. A., Springer, P. A., Loeb, G. E., Berwald-Netter, Y., and Okun, L. M. (1972). A miniature microeletrode array to monitor the bioelectric activity of cultured cells. *Exp. Cell Res.* **74**, 61–66.
2. Quinn, C. P., Pathak, C. P., Heller, A., and Hubbell, J. A. (1995). Photo-crosslinked copolymers of 2-hydroxyethyl methacrylate, poly(ethyleneglycol) tetra-acrylate and ethylene dimethacrylate for improving biocompatiblity of biosensors. *Biomaterials* **16**, 389–396.
3. Sagen, J., Wang, H., Tresco, P. A., and Aebischer, P. (1993). Transplants of immunologically isolated xenogeneic chromaffin cells provide a long-term source of pain-reducing neuroactive substances. *J. Neurosci.* **13**, 2415–2423.
4. Tresco, P. A., Winn, S. R., Tan, S., Jaeger, C. B., Greene, L. A., and Aebischer, P. (1992). Polymer-encapsulated PC12 cells: Long-term survival and associated reduction in lesion-induced rotational behavior. *Cell Transplant.* **1**, 255–264.
5. Winn, S. R., Tresco, P. A., Zielinski, B., Sagen, J., and Aebischer, P. (1992). Microencapsulated bovine chromaffin cells in vitro: Effect of density and coseeding with a NGF-releasing cell line. *J. Neural Transplant. Plast.* **3**, 115–124.
6. Aebischer, P., Wahlberg, L., Tresco, P. A., and Winn, S. R. (1991). Macroencapsulation of dopamine-secreting cells by coextrusion with an organic polymer solution. *Biomaterials* **12**, 50–56.
7. Bonadio, J., Smiley, E., Patil, P., and Goldstein, S. (1999). Localized, direct plasmid–gene delivery in vivo: Prolonged therapy results in reproducible tissue regeneration. *Nat. Med.* **5**, 753–759.
8. Thomas, M., Yang, L., and Hornsby, P. J. (2000). Formation of functional tissue from transplanted adrenocortical cells expressing telomerase reverse transcriptase. *Nat. Biotechnol.* **18**, 39–42.
9. Lanza, R. P., Langer, R., and Vacanti, J. P., eds. (2000). "Principles of Tissue Engineering," 2nd ed. Academic Press, San Diego, CA.
10. Langer, R., and Vacanti, J. P. (1993). Tissue engineering. *Science* **260**, 920–926.
11. Hubbell, J. A. (1995). Biomaterials in tissue engineering research. *Bio/Technology* **13**, 565–576.
12. Pierschbacher, M. D., Polarek, J. W., Craig, W. S., Tschopp, J. F., Sipes, N. J., and Harper, J. R. (1994). Manipulation of cellular interactions with biomaterials toward a therapeutic outcome: A perspective. *J. Cell. Biochem.* **56**, 150–154.
13. Williams, S. K., Jarrell, B. E., Friend, L., Radomski, J. S., Carabasi, R. A., Koolpe, E., Mueller, S. N., Thornton, S. C., Marinucci, T., and Levine, E. (1985). Adult human endothelial cell compatibility with prosthetic graft material. *J. Surg. Res.* **38**, 68–629.

14. Horbett, T. A. (1994). The role of adsorbed proteins in animal cell adhesion. *Colloids Surf. B: Biointerfaces* **2**, 225–240.

15. Norde, W., and Lyklema, J. (1991). Why proteins prefer interfaces. *J. Biomater. Sci. Polymer Edn* **2**, 183–202.

16. Horbett, T. (1993). Principles underlying the role of adsorbed plasma proteins in blood interactions with foreign materials. *Cardiovasc. Pathol.* **2**, 1375–1485.

17. Horbett, T. (1982). Protein adsorption on biomaterials. *In* "Biomaterials: Techniques of Biocompatibility Testing" (D. F. Williams, ed.), pp. 183–214. CRC Press, Boca Raton, FL.

18. Tang, L., Jennings, T. A., and Eaton, J. W. (1998). Mast cells mediate acute inflammatory responses to implanted biomaterials. *Proc. Natl. Acad. Sci. U.S.A.* **95**, 8841–8846.

19. McClary, K. B., Ugarova, T., and Grainger, D. W. (2000). Modulating fibroblast adhesion, spreading, and proliferation using self-assembled monolayer films of alkylthiolates on gold. *J. Biomed. Mater. Res.* **50**, 428–439.

20. McClary, K. B., and Grainger, D. W. (1999). RhoA-induced changes in fibroblasts cultured on organic monolayers. *Biomaterials* **20**, 2435–2446.

21. Lopez, G., Albers, M., Schreiber, S., Carroll, R., Reralta, E., and Whitesides, G. (1993). Convenient methods for patterning the adhesion of mammalian cells to surfaces using self-assembled monolayers of alkanethiols on gold. *J. Am. Chem. Soc.* **115**, 5877–5888.

22. Webb, K., Hlady, V., and Tresco, P. A. (1998). Relative importance of surface wettability and charged functional groups on NIH 3T3 fibroblast attachment, spreading, and cytoskeletal organization. *J. Biomed. Mater. Res.* **41**, 422–430.

23. Chen, C. S., Mrksich, M., Huang, S., Whitesides, G. M., and Ingber, D. E. (1998). Micropatterned surfaces for control of cell shape, position, and function. *Biotechnol. Prog.* **14**, 356–363.

24. Mrksich, M. (1998). Tailored substrates for studies of attached cell culture. *Cell Mol. Life Sci.* **54**, 653–662.

25. Mrksich, M., and Whitesides, G. M. (1996). Using self-assembled monolayers to understand the interactions of man-made surfaces with proteins and cells. *Annu. Rev. Biophys. Biomol. Struct.* **25**, 55–78.

26. Kane, R. S., Shuichi, T., Ostuni, E., Ingber, D. E., and Whitesides, G. M. (1999). Patterning proteins and cells using soft lithography. *Biomaterials* **20**, 2363–2376.

27. Bhatia, S. N., Yarmush, M. L., and Toner, M. (1997). Controlling cell interactions by micropatterning in co-cultures: Hepatocytes and 3T3 fibroblasts. *J. Biomed. Mater. Res.* **34**, 189–199.

28. Elbert, D. L., and Hubbell, J. A. (1996). Surface treatments of polymers for biocompatibility. *Annu. Rev. Mater. Sci.* **26**, 365–394.

29. Bearinger, J. P., Castner, D. G., Golledge, S. L., Rezania, A., Hubchak, S., and Healy, K. E. (1997). P(Amm-*co*-EG) interpenetrating polymer networks grafted to oxide surfaces: Surface characterization, protein adsorption, and cell attachment studies. *Langmuir* **13**, 5175–5183.

30. Healy, K. E., Thomas, C. H., Rezania, A., Kim, J. E., McKeown, P. J., Lom, B., and Hockberger, P. E. (1996). Kinetics of bone cell organization and mineralization on materials with patterned surface chemistry. *Biomaterials* **17**, 195–208.

31. Matsuda, T., and Sugawara, T. (1995). Development of surface photochemical modification method for micropatterning of cultured cells. *J. Biomed. Mater. Res.* **29**, 749–756.

32. Ratner, B. D., Chilkoti, A., and Lopez, G. P. (1990). Plasma deposition and treatment for biomaterial applications. *In* "Plasma Deposition, Treatment, and Etching of Polymers" (R. d'Agostino, ed.), pp. 468–474. Academic Press, San Diego, CA.

33. Ertel, S. I., Ratner, B. D., and Horbett, T. A. (1991). The adsorption and elutability of albumin, IgG, and fibronectin on radio-frequency plasma deposited polystyrene. *J. Colloid Interface Sci.* **147**, 433–442.

34. Ertel, S. I., Chilkoti, A., Horbett, T. A., and Ratner, B. D. (1991). Endothelial cell growth on oxygen-containing films deposited by radio-frequency plasmas: The role of surface carbonyl groups. *J. Biomater. Sci. Polymer Edn* **3**, 163–183.

35. Van Wachem, P. B., Mallens, B. W. L., Dekker, A., Beugeling, T., Feijen, J., Bantjes, A., Detmers, J. P., and van Aken, W. G. (1987). Adsorption of fibronectin derived from serum and from human endothelial cells onto tissue culture polystyrene. *J. Biomed. Mater. Res.* **21**, 1317–1327.

36. Van Wachem, P. B., Beugeling, T., Mallens, B. W. L., Dekker, A., Feijen, J., Bantjes, A., and van Aken, W. G. (1988). Deposition of endothelial fibronectin on polymeric surfaces. *Biomaterials* **9**, 121–123.

37. Schakenraad, J. M., Arends, J., Busscher, H. J., Dijk, F., Van Wachem, P. B., and Wildevuur, Ch. R. H. (1989). Kinetics of cell spreading on protein pre-coated substrata: A study on interfacial aspects. *Biomaterials* **10**, 43–50.

38. Andrade, J. D., ed. (1985). "Principles of Protein Adsorption." Plenum, New York.

39. Edelman, G. M. (1983). Cell adhesion molecules. *Science* **219**, 450–457.

40. Rouslahti, E., and Pierschbacher, M. D. (1987). New perspectives in cell adhesion: RGD and integrins. *Science* **238**, 491–497.

41. Clark, E. A., and Brugge, J. S. (1995). Integrins and signal transduction pathways: The road taken. *Science* **69**, 11–25.

42. Kobayashi, H., and Ikada, Y. (1991). Corneal cell adhesion and proliferation on hydrogel sheets bound with cell-adhesive proteins. *Curr. Eye Res.* **10**, 899–908.

43. Mooney, D., Hansen, L., Vacanti, J., Langer, R., Farmer, S., and Ingber, D. (1992). Switching from differentiation to growth in hepatocytes: Control by extracellular matrix. *J. Cell. Physiol.* **151**, 497–505.

44. LeBaron, R. G., and Athanasiou, K. A. (2000). Extracellular matrix cell adhesion peptides: Functional applications in orthopaedic materials. *Tissue Eng.* **6**, 85–103.

45. McFarland, C. D., Mayer, S., Scotchford, C., Dalton, B. A., Steele, J. G., and Downes, S. (1999). Attachment of cultured human bone cells to novel polymers. *J. Biomed. Mater. Res.* **44**, 1–11.

46. Lewandowska, K., Balachander, N., Sukenik, C. N., and Culp, L. A. (1989). Modulation of fibronectin adhesive functions for fibroblasts and neural cells by chemically derivatized substrata. *J. Cell. Physiol.* **141**, 334–345.

47. Staatz, D. W., Fok, K. F., Zutter, M. M., Adams, S. P., Rodriguez, B. A., and Santoro, S. A. (1991). Identification of a tetrapeptide recognition sequence for the $\alpha_2\beta_1$ integrin in collagen. *J. Biol. Chem.* **266**, 7363.

48. Ingber, D. E. (1998). The architecture of life. *Sci. Am.* **278**, 48–57.

49. Grainger, D. W., Pavon-Djavid, G., Migonney, V., and Josefowicz, M. (2001). Assessment of fibronectin conformation adsorbed to polytetrafluoroethylene surfaces from serum protein mixtures and correlation to support of cell attachment in culture. *J. Biomater. Sci. Polymer Edn.*

50. Grinnell, F., and Feld, M. K. (1981). Adsorption characteristics of plasma fibronectin in relationship to biological activity. *J. Biomed. Mater. Res.* **15**, 363–381.

51. Grinnell, F., and Feld, M. K. (1982). Fibronectin adsorption on hydrophilic and hydrophobic surfaces detected by antibody binding and analyzed during cell adhesion in serum-containing medium. *J. Biol. Chem.* **257**, 4888–4893.

52. Grinnell, F., and Phan, T. V. (1983). Deposition of fibronectin on material surfaces exposed to plasma: Quantitative and biological studies. *J. Cell. Physiol.* **116**, 289–296.

53. Craig, W. S., Cheng, S., Mullen, D. G., Blevitt, J., and Pierschbacher, M. D. (1995). Concept and progress in the development of RGD-containing peptide pharmaceuticals. *Biopolymers* **37**, 157–175.

54. Dillow, A. K., and Tirrell, M. (1998). Targeted cellular adhesion at biomaterial interfaces. *Curr. Opin. Solid State Mater. Sci.* **3**, 252–259.

55. Hammer, D. A., and Tirrell, M. (1996). Biological adhesion at interfaces. *Annu. Rev. Mater. Sci.* **26**, 651–691.

56. Rouslahti, E. (1988). Fibronectin and its receptors. *Annu. Rev. Biochem.* **57**, 375–413.

57. Pierschbacher, M. D., Rouslahti, E., Sundelin, J., Lind, P., and Peterson, P. A. (1982). The cell attachment domain of fibronectin. Determination of the primary structure. *J. Biol. Chem.* **257**, 9593–9597.

58. Akiyama, S. K., and Yamada, K. M. (1987). Fibronectin. *Adv. Enzymol.* **59**, 1–57.

59. Hayman, E. G., Pierschbacher, M. D., Suzuki, S., and Rouslahti, E. (1985). Vitronectin—a major cell attachment-promoting protein in fetal bovine serum. *Exp. Cell Res.* **160**, 245–258.

60. Myles, J. L., Burgess, B. T., and Dickenson, R. B. (2000). Modification of the adhesive properties of collagen by covalent grafting with RGD peptides. *J. Biomater. Sci. Polymer Edn* **11**, 69–86.

61. Fields, G. B., Lauer, J. L., Dori, Y., Forns, P., Yu, Y. C., and Tirrell, M. (1998). Protein-like molecular architecture: Biomaterial applications for inducing cellular receptor binding and signal transduction. *Biopolymers* **47**, 143–151.

62. Yu, Y. C., Roontga, V., Daragan, V. A., Mayo, K. H., Tirrell, M., and Fields, G. B. (1999). Structure and dynamics of peptide-amphiphiles incorporating triple-helical protein-like molecular architecture. *Biochemistry* **38**, 1659–1668.

63. Kielty, C. M., Hopkinson, I., and Grant, M. E. (1992). Collagen: The collagen family—Structure, assembly, and organization in the extracellular matrix. *In* "Connective Tissue and Its Heritable Disorders: Molecular, Genetic, and Medical Aspects" (P. M. Royce and B. Steinmann, eds.), pp. 103–147. Wiley, New York.

64. Boskey, A. L. (1989). Noncollagenous matrix proteins and their role in mineralization. *Bone Miner.* **6**, 111–123.

65. Stanford, C., and Keller, J. (1991). The concept of osseointegration and bone matrix expression. *Crit. Rev. Oral Biol. Med.* **2**, 83–101.

66. Savage, B., and Ruggeri, Z. M. (1991). Selective recognition of adhesive sites in surface-bound fibrinogen by glycoprotein IIb–IIIa on nonactivated platelets. *J. Biol. Chem.* **266**, 11227–11233.

67. Johnson, P. C., Shepeck, R. A., Hribar, S. R., Bentz, M. L., Janosky, J., and Dickson, C. S. (1991). Inhibition of platelet retention on artificial microvascular grafts with monoclonal antibodies and a high-affinity peptide directed against platelet membrane glycoproteins. *Arteriosclerosis Thromb.* **11**, 552–560.

68. Massia, S. P., Rao, S. S., and Hubbell, J. A. (1993). Covalently immobilized laminin peptide Tyr-Ile-Gly-Ser-Arg (YIGSR) supports cell spreading and co-localization of the 67-kilodalton laminin receptor with alpha-actinin and vinculin. *J. Biol. Chem.* **268**, 8053–8059.

69. Tong, Y. W., and Shoichet, M. S. (1998). Enhancing the interaction of central nervous system neurons with poly(tetrafluoroethylene-*co*-hexafluoropropylene) via a novel surface amine-functionalization reaction followed by peptide modification. *J. Biomater. Sci. Polymer Edn* **9**, 713–729.

70. Dekker, A., Poot, A. A., van Mourik, J. A., Workel, M. P., Beugeling, T., Bantjes, A., Feijen, J., and van Aken, W. G. (1991). Improved adhesion and proliferation of human endothelial cells on polyethylene precoated with monoclonal antibodies directed against cell membrane antigens and extracellar matrix proteins. *Thromb. Haemostasis* **66**, 715–724.

71. Ruiz, L., Fine, E., Voeroes, J., Makohliso, S. A., Leonard, D., Johnston, D. S., Textor, M., and Mathieu, H. J. (1999). Phosphorylcholine-containing polyurethanes for the control of protein adsorption and cell attachment via photoimmobilized laminin oligopeptides. *J. Biomater. Sci. Polymer Edn* **10**, 931–955.

72. Ruoshlahti, E. (1996). RGD and other recognition sequences for integrins. *Annu. Rev. Cell Dev. Biol.* **12**, 697–715.

73. Hynes, R. O. (1990). "Fibronectins." Springer, New York.

74. Nagai, T., Yamakawa, N., Aota, S., Yamada, S. S., Akiyama, S. K., Olden, K., and Yamada, K. M. (1991). Monoclonal antibody characterization of two distant sites required for function of the central cell-binding domain of fibronectin in cell adhesion, cell migration, and matrix assembly. *J. Cell Biol.* **114**, 1295–1305.

75. Aota, S., Nagai, T., and Yamada, K. M. (1991). Characterization of regions of fibronectin besides the arginine-glycine-aspartic acid sequence required for adhesive function of the cell-binding domain using site-directed mutagenesis. *J. Biol. Chem.* **266**, 15938–15943.

76. Manabe, R., Ohe, N., Maeda, T., Fukuda, T., and Sekiguchi, K. (1997). Modulation of cell-adhesive activity of fibronectin by the alternatively spliced EDA segment. *J. Cell Biol.* **139**, 295–307.

77. Klebe, R. J., Bentley, K. L., and Schoen, R. C. (1981). Adhesive substrates for fibronectin. *J. Cell. Physiol.* **109**, 481–488.

78. Schwartz, M. A., and Juliano, R. L. (1984). Surface activation of the cell adhesion fragment of fibronectin. *Exp. Cell Res.* **153**, 550–555.

79. Grinnell, F. (1982). Migration of human neutrophils in hydrated collagen lattices. *J. Cell Sci.* **58**, 95–108.

80. Grinnell, F. (1978). Cellular adhesiveness and extracellular substrata. *Int. Rev. Cytol.* **53**, 65–144.

81. Cadroy, Y., Houghten, R. A., and Hanson, S. R. (1987). RGDV peptide selectively inhibits platelet-dependent thrombus formation in vivo. Studies using a baboon model. *J. Clin. Invest.* **84**, 939–944.

82. Ferris, D. M., Moodie, G. D., Dimond, P. M., Gioranni, C. W. D., Ehrlich, M. G., and Valentini, R. F. (1999). RGD-coated titanium implants stimulate increased bone formation in vivo. *Biomaterials* **20**, 2323–2331.

83. Moodie, G. D., Ferris, D. M., Hertzog, B. A., Wimmer, N., Morgan, H., Chen, C. Y., Mathiowitz, E., and Valentini, R. F. (1998). Early osteoblast attachment, spreading, and focal adhesions on RGD coated surfaces. *Mater. Res. Soc. Symp. Proc.* **550**, 207–214.

84. Dee, K. C., Andersen, T. T., and Bizios, R. (1993). Cell function on substrates containing immobilized bioactive peptides. *Mater. Res. Soc. Symp. Proc.* **331**, 115–119.

85. Drumheller, P. D., and Hubbell, J. A. (1994). Polymer networks with grafted cell adhesion peptides for highly biospecific cell adhesive substrates. *Anal. Biochem.* **222**, 380–388.

86. Aigner, J., Tegeler, J., Hutzler, P., Campoccia, D., Pavesio, A., Hammer, C., Kastenbauer, E., and Naumann, A. (1998). Cartilage tissue engineering with novel nonwoven structural biomaterial based on hyaluronic acid benzyl ester. *J. Biomed. Mater. Res.* **42**, 172–181.

87. Tona, A., and Valentini, R. F. (1996). Derivatized hyaluronic acid films support mesenchymal stem cell attachment and proliferation. *Soc. Biomater.*, p. 849.

88. McFarland, C. D., Thomas, C. H., DeFilippis, C., Steele, J. G., and Healey, K. E. (2000). Protein adsorption and cell attachment to patterned surfaces. *J. Biomed. Mater. Res.* **49**, 200–210.

89. Grzesiak, J. J., Pierschbacher, M. D., Amodeo, M. F., Malaney, T. I., and Glass, J. R. (1997). Enhancement of cell interactions with collagen/glycosaminoglycan matrices by RGD derivatization. *Biomaterials* **18**, 1625–1632.

90. Yoneyama, T., Ito, M., Sugihara, K., Ishihara, K., and Nakabayashi, N. (2000). Small diameter vascular prosthesis with a nonthrombogenic phospholipid polymer surface: Preliminary study of a new concept for functioning in the absence of pseudo- or neointima formation. *Artif. Organs* **24**, 23–28.

91. Hermanson, G. T. (1996). "Bioconjugate Techniques." Academic Press, Inc., San Diego, CA.

92. Healy, K. E. (1999). Molecular engineering of materials for bioreactivity. *Curr. Opin. Solid State Mater. Sci.* **4**, 381–387.

93. Ingber, D. E. (1994). Integrating with integrins. *Mol. Biol. Cell* **5**, 389–393.

94. Hynes, R. O. (1992). Integrins: Versatility, modulation, and signaling in cell adhesion. *Cell (Cambridge, Mass.)* **69**, 11–25.

95. Ruoslahti, E. (1991). Integrins. *J. Clin. Invest.* **87**, 1–5.

96. Albelda, S. M., and Buck, C. A. (1990). Integrins and other cell adhesion molecules. *FASEB J.* **4**, 2868–2880.

97. Hynes, R. O. (1987). Integrins: A family of cell surface receptors. *Cell (Cambridge, Mass.)* **48**, 549–554.

98. Bowditch, R. D., Halloran, C. E., Aota, S., Obara, M., Plow, E. F., Yamada, K. M., and Ginsberg, M. H. (1991). Integrin $\alpha_{IIb}\beta_3$ (platelet GPIIb-IIIa) recognizes multiple sites in fibronectin. *J. Biol. Chem.* **266**, 23323–23328.

99. Werb, Z., Tremble, P. M., Behrendtsen, O., Crowley, E., and Damsky, C. H. (1989). Signal transduction through the fibronectin receptor induces collagenase and stromelysin gene expression. *J. Cell Biol.* **109**, 877–889.

100. Damsky, C., Tremble, P., and Werb, Z. (1992). Signal transduction via the fibronectin receptor: Do integrins regulate matrix remodeling? *Matrix Suppl.* **1**, 184–191.

101. Sakiyama, S. E., Schense, J. C., and Hubbell, J. A. (1999). Incorporation of heparin-binding peptides into fibrin gels enhances neurite extension: An example of designer matrices in tissue engineering. *FASEB J.* **13**, 2214–2224.

102. Hubbell, J. A., Massia, J. P., Desai, N. P., and Drumheller, P. D. (1991). Endothelial cell-selective materials for tissue engineering in the vascular graft via a new receptor. *Bio/technology* **9**, 568–572.

103. Yamamoto, K., and Yamamoto, M. (1994). Cell adhesion receptors for native and denatured type I collagens and fibronectin in rabbit arterial smooth muscle cells in culture. *Exp. Cell Res.* **214**, 258.

104. Cardarelli, P. M., Yamagata, S., Taguchi, I., Gorscan, F., Chiang, S. L., and Lobl, T. (1992). The collagen receptor $\alpha_2\beta_1$, from MG-63 and HT-1080 cells, interacts with a cyclic RGD peptide. *J. Biol. Chem.* **267**, 23159.

105. Ito, Y., Zheng, J., Imanishi, Y., Yonezawa, K., and Kasuga, M. (1996). A protein-free cell culture on an artificial substrata covalently immobilized with insulin. *Proc. Natl. Acad. Sci. U.S.A.* **93**, 3598–3601.

106. Ito, Y., Chen, G., and Imanishi, Y. (1996). Photo-immobilization of insulin onto polystyrene dishes for protein-free cell culture. *Biotechnol. Prog.* **12**, 700–703.

107. Ito, Y., Kondo, S., Chen, G., and Imanishi, Y. (1997). Patterned artificial juxtacrine stimulation of cells by covalently immobilized insulin. *FEBS Lett.* **403**, 159–162.

108. Kuhl, P. R., and Griffith-Cima, L. G. (1996). Tethered epidermal growth factor as a paradigm for growth factor-induced stimulation from the solid phase. *Nat. Med.* **2**, 1022–1027.

109. Ito, Y., Li, J. S., Takahashi, T., Imanishi, Y., Okabayashi, Y., Kido, Y., and Kasuga, M. (1997). Enhancement of the mitogenic effect by artificial juxtacrine stimulation using immobilized EGF. *J. Biochem. (Tokyo)* **121**, 514–520.

110. Ito, Y., Chen, G., and Imanishi, Y. (1998). Artificial juxtacrine stimulation for tissue engineering. *J. Biomater. Sci. Polymer Edn* **9**, 879–890.

111. Ito, Y., Chen, G., and Imanishi, Y. (1998). Micropattern immobilization of epidermal growth factor to regulate cell functions. *Bioconjugate Chem.* **9**, 277–282.

112. Dee, K. C., Anderson, T. T., and Bizios, R. (1998). Design and function of novel osteoblast-adhesive peptides for chemical modification of biomaterials. *J. Biomed. Mater. Res.* **40**, 371–377.

113. Sakiyama, S. E., Schense, J. C., and Hubbell, J. A. (1999). Incorporation of heparin-binding peptides into fibrin gels enhances neurite extension: An example of designer matrices in tissue engineering. *FASEB J.* **13**, 2214–2224.

114. Griffith, L. G., and Lopina, S. (1998). Microdistribution of substratum-bound ligands affects cell function: Hepatocyte spreading on PEO-tethered galactose. *Biomaterials* **19**, 979–986.

115. Franks, L. M., and Teich, N. M. (1991). "Introduction to the Cellular and Molecular Biology of Cancer." Oxford University Press, Oxford, UK.

116. Faustman, D., and Coe, C. (1991). Prevention of xenograft rejection by masking donor HLA class I antigens. *Science* **252**, 1700–1702.

117. Kao, N. J., Hubbell, J. A., and Anderson, J. M. (1999). Protein-mediated macrophage adhesion and activation on biomaterials: A model for modulating cell behavior. *J. Mater. Sci.: Mater. Med.* **10**, 601–605.

118. Yun, J. K., DeFife, K., Colton, E., Stack, S., Azeez, A., Cahalan, L., Verhoeven, M., Cahalan, P., and Anderson, J. M. (1995). Human monocyte/macrophage adhesion and cytokine production on surface-modified poly (tetrafluoroethylene/hexafluoropropylene) polymers with and without protein preadsorption. *J. Biomed. Mater. Res.* **29**, 257–268.

119. Smetana, K., Lukas, J., Paleckova, V., Bartunkova, J., Liu, F. T., Vacik, J., and Gabius, H. J. (1997). Effect of chemical structure of hydrogels on the adhesion and phenotypic characteristics of human monocytes such as the expression of galectins and other carbohydrate-binding sites. *Biomaterials* **18**, 1009–1014.

120. Daeschel, M. A., and McGuire, J. (1998). Interrelationships between protein surface adsorption and bacterial adhesion. *Biotechnol. Genet. Eng. Rev.* **15**, 413.

121. Desai, N. P., Hossainy, S. F., and Hubbell, J. A. (1992). Surface-immobilized polyethylene oxide for bacterial repellence. *Biomaterials* **13**, 417–420.

122. Carr, P. W., and Bowers, L. D. (1980). "Immobilized Enzymes in Analytical and Clinical Chemistry: Fundamentals and Applications," pp. 172–173. Wiley, New York.

123. Amos, R. A., Anderson, A. B., and Clapper, D. L. (1995). Biomaterial surface modification using photochemical coupling technology. *In* "Encyclopedia Handbook of Biomaterials and Bioengineering. Part A: Materials" (D. L. Wise, D. J. Trantolo, and D. E. Altobelli, eds.), pp. 895–926. Dekker, New York.

124. von Recum, A. F., Shannon, C. E., Cannon, C. E., Long, K. J., van Kooten, T. G., and Meyle, J. (1996). Surface roughness, porosity, and texture as modifiers of cellular adhesion. *Tissue Eng.* **2**, 241–253.

125. Deutsch, J., Motlagh, D., Russell, B., and Desai, T. A. (2000). Fabrication of microtextured membranes for cardiac myocyte attachment and orientation. *J. Biomed. Mater. Res.* **53**, 267–275.

126. Kapur, R., Spargo, B. J., Chen, M. S., Calvert, J. M., and Rudolph, A. S. (1996). Fabrication and selective surface modification of 3-dimensionally textured biomedical polymers from etched silicon substrates. *J. Biomed. Mater. Res.* **33**, 205–216.

127. Den Braber, E. T., de Ruijter, J. E., Ginsel, L. A., von Recom, A. F., and Jansen, J. A. (1998). Orientation of ECM protein deposition, fibroblasts, cytoskeleton, and attachment complex components on silicone microgrooved surfaces. *J. Biomed. Mater. Res.* **40**, 291–300.

128. Mertz, P. M., Davis, S. C., Franzen, L., Uchima, F. D., Pickett, M. P., Pierschbacher, M. D., and Polerek, J. W. (1996). Effects of arginine-glycine-aspartic acid peptide-containing artificial matrix on epithelial migration in vitro and experimental second-degree burn wound healing in vivo. *J. Burn Care Rehabil.* **17**, 199–206.

129. Puleo, D. A., Preston, K. E., Shaffer, J. B., and Bizios, R. (1993). Examination of osteoblast-orthopaedic biomaterial interactions using molecular techniques. *Biomaterials* **14**, 111–114.

130. Dee, K. C., Rueger, D. C., Andersen, T. T., and Bizios, R. (1996). Conditions which promote mineralization at the bone–implant interface: A model in vitro study. *Biomaterials* **17**, 209.

131. Suzuki, T., Tanihara, M., Suzuki, K., Saitou, A., Sufan, W., and Nishimura, Y. (2000). Alginate hydrogel linked with synthetic oligopeptide derived from BMP-2 allows ectopic oeteoinduction in vivo. *J. Biomed. Mater. Res.* **50**, 405–409.

132. Borkenhagen, M., Clemence, J.-F., Sigrist, H., and Aebischer, P. (1998). Three-dimensional extracellular matrix engineering in the nervous system. *J. Biomed. Mater. Res.* **40**, 392–400.

133. Biran, R., Noble, M. D., and Tresco, P. A. (1999). Characterization of cortical astrocytes on materials of differing surface chemistry. *J. Biomed. Mater. Res.* **46**, 150–159.

134. Kam, L., Shain, W., Turner, J. N., and Bizios, R. (1999). Correlation of astroglial cell function on micro-patterned surfaces with specific geometric parameters. *Biomaterials* **20**, 2343–3250.

135. Elbert, D. L., and Hubbell, J. A. (1998). Reduction of fibrous adhesion formation by a copolymer possessing an affinity for anionic surfaces. *J. Biomed. Mater. Res.* **42**, 55–65.

136. Jaeger, C. B., Aebischer, P., Tresco, P. A., Winn, S. R., and Greene, L. A. (1992). Growth of tumour cell lines in polymer capsules: Ultrastructure of encapsulated PC12 cells. *J. Neurocytol.* **21**, 469–480.

137. Tresco, P. A., Winn, S. R., and Aebischer, P. (1992). Polymer encapsulated neurotransmitter secreting cells. Potential treatment for Parkinson's disease. *ASAIO J.* **38**, 17–23.

138. Lysaght, M., Gentile, F., Li, R., Rein, D., and Schoichet, M. (1994). Implants containing living cells for CNS therapy. *Artif. Cells, Blood Substitutes, Immobilization Biotechnol.* **22**, 1073–1199.

139. Lysaght, M., Gentile, F., Li, R., Rein, D., and Schoichet, M. (1994). Implants containing living cells for CNS therapy. *Artif. Cells, Blood Substitutes, Immobilization Biotechnol.* **22**, 1073–1199.

140. Kao, W. J., and Hubbell, J. A. (1998). Murine macrophage behavior on peptide-grafted polyethyleneglycol-containing networks. *Biotechnol. Bioeng.* **59**, 2–9.

141. Silver, J. H., Hergenrother, R. W., Lin, J.-C., Lim, F., Lin, H.-B., Okada, T., Chaudhury, M. K., and Cooper, S. L. (1995). Surface and blood-contacting properties of alkylsiloxane monolayers supported on silicone rubber. *J. Biomed. Mater. Res.* **29**, 535–548.

142. Desai, N. P., and Hubbell, J. A. (1989). The short-term blood biocompatibility of poly(hydroxyethyl methacrylate-*co*-methyl methacrylate) in an in vitro flow system measured by digital videomicroscopy. *J. Biomater. Sci., Polymer Edn* **1**, 123–146.

143. Wagner, W. R., and Hubbell, J. A. (1990). Local thrombin synthesis and fibrin formation in an in vitro thrombosis model result in platelet recruitment and thrombus stabilization on collagen in heparinized blood. *J. Lab. Clin. Med.* **116**, 636–650.

144. Cruise, G. M., Hegre, O. D., Scharp, D. S., and Hubbell, J. A. (1998). A sensitivity study of the key parameters in the interfacial photopolymerization of poly(ethylene glycol) diacrylate upon porcine islets. *Biotechnol. Bioeng.* **57**, 655–665.

145. McKay, R. (1997). Natural and engineered stem cells in the central nervous system. *Science* **276**, 66–71.

146. Stocum, D. L. (1998). Regenerative biology and engineering: Strategies for tissue restoration. *Wound Rep. Regeneration* **6**, 276–290.

147. Mayer, J., Karamuk, E., Akaike, T., and Wintermantel, E. (2000). Matrices for tissue engineering-scaffold structure for a bioartificial liver support system. *J. Controlled Release* **64**, 81–90.

148. Kamamuk, E., Mayer, J., Wintermantel, E., and Akaike, T. (1999). Partially degradable film/fabric composites: Textile scaffolds for liver cell culture. *Artif. Organs* **23**, 881–884.

149. Wintermantel, E., Cima, L., Schoo, B., and Langer, R. (1992). Angiopolarity of cell carriers: Directional angiogenesis in resorbable liver cell transplantation devices. *Exs* **61**, 331–334.

150. Griffith, L. G., Wu, B., Cima, M. J., Powers, M. J., Chaignaud, B., and Vacanti, J. P. (1997). In vitro organogenesis of liver tissue. *Ann. N.Y. Acad. Sci.* **831**, 382–397.

151. Corwin, J. T., and Cotanche, D. A. (1988). Regeneration of sensory hair cells after acoustic trauma. *Science* **240**, 1774–1776.

152. Cruise, G. M., Hegre, O. D., Lamberti, F. V., Hager, S. R., Hill, R., Scharp, D. S., and Hubbell, J. A. (1999). In vitro and in vivo performance of porcine islets encapsulated in interfacially photopolymerized poly(ethylene glycol) diacrylate membranes. *Cell Transplant.* **8**, 293–306.

153. Hill-West, J. L., Chowdhury, S. M., Dunn, R. C., and Hubbell, J. A. (1994). Efficacy of a resorbable hydrogel barrier, oxidized regenerated cellulose, and hyaluronic acid in the prevention of ovarian adhesions in a rabbit model. *Fertil. Steril.* **62**, 630–634.

154. Border, W. A., Noble, N. A., Yamamoto, T., Harper, J. R., Yamaguchi, Y., Pierschbacher, M. D., and Rouslahti, E. (1992). Natural inhibitor of transforming growth factor-beta protects against scarring in experimental kidney disease. *Nature (London)* **360**, 361–364.

155. Massia, S. P., and Hubbell, J. A. (1991). An RGD spacing of 44 nm is sufficient for integrin $\alpha_V \beta_3$-mediated fibroblast spreading and 140 nm for focal contact and stress fiber formation. *J. Cell Biol.* **114**, 1089–1100.

156. Bunge, R. P. (1994). The role of Schwann cells in trophic support and regeneration. *J. Neurol.* **242**, S19–S21.

157. Tong, Y. W., and Shoichet, M. S. (1998). Peptide surface modification of poly(tetrafluoroethylene-co-hexafluoropropylene) enhances its interaction with central nervous system neurons. *J. Biomed. Mater. Res.* **42**, 85–95.

158. Steele, J. G., Johnson, G., Jenkins, M., McLean, K. M., Griesser, H. J., Kwak, J., and Goodman, M. (2000). Peptoid-containing collagen mimetics with cell binding activity. *J. Biomed. Mater. Res.* **51**, 612–624.

159. Schroeder, A., Francz, G., Bruinink, A., Hauert, R., Mayer, J., and Wintermantel, E. (2000). Titanium containing amorphous hydrogenated carbon films (a-C: H/Ti): Surface anaylsis and evaluation of cellular reactions using bone marrow cell cultures in vitro. *Biomaterials* **21**, 449–456.

160. Xiao, S.-J., Textor, M., Spencer, N. D., and Sigrist, H. (1998). Covalent attachment of cell-adhesive, (Arg-Gly-Asp)-containing peptides to titanium surfaces. *Langmuir* **14**, 5507–5516.

161. Rezania, A., and Healy, K. E. (1999). Biomimetic peptide surfaces that regulate adhesion, spreading, cytoskeletal organization, and mineralization of the matrix deposited by osteoblast-like cells. *Biotechnol. Prog.* **15**, 19–32.

162. Han, D. K., Park, K. D., Hubbell, J. A., and Kim, Y. H. (1998). Surface characteristics and biocompatibility of lactide-based poly(ethylene glycol) scaffolds for tissue engineering. *J. Biomater. Sci. Polymer Edn* **9**, 667–680.

163. Hern, D. L., and Hubbell, J. A. (1998). Incorporation of adhesion peptides into nonadhesive hydrogels useful for tissue resurfacing. *J. Biomed. Mater. Res.* **39**, 266–276.

164. Drumheller, P. D., and Hubbell, J. A. (1994). Polymer networks with grafted cell adhesion peptides for highly biospecific cell adhesive substrates. *Anal. Biochem.* **222**, 380–388.

165. Brandley, B., and Schnaar, R. (1988). Covalent attachment of an Arg-Gly-Asp sequence peptide to derivatizable polyacrylamide surfaces: Support of fibroblast adhesion and long-term growth. *Anal. Biochem.* **172**, 270–278.

166. Massia, S. P., and Hubbell, J. (1991). Human endothelial cell interactions with surface-coupled adhesion peptides on a nonadhesive glass substrate and two polymeric biomaterials. *J. Biomed. Mater. Res.* **25**, 223–242.

167. Cook, A. D., Hrkach, J. S., Gao, N. N., Johnson, I. M., Pajvani, U. B., Cannizzaro, S. M., and Langer, R. (1997). Characterization and development of RGD-peptide-modified poly(lactic acid-co-lysine) as an interactive, resorbable biomaterial. *J. Biomed. Mater. Res.* **35**, 513–523.

168. Holt, D. B., Eberhart, R. C., and Prager, M. D. (1994). Endothelial cell binding to Dacron modified with polyethylene oxide and peptide. *ASAIO J.* **40**, M858–M863.

169. Jo, S., and Mikos, A. G. (1999). Modification of poly(ethylene glycol)-tethered poly(propylene-co-fumarate) with RGD peptide. *Polym. Prepr., Am. Chem. Div. Pol. Chem.* **40**, 183–184.

170. Lin, H. B., Sun, W., Mosher, D. F., Garcia-Echevierria, C., Schaufelberger, K., Lelkes, P. I., and Cooper, S. L. (1994). Synthesis, surface, and cell-adhesion properties of polyurethanes containing covalently grafted RGD-peptides. *J. Biomed. Mater. Res.* **28**, 329–342.

171. Lin, H. B., Garcia-Echeverria, C., Asakura, S., Sun, W., Mosher, D. F., and Cooper, S. L. (1992). Endothelial cell adhesion on polyurethanes containing covalently attached RGD–peptides. *Biomaterials* **13**, 905–914.

172. Lee, K. Y., and Mooney, D. T. (2001). Hydrogels for tissue engineering. *Chem. Rev.* **101**, 1869–1879.

173. Winger, T. M., and Ludovice, P. J. (1999). Formation and stability of complex membrane-mimetic monolayers on solid supports. *Langmuir* **15**, 3866–3874.

174. Massia, S. P., and Hubbell, J. A. (1990). Covalent surface immobilization of Arg-Gly-Asp- and Tyr-Ile-Gly-Ser-Arg-containing peptides to obtain well-defined cell-adhesive substrates. *Anal. Biochem.* **187**, 292–301.

175. Roberts, C., Chen, C. S., Mrksich, M., Martichonok, V., Ingber, D. E., and Whitesides, G. M. (1998). Using mixed self-assembled monolayers presenting RGD and (EG)$_3$OH groups to characterize long-term attachment of bovine capillary endothelial cells to surfaces. *J. Am. Chem. Soc.* **120**, 6548–6555.

176. Sugawara, T., and Matsuda, T. (1995). Photochemical surface derivatization of a peptide containing Arg-Gly-Asp (RGD). *J. Biomed. Mater. Res.* **29**, 1047–1052.

177. Bearinger, J. P., Castner, D. G., and Healy, K. E. (1998). Biomolecular modification of p(Amm-*co*-EG/AA) IPNs supports osteoblast adhestion and phenotypic expression. *J. Biomater. Sci. Polymer Edn.* **9**, 629–652.

178. Patel, N., Bhandari, R., Shakesheff, K. M., Cannizzaro, S. M., Davies, M. C., Langer, R., Roberts, C. J., Tendler, S. J. B., and Williams, P. M. (2000). Printing patterns of biospecifically-adsorbed protein. *J. Biomater. Sci. Polymer Edn.* **11**, 319–331.

179. Lom, B., Healy, K. E., and Hockberger, P. E. (1993). A versatile technique for patterning biomolecules onto glass coverslips. *J. Neurosci. Methods* **50**, 385–397.

180. Thomas, C. H., Lhoest, J. B., Castner, D. G., McFarland, C. D., and Healy, K. E. (1999). Surfaces designed to control the projected area and shape of individual cells. *J. Biomech. Eng.* **121**, 40–48.

181. Kim, S.-H., and Chu, C.-C. (2000). Pore structure analysis of swollen dextran–methacrylate hydrogels by SEM and mercury intrusion porosimetry. *J. Biomed. Mater. Res.* **53**, 258–266.

182. Hennink, W. E., Talsma, H., Borchert, J. C. H., Smedt, S. C. D., and Demeester, J. (1996). Controlled release of proteins from dextran hydrogels. *J. Controlled Release* **39**, 47–55.

183. Pierschbacher, M. D., and Ruoslahti, E. (1984). Cell attachment activity of fibronectin can be duplicated by small synthetic fragments of the molecule. *Nature (London)* **309**, 30–33.

184. Bhat, V. D., Truskey, G. A., and Reichert, W. M. (1998). Fibronectin and avidin–biotin, a heterogeneous ligand system for enhanced endothelial cell adhesion. *J. Biomed. Mater. Res.* **41**, 377–385.

185. Green, N. M. (1975). Avidin. *Adv. Protein Chem.* **29**, 85–133.

186. Shakesheff, K., Cannizzaro, S., and Langer, R. (1998). Creating biomimetic microenvironments with synthetic polymer–peptide hybrid molecules. *J. Biomater. Sci. Polymer Edn.* **9**, 507–518.

187. Huebsch, J. B., Fields, G. B., Triebes, T. G., and Mooradian, D. L. (1996). Photoreactive analog of peptide FN-C/H-V from the carboxy-terminal heparin-binding fragment of fibronectin involves heparin-dependent and -independent activities. *J. Cell Biol.* **110**, 777–787.

188. Urry, D. W., Pattanaik, A., Xu, J., Woods, T. C., McPherson, D. T., and Parker, T. (1998). Elastic protein-based polymers in soft tissue augmentation and generation. *J. Biomater. Sci. Polymer Edn.* **9**, 1015–1048.

189. Walsh, W. R., Kim, H. D., Jong, Y. S., and Valentini, R. F. (1995). Controlled release of platelet-derived growth factor using ethylene vinyl acetate copolymer (EVAc) coated on stainless-steel wires. *Biomaterials* **16**, 1319–1325.

190. Kim, H. D., and Valentini, R. F. (1997). Human osteoblast response in vitro to platelet-derived growth factor and transforming growth factor-β; delivered from controlled-release polymer rods. *Biomaterials* **18**, 1175–1184.

191. Cruise, G. M., Scharp, D. S., and Hubbell, J. A. (1998). Characterization of permeability and network structure of interfacially photopolymerized poly(ethylene glycol) diacrylate hydrogels. *Biomaterials* **19**, 1287–1294.

192. Kushida, A., Masayuki, Y., Konno, C., Kikuchi, A., Sakurai, Y., and Okano, T. (1999). Decrease in culture temperature releases monolayer endothelial cell sheets together with deposited fibronectin matrix from temperature-responsive culture surfaces. *J. Biomed. Mater. Res.* **45**, 355–362.

193. Haugland, R. (1998). "Handbook of Fluorescent Probes and Research Chemicals," 6th ed. Molecular Probes, Eugene, OR. Available at: (www.probes.com).

194. Laemmli, U. K. (1970). Cleavage of structural proteins during the assembly of the head of bacteriophage T4. *Nature (London)* **277**, 680–685.

195. Kishida, A., Kato, S., Ohmura, K., Sugimura, K., and Akashi, M. (1996). Evaluation of biological responses to polymeric biomaterials by RT-PCR analysis. I. Study of IL-1 beta mRNA expression. *Biomaterials* **17**, 1301–1305.

196. Kato, S., Akagi, T., Kishida, A., Sugimura, K., and Akashi, M. (1997). Evaluation of biological responses to polymeric biomaterials by RT-PCR analysis II: Study of HSP 70 mRNA expression. *J. Biomater. Sci. Polymer Edn* **8**, 809–814.

197. Sambrook, J., Fritsch, E. F., and Maniatis, T. (1989). "Molecular Cloning: A Laboratory Manual," 3 vols. Cold Spring Harbor Lab. Press, Cold Spring Harbor, NY.

198. Garcia, A. J., Ducheyne, P., and Boettinger, D. (1997). Cell adhesion strength increases linearly with adsorbed fibronectin surface density. *Tissue Eng.* **3**, 197–203.

199. Belcheva, N., Baldwin, S. P., and Saltzman, W. M. (1998). Synthesis and characterization of polymer-(*multi*-)peptide conjugates for control of specific cell aggregation. *J. Biomater. Sci. Polymer Edn.* **9**, 207–226.

200. Yarema, K. J., Mahal, L. K., Bruehl, R. E., Rodriguez, E. C., and Bertozzi, C. R. (1998). Metabolic delivery of ketone groups to sialic acid residues. Applications to cell surface glycoform engineering. *J. Biol. Chem.* **273**, 31168–31179.

CHAPTER 67

CELL–SYNTHETIC SURFACE INTERACTIONS: PHYSICOCHEMICAL SURFACE MODIFICATION

Gilson Khang and Hai Bang Lee

The chapter presents physicochemical treatments for improving the wettability and hydrophilicity of poly(L-lactic-co-glycolic acid) (PLGA, 75:25 mole ratio of lactide to glycolide) film and porous scaffold materials fabricated by the emulsion freeze-drying method. Chemical treatments were 70% perchloric acid, 50% sulfuric acid, and 0.5 N sodium hydroxide solution, and physical methods were corona and plasma treatments generated by radio frequency glow discharge. Also, we prepared a wettability gradient on the PLGA films by treating them in air with corona from a knife-type electrode whose power increases gradually along the sample length. The PLGA surfaces oxidized gradually with increasing corona power, and a wettability chemogradient was created on the surfaces. The water contact angle of PLGA film that had been surface-treated with various methods decreased from 73° to 50–60°; that is, the introduction of an oxygen-containing functional group onto the PLGA backbone served to increase hydrophilicity as measured by electron spectroscopy for chemical analysis (ESCA). It could be observed that the adhesion and growth of the cell on physicochemically treated PLGA surfaces, especially those treated with perchloric acid, occurred more actively than on the control.

The wettability of chemically treated PLGA scaffolds was ranked in the order of perchloric acid, sulfuric acid, and sodium hydroxide solution by blue dye intrusion experiment, whereas physical methods had no effects. It is suggested that chemical treatment method may be useful in uniformly seeding porous biodegradable PLGA scaffolds. It seems that wettability plays important roles in cell adhesion, spreading, and growth on PLGA surfaces and in facilitating the intrusion of nutrient media into the PLGA scaffold. The surface modification technique used in this chapter may be applicable to the area of tissue engineering for the improvement of tissue compatibility of film- or scaffold-type substrates.

INTRODUCTION

Since the mid-1990s, poly(L-lactide-co-glycolides) have been extensively used or tested for a wide range of medical applications as bioerodible materials owing to their good biocompatibility, controllable biodegradabilitiy, and relatively good processibility [1]. Poly(L-lactic-co-glycolic acid) (PLGA) is a bioresorbable polyester belonging to the group of poly(α-hydroxy acids). This polymer and its homopolymers [polylactide (PLA) and polyglycolide (PGA)] degrade by nonspecific hydrolytic scission of their ester bonds. The hydrolysis of PLA yields lactic acid, a normal by-product of anaerobic metabolism in the human body, which is incorporated into the tricarboxylic acid (TCA) cycle and finally excreted as carbon dioxide and water. PGA degrades by a combination of hydrolytic scission and enzymatic (esterase) action to produce glycolic acid, which can either enter the TCA cycle or be excreted in urine and eliminated as carbon dioxide and water [2]. The degradation time of PLGA can be controlled from weeks to over a year by varying both the ratio of monomers and the processing conditions. It might be a suitable biomaterial for use in tissue-engineered repair systems [3] in which cells are cultured within PLGA films or scaffolds and in drug delivery systems [4] in which drugs are loaded within PLGA microspheres.

On the other hand, the interaction of cells with polymer surfaces is important for a variety of biomedical and biotechnology applications. It is generally recognized that adhesion and proliferation of different polymeric materials on cells of different types depend largely on surface characteristics such as wettability (hydrophilicity/hydrophobicity of surface free energy), chemistry, charge, roughness, and rigidity [5]. A large number of research groups, including our laboratories, have modified polymer surfaces and investigated the interaction of cultured cells and polymers with different surface properties.

For biomedical applications of the poly(α-hydroxy acid) family, it is more desirable to change the hydrophobic PGA, PLA, and PLGA surface to make it hydrophilic [6]. Hydrophobic surfaces in aqueous solutions possess high interfacial free energy, which tends to unfavorably influence the surfaces cell, tissue, and blood compatibility in the initial stage of contact [7]. Also, for tissue engineering applications, a cell suspension or culture medium plated on a porous scaffold, should not contact PLGA: the hydrophobic PLGA surface would result in the soaking of the cell suspension into the air-filled porous scaffold. To overcome this problem, Mikos *et al.* [8] proposed prewetting by ethanol. However, this method not only has the drawback of the toxicity of the remaining ethanol, but the surface is still hydrophobic, as well.

This chapter presents a physicochemical treatment for improving the wettability and hydrophilicity of PLGA films and porous scaffolds fabricated by the emulsion freeze-drying method. Chemical treatments are 70% perchloric acid, 50% sulfuric acid, and 0.5 N sodium hydroxide solution, and physical methods are corona and plasma treatments generated by radio frequency glow discharge (RFGD) [9]. Also, we have illustrated a method of preparation a wettability gradient on PLGA films by treating them in air with corona from a knife-type electrode whose power increases gradually along the sample length [10]. The PLGA surfaces oxidize gradually with increasing corona power, and a wettability gradient is created on the surfaces.

Physicochemically modified PLGA surfaces are characterized by means of a water contact angle goniometer and by ESCA. NIH/3T3 fibroblast cells are cultured on surface-modified PLGA for the evaluation of cell attachment and proliferation in terms of surface hydrophilicity/hydrophobicity. Also, the wettability of physicochemically treated PLGA scaffolds is characterized by blue dye intrusion.

MATERIALS
POLYMER PROCESSING
Chemicals
1. *Monomers*: L-lactide and glycolide (Boehringer-Ingelheim, Germany)
2. *Initiator*: stannous 2-ethylhexanoate (Wako Chemical Co., Japan)
3. *Solvents*: toluene (Junsei Chemical Co., Japan); methylene chloride (MC, Tedia. Co. U.S.A.); methyl alcohol (Junsei); sulfuric acid (H_2SO_4, Jin Chemical & Pharmaceutical Co., Korea); hyperchloric acid (70% $HClO_4$, Junsei); potassium chlorate ($KClO_3$, Junsei)
4. *Dye*: Coomassie brilliant blue dye R-250 (Bio-Rad Lab, Richmond, VA)

Laboratory Supplies
1. Polymerization reactor with Pyrex glass (30 mm × 35 cm).
2. Silicon oil bath (custom-designed)
3. Pyrex petri dishes (100-mm diameter) with cover (Fisher Scientific, Pittsburgh, PA)
4. Rectangular Teflon molds (custom-designed)

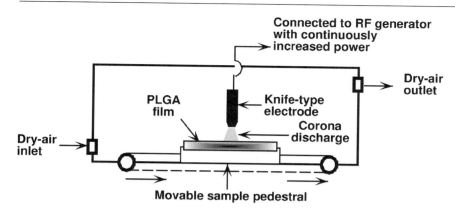

Fig. 67.1. Schematic diagram of the corona discharge apparatus for the preparation of a wettability chemogradient.

5. Corona discharge apparatus with wettability gradient (custom-designed, 100 kHz, knife-type electrode) (Fig. 67.1).
6. Plasma discharge apparatus (custom-designed, 100 kHz, bell jar type) (Fig. 67.2)

Instruments for Analysis
1. Gel permeation chromatography (GPC, Waters Chromatograph 200 Series, Delaware)
2. Contact angle goniometer (model 100-0, Rame-Hart, Inc., U.S.A.)
3. ESCA (ESCALAB MK II, V. G. Scientific Co., U.K.)
4. Scanning electron microscopy (Jeol Co., Japan, model JSM-840-A)
5. Plasma sputter (Emscope, U.K., model SC 500K)
6. Mercury porosity meter (Micrometrics Co., U.S.A., AutoPore II 9220)
7. Video high scope (Airox, Tokyo, KH-2200 Hi-Scope)
8. Freeze-dryer (FDU-540, EYELA, Tokyo)

CELL CULTURE
1. Tissue culture polystyrene flasks (25 and 75 cm^2, Corning, U.S.A.)
2. 0.25% trypsin–EDTA in phosphate-buffered saline (PBS) (Gibco Laboratories, U.S.A.)
3. Custom-made cell culture chamber with Teflon block (1.0 cm × 5.0 cm)

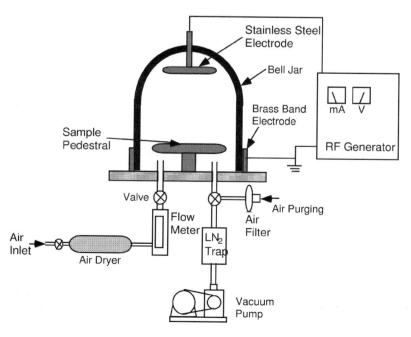

Fig. 67.2. Schematic diagram of the plasma discharge apparatus.

4. Dulbecco's phosphate-buffered saline (PBS, pH 7.3–7.4; Sigma, U.S.A.), free of Ca^{2+} and Mg^{2+}

5. *Culture media*: RPMI-1640 nutrient mixture (Gibco) containing 10% fetal bovine serum (FBS, Gibco), 100 units/ml penicillin (Sigma), and 100 μg/ml gentamicin sulfate (Sigma)

6. *Fixation medium*: 2.5% glutaraldehyde (Gibco)

7. *Dehydrated solution*: ethanol, graded series (50, 60, 70, 80, 90, and 100% with deionized water)

METHODS

POLYMER PROCESSING

PLGA Synthesis and Characterizations

Details of synthesis and characterization of the PLGA polymer described here can be found elsewhere [6,9,10].

1. A 30-g mixture of the L-lactide (75 mol %) and glycolide (25 mol %) is preheated in an evacuated flask at 60°C for 2 h to remove water.*

2. Stannous 2-ethylhexanoate in toluene (150 ppm) is added to the polymerization reactor with agitation at 100 rpm. Dry nitrogen gas is flushed through the reactor during the whole process.

3. After the catalyst has been added, the copolymerization reaction is carried out at 165°C for 4.5 h. The light brownish PLGA obtained is purified by dissolving in MC, followed by slow precipitation in excess methanol to remove unreacted monomers and other impurities. The molecular weight of PLGA can be controlled by controlling the reaction time. The longer reaction time, the higher molecular weight that can be obtained.†

4. The polymer, which very sensitive to moisture in the air and thus at risk of premature hydrolytic degradation, is dried *in vacuo* at room temperature for 7 days and kept until use.

5. The synthesized PLGA is characterized by means of gel permeation chromotograph (GPC). Measurement is carried out on a Waters Chromatograph 200 series instrument equipped with 6-μm Styragel columns; pore sizes are 10^5, 10^4, 10^3, and 500 Å.

6. Tetrahydrofuran is used as an eluent solvent; temperature, flow rate, and injection volume were 30°C, 1 ml/min, and 15 μl, respectively.

7. A polystyrene monodisperse standard series is used to calibrate the molecular weight. The average molecular weight (MW) and molecular weight distribution (MWD) of purified samples are in the range of 50,000–70,000 g/mol and 1.5–1.9, respectively, for the 4.5-h reaction time, with good reproducibility.

Preparation of PLGA Films

1. Ten percent (w/v) of PLGA (MW 55,000 g/mol, and MWD 1.79) is dissolved in MC.

2. Two grams of PLGA solution is cast onto 100-mm-diameter Pyrex petri dishes with a horizontal level to get PLGA films 400–600 μm. After evaporation of the MC at room temperature, the films are cut into rectangles 70 cm × 5.0 cm.

3. After drying *in vacuo* overnight to remove residual methylene chloride and ultrasonically washed in ethanol, the PLGA films are kept under vacuum until use.

Preparation of PLGA Scaffolds by Emulsion Freeze-Drying Methods

The following preparation of PLGA scaffolds is described by Khang *et al.* [11] and Whang *et al.* [12] in detail.

1. Ten percent (w/v) of PLGA is dissolved in methylene chloride.

---

*If even a trace of water remains, the PLGA will not reach higher molecular weights.

†If unreacted monomers and other impurities are allowed to remain in the polymerized PLGA, the properties of the PLGA films and scaffolds will not be suitable for the intended end use. For example, the PLGA films will be opaque and scaffolds will tend to shrink.

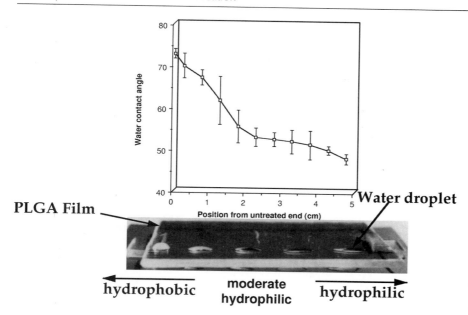

PLGA Film

Water droplet

hydrophobic   moderate   hydrophilic
                hydrophilic

Fig. 67.3. Changes in the water contact angle of a corona-treated PLGA surface and morphology of a water droplet along the sample length.

2. In a 10-W sonicator, 3 g of PLGA solution is thoroughly mixed with 7 g of deionized water in a sample vial for 30 s. The pore diameter obtained ranges 30 to 300 μm. Scaffold porosity from can be controlled by controlling the ratio of the concentration of PLGA solution to deionized water. The higher amount of deionized water, the larger pore size, resulting in the faster promotion of vascular penetration into the scaffold interior. However, over 50% of the deionized water did not form a microporous structure. It is noted as well that this method can be used to fabricate controlled-release scaffolds of bioactive molecules by dissolving cytokines or another bioactive molecules in the water phase.

3. The emulsion obtained in step 3 is poured into a rectangular (or cylindrical or tubular) Teflon mold and then quenched by quickly immersing it in a liquid nitrogen bath at $-196°C$.

4. The frozen samples are freeze-dried at 30 mtorr and $-55°C$.

5. Samples are placed in a vacuum desiccator at room temperature for at least 7 days to remove any residual solvent.

6. A sharp razor blade is used to cut the microporous scaffolds to 2-mm thickness (1.0×1.0 cm^2 size).

7. The microporous PLGA scaffolds are characterized by scanning electron microscope (SEM) and mercury porosity meter to determine pore size and pore size distribution.

Physicochemical Treatments of PLGA Films and Scaffolds

1. Chemical treatment [9].

   a. The following chemicals are prepared: 50% H_2SO_4, 0.5 N NaOH, and 70% perchloric acid. (*Note*: These chemicals are highly toxic and harmful.)
   b. The PLGA film samples are immersed for 30 min in an orbital shaker containing the chemical solution at 30 rpm and 5° angle.
   c. The PLGA scaffolds receive the same treatment outlined in step b, but the PLGA surface is damaged (e.g., the transparent PLGA surfaces become opaque). Also, the water contact angle of the chemically modified PLGA decreases from 73° to 50–60° (i.e., increased hydrophilicity due to the chemical oxidation).
   d. The surface-treated PLGA films and scaffolds are thoroughly washed five times with excess PBS to remove toxicity due to residual chemicals.

2. Corona treatment with wettability gradient as physical method [5,9,10]

1 DAY   **2 DAYS**

0.5 CM

1.5 CM

2.5 CM

4.5 CM

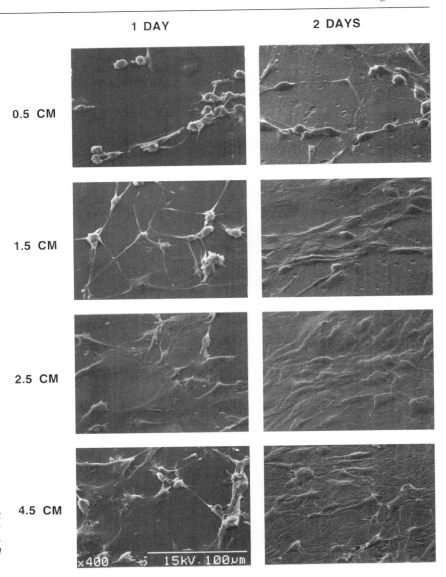

Fig. 67.4. SEM photomicrographs of fibroblast cells attached on corona-treated PLGA surfaces with wettability chemogradient along the sample length after 1- and 2-day cultures. Original magnification, ×400.

a. A knife-type electrode is connected to the RFGD generator, whose power is gradually increased by a motorized drive as shown in Fig. 67.1.*

b. The cleaned PLGA sample is placed on the sample bed and dry air is purged through the apparatus at a flow rate of 20 liters/min. The electrode is 1.5 mm away from the PLGA surface.

c. At the same time as the sample bed is being translated at a constant speed, 1.0 cm/s, the corona from the electrode is discharged onto the sample with gradually increasing power (from 10 W to 50 W at 100 kHz).

d. The sample sheet (7.0 cm × 5.0 cm) is treated for 5 s. By this treatment, the sample PLGA surface is continuously exposed to the corona with increasing power.

3. Plasma treatment as physical method [6]

*The corona discharge treatment did not show any visible changes on the PLGA surface, but the water contact angles of the PLGA surface gradually decreased (from 75° to 47°) along the sample length with increasing corona power (Fig. 67.3). The decrease in contact angles (and thus the increase in wettability) along the sample length may be due to the oxygen-based polar functionalities incorporated onto the surface by the ESCA analysis. The possible mechanism of oxidation on the surface may involve the decomposition and re-formation of free radicals of the PLGA molecular chain backbone by corona discharge [13–15].

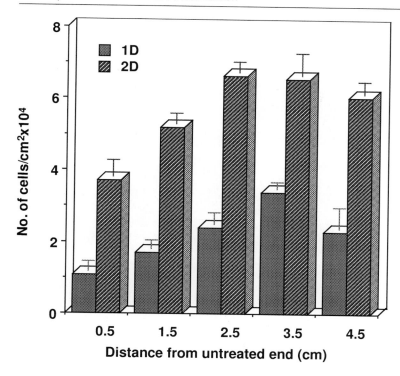

Fig. 67.5. Plot of fibroblast cell adhesion (1 day) and growth (2 days) on corona-treated PLGA surfaces versus wettability chemogradient along the sample length (number of seeded cells; 4 × 10⁴/cm²), n = 3.

a. The PLGA samples are treated with a custom-designed RFGD plasma-generating apparatus, which consists of a Pyrex chamber (~6-liter volume) with an upper stainless steel electrode and a lower brass band electrode in dry air as shown Fig. 67.2.

b. The power supply of the RFGD generator is 200 V, 20 mA, at 100 kHz. The plasma treatment is performed in a reactor at 0.1-torr vacuum.

c. The cleaned PLGA film is located on the lower plate of the reactor chamber. The samples in the reactor chamber are exposed to the plasma for 1–60 s. Wettability or hydrophilicity can be controlled by plasma exposed time. Note that the contact angles on the plasma-treated PLGA surfaces decreased with increasing plasma exposure time from 92° to about 30° (i.e., increased hydrophilicity). Beyond 25 s, the water contact angle did not change even at 60 s [6].

d. After the chamber has been degassed for 10 min, the surface-treated PLGA sample is removed from the Pyrex chamber and used for surface characterization and observation of cell and platelet adhesion.

Surface Characterization of Physicochemically Modified PLGA Surfaces

1. The water contact angle, an indicator of the wettability of surfaces, is measured with an optical bench-type contact angle goniometer, using a sessile drop method at room temperature. The more hydrophilic the sample, the lower the water contact angle observed.

2. To identify the functional groups introduced on the PLGA films, the physicochemically treated PLGA surfaces are analyzed by ESCA with $A1K\alpha$ at 1487 eV and 300 W power anode (incidence angle 30°). Survey scan and carbon 1S core-level scan spectra are taken to analyze each sample (analysis area, ~5 mm²).

MEASUREMENT OF WETTING PROPERTY OF PHYSICOCHEMICALLY TREATED PLGA SCAFFOLDS

1. Immerse physicochemically treated PLGA scaffolds (1.0 cm ×1.0 cm ×2.0 mm thick) in a vial of 0.05% w/v blue dye solution with orbital shaking (30 rpm and 5° angle) and remove at scheduled times, such as 0.5, 1, 2, 4, 6, and 24 h.

2. Use the freeze-dryer to dry the samples, which have taken up some blue strain, and use a video camera to observe the extent of water soaking in photos of cross sections of

1 Day   2 Days

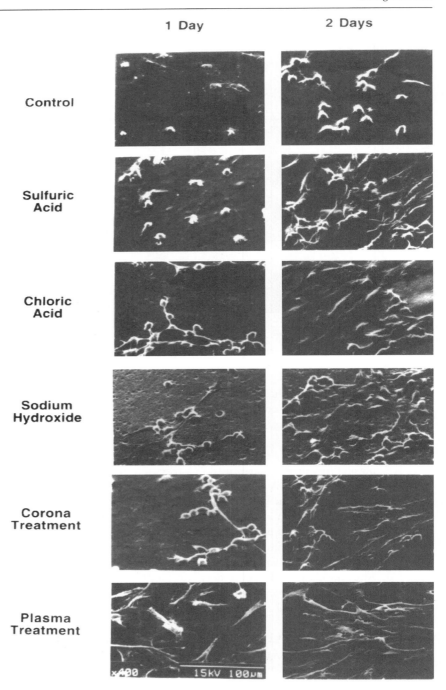

Control

Sulfuric
Acid

Chloric
Acid

Sodium
Hydroxide

Corona
Treatment

Plasma
Treatment

Fig. 67.6. SEM photomicrographs of fibroblast cells attached to physicochemically treated PLGA surfaces after 1- and 2-day cultures. Original magnification, ×400.

the scaffold. The wettability of the chemically treated PLGA scaffolds was ranked in the following order: perchloric acid, sulfuric acid, and sodium hydroxide solution, according to blue dye intrusion, whereas physical methods had no effect. It can be suggested that the chemical treatment method may be useful in uniformly seeding porous biodegradable PLGA scaffolds.

CELL CULTURE ON PHYSICOCHEMICALLY TREATED PLGA SURFACES

1. NIH/3T3 fibroblast cells, routinely cultured in tissue culture polystyrene flasks at 37°C under 5% CO_2 atmosphere, are harvested after treatment with 0.25% trypsin.

2. The physicochemically modified PLGA surfaces (culture area 1.0 cm ×5.0 cm) are placed in a custom-made cell culture chamber.

Control	Chloric Acid	H₂SO₄	NaOH	Plasma

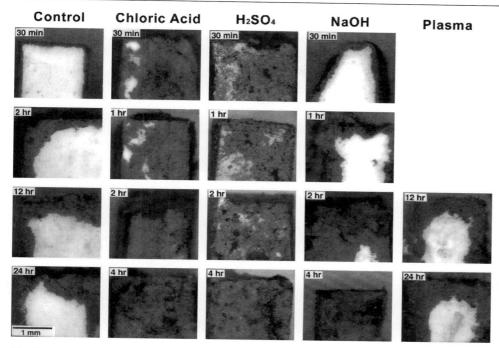

Fig. 67.7. *Wetting property of physicochemically treated porous PLGA scaffolds as demonstrated by the blue dye intrusion methods for 0.5, 1, 2, 4, 12, and 24 h.*

3. The PLGA film surfaces placed on the culture plate are equilibrated for 30 min with PBS (pH 7.3–7.4) that is free of Ca^{2+} and Mg^{2+}.

4. After the PBS solution has been removed from the chambers by pipeting, the cells ($4 \, cm \times 10^4/cm^2$) are seeded to the surfaces.

5. The culture medium used is RPMI-1640 nutrient mixture containing 10% FBS, 100 units/ml penicillin, and 100 μg/ml gentamicin sulfate.

6. The cell culture is allowed to grow for up to 2 days on the surface-modified PLGA surfaces. The culture medium changed once after 1 day.

7. After incubation at 37°C under 5% CO_2 atmosphere, the surfaces are washed with PBS and the cells attached on the surfaces are fixed with 2.5% glutaraldehyde in PBS for 24 h at room temperature. After thorough washing with PBS, the cells on the surfaces are dehydrated in ethanol, graded series (50, 60, 70, 80, 90, and 100%), for 10 min each and allowed to dry on a clean bench at room temperature.

8. The cell-attached PLGA surfaces are gold-deposited in vacuum and examined by SEM with a tilt angle of 45°.

9. The cell density on the surfaces is estimated by counting the number of attached cells on the PLGA surfaces. Different fields for each sample are randomly counted, and the results are expressed in terms of the number of cells attached per square centimeter.

CONCLUSIONS

It has been shown that a physicochemical treatment will improve the wettability and hydrophilicity of PLGA films and porous scaffolds fabricated by the emulsion freeze-drying method. Figure 67.4 shows that the fibroblast cells were more adherent-positions with moderate hydrophilicity on the wettability chemogradient PLGA surface than to the more hydrophobic or hydrophillic positions. Maximum adhesion of the cells appeared at around 2.5–3.5 cm (water contact angle ∼53–55°), as shown in Fig. 67.5. The cell morphology was also changed along the wettability gradient, as observed by SEM. The cells were spread and flattened more on the moderately hydrophilic position of the wettability gradient PLGA surface than on the more hydrophobic or even hydrophilic ones after a 1-day culture. The cells after a 2-day culture were almost flattened on all positions of the gradient except the hydrophobic one at 0.5 cm. Figure 67.6 shows more attachment and spreading of the cells on the physicochemically treated PLGA surface than on the control. For wettability, the chemically treated PLGA scaffolds ranked in the following order: perchloric acid, sulfuric acid, and sodium hydroxide solution, according to the blue dye intrusion method, whereas physical methods were of no effect (Fig. 67.7).

It is suggested that the chemical treatment method may be useful in uniformly seeding porous biodegradable PLGA scaffolds. It seems that wettability plays important roles in cell adhesion, spreading, and growth on the PLGA surface, as well as the intrusion of nutrient media into the PLGA scaffold. The surface modification technique described in this chapter, especially the chemical treatment methods, may be applicable to the area of tissue engineering for the improvement of the tissue compatibility of film- or scaffold-type substrates.

ACKNOWLEDGMENTS

This work was supported by grants HMP-97-E-0016 from the Korea Ministry of Health and Welfare and 97-N1-02-05-A-02 from the Korea Ministry of Science and Technology.

REFERENCES

1. Agrawal, C. M., Niederauer, G. G., Micallef, D. M., and Athanasiou, K. A. (1995). The use of PLA-PGA polymers in orthopedics. *In* "Encyclopedic Handbook of Biomaterials and Bioengineering" (D. L. Wise, D. J. Trantolo, D. E. Altobelli, M. J. Yasemski, J. D. Gresser, and E. R. Schwartz, eds.), Part A, Vol. 2, pp. 1055–1089. Dekker, New York.
2. Holland, S. J., Tighe, B. J., and Gould, P. L. (1986). Polymers for biodegradable medical devices, 1. The potential of polyesters as controlled macromolecular release systems. *J. Controlled Release* **4**, 155–180.
3. Langer, R., and Vacanti, J. P. (1993). Tissue Engineering. *Science* **260**, 920–926.
4. Lee, H. B., Khang, G., Cho, J. C., Rhee, J. M., and Lee, J. S. (1999). Fentanyl-loaded PLGA microspheres for local anesthesia. *Polym. Prepr.* **40**, 288–299.
5. Lee, H. B., and Lee, J. H. (1995). Biocompatibility of solid substrates based on surface wettability. *In* "Encyclopedic Handbook of Biomaterials and Bioengineering" (D. L. Wise, D. J. Trantolo, D. E. Altobelli, M. J. Yasemski, J. D. Gresser, and E. R. Schwartz, eds.), Part A, Vol. 1, pp. 371–398. Dekker, New York.
6. Khang, G., Lee, J. W., Jeon, J. H., Cho, S. C., and Lee, H. B. (1997). Cell and platelet adhesion on plasma glow discharge treated poly(lactide-*co*-glycolide). *Biomed. Mater. Eng.* **7**, 357–368.
7. Khang, G., Lee, H. B., and Park, J. B. (1995). Biocompatibility of polysulfone. II. Platelet adhesion and CHO cell growth on surface modified polysulfone. *Biomed. Mater. Eng.* **5**, 259–273.
8. Mikos, A. G., Lyman, M. D., Freed, L. E., and Langer, R. (1994). Wetting of poly(L-lactic acid) and poly(DL-lactic-*co*-glycolic acid) foams for tissue culture. *Biomaterials* **15**, 55–58.
9. Khang, G., Jeon, J. H., Cho, J. C., Rhee, J. M., and Lee, H. B. (1999). Improvement of wetting property for porous PLGA scaffold by physicochemical treatment. *Polymer (Korea)* **23**, 861–868.
10. Khang, G., Lee, S. J., Lee, J. H., and Lee, H. B. (1999). Interaction of fibroblast cells on poly(L-lactide-*co*-glycolide) surfaces with wettability chemogradient. *Biomed. Mater. Eng.* **9**, 179–189.
11. Khang, G., Jeon, J. H., Cho, J. C., and Lee, H. B. (1999). Fabrication of tubular porous PLGA scaffolds by emulsion freeze drying methods. *Polymer (Korea)* **23**, 471–477.
12. Whang, K., Thomas, C. H., and Healy, K. E. (1995). A novel method to fabricate bioresorbable scaffolds. *Polymer* **36**, 837–842.
13. Lee, J. H., Kim, H., Khang, G., Lee, H. B., and Jhon, M. S. (1992). Characterization of wettability gradient surfaces prepared by corona discharge treatment. *J. Colloid Interface Sci.* **152**, 97–104.
14. Lee, J. H., Khang, G., Lee, J. W., and Lee, H. B. (1997). Interaction of cells on chargeable functional group gradient surfaces. *Makromol. Chem., Macromol. Symp.* **118**, 571–576.
15. Khang, G., Lee, J. H., Lee, I., Rhee, J. M., and Lee, H. B. (2000). Interaction of different types of cells on poly(L-lactide-*co*-glycolide) surfaces with wettability chemogradient. *Korea Polymer J.* **8**, 276–284.

CHAPTER 68

Microencapsulation Methods: Agarose–PSSa

Hiroo Iwata

INTRODUCTION

Enclosure of islets of Langerhans (islets) into a semipermeable membrane prior to transplantation, that is, immunoisolation, is a promising approach to the problem of overcoming adverse host immune reactions against the transplanted islets. The semipermeable membrane permits the passage of oxygen, nutrients, and insulin secreted by the islets, but inhibits contacts of immunocompetent cells and molecules with the islets. Various devices—diffusion chambers, tubular intravascular units, and microcapsules—have been developed [1]. Each of these has its own advantages and disadvantages. Microcapsules can be easily implanted without major surgical procedures. Each microcapsule contains a small number of islets. This allows sufficient oxygen and nutrient supply to reach the islets.

Since 1980, when Lim and Sun [2] introduced the alginate–poly(L-lysine) microcapsule to enclose living islets, polyion complex membranes made of various combinations polyanions and polycations have been examined, and superior characteristics have been claimed for them. The viability of enclosed islet cells is high. They do not induce inflammatory reactions, and their cutoff molecular weights can be easily controlled by changing the molecular weights of the polyelectrolytes and the time for polyion complex formation. One major drawback of this type of microcapsule, however, is its low mechanical strength. When implanted into recipients, the devices are seriously deformed and sometimes are ruptured even in the peritoneal cavity.

To overcome this problem, we developed microbeads made of agarose hydrogel [3,4]. Agarose is soluble in hot water. Its solution congeals between 20 and 40°C to form a firm, resilient gel, which does not melt below 60°C. The large hysteresis observed in gelling and melting temperatures is suitable for microencapsulation islets. Islets can be microencapsulated below body temperature, and the microbeads are stable after implantation. Firm agarose microbeads resist mechanical force exerted after implantation, and agarose gels are fairly stable. Although a few microorganisms metabolize agarose or elaborate enzymes that degrade it, mammals do not. The microbeads can exist for a long time in the body without deterioration. In addition, agarose is compatible with various polysaccharides and proteins, and some synthetic polymers. Flocculation or marked degradation does not occur when their dispersion is mixed. This property is useful for modification of the characteristics of agarose microbeads. Membrane permeability and molecular weight cutoff can be modulated by the addition to agarose hydrogel of other polymers, and biological active substances can be added to agarose microbeads [5].

ENCAPSULATION OF ISLETS IN AGAROSE MICROBEADS

INSTRUMENTS AND EQUIPMENT

Autoclaved/sterile surgical instruments: two pairs of scissors, straight and curved tweezers, a razor

Cork plate, 20×30 cm^2

Glass centrifugal tubes, 50-ml, with a deep screw cap

Water bath at 37°C

Laminar flow hood

Stereoscopic microscope

DISPOSABLE ITEMS

Petri dishes 60-mm diameter for bacteria culture (Falcon, Becton Dickinson, Lincoln Park, NJ).[*]

Plastic pipettes, 10 ml

Pasteur pipettes coated with bovine serum albumin (BSA) (Sigma).[†]

Pasteur pipettes with a capillary tip, also coated with BSA

Plain bulbs for Pasteur pipettes and 10-ml pipettes

Centrifuge plastic tubes, 15 ml and 50 ml

SOLUTIONS

1. *Buffers*: Ca^{2+}- and Mg^{2+}-free Hanks' Balanced Salt Solution (HBSS; Nissui, Tokyo, Japan) and Ca^{2+}- and Mg^{2+}-free Dulbecco's phosphate-buffered saline (PBS; Nissui). Serum-free Eagle's Minimum Essential Medium (MEM, autoclavable, Nissui) containing 60 mg/liter kanamycin and 0.15% sodium bicarbonate. Filter all these solutions through a 0.22-μm sterile filtration unit.

2. *Collagenase solution*: Dissolve collagenase (type S-1, Nitta Gelatin, Osaka, Japan) in HBSS to 800 mg/dl. Filter the solution through a 0.22-μm sterile filtration unit. Pipette 5 ml of the collagenase solution into 50-ml plastic centrifugal tubes and store them at –20°C until use. Thaw the solution at 37°C just before use.

3. *Primary culture medium*: Eagle's MEM (Nissui) containing 60 mg/liter kanamycin, 0.292 g/liter L-glutamine, 0.15% sodium bicarbonate, and 5% heat-inactivated fetal bovine serum (FBS) (HyClone Laboratory, Logan, UT).

4. *Agarose solution*: Suspend agarose powder (0.15 g) (low gelling temperature, Nacalai Tesque, Kyoto, Japan) into serum-free MEM (3 ml) in 50-ml glass centrifugal tubes.[‡] Autoclave the glass tubes for sterilization and to dissolve agarose powder. Use a vortex mixer to mix the solution well at high temperature and store at –20°C until use.

5. *Microencapsulation medium*: Mix 150 ml of paraffin oil (7174, Merck, Darmstadt, Germany) vigorously with 30 ml of PBS in a 200-ml glass bottle. Leave the mixture for 20 min at room temperature, and remove the PBS layer by suction. Repeat this procedure three times to extract water-soluble toxic substances from the paraffin oil. Autoclave the glass bottle, and store it at room temperature until use.

ISLET ISOLATION[§]

1. Kill a golden hamster by deep anesthesia (inhalation of ether) and fix it on a cork plate. Swab the ventral surface with 70% ethanol and shave the ventral hair. Cut the ventral skin along and perpendicular to the median line and remove the skin to expose the untouched ventral surface of the abdominal wall.

---

[*]Untreated petri dishes should be used to inhibit islet adherence to a substrate.

[†]Islets in serum-free medium easily stick to glassware. Pasteur pipettes should be coated with BSA to prevent loss of islets when islets in serum-free medium are handled. Pipettes can be coated with BSA by simply rinsing them with 1% BSA solution in phosphate-buffered saline.

[‡]A glass tube should be used to prepare agarose microbeads. If a plastic tube is used, the agarose solution cannot be well suspended into paraffin oil.

[§]Since the report of islet isolation from the rat pancreas in 1967 by Lacy and Kostianovsky [6], several modifications and standardization have been introduced for the purpose of increasing the yield and the reproducibility of the isolation. However, the procedure remains difficult for beginners. Islets are most easily isolated from the hamster pancreas.

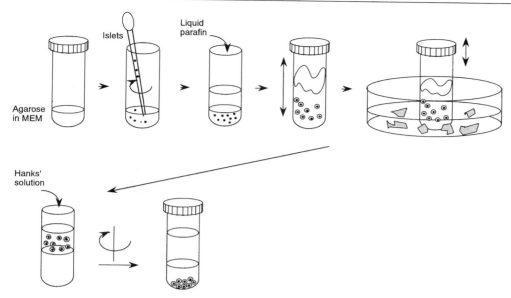

Fig. 68.1. Procedure for encapsulating islets in agarose microbeads.

2. Cut longitudinally along the median line with sterile scissors, revealing the viscera. Push the intestines a side. Aseptically remove the entire pancreas and place it on a petri dish kept on ice. Collect pancreatic tissue from six hamsters.

3. Trim away mesenteric membranes and other extraneous matter, and use sharp scissors to mince the pancreatic tissue into pieces to approximately 1 mm³.

4. Transfer the minced tissue to a 50-ml plastic centrifuge tube containing 5 ml of the collagenase solution, and cap the tube. Agitate the tube vigorously in the longitudinal direction with manual shaking for 5–7 min until the solutions becomes milky and small cell clusters stick to the wall of the tube. *

5. Add cold HBSS to the tube to 45 ml and use a 10-ml pipette to draw the solution up and down gently, to disperse the tissue. Leave the tube for 3 min to allow islets to settle. Remove 20 ml of the supernatant fluid. Repeat these washing steps eight times with fresh HBSS.

6. Resuspend the digested tissue in 50 ml of HBSS. Centrifuge at 140*g* for 5 min, discard the supernatant, and resuspend the tissue pellet into 50 ml of MEM supplemented with 5% FBS.

7. Transfer the suspension to 60-mm petri dishes, 5 ml per dish.

8. Use a Pasteur pipette with a fine tip to handpick islets under a stereoscopic microscope. Resuspend the islets into 5 ml of MEM supplemented with 5% FBS in a 15-ml plastic centrifuge tube. Generally, 1500–2000 islets are obtained at this stage.

9. Encapsulate the islets in agarose hydrogel microbeads immediately.

Encapsulation of Islets (see Fig. 68.1)

1. Keep a glass bottle containing paraffin oil at 37°C for at least 30 min until use.

2. Microwave a glass tube containing agarose gel to melt the gel and keep it in a 37°C water bath for 8 min until use. †

3. Sediment the islets by centrifugation at 140*g* for 5 min.

4. Resuspend the islets in 5 ml of HBSS and then pellet by centrifugation at 140*g* for 5 min. Remove the supernatant as carefully as possible. Keep the tube containing the islet pellet at 37°C for 2 min.

---

*Minced pancreatic tissue is digested by collagenase to dissociate exocrine tissue. Without vigorous shaking, islets remain trapped in a viscous liquid. The test tube should thus be vigorously agitated.

†While a tube is microwaved, it should be immersed in hot water in a 100 ml beaker, and the screw cap should be released.

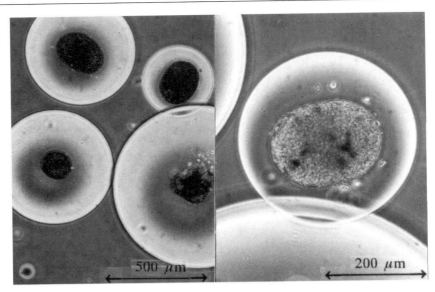

Fig. 68.2. Agarose microbeads containing hamster islets.

5. Use a Pasteur pipette to transfer the islet pellet to a glass tube containing 3 ml of an agarose solution kept at 37°C. * Then use the Pasteur pipette to suspend the islets well into the agarose solution.

6. Add approximately 20 ml of paraffin oil to the glass tube. Suspend the agarose solution in the paraffin oil with manual shaking at room temperature for about 5 min, until the desired droplet size is obtained.

7. Immerse the tube in an ice bath for 5 min with agitation to encourage gelation of agarose solution droplets. †

8. Add 25 ml of cold HBSS to the tube. Centrifuge the tube at 800g for 15 min to spin off the agarose microbeads to the bottom of the tube. Remove the paraffin oil and HBSS layers by suction.

9. Resuspend the microbeads into 45 ml of cold HBSS. Transfer the suspension to a 50-ml plastic centrifugal tube. Centrifuge the tube at 800g for 15 min and remove supernatant by suction. This procedure is repeated twice.

10. Resuspend the microbeads in 50 ml of MEM containing 5% FBS. Transfer the suspension into 10 petri dishes, 5 ml per dish.

11. Using a Pasteur pipette under a stereoscopic microscope, handpick microbeads containing islets and transfer them to petri dishes 6 cm in diameter, containing 5 ml of MEM supplemented with 5% FBS (Fig. 68.2).

ENCAPSULATION OF ISLETS IN AGAROSE–POLY(STYRENE SULFONIC ACID) (PSSA) MICROBEADS

All procedures and solutions are the same as described in connection with encapsulation in agarose microbeads, except for the following.

SOLUTIONS

Agarose–PSSa solution: Suspend agarose powder (0.15 g) (low gelling temperature, Nacalai, Tesque, Kyoto, Japan) and PSSa powder (0.15 g) (average molecular weight 50 kDa, Toso, Tokyo) in serum-free MEM (3 ml) in a 50-ml glass centrifuge tube. Autoclave the glass tube for sterilization, and to dissolve the agarose and PSSa. Use a vortex mixer at high temperature to mix the solution well and store at −20°C until use.

---

* A Pasteur pipette coated with BSA should be used to prevent loss of islets.
† Agarose gel film is concomitantly formed on the glass test tube. Some islets are occluded in the gel film.

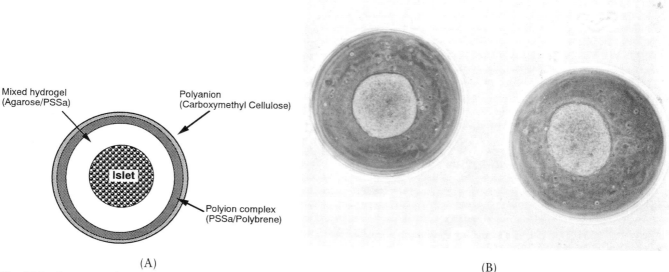

Fig. 68.3. Agarose–poly(styrenesulfonic acid) microbeads. (A) Schematic representation of agarose–PSSa microbeads. (B) Photomicrographs of microbeads.

Polybrene solution: Dissolve hexadimethrine bromide (Polybrene) (molecular weight 5–7 kDa, Aldrich Chemical Co.) in 50 ml of serum-free MEM to 1%. Filter the solution through a 0.22-μm sterile filtration unit into 50-ml plastic centrifuge tubes. Keep the tubes on ice until use.

Solution for coating agarose–PSSa microbeads: Dissolve polyanion, carboxymethyl-cellulose (MW 210–250 kDa, Wako Pure Chemicals, Osaka, Japan), poly(acrylic acid) (MW 200–500 kDa, Wako Pure Chemicals) or chondroitin sulfate (MW 40–80 kDa, Wako Pure Chemicals), in 30 ml of serum-free MEM to 1% in a 50-ml glass centrifuge tube. Autoclave the glass tube for sterilization, and to dissolve the polyanion. Use a vortex mixer to mix the solution well. Keep the tube on ice until use.

ENCAPSULATION OF ISLETS

1. Microwave a glass tube containing 3 ml of agarose–PSSa mixed gel to melt the gel, and keep the tube in a 40°C water bath for 8 min.[*]

2. Sediment the islets by centrifugation at 140g for 5 min. Resuspend the islets in 5 ml of HBSS and then pellet by centrifugation at 140g for 5 min. Remove the supernatant as carefully as possible. Keep the tube containing the islet pellet at 40°C for 2 min.

3. Use a Pasteur pipette to transfer the pellet to the tube and then mix the islets with the agarose–PSSa solution well, again using the Pasteur pipette.[†]

4. Add approximately 20 ml of paraffin oil to the tube. Suspend the liquids by manual shaking to form agarose–PSSa droplets of the appropriate size at room temperature for about 5 min.

5. Immerse the tube in an ice bath for 5 min with agitation to induce gelation of agarose–PSSa droplets.

6. Add 25 ml of a 1% Polybrene solution to the tube and gently agitated in an ice bath for 5 min.[‡]

---

[*] An agarose–PSSa solution becomes viscous at 37°C during incubation. The tube should be kept at 40°C. While being microwaved, it should be immersed in hot water in a 100-ml beaker, and the screw cap should be released.

[†] Islet suspension should be quickly mixed with the agarose–PSSa solution at 40°C.

[‡] Microbeads and highly viscous substances are released from paraffin oil in this step. Be careful not to occlude the microbeads into the viscous substance.

7. Spin the microbeads into the Polybrene solution by centrifugation at 800g for 10 min. Remove the paraffin oil and Polybrene solution layers by suction.

8. Resuspend the microbeads into 25 ml of 1% Polybrene solution; transfer the suspension to a plastic centrifuge tube, and leave it on ice with gentle agitation for 15 min. After this period, settle the microbeads by centrifuge at 140g for 5 min. Discard the supernatant.

9. Resuspend the microbeads in HBSS. Centrifuge the tube at 800g and discard the supernatant. This procedure is repeated twice.

10. Resuspend the microbeads into 45 ml of 1% polyanion solution and leave on ice for 30 min with gentle agitation. After this period, settle the microbeads by centrifuge at 140g for 5 min. Discard the supernatant.

11. Resuspend the microbeads in 45 ml of MEM supplemented with 5% FBS. Transfer the suspension into 10 petri dishes 6 cm in diameter, 5 ml per dish.

12. Use a Pasteur pipette under a stereoscopic microscope to handpick microbeads containing islets and transfer them to petri dishes 6 cm in diameter and containing 5 ml of MEM supplemented with 5% FBS (Fig. 68.3).

ACKNOWLEDGMENTS

The author's work is supported by grant JSPS-RFTF 96I 00204 and by Grant-in-Aid for Scientific Research (C) 10838016 from the Japan Society for the Promotion of Science.

REFERENCES

1. Kühtreiber, W. M., Lanza, R. P., and Chick, W., eds. (1998). "Cell Encapsulation Technology and Therapeutics." Birkhäuser, Boston.

2. Lim, F., and Sun, A. M. (1980). Microencapsulated islets as bioartificial endocrine pancreas. *Science* **210**, 908–910.

3. Iwata, H., Amemiya, H., Matsuda, T., Takano, H., Hayashi, R., and Akutsu, T. (1989). Evaluation of microencapsulated islets in agarose gel as bioartificial pancreas by studies of hormone secretion in culture and by xenotransplantation. *Diabetes* **38**(Suppl. 1), 224–225.

4. Iwata, H., Takagi, T., Amemiya, H., Shimizu, H., Yamashita, K., Kobayashi, K., and Akutsu, T. (1992). Agarose for a bioartificial pancreas. *J. Biomed. Mater. Res.* **26**(7), 967–977.

5. Iwata, H., Takagi, T., Kobayashi, K., Oka, T., Tsuji, T., and Ito, F. (1994). Strategy for developing microbeads applicable to islet xenotransplantation into a spontaneous diabetic NOD mouse. *J. Biomed. Mater. Res.* **28**(10), 1201–1207.

6. Lacy, P. E., and Kostoanovsky, M. (1967). Method for the isolation of intact islets of Langerhans from the rat pancreas. *Diabetes* **16**(1), 35–39.

CHAPTER 69

MICROENCAPSULATION METHODS: ALGINATE (CA^{2+}-INDUCED GELATION)

Hua Yang and James R. Wright Jr.

INTRODUCTION

Alginate, a naturally occurring copolymer, has long been used in the food and pharmaceutical industries as a thickener to enhance viscosity, and to stabilize aqueous mixtures, dispersions, and emulsions [1]. For these initial applications, alginates were exploited largely on the basis of empirical knowledge. After 1980, the year that Lim and Sun reported that an alginate–polylysine microencapsulation system permitted islet transplantation in diabetic rodents [2], attention began to focus on the chemical and physical properties of alginate, primarily because others had difficulty replicating the results reported. In the following years, significant amounts of literature were devoted to the characterization of alginates as an encapsulation material. Today, it is fair to say that calcium alginate is the most widely used and studied encapsulation material [3–5].

Traditionally, most of the procedures for producing alginate-based microcapsules involve extruding a mixture of cells and sodium alginate into a divalent cation solution, such as calcium chloride. Calcium ions diffuse into the alginate droplets, forming alginate gel spheres consisting of an ionically cross-linked three-dimensional lattice. In most cases the negatively charged gel spheres are then coated with positively charged polymers, such as poly(L-lysine) (PLL), to increase strength and modulate the permeability. Often, in accordance with Lim and Sun's initial microencapsulation system, the calcium alginate core is liquefied with citrate buffer after PLL coating in an attempt to optimize the microenvironment for cell survival. To optimize diffusion of oxygen and nutrients, the diameter of microcapsules is generally maintained between 300 and 800 μm. Many authors limit the size to less than 500 μm.

In 1995, Lanza et al. reported an unexpected finding [6]. By simply increasing the alginate gel sphere diameter to 3700 μm (i.e., without PLL coating), they found that porcine and bovine islet xenograft survival could be significantly prolonged in diabetic mice without any immunosuppression. The immunoprotective effect of this large-diameter, uncoated alginate gel encapsulation method compared favorably with data from earlier of work with traditional PLL-coated alginate microcapsules [6,7]. We applied this method to an even more discordant xenograft model, teleost fish islets transplanted into streptozotocin (STZ)-diabetic rats and mice. In our laboratory, fish islet xenograft survival was significantly prolonged in nonimmunosuppressed rats and mice [8,9]. In combination with low-dose immunosuppression, large-diameter uncoated alginate gel capsules provided long-term fish islet xenograft survival in STZ-diabetic mice and rats [8,9] and long-term islet allograft survival in spontaneously diabetic dogs [7]. The finding that uncoated alginate gel sphere encapsulation could markedly prolong islet xenograft survival between such highly discordant species was particularly surprising, since alginate gel has sufficient porosity to permit the entry of both antibodies and complement [6,10,11]. Clearly, the mechanical barrier cannot fully explain the immunoprotective effect provided by large-diameter uncoated alginate spheres. Therefore, it seems likely that other mechanism(s) must also contribute to this effect.

ALGINATE CHEMICAL STRUCTURE AND GEL-FORMING KINETICS

Alginates are polysaccharides found in all species of brown marine algae (Phaeophyceae) and in some microorganisms [1,12]. They are linear copolymers of 1,4-linked α-L-guluronic acid (G) and β-D-mannuronic acid (M) arranged as chains in alternating blockwise patterns (Fig. 69.1). Depending on the alginate source, the polymers are composed of three different block types: polymannuronate (MM), polyguluronate (GG), and mixed mannuronate and guluronate blocks (MG). Depending upon the chemical composition, the molecular weight of alginates can vary from approximately 50 to 500 kDa [13]. Sodium alginate extracted from stipes of *Laminaria hyperborea* (Pronova LVG, Pronova Biopolymer, Drammen, Norway) is rich in guluronic acid (high G) and has a molecular weight of 277 kDa, while sodium alginate isolated from *Macrocystis pyrifera* (Kelco Division of Merck, San Diego, CA) is high in mannuronic acid (high M) and has a molecular weight of 210 kDa.

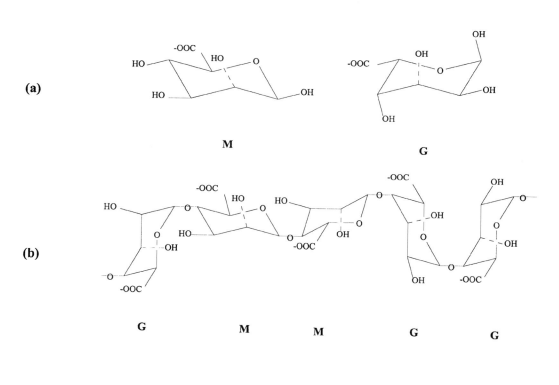

Fig. 69.1. Structure of alginate: (a) monomer conformation (M, mannuronate; G, guluronate), (b) chain conformation, and (c) schematic alginate chain sequence. Reprinted with permission of Birkhäuser from Yang and Wright [44, p. 80].

The monomeric composition, sequential structure, molecular size, and gel-forming kinetics of an alginate determine its diffusion coefficient (i.e., porosity), swelling behavior, stability, biodegradability, gel strength, and biocompatibility. Capsules made from alginates with high G composition and longer G blocks have higher mechanical strength, lower shrinkage, more stability toward monovalent cations, and higher porosity [4,5,11,13]. Diffusion also depends upon the alginate concentration, the M/G ratio, and the gel-forming method. Upon forming a gel in the presence of a charged osmolyte such as NaCl, the alginate gel sphere will become homogeneous (i.e., the distribution of alginate within the sphere is isotropic), while gels formed in the absence of a charged osmolyte tend to be inhomogeneous, with a higher alginate concentration near the surface of the gel sphere. Entrapped bovine serum albumin (BSA) is diffused more quickly from homogeneous spheres than from inhomogeneous spheres [10]. Because the alginate gel is negatively charged, pH also affects the diffusion of charged substrates and products; for instance, the rate of diffusion of BSA from alginate spheres increases with pH because of the increasing negative charge on the protein. The temperature dependence of the diffusion coefficient follows the Arrhenius equation. It is worth mentioning that one disadvantage of a calcium alginate gel is its sensitivity to calcium chelators, such as phosphates, lactate, or EDTA, and to cations such as sodium or magnesium, which are able to displace calcium. This sensitivity is problematic because these compounds are often found in culture media or in fixatives. On the other hand, exposure to citrate can be used to liquefy the microcapsule's alginate core or to recover encapsulated cells.

Alginates possess a number of favorable characteristics. The conditions required for gelation are very mild. Gelation does not require pH changes, temperature changes, or toxic reagents, and the process produces no residues. The gelation procedure is simple and low in cost. Therefore, alginate gel has long been used in the food and pharmaceutical industries [1,14] with only a few reports of complications [15–17]. In recent years, alginate has been used in a wide range of biomedical fields [18] including controlled-release systems for drugs [19–23], somatic gene therapy [24–27], and oral vaccines [28–30]. Alginate-encapsulated cell lines have been used for the *in vivo* delivery of monoclonal antibodies and recombinant products [31,32]. Alginates have also been used for scaffolding to support cell proliferation to promote rebuilding of cartilage [33] and wound healing [14,34–37], as well as for hemostatic agents and swabs [14,16,17]. Alginate has even been used to encapsulate embryos for *in vitro* fertilization [38]. However, most investigators have focused on the use of alginate gels for immunoisolation systems to facilitate cell transplantation.

LARGE-DIAMETER, UNCOATED ALGINATE GEL SPHERE ENCAPSULATION TECHNIQUE

In our laboratory, we use tilapia, a tropical teleost fish with large, anatomically discrete islet organs called Brockmann bodies (BBs), as islet donors for xenotransplantation studies [39–42], and we are currently producing transgenic tilapia with a "humanized" tilapia insulin gene, which we hope may provide an inexpensive future source of donor islets for clinical islet xenotransplantation [43]. Using tilapia islet donors, we have examined the protective effect provided by uncoated alginate gel spheres on islet xenograft survival in several different strains of diabetic mouse and rat recipients [9,44].

Figure 69.2 shows our encapsulation procedure [9], which is based on the method described by Lanza *et al.* [7]. The initial step is the formation of an islet–alginate suspension. BB fragments are collected [41] and mixed into a 1.5% LVG sodium alginate (Pronova Biopolymer, Drammen, Norway) solution. Islet tissue from 1000 g of donor fish is suspended in 1 ml of sodium alginate for transplantation into murine recipients. This volume of fish islet tissue is the functional equivalent of roughly 600–650 handpicked rodent islets or several thousand human or porcine islet equivalents [45]. For rat recipients, islet tissue from 7000 g of donor fish is suspended in 7 ml of sodium alginate. The next step is the formation of islet-containing calcium alginate gel spheres. The islet–alginate mixture is placed in a syringe and then extruded dropwise through a 16-gauge intravenous catheter (Critikon, Tampa, FL) into a beaker containing a stirred 1.5% $CaCl_2$ solution (Fig. 69.2). This procedure generates alginate gel spheres measuring 4.5 mm in diameter (Fig. 69.3).

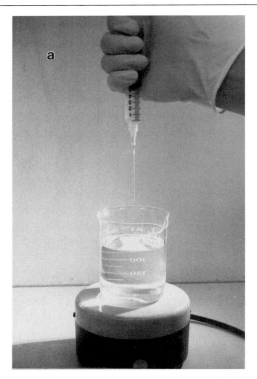

Fig. 69.2. Encapsulation process: (a) islet fragments suspended in 1.5% LVG sodium alginate saline solution are extruded through a 16-gauge intravenous catheter (Critikon, Tampa, FL) into 1.5% CaCl₂ solution to form (b) islet-containing calcium alginate gel spheres. Reprinted with permission of Birkhäuser from Yang and Wright [44, p. 81].

A third step, fine coating, is optional. After gelation, the gel spheres are washed with normal saline and resuspended in a beaker containing a 0.7% solution of UP (ultrapure) LVG sodium alginate (Pronova); the suspension is slowly stirred with a stirring bar for 5 min. Finally, the capsules are washed with Hanks's Balanced Salt Solution (HBSS) and cultured overnight ($37°$C, 5% CO_2) in CMRL-1066 culture medium with 10% fetal calf serum, 100 U/ml penicillin, and 100 µg/ml streptomycin sulfate before transplantation. In our initial studies, the LVG alginate gel spheres were fine-coated with pure UP LVG alginate gel to improve capsule biocompatibility [9]. Subsequently, we determined that fine coating can be omitted without significant loss in biocompatibility.

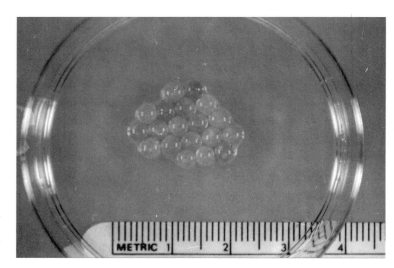

Fig. 69.3. Encapsulated fish islet grafts recovered from diabetic recipient mice over 50 days post-transplantation. Reprinted with the permission of Lippincott Williams & Wilkins from Yang et al. [9].

UNCOATED ALGINATE GEL FOR ISLET XENOTRANSPLANTATION

In 1995 Lanza *et al.* reported that uncoated alginate gel spheres permitted survival of porcine or bovine islet xenografts and reversed hyperglycemia in STZ-diabetic mice for at least 30 and 43 days, respectively, without any immunosuppression, while nonencapsulated grafts survived less than 4 days [6]. The same group subsequently reported that uncoated alginate gel encapsulation, combined with 10–20 mg/kg/day cyclosporin A (CsA) treatment, permitted long-term function of both porcine and bovine islet xenografts in STZ-diabetic Lewis rats, while unencapsulated controls experienced primary nonfunction [7].

In our laboratory, encapsulation with large-diameter uncoated alginate gel spheres markedly prolonged mean graft survival time (mGST) of fish islets transplanted into STZ-diabetic Balb/c mice [9]. Unencapsulated fish islets transplanted intraperitoneally (IP) into euthymic Balb/c mice uniformly failed within 5 days; however, survival was significantly prolonged to over 27 days by encapsulation ($p < 0.005$), with one of seven mice achieving long-term function. Immunosuppression with either CsA (10 mg/kg/day) or 15-deoxyspergualin (DSG, 2.5 mg/kg/day) minimally prolonged unencapsulated fish islet graft survival ($p < 0.05$, and n.s.); however, when combined with encapsulation, mGSTs were markedly prolonged relative to both control ($p < 0.001$) and drug-only treatment groups ($p < 0.005$). Long-term graft function was achieved in 4 of 6 mice treated with CsA and in all mice treated with DSG. We also confirmed biocompatibility by transplanting encapsulated fish islets into STZ-diabetic athymic nude mice, GSTs were uniformly over 50 or over 100 days in STZ-diabetic nude mice. In all instances of long-term graft function in either Balb/c mice or nude mice, graft dependency was confirmed by graft retrieval from each recipient's peritoneal cavity 50–100 days post-transplantation. Histological examination of the explanted encapsulated islets revealed numerous viable, well-granulated islets (Fig. 69.4a, b) and occasional necrotic islets. The explanted capsules did not show gross (Fig. 69.3) or histological evidence of fibroplasia (Fig. 69.4a, b).

Large-diameter alginate gel spheres also prolonged fish islet xenograft function in STZ-diabetic rats [44]. Unencapsulated fish islets transplanted into STZ-diabetic Lewis rats experienced primary nonfunction, even with adjuvant immunosuppression. However, mGST of alginate-encapsulated fish islet grafts was 13 days in untreated Lewis rats, over 46 days in Lewis rat recipients receiving 2.5 mg/kg/day DSG (3/6 rats achieved >50 day function), and 27 days in Lewis rats receiving 10 mg/kg/day CsA. Histological examination of explanted grafts retrieved from rats with 50-day function confirmed viable, well-granulated islets (Fig. 69.4c) and good capsule biocompatibility.

UNCOATED ALGINATE GEL FOR ISLET ALLOTRANSPLANTATION

The majority of work with uncoated alginate spheres has pertained to experimental islet xenotransplantation. However, Lanza and Chick have performed some studies using a large-animal islet allograft model [3]. They demonstrated that in spontaneously diabetic dogs, encapsulated canine islets, when combined with low-dose CsA, resulted in normoglycemia without exogenous insulin treatment for 60 to over 175 days. In our laboratory, encapsulated Wistar-Furth (WF) rat islet allografts survived uniformly over 50 days in diabetic Lewis rat recipients ($n = 3$).

CO-ENCAPSULATION METHOD

Although, we were able to achieve significant prolongation of fish islet graft survival in diabetic euthymic mouse and rat recipients by transplanting large-diameter alginate gel encapsulated fish islets [9,44], long-term graft survival required supplemental systemic immunosuppression. To date, we are unaware of any encapsulation system that provides complete protection for xenogenic islet grafts between discordant species. There are two conflicting issues involved in the design of encapsulation systems. Smaller membrane "pore" sizes provide better protection from rejection, but compromise permeability for oxygen, nutrients, and functional products. Xenograft rejection involves not only large factors such as cells and immunoglobulin, but also many factors with very small molecular weights.

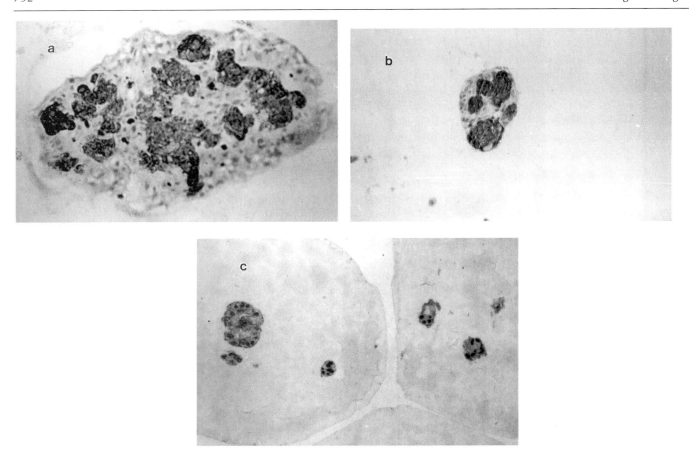

Fig. 69.4. Histological sections of encapsulated fish islet grafts removed from (a) a diabetic nude mouse at day 55, (b) a diabetic Balb/c mouse receiving DSG treatment at day 50, and (c) a diabetic Lewis rat receiving DSG treatment at day 50 post-transplantation. Islet tissue is viable and contains both β and non-β cells [stained with immunoperoxidase for fish insulin, original magnification (a) ×200; (b) ×100; and (c) ×40. (a) and (b) reprinted with the permission of Lippincott Williams & Wilkins from Yang et al. [9]. (c) reprinted with permission of Birkhäuser from Yang and Wright [44, p. 84].

Over the past 20 years, much effort has been devoted to controlling capsule porosity [3,44]; however, many of these efforts such as further coating with synthetic materials or increasing the density of the encapsulation materials had deleterious effects, such as stimulating fibroplasia and graft hypoxia. Therefore, engineering capsule porosity alone may not be sufficient to meet the demand of a clinically applicable encapsulation system. One possible method of circumventing this capsule engineering obstacle is "co-encapsulation" of the islets with cells that produce immunomodulating substances, which in turn supplement the capsule's immunoprotective effect (Fig. 69.5). In our laboratory, Sertoli-enriched testicular cell fractions have been used as a source of immunomodulating Sertoli cells to facilitate fish islet xenograft survival in rodent models [46].

The rationale for this approach is based both on Selawry's observations that the cryptorchid testis provides an immune privileged site for islet allo- and xenografts in rats [47,48] and on the belief that testicular Sertoli cells are responsible for this immune protective effect [49,50]. When co-transplanted with islet grafts in an ectopic site, such as under the renal capsule, Sertoli cells markedly prolong islet allograft survival [50–52]. However, the islet grafts must be in close association with the Sertoli cells to receive bystander immune privileges. Because the human renal subcapsular space is not suitable for islet transplantation and because the intraportal route does not achieve close association, Korbutt *et al.* proposed that co-encapsulation, a system that combines the benefits of encapsulation and closely associates Sertoli cells and islet grafts, could have great future clinical ap-

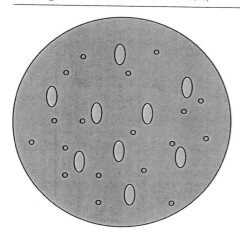

Fig. 69.5. Schematic diagram showing co-encapsulation model of fish islets and rodent Sertoli cell aggregates.

◦ Sertoli cell aggregates

◉ Fish islet

plicability in the treatment of diabetes. Korbutt *et al.* presented preliminary data showing that co-encapsulation of rat islets and Sertoli cells extends allogenic islet allograft survival in a rat model [53]. In our laboratory, Sertoli-enriched testicular cell fractions were enzymatically harvested from adult Balb/c mice or WF rats. After 48 h of culture, Sertoli cell aggregates were co-encapsulated with fish islet fragments in large-diameter uncoated alginate gel spheres. Islets encapsulated alone or islets co-encapsulated with Sertoli cells were transplanted IP into STZ-diabetic Balb/c mice. Our results show that co-encapsulation with testicular cell fractions enriched by Sertoli cells further prolonged fish islet graft survival time from 21 ± 6.7 days (encapsulated alone) to over 46 ± 6.3 days (co-encapsulated with syngeneic murine Sertoli cells), without additional systemic immunosuppression. Testicular cells harvested from xenogenic WF rats produced similar protective effects ($> 46 \pm 10.9$ days). Immunoperoxidase staining demonstrated that insulin-positive islets and vimentin-positive Sertoli cells coexist in the alginate gel spheres harvested 50 days after implantation in Balb/c mouse recipients. When diabetic WF rats were used as recipients, co-encapsulation of WF rat testicular cells further prolonged mean fish islet graft survival time from 13 ± 3 days (i.e., encapsulated alone) to 31 ± 9.2 days. At present, although the responsible mechanisms have not been elucidated, we have demonstrated that co-encapsulated syngeneic or xenogeneic Sertoli-enriched testicular cell fractions can provide additional immune protection to fish islet grafts in the absence of systemic immune suppression. Our results support the theory that Sertoli cells produce local immunosuppressive factor(s), which supplement the immune protective features of alginate microcapsules.

ENCAPSULATION OF HYBRIDOMA CELLS

Another valuable application for the alginate gel sphere encapsulation technique is to permit transplantation of hybridoma cells or gene transfected cells lines for the *in vivo* delivery of either monoclonal antibodies or recombinant products, a technique developed by Savelkoul *et al.* [31,54]. In this system, the alginate gel spheres implanted into the abdominal cavity of a recipient animal protect the encapsulated cells from host rejection while allowing these cells to produce and secrete products *in vivo*. This system had been applied in a number of *in vivo* studies, including those involving infection [55,56], cytokine modulation [54], asthma [57], and cancer [58]. In 1999, Dickson [32], in our laboratory, was able to rapidly and inexpensively screen the effects of a large number of monoclonal antibodies directed at lymphocyte subsets, adhesion molecules, and various cytokines in our fish-to-mouse islet xenotransplantation model.

The alginate encapsulation technique used in our laboratory is a merger of the "alginate-entrapping" method of Savelkoul [31] and the technique developed in our laboratory for islet transplantation (described in detail earlier). Briefly, hybridoma cells are maintained according to standard cell culture procedures provided by the hybridoma sup-

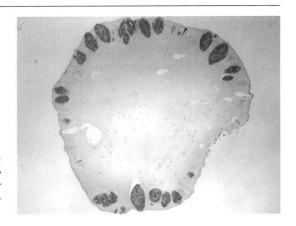

Fig. 69.6. Histological section showing alginate encapsulated hybridoma cells (GK1.5) after 25 days of intraperitoneal implantation in a Balb/c mouse; note that the colonies of proliferated hybridoma cells are preferentially found near the periphery of the capsules. Hematoxylin–Iosin (HE) stain; original magnification 40×.

pliers. One day before transplantation, cells are subcultured 1:5 to stimulate growth. On the day of transplantation, hybridoma cells are harvested and mixed into 1.5% sodium alginate (LVG, from Pronova) at a concentration of about 2×10^6 cells/ml. The cell-containing gel mixture is then extruded dropwise by syringe through a 16-gauge needle into a 1.5% calcium chloride solution. Those gel drops are allowed to solidify into gel spheres for 5 min. After washing twice with physiological saline followed by overnight culture, they are ready for transplantation. Our final cell-containing gel spheres have a mean diameter of about 4.5 mm. For each recipient mouse a total number of 2×10^6 cells are implanted IP either on the day of islet transplantation or up to 7 days before islet transplantation, depending upon the titers of monoclonal antibodies desired at the time of islet transplantation. Dickson [32] was able to demonstrate the survival of alginate-encapsulated hybridoma cells for over 25 days after implantation (Fig. 69.6), as well as high titers of circulating monoclonal antibodies in recipient mice. We believe that the high-quality alginate products and larger sphere diameters used in our laboratory, which differ from those described in earlier publications [31,54], provided better protection for encapsulated hybridoma cells and prolonged their function. In summary, this technology can be extremely valuable for preliminary screening of the effects of cell secretory products or monoclonal antibodies in "*in vivo*" models.

SPHERE DIAMETER

One of the unique features of this encapsulation system is the relatively large diameter of the capsules. Using a discordant porcine-to-rat islet xenograft model, Lanza *et al.* [6] compared the viability of islet grafts immobilized in uncoated alginate gel spheres varying from 800 μm to 3.7 mm in diameter. They found that graft survival correlated with capsule diameter. Grafts transplanted in capsules of less than 1.6 mm did not even survive 2 weeks based on functional and histological parameters. However, graft viability increased up to 85% for islets entrapped in 3.7-mm-diameter gel spheres. The authors of this study did not report whether they have ever tried using capsules with even larger diameters.

We likewise tested the effect of uncoated alginate gel sphere diameter on graft survival using our fish-to-Balb/c mouse islet xenotransplantation model. Balb/c mice that received tilapia islet grafts in 4.5-mm gel spheres (45 spheres per milliliter of gel) had a mean graft survival time (mGST) of $> 27 \pm 13$ days ($n = 10$), while in those that received 2-mm gel spheres (100 spheres/ml) or 6-mm gel spheres (15 spheres/gel), the mGSTs were reduced to 14 ± 7 and 13 ± 2 days, respectively.

By transplanting 4.5-mm gel spheres and 6-mm gel spheres into STZ-diabetic nude mice, we obtained results that are similar to that achieved in diabetic Balb/c mice, indicating that graft failure is due to non-immune-mediated mechanisms. Islet grafts embedded in 4.5-mm alginate gel spheres functioned uniformly over 50 days or over 100 days, while those embedded in 6-mm gel spheres had an mGST of 15 ± 6.2 days. Even though tilapia islets are markedly more resistant to hypoxia than mammalian islets [59], oxygen diffusion in 6-mm spheres was likely insufficient to maintain their viability. It seems likely that 3.7 and 4.5 mm represent the approximate upper limit of alginate gel sphere diameter for mammalian and fish islets, respectively.

STRAIN AND SPECIES DIFFERENCES

Several donor–recipient species combinations have been used to examine the immuno-protective effects of large-diameter uncoated alginate gel spheres. Lanza *et al.* found that uncoated alginate spheres containing either porcine or bovine islet xenografts consistently provided long-term normoglycemia in diabetic C57BL/6J mice. However, porcine islet xenograft survival was only 13 days in a STZ-diabetic Lewis rat recipient. With adjuvant CsA treatments, long-term survival of porcine and bovine islets in Lewis rats, as well as long-term survival of canine islets in spontaneously diabetic dogs, was achieved. On the other hand, uncoated alginate spheres did not maintain bovine islet graft survival in diabetic dogs even when they received triple immunosuppressive therapy [3]. In our laboratory, tilapia islet GSTs differed significantly between rat and mouse recipients and even between different strains of recipient mice. The mGST of alginate-encapsulated tilapia islet grafts was over 49 days in STZ-diabetic C57 BL/6J mouse recipients, with 5 of 6 mice remaining normoglycemic for greater than 50 days, while mGST in Balb/c mice was over 27 days, with only one mouse remaining normoglycemic for greater than 50 days ($t = 4.35$, $df = 11$, $p < 0.005$). Mean GST in Lewis rats was only 13 days. Clearly, the immunoprotective effects of large-diameter uncoated alginate gel spheres vary between recipient strains and species.

BIOCOMPATIBILITY

For any encapsulation material, biocompatibility is critical. Considerable evidence suggests that alginate gels produced from a variety of sources are nontoxic to encapsulated cells; encapsulated islets uniformly maintain normal morphology, insulin content, and glucose responsiveness during long-term *in vitro* culture [60–62]. However, after transplantation, the results achieved with alginate-encapsulated islets have been quite variable, primarily because of host "foreign-body" reaction (i.e., fibroplasia or fibrosis), which prevents diffusion of nutrients and oxygen. Numerous early studies using PLL-coated alginate microcapsules produced from poorly characterized alginates yielded contradictory results, primarily because of variable degrees of fibroplasia. As mentioned earlier, this sparked considerable interest in the relationship between alginate chemical structure and biocompatibility. This marked variability was eventually explained, at least in part, by Soon-Shiong *et al.* [63,64], who discovered that alginates with low M content evoked minimal cytokine responses and fibroplasia. In our laboratory, purified low-M sodium alginate (LVG, Pronova) has consistently provided excellent biocompatibility in both STZ-diabetic and spontaneously diabetic nonobese diabetic rodents. At 50–100 days after intraperitoneal transplantation, we consistently find over 80% of alginate gel spheres containing fish islet xenografts to be free floating within the abdominal cavity. Microscopic examination of these encapsulated islets consistently reveals little or no inflammation or fibroplasia (Fig. 69.4). In general, fewer than 20% of the capsules are adherent to connective tissue; even in these instances, the capsules have not precipitated fibroblastic overgrowth, but rather have been surrounded by adipose tissue [9]. Lanza *et al.* [6,7] have reported similar biocompatibility in work with Pronova LVG alginate. In contrast, when we have used different, poorly characterized alginates to produce uncoated alginate capsules, GST has been diminished by fibrotic overgrowth (Fig. 69.7).

We have also examined the biocompatibility of Pronova LVG alginate gel toward embedded islet grafts *in vivo*. Figure 69.8b–f shows an alginate encapsulated WF rat islet allograft removed from a diabetic Lewis rat recipient 50 days after intra-abdominal implantation. Serial sections stained by the immunoperoxidase method for insulin, glucagon, somatostatin, and pancreatic peptide show that encapsulated islets maintain a normal topographical distribution of all four endocrine cells types. However, compared with alginate-encapsulated WF rat islets examined before transplantation (Fig. 69.8a), the overall islet size appeared to decrease somewhat over time, probably secondary to β cell attrition due to central necrosis. One interesting finding was that macrophage-like cells with ballooned

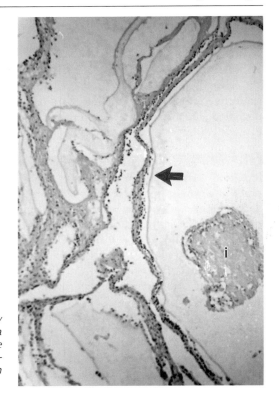

Fig. 69.7. Histological section showing fibrotic reaction and mild inflammatory response at 20 days post-transplantation to fish islet grafts encapsulated using a poorly characterized alginate. The arrow shows the edge of an alginate gel sphere containing a necrotic islet (i). Several other capsules (left) collapsed during histological processing HE stain; original magnification 40×. Reprinted with permission of Birkhäuser from Yang and Wright [44, p. 86].

granules migrated out of the islets and occupied the space between the islet surface and the alginate gel (Fig. 69.8, arrows); upon examination by electron microscopy, the ballooned granules in the cytoplasm appeared to contain phagocytized loosed alginate gel (Fig. 69.9). This finding correlates with the results of an earlier publication studying the ultrastructural appearance of alginate–PLL encapsulated islets maintained in tissue culture from 0 to 100 days [61]. Fraser *et al.* showed that phagocytic cells having macrophage-like ultrastructural features migrated from encapsulated islets and attempted to ingest the encapsulation materials from within; however, these cells did not break through the microcapsule. Whether these cells represent "passenger leukocytes" or contribute to "shed antigen" is still under investigation [61].

Although our results are consistent with the contention by others that alginate gel's *in vivo* biocompatibility depends upon its monomeric composition and sequential arrangements [4,5,11,13,63,64], several recent publications suggest that alginate purity is also an important factor. Removal of impurities, even from alginates high in mannuronic acid, can improve *in vivo* biocompatibility [65,66]. These studies suggest that high M content is not solely responsible for alginate-induced fibroplasia. In fact, at least 10–20 bioactive, immune stimulatory fractions from commercial alginates have been characterized by means of free-flow electrophoretic separation [67]. The number and distribution of those impurities varied with the source of the alginate and even within different batches of the same product. Nevertheless, a study by De Vos *et al.* [66] found that even after the alginate had been purified and the alginate–polylysine microcapsule's biocompatibility improved, graft survival was limited, probably because of hypoxia secondary to the lack of blood supply to encapsulated islets.

RETRIEVABILITY AND BIODEGRADATION OF ALGINATE GEL CAPSULES

Given attrition over time, periodic replacement of encapsulated cells appears to be inevitable, raising the issues of retrievability and biodegradation. In our rodent studies, alginate gel capsules were highly biocompatible and remained free floating within the peritoneal cavity over time. If this proves true in man, the majority of nonfunctioning capsules

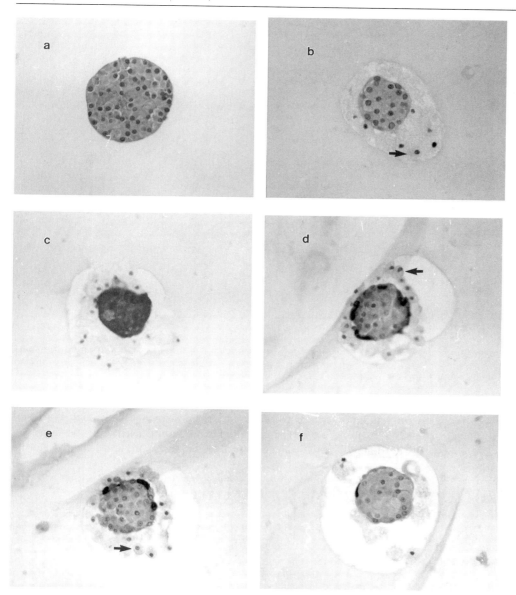

Fig. 69.8. Histological sections showing alginate encapsulated WF rat islets. (a) Alginate encap-sulated WF islet before transplantation (i.e., day 0) and stained with HE. (b)–(f) Serial histological sections of an alginate encapsulated WF rat islet graft removed from a diabetic Lewis rat recipient 50 days after intraperitoneal transplantation. HE stain (b) and immunoperoxidase staining for (c) insulin, (d) glucogon, (e) somatostatin, and (f) pancreatic peptide. Note also that macrophage-like cells with ballooned granules have migrated out of the islets and occupied the space between the islet surface and the alginate gel (arrows). Original magnification 200×.

could be laparoscopically removed by lavage prior to retransplantation. Although alginate gel spheres usually appeared to be intact after retrieval from the peritoneal cavities of small- and large-animal recipients [3,6,7,9], microscopic chipping and heterophagocytosis were observed [6]. Most evidence suggests that alginate is biodegradable. Al-Shamkhani and Duncan [68], who studied clearance of sodium alginate solution after intravenous adminis-tration in rats, found that the low molecular weight fraction (<48 Da) was excreted in the urine within 24 h, while the larger fraction remained in the circulation and did not read-ily accumulate in any of the tissues. These investigators also found that sodium alginate solution administered intraperitoneally was transferred to the blood within 24 h. Other studies suggest that alginate administered intraperitoneally or subcutaneously reaches the

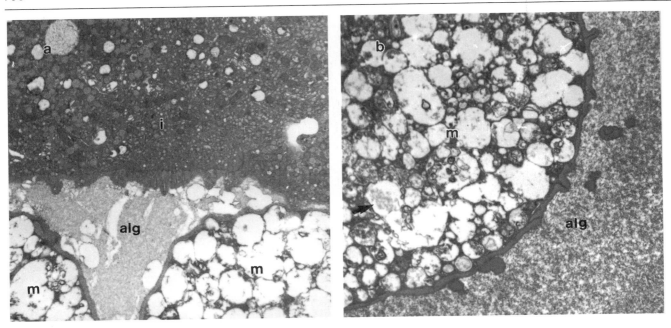

Fig. 69.9. (a) Transmission electron micrograph showing macrophage-like cells migrating out of an islet graft and filling the space between the islet surface and the alginate material. The ballooned cytoplasmic granules (b) are filled with loose alginate (arrow). Original magnification 9800×. Alg, alginate; i, islet cell; m, macrophage-like cells.

bloodstream via the lymphatic system [69–71]. Therefore, it seems likely that any residual alginate spheres remaining after laparoscopic lavage would eventually be degraded by chipping and heterophagocytosis, transferred to the bloodstream, and excreted via the urine.

PROTECTIVE MECHANISM(S) OF UNCOATED ALGINATE GEL

The mechanism(s) responsible for graft rejection after transplantation of unencapsulated islets between discordant species have not been fully elucidated, but most investigators working in this field attribute rejection to a combination of nonspecific and specific immune responses, including both cellular and humoral responses. Clearly, T cells play a critical role, since athymic nude mice do not reject islet xenografts when transplanted under the kidney capsule [39,45]. Various small molecular weight cytokines produced by immune effector cells are directly toxic to islet cells [72] and are presumed to play important roles. The relative strength of the humoral response and the role of complement activation in the rejection process are matters that remain to be clasified. Therefore, it is unwise to make dogmatic statements pertaining to the protective mechanism(s) afforded by alginate encapsulation except to say that encapsulation does prevent direct immune effector cell access to the graft.

The mechanism by which large-diameter uncoated alginate gel spheres prolong discordant islet xenograft survival is unclear, especially because alginate gel has high porosity. Calcium alginate gel is generally permeable to molecules with a molecular weight lower than 600 kDa [6,7]. Preformed "natural" antibodies, which are in large part responsible for hyperacute whole-organ xenograft rejection, are IgM class (M_w 900 kDa) and thus would be excluded; however, it is believed that this mechanism of rejection does not occur with nonvascularized cellular xenografts [73]. IgG ($M_w = 154$ kDa), components of the complement system ($M_w = 24$–570 kDa), and cytokines ($M_w = 10$–30 kDa), which are thought to play more important roles, would be expected to have free access to the encapsulated grafts. One hypothesis is that the negatively charged molecular network in the alginate gel prevents the entry of negatively charged immunoglobulins and complement components. At first glance, this seems like an unlikely explanation, since IgG class monoclonal antibodies readily pass through large-diameter uncoated alginate gel spheres when hybridoma cells are encapsulated [32]. However, Tanaka *et al.* [10] and Yuet *et al.* [74] found that although negatively charged large proteins diffused rapidly out of alginate gel spheres, diffusion of the same proteins into the gel spheres was inhibited. Conversely, Thu *et al.* [5] found that positively charged lysozyme molecules rapidly diffused into alginate gel spheres. Zekorn *et al.* [75] have suggested that nonspecific coating of the pores of encapsulation materials by negatively charged serum proteins may protect the encapsulated islets from damage by small molecular weight cytokines. Another hypothesis is that the negatively charged alginate gel alters the immunoglobulins and complement components via electrostatic interactions as they diffuse into and while they are within the capsule, thus disrupting their function. Ionic interactions between the gel matrix and proteins could also explain why larger gel spheres provide better protection than smaller gel spheres, since larger spheres would be expected to have a more negatively charged core.

SUMMARY

Microencapsulation via large diameter alginate gel spheres is probably the simplest immunoisolation system available. Although providing effective immunoisolation of highly discordant islets in rodent models, it remains to be determined whether this method will prove equally effective in large animals and, ultimately, in man. There are also still many unanswered questions related to the mechanism(s) responsible for the immunoprotective effect. However, future studies will undoubtedly further elucidate these mechanism(s) and will facilitate the development of better immunoisolation systems.

REFERENCES

1. McNeely, W. H., and Pettitt, D. J. (1973). Algin. *In* "Industrial Gums—Polysaccharides and their Derivatives" (R. L. Whistler, ed.), pp. 49–81. Academic Press, New York.
2. Lim, F., and Sun, A. M. (1980). Microencapsulated islets as bioartifical endocrine pancreas. *Science* 210, 908–910.
3. Lanza, R. P., and Chick, W. L. (1997). Transplantation of encapsulated cells and tissues. *Surgery* 121, 1–9.
4. Thu, B., Bruheim, P., Espevik, T., Smidsrod, O., Soon-Shiong, P., and Skjak-Braek, G. (1996). Alginate polycation microcapsules. 1. Interaction between alginate and polycation. *Biomaterials* 17, 1031–1040.
5. Thu, B., Bruheim, P., Espevik, T., Smidsrod, O., Soon-Shiong, P., and Skjak-Braek, G. (1996). Alginate polycation microcapsules. 2. Some functional properties. *Biomaterials* 17, 1069–1079.
6. Lanza, R. P., Kühtreiber, W. M., Ecker, D., Staruk, J. E., and Chick, W. L. (1995). Xenotransplantation of porcine and bovine islets without immunosuppression using uncoated alginate microspheres. *Transplantation* 59, 1377–1384.
7. Lanza, R. P., Kühtreiber, W. M., Ecker, D., Staruk, J. E., Marsh, J., and Chick, W. L. (1995). A simple method for transplanting discordant islets into rats using alginate gel spheres. *Transplantation* 59, 1486–1487.
8. Yang, H., O'Hali, W., Kearns, H., and Wright, J. R., Jr. (1996). Reversal of diabetes in rodents by encapsulated fish islets. *Abstr., No. 7.08, 3rd Int. Congr. Cell Transplant Soc.*, Miami, FL.
9. Yang, H., O'Hali, W., Kearns, H., and Wright, J. R., Jr. (1997). Long-term function of fish islet xenografts in mice by alginate encapsulation. *Transplantation* 64, 28–32.
10. Tanaka, H., Matsumura, M., and Veliky, I.A. (1984). Diffusion characteristics of substrates in Ca–alginate gel beads. *Biotechnol. Bioeng.* 26, 53–58.
11. Martinsen, A., Storro, I., and Skjak-Braek, G. (1992). Alginate as immobilization material. III. Diffusional properties. *Biotechnol. Bioeng.* 39, 186–194.
12. Morris, V. J. (1986). Gelation of polysaccharides. *In* "Functional Properties of Food Macromolecules" (J. R. Mitchell and D. A. Ledward, eds.), pp. 121–128. Elsevier, New York.
13. Martinsen, A., Skjak-Braek, G., and Smidsrod, O. (1989). Alginate as immobilization material. I. Correlation between chemical and physical properties of alginate gel beads. *Biotechnol. Bioeng.* 33, 79–89.
14. Sandford, P.A. (1992). High purity chitosan and alginate: Preparation, analysis and applications. *Front. Carbohydr. Res.* 2, 250–269.
15. Levin, D. M. (1985). Alginate powder hazards: Are you safe? *Dent. Assist.* 4, 30–31.
16. Matthew, I. R., Brownwe, R. M., Frame, J. W., and Millar, B. G. (1993). Tissue response to a haemostatic alginate wound dressing in tooth extraction pockets. *Br. J. Oral Maxillary Surg.* 31, 165–169.
17. Odell, E. W., Oades, P., and Lombardi, T. (1994). Symptomatic foreign body reaction to haemostatic alginate. *Br. J. Oral Maxillary Surg.* 32, 178–179.
18. Smidsrod, O., and Skjak-Braek, G. (1990). Alginate as immobilization matrix for cells. *Trends Biotechnol.* 8, 71–78.
19. Bodmeier, R., and Paertakul, O. (1989). Spherical agglomerates of water insoluble drugs. *J. Pharm. Sci.* 78, 964–967.
20. Downs, E. C., Robertson, N. E., Riss, I. L., and Plunkett, M. L. (1992). Calcium alginate beads as a slow-release system for delivering angiogenic molecules in vivo and in vitro. *J. Cell. Physiol.* 152, 422–429.
21. Prisant, L. M., Bottini, B., DiPiro, J. T., and Carr, A. A. (1992). Novel drug-delivery systems for hypertension. *Am. J. Med.* 93, 459–559.

22. Takka, S., and Acartuck, F. (1999). Calcium alginate microparticles for oral administration. I. Effect of sodium alginate type on drug release and drug entrapment efficiency. *J. Microencapsul.* **16**, 275–290.

23. Acartuck, F., and Takka, S. (1999). Calcium alginate microparticles for oral administration. II. Effect of formulation factors on drug release and drug entrapment efficiency. *J. Microencapsul.* **16**, 291–301.

24. Chang, P. L. (1997). Microcapsules as bio-organs for somatic gene therapy. *Ann. N.Y. Acad. Sci.* **831**, 461–473.

25. Hortelano, G., Al-Hendy, A., Ofosu, F. A., and Chang, P. L. (1996). Delivery of human factor IX in mice by encapsulated recombinant myoblasts: A novel approach towards allogeneic gene therapy of hemophilia B. *Blood* **87**, 5095–5103.

26. Al-Hendy, A., Hortelano, G., Tannenbaum, G. S., and Chang, P. L. (1995). Correction of the growth defect in dwarf mice with nonautologous microencapsulated myoblasts—an alternate approach to somatic gene therapy. *Hum. Gene Ther.* **6**, 165–175.

27. Tai, I. T., and Sun, A. M. (1993). Microencapsulation of recombinant cells: A new delivery system for gene therapy. *FASEB J.* **7**, 1061–1069.

28. Pier, G. B. (1991). Vaccine potential of *Pseudomonas aeruginosa* mucoid exopolysaccharide (alginate). *Antibiot. Chemother.* **44**, 134–142.

29. Bowersock, T. L., HogenEsch, H., Suckow, M., Porter, R. E., Jackson, R., Park, H., and Park, K. (1996). Oral vaccination with alginate microsphere systems. *J. Controlled Release* **39**, 209–220.

30. Bowersock, T. L., HogenEsch, H., Suckow, M., Guimond, P., Martin, S., Borie, D., Torregrosa, S., Park, H., and Park, K. (1999). Oral vaccination of animals with antigens encapsulated in alginate microspheres. *Vaccine* **17**, 1804–1811.

31. Savelkoul, H. F. J. (1994). Alginate encapsulation of cytokine gene-transfected cells. *In* "Immunology Methods Manual: The Comprehensive Sourcebook of Techniques" (I. Lefkovits, ed.), pp. 1130–1135. Academic Press, San Diego, CA.

32. Dickson, B. (1999). Characterization of host immune cells contributing to the rejection of tilapia-to-mouse islet xenografts. M.Sc. Thesis, Dalhousie University, Halifax, N.S.

33. Paige, K. T., Cima, L. G., Yaremchuk, M. J., Schloo, B. L., Vacanti, J. P., and Vacanti, C. A. (1996). De novo cartilage generation using calcium alginate–chondrocyte constructs. *Plast. Reconstr. Surg.* **97**, 168–178.

34. Piacquadio, D., and Nelson, D. B. (1992). Alginates. A "new" dressing alternative. *J. Dermatol. Surg. Oncol.* **18**, 992–995.

35. Wilson, P. R. (1996). Dressed to heal: New options for graft site dressing. *Australas. J. Dermatol.* **37**, 157–158.

36. Jankovsky, M., and Vasakova, L. (1996). Immobilization in alginate gels. *Vet. Med. (Prague)* **41**, 159–164.

37. Lin, S. S., Ueng, S. W., Lee, S. S., Chan, E. C., Chen, K. T., Yang, C. Y., Chen, C. Y., and Chan, Y. S. (1999). In vitro elution of antibiotic from antibiotic-impregnated biodegradable calcium alginate wound dressing. *J. Trauma* **47**, 136–141.

38. Adaniya, G. K., Roblero, L., Rawlins, R. G., Miller, I. F., Quigg, J. M., and Zaneveld, L. J. D. (1993). First pregnancies and live births from transfer of sodium alginate encapsulated embryos in a rodent model. *Fertil. Steril.* **59**, 652–655.

39. Wright, J. R., Jr., Polvi, S., and MacLean, H. (1992). Experimental transplantation using principal islets of teleost fish (Brockmann bodies): Long-term function of tilapia islet tissue in diabetic nude mice. *Diabetes* **41**, 1528–1532.

40. Wright, J. R., Jr. (1994). Procurement of fish islets (Brockmann bodies). *In* "Pancreatic Islet Transplantation Series" (R. P. Lanza and W. L. Chick, eds.), Vol. 1, pp. 123–133. R. G. Landes, Austin, TX.

41. Yang, H., and Wright, J. R., Jr. (1995). A method for mass harvesting islets (Brockmann bodies) from teleost fish. *Cell Transplant.* **4**, 621–628.

42. Wright, J. R., Jr., and Yang, H. (1998). Tilapia Brockmann bodies: An inexpensive, simple model for discordant islet xenotransplantation. *Ann. Transplant.* **2**, 72–76.

43. Wright, J. R., Jr., and Pohajdak, B. (2001). Cell therapy for diabetes using piscine islet tissue. *Cell Transplant.* **10**, 125–143.

44. Yang, H., and Wright, J. R., Jr. (1999). Calcium alginate encapsulation. *In* "Handbook of Encapsulated Cell Technology and Therapeutics" (W. M. Kühtreiber, R. P. Lanza, and W. L. Chick, eds.), pp. 79–89. Birkhäuser, Boston.

45. Yang, H., Dickson, B. C., O'Hali, W., Kearns, H., and Wright, J. R., Jr. (1997). Functional comparison of mouse, rat, and fish islet grafts transplanted into diabetic nude mice. *Gen. Comp. Endocrinol.* **106**, 384–388.

46. Yang, H., and Wright, J. R., Jr. (1999). Co-encapsulation of sertoli enriched testicular cell fractions further prolongs fish-to-mouse islet xenograft survival. *Transplantation* **67**, 815–820.

47. Selawry, H. P., and Whittington, K. (1984). Extended allograft survival of islets grafted into the intra-abdominally placed testis. *Diabetes* **33**, 405–406.

48. Bellgrau, D., and Selawry, H. (1990). Cyclosporine-induced tolerance to intratesticular islet xenografts. *Transplantation* **50**, 654–657.

49. Selawry, H., Kotb, M., Herrod, H., and Lu, Z. N. (1991). Production of a factor, or factors, suppressing IL-2 production and T cell proliferation by Sertoli cell-enriched preparations. *Transplantation* **52**, 846–850.

50. Selawry, H. P., and Cameron, D. F. (1993). Sertoli cell-enriched fractions in successful islet cell transplantation. *Cell Transplant.* **2**, 123–129.

51. Takeda, Y., Gotoh, M., Dono, K., Nishihara, M., Grochowiechi, T., Kimura, F., Yoshida T., Ohta, Y., Ota, H., Ohzato, H., Umeshita, K., Takeda, T., Matsuura, N., Sakon, M., Kayagaki, N., Yagita, H., Okumura, K., Miyasaka, M., and Monden, M. (1998). Protection of islet allografts transplanted together with Fas ligand expressing testicular allografts. *Diabetologia* **41**, 315–321.

52. Korbutt, S. G., Elliott, F. J., and Rajotte, R. V. (1997). Co-transplantation of allogeneic islets with allogeneic testicular cell aggregates allows long-term graft survival without systemic immunosuppression. *Diabetes* **46**, 317–322.

53. Korbutt, S. G., Ao, Z., Flashner, M., Elliot, F. J., and Rajotte, R. V. (1998). Co-encapsulation of allogenic islets with allogenic Sertoli cells prolongs graft survival without systemic immunosuppression. *Transplant. Proc.* **30**, 419.

54. Savelkoul, H. F. J., van Ommen, R., Vossen, A. C. T. M., Breedland, E. G., Coffman, R. L., and van Oudenaren, A. (1994). Modulation of systemic cytokine levels by implantation of alginate encapsulated cells. *J. Immunol. Methods* **170**, 185–196.

55. Bianchi, A. T., Moonen-Lausen, H. W., van Milligen, F. J., Savelkoul, H. F., Zwart, R. J., and Kimman, T. G. (1998). A mouse model to study immunity against pseudorabies virus infection: significant of CD4$^+$ and CD8$^+$ cells in protective immunity. *Vaccine* **16**, 1550–1558.

56. Reddy, L. V., Ring, C., Kumar, A., and Kurup, V. P. (1993). Depletion of CD4 cells in mice with intraperitoneal injection of alginate-encapsulated GK1.5 hybridoma cells: A potential use in development of animal models of infections disease. *J. Infect. Dis.* **168**, 1082–1083.

57. VanOosterhout, A. J., VanArk, I., Hofman, G., Savelkoul, H. F., and Nijkamp, F. P. (1993). Recombinant interleukin-5 induces in vivo airway hyperresponsiveness to histamine in guinea pigs. *Eur. J. Pharmacol.* **239**, 277.

58. Lang, M. S., Hovenkamp, E., Savelkoul, H. F., Knegt, P., and van Ewijk, W. (1995). Immunotherapy with monoclonal antibodies directed against the immunosuppressive domain of p15E inhibits tumour growth. *Clin. Exp. Immunol.* **102**, 468–475.

59. Wright, J. R., Jr., Yang, H., and Dooley, K. C. (1997). Tilapia—A source of hypoxia-resistant islets for encapsulation. *Cell Transplant.* **9**, 299–307.

60. Lanza, R. P., Butler, D. H., Borland, K. M., Staruk, J. E., Faustman, D. L., Solomon, B. A., Muller, T. E., Rupp, R. G., Maki, T., Monaco, A. P., and Chick, W. L. (1991). Xenotransplantation of canine, bovine, and porcine islets in diabetic rats without immunosuppression. *Proc. Natl. Acad. Sci. U.S.A.* **88**, 11100–11104.

61. Fraser, R. B., MacAulay, M. A., Wright, J. R., Jr., Sun, A. M., and Rowden, G. (1995). Migration of macrophage-like cells within encapsulated islets of Langerhans maintained in tissue culture. *Cell Transplant.* **4**, 529–534.

62. Falorni, A., Basta, G., Santeusanio, F., Brunetti, P., and Calafiore, R. (1996). Culture maintenance of isolated adult porcine pancreatic islets in three-dimensional gel matrices: Morphologic and functional results. *Pancreas* **12**, 221–229.

63. Soon-Shiong, P., Otterlie, M., Skjak-Braek, G., Smidsrod, O., Heintz, R., Lanza, R. P., and Espevik, T. (1991). An immunological basis for the fibrotic reaction to implanted microcapsules. *Transplant. Proc.* **23**, 758–759.

64. Soon-Shiong, P., Feldman, E., Nelson, R., Komtebedde, J., Smidsrod, O., Skjak-Braek, G., Espevik, T., Heintz, R., and Lee, M. (1992). Successful reversal of spontaneous diabetes in dogs by intraperitoneal microencapsulated islets. *Transplantation* **5**, 769–774.

65. Klöck, G., Pfeffermann, A., Ryser, C., Gröhn, P., Kuttler, P., Hanh, H. J., and Zimmermann, U. (1997). Biocompatibility of mannuronic-acid rich alginate. *Biomaterials* **18**, 707–713.

66. De Vos, P., De Haan, B. J., Wolters, G. H. J., Strubble, J. H., and van Schilfgaarde, R. (1997). Improved biocompatibility but limited graft survival after purification of alginate for microencapsulation of pancreatic islets. *Diabetologia* **40**, 262–270.

67. Zimmermann, U., Klöck, G., Federlin, K., Hannig, K., Kowalski, M., Brezel, R. G., Horcher, A., Entenmann, H., Siebers, U., and Zekorn, T. (1992). Production of mitogen contamination free alginates with variable ratios of mannuronic to guluronic acid by free flow electrophoresis. *Electrophoresis* **13**, 269–274.

68. Al-Shamkhani, A., and Duncan, R. (1995). Radioiodination of alginate via covalently bound tyrosinamide allows monitoring of its fate in vivo. *J. Bioact. Compat. Polym.* **10**, 4–13.

69. Flessner, M. F., Dedrick, R. L., and Schultz, J. S. (1985). Exchange of macromolecules between peritoneal cavity and plasma. *Am. J. Physiol.* **248**, H15–H25.

70. Seymour, L. W., Duncan, R., Strohalm, J., and Kopecek, J. (1987). Effect of molecular weight (M_w) of N-(2-hydroxypropyl) methacrylamide copolymers on body distribution and rate of excretion after subcutaneous, intraperitoneal, and intravenous administration to rats. *J. Biomed. Mater. Res.* **21**, 1341–1358.

71. Tomlinson, E. (1986). Site-specific drug carriers. *Eng. Med. (Berlin)* **15**, 197–202.

72. Rabinovitch, A. (1998). An update on cytokines in the pathogenesis of insulin-dependent diabetes mellitus. *Diabetes Metab. Rev.* **14**, 129–151.

73. Yu, W., and Wright, J. R., Jr. (1999). Heterotopic cardiac xenotransplantation: Fish-to-rat. *Xenotransplantation* **6**, 213–219.

74. Yuet, P. K., Kwok, W. Y., Harries, T. J., and Goosen, M. F. A. (1993). Mathematical modelling of protein diffusion and cell growth in microcapsules. In "Fundamentals in Animal Cell Encapsulation and Immobilization" (M. F. A. Goosen, ed.), pp. 79–111. CRC Press, Boca Raton, FL.

75. Zekorn, T., Siebers, U., Bretzel, R. G., Renardy, M., Planck, H., Zschocke, P., and Federlin, K. (1990). Protection of islets of Langerhans from interleukin-1 toxicity by artificial membranes. *Transplantation* **50**, 391–394.

CHAPTER 70

MICROENCAPSULATION METHODS: ALGINATE–POLY(L-LYSINE)

Elliot L. Chaikof, Susan Safley, and Colin J. Weber

O ne of the most promising cell encapsulation methods has been based upon the the use of the anionic acidic polysaccharide alginic acid, which gels on contact with a solution of calcium chloride, initially reported by Lim and Sun [1]. In fact, Hunkeler [2] has estimated that over 85% of all articles dealing with cell encapsulation and published since Sun's paper have involved modifications of the alginate–calcium chloride–poly(L-lysine) system. The use of the alginate–poly(L-lysine) encapsulation system has been most commonly applied to pancreatic islet transplantation for the treatment of diabetes. Specifically, a large number of studies have shown that intraperitoneal xenografts of alginate–poly(L-lysine)-encapsulated rat, dog, pig, or human islets into streptozotocin-diabetic mice or rats can normalize blood glucose for over 100 days [3–5]. Moreover, long-term normalization of hyperglycemia by microencapsulated canine islet allografts and pig islet xenografts, and early function of one human islet allograft, have been reported [6–8].

It has been postulated by several investigators that alginate–poly(L-lysine) microcapsules promote survival of xenogeneic and allogeneic cell transplants by either preventing or minimizing release of donor antigen(s), thereby reducing host sensitization, and/or by preventing or reducing exposure of the donor cells to host immunological effector mechanisms [9–12]. Recently, several new approaches to modify xenogeneic donors have been reported, including the generation of cells that do not express major histocompatibility antigen. Nonetheless, it is unlikely that all immunogenic proteins or carbohydrates can be eliminated through genetic strategies. Thus, even with the advent of donor cells from porcine or other sources that have been genetically engineered to reduce immunogenicity, mechanical devices such as microcapsules will remain an important tool for protecting xenogeneic cells and providing delivery of these cells in an optimal, physiological manner.

PRINCIPLE OF DIVALENT CATION MEDIATED ALGINATE CROSS-LINKING

Alginate is a family of non-branched binary copolymers of 1-4-glycosidically linked β-D-mannuronic acid (M) and α-L-guluronic acid (G) residues extracted from seaweed. The relative amounts of the two uronic acid monomers and their sequential arrangement along the polymer chain vary widely, depending on the origin of the alginate. The uronic acid residues are distributed along the polymer chain in a pattern of blocks, where homopolymeric blocks of G residues (G blocks), homopolymeric blocks of M residues (M blocks), and blocks with alternating sequence of M and G units (MG blocks) coexist [13–15]. Thus, the alginate molecule cannot be described by the monomer composition alone. Rather, NMR characterization of the sequence of M and G residues in the alginate chain is needed to calculate M/G ratio, G content, consecutive number of G monomers (i.e., G > 1), and average length of blocks of consecutive G monomers. Characteristically, commercial alginates have a polydispersity index in the range of 1.5–3.0.

Current models of alginate cross-linking suggest that junctions are created by the side-by-side alignment of oligo G blocks that surround an appropriate counterion, such as calcium. These junctions, in turn, link the elastic segments of the alginate chains, composed largely of alternating G and M residues. It is also recognized that G-block length and number within a chain are critical to the gelling process [16]. While this provides a general framework for understanding gel formation, the location, length, and composition of the homo- and copolymeric blocks are variable, depending on the seaweed and kelp species from which the alginate is extracted. Fundamentally, it remains unclear on a molecular level how these features influence the formation of junction zones and, in turn, how cross-link density in association with block structure dictates membrane mechanical and transport properties.

THE PRINCIPLES OF ALGINATE–POLY(L-LYSINE) MICROCAPSULE FORMATION

In the alginate system, the process of microcapsule formation involves the gelation of an alginate–cell suspension in a calcium chloride bath, coating of the outer membrane surface with poly(L-lysine) (PLL), and the subsequent liquefaction of the droplet with sodium citrate. Permeability of the membrane is controlled, in part, by PLL molecular weight. Specifically, a decrease in the molecular weight of PLL reduces permeability, presumably because the PLL chains penetrate the alginate membrane, yielding a more compact membrane owing to enhanced intermolecular interactions [17]. Membrane strength may be increased by reducing polylysine molecular weight, by altering alginate concentration and composition [18,19], or by repeating the membrane-producing steps to create double-walled or multilayered capsules [3]. However, these maneuvers often have a secondary effect on diffusive transport.

LIMITATIONS OF THE ALGINATE–POLY(L-LYSINE) MICROCAPSULE STRATEGY

PURITY AND BIOCOMPATIBILITY

As a natural product, alginate is readily available and relatively inexpensive. However, its use is associated with a number of limitations common to other naturally occurring polysaccharides that have been used in other cell encapsulation systems, including chitosan and κ-carrageenan. For example, extraneous substances and materials may be present in the alginate. There impurities may be processing aids related to the manufacture of alginate such as sodium carbonate, heavy metals, formaldehyde, which is used by some manufacturers to conserve harvested seaweed, and polyphenolic compounds that may originate from the algae itself. In addition, bacteria, yeast, and mold are also impurities, that can arise in an alginate sample. Significantly, the presence of bacteria may contribute to the presence of endotoxins. Elimination of contaminating pyrogens or mitogens and the observed lot-to-lot variability for all these polysaccharides is related, in part, to their structural diversity and to the problems inherent in the purification process. Descriptions for removing these contaminants have been detailed elsewhere [20].

NATURAL PRODUCT CHARACTERIZATION AND EVOLVING KNOWLEDGE OF STRUCTURE–FUNCTION RELATIONSHIPS

It is anticipated that our ability to optimize the selection process for naturally occurring alginates will continue to improve in parallel with the refinement of analytical tools for the structural characterization of complex carbohydrates, such as matrix-assisted laser desorption mass spectroscopy and two-dimensional NMR, particularly when combined with computational molecular models [21,22]. However, our understanding of alginate structure–function relationships remains limited. For example, it is known that alginate gels with high G content alginate are generally more rigid and brittle, whereas those from high M content alginate are more elastic [23]. However, while some investigators claim that mechanical strength is improved upon use of alginate with an L-guluronic acid content exceeding 70% [24], others cite as critical the alginate purity and the M/G ratio [8]. Moreover, the role of counterion type may further complicate this analysis, since membrane elasticity, as measured by Young's modulus, apparently increases with the affinity of the cation for the alginate ($Cd^{2+} > Ba^{2+} > Ca^{2+}$) [2].

CONTROL OF MEMBRANE PERMEABILITY

Membrane permeability is a function of both transport and thermodynamic properties, which are dependent upon the molecular characteristics of both the membrane and solute population [25]. With the exception of perfusion-based devices for cell transplantation, in which convective flux may play a significant role in transmembrane molecular transport, transport and thermodynamic properties are defined experimentally by diffusion and partition coefficients, respectively. Although both parameters are strongly dependent upon the molecular size of a given marker, even solutes of identical molecular weight will vary in their shape or conformation, as well as in the nature and surface distribution of charged, hydrophobic, and polar sites. All these properties will usually influence interactions relevant to both parameters and limit ones's ability to extrapolate experimental findings from one solute to another.

All this suggests that the description of the permeability of alginate–poly(L-lysine) membranes in terms of a molecular weight cutoff (MWCO) is limited. Use of this term assumes that molecular partitioning and diffusion are predominantly a size exclusion effect and independent of electrostatic, hydrophobic, biospecific, or conformational features of the solute. However, even if one excludes these effects and applies ideal size exclusion theories, Gehrke *et al.* [25–27] have demonstrated that sharp molecular weight cutoffs for either partition or diffusion coefficients are not predicted in swollen gels. Thus, while the experimentally determined rejection coefficient of a membrane may be 1.0 for a given solute, indicating that it does not pass convectively through the membrane, diffusive parameters are often nonzero. This is the case even for larger molecules, such as immunoglobulins, in which slow diffusive transport through the membrane remains in effect, albeit on a time scale that may not be clinically relevant [28]. Recent efforts at defining the membrane permeability of biologically relevant proteins, rather than the use arbitrary markers of varying molecular size, will likely have greater a predictive capacity with respect to *in vivo* performance [29]. In the interim, the permeability of the alginate–poly(L-lysine) system remain incompletely defined.

INTERDEPENDENCE OF PERMEABILITY AND MECHANICAL PROPERTIES

An additional limitation of the alginate–poly(L-lysine) system is the inability to achieve independent control of permeability and mechanical strength. As noted earlier, membrane strength may be increased by reducing polylysine molecular weight, by altering alginate concentration and composition [18,19], or by repeating the membrane-producing steps to create double-walled or multilayered capsules [3]. However, these maneuvers often have a secondary effect on diffusive transport. One solution has been the addition of one or more postcoating steps with oppositely charged polymers, to further modify biocompatibility, reduce membrane permeability, or improve mechanical properties. For example, to reduce protein adsorption and the associated cellular response to a polymeric membrane, several reports have documented the utility of coating membrane surfaces with poly(ethylene oxide) (PEO). For example, Sawhney and Hubbell synthesized a graft copolymer having PLL as the backbone and monomethoxy poly(ethylene glycol) as pendent chains [30]. This polycationic copolymer was used to form with sodium alginate microcapsules that demonstrated reduced protein adsorption, complement binding, and cell adhesion. In an extension of this effort, a cross-linked film of PEO was formed on the surface of alginate–PLL microcapsules by interfacial photopolymerization of acrylate-functionalized PEO [31].

IMPROVEMENTS IN ALGINATE–POLY(L-LYSINE) MICROCAPSULE DESIGN

Our group has characterized an improved formulation of poly(L-lysine)–alginate microencapsulation that involves forming a "double-wall" membrane and increasing alginate concentration. The "double-walled" microcapsule is more durable than conventional microcapsules, in part, because the likelihood of islets protruding from the capsule surface is reduced. Furthermore, we have found that increasing alginate from 1.85% to 2.0% reduces membrane permeability to approximately 100,000 kDa with apparent exclusion of IgG [3]. To date, this improved formulation yields prolonged and occasionally indefinite survival of rat islet xenografts in nondoese diabetic (NOD) mice, with no additional immunosuppression required.

The technique of microencapsulation employed in our laboratory utilizes alginate (low viscosity, bacterial count <25/g, from ISP Alginates, San Diego, CA) in aqueous phase, under conditions that are physiological with regard to pH and temperature [3,4]. Briefly, isolated islets are suspended 1:10 (v/v) in 2.0% sodium alginate in 0.9% saline. Droplets containing islets in alginate are produced by extrusion (1.7 ml/min) through a 22-gauge airjet needle (airflow 5 liters/min). Droplets fall 2 cm in to a 20-ml beaker containing 10 ml of 1.1% $CaCl_2$ in 0.9% saline, 25 mM HEPES-buffered, pH 7.1. Gelled droplets are decanted, transferred to a 50-ml centrifuge tube, and filled completely with 1.1% $CaCl_2$ for each 2–4 ml of microcapsules; the tube is then rotated gently, end over end, one revolution each 10 s, for 10 min. Microcapsules are allowed to settle; supernatant is aspirated; and then microcapsules are washed in 0.55% $CaCl_2$, followed by 0.28% $CaCl_2$ (in 0.9% saline, 25 mM HEPES, pH 7.1). After a final wash in 0.9% saline, poly(L-lysine) (MW 18,000) 0.5 mg/ml in saline, is added to completely fill the tube; and the tube is rotated for 6 min. Microcapsules are allowed to settle and then are washed in 0.1% CHES buffer (cyclohexylaminaethanesulfonic acid) in saline, pH 8.2, followed by another wash in 0.9% saline. Subsequently, 0.2% sodium alginate is added and the microcapsules are rotated again for 4 min. Thereafter, alginate is aspirated and discarded and microcapsules washed in 0.9% saline.

A final wash in 55 mM sodium citrate is used to solubilize any alginate not reacted with poly(L-lysine). At this stage, capsules may be implanted or maintained in petri dish culture. This technique produces "single-walled" microcapsules, which are similar to those reported from several other laboratories. As noted, we have found that "single-walled" capsules are fragile and, as a consequence, have added several steps to the our protocol prior to final the sodium citrate step. In essence, "double-walled" capsules are produced by reincubation of capsules with both poly(L-lysine) and dilute alginate. Specifically, poly(L-lysine) (0.5 mg/ml in saline) is added and the microcapsule suspension rotated for an additional 6 min. The microcapsules are then washed in 0.1% CHES (0.9% saline, pH 8.2) and saline, followed by reincubation in dilute 0.2% sodium alginate for 4 min. Microcapsules are then washed in 0.9% saline, followed by incubation in 55 mM sodium citrate (pH 7.4) for an additional 10 min. The capsules are washed three times in 0.9% saline, and then transferred to conventional tissue culture medium.

"Single-walled" microcapsules are translucent, measuring approximately 700 μm, while "double-walled" microcapsules are slightly smaller (~500 μm) and are somewhat opaque. When implanted in NOD mice, neither "single-wall" nor a "double-wall" elicits a cellular reaction in the absence of donor cells, if otherwise intact. After 6 months *in vivo*, "double-walled" microcapsules demonstrate better capsule integrity and greater capsule wall thickness than has been the case with "single-walled" microcapsules.

With these refinements in microcapsule preparation, we have observed an associated increase in encapsulated islet xenograft survival. For exmple, rat-to-NOD islet survival increased from 19 ± 13 days for "double-walled" capsules composed of 1.85% alginate to 86 ± 24 days for similar capsules composed of 2.0% alginate ($p < 0.001$). The responsible mechanisms for these observed changes in graft survival remain under active investigation.

ALGINATE–POLY(L-LYSINE)
MICROCAPSULE BIOCOMPATIBILITY

Empty microcapsules that have been formulated as described earlier have been recovered intact, free of any host reaction, from the peritoneal cavity of dozens of NOD mice and Lewis rats after more than 180 days in multiple experiments done over 10 years [3,4,32]. Similar observations have been made by other investigators [8,33–37]. Of note, we have observed a limited cellular reaction to fractured microcapsules and capsule fragments.

Reports by other groups [38,39] of fibrotic reactions surrounding empty microcapsules may be related to impurities in reagents, as described earlier. From a practical point of view, we have found it quite acceptable to select low viscosity, low bacterial count alginate (ISP Alginate, San Diego, CA), with screening of empty capsules in NOD recipients for 2–4 months prior to the use of encapsulated islets.

REFERENCES

1. Lim, F., and Sun, A. M. (1980). Microencapsulated islets as a bioartificial endocrine pancreas. *Science* 210, 908.
2. Hunkeler, D. (1997). Polymers for bioartificial organs. *Trends Pharmacol.* 5, 286–293.
3. Weber, C. J., and Reemstma, K. (1994). Microencapsulation in small animals—Xenografts. *In* "Pancreatic Islet Transplantation" (R. P. Lanza and W. L. Chick, eds.), pp. 59–79. R.G. Landes, Austin, TX.
4. Weber, C., Zabinski, S., Kosschitzky, T., *et al.* (1990). The role of CD4+ helper T cells in destruction of microencapsulated islet xenografts in NOD mice. *Transplantation* 49, 396–404.
5. Siebers, U., Horcher, A., Brandhorst, H., Brandhorst, D., Federlin, K., Bretzel, R. G., *et al.* (1998). Time course of the cellular reaction toward microencapsulated xenogeneic islets in the rat. *Transplant. Proc.* 30(2), 494–495.
6. Soon-Shiong, P., Feldman, E., Nelson, R., Heintz, R., Yao, Q., Yao, Z., *et al.* (1993). Long-term reversal of diabetes by the injection of immunoprotected islets. *Proc. Natl. Acad. Sci. U.S.A.* 90(12), 5843–5847.
7. Soon-Shiong, P., Heintz, R. E., Merideth, N., Yao, Q. X., Yao, Z., Zheng, T., *et al.* (1994). Insulin independence in a type 1 diabetic patient after encapsulated islet transplantation. *Lancet* 343, 950–951.
8. Sun, Y., Ma, X., Zhou, D., Vacek, I., and Sun, A. M. (1996). Normalization of diabetes in spontaneously diabetic cynomolgus monkeys by xenografts of microencapsulated porcine islets without immunosuppression. *J. Clin. Invest.* 98(6), 1417–1422.
9. Colton, C. K., and Avgoustiniatos, E. S. (1991). Bioengineering in development of the hybrid artificial pancreas. *J. Biomech. Eng.* 113(2), 152–170.
10. Halle, J., Bourassa, S., Leblond, F., *et al.* (1993). Protection of islets of Langerhans from antibodies by microencapsulation with alginate–poly(l-lysine) membranes. *Transplantation* 44, 350–354.
11. Iwata, H., Morikawa, N., Fujii, T., Takagi, T., Samejima, T., and Ikada, Y. (1995). Does immunoisolation need to prevent the passage of antibodies and complements? *Transplant. Proc.* 27(6), 3224–3226.
12. Zekorn, T., Endl, U., Horcher, A., Siebers, U., Bretzel, R. G., and Federlin, K. (1995). Mixed lymphocyte islet culture for assessment of immunoprotection by islet microencapsulation. *Transplant. Proc.* 27(6), 3362–3363.
13. Skjak-Braek, G. (1992). Alginates: Biosyntheses and some structure–function relationships relevant to biomedical and biotechnological applications. *Biochem. Soc. Trans.* 20(1), 27–33.
14. Draget, K. I., Skjak-Braek, G., and Smidsrod, O. (1997). Alginate based new materials. *Int. J. Biol. Macromolecules* 21(1–2), 47–55.
15. Gacesa, P. (1998). Bacterial alginate biosynthesis—Recent progress and future prospects. *Microbiology* 144(pt. 5), 1133–1143.
16. Stokke, B. T., Smidsrod, O., Zanetti, F., Strand, W., and Skjakbraek, G. (1993). Distribution of uronate residues in alginate chains in relation to alginate gelling properties 2. Enrichment of beta-d-mannuronic acid and depletion of alpha-l-guluronic acid in sol fraction. *Carbohydr. Polym.* 21, 39–46.
17. King, G. A., Daugulis, A. J., Faulkner, P., and Goosen, M. F. A. (1987). Alginate–polylysine microcapsules of controlled membrane molecular weight cutoff for mammalian cell culture engineering. *Biotechnol. Prog.* 3, 231–240.
18. Thu, B., Bruheim, P., Espevik, T., Smidsrod, O., Soon-Shiong, P., and Skjak-Braek, G. (1996). Alginate polycation microcapsules. II. Some functional properties. *Biomaterials* 17(11), 1069–1079.
19. Thu, B., Bruheim, P., Espevik, T., Smidsrod, O., Soon-Shiong, P., and Skjak-Braek, G. (1996). Alginate polycation microcapsules. I. Interaction between alginate and polycation. *Biomaterials* 17(10), 1031–1040.
20. Prokop, A., and Wang, T. G. (1997). Purification of polymers used for fabrication of an immunoisolation barrier. *Ann. N.Y. Acad. Sci.* 831, 223–231.
21. Mulloy, B. (1996). High-field NMR as a technique for the determination of polysaccharide structures. *Mol. Biotechnol.* 6(3), 241–265.
22. Peters, T., and Pinto, B. M. (1996). Structure and dynamics of oligosaccharides: NMR and modeling studies. *Curr. Opin. Struct. Biol.* 6(5), 710–720.
23. Gombotz, W. R., and Wee, S. F. (1998). Protein release from alginate matrices. *Adv. Drug Delivery Rev.* 31, 267–285.
24. Martinsen, A., Skjak-Braek, G., and Smidsrod, O. (1989). Alginate as immobilization material. 1. Correlation between chemical and physical properties of alginate gel beads. *Biotechnol. Bioeng.* 33, 79–89.
25. Gehrke, S. H., Fisher, J. P., Palasis, M., and Lund, M. E. (1997). Factors determining hydrogel permeability. *Ann. N.Y. Acad. Sci.* 831, 179–207.

26. Gehrke, S. H. (1993). Synthesis, equilibrium swelling, kinetics, permeability and applications of environmentally responsive gels. *Adv. Polym. Sci.* **110**, 81–144.

27. Palasis, M., and Gehrke, S. H. (1992). Permeability of responsive poly(N-isopropylacrylamide) gel to solutes. *J. Controlled Release* **18**, 1–11.

28. Dionne, K. E., Cain, B. M., Li, R. H., Bell, W. J., Doherty, E. J., Rein, D. H., *et al.* (1996). Transport characterization of membranes for immunoisolation. *Biomaterials* **17**(3), 257–266.

29. Brissova, M., Lacik, I., Powers, A. C., Anilkumar, A. V., and Wang, T. (1998). Control and measurement of permeability for design of microcapsule cell delivery system. *J. Biomed. Mater. Res.* **39**(1), 61–70.

30. Sawhney, A. S., and Hubbell, J. A. (1992). Poly(ethylene oxide)-*graft*-poly(L-lysine) copolymers to enhance the biocompatibility of poly(L-lysine)–alginate microcapsule membranes. *Biomaterials* **13**, 863–870.

31. Sawhney, A. S., Pathak, C. P., and Hubbell, J. A. (1993). Interfacial photopolymerization of poly(ethylene glycol)-based hydrogels upon alginate–poly(L-lysine) microcapsules for enhanced biocompatibility. *Biomaterials* **14**, 1008–1016.

32. Weber, C., Ayres-Price, J., Costanzo, M., and Stall, A. (1994). NOD mouse peritoneal cellular response to poly(L-lysine)–alginate microencapsulated rat islets. *Transplant. Proc.* **26**, 1116–1119.

33. Soon-Shiong, P., Feldman, E., Nelson, R., Komtebedde, J., Smidsrod, O., Skjak-Braek, G., *et al.* (1992). Successful reversal of spontaneous diabetes in dogs by intraperitoneal microencapsulated islets. *Transplantation* **54**(5), 769–774.

34. Wang, T., Lacik, I., Brissova, M., Anilkumar, A. V., Prokop, A., Hunkeler, D., *et al.* (1997). An encapsulation system for the immunoisolation of pancreatic islets. *Nat. Biotechnol.* **15**(4), 358–362.

35. Lanza, R. P., Kuhtreiber, W. M., Ecker, D., Staruk, J. E., and Chick, W. L. (1995). Xenotransplantation of porcine and bovine islets without immunosuppression using uncoated alginate microspheres. *Transplantation* **59**(10), 1377–1384.

36. Goosen, M. F. A., O'Shea, G. M., Gharapetian, H., Chou, S., and Sun, A. M. (1985). Optimization of microencapsulation parameters: Semipermeable microcapsules as a bioartificial pancreas. *Biotechnol. Bioeng.* **27**, 146–150.

37. Horcher, A., Siebers, U., Bretzel, R. G., Federlin, K., and Zekorn, T. (1995). Transplantation of microencapsulated islets in rats: Influence of low temperature culture before or after the encapsulation procedure on the graft function. *Transplant. Proc.* **27**(6), 3232–3233.

38. De Vos, P., De Haan, B., Wolters, G. H., and Van Schilfgaarde, R. (1996). Factors influencing the adequacy of microencapsulation of rat pancreatic islets. *Transplantation* **62**(7), 888–893.

39. Wijsman, J., Atkison, P., Mazaheri, R., Garcia, B., Paul, T., Vose, J., *et al.* (1992). Histological and immunopathological analysis of recovered encapsulated allogeneic islets from transplanted diabetic BB/W rats. *Transplantation* **54**(4), 588–592.

CHAPTER 71

MICROENCAPSULATION METHODS: ALGINATE–POLY(LYSINE)–POLY(ETHYLENEIMINE)–PROTAMINE SULFATE–HEPARIN

Tadeusz Orlowski

In the past, several compounds were used for the immunoisolation of islets. Now capsules consisting of alginate–poly(L-lysine)–alginate (APA) membranes, which are less resistant but more compatible than alginate—(poly-L-lysine)–poly(ethyleneimine) (APPi) are commonly used.

To improve the biocompatibility of APPi capsules, Tatarkiewicz covered them with an additional layer consisting of heparin–protamine complex (APiPrH). Apropriate studies show that in long-term experiments, immunoisolation with APiPrH microcapsules is less effective than that with APA beads. Nevertheless, in situations in which the immunoisolated material is exposed to intensive mechanical and/or chemical stress APiPrH microcapsules are preferable.

INTRODUCTION

There are many applications of semipermeable membranes in medicine [1,2]. The most promising is immunoisolation of transplanted islets of Langerhans in treating diabetes [3–9]. Diabetes mellitus is a chronic disease, leading potentially to severe, often fatal, complications, such as micro- and macroangiopathy, end-stage renal failure, neuropathy, and blindness. Adequate control of glycemia can prevent the appearance, of these complications, but even frequently applied exogenous insulin fails to provide perfect glucose homeostasis. Only transplantation of whole pancreas or separated insulin-secreting islets of Langerhans can assure permanent normalization of glucose metabolism and prevent development of late diabetic complications [10–13]. Since, however, allografts and xenografts are highly antigenic, post-transplant graft immunoprotection is mandatory [14–24].

To eliminate the need for potentially dangerous immunosuppressive therapy, immunoisolation techniques were developed in which the graft is encapsulated within a membrane that exhibits an appropriate degree of selective permeability. The membrane used for microencapsulation must be biocompatibile, nondegradable, resistant to mechanical stress and as indicated, selectively permeable. It must allow the diffusion into the capsule of oxygen and low molecular weight substances, such as glucose, nutrients, and insulin, and the expulsion of waste products. At the same time it should be impermeable to the cellular and higher molecular weight components of the immune system [20–26].

Table 71.1. The Tatarkiewicz method for the Encapsulation of Islets of Langerhans

1. Prepare a suspension of islets (1000–2000/ml) in 1.2% w/v sodium alginate (Sigma) in PBS at pH 7.4
2. Use an airjet–syring pump generator to extrude the suspension into a solution of 1.5% $CaCl_2$
3. Decant the $CaCl_2$ and wash the microcapsules ($\sim$ 0.6 μm in diameter) successively:
   a. In 0.2% poly(ethyleneimine) in PBS for 4 min
   b. In PBS several times
   c. In 0.1% protamine sulfate in PBS for 4 min
   d. In heparin (25 U/ml) in PBS to neutralize protamine for 2 min
4. Afterward allow the capsules to react for 6–10 min with 0.1% sodium citrate in PBS to liquiefy the alginate.
5. Wash the beads several times in PBS and culture medium prior to incubation.

Since 1980, when Lim and Sun demonstrated that islets encapsulated in alginate–poly(L-lysine)–poly(ethyleneimine) (APPi) can be successfully grafted [27] into streptozotocin-diabetic rats, transplantation of immunoisolated islets has developed as an attractive method of diabetes therapy [6–10].

Several compounds were applied for the encapsulation of islets [9–19]. Now capsules consisting of alginate–poly(L-lysine)–alginate (APA) membranes, which are less resistant but more compatible than APPi, are usually used [28–37].

To improve the biocompatibility of capsules, Tatarkiewicz [38–40] covered them with additional layer consisting of heparin–protamine complex (APiPrH). Protamine, a strongly basic low molecular weight protein, combines with the strongly acid heparin to form a stable salt without anticoagulant activity [40]. Both protamine and heparin are nonantigenic.

To assess the value of APiPrH microcapsules appropriate studies, were performed.

MATERIALS AND METHODS

Islets were isolated from male WAG rats by means of the three-step digestion modification [41] of the Lacy–Kostianovsky method [42] and purification by Ficoll discontinuous gradient centrifugation. Lim and Sun's original methods [27] was control APA microcapsules, and APiPrH beads were prepared according to the Tatarkiewicz prescription (Table 71.1) [9,38,39]. In brief, day after isolation islets are suspended in 1.2% w/v sodium alginate (1000–2000/ml) and spherical droplets are formed by an air jet–syringe pump generator and gelled in a 1.5% $CaCl_2$ solution. The beads that are formed ($\sim$0.6 μm in diameter) are washed successively in 0.2% aqueous poly(ethyleneimine) (Pi), 1% calcium chloride, 0.1% protamine sulfate (Pr), and heparin (H) in amounts adjusted to neutralize the protamine. Afterward they are washed–several times in phosphate-buffered saline (PBS) [pH 7.4]. The capsules are then allowed to react with 0.1% sodium citrate in PBS (pH 7.4) to liquefy the alginate, and finally, before incubation, they are washed several times in 0.9% PBS and culture medium.

In *in vitro* studies standard methods were used [10,15,39]. In *in vivo* experiments three sets of 1000 rat islets were transplanted intraperitoneally into streptozotocin-induced diabetic BALB/c mice (islets were free encapsulated with APL or APiPrH). The blood glucose concentration of the recipients was measured by radioimmunological methods, with rat insulin as the standard and double antibody separation. For statistical analysis, Student's t test was used. Differences with p values less than 0.05 were considered to be significant.

COMPARISON OF APIPRH AND APA CAPSULES

Several features of APiPrH and APA capsules are comparable. Both APA and APiPrH microcapsules are spherical and smooth, with reproducible diameters of approximately 600 μm (Fig. 71.1).

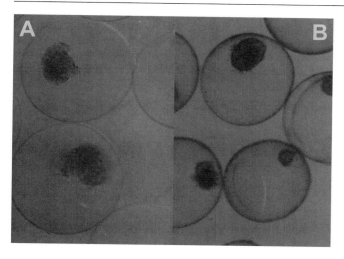

Fig. 71.1. Photomicrographs of APA and APiPrH microencapsulated rat islets.

As indicated in Table 71.2, empty APiPrH capsules are more resistant *in vitro* than APA to ultrasonic damage, alkalization, several passages through the syringe needles, osmotic stress, and freezing. Beads of both types are very resistant to acidification [10,15,39].

Both types of capsule wall exhibit selective permeability [9,39]. The capsule wall is impermeable to albumins and globulins, as well as to the cells. The diffusion of glucose and insulin throughout the capsular wall is excellent; the lag time in response to glucose challenge is short (<5 min) [9].

In culture, the APiPrH-microencapsulated rat islets remain viable, morphologically and functionally intact, for at least one month. During this time they secrete insulin under standard culture conditions and after varying stimulation with glucose as good as free islets. After 2–3 days, secretion of insulin (into the culture medium) from both types of encapsulated islet was comparable with that from the control islets. In magnitude of insulin secretion in response to glucose challenge during low (1.67 mM), high (16.7 mM), and again low (1.67 mM) glucose challenge, both during perifusion and during static incubation, free, APA- and APiPrH-encapsulated islets are comparable. *In vitro* mitogenic action on mouse splenocytes of particular APA and APiPrH capsule wall components is negligible in comparison to such potent mitogens as Concanavolin A (Con A) (Table 71.3). *In vivo* empty APA and APiPrH capsules grafted into the peritoneal cavity or under the renal capsule provoke no inflammatory reaction and are free from cellular and/or fibrotic overgrowth. They floot freely in the peritoneal cavity and can be removed easily even more than 4 weeks af-

Table 71.2. Responses of Microcapsules to Different Kinds of Mechanical and Chemical Stress

	Type of capsules	
Kind of stress	APiPrH	APA
Suspension in distilled water:	Some swelling	Got wrinkles
in 1 N HCl [pH=3.54]	Resistant	Resistant
in <0.003 N NaOH (pH=8.4)	Resistant	Resistant
in >0.025 N NaOH (pH=10.84)	Resistant	Dissolved
Cryopreservation	Resistant	Got wrinkles
Ultrasonic homogenization with "Labsonic U" power	Resistant to 3 attempts	Destroyed after first attempt
Repeating duty maximal	Resistant	Destroyed
Repeating duty cycle 0.9 of 6' each	Resistant	Destroyed

Table 71.3. *In Vitro Mitogenic Activity of Mouse Splenocytes after Stimulation by Components of the Capsular Wall: Presented as the Ratio of Particular Response to the Response of Resting Cells*

Components	Days of culture				
	1	2	3	4	5
Splenocytes	1	1	1	1	1
Splenocytes + Con A	77–145	167–610	15–156	3.4	0.6
Splenocytes + protamine sulfate	0.3–1.3	0.3–0.9	0.2–1.6	1.3	0.3
Splenocytes + heparin	0.5–0.8	0.4–1.0	0.5–2.2	0.9	0.7
Splenocytes + poly(ethyleneimine)	0.6–1.1	0.5–3.0	0.2–2.4	1.4	1.1
Splenocytes + protamine sulfate + heparin	1.6	1.5	0.3	1.0	1.3
Splenocytes + poly(L-lysine)	0.3–0.6	0.3–1.5	0.1–2.3	1.2	1.1
Splenocytes + sodium alginate (Sigma)	1.4–2.1	1.1–5.1	1.5–2.0	4.4	3.4
Splenocytes + sodium alginate (Manugel, Kelco)	2.1–7.4	0.8–9.3	1.4–2.9	7.6	3.2

ter implantation. (Fig. 71.2) Capsules implanted retroperitoneally provoke strong fibrotic reaction [10,15].

Whereas islets encapsulated with APA were clearly visible even after more than 6 months, after several weeks APiPrH beads were already opaque and firm, and the and enclosed islets were invisible [10].

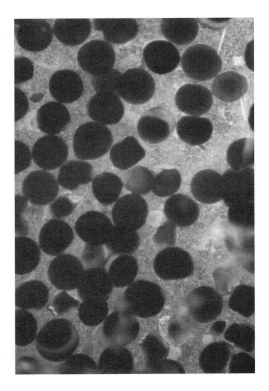

Fig. 71.2. *APiPrH capsules removed 4 weeks after intraperitoneal transplantation.*

Intraperitoneal transplantation int. streptozotocin-diabetic mice of 1000 rat islets immunoisolated in APiPrH can reverse hypoglycemia for more than 6 weeks (46 ± 15 days), whereas APA-isolated grafts can secrete appropriate amounts of insulin much longer (>4 months) [10,15]. Because there are no signs of local inflammatory reaction around the APiPrH capsules, it is probable that the permeability of their walls will decrease with time.

CONCLUSION

Long-term experiments indicate that immunoisolation with APiPrH microcapsules is less effective than immunoisolation with APA beads. Nevertheless, in situations in which the immunoisolated material is exposed to intensive mechanical and/or chemical stress, APiPrH microcapsules are preferable.

REFERENCES

1. Lanza, R. P., Beyer, A. M., Staruk, J. E., et al. (1993). Biohybrid artificial pancreas. Transplantation 56, 1067–1072.
2. Lanza, R. P., Chick, W. L. (1997). Transplantation of encapsulated cells and tissues. Surgery 121, 1–9.
3. Jain, K., Yang, H., Cai, B. R., et al. (1995). Retrievable replaceable, macroencapsulated pancreatic islet xenograft. Transplantation 59, 319–324.
4. Chen, Ch. F., Chern, H. T., Leu, F. J. (1994). Xenotransplantation of microencapsulated canine islets into diabetic rats. Art. Organs. 18, 193–197.
5. Lanza, R. P., Kuhtreiber, W. M., Ecker, D. et al. (1995). Xenotransplantation of porcine and bovine islets without immunosuppression using uncoated alginate microspheres. Transplantation 59, 1377–1384.
6. Weiner, C. P., et al. (1997). Encapsulated beta- islet cells as a bioartificial pancreas to treat insulin-dependent 197 diabetes during pregnancy. Am. J. Obstet. Gynecol. 177, 746–752.
7. Kawakami, Y., Inoue, K., Hayashi, H., et al. (1997). Subcutaneous xenotransplantation of hybrid artificial pancreas encapsulating pancreatic B cell line. Cell Transplant. 6, 541–545.
8. Cohrum, K., Jemtrud, S., and Dorian, R. (1995). Successful xenografts in mice with microencapsulated dog islets. Transplant. Proc. 27, 3297–3301.
9. Tatarkiewicz, K., Sitarek, E., Sabat, M., and Orlowski, T. (1997). Reversal of hyperglycemia in streptozotocin-diabetic mice by xenotransplantation of microencapsulated rat islets. Ann. Transplant. 2, 20–23.
10. Orlowski, T., Tatarkiewicz, K., Sitarek, E., Sabat, M., Fiedor, P., and Samsel, R. (1996). Experience with pancreas islets separation, immunoisolation and cryopreservation, Ann. Transplant. 1, 54–58.
11. Garfinkel, M. R., Harland, R. C., and Opara, O. C. (1998). Optimalization of the macroencapsulated islet for transplantation. Surg. Res. 76, 7–10.
12. Ohyama, T., Nakajima, Y., Kenehiro, H., et al. (1998). Long-term normalization of diabetes by xenotransplantation of newly developed encapsulated pancreatic islets. Transplant. Proc. 30, 3433–3435.
13. Cohrum, K., Jemtrud, S., and Dorian, R.(1995). Successful xenografts in mice with microencapsulatrd dog islets. Transplant. Proc. 27, 3297–3301.
14. Sun, Y. L., Ma, X., Zhou, D., Vacek, I., and Sun, A. M. (1993). Porcine pancreatic islets: Isolation, microencapsulation, and xenotransplantation. Artif. Organs 17, 727–733.
15. Orlowski, T., Sitarek, E., Tatarkiewicz, K., Sabat, M., and Antosiak, M. (1997). Comparison of two methods of pancreas islets immunoisolation. Int. J. Artif. Organs 20, 701–703.
16. Lanza, R. P., and Chick, W. L. (1997). Immunoisolation: At a turning point. Immunol. Today 16, 135–138.
17. Lanza, R. P., Hajes, J. L., and Chick, W. L. (1996). Encapsulated cell technology. Natur. Biotechnol. 14, 107–111.
18. Jain, K., Yang, H., Asina, S. K. et al. (1996). Long term preservation of islets of Langerhans in hydrophilic macrobeads. Transplantation 59, 1377–1384.
19. Huguet, M. L., and Dellacheria, E. (1996). Calcium alginate beads coated with chitosan. Proc. Biochem. 31(8), 745–751.
20. O'Shea, G. M., Goosea, M. F. A., and Sun, A. M. (1984). Prolonged survival of transplanted islets of Langerhans encapsulated in a biocompatible membrane. Biochim. Biophys. Acta 804, 133–136.
21. Calafiore, R., Basta, G., Falorni, A. et al. (1992). Fabrication of high performance microcapsules for pancreatic islets transplantation. Diabetes Nutr. Metab. 5(Suppl. 1), 173–176.
22. Robitaille, R., Pariseau, J. F., Leblond, F. A., et al. (1999). Studies on small [<350 microm.] alginate-poly-L-lysin microcapsules. J. Biomed. Mater. Res. 44, 116–120.
23. Lanza, R. P., Ecker, D., Kuhtreiber, W. M., et al. (1999). Transplantation of islets using microencapsulation. J. Mol. Med. 77, 206–210.
24. Olivers, E., Nilsson,K., and Malaisse, W. J. (1999). Microencapsulation of single rat pancreatic islets with preserved secretory potential. Med. Sci. Res. 27, 411–413.
25. De Vos, P., De Haan, B. J., and Van Schilfgaarde, R. (1998). Factors causing failure of islets in nonovergrown capsules. Transplant. Proc. 30, 496–497.
26. De Vos, P., De Haan, B., Wolters, G. H. J., et al. (1996). Factors Influencing the adequacy of microencapsulation of rat pancreatic islets. Transplantation 62, 888–893.
27. Lim, F., and Sun, A. M. (1980). Microencapsulated islets as a bioartificial pancreas. Science 210, 908–910.
28. Kawakami, Y., Inoue, K., and Hayashi, H., et al. (1997). Subcutaneous xenotransplantation of hybrid artificial pancreas encapsulating pancreatic B cell line. Cell. Transplant. 6, 541–545.

29. Constantinidis, I., Rask, I., Long, R. C., *et al.* (1999). Effects of alginate composition on the metabolic, secretory, and growth characteristics of entrapped beta TC3 mouse insulinoma cells. *Biomaterials* **20**, 2019–2027.

30. De Vos, P., De Haan, B., and Van Schilfgaarde, R. (1997). Effect of the alginate composition on the biocompatibility of alginate–polylysine microcapsules. *Biomaterials* **18**, 273–278.

31. Clayton, H. A., London, N. J. M., Colloby, P. S. et al. (1992). The effect of capsule composition on the viability and biocompatibility of sodium alginate/poly-L-lysine encapsulated islets. *Transplant. Proc.* **24**, 956.

32. Lanza, R. P., Beyer, A. M., Staruk, J. E., *et al.* (1993). Biohybrid artificial pancreas. *Transplantation* **56**, 1067–1072.

33. Lanza, R. P., Kuhtreiber, W. M., Ecker, D., *et al.* (1995). Xenotransplantation of porcine and bovine islets without immunosuppression using uncoated alginate microspheres. *Transplantation* **59**, 1377–1384.

34. De Vos, P., De Haan, B., and Van Schilfgaarde, R. (1998). Is it possible to use the standard alginate-PLL procedure for production of small capsules. *Transplant. Proc.* **30**, 492–493.

35. Lanza, R. P., Beyer, A. M., Staruk, J. E., *et al.* (1993). Biohybrid artificial pancreas. *Transplantation* **56**, 1067–1072.

36. Lanza, R. P., Kuhtreiber, W. M., Ecker, D., *et al.* (1995). Xenotransplantation of porcine and bovine islets without immunosuppression using uncoated alginate microspheres. *Transplantation* **59**, 1377–1384.

37. Kulseng, B., Thu, B., Espevik, T., *et al.* (1997). Alginate poli-L-lysine microcapsules as immune barrier. *Cell Transplant.* **6**, 387–394.

38. Tatarkiewicz, K. (1988). New membrane for cell encapsulation. *Artif. Organs* **12**, 446–448.

39. Tatarkiewicz, K. (1994). Ocena biozgodnosci blony protaminowo-heparynowej i wplyw mikrooplaszczania na czynności życiowe wysp trzustkowych in vitro. *Proc. Inst. Biocybrn. Biomed. Engin.* **37**, 1–86.

40. Levine, W. G. (1970). Anticoagulants. *In* "Pharmacological Basis of Therapeutics" (L. S. Goodman and A. Gilman, eds.), 4th ed., pp. 1445–1463. Macmillan, London.

41. Sitarek, E., Tatarkiewicz, K., Sabat, M., and Orlowski, T. (1993). Methods of rat langerhans islets isolation. *Bull. Ac. Pol. Sci. Scis. Biol.* **41**, 117–121.

42. Lacy, P. E., and Kostanovsky, M. (1967). Method for the isolation of intact islets of Langerhans from the rat pancreas. *Diabetes* **16**, 35–39.

CHAPTER 72

MICROENCAPSULATION METHODS: GLYCOSAMINOGLYCANS AND CHITOSAN

Victor S. Lin and Howard W. T. Matthew

In this chapter we describe principles of direct microencapsulation of cells using ionic interactions between oppositely charged, water-soluble polymers. The particular polymers of interest are the extracellular matrix polysaccharides termed glycosaminoglycans and the chitin derivative chitosan. These materials possess biological activity that may be particularly useful from a tissue engineering standpoint. Protocols are presented for microencapsulation in microcapsules composed of carboxymethylcellulose, chondroitin sulfate, and chitosan with or without an interior gel. The physical and biological properties of the microcapsules are discussed, together with mechanisms for modifying these properties.

INTRODUCTION

Microencapsulation is a well-established technology that has been applied to culturing animal and microbial cells for a variety of applications. The need for continued improvement in cell culture systems fueled the development of a variety of encapsulation methods, each of which has its own strengths. The advent of tissue engineering as a distinct discipline provides new opportunities for the application of microencapsulation technology. In this chapter we describe protocols for microencapsulation of cells and cellular aggregates via complex formation between the polycation chitosan and a number of polyanionic species, most notably the glycosaminoglycans. Microencapsulation provides a number of clear advantages for *in vitro* cultures and also provides a useful mechanism for effecting cell transplantation for tissue engineering applications. Microencapsulated proliferating cells can produce very high density cultures and may be able to generate a particularly favorable microenvironment. Cells within capsule membranes are protected from shear damage in flow or stirred systems.

Microencapsulation also provides a convenient method of cell handling, since encapsulated cells can be readily harvested from culture systems, or injected/implanted into specific physiological locations. With a capsule membrane of appropriate permeability, transplanted allogeneic or xenogeneic cells can be protected from the host's immune system. In addition, capsule membranes may serve to retain and sequester potentially allergenic cellular products.

Yoshioka *et al.* [1] described a microencapsulation method based on the formation of ionic complexes between cationic chitosan and anionic carboxymethylcellulose (CMC). In that procedure, microcapsules were formed by deposition of an insoluble ionic complex at the interface between droplets of CMC solution suspended in a chitosan solution. The formation of an insoluble complex by ionic interaction of oppositely charged soluble polymers is termed complex coacervation. The microcapsules formed in the original procedure were fragile and exhibited a high degree of swelling (>300% increase in volume). In spite of this drawback, the complex coacervation concept is attractive because of its simplicity, its gentle chemistry, and the availability of a large number of biocompatible, ionic polymers and polysaccharides capable of interacting to form coacervates [2–15]. The ionic polysaccharides termed glycosaminoglycans (GAGs) are of particular interest because of their broad range of biological activity [13,15–22]—activity that can be harnessed to enhance the effectiveness of tissue-engineered systems. The potential contributions of a GAG capsule component include modulating cell behavior (e.g., inducing aggregation) within the microcapsules, altering cell adhesion, proliferation, and migration around microcapsules, inhibition of blood coagulation, and enhancing angiogenesis in the vicinity of implanted microcapsules.

ENCAPSULATION BY IONIC COMPLEX FORMATION

In principle, any pair of oppositely charged, water-soluble macromolecules of sufficiently high molecular weight can be used to form microcapsules by ionic complex formation. In practice, many such ionic interactions produce amorphous precipitates or soluble complexes and are thus not suitable for encapsulation. The reasons for such failures stem from variations in molecular properties such as charge density, degree of ionization, and chain conformations in solution. As a result, predicting the form of such polyelectrolyte complexes is difficult and is beyond the scope of this chapter.

The fundamental principle of this encapsulation approach involves first suspending the cells to be encapsulated in a viscous solution of the polyanionic components of the complex. Gel-forming agents such as type I collagen may also be included if desired. Droplets of this cell suspension are then generated by extrusion through a needle or catheter. The droplets are collected into a rapidly stirred solution of a polycation, where interactions between the positively charged polycation chains and the negatively charged polyanions forms an ionically cross-linked membrane around the droplet. The initial complex membrane formed is very thin (<1 μm) but rapidly thickens as the lower molecular weight polymers diffuse across the membrane and react to create additional layers of complex. This process is self-limiting, as membrane thickening rapidly decreases the transmembrane diffusion rate. At this point, the formed microcapsules are ready for washing to remove excess polymer. Washing is followed by surface stabilization, and any other required manipulations.

Surface stabilization involves complexing excess cationic polymer chains on the external surfaces with a polyanion. If this is not done, residual anion diffusing from the microcapsules can complex with surface polycation on adjacent microcapsules, resulting in extensive intercapsule adhesion. The stabilizing polyanion can be the same as that in the intracapsule solution, but it may also be chosen on the basis of imparting particularly desirable surface properties; for example, heparin may be used to confer anticoagulant activity. We have reported on the use of poly(galacturonic acid) (PGA) as a stabilizing polyanion to minimize capsule swelling after formation [10].

As mentioned earlier, the original procedure of Yoshioka et al. [1] employed CMC as the sole polyanion and generated very thin membranes. Addition to the CMC solution of the GAG chondroitin 4-sulfate (chondroitin sulfate A, CSA) produced a much more sturdy capsule membrane with a significantly higher permeability [10]. Further experiments demonstrated that viable microcapsules could be formed by chitosan interacting with most GAGs without a need for CMC. The strength and permeability of some of these formulations were examined, and while the CMC/CSA–chitosan–PGA formulation proved to be the most sturdy, other formulations may have greater tissue engineering potential owing to the biological activity of specific GAG species.

The chitosan–CMC-GAG microcapsules are hollow and permeable to macromolecules, and they ultimately equilibrate completely with the exterior medium. As a result, the encapsulated cells settle and may attach directly to the capsule wall. With some cell types, such as hepatocytes, the settled cells preferentially aggregate into spheroids. If the number of cells in the capsule is large enough, this behavior can lead to the formation of large (e.g., 400-μm-diameter) spheroids with necrotic centers. Similarly, if preformed aggregates are encapsulated, aggregate fusion can occur, again leading to large nutrient-limited spheroids with necrotic regions. These problems can be abrogated by coencapsulating a gel-forming agent. After capsule formation, the gel serves to immobilize the cells or spheroids in a dispersed pattern and prevents fusion into superspheroids with greater diffusion limitations. Furthermore, biologically active gels such as type I collagen and solubilized Engelbreth-Holm-Swarm (EHS) matrix (Matrigel, Collaborative Biomedical, Bedford, MA) may serve to enhance the survival, growth, and function of encapsulated cells and cellular aggregates.

MICROCAPSULE PHYSICAL PROPERTIES

Optimization of a microencapsulation procedure, or a tissue engineering application based on such a procedure, requires some detailed knowledge of the physical properties of the capsule membranes. Capsule burst strength, macromolecular permeability, and wall composition are the three most important parameters from a practical standpoint. Burst strength determines factors such as limitations on handling and physiological placement. Permeability determines the level of immunological isolation/protection of encapsulated cells, as well as the utility of a given capsule formulation when diffusion of high molecular weight factors into or out of the microcapsules is essential. Capsule wall composition may be the determining factor in interactions such as blood compatibility and the tissue response to implanted microcapsules. For example, fibrotic responses to implanted microcapsules may result in deposition of a thick collagenous "capsule" and implant failure due to the associated diffusion limitations.

The thickness and strength of the capsule wall can be controlled to some extent by varying the molecular weights of the constituent polymers. Lower molecular weight forms diffuse through the capsule membrane more rapidly and thus produce thicker walls before being hampered by the diffusion distance. For example, lower molecular weight chitosan produces capsule walls that are both thicker and less elastic than those made with higher molecular weight forms. Both chitosan and GAGs in solution can be stoichiometrically depolymerized by means of the sodium nitrite–nitrous acid procedure [23,24]. Reductions in capsule wall permeability can also be expected with reduced molecular weight polymers.

In 1993 we reported on the albumin permeability of the chitosan–CMC/CSA microcapsules and related capsule formulations [10]. All the complex coacervate microcapsules proved to be permeable to bovine albumin. Results of a more detailed kinetic analysis of diffusion into microcapsules are presented later.

Because the ionization level of the chitosan, and to a lesser extent that of the GAG, is pH dependent, the strength of the chitosan–GAG ionic bonds changes with pH and actually decreases as the acidic chitosan solution used for capsule formation goes to the physiological solutions used before culture and implantation. As a result, a significant percentage of the initially complexed GAG is desorbed from the capsule membrane. This desorption has little effect on membrane physical properties but may significantly alter such membrane biological properties as cell adhesion, protein adsorption, and blood compatibility.

MATERIALS
CMC/CSA–CHITOSAN–PGA MICROCAPSULES

1. *Polyanion solution:* This solution is used to suspend the cells to be encapsulated. The volume required approximately equals the volume of microcapsules to be made. Medium-viscosity CMC and CSA are dissolved in an appropriate volume of a sorbitol–HEPES buffer to give final concentrations of 1.5 wt% CMC and 4 wt% CSA. The sorbitol–HEPES buffer contains 0.4 g/liter KCl, 0.5 g/liter NaCl, 3.0 g/liter HEPES·Na, and 36 g/liter sorbitol in water. The pH is adjusted to 7.4 with 0.1 M NaOH. The CMC and CSA powders should be mixed and added slowly to the rapidly stirred buffer to avoid excessive clumping. The solution is then sterilized by autoclaving at 121°C for 30 min. Note that because of the rapid increase in solution viscosity, dissolution of the solids may be time-consuming at room temperature. Therefore, the suspension can be autoclaved with suspended solids present. Complete dissolution of suspended solids will occur during autoclaving.

2. *Chitosan solution:* Since chitosan in solution decomposes upon heating, the polymer is sterilized before dissolution. Thus, 3 g of chitosan powder is suspended in 250 ml

of water with a magnetic stir bar. Similarly, 19 g of sorbitol is separately dissolved in an additional 250 ml of water. The two volumes are sterilized by autoclaving (121°C, 30 min). Under sterile conditions, 0.6 ml of glacial acetic acid is added to the chitosan suspension, and the mixture is stirred for at least 4 h to partially dissolve the chitosan. The sorbitol and chitosan solutions are then mixed. The solution may then be centrifuged or filtered through sterile gauze to remove undissolved particles of excess chitosan.

3. *Surface stabilization solution*:   Sodium polygalacturonate (PGA) is dissolved in sorbitol–HEPES buffer to give a concentration of 0.1 wt%. The solution is then sterilized by autoclaving (121°C, 30 min).

4. Tween 20 or Tween 80, sterilized by filtration or autoclaving.

5. Phosphate-buffered saline for capsule washing steps, sterilized by filtration or autoclaving.

6. *Droplet forming apparatus*:   A variety of atomization devices for forming liquid droplets of controlled size are commercially available or can be easily fabricated. One such simple device is designed to blow compressed air coaxially around a needle or catheter. The suspension of cells in polyanion solution is extruded through a 22-gauge, blunt needle or preferably a 24-gauge Teflon catheter. Compressed air is blown coaxially around the catheter tip, thus shearing droplets away before they reach full size. Droplet size can be controlled by varying the airflow rate to remove liquid drops from the catheter tip at some predetermined size.

MICROCAPSULES WITH AN INTERIOR EXTRACELLULAR MATRIX (ECM) GEL

When microcapsules with an internal ECM gel are desired, the following components are used in lieu of the regular polyanion solution. Gels may be composed of collagen or Matrigel. Other agents that exhibit gel formation at physiological temperatures may also be used if they are miscible with the membrane forming polyanions.

1. *Gel-forming solution*:   either type I collagen (bovine or rat tail) dissolved in 1 mM HCl at a concentration of 2 mg/ml or solubilized EHS extracellular matrix (Matrigel, Collaborative Biomedical).

2. *10× DMEM*:   for use with the collagen solution. Commercial Dulbecco's Modified Eagle's Medium powder and the required amount of $NaHCO_3$ are dissolved in 10% of the recommended water volume. For example, powder for 1000 ml of culture medium is dissolved in 100 ml of water.

3. *Double-strength polyanion solution*:   medium-viscosity CMC and CSA are dissolved in sorbitol–HEPES buffer to give final concentrations of 3 wt% CMC and 8 wt% CSA. The pH is adjusted to 7.4 with 0.1 M NaOH, and the solution is sterilized by autoclaving (121°C, 30 min).

METHODS

The protocols that follow can be used to form general-purpose microcapsules composed of chondroitin 4-sulfate (CSA), carboxymethylcellulose (CMC), chitosan, and polygalacturonate (PGA). This basic protocol may also be modified as indicated later to form microcapsules with cells or cell spheroids embedded in internal gels of collagen or Matrigel. The basic protocols may also be modified to obtain application-specific properties by changing the type of GAG used, the GAG concentration, the GAG:CMC ratio, or the polymer molecular weights.

MICROENCAPSULATION WITHIN CMC/CSA–CHITOSAN–PGA MEMBRANES

The microencapsulation should be conducted with solutions chilled to 4°C to reduce cell metabolism and minimize adverse effects on cell viability. Likewise, tubes containing cells or cell suspensions should be kept on ice until actually used.

1. The cells or cellular spheroids to be encapsulated should be prepared as a suspension in a 15-ml centrifuge tube and then allowed to settle from suspension under either unit gravity or at low-speed centrifugation (<300 rpm).

2. The supernatant is carefully removed by aspiration or pipetting. Polyanion solution (CMC/CSA) is then added to the cell pellet to give a final volume based on the encapsulation density or total capsule volume desired. A high encapsulation density uses 10–15 million cells per milliliter of total (capsule) volume. For example, 30 million rat hepatocytes in a cell pellet are made up to a total volume of 3 ml with polyanion solution. The cells or spheroids are then suspended in the polyanion solution by repeatedly inverting and tapping the tube.

3. A 30-ml volume of chitosan solution is transferred to a sterile 50-ml beaker containing a 2.5-cm magnetic stirring bar. The solution is stirred at a speed high enough to generate a vortex about 2.5 cm deep (~250 rpm). Two or three drops of sterile Tween 20 or Tween 80 are then added.

4. A 5-ml sterile disposable syringe is fitted with an 18-gauge needle, and 3 ml of the cell–polyanion suspension is carefully aspirated into the syringe. After mixing briefly by inversion to ensure uniformity, the 18-gauge needle is replaced with a 24-gauge catheter or a 22-gauge needle and the syringe is transferred to the droplet-generating apparatus.

5. The air supply to the apparatus is connected, and the airflow is started. Several drops of the cell–polyanion suspension are dispensed from the syringe by depressing the plunger. The droplet size is assessed, and the airflow is adjusted upward (for smaller droplet size) or downward (for larger droplet size), as desired. Excellent results are obtained with droplets 0.3–1.0 mm in diameter.

6. Once the airflow has been adjusted to a suitable level, the droplet-generating apparatus is positioned over the stirring chitosan solution and the entire 3 ml of suspension is dispensed carefully into the stirring chitosan solution. While the suspension can be dispensed rapidly, care should be taken to ensure that it is dispensed as discrete droplets, as opposed to a jet or continuous stream. The apparatus should also be positioned so that droplets enter the chitosan solution approximately midway between the beaker wall and the vortex center. Capsule membranes are generated at the droplet–chitosan solution interface by complex formation between the CMC/CSA and chitosan.

7. After the entire aliquot in the syringe has been dispensed into the chitosan solution, the stirring speed is reduced and 20 ml of phosphate-buffered saline (PBS) is poured into the chitosan–capsule suspension. After an additional 2–3 min of stirring, this mixture is rapidly poured into a 250-ml conical flask containing 150 ml of PBS. As soon as the microcapsules have settled, about 180 ml of the supernatant is rapidly aspirated off without tilting the flask or disturbing the microcapsules. The flask with the remaining 20 ml of liquid is vigorously shaken to disrupt any adhesion between microcapsules, and another 150 ml of PBS is poured into the flask. After the microcapsules have settled, the supernatant is again aspirated.

8. The microcapsules are surface-stabilized by adding 10 ml of the PGA surface stabilizing solution and swirling briefly. The liquid is then completely removed by aspiration, and the microcapsules are rinsed twice with PBS.

9. At this point the microencapsulated cells can be equilibrated at 37°C with culture medium or a balanced salt solution prior to culture or *in vivo* implantation. Since the interior CSA and CMC do diffuse out of the capsules over time, equilibration with culture medium for at least 30 min is recommended to allow removal of the excess polyanions.

MICROENCAPSULATION WITH AN INTERNAL CAPSULE GEL

As mentioned earlier, the microencapsulation should be conducted with solutions chilled to 4°C. When gel-forming agents are employed, low temperature is particularly important to prevent premature gelation of the collagen or Matrigel. With both types of gelling agent, gel formation occurs within the capsules during equilibration with culture medium.

Matrigel Interior Gel

Care must be taken to ensure that all solutions, pipettes, and containers are prechilled and not allowed to warm significantly while Matrigel or its mixtures are being handled. A 1-ml aliquot of Matrigel is thawed at 4°C and mixed with an equal volume of cold, double-strength polyanion solution (3% CMC, 8% CSA). This mixture is then used (instead of the regular polyanion solution) to suspend the cells in step 2 of the microencapsulation protocol.

Collagen Interior Gel

A 4.5-ml volume of cold collagen solution is transferred to a chilled 15-ml centrifuge tube. A 0.5-ml volume of cold 10× DMEM is then added to the collagen solution and rapidly mixed by inversion. A 5-ml volume of cold, double-strength polyanion solution is then added to the collagen DMEM mixture and again mixed by inversion. This collagen–DMEM–polyanion mixture is used (instead of the regular polyanion solution) to suspend the cells in step 2 of the microencapsulation protocol.

APPLICATION-SPECIFIC MODIFICATIONS TO THE ENCAPSULATION PROTOCOLS

The protocols just presented describe general-purpose microcapsules, which may be suitable for most *in vitro* applications. However, *in vivo* or *ex vivo* applications may require modifications in either capsule composition or physical properties. In this section we describe options for modifying the general protocols to achieve application-specific microcapsule performance.

Biological Activity

Microcapsule biological activity can be modified by changing the macromolecular components of the polyanion solution and the surface-stabilizing solution. For example, using another GAG (e.g., heparin or hyaluronic acid) as a component of the polyanion solution may alter GAG-dependent proliferative or phenotype changes in the encapsulated cells [19, 21,22]. Such changes could be mediated either by GAG-modulated changes in the activity of so-called heparin binding growth factors (e.g., FGF, EGF, VEGF, HGF, etc.) or by direct GAG–receptor interactions [13,14,16–18]. Related effects may be produced in tissues surrounding a microcapsule implantation site and could include enhanced angiogenesis and vascularization of the implant. Similary, replacing PGA with heparin in the surface stabilization solution may generate microcapsules with improved blood compatibility.

Microcapsule Permeability

Earlier work demonstrated that the macromolecular permeability of the capsule membrane is a function of both the microcapsule composition and the molecular weights of the polymers used. The highest permeability to bovine albumin was obtained with a hyaluronate–chitosan–hyaluronate microcapsule formulation. Likewise, the lowest permeability was obtained with the original CMC–chitosan–CMC formulation. In principle, the permeability of any given formulation can be reduced by reducing the molecular weights of the constituent polymers. This can be done with both chitosan and GAGs by reacting them in solution with nitrous acid. In particular, the reaction of chitosan with nitrous acid produces a stoichiometric cleavage of the chitosan chains. Thus fairly precise predictions of the final molecular weight (average) can be made, even if the initial molecular weight is unknown [23].

Microcapsule Strength

Microcapsule strength is a function of microcapsule composition and wall thickness. Since the wall thickness is also a function of how rapidly the component polymers diffuse across the initial capsule membrane, reducing the molecular weight of a given polymer will also tend to increase wall thickness and capsule strength.

MEASUREMENT OF MICROCAPSULE PERMEABILITY

To ensure survival and function of encapsulated cells, nutrients and wastes must be able to diffuse into and out of the microcapsules at an adequate rate. Since the capsule membrane and interior gel are hydrogel materials, passage of substances into and within the microcapsules is almost exclusively by molecular diffusion. Therefore experiments were performed to assess the permeability of microcapsules (CMC/CSA/collagen gel–chitosan–PGA) by measuring the diffusivity of various substances within the microcapsules.

For each diffusion experiment, a 4-ml mixture of 0.9% NaCl solution (saline) and saline-equilibrated microcapsules was placed in a 15-ml centrifuge tube. The volume ratio of saline to microcapsules was 2.3:1. Capsule volume was precisely ascertained by determining an average diameter for the capsule population via microscopic measurement, and then counting out a specific number of capsules. To this mixture was added a concentrated tracer solution of resorufin (sodium salt, Molecular Probes, Eugene, OR), insulin [mixture, with and without fluorescein isothiscyanate (FITC)-labeling], or bovine albumin (mixture, with and without FITC labeling). Thus the final tracer concentration in the extracapsular solution was 250 ng/ml resorufin, 1 mg/ml insulin (with 5 µg/ml FITC-labeled insulin), or 1 mg/ml albumin (with 5 µg/ml FITC-labeled albumin).

After the addition of the tracer solutions, the tubes were placed on an orbital shaker and mixed at 100 rpm to allow the tracer molecules in the extracapsular fluid to diffuse into the microcapsules over time. Samples (0.2 ml) of the extracapsular fluid were periodically withdrawn so their concentrations could be measured in a fluorometer. The resorufin concentration was measured with excitation and emission wavelengths of 571 and 585 nm, respectively. The insulin and albumin concentrations were measured at excitation (emission) wavelengths of 488 (518) nm. Samples were withdrawn for measurements until the concentration reached a steady state plateau. Figure 72.1 plots tracer fractional uptake versus time. Fractional uptake M/M_f is the amount of substance at any time divided by the final equilibrium amount.

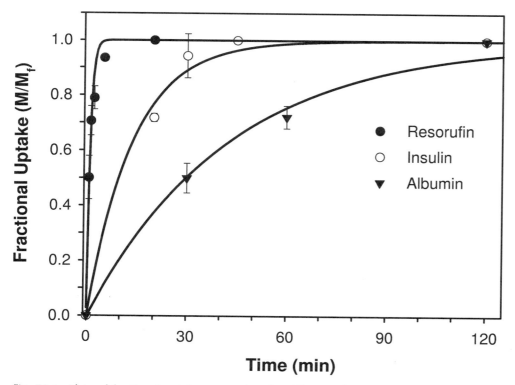

Fig. 72.1. Plots of fractional uptake versus time for diffusion of tracer molecules from a well-mixed solution of finite volume into CMC/CSA/collagen gel–chitosan–PGA microcapsules. Curve fits were obtained using the approximate solution of Crank [25], described in the text.

Table 72.1. Effective Diffusivities of Tracer Molecules

Tracer	Molecular weight	Diffusivity (cm²/s)
Resorufin	235	$4.34 \pm 1.49 \times 10^{-6}$
Insulin	5,734	$5.57 \pm 1.86 \times 10^{-7}$
Albumin	67,000	$2.14 \pm 0.25 \times 10^{-7}$

Non-steady-state diffusion in a sphere is governed by the following diffusion equation:

$$\frac{\partial C}{\partial t} = D\left(\frac{\partial^2 C}{\partial r^2} + \frac{2\partial C}{r\partial r}\right)$$

which relates the tracer concentration C at various distances r from the center of the sphere to the time t. The diffusivity D is determined by fitting the mathematical solution of the equation to the data of concentration versus time.

Several simplifying assumptions were made in applying this diffusion model. The diffusivity within a microcapsule containing a collagen gel was treated as a constant throughout the microcapsule. The extracapsular fluid was assumed to be well mixed, and the microcapsules were assumed to be homogeneous spheres. The solution used was modified from the one derived by Crank [25] for diffusion from a well-stirred solution of limited volume. Therefore, the fractional uptake M/M_f is approximately related to time by

$$\frac{M}{M_f} = 1 - \frac{6\alpha(\alpha + 1)\exp(-DB^2t/R^2)}{9 + 9\alpha + B^2\alpha^2}$$

where α is the solution-to-capsule ratio and B, a constant that corresponds to α, is 3.465 for our case (α = 2.3) [25]. D is the diffusivity (cm²/s), t is the elapsed time (s), and R is the radius of the microcapsules (0.085 cm for the data in Fig. 72.1). This approximate solution was reported to work well for values of M/M_f above 0.4. Table 72.1 shows the diffusivities determined for resorufin, insulin, and albumin. Although several times smaller than values for molecules of similar molecular weight diffusing in water, these values are still considerably larger than those for molecules diffusing in alginate or agarose gels [26].

REFERENCES

1. Yoshioka, T., Hirano, R., Shioya, T., and Kako, M. (1990). Encapsulation of mammalian-cell with chitosan CMC capsule. *Biotechnol. Bioeng.* **35**, 66–72.
2. Chavasit, V., Kienzlesterzer, C., and Torres, J. A. (1988). Formation and characterization of an insoluble poly-electrolyte complex–chitosan-polyacrylic acid. *Polym. Bull.* **19**, 223–230.
3. Chavasit, V., and Torres, J. A. (1990). Chitosan poly(acrylic acid)—mechanism of complex-formation and potential industrial applications. *Biotechnol. Prog.* **6**, 2–6.
4. Takahashi, T., Takayama, K., Machida, Y., and Nagai, T. (1990). Characteristics of polyion complexes of chitosan with sodium alginate and sodium polyacrylate. *Inter. J. Pharmac.* **61**, 35–41.
5. Daniels, R., and Mittermaier, E. M. (1995). Influence of pH adjustment on microcapsules obtained from complex coacervation of gelatin and acacia. *J. Microencapsul.* **12**, 591–599.
6. Duquemin, S. J., and Nixon, J. R. (1986). The effect of surfactants on the microencapsulation and release of phenobarbitone from gelatin–acacia complex coacervate microcapsules. *J. Microencapsul.* **3**, 89–93.
7. Koh, G. L., and Tucker, I. G. (1988). Characterization of sodium carboxymethylcellulose–gelatin complex coacervation by chemical analysis of the coacervate and equilibrium fluid phases. *J. Pharm. Pharmacol.* **40**, 309–312.
8. Koh, G. L., and Tucker, I. G. (1988). Characterization of sodium carboxymethylcellulose–gelatin complex coacervation by viscosity, turbidity and coacervate wet weight and volume measurements. *J. Pharm. Pharmacol.* **40**, 233–236.
9. McMullen, J. N., Newton, D. W., and Becker, C. H. (1982). Pectin–gelatin complex coacervates I. Determinants of microglobule size, morphology, and recovery as water-dispersible powders. *J. Pharm. Sci.* **71**, 628–633.
10. Matthew, H. W., Salley, S. O., Peterson, W. D., and Klein, M. D. (1993). Complex coacervate microcapsules for mammalian cell culture and artificial organ development. *Biotechnol. Prog.* **9**, 510–519.
11. Tsung, M., and Burgess, D. J. (1997). Preparation and stabilization of heparin/gelatin complex coacervate microcapsules. *J. Pharm. Sci.* **86**, 603–607.
12. Chavasit, V., and Torres, J. A. (1990). Chitosan–poly(acrylic acid): Mechanism of complex formation and potential industrial applications. *Biotechnol. Prog.* **6**, 2–6.

13. Denuziere, A., Ferrier, D., and Domard, A. (2000). Interactions between chitosan and glycosaminoglycans (chondroitin sulfate and hyaluronic acid): Physicochemical and biological studies. *Ann. Pharm. Fr.* **58**, 47–53.

14. Denuziere, A., Ferrier, D., Damour, O., and Domard, A. (1998). Chitosan–chondroitin sulfate and chitosan–hyaluronate polyelectrolyte complexes: Biological properties. *Biomaterials* **19**, 1275–1285.

15. Hamano, T., Teramoto, A., Iizuka, E., and Abe, K. (1998). Effects of polyelectrolyte complex (PEC) on human periodontal ligament fibroblast (HPLF) function. I. Three-dimensional structure of HPLF cultured on PEC. *J. Biomed. Mater. Res.* **41**, 257–269.

16. Lindahl, U., and Hook, M. (1978). Glycosaminoglycans and their binding to biological macromolecules. *Annu. Rev. Biochem.* **47**, 385–417.

17. Lindahl, U., and Kjellen, L. (1991). Heparin or heparan sulfate—What is the difference? *Thromb. Haemostasis* **66**, 44–48.

18. Lindahl, U., Lidholt, K., Spillmann, D., and Kjellen, L. (1994). More to "heparin" than anticoagulation. *Thromb. Res.* **75**, 1–32.

19. Spray, D. C., Fujita, M., *et al.* (1987). Proteoglycans and glycosaminoglycans induce gap junction synthesis and function in primary liver cultures. *J. Cell Biol.* **105**, 541–551.

20. Sechriest, V. F., Miao, Y. J., *et al.* (2000). GAG-augmented polysaccharide hydrogel: A novel biocompatible and biodegradable material to support chondrogenesis. *J. Biomed. Mater. Res.* **49**, 534–541.

21. Madihally, S. V., Flake, A. W., and Matthew, H. W. (1999). Maintenance of CD34 expression during proliferation of CD34+ cord blood cells on glycosaminoglycan surfaces. *Stem Cells* **17**, 295–305.

22. Fujita, M., Spray, D. C., *et al.* (1987). Glycosaminoglycans and proteoglycans induce gap junction expression and restore transcription of tissue-specific mRNAs in primary liver cultures. *Hepatology* **7**, 1S–9S.

23. Allan, G. G., and Peyron, M. (1995). Molecular weight manipulation of chitosan. I. Kinetics of depolymerization by nitrous acid. *Carbohydr. Res.* **277**, 257–272.

24. Bosworth, T. R., and Scott, J. E. (1994). A specific fluorometric assay for hexosamines in glycosaminoglycans, based on deaminative cleavage with nitrous acid. *Anal. Biochem.* **223**, 266–273.

25. Crank, J. (1975). "Mathematics of Diffusion," 2nd ed., pp. 93–96. Oxford University Press, London.

26. Li, R. H., Altreuter, D. H., and Gentile, F. T. (1996). Transport characterization of hydrogel matrices for cell encapsulation. *Biotechnol. Bioeng.* **50**, 365–373.

CHAPTER 73

MICROENCAPSULATION METHODS: POLYACRYLATES

Minhua Feng and Michael V. Sefton

This chapter reviews the basic principles and laboratory methods of cell microencapsulation with polyacrylates. This review is focused on the use of synthetic polyacrylates for cell microencapsulation, the microencapsulation protocols with typical synthetic polymers, and the methods for the evaluation of cell capsules. HEMA-MMA (hydroxyethyl methacrylate–methyl methacrylate) copolymers are the most common synthetic polymers for cell microencapsulation, and microencapsulations with water-insoluble polymers are usually performed by interfacial precipitation to minimize the contact time and toxicity of the solvents.

INTRODUCTION

Cell microencapsulation is the process of enclosing cells (usually as a cell suspension) within a polymer membrane. The microcapsules are, in most cases, spherical, resulting in an implant having a high surface-to-volume ratio with sizes ranging from a few hundred to roughly a thousand micrometers. The membrane allows small-molecule nutrients from the external environment to enter the capsule and small metabolic products to leave the capsule, while acting as a barrier to high molecular weight antibodies and complement substances from host tissues [1–3]. The therapeutic purposes of encapsulated cells are reviewed elsewhere [4–15].

Living cells are sensitive biological materials that will lose viability if they are exposed to most solvents, acids, and bases, or to temperature extremes or energy sources. Therefore, most of the traditional chemical, physicochemical, or mechanical processes for the microencapsulation of materials such as dyes, agricultural chemicals, drugs, proteins, food additives, and fragrances are not suitable for the encapsulation of living cells. In addition, to keep the encapsulated cells viable after implantation, the polymer walls must be not only semipermeable and protective, but also biocompatible with both the cells and the host tissues.

Chang at McGill University pioneered the microencapsulation of proteins and cells with nitrocellulose or polyamide (nylon) as artificial cells in 1960s [16]. However, microencapsulation was not carried out under physiological conditions, and the artificial cells were not viable [17]. Microencapsulation of living cells under physiological conditions was reported in 1980 by Lim and Sun, who encapsulated pancreatic islets with calcium alginate, followed by a series of post-treatment steps [18]. Since then alginate, a naturally occurring copolymer of guluronic acid and mannuronic acid, has been widely used in the microencapsulation of living cells under mild conditions, in conjunction with divalent cations or positively charged polymers [19–22]. However, reliance on capsule membranes with alginate has been questioned because of perceived limitations in biocompatibility, reproducibility, and physicochemical stability [23–26].

Since the late 1970s our group has been working on the microencapsulation of cells with synthetic polymers as an alternative approach [27,28]. Synthetic polymers were used for cell microencapsulation because it is relatively straightforward to optimize the properties of synthetic polymers for specific applications by controlling and/or adjusting their structures, compositions, purity, and so on. Several types of polymers with desirable properties have been synthesized, and corresponding processes of microencapsulating cells under mild conditions were devised. Promising results have been obtained both *in vitro* and *in vivo*.

GENERAL MICROENCAPSULATION PROCESSES

Microencapsulation first achieved significant recognition in 1954, when carbonless copy paper was introduced [29]. The microcapsules of color-reactive oils coated on to paper were formed by simple coacervation of gelatin with sodium sulfate [30] or complex coacervation of gelatin with gum arabic [31]. Carbonless copy paper, the first commercial application of microcapsules, is now a $6 billion business annually worldwide, and microcapsules of other materials, such as pesticides, chemical additives, drugs, vaccines, proteins, and enzymes, have been produced on industrial or laboratory scales.

There are more than 200 microencapsulation methods, most of them patented, but all include three basic steps: enclosure of the core, formation of the microcapsule, and hardening of the microcapsule. The methods are generally classified as: chemical, physicochemical, or mechanical processes based on the mechanisms of capsule formation. For detailed information about general microencapsulation processes, readers are referred to a 1999 book edited by Arshady [32].

A chemical process is used to form microcapsules by reaction of wall-forming substances, usually polymerization or polycondensation of monomers. The first step core closure, is achieved by means of dispersion, suspension, or emulsion technology. Capsule formation and hardening happen at the same time, owing to the polymerization or polycondensation of monomers. Although a chemical process is widely employed for the formation of microcapsules of various types, it is apparently not suitable for the microencapsulation of cells because of the toxic monomers, solvents, free radicals, or high temperatures usually involved in the process. In the 1960s Chang used an interfacial polycondensation protocol to form microcapsules of proteins or cells as artificial cells with a polyamide (nylon) shell [33], but viable cells were not obtained. To limit the toxic effects of reactive molecules to cells, viable cells have been embedded in a gel and then encapsulated by polymerization [34,35].

In a physicochemical process, preformed polymers are used instead of monomers, and microcapsule formation results from the physicochemical changes of the system due to solvent removal (extraction/evaporation), cross-linking, or coacervation. The polymers dissolved in the solutions become membranes that are ultimately not soluble in the systems used. Solvent extraction and evaporation methods generally involve the dissolution or dispersion of the core materials in a polymer solution, and the droplets are made by suspension in an immiscible medium (or solvent) or extrusion from orifices. Subsequently, the capsules are formed by means of solvent extraction—for example, by adding the suspension mixture to a polymer precipitant. Solvent extraction is an important method for the microencapsulation of living cells as long as the polymers used are soluble in low-toxicity solvents and can be precipitated in aqueous solutions or other nontoxic precipitants. The polymers and protocols for cell microencapsulation with this method are discussed in the sections that follow.

Microencapsulation by cross-linking is similar to solvent extraction/evaporation in the core closure and droplet formation steps, but the conversion of liquid droplets to the corresponding solid microcapsules results from cross-linking of the polymers. Cross-linking can be achieved by using a cross-linker for reactive polymers, by increasing the temperature for thermally sensitive polymers, or by a combination of both. Although some polymers gel in a physiologically suitable temperature range or can be cross-linked by ions or other suitable agents, the cross-linking method is not very suitable for cell microencapsulation owing to the chemical reaction or temperature. The coacervation method is widely used in microencapsulation, but the actual mechanism is complex. Coacervation is usually accomplished by introducing into the polymer solution a miscible solvent (but a nonsolvent for the polymer) or by mixing a solution of one charged polymer with a solution of an oppositely charged polymer. Since coacervation can be carried out in aqueous solutions under mild conditions, microcapsules of cells are commonly prepared with this method. In fact, cell microencapsulation with alginate is a typical example.

In a mechanical process, microcapsules are formed by methods such as extrusion, spray drying, fluidized-bed coating, and micronization. Both liquids and solid can be microencapsulated mechanically, but the process is mostly suitable for the solid microparticles. Preformed polymers or other materials are used as wall materials. Mechanical processes are of limited suitability in cell microencapsulation because of the relatively drastic conditions (e.g., high temperature, pressure, or vacuum, mechanical force, solvents) usually involved.

SYNTHETIC POLYMERS FOR CELL MICROENCAPSULATION

While the success of microencapsulation is dependent on the process, the properties of a microcapsule are mainly determined by wall material. Since the structures as well as the properties of synthetic polymers are controllable, it is easier to devise synthetic polymers than natural polymers with desirable properties for cell microencapsulation. The polymers for cell microencapsulation must be easily processed under mild conditions, and they must be semipermeable, biocompatible, and stable. Biodegradability is unlikely to be useful for immunoisolation devices. Ideally, the following requirements need to be met:

1. Formed polymers (or monomers) are nontoxic, and soluble in aqueous solutions or low-toxicity solvents.

2. Microencapsulation is carried out under mild conditions that do not involve toxic chemicals, solvents, high or low temperatures, high energies, or physical forces.

3. The resulting capsule walls are permeable to low molecular weight nutrients or metabolized products, and impermeable to high molecular weight antibodies or complement substances (or at least the cells of the host immune system).

4. The polymer walls are compatible with the host tissues and encapsulated cells.

5. The walls are stable after the capsules are implanted.

Table 73.1 lists the synthetic polymers that have been used for cell microencapsulation. Like naturally occurring polyelectrolytes such as alginate, synthetic polyelectrolytes (charged polymers) have been used for cell microencapsulation by coacervation. Charged polymers are usually water soluble and can form coacervates when exposed to the aqueous solutions of oppositely charged polymers. The microencapsulation of cells with synthetic polyelectrolytes was initiated by Gharapetian *et al.* to improve capsule strength and other properties relative to what was then obtainable using alginate–polylysine [36]. Systematic studies with (meth)acrylate-based polyelectrolytes were reported by Stevenson and cowork-

Table 73.1. Synthetic Polymers Used for Cell Microencapsulation

Polymer	Cells encapsulated initially	Year	Ref.
Polyamide	Erythrocytes	1964	[16]
Eudragit polymers	Red blood cells	1983	[45]
HEMA-MMA (75:25) copolymer	CHO cells	1987	[55]
Polyacrylamide	Islet cells	1988	[34]
DMAEMA-MMA polymers	Diploid fibroblasts	1989	[49]
(Meth)acrylate-based polyelectrolytes	Red blood cells	1991	[37]
N-Isopropylacrylamide-based polymers	Islet cells	1996	[44]
HEMA-MMA (90:10) copolymer	CHO cells	2000	[65]
HEMA-BHEAA polymers	CHO cells	2000	[67]

ers [37–40]. It was indicated that synthetic polyelectrolytes were generally more toxic than natural polyelectrolytes, but the toxicity of the former was dependent on molecular weight and could be reduced after coacervation. In the work of Prokop, Hunkeler, and coworkers, 1235 polyelectrolyte combinations were tested, and 47 combinations were found to form sufficiently stable capsules [41–43]. The entrapment of cells in aqueous media under mild conditions is the key advantage of using polyelectrolytes. However, the water solubility of polyelectrolytes may have drawbacks in long-term protection *in vivo* (owing to instability in aqueous media) and with respect to biocompatibility (owing to their surface charge).

Some water-soluble polymers below the critical solution temperature (LCST) become water insoluble upon heating. *N*-Isopropylacrylamide-based polymers, which are thermally sensitive, have been used for the encapsulation of cells [44], but their biocompatability has not been demonstrated.

Water-insoluble polyacrylates have been studied for over 20 years to improve on biocompatibility and potential instability problems posed by water-soluble polymers. Our initial cell microencapsulation was carried out with Eudragit RL (Rohm Pharma), a commercially available copolymer of acrylate and methacrylate monomers; it has 5% quaternary ammonium groups and is soluble in diethyl phthalate [45–47]. The Eudragit RL capsules had a limited permeability and were not biocompatible, presumably owing to their cationic nature [48]. To encapsulate anchorage-dependent cells for *in vitro* purposes, copolymers of dimethylaminoethyl methacrylate (DMAEMA) and methyl methacrylate (MMA) were used [49], since surface charge plays a major role in cell proliferation [50, 51]. Microcapsules of human diploid fibroblasts (HDF) were prepared with the DMAEMA-MMA copolymers dissolved in diethyl phthalate. While cell growth did not occur in intact capsules, cells that leaked from the capsules grew well on the bottom of the culture dish. This observation suggested that microencapsulation did not damage the cells but that the copolymers were impermeable to essential nutrients, presumably because of their low water content (10% in phosphate-buffered saline). The water content of the polymers was increased by introducing methacrylic acid (MAA) into the polymers, and the terpolymer of DMAEMA-MMA-MAA (4.4% MAA) was synthesized with a water content of 27%. A small improvement in cell viability was achieved with the terpolymer, but the terpolymer was not a good substrate for HDF growth (the surface charge had changed). Cell growth in intact capsules of the terpolymer was not observed.

To obtain water-insoluble polyacrylates with better permeability and biocompatibility, the thermoplastic copolymers of hydroxyethyl methacrylate (HEMA) and methyl methacrylate (MMA) with different compositions were synthesized [52]. HEMA and MMA were chosen because poly(HEMA) is widely used as contact lenses and artificial corneas, and poly(MMA) is used as bone cement. We found that the copolymer with the mole fraction of 75% HEMA and 25% MMA was optimal in overall properties [e.g., water insolubility, relatively high water uptake (20–24%), and solubility in poly(ethylene glycol 200), a solvent low in toxicity [53,54]. The copolymer of HEMA-MMA (75:25) has been extensively used for the microencapsulation of various cells with great success [55–62], at least *in vitro*.

The HEMA-MMA capsules were soft, elastic, and tough, and only slight fibrosis of the capsules was observed. However, these polymers have the potential drawback of low permeability to water-soluble nutrients and the added challenge of maintaining cell viability in the presence of nonaqueous solvents. To increase the permeability of the HEMA-MMA copolymers, a new 75:25 copolymer was synthesized from 100% pure HEMA monomer without contaminating cross-linking agents [63]. Such contaminants lead to branched polymers with more limited properties. The purer polymer has higher water content (29%) than the corresponding polymer (21%) made from incompletely purified monomers. In addition, smaller cell microcapsules could be prepared with the purer polymer at the same conditions (because of viscosity differences). Other copolymers that are soluble in iodinated X-ray contrast agents (aqueous organic solutions) have also been synthesized [64,65].

The concerns over diffusion limitations (and potentially insufficient biocompatibility) of HEMA-MMA copolymers had led us to explore other water-insoluble polymers with higher water contents. For example, a PMMA-PEG-PMMA triblock copolymer with expected high hydrophilicity was synthesized, but the difficulties in isolating the polymer prevented us from testing it [66]. Copolymers of HEMA and *N*-benzyl-*N*-hydroxylethyl acrylamide (BHEAA) have been synthesized, and promising results were obtained [67]. In addition, water-insoluble polymers with water contents over 60% were recently invented in our laboratories, and preliminary results show that the polymers are more biocompatible and permeable than HEMA-MMA copolymers. Other features of the polymers include solubility in solvents low in toxicity, nonstickiness, and transparency.

CELL MICROENCAPSULATION PROTOCOLS
HEMA-MMA COPOLYMERS

The protocols of cell microencapsulation with HEMA-MMA copolymers were initially devised in 1987 and modified in later practice [68]. The principle of the method is solvent extraction, in which the solvents for the HEMA-MMA polymers are extracted with water or aqueous solutions to leave the water-insoluble polymers as membranes around cells. HEMA-MMA (75:25) polymers are soluble in PEG-200 (MW 200), a water-soluble, relatively low-toxicity solvent, while HEMA-MMA (90:10) polymers are soluble in the aqueous solutions of X-ray contrast agents such as iopamidol [65]. Cells are usually dispersed in the culture medium with or without an immobilization gel. Coaxial double needles are used to extrude the cell suspension and the polymer solutions. The cells extruded with the inner needle are enclosed within the droplets of the polymer solutions at the tips of the needles, so that the contact time between the cells and the polymer solutions is minimized. When the droplets of the polymer solutions with cells inside reach water or an aqueous buffer solution such as phosphate-buffered saline (PBS) as a nonsolvent for the polymers, the solvents (PEG-200 or iopamidol solution) are extracted into the nonsolvent, leaving the water-insoluble polymers as capsule walls. There are a few ways to shear off the polymer droplets, all of which influence the shape, size, and even the properties of the microcapsules. Initially, the droplets were sheared off from the tips of the needles with a coaxial airstream [55], but this method was replaced by the practice of oscillating the coaxial needle assembly in the vertical plane at a hexadecane liquid–air interface [56]. Now the flow of hexadecane through a coaxial nozzle (a third coaxial channel) is used to shear off the droplets with a higher shearing force [69]. In the present setup, smaller and more uniform capsules are prepared than with the earlier methods. A typical protocol of cell microencapsulation with HEMA-MMA copolymers follows.

HEMA-MMA copolymers were synthesized by solution polymerization, and a polymer solution (10%, w/v) was prepared by dissolving the polymer in PEG-200. The cells [e.g., Chinese hamster ovary (CHO)] were suspended in minimum essential medium (α–MEM) supplemented with 10% fetal bovine serum (FBS) and 15% Ficoll-400.

The microencapsulation was done in a flow hood under sterile conditions. The needle assembly (Fig. 73.1) was placed over a receiving dish containing 500 ml of sterile PBS with 150 ppm Pluronic L101 (surfactant) and about 500 ml of hexadecane overlaid on the PBS. An overhead magnetic stirring bar was operated in the receiving dish, and the tip of needle assembly just touched the upper level of the hexadecane. All the materials were autoclaved or gas-sterilized, or used as sterile. The cell solution–suspension and the polymer solution were coextruded through the coaxial needle assembly, and the droplets were sheared off by the flow of hexadecane. The flow rates of core solution–suspension and polymer solution were 5.5 and 12.5 μl/min, respectively. Hexadecane flowed at the rate of 120 ml/min. About 60 droplets were produced per minute with these flow rates. The capsules were further stirred for 30 min in the receiving dish and transferred into fresh PBS for washing for another half-hour. The encapsulated cells were transferred into and maintained in the α-MEM medium in a 100-mm tissue culture dish at 37°C, and the medium was changed once a week. For more detailed information, we recommend that readers refer to an earlier review [68], which also details the methods for using agarose, collagen, or Matrigel as a cell immobilization material within HEMA-MMA.

MICROENCAPSULATION WITH OTHER POLYACRYLATES

Encapsulation of cells with (meth)acrylate-based polyelectrolytes was accomplished through complex coacervation [37]. Cells were suspended in a polyelectrolyte polymer so-

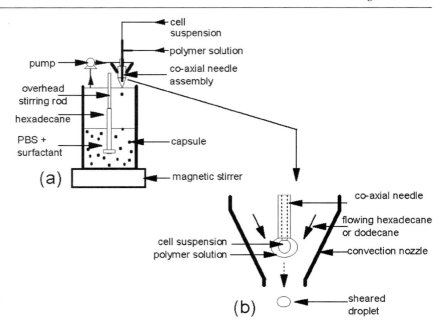

Fig. 73.1. (a) Microencapsulation with (b) the needle assembly.

lution, and the mixture was extruded into a receiving bath containing the solution of an oppositely charged polymer. Capsules were formed and fabricated in the receiving solution by coacervation for 10–30 min, then transferred into a medium for maintenance or tests. Detailed information can be found elsewhere [37–40].

Besides HEMA-MMA polymers, other water-insoluble polyacrylates such as Eudragit and DMAEMA-MMA have been used in the microencapsulation of cells. The process with these polymers was quite similar to that with HEMA-MMA, which was described earlier. The differences for DMAEMA-MMA, for example, included the solvents, the means of shearing the polymer droplets, and the nonsolvents. Detailed protocols for microencapsulation with these particular polymers are given elsewhere [45,49].

EVALUATION OF CELL CAPSULES

CAPSULES

Cell capsules are in the range of 100–2000 μm (in most cases, 300–1000 μm). A microscope with eyepiece reticules (after calibration) is usually used to determine capsule size. Capsule morphology is normally assessed with scanning electron microscopy (SEM), which can provide information about the porosity and thickness of a capsule wall as well as the eccentricity [70].

Mechanical properties including toughness, brittleness, elasticity, and stickiness are important but difficult to measure, and there are few standard methods for determining these [71,72]. While force–displacement curves can be determined under compression, these curves cannot be related to fundamental capsule properties.

CAPSULE WALLS

For capsule permeability, a particular solute is enclosed in the capsules, and the amount of the solute diffusing from the capsules is measured at different times to calculate the mass diffusion rate [56]. Alternatively, empty capsules can be loaded with the test solute and its disappearance from the incubation solution or its leakage back into an initially solute-free solution can be measured. In another approach, proteins with different sizes are encapsulated and the amount of proteins released at equilibrium is determined to estimate the molecular weight cutoff [73].

Assessment of biocompatibility is obviously crucial, and implant studies are critical [74–76]. Histopathological evaluation of the tissue surrounding the explant is the standard method of evaluation, although there are also several *in vivo* markers and *in vitro* tests that can be done. Result interpretation is very important, but oversimplistic analysis is typical. For more information, readers are directed to the extensive literature on biocompatibility and a late—1990s review on host response to tissue-engineered devices [77].

Physicochemical stability refers to the overall stability of capsules *in vivo*. The capsule walls must be mechanically strong, water insoluble, not hydrolyzable, and not (bio)degradable with chemicals and enzymes.

Encapsulated Cells

The MTT assay is one of many used to examine the viability of encapsulated cells. Only living cells are able to convert MTT (3-(4,5-dimethylthiazol-2-yl)-2,5-diphenyltetrazolium bromide) to formazan by the action of succinate dehydrogenase. The amount of formazan formed is related to the biological activity of the cells and their number [78].

Fluorescent viability can be used to identify living or dead encapsulated cells with calcein-AM and ethidium homodimer **1** stains. While red fluorescent ethidium homodimer **1** permeates only the compromised membrane of unhealthy cells and stains their nuclear DNA, calcein-AM is converted by living cells into a product that fluoresces green [79].

Proliferation of encapsulated cells is examined by determining the average number of cells in the capsules at different times. Capsules are cut open to release/dissociate the inside cells, and the cell numbers are counted with a hemocytometer [80].

Functional ability is determined by measuring the amount of specific products metabolized by particular cells. For example, insulin is measured with encapsulated islets [81].

CONCLUDING REMARKS

Because of the diversity of structures, as well as procedural flexibility, several synthetic polymers have been developed for cell microencapsulation. Although the copolymers of HEMA-MMA (hydroxymethyl methacrylate–methyl methacrylate) are the most common synthetic polymers for cell microencapsulation, other water-insoluble, uncharged polymers have also been synthesized. Microencapsulation of cells with synthetic polymers is usually performed by interfacial precipitation to minimize the problems associated with toxic solvents or chemicals, while cell microencapsulation with some synthetic polymers can also be done in aqueous solutions.

REFERENCES

1. Uludag, H., Kharlip, L., and Sefton, M. V. (1993). Protein delivery by microencapsulated cells. *Adv. Drug Delivery Rev.* **10**, 115–130.
2. Babensee J., and Sefton, M. V. (1996). Polyacrylate microcapsules for cell delivery. *In* "Microparticulate Systems for the Delivery of Proteins and Vaccines" (S. Cohen and H. Bernstein, eds.), pp. 477–519. Dekker, New York.
3. Li, R. H. (1998). Materials for immunoisolated cell transplantation. *Adv. Drug Delivery Rev.* **33**, 87–109.
4. Soon-Shiong, P., Heintz, R., Merideth, N., Yao, Q., Yao, Z., Zheng, T., Murphy, M., Moloney, M., Schmehl, M., Harris, M., Mendez, R., and Sandford, P. (1994). Insulin independence in type 1 diabetic patient after encapsulated islet transplantation. *Lancet* **343**, 950–951.
5. Sun, A. M., Lim, F., Van Rooy, H., and O'Shea, G. (1981). Long-term studies of microencapsulated islets of Langerhans: A bioartificial endocrine pancreas. *Artif. Organs* **5**, 784.
6. Roberts, T., De Boni, U., and Sefton, M. V. (1996). Dopamine secretion by PC12 cells microencapsulated in a hydroxyethyl methacrylate–methyl methacrylate copolymer. *Biomaterials* **17**, 267–275.
7. Aebischer, P., Goddard, M., Signore, A., and Timpson, R. (1994). Functional recovery in hemiparkinsonian primates transplanted with polymer-encapsulated PC12 cells. *Exp. Neurol.* **126**, 151–158.
8. Winn, S. R., Hammang, J. P., Emerich, D. F., Lee, A., Palmiter, R. D., and Baetge, E. E. (1994). Polymer-encapsulated cells genetically modified to secrete human nerve growth factor promote the survival of axotomized septal cholinergic neurons. *Proc. Natl. Acad. Sci. U.S.A.* **91**, 2324–2328.
9. Emerich, D., Lindner, M., Winn, S., Chen, E., Frydel, B., and Kordower, J. (1996). Implants of encapsulated human CNTF-producing fibroblasts prevent behavioral deficits and striatal degeneration in a rodent model of Huntington's disease. *J. Neurosci.* **16**, 5168–5181.
10. Fu, X. W., and Sun, A. M. (1989). Microencapsulated parathyroid cells as a bioartificial parathyroid. In vivo studies. *Transplantation* **47**, 432–435.
11. Hasse, C., Schrezenmeir, J., Stinner, B., Schark, C., Wagner, P. K., Neumann, K., and Rothmund, M. (1994). Successful allotransplantation of microencapsulated parathyroids in rats. *World J. Surg.* **18**, 630–634.
12. Dixit, V., Darvasi, R., Arthur, M., Brezina, M., Lewin, K., and Gitnick, G. (1990). Restoration of liver function in Gunn rats without immunosuppression using transplanted microencapsulated hepatocytes. *Hepatology* **12**, 1342–1349.
13. Dixit, V., and Gitnick, G. (1995) Transplantation of microencapsulated hepatocytes for liver function replacement. *J. Biomater. Sci. Polymer Edn* **7**, 343–357.

14. Sagen, J., Wang, H., Tresco, P., and Aebischer, P. (1993). Transplants of immunologically isolated xenogeneic chromaffin cells provide a long-term source of pain-reducing neuroactive substances. *J. Neurosci.* **13**, 2415–2423.

15. Liu, H. W., Ofosu, F. A., and Chang, P. L. (1993). Expression of human factor IX by microencapsulated recombinant fibroblasts. *Hum. Gene Ther.* **4**, 291–301.

16. Chang, T. M. S. (1964). Semipermeable microcapsules. *Science* **146**, 524–525.

17. Chang, T. M. S. (1965). Semipermeable microcapsules. Ph.D. Thesis, McGill University, Montreal, Canada.

18. Lim, F., and Sun, A. M. (1980). Microencapsulated islets as a bioartificial pancreas. *Science* **210**, 908.

19. Winn, S. R., and Tresco, P. A. (1994). Hydrogel applications for encapsulated cellular transplants. *Methods Neurosci.* **21**, 387–402.

20. Grohn, P., Klock, G., Schmitt, J., Zimmerman, U., Horcher, A., Bretzel, R. G., Hering, B. J., Brandhorst, H., Zekorn, T., and Federlin, K. (1994). Large-scale production of Ba^{2+} alginate coated islets of Langerhans for immunoisolation. *Exp. Clin. Endocrinol.* **102**, 380–387.

21. Peirone, M., Ross, C. J. D., Hortelano, G., Brash, J. L., and Chang, P. L. (1998) Encapsulation of various recombinant mammalian cell types in different alginate microcapsules. *J. Biomed. Mater. Res.* **42**, 587–596.

22. Chang, S. J., Lee, C. H., and Wang, Y. J. (1999). Microcapsules prepared from alginate and a photosensitive poly(L-lysine). *J. Biomater. Sci., Polymer Edn* **10**, 531–542.

23. Weber, C. J., Zabinski, S., Koschitzky, T., Wicker, L., Rajotte, R., D'Agati, V., Peterson, L., Norton, J., and Reemtsma, K. (1990). The role of CD4+ helper T cells in the destruction of microencapsulated islet xenografts in NOD mice. *Transplantation* **49**, 396–404.

24. Wijsman, J., Atkinson, P., Mazaheri, R., Garcia, B., Paul, T., Vose, J., O'Shea, G., and Stiller, C. (1992). Histological and immunopathological analysis of recovered encapsulated allogenic islets from transplanted diabetic BB/W rats. *Transplantation* **54**, 588–592.

25. Lum, Z. P., Krestow, M., Tai, I. T., Vacek, I., and Sun, A. M. (1992). Xenografts of rat islets into diabetic mice: An evaluation of new smaller capsules. *Transplantation* **53**, 1180–1183.

26. Dixit, V., and Gitnick, G. (1995). Transplantation of microencapsulated hepatocytes for liver function replacement. *J. Biomater. Sci. Polymer Edn* **7**, 343–357.

27. Stevenson, W. T. K., and Sefton, M. V. (1993). Development of polyacrylate microcapsules. *In* "Fundamentals of Animal Cell Encapsulation and Immobilization" (M. F. A. Goosen, ed.), pp. 143–181. CRC Press, Boca Raton, FL.

28. Sefton, M. V. (1982). Encapsulation of live animal cells. U.S. Pat. 4,353,888.

29. Green, B. K. (1979). History and principles of microencapsulation. *In* "Microencapsulation: New Techniques and Application" (T. Kondo, ed.), pp. 1–9. Techno Inc., Tokyo.

30. Green, B. K. (1957). Oil containing microscopic capsule and method for making them. U.S. Pat. 2,800,458.

31. Green, B. K., and Schleicher, L. (1956). Manifold record materials. U.S. Pat. 2,730,456.

32. Arshady, R. (1999). Manufacturing methodology of microcapsules. *In* "Microspheres, Microcapsules and Liposomes" (R. Arshady, ed.), Vol. 1, pp. 279–326. Citus Books, London.

33. Chang, T. M. S. (1985). Biomedical applications of artificial cells containing immobilized enzymes, proteins, cells, and other biologically active materials. *In* "Enzymes and Immobilized Cells in Biotechnology" (A. L. Laskin, ed.), Chapter 11. Benjamin/Cummins, Menlo Park, CA.

34. Dupuy, B., Gin, H., Baquey, C., and Ducassou, D. (1988). In situ polymerization of a microencapsulating medium round living cells. *J. Biomed. Mater. Res.* **22**, 1061–1070.

35. Dupuy, B., Cadic C., Gin, H., Baquey, C., Dufy, B., and Ducassou, D. (1991). Microencapsulation of isolated pituitary cells by polyacrylamide microlatex coagulation on agarose beads. *Biomaterials* **12**, 493–496.

36. Gharapetian, H., Davies, N. A., and Sun, A. M. (1986). Encapsulation of viable cells within polyacrylate membranes. *Biotechnol. Bioeng.* **28**, 1595–1600.

37. Wen, S., Yin, X., and Stevenson, W. T. K. (1991). Microcapsules through polymer complexation. I. By complex coacervation of polymers containing a high charge density. *Biomaterials* **12**, 374–384.

38. Wen, S., Yin, X., Stevenson, W. T. K., and Alexander, H. (1991). Microcapsules through polymer complexation. II. By complex coacervation of polymers containing a low charge density. *Biomaterials* **12**, 479–488.

39. Wen, S., Alexander, H., Inchikel, A., and Stevenson, W. T. K. (1995). Microcapsules through polymer complexation. Part III. Encapsulation and culture of human Burkitt lymphoma cells in vitro. *Biomaterials* **16**, 325–335.

40. Dyakonov, T. A., Zhou, L., Wan, Z., Huang, B., Meng, Z., Guo, X., Alexander, H., Moore, W. V., and Stevenson, W. T. K. (1997). Synthetic strategies for the preparation of precusor polymers and of microcapsules suitable for cellular entrapment by polyelectrolyte complexation of those polymers. *Ann. N.Y. Acad. Sci.* **381**, 72–85.

41. Hunkeler, D., Prokop, A., Power, A., Haralson, M., Dimari, S., and Wang, T. G. (1997). A screening of polymers as biomaterials for cell encapsulation. *Polymer News* **22**, 232–240.

42. Prokop, A., Hunkeler, D., DiMari, S., Haralson, M. A., and Wang, T. G. (1998). Water soluble polymers for immunoisolation. I. Complex coacervation and cytotoxicity. *Adv. Polym. Sci.* **136**, 1–52.

43. Prokop, A., Hunkeler, D., Power, A., Whiteesell, R. R., and Wang, T. (1998). Water soluble polymers for immunoisolation II: evaluation of multicomponent microencapsulation systems. *Adv. Polym. Sci.* **136**, 53–73.

44. Shimizu, S., Yamazaki, M., Kubota, S., Ozasa, T., Moriya, H., Kobayashi, K., Mikami, M., Mori, Y., and Yamaguchi, S. (1996). In vitro studies on new method for islet microencapsulation using a thermoreversible gelation polymer, N-iso-propylacrylamide-based copolymer. *Artif. Organs* **20**, 1232–1237.

45. Sefton, M. V., and Broughton, R. L. (1983). Microencapsulation of erythrocytes. *Biochim. Biophys. Acta* **717**, 473–477.

46. Sefton, M. V., Dawson, R. M., Broughton, R. L., Blysniuk, J., and Sugamori, M. E. (1987). Microencapsulation of mammalian cells in a water insoluble polyacrylate by coextrusion and interfacial precipitation. *Biotechnol. Bioeng.* **29**, 1135–1143.

47. Sugamori, M., and Sefton, M. V. (1989). Microencapsulation of pancreatic islets in a water insoluble polyacrylate. *Trans. ASAIO* **35**, 791–799.

48. Broughton, R. L., and Sefton, M. V. (1989). Effect of capsule permeability on growth of CHO cells in Eudragit RL microcapsules: Use of FITC dextran as a marker of capsule quality. *Biomaterials* **10**, 462–465.

49. Mallabone, C. L., Crooks, C. A., and Sefton, M. V. (1989). Microencapsulation of human diploid fibroblasts in cationic polyacrylates. *Biomaterials* **10**, 380–386.

50. Rosen, J. J., Gibbons, D. F., and Culp, L. A. (1975). Fibroblast interactions with hydrogel surfaces. *Polym. Prepr. Am. Chem. Soc., Dis. Polym. Prepr.* **21**, 553.

51. Reuveny, S., Silberstein, L., Shahar, A., Freeman, E., and Mizraki, A. (1982). DE-52 and DE-53 cellulose microcarriers. I. Growth of primary and established anchorage-dependent cells. *In Vitro* **18**(2), 92–99.

52. Stevenson, W. T. K., Evangelista, R., and Sefton, M. V. (1987). Preparation and characterization of thermoplastic polymers from hydroxyalkyl methacrylates. *J. Appl. Polym. Sci.* **34**, 65–83.

53. Douglas, J. A., and Sefton, M. V. (1990). The permeability of Eudragit RL and HEMA-MMA microcapsules to glucose and insulin. *Biotechnol. Bioeng.* **36**, 653–664.

54. Stevenson, W. T. K., and Sefton, M. V. (1988). The equilibrium water content of some thermoplastic hydroxyalkyl methacrylate polymers. *J. Appl. Polym. Sci.* **36**, 1541–1553.

55. Dawson, R. M., Broughton, R. L., Stevenson, W. T. K., and Sefton, M. V. (1987). Microencapsulation of CHO cells in a hydroxyethyl methacrylate–methyl methacrylate copolymer. *Biomaterials* **8**, 360–366.

56. Crooks, C. A., Douglas, J. A., Broughton, R. L., and Sefton, M. V. (1990). Microencapsulation of mammalian cells in a HEMA-MMA copolymer: Effects on capsule morphology and permeability. *J. Biomed. Mater. Res.* **24**, 1241–1262.

57. Sefton, M. V., Uludag, H., Babensee, J., Roberts, T., Horvath, V., and De Boni, U. (1994). Microencapsulation of cells in a thermoplastic copolymer (hydroxyethyl methacrylate–methyl methacrylate). *J. Neurosci.* **21**, 371–386.

58. Tse, M., Uludag, H., Sefton, M. V., and Chang, P. L. (1996). Secretion of recombinant proteins from hydroxyethyl methacrylate–methyl methacrylate capsules. *Biotechnol. Bioeng.* **51**, 271–280.

59. Babensee, J. E., Cornelius, R. M., Brash, J. L., and Sefton, M. V. (1998). Immunoblot analysis of proteins associated with HEMA-MMA microcapsules: Human serum proteins in vitro and rat proteins following implantation. *Biomaterials* **19**, 839–849.

60. May, M. H., and Sefton, M. V. (1999). Conformal coating of small particles and cell aggregates at a liquid–liquid interface. *Ann. N.Y. Acad. Sci.* **875**, 126–134.

61. Sefton, M. V., May, M. H., Lahooti, S., and Babensee, J. E. (2000). Making microencapsulation work: Conformal coating, immobilization gels and in vivo performance. *J. Controlled Release* **65**, 173–186.

62. Lahooti, S., and Sefton, M. V. (2000). Effect of an immobilization matrix and capsule membrane permeability on the viability of encapsulated HEK cells. *Biomaterials* **21**, 987–95.

63. Feng, M., and Sefton, M. V. (2000). Hydroxyethyl methacrylate–methyl methacrylate (HEMA-MMA) copolymers for cell microencapsulation: Effect of HEMA purity. *J. Biomater. Sci. Polymer Edn* **11**, 537–545.

64. Morikawa, N., Iwata, H., Matsuda, S., Miyazaki, J., and Ikada, Y. (1997). Encapsulation of Mammalian cells into synthetic polymer membranes using least toxic solvents. *J. Biomater. Sci. Polymer Edn* **8**, 575–586.

65. Feng, M., and Sefton, M. V. (2000). Polymers that are soluble in X-ray contrast agents and their use in cell immunoisolation. *6th World Biomater. Congr.*, Hawaii, *2000*, Abstr. 566.

66. Eisa, T., and Sefton, M. V. (1993). Towards the preparation of a MMA-PEO block copolymer for the microencapsulation of mammalian cells. *Biomaterials* **14**, 755–761.

67. Feng, M., and Sefton, M. V. (2000). Development of a novel copolymer (HEMA-BHEAA) for cell microencapsulation. *6th World Biomater. Congr.*, Hawaii, *2000*, Abstr. 1419.

68. Lahooti, S., and Sefton, M. V. (1998). Methods for microencapsulation with HEMA-MMA. *Methods Mol. Med.* **18**, 331–348.

69. Uludag, H., Horvath, V., Black, J. P., and Sefton, M. V. (1994). Viability and protein secretion from human hepatoma (HepG2) cells encapsulated in 400 μm polyacrylate microcapsules by submerged nozzle-liquid jet extrusion. *Biotechnol. Bioeng.* **44**, 1199–1204.

70. Babensee, J., and Sefton, M. V. (1997). Protein delivery by microencapsulated cells. *In* "ACS Professional Reference Book. Controlled Drug Delivery Challenges and Strategies" (K. Park, ed.), pp. 311–332. American Chemical Society, Washington, DC.

71. Lee, C. S., and Chu, I. M. (1997). Characterization of modified alginate–poly-L-lysine microcapsules. *Art. Organs* **21**, 1002–1006.

72. Matthew, H. W., Sally, S. O., Peterson, W. D., and Klein, M. D. (1993). Complex coacervate microcapsules for mammalian cell culture and artificial organ development. *Biotechnol. Prog.* **9**, 510–519.

73. Sefton, M. V., Hwang, J. R., and Babensse, J. E. (1997). Selected aspects of the microencapsulation of mammalian cells in HEMA-MMA. *Ann. NY. Acad. Sci.* **831**, 260–270.

74. De Vos, P., Wolters, G. H., Fritschy, W. M., and Van Schilfgaarde, R. (1993). Obstacles in the application of microencapsulation in islet transplantation. *Int. J. Artif. Organs* **16**, 205–212.

75. Soon Shiong, P., Otterlei, M., Skjak-Brek, G., Smidsrod, O., Heintz, R., Lanza, R. P., and Espevik, T. (1991). An immunological basis for the fibrotic reaction to implanted microcapsules. *Transplant. Proc.* **23**, 758–759.

76. Campioni, E. G., Nobrega, J. N., and Sefton, M. V. (1998). HEMA/MMA microcapsule implants in hemiparkinsonian rat brain: Biocompatibility assessment using [^{3}H]PK11195 as a marker for gliosis. *Biomaterials* **19**, 829–837.

77. Babensee, J. E., Anderson, J. M., McIntire, L. V., and Mikos, A. G. (1998). Host response to tissue engineered devices. *Adv. Drug Delivery Rev.* **33**, 111–139.

78. Uludag, H., and Sefton, M. V. (1993). Metabolic activity and proliferation of CHO cells in hydroxyethyl methacrylate–methyl methacrylate (HEMA-MMA) microcapsules. *Cell Transplant.* **2**, 175–182.

79. Lahooti, S. (1999). In vitro and in vivo characterization of genetically engineering cells microencapsulated in a HEMA-MMA copolymer. Ph.D. Thesis, University of Toronto, Toronto, Canada.

80. Uludag, H., and Sefton, M. V. (1993). Microencapsulated human hepatoma (HEPG2) cells: In vitro growth and protein release. *J. Biomed. Mater. Res.* **27**, 1213–1224.

81. Sefton, M. V., and Kharlip, L. (1994). Insulin release from rat pancreatic islets microencapsulated in a HEMA-MMA polyacrylate. *In* "Pancreatic Islet Transplantation" (R. P. Lanza and W. L. Chick, eds.), Vol. 3, pp. 107. R. G. Landes, Austin, TX.

CHAPTER 74

MICROENCAPSULATION METHODS: POLY(VINYL ALCOHOL) (PVA)

Brian C. Anderson, Aurelia Amanda, and Surya K. Mallapragada

The use of poly(vinyl alcohol) (PVA) for microencapsulations related to tissue engineering applications is potentially very promising. The biocompatibility of PVA has already made this polymer the material of choice for several biological applications. The slow dissolution of the polymer has served well in the development of delivery devices for sustained release of bioactive agents. Because of the harsh chemicals or low temperatures required to cross-link PVA, its use in the microencapsulation of living cells has been rather restrained. In several thoroughly investigated applications, however, a previously cross-linked microparticle can be loaded with a bioactive molecule. PVA has also been used in microencapsulation technology as an emulsifier or stabilizer at low concentrations for microparticles of several polymers.

BACKGROUND

Poly(vinyl alcohol) (PVA) was first synthesized by Herrmann and Haehnel in 1924 [1]. It was synthesized by polymerization of vinyl acetate monomers into poly(vinyl acetate), followed by the hydrolysis of the acetyl groups into poly(vinyl alcohol). The applications of PVA in the biomedical area include drug delivery devices, hemodialysis membranes, soft contact lenses, and membranes for bioseparations. The use of PVA in drug delivery devices has been studied extensively [2]. PVA is hydrophilic, nontoxic, and easy to process. It has good pH stability and excellent mechanical properties, and it exhibits low protein and cell adhesion. PVA is of particular interest for biomedical applications because of its semipermeability and biocompatibility. The semipermeability of PVA hydrogels allows transport of oxygen and nutrients that are necessary for cell survival as well as for the removal of wastes secreted by the cells. Meanwhile, this semipermeable membrane also creates a barrier for macromolecules from the recipient immune system such as immunoglobulins and immune cells [3]. PVA is also used as an emulsifier in various microcapsule preparations [4], and it has a slow dissolution rate for long-term applications. Some of the disadvantages of PVA are the slow dissolution rate for short-term applications and the need to use cross-linking agents to improve its mechanical strength. The use of cross-linking agents has restrained the applications of PVA for cell microencapsulation because most cross-linking agents are hazardous to living cells.

BIOCOMPATIBILITY

Biocompatibility is one of the most important factors to be considered in cell encapsulation. PVA is a good choice for microencapsulation because of its biocompatibility. PVA loaded with heparin was used as antithrombin coating on biomaterials, and the device was shown to be capable of preventing blood clotting [5,6]. Noguchi and coworkers implanted

PVA hydrogels as artificial articulate cartilage in the knee joints of male adolescent white rabbits [7]. The implantation caused only mild inflammation, which stopped a couple of weeks after the surgery. PVA was grafted with methyl methacrylate and acrylamide, and it was found to have lower thrombogenicity and better cell attachment than unmodified PVA [8]. PVA was also studied for potential use in soft contact lenses [9]. It was claimed that PVA had better mechanical strength and less protein adsorption than the commercial soft contact lenses, and there was no change in the corneal epithelium of rabbit eyes wearing the soft contact lenses. The platelet adhesion on PVA was increased as the water content of the hydrogels decreased by heat treatment [10]. *In vitro* and *ex vivo* studies by Fujimoto and coworkers showed lower protein adsorption and platelet adhesion for PVA heat-treated in the presence of glycerol [11]. Membranes for hemodialysis made of nylon 4 were coated with PVA to improve the blood compatibility of the membranes [12]. Hemodialysis PVA membranes loaded with acetylsalicylic acid also showed good resistance to platelet adhesion [13]. Burczak and coworkers used PVA to encapsulate islets of Langerhans in macro bags to make a hybrid-type artificial pancreas [14]. Long-term *in vivo* studies showed that slower solute diffusion resulted from modifications of the hydrogels. They also showed that a chronic inflammatory response from long-term implantation was induced by cellular enzyme activities.

CELLULAR ENCAPSULATION

Although PVA is ruled out for some applications because of the harsh chemical cross-linking agents, often aldehydes, used in making it into microcapsules for cellular encapsulation, there is still a place for PVA in encapsulation technologies. The biocompatibility of PVA and a slow dissolution rate make it an ideal candidate for applications in which living cells need to be trapped inside an encapsulation matrix to secrete proteins *in vivo*. Applications where the cells can be incorporated into a polymer matrix after cross-linking avoid contact between of aldehydes and living cells. Two such uses discussed here are dopamine secretion for combating Parkinson's disease and the encapsulation of islets of Langerhans for artificial pancreases [14–17].

The use of a PVA matrix has been tested and found to be an efficient vehicle for cate-cholamine release *in vivo* [15]. A PVA foam scaffold was used in hollow-fiber membranes to trap dopamine-secreting PC12 cells. A scaffold is needed to prevent the cells from aggregating into large clusters, thus depriving the center cells of necessary nutrients. Commercial PVA sponges, formed by cross-linking with formaldehyde vapor aeration and having an average pore size of about 150 μm, were used under aqueous swollen conditions. The sponges were seeded with collagen and cut into the appropriate size and shape. The PC12 cells were then seeded after sterilization by injection into the polymer matrix. *In vitro* and *in vivo* studies were performed to evaluate the release of basal L-dopa versus a similar matrix made with chitosan. The release of basal L-dopa was much faster with the PVA matrix (Fig. 74.1).

Another *in vivo* application that has been investigated is the use for entrapment of islets of Langerhans. Several methods have been used to investigate entrapment matrices of PVA for this application. Burczak *et al.* have formed a macrocapsule of PVA by cross-linking the polymer with glutaraldehyde in the presence of hydrochloric acid as a catalyst [14]. The entrapment matrix was then entrapped in a poly(ethylene terephthalate) coating. The transport of large molecules through the matrix was found to be hindered over time when implanted in rats, but the transport of smaller molecules, like glucose, was not affected by implantation. Tamura *et al.* have produced a similar macrocapsule using a freeze–thaw technique to cross-link the polymer without irradiation [16]. Lozinsky *et al.* have also used a freeze–thaw technique to make polymer entrapment cylinders that were examined with several model cells and microorganisms [17]. The primary focus of this study was the mechanical properties of the matrix.

BIOACTIVE MOLECULE RELEASE

In another use of PVA encapsulation in bioengineering, polymer microcapsules deliver bioactive molecules into a specific area over time. This release can be achieved by loading

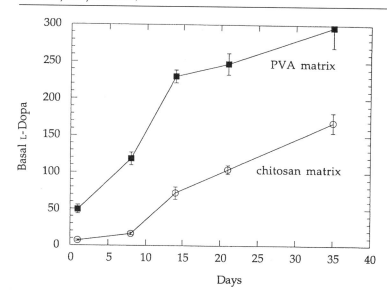

PVA microspheres with the appropriate proteins or hormones and allowing them to release the molecules at the implantation site. This approach could be used for tissue engineering, where growth factors or hormones must be applied to insure growth or biocompatibility. Both cross-linked and non-cross-linked PVA microcapsules have been investigated.

An example of cross-linked PVA microspheres for enzyme delivery is found in study by Bachtsi and Kiparissides [18]. PVA microspheres were made by cross-linking with glutaraldehyde in the catalytic presence of sulfuric acid. The microspheres were washed thoroughly and loaded with protease by soaking the particles in a protease solution for 2 days. The loaded microspheres were then tested for their release kinetics (Fig. 74.2). A higher degree of cross-linking (10–20%) led to a slower, more uniform release, while a lower degree of cross-linking (~1%) led to more enzyme delivery.

Some efforts have been made to form microspheres without cross-linking agents, to avoid the use of these toxic materials [19]. The particles were formed by means of repeated freezing and thawing of a two-phase oil–water mixture in the presence of sodium lauryl sulfate. Control of the particle size was attained by controlling the ratio of oil phase and the amount of PVA. The particles were loaded with bovine serum albumin by adding BSA to the PVA solution before the particle formation. The absence of an aldehyde cross-linking agent allows for the addition of the enzyme before the particles are formed. The release of BSA from these microspheres was then studied (Fig. 74.3).

Microspheres have also been made in which the PVA is a thin membrane coating a small amount of a drug [20]. A PVA coating was spray-dried onto microspheres of indomethacin

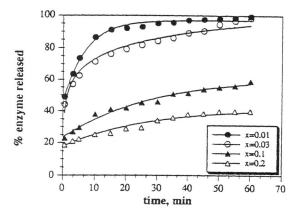

Fig. 74.2. Enzyme release from cross-linked PVA microparticles with different degrees of crosslinking. From J. Microencapsulation **12**(1), 23–35, Bachtsi, A. R., and Kiparissides, C., Taylor & Francis, Philadelphia, PA [18]. Reproduced with permission. All rights reserved.

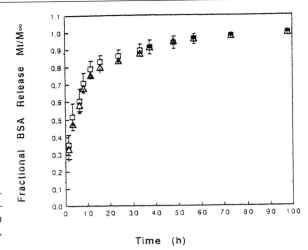

Fig. 74.3. Fractional release of bovine serum albumin from PVA microparticles prepared by freezing and thawing technique with diameters of 300–500 μm (□), 500–710 μm (△), and 850–1400 μm (+). Reprinted from Journal of Controlled Release **27**, Ficek and Peppas [19]. Copyright 1993, with permission from Elsevier Science.

to increase the dissolution of the drug into an aqueous solution. Because indomethacin is relatively insoluble in water, it release rate into an aqueous solution is slow. The addition of a PVA coating greatly accelerated this release (Fig. 74.4).

USE AS A STABILIZER

Probably the most widely used application of PVA in encapsulation technology is as a stabilizer or emulsifier. Many researchers have used PVA for this purpose in making microparticles of poly(D,L-lactic acid–co-glycolic acid) [4,21,22], locust bean gum [23], polystyrene [24,25], ethyl cellulose [26], polycarbonate [27], and polyterephthalamide [28]. Although PVA is only an indirect participant in these encapsulations, its use is quite advantageous to the overall encapsulation procedure because of its biocompatibility and nontoxic nature.

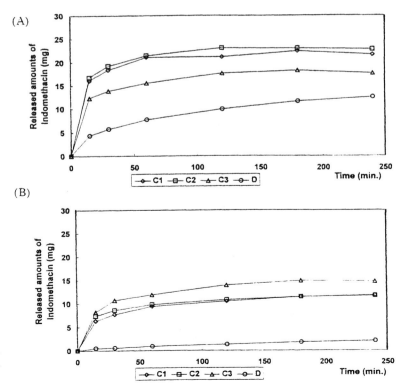

Fig. 74.4. Indomethacin release from microparticles uncoated and coated with PVA prepared by spray-drying technique: (A) in phosphated buffer, pH 7.4, and (B) in distilled water. C1, C2, C3, and D are 0.1, 0.2, 0.5, and 0% PVA solution, respectively. From J. Microencapsulation **16**(3), 315–324, Ermis, D., and Yuksel, A., Taylor & Francis, Philadelphia, PA [20]. Reproduced with permission. All rights reserved.

REFERENCES

1. Winkler, H. (1973). Historical development of polyvinyl alcohol. *In* "Polyvinyl Alcohol" (C. A. Finch, ed.), Vol. 1, pp. 1–15. Wiley, London.

2. Mallapragada, S. K., and McCarthy-Schroeder, S. (2000). Poly(vinyl alcohol) as a drug delivery carrier. *In* "Handbook of Pharmaceutical Controlled Release Technology" (D. L. Wise, ed.), pp. 31–47. Dekker, New York.

3. Burczak, K., Fujisato, T., Hatada, M., and Ikada, Y. (1994). Protein permeation through poly(vinyl alcohol) hydrogels membranes. *Biomaterials* 15(3), 231–238.

4. Mandal, T. K., Shekleton, M., Onyebueke, E., Washington, L., and Penson, T. (1996). Effect of formulation and processing factors on the characteristics of biodegradable microcapsules of zidovudine. *J. Microencapsul.* 13(5), 545–557.

5. Smith, B. A. H., and Sefton, M. V. (1988). Permeability of a heparin–polyvinyl alcohol hydrogel to thrombin and antithrombin III. *J. Biomed. Mater. Res.* 22, 673–685.

6. Sefton, M. V. (1989). Blood, guts, and chemical engineering. *Can. J. Chem. Eng.* 67, 705–712.

7. Noguchi, T., Yamamuro, T., Oka, M., Kumar, P., Kotoura, Y., Hyon, S. H., and Ikada, Y. (1991). Poly(vinyl alcohol) hydrogel as an artificial articular cartilage: Evaluation of biocompatibility. *J. Appl. Biomater.* 2, 101–107.

8. Mathew, J., and Kodama, M. (1992). Study of blood compatible polymers. I. Modification of poly(vinyl alcohol). *Polym. J.* 24(1), 31–41.

9. Hyon, S. H., Cha, W. I., Ikada, Y., Kita, M., Ogura, Y., and Honda, Y. (1994). Poly(vinyl alcohol) hydrogels as soft contact lens material. *J. Biomater. Sci. Polymer Edn* 5(5), 397-406.

10. Ko, J. H., Ericson, D., Tucker, J., and Walker, R. (1994). Effect of annealing on the physical properties and blood compatibility of poly(vinyl alcohol) hydrogels. *ANTEC*, pp. 2651–2655.

11. Fujimoto, K., Minato, M., and Ikada, Y. (1994). Poly(vinyl alcohol) hydrogels prepared under different annealing conditions and their interactions with blood components. *ACS Symp. Ser.* 540, 229–242.

12. Shih, C. Y., and Lai, J. Y. (1993). Polyvinyl alcohol plasma deposited nylon 4 membrane for hemodialysis. *J. Biomed. Mater. Res.* 27, 983–989.

13. Paul, W., and Sharma, C. P. (1997). Acetylsalicylic acid loaded poly(vinyl alcohol) hemodialysis membranes: Effect of drug release on blood compatibility and permeability. *J. Biomater. Sci., Polymer Edn* 8(10), 755–764.

14. Burczak, K., Gamian, E., and Kochman, A. (1996). Long-term *in vivo* performance and biocompatibility of poly(vinyl alcohol) hydrogel macrocapsules for hybrid-type artificial pancreas. *Biomaterials* 17(24), 2351–2356.

15. Li, R. H., White, M., Williams, S., and Hazlett, T. (1998). Poly(vinyl alcohol) synthetic scaffolds for cell encapsulation. *J. Biomater. Sci., Polymer Edn* 9(3), 239–258.

16. Tamura, K., Ike, O., Hitomi, S., Isobe, J., Shimizu, Y., and Nambu, M. (1986). A new hydrogel and its medical application. *Trans Am. Soc. Artif. Intern. Organs* 23, 605–608.

17. Lozinsky, V. I., Zubov, A. L., and Titova, E. F. (1997). Poly(vinyl alcohol) cryogels employed as matrices for cell immobilization. 2. Entrapped cells resemble porous fillers in their effects on the properties of PVA–cryogel carrier. *Enzyme Microb. Technol.* 20, 182–190.

18. Bachtsi, A. R., and Kiparissides, C. (1995). An experimental investigation of enzyme release from poly(vinyl alcohol) crosslinked microspheres. *J. Microencapsul.* 12(1), 23–35.

19. Ficek, B. J., and Peppas, N. A. (1993). Novel preparation of poly(vinyl alcohol) microparticles without crosslinking agent for controlled drug delivery of proteins. *J. Controlled Release* 27, 259–264.

20. Ermis, D., and Yuksel, A. (1999). Preparation of spray-dried microspheres of indomethacin and examination of the effects of coating on dissolution rates. *J. Microencapsul.* 16(3), 315–324.

21. Torres, A. I., Boisdron-Celle, M., and Benoit, J. P. (1996). Formulation of BCNU-loaded mcrospheres: Influence of drug stability and solubility on the design of the microencapsulation procedure. *J. Microencapsul.* 13(1), 41–51.

22. Erden, N., and Celebi, N. (1996). Factors influencing release of salbutamol sulphate from poly(lactide-*co*-glycolide) microspheres prepared by water-in-oil-in-water emulsion technique. *Int. J. Pharm.* 137, 57–66.

23. Suzuki, S., and Lim, J. K. (1994). Microencapsulation with carrageenan–locust bean gum mixture in a multiphase emulsification technique for sustained drug release. *J. Microencapsul.* 11(2), 197–203.

24. Chatzi, E. G., Boutris, C. J., and Kiparissides, C. (1991). On-line monitoring of drop size distributions in agitated vessels. 1. Effects of temperature and impeller speed. *Ind. Eng. Chem. Res.* 30, 536–543.

25. Mahabadi, H. K., Ng, T. H., and Tan, H. S. (1996). Interfacial/free radical polymerization microencapsulation: Kinetics of particle formation. *J. Microencapsul.* 13(5), 559–573.

26. Arabi, H., Hashemi, S. A., and Fooladi, M. (1996). Microencapsulation of allopurinon by solvent evaporation and controlled release investigation of drugs. *J. Microencapsul.* 13(5), 527–525.

27. Thanoo, B. C., Sunny, M. C., and Jayakrishnan, A. (1993). Oral sustained-release delivery systems using polycarbonate microspheres capable of floating on the gastric fluid. *J. Pharm. Pharmacol.* 45, 21–24.

28. Alexandridou, S., and Kiparissides, C. (1994). Production of oil-containing polyterephthalamide microcapsules by interfacial polymerization. An experimental investigation of the effect of process variables on the microcapsule size distribution. *J. Microencapsul.* 11(6), 603–614.

CHAPTER 75

MICROENCAPSULATION METHODS: PMCG CAPSULES

Taylor G. Wang

INTRODUCTION

Transplantation of cells to treat a variety of human diseases such as hormone or protein deficiencies is limited because the cells are quickly destroyed by the recipient's immune system. To overcome this limitation, it was proposed that hormone- or protein-secreting cells be enclosed in a semipermeable membrane that would protect cells from immune attack and yet allow the influx of molecules important for cell function/survival and the efflux of such desired cellular products as hormones [1–4]. The principle of immunoisolation or immunoprotection of cells for transplantation has two great potentials: cell transplantation without the need for immunosuppression and its accompanying side effects, and transplantation of cells from nonhuman species (xenograft) to compensate for the limited supply of donor cells for such diseases as diabetes [5–7]. Since any disease that is best treated by the regulated release of a cellular product (hormone, protein, neurotransmitter, etc.) is a candidate for transplantation of immunoisolated cells, a variety of cell types including pancreatic islets, hepatocytes, neurons, parathyroid cells, and cells secreting various clotting factors, can be considered for enclosure within such devices.

To achieve this promise, a capsule must be developed to satisfy a set of stringent and dichotomous requirements. The capsule membrane, to serve as an immunoisolation device, must be able to keep the immune system away from the living cells, yet allow nutrients, oxygen, and proteins to pass through without much impediment. The membrane must be biocompatible to the host and to the cells it encloses. It also needs to be strong enough to survive handling, transplantation, and the hostile environment inside the human body.

The membranes of choice today are polymer systems of both natural and synthetic origin. Polymers offer a large variety of physical, chemical, and structural properties that make them highly favorable for various biomedical applications. When the polymeric material is, as in this case, aimed at the immunoprotection of living cells, the optimization of the polymer system demands the ability to meet the following requirements:

1. *Engineering*: mechanical stability, thickness, porosity, homogeneity, geometry, purification, and sterilization
2. *Biological*: biocompatibility, resistance to the biodegradation, nontoxicity, retrievability, and permeability adjustment

All these requirements determine the number of polymer candidates that can be utilized in immunoisolation devices of various types. They are represented by the membranes in a form of the hollow fibers (copolymer of acrylonitrile and vinyl chloride), tubular diffusion chambers (polyacrylates, sodium alginate) [7,8] vascularized perfusion devices (polyacrylates) [9], and microcapsules: polyelectrolyte complexes [10], agarose [11] and poly(hydroxyethyl methacrylate-*co*-methyl methacrylate) [12]. Among them, however, microcapsules seem to exhibit superior insulin response and long-term islet viability and function [13]. The microcapsule system of sodium alginate and poly(L-lysine), originated by Lim and Sun, has demonstrated diabetes reversal in animals and in a small-scale clinical trial [14–16]. However, the inability to provide independent adjustment of capsule parameters (e.g., mechanical strength or permeability) has limited the success of this system [17]. Other approaches utilizing different principles of capsule formation, such as temperature-induced sol–gel transition [11] solvent extraction, and precipitation [12], suffer similar limitations.

Although fundamental advances in our understanding of the human immune system and the immune rejection process may eventually lead to means of developing technologies that will overcome the vigorous humoral and cellular immune responses associated with the transplantation of xenogeneic tissues, these solutions are likely to be decades in the future. In the meantime, encapsulation may be the only way to establish prolonged survival of cell and tissue transplants in patients with diseases caused by the loss of specific vital metabolic functions.

In addition to diabetes, encapsulation has broad application to the treatment of major diseases such as cancer and AIDS and a wide range of other disorders resulting from functional defects of native cell systems [18–25]. Such application include the use of hepatocytes for the treatment of liver failure and enzymatic defects [19–25], adrenal cells in Parkinson's disease [26–28], cells that produce nerve growth factors in Alzheimer's disease [29,30], epilepsy, amyotrophic lateral sclerosis (Lou Gehrig's disease) and Huntington's disease [31–33], spinal cord injuries [34], and strokes, as well as cells that produce clotting factors VIII and IX in hemophilia [35,36] and endocrine cells in other disorders resulting from hormone deficiency [25,37]. Moreover, by using recombinant DNA and cell engineering technologies in conjunction with encapsulation technologies, it should also prove possible to treat patients suffering from chronic pain [27], Kaposi's sarcoma, and various hematological disorders.

To date, most of the research in the area of encapsulated cell transplantation has been carried out with pancreatic islets. There are several reasons for this.

1. Diabetes is a leading cause of morbidity and mortality in the world, at an annual cost of approximately $134 billion in the United States alone.
2. Pancreatic islets, which comprise only 1–2% of the human adult pancreatic volume, can be isolated from animal sources.
3. The quantity of differentiated islet tissue to be transplanted is within a reasonable range.

Like other human organs, donor pancreases are in very short supply. In the United States, it is estimated that only approximately 1000 pancreases are recovered each year. Even with improved procurement of human organs, the supply of donor tissue would remain quite inadequate if pancreatic islet transplantation were to be developed as an effective therapy. For the immediate future, the logical alternative is to use nonhuman donor islets. For example, methods have been developed to isolate pancreatic islets from porcine and bovine glands; insulins from these animals are fully active in man and have been used to treat diabetics for over 70 years. Pig and cow islets are an attractive option because they are readily available and the amino acid sequence of these insulins is similar to that of human insulin. Further, herds of specific-pathogen-free (SPF) animals would be available as a potential safeguard against the transfer of infectious organisms to human islet recipients.

THEORETICAL MODEL

The details of this calculation can be found elsewhere [38]. Only key points are included here, for of purpose clarification.

IMMUNOISOLATION

A new physical picture of immunoisolation of living cells has been proposed [38]. In this new model, the large pores will allow the nutrients, proteins as well as immune systems, to enter the membrane. However, with proper selection of wall thickness, the small pores inside the membrane will prevent or delay most of the immune system from passing all the way through to the inner volume of the capsule where the cells reside. The capsule membrane operates more as an immune system *barrier* (or entrapper) than as a *gatekeeper*. Thus we write

$$\Gamma = 3\frac{d^2}{R^2}f\tau \tag{1}$$

where Γ is the time required for for immune system to pass through the membrane, f is the ratio between surface areas of small pores and large pores, d is the membrane thickness, R is the pore size, and τ is the time during which the immune system can be trapped by the small pores in each collision.

MASS TRANSPORT

To maintain the islets viable inside an immunoisolation capsule, nutrient and oxygen must be allowed in, and insulin out. This is a diffusive process. The mass transport across a random network structure can be approximated by

$$Q \sim \frac{R^4 T}{r\eta d}\Delta CN \tag{2}$$

where R is the pore size, d is the thickness of the membrane, ΔC is the mass concentration difference, T is the temperature, and N is related to the effective capsule surface area that participates in mass transport. The diffusion of biomaterial in liquids is inverse proportional to the fluid viscosity η and particle radius r. This is usually called Stokes velocity. Equation (2) is phenomenological. However, it provides some valuable insight into the capsule design optimization.

MECHANICAL STRENGTH

Many immunoisolation devices fail inside their animal or human host as a result of capsule breakage, which exposes unprotected living cells to the host's immune system. At best, the cells are destroyed; at worst, an immunoreaction can be triggered. Therefore, if the immunoisolation devices are to be used for the long-term cure of hormone deficiency diseases in humans, the capsules must have sufficient mechanical strength to survive the constant pressure exerted on them. A spherical capsule with a resting radius of R_0, distorts the shape, δr, under an unidirectional pressure P_x.

$$\frac{\delta r}{R_0} = \frac{R_0}{4Yd}P_X \tag{3}$$

where Y is Young's modulus and d the thickness. Equations (1), (2), and (3) clearly show that if the capsule parameters cannot be adjusted independently, it will be very difficult to satisfy all those dichotomy requirements on immunoisolation, mass transport, and mechanical strength.

PMCG CAPSULES

Most cell encapsulation devices today utilize modifications of the procedure originated by Lim and Sun in which the encapsulant is suspended in a polyanionic aqueous solution and extruded by an air jet–syringe pump droplet generator into calcium ions. Poly(L-lysine), which is a cationic macromolecule, is then mixed with the hardened polyanionic gel, and a membrane is formed at the interface as a result of the ionic interaction. However, because this is a binary system, all membrane parameters are tied to a single chemical complex, and attempts to optimize one parameter will affect all other parameters. The inability to optimize the capsule functions independently (e.g., immunoisolation vs mass transport) has limited the success of this system. Capsules made with other binary polymer systems, such as HEMA/MMA, suffer similar limitations [39].

CAPSULE DEVELOPMENT

Polymer Screening

After the encapsulant has been suspended in a polyanionic aqueous solution and extruded into calcium ions, poly(L-lysine) is mixed with the hardened polyanionic gel, and a membrane is formed at the interface. In this a binary system, since all membrane parameters are tied to a single chemical complex, attempts to optimize one parameter will affect all other parameters. For example, increasing membrane thickness to improve mechanical strength will hamper mass transport, and vice versa. The success of this system has been limited by the inability to optimize the capsule functions independently, together with the erroneous assumption of uniform capsule pore size. Capsules made with other binary polymer systems suffer similar limitations.

To overcome these limitations, Vanderbilt University investigators developed a new multicomponent capsule. Over a thousand combinations of polyanions and polycanions were studied in search of polymer candidates that would be suitable for barrier capsules (Table 75.1) [40,41]. The combination of sodium alginate (SA), cellulose sulfate (CS), poly(methylene-co-guanidine) (PMCG), calcium chloride (CaCl$_2$), and sodium chloride

Table 75.1. The 35 Polycations and 35 Polyanions That Were Screened to Determine the Interactions of Different Pairs

Polycation group	Polyanion group
Chitosan	Sodium alginate
Glycol chitosan	PG alginate
DEAE–dextran	Carboxymethyl cellulose
Quaternized hydroxymethyl cellulose, JR-125	Cellulose sulfate
Poly(dimethylamino-co-epichlorhydrin), quaternized	Gellan gum
Poly(ethyleneimine)	Gum arabic
Poly(ethyleneimine/hydroxyethyl)	Maleic anhydride
Gelatin A, pH 4	p-acrylamide carboxyl-mod.
Qua ternized polyamine, B50	p-AA/MPSA
Poly(L-lysine)	p-maleic acid
Poly(allylamine)	Polyphosphate
Poly(diallyl dimethylammonium chloride)	Poly(vinylsulfone)
DADMAC, acrylamide, C3204/C505	Poly(vinylsulfonic acid)
Cationic polyacrylamide, Jayfloc 3466	Poly(tyrene sulfate)
Poly(2-DMAE-methacrylate)	Poly(vinylphosphonic acid)
Poly(acrylamide methacryloxyethyl–trimethyl ammonium brimide)	Poly(vinyl phosphate)
Poly(methacryloxyethyl–trimethyl ammonium bromide)	Poly(acrylic acid)
Poly(2-hydroxy-3-methacryloxypropyl–trimethyl ammonium chloride)	p-methacrylic acid
Poly(3-chloro-2-hydroxypropyl-methacryloxyethyl–dimethyl ammonium chloride)	p-acrylamide acrylic acid
Poly(butylacrylate-methacryloxyethyl–trimethyl ammonium bromide)	Anionic acrylamide
Copolymer DMAE–methacrylate-acrylamide (Betz no. 1158)	Gelatin A
Copolymer DMAE–acrylate-acrylamide (Betz nos. 1 to 5)	Gelatin B pH 6.5
Copolymer dimethylaminaethyl–acrylate acrylamide (Hunk no. 5)	p-glutamic acid
Copolymer MDEAE–acrylate acrylamide (Betz no. 1160P)	p-lacturonic acid
Poly(methylene-co-guanidine), plus 0.2 g/liter CaCl$_2$	CM-dextran
Cationic polyacrylamide proprietary (496C/492C)	CM-amylose
Poly(vinyl pyrrolidone–DMAE–methacrylate)-(quaternized)	Dextran sulfate
Poly(ethyleneimine), epichlorhydrin modified	Heparin
Poly(1-methyl-2-vinylpyridinium bromide)	Chondroitin 4-sulfate
Poly(1-methyl-4-vinylpyridinium bromide)	Chondroitin 6-sulfate
Polyamine 4030	Byaluronic acid
Polyamide 5087/792a	Xanthan
Poly-DADMAC/N-isopropyl–acrylamide(pDADMAC/NIPAAM)	Carrageenan
Polymidazoline, quaternized	Methylvinyl ether–maleic acid
Polyvinylamine	Methylvinyl-ether–maleic anhydride

(NaCl) was found to be most promising based on extensive *in vitro* studies of the roles of various components of the capsule (Table 75.2) [6,42].

From the point of view of polymer chemistry, the polyelectrolyte complex is formed by mixing two polymer solutions, prepared in the same solvent, leading to a two-phase system in which both polymers are concentrated in one of the phases (the "precipitate"), while the other phase (the "supernatant") is essentially polymer free [43]. This is also termed complex coacervation. The character of a polyelectrolyte complex is generally governed by the long-range electrostatic interactions and the presence of small-molecule counterions. The effects

Table 75.2. Summary of Capsule Formation Process and the Roles of Individual Components

Exper- iment	Components				Appearance at[a]		
	SA-HV 1 wt% PBS	CS, 1 wt% PBS	$CACl_2$ 1 wt% H_2O	PGMG, 0.9 wt% $NaCl_5$ pH 7.5	First wash in PBS	Sodium citrate 50 mM, 5 min	Storage in PBS 24 h
1	+	−	+	−	T, beads slowly dissolved	Quickly dissolved	−
2	+	−	−	+	Partially shrunk		Swelling ruptured
3[b]	+	−	+	+	Beads	Swelling	T, swollen, stable, fragle
4[c]	+	−	+	+	Beads	Swelling	T, swollen, stable, fragle
5	−	+	+	−	Shrunk, slowly dissolved	Quickly dissolved	−
6	−	+	−	+	Shrunk, stable	−	NT, shrunk, stable
7[b]	−	+	+	+	Shrunk, stable	Shrunk, stable	ST, shrunk, stable, fine, spongelike
8[c]	−	+	+	+	Shrunk, stable	Shrunk, stable	NT, shrunk, stable, carse, spongelike
9[d]	+	+	+	−	Beads slowly dissolved	Quickly dissolved	−
10[d]	+	+	−	+	Irregular shape	−	Swelling rupture
11[b,d]	+	+	+	+	ST beads	ST capsules, stable	ST, strong, stable, visible ST membrane
12[c,d]	+	+	+	+	ST beads	ST capsules, stable	ST, strong, stable, visible membrane, precipitated core

[a]T, transparent; ST, semitransparent; NT, nontransparent.
[b]Poly(methylene-*co*-guariidine)–CaCl mixture: 1 wt% PMCG, 1 wt% $CaCl_2$ in 0.9 wt% NaCl at pH 7.5.
[c]PMCG reaction is a second step following the first step of bead formation via $CaCl_2$.
[d]The SA-HV/CS mixture contained 1 wt% sodium alginate and 1 wt% cellulose sulfate in phosphate buffered soline (PBS).

of polymer architecture on the character of the resulting polyelectrolyte complex [44] are given by cooperative interactions (affecting the entropy of complementary polymers), concerted interactions (coexistence of different kinds of interaction forces), degree of matching distribution of active sites, and appearance of specific microdomains (hydrophobic and electrostatic effects). Besides these, other factors such types of interaction force, ionic strength, pH, temperature, solvent, steric conformity and polymer chain length may also play important roles in the complex formation process. These factors point out the complexity of the conditions involved in the capsule formation protocol. Furthermore, the capsule formation must proceed under physiological conditions with respect to ionic strength, pH, and temperature.

Capsule Processing

Capsules are made in a novel chemical reaction chamber developed in our laboratory. The heart of this apparatus is a multiloop chamber reactor filled with cation solution. This cation solution bath is fed by a cation stream, which continuously replenishes the solution and carries away the anion drops being introduced into the chamber. SA-CS droplets, with pancreatic islets enclosed, enter the PMCG-CaCl$_2$-NaCl stream at selected impact parameters to promote the islet centering [45–50]. The droplets are then carried into the multiloop reactor by the polycation stream. At the upper half of the loop, the encapsulated islets fall toward the center of the loop, and at the lower half of the loop, the encapsulated islets fall away from the center of the loop. Since the droplets alternate falling toward and away from the center, they can be processed without impacting the inner wall. The reactor facilitates tight control of capsule sphericity, membrane thickness, and uniformity, as well as negating gravitational effects. We have produced capsules with diameters from 0.5 to 3.0 mm and membrane thicknesses from 6 to 200 μm (Fig. 75.1).

Mechanical Strength

An apparatus developed in our laboratory is used to measure mechanical strength by placing an increasing uniaxial load on the capsule until bursts. The capsule mechanical strength, a function of membrane thicknesses, can be adjusted anywhere from a load of a fraction of a gram to decagrams, to meet the needs of transplantation without altering the permeability of the capsule. Figure 75.2, a plot of rupture load versus product of membrane thickness and capsule size, demonstrates that this new capsule has much greater flexibility in mechanical strength than capsules formulated with the widely used sodium alginate–polylysine system. The slope of the curve represents the rupture stress and thereby, indirectly, the inherent strength of the capsular membrane [51].

Permeability

Measurement of capsule permeability was improved by utilizing two complementary methods [52] size exclusion chromatography (SEC) with dextran molecular weight standards and a newly developed method of assessing the permeability of a series of biologically relevant proteins by means of encapsulated protein A–Sepharose (PAS). By combining

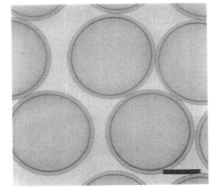
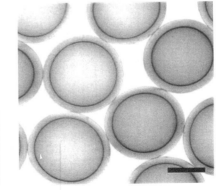

Fig. 75.1. Empty PMCG capsules. Capsules prepared under identical processing steps were transplanted intraperitoneally into normal C57/Bl6 mice. Capsules were retrieved 30 days later and photographed. Left: capsules prior to transplantation (bar = 340 μm); right, capsules retrieved from normal mice (bar = 219 μm).

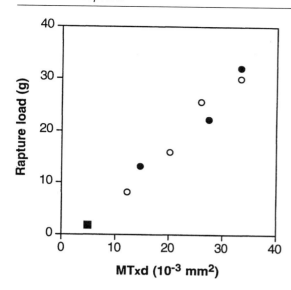

Fig. 75.2. Capsule mechanical strength. The mechanical burst strength of the capsules as a function of their wall thickness: open circles, 0.6% CS–0.6% alginate system; solid circles, 0.9% CS–0.9% alginate system; square, SA-PLL system.

these methodologies to measure permeability and component concentration manipulations to control capsule permeability, a series of capsules with a range of permeabilities (40–230 kDa, based on dextran exclusion measurement) was developed and characterized. The apparent pore size of the capsular membrane was determined by size exclusion chromatography, which measures the exclusion of dextran solutes from the column packed with microcapsules. The use of neutral polysaccharide molecular weight standards makes it possible to evaluate the membrane properties when solute diffusion is controlled only by its molecular dimension. Based on the measured values of solute size exclusion coefficients ($KSec$) and known size of solute molecules, one can estimate the membrane pore size distribution (PSD). The differential PSD for the microcapsule with a 230-kDa exclusion limit showed a maximum at a viscosity radius R_μ equal to 50 Å, and a cutoff radius R_μ^3 117 Å (hatched area in Fig. 75.3).

To assess capsule permeability to immunologically relevant proteins, a new method involving encapsulated protein A–Sepharose (PAS) was developed. IgG readily entered capsules with the 230-kDa exclusion limit, but not the one at 40 kDa (data not shown). The SA-PLL capsules used by many laboratories have an exclusion limit of 110 kDa based on our dextran measurements, which corresponds to an R_μ of 78 Å. It is noteworthy that for 52 Å, the molecular size of IgG reported in the literature, IgG would expected to pass a

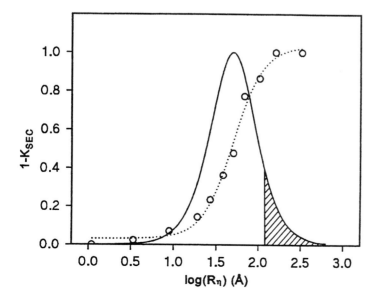

Fig. 75.3. Pore size distribution (PSD). Apparent PSD of PMCG capsules estimated based on dextran viscosity radius R_0^η The cumulative PSD (dotted curve) and the differential PSD (solid curve) are shown. The exclusion limit of this capsule is 230 kDa.

capsular membrane a 110-kDa exclusion limit. Even though IgG enters the capsules, islet destruction does not always occur *in vivo*; the entry of complement for cell lysis by IgG could be prevented or delayed.

Our mathematical model shows that it is not solely the nominal cutoff value that determines whether immunoglobulin or the complement system is or is not kept out, but rather, the probability of penetration all the way through of immunoglobulin or the complement system in a random network system. Our capsule design is flexibe enough to keep the biologically relevant molecular system out, as demonstrated in our positive results in non obese diabetic (NOD) mice with the capsule of reported cutoff values.

Biocompatibility

Extensive biocompatibility testing of empty capsules was performed prior to transplantation of encapsulated islets into animals. Capsules transplanted into the peritoneal cavity of healthy mice and dogs to study biocompatibility were later retrieved to study the tissue reaction and fibrosis on the capsule surface. Experimental results have shown that capsules similar in physical appearance, mechanical strength, and permeability may exhibit drastic differences in biocompatibility. These data suggest that there is a set of operational parameters in chemistry and capsule processing and that minor variations may have strong impact on biocompatibility.

Blood Glucose Control

The purpose of islet transplantation is to provide physiological control of blood glucose concentration. Perifusion of encapsulated islets with glucose elicited an approximately fourfold average increase from basal insulin secretion [6,8]. There was a delay of 7 ± 1 min before the insulin concentration in the perfusate began to increase (Fig. 75.4). This response is well within a time frame compatible with closed-loop insulin delivery (pharmacokinetic modeling of glucose homeostasis in humans suggests that the lag time of the increase in insulin delivery by an artificial pancreas must be <15 min to avoid the overexcursion of postprandial blood glucose) [53]. Immunoisolated islets lack intimate vascular access and must be supplied with oxygen and nutrients by diffusion from the nearest blood vessels over distances greater than those normally encountered. Only a rim of islets remained viable from the capsules that were retrieved [5]. Similar results were obtained with canine islet implants into rats. These findings are in agreement with the theoretical model.

This breakthrough in encapsulation technology and capsule science offers distinct advantages for the task of immunoisolating cells for transplantation. It allows us to customize

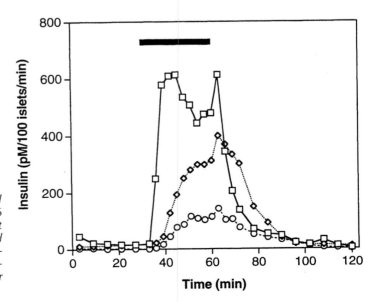

Fig. 75.4. Viability of encapsulated islets. Capsules retrieved one year following transplantation into diabetic C57/B16 mice were assessed in the cell perifusion apparatus. Bar at top shows a stimulation period with 20 mM glucose and 0.045 mM isobutyl methyl xanthine. Insulin secretion of unencapsulated islets (squares), freshly encapsulated islets (diamonds), and retrieved encapsulated islets (circles) shown, for comparison.

capsules for biocompatibility performance in a way that addresses the specific biological requirements of the transplantation hosts.

ANIMAL STUDIES

Extensive animal studies have been conducted to test the functions of PMCG capsule designs.

RODENT STUDIES

Encapsulated rat islets readily reversed chemically induced (by streptozocin) diabetes in mice, and islet function was maintained for at least 4–6 months (Fig. 75.5). Similar studies were performed with NOD female mice, and diabetes was reversed for up to 241 days. Encapsulated islets have long-term function as assessed by their ability to secrete insulin following glucose stimulation. Recently, Wang and Elliott of New Zealand, without the use of any immunosuppression, successfully used encapsulated porcine islets to reverse diabetes (14 weeks) in a NOD mouse model [54] (Fig. 75.6). The eventual failure of encapsulated islets transplanted into animals with chemically induced diabetes did not result from capsule rupture or from an immune attack. Capsules retrieved from those animals were free floating, and their surface was free of fibrosis. Thus, the failure likely resulted from islet death within the capsule, due either to islet injury or to nutrient deficiency or islet toxicity caused by soluble immune factors (cytokines). In contrast, the failure of encapsulated islets transplanted into NOD mice may have resulted from an immune or autoimmune attack. Capsules retrieved from those animals were clumped and demonstrated marked fibrosis around the capsule surface.

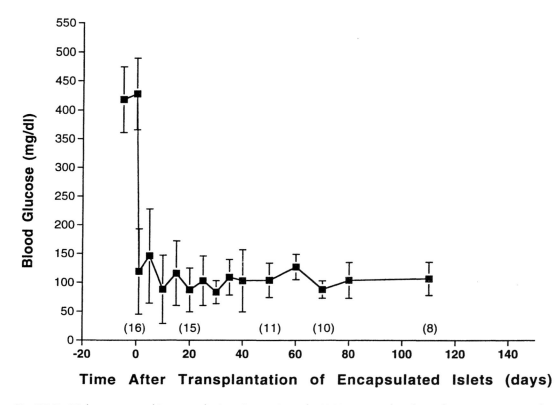

Fig. 75.5. Diabetes reversal in normal mice. Approximately 1000 encapsulated rat islets (0.8-mm capsule with 0.1-mm wall thickness and exclusion limit of 230 kDa) were transplanted intraperitoneally into diabetic C57/Bl6 mice (strepzotocin-induced). The mean and standard deviation of blood sugars of nondiabetic animals (measured with a One-Touch Glucometer) are plotted against the number of days following transplantation. Of the 16 diabetic mice that received transplants, the number of animals that were nondiabetic at selected time points is shown in parentheses. Of the 8 animals with normal blood sugar 110 days following transplantation, 2 later became diabetic, while 6 were still normoglycemic 300 days following transplantation.

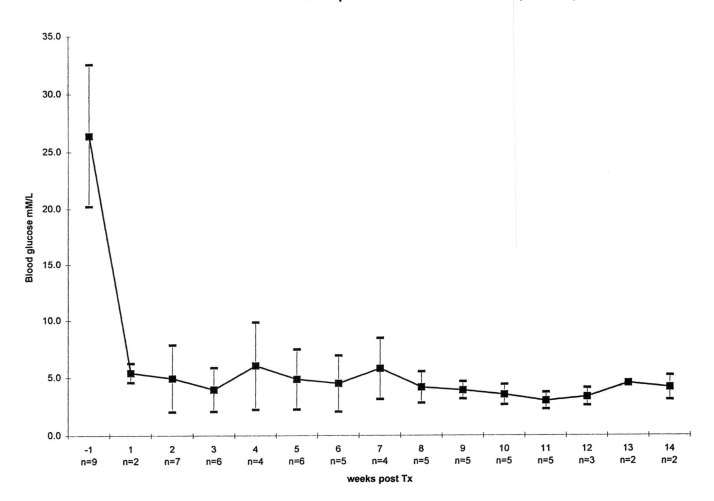

Fig. 75.6. *NOD diabetes reversal with encapsulated porcine islets. Encapsulated porcine islets were transplanted into the peritoneal cavity of female NOD mice that had developed spontaneous diabetes by the age of 20 weeks. Of the diabetic mice that received transplants, the number of animals that were nondiabetic at selected time points is shown in parentheses. This experiment is part of the results of a collaborative effort between T. Wang and R. Elliott of New Zealand [54].*

LARGE ANIMAL STUDIES

Encapsulated porcine islets were transplanted into peritoneal cavity of dogs. The PMCG capsules for the porcine C-peptide studies (Fig. 75.7) were of the same design as the ones used in rodent diabetes reversal, with one exception: ultrapurified polymer coatings (Monsanto UPA) were used instead of industrial grade. No insulin response from the neonatal porcine islets was observed in the first month of the transplantation. Porcine C-peptide was measured from day 30 to day 100 before the experiment was terminated for autopsy studies. The detection of porcine C-peptide indicated that the encapsulated porcine islets in canine were viable and functioning. This clearly demonstrated that our capsules were able to provide immunoisolation protection in a large-animal model. At the end of the experiment (> 100 days), the histology showed all the capsules to be broken and clumped. The animal studies show us that capsule design must be tailored specifically to the transplantation host; one design does not fit all. We need to systematically optimize the capsule for human trial.

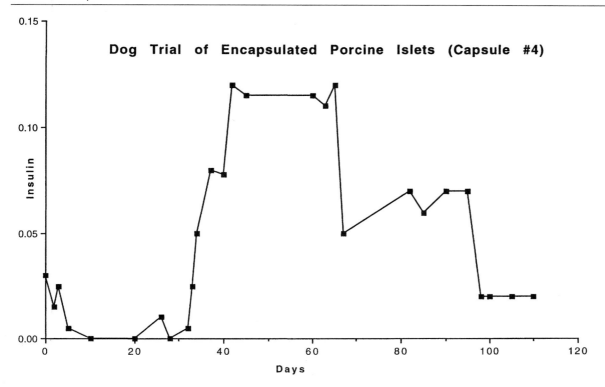

Fig. 75.7. The porcine C-peptide studies. Porcine C-peptide secreted by encapsulated neonatal porcine islets in the peritoneal cavity of dog was measured. Minimal porcine insulin responses were observed in the first month of transplantation. The data suggest islet maturation and differentiation. Loss of the porcine insulin secretion after 100 days is the result of capsule clumping and breakage.

OPTIMIZATION STUDIES

Our data suggest [38] that there exists a window of operation for the composition of polycation solutions at given conditions under which the capsules elicit little or no fibrosis stimulation. Therefore, our work in this direction will be aimed at studying the operating windows for the compositions of polyelectrolyte solutions. Our knowledge of the PMCG capsule formation process is comprehensive enough to permit us to understand the role of each individual component. This background has been applied in the up-to-date capsule development stage. Nevertheless, we lack the quantitative information on how the modification of the recipe will influence membrane composition and stability in humans. The experimental design involves forming capsules under different conditions, with the goal of completing the polyelectrolyte reaction (as determined by spectroscopic analysis). This work is combined with long-term *in vitro* studies of capsules incubated in human or animal serum at 37°C, and measurements of their mechanical strength and permeability as a function of time.

Additional experiments concentrate on the detection of membrane components leached out of the capsule with time under conditions favoring instability of the polyelectrolyte complex. Membrane composition uses is analyzed by means of infrared and elemental techniques to assess the saturated concentrations of PMCG in a polycation solution for a given concentration of SA-CS. The leached compounds will be detected by spectroscopic techniques. As a result, comprehensive knowledge about the most favorable conditions for stabilizing the capsule for *in vivo* tests should be obtained.

PRELIMINARY OPTIMIZATION RESULTS

In this experiment, we modified the capsules used in large-animal studies by increasing the membrane polyanion compositions concentration by 1% and reducing the gelling time by a factor of 2. The stability of the modified capsules (Fig. 75.8) was tested with the procedure given earlier. Figure 75.9 shows the capsules (0.5 min gelling time) on the omentum of

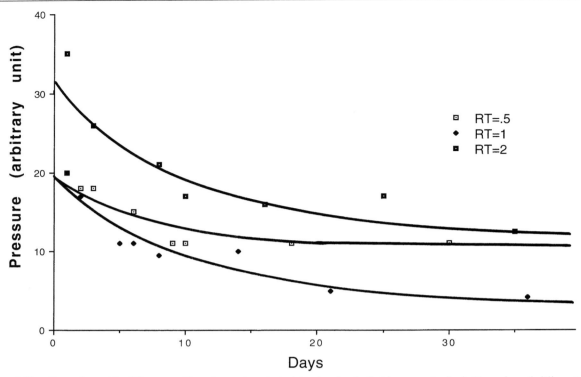

Fig. 75.8. Capsule stability. Capsules with different gelling times, incubated in dog body fluid, were studied. Capsules of different composition and different reaction time reached different steady state values. This suggests that capsule stability inside the transplantation hosts can be optimized.

a dog at the one-month examination. The capsules were free to roll back and forth when the omentum was tilted. The histology shows that all the capsules were free-floating and intact. The surface of the vast majority of the capsules is clean. A few capsules (< 1%, probably) have a scant amount of fibrin and rare mononuclear cells adherent to the surface. There was no host inflammation. This suggests that the stability of capsules inside a transplantation host can be optimized. These data strongly suggest that optimization experiments will yield improvements in graft performance and survival in humans. This is a significant advance over traditional capsules.

DISCUSSION

Organ transplantation is one of the miracles of the twentieth century. Polymer membranes are the material of choice today for encapsulating living cells for immunoisolation. Recent data have shown some exciting success in animal models. A new physical picture and a better understanding of mechanism of immunoisolation have advanced systematic studies in preparation for human clinical trials.

The next few years should prove to be a period of great challenge and, hopefully, of great excitement and promise to those involved in islet transplantation. The first clinical trials utilizing encapsulated islet to treat diabetic patients will likely occur in 2002. This in turn will serve as an important starting point for developing living drug delivery systems for treating a wide range of additional disorders. For patients with diabetes, as well as those with other diseases (e.g., cancer, hemophilia, liver failure, Parkinson's), the coming years hold promise for greatly improved therapy.

The capsule described in this chapter is based on a mixture of polyanions reacting with an oligomeric polycation in the presence of the low molecular weight gelling agent $CaCl_2$ and antigelling NaCl. Multicomponent polyelectrolyte systems have often been studied to elucidate the preferential binding of the individual components and, by combining their different features, to achieve the required complex properties.

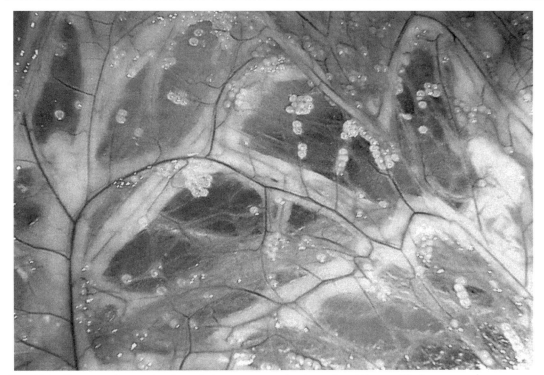

Fig. 75.9. One-month examination of the new capsules in a dog. We modified our membrane compositions to improve capsule stability and transplanted the new capsules into a healthy dog. At the one-month examination, these capsules were free to roll back and forth when the omentum was tilted. There was no host inflammation.

When all these components are involved in the complexation reaction, the cation compounds compete according to their mobility and affinity for the anionic reactive sites of the polymer matrix. Obviously, individual interactions are influenced by each component of the reaction mixture. Nevertheless, one can identify the dominant interactions that govern capsule formation. Calcium ions diffuse faster than PMCG molecules and occupy more reactive sites. The capsule retains its spherical shape, which can be related to the interactions of Ca^{2+} with SA-HV (high viscosity) at the given concentration ratio of SA-HV to CS. Owing to its higher charge density and its flexible molecule, cellulose sulfate is more likely to form a stable complex with PMCG than SA. The ratio of SA-HV to CS determines the character of the membrane; it becomes more precipitate-like with increasing CS concentration. After calcium removal, SA-HV may act as a adiluent in the CS-PMCG network because weaker interactions with PMCG provide hydrophilic domains and, thus, prevent the network from losing water and collapsing.

In addition to the processes just described, sodium chloride significantly affects capsule formation. We found that NaCl is one of the factors determining membrane permeability as well as surface topology. In this respect, sodium ions control the initial stage of complex formation. They screen the negative charges of the reacting polyelectrolytes, hence suppress the reaction of calcium ions with the polyanion matrix. By increasing the ratio of $[Na^+]$ to $[Ca^{2+}]$ for a given polyanion mixture, it is possible to extend the time scale for freezing the capsule shape and obtain a wrinkle-free surface. This is in accordance with our observations and literature data.

We have found that the molecular weights of sodium alginate and cellulose sodium sulfate are also significant factors that have an impact on the stability and mechanical properties of this capsule. When high molecular weight sodium alginate, in a mixture of 0.6 wt% SA and 0.6 wt% CS, was replaced by lower molecular weight sodium alginate (SA-LV, $M_w = 2.5 \times 10^5$ Da), the resulting membrane became weaker, exhibiting more surface wrinkles and irregularity in the capsular shape.

Acknowledgments

The Vanderbilt University studies were supported by grants from Juvenile Diabetes Foundation, Evans and Gilruth Foundation, Vanderbilt University School of Engineering, and the Vanderbilt Diabetes Research and Training Center (NIH DK20593). The author wishes to express his gratitude for the valuable contribution provided to him by Robert Lanza M. D. Advanced Cell Technology, Igor lacik, Ph.D., Polymer Institute, Slovak Academy of Sciences., A. V. Anilkumar, Ph.D., M. Brissova Ph.D., D. Hunkler Ph.D., A. Prokop Ph.D., K. Shahrokhi M. S., and Paul LeMaster, M. S., of Vanderbilt University, and A. C. Powers, M. D., Phil Williams B.S., Kay Washington, M. D. Alan Cherrington Ph.D, of Vanderbilt University Medical School/VA Hospitals. The author also acknowldges the excellent support provided to him by his New Zealand colleagues in particular, R. Elliot, MD., and E. Livia, MD.

References

1. Chang, T. M. (1992). Hybrid artificial cells: Microencapsulation of living cells. *ASAIO J.* **38**, 128–130.
2. Lacy, P. E. (1995). Treating diabetes with transplanted cells. *Sci. Am.* **273**, 50–58.
3. Lanza, R. P., Sullivan, S. J., and Chick, W. L. (1992). Islet transplantation with immunoisolation. *Diabetes* **41**, 1503.
4. Colton, C. K. and Avgoustiniatos, E. S. (1991). Bioengineering in development of the hybrid artificial pancreas. *J. Biomech. Eng.* **113**, 152–170.
5. Wang, T. G., and Lanza, R. P. (2000). "Bioartificial Pancreas: Principles of Tissue Engineering," (2nd ed.), Vol. 36, pp. 495–507. Academic Press, San Diego, CA.
6. Wang, T. G., Lacik, I., Brissova, M., Anilkumar, A. V., Prokop, A., Hunkeler, D., Green, R., Shahrokhi, K., and Powers, A. C. (1997). An encapsulation system for the immunoisolation of pancreatic islets. *Nat. Biotechnol.* **15**, 358–362.
7. Lanza, R. P., Kuhtreiber, W. M., Ecker D., *et al.* (1995). Xenotransplantation of porcine and bovine islets without immunosuppression using uncoated alginate microspheres. *Transplantation* **59**, 1377.
8. Lanza, R. P., Butler, D. H., Borland, K. M., *et al.* (1991). Xenotransplantation of canine, bovine, and porcine islets in diabetic rats without immunosuppression. *Proc. Natl. Acad. Sci. U.S.A.* **88**, 11100.
9. Maki, T., Lodge, J. P. A., Carretta, M., Ohzato, H., Borland, K. M., Sullivan, S. J., Staruk, J., Muller, T. E., Solomon, B. A., Chick, W. L., and Monaco, A. P. (1993). Treatment of severe diabetes mellitus for more than one year using a vascularized hybrid artificial pancreas. *Transplantation* **55**, 713–718.
10. Lim, F., and Sun, A. M. (1980). Microencapsulated islets as bioartificial endocrine pancreas. *Science* **210**, 908–910.
11. Iwata, H., Takagi, T., Kobayashi, K., Oka, T., Tsuki, T., and Ito, F. (1994). Strategy for developing microbeads applicable to islet xenotransplantation into a spontaneous diabetic NOD mouse. *J. Biomed. Mat. Res.* **28**, 1201–1207.
12. Crooks, C. A., Douglas, J. A., Broughton, R. L., and Sefton, M. V. (1990). Microencapsulation of mammalian cells in a HEMA-MMA copolymer: Effects on capsule morphology and permeability. *J. Biomed. Mat. Res.* **24**, 1241–1262.
13. Mikos, A. G., Papadaki, M. D., Kouvroukoglou, S., Ishaoug, S. L., and Thomson, R. C. (1994). Mini-review: Islet transplantation to create a bioartificial pancreas. *Biotechnol. Bioeng.* **43**, 673–677.
14. Soon-Shiong, P., Feldman, E., Nelson, R., Heintz, R., Yao, Q., Yao, Z., *et al.* (1993). Long-term reversal of diabetes by the injection of immunoprotected islets. *Proc. Nat. Acad. Sci. U.S.A.* **90**, 5843–5847.
15. Soon-Shiong, P., Heintz, R. E., Merideth, N., Yao, Q. X., Yao, Z., Zheng, T., *et al.* (1994). Insulin independence in a type 1 diabetic patient after encapsulated islet transplantation. *Lancet* **343**, 950–951.
16. Sun, Y., Ma, X., Zhou, D., Vacek, I., and Sun, A. M. (1996). Normalization of diabetes in spontaneously diabetic cynomologus monkeys by xenografts of microencapsulated porcine islets without immunosuppression. *J. Clin. Invest.* **98**, 1417–1422.
17. De Vos, P., Wolters, G. H. J., Fritschy, W. M., and Van Schilfgaarde, R. (1993). Obstacles in the application of microencapsulation in islet transplantation. *Int. J. Artif. Organs* **16**, 205–212.
18. Lanza, R. P., and Chick, W. L. (1995). Encapsulated cell therapy. *Scien. Am. Sci. Med.* **2**, 16.
19. Cai, Z., Shi, Z., Sherman, M., *et al.* (1989). Development and evaluation of a system of microencapsulation of primary rat hepatocytes. *Hepatology* **10**, 855.
20. Eguchi, S., Chen, S., Rozga, J., and Demetriou, A. A., (1996). Tissue engineering/hybrid tissues: Liver, Yearbook of Cell and Tissue Transplantation, (R. P. Lanza and W. L. Chick, eds.). Kluwer Academic Press, Dordrecht, The Netherlands.
21. Hori, M. (1988). Will artificial liver therapy ever become a reality? *Artif. Organs* **12**, 293.
22. Makowka, L., Falk, R. E., Rostein, L. E., *et al.* (1980). Cellular transplantation in the treatment of experimental hepatic failure. *Science* **210**, 901.
23. Mito, M., and Sawa, M. (1997). Hepatocyte transplantation. *In* "Yearbook of Cell and Tissue Transplantation" (R. P. Lanza and W. L. Chick, eds.). Kluwer Academic Press, Dordrecht, The Netherlands.
24. Sun, A. M., Cai, Z., Shi, Z., *et al.* (1986). Microencapsulated hepatocytes as a bioartificial liver. *Trans. Am. Soc. Artif. Intern. Organs* **32**, 39.
25. Sun, A. M., Cai, Z., Shi, Z., *et al.* (1987). Microencapsulated hepatocytes: An in vitro and in vivo study. *Biomater. Art. Cells Art. Organs.* **15**, 1483.
26. Freed, W. J. (1997). Neural transplantation Brain. *In* "Yearbook of Cell and Tissue Transplantation" (R. P. Lanza and W. L. Chick, eds.). Kluwer Academic Press, Dordrecht, The Netherlands.
27. Sagen, J. (1997). Chromaffin cell transplantation. *In* "Yearbook of Cell and Tissue Transplantation" (R. P. Lanza and W. L. Chick, eds.). Kluwer Academic Press, Dordrecht, The Netherlands.

28. Widner, H., Brundin, J., Rehncroma, S., *et al.* (1991). Transplanted allogeneic fetal dopamine neurons survive and improve motor function in idiopathic Parkinson's disease. *Transplant. Proc.* **23**, 793.

29. Hefti, F., and Knusel, B. (1988). Chronic administration of nerve growth factor and other neurotropic factors to the brain. *Neurobiol. Aging* **9**, 689.

30. Phelps, C. H., Gage, F. H., Growdon, J. H., *et al.* (1989). Potential use of nerve growth factor to treat Alzheimer's disease. *Neurobiol. Aging* **10**, 205.

31. Frim, D. M., Short, M. P., Rosenberg, W. S., *et al.* (1993). Local protective effects of nerve growth factor-secreting fibroblasts against excitotoxic lesions in the rat striatum. *J. Neurosurg.* **78**, 267.

32. Schumacher, J. M., Short, M. P., Hyman, B. T., *et al.* (1991). Intracerebral implantation of nerve growth factor–producing fibroblasts protects striatum against neurotoxic levels of excitatory amino acids. *Neuroscience* **45**, 561.

33. Winn, S. R., Hammang, J. P., Emerich, D. F., *et al.* (1994). Polymer-encapsulated cells genetically modified to secrete human growth factor promote the survival of axotomized septal cholinergic neurons. *Proc. Natl. Acad. Sci. U.S.A.* **91**, 23.

34. Tessler, A. (1997). Neural transplantation: Spinal cord. *In* "Yearbook of Cell and Tissue Transplantation" (R. P. Lanza and W. L. Chick, eds.). Kluwer Academic Press, Dordrecht, The Netherlands.

35. Hellman, L., Smedsrod, B., Sandberg, H., *et al.* (1989). Secretion of coagulant factor VIII activity and antigen by in vitro cultivated rat liver sinusoidal endothelial cells. *Br. J. Haematol.* **73**, 348.

36. Roberts, H. R. (1989). The treatment of hemophilia: Past tragedy and future promise. *N. Engl. J. Med.* **321**, 1188.

37. Fu, X. W., and Sun, A. M. (1989). Microencapsulated parathyroid cells as a bioartificial parathyroid: In vivo studies. *Transplantation* **47**, 432.

38. Wang, T. G. (1997). New technologies for artificial cells. *Artif. Organs* **21**(6).

39. Sefton, M. (2000). A status report on microencapsulation of cells for immunoisolation. *E-biomed* **1**, 49–54.

40. Prokop, A., Hunkeler, D., DiMari, S., Haralson, M., and Wang, T. G. (1998). Water soluble polymer for immunoisolation. I. Complex coacervation and cytotoxicity. *Adv. Polym. Sci.* **136**, 1–53.

41. Prokop, A., Hunkeler, D., Powers, A. C., Whitesell, R., and Wang, T. G. (1998). Water soluble polymer for immunoisolation. II. Evaluation of multicomponent microencapsulation systems. *Adv. Polym. Sci.* **136**, 54–73.

42. Lacik, I., Brissova, M., Anilkumar, A. V., Powers, A. C., and Wang T. G. (1998). New capsule with tailored properties for the encapsulation of living cells. *J. Biomed. Mat. Res.* **39**, 52–60.

43. Molyneux, P. (1984). Interactions between polymers. *In* "Water-Soluble Synthetic Polymers: Properties and Behavior," Vol. II, p. 159. CRC Press, Boca Raton, FL.

44. Tsuchida, E. (1986). Polyelectrolyte complexes. *In* "Developments in Ionic Polymers-2" (A. D. Wilson and H. J. Prosser, eds.), p. 191. Elsevier, Amsterdam.

45. Lee, M. C., Feng, I.-A., Elleman, D. D., Wang, T. G., and Young, A. T. (1982). Generation of a strong core centering force submillimeter compound droplet system. *Proc. 2nd Int. Colloq. Drops Bubbles.* (D. Lecroissate ed.), p. 107.

46. Wang, T. G., Anilkumar, A. V., Lee, C. P., and Lin, K. C. (1994). Core-centering of compound drops in capillary oscillations: Observations on USML-1 experiments in space. *J. Colloid Interface Sci.* **165**, 19–30.

47. Wang, T. G. (1986). "Spherical Shell Technology and Science: Microgravity Science and Applications." National Academy Press, Washington, DC.

48. Kendall, J. M., Chang, M., and Wang, T. G. (1988). Fluid and chemical dynamics relating to encapsulation technology. *Am. Proc.* **197**, 487–495.

49. Lee, C. P., and Wang, T. G. (1988). The centering dynamics of a thin liquid shell in capillary oscillations. *J. Fluid Mech.* **188**, 411–435.

50. Liu, K. C., and Wang, T. G. (1992). A novel method for producing microspheres with semipermeable polymer membranes. *AIAA*, **92**, 118.

51. Lardner, T., and Pujara, P. (1980). Compression of spherical cells. *Mechan. Today*, pp. 161–176.

52. Brissova, M., Lacik, I., Powers, A. C., Anilkumar, A. V., and Wang, T. G. (1998). Control and measurement of permeability for design of microcapsule cell delivery system. *J. Biomed. Mat. Res.* **39**, 61–70.

53. Kraegen, E. W., Chisholm, D. J., and MacNamara, M. E. (1981). Timing of insulin delivery with meals. *Horm. Metab. Res.* **13**, 365.

54. Wang, T. G., and Elliott, R. (1999). NOD mice reversal with encapsulated neonatal porcine islets without immunosuppressive drug (private communication).

CHAPTER 76

MICROENCAPSULATION METHODS: CHITOSAN AND ALGINATE

Mattheus F. A. Goosen

T he applications of two inexpensive polymers, chitosan and alginate, are reviewed with respect to the encapsulation of cells. Membrane permeability control techniques are also considered. Electrostatic droplet generation is presented, as well as modeling of the droplet formation process. The chapter closes with a report on a study in which electrostatics was used to investigate the growth of somatic tissue encapsulated in alginate.

INTRODUCTION

The development of technically effective as well as commercially viable microencapsulation systems requires droplet generation processes that can be easily scaled up, as well as inexpensive, biocompatible polymers [1,2]. A variety of encapsulation methods have evolved over the past few decades, including, for example, emulsion, interfacial polymerization, and extrusion [3,4]. Numerous polymers have also been employed for cell encapsulation, such as agarose, alginate, polyacrylates, and poly(vinyl alcohol) [5,6]. The significant advances in technical development, as well as the gaining of a more fundamental understanding of immobilized cell culture processes, have in part contributed to successes in the application of tissue engineering to the solution of biomedical problems [2,7]. This chapter reviews the application of two inexpensive polymers, chitosan and alginate, with respect to the encapsulation of cells. Electrostatic droplet generation is also presented, as well as modeling of the droplet formation process.

CHITOSAN MEMBRANES AND CELL ENCAPSULATION

Chitosan is a polysaccharide obtained by deacetylating chitin, which is the major constituent of the exoskeleton of crustaceous water animals. Chitosan was reportedly discovered by in 1859 when Rouget [8] boiled chitin in a concentrated potassium hydroxide solution, there by deacetylating the chitin. Fundamental research on chitosan did not start in earnest until about a century later. In 1934 two patents, one for producing chitosan from chitin and the other for making films and fibers from chitosan, were obtained by Rigby [9,10].

The main driving force in the development of new applications for chitosan lies in the fact that the polysaccharide is not only naturally abundant, but is also nontoxic and biodegradable. Most chitosan in practical and commercial use comes from the production of deacetylated chitin, with the shells of crab, shrimp, and krill (the major waste by-product of the shellfish processing industry) being the most available sources of chitosan [11,12].

Chelation takes advantage of one of the most useful properties of chitosan. Chitosan can selectively bind desired materials such as cholesterol, fats, metal ions, proteins, and tumor cells. Other properties that make chitosan very useful include inhibition of tumor cells [13], antifungal effects [14], acceleration of wound healing [15,16], stimulation of the immune system [17–19], and acceleration of plant germination [20].

Chitosan is a good cationic polymer for membrane formation. Chitosan membranes may be prepared in various ways: evaporation of chitosan solvents, cross-linking with bifunctional reagents, chelating with anionic counterions, or complexing with polymers and proteins. In early research it was shown that membranes formed from the polymer could be exploited for water clarification, filtration, fruit coating, surgical dressing, and controlled release. In 1978, for example, Hirano showed that N-acetyl chitosan membranes were ideal for controlled agrochemical release [21]. Later, he found that a semipermeable membrane with a molecular weight cutoff ranging from 2900 to 13,000 could be formed [22] from chitosan. In 1984 Rha *et al.* first documented a procedure for preparing chitosan capsules for cell encapsulation [23]. The chitosan–alginate capsules had a liquid alginate core. Since then, several other studies have been reported on the use of chitosan copolymers for the immobilization of hybridoma cells and plant cells [24,25]. However, Smith *et al.* [26] and McKnight *et al.* [27] indicated the apparently poor biocompatibility of chitosan with hybridoma and insect cells.

Ionotropic gelation for the formation of chitosan membranes is a very mild process. Chitosan membranes have been formed with a variety of counterions or polymers, such as pyrophosphate, octylsulfate, and alginate. Among the counterion polymers that gelate with chitosan, alginate is the most widely used. Usually, alginate gels are prepared by reacting the polymer with divalent ions such as calcium. The encapsulation of living cells by using alginate polycations was developed by Lim and Sun [28] for transplantation purposes (bioartificial pancreas). In their process, a solution of sodium alginate and suspended cells was extruded into a calcium chloride bath. In the reaction that followed, sodium ions in the alginate solution were exchanged for calcium ions. The resultant gel beads were then reacted with a polycationic polymer [poly(L-lysine)] to form a membrane and hardened with poly(ethyleneimine) (to reduce digestibility *in vivo* and capsule clumping). The hard inner core could be reliquefied with sodium citrate to increase the mass transfer rate [29]. In the membrane formation step, a salt bond formed between the positive amine on the polycationic polymer and the negative carboxylic acid on the alginate. Several modifications of this procedure were attempted to increase the capsule's useful *in vivo* lifetime and membrane strength. Rha *et al.* [23] simplified the procedure of capsule formation and replaced poly(L-lysine) with chitosan. The formation of salt bonds between positively and negatively charged functional groups was used to explain the formation of the chitosan–alginate membrane. One advantage of Rha and Rodriguez-Sanchez's procedure [30] was direct membrane formation, which facilitated the free orientation of the polymer chains and increased the membrane strength.

The earliest industrial applications of encapsulation, in the 1950s, entailed the preparation of microcapsules containing ink for carbonless copy paper [31]. More recently, the technique has attracted interest because of its potential for providing higher cell densities and product concentrations. Chitosan–copolymer capsules have been successfully used to culture *Bacillus* [23], plant cells [24], and hybridoma cells [32,33]. The capsule membrane permeability and molecular weight cutoff can be controlled by modulating the viscosity-average molecular weight and concentration of the membrane-forming polymers, the pH, and the ionic strength, as well as the reaction time [34–38]. The authors found that the durability of chitosan–alginate capsule membranes depended on the chitosan molecular weight: the lower the chitosan molecular weight, the stronger and thicker the membranes. They tentatively attributed this effect to the limitation of the molecular size with respect to mass diffusion of chitosan through the pores in the alginate gel matrix.

MEMBRANE PERMEABILITY CONTROL TECHNIQUES

In practice, several factors can be manipulated to control membrane transport in microcapsules. In general, polymer permeability decreases as the molecular weight of nonionic polymers increases. Such intrinsic factors as the tightness of the polymeric chain packing, the rigidity of the polymer chains, and the degree of polymer crystallinity are known to

Table 76.1. Effect of Membrane Polymer Molecular Weight on Capsule Permeability and Strength [40]

Membrane polymer	Permeant[a]	Effect of polymer Mw increase on[b]		Ref.
		Permeability	**Strength**	
PLL	BSA	↑		[42]
PLL	BSA	↑	↓	[43]
Chitosan	BSA	↑		[44]
Chitosan		↑		[44]
Chitosan	BSA	No change	↓	[45]
Chitosan	BSA	↓		[46]

[a]BSA, bovine serum albumin.
[b]↑, increase; ↓, decrease.

exert varying influences on solute diffusivity across polymer membranes [39–41]. It is believed that as polymer molecular weight increases, the degree of molecular chain packing, and consequently the network of interchain bonding and polymer chain stiffness, also rises. Thus, the free volume (hence the diffusion channels) of the amorphous fraction decreases, and so also diffusivity.

However, permeation studies on polymers used in cell encapsulation indicate that the relationship between polymer molecular weight and solute diffusivity elucidated for non-ionic polymers does not hold for polyelectrolytes. Table 76.1 summarizes the findings of various workers using either poly(L-lysine) (PLL) or chitosan as the membrane polymer [42–46]. A careful analysis suggests that polymer membrane diffusivity, hence permeability, appears to be mostly dependent on the extent of ionic interaction between the cationic membrane and the anionic core. A decrease in polymer molecular weight means reductions in polymer chain length and complexity, as well as lower solution viscosity. Therefore, the membrane solution is afforded a greater capacity to penetrate the capsule core and react by forming thicker, more compact membranes with fewer defects. Thus, polyelectrolytes differ from nonionic polymers in that in the former, a fall in molecular weight decreases membrane permeability, provided molecular weight does not fall below a critical level.

The physical strength of a capsule membrane is a function of several factors. Rowe [47,48] observed that the mechanical strength of hydroxypropylmethylcellulose (HPMC) films rose, while their flexibility and incidence of physical defects declined, as polymer molecular weight increased. Sato [49] proposed that cracks or physical defects will occur in a polymeric membrane if the internal stress is greater than or equal to the tensile strength of the membrane.

An effective approach to minimizing internal stress is to plasticize the membrane polymers with noncytotoxic compounds such as the poly(ethylene glycols) and citric acid [50,51].

Applying relationships established earlier, Okhamafe and York [52] showed that the toughness of a membrane, hence its stress crack resistance, is dependent on its strength and flexibility. Internal flaws or incipient cracks in the membrane constitute stress locations that may subsequently propagate, leading to physical defects such as rupture. This finding is significant because defects such as ruptures and cracks severely impair the immunoprotective capacity of a membrane and dislocate transmembrane transport processes.

Even where defects are not manifest at the time of capsule production, some encapsulated systems, upon incubation in a culture medium, show volume expansion due to liquefication and swelling of the capsule core [42,43,53]. If the membrane is not sufficiently tough or resistant to the exerted stress, cracks, and even rupture, may ensue. Furthermore, the swelling-induced stretching of the membrane could widen its diffusion pores and pathways, thus leading to increased solute diffusivity. For thermoplastic polymers, Gentile *et al.* [51]

have noted that increasing membrane composition and structure through the application of techniques that increase the isoreticulated structure ultrafiltration and microfiltration properties of membranes can enhance tensile strength. So also does a reduction in membrane porosity.

The work of King *et al.* [43] and McKnight [45] demonstrated that the higher the molecular weight of PLL and chitosan, the lower the strength, flexibility, and thickness of the membrane, and the higher the incidence of physical defects. An increase in molecular weight (which translates to longer polymer chain length) reduces the capacity of the membrane polymer to penetrate and interact with the core polymer.

Membrane thickness is closely linked to membrane strength and polymer molecular weight, as already outlined. Any measures taken to enhance membrane strength also increase thickness. Since a direct correlation exists between membrane strength and thickness in encapsulated cell systems, efforts to control these factors often exert a profound impact on membrane permeability.

Alteration of the chemical structure of an existing membrane polymer is perhaps one of the least favored approaches to the control of membrane permeability. For one thing, the outcome of such an effort is often very uncertain. Even where the desired membrane permeation characteristics are achieved, the problems of cytotoxicity and biocompatibility may emerge. Furthermore, because the polymer is now an entirely new material, the potential cost of carrying out all the tests for implantable systems required by the relevant regulatory authorities, as well as the attendant delays, generally makes the approach unattractive. Although the success rate appears low, Gharapetian *et al.* [55] have reported that modulation of polyacrylate membrane permeability is possible by modifying the copolymer structure.

There is hardly any doubt now that formation around a capsule of one or more membranes (or, as it seems in some cases, a multilayered membrane) provides an effective tool for modifying membrane transport properties. King *et al.* [43] achieved a threefold increase in intracapsular monoclonal antibody production in a PLL–alginate system consisting of two PLL membranes (molecular weight 2.0×10^5 and 2.2×10^4), compared with a capsule system with a single membrane. Encapsulated cell systems employing composite membranes reportedly have been used to treat type I diabetes [54].

Chang [56] demonstrated that using the standard PLL–alginate method to encapsulate hepatocytes leaves a few of the cells trapped in the membrane matrix, with the result that the membrane is thin and malformed at the points of entrapment. When such capsules were implanted peritoneally in mice, a severe cell-mediated immune response resulted not only in the formation of fibrous tissue around the capsule but also in penetration of the membrane at the protruding sites by elements of the immune system. Although these authors also devised a novel technique whereby the cells are initially entrapped in alginate microspheres, which are then embedded in larger alginate beads and finally enclosed within a PLL membrane, it seems that application of multiple membranes could also solve this problem.

Several studies have examined the effect of polymer solution concentration on the membrane characteristics of encapsulated systems. Goosen *et al.* [42] and King *et al.* [43] found that although membrane thickness remained unchanged when the PLL solution concentration used for preparing PLL–alginate capsules was increased, membrane strength rose while permeability decreased. These and other findings are summarized in Table 76.2 [57–59]. It would appear that for most polyelectrolytes, a rise in mechanical strength and a fall in membrane permeation accompany an increase in membrane polymer concentration. This is because of the greater number of interactions at the membrane–core interface, which produces more compact membranes. There is, however, a threshold solution viscosity beyond which penetration of, and interaction with, the core is inhibited.

The reaction time denotes the period when the membrane solution is allowed to react or interact with the capsule core before the capsules are separated from the polymer solution and washed. This factor has been shown to be of great importance in the PLL–alginate system, where prolongation of reaction time from 3 min to 40 min produced thicker and more compact membranes with correspondingly lower permeability [43]. Apparently, the higher the reaction time, the greater the opportunity afforded the membrane solution to penetrate and interact with the core. However, reaction time has no influence on chitosan–

Table 76.2. Effect of Polymer Solution Concentration on Membrane Permeability, Strength, and Thickness [40]

Membrane polymer	Permeant	Effect of solution concentration increase on			Ref.
		Permeability	**Strength**	**Thickness**	
PLL			↑	No change	[42]
PLL	Carbonic anhydrase	↓	↓	↑	[43]
Chitosan	BSA	↓ (small)			[57]
Chitosan	BSA	↓			[46]
Alginate			↓		[58]
Polyacrylate			↑		[59]

alginate system [44,46]. On the other hand, the alginate membranes of alginate–chitosan capsules exhibited improved strength when reaction time was increased [58] but the authors did not report any permeation data.

The pH of a solution is a handy tool for modulating polyelectrolyte membranes. The cationic polymer chitosan appears to have attracted most attention. This polymer is soluble at low pH, but its solubility decreases as alkaline pH is approached. Furthermore, the micropores of chitosan membranes assume minimum dimensions at low pH, but at about pH 6 and higher, repulsive forces cause the pores to open up, and consequently, membrane permeation rises [57,60]. Polk reported lower membrane strength [46] and flexibility as chitosan pH was lowered from 5.5, but the work of Kim and Rha [57] shows that this effect notwithstanding, membrane permeability decreased as pH was lowered from 6.0 to 3.2.

In polymer science, the use of additives is undoubtedly one of the most convenient techniques for modulating the permeation and mechanical properties of polymeric membranes. Additives, however, have enjoyed only limited applications in encapsulated cell studies, presumably because of uncertainties regarding their biocompatibility and effect on cell viability.

Two main types of additive have attracted the attention of investigators. First are selected ionic additives, which can be employed to modify the power dimensions and frequency of polyelectrolyte membranes, as well as the charge density. Sodium chloride was found to enhance the permeability of chitosan membranes to bovine serum albumin (BSA), but the effect was greater for the low molecular weight chitosan than for the higher molecular weight type. The effect is thought to be largely due to the anion (Cl$^-$), which neutralizes some of the cationic charge on chitosan, thus limiting its capacity to interact with the anionic core to form compact, thick, and strong membranes. Another probable factor is that reducing the cationic charge density in the chitosan membranes actually facilitates protein transport across the membrane because of lower repulsive forces along the diffusion pores and pathways.

The second type of additive is plasticizers. Plasticization results in the severance of the polymer chain segments of a membrane as the plasticizer molecules become sandwiched between the segments. Consequently, polymer chain or segmental mobility increases, a change that is manifested as larger and/or more diffusion pathways as well as increased flexibility [50,52,61–65]. Plasticization has generally been found to cause increased permeability [51] while minimizing coating/membrane defects [52]. The polyols [e.g.. poly(ethylene glycols), propylene glycol, glycerol], are suitable plasticizers for cellulosic polymers [64] and are generally biocompatible and noncytotoxic. Table 76.3 summarizes the influences of certain additives on capsule membrane permeability and mechanical properties [66,67].

MATHEMATICAL MODELING OF POLYMER DROPLET FORMATION USING ELECTROSTATICS

Droplet formation in the presence of an electric field has been analyzed elsewhere [68–73]. If gravity were the only force acting on the meniscus of a droplet attached to the

Table 76.3. Influence of Additives on the Permeability and Mechanical Properties of Capsule Membranes [40]

Membrane polymer	Additive	Permeant	Effect of additive on		Ref.
			Permeability	Strength	
Chitosan	NaCl	BSA	↑		[44]
	Na$_2$PO$_4$	BSA	↑		[44]
Chitosan	NaCl	BSA	↑		[57]
Chitosan	CaCl			↑	
Alginate	Glucose			↑	[58]
Methacrylates	PEG 200		↑		[66]
	Glyerol		↑		[66]
Alginate	Trizma base buffer	BBT^a	↑	↑	[67]
Alginate	Borax			↑	[67]

^{a}Brilliant blue tartrazine.

end of a tube, large uniformly sized droplets would be produced. The gravitational force F_g pulling the droplet from the end of the tube is given by

$$F_g = \frac{4}{3}\pi r^3 \rho g \qquad (1)$$

where ρ is the density of the polymer solution, r is the droplet radius, and g is the acceleration due to gravity. The capillary surface force F_γ, which holds the droplet to the end of the tube, is given by

$$F_\gamma = 2\pi r_0 \gamma \qquad (2)$$

where r_0 is the internal radius of the tube and γ is the surface tension.

Equating the gravitational force on the droplet to the capillary surface tension force holding the droplet to the tube (i.e., the extrusion orifice) gives

$$r = \left(\frac{3r_0\gamma}{2\rho g}\right)^{1/3} \qquad (3)$$

In the presence of an applied voltage, the electric force F_e acting along with the gravitational force F_g would reduce the critical volume for drop detachment, resulting in a smaller droplet diameter. Equating the gravitational and electrical forces on the droplet to the capillary surface force F_γ yields

$$F_\gamma = F_g + F_e \qquad (4)$$

In the case of a charged needle, the stress produced by the external electric field at the needle tip is obtained by using a modified expression developed by De Shon and Carlson [74]:

$$F_e = 4\pi\varepsilon_0\left[\frac{V}{\ln(4H/r_0)}\right]^2 \qquad (5)$$

where H is the distance between the needle tip and collecting solution, V is the applied voltage, and ε_0 is the permittivity of the air.

The effect of applied potential on the droplet radius for a charged needle arrangement can be derived by substituting Eqs. (1), (2), and (5) into Eq. (4):

$$r = \left\{\left[\frac{3}{2\rho g}\right]\left\{r_0\gamma - 2\varepsilon_0\left[\frac{V}{\ln(4H/r_0)}\right]^2\right\}\right\}^{1/3} \qquad (6)$$

Table 76.4. Comparison of Experimental and Calculated Microbead Diameter as a Function of Extrusion Orifice Size and Applied Potential [73]

Extrusion orifice diameter (μm)	Applied potential (kV)	Microbead diameter (μm)	
		Calculated[a]	Experimental[b]
400	0	2600	2000
1000	0	3500	2800
1900	0	4400	3700
1900	5	4018	3500
1900	10	1680	1700

[a]Bead size at 0 kV was determined by using Eq. (3) and at 10 kV by using Eq. (6).

[b]Bead size using 4% w/v nonautoclaved sodium alginate in water with an electrode distance of 6 cm and a 22-gauge needle (r_0 = 500 μm).

Equation (3) was employed to calculate the microbead diameter in the absence of an applied voltage (i.e., 0 kV). The surface tension γ of the alginate solution was assumed to be 73 g/s^2, which is the value for water against air [75]. The density of the polymer solution was taken as 1 g/cm^3. In the presence of an applied voltage, Eq. (6) was used to determine microbead size. The permittivity of air ε_0, used in calculations was 1.0 g cm/s^2 kV^2. This value was estimated based on earlier studies [68].

Reasonably good agreement was obtained between calculated and experimental values of microbead diameter (Table 76.4) [76]. For example, when the extrusion orifice diameter decreased from 1900 μm to 400 μm, the calculated bead diameters decreased from 4400 μm to 2600 μm, and the experimental values decreased from 3700 μm to 2000 μm. When the extrusion orifice diameter was kept constant at 1900 μm and the applied voltage was increased from 0 to 10 kV, there was also a similar decrease in bead size from 4400 μm to 1690 μm for the calculated values and from 3700 μm to 1700 μm for the experimental values.

USE OF ELECTROSTATICS TO STUDY THE GROWTH OF SOMATIC TISSUE ENCAPSULATED IN ALGINATE

Several investigators [69–72] have studied the mechanism of alginate droplet formation by means of an electrostatic droplet generator attached to a variable-voltage power supply (Fig. 76.1). Animal cell suspensions were successfully extruded from the electrostatic droplet generator. The application of this technology to plant cell immobilization was reported in the late 1990s [3,72,76]. A major concern in cell and bioactive agent immobilization has been the production of very small microbeads, to minimize the mass transfer resistance problem associated with large-diameter beads (>1000 μm). Klein *et al.* [77] reported production of alginate beads with diameters of 100–400 μm where compressed air was used to quickly pass the cell–gel solution through a nozzle. Few attempts have been made in the application of electric fields to the production of micrometer-size polymer beads for cell immobilization [73].

Somatic embryogenesis is a new plant tissue culture technology in which somatic cells (i.e., any cell except a germ or seed cell) are used to produce an embryo (i.e., plant in early state of development) [78]. The technique of somatic embryogenesis in liquid culture, which is believed to be an economical way for future production of artificial seeds, may benefit from cell immobilization technology. Encapsulation may aid in the germination of somatic embryos by allowing for higher cell densities, protecting cells from shear damage in suspension culture, allowing for surface attachment in the case of anchorage-dependent cells, and being very suitable for scale-up in bioreactors.

The long-term objective of the project reported in this section is the development of an effective method for the large-scale production of artificial seeds by means of electrostatic droplet formation and immobilized cell culture technology. The short-term aim was to use

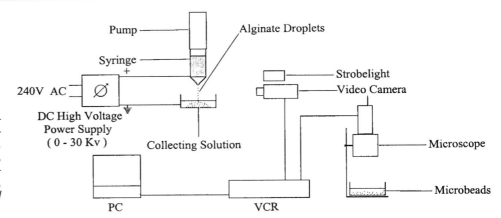

an electrostatic droplet generator to investigate the production of small alginate microbeads. Callus tissue from African violets was also immobilized and cultured.

MICROBEAD SIZE AND MECHANISM OF ALGINATE DROPLET FORMATION

To be able to encapsulate callus tissue aggregates 1–2 mm in size without having the tissue clogg the extrusion orifice, a syringe (1900-μm-i.d. outlet) had to be used without a needle. In addition, alginate was dissolved directly in the medium, instead of pure water or saline solution, to help keep the plant tissue viable. The effect of these changes on electrostatic droplet generation was assessed. There was a significant difference, for example, in alginate bead size resulting from extrusion of nonautoclaved 2% alginate dissolved in the medium and from extrusion of the polysaccharide dissolved in water (Fig. 76.2). At 5 kV, for instance, 2% alginate in medium produced gel beads 4300 μm in diameter, while 2% alginate in water gave beads 2300 μm in diameter. The effect of the medium on bead size, which was observed only at 2% alginate concentration and at an applied voltage of less than 10 kV, may have been due to a change in the liquid surface tension as a result of amino acids in the medium.

With 4% alginate in water, the bead size decreased from 3700 μm to 1200 μm, when the applied voltage was increased from 0 to 20 kV. With 2% alginate at 20 kV, microbeads as small as 500 μm could be produced. The sharpest drop occurred between 5 and 10 kV, probably because the electrostatic force had overcome the surface tension force holding the droplet to the syringe orifice. The decrease in bead size with increasing applied potential (Fig. 76.2) for extrusion directly from a syringe observed in the present investigation is similar to what was reported by Bugarski *et al.* [69,70] and Goosen *et al.* [72] when a stainless steel needle was employed as the extrusion orifice. Smaller bead sizes, however, would decrease the effective plating density of cells or aggregates per drop.

Video photography and digital image processing of the mechanism of alginate droplet formation directly from a syringe, in the absence of a needle, revealed the following sequence of events. With the power switched off (i.e., 0 kV applied potential), a large 4000-μm droplet (2% alginate in water) formed at the tip of the syringe, held on by surface tension (Fig. 76.3A). When the power was switched on (10 kV), the electrostatic force produced a Taylor cone–like droplet (Fig. 76.3B). The alginate cone extended to form a thin strand (Fig. 76.3C). Finally, the fine alginate strand broke away from the orifice and formed small alginate droplets (Fig. 76.3D). Note that the positive electrode was inserted directly into the syringe (top right). A fine alginate stream was also observed when a needle was attached to the syringe. In a similar study, Stegemann and Sefton used video and digital image analysis methods[5] in an assessment of a submerged jet microencapsulation process with HEMA-MMA (i.e., hydroxymethyl methacrylate and methyl methacrylate).

Good agreement was obtained between calculated and experimental values of microbead diameter (Table 76.4). For example, when the extrusion orifice diameter decreased from 1900 μm to 400 μm at o kV, the calculated bead diameters decreased from 4400 μm

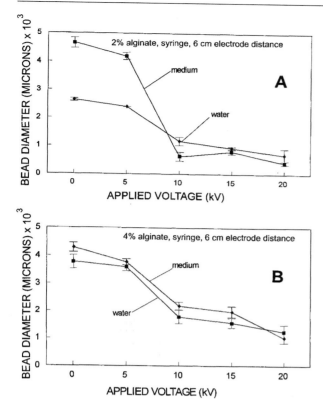

Fig. 76.2. Effect of medium, alginate concentration, and applied voltage on microbead diameter when direct syringe extrusion was used without a needle: (A) 2% alginate and (B) 4% alginate. Error bars refer to the standard deviation based on a sample size of 30 beads. Reprinted with permission from Al-Hajry et al. [76]. Copyright 1999, American Chemical Society.

to 2600 μm and the experimental values decreased from 3700 μm to 2000 μm, respectively. When the extrusion orifice diameter (i.e., syringe) was kept constant at 1900 μm microns and the applied voltage was increased from 0 to 10 kV, there was also a similar decrease in bead size from 4400 μm to 1690 μm for the calculated values and from 3700 μm to 1700 μm for the experimental values, respectively. The original paper by DeShon and Carlson [74] noted that the expression for the stress produced by the electric field, though valid in the case of a metal sphere–plate electrostatic system, leads to errors of 20% when used for the prediction of drop formation.

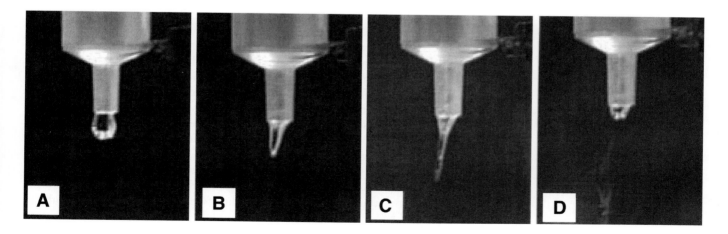

Fig. 76.3. Mechanism of alginate droplet formation at syringe orifice. (A) Droplet formation in the absence of applied potential. (B) Power switched on, Taylor cone–like droplet forms. (C) Cone extends to form thin alginate strands. (D) Alginate strand breaks away from orifice and breaks up to form droplets. Note the positive electrode inserted directly into the syringe (top right). Extrusion parameters: 10 kV applied potential, 6-cm electrode distance, 2% alginate in water; syringe orifice, 1.9 mm i.d., 4.0 mm o.d. Reprinted with permission from Al-Hajry et al. [76]. Copyright 1999, American Chemical Society.

Deformed alginate microbeads were obtained in our study with 1% alginate solution, whereas proper beads were produced with 2 and 4% alginate. This was slightly different from the teardrop-shaped bead found in the case of extrusion through a needle [68,79]. We can speculate that at 1% alginate, the surface tension is not strong enough to cause the alginate–medium droplet to form into a perfect sphere before impacting the surface of the $CaCl_2$ hardening solution. Internal channels/striations could also just be seen at the lowest alginate concentration. This of course was not the case at the 2 and 4% alginate concentrations. Striations or interstitial spaces are due to the mechanism of alginate gel formation, which has been well documented [80,81]. Upon addition of sodium alginate solution to a calcium chloride solution, interfacial polymerization is instantaneous, with precipitation of calcium alginate followed by a more gradual gelation of the interior as calcium ions permeate through the alginate. Block copolymer segments of the neutralized guluronic acid residues in alginate assemble into an "eggbox" type of complex around an array of calcium ions. The mannuronic acid residues of alginate, on the other hand, form a random agglomeration of chains that act as complex breakers. These residues randomize the gelatin reaction product into a three-dimensional configuration that, at low alginate concentrations, contains interstitial spaces or striations.

Autoclaving had an appreciable effect on the viscosity of the alginate solution. For example, on the basis of the viscometric method, the viscosity of a 4% sodium alginate in water solution decreased from 655 cP to 343 cP as a result of autoclaving for 20 min at 120°C and 15 psi. At 6% alginate the viscosity of the polysaccharide in water solution decreased from 2447 cP to 628 cP, while that of the alginate in medium solution dropped from 2558 cP to 649 cP, as a result of autoclaving. Investigators need to be aware that if an alginate solution must be autoclaved, its viscosity should be checked before and after sterilization treatment. If the alginate viscosity due to autoclaving significantly drops, the initial alginate concentration should be increased to counteract this decrease in viscosity. It is the final viscosity of the alginate solution that will determine microbead morphology.

CULTURE AND GROWTH OF IMMOBILIZED CALLUS TISSUE

A combination of needle (21-gauge) and syringe was employed in a preliminary cell–tissue immobilization study. The plant tissue clogged the needle, however and it was found necessary to switch to 19-gauge needle and finally to no needle at all (i.e., syringe only). The positive electrode in the last case was inserted directly into the side of the plastic syringe, close to the syringe outlet as depicted in Fig. 76.3. With this modified droplet generation technique, the problem of tissue segments/aggregates clogging the syringe outlet was solved. Even with a relatively wide syringe orifice of 1900 μm, the use of electrostatics gave bead diameters as low as 500 μm (Fig. 76.2).

Three tissue preparation methods were assessed in the growth of immobilized callus. These included cutting of callus to produce tissue segments 1–2 mm in size, sieving to give cell aggregates, and using a very fine sieve to produce dissociated individual cells. The tissue/cells were immobilized in 2, 4, and 6% alginate prepared with MS medium. Overall, the tissue segments immobilized in 4% alginate in MS medium and cultured on agar gel grew best (Fig. 76.4). Immobilized single cells were the most difficult to culture and rarely resulted in cell growth. Tissue immobilized in 6% alginate showed poor cell growth on agar and no cell growth in liquid (suspension) culture. This was presumable due to mass transfer limitations. African violet tissue immobilized in 4% alginate grew better than that grown in 2% alginate. This was unexpected, since the latter gel structures should be more porous. With the alginate dissolved in MS medium, some of the key higher molecular weight nutrients may have been retained by the gel bead at the 4% alginate concentration, with the result that they could be used by the plant cells. Immobilized cell culture on agar gel was more effective than suspension culture in liquid medium in E-flasks.

Tissue segments entrapped in 2.0% alginate microbeads and cultured on agar gave good cell growth, with 70% of the encapsulated tissue producing shoots. On the other hand, cell aggregates under the same immobilization/culture conditions produced poor cell growth, with only 5% of the encapsulated tissue forming callus. Why were shoots produced only from tissue segments and not from cell aggregates or individual cells? Shoot

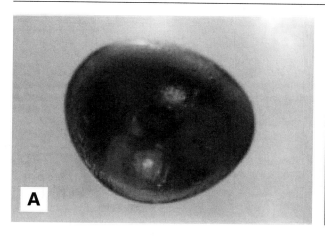
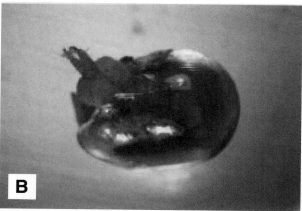
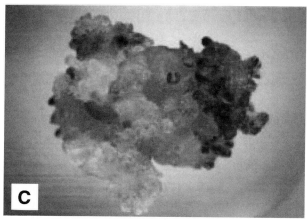

Fig. 76.4. Callus tissue from African violets encapsulated in 4% (w/v) sodium alginate in medium using electrostatics at 10 kV, 4-cm electrode distance, and with a syringe: (A) day 0, (B) day 15, and (C) day 30. Immobilized tissue cultures on agar gel (microbead size: 2000 μm). Reprinted with permission from Al-Hajry et al. [76]. Copyright 1999, American Chemical Society.

development depends on the type of callus tissue formed, which in turn is determined by the direction of cell differentiation. Cells that are preorganogenically determined, as was the case for the cell aggregates growing in 2.0 and 4.0% alginate on agar, cannot change their direction to the next stage. They are preembryonically determined even in the presence of shoot growth promoters (IAA and BA). Adventitious shoots, though, were formed from tissue segments entrapped in 2.0 and 4.0% alginate microbeads and cultures on agar. We can speculate that perhaps a critical cell density may need to be reached before callus cells can change their direction to form organized cell structures (i.e., shoots).

Plantlets obtained from 4% alginate beads on agar, originally immobilized at 10 kV, 6 cm distance, were transferred to sterilized potting mixture at 2 months culture. The plantlets grew well and showed complete leaf and root development by 4 months (Fig. 76.5). Suspension culture of encapsulated callus tissue was less successful. Piccioni [82] in 1997 investigated the growth of plantlets from alginate-encapsulated buds of M.26 apple rootstocks. He showed that the addition to the somatic tissue culture of growth regulators (e.g., indolebutyric acid) several days prior to the encapsulation, as well as the addition of the same regulators to the encapsulation matrix, improved the production of plantlets in suspension culture from 10% to more than 60%. We can speculate that culturing the African violet somatic tissue in the presence of growth regulators prior to and during encapsulation may enhance the production of plantlets from suspension culture. Tissues from plants of various types, though, behave quite differently. For example, mulberry and banana plantlets were attained from alginate-encapsulated tissue without any specific root induction treatment [83,84], while Piccioni's apple rootstock bud required growth regulators.

Fig. 76.5. African violet plantlets grown from encapsulated somatic leaf tissue at 4 months. (A) Plantlets in sterilized potting mix composed of 50% peat moss and 50% vermiculite. (B) Complete plant. Reprinted with permission from Al-Hajry et al. [76]. Copyright 1999, American Chemical Society.

Electrostatic droplet generation does not appear to have a negative impact on somatic tissue viability, since cell growth and plantlet formation were observed. This is in agreement with similar studies reported for insect cells and mammalian cells [73,79] which showed that high electrostatic potentials did not affect cell viability.

CONCLUDING REMARKS

The ability to develop successful and well-understood microencapsulation systems will necessitate close collaboration among scientists with different areas of expertise such as engineering, microbiology, biochemistry, and medicine. Over the next decade, as our fundamental knowledge of encapsulated cell systems increases, we can expect to see many new areas of application in medicine, in agriculture, and in the bioprocess industries.

ACKNOWLEDGMENT

The financial support of the Natural Sciences and Engineering Research Council of Canada, and of Sultan Qaboos University (grant AGBIOR 9505 to the author) is gratefully acknowledged. The membrane permeability control experiments were performed and analyzed by Dr. A. Okhamafe of Benin University. The plant cell immobilization study was performed by H. Al-Hajry, S. Al-Maskari, and L. Al-Kharousi (SQU). The assistance of Dr. O. El-Mardi in the plant tissue culture is also acknowledged.

REFERENCES

1. Colton, C. K. (1996). Engineering challenges in cell encapsulation technology. *Trends Biotechnol.* **14**, 158–162.
2. Lanza, R. P., Hayes, J. L., and Chick, W. L. (1996). Encapsulated cell technology. *Nat. Biotechnol.* **14**, 1107–1111.
3. Goosen, M. F. A. (1999). Mass transfer in immobilized cell systems. *In* "Cell Encapsulational Technology and Therapeutics" (W. M. Kuhtreiber, R. P. Lanza, and W. L. Chick., eds.), Chapter 3, pp. 18–28. Birkhaeuser and Springer-Verlag, Berlin.
4. Li, Q., Goosen, M. F. A., Grandmaison, E. W., and Taylor, D. (2000). Convection-enhanced mass transfer in aggregated beads for gel chromatography. *AIChE J.* **50** (in press).
5. Stegemann, J. P., and Sefton, M. V. (1996). Video analysis of submerged jet microencapsulation using HEMA-MMA. *CJChE* **74**, 518–525.
6. De Vos, P. (1996). Association between capsule diameter, adequacy of encapsulation, and survival of microencapsulation rat islet allografts. *Transplantation* **62**, 893–899.
7. Hunkeler, D. (1999). Bioartificial organs: Risks and requirements. *Ann. N.Y. Acad. Sci.* **875**, 1–6.

8. Muzzarelli, R. A. A. (1977). "Chitin." Pergamon of Canada Ltd., Toronto.

9. Rigby, G. W. (1934). Substantially undergraded deacetylated chitin and process for producing the same. U.S. Pat. 2,040,879.

10. Rigby, G. W. (1934). Process for the preparation of films and filaments and products thereof. U.S. Pat. 2,040,880.

11. Brine, C. J. (1984). *In* "Chitin, Chitosan and Related Enzymes" (J. P. Zikakis, ed.), Chapters 17–23. Academic Press, New York.

12. Li, Q., Dunn, E. T., Grandmaison, E. W., and Goosen, M. F. A. (1997). Applications and properties of chitosan. *In* "Applications of Chitin and Chitosan" (M. F. A. Goosen, ed.), pp. 1–29. Technomic Publ., NJ.

13. Sirca, A. E., and Woodman, R. J. (1971). *JNCI, J. Natl. Cancer Inst.* **47**(2), 377–388.

14. Allan, C. R., and Hadwiger, L. A. (1979). *Exp. Mycol.* **3**(3), 285–287.

15. Balassa, L. L., and Prudden, J. F. (1978). MIT sea grant Rep. mITSG, 78-7. *Proc. Int. Conf. Chitin/Chitosan, 1s,* PB 285 640, pp. 296–305.

16. Malette, W. G., Jr., Quigley, H. J., and Adiches, E. D. (1986). *In* "Chitin in Nature and technology" (R. Muzzarelli, C. Jeuniaux, and G. W. GooDay, eds.), pp. 435–442. Plenum, New York.

17. Suzuki, K., Ogawa, Y., Hashimoto, K., Suzuki, S., and Suzuki, M. (1984). *Microbiol. Immunol.* **28**, 903–912.

18. Nishimura, K., Nishimura, S., Nishi, N., Tokura, S., and Azuma, I. (1986). *In* "Chitin in Nature and Technology" (R. Muzzarelli, C. Jeuniaux, and G. W. GooDay, eds.), pp. 477–483. Plenum, New York.

19. Eida, T., and Hidaka, H. (1988). *Jpn. Fudo Saiensu* **27**(12), 56–63.

20. Yano, S., and Tsugita, T. (1988). Chitosan-containing seed coatings for yield enhancement, Jpn. Kokai Tokkyo Koho JP 63,139,102 [88,139,102].

21. Hirano, S. (1978). *Agric. Biol. Chem.* **42**(10), 1939–1940.

22. Hirano, S., Tobetto, K., Hasegawa. M., and Matsuda, N. (1980). *J. Biomed. Mater. Res.* **14**, 477–486.

23. Rha, C., Rodriguez-Sanchez, D., and Kienzle-Sterzer, C. (1984). *In* "Biotechnology of Marine Polysaccharides" (R. R. Colwell, E. R. Pariser, and A. J. Sinskey, eds.), pp. 283–311. Hemisphere Publishing, Washington, DC.

24. Knorr, D., Beaumont, M. S., and Pandya, Y. (1987). *Biotechnol. Food Ind., Proc. Int. Symp.*, pp. 389–400.

25. Shiotani, T., and Shiiki, Y. (1986). Preparation of capsules using phosphates and chitin derivatives. Jpn. Kokai Tokkyo Koho JP 61,153,135 [86,153,135].

26. Smith, N. A., Goosen, M. F. A., King, G. A., Faulkner, P., and Daugulis, A. J. (1989). Toxicity analysis of encapsulation solutions and polymers in the cultivation of insect cells. *Biotechnol. Lett.* **3**(1), 61–66.

27. McKnight, C. A., Ku, A., Goosen, M. F. A., Sun, D., and Penney, C. (1987). Synthesis of chitosan–alginate microcapsule membranes. *J. Bioact. Compat. Polym.* **3**, 334–355.

28. Lim, F., and Sun, A. M. (1980). *Science* **210**, 908–910.

29. Lim, F. (1982). Encapsulation of biological materials. U.S. Pat. 4,352.883.

30. Rha, C., and Rodriguez-Sanchez, D. (1988). Encapsulated active material system. U.S. Pat. 4,749,620.

31. Green, B. K. (1955). Pressure-sensitive record materials. U.S. Pat. 2,712,507.

32. Yoshioka, T., Hirano, R., Shioya, T., and Kako, M. (1990). *Biotechnol. Bioeng.* **35**, 66–72.

33. Kim, S. K., and Rha, C. (1989). *In* "Chitin and Chitosan" (G. Skjak-Braek, T. Anthonsen, and P. Sandford, eds.), pp. 617–626. Elsevier, New York.

34. Goosen, M. F. A., King, G. A., McKnight, C. A., and Marcotte, N. (1989). Animal cell culture engineering using alginate polycation microcapsules of controlled membrane molecular weight cut-off. *Membr. Sci.* **41**, 323–343.

35. Goosen, M. F. A. (1987). Insulin delivery systems and the encapsulation of cells for medical and industrial use. *CRC Crit. Rev. Biocompat.* **3**(1), 1–24.

36. Daly, M. M., Keown, R. W., and Knorr, D. W. (1989). Chitosan alginate capsules. U.S. Pat. 4,808,707.

37. Shioya, T., and Rha, C. (1989). *In* "Chitin and Chitosan" (G. Skjak-Braek, T. Anthonsen, and P. Sandford, eds.), pp. 627–634. Elsevier, New York.

38. Kim, S. K., and Rha, C. (1989). *In* "Chitin and Chitosan" (G. Skjak-Braek, T. Anthonsen, and P. Sandford, eds.), pp. 627–642. Elsevier, New York.

39. Okhamafe, A. O., and York, P. (1987). Interactive phenomenon in pharmaceutical film coatings and testing methods. *Int. J. Pharm.* **39**, 1–21.

40. Okhamafe, A. O., and Goosen, M. F. A. (1999). Modulation of membrane permeability. *In* "Cell Encapsulation Technology and Therapeutics" (W. M. Kuhtreiber, A. P. Lanza, and W. L. Chick, eds.), Chapter 5, pp. 53–63. Birkhaeuser and Springer-Verlag, Berlin.

41. Rogers, C. E. (1976). Structural factors governing controlled release. *ACS Symp. Ser.* **33**, 15–25.

42. Goosen, M. F. A., O'Shea, G. M., Gharapetian, H. M., Chou, S., and Sun, A. M. (1985). Optimization of microencapsulation parameters; semipermeable microcapsules as a bioartificial pancreas. *Biotechnol. Bioeng.* **27**, 146–150.

43. King, G. A., Daugulis, A. J., Faulkner, P., and Goosen, M. F. A. (1987). Alginate–polylysine microcapsules of controlled membrane molecular weight cut-off for mammalian cell culture engineering. *Biotechnol. Prog.* **3**, 231–240.

44. Shioya, T., and Rha, C. (1989). Transmembrane permeability of chitosan/carboxymethyl cellulose capsule. *In* "Chitin and Chitosan" (G. Skjaek-Brake, T. Anthonsen, and P. Sandford, eds.), pp. 627–634. Elsevier, London.

45. McKnight, C. A. (1987). Chemical modification of chitosan for the microencapsulation of mammalian cells. M.Sc. Thesis, Queen's University, Kingston, Canada.

46. Polk, A. E. (1990). Development of chitosan–alginate microcapsules for the oral delivery of vaccines in aquaculture. M.Sc. Thesis, Queen's University, Kingston, Canada.

47. Rowe, R. C. (1976). The effect of molecular weight on the properties of films prepared from hydroxypropyl methylcellulose. *Pharm. Acta Helv.* **51**, 330–334.

48. Rowe, R. C., and Forse, S. F. (1980). The effect of polymer molecular weight on the incidence of film cracking and splitting in film-coated tablets. *J. Pharm. Pharmacol.* **32**, 583–584.

49. Sato, K. (1980). The internal stress of coating films. *Prog. Org. Coat.* **8**, 143–148.

50. Okhamafe, A. O., and York, P. (1985). Interaction phenomena in some aqueous-based tablet coating polymer systems. *Pharm. Res.* **2**, 19–23.

51. Okhamafe, A. O., and York, P. (1988). Studies of interaction phenomena in aqueous-based film coatings containing soluble additives using thermal analysis techniques. *J. Pharm. Sci.* **77**, 435–444.

52. Okhamafe, A. O., and York, P. (1985). Stress crack resistance of some pigments and unpigmented tablet film coating systems. *J. Pharm. Pharmacol.* **37**, 449–454.

53. Yoshioka, T., Hirano, R., Shioya, T., and Kako, M. (1990). Encapsulation of mammalian cells with chitosan–CMC capsule. *Biotechnol. Bioeng.* **35**, 66–72.

54. Gentile, F. T., Doherty, E. J., Rein, D. H., Shoichet, M. S., and Winn, S. H. (1995). Polymer science for macroencapsulation of cells for central nervous system transplantation. *React. Polym.* **25**, 207–227.

55. Ggharapetian, H., Maleki, M., O'Shea, G. M., Carpenter, R. C., and Sun, A. M. (1987). Polyacrylate microcapsules for cell encapsulation: Effects of copolymer structure on membrane properties. *Biotechnol. Bioeng.* **30**, 775–779.

56. Chang, T. M. S. (1993). Living cells and microorganisms immobilized by microencapsulation inside artificial cells. *In* "Fundamentals of Animal Cell Encapsulation and Immobilization" (M. F. A. Goosen, ed.), pp. 183–196. CRC Press, Boca Raton, FL.

57. Kim, S. K., and Rha, C. (1989). Transmembrane permeation of proteins in chitosan capsules. *In* "Chitin and Chitosan" (G. Skjaek-Braek, T. Anthonsen, and P. Sandford, eds.), pp. 635–642. Elsevier, London.

58. Daly, M. M., and Knorr, D. (1988). Chitosan–alginate complex coacervate capsules: Effect of calcium chloride, plasticizers, and polyelectrolytes on mechanical stability. *Biotechnol. Prog.* **4**, 76–83.

59. Sefton, M. V., Blysniuk, J., Broughton, R. L., Dawson, R. M., and Sugamori, M. E. (1987). Microencapsulation of mammalian cells in a water insoluble polyacrylate by coextrusion and interfacial precipitation. *Biotechnol. Bioeng.* **29**, 1135–1143.

60. Okhamafe, A. O., Amsden, B., Chu, W., and Goosen, M. F. A. (1996). Modulation of protein release from chitosan–alginate microcapsules using the pH-sensitive polymer hydroxypropyl methylcellulose acetate succinate. *J. Microencapsul.* **13**, 497–508.

61. Deschesne, J. P., Delporte, J. P., Jaminet, F., and Venturas, K. (1982). Influence of plasticizers on properties of applied films of Eudragjit L30D. *J. Pharm. Belg.* **37**, 283–286.

62. Okhamafe, A. O., and Iwebor, H. U. (1987). Moisture permeability mechanisms of some aqueous-based tablet film coatings containing soluble additives. *Pharmazie* **42**, 611–613.

63. Porter, S. C. (1982). The practical significance of the permeability and mechanical properties of polymer films used for the coating of pharmaceutical dosage forms. *Int. J. Pharm. Technol. Prod. Manuf.* **3**, 21–25.

64. Sakellarious, P., Rowe, R. C., and White, E. F. T. (1986). An evaluation of the interaction and plasticizing efficiency of the polyethylene glycols in ethylcellulose and hydroxypropyl methycellulose films using torsional braid pendulum. *Int. J. Pharm.* **3**, 55–64.

65. Yuet, P., Harris, T. J., and Goosen, M. F. A. (1995). Mathematical modelling of immobilized animal cell growth. *Int. J. Artif. Cells, Blood Substitutes Immobilization Biotechnol.* **23**(1), 109–133.

66. Crooks, C. A., Douglas, J. A., Broughton, R. L., and Sefton, M. V. (1990). Microencapsulation of mammalian cells in HEMA-MMA copolymer effects on capsule morphology and permeability. *J. Biomed. Mater. Res.* **24**, 1241–1247.

67. Knorr, D., and Daly, M. (1988). Mechanics and diffusional changes observed in multilayer chitosan/alginate coacervate capsule. *Process Biochem.* **23**, 48–50.

68. Bugarski, B., Amsden, B., Neufeld, R. J., Poncelet, D., and Goosen, M. F. A. (1994). Effect of electrode geometry and charge on the production of polymer microbeads by electrostatics. *CJChE* **72**, 517–521.

69. Bugarski, B., Smith, J., Wu, J., and Goosen, M. F. A. (1993). Methods of animal cell immobilization using electrostatic droplet generation. *Biotechnol. Tech.* **6**(9), 677–682.

70. Bugarski, B., Li, Q., Goosen, M. F. A., Poncelet, D., Neufeld, R., and Vunjak, G. (1994). Electrostatic droplet generation: Mechanism of polymer droplet formation. *AIChE J.* **440**(6), 1026–1031.

71. Goosen, M. F. A., Mahmoud, E. S. E., Al-Ghafri, A. S., Al-Hajri, H. A., Al-Sinani, Y. S., and Bugarski, B. (1996). Immobilization of cells using electrostatic droplet generation. *In* "Methods in Molecular Biology: Immobilization Enzymes and Cells" (G. Bickerstaff, ed.), Humana Press, Totowa, NJ.

72. Goosen, M. F. A., Al-Ghafri, A. S., El-Mardi, O., Al-Belushi, M. I. J., Al-Hajri, H. A., Mahmoud, E. S. E., and Consolacion, E. C. (1997). Electrostatic droplet generation for encapsulation of somatic tissue: Assessment of high voltage power suller. *Biotechnol. Prog.* **13**(4), 497–502.

73. Goosen, M. F. A., O'Shea, G. M., Gharapetian, M. M., and Sun, A. M. (1986). Immobilization of living cells in biocompatible semipermeable microcapsules: Biomedical and potential biochemical engineering applications. *In* "Polymers in Medicine" (E. Chielline, ed.), p. 235. Plenum, New York.

74. De Shon, E. W., and Carlson, R. (1968). Electric field and model for electrical liquid spraying. *J. Colloid Sci.* **28**, 161–166.

75. Weast, R. C. (1979). "Handbook of Chemistry and Physics," 59th ed., p. F45. CRC Press, Boca Raton, FL.

76. Al-Hajry, H. A., Al-Maskary, A. A., Al-Kharousi, L. M., El-Mardi, O., Shayya, W. H., and Goosen, M. F. A. (1999). Electrostatic encapsulation and growth of plant cell cultures in alginate. *Biotechnol. Prog.* **15**, 768–774.

77. Klein, J., Stock, J., and Vorlop, D. K. (1993). Pore size and properties of spherical calcium alginate biocatalysts. *Eur. J. Appl. Microb. Biotechnol.* **18**, 86.

78. Teng, W. L., Liu, Y. J., Tsai, V. C., and Soong, T. S. (1994). Somatic Embryogenesis of carrot in bioreactor culture systems. *Hortic. Sci.* **29**(11), 1349–1352.

79. King, G. A., Daugulis, A. J., Faulkner, P., Bayly, D., and Goosen, M. F. A. (1989). Alginate concentration: A key factor in growth of temperature-sensitive insect cells in microcapsules. *Biotechnol. Bioeng.* **34**, 1085–1091.

80. Stevenson, W. T. K., and Sefton, M. F. (1993). Development of polyacrylate microcapsules. *In* "Fundamentals of Animal Cell Encapsulation and Immobilization" (M. F. A. Goosen, ed.), pp. 143–182. CRC Press, Boca Raton, FL.

81. Fraser, J. E., and Bickerstaff, G. F. (1997). Entrapment in calcium alginate. *In* "Methods in Microbiology: Immobilization of Enzymes and Cells" (G. F. Bicker Staff, ed.), pp. 61–66. Humana Press, Totowa, NJ.

82. Piccioni, E. (1997). Plantlets from encapsulated micropropagated buds of m. 26 apple rootstock. *Plant Cell, Tissue Organ Cult.* **47,** 255–260.

83. Bapat, V. A., and Rao, P. S. (1990). In-vivo growth of encapsulated auxilliary buds of mulberry (*Morus indica*, L). *Plant Cell, Tissue Organ Cult.* **20,** 60–70.

84. Ganapathi, T. R., Suprasanna, P., Bapat, V. A., and Rao, P. S. (1992). Propagation of banana through encapsulated shoot tips. *Plant Cell Rep.* **11,** 571–575.

SECTION III

METHODS FOR ENGINEERING CELLS AND TISSUES

CHAPTER 77

FETAL CELL CULTURE

Paolo De Coppi, Gunter Schuch, and Anthony Atala

INTRODUCTION

The use of fetal cells has a number of theoretical advantages that make it attractive for tissue engineering and transplantation research [1].

With an ability to differentiate into multiple cell types, fetal cells can often differentiate in response to environmental cues or according to an intrinsic program. This plasticity means that such cells may migrate, proliferate, differentiate, and establish functional connections with other cells.

Fetal cells usually proliferate more rapidly than mature, fully differentiated cells. They can generally survive at lower oxygen levels than mature cells. They are, therefore, more resistant to ischemic damage during *in vitro* manipulation or after transplantation, and they display better survival after refrigeration and cryopreservation protocols [2]. Fetal cells also produce high levels of angiogenic and neurotrophic factors, which enhance their ability to grow once grafted and facilitate the regeneration of surrounding host tissue [3]. In addition, antigens are expressed at lower levels in some fetal tissues than in corresponding adult tissues, thus making fetal tissues less susceptible to rejection [4,5].

SOURCES FOR FETAL CELLS

For decades, fetal cells of several types have been used for transplantation, either experimentally or clinically. There are two principal sources from which fetal cells can be isolated: umbilical cord blood and the fetus itself.

Elective abortions remain the preferred means for obtaining human fetal cells. Tissues obtained from spontaneously aborted fetuses are usually nonviable and may not be available in a predictable manner [6].

Minimally invasive harvest of fetal tissue can be performed *in utero*. Tissue can be engineered *in vitro* in parallel to the remainder of gestation and may be available for surgical implantation *in utero*, at birth, or later in life.

FETAL CELL ISOLATION
HEMATOPOIETIC STEM CELLS FROM HUMAN FETAL BLOOD

Hematopoietic stem cells circulate in the fetal blood and may be isolated from the placenta through the umbilical cord after birth (after delivery of the newborn, up to 150 ml of fetal blood remains in the placental blood vessels) [7,8]. These cells have been safely isolated, then transplanted into recipients to treat malignancies and bone marrow failure [9, 10].

Isolation of Hematopoietic Stem Cells from Umbilical Cord Blood (UCB) [11,12]
1. The umbilical vein is punctured after the umbilical cord had been clamped, cut, and cleaned with an antiseptic solution. Early clamping of the cord must be avoided to prevent neonatal anemia later.

2. Blood is collected by gravity into a standard donation blood bag that contains 25 ml of acid citrate dextrose (anticoagulant) and stored at 4°C until processing (within 24 h).

3. Cells are separated by hydroxyethyl starch (HES) after standing for a minimum of 1 h at 22°C.

   a. The UCB is diluted 4:1 with Hanks's balanced salt solution.
   b. HES (molecular weight 450,000, 6% w/v in 0.9% NaCl) is added to the UCB suspension, for a final concentration of 0.75%.
   c. The bag is hung in an inverted position at 22°C for gravity sedimentation of red blood cells (RBC's).
      When a clear demarcation between RBCs and leukocyte-rich plasma is visible, drain the RBCs into a second bag, taking care not to disturb the interface, until the separation line is 0.5–1 cm above the outlet.
   d. The bag containing the leukocyte-rich plasma is centrifuged in the upright position at 22°C and 800 g for 10 min.
   e. The supernatant plasma is removed and mixed with equal volumes of cell suspension and cryoprotectant solution containing 20% dimethyl sulfoxide (DMSO) in tissue culture medium M-199. The samples, are stored in 4.5-ml cryotubes (Nunc Inc., Naperville, IL) at a cooling rate of 1°C/min in a programmable cell freezer.

TRANSPLANTATION OF FETAL PANCREATIC ISLET CELLS

Standard insulin replacement therapy often cannot prevent significant and life-shortening complications, including kidney disease, cardiovascular disease, and blindness. These problem could be prevented by the more precise regulation of glucose levels resulting from transplantation of endocrine pancreatic cells. Fetal pancreatic tissue may be preferable to adult tissue because of its high ratio of endocrine to exocrine tissue and its relative lack of highly antigenic passenger cells, which provoke graft rejection [4,13].

It has been proven that enzymatic digestion and culture of human fetal pancreas leads to the formation of isletlike cell clusters (ICCs) containing a high proportion of undifferentiated cells, putatively containing a high proportion of undifferentiated cells, precursors of the hormone-producing cells found in mature islets [14].

Isolation of Human Fetal Pancreatic Cells [15]

1. Human fetal pancreas is obtained between 18 and 24 weeks of gestation.
2. The extraneous material is resected and the pancreas is minced into 1 mm pieces in cold Hanks's Buffer Balanced Solution (HBBS).
3. The fragments are digested with 5.5 mg/ml collagenase in a 50-ml tube, with vigorous shaking for 8–12 min in a water bath at 37°C.
4. After several washes at 4°C with HBBS, the digest is divided among several petri dishes that discourage cell attachment in RPMI-1640 medium containing 10% pooled human serum and antibiotics (100 U/ml penicillin, 0.1 mg/ml streptomycin, and 1 μg/ml amphotericin B).
5. Free-floating ICCs (50–150 μm in diameter), visible under the stereomicroscope after 24 h, can be picked up with a positive displacement micropipettor after 5 days in culture.

Storage of ICCs [16]

ICCs are incubated in 0.2 ml of medium containing 10% human serum in DMSO at increasing concentrations (0.66, 1, and 1.5 mol/liter) for periods of 5, 25, and 15 min, respectively, then supercooled to –7.5°C for 5 min. The samples are cooled to –40°C in a controlled fashion of 0.25°C/min in a calibrated glass container in a bath of liquid nitrogen. The samples are stored in liquid nitrogen.

TRANSPLANTATION OF FETAL LIVER CELLS

In hepatic storage disorders, the absence of functional enzymes leads to the buildup of unmetabolized substrates and to illness. Fetal liver cells, transplanted to these patients, may secrete the missing enzymes, which could then be taken up by the host cells, correcting the defective metabolism [17,18].

From 4 to 18 weeks of gestation the fetal liver is a concentrated source of pluripotent hematopoietic stem cells. The limited supply of histocompatible bone marrow for transplantation may be offset by the availability of fetal liver tissue [19,20]. The fetal liver can provide lifesaving cells that may be lacking in patients with many hematopoietic disorders [4].

Isolation of Fetal Hepatocytes [21,22]

1. The liver is perfused through the portal or umbilical vein, with 1 liter of calcium-free HEPES buffer solution (pH 7.6, 37°C) for 20 min and with HEPES buffer solution containing 0.05% collagenase and 10 mmol/liter calcium for 25 min.

2. The dissociated cells are suspended in hormonally defined medium (HDM), consisting of Dulbecco's Modified Eagle's Medium and Hanis F-1: nutrient medium (DMEM/F-12), containing 15% Myoclone Super Plus fetal calf serum (FCS), 0.1% bovine serum albumin (BSA), 10^{-8} M insulin, 2 mM glutamine, and antibiotics.

3. The cells are washed and separated by two successive centrifugations at 50 g for 5 min.

4. The cells are resuspended in HDM and seeded at a density of $2-5 \times 10^5$ cells per 6-cm dish. The medium is changed 4–6 h after seeding, and daily thereafter, with serum-free HDM, containing 10^{-8} M insulin, 10^{-8} M dexamethasone, 10^{-8} M 3-3′-triiodo-L-thyronine, 10^{-2} M lactic acid plus 1% (v/v) of a mixture of albumin and linoleic acid (1%).

5. The culture is maintained at 37°C in 5% CO_2.

TRANSPLANTATION OF FETAL MUSCLE CELLS

Certain myopathic conditions may be improved through fetal tissue transplantation. In animal models of muscular dystrophy, transplanted myoblasts have fused with degenerating muscle fibers, supplying sufficient numbers of normal genes or gene products to rescue the muscle fibers of the host [4].

Isolation of Fetal Satellite Muscle Cells by Enzymatic Digestion [23]

1. The muscle bundles are handled gently, to minimize damage to the fibers.

2. The muscle is rinsed in Dulbecco's phosphate-buffered saline (PBS), placed into a 50-mm diameter plastic petri dish containing 8 ml of 0.2% type I collagenase in DMEM and incubated in a shaker water bath at 37°C for 1.5–2 h.

3. Following digestion, the muscle is transferred to the first of a series of 50-mm dishes prerinsed with undiluted horse serum (HS) to prevent adhesion of the fibers. Under a transilluminating stereo dissecting microscope, the muscle is gently sucked in and out of the wide-mouthed Pasteur pipette to release single fibers from its surface. As the muscle diameter decreases with increasing loss of fibers, progressively narrower Pasteur pipettes are used. Once 20–30 fibers have been released, the mass of muscle is removed to a second similar petri dish and the procedure is repeated until sufficient single muscle fibers have been released.

4. The muscle fibers are individually picked with a Pasteur pipette, placed in single wells of 24-well tissue culture plates precoated with 10% Matrigel, and incubated for 30–90 min. The plated fibers are allowed to settle and attach for 3 min to a Matrigel substrate with 0.5 ml of 10% horse serum and 0.5% chick embryo extract in DMEM. The plate is incubated at 37°C and 5% CO_2.

5. The cultures are examined daily for the duration of the experiment. Putative satellite cells, dissociated from their fibers, appear 12–24 h after plating. By 3–4 days after plating, 50–300 cells should surround each fiber. At this time the fibers are removed and proliferative medium (20% FBS, 10% HS, and 1% chick embryo extract in DMEM) is added. When the cultures become dense, the cells may be trypsinized and subcultured or allowed to differentiate.

TRANSPLANTATION OF FETAL NEURAL TISSUE

Brain diseases such as Parkinson's represent an important target for tissue engineering [24,25]. Fetal nerve cell grafts have been performed successfully in animal models of neurodegenerative disease. Fetal neurons, especially monoamine neurons, express a high degree of neural plasticity and regenerative capacity. These neurons establish extensive efferent synaptic connections with previously denervated or neuron-depleted host brain regions. Thus, the grafts partially restore some form of neural circuitry through the development of dendritic processes from the transplant to the host neurons [4,6].

Isolation of Fetal Neuron Cells

1. The tissue is placed onto 100-mm petri dishes on a bed of ice, and 6 ml of Ca^{2+}-Mg^{2+}-free Hanks's solution (CMFH) is added. The tissue is minced into small pieces (= 0.5 cm) and transferred with a 10 ml wide-tipped pipette to a 125-ml container.

2. The jar is sealed (to minimize changes in medium pH) and the tissue is allowed to settle on ice ($\approx$3 min). The volume of the CMFH liquid should be about double that of the settled cut tissue.

3. The medium is removed, 10 ml of CMFH is added, and the tissue is again allowed to settle. This wash is repeated two times, and the jar is sealed tightly.

4. The jar containing the washed tissue is placed in a water bath at 37°C and incubated for 20 min.

5. In the hood, the medium is carefully removed and 6 ml of trypsin solution is added.

6. The suspension is incubated at 37°C for 20 min.

7. To the jar containing the trypsinized tissue is added 18 ml of DNase solution. The jar is gently swirled.

8. Another 18 ml of DNase solution is added to the jar. The jar is swirled and the contents carefully divide into three tubes. The tissue is dissociated in each tube by gently pipetting up and down with a 10-ml standard-tip pipette. After being allowed to settle, the undissociated tissue fragments are transferred to a new test tube.

9. Fresh DNase is added to the undissociated fragments, 3 ml to each tube. The tissue in each tube is dissociated by pipetting. The undissociated fragments are allowed to settle, and the tissue is transferred to a new test tube. This step is repeated.

10. The tubes containing the cell suspensions are centrifuged at room temperature at 1000 rpm for 7 min and the supernatant fluid is discarded.

11. The pellets are resuspended with 50 ml insulin–transferrin–selenium supplement [100 ml of DMEM, 1.0 ml of ITS+ (content per liter: 6.25 mg of insulin, 6.25 mg of transferrin, 6.25 µg of selenous acid, 1.25 g BSA, and 5.35 mg linoleic acid), 0.5 ml of antibiotic/antimycotic, 0.1 ml of T_3 with cortisol, 0.6 ml of 45% D-glucose] solution and transferred to a 125-ml jar.

12. An aliquot of 25-µl cell suspension is mixed with 475 µl of trypan blue, and a cell count is performed with a hemocytometer. The yield of viable cells should be between 85 and 95%.

REFERENCES

1. Edwards, R. (1992). "Fetal Tissue Transplants in Medicine." Cambridge University Press, Cambridge, England.
2. Council on Scientific Affairs and Council on Ethical and Judicial Affairs (1990). Medical applications of fetal tissue transplantation *Jama, J. Am. Med. Assoc.* **263**, 565–570.
3. Bjorklund A., Lindvall O., Isacson O., *et al.* (1987). Mechanisms of action of intracerebral neural implants: Studies on nigral and striatal grafts to the lesioned striatum. *Trends Neurosci.* **10**, 509–516.
4. Fine, A. (1994). Transplantation of fetal cells and tissue: An overview. *Can. Med. Assoc. J.* **151**, 1261–1268.
5. Rojansky, N., and Schenker, J. G. (1993). The use of fetal tissue for therapeutic applications. *Int. J Gynecol. Obstet.* **41**, 233–240.
6. Tabbal, S., Fahn, S., and Frucht, S. (1998). Fetal tissue transplantation [correction of transplanation] in Parkinson's disease. *Curr. Opin. Neurol.* **11**, 341-349.
7. Craven, C. M., and Ward, K. (1999). Transfusion of fetal cord blood cells: An improved method of hematopoietic stem cell transplantation? *J. Reprod. Immunol.* **42**, 59–77.
8. Shaaban, A. F., and Flake, A. W. (1999). Fetal hematopoietic stem cell transplantation. *Semin. Perinatol.* **23**, 515–523.
9. Gluckman, E., Rocha, V., and Chastang, C. (1998). European results of unrelated cord blood transplants. Eurocord group. *Bone Marrow Transplant.* **21**, S87–S91.

10. Kawada, H., *et al.* (1999). Rapid ex vivo expansion of human umbilical cord hematopoietic progenitors using a novel culture system. *Exp. Hematol.* **27**, 904–915.

11. Regidor, C., *et al.* (1999). Umbilical cord blood banking for unrelated transplantation: Evaluation of cell separation and storage methods. *Exp. Hematol.* **27**, 380–385.

12. Elchalal, U., *et al.* (2000). Postpartum umbilical cord blood collection for transplantation: A comparison of three methods. *Am. J. Obstet. Gynecol.* **182**, 227–232.

13. Crombleholme, T. M., Langer, J. C., Harrison, M. R., and Zanjani, E. D. (1991). Transplantation of fetal cells. *Am. J. Obstet. Gynecol.* **164**, 218–230.

14. Tuch, B. E., Monk, R. S., and Beretov, J. (1991). Reversal of diabetes in athymic rats by transplantation of human fetal pancreas. *Transplantation* **52**, 172–175.

15. Beattie, G. M., Butler, C., and Hayek, A. (1994). Morphology and function of cultured human fetal pancreatic cells transplanted into athymic mice: A longitudinal study. *Cell Transplant.* **3**, 421–425.

16. Beattie, G. M., Otonkoski, T., Lopez, A. D., and Hayek, A. (1993). Maturation and function of human fetal pancreatic cells after cryopreservation. *Transplantation* **56**, 1340–1343.

17. Sharma, S., Pati, H. P., and Mohanty, A. (1997). In vitro colony formation from fetal liver cells and their infusions in patients of aplastic anaemia. *Med. Oncol.* **14**, 137–139.

18. Thomas, D. B. (1993). The infusion of human fetal liver cells. *Stem Cells* **11**(Suppl 1), 66–71.

19. Touraine, J. L. (1991). In utero transplantation of fetal liver stem cells in humans. *Blood Cells* **17**, 379–387.

20. Touraine, J. L. (1996). In utero transplantation of fetal liver stem cells into human fetuses. *J. Hematother.* **5**, 195–199.

21. Vons, C., *et al.* (1991). Regulation of fatty-acid metabolism by pancreatic hormones in cultured human hepatocytes. *Hepatology* **13**, 1126–1130.

22. Andreoletti, M., *et al.* (1997). Preclinical studies for cell transplantation: Isolation of primate fetal hepatocytes, their cryopreservation, and efficient retroviral transduction. *Hum. Gene Ther.* **8**, 267–274.

23. Rosenblatt, J. D., Lunt, A. I., Parry, D. J., and Partridge, T. A. (1995). Culturing satellite cells from living single muscle fiber explants. *In Vitro Cell. Dev. Biol. Anim.* **31**, 773–779.

24. Helmuth, L. (1999). Fetal cells help Parkinson's patients. *Science* **286**, 886–887.

25. Koller, W. C., Pahwa, R., Lyons, K. E., and Albanese, A. (1999). Surgical treatment of Parkinson's disease. *J. Neurol. Sci.* **167**, 1–10.

CHAPTER 78

BREAST RECONSTRUCTION

Geoffrey L. Robb, Michael J. Miller, and Charles W. Patrick Jr.

INTRODUCTION

Breast cancer continues to be the most common cancer among women, other than cancers of the skin, with an incidence of nearly one of every three cancers occurring in American women. Approximately one in nine women in this country will develop breast cancer by 85 years of age. The United States has the highest incidence of breast cancer in the world, with 110.6 cases per 100,000 women. In the year 2000, roughly 182,800 new cases of invasive breast cancer were diagnosed.

From a surgical therapeutic perspective, 69,683 women underwent breast reconstruction in 1998, a 135% increase since 1992. This represents 3.1% of all plastic surgery procedures for that year. The majority of patients, 49%, were in the 35- to 50-year-old age group, while 36% were in the 51- to 64-year-old age group. Thirty-nine percent of the reconstructions were performed at the same time as the mastectomy, which represents the immediate use of both implants in 46% of patients as well as autogenous tissues in 37% of patients for breast reconstruction [1].

These statistics underscore the growing importance of cancer rehabilitation in the form of breast reconstruction for women affected by breast cancer in all age groups. As opposed to being an issue of mere vanity, restoration of the breast form and contour is valued for the necessary maintenance of self-esteem and body image. Even limited excisions of the breast for the eradication of cancer can produce permanent breast deformities and breast asymmetry. The relevant importance of this "woman's issue" was recently supported by the passage by both houses of Congress of the Omnibus Budget Bill. In this bill, insurance companies are required not only to underwrite reconstructive breast surgery following cancer treatment, but also to cover the additional procedures necessary to maintain symmetry with the opposite normal breast. There is a clear ethical and personal mandate to support the reconstruction of breast deformity, whether congenital, secondary to trauma, or, in particular, following cancer treatment.

TYPES OF BREAST RECONSTRUCTION

Plastic surgery for the breast may be broadly classified as either primarily reconstructive or aesthetic depending on the nature of the deformity. This distinction is primarily one of degree. Cosmetic procedures address deformities that are anatomically within normal limits but nevertheless present an appearance that is unsatisfactory to the patient. Aesthetic breast enlargement surgery, or augmentation mammaplasty, is one of the most common aesthetic procedures. Several hundred cubic centimeters of additional soft tissue or tissue equivalent may be required. Usually this is supplied by breast implants consisting of an envelope made of silicone elastomer filled with either saline solution or silicone gel. The ability to engineer additional fat would potentially eliminate the need for artificial breast implants. Reconstructive operations correct more extreme problems; however, they must still follow proper aesthetic principals. After all, the most sophisticated breast reconstruction that does not look like a normal breast will not be well accepted. Both reconstructive and aesthetic operations may require tissue replacement and therefore may be influenced by developments in tissue engineering.

Breast reconstruction is one of the most common reconstructive procedures. The usual indication is to restore the breast following complete removal (i.e., total mastectomy) performed for cancer treatment. It has been shown that women with breast cancer must deal with two separate emotional issues, the reality of a life-threatening disease and the possibility of losing a breast. Loss of the breast is not life-threatening, but many women find the deformity emotionally and psychologically disturbing [2]. A mastectomy causes a significant functional and cosmetic deformity, replacing the soft, projecting breast with a long, flat scar. The breast is a significant part of female body image and sense of femininity. The patient is reminded of her cancer experience every time she looks in a mirror. It can be difficult to find clothing. For some women, an external prosthesis may be satisfactory, but many find such devices cumbersome and unacceptable. Breast reconstruction is intended to overcome these problems and enhance the quality of life for women following mastectomy for breast cancer. The structure of the female breast consists of a container made of skin filled with soft glandular and fatty tissue.

Postmastectomy breast reconstruction involves replacing missing skin and soft-tissue volume to recreate the appearance of the breast. Current methods rely on breast implants [3], soft-tissue flaps [4], or a combination of these [5]. Each technique offers certain advantages depending on the patient. Reconstruction based primarily on artificial devices uses a process known as tissue expansion to create additional skin. This involves placing an inflatable silicone device called a tissue expander beneath the tissues. The expander is gradually inflated with physiological saline solution injected over several weeks. When the expansion is complete, the device is removed and the tissue is ready to use. This process has been shown to increase the amount of tissue and improve the blood supply. It requires up to 6 months to complete. The tissue expander is then replaced with a permanent implant to provide the necessary volume for the completed reconstruction.

Autologous tissue reconstruction is most often performed by means of skin and fat obtained from the lower anterior abdominal wall as a flap, called a transverse rectus abdominis musculocutaneous (TRAM) flap (Fig. 78.1). The tissue may be transferred by either keeping the rectus abdominis muscle attached superiorly or by performing a microvascular transfer [6]. The skin and fat may then be shaped to simulate the appearance of the breast. The tissue is a similar in consistency to breast tissue and provides a reconstructed breast that looks and feels the most natural. The transverse scar at the donor site, located midway between the umbilicus and the pubic area, is easily hidden by clothing. When there is inadequate tissue on the lower abdomen, the procedure can be combined with placement of a breast implant. In such cases, tissue may be harvested from either the back or the abdomen. If tissue is used, however, the results tend to be more natural and long-lasting. The disadvantages of these operations are that they are more time-consuming, have greater risks, and cause more scarring than other techniques.

There are advantages minimizing dependence on permanent breast implants for breast reconstruction. Implants can erode through the skin, become infected, and form deforming scars. They are more difficult to control during the shaping and contouring and can create unnatural surface contours over time. Breast implants are not a good option in patients who have been treated with radiation because of a tendency for firm scars to form around the implant. Reconstruction based entirely on tissue avoids these problems, but requires surgery other sites on the patient, resulting in alteration of normal areas. It is the opportunity to achieve a natural tissue reconstruction without donor site problems that provides the incentive to develop fat-tissue approaches to breast reconstruction. A tissue-engineered soft-tissue alternative would have wide application in postmastectomy breast reconstruction.

SURGICAL PRINCIPLES AND TISSUE ENGINEERING

To envision how tissue engineering methods might be applied to breast reconstruction, it is helpful to consider some surgical principles. The use of tissue in reconstructive surgery involves a two-step process of *transfer* from an uninjured location (donor site) and *modification* to replace or simulate the breast tissue that which has been lost (Fig. 78.1).

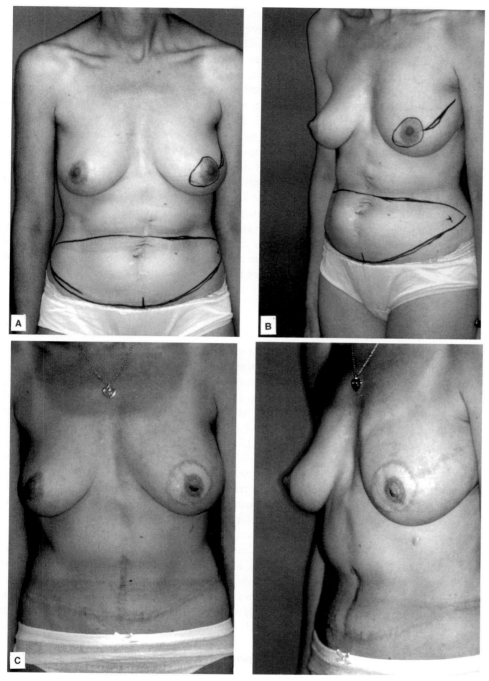

Fig. 78.1. *(A) Preoperative skin-sparing mastectomy utilizing larger TRAM flap design owing to midline lower abdominal scar; (B) preoperative free TRAM reconstruction using skin-sparing mastectomy; (C) postoperative free TRAM reconstruction, nipple–areolar reconstruction completed; and (D) postoperative free TRAM reconstruction.*

TISSUE TRANSFER

Tissue transfer methods may be classified as either tissue grafts or flaps. A graft is any tissue transferred without its blood supply. Graft healing depends upon nutritive support passively available in the tissues surrounding the defect. Small amounts of skin, dermis, and fat may be transferred in this way. In breast reconstruction, these tissues are autologous, or obtained from an uninjured location on the same patient. The volume of tissue required for breast reconstruction is sufficiently large that it cannot survive transfer as a graft.

When a large amount of tissue is required, it must be transferred with a blood supply that originates from outside the zone of injury. This is the definition of a surgical "flap," the traditional term for a unit of tissue moved to another location with preservation of its blood supply. Tissue transferred as flaps may be moved into a compromised area because they will not depend on the ability of the surrounding tissues to supply nutritive support. A variety of surgical flaps have been described that provide skin and fat suitable for breast reconstruction. The most common, however, is located on the lower abdomen. The blood supply to this area passes through the rectus abdominis muscle. It is therefore possible to move the skin and fat of the lower abdomen to the chest for breast reconstruction, using the rectus muscle as a conduit for the blood supply. The most advanced transfer technique is a microvascular transfer. This method involves isolating the tissue unit on its primary vascular supply and temporarily dividing the blood vessels, cutting it "free" from the patient. Tissues transferred in this way are often called "free flaps." The vessels supplying the flap, usually 1–3 mm in diameter, are sewn with extremely fine suture materials to other vessels near the defect; the surgeon is aided by an operating microscope. Usually, two microvascular anastomoses are required, one for the artery and one for the vein. Skin and fat transferred in this way heal in the normal way with little contracture or loss of substance. Most of the history of plastic surgery consists of advances in techniques to transfer tissue.

TISSUE MODIFICATION

Tissue modification is the second step of the reconstruction. After transfer, tissues must be reshaped to simulate missing structures. In contrast to tissue transfer techniques, tissue modification methods have changed little through the centuries. Surgeons still learn to manually alter tissues at the time of surgery, an often difficult and time-consuming process. The results are never exact. Original breast is approximated to a degree that varies depending on the nature of the tissues and the personal skill of the surgeon. The primary factor limiting these methods is the need to preserve adequate blood supply to all portions of the tissue. Efforts to overcome these limitations have focused improving the blood supply and "prefabricating" structures prior to transfer [7–14].

More elaborate methods of tissue modification leading up to tissue engineering have been described experimentally and used in selected patients. As early as 1963, attempts were made to revascularize tissues by direct transfer of blood vessels [15–17]. These techniques are based upon rearranging mature tissue elements into useful configurations prior to transfer. Even more advanced than simple rearrangement is direct modification of the tissue elements by using implants and induction factors to mold and transform. Hollow molding chambers made from inert titanium or silicone have been used in laboratory animals to create tissue flaps of different shapes [18,19]. Khouri *et al.* added a potent growth factor to control the differentiation of soft tissue inside molding chambers placed inside laboratory animals [20]. Despite these laboratory studies that demonstrate the feasibility of fabricating surgical flaps into different shapes, clinically useful techniques have yet to emerge. Only small amounts of tissue have been produced, and the proper shape has not been retained after removal of the mold.

The goal of tissue engineering is to improve our ability to modify tissues by shifting from working with whole tissues to more fundamental levels. From a surgeon's viewpoint, tissue engineering is modification of existing tissues at the cellular or molecular level to fabricate new tissues for reconstructive surgery.

TISSUE ENGINEERED BREAST EQUIVALENT
PREFACE

The application of tissue engineering to breast fabrication is a relatively new effort. The following sections give an overview of the state of the art and the preliminary attempts in this new venue of tissue engineering. The guiding concept is to develop a vascular construct to restore the breast mound and provide optimum cosmesis such that the limitations with tissue transfer and breast implants are abrogated. Strategies for developing tissue constructs within the breast envelope and *ex vivo* followed by subsequent implantation into the breast envelope are being investigated [21–24]. Restoring functional aspects of the breast, such as lactation and tactile stimulation, are beyond the scope of current strategies. However, investigators have utilized tissue engineering strategies for nipple reconstruction [25].

To be sure, the development of a breast equivalent is particularly challenging. Unlike most other tissues and organs, breast tissue is highly variable among patients with respect to volume, composition, shape, soft-tissue biomechanics, ethnicity, age, and hormonal environment (i.e., pre-/postmenopause, pregnancy). Moreover, the final aesthetic outcome of a breast strongly affects the emotional well-being of a patient. In addition, breast aesthetics truly follow the platitude "beauty is in the eye of the beholder," and patient expectations often overrule a surgeon's concept of the individual's optimum breast.

Tissue engineering modalities can be segregated into four fundamental components, namely, cells, scaffold, microenvironment, and elucidation of patient-specific design parameters. Each is discussed here under the aegis of breast tissue engineering.

CELLS FOR A BREAST EQUIVALENT
Adipose

A breast largely consists of adipose tissue setting in the skin envelope against the pectoral muscles of the chest wall. Naturally, the development of a tissue equivalent for breast restoration has focused on human adipose tissue. Adipose tissue is ubiquitous, the largest tissue in the body, uniquely expendable, and most patients possess excess that can be harvested without creating contour deformities. Autologous fat transplantation gives poor results, with 40–60% reduction in graft volume [22,26,27]. The reduction in adipose volume is postulated to be related to insufficient revascularization. The advent of liposuction led investigators to attempt using single-cell suspensions of mature adipocytes. However, since adipocytes possess a cytoplasm composed of 80–90% lipid, they readily tend, upon aspiration, to be traumatized by the mechanical forces of liposuction, resulting in about 90% damaged cells. The remaining 10% tend to form cysts or localized necrosis postinjection. Moreover, mature adipocytes cannot be expanded *ex vivo* because they are terminally differentiated.

Recent progress has been made by using preadipocytes, precursor cells that differentiate into mature adipocytes. Preadipocytes are fibroblast-like cells that uptake lipid during differentiation (see Chapter 10). They grow easily with standard cell culture technologies, they can be expanded *ex vivo*, and the molecular biology involved in differentiation has largely been elucidated through research in the obesity and diabetes areas [28,29]. However, much of the application-based biology of preadipocytes remains unknown (e.g., cell adhesion, cell motility, response to various microenvironments). Human, rat, and swine preadipocytes have been routinely cultured [30–35]. Preadipocytes are normally isolated from enzyme-digested adipose tissue or liposuction material [22]. Alternatively, adipocyte stem cells may potentially allow one to develop cultures of preadipocytes. Researchers are predominantly focusing on using subcutaneous preadipocytes for tissue engineering strategies. It is known that fat depots at different anatomical locations behave differently [36–39]. Hence, it remains to be seen if subcutaneous preadipocytes can adequately replace mammary adipose.

Microvascular Network

Any potential clinically translatable tissue engineering modality must consider the microvasculature. Adipose tissue is unique in that it has the capacity to continue to grow and its vascular network grows in tandem (i.e., *de novo* angiogenesis) [40]. Adipose tissue is highly vascular. The capillary density of adipose is approximately one-third that of muscle. However, from a metabolism standpoint and correcting for active protoplasm (i.e., since an adipocyte is largely lipid within its cytoplasm) the capillary bed of adipose is far richer ($\sim$ two or three times) than that of muscle. Adipose tissue is also known to enhance angiogenesis through the secretion of growth factors extracellular matrices (ECMs) [41–43].

Of the three biological mechanisms available to vascularize a tissue equivalent, only two are available to adults, namely, revascularization and inosculation. Revascularization denotes the growth of capillaries from a host site or tissue into a tissue equivalent. Except for relatively thin constructs, which can survive by diffusion, the slow kinetics (on the order of weeks) of this process typically abrogates it use for large constructs. It has

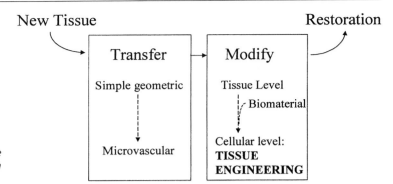

Fig. 78.2. Tissue engineering viewed in the context of the process of reconstructive surgery may be considered an advance in the step of tissue modification.

been proposed to use the highly vascular and adipocyte-rich omentum to encase constructs [44–46]. Inosculation is the process of two capillaries or capillary networks fusing together. The kinetics of inosculation occurs on the order of hours and is the predominant factor that allows plastic surgeons to transfer tissue from a donor site to a recipient site. The capillary networks of the recipient site and the graft fuse together, thus forming a patent vascular network throughout the graft. The use of inosculation in a tissue engineering strategy requires either the seeding of microvascular endothelial cells into or the *ex vivo* or *in situ* development of capillary networks within a tissue equivalent. Both modalities are being investigated but are currently hindered by the lack of understanding of the biological mechanisms that control inosculation and of understanding of capillary formation and cell culture technology of microvascular endothelial cells. Knowledge gained by using vein- or artery-derived endothelial cells cannot be directly translated to capillary endothelial cells.

SCAFFOLDS

A support structure is required for anchorage-dependent cells to migrate and proliferate and to give a tissue equivalent the boundary conditions for final overall tissue shape. Implantable materials utilized have predominantly been porous biodegradable polymer foams [21–23]. For instance, poly(actic-*co*-glycolic acid) (PLGA) scaffolds preseeded with preadipocytes have demonstrated adipose tissue formation [21]. Polymer foams, however, will probably not be the optimum choice for breast scaffolds: they are too rigid for the breast envelope and would be uncomfortable for the patient. In 1999 Kral and Crandall used a non biodegradable scaffold to demonstrate the attachment and proliferation of preadipocytes on Fluorotex monofilament-expanded poly(tetrafluoroethylene) scaffolds coated with various ECMs [47]. Injectable materials, such as hydrogels, inherently possess optimum properties for use in the breast envelope. Both alginate and hyaluronic acid gels have been investigated [23,48,49]. In addition, preadipocytes successfully proliferated and differentiated within fibrin gels.

Finally, adjustable implants have been proposed. Vacanti and colleagues have conceptualized serial injections of a cell-seeded hydrogel within a tissue expander device, with the tissue expander being decreased in size each time an injection is conducted [50]. The optimum scaffold for breast tissue engineering remains elusive. Derivatizing polymers with adhesion molecules can potentially optimize scaffolds. However, this strategy is complicated by the variation of the constitution and distribution of the ECM during adipocyte differentiation [51]. Although short-term studies have demonstrated adipose formation within biodegradable polymers, it remains to be determined whether the formed adipose tissue resorbs over the long term. Investigators are involved in a year-long study to determine the sustainability of tissue engineered adipose [C. W. Patrick, Jr., unpublished data].

ADIPOSE MICROENVIRONMENT

The microenvironment surrounding a tissue construct affects its differentiation and rate of tissue formation. Adipogenesis can be affected, in part, by growth factors (endogenous and exogenous), pO_2 (normoxia vs hypoxia), pH, adhesion molecule on ECM and support cells, and micromotion. Kawaguchi *et al.* demonstrated *de novo* adipogenesis following in-

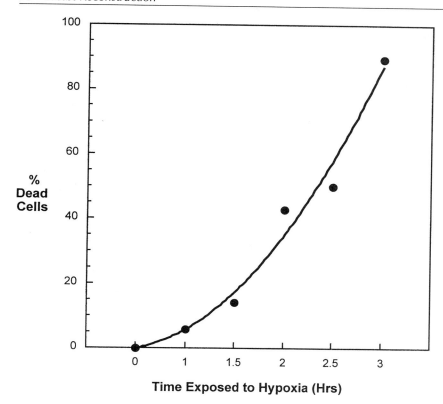

Fig. 78.3. Viability of rat preadipocytes exposed to a hypoxic environment (0% O$_2$).

jection of Matrigel and basic fibroblast-like growth factor (bFGF) in mice [52]. Yuksel and colleagues have used biodegradable microspheres loaded with insulin, bFGF, and insulin-like growth factor 1 to differentiate preadipocytes to mature adipocytes *in vivo* [53,54]. In addition, both epidermal growth factor and tumor necrosis factor α inhibit adipose differentiation [55,56]. Preadipocytes are extremely sensitive to hypoxic environments (Fig. 78.3). This is not surprising based on the historical results of free fat grafting. It is a major design constraint, however, insofar as it limits the time preadipocytes can be placed in a breast envelope without an adequate microvascular network. In contrast, microvascular endothelial cells have been shown to survive hypoxic conditions for 5–7 days.

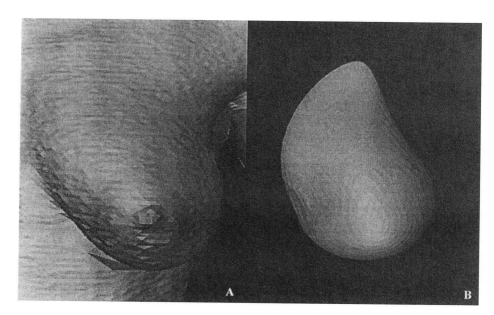

Fig. 78.4. Virtual reality breast simulator. (A) Range data of a patient's breast (three-dimensional surface scan). (B) Fitted virtual breast model, resulting in a volume of 987 ml and surface area of 453 cm^2.

PATIENT-SPECIFIC DESIGN PARAMETERS

To be truly clinically translatable a breast tissue engineering strategy must be patient specific. Unlike strategies for organs that can largely be grown as "one size fits all," breast shape and volume vary widely among the patient population. Breast implants, for instance, range from 100 ml to 2 liters. Hence, methods must exist to predetermine design parameters preoperatively such that the final outcome is known *a priori*. To accomplish this goal, bioengineers, physicians, and computer scientists have combined skill sets to develop a virtual reality breast simulator. A first-generation VR model of the female breast has been developed (Fig. 78.4) The system uses a global parametric deformable model of an ideal breast and allows the surgeon to manipulate the shape of the breast by varying five key shape variables, analogous to the aesthetic and structural elements surgeons inherently vary manually during breast reconstruction. The variables are ptosis (sagging of the breast), top-shape (top's concavity/convexity), turn-top (orientation of top half of the breast with respect to the shoulders), flatten-side (side's concavity/convexity), and turn (deflection of nipple orientation from a perpendicular axis originating at the chest wall). The second generation of the VR model is being developed to be patient specific by importing three-dimensional measurements of the surface of a patient's breast obtained via surface scanning.

REFERENCES

1. Robb, G. L. (2000). Breast reconstruction. Internet Site: YourDoctor.com.
2. Gilboa, D., Borenstein, A., Floro, S., *et al.* (1990). Emotional and psychological adjustment of women to breast reconstruction and detection of subgroups at risk for psychological morbidity. *Ann. Plast. Surg.* **25**, 397–401.
3. Cohen, B. E., Casso, D., and Whetstone, M. (1992). Analysis of risks and aesthetics in a consecutive series of tissue expansion breast reconstructions. *Plast. Reconstr. Surg.* **89**, 840–843.
4. Bostwick, J., and Jones, G. (1994). Why I choose autogenous tissue in breast reconstruction. *Clin. Plast. Surg.* **21**, 165.
5. Fisher, J., and Hammond, D. (1994). The combination of expanders with autogenous tissue in breast reconstruction. *Clin. Plast. Surg.* **21**, 309.
6. Schusterman, M. A., Kroll, S. S., and Weldon, M. E. (1992). Immediate breast reconstruction: Why the free TRAM over the conventional TRAM? *Plast. Reconstr. Surg.* **90**, 255–261.
7. Özgenta, H. E., Shenaq, S., and Spira, M. (1995). Prefabrication of a secondary TRAM flap. *Plast. Reconstr. Surg.* **95**, 441–449.
8. Itoh, Y. (1992). An experimental study of prefabricated flaps using silicone sheets, with reference to the vascular patternization process. *Ann. Plast. Surg.* **28**, 140–146.
9. Mulliken, J. B., and Glowacki, J. (1980). Induced osteogenesis for repair and construction in the craniofacial region. *Plast. Reconstr. Surg.* **65**, 553–560.
10. Hirase, Y., Valauri, F. A., and Buncke, H. J. (1988). Prefabricated sensate myocutaneous and osteomyocutaneous free flaps: An experimental model. Preliminary report. *Plast. Reconstr. Surg.* **82**, 440–445.
11. Stark, G. B., Hong, C., and Futrell, J. W. (1987). Enhanced neovascularization of rat tubed pedicle flaps with low perfusion of the wound margin. *Plast. Reconstr. Surg.* **80**, 814–824.
12. Hussl, H., Russell, R. C., Zook, E. G., and Eriksson, E. (1986). Experimental evaluation of tissue Revascularization using a transferred muscular–vascular pedicle. *Ann. Plast. Surg.* **17**, 299–305.
13. Hyakusoku, H., Okubo, M., Umeda, T. A., and Fumiiri, M. (1987). A prefabricated hair-bearing island flap for lip reconstruction. *Br. J. Plast. Surg.* **40**, 37–39.
14. Khouri, R. K., Tark, K. C., and Shaw, W. W. (1992). Prefabrication of flaps using an arteriovenous bundle and angiogenesis factors. *Surg. Forum* pp. 597–599.
15. Dickerson, R. C., and Duthie, R. B. (1963). The diversion of arterial blood flow to bone. *J. Bone J. Surg. Am. Vol.* **45A**, 356.
16. Woodhouse, C. F. (1963). The transplantation of patent arteries into bone. *J. Int. Coll. Surg.* **39**, 437.
17. Orticochea, M. (1971). A new method for total reconstruction on the nose: The ears of donor areas. *Br. J. Plast. Surg.* **24**, 225.
18. Albrektsson, T., Branemark, P. A., Erikson, A., and Lindstrom, J. (1978). The preformed autologous bone graft. An experimental study in the rabbit. *Scand. J. Plast. Reconstr. Surg.* **12**, 215–223.
19. Fisher, J., and Yang, W. Y. (1988). Experimental tissue molding for soft tissue reconstruction: A preliminary report. *Plast. Reconstr. Surg.* **82**, 857–864.
20. Khouri, R. K., Koudsi, B., and Reddi, H. (1991). Tissue transformation into bone in vivo. *J. Am. Med. Assoc.* **266**, 1953–1955.
21. Patrick, C. W., Jr., Chauvin, P. B., *et al.* (1999). Preadipocyte seeded PLGA scaffolds for adipose tissue engineering. *Tissue Eng.* **5**, 139–151.
22. Patrick, C. W., Jr., Chauvin, P. B., *et al.* (1998). Tissue engineered adipose. *In* "Frontiers in Tissue Engineering" (C. W., Patrick, Jr., A. G. Mikos and L. V. McIntire, eds.), pp. 369–382. Elsevier, Oxford, UK.
23. Lee, K. Y., Halberstadt, C. R., *et al.* (2000). Breast reconstruction. *In* "Principles of Tissue Engineering" (R. P. Lanza, R. Langer and J. Vacanti, eds.), pp. 409–423. Academic Press, San Diego, CA.
24. Katz, A. J., Llull, R., *et al.* (1999). Emerging approaches to the tissue engineering of fat. *Clin. Plast. Surg.* **26**, 587–603.

25. Cao, Y. L., Lach, E., *et al.* (1998). Tissue engineered nipple reconstruction. *Plast. Reconstr. Surg.* **102**, 2293–2298.

26. Billings, E., Jr., and May, J. W., Jr. (1989). Historical review and present status of free fat graft autotransplantation in plastic and reconstructive surgery. *Plast. Reconstr. Surg.* **83**, 368–381.

27. Ersek, R. A. (1991). Transplantation of purified autologous fat: A 3-year follow-up disappointing. *Plast. Reconstr. Surg.* **87**, 219–227.

28. Mandrup, S., and Lane, D. (1997). Regulating adipogenesis. *J. Biol. Chem.* **272**, 5367–5370.

29. Loftus, T. M., and Lane, M. D. (1997). Modulating the transcriptional control of adipogenesis. *Curr. Opin. Genet. Dev.* **7**, 603–608.

30. Entenmann, G., and Hauner, H. (1996). Relationship between replication and differentiation in cultured human adipocyte precursor cells. *Am. J. Physiol.* **270**, C1011–C1016.

31. Novakofski, J. E. (1987). Primary cell culture of adipose tissue. *In* "Biology of the Adipocyte: Research Approaches" (G. J. Hausman and R. J. Martin, eds.), pp. 160–197. Van Nostrand-Reinhold, New York.

32. Strutt, B., Khalil, W., *et al.* (1996). Growth and differentiation of human adipose stromal cells in culture. *In* "Methods in Molecular Medicine: Human Cell Culture Protocols" (G. E. Jones, ed.), pp. 41–51. Humanaa Press, Totowa, NJ.

33. Hausman, G. J., and Richardson, R. L. (1998). Newly recruited and pre-existing preadipocytes in cultures of porcine stromal–vascular cells: Morphology, expression of extracellular matrix components, and lip accretion. *Anim. Sci.* **76**, 48–60.

34. Shillabeer, G. Z., Li, Z.-H., *et al.* (1996). A novel method for studying preadipocyte differentiation in vitro. *Int. J. Obes.* **20**(3), S77–S83.

35. Van, R. L. R., and Roncari, D. A. K. (1977). Isolation of fat cell precursors from adult rat adipose tissue. *Cell Tissue Res.* **181**, 197–203.

36. Dijan, P., Roncari, D. A. K., *et al.* (1983). Influence of anatomic site and age on the replication and differentiation of rate adipocyte precursors in culture. *J. Clin. Invest.* **72**, 1200–1208.

37. Kirkland, J. L., Hollenberg, C. H., *et al.* (1996). Effects of fat depot site on differentiation-dependent gene expression in rate preadipocytes. *Int. J. Obes. Relat. Metab. Disord.* **20**(Suppl 3), S102–S107.

38. Kirkland, J. L., Hollenberg, C. H., *et al.* (1994). Effects of age and anatomic site on preadipocyte number in rat fat depots. *J. Gerontol.* **49**, B31–35.

39. Hauner, H., and Entenmann, G. (1991). Regional variation of adipose differentiation in cultured stromal–vascular cells from the abdominal and femoral adipose tissue of obese women. *Int. J. Obes.* **15**, 121–126.

40. Crandall, D. L., Hausman, G. J., *et al.* (1997). A review of the microcirculation of adipose tissue: Anatomic, metabolic, and angiogenic perspectives. *Microcirculation* **4**, 211–232.

41. Silverman, K. J., Lund, D. P., *et al.* (1988). Angiogenic activity of adipose tissue. *Biochem. Biophys. Res. Commun.* **153**, 347–352.

42. Zhang, Q. X., Magovern, C. J., *et al.* (1997). Vascular endothelial growth factor is the major angiogenic factor in omentum: Mechanism of the omentum-mediated angiogenesis. *J. Surg. Res.* **67**, 147–154.

43. Varzaneh, F. E., Shillabeer, G., *et al.* (1994). Extracellular matrix components secreted by microvascular endothelia cells stimulate preadipocyte differentiation in vitro. *Metab. Clin. Exp.* **43**(7), 906–912.

44. Erol, O. O., and Spira, M. (1990). Reconstructing the breast mound employing a secondary island omental skin flap. *Plast. Reconstr. Surg.* **86**, 219–227.

45. Marschall, M. A., Gigas, E. G., *et al.* (1990). The omentum in reconstructive surgery. *In* "Mastery of Plastic & Reconstructive Surgery" (M. Cohn and R. M. Goldwyn, eds.), Vol. 1, pp. 95–101. Little, Brown, Boston.

46. Sugawara, Y., Harii, K., *et al.* (1998). Reconstruction of skull defects with vascularized omentum transfer and split calvarial bone graft: Two case reports. *J. Reconstr. Microsurg.* **13**, 101–108.

47. Kral, J. G., and Crandall, D. L. (1999). Development of a human adipocyte synthetic polymer scaffold. *Plast. Reconstr. Surg.* **104**, 1732–1738.

48. Marler, J. J., Guha, A., *et al.* (2000). Soft-tissue augmentation with injectable alginate and syngeneic fibroblasts. *Plast. Reconstr. Surg.* **105**, 2049–2058.

49. Duranti, F., Salti, G., *et al.* (1998). Injectable hyaluronic acid gel for soft tissue augmentation. *Dermatol. Surg.* **24**, 1317–1325.

50. Vacanti, J. P., Atala, A., *et al.* (1998). Breast tissue engineering. U.S. Pat. 5,716,404.

51. Kubo, Y., Kaidzu, S., *et al.* (2000). Organization of extracellular matrix components during differentiation of adipocytes in long-term culture. *In Vitro Cell. Dev. Biol.* **36**, 38–44.

52. Kawaguchi, N., Toriyama, K., *et al.* (1998). *De novo* adipogenesis in mice at the site of injection of basement membrane and basic fibroblast growth factor. *Cell Biol.* **95**, 1062–1066.

53. Yuksel, E., Weinfeld, B., *et al.* (2000). Increased free fat-graft survival with the long-term, local delivery of insulin, insulin-like growth factor-I, and basic fibroblast growth factor by PlGA/PEG microspheres. *Plast. Reconstr. Surg.* **105**, 1712–1720.

54. Yuksel, E., Weinfeld, A. B., *et al.* (2000). *De novo* adipose tissue generation through long-term, local delivery of insulin and insulin-like growth factor-1 by PLGA/PEG microspheres in an *in vivo* rat model: A novel concept and capability. *Plast. Reconstr. Surg.* **105**, 1721–1729.

55. Serrero, G. (1987). EGF inhibits the indifference of adipocyte precursors in primary cultures. *Biochem. Biophys. Res. Commun.* **146**, 194–202.

56. Kras, K. M., Hausman, D. B. *et al.* (2000). Tumor necrosis factor stimulates cell proliferation in adipose tissue-derived stromal–vascular cell culture: Promotion of adipose tissue expansion by paracrine growth factors. *Obes. Res* **8**, 186–193.

CHAPTER 79

BLOOD VESSEL SUBSTITUTE

Dror Seliktar and Robert M. Nerem

The development of a tissue-engineered blood vessel substitute is being pursued by means of a variety of different approaches. These include techniques based on collagen, synthetic polymers, and cell self-assembly, as well as acellular techniques. The 30-year history of this technology has shown many promising developments, and in particular, since 1998 there has been significant progress toward making these tissue analogues more functional. Tissue-engineered constructs now exhibit characteristics remarkably similar to those of natural arteries, such as highly organized histological structure and burst strengths in excess of 2000 mmHg. Short-term feasibility studies in various animal models have demonstrated the potential for clinical application and the functionality of these substitute vessels for replacement procedures. There are still some issues regarding this research that must be addressed; however, it is likely that with the help of further research, the technologies now being developed will make their way to the clinic in the coming years. In this chapter, we review the history of the field and the requirements of a tissue-engineered blood vessel substitute. Furthermore, we describe, in detail, four different technologies for constructing engineered blood vessel analogues, including a discussion of their advantages and limitations. The chapter concludes with a brief perspective on future trends in the tissue engineering of blood vessel substitutes.

INTRODUCTION

The production of physiologically functional tissue constructs from biological and/or synthetic components for cardiovascular replacement therapies is an attractive alternative to autologous, native tissue approaches. Hence, tissue engineering methodologies have emerged as a common strategy for constructing blood vessels in the laboratory. There are at least four different in the methods for engineering these implants, but the fundamental concept that ties them together is their integration of a living (cellular) component into the construct. Such tissue constructs can be made from a variety of biomaterial scaffolds, which include resorbable synthetics, nondegradable synthetics, purified biologicals, and wholly biological matrices, provided cell function can be tailored (via the matrix) to the specific function of the replacement organ.

Early attempts at the engineering of a blood vessel substitute involved either seeding or recruiting endothelial cells onto a synthetic graft surface. The purpose was to provide a nonthrombogenic lining to interface with the flowing blood, and much of the focus of this research has been on modifications of synthetic surfaces for enhanced endothelial cell attachment or migration. While these methods are certainly worth mentioning, they are not covered in any detail here because such a blood vessel substitute does not have the full functionality of a native vessel. For further information about these efforts, see Zilla and Greisler [1].

In recent years, the focus of blood vessel tissue engineering has been the *in vitro* development of three-dimensional constructs. There are four main methodologies used to develop these complex tissue analogues. Before describing these in detail, it is worth reviewing what might be the desired characteristics of a functional, tissue-engineered blood vessel substitute.

REQUIREMENTS FOR A BLOOD VESSEL SUBSTITUTE

We begin by highlighting some of the critical requirements for a blood vessel substitute. The complicated structure and function of a blood vessel necessitate a somewhat intricate structure of the tissue analogue. To this end, tissue-engineered vascular constructs are typically made as a coculture of vascular smooth muscle cells (SMCs) and endothelial cells (ECs). The SMCs function to provide the construct with vasoactivity, a critical element for short-term and long-term adaptation to varying hemodynamic conditions, while the ECs provide a nonthrombogenic lining of the conduit for interfacing with the flowing blood. The presence of both SMCs and ECs is essential for proper physiological function after implantation; whether the cells must be incorporated into the construct prior to implantation or can be recruited after implantation, however, is still a matter of debate. It is not clear which approach will ultimately gain clinical acceptance.

The presence of an adherent endothelium is required for a proper interfacing between the flowing blood in the vessel and the underlying SMCs. The ECs must therefore be completely confluent to ensure a nonthrombogenic surface coverage. The ECs should also exhibit a quiescent phenotype to prevent the activation of platelets and other clotting factors. A fully functional endothelium would also enable EC-mediated vasoactivity by acting as a signaling interface between the blood flow in the vessel lumen and the underlying SMCs.

The SMCs of the medial layer also should be organized circumferentially and should express a specific phenotype to function properly. The cellular morphology must enable an efficient contraction and dilation of the vessel in response to vasoactive signals. Thus, the SMCs need to be elongated circumferentially in the direction of the collagen and elastin fibers. Furthermore, the SMCs should exhibit contractile behavior within the vessel wall (i.e. a contractile phenotype). An SMC population expressing a synthetic phenotype would result in excess extracellular matrix (ECM) production that could lead to intimal thickening and occlusion of the engineered vessel.

In addition to exhibiting complex biological functionality, the blood vessel wall exhibits very distinct biomechanical characteristics. The hemodynamic environment imposes a considerable load on the blood vessel, with pulsatile pressure causing cyclic distensions of nearly 10% radial strain in certain locations of the vasculature. To endure this repeated loading, the structure of the artery wall is designed to withstand the pressure by organizing collagen and elastin fibers concentrically in a meshlike pattern [2]. This organization is difficult to imitate in a tissue-engineered construct, but in developing a functional analogue, one must be aware of the importance of two components of blood vessel integrity: mechanical strength and proper viscoelasticity. That is, a vascular construct must be able to easily maintain the operating pressures of the vasculature and also must have an elastic recovery very similar to that of the native artery. To this end, a matched impedance of the engineered blood vessel to the native system is believed to be a critical factor for long-term *in vivo* success.

Elastin provides the distinctive mechanical properties enabling the structure of a natural artery to cyclically recover after each cardiac cycle. The presence of mature elastic fibers in a tissue-engineered blood vessel construct may therefore significantly improve the viscoelastic response of the tissue analogue. Furthermore, elastin may be necessary for tissue-engineered constructs to endure long-term repeated cyclic loading without developing aneurysms. There is evidence to suggest that instabilities in the blood vessel leading to aneurysms are linked to alterations in the elastin of the artery wall [3–6]. Furthermore, *in vitro* degradation of elastin causes aneurysmal dilation in canine carotid arteries, though not to the extent of aneurysmal arteries *in vivo* [7]. While there is still no clear consensus on how changes in elastin impact aneurysm formation, it is clear that elastin should be present in the design of a tissue-engineered blood vessel for long-term *in vivo* success.

Developing constructs with elastin in a structurally stable configuration has also proven to be a significant challenge. Therefore, some methods rely on *in vivo* remodeling to provide the construct with the elastic component. This entails developing constructs that initially lack elastin but have adequate mechanical stability for implantation. The elastic fibers would then be integrated into the tissue analogue through a process of cell-mediated remodeling to reinforce the existing construct architecture. This approach can work only if, in the interim, constructs do not develop aneurysms, since structurally stable elastic fibers can take up to 24 weeks to develop *in vivo* [8].

Immune acceptance in vascular tissue engineering has always been a considerable hurdle because both ECs and SMCs are nonimmunoprivileged cell types. Currently, construct technologies rely on autologous strategies to avert this issue. However, for tissue engineering approaches to truly gain acceptance at a clinically relevant scale, allogeneic cells would have to be used in place of autologous cells. Using up-to-date cell expansion technologies would offer large supplies of allogeneic cells for tissue engineered vascular constructs, provided immune acceptance can be achieved. One possible solution is to engineer the immune response by genetically modifying these cells not to express major histocompatibility complex (MHC) antigens. Another possibility is to use a co-stimulatory blockade to prolong the survival of engineered grafts [9]. While there has been some progress in this field, we will not focus our attention on immunogenicity.

We have thus detailed a number of characteristics that a living blood vessel substitute should exhibit prior to implantation into the host system. Our current knowledge of engineering tissues does not, however, enable a truly comprehensive design, one inclusive of all these requirements without *in vitro* cell-mediated construct reorganization. Therefore, most of the methodologies reviewed rely heavily on an *in vitro* maturation period to improve construct integrity, vasoactive response, and EC stabilization. It is important to also realize that cell-mediated remodeling in tissue engineering is a fundamental requirement, and that understanding and manipulating the development process can ultimately produce a superior blood vessel construct.

COLLAGEN-BASED CONSTRUCT TECHNOLOGIES

The earliest three-dimensional vascular construct produced from living cells was the collagen-based construct model (Fig. 79.1). Weinberg and Bell first reported on the reconstruction of a vascular wall with a reconstituted collagen scaffold in the mid-1980s [10]. They demonstrated the feasibility of creating an adventitia-like structure made from fibroblasts and collagen, a media-like structure made from SMCs and collagen, and an intima-like monolayer of ECs, molded into a tubular configuration. Prior to this, the collagen gel approach was used mainly for skin substitutes and *in vitro* cell studies [11]. Since then, extensive research by a number of groups has built on this initial design, which mainly lacked the mechanical integrity required to withstand physiological blood pressure. The appeal of using a reconstituted collagen scaffold lies in the ease with which cells are able to reorganize the collagenous matrix. Thus the focus of the ensuing research has been *in vitro* cell-mediated construct remodeling.

To improve the mechanical properties of the construct, L'Heureux *et al.* introduced a culturing technique that allowed the cells to contract the collagen gel over a central mandrel, without adhering to it [12]. Constructs developed in this manner exhibited significant increases in material modulus compared with constructs cultured without a central man-

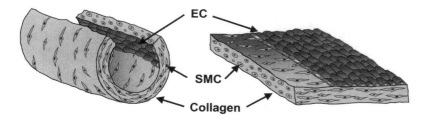

Tubular Construct   **Slab Construct**

Fig. 79.1. Illustration of a coculture collagen-based construct model used by Nerem and coworkers [17] to develop constructs in a tubular configuration and also in a slab configuration. The construct is made with reconstituted type I rat tail collagen and vascular smooth muscle cells (SMC). Endothelial cells (EC) are subsequently seeded onto the lumen of the vessel to form a confluent monolayer.

drel [13]; however, they still were not strong enough to withstand arterial pressure. Hirai *et al.* varied the initial collagen concentration to improve the burst strength of collagen constructs [14]. Tranquillo *et al.* introduced the concept of magnetic prealignment during fibrillogenesis in an effort to induce a favorable realignment of the collagen fibrils in the circumferential direction [15]. This approach proved to be successful with constructs cultured without a central mandrel; however, mandrel compaction was more beneficial to construct integrity than magnetic prealignment. Girton *et al.* demonstrated that significant stiffening and strengthening of collagen gel constructs could be achieved through glycation of the collagenous matrix [16]. This was done by culturing constructs in a medium containing elevated levels of glucose or ribose. Their approach also illustrates the potential of an appropriate biochemical environment in construct development. Finally, Seliktar *et al.* elicited favorable remodeling by the embedded SMCs by exposing cell-seeded collagen constructs to cyclic distensions of 10% radial strain [17]. This mechanical conditioning approach resulted in a significant enhancement of the mechanical properties of the tissue analogue as well as an improved histological organization of the SMCs and collagen.

METHODOLOGY

The formation of a collagen gel construct begins with a solution of solubilized collagen, suspended cells, and growth supplements. The solubilized collagen spontaneously polymerizes at neutral pH and 37°C. Neutralization is achieved with a correct proportion of a mild base such as sodium hydroxide. The polymerization step, referred to here as gelation, takes on the order of minutes. Once polymerized, the collagen forms a stable hydrogel network, with the cells being entrapped in the interstices of the highly entangled array of collagen fibrils. The cellular solution can therefore be molded into the desired geometry by initiating collagen polymerization in the proper geometric configuration. The cells, which initially have a rounded morphology, spread out in three dimensions as they become anchored to the surrounding fibrils. Table 79.1 summarizes the formulas for calculating the correct amount of constituents in the making of a collagen gel construct as reported by Seliktar *et al.* [17].

During the initial 48 h after gelation, the cells extensively rearrange the collagen fibrils by bringing them in close proximity to one another [18–20]. This process of "densification" or "contraction" causes a marked alteration in the appearance of the collagen gel. The mechanical properties are also changing with densification, and in general, the rearrangement of fibrils works to improve the mechanical stability of the gel (see review by Tranquillo [21]). The densification is considerably reduced as the gel reaches a limiting density of collagen fibrils and eventually levels off after about 10–15 days in culture, depending on cell type, cell

Table 79.1. Volume Calculations for Making a Collagen Gel

Total volume (V_T)	= 4.5 ml × no. gels	V_T depends on the number of gels made per batch
FBS (V_F)	= 0.1 × V_T	10% of V_T
Collagen (V_C) [C] = concentration	= V_T × 2/[C]	V_C depends on the concentration of collagen stock solution
0.1 N NaOH (V_N)	= 0.2 × V_C	0.02 M acid divided by 0.1 N base equals 0.2
5 × Medium (V_M)	= $V_T - V_F - V_C - V_N$	Remaining volume is concentrated medium

V_T, total volume; V_S, serum volume; V_C, collagen volume; V_N, NaOH volume; V_M, 5× concentrated medium volume; [C], solubilized collagen concentration (order of mixing: V_M, V_S, cells, V_N, V_C).

concentration, and collagen concentration. These parameters can also be varied to control the amount of contraction and the final density of the collagen construct.

There have been several attempts to enhance the cell-mediated alterations of the collagen gels by adding biochemical supplements and/or modifying the mechanical environment of the system. For example, mechanically conditioning collagen constructs with cyclic strain result in cell-mediated modifications of their histological organization as well as improved mechanical properties of the collagen gel [17,22]. Similarly, adding high amounts of ribose to the culture medium (30 mM) also enhances the glycation of proteins within the gel network and results in improved mechanical properties of the construct [16,23]. The orientation of cells can also be significantly enhanced when gel contraction occurs over a central mandrel [12,13]; the alterations in stress distribution within the construct wall apparently have a favorable impact on the orientation of the embedded cells. These results indicate the importance of the mechanical and biochemical environments in the development of vascular constructs.

Our laboratory has focused its attention on understanding the *in vitro* development process of collagen-based constructs in a unique mechanical and biochemical environment. We have developed a bioreactor system that enables us to expose SMC-seeded constructs to a prescribed cyclic strain environment during their maturation [17]. This mechanical environment affects the expression of remodeling proteins produced in the constructs, including the production of matrix-degrading enzymes called matrix metalloproteinases (MMPs) [24]. The MMPs in part regulate the construct structural development in response to the dynamic mechanical queues. We found significant increases in the mechanical properties of constructs conditioned with cyclic strain, compared with unconditioned controls, and this correlated well to the increased expression of MMPs in the conditioned construct [25]. Altering the biochemical environment in the bioreactor to reduce MMP expression resulted in a reduction in the strain-stimulated enhancement of mechanical properties. We also found that SMCs increase their expression of elastin mRNA in response to mechanical stimulation [26]. Thus, we can stimulate remodeling of vascular constructs by regulating the mechanical and biochemical environment inside the bioreactor.

LIMITATIONS

Collagen-based constructs are still limited by the poor mechanical integrity of the reconstituted collagen gel; despite the promise of *in vitro* manipulations, the strength and viscoelastic properties are not improved enough to make them suitable for implantation. The burst strengths reported for collagen-based constructs range from as low as 50 mmHg up to 225 mmHg [14,23], with the latter being marginally acceptable for implantation. Furthermore, there is no evidence in the literature that elastin is integrated into the collagen constructs [23]. Therefore, it is unlikely that the constructs would be able to endure repeated cyclic loading for a prolonged duration without developing aneurysms. Without the aid of a synthetic support sleeve, these constructs appear to be able to support venous loading only [14,27].

For this approach to be successful, the construct's *in vitro* development will need to foster a more favorable reorganization of the collagen matrix and thus enhance its mechanical properties and suturability. It is also not clear whether the *in vivo* remodeling that takes place upon implantation will favor construct integration with the host tissue. Furthermore, while this model provides the opportunity for functional vasoactivity, it has yet to be demonstrated by *in vitro* testing methods. Finally, there remain a number of immune acceptance issues related to the use of nonhuman collagen that must also be addressed before these grafts can be used successfully. Fortunately, human collagen is becoming more readily available.

CELL-SEEDED POLYMERIC SCAFFOLD TECHNOLOGIES

Another promising new development was reported in 1999 by Niklason *et al.* [28], who describe a cell-seeded polymeric scaffold model of a blood vessel substitute. This method employs a scaffold made of a biodegradable poly(glycolic acid) (PGA) mesh and subcultured vascular cells (Fig. 79.2). Advanced Tissue Sciences, (La Jolla, CA) also has reported

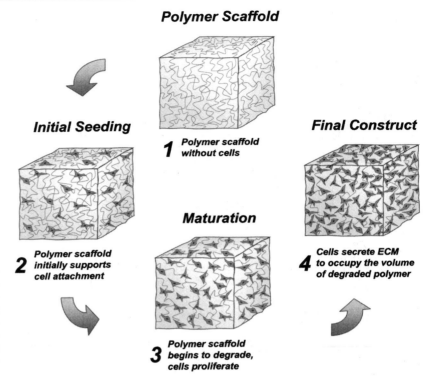

Polymer Scaffold

Initial Seeding

Final Construct

1 *Polymer scaffold without cells*

2 *Polymer scaffold initially supports cell attachment*

Maturation

4 *Cells secrete ECM to occupy the volume of degraded polymer*

3 *Polymer scaffold begins to degrade, cells proliferate*

Fig. 79.2. The development process of a cell-seeded polymeric scaffold construct model. During the course of development, the cells are initially seeded into a highly porous PGA mesh; as the cells attach and proliferate, the PGA hydrolyzes and degrades. After several weeks in culture, the construct is composed primarily of cells and extracellular matrix; very little polymer remains in the construct.

on a similar model of tissue-engineered blood vessel substitutes [29,30]. These are the first reports of a successful blood vessel construct developed from a cell-seeded polymeric scaffold, despite the remarkable success of this methodology in tissue-engineered skin and its use in cartilage substitutes [31,32]. Nevertheless, this approach has been extensively studied with many cell types, including osteoblasts, chondrocytes, fibroblasts, smooth muscle cells, and hematopoietic cells. The main premise is to seed a population of cells within the intricate structure of a uniquely designed biodegradable polymeric scaffold. As the scaffold degrades, the cell population proliferates and produces its own ECM proteins to occupy the void volume of the degrading polymer. Over the course of days and weeks in culture, the construct progressively transforms from a synthetic to a biological tissue analogue.

Despite the simple concept, the implementation of this methodology in cardiovascular tissue engineering is clearly nontrivial. Success depends on many factors, not the least of which is a proper *in vitro* culture environment for the maturation of the construct. The cells must receive the appropriate environmental stimuli to produce ECM and must be able to properly organize this matrix into a mechanically enduring tissue analogue. Furthermore, cell phenotype is regulated in part by environmental stimuli. Therefore, care must be taken not to cause fundamental changes in cellular behavior that would be adverse to tissue functionality. These complications not withstanding, the work of Niklason *et al.* and Advanced Tissue Sciences demonstrates the progress being made in this field. While these are the first groups to report on cardiovascular constructs developed by cell-seeded polymeric scaffold technologies, future advances are sure to come.

METHODOLOGY

The methods for growing vascular constructs with a polymeric scaffold are understandably more complicated than those of the collagen-based approach. The methods of Niklason *et al.*, for example, involve several steps, described shortly. First, the scaffolding material, PGA, was chemically modified with sodium hydroxide to improve hydrophilicity, protein adsorption, and cell attachment in the construct. The modified PGA scaffold was then seeded for 30 min with SMCs (a typical seeding density per vessel is 7.5×10^6 cells). A special culture chamber was used to provide the constructs with the nutrients and environmental stimuli required for proper tissue maturation. The maturation process involved 8 weeks of *in vitro* culture. After this maturation period, the constructs exhibited the desired characteristics, including sufficient mechanical integrity and a proper histological organization. During this time, cells produced and organized ECM proteins to reinforce the degrading synthetic matrix as both the mechanical stability and histological organization were progressively improving. However, the construct mechanical strength began to build only after the fifth week in culture. Consequently, the synthetic polymer was almost completely hydrolyzed by the fifth week (only 15% of the initial polymer remains).

Much of the novelty in the cell-seeded polymeric scaffold approach for engineering vascular constructs lies in the determination of the culture conditions. Finding the correct combination of mechanical and biochemical signals for tissue development requires careful consideration. For example, Niklason *et al.* reported the use of a biomimetic system that imparts pulsatile radial stresses of 165 beats per minutes and 5% radial strain to the developing constructs. Furthermore, they also reported a culture medium containing growth factors and supplements such as 20% fetal bovine serum, ascorbic acid, copper sulfate, proline, alanine, and glycine. These supplements were added to ensure the availability of the biochemical signaling and building blocks needed for cells to produce and cross-link matrix proteins as the polymer degrades. Consequently, construct development was sensitive to changes in culture conditions, and removal of mechanical stimulation or ascorbic acid, for example, resulted in a significant reduction in the quality of the final construct (i.e., poorer mechanical properties and histological organization).

After 8 weeks of construct maturation, the final stage of the construct development involved seeding ECs onto the fully matured vessel lumen. Endothelialization was done by injecting a bolus of ECs in culture medium (3×10^6 cells/ml) into the lumen of the construct. The cells then formed attachments to the lumen surface under static conditions for 90 min. The vessel was then exposed to shear stress by flowing culture medium through the lumen of the tube at a predetermined rate. The mechanical signals generated by the imposition of fluid flow may have contributed to the stabilization of the developing endothelium. After 3 days of shear stress, when a confluent monolayer of ECs was formed, the vessel was considered to be fully matured and ready for *in vivo* testing.

Considering the complexity of developing a mature blood vessel from a synthetic starting material, the results obtained to date are promising. When cultured under pulsatile conditions the engineered vessels have a histological appearance similar to that of native arteries. Pulsatile conditions also stimulated cells to form mechanically stable constructs, as illustrated by the very high rupture strength of pulsed vessels after 8 weeks (>2000 mmHg) in comparison to nonpulsed vessels (< 300 mmHg). Furthermore, since the majority of the PGA is hydrolyzed by the fifth week in culture, it is unlikely that the synthetic material contributed to the mechanical stabilization of the fully matured construct. One possible explanation for the high rupture strengths is the high collagen content of pulsed constructs (50% after 8 weeks), which is significantly less in nonpulsed constructs. Another functional attribute of these tissue analogues is their ability to contract in response to known vasoactive substances such as serotonin and endothelin-1. However, the constructs did not demonstrate full vasoactivity. Further evidence of SMC contractile properties was demonstrated by the presence of smooth muscle α-actin and calponin, as well as myosin heavy chain. Initial *in vivo* studies in Yucatan miniature swine demonstrated the feasibility of implanting these engineered blood vessel constructs into the arterial system in the right saphenous artery location. The results did not indicate bleeding at the anastomoses or mechanical breakdown after 4 weeks, suggesting that the matured constructs could endure the demanding mechanical environment of the arterial system.

LIMITATIONS

The cell-seeded polymeric scaffold method has potential, given the results obtained thus far. There are, however, several issues that still must be addressed. For example, this method involves a long and very complicated *in vitro* development process that may be costly from a manufacturing standpoint. The maturation process would therefore need to be accelerated considerably to reduce culture time and manufacturing costs. A further assessment of the construct material properties should also answer questions of impedance mismatching and viscoelasticity. There is also no evidence that mature elastic fibers are formed during *in vitro* maturation. Hence, this method relies on *in vivo* elastin production (and organization) for long-term success. While these constructs appear to endure *in vivo* loading for up to 4 weeks, longer implantation studies are required to investigate elastic fiber development and possible aneurysm formation.

It is not entirely clear whether such a construct is able to support a stable EC monolayer under physiological flow, an issue deserving much attention because exposed subendothelial tissue may accelerate thrombosis and cause vessel occlusion. Finally, the presence of a hydrolized polymer within the vessel wall may be detrimental to graft survival, inasmuch as such foreign objects may lead to immunological infiltration. While the latter did not present a problem in the miniature swine study, a human model may be significantly more challenging in this respect. Furthermore, all the data reported thus far have been of animal cell constructs; the successful implementation of this technology with human cells might prove to be quite difficult, considering the differences between the behavior of animal cells and human cells in culture.

CELL SELF-ASSEMBLY TECHNOLOGIES

A third and somewhat unique approach to vascular construct development is the cell self-assembly method described by Auger and coworkers (Fig. 79.3) [33]. Here, intact layers of human vascular cells are grown to overconfluence, forming viable sheets of cells and ECM. These sheets are made from either SMCs or fibroblasts. After some time in culture, the sheets are rolled over a mandrel to assume a tubular configuration and then additionally cultured until fully matured. This methodology was adapted from work by Auger and coworkers with skin equivalents, which are also produced by cell self-assembly [34]. The cell self-assembly model is unique because human cells make the construct completely from secreted human proteins. Unlike the collagen-based approach, which to date has used reconstituted animal proteins, this method eliminates immunological mismatch, provided autologous cells are used. Furthermore, unlike the case of polymer scaffold methodology, there is no concern about remaining synthetic polymer in the vessel wall. Hence, the final construct is not only "wholly" biological, but also free of mismatched constituents.

The underlying premise of the cell self-assembly methodology is relatively simple: that is, to culture a sheet of vascular cells to confluence and tubularize it over a mandrel. However, the successful implementation of such a strategy has proven to be a considerable challenge. This is mainly because of the complicated behavior of vascular cells in a tissue culture environment, particularly when there are stringent requirements for the final product. Thus, understanding what regulates the synthetic activities of vascular cells *in vitro* is a fundamental requirement for this approach. Building on their highly studied skin substitute model, Auger and coworkers were able to establish a set of culture conditions that stimulate vascular cells to produce tissue analogues with functional characteristics. The main drawback to this method is the length of time required for cultured constructs to fully mature (>13 weeks). Nonetheless, as the culture conditions are further refined, construct maturation quite possibly could be considerably accelerated.

METHODOLOGY

The first step in this development process was the initial cell culture work. Human umbilical vein SMCs were grown in large rectangular flasks in culture medium supplemented

Fig. 79.3. An illustration of a cell self-assembly construct model similar to the model described by Auger and coworkers [33]. Initially, mature sheets of cells are rolled over a central mandrel. The rolled tubes are cultured for 8 weeks to form mature laminated tubular constructs.

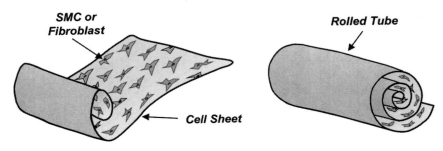

with 10% bovine serum and 50 μg/ml sodium ascorbate for as long as 5 weeks. Likewise, human skin fibroblasts were cultured under identical conditions to form a cell–ECM sheet. After sufficient time in culture, the SMC and fibroblast sheets were manually peeled off the culture flask and rolled over a permeable mandrel to assume a tubular configuration.

To assemble the tubular construct, the following steps were carried out under sterile culture conditions. First, an acellular inner membrane was made from a mature sheet of fibroblasts rolled around a central mandrel and then dehydrated. The inner membrane was passed over a perforated tubular mandrel (3.0 mm o.d.). Next, a sheet of SMCs was rolled over the top of the membrane and the entire construct was placed into a bioreactor for one week of maturation. The next step was to role a sheet of fibroblasts over the construct, to create an adventitial layer. The tube was then cultured to maturity over a time span of 8 weeks and under prescribed culture conditions. During this time, the medial cells assumed a circumferential orientation and produced large amounts of ECM. The fibroblasts also produced ECM and enhanced the structural characteristic of the adventitial layer. The bioreactor system provided both luminal flow of culture medium and mechanical support for the developing tubular constructs. After the final maturation process was complete, the constructs were seeded with ECs and cultured additionally to allow stable endothelial coverage of the lumen.

The methods for developing these constructs are time-consuming and very complicated. The average time to produce one graft is approximately 3 months. Furthermore, proper execution of the techniques described is by no means trivial, and considerable effort is required to attain successful results. Despite these difficulties, however, the results obtained by Auger and coworkers are promising. The fully matured construct, for example, can withstand more than 2000 mmHg pressure before bursting, which would suggest sufficient construct mechanical integrity. Surprisingly, the adventitial layer exhibited the highest mechanical strength of all the layers, and accounted for the superior integrity of the construct. In addition, the adventitial layer exhibited high amounts of elastin assembled in small fibers, organized in circular arrays. Furthermore, histological evaluation of the construct revealed elongated SMCs with circumferential and longitudinal orientations. The SMCs also stained positive for smooth muscle α-actin and desmin, two markers of the smooth muscle phenotype.

LIMITATIONS

While the cell self-assembly method has potential for cardiovascular tissue engineering, there are also some issues that must be addressed. For example, it is not evident from initial reports whether the construct material exhibits viscoelastic response under physiological loading, hence whether such a construct would experience compliance mismatch upon implantation. It is important in the future for more comprehensive testing of mechanical properties to be done on these constructs. Furthermore, these constructs derive most of their structural integrity from their adventitial layer, whereas the medial layer provides integrity for natural arteries under physiological loading, and this discrepancy may result in long-term maladaptation of the tissue analogue. The long culture time required to produce each construct by means of the cell self-assembly method is also a considerable limitation. In addition to the 5-week culture period for the cellular sheets, the construct is matured for another 8 weeks prior to implantation. Such a long manufacturing process can be extremely expensive. However, it is possible that the maturation process can be accelerated with further optimization of culture conditions. Optimized culture conditions may include mechanical stimulation, which has been shown in other engineered blood vessel models to enhance ECM production and structural development [17,28,35]. Even with these apparent limitations, there are high hopes that this approach may be used in the tissue engineering of a blood vessel substitute.

ACELLULAR TECHNOLOGIES

The methods described thus far integrate living cells into the design of a vascular construct prior to implantation. Another approach, however, is to use a noncellular construct that once implanted, recruits cells from the surrounding host tissue. Such an acellular methodology offers some advantages over cellular approaches because of the difficulties of isolating and expanding autologous human cells and because of immune acceptance issues with allogeneic cells. Integrating cells *in vitro* can also increase the risk of contamination, hence complicating the implantation process; however, recruiting cells into an acellular vascular construct can be a daunting task. The process of *in vivo* vascular cell migration is poorly understood at best, and engineering this complicated response into a vascular construct remains a major challenge. Nonetheless, there are a number of reports that illustrate the potential impact acellular methods may have on the field.

One concept that has been investigated is the use of a rolled small intestinal submucosa (SIS) as a small-diameter vascular graft [36–39]. The SIS is a cell-free, 100-μm-thick collagen layer derived from small intestine. The mechanical properties of the SIS make it a suitable candidate for vascular implantation [40,41]. The mean burst pressure of a 5.5-mm-diameter graft is 3517 mmHg. The SIS graft is approximately half as compliant as a canine carotid artery, but nearly four times as compliant as a typical vein graft [40]. The healing characteristics of the small diameter (3.5–5 mm) SIS grafts also make them a suitable vascular substitute. When implanted in the location of the canine carotid artery, for example, seven out of eight grafts remained patent after 120 days. This compares to two out of eight for similar extended poly(tetrafluorothylene) (ePTFE) grafts implanted as controls. The formation of a cellularized neointima and confluent monolayer of ECs was observed as early as 28 days after implantation [38].

A recent report by Sullivan and coworkers demonstrated the long-term feasibility of a similar acellular approach with a rabbit arterial bypass model [42]. They used the intestinal collagen layer (ICL) of the SIS, isolated from porcine tissue and treated with a nondetergent, chemical cleaning process that removed most of the cellular materials without compromising the native architecture of the collagen. The result was an ICL comprising mostly of type I collagen and essentially free of other proteins, lipids, and nucleic acids. A thin layer of bovine type I collagen was also placed on the luminal surface of the graft to improve the graft patency. These grafts demonstrated sufficient mechanical integrity for implantation; the burst pressure was approximately 1000 mmHg, and the suture retention strength was twice the surgical requirement. Furthermore, the layer of dense fibrillar bovine collagen (DFC) on the lumen significantly improved graft patency in comparison to uncoated grafts in the rabbit model.

METHODOLOGY

The method described here is an example of an acellular technology used by Sullivan and coworkers [42]. The small intestine of pigs was manually denuded of the mesenteric layer and mechanically stripped of the mucosal and membranous layers with a commercially available gut cleaner. The remaining collagen submucosa was chemically processed with EDTA and salt solutions to remove cellular debris and then tested for sterility prior to use. Under sterile conditions, the ICL was wrapped twice around a 4-mm mandrel to create a two-layered tube that was then filled with acid-solubilized bovine tendon type I collagen. Following a 30-min incubation, the graft was cross-linked in 1 mM 1-ethyl-3-(3-dimethylaminopropyl) carbodiimide hydrochloride (EDC) and sterilized with 0.1% peracetic acid. To reduce acute thrombogenicity, the final construct was coated with heparin–benzalkonium chloride complex (HBAC) prior to implantation.

Chemically cross-linking the ICL construct with EDC provided the mechanical properties needed to operate under physiological loading without compromising the porosity of the material. Thus, the graft endured cyclic stresses and enabled cellular infiltration into the vessel wall. Furthermore, the EDC reduced antigenicity and eliminated the residual toxicity associated with other forms of aldehyde fixation. The DFC layer proved to be effective as a better substrate for endothelial cell attachment, since grafts remained patent after 90 days of implantation and were effectively endothelialized at the time of explant; not modifying the lumen resulted in significantly lower patency rates for these grafts.

The ICL has proven that it can perform as a vascular graft and eventually be integrated into the host tissue through cellular infiltration and vascular remodeling. These grafts were implanted into male New Zealand white rabbits for up to 90 days, and all the grafts remained patent. Histological evaluation revealed infiltration by SMCs, fibroblasts, and a modest inflammatory response within the wall of the construct. There was no indication of immunological response to the porcine or bovine collagen. By 90 days, it was difficult to distinguish the original ICL starting material from the neoartery that had formed. There was, however, a thickening of the vessel wall as a result of the cellular infiltration, similar to the hyperplastic response observed with vein grafts in similar experimental models. Nevertheless, this wall thickening did not appear to change the vessel diameter. A direct result of the cellular infiltration was illustrated by the vasomotor responsiveness of these grafts when stimulated with vasoactive agonists such as norepinephrine, serotonin, and bradykinin. The ICL grafts demonstrated contractile properties after 2 months of implantation, suggesting that the graft was becoming a functional neovessel *in vivo*.

Inoue *et al.* reported on a similar technique that employed rolled human acellular dermal (ACD) matrix as a small-diameter vascular graft [43]. The patency rates of these ACD grafts were superior to ePTFE grafts upon implantation into the rat femoral artery for 28 days; however, they exhibited false aneurysms around the longitudinal suture line. A different approach taken by Wilson *et al.*, who used a detergent–enzymatic extraction process for decellularizing bypass allografts [44]. This process also removes the cells without disturbing the arrangement of the structural proteins. When implanted into a canine model, the grafts showed no inflammation, but only minimal cellular repopulation. Clearly, this model needs further improvement prior to clinical consideration.

Another acellular tissue engineering approach involves the use of modified hydrogels that coat the vascular wall to recruit and accelerate EC coverage [45]. For the surface to become more attractive for migrating ECs, the hydrogel was modified with synthetic adhesion ligands or genetically engineered adhesion proteins. The subsequent interactions between ECs and the modified surface promoted both EC migration and anchorage.

LIMITATIONS

The acellular technologies have one considerable drawback: they lack an initial non-thrombogenic endothelial cell lining on the lumen of the graft. Therefore, for this approach to be successful, endothelial cells must be recruited to cover the luminal surface after implantation, and this recruitment must take place at a reasonably rapid rate. The work of Sullivan and coworkers demonstrated that this is possible in a rabbit model with the inclusion of a layer of collagen on the graft surface that provides a superior anchoring substrate for ECs. In this case, ECs can migrate from the anastomoses onto the graft surface and form a confluent layer. This task, however, is considerably more difficult with human models, and it remains to be seen whether human ECs will occupy the graft surface in its entirety.

SUMMARY AND CONCLUSIONS

There has been much progress in the area of vascular tissue engineering over the past three decades. This chapter highlighted four very different approaches toward the common goal of engineering a substitute blood vessel, and each has distinct advantages and disadvantages. The collagen-based approach, which offers the advantages of relatively fast *in vitro* development and simple handling, is still limited by poor mechanical integrity. The synthetic polymer approach, which has proven to be very successful with an animal model, is still limited by long culture times and still needs to be tested and optimized using human cells. The cell self-assembly approach, which offers few immunogenic barriers, requires a very long and complicated manufacturing process. And finally the acellular approach, which offers the advantages of simple manufacturing and ease of handling and is immune acceptable, requires further refinement to gain clinical acceptance.

The future of tissue engineering research is undoubtedly coupled to its current limitations. Engineered constructs will need to be sufficiently tested for their mechanical properties, which include visocelastic response under physiological loading. Furthermore, the vasoactive response of the constructs would require comprehensive characterization to validate a functional SMC–medial layer. The endothelium of blood vessel constructs would require *in vitro* assessment to confirm a quiescent and viable monolayer that is firmly anchored to the subendothelial surface, even in the presence of physiological fluid flow. The culture environment will require further optimization, including biochemical and mechanical stimulation, to reduce *in vitro* maturation times and enhance construct development. The integration of construct with host is also a topic worthy of investigation in particular with respect to the questions of impedance mismatch and immune response for tissue engineering technologies. In the context of host integration, there remains the question of long-term adaptability of the tissue analogue. Particularly, it will be necessary to determine how much the tissue must resemble a native artery prior to implantation and how much it may be remodeled after implantation. The control of intimal thickening and thrombosis is paramount to *in vivo* success with this technology, and methods that allow such control will undoubtedly take center stage in future *in vivo* testing. And finally, only the acellular approach is not limited by the use of autologous cells. For the other approaches, the use of nonautologous cells for the final product would certainly be required for the technology to have a significant impact. Thus, the engineering of immune acceptance in these constructs will certainly need to be the focus of future research. While there are likely other issues deserving of attention, we have listed a few that are relevant.

The research summarized in this chapter offers new insights in to construct maturation and construct integration that will undoubtedly pave the way toward the development of a blood vessel substitute that has the desired characteristics, a simple manufacturing process, and relatively fast development. Furthermore, in the near future, we should know more about such critical issues as immunogenicity, postoperative construct stability, and long-term integration with the host. As we begin to understand the structure–function relationship of substitute blood vessels—with regard to both the *in vivo* system and the *in vitro* models—we can design blood vessel analogues that bring this benchtop technology much closer to clinical application.

REFERENCES

1. Zilla, P., and Greisler, P. H. (1999). "Tissue Engineering of Vascular Prosthetic Grafts." R.G. Landes, Austin, TX.
2. Fung, Y. (1981). "Biomechanics: Mechanical Properties of Living Tissues," Chapter 8. Springer-Verlag, New York.
3. Rizzo, R. J., McCarthy, W. J., Dixit, S. N., Lilly, M. P., Shively, V. P., *et al.* (1989). Collagen types and matrix protein content in human abdominal aortic aneurysms. *J. Vasc. Surg.* 10, 365–373.
4. Minion, D. J., Davis, V. A., Nejezchleb, P. A., Wang, Y., McManus, B. M., and Baxter, B. T. (1994). Elastin is increased in abdominal aortic aneurysms. *J. Surg. Res.* 57, 443–446.
5. Grange, J. J., Davis, V., and Baxter, B. T. (1997). Pathogenesis of abdiminal aortic aneurysm: An update and look toward the future. *Cardiovasc. Surg.* 5, 256–265.
6. Gandhi, R. H., Irizarry, E., Cantor, J. O., Keller, S., Nackman, G. B., *et al.* (1994). Analysis of elastin cross-linking and the connective tissue matrix of abdominal aortic aneurysms. *Surgery* 115, 617–620.
7. Dobrin, P. B., Schwarcz, T. H., and Baker, W. H. (1988). Mechanisms of arterial and aneurysmal tortuosity. *Surgery* 104, 568–571.
8. Hirai, J., and Matsuda, T. (1996). Venous reconstruction using hybrid vascular tissue composed of vascular cells and collagen: Tissue regeneration process. *Cell Transplant.* 5, 93–105.
9. Larsen, C. P., Elwood, E. T., Alexander, D. Z., Ritchie, S. C., Hendrix, R., *et al.* (1996). Long-term acceptance of skin and cardiac allografs after blocking CD40 and CD28 pathways. *Nature* (*London*) 381, 434–438.
10. Weinberg, C. B., and Bell, E. (1986). A blood vessel model constructed from collagen and cultured vascular cells. *Science* 231, 397–400.
11. Bell, E., Ivarsson, B., and Merrill, C. (1979). Production of a tissue-like structure by contraction of collagen lattices by human fibroblasts of different proliferative potential in vitro. *Proc. Nat. Acad. Sci. U.S.A.* 76, 1274–1278.
12. L'Heureux, N., Germain, L., L'abbé, R., and Auger, F. A. (1993). In vitro construction of a human blood vessel from cultured vascular cells: A morphologic study. *J. Vasc. Surg.* 17, 499–509.
13. Barocas, V. H., Girton, T. S., and Tranquillo, R. T. (1998). Engineered alignment in media equivalents: Magnetic prealignment and mandrel compaction. *J. Biomech. Eng.* 120, 660–666.
14. Hirai, J., Kanda, K., Oka, T., and Matsuda, T. (1994). Highly oriented, tubular hybrid vascular tissue for a low pressure circulatory system. *ASAIO J.* 40, M383–M388.
15. Tranquillo, R. T., Girton, T. S., Bromberek, B. A., Triebes, T. G., and Mooradian, D. L. (1996). Magnetically orientated tissue-equivalent tubes: Application to a circumferentially oriented media-equivalent. *Biomaterials* 17, 349–357.
16. Girton, T. S., Oegema, T. R., and Tranquillo, R. T. (1999). Exploiting glycation to stiffen and strengthen tissue equivalents for tissue engineering. *J. Biomed. Mater. Res.* 46, 87–92.
17. Seliktar, D., Black, R. A., Vito, R. P., and Nerem, R. M. (2000). Dynamic mechanical conditioning of collagen–gel blood vessel constructs induces remodeling in vitro. *Ann. Biomed. Eng.* 28, 351–362.
18. Ehrlich, H. P., Borland, K. M., Muffly, K. E., and Hall, P. F. (1986). Contraction of collagen lattice by peritubular cells from rat testis. *J. Cell Sci.* 82, 281–294.

19. Stopak, D., and Harris, A. K. (1982). Connective tissue morphogenesis by fibroblast traction. I. Tissue culture observations. *Dev. Biol.* **90**, 383–398.

20. Yamato, M., Adachi, E., Yamamoto, K., and Hayashi, T. (1995). Condensation of collagen fibrils to the direct vicinity of fibroblasts as a cause of gel contraction. *J. Biochem. (Tokyo)* **117**, 940–946.

21. Tranquillo, R. T. (1999). Self-organization of tissue-equivalents: The nature and role of contact guidance. *Biochem. Soc. Symp.* **65**, 27–42.

22. Kanda, K., and Matsuda, T. (1994). Mechanical stress-induced orientation and ultrastructural change of smooth muscle cells cultured in three-dimensional collagen lattices. *Cell Transplant.* **3**, 481–492.

23. Girton, T. S., Oegema, T. R., Grassl, E. D., Isenberg, B. C., and Tranquillo, R. T. (2001). Mechanisms of stiffening and strengthening in media-equivalents fabricated using glycation. *J. Biomech. Eng.* (to be published).

24. Seliktar, D., Galis, Z. G., and Nerem, R. M. (1999). Cyclic mechanical strain stimulated in vitro remodeling of tissue engineered vascular grafts. *BMES-EMBS Ann. Meet.* Atlanta, GA.

25. Seliktar, D., Galis, Z. G., and Nerem, R. M. (2000). Matrix metalloproteinase-2 regulates tissue remodeling of smooth muscle cell-seeded collagen constructs subjected to cyclyc mechanical strain. *ISACB 7th Bien. Meet.* Tucson, AZ.

26. Seliktar, D., Galis, Z. G., and R. M. N. (1999). In vitro remodeling of cell-seeded collagen–gel vascular constructs in response to a uniquely defined mechanical environment. *4th Int. Conf. Cell. Eng.* Nara, Jpn.

27. Hirai, J., and Matsuda, T. (1995). Self-organized, tubular hybrid vascular tissue composed of vascular cells and collagen for low-pressure-loaded venous system. *Cell Transplant.* **4**, 597–608.

28. Niklason, L. E., Gao, J., Abbott, W. M., Hirschi, K. K., Houser, S., *et al.* (1999). Functional arteries grown in vitro. *Science* **284**, 489–493.

29. Landeen, L. K., Graham, D. A., Alexander, H. G., Garcia, A., Fino, M. R., *et al.* (1999). Effects of scaffold design on tissue formation in tissue-engineered vascular grafs. *3rd Annu. Hilton Head Workshop*, Hilton Head, SC.

30. Landeen, L. K., Alexander, H. G., Graham, D. A., Garcia, A., Chen, P., *et al.* (2000). Biochemical and biomechanical characterization of a tissue engineered vascular graft. *ISACB 7th Bienn. Meet.* Tucson, AZ.

31. Cima, L. G., Langer, R., and Vacanti, J. P. (1991). Polymers for tissue and organ culture. *J. Bio. Compat. Polym.* **6**, 232-0239.

32. Langer, R., and Vacanti, J. P. (1993). Tissue engineering. *Science* **260**, 920–926.

33. L'Heureux, N., Paquet, S., L'abbé, R., Germain, L., and Auger, F. A. (1998). A completely biological tissue-engineered human blood vessel. *FASEB J.* **12**, 47–56.

34. Michel, M., L'Heureux, N., Pouliot, R., Xu, W., Auger, F. A., and Germain, L. (1999). Characterization of a new tissue-engineered human skin equivalent with hair. *In Vitro Cell. Dev. Biol. Anim.* **35**, 318–326.

35. Kim, B. S., Nikolovski, J., Bonadio, J., and Mooney, D. J. (1999). Cyclic mechanical strain regulates the development of engineered smooth muscle tissue. *Nat. Biotechnol.* **17**, 979–983.

36. Lantz, G. C., Badylak, S. F., Hiles, M. C., Coffey, A. C., Geddes, L. A., *et al.* (1993). Small intestinal submucosa as a vascular graft: A review. *J. Invest. Surg.* **6**, 297–310.

37. Badylak, S. F., Lantz, G. C., Coffey, A., and Geddes, L. A. (1989). Small intestinal submucosa as a large diameter vascular graft in the dog. *J. Surg. Res.* **47**, 74–80.

38. Sandusky, G. E., Lantz, G. C., and Badylak, S. F. (1995). Healing comparison of small-intestine submucosa and ePTFE grafts in the canine carotid artery. *J. Surg. Res.* **58**, 415–420.

39. Sandusky, G. E., Jr., Badylak, S. F., Morff, R. J., Johnson, W. D., and Lantz, G. (1992). Histologic findings after in vivo placement of small intestine submucosal vascular grafts and saphenous vein grafts in the carotid artery in dogs. *Am. J. Pathol.* **140**, 317–324.

40. Roeder, R., Wolfe, J., Lianakis, N., Hinson, T., Geddes, L. A., and Obermiller, J. (1999). Compilance, elastic modulus, and bust pressure of small-intestine submucosa (SIS), small-diameter vascular grafts. *J. Biomed. Mater. Res.* **47**, 65–70.

41. Hiles, M. C., Badylak, S. F., Lantz, G. C., Kokini, K., Geddes, L. A., and Morff, R. J. (1995). Mechanical properties of xenogeneic small-intestinal submucosa when used as an aortic graft in the dog. *J. Biomed. Mater. Res.* **29**, 883–891.

42. Huynh, T., Abraham, G., Murray, J., Brockbank, K., Hagen, P. O., and Sullivan, S. (1999). Remodeling of an acellular collagen graft into a physiologically responsive neovessel. *Nat. Biotechnol.* **17**, 1083–1086.

43. Inoue, Y., Anthony, J. P., Lleon, P., and Young, D. M. (1996). Acellular human dermal matrix as a small vessel substitute. *J. Reconstr. Microsurg.* **12**, 307–311.

44. Wilson, G. J., Courtman, D. W., Klement, P., Lee, J. M., and Yeger, H. (1995). Acellular matrix: A biomaterials approach for coronary artery bypass and heart valve replacement. *Ann. Thoracic Surg.* **60**, S353–S358.

45. Hubbell, J. A. (1995). Biomaterials in tissue engineering. *Bio/Technology* **13**, 565–576.

CHAPTER 80

SMALL-DIAMETER VASCULAR GRAFTS

Laura E. Niklason and Mitchel Seruya

INTRODUCTION

Atherosclerotic vascular disease, in the form of coronary artery and peripheral vascular disease, is the largest cause of mortality in the United States [1]. The ubiquity of atherosclerosis and its related diseases means that diseased muscular arteries are replaced far and away more frequently than any other tissues in the body. Replacement of diseased arteries, as in cardiac bypass or peripheral bypass surgery, is generally with autologous vein or artery. Vein grafts, however, have thin walls that are damaged when transplanted into the arterial system [2], and suitable veins are not available in all patients. Internal mammary arteries, which comprise the majority of arterial grafts and function better than veins, are useful only in the coronary circulation. Therefore, many patients who are in need of bypass surgery do not possess adequate conduits to perform the job. These patients then face palliative medical therapy, and often suffer myocardial infarctions (heart attacks) or endure limb amputation. These medical realities have spurred many investigators to attempt to develop artificial and biological replacements for small-caliber arteries, to enable the treatment of the many thousands of patients with surgically correctable atherosclerosis but no suitable autologous conduit.

GENERAL CONSIDERATIONS

Poly(ethyleneterephthalate) (Dacron) and expanded poly(tetrafluoroethylene) (ePTFE) are the materials most widely used for artificial vascular grafts. Though fairly inert, these substances do evoke cellular and humoral foreign body responses, including the deposition of plasma proteins and platelets, the infiltration of neutrophils and monocytes, and the migration of endothelial and smooth muscle cells. These interactions are amplified in low flow rate, high-resistance, small-caliber grafts, resulting in graft occlusion and subsequent failure [3]. Indeed, artificial grafts used to bypass arteries that are less than 6 mm in diameter suffer thrombosis rates greater than 40% after 6 months [4].

The poor biocompatibility associated with synthetic vascular grafts has motivated researchers to develop more nativelike conduits. Normal muscular arteries have a trilamellar structure (Fig. 80.1), with each layer conferring specific functional properties [5]. The endothelium is a single-cell layer that functions to prevent spontaneous blood clotting in the vessel. The medial layer is composed of smooth muscle cells (SMCs) and extracellular matrix components such as collagen, elastin, and proteoglycans. The media confers mechanical strength to the vessel, as well as its native ability to contract or relax in response to external stimuli. The adventitia, composed primarily of fibroblasts and extracellular matrix, also harbors the microscopic blood supply of the artery as well as its innervation. Mimicry of some or all of these three layers of native arteries has been the underlying strategy in all tissue engineering approaches to arterial replacement.

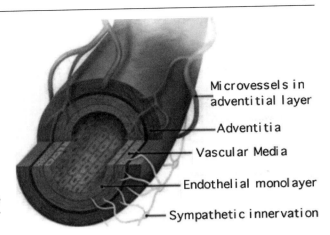

Microvessels in adventitial layer

Adventitia

Vascular Media

Endothelial monolayer

Sympathetic innervation

Fig. 80.1. Native muscular artery. Normal arteries divide into three component layers: the endothelial layer, the medial layer, and the adventitial layer.

APPROACHES TO THE PROBLEM
COLLAGEN GEL BASED GRAFTS

The first attempts to develop tissue-engineered arteries began in the 1980s, when Weinberg and coworkers [6,7] described techniques for culturing arterial grafts using cells grown in preformed, tubular collagen gels. Because grafts based on collagen gels are composed primarily of non-cross-linked fibrils and have rupture strengths of 100–300 mmHg, these vessels proved too weak to be used as arterial grafts *in vivo* [8]. In an effort to improve construct structural integrity without the use of synthetic support, L'Heureux and coworkers [9] introduced a cell culture technique in which a collagen gel developed over a central mandrel. Though mandrel compaction produced significant increases in material modulus, constructs were still not strong enough to withstand arterial pressure. To increase circumferential tensile strength, Tranquillo and coworkers [10,11] have explored the use of glycation, which is the nonenzymatic cross-linking of collagen fibrils by reducing sugars, as well as magnetic prealignment of collagen fibrils. Glycated collagen gel constructs exhibit compliance values similar to those of native arteries, but they possess less tensile strength and far less burst strength. Nevertheless, this approach has proven to be beneficial to construct integrity. Finally, Nerem and coworkers [12,13] have made an extensive study of the interactions between cultured vascular cells and collagen gel matrix in both dynamic and static culture environments. In comparison to statically conditioned constructs, dynamically conditioned constructs possess higher mechanical strength and increased tissue organization. Collagen gel techniques have been well studied and continue to hold promise, but a graft with mechanical strength comparable to that of native vessels has not yet been demonstrated with these approaches.

DECELLULARIZED GRAFTS

A different approach to arterial replacement utilizes tubes of submucosal collagen from porcine small intestine [14]. The submucosal collagen layer is coated with bovine fibrillar collagen and then is complexed with heparin–benzalkonium chloride on the luminal surface. These nonliving vascular grafts are prepared with a minimum of chemical cross-linking, thereby preserving their biocompatibility and capacity for cellular infiltration *in vivo*. The cross-linked porcine collagen layer provides mechanical strength, while the luminally complexed heparin reduced its tendency for thrombosis. These grafts have been implanted into rabbits and followed for up to 13 weeks, during which all the implanted grafts remained patent. This impressive result was accompanied by the observation that the initially nonviable grafts became invested with smooth muscle cells and endothelial cells (ECs) during implantation. The implanted grafts incorporated cells to an extent that allowed them to exhibit small contractile responses to pharmacological agents after explantation. However, the response of the human cardiovascular system to vascular grafts that are made from animal collagens remains unknown. Whether these grafts will stimulate an inflammatory response in humans, and whether they will develop the endothelial layer that may have contributed to their patency in animals, are unanswered questions.

CELL–POLYMER CONSTRUCTS

Biodegradable polymers have also been exploited to support vascular cell growth for tissue engineering. These materials provide a three-dimensional template for tissue formation, while degrading to biocompatible subunits over time. To tissue-engineer pulmonary artery grafts, Vacanti and coworkers [15] have seeded a mixed population of vascular cells on polyglactin–poly(glycolic acid) tubular scaffolds, 15 mm in diameter. Autologous grafts were grown for 7 days in culture and then implanted autologously into lambs. All reported grafts remained patent for up to 24 weeks, and histological analysis showed an absence of residual scaffold fragments after 11 weeks *in vitro*. The presence of elastic and collagen fibers was demonstrated in the medial layer, and ECs were identified on the luminal surface. The early results with these tissue-engineered pulmonary conduits seem promising, yet mechanical and pharmacological assessments still need to be performed.

As an approach to culturing small-diameter vascular conduits, we [16,17] have seeded vascular SMCs and ECs on highly porous, degradable polymer scaffolds made from poly(glycolic acid). During culture, scaffolds with growing vascular cells are subjected to pulsatile radial distensions that in some ways mimic the pulsatile strains in the human cardiovascular system. After approximately 8 weeks, the polymer scaffold substantially degrades and is replaced by a dense smooth muscle cell medial layer and an inner endothelial lining. Engineered vessels cultured under pulsatile conditions possess many of the physiological and mechanical characteristics of native arteries, including burst pressures exceeding 2000 mmHg. These techniques were used to implant autologous tissue-engineered arteries in miniature swine, where they were observed to be functional for up to 4 weeks. However, further work is required to understand the biological effects of various culture conditions and cell–polymer interactions on the cultured vascular cells both *in vitro* and *in vivo*.

COMPLETELY BIOLOGICAL GRAFTS

Exogenous extracellular matrices such as submucosal collagen, collagen gels, and biodegradable polymers have been used to support the growth of vascular cells. In contrast, L'Heureux and colleagues reported in 1998 an alternative approach to arterial replacement. This method is unique in that it incorporates cultured cells without any type of exogenous extracellular matrix [9,18]. These investigators cultured skin fibroblasts and human umbilical vein SMCs as flat sheets. The cellular sheets were then rolled around a mandrel in sequential layers to form a tubular vessel construct. After a suitable culture period, they coated the inner lumen of the tubular construct with ECs. In this way, an engineered tissue was created *in vitro* that comprised all three layers of a native artery. These artificial vessels could withstand relatively high-pressure stresses, displaying rupture strengths of greater than 2000 mmHg. Although these human-engineered vessels suffered a 50% thrombosis rate after one week of implantation in a dog model, this work showed that the creation of a mechanically robust and functional tissue could be achieved from biological components alone.

DESIGN GOALS FOR ENGINEERED ARTERIES

All strategies for engineered graft development share the common goals of producing sufficient mechanical integrity and elastic moduli to allow long-term tolerance of systemic arterial pressures, a differentiated smooth muscle cell wall, and a confluent and quiescent endothelium [19]. The fundamental requirement of mechanical strength is perhaps the most important of these attributes, since a biological graft that suffers dilatation or rupture under hemodynamic forces cannot function adequately as an arterial conduit. All biological tissues that serve a mechanical function (including bone, skin, ligament, blood vessel, etc.) derive their tensile strength primarily from extracellular matrix proteins that are produced by cells in the tissue [20]. The elastic modulus of native arteries at intraluminal pressures near the normal blood pressure is largely a function of the elastin content and the degree of smooth muscle cell activation (Fig. 80.2). Studies have shown that elastin may also play a

A

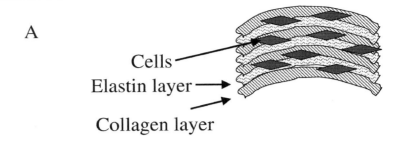

Cells
Elastin layer →
Collagen layer

B

High Pressure
Range

E = Elastic Modulus

$= \dfrac{\text{Stress}}{\text{Strain}}$

= Slope of curve

Physiologic Range

Stress

Strain

Fig. 80.2. (A) Schematic of medial layer. Within the medial layer, vascular smooth muscle cells synthesize layers of collagen and elastin fibers. (B) Stress–strain behavior of native vessels. The elastic modulus is taken as the slope of the stress–strain curve. In the physiological range, the elastic modulus is representative of the contribution of elastin fibers and level of smooth muscle cell activation. At intraluminal pressures exceeding the normal blood pressure, content and recruitment of collagen fibrils account for the bulk of the elastic modulus.

fundamental role in the embryological development of muscular arteries, and in the maintenance of smooth muscle cell quiescence *in vivo* [21]. In contrast, at intraluminal pressures that greatly exceed the normal blood pressure, the maximal elastic modulus or the "burst strength" of both native and engineered vessels is related to both the content and recruitment of collagen fibrils (Table 80.1) [16,22,23].

To develop mechanically robust and functional engineered arteries, controlling the processes that regulate the synthesis, deposition, and remodeling of extracellular matrix proteins by cells in the engineered vessel wall is of fundamental importance. Extracellular matrix deposition and remodeling are dependent upon cell sourcing, culture scaffold, bioreactor design, nutrient medium composition, growth factor supplementation, mass transfer considerations, and required culturing time. This chapter discusses in depth the scientific and practical issues accompanying such parameters, as encountered in our laboratory.

Table 80.1. Modified Constitutive Equation of the Vessel Wall

$$E = E_E + E_C f_C + E_{SMC} f_A$$

where
$E_{SMC} = 8 \pm 7 \times 10^6$ dyn/cm^2
$E_E = 3 \pm 2 \times 10^6$ dyn/cm^2
$E_C = 1 \pm 0.5 \times 10^9$ dyn/cm^2
f_C is the fraction of total collagen fibers that support wall stress at a given strain
f_A is the degree of SMC activation

METHODS FOR VASCULAR TISSUE ENGINEERING
CELL ISOLATION AND CULTURE

Adequate tissue development requires cells with high proliferative capacity. For this reason, our lab uses SMCs and ECs that are derived from vascular biopsy samples taken from young animals. To isolate ECs [24], an arterial biopsy sample is incised longitudinally and the luminal surface is scraped with a scalpel blade. EC scrapings are then incubated in phosphate-buffered saline (PBS) and are intermittently aspirated up and down with a pipette, to break up clumps of cells. ECs are then centrifuged and resuspended in DMEM–10% CS: Dulbecco's Modified Eagle's Medium and 10% calf serum, supplemented with 100 U/ml penicillin, 100 μg/ml streptomycin, and 125 μg/ml heparin U/ml. Cells are normally grown at 37°C in a humid atmosphere with 10% CO_2, and are passaged at subconfluence by means of a 0.25% solution of trypsin–EDTA. EC identity is confirmed by a cobblestone appearance and by positive staining for von Willebrand factor. To avoid possible dedifferentiation or senescence associated with successive EC passaging, ECs are generally used at or below passage 4 for engineered vessels.

SMCs are isolated from the medial layer of an arterial biopsy sample. After EC isolation, the medial layer of the vessel is dissected free of the endothelial and adventitial layers, thereby ensuring as pure a population of SMCs as possible. The tissue is cut into segments approximately 1 cm square and placed intimal-side-down in tissue culture dishes to allow SMC migration from the explant specimens. Several milliliters of culture medium [DMEM–20% FBS: DMEM with 20% fetal bovine serum (FBS), penicillin, and streptomycin] is added to the specimens, which are then placed in an incubator at 37°C at 10% CO_2 overnight. The following day, excess DMEM–20% FBS is added to the dishes, and explant cultures are maintained for 10 days, after which time SMCs have migrated off the specimens and become established in a two-dimensional culture. The specimens are then removed, and the SMCs maintained in culture until confluence at which time they are passaged. SMC identity is confirmed by their characteristic "hill and valley" appearance, by positive staining for SM α-actin, and by subsequent positive staining for calponin and myosin heavy chain in engineered vessel walls. SMCs are generally used at or below passage 4 for engineered vessels.

POLYMER SCAFFOLDS

An ideal three-dimensional culture scaffold should secure spatially uniform cell attachment in conjunction with the maintenance of cell phenotype, permit sufficient mass transfer of gases and nutrients, and degrade in synchrony with the formation of tissue components. Poly(glycolic acid) (PGA), an FDA-approved polyester that degrades by hydrolysis of ester linkages in the polymer backbone, has been successfully used as a biocompatible, biodegradable polymer for tissue engineering [25]. PGA scaffolds are manufactured as described elsewhere [25] and processed into nonwoven sheets of fibers 13 μm in diameter to enable cellular attachment, cross-talk, and minimal resistance to nutrient transfer. We perform surface treatment of PGA meshes, to increase hydrophilicity and SMC seeding, as described elsewhere [26]. Briefly, meshes are washed to remove possible surface contaminants, immersed in a 1.0 N solution of NaOH for one minute, washed in distilled water, and dried overnight under vacuum. This treatment hydrolyzes ester bonds on the PGA fiber surface, creating hydroxyl and carboxylic acid groups, thereby increasing surface hydrophilicity. To form tubular scaffolds for vessel culture, treated sheets of PGA mesh approximately 8×1.5 cm² are sewn with 6-0 PGA suture into tubes of 3.1 mm internal diameter. Scaffolds are sewn at either end to short sleeves of Dacron vascular graft material (5 mm long × 5 mm i.d.) to facilitate attachment of the PGA scaffold to the glass bioreactors.

BIOREACTOR SYSTEM AND TISSUE CULTURE

In general, a bioreactor system should maintain spatial uniformity of cell seeding, ensure sufficient mass transfer, and supply appropriate biochemical and mechanical stimuli to growing tissues [27,28]. Motivated by the exposure of vascular cells to pulsatile physical forces during most of vasculogenesis [29] and throughout life, the bioreactor system in our laboratory (Fig. 80.3) is designed to apply a pulsatile physical stress to vascular cells

Bioreactor with Seeded Scaffold

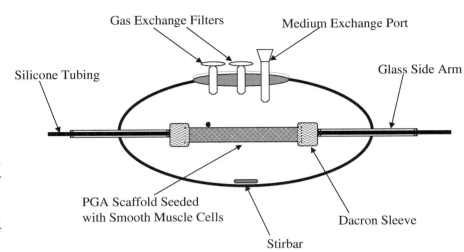

Fig. 80.3. Bioreactor system. Bioreactors are assembled with PGA scaffold and Dacron sleeves, allowing the formation of a fluid-tight connection between the vascular tissue and the bioreactor. Silicone tubing extends through the lumen of the vessel and through the glass sidearms of the bioreactor.

throughout the culturing period [16,17]. Inside the bioreactors, the tubular PGA scaffolds are threaded over medical grade silicone tubing having a known compliance and an outer diameter of 3.1 mm. The silicone tubing exits through the sidearms of each bioreactor and connects to the pulsatile perfusion system. After assembly, the complete bioreactor with PGA scaffold is sterilized prior to cell seeding. Each bioreactor contains a stir bar to allow continuous stirring and improve nutrient convection, a lid that is fitted for gas exchange, and a rubber membrane for culture medium exchange.

After sterilization, PGA scaffolds in bioreactors are seeded with SMCs. Vascular SMCs, passage 2 or 3, are grown to confluence in culture flasks, trypsinized, centrifuged, and resuspended to a density of 5×10^6 cell/ml in culture medium. For each engineered vessel, 1.5 ml of the SMC cell suspension is pipetted onto the scaffold, and the bioreactor is capped and removed to a tissue culture incubator. Bioreactors are slowly rotated for 30 min to facilitate uniform cell seeding around the scaffold, and then the bioreactors are filled with enhanced culture medium. This enhanced medium consists of enhanced DMEM: 20% FBS, growth factors, penicillin G (100 U/ml), HEPES (5 mM), $CuSO_4$ (3 ng/ml), proline, glycine, ascorbic acid (50 µg/ml), and alanine 20 (µg/ml). Growth factors and serum support cell proliferation, while proline, alanine, and glycine serve as the building blocks for procollagen and tropoelastin production. Ascorbic acid (vitamin C) supports hydroxyproline formation and has been shown to increase the production of extracellular matrix proteins [18,30], while copper ion has been shown to be important for collagen and elastin biosynthesis and cross-link formation [31–33]. It must be stressed that the quality of collagen content, namely, the degree of fibrillar cross-linking, is just as important as the amount of collagen production. Normal cross-linking is essential to provide resistance to elastolysis and collagenolysis by nonspecific proteinases in the vessel wall [34]. The degree of cross-linked collagen fibrils may also regulate the extent of fiber recruitment at high stresses and affect the "burst strength" of native and engineered vessels [16,22,23].

Filled with medium, bioreactors are transported to an incubator that is maintained at 10% CO_2 in air, 100% humidity, and 37°C. Bioreactors are placed on a stir plate that mixes the culture medium continuously for the duration of the experiment. The silicone tubing in each bioreactor is connected to a pulsatile perfusion system, which incorporates a bellows-style pump that operates a range of pulse rates from 60 to 165 beats per minute. The pump exerts pressures up to 270 ± 30 mmHg with a variable stroke volume of 0–10 ml/stroke. The perfusion system provides a pulsatile flow of sterile PBS through the silicone tubing in the lumen of each engineered vessel and a cyclic distension of up to 5% with each cycle. Bioreactors are fed with fresh enhanced DMEM for half the bioreactor volume twice per week. Vessel culture continues for up to 8 weeks, at which time the vessels are seeded with luminal ECs.

To seed an endothelial layer in an engineered vessel, the bioreactor is detached from the perfusion system and the silicone tubing is carefully removed from the vessel lumen through the bioreactor sidearms in sterile fashion. ECs that are grown in DMEM–10% CS, passage 4 or below, are trypsinized and resuspended to a density calculated to achieve twice confluence on the luminal surface of the vessel, or approximately 3×10^6 cell/ml. ECs are injected into the vessel lumen via one sidearm of the bioreactor. The ends of the bioreactor are capped, and the bioreactor is placed in the incubator and rotated about the vessel axis for 90 min to assure circumferentially uniform luminal EC seeding. The bioreactors are then returned to the incubator and connected to a modified perfusion system. In the modified perfusion system, the bellows pump is replaced with a continuous roller pump, and the sterile buffer is exchanged for enhanced DMEM with 5% FBS. This modified system allows continuous perfusion of the endothelialized vessel lumen with culture medium at flow rates that are increased gradually from 0.03 ml/s to 0.1 ml/s over 3 days of culture. After completion of endothelialization, the tissue-engineered graft possessing smooth muscle cells and endothelial layers of native artery may be either removed for *in vitro* analysis or implanted into an autologous donor animal [16].

ASSESSMENT OF GRAFT PROPERTIES

The fabrication of an engineered vessel that appears grossly similar to a native artery is not sufficient evidence of a clinically useful conduit. Histological, ultrastructural, biochemical, immunocytochemical, pharmacological, mechanical, and *in vivo* analyses must be performed to fully assess the degree of organization and functionality of engineered vessels. Information regarding vessel architecture, including the presence and orientation of extracellular matrix proteins, can be derived from histological and ultrastructural analyses. Biochemical analyses, including DNA and collagen assays, quantify cellular and collagen content, respectively, thereby providing an objective means of comparing the properties of native and engineered tissues.

Immunocytochemical staining for SM α-actin, calponin, and myosin aids in the evaluation of the differentiation state of the SMCs in the vessel wall. In addition, staining for proliferating cell nuclear antigen (PCNA) may indicate the likelihood of SMC hyperplasia and consequent luminal occlusion following implantation. Pharmacological testing—specifically, the response of SMCs to vasoconstrictive and vasodilatory agents—can further demonstrate the functional status of engineered vascular cells.

Expression of von Willebrand factor, platelet endothelial cell adhesion molecule (PECAM), and tissue factor receptor molecules confirms EC identity, but may indicate an activated procoagulant phenotype [35]. In contrast, detection of thrombomodulin or prostacyclin production would be indicative of a less prothrombotic state [35]. Ultimately, endothelial characterization helps predict the risk of a thrombogenic response following implantation and the likelihood of long-term graft patency.

Mechanical testing and *in vivo* implantation studies ultimately assess the graft's ability to withstand long-term exposure to systemic arterial pressures and hemodynamic forces. Stress–strain analysis provides the elastic modulus and compliance data, which can predict the graft's resistance to dilatation and aneurysm formation *in vivo*, while measurements of suture retention strength help predict the feasibility of clinical implantation [16,17]. *In vivo* implantation studies also provide a definitive means of evaluating the biocompatibility of engineered vessels and their practical utility as small-diameter arterial replacements.

FUTURE DIRECTIONS

What will the replacement arteries of the future look like? Although the final solution is not clear, reports in the last several years have shown us several exciting possibilities. Nonliving tubular grafts have been produced from natural biological materials that can become functioning neoarteries after implantation *in vivo*. Living vascular grafts possessing excellent mechanical properties have been cultured *in vitro* both from autologous vascular cells and from nonautologous neonatal human cells. Although xenogeneic or "immunologically neutral" cells in living vascular grafts are a long-term possibility, it is likely that any cell-based vascular grafts in the near future will incorporate autologous vascular cells. The eventual solution for living vascular grafts may incorporate cells cultured alone, or it may include cells in combination with synthetic or natural materials.

The challenging hurdles still remain of creating a functional nerve supply and microvasculature *in vitro* to support living vascular tissues. While innervation and microcirculation may not be essential requirements for vascular surgical conduits, their production in any cultured tissue would constitute a giant leap forward for the field of tissue engineering. Functional engineered arteries may one day revolutionize the treatment of vascular disease. They will also be a crucial enabling technology for the development and implantation of other engineered tissues, all of which will require a blood supply for their function in sustaining human life.

REFERENCES

1. Ross, R. (1993). The pathogenesis of atherosclerosis: A perspective for the 1990's. *Nature (London)* **362**, 801–809.
2. Davies, M. G., and Hagen, P.-O. (1994). Structural and functional consequences of bypass grafting with autologous vein. *Cryobiology* **31**, 63–70.
3. Zarge, J. I., Huang, P., and Greisler, H. P. (1997). Blood vessels. *In* "Principles of Tissue Engineering" (R. P. Lanza, R., Langer, and W. L., Chick, eds.), pp. 349–364. Academic Press, San Diego, CA.
4. Sayers, R. D., Raptis, S., Berce, M., and Miller, J. H. (1998). Long-term results of femorotibial bypass with vein or polytetrafluoroethylene. *Br. J. Surg.* **85**, 934–938.
5. Kelly, D. E., Wood, R. L., and Enders, A. C. (1984). "Bailey's Textbook of Microscopic Anatomy." Williams & Wilkins, Baltimore.
6. Chen, J.-K., Haimes, H. B., and Weinberg, C. B. (1991). Role of growth factors in the contraction and maintenance of collagen lattices made with arterial smooth muscle cells. *J. Cell. Physiol.* **146**, 110–116.
7. Weinberg, C. B., and Bell, E. (1986). A blood vessel model constructed from collagen and cultured vascular cells. *Science* **231**, 397–400.
8. Kobashi, T., and Matsuda, T. (1999). Fabrication of branched hybrid vascular prostheses. *Tissue Eng.* **5**, 515–524.
9. L'Heureux, N., Germain, L., Labbe, R., and Auger, F. A. (1993). In vitro construction of a human blood vessel from cultured vascular cells: A morphologic study. *J. Vasc. Surg.* **17**, 499–509.
10. Girton, T. S., Oegema, T. R., Grassi, E. D., Isenberg, B. C., and Tranquillo, R. T. (2000). Mechanisms of stiffening and strengthening in media-equivalents fabricated using glycation. *J. Biomech. Eng.* **122**, 216–223.
11. Barocas, V. H., Girton, T. S., and Tranquillo R. T. (1998). Engineered alignment in media equivalents: Magnetic pre-alignment and mandrel compaction. *J. Biomech. Eng.* **120**, 660–666.
12. Ziegler, T., and Nerem, R. M. (1994). Tissue engineering a blood vessel: Regulation of vascular biology by mechanical stresses. *J. Cell. Biochem.* **56**, 204–209.
13. Seliktar, D., Black, R. A., Vito, R. P., and Nerem, R. M. (2000). Dynamic mechanical conditioning of collagen–gel blood vessel constructs induces remodeling in vitro. *Ann. Biomed. Eng.* **28**, 351–362.
14. Huynh, T., Abraham, G., Murray, J., Brockbank, K., Hagen, P.-O., and Sullivan, S. (1999). Remodeling of an acellular collagen graft into a physiologically responsive neovessel. *Nat. Biotechnol.*
15. Shinoka, T., Shum-Tim, D., Ma, P. X., Tanel, R. E., Isogai, N., Langer, R., Vacanti, J. P., and Mayer, J. E., Jr. (1998). Creation of viable pulmonary autografts through tissue engineering. *J. Thorac. Cardiovasc. Surg.* **115**, 536–546.
16. Niklason, L. E., Gao, J., Abbott, W. M., Hirschi, K., Houser, S., Marini, R., and Langer, R. (1999). Functional arteries grown in vitro. *Science* **284**, 489–493.
17. Niklason, L. E., Abbott, W. A., Gao, J., Klagges, B., Conroy, N., Jones, R., Vasanawala, A., and Langer, R. (2001). Morphologic and mechanical characteristics of bovine engineered arteries. *J. Vasc. Surg.* **33**, 628–638.
18. L'Heureux, N., Paquet, S., Labbe, R., Germain, L., and Auger, F. A. (1998). A completely biological tissue-engineered human blood vessel. *FASEB J.* **12**, 47–56.
19. Cox, J. L., Chiasson, D. A., and Gotlieb, A. I. (1991). Stranger in a strange land: The pathogenesis of saphenous vein graft stenosis with emphasis on structural and functional differences between veins and arteries. *Progr. Cardiovasc. Dis.* **34**, 45–68.
20. Cox, R. H. (1978). Passive mechanics and connective tissue composition of canine arteries. *Am. J. Physiol.* **234**, H533–H541.
21. Li, D. Y., Brooke, B., Davis, E. C., Mecham, R. P., Sorensen L. K., Boak, B. B., Eighwald E., and Keating, M. T. (1998). Elastin is an essential determinant of arterial morphogenesis. *Nature (London)* **393**, 276–280.
22. Armentano, R. L., Levenson, J., Barra, J. G., Cabrera Fischer, E. I., Breitbart, G. J., Pichel, R. H., and Simon, A. (1991). Assessment of elastin and collagen contribution to aortic elasticity in conscious dogs. *Am. J. Physiol.* **260**, H1870–H1877.
23. Armentano, R. L., Barra, J. G., Levenson, J., Simon, A., and Pichel, R. H. (1995). Arterial wall mechanics in conscious dogs: Assessment of viscous, inertial, and elastic moduli to characterize aortic wall behavior. *Circ. Res.* **76**, 468–478.
24. D'Amore, P.A., and Smith, S.R. (1993). Growth factor effects on cells of the vascular wall: A survey. *Growth Factors* **8**, 61–75.
25. Freed, L. E., Vunjak-Novakovic, G., Biron, R. J., Eagles, D. B., Lesnoy, D. C., Barlow, S. K., and Langer, R. (1994). Biodegradable polymer scaffolds for tissue engineering. *Bio/Technology* **12**, 689–693.
26. Gao, J., Niklason, L. E., and Langer, R. S. (1998). Surface modification of polyglycolic acid meshes increases the seeding density and spreading of smooth muscle cells. *J. Biomed. Mater. Res.* **42**, 417–424.

27. Freed, L. E., Vunjak-Novakovic, G., and Langer, R. (1993). Cultivation of cell-polymer cartilage implants in bioreactors. *J. Cell. Biochem.* **51**, 257–264.

28. Freed, L. E., Hollander, A. P., Martin, I., Barry, J. R., Langer, R., and Vunjak-Novakovic, G. (1998). Chondrogenesis in a cell–polymer–bioreactor system. *Exp. Cell Res.* **240**, 58–65.

29. Risau, W., and Flamme, I. (1995). Vasculogenesis. *Annu. Rev. Cell Dev. Biol.* **11**, 73–91.

30. Phillips, T. M., and Dickens, B. F. (1998). Analysis of recombinant cytokines in human body fluids by immunoaffinity capillary electrophoresis. *Electrophoresis* **19**, 2991–2996.

31. Tinker, D., and Rucker, R. B. (1985). Role of selected nutrients in synthesis, accumulation, and chemical modification of connective tissue proteins. *Physiol. Rev.* **65**, 607–657.

32. Hill, K. E., and Davidson, J. M. (1986). Induction of increased collagen and elastin biosynthesis in copper-deficient pig aorta. *Arteriosclerosis* **6**, 98–104.

33. Vadlamudi, R. K., McCormick, R. J., Medeiros, D. M., Vossoughi, J., and Failla, M. L. (1993). Copper deficiency alters collagen types and covalent cross-linking in swine myocardium and cardiac valves. *Am. J. Physiol.* **264**, H2154–H2161.

34. Rucker, R. B., Kosonin, T., Clegg, M. S., Mitchell, A. E., Rucker, B. R., Uriu-Hare, J. Y., and Keen, C. L. (1998). Copper, lysyl oxidase, and extracellular matrix protein cross-linking. *Am. J. Clin. Nutr.* **67**(Suppl.), 996S–1002S.

35. Rosenberg, R. D., and Aird, W. C. (1999). Vascular-bed-specific hemostasis and hypercoagulable states. *N. Engl. J. Med.* **340**, 1555–1564.

CHAPTER 81

CARDIAC TISSUE

Robert E. Akins

This chapter presents a method for preparing three-dimensional constructs of mammalian cardiac ventricle cells. The method takes advantage of the innate capacity of isolated cardiac cells to reestablish complex, three-dimensional tissue organization *in vitro* and is performed in the absence of scaffold materials, exogenous growth factors, or serum. Only orienting surfaces are used to initiate the organizational process. Underlying the method is the concept that the intimate cell–cell interactions seen in intact cardiac tissue should be established in the earliest phases of culture. To this end, constructs are prepared in a low-shear suspension culture environment that allows the colocalization of cells in three dimensions for extended periods of time. Cells are inoculated into this environment at a relatively high density to promote interactions [1–3].

The constructs prepared by means of the present method share many attributes with intact cardiac tissue, including distinctive multicellular organization and oriented contractile function. The constructs formed may provide material for initial clinical applications. In addition, the scaffold-free nature of this approach may prove especially useful in the study of cell–cell interactions in the establishment and maintenance of cardiotypic architectures. It is hoped that the method will provide useful models for studying diverse aspects of cardiac tissue engineering and cardiac cell biology.

INTRODUCTION

THE NEED FOR CARDIAC TISSUE ENGINEERING

Congenital and acquired heart diseases are substantial health problems. Congenital heart defects occur in nearly 14 of every 1000 newborns [4]. In fact, of all congenital defects, those in the heart are the most common and are the leading cause of death in the first year of life [5]. In addition to congenital defects, many people born with structurally normal hearts suffer cardiac malfunctions at some point during their lives owing to disease, infection, or poor coronary circulation [4,6]. According to U.S. government statistics, approximately 33% of all deaths are related to some type of heart disease.

Treatments for severe cardiac problems have become highly advanced, but when they fail, organ transplantation remains the only other option. Unfortunately, there is a profound shortage of donor hearts, and many patients do not receive the necessary procedures [7]. Clearly there is a need for new approaches to treat profound heart disease, and the possibility that diminished cardiac function may be recovered through the implantation of biosynthetic constructs has great appeal. The ability to reestablish cardiac tissue architecture and function *de novo* from isolated cells may make it possible to specifically "tissue-engineer" constructs for implantation. Such methods would have potential application in a wide variety of debilitating conditions.

GENERAL APPROACHES TO CARDIAC TISSUE ENGINEERING

The first demonstration of *in vitro* cardiac cell culture was provided in the early 1900s [8,9]. Over the years, *in vitro* culture models have been used to perform a substantial amount of research in the arena of cardiac cell biology. This has included research into the reassembly of contractile, three-dimensional assemblages [10]. The application of this wealth of information to potential regenerative therapies has only recently begun. The field of cardiac tissue engineering is in its infancy, and it is still unclear which culture approaches or specific methods may be fruitful in the long run. The method provided here gives one set of conditions under which isolated cells reestablish aspects of cardiac structure; however, other approaches to cardiac tissue engineering have also been attempted.

Three general tissue engineering approaches have been attempted thus far. The first approach involves the implantation of cell suspensions directly into the heart. A number of cell implantation studies have been carried out with a variety of cell types [11–15]. Results from these studies show that implanted cells can become incorporated into existing cardiac structures; however, this approach may be of little clinical benefit when local cardiac structure is missing or seriously damaged [16,17]. The second approach involves culturing cardiac cells with three-dimensional mesh, foam, or ceramic scaffolds [18–21]. Highly engineered scaffolds could be used to direct the gross conformation of a construct, to influence the phenotype of the cellular components of a construct, or to direct particular cells to specific sites within a construct. Although this approach is very new, it may prove immensely powerful in the eventual production of large implants. The third approach involves the culture of cells in suspension with [2,22] or without [10] orienting surfaces and cell adhesion molecules. The establishment of structure through this last general approach is primarily "cell-directed" rather than "scaffold-directed," and the method may be best suited for studying tissue engineering concepts or for producing small implants.

The method described herein is an example of a "cell-directed" approach. It is presented in four sections: Cell Isolation, Selection of Medium, Preparation of Cell Support Materials, and Bioreactor Setup and Use.

CELL ISOLATION

For use in eventual clinical applications, human cardiac cells may be the most desirable. Autologous or heterologous cells collected from biopsies, surgical remnants, donor organs, or unwanted tissues may have practical use in the construction of biosynthetic implants. Alternatively, there is the possibility that cell lines, embryonic stem cells, or even nonhuman cells may prove useful. In the early stages of tissue engineering research, however, animal models are most commonly used, and neonatal rat ventricular cells are among the most common of animal models. The cell isolation method below is designed for 2-day-old rat ventricular cells; however, the method can be easily adapted to other systems. The central concept to be retained in any adaptations is that specific combinations of purified proteolytic enzymes result in the highest quality of cells and the best overall yield.

MATERIALS AND REAGENTS

NB: Tissue dissections and cell isolations should be carried out under sterile conditions to avoid possible bacterial or fungal contamination. Reagents and materials from a variety of commercial sources have been used successfully, and vendors not cited unless only one source has been used. MSDS sheets should be consulted for each of the reagents to determine safe handling procedures.

1. *Animals*: 2-day old rat pups. (Comments on determining the number of rats needed follow the section on procedure).
2. Magnetic stir plate.
3. Water-jacketed 25- to 50-ml sterile spinner flask, hooked up to a circulating water bath set at 36 ($\pm$0.5)°C.
4. Bacteriological grade culture dishes, 100 mm.
5. Sterile dissecting instruments (forceps and scissors).
6. Sterile Ca^{2+}-, Mg^{2+}-free Hanks's Balanced Salt Solution (HBSS) buffered with 10 mM HEPES to pH 7.4 (HBSSH).
7. Growth medium (see later section, Selection of Medium).
8. Two 250-ml beakers plus one squeeze bottle of 70% ethanol.

9. Proteolytic enzyme solution prepared in HBSSH. This solution should be prepared fresh just before use. (See later comments regarding the use of purified enzymes.)

    a. Trypsin (Worthington Biochemical Corporation, Freehold, NJ, cat. no. LS03707) to total activity of 238 U/100 ml

    b. Chymotrypsin (Worthington cat. no. LS01450) to a total activity of 278 U/100 ml

    c. Elastase (Calzyme Laboratories, San Luis Obispo, CA, product no. 2-1-8) to a total activity of 5.32 U/100 ml

10. Protease inhibitor solution. (See later comments on the need for protease inhibitors.)
11. Sterile cotton balls or small gauze pads.
12. Sterile 50-ml centrifuge tubes.
13. Cell support materials [e.g., culture dishes, microcarrier beads, collagen or poly(glycolic acid) sutures, coverslips, etc.] precoated the previous day with fibronectin as explained later (Preparation of Cell Support Materials)
14. Sterile pipettes, 1, 10, and 25 ml.
15. Hemocytometer or other cell counter (see Chapter 4 in this volume for details on cell counting).
16. Centrifuge for spinning 50-ml tubes.
17. Falcon cell strainers (Becton Dickinson, Franklin Lakes, NJ, cat. no. 2350).

PROCEDURE

NB: Animals should be handled in accordance with the guidelines of the local Institutional Animal Care and Use Committee (IACUC).

1. Fill one of the 100-mm culture dishes with about 25 ml of sterile HBSSH. This is where the dissected hearts will be placed.
2. Place dissecting instruments into a beaker of 70% ethanol.
3. Sterilize the work area inside a laminar flow hood by swabbing with cotton or gauze soaked in 70% ethanol. (See later comment on laminar flow hoods.)
4. Anesthetize and euthanize an animal per local IACUC guidelines. It is important to perform the dissection as soon as possible after euthanasia.
5. Swab the ventral surface of the animal (neck through abdomen) with cotton or gauze soaked in 70% alcohol.
6. Using sharp scissors, make a lateral incision across the chest just under the rat's "armpits."
7. Remove the heart rapidly, but carefully, with forceps and transfer to the culture dish containing HBSSH.
8. Dispose of the carcass.
9. Rinse the dissecting instruments in sterile water and place them back in the 70% ethanol.
10. Repeat steps 3–9 until all the hearts have been collected. Maintain sterility throughout the procedure.
11. Examine each heart and remove any pulmonary or other noncardiac tissue that is present.
12. Transfer the tissue to a clean, dry 100-mm dish, and mince the tissue into pieces of about 1–2 mm^3. This step is critical: too fine a mince will result in the loss of cells in the subsequent wash steps, and too coarse a mince will result in cells being left in the tissue remnants after digestion.
13. Using the sterile scraper, transfer the minced tissue to the water-jacketed spinner flask.
14. Add the remainder of the sterile HBSSH to the tissue to rinse away excess blood cells and debris. Carefully decant the rinse solution, leaving the tissue fragments in the flask.
15. Add proteolytic enzyme solution to the minced tissue. Use 1 ml per heart equivalent. Place the stirrer on the magnetic stir plate, and set the rate of revolution to just suspend the fragments (<200 rpm). Incubate for 15 min, then decant the enzyme solution into a waste beaker.

16. Repeat step 15 once. The first two enzyme treatments are primarily blood cells and are discarded.

17. Add proteolytic enzyme solution to the minced tissue; again use 1 ml per heart equivalent. Place the stirrer on the stir plate, and suspend the fragments. Incubate for 20 min, then decant the enzyme solution into a 50-ml centrifuge tube containing 15 ml of growth medium. (See later comment on the need for protease inhibitors.)

18. Collect the cells by gentle centrifugation at approximately 250g for 10 min. When centrifugation is complete, decant the solution from the pellet into the waste beaker. Be careful not to dislodge a soft pellet. Resuspend the pellets in 5 ml of growth medium, pooling them as you go.

19. Repeat steps 17 and 18 until no more cells come out of the tissue fragments (i.e., the enzyme solution is clear after the 20-min incubation).

20. Gently triturate the cell suspension and strain it through the Falcon cell strainers to remove any undissociated chunks of tissue.

21. Use a hemocytometer or other cell counter to count the trypan-blue-excluding cells (see Chapter 4 in this volume for details).

21. Dilute the cells with growth medium to a density of 1×10^6 cells/ml.

COMMENTS AND SUGGESTIONS
Determining the Number of Rats Needed

Expect a yield of about 5×10^6 viable cells per rat. Cells are inoculated at a density of 1×10^6 cells/ml. So, a 50-ml bioreactor would require 10 neonatal rats. To estimate the degree of confluence expected upon inoculation, a general rule of thumb can be used that relates the volume of medium to the number of cells to the surface area of attachment material: 1 ml of medium per million cells per 5 cm^2 of surface. It should also be noted that although neonatal rat cardiomyocytes have very little proliferative activity, other cells from the ventricle (e.g., endocardial cells or fibroblasts) do proliferate. Inoculating at too low a density under conditions that allow nonmyocyte proliferation may result in the underrepresentation of myocytes in the final construct.

Use of Purified Enzyme

The use of a mixture of purified enzymes is based on published analysis of enzyme mixtures for tissue disaggregation [23]. It should be noted that alternative digestion methods for the preparation of cardiac cells are possible. The degree of tissue digestion, the speed with which the procedure may be completed, and the quality of the resulting cells all must be considered when one is choosing a dissociation method. The method presented here makes it possible to reproducibly prepare high-quality neonatal rat heart cells that resume contractile activity within 24 h after plating. In addition, this method routinely yields a high proportion of adherent cells and gives a very good yield of cardiomyocytes as a proportion of the entire cell isolate.

Need for Protease Inhibitors

It is helpful to neutralize the proteolytic enzymes prior to culture, especially when a serum-free medium will be used. Adding ovomucoid or soybean trypsin inhibitor to the freshly isolated cells prior to centrifugation works well. Alternatively, additional rinsing can be done prior to plating the cells.

Laminar Flow Hoods

Laminar flow hoods provide sterile fields for dissection and culture work. In general, both vertical and horizontal flow hoods are usable for the preparation of animal cells; for work with human material, however, vertical flow hoods should be used to limit potential exposure to infectious agents. To keep the work surface of the hood clean, the hood can be lined with aluminum foil sheets, sterilized with 70% ethanol. In general, flame sterilization techniques should not be used inside laminar flow hoods because the heat generated disrupts the airflow and may damage the hood.

SELECTION OF MEDIUM

There are numerous nutritive media available for mammalian cell culture [24]. The choice of medium for use in tissue engineering experiments must be based on a variety of considerations; however, serum-free media have a great many advantages over serum-containing media (e.g., see Chapter 1 in this volume). For tissue engineering, the long-term goal is to reproducibly generate implantable constructs, and the use of an undefined component like serum probably should be avoided. The selection of a medium must be based on overall experimental design and research goals, and serum-containing media can be used in the current method.

A few serum-free media have been described for the culture of mammalian cardiac cells. As a rule, these formulations are basal media supplemented with specific hormones or factors like insulin, thyroid hormone, holotransferrin, and steroids [25–29]. Other researchers have used combinations of serum-free medium and fibroblast-conditioned medium for cardiac cell culture [30–32]. Any of these media can be used for cardiac tissue engineering with the current method. The author has successfully used a previously described serum-free and hormone-free medium (SF-HF) [28] for the culture of cardiac cells [1–3]. SF-HF is based on a 1:1 mixture of Ham's F-12 and Dulbecco's Modified Eagle's Medium and works well for the establishment and maintenance of three-dimensional cardiac constructs from rat heart cells. More recently, an improved serum-free medium has been developed and is commercially available (AI-1 Medium, KD Medical, Columbia, MD).

PREPARATION OF CELL SUPPORT MATERIALS

Cardiac cells are attachment dependent and contact inhibited (see Chapter 3 in this volume). To initiate cell culture in the present method, surfaces onto which the cells can attach are provided. Standard support materials include Biosilon microcarrier beads (Nunc/InterMed, Roskilde, Denmark), CytoDex microcarrier beads (Sigma Chemical Company, St. Louis, MO), non-oriented collagen cultispheres (HyClone Laboratories, Logan, UT), oriented-collagen sutures (Organogenesis, Canton, MA), and various other suture materials [e.g., poly(glycolic acid), Davis & Geck, Manati, Puerto Rico]. If not shipped sterile, these support materials must be sterilized prior to use. Also, if a serum-free medium is to be used, the material should be coated with fibronectin (or laminin) prior to culture to encourage cell attachment. These steps can be carried out as follows.

PROCEDURE

1. Rinse material in sterile, deionized water.
2. Place material into a solution of 70% ethanol for 24 h.
3. Rinse material in sterile, deionized water.
4. Store material in sterile Dulbecco's phosphate-buffered saline.
5. Prepare a stock solution of fibronectin (Collaborative Research, Waltham, MA, cat. no. 35-4008) by adding 10 ml of sterile deionized water to a 1-mg vial of fibronectin. This is done several hours before use to assure that the fibronectin is solubilized. Note that a laminin stock solution can be prepared in a similar manner.
6. Place the support material into the fibronectin stock solution for 24 h at 4°C.
7. Rinse the support material in cell culture medium just prior to use.

BIOREACTOR SETUP AND USE

There are many types of bioreactor available for cell culture, and the criteria by which a bioreactor may be judged useful or not vary depending on the application. Any of several bioreactors could be used with the methodology presented here as long as the selection criteria include the minimization of damaging fluid shear and the ability for cells to colocalize in a fluid environment. Low fluid shear is a desirable characteristic because cardiac cells are known to be mechanically sensitive [33–37], and colocalization allows the desired cell–cell contacts to form. In addition, the bioreactor should be able to support the metabolic requirements of the highly active cardiac cells. Two related types of bioreactor that meet these criteria have been used in our early tissue engineering efforts.

The first type of bioreactor we have used is a high-aspect-ratio vessel (HARV; Synthecon, Inc., Friendswood, TX). HARV bioreactors provide a low-shear, fluid environment (<0.52 dyn/cm^2) [38], allow cells to colocalize, and support the metabolic requirements of neonatal rat cardiac cells [2]. The vessel portion of the HARV is a machined 10-ml Lexan culture dish that is operated in the upright position. There are two halves to the vessel. One half contains several small pores that are covered by a silicone membrane, which provides a gas exchange surface for the system. The other half of the vessel has a circular recessed area and two Luer-locking ports through which cells, constructs, and media can be accessed. When assembled and loaded, the HARV is a closed, fluid-filled system that can be mounted onto a motorized base and rotated.

The second type of bioreactor that we have used is a simple VueLife fluoropolymer (PTFE) bag (American Fluoroseal Corporation, Gaithersburg, MD, cat. no. 1P-0007). VueLife bags are single 7-ml units with a single access port. The bags are optically transparent, and the contents can be visualized on a standard light microscope without sampling. The PTFE material itself is gas permeable, and the entire bag serves as a gas exchange surface. When assembled and loaded, a VueLife bag is a closed, fluid-filled system that can be rotated to suspend the contents.

The key to the utility of these bioreactors has been discussed elsewhere [1] (also see Chapter 6 in this volume). Briefly, the sedimentation of cells (or support material, etc.) in a bioreactor combined with the rotational movement of the medium results in the cell taking an elliptical path relative to the observer. By adjusting the rotation rate of the bioreactor, the dimensions of this elliptical path can be reduced so that vessel, medium, and contents appear to maintain their relative relationships in a low-shear environment.

A method for the setup and use of the HARV and VueLife bioreactors follows. Manufacturers' recommendations should be consulted regarding specific details for the handling of each type of bioreactor.

PROCEDURE

1. Rinse the bioreactor thoroughly with sterile, deionized water. The HARV should be disassembled for this step; the VueLife is a single unit. Care should be taken with both bioreactors to assure that foreign materials are not introduced. Surfactants should be avoided.

2. Sterilize the bioreactor. For the HARV, ethylene oxide gas or 70% ethanol is appropriate; autoclaving is not. For the VueLife bags, autoclaving or 70% ethanol is appropriate; ethylene oxide should not be used.

3. Rinse the bioreactor again in sterile, deionized water just prior to use. A HARV bioreactor should be checked for potential leaks through the fittings or through the silicone gas exchange membrane at this point and any defects corrected. Empty the bioreactor.

4. Use the available access port(s) to load the bioreactor with support materials, cells suspension, and medium. Remove any air bubbles that may be present. The inoculation density will depend on the desired results. Good results may be obtained using the density specified in the rule of thumb given earlier: 1 ml of medium for 1×10^6 cells on 5 cm^2 of surface. Changes to this ratio will alter the size and cellular thickness of the constructs prepared.

5. Attach the bioreactor to the rotating platform. The HARV bioreactor comes with a rotating platform and a controller. A rotator for the VueLife bags can be devised by using a small variable speed motor (0–100 rpm) and a simple mounting platform.

6. Adjust the rotation rate of the bioreactor to keep the support material and cells in static suspension. As the bioreactor rotates, it should be possible to use oblique illumination from a penlight to visualize the elliptical path traced by the included support surfaces (the cells will be too small to be seen easily). As the rate of rotation is increased, the elliptical paths will get smaller until the supports appear to remain stationary relative to the wall of the bioreactor. Note that if the rate of rotation is set too high, an undesirable centrifugal effect may be seen.

7. Place the bioreactor into an incubator at 37°C in 95–100% humidity with CO_2 (if required). Note that the level of CO_2 depends on the medium chosen, especially the sodium bicarbonate level in the medium. For example, SF-HF contains 2.45 mg/ml sodium

bicarbonate and is used in a 5% CO_2 environment. The final pH of the medium should be about 7.2.

8. Assess the attachment and contractile effort of the culture. Take a sample of cells from the HARV for microscopic observation, or place a VueLife bag directly on the microscope stage. Cells should attach to the support materials within 16 h and should be spontaneously beating within 24 h of inoculation into the bioreactor. Delays in this time frame may indicate poor preparation of the bioreactor, the support materials, the medium, or the cells.

9. Change the medium at 24 h to remove any unattached cells. The exchange of medium, addition of compounds, and sampling of the vessels can be carried out through the available access ports.

10. Feed the cultures with a complete change of medium at 48-h intervals.

REFERENCES

1. Akins, R. E. (2000). Prospects for the use of cell implantation, gene therapy, and tissue engineering in the treatment of myocardial disease and congenital heart defects. *In* "Medizinische Regeneration and Tissue Engineering" (K. Sames, ed.), Chapter XI, pp. 1–12. EcoMed, Landsberg, Germany.

2. Akins, R. E., Boyce, R. A., Madonna, M. L., Schroedl, N. A., Gonda, S. R., McLaughlin, T. A., and Hartzell, C. R. (1999). Cardiac organogenesis in vitro: Reestablishment of three-dimensional tissue architecture by dissociated neonatal rat ventricular cells. *Tissue Eng.* 5, 103–118.

3. Akins, R., Schroedl, N., Gonda, S., and Hartzell, C. (1997). Neonatal rat heart cells cultured in simulated microgravity. *In Vitro Cell. Dev. Biol.—Anim.* 33, 337–343.

4. Gillum, R. F. (1994). Epidemiology of congenital heart disease in the United States. *Am. Heart J.* 127, 919–927.

5. Hoffman, J. I. (1995). Incidence of congenital heart disease. I. Postnatal incidence. *Pediatr. Cardiol.* 16, 103.

6. McGovern, P. G., Pankow, J. S., Shahar, E., Doliszny, K. M., Folsom, A. R., Blackburn, H., and Luepker, R. V. (1996). Recent trends in acute coronary heart disease—mortality, morbidity, medical care, and risk factors. *N. Engl. J. Med.* 334, 884–890.

7. Stevenson, L. W., Warner, S. L., Steimle, A. E., Fonarow, G. C., Hamilton, M. A., Moriguchi, J. D., Kobashigawa, J. A., Tillisch, J. H., Drinkwater, D. C., and Laks, H. (1994). The impending crisis awaiting cardiac transplantation. Modeling a solution based on selection. *Circulation* 89, 450–457.

8. Burrows, M. T. (1912). Rhythmical activity of isolated heart muscle cells in vitro. *Science* 36, 90–92.

9. Burrows, M. T. (1910). The cultivation of tissues of the chick embryo outside the body. *JAMA J. Am. Med. Assoc.* 55, 2057–2058.

10. Sperelakis, N. (1978). Cultured heart cell reaggregate model for studying cardiac toxicology. *Environm. Health Perspect.* 26, 243–267.

11. Koh, G. Y., Soonpaa, M. H., Klug, M. G., and Field, L. J. (1993). Long-term survival of AT-1 cardiomyocyte grafts in syngeneic myocardium. *Am. J. Physiol.* 264, H1727–H1733.

12. Koh, G. Y., Klug, M. G., Soonpaa, M. H., and Field, L. J. (1993). Differentiation and long-term survival of C2C12 myoblast grafts in heart. *J. Clin. Invest.* 92, 1548–1554.

13. Soonpaa, M. H., Koh, G. Y., Klug, M. G., and Field, L. J. (1994). Formation of nascent intercalated disks between grafted fetal cardiomyocytes and host myocardium. *Science* 264, 98–101.

14. Chiu, R. C., Zibaitis, A., and Kao, R. L. (1995). Cellular cardiomyoplasty: Myocardial regeneration with satellite cell implantation. *Ann. Thorac. Surg.* 60, 12–18.

15. Klug, M. G., Soonpaa, M. H., Koh, G. Y., and Field, L. J. (1996). Genetically selected cardiomyocytes from differentiating embryonic stem cells form stable intracardiac grafts. *J. Clin. Invest.* 98, 216–224.

16. Aoki, M., Morishita, R., Higaki, J., Moriguchi, A., Hayashi, S., Matsushita, H., Kida, I., Tomita, N., Sawa, Y., Kaneda, Y., and Ogihara, T. (1997). Survival of grafts of genetically modified cardiac myocytes transfected with FITC-labeled oligodeoxynucleotides and the beta-galactosidase gene in the noninfarcted area but not the myocardial infarcted area. *Gene Ther.* 4, 120–127.

17. Watanabe, E., Smith, D. M., Jr., Delcarpio, J. B., Sun, J., Smart, F. W., Van Meter, C. H., Jr., and Claycomb, W. C. (1998). Cardiomyocyte transplantation in a porcine myocardial infarction model. *Cell Transplant.* 7, 239–246.

18. Bursac, N., Papadaki, M., Cohen, R. J., Schoen, F. J., Eisenberg, S. R., Carrier, R., Vunjak-Novakovic, G., and Freed, L. E. (1999). Cardiac muscle tissue engineering: Toward an in vitro model for electrophysiological studies. *Am. J. Physiol.* 277, H433–H444.

19. Carrier, R. L., Papadaki, M., Rupnick, M., Schoen, F. J., Bursac, N., Langer, R., Freed, L. E., and Vunjak-Novakovic, G. (1999). Cardiac tissue engineering: Cell seeding, cultivation parameters, and tissue construct characterization. *Biotechnol. Bioeng.* 64, 580–589.

20. Polonchuk, L., Elbel, J., Eckert, L., Blum, J., Wintermantel, E., and Eppenberger, H. M. (2000). Titanium dioxide ceramics control the differentiated phenotype of cardiac muscle cells in culture. *Biomaterials* 21, 539–550.

21. Li, R. K., Yau, T. M., Weisel, R. D., Mickle, D. A., Sakai, T., Choi, A., and Jia, Z. Q. (2000). Construction of a bioengineered cardiac graft. *J. Thorac. Cardiovasc. Surg.* 119, 368–375.

22. Fink, C., Ergun, S., Kralisch, D., Remmers, U., Weil, J., and Eschenhagen, T. (2000). Chronic stretch of engineered heart tissue induces hypertrophy and functional improvement. *FASEB J.* 14, 669–679.

23. Speicher, D. W., and McCarl, R. L. (1978). Evaluation of a proteolytic enzyme mixture isolated from crude trypsins in tissue disaggregation. *In Vitro* **14**, 849–853.

24. Mather, J. P. (1998). Making informed choices: Medium, serum, and serum-free medium. *In* "Animal Cell Culture Methods" (J. P. Mather and D. Barnes, eds.), Vol. 57, pp. 19–30. Academic Press, San Diego, CA.

25. Karliner, J. S., Simpson, P. C., Taylor, J. E., Honbo, N., and Woloszyn, W. (1985). Adrenergic receptor characteristics of cardiac myocytes cultured in serum-free medium: Comparison with serum-supplemented medium. *Biochem. Biophys. Res. Commun.* **128**, 376–382.

26. Libby, P. (1984). Long-term culture of contractile mammalian heart cells in a defined serum-free medium that limits non-muscle cell proliferation. *J. Mol. Cell. Cardiol.* **16**, 803–811.

27. Kessler-Icekson, G., Sperling, O., Rotem, C., and Wasserman, L. (1984). Cardiomyocytes cultured in serum-free medium. Growth and creatine kinase activity. *Exp. Cell Res.* **155**, 113–120.

28. Mohamed, S. N., Holmes, R., and Hartzell, C. R. (1983). A serum-free, chemically-defined medium for function and growth of primary neonatal rat heart cell cultures. *In Vitro Cell. Dev. Biol.* **19**, 471–478.

29. Freerksen, D. L., Schroedl, N. A., and Hartzell, C. R. (1984). Control of enzyme activity levels by serum and hydrocortisone in neonatal rat heart cells cultured in serum-free medium. *J. Cell. Physiol.* **120**, 126–134.

30. Suzuki, T., Hoshi, H., and Mitsui, Y. (1990). Endothelin stimulates hypertrophy and contractility of neonatal rat cardiac myocytes in a serum-free medium. *FEBS Lett.* **268**, 149–151.

31. Suzuki, T., Tsuruda, A., Katoh, S., Kubodera, A., and Mitsui, Y. (1997). Purification of endothelin from a conditioned medium of cardiac fibroblastic cells using a beating rate assay of myocytes cultured in a serum-free medium. *J. Mol. Cell. Cardiol.* **29**, 2087–2093.

32. Suzuki, T., Hoshi, H., Sasaki, H., and Mitsui, Y. (1991). Endothelin-1 stimulates hypertrophy and contractility of neonatal rat cardiac myocytes in a serum-free medium. II. *J. Cardiovasc. Pharmacol.* **17**, S182–S186.

33. Andries, L. J., Sys, S. U., and Brutsaert, D. L. (1995). Morphoregulatory interactions of endocardial endothelium and extracellular material in the heart. *Herz* **20**, 135–145.

34. Yamazaki, T., Komuro, I., Kudoh, S., Zou, Y., Shiojima, I., Hiroi, Y., Mizuno, T., Maemura, K., Kurihara, H., Aikawa, R., Takano, H., and Yazaki, Y. (1996). Endothelin-1 is involved in mechanical stress-induced cardiomyocyte hypertrophy. *J. Biol. Chem.* **271**, 3221–3228.

35. Swynghedauw, B. (1986). Developmental and functional adaptation of contractile proteins in cardiac and skeletal muscles. *Physiol. Rev.* **66**, 710–771.

36. Sadoshima, J., and Izumo, S. (1993). Mechanical stretch rapidly activates multiple signal transduction pathways in cardiac myocytes: Potential involvement of an autocrine/paracrine mechanism. *EMBO J.* **12**, 1681–1692.

37. Dzau, V. J. (1993). The role of mechanical and humoral factors in growth regulation of vascular smooth muscle and cardiac myocytes. *Curr. Opin. Nephrol. Hypertens.* **2**, 27–32.

38. Goodwin, T. J., Prewett, T. L., Wolf, D. A., and Spaulding, G. F. (1993). Reduced shear stress: A major component in the ability of mammalian tissues to form three-dimensional assemblies in simulated microgravity. *J. Cell. Biochem.* **51**, 301–311.

CHAPTER 82

CARDIAC VALVES AND ARTERIES

John E. Mayer Jr., Kristine Guleserian, Fraser Sutherland, and Tjorvi Perry

Tissue engineering is a developing science that brings together engineering and biology in an attempt to develop replacement tissues, beginning with their individual cellular components. Much of the strength and flexibility in normal tissues is due to the specialized proteins and polysaccharide–protein complexes that form the matrix of the normal heart valve and vascular wall. This matrix is normally produced by the cells in the valve and vascular tissue, and there is active production and turnover of the matrix. Although it is has been possible for some time to grow cells of individual types in culture, it is more difficult to induce these cells to assemble or organize into the more complex structural arrangements, are found in normal tissues or to produce normal structural proteins in an organized fashion. Our laboratory efforts to develop heart valves and large conduit arteries have been based on the concept of using biodegradable polymers as scaffolds upon which to have normal cells grow and proliferate. The scaffold also temporarily provides the structure and the mechanical stability, which are necessary for "tissues" to develop from their individual cellular components. Ideally, these polymer scaffolds then degrade during the time that the cells in the developing "tissue" are producing their own normal structural proteins and are becoming organized and oriented to replicate normal tissue structure.

INTRODUCTION

Diseases of the heart valves and large "conduit" arteries account for approximately 60,000 cardiac surgical procedures each year in the United States but the currently available replacement devices have well-known significant limitations. Ideally, any valve or artery substitute would function similarly to the normal valve or artery to allow blood to pass through it without stenosis or regurgitation, but would also have the following characteristics: durability, growth potential, compatibility with blood (so that no thrombus and emboli would form on its surface), and resistance to infection. None of the currently available devices constructed from prosthetic or biological materials meet these criteria. Our concept was that if new valves or arteries could be made from individual cells, then the new living tissues could potentially have these desirable characteristics. The potential for growth is of particular importance to children with malformed or diseased valves or arteries.

LABORATORY PROJECTS

Several projects have been undertaken in our laboratory to initially construct a heart valve leaflet and large arteries using the tissue engineering approach. In the initial studies [1–5], a poly(glycolic acid)–poly(lactic acid) composite polymer (PGA/PLA) served as the scaffold on which the cells were grown. The cells came from normal arteries, which could be removed from the animal and separated into the various cell components. We found that it was important to use autologous tissue as the source of the cells, thereby eliminating the possible of immune rejection of the tissues once they were reimplanted [3]. The number of cells were "expanded" in cultures by allowing them to divide, and then suspensions of the cells were added to the polymer scaffolds. These cell–polymer constructs were then incubated in culture for several more days before they were implanted to replace a valve or an artery in the same animal from which the cells had been originally removed.

Valve leaflets [2–4] and segments of pulmonary artery [5] functioned well for periods of up to 4 months without structural failure and without thrombus formation on the surfaces. Importantly, when these structures were implanted into growing animals, they demonstrated growth and complete degradation of the polymer by 6 weeks after implantation [4,5]. We were also able to demonstrate that cells from the wall of a systemic vein could be used to form a conduit artery for the pulmonary circulation [5]. The tissues appeared to have relatively normal structure, and they produced the normal matrix proteins.

Despite these encouraging results, a significant number of problems remain to be resolved prior to application of these techniques for clinical use. First, all our published experiments to date have been carried out in animals, and it remains to be determined whether human cells can be used to develop tissues in the same way. We are engaged in studies with human cells to address this question. Second, the identification of an "ideal" polymer scaffold has not been accomplished. The PGA-PLA polymer used as the scaffold in the initial experiments is stiff and difficult to work with surgically. Our (unreported) attempts to create a three-leaflet valve yielded valve leaflets that were quite thick, and implantation of the valve was associated with intolerable pulmonary valvar stenosis. Experience with a much more flexible polymer, polyhydroxyoctanoate (PHO), revealed that it is possible to construct a trileaflet, valved conduit that has good early function *in vivo* [6,7]. However, the prolonged degradation time of this material was associated with a less favorable outcome of the engineered tissue in both the leaflet and the wall of the conduit [6], despite modification of the PHO by salt leaching to make it more porous [7]. Based on these experiments, we have concluded that a short degradation time for the polymer scaffold is preferable. More recently, we have used a composite material formed from PGA and poly(4-hydroxybutyrate) (P4HB) with better early and intermediate-term results [8]. This composite polymer appears to have been completely degraded within 6 weeks of implantation, but is more flexible than the PGA native polymer. Others have taken an approach in which native tissues are "decellularized" by incubation in trypsin–EDTA for 48 h, leaving only the matrix proteins as a substrate to which donor cells are then added [9]. Steinhoff and colleagues report that the recellularized valves showed "complete histological restitution of valve tissue and confluent endothelial surface coverage in all cases" [9]. The interstitium of the valve was populated by myofibroblast-like cells, which stained for smooth muscle α-actin [9]. They do report that the valve leaflets were thicker than normal. We are continuing to search for a biodegradable polymer or native scaffolding material that will have more acceptable initial strength and flexibility characteristics while still providing a hospitable environment for the cells to develop into tissues.

The ideal source of the cells for the developing "tissues" has not been determined. In patients, if it is necessary to use vascular wall cells for the engineered tissue, it would be preferable to use veins rather than arteries as the initial source of the cells, since veins are more plentiful and removal is less likely to compromise the blood supply to normal tissues. We have some evidence that heart valves developed by using cells from the skin do not function as well as those developed by using cells from the wall of the artery [4], but vein wall cells seem to work reasonably well in conduit arteries [5]. We have begun to explore the use of circulating endothelial precursor cells as a potential source of cells for tissue-engineered cardiovascular structures [10]. These cells are isolated from the buffy coat based on their adherence to fibronectin, and we have added them to decellularized porcine iliac artery scaffolds. This approach has allowed us to create tissue-engineered small-caliber arterial grafts that remained patent in sheep for periods of up to several months [10]. Histologically, these tissue-engineered vessels appear to remain patent and viable, and they seem to form new matrix components in the wall of the neovessel. If similar results are obtained in creating tissue-engineered cardiac valves from these circulating endothelial cells (ongoing experiments), the sacrifice of native blood vessels for cells to seed scaffolds can be avoided.

All our experiments to date have been carried out in immature growing animals, and it is not clear whether the cells that are used to form these "tissues" must be from immature animals. There is some reason to believe that fetal sources of cells would be preferable, and it has been our observation that fetal cells proliferate much more rapidly under tissue culture conditions. We believe that we have successfully isolated circulating endothelial cells from human cord blood [K. Guleserian, unpublished observations]. Since it is now possible to diagnose many forms of congenital heart disease in fetal life, one might imagine using fetal cells to develop replacement valves or arteries while the gestation is continuing. At birth these replacement valves or arteries, developed from the child's own cells, could be ready for implantation.

FUTURE PROSPECTS

Finally, a number of questions remain regarding optimal *in vitro* conditions under which the tissue-engineered structures should be grown. One important area is that of the physical signals that are delivered to the developing tissues. We have recently completed initial studies that assessed the effects of subjecting the developing valve leaflet tissue to physical forces (shear stress and hydrostatic pressure) provided by a pulse duplicator system [8]. These studies have shown that the developing tissue, which was grown on the PGA-P4HB polymer scaffold, does respond to imposed physical "signals" by increasing the amount of collagen produced and increasing the number of cells in the tissue [8]. The early function and the gross and histological appearance of the tissue valves produced in this way are superior to those resulting from the use of prior techniques, although there was some central valvar regurgitation after 4 months *in vivo* [8]. A variety of chemical or cytokine signals will likely also have an impact on the development of these tissues *in vitro*, and we may be able to alter the chemical and/or cytokine signals to the developing tissue to further "engineer" the tissue that is formed. Our understanding of how these developing tissues will respond to any number of physical and chemical signals remains limited. However, we anticipate that by controlling both the physical and biochemical environment of the developing tissue, it will be possible to more precisely guide the development of valve and arterial conduit tissues prior to implantation into the blood-stream.

Tissue engineering is a new approach to solving the problem of creating replacement tissues for use as heart valves or arteries. Although initial animal studies have been encouraging, numerous questions must be resolved before clinical trials can begin.

REFERENCES

1. Breuer, C., Shin'oka, T., Tanel, R. E., Zund, G., Money, D. J., Ma, P. X., Langer, R., Vacanti, J. P., and Mayer, J. E. (1996). Tissue engineering lamb heart valve leaflets. *Biotechnol. Bioeng.* **50**, 562–567.
2. Shin'oka, T., Ma, P. X., Shum-Tim, D., Breuer, C. K., Cusick, R. A., Zund, G., Langer, R., Vacanti, J. P., and Mayer, J. E. (1996). Tissue engineered heart valves—autologous valve leaflet replacement study in a lamb model. *Circulation* **94**, III64–III68.
3. Shin'oka, T., Breuer, C. K., Tanel, R. E., Zund, G., Miura, T., Ma, P. X., Langer, R., Vacanti, J. P., and Mayer, J. E. (1995). Tissue engineering heart valves: Valve leaflet replacement study in a lamb model. *Ann. Thorac. Surg.* **60**, S-513–S-516.
4. Shin'oka, T., Shum-Tim, D., Ma, P. X., Tanel, R. E., Langer R., Vacanti, J. P., and Mayer J. E. (1997). Tissue engineered valve leaflets: Does cell origin affect outcome? *Circulation* **96**(Suppl. II), II-102–II107.
5. Shin'oka, T., Shum-Tim, D., Ma, P. X., Tanel, R. E., Isogai, N., Langer, R., Vacanti, J. P., and Mayer, J. E. (1998). Creation of viable pulmonary artery autografts through tissue engineering. *J. Thorac. Cardiovasc. Surg.* **115**, 536–546.
6. Stock, U., Nagashima, M., Khalil, P. N., Nollert, G. D., Herden, T., Sperling, J. S., Moran, A. M., Lien, J., Martin, D. P., Schoen, F. J., Vacanti, J. P., and Mayer, J. E. (2000). Tissue engineered valved conduits in the pulmonary circulation. *J. Thorac. Cardiovasc. Surg.* **119**, 732–740.
7. Sodian, R., Hoerstrup, S. P., Sperling, J. S., Daebritz, S., Martin, D. P., Moran, A. M., Kim, B. S., Schoen, F. J., Vacanti, J. P., and Mayer, J. E., Jr. (2000). Early in vivo experience with tissue engineered trileaflet heart valves. *Circulation* **102**, III-22–III-29.
8. Hoerstrup, S., Sodian, R., Daebritz, S., Wang, J., Bacha, E. A., Martin, D. P., Moran, A. M., Guleserian, K. J., Sperling, J. S., Kaushal, S., Vacanti, J. P., Schoen, F. J., and Mayer, J. E. (2000). Functional living trileaflet heart valves grown in vitro, *Circulation*, **102**(Suppl. III), III-44–III-49.
9. Steinhoff, G., Stock, U., Karim, N., Bertsching, H., Timke, A., Meliss, R., Pethig, K., Haverich, A., and Bader, A. (2000). Tissue engineering of pulmonary heart valves on allograft acellular matrix conduits. *Circulation* **102**(Suppl. III), III-50–III-55.
10. Kaushal, S., Amiel, G., Guleserian, K., Perry, T., Soker, S., Roth, S., Atala, A., Bischoff, J., and Mayer, J. E. (2000). Circulating endothelial cells for tissue-engineered small diameter vascular grafts. *Circulation* **102**(Suppl. II), II-766 (abstr.).

CHAPTER 83

CORNEA

May Griffith, Vickery Trinkaus-Randall, Mitchell A. Watsky, Chia-Yang Liu, and
Heather Sheardown

THE CORNEA

The cornea provides two major functions, protecting the intraocular contents and serving as major optical element of the eye. In fact 75% of the diopteric power of the eye depends on the interface of the cornea and air [1]. Any injury, disease, or cellular failure that causes opacification of the cornea causes visual impairment and in serious cases, corneal blindness. Affecting more than 10 million patients worldwide, corneal opacification is often treatable by transplantation with human donor tissue (estimates from U.S. Vision Share Eye Bank Consortium). However, for patients afflicted with specific disorders such as autoimmune conditions or alkali burns, this procedure has a poor success rate (Dohlman, personal communications). Furthermore, in developing countries where the number of cornea blindness cases is increasingly problematic, healthy donor tissue is rare. An alternative for these patients is replacement of the damaged cornea with an artificial substitute.

CORNEAL BIOCHEMISTRY AND IDENTIFYING MARKERS
Corneal Epithelium

The corneal epithelium is a stratified, nonkeratinized, noncornified, epithelium comprising five to seven cell layers. A prominent feature of the corneal epithelium is an abundance of intracellular intermediate filaments that are rich in the K3/K12 keratin pair and hemidesmosomes that are present in the basal cells and play a role in signaling [1]. Like other keratin pairs, K3/K12 contributes towards the cellular integrity of corneal epithelium. Loss of the K12 gene-by-gene targeting results in fragile corneal epithelium [2]. Likewise, mutations in K3 or K12 genes cause Meesmann's corneal dystrophy [3]. Chapter 9 and Ref. [1] provide more details on the corneal epithelium and its culture.

Corneal Stroma

The corneal stroma comprises 85–90% of the corneal thickness and has a pivotal role in the maintenance of corneal transparency. It is composed mainly of hydrated extracellular matrix (ECM) with a population of fibroblastic-like cells known as keratocytes. The stromal matrix is composed of collagens (70% type I, 20% type V, along with minor collagens), proteoglycans, and other minor proteins. Type I collagen, which forms the bulk of the ECM, is composed of regularly packed collagen fibrils that are organized into lamellae. It has been suggested that the stoichiometry and interaction of different collagen types play important roles in modulating collagen fibril diameter [5,6]. It is also believed that various proteoglycans within the stroma interact with the collagen fibrils to regular interfibrillar spacing [7,8]. Keratan sulfate proteoglycans (KSPG) are the predominant proteoglycan in the corneal stroma, followed by dermatan sulfate proteoglycans (DSPG) and chondroitin sulfate proteoglycans (CSPG). The ratios of various proteoglycans govern stromal water distribution [9–11].

Keratocan, lumican, and mimecan are three major KSPGs in the corneal stroma. The role of KSPG in regulating collagen fibrillogenesis has been demonstrated in lumican-null mice via gene targeting. Lumican-null mice develop cloudy corneas and unorganized collagen fibers [12,13]. In humans, mutations of the keratocan gene have been associated with corneal plana [14], indicating a major role in corneal transparency. It is interesting that keratocan gene expression tracks corneal stromal morphogenesis and is restricted to keratocytes in adult mice [15], making it a potential definitive marker for stromal keratocytes.

Corneal Endothelium

The human corneal endothelium is a single-layer of flattened hexagonal cells. It is not a true endothelium (i.e., lining a vascular or lymphatic channel), but a mesothelium (i.e., resembling the cells that line such body cavities as peritoneum and pleural, or pericardial clefts or spaces). Gap junctions are found along the lateral membranes of cells, and tight junctions are found apically, forming incomplete seals [16]. Along with nutrients, water leaks into the stroma across the endothelium, and subsequently this fluid is "pumped" back in the anterior chamber — hence the pump–leak theory of corneal hydration and nutrition. Disease or trauma resulting in any significant imbalance of this pump–leak cycle will result in corneal edema and loss of transparency. In the human, corneal endothelial cells have little potential for cell division, and significant cell death is often a major factor contributing to an imbalance of the pump–leak system. Patients with severe endothelial damage are treated with corneal transplants (penetrating keratoplasty), which supply a new population of endothelial cells to the cornea, allowing for restoration of the pump–leak machinery. A nonswelling corneal prosthesis or artificial cornea would potentially not require an endothelium as long as sufficient movement of aqueous humor could occur through the implant to provide nutrition to resident cells. In addition, the posterior surface would need to be treated to prevent formation synechiae.

CORNEAL SUBSTITUTES AND EQUIVALENTS

Corneal substitutes or artificial corneas can be broadly divided into two types: (1) keratoprostheses and (2) corneal equivalents. Keratoprostheses that have been developed to date have been primarily composed of biocompatible synthetic polymers. Most comprise a solid, transparent optic core, designed to allow the passage of light to the posterior of the eye, and a surrounding porous skirt that allows cell infiltration from the host to anchor the prosthesis into place [reviewed in Ref. [1]].

There have also been various attempts at fabrication of tissue equivalents. These range from use of purely biological materials synthesized by corneal cells in culture to use of noncorneal tissues as substitutes and combinations of biological and synthetic materials. Unlike the keratoprostheses models that exist, tissue-engineered corneal equivalents are cell-based materials.

HANDLING CORNEAL CELLS
CELL ISOLATION AND PROPAGATION

Chapter 9 gives various protocols for isolation and maintenance of human corneal epithelial cells. Protocols for developing continuous cell lines are also given. These protocols can be easily adapted for the other two main cell types within the cornea: stromal fibroblasts (keratocytes) and endothelium. The following protocols for handling human corneal fibroblasts and endothelial cells can be used when there are differences in methodology.

Isolation of Corneal Endothelial Cells

MATERIALS: (STERILE)

- *Culture medium*: Medium 199 + 10% fetal bovine serum (FBS) + insulin–transferrin–selenium (ITS Premix, Becton-Dickinson: 5 mg/liter insulin, 5 mg/liter transferrin, and 5 μg/liter selenus acid), pH 7.2–7.4
- OptiSol (Chiron)
- Phosphate-buffered saline (PBS), Hanks's Balanced Saline Solution (HBSS), or culture medium without serum

- Corneal gill knife (Storz Instruments)
- Tissue culture dishes, 35 mm, precoated with 1% aqueous solution of gelatin

The isolation proceeds as follows.

1. Wash each donor cornea thoroughly in several washes of PBS, HBSS, or serum-free medium. Check under a dissecting microscope to ensure that the cornea is free from contaminating debris or cells.

2. Remove the epithelium (by scraping with a gill knife as described in Chapter 9 and place the cornea into a 1% gelatin-precoated dish containing culture medium comprising a equal parts of Medium 199 and OptiSol. Medium 199 can be used alone, but this is suboptimal.

3. Using a corneal gill knife and a dissecting microscope, with the endothelial side (concave side) facing upward, very gently scrape off the endothelium. The endothelium is a single layer of cells sitting on an acellular Descemet's membrane—excessive pressure will penetrate the stroma and liberate underlying stromal fibroblasts that will very quickly overgrow an endothelial culture.

4. Make sure that the cells have settled onto the plastic dish and have spread before replacing the medium (allow up to 2 weeks). If required, feed cultures by adding fresh medium. You may need to subculture or clonally select your cells to remove contaminating fibroblasts (if present). If the endothelial cells fail to adequately settle and adherie to the dish, add a solution of fibronectin (Sigma) to the culture medium (0.01% v/v).

Note: Pretreatment of the cornea in a dilute solution of collagenase or Dispase can help loosen the basement membrane, and the endothelium can be lifted off as a sheet with Descemet's membrane. To prevent contamination with epithelial cells, however, the epithelium should be removed first. Additionally, endothelial-like cells are also found in the surrounding trabecular meshwork although their exact identity is not known. Most importantly, corneas from younger donors and those processed within hours of death and in storage for minimal lengths of time (within 24 or at most 48 h) work best.

Isolation of Corneal Stromal Fibroblasts

MATERIALS: (STERILE)

- *Culture medium*: DMEM + 10% fetal bovine serum (FBS) + insulin–transferrin–selenium (ITS Premix, Becton-Dickinson: 5 mg/liter insulin, 5 mg/liter transferrin, and 5 μg/l selenous acid), pH 7.2–7.4
- Phosphate-buffered saline (PBS), Hanks's Balanced Salt Solution (HBSS), or serum-free culture medium
- Collagenase I (Life Technologies), 0.25%, in calcium- and magnesium-free PBS
- Tissue culture dishes, 35 mm, precoated with 1% aqueous solution of gelatin

The isolation proceeds as follows.

1. Remove all epithelium and endothelium from each donor cornea, either mechanically by scraping or with the help of enzymatic treatment. Cut off any surrounding sclera. Check under a dissecting microscope to make sure that there is no remaining epithelium or endothelium.

2. Cut the stroma into pieces and digest with 0.25% collagenase at 37°C (a cell culture incubator works well). The collagenase can be prewarmed to start the reaction. The amount of time needed to digest the pieces will depend upon the age of the donor cornea. Digest for 1–2 h.

3. Remove the pieces of stroma, place into gelatin-precoated tissue culture dishes, tease the pieces apart with forceps, and allow the stromal fibroblasts to move out onto the dish. For animal stromas that are less dense or for more digested samples of human stromas, the fibroblasts released by the enzymatic treatment can be separated from the matrix debris by filtering through a Nitex membrane with or a plastic cell sieve with 20-μm pores.

4. Allow the cells to settle before changing the culture medium. In the meantime, feed the cells by adding more medium to the culture dish.

Stromal Keratocytes vs Fibroblasts

Current thinking has demonstrated that there are several phenotypes of stromal cells, which can be stimulated by the addition of specific factors to serum-free medium. Evidence of the phenotypes has been demonstrated by Beales [17] with increased keratan sulfate synthesis. Others have reported that cells differ in their ability to contract. We have shown that cells differ in their ability to migrate in migration chambers. We have also shown in preliminary studies that cellular phenotypes differ in their ability to translocate heparan sulfate proteoglycans.

To isolate corneal stromal keratocytes (which are defined as the cells that synthesize keratan sulfates, specifically, keratocan [15] and have dendrite-looking processes instead of being spindle shaped like typical fibroblasts), corneas must be removed and must be devoid of sclera. Intact corneas should then be incubated in collagenase for 45 min to remove the epithelium, endothelium, and external stromal cells that were injured by corneal removal. Incubate further in fresh collagenase until cells are released. Plate the stromal cells in serum-free Dulbecco's Modified Eagle's Medium (DMEM) containing antibiotics and antimycotics, and incubate for the desired time period. Stromal cells are avascular naturally and have been shown to survive for extended periods of time in serum-free medium. Beales *et al.* [17] reported a similar methodology and demonstrated that culturing the cells in serum resulted in a decreased amount of keratan sulfate synthesis. Earlier Brown *et al.* [18] had reported that cells cultured in medium containing 1% platelet-poor horse serum had greater levels of keratan sulfate synthesis; however, the results were inconsistent with different batches of serum.

Propagation of Cells in Culture

MATERIALS

- *Culture medium for endothelium*: Medium 199 + 10% fetal bovine serum (FBS) + insulin–transferrin–selenium (ITS Premix, Becton-Dickinson: 5 mg/liter insulin, 5 mg/liter transferrin, and 5 μg/liter selenous acid)
- *Culture medium for stromal fibroblasts*: DMEM (or Medium 199) + 10% FBS + ITS Premix
- *Culture medium for stromal keratocytes*: DMEM without serum
- Trypsin–ethylenediaminetetraacetic acid (trypsin-EDTA, 0.05% Life Technologies)
- Calcium- and magnesium-free PBS (Life Technologies)
- Tissue culture plasticware, precoated with 1% (w/v) aqueous gelatin

The propagation proceeds as follows.

1. Once the cells have settled onto the tissue culture plastic and have spread, replace the growth medium with fresh medium every 2–3 days.

2. Split the cells while they are still in the growth phase (at about 65–70% confluence to be safe). To split the cultures, aspirate the medium, wash the cells with PBS (Ca^{2+}, Mg^{2+} free).

3. Aspirate PBS and add prewarmed trypsin containing 0.05% EDTA. Add 1 ml to 100-ml tissue culture dishes or 750-ml flasks.

4. Place in 37°C incubator for 1–3 min. Check repeatedly under the microscope for action of trypsin, as cells will round up.

5. Inactivate trypsin with an excess of serum-containing culture medium or soybean trypsin inhibitor. Cells can now be split into aliquots and reseeded. If serum-free medium is used, inhibit the action of the trypsin with soybean trypsin inhibitor to prevent the binding of serum to the cells, which would influence expression or adherence. At this stage, the cells can be prepared for cryogenic storage as described in Chapter 9 for corneal epithelial cells.

Preparation of Cell Lines with Extended Life Spans

Lines of cells with extended life spans are useful for the early stages of testing polymers used in the development of keratoprostheses and also in the development of cell-based tissue engineered corneas. The methods used for corneal epithelial cells (Chapter 9) can be applied to stromal fibroblasts and endothelial cells. Selection of cell lines that approximate the low-passage primary cells are similar. As with epithelial cells, electrophysiology can be very useful in differentiating the immortalized cells from cells that have transformed phenotypes, prior to applying other screening tests.

Figure 83.1 shows representative current tracings and current–voltage (*I-V*) relationships of cultured human stromal cells. Figure 83.2 shows this information for cultured human stromal cells lines genetically altered to have extended life spans (immortalized). A tail current protocol (tracings not shown) was used to determine reversal potentials (E_{rev}). To obtain tail currents, cells were held at 0 mV, depolarized to 100 mV, and then hyperpolarized to the different voltages shown on the *x* axis. E_{rev} was measured with cells bathed in a NaCl Ringer's bath and a K^+ methanesulfanate–amphotericin pipette solution. Offset potentials were corrected for by subtracting the E_{rev} obtained in a symmetric KCl Ringer's bath. While the cultured and immortalized ion currents are not identical, the similarities between the have been used to select cells for use in constructing corneal tissue equivalents [19]. Although not shown here, stromal cells deemed "transformed" following additional screening assays have a relatively unique current, as was found in the other transformed corneal cell types [19].

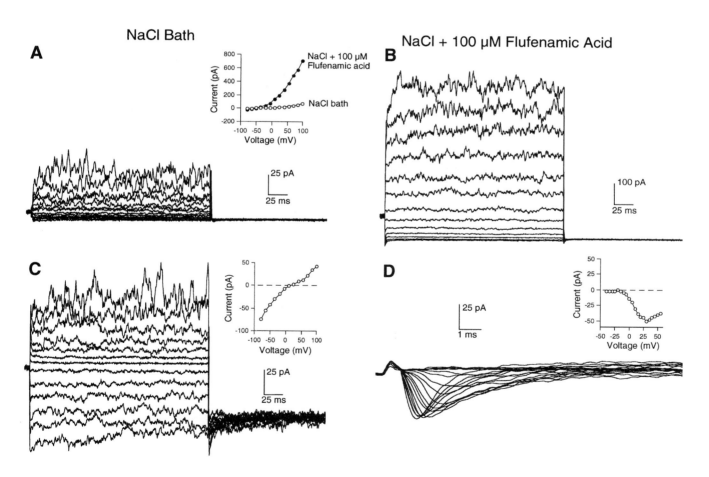

Fig. 83.1. Representative current tracings and current–voltage (I-V) relationships of cultured human stromal cells. (A) and (B) Current and I-V relationships for nonselective cation ($E_{rev} = -13$ mV) and fenamate-activated K^+ current ($E_{rev} = -40$ mV). (C) Representative inwardly rectifying K^+ current (KCl bath). (D) Na^+ current.

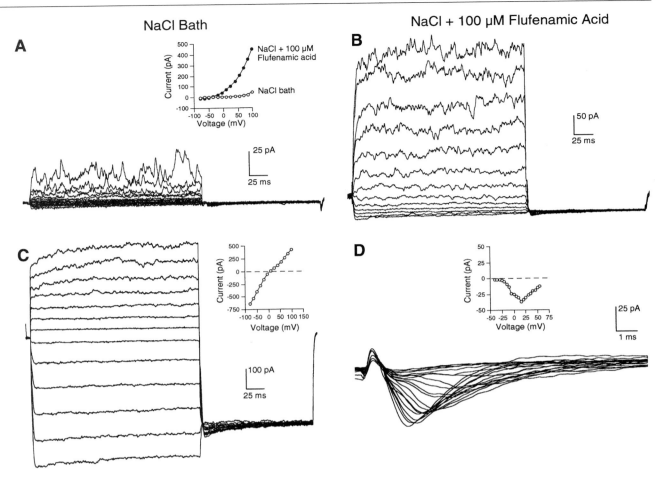

Fig. 83.2. Representative current tracings and current–voltage (I-V) relationships of a cultured human stromal cell line from of 12-year-old donor, genetically modified to have extended life spans (immortalized). (A) and (B) Current and I-V relationships for nonselective cation ($E_{rev} = -18$ mV) and fenamate-activated K^+ current ($E_{rev} = -48$ mV), revealing similarity to the cells described in Fig. 83.1. (C) and (D) demonstrate inwardly rectifying K^+ current and Na^+ current, respectively, as in Fig. 83.1.

Griffith *et al.* [19] showed that corneal tissue equivalents had the most *in vivo*–like morphological appearance and physiological function when cell lines with currents closest to normal cells were utilized. On the contrary, when cells expressing the electrophysiological characteristics of transformed cells were used in the reconstruction, the reconstruction was unsuccessful. This was because cells that expressed the current indicative of transformed cells invaded the matrix and did not differentiate like the immortalized cells.

KERATOPROSTHESES

DESIGNING A KERATOPROSTHESES

The ideal keratoprosthesis has been suggested to be an epithelialized artificial donor button suitable for implantation in a similar fashion to a penetrating keratoplasty [20]. It would comprise a flexible, transparent optical core surrounded by a porous skirt of similar chemical composition and flexibility. The core and skirt should be permanently joined in a manner that prevents loosening or leakage at the interface of the materials. Furthermore, the skirt would permit fibroblast ingrowth and collagen deposition sufficient to provide firm anchorage and complete integration with the surrounding host tissues without the need for additional mechanical features. The skirt should also be strong enough for suture placement. The anterior surface of the device should allow for epithelialization to occur. The posterior surface should inhibit cellular attachment and proliferation to prevent the formation of opaque retroprosthetic membranes. The material used in the fabrication of the optic portion of the device, aside from being transparent, should provide a refractive power similar to that of the normal cornea, and the diameter of the optic should be sufficient to allow posterior segment visualization and a reasonable field of view. None of the materials used in device fabrication should elicit any adherent biological response, including inflammation or mutagenesis.

SELECTION OF MATERIALS

A range of polymers with varying degrees of ophthalmic compatibility that have been used for keratoprosthesis development, used both as the optic and the haptic. For a review, see Trinkaus-Randall [1]. A few of the more promising materials (e.g., pHEMA, hydrogel, silicone rubber) discussed. A method for preparing pHEMA is provided, along with suggested methods for surface modification. Silicon rubber, from Dow Corning, can be prepared according to manufacturer's instructions.

Silicon Rubber and Polyurethane

Silicone rubber [20–28], a hydrophobic elastomer, has shown promise as a possible optic material in terms of tensile strength, elongation at break, and optical properties. In fact a poly(vinylpyrrolidone)-coated medical grade poly(dimethylsiloxane) (PDMS) optic has been used in a keratoprosthesis that was tested on human patients [29]. Silicone rubber membranes can be prepared by using a monomer and a curing cross-linking agent available in a kit from Dow Corning. Lee *et al.* [30] spin-cast their membranes to obtain membranes of the desired thickness. Polyurethanes, widely used as biomaterials in blood contacting applications, have also been suggested as potential optic materials [31].

Hydrogels and pHEMA

Hydrogels are hydrophilic, water-swellable polymers with good flexibility, elasticity, and optical transparency. They have been widely studied as potential keratoprosthetic materials, both for the transparent optic and the surrounding haptic. The structure of these polymers allows for the diffusion of nutrients, and therefore there is the possibility that these polymers can be fabricated to support an overlying epithelial layer.

Poly(hydroxyethyl methacrylate) (pHEMA), widely used as a contact lens material, has also been studied as a potential keratoprosthetic material. Chirila *et al.* have developed a system consisting of a central optical zone formed from homogeneous transparent pHEMA and a peripheral skirt made from spongy pHEMA [32]. The boundary between the two forms of pHEMA has been shown to fulfill the requirements for an interpenetrating network (IPN), which the authors believe confers an advantage to their device. Interpenetrating networks of the porous and nonporous materials are formed by first polymerizing the porous outer ring in a doughnut-shaped mold for 5 h at 60°C, and then subsequently doing the haptic polymerization in the inner ring. In this way, the monomers from the second polymerization penetrate the spongy outer region. The use of similar polymers for both optic and skirt eliminates the point of weakness that is found in other models that have used chemically dissimilar polymers in the formation of the optic and skirt. Polymerizing with divinyl glycol as the cross-linking agent somewhat enhanced the poor mechanical strength associated with the pHEMA sponges [33].

Following implantation of the sponges into rabbit eyes, cellular invasion was noted, particularly at 12 weeks postimplantation [34]. Newly synthesized type III collagen was produced within the sponge within one month of implantation [35]. Problems associated with this material in the current application include its poor mechanical strength, the deposition of calcium [20], and the potential for degradation [36].

pHEMA Preparation

Similar techniques are used in the fabrication of pHEMA sponges and sheets [32,37]. Variations in the water content result in the formation of the porous sponge materials suitable for stromal ingrowth or optically transparent haptic materials.

1. Mix appropriate amounts of purified HEMA monomer with water: use 20:80 wt/wt for porous materials and 70:30 for optically clear materials. Cross-link by mixing in 0.5% (by weight) divinyl glycol or ethylene glycol dimethacrylate.

2. Add a 0.12% (equimolar mixture, by weight) aqueous solution of ammonium persulfate and sodium metabisulfite, or 0.12% (by weight) of benzoyl peroxide to initiate the reaction. Mix thoroughly with the monomers.

3. Pour solutions into moulds of the desired shape.

4. Use a vacuum for 5 min to evacuate air from the system. Then purge the system with nitrogen.

5. Place the mold in a constant temperature water bath and run a temperature program consisting of three 10-h steps at 30, 40, and 50°C. This polymerizes and cures the samples, which can then be removed from the molds for use.

MODIFICATION OF POLYMERS FOR IMPROVED
CELLULAR INTERACTION
Modification of Epithelial Interface

It is generally agreed that formation of a corneal epithelial layer is desirable and possibly necessary for device success in many cases [20,38]. The corneal epithelium maintains a connection to its basement membrane through adhesive interactions between the surface integrins on the cells of the basal layer and adhesive peptide sequences within surrounding extracellular matrix proteins such as collagens (types I, III, IV), laminin, and fibronectin. Therefore, surface modification with intact extracellular matrix proteins as well as the shorter binding peptide sequences can be used to enhance corneal epithelial cell interactions with polymeric materials. Following, are several selected protocols for producing hydroxyl groups on polymeric surfaces (silicon rubber membranes) and for attachment of adhesive peptides onto modified surfaces.

Method for Surface Modification to Generate Hydroxyl Groups

The modifications are performed according to the protocol of Wickson and Brash [39] as follows.

1. Place silicone rubber (or other polymeric) membranes in a radio frequency glow discharge (RFGD) reactor under a vacuum exceeding 50 µmHg and exposed to argon plasma (50 psig) for 5 min. Forward voltage, 20 W; reflected voltage, 2 W.

2. Introduce allyl alcohol into the reactor at 50 psig and expose surfaces for 10 min.

3. Expose surfaces to $NaBH_4$ for 24 h to complete conversion of hydroxyl groups.

Peptide Attachment of Hydroxylated Surfaces via Tresyl Chloride Reaction [40]

1. Dry hydroxylated surfaces and expose to a solution of acetone (5 ml), pyridine (400 µl), tresyl chloride (trifluoroethanesulfonyl chloride, 200 µl) for 1 h at room temperature.

2. Rinse surfaces in 1 mM HCl, followed by 0.2 M $NaHCO_3$.

3. After the rinses, expose surfaces to peptides at a concentration of 100 ng/ml in 0.2 M $NaHCO_3$ for 24 h.

4. Rinse surfaces with 0.2 M $NaHCO_3$.

5. Rinsed surfaces with β-mercaptoethanol.

Other modification methods have been used to attach biological moieties to a polymeric surface. Radio frequency glow discharge with poly(acrylic acid) [31] and heptylamine [41,42] has been used to generate different reactive groups at the surface. Linkage of the bioactive moiety has been accomplished by using various linkers including diisocyanate, polyisocyanate, and cyanogen bromide [43,44], as well as by N-(3-dimethylaminopropyl)-N'-ethylcarbodiimide (EDC) and N-hydroxysuccinimide (NHS) chemistry. Coupling to a perfluorinated ethylene–propylene copolymer (FEP) surfaces was accomplished by mean of RFGD deposition of thin aldehyde layers followed by reductive amination [45].

Effects of Modification

A number of studies have shown that collagen and fibronectin modification can enhance corneal epithelial cell growth. As well, enhanced corneal epithelial tissue outgrowth was noted on a Vitrogen-modified surface. Modification with cell adhesion peptides also results in enhanced corneal epithelial cell attachment and growth. While RGDS [44] has been shown to have an effect, Merrett et al. [46] found that surface modification with the laminin-binding peptide YIGSR (Tyc-Ile-Gly-Ser-Arg) leads to a significant increase in the initial adhesion of human corneal epithelial cells in comparison. Steele et al. [47] found that collagen based peptides containing the –Gly-Pro-Nleu sequence also supported attachment and growth of corneal epithelial cells. Furthermore, attachment via a carboxymethyl dextran spacer resulted in an even greater degree of cell outgrowth [48].

Surface modification with moieties that are not necessarily components of the normal corneal epithelium can also result in an enhancement in corneal epithelial cell growth. A simple RFGD argon plasma treatment of poly(vinyl alcohol) rendered the surfaces more conducive to corneal epithelial cell growth [49], with a full confluent epithelium detected after 3 weeks of in vivo implantation. Corneal epithelial cells in vitro demonstrated a high affinity for silicone rubber surfaces modified by RGFD attachment of hydrophilic monomers including 2-hydroxyethyl methacrylate (HEMA) [22,23]. This study demonstrates the obvious effect of the base polymer selected, since the surface of pHEMA implanted into rabbit corneas showed no epithelialization [50,51]. Attachment of polysaccharides to an FEP polymeric substrate led to either enhancement or inhibition of corneal epithelial cell growth [52] depending on the polysaccharide used.

Additional Factors Affecting Corneal Epithelial Cell Growth

Surface topography and porosity have also been shown to play a significant role in determining the extent of corneal epithelial cell attachment to polymeric surfaces. On polycarbonate membranes, surfaces with pores in the diameter range of 0.1–0.8 μm supported superior stratification and protein deposition compared with containing pores greater than 1.0 μm [53]. The assembly of adhesive structures on surfaces with pore diameters between 0.4 and 2.0 μm was restricted to regions of the polymer between the pores, while no adhesive structures assembled on the nonporous or the 3.0-μm surface [54,55]. To overcome diffusional limitations present in a number of polymers that are used as optic materials, alternative delivery methods for the nutrients may be possible. Kuhl and Griffith-Cima have developed a method of tethering growth factors such as epidermal growth factor to polymeric substrates [56] and demonstrated that the tethered growth factor was as effective as soluble epidermal growth factor in eliciting a response with hepatocytes. Such an approach may prove useful in artificial corneal development.

Modifications for Stromal Keratocyte Ingrowth

Porous materials have been widely used to anchor the keratoprosthesis to the existing ocular tissue by stromal invasion. In the studies of Trinkaus-Randall et al. and Legeais et al., pore diameters of at least 20 μm are required for significant cellular infiltration. Furthermore, Trinkaus-Randall's studies suggested that a fiber diameter between 2 and 12 μm resulted in more extensive cellular invasion. However, likely because of the relatively robust nature of stromal keratocytes, very little has been done to enhance the infiltration of these cells. Trinkaus-Randall et al. evaluated four methods of pretreating the porous disks prior to placement in the cornea [57]. These modifications included preincubation in tissue culture medium, preincubation with a type I collagen solution, preculture with stromal keratocytes prior to implantation, and preincubation with type I collagen followed by preincubation with stromal keratocytes. Overall, the results suggest that preseeding the cells without the additional collagen preincubation gave the best results. Chirila et al. [35] also preincubated their devices in collagen prior to implantation to facilitate stromal invasion.

CORNEAL TISSUE EQUIVALENTS

Various methods have been tried for reconstruction of three-dimensional split and full-thickness corneal equivalents. We list several examples of methods used in the in vitro reconstruction of human corneas. Methods for construction of three-dimensional corneal equivalents for bovine, and rabbit plus mouse cells can be found in Minami *et al.* [58] and Zieske *et al.* [59], respectively.

TOTALLY CELL-BASED METHODS

Treatment of corneal stromal keratocytes with ascorbic acid (vitamin C) will cause an increase in the synthesis of collagen, the main noncellular component of the cornea [60,61]. Germain *et al.* [62] have very elegantly adopted this information to develop corneal constructs from corneal epithelium and keratocytes without the addition of extracellular matrix proteins. Essentially, keratocytes were cultured in the presence of ascorbic acid to produce thick fibrous sheets *in vitro*. Two sheets were then superimposed to form a stromal equivalent, after which corneal epithelial cells were seeded on top.

CELLS AND EXTRACELLULAR MATRIX SCAFFOLDING

We have modified the following protocol from Zieske *et al.* [59] for use with human corneal cells instead of animal cells. An alternative matrix that is chemically cross-linked for strength [19] is described subsequently.

METHODS

- Six-well culture inserts coated with collagen type I (e.g., precoated from Becton-Dickinson) or with collagen
- Deep-well plate (Biocoat cell environments, cat. no. 05467) (optional)
- Culture media

Media with Serum

> Medium 199 + 10% fetal bovine serum (FBS) + insulin–transferrin–selenium (ITS Premix, Becton-Dickinson: 5 mg/liter insulin, 5 mg/liter transferrin, and 5 μg/liter selenium), pH 7.2–7.4 + antibiotics, if needed
> *SHEM medium*: 1:1 mixture of Dulbecco's Modified Eagle's Medium (DMEM) and Ham's F-12 nutrient, 15% FBS, 10 ng/ml epidermal growth factor (EGF), 5 μg/ml insulin, 0.1 μg/ml cholera toxin, α subunit, 5 mM L-glutamine, 0.5% dimethyl sulfoxide (DMSO), and antibiotics, if needed

Serum-Free Media

Keratinocyte serum-free medium (Life Technologies)

- Type I collagen from bovine skin (Becton-Dickinson)

The procedure is as follows.

1. Seed corneal endothelial cells onto the bottom of a collagen-coated culture insert. Supplement with serum-containing DMEM both inside (2 ml) and outside (2 ml) the culture insert and grow till confluence is achieved.
2. Add 1 ml of ice-cold, neutralized collagen containing 5×10^4 human corneal fibroblasts ml onto the insert and allow to gel at room temperature. Feed with DMEM-containing serum. The gel is allowed to contract submerged in culture medium for 5–7 days.
3. A suspension of human corneal epithelial cells (1.8×10^5 cells/insert) is then seeded onto the contracted gel. SHEM or serum-free keratinocyte medium is added to the insert of the culture.
4. The culture is grown submerged in serum-free keratinocyte medium for 3–7 days. Change the medium to keratinocyte–serum-for-medium supplemented with 0.3% FBS and 1.8 mM $CaCl_2$ or SHEM; after which the cultures should be raised to a "moist" air–liquid interface. This is done by removing the medium from within the insert but adding an extra milliliter of medium outside the insert.

Alternative Protocol for Cross-Linked Stromal Matrix

The following alternative formulation for a stroma that does not contract is adapted from a protocol we have used to construct human corneal equivalents for *in vitro* use [19]. The materials listed are sufficient for six 30-mm-diameter culture insert.

We formerly used type I bovine collagen (Becton-Dickinson) and found that it produced a much more transparent corneal equivalent than that obtained from using rat tail collagen. However, because of supplier sourcing changes and difficulty in obtaining commerciall bovine dermal collagen that would form suitable gels, we began to use rat tail collagen as a starting material. This material, however, requires considerable strengthening before it can be used in transplantations.

MATERIALS (ALL STERILE)

- Collagen buffer
- Type I collagen from rat tendon (Becton-Dickinson)
- 1 N sodium hydroxide
- Chondroitin sulfate C (Sigma), 20% w/v solution in PBS or DMEM without additives
- Dextran (Sigma)
- Glutaraldehyde, purified (Sigma, ICN)
- Glycine, tissue culture grade

The matrix is is prepared as follows.

1. Using precooled 50-ml acid-cleaned glass or disposable Falcon centrifuge tubes, mix 6 ml of type I collagen with collagen buffer, stirring very gently to avoid bubble formation.

2. Neutralize the collagen solution by adding sterile 1 N sodium hydroxide by titrating with 10 μl at a time. The final pH should be about 7.5.

3. Add and simultaneously mix in by stirring very gently but thoroughly, 520 μl of the 20% chondroitin sulfate solution. If the mixture is not ice cold, there may be slight cloudiness. However, this will become clear as the chondroitin sulfate becomes homogeneously mixed into the neutralized collagen. (Ice-cold collagen will not go through a cloudy stage.)

4. In a separate Eppendorf tube, make up a 1.5% glutaraldehyde solution in 20% dextran in DMEM. Make sure that the glutaraldehyde solution is prepared just before use. Aliquot the glutaraldehyde and then make sure each aliquot is kept sealed tight since exposure to air will cause oxidation, rendering the material useless for cross-linking.

5. Add the 1.5% glutaraldehyde mixture to the collagen matrix to give a final concentration of 0.02% glutaraldehyde. Mix thoroughly, but gently, avoiding bubble formation.

6. Cross-link on ice for 2 h—stir a few times to make sure that everything is well mixed.

7. Unreacted glutaraldehyde is bound up and rendered "non toxic" by adding 500 μl of 25% glycine in DMEM (store this at room temperature in small aliquots). Mix thoroughly, and let the mixture react for at least 1 h on ice.

8. While waiting, trypsinize the stromal fibroblast cells. Neutralize the trypsin by adding serum-containing culture medium. Centrifuge at 1500 rpm for 5 min at 18°C. There should be a pellet of cells at the bottom. Aspirate off the supernatant.

9. Resuspend the cells in a minimal amount of medium (100 μl) and monodisperse the cells. Add 5×10^4 cells per milliliter to the collagen–chondroitin sulfate mixture and mix to ensure even dispersion of the cells.

10. Depending on the desired thickness of cornea, add 0.7–1 ml of matrix containing monodispersed cells to each insert. Incubate at 37°C to gel. This will take about 15–20 min for rat tail collagen. Epithelial cells can then be seeded on top of the gel.

ASSAYS OF CELLULAR DIFFERENTIATION

MORPHOLOGICAL AND BIOCHEMICAL MARKERS

Morphological and biochemical markers for corneal epithelium in three-dimensional constructs were discussed in Chapter 9. Similar methods can be used to evaluate epithelial

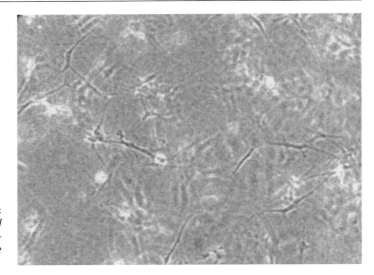

Fig. 83.3. Phase contrast micrograph of corneal keratocytes within a three-dimensional collagen–glycosaminoglycan gel that was cross-linked with 0.02% glutaraldehyde and post-treated to remove unbound glutaraldehyde. These cells are fibroblast-like, with dendritic processes.

differentiation on top of keratoprostheses. Zieske *et al.* [59] examined their constructs for expression of (1) differentiation markers using anti-keratin 3 (ICN Biomedicals, Inc.) and anti-keratin 12 [15]; (2) basement membrane formation using anti-laminin (Chemicon), and (3) tight junction formation using anti-ZO-1 (Zymed).

Stromal keratocytes in native corneas appear in cross section as elongated cells sandwiched between lamellae of collagen. Detailed microscopical observations indicate that these cells are linked to form a network, contacting each other through gap junctions [1]. Figure 83.3 shows a phase contrast image of keratocytes within a glutaraldehyde cross-linking matrix (the alternative stroma matrix just described). Keratocan is reportedly one of the most corneal-keratocyte-specific proteoglycans described to date [15]. This observation indicates that keratocan can serve as a specific marker for keratocytes.

Endothelial cells form a single layer of low cuboidal cells. They have an active sodium–potassium ATPase pump for osmoregulation that can be visualized by immunohistochemistry. The cells also express type VIII collagen, which can be used as a marker, by *in situ* hybridization (Fig. 83.4).

FUNCTIONAL MARKERS

Griffith *et al.* [19] used a selection of functional markers to assess the biological activity of the cells within the whole three-dimentional construct. In the keratoprosthesis, however, specific cellular functioning is not as important as overall biocompatibility and the abil-

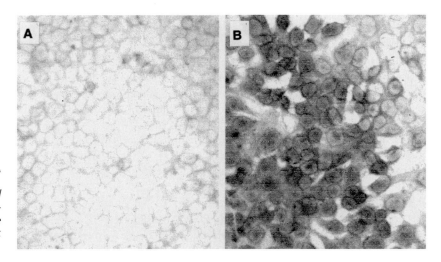

Fig. 83.4. In situ hybridization using an RNA probes conjugated to digoxigenin (DIG). (A) Control: corneal endothelial cells hybridized with a sense probe remain unstained. (B) Endothelial cells hybridized with an antisense probe show expression of type VIII collagen (dark staining).

ity of the prosthesis to remain transparent while integrated within the host tissue without provoking immune or inflammatory responses.

For both tissue equivalents and keratoprostheses, ensuring that the corneal replacement is transparent and will not scatter light significantly is key to ensuring good quality of vision.

REFERENCES

1. Trinkaus-Randall, V. (2000). Cornea: Biological responses. *In* "Principles of Tissue Engineering" (R. Lanza, R. Langer, and E. Chick, eds.), 2nd ed., Chapter 35, pp. 471–491. Academic Press, San Diego, CA.
2. Kao, W. W.-Y., Liu C.-Y., Converse, R. L., Shiraishi A., Kao, C. W.-C., Ishizaki, M., Doetschman, T. C., and Duffy, J. (1996). Keratin 12 deficient mice with epithelial erosion. *Invest. Ophthalmol. Visual Sci.* **37**, 2572–2584.
3. Irvine, A. D., Corden, L. D., Swensson, O., Swensson, B., Moore, J. E., Frazer, D. G., Smith, F. J. D., Knowlton, R. G., Christophers, E., Rochels, R., Uitto, J., and McLean, W. H. I. (1997). Mutations in cornea-specific keratin K3 or K12 genes cause Meesmann's corneal dystrophy. *Nat. Genet.* **16**, 184–187.
4. Linsenmayer, T. F., Fitch, J. M., and Birk, D. E. (1990). Heterotypic collagen fibrils and stabilizing collagens. Controlling elements in corneal morphogenesis. *Ann. N. Y. Acad. Sci.* **580**, 143–160.
5. Doane, K. J., Babiarz, J. P., Fitch, J. M., Linsenmayer, T. F., and Birk, D. E. (1992). Collagen fibril assembly by corneal fibroblasts in three-dimensional collagen gel cultures: Small-diameter heterotypic fibrils are deposited in the absence of keratan sulfate proteoglycan. *Exp. Cell Res.* **202**, 113–124.
6. Rada, J. A., Cornuet, P. K., and Hassell, J. R. (1993). Regulation of corneal collagen fibrillogenesis *in vitro* by corneal proteoglycan (lumican and decorin) core proteins. *Exp. Eye Res.* **56**, 635–648.
7. Scott, J. E. (1996). Proteodermatan and proteokeratan sulfate (decorin, lumican/fibromodulin) proteins are horseshoe shaped. Implications for their interactions with collagen. *Biochemistry* **35**, 8795–8799.
8. Weber, I. T., Harrison, R. W., and Iozzo, R. V. (1996). Model structure of decorin and implications for collagen fibrillogenesis. *J. Biol. Chem.* **271**, 31767–31770.
9. Axelsson, I. (1984). Heterogeneity, polydispersity, and physiological role of corneal proteoglycan. *Acta Ophthalmol.* **62**, 25–38.
10. Poole, A. R. (1986). Proteoglycan in health and disease: Structures and functions. *Biochem. J.* **236**, 1–14.
11. Hassell, J. R., Newsome, D. A., Krachmer, J. H., and Rodrigues, M. M. (1980). Macular corneal dystrophy: Failure to synthesize a mature keratan sulfate proteoglycan. *Proc. Natl. Acad. Sci. U.S.A.* **77**, 3705–3709.
12. Charkravarti, S., Magnuson, T., Lass, J. H., Jepsen, K. J., LaMantia, C., and Carroll, H. (1998). Lumican regulates collagen fibril assembly: Skin fragility and corneal opacity in the absence of lumican. *J. Cell Biol.* **141**, 1277–1286.
13. Saika, S., Shiraishi, A., Liu, C. Y., Funderburgh, J. L., Kao, C. W., Converse, R. L., and Kao, W. W. (2000). Role of lumican in the corneal epithelium during wound healing. *J. Biol. Chem.* **275**, 2607–2612.
14. Pellegata, N. S., Dieguez-Lucena, J. L., Joensuu, T., Lau, S., Montgomery, K. T., Krahe, R., Kivela, T., Kucherlapati, R., Forsius, H., and de la Chapelle, A. (2000). Mutations in KERA, encoding keratocan, cause cornea plana. *Nat. Genet.* **25**, 91–95.
15. Liu , C.-Y., Shiraishi, A., Kao, C. W.-C., Converse, R. L., Funderburgh, J. L., Corpuz, L. M., Conrad, G. W., and Kao, W. W.-Y. (1998). The cloning of mouse keratocan cDNA and genomic DNA and the characterization of its expression during eye development. *J. Biol. Chem.* **273**, 22584–22588.
16. Waring, G. O., Bourne, W. M., Edelhauser, H. F., and Kenyon, K. R. (1982). The corneal endothelium: Normal and pathological structure and function. *Ophthalmology (Rochester, Minn.)* **89**, 531–590.
17. Beales, M. P., Funderburgh, J. L., Jester, J. V., and Hassell, J. R. (1999). Proteoglycan synthesis by bovine keratocytes and corneal fibroblasts: Maintenance of the keratocyte phenotype in culture. *Invest. Ophthalmol. Visual Sci.* **40**, 1658–1663.
18. Brown, C. T., Nugent, M. A., and Trinkaus-Randall, V. (1996). Altered glycosaminoglycan synthesis in platelet poor serum. *Mol. Biol. Cell* **7**, 55a.
19. Griffith, M., Osborne, R., Munger, R., Xiong, X., Doillon, C., Laycock, N. L. C., Hakim, M., Song, Y., and Watsky, M. A. (1999). Functional human corneal equivalents from cell lines. *Science* **286**, 2169–2172.
20. Hicks, C. R., Fitton, J. H., Chirila, T. V., Crawford, G. J., and Constable, I. J. (1997). Keratoprosthesis: Advancing toward a true artificial cornea. *Surv. Ophthalmol.* **42**, 175–189.
21. Kain, H. L. (1990). A new concept for keratoprosthesis. *Klin. Monatsbl. Augenheilkd.* **197**, 386–392.
22. Chang, P. C., Lee, S. D., and Hsiue, G. H. (1998). Heterobifunctional membranes by plasma induced graft polymerization as an artificial organ for penetration keratoprosthesis. *J. Biomed. Mater. Res.* **39**, 380–389.
23. Hsiue, G. H., Lee, S. D., and Chang, P. C. (1996). Surface modification of silicone rubber membrane by plasma induced graft copolymerization as artificial cornea. *Artif. Organs* **20**, 1196–1207.
24. Lee, S. D., Hsiue, G. H., Chang, P. C., and Kao, C. Y. (1996). Plasma-induced graft polymerization of acrylic acid and subsequent grafting of collagen onto polymer film as biomaterials. *Biomaterials* **17**, 1599–1608.
25. Lee, S. D., Hsiue, G. H., Kao, C. Y., and Chang, P. C. (1996). Artificial cornea: Surface modification of silicone rubber membrane by graft polymerization of pHEMA via glow discharge. *Biomaterials* **17**, 587–595.
26. von Fischern, T., Langefeld, S., Yuan, L., Volcker, N., Reim, M., Kirchhof, B., and Schrage, N. F. (1998). Development of a surface modified silicone-keratoprosthesis with scleral fixation. *Acta Chir. Hung.* **37**, 219–225.
27. von Fischern, T., Langefeld, S., Yuan, L., Volcker, N., Reim, M., Kirchhof, B., and Schrage, N. F. (1999). The "Aachen" keratoprosthesis: A new approach towards successful keratoprosthesis surgery. *Int. J. Artif. Organs* **22**, 52–57.
28. Langefeld, S., Volcker, N., Kompa, S., von Fischern, T., Bienert, H., Klee, D., Hocker, H., Reim, M., and Schrage, N. F. (1999). Functionally adapted surfaces on a silicone keratoprosthesis. *Int. J. Artif. Organs* **22**, 235–241.

29. Legeais, J. M., and Renard, G. (1998). A second generation of artificial cornea (Biokpro II). *Biomaterials* **19**, 1517–1522.

30. Lee, S. D., Hsiue, G. H., Kao, C. Y., and Chang, P. C. T. (1996). Artificial cornea: Surface modification of silicone rubber membrane by graft polymerization of pHEMA via glow discharge. *Biomaterials* **17**, 587–595.

31. Bruin, P., Meeuwsen, E. A., van Andel, M. V., Worst, J. G., and Pennings, A. J. (1993). Autoclavable highly cross-linked polyurethane networks in ophthalmology. *Biomaterials* **14**, 1089–1097.

32. Chirila, T. V., Vijayasekaran, S., Horne, R., Chen, Y.-C., Dalton, P. D., Constable, I. J., and Crawford, G. J. (1994). Interpenetrating polymer network (IPN) as a permanent joint between the elements of a new type of artificial cornea. *J. Biomed. Mater. Res.* **28**, 745–753.

33. Chirila, T. V., Yu, D.-Y., Chen, Y.-C., and Crawford, G. J. (1995). Enhancement of mechanical strength of poly(2-hydroxyethyl methacrylate) sponges. *J. Biomed. Mater. Res.* **29**, 1029–1032.

34. Vijayasekaran, S., Fitton, J. H., Hicks, C. R., Chirila, T. V., Crawford, G. J., and Constable, I. J. (1998). Cell viability and inflammatory response in hydrogel sponges implanted in the rabbit cornea. *Biomaterials* **19**, 2255–2267.

35. Chirila, T. V., Thompson-Wallis, D. E., Crawford, G. J., Constable, I. J., and Vijayasekaran, S. (1996). Production of neocollagen by cells invading hydrogel sponges implanted in the rabbit cornea. *Graefe's Arch. Clin. Exp. Ophthalmol.* **234**, 193–198.

36. Jeyanthi, R., and Pandauanga, R. (1990). In vivo biocompatibility of collagen poly(hydroxyethyl methacrylate) hydrogels. *Biomaterials* **11**, 238–243.

37. Crawford, G. J., Constable, I. J., Chirila, T. V., Vijayasekaran, S., and Thompson, D. E. (1993). Tissue interaction with hydrogel sponges implanted in the rabbit cornea. *Cornea* **12**, 348–357.

38. Leibowitz, H., Trinkaus-Randall, V., Tsuk, A. G., and Franzblau, C. (1994). Progress in the development of a synthetic cornea. *Prog. Retinal Eye Res.* **13**, 605–621.

39. Wickson, B. M., and Brash, J. L. (1999). Surface hydroxylation of polyethylene by plasma polymerization of allyl alcohol and subsequent silylation. *Colloids Surf.* **156**, 201–213.

40. Merrett, K., McGuire, A., Griffith, C. M., Deslandes, Y., Plezier, G., and Sheardown, H. (2000). Optimization of surface modification of pHEMA with cell adhesion peptide RGDS and YIGSR for interaction with corneal epithelial cells. *J. Biomater. Sci. Polym. Ed.* (in press).

41. Xie, R. Z., Sweeney, D. F., Beumer, G. J., Johnson, G., Griesser, H. J., and Steele, J. G. (1997). Effects of biologically modified surfaces of synthetic lenticles on corneal epithelialization in vivo. *Aust. N. Z. J. Ophthalmol.* **25**, S46–S49.

42. Thissen, H., McLean, K., Johnson, G., Steele, J. G., and Griesser, H. J. (1999). Covalent immobilization of Vitrogen to improve corneal epithelial tissue outgrowth and adhesion. *Trans. Soc. Biomater. Meet.* **22**, 450.

43. Koybayshi, H., and Ikada, Y. (1991). Covalent immobilization of proteins on to the surface of poly(vinyl alcohol) hydrogel. *Biomaterials* **12**, 747–751.

44. Koyayashi, H., and Ikada, Y. (1991). Corneal cell adhesion and proliferation in hydrogel sheets bound with cell-adhesive proteins. *Curr. Eye Res.* **10**, 899–908.

45. Dai, L., St. John, H. A. W., Bi, J., Zientek, P., Chatelier, R. C., and Griesser, H. J. (2000). Biomedical coatings by the covalent immobilization of polysaccharides onto gas plasma activated surfaces. *SIA, Surf. Interface Anal.* **29**, 46–55.

46. Merrett, K., Deslandes, Y., Griffith, C. M., and Sheardown, H. (1999). Attachment of cell adhesion peptides to polymer surfaces via hydroxyl groups: Effect on the growth of human corneal epithelial cells. *Trans. Soc. Biomater. Meet.* **22**, 449.

47. Johnson, G., Jenkins, M. L., McLean, K. M., Griesser, H. J., Kwak J., Goodman, M., and Steele, J. G. (2000). Peptoid containing collagen mimetics with cell binding activity. *J. Biomed. Mater. Res.* **51**, 612–624.

48. McLean, K. M., Gong, X., Kingshott, P., Johnson, G., Jenkins, M., Steele, J. G., Goodman, M., and Griesser, H. J. (1999). Surface immobilization of synthetic collagen-like molecules containing peptoids and interactions with cells. *Trans. Soc. Biomater. Meet.* **22**, 241.

49. Latkany, R., Tsuk, A., Sheu, M.-S., Loh, I.-H., and Trinkaus-Randall, V. (1997). Plasma surface modification of artificial corneas for optimal epithelialization. *J. Biomed. Mater. Res.* **36**, 29–37.

50. Hicks, C. R., Chirila, T. V., Dalton, P. D., Clayton, A. B., Vijayasekaran, S., Crawford, G. J., and Constable, I. J. (1996). Keratoprosthesis: Preliminary results of an artificial corneal button as a full thickness implant in the rabbit model. *Aust. N. Z. J. Ophthalmol.* **24**, 297–303.

51. Crawford, G. J., Chirila, T. V., Vijayasekaran, S., Dalton, P. D., and Constable, I. J. (1996). Preliminary evaluation of a hydrogel core-and-skirt keratoprosthesis in the rabbit cornea. *J. Refract. Surg.* **12**, 526–529.

52. McLean, K. M., Chatelier, R. C., Beumer, G. J., Brack, N., Johnson, G., Jenkins, M., Steele, J. G., and Griesser, H. J. (1999). Corneal epithelial tissue migration on different polysaccharide coatings. *Trans. Soc. Biomater. Meet.* **22**, 329.

53. Evans, M. D. M., Dalton, B. A., and Steele, J. G. (1999). Persistent adhesion of epithelial tissue is sensitive to polymer topography. *J. Biomed. Mater. Res.* **46**, 485–493.

54. Fitton, J. H., Dalton, B. A., Beumer, G., Johnson, G., Griesser, H. J., and Steele, J. G. (1998). Surface topography can interfere with epithelial tissue migration. *J. Biomed. Mater. Res.* **42**, 245–257.

55. Dalton, B. A., Evans, M. D. M., McFarland, G. A., and Steele, J. G. (1999). Modulation of corneal epithelial stratification by polymer surface topography. *J. Biomed. Mater. Res.* **45**, 384–394.

56. Kuhl, P. R., and Griffith-Cima, L. G. (1996). Tethered epidermal growth factor as a paradigm for growth factor induced stimulation from the solid phase. *Nat. Med.* **2**, 1022–1027.

57. Trinkaus-Randall, V., Banwatt, R., Wu, X. Y., Leibowitz, H. M., and Franzblau, C. (1994). Effect of pretreating porous webs on stromal fibroplasia in vivo. *J. Biomed. Mater. Res.* **28**, 195–202.

58. Minami, Y., Sugihara, H., and Oono, S. (1993). Reconstruction of cornea in three-dimensional collagen gel matrix culture. *Invest. Ophthalmol. Visual Sci.* **34**, 2315–2324.

59. Zieske, J. D., Mason, V. S., Wasson, M. E., Meunier, S. F., Nolte, C. J. M., Fukai, N., Olsen, B. R., and Parenteau, N. L. (1994). Basement membrane assembly and differentiation of cultured corneal cells: Importance of culture environment and endothelial cell interaction. *Exp. Cell Res.* **214**, 621–633.

60. Grinnell, F., Fukamizu, H., Pawelek, P., and Nakagawa, S. (1989). Collagen processing, cross-linking, and fibril bundle assembly in matrix produced by fibroblasts in long-term cultures supplemented with ascorbic acid. *Exp. Cell Res.* **181**, 483–491.

61. Geesin, J. C., Hendricks, L. J., Gordon, J. S., and Berg, R. A. (1991). Modulation of collagen synthesis by growth factors: The role of ascorbate-stimulated lipid peroxidation. *Arch. Biochem. Biophys.* **289**, 6–11.

62. Germain, L., Carrier, P., Giasson, M., Grandbois, E., Auger, F. A., and Guérin, S. L. (1999). A new method for in vitro production of human reconstructed cornea by tissue engineering. *Invest. Ophthalmol. Visual Sci.* **S329**(Abstr. No. 1745).

CHAPTER 84

ALIMENTARY TRACT

Shojiro Matsuda and Yoshito Ikada

The alimentary tract has chemical functions such as digestion and internal secretion. Cells are essential to regenerate these chemical functions. When the alimentary tract is made from a long artificial tube, it is important to anastomose the artificial tube without leakage and stenosis. A cell covering the inner wall of the connection site between the native and the artificial organ seems to be an effective means of overcoming these problems. This chapter describes recent work on the tissue engineering of esophagus and small intestine.

For tissue-engineered esophagus, two types of scaffold were made. One was a tube of poly(glycolic acid) mesh–collagen complex, with the inner side covered by esophageal epithelial cells; the other was a bilayered construct made of porous collagen sponge and silicone. Intestinal epithelial organoid units were seeded onto highly porous poly(glycolic acid) tubes and anastomosed to the native intestine with good results. Collagen sponges and the porous polymer tubes proved to be good scaffolds for mucous cell growth and differentiation. The scaffolds degraded, resulting in regeneration of esophagus or small intestine.

INTRODUCTION

Although a large number of tissues are targets of tissue engineering, very few studies have been devoted to the tissue engineering of the alimentary tract, probably because it is one of the most complex organs in the body. The alimentary tract, a musculomembranous tube that extends from the mouth to the anus, is lined throughout its entire extent by mucous membrane. At its commencement is the mouth, where provision is made for the mechanical division of the food and for its admixture with a fluid secreted by the salivary glands. Beyond this are the pharynx and esophagus, which convey food into the stomach, where it is stored for a time and where the first stages of the digestive process take place. The stomach is followed by the small intestine, which is divided for purposes of description into three parts: the duodenum, the jejunum, and the ileum. In the small intestine the process of digestion is completed, and the resulting products are absorbed into the blood and lacteal vessels. Finally, the small intestine ends in the large intestine, made up of cecum, colon, rectum, and anal canal, the last terminating on the surface of the body at the anus.

ESOPHAGUS REPLACEMENT BY TISSUE ENGINEERING

A variety of artificial esophagi have been designed for patients who have had gastrectomy or colectomy [1–4]. However, it is impossible to prevent complications, even with an artificial esophagus made of materials high in tissue compatibility. There are two different mechanisms responsible for the occurrence of complications. One is leakage and infection in the early stage, and the other is stenosis after displacement of the prosthesis in the late stage. The former is attributable to the materials used to prepare the artificial esophagus, that is, to biocompatibility problems between the prosthesis and the host tissue. The latter is caused by the regenerated tissue itself and is, therefore, a problem associated with the immaturity of the regenerated tissue, especially the submucosal tissue. Thus, it seems probable that tissue engineering is an alternative for replacement of the lost esophageal tissue. Two approaches attempted by two different research groups in Japan for the tissue engineering of the esophagus are represented.

USE OF ESOPHAGEAL EPITHELIAL CELLS FOR TISSUE ENGINEERING

Sato *et al.* proposed the harvesting of a small amount of normal esophageal tissue from a patient with esophageal cancer before surgery, followed by culture of the esophageal cells and the construction *in vitro* of an artificial esophagus, which would be then grafted into the same patient to replace the cancerous esophagus [5,6]. For this purpose the investigators made a tubular artificial esophagus from a poly(glycolic acid) (PGA) mesh–collagen complex tube whose inner side was covered *in vitro* by cultured human esophageal epithelial cells. After covering, the tube was grafted in latissimus dorsi muscle flaps of athymic rats.

The PGA mesh–collagen complex was prepared as follows. An acid-soluble type I collagen solution, made from pig tendon, was mixed with Ham's F-12 medium and HEPES, supplemented with $NaHCO_3$ and NaOH. Three milliliters of this mixture was prepared in a 60-mm culture dish and a PGA mesh (1.5 cm × 4 cm) was embedded in the solution. After warming to 37°C, the collagen solution became a gel, completing the PGA mesh–collagen complex. To reconstruct a mucosal tube, samples of normal mucosa (10 cm^2) were resected immediately after surgical removal from specimens of esophageal cancer patients. They were treated with Dispase and trypsin, and disaggregated epithelial cells were pelleted by centrifugation. Approximately 2×10^6 cells were inoculated on the surface of each PGA mesh–collagen complex and cultured in keratinocyte growth medium. After the cultures reached confluence, in 5–7 days, the medium was changed to keratinocyte growth medium supplemented with 20% fetal calf serum, and the cultures were maintained for more 3–5 days. At this time, the edges of the PGA mesh were sutured to create a tube, 0.5 cm in diameter and 4 cm long. The backs of 10-week-old athymic rats were incised and the latissimus dorsi muscles were exposed and dissected to make muscle flaps with pedicles. The restructured mucosal tubes were wrapped in the flaps, using a fibrin glue. The animals were killed at 4, 8, 20, and 28 days after grafting, and the muscle tubes were excised for macroscopic and microscopic examination. All the tubes maintained a tubular shape, and the inside was covered with a white translucent membranelike epithelium. The controls, PGA mesh–collagen complex tubes without an epithelial cell coating, collapsed and did not maintain a tubular shape. The microscopic appearance of the tube's inner surface showed growth of grafted epithelia. At 8 days after grafting, the epithelium was about 10 layers thick. Connective tissue infiltrated the collagen layer from muscle in a time-dependent fashion, and neovascularization appeared. At 20 days after grafting, there were more cell layers and the epithelium was thicker than at 8 days. There was neither stenosis of the lumen due to overgranulation nor contraction by means of scar. The wall structure, from the outside in, consisted of a complex and multilayered region of human esophageal epithelial cells. Ultimately it was similar to a normal esophageal wall. At 8 days after grafting, the junction between the epithelium and the collagen layer stained discontinuously and vaguely, but at 20 days after grafting this zone became continuous and thicker, and it appeared to be a basement membrane.

Human esophageal epithelial cells from a small amount of mucosa could be cultured to sufficient quantities to cover the entire inside of an artificial esophagus. During a 2-week culture period, a 250-cm^2 sheet of cultured epithelium could be produced from about 2.5 cm^2 of mucosa, and the cultured cells could be cryopreserved.

USE OF A COLLAGEN SCAFFOLD FOR ESOPHAGEAL TISSUE REGENERATION

Natsume *et al.* developed a new type of artificial esophagus with bilayered structure made of porous collagen sponge and silicone [7–10]. The bilayered structure of the tube is depicted in Fig. 84.1. The outer porous collagen sponge layer serves as a template for the construction of the neoesophagus, and the inner silicone tube layer imparts rigidity to the artificial esophagus temporarily and protects the anastomotic sites from infection, leakage, collapse, and dislocation until neoesophagus formation is complete. The tube was prepared as follows. A silicone tube, 5 cm long with an internal diameter of 2.5 cm, was made of a medical grade 1-mm-thick silicone sheet, which was reinforced by a nylon mesh embedded with silicone glue. The outer surface of the tube was exposed to corona discharge to make

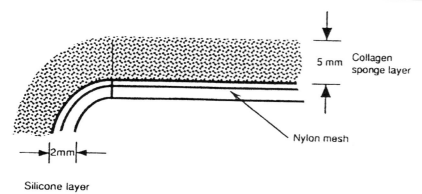

Fig. 84.1. Schematic representation of the artificial esophagus. From Takimoto, Y., et al. (1995). ASAIO J. **41**, M605–M608 [10], with permission.

the surface hydrophilic and then placed in a poly(tetrafluoroethylene) (PTFE) tube with an internal diameter of 3 cm. The collagen solution, bubbled under stirring in a refrigerated homogenizer, was poured between the PTFE tube and the silicone tube and then freeze-dried to form a porous sponge (5 mm thick). The sponge was heated at 105°C under vacuum to introduce cross-linking in the collagen molecules. This process was necessary to maintain the highly porous structure during suturing and handling. The pore sizes of the sponge were always controlled to be in the range of 100–500 μm.

The bilayered tubes were implanted into adult mongrel dogs. The trachea and the cervical esophagus were carefully isolated, taking care to preserve the vagus nerves. A 5-cm length of the cervical esophagus was resected and replaced with an artificial esophagus. The edges (2–3 mm) of the prosthesis were invaginated into the inside of the esophagus so that the resected edge of the esophagus was only in contact with the outer collagen sponge layer of the prosthesis, as shown in Fig. 84.2 and anastomosed with interrupted sutures, using an absorbable suture (3-0 Vicryl). After esophageal replacement, an intravenous hyperalimentation (IVH) tube was inserted into the left femoral vein. Dogs received no food orally but were fed by IVH alone supplied through an infusion pump, for at least 2 weeks after implantation. Penicillin and streptomycin were given intravenously during this period. After termination of IVH, animals were given standard dry dog food.

All dogs survived for more than 7 days as summarized in Table 84.1, and esophageal tissue regeneration was evident in all of them. Four died of unknown causes, but neither breakdown of prosthesis nor leakage in anastomotic sites was observed in those animals. Microscopic studies confirmed that there was no local infection associated with the implant. The esophageal regeneration was comparable to that observed in sacrificed animals. After 2 weeks, substantial neoesophageal tubes had been re-formed and residual collagen layers of the implants could not be observed macroscopically. The silicone tube had remained in place, while epithelial regeneration had started already from both sides of the anastomotic edges beneath the silicone layer. The maximum distance of the epithelialization migration front was 2.5 cm, and the minimum was 0.5 cm. At 3 weeks after surgery, epithelial continuity was almost restored, although the regenerated epithelial sheet was thin and immature.

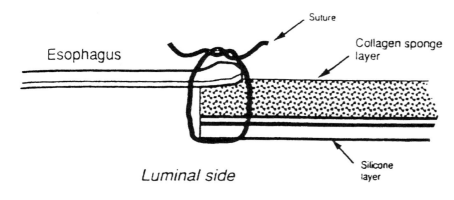

Fig. 84.2. Anastomotic technique to secure the implant. From Takimoto, Y., et al. (1995). ASAIO J. **41**, M605–M608 [10], with permission.

Table 84.1. Postoperative Resuts of Implanted Artificial Esophagus

| Survival time (days) | Cause of death | Complications | | | Silicone tube in replacement site | Epithelial regeneration[c] |
		Leakage	Granulation[a]	Stenosis[b]		
9	Accident	–	–	–	Present	–
12	Uncertain	–	–	–	Present	–
14	Sacrificed	–	–	–	Present	–
14	Sacrificed	–	–	–	Present	–
14	Sacrificed	–	–	–	Present	–
14	Uncertain	–	–	–	Present	–
17[c]	Sacrificed	–	–	+–	Present	+
21(15)	Sacrificed	–	+	+–	Not present	–
21	Sacrificed	–	–	++	Not present	+
21	Sacrificed	–	–	–	Present	+
21	Sacrificed	–	–	–	Present	–
21	Sacrificed	–	–	–	Present	–
21[c]	Uncertain	–	–	–	Present	–
23(21)	Uncertain	–	–	+	Not present	+
27	Sacrificed	–	–	–	Present	+
30	Sacrificed	–	–	–	Present	+
36(21)	Sacrificed	–	–	++	Not present	+
70(23)	Alive	–	–	+	Not present	+
370(22)	Alive	–	–	+	Not present	+

[a]Exuberant granulated tissue development: +, present; –, not present.
[b]Degree of stenosis: –, not observed; +–, low grade; +, moderate grade; ++ severe.
[c]Continuity of the regenerated epithelial lining was judged by histological examination: +, continuous; –, not continuous.

From Takimoto, Y., et al. (1993). *ASAIO J.* **39**, M736–M739 [8], with permission.

After 4–5 weeks, the maturation of the newly formed epithelium had progressed and become firm. In five dogs, immediately after IVH was terminated and oral feeding started, the silicone tube dropped into the stomach spontaneously. These dogs could take the standard dry dog food, but a tendency toward stenosis became apparent. One dog started oral feeding on day 21 but vomited severely and was sacrificed on day 36 because of malnutrition. However, the entire neoesophageal luminal surface was covered with mature mucosa, the transition from the uninjured intact portion to the regenerated area was smooth, and no granulation tissue had formed on the luminal surface. Two dogs that were not sacrificed underwent a long-term survival study (>370 and >70 days) and vomited for several weeks after starting oral feeding (on days 23 and 25, respectively) but improved gradually and later took food easily; no weight reduction was observed. Esophagography was performed in one dog at 6 and 12 months after implantation. After implantation, an acute inflammatory response was elicited and lasted for 1–2 weeks. The major inflammatory cells were neutrophils and histocytes, and occasionally small lymphocytes. After 2 weeks, the number of neutrophils declined gradually, and there was a compensatory increase in mononuclear cells. Within 2 weeks, the replacement site had been massively infiltrated with fibroblasts and endothelial cells. The extracellular matrix, consisting of disorganized fine fibers around the fibroblasts, was strongly stained with Azan, which stains collagen. After 2 weeks, the epithelial cells advanced from the anastomotic lines of the neoesophageal lumen in all the animals sacrificed on day 14, and the epithelial cell migration front was clearly observed. The fronts of the extending epithelial sheets consisted exclusively of poorly differentiated, spindle-shaped cells. However, the migrating epithelial cells adjacent to the anastomotic line had differentiated, with appearances identical to those of the uninjured intact epithelium. The cells lying on the boundary between the submucosal connective tissue stratified into a cuboidal multilayered formation and the cell layers at the luminal surface showed terminal

Table 84.2. Late Postoperative Results of Long-Term Follow-up

Animal no.	Survival time (days)	Duration of stent (days)	Cause of death	Complication: Stenosis[a]	Epithelial Regeneration
18	730	22	Accident	+	Complete
19	479	24	Alive	–	Complete
23	136	30	Pneumonia	–	Complete
20	239	33	Pneumonia	–	Complete
21	322	36	Alive	–	Complete
22	224	39	Alive	–	Complete
24	441	59	Alive	–	Complete
25	41	30	Alive	–	Complete

[a]+, present; –, not present.
From Takimoto, Y., et al. (1995). *ASAIO J.* **39**, M736–M739 [8], with permission.

differentiation (i.e., flattened and keratinized cells). In the center of the neoesophageal tissue, which was not yet covered with epithelial cells, small residues of implanted collagen sponge were noticed near the luminal surface. After 4–5 weeks, the neoesophageal lumen was covered completely with a mature epithelial sheet, consisting of well-differentiated, multilayered (10–20 layers) stratified cells and almost as thick as the normal esophagus.

As described in Table 84.1, the tendency for stenosis was only one complication observed with this artificial esophagus. To investigate whether stenosis occurred because of poor submucosal tissue formation, the stenting time was changed to examine the relationship between stenting time and stenosis in the late postoperative period. After surgery, the silicone tube was removed in selected animals every week from 2–8 weeks, when oral feeding was again initiated. In 19 cases in which the stent was dislodged within 3 weeks, stenosis developed rapidly and the dog became unable to swallow within 3 days. In two cases in which the stent moved between the third and fourth weeks, stenosis developed gradually after dislodgement. In four cases in which the stents remained in place for more than 4 weeks, stenosis did not occur, and subsequent oral feeding remained possible without weight loss, even at 12 months after surgery (Table 84.2). Endoscopic and macroscopic examination of the postoperative esophagus revealed that no granulation tissue had formed and no infection or anastomotic leakage had occurred at the junction of the prosthesis and esophagus at time of death. In all dogs, neoesophageal epithelialization was complete in the regenerated esophagus. In cases in which the stent moved 4 or more weeks after surgery and the animals survived for more than 106 days, the neoesophagus was covered with a polylayer of squamous epithelium and had normal esophageal glands and a muscle layer.

The studies just described were concerned with short segment replacement using a 5-cm-long bilayered tube. To confirm that longer segments of esophagus can be replaced safely with this type of artificial esophagus, an artificial esophagus was fabricated to replace a 10-cm-long gap. This was similar in structure and made of a silicone tube reinforced with a nylon cloth coated on the outside with 5-mm-thick freeze-dried collagen sponge. Seven adult mongrel dogs were used for the implantation. Five of the seven dogs could be fed orally, and their body weight was stable. These five dogs were allowed to live for more than 6 months. The other two dogs died of anesthetic accidents when the silicone stent was removed. Radiographic examination using radiopaque medium showed no stenosis or shortening of the reconstructed esophagus, and peristalsis was observed in the neoesophagus in all five dogs. Peristaltic waves were transmitted in succession from the oral side of the host esophagus to the anal side through the regenerated segment. The results of postoperative findings are tabulated in Table 84.3. Neoesophagus was regenerated promptly in all the animals, and the luminal surface was covered with a polylayer of squamous epithelium. The regenerated esophagus had normal esophageal glands and immature muscle tissue.

Table 84.3. Postoperative Results

Animal no.	Survival time (days)	Duration of stent (days)	Prognosis and death	Complications[a]		
				Leakage	Infection	Stenosis
1	362	42	Alive	–	–	–
2	42	42	Death	–	–	–
3	231	42	Alive	–	–	–
4	216	42	Alive	–	–	–
5	42	42	Death	–	–	–
6	204	42	Alive	–	–	–
7	182	42	Alive	–	–	–

[a]–, not present.

From Takimoto, Y., *et al.* (1995). *ASAIO J.* **41**, M606–M608 [10], with permission.

TISSUE ENGINEERING OF THE SMALL INTESTINE

Vacanti *et al.*, who used intestinal epithelial organoid units seeded onto highly porous biodegradable polymer tubes to investigate the tissue engineering of small intestine, anastomosed the tissue-engineered intestine to native small bowel alone or combined with small-bowel resection on neointestinal regeneration. Their methods for tube fabrication, organoid seeding, and anastomosis of the tissue-engineered intestine to native small bowel are as follows [11,12].

Microporous biodegradable polymer tubes, 10 mm in length, and 5 mm in outer diameter, with an internal diameter of 2–3 mm, were created from sheets of nonwoven mesh of poly(glycoric acid) (PGA) fibers sprayed on the outer surface with 5% poly(L-lactic acid). Intestinal epithelial organoid units were harvested from male and female 7-day-old Lewis rats. In brief, total small bowel was harvested and cut into small pieces. The pieces were digested with an enzyme solution containing 300 U/ml collagenase XI and 0.1 mg/ml Dispase for 25 min. After mechanical agitation and purification, isolated organoid units were seeded onto polymer at a density of $7.9–9.0 \times 10^4$ u/polymer. These unit–polymer constructs were implanted into the abdominal cavity of male Lewis rats wrapped with omentum. Three weeks after the initial surgery, the unit–polymer constructs were opened longitudinally and anastomosed with interrupted 6-0 silk to the native jejunum in a side-to-side fashion. Ten weeks after the initial surgery, all rats were sacrificed and the tissue-engineered constructs were harvested. Size of the tissue-engineered intestine was measured 3 and 10 weeks after implantation. Upper gastrointestinal study was performed in four randomly selected rats before sacrifice. All specimens were stained with hematoxylin–eosin for histological assessment.

All rats survived this study, and cyst development was noted in all animals. All anastomoses were patent at 10 weeks. Histology revealed the development of a vascularized tissue with a neomucosa lining the lumen of the cyst with invaginations resembling crypt-villus structures. As shown in Table 84.4, morphometric analysis demonstrated significantly greater villus number, villus height, crypt number, crypt area, and mucosal surface length in groups 2 and 3 compared with group 1, and significantly greater villus number, villus height, crypt area, and mucosal surface length in group 3 compared with group 2. (Group 1, implantation alone, $n = 9$; group 2, implantation followed by anastomosis to native small bowel at 3 weeks, $n = 11$; group 3, implantation following 75% small-bowel resection and subsequent anastomosis to native small bowel at 3 weeks, $n = 8$.) All rats increased their body weights significantly after anastomosis, and upper GI study revealed that there was no evidence of either bowel stenosis or obstruction at the anastomotic site. The patency of the anastomosis was 90% (18 of 20 rats), and in some cases, the lumen of the cyst was visualized by the upper GI study. Histology showed that the cysts were lined by a neomucosal layer characterized by crypt–villus structures and had continuity to the native mucosa.

Table 84.4. Morphometric Analysis of the Neomucosa

Group	Number of villi	Villus height (μm)	Number of crypts	Crypt area (μm^2)	Mucosal length (μm)
1	14 ± 5	105 ± 31	25 ± 11	$53{,}355 \pm 25{,}706$	$7{,}277 \pm 1{,}606$
2	18 ± 4^a	256 ± 51^a	81 ± 74^a	$167{,}129 \pm 87{,}362^a$	$13{,}138 \pm 3{,}697^a$
3	$22 \pm 2^{b,c}$	$472 \pm 65^{b,c}$	93 ± 30^b	$543{,}565 \pm 144{,}915^{b,c}$	$22{,}254 \pm 2{,}427^{b,c}$

[a]Group 2 versus group 1.

[b]Group 3 versus group 1.

[c]Group 3 versus group 2.

Note: $p < 0.05$, analysis of variance; Tukey test. Reproduced with permission from Ref. [11].

REFERENCES

1. Morfit, H. M., and Karmish, D. (1962). Long-term end results in bridging esophageal defects in human beings with Teflon prostheses. *Am. J. Surg.* **104**, 756–760.

2. Barnes, W. A., Redo, S. F., and Ogata, K. (1972). Replacement of portion of canine esophagus with composite prosthesis and greater omentum. *J. Thorac. Cardiovasc. Surg.* **64**, 892–896.

3. Feng-Lin, W., Nieuwenhuis, P., Gogolewski, S., Pennings, A. J., and Wildevuur, C. R. H. (1984). "Oesophageal Prosthesis," pp. 317–332. Elsevier, New York.

4. Natsume, T., Ike, O., Okada, T., Shimizu, Y., Ikada, Y., and Tamura, K. (1990). Experimental studies of a hybrid artificial esophagus combined with autologous mucosal cells. *ASAIO Trans.* **36**, M435–M437.

5. Sato, M., Ando, N., Ozawa, S., Nagashima, A., and Kitajima, M. (1993). A hybrid artificial esophagus using cultured human esophageal epithelial cells. *ASAIO J.* **39**, M554–M557.

6. Sato, M., Ando, N., Ozawa, S., Miki, H., and Kitajima, M. (1994). An artificial esophagus consisting of cultured human esophageal epithelial cells, polyglycolic acid mesh, and collagen. *ASAIO J.* **40**, M389–M392.

7. Natsume, T., Ike, O., Okada, T., Takimoto, N., Shimizu, Y., and Ikada, Y. (1993). Porous collagen sponge for esophageal replacement. *J. Biomed. Mater. Res.* **27**, 867–875.

8. Takimoto, Y., Okumura, N., Nakamura, T., Natsume, T., and Shimizu, Y. (1993). Long-term follow-up of the experimental replacement of the esophagus with a collagen–silicone composite tube. *ASAIO J.* **39**, M736–M739.

9. Takimoto, Y., Teramachi, M., Okumura, N., Nakamura, T., and Shimizu, Y. (1994). Relationship between stenting time and regeneration of neoesophageal submucosal tissue. *ASAIO J.* **40**, M793–M797.

10. Takimoto, Y., Nakamura, T., Teramachi, M., Kiyotani, T., and Shimizu, Y. (1995). Replacement of long segments of the esophagus with a collagen–silicone composite tube. *ASAIO J.* **41**, M605–M608.

11. Kim, S. S., Kaihara, S., Benvenuto, M. S., Choi, R. S., Kim, B. S., Mooney, D. J., and Vacanti, J. P. (1999). Effects of anastomosis of tissue-engineered neointestine to native small bowel. *J. Surg. Res.* **87**, 6–13.

12. Kaihara, S., Kim, S. S., Benvenuto, M. S., Choi, R. S., Kim, B. S., Mooney, D. J., Tanaka, K., and Vacanti, J. P. (1999). Anastomosis between tissue-engineered intestine and native small bowel. *Transplant. Proc.* **31**, 661–662.

CHAPTER 85

MONITORING METABOLIC ACTIVITY AND DIFFERENTIATED FUNCTION IN A BIOARTIFICIAL LIVER DEVICE

Susan Fugett Abu-Absi, Donavon J. Hess, Kristine E. Groehler, Rory P. Remmel, Timothy D. Sielaff, and Wei-Shou Hu

INTRODUCTION

The liver is the largest gland in the body, weighing about 1–1.5 kg in the adult human. It receives 28% of the total amount of blood flow through the body and uses 20% of the oxygen supply. This singular organ performs several diverse functions that are crucial for the normal operation of many other organs in the healthy mammal.

The liver takes up amino acids, carbohydrates, lipids, and vitamins from blood circulation, stores them, metabolizes them, and then returns the products to the bloodstream. The liver also plays a biosynthetic role in producing plasma proteins involved with blood coagulation, complement systems, and proteins such as albumin that transport nutrients and wastes through the bloodstream [1].

In addition, the liver plays a vital role in detoxification by transforming harmful substances into safe, excretable forms. The liver detoxifies endogenous and exogenous compounds by two means. Phase I biotransformations involve oxidation, reduction, or hydrolysis of a compound, and phase II biotransformations involve conjugation of a compound with a small molecule. Phase II reactions often follow modification by phase I reactions.

Because the liver performs so many diverse functions, liver failure is a serious concern and is associated with approximately 30,000 deaths per year in the United States [2]. Viral infection and poisonings by drug overdoses and other substances are common causes of fulminant hepatic failure, which results in coma within 8 weeks of onset. Treatments for acute liver failure remain elusive, however. Over the years, mechanical systems such as charcoal hemoperfusion [3] and hemodialysis [4] have not been shown to improve patient survival. Modest successes with whole-organ perfusion [5] and the success of liver transplantation have led to the idea that a biological component may be more effective in supporting the failing liver. Most patients who die of acute liver failure die of the complications of intracranial hypertension. The immediate aim of artificial liver support is to delay this lethal complication of liver failure until a donor liver becomes available for transplantation or until the patient's own liver regenerates.

BIOARTIFICIAL LIVER DESIGNS

One consideration in developing a bioartificial liver (BAL) device is choice of cellular component—usually either primary porcine or rat hepatocytes or transformed hepatic cell lines. Cultured primary hepatocytes retain a higher percentage of liver-specific activity than immortalized cell lines [6–8], and are therefore used in many BAL devices. In addition to losing differentiated function upon transformation, cell lines are also potentially tumorigenic [9,10] and thus may pose an additional threat to the patient. However, the isolation of primary hepatocytes from animals is labor intensive and must be performed often to supply cells to load into the device. The easy maintenance of continuous cell lines and their ability to proliferate to fill the volume of a bioreactor make them attractive for use in bioartificial liver support devices. Therefore, a few designs employ HepG2 and C3A cells.

The other major BAL design variable is the configuration of the device. In a few BAL designs, hepatocytes are cultured in suspension. However, for the maintenance of liver-specific functions for more than a few hours postharvest, hepatocytes must be cultured on a substrate to which they can attach. Therefore, many BAL devices employ microcarriers, membranes, or gel matrices, or a combination thereof. Membranes and microencapsulation provide the added benefit of immunoprotection by keeping the cells sequestered from most of the patient's immunoglobulins and complement proteins. Hollow-fiber bioreactors are utilized in a majority of designs for several reasons. They offer a convenient means of compartmentalizing the cellular component from the patient's blood, and their large surface area provides good mass transfer characteristics. Various designs seed the cells either in the interior of the hollow fibers or in the extracapillary space. In addition, the choice of membrane material confers different characteristics on the device. Several publications provide a more complete description of current hollow-fiber, microencapsulation, and packed-bed reactor BAL designs and discussions of the advantages and disadvantages of the design choices [11–15].

Several bioartificial liver (BAL) devices are in the midst of human clinical trials to evaluate their safety and efficacy in treating liver failure. Over the years, many other designs have been developed and tested in *in vitro* studies. This chapter outlines procedures for monitoring metabolic activity and liver-specific functions in BAL devices. Oxygen consumption, urea production, and albumin synthesis can be used to measure the metabolism and biosynthetic activity of the hepatocytes. In addition, biotransformation can be monitored by analyzing both phase I (oxidation) and phase II (conjugation) drug metabolism. Methods of examining the phase I activity of cytochrome P450 enzymes are described, and monitoring the conjugation of 4-methylumbelliferone is discussed as one way to assess phase II activity. In addition, a convenient method of inspecting the viability of the cells at the end of the culture period is discussed. To interpret concentration data obtained from these techniques, it is necessary to complete a material balance over the system to calculate activities. A thorough discussion of the material balance calculations is provided in addition to pertinent background information on the techniques discussed.

The methods described in this chapter are explained in reference to a gel entrapment bioartificial liver, although many can be modified and applied to most hepatocyte culture systems.

Gel Entrapment Bioartificial Liver

The example system described in the remainder of this chapter is a BAL that utilizes primary hepatocytes entrapped in collagen gel in the luminal spaces of hollow-fiber cartridges. It was developed at the University of Minnesota by a team of surgeons, engineers, and biologists [16]. A suspension of cells in a collagen solution is injected into the lumen of the hollow-fiber cartridge. Warm culture medium is circulated in the extracapillary space over the following 24 h to allow the collagen to gel. After gelation occurs, the cells pull on the collagen to contract it to approximately 60% of the fiber diameter. This allows room to perfuse culture medium in the intracapillary space as the patient's blood is passed through the shell side. Use of this device containing $6–7 \times 10^9$ porcine hepatocytes in cellulose diacetate cartridges significantly improved the survival rate in a canine model of fulminant hepatic failure [17].

Culturing hepatocytes inside a collagen gel matrix was shown to extend the length of the culture period and to promote sustained viability and function [18,19]. The collagen matrix enables the cells to retain a rounded morphology with close cell–cell interaction. This method of gel entrapment has also been modified for use in *in vitro* studies [20].

Cells are entrapped in a mixture of cell culture medium and collagen. One volume of fourfold concentrated culture medium is mixed with three volumes of type I collagen to produce a final solution containing the correct concentration of nutrients. To promote function and viability of the hepatocytes, growth factors are added to the gel mixture.

Entrapped hepatocytes can be cultured in two basic configurations for cultures independent of the hollow-fiber system. Hepatocytes in a collagen solution can be injected into the lumen of silicon tubing to form cylindrical gels or transferred to tissue culture dishes to form discs. *In vitro* studies with collagen cylindrical gels and discs are valuable for developing techniques and assays without the expense and trouble of a hollow-fiber cartridge and on a smaller scale requiring fewer cells. The protocols in the remainder of this chapter, although discussed in relation to a BAL device, can easily be adapted for use in gel disc and cylindrical gel cultures.

Protocol for Collagen Gel Entrapment

MATERIALS

Freshly isolated primary hepatocytes. Keep on ice. Methods of hepatocyte isolation have been published elsewhere [21,22].

Hepatocyte culture medium. (Researchers have developed many formulations of cell culture medium to support hepatocyte viability and growth. These recipes can be found in the literature.)

Vitrogen 100, typically 3 mg of collagen per milliliter (Collagen type I, Cohesion, Palo Alto, CA). Keep on ice.

Fourfold concentrated ($4\times$) hepatocyte culture medium. Keep on ice.

1 N NaOH.

12-well tissue culture plates.

35-mm tissue culture dishes.

100-mm tissue culture dishes.

12-ml sterile syringe with 14-gauge needle.

Silastic medical- grade tubing, 0.025 in. i.d., sterile (Dow Corning, Midland, MI).

Scalpel, sterile.

Forceps, sterile.

PROCEDURE

1. Spin the hepatocyte cell suspension at 30g for 2 min to reduce the volume. Aspirate medium from the cell pellet. Set on ice.
2. Combine 3 volumes of Vitrogen with 1 volume of $4\times$ medium. Adjust pH to 7.4 with 1 N NaOH.
3. Add hepatocytes to a concentration of 10^6/ml and swirl gently to mix.
   The following steps are different for discs and cylindrical gels and are discussed separately.
4a. *Preparation of discs*
   i. Place 1 ml of this collagen–cell mixture into each well of a 12-well tissue culture plate. Swirl gently to ensure homogeneous coating and cell concentration.
   ii. Promptly place in 37°C humidified incubator to gel.
4b. *Preparation of cylindrical gels*
   i. Use a 14-gauge sterile needle to gently take up the rest of the collagen–cell mixture into a sterile 12-ml syringe.
   ii. Slowly push the collagen–cell mixture into the lumen of the sterile silicone tubing. Performing this step in a rapid fashion applies unnecessary shear to the cells and may decrease the viability.
   iii. Place tubing into a 37°C humidified incubator to gel. Maintain sterility by wrapping in sterile foil or placing in a 100-mm tissue culture dish.

5. After 30 min check to see if the gelation of the discs and cylindrical gels is complete. *The following steps are different for discs and cylindrical gels and are discussed separately.*

6a. *For discs*

    i. Add 2 ml of culture medium to each well and use the dull end of a sterile scalpel to gently release the gel discs from the sides of the wells. This step allows the gel to contract. Return to incubator.

    ii. About 3 h later, with sterile forceps, carefully remove the gel discs from the 12-well plate and transfer them into 35-mm tissue culture dishes containing 2 ml of fresh culture medium. This step allows the bottom side of the discs access to medium.

    iii. Culture at 37°C, replacing the medium daily.

6b. *For cylindrical gels*

    i. Release the collagen cylindrical gels from the tubing by gently pushing air through the tubing with a sterile syringe.

    ii. Place 2 ml of culture medium into 35-mm tissue culture dishes and transfer the cylindrical gels to the dishes.

    iii. Culture at 37°C, replacing the medium daily.

Description of Culture System

The foregoing technique for collagen entrapment in the lumen of silicon tubing can be adapted to load a hepatocyte–collagen mixture into the lumen of hollow-fiber cartridges. Most hollow-fiber cartridges are configured so that luminal access takes place through end caps. After loading, however, the end caps of the cartridge must be flushed with cell culture medium to remove collagen solution so that access to the lumen is not restricted. For small-scale metabolic studies, miniature hollow-fiber cartridges varying in fiber diameter, fiber material, and molecular weight cutoff can be purchased from Amicon (Beverly, MA), Minntech Corporation (Minneapolis, MN), and Spectrum Laboratories (Laguna Hills, CA).

After loading, the collagen suspension will set and hepatocytes will contract the gel over the following 24 h. During the gelation and contraction stages, oxygenated culture medium at 37°C should be recirculated through the extracapillary space to provide the cells with nutrients and oxygen. The flow rate is typically chosen to avoid limiting the delivery of oxygen to the cells. The temperature can be maintained by culturing the entire system in a 37°C incubator or by controlling the temperature of the culture medium. After 24 h, the gel will have contracted to fill only about 60% of the internal diameter of the fibers, leaving room for hormonally supplemented culture medium to be perfused through the lumen at a slow rate that will not disturb the gels. Growth factors and other supplements necessary for sustained hepatocyte function should be included in the lumen medium.

Because of their excellent mass transfer properties, hollow fibers may be loaded with a very high density of cells. However, with increasing cell number, the problems of oxygen supply, waste removal, and pH control become increasingly significant. Most cell culture systems of this type rely on a gas exchanger to supply oxygen and carbon dioxide (for pH control) to the culture medium, and more sophisticated systems employ pH control algorithms to supply air, O_2, and CO_2 in the proper ratio. It is also convenient to have in-line pH and dissolved oxygen (DO) probes to continuously monitor the culture conditions. Since the culture medium that flows through the shell side of the hollow-fiber cartridge is typically recirculated, it is often necessary to supply fresh nutrients and remove spent medium from the system. A general guideline is to renew the entire volume of culture medium every 24 h to prevent glucose depletion and keep the lactic acid concentration below 1 g/liter. Each system should be designed to include convenient sampling ports on both the shell and lumen sides. Figure 85.1 shows a schematic of a standard hollow-fiber bioreactor system.

Since the precise factors required for effective artificial liver support are unknown, markers of hepatic function must suffice to monitor the cells within a bioartificial liver device. Performance variables that are often monitored include oxygen consumption, ureagenesis, albumin production, viability, and drug metabolism. The following sections describe

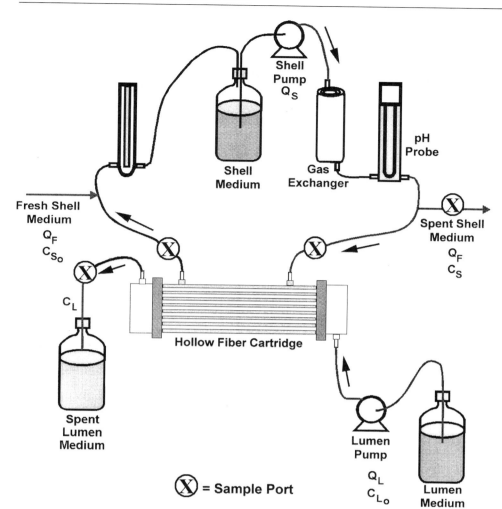

Fig. 85.1. Schematic of gel entrapment hollow-fiber bioreactor. Shell medium recirculates at a flow rate of Q_S cocurrent to the flow of lumen media at Q_L. Fresh medium is continually added and spent medium removed at a flow rate of Q_F. A gas exchanger is used to oxygenate the shell medium and a pH and DO probe are used to monitor the culture environment. Sample ports are included on both the shell and lumen sides.

protocols for these assays, and the last section details the calculations necessary to interpret the information obtained from the functional assays.

OXYGEN CONSUMPTION
OXYGEN CONSUMPTION IN REACTOR CONFIGURATIONS

Monitoring oxygen consumption is useful as an instantaneous indication of cell viability and metabolic activity. Measurement of the oxygen drop across the reactor can be taken with in-line DO probes, or off-line sample measurement can be taken with a clinical blood gas analyzer or by DO probe. Bubble-free samples can be removed with a sterile Luer-Lock syringe and promptly analyzed.

Once the samples have been analyzed, the drop in oxygen tension, ΔpO_2 (mmHg), can be calculated by subtracting the reactor outlet oxygen tension from the inlet oxygen tension. The specific rate of oxygen consumption in cell culture medium can then be estimated by the following equation

$$q_{O_2} = 1.29 \times 10^{-9} \frac{\text{mol } O_2}{\text{mmHg} \cdot \text{ml}} \left(\frac{Q_S}{X} \right) (\Delta pO_2) \tag{1}$$

which assumes an oxygen solubility constant of 1.29×10^{-9} mol O_2/(mmHg·ml). This constant was derived by assuming a Henry's law constant of 31.9 (m^3·atm)/kg, which is applicable for cell culture medium at 37°C. In Eq. (1), X is the number of cells in the culture and Q_s is the flow rate of medium through the shell space.

OXYGEN CONSUMPTION IN DISCS AND CYLINDRICAL GELS

Measuring oxygen consumption in collagen gel discs and cylinders is more difficult. The cells must first be confined to a closed system. Then the dissolved oxygen in the medium can be measured over a period of time. A simple way to carry out these measurements is to seal the disc or cylindrical gel in culture medium inside an airtight glass syringe maintained at 37°C. A blood gas analyzer can be used to take the DO measurement directly from the syringe at regular intervals. The calculation is slightly different.

$$q_{O_2} = 1.29 \times 10^{-9} \frac{\text{mol O}_2}{\text{mmHg} \cdot \text{mL}} \left(\frac{V}{X}\right) \left(\frac{\Delta pO_2}{\Delta t}\right) \tag{2}$$

Here, V denotes the volume of the culture medium in the syringe, and the oxygen drop is reported per unit time ($\Delta pO_2/\Delta t$).

UREAGENESIS AND ALBUMIN PRODUCTION

Ammonia is a by-product of cellular metabolism and is toxic to cells. It is metabolized to urea exclusively in the liver in the Krebs cycle [23]. Because this metabolic function is the exclusive province of the hepatocyte, and because it is measured with relative ease, urea production is a mainstay of tests used to monitor BAL function. Together with oxygen consumption data, this assay confirms both cell viability and the presence of differentiated hepatocytes. A spectrophotometric kit to detect urea is available from Sigma (St. Louis: MO, Urea Nitrogen Diagnostics kit, Procedure 640). The kit was developed according to the methods of Fawcett and Scott [24] and Chaney and Marback [25]. Since this type of assay is routinely performed in hospital laboratories and is well illustrated elsewhere, details are not provided here.

As urea production serves as a marker for hepatic metabolic function, albumin production serves as a marker for hepatic protein synthetic function. Albumin, like urea, is produced exclusively in the liver [26]. It accounts for 60% of plasma protein, and is important in providing colloid osmotic pressure and in serving as a carrier for small molecules, including unconjugated bilirubin and hormones. Measuring the amount of albumin produced during hepatocyte culture is much more labor intensive than the measurement of ureagenesis. However, albumin production is a more reliable indicator of hepatocyte function because it is relatively constant [27], whereas the levels of enzymes involved in ureagenesis can vary greatly with protein uptake [28].

Albumin concentrations can be measured by an enzyme-linked immunosorbent assay (ELISA). A common procedure for measuring albumin concentrations is sandwich ELISA, which takes place in three steps. The first step requires that an antibody-coated solid be incubated with the sample solution, allowing the albumin in the test solution to bind to the immobilized antibody. Second, an antibody–enzyme conjugate is added to the reaction mixture and binds to any albumin that had been bound to the surface, creating an antibody–antigen–antibody complex, hence the term "sandwich ELISA." Last, any unbound conjugate is washed off, a substrate solution is added, and time is allowed for incubation. Once the incubation is over, a spectrophotometer can be used to determine the concentration of colored product formed. The concentration of colored product is proportional to the concentration of albumin in the sample [29]. The sandwich ELISA method has been used to quantify porcine albumin in BAL devices [19]. Another way to assay for albumin is competitive ELISA. A complete protocol was published in 1999 [30].

The ELISA can detect antigen concentrations to about 5 µg/liter, but to achieve this detection level, the antibodies must be of the highest quality [31]. The antibody–enzyme conjugate gives the least amount of background if it is affinity-purified and is free of aggregates and soluble complexes. It should also be specific to the species and the class of the intermediate antibody in the case of the sandwich ELISA. To reduce the nonspecific binding of later layers of antibody, antibodies used to coat the solid should be affinity-purified and from the same species as the enzyme-conjugated antibody. The use of antibodies of these types should reduce the possibility of cross-species contamination. This is especially important if bovine serum albumin (BSA) is added as a culture supplement. Once an albumin ELISA protocol has been developed, a test should be run with blank media samples containing BSA to evaluate the amount of cross-species interference in the assay.

DRUG METABOLISM

BACKGROUND

The liver has two primary means of metabolizing xenobiotic compounds: phase I biotransformations (oxidation, reduction, and hydrolysis) and conjugative, or phase II, biotransformations (sulfation, glucuronidation, methylation, acetylation, glutathione transfer, and amino acid conjugation). Phase I biotransformations are generally mediated by hepatic microsomal enzyme systems, especially the cytochrome P450 enzyme systems. By definition, phase II biotransformations follow phase I reactions in the process of detoxification, but direct conjugation (e.g., 4-methylumbelliferone glucuronidation) may also occur for some compounds. In some experimental situations, it is useful to measure the detoxification functions of hepatocytes. Effective detoxification of endogenous and exogenous compounds is likely to be vital to the efficacious performance of liver support devices. Additionally, detoxification functions are further indicators of higher liver function.

To measure oxidative function (phase I activity), the conversion of lidocaine to monoethylglycine xylidide (MEGX), its primary metabolite, and other metabolites can be measured by either high-performance liquid chromatography (HPLC) or gas chromatography. A fluorescence polarization immunoassay for MEGX is also available [32]. MEGX formation, the result of an oxidative N-deethylation, is catalyzed by CYP1A2, CYP2C11, CYP2B1, and CYP3A2 in rats and CYP3A4 in humans [33]. In untreated rats and humans, CYP3A appears to be the primary enzyme responsible for MEGX formation [34]. Other lidocaine metabolites (3-OH-lidocaine, 2-hydroxymethyl-lidocaine) are formed by different P450 enzymes. MEGX is further metabolized by oxidation to 3-OH-MEGX and amide hydrolysis to 2,6-dimethylaniline, especially after long incubation times [35].

4-Methylumbelliferone (4-MU, Hymecromone) clearance is used to measure conjugation (phase II) activity. 4-MU is metabolized primarily by glucuronidation, with some of the compound undergoing sulfation in rats [36,37]. The concentration of 4-MU and its conjugates (4-MU glucuronide and 4-MU sulfate) in media can be measured by means of HPLC [38]. In addition, 4-MU disappearance can also be monitored by fluorometry (excitation and emission wavelengths of 365 and 445 nm, respectively).

A major consideration of drug metabolism studies *in vitro* is controlling for drug adsorbance to the cultureware (tubing and gas exchanger, as well as hollow-fiber membranes). Therefore, it is also important to monitor the appearance of drug metabolites or conjugates, for this allows one to rule out the possibilities that the drug disappearance is due to binding to the cultureware or that the drug is being taken up into the cells but not metabolized. The methods described here utilize HPLC to analyze the media samples. Lovdahl *et al.* [38] and Nyberg *et al.* [39] provide more detailed information on HPLC methods.

For standard curves, it is necessary to analyze known concentrations of the drugs and their metabolites. Many of these substances are commercially available, but in some cases must be synthesized *a priori* or obtained from other laboratories or pharmaceutical companies. The metabolites of lidocaine can be synthesized by following procedures developed by Nelson *et al.* [40] or Keenaghan and Boyes [35].

MEASUREMENT OF LIDOCAINE AND ITS METABOLITES AND 4-MU AND ITS CONJUGATES

MATERIALS

Lidocaine (Sigma, St. Louis, MO)
Monoethylglycine xylidide (MEGX)
Glycine xylidide (GX)
3-OH-Lidocaine
2-Hydroxy-methyl-lidocaine (2-MeOH-lidocaine)
3-OH-MEGX

4-Methylumbelliferone (4-MU) (Sigma)
4-Methylumbelliferyl glucuronide (4-MUG) (Sigma)
4-Methylumbelliferyl sulfate (4-MUS) (Sigma)
Chemicals and apparatus for extraction and HPLC analysis

PROCEDURE

1. Prepare shell media containing 12 µg/ml lidocaine or 10 µg/ml 4-MU.
2. Take shell samples at 4- to 6-h intervals prior to cell loading to determine rate of drug adsorbance to cultureware.
3. Inoculate cells and take shell media samples at t = 0, 0.5, 2, 4, 8, 20, and 24 h, etc. Extending this experiment to longer times may require pulsing the shell media with additional drugs. Also, lumen samples must be taken at time points after the initiation of luminal flow so that calculations can include that information.

HPLC Analysis

Briefly, the drugs and their metabolites must first be isolated by solid phase extraction or liquid–liquid extraction, and then separated by HPLC [38,39]. Detection for lidocaine, MEGX, GX, 3-OH-lidocaine, 2-MeOH-lidocaine, and 3-OH-MEGX is performed at 214 nm; for 4-MU, 4-MUG, and 4-MUS detection, 314 nm is used. Standard solutions of all substances must also be analyzed to produce a standard curve. Samples are then quantified by comparing the peak area ratio of the component of interest to that of an internal standard. The concentrations of the drugs and metabolites are determined by extrapolating the peak height ratios from the samples onto the standard curves generated.

CONFOCAL MICROSCOPY

The confocal microscope is a powerful tool for imaging through tissues and three-dimensional samples. It is very useful for studies of gel-entrapped hepatocytes because it allows one to optically section through the gel disc or cylindrical gel to assay the viability and other characteristics of the cells contained therein. We describe methods for determining viability and assaying for cytochrome P450 function *in situ*.

IN SITU ASSAY FOR CYTOCHROME P450 ACTIVITY

In addition to lidocaine clearance, several other substrates are available to detect the activity of various cytochrome P450 isozymes. Fluorogenic assays with alkoxyresorufin substrates are convenient and sensitive. Pentoxyresorufin O-dealkylation (PROD) and ethoxyresorufin-O-dealkylation (EROD) are measures of rat CYP2B1/2 and CYP1A1/2 function, respectively. CYP1A1 and CYP2B1 are not constitutively expressed in liver but are highly inducible by aromatic hydrocarbons (such as β-naphthoflavone) and phenobarbital, respectively [41]. PROD activity, although inducible by phenobarbital treatment [42], is detectable without drug inducement, owing to the contribution of several different P450 enzymes, including constitutive CYP1A2. Several helpful reviews of cytochrome P450 functions are available [41,43,44].

During the EROD and PROD assays, CYP1A1/2 and CYP2B1/2 isozymes in rat hepatocytes convert the alkoxyresorufin to a fluorescent product, resorufin. In addition to the substrate, probenecid and dicumarol are added to the culture medium to restrict the transport of resorufin from the cells and to prevent it from being processed into a nonfluorescent derivative. By carefully controlling the assay conditions and settings of the confocal microscope, this technique can be used to qualitatively compare the cytochrome P450 activity in different cultures and, more importantly, to examine the spatial distribution of CYP450 within a sample. Aside from confocal microscopy, cytochrome P450 activity can be detected by measuring the amount of resorufin in the culture medium after addition of pentoxyresorufin or ethoxyresorufin. This can be accomplished easily by using a fluorescence plate reader.

MATERIALS

Dicumarol (Sigma), 10 mM stock solution in 0.5 N NaOH

Probenecid (Sigma), 0.16 M stock solution in 0.5 N NaOH

Na phenobarbital, 0.4 M stock solution in distilled water, or β-naphthoflavone (Aldrich, Milwaukee, WI), 100 mM stock solution in dimethyl sulfoxide (DMSO).

Clear Williams E medium containing no phenol red (Life Technologies, Rockville, MD). Adjust pH to 7.2 with HEPES buffer.

Ethoxy- or pentoxyresorufin (Molecular Probes, Eugene, OR), 1 mM stock solution in DMSO. Store in the dark.

Confocal microscope with filter set to detect rhodamine fluorescence.

PROCEDURE

1. The culture should be induced for at least 24 h before the assay by switching to culture medium containing 0.5–2 mM Na phenobarbital (for PROD) or 50 μM β-naphthoflavone (for EROD).

2. Expel collagen cylindrical gels from the hollow-fiber cartridge by blowing air through the lumen. Remove as much liquid as possible. If discs or cylindrical gels are used, remove culture medium from the dish. Wash two times with Williams E medium containing no phenol red.

3. Prepare incubation medium (clear Williams E containing 25 μM dicumarol and 2 mM probenecid). Adjust pH to 7.2 with 1 M HEPES buffer.

4. Add incubation medium to sample and incubate at 37°C.

5. Add 20 μl of pentoxyresorufin (for PROD) or ethoxyresorufin (for EROD) per milliliter of incubation medium and incubate at 37°C.

6. Assay for resorufin formation within 15 min (EROD) or 30 min (PROD) using confocal microscope set to measure rhodamine fluorescence.

VIABILITY ASSAY

Prior to culture, hepatocyte viability is tested by trypan blue exclusion to assure an adequate viable cell mass. It is also informative to assess the viability of the hepatocytes at the end of the culture. The fluorescence-based method described here to assay viability is a powerful tool for three-dimensional culture systems. It allows the investigator to obtain viability information with reference to position in the culture. Fluorescein diacetate is cleaved by esterases in live cells to the fluorescent molecule fluorescein, while ethidium bromide passes through the membranes of dead cells and binds to nucleic acids, staining the cell nuclei red. Dual scanning on the confocal microscope can be used to detect both fluorophores simultaneously, and the technique can be used successfully with a fluorescence microscope as well.

MATERIALS

Fluorescein diacetate (FDA) (Sigma), 5 mg/ml stock solution in DMSO. Store at 4°C in the dark.

Ethidium bromide (EB) (Sigma), 40 μg/ml stock solution in phosphate-buffered saline. Store at 4°C in the dark.

Phosphate-buffered saline (PBS).

Confocal microscope with filter sets for fluorescein and rhodamine.

PROCEDURE

1. Prepare staining solution by mixing 10 ml of EB (40 μg/ml) with 20 μl of FDA (5 mg/ml). The resulting solution will contain 10 μg/ml FDA and 40 μg/ml EB.

2. Expel collagen cylindrical gels from the hollow-fiber cartridge by blowing air through the lumen. Remove as much liquid as possible. If discs or cylindrical gels are used, remove culture medium from the dish.
3. Add staining solution and incubate for 2 min at room temperature.
4. Aspirate off staining solution and rinse 3 times with PBS.
5. View on a confocal microscope. Red cells are dead and live cells are green.

MATHEMATICAL ANALYSIS OF REACTOR PERFORMANCE

To complete the mass balance for the system in Fig. 85.1, one would need to solve a system of coupled first-order partial differential equations. However, by assuming that both the lumen and shell streams are well mixed, so that there is no concentration gradient in either the axial or radial direction, a more simplified material balance can be written and used for analysis of the reactor performance. A mass balance over both streams produces the following equations

$$\text{Lumen:} \quad \frac{d(C_L V_L)}{dt} = qX + Q_L(C_{Lo} - C_L) - R_M \tag{3}$$

$$\text{Shell:} \quad \frac{d(C_S V_S)}{dt} = Q_F(C_{So} - C_S) + R_M - R_S \tag{4}$$

where the subscripts L, S, and F denote lumen, shell, and feed, Q is the volumetric flow rate, C is the mass concentration of a product or substrate, V is volume, q represents the specific production or clearance rate of a substance in units of mass/(time · cell), X represents the cell number inside the reactor, R_M represents the rate of transfer of a substance across the membrane from the lumen to shell, and R_S is the rate of removal by sampling. Integrating Eqs. (3) and (4), adding them together, and rearranging gives the following equation:

$$qX\Delta t = \Delta(C_L V_L) + \Delta(C_S V_S) - [Q_F(C_{So} - C_S) + Q_L(C_{Lo} - C_L)]\Delta t + (VC)_{\text{sample}} \tag{5}$$

which can easily be used to calculate specific production rates and the total amount of product. Here $(VC)_{\text{sample}}$ represents the product of the volume of sample taken over the interval Δt, and the concentration, C, in the sample.

Once the concentration of albumin, urea, lidocaine and its metabolites, or 4-MU and its conjugates has been determined in each sample, cumulative amounts of product formed or substrate consumed and the specific production rate or clearance rate must be calculated. Equation (5) can be used to compute the amount of product formed in the time interval Δt. A negative value of $qX\Delta t$ represents a decrease in the amount of substrate and indicates a negative value of q. In such a case, q would represent a specific clearance rate. The cumulative amount of substrate or product $(M_{P/S})_j$ in the system up to time t_j is the sum of $qX\Delta t$ from time t_0 through time t_j or

$$\left(M_{P/S}\right)_j = \sum_{i=0}^{j}(qX\Delta t)_i \tag{6}$$

For albumin and urea one can assume zero-order kinetics, and linear regression can be performed to fit cumulative mass calculated with Eq. (6) and time data to a line. Dividing the slope of this line by the total number of cells in the reactor yields the specific production rate, q. Figure 85.2a illustrates this procedure.

Equation (5) can also be used to calculate the total amount of drug consumed or metabolite produced during a time interval when drug is present in the culture system. It includes contributions from both the lumen and shell sides. For clearance of lidocaine and 4-MU, experimental results have indicated that the specific rate of clearance does not stay constant over the course of the culture, but is a function of drug concentration. Therefore, clearance data can be fit to a first-order rate equation to calculate the rate constant. This can be done by plotting the ratio of the initial mass of drug and the mass of drug remaining in the culture as calculated from Eq. (6) ($M_{P/S_0}/M_{P/S}$) versus time on a semilog plot and fitting with an exponential curve. For simplification, the same result can be obtained by plotting

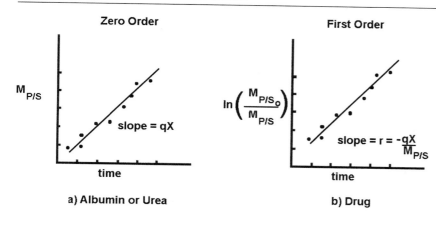

Zero Order

$M_{P/S}$

slope = qX

time

a) Albumin or Urea

First Order

$\ln\left(\dfrac{M_{P/S_0}}{M_{P/S}}\right)$

slope = r = $-\dfrac{qX}{M_{P/S}}$

time

b) Drug

Fig. 85.2. Calculation of (a) the rate of production of albumin and urea and (b) the clearance rate of lidocaine and 4-MU. Plotting the data as shown will allow linear regression to be performed on the data.

the natural log of $M_{P/S_0}/M_{P/S}$ versus time and fitting with a line (Fig. 85.2b). The slope of the line obtained is r, where r is the clearance rate in units of time^{-1}. The specific clearance rate, q, will have a negative value. The relationship between r and the specific clearance rate, q, is as follows

$$r = -\frac{qX}{M_{P/S}} \qquad (7)$$

As mentioned earlier, drug metabolism studies in BAL systems are significantly complicated by drug adsorbance to cultureware, and this can introduce serious error into the calculations just described. For this reason, it is crucial to assay for the appearance of metabolites in the culture in addition to the disappearance of lidocaine or 4-MU. The appearance of metabolites confirms that the drug is actually being metabolized and that disappearance is not merely due to drug binding to tubing and cultureware components. Calculations of the specific production rate for drug metabolites can also assume a first-order dependence on the concentration of drug. The classic textbook on pharmacokinetics by Gibaldi and Perrier should be consulted for more information on the analysis of drug metabolism studies [45].

The calculations are much simpler for cultures of discs and cylindrical gels. Since these are simply batch cultures, the raw concentration data can easily be converted to mass and summed to give the cumulative mass for each measurement point. The cumulative mass over time can be plotted as in Fig. 85.2 and analyzed. Complications arise, however, because the concentration of metabolites may not be the same in the gel and in the culture medium. Such a discrepancy can introduce serious error into the calculations, and the posibility must be considered.

Using primary cells to interpret the results of an experiment may be difficult because there is often variation in cell performance from animal to animal. Many researchers find it useful to run simultaneous batch control cultures in petri dishes or spinner flasks. Comparing the performance of cells in the control culture to the BAL is an easy and reliable way of detecting problems or advantages specific to the BAL and not caused by cell variation. It is also useful to compare performance to the literature, keeping in mind that mass transfer characteristics, cell source, and media components may all affect specific production rates and drug clearance activities.

CONCLUSIONS

The bioartificial liver may prove to be effective at increasing survival in patients waiting for liver transplantation, and possibly in promoting liver regeneration in some cases. Several clinical trials are under way to examine the safety and efficacy of BAL devices in human treatment. In addition, BAL systems are a valuable tool for *in vitro* investigations of drug metabolism and hepatocyte function. This chapter has provided information useful for developing an *in vitro* hepatocyte cell culture system and monitoring its metabolic activity and differentiated functions. The methods discussed enable quantitative comparison among different hepatocyte cell culture systems and conditions. In addition, the protocols for collagen disc and cylindrical gel culture are useful for developing new techniques and assays for use in BAL devices.

REFERENCES

1. Arias, I. M., Boyer, J. L., Fausto, N., Jakoby, W. B., Schachter, D., and Shafritz, D. A., eds. (1994). "The Liver: Biology and Pathobiology," 3rd ed. Raven Press, New York.
2. National Center for Health Statistics. (1993). "Vital Statistics of the United States." National Center for Health Statistics, Division of Vital Statistics, Hyattsville, MD.
3. O'Grady, J., Gimson, A., O'Brien, C., Pucknell, A., Hughes, R., and Williams, R. (1988). Controlled trials of charcoal hemoperfusion and prognostic factors in fulminant hepatic failure. *Gastroenterology* **94**, 1186–1192.
4. Opolon, P., Rapin, J. R., Huguet, C., Granger, A., Delorme, M. L., Boschat, M., and Sausse, A. (1976). Hepatic failure coma treated by polyacrylonitrile membrane hemodialysis. *Trans. Am. Soc. Artif. Int. Organs* **22**, 701–710.
5. Fox, I. J., Langnas, A. N., Fristoe, L. W., Shaefer, M. S., Vogel, J. E., Antonson, D. L., Donovan, J. P., Heffron, T. G., Markin, R. S., and Sorrell, M. F. *et al.* (1993). Successful application of extracorporeal liver perfusion: A technology whose time has come. *Am. J. Gastroenterol.* **88**, 1876–1881.
6. Nyberg, S. L., and Misra, S. P. (1998). Hepatocyte liver-assist systems—A clinical update. *Mayo Clin. Proc.* **73**, 765–771.
7. Nyberg, S. L., Remmel, R. P., Mann, H. J., Peshwa, M. V., Hu, W.-S., and Cerra, F. B. (1994). Primary hepatocytes outperform Hep G2 cells as the source of biotransformation functions in a bioartificial liver. *Ann. Surg.* **220**, 59–67.
8. Shiraha, H., Koide, N., Hada, H., Ujike, K., Nakamura, M., Shinji, T., Gotoh, S., and Tsuji, T. (1996). Improvement of serum amino acid profile in hepatic failure with the bioartificial liver using multicellular hepatocyte spheroids. *Biotechnol. Bioeng.* **50**, 416–421.
9. Stange, J., and Mitzner, S. (1996). Cell sources for bioartificial liver support. *Int. J. Artif. Organs* **19**, 14–17.
10. Cao, S., Esquivel, C. O., and Keeffe, E. B. (1998). New approaches to supporting the failing liver. *Annu. Rev. Med.* **49**, 85–94.
11. Tzanakakis, E. S., Hess, D. J., Sielaff, T., and Hu, W.-S. (2000). Extracorporeal tissue engineered liver-assist devices. *Annu. Rev. Biomed. Eng.* **2**, 607–632.
12. Busse, B., Smith, M. D., and Gerlach, J. C. (1999). Treatment of acute liver failure: Hybrid liver support a critical overview. *Lagenbeck's Arch. Surg.* **384**, 588–599.
13. Friend, J. R., and Hu, W.-S. (1998). Engineering a bioartificial liver support. *In* "Frontiers in Tissue Engineering" (C. W. Patrick, A. G. Mikos, and L. V. McIntire, eds.), pp. 678–695. Elsevier, New York.
14. Ohshima, N. (1997). Tissue engineering aspects of the development of bioartificial livers. *J. Chin. Inst. Chem. Eng.* **28**, 441–453.
15. Dixit, V., and Gitnick, G. (1996). Artificial liver support: State of the art. *Scand. J. Gastroenterol., Suppl.*, pp. 101–114.
16. Nyberg, S. L., Shatford, R. A., Peshwa, M. V., White, J. G., Cerra, F. B., and Hu, W.-S. (1993). Evaluation of a hepatocyte-entrapment hollow fiber bioreactor. A potential bioartificial liver. *Biotechnol. Bioeng.* **41**, 194–203.
17. Sielaff, T. D., Hu, M. Y., Amiot, B., Rollins, M. D., Rao, S., McGuire, B., Bloomer, J. R., Hu, W.-S., and Cerra, F. B. (1995). Gel-entrapment bioartificial liver therapy in galactosamine hepatitis. *J. Surg. Res.* **59**, 179–184.
18. Koebe, H. G., Pahernik, S., Eyer, P., and Schildberg, F. W. (1994). Collagen gel immobilization: A useful cell culture technique for long-term metabolic studies on human hepatocytes. *Xenobiotica* **24**, 95–107.
19. Lazar, A., Mann, H. J., Remmel, R. P., Shatford, R. A., Cerra, F. B., and Hu, W.-S. (1995). Extended liver-specific functions of porcine hepatocyte spheroids entrapped in collagen gel. *In Vitro Cell. Dev. Biol.—Anim.* **31**, 340–346.
20. Sielaff, T. D., Hu, M. Y., Rao, S., Groehler, K., Olson, D., Mann, H. J., Remmel, R. P., Shatford, R. A., Amiot, B., Hu, W.-S., and Cerra, F. B. (1995). A technique for porcine hepatocyte harvest and description of differentiated metabolic functions in static culture. *Transplantation* **59**, 1459–1463.
21. Seglen, P. O. (1976). Preparation of isolated rat liver cells. *Methods Cell Biol.* **13**, 29–83.
22. Hansen, L. K., Friend, J. R., Remmel, R., Cerra, F. B., and Hu, W.-S. (1999). Development of a bioartificial liver device. *Methods Mol. Med.* **18**, 423–431.
23. Jones, M. E., Anderson, A. D., Anderson, C., and Hodes, S. (1961). Citrulline synthesis in rat tissues. *Arch. Biochem. Biophys.* **95**, 499–507.
24. Fawcett, J., and Scott, J. (1960). A rapid and precise method for the determination of urea. *J. Clin. Pathol.* **13**, 156–160.
25. Chaney, A., and Marback, E. (1962). Modified reagents for determination of urea and ammonia. *Clin. Chem. (Winston-Salem, N.C.)* **8**, 130–132.
26. Rothschild, M. A., Oratz, M., and Schreiber, S. S. (1972). Albumin synthesis. *N. Engl. J. Med.* **286**, 748–757.
27. Pietrangelo, A., and Shafritz, D. (1994). Gene expression during pathophysiologic states. *In* "The Liver: Biology and Pathobiology" (I. M. Arias, J. L. Boyer, N. Fausto, W. B. Jakoby, D. Schachter, and D. A. Shafritz, eds.), pp. 85–98. Raven Press, New York.
28. Ampola, M. G. (1994). The urea cycle enzymes and defects. *In* "The Liver: Biology and Pathobiology" (I. M. Arias, J. L. Boyer, N. Fausto, W. B. Jakoby, D. Schachter, and D. A. Shafritz, eds.), pp. 365–378. Raven Press, New York.
29. Kimball, J. W. (1990). "Introduction to Immunology," 3rd ed. Macmillan, New York.
30. Friend, J. R., Wu, F. J., Hansen, L. K., Remmel, R. P., and Hu, W.-S. (1999). Formation and characterization of hepatocyte spheroids. *Methods Mol. Med.* **18**, 245–252.
31. Catty, D., Ling, N. R., Lowe, J. A., and Raykundalia, C. (1981). Antisera in immunoassays with special reference to monoclonal antibodies to human immunoglobulins. *In* "Immunoassays for the 80s" (A. Voller, A. Bartlett, and D. Bidwell, eds.), pp. 133–153. MTP Press, Lancaster, England.

32. Shatford, R. A., Nyberg, S. L., Meier, S. J., White, J. G., Payne, W. D., Hu, W.-S., and Cerra, F. B. (1992). Hepatocyte function in a hollow fiber bioreactor: A potential bioartificial liver. *J. Surg. Res.* **53**, 549–557.

33. Imaoka, S., Enomoto, K., Oda, Y., Asada, A., Fujimori, M., Shimada, T., Fujita, S., Guengerich, F. P., and Funae, Y. (1990). Lidocaine metabolism by human cytochrome P-450s purified from hepatic microsomes: Comparison of those with rat hepatic cytochrome P-450s. *J. Pharmacol. Exp. Ther.* **255**, 1385–1391.

34. Nakomoto, T., Oda, Y., Imaoka, S., Funae, Y., and Fujimori, M. (1996). Effect of phenobarbital on the pharmacokinetics of lidocaine, monoethylglycinexylidide and 3-hydroxylidocaine in the rat: Correlation with P450 isoform levels. *Drug Metab. Dispos.* **25**, 296–300.

35. Keenaghan, J. B., and Boyes, R. N. (1972). The tissue distribution, metabolism and excretion of lidocaine in rats, guinea pigs, dogs and man. *J. Pharmacol. Exp. Ther.* **180**, 454–463.

36. Mulder, G. J. (1973). The rate-limiting step in the biliary elimination of some substrates of uridine diphosphate glucuronyltransferase in the rat. *Biochem. Pharmacol.* **22**, 1751–1763.

37. Mulder, G. J., Brouwer, S., Weitering, J. G., Scholtens, E., and Pang, K. S. (1985). Glucuronidation and sulfation in the rat in vivo. *Biochem. Pharmacol.* **34**, 1325–1329.

38. Lovdahl, M. J., Reher, K. J., Mann, H. J., and Remmel, R. P. (1994). Determination of 4-methylumbelliferone and its metabolites in William's E Media and dog plasma. *J. Liq. Chromatogr.* **17**, 1795–1809.

39. Nyberg, S. L., Shirabe, K., Peshwa, M. V., Sielaff, T. D., Crotty, P. L., Mann, H. J., Remmel, R. P., Payne, W. D., Hu, W.-S., and Cerra, F. B. (1993). Extracorporeal application of a gel-entrapment, bioartificial liver: Demonstration of drug metabolism and other biochemical functions. *Cell Transplant.* **2**, 441–452.

40. Nelson, S. D., Garland, W. A., Breck, G. D., and Trager, W. F. (1977). Quantification of lidocaine and several metabolites utilizing chemical-ionization mass spectrometry and stable isotope labeling. *J. Pharm. Sci.* **66**, 1180–1190.

41. Soucek, P., and Gut, I. (1992). Cytochromes P-450 in rats: Structures, functions, properties and relevant human forms. *Xenobiotica* **22**, 83–103.

42. Waxman, D. J., and Azaroff, L. (1992). Phenobarbital induction of cytochrome P-450 gene expression. *Biochem. J.* **281**, 577–592.

43. Omiecinski, C. J., Remmel, R. P., and Hosagrahara, V. P. (1999). Concise review of the cytochrome P450s and their roles in toxicology. *Toxicol. Sci.* **48**, 151–156.

44. Pelkonen, O., Maenpaa, J., Taavitsainen, P., Rautio, A., and Raunio, H. (1998). Inhibition and induction of human cytochrome P450 (CYP) enzymes. *Xenobiotica* **28**, 1203–1253.

45. Gibaldi, M., and Perrier, D. (1982). "Pharmacokinetics," 2nd ed. Dekker, New York.

CHAPTER 86

BLOOD CELL SUBSTITUTES

Thomas Ming Swi Chang

B asic research on polyhemoglogin was started sometime ago [1]. However, it is the HIV in donor blood that stimulated the development of blood substitutes for clinical use. Soluble polymerized hemoglobin (polyhemoglobin) is now in phase III clinical trials. Patients have received up to 20 units (10 liters) in trauma surgery and other surgery. Polyhemoglobin can be stored for more than a year. It has no blood group antigen and can be used as a universal donor. It can also be sterilized. A typical method for the laboratory preparation of polyhemoglobin is described in detail in this chapter. The use of *in vitro* complement activation of human plasma as a screening test also is described. With a circulation half-time of 24 h, polyhemoglobin is being used mainly for perioperative uses.

Future generations of blood substitutes are being studied. For example, under conditions with potential for ischemia–reperfusion injuries, a new polyhemoglobin–superoxide dismutase–catalase that can reduce oxygen radicals is being developed. To increase the circulation time, artificial red blood cells are prepared with a bilayer lipid membrane (hemoglobin liposome) or with biodegradable polymer membranes like polylactide (hemoglobin nanocapsules). The preparation of biodegradable polymeric hemoglobin nanocapsules optimized for *in vitro* properties is described in this chapter.

BACKGROUND AND PRINCIPLES

Hemoglobin is the red blood cell protein that specializes in transport, especially of oxygen. Unfortunately, it cannot be used for transfusion into the body. Each hemoglobin molecule, consisting of four subunits, or tetramers, breaks down into two toxic dimers when infused into the body. As a result, hemoglobin must be modified before it can be used [2,3]. The first attempt to overcome this problem was carried out by Chang, who tried to prepare artificial red blood cells by microencapsulating hemoglobin [1]. This attempt originally proved be too ambitious but this is being developed as a third-generation blood substitute. Simpler methods of modification of hemoglobin also are being developed for use in clinical trials. For example, in earlier studies to improve the circulation time of artificial red blood cells, Chang used cross-linked hemoglobin [1,4]. Initially, the goal was to form cross-linked hemoglobin membranes for artificial red blood cells. When Chang kept decreasing the diameter to about 1 μm, all the hemoglobin was cross-linked into polyhemoglobin [1,4].

SOLUBLE MODIFIED HEMOGLOBIN

There are two classes of soluble modified hemoglobin [2,3]. The first class consists of three types of cross-linked hemoglobin: polyhemoglobin, conjugated hemoglobin, and intramolecularly cross-linked hemoglobin. The second type consists of recombinant human hemoglobin.

Chang published the first report in 1964 on the use of a bifunctional agent, sebacyl chloride, to cross-link hemoglobin [1,4]. This compound was used first to form cross-linked hemoglobin membranes for artificial red blood cells. Chang found that with decreasing size of the artificial cells, all the hemoglobin molecules are cross-linked into polyhemoglobin [1,4]. At that time he proposed that the bifunctional agent cross-linked some of the many amino groups on the surface of the hemoglobin molecule. Cross-linking with a bifunctional agent prevents the breakdown of hemoglobin tetramers into dimers. The reaction is as follows:

$$Cl-CO-(CH_2)_8-CO-Cl \; + \; HB-NH_2 \; \rightarrow \; HB-NH-CO-(CH_2)_8-CO-NH-HB$$

Sebacyl chloride   Hemoglobin   Cross-linked hemoglobin

$$+ \; Cl_2$$

In 1971 Chang reported the use of another bifunctional agent, glutaraldehyde, to cross-link hemoglobin and the red blood cell enzyme catalase [5]. The reaction is as follows:

$$H-CO-(CH_2)_3-CO-H \; + \; HB-NH_2 \; \rightarrow \; HB-NH-CO-(CH_2)_3-CO-NH-HB$$

Glutaraldehyde   Hemoglobin   Cross-linked hemoglobin

$$+ \; H_2$$

In this 1971 procedure, catalase was added to 10 g/dl hemoglobin and enclosed in 1.5 ml of microcapsules. To this was added 100 ml of a solution (0.1 mol/liter metaborate and 0.001 mol/liter benzamideine HCl). Then, 200 μl of 50% glutaraldehyde (5.56 mol/liter) was added, and the mixture was kept slightly agitated at 20°C for 1 h. Then 100 of 0.05 mol/liter sodium borohydride was added and left at 4°C for 20 min, then washed twice with 200 ml of 0.154 mol/liter sodium chloride solution. Cross-linking hemoglobin and catalase with glutaraldehyde in this way increases the stability of the enzyme [5]. In this approach cross-linking was adjusted so that the cross-linked proteins are in a soluble state. This results in less steric hindrance and greater ease of substrate diffusion. This procedure is the first reported study of the use of glutaraldehyde to cross-link protein molecules to one another (intermolecular). It can also cross-link the protein internally (intramolecular). In 1973 Payne used an extension of this glutaraldehyde method to cross-link protein to form soluble molecular weight markers for separation studies [6].

Glutaraldehyde as Cross-Linker for Polyhemoglobin

The 1971 glutaraldehyde approach of Chang [5] was developed as pyridoxalated polyhemoglobin blood substitutes by Dudziak and Bonhard in 1976 [7], Moss *et al.* in 1980 [8], DeVenuto and Zegna in 1982 [9], Chang's group in 1982 [10], Stabilini *et al.* in 1983 [11], and Hobbhahn *et al.* in 1985 [12]. This type of polyhemoglobin requires the use of Benesch's pyridoxal phosphate to substitute for the 2,3-diphosphoglycerate [13]. The group of Gould and Moss [8,14] has developed this very extensively to its present refined state for clinical trials. The product from this group was effective when tested in clinical trials in acute trauma surgical patients [15]. Phase III clinical trials using up to 10,000 ml in acute trauma surgical patients are in progress. This approach has also been developed extensively by the Biopure group, using bovine hemoglobin to form polyhemoglobin [16,17]. These investigators have reported the effectiveness of the product in aortic surgery [18]. They are now in phase III clinical trials, using about 10,000 ml in surgical patients.

Other Cross-Linking Agents

Basic research has resulted in increasing understanding of the structural–functional relationship of hemoglobin. This has made it possible for many groups to carry out excellent in-depth studies to design other specific cross-linkers and modifiers. Many have investigated in depth the use of very specific cross-linking agents and modifiers. This has resulted in exciting improvements in the specificity of cross-linkers that have the dual function of modifying the hemoglobin to improve its oxygen release properties.

Walder *et al.* in 1979 reported the use of a 2,3-DPG pocket modifier, bis(3,5-dibromosalicyl)fumarate (DBBF) to intramolecularly cross-link the two α subunits of the hemoglobin molecule [19]. This prevents dimer formation and improves oxygen affinity. Estep and collaborators extensively developed this modification [20,21] for clinical trials. The clinical results, however, led Baxter to discontinue this approach and to develop another modified hemoglobin [21]. Hsia originated a dialdehyde prepared from open-ring sugars to form polyhemoglobin [22]. This has been developed and extended [23], and a phase III clinical trial in which polyhemoglobin was used in coronary bypass surgery has been completed.

Conjugated hemoglobin formed by cross-linking hemoglobin to soluble polymer is also in clinical trial [24,25].

RECOMBINANT HEMOGLOBIN

Another very exciting area is the use of recombinant technology to produce recombinant human hemoglobin from *E. coli* [26,27]. Clinical results showing vasoactivity led these investigators to develop a second-generation recombinant hemoglobin [28]. A second-generation recombinant human hemoglobin with the site for nitric oxide blocked by a tryptophan group has resulted in a preparation with no vasoactivity [29]. This is now being developed by Baxter.

LABORATORY PROCEDURE FOR THE PREPARATION OF POLYHEMOGLOBIN

Since glutaraldehyde-cross-linked polyhermoglobins are now in the most advanced stage of clinical trials, the following is a simplified laboratory method for using glutaraldehyde to prepare polyhemoglobin. This method, which incorporates most of the basic laboratory principles, can form the basis of extension into new approaches. It is also good for preparing basic polyhemoglobin for use in different areas of laboratory research. However, it is not for use in preclinical or clinical application. For this, other elaborated and detailed procedures more suitable for industrial scale-up are required. Details of tests for endotoxin, complement activation, contaminants, and sterility are available elsewhere [2,3].

PREPARATION OF STROMA-FREE HEMOGLOBIN SOLUTION

Stroma-free hemoglobin solution is the most important starting material and must be very carefully prepared. The easiest way is to buy the solution when it is available (e.g., from Biopure) in the form of purified bovine hemoglobin. Otherwise, it can be prepared in the laboratory as follows.

Blood (either outdated whole blood obtained from a blood bank or fresh animal blood) is centrifuged at 6000 rpm (4000g) for 20 min at 4°C. The plasma and the buffy coat containing the white blood cells and platelets are removed by aspiration. The sedimented red blood cells are washed four times as follows. In each washing, three times the red blood cells volume of ice-cold, sterile isotonic saline was added and mixed, then centrifuged to remove the saline. The red blood cells were then lysed by the addition of two volumes of hypotonic, 15 mls M phosphate buffer, pH 7.4, to one volume of packed cells. This is mixed by repeated inversion and swirling for 2–3 min. The suspension is allowed to stand for 20–30 min, then is placed into a large separation funnel and 0.5 volume of cold, reagent grade toluene is added. This is mixed by shaking vigorously for 5 min and then is left standing for 3 h at 4°C. The top layer, consisting of toluene with extracted stroma lipid and cellular debris, is aspirated. The solution is then centrifuged at 25,000g at 4°C for 1 h. This is followed by a second toluene extraction, but instead of standing for 3 h, the funnel is left overnight in a cold room or refrigerator at 4°C. The lower layer of hemoglobin solution is then separated and centrifuged at 25,000g at 4°C for 1 h.

Although not absolutely required, a crystallization step can be carried out to improve purity. Here, the hemoglobin is crystallized in a 2.8 M phosphate buffer as follows. The stroma-free hemoglobin is dialyzed (1:5 v/v) against 2.8 M potassium phosphate (pH 7.0) at 37°C for 5 h, then for 18 h against a fresh 2.8 M potassium phosphate buffer. The crystals from inside the dialysis fiber are redissolved for the next step.

The final step is to dialyze the stroma-free hemoglobin solution against physiological Ringer s lactate solution at 4°C. (An isotonic, buffered dialysate solution, pH 7.35, can also be prepared from NaCl, 5.85 g/liter; NaHCO$_3$, 3.61/liter; NaH$_2$PO$_4$·H$_2$O, 0.44 g/liter; KCl, 0.37 g/liter; CaCl$_2$·2H$_2$O, 0.22 g/liter; and MgCl$_2$·5H$_2$O 0.08 g/liter.) For a small volume, dialysis can be easily carried out by placing the solution in dialysis tubing, both ends of which carefully tied. The dialysis tubing is then suspended in the Ringer's lactate solution that is kept stirred with a magnetic stirrer. For larger volumes, a standard hollow-fiber dialyzer is required.

PYRIDOXYLATION OF HEMOGLOBIN

Pyridoxylation is needed only when human hemoglobin is used. This step is not needed with bovine hemoglobin, which has a much higher starting oxygen affinity value (P_{50}) Stroma-free hemoglobin is deoxygenated under continuous nitrogen bubbling for 1–2 h at 4°C. Pyridoxal 5′-phosphate (Sigma Chemical Co.) in Tris-HCl is added in a 4 : 1 molar ratio, then reduced with excess $NaBH_4$ under nitrogen for 18 h. Excess reagents are removed by dialysis against Ringer's lactate. An isotonic, buffered dialysate solution (pH 7.35) can also be prepared in the laboratory from NaCl, 5.85 g/liter; $NaHCO_3$, 3.61/liter; $NaH_2PO_4·H_2O$, 0.44 g/liter; KCl, 0.37 g/liter; $CaCl_2·2H_2O$, 0.22 g/liter; and $MgCl_2·5H_2O$, 0.08 g/liter.

PREPARATION OF POLYHEMOGLOBIN

In a cold room at 4°C, one adds 0.4 ml of 1.3 M lysine monohydrochloride in a 0.1 M phosphate buffer to 30 ml of pyridoxlated hemoglobin (10 g/dl). Next, 3 ml of ice-cold, degassed 0.25 M glutaraldehyde in 0.1 M phosphate buffer is added slowly. The reacting mixture is left rotating in a mixer in the cold room. Methemoglobin formation will be marked if reaction is not carried out at 4°C. The progress of the reaction can be monitored either by agarose gel chromatography columns (Sepharose 4B, 60,000–2 × 10⁶ Da and Sepharose 6B, 10,000–2 × 10⁶ Da or by a colloid osmometer. This gives information on the molecular distribution of the polyhemoglobin. A rough estimate of the reaction can be done by ultrafiltering a small sample in a small 100-Da cutoff centrifuge-tube-type ultrafiltrator. The amount of tetramers ultrafiltered would give a quick estimate of the degree of polymerization of the hemoglobin preparation. The molecular weight range of the polyhemoglobin can be controlled by adjusting the reaction time, the ratio of glutaraldehyde to hemoglobin, and other reaction parameters.

When the desired molecular weight range of polyhemoglobin is obtained, the cross-linking is quenched. This is done by adding 40 ml of the 1.3 M lysine monohydrochloride in a 0.1 M phosphate buffer. The solution is then centrifuged at $25,000g$ for 1 h. The supernatant is dialyzed at 4°C for 3 h against a Ringer's lactate solution or an isotonic, buffered dialysate solution (pH 7.35) as described earlier. Dialysis removes free glutaraldehyde and excess lysine. It also allows the electrolytes to be adjusted to physiological concentrations. Removal of unreacted reagents is very important. To allow for better a surface–volume relationship, important for adequate dialysis, only 10 ml is placed in each only partially filled section of tubing. Dialysis is completed when the electrolyte composition of the final product is the same as Ringer's lactate.

For better final purification (not needed for laboratory preparation), all hemoglobin preparations are passed through a composite artificial kidney consisting of a hollow-fiber dialysis unit in series with a 100-g column of collodion-coated activated charcoal.

The solution is ultrafiltered to the desired hemoglobin concentration. The solution is then filtered through a 0.45-μm sterilization filter unit. It is stored under sterile condition 4°C to slow the formation of methemoglobin.

CHARACTERIZATION

COMPOSITION

Hemoglobin preparations are modified to have electrolyte composition similar to that of plasma. It is also important for the oncotic pressure (colloid osmotic pressure) to be close to that of plasma. Most modified hemoglobin solutions are similar in electrolyte composition and oncotic pressures. However, there are variations in the types of other chemical added to the preparation to slow methemoglobin (MetHb) formation, to achieve antioxidant effects, and for other actions.

Table 86.1 gives as an example the composition of pyridoxylated polyhemoglobin prepared from the laboratory procedure just described. It must be remembered that this is a laboratory procedure, more complicated methods of optimization to improve P_{50} or to remove tetrmeric hemoglobin are not included.

Table 86.1. Composition of Pyridoxylated Polyhemoglobin Solution

Assay	Mean ± SD
Hemoglobin	13.5 ± 1.0 g/dl
MetHb	0.4–1.0 ± 0.12 g/dl
Na$^+$	145.0 ± 5.2 mequiv/liter
K$^+$	4.9 ± 0.3 mequiv/liter
Cl$^-$	113.6 ± 2.6 mequiv/liter
Ca^{2+}	11.2 ± 1.0 g/dl
Mg^{2+}	2.0 ± 0.2 g/dl
Phosphate	3.2 ± 0.1 g/dl
Osmolality	312.4 ± 9.9 mOsm/kg
pH (at 37°C)	7.37 ± 0.05

MOLECULAR WEIGHT DISTRIBUTION

Molecular weight distribution varies widely among the different types of cross-linked hemoglobin blood substitute. All the intramolecularly cross-linked hemoglobins are in the tetrameric form, with a molecular weight of about 68,000 Da. In polyhemoglobin, the molecular weights follow a distribution curve from tetrameric to much larger molecular weight. Some groups prefer to use a larger mean molecular weight. Other groups prefer a smaller mean molecular weight. Blood substitues with lower mean molecular weight also have more tetrameric hemoglobin. Those with higher mean molecular weight have less tetrameric hemoglobin because of the higher degree of polymerization. Chang [1,2] and Gould [14] have emphasized the need to remove as much tetrameric hemoglobin as possible. Thus, after the preparation of polyhemoglobin, Gould's group carried out another step to remove much of the tetrameric hemoglobin. This seems to have prevented vasoactivity even when large volumes (≤10,000 ml) were infused. The pyridoxylated polyhemoglobin prepared from the laboratory procedure described earlier can be used as an example. Even with this procedure, the molecular weight distribution can be varied at will by changing the reaction time, the ratio of glutaraldehyde to hemoglobin, and other reaction parameters.

For molecular weight distribution, pyridoxylated polyhemoglobin (PP-PolyHb) was run on Sephadex G-200 (1.6 × 70 cm column) in 0.1 M Tris-HCl (pH 7.5) at 12 ml/h. A selectivity curve (K_{av} vs log MW) generated from elution volumes of three protein standards (thyroglobulin, catalase, and aldolase from Pharmacia Fine Chemicals) was used to approximate molecular weights of the different fractions comprising the PP-PolyHb preparation. The K_{av} values for each peak enabled an approximation of the molecular weight of each peak. Percent contribution of these fractions to overall composition of the PP-PolyHb solution in Table 86.2 was estimated from the area under each peak. The fraction of polymerized hemoglobin was 93%, with 60% falling in the 130,000 to 350,000 Da range and 33% ranging in molecular weight from 3 50,000 to 750,000 DA. Only 7% was in the tetrameric form.

COLLOID PRESSURE OF POLYHEMOGLOBIN

Modified hemoglobin solutions are usually prepared to have the same oncotic pressure [colloid osmotic pressure(COP)] as plasma due to plasma proteins. This can be measured by using an oncometer (e.g., IL 186 Weil Oncometer calibrated with 5 g% albumin solution). For example, both pyridoxylated polyhemoglobin (PP-PolyHb) and stroma-free hemoglobin (SFHb) have been measured to ascertain COP at various Hb levels. To attain isooncotic pressures (COP 25 mmHg) SFHb had to be diluted to 7 g/dl, compared with cross-linked PP-PolyHb at 14 g/dl. Thus, the COP values for SFHb were at least double that of PP-PolyHb at similar concentrations.

Table 86.2. Molecular Weight Distribution of Pyridoxylated Polyhemoglobin

Peak	V_e (ml)	$K_{av}{}^a$	Molecular weight (Da)	Percent
1st	37.7	0.02	750,000	10
2nd	45.5	0.11	470,000	23
3rd	54.5	0.22	260,000	39
4th	65.4	0.35	130,000	21
5th	77.2	0.49	66,000	7

$^a K_{av} = (V_e - V_0)/(V_t - V_0)$, where V_e is elution volume, V_0 is void volume, and V_t is bed volume.

VISCOSITIES

Viscosity varies with the type of modified hemoglobin, the concentration of the modified hemoglobin and, for polyhemoglobin, the molecular size of the polyhemoglobin. This property can be measured in an Ubbelohde Viscometer (Canon Instrument Co.) equilibrated at 25 and 37°C. Multiplying kinematic viscosity times the measured density of each solution yielded values for dynamic viscosity (centipoises: cP). For the 14-g/dl pyridoxylated polyhemoglobin prepared with the foregoing molecular weight distribution, the intrinsic (dynamic) viscosity is 3.83 cP at 37°C, which is similar to that of whole blood. Lower molecular weight distribution would result in lower viscosity.

OXYGEN AFFINITY (P_{50})

Oxygen affinity is a measure of the ease of unloading oxygen to the tissue as the gas passes through the capillaries. Hemoglobin binds oxygen from the lung. As it reaches the tissues, hemoglobin releases oxygen to the tissue depending on the oxygen tension in the tissue. Oxygen affinity is a measurement of how easily oxygen can be released from hemoglobin when required by the tissues. Inside human red blood cells, the presence of 2,3-DPG results in hemoglobin being able to readily release oxygen when required. Pyridoxal phosphate (PLP) applied to polyhemoglobin can replace 2,3-DPG.

Oxygen dissociation curves are measured under standard conditions (pH 7.4, pCO_2 40 mmHg, at 37°C) by using tonometry, an IL 282 Hb Co-oximeter, and a Corning 175 pH/Blood Gas Analyzer. As discussed earlier, P_{50} measures the ease with which hemoglobin is able to unload oxygen at a given oxygen tension. P_{50} is the oxygen tension at which fully oxygen-saturated hemoglobin releases 50% of the oxygen it carries.

Whole blood in this case has a P_{50} of 24–26 mmHg. Stroma-free hemoglobin has a lower P_{50} (12–14 mmHg) because of the absence of 2,3-DPG. For pyridoxylated hemoglobin before cross-linking into polyhemoglobin, the P_{50} is 28 torr. Here, the pyridoxal 5′-phosphate, acting as a 2,3-DPG analogue, effectively increased the P_{50} of stroma-free hemoglobin. After polymerization, pyridoxylated polyhemoglobin has a P_{50} of only 20 mmHg. More complicated special reaction conditions can improve the P_{50} to 28 mmHg.

ANIMAL SAFETY STUDIES ARE NOT NECESSARILY VALID FOR HUMANS

Response in animals is not always the same as for human. This is especially true in tests for hypersensitivity, complement activation, and immunology. We have designed a screening test to bridge the gap between animal safety studies and use in humans based on *in vitro*, detection of complement activation [30–32]. Complement activation is important in many adverse reactions of humans to modified hemoglobin. Modified hemoglobin may be contaminated with trace amounts of blood group antigen that can form an antigen–antibody complex. This antigen–antibody complex can be detected by complement activation. Other potential materials can also cause complement activation. These include endotoxin, microorganisms insoluble immune complexes, chemicals, polymers and organic solvents.

In Vitro Screening Test Using Human Plasma

We have devised an *in vitro* test tube screening test using human plasma or blood [31–33]. The use of human plasma or blood gives the closest response to that of a person can be obtained without an actual injection. If we want to be more specific, we can use the plasma of the patient who is to receive the blood substitute. Many components of human blood or plasma can be used for this *in vitro* screening test. If one were to select only one test, perhaps the most useful one would be the effect of modified hemoglobin on complement activation (C3a) when added to human plasma or blood. This simple test consists of adding 0.1 ml of modified hemoglobin to a test tube containing 0.4 ml of human plasma or blood. The plasma is incubated for an hour and then analyzed for complement activation.

Laboratory Procedure Based on Human Plasma [31,21]

Obtained blood by clean venous puncture from human volunteers into 50-ml polypropylene (Sastedt) heparinized tubes (10 IU heparin per milliliter of blood). Immediately separate the plasma by centrifugation at $5500g$ at 2°C for 20 min and freeze the plasma in separate portions at −70°C. Do not use serum: coagulation initiates complement activation. EDTA should not be used as an anticoagulant because it interferes with complement activation.

Immediately before use, the plasma sample is thawed and 0.4 ml of the plasma are pipetted into 4-ml sterile polypropylene tubes (Fisher). Then 0.1 ml of pyrogen-free saline (or Ringer's lactate) for injection is added to the 0.4 ml of human plasma as a control. The amount of 0.1 ml of hemoglobin or modified hemoglobin is added to each of the other tubes containing 0.4 ml of human plasma. The reaction mixtures are incubated at 37°C at 60 rpm for 1 h in a Lab-Line Orbit Environ Shaker (Fisher Scientific, Montreal, Canada). After 1 h the reaction is quenched by adding 0.4 ml of this solution to a 2-ml EDTA sterile tube containing 1.6 ml of sterile saline. The samples are immediately stored at −70°C until analyzed.

The analytical kit for human complement C3a can be purchased from Amersham Canada. The method of analysis is the same as the instructions in the kit with two minor modifications. Centrifugation is carried out at $10,000g$ for 20 min. After the final step of inversion, the inside walls the tubes are carefully blotted with Q-Tips.

The exact procedure and precise timing described here are important in obtaining reproducible results. The baseline control level of C3a complement activation will vary with the source and procedure of obtaining the plasma. Therefore, a control baseline level must be used for each analysis. Furthermore, all control and test studies should be carried out in triplicate. Reproducibility must be established before this test can be used for scientific analysis, and much practice is needed to acquire the proper techniques.

Procedure Based on Blood from Finger Pricks [33]

Instead of using plasma and having to withdraw blood with a syringe, a simpler procedure involves obtaining blood from finger pricks. Sterile methods are used to prick a finger, and blood is collected in heparinized microhematocrit tubes. The tubes are kept at 4°C, then used immediately as follows. Each blood sample used in testing contains 80 µl of whole blood and 20 µl of saline. Each test solution is added to a blood sample. Test solutions include saline (negative control), Zymosan (positive control), or hemoglobin. This is incubated at 37°C at 60 rpm. After 1 h of incubation, EDTA solution is added to the sample to stop the reaction. The analysis for C3a is then carried out as described. The test kit is based on the ELISA C3a Enzyme Immunoassay from Quidel Co. (San Diego, CA).

Areas of Application

In research, development, or industrial production of blood substitute, different chemicals, reagents, and organic solvents, including cross-linkers, lipids, solvents, chemicals, and polymers. Some of these materials can potentially result in complement activation and other reactions in humans. Other potential sources of problems include trace contaminants from ultrafilters, dialyzers, and chromatography.

USE IN RESEARCH

Why wait for the completion of research, industrial production, and preclinical animal studies? Why not do this test right at the beginning, during the research stage? If a new system is found to cause complement activation at this stage, tremendous wastes of time and money—in further development, industrial production, and preclinical animal study—could be avoided [34].

In our ongoing study of hemoglobin nanocapsules, different polymers, lipids, reagents, and solvents are used. We therefore analyzed their effects on complement activation of human plasma *in vitro* [34]. One type of L-phosphatidylcholine caused marked increases in complement activation. Another type of L-phosphatidylcholine did not result in marked increases in complement activation. Some polymers tested, such as poly(actic acids) and ethylcellulose, did not result in this degree of complement activation. After repeated washing, there was no longer any significant complement activation. Another polymer, isobutyl 2-cyanoacrylate, resulted in a C3a level lower than that of the control. The reason for this is that the polymer does not cause complement activation, while it adsorbs C3a. The emulsifying agent Tween 20 also did not result in complement activation. Other groups are now starting to use this idea in research [35].

USE IN INDUSTRIAL PRODUCTION

We have also used this preclinical test to help others in industrial production [34]. Thus in one of the earlier industrial scale-ups of polyhemoglobin, the *in vitro* screening test showed that certain batches caused complement activation. Further use of this test showed that activation was the result of the use of new ultrafiltrators. Reused or washed ultrafiltrators did not cause complement activation. Without this test, some batches could have resulted in adverse effects of "unknown causes" in humans. Chromatography, ultrafiltrators, dialysis membranes, and other separation systems are used extensively in the preparation of different types of blood substitute. It is therefore important to screen for the possibility of trace contaminants that could cause complement activation. In the same way, different chemical agents and different reactants used in industrial production could be similarly tested.

CLINICAL TRIALS AND USE IN HUMANS

The *in vitro* test may be useful in large-scale screening for human response [34]. For instance, it could be used to study variation in production batches. It could also be used to study individual variations. Furthermore, it could also be used to analyze the response of different human populations, especially with different disease conditions. It is important to note that all these tests could be done without ever introducing any blood substitute into a person.

NEW GENERATIONS
POLYHEMOGLOBIN–SOD–CATALASE

Although first-generation blood substitutes in the form of modified hemoglobin have important potential for some clinical uses, they would not be applicable to all clinical conditions. In sustained ischemia in strokes, in myocardial infarction, and in severe and sustained hemorrhagic shock reperfusion with oxygen-carrying fluids can result in the release of oxygen radicals producing ischemia–reperfusion injuries [36,37]. The present first-generation modified hemoglobins are prepared from ultrapure hemoglobin and do not contain any red blood cell enzymes. We are studying the cross-linking of trace amounts of catalase and superoxide dismutase (SOD) to hemoglobin (Hb) to form PolyHb–SOD–catalase [36,37]. Our studies show that under conditions of sustained hypoxia, perfusion using this preparation significantly reduces the increase in oxygen radicals in comparison polyhemoglobin.

ARTIFICIAL RED BLOOD CELLS

Since modified hemoglobin in solution is not covered, it must be ultrapure to avoid adverse reactions. Furthermore, the circulation half-time is only 24 h. As a result, the original idea of a complete artificial red blood cell [1] is now being developed in the form of third-generation blood substitutes. The most recent developments include two approaches. One is based on the use of submicrometer lipid membrane microencapsulated hemoglobin [38–40]. This option is being explored especially more recently by the groups of Tsuchida in Japan [38] and Rudolph in the United State [39]. Modification of surface properties has resulted in a circulation half-time of about 50 h [40]. We are developing another approach based on our earlier preparation of biodegradable polymer encapsulated biologically active materials [41] and combining it with nanotechnology. The result is biodegradable polymeric hemoglobin nanocapsules having a diameter of about 150 nm [42–44]. Unlike lipid, polylactide and other biodegradable polymer membranes can be readily converted to water and carbon dioxide after use and therefore the polymer membrane does not accumulate in the body. We have included superoxide dismutase, catalase, and carbonic anhydrase, and in addition, multienzyme systems in the nanocapsules [43,44]. Having optimized the *in vitro* characteristics, we are at present successfully developing and optimizing their *in vivo* characteristics, especially for increase in circulation tissue. The following methods are those we have used in optimizing the *in vitro* characteristics.

Materials

Poly(lactic acid) (PLA) was obtained from Polysciences Inc. Isobutyl 2-cyanoacrylate (IBCA), the surfactants Tween 20, Span 85, Triton X-100, and Pluronic F68, ethyl cellulose, and phosphatidylcholine (hydrogenated) and other phospholipids, such as distearoyl phosphatidylcholine (DSPC) or distearoyl phosphatidylglycerol (DSPG) and tocopherol acetate were obtained from Sigma Chemical Co. Dialysis membrane (Spectrapor 5) was purchased from Fisher Scientific Co. Diethyl ether, ethyl acetate, cyclohexane, and chloroform were purchased from BDH Chemical (Canada). All the other chemicals were of reagent grade.

Preparation of Hemoglobin Solution

Stroma-free hemoglobin was prepared as described earlier in connection with polyhemoglobin.

Preparation of Biodegradable Polymer Membrane Nanocapsules Containing Hemoglobin and Enzymes [44]

Preparation with D,L-PLA

First 100–150 mg of PLA and 50–100 mg of phospholipid and 5 mg of tocopherol are dissolved in a mixed solution of ethanol (2–5 ml) and acetone (5–10 ml). This solution is slowly injected into 10–30 ml of 0.5–10% hemoglobin solution containing 0.1–1.0%. Tween 20 under constant magnetic stirring. Diffusion of ethanol and acetone into the aqueous phase results in particle formation. The ethanol and acetone in the aqueous phase can be easily eliminated by dialysis against saline solution or phosphate buffer at 4°C. The hemoglobin can be separated from polyhemoglobin by high-speed centrifugation (30, 000g) or by ultrafiltration.

Preparation with Isobutyl Cyanoacrylate (IBCA)

First 50–100 mg of IBCA, 35–50 mg of phospholipid, and 5 mg of tocopherol are dissolved in 5–10 ml of ethanol. This solution is slowly injected into 10–25 ml of 0.5–10% hemoglobin solution containing 0.1–1.0%. Tween 20 under constant magnetic stirring. The polymer membrane particles containing hemoglobin are simultaneously formed. The ethanol is eliminated by dialysis against phosphate buffer (pH 7.4) or saline solution at 4°C. The particles may be separated from hemoglobin by high-speed centrifugation or ultrafiltration.

Preparation with L-Polylactide (L-PLA)

First 150 mg of PLA and 50 mg of phosphatidylcholine are dissolved in 10 ml of chloroform containing mixed surfactants. Then 0.5 ml of stroma-free hemoglobin solution is emulsified in the solution and 25 ml of diethyl ether is injected into the emulsion under magnetic stirring. The biodegradable polymer membrane nanocapsules containing hemoglobin are simultaneously formed by interfacial polymer deposition. The particles are solidified by adding cyclohexane. The solidified particles are then separated by centrifugation and suspended in saline containing surfactant. The mixed suspensions were dialyzed against phosphate buffer (pH 7.4) at 4°C.

Preparation with D,L-PLA and Ethyl Cellulose

First 100 mg of PLA, 50 mg of ethyl cellulose, and 50 mg of phosphatidylcholine are dissolved in 10 ml of ethyl acetate containing mixed surfactants. Then 0.5 ml of hemoglobin solution is emulsified in this solution and 25 ml of cyclohexane is injected into the ethyl acetate–hemoglobin emulsion under magnetic stirring. The biodegradable polymer membrane containing hemoglobin is simultaneously formed by interfacial polymer deposition. The biodegradable polymer membrane containing hemoglobin is solidified by adding another 100 ml of cyclohexane. The solidified particles are then separated by centrifugation and suspended in saline-containing surfactant. The suspensions are dialyzed against phosphate buffer (pH 7.4) at 4°C.

Preparation with Poly(lactide-co-glycolide)

First 100 mg of poly(actide-*co*-glycolide), 10–50 mg of phospholipid, and 5 mg of tocopherol are dissolved in a 5–10 ml of acetone with the help of water bath sonication. This solution is slowly injected into 10–30 ml of 0.5–10% hemoglobin solution containing 0.1–1.0% Tween 20 under constant magnetic stirring. The polymer membrane particles containing hemoglobin are formed after the diffusion of acetone into the aqueous phase. The acetone can be removed by dialysis against saline solution or phosphate buffer (pH 7.4) or saline solution at 4°C. The particles can be separated by high-speed centrifugation or ultrafiltration.

REFERENCES

1. Chang, T. M. S. (1964). Semipermeable microcapsules. *Science* **146**, 524.
2. Chang, T. M. S. (1997). "Blood Substitutes: Principles, Methods, Products and Clinical Trials," Vol. 1 (monogr.) Karger, Basel/R.G. Landes, Austin, TX.
3. Chang, T. M. S., ed., (1998). "Blood Substitutes: Principles, Methods, Products and Clinical Trials," Vol. 2. Karger, Basel.
4. Chang, T. M. S. (1972). "Artificial Cellls" (monogr.) Thomas, Springfield, IL.
5. Chang, T. M. S. (1971). Stabilization of enzyme by microencapsulation with a concentrated protein solution or by crosslinking with glutaraldehyde. *Biochem. Biophys. Res. Commun.* **44**, 1531–1533.
6. Payne, J. W. (1973). Glutaraldehyde crosslinked protein to form soluble molecular weight markers. *Biochem. J.* **135**, 866–873.
7. Dudziak, R., and Bonhard, K. (1980). The development of hemoglobin preparations for various indications. *Anesthesist* **29**, 181.
8. Sehgal, I. R., Rosen, A. L., Gould, S. A., Sehgal, H. L., Dalton, L., Mayoral, J., and Moss, G. S. (1980). In vitro and in vivo characteristics of polymerized pyridoxalated hemoglobin solution. *Fed. Proc., Fed. Am. Soc. Exp. Biol.* **39**, 2383.
9. DeVenuto, F., and Zegna, A. I. (1982). Blood exchange with pyridoxalated-polymerized hemoglobin. *Surg., Gynecol. Obstet.* **155**, 342.
10. Keipert, P. E., Minkowitz, J., and Chang, T. M. S. (1982). Cross-linked stroma-free polyhemoglobin as a potential blood substitute. *Int. J. Artif. Organs* **5**(6), 383.
11. Stabilini, R., Palazzini, G., Pietta, G. P., Pace, M., Calatroni, A., Raffaldoni, E., Ghessi, A., Aguggini, G., and Agoston, A. (1983). A pyridoxalated polymerized hemoglobin solution as oxygen carrying substitute. *Int. J. Artif. Organs* **6**, 319.
12. Hobbhahn, J., Vogel, H., Kothe, N., Brendel, W., Keipert, P., and Jesch, F. (1985). Hemodynamics and transport after partial and total blood exchange with pyridoxalated polyhemoglobin in dogs. *Acta Anaesth. Scand.* **29**, 537–543.
13. Benesch, R., Benesch, R. E., Yung, S., and Edalji, R. (1975). Hemoglobin covalently bridged across the polyphosphate binding site. *Biochem. Biophys. Res. Commun.* **63**, 1123.
14. Gould, S. A., Sehgal, L. R., Sehgal, H. L., DeWoskin, R., and Moss, G. S. (1998). The clinical development of human polymerized hemoglobin. *In* "Blood Substitutes: Principles, Methods, Products and Clinical Trials" (T. M. S. Chang, ed.), Vol. 2, pp. 12–28. Karger, Basel/R.G. Landes, Austin, TX.
15. Gould, S. A., Moore, E. E., Hoyt, D. B., Burch, J. M., Haenel, J. B., Garcia, J., DeWoskin, R., and Moss, G. S. (1998). The first randomized trial of human polymerized hemoglobin as a blood substitute in acute trauma and emergent surgery. *J. Am. Coll. Surg.* **187**, 113–120.
16. Hughes, G. S., Francom, S. F., Antal, E. J. *et al.* (1995). Hematologic effects of a novel hemoglobin-based oxygen carrier in normal male and female subjects. *J. Lab. Clin. Med.* **126**, 444–451.
17. Pearce, L. B., and Gawryl, M. S. (1998). Overview of preclinical and clinical efficacy of Biopure's HBOCs. *In* "Blood Substitutes: Principles, Methods, Products and Clinical Trials" (T. M. S. Chang, ed.), Vol. 2, pp. 82–98. Karger, Basel/R.G. Landes, Austin, TX.

18. LaMuraglia, G. M., *et al.* (2000). The reduction of the allogenic transfusion requirement in aortic surgery with a hemoglobin-based solution. *J. Vasc. Surg.* **31**(2), 299–308.

19. Walder, J. A., Zaugg, R. H., Walder, R. Y., Steele, J. M., and Klotz, I. M. (1979). Diaspirins that cross-link alpha chains of hemoglobin: Bis(3,5-dibromosalicyl) succinate and bis(3,5-dibormosalicyl)fumarate. *Biochemistry* **18**, 4265–4270.

20. Estep, T. N., Gonder, J., Bornstein, I., Young, S., and Johnson, R. C. (1992). Immunogenicity of diaspirin crosslinked hemoglobin solutions. *Biomater. Artif. Cells, Immobil. Biotechnol.* **20**, 603–610.

21. Nelson, D. J. (1998). Blood and HemAssistTM (DCLHb): Potentially a complementary therapeutic team. *In* "Blood Substitutes: Principles, Methods, Products and Clinical Trials" (T. M. S. Chang, ed.), Vol. 2, pp. 39–57. Karger, Basel/R.G. Landes, Austin, TX.

22. Hsia, J. C. (1991). *o*-Raffinose polymerized hemoglobin as red blood cell substitute. *Biomater. Artif. Cells, Immobil. Biotechnol.* **19**, 402.

23. Adamson, J. G., Moore, C., and Hemolink, T. M. (1998). An *o*-raffinose crosslinked hemoglobin-based oxygen carrier. *In* "Blood Substitutes: Principles, Methods, Products and Clinical Trials" (T. M. S. Chang, ed.), Vol. 2, pp. 62–79. Karger, Basel/R.G. London, Austin, TX.

24. Shorr, R. G., Viau, A. T., and Abuchowski, A. (1996). Phase 1B safety evaluation of PEG hemoglobin as an adjuvant to radiation therapy in human cancer patients. *Artif. Cells, Blood Substitutes, Immobil.. Biotechnology, Int. J.* **24** (abstr. issue), 407.

25. Privalle, C., *et al.* (2000). Pyridoxated hemaglobin polyoxyethylene: A nitric oxide scavenger. *Free Radical Biol. Med.* **28**, 1507–1517.

26. Hoffman, S. J., Looker, D. L., Roehrich, J. M. *et al.* (1990). Expression of fully functional tetrameric human hemoglobin in *Escherichia coli. Proc. Natl. Acad. Sci. U.S.A.* **87**, 8521–8525.

27. Looker, D., Durfee, S., Shoemaker, S., Mathews, A., Kiyoshi, N. and Stetler, G. (1991). Production of recombinant hemoglobin specifically engineered to enhance delivery and circulating half-life: A recombinant cell-free blood substitute. *Biomater. Artif. Cells, Immobil. Biotechnol.* **19**(2), 418.

28. Freytag, J. W., and Templeton, D. (1997). Optro TM (recombinant human hemoglobin): A therapeutic for the delivery of oxygen and the restoration of blood volume in the treatment of acute blood loss in trauma and surgery. *In* "Red Cell Substitutes; Basic Principles and Clinical Application" (A. S. Rudolph, R. Rabinovici, and G. Z. Feuerstein, eds.), pp. 325–334. Dekker, New York.

29. Doherty, D. H., Doyle, M. P., Curry, S. R., Vali, R. L., Fattor, T. J., Olson, J. S., and Lemon, D. D. (1998). Rate of reaction with nitric oxide determines the hypertensive effect of cell-free hemoglobin. *Nat. Biotechnol.* **16**, 672–676.

30. Chang, T. M. S. (1997). Recent and future developments in modified hemoglobin and microencapsulated hemoglobin as red blood cell substitutes. *Artif. Cells, Blood Substitutes, Immobil. Biotechnol. Int. J.* **25**, 1–24.

31. Chang, T. M. S., and Lister, C. (1990). A screening test of modified hemoglobin blood substitute before clinical use in patients–based on complement activation of human plasma. *Biomat. Artif. Cells, Organs* **18**(5), 693–702.

32. Chang, T. M. S., and Lister, C. W., (1993). Screening test for modified hemoglobin blood substitute before use in human. U.S. Pat. 5,200,323.

33. Chang, T. M. S., and Lister, C. W. (1993). Use of finger-prick human blood samples as a more convenient way for in vitro screening of modified hemoglobin blood substitutes for complement activation: A preliminary report. *Biomater. Artif. Cells, Immobil. Biotechnol.* **21**, 685–690.

34. Chang, T. M. S., and Lister, C. W. (1994). Assessment of blood substitutes: II. In vitro complement activation of human plasma and blood for safety studies in research, development, industrial production and preclinical analysis. *Artif. Cells, Blood Substitutes, Immobil. Biotechnol. Int. J.* **22**, 159–170.

35. Szebeni, J., Wassef, N. M., Rudolph, A.S., and Alving, C. R. (1996). Complement activation in human serum by liposome-encapsulated hemoglobin: The role of natural anti-phospholipid antibodies. *Biochim. Biophys. Acta* **1285** 127–130.

36. D'Agnillo, F., and Chang, T. M. S. (1998). Polyhemoglobin–superoxide dismutase—catalase as a blood substitute with antioxidant properties. *Nat. Biotechnol.* **16**(7), 667–671.

37. Razack, S., D'Agnillo, F., and Chang, T. M. S. (1997). Crosslinked hemoglobin–superoxide dismutase–catalase scavenges free radicals in a rat model of intestinal ischemia–reperfusion injury. *Artif. Cells, Blood Substitutes, Immobil. Biotechnol. Int. J.* **25**, 181–192.

38. Tsuchida, E., ed. (1998). "Blood Substitutes: Present and Future Perspectives." Elsevier, Amsterdam.

39. Rudolph, A. S., Rabinovici, R., and Feuerstein, G. Z., eds. (1997). "RBC Substitutes." New York.

40. Philips, W. T., Klpper, R. W., Awasthi, V. D., Rudolph, A. S., Cliff, R., Kwasiborski, V. V., and Goins, B. A. (1999). Polyethylene glyco-modified liposome-encapsulated hemoglobin: A long circulating red cell substitute. *J. Pharmacol. Exp. Ther.* **288**, 665–670.

41. Chang, T. M. S. (1976). Biodegradable semipermeable microcapsules containing enzymes, hormones, vaccines, and other biologicals. *J. Bioeng.* **1**, 25–32.

42. Yu, W. P., and Chang, T. M. S. (1996). Submicron polymer membrane hemoglobin nanocapsules as potential blood substitutes: Preparation and characterization. *Artif. Cells, Blood Substitutes, Immobil. Biotechnol. Int. J.* **24**, 169–184.

43. Chang, T. M. S., and Yu, W. P. (1998). Nanoencapsulation of hemoglobin and rbc enzymes based on nanotechnology and biodegradable polymer. *In* "Blood Substitutes: Principles, Methods, Products and Clinical Trials" (T. M. S. Chang, ed.), Vol. 2, pp. 216–231. Karger, Basel.

44. Chang, T. M. S., and Yu, W. P. (1997). Biodegradable polymer membrane containing hemoglobin for blood substitutes. U.S. Pat. 5,670,173.

CHAPTER 87

LIVER

Kohei Ogawa and Joseph P. Vacanti

Orthotopic liver transplantation is currently the only established successful treatment for end-stage liver disease. However, the difficulties in applying this procedure due to organ shortages, technical complexity, and cost have increased interest in hepatocyte transplantation as an alternative treatment. Tissue-engineered liver represents an improvement over the traditional approach to hepatocyte transplantation. Because the liver is one of the most sophisticated organs in the human body, tissue engineering liver is complex and difficult. However, this technique is very attractive because it has the promise of providing a permanent treatment for end-stage liver and enzyme deficiency diseases, thus becoming a viable alternative for patients. This chapter discusses the basic methods of engineering liver tissue.

INTRODUCTION

For those suffering from end-stage or metabolic liver diseases, transplantation is often the only hope for survival and for leading a normal life. However, such treatments are not ideal, and severe donor organ shortages and the technical complexity and high costs of the procedure can make them a source of frustration to doctors and patient alike. In fact, despite major advances in the fields of immunology and transplantation, 26,000 people die of end-stage liver disease each year in the United States, costing an estimated annual value of $9 billion [1]. Even following successful procedures, debilitating immunosuppressive drugs are often used for the rest of a patient's life to control allograft rejection.

These difficulties with orthotopic liver transplantation have increased interest and research into alternative treatments such as hepatocyte transplantation, including liver tissue engineering and bioartificial liver-assist devices. Artificial assist devices are intended to be only temporary supplements to the patient's own liver until a donor organ becomes available for transplant. On the other hand, by transplanting liver cells separately, especially hepatocytes (liver parenchymal cells that perform most of the organ functions and make up 80% of cytoplasmic mass), there is the potential for long-term treatment. Hepatocyte transplantation can utilize cells from one donor for multiple recipients, thus decreasing the risk and expense associated with major surgical procedures; if autologous cells can be used, the need for immunosuppression is obviated. In addition, there is the capacity for liver-directed gene therapy to treat inborn errors of metabolism due to single gene defects [2].

A variety of approaches have been used to transplant hepatocytes into recipient animals, including injection into the spleen or liver directly or through the portal vein [3,4]. In addition, liver cells have been encapsulated in biocompatible membranes to prevent immune reactions, or attached to microcarrier beads, and injected into the abdominal cavity [5–7]. However, these approaches have met with only limited success because only a small number of cells can be transplanted, and such cells are not able to keep their liver-specific function over long periods of time.

Tissue-engineering of liver represents an improvement over the traditional approaches to hepatocyte transplantation [8,9]. Polymer scaffolds have a large surface area and the potential to accommodate large numbers of hepatocytes. Moreover, by using biodegradable polymers, only the newly created tissue will remain following polymer resorption. The ul-

Methods of Tissue Engineering

timate goal of this approach is to create implantable liver tissue that can provide long-term hepatic support.

ENGINEERING LIVER TISSUES

The liver has an extremely complicated structure because it is a highly vascularized organ, comprising multiple cell types. The highly metabolic hepatocytes require rapid access to oxygen and nutrition supply after transplantation. Many critical functions are performed by the liver, which means that transplanted cells must keep a high degree of liver-specific function. These elements have made liver tissue engineering complicated and difficult.

Engineering liver tissue has some fundamental steps. First, hepatocytes are seeded onto a polymer scaffold. After seeding, the cell–polymer construct is cultured for varying periods of time before being implanted *in vivo*. Some type of hepatotrophic stimulation usually accompanies this implantation, and the transplanted hepatocytes can be analyzed both histologically and functionally at progressive time points (Fig. 87.1).

TISSUE ENGINEERING MATRICES

Hepatocytes are anchorage-dependent cells and require an insoluble extracellular matrix for survival, reorganization, proliferation, and function. These extracellular matrices not only provide a surface for cell adhesion, but also have profound influences on modulating cell shape and gene expression related to cell growth and liver-specific function [10]. While other cell transplantation techniques such as hepatocyte injection utilize existing stromal tissue such as the spleen or liver as a vascular extracellular matrix, the space required to accommodate the transplanted hepatocytes is limited. A variety of synthetic polymers, both degradable and nondegradable, have been utilized to fabricate tissue-engineered matrices [9,11]. These polymer scaffolds serve as a template to guide cell organization and

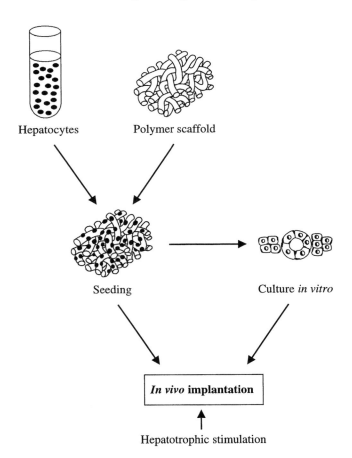

Fig. 87.1. One basic method of engineering liver tissue.

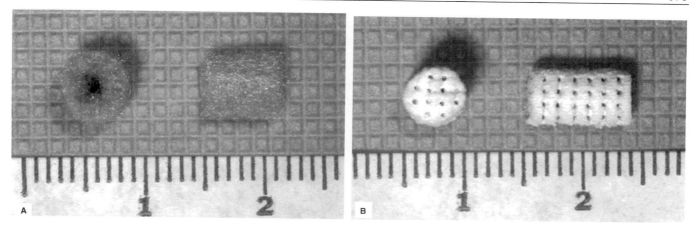

Fig. 87.2. (A) PGA polymer tube. (B) Three-dimensional printed (3DP) polymer.

growth. The microporous structure allows diffusion of oxygen and nutrients to and removal of waste from the implanted cells, and also provides a space for neovascularization from surrounding tissue and hepatocyte reorganization. Owing to the high metabolic demands of hepatocytes, the polymer devices need to be at least 95% porous to allow vascular ingrowth and exchange of nutrients and wastes [8,12]. Among the matrices used to study prevascularization into the polymer devices for cell attachment *in vivo* have been poly(vinyl alcohol) (PVA) sponges [13]. However, these nondegradable sponges can act as a nidus for infection and chronic inflammation; thus poly(glycolic acid) (PGA) and poly(L-lactic acid) (PLLA), which are biodegradable, are mainly used at present in this field (Fig. 87.2A) [14, 15]. Biodegradable polymers are especially attractive for applications in tissue engineering because after polymer erosion, natural tissue will result. Different extracellular matrices can be incorporated or coated on these polymer scaffolds before cell seeding to enhance cell attachment, function, and reorganization. In the late 1990s complex, three-dimensional, synthetic, biodegradable polymer scaffolds were developed by means of a technique called three-dimensional printing (3DP) [16]. These polymers have large surface areas for cell attachment and an intrinsic network of interconnected vascular channels. This technique creates scaffolds of any size or shape with a high degree of macroarchitectual and microarchitectural complexity (Fig. 87.2B).

CELLS

Hepatocytes

Differentiated hepatocytes are difficult to culture *in vitro* because they do not proliferate well and quickly lose viability. Thus, freshly isolated cells are generally harvested to seed large numbers of viable hepatocytes onto the polymer scaffold. Techniques for harvesting hepatocytes from the rat liver have been described by Berry and Friend [17] and Seglen [18], who uses a two-step collagenase digestion. The parenchymal hepatocyte fraction is isolated from the nonparenchymal cell fraction after low-speed centrifuging, and cell suspension for seeding is prepared. This enzymatic digestion technique has also been applied to other animals [19–21]. The concentration of the hepatocyte suspension can be varied depending on the need.

Recently, there have been numerous reports of the existence of highly proliferative hepatocytes in the nonparenchymal cell fraction of rat livers [22,61,62]. Since these hepatocytes are smaller than those in the parenchymal cell fraction, they are called small hepatocytes (SHCs) [22]. Using this type of cell might offer a possibility of better engraftment and proliferation on the polymer scaffold than is obtainable with mature hepatocytes.

Cocultures with Other Cell Types

It is well known that hepatocytes are highly differentiated, and while capable of repeatedly regenerating *in vivo*, they typically differentiate rapidly and do not grow in monolayer culture. In addition, they express liver-specific function only for relatively short periods, and some functions are hardly ever expressed at all. Many methods for improving and overcoming these disadvantages have been reported, including the addition of various extracellular matrix (ECM) components, specific soluble factors (dimethyl sulfoxide, nicotinamide, or phenobarbital) and cocultures with various other cell types [23,24]. These studies indicate that to retain their liver-specific functions, hepatocytes in culture should be maintained in contact with either ECM or homo- or heterotypic cells.

Hepatocytes cocultured with biliary epithelial cells have shown ductular formation *in vitro* and *in vivo* [25], and vestiges of ductular formation in hepatocyte–polymer constructs have been histologically observed [26]. Cocultures of hepatocytes with a nonparenchymal liver cell lines that have endothelial characteristics demonstrated ECM production between the two cell types [27]. In addition, cocultures of hepatocytes and pancreatic islet cells have been studied to aid in hepatocyte survival, growth, and maintenance. The pancreas is an important source of hepatotrophic stimuli found in portal venous circulation [28], and the cotransplantation of hepatocytes with islets of Langerhans as free grafts effectively delivered hepatotrophic stimuli that improved the growth and function of transplanted hepatocytes [29,30]. A similar series of cotransplantation experiments upon polymer matrices showed improvements in hepatocyte survival [31].

CELL SEEDING ON A POLYMER SCAFFOLD

To successfully replace liver function, an estimated 10–20% of the liver mass needs to be implanted [32]. Thus a large number of hepatocytes must be seeded onto the polymer and delivered into the body. A variety of seeding methods have been utilized, the two most significant being static and dynamic seeding. Static seeding uses a pipette and capillary action to deliver cells into the polymer [16,33], but dynamic seeding is generally used to create more uniform cell attachment by moving cells through and within the spaces of the polymer. Studies have shown that various types of cell used in tissue engineering, including chondrocytes, do attach more uniformly when dynamic seeding techniques are used [34], and preliminary studies have also demonstrated this to be true of hepatocytes [35]. Some methods include flow through the polymer with a peristaltic pump (Fig. 87.3A), or using a spinner flask culture system, where the polymer is hung in a cell suspension that is stirred at 50 rpm at 37°C with 5% CO_2 (Figs. 87.3B and 87.4).

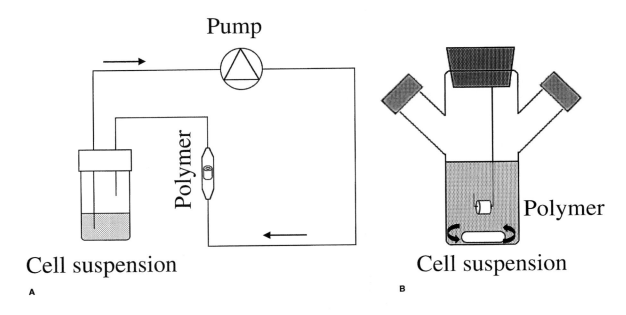

Fig. 87.3. The schema of seeding methods: (A) flow seeding and (B) spinner flask seeding.

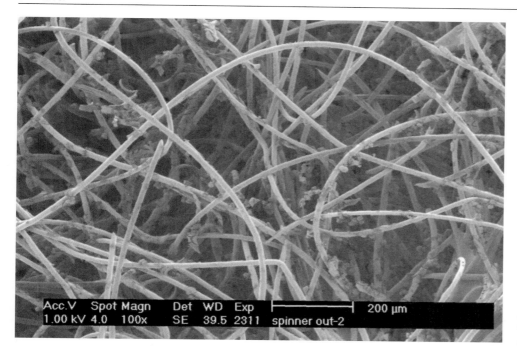

Fig. 87.4. Scanning electron micrograph of cells seeded onto a PGA tube with a spinner flask.

IMPLANTATION OF CELL–POLYMER CONSTRUCTS AND VASCULARIZATION INTO THE POLYMER SCAFFOLD

After cell seeding, the polymer–cell constructs can be implanted *in vivo*. The site of implantation must be rich in microvasculature and must have wide spaces. Omentum, small intestinal mesentery, retroperitoneal space, and subcutaneous tissue have all been used, but early studies have shown that in many of these sites, only a small percentage of cells are viable one week after implantation [26]. Neovascularization usually occurs only in the surface area of the polymer, close to the host tissue, shortly after implantation, and many cells die within a few days. The limitations of oxygen–nutrient diffusion and waste exchange are critical factors for the viability of these highly metabolically active cells during the initial post-transplantation period [16].

To address this issue, various approaches to the induction of vascularization into the polymer scaffold have been investigated. One approach was to create a time lag of 5–7 days between polymer implantation and hepatocyte introduction [13]. The polymer devices were preimplanted into the recipient animals, but while the devices were well vascularized, the potential space for introduction of cells was essentially eliminated. Prevascularization of these devices before cell transplantation resulted in a considerable improvement in cell survival over the first day or two. Thompson *et al.* reported that they could achieve the formation of organoid neovascular structures when angiogenic growth factors were incorporated into the devices to promote neovascularization of the implanted structure [36]. In 1999 Kim *et al.* created an animal model for the implantation of hepatocytes seeded onto biodegradable polymer tubes directly into the portal circulation, using small intestinal submucosa as a small-caliber venous graft between the portal vein and the inferior vena cava. Preliminary experiments demonstrated that hepatocytes survived in portal blood 2 days after implantation [37].

HEPATOTROPHIC STIMULATION

Upon clinical damage or after surgical resection, different factors that regulate liver regeneration, such as hepatocyte growth factor (HGF), become recognized. In general, hepatocytes in a healthy liver have a tremendous capacity to proliferate following hepatotrophic stimulation. However, it is difficult to provoke transplanted hepatocytes to grow and maintain their liver-specific function without stimulation *in vivo*. This is because they need certain growth factors such as HGF to proliferate and can easily be defeated in competition with native liver tissue. To address this issue, special efforts have been made to provide hepatotropic stimulation only to transplanted hepatocytes. Incompletely characterized trophic stimuli in the portal venous circulation are known to contribute to liver regeneration [38, 39]. For this reason the porta-caval shunt, a surgical technique developed to drain portal venous flow into systemic circulation, has been used to prevent first-pass clearance of hepatotrophic stimuli by the native liver and to provide transplanted hepatocytes with increased access to these stimuli [13,14]. In addition, a partial hepatectomy to remove a portion of the native liver can also be used to provide a proliferative stimulus [40–42]. Use of a partial hepatectomy alone is not effective however, since only one or two rounds of cell division are needed to restore the liver mass to its original size [43,44], and both native liver and transplanted hepatocytes receive this stimulus. Therefore, a combination of a porta-caval shunt and partial hepatectomy is a more effective method of hepatotrophic stimulation [13,14].

Some chemicals such as carbon tetrachloride and D-galactosamine are known to induce liver injury [2,9] and can be used to inhibit the proliferation of native liver cells following a partial hepatectomy [45,46]. Another drug, called retrorsine, is known to inhibit the regeneration of the normal liver by producing a block in the hepatocyte cell cycle, causing an accumulation of cells in late S and/or G2 phase [47]. Retrorsine treatment before hepatocyte transplantation may effectively inhibit the proliferation of the native liver, thus helping the proliferation of transplanted hepatocytes. In 1998 an animal model of hepatocyte transplantation incorporating a combination of retrorsine treatments and a partial hepatectomy was reported with successful results [48].

Besides using endogenous hepatotrophic stimulating factors, the administration of exogenous stimulating factors, such as HGF, hepatic stimulatory substance (HSS), and heparin-binding epidermal growth factor–like growth factor (HB-EGF), have been reported to be effective for cell proliferation after transplantation [2,46,47].

IN VITRO EXPERIMENTS

To create organized and vascularized liver tissue *in vitro* before *in vivo* implantation, cell–polymer constructs can be cultured *in vitro* after cell seeding. Culture systems used to grow the engineered tissues are called bioreactors and generally provide more efficient O_2 delivery and CO_2 removal, a more physiological pH, and an environment more favorable for cell metabolism and function than static conditions alone. For example, one continuous flow system circulates growth medium through the construct, as illustrated in Fig. 87.5. The medium is pumped from the reservoir, through the oxygenation tube, air trap, and cell–polymer housing units, and recirculated back to the reservoir. The entire unit is maintained at 37°C with 5% CO_2. Flow rates for this system have been described by Kim *et al.* [16],

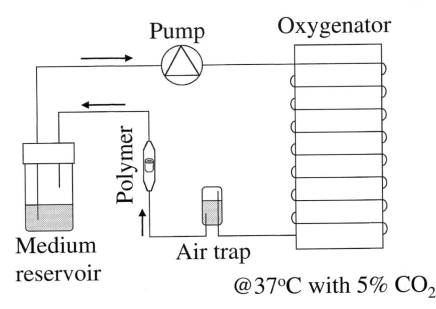

Fig. 87.5. The schema of a flow condition culture system.

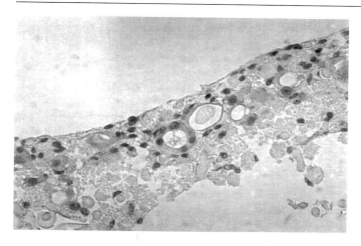

Fig. 87.6. Histological appearance of new tissue within the polymer scaffold after 3 weeks of culture in a bioreactor. Original magnification ×400.

and the volume oxygen consumption rate (QO_2) was approximately $2-6 \times 10^{-5}$ mmol/cm^3 cell mass-second for metabolically active cells. The polymer scaffold contained less than 20% by volume cells after cell seeding, and the scaffold void volume was 0.11 cm^3. To maintain cell viability, the specified oxygen concentration should be above 50% from the inflow to the outflow of the cell–polymer housing unit. Thus, the flow rate F needed to maintain the specified oxygen concentration after cell seeding can be estimated by:

$$F = \frac{0.2 \times V \times QO_2}{0.5 \times CO_2 \text{ inflow}}$$

for a given value of inflow oxygen concentration (CO_2 inflow). From this formula, the flow rate needed for the survival of the initial cell mass is estimated to be at most 1.1 ml/min.

This culture system is also useful for assaying the cell function and proliferative capacity within the polymer scaffold. In fact, hepatocytes–polymer constructs in a flow condition culture system synthesized significantly higher levels of albumin than those in static conditions after 2 days [16]. This culture system is now being used to study the effects of cocultured nonparenchymal cells and SHCs from adult rats. Recently, there has been success this bioreactor system has been used successfully to culture SHCs for up to 3 weeks while maintaining high levels of albumin production. In addition, it has been found that these constructs formed new tissue that included tubular or cystic formation in the polymer scaffold (Fig. 87.6) [unpublished data].

FUTURE DIRECTIONS

Studies in tissue engineering have made great progress in recent years. For further advances, however, especially with tissue engineering liver, there are two major obstacles to overcome. First, neither hepatocytes in culture nor hepatocytes transplanted *in vivo* have a high proliferative potential while maintaining their liver-specific function as native liver in an appropriate environment. Although a variety of methods (coculturing, provision of hepatotrophic stimulation, and improvement of extracellular matrices including polymer scaffolds) have been investigated, these techniques have not been able to cause hepatocyte proliferation and maintenance of specific liver functions. One possible solution may be the use of highly proliferative cells such as liver progenitor cells or liver stem cells instead of mature hepatocytes. Oval cells are known to appear during hepatocarcinogenesis and chemically induced severe injury [25,49–52]. These cells are considered to be a potential candidate for progenitor cells because they can differentiate into both hepatocytes and biliary epithelial cells, and because they appear to have a higher replicative potential than mature hepatocytes. Small hepatocytes, which are considered to be "committed progenitor cells" because they can further differentiate into mature hepatocytes [53], also have this proliferative potential. Special efforts have been made to detect the liver stem cells in many institutes [54–57]. These cell types may prove to be better candidates for transplantation than mature hepatocytes.

A second obstacle that needs to be addressed is the poor diffusion of nutrients waste and within the engineered constructs. Past experiments have sought to induce the ingrowth of blood vessels into tissue-engineered devices to archive permanent vascularization; however, these approaches have been met with only limited success. More recently, an attempt to use microfabrication technology and cell biology to fabricate capillary vessels *in vitro* was reported [58]. If successful, this method may accelerate the progress of tissue engineering not only for liver, but also many other large and complex tissues such as the kidney or the heart.

Hepatocyte transplantation has already been applied to clinical use as a bridge to orthotopic liver transplantation and for some kinds of metabolic liver disease [59,60]. Tissue-engineered livers have the promise of providing a permanent treatment for end-stage liver and enzyme deficiency diseases, thus becoming a viable alternative for patients.

REFERENCES

1. American Liver Foundation (1996). "Fact Sheet: Hepatitis, Liver and Gallbladder Disease in the United States." American Liver Foundation, Cedar Grove, NJ.
2. Raper, S., and Wilson, J. M. (1993). Cell transplantation in liver-directed gene therapy. *Cell Transplant.* **2**, 381–400.
3. Matas, A. J., Surtherland, D. E. R., Steffes, M. W., Mauer, S. M., Sowe, A., Simmons, R. L., and Najarian, J. S. (1976). Hepatocellular transplantation for metabolic deficiencies: Decrease of plasma bilirubin in Gunn rats. *Science* **192**, 892–894.
4. Ponder, K. P., Gupta, S., Leland, F., Darlington, G., Finegold, M., DeMayo, J., Ledley, F. D., Chowdhury, J. R., and Woo, S. L. (1991). Mouse hepatocytes migrate to liver parenchyma and function indefinitely after intrasplenic transplantation. *Proc. Natl. Acad. Sci. U.S.A.* **88**, 1217–1221.
5. Wong, H., and Chang, T. M. (1986). Bioartificial liver: Implanted artificial cells microencapsulated living hepatocytes increases survival of liver failure rats. *Int. J. Artif. Organs* **9**, 335–336.
6. Dixit, V., Darvasi, R., Arthur, M., Brezina, M., Lewin, K., and Gitnick, G. (1990). Restoration of liver function in Gunn rats without immunosuppression using transplanted microencapsulated hepatocytes. *Hepatology* **12**, 1342–1349.
7. Lacy, P. E., Hegre, O. D., Gerasimidi-Vazeou, A., Gentile, F. T., and Dionne, K. E. (1990). Maintenance of normoglycemia in diabetic mice by subcutaneous xenografts of encapsulated islets. *Science* **254**, 1782–1784.
8. Vacanti, J. P., Morse, M. A., Saltzman, W. M., Domb, A. J., Perez-Atayde, A., and Langer, R. (1988). Selective cell transplantation using bioabsorbable artificial polymers as matrices. *J. Pediatr. Surg.* **23**, 3–9.
9. Langer, R., and Vacanti, J. P. (1993). Tissue engineering. *Science* **260**, 920–926.
10. Mooney, D., Hansen, L., Vacanti, J., Langer, R., Farmer, S., and Ingber, D. (1992). Switching from differentiation to growth in hepatocytes: Control by extracellular matrix. *J. Cell. Physiol.* **151**, 497–450.
11. Peppas, N. A., and Langer, R. (1994). New challenges in biomaterials. *Science* **263**, 1715–1720.
12. Mikos, A. G., Sarakinos, G., Lyman, M. D., Ingber, D. E., Vacanti, J. P., and Langer, R. (1993). Prevascularization of porous biodegradable polymers. *Biotechnol. Bioeng.* **42**, 716–723.
13. Uyama, S., Kaufmann, P. M., Takeda, T., and Vacanti, J. P. (1993). Delivery of whole liver-equivalent hepatocyte mass using polymer devices and hepatotrophic stimulation. *Transplantation* **55**, 932–935.
14. Sano, K., Cusick, R. A., Lee, H., Pollok, J. M., Kaufmann, P. M., Uyama, S., Mooney, D., Langer, R., and Vacanti, J. P. (1996). Regenerative signals for heterotopic hepatocyte transplantation. *Transplant. Proc.* **28**, 1857–1858.
15. Kaihara, S., and Vacanti, J. P. (1999). Tissue engineering: Toward new solution for transplantation and reconstructive surgery. *Arch. Surg. (Chicago)* **134**, 1184–1188.
16. Kim, S. S., Utsunomiya, H., Koski, J. A., Wu, B. M., Cima, M. J., Sohn, J., Mukai, K., Griffith, L. G., and Vacanti, J. P. (1998). Survival and function of hepatocytes on a novel three-dimensional synthetic biodegradable polymer scaffold with an intrinsic network of channels. *Ann. Surg.* **228**(1), 8–13.
17. Berry, M. N., and Friend, D. S. (1969). High yield preparation of isolated rat liver parenchymal cells. *J. Cell Biol.* **43**, 506–520.
18. Seglen, P. O. (1976). Preparation of isolated rat liver cells. *Methods Cell Biol.* **13**, 29–83.
19. Naik, S., Trenkler, D., Santagini, H., Pan, J., and Jaureguni, H. O. (1996). Isolation and culture of porcine hepatocytes for artificial liver support. *Cell Transplant.* **5**, 107–115.
20. Forsell, J. H., Jesse, B. W., and Shll, L. R. (1985). A technique for isolation of bovine hepatocytes. *J. Anim. Sci.* **60**, 1597–1609.
21. Strom, S. C., Jirtle, R. L., Novicki, D. L., Rosenberg, M. R., Novotny, A., Irons, G., and McLain, J. R. (1982). Isolation, culture, and transplantation of human hepatocytes. *JNCI, J. Natl. Cancer Inst.* **68**, 771–778.
22. Mitaka, T., Mizuguchi, T., Sato, F., Mochizuki, C., and Mochizuki, Y. (1998). Growth and maturation of small hepatocytes. *J. Gastroenterol. Hepatol.* **13**, S70–S77.
23. Guguen-Guillouso, C. (1986). Role of homotypic cell interactions in expression of specific functions by cultured hepatocytes. *In* "Research in Isolated and Cultured Hepatocytes" (A. Guillouzo and C. Guguen-Guillouzo, eds.), pp. 259–284. John Libbey Eurotext/INSERUM, London.
24. Strain, A. J. (1994). Isolated hepatocytes: Use in experimental and clinical hepatology. *Gut* **35**, 433–436.
25. Sirica, A. E., Mathis, G. A., Sano, N., and Elmore, L. W. (1990). Isolation, culture, and transplantation of intrahepatic biliary epithelial cells and oval cells. *Pathobiology* **58**, 44–64.

26. Hansen, L. K., and Vacanti, J. P. (1992). Hepatocyte transplantation using artificial biodegradable polymers. *In* "Current Controversies in Biliary Atresia" (M. A. Hoffman, R. Austin, and G. Landes, eds.), pp. 96–106. CRC Press, Boca Raton, FL.

27. James, N. H., Molloy, C. A., Soames, A. R., French, N. J., and Roberts, R. A. (1992). An in vitro model of rodent nongenotoxic hepatocarcinogenesis. *Exp. Cell Res.* **203,** 407–419.

28. Starzl, T. E., Francavilla, A., Halgrimson, C. G., Francavilla, F. R., Porter, K. A., Brown, T. H., and Putnam, C. W. (1973). The origin, hormonal nature, and action of hepatotrophic substances in portal venous blood. *Surg. Gynecol. Obstet.* **137,** 179–199.

29. Ricordi, C., Lacy, P. E., Callery, M. P., Park, P. W., and Flye, M. W. (1989). Trophic factors from pancreatic islets in combined hepatocyte–islet allografts enhance hepatocellular survival. *Surgery* **105,** 218–223.

30. Kamei, T., Falqui, L., Lacy, P. E., Callery, M. P., and Flye, M. W. (1990). Allogeneic islets transplanted intraportally maintain hepatic integrity following portacaval shunt in rats. *Transplant. Proc.* **22,** 810–813.

31. Kaufmann, P. M., Sano, K., Uyama, S., Schloo, B., and Vacanti, J. P. (1994). Heterotopic hepatocyte transplantation using three-dimensional polymers: Evaluation of the stimulatory effects by portacaval shunt or islet cell cotransplantation. *Transplant. Proc.* **26,** 3343–3345.

32. Asonuma, K., Gilbert, J. C., Stein, J. E., Takeda, T., and Vacanti, J. P. (1992). Quantitation of transplanted hepatic mass necessary to cure the Gunn rat model of hyperbilirubinemia. *J. Pediatr. Surg.* **27,** 298–301.

33. Freed, L. E., Marquis, J. C., Nohria, A., Emmanual, J., Mikos, A. G., and Langer, R. (1993). Neocartilage formation in vitro and in vivo using cells cultured on synthetic biodegradable polymers. *J. Biomed. Mater.* **27,** 11–23.

34. Freed, L. E., and Vunjak-Novakovic, G. (1995). Tissue engineering of cartilage. *In* "The Biomedical Engineering Handbook" (J. D. Bronzino, ed.), pp. 1788–1806. CRC Press, Boca Raton, FL.

35. Kim, S. S., Sundback, C. A., Kaihara, S., Benvenuto, M. S., Kim, B. S., Mooney, D. J., and Vacanti, J. P. (2000). Dynamic seeding and *in vitro* culture of hepatocytes in a flow perfusion system. *Tissue Eng.* **6,** 39–44.

36. Thompson, J. A., Haudenschild, C. C., Anderson, K. D., DiPietro, J. M., Anderson, W. F., and Maciag, T. (1989). Heparin-binding growth factor 1 induces the formation of organoid neovascular structures in vivo. *Proc. Natl. Acad. Sci. U.S.A.* **86,** 7928–7932.

37. Kim, S. S., Kaihara, S., Benvenuto, M. S., Kim, B. S., Mooney, D. J., and Vacanti, J. P. (1999). Small intestinal submucosa as a small-caliber venous graft: A novel model for hepatocyte transplantation on synthetic biodegradable polymer scaffolds with direct access to the portal venous system. *J. Pediatr. Surg.* **34**(1), 124–128.

38. Starzl, T. E., Porter, K. A., and Kashiwagi, N. (1975). Portal hepatotrophic factors, diabetes mellitus and acute liver atrophy, hypertrophy and regeneration. *Surg. Gynecol. Obstet.* **141,** 843–858.

39. Jaffe, V., Darby, H., Bishop, A., and Hodgson, H. J. (1991). The growth of liver cells in the pancreas after intra-splenic implantation: The effects of portal perfusion. *Int. J. Exp. Pathol.* **72,** 289–299.

40. Zhang, H., Miescher-Clemens, E., Drugas, G., Lee, S. M., and Colombani, P. (1992). Intrahepatic hepatocyte transplantation following subtotal hepatectomy in the recipient: A possible model in the treatment of hepatic enzyme deficiency. *J. Pediatr. Surg.* **27,** 312–315.

41. Kokudo, N., Ohashi, K., Takahashi, S., Bandai, Y., Sanjo, K., Idezuki, Y., and Nozawa, M. (1994). Effect of 70% hepatectomy on DNA synthesis in rat hepatocyte isograft into the spleen. *Transplant. Proc.* **26,** 3464–3465.

42. Borel Rinkes, I. H., Bijma, A., Kazemier, G., Sinaasappel, M., Valerio, D., and Terpstra, O. T. (1994). Proliferative response of hepatocytes transplanted into spleen or solid support. *J. Surg. Res.* **56,** 417–423.

43. Bucher, N. L. R., and Swaffield, M. N. (1964). The rate of incorporation of labeled thymidine into the deoxyribonucleic acid of regenerating rat liver in relation to the amount of liver excised. *Cancer Res.* **240,** 1611–1625.

44. Rabes, H. M., Wirsching, R., Tuczek, H. V., and Iseler, G. (1976). Analysis of cell cycle compartments of hepatocytes after partial hepatecomy. *Cell Tissue Kinet.* **9,** 517–532.

45. Gagandeep, S., Rajvanshi, P., Sokhi, R. P., Slehria, S., Palestro, C. J., Bhargava, K. K., and Gupta, S. (2000). Transplanted hepatocytes engraft, survive, and proliferate in the liver of rats with carbon tetrachloride-induced cirrhosis. *J. Pathol.* **191,** 78–85.

46. Gupta, S., Rajvanshi, P., Aragona, E., Lee, C. D., Yerneni, P. R., and Burk, R. D. (1999). Transplanted hepatocytes proliferate differently after CCl₄ treatement and hepatocyte growth factor infusion. *Am. J. Physiol.* **276,** G629–G638.

47. Peterson, J. E. (1965). Effects of the pyrrolizidine alkaloid lasiocarpine-*N*-oxide on nuclear and cell division in the liver of rats. *J. Pathol. Bacteriol.* **89,** 153–171.

48. Laconi, E., Oren, R., Mukhopadhyay, D. K., Hurston, E., Laconi, S., Pani, P., Dabeva, M. D., and Shafritz, D. A. (1998). Long-term, near-total liver replacement by transplantation of isolated hepatocytes in rats treated with retrorsine. *Am. J. Pathol.* **153,** 319–329.

49. Grisham, J. W. (1980). Cell types in long-term propagable cultures of rat liver. *Ann. N.Y. Acad. Sci.* **349,** 128–137.

50. Sell, S. (1990). Is there a liver stem cell? *Cancer. Res.* **50,** 3811–3815.

51. Thorgeirsson, S. S., and Evarts, R. P. (1992). Growth and differentiation of stem cells in adult rat liver. *In* "The Role of Cell Types in Hepatocarcinogenesis" (A. E. Sirica, ed.), pp. 109–120. CRC Press, Boca Raton, FL.

52. Sigal, S. H., Brill, S., Fiorino, A. S., and Reid, L. M. (1992). The liver as a stem cell and lineage system. *Am. J. Physiol.* **263,** G139–G148.

53. Mitaka, T. (1998). The current status of primary hepatocyte culture. *Int. J. Exp. Pathol.* **79,** 393–409.

54. Alison, M. (1998). Liver stem cells: A two-compartment system. *Curr. Opin. Cell Biol.* **10,** 710–715.

55. Werlich, T., Stiller, K. J., and Machnik, G. (1998). Experimental studies on the stem cell concept of liver regeneration. I. *Exp. Toxicol. Pathol.* **50,** 73–77.

56. Taniguchi, H., Kondo, R., Suzuki, A., Zheng, Y., Ito, S., Takada, Y., Fukunaga, K., Seino, K., Yuzawa, K., Otsuka, M., Fukao, K., Yoshiki, A., Kusakabe, M., and Nakauchi, H. (1999). Evidence for the presence of hepatic stem cells in the murine fetal liver. *Transplant. Proc.* **31,** 454.

57. Taniguchi, H., Suzuki, A., Zheng, Y., Kondo, R., Takada, Y., Fukunaga, K., Seino, K., Yuzawa, K., Otsuka, M., Fukao, K., and Nakauchi, H. (2000). Usefulness of flow-cytometric cell sorting for enrichment of hepatic stem and progenitor cells in the liver. *Transplant. Proc.* **32**, 249–251.

58. Kaihara, S., Borenstein, J., Koka, R., Lalan, S., Ochoa, E. R., Ravens, M., Pien, H., Cunningham, B., and Vacanti, J. P. (2000). Silicon micromachining to tissue engineer branched vascular channels for liver fabrication. *Tissue Eng.* **6**, 105–117.

59. Strom, S. C., Fisher, R. A., Thompson, M. T., Sanyal, A. J., Cole, P. E., and Posner, M. P. (1997). Hepatocyte transplantation as a bridge to orthotopic liver transplantation in terminal liver failure. *Transplantation* **63**, 559–569.

60. Fox, I. J., Chowdhury, J. R., Kaufman, S. S., Goertzen, T. C., Chowdhury, N. R., Warkentin, P. I., Dorko, K., Sauter, B. V., and Storm, C. S. (1998). Treatment of the Crigler–Najjar syndrome type I with hepatocyte transplantation. *N. Engl. J. Med.* **338**, 1422–1426.

61. Mitaka, T., Mikami, M., Sattler, G. L., Pitot, H. C., and Mochizuki, Y. (1992). Small cell colonies appear in the primary culture of adult rat hepatocytes in the presence of nicotinamide and epidermal growth factor. *Hepatology* **16**, 440–447.

62. Tateno, C., and Yoshizato, K. (1996). Growth and differentiation in culture of clonogenic hepatocytes that express both phenotypes of hepatocytes and biliary epithelial cells. *Am. J. Pathol.* **149**, 1593–1605.

CHAPTER 88

Extracorporeal Kidney

Evangelos Tziampazis and H. David Humes

INTRODUCTION

The kidney was the first major organ for which a successful extracorporeal replacement therapy has been developed and applied extensively to restore partially the lost tissue function. Indeed, following their integration into mainstream clinical practice during the last three decades, dialysis and derivative blood purification treatments (based on artificial semipermeable membranes) have extended the life of millions of people suffering from renal failure. Still, kidney failure patients undergoing dialysis treatment suffer unacceptably high mortality and morbidity rates. Thus, for example, the expected remaining years of life of end-stage renal disease (ESRD) patients on dialysis is 2.7- to 6.4-fold lower than that of corresponding age-sex-race groups in the overall U.S. population. As an indication of the impact of renal disease on society, direct expenditures for ESRD treatment exceed $15 billion in the United States alone [1].

Despite advances in the treatment of kidney failure through the use of artificial membranes [2–5], improvements in therapeutic outcomes are slow, and all such treatments are limited by the inherent inability of inanimate membranes to replicate the regulatory and metabolic functionality provided by the healthy kidney. As a result, unlike kidney transplantation, these renal replacement therapies provide only partial substitution for the lost renal function and thus lead to long-term complications and reduced duration and quality of life [1,6]. However, recent advances in tissue engineering promise to lead to the development of bioartificial constructs that can approximate, and eventually may replicate completely, the functionality of the healthy kidney. The early development of these bioartificial renal constructs has focused on their use in extracorporeal hemoperfusion circuits, but it is expected that once therapeutic benefits have been established and the fundamental technical difficulties addressed, the natural evolution of the technology will lead to the development of an implantable bioartificial kidney. Understanding of kidney physiology and functionality is a prerequisite for the development of successful bioartificial renal tissues, whether extracorporeal or implantable.

KIDNEY FUNCTION

The functional element of the kidney is the nephron; there are one million such elements in each kidney. The nephron has three major parts: the *glomerulus*, the *tubule*, and the *collecting duct* (Fig. 88.1). The glomerulus, responsible for the removal of toxic substances from the bloodstream, consists of a tuft of small, highly permeable blood vessels located inside a hollow structure called Bowman's capsule. Together, these two elements make up the renal corpuscle. Bowman's capsule opens up into the tubule, the part that carries out most of the metabolic, transport, and regulatory functions of the kidney. The tubule portion of the nephron can be subdivided into different segments based on their functional and morphological characteristics. Functionally, there are three major segments: the proximal tubule, the loop of Henle, and the distal tubule, with each segment playing a special role in ensuring proper kidney function.

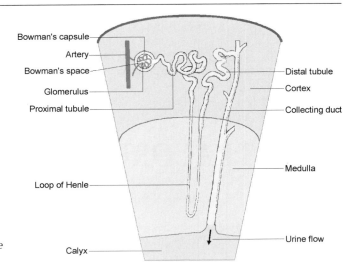

Bowman's capsule
Artery
Bowman's space
Glomerulus
Proximal tubule
Distal tubule
Cortex
Collecting duct
Medulla
Loop of Henle
Urine flow
Calyx

Fig. 88.1. Schematic of a nephron, the functional element of the kidney.

Unlike typical blood vessels, the glomerulus is characterized by very high hydraulic permeability, one to two orders of magnitude larger than that of capillaries in other tissues [7]. The high hydraulic permeability of the glomerulus enables the filtration of circulating toxins by means of convection rather than diffusion. As a result, toxins are removed rapidly, regardless of concentration and size (up to a certain molecular weight cutoff), leaving behind practically toxin-free blood. In addition to supporting very high glomerular ultrafiltration rates (around 130 liters/day in an average size human), kidney evolution ensured that toxin removal does not result in excessive loss of water, nutrients, salts, proteins, and other important blood constituents. Circulating blood cells and large molecular weight components, including albumin and other large proteins, cannot permeate the glomerular membrane and are retained in the filtered blood, while essential small-sized molecules that filter through are reclaimed and regulated in the tubule. To achieve the required semipermeability characteristics and extremely high ultrafiltration rates, the glomerular membrane evolved a complex structure comprising a specialized fenestrated endothelium, a charged basement membrane, and highly differentiated epithelial cells. Together, these components impart to the glomerular membrane its extraordinary characteristics, clearly unmatched by purely artificial membranes.

The vast majority (>98%) of the filtrate volume that drains into the tubule is eventually reabsorbed back into the bloodstream. Approximately half the recovered water, along with the majority of glucose, sodium, and other filtered compounds, is reabsorbed in the proximal tubule; the rest is processed and regulated in more distal parts of the nephron. Unlike glomerular filtration, transport through the tubule wall is driven by active solute transfer, from the apical to the basal side, across membrane channels in the specialized epithelial cells lining the tubule. The resulting osmotic pressure gradients drive the transport of water in the same direction. Besides the reclamation of important blood components, tubule epithelial cells are also responsible for the metabolic degradation of certain filtered solutes, such as β_2 microglobulin (β_2M). It should be noted that insufficient removal of β_2M during traditional hemodialysis has been associated with the development of dialysis-related amyloidosis. Active transport is also used to remove from the bloodstream certain compounds, such as penicillin and other drugs, that are too large to be removed in the glomerulus.

Expansive vascularization networks in the kidney facilitate the extensive mass transport taking place in the organ. In addition, the orientation of nephrons in the kidney, whereby the loop of Henle is toward the interior part of the organ, plays an important role in the regulation of the final steps of urine formation. Indeed, the anatomic and functional characteristics of the various tubule segments of the nephron lead to very different solute concentrations for the inner (medulla) and outer (cortex) portions of the kidney. The very high salinity of the medulla is the osmotic force behind the final removal of water in the collecting duct and the formation of urine with high salt and toxin concentrations; to ensure appropriate volume homeostasis, this process is further regulated by the antidiuretic hormone ADH.

When functioning properly, all kidney components work together to ensure the removal of toxic compounds from plasma, while retaining the useful components (e.g., glucose and proteins), maintaining appropriate fluid volumes, controlling the level of ions, and regulating the function of other organs in the body. It is also important to note that evolution ensured that the kidney fulfills its role even under stress, during conditions of high toxin load. As a result, only a small portion of the overall capacity of the kidneys is required under normal conditions, a helpful characteristic for the effort to develop a bioartificial kidney.

ENGINEERING A BIOARTIFICIAL KIDNEY
REQUIREMENTS

From the description of the functions performed by a healthy kidney, it is evident that current kidney treatments are not capable of providing an adequate kidney replacement. Complete replacement of kidney functionality requires the development of bioartificial constructs capable of replicating the filtration, transport, metabolic, and regulatory functions of the organ. In turn, this requires the presence not only of a filtration unit but also of units that provide the functions of proximal tubule, loop of Henle, distal tubule, and even collecting duct.

It is apparent that the resultant complexity makes the development of such a system impractical, even though the underlying technological principles are fairly well understood. However, compared with the current treatments based on artificial membrane, substantial clinical benefits may be achieved even without incorporating the functionality of the loop of Henle and the distal nephron. This notion is supported by clinical experience with patients who suffer moderate renal insufficiency and, even though their kidneys cannot provide the fine control of water and salt homeostasis associated with the healthy kidney, are able to maintain, with reasonable success, fluid and electrolyte homeostasis [8]. Thus, we believe that the development of a relative simple system combining filtration and proximal tubule functionality can improve considerably the prognosis of patients with renal disease.

HEMOFILTER

To replace successfully the filtration function of the nephron, the hemofilter must be able to replicate the ability of the kidney glomerulus to efficiently remove blood toxins, irrespective of their molecular size and concentration. This requirement can be satisfied if the membrane permeability is high enough to permit pressure-driven ultrafiltration fluxes instead of relying on diffusion alone. Ultrafiltration removes water and along with it clears effectively all solutes up to the characteristic molecular weight cutoff of the membrane. In contrast, clearance by diffusion depends on concentration gradients and solute mobility; as a result, diffusion is ineffective for the removal of large and/or highly dilute solutes.

The recognition of these inherent limitations in diffusive transport has led to the development of membranes with high hydraulic permeabilities and of therapeutic modalities in which blood cleansing is achieved via convective hemofiltration rather than diffusive hemodialysis [9,10]. While hemofiltration can be achieved today successfully with purely artificial membranes, new research in biomaterials and tissue engineering holds promise for significant future improvements.

Artificial Membranes

Modern membranes are used reliably all over the world in countless hemodialysis and hemofiltration procedures. Typically, membranes are used in the form of hollow-fiber modules and are based on cellulose or synthetic polymers (including polysulfone, polyacrylonitrile, and polyamide), used as such or modified with a variety of agents [11–16]. In particular, polysulfone hollow-fiber membranes have been used successfully as blood hemofilters in extracorporeal circuits, maintaining ultrafiltration functionality over several weeks with a single device [17,18]. Moreover, today's high-flux membranes can provide effective filtration of small protein molecules, such as β_2 microglobulin, and thus may help reduce the side effects of chronic hemodialysis treatment, such as amyloidosis [19–22]. However, despite considerable advances in a number of areas, such as biocompatibility and progressive fouling by protein deposition, the performance of current membranes is still far from optimal.

Membrane incompatibility is the most critical shortcoming of artificial membranes. Indeed, current membranes have been engineered from off-the-shelf polymeric materials, with development focusing primarily on improving membrane transport characteristics, rather than on controlling blood–material interactions and optimizing biocompatibility [2]. As a result, current membranes induce a cascade of biochemical events leading to blood coagulation and thrombus formation and, frequently, to complement activation and inflammation [23].

In clinical practice biocompatibility problems are addressed with anticoagulation therapy, even though this offers only a partial solution and may lead to additional complications, such as hemorrhage. Systemic administration of heparin, the most commonly prescribed anticoagulation therapy, helps reduce but does not fully eliminate thrombus formation [24]; moreover, in addition to bleeding, heparin may induce negative immune responses and under certain conditions may actually initiate the blood coagulation cascade [25,26]. Furthermore, protamine, which is used frequently to neutralize the free circulating heparin, may also lead to complement activation and adverse hemodynamic responses [27]. In recognition of these limitations, researchers are trying to improve anticoagulation therapy by replacing heparin with alternative thrombin inhibitors that evolved in bloodsucking animals [28]. Most attention has focused on hirudin, a leech-derived polypeptide that forms a practically irreversible bond with thrombin and exhibits exceptional antithrombotic action [29,30]. Recombinant hirudin and synthetic antithrombotic agents based on the structure of hirudin have been synthesized and are being evaluated as heparin replacements [31–33], including use in hemoperfusion circuits [34,35]. As in the case of heparin [36], these antithrombotic agents may be administered into the systemic circulation [35] or may be incorporated directly onto the polymeric membrane [37].

The latter approach is part of broader efforts to improve biomaterial–blood compatibility by modifying the membrane material itself rather than by altering the systemic response of the host. In addition to using active anticoagulants like heparin and hirudin, which act on the thrombotic cascade, membrane surfaces can be modified by the adsorption and grafting of more passive antithrombotic agents [e.g., poly(ethylene glycol) [38]], which alter the nature of protein–biomaterial interactions and result in an improved adsorbed protein state. Such modifications have been applied successfully to improve the blood compatibility of membranes based both on synthetic and biological polymers [39,40]. In addition to coating, antithrombotic agents may be incorporated throughout the membrane, normally by block copolymerization as in the case of polysulfone–poly(ethylene oxide) membranes [16]; bulk incorporation may be advantageous because antithrombotic moieties are far less likely to be removed by erosion or by scratching of the surface.

It is evident from the foregoing discussion that over the past six decades blood purification membranes have evolved from their lowly beginnings as sausage casings [41]. Even though progress has been rather slow by modern technical standards [42], membrane efficiency and filtration characteristics have improved significantly. Moreover, as our understanding of how material chemistry and microstructure define host–biomaterial interaction improves further, so will the potential of developing inherently blood-compatible membranes [12,43]. It will be many years, however, before the blood compatibility of artificial membranes can approach that of the natural endothelium, the gold standard for blood compatibility [44].

Bioartificial Filter

The incorporation of a confluent endothelial monolayer on a porous membrane may provide a solution to the membrane–blood compatibility problems and form the basis for the development of a bioartificial hemofilter (BAHF). In addition to improved biocompatibility, a bioartificial membrane is inherently better suited to replicate glomerular filtration and thus provide a complete functional replacement for the glomerulus. However, owing to the relative clinical success of hemodialysis treatments and the complexity inherent in tissue engineering kidney replacements [45], the literature is devoid of promising efforts to develop a BAHF. Nevertheless, the feasibility of the BAHF concept is supported strongly by a series of related developments.

Thus, animal studies have demonstated that endothelial seeding on synthetic vascular grafts inhibits the long-term deposition of platelets and formation of thrombus [46,47]. Furthermore, implantation of synthetic grafts endothelialized *in vitro* with autologous endothelial cells (EC) was also very promising in humans, replicating the positive effects exhibited in animal studies and improving the long-term patency of the grafts [48].

The antithrombotic properties of endothelialized membranes could be further improved if the ECs are engineered to secrete anticoagulation factors [49]. The ability to produce vascular grafts lined with genetically engineered endothelial cells capable of producing high amounts of active recombinant proteins has already been demonstrated [50,51]. Accordingly, endothelial cells engineered to express anticoagulant factors could be utilized to coat vascular grafts and stents as well as semipermeable membranes and improve their long-term performance *in vivo* [52]. The same approaches can be extended to improve the patency of endothelialized hemofiltration membranes as well [53]. The isolation and cloning of anticoagulation proteins [54,55] has facilitated these developments, and to date researchers have engineered ECs that secrete [56] anticoagulation proteins or, alternatively, incorporate them onto the plasma membrane [49,57]. Moreover, the expression of anticoagulation proteins presented by the plasma membrane can be regulated by the activation of the endothelium [58], thus giving rise to the potential for local, on-demand inhibition of the thrombotic cascade.

Even though endothelial cells can be isolated from both animal and human glomeruli and propagated *in vitro* [59], financial and technical considerations limit the application potential of this autologous cell source for the endothelialization of the hemofilter membrane. Alternatively, autologous endothelial cells can be isolated from peripheral blood vessels following standard enzymatic procedures but can also be derived from progenitor endothelial cells isolated from bone marrow [60] or, preferentially, from peripheral blood and propagated *in vitro* [61,62]. Moreover, current cell culture protocols are very effective in utilizing appropriate extracellular matrix (ECM) components and soluble growth factors to support endothelial cell adhesion and growth into endothelium monolayers *in vitro* [63–65]. Of note, ECM components and soluble factors can be used to induce morphogenic changes on cultured endothelia, including the formation of fenestrae [66–68].

It is thus evident that all fundamentals are in place for the development of a hybrid hemofilter comprising a semipermeable inanimate membrane that provides the desired filtration characteristics and a confluent endothelial monolayer that imparts superior blood compatibility. It may also be possible to develop a bioartificial hemofilter in which living cells and their ECM provide filtration functionality mimicking the glomerulus. Even though the protocols required for this step are not yet fully established, continuous advances in our understanding of the *in vitro* and *in vivo* interactions between different glomerular cell types, as well as between cells and ECM or growth factors [69–73], suggest that such a development is forthcoming.

BIOARTIFICIAL TUBULE

Unlike glomerular filtration, which is based primarily on physicochemical principles, tubule functions are controlled by biological processes. As a result, inanimate, purely artificial devices cannot be used to replicate, to any significant measure, the functionality of the tubule. Instead, replacement of the multivariate tubular functions of the kidney requires the use of the biological components that have evolved specifically to carry out those functions: the renal tubule epithelium.

Thus, according to the principles of tissue engineering, a tubule replacement can be conceived to consist of functional living tubule epithelial cells, grown in the form of monolayers onto scaffolds made of appropriate synthetic and/or biological materials [74–76]. Because the scaffolding material provides the required physical support for the tubule cells, it should not degrade considerably over the life of the bioartificial construct; furthermore, it should permit the unhindered transport of solutes reclaimed by the tubule cells on one side to the filtered blood circulating at the other side, provide large surface area to facilitate this mass transport, be compatible with blood, and not leach cytotoxic products to the circulation. Semipermeable hollow-fiber membranes such as those employed in hemodialysis circuits meet all these requirements and, in addition, provide an immunoprotective barrier that enables the use of nonautologous tubule epithelial cells [77–79].

In the late 1980s Patrick Aebischer and coworkers [74,75,79,80] demonstrated first the feasibility of constructing a bioartificial tubule by growing confluent tubule epithelial cell monolayers on synthetic hollow fibers. However, they employed permanent renal cell lines, derived from proximal and distal tubule segments, which do not exhibit fully the differentiated transport properties that are characteristic of normal tubule epithelia [8]. As a result, even though these constructs exhibited certain aspects of tubule function, like the ability to reabsorb glucose and secrete tetraethylammonium under uremic conditions, they were not able to replicate other specific functions, such as secretion of para-aminohippurate (PAH) [75,79]. Thus, use of differentiated cells is necessary if the majority of the specific tubule functions are to be maintained.

The ability to isolate and propagate *in vitro* primary cells from renal epithelia, as well as from other kidney tissue, has been established for a long time [81,82]. However, such cells have limited capacity for replication and can maintain their differentiated properties for only a limited number of passages. On the other hand, stem cells that maintain large capacity for self-renewal and ability to differentiate under defined conditions provide the optimal source of functional tubule epithelia. Embryonic stem cell research has verified the existence of pluripotent renal stem cells [83], while clinical observations of the ability of renal tubules to regenerate after acute ischemic or nephrotoxic injury [84–86] suggest that tubule progenitor or stem cells do exist in the adult mammalian kidney. Pluripotent stem cells have not been isolated to date, but research in our laboratory has demonstrated that renal epithelial progenitor cells can be isolated from mammalian adult kidneys and grown *in vitro* [87,88]. In accordance with the fundamental role of growth factors [86,89–91] and extracellular matrix proteins [92–94] on renal epithelial cell function, appropriate combinations of soluble and insoluble factors can be used to induce and maintain morphogenic differentiation of these cells *in vitro* [53,87]. Even though successful *in vitro* expansion has been achieved with renal tubule progenitor cells isolated from both animal and human adult kidneys, use of animal tissue offers certain advantages in terms of procurement and transfer of infectious pathogens. Pig donors in particular are the preferred source of xenogeneic tissue, in part because of their general anatomic and physiological similarities with humans and the relative ease with which they can be bred and raised in controlled herds [95].

With the availability of large amounts of differentiated proximal tubule epithelial cells warranted by these developments, our laboratory has proceeded to develop a bioartificial proximal tubule [77,78,96–98]. The first stage of our effort has focused on developing a bioartificial renal tubule assist device (RAD) suitable for use in an extracorporeal hemoperfusion circuit in series with a hemofilter. As discussed in more detail earlier, the addition of proximal tubule functionality to a hemofilter provides the first realistic approximation of kidney function and can improve substantially the therapeutic outcomes of patients with kidney failure. Accordingly, a renal tubule assist device based on porcine proximal tubule progenitor cells was developed and tested extensively *in vitro* for differentiated proximal tubule functions [77,98]. Proximal tubule progenitor cells were seeded onto the ECM-coated lumina of high-flux polysulfone hollow-fiber cartridges and were cultured *in vitro* for a minimum of 14 days. The resulting confluent monolayers, as many as 2.5×10^9 cells for 0.7 m^2, exhibited morphological characteristics (e.g., apical microvilli, endocytic vesicles, tight junctional complexes) typical for differentiated tubule epithelia as well as differentiated metabolic, secretory, and vectorial transport activities [77]. In particular, these *in vitro* studies demonstrated the ability of RADs to actively transport sodium, glucose, bicarbonate, and organic anions, such as PAH; they also reabsorbed significant amounts of water in response to oncotic and osmotic pressure gradients. In contrast, early bioartificial tubule constructs based on permanent cell lines exhibited incomplete vectorial active transport behavior, reabsorbing glucose but failing to transport PAH [75]. In addition, renal proximal tubule cells cultured in RADs or in culture plates maintained their ability to synthesize and excrete ammonia (a function critical for the maintenance of acid–base homeostasis in the body), produce $1,25\text{-}(OH)_2D_3$ (the most active form of vitamin D), and, through metabolic degradation and transport of its constituent amino acids, remove glutathione from the perfusate [77]. It is thus evident that RADs possess the ability to replicate the major differentiated transport, metabolic, and secretory functions performed by the healthy renal proximal tubule. Furthermore, in accordance with earlier results with LLC-PK1 cells [99], a permanent cell line derived from porcine proximal renal tubules, we believe that cells in the RAD may be successful in metabolizing and removing β_2 microglobulin from the bloodstream, thereby helping to ameliorate dialysis-caused amyloidosis [100,101].

BIOARTIFICIAL EXTRACORPOREAL KIDNEY

The successful application of tissue engineering principles to the development of a bioartificial renal proximal tubule was followed by animal studies in which the bioartificial tubule construct was connected in an extracorporeal hemoperfusion circuit in series with a hemofilter [78]. The primary objectives of these studies were to evaluate the bioartificial extracorporeal kidney in large animals (dogs) and under physiological conditions and to optimize the system in terms of its functionality and ease of use. Animals became uremic after removal of both kidneys, and following 24 h of postoperative recovery they were connected to an extracorporeal hemoperfusion circuit for treatment either with venovenous hemofiltration incorporating a RAD cartridge or with hemofiltration with a sham RAD cartridge containing no cells. Animals were treated for either 7–9 h/day for 3 consecutive days or for 24 h continuously.

Figure 88.2 illustrates the extracorporeal bioartificial kidney circuit used. Briefly, venous blood was pumped into the luminal space of a conventional polysulfone hemofilter, where it was split into two distinct fluid streams. Posthemofilter blood was directed to RAD extracapillary space (ECS), while the ultrafiltrate collected into the ECS of the hemofilter was pumped into the luminal space of the RAD. This way, proximal tubule cells in the RAD were exposed to blood ultrafiltrate, much as in the *in vivo* situation in the nephron, while being isolated from the host's blood and immune systems. Following tubule epithelial cell processing in the RAD, blood was returned to the host animal while the processed ultrafiltrate was discarded as urine. This extracorporeal circuit enabled the administration of replacement fluids and heparin, the sampling of perfused fluids for off-line analysis, as well as on-line monitoring and regulation of pressures, temperatures, and flow rates.

The bioartificial extracorporeal kidney resulting from the combination of filtration and functional renal epithelial processing units constituted the first man-made system to provide a realistic approximation of kidney function, including transport, metabolic, and endocrinologic activity [78,102]. Cells in the bioartificial tubule cartridge retained their viability and, despite prolonged hemoperfusion, cell loss due to shearing was insignificant, with less than one cell in 10,000 lost after 24 h perfusion. The rate of ultrafiltration through the synthetic hemofilter was 5–7 ml/min, of which 40–50% was reclaimed through reabsorption by the RAD.

It is interesting to note that the proximal tubule absorbs *in vivo* a similar percentage of glomerular ultrafiltrate. Compared with the sham control, treatment with the extracor-

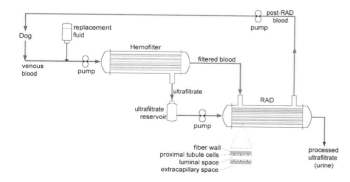

Fig. 88.2. Schematic of the hemoperfusion circuit utilized to connect the extracorporeal bioartificial kidney to the uremic animal model (dog). The bioartificial organ comprises two functional units in series: a synthetic hemofilter and a bioartificial renal tubule assist device (RAD).

poreal bioartificial kidney provided substantial improvements to the values of a multitude of renal function indicators. In particular, the RAD was successful in removing substantial amounts (>50%) of the glutathione present in the ultrafiltrate, resulting in plasma glutathione levels that were threefold higher than those achieved with sham controls, and performing active transport of glucose, potassium, and bicarbonate. Treatment with the extracorporeal bioartificial kidney was also more successful than hemoperfusion alone (sham control) in maintaining blood urea nitrogen levels. Furthermore, RADs excreted considerable amounts of NH_3 into the ultrafiltrate, up to 100 μmol/h, compared with no secretion in the controls. Finally, in animals treated with the RAD 1,25-$(OH)_2D_3$ values attained their prenephrectomy levels, while 1,25-$(OH)_2D_3$ levels continued dropping from the already low levels of the uremic animals in the absence of RAD.

These animal studies have shown clearly that RADs maintained their functionality when used *ex vivo* to improve the renal therapy of large uremic animals. Moreover, these studies have shown that a bioartificial extracorporeal kidney comprising a RAD in series with a hemofilter can replace the filtration, transport, metabolic, and secretory functions lost in renal failure. As such, these exciting results set the stage for pilot human trials, in which the whole bioartificial extracorporeal kidney is used to treat acute tubule necrosis.

If, as predicted by all available data, human trials prove successful, we envision a number of further research directions, explored in series or independently, that will lead to further improvements in the functionality and application potential of bioartificial kidney constructs. For example, utilization of a bioartificial hemofilter and further improvements in biomaterial biocompatibility may enable the long-term use of the extracorporeal bioartificial kidney without anticoagulation therapy, while additional tubule functionality may be added by incorporating into the circuit hollow-fiber cartridges lined with differentiated epithelial cells from the loop of Henle and the distal tubule. Similarly, improving the performance efficiency of the circuit will result in size reductions and could thus lead to the progressive evolution of the bioartificial kidney from bedside to wearable and, eventually, implantable devices.

SUMMARY

The kidney was the first major organ for which successful therapeutic modalities were developed and used extensively to replace a portion of the functions performed by the healthy tissue. Thus, thanks to regular dialysis treatments, millions of renal failure patients have been sustained to life. However, because current dialytic therapy is based on purely artificial (acellular) systems, it cannot provide complete replacement of renal function and results in unacceptably high rates of mortality and morbidity. Only living cells are capable of providing the metabolic, endocrinological, and active transport functions of the kidney lost in renal failure. To that extent, tissue engineering principles are being applied to develop a bioartificial kidney capable of replicating the functionality of the healthy nephron.

To achieve this, the bioartificial kidney must become an adequate substitute for the filtration function of the glomerulus and the epithelial processing performed in the renal tubule. In principle, filtration substitution may be accomplished either by inanimate or bioartificial membranes, but only the former are available today. Even though current acellular membranes are used successfully in extracorporeal hemofiltration circuits, there is considerable potential for improvement, especially in terms of blood compatibility and filtration efficiency. Advances in anticoagulation therapy, biomaterials science, and tissue engineering have already generated the technical foundation required for such improvements. The development of the bioartificial tubule and the resultant potential for more complete renal replacement therapy, first extracorporeally and eventually through an implantable bioartificial kidney, will undoubtedly provide the impetus for the development of improved hemofilters that mimic the healthy glomerulus.

Replacement of the epithelial processing of the tubule requires the combination of differentiated cells and a synthetic support structure. In general, this bioartificial tubule may incorporate different segments to substitute the major functional parts of the renal tubule. However, clinical experience suggests that incorporation of proximal tubule functionality alone could improve substantially the therapeutic outcomes of renal failure patients. Consistent with the concept of staged development, our laboratory has developed an extracorporeal bioartificial proximal renal tubule, suitable for use in a hemoperfusion circuit. Connected in series with a hemofilter, the engineered tissue can thus function as a renal tubule assist device (RAD), providing proximal tubule functionality to standard arteriovenous hemofiltration. Extensive testing *in vitro* illustrated the capability of the engineered tissue to replicate the major transport, metabolic, and endocrinological functions of the proximal tubule, and thus its potential to form the basic building block for a bioartificial kidney. This potential was verified in studies with uremic dogs, whereby the combination of RAD and synthetic hemofilter provided the first ever realistic replacement of lost kidney function, including filtration, active transport, metabolic, and endocrinological activity, and has set the stage for the initiation of human trials.

Should these trials prove successful, the combination of hemofilter and RAD will provide immediate therapeutic benefits to a large number of patients and will form the impetus for further developments. These may involve not only the incorporation of additional bioartificial units—including a bioartificial hemofilter, distal tubule, and loop of Henle, or collecting duct—but also the optimization [98,103,104] of the long-term efficiency and functionality of the extracorporeal bioartificial kidney and the move from bedside to wearable and eventually implantable systems. These are clearly exciting times for renal replacement therapy!

REFERENCES

1. U.S. Renal Data System (1999). "USRDS 1999 Annual Data Report." National Institutes of Health, and National Institute of Diabetes and Digestive and Kidney Diseases, Bethesda, MD.
2. Cheung, A. K., and Leypoldt, J. K. (1997). The hemodialysis membranes: A historical perspective, current state and future prospect. *Semin. Nephrol.* **17**, 196–213.
3. Grootendorst, A. F., and van Bommel, E. F. (1993). The role of hemofiltration in the critically-ill intensive care unit patient: Present and future. *Blood Purif.* **11**, 209–223.
4. Leber, H. W., Wizemann, V., Goubeaud, G., Rawer, P., and Schutterle, G. (1978). Hemodiafiltration: A new alternative to hemofiltration and conventional hemodialysis. *Artif. Organs* **2**, 150–153.
5. Ronco, C., Ghezzi, P. M., and La Greca, G. (1999). The role of technology in hemodialysis. *J. Nephrol.* **12**(Suppl. 2), S68–S81.
6. Simmons, R. G., Anderson, C., and Kamstra, L. (1984). Comparison of quality of life of patients on continuous ambulatory peritoneal dialysis, hemodialysis, and after transplantation. *Am. J. Kidney Dis.* **4**, 253–255.
7. Brenner, B. M., and Humes, H. D. (1977). Mechanics of glomerular ultrafiltration. *N. Engl. J. Med.* **297**, 148–154.
8. Cieslinski, D., and Humes, H. D. (1994). Tissue engineering of a bioartificial kidney. *Biotechnol. Bioeng.* **43**, 678–681.
9. Colton, C., Henderson, L., Ford, C., and Lysaght, M. J. (1975). Kinetics of hemodiafiltration. I. In vitro transport characteristics of a hollow-fiber blood ultrafilter. *J. Lab. Clin. Med.* **85**, 355–371.
10. Henderson, L., Colton, C., and Ford, C. (1975). Kinetics of hemodiafiltration. II. Clinical characterization of a new blood cleansing modality. *J. Lab. Clin. Med.* **85**, 372–391.
11. Vienken, J., Diamantoglou, M., Henne, W., and Nederlof, B. (1999). Artificial dialysis membranes: From concept to large scale production. *Am. J. Nephrol.* **19**, 355–362.
12. Deppisch, R., Gohl, H., and Smeby, L. (1998). Microdomain structure of polymeric surfaces—Potential for improving blood treatment procedures. *Nephrol. Dial. Transplant.* **13**, 1354–1359.
13. Ronco, C., Brendolan, A., Cappelli, G., Ballestri, M., Inguaggiato, P., Fortunato, L., Milan, M., Pietribiasi, G., and La Greca, G. (1999). In vitro and in vivo evaluation of a new polysulfone membrane for hemodialysis. Reference methodology and clinical results. (Part. 2: In vivo study). *Int. J. Artif. Organs* **22**, 616–624.
14. Fushimi, F., Nakayama, M., Nishimura, K., and Hiyoshi, T. (1998). Platelet adhesion, contact phase coagulation activation, and C5a generation of polyethylene glycol acid–grafted high flux cellulosic membrane with varieties of grafting amounts. *Artif. Organs* **22**, 821–826.
15. Hoenich, N., Stamp, S., and Roberts, S. (2000). A microdomain-structured synthetic high-flux hollow-fiber membrane for renal replacement therapy. *ASAIO J.* **46**, 70–75.
16. Hancock, L., Fagan, S., and Ziolo, M. (2000). Hydrophilic, semipermeable membranes fabricated with poly(ethylene oxide)–polysulfone block copolymer. *Biomaterials* **21**, 725–733.
17. Golper, T. (1986). Continuous arteriovenous hemofiltration in acute renal failure. *Am. J. Kidney Dis.* **6**, 373–381.
18. Kramer, P., Wigger, W., Rieger, J., Matthaei, D., and Scheler, F. (1977). Arteriovenous haemofiltration: A new and simple method for treatment of overhydrated patients resistant to diuretics. *Klin. Wochenschr.* **55**, 1121–1122.
19. Campistol, J. M., Torregrosa, J. V., Ponz, E., and Fenollosa, B. (1999). Beta2-microglobulin removal by hemodialysis with polymethylmethacrylate membranes. *Contrib. Nephrol.* **125**, 76–85.
20. Clark, W. R., Macias, W. L., Molitoris, B. A., and Wang, N. H. (1994). Membrane adsorption of beta2-microglobulin: Equilibrium and kinetic characterization. *Kidney Int.* **46**, 1140–1146.
21. Floege, J., and Koch, K. M. (1994). Beta 2-microglobulin associated amyloidosis and therapy with high flux hemodialysis membranes. *Clin. Nephrol.* **42**(Suppl. 1), S52–S56.

22. Schaefer, R., Gilge, U., Goehl, H., and Heidland, A. (1990). Evaluation of a new polyamide membrane (Polyflux 130) in high-flux dialysis. *Blood Purif.* **8**, 23–31.

23. Himmelfarb, J., Tolkoff Rubin, N., Chandran, P., Parker, R. A., Wingard, R. L., and Hakim, R. (1998). A multicenter comparison of dialysis membranes in the treatment of acute renal failure requiring dialysis. *J. Am. Soc. Nephrol.* **9**, 257–266.

24. Eberhart, R. C., and Clagett, C. P. (1991). Catheter coatings, blood flow, and biocompatibility. *Semin. Hematol.* **28**, 42–48.

25. Mammen, E. (1999). Low molecular weight heparins and heparin-induced thrombocytopenia. *Clin. Appl. Thromb. Hemostasis* **5**(Suppl. 1), S72–S75.

26. Blezer, R., Fouache, B., Willems, G., and Lindhout, T. (1997). Activation of blood coagulation at heparin-coated surfaces. *J. Biomed. Mater. Res.* **37**, 108–113.

27. Kim, J. S., Vincent, C., Teng, C. L., Wakefield, T. W., and Yang, V. C. (1989). A novel approach to anticoagulation control. *ASAIO Trans.* **35**, 644–646.

28. Dodt, J., Otte, M., Strube, K. H., and Friedrich, T. (1996). Thrombin inhibitors of bloodsucking animals. *Semin. Thromb. Hemostasis* **22**, 203–208.

29. Beretz, A., and Cazenave, J. P. (1991). Old and new natural products as the source of modern antithrombotic drugs. *Planta Med.* **57**, S68–S72.

30. Heras, M., Chesebro, J., Webster, M., Mruk, J., Grill, D., Penny, W., Bowie, E., Badimon, L., and Fuster, V. (1990). Hirudin, heparin, and placebo during deep arterial injury in the pig. The in vivo role of thrombin in platelet-mediated thrombosis. *Circulation* **82**, 1476–1484.

31. Fareed, J., Lewis, B., Callas, D., Hoppensteadt, D., Walenga, J., and Bick, R. (1999). Antithrombin agents: The new class of anticoagulant and antithrombotic drugs. *Clin. Appl. Thromb. Hemostasis* **5**, S45–S55.

32. Lombardi, A., De Simone, G., Galdiero, S., Staiano, N., Nastri, F., and Pavone, V. (1999). From natural to synthetic multisite thrombin inhibitors. *Biopolymers* **51**, 19–39.

33. Maraganore, J., Chao, B., Joseph, M., Jablonski, J., and Ramachandran, K. (1989). Anticoagulant activity of synthetic hirudin peptides. *J. Biol. Chem.* **264**, 8692–8698.

34. Bucha, E., Nowak, G., Czerwinski, R., and Thieler, H. (1999). r-Hirudin as anticoagulant in regular hemodialysis therapy: Finding of therapeutic r-hirudin blood/plasma concentrations and respective dosages. *Clin. Appl. Thromb. Hemostasis* **5**, 164–170.

35. Fischer, K., van de Loo, A., and Bohler, J. (1999). Recombinant hirudin (lepirudin) as anticoagulant in intensive care patients treated with continuous hemodialysis. *Kidney Int., Suppl.* **72**, S46–S50.

36. Edmunds, L. J. (1994). Surface-bound heparin—Panacea or peril? *Ann. Thorac. Surg.* **58**, 285–286.

37. Seifert, B., Romaniuk, P., and Groth, T. (1997). Covalent immobilization of hirudin improves the haemocompatibility of polylactide–polyglycolide in vitro. *Biomaterials* **18**, 1495–1502.

38. Mori, Y., Nagaoka, S., Takiuchi, H., Kikuchi, T., Noguchi, N., Tanzawa, H., and Noisshiki, Y. (1982). A new antithrombogenic material with long polyethylene oxide chains. *Trans. Am. Soc. Artif. Intern. Organs* **28**, 459–463.

39. Amiji, M., and Park, K. (1993). Surface modification of polymeric biomaterials with poly(ethylene oxide), albumin, and heparin for reduced thrombogenicity. *J. Biomater. Sci., Polym. Ed.* **4**, 217–234.

40. Ishihara, K., Takayama, R., Nakabayashi, N., Fukumoto, K., and Aoki, J. (1992). Improvement of blood compatibility on cellulose dialysis membranes. 2. Blood compatibility of phospholipid polymer grafted cellulose membrane. *Biomaterials* **13**, 235–239.

41. Kolff, W. J. (1998). The early years of artificial organs at the Cleveland Clinic. Part I: Artificial kidney and kidney transplantation. *ASAIO J.* **44**, 3–11.

42. Lysaght, M. J. (1995). Evolution of hemodialysis membranes. *Contrib. Nephrol.* **113**, 1–10.

43. Ishihara, K. (1993). Blood compatible polymers. In "Biomedical Applications of Polymeric Materials" (T. Tsuruta *et al.*, eds.), pp. 89–115. CRC Press, Boca Raton, FL.

44. Kim, S. W., and Jacobs, H. (1996). Design of nonthrombogenic polymer surfaces for blood-contacting medical devices. *Blood Purif.* **14**, 357–372.

45. Vanholder, R. (1993). Hybrid bioartificial kidney: Pitfalls and present state of the art. *Contrib. Nephrol.* **103**, 160–167.

46. Schneider, P. A., Hanson, S. R., Price, T. M., and Harker, L. A. (1988). Preformed confluent endothelial cell monolayers prevent early platelet deposition on vascular prostheses in baboons. *J. Vasc. Surg.* **8**, 229–235.

47. Shepard, A. D., Eldrup-Jorgensen, J., Keough, E. M. *et al.* (1986). Endothelial cell seeding of small-caliber synthetic grafts in the baboon. *Surgery* **99**, 318–326.

48. Kadletz, M., Magometschnigg, H., Minar, E., Konig, G., Grabenwoger, M., Grimm, M., and Wolner, E. (1992). Implantation of in vitro endothelialized polytetrafluoroethylene grafts in human beings. A preliminary report. *J. Thorac. Cardiovasc. Surg.* **104**, 736–742.

49. Chen, D., Riesbeck, K., Kemball-Cook, G., McVey, J. H., Tuddenham, E. G., Lechler, R. I., and Dorling, A. (1999). Inhibition of tissue factor-dependent and -independent coagulation by cell surface expression of novel anticoagulant fusion proteins. *Transplantation* **67**, 467–474.

50. Wilson, J. M., Birinyi, L. K., Salomon, R. N., Libby, P., Callow, A. D., and Mulligan, R. C. (1989). Implantation of vascular grafts lined with genetically modified endothelial cells. *Science* **244**, 1344–1346.

51. Zwiebel, J., Freeman, S., Kantoff, P., Cornetta, K., Ryan, U., and Anderson, W. (1989). High-level recombinant gene expression in rabbit endothelial cells transduced by retroviral vectors. *Science* **243**, 220–222.

52. Dichek, D., Neville, R., Zwiebel, J., Freeman, S., Leon, M., and Anderson, W. (1989). Seeding of intravascular stents with genetically engineered endothelial cells. *Circulation* **80**, 1347–1353.

53. Humes, H. D., and Cieslinski, D. A. (1996). Methods and compositions of a bioartificial kidney suitable for use in vivo and ex vivo. U.S. Pat. 5,549,674.

54. Scacheri, E., Nitti, G., Valsasina, B., Orsini, G., Visco, C., Ferrera, M., Sawyer, R. T., and Sarmientos, P. (1993). Novel hirudin variants from the leech *Hirudinaria manillensis*. Amino acid sequence, cDNA cloning and genomic organization. *Eur. J. Biochem.* **214**, 295–304.

55. Strube, K. H., Kroger, B., Bialojan, S., Otte, M., and Dodt, J. (1993). Isolation, sequence analysis, and cloning of haemadin. An anticoagulant peptide from the Indian leech. *J. Biol. Chem.* **268**, 8590–8595.

56. Rade, J. J., Cheung, M., Miyamoto, S., and Dichek, D. A. (1999). Retroviral vector-mediated expression of hirudin by human vascular endothelial cells: Implications for the design of retroviral vectors expressing biologically active proteins. *Gene Ther.* **6**, 385–392.

57. Riesbeck, K., Chen, D., Kemball-Cook, G., McVey, J. H., George, A. J., Tuddenham, E. G., Dorling, A., and Lechler, R. I. (1998). Expression of hirudin fusion proteins in mammalian cells: A strategy for prevention of intravascular thrombosis. *Circulation* **98**, 2744–2752.

58. Chen, D., Riesbeck, K., McVey, J. H., Kemball-Cook, G., Tuddenham, E. G., Lechler, R. I., and Dorling, A. (1999). Regulated inhibition of coagulation by porcine endothelial cells expressing P-selectin-tagged hirudin and tissue factor pathway inhibitor fusion proteins. *Transplantation* **68**, 832–839.

59. Green, D. F., Hwang, K. H., Ryan, U. S., and Bourgoignie, J. J. (1992). Culture of endothelial cells from baboon and human glomeruli. *Kidney Int.* **41**, 1506–1516.

60. Masek, L. C., and Sweetenham, J. W. (1994). Isolation and culture of endothelial cells from human bone marrow. *Br. J. Haematol.* **88**, 855–865.

61. Boyer, M., Townsend, L., Vogel, L., Falk, J., Reitz-Vick, D., Trevor, K., Villalba, M., Bendick, P., and Glover, J. (2000). Isolation of endothelial cells and their progenitor cells from human peripheral blood. *J. Vasc. Surg.* **31**, 181–189.

62. Asahara, T., Murohara, T., Sullivan, A., Silver, M., van der Zee, R., Li, T., Witzenbichler, B., Schatteman, G., and Isner, J. M. (1997). Isolation of putative progenitor endothelial cells for angiogenesis. *Science* **275**, 964–967.

63. Lin, H. B., Garcia-Echeverria, C., Asakura, S., Sun, W., Mosher, D. F., and Cooper, S. L. (1992). Endothelial cell adhesion on polyurethanes containing covalently attached RGD-peptides. *Biomaterials* **13**, 905–914.

64. Nikolaychik, V. V., Samet, M. M., and Lelkes, P. I. (1994). A new, cryoprecipitate based coating for improved endothelial cell attachment and growth on medical grade artificial surfaces. *ASAIO J.* **40**, M846–M852.

65. Terramani, T. T., Eton, D., Bui, P. A., Wang, Y., Weaver, F. A., and Yu, H. (2000). Human macrovascular endothelial cells: Optimization of culture conditions. *In Vitro Cell. Dev. Biol.—Anim.* **36**, 125–132.

66. Carley, W., Milici, A., and Madri, J. (1988). Extracellular matrix specificity for the differentiation of capillary endothelial cells. *Exp. Cell Res.* **178**, 426–434.

67. Esser, S., Wolburg, K., Wolburg, H., Breier, G., Kurzchalia, T., and Risau, W. (1998). Vascular endothelial growth factor induces endothelial fenestrations in vitro. *J. Cell Biol.* **140**, 947–959.

68. Hammes, M., and Singh, A. (1994). Effect of polycations on permeability of glomerular epithelial cell monolayers to albumin. *J. Lab. Clin. Med.* **123**, 437–446.

69. Ballermann, B. J. (1989). Regulation of bovine glomerular endothelial cell growth in vitro. *Am. J. Physiol.* **256**, C182–C189.

70. Carey, D. (1991). Control of growth and differentiation of vascular cells by extracellular matrix proteins. *Annu. Rev. Physiol.* **53**, 161–177.

71. Grant, D., Tashiro, K., Segui-Real, B., Yamada, Y., Martin, G., and Kleinman, H. (1989). Two different laminin domains mediate the differentiation of human endothelial cells into capillary-like structures in vitro. *Cell (Cambridge, Mass.)* **58**, 933–943.

72. Liu, A., Dardik, A., and Ballermann, B. J. (1999). Neutralizing TGF-beta1 antibody infusion in neonatal rat delays in vivo glomerular capillary formation. *Kidney Int.* **56**, 1334–1348.

73. Takahashi, T., Huynh-Do, U., and Daniel, T. O. (1998). Renal microvascular assembly and repair: Power and promise of molecular definition. *Kidney Int.* **53**, 826–835.

74. Aebischer, P., Ip, T. K., Panol, G., and Galletti, P. M. (1987). The bioartificial kidney: Progress towards an ultrafiltration device with renal epithelial cells processing. *Life Support Syst.* **5**, 159–168.

75. Ip, T. K., Aebischer, P., and Galletti, P. M. (1988). Cellular control of membrane permeability. Implications for a bioartificial renal tubule. *ASAIO Trans.* **34**, 351–355.

76. Humes, H. D. (1996). The bioartificial renal tubule: Prospects to improve supportive care in acute renal failure. *Renal Failure* **18**, 405–408.

77. Humes, H. D., MacKay, S. M., Funke, A. J., and Buffington, D. A. (1999). Tissue engineering of a bioartificial renal tubule assist device: In vitro transport and metabolic characteristics. *Kidney Int.* **55**, 2502–2514.

78. Humes, H. D., Buffington, D. A., MacKay, S. M., Funke, A. J., and Weitzel, W. F. (1999). Replacement of renal function in uremic animals with a tissue-engineered kidney. *Nat. Biotechnol.* **17**, 451–455.

79. Uludag, H., Ip, T. K., and Aebischer, P. (1990). Transport functions in a bioartificial kidney under uremic conditions. *Int. J. Artif. Organs* **13**, 93–97.

80. Ip, T. K., and Aebischer, P. (1989). Renal epithelial-cell-controlled solute transport across permeable membranes as the foundation for a bioartificial kidney. *Artif. Organs* **13**, 58–65.

81. Horster, M. (1979). Primary culture of mammalian nephron epithelia: Requirements for cell outgrowth and proliferation from defined explanted nephron segments. *Pfluegers Arch.* **382**, 209–215.

82. Horster, M. (1980). Tissue culture in nephrology: Potential and limits for the study of renal disease. *Klin. Wochenschr.* **58**, 965–973.

83. Qiao, J., Cohen, D., and Herzlinger, D. (1995). The metanephric blastema differentiates into collecting system and nephron epithelia in vitro. *Development (Cambridge, UK)* **121**, 3207–3214.

84. Venkatachalam, M. A., Bernard, D. B., Donohoe, J. F., and Levinsky, N. G. (1978). Ischemic damage and repair in the rat proximal tubule: Differences among the S1, S2, and S3 segments. *Kidney Int.* **14**, 31–49.

85. Cuppage, F. E., and Tate, A. (1967). Repair of the nephron following injury with mercuric chloride. *Am. J. Pathol.* **51**, 405–429.

86. Coimbra, T. M., Cieslinski, D. A., and Humes, H. D. (1990). Epidermal growth factor accelerates renal repair in mercuric chloride nephrotoxicity. *Am. J. Physiol.* **259**, F438–F443.

87. Humes, H. D., and Cieslinski, D. A. (1992). Interaction between growth factors and retinoic acid in the induction of kidney tubulogenesis in tissue culture. *Exp. Cell Res.* **201**, 8–15.

88. Humes, H. D., Krauss, J. C., Cieslinski, D. A., and Funke, A. J. (1996). Tubulogenesis from isolated single cells of adult mammalian kidney: Clonal analysis with a recombinant retrovirus. *Am. J. Physiol.* **271**, F42–F49.

89. Humes, H. D., Beals, T. F., Cieslinski, D. A., Sanchez, I. O., and Page, T. P. (1991). Effects of transforming growth factor-beta, transforming growth factor-alpha, and other growth factors on renal proximal tubule cells. *Lab. Invest.* **64**, 538–545.

90. Schena, F. P. (1998). Role of growth factors in acute renal failure. *Kidney Int., Suppl.* **66**, S11–S15.

91. Humes, H. D., MacKay, S. M., Funke, A. J., and Buffington, D. A. (1997). Acute renal failure: Growth factors, cell therapy, and gene therapy. *Proc. Assoc. Am. Physicians* **109**, 547–557.

92. Miner, J. H. (1999). Renal basement membrane components. *Kidney Int.* **56**, 2016–2024.

93. Kanai, N., Fujita, Y., Kakuta, T., and Saito, A. (1999). The effects of various extracellular matrices on renal cell attachment to polymer surfaces during the development of bioartificial renal tubules. *Artif. Organs* **23**, 114–118.

94. Tam, V. K., Clemens, T. L., and Green, J. (1998). The effect of cell–matrix interaction on parathyroid hormone (PTH) receptor binding and PTH responsiveness in proximal renal tubular cells and osteoblast-like cells. *Endocrinology (Baltimore)* **139**, 3072–3080.

95. Cozzi, E., and White, D. (1995). The generation of transgenic pigs as potential organ donors for humans. *Nat. Med.* **1**, 964–966.

96. Humes, H. D., MacKay, S. M., Funke, A. J., and Buffington, D. A. (1997). The bioartificial renal tubule assist device to enhance CRRT in acute renal failure. *Am. J. Kidney Dis.* **30**, S28–S31.

97. MacKay, S. M., Funke, A. J., Buffington, D. A., and Humes, H. D. (1998). Tissue engineering of a bioartificial renal tubule. *ASAIO J.* **44**, 179–183.

98. Nikolovski, J., Gulari, E., and Humes, H. D. (1999). Design engineering of a bioartificial renal tubule cell therapy device. *Cell Transplant.* **8**, 351–364.

99. Sanaka, T., Agishi, T., Hayashi, T., Hayasaka, Y., Yasuo, M., Fukui, K., Ota, K., and Sugino, N. (1989). Extracorporeal hybridization of proximal renal tubular cells and an artificial membrane for the purpose of beta2 microglobulin removal. *ASAIO Trans.* **35**, 527–530.

100. Gejyo, F., Homma, N., Hasegawa, S., and Arakawa, M. (1993). A new therapeutic approach to dialysis amyloidosis: Intensive removal of beta2-microglobulin with adsorbent column. *Artif. Organs* **17**, 240–243.

101. Drueke, T., Touam, M., and Zingraff, J. (1995). Dialysis-associated amyloidosis. *Adv. Renal Replace. Ther.* **2**, 24–39.

102. Humes, H. D. (2000). Bioartificial kidney for full renal replacement therapy. *Semin. Nephrol.* **20**, 71–82.

103. Moussy, Y. (2000). Bioartificial kidney. I. Theoretical analysis of convective flow in hollow fiber modules: Application to a bioartificial hemofilter. *Biotechnol. Bioeng.* **68**, 142–152.

104. Moussy, Y. (2000). Bioartificial kidney. II. A convective flow model of a hollow fiber bioartificial renal tubule. *Biotechnol. Bioeng.* **68**, 153–159.

CHAPTER 89

INTRACORPOREAL KIDNEY

James J. Yoo, Tae-Gyun Kwon, and Anthony Atala

INTRODUCTION

End-stage renal failure is a devastating disease that involves multiple organs in affected individuals. Although currently available treatment modalities, including dialysis and transplantation, can prolong survival for many patients, problems such as donor shortage, complications, and graft failure remain a continuing concern. Numerous investigative efforts have been attempted to improve, restore, or replace renal function. Tissue engineering approaches using cell therapies have been proposed as an alternative method. In this chapter we describe various engineering approaches to the achievement of functional intracorporeal kidney support.

Owing to its complex structure and function, the kidney is the most challenging organ to reconstruct in the genitourinary system. Normal renal function includes synthesis of 1,25-vitamin D_3, erythropoietin, glutathione, and free radical scavenging enzymes. The kidney also participates in the catabolism of low molecular weight proteins [1] and in the production and regulation of cytokines [2,3]. Because these functions could not be replaced with dialysis therapy, long-term problems, such as anemia and malnutrition, are prevalent in such patients.

Renal replacement therapies now available are limited owing to the inherent shortage of donor organs, as well as to associated complications involving dialysis, transplantation, and immunosuppressive therapies [4–7]. Limitation of current therapies for renal functional augmentation has led investigators to pursue alternative therapeutic modalities.

The concept of using tissue engineering techniques for cell transplantation has been proposed as a method of improving, restoring, or replacing renal function [8–11]. The emergence of tissue engineering strategies has presented unlimited possibilities for the management of pathological renal conditions. Augmentation of either isolated or total renal function with kidney cell expansion may be a feasible solution. We have followed an approach that involves the development of intracorporeal support systems for renal functional replacement.

COMPONENTS REQUIRED FOR THE ENGINEERING OF KIDNEY TISSUE

Components required to achieve partial or total renal function are renal cells, three-dimensional scaffolds, and an *in vivo* environment. The challenge associated with renal cell culture is due to the unique structural and cellular heterogeneity present within the kidney. The system of nephrons and collecting ducts is composed of multiple functionally and phenotypically distinct segments. For this reason, appropriate conditions need to be provided for long-term survival, differentiation, and growth of the cells. Extensive research has been performed to determine optimal growth conditions for renal cell enrichment [12–17]. Based on the literature and our experience, we were able to obtain optimal growth conditions for a stable cell culture system.

Although isolated renal cells retain their phenotype and function in culture, transplantation of these cells *in vivo* may not result in structural remodeling. In addition, the limited diffusion may prevent the implantation of cell or tissue components in large volumes [18]. Thus, a cell support matrix is necessary to allow diffusion of nutrients across the entire implant. A variety of synthetic materials were examined to determine ideal support structures for the regeneration of urolqgic tissue [19,20]. Biodegradable materials, such as poly(lactic acid) and poly(glycolic acid) polymers, and collagen scaffolds, have been preferred for their biocompatibility and ease of processing [21–25].

ENGINEERING RENAL STRUCTURES *IN VIVO*

The kidney is responsible not only for urine excretion but for several other important metabolic functions. Our initial study involved the use of tissue engineering techniques to investigate the feasibility of achieving renal cell growth, expansion, and *in vivo* reconstitution [19]. New Zealand white rabbits underwent nephrectomy and renal artery perfusion with a nonoxide solution that promoted iron particle entrapment in the glomeruli. Homogenization of the renal cortex and fractionation in 83- and 210-μm sieves with subsequent magnetic extraction yielded three separate purified suspensions of distal tubules, glomeruli, and proximal tubules. The cells were plated separately *in vitro* and seeded onto biodegradable poly(glycolic acid) scaffolds. Polymer scaffolds were implanted subcutaneously into host athymic mice. Included were implants of proximal tubular cells, glomeruli, distal tubular cells, and a mixture of all three cell types. Polymers alone served as controls. Animals were sacrificed at 1 week, 2 weeks, and 1 month after implantation, and the retrieved scaffolds were analyzed. An acute inflammatory phase and a chronic foreign body reaction were seen, accompanied by vascular ingrowth by 7 days after implantation. Histological examination demonstrated progressive formation and organization of the nephron segments within the polymer fibers with time. Renal cell proliferation in the cell–polymer scaffolds was detected by *in vivo* labeling of replicating cells with the thymidine analogue bromodeoxyuridine (BrdU). BrdU incorporation into renal cell DNA was identified immunocytochemically with monoclonal anti-BrdU antibodies. These results demonstrated that renal-specific cells can be successfully harvested, survive in culture, and attach to artificial biodegradable polymers. The renal cell–polymer scaffolds can be implanted into host animals, where the cells replicate and organize into nephron segments as the polymer, which acts as a cell delivery vehicle, undergoes biodegradation.

The initial experiments demonstrated that implanted cell–polymer scaffolds gave rise to renal tubular structures. However, it was unclear whether the tubular structures had reconstituted *de novo* from dispersed renal elements or merely represented fragments of donor tubules that had survived intact. Further investigation was conducted to examine the tubular reconstitution process [26]. Mouse renal cells were harvested, grown, and expanded in culture. Subsequently, single isolated cells were seeded on biodegradable polymers and implanted into syngeneic hosts. Renal epithelial cells were observed to reconstitute into tubular structures *in vivo*. Sequential analyses of the retrieved implants over time demonstrated that renal epithelial cells first organized into a cordlike structure with a solid center. Subsequent canalization into a hollow tube could be seen by 2 weeks. Histological examination with nephron-segment-specific lactins showed successful reconstitution of proximal tubules, distal tubules, loops of Henle, collecting tubules, and collecting ducts. These results showed that single suspended cells are capable of reconstituting into tubular structures, with homogeneous cell types in each tubule.

CREATION OF FUNCTIONING RENAL UNITS *IN VIVO*

The kidneys are critical to body homeostasis because of their excretory, regulatory, and endocrinological functions. The excretory function is initiated by filtration of blood at the glomerulus, and the regulatory function is provided by the tubular segments. Although our prior studies demonstrated that renal cells seeded on biodegradable polymer scaffolds are able to reconstitute into renal structures *in vivo*, the type and structural configuration of polymers used prevented the achievement of complete renal function. In our subsequent study we explored the feasibility of creating a functional artificial renal unit, wherein urine production could be achieved [27]. Mouse renal cells were harvested and expanded in culture. The cells were seeded onto a tubular device constructed from polycarbonate (4 μm pore size), connected at one end with a Silastic catheter that terminated in a reservoir. The device was implanted in the subcutaneous space of athymic mice. Animals were sacrificed at 1, 2, 3, 4, and 8 weeks after implantation, and the retrieved specimens were examined histologically and immunocytochemically. Fluid was collected from inside the implant, and uric acid and creatinine levels were determined.

Histological examination of the implanted device revealed extensive vascularization, as well as formation of glomeruli and highly organized tubule-like structures. Immunocytochemical staining with anti-osteopontin antibody, which is secreted by proximal and distal tubular cells and the cells of the thin ascending loop of Henle, stained the tubular sections.

Immunohistochemical staining for alkaline phosphatase stained proximal tubule–like structures. Uniform staining for fibronectin in the extracellular matrix of newly formed tubes was observed. The fluid collected from the reservoir was yellow and contained 66 mg/dl uric acid (vs 2 mg/dl in plasma), suggesting that these tubules are capable of unidirectional secretion and concentration of uric acid. The creatinine assay performed on the collected fluid showed an 8.2-fold increase in concentration, in comparison to serum. These results demonstrated that single cells form multicellular structures and become organized into functional renal units that are able to excrete high levels of solutes through a urine-like fluid.

USING A COLLAGEN-BASED KIDNEY MATRIX FOR RENAL TISSUE ENGINEERING

In our earlier study, we showed that renal cells seeded on synthetic renal devices with a collecting system are able to form functional renal structures with urinelike fluid excretion. However, a naturally derived tissue matrix with an existing three-dimensional kidney architecture would be preferable, in that it would allow for transplantation of a large number of cells for the creation of greater renal tissue volumes.

We developed an acellular collagen-based kidney matrix, which is identical to the native renal architecture; the protocol is given shortly. In a subsequent study we investigated whether the collagen-based matrices could accommodate large volumes of renal cells that could proliferate and form kidney structures *in vivo* [28].

Acellular collagen matrices, derived from porcine kidneys, were obtained through a multiple-step decellularization process. Serial evaluation of the matrix for cellular remnants was performed by using histochemistry, scanning electron microscopy (SEM), and reverse transcription–polymerase chain reaction analysis (RT-PCR). Mouse renal cells were harvested, grown, and seeded on 80 collagen matrices at a concentration of 30×10^6 cells/ml. Forty cell–matrix constructs grown *in vitro* were analyzed 3 days, 1, 2, 4, and 6 weeks after seeding. The remaining 40 cell–matrices were implanted in the subcutaneous space of 20 athymic mice. The animals were sacrificed 3 days, 1, 2, 4, 8, and 24 weeks after implantation for analyses. Gross, SEM, histochemical, immunocytochemical, and biochemical analyses were performed.

Scanning electron microscopy and histological examination confirmed the acellularity of the processed matrix. RT-PCR analysis performed on the kidney matrices demonstrated the absence of any RNA residues. Renal cells seeded on the matrix adhered to the inner surface and proliferated to confluency 7 days after seeding, as demonstrated by SEM. Histochemical and immunocytochemical analyses performed using with hematoxylin and eosin, periodic acid Schiff, alkaline phosphatase, anti-osteopontin, and anti-CD-31 identified stromal, endothelial, and tubular epithelial cell phenotypes within the matrix. Renal tubular and glomerulus-like structures were observed 8 weeks after implantation. 3-(4,5-dimethylthiazol-2-yl)-2,5-diphenyltetrazolium-bromide (MTT) proliferation and radioactive thymidine incorporation assays performed 6 weeks after cell seeding demonstrated a cell population increase of 116 and 92%, respectively, in comparison to the 2-week time points. This study demonstrates that renal cells are able to adhere and proliferate on the collagen-based kidney matrices. The renal cells reconstitute renal tubular and glomerulus-like structures. The collagen-based kidney matrix system seeded with renal cells may be useful in the future for augmenting renal function.

Several challenges remain in engineering renal tissue for intracorporeal use. The kidney mass needs to be large enough for functional replacement while allowing for adequate vascularization. In addition, efficient systems for urine excretion need to be optimized. Work in our laboratory is aimed at addressing these challenges.

PROTOCOL

Preparation of Acellular Collagen-Based Matrix

1. Prepare kidney tissue.
   a. Surgically remove fresh kidneys.
   b. Sagittally slice each kidney with a scalpel.
   c. Remove collecting system and renal capsule.
   d. Rinse tissue thoroughly.
2. Treat tissue with distilled water in a magnetic stirring flask (moderate speed) overnight at 4°C. This step results in cell membrane rupture and removes cellular debris.
3. Remove distilled water and treat with decellularizing solution.
   a. To 1 liter of isotonic solution (phosphate-buffered saline: 1× PBS), add the following reagents: 10 ml of Triton X-100 and 1 ml of 1% ammonium hydroxide.
   b. Agitate for 15 min to dissolve Triton X-100.
   c. Place tissue slices into a flask containing decellularizing solution.
   d. Place the container securely on the agitator at 4°C (~120 rpm).
4. Replace solution after 48 h and repeat agitation.
5. Rinse the tissue slices thoroughly when replacing the solution (step 3).
6. Wash vigorously with distilled water.
   a. Place the kidney matrix in a clean flask containing 1× PBS and securely place on the rotator at 4°C.
   b. Change solution (1× PBS) every 4 h three times.
   c. Leave the flask in the rotator with 1× PBS overnight.
7. Lyophilize the tissue matrix samples.
8. Pack and sterilize in ethylene oxide (cold cycle recommended).
9. Store until used.
10. When ready to use, equilibrate the tissue in medium overnight prior to cell seeding.

REFERENCES

1. Maack, T. (1992). Renal handling of proteins and polypeptides. *In* "Handbook of Physiology" (E. E. Windhager, ed.), pp. 2039–2118. Oxford University Press, New York.
2. Engler, F. J., Blum, G., Rodemann, H. P., *et al.* (1993). Human renal tubular cells as a cytokine source: PDGF-B, GM-CSF AND IL-6 mRNA expression in vitro. *Exp. Nephrol.* **1**, 26–35.
3. Stadnyk, A. W. (1994). Cytokine production by epithelial cells. *FASEB J.* **8**, 1041–1047.
4. Amiel, G., Yoo, J. J., and Atala, A. (2000). Renal therapy using tissue engineered constructs and gene therapy. *World J. Urol.* **18**, 71–79.
5. Chazan, J. A., Libbey, N. P., London, M. R., *et al.* (1991). The clinical spectrum of renal osteodystrophy in 57 hemodialysis patients: A correlation between biochemical parameters and bone pathology findings. *Clin. Nephrol.* **35**, 78–85.
6. Cohen, J., Hopkin, J., and Kurtz, J. (1994). Infectious complications after renal transplantation. *In* "Kidney Transplantation: Principles and Practice" (P. J. Morris, ed.), pp. 364–389. Saunders, Philadelphia.
7. Feldman, H. I., Kobrin, S., and Wasserstein, A. (1996). Hemodialysis vascular access morbidity. *J. Am. Soc. Nephrol.* **7**, 523–535.
8. Amiel, G. E., and Atala, A. (1999). Current and future modalities for functional renal replacement. *Urol. Clin. North Am.* **26**, 235–246.
9. Atala, A. (1999). Future perspectives in reconstructive surgery using tissue engineering. *Urol. Clin. North Am.* **26**, 157–165.
10. Atala, A. (1997). Tissue engineering in the genitourinary system. *In* "Tissue Engineering" (A. Atala and D. Mooney, eds.), Chapter 8. Birkhaeuser, Boston.
11. Humes, D. H., Buffington, D. A., MacKay, S. M., Funke, A. J., and Weitzel, W. F. (1999). Replacement of renal function in uremic animals with a tissue engineered kidney. *Nat. Biotechnol.* **17**, 451–455.
12. Brezis, M., and Rosen, S. (1995). Hypoxia of the renal medulla—Its implications for disease. *N. Engl. J. Med.* **332**, 647–655.
13. Schena, F. P. (1998). Role of growth factors in acute renal failure. *Kidney Int.* **53**(Suppl. 66), S-11–S-15.

14. Milici, A. J., Furie, M. B., and Carley, W. W. (1985). The formation of fenestrations and channels by capillary endothelium in vitro. *Proc. Natl. Acad. Sci. U.S.A.* **82**, 6181–6185.

15. Carley, W. W., Milici, A. J., and Madri, J. A. (1988). Extracellular matrix specificity for the differentiation of capillary endothelial cells. *Exp. Cell Res.* **178**, 426–434.

16. Humes, H. D, and Ceileski, D. A. (1992). Interaction between growth factors and retinoic acid in the induction of key tubulogenesis in tissue culture. *Exp. Cell Res.* **201**, 8–15.

17. Hirokoshi, S., Koide, H., and Shirai, T. (1988). Monoclonal antibodies against laminin A chain and B chain in the human and mouse kidneys. *Lab. Invest.* **58**, 532–538.

18. Folkman, J., and Hochberg, M. M. (1973). Self-regulation of growth in three dimensions. *J. Exp. Med.* **138**, 745–749.

19. Atala, A., Schlussel, R. N., and Retik, A. B. (1995). Renal cell growth in vivo after attachment to biodegradable polymer scaffolds. *J. Urol.* **153**, 4.

20. Tachibana, M., Nagamatsu, G. R., and Addonizio, J. C. (1985). Ureteral replacement using collagen tube sponge graftS. *J. Urol.* **133**, 866–869.

21. Mooney, D. J., Breuer, C., McNamara, K., *et al.* (1996). Fabricating tubular devices from polymers of lactic and glycolic acid for tissue engineering. *Tissue Eng.* **1**, 107–118.

22. Mikos, A. G., Sarakinos, G., Lyman, M. D., *et al.* (1993). Prevascularization of porous biodegradable polymers. *Biotechnol. Bioeng.* **42**, 716–723.

23. White, R. A., Hirose, F. M., Sproat, R. W., *et al.* (1981). Histologic observations after short-term implantation of two porous elastomers in dogs. *Biomaterials* **2**, 171–176.

24. Freed, L. E., Vunjak-Novakovic, G., Biron, R. J., Eagles, D. B., Lesnoy, D. C., Barlow, S. K., and Langer, R. (1994). Biodegradable polymer scaffolds for tissue engineering. *Bio/Technology* **12**, 689–693.

25. Hubbel, J. A., Massia, S. P., *et al.* (1991). Endothelial cell-selective materials for tissue engineering in the vascular graft via a new receptor. *Bio/Technology* **9**, 568–572.

26. Fung, L. C. T., Elenius, K., Freeman, M., Donovan, M. J., and Atala, A. (1996). Reconstitution of poor EGFr-poor renal epithelial cells into tubular structures on biodegradable polymer scaffold. *Pediatrics* **98**(Suppl.), S631.

27. Yoo, J. J., Ashkar, S., and Atala, A. (1996). Creation of functional kidney structures with excretion of kidney-like fluid in vivo. *Pediatrics* **98S**, 605.

28. Amiel, G. E., Yoo, J. J., and Atala, A. (2000). Renal tissue engineering using a collagen-based kidney matrix. *Tissue Eng.*, Suppl. **6**, 675.

CHAPTER 90

URETHRAL TISSUE REPAIR

James J. Yoo, Shay Soker, and Anthony Atala

INTRODUCTION

When there is limited urethral mucosa for adequate reconstruction, tissues from other sources have been used, such as genital and extragenital skin flaps or grafts, mucosal grafts from the bladder or buccal regions, tunica vaginalis, and peritoneal grafts [1–3]. However, problems related to these materials have been reported. Complications such as hair growth, graft shrinkage, stricture, stone formation, and diverticuli have been associated with skin grafts [4–8]. Bladder mucosa free grafts have been used for urethral reconstruction; however, problems with mucosal glandular protrusion and donor site morbidity have limited their applicability [2]. Other types of mucosal tissue (e.g., buccal, labial, rectal grafts) have also been tried clinically for urethral repair, with various results [1,9]. A free graft or flap derived from tunica vaginalis has been tried clinically in hypospadiac patients, but all developed meatal stenosis [10].

A variety of synthetic grafts composed of silicone, Teflon, and Dacron have been proposed for urethral reconstruction. These materials have been associated with erosion, dislodgment, fistula, stenosis, extravasation, or calcification [11–13]. The problems associated with the use of nondegradable materials led to the investigation of biodegradable substitutes. A polyglactin fiber mesh tube coated with poly(hydroxybutyric acid) and hyaluronan benzyl ester has been used experimentally. Complete regeneration of the urethral epithelium and the adjacent connective tissue was achieved. The biodegradable polymer meshes served as scaffolds that guided urothelial and connective tissue regeneration [14,15].

Although several innovative tissues have been proposed as possible free grafts for urethral repair, it is evident that all have specific advantages and disadvantages. The use of these tissues may be associated with additional procedures for graft retrieval, prolonged hospitalization, and donor site morbidity. For these reasons, alternate materials have been sought for urethral repair. Urethral tissue substitutes can be obtained through several tissue engineering approaches. These include the use of autologous urologic cells to create urethral tissue for subsequent repair and tissue regeneration inducing biomaterials. Each approach has its own advantages and indications for clinical utility.

AUTOLOGOUS UROLOGIC CELLS FOR URETHRAL TISSUE REPAIR

Our initial experiment involving urethral tissue engineering was directed toward creating urethral tissue composed of urologic cells. Autologous bladder urothelial and muscle cells were harvested, grown, and expanded in culture (see Protocols I.A and I.B). The cells were seeded onto nonwoven meshes of poly(glycolic acid) [16]. Partial urethrectomies were performed in rabbits, and a segment of the polymer mesh of the appropriate diameter was interposed to form the neourethra in each animal. There was no evidence of voiding difficulties or any other complications. Retrograde urethrograms showed no evidence of stricture formation. Histological examination of the neourethras demonstrated complete reepithelialization of the polymer mesh implanted sites by day 14, and reepithelialization continued for the entire duration of the study. Polymer fiber degradation was evident 14 days after implantation.

PROTOCOL I.A: UROTHELIAL CELL CULTURE

1. Materials and medium

   a. *Tissue source*: bladder tissue.
   b. *Medium*: keratinocyte and serum-free medium (Gibco/BRL), bovine pituitary extract (25 mg/500 ml medium), and recombinant epidermal growth factor (2.5 µg/ 500 ml medium).

2. Tissue harvest

   a. Obtain bladder specimen.
   b. Gently rinse the specimen several times with medium in culture plates.
   c. Mechanically scrape urothelial surface gently with a no. 10 scalpel blade. Be sure to use gentle short strokes and avoid cutting into the specimen. Urothelial cell clumps should be visible as tiny opaque material dispersing into the medium.
   d. Aspirate urothelial cell–medium suspension and plate the cells in a 24-well cell culture plate with approximately 0.5 ml of the suspension in each well. Add an additional 0.5 ml to make a final volume of 1 ml. Incubate cells at 37°C with 5% CO_2.
   e. On the following day, aspirate the medium from the wells and replace with fresh medium.
   f. Centrifuge the cells in the aspirated medium at 1000 rpm for 4 min.
   g. Remove the supernatant and resuspend the cells in 3–4.5 ml of fresh medium. Replate the cells in new wells.

3. Maintenance of urothelial cells

   a. Replace the medium with fresh warm (37°C) medium every 3 days, depending on the cell density.
   b. Trypsinize cells when they are 80–90% confluent.

4. Subculture of corporal smooth muscle cells

   a. Remove medium and add 1 ml of PBS-EDTA (0.5 M) to each well or 10 ml to each 10-cm culture plate. Observe the cells under a phase contrast microscope.
   b. When cell–cell junctions are separated for the majority of the cells, remove PBS-EDTA and add 300 µl of trypsin–EDTA to each well or 5 ml to each 10-cm culture plate.
   c. Periodically agitate the plates. When 80–90% of the cells are detached, add 30 µl of soybean trypsin inhibitor (Gibco/BRL, 294 mg of inhibitor) to 20 ml of PBS to each well, or 700 µl to each 10-cm plate. Add 0.5 ml of medium to each well or 3 ml to each 10-cm plate.
   d. Aspirate and centrifuge the cell suspension at 1000 rpm for 4 min, and remove the supernatant.
   e. Resuspend cells and count the number of viable cells by means of trypan blue exclusion.
   f. Aliquot the desired number of cells on the plate and place the cells in the incubator.

PROTOCOL I.B: BLADDER SMOOTH MUSCLE CELL CULTURE

1. Materials and medium

   a. *Tissue source*: bladder tissue.
   b. *Medium*: Dulbecco's Modified Eagle's Medium (DMEM), 10% fetal bovine serum (FBS), and antibiotic [penicillin (100 U/ml)–streptomycin (100 µg/ml), and amphotericin B (0.25 µg/ml)].

2. Tissue harvest

    a. Obtain fresh bladder tissue specimen.

    a. Use sharp tenotomy scissors to cut muscle tissue into small fragments (2–3 mm).

    c. Space muscle fragments evenly onto a cell culture plate (100 mm).

    d. Allow muscle fragments to dry and adhere to the plate (5–10 min).

    e. Add 15 ml of DMEM and incubate for 5 days.

    f. Change medium on the sixth day and remove nonadherent tissue fragments.

    g. When small islands of cell colony are formed, remove the tissue fragments and change medium.

    h. When sufficient cells are grown, trypsinize, count, and plate them onto new plates.

3. Maintenance of bladder smooth muscle cells

    a. Feed cells every 3 days, depending on the cell density.

    b. Trypsinize cells when they are 80–90% confluent.

4. Subculture of bladder smooth muscle cells

    a. Remove medium and add 10 ml of PBS-EDTA (0.5 M) for 4 min. Confirm the separation of cell junction under a phase contrast microscope.

    b. Remove PBS-EDTA and add 5 ml of trypsin–EDTA.

    c. Add 5 ml of medium when 80–90% of the cells lift under microscope.

    d. Aspirate the cell suspension into a 15-ml test tube.

    e. Centrifuge the cells at 1000 rpm for 4 min and remove the supernatant.

    f. Resuspend cells and count the number of viable cells by means of trypan blue exclusion.

    g. Aliquot the desired number of cells on the plate and make the volume of medium to a total of 10 ml.

    h. Place the cells in the incubator.

URETHRAL TISSUE REGENERATION

Several naturally derived tissue substitutes have been tried in urethral reconstructive procedures. Lyophilized human dura and vein homografts were proposed earlier to guide transitional epithelium ingrowth. Upon *in vivo* implantation, the luminal surface of lyophilized human dura was entirely epithelialized, and the material was absorbed and replaced by granulation tissue [17]. Urethral tissue replacement with lyophilized vein homografts resulted in the ingrowth of transitional epithelium [18]. Although urothelial cell epithelialization was observed in these substitute materials, none of them demonstrated urethral smooth muscle regeneration.

Free grafts of tubularized peritoneum were used as urethral tissue substitutes experimentally in rabbits. Organized multilayered graft epithelialization occurred; however, fistulas formed in two of the animals [19]. Later, porcine small intestinal submucosa (SIS) was used for urethral repair in a rabbit model to determine whether this material can evoke urethral regeneration. The SIS onlay grafts were shown to promote regeneration of the normal rabbit epithelium supported by a vascularized collagen and smooth muscle backing. Although the results appear promising, clinical studies had not been reported as of 1998 [20].

In our laboratory, a naturally derived acellular collagen based tissue substitute was developed from donor porcine bladder (see Protocol II). The acellular collagen matrix had been initially developed in our laboratory as a biomaterial for bladder augmentation. Studies performed using this material for bladder augmentation had showed that the acellular matrix was biocompatible and was able, upon *in vivo* implantation, to form bladder tissue similar to the native bladder [21]. In a subsequent study we investigated whether the acellular collagen matrix would be suitable for urethral reconstruction [22].

A ventral urethral defect measuring 1×0.7 cm^2 (approximately half of the urethral circumference) was created in 10 male rabbits. The acellular collagen matrix was trimmed and used to replace the urethral defect in an onlay fashion. Serial urethrography was performed pre- and postoperatively at 0.5, 1, 2, 3, and 6 months. Animals were sacrificed at 0.5, 1, 2, 3, and 6 months after surgery. All animals survived until sacrifice without any noticeable voiding dysfunction. Serial urethrograms confirmed the maintenance of a wide urethral caliber without any signs of strictures. Gross examination at retrieval showed normal-appearing tissue without any evidence of fibrosis. At retrieval, the distances between the marking sutures placed at the anastomotic margins remained stable, with no distance varying more than 10% in any axis, indicating the maintenance of the initial implant diameter. Histologically, the implanted matrices contained host cell infiltration and generous angiogenesis by 2 weeks after surgery. There was no evidence of fibrosis or scarring in the urethras at any of the retrieval time periods.

The presence of a complete transitional cell layer over the graft was confirmed 2 weeks after the repair, and this was consistent throughout the study. The urothelial cell layers stained positively with the broadly reacting anti-pancytokeratins AE1/AE3 in all implants. Normal-appearing organized muscle fiber bundles were evident 6 months after implantation. These results demonstrated that the acellular collagen matrix could be a useful material for urethral repair in the rabbit.

PROTOCOL II: ACELLULAR COLLAGEN MATRIX PREPARATION

1. Obtain donor bladder tissue.
2. Isolate the submucosa from the muscular and serosal layers means of microdissection techniques.
3. Treat tissue with distilled water in a magnetic stirring flask set at moderate speed for 24–48 h at 4°C.
4. Remove distilled water and treat with Triton X-100 (0.5%) and ammonium hydroxide (0.05%) in fresh distilled water for 72 h in a stirring flask at 4°C.
5. Wash with distilled water in a stirring flask for 24–48 h at 4°C. After this washing step, take a small piece of tissue for histological analysis to confirm any cellular remnants. Tissue matrix is usually decellularized at this time.
6. After confirmation of decellularization, wash with distilled water in a stirring flask for 24–48 h at 4°C. Tissues retaining cellular components should undergo an additional cycle of treatment. Repeat steps 4 and 5, and perform another histological analysis.
7. After the washing cycle with distilled water, rinse with 1 × PBS overnight.
8. Freeze-dry the tissue samples overnight.
9. Pack the samples and sterilize in ethylene oxide.
10. Store until used. When ready to use, equilibrate the tissue in 1× PBS or normal saline.

Following our experimental experience with the collagen-based acellular matrix, we used the material clinically for urethral reconstruction [23]. Four patients with a history of hypospadias surgery underwent reoperative repair in which the collagen-based matrix was used for urethral reconstruction. The collagen matrix, obtained from donor cadaver bladder, was processed and trimmed to size as needed for each individual patient (see Protocol II). The neourethras were created by anastomosing the matrix to the urethral plate in an onlay fashion. The size of the created neourethra ranged from 5 to 15 cm. After a 22-month follow-up, three of the four patients had a successful outcome in regard to cosmetic appearance and function. One patient who had a 15-cm neourethra created developed a subglanular fistula. Thus the use of a collagen-based acellular matrix appears to be beneficial for patients with prior hypospadias repair who may lack sufficient genital skin for reconstruction.

In a subsequent study, 26 patients with a diagnosis of urethral stricture underwent reconstructive surgery in which a collagen-based inert matrix was used for urethral reconstruction [24]. The inert collagen matrix was trimmed to size as needed for each patient, and the neourethras were created by anastomosing the matrix to the urethral plate in an onlay fashion. The size of the created neourethra ranged from 1.5 to 16 cm. Urethrograms were performed routinely 4 months postoperatively. Cystoscopic studies with urethral biopsies were also performed.

After an 18-month follow-up, 23 of the 26 patients had a successful outcome in regard to function. Two patients had a recurrent anastomotic stricture, and one patient developed a subcoronal fistula, which closed spontaneously after one year of repair. Cystoscopic studies showed adequate caliber conduits and normally appearing urethral tissues. Histological examination of the biopsy specimens showed the typical urethral stratified epithelium.

BIOCOMPATIBLE URETHRAL STENTS FOR STRICTURE DISEASES

Current treatment modalities for urethral strictures, such as ablation and the placement of synthetic stents, are suboptimal, and recurrence rates are high. The use of a natural urethral stent made of autologous tissue would be advantageous by reason of its biocompatibility. We investigated the feasibility of engineering *in vitro* tubular cartilaginous stents, which could be used as permanent urethral stents in stricture disease [25]. Thirty cylindrical tubes (10 mm × 9 mm) were fabricated from poly(L-lactic acid)-coated, poly(glycolic acid) polymer scaffolds. Chondrocytes, harvested and isolated from the articular surface of calf shoulder, were seeded onto the tubular scaffolds at a density of 60×10^6 cells/ml (see Protocol III). Fifteen stents were placed in a stirring bioreactor in culture. The remaining 15 stents were implanted in the subcutaneous space of athymic mice for comparison. Gross and histological analyses and biomechanical studies were performed at 4 and 10 weeks after cell seeding.

Scanning electron microscopy showed adequate chondrocyte attachment to the matrix. Gross examination of the stents engineered *in vivo* or *in vitro* showed the presence of well-formed, milky white tubular cartilage structures without any evidence of tissue ingrowth into the lumen. Histological analyses confirmed the presence of mature cartilage and the deposition of collagen and glycosaminoglycan in both groups. Biomechanical studies demonstrated that the engineered stents, formed in both conditions, were readily elastic and could withstand high degrees of pressure. This study demonstrates the feasibility of engineering cartilaginous stents *in vitro* using a bioreactor. The *ex vivo* engineered cartilage stents composed of autologous chondrocytes may be useful clinically.

PROTOCOL III: CARTILAGE TISSUE HARVEST

1. Materials

    a. *Medium*: Ham's F-12 (Gibco, Grand Island, NY), 10% fetal bovine serum, vitamin C (5 μg/ml), and antibiotic [penicillin (100 U/ml)–streptomycin (100 μg/ml) amphotericin B (0.25 μg/ml)].

    b. Digestion solution: 3% collagenase type II (Worthington Biochemical, Lakewood, NJ)

2. Tissue harvest

    a. Obtain cartilage tissue in a sterile manner.

    b. Use either a surgical blade or sharp tenotomy scissors to cut the tissue into small fragments (2–3 mm).

    c. Digest cartilage tissue fragments in 3% collagenase type II solution.

    d. Place the tube containing the digesting solution and cartilage fragments in a 37°C agitating incubator for 6–8 h. Be sure to check the tissue fragments periodically for overtreatment.

    e. When digesting step is complete, filter solution through a nylon mesh to remove undigested cartilage tissue.

    f. Wash the filtered cells with 1× PBS two times.

    g. Centrifuge the cells at 1200 Rpm for 10 min.

    h. Count the viable cells on a hemocytometer by means of trypan blue exclusion. Plate the cells in culture dishes at a desirable density.

    i. Incubate the cells at 37°C in the presence of 5% CO_2 and maintain the cells in a routine manner.

REFERENCES

1. Dessanti, A., Rigamonti, W., and Merulla, V. (1992). Autologous buccal mucosa graft for hypospadias repair: An initial report. *J. Urol.* **147**, 1081–1084.

2. Ehrlich, R. M., Reda, E. F., and Koyle, M. A. (1989). Complications of bladder mucosal graft. *J. Urol.* **142**, 626–627.

3. Humby, G. (1941). A one-stage operation for hypospadias. *Br. J. Surg.* **29**, 84–92.

4. Brannan, W., Ochsner, M. G., Fuselier, H. A., and Goodlet, J. S. (1976). Free full thickness skin graft urethroplasty for urethral stricture: Experience with 66 patients. *J. Urol.* **115**, 677–680.

5. Devine, C. J., Jr., and Horton, C. E. (1961). A one-stage hypospadias repair. *J. Urol.* **85**, 166–172.

6. Hendren, W. H., and Reda, E. F. (1986). Bladder mucosa graft for construction of male urethra. *J. Pediatr. Surg.* **21**, 189–192.

7. Li, Z. C., Zheng, Y. H., Shey, Y. X., and Cao, Y. F. (1981). One-stage urethroplasty for hypospadias using a tube constructed with bladder mucosa: A new procedure. *Urol. Clin. North Am.* **8**, 463–470.

8. Ozcan, M., and Kahveci, R. (1987). One-stage repair of distal and midpenile hypospadias by a modified Hodgson III technique. *Eur. J. Plast. Surg.* **10**, 159–163.

9. Guzman, L. (1999). Neourethra with rectum, posterior sagittal approach. *In* "Reconstructive and Plastic Surgery of the External Genitalia: Adult and Pediatric" (R. M. Ehrlich and G. J. Alter, eds.), pp. 101–108. Saunders, Philadelphia.

10. Snow, B. W., and Cartwright, P. C. (1992). Tunica vaginalis urethroplasty. *Urology* **40**, 442–445.

11. Anwar, H., Dave, B., and Seebode, J. J. (1984). Replacement of partially resected canine urethra by polytetrafluoroethylene. *Urology* **24**, 583–586.

12. Hakky, S. I. (1976). Urethral replacement by Dacron mesh. *Lancet* **2**, 1192.

13. Hakky, S. I. (1977). The use of fine double siliconised Dacron in urethral replacement. *Br. J. Urol.* **49**, 167–172.

14. Italiano, G., Abatangelo, G., Jr., Calabro, A., Abatangelo, G., Sr., Zanoni, R., O'Regan, M., and Glazel, G. P. (1997). Reconstructive surgery of the urethra: A pilot study in the rabbit on the use of hyaluronan benzyl ester (Hyaff-11) biodegradable grafts. *Urol. Res.* **25**, 137–142.

15. Olsen, L., Bowald, S., Busch, C., Carlsten, J., and Eriksson, I. (1992). Urethral reconstruction with a new synthetic absorbable device. *Scand. J. Urol. Nephrol.* **26**, 323–326.

16. Cilento, B. G., Retik, A. B., and Atala, A. (1995). Urethral reconstruction using a polymer mesh. *J. Urol.* **153**, 371A.

17. Kelami, A., Korb, G., Ludtke-Handjery, A., Rolle, J., Schnell, J., and Lehnhardt, F. H. (1971). Alloplastic replacement of the partially resected urethra on dogs. *Invest. Urol.* **9**, 55–58.

18. Kjaer, T. B., Nilsson, T., and Madsen, P. O. (1976). Total replacement of part of the canine urethra with lyophilized vein homografts. *Invest. Urol.* **14**, 159–161.

19. Shaul, D. B., Xie, H. W., Diaz, J. F., Mahnovski, V., and Hardy, B. E. (1996). Use of tubularized peritoneal free grafts as urethral substitutes in the rabbit. *J. Pediatr. Surg.* **31**, 225–228.

20. Kropp, B. P., Ludlow, J. K., Spicer, D., Rippy, M. K., Badylak, S. F., Adams, M. C., Keating, M. A., Rink, R. C., Birhle, R., and Thor, K. B. (1998). Rabbit urethral regeneration using small intestinal submucosa onlay grafts. *Urology* **52**(1), 138–142.

21. Yoo, J. J., Meng, J., Oberpenning, F., and Atala, A. (1998). Bladder augmentation using allogenic bladder submucosa seeded with cells. *Urology* **51**, 221–225.

22. Chen, F., Yoo, J. J., and Atala, A. (1999). Acellular collagen matrix as a possible "off the shelf" biomaterial for urethral repair. *Urology* **54**, 407–410.

23. Atala, A., Guzman, L., and Retik, A. B. (1999). A novel inert collagen matrix for hypospadias repair. *J. Urol.* **162**, 1148–1151.

24. Kassaby, E. A., Yoo, J. J., Retik, A. B., and Atala, A. (2000). A novel inert collagen matrix for urethral stricture repair. *J. Urol.* **163**, 308A.

25. Amiel, G. E., Kim, B., Yoo, J. J., and Atala, A. (2000). Ex vivo engineered stents for urethral strictures. *J. Urol.* **163**, 319A.

CHAPTER 91

PENIS

Heung Jae Park, DeZheng Dong, and Anthony Atala

INTRODUCTION

Male genitalia having abnormalities due to congenital anomalies or acquired conditions require surgical correction. Owing to the shortage of autologous penile tissue, multiple staged surgeries using nongenital tissues and silicone prostheses have been the mainstay in phallic reconstruction. However, graft failure and prosthesis-related complications remain a problem. Creation of penile structures composed of autologous tissue would be a preferable treatment approach for these patients. In this chapter we describe tissue engineering methods that may be applicable to genital reconstruction.

One of the major limitations of penile tissue reconstruction is the availability of sufficient autologous tissue. Nongenital tissue sources have been used over the years; however, complications such as infection, graft failure, and donor site morbidity have posed continuing problems [1–6]. The ability to engineer penile tissue composed of autologous cells would be beneficial.

CORPORAL TISSUE RECONSTITUTION

In attempting to engineer autologous penile tissue, our initial effort was focused on the formation of corporal tissue, since corpus cavernosum is one of the major tissue components of the phallus [7]. Human corporal smooth muscle cells were isolated, grown, and expanded in culture (see Protocol I.A). The cells were seeded on biodegradable poly(glycolic acid) polymers for implantation. Multilayers of corporal smooth muscle cells were identified grossly and histologically. Smooth muscle phenotype was confirmed immunocytochemically and by Western blot analyses. This study provided the evidence that cultured human corporal smooth muscle cells could be used in conjunction with biodegradable polymers to create cavernosal smooth muscle tissue *in vivo*.

PROTOCOL I.A:
CORPUS CAVERNOSAL SMOOTH MUSCLE CELL CULTURE

1. Materials and medium

   a. *Tissue source*: human corpus cavernosum.
   b. *Medium*: Dulbecco's Modified Eagle's Medium (DMEM), 10% fetal bovine serum (FBS), and antibiotic [penicillin (100 U/ml)–streptomycin (100 µg/ml), amphotericin B (0.25 µg/ml)].

2. Tissue harvest

   a. Obtain fresh cavernosal tissue specimen.
   b. Use sharp tenotomy scissors to cut muscle tissue into small fragments (2–3 mm).
   c. Space muscle fragments evenly onto a cell culture plate (100 mm).
   d. Allow muscle fragments to dry and adhere to the plate (5–10 min).
   e. Add 15 ml of DMEM and incubate for 5 days.
   f. Change medium on the sixth day and remove nonadherent tissue fragments.

g. When small islands of cell colony are formed, remove the tissue fragments and change the medium.

h. When sufficient cells are grown, trypsinize, count, and plate the cells onto new plates.

3. Maintenance of corporal smooth muscle cells

  a. Feed cells every 3 days, depending on the cell density.
  b. Trypsinize cells when they are 80–90% confluent.

4. Subculture of corporal smooth muscle cells

  a. Remove medium and add 10 ml of PBS-EDTA (0.5 M) over 4 min. Confirm the separation of cell junction under a phase contrast microscope.
  b. Remove PBS-EDTA and add 5 ml of trypsin–EDTA.
  c. Add 5 ml of medium when 80–90% of the cells lift under the microscope.
  d. Aspirate the cell suspension into a 15-ml test tube.
  e. Centrifuge the cells at 1000 rpm for 4 min and remove the supernatant.
  f. Resuspend cells and use trypan blue exclusion to count the number of viable cells.
  g. Aliquot the desired number of cells on the plate and make the volume of medium to a total of 10 ml.
  h. Place the cells in the incubator.

The main cellular components of corporal tissue consist of cavernosal smooth muscle and endothelial cells. In a subsequent study we investigated the possibility of developing corporal tissue by combining smooth muscle and endothelial cells. Normal human cavernosal smooth muscle cells and ECV 304 human endothelial cells were seeded on biodegradable polymers for implantation [8]. ECV 304 endothelial cells were used in the study, to allow the investigators to distinguish the implanted cells from the host endothelial cells. The retrieved structures showed formation of distinct tissue structures, consisting of organized smooth muscle tissue adjacent to endothelial cells. Presence of vascular structures was evident. Each cell type was confirmed by means of various assessment methods. This study showed that human corporal smooth muscle and endothelial cells seeded on biodegradable polymer scaffolds are able to form vascularized cavernosal tissue when implanted *in vivo*. Endothelial cells can act in concert with the native vasculature. These results suggest that the creation of well-vascularized autologous corpuslike tissue consisting of smooth muscle and endothelial cells may be possible.

Although we are able to form tissue consisting of corporal smooth muscle and endothelial cells *in vivo*, three-dimensional corporal structures could not be achieved because of the type of polymer matrix used. We developed a naturally derived collagen matrix, which is structurally similar to the native corporal architecture [9]. Acellular collagen matrices, derived from rabbit corpora, were obtained by means of cell lysis techniques (see: Protocol I.B). Human corpus cavernosal muscle and endothelial cells were grown and expanded in culture (see later: Protocol I.C). We have used human capillary endothelial cells, isolated from newborn foreskin via *Ulex europaeus* I (UEA I)-coated Dynabeads [10,11]. However, the source of endothelial cells could also be corpus cavernosum.

Primary human cavernosal smooth muscle and endothelial cells were seeded in a stepwise fashion. Cavernosal smooth muscle cells were initially seeded on the collagen matrices at a concentration of 30×10^6 cells/ml. The cells were allowed to attach and grow for 3 days in culture. Endothelial cells were then seeded at a concentration of 3×10^6 cells/ml. Cell matrices seeded with corporal cells were implanted *in vivo*. The implanted cell matrices showed neovascularity into the sinusoidal spaces by 1 week after implantation. Increased organization of smooth muscle and endothelial cells lining the sinusoidal walls was observed at 2 weeks and continued with time. The matrices were covered with the appropriate cell architecture 4 weeks after implantation [9,12]. This study demonstrates that human cavernosal smooth muscle and endothelial cells seeded on three-dimensional acellular collagen matrices derived from donor corpora are able to form a well-vascularized corporal architecture *in vivo*.

PROTOCOL I.B: ACELLULAR COLLAGEN MATRIX PREPARATION

1. Obtain corpus cavernosum from donor rabbits.
2. Take cross-sectional corporal fragments 0.5 cm in thickness.
3. Treat tissue with distilled water in a magnetic stirring flask (moderate speed) for 24–48 h at 4°C.
4. Remove distilled water and treat with Triton X-100 (0.5%) and ammonium hydroxide (0.05%) in fresh distilled water for 72 h in a stirring flask at 4°C.
5. Wash with distilled water in a stirring flask for 24–48 h at 4°C. After this washing step, take a small piece of tissue for histology to confirm any cellular remnants.
6. A small tissue mass is usually decellularized at this time. After confirmation of decellularization, repeat step 5. Dense tissue may require another cycle of treatment. Repeat steps 4 and 5, and perform another histological analysis.
7. After washing the tissue with distilled water, rinse with 1 × PBS overnight.
8. Freeze-dry the tissue samples overnight.
9. Pack and sterilize in ethylene oxide.
10. Store until used. When ready to use, equilibrate the tissue in culture medium overnight prior to cell seeding.

PROTOCOL I.C:
HUMAN ENDOTHELIAL CELL CULTURE FROM FORESKIN

1. Materials and media

   a. *Medium A* (for primary culture and first passage after UEA-I bead selection): 38.5 ml of endothelial basal medium 131 (Clonetics Corp., cat. no. CC 3121), 10 ml of 20% fetal bovine serum (FBS) (Hyclone, cat. no. 17-1111-L), 0.5 ml (2 mM) L-glutamine (Irvine Scientific, cat. no. 9317, 100× stock), 0.5 ml of PSF (antibiotic–antimycotic) [Gibco, cat. no. 600-5240AG: 100 U/ml penicillin G sodium (GPS), 100 μg/ml of streptomycin sulfate, 0.25 μg/ml amphotericin B (fungizone)], 0.5 ml (0.5 mM) dibutyryl cyclic AMP (Sigma, cat. no. D-0627), and 50 μl (1 μg/ml) hydrocortisone (Sigma, cat. no. H-0888).
   b. *Medium B* (for passage 2 and all following passages): Endothelial basal medium 131, 1× GPS, 10% FBS, and 2 μg/ml basic fibroblast growth factor (25 μg/ml stock solution) (Scios Nova).
   c. *Gelatin coating* (1% Difco Bacto Gelatin in PBS): Dissolve gelatin in PBS; autoclave to sterilize, and filter to remove particles.

2. Processing foreskin

   a. *Preapare foreskin* collecting medium: 450 ml of DMEM, 25 ml of FBS (5%), 20 ml of antibiotic–antimycotic (400 U/ml penicillin, 400 μg/ml streptomycin, 1 μg/ml fungizone), 5 ml of L-glutamine (2 mM), and 1 ml of gentamicin sulfate (100 μg/ml).
   b. Place the collecting medium with the foreskin in a culture plate (100 mm) in a tissue culture hood.
   c. Rinse two or three times with the collecting medium.
   d. Add 30 ml of collecting medium to a new 50-ml Falcon tube. Add an additional 2 ml of antibiotic–antimycotic.
   e. Separate the skin and subcutaneous tissue with a sterile scalpel blade and transfer the segments into the collecting medium in a 50-ml tube.
   f. Agitate the segments in the collecting medium at room temperature for at least 4–5 h to kill bacteria and spores that reside on the skin.

3. Isolation of endothelial cells

a. *Prepare digestion solution*: 7.5 ml of 1: 250 trypsin (40× trypsin, Sigma cat. no. T-0511), 2.7 ml of 0.5 M EDTA, pH 8.0, and 40 ml of Hanks' Balanced Salt Solution (HBSS).

b. *Prepare 10× HBSS without Ca^{2+} and Mg^{2+}*: 40 g of NaCl, 2 g of KCl, 240 mg of Na_2HPO_4, 300 mg of KH_2PO_4, 1750 mg of $NaHCO_3$, 5 g of glucose, and 100 mg of phenol red.

c. *Prepare wash solution (HBBS with 1× Ca^{2+} and Mg^{2+})*: 50 ml of 10× HBSS, 92.7 mg of $CaCl_2 \cdot 2H_2O$ (1.26 mM final), 100 mg of $MgSO_4 \cdot 7H_2O$ (0.8 mM final), 25 ml of FBS (5% final), and 5 ml of PSF (antibiotic/antimycotic).

d. Coat a petri dish (100 mm) for each one or two foreskins with 8.0 ml of 1% gelatin–PBS. Remove excessive gelatin before plating.

e. Autoclave a Teflon homogenizer (2.5 cm diameter) and gauze.

f. Remove the collecting medium from the foreskin segments.

g. Transfer the tissue segments into a sterile culture plate (100 mm).

h. Cut the foreskin segments into 4-mm² fragments with a sterile scalpel blade.

i. Transfer the tissue fragments to a sterile 50-ml Falcon tube and add 6.0 ml of digestion solution for 1–2 foreskins. Agitate vigorously at 37°C for 10 min.

j. Allow the skin fragments to sediment by gravitational force and aspirate the digestion medium. Wash once with 20 ml of wash solution, swirl vigorously, and remove the wash solution.

k. Add 10 ml of fresh wash solution and squeeze the fragments with the homogenizer.

l. Filter through 8–10 layers of sterile gauze into a 50-ml Falcon tube (mesh filter).

m. Repeat steps k and l, and collect the expelled cells into the same Falcon tube.

n. Centrifugate cells at 1000 rpm for 10 min at room temperature.

o. Aspirate the supernatant and plate the cells with 10 ml of EBM 131 (culture medium A) in a gelatin-coated culture dish (100 mm). Place the cells in an incubator overnight with 5% CO_2.

p. Wash the cells vigorously three or four times with PBS. Feed the cells with 10 ml of culture medium A.

q. Change medium every 2 days. The primary culture will be subconfluent after 7–8 days. They will be ready for the UEA-I isolation procedure at this point.

4. *Ulex europaeus* (UEA) I selection of endothelial cells

a. *Coating of Dynabeads with UEA*: Mix together 250 µl of Dynabeads (4×10^8 beads/ml) (Dynal, cat. no. 140.03), M-450, tosylactivated, 50 µg of unconjugated UEA-I (Vector, cat no. L-1060), and 225 µl of 0.5 M boric acid, pH 9.5. The bead/lectin ratio should be 2.0×10^6 beads per microgram of lectin. The volume ratio of Dynabeads to boric acid with lectin should be 1:1.

b. Reconstitute the UEA-I with 1.0 ml of sterile PBS–0.1 mM $CaCl_2$ to 2 mg/ml and store at 4°C (UEA-I is quite stable); 50 µg = 25 µl.

c. Mix Dynabeads, lectin, and boric acid in a sterile 2.0-ml screw-cap tube and agitate on a rotator at room temperature overnight.

d. Pipette the bead–lectin mixture (in 10 ml of HBSS) into a 15-ml Falcon tube. Wash with 10 ml of HBSS (plus Ca^{2+}/Mg^{2+}, 1% BSA) on the rotator for 15 min at room temperature.

e. Place the tube in a magnetic particle concentrator (MCP-1, Dynal, cat. no. 12001) and wait one minute for the beads to be collected onto the magnet. Aspirate the supernatant with a Pasteur pipette. Take the tube out of the MCP, rinse three times at room temperature for 15 min, and once overnight at 4°C.

f. Resuspend the beads in 250 µl of HBSS (plus Ca^{2+}/Mg^{2+}, 5% FBS, 1× PBS) and store at 4°C in a sterile 2.0-ml screw-cap tube. The beads will be stable for several months.

5. Purification of endothelial cells from primary cultures

a. Trypsinize subconfluent cell cultures (7–8 days) with $1\times$ trypsin–EDTA.

b. Centrifuge the trypsinized cells at $208g$ (1000 rpm) for 10 min.

c. Resuspend the cell pellet from one 100-mm petri dish in 190 μl of HBSS buffer. Pipette up and down several times with a 200 μl Pipetman to break up the cell clusters. Transfer the cell suspension into a sterile 2-ml screw-cap tube and add 5 μl UEA-I-coated Dynabeads.

d. Incubate cells and the beads for 3–5 min. Hold the tube in your hand and roll it between your palms gently to keep the beads in suspension. Endothelial cells and beads will form visible tiny clusters.

e. Transfer the cell–bead mixture to a 15-ml Falcon tube. Add 5 ml of HBSS buffer and pipette the cells several times up and down with the buffer. Place the Falcon tube into the MCP and collect the beads onto the magnet for about 1 min. Aspirate the wash solution with a Pasteur pipette while the tube is in the MCP. Take the tube out of the MCP. Repeat this wash 4 times with 5 ml HBSS wash buffer.

f. Resuspend the cells in 6 ml of EBM 131 growth medium A and place 3 ml onto each gelatin-coated 60-mm petri dish. This passage is designated as passage 1. Let the cells grow to confluence at 37°C and 5% CO_2. Change the medium every 3–4 days or twice a week.

g. When endothelial cells become confluent, trypsinize and split the cells 1:3 to 1:4. From now on (passage 2 and all the following passages), endothelial cells are cultured in growth medium B.

h. The endothelial cells should be fed every 2–3 days and split every 5–7 days (at least once a week).

PENILE PROSTHESES FOR RECONSTRUCTION

Penile reconstruction was attempted several decades ago with rib cartilage used as a stiffener. However, unsatisfactory functional and cosmetic results, due to curvature, discouraged its use [3,13]. Silicone prostheses were popularized owing to their unique mechanical properties. Although silicone is an acceptable biomaterial, biocompatibility is a concern for selected patients [14,15]. The use of a natural prosthesis composed of autologous cells may be beneficial.

Of the tissues existing in the human body, cartilage would serve as an ideal prosthesis for penile reconstruction, owing to its biomechanical properties [16,17]. Initial studies performed in our laboratory showed that chondrocytes suspended in biocompatible polymers form cartilage structures when implanted *in vivo* [18]. A feasibility study of engineering natural penile prostheses made of cartilage was attempted. Chondrocytes, harvested from bovine articular cartilage tissue, were grown, expanded and seeded onto preformed cylindrical poly(glycolic acid) polymer rods for implantation *in vivo* (see Protocol II.A) [16]. Chondrocytes were seeded onto preformed cylindrical poly(glycolic acid) polymer rods at a concentration of 50×10^6 chondrocytes/cm^3. The cell–polymer rods were implanted *in vivo*.

The retrieved implants formed milky white rod-shaped cartilaginous structures, maintaining their preimplantation size and shape. Biomechanical properties of the engineered cartilage rods, including compression, tension, and bending, showed that the cartilage tissues were readily elastic and could withstand high degrees of pressure. These results indicate that the engineered cartilage rods possessed the mechanical properties required to maintain penile rigidity. Histomorphological analyses confirmed the presence of mature and well-formed cartilage in all the cell-seeded implants.

PROTOCOL II.A: CARTILAGE TISSUE HARVEST

1. Materials and medium

    a. *Medium*: Ham's F-12 nutrient medium (Gibco, Grand Island, NY), 10% fetal bovine serum, vitamin C (5 μg/ml), and antibiotic [penicillin (100 U/ml)–streptomycin (100 μg/ml), amphotericin B (0.25 μg/ml)].

    b. *Digestion solution*: 3% collagenase type II (Worthington Biochemical, Lakewood, NJ).

2. Obtain cartilage tissue in a sterile manner.
3. Use either a surgical blade or sharp tenotomy scissors to cut the tissue into small fragments (2–3-mm).
4. Digest cartilage tissue fragments in 3% collagenase type II solution.
5. Place the tube containing the digesting solution and cartilage fragments in a 37°C agitating incubator for 6–8 h. Be sure to check the tissue fragments periodically for overtreatment.
6. When digesting step is complete, filter through a nylon mesh to remove undigested cartilage tissue.
7. Wash the filtered cells twice with $1\times$ phosphate-buffered saline.
8. Centrifuge the cells at 1200 rpm for 10 min.
9. Use a hemocytometer to count the viable cells by means of trypan blue exclusion. Plate the cells in culture dishes at a desirable density.
10. Incubate the cells at 37°C in the presence of 5% CO_2 and maintain the cells in a routine manner.

In a subsequent study using an autologous system, the feasibility of applying the engineered cartilage rods *in situ* was investigated [19]. Autologous cartilages harvested from rabbit ear were dissected into small fragments (2×2 mm^2). The technique describe in Protocol II.A was used to harvest chondrocytes under sterile conditions [18,20]. The chondrocytes were expanded until sufficient cell quantities were available. The cells were trypsinized, collected, washed, and counted for seeding. Chondrocytes were seeded onto preformed poly(L-lactic acid) coated poly(glycolic acid) polymer rods at a concentration of 50×10^6 chondrocytes/cm^3. The chondrocyte–polymer scaffolds were implanted in the corporal spaces of rabbits. Bilateral intracorporal implantation of the cell–polymer scaffolds were performed. The implants were retrieved and analyzed grossly and histologically 1, 2, 3, and 6 months after surgery.

Gross examination at retrieval showed the presence of well-formed milky white cartilage structures within the corpora at 1 month. There was no evidence of erosion or infection in any of the implant sites. Histological analyses demonstrated the presence of mature and well-formed chondrocytes in the retrieved implants. Autologous chondrocytes seeded on preformed biodegradable polymer structures are able to form cartilage structures within the rabbit corpus cavernosum. This technology appears to be useful for the creation of autologous penile prostheses.

REFERENCES

1. Puckett, C. L., and Montie, J. E. (1978). Construction of the male genitalia in the transsexual using a tubed groin flap for the penis and a hydraulic inflation device. *Plast. Reconstr. Surg.* **61**, 523–530.
2. Chang, T. S., and Hwang, W. Y. (1984). Forearm flap in one-stage reconstruction of the penis. *Plast. Reconstr. Surg.* **74**, 251–258.
3. Gilbert, D. A., Williams, M. W., Horton, C. E., *et al.* (1988). Phallic reinnervation via the pudendal nerve. *J. Urol.* **140**, 295–299.
4. Goodwin, W. E., and Scott, W. W. (1952). Phalloplasty. *J. Urol.* **68**, 903.
5. Horton, C. E., and Dean, J. A. (1990). Reconstruction of traumatically acquired defects of the phallus. *World J. Surg.* **14**, 757.
6. Sharaby, J. S., Benet, A. E., and Melman, A. (1995). Penile revascularization. *Urol. Clin. North Am.* **22**(4), 821–832.
7. Kershen, R. T., Yoo, J. J., Moreland, R. B., *et al.* (1998). Novel system for the formation of human corpus cavernosum smooth muscle tissue in vivo. *J. Urol.* **159**(Suppl.), 156.
8. Park, H. J., Yoo, J. J., Kershen, R. T., *et al.* (1999). Reconstruction of human corporal smooth muscle and endothelial cells in vivo. *J. Urol.* **162**, 1106–1109.
9. Falke, G., Yoo, J. J., Machado, M. G., Moreland, R., and Atala, A. (2000). Penile reconstruction using engineered corporal tissue. *J. Urol.* **163**(Suppl.), 980.
10. Jackson, C. J., Garbett, P. K., Nissen, B., *et al.* (1990). Binding of human endothelium to *Ulex europaeus* I-coated Dynabeads: Application to the isolation of microvascular endothelium. *J. Cell Sci.* **96**, 257–262.
11. Kraling, B. M., and Bischoff, J. (1998). A simplified method for growth of human microvascular endothelial cells results in decreased senescence and continued responsiveness to cytokeratines and growth factors. *In Vitro Cell. Dev. Biol.* **34**(4), 308–315.
12. Atala, A. (1999). Tissue engineering applications for erectile dysfunction. *Int. J. Impotent Res.* **11**(Suppl. 1), s41–s47.
13. Frumpkin, A. P. (1944). Reconstruction of male genitalia. *Am. Rev. Sov. Med.* **2**, 14.

14. Nukui, F., Okamoto, S., Nagata M., *et al.* (1997). Complications and reimplantation of penile implants. *Int. J. Urol.* 4(1), 52–54.

15. Kardar, A., and Pettersson, B. A. (1995). Penile gangrene: A complication of penile prosthesis. *Scand. J. Urol. Nephrol.* 29(3), 355–356.

16. Yoo, J. J., Lee, I., and Atala, A. (1998). Cartilage rods as a potential material for penile reconstruction. *J. Urol.* 160, 1164–1168.

17. Yoo, J. J., Park, H. J., and Atala, A. (2000). Tissue engineering applications for phallic reconstruction. *World J. Urol.* 18, 62–66.

18. Atala, A., Cima, L. G., Kim, W., Paige, K. T., Vacanti, J. P., Retik, A. B., and Vacanti, C. A. (1993). Injectable alginate seeded with chondrocytes as a potential treatment for vesicoureteral reflux. *J. Urol.* 150, 745–747.

19. Yoo, J. J., Park, H. J., Lee, I., *et al.* (1999). Autologous engineered cartilage rods for penile reconstruction. *J. Urol.* 162, 1119–1121.

20. Atala, A., Kim, W., Paige, K. T., Vacanti, C. A., and Retik, A. B. (1994). Endoscopic treatment of vesicoureteral reflux with a chondrocyte–alginate suspension. *J. Urol.* 152, 641–643.

CHAPTER 92

TESTES

Marcelle Machluf and Anthony Atala

INTRODUCTION

The human testis is a complex organ comprising germ cells and a variety of somatic cells such as Sertoli, Leydig, endothelial, fibroblast, macrophage, and peritubular myoid cells.

The testis has two function; spermatogenesis, which occurs in the seminiferous tubules, and secretion of steroid hormones (androgens) by the Leydig cells in the interstitial tissue. These testicular functions are intimately related because testosterone synthesis is required for the production of sperm as well as for the development of secondary sexual characteristics and normal sexual behavior.

Overall the functionality of this organ is controlled and regulated by the hypothalamic–pituitary system by way of luteinizing hormone (LH), which acts on Leydig cells and regulates androgen secretion, whereas spermatogenesis is controlled by follicle-stimulating hormone (FSH), which acts on Sertoli cells and locally produced androgens.

In males, androgens, in particularly testosterone, are known to have many important physiological actions, including effects on muscle, bone, central nervous system, prostate, bone marrow, and sexual function. Testicular dysfunction and hypogonadal disorders evolve from different pathophysiological conditions such as Klinefelter's syndrome, bilateral mump orchitis, toxic damage from alcohol or chemotherapy, and orchiectomy [1]. Patients with such conditions require lifelong androgen replacement therapy to maintain physiological levels of serum testosterone. Such therapy may increase muscle strength, stabilize bone density, improve osteoporosis, and restore secondary sexual characteristics, including libido and erectile function [2].

To date different approaches have been attempted to treat testicular dysfunction and hypogonadal disorders. This chapter briefly describes these approaches, while focusing on a new tissue-engineered system for androgen replacement therapy.

TRANSPLANTATION OF TESTES

The first authenticated record of gonadal transplantation is attributed to an eighteenth-century Scottish anatomist and surgeon, John Hunter, who grafted chicken testes to the body cavity of both male and female hosts. Full details of this work have not survived, and it is difficult to evaluate its outcome.

Berthold (1849) was the first to report on a successful testicular transplant, since he used autografts and avoided the risk of rejection. When he replaced the testes of capons in their own body cavity, he found that the growth of comb and plumage, and courting behavior, all of which are androgen dependent, were maintained [3]. A century later, interest in testicular transplant increased as a result of the misapprehension that somatic aging is caused by withdrawal of sex hormones. Lydston (1916) had publicized a series of testicular transplantation experiments performed in his patients [4]. He and others following him believed that transplanted sex glands produced a hormone that was a "cell stimulant, nutrient and regenerator" capable of prolonging life and restoring waning sexual functions, arteriosclerosis and other infirmities of midlife and later.

Voronoff (1923) was the first to use chimpanzee and baboon organs for treating patients [5]. This approach was taken by other surgeons, but none of them used microsurgery to join blood vessels of the graft to the host's circulation, resulting in ischemic necrosis preceded by organ rejection. Later, successful testicular transplantation could be achieved when ischemia time was reduced to less than an hour by using vascular anastamoses in dogs [6] and inbred rats [7,8]. The first convincing human testicular transplant was published by Silber (1978), who grafted a patient with a testis from the patient's genetically identical twin brother [9]. However, with time the stringent requirements for success have precluded a surge in demand for this operation. Moreover, carefully conducted grafting trials failed to confirm former claims; the new synthetic sex steroids were shown not to affect the life span of experimental animals [10]. Nevertheless, testicular transplantation may still be regarded as having clinical potential—for example, who in carriers of genetic disease, who can receive normal germ cells from donors.

TRANSPLANTATION OF TESTICULAR TISSUES

The problems arising from the size of the testis and its fibrous capsule led some transplanters to use sliced or minced organs. Kearns (1941), who reimplanted testicular tissue subcutaneously in a victim of accidental castration, reported the most plausible case [11]. According to this report, testosterone was being produced by the autograft, but without the normal architecture of the seminiferous epithelium, it hard to understand how germ cell transfer could have restored spermatogenesis. Furthermore, injecting spermatogonial stem cells from donor testes into atrophic tubules is daunting the testes must produce millions of spermatozoa per day to be fertile. Therefore, efforts to develop tissue grafting for the purpose of improving testosterone levels in hypogonadal men are more likely to succeed than are attempts at restoring fertility. The former goal appears to be simple, requiring the transfer of interstitial cells (Leydig cells), which are readily isolated from the donor testes by means of collagenase. Interstitial cells grafted in castrated rodents resulted in partial restoration of body weight, and testosterone levels above those of controls [12–14]. A number of vehicles and several implantation sites for interstitial cells have been tried, but none fully replaced testicular androgen production.

TESTOSTERONE DELIVERY SYSTEMS

The main goal of androgen replacement therapy is to maintain physiological levels of serum testosterone and also its metabolites, dihydrotestosterone and estradiol. Hypogonadal states secondary to hypothalamic–pituitary disorders, gonadal abnormalities, and defects in androgen action or secretion may benefit from androgen replacement.

Androgen replacement modalities include oral administration of testosterone tablets or capsules [15–17], depot injections [18–20], sublingual treatment [21], and skin patches [22–24]. When taken orally, testosterone preparations are largely rendered metabolically inactive during the "first pass" through the liver. This metabolic inactivation requires large oral doses of testosterone (>200 ng/day) to reach normal serum levels. These large doses of testosterone may be toxic to the liver and may lead to hepatitis, hepatoma, or hepatocarcinoma [19,25,26].

Parenteral depot preparations include testosterone enanthante (Delatestryl) and testosterone cypionate (Depot testosterone cypionate). These preparations are based on 17β-hydroxyl esters, which are given intramuscularly (IM), with slow-release, oil-based injection vehicles every 10–21 days. Testosterone levels with these preparations rise to supernormal levels for 1 or 2 days, after which they gradually fall within the normal range for 10–12 days, reaching baseline at approximately 21 days. This fluctuation in testosterone levels may produce significant swings in mood, libido, and sexual function [2,20].

Transdermal testosterone therapy includes both scrotal and nonscrotal patches. Testoderm and Androderm are multilayered skin patches that deliver measured doses of testosterone across the skin, with doses ranging from 4 to 6 mg/day. The scrotal skin is used as a delivery target because of the 5α-reductase activity present within this site. Androderm comes as a patch, which delivers 2.5 mg of testosterone in an alcohol-based gel reservoir. The gel enhances drug penetration by contributing to the breakage of the epidermal barrier.

When used in nonscrotal skin, the patch has to be applied twice daily. Advantages of these systems include a reduced frequency of administration, thereby increasing patient compliance, avoidance of gastric and hepatic first-pass metabolism, and achievement of steady state plasma concentration levels of testosterone. However, despite these advantages, the transdermal systems have been associated with adverse effects, such as transient erythema, pruritis, induration, burning, rash, and skin necrosis [23,24,27].

CELL ENCAPSULATION FOR TESTOSTERONE THERAPY

Cell transplantation has long been proposed as a treatment for several diseases involving hormone or protein deficiencies. Cell rejection by the host immune system, however, has limited the use of this strategy. Encapsulation of living cells in a protective, biocompatible, and semipermeable polymeric membrane has been proven to be an effective method of immunoprotection of the desired cells, regardless of the type of recipient (allograft, xenograft) [28]. A majority of the implantation work using microencapsulated cells as delivery vehicles employs two polymers; sodium alginate and poly(L-lysine) (PLL) [29]. Alginate microcapsules have been used for various applications [30,31], particularly for the encapsulation of pancreatic islet cells/or insulin delivery [29,32], and recombinant cells have served for the delivery of therapeutic gene products [33].

The Leydig cells of the testes are the major source of testosterone in men (95%). Implantation of heterologous Leydig cells has been proposed as a method for chronic testosterone replacement. However, these approaches were limited by tissue and cell failure to produce long-term testosterone and dissemination of the implanted cells. Therefore, encapsulation of Leydig cells might be useful for testosterone replacement therapy. Such a system might be able to simulate the normal diurnal pattern of testosterone release by the testes, thereby avoiding side effects such as those associated with chemically modified testosterone administration. Leydig cell transplantation may be also beneficial not only for testosterone replacement but also for the secretion of other associated hormones and growth factors such as melanocytes, β-andorpilin, prostaglandins, insulin-like growth factor 1 (IGF-1), and interleukins [34].

Studies in our laboratory have been focused on the encapsulation and implantation of isolated Leydig cells for long-term testosterone delivery. Leydig cells were isolated from male Sprague-Dawley rats, 56–70 days old, by means of collagenase and Percoll gradient separation. The isolated Leydig cells were encapsulated within microspheres composed of calcium–alginate, coated with the positively charged polyelectrolyte PLL, and recoated with alginate (Fig. 92.1). Based on the molecular weight of testosterone (300 Da), PLL having

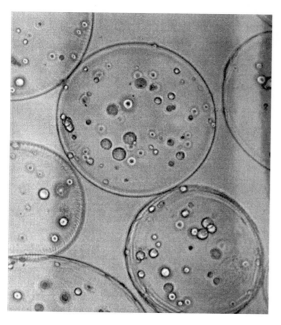

Fig. 92.1. Light microscopy of the encapsulated Leydig cells. Original magnification ×25, 0.7 mm in diameter.

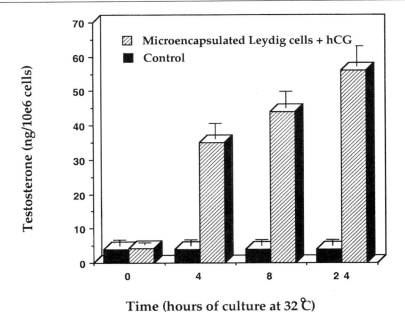

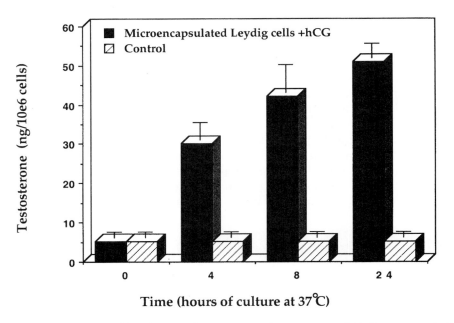

Fig. 92.2. The response of cultured microencapsulated rat Leydig cells to human chorionic gonadotropin (0.06 IU/ml) over a period of 24 h. The experiments were performed at 37°C (top) and at 32°C (bottom).

a molecular weight of 21 kDa and 1.2 % sodium alginate with a high guluronic acid content (> 65%) were chosen. PLL with a molecular weight ranging between 16 and 22 kDa produced a semipermeable membrane with a molecular weight cutoff of 70 kDa (i.e., preventing the diffusion of cells and antibodies). Methyl(this) tetrazole assays performed daily on the microencapsulated cells showed that the cells remained viable during the experiments. Testosterone secretion from cultured encapsulated Leydig cells in response to human chorionic gonadotropin (hCG) was the highest 24 h after hCG stimulation (0.06 IU/ml) (Fig. 92.2). There was no significant difference in testosterone secretion when cells were cultured at either 32 or 37°C (Fig. 92.2(top), 92.2(bottom)). The encapsulated and nonencapsulated Leydig cells were found to be resistant to temperature changes. This finding

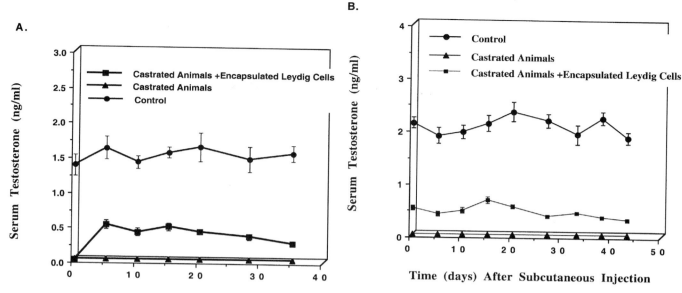

Fig. 92.3. Serum testosterone levels in normal rats, castrated rats without Leydig cell implantation, and castrated rats, which received 5×10^6, encapsulated Leydig cells intraperitoneally (A) and subcutaneously (B).

broadens the list of possible *in vivo* sites for Leydig cell transplantation. The intraperitoneal cavity was the first site chosen for cell implantation because of its generous vascular and nutritional capabilities. *In vivo* studies performed in castrated rats showed that the total testosterone levels measured in the serum of castrated rats that were injected intraperitoneally with 5×10^6 encapsulated Leydig cells were between 0.23 to 0.51 ng/ml for more than 3 weeks (Fig. 92.3A). These animals did not receive any exogenous hCG stimulation. Similar testosterone levels (0.24–0.48 ng/ml) were obtained when encapsulated Leydig cells were injected subcutaneously (Fig. 92.3B). However, testosterone was detected for a longer time period in the subcutaneous group (43 days) than in the intraperitoneal group (35 days). These testosterone levels were lower than the ones detected in the control rats (1.7 ng/ml), which were not castrated. However, only 5×10^6 microencapsulated cells were implanted in each animal, representing only 10% of the normal adult rat Leydig cell population.

METHODS FOR ENCAPSULATION

Microencapsulation is currently the optimal immunoisolation technique. Different approaches and polymers are being used for encapsulating cells and tissue for therapeutic applications. The technique of microencapsulation used by our laboratory utilizes two polymers: highly purified calcium-alginate (Pronova, Norway) and low molecular weight (23.6 kDa, Sigma) poly(L-lysine). This procedure is described as follows.

METHOD FOR CELL ENCAPSULATION

1. Isolated cells are suspended in sodium alginate (1.2%) (60% guluronic acid content) in 0.9% saline for 5 min.
2. The cell-alginate suspension is extruded through a 22-gauge airjet–needle into a CaCl$_2$-HEPES solution (1.5%).
3. The beads are stirred for 20 min in the CaCl$_2$-HEPES solution.
4. Gelled droplets are transferred to ecno-colums (Bio-Rad) and decanted.
5. The columns are filled with 15 ml of poly(L-lysine) solution in 0.9% saline, sealed, and rotated gently for 12 min.
6. The poly(L-lysine) solution is decanted from the columns and washed three times with HEPES solution.

7. A 0.125% alginate solution is added, and the mixture is rotated for 10 min. Then the alginate solution is decanted and the supernatant is washed three times with HEPES prior to culturing.

CONCLUDING REMARKS

This preliminary study shows that encapsulated Leydig cells are able to secrete testosterone both *in vitro* and *in vivo*. Nevertheless, further studies regarding the ideal *in vivo* setting for Leydig cell implantation is required. In addition, higher levels of testosterone are required for this system to be useful therapeutically. This may be achieved by implanting a larger number of Leydig cells. Encapsulation of cocultured Leydig cells with Sertoli cells may also improve the time period and levels of testosterone secretion. Stable levels of testosterone were achieved when thus system was used, and this the system may be helpful in preventing the side effects associated with testosterone fluctuations in conventional therapy.

REFERENCES

1. Griffin J. E., and Wilson, J. D. (1998). Disorders of the testes and the male reproductive tract. *In* "William's Textbook of Endocrinology" (J. D. Wilson, *et al.*, eds.), pp. 819–875. Saunders, Philadelphia.
2. Bhasin S., and Bremner, W. J. (1997). Clinical review: 85 emerging issues in androgen replacement therapy. *J. Clin. Endocrinol. Metab.* **82**, 3–7.
3. Berthold, A. A. (1849). Transplantation der hoden. *Arch. Anat. Physiol. Wiss. Med.* **16**, 42–46.
4. Lydston, G. F. (1916). Sex gland implantation. Additional cases and conclusions to date. *JAMA, J. Am. Med. Assoc.* **66**, 1540–1543.
5. Brinster, R. L., and Z. J. W. (1960). Spermatogenesis following male germ-cell transplantation. *Proc. Natl. Acad. Sci. U.S.A.* **91**, 11298–11302.
6. Attaran, S. E., H. C. V., Crary, L. S., *et al.* (1966). Homotransplnat of the testis. *J. Urol.* **95**, 387–389.
7. Gittes, R. F., A. J. E., Yen, S. S. C., and Lee, S. (1972). Testicular transplantation in the rat: Long-term gonadotropin and testosterone radioimmunoassays. *Surgery* **72**, 187–192.
8. Lee, S., T. K., and Orloff, M. J. (1971). Testicular transplantation in the rat. *Transplant. Proc.* **3**, 586–590.
9. Silber, S. J. (1978). Transplantation of a human testis for anorchia. *Fertil. Steril.* **30**, 181–187.
10. Parkes, A. S. (1966). The rise of reproductive endocrinology, 1920-1940. *J. Endocrinol.* **34**, 20–32.
11. Kearns, W. (1941). Testicular transplantation. Successful autlplastic graft following accidental castration. *Ann. Surg.* **114**, 886–890.
12. Fox, M., B. P. E., and Hammonds, J. C. (1973). Transplantation of interstitial cells of the testis. *Br. J. Urol.* **45**, 696–701.
13. Tai, J., J. H., and Tze, W. J. (1989). Successful transplantation of Leydig cells in castrated inbred rats. *Transplantation* **47**, 1087–1089.
14. Boyle, P. F., Fox, M., and Slater, D. (1976). Transplantation of interstitial cells of the testis: Effect of implant site, graft mass and ischaemia. *Br. J. Urol.* **47**, 891–898.
15. Canteril, J., *et al.* (1984). Which testosterone therapy? *Clin. Endocrinol.* **21**, 97–107.
16. Ferrini, R. L., and Barrett-Connor, E. (1998). Sex hormones and age: A cross-sectional study of testosterone and estradiol. *Am. J. Epidemiol.* **147**, 750–755.
17. Franchimont, P., *et al.* (1978). Effect of oral testosterone undecanoate in males. *Clin. Endocrrinol.* **9**, 313–320.
18. Fujioka, M., *et al.* (1986). Pharmacokinetic properties of testosterone propionate in normal men. *J. Clin. Endocrinol. Metab.* **63**, 1361–1364.
19. Snyder, P. J., and Lawrence, D. A. (1980). Treatment of male hypogonadism with testosterone enanthate. *J. Clin. Endocrinol. Metab.* **51**, 1335–1339.
20. Sokol, R. Z., *et al.* (1982). Comparison of the kinetics of injectible testosterone in eugonadal and hypogonadal men. *Fertil. Steril.* **37**, 425–430.
21. Stuendle, C. S., and Dudlly, R. E. (1990). Sublingual testosterone simulated episodic androgen release. *Proc. 73rd Ann. Meet Endocr. Soc.* **426**.
22. Wilson, D. E., *et al.* (1998). Bioequivalence assessment of a single 5 mg/day testosterone transdermal system versus two 2.5 mg/day systems in hypogonadal men. *J. Clin. Pharmacol.* **38**, 54–59.
23. McClellan, K. J., and Goa, K. L. (1998). Transdermal testosterone. *Adis New Drug Profile* **55**, 253–258.
24. Bennett, N. J. (1998). A burn-like lesion caused by testosterone transdermal system. *Burns* **24**, 478–480.
25. Bagatell, C. J., and Bremner, W. J. (1996). Drug therapy: Androgen in men, uses and abuses. *N. Eng. J. Med.* **334**, 707–710.
26. Gooren, L. J. G. (1994). A ten year safety study of the oral androgen testosterone undecanoate. *Andrology* **15**, 212–216.
27. Hogan, D. J., and Maibach, H. I. (1990). Adverse dermatological reactions to transdermal drug delivery systems. *J. Am. Acad. Dermatol.* **22**, 811–814.
28. Chang, T. M. S. (1993). Bioencapsulation in biotechnology. *Biomater. Artif. Cells, Immobil. Biotechnol.* **21**, 291–293.
29. Lim, F., and Sun, A. M. (1980). Microencapsulated islets as bioartificial endocrine pancreas. *Science* **210**, 908–912.
30. Chang, T. M. S. (1998). Pharmaceutical applications of artificial cells including microencapsulation. *Eur. J. Pharm. Biopharm.* **45**, 3–8.

31. Machluf, M., J. T., Atala A., *et al.* (2001). Continuous release of endostatin from microencapsulated engineered cells for tumor therapy. *Nat. Biotechnol.* (in press).

32. Wang, T., *et al.* (1997). An encapsulation system for the immunoisolation of pancreatic islets. *Nat. Biotechnol.* **15,** 358–362.

33. Tai, I. T., and Sun, A. M. (1993). Microencapsulation of recombinant cells: A new delivery system for gene therapy. *FASEB J.* **7,** 1061–1066.

34. Verhoeven, G. (1992). Local control system within the testis. *In* "Bailliere's Clinical Endocrinology and Metabolism" (B. Tindall, ed.), pp. 313–333.

CHAPTER 93

CARTILAGE RECONSTRUCTION

Lawrence J. Bonassar

CARTILAGE PHYSIOLOGY
ANATOMY, FUNCTION, AND PATHOLOGY

Cartilage is a highly specialized, avascular, aneural, connective tissue found throughout the body. The primary function of this tissue in all its locations is to provide structural support for surrounding tissues, such as in the ear or nose, or to cushion or resist externally applied stresses, such as in the trachea or articular joints. There are four major types of cartilage, each distinguished by specific constitutive or organizational components and by subspecialized functions of the chondrocyte phenotype: hyaline cartilage, elastic cartilage, costochondral cartilage, and fibrocartilage (Table 93.1).

Hyaline cartilage is the most prevalent type in the body and is distinguished by a shiny white appearance. The chondrocytes in hyaline cartilage are generally round or slightly triangular and tend to be spaced evenly throughout the extracellular matrix (ECM). The ECM is composed primarily of type II collagen and the large aggregating proteoglycan, aggrecan. Elastic cartilage is similar in appearance to hyaline cartilage but is distinguished from it by the presence of elastin in the ECM, which gives the tissue large extensibility. Costochondral cartilage is found in direct apposition with bone, either in the ribs or in the growth plate. It is distinguished from other types of cartilage by the presence of hypertrophic chondrocytes, which appear large and square, and by the presence of type X collagen in the ECM. Fibrocartilage, as the name implies, is essentially a hybrid of cartilage and fibrous tissue. The chondrocytes in fibrocartilage are less round and more elongated than other chondrocytes and resemble fibroblasts histologically or in culture. The ECM of fibrocartilage contains large amounts of type I collagen and reduced amounts of proteoglycans, giving it mechanical properties that are inferior to those of other types of cartilage.

There are many causes and forms of cartilage pathology and dysfunction, all of which are relevant to the discussion of strategies for engineering replacement tissues. The most universal cause necessitating cartilage repair is trauma, which can affect articular, nasal, auricular, or rib cartilage, as well as intervertebral disc and meniscus. Surgical treatment of congenital defects such as microtia [1] or cleft palate [2] often involve use of cartilage transplantation in the reconstruction of craniofacial features. Degenerative diseases such as arthritis are the major cause of loss of articular cartilage [3] and intervertebral disk [4]. Cartilage cancers are rare, but their removal from sites such as articular joints [5] and trachea [6] results in significant loss of tissue.

Loss of cartilage tissue, regardless of cause, is particularly problematic, given the extremely limited repair capacity of the tissue [7]. Cartilage transplantation has never met with much success, owing to scarcity of donor sites and associated morbidity in harvest procedures as well as resorption of transplanted cartilage from the implant site. This indeed has provided the major motivation behind the field of cartilage tissue engineering. The recreation of tissue from its component parts, cells and matrix, has been the focus of much research in the past decade. This chapter reviews earlier work and summarizes state-of-the-art techniques used to solve this critical problem.

Table 93.1. Summary of Types and Locations of Cartilage throughout the Body

Hyaline	Fibrocartilage	Costochondral	Elastic
Articular joints	Tendon insertion site	Rib	Ear
Nose	Ligament insertion site	Growth plate	
Trachea	Meniscus		
Intervertebral disk (NP)	Intervertebral disk (AP)		
Vertebral end plate			

COMPOSITION AND STRUCTURE

A reasonable attempt at generation or regeneration of a tissue cannot be undertaken without an in-depth understanding of the composition and function of the tissue at hand. Cartilage, in all forms, is a highly hydrated tissue, with a water content of approximately 75–80%, the remaining 20–25% being the solid portion of the ECM. The chemical constituents of the cartilage matrix are of extreme importance when one is discussing tissue function [8]. These macromolecules include proteoglycans, collagens, other proteins such as link protein, and hyaluronic acid, which are connected by chemical and mechanical crosslinks to form an effectively continuous three-dimensional network throughout the tissue. In addition, each type of matrix macromolecule has its own distinct chemical nature that contributes to the bulk properties of the material.

Proteoglycans

As a group, all proteoglycans in cartilage have some structural similarities. As the name implies, proteoglycans (PGs) are macromolecules composed of proteinaceous regions and glycosylated regions. Cartilage PGs consist of glycosaminoglycan (GAG) carbohydrate chains, which are covalently bonded to a protein core. The nature of both the core protein and the chains varies with PG type. For the purposes of this discussion, cartilage PGs are divided into two types: large aggregating PGs (aggrecan) and small PGs (decorin, biglycan, and fibromodulin).

The most abundant proteoglycan in cartilage by weight is the large aggregating proteoglycan, aggrecan [9]. Aggrecan is a macromolecule of considerable size, with total relative molecular mass M_r of $1-3 \times 10^6$ Da. The core protein M_r is approximately 300,000 Da and extends to a length of approximately 300 nm [8]. In addition, there are approximately 100 chondroitin sulfate (CS) chains and 30 keratan sulfate (KS) chains bound to the protein core, which provide the balance of molecular mass.

The small proteoglycans of primary interest in cartilage are decorin and biglycan. Decorin and biglycan have similar, but nonidentical structures. Both have a core protein M_r of 40 kDa. Decorin, with a single CS chain, has a total M_r of approximately 85 kDa, while biglycan, with 2 CS chains, has a total M_r of 130 kDa [8,9]. Decorin is known to bind to type IX collagen fibrils during formation and may inhibit the production of such fibrils. Though structurally similar, biglycan appears to have no such function. Additionally, it has been observed that both decorin and biglycan have the ability to immobilize certain growth factors, particularly basic fibroblast growth factor (bFGF), and thus may play a role in regulating matrix assembly [9].

Hyaluronic Acid

Hyaluronan is the long chain backbone macromolecule to which aggrecan molecules bind to form aggregates. Hyaluronate molecules are extremely large, with M_r up to $5-6 \times 10^6$ Da and a length of up to 10 μm. Typical sizes for hyaluronan are an M_r of $3-6 \times 10^5$ Da and a length of 0.5–1 μm [10]. As expected, the size of the hyaluronic acid chain determines how many proteoglycans can bind to the molecule. Since GAG chains typically extend from the core proteins of proteoglycans, steric hindrance is an important determinant of molecular packing. It is estimated that these hindrances allow for one aggrecan molecule per 7 kDa of hyaluronate chain [10]. As such, the largest of the hyaluronate molecules can bind from 400 to 800 aggrecan molecules. The M_r of such aggregates can be as much as 100×10^6 kDa [9].

As mentioned, proteoglycans are held in the matrix by means of an attachment between the G1 domain of aggrecan and the hyaluronate backbone. This attachment is further enhanced by a small polypeptide known as link protein, which binds both to the G1 domain of aggrecan and to hyaluronate. Link protein is similar in size and composition to the G1 domain of aggrecan. Like the G1 domain, it consists of an Ig fold and a protein tandem repeat sequence. It is postulated that the primary role of link protein may be to increase the stability of aggrecan–hyaluronan complexes in the first stages of aggregation, when the binding of proteoglycan to hyaluronate may be incomplete.

Collagens

Collagens play a key role in the matrix structure of cartilage. Collagen fibers are very strong in tension, and in large part their function is to resist the osmotic swelling pressure created by the highly charged proteoglycans. This structure allows the collagen to provide strength when the tissue is exposed to tensile and shear forces. The collagens most prevalent in articular cartilage are types II, IX, and XI. Together, these three types make up 99% by weight of all the collagen present in articular cartilage [11]. Of the three, type II is by far the most common, constituting 90–95% of all the collagen in the tissue.

Type II Collagen

Type II collagen is found throughout the cartilage ECM, forming the framework for the tissue. The type II network serves to entrap molecules within the tissue and restrain tissue swelling due to osmotic pressure of the charged proteoglycans. The type II molecule is composed of three identical $\alpha 1(II)$ chains wound together and consists of a single continuous triple helical domain, with unwound telopeptide regions on both the C- and N-termini. Type II molecules assemble into fibrils by means of cross-linking hydroxylysyl pyridinoline residues in telopeptide regions. This structure gives type II collagen a distinctive banded structure, which repeats every 67 nm [11]. The continuous assembly of type II collagen molecules into fibrils results in an increase in fibrillar diameter with age, from less than 20 nm in fetal tissue to 50–100 nm in adult tissue. The type II fibril orientation varies with location in the tissue. In the uppermost regions of the tissue, near the articular surface, the orientation of the fibrils is tangential to the surface, while deeper in the tissue, the fibrils are essentially oriented randomly.

Type XI Collagen

Type XI collagen constitutes approximately 3% by weight of all the collagen in the cartilage matrix [11]. Biochemical and immunochemical data suggest that the primary function of type XI collagen is to act as the core for type II fibrils. Unlike type II, type XI collagen is a heterogeneous trimer, composed of three chains, $\alpha 1(XI)$, $\alpha 2(XI)$, and $\alpha 3(XI)$. The $\alpha 3(XI)$ chain is indistinguishable from $\alpha 1(II)$, which may aid in its function as the core molecule of type II collagen. The $\alpha 1(XI)$ and $\alpha 2(XI)$ chains are composed of separate and distinct amino acid sequences [11].

Type IX Collagen

Type IX collagen represents 1% of the collagenous protein in mature cartilage. There is evidence that as much as 10% of the collagen in fetal cartilage is type IX [11], implying that it may play an important role in matrix assembly. Type IX collagen is a heterotrimer of chains $\alpha 1(IX)$, $\alpha 2(IX)$, and $\alpha 3(IX)$. The molecule consists of three collagenous domains, COL1, COL2, and COL3, which are woven in a triple helix, and four globular domains, NC1, NC2, NC3, and NC4, which are located in between the collagenous domains and at the ends of the molecule. In the NC3 domain there is a site of attachment for a chondroitin sulfate chain that has ionizable charge groups. The N-terminal region of the COL2 domain appears to contain the sites of attachment that cross-link type IX collagen to type II collagen. All three type IX chains, $\alpha 1(IX)$, $\alpha 2(IX)$, and $\alpha 3(IX)$, contain the hydroxylysine residues that covalently bind to the α1(II) chain on the type II molecule [12,13]. This method of attachment results in the orientation of the type II and type IX molecules in an antiparallel fashion, as observed by electron microscopy [14].

MECHANICAL PROPERTIES

The mechanical properties of cartilage in all forms are related to its complex physical structure. In articular cartilage, growth plate cartilage, and intervertebral disk, the primary loads experienced by the tissues are compressive. The compressive properties of the tissue arise primarily from the proteoglycan component of the ECM. The large proteoglycan, aggrecan, associates noncovalently with hyaluronan to form a network of large aggregates throughout the tissue. Branching off from the aggrecan core protein are many GAG chains containing sulfate and carboxylic acid charge moities, which endow the tissue with a significant charge density of approximately 0.1–0.2 mol of charge per liter of fluid [15]. Under physiological conditions, all these charge groups are effectively ionized. The attraction of positive counterions from the surrounding fluid to the negative charges fixed on the matrix creates an electrical double layer at the surface of the GAG chains. When compressive strains are applied to the tissue, the effective distance between double layers is decreased. As the double layers overlap, significant electrostatic repulsion develops, which is translated into a marked increase in tissue stiffness [16]. The equilibrium modulus of cartilage in compression can range from as high as 0.5–1.0 MPa [17] in articular cartilage to as low as 0.05 MPa in nucleus pulposus. It has been estimated that as much as half the stiffness of the tissue in compression is attributable to electrostatic repulsion of GAG chains [18]. Given that the electrostatic repulsion forces are the same in all directions, tissues such as midzone articular cartilage and nucleus pulposus tend to be isotropic in compression.

The fluid phase of the highly hydrated matrix also has a significant impact on the physical properties of articular cartilage. As the tissue is compressed, fluid leaves the tissue. The frictional forces that result from fluid movement relative to the solid phase tend to stiffen the tissue. As loading time is decreased (or loading frequency is increased), the velocity of the fluid relative to the matrix increases, and thus frictional forces increase as well. This results in a dynamic tissue stiffness that is highly frequency dependent, with values up to 20 MPa at 1 Hz.

In contrast to articular cartilage, growth plate cartilage, and intervertebral disk, the primary mechanical requirements of auricular and tracheal cartilages involve bearing loads in tension and shear. The tensile and shear properties of cartilage are primarily derived from the collagenous portion of the ECM. The tensile modulus of cartilage ranges from 10 to 30 MPa [19], while the shear modulus ranges from 5 to 20 MPa [20]. Similar to the compressive properties, the tensile properties of cartilage show a marked time dependence, with the tissue becoming significantly stiffer at higher loading rates. While fluid flows do occur during tensile deformation, the time-varying and frequency-dependent properties in this case are due to frictional interaction between solid components of the ECM and not to solid–fluid interaction. As in other soft tissues, the arrangement of collagen fibers plays a critical role in determining the tensile properties of cartilage [21]. As such, the tensile properties of cartilage can be highly anisotropic. This is particularly evident in surface zone articular cartilage, annulus fibrosus, and tracheal cartilage.

ENGINEERING OF CARTILAGENOUS TISSUES
GENERAL PRINCIPLES

The lack of innate reparative capacity in all forms makes cartilage a likely candidate for tissue engineering applications. As in most tissue engineering applications, the generation of cartilage involves one or more of three critical components: cells, matrix, and growth factors. Cell transplantation and growth factor administration have both been used individually in attempts at site-specific cartilage repair. The use of matrix scaffolds alone to conduct ingrowth from adjacent tissue has met with much success in bone repair but has not been successful in cartilage repair, likely because of the very limited proliferative and migrational capacity of chondrocytes. As such, the vast majority of cartilage tissue engineering approaches use scaffolds in combination with cells and/or growth factors.

CELLS

Chondrocyte Transplantation

In any approach involving cell transplantation, identification of a suitable population of cells is the first critical step. While the demand for engineered cartilage for replacement of tissue lost to disease or trauma is great, rarely are such situations life threatening. The use of allogenic cells for cartilage therapies is of limited feasibility, given the risks for disease transmission and the necessity of long-term immunosuppression in the recipient. As such, autologous chondrocyte transplantation has been the subject of much study. Methods for isolation of chondrocytes from cartilage have been widely reported, with most involving digestion of the tissue in collagenase in concentrations ranging from 0.08 to 0.3%.

While obtaining chondrocytes for tissue engineering studies is relatively routine, obtaining sufficient quantities of cells for clinically relevant implants can be challenging. Typical cell density of tissue-engineered cartilage constructs ranges from 25 to 50×10^6 cells/ml [22]. With reconstructive surgery applications that may require several cubic centimeters of tissue, the number of cells required for construct assembly could approach $100-200 \times 10^6$ cells [23]. Given the inherent inefficiency of collagenase digestion, this might require several grams of cartilage from the donor site to achieve this cell number, which is not clinically feasible.

The challenge then becomes starting with a relatively small number of chondrocytes and expanding this population in monolayer culture to a clinically feasible level. This is particularly problematic, since chondrocytes are known to dedifferentiate during monolayer culture [24]. Indeed, studies using articular chondrocytes have demonstrated that the quality of engineered tissue decreases dramatically as the cells are passaged in extended monolayer culture [25]. Attempts have been made to circumvent this problem by using insulin-like growth factor (IGF-I) and transforming growth factor β (TGF-β) to induce redifferentiation to the chondrocyte phenotype [26]. Other studies have demonstrated that returning dedifferentiated chondrocytes to three-dimensional culture can induce redifferentiation as well [24]. Each of these approaches has considerable merit, but tissue quality remains a persistent obstacle in bringing this technology to practice.

The greatest amount of data in engineering of human cartilage tissue has likely been generated in studies involving autologous chondrocyte transplantation for repair of articular cartilage defects [27]. This technique involves harvest of chondrocytes from non-weight-bearing regions of the femoral condyles, minimally expanding the cell population in culture, and delivering these cells to a small-volume, partial-thickness defect in the load-bearing region of the knee (Fig. 93.1A). To prevent cells from leaking into and being cleared from the synovial fluid, a patch of periosteal tissue is sutured over the defect. While initially quite encouraging, the results from this technique in human patients have been highly variable [27,28]. Further, there are no definitively successful results in any animal model, including dogs [29] and rabbits [30]. Interpretation of results from these studies is often complicated, given that the role of the periosteal flap is difficult to assess. In this, as in all approaches to articular cartilage repair, the integration of new tissue with existing tissue is of critical importance. Several studies have attempted to explore the mechanisms by which pieces of cartilage tissue can integrate [31,32] and the mechanics of this cell–tissue [33] and tissue–tissue interface [34].

Many additional studies have investigated the feasibility of chondrocyte transplantation for the regeneration of other types of cartilage. Reconstruction of the ear with tissue-engineered cartilage has been explored through xenogenic transplantation into athymic mice [35] and in autologous transplants in pigs [36]. The feasibility of cartilage implants for nasal [37], temporomandibular disk [38], tracheal [39], intervertebral disk [40], and meniscal cartilage [41] has been demonstrated in implantation in athymic mice. The use of autologous transplanted chondrocytes as a means of tissue bulking has been explored in applications of treating urinary incontinence [42] and in nipple reconstruction [43].

Mesenchymal Stem Cells

Donor site morbidity and tissue scarcity are issues that confound the use of autologous chondrocytes as a source of cells for cartilage tissue engineering. An alternative approach

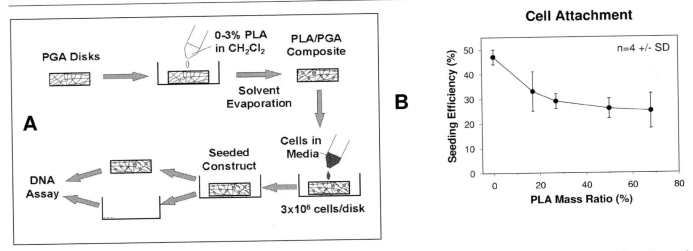

Fig. 93.1. (A) Schematic illustration of the process of polymer coating by solvent evaporation and the evaluation of the effects of the process on cell adhesion. The coating polymer (PLA) is dissolved in a suitable solvent (CH_2Cl_2), and the solution is poured onto the polymer scaffold to be coated (PGA). As the solvent is removed, PLA is deposited on PGA fibers, strengthening the composite and allowing it to retain shape. (B) To evaluate effect of this process on cell adhesion, cells are pipetted onto scaffolds of varying PGA/PLA ratios. As more PLA is added to the composite, the efficiency of the seeding process decreases significantly.

has been proposed involving the use of marrow-derived mesenchymal stem cells (MSCs). MSCs account for only one of every million cells found in the bone marrow [44], but purified cultures can be obtained from bone marrow aspirates by means of antibody-based isolation techniques. These cells have the capacity to proliferate extensively in culture and to differentiate into several musculoskeletal phenotypes, including bone, tendon, ligament, and cartilage [45]. Guiding MSCs into the chondrocyte phenotype typically involves extended time in pellet culture in the presence of TGF-β [46]. The development of cartilage-like tissue from *in vitro* cultures of MSCs on collagen scaffolds has been demonstrated [47], and there are preliminary investigations into the use of MSCs for repair of articular cartilage defects [48].

Dermal Fibroblasts

Another alternative to the use of primary autologous chondrocytes as the source of cells for tissue-engineered cartilage involves the complex manipulation of dermal fibroblasts. In the late 1990s, studies demonstrated that primary dermal fibroblasts placed in hypoxic micromass culture (200×10^6 cells/ml) exhibit behavior typical of chondrocytes, including production of type II collagen and proteoglycans [49]. The important step in this process involves the use of staurisporine, an agent that specifically blocks the production of type I collagen without affecting type II collagen production. Interestingly, the process of chondrocyte redifferentiation from a fibroblast phenotype has been shown to occur more efficiently in hypoxic environments [50]. Further emphasizing the potential of this technology is the fact that this process has been completed using neonatal foreskin fibroblasts as source cells, which may have the potential to act as universal donor cells because at an early stage of development these cells lack MHC class antigens.

MATRIX MATERIALS

A variety of matrix materials have been used to maintain cells in three-dimensional culture and to provide biochemical and organizational cues to these cells. Both the chemical identity and the way in which the materials are processed play important roles in determining the physical properties of the materials, as well as the way in which cells interact with these materials. While there are literally dozens of types of scaffold used for cartilage tissue engineering, these can be categorized in three convenient groups: synthetic scaffolds, hydrogels, and natural polymers.

Synthetic Scaffolds

Many of the synthetic scaffold materials used for cartilage tissue engineering have documented histories in other biomedical applications, particularly in drug delivery. Particularly, the poly(α-hydroxy acids), including poly(glycolic acid) (PGA), poly(lactic acid) (PLA), and poly(lactic-co-glycolic acid) (PLGA) copolymers, have figured prominently in the cartilage tissue engineering literature. This stems from their ease of use and their long history with the FDA, which dates back for more than 20 years with their use in biodegradable sutures.

These materials degrade through hydrolysis of the ester bond in the polymer backbone. The degradation rates of scaffolds made from these materials depend not only on the materials, but on the physical characteristics of the scaffolds, such as fiber diameter, pore size, and polymer crystallinity. Control of these parameters allows for regulation of overall scaffold degradation rates, which can range from 6–8 weeks in the case of highly porous PGA fibrous mesh [51] to 6–18 months in the case of a highly crystalline PLA [52]. Similarly, the mechanical properties of these scaffolds can be regulated and have been shown to range from 5 kPa [53] to 1 GPa [54].

Processing of these materials is an important component of the scaffold fabrication process and often contributes to the difficulty and expense of the use of these materials. The most common material used for cartilage tissue engineering has been nonwoven PGA mesh. This has been used in the engineering of articular [55], auricular [35], nasal [22], tracheal [39], and intervertebral disk [40] cartilages. As a source material, nonwoven PGA mesh is quite expensive, since the elaborate production process involves either extrusion through 15-μm holes or spinning to produce fibers. Owing to lack of mechanical stiffness, meshes are often mechanically cross-linked by coating with other, stiffer polymers. This measure has been shown to enhance the mechanical properties of the meshes, but is known to inhibit the process of cell attachment to these scaffolds [53] (Fig. 93.1B).

PLA and PLGA sponges and foams are inherently stiffer than PGA scaffolds and are in general easier to process than PGA owing to their solubility in common organic solvents. PLA and PLGA copolymers with more than 50% lactic acid content are soluble in chloroform, methylene chloride, and carbon tetrachloride, while PGA is not. Many processing techniques rely on this solvent susceptibility, including freeze-drying [56] and salt-leaching [57]. Freeze-drying is a multistep process involving dissolving the polymer in a suitable solvent, bringing the solution to a temperature below that of the melting point of the solvent, and evaporating the solvent by placing the frozen sample in a vacuum while maintaining a low temperature. This technique can produce scaffolds very high in porosity, which is controlled by regulating the solvent-to-polymer ratio. However, the pore size of these samples may be influenced by the rate at which the polymer solution is frozen and the temperature at which solvent evaporation takes place. Salt-leaching techniques (Fig. 93.2) provide an option for more precise control of pore dimension through the use of salt crystals as porogens [57]. This technique involves inclusion of salt crystals (typically NaCl) of known size in the polymer solution. After evaporation of the solvent, salt crystals remain inside the polymer and are removed by successive washes in water. The pores that remain after washing are the same size as the included salt crystals but can be modified by heat treatment.

While this technology has been used in many applications for cartilage tissue engineering, many challenges remain in bringing this technology to clinical practice or commercialization. While cells adhere to these polymers, the seeding efficiency of these scaffolds is often not very high. Further, there is a significant foreign body response associated with these systems. This is particularly true in applications for reconstructive surgery, where acute reactions to subcutaneous implants involve a significant inflammatory response [36]. Inflammatory agents such as interleukin 1 (IL-1) have been shown to inhibit cartilage formation [58] and degrade cartilage ECM [59]. Even in a noninflammatory environment, the natural breakdown products of these materials are acidic and will lower local pH [60]. Chondrocytes are known to be acutely sensitive to changes in environmental pH [61], and thus controlling the fate of polymer degradation fragments becomes an important consideration.

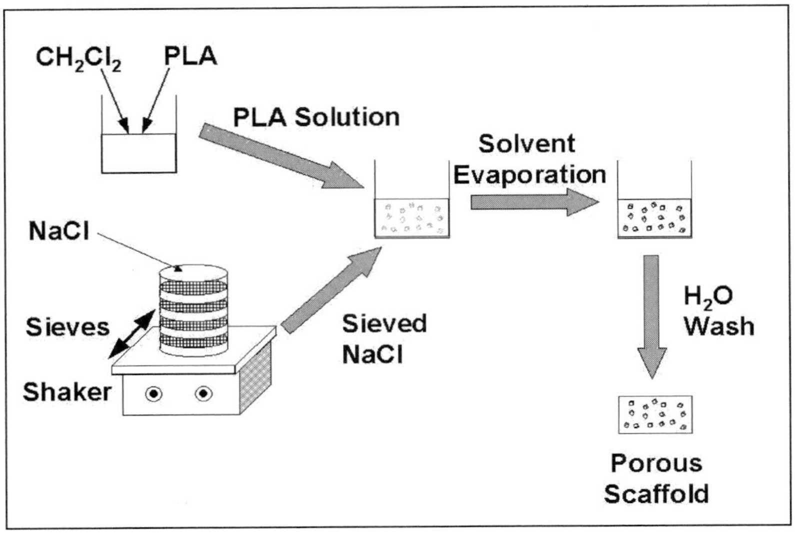

Fig. 93.2. Graphical description of the production of porous scaffolds by salt leaching. The desired polymer is dissolved in a suitable solvent and combined with NaCl particles of a desired size, obtained from passage through an arrangement of sieves. The solvent is removed by evaporation, leaving a two-phase solid composed of polymer and salt domains. The NaCl is removed by successive washes in water, producing a porous scaffold with pore of the same size as the salt crystals.

Hydrogels

Hydrogels are single-phase materials composed of highly hydrated polymer networks. They offer many specific advantages for tissue engineering, since their high water content (70–99%) typically endows these materials with high biocompatibility. Such materials are usually formed by dissolving the solid polymer in an aqueous buffer, then initiating gelation via changes in temperature [43,62], ionic content [63], or pH [64]. If cells are included in the buffer solution, these techniques are efficient methods for cell encapsulation.

There is a significant body of literature documenting the use of hydrogels as model systems for chondrocyte culture. Both alginate [62] and agarose [63] have been used in such applications, and both are known to support the chondrocyte phenotype and the assembly of a cartilage-like extracellular matrix. Alginate is an anionic polysaccharide derived from seaweed. It is a linear copolymer of mannuronic and guluronic acids, with segments composed of mannuronic acid (M) blocks, guluronic acid (G) blocks, and alternating M-G blocks. In the presence of multivalent cations, such as Ca^{2+} or Mg^{2+}, alginate chains are cross-linked by ionic bonding through G segments. Typical methods for *in vitro* chondrocyte culture in alginate involve encapsulation of cells in beads made by dropping cells suspended in 1–2% alginate solution into a bath of $CaCl_2$. While this is a useful experimental system, the tissue engineering applications of chondrocyte–alginate beads are minimal. As a result, methods have been developed using $CaSO_4$ as a cross-linking agent to lengthen the time required for cross-linking. This allows for injectable delivery of chondrocytes in alginate and results in *in situ* gelation [65]. These techniques have been extended to enable production of shaped implants by injection molding [66] (Fig. 93.3).

Agarose is a modified form of the agar material used for bacterial cell culture. Unlike alginate, encapsulation techniques using agarose involve temperature-induced phase transitions. Typically, agarose is heated in aqueous solution to above 70°C to maximize solubility, then mixed with cold media containing cells [67]. The mixture forms a single-phase hydrogel when cooled below 40°C, effectively encapsulating cells. Extended cultures of chondrocytes in alginate have demonstrated the production of significant amounts of cartilage ECM and increases in the mechanical properties of the constructs. Further, the response of cells to mechanical stimuli in these systems is similar to that of chondrocytes in native tissue [68]. There has been little investigation of *in vivo* agarose implants, likely because agarose is not a biodegradable polymer.

The Pluronics or polaxomers are polymers composed of blocks of poly(ethylene oxide) (PEO) and poly(propylene oxide) (PPO) that also exhibit thermally induced gelation [69]. In these systems gelation occurs at higher temperatures, when thermal energy exceeds the energy of hydrogen bonding that occurs between water and the hydrophilic PEO blocks. This phenomenon is dependent on both the concentration and the molecular weight of the poly-

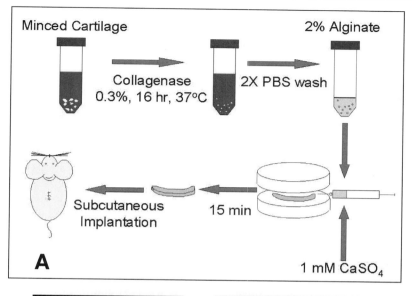
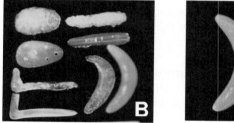
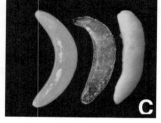

Fig. 93.3. (A) Schematic of the injection molding process. Cells obtained from cartilage via collagenase digestion are suspended in a solution of 2% alginate. Immediately before injection into the mold, sterilized CaSO₄ is mixed with chondrocytes in alginate to initiate gel formation. The chondrocyte–alginate–CaSO₄ mixture is injected into the sterilized mold by means of a syringe and needle. Formed shapes are removed from molds 15 min after injection. (B) Examples of a silicon nose bridge, chin, malar, and nasal septum implant and corresponding alginate constructs immediately after molding. (C) Silicon chin implant (left), corresponding molded alginate construct (middle), and resultant cartilage (right) removed 30 weeks after implantation into the subcutaneous space of a nude mouse.

mer in question, with higher molecular weights and higher concentrations resulting in gelation at lower temperatures. At 37°C, Pluronic (F68), having a number-average molecular weight M_n of 168,000 Da gels in concentrations in excess of 60%, while a heavier Pluronic (F127, M_n 127,000 Da), gels at 37°C in concentrations in excess of 17%, with concentrations of 30% producing gelation at 8°C. Gelation at temperatures near body temperature make Pluronic an ideal candidate for *in vivo* cell delivery. Indeed, the use of Pluronic F127 to deliver articular and auricular chondrocytes has been successful in promoting cartilage formation [36]. However, the rapid degradation times of these polymers make this a suboptimal delivery choice for cells with low biosynthetic rates, such as the nucleus pulposus [40]. Recently, other polymer systems have been developed that exhibit similar thermal gelation characteristics. The KLD12 peptide [70] is a polymer made from block repeats of a tripeptide sequence that forms β sheets in aqueous solution. Much as in the Pluronic system, as the temperature is raised, hydrogen bonding is disrupted and the polymer falls out of solution, forming a hydrated gel. Preliminary studies have demonstrated that chondrocytes cultured in KLD12 gels produce proteoglycans and collagen similar to those of cartilage.

The use of hydrogels has been highly successful as a means of producing three-dimensional chondrocyte cultures and allowing *in vivo* tissue formation. The main shortcomings of these systems are in the mechanical properties of such gels, which are normally quite weak. Further, different hydrogel systems have distinct degradation times, but control of degradation rates for individual systems has not been demonstrated. Thus considerable challenges remain in bringing hydrogel technology to clinical practice in cartilage tissue engineering.

Natural Polymers

The prospect of using naturally derived polymers for support of chondrocyte growth is appealing because these are materials to which cells are known to react well. The classic example of a natural polymer used as a scaffold material is type I collagen. This is a system of choice for several applications, largely because of the ease of cell delivery. Many preparations of collagen type I are soluble under acidic conditions. The neutralization of collagen solutions results in the formation of a hydrated collagen gel. When cells are included in the neutralizing solution, they are effectively encapsulated in the collagen gel. This technology has been used for the delivery of chondrocytes for articular cartilage repair [71] and for the engineering of intervertebral disk [72] and meniscus [73]. One of the main obstacles in using these systems in clinical practice is the mechanical performance of the gels, which have moduli on the order of 0.1–1 kPa. An additional concern associated with the use of type I collagen as a scaffold material is that most cartilages do not contain type I collagen. Therefore scaffolds composed of type II collagen have been constructed from native cartilage matrix [74]. However the purity of the type II collagen in such preparations is a concern.

The other natural polymer based system used for a variety of cartilage applications is fibrin. This technology is based on knowledge of hemostasis and the clotting process. Fibrin clots or fibrin glues form from the reaction of fibrinogen and thrombin. Cells present in either component during reaction are encapsulated in the resulting fibrin structure. The formation of a fibrin network takes place over the course of 10–15 min, which leaves enough time for injection of the mixture into the body for *in situ* gelation. This method has been successful in the formation of cartilage in subcutaneous models [75] and in cartilage interface repair models [76–78]. The most difficult obstacle in this area is the acquisition of purified materials, since the source of the reagents is typically blood cryoprecipitate.

GROWTH FACTORS

The treatment of cartilage defects with peptide growth factors alone would be of great advantage for commercialization of new products. As such, several studies have investigated the potential for growth factors such as IGF-1, TGF-β, and bone morphogenetic protein 2 (BMP-2) to repair cartilage lesions. This is a particularly challenging task, since cartilage defects are by definition devoid of cells, and chondrocytes are quite slow to migrate. Thus, many of these techniques rely on other cell sources in the body, such as stem cells from the blood.

The use of IGF-1 in an articular cartilage defect model demonstrated a slight augmentation to the repair process, as indicated by slight increases in type II collagen and proteoglycans in the defect site [79]. The use of liposome-encapsulated TGF-β to treat articular cartilage defects has also shown promise, and it is suggested that synovial fibroblasts may play a role in this process. Collagen sponges impregnated with BMP-2 and implanted in articular cartilage defects showed enhanced cartilage formation and integration with surrounding tissue [80]. All applications of growth factors for cartilage tissue engineering thus far have been focused in the area of articular cartilage defects. These technologies may be applicable to intervertebral disk and meniscal repair, but they are likely to be of limited use in nasal or auricular cartilage repair.

FUTURE DIRECTIONS

Great strides have been taken toward the goal of using tissue engineering to achieve functional cartilage replacement. The development of technologies for the delivery of chondrocytes to generate cartilage in joints, spine, ear, nose, and other structures has demonstrated the potential to replace or regrow a tissue with little innate reparative capacity. However, many challenges remain in all areas of cartilage tissue engineering research. Procedures for obtaining sufficient numbers of chondrocytes to generate large amounts of tissue are not currently available. The improvement of methods for culture of primary chondrocytes and the demonstration of feasibility of large-scale chondrogenesis from MSCs or dermal fibroblasts remain worthy goals. Similarly, the improvement of materials available for cell transplantation is a considerable challenge.

Understanding and regulating the processes of cell adhesion and proliferation on synthetic matrices is a critical issue. Indeed, progress has been made in this area through grafting of cell adhesion ligands [81] and controlling pore structure through techniques such as three-dimensional printing [82]. In addition, the control of the mechanical environment [60] and the nutrient supply [83] to growing constructs may allow for more precise regulation of cartilage matrix assembly on material scaffolds. With these great challenges ahead, the field of cartilage tissue engineering will continue to be an area of active research for many years to come.

REFERENCES

1. Harris, J., Kallen, B., and Robert, E. (1996). The epidemiology of anotia and microtia. *J. Med. Genet.* **33**, 809–813.

2. Lorenz, H. P., Hedrick, M. H., Chang, J., Mehrara, B. J., and Longaker, M. T. (2000). The impact of biomolecular medicine and tissue engineering on plastic surgery in the 21st century. *Plast. Reconstr. Surg.* **105**, 2467–2481.

3. Buckwalter, J. A., and Mankin, H. J. (1998). Articular cartilage: Degeneration and osteoarthritis, repair, regeneration, and transplantation. *Instruc. Course Lect.* **47**, 487–504.

4. Connell, M. D., and Wiesel, S. W. (1992). Natural history and pathogenesis of cervical disk disease. *Orthop. Clin. North Am.* **23**, 369–380.

5. Lewis, M. M., Sissons, H. A., Norman, A., and Greenspan, A. (1987). Benign and malignant cartilage tumors. *Instruc. Course Lect.* **36**, 87–114.

6. Vinod, S. K., Macleod, C. A., Barnes, D. J., and Fletcher, J. (1999). Malignant fibrous histiocytoma of the trachea. *Respirology* **4**, 271–274.

7. Buckwalter, J. A., and Mankin, H. J. (1998). Articular cartilage repair and transplantation. *Arthritis Rheum.* **41**, 1331–1342.

8. Hardingham, T. E., and Fosang, A. J. (1992). Proteoglycans: Many forms, many functions. *FASEB J.* **6**, 861–870.

9. Heinegaard, D. K., Franzen, A., Hedbom, E., and Sommarin, Y. (1986). Common structures of the core proteins of interstitial proteoglycans. *In* "Functions of the Proteoglycans" (V. C. Hascall, ed.), pp. 69–88. Wiley, New York.

10. Hardingham, T. E., Fosang, A. J., and Dudhia, J. (1992). "Aggrecan, the Chondroitin Sulfate/Keratan Sulfate Proteoglycan from Cartilage," pp. 5–20. Raven Press, New York.

11. Eyre, D. R., Dickson, I. R., and van Ness, K. (1991). The collagens of articular cartilage. *Semin. Arthritis Rheum.* **21**(S2), 2–11.

12. Eyre, D. R., Dickson, I. R., and van Ness, K. (1988). Collagen cross-linking in human bone and articular cartilage: Age-related changes in the content of mature hydroxypyrindinium residues. *Biochem. J.* **252**, 495–500.

13. Wu, J. J., Woods, P. E., and Eyre, D. R. (1992). Identification of cross-linking sites in bovine articular cartilage reveals an antiparallel type II–type IX molecular relationship and type IX to type II bonding. *J. Biol. Chem.* **267**, 23007–23014.

14. Vaughn, L. (1988). Periodic distribution of type IX collagen over cartilage fibrils. *J. Cell Biol.* **106**, 991–997.

15. Maroudas, A. (1980). Physical chemistry of articular cartilage and the intervertebral disc. *In* "The Joints and Synovial Fluid" (L. Sokoloff, ed.), Vol. 2, pp. 239–291. Academic Press, New York.

16. Grodzinsky, A. J. (1983). Electromechanical and physicochemical properties of connective tissue. *CRC Crit. Rev. Bioeng.* **9**, 133–199.

17. Mow, V. C., Kuei, S. C., Lai, W. M., and Armstrong, C. G. (1980). Biphasic creep and stress relaxation of articular cartilage in compression? Theory and experiments. *J. Biomech. Eng.* **102**, 73–84.

18. Frank, E. H., and Grodzinsky, A. J. (1987). Cartilage electromechanics. I. Electrokinetic transduction and the effects of electrolyte pH and ionic strength. *J. Biomech.* **20**, 615–627.

19. Akizuki, S., Mow, V. C., Muller, F., Pita, J. C., Howell, D. S., and Manicourt, D. H. (1986). Tensile properties of human knee joint cartilage. I. Influence of ionic conditions, weight bearing, and fibrillation on the tensile modulus. *J. Orthop. Res.* **4**, 379–392.

20. Setton, L. A., Mow, V. C., Muller, F. J., Pita, J. C., and Howell, D. S. (1994). Mechanical properties of canine articular cartilage are significantly altered following transection of the anterior cruciate ligament. *J. Orthop. Res.* **12**, 451–463.

21. LeRoux, M. A., Arokoski, J., Vail, T. P., Guilak, F., Hyttinen, M. M., Kiviranta, I., and Setton, L. A. (2000). Simultaneous changes in the mechanical properties, quantitative collagen organization, and proteoglycan concentration of articular cartilage following canine meniscectomy. *J. Orthop. Res.* **18**, 383–392.

22. Puelacher, W. C., Kim, S. W., Vacanti, J. P., Schloo, B., Mooney, D., and Vacanti, C. A. (1994). Tissue-engineered growth of cartilage: The effect of varying the concentration of chondrocytes seeded onto synthetic polymer matrices. *Int. J. Oral. Maxillofacal Surg.* **23**, 49–53.

23. Chang, S. C. N., Rowley, J. A., Tobias, G., Genes, N. G., Roy, A., Mooney, D. J., Vacanti, C. A., and Bonassar, L. J. (2001). Injection molding of chondrocyte/alginate constructs in the shape of facial implants. *J. Biomed. Mater. Res.* **55**, 503-511.

24. Benya, P. D., and Shaffer, J. D. (1982). Dedifferentiated chondrocytes reexpress the differentiated collagen phenotype when cultured in agarose gels. *Cell (Cambridge, Mass.)* **30**, 215–224.

25. Vacanti, C. A., and Upton, J. (1994). Tissue-engineered morphogenesis of cartilage and bone by means of cell transplantation using synthetic biodegradable polymer matrices. *Clin. Plast. Surg.* **21**, 445–462.

26. Yaeger, P. C., Masi, T. L., de Ortiz, J. L., Binette, F., Tubo, R., and McPherson, J. M. (1997). Synergistic action of transforming growth factor-beta and insulin-like growth factor-I induces expression of type II collagen and aggrecan genes in adult human articular chondrocytes. *Exp. Cell Res.* **237**, 318–325.

27. Brittberg, M., Lindahl, A., Nilsson, A., Ohlsson, C., Isaksson, O., and Peterson, L. (1994). Treatment of deep cartilage defects in the knee with autologous chondrocyte transplantation. *N. Engl. J. Med.* **331**, 889–895.

28. Nehrer, S., Spector, M., and Minas, T. (1999). Histologic analysis of tissue after failed cartilage repair procedures. *Clin. Orthop. Relat. Res.* **365**, 149–162.

29. Breinan, H. A., Minas, T., Hsu, H. P., Nehrer, S., Sledge, C. B., and Spector, M. (1997). Effect of cultured autologous chondrocytes on repair of chondral defects in a canine model. *J. Bone Jt. Surg., Am. Vol.* **79**, 1439–1451.

30. Brittberg, M., Nilsson, A., Lindahl, A., Ohlsson, C., and Peterson, L. (1996). Rabbit articular cartilage defects treated with autologous cultured chondrocytes. *Clin. Orthop. Relat. Res.* **326**, 270–283.

31. Peretti, G. M., Randolph, M. A., Caruso, E. M., Rossetti, F., and Zaleske, D. J. (1998). Bonding of cartilage matrices with cultured chondrocytes: An experimental model. *J. Orthop. Res.* **16**, 89–95.

32. Schinagl, R. M., Kurtis, M. S., Ellis, K. D., Chien, S., and Sah, R. L. (1999). Effect of seeding duration on the strength of chondrocyte adhesion to articular cartilage. *J. Orthop. Res.* **17**, 121–129.

33. Peretti, G. M., Bonassar, L. J., Caruso, E. M., Randolph, M. A., Trahan, C. A., and Zaleske, D. J. (1999). Biomechanical analysis of a chondrocyte-based repair model of articular cartilage. *Tissue Eng.* **5**, 317–326.

34. Reindel, E. S., Ayroso, A. M., Chen, A. C., Chun, D. M., Schinagl, R. M., and Sah, R. L. (1995). Integrative repair of articular cartilage in vitro: Adhesive strength of the interface region. *J. Orthop. Res.* **13**, 751–760.

35. Cao, Y., Vacanti, J. P., Paige, K. T., Upton, J., and Vacanti, C. A. (1997). Transplantation of chondrocytes utilizing a polymer–cell construct to produce tissue-engineered cartilage in the shape of a human ear. *Plast. Reconstr. Surg.* **100**, 297–302.

36. Cao, Y., Rodriguez, A., Vacanti, M., Ibarra, C., Arevalo, C., and Vacanti, C. A. (1998). Comparative study of the use of poly(glycolic acid), calcium alginate and Pluronics in the engineering of autologous porcine cartilage. *J. Biomater. Sci., Polym. Ed.* **9**, 475–487.

37. Puelacher, W. C., Mooney, D., Langer, R., Upton, J., Vacanti, J. P., and Vacanti, C. A. (1994). Design of nasoseptal cartilage replacements synthesized from biodegradable polymers and chondrocytes. *Biomaterials* **15**, 774–778.

38. Puelacher, W. C., Wisser, J., Vacanti, C. A., Ferraro, N. F., Jaramillo, D., and Vacanti, J. P. (1994). Temporomandibular joint disc replacement made by tissue-engineered growth of cartilage. *J. Oral Maxillofacial Surg.* **52**, 1172–1177.

39. Sakata, J., Vacanti, C. A., Schloo, B., Healy, G. B., Langer, R., and Vacanti, J. P. (1994). Tracheal composites tissue engineered from chondrocytes, tracheal epithelial cells, and synthetic degradable scaffolding. *Transplant. Proc.* **26**, 3309–3310.

40. Kusior, L., Vacanti, C. A., Bayley, J., and Bonassar, L. J. (1998). Tissue engineering of bovine nucleus pulposus in nude mice. *Tissue Eng.* **4**, S468.

41. Ibarra, C., Jannetta, C., Vacanti, C. A., Cao, Y., Kim, T. H., Upton, J., and Vacanti, J. P. (1997). Tissue engineered meniscus: A potential new alternative to allogeneic meniscus transplantation. *Transplant. Proc.* **29**, 986–988.

42. Atala, A., Cima, L. G., Kim, W., Paige, K. T., Vacanti, J. P., Retik, A. B., and Vacanti, C. A. (1993). Injectable alginate seeded with chondrocytes as a potential treatment for vesicoureteral reflux. *J. Urol.* **150**, 745–747.

43. Cao, Y. L., Lach, E., Kim, T. H., Rodriguez, A., Arevalo, C. A., and Vacanti, C. A. (1998). Tissue-engineered nipple reconstruction. *Plast. Reconstr. Surg.* **102**, 2293–2298.

44. Caplan, A. I. (1991). Mesenchymal stem cells. *J. Orthop. Res.* **9**, 641–650.

45. Pittenger, M. F., Mackay, A. M., Beck, S. C., Jaiswal, R. K., Douglas, R., Mosca, J. D., Moorman, M. A., Simonetti, D. W., Craig, S., and Marshak, D. R. (1999). Multilineage potential of adult human mesenchymal stem cells. *Science* **284**, 143–147.

46. Johnstone, B., Hering, T. M., Caplan, A. I., Goldberg, V. M., and Yoo, J. U. (1998). In vitro chondrogenesis of bone marrow-derived mesenchymal progenitor cells. *Exp. Cell Res.* **238**, 265–272.

47. Worster, A. A., Nixon, A. J., Brower-Toland, B. D., and Williams, J. (2000). Effect of transforming growth factor beta1 on chondrogenic differentiation of cultured equine mesenchymal stem cells. *Am. J. Vet. Res.* **61**, 1003–1010.

48. Ponticiello, M. S., Schinagl, R. M., Kadiyala, S., and Barry, F. P. (2000). Gelatin-based resorbable sponge as a carrier matrix for human mesenchymal stem cells in cartilage regeneration therapy. *J. Biomed. Mater. Res.* **52**, 246–255.

49. Nicoll, S. B., Wedrychowska, A., and Bhatnagar, R. S. (1998). Induction of a chondrocyte-like phenotype in human dermal fibroblasts by micromass culture in a functionally hypoxic environment: Regulation by protein kinase C. *Trans. Orthop. Res. Soc.* **23**, 36.

50. Domm, C., Schunke, M., Fay, J., and Kurz, B. (1998). Influence of oxygen on the redifferentiation of dedifferentiated bovine articular chondrocytes in the 3D-alginate system. *Trans. Orthop. Res. Soc.* **25**, 57.

51. Kusior, L., Vacanti, C. A., Bayley, J., and Bonassar, L. J. (1999). Tissue engineering of bovine nucleus pulposus in nude mice. *Trans. Orthop. Res. Soc.* **24**, 807.

52. Hooper, K. A., Macon, N. D., and Kohn, J. (1998). Comparative histological evaluation of new tyrosine-derived polymers and poly(L-lactic acid) as a function of polymer degradation. *J. Biomed. Mater. Res.* **41**, 443–454.

53. Moran J. M., and Bonassar, L. J. (1998). Fabrication and characterization of PLA/PGA composites for cartilage tissue engineering. *Tissue Eng.* **4**, S498.

54. Daniels, A.U., Andriano, K. P., Smutz, W. P., Chang, M. K., and Heller, J. (1994). Evaluation of absorbable poly(ortho esters) for use in surgical implants. *J. Appl. Biomater.* **5**, 51–64.

55. Freed, L. E., Grande, D. A., Lingbin, Z., Emmanual, J., Marquis, J. C., and Langer, R. (1994). Joint resurfacing using allograft chondrocytes and synthetic biodegradable polymer scaffolds. *J. Biomed. Mater. Res.* **28**, 891–899.

56. Thomson, R. C., Yaszemski, M. J., Powers, J. M., and Mikos, A. G. (1995). Fabrication of biodegradable polymer scaffolds to engineer trabecular bone. *J. Biomater. Sci. Polym. Ed.* **7**, 23–38.

57. Lu, L., Peter, S. J., Lyman, M. D., Lai, H. L., Leite, S. M., Tamada, J. A., Uyama, S., Vacanti, J. P., Langer, R., and Mikos, A. G. (2000). In vitro and in vivo degradation of porous poly(DL-lactic-*co*-glycolic acid) foams. *Biomaterials* **21**, 1837–1845.

58. Rayan, V., and Hardingham, T. (1994). The recovery of articular cartilage in explant culture from interleukin-1 alpha: Effects on proteoglycan synthesis and degradation. *Matrix Biol.* **14**, 263–271.

59. Bonassar, L. J., Sandy, J. D. Lark, M. W., Plaas, A. H. K., Frank, E. H., and Grodzinsky, A. J. (1997). Inhibition of cartilage degradation and changes in physical properties induced by IL-1β and retinoic acid. *Arch. Biochem. Biophys.* **344**, 325–334.

60. Pazzano, D., Mercier, K. A., Moran, J. M., Fong, S. S., DiBiasio, D. D., Rulfs, J. X., Kohles, S. S., and Bonassar, L. J. (2000). Comparison of chondrogenesis in static and perfused bioreactor culture. *Biotechnol. Prog.* **16**, 893–896.

61. Gray, M. L., Pizzanelli, A. M., Grodzinsky, A. J., and Lee, R. C. (1988). Mechanical and physiochemical determinants of the chondrocyte biosynthetic response. *J. Orthop. Res.* **6**, 777–792.

62. Sun, D., Aydelotte, M. B., Maldonado, B., Kuettner, K. E., and Kimura, J. H. (1986). Clonal analysis of the population of chondrocytes from the Swarm rat chondrosarcoma in agarose culture. *J. Orthop. Res.* **4**, 427–436.

63. Guo, J. F., Jourdian, G. W., and MacCallum, D. K. (1989). Culture and growth characteristics of chondrocytes encapsulated in alginate beads. *Connect. Tissue Res.* **19**, 277–297.

64. Okude, M., Yamanaka, A., and Akihama, S. (1995). The effects of pH on the generation of turbidity and elasticity associated with fibrinogen–fibrin conversion by thrombin are remarkably influenced by sialic acid in fibrinogen. *Biol. Pharm. Bull.* **18**, 203–207.

65. Paige, K. T., Cima, L. G., Yaremchuk, M. J., Vacanti, J. P., and Vacanti, C. A. (1995). Injectable cartilage. *Plast. Reconstr. Surg.* **96**, 1390–1398.

66. Chang, S. C. N., Rowley, J. A., Tobias, G., Genes, N. G., Roy, A., Mooney, D. J., Vacanti, C. A., and Bonassar, L. J. (2001). Injection molding of chondrocyte/alginate constructs in the shape of facial implants *J. Biomed. Mater. Res.* **55**, 503–511.

67. Buschmann, M. D., Gluzband, Y. A., Grodzinsky, A. J., Kimura, J. H., and Hunziker, E. B. (1992). Chondrocytes in agarose culture synthesize a mechanically functional extracellular matrix. *J. Orthop. Res.* **10**, 745–758.

68. Buschmann, M. D., Gluzband, Y. A., Grodzinsky, A. J., and Hunziker, E. B. (1995). Mechanical compression modulates matrix biosynthesis in chondrocyte/agarose culture. *J. Cell Sci.* **108**, 1497–1508.

69. Miyazaki, S., Tobiyama, T., Takada, M., and Attwood, D. (1995). Percutaneous absorption of indomethacin from Pluronics F127 gels in rats. *J. Pharm. Pharmacol.* **47**, 455–457.

70. Kisiday, J. (2000). Cartilage tissue engineering using a new self-assembling peptide gel. *Ann. Biomed. Eng.* **28S**, S121.

71. Nehrer, S., Breinan, H. A., Ramappa, A., Hsu, H. P., Minas, T., Shortkroff, S., Sledge, C. B., Yannas, I. V., and Spector, M. (1998). Chondrocyte-seeded collagen matrices implanted in a chondral defect in a canine model. *Biomaterials* **19**, 2313–2328.

72. Schneider, T. O., Mueller, S. M., Shortkroff, S., and Spector, M. (1999). Expression of alpha-smooth muscle actin in canine intervertebral disc cells in situ and in collagen–glycosaminoglycan matrices in vitro. *J. Orthop. Res.* **17**, 192–199.

73. Mueller, S. M., Shortkroff, S., Schneider, T. O., Breinan, H. A., Yannas, I. V., and Spector, M. (1999). Meniscus cells seeded in type I and type II collagen–GAG matrices in vitro. *Biomaterials* **20**, 701–709.

74. Nehrer, S., Breinan, H. A., Ramappa, A., Shortkroff, S., Young, G., Minas, T., Sledge, C. B., Yannas, I. V., and Spector, M. (1997). Canine chondrocytes seeded in type I and type II collagen implants investigated in vitro. *J. Biomed. Mater. Res.* **38**, 95–104.

75. Silverman, R. P., Passaretti, D., Huang, W., Randolph, M. A., and Yaremchuk, M. J. (1999). Injectable tissue-engineered cartilage using a fibrin glue polymer. *Plast. Reconstr. Surg.* **103**, 1809–1818.

76. Peretti, G. M., Randolph, M. A., Caruso, E. M., Rossetti, F., and Zaleske, D. J. (1998). Bonding of cartilage matrices with cultured chondrocytes: An experimental model. *J. Orthop. Res.* **16**, 89–95.

77. Peretti, G. M., Bonassar, L. J., Caruso, E. M., Randolph, M. A., Trahan, C. A., and Zaleske, D. J. (1999). Biomechanical analysis of a chondrocyte-based repair model of articular cartilage. *Tissue Eng.* **5**, 317–326.

78. Silverman, R. P., Bonassar, L., Passaretti, D., Randolph, M. A., and Yaremchuk, M. J. (2000). Adhesion of tissue-engineered cartilage to native cartilage. *Plast. Reconstr. Surg.* **105**, 1393–1398.

79. Nixon, A. J., Fortier, L. A., Williams, J., and Mohammed, H. (1999). Enhanced repair of extensive articular defects by insulin-like growth factor-I-laden fibrin composites. *J. Orthop. Res.* **17**, 475–487.

80. Sellers, R. S., Zhang, R., Glasson, S. S., Kim, H. D., Peluso, D., D'Augusta, D. A., Beckwith, K., and Morris, E. A. (2000). Repair of articular cartilage defects one year after treatment with recombinant human bone morphogenetic protein-2 (rhBMP-2). *J. Bone Jt. Surg. Am. Vol.* **82**, 151–160.

81. Rowley, J. A., Madlambayan, G., and Mooney, D. J. (1999). Alginate hydrogels as synthetic extracellular matrix materials. *Biomaterials* **20**, 45–53.

82. Park, A., Wu, B., and Griffith, L. G. (1998). Integration of surface modification and 3D fabrication techniques to prepare patterned poly(L-lactide) substrates allowing regionally selective cell adhesion. *J. Biomater. Sci. Polym. Ed.* **9**, 89–110.

83. Freed, L. E., Hollander, A. P., Martin, I., Barry, J. R., Langer, R., and Vunjak-Novakovic, G. (1998). Chondrogenesis in a cell–polymer–bioreactor system. *Exp. Cell Res.* **240**, 58–65.

CHAPTER 94

PHALANGES AND SMALL JOINTS

Noritaka Isogai and William J. Landis

BACKGROUND

The repair of severely traumatized or congenitally deficient joints requires multiple complicated reconstructive procedures, and recognition of the difficulties associated with such approaches has led to investigations of new therapeutic strategies. Free whole-joint transfer with a microvascular anastomosis technique was developed and has been performed clinically as a useful procedure to replace the function of a defective joint [1–7]. However, a vascularized autogenous whole-joint transfer is limited by donor site morbidity and availability of tissue. Also, sacrifice of the big toe or the second toe makes this surgical procedure aesthetically unacceptable. These current limitations in replacing organ/tissue function by whole-organ/tissue transplantation and artificial prostheses have fostered interest in the possibility of novel engineering of tissues and organs. In this context, initial tissue engineering experiments have demonstrated that the formation of a model of human phalanges and small joints can be accomplished by a selective placement of bovine periosteum, chondrocytes, and tenocytes into a biodegradable synthetic polymer scaffold [8]. In this approach directed ultimately toward regenerating phalanges and joints, three fundamental tissue engineering steps are involved: (1) fabrication of a three-dimensional biodegradable synthetic polymer serving as a framework for the seeding and growth of cells, (2) isolation and delivery of specific bovine cells to the polymer scaffolds, and (3) transplantation of cultured cell–polymer constructs to induce structural, biochemical, and biological development of the tissues in athymic nude mice. This chapter outlines the experimental procedures used for the formation of the model tissue and briefly describes the morphology of the resulting composite grown in the shape and dimensions of a human phalanx and joint. More complete details concerning the model phalanges and small joints were published in 1999 [8].

LABORATORY PROTOCOL
FABRICATION OF THE THREE-DIMENSIONAL BIODEGRADABLE SYNTHETIC POLYMER
Materials Required
1. Poly(glycolic acid) (PGA) fiber (Albany International, Mansfield, MA). A nonwoven mesh of PGA fiber having a diameter of 15 μm and interfiber spaces averaging 75–100 μm is used (Fig. 94.1). The fibers provide mechanical strength, while the interfiber spaces provide nutritional pathways for the selectively seeded cells. The PGA scaffold permits cell processes to anchor to the framework and thereby allows the cells to become securely integrated over and within the mesh.
2. Poly(L-lactic acid) (PLLA) (Polysciences, Warrington, PA)
3. Methylene chloride (Sigma, St. Louis, MO)
4. Impression material putty (vinyl polysiloxane) (3M Dental Products, St. Paul, MN)

Fig. 94.1. Scanning electron micrograph of the PGA mesh prior to cell seeding. Fibers of uniform diameter are randomly oriented and maintain open interfiber spaces throughout the scaffold. Bar = 100 µm.

Procedure for Copolymer Formation (Fig. 94.2)

1. Immerse the PGA mesh in a 2% PLLA solution dissolved in methylene chloride. Use gauze to absorb excessive solution.
2. Transfer the wet mesh into negative impression molds of a human phalanx, modeled from cadaveric human distal and middle phalangeal bones.
3. After spontaneous evaporation of the solvent in a hood, remove the scaffold (the PGA-PLLA copolymer shaped into the form of a human phalanx) from the mold with clean forceps.
4. Place the scaffold into a dry sterilizer and heat-treat it at 195°C for 90 min.
5. Sterilize the polymer with ethylene oxide or gamma irradiation. The process of immersion and heating results in the formation of covalent cross-links in the copolymer and generates a three-dimensional scaffold for use as the mold for the phalanx.

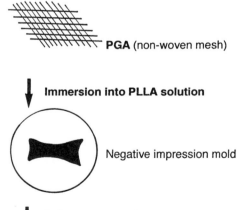

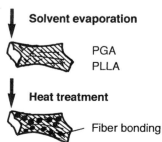

Fig. 94.2. Schematic illustrating the copolymer fabrication process. PGA is immersed in a petri dish containing PLLA in methylene chloride. The wet mesh is placed into a negative impression mold of an appropriate cadaveric bone, and the solvent evaporates. The resulting PGA-PLLA copolymer is heated to 195°C for 90 min to induce bonding of composite fibers. Diagram adapted from Isogai et al. [8] and used with permission from the Journal of Bone and Joint Surgery.

ISOLATION AND DELIVERY OF BOVINE CELLS TO THE BIODEGRADABLE POLYMER SCAFFOLDS

Materials Required
1. Calf shoulder (newborn), harvested and kept on ice (4°C)
2. Sterile surgical instruments (forceps, scissors, elevator, scalpel)
3. Biodegradable polyglactin sutures (5-0 Vicryl, Ethicon, Somerville, NJ)
4. 50-ml Falcon conical centrifuge tubes (Becton Dickinson Labware, Lincoln Park, NJ)
5. COSTAR 6-well polystyrene cell culture plates (Corning, Corning, NY)
6. 0.3% type II collagenase (Worthington, Freehold, NJ)
7. Sterile magnesium- and calcium-free phosphate-buffered saline (Dulbecco's PBS, Gibco, Grand Island, NY)
8. Nylon mesh filter (150-μm pore size)
9. Hemocytometer

Preparation of the Media
1. Medium for the periosteum: Medium 199 (M-199) (Gibco, Grand Island, NY)
2. Medium for chondrocytes and tenocytes: Ham's F-12 medium (Gibco)
3. Media supplementation

   a. 10% fetal bovine serum (Gibco)
   b. L-Glutamine (292 μg/ml)
   c. Penicillin (100 U/ml)
   d. Streptomycin (100 μg/ml)
   e. L-Ascorbic acid (50 μg/ml)

Procedure (Fig. 94.3)
1. Preparation of the periosteum–copolymer construct

   a. With the surgical elevator, remove in sterile fashion a section of the full thickness of the periosteum from the radial diaphysis of the harvested calf shoulder.
   b. Wrap the copolymer scaffold with the fresh periosteum, and in doing so place the cambium layer in contact with the copolymer surface.
   c. Suture (5-0 Vicryl) the two opposing edges of the periosteum as the tissue is wrapped about the copolymer.
   d. Place the periosteum–copolymer construct in M-199 medium and incubate it at 37°C for 1 week (5% CO_2 atmosphere).

2. Preparation of the chondrocyte–polymer construct

   a. With the scalpel and in sterile fashion, obtain thin slices of cartilage from the articular surfaces of the fresh calf shoulder.
   b. Collect the cartilage slices in a 50-ml Falcon tube and rinse with sterile PBS.
   c. Centrifuge the collected cartilage for 10 min at 2000 rpm and 4°C.
   d. Aspirate the supernatant and add 30 ml of collagenase solution.
   e. Place the cartilage–collagenase solution in an incubated shaker bath for 12 h at 37°C.
   f. Filter the enzymatically digested solution through sterile nylon mesh and wash the resulting population of cells three times with M-199 medium.
   g. Remove a sample of the washed cells to determine cell number with the hemocytometer.
   h. Resuspend the cells to a final concentration of 150×10^6 cells/ml.
   i. Using 100 μl of the concentrated cell suspension, seed cells by pipette delivery onto the PGA polymer mesh (individual sheets 10×10 mm^2 in size and 2 mm thick) separately placed in 6-well culture plates.
   j. One hour after seeding, add F-12 medium to cover the mesh.

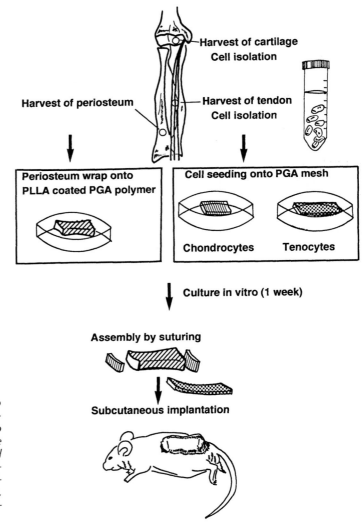

Fig. 94.3. Diagram of the general experimental design used to form models of human phalanges and small joints. Bovine material is harvested and periosteum is wrapped directly onto a PGA-PLLA copolymer molded to cadaveric bone. Bovine chondrocytes and tenocytes are isolated biochemically and separately seeded onto PGA polymer mesh sheets. All materials are cultured and composite tissue structures are then fabricated to create models of a distal phalanx, a middle phalanx, and a distal interphalangeal joint. The sutured tissues are individually implanted in athymic mice.

    k. Change medium every 2–3 days for 1 week. Culturing for 1 week resulted in the ingrowth of chondrocytes throughout the porous spaces of the PGA polymer mesh. Continued medium exchange ensures a nutrient supply for extended survival of the cultured cells and elaboration of a cartilage extracellular matrix (Fig. 94.4).

3. Preparation of the tenocyte–polymer construct

    a. Dissect the flexor tendons from the calf forelimbs and dice them into pieces approximately 5 mm³ in size.

    b. Collect the tendon pieces in a 50-ml Falcon tube and rinse with sterile PBS.

    c. Repeat steps 2c–2k.

Transplantation of Cultured Cell–Polymer Constructs
Materials Required

    1. Nude mice (homozygous, Balb/c, 4 to 6-week-old animals)

    2. Sterile surgical instruments (forceps, scissors, elevator, scalpel)

    3. Biodegradable polyglactin sutures (5-0 Vicryl, Ethicon, Somerville, NJ)

    4. Nylon sutures (5-0, Ethicon)

    5. Silicone sheet (0.5-mm thickness)

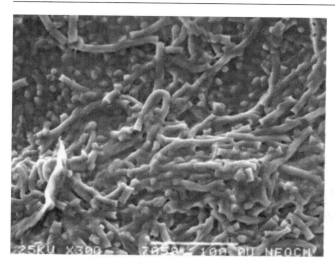

Fig. 94.4. Scanning electron micrograph of PGA mesh seeded with isolated chondrocytes and cultured for 10 days. The cells appear as smaller polygonal structures along the PGA fibers and throughout the polymer scaffold. They produce sheetlike extracellular matrix that covers the fibers and fills the interfiber spaces of the mesh. By 10 days in culture, hydrolysis has reduced both the length and diameter of the PGA fibers (see Fig. 94.1 at the same magnification). Bar = 100 μm.

Procedure

1. Under sterile conditions, attach the periosteum–copolymer construct and chondrocyte-seeded PGA polymer mesh by suturing with biodegradable polyglactin. This creates either the distal phalanx model with the periosteal construct and one articular surface or the middle phalanx model with the periosteal construct and two articular surfaces (Fig. 94.5).

2. Using polyglactin, suture the tenocyte-seeded PGA polymer mesh to the chondroosseous junction of the middle phalanx model to approximate an insertion site of a phalangeal joint tenocapsule (Fig. 94.5).

3. Without suturing, insert the 0.5-mm-thick silicone sheet (cut to size) between the two adjacent PGA polymers seeded with chondrocytes (belonging to the distal phalanx model and the middle phalanx model). The silicone placement prevents direct contact of the PGA sheets in the formation of the complete phalangeal joint.

4. Wrap the tenocyte–polymer construct about the apposing chondrocyte-seeded PGA sheets to create a model of the distal interphalangeal joint of the human finger (Fig. 94.5).

5. Implant the various cultured periosteum–cell–polymer constructs into the dorsal subcutaneous space of the nude mice (one construct/animal) and suture the back with 5-0 nylon (Fig. 94.3).

OBSERVATION OF THE TISSUE-ENGINEERED CONSTRUCTS

Subcutaneous implantation of the periosteum–cell–polymer constructs into athymic nude mice resulted in the formation of new tissues having the shape and dimensions of

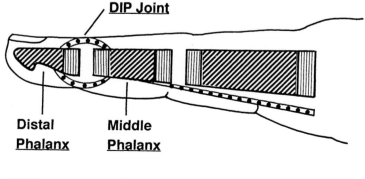

Fig. 94.5. Schematic of the three types of tissue-engineered constructs modeled to a human finger. The various polymer scaffolds wrapped with periosteum or seeded with either chondrocytes or tenocytes are sutured to form a distal or middle phalanx or a distal interphalangeal joint (DIP).

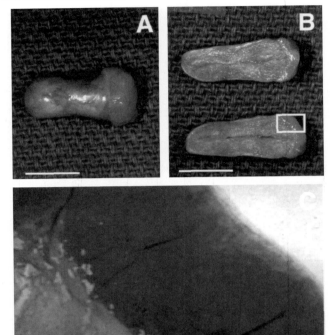

Fig. 94.6. (A) Representative distal phalanx model photographed directly after 20 weeks of implantation in an athymic mouse. Bar = 1.2 cm. The tissue-engineered construct has become vascularized, and an articular cartilage surface has developed. (B) In longitudinal section, the articular cartilage can be distinguished from new bone also comprising the construct. Bar = 1.2 cm. (C) Enlargement of the articular region (boxed area in B) shows cartilage as a layer of tissue, darkly stained here with Safranin-O, and an intact chondro-osseous junction with newly formed subchondral bone (lower left). Details of Figs. 94.6–94.8 may be found in Isogai et al. [8]. (A) and (B) are from Isogai et al. [8] and are used with permission from the Journal of Bone and Joint Surgery.

a human distal phalanx (Fig. 94.6), middle phalanx (Fig. 94.7), and distal interphalangeal joint (Fig. 94.8). Histological examination revealed articular cartilage formation that was clearly distinguishable from subchondral bone that formed in both distal (Fig. 94.6C) and middle (Fig. 94.7B) phalanx models. Neotendon developed in the middle phalanx model (Fig. 94.7C, D). At the tenochondral junction (Fig. 94.7C), fibrocartilage was present and indicates a normal biological response in bone growth and development. In the distal interphalangeal joint model, a tenocapsule formed with articular cartilage along the joint cavity created by the presence of the inserted silicone sheet (Fig. 94.8C, D).

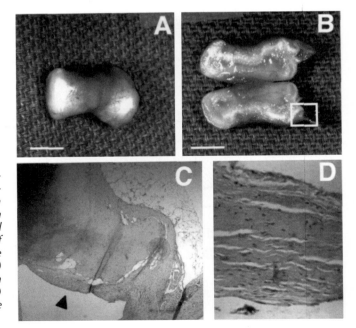

Fig. 94.7. (A) A middle phalanx model after 20 weeks of implantation, showing two articular cartilage surfaces and vascularity surrounding the central shaft. Bar = 1.0 cm. (B) A longitudinal section demonstrates ingrowth of a vascular supply to the middle portion of the construct and integration of a developing neotendon (boxed area). Bar = 1.0 cm. (C) Enlargement of the tenochondral region of the construct reveals an intact junction (arrowhead) between the two cell types. (D) Enlarged portion (original magnification 650×) of (C) in which tenocytes and tendon microfibrils are aligned in linear and parallel fashion as observed in vivo. (A), (B), and (D) are from Isogai et al. [8] and are used with permission from the Journal of Bone and Joint Surgery.

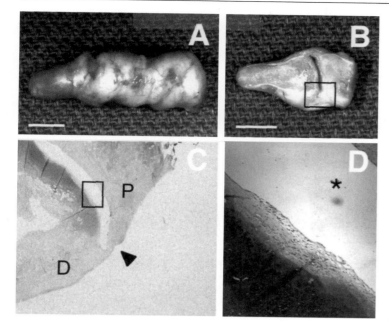

Fig. 94.8. (A) A distal interphalangeal joint 20 weeks following implantation. The construct is adequately supplied with vascularity. Bar = 1.0 cm. (B) Longitudinal profile of the distal phalanx and a portion of the middle phalanx of the model shows a well-formed tenocapsule, the joint cavity created by insertion of a silicone sheet, and newly developing bone. Bar = 1.0 cm. (C) Enlargement of boxed area in (B) demonstrates cartilage, bone, and the intact tenocapsule (arrowhead) between the distal (D) and middle phalanges of the construct as well as remnants of polymer (P) in the middle phalanx. (D) Enlargement of the boxed region of (C) reveals Safranin-O staining of articular surfaces along the joint cavity (asterisk) with a relatively lightly staining tangential cartilage zone above a richer staining transitional zone as found in vivo. (A) and (B) are from Isogai et al. and are used with permission from the Journal of Bone and Joint Surgery.

REFERENCES

1. Buncke, H. J., Jr., Daniller, A. I., Schulz, W. P., and Chase, R. A. (1967). The fate of autogenous whole joints transplanted by microvascular anastomoses. *Plast. Reconstr. Surg.* **39,** 333–341.
2. Goldberg, V. M., Porter, B. B., and Lance, E. M. (1980). Transplantation of the canine knee joint on a vascular pedicle. A preliminary study. *J. Bone Jt. Surg., Am.* **62-A,** 414–424.
3. Hurwitz, P. J. (1979). Experimental transplantation of small joints by microvascular anastomoses. *Plast. Reconstr. Surg.* **64,** 221–231.
4. Mathes, S. J., Buchannan, R., and Weeks, P. M. (1980). Microvascular joint transplantation with epiphyseal growth. *J. Hand Surg.* **5,** 586–589.
5. O'Brien, B. M. (1977). "Microvascular Reconstructive Surgery." Churchill-Livingstone, New York.
6. Tsai, T. M., Jupiter, J. B., Kutz, J. E., and Kleinert, H. E. (1982). Vascularized autogenous whole joint transfer in the hand—A clinical study. *J. Hand Surg.* **7,** 335–342.
7. Wray, R. C., Jr., Mathes, S. M., Young, V. L., and Weeks, P. M. (1981). Free vascularized whole-joint transplants with ununited epiphyses. *Plast. Reconstr. Surg.* **67,** 519–525.
8. Isogai, N., Landis, W., Kim, T. H., Gerstenfeld, L. C., Upton, J., and Vacanti, J. P. (1999). Formation of phalanges and small joints by tissue engineering. *J. Bone Jt. Surg., Am.* **81-A,** 306–316.

CHAPTER 95

MENISCUS

Andrew J. Carter and James W. Huckle

INTRODUCTION

Prior to the 1970s the meniscus was thought to be a vestigial organ, and thus the chosen treatment for a torn meniscus was complete resection. However, in patients who had undergone meniscectomy there was a high incidence of degenerative joint disease, which alluded to the important role of the menisci in joint function [1–3]. It is now widely accepted that menisci perform a number of important functions, including tibiofemoral load transmission, shock absorption, lubrication, passive stabilization, and proprioception. Mow *et al.* [4], O'Connor and McConnaughey [5], and Baratz *et al.* [6] demonstrated that the contact stresses on the articular cartilage tend to increase in direct proportion to the amount of meniscus removed.

Surgeons now recognize the need to minimize the amount of tissue resected, and partial meniscectomy has replaced complete excision as the treatment of choice for dealing with meniscal injuries that are not amenable to repair. The goal of partial meniscectomy is to remove as little tissue as possible, to minimize the increase in compartment pressure within the knee. Although there is strong evidence to support partial meniscectomy over total meniscectomy, degenerative changes still occur to a lesser degree in individuals who have undergone partial excision of the meniscus [7]. Moreover, in a follow-up study of 91 soccer players, Muckle [8] determined that all 91 had developed radiographic degenerative changes in the knee within 10 years of surgery regardless of whether a partial or complete meniscectomy had been performed. There has also been a growing interest in the fields of meniscal repair and regeneration.

Tissue engineering approaches have shown promise for the successful resurfacing of articular cartilage defects, and similar technology is being developed for meniscal cartilage regeneration. The creation of a tissue-engineered meniscus requires a multidisciplined approach utilizing various aspects of biology and engineering. The factors to be considered include selection, isolation, and maintenance of the appropriate cells, composition of the growth media for cell expansion and neotissue formation, selection and design of the appropriate matrix scaffold, and the design of bioreactors capable of providing an environment that will stimulate meniscal cartilage production. This chapter describes some of the key aspects involved in the tissue engineering process that could be utilized to generate implants that may prove to be the ideal treatment for meniscal injuries.

MENISCAL DEVELOPMENT

The embryonic development of the meniscus has been studied in the rabbit by Bland and Ashhurst [9]. The formation of the meniscus is initiated via a mesenchymal aggregation of cells, which can be seen in the developing fetus at 20–25 days. At this developmental stage a matrix is formed that contains collagen types I, III, and V. The same collagens are also present in the meniscus of a 1-week neonatal rabbit, but by 3 weeks type II collagen is also present. Fibrochondrocytes in lacunae are seen at 12–14 weeks, and by 2 years these cells occupy the central region of the inner two-thirds of the meniscus. At 12–14 weeks there are distinct regions of fibrocartilage containing collagen type I, and a more hyaline-

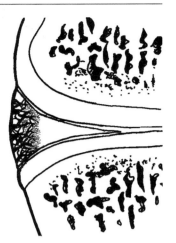

Fig. 95.1. Diagrammatic representation of a cross section through the media compartment of the human knee. The vascular supply branching from the radial vessels of the perimeniscal capillary plexus penetrates into the outer third of the meniscus (shaded).

like tissue containing collagen type II. As the fetus develops, the joints begin to move and the collagen begins to align, giving rise to oriented collagen fiber bundles in response to this loading.

During development, the knee meniscus has blood vessels throughout; however, postnatally, the inner portion (66–75%) becomes avascular (white zone) [10], as shown in Fig. 95.1. As an individual matures, the blood supply retreats such that vascularization in adults is limited to the peripheral (10–30%) region.

The technique of meniscal repair was first reported by Annandale [11] in 1885. However, it was not until the past decade that this surgery has been extensively performed. It is now accepted that tears in the vascular periphery of the menisci should heal (or be repaired) with a retention of functional properties [12]. However, questions still remain in relation to the clinical effectiveness of repairing tears within the avascular region. Estimates (Medical Data International, 1997) indicate that meniscal repair is attempted in only 10% of meniscal injuries, either by suturing or by the use of other fixation devices such as meniscal arrows or tacks. Various meniscal replacement strategies utilizing synthetic prostheses [13–16] have been attempted with limited success. Problems associated with the biocompatibility and mechanical performance of these synthetic implants suggest the importance of devising a more biological approach to meniscal replacement.

Several studies demonstrating the use of meniscal allografts to heal damaged menisci have been performed in animal models [17–20]. However, meniscal allografts are not widely used owing to the risk of transmission of disease, the sizing of the graft, and the limited availability of the grafting material. Earlier studies suggest that an allogeneic tissue-engineered approach to meniscal repair is feasible. The study by Ertl *et al.* [21] involved the *in vitro* culture of allogeneic meniscal implants that were subsequently used for the repair of defects in a lapine model. Cardinnel *et al.* [22] used an alginate delivery system to implant allogeneic meniscal fibrochondrocytes, in conjunction with growth factors, into the avascular zone of the menisci in a lapine model. The authors reported healing of avascular defects with no evidence of an immune or inflammatory response. Our belief is that the *in vitro* culture of meniscal tissue from allogeneic cell lines that have been screened for the absence of disease would provide an alternative source of grafting material.

MENISCAL TISSUE ENGINEERING
CELL SOURCE AND ISOLATION

There are several sources of cells that could potentially be used for meniscal tissue engineering including mesenchymal stem cells, chondrocytes, synoviocytes, dermal fibroblasts, and fibrochondrocytes. The most straightforward choice of cell source comprises cells within the menisci, since these are already committed to producing the desired tissue type. The challenge is to isolate and culture these cells without loss of phenotype. A further issue is whether to use an allogeneic or autologous approach. There are certain advantages to the use of an allogeneic approach in that the product would be available off the shelf, with no need to harvest the initial biopsy sample from the patient and wait several weeks for the implant to be cultured.

MENISCAL CELL ISOLATION METHOD

Webber *et al.* [23] successfully isolated rabbit meniscal fibrochondrocytes via sequential enzymatic digestion, using a method essentially as described by Green [24]. Various methods have been evaluated in our laboratory. For all the methods, menisci were aseptically dissected from ovine hind limbs and any adherent tissues detached. The inner (vascular) and outer (avascular) zones were separated before digestion. Except for the first of the five methods described here, tissue was minced into pieces of 2–3 mm^3 before tissue digestion. During digestion, the tissue was incubated at 37°C and gently agitated on an orbital shaker. The viable cell yields were then estimated by means of standard cell counts and trypan blue exclusion.

The following enzymatic digestion methods were evaluated.

A. *Method of Green* [24]: Whole menisci were incubated in 0.05% w/v hyaluronidase in phosphate-buffered saline (PBS) for 3 min. Tissue was then minced and incubated in 0.2% v/v trypsin in PBS for 30 min. The partially digested tissue was then rinsed in PBS, followed by digestion in 0.2% w/v type 2 collagenase (262 U/mg) in medium overnight.

B. *Adapted (no hyaluronidase step) method of Green* [24]: The minced tissue was incubated in 0.2% v/v trypsin in PBS for 30 min. The resulting partially digested tissue was rinsed in PBS, followed by 0.2% w/v digestion with type II collagenase (262 U/mg) in medium overnight.

C. *Method of Webber et al.* [23]: The minced tissue was digested in 0.05% w/v hyaluronidase in PBS for 30 min followed by 0.2% v/v trypsin in PBS for 30 min. The resulting partially digested tissue was rinsed in PBS, followed by digestion in 0.2% w/v type II collagenase (262 U/mg) in medium overnight.

D. *Adapted method of Freed et al.* [25]: The minced tissue was digested in 0.2% w/v type 2 collagenase (262 units/mg) in medium overnight.

E. *Adapted method for the isolation of articular chondrocytes by Kuettner et al.* [26]: The minced tissue was digested in 0.1% w/v Pronase in medium for 3 h. The partially digested tissue was then rinsed in PBS, followed by digestion in 0.2% w/v type 2 collagenase in medium overnight.

From the results shown in Fig. 95.2, it can be seen that method E gave the best yield of viable cells.

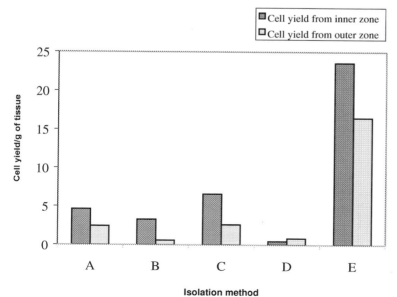

Fig. 95.2. Cell yields obtained from meniscal tissue by various isolation methods.

CULTURE OF MENISCAL FIBROCHONDROCYTES

The first meniscal fibrochondrocytes cultured under monolayer conditions were isolated from rabbit menisci by Webber *et al.* [23]. Morphologically these fibrochondrocytes took on the typical flattened appearance of cells in monolayer culture. However, two subpopulations of cells could be grown, depending on the type of basal nutrient medium used for primary culture. They could be distinguished by morphological, proliferative, and synthetic characteristics, one being more elongated and the other more polygonal. It was suggested that these two cell types represent the superficial and deep fibrochondrocytes, respectively, as seen in electron microscopic studies of rabbit menisci [27]. Population doubling times were within the range (15–20 h) reported for hyaline cartilage chondrocytes [28, 29].

Rabbit fibrochondrocytes synthesize sulfated glycosaminoglycans (GAGs) in monolayer culture [30]. The newly synthesized GAGs contain approximately 87% chondroitin sulfate and 5% dermatan sulfate. The amount and types of GAGs synthesized by these cells in monolayer are similar to values produced by this cell type in organ culture [31]. Verbruggen *et al.* [32] studied the effect of long-term culture of human menisci on matrix metabolism. The bulk of the synthesis of proteoglycans was reported in the first 2 weeks, declining dramatically after 4 weeks, with increased proportions in the incubation media. Kumagee [33] cultured human meniscal fibrochondrocytes and observed no morphological difference in cells from the different regions; however, proteoglycan synthesis from the inner two-thirds was higher than that by cells from the outer third of the meniscus. There was no difference in collagen synthesis.

The bioreactor system described in 1995 by Dunkelman *et al.* [34] has been used to evaluate various cell sources for their ability to secrete a fibrocartilage matrix *in vitro*. A poly(glycolic acid) scaffold (described in more detail shortly), 10 mm in diameter and 2 mm thick, with a density of 45 mg/cm^3 was used. Four million cells of each cell type were seeded into each scaffold by means of the method described by Dunkelman *et al.* [34], and the scaffolds were transferred into the culture system. The cultures were maintained in Dulbecco's Modified Eagle's Medium (DMEM) with 10% fetal calf serum (FCS: 50 ml/scaffold), and the medium was changed (50% split feed) once a week for a 4-week culture period. A method adapted from Woessner [35] was used to analyze the resulting constructs for total collagen content. Primary isolated meniscal cells from the entire meniscus (Entire men in Fig. 95.3), primary isolated cells from the avascular region of the meniscus (Inner men), primary isolated cells from the vascular region (Outer men), human neonatal dermal fibroblasts (HNDF), passage-3 articular chondrocytes (P3 AC), and cocultures of neonatal dermal fibroblasts, and primary isolated articular chondrocytes (P0 AC

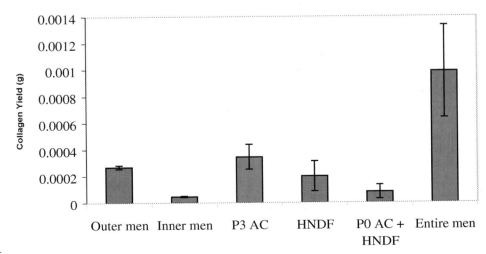

Fig. 95.3. Collagen yield of constructs grown from various cell types.

and HNDF) have been evaluated. Cells isolated from the entire meniscus produced the best constructs with greatest collagen content, as is shown in Fig. 95.3.

SCAFFOLD DESIGN

Our strategy for the culture of implants with a three-dimensional shape is to use a fibrous bioresorbable scaffold. This approach has been utilized by other workers for the growth of tissue-engineered articular cartilage [36,37]. The scaffold is required to fulfill several roles:

Define a three-dimensional shape
Allow anchorage of the cells in the required three-dimensional space
Provide access for nutrients to reach the cells and for removal of waste products
Provide mechanical support to maintain space for tissue to form
Support differentiated cell/tissue growth

The use of scaffolds manufactured by felting, a simple textile process, has been successful for tissue engineering of articular cartilage implants [34,36]. Various fiber types can be used, giving rise to scaffolds with controllable parameters. For example, poly(lactic acid) (PLA) and poly(glycolic acid) (PGA) fibers hydrolyze into naturally occurring metabolic intermediates, and degradation times can be manipulated from days to years depending on the ratio of the two fiber types within the scaffold.

FELT CHARACTERIZATION

The felt is produced by means of a needling process that essentially creates a randomized entanglement of filaments. A typical structure is shown in Fig. 95.4.

A more detailed analysis of the structural properties of the felt was obtained by image analysis of sections taken through the felt. The felt is embedded in epoxy resin prior to sectioning and surface polishing with decreasing grades of alumina suspension to a final grade of 0.5 μm. The surface of the sample is etched to achieve adequate contrast between the fibers and the resin. Then the fiber distribution and orientation can be viewed under a light microscope and captured for image analysis (Fig. 95.5).

The sectioned fibers appear as ellipses, and therefore the three-dimensional orientation of the fibers may be determined by the direction of the major ellipse and its ratio to the minor axis. The example of the frequency distribution of fiber spacing presented in Fig. 95.6 shows the results from two separate manufacturing runs, illustrating the reproducibility of the process.

When the felt process is used, with PGA fibers typically 12–20 μm in diameter, scaffolds 1–8 mm thick, having densities ranging from 15 to 80 mg/cm^3, can be produced.

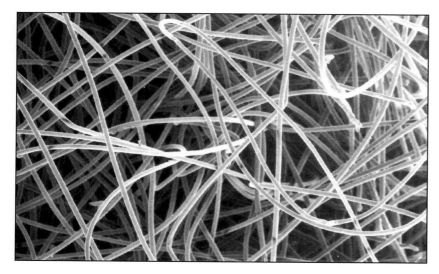

Fig. 95.4. Electron micrograph of a PGA felt scaffold (45 mg/cm^3).

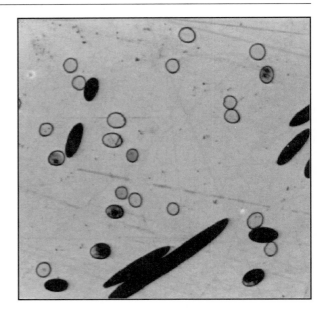

Fig. 95.5. Image-enhanced view of the section through a PGA felt.

EVALUATION OF SCAFFOLD POLYMER TYPE

The bioreactor system described by Dunkelman *et al.* [34] was used to compare the collagen deposition of meniscal fibrochondrocytes on three different scaffolds made from PLA, PGA, and PET [a nonresorbing poly(ethylene terephthalate)]. In all the scaffolds dimensions (10 mm in diameter × 2 mm thick), the fiber diameter (13 μm), and density (45 mg/cm^3) were constant.

Four million meniscal cells were seeded into each scaffold, and the scaffolds transferred into the culture system. The cultures were maintained in DMEM with 10% FCS (50 ml/scaffold) and the media changed (50% split feed) once a week for a 4-week culture period. A method adapted from Woessner [35] was used to analyze the resulting constructs for total collagen content (Fig. 95.7).

The PLA- and PET-cultured constructs had comparable collagen levels, which were much superior to the levels in the PGA constructs (Fig. 95.7). The reason for the differences observed is not known. A high concentration of glycolic acid monomer in the medium, released upon degradation of the PGA felt, may be exerting an adverse effect. Sittinger

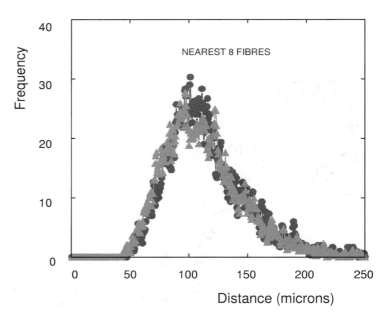

Fig. 95.6. Average spacing to 8 nearest fibers in a PGA felt scaffold by image analysis. Felt density, 45 mg/cm^3. Results shown from two separate batches (circles and triangles).

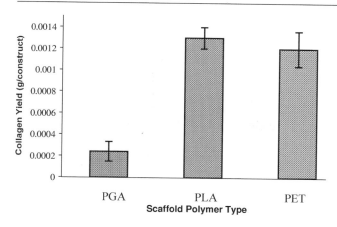

Fig. 95.7. Collagen yield of constructs cultured on scaffolds fabricated from different polymers.

et al. [38] compared the biocompatibility of lactic acid and glycolic acid with that of chondrocytes and concluded that "regarding the degradation products, glycolic acid and lactic acid, non-wovens of PLA are more biocompatible with chondrocytes than non-wovens of PGA." For systems in which a large amount of scaffold degradation is expected, it may be necessary to increase the amount of media used to dilute the degradation products. It may also be necessary to optimize the mass transport to ensure that degradation products are removed from the growing tissue-engineered construct.

GROWTH OF SHAPED CONSTRUCTS

Meniscal-shaped PLA scaffolds have been produced by heat-pressing a nonwoven PLA felt (45 mg/cm^3) into the shape of the meniscus. Figure 95.8 shows photographs of an ovine native meniscus (A) and a heat-pressed, meniscal-shaped scaffold (B) in plan view (top) and in cross section (bottom). Meniscal-shaped tissue engineered implants (C) have been produced by means of the shaped PLA scaffolds and a modified bioreactor system for construct culture. The construct has formed in the shape of the original scaffold, and the voids in between the scaffold fibers have filled with neotissue. Thus it can be seen that the scaffold can be used to define the shape of the resultant tissue-engineered construct.

BIOREACTOR DESIGN

Identification of the key environmental/bioreactor features that stimulate meniscal cartilage formation is essential to the process of meniscal tissue engineering. The fluid flow and the mixing of the media within the bioreactor have been shown to be important in the process of articular cartilage tissue engineering [39,40]. Moreover, Freed and Vunjak-

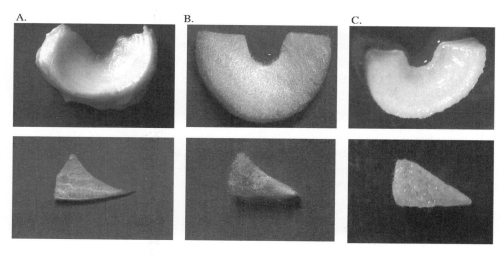

Fig. 95.8. Production of meniscal-shaped constructs.

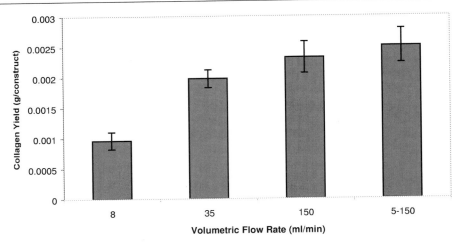

Fig. 95.9. Collagen yield of constructs cultured under various volumetric flow rates.

Novakovic [40] suggest that the fluid flow effects cartilage tissue production via changes in mass transfer rates and by direct hydrodynamic effects on cell shape.

The effect of increasing the volumetric flow rate on meniscal tissue formation has been studied by means of a custom-made tubular fixed-bed bioreactor with an open channel bypass. Meniscal fibrochondrocytes were seeded on to PLA felts and cultured in the bioreactor at various flow rates: 8, 35, and 150 ml/min and a ramped flow rate (increasing the flow rate incrementally every second day during the culture period) from 5 ml/min to 150 ml/min. The constructs were analyzed for collagen content after 2 weeks of culture (Fig. 95.9). This study indicates that optimization of fluid flow conditions is important in the meniscal tissue engineering process.

CONCLUSIONS

Various methods have been evaluated for the isolation of meniscal cells. We have shown that a digestion procedure with a combination of pronase and collagenase give the greatest yield of viable cells. Various cell sources were evaluated for their ability to secrete a collagenous matrix within a three-dimensional culture system. Meniscal cells isolated from the entire meniscus produced constructs with collagen levels superior to these of the other cell types, which included meniscal cells isolated from the avascular region or the vascular region of the meniscus. Thus it may be a requirement to isolate cells from the entire meniscal population, to be able to take advantage of a possible beneficial interaction between the various meniscal cell types.

These studies indicate that the scaffold material (PLA or PGA) on which the cells are cultured has an effect on the shape and yield of the tissue generated in bioreactor culture, the latter possibly being an effect of the concentration of the glycolic acid monomer in the media or a local mass transport effect in the growing construct. Fluid flow conditions have been shown to affect tissue production in our bioreactor systems, and optimization of the fluid dynamics is critical to the process of meniscal tissue engineering. We are investigating the ability of tissue-engineered implants to stimulate the repair of meniscal defects *in vivo*. This tissue engineering approach may in the future provide an ideal alternative to meniscectomy and meniscal allografts in the treatment of meniscal defects.

ACKNOWLEDGMENTS

This work was carried out as part of a joint venture between Smith & Nephew and Advanced Tissue Sciences. We thank Dr. Emma Robinson, Dr. Mark Smith, and Mr. Nick Medcalf for their help in the preparation of this manuscript. The image analysis methodology for felt characterization was developed by the University of Leeds.

REFERENCES
1. King, D. (1936). The healing of semilunar cartilages. *J. Bone Jt. Surg.* **18**, 333.
2. Lagergren, M. (1953). Meniscus operations in secondary arthrosis defomans. *Acta Orthop. Scand.* **14**, 280.

3. Fox, J. M., Blazina, M. E., and Carlsen, G. T. (1971). Multiphasic view of medial meniscectomy. *Am. J. Sports Med.* **7**, 161–164.

4. Mow, V. C., Ratcliffe, A., Chern, K. Y., and Kelly, M. A. (1993). Structure and function relationships of the menisci of the knee. *In* "Knee Meniscus: Basic and Clinical Foundations" (V. C. Mow, S. P. Arnoczky, and W. J. Douglas, eds.).

5. O'Connor, B. L., and McConnaughey, J. S. (1978). The structure and innervation of cat knee meniscus, and their relation of a 'sensory hypothesis' of meniscal function. *Am. J. Anat.* **153**, 431–442.

6. Baratz, M. E., Fu, F. H., and Mengato, R. (1986). Meniscal tears: The effect of meniscectomy and repair on intraarticular contact areas and stress in the human knee. A preliminary report. *Am. J. Sports Med.* **14**, 270–275.

7. Cox, J. S., Nye, C. E., Schaefer, W. W., and Woodstein, I. J. (1975). The degenerative effects of partial and total resection of the medial meniscus in dog's knees. *Clin. Orthop. Relat. Res.* **109**, 178–183.

8. Muckle, D. S. (1983). Factors affecting the prognosis of meniscectomy in soccer players. *Br. J. Sports Med.* **17**(20), 88–90.

9. Bland, Y. S., and Ashhurst, D. E. (1996). Changes in the content of the fibrillar collagens and the expression of their mRNAs in the menisci of the rabbit knee joint during development and ageing. *Histochem. J.* **28**(4), 265–274.

10. O'Meara, P. M. (1993). The basic science of meniscal repair. *Arthroscopy* **7**, 120–125.

11. Annandale, T. (1885). An operation for displaced seminular cartilage. *Br. Med. J.* **1**, 779.

12. Newman, A. P., Daniels, A. U., and Burks, R. T. (1993). Principles and decision making in meniscal surgery. *J. Arthrop. Relat. Surg.* **9**, 33–51.

13. Veth, R. P. H., Jansen, H. W. B., Leenslag, J. W., Pennings, A. J., Hartel, R. M., and Nielsen, H. K. L. (1986). Experimental meniscal lesions with a carbon fiber–polyurethane–poly(lactide) graft. *Clin. Orthop. Relat. Res.* **202**, 286–293.

14. Wood, D. J., Minns, R. J., and Strover, A. (1990). Replacement of the rabbit medial meniscus with a polyester–carbon fiber bioprosthesis. *Biomaterials* **11**, 13–16.

15. Messner, K. (1994). The concept of a permanent synthetic prosthetic meniscus: A critical discussion after 5 years of experimental investigations used in Dacron and Teflon implants. *Biomaterials* **15**, 243–250.

16. De-Groot, J. H., De-Vrijer, R., Pennings, A. J., Klompmaker, J., Veth, R. P. H., and Jansen, H. W. B. (1996). Use of porous polyurethanes for meniscal reconstruction and meniscal prostheses. *Biomaterials* **17**(2), 163–173.

17. Arnoczky, S. P., Warren, R. F., and McDevitt, C. A. (1990). Meniscal replacement using a cryopreserved allograft—An experimental study in the dog. *Clin. Orthop. Relat. Res.* **252**, 121–128.

18. Canham, W., and Stanish, W. (1986). A study of the biological behaviour of the meniscus as a transplant in the medial compartment of a dog's knee. *Am. J. Sports Med.* **14**, 376–379.

19. Milachowski, K. A., Weismeier, K., and Wirth, C. J. (1989). Homologous meniscus transplantation. *Int. Orthop.* **13**, 121–128.

20. Veltri, D. M., Warren, R. F., Wickiewicz, T. L., and O'Brien, S. J. (1994). Current status of allograft meniscal transplantation. *Clin. Orthop. Relat. Res.* **303**, 44–55.

21. Ertl, W., Schwartz, R., Manji, R., Paulino, C., and Grande, D. (1995). Tissue engineering approaches for repair of meniscal injuries. *Orthop. Res. Soc.* **20**(2), 13–15.

22. Cardinnel, L., Wezeman, F., Patwardhan, A., and Light, T. (1998). Fibrochondrocyte allografts promote healing of standardised meniscal defects in the avascular zone. *Orthop. Res. Soc. Trans.* **44**(23/1), 306.

23. Webber, R. J., Harris, M., and Hough, A. J. (1985). Cell culture of rabbit meniscal fibrochondrocytes: Proliferative and synthetic response to growth factors and ascorbate. *J. Orthop. Res.* **3**, 36.

24. Green, W. T. (1971). Behaviour of articular chondrocytes in cell culture. *Clin. Orthop. Relat. Res.* **75**, 248–260.

25. Freed, L. E., Marquis, J. C., Norhria, A., Emmanual, J., Mikos, A. G., and Langer, R. (1993). Neocartilage formation *in vitro* and *in vivo* using cells cultured on biodegradable polymers. *J. Biomed. Mater. Res.* **27**, 11–23.

26. Kuettner, K. E., Pauli, B. U., Gall, G., Memoli, V. A., and Schenk, R. K. (1982). Synthesis of cartilage matrix by mammalian cells in vitro. Isolation, culture characteristics, and morphology. *J. Cell Biol.* **93**, 742–750.

27. Ghadially, F. N., Thomas, I., Yong, N. K., and La-Londe, J. M. (1978). Ultrastructure of rabbit semilunar cartilages. *J. Anat.* **125**, 499.

28. Webber, R. J., Malemud, C. J., and Sokoloff, L. (1977). Species differences in cell culture of mammalian articular chondrocytes *Calcif. Tissue Res.* **23**, 61.

29. Webber, R. J., and Sokoloff, L. (1981). *In vitro* culture of rabbit growth plate chondrocytes: Age dependence of response to fibroblast growth factor and chondrocyte growth factor. *Growth* **45**, 252.

30. Webber, R. J., and Hough, A. J. (1988). Cell culture of rabbit meniscal fibrochondrocytes. II. Sulphated proteoglycan synthesis. *Biochemistry* **70**, 193.

31. Webber, R. J., Norby, D. P., Malemud, C. J., Goldberg, V. M., and Moskowitz, R. W. (1984). Characterisation of newly synthesised proteoglycans from rabbit menisci in organ culture. *Biochem. J.* **221**, 875.

32. Verbruggen, G., Verdonk, R., Veys, E. M., Van-Daele, P., De-Smet, P., Van-den-Abbeele. K., Claus. B., and Baeten, D. (1996). Human Meniscal proteoglycan metabolism in long-term tissue culture. *Knee Surg. Sports Traumatol. Arthroscopy* **1**, 57–63.

33. Kumagee, Y. (1994). Proteoglycan and collagen synthesis of cultured fibrochondrocytes from the human knee joint meniscus. *Nippon Seikeigeka Gakkai Zasshi* **68**(10), 885–894.

34. Dunkelman, N. S., Zimber, M. P., LeBaron, R. G., Pavelec, R., Kwan, M., and Purchio, A. F. (1995). Cartilage production by rabbit articular chondrocytes on polyglycolic acid scaffolds in a closed bioreactor system. *Biotechnol. Bioeng.* **46**(4), 299–305.

35. Woessner, J. F. (1961). The determination of hydroxyproline in tissue and protein samples containing small proportions of this amino acid. *Arch. Biochem. Biophys.* **93**, 440–447.

36. Vacanti, C. A., Langer, R., Schloo, B., and Vacanti, J. P. (1991). Synthetic polymers seeded with chondrocytes provide a template for new cartilage formation. *Plast. Reconst. Surg.* **88**, 753–759.

37. Freed, L. E., Vunjak-Novakovic, G., Biron, R., Eagles, D., Lesony, D., Barlow, S., and Langer, R. (1994). Biodegradable polymer scaffolds for tissue engineering. *Bioshill Technology* **20**, 689–693.

38. Sittinger, M., Reitzel, D., Dauner, M., Hierlemann, H., Hammer, C., Kastenbauer, E., Planck, H., Burmester, G. R., and Bujia, J. (1996). Resorbable polyesters in cartilage engineering: Affinity and biocompatibility of polymer fiber structures to chondrocytes. *J. Biomed. Mater. Res.* **33**(2), 57–63.

39. Vunjak-Novakovic, G., Freed, L. E., Biron, R., Eagles, D., and Langer, R. (1996). Effects of mixing on the composition and morphology of tissue engineered cartilage. *Bioeng. Food Nat. Prod.* **42**(3), 850–860.

40. Freed L. E., and Vunjak-Novakovic, G. (1977). Tissue culture bioreactors: Chondrogenesis as a model system. *In* "Principles in tissue engineering" (R. Lanza, R. Langer and W. Chick, eds.), R.G. Landes, Austrin, TX.

CHAPTER 96

CELL-BASED THERAPIES FOR THE TREATMENT OF ARTICULAR CARTILAGE INJURY

Bernd Kinner and Myron Spector

INTRODUCTION

The limited healing potential of cartilage has been reported for more than 200 years [1,2]. As with other tissues, cartilage repair depends on fibrin clot formation and the subsequent cascade of chondroprogenitor cells derived from adjacent cartilage, underlying marrow, or synovium that eventually leads to the formation of reparative tissue in the defect. The rather unique anatomical and physiological characteristics of cartilage (viz., avasularity, low mitotic activity of chondrocytes, and the chondrocytic release of degrading enzymes) prevent the initialization of this process if the defect does not penetrate the subchondral bone. Even if the defect extends into the subchondral bone, patients are more likely to develop fibrocartilage of questionable functional value instead of the articular form of hyaline cartilage [3].

Several traditional techniques of cartilage repair are based on accessing the subchondral vascularity and marrow either by drilling of the subchondral bone [4], spongialization [5], or microfracturing [6–8]. These procedures are usually combined with debriding of loose cartilage particles, abrasion arthroplasty [9,10], and lavage of the joint. However, the fibrocartilage-like repair tissue that results from these techniques lacks the composition, structure, and mechanical properties of normal cartilage, and the long-term clinical outcomes are unpredictable [11]. Patients treated with these techniques often experience short-term pain relief but develop progressive symptoms when the repair tissue fails [12].

As an alternative approach, novel techniques are being employed to deliver chondrogenic cells to the cartilage defect, either in form of autologous chondrocytes isolated from a biopsy of healthy cartilage [13] or in form of precursor cells within the periosteum [14] or the perichondrium [15] that will eventually undergo terminal differentiation to chondrocytes.

LABORATORY BASIS
EXPANSION OF CELLS IN VITRO
Autologous Articular Chondrocytes

The rationale for using articular chondrocytes for a cell-based therapy is that they already possess the desired phenotype [16]. Chondrocytes comprise the single cellular component of adult hyaline cartilage and are considered to be terminally differentiated, thus being highly specialized. Their main function is to maintain the cartilage matrix, synthesizing types II, IX, and XI collagen; the large aggregating proteoglycan, aggrecan; the smaller proteoglycans biglycan and decorin; and several specific and nonspecific matrix proteins that are expressed at defined stages during growth and development. Freshly isolated articular chondrocytes continue to exhibit their specific phenotype in culture for at least several days to weeks. This makes them a suitable cell type for a cell-based treatment of chondral defects.

However, articular chondrocytes do not readily proliferate *in vitro*. Cells from a younger population (e.g., third and fourth decades of life) have been found to undergo 0.3 doubling per day; a standardized and validated approach for culturing cells for later implantation was used [17]. Even lower proliferation rates are obtained in elderly patients and with arthritic cartilage [18]. When chondrocytes are deprived of their three-dimensional environment, their phenotype switches to a more fibroblastic cell form, expressing types I and III collagen instead of the cartilage-specific type II collagen [19–21].

The process of "dedifferentiation" is known to be dependent upon culture conditions. Expression of fibroblast-like proteins and morphology can be promoted by seeding at low density and treatment with certain cytokines [20]. On the other hand "redifferentiation," or reexpression of cartilage-specific behavior, can be accomplished by using selected culture systems, including spinner flasks [22] and dishes coated with materials that prevent cell adherence such as agarose or collagen gels [23]; by seeding in high-density micromass cultures [24,25]; by using hypoxic culture conditions [26]; and by embedding the cells in solid matrices that do not allow adherence, such as agarose [27], collagen, or alginate gels [28,29]. The chondrocytic phenotype can also be maintained when the cells are seeded in certain sponge-like scaffolds used for tissue engineering [30–32]. Additionally, different cytokines have influence on the degree of expression of cartilage specific molecules. Members of the tranforming growth factor β (TGF-β) superfamily can trigger the expression of the chondrocytic phenotype [33]. Additionally, staurosporine, a protein kinase C inhibitor [25,26,34], insulin-like growth factor (IGF) with or without addition of insulin [35–37], hepatocyte growth factor (HGF), and fibroblast growth factor (FGF) [32,38] have been shown to increase the expression of cartilage-specific matrix products.

Other studies propose that chondrocytes, even if dedifferentiated after a extended time in monolayer culture, reexpress their chondrocytic phenotype if implanted into a cartilage defect *in vivo*, probably owing to the release of growth and differentiation factors from the adjacent host tissue [39]. However, more recent studies suggest that this process might not be occurring in every case [31].

Another issue in employing autologous articular chondrocytes relates to donor site morbidity. To obtain autologous chondrocytes, healthy cartilage must be harvested from uninvolved regions of the joint. Although it has been assumed that there are no major problems associated with the harvest of cartilage, evidence is scarce. There are no published studies focusing on the harvest site. Harvesting cartilage for autologous chondrocyte implantation (ACI) makes a second operation necessary, with all theoretical possibilities of complication (e.g., infection). Moreover, harvesting cartilage might initiate additional damage to the joint. In a canine investigation reported in 2000 [40], up to threefold changes in certain mechanical properties were recorded for articular cartilage samples taken away from the harvest site 18 weeks after surgery. These alterations were consistent with hypertrophy that has been found to precede osteoarthritic changes in joints. Clearly, additional studies are necessary to determine the consequences of harvest of articular cartilage for the isolation of chondrocytes for cell-based therapies.

Other recent studies have demonstrated that as articular chondrocytes are expanded in monolayer culture, a greater percentage of cells express the gene that encodes for a contractile muscle actin, smooth muscle α-actin (SMA). Moreover, these studies have demonstrated that SMA-expressing cells derived from canine [41] and human [42] articular cartilage can contract a collagen–glycosaminoglycan analogue of extracellular matrix *in vitro*. This work raises the question of the role that SMA expression may play once the cells have been injected into a cartilage defect.

Mesenchymal Stem Cells

The difficulty of obtaining chondrocytes and maintaining differentiated cell cultures has led to research on other cell types for cell-based therapies for cartilage repair. Several studies have shown that autologous bone marrow stromal cells and periosteum-derived cells are able to exhibit a chondrocytic phenotype *in vivo* [43] and *in vitro* under certain conditions [44]. Culturing cells at high density or under hypoxic conditions promotes differentiation toward a chondrocytic phenotype [24,45,46]. This process may be facilitated by addition of selected cytokines including TGF-β and bone morphogenetic protein 2 (BMP-2), and perhaps by merely adding dexamethasone to the culture medium [45]. The principal advantages associated with the use of these cells over autologous articular chondrocytes for cartilage repair procedures are ready availability and minimal donor site morbidity.

Johnstone and Yoo [45] developed a culture system that facilitates the chondrogenic differentiation of rabbit bone marrow derived mesenchymal progenitor cells. Cells obtained in bone marrow aspirates were first isolated by monolayer culture and then transferred into tubes and allowed to form three-dimensional aggregates in a chemically defined medium that included dexamethasone and/or TGB-β1. The chondrogenic differentiation of cells within the aggregate was evidenced by the appearance of toluidine blue metachromasia and the immunohistochemical detection of type II collagen [44]. This approach seemed to recapitulate embryonic chondrogenesis. One of the major drawbacks is that these cells may further differentiate into hypertrophic chondrocytes, eventually favoring apatite mineral deposition and endochondral ossification.

Fibroblasts

Nicoll *et al.* [26] described a model for the conversion of human dermal fibroblasts to chondrocyte-like cells and the potential application of this methodology to cartilage tissue engineering. Human neonatal foreskin fibroblasts were seeded in two-dimensional high-density micromass cultures in the presence of staurosporine and lactic acid to induce functional hypoxia. Cells were also seeded into three-dimensional polymer scaffolds. Northern analysis revealed aggrecan core protein expression in lactate-treated micromass cultures, and type I collagen gene expression was virtually abolished in all cultures supplemented with staurosporine. Moreover, the cells in these cultures displayed a rounded, cobblestone-shaped morphology typical of differentiated chondrocytes and were organized into nodules that stained positive with Alcian blue. When these cells were seeded on matrices of poly(glycolic acid)–poly(L-lactic acid) (PGA-PLLA), a chondrocyte-like morphology was observed in cultures treated with lactate and staurosporine, in contrast to the flattened sheets of fibroblast-like cells seen in untreated controls. This approach holds promise for the use of readily accessible nonchondrocytic autologous cells for cartilage repair procedures.

Allogeneic Chondrocytes

Transplantation of allogeneic osteochondral grafts has been used clinically for many years [47–49], and several investrgations have used allogeneic cells to study tissue engineering [50,51]. Moreover, it has been proposed [52] that allogeneic chondrocytes from amputated limbs or joint arthoplasties might play a major role in the future. However, this approach is less attractive for cell therapy because issues related to immune response and transmission of disease must be taken into account.

ANIMAL STUDIES

We focus here on autologous chondrocyte implantation (ACI) because it is the most widely employed cell-based therapy for cartilage repair. Animal investigation of cartilage repair by means of articular cnondrocytes (viz., allogeneic cells), expanded *in vitro*, dates back more than 20 years [53].

Experiments evaluating the possibilities of ACI were first reported in 1987. In a study in rabbits, Grande *et al.* [54] showed that defects that had received transplants had significantly more cartilage reconstituted (82%) than ungrafted controls (18%). Brittberg *et al.* [55] later obtained similar results upon treating 51 New Zealand white rabbits. ACI significantly increased the amount of newly formed repair tissue up to 52 weeks, in

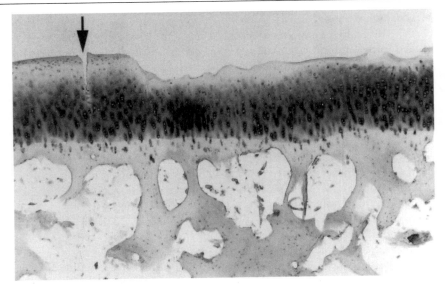

Fig. 96.1. ACI-treated canine articular cartilage defect after 6 months, showing hyaline and articular cartilage extending over much of the lesion but with an irregular surface comprising fibrocartilage and fibrous tissue. The regenerated cartilage is well integrated with the calcified cartilage. Hematoxylin–eosin stain; original magnification, 60×.

contrast to the lack of intrinsic repair or the absence of repair with periosteal grafts alone. However, Brittberg and colleagues also noted that repair tissue tended to be incompletely bonded to the adjacent cartilage. Subsequently, Breinan *et al.* [56] repeated these experiments in a canine model. They found no significant differences among the treated and control (periosteum alone and nontreated defects) groups after one year. By 18 months, neither a complete filling nor the restoration of the architecture was complete (Figs. 96.1 and 96.2) [57].

Moreover, cartilage surrounding the defect showed degenerative changes, which seemed to be related to suturing of the periosteal flap. These contradicting results might have been due in part to differences in the animal models. Dogs have a thin subchondral bone plate

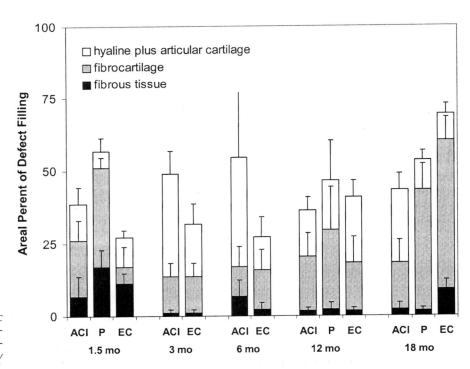

Fig. 96.2. Results of histomorphometric evaluations of repair tissue in a canine articular cartilage defect model: ACI, cultured autologous chondrocytes; EC empty (untreated) control, periosteum control. Mean ± SEM.

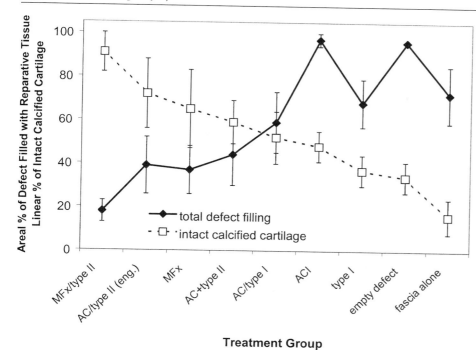

Fig. 96.3. Graph depicting the inverse relationship between total filling and remaining intact calcified cartilage for 15-week canine implant groups and controls. Linear regression analysis of the correlation of total filling with intact calcified cartilage yielded $R^2 = 0.61$: MFx, microfracture, AC, autologous cultured chondrocytes; ACI, autologous chondrocyte implantation; type I and type II refer to the applied collagen matrices.

that can easily be damaged. As a consequence, mesenchymal stem cells can get access to the defect and mix with the implanted chondrocytes.

Another interesting finding we recently made, using the same canine model, in which a defect was made to the level of the tidemark, demonstrated a significant correlation between the degree to which the calcified cartilage layer and subchondral bone were disturbed and the amount of filling [58]. Moreover, when it forms, hyaline cartilage most frequently occurs superficial to intact calcified cartilage. This work stresses the importance of an intact calcified cartilage layer for obtaining repair tissue composed mainly of artilcular cartilage. However, the amount of filling repair tissue is inversely related to the remaining intact calcified cartilage (Fig. 96.3).

Additionally, the observation that some spontaneous regeneration can occur in a canine model [59] raised the question of the degree to which such regeneration can occur in humans. In a surgically created full-thickness cartilage defect (to the tidemark) in 20 adult mongrel dogs, 40% of the created defects were filled with reparative tissue, 19% of which were found to be hyaline-like articular cartilage [59]. Therefore investigations of new modalities of treating lesions in articular cartilage must acknowledge, through careful design of controls, the potential for spontaneous regeneration [59].

CLINICAL APPLICATIONS
AUTOLOGOUS CHONDROCYTE IMPLANTATION (ACI)

On the basis of promising animal studies (rabbit), ACI was introduced into the clinic. Brittberg *et al.* [13] were the first to publish their results on 23 patients treated in Sweden for symptomatic cartilage defects. Thirteen patients had femoral condylar defects, ranging in size from 1.6 to 6.5 cm^2, due to trauma or osteochondritis dissecans. Seven patients had patellar defects. Ten patients had been treated with shaving and debridement of unstable cartilage. The results were very promising for the condylar defects. Patients were followed for 16–66 months (mean, 39 months). Initially, the transplants eliminated knee locking and reduced pain and swelling in all patients. After three months, arthroscopy showed that the transplants were level with the surrounding tissue and spongy when probed, with visible borders. A second arthroscopic examination showed that in many instances the transplants had the same macroscopic appearance as earlier but were firmer when probed and similar in appearance to the surrounding cartilage. Two years after transplantation, 14 of the 16 patients with femoral condylar transplants had good-to-excellent results. Two patients required a second operation because of severe central wear in the transplants, with locking and pain. A mean of 36 months after transplantation, the results were excellent or good in two of the seven patients with patellar transplants, fair in three, and poor in two; two patients required a second operation because of severe chondromalacia. Biopsies showed that 11 of the 15 femoral transplants and 1 of the 7 patellar transplants had the appearance of "hyaline-like" cartilage. These results and the establishment of a commercial service for culturing autologous chondrocytes led to a dramatic increase in the use of this cell-based therapy for cartilage repair.

While there appear to be general similarities in the procedures used by various commercial and academic laboratories for the isolation and expansion of articular chondrocytes for ACI, there may be important differences. One such difference is the use of the patient's own serum for culturing the cells, as described in the original method by Brittberg *et al.* [13]. One commercial enterprise, Genzyme Biosurgery uses approved and validated fetal bovine serum instead of the patient's serum in the culture media. Another potentially important difference is that Genzyme needs to freeze and store the isolated cells to allow for verification of adequate insurance coverage prior to the implantation procedure. A study reported in 2000, however, has indicated that this freeze–thaw cycle may adversely affect the outcome of the procedure. Perka *et al.* [60] demonstrated that cryopreserved chondrocytes seeded into polymer scaffolds yielded a 85% repair of an osteochondral defect in rabbits, whereas 100% of the defects treated with noncryopreserved cells were filled. Additional work is necessary to more fully explore the effects of certain handling and culture procedures on the performance of monolayer-expanded chondrocytes in ACI.

Techniques of cell isolation, expansion in culture, and implantation have remained essentially the same since first published in 1994. Cartilage (150–300 mg) is harvested arthroscopically from a minimally load-bearing area of the upper aspect or the medial condyle of the affected knee. The biopsy specimen is then transported to a laboratory facility in transport media. Chondrocytes are isolated by means of standard techniques as outlined in Chapter 25. After a certain period of cell expansion (11–21 days [18], depending upon the growth kinetics) a certain number of cells (e.g., 12×10^6 for Genzyme's Carticel procedure) are provided in a serum-free and gentamicin-free transport medium.

A medial or lateral parapatellar incision is used to debride the defect. The integrity of the tidemark needs to be maintained to avoid infiltration of undifferentiated mesenchymal stem cells, which could contribute to the formation of fibrocartilagenous repair tissue [61]. A periosteal flap is harvested from the anterior aspect of the proximal tibia or femur, formed to the shape of the lesion, and sutured to the rim of the defect. Then fibrin glue is used to seal the border of periosteal cover. The chondrocyte suspension is subsequently injected under the periosteal flap. Postoperative rehabilitation protocols generally involve continuous passive motion and limited weight bearing.

Clinical Results

Follow-up investigations of the two largest patient groups have been reported: patients in Sweden treated predominantly by L. Peterson, M.D., and M. Brittberg, M.D., and patients treated by the surgeons participating in the Genzyme Cartilage Repair Registry. In 2000 Peterson *et al.* [18] reported their 2- to 9-year results, including clinical, arthroscopic, and histological evaluations of 101 patients. In this retrospective study ACI yielded good results in 92% for isolated femoral lesions, 67% for multiple lesions, 89% for osteochondritits dissecans, and 65% for patella defects after an average follow-up of 4.2 years. Arthroscopic follow-up in 65 of 93 patients showed slow maturing of the tissue during the first year, but repair tissue at the subsequent follow-ups was as firm as the adjacent tissue. Histopathological analysis in 21 patients revealed a homogeneous matrix with low cellularity considered to be "hyaline-like" in 17 patients, whereas 4 patients showed fibrous repair tissue. Immunohistochemistry for collagen type II was positive in all the patients with "hyaline-like" repair tissue and negative in the fibrous repair tissue. Adverse events were reported in 51% of the patients, including 7 graft failures (7%) and 10 adhesions, which needed arthroscopic intervention. Graft hypertrophy, attributed to the periosteal flap, was seen in 26 patients.

The last volume of the Genzyme Patient Registry Report included 4-year patient outcomes for individuals treated outside Sweden. Progressive improvement in the overall condition of the patients and in symptomatology could be demonstrated at 24, 36, and 48 months. Improvement in comparison to baseline was 85% for the femoral condyle as rated by the clinician and 81% as rated by the patient. Adverse events were reported in 8% of the patients. These included adhesions or fibroarthorsis in 1.9% and hypertropohic changes in 1.3%. The cumulative incidence of treatment failure was estimated as 3.2% at 48 months, and 6.3% of the patients reported an operation following implantation. Updated information is available on the World Wide Web (http://www.genzyme.com/carticel/patreg.htm).

Regulations

Much attention has been paid to the standardization and validation of the cell culture procedures since their clinical introduction. In August 1997 the U.S. Food and Drug Administration granted accelerated approval for Genzyme's Carticel procedure for the repair of symptomatic, cartilaginous defects of the femoral condyle, including both first- and second-line repairs (FDA Talk paper, T97-38). As a condition of this approval, Genzyme agreed to conduct two studies, including a multicenter, randomized controlled trial in 300 patients comparing Carticel with other primary repair techniques, to confirm the benefits of Carticel in this setting.

Genzyme was unable, however, to enroll an adequate number of patients in either of the planned studies. The company subsequently changed the product labeling, narrowing the indications for Carticel to second-line therapy for the repair of cartilage defects of the femoral condyle.

Risks and Quality Control

The risks associated with the currently used cell-based products are unknown but might include the following: adverse effects of the harvest procedure to procure the articular cartilage tissue for cell isolation, effects of the arthrotomy currently required for the open implantation of the cells, degeneration of the adjacent articular cartilage related to the damage due to suturing, flap detachment, and malignant transformation of cells in culture.

Fortunately, while there is a finite possibility (albeit small) for malignant or dysplastic transformation of cells during their *in vitro* expansion, no such occurrence has been reported. In addition, implantation of normal autologous chondrocytes could potentially stimulate growth of malignant cells in the area of the implant, although there have been no reported incidents in humans.

Another problem in using autologous articular chondrocytes arises on the harvest site. To obtain autologous chondrocytes, healthy cartilage must be harvested from uninvolved and principal "unloaded" areas of the femoral side of the knee. Published data on the healing of the harvest sites is scarce. For all these techniques, it is assumed that cartilage harvest does not affect overall outcome. There are no studies, however, focusing on the harvest site. Harvesting cartilage for ACI makes a second operation necessary with all theoretical possibilities of complication (e.g., infection). Moreover, as described earlier in this chapter, the harvest procedure may place the other cartilage in the joint at risk of degeneration.

Although autologous cell therapies circumvent complications like graft rejection or viral transmission, significant challenges exist in assuring a safe and reproducible product. Genzyme established a quality assurance program based on U.S. FDA good manufacturing practice regulations, which was reviewed in 1998 [17]. Process variables have to be controlled rigorously, and testing performed for sterility and the presence of endotoxin. Moreover, assessment of cell viability and growth kinetics is a crucial part of nonconformance reporting.

According to Genzyme data, 1.64% of the cartilage biopsy samples received were contaminated. Contamination was recorded only for 0.03% during processing and in 0.16% at release. Endotoxin content ranged between less than 0.15 and 0.5 EU/ml (allowable limit 82.5 EU/ml), and cell viability was $90.9 \pm 4.06\%$ at release. Measurement of growth kinetics revealed 0.311 doubling per day. Out of 1377 cartilage specimens, 86 nonconformances were identified related to biopsy quality; only 12 were related to cell processing.

AUTOLOGOUS PERIOSTEUM GRAFTS

The use of autologous periosteum alone [14,62–67] has also been investigated for cartilage repair. Animal studies by several groups showed that neochondrogenesis of hyaline cartilage is also possible with this method [68–70].

Periosteal transplantation was initially described by Rubak [71]. He used a periosteal flap to treat cartilage defects in rabbits. The results revealed that the defects were repaired and filled after 4 weeks with a hyaline-like cartilage, whereas the empty control defect showed fibrocartilage-like repair tissue. The first trial in humans was published 3 years later by Niedermann *et al.* [72]. They reported successful results in all their four initially treated patients. Hoikka *et al.* [73] treated 13 patients: 8 had a good results, 4 a fair outcome, and 1 a poor result. O'Driscoll contributed in the following years valuable basic findings. It was found that orientation of the periosteal flap (cambium layer facing up), postoperative factors such as the use of continuous passive motion [62], and the age and maturity of the experimental animal were of importance when dealing with periosteal transplants. O'Driscoll [14] himself treated approximately 40 patients, 20 of whom had defects in the knee and were followed up by the author: 12 individuals had a good outcome, 4 had poor results, and 4 patients had inadequate length of follow-up.

Altogether the results are promising, but as with the other techniques, there are no controlled studies, which would allow for comparison of various methods.

PERICHONDRIUM

Autologous perichondrium has also been employed for cartilage repair [15,74–76]. Perichondrium, taken from the cartilaginous covering of the rib, is placed into the chondral defect of the affected joint. Homminga *et al.* [15] published the first clinical data obtained from this approach. Twenty-five patients with 30 symptomatic chondral lesions received autologous perichondrial grafts taken from the costal arch and fixed to the subchondral bone with human fibrin glue. The majority of the lesions were graded Outerbridge grade III/IV, and half were located on the medial femoral condyle. The opposing articulating surface had no greater than Outerbridge II changes. Techniques very similar to those employed for ACI were used in preparing the defects.

Postoperatively, continuous passive motion was started 2 weeks after surgery and non-weight-bearing exercise was continued for 3 months. All patients were examined arthroscopically at an average of 10 months after implantation. Of the 30 grafted defects, 27 had completely filled with tissue resembling cartilage. In two cases the defect was unchanged, and one patella graft was covered with white tissue with fibrillated surface.

Three biopsy samples were taken at one year, two of which showed disruption of the cartilage–bone junction. Histologically, the regenerated cells appeared to be chondrocytes. Clinically at 1 year, the mean clinical improvement was 80%, as rated by the Hospital for Special Surgery grading scheme; 18 of the 25 patients were completely symptom free. These results remained stable even after 23.5 months, but at 5 years 20 of the 27 grafts were associated with pain and degeneration. A total of 88 patients were treated between 1986 and 1992 in one series [77]. After a mean follow-up of 4 years, only 38% showed a good and 8% a fair outcome; 55% had a poor result. Graft failure ranged between 33 and 62% according to the location of the defect. Only a small, carefully selected group of patients (isolated defects) showed mostly (91%) good results.

Animal experiments subsequently demonstrated an increased calcification of the basal layer of the repair tissue [76], and this was confirmed clinically. Twenty-five of 47 patients displayed calcification of the graft, reason for the authors to start a treatment with with indomethacin [77].

One of the main shortcomings of perichondrial grafting is the limited availability of large grafts. Graft size is limited to the rib size, with the result that several rib perichondrial grafts have to been harvested to fill a large defect. Additionally, endochondral ossification and delamination of the cartilage from the subchondral bone plate are potentially significant limitations to the long-term efficacy of this repair.

CURRENT AND FUTURE INVESTIGATIONS

Current investigations are addressing the following issues:

1. Refinement of surgical procedure (viz., arthroscopic delivery of cells)
2. Alternative cell sources
3. Optimization of culture conditions
4. Engineered scaffolds for cell delivery
5. Genetically modified cells

As outlined earlier, different cell types are under investigation (Table 96.1). To date the committed chondrocytes seem to be the most promising cell for cartilage repair, despite undergoing phenotypic changes in monolayer culture. However, a final judgment of the best cell type is impossible because of lack of information. Further studies are needed to show clearly the advantages and disadvantages of each cell type, and whether the chosen cell type realizes its potential clinically.

Basic investigations have to show more clearly the phenotypic changes, on an RNA and protein level, involved in expanding chondrocytes in monolayer culture and in three-dimensional systems. Culture conditions have a profound effect on cell growth–kinetics, phenotypic stability, and matrix production, but the optimal conditions have yet to be defined. Addition of growth factors is under intensive investigation [35,36,38,78,79]. Techniques for the culture of cells under hydrodynamic pressure [80], continuous medium perfusion [81], and using microgravity in bioreactors [82–88] have been implemented and shown promise.

MATRICES

Phenotypic changes of chondrocytes in monolayer culture have been shown in many studies, and thus interest has focused on three-dimensional systems to culture and deliver the cells to the defect. These systems can act as templates for growth, hence contributing to the phenotypic stability of the chondrocytes.

The matrix, or scaffold, can have several roles in the process of tissue engineering. These roles include structural support for the defect site, barrier to the ingrowth of undesirable cell and tissue types, scaffold for cell migration and proliferation, and carrier or reservoir of cells or regulators (e.g., growth factors). A variety of scaffolds have shown promise thus far. Table 96.2 summarizes the matrices under investigation for cartilage repair.

Table 96.1. Summary of Cells Used in Cell-Based Cartilage Repair Procedures

		Ref.	
Cell type	*In vitro*	*In vivo*	Clinical application
Autologous chondrocytes		[55][a], [56][b]	[13,18], [89,90]
Allogeneic chondrocytes		[52][a], [91][a], [92][a]	
Autologous rib/ear perichonrial cells	[93]	[94][c] [76][a]	[77]
Bone marrow stromal cells	[44]	[95][a,d]	
Neonatal foreskin fibroblasts	[26,96]		

[a]Rabbit.
[b]Dog.
[c]Sheep.
[d]Mouse.

Table 96.2. Summary of Matrices Used for Cartilage Tissue Engineering

Matrix composition	Ref.
Collagen, type I	[30,52,98,99]
Collagen, type II	[30]
Devitalized articular cartilage	[100–102]
Agarose	[103]
Cellulose	[103,104]
Coral, hydroxyapatite	[105]
Poly(lactic acid) (PLA)	[106,107]
Poly(glycolic acid) (PGA)	[98]
PLA-PGA composite	[98,108,109]
Polypropylene	[104]
PEMA/THEMA	[110]

Bell [97] described the ideal scaffold for tissue engineering as one that provides a transitional framework whereby the cells populating it create a replacement tissue as the scaffolding material disappears. Ideally this scaffold would be degraded at the same speed with which the chondrocytes produce their own framework. Other studies have demonstrated that matrix composition affects cell viability, cell attachment, morphology, and synthesis of matrix components [98–110]. For example, chondrocytes seeded into agarose proliferated and synthesized matrix molecules over a period of weeks [104], whereas cellulose matrix was found to be toxic. Grande [98] found that chondrocytes seeded onto type I collagen matrices were more strongly attached, more spherical, and able to produce more type II collagen and higher levels of proteoglycan than chondrocytes seeded onto PGA or Vicryl (PLA-PGA copolymer) matrices.

Although type I collagen matrices are the more widely researched scaffold, we have shown that a type II collagen scaffold supported a more chondrocytic morphology and bio synthetic activity [30,111]. In addition to chemical composition, matrix geometry can influence the performance of the implant. Collagen can be employed as gel [52] or as a sponge with varying degrees of porosity and a range of pore diameters [31,98,99].

GENETICALLY MODIFIED CELLS

Interest in combining tissue engineering and gene therapy approaches has been growing. Principal applications of genetically augmented engineered tissues of the musculoskeletal system were reviewed in 1999 [112]. A target gene, encoding a specific protein molecule, can be introduced into the cell by means of different vectors. There are several advantages to delivering genes, rather than the gene products, to patients, including the ability to achieve high concentrations of the gene products locally in a sustained manner for extended times. Such capabilities are likely to be especially valuable in orthopedic tissue engineering and tissue repair, where it may be necessary to expose discrete populations of cells to various growth factors in precise anatomic locations for lengths of time that go beyond conventional means of delivery [113]. Also, endogenously synthesized proteins might have greater biological activities than exogenously administered recombinant proteins [114].

Cell therapies and tissue engineering provide the ideal partner for *ex vivo* gene therapy, since *in vivo* gene transfer in humans is constrained by safety concerns and the lack of suitable vectors [113]. *Ex vivo* gene transfer is safer because vectors are not introduced directly into the patient, and genetically modified cells can be screened extensively before they are implanted.

Moreover, gene delivery to chondrocytes may require *ex vivo* techniques because the dense cartilaginous matrix is likely to restrict access of vectors to the cells by *in vivo* delivery. Several feasibility studies have investigated the delivery of marker genes like *lacZ*—encoding for the enzyme β-galactosidase—to different musculoskeletal tissues [115–119]. Genetically modified mesenchymal stem cells may also prove to be useful agents of tissue repair in orthopedics [112]. Moreover, recent studies could show an increased matrix synthesis after adenoviral transfer of TGF-β1 gene to canine and human meniscal cells [119] and disc cells [120].

DISCUSSION

Biological or tissue-engineered therapies for cartilage defects have progressed significantly and are becoming important modalities of treatment in orthopedic surgery. Long-term outcome is unknown for all these therapies, however, and there is a lack of controlled studies comparing the different treatment options. Prospective studies are needed to better understand which of the different options will be the most suitable for specific indications.

Nehrer *et al.* [121] compared the histological outcomes of failed cartilage repair procedures. The reparative tissue was retrieved during revision surgery from full-thickness chondral defects in 18 patients in whom abrasion arthroplasty ($n = 12$), grafting of perichondrial flaps ($n = 4$), and periosteal patching augmented by autologous chondrocyte implantation in cell suspension ($n = 6$) had failed to provide lasting relief of symptoms. Histological and immunohistochemical investigations showed fibrous, spongiform tissue comprising type I collagen in $22 \pm 9\%$ (mean $\pm$ standard error of the mean) of the cross-sectional area, and degenerating hyaline tissue ($30 \pm 10\%$) and fibrocartilage ($28 \pm 7\%$) with positive type II collagen staining. Three of four specimens obtained after implantation of perichondrium failed as a result of bone formation that was found in $19 \pm 6\%$ of the cross-sectional area, including areas staining positive for type X collagen, as an indicator for hypertrophic chondrocytes. Revision after autologous chondrocyte implantation was associated with partial displacement of the periosteal graft from the defect site because of insufficient ongrowth or early suture failure. When the graft edge displaced, repair tissue was fibrous ($55 \pm 11\%$), whereas graft tissue attached to subchondral bone displayed hyaline tissue (to 6%) and fibrocartilage (to 12%) comprising type II collagen at 3 months after surgery.

Competing and alternative techniques like osteochondral autograft transfer [122,123], mosaicplasty [124], or osteochondral paste must be considered or included into a therapeutical-algorithm [125]. The quality of the provided cells and the technical procedure of implanting the cells are not the only factors to affect the outcome. Correction of predisposing factors like malalignment of the knee or maltracking of the patella have been shown to be necessary for a good clinical outcome [126].

REFERENCES

1. Hunter, W. (1743). Of the structure and diseases of articulating cartilages. *Philos. Trans. R. Soc. London, Ser. B* **42**, 514–522.
2. Paget, J. (1853). Healing of injuries in various tissues. *Lect. Surg. Pathol.* **1**, 262–270.
3. Wirth, C. I., and Rudert, M. (1996). Techniques of cartilage growth enhancement: A review of the literature. *Arthroscopy* **12**, 300–308.
4. Pridie, K. H. (1959). A method of resurfacing osteoarthritic knee joints. *J. Bone Jt. Surg.* **41**, 618–619.
5. Ficat, R. P., Ficat, C., Gedeon, P. K. and Toussaint, J. F. (1984). Spongialization: A new treatment for diseased patella. *Clin. Orthop. Relat. Res.* **182**, 200–205.
6. Frisbie, D. D., Trotter, G. W., Powers, B. E., Rodkey, W. G., Steadman, J. R., Howard, R. D., Park, R. D. and McIlwraith, C. W. (1999). Arthroscopic subchondral bone plate microfracture technique augments healing of large chondral defects in the radial carpal bone and medial femoral condyle of horses. *Vet. Surg.* **28**, 242–255.
7. Steadman, J. R., Rodkey, W. G., Briggs, K. K., and Rodrigo, J. J. (1999). [The microfracture technic in the management of complete cartilage defects in the knee joint]. *Orthopade* **28**, 26–32.
8. Steadman, J. R. (1998). Long-term clinical results with microfracture (MCFR) and debridement for treatment of full-thickness chondral defects. *65th Ann. Mtg. Amer. Acad. Orthop. Surg.* p. 214.
9. Friedmann, M. J., Berasi, D. O., Fox, J. M., Pizzo, W. D., Snyder, S. J., and Ferkel, R. D. (1984). Peliminary results with abrasion arthroplasty in the osteoarthritic knee. *Clin. Orthop. Relat. Res.* **182**, 200–205.
10. Ewing, J. W. (1990). Arthroscopic treatment of degenerative meniscal lesion and early degenerative arthritis of the knee. In "Articular Cartilage and Knee Joint Function: Basic Science and Arthroscopy" (Ewing, J. W., ed.), pp. 137–145. Raven Press, New York.
11. Johnson, L. L. (1991). Characteristics of the immediate postarthroscopic blood clot formation in the knee joint. *Arthroscopy* **7**, 14–23.
12. Johnson, L. L. (1986). Arthroscopic abrasion arthroplasty, historical and pathologic perspective: Present status. *Arthroscopy* **2**, 54–69.

13. Brittberg, M., Lindahl, A., Nilsson, A., Ohlsson, C., Isaksson, O., and Peterson, L. (1994). Treatment of deep cartilage defects in the knee with autologous chondrocyte transplantation. *N. Engl. J. Med.* **331**, 889–995.

14. O'Driscoll, S. W. (1999). Articular cartilage regeneration using periosteum. *Clin. Orthop. Relat. Res.* 367S, S186–S203.

15. Homminga, G. N. Boulstra, S. K., Bouwmeester, P. S. M., and van der Linden, A. J. (1990). Perichondral grafting for cartilage lesions of the knee. *J. Bone Jt. Surg. Br. Vol.* **72B** 1003–1007.

16. Lee, C. R. and Spector, M. (1998). Status of articular cartilage tissue engineering. *Curr. Opin. Orthop.* **9**, 88–93.

17. Mayhew, T. A., Williams, G. R., Senica, M. A., Kuniholm, G., and Du Moulin, G. C. (1998). Validation of a quality assurance program for autologous cultured chondrocyte implantation. *Tissue Eng.* **4**, 325–334.

18. Peterson, L., Minas, T., Brittberg, M., Nilsson., A., Sjögren-Jansson, E., and Lindahl, A. (2000). Two- to 9-year outcome after autologous chondrocyte transplantation of the knee. *Clin. Orthop. Relat. Res.* **374**, 212–234.

19. Goldring, M. B., Sandell, L. J., Stephenson, M. L., and Krane, S. M. (1986). Immune interferon suppresses levels of procollagen mRNA and type II collagen synthesis in cultured human articular and costal chondrocytes. *J. Biol. Chem.* **261**, 9049–9055.

20. Goldring, M. B., Birkhead, J., Sandell, L. J., Kimura, T., and Krane, S. M. (1988). Interleukin 1 suppresses expression of cartilage-specific types II and IX collagens and increases types I and III collagens in human chondrocytes. *J. Clin. Invest.* **82**, 2026–2037.

21. Saadeh, P. B., Brent, B., Mehrara, B. J., Steinbrech, D. S., Ting, V., Gittes, G. K., and Longaker, M. T. (1999). Human cartilage engineering: Chondrocyte extraction, proliferation, and characterization for construct development. *Ann. Plast. Surg.* **42**, 509–513.

22. Norby, D. P., Malemud, C. J. and Sokoloff, L. (1977). Differences in the collagen types synthesized by lapine articular chondrocytes in spinner and monolayer culture. *Arthritis Rheum.* **20**, 709–716.

23. Watt, F. M., and Dudhia, J. (1988). Prolonged expression of differentiated phenotype by chondrocytes cultured at low density on a composite substrate of collagen and agarose that restricts cell spreading. *Differentiation (Berlin)* **38**, 140–147.

24. Denker, A. E., Nicoll, S. B., and Tuan, R. S. (1995), Formation of cartilage-like spheroids by micromass cultures of murine C3H10T1/2 cells upon treatment with transforming growth factor-beta 1. *Differentiation, (Berlin)* **59**, 25–34.

25. Kulyk, W. M., and Reichert, C. (1992). Staurosporine, a protein kinase inhibitor, stimulates cartilage differentiation by embryonic facial mesenchyme. *J. Craniofacial Genet. Dev. Biol.* **12**, 90–97.

26. Nicoll, S. B., Wedrychowska, A., Smith, N., and Bhatnage, R. S. (1998). A new approach to cartilage tissue engineering using human dermal fibroblasts seeded on three-dimmensional polymer scaffolds. *Proc. Mater. Res. Soc.* **530**.

27. Benya, P. D., and Shaffer, J. D. (1982). Dedifferentiated chondrocytes reexpress the differentiated collagen phenotype when cultured in agarose gels. *Cell (Cambridge, Mass.* **30**, 215–224.

28. Guo, J. F., Jourdian, G. W., and MacCallum, D. K. (1989). Culture and growth characteristics of chondrocytes encapsulated in alginate beads. *Connect Tissue Res.* **19**, 277–297.

29. Häuselmann, H. J., Masuda, K., Hunziker, E. B., Neidhardt, M., Mok, S. S., Beat, A., and Thonar, E. J. (1996). Adult human chondrocytes cultured in alginate form a matrix similar to native human articular cartilage. *Am. J. Physiol.* **271**, C742–C752.

30. Nehrer, S., Breinan, H. A., Ramappa, A., Young, R. G., Shortkroff, S., Louie, L. B., Sledge, C. B., Yannas, C. B., and Spector, M. (1997). Matrix collagen type and pore size influence behavior of seeded canine chondrocytes. *Biomaterials* **18**, 769–776.

31. Nehrer, S., Breinan, H. A., Ramappa, A., Hsu, H.-P., Minas, T., Shortkroff, S., Sledge, C. B., Yannas, I. V., and Spector, M. (1998). Chondrocyte-seeded collagen matrices implanted in a chondral defect in a canine model. *Biomaterials* **19**, 2313–2328.

32. Martin, I., Vunjak-Novakovic G., Yang, J., Langer, R., and Freed, L. E. (1999). Mammalian chondrocytes expanded in the presence of fibroblast growth factor 2 maintain the ability to differentiate and regenerate three-dimensional cartilaginous tissue, *Exp. Cell. Res.* **253**, 681–688.

33. Frenkel, S. R., Saadeh, P. B., Mehrara, B. J., Chin, G. S., Steinbrech, D.S., Brent, B., Gittes, G. K., and Longaker, M. T. (2000). Transforming growth factor beta superfamily members: Role in cartilage modeling. *Plast. Reconstr. Surg.* **105**, 980–990.

34. Kulyk, W. M., Franklin, J. L., and Hoffman, L. M. (2000). Sox9 expression during chondrogenesis in micromass cultures of embryonic limb mesenchyme. *Exp. Cell Res.* **255**, 327–332.

35. Chopra, R. and Anastassiades, T. (1998). Specificity and synergism of polypeptide growth factors in stimulating the synthesis of proteoglycans and a novel high molecular weight anionvc glycoprotein by articular chondrocyte cultures. *J. Rheumatol.* **25**, 1578–1584.

36. Dunham, B. P., and Koch, R. I. (1998). Basic fibroblast growth factor and insulinlike growth factor I support the growth of human septal chondrocytes in a serum-free environment. *Arch. Otolatyngol. Head Neck Surg.* **124**, 1325–1330.

37. Goto, K., Yamazaki, M., Tagawa, M., Goto, S., Kon, T., Moriya, H., and Fujimura, S. (1998). Involvement of insulin-like growth factor I in development of ossification of the posterior longitudinal ligament of the spine. *Calcif. Tissue Int.* **62**, 158–165.

38. Fujisato, T., Sajiki, T., Liu, Q., and Ikada, Y. (1996). Effect of basic flbroblast growth factor on cartilage regeneration in chondrocyte-seeded collagen sponge scaffold. *Biomaterials* **17**, 155–162.

39. Shortkroff, S., Barone, L., Hsu, H. P., Wrenn, C., Gagne, T., Chi, T., Breinan, H., Minas, T., Sledge, C. B., Tubo, R., and Spector, M. (1996). Healing of chondral and osteochondral defects in a canine model: The role of cultured chondrocytes in regeneration of articular cartilage. *Biomaterials* **17**, 147–154.

40. Lee, C. R., Grodzinsky, A. J., Hsu, H.-P., Martin, S. D., and Spector, M. (2000). Effects of harvest and selected cartilage repair procedures on the physical and biochemical properties of articular cartilage in the canine knee. *J. Orthop. Res.* **18**, 790–799.

41. Lee, C. R., Breinan, H. A., Nehrer, S. and Spector, M. (2000). Articular cartilage chondrocytes in type I and type II collagen–GAG matrices exhibit contractile behavior in vitro. *Tissue Eng.* **6**, 555–565.

42. Kinner, B., and Spector, M. (2001). Smooth muscle actin expression by human articular chondrocytes and their contraction of a collagen–glycosaminoglycan matrix in vitro. *J. Orthop. Res.* **19**, 233–241.

43. Wakitani, S., Goto, T., Pineda, S. J., Young, R. G., Mansour, J. M., Caplan, A. I., and Goldberg, V. M. (1994). Mesenchymal cell–based repair of large, full-thickness defects of articular cartilage. *J. Bone Jt. Surg.* **76**, 597.

44. Johnstone, B., Hering, T. M., Caplan, A. I., Goldberg, V. M., and Yoo, J. U. (1998). In vitro chondrogenesis of bone marrow–derived mesenchymal progenitor cells. *Exp. Cell Res.* **238**, 265–272.

45. Johnstone, B., and Yoo, J. U. (1999). Autologous mesenchymal progenitor cells in articular cartilage repair. *Clin. Orthop. Relat. Res.* **367S**, S156–S162.

46. Denker, A. E., Haas, A. R., Nicoll, S. B., and Tuan, R. S. (1999). Chondrogenic differentiation of murine C3H10T1/2 multipotential mesenchymal cells: I. Stimulation by bone morphogenetic protein-2 in high-density micromass cultures. *Differentiation (Berlin)* **64**, 67–76.

47. Ghazavi, M. T., Pritzker, K. P., Davis, A. M., and Gross, A. E.(1997). Fresh osteochondral allografts for post-traumatic osteochondral defects of the knee. *J. Bone Jt. Surg. Br. Vol* **79**, 1008–1013.

48. Gross, A. E., Silverstein, E. A., Falk, J., Falk, R., and Langer, F. (1975). The allotransplantation of partial joints in the treatment of osteoarthritis of the knee. *Clin. Orthop. Relat. Res.* **108**, 7–14.

49. McDermott, A. G., Langer, F., Pritzker, K. P., and Gross, A. E. (1985). Fresh small-fragment osteochondral allografts. Long-term follow-up study on first 100 cases. *Clin. Orthop. Relat. Res.* **197**, 96–102.

50. Rahfoth, B., Weisser, J., Sternkopf, F., Aigner, T., von der Mark, K., and Brauer, R. (1998). Transplantation of allograft chondrocytes embedded in agarose gel into cartilage defects of rabbits. *Osteoarthritis Cartilage* **6**, 50–65.

51. Wakitani, S., Goto, T., Young, R. G., Mansour, J. M., Goldberg, V. M., and Caplan, A. I. (1998). Repair of large full-thickness articular cartilage defects with allograft articular chondrocytes embedded in a collagen gel. *Tissue Eng.* **4**, 429–444.

52. Kawamura, S., Wakitani, S., Kimura, T., Maeda, A., Caplan, A. I., Shino, K., and Ochi, T. (1998). Articular cartilage repair. Rabbit experiments with a collagen gel–biomatrix and chondrocytes cultured in it. *Acta Orthop. Scand.* **69**, 56–62.

53. Green, W. T., Jr. (1977). Articular cartilage repair. Behavior of rabbit chondrocytes during tissue culture and subsequent allografting. *Clin. Orthop. Relat. Res.* **124**, 237–250.

54. Grande, D. A., Singh, I. J., and Pugh, J. (1987). Healing of experimentally produced lesions in articular cartilage following chondrocyte transplantation. *Anat. Rec.* **218**, 142–148.

55. Brittberg, M., Nilsson, A., Lindahl, A., Ohlsson, C., and Peterson, L. (1996). Rabbit articular cartilage defects treated with autologous cultured condrocytes. *Clin. Orthop. Relat. Res.* **326**, 270–283.

56. Breinan, H. A., Minas, T., Hsu, H.-P., Neher, S., Sledge, C. B., and Spector, M. (1997). Effect cultured autologous chondrocytes on repair of chondral defects in a canine model. *J. Bone Jt. Surg., Am. Vol.* **79-A**, 1439–1451.

57. Breinan, H. A., Minas, T., Hsu, H.-P., Nehrer, S., Shortkroff, S., and Spector, M. (2001). Autologous chondrocyte implantation in a canine model: Change in composition of reparative tissue with time. *J. Orthop. Res.* **19**, 282–292.

58. Breinan, H. A., and Spector, M. (2001). Chondral defects in animal models: Effects of selected repair procedures in canines. *Clin. Orthop. Relat. Res.* (in press).

59. Wang, Q., Breinan, H. A., Hsu, H. P., and Spector, M. (2000). Healing of defects in canine articular cartilage: Distribution of nonvascular alpha-smooth muscle actin-containing cells. *Wound Repair Regeneration* **8**, 145–158.

60. Perka, C.,Sittinger, M., Schultz, O., Spitzer, R.-S., Schlenzka, D., and Burmester, G. R. (2000). Tissue engineered cartilage repair using cryopreserved and noncryopreserved chondrocytes. *Clin. Orthop. Res.* **378**, 245–254.

61. Brittberg, M. (1999). Autologous chondrocyte transplantation. *Clin. Orthop. Res.* **367S**, 147–155.

62. O'Driscoll, S. W., Keeley, F. W., and Salter, R. B. (1986). The chondrogenic potential of free autogenous periosteal grafts for biological resurfacing of major full-thickness defects in joint surfaces under the influence of continuous passive motion. An experimental investigation in the rabbit. *J. Bone Jt. Surg., Am. Vol.* **68-A**, 1017–1035.

63. O'Driscoll, S. W., and Salter, R. B. (1986). The repair of major osteochondral defects in joint surfaces by neochondrogenesis with autogenous osteoperiosteal grafts stimulated by continuous passive motion. An experimental investigation in the rabbit. *Clin. Orthop. Relat. Res.* **208**, 131–140.

64. O'Driscoll, S. W., Keeley, F. W. (1988). Durability of regenerated articular cartilage produced by free autogenous periostal grafts in major full thickness defects in joint surface under the influence of continous passive motion. *J. Bone Jt. Surg., Am. Vol.* **70-A**, 595–606.

65. O'Driscoll, S. W., Recklies, A. D., and Poole, A. R. (1994). Chondrogenesis in periosteal explants. An organ culture model for in vitro study. *J. Bone Jt. Surg., Am. Vol.* **76-A**, 1042–1051.

66. O'Driscoll, S. W., Fitzsimmons, J. S., and Commisso, C. N. (1997). Role of oxygen tension during cartilage formation by periosteum. *J. Orthop. Res.* **15**, 682–687.

67. O'Driscoll, S. W., Meisami, B., Miura, Y., and Fitzsimmons, J. S. (1999). Viability of periosteal tissue obtained postmortem. *Cell Transplant* **8**, 611–616.

68. O'Driscoll, S. W., and Salter, R. B. (1984). The induction of neochondrogenesis in free intra-articular periostal autografts under the influence of continuous passive motion. *J. Bone Jt. Surg., Am. Vol.* **66-A**, 1248–1257.

69. Carranza-Bencana, A., Garcia-Paino, L., Armas Padron, J. R., and Cayuela Dominguez, A. (2000). Neochondrogenesis in repair of full-thickness articular cartilage defects using free autogenous periosteal grafts in the rabbit. A follow-up in six months. *Osteoarthritis Cartilage* **8**, 351–358.

70. Zarnett, R., and Salter, R. B. (1989). Periosteal neochondrogenesis for biologically resurfacing joints: Its cellular origin. *Can. J. Surg.* **32**, 171–174.

71. Rubak, J. M. (1982). Reconstruction of articular cartilage defects with free periosteal grafts. An experimental study. *Acta Orthop. Scand.* **53**, 175–180.

72. Niedermann, B., Boe, S., Lauritzen, J., and Rubak, J. M. (1985). Glued periosteal grafts in the knee. *Acta Orthop. Scand.* **56**, 457–460.

73. Hoikka, V. E., Jaroma, H. J., and Ritsila, V. A. (1990). Reconstruction of the patellar articulation with periosteal grafts. 4- year follow-up of 13 cases. *Acta Orthop. Scand.* **61**, 36–39.

74. Homminga, G. N., van der Linden, A. J., Terwindt-Rouwenhorst, W. A. W., and Drukker, J. (1989). Repair of articular cartilage defects by perichondrial grafts. Experiments in the rabbit. *Acta Orthop. Scand.* **60**, 326–329.

75. Homminga, G. N., van der Linden, T. J., Terwindt-Rouwenhorst, E. A., and Drukker, J. (1989). Repair of articular defects by perichondrial grafts. Experiments in the rabbit. *Acta Orthop. Scand.* **60**, 326–329.

76. Homminga, G. N., Bulstra, S. K., Kuijer, R., and van der Linden, A. J. (1991). Repair of sheep articular cartilage defects with a rabbit costal perichondrial graft. *Acta Orthop. Scand.* **62**, 415–418.

77. Bouwmeester, S. J., Beckers, J. M., Kuijer, R., van der Linden, A. J., and Bulstra, S. K. (1997). Long-term results of rib perichondrial grafts for repair of cartilage defects in the human knee. *Int. Orthop.* **21**, 313–317.

78. van Susante, J. L., Buma, P., van Beuningen, H. M., van den Berg, W. B., and Veth, R. P. (2000). Responsiveness of bovine chondrocytes to growth factors in medium with different serum concentrations. *J. Orthop. Res.* **18**, 68–77.

79. de Haart, M., Marijnissen, W. J., van Osch, G. J., and Verhaar, J. A. (1999). Optimization of chondrocyte expansion in culture. Effect of TGF beta-2, bFGF and L-ascorbic acid on bovine articular chondrocytes. *Acta Orthop. Scand.* **70**, 55–61.

80. Klein-Nulend, J., Veldhuijzen, J. P., van de Stadt, R. J., van Kampen, G. P., Kuijer, R., and Burger, E. H. (1987). Influence of intermittent compressive force on proteoglycan content in calcifying growth plate cartilage in vitro. *J. Biol. Chem.* **262**, 15490–15495.

81. Sittinger, M., Schultz, O., Keyszer, G., Minuth, W. W., and Burmester, G. R. (1997). Artificial tissues in perfusion culture. *Int. J. Artif. Organs* **20**, 57–62.

82. Freed, L. E., Vunjak-Novakovic, G., and Langer, R. (1993). Cultivation of cell–polymer cartilage implants in bioreactors. *J. Cell. Biochem.* **51**, 257–264.

83. Freed, L. E., Langer, R., Martin, I., Pellis, N. R., and Vunjak-Novakovic, G. (1997). Tissue engineering of cartilage in space. *Proc. Natl. Acad. Sci. U.S.A.* **94**, 13885–13890.

84. Freed, L. E., and Vunjak-Novakovic, G. (1997). Microgravity tissue engineering. *In Vitro Cell. Dev. Biol.—Anim.* **33**, 381–385.

85. Freed, L. E., Hollander, A. P., Martin, I., Barry, J. R., Langer, R., and Vunjak-Novakovic, G. (1998). Chondrogenesis in a cell–polymer–bioreactor system. *Exp. Cell Res.* **10**, 58–65.

86. Baker, T. L., and Goodwin, T. J. (1997). Three-dimensional culture of bovine chondrocytes in rotating-wall vessels. *In Vitro Cell. Dev. Biol.—Anim.* **33**, 358–365.

87. Martin, I., Obradovic, B., Treppo, S., Grodzinsky, A. J., Langer, R., Freed, L. E., and Vunjak-Novakovic, G. (2000). Modulation of the mechanical properties of tissue engineered cartilage. *Biorheology* **37**, 141–147.

88. Vunjak-Novakovic, G., Martin, I., Obradovic, B., Treppo, S., Grodzinsky, A. J., Langer, R., and Freed, L. E. (1999). Bioreactor cultivation conditions modulate the composition and mechanical properties of tissue-engineered cartilage. *J. Orthop. Res.* **17**, 130–138.

89. Wright, J. G. (1995). Autologous chondrocyte transplantation. *N. Engl. J. Med.* **332**, 540.

90. Minas, T. (1998). Chondrocyte implantation in the repair of chondral lesions of the knee: Economics and quality of life. *Am. J. Orthop.* **27**, 739–744.

91. Rahforth, B., Weisser, J., Sternkopf, F., Aigner, T., von der Mark, K., and Bräuer, R. (1998). Transplantation of allograft chondrocytes embedded in agarose gel into carilage defects of rabbits. *Osteoarthritis Cartilage* **6**, 50–65.

92. Baragi, V. M., Renkiewicz, R. R., Qiu, L., Brammer, D., Riley, J. M., Sigler, R. E., Frenkel, S. R., Amin, A., Abramson, S. B., and Roessler, B. J. (1997). Transplantation of adenovirally transduced allogeneic chondrocytes into articular cartilage defects in vivo. *Osteoarthritis Cartilage* **5**, 275–282.

93. Bulstra, S. K., Homminga, G. N., Buurman, W. A., Terwindt-Rouwenhors, E. and van der Linden, A. J. (1990). The potential of adult human perichondrium to form hyalin cartilage in vitro. *J . Orthop. Res.* **8**, 328–335.

94. Chu, C. R., Dounchis, J. S., Yoshioka, M., Sah, R. L., Coutts, R. D., and Amiel, D. (1997). Osteochondral repair using perichondrial cells. A 1-year study in rabbits. *Clin Orthop. Relat. Res.* **340**, 220–229.

95. Sittinger, M., Perka, C., Schultz, O., Haupl, T., and Burmester, G. R. (1999). Joint cartilage regeneration by tissue engineering. *Z Rheumatol.* **58**, 130–135.

96. Mizuno, S. and Glowacki, J. (1996). Chondroinduction of human dermal fibroblasts by demineralized bone in three-dimensional culture. *Exp. Cell Res.* **227**, 89–97.

97. Bell, E. (1995). Strategy for the selection of scaffolds for tissue engineering. *Tissue Eng.* **1**, 163–179.

98. Grande, D. A., Halberstadt, C., Naughton, G., Schwartz, R., and Manji, R. (1997). Evaluation of matrix scaffolds for tissue engeneering of articular cartilage grafts. *J. Biomed. Mater. Res.* **34**, 211–220.

99. Frenkel, S. R., Toolan, B., Menche, D., Pitman, M. I., and Pachence, J. M. (1998). Chondrocyte transplantation using a collagen bilayer matrix for cartilage repair. *J. Bone Jt. Surg., Br. Vol.* **79B**, 831–636.

100. Peretti, G. M., Bonassar, L. J., Caruso, E. M., Randolph, M. A., Trahan, C. A., and Zaleske, D. J. (1999). Biomechanical analysis of a chondrocyte-based repair model of articular cartilage. *Tissue Eng.* **5**, 317–326.

101. Peretti, G. M., Randolph, M. A., Caruso, E. M., Rossetti, F. and Zaleske, D. J. (1998). Bonding of cartilage matrices with cultured chondrocytes: an experimental model. *J. Orthop. Res.* **16**, 89–95.

102. Toolan, B. C., Frenkel, S. R., Pereira, D. S., and Alexander, H. (1998). Development of a novel osteochondral graft for cartilage repair. *J. Biomed. Mater. Res.* **41**, 244–250.

103. Matthew, I. R., Browne, R. M., Frame, J. W., and Millar, B. G. (1995). Subperiosteal behaviour of alginate and cellulose wound dressing materials. *Biomaterials* **16**, 275–287.

104. Cook, J. L., Kreeger, J. M., Payne, J. T., and Tomlinson, J. L. (1997). Three-dimensional culture of canine articular chondrocytes on multiple transplantable substrates. *Am. J. Vet. Res.* **58**, 419–424.

105. van Susante, J. L., Buma, P., Homminga, G. N., van den Berg, W. B., and Veth, R. P. (1998). Chondrocyte-seeded hydroxyapatite for repair of large articular cartilage defects. A pilot study in the goat. *Biomaterials* **19**, 2367–2374.

106. Chu, C. R., Monosov, A. Z., and Amiel, D. (1995). In situ assessment of cell viability within biodegradable polylactic acid polymer matrices. *Biomaterials* **16**, 1381–1384.

107. von Schroeder, H. P., Kwan, M., Amiel, D., and Coutts, R. D. (1991). The use of polylactic acid matrix and periosteal grafts for the reconstruction of rabbit knee articular defects. *Biomed. Mater. Res.* **25**, 329–339; erratum: *Ibid.* **26**(4), 553 (1992).

108. Rotter, N., Aigner, I., Naumann, A., Planck, H., Hammer, C., Burmester, G., and Sittinger, M. (1998). Cartilage reconstruction in head and neck surgery: Comparison of resorbable polymer scaffolds for tissue engineering of human septal cartilage. *J. Biomed. Mater. Res.* **42**, 347–356.

109. Athanasiou, K. A., Agrawal, C. M., Barber, F. A., and Burkhart, S. S. (1998). Orthopaedic applications for PLA-PGA biodegradable polymers. *Arthroscopy* **14**, 726–737.

110. Wyre, R. M., and Downes, S. (2000). An in vitro investigation of the PEMA/THFMA polymer system as a biomaterial for cartilage repair. *Biomaterials* **21**, 335–343.

111. Nehrer, S., Breinan, H. A., Ramappa, A., Hsu, H.-P., Minas, T., Shortkroff, S., Sledge, C. B., Yannas, C. B., and Spector, M. (1998). Chondrocyte-seeded collagen matices implanted in a chondral defect in a canine model. *Biomaterials* **19**, 2313–2328.

112. Evans, C. H., and Robbins, P. D. (1999). Genetically augmented tissue engineering of the musculoskeletal system. *Clin. Orthop. Relat. Res.* **367S**, S410–S418.

113. Kang, R., Ghivizzani, S. C., Muzzonigro, T. S., Herndon, J. H., Robbins, P. D., and Evans, C. H. (2000). Orthopaedic applications of The Marshall R. Urist Young Investor Award. Orthopaedic applications of gene therapy. From concept to clinic. *Clin. Orthop. Relat. Res.* **375**, 324–337.

114. Sandhu, J. S., Gorczynski, R. M., Waddell, J., Nguyen, H., Squires, J., Boynton, E. L., and Hozumi, N. (1999). Effect of interleukin-6 secreted by engineered human stromal cells on osteoclasts in human bone. *Bone* **24**, 217–227.

115. Kang, R., Marui, T., Chivizzani, S. C., Nita, I. M., Georgescu, H. I., Suh, J. K., Robbins, P. D., and Evans, C. H. (1997). Ex vivo gene transfer to chondrocytes in full-thickness articular cartilage defects: A feasibility study. *Osteoarthritis Cartilage* **5**, 139–143.

116. Gerich, T. G., Fu, F. H., Robbins, P. D., and Evans, C. H. (1996). Prospects for gene therapy in sports medicine. *Knee Surg. Sports Traumatol. Arthroscopy* **4**, 180–187.

117. Lou, J., Tu, Y. Ludwig, F. J., Zhang, J., and Manske, P. R. (1999). Effect of bone morphogenetic protein-12 gene transfer on mesenchymal progenitor cells. *Clin. Orthop. Relat. Res.* **369**, 333–339.

118. Nakamura, N., Timmermann, S. A., Hart, D. A., Kaneda, Y., Shrive, N. G., Shino, K., Ochi, T., and Frank, C. B. (1998). A comparison of in vivo gene delivery methods for antisense therapy in ligament healing. *Gene Ther.* **5**, 1455–1461.

119. Goto, H., Shuler, F. D., Lamsam, C., Moller, H. D., Niyibizi, C., Fu, F. H., Robbins, P. D., and Evans, C. H. (1999). Transfer of *lacZ* marker gene to the meniscus. *J. Bone Jt. Surg., Am. Vol.* **81-A**, 918–925.

120. Nishida, K., Kang, J. D., Gilberston, L. G., Moon, S. H., Suh, J. K., Vogt, M. T., Robbins, P. D., and Evans, C. H. (1999). Modulation of the biologic activity of the rabbit intervertebral disc by gene therapy: An in vivo study of adenovirus-mediated transfer of the human transforming growth factor beta 1 encoding gene. *Spine* **24**, 365, 2419–2425.

121. Nehrer, S., Spector, M., and Minas, T. (1999). Histological analysis of tissue after failed cartilage repair procedures. *Clin. Orthop. Relat. Res.* **365**, 149–162.

122. Matsusue, Y., Yamamuro, T., and Hama, H. (1993). Arthroscopic multiple osteochondral transplantation to chondral defect in the knee associated with anterior cruciate ligament disruption. *Arthroscopy* **9**, 318–321.

123. Bobic, V. (1996). Arthroscopic osteochondral autograft transplantation in anterior cruciat ligament reconstruction: A preliminary clinical study. *Knee Surg., Sports Traumatol. Arthroscopy* **3**, 262–264.

124. Hangody, L., Kish, G., Karpati, Z., Udvarhelyi, I., Szigeti, I., and Bely, M. (1998). Mosaicplasty for the treatment of articular cartilage defects: Application in clinical practice. *Orthopaedics* **21**, 751–765.

125. Minas, T. (1999). The role of cartilage repair techniques, including chondrocyte transplantation, in focal chondral knee damage. *Instruct. Course Lect.* **48**, 629–643.

126. Minas, T., and Neher, S. (1997). Current concepts in the treatment of articular cartilage defects. *Orthopaedics* **20**, 525–538.

CHAPTER 97

CELL-BASED THERAPIES FOR BULKING AGENTS

Hans G. Pohl and Anthony Atala

Open surgical correction of primary vesicoureteral reflux and of urinary incontinence that results from intrinsic sphincteric dysfunction is almost uniformly successful. Yet despite surgical cure in 99% of children following ureteral reimplantation and in 80% of women following placement of a urethral sling, further surgical innovation may afford similar success rates with less morbidity. The rationale for the endoscopic management of reflux and urinary incontinence relies on shorter operative times, reduced procedural costs, ease of surgical technique, and reduced patient morbidity. Over the last several decades, mounting clinical experience has demonstrated that the endoscopic correction of primary vesicoureteral reflux and urinary incontinence is both possible and effective. Most recently attempts have also been made to correct dysphonia secondary to vocal cord scarification via endoscopic means. The results of two limited trials are encouraging and lend further support to the rationale of using autologously derived cells as bulking agents.

Although reflux, intrinsic sphincteric incontinence, and dysphonia differ in their underlying pathophysiology, the basic principle behind the correction of each incorporates reliance on the provision of additional submucosal bulk. Ideally, the resultant mucosal coaptation produces an increase in resistance that is sufficient to prevent the inappropriate flow of urine either retrograde (to the kidneys) or antegrade (out the urethra) or sufficient vocal fold bulk to provide resonance. Since the introduction of injectable poly(tetrafluoroethylene) (Teflon) paste for use in the treatment of urinary incontinence in the early 1970s, many novel substances have been developed and promoted for use as injectable endoscopic bulking agents (Table 97.1). Teflon and purified glutaraldehyde-cross-linked bovine collagen (GAX–collagen) remain the most widely used and accepted agents worldwide. Despite their therapeutic efficacy, however, both these substances have had drawbacks that limit and/or restrict their clinical utility. In the case of Teflon, reports of particulate migration, embolization, and granuloma formation have led to questions of the materials safety and curtailed its use in the United States [1,2]. GAX–collagen, though not prone to migration or granuloma formation, fails to conserve its volume over time, resulting in the need for periodic re-treatment [3]. In addition, its administration has been implicated in acute hypersensitivity reactions, and in the sporadic development of connective tissue diseases post-therapy [4].

The absence of an ideal bulking agent has resulted in a continued research effort directed toward the development of new injectable substances that may be used for the endoscopic treatment of vesicoureteral reflux, incontinence, and dysphonia. Theoretically, an injectable substance used for such purposes should adhere to certain ideal material characteristics. The first, "anatomical integrity," is the ability of the material to be delivered endoscopically and to conserve its volume over time; for the achievement of the second, "material safety," a material would have to be biocompatible, nonantigenic, noncarcinogenic, and nonmigratory. To date, no substance has successfully met all these criteria, therefore limiting the implementation of these cost-effective, low-morbidity procedures.

Table 97.1. Nonautologous and Autologous Agents That Have Been Used for the Endoscopic Correction of Vesicoureteral Reflux

Nonautologous agents	Autologous agents
Poly(tetrafluoroethylene) paste (Teflon)	Autologous fat
Glutaraldehyde-cross-linked (GAX) bovine collagen	Autologous collagen
Particulate silicone microimplants (Macroplastique)	Tissue–engineered therapies
Poly(vinyl alcohol) foam	
Injectable Bioglass	
Dextranomer microspheres	
Detachable self-sealing membrane system	

The goal of tissue engineering is to use selective, autologous cell transplantation to create new, functional, nonimmunogenic tissue that can survive *in vivo*. Donor tissue is harvested, dissociated into individual cells or small tissue fragments, and reimplanted into the autologous host, alone or attached to a support matrix. Much of the power of this technique lies in the fact that cultured cells can be expanded to great numbers prior to reimplantation, permitting their use in surgical reconstruction. Thus the physician has a unique opportunity to surgically reestablish near-normal anatomy with a physiologically functional tissue. This approach has been used in the genitourinary tract to engineer new, functional urological structures [5]. A living, injectable, tissue-engineered autologous bioimplant for use in the correction of urinary incontinence and vesicoureteral reflux would be ideal. Such an implant would optimally satisfy all the criteria in terms of anatomical integrity and material safety. Toward this goal, the suitability of implementing injectable tissue-engineered, autologous chondrocytes and smooth muscle cells for the endoscopic treatment of urinary incontinence and vesicoureteral reflux has been investigated.

AUTOLOGOUS CHONDROCYTES

Because chondrocytes posse the ability to form viable cartilage, they have good potential as a stable submucosal bulking material. Mammalian chondrocytes have been shown to grow readily in culture and can survive and synthesize cartilage matrix *in vitro* [6,7]. Once seeded onto preformed biodegradable polymer templates, these cells can also survive *in vivo* and synthesize new cartilaginous structures. Initial investigative efforts were directed at determining the viability of injected chondrocytes transplanted *in vivo* via an alginate–polymer delivery vehicle [8]. Alginate, a liquid copolymer of glucuronic and mannuronic acids, seemed to be an excellent synthetic substrate: it solidifies by gelation once injected *in vivo* and undergoes hydrolytic biodegradation over time [9]. Such a system would serve as a synthetic substrate for injectable delivery and initial maintenance of cartilage architecture *in vivo*. Concentrated chondrocyte suspensions harvested from bovine articular cartilage were mixed with dry alginate powder, forming a gel. This chondrocyte–alginate gel suspension was injected subcutaneously into athymic mice. Histological analyses of retrieved specimens demonstrated new cartilage formation in all of the experimental subjects (Fig. 97.1). Gross examination of the injection sites over time demonstrated progressive replacement of the polymer gels with new cartilage. The newly formed cartilage retained its volume over time, which appeared to be related to the volume of chondrocyte–alginate gel suspension initially injected. The lack of cartilage formation at sites where either alginate alone or free chondrocytes were injected supported the necessity of a cell–polymer construct.

With these encouraging results, a system was envisioned under which autologous chondrocytes harvested from a patient could be expanded *in vitro*, combined with an alginate copolymer, and injected cystoscopically. Working toward this goal, the earlier strategy was implemented in experiments involving the endoscopic treatment of vesicoureteral reflux in a porcine model [10]. Bilateral vesicoureteral reflux was created in four Hanford minipigs via an open-bladder technique and confirmed radiologically [11]. Chondrocytes were

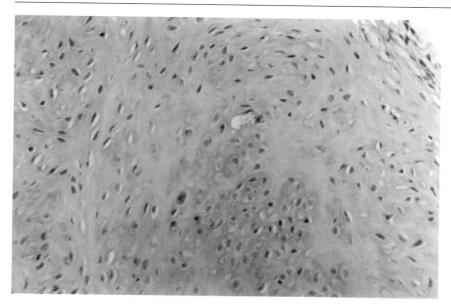

Fig. 97.1. Experimental cartilagenous structure retrieved from an a thymic mouse 12 weeks after initial injection; hematoxylin–eosin stain. Homogeneous plate of cartilage is evident with lacunar formation in which chondrocytes are enclosed. Reduced from original magnification of 20×.

harvested from hyaline cartilage obtained from the auricular surfaces of each miniswine. The isolated cells were grown and expanded *in vitro* for 5–8 weeks. Chondrocyte cell suspensions in sodium alginate were prepared and stored at 32°C until use. All four minipigs underwent unilateral endoscopic injection of an autologous chondrocyte–alginate suspension that was evoked to initiate gelification just prior to use by the addition of calcium sulfate. A total of approximately 2–3 ml of chondrocyte–alginate suspension (40–60×10^6 chondrocytes) was delivered to the subureteral region of each animal via a 22-gauge cystoscopic needle. The opposite ureteral orifice remained untreated and served as an internal control. Serial evaluation with cystoscopy, cystography, and excretory urography was performed at 2, 4, and 6 months after treatment. Cystoscopy showed a smooth bladder wall in all animals throughout the study period. Cystography demonstrated complete resolution of reflux on the treated side of all animals, while reflux persisted on the untreated (control) side. Excretory urography failed to demonstrate hydroureteronephrosis in any of the animals, suggesting the lack of significant obstruction related to the newly formed cartilage. At the time of sacrifice, gross examination of the bladder injection site showed a well-defined, rubbery-to-hard cartilage structure in the subureteral region. The presence of cartilage in all the retrieved implants was confirmed with hematoxylin and eosin staining, as well as aldehyde fuchsin–Alcian blue staining for chondroitin sulfate. Histological examination of the bladder failed to demonstrate any inflammation near the injection sites. In addition, there was no evidence of chondrocyte or alginate migration nor granuloma formation in any of the retrieved organs, including bladder, ureters, lymph nodes, kidneys, lungs, liver, and spleen.

Clinical trials for the use of this substance in humans have been approved by the U.S. Food and Drug Administration and are under way. Diamond and Caldamone have reported the preliminary results of a study directed toward the correction of vesicoureteral reflux [12]. Twenty-nine children with varying grades of reflux underwent subureteric endoscopic injection of autologous chondrocytes into 50 ureters. The chondrocytes were grown from a small cartilage specimen harvested from each patient's left ear at the time of initial diagnostic cystoscopy. Three months after 60% of the ureters were free of reflux after a single injection, as demonstrated by radionuclide cystogram. Overall, 79% of the ureters were cured at 3 months after one or two injections. These initial results suggest that the use of autologous chondrocytes for the endoscopic correction of reflux is both safe and effective. Long-term studies directed at confirming these results are in progress (Fig. 97.2).

Phase I clinical trials implementing this technology for the endoscopic correction of urinary incontinence due to intrinsic sphincteric deficiency are also in progress. *In vitro* expanded autologous chondrocytes combined with the alginate copolymer are injected sub-

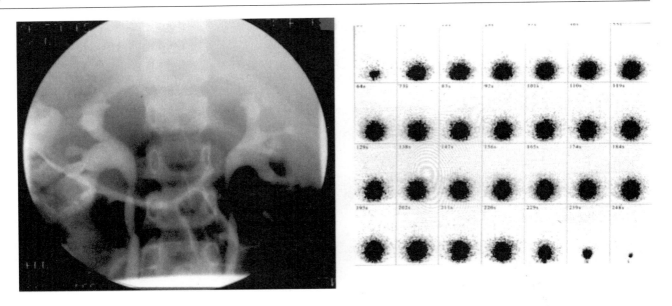

Fig. 97.2. (Left) Cystogram of a patient before endoscopic treatment with autologous chondrocytes showing bilateral reflux. (Right) Cystograms of a patient 3 months after endoscopic treatment with autologous cells showing no further evidence of reflux.

mucosally into the proximal urethra. The implant remains within the submucosa and results in increased urethral resistance to the passive outflow of urine. Clinical results reported in 1999 by Bent *et al.* showed this technique also to be safe and effective for the treatment of women with intrinsic sphincter deficiency [13]. Autologous chondrocytes were injected endoscopically to occlude the bladder neck in 32 patients. Six months after a single injection, 17 women were dry and 10 showed marked improvement, for an overall success rate of 84%.

PREPARATION OF ALGINATE POLYMER

Dry alginate impression powder is used as a delivery vehicle for autologous chondrocytes harvested from a limited biopsy under sterile conditions (see Chapter 96). In pure form, alginate is packaged with calcium phosphates and sulfates that control the gelation reaction. A 2% w/v sodium alginate suspension (0.1 M potassium phosphate, 0.135 M sodium chloride, pH 7.4) is made and sterilized in ethylene oxide. A 1.5-ml aliquot of 40×10^6 cells/ml chondrocyte suspension is added to an equal volume of the alginate solution for a final concentration of alginate of 1%. The cell–alginate suspension is maintained at 32°C until injection. Immediately prior to injection, 0.2 g/ml calcium sulfate is added to initiate the gelation process. The chondrocyte–alginate suspension spiked with calcium sulfate is vortexed and stored on ice until application. Gelation is a relatively slow process, thus the chondrocyte–alginate suspension remains in a liquid state for approximately 40 min.

AUTOLOGOUS BLADDER MUSCLE CELLS

The success with chondrocyte injection encouraged the investigation of another possible cell type for use as an autologous transplantable bulking agent. Bladder smooth muscle cells seemed ideal; besides being nonimmunogenic, derived from the bladder itself, these cells would retain their normal program of differentiation and act as a native tissue once transplanted *in situ*. To explore the feasibility of using these cells, a human bladder smooth muscle cell–alginate gel complex was injected subcutaneously in athymic nude/nude mice [14]. A 22-gauge needle was used to deliver 5 million cultured cells to each of the experimental injection sites in a total volume of 0.5 ml of gel. Injections of alginate alone, or bladder muscle cells alone, served as controls. Histological examination of the injection areas up to 2 months after delivery demonstrated evidence of *de novo* muscle formation.

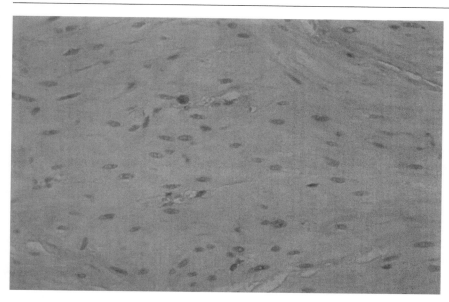

Fig. 97.3. Autologous bladder smooth muscle cell bioimplant retrieved from the treated subureteral region of a Hanford minipig demonstrating the presence of viable smooth muscle cells; hematoxylin–eosin stain. Reduced from original magnification of 20×.

Analysis of each implant revealed the progressive replacement of alginate with muscle over time, with the size of each implant remaining uniform and stable. Immunohistochemical staining with anti-desmin antibodies suggested that the cells maintained a normal program of muscular differentiation. There was no evidence of muscle growth in either of the control groups (alginate alone or muscle cells alone).

Further *in vivo* experiments were performed in 5 Hanford minipigs to determine whether reflux could be corrected by the endoscopic delivery of the bladder muscle cell–alginate gel complex [15]. As in the chondrocyte experiment, bilateral reflux was created in each animal by unroofing the intravesical ureters in open surgery. Three months postoperatively, the presence of bilateral reflux was confirmed by cystography. Muscle cells were harvested from each animal, then cultured and expanded *in vitro* by means of tissue engineering techniques [16]. Prior to injection, the autologous cells of each animal were pooled in a muscle cell–alginate suspension to a final concentration of 20×10^6 cells/ml. Each pig was then injected unilaterally with its own cells in the subureteral region, using a 21-gauge needle through a cystoscope. Follow-up cystograms demonstrated complete resolution of reflux on the treated side of all animals, while reflux persisted in the untreated, contralateral control ureter. Excretory urography confirmed the absence of ureteral obstruction in the treated side. At the time of sacrifice, histological examination revealed viable smooth muscle cells present in the subureteric region (Fig. 97.3). It appeared that the injected muscle cells replaced the alginate copolymer over time as observed earlier. The volume of the implant, however, remained stable. Examination of distant organs failed to demonstrate migration of bladder muscle cells or alginate, or evidence of granuloma formation. The bladder muscle cell–alginate complex, like the chondrocyte–alginate complex, corrected reflux and was nonimmunogenic, nonmigratory, and volume stable. The same system can be applied toward the endoscopic correction of urinary incontinence.

AUTOLOGOUS PRIMARY MYOBLASTS

The potential use of injectable, cultured myoblasts for the treatment of stress urinary incontinence has been investigated in preliminary experiments by Yokoyama *et al.* [17] and Huard *et al.* [18]. Primary myoblasts obtained from mouse skeletal muscle were transduced *in vitro* to carry the β-galactosidase reporter gene and were then incubated with fluorescent microspheres, which would serve as markers for the original cell population. Cells were directly injected into the proximal urethra and lateral bladder walls of nude mice with a microsyringe in an open surgical procedure. Tissue was harvested up to 35 days postinjection, analyzed histologically, and assayed for β-galactosidase expression. Myoblasts expressing β-galactosidase and containing fluorescent microspheres were found at each of the retrieved time points. In addition, regenerative myofibers expressing β-galactosidase were identified within the bladder wall. By 35 days postinjection, some of the injected cells expressed the contractile filament smooth muscle α-actin, suggesting the possibility of myoblastic differentiation into smooth muscle. The authors reported that a significant portion of the injected myoblast population persisted *in vivo*. The survival of the cells and their apparent commencement of the process of myogenic differentiation, further support the feasibility of using cultured cells of muscular origin as an injectable bioimplant.

Similar investigations were performed with autologous primary skeletal muscle cells in an effort to determine the feasibility of using skeletal muscle as a source of myogenic precursors. The authors demonstrated significantly greater persistence of submucosal bulk following injection of skeletal muscle cells than when bovine collagen was used [19].

AUTOLOGOUS FAT CELLS

The rationale for the use of autologous fat as a bulking agent is based on the successful use of autologous fat free flap transplants by cosmetic surgeons. Variable results are reported owing to loss of tissue bulk over time, yet it appears that smaller volumes of injectable adipocytes are more likely to be successful [20]. A limited number of attempts have been made to use autologous fat cells as a bulking agent both experimentally and clinically. However, these investigations have been followed by disappointing results that militate against the widespread clinical application of adipocytes in the treatment of reflux and incontinence.

Canning *et al.* have found that autologous fat injected submucosally into rabbit bladders remains viable for up to 11 months, however significant absorption does occur [21]. Furthermore, experimental data in rabbits demonstrated that injectable adipocytes can provide sufficient resistance at the bladder neck to permit consideration of their use for the treatment of incontinence [22]. However, histological analysis following injection of autologous adipocytes has identified inflammation at the implantation site [23]. As inflammation subsides, tissue fibrosis prevails rather than permanence of the injectable autologous adipocytes. Ultimately, only approximately 10% of injected adipocytes survive, the remaining bulk being composed of postinflammatory fibrosis [24]. Similar disappointing results were obtained following injection of autologous adipocytes into rat bladder neck submucosa.

Clinical investigations with autologous adipocytes injected as a bulking agent to treat reflux or urinary incontinence reflect the unstable nature of the implant. Trials of injectable autologous fat have been wholly unsuccessful in the treatment of vesicoureteral reflux, either primary or secondary to neurogenic bladder, thus leading many to abandon its use in this condition. Clinical studies have also attempted to use autologous fat injections for the treatment of incontinence with initially good results. In general, initial cure rates range from approximately 30 to 50% after one or two injections, however most patients relapse after 36 months. Results of a prospective trial comparing autologous fat injection to periurethral collagen injection, published in 1997, demonstrated a twofold greater improvement in incontinence among patients receiving collagen [25].

Endoscopic injection of autologous adipocytes into the vocal cords of patients with dysphonia has achieved notable success. In a dog model, Woo *et al.* implanted fat cells into Reinke's pouch and noted that 6 weeks following implantation, sufficient vocal cord bulk was present to allow phonation [26]. Clinical studies in patients with severe dysphonia have similarly supported the hope that a cell-based endoscopic approach will benefit patients with even severely scarified cords [27].

CONCLUSIONS

The endoscopic approach of using bulking agents for several medical conditions is attractive at several levels. In general, these procedures are shorter, more cost-effective, and more easily performed than their open counterparts, often in an outpatient setting. These factors are of particular significance in the management of vesicoureteral reflux, where treatment decisions are often dictated by the balance between the morbidity of corrective surgery versus the uncertainty of spontaneous resolution. A significant reduction in the former would very likely shift the balance in favor of surgical intervention in selected patients [28]. In addition, many patient who suffer from urinary incontinence are elderly and/or unfit for major surgery; these procedures, which can be performed under local anesthesia, afford them the opportunity to receive a much-desired treatment for which they would have otherwise been excluded. Though currently available bulking agents have been shown to be effective, they have had significant drawbacks in terms of material safety and anatomical integrity. The development of an injectable substance that demonstrated long-term clinical efficacy without the risk of particle migration or bioincompatibility would make these therapies more appealing to both clinicians and patients. The application of tissue engineering techniques to the development of such a substance is logical. A tissue-engineered bioimplant would be autologous, regarded as self, and therefore nonimmunogenic and completely biocompatible. Autologous therapies are safe, and to date there have been no reported cases of disease transmission. Initial experimental results with autologous chondrocytes and muscle cells have been promising. The long-term clinical results of these tissue-engineered injectable substances, once obtained, will require critical review. Perhaps the near future will indicate which bulking agent(s) are the most efficacious and safest for the treatment of vesicoureteral reflux and/or incontinence.

REFERENCES

1. Malizia, A. A., Jr., Reiman, H. M., Myers, R. P., Sande, J. R., Barham, S. S., Benson, R. C., Jr., Dewanjee, M. K., and Utz, W. J. (1984). Migration and granulomatous reaction after periurethral injection of Polytef (Teflon). *JAMA, J. Am. Med. Assoc.* **251**, 3277–3281.
2. Mittleman, R. E., and Marraccini, J. V. (1983). Pulmonary Teflon granulomas following periurethral Teflon injection for urinary incontinence. *Arch. Pathol. Lab. Med.* **107**, 611–612.
3. Ang, L. P., Tay, K. P., Lim, P. H., and Chang, H. C. (1997). Endoscopic injection of collagen for the treatment of female urinary stress incontinence. *Int. J. Urol.* **3**, 254–258.
4. Cukier, J., Beauchamp, R. A., Spindler, J. S., Spindler, S., Lorenzo, C., and Trentham, D. E. (1993). Association between bovine collagen dermal implants and a dermatomyositis or a polymyositis-like syndrome. *Ann. Intern. Med.* **118**, 920–928.
5. Atala, A. (1998). Tissure engineering in urologic surgery. *Urol. Clin. North Am.* **25**, 39–50.
6. Klagsbrun, M. (1979). Large-scale preparation of chondrocytes. *Methods Enzymol.* **58**, 560–564.
7. Guo, J. F., Jourdian, G. W., and MacCallum, D. K. (1989). Culture and growth characteristics of chondrocytes encapsulated in alginate beads. *Connect. Tissue Res.* **19**, 277–297.
8. Atala, A., Cima, L. G., Kim, W., Paige, K. T., Vacanti, J. P., Retik, A. B., and Vacanti, C. A. (1993). Injectable alginate seeded with chondrocytes as a potential treatment for vesicoureteral reflux. *J. Urol.* **150**, 745–747.
9. Yasin, M., Holland, S. J., Jolly, A. M., and Tighe, B. J. (1989). Polymers for biodegradable medical devices. VI. Hydroxybutyrate–hydroxyvalerate copolymers: Accelerated degradation of blends with polysaccharides. *Biomaterials* **10**, 400–412.
10. Atala, A., Kim, W., Paige, K. T., Vacanti, C. A., and Retik, A. B. (1994). Endoscopic treatment of vesicoureteral reflux with a chondrocyte–alginate suspension. *J. Urol.* **152**, 641–643.
11. Atala, A., Peters, C. A., Retik, A. B., and Mandell, J. (1992). Endoscopic treatment of vesicoureteral reflux with a self-detachable balloon system. *J. Urol.* **148**, 724–727.
12. Diamond, D. A., and Caldamone, A. A. (1998). Endoscopic treatment of vesicoureteric reflux in children using autologous chondrocytes—preliminary results. *Pediatrics* **102**, 107A.
13. Bent, A. E., Tutrone, R. T., Lloyd, L. K., Badlani, G., and Kennelly, M. J. (1999). Treatment of intrinsic sphincter deficiency using autologous ear cartilage as a periurethral bulking agent. *Continence Forum*, Denver, CO, 1999.
14. Atala, A., Cilento, B. G., Paige, K. T., and Retik, A. B. (1994). Injectable alginate seeded with human bladder muscle cells as a potential treatment for vesicoureteral reflux. *J. Urol.* **151**, 362A.
15. Cilento, B. G., and Atala, A. (1995). Treatment of reflux and incontinence with autologous chondrocytes and bladder muscle cells. *Dial. Pediatr. Urol.* **18**, 11.
16. Atala, A., Freeman, M. R., Vacanti, J. P., Shepard, J., and Retik, A. B. (1993). Implantation in vivo and retrieval of artificial structures consisting of rabbit and human urothelium and human bladder muscle. *J. Urol.* **150**, 608–612.
17. Yokoyama, T., Chancellor, M. B., Watanabe, T., Ozawa, H., Yoshimura, N., de Groat, W. C., Qu, Z., and Huard, J. (1999). Primary myoblasts injection into the urethra and bladder as a potential treatment of stress urinary incontinence and impaired detrusor contractility; long term survival without significant cytotoxicity. *J. Urol.* **161**, 307.
18. Huard, J., Yokoyama, T., Lavelle, J., Teahan, S., Watanabe, T., Ozawa, H., Yoshimura, N., de Groat, W. C., Qu, Z., and Chancellor, M. B. (1999). Differentiation of primary myoblast injection into the lower urinary tract; creation of detrusor cellular myoplasty. *J. Urol.* **161**, 66.
19. Yokoyama, T., Hauck, J., and Chancellor, M. B. (2000). Myoblast therapy for stress urinary incontinence and bladder dysfunction World. *J. Urol.* **18**(1), 56.
20. Canning, D. A. (1995). The future of fat tissue transplantation for the correction of incontinence or vur in children. *Dial. Pediatr. Urol.* **18**, 11.

21. Mathews, R. D., Christensen, J. P., and Canning, D. A. (1994). Persistence of autologous free fat transplant in bladder submucosa of rats. *J. Urol.* **152**(92, pt. 2), 819.

22. Canning, D. A., Seibold, J., Saito, M., *et al.* (1995). Use of injectable fat to obstruct the urethra in rabbits. *Neurourol. Urodyn.* **14**, 259.

23. Nguyen, A., Pasyk, K. A., Bouvier, T. N., Hassett, C. A., and Argenta, L. K. (1990). Comparative study of survival of autologous adipose tissue taken and transplanted by different techniques. *Plast. Reconstr. Surg.* **83**(3), 378.

24. Olson, M. E., Morck, D. W., Ceri, H., Lee, C. C., and Chancellor, M. B. (1998). Evaluation of autologous fat implantation in the rat urinary bladder submucosa. *Urology* **52**(2), 915.

25. Haab, F., Zimmern, P. E., and Leach, G. E. (1997). Urinary stress incontinence due to intrinsic sphincteric deficiency: Experience with fat and collagen periurethral injections. *J. Urol.* **157**, 1287.

26. Woo, P., Rahbar, R., and Wang, Z. (1999). Fat implantation into Reinke's space: A histologic and stroboscopic study in the canine. *Ann. Otol. Rhinol. Laryngol.* **108**(8), 738.

27. Sataloff, R. T., Spiegel, J. R., and Hawkshaw, M. J. (1997). Vocal fold scar. *Ear Nose Throat J.* **76**(11), 776.

28. Atala, A., and Casale, A. J. (1990). Management of primary vesicoureteral reflux. *Inf. Urol.* **2**, 39.

CHAPTER 98

MYOBLAST TRANSPLANTATION

Daniel Skuk and Jacques P. Tremblay

The *in vivo* implantation of normal or genetically modified myogenic cells, following culture and expansion *in vitro*, is referred as "myoblast transplantation" or "myoblast transfer therapy." In its original concept, myoblast transplantation is an experimental approach to treat severe genetic muscle diseases. Myoblasts are injected in the affected muscles to fuse with mature fibers, incorporating the normal genome to correct the disease. In addition, implantation of myoblasts into muscle and other organs seems to be promising for the treatment of nonmyopathic diseases. The chapter reviews the techniques currently used for myoblast transplantation research and proposes a scope of the techniques that could be used in humans.

SEARCHING FOR A TREATMENT OF GENETIC MYOPATHIES

Genetic myopathies are characterized by muscle weakness determined by molecular defects. The severity of muscle weakness depends on the particular entity, ranging from severe hypotonia at birth to moderate muscle weakness beginning in adult life. Some myopathies cause death in the early childhood, while others do not significantly compromise life expectancy. The most frequent genetic myopathy, and the principal target of myoblast transplantation (MT), is Duchenne muscular dystrophy (DMD). DMD is characterized by progressive muscle wasting beginning in early childhood, the patient becoming nonambulatory around the age of 10, and dying by the age of 20. DMD is caused by a deficiency of dystrophin, a subsarcolemmal fibrillar protein implicated in the connection between the intracellular contractile proteins and the extracellular basal lamina [1]. To date, there is no rational treatment for most genetic myopathies. Although gene therapy strategies seem promising [2], they cannot provide hope for patients at advanced stages of disease, since lost muscle fibers cannot be restored by replacing simply genes. On the other hand, MT proposes the intramuscular implantation of normal or genetically corrected myogenic cells [3]. This therapeutic approach allows the introduction of the normal genome and also new myogenic cells into wasted muscles. This is the only possibility by which new muscle fibers can be made or existing ones enlarged, thus increasing strength in myopathic patients.

MUSCLE CHARACTERISTICS CONDITIONING MT
MUSCLE STRUCTURE

As for other tissues, understanding the specific characteristics of muscle is fundamental to the design of a strategy of muscle engineering *in vivo*. Skeletal muscles are composed of long syncytia (myofibers), highly specialized to generate mechanical work. Myofibers are bundles of contractile filaments enveloped by a membrane (the sarcolemma), with a system allowing the conversion of the electrical stimuli into changes in the myofiber length. Myofibers can reach many centimeters in length and are disposed parallel to each other in a connective tissue matrix. Individual myofibers are grouped into fascicles and are joined by a delicate network of connective tissue, the endomysium. Fascicles are joined by the perimysium, and the fascicle ensemble is surrounded by the epimysium. This connective network ensures the coordinated mechanical work of the muscle. Thus, muscle tissue engineering must be capable of forming long contractile syncytia, parallel and longitudinally disposed into an organized matrix of connective tissue. Moreover, the muscles must be innervated by motor neurons and may contain proprioceptive elements.

THE SATELLITE CELL

Since myofibers are highly specialized syncytia, they cannot enter mitosis. For muscle tissue engineering, a precursor myogenic cell capable of proliferating and of later differentiating into mature myofibers is needed. During embryonic development, myofibers are formed by the fusion of mononucleated precursors called myoblasts. In postnatal life, muscle growth during the hypertrophic phase is done by mononucleated precursor cells that are present in the periphery of myofibers [4,5]. These precursor cells (named "satellite cells") remain at the periphery of myofibers throughout life. They occupy a depression on the mature fiber, lying between the sarcolemma and the basal lamina (Fig. 98.1). Satellite cells can be enzymatically isolated, and they can be proliferated *in vitro*, maintaining their capacity to fuse into myotubes and to differentiate into myofibers. Moreover, cultured myogenic cells (called also "myoblasts") can be reintroduced in muscles, where they can be integrated into the host myofibers [6].

MUSCLE REGENERATION

For MT purposes, myoblasts introduced into muscles must fuse with preexisting myofibers or must form new myofibers. Because mature myofibers do not automatically incorporate new progenitor cells, myoblasts introduced into a muscle do not necessarily fuse with

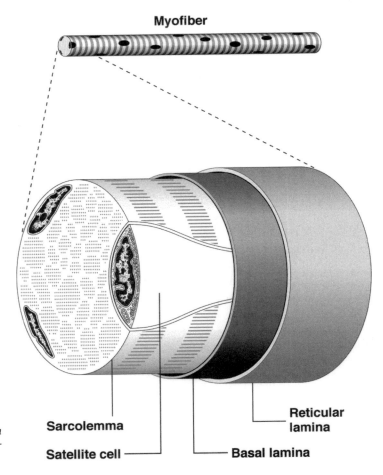

Fig. 98.1. Schematic representation of the location of a satellite cell relative to the myofiber and the extracellular matrix.

myofibers if not stimulated to do it. A mechanism that efficiently allows the incorporation of myogenic precursors into mature myofibers is muscle regeneration [7]. After total or segmental necrosis, satellite cells are drawn out from their quiescent state, becoming activated to proliferate, increasing in number, and fusing to form myotubes in a process that mimics muscle histogenesis in the embryo. Myotubes synthesize contractile proteins and develop into a mature myofiber. Preservation of basement membrane and endomysium is important to provide a scaffold for the regenerating fibers [8], maintaining the parallel orientation and links necessaries for a coordinated mechanical performance. By this mechanism, activated satellite cells can fill the defect produced by segmental necrosis or produce a new fiber after total necrosis.

ANIMAL MODELS FOR MT STUDIES
MOUSE MODELS

Although many MT experiments can be performed in normal mice or rats, myopathic mice provide good models for studying the capacity of MT to correct muscle diseases. Spontaneous mutants of C57BL mice [9], *mdx* mice lack dystrophin as a result of a point mutation in the dystrophin gene [10]. They were extensively used for MT experiments [11–18], although progressive weakness developing only late in life hinders the functional evaluation of the MT treatment. The *dy/dy* mouse, a model for congenital dystrophy with merosin absence [19], exhibits severe and progressive dystrophy and was also a target of MT experiments [20,21]. The possibility of producing knockout mice for any known protein associated with a myopathy [22,23] opens a large spectrum of possibilities for MT experiments. Severe combined immunodeficient (SCID) mice, which are useful recipients for xenotransplantation, provide an important tool for readily testing *in vivo* the myogenic capacity of myoblasts from humans or large animals [24,25]. Implantation of myoblasts from myopathic patients in SCID mice can also be a model for the creation of myopathic human muscle *in vivo* [26].

LARGE ANIMALS

The conditions present in mouse MT are not representative of those of humans [27], the size of the muscles being at least one obvious difference. Indeed, skeletal muscle regeneration may be different in mice and humans [28]. For this reason, experiments in large animals are crucial to develop strategies applicable to patients. Dystrophin-deficient dogs exhibit histological and clinical features similar to those of DMD patients [29,30], providing an ideal model for MT experiments [31,32]. Nevertheless, MT experiments in dogs have thus far have shown limited results [25,32]. Monkeys are important models for MT [33–36] because their transplantation immunology is similar to that of humans [37] and because their muscle size and anatomy are more appropriate for defining MT techniques applicable to humans [34].

CULTURING AND TRANSPLANTING MUSCLE CELLS
MYOBLAST PRODUCTION
Primary Cultures
Mice

Most mouse myoblast cultures in our laboratory are from newborns 1–3 days old. Myogenic cells are released from the limb muscles by a serial enzyme treatment modified from Cossu *et al.* [38]. After newborn mice have been killed, their limbs are collected, the skin is peeled away, and the muscles are dissected and chopped. Thereafter, muscles are submitted to a 1-h digestion at 37°C in 600 U/ml collagenase, followed by a 30-min incubation at 37°C in Hanks's Balanced Salt Solution (HBSS) containing 0.1% trypsin. The cell suspension obtained after both digestions is plated on culture flasks, with the growth medium (M-199) supplemented with 15% fetal bovine serum.

Although very efficient MT can be done with cells obtained this way [11,12,39], primary cultures are composed predominantly by fibroblasts [40–42]. Preplating is a simple method that takes advantage of the different attachment timings of fibroblasts and myoblasts to eliminate fibroblasts. The original method of Yaffe [43] consists of placing the cell suspension into a culture flask for 40–60 min, after which the medium containing cells that did not plate is collected by aspiration and pooled into another culture flask. Most cells attached to the substratum during that period are fibroblasts and epithelial cells; most myoblasts remain unattached. Nonmyogenic cells are removed from the culture with several preplates [40–44].

Large Animals

In monkeys and dogs, primary cultures are obtained from muscle biopsys samples taken at open surgery. Our muscles of choice are the quadriceps femoris in monkeys and the triceps femoris in dogs, because it is easy to obtain a long but narrow sample, providing abundant tissue without too much transverse damage to the muscle fibers. Samples are chopped and digested for 1 h at 37°C in collagenase followed by 30 min at 37°C in HBSS containing 0.1% trypsin. After dissociation, cells are grown in MCDB 120 medium [45] with 15% fetal calf serum and 10 ng/ml basic fibroblast growth factor (bFGF). The percentage of myogenic cells (detected by the expression of *NKH-1*) in those primary cultures is normally high: 90–100% in monkeys [34] and 50–100% in dogs [25].

Cell Lines

Since the proliferative capacity of primary myoblasts is finite, some MT studies use established myogenic cell lines such as C2 in mice [46–48] and L6 in rats [49,50]. Nevertheless, established myogenic cell lines may form tumors *in vivo* [46,51,52]. To generate myogenic cell lines with the ability to proliferate indefinitely *in vitro* while maintaining the capacity of normal differentiation *in vivo*, some studies use myoblasts isolated from the H-$2K^b$-$tsA58$ transgenic mice [20,53–55]. These mice carry the thermolabile *tsA58* mutant of SV40 large-T antigen under the control of the H-$2K^b$ promoter [56]. The thermolabile *tsA58* protein is functionally active at 33°C (but not a 39°C), and the promoter is activated by interferon γ, those being the permissive conditions for the indefinite proliferation of these cells *in vitro*. Nevertheless, some observations suggest that the success of transplanting T-antigen immortalized cells can be limited by a higher mortality rate of these cells (unpublished observations).

MYOBLAST IMPLANTATION
Intramuscular Myoblast Implantation

Direct injection into muscles is by far the most common means of introducing cultured myoblasts into muscles.

Muscles of Choice

The tibialis anterior is the muscle currently used for mouse MT in our laboratory: it is a medium-sized muscle, easily identified, close and external to the anterior edge of the tibia, superficially located under a thin skin, and easy to collect by identifying the distal tendon. The extensor digitorum longus, a fusiform thin muscle with two tendons, is ideal for force measuring *in vitro* [57] and for special techniques like orthotopic muscle autografting [57,58]. The soleus is also used by other authors [40,59–61]. In monkeys, our muscles of choice are the biceps brachii and the quadriceps femoris. The biceps brachii is easy to delimit in surface and can be entirely injected percutaneously. In dogs, we choose the deltoid, because it is one of the more severe affected muscles in dystrophic dogs [30]. Also, it is easy to identify, close to the clavicle, and allows one to work with the animal lying laterally. Kornegay *et al.* [32] used the cranial sartorius for MT experiments because it is also markedly involved in dystrophic dogs.

Myoblast Injection

For transplantation, cells are detached from culture flasks by means of 0.1% trypsin in HBSS, followed by three rinses in HBSS. The final cell pellet is suspended in HBSS into a small tube and stocked on ice until transplantation. Some authors propose including bovine serum albumin in the saline to keep cells before MT improved their viability [42,62]. Cells can be sedimented by centrifugation before stocking on ice, or immediately before transplantation. The number of myoblasts injected into a mouse muscle varies with the need of the study, being normally between 5×10^5 and 5×10^6. For injection, the supernatant is removed and the final pellet resuspended in 10–20 μl of HBSS. Considering the small size of mouse muscles, we perform cell injections using a glass micropipette with a 50-μm tip. Injections can be performed percutaneously or following exposure of the muscle belly by opening the skin.

For MT in large animals, a Hamilton syringe of 50–100 μl allows precision delivery of small quantities of cell suspension (5–10 μl) through a long trajectory of injection (≥1 cm). In monkeys, MT into an entire muscle can be performed percutaneously. For small volumes of muscle, MT can be also performed percutaneously, identifying the site of injection by marks in the skin (two superficial incisions comprising the site ensure indelible marks), or by an open surgery, identifying the site between two nonabsorbable sutures. We prefer the last method for dogs, because the skin and subcutaneous tissue are thicker than in monkeys. If a precise interinjection distance is critical for an experiment, a sterile transparent dressing with a grid should be placed on the skin.

Other Routes of Myoblast Implantation
Injection of Myoblasts into the Bloodstream
Direct injection into the bloodstream was proposed to reach muscles that are relatively inaccessible for direct injection, such as the diaphragm, either by direct intra-aortal infusion [63] or by extracorporeal circulation [50]. Nevertheless, the efficacy of this method is very limited.

Extramuscular Myoblast Implantation
Myoblasts can be implanted under the skin to create ectopic muscles [52].

IMMUNOLOGICAL ASPECTS OF MYOBLAST IMPLANTATION
The major histocompatibility complex (MHC) is expressed in myoblasts and myotubes [64–66] and, after MT, in myofibers [61,67]. Consequently, cellular rejection is present following myoblast allotransplantation in nonimmunosuppressed animals [15,60, 61,68–70]. Thus, the immunology of MT must be always considered in MT experiments. Autologous [58] or syngenic [17,71] transplantation avoids rejection. Although syngenic male MT into female in nonimmunosuppressed mice was frequently used [40,59,72] some results suggest that antigens linked to the Y chromosome are enough to trigger rejection [71]. Rejection can also be avoided by using SCID [24,57] or nude mice [18] as hosts. Immunosuppression may also be required for allotransplantation or in autologous and syngenic transplantation of transfected myoblasts (depending on the antigenicity of the transgene proteins for the last case). Not all immunosuppressive agents are equally effective for MT experiments [73,74]. Cyclophosphamide is not a good agent for MT, probably because of its cytotoxic effects on the transplanted cells [75]. Cyclosporine A effectively controlled rejection in MT mouse experiments at 50 mg/kg administered by daily injections [14,60,61]. Rapamycin [74] was effective for MT mouse experiments when injected daily intraperitoneally at 5 mg/kg. Rapamycin is more effective than cyclosporine A at controlling the humoral immune response [73,74]. According to our experiments, the best agent for mouse MT is FK506 [39], administered intramuscularly once a day at 2.5 mg/kg. In monkeys, FK506 is administered intramuscularly once a day at 0.3–0.5 mg/kg [33,34], and its efficacy was confirmed up to one year after MT [34]. Lower FK506 doses do not control cellular rejection in monkeys [33]. FK506 toxicity is more prominent in dogs than in other species, and for this reason, we used low daily oral doses of FK506 (0.3 mg/kg) combined with RS-61443 (20 mg/kg) and cyclosporine A (20 mg/kg) [25]. This protocol provided better immunosuppression for dog MT than FK506 alone or RS-61443 combined with cyclosporine A [25].

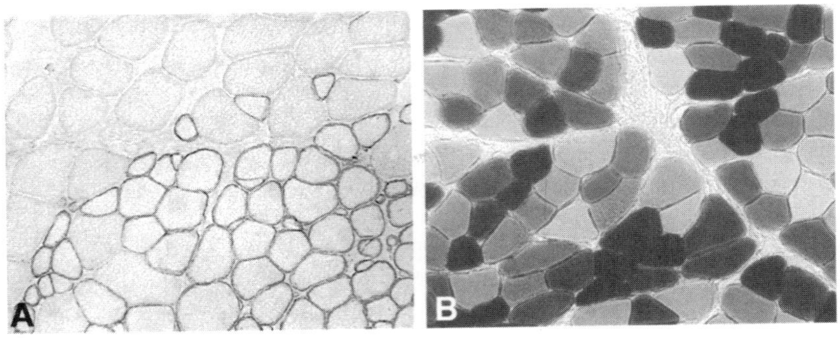

Fig. 98.2. Identification of fibers that incorporate transplanted myoblasts: (A) dystrophin expression in an mdx mouse muscle after injection of normal myoblasts and (B) β-Gal expression in a monkey muscle injected with LacZ-transgenic myoblasts.

MONITORING THE SUCCESS OF MT
ANIMAL MODELS
Histological Methods

The parameter most frequently used to quantify MT success is the number of myofibers that incorporate the transplanted myoblasts, observed in muscle cross sections. When the host is deficient in a protein present in the normal transplanted myoblasts, MT success can be evaluated by the immunohistochemical detection of this protein. Transplantation of normal myoblasts leads to the formation of dystrophin-positive fibers in *mdx* muscles (Fig. 98.2) [11,16,48,76], as well as merosin-positive fibers in *dy/dy* mice [20,21]. Myofibers formed by the fusion of human myoblasts implanted in SCID mice can be identified with antibodies that recognize epitopes present in human dystrophin but not in mouse dystrophin [24,26,77]. If normal animals are used as MT hosts, the injected myoblasts must be labeled. The most common label is β-galactosidase (β-Gal), using cells genetically modified by incorporating the LacZ gene. When LacZ-expressing myoblasts form new myofibers or fuse with preexisting myofibers, β-Gal is detected histochemically into the sarcoplasm (Fig. 98.2). LacZ-expressing mouse myoblasts can be obtained from transgenic animals [78]. Otherwise, myoblasts obtained from normal animals can be transfected *in vitro* with the LacZ gene [25,33,36]. To detect β-Gal, the muscle sections are fixed in 0.25% glutaraldehyde for 5 min, followed by overnight incubation at room temperature in a solution containing 0.4 mM 5-bromo-4-chloro-3-indolyl-β-D-galactopyranoside (from a 20 mg/ml stock solution in dimethylformamide), 3 mM potassium ferrocyanide, 3 mM potassium ferricyanide, and 1 mM magnesium chloride in phosphate-buffered saline at pH 7.4. Transplantation of male myoblasts into females allows us to identify the hybrid fibers by *in situ* hybridization with Y-chromosome-specific DNA probes [40,79].

Molecular and Biochemical Methods

MT success can be quantified by measuring in host muscle homogenates the expression of a transgene or a normal gene present exclusively in the donor myoblasts. Partridge's team evaluates the success of MT by injecting myoblasts from *Gpi1-s^b* allotype mice, which produced the BB isoenzyme of dimeric glucose 6-phosphate isomerase (GPI), into *Gpi1-s^a* allotype mice, which produced the AA isoenzyme [18,46,80,81]. Rando *et al.* [62] analyzed β-Gal activity in muscles transplanted with β-Gal-expressing myoblasts. Assays by means of the polymerase chain creation (PCR) quantify the expression of donor dystrophin in *mdx* muscles injected with normal myoblasts [13,14].

Combined Methods

The same muscle can be analyzed by different methods by sectioning the frozen sample in a cryostat. Serial sections are collected alternatively onto glass slides for histology and in a tube for biochemical or molecular analysis [26,62].

Physiological Methods

Muscles injected with myoblasts can be studied functionally, to evaluate whether MT modifies the capacity of muscles to generate mechanical work. We measure the contractile properties of the tibialis anterior in mice anesthetized with 80 mg/kg pentobarbitone sodium, administered intraperitoneally. With the animal placed on a temperature-controlled platform, an incision is made over the experimental muscle, and the distal tendon is sectioned and attached to a hook linked to a force transducer. The motor nerve is isolated and stimulated with a pair of cuff electrodes. Both muscle and nerve are bathed regularly with saline at $\sim 37°C$. Muscle length is adjusted until the force developed during a single twitch reaches a maximum and the optimum muscle length (L_o) is defined. Subsequent measurements of isometric force are made with the muscles at L_o. The maximum isometric force (P_o) is measured at a stimulation frequency between 60 and 120 Hz and a train duration of 300 ms. Other force measuring methods for MT experiments can be found in other studies [40,57,58,60].

DYSTROPHIC HUMAN PATIENTS

Tests similar to those used in animal models were also applied during earlier clinical trials [67,82–88].

Histological Evaluation

MT clinical trials quantified dystrophin-positive fibers in muscle cross sections. Nevertheless, the presence of revertant fibers in DMD patients [89] can be a cause of misinterpretation [90]. In revertant fibers, a second mutation has restored the in-frame traduction of dystrophin, allowing the synthesis of a modified but functional protein that is detected in the sarcolemma. To avoid confusion with revertant fibers, the presence of donor dystrophin must be immunohistochemically demonstrated with antibodies that recognize exons absent from the patient genome [84,85,90]. Another way of detecting the fusion of donor myoblasts with the host fibers is to use fluorescence *in situ* hybridization (FISH) to detect the presence of donor nuclei [91,92]. For this, the patients must have characterized deletions in the dystrophin gene, enabling the design of DNA probes that hybridize to donor nuclei only [92].

Molecular Methods

Since dystrophin is not expressed in myoblasts but in myotubes and myofibers, the presence of donor-derived dystrophin mRNA in muscle homogenates was used to monitor the fusion of transplanted normal myoblasts into the muscles of DMD patients [67,82,83]. Immunoblot analysis of dystrophin was also done in some clinical trials [67,83,86,88].

Functional Methods

The capacity of MT to improve strength is the fundamental issue of an MT clinical trial. For measurements of voluntary strength, placebo injection of saline in the contralateral muscle by a double-blind procedure is necessary to avoid subjective conditioning. In earlier studies, we evaluated static voluntary strength for elbow flexion and extension after MT in the biceps brachii four times before and four times after MT [93]. Measurements were made with a computer-controlled dynamometer on patients who were seated, with shoulders and trunk stabilized, one arm being attached to a support. Force was recorded during maximal isometric elbow flexions and extensions, with the elbow at 90°. Variants in the protocol of force measuring can be found in other clinical studies [67,84–86].

IMPROVING THE SUCCESS OF MT

INDUCTION OF MUSCLE NECROSIS

Inducing necrosis in the host tissue where myoblasts are implanted increases the number of regenerating fibers that can incorporate the injected myoblasts.

Injection of Myotoxic Substances

Specific Myotoxins

Notexin is a potent myotoxic phospholipase purified from the venom of the Australian tiger snake *Notechis scutatus scutatus* [94,95]. Notexin induces the necrosis of the skeletal muscle fibers, preserving the satellite cells, nerves, and vascular elements [96]. This specific effect allows complete and rapid muscle regeneration [97]. In mice, we use notexin at

5–10 μg/ml, diluted in HBSS. Normally, 10–20 μl is injected in a tibialis anterior with a glass micropipette, 24 h before MT [24,39]. In large muscles of dogs and monkeys, it is better to inject myoblasts resuspended in the notexin solution, to ensure that the implanted myoblasts are placed exactly in the areas of necrosis [32,33]. Cardiotoxin, another myotoxin purified from snake venoms, is used to trigger a degeneration–regeneration process in muscles [98–100].

Local Anesthetics

Local anesthetics cause myonecrosis after intramuscular injection [101]. Bupivacaine hydrochloride was used to induce muscle degeneration–regeneration in some MT experiments, injected at 0.5% before [102] or concomitantly [49] to MT.

Orthopic Muscle Autotransplantation

Orthotopic muscle autotransplantation is an experimental method in which the extensor digitorum longus is completely removed from its bed and returned to its original site before MT, producing devascularization and necrosis [58].

INHIBITION OF HOST MYOGENIC CELL PROLIFERATION

Inhibiting the participation of host myogenic cells in the process of muscle regeneration [103] improves the participation of the transplanted myoblasts [104]. The most frequent method for this is host muscle irradiation with a single high dose of X [26,55,58,59] or gamma rays [20,39,76]. An electron accelerator (X-ray) is preferred to a cobalt machine (gamma-ray), because the former has the narrower penumbra. This is a critical factor when the small posterior limbs of mice are being irradiated, since a wide penumbra exposes the lower abdomen to important quantities of radiation.

COMBINATION OF NECROSIS WITH HOST CELL INHIBITION

Irradiation and notexin are frequently combined for mouse MT [24,26,39]. Alternatively, freezing (cryodamage) destroys all muscle cells, combining the induction of muscle necrosis with the elimination of host satellite cells [40,48,60]. Irintchev *et al.* [40] apply the flat end of a copper rod precooled in liquid nitrogen onto the surface of the soleus, midway between the tendons. Myoblasts are implanted immediately after thawing. However, killing blood vessel cells, fibroblasts, and intramuscular nerves can lead to less efficient regeneration than the use of specific myotoxins [105].

MYOBLAST STIMULATION *In Vitro*

The addition of basic fibroblast growth factor to the growth medium quadrupled MT success in mice [12]. To obtain this effect, primary cultured myoblasts were incubated in medium containing 100 ng/ml human recombinant basic fibroblast growth factor for 48 h before MT. Adding 20 μg/ml concanavalin A to the growth medium 48 h before MT increased more than three fold the area of dispersion of the injected myoblasts [106]. This effect was attributed to an increase in myoblast migration, since concanavalin A increases the synthesis of collagenase [107] and plasminogen activator [108], while reducing the expression of the tissue inhibitor metalloproteinases [109].

FOLLOWING THE EARLY FATE OF IMPLANTED MYOBLASTS

The early behavior of myoblasts implanted into muscles is an important issue in MT research. Many experiments were undertaken to analyze the survival of cells after implantation [41,54,55,72,110–113], as well as the capacity of these cells to migrate [72,106,114].

DETECTION OF MARKERS IN MUSCLE HOMOGENATES

The intensity of a label present in the transplanted myoblasts and detected in host muscle homogenates allows an estimation of the quantity of transplanted cells present in the host tissue.

β-Galactosidose Assay

Frequently the percentage of transplanted cells remaining at different periods post-MT was evaluated by quantifying β-Gal activity in muscles implanted with LacZ myoblasts [41,54,62,110,111]. The muscles are homogenized in 800 μl of Tris–hydroxymethylamine buffer, pH 8.0, followed by a 5-min centrifugation at 6500 rpm. From the supernatant, 100 μl is taken and mixed with 3 μl of a solution 0.1 M MgCl$_2$ and 4.5 M β-mercaptoethanol plus 66 μl of a solution of 4 mg/ml *o*-nitrophenyl-β-D-galactopyranoside diluted in 0.1 M sodium phosphate, pH 7.5. Finally, 131 μl of the same buffer is added. The samples are incubated for 30 min at 37°C, and the reaction is stopped by adding 150 μl of 1 M Na$_2$CO$_3$. The optical density is read at 420 nm on a spectrophotometer.

Y Chromosome

Since the level of β-Gal expression can be affected (e.g., promoter shut off) by external factors [115], other quantitative methods were also used for myoblast quantification. Male myoblasts implanted into muscles of females can be quantified by detecting the Y chromosome in muscle homogenates [55,112,116]. We developed a quantitative competitive PCR-based assay, modifying a method described earlier [117] by introducing a competitor plasmid containing the priming sites for a sequence of the mouse Y chromosome [116].

Radioactive Labels

Detection of β-Gal or Y chromosome quantifies not only surviving cells but also cell proliferation. In contrast, if myoblasts are labeled with [^{14}C]thymidine, the marker will be divided equally between daughter cells during proliferation and will not increase with cell proliferation [55,112]. Radiolabeling of myoblasts is performed by adding 0.25 μCi/ml methyl-[^{14}C]thymidine to the growth medium for 16 h. The radiolabeled myoblasts are implanted in muscles, and these muscles are dissected after different periods of time. To measure the amount of radiolabel, the muscles are digested for 16 h at 50°C in 50 mM Tris–HCl, 0.1 M EDTA, 0.1 M NaCl, pH 8.0, containing 0.5% sodium dodecyl sulfate and 500 μg/ml proteinase K. An aliquot is mixed with 10 ml of a liquid scintillation counting cocktail, and the radiolabel is measured in a counter system. Beauchamp *et al.* [55,112] combined Y-chromosome and radiolabel detection to evaluate both cell death and cell proliferation in a given muscle.

DETECTING IMPLANTED CELLS IN MUSCLE SECTIONS

To identify the donor myoblasts in the host tissue, labels must be detected in the mononuclear cells. This excludes proteins that are expressed only after fusion, such as dystrophin (for transplantation in *mdx* mice) or β-Gal under the control of promoters activated after differentiation (i.e., Tn*I*-LacZ myoblasts [78]). Methods of identifying the implanted myoblasts include transgene myoblasts carrying the LacZ gene under constitutive promoters [54,113], myoblasts stained with PKH26 (a fluorescent dye that incorporates into the membrane of cells) [114], and fluorescent latex microspheres incorporated into myoblasts [118]. Transplantation of male myoblasts into females allows their identification by *in situ* hybridization with Y-chromosome-specific DNA probes [72].

IN VITRO GENETIC MODIFICATION OF MYOBLASTS

The possibility of proliferating myogenic cells in culture allows the genetic manipulation of these cells *in vitro* before transplantation. Incorporation of foreign genes (as β-Gal) is a useful tool for identifying the transplanted cells in basic research. The genetic correction of autologous myoblasts *ex vivo* is proposed to avoid the immune response against allogeneic myoblasts, thus avoiding the complications of immunosuppression in human patients. In this case, it was demonstrated that both the human dystrophin minigene and the full-length dystrophin gene can be transferred with adenoviral vectors to *mdx* and DMD myoblasts, and these corrected myoblasts were successfully transplanted respectively in *mdx* and SCID mice [81,119,120]. Myoblasts can also be vehicles for the delivery of factors and hormones for therapeutic purposes [121]. Myoblasts can be genetically manipulated to obtain systemic secretion of substances like coagulation factor IX [122,123], erythropoietin [124–126], growth hormone [127], proinsulin [128], and granulocyte colony stimulating factor 1 [129]. Transgenic myoblasts can also deliver factors locally, such as antiarthritic proteins into joints [130] and neurotrophic factors in the central nervous system [131].

PERSPECTIVES FOR HUMAN USE

Developing MT strategies for human applications implies an existing notion of the objective to be accomplished and the best method of achieving this objective. MT aims to improve the muscle function of severe myopathic patients to ensure acceptable quality of life. For DMD, the question is: How many dystrophin-positive fibers must be produced by MT in a muscle to improve the condition of these patients? Although speculative, an estimation could be obtained by the observations in woman DMD carriers, who exhibit a mosaic of dystrophin-positive and dystrophin-negative fibers and either have mild myopathy or are asymptomatic. Asymptomatic carriers reportedly exhibited 81–97% dystrophin-positive fibers in samples from muscle biopsies [132]. Mild symptomatic DMD carriers

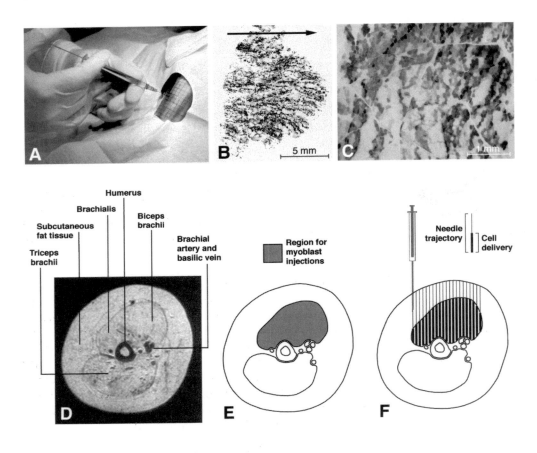

Fig. 98.3. MT experiments in monkeys (A–C) help to identify the potential conditions for MT in humans. For homogeneous incorporation of donor myoblasts through an entire monkey biceps brachii (A), close parallel injections are performed (a transparent dressing with a grid is used to control the distance between injections), and small quantities of cell suspension are homogeneously delivered during the needle withdrawal (5–10 μl/cm). (B) Cross section of a biceps brachii injected this way with β-Gal-labeled myoblasts illustrates how ß-Gal-positive fibers are disposed in parallel bands following the pattern of injections (arrow indicates orientation of injections). Biopsies performed in these muscles exhibit 50–75% β-Gal-positive fibers (C), values that may be therapeutic if reproduced as dystrophin-positive fibers in a DMD muscle (see text). Injecting many human muscles by this method is a technical challenge that should be resolved by developing an automated device coordinated by image scanning. We envision a procedure of injections in three steps (D–F). First, to identify the anatomical structures, the limb is fixed and scanned, giving a magnetic resonance image of a DMD patient's arm as shown (D). Second, the operator selects the region to be injected with myoblasts (E). Third, a robot system performs injections, delivering cells in the selected regions (brachialis and biceps brachii in this case), without reaching the humerus, brachial artery, and basilic vein, and avoiding loss of donor-cells in the subcutaneous tissue (F).

exhibited 68–82% [133] and 51–85% [132] dystrophin-positive fibers. Finally, DMD-like girls presented 22 and 6% dystrophin-positive fibers [132]. These observations show that the severity of myopathy depends on the percentage of dystrophin-positive fibers. Patients with more than 80% of dystrophin-positive fibers either are asymptomatic or develop mild symptoms as they get older. With 50–80% of dystrophin-positive myofibers, there is mild muscle weakness. These last percentages are a realistic goal for MT, since we demonstrated that it is possible to obtain 50–70% of β-Gal-positive myofibers after injecting β-Gal-labeled myoblasts in the whole-biceps brachii of a monkey (Fig. 98.3) [34]. Nevertheless, the only way to ensure the fusion of transplanted myoblasts in at least 50% of the fibers is to distribute them by multiple close injections throughout the entire muscle. Injecting myoblasts in a whole large muscle by close injections takes much time and requires great precision to completely inject all regions of some muscles. Moreover, to provide functional motility to a patient, many groups of muscles must be injected in the limbs, trunk, and neck. The complexity of this task is evident when one remembers that skeletal muscle tissue accounts for 50% of the total body weight. Considering the topographical precision needed, it appears logical that the use of an automated device would be more effective than manual injections. The ideal device whould be a robot able to perform rapid and precise parallel close injections. Injections could be coordinated by a system of image scanning, able to coordinate the injection trajectories with the muscle anatomy. This automated technique would avoid cell loss in extramuscular tissues, damage to bones, and lesions of peripheral nerves and large vessels (Fig. 98.3).

On the other hand, MT as a strategy for secreting hormones or factors into the bloodstream needs only to incorporate a relatively small number of "secreting" myoblasts into the host tissue; there is no requirement for precise distribution throughout the whole muscle. Therefore, the injection of myoblasts genetically engineered to secrete substances could be performed manually, by multiple injections in a reduced volume of muscle.

References

1. Watkins, S. C., Cullen, M. J., Hoffman, E. P., and Billington, L. (2000). Plasma membrane cytoskeleton of muscle: A fine structural analysis. *Microsc. Res. Tech.* **48**, 131–141.
2. Hartigan-O'Connor, D., and Chamberlain, J. S. (2000). Developments in gene therapy for muscular dystrophy. *Microsc. Res. Tech.* **48**, 223–238.
3. Skuk, D., and Tremblay, J. P. (2000). Progress in myoblast transplantation: a potential treatment of dystrophies. *Microsc. Res. Tech.* **48**, 213–222.
4. Ishikawa, H. (1966). Electron microscopic observations of satellite cells with special reference to the development of mammalian skeletal muscles. *Z. Anat. Entwicklungsgesch.* **125**, 43–63.
5. Church, J. C. (1969). Satellite cells and myogenesis; a study in the fruit-bat web. *J. Anat.* **105**, 419–438.
6. Lipton, B. H., and Schultz, E. (1979). Developmental fate of skeletal muscle satellite cells. *Science* **205**, 1292–1294.
7. Grounds, M. D. (1999). Muscle regeneration: Molecular aspects and therapeutic implications. *Curr. Opin. Neurol.* **12**, 535–543.
8. Vracko, R., and Benditt, E. P. (1972). Basal lamina: The scaffold for orderly cell replacement. Observations on regeneration of injured skeletal muscle fibers and capillaries. *J. Cell Biol.* **55**, 406–419.
9. Bulfield, G., Siller, W. G., Wight, P. A., and Moore, K. J. (1984). X chromosome-linked muscular dystrophy (*mdx*) in the mouse. *Proc. Natl. Acad. Sci. U.S.A.* **81**, 1189–1192.
10. Sicinski, P., Geng, Y., Ryder-Cook, A. S., Barnard, E. A., Darlison, M. G., and Barnard, P. J. (1989). The molecular basis of muscular dystrophy in the *mdx* mouse: A point mutation. *Science* **244**, 1578–1580.
11. Kinoshita, I., Vilquin, J. T., Guerette, B., Asselin, I., Roy, R., Lille, S., and Tremblay, J. P. (1994). Immunosuppression with FK 506 insures good success of myoblast transplantation in MDX mice. *Transplant. Proc.* **26**, 3518.
12. Kinoshita, I., Vilquin, J. T., and Tremblay, J. P. (1995). Pretreatment of myoblast cultures with basic fibroblast growth factor increases the efficacy of their transplantation in *mdx* mice. *Muscle Nerve* **18**, 834–841.
13. Asselin, I., Tremblay, M., Vilquin, J. T., Guerette, B., and Tremblay, J. P. (1994). Polymerase chain reaction–based assay to assess the success of myoblast transplantation in *mdx* mice. *Transplant. Proc.* **26**, 3389.
14. Asselin, I., Tremblay, M., Vilquin, J. T., Guerette, B., Roy, R., and Tremblay, J. P. (1995). Quantification of normal dystrophin mRNA following myoblast transplantation in *mdx* mice. *Muscle Nerve* **18**, 980–986.
15. Guerette, B., Asselin, I., Vilquin, J. T., Roy, R., and Tremblay, J. P. (1995). Lymphocyte infiltration following allo- and xenomyoblast transplantation in *mdx* mice. *Muscle Nerve* **18**, 39–51.
16. Partridge, T. A., Morgan, J. E., Coulton, G. R., Hoffman, E. P., and Kunkel, L. M. (1989). Conversion of *mdx* myofibres from dystrophin-negative to -positive by injection of normal myoblasts. *Nature* (London) **337**, 176–179.
17. Vilquin, J. T., Wagner, E., Kinoshita, I., Roy, R., and Tremblay, J. P. (1995). Successful histocompatible myoblast transplantation in dystrophin-deficient *mdx* mouse despite the production of antibodies against dystrophin. *J. Cell Biol.* **131**, 975–988.

18. Morgan, J. E., Pagel, C. N., Sherratt, T., and Partridge, T. A. (1993). Long-term persistence and migration of myogenic cells injected into pre-irradiated muscles of *mdx* mice. *J. Neurol. Sci.* **115**, 191–200.

19. Sunada, Y., Bernier, S. M., Kozak, C. A., Yamada, Y., and Campbell, K. P. (1994). Deficiency of merosin in dystrophic *dy* mice and genetic linkage of laminin M chain gene to *dy* locus. *J. Biol. Chem.* **269**, 13729–13732.

20. Vilquin, J. T., Kinoshita, I., Roy, B., Goulet, M., Engvall, E., Tome, F., Fardeau, M., and Tremblay, J. P. (1996). Partial laminin alpha2 chain restoration in alpha2 chain-deficient *dy/dy* mouse by primary muscle cell culture transplantation. *J. Cell Biol.* **133**, 185–197.

21. Vilquin, J. T., Guerette, B., Puymirat, J., Yaffe, D., Tome, F. M., Fardeau, M., Fiszman, M., Schwartz, K., and Tremblay, J. P. (1999). Myoblast transplantations lead to the expression of the laminin alpha 2 chain in normal and dystrophic (*dy/dy*) mouse muscles. *Gene Ther.* **6**, 792–800.

22. Hack, A. A., Ly, C. T., Jiang, F., Clendenin, C. J., Sigrist, K. S., Wollmann, R. L., and McNally, E. M. (1998). gamma-Sarcoglycan deficiency leads to muscle membrane defects and apoptosis independent of dystrophin. *J. Cell Biol.* **142**, 1279–1287.

23. Duclos, F., Straub, V., Moore, S. A., Venzke, D. P., Hrstka, R. F., Crosbie, R. H., Durbeej, M., Lebakken, C. S., Ettinger, A. J., van der Meulen, J., Holt, K. H., Lim, L. E., Sanes, J. R., Davidson, B. L., Faulkner, J. A., Williamson, R., and Campbell, K. P. (1998). Progressive muscular dystrophy in alpha-sarcoglycan-deficient mice. *J. Cell Biol.* **142**, 1461–1471.

24. Huard, J., Verreault, S., Roy, R., Tremblay, M., and Tremblay, J. P. (1994). High efficiency of muscle regeneration after human myoblast clone transplantation in SCID mice. *J. Clin. Invest.* **93**, 586–599.

25. Ito, H., Vilquin, J. T., Skuk, D., Roy, B., Goulet, M., Lille, S., Dugre, F. J., Asselin, I., Roy, R., Fardeau, M., and Tremblay, J. P. (1998). Myoblast transplantation in non-dystrophic dog. *Neuromuscular. Disord.* **8**, 95–110.

26. Skuk, D., Furling, D., Bouchard, J. P., Goulet, M., Roy, B., Lacroix, Y., Vilquin, J. T., Tremblay, J. P., and Puymirat, J. (1999). Transplantation of human myoblasts in SCID mice as a potential muscular model for myotonic dystrophy. *J. Neuropathol. Exp. Neurol.* **58**, 921–931.

27. Grounds, M. D. (1996). Commentary on the present state of knowledge for myoblast transfer therapy. *Cell Transplant.* **5**, 431–433.

28. Borisov, A. B. (1999). Regeneration of skeletal and cardiac muscle in mammals: Do nonprimate models resemble human pathology? *Wound Repair Regeneration* **7**, 26–35.

29. Cooper, B. J. (1989). Animal models of Duchenne and Becker muscular dystrophy. *Br. Med. Bull.* **45**, 703–718.

30. Valentine, B. A., Cooper, B. J., Cummings, J. F., and de Lahunta, A. (1990). Canine X-linked muscular dystrophy: Morphologic lesions. *J. Neurol. Sci.* **97**, 1–23.

31. Bartlett, R. J., Sharp, N. J., Hung, W. Y., Kornegay, J. N., and Roses, A. D. (1990). Molecular markers for myoblast transplantation in GRMD. *Adv. Exp. Med. Biol.* **280**, 273–278.

32. Kornegay, J. N., Prattis, S. M., Bogan, D. J., Sharp, N. J. H., Bartlett, R. J., Alameddine, H. S., and Dykstra, M. J. (1992). Results of myoblast transplantation in a canine model of muscle injury. *In* "Duchenne Muscular Dystrophy: Animal Models and Genetic Manipulation" (B. A. Kakulas, J. M. Howell, and A. D. Roses, eds.), pp. 203–212. Raven Press, New York.

33. Skuk, D., Roy, B., Goulet, M., and Tremblay, J. P. (1999). Successful myoblast transplantation in primates depends on appropriate cell delivery and induction of regeneration in the host muscle. *Exp. Neurol.* **155**, 22–30.

34. Skuk, D., Goulet, M., Roy, B., and Tremblay, J. P. (2000). Myoblast transplantation in whole muscle of nonhuman primates. *J. Neuropathol. Exp. Neurol.* **59**, 197–206.

35. Kinoshita, I., Vilquin, J. T., Gravel, C., Roy, R., and Tremblay, J. P. (1995). Myoblast allotransplantation in primates. *Muscle Nerve* **18**, 1217–1218.

36. Kinoshita, I., Roy, R., Dugre, F. J., Gravel, C., Roy, B., Goulet, M., Asselin, I., and Tremblay, J. P. (1996). Myoblast transplantation in monkeys: Control of immune response by FK506. *J. Neuropathol. Exp. Neurol.* **55**, 687–697.

37. Hamawy, M. M., and Knechtle, S. J. (1998). Strategies for tolerance induction in nonhuman primates. *Curr. Opin. Immunol.* **10**, 513–517.

38. Cossu, G., Zani, B., Coletta, M., Bouche, M., Pacifici, M., and Molinaro, M. (1980). In vitro differentiation of satellite cells isolated from normal and dystrophic mammalian muscles. A comparison with embryonic myogenic cells. *Cell Differ.* **9**, 357–368.

39. Kinoshita, I., Vilquin, J. T., Guerette, B., Asselin, I., Roy, R., and Tremblay, J. P. (1994). Very efficient myoblast allotransplantation in mice under FK506 immunosuppression. *Muscle Nerve* **17**, 1407–1415.

40. Irintchev, A., Langer, M., Zweyer, M., Theisen, R., and Wernig, A. (1997). Functional improvement of damaged adult mouse muscle by implantation of primary myoblasts. *J. Physiol. (London)* **500**, 775–785.

41. Qu, Z., Balkir, L., van Deutekom, J. C., Robbins, P. D., Pruchnic, R., and Huard, J. (1998). Development of approaches to improve cell survival in myoblast transfer therapy. *J. Cell Biol.* **142**, 1257–1267.

42. Rando, T. A., and Blau, H. M. (1994). Primary mouse myoblast purification, characterization, and transplantation for cell-mediated gene therapy. *J. Cell Biol.* **125**, 1275–1287.

43. Yaffe, D. (1968). Retention of differentiation potentialities during prolonged cultivation of myogenic cells. *Proc. Natl. Acad. Sci. U.S.A.* **61**, 477–483.

44. Richler, C., and Yaffe, D. (1970). The in vitro cultivation and differentiation capacities of myogenic cell lines. *Dev. Biol.* **23**, 1–22.

45. Ham, R. G., St. Clair, J. A., Webster, C., and Blau, H. M. (1988). Improved media for normal human muscle satellite cells: Serum-free clonal growth and enhanced growth with low serum. *In Vitro Cell. Dev. Biol.* **24**, 833–844.

46. Morgan, J. E., Moore, S. E., Walsh, F. S., and Partridge, T. A. (1992). Formation of skeletal muscle in vivo from the mouse C2 cell line. *J. Cell Sci.* **102**, 779–787.

47. Watt, D. J., Karasinski, J., and England, M. A. (1993). Migration of *lacZ* positive cells from the tibialis anterior to the extensor digitorum longus muscle of the X-linked muscular dystrophic (*mdx*) mouse. *J. Muscle Res. Cell Motil.* **14**, 121–132.

48. Wernig, A., and Irintchev, A. (1995). "Bystander" damage of host muscle caused by implantation of MHC-compatible myogenic cells. *J. Neurol. Sci.* **130**, 190–196.

49. Pin, C. L., and Merrifield, P. A. (1997). Developmental potential of rat L6 myoblasts in vivo following injection into regenerating muscles. *Dev. Biol.* **188**, 147–166.

50. Torrente, Y., D'Angelo, M. G., Del Bo, R., DeLiso, A., Casati, R., Benti, R., Corti, S., Comi, G. P., Gerundini, P., Anichini, A., Scarlato, G., and Bresolin, N. (1999). Extracorporeal circulation as a new experimental pathway for myoblast implantation in *mdx* mice. *Cell Transplant.* **8**, 247–258.

51. Wernig, A., Irintchev, A., Hartling, A., Stephan, G., Zimmermann, K., and Starzinski-Powitz, A. (1991). Formation of new muscle fibres and tumours after injection of cultured myogenic cells. *J. Neurocytol.* **20**, 982–997.

52. Irintchev, A., Rosenblatt, J. D., Cullen, M. J., Zweyer, M., and Wernig, A. (1998). Ectopic skeletal muscles derived from myoblasts implanted under the skin. *J. Cell Sci.* **111**, 3287–3297.

53. Morgan, J. E., Beauchamp, J. R., Pagel, C. N., Peckham, M., Ataliotis, P., Jat, P. S., Noble, M. D., Farmer, K., and Partridge, T. A. (1994). Myogenic cell lines derived from transgenic mice carrying a thermolabile T antigen: A model system for the derivation of tissue-specific and mutation-specific cell lines. *Dev. Biol.* **162**, 486–498.

54. Guerette, B., Skuk, D., Celestin, F., Huard, C., Tardif, F., Asselin, I., Roy, B., Goulet, M., Roy, R., Entman, M., and Tremblay, J. P. (1997). Prevention by anti-LFA-1 of acute myoblast death following transplantation. *J. Immunol.* **159**, 2522–2531.

55. Beauchamp, J. R., Morgan, J. E., Pagel, C. N., and Partridge, T. A. (1999). Dynamics of myoblast transplantation reveal a discrete minority of precursors with stem cell–like properties as the myogenic source. *J. Cell Biol.* **144**, 1113–1122.

56. Jat, P. S., Noble, M. D., Ataliotis, P., Tanaka, Y., Yannoutsos, N., Larsen, L., and Kioussis, D. (1991). Direct derivation of conditionally immortal cell lines from an *H-2K^b-tsA58* transgenic mouse. *Proc. Natl. Acad. Sci. U.S.A.* **88**, 5096–5100.

57. Arcila, M. E., Ameredes, B. T., DeRosimo, J. F., Washabaugh, C. H., Yang, J., Johnson, P. C., and Ontell, M. (1997). Mass and functional capacity of regenerating muscle is enhanced by myoblast transfer. *J. Neurobiol.* **33**, 185–198.

58. Alameddine, H. S., Louboutin, J. P., Dehaupas, M., Sebille, A., and Fardeau, M. (1994). Functional recovery induced by satellite cell grafts in irreversibly injured muscles. *Cell Transplant.* **3**, 3–14.

59. Wernig, A., Zweyer, M., and Irintchev, A. (2000). Function of skeletal muscle tissue formed after myoblast transplantation into irradiated mouse muscles. *J. Physiol. (London)* **522**, 333–345.

60. Wernig, A., Irintchev, A., and Lange, G. (1995). Functional effects of myoblast implantation into histoincompatible mice with or without immunosuppression. *J. Physiol. (London)* **484**, 493–504.

61. Irintchev, A., Zweyer, M., and Wernig, A. (1995). Cellular and molecular reactions in mouse muscles after myoblast implantation. *J. Neurocytol.* **24**, 319–331.

62. Rando, T. A., Pavlath, G. K., and Blau, H. M. (1995). The fate of myoblasts following transplantation into mature muscle. *Exp. Cell Res.* **220**, 383–389.

63. Neumeyer, A. M., DiGregorio, D. M., and Brown, R. H., Jr. (1992). Arterial delivery of myoblasts to skeletal muscle. *Neurology* **42**, 2258–2262.

64. Mantegazza, R., Hughes, S. M., Mitchell, D., Travis, M., Blau, H. M., and Steinman, L. (1991). Modulation of MHC class II antigen expression in human myoblasts after treatment with IFN-gamma. *Neurology* **41**, 1128–1132.

65. Michaelis, D., Goebels, N., and Hohlfeld, R. (1993). Constitutive and cytokine-induced expression of human leuko-cyte antigens and cell adhesion molecules by human myotubes. *Am. J. Pathol.* **143**, 1142–1149.

66. Cifuentes-Diaz, C., Delaporte, C., Dautreaux, B., Charron, D., and Fardeau, M. (1992). Class II MHC antigens in normal human skeletal muscle. *Muscle Nerve* **15**, 295–302.

67. Karpati, G., Ajdukovic, D., Arnold, D., Gledhill, R. B., Guttman, R., Holland, P., Koch, P. A., Shoubridge, E., Spence, D., Vanasse, M., Watters, G. V., Abrahamowicz, M., Duff, C., and Worton, R. G. (1993). Myoblast transfer in Duchenne muscular dystrophy. *Ann. Neurol.* **34**, 8–17.

68. Guerette, B., Roy, R., Tremblay, M., Asselin, I., Kinoshita, I., Puymirat, J., and Tremblay, J. P. (1995). Increased granzyme B mRNA after alloincompatible myoblast transplantation. *Transplantation* **60**, 1011–1016.

69. Guerette, B., Tremblay, G., Vilquin, J. T., Asselin, I., Gingras, M., Roy, R., and Tremblay, J. P. (1996). Increased interferon-gamma mRNA expression following alloincompatible myoblast transplantation is inhibited by FK506. *Muscle Nerve* **19**, 829–835.

70. Pavlath, G. K., Rando, T. A., and Blau, H. M. (1994). Transient immunosuppressive treatment leads to long-term retention of allogeneic myoblasts in hybrid myofibers. *J. Cell Biol.* **127**, 1923–1932.

71. Boulanger, A., Asselin, I., Roy, R., and Tremblay, J. P. (1997). Role of non–major histocompatibility complex antigens in the rejection of transplanted myoblasts. *Transplantation* **63**, 893–899.

72. Fan, Y., Maley, M., Beilharz, M., and Grounds, M. (1996). Rapid death of injected myoblasts in myoblast transfer therapy. *Muscle Nerve* **19**, 853–860.

73. Vilquin, J. T., Asselin, I., Guerette, B., Kinoshita, I., Lille, S., Roy, R., and Tremblay, J. P. (1994). Myoblast allo-transplantation in mice: Degree of success varies depending on the efficacy of various immunosuppressive treatments. *Transplant. Proc.* **26**, 3372–3373.

74. Vilquin, J. T., Asselin, I., Guerette, B., Kinoshita, I., Roy, R., and Tremblay, J. P. (1995). Successful myoblast allotrans-plantation in *mdx* mice using rapamycin. *Transplantation* **59**, 422–426.

75. Vilquin, J. T., Kinoshita, I., Roy, R., and Tremblay, J. P. (1995). Cyclophosphamide immunosuppression does not permit successful myoblast allotransplantation in mouse. *Neuromuscular Disord.* **5**, 511–517.

76. Vilquin, J. T., Wagner, E., Kinoshita, I., Roy, R., and Tremblay, J. P. (1995). Successful histocompatible myoblast transplantation in dystrophin-deficient *mdx* mouse despite the production of antibodies against dystrophin. *J. Cell Biol.* **131**, 975–988.

77. Huard, J., Tremblay, G., Verreault, S., Labrecque, C., and Tremblay, J. P. (1993). Utilization of an antibody specific for human dystrophin to follow myoblast transplantation in nude mice. *Cell Transplant.* **2**, 113–118.

78. Kinoshita, I., Huard, J., and Tremblay, J. P. (1994). Utilization of myoblasts from transgenic mice to evaluate the efficacy of myoblast transplantation. *Muscle Nerve* **17**, 975–980.

79. Grounds, M. D., Lai, M. C., Fan, Y., Codling, J. C., and Beilharz, M. W. (1991). Transplantation in the mouse model—The use of a Y-chromosome-specific DNA clone to identify donor cells in situ. *Transplantation* **52**, 1101–1105.

80. Gross, J. G., and Morgan, J. E. (1999). Muscle precursor cells injected into irradiated *mdx* mouse muscle persist after serial injury. *Muscle Nerve* **22**, 174–185.

81. Moisset, P. A., Skuk, D., Asselin, I., Goulet, M., Roy, B., Karpati, G., and Tremblay, J. P. (1998). Successful transplantation of genetically corrected DMD myoblasts following ex vivo transduction with the dystrophin minigene. *Biochem. Biophys. Res. Commun.* **247**, 94–99.

82. Gussoni, E., Pavlath, G. K., Lanctot, A. M., Sharma, K. R., Miller, R. G., Steinman, L., and Blau, H. M. (1992). Normal dystrophin transcripts detected in Duchenne muscular dystrophy patients after myoblast transplantation. *Nature (London)* **356**, 435–438.

83. Huard, J., Bouchard, J. P., Roy, R., Malouin, F., Dansereau, G., Labrecque, C., Albert, N., Richards, C. L., Lemieux, B., and Tremblay, J. P. (1992). Human myoblast transplantation: Preliminary results of 4 cases. *Muscle Nerve* **15**, 550–560.

84. Mendell, J. R., Kissel, J. T., Amato, A. A., King, W., Signore, L., Prior, T. W., Sahenk, Z., Benson, S., McAndrew, P. E., Rice, R., Nagaraja, H., Stephens, R., Lantry, L., Morris, G. E., and Burghes, A. H. M. (1995). Myoblast transfer in the treatment of Duchenne's muscular dystrophy. *N. Engl. J. Med.* **333**, 832–838.

85. Miller, R. G., Sharma, K. R., Pavlath, G. K., Gussoni, E., Mynhier, M., Lanctot, A. M., Greco, C. M., Steinman, L., and Blau, H. M. (1997). Myoblast implantation in Duchenne muscular dystrophy: The San Francisco study. *Muscle Nerve* **20**, 469–478.

86. Neumeyer, A. M., Cros, D., McKenna-Yasek, D., Zawadzka, A., Hoffman, E. P., Pegoraro, E., Hunter, R. G., Munsat, T. L., and Brown, R. H., Jr. (1998). Pilot study of myoblast transfer in the treatment of Becker muscular dystrophy. *Neurology* **51**, 589–592.

87. Tremblay, J. P., Bouchard, J. P., Malouin, F., Theau, D., Cottrell, F., Collin, H., Rouche, A., Gilgenkrantz, S., Abbadi, N., Tremblay, M., *et al.* (1993). Myoblast transplantation between monozygotic twin girl carriers of Duchenne muscular dystrophy. *Neuromuscular Disord.* **3**, 583–592.

88. Morandi, L., Bernasconi, P., Gebbia, M., Mora, M., Crosti, F., Mantegazza, R., and Cornelio, F. (1995). Lack of mRNA and dystrophin expression in DMD patients three months after myoblast transfer. *Neuromuscular Disord.* **5**, 291–295.

89. Nicholson, L. V., Davison, K., Johnson, M. A., Slater, C. R., Young, C., Bhattacharya, S., Gardner-Medwin, D., and Harris, J. B. (1989). Dystrophin in skeletal muscle. II. Immunoreactivity in patients with Xp21 muscular dystrophy. *J. Neurol. Sci.* **94**, 137–146.

90. Partridge, T., Lu, Q. L., Morris, G., and Hoffman, E. (1998). Is myoblast transplantation effective? *Nat. Med.* **4**, 1208–1209.

91. Gussoni, E., Wang, Y., Fraefel, C., Miller, R. G., Blau, H. M., Geller, A. I., and Kunkel, L. M. (1996). A method to codetect introduced genes and their products in gene therapy protocols. *Nat. Biotechnol.* **14**, 1012–1016.

92. Gussoni, E., Blau, H. M., and Kunkel, L. M. (1997). The fate of individual myoblasts after transplantation into muscles of DMD patients. *Nat. Med.* **3**, 970–977.

93. Tremblay, J. P., Malouin, F., Roy, R., Huard, J., Bouchard, J. P., Satoh, A., and Richards, C. L. (1993). Results of a triple blind clinical study of myoblast transplantations without immunosuppressive treatment in young boys with Duchenne muscular dystrophy. *Cell Transplant.* **2**, 99–112.

94. Harris, J. B., and Johnson, M. A. (1978). Further observations on the pathological responses of rat skeletal muscle to toxins isolated from the venom of the Australian tiger snake, *Notechis scutatus scutatus*. *Clin. Exp. Pharmacol. Physiol.* **5**, 587–600.

95. Pluskal, M. G., Harris, J. B., Pennington, R. J., and Eaker, D. (1978). Some biochemical responses of rat skeletal muscle to a single subcutaneous injection of a toxin (notexin) isolated from the venom of the Australian tiger snake *Notechis scutatus scutatus*. *Clin. Exp. Pharmacol. Physiol.* **5**, 131–141.

96. Harris, J. B., and Cullen, M. J. (1990). Muscle necrosis caused by snake venoms and toxins. *Electron Microsc. Rev.* **3**, 183–211.

97. Sharp, N. J., Kornegay, J. N., Bartlett, R. J., Hung, W. Y., and Dykstra, M. J. (1993). Notexin-induced muscle injury in the dog. *J. Neurol. Sci.* **116**, 73–81.

98. Couteaux, R., Mira, J. C., and d'Albis, A. (1988). Regeneration of muscles after cardiotoxin injury. I. Cytological aspects. *Biol. Cell* **62**, 171–182.

99. d'Albis, A., Couteaux, R., Janmot, C., Roulet, A., and Mira, J. C. (1988). Regeneration after cardiotoxin injury of innervated and denervated slow and fast muscles of mammals. Myosin isoform analysis. *Eur. J. Biochem.* **174**, 103–110.

100. Kherif, S., Lafuma, C., Dehaupas, M., Lachkar, S., Fournier, J. G., Verdiere-Sahuque, M., Fardeau, M., and Alameddine, H. S. (1999). Expression of matrix metalloproteinases 2 and 9 in regenerating skeletal muscle: A study in experimentally injured and *mdx* muscles. *Dev. Biol.* **205**, 158–170.

101. Nonaka, I., Takagi, A., Ishiura, S., Nakase, H., and Sugita, H. (1983). Pathophysiology of muscle fiber necrosis induced by bupivacaine hydrochloride (Marcaine). *Acta Neuropathol.* **60**, 167–174.

102. Cantini, M., Massimino, M. L., Catani, C., Rizzuto, R., Brini, M., and Carraro, U. (1994). Gene transfer into satellite cell from regenerating muscle: Bupivacaine allows beta-Gal transfection and expression in vitro and in vivo. *In Vitro Cell. Dev. Biol.—Anim.* **30A**, 131–133.

103. Wakeford, S., Watt, D. J., and Partridge, T. A. (1991). X-irradiation improves *mdx* mouse muscle as a model of myofiber loss in DMD. *Muscle Nerve* **14**, 42–50.

104. Morgan, J. E., Hoffman, E. P., and Partridge, T. A. (1990). Normal myogenic cells from newborn mice restore normal histology to degenerating muscles of the *mdx* mouse. *J. Cell Biol.* **111**, 2437–2449.

105. Manor, D., and Sadeh, M. (1989). Muscle fibre necrosis induced by intramuscular injection of drugs. *Br. J. Exp. Pathol.* **70**, 457–462.

106. Ito, H., Hallauer, P. L., Hastings, K. E., and Tremblay, J. P. (1998). Prior culture with concanavalin A increases intramuscular migration of transplanted myoblast. *Muscle Nerve* **21**, 291–297.

107. Wang, H. M., Hurum, S., and Sodek, J. (1983). Con A stimulation of collagenase synthesis by human gingival fibroblasts. *J. Periodontal Res.* **18**, 149–155.

108. Vassalli, J. D., Hamilton, J., and Reich, E. (1977). Macrophage plasminogen activator: Induction by concanavalin A and phorbol myristate acetate. *Cell (Cambridge, Mass.)* **11**, 695–705.

109. Overall, C. M., and Sodek, J. (1990). Concanavalin A produces a matrix-degradative phenotype in human fibroblasts. Induction and endogenous activation of collagenase, 72-kDa gelatinase, and *Pump-1* is accompanied by the suppression of the tissue inhibitor of matrix metalloproteinases. *J. Biol. Chem.* **265**, 21141–21151.

110. Guerette, B., Asselin, I., Skuk, D., Entman, M., and Tremblay, J. P. (1997). Control of inflammatory damage by anti-LFA-1: Increase success of myoblast transplantation. *Cell Transplant.* **6**, 101–107.

111. Merly, F., Huard, C., Asselin, I., Robbins, P. D., and Tremblay, J. P. (1998). Anti-inflammatory effect of transforming growth factor-beta1 in myoblast transplantation. *Transplantation* **65**, 793–799.

112. Beauchamp, J. R., Pagel, C. N., and Partridge, T. A. (1997). A dual-marker system for quantitative studies of myoblast transplantation in the mouse. *Transplantation* **63**, 1794–1797.

113. Skuk, D., and Tremblay, J. P. (1998). Complement deposition and cell death after myoblast transplantation. *Cell Transplant.* **7**, 427–434.

114. Torrente, Y., El Fahime, E., Caron, N. J., Bresolin, N., and Tremblay, J. P. (2000). Intramuscular migration of myoblasts transplanted after muscle pretreatment with metalloproteinases. *Cell Transplant.* **9**, 539–549.

115. Qin, L., Ding, Y., Pahud, D. R., Chang, E., Imperiale, M. J., and Bromberg, J. S. (1997). Promoter attenuation in gene therapy: Interferon-gamma and tumor necrosis factor-alpha inhibit transgene expression. *Hum. Gene. Ther.* **8**, 2019–2029.

116. Caron, N. J., Chapdelaine, P., and Tremblay, J. P. (1999). Male-specific competitive PCR for the quantification of transplanted cells in mice. *BioTechniques* **27**, 424–426, 428.

117. Ogata, H., Bradley, W. G., Inaba, M., Ogata, N., Ikehara, S., and Good, R. A. (1995). Long-term repopulation of hematolymphoid cells with only a few hemopoietic stem cells in mice. *Proc. Natl. Acad. Sci. U.S.A.* **92**, 5945–5949.

118. Satoh, A., Huard, J., Labrecque, C., and Tremblay, J. P. (1993). Use of fluorescent latex microspheres (FLMs) to follow the fate of transplanted myoblasts. *J. Histochem. Cytochem.* **41**, 1579–1582.

119. Moisset, P. A., Gagnon, Y., Karpati, G., and Tremblay, J. P. (1998). Expression of human dystrophin following the transplantation of genetically modified *mdx* myoblasts. *Gene Ther.* **5**, 1340–1346.

120. Floyd, S. S., Jr., Clemens, P. R., Ontell, M. R., Kochanek, S., Day, C. S., Yang, J., Hauschka, S. D., Balkir, L., Morgan, J., Moreland, M. S., Feero, G. W., Epperly, M., and Huard, J. (1998). Ex vivo gene transfer using adenovirus-mediated full-length dystrophin delivery to dystrophic muscles. *Gene Ther.* **5**, 19–30.

121. Blau, H. M., and Springer, M. L. (1995). Muscle-mediated gene therapy. *N. Engl. J. Med.* **333**, 1554–1556.

122. Roman, M., Axelrod, J. H., Dai, Y., Naviaux, R. K., Friedmann, T., and Verma, I. M. (1992). Circulating human or canine factor IX from retrovirally transduced primary myoblasts and established myoblast cell lines grafted into murine skeletal muscle. *Somatic Cell Mol. Genet.* **18**, 247–258.

123. Wang, J. M., Zheng, H., Blaivas, M., and Kurachi, K. (1997). Persistent systemic production of human factor IX in mice by skeletal myoblast-mediated gene transfer: Feasibility of repeat application to obtain therapeutic levels. *Blood* **90**, 1075–1082.

124. Hamamori, Y., Samal, B., Tian, J., and Kedes, L. (1994). Persistent erythropoiesis by myoblast transfer of erythropoietin cDNA. *Hum. Gene Ther.* **5**, 1349–1356.

125. Hamamori, Y., Samal, B., Tian, J., and Kedes, L. (1995). Myoblast transfer of human erythropoietin gene in a mouse model of renal failure. *J. Clin. Invest.* **95**, 1808–1813.

126. Bohl, D., Naffakh, N., and Heard, J. M. (1997). Long-term control of erythropoietin secretion by doxycycline in mice transplanted with engineered primary myoblasts. *Nat. Med.* **3**, 299–305.

127. Dhawan, J., Pan, L. C., Pavlath, G. K., Travis, M. A., Lanctot, A. M., and Blau, H. M. (1991). Systemic delivery of human growth hormone by injection of genetically engineered myoblasts. *Science* **254**, 1509–1512.

128. Simonson, G. D., Groskreutz, D. J., Gorman, C. M., and MacDonald, M. J. (1996). Synthesis and processing of genetically modified human proinsulin by rat myoblast primary cultures. *Hum. Gene Ther.* **7**, 71–78.

129. Dhawan, J., Rando, T. A., Elson, S. E., Lee, F., Stanley, E. R., and Blau, H. M. (1996). Myoblast-mediated expression of colony stimulating factor-1 (CSF-1) in the cytokine-deficient op/op mouse. *Somatic Cell Mol. Genet.* **22**, 363–381.

130. Day, C. S., Kasemkijwattana, C., Menetrey, J., Floyd, S. S., Jr., Booth, D., Moreland, M. S., Fu, F. H., and Huard, J. (1997). Myoblast-mediated gene transfer to the joint. *J. Orthop. Res.* **15**, 894–903.

131. Deglon, N., Heyd, B., Tan, S. A., Joseph, J. M., Zurn, A. D., and Aebischer, P. (1996). Central nervous system delivery of recombinant ciliary neurotrophic factor by polymer encapsulated differentiated C2C12 myoblasts. *Hum. Gene Ther.* **7**, 2135–2146.

132. Di Blasi, C., Morandi, L., Barresi, R., Blasevich, F., Cornelio, F., and Mora, M. (1996). Dystrophin-associated protein abnormalities in dystrophin-deficient muscle fibers from symptomatic and asymptomatic Duchenne/Becker muscular dystrophy carriers. *Acta Neuropathol.* **92**, 369–377.

133. Bonilla, E., Schmidt, B., Samitt, C. E., Miranda, A. F., Hays, A. P., de Oliveira, A. B., Chang, H. W., Servidei, S., Ricci, E., Younger, D. S., *et al.* (1988). Normal and dystrophin-deficient muscle fibers in carriers of the gene for Duchenne muscular dystrophy. *Am. J. Pathol.* **133**, 440–445.

CHAPTER 99

SKELETAL MUSCLE

Jennifer J. Marler

T his chapter provides a cursory review of muscle cell biology, followed by a description of different tissue engineering approaches—myoblast transplantation for gene therapy, myoblast transplantation for structural reanimation of skeletal and cardiac muscle, and the development of three-dimensional engineered skeletal muscle from a combination of myoblasts and polymers. It is intended as an overview of the diverse methods used for skeletal muscle cells in tissue engineering rather than an exhaustive list of experiments.

MUSCLE CELL BIOLOGY

MUSCLE DEVELOPMENT

The formation of skeletal muscle during embryogenesis involves the commitment of mesodermal stem cells to the skeletal muscle lineage and the subsequent differentiation of embryonic myoblasts (muscle stem cells) to form multinucleate myotubes, which mature into skeletal muscle fibers. Skeletal muscle is derived from the somites, which form in a rostral-to-caudal gradient by segmentation of the paraxial mesoderm along the neural tube. These somites become compartmentalized to form the sclerotome, giving rise to the ribs and vertebrae, and the dermatomyotome, which divides into the myotome, giving rise to the axial muscles of the back, and the dermatome, forming the dermis of the axial back. While the axial muscles of the back are formed from the myotomal compartment of the somites, the limb muscles are derived from myogenic precursor cells that migrate from the ventrolateral edge of the dermatomyotome.

SKELETAL MUSCLE HISTOARCHITECTURE

Mature striated skeletal muscle consists of bundles of very long (≤4 cm) myofibers—cylindrical multinucleated cells with a diameter of 10–100 μm. Multinucleation results from the fusion of the embryonic mononucleated myoblasts. The oval nuclei are usually found at the periphery of the cell under the cell membrane.

The myofibers that constitute muscle are arranged in regular bundles surrounded by an epimysium—an external sheath of dense connective tissue. From the epimysium, thin septa of connective tissue, the perimysia, extend inward, dividing the myofibers into fascicles. Each muscle fiber is surrounded by an endomysium. Blood vessels penetrate the muscle with the connective tissue septa and form a rich capillary network that runs between and parallel to the muscle fibers. Lymphatics are found in the connective tissue.

The sarcoplasm of each muscle fiber is filled with long cylindrical filamentous bundles called myofibrils, which have a diameter of 1–2 μm, run parallel to the long axis of the muscle, and are composed of an end-to-end chainlike arrangement of sarcomeres. As a consequence of the arrangement of sarcomeres in adjacent myofibrils, the entire muscle fiber exhibits a characteristic pattern of transverse striations. Electron microscopic studies reveal that the sarcomere pattern is due to the presence of alternating thick and thin filaments that lie parallel to the long axis of the muscle in an alternating pattern. Thin filaments are composed primarily of actin, tropomyosin, and troponin. Thick filaments consist primarily of myosin. These filaments constitute the contractile apparatus of skeletal muscle. Other contractile proteins, such as α-actinin and β-actinin are also present, in smaller amounts.

To contract, each myofiber requires innervation by a motor nerve branch. Myelinated motor nerves branch out within the perimysial connective tissue, where each nerve gives rise to several terminal twigs. At the site of innervation, the nerve loses its myelin sheath and forms a dilated termination that sits within a trough at the myofiber's surface, forming a motor end plate or neuromuscular junction. A single nerve fiber may innervate one muscle fiber or may branch and innervate multiple fibers, up to 160 or more. A single nerve fiber and all the fibers it innervates are termed the motor unit. In vertebrates, each myofiber has only one single neuromuscular junction.

Muscle fibers are of two major physiological types. Type I fibers are slow twitch, and their contraction is dependent on aerobic metabolism. Type II fibers are fast twitch, and their contraction depends primarily on glycogenolytic (anaerobic) metabolism. There are many differences in gene expression between the two fiber types, including metabolic enzymes, membrane channel protein isoforms, and myofibrillar protein isoforms. Most skeletal muscle in humans contain mixtures of the two fiber types. Some fibers show intermediate properties and represent minor subtypes. Developmental studies using retroviral cell lineage marking have revealed that fetal myoblasts can migrate across basal laminae during development and can contribute nuclei to both fast and slow muscle fibers [1,2].

Myoblasts, also known as "satellite cells," remain closely associated with myofibers during the course of their life cycle. They lie nestled beneath the basement membrane, closely apposed to each fiber. If the fiber becomes injured, they remain capable of dividing and repopulating it [3–5]. They can be recruited locally at sites of focal muscle injury or from undamaged sites in the same muscle; myoblasts appear capable of migrating between adjacent regenerating muscles in a mouse model [6]. Thus, a source of "muscle stem cells" exists from fetal life through old age. These cells can be liberated from tissue by enzymatic digestion and grown *in vitro*.

Myoblast Tissue Culture

Myoblasts can be cultivated in tissue culture. Their behavior can be directed by the *in vitro* milieu. Under high-serum conditions, they continue to proliferate, up to 30 doublings [7]. Under low-serum conditions, they exit the cell cycle and fuse to form multinucleated myotubes that will become contractile.

There is evidence for functional heterogeneity of myoblasts. Myoblasts isolated from masseter muscles grow more slowly than those from the limb; the masseter muscle also regenerates much less effectively than limb muscle [8].

TISSUE ENGINEERING APPROACHES

To date, tissue engineering with myoblasts has fallen into three principal categories:

Myoblast transplantation for gene therapy (delivery of specific proteins) or for structural "reanimation" of skeletal muscle itself or of the myocardium
Transplantation of encapsulated myoblasts
Creation of three-dimensional tissue from cultured myoblasts and polymers (either solid scaffolds or hydrogels)

Myoblast Transplantation
Gene and Cell Therapy

Myoblasts are ideal cells for gene therapy for at least four reasons: (1) skeletal muscle is easily accessible for the delivery of cells or recombinant genes, making up 10% of the total human body mass, (2) myofibers are long-lived, providing a stable environment for long-term expression of recombinant transgenes, (3) myofibers are multinucleated, permitting delivery of multiple genes encoding products that can interact inside the cell, and (4) genetically altered skeletal muscle is efficient at producing and delivering recombinant gene products to the circulation [9].

The strategies for gene delivery into muscle fall into two categories: *in vivo* approaches, where a vector (either viral or nonviral) harboring a therapeutic transgene is introduced directly into muscle tissue, and *ex vivo* approaches, in which myoblasts isolated from tissue (either autologous or allogeneic), or genetically engineered *in vitro*, are introduced into a recipient muscle, where they become integrated into preexisting tissue. This chapter focuses on *ex vivo* approaches.

Demonstrations that myoblasts genetically manipulated in tissue culture could express biologically active peptides *in vivo* included experiments introducing genes for human growth hormone, human and canine recombinant factor IX, and human erythropoietin into C2C12 cells, a myoblast cell line [10–14]. In all cases, stable expression was noted in the serum of transplanted mice. A *lacZ* reporter gene introduced by retroviral transfection confirmed that the C2C12 cells formed muscle upon transplantation [12].

One potential shortcoming of introducing C2C12 cells is the formation of tumors by these cells in mice [15,16]. However, subsequent experiments have confirmed stable long-term expression of factor IX in engineered primary myoblasts [17]. Repeated transplantation of engineered primary myoblasts was necessary to achieve therapeutic levels of human factor IX [18]. Transplanted primary myoblasts have been shown to survive as muscle precursor cells, in addition to fusing with existing myofibers [19].

Myoblast transplantation has been identified as a promising therapeutic avenue for patients with Duchenne muscular dystrophy (DMD). DMD is a devastating X-linked neuromuscular disease caused by mutations in the gene coding for dystrophin, an essential sarcolemma-associated cytoskeletal protein [20]. The absence of dystrophin causes recurrent cycles of muscle fiber necrosis that lead to progressive muscle fiber loss [21]. Clinically, this progressive muscle degeneration leads to wheelchair confinement at around age 10 and death by the early twenties.

Most of the experimental approaches for the treatment of DMD aim to introduce the gene of normal dystrophin into muscle fibers [22]. Myoblast transplantation (MT) is one of these strategies [23]. The rationale of this method is the incorporation of myoblasts obtained from a healthy donor into the diseased muscles of a patient. Myoblasts are obtained from a biopsy sample of healthy donor muscle that is dissociated *in vitro* with cultivation and expansion of the liberated cells.

MT experiments in mice have shown successful results. MT could efficiently restore dystrophin expression in *mdx* mouse muscles, an animal model of DMD [24,25]. MT restored mass and force in muscles that had been irreversibly damaged, as well as increased twitch and tetanic tension in regenerating muscles [26–28]. In *mdx* mice immunosuppressed with FK506 and transplanted with normal myoblasts expressing the β-galactosidase (β-Gal) gene, up to 95% of the muscle fibers in *mdx* mouse expressed both β-Gal and dystrophin [29]. In addition, dystrophin produced by the donor myoblasts protected the *mdx* muscle against the mechanical stress that is responsible for fiber necrosis in DMD [30,31].

This initial success led to the implementation of clinical trials with MT [32–37]. These have not yet achieved widespread success in the treatment of DMD patients. Three types of problems have been identified as limiting factors in the success of clinical trials [38]: the specific immune response, limited diffusion and proliferation of transplanted myoblasts, and an inflammatory response responsible for a rapid death of transplanted cells. Recently, progress has been made in understanding and addressing these limitations.

The specific immune response toward the transplanted myoblasts, as well as to the muscle fibers created by their fusion, can be controlled with immunosuppression [39]. Several approaches have yielded means that more effectively deliver myoblasts throughout a muscle. Intramuscularly injected myoblasts tend to remain near sites of injection without spreading. Notexin, a potent myotoxic phospholipase purified from the venom of the Australian tiger snake *Notechis scutatus scutatus*, induces the necrosis of muscle fibers (preserving the satellite cells, nerves, and vessels) followed by complete regeneration of muscle fibers [40–43]. This effect has been used to improve the success of MT in both mice [44,45] and monkeys in which the number of β-Gal-positive fibers after MT in small sites is doubled [46]. Myoblasts that are precultured with concanavalin A or express matrilysin, a matrix metalloproteinase, demonstrate increased myogenic potential and fusion in mice [38,47]. DMD myoblasts transformed by SV40 large-T antigen and then infected with a telomerase retrovirus demonstrate a tremendous increase in proliferative capacity, suggesting that these cells may constitute a source of autologous cells for DMD patients [48,49].

Recently, successful myoblast transplantation has been reported in whole muscle of immunosuppressed macaque monkeys. Allogeneic myoblasts transduced with the β-Gal gene were injected into whole biceps brachii muscles; β-Gal-positive fibers were evident up to one year after MT [50]. Withdrawal of immunosuppression led to histological evidence of cellular rejection. The highest numbers of β-Gal-positive fibers achieved at one year were 50 and 20%.

MT has not yet reached a level of efficiency suitable for clinical implementation in DMD patients. It has been estimated, based on the relationship of dystrophin-positive fibers in muscle biopsy samples to the symptomatology of specific patients, that greater than 80% hybrid fibers will be required to completely alleviate symptoms in patients [50]. It is likely that *in vivo* gene therapy approaches will be more successful in delivering genes diffusely to large muscle masses. Nevertheless, MT done for DMD has yielded insight into myoblast biology and may still prove to be a valuable therapeutic modality for these patients.

Structural Reanimation

Myoblasts transplantation to "irreversibly" injured rat extensor digitorum longus muscle results in myofiber regeneration at the grafting site and enhances functional recovery [26]. Similar results have been reported for cryodamaged mouse soleus muscles, though the implanted cells significantly contributed to muscle regeneration only when the damage was extensive [27].

Myoblasts have been used as a source of cells for the "rejuvenation" of injured myocardium, which lacks the stem cells necessary for self-repair. Injections of myoblasts labeled with a *lacZ* gene into the cryoinjured canine myocardium revealed the differentiation of these cells into "cardiac-like" muscle cells within a dense scar [51,52]. These cells had centrally located nuclei and intercalated disks, both features of cardiac myocytes. Further evidence for the transformation of implanted cells was provided by a study using 4',6-diamino-2-phenylindole labeling of myoblasts injected into rat myocardium [53].

Other work has demonstrated that the contractility of implanted myoblasts improved the physiological performance of the recipient, regenerating cryoinfarcted myocardium [54–56]. The transition of the implanted cells to the "cardiac-like" phenotype is not yet understood. Two different cell types, in fact, have been noted after injection of cells into the cryoinjured rabbit myocardium—multinucleated cells within the cryolesion and isolated clusters of nonskeletal muscle cells that resembled immature cardiocytes at the periphery of the cryolesion [57]. Coronary infusion of rabbit myoblasts transduced with adenovirus to express β-Gal demonstrated that intravascular delivery of these cells is feasible; this resulted in distribution of β-Gal-positive cells to all cardiac layers, whereas direct injection resulted in clustering of cells principally in the myocardium [58,59]. A report of the use of marrow stromal cells, mesenchymal stem cells (MSCs), for cellular cardiomyoplasty appeared in 2000 [60]. Rat MSCs labeled with 4',6-diamino-2-phenylindole injected into isogeneic rat myocardia demonstrated myogenic differentiation, expressing sarcomeric myosin heavy chain and organized contractile proteins [60]. This suggests that MSCs may be a useful cell source in the future for cellular cardiomyoplasty.

ENCAPSULATION OF MYOBLAST FOR TRANSPLANTATION

Myoblasts have also been used as a cell source in encapsulated cell therapy. Encapsulation is based on the premise that cells, once sequestered within a semipermeable membrane, are isolated from the immune system and therefore cannot be recognized or damaged by normal host defenses. In addition, the membrane prevents any potential outgrowth of cells into the host parenchyma, allowing the use of mitotically active cells. This represents an emerging technology for the treatment of a number of diseases, such as neurodegenerative diseases. C2C12 cells transfected to secrete ciliary neurotrophic factor (CNTF), immobilized in matrices and encapsulated into hollow-fiber membrane devices, demonstrate *in vitro* viability and secretion of CNTF [61]. Dose control has been achieved with such cells by encapsulating discrete numbers of cell-containing microcarriers embedded in nonmitogenic hydrogels [62]. Encapsulated myoblasts have not yet been used in any clinical trials.

ENGINEERING THREE-DIMENSIONAL SKELETAL MUSCLE TISSUE

Implantation of myoblast–polymer constructs has yielded variable results. Syngeneic myoblasts from Oxford-albino rats seeded on collagen matrices failed to demonstrate muscle formation when implanted in the rectus abdominis muscle [63]. However, myoblasts from neonatal Fisher rats seeded onto poly(glycolic acid) meshes and implanted into the omentum of syngeneic adults formed histologically organized neomuscle [64]. This may represent an effect of the recipient environment.

One disadvantage of rigid polymer scaffolds is the limited ability of such frameworks to transmit dynamic environmental forces to cells, which generally experience a wide range of tissue forces. The most promising results to date for tissue-engineered muscle have combined myoblasts with extracellular matrix molecules, either as a substrate or as a three-dimensional hydrogel matrix. "Kinetically engineered" skeletal muscle was produced by culturing embryonic avian skeletal muscle myoblasts on a collagen-coated elastic substratum, either quiescently or under tension [65].

Amplitude, direction, and time history of the applied forces were designed to recapitulate those during *in vivo* embryogenesis, that is, static tension due to skeletal growth and dynamic tension due to muscular contraction. Cultured cells were first subject to unidirectional stretch, then constant tension and, finally, to unidirectional stretch with superimposed dynamic stretch/relaxation. This resulted in formation of a monolayer of multinucleated myotubes that were oriented in the direction of stretch and formed a rodlike "organoid." Organoids grown in the presence of mechanical forces were 1–2 mm in diameter by 30–35 mm long and contained parallel networks of unbranched fibers, a well-defined epimysium, and developing tendons [65]. The engineered muscle contracted and increased its axial tension by 91% in response to potassium. In contrast, organoids grown without kinetic manipulation were only 0.5–1 mm long, consisted of disorganized, branching fibers, and could not perform directed work.

In a subsequent study, cells from the myogenic C2C12 line were used to create organoids by mixing them with collagen and Matrigel, casting them in semicircular molds, and cultivating them under tension [66]. The C2C12 cells were genetically modified to secrete recombinant human growth hormone (rhGH), rendering these organoids a potential vehicle for gene therapy. When implanted subcutaneously in syngeneic mice, they secreted physiological doses of rhGH, which attenuated skeletal muscle disuse atrophy in mice that underwent hindlimb unloading [67]. Subsequently, human adult primary myoblasts were transduced *ex vivo* with replication-deficient retroviral expression vectors to secrete rhGH; these cells were used to form bioartificial muscle that continued to express the protein [68]. Histological examination of C2C12 myoblasts suspended in a collagen gel and implanted subcutaneously in nude mice revealed the formation of multinucleated myotubes with a rich capillary supply permeating the construct by 2 weeks [69].

Fetal myoblasts suspended in a collagen hydrogel have been used to create cellular constructs for surgical repair. Cells isolated from a fetal lamb biopsy sample were expanded in tissue culture, and suspended in a collagen hydrogel to create a circular piece of tissue for diaphragmatic repair in the same animal postnatally [70]. Myoblast-containing repair sites demonstrated a smooth, glistening surface with minimal eventration and a myofibrous architecture. By contrast, diaphragmatic defects repaired with cell-free collagen hydrogels had an irregular, distorted appearance, eventration in all cases, and consisted histologically of a thin fibrous septum [70]. This is the first study demonstrating the surgical application of bioartificial muscle for reconstruction.

SUMMARY AND FUTURE PROSPECTS

Myoblasts have been used in a wide array of tissue engineering approaches, employing diverse methodologies. Exciting new developments in the understanding of myoblast biology will undoubtedly contribute new avenues for tissue engineering. The myogenic potential of circulating bone marrow derived mesenchymal cells [71] and neural stem cells [72] may yield new sources of myoblast precursors. Clonally derived neural stem cells from mice and humans can be directed to produce skeletal myotubes, both *in vivo* and *in vitro*; in this extraordinary demonstration, these neural stem cells are generating cells that normally originate from a different germ cell layer [72].

REFERENCES

1. Hughes, S. M., and Blau, H. M. (1990). Migration of myoblasts across basal lamina during skeletal muscle development. *Nature (London)* **345**, 350–353.
2. Hughes, S. M. (1999). Fetal myoblast clones contribute to both fast and slow fibres in developing rat muscle. *Int. J. Dev. Biol.* **43**, 149–155.
3. Snow, M. H. (1977). Myogenic cell formation in regenerating rat skeletal muscle injured by mincing. I. A fine structural study. *Anat. Rec.* **188**, 181–199.
4. Lipton, B. H., and Schultz, E. (1979). Developmental fate of skeletal muscle satellite cells. *Science* **205**, 1292–1294.
5. Robertson, T. A., Grounds, M. D., Mitchell, C. A., and Papadimitriou, J. M. (1990). Fusion between myogenic cells in vivo: An ultrastructural study in regenerating murine skeletal muscle. *J. Struct. Biol.* **105**, 170–182.
6. Watt, D. J., Morgan, J. E., Clifford, M. A., and Partridge, T. A. (1987). The movement of muscle precursor cells between adjacent regenerating muscles in the mouse. *Anat. Embryol.* **175**, 527–536.
7. Rando, T. A., and Blau, H. M. (1994). Primary mouse myoblast purification, characterization, and transplantation for cell-mediated gene therapy. *J. Cell Biol.* **125**, 1275–1287.
8. Pavlath, G. K., Thaloor, D., Rando, T. A., Cheong, M., English, A. W., and Zheng, B. (1998). Heterogeneity among muscle precursor cells in adult skeletal muscles with differing regenerative capacities. *Dev. Dyn.* **212**, 495–508.
9. Ozawa, C. R., Springer, M. L., and Blau, H. M. (2000). A novel means of drug delivery: Myoblast-mediated gene therapy and regulatable retroviral vectors. *Annu. Rev. Pharmacol. Toxicol.* **40**, 295–317.
10. Barr, E., and Leiden, J. M. (1991). Systemic delivery of recombinant proteins by genetically modified myoblasts. *Science* **254**, 1507–1509.
11. Dhawan, J., Pan, L. C., Pavlath, G. K., Travis, M. A., Lanctot, A. M., and Blau, H. M. (1991). Systemic delivery of human growth hormone by injection of genetically engineered myoblasts. *Science* **254**, 1509–1512.
12. Yao, S. N., and Kurachi, K. (1992). Expression of human factor IX in mice after injection of genetically modified myoblasts. *Proc. Natl. Acad. Sci. U.S.A.* **89**, 3357–3361.
13. Dai, Y., Roman, M., Naviaux, R. K., and Verma, I. M. (1992). Gene therapy via primary myoblasts: Long-term expression of factor IX protein following transplantation in vivo. *Proc. Natl. Acad. Sci. U.S.A.* **89**, 10892–10895.
14. Hamamori, Y., Samal, B., Tian, J., and Kedes, L. (1995). Myoblast transfer of human erythropoietin gene in a mouse model of renal failure. *J. Clin. Invest.* **95**, 1808–1813.
15. Morgan, J. E., Moore, S. E., Walsh, F. S., and Partridge, T. A. (1992). Formation of skeletal muscle in vivo from the mouse C2 cell line. *J. Cell Sci.* **102**, 779–787.
16. Wernig, A., Irintchev, A., Hartling, A., Stephan, G., Zimmermann, K., and Starzinski-Powitz, A. (1991). Formation of new muscle fibres and tumours after injection of cultured myogenic cells. *J. Neurocytol.* **20**, 982–997.
17. Yao, S. N., Smith, K. J., and Kurachi, K. (1994). Primary myoblast-mediated gene transfer: Persistent expression of human factor IX in mice. *Gene Ther.* **1**, 99–107.
18. Wang, J. M., Zheng, H., Blaivas, M., and Kurachi, K. (1997). Persistent systemic production of human factor IX in mice by skeletal myoblast-mediated gene transfer: Feasibility of repeat application to obtain therapeutic levels. *Blood* **90**, 1075–1082.
19. Yao, S. N., and Kurachi, K. (1993). Implanted myoblasts not only fuse with myofibers but also survive as muscle precursor cells. *J. Cell Sci.* **105**, 957–963.
20. Hoffman, E. P., Brown, R. H., and Kunkel, L. M. (1992). Dystrophin: The protein product of the Duchenne muscular dystrophy locus. *Bio/Technology* **24**, 457–466.
21. Ervasti, J. M., and Campbell, K. P. (1993). Dystrophin-associated glycoproteins: Their possible roles in the pathogenesis of Duchenne muscular dystrophy. *Mol. Cell Biol. Hum. Dis. Ser.* **3**, 139–166.
22. Kakulas, B. A. (1997). Problems and potential for gene therapy in Duchenne muscular dystrophy. *Neuromuscular Disord.* **7**, 319–324.
23. Partridge, T. A. (1991). Invited review: Myoblast transfer: A possible therapy for inherited myopathies? *Muscle Nerve* **14**, 197–212.
24. Dunckley, M. G., Wells, D. J., Walsh, F. S., and Dickson, G. (1993). Direct retroviral-mediated transfer of a dystrophin minigene into *mdx* mouse muscle in vivo. *Hum. Mol. Genet.* **2**, 717–723.
25. Vincent, N., Ragot, T., Gilgenkrantz, H., Couton, D., Chafey, P., Gregoire, A., Briand, P., Kaplan, J. C., Kahn, A., and Perricaudet, M. (1993). Long-term correction of mouse dystrophic degeneration by adenovirus-mediated transfer of a minidystrophin gene. *Nat. Genet.* **5**, 130–134.
26. Alameddine, H. S., Louboutin, J. P., Dehaupas, M., Sebille, A., and Fardeau, M. (1994). Functional recovery induced by satellite cell grafts in irreversibly injured muscles. *Cell Transplant.* **3**, 13–14.
27. Irintchev, A., Langer, M., Zweyer, M., Theisen, R., and Wernig, A. (1997). Functional improvement of damaged adult mouse muscle by implantation of primary myoblasts. *J. Physiol. (London)* **500**, 775–785.
28. Arcila, M. E., Ameredes, B. T., DeRosimo, J. F., Washabaugh, C. H., Yang, J., Johnson, P. C., and Ontell, M. (1997). Mass and functional capacity of regenerating muscle is enhanced by myoblast transfer. *J. Neurobiol.* **33**, 185–198.
29. Karpati, G., Ajdukovic, D., Arnold, D., Gledhill, R. B., Guttmann, R., Holland, P., Koch, P. A., Shoubridge, E., Spence, D., Vanasse, M., et al. (1993). Myoblast transfer in Duchenne muscular dystrophy. *Ann. Neurol.* **34**, 8–17.
30. Morgan, J. E., Pagel, C. N., Sherratt, T., and Partridge, T. A. (1993). Long-term persistence and migration of myogenic cells injected into preirradiated muscles of *mdx* mice. *J. Neurol. Sci.* **115**, 191–200.

31. Brussee, V., Merly, F., Tardif, F., and Tremblay, J. P. (1998). Normal myoblast implantation in *mdx* mice prevents muscle damage by exercise. *Biochem. Biophys. Res. Commun.* **250**, 321–327.

32. Gussoni, E., Pavlath, G. K., Lanctot, A. M., Sharma, K. R., Miller, R. G., Steinman, L., and Blau, H. M. (1992). Normal dystrophin transcripts detected in Duchenne muscular dystrophy patients after myoblast transplantation. *Nature* (*London*) **356**, 435–438.

33. Huard, J., Bouchard, J. P., Roy, R., Malouin, F., Dansereau, G., Labrecque, C., Albert, N., Richards, C. L., Lemieux, B., and Tremblay, J. P. (1992). Human myoblast transplantation: Preliminary results of 4 cases. *Muscle Nerve* **15**, 550–560.

34. Law, P. K., Goodwin, T. G., Fang, Q., Hall, T. L., Quinley, T., Vastagh, G., Duggirala, V., Larkin, C., Florendo, J. A., Li, L., Jackson, T., Yoo, T. J., Chase, N., Neel, M., Krahn, T., and Holcomb, R. (1997). First human myoblast transfer therapy continues to show dystrophin after 6 years. *Cell Transplant.* **6**, 95–100.

35. Tremblay, J. P., Malouin, F., Roy, R., Huard, J., Bouchard, J. P., Satoh, A., and Richards, C. L. (1993). Results of a triple blind clinical study of myoblast transplantations without immunosuppressive treatment in young boys with Duchenne muscular dystrophy. *Cell Transplant.* **2**, 99–112.

36. Tremblay, J. P., Bouchard, J. P., Malouin, F., Theau, D., Cottrell, F., Collin, H., Rouche, A., Gilgenkrantz, S., Abbadi, N., Tremblay, M., *et al.* (1993). Myoblast transplantation between monozygotic twin girl carriers of Duchenne muscular dystrophy. *Neuromuscular Disord.* **3**, 583–592.

37. Morgan, J. E. (1994). Cell and gene therapy in Duchenne muscular dystrophy. *Hum. Gene Ther.* **5**, 165–173.

38. Caron, N. J., Asselin, I., Morel, G., and Tremblay, J. P. (1999). Increased myogenic potential and fusion of matrilysin-expressing myoblasts transplanted in mice. *Cell Transplant.* **8**, 465–476.

39. Kinoshita, I., Roy, R., Dugre, F. J., Gravel, C., Roy, B., Goulet, M., Asselin, I., and Tremblay, J. P. (1996). Myoblast transplantation in monkeys: Control of immune response by FK506. *J. Neuropathol. Exp. Neurol.* **55**, 687–697.

40. Harris, J. B., and Johnson, M. A. (1978). Further observations on the pathological responses of rat skeletal muscle to toxins isolated from the venom of the Australian tiger snake, *Notechis scutatus scutatus*. *Clin. Exp. Pharmacol. Physiol.* **5**, 587–600.

41. Pluskal, M. G., Harris, J. B., Pennington, R. J., and Eaker, D. (1978). Some biochemical responses of rat skeletal muscle to a single subcutaneous injection of a toxin (notexin) isolated from the venom of the Australian tiger snake *Notechis scutatus scutatus*. *Clin. Exp. Pharmacol. Physiol.* **5**, 131–141.

42. Harris, J. B., and Cullen, M. J. (1990). Muscle necrosis caused by snake venoms and toxins. *Electron Microsc. Rev.* **3**, 183–211.

43. Sharp, N. J., Kornegay, J. N., Bartlett, R. J., Hung, W. Y., and Dykstra, M. J. (1993). Notexin-induced muscle injury in the dog. *J. Neurol. Sci.* **116**, 73–81.

44. Huard, J., Roy, R., Guerette, B., Verreault, S., Tremblay, G., and Tremblay, J. P. (1994). Human myoblast transplantation in immunodeficient and immunosuppressed mice: Evidence of rejection. *Muscle Nerve* **17**, 224–234.

45. Vilquin, J. T., Guerette, B., Kinoshita, I., Roy, B., Goulet, M., Gravel, C., Roy, R., and Tremblay, J. P. (1995). FK506 immunosuppression to control the immune reactions triggered by first-generation adenovirus-mediated gene transfer. *Hum. Gene Ther.* **6**, 1391–1401.

46. Skuk, D., Roy, B., Goulet, M., and Tremblay, J. P. (1999). Successful myoblast transplantation in primates depends on appropriate cell delivery and induction of regeneration in the host muscle. *Exp. Neurol.* **155**, 22–30.

47. Ito, H., Hallauer, P. L., Hastings, K. E., and Tremblay, J. P. (1998). Prior culture with concanavalin A increases intramuscular migration of transplanted myoblast. *Muscle Nerve* **21**, 291–297.

48. Seigneurin-Venin, S., Bernard, V., and Tremblay, J. P. (2000). Telomerase allows the immortalization of T antigen-positive DMD myoblasts: A new source of cells for gene transfer application. *Gene Ther.* **7**, 619–623.

49. Seigneurin-Venin, S., Bernard, V., Moisset, P. A., Ouellette, M. M., Mouly, V., Di Donna, S., Wright, W. E., and Tremblay, J. P. (2000). Transplantation of normal and DMD myoblasts expressing the telomerase gene in SCID mice. *Biochem. Biophys. Res. Commun.* **272**, 362–369.

50. Skuk, D., Goulet, M., Roy, B., and Tremblay, J. P. (2000). Myoblast transplantation in whole muscle of nonhuman primates. *J. Neuropathol. Exp. Neurol.* **59**, 197–206.

51. Marelli, D., Desrosiers, C., el-Alfy, M., Kao, R. L., and Chiu, R. C. (1992). Cell transplantation for myocardial repair: An experimental approach. *Cell Transplant.* **1**, 383–390.

52. Chiu, R. C., Zibaitis, A., and Kao, R. L. (1995). Cellular cardiomyoplasty: Myocardial regeneration with satellite cell implantation. *Ann. Thorac. Surg.* **60**, 12–18.

53. Dorfman, J., Duong, M., Zibaitis, A., Pelletier, M. P., Shum-Tim, D., Li, C., and Chiu, R. C. (1998). Myocardial tissue engineering with autologous myoblast implantation. *J. Thorac. Cardiovasc. Surg.* **116**, 744–751.

54. Murry, C. E., Wiseman, R. W., Schwartz, S. M., and Hauschka, S. D. (1996). Skeletal myoblast transplantation for repair of myocardial necrosis. *J. Clin. Invest.* **98**, 2512–2523.

55. Taylor, D. A., Atkins, B. Z., Hungspreugs, P., Jones, T. R., Reedy, M. C., Hutcheson, K. A., Glower, D. D., and Kraus, W. E. (1998). Regenerating functional myocardium: Improved performance after skeletal myoblast transplantation. *Nat. Med.* **4**, 929–933; erratum: *Ibid.* **4**(10), 1200.

56. Atkins, B. Z., Hueman, M. T., Meuchel, J. M., Cottman, M. J., Hutcheson, K. A., and Taylor, D. A. (1999). Myogenic cell transplantation improves in vivo regional performance in infarcted rabbit myocardium. *J. Heart Lung Transplant.* **18**, 1173–1180.

57. Atkins, B. Z., Lewis, C. W., Kraus, W. E., Hutcheson, K. A., Glower, D. D., and Taylor, D. A. (1999). Intracardiac transplantation of skeletal myoblasts yields two populations of striated cells in situ. *Ann. Thorac. Surg.* **67**, 124–129.

58. Robinson, S. W., Cho, P. W., Levitsky, H. I., Olson, J. L., Hruban, R. H., Acker, M. A., and Kessler, P. D. (1996). Arterial delivery of genetically labelled skeletal myoblasts to the murine heart: Long-term survival and phenotypic modification of implanted myoblasts. *Cell Transplant.* **5**, 77–91.

59. Taylor, D. A., Silvestry, S. C., Bishop, S. P., Annex, B. H., Lilly, R. E., Glower, D. D., and Kraus, W. E. (1997). Delivery of primary autologous skeletal myoblasts into rabbit heart by coronary infusion: A potential approach to myocardial repair. *Proc. Assoc. Am. Physicians* 109, 245–253.

60. Wang, J. S., Shum-Tim, D., Galipeau, J., Chedrawy, E., Eliopoulos, N., and Chiu, R. C. (2000). Marrow stromal cells for cellular cardiomyoplasty: Feasibility and potential clinical advantages. *J. Thorac. Cardiovasc. Surg.* **120**, 999–1006.

61. Li, R. H., Williams, S., Burkstrand, M., and Roos, E. (2000). Encapsulation matrices for neurotrophic factor-secreting myoblast cells. *Tissue Eng.* **6**, 151–163.

62. Li, R. H., Williams, S., White, M., and Rein, D. (1999). Dose control with cell lines used for encapsulated cell therapy. *Tissue Eng.* **5**, 453–466.

63. van Wachem, P. B., Brouwer, L. A., and van Luyn, M. J. (1999). Absence of muscle regeneration after implantation of a collagen matrix seeded with myoblasts. *Biomaterials* **20**, 419–426.

64. Saxena, A. K., Marler, J., Benvenuto, M., Willital, G. H., and Vacanti, J. P. (1999). Skeletal muscle tissue engineering using isolated myoblasts on synthetic biodegradable polymers: Preliminary studies. *Tissue Eng.* **5**, 525–532.

65. Vandenburgh, H. H., Swasdison, S., and Karlisch, P. (1991). Computer-aided mechanogenesis of skeletal muscle organs from single cells in vitro. *FASEB J.* **5**, 2860–2867.

66. Vandenburgh, H., Del Tatto, M., Shansky, J., Lemaire, J., Chang, A., Payumo, F., Lee, P., Goodyear, A., and Raven, L. (1996). Tissue-engineered skeletal muscle organoids for reversible gene therapy. *Hum. Gene Ther.* **7**, 2195–2200.

67. Vandenburgh, H., Del Tatto, M., Shansky, J., Goldstein, L., Russell, K., Genes, N., Chromiak, J., and Yamada, S. (1998). Attenuation of skeletal muscle wasting with recombinant human growth hormone secreted from a tissue-engineered bioartificial muscle. *Hum. Gene Ther.* **9**, 2555–2564.

68. Powell, C., Shansky, J., Del Tatto, M., Forman, D. E., Hennessey, J., Sullivan, K., Zielinski, B. A., and Vandenburgh, H. H. (1999). Tissue-engineered human bioartificial muscles expressing a foreign recombinant protein for gene therapy. *Hum. Gene Ther.* **10**, 565–577.

69. Okano, T., and Matsuda, T. (1998). Muscular tissue engineering: Capillary-incorporated hybrid muscular tissues in vivo tissue culture. *Cell Transplant.* **7**, 435–442.

70. Fauza, D. O., Marler, J. J., Koka, R., Forse, R. A., Mayer, J. E., and Vacanti, J. P. (2001). Fetal tissue engineering: Diaphragmatic replacement. *J. Pediatr. Surg.* **36**, 146–151.

71. Ferrari, G., Cusella-De Angelis, G., Coletta, M., Paolucci, E., Stornaiuolo, A., Cossu, G., and Mavilio, F. (1998). Muscle regeneration by bone marrow-derived myogenic progenitors. *Science* **279**, 1528–1530; erratum: *Ibid.* **281**, 923.

72. Galli, R., Borello, U., Gritti, A., Minasi, M. G., Bjornson, C., Coletta, M., Mora, M., De Angelis, M. G., Fiocco, R., Cossu, G., and Vescovi, A. L. (2000). Skeletal myogenic potential of human and mouse neural stem cells. *Nat. Neurosci.* **3**, 986–991.

CHAPTER 100

VISION ENHANCEMENT SYSTEMS

Gislin Dagnelie, Kah-Guan Au Eong, James D. Weiland, and Srinivas R. Sadda

INTRODUCTION

The long-term goal of vision enhancement is to compensate for loss of optical imaging quality in the eye and for loss of transduction capability of the retina and visual pathway caused by injury or disease. Compensation methods for optical vision loss are presented elsewhere in this volume (Chapter 83). Visual impairment and blindness caused by neural vision loss may arise at any anatomical site along the visual pathway. Among the causes are conditions such as age-related macular degeneration (AMD) and retinitis pigmentosa (RP), which damage the outer retina; diabetic eye disease, which damages the inner retina; glaucoma, which affects the optic nerve, and a host of intracranial lesions that affect the higher visual pathways and brain areas.

There is currently no effective treatment to reverse neural blindness. Advances in microtechnology, computer science, optoelectronics, and microsurgery have increased the feasibility of improving and restoring functional vision through enhancement of the image being presented, through prosthetic devices that directly stimulate the remaining neural tissues proximal to the site of damage, and through cell transplantation and differentiation leading to functional restoration of lesioned tissue. Each of these strategies has made great strides in the past few years, but all remain faced with numerous and formidable challenges. In this chapter we discuss some of the necessary and most promising building blocks for vision enhancement and restoration through these three approaches.

OPTOELECTRONIC VISION ENHANCEMENT

The loss of the eye's optical imaging quality, the loss of phototransduction in the retinal photoreceptors, and the loss of signal transduction in neural elements along the visual pathways can all be modeled as a loss in the transfer characteristic of the visual system. To a limited extent, one can compensate for such losses by preprocessing the images presented to the eyes. In the case of optical (refractive) errors of the ocular media, the necessary inverse filtering to restore the optical wavefront reaching the photoreceptors can be performed with almost unlimited precision. This is not true for neural losses, for which the required inverse filters are not physically realizable. Expressed in terms of spatial frequencies, the loss of high-frequency information in the image is inherently accompanied by phase shifts, for which no compensation is possible; in spatial terms, there is no physical deconvolution that can compensate for the broadening of the point spread function arising in the image wherever neural elements are degenerating.

The traditional way to compensate for degraded image information is the use of magnification aids such as loupes and telescopes. Within a limited range, and at the expense of a reduced field of view, such devices provide the most straightforward and affordable compensation for loss of perceived image detail. In more severe cases, two additional approaches can be called upon: preemphasis filtering and feature extraction.

Methods of Tissue Engineering

PREEMPHASIS FILTERING

In cases of relatively mild vision loss (visual acuity > 20/400), a close approximation to restoration of the original image may be provided by image processing techniques such as preemphasis filtering. These techniques attempt to restore the image by boosting high spatial frequencies in a Fourier transformation of the image or, equivalently, through deconvolution in the space domain. A number of image processing software packages available commercially or as freeware use these approaches, and depending on the processing power available, close to real-time performance (with frame rates well in excess of 10 per second, although sometimes with a time lag of 1–2 frames) may be obtained.

Figure 100.1 shows an example of the degradation of facial features by a low-pass filter, symbolizing a mild degree of visual impairment. The Sobel filter and the Peli adaptive filter (Fig. 100.1c and 100.1d, respectively) represent two attempts at restoring visibility of the facial features. Peli *et al.* [1] demonstrated that enhancement filters that render faces more recognizable to low vision patients may have the opposite effect in normally sighted observers, thus demonstrating that the effectiveness of these filters can be tested only in visually impaired observers.

In 2000 a conference presentation [2] gave results of a preference test by low vision observers among different image enhancement algorithms in the ImageLab software package, developed at the University of Waterloo, Ontario [3], and applied to stationary colored images such as faces and outdoor scenes. Both deconvolution (with varying kernel size

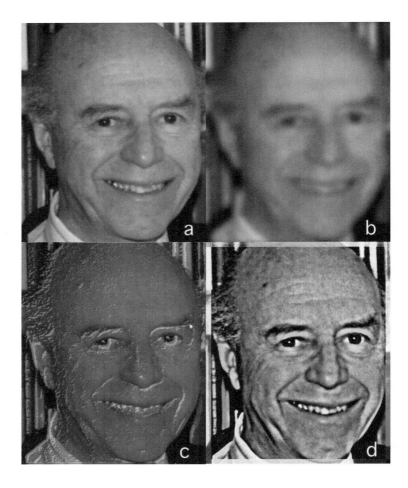

Fig. 100.1. Application of spatial and spatial–frequency domain enhancement filters to a photograph (a), which would be perceived as blurred by a low vision observer (b). To enhance the image, Sobel filtering can be carried out as a real-time convolution and added to the original (c). The Peli adaptive filter (d) uses spatial frequency enhancement. Figure based on Omoruyi and Leat [2], and on images made available by Drs. Gloria I. Omoruyi and Susan J. Leat.

and complexities) and inverse filtering in the spatial frequency domain (with varying bandwidths, center frequencies, and steepness) were presented, and preferences were attributed almost equally to filters in either domain. Common properties of preferred filters were their relatively low order and high bandwidth (i.e., larger and smoother deconvolution kernels), while more enhancing algorithms that emphasized the restoration of edges were rejected as exaggerated.

Unfortunately, larger kernels are computationally much more intensive, as is the use of frequency domain filtering. A compromise would be the use of smaller kernels such as a Sobel filter [4] followed by mixing of filtered and original images to de-emphasize gradients and edges; only if real-time frequency domain filtering is available, however, will it be feasible to apply dedicated algorithms such as the Peli adaptive filters [1] to real-time vision enhancement for visually impaired individuals. With increasingly powerful image processing capability becoming more affordable and portable, implementation of many satisfactory image enhancement algorithms in portable optoelectronic low vision aids can be anticipated.

FEATURE EXTRACTION

Magnification and preemphasis filtering may not permit patients with severe vision loss to recognize real-life scenes (e.g., finding a familiar person or the exit door in a large gathering) and achieve visual support for complex motor tasks (e.g., safely crossing a busy intersection). In situations like these, nonlinear filtering techniques developed for robotics and machine vision—such as contour extraction and template matching—can be called upon to assist the visually impaired individual (in addition to nonvisual approaches, such as voice recognition and cane travel). Thus, a template-matching algorithm can be invoked to find a doorway, the desired face (assuming the algorithm allows one to choose from among a collection of familiar faces), the curbside, or any approaching vehicle that is not coming to a stop. The feedback could be visual (in the form of highlights in a head-worn display, if the wearer's vision is adequate), auditory, or tactile.

IMAGE REMAPPING

Images presented to a visually impaired or blind person may undergo distortion in the perceptual process. The most dramatic example of this will be encountered in visual prostheses based on cortical stimulation (see later), where the perceived locations of phosphenes, created by electrical stimulation of a regular array of implanted electrodes, will not correspond to the same regular pattern, but rather, will be determined by the retinocortical projection map. Dagnelie and Vogelstein [5], through simulations in normally sighted observers, have demonstrated that an accurate map of the perceived phosphenes can be constructed through a combination of three techniques—notably the placement of a finger on a touchscreen, a gaze shift following a briefly presented phosphene, and distance comparisons in phosphene triads. With current digital signal processors (DSPs) the inverse point-to-point projection can be implemented in real time, albeit with a one-frame delay.

As another example, patients with AMD or other retinal diseases involving the macula often complain of distortions of their visual perception, in which straight lines appear crooked. If this distortion is stable enough to be mapped, through a detailed procedure similar to those proposed by Dagnelie and Vogelstein [5], an inverse transformation can be applied to the image. Thus, the visual image presented to a low vision patient or cortical prosthesis wearer can be made to appear close to normal, and habituation (presumably a cortical "learning" process) is expected to further reduce any remaining perceived distortions.

CURRENT AND PROSPECTIVE DEVICES

Although the building blocks for head-worn, optoelectronic low vision devices capable of preemphasis filtering and feature extraction exist, real-time, general-purpose, and wearable implementations may still be out of reach of, and too costly for, most potential beneficiaries. However, several optoelectronic low vision headsets already use DSPs to control zoom and autofocus functions for their image acquisition cameras, and as DSPs of increasing power become available, at least some of the image processing methods discussed are likely to be implemented in low vision aids within the next few years.

VISUAL PROSTHETIC DEVICES BASED ON ELECTRICAL STIMULATION

Electrical stimulation has been prescribed as a therapy for a variety of conditions for literally hundreds of years. Roman physicians used electric eels from the Mediterranean as a therapy for headaches [6]. At the turn of the eighteenth century, English surgeon C. P. Wilkinson advertised his Medico-electric rooms for treating hemorrhoids, deafness, and "female complaints" [7]. Volta placed either end of a primitive battery in each of his own ears and subsequently reported a bubbling sound [6]. While the potential therapeutic effects of electricity were suspected by these early physical scientists, our improved understanding of the human body and the technologies that allow us to control electricity have enabled the implementation of sophisticated electronic implants in the human body to replace lost function. The reader is referred to Hambrecht [6] for a comprehensive review of the history of neural stimulation.

ELECTRICAL STIMULATION OF NEURAL TISSUE

Nerve cells communicate via electrochemical signals. Neurotransmitter released from a cell binds to membrane proteins in an adjacent cell, opening the chemically activated ion channels in the membrane and thereby changing the electric properties of the membrane. This results in a transient change in the electrical potential of the membrane, which in turn opens voltage-sensitive membrane ion channels, resulting in a continuation of the membrane potential change. Equilibrium potential is eventually reestablished as the ion channels again close, but if the stimulus is sufficiently strong, the cell will have remained depolarized long enough to transmit the signal to the next cell in the signal path.

Electrical stimulation elicits a neural response by "turning on" the voltage-sensitive ion channels, bypassing the chemically gated channels in the stimulated cell. Nerve cells have a transmembrane equilibrium potential of approximately −60 mV; that is, the intracellular potential is negative with respect to extracellular potential. An electrical signal applied with an extracellular stimulating electrode imposes an electric field on the cell, depolarizing the cell by making the extracellular potential more negative—and making the intracellular and extracellular potentials closer in magnitude—when negative current is applied. Electric field and current density are directly proportional, so it is important to note that the size of the electrode matters as much as total current in any attempt to achieve threshold depolarization (e.g., 100 µA applied through a 100,000-µm^2 electrode will have less stimulus strength than the same stimulus applied through a 1000-µm^2 electrode). In theory, current density should be the only factor determining stimulus threshold. However, in practice, the relationship is not directly proportional [8]: a small electrode will require a higher current density than a larger electrode. Also, since electrical current requires both a source and a sink, negative current in one place implies positive current in another place. However, if the current sink (current return) is large and placed far away from the current source (stimulating electrode), then the current return has a low current density and does not stimulate neural tissue.

The ideal electrical stimulus would be a single cathodic (i.e., negative) current pulse. It would require the least amount of power and would result in depolarization of the cell membrane under the electrode. However, current pulses are typically applied in trains to convey a continuous stimulus to the cell as well as to the prosthesis wearer's perception. If a stimulus waveform consisted of a repeated, cathodic pulse, residual negative charge would build up at the electrode–tissue interface, eventually resulting in the generation of gaseous hydrogen or oxygen (gassing or bubbling) [8,9]. To avoid such charge accumulation, a stimulus pulse must be charge balanced. This can be accomplished by capacitively coupling the electrode—at the end of the current pulse, the capacitor discharges so that no net current is applied to the electrode—or by using a charge-balanced stimulus pulse—the charge is reversed by applying a positive current pulse after the negative pulse, again resulting in no net charge transfer.

An important consideration for a stimulating electrode is the material that forms the interface to the tissue. Since the electrode must transfer a substantial charge, metals are best suited for this purpose. Obviously the electrode surface should not be corroded by the physiological milieu, nor should it be neurotoxic. The noble metals (gold, platinum, iridium) satisfy these constraints. Also, the surface must be able to transfer large charges without undergoing physical changes. Gold dissolves when stimulating currents are applied, but platinum and iridium can withstand large stimulating currents. A stable oxide of iridium, $IrO_x(OH)_x$, has superior charge-carrying capacity owing to its large and irregular surface structure, and its ability to pass charge through reversible faradaic reactions [10]. In these reactions, electrical charge crosses from electrode to electrolyte. In general, such reactions introduce harmful products into the tissue, but both iridium oxide and platinum possess the ability to safely transfer charge through these reactions.

A second mechanism of transferring electrical charge is through double-layer charging. This occurs at the electrode–tissue interface, since ions in solution are hydrated by the oppositely charged end of the polar water molecules. Therefore, since the water acts as a dielectric layer, separating ionic and electron charge, resulting in a capacitance, when an ion is attracted to the oppositely charged electrode, it need not directly contact the electrode. This type of current flow is actually a rearrangement of charge, since no charge crosses the interface, but the effect is the same. It would seem that this capacitive stimulation would be ideal for safety purposes, but in practice, physical limitations in the amount of charge that can be stored (due to the low breakdown voltage of the water dielectric layer), prevent purely capacitive electrodes from being much different from platinum [11], leaving iridium oxide as the best electrode material.

Electrochemical limits are only one consideration when one is determining the safety of neural stimulation. It has been determined empirically that neural damage arises as a function of both charge density and total charge. For small electrodes (<10,000 μm^2), a greater charge density will be tolerated by neural tissue (1 mC/cm^2) [12]. To explain this, it has been surmised that since the total charge of the pulse through a microelectrode is small, some local pH change or gas bubble formation can be buffered by the bulk tissue.

RETINAL PROSTHESIS

The success of the cochlear implant in bypassing damaged or absent hair cell receptors and electrically stimulating more proximal neurons [13,14] has prompted investigators to embark on the idea of the retinal prosthesis [15]. Working within the eye presents formidable challenges. In comparison to the inner ear, the eye is prone to chronic inflammation, its neural tissues (retina) can easily detach, and the potential for a devastating infection makes the use of a design in which wires penetrate the wall of the eye highly undesirable.

A diagram of a retinal prosthesis as envisioned by Humayun and associates is shown in Fig. 100.2. The system consists of several components.

1. External to the eye and body, a video camera captures the visual environment, and electronic image processing circuitry reduces the resolution and complexity of the image. Both components are mounted on an eyeglass frame worn by the subject.

2. The image data are digitally encoded and fed via a telemetry link (laser or radio frequency modulated signal) to a decoder chip implanted in the eye. In addition to the image data, the transmission beam will be used to supply power to the implanted circuitry. The advantages of using laser include low heat dissipation, good transmission through the ocular media, small size, and low cost. However, there is a need to target the laser delivery accurately onto the implant. In addition, the laser transmission can be interrupted by media opacities (such as cataract and vitreous hemorrhage) and blinking. Accurate targeting and clear media are not necessary for radio frequency transmission, but interference from other electronic sources may occur.

3. The decoder chip inside the eye converts the transmitted image data and produces the necessary pattern of small electrical currents to be applied to the retina through a two-dimensional array of electrodes positioned at the inner retinal surface. Each individual electrode directly stimulates the underlying retinal neurons, which then relay this information

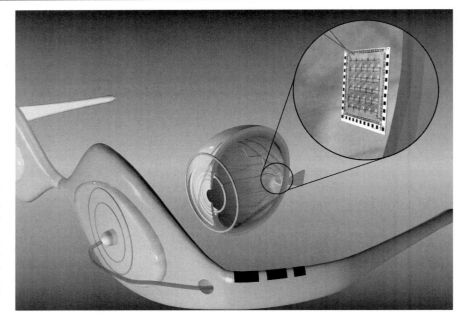

Fig. 100.2. Schematic representation of the visual prosthesis proposed by the Intraocular Retinal Prosthesis group at the Johns Hopkins University and North Carolina State University. Image data from an external video camera and signal preprocessor are transmitted to an intraocular decoder chip and electrode array, which convey localized stimulation to the underlying retinal tissue.

to the visual cortex, resulting in perception of a dot of light at a point in the visual field corresponding to the retinal location.

Simultaneous activation of multiple electrodes in the array will create a pattern of individual dots of light, not unlike the formation of a letter from single dots on a dot-matrix printer, or an image on a stadium scoreboard. The engineering of the system is considered to be feasible with available technology, much of which is currently used in cochlear implants for the hearing impaired.

At first glance, it may appear preferable to engineer a single implantable prosthesis with all system components for light detection, image processing, current generation, and electrode stimulation implemented as a layered structure. In such a device, the image would be formed on a light detector by the optics of the eye. However, a prototype device with discrete subsystems with a majority of the electronics outside the eye would reduce the size of, and heat generated by, the intraocular components, and would allow the external components to be repaired, modified, or upgraded without additional surgery.

The prerequisites that are paramount to the success of the proposed retinal prosthesis include the following:

Positional stability of the prosthesis after implantation
Biocompatibility of prosthesis materials inside the eye
Specificity of the stimulated cell population
Functionality of the visual percept

Positional Stability of the Prosthesis after Implantation

The greatest challenge to the success of a retinal prosthesis is the integration of electronic technology with the nervous system of the visual pathway. The stimulating electrodes could be placed in the subretinal [16,17] or epiretinal location [18–24].

One potential advantage of a prosthesis in the subretinal space over an epiretinal device is that the electrodes may stimulate more distal neural elements in the afferent visual pathway. This has the theoretical advantage of capturing some of the early neural processing that occurs in the bipolar cells of the inner retina (as discussed shortly in the section on specificity). Also, with current vitreoretinal techniques, it is easier to place a prosthesis in the subretinal space than to fix it onto the epiretinal surface. However, a subretinal prosthesis is a highly unnatural substrate for the overlying neural elements. Exchange of nutrients, oxygen, and waste materials between the retina and the underlying retinal pigment epithelium and choroidal circulation may be disrupted or impaired by the subretinal prosthesis. Moreover, the pressure exerted on the retina may lead to necrosis. While the former issue (metabolic exchange between retina and choroid) may be of minor importance in a retina whose outer layers are degenerated by AMD or RP—note that the inner retina in humans has its own circulatory supply—the risk of pressure necrosis will require long-term studies in animals and blind patients. This issue can be considered to be one of the major safety aspects covered by the phase I clinical trials of subretinal implantation announced by Chow and colleagues in Chicago (AP news release, June 29, 2000).

The epiretinal approach places the electrode array in contact with the internal limiting membrane, in closest proximity to the ganglion cell axons in the nerve fiber layer and to the ganglion cell bodies. Although free of some of the potential problems associated with the subretinal prosthesis, this approach has its own challenges. The prosthesis must remain in its position to ensure stable signal transduction. Moreover, there is no generally accepted technique for long-term fixation of an electrode array at a predetermined distance from the retina. This is of great importance because the inertia of the array and the drag of the intraocular fluid due to angular acceleration during saccadic movements would tend to shear the array off the retinal surface. It is possible that stimulation of Mueller and other cells could lead to the formation of epiretinal membranes and fibrous "cocoons" surrounding the electrode array, not unlike that seen in chronic intraocular foreign bodies. On the other hand, such membranes could interfere with the function of the prosthesis by becoming a barrier of high resistance between the electrodes and the inner retina.

Walter et al. reported several fixation techniques for epiretinal microelectrode arrays in rabbit eyes [25,26]. First, without much success, they used fluid–air exchange and fibrin glue for retinal fixation in seven rabbit eyes. They began by performing cataract extraction, vitrectomy, and 360° endolaser photocoagulation to the peripheral retina. Two weeks later, a second vitrectomy was performed to remove the residual cortical vitreous humor and the microelectrode array was placed on the visual streak and pressed against the retinal surface. A fluid–air exchange was performed in four eyes, and a fibrin glue (Tissue-Col) was used to cover the implant in three eyes. Of the four eyes with no fibrin glue, the microelectrode array dislocated into the anterior chamber by the first postoperative day in one eye and the eye developed a tractional retinal detachment. The other three prostheses remained in position for 1 week, but all were dislocated into the inferior half of the vitreous cavity after 1–2 weeks. In the three eyes in which the microelectrode array was covered with fibrin glue, the prosthesis remained in the original position throughout the 2- to 4-week follow-up period but became embedded in a fibrous network.

Margalit et al. [27] examined nine commercially available compounds for their suitability as intraocular adhesives for attaching an electrode array to the retina, studying their short-term biocompatilibity, adhesion, consistency, and stability. The adhesives tested included commercially available fibrin sealant (Haemacure Co.), autologous fibrin, Cell-Tak (Becton Dickinson), three different photocurable glues (Star Technology, Lightwave Energy Systems Co.—LESCO, and Loctite), and three different poly(ethylene glycol) hydrogels (Shearwater Polymers, Cohesion Technologies Inc.). Margalit and colleges found that poly(ethylene glycol) hydrogels appear to be excellent candidates for use as an adhesive in the eye, based on their adhesive strength, consistency, and minimal acute toxicity. However, further studies are necessary to evaluate the long-term adhesion and biocompatibility of these compounds in the eye. The finding of these investigators that the sea mussel adhesive Cell-Tak causes acute tissue damage when used intraocularly concurred with results of Liggett et al. [28].

Walter et al. more recently reported successful long-term implantation of electrically inactive epiretinal microelectrode arrays in rabbit eyes using retinal tacks [26]. During an initial procedure, a lens-sparing three-port core vitrectomy was performed, and the prospective fixation area inferior to the optic nerve was coagulated with an infrared diode endolaser. Three weeks later, a second vitrectomy was performed to remove residual cortical vitreous humor; the microelectrode array was implanted, and a retinal tack made of titanium was used to fixate the implant by penetrating the area of the laser scar. In one case, a total retinal detachment with dense cataract formation occurred after implantation. In the remaining nine cases, the implant remained in its original location throughout 6 months of follow-up. The retina remained attached in these eyes, but in two cases epiretinal membranes were seen around the tack.

Biocompatibility of Prosthesis Materials inside the Eye

A number of biocompatibility issues must be addressed before any retinal prosthesis can be implanted routinely in patients for any length of time. Part of the difficulty in resolving these issues may be related to subtle differences between materials and species used. Thus, in the initial experiments by Walter *et al.* [25], a prosthesis consisting of a foil of silicon, weighing 9.4 mg and embedded in a soft carrier of silicone rubber (Fraunhofer Institute for Microelectronic Circuits and Systems, Duisburg, Germany), initially did not reveal any sign of toxicity in the electroretinogram of rabbits with successful prosthesis implantation. However, the visually evoked potential, although within normal limits shortly after implantation, showed a considerable reduction of the P2 component after 10 weeks, indicating a toxic effect of the device on the ganglion cells or their axons. This interpretation was consistent with the observed narrowing of the optic nerve vessels in the papilla of two rabbits with a follow-up of 20 weeks after implantation. Edell *et al.* [29], on the other hand, found that silicon electrode shafts in brain tissue do not cause tissue damage, except around the electrode tip. Majji *et al.* [30] found no sign of adverse effects using a grid of platinum electrodes embedded in a silicone matrix similar to that of Walter *et al.*

While some electrode and insulation materials may thus have toxic effects on intraocular tissues, problems for active electronics and substrate materials posed by the physiological milieu inside the eye are equally serious. A variety of studies have indicated the need for hermetic sealing of microelectronics to be implanted in the human body [31]. Even materials such as polyimides, thought to be suitable electrode substrates several years ago [32], have been found to swell, crack, and lose their insulating properties after prolonged immersion in solutions similar to aqueous and vitreous humor [J.D. Weiland, unpublished observations].

There is significant controversy about the ability of silicon and related materials to withstand long-term exposure to the intraocular milieu. At a meeting of intraocular prosthesis researchers in June 2000, the group of Chow reported having found no adverse effects on their photodiode arrays, while the groups of Zrenner and Humayun did find deterioration of silicon structures; Zrenner's group reported that silicon nitride coatings appear to provide adequate protection for periods of at least several months [M. S. Humayun, personal communication]. Amorphous silicon carbide has been reported to be a superior electrode insulator over prolonged implantation periods [33].

Specificity of the Stimulated Cell Population

Whether the stimulating electrode array is placed on the vitreal side of the retina, close to the nerve fiber and ganglion cell layers, or subretinally, close to any remaining photoreceptors and horizontal cells, the charge released by the electrodes will spread into deeper retinal layers, especially if the reference electrode is located at some distance, or even outside the eye. Therefore, one cannot assume that only cellular segments or even entire cells in the immediate vicinity of the electrodes will be stimulated, inasmuch as the electrical properties and the actual activation/depolarization process for each cell type need to be taken into account. Greenberg *et al.* [34], in a model study of retinal ganglion cells, concluded that activation of the soma and dendrites would have lower stimulation thresholds, as long as the duration of the stimulating pulse was well in excess of 1 ms. Below 1 ms, on the other hand, thresholds for depolarization—and thus activation of the cell—are lowest for ganglion cell axons, owing to their different membrane properties and aided by their closer vicinity to an epiretinal electrode.

Support for this model was obtained in human intraocular stimulation experiments. Weiland *et al.* [35] stimulated normal retina as well as two areas of laser-induced retinal damage in one eye of two subjects scheduled for exenteration due to recurrent cancer near that eye. Electrical stimulation of normal and laser-damaged retinal areas resulted in significantly different visual percepts. In particular, both patients reported seeing light "threads" in areas where all retinal cell types had been destroyed by argon green laser. The same percept, consistent with stimulation of a bundle of ganglion cell axons directly under the electrode, was reported by RP patients during local electrical stimulation of the retina with pulse durations shorter than 1 ms [19].

Evidence for the ability to selectively stimulate cells in different retinal layers has also been provided by *in vitro* stimulation of amphibian retina [36]. Strength–duration curves for photoreceptors, bipolar cells, and ganglion cells were determined by stimulating the retina with local light or electrical stimuli and recording from the optic nerve, in dark-adapted, light-adapted, and "synaptic block" (i.e., a chemical agent such as Co^{2+} is used to block all signal transmission from bipolar to ganglion cells) conditions. Electrical stimulation thresholds were lowest in the dark-adapted (i.e., maximum photoreceptor gain) condition, and chronaxie and rheobase values were consistent with those published for photoreceptors. Thresholds in light-adapted retina were lower than in synaptically blocked retina, except for stimulus durations below 1 ms, suggesting that bipolar cells have lower activation thresholds than ganglion cells, but only for longer stimulus durations.

Finally, in intraocular human stimulation experiments, the visual percepts of a "dark spot" in intact retinas [35] and in some RP eyes with minimal remaining sight [19] suggest mediation through photoreceptor stimulation—note that photoreceptors are hyperpolarized by light, and electrical depolarization would thus correspond to a dark stimulus—while the percepts of a "bright spot" in a retinal area where photoreceptors were destroyed by a krypton red laser [35] and in RP eyes without remaining vision [19] suggest mediation through electrical activation of bipolar cells.

It is preferable to stimulate early rather than late stages along the retinal processing pathway, to maximize the chances that the percept resulting from stimulation across an array of electrodes will resemble images produced by normal vision, and thus be easily understandable to future prosthesis wearers. Thus, the ability to use stimulus pulse durations well over 1 ms to perform selective stimulation of the bipolar cells in a degenerated retina appears to be an important step forward toward prosthetic vision.

Functionality of the Visual Percept

Studies in volunteers have demonstrated the ability to convey a spot of light by electrically stimulating a retinal location that contains no photoreceptors [35,37], as well as the ability to recognize crude shapes [37]. The percept corresponds with the stimulus time and location, suggesting that the brain, without any training, is capable of responding to signals from a retina that has lost photoreceptor function years earlier. Resolution obtained with handheld electrode arrays over the human retina is on the order of 1°—adequate for mobility purposes—and permanently implanted arrays are likely to provide slightly better resolution. In addition, work by Cha *et al.* [38] and Dagnelie and Vogelstein [5] in normally sighted human subjects has shown that reduction of visual input to a 25 × 25 array of dots in the central visual field with resolutions between 0.25 and 1.25° can provide useful visually guided mobility face recognition, and reading. Even the mere ability to perceive light may be useful to some totally blind subjects [39].

CORTICAL PROSTHESIS

Electrical stimulation of the human visual cortex produces the sensation of phosphenes [40–42]. Following several early reports related to the development of a cortical visual prosthesis in the 1970s [43–46], Dobelle in 2000 reported the first visual prosthesis providing useful artificial vision to a volunteer, blind from trauma in both eyes, by connecting a video camera, image processing, and stimulating electronics to the visual cortex [47]. The subject, now 66 years old, lost his vision from trauma in one eye at age 22 and in the fellow eye at age 36. In 1978 an intracranial electrode array was implanted under local anesthesia on the medial side of his right occipital cortex. Equipped with the new electronics and software, he is able to scan and recognize a 6-inch-square "tumbling E" at 5 ft—corresponding to a visual acuity of approximately 20/1200—and to travel alone using public transport. The low resolution of these electrodes illustrates an important complication that has been dogging researchers ever since electrodes were chronically implanted into superficial brain layers: glial tissue starts forming around an electrode at the site of insertion, resulting either in extrusion of the electrode or at least in increased distance between electrode and nerve cell being monitored or stimulated and, in the case of stimulation, to increased threshold currents [R. A. Normann, personal communication].

Two other important efforts in this area have been going on for a number of years. Researchers at the National Institute for Neurological Disorders and Stroke published results on mapping and perception experiments in a subject implanted with 38 (13 pairs and 12 single electrodes) 2-mm-long iridium wires over the occipital cortex [48]. Since then, these investigators have developed arrays of implantable electrodes, which they have embedded in the primary visual cortex of primates. High-quality recordings from up to 64 electrodes have been obtained, and behavioral stimulation experiments are in progress [C. V. Kufta, personal communication]. At the University of Utah, Normann *et al.* [49] have developed square arrays of 64 and 100 amorphous silicon electrode shafts up to 1.5 mm long. These electrodes have been used for recording in primates and cats, and for functional stimulation in behaving cats, who were trained to respond to an auditory stimulus and then produced highly reliable responses to electrical stimulation of the auditory cortex.

Biocompatibility issues are an important factor in the development of materials for a cortical visual prosthesis. While the researchers in Utah have reported ways of preventing adhesion of the dura mater to implanted electrode arrays [50], the problems of glial tissue proliferation (mentioned earlier) still await an effective solution.

A final issue of importance for the development of a functional cortical visual prosthesis is the need to disentangle the complex projection from the retina to the primary, and possibly also higher areas of the occipital cortex. As reported by Schmidt *et al.* [48], this mapping can be highly irregular, and even with small electrode arrays over the foveal projection in primary visual cortex there is no guarantee that the projection will be regular [51], although it is possible that some of these perceived irregularities can be "unlearned." Dagnelie and Vogelstein [5] have developed several approaches that will allow precise cartography of the phosphene map perceived by a cortical implant wearer, thus allowing use of the inverse transformation in an image preprocessor to render the images intelligible to the prosthesis wearer.

TISSUE TRANSPLANTATION

Success in transplantation of other organs and tissues has encouraged some researchers to pursue tissue transplantation to treat neural blindness [52,53]. Replacement of the entire retina is generally regarded as beyond current capabilities, primarily because it is impossible to reconnect the retina to the brain via outgrowth of new optic nerve fibers. Two distinct forms of tissue transplantation—namely, transplantation of neural retinal cells or retinal stem cells and transplantation of retinal pigment epithelium cells—are thought to be more promising. Numerous advances in tissue transplantation have been made in recent years, yet even basic issues such as donor cell source, optimal age for transplantation, transplant technology, and rejection have yet to be fully explored and harnessed.

TRANSPLANTATION OF NEURAL RETINAL CELLS

In eyes with retinitis pigmentosa (RP) and age-related macular degeneration (AMD), the inner retina is relatively preserved despite death of photoreceptors in the outer retina. In theory, healthy photoreceptor cells, if successfully transplanted, might be able to form functional connections with the inner retina and thus restore vision. Another mechanism by which photoreceptor transplantation may be beneficial in RP patients is through the rescue of remaining host photoreceptors. Photoreceptor transplants have been shown to increase host cone survival in a mouse model of hereditary photoreceptor cell degeneration [54,55].

Neural retinal transplantation was first performed by Tansley in 1946, but significant interest did not develop until the mid-1980s. Del Cerro *et al.* [56] performed transplants of full-thickness retina into the anterior chamber of mice. Turner and Blair [57] transplanted neonatal rat retina into the subretinal space via a transscleral approach and showed survival and differentiation into retinal layers. Subsequent studies (in various animal models of retinal degeneration) by a number of investigators have shown that the transplants can differentiate into photoreceptors that appear to be functionally active, as evidenced by electrophysiolgical activity [58] and light-dependent shifts in phototransduction proteins [59].

A number of factors have been identified that limit the potential success of neural retinal cell transplantation. Aramant *et al.* [60] observed decreased survival of the transplant with increasing age of the donor tissue. Consequently, fetal tissue was suggested as the optimal source of donor tissue. In addition, the organization of the donor tissue was found to be important. Initially, investigators prepared microaggregate suspensions of retinal tissue by taking small retinal fragments, which were then dissociated into smaller clusters of cells by repeated aspiration and extrusion from a microsyringe. These preparations were easy to manipulate and could be injected into the subretinal space with a small-diameter cannula. However, microaggreate suspensions frequently formed rosettes [61,62], an anatomic arrangement limiting the cells' ability to form connections with surrounding host tissue. Subsequently, investigators utilized sheets of retinal tissue, whereupon they observed a reduced incidence of rosette formation. Sheets, however, are more difficult to manipulate and insert into the subretinal space. Hence, some investigators [63] developed techniques to embed the retinal sheet in gelatin, thereby obtaining a firm substrate that could be rolled into a special applicator for insertion into the subretinal space, where it unfolds and can be positioned.

Some investigators have observed a possible glial cell barrier that develops between the transplant and the host retinal tissue [64]. Others have suggested that inner retinal neurons included in the transplant may impair the formation of connections between the transplanted cells and the remaining host retinal neurons. Consequently, investigators have minimized the presence of inner retinal elements in the graft by vibratome sectioning [65,66] or excimer laser ablation [67].

The current view, based on these studies, is that a fetal retinal sheet with removal of inner retinal elements is the optimal source of tissue for transplantation. Despite the apparent anatomic success of transplants in animal models of retinal degeneration, improvement in visual function or formation of functional connections between host and transplant cells, has not been conclusively demonstrated. In 2000 McCall *et al.* [68] were able to record light responses from the superior colliculus following cotransplantation of neural retinal sheets and rational pigment epithelium (RPE) in RCS rats. Once again, however, a rescue effect on remaining host photoreceptors cannot be entirely ruled out.

Despite the lack of definite "proof of principle" in animal models, a few small pilot studies of neural retinal transplantation in human patients—all with advanced RP, with the exception of one patient with advanced neovascular AMD—have been undertaken [69–72]. Although temporary objective improvements were demonstrated in some patients [72], and other studies noted subjective mild, transient improvement in some patients, no sustained benefits were observed. At the same time, no patient experienced rejection, and adverse outcomes were rare (one case of retinal detachment in Das's series [70]). In one patient in Radtke's study [71], there was a transient improvement in the multifocal electroretinogram (ERG) results over the graft. Although the ERG suggests that the transplant was functional, it does not establish the presence of graft–host connections. In Humayun's series, the patient with AMD died 3 years after transplantation. Histological exam [73] disclosed survival of the transplanted tissue, but no photoreceptor differentiation (possibly due to the lack of functional RPE).

Although the precise reason for failure of neural retinal transplantation in human patients is not known, the prevailing view is that poor graft–host connectivity is the major problem. Consequently, future efforts in this field are being directed toward the enhancement of graft–host functional integration. One possible approach is the use of retinal stem cells, which may have a greater capacity to form new synapses.

TRANSPLANTATION OF RETINAL STEM CELLS

Until recently, notwithstanding remarkable findings of pluripotential cells in various organ systems, the adult retina was believed to lack regenerative capacity. Tropepe *et al.* [74], however, identified stem cells (self-renewing and multipotential cells) in the pigmented ciliary margin of adult mouse eyes, which were capable of clonal proliferation *in vitro*. Stem cells were cultured by incubation of isolated pigmented ciliary margin cells in serum-free media containing epidermal growth factor (EGF) or basic fibroblast growth factor (FGF-2). In the presence of appropriate growth factors, these cells were found to be capable of differentiating into cells expressing markers for various retinal cell types, including photoreceptors, bipolar neurons, and glial cells. In addition, other investigators [75] have isolated neural stem cells from the hippocampus of adult rats. When these adult hippocampus-derived neural progenitor cells (AHPC) were transplanted into the vitreous cavity of adult rat eyes, they migrated into the retina and adopted morphologies similar to Mueller, astroglial, bipolar, horizontal, amacrine, and photoreceptor cells. None of the cells, however, developed end-stage markers specific for retinal neurons. Chacko *et al.* [76] performed transplants of cultured retinal progenitor cells (derived from embryonic rat retina) into the subretinal space of adult rats. Some cells expressed certain photoreceptor-specific markers, but typical photoreceptor morphology and outer segment formation was not observed.

In summary, the discovery of retinal stem cells is a promising finding for retinal transplantation for a number of reasons. First, stem cells represent a potentially unlimited source of tissue for transplantation. Second, as stem cells can be harvested from adult eyes, the ethical concerns raised by fetal retinal transplantation are avoided. Finally, stem cells, by virtue of their immaturity, may be more capable of forming synaptic connections. Before retinal stem cells can enable a viable treatment for retinal degenerations, however, significant problems must be overcome. In particular, techniques must be devised to reproducibly direct stem cells to differentiate (and differentiate completely) into the desired phenotype(s).

TRANSPLANTATION OF RETINAL PIGMENT EPITHELIUM (RPE)

The functional and morphological integrity of the photoreceptor–RPE complex is crucial to the normal function of the retina [77]. In age-related macular degeneration, the RPE becomes dysfunctional, leading to degeneration of the overlying photoreceptors. Gouras *et al.* [78] postulated that RPE cell transplantation to the area of RPE degeneration would reestablish functional and morphological integrity of the photoreceptor–RPE complex, improving or restoring visual function. Most investigators now involved in RPE transplantation in AMD patients and animal models of AMD are using either a sheet of freshly harvested RPE from an adult donor, a patch of fetal RPE cells cultured *in vitro* prior to transplantation, or RPE cell suspensions [79–84]. More recently, some have grown human fetal RPE cells as microspheres on cross-linked fibrinogen for transplantation in rabbit eyes [85,86].

Although it is traditionally thought that the eye is an immunologically privileged site, that privilege is not absolute and can be broken [87]. Zhang and Bok [88] showed that homologous RPE cells injected into the subretinal space in rats caused chronic rejection despite the absence of acute immune rejection. Studies in rabbits by Crafoord *et al.* [89] on allotransplantation of RPE cell suspensions into the subretinal space showed that although the allograft persisted at 6 months, its survival was limited. The cells generally formed a monolayer, but focal fragmentation and disruption with dispersion of melanin pigment occurred. Focal multilayers of cells containing large macrophages were also present, with associated overlying photoreceptor damage.

The cellular response to the allografts in the subretinal space does not involve the invasion of lymphocytes as in classical host–graft rejection. Crafoord *et al.* [90] also showed that systemic cyclosporine A did not prevent RPE allograft destruction in rabbits. When transplantation into the subretinal space of human RPE allografts, even fetal cells, has been attempted without immunosuppression, rejection also has resulted [79,81,82].

Several attempts at RPE transplantation in humans with advanced AMD have been reported [79–82,91,92]. The technique of Peyman *et al.* [91] involved raising a large retinal flap encompassing the macula and the temporal vascular arcades, removal of submacular scar, and replacement of retinal pigment epithelial cells. These researchers used an autologous pedicle graft in one patient and homologous RPE with Bruch's membrane in another. Fourteen months after the operation, the visual acuity of the patient with a pedicle graft had improved from counting fingers to 20/400 and the patient fixated over the transplanted RPE cells. The homologous graft became encapsulated with a fine subretinal membrane

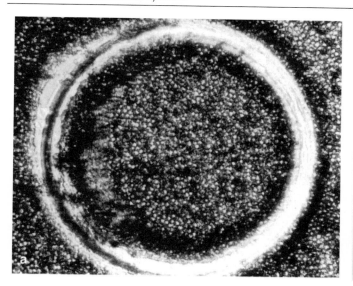
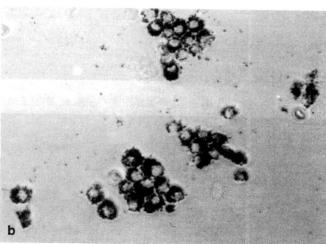

Fig. 100.3. (a) Circular human fetal RPE patch transplant trephined in the culture plate prior to loading into micropipette: patch diameter 0.6 mm. (b) Suspension of RPE cells, prior to transplantation, shows single cells or clusters, most often in the range of 5–10 cells. (Figure 1 in Algvere et al. [81]; reproduced with permission of the copyright holders Springer-Verlag and the author.)

without neovascular tissue after 10 months, and the visual acuity remained counting fingers at 2 feet.

Algvere *et al.* [79,81,82] and Gouras and Algvere [80] have reported on two different techniques for RPE transplantation in humans, both using fetal human RPE (Fig. 100.3). One technique involves harvesting small sheets of RPE from eyes of human fetuses at gestational age 13–20 weeks. Before being dissected, the fetal eyes are kept in a refrigerator for 1–2 days in Earl's Balanced Salt Solution without Ca and Mg and supplemented with penicillin (50 U/ml) and streptomycin (50 μm/ml). The eyes are dissected under sterile conditions in a laminar flow hood with the aid of an operating microscope. The anterior segment of the eye, vitreous humor and neurosensory retina are removed. The remaining posterior segment is cut into quarters and placed in Dulbecco's Medium in a culture plate. The sclera is removed from each quadrant, and fine forceps are used to dissect small sheets of RPE, 5–10 mm², away from Bruch's membrane and the choroid. Each sheet of RPE is kept in a separate culture plate and maintained in Dulbecco's Medium supplemented with 20% newborn calf serum and 0.45% glucose in a humidified incubator at 37°C with 5% CO_2 and air mixture for 2–7 days before transplantation. By the time of the transplantation, a large and broad zone of monolayered, less pigmented, RPE cells has grown out in the periphery but only the original darkly pigmented patch is selected for transplantation.

The second technique involves microdissection: RPE cells are harvested from human fetal eyes by injecting Earl's Balanced Salt Solution under the RPE and mechanically dissociating the loosened cells into single cells or clusters by gentle trituration, using a fire-polished Pasteur pipette. After centrifugation and removal of the supernatant, the cells are resuspended in 10–20 μl of Earl's Balanced Salt Solution. Enzymes are not used for dissociation. The cells are not cultured but are transplanted within 2 h. The number of cells used in each transplant is calculated to be approximately 4×10^4 to 6×10^6.

RPE transplantation is performed using pars plana vitrectomy techniques [79–81]. In cases of neovascular AMD, following vitrectomy and peeling of the posterior hyaloid face from the posterior pole of the eye, the choroidal neovascular membrane (CNVM) is removed through a tiny diathermy-induced retinotomy in the macula. The RPE transplant is then injected into the subretinal space through a glass micropipette with an outer tip diameter of 100–150 μm. A fluid–air exchange is performed in the vitreous cavity before it is filled with silicone oil. In eyes with geographic atrophy (no CNVM), a small retinal bleb is formed by injecting balanced salt solution through a retinotomy (Fig. 100.4). The patch or suspension of RPE cells is injected into the subretinal space through the retinotomy. Fol-

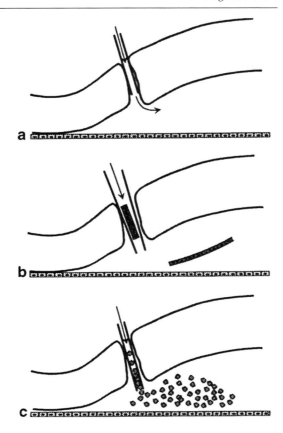

Fig. 100.4. Illustration of the method of producing a neurosensory detachment with a jet stream of fluid through the neural retina (a), the introduction of RPE patch transplants (b), and the injection of a suspension of RPE cells (c) into the subretinal space. The orifice of the pipette required to inject cell suspensions is smaller than that for transplanting an RPE patch. (Figure 2 in Algvere et al. [81]; reproduced with permission of the copyright holders Springer-Verlag and the author.)

lowing fluid–air exchange, the vitreous cavity is filled with either sulfur hexafluoride (SF_6) or perfluoropropane (C_3F_8) gas for internal tamponade. Postoperatively, the patient is positioned facing down for 2–4 days.

The long-term outcome of RPE allografts in 16 nonimmunosuppressed patients with AMD reported by Algvere et al. [82] indicates that rejection is a major problem for both patch grafts and cell suspensions. Allografts transplanted into the subretinal space of eyes with neovascular AMD were rejected within 3 months, while those transplanted in nonneovascular AMD were rejected more slowly over 12–20 months. Rejection was evidenced by the presence of retinal edema, fluorescein leakage, and fragmentation, disruption, or disappearance of the graft, with loss of overlying visual function. Four of nine transplants in eyes with nonneovascular AMD did not show clinical signs of rejection. The investigators concluded that the presence of an intact blood–retinal barrier tends to prevent the development of rejection of these transplants. No visual improvement was observed in any of the 16 patients who had RPE allograft transplantation.

Weisz et al. [92] reported a case of allogeneic fetal RPE transplantation in a patient with severe atrophic AMD followed over 24 months. After pars plana vitrectomy, a retinotomy was made along the superotemporal vascular arcade at the edge of an area of geographic atrophy. A suspension of RPE cells was then infused into the subretinal space through the retinotomy. Visual acuity remained unchanged at 20/80 following surgery, but 5 months later it had dropped to 20/267. By 14 months after surgery, the visual acuity was 20/500 and remained stable from that time on. Because the type and pattern of progressive visual loss fell within the observed findings of the natural history of geographic atrophy [93], the visual loss was believed to be due to progression of the geographic atrophy rather than to the surgery. Although the allogeneic fetal tissue remained identifiable during the 24-month study period without immunosuppression, fluorescein angiography demonstrated early RPE–choroidal leakage and subsequent vasculitis in the area including and surrounding the graft, indicative of graft rejection. There was also progressive subretinal fibrosis in the area of transplant. Moreover, the patient developed a weakly positive cellular response to phosducin and rhodopsin.

TRANSPLANTATION OF AUTOLOGOUS IRIS PIGMENT EPITHELIUM (IPE)

Theoretically, transplantation of autologous RPE would circumvent the problem of rejection. However, because AMD is frequently bilateral and the harvesting of RPE from an intact eye is difficult, healthy autologous RPE cells are not readily available for transplantation. On the other hand, autologous iris pigment epithelial (IPE) cells, which have the same embryonic origin as RPE, can be readily harvested by simple surgical procedures. Since both IPE and RPE cells are of neuroepithelial origin, are pigmented, exhibit limited replicative capacity *in vivo* but show some plasticity *in vitro*, contain cathepsin D, and are phagocytic, some investigators have reasoned that it may be possible to use IPE cells as a replacement for diseased RPE in an autograft procedure [87,94]. Transplantation of autologous IPE into the subretinal space has therefore been attempted as one approach for the treatment of age-related macular degeneration.

Thumann *et al.* [94] harvested IPE cells by iridectomy and used the pars plana approach to transplant the cells directly into the subretinal space of the fellow eye in 25 rabbits. They found that the autologous IPE survived and formed a polarized monolayer above the retinal pigment epithelium in the subretinal space, with apical microvilli adjacent to the photoreceptors. In addition, they observed fragments of phagocytosed photoreceptor rod outer segments in phagosomes within the cytoplasm of the transplanted IPE cells. The rod outer segments adjacent to the transplanted cells remained healthy throughout the experimental period of 5 months, and no signs of a cell-mediated immunological response were observed.

The short-term visual outcome of 21 patients who had autologous IPE transplantation for neovascular AMD was reported in 2000 [95,96]. The IPE cells were first obtained by peripheral iridectomy and cultured using autologous serum. The cultured cells were then transplanted into the subretinal space after CNVM removal. Postoperative follow-up ranged from 1 to 16 months. Visual acuity improved in 9 patients, was unchanged in 2 patients, and worsened in 10 patients. The postoperative visual acuity tended to decrease initially and then to improve after 3 months. For 13 patients with a follow-up of 6 months or longer, the visual acuity improved in 7, was unchanged in 1, and decreased in 5 eyes. Interestingly, the investigators observed that eyes with a greater number of transplanted cells had better visual acuity than those with a smaller number of transplanted cells. However, pre- and postoperative retinal sensitivities as tested by electroretinography and Humphrey visual field examination were unchanged.

CONCLUSIONS

In this chapter we have reported on some of the methods and approaches by which tissue engineering and related disciplines may lend a hand to vision preservation and restoration. This is a complex field, which may still be years away from routine application in patients, but rapid progress is being made in many laboratories, and the interdisciplinary nature of this work offers fascinating challenges as well as rewarding opportunities for young researchers.

REFERENCES

1. Peli, E., Lee, E., Trempe, C. L., and Buzney, S. (1994). Image enhancement for the visually impaired: The effects of enhancement on face recognition. *J. Opt. Soc. Am.* **11**, 1929–1939.
2. Omoruyi, G. I., and Leat, S. J. (2000). Selection of potential image processing algorithms for improving the visibility of images for low vision observers (ARVO abstract). *Invest. Ophthalmol. Visual Sci.* **41**, S431.
3. Kennedy, A. J., Leat, S. J., and Jernigan, M. E. (1998). A generalized image processing application for the enhancement of images for the visually impaired. *In* "Vision Science and its Applications," OSA Tech. Dig., pp. 124–127. Optical Society of America, Washington, DC.

4. Sobel, I. (1970). "Camera Models and Machine Perception," AIM-21. Stanford Artificial Intelligence Lab, Palo Alto, CA.

5. Dagnelie, G., and Vogelstein, J. V. (1999). Phosphene mapping procedures for prosthetic vision. *In* "Vision Science and its Applications," OSA Tech. Dig., pp. 294–297. Optical Society of America, Washington, DC.

6. Hambrecht, F. T. (1990). The history of neural stimulation and its relevance to future neural prostheses. *In* "Neural Prostheses: Fundamental Studies" (W. F. Agnew and D. B. McCreery, eds.), 1st ed., pp. 1–23. Prentice-Hall, Englewood Cliffs, NJ.

7. Wilkinson, C. P. (1799). "The Effects of Electricity…" M. Allen, London.

8. Lilly, J. C., Hughes, J. R., Alvord, E. R., and Galkin, T. W. (1955). Brief, noninjurious electric waveform for stimulation of the brain. *Science* **121**, 468–469.

9. Brummer, S. B., and Turner, M. J. (1975). Electrical stimulation of the nervous system: The principle of safe charge injection with noble metal electrodes. *Bioelectrochem. Bioenerg.* **2**, 13–25.

10. Robblee, L. S., Lefko, J. L., and Brummer, S. B. (1983). Activated Ir: An electrode suitable for reversible charge injection in saline solution. *J. Electrochem. Soc.* **130**(3), 731–732.

11. Rose, T. L., Kelliher, E. M., and Robblee, L. S. (1985). Assessment of capacitor electrodes for intracortical neural stimulation. *J. Neurosci. Methods* **12**(3), 181–193.

12. McCreery, D. B., Agnew, W. F., Yuen, T. G. H., and Bullara, L. (1990). Charge density and charge per phase as cofactors in neural injury induced by electrical stimulation. *IEEE Trans. Biomed. Eng.* **37**(10), 996–1001.

13. Balkany, T., Hodges, A., and Luntz, M. (1996). Update on cochlear implantation. *Otolaryngol. Clin. North Am.* **29**, 277–289.

14. Clark, G. M., Tong, Y. C., and Patrick, J. F. (1990). Introduction. *In* "Cochlear Prostheses" (G. M. Clark, Y. C. Tong and J. F. Patrick, eds.), pp. 1–14. Churchill-Livingstone, Melbourne, Australia.

15. Humayun, M. S., and de Juan, E., Jr. (1998). Artificial vision. *Eye* **12**, 605–607.

16. Chow, A., and Chow, V. (1997). Subretinal electrical stimulation of the rabbit retina. *Neurosci. Lett.* **225**, 13–16.

17. Zrenner, E., Miliczek, K. D., Gabel, V. P., Graf, H. G., Guenther, E., Haemmerle, H., Hoefflinger, B., Kohler, K., Nisch, W., Schubert, M., Stett, A., and Weiss, S. (1997). The development of subretinal microphotodiodes for replacement of degenerated photoreceptors. *Ophthalmic Res.* **29**, 269–280.

18. Humayun, M. S., Propst, R., de Juan, E., Jr., McCormick, K., and Hickingbotham, D. (1994). Bipolar surface electrical stimulation of the vertebrate retina. *Arch. Ophthalmol.* **112**, 110–116.

19. Humayun, M. S., de Juan, E., Jr., Greenberg, R. J., Dagnelie, G., Rader, R. S., and Katona, S. J. (1997). Electrical stimulation of the retina in patients with photoreceptor loss (ARVO abstract). *Invest. Ophthalmol. Visual Sci.* **38**, S39.

20. Eckmiller, R. (1997). Learning retina implants with epiretinal contacts. *Ophthalmic Res.* **29**, 281–289.

21. Wyatt, J. L., Rizzo, J. F., Grumet, A., Edell, D., and Jensen, R. J. (1994). Development of a silicon retinal implant: Epiretinal stimulation of retinal ganglion cells in the rabbit (ARVO abstract). *Invest. Ophthalmol. Visual Sci.* **35**, 1380.

22. Narayanan, M. V., Rizzo, J. F., Edell, D., and Wyatt, J. L. (1994). Development of a silicon retinal implant: Cortical evoked potentials following focal stimulation of the rabbit retina with light and electricity (ARVO abstract). *Invest. Ophthalmol. Visual Sci.* **35**, 1380.

23. Mann, J., Edell, D., Rizzo, J. F., Raffel, J., and Wyatt, J. L. (1994). Development of a silicon retinal implant: Micro-electronic system for wireless transmission of signal and power (ARVO abstract). *Invest. Ophthalmol. Visual Sci.* **35**, 1380.

24. Wyatt, J., and Rizzo, J. (1996). Ocular implants for the blind. *IEEE Spectrum* **33**(5), 47–53.

25. Walter, P., Szurman, P., Krott, R., Baum, U., Bartz-Schmidt, K.-U., and Heimann, K. (1997). Experimental implantation of devices for electrical retinal stimulation in rabbits: Preliminary results. *In* "Advances in ocular toxicology" (K. Green, H. F. Edelhauser, R. B. Hackett, D. S. Hull, D. E. Potter and R. C. Tripathi, eds.), pp. 113–120. Plenum, New York.

26. Walter, P., Szurman, P., Vobig, M., Berk, H., Ludtke-Handjery, H.-C., Richter, H., Deng, Mittermayer, C., Heimann, K., and Sellhaus, B. (1999). Successful long-term implantation of electrically inactive epiretinal microelectrode arrays in rabbits. *Retina* **19**, 546–552.

27. Margalit, E., Fujii, G. Y., Lai, J. C., Gupta, P., Chen, S. J., Shyu, J. S., Piyathaisere, D. V., Weiland, J. D., de Juan, E., Jr., and Humayun, M. S. (2000). Bioadhesives for intraocular use. *Retina* **20**, 469–477.

28. Liggett, P. E., Cano, M., Robin, J. B., and et al. (1990). Intravitreal biocompatibility of mussel adhesive protein. A preliminary study. *Retina* **10**, 144–147.

29. Edell, D. J., Toi, V. V., McNeil, V. M., and Clark, L. D. (1992). Factors influencing the biocompatibility of insertable silicon microshafts in cerebral cortex. *IEEE Trans. Biomed. Eng.* **39**, 635–643.

30. Majji, A. B., Humayun, M. S., Weiland, J. D., Suzuki, S., D'Anna, S. A., and de Juan, E., Jr. (1999). Long-term histological and electrophysiological results of an inactive epiretinal electrode array implantation in dogs. *Invest. Ophthalmol. Visual Sci.* **40**, 2073–2081.

31. Lund, J. L., and Wise, K. D. (1994). Chip-level encapsulation of implantable CMOS microelectrode arrays. *Dig. Solid State Sensor Actuator Workshop 94*, 29–32.

32. Richardson, R. R., Jr., Miller, J. A., and Reichert, W. M. (1993). Polyimides as biomaterials: Preliminary biocompatibility testing. *Biomaterials* **14**, 627–635.

33. Cogan, S. F., Liu, Y. P., Jones, R. B., and Edell, D. J. (1998). Silicon carbide coated electrodes for neural prostheses. *NINDS 29th Neural Prosthesis Workshop*, Bethesda, MD.

34. Greenberg, R. J., Velte, T. J., Humayun, M. S., Scarlatis, G. N., and de Juan, E., Jr. (1999). A computational model of electrical stimulation of the retinal ganglion cell. *IEEE Trans. Biomed. Eng.* **46**, 505–514.

35. Weiland, J. D., Humayun, M. S., Dagnelie, G., de Juan, E., Jr., Greenberg, R. J., and Iliff, N. T. (1999). Understanding the origin of visual percepts elicited by electrical stimulation of the human retina. *Graefe's Arch. Clin. Exp. Ophthalmol.* **237**, 1007–1013.

36. Greenberg, R. J. (1998). Analysis of electrical stimulation of the vertebrate retina — work towards a retinal prosthesis. Ph.D. Dissertation, Johns Hopkins University, Baltimore, MD.

37. Humayun, M. S., de Juan, E., Jr., Weiland, J. D., Dagnelie, G., Katona, S., Greenberg, R. J., and Suzuki, S. (1999). Pattern electrical stimulation of the human retina. *Vision Res.* **39**, 2569–2576.

38. Cha, K., Horch, K. W., and Normann, R. A. (1992). Simulation of a phosphene-based visual field: Visual acuity in a pixelized vision system. *Ann. Biomed. Eng.* **20**, 439–449.

39. Ross, R. D. (1998). Is perception of light useful to the blind patient? *Arch. Ophthalmol.* (*Chicago*) **116**, 236–238.

40. Brindley, G. S., and Lewin, W. S. (1968). The sensations produced by electrical stimulation of the visual cortex. *J. Physiol.* (*London*) **196**, 479–493.

41. Dobelle, W. H., and Mladejovsky, M. G. (1974). Phosphenes produced by electrical stimulation of human occipital cortex and their application to the development of a prosthesis for the blind. *J. Physiol.* (*London*) **245**, 553–576.

42. Bak, M., Girvin, J. P., Hambrecht, F. T., Kufta, C. V., Loeb, G. E., and Schmidt, E. M. (1990). Visual sensations produced by intracortical microstimulation of the human occipital cortex. *Med. Biol. Eng. Comput.* **28**, 257–259.

43. Dobelle, W. H., Mladejovsky, M. G., and Girvin, J. P. (1974). Artificial vision for the blind: Electrical stimulation of visual cortex offers hope for a functional prosthesis. *Science* **183**, 440–444.

44. Dobelle, W. H., Mladejovsky, M. G., Evans, J. R., Roberts, T. S., and Girvin, J. P. (1976). "Braille" reading by a blind volunteer by visual cortex stimulation. *Nature* (*London*) **259**, 111–112.

45. Dobelle, W. H., Quest, D., Antunes, J., Roberts, T., and Girvin, J. P. (1979). Artificial vision for the blind by electrical stimulation of the visual cortex. *Neurosurgery* **5**, 521–527.

46. Klomp, G. F., Womack, M. V. B., and Dobelle, W. H. (1977). Fabrication of large arrays of cortical electrodes for use in man. *J. Biomed. Mater. Res.* **11**, 347–364.

47. Dobelle, W. H. (2000). Artificial vision for the blind by connecting a television camera to the visual cortex. *ASAIO J.* **46**, 3–9.

48. Schmidt, E. M., Bak, M. J., Hambrecht, F. T., Kufta, C. V., O'Rourke, D. K., and Vallabhanath, P. (1996). Feasibility of a visual prosthesis for the blind based on intracortical microstimulation of the visual cortex. *Brain* **119**, 507–522.

49. Normann, R. A., Maynard, E. M., Rousche, P. J., and Warren, D. J. (1999). A neural interface for a cortical vision prosthesis. *Vision Res.* **39**, 2577–2587.

50. Maynard, E. M., Fernandez, E., and Normann, R. A. (2000). A technique to prevent dural adhesions to chronically implanted microelectrode arrays. *J. Neurosci. Methods* **97**, 93–101.

51. Normann, R. A., Maynard, E. M., Guillory, K. S., and Warren, D. J. (1996). Cortical implants for the blind. *IEEE Spectrum* **33**(5), 54–59.

52. Del Cerro, M., Lazar, E. S., and DiLoreto, D. (1997). The first decade of continuous progress in retinal transplantation. *Microsc. Res. Tech.* **36**, 130–141.

53. Litchfield, T. M., Whiteley, S. J. O., and Lund, R. D. (1997). Transplantation of retinal pigment epithelial, photoreceptor and other cells as treatment for retinal degeneration. *Exp. Eye Res.* **64**, 655–666.

54. Mohand-Said, S., Hicks, D., Simonutti, M., Tran-Minh, D., Deudon-Combe, A., Dreyfus, H., Silverman, M. S., Ogilvie, J. M., Tenkova, T., and Sahel, J. (1997). Photoreceptor transplants increase host cone survival in the retinal degeneration (*rd*) mouse. *Ophthalmic Res.* **29**, 290–297.

55. Mohand-Said, S., Deudon-Combe, A., Hicks, D., Simonutti, M., Forster, V., Fintz, A. C., Leveillard, T., Dreyfus, H., and Sahel, J. A. (1998). Normal retina releases a diffusible factor stimulating cone survival in the retinal degeneration mouse. *Proc. Natl. Acad. Sci. U.S.A.* **95**, 8357–8362.

56. Del Cerro, M., Gash, D. M., Rao, G. N., Notter, M. F., Wiegand, S. J., and Gupta, M. (1985). Intraocular retinal transplants. *Invest. Ophthalmol. Visual Sci.* **26**, 1182–1185.

57. Turner, J. E., and Blair, J. R. (1986). Newborn rat retinal cells transplanted into a retinal lesion site in adult host eyes. *Dev. Brain Res.* **26**, 91–104.

58. Adolph, A. R., Zucker, C. L., Ehinger, B., and Bergstrom, A. (1994). Function and structure in retinal transplants. *J. Neural Transplant. Plast.* **5**, 147–161.

59. Seiler, M. J., Aramant, R. B., and Ball, S. L. (1999). Photoreceptor function of retinal transplants implicated by light-dark shift of S-antigen and rod transducin. *Vision Res.* **39**, 2589–2596.

60. Aramant, R., Seiler, M., and Turner, J. E. (1988). Donor age influences on the success of retinal grafts to adult rat retina. *Invest. Ophthalmol. Visual Sci.* **29**, 498–503.

61. Del Cerro, M., Ison, J. R., Bowen, G. P., Lazar, E., and del Cerro, C. (1991). Intraretinal grafting restores visual function in light-blinded rats. *NeuroReport* **2**, 529–532.

62. Seiler, M. J., and Aramant, R. B. (1994). Photoreceptor and glial markers in human embryonic retina and in human embryonic retinal transplants to rat retina. *Brain Res. Dev. Brain Res.* **80**, 81–95.

63. Seiler, M. J., and Aramant, R. B. (1998). Intact sheets of fetal retina transplanted to restore damaged rat retinas. *Invest. Ophthalmol. Visual Sci.* **39**, 2121–2131.

64. Gouras, P., Du, J., Kjeldbye, H., Yamamoto, S., and Zack, D. J. (1994). Long-term photoreceptor transplants in dystrophic and normal mouse retina. *Invest. Ophthalmol. Visual Sci.* **35**, 3145–3153.

65. Wang, X., and Silverman, M. (1996). A new technique for isolation of photoreceptor layer (in Chinese). *Chung-hua Yen K'o Tsa Chih* **32**, 304–306.

66. Ghosh, F., Juliusson, B., Arner, K., and Ehinger, B. (1999). Partial and full-thickness neuroretinal transplants. *Exp. Eye Res.* **68**, 67–74.

67. Huang, J. C., Ishida, M., Hersh, P., Sugino, I. K., and Zarbin, M. A. (1998). Preparation and transplantation of photoreceptor sheets. *Curr. Eye Res.* **17**, 573–585.

68. McCall, M. A., Woch, G., Aramant, R. B., and Seiler, M. J. (2000). Transplanted fetal retina can restore visual responses in a rat model of retinal degeneration (ARVO abstract). *Invest. Opthalmol. Visual Sci.* **41**, S855.

69. Kaplan, H. J., Tezel, T. H., Berger, A. S., Wolf, M. L., and Del Priore, L. V. (1997). Human photoreceptor transplantation in retinitis pigmentosa: A safety study. *Arch. Ophthalmol.* **115**, 1158–1172.

70. Das, T., del Cerro, M., Jalali, S., Rao, V. S., Gullapalli, V. K., Little, C., Loreto, D. A., Sharma, S., Sreedharan, A., del Cerro, C., and Rao, G. N. (1999). The transplantation of human fetal neuroretinal cells in advanced retinitis pigmentosa patients: Results of a long-term safety study. *Exp. Neurol.* **157**, 58–68.

71. Radtke, N. D., Aramant, R. B., Seiler, M., and Petry, H. M. (1999). Preliminary report: Indications of improved visual function after retinal sheet transplantation in retinitis pigmentosa patients. *Am. J. Ophthalmol.* **3**, 384–387.

72. Humayun, M. S., de Juan, E. Jr., Del Cerro, M., Dagnelie, G., Radner, W., Sadda, S. R., and del Cerro, C. (2000). Human neural retinal transplantation. *Invest. Ophthalmol. Visual Sci.* **41**, 3100–3106.

73. Del Cerro, M., Humayun, M. S., Sadda, S. R., Cao, J., Hayashi, N., Green, W. R., del Cerro, C., and de Juan, E., Jr. (2000). Histologic correlation of human neural retinal transplantation. *Invest. Ophthalmol. Visual Sci.* **41**, 3142–3148.

74. Tropepe, V., Coles, B. L., Chiasson, B. J., Horsford, D. J., Elia, A. J., McInnes, R. R., and van der Kooy, D. (2000). Retinal stem cells in the adult mammalian eye. *Science* **287**, 2032–2036.

75. Takahashi, M., Palmer, T. D., Takahashi, J., and Gage, F. H. (1998). Widespread integration and survival of adult-derived neural progenitor cells in the developing optic retina. *Mol. Cell. Neurosci.* **12**, 340–348.

76. Chacko, D. M., Rogers, J. A., Turner, J. E., and Ahmad, I. (2000). Survival and differentiation of cultured retinal progenitors transplanted in the subretinal space of the rat. *Biochem. Biophys. Res. Commun.* **268**, 842–846.

77. Kaplan, H. J., Tezel, T. H., and Del Priore, L. V. (1998). Retinal pigment epithelial transplantation in age-related macular degeneration. *Retina* **18**, 99–102.

78. Gouras, P., Flood, M. T., Kjeldbye, H., Bilek, M. K., and Eggers, H. (1985). Transplantation of cultured human retinal epithelium to Bruch's membrane of the owl monkey's eye. *Curr. Eye Res.* **4**, 253–265.

79. Algvere, P. V., Berglin, L., Gouras, P., and Sheng, Y. (1994). Transplantation of fetal retinal pigment epithelium in age-related macular degeneration with subfoveal neovascularization. *Graefe's Arch. Clin. Exp. Ophthalmol.* **232**, 707–716.

80. Gouras, P., and Algvere, P. V. (1996). Retinal cell transplantation in the macula: New techniques. *Vision Res.* **36**, 4121–4125.

81. Algvere, P. V., Berglin, L., Gouras, P., Sheng, Y., and Dafgard Kopp, E. (1997). Transplantation of RPE in age-related macular degeneration: Observations in disciform lesions and dry RPE atrophy. *Graefe's Arch. Clin. Exp. Ophthalmol.* **235**, 149–158.

82. Algvere, P. V., Gouras, P., and Dafgard Kopp, E. (1999). Long-term outcome of RPE allografts in non-immunosuppressed patients with AMD. *Eur. J. Ophthalmol.* **9**, 217–230.

83. Tezel, T. H., Del Priore, L. V., and Kaplan, H. J. (1997). Harvest and storage of adult human retinal pigment epithelial sheets. *Curr. Eye Res.* **16**, 802–809.

84. Thumann, G., Schraermeyer, U., Bartz-Schmidt, K. U., and Heimann, K. (1997). Descemet's membrane as membranous support in RPE/IPE transplantation. *Curr. Eye Res.* **16**, 1236–1238.

85. Oganesian, A., Gabrielian, K., Ernest, J. T., and Patel, S. C. (1999). A new model of retinal pigment epithelium transplantation with microspheres. *Arch. Ophthalmol. (Chicago)* **117**, 1192–1200.

86. Gabrielian, K., Oganesian, A., Farrokh-Siar, L., Rezai, K. A., Verp, M. S., Patel, S. C., and Ernest, J. T. (1999). Growth of human fetal retinal pigment epithelium as microspheres. *Graefe's Arch. Clin. Exp. Ophthalmol.* **237**, 241–248.

87. Williams, K. A. (1999). Transplantation of autologous iris pigment epithelial cells as a treatment for age-related macular degeneration? *Transplantation* **68**, 171–172.

88. Zhang, X., and Bok, D. (1998). Transplantation of retinal pigment epithelial cells and immune response in the subretinal space. *Invest. Ophthalmol. Visual Sci.* **39**, 1021–1027.

89. Crafoord, S., Algvere, P. V., Seregard, S., and Dafgard Kopp, E. (1999). Long-term outcome of RPE allografts to the subretinal space of rabbits. *Acta Ophthalmol. Scand.* **77**, 247–254.

90. Crafoord, S., Algvere, P. V., Dafgard Kopp, E., and Seregard, S. (2000). Cyclosporine treatment of RPE allografts in the rabbit subretinal space. *Acta Ophthalmol. Scand.* **78**, 122–129.

91. Peyman, G. A., Blinder, K. J., Paris, C. L., Alturki, W., Nelson, N. C., and Desai, U. (1991). A technique for retinal pigment epithelium transplantation for age-related macular degeneration secondary to extensive subfoveal scarring. *Ophthalmic Surg.* **22**, 102–108.

92. Weisz, J. M., Humayun, M. S., de Juan, E., Jr., Del Cerro, M., Sunness, J. S., Dagnelie, G., Soylu, M., Rizzo, L., and Nussenblatt, R. B. (1999). Allogenic fetal retinal pigment epithelial cell transplant in a patient with geographic atrophy. *Retina* **19**, 540–545.

93. Sunness, J. S., Gonzalez-Baron, J., Applegate, C. A., Bressler, N. M., Tian, Y., Hawkins, B., Barron, Y., and Bergman, A. (1999). Enlargement of atrophy and visual acuity loss in the geographic atrophy form of age-related macular degeneration. *Ophthalmology* **106**, 1768–1779.

94. Thumann, G., Bartz-Schmidt, K. U., El Bakri, H., Schraermeyer, U., Spee, C., Cui, J. Z., Hinton, D. R., Ryan, S. J., and Heimann, K. (1999). Transplantation of autologous iris pigment epithelium to the subretinal space in rabbits. *Transplantation* **68**, 195–201.

95. Kano, T., Yoshida, M., Abe, T., Tomita, H., Akasaka, S., and Tamai, M. (2000). The visual functions of cultured auto iris pigment epithelial cell transplantation in patients with age-related macular degeneration in short time periods (ARVO abstract). *Invest. Ophthalmol. Visual Sci.* **41**, S853.

96. Yoshida, M., Kano, T., Abe, T., Tomita, H., Akasaka, S., and Tamai, M. (2000). Macular changes and visual outcome of cultured auto iris pigment epithelial cells transplantation in age-related macular degeneration (ARVO abstract). *Invest. Ophthalmol. Visual Sci.* **41**, S853.

CHAPTER 101

CNS GRAFTS FOR TREATMENT OF NEUROLOGICAL DISORDERS

Jonathan H. Dinsmore, Jerri Martin, Julie Siegan, Jody Pope Morrison,
Charles Lindberg, Judson Ratliff, and Douglas Jacoby

Cell transplantation for the treatment of neurological disorders is an emerging therapeutic approach. Extensive studies have been performed in animal models and in patients for treatment of Parkinson's disease and Huntington's disease. Less exhaustive studies have been performed for the treatment of other neurological disorders, but there is a growing body of evidence to support the utilization of cell therapy for additional applications. Many of the human studies have relied on the use of human fetal tissue, which has presented limitations due to the difficulty in obtaining human fetal tissue. We have developed the use of fetal pig tissue as an alternative to human fetal tissue. Pig fetal neural cells have proven to perform similarly to human cells in both animal and human studies. Data are presented describing the utilization of pig fetal cells for the treatment of two conditions not previously tested in humans, stroke and chronic intractable pain. Both applications are being tested in initial human safety studies. The applicability of fetal pig cell grafts for multiple neurological disorders is discussed.

INTRODUCTION

Neural cell transplantation shows promise as a therapeutic intervention for neurological disorders in both animal studies and in patients. One limitation to the broad application of neural cell transplants has been the limited access to human fetal tissue. We have circumvented the limitations on human fetal tissue through use of an alternative tissue source, fetal pig neuronal cells. Much work and much attention have been focused on transplantation of fetal neural tissue for the treatment of Parkinson's disease and Huntington's disease. Yet a substantial amount of research shows that other types of neurological disorder (stroke, epilepsy, Alzhiemer's disease, spinal cord injury, and chronic intractable pain) may be equally well treated with neural cell transplants [1]. Clinical studies with fetal pig cell transplants have been performed or are ongoing for Parkinson's disease, Huntington's disease, stroke, and epilepsy and will soon commence for the treatment of spinal cord injury and chronic intractable pain. The most extensive clinical data are available for the treatment of Parkinson's disease, and much of that work is described elsewhere [2–4]. Further data for transplantation of fetal pig cells for Huntington's disease [5,6] and epilepsy [7] have also been reported. Therefore, we focus here on work conducted on the feasibility of using fetal pig neuronal cells for two new treatment applications, stroke and chronic intractable pain.

One of the major challenges associated with the use of pig fetal tissue is the prevention of immune rejection of the grafted cells. Rejection of central nervous system (CNS) grafts can be overcome with standard systemic immunosuppression (i.e., cyclosporine A) or by use of an alternative immunosuppression strategy that involves treatment of cells prior to transplantation with a F(ab')$_2$ fragment of a monoclonal antibody directed against class I major histocompatibility complex (MHC-I) on the surface of the donor pig cells [8]. The presence of F(ab')$_2$ bound to the donor MHC-I antigens alters the process by which T cells recognize and destroy engrafted cells, thus inhibiting graft rejection without compromising the recipient's immune system [9]. Grafted cells survive indefinitely after F(ab')$_2$ treatment, without any further immunosuppression. Thus, transplants of fetal pig neuronal cells can be successfully performed and the functional consequences of the grafts can be assessed.

Depending on the specific disease application that is being addressed, there are various rat animal models and various fetal brain regions that can be utilized to test the ability of cells to restore appropriate function. Specifically, for the application of pig cells for treatment of stroke or chronic intractable pain, we utilized a single cell source, fetal pig lateral ganglionic eminence (LGE) cells. The LGE is the anlage to the adult striatum and contains an abundance of γ-aminobutyric acid (GABA) producing neurons.

GABAergic neurons are abundant throughout the nervous system and have been identified as key regulatory neurons in spinal cord pain pathways. The LGE cells, being rich in GABA neurons, could potentially be utilized as a means of delivering GABA at a focal location. This would not be diffuse release of GABA. Rather, the differentiated GABAergic neurons would supply GABA via synaptic release. Because neurotransmitters are normally released via synapses, the transplanted cells would be expected to more naturally mimic the missing or damaged host cells. We showed earlier that when LGE cells are placed into appropriate striatal target locations, the cells engraft and differentiate much as they do when transplanted to ectopic sites, hippocampus or frontal cortex. Thus, the expectation was that LGE cells transplanted to the spinal cord would behave similarly.

For stroke, we also used LGE cells, although not solely as sources of GABA, but also as a potential means of reconstructing damaged striatal pathways. The LGE cells are striatal precursors, and repeated studies have shown that fetal LGE cells transplanted to the adult brain can reform proper striatal signaling [10]. These studies suggested that striatal damage could be repaired in part through the reestablishment of anatomical connections and appropriate release of GABA. The striatal grafts developed an internal organization that contained areas derived from both glial and neuronal cells [5,6,8]. Striatal grafts became incorporated into host neural circuitry and provided functional effects that appeared to be dependent upon the reconstruction of reciprocal circuits in the host brain. Studies have shown that inputs to and outputs from the striatum are reformed to some extent by striatal grafts, and this was observed for both allografts and xenografts [5,6,8,10].

We describe next the utilization of fetal pig LGE cells to treat stroke (i.e., to repair damaged striatal functions) and to treat chronic intractable pain (i.e., to act as a localized source of GABA). Striatal grafts survived both in the ischemic brain and in the spinal cord at high frequency, and equivalent graft survival was observed with either cyclosporine A or F(ab')$_2$ treatment. Grafts either in the brain or in the spinal cord gray matter differentiated into GABAergic neurons and formed synaptic connections.

METHODS
FETAL PORCINE CELL ISOLATION

Donor gilts were naturally bred, and at 34–38 days of gestation, uteri were harvested and cells prepared as described elsewhere [5,7,8]. Prior to transplantation, approximately half the LGE cells obtained from a fetal harvest were immunomodulated with an anti-MHC Class I F(ab')$_2$ fragment to prevent rejection. Cells were resuspended to 1×10^7 cells/ml in Hanks's Balanced Salt Solution (HBSS) containing 10 µg/ml anti-MHC class I F(ab')$_2$ fragment (PT85, Veterinary Medicine Research and Development, Pallman, WA). Following a 1-h incubation period at 4°C, the cells were rinsed with ice-cold HBSS to remove unbound F(ab')$_2$ fragment and resuspended at 4°C in transplantation media containing 1 µg/ml masking F(ab')$_2$ fragment.

ANIMALS AND EXPERIMENTAL DESIGN

Sprague-Dawley (spinal cord and cerebral ischemia) and male spontaneous hypertensive (SHR) rats (Taconic Farms, Germantown, NY, or Charles River Laboratories, Wilmington, MA) weighing 250–340 g at the start of the study were used. To suppress xenograft rejection, two methods were employed:

1. General immunosuppression of animals by daily injection of cyclosporine A as described elsewhere [7]: subcutaneous dose of 10 mg/kg, for intracerebral transplants and 15 mg/kg for grafts in the spinal cord was used [11]
2. F(ab')$_2$ treatment of cells prior to transplant

Lesion and Transplantation Surgery

Published methods were used to create stroke damage in the brain, and the middle cerebral artery was occluded via an embolus [12]. The embolus was left in place for 1–1.5 h prior to reperfusion of the affected area. A microcannula approach was used for all stereotaxic surgeries [13,14]. For the stroke rats, transplants were performed at 3, 7, 14, or 28 days after occlusion surgery. Briefly, animals were anesthetized with a Rompum/Ketamine cocktail (1:7 mixture of 100 mg/ml solution of each; administered intraperitoneally at 1 ml/kg) and placed in a stereotaxic frame with incisor bar set at –2.5. A small incision was made to expose the skull, and a 1-mm burr hole drilled +5 mm lateral to the intersection of the bregma and the midline. LGE cells or saline or vehicle (transplantation media) were then drawn into a 5-μl Hamilton syringe bearing a pulled glass micropipette tip (outer diameter: approx. 0.16 mm, inner diameter: 0.1 mm). The syringe was set at an angle of +15° from vertical and the glass tip was inserted 5.5 mm into the brain as measured from the dura mater. Boluses of cells or vehicle were injected at the following depths and amounts at a rate of 1 μl/min: at –5.5 mm, 1 μl; at –4 mm, 1.5 μl, and at –2 mm, 1 μl. The first two boluses totaling 250,000 cells were deposited in the striatum and the last bolus of 100,000 cells was placed in the cortex. This made a total of 350,000 LGE cells injected in a 3.5-μl volume. Between depths, the syringe was left in place for 1 min before being slowly drawn upward to the next injection site. A 5-min wait was included after the final injection before the micropipette was slowly withdrawn. The incision was closed, and the animals were coded for blinded behavior studies.

Intraspinal transplantation was achieved by means of familiar methods [15,16]. Rats were anesthetized with ketamine (87 mg/ml) and xylazine (13 mg/ml) for all surgical procedures. Several spinal segments (T12-L2) of the spinal cord were exposed via drilling or laminectomy, and the cord was immobilized on a frame to assure precise stereotaxic implantation. Engraftment was accomplished by injecting 1 μl of 1×10^5 cells through a glass micropipette attached to a 1-μl Hamilton syringe into the intact spinal cord. The pulled micropipettes have an outside diameter of 50–100 μm, thus increasing the accuracy of delivering cells to small targets and decreasing the damage caused by the needle. Two discrete deposits of 0.5 μl of either saline or 0.5×10^5 of porcine cells were injected each over a 1-min period within the same needle track at the following coordinates relative to the dorsal spinal artery: lateral, –0.7 mm; ventral, –0.7 and –0.5 mm. There was a 1-min wait between injections. In addition, the glass tip remained in the spinal cord for 5 min following the –0.5-mm injection to prevent back-flow out of LGE cells upon removal of the micropipette tip. The musculature was sutured (4-0 silk), and the skin closed with autoclips.

Behavioral Assessments

Rats with cells transplanted to the spinal cord were observed weekly for a series of 12 weeks. Responses to noxious and innocuous stimuli were determined from three tests, used sequentially: a copper plate chilled over ice to 4°C for cold allodynia, von Frey thresholds for tactile allodynia, and a pin prick test for mechanical hyperalgesia. Locomotion was assessed using the modified locomotion open-field test of Basso *et al.* [17].

The behavior of rats with middle cerebral artery (MCA) occlusion was assessed by means of a battery of behavioral tests designed to assess both sensory and motor effects. Animals were trained several times and tested once before lesioning; they were tested every week following MCA occlusion surgery until transplant, and every 2–4 weeks following cell transplantation. The following sensory tests were used to assess the degree of stroke-related impairment: visual placing test [18], vibrissae tactile test [19], forelimb tactile test ("Bilateral Asymmetry test") [20], and hindlimb tactile test [21]. The following **motor** tests were used to assess the degree of stroke-related impairment: elevated body swing test (EBST) [22], posture reflex test [23], outreach test [24], and balance beam (forelimb/hindlimb motor control) [25].

HISTOLOGICAL PROCEDURES

Transcardial perfusion was performed on each animal with 0.9% heparinized saline followed by Zamboni's fixative. Spinal cords were removed intact with the vertebral column. The vertebral column and spinal cord were postfixed for 24 h prior to identification and removal of the transplanted area. Spinal cord tissue was paraffin embedded and sectioned at 5–7 μm. Serial sections were stained with hematoxylin and eosin to appropriately identify the transplanted area prior to performing additional immunohistochemical tests.

Brains from animals in which cells were transplanted into the ischemic area were removed intact and placed in 10% buffered formalin for 2–4 h. Brains were then cut into slices 2–3 mm wide and returned to 10% buffered formalin to fix for 12–16 h. After fixation, tissue was processed for paraffin and thin sections were cut at 6–10 μm. Sections mounted on slides were stained with hematoxylin and eosin for identification of the graft site and to monitor lymphocytic infiltration. Brain sections were immunostained with antibodies [5,7] to identify specific neuronal and glial cells.

RESULTS

STROKE STUDIES

The two strains of rat were divided into four groups to receive LGE cell implants at either 3, 7, 14, or 28 days following ischemic injury. Control animals for each group received vehicle instead of LGE cells. Animals were assessed every 2–4 weeks by means of the battery of sensory and motor behavior tests described earlier and either were immunosuppressed by daily cyclosporine A administration or received cells treated with $F(ab')_2$. All animals were necropsied at 12 weeks following transplantation surgery, and the brains were processed for histological examination of the graft and host tissue. Our studies showed that LGE cells survived equally well at all postischemia test times and in all rat strains tested. Further, survival was equivalent between animals given cells treated with cyclosporine A or $F(ab')_2$ (Table 101.1).

Grafts survived well within ischemic zones, showing organotypic organization typical of striatal grafts. The cells formed solid grafts that filled from 30–100% of the ischemic area (Fig. 101.1) and integrated with surrounding host tissue (Figs. 101.2 and 101.3). Grafts integrated into the host brain with such minimal disruption that the border areas between host and graft were difficult to discern without the assistance of species-specific staining (Figs. 101.2 and 101.3). Pig neurons were shown to send neuronal projections from the graft to host structures (Fig. 101.3). Additionally, a pig-specific synaptic marker, synaptobrevin, revealed pig synapses that formed both within the graft and within the host brain (Fig. 101.2). Pig glial cells were also distributed throughout the graft, with some glial fibers extending into host brain (Fig. 101.2). Behavioral improvements were also shown to occur in grafted rats. The behavioral recovery seen in transplanted rats did not differ in total extent from that seen in control rats, but the rate of initial recovery was more rapid in the transplant group. The spontaneous recovery of function in rats following ischemic damage

Table 101.1. Comparison of Graft Survival Rates of Transplanted Porcine LGE Cells[a]

Target location	Cyclosporine A cells		$F(ab')_2$-treated cells	
	Graft survival (%)	*n*	Graft survival (%)	*n*
Ischemic brain	86	43	83	30
Spinal cord	73	26	77	13

[a]Survival rates for each experiment listed and for the total of pooled experiments were compiled, and a chi-squared test performed to determine if survival rates were equivalent.

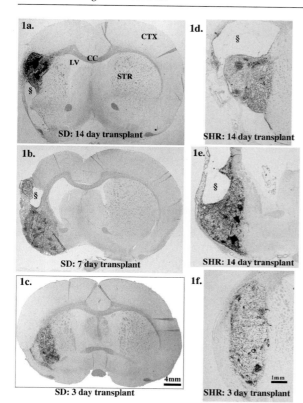

Fig. 101.1. In situ hybridization for pig DNA in stroke grafts: micrographs taken from different rats. The grafts were stained by in situ hybridization for pig DNA. Graft areas appear darkly stained next to the adjacent host brain. The infarct cavity is apparent in rats where the graft did not fill the entire infarct (a, b, d, and e) and nonexistent in rats where the graft completely filled the infarct (c, f). SD, Sprague-Dawley; SHR, spontaneous hypertensive rat; LV, lateral ventricle, CC, corpus callosum, STR, striatum, CTX, cortex, §, infarct cavity.

has been shown by others [26,27], and is a weakness of the rat stroke model. However, the more rapid behavioral recovery in transplanted rats suggests that cells have a beneficial effect on behavioral recovery. Furthermore, the results clearly demonstrated cells could be administered any time after a stroke without adverse effects.

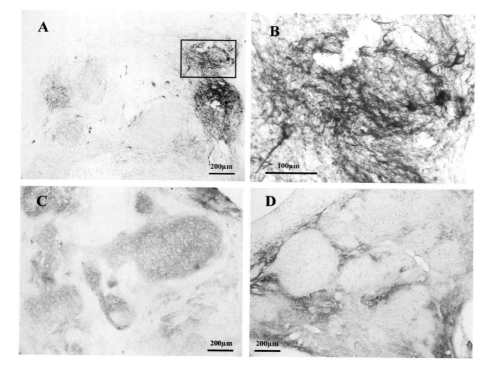

Fig. 101.2. Pig-specific neuronal and glial staining of grafts: micrographs taken from the same adjacent sections from the same animal. (A) and (B) were stained for pig neurofilament, (C) for pig synaptobrevin, and (D) for pig glia. (A), (C), and (D) are at the same magnification, (B) is at a higher magnification. Boxed region in (A) is the area shown at higher magnification in (B). Note the pattern of staining. Neuronal cell staining (A)–(C) is segregated from glial staining (D). Neuronal cells occur as aggregates of cells surrounded by areas rich in glial cells. Synaptic staining is concentrated in areas occupied by neuronal cells (C), as is appropriate because neuronal cells do not form synapses with glial cells.

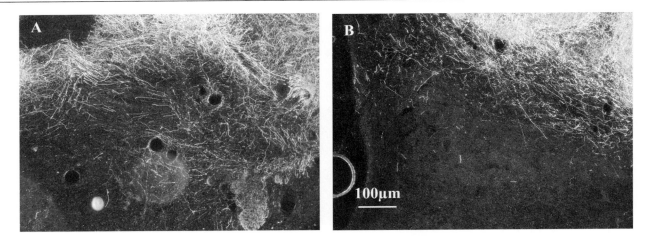

Fig. 101.3. Darkfield imaging of pig neurofilament staining in grafts from two different rats. The sections were stained with an antibody to pig neurofilament and were viewed by dark-field optics to more sensitively reveal the neuronal projections form grafted pig cells. There is a dense plexus of neuronal projections within the graft (upper right region in both micrographs) accompanied by an extensive network of projections penetrating the host rat brain.

PAIN STUDIES

Rats were divided between those that received porcine LGE cells and were immuno-suppressed with daily injections of cyclosporine A and those that received LGE cells treated with F(ab')$_2$. As with the transplants for stroke, the survival of grafts was equivalent for the two groups (Table 101.1). Grafts were found within the target sites in the spinal gray matter and stained for specific pig neuronal and glial markers (Fig. 101.4). Although cells were transplanted into the normal spinal cord, there were no associated adverse effects of the transplant at either the behavioral or the histological level. Transplanted rats were tested for several abnormal pain (cold allodynia, tactile allodynia, and mechanical hyperalgesia) or motor behaviors, and none were observed (data not shown).

Within the graft site, the transplants were highly integrated with the host gray matter structure. Borders between graft and host were continuous, with little evidence of any

Fig. 101.4. Neurofilament and glutamic acid decarboxylase (GAD) staining of grafts in the spinal cord. (A)–(D) Staining of graft to the spinal cord from adjacent sections taken from the same rat: (A) and (C) are low- and high-magnification views, respectively, of the graft stained for pig-specific neurofilament; (B) and (D) are low- and high-magnification views, respectively, the graft stained for GAD. GAD staining reveals the GABAergic neurons within the graft, and the neurofilament reveals the neuronal projections from the neurons in the graft. The graft boundaries are marked by arrows in panels (A) and (B). Boxed regions in (A) and (B) are the areas shown at higher magnification in (C) and (D). Arrows in (C) highlight neuronal projections extending from the graft to the host spinal cord, and the dashed line represents the graft–host border. Arrows in (D) point to GAD-positive cell bodies within the graft.

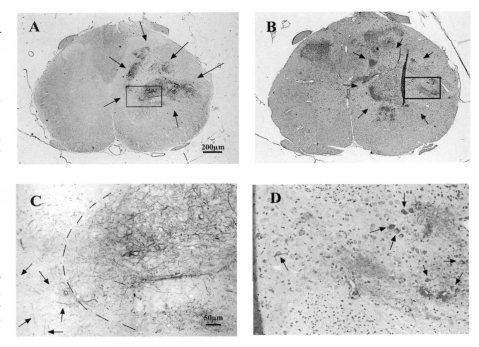

host gliotic response (Fig. 101.4). Cells within the graft expressed both glial and neuronal markers and were organized within the graft in a manner highly reminiscent of that seen when these cells were transplanted into the striatum. Thus, fetal striatal precursor (LGE) cells behave similarly regardless of whether they are transplanted into their natural environment in the striatum or into a foreign environment in the spinal cord gray matter. This feature of the grafts is extremely important for the proper utilization of these grafts to deliver GABA to the spinal cord in a site-specific and neuronally regulated manner. Since neurons from the grafts send neuronal extensions into the host neuronal tissue, the delivery of GABA by the grafts is not limited to the region immediately adjacent to the graft (Fig. 101.4). These neurite projections can extend for many millimeters away from the graft. GABA is released at the nerve terminal; thus, the effective region over which GABA can be released is greatly increased by the extent of neurite outgrowth. Neurons from the graft do form synapses and was demonstrated by staining with pig-specific antibodies directed against the synapse-specific protein synaptobrevin (data not shown).

CONCLUSIONS

Fetal pig striatal (LGE) cells provide a novel means of reconstructing neuronal defects in two dissimilar conditions, stroke and chronic intractable pain. The ultimate benefits of these cell transplants must be tested in patients. The positive demonstration of graft survival with appropriate structural and functional elements is the first step to applying cells in patients. The future will be exciting as more experience is gained with cellular therapy in patients.

REFERENCES

1. Edge, A. S., Gosse, M. E., and Dinsmore, J. (1999). Xenogeneic cell therapy: Current progress and future developments in porcine cell transplantation. *Cell Transplant* 7(6), 525–539.
2. Deacon, T. *et al.* (1997). Histological evidence of fetal pig neural cell survival after transplantation into a patient with Parkinson's disease. *Nat. Med.* 3, 350–353.
3. Fink, J. S., *et al.* (2000). Porcine xenografts in Parkinson's disease and Huntington's disease patients: Preliminary results. *Cell Transplant* 9(2), 273–278.
4. Schumacher, J. M., *et al.* (2000). Transplantation of embryonic porcine mesencephalic tissue in patients with PD. *Neurology* 54(5), 1042–1050.
5. Deacon, T. W., Pakzaban, P., Burns, L. H., Dinsmore, J., and Isacson, O. (1994). Cytoarchitectonic development, axon–glia relationships, and long distance axon growth of porcine striatal xenografts in rats. *Exp. Neurol.* 130(1), 151–167.
6. Isacson, O., Deacon, T. W., Pakzaban, P., Galpern, W. R., Dinsmore, J., and Burns, L. H. (1995). Transplanted xenogeneic neural cells in neurodegenerative disease models exhibit remarkable axonal target specificity and distinct growth patterns of glial and axonal fibres. *Nat. Med.* 1(11), 1189–1194.
7. Jacoby, D. B., Lindberg, C., Cunningham, M. G., Ratliff, J., and Dinsmore, J. (1999). Long-term survival of fetal porcine lateral ganglionic eminence cells in the hippocampus of rats. *J. Neurosci. Res.* 56(6), 581–594.
8. Pakzaban, P., Deacon, T. W., Burns, L. H., Dinsmore, J., and Isacson, O. (1995). A novel mode of immunoprotection of neural xenotransplants: Masking of donor major histocompatibility complex class I enhances transplant survival in the central nervous system. *Neuroscience* 65, 983–996.
9. DerSimonian, H., Pan, L., Yatko, C., Rodrigue-Way, A., Johnson, E., and Edge, A. S. (1999). Human anti-porcine T cell response: Blocking with anti-class I antibody leads to hyporesponsiveness and a switch in cytokine production. *J. Immunol.* 162(12), 6993–7001.
10. Dunnett, S. B. (1995). Functional repair of striatal systems by neural transplants: Evidence for circuit reconstruction. *Behav. Brain Res.* 66(1-2), 133–142.
11. Giovanini, M. A., Reier, P. J., Eskin, T. A., Wirth, E., and Anderson, D. K. (1997). Characteristics of human fetal spinal cord grafts in the adult rat spinal cord: Influences of lesion and grafting conditions. *Exp. Neurol.* 148(2), 523–543.
12. Zea Longa, E., Weinstein, P. R., Carlson, S., and Cummins, R. (1989). Reversible middle cerebral artery occlusion without craniectomy in rats. *Stroke* 20(1), 84–91.
13. Nikkhah, G., Olsson, M., Eberhard, J., Bentlage, C., Cunningham, M. G., and Bjorklund, A. (1994). A microtransplantation approach for cell suspension grafting in the rat Parkinson model: A detailed account of the methodology. *Neuroscience* 63(1), 57–72.
14. Nikkhah, G., Cunningham, M. G., Jodicke, A., Knappe, U., and Bjorklund, A. (1994). Improved graft survival and striatal reinnervation by microtransplantation of fetal nigral cell suspensions in the rat Parkinson's model. *Brain Res.* 633, 133–143.
15. Reier, P. J., Bregman, B. S. and Wujek, J. R. (1986). Intraspinal transplantation of embryonic spinal cord tissue in neonatal and adult rats. *J. Comp. Neurol.* 247(3), 275–296.
16. Stokes, B. T., and Reier, P. J. (1992). Fetal grafts alter chronic behavioral outcome after contusion damage to the adult rat spinal cord. *Exp. Neurol.* 116(1), 1–12.

17. Basso, D. M., Beattie, M. S., and Bresnahan, J. C. (1995). A sensitive and reliable locomotor rating scale for open field testing in rats. *J. Neurotrauma* **12**(1), 1–21.

18. De Ryck, M., Van Reempts, J., Duytschaever, H., Van Deuren, B., and Clincke, G. (1992). Neocortical localization of tactile/proprioceptive limb placing reactions in the rat. *Brain Res.* **573**, 44–60.

19. Tucker, J. C., McDaniel, W. F., and Smith, S. R. (1992). A behavioral study of bilateral middle cerebral artery hemorrhagic ischemia in rats. *NeuroReport* **3**(8), 725–728.

20. Markgraf, C. G., Green, E. J., Hurwitz, B. E., Morikawa, E., Deitrich, W. D., McCabe, M., Ginsberg, M. D., and Schneiderman, N. (1992). Sensorimotor and cognitive consequences of middle cerebral artery occlusion in rats. *Brain Res.* **575**, 238–246.

21. Kawamata, T., Alexis, N. E., Dietrich, W. D., and Finklestein, S. P. (1996). Intracisternal basic fibroblast growth factor (bFGF) enhances behavioral recovery following focal cerebral infarction in the rat. *J. Cereb. Blood Flow Metab.* **16**, 542–547.

22. Borlongan, C. V., and Sanberg, P. R. (1995). Elevated body swing test: A new behavioral parameter for rats with 6-hydroxydopamine-induced hemiparkinsonism. *J. Neurosc.* **15**(7), 5372–5378.

23. Bederson, J. B., Pitts, L. H., Tsuji, M., Nishimura, M. C., Davis, R. L., and Bartkowski, H. (1986). Rat middle cerebral artery occlusion: Evaluation of the model and development of a neurologic examination. *Stroke* **17**(3), 472–476.

24. Aronowski, J., Samways, E., Strong, R., Rhoades, H. M., and Grotta, J. C. (1996). An alternative method for the quantitation of neuronal damage after experimental middle cerebral artery occlusion in rats: Analysis of behavioral deficit. *J. Cereb. Blood Flow Metab.* **16**(4), 705–713.

25. Alexis, N. E., Dietrich, W. D., Green, E. J., Prado, R., and Watson, B. D. (1995). Nonocclusive common carotid artery thrombosis in the rat results in reversible sensorimotor and cognitive behavioral deficits. *Stroke* **26**, 2338–2346.

26. Grabowski, M., Brundin, P., and Johansson, B. B. (1993). Paw-reaching, sensorimotor, and rotational behavior after brain infarction in rats. *Stroke* **24**, 889–895.

27. Markgraf, C. G., Johnson, M. P., Braun, D. L., and Bickers, M. V. (1997). Behavioral recovery patterns in rats receiving the NMDA receptor antagonist MDL 100,453 immediately post-stroke. *Pharmacol., Biochem. Behav.* **56**(3), 391–397.

CHAPTER 102

PERIPHERAL NERVE REGENERATION

Byung-Soo Kim and Anthony Atala

INTRODUCTION

The usual method of repairing a severed nerve involves direct suturing of the proximal and distal stumps. When the gap of a transected nerve is large, it cannot be repaired by direct anastomosis without tension, and often interposition of autologous donor nerve grafts is required. However, the autograft method may cause donor site morbidity and does involve a secondary surgery.

A number of methods have used in attempts to repair peripheral nerve defects that are too lengthy to be repaired with direct anastomosis in animal models. Tubular conduits have been employed to bridge defective nerve gaps for guided nerve regeneration. These conduits provide a nerve guidance channel through which axons can sprout from the proximal nerve stump, thus reducing the infiltration of fibrous tissue. Several studies have demonstrated that the use of such conduits can repair nerve injury without the use of autografts [1–5]. The tubular conduits have been fabricated from various materials, including nondegradable synthetic materials such as silicone [1–4], polycarbonate [6], and poly(acrylonitrile-co-vinyl chloride) [7], as well as biodegradable polymers such as poly(lactic acid) [8,9], poly(lactic-co-glycolic acid) (PLGA) [9,10], poly(glycolic acid) [11], poly(lactic acid-co-ε-caprolactone) [12–14], and polyhydroxybutyrate [15]. Conduits fabricated from nondegradable synthetic materials remain *in situ* as a foreign body and may elicit an inflammatory response, limiting nerve regeneration. In contrast, conduits fabricated from biodegradable scaffolds offer the advantage of resorption when nerve regeneration is complete and reduce the chance of inflammatory responses.

It has been also shown that loading nerve grafts with neurotrophic factors stimulates peripheral nerve regeneration [16–20]. The soluble factors beneficial for nerve regeneration include acidic fibroblast growth factor (aFGF) [16], basic fibroblast growth factor (bFGF) [19], nerve growth factor (NGF) [17], glial growth factor (GGF) [18], brain-derived neurotrophic factor (BDNF) [20], ciliary neurotrophic factor (CNTF) [20], platelet-derived growth factor (PDGF) [21], and insulin-like growth factor 1 (IGF-1) [21]. These soluble factors play important roles in the differentiation, maintenance, and survival of neurons.

Loading nerve grafts with extracellular matrices (ECMs) has been found to enhance peripheral nerve regeneration. These ECM molecules include collagen [11,22], laminin [23,24], glycosaminoglycan [22], and fibronectin [23]. Nerve grafts loaded with laminin and fibronectin showed improved axonal growth and regenerated nerve function [23]. In another study, tubular conduits with lumina filled with collagen and glycosaminoglycan resulted in improved peripheral nerve regeneration [22]. Reports that, alginate [24] and fibrin [25] also promote peripheral nerve regeneration appeared in 2000.

Several studies have demonstrated that seeding of Schwann cells into nerve grafts promotes peripheral nerve regeneration [26,27]. Schwann cells, which ensheath axons, are known to secrete neurotrophic factors (e.g., NGF, BDNF, CNTF, IGF-1) and ECM molecules (e.g., laminin, collagen type IV) and to guide the growth of regenerating axons [28–30]. Thus, in the process of peripheral nerve regeneration, Schwann cells may play a critical role, serving as both a source of various neurotrophic factors and a substrate that guides axonal growth.

Recent experiments in our laboratory indicate that acellular nerve matrix allografts may be a useful biomaterial for functional peripheral nerve regeneration. The acellular nerve matrix allografts represent the ECM of peripheral nerves in its native structure, composition, and mechanical properties. This preservation of complex features of native ECM may be physiologically relevant in the reconstruction of functional peripheral nerves. In this chapter, we describe the preparation of biodegradable polymer conduits and acellularized nerve allografts, the loading of grafts with growth factors, ECMs, or Schwann cells, and the assessment of nerve functional recovery in animal models.

NERVE GRAFTING WITH ACELLULAR NERVE MATRIX
ACELLULAR NERVE MATRIX PREPARATION
To obtain acellularized nerve allografts for nerve repair, sciatic nerves are taken from cadaveric donors and processed by means of a multiple-step decellularization technique.

Materials
    500 ml flask
    Orbital shaker
    Lyophilization chamber and lyophilizer
    Ethylene oxide sterilization unit
    Triton X-100 (Sigma, St. Louis, MO)
    Ammonium hydroxide

Methods
1. Immerse the isolated nerve segments in distilled water at 4°C for 24 h. Replace distilled water several times during this period.
2. Treat the nerve segments with 0.5% (v/v) Triton X-100 and 47.6 mM ammonium hydroxide in a 500-ml flask, and stir the flask at 4°C for 72 h for cell lysis.
3. Wash the nerve segments thoroughly with distilled water several times. Repeat step 2.
4. Lyophilize the nerve segments for 24 h.
5. Cut into 1-cm lengths.
6. Sterilize with ethylene oxide and leave for a sufficient period (>48 h) for the release of ethylene oxide after sterilization.

PROCEDURES FOR GRAFTING
The processed nerve allografts can be placed across the nerve defect gap and anastomosed microsurgically.

Materials
Surgical instruments should be washed and autoclaved.

    Sterile gauze
    Microsurgical scope
    Acellular nerve matrix grafts
    9-0 nylon sutures (Ethicon, Somerville, NJ)

Methods
1. Expose the defective nerve. Confirm functional status and excise defective segment. The nerve should be handled gently, to prevent axonal damage.
2. Place the nerve allograft across the gap.

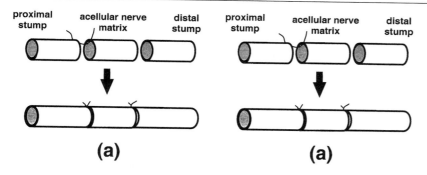

Fig. 102.1. Schematics showing surgical placement of (a) acellular nerve matrix graft and (b) tubular conduit.

3. Suture the graft microsurgically to the epineurium of the proximal and distal nerve stumps with two or three sutures (9-0 nylon) at each junction (Fig. 102.1a).

4. Close the wound.

NERVE GRAFTING WITH TUBULAR CONDUITS LOADED WITH ECMS, SCHWANN CELLS, OR GROWTH FACTORS

Tubular conduits can be fabricated from biodegradable polymers, such as PLGA. To enhance the regeneration of injured nerve, the tubular conduits can be loaded with ECM, such as collagen and laminin, growth factors, such as aFGF, NGF, and GGF, and/or Schwann cells.

BIODEGRADABLE POLYMER CONDUIT PREPARATION

Materials

Thin-walled Polytef sheath (i.d. 2.0 mm), glass capillary (o.d. 1.4 mm)
Fume hood
85:15 PLGA (Birmingham Polymers, Birmingham, AL)
Chloroform ($CHCl_3$) or methylene chloride (CH_2Cl_2)
Vacuum chamber
Ethylene oxide sterilization unit

Methods

1. A thin-walled Polytef sheath (i.d. 2.0 mm) is passed over a glass capillary (o.d. 1.4 mm).

2. Inside a fume hood, the tube is dipped for 15 s into a 5% (w/v) polymer solution of 85:15 PLGA dissolved in either chloroform or methylene chloride.

3. The tube is removed the polymer solution and suspended in air for 15 s for solvent evaporation.

4. Repeat 20 cycles of steps 2 and 3.

5. After the dipping procedure, the tube remains in the fume food overnight and under vacuum for 48 h for residual solvent evaporation.

6. The formed 85:15 PLGA tubular conduit is removed from the tube and cut into 12-mm-long conduits.

7. The polymer conduits are sterilized with ethylene oxide before use.

ECM PREPARATION

Materials

Laminin (Biomedical Technologies, Stoughton, MA)
Vitrogen 100 (rat tail collagen, type I, Collagen Corp., Palo Alto, CA)
Dulbecco's Modified Eagle's Medium (DMEM)
1 N NaOH

Methods

All pipettes and tissue culture materials that contact laminin solution and collagen solution must be cold before use to prevent gelation.

1. To make laminin solution, dissolve laminin in DMEM at a concentration of 1 mg/ml. The laminin solution must be stored at 4°C until use because it forms a gel at room temperature.

2. To make collagen solution, mix 300 μl of Vitrogen 100 with ice-cold DMEM (600 μl) and 1 N NaOH (7μl). The collagen final concentration is 1.1 mg/ml. The collagen solution must be stored at 4°C until use because it forms a gel at room temperature.

ISOLATION AND CULTURE OF SCHWANN CELLS
Materials

    Vitrogen 100
    0.02 N acetic acid
    Culture dishes
    Phosphate-buffered saline (PBS), DMEM, fetal bovine serum (FBS), penicillin–
        streptomycin (1000 U/ml)
    No. 15 scalpel blade, forceps
    Trypsin, collagenase, and hyaluronidase (Sigma)
    Microsurgical scope

Methods

1. Dilute Vitrogen 100 to 50 μg/ml with 0.02 N acetic acid.

2. Add 1–2 ml of this solution to 35-mm culture dish and incubate at room temperature for 1 h.

3. Carefully aspirate remaining solution, and rinse well with PBS to remove acid. Store the collagen-coated dishes at 4°C until use.

4. Donor peripheral nerves are obtained.

5. The epineurial tissues are microscopically peeled away, and the remaining tissues are cut into 1-mm-long segments.

6. The nerve segments are placed on collagen-coated culture dishes in DMEM supplemented with 10% (v/v) FBS and penicillin–streptomycin (1000 U/ml).

7. Every 5 days, the nerve segments are transferred to new collagen-coated culture dishes, leaving behind the fibroblasts that migrate out onto the collagen. After four such transfers, almost all the fibroblasts will have left the nerve segments.

8. The nerve segments are digested with 0.3% (w/v) trypsin, 0.1% (w/v) collagenase, and 0.1% (w/v) hyaluronidase in DMEM at 37°C for 2 h.

9. The cell suspension is centrifuged at 200g for 5 min, and the supernatant is discarded.

10. Add fresh DMEM, resuspend the cells, and repeat step 9.

11. Resuspend the cell peallet in DMEM supplemented with 10% (v/v) FBS, 2 mM forskolin, and 10 mg/ml pituitary extract, and culture at 37°C in 100% humidified atmosphere with 5% (v/v) CO_2.

PROCEDURES FOR GRAFTING
Materials

    Microsurgical scope
    85:15 PLGA tubular conduits (i.d. 1.4 mm, length 16 mm)
    Collagen or laminin solution as prepared in connection with ECM
    Schwann cells
    NGF, aFGF, bFGF, IGF-1, BDNF, CNTF, PDGF-BB, and/or GGF (all soluble factors
        from Sigma, except GGF from Cambridge Neurosciences, Cambridge, MA)
    9-0 nylon sutures
    Fibrin glue (Vitex Technologies, New York)

Methods

1. Place the tubular conduit (fabricated from 85:15 PLGA) across the nerve gap.

2. A suture (9-0 nylon) is (a) placed 1 mm in from the end of conduit and pulled through the lumen of the conduit, (b) placed through the epineurium 1 mm in from the proximal nerve stump, (c) placed back to the lumen 1 mm in from the end of the conduit, and (d) carried through to the outer wall of the conduit (Fig. 102.1b).

3. The proximal nerve stump is pulled into the conduit by gently pulling on the suture/nerve until it moves into the conduit. The knot is tied using three ties.

4. The junction of proximal nerve stump and conduit is sealed with fibrin glue.

5. A 50-μl syringe is used to fill the conduit with 20 μl of either collagen or laminin solution. Also, growth factor(s), such as NGF (100 ng/ml), aFGF (1000 U/ml), bFGF (10 μg/ml), IGF-1 (30 μg/ml), BDNF (10 μg/ml), CNTF (10 μg/ml), PDGF-BB (60 μg/ml), and/or GGF (10 μg/ml), and/or Schwann cells (5×10^7 cells/ml) can be mixed with the ECM and injected into the lumen of the conduit. The ECM forms a gel at body temperature.

6. The same procedure is used to position and anchor the distal stump to the conduit. Fibrin glue is applied to seal the junction of distal nerve stump and conduit.

ASSESSMENTS OF NERVE FUNCTION RECOVERY

NEUROBEHAVIORAL TESTS

Motor function in animals can be examined by measuring the extensor postural thrust (EPT). Proprioceptive functions can be evaluated by measuring the tactile placing response (TPR) and the hopping response (HR). Sensory functions can be evaluated by measuring the withdrawal reflux latency (WRL) of a single limb to a hot plate surface [31–33].

1. Extensor postural thrust

   a. Hold the animal upright with the hindlimb extended so that the body's weight is supported by the distal metatarsus and toes.

   b. EPT is measured as the force, applied to a distal platform balance (Ohaus LoPro; Fisher Scientific, Floral Park, NJ), that resists contact of the platform by the heel.

2. Tactile placing response

   a. Hold the animal upright on the test surface.

   b. Flex the toes back, bringing the dorsum of the foot into contact with the testing surface. Ordinarily, a prone animal will respond to having its hindpaw pulled back by returning it to a position alongside its flank.

   c. Measure the ability to reposition the toes to a normal position (TPR) by grading on a scale from 1 to 4 (1, normal return of position: 2, slightly impaired; 3, greatly impaired, and 4, completely absent).

3. Hopping response

   a. Hold the animal upright with the hindlimb on a supporting surface.

   b. One hindlimb at a time is lifted off the ground, and the animal's body is moved laterally. Ordinarily this stimulation of the lateral surface of the foot elicits a hopping reflex, to prevent the animal from falling over. The HR is recorded as either present or absent.

4. Thermal nociception

   a. Hold the animal with a cloth gently wrapped above its waist to restrain the upper extremities and obstruct vision.

   b. Position the animal to stand with one hind paw on a 56°C hot plate (model 35-D, IITC Life Science Instruments, Woodland Hills, CA) and the other on a room temperature plate.

   c. WRL is measured using a stopwatch or a computer data collection system (PC with a footpad switch, Cupertino, CA). If no withdrawal occurred from the hot plate within 12 s, terminate the trial to prevent injury.

ELECTROPHYSIOLOGICAL TEST

To evaluate recovery of motor function, the contractile response of reinnervated muscle upon the sciatic nerve stimulation is tested by recording electromyograms (TE42 machine, TECA Corporation, Pleasantville, NY).

1. Anesthetize the animal (200 g) by a 0.7-ml injection of ketamine (33 mg/ml) and xylazine (6.7 mg/ml) in a 1:1 volume proportion.
2. Expose the peripheral nerve of interest.
3. Place a recording needle in the innervated muscle of interest.
4. Stimulate the nerve with two silver wire electrodes (0.01 ms duration, 1 Hz, 2 V).
5. Record compound action potentials.
6. Measure the early latency and peak amplitude.

IMMUNOHISTOCHEMISTRY

The axon regeneration and Schwann cell ingrowth can be examined by immunohistochemistry.

Materials

PBS-buffered 10% (v/v) formalin
50, 70, 95, and 100% (v/v) ethanol, xylene
Histology cassettes and molds
Water bath
65°C hot plate
Microtome
Glass slides and coverslips
Blocking serum prepared from the species in which the secondary antibody is made
Primary antibodies: NF 200 (Sigma) for axon, S-100 (DAKO, Denmark) for Schwann cells
Vectastain ABC kit (Vector Laboratories, Burlingame, CA)
Diaminobenzidine tetrahydrochloride (DAB) kit (Vector Laboratories)
Xylene-based mounting solution (Cytoseal, Stephens Scientific, Riverdale, NJ)

Methods

1. Dissect the nerve grafts and fix the samples in PBS-buffered 10% (v/v) formalin solution overnight at room temperature.
2. Pour off the formalin solution and replace with 50% (v/v) ethanol. Incubate for 15 min at room temperature.
3. Continue dehydration by incubating three times in a series of 70, 95, and 100% (v/v) ethanol for 15 min each time at room temperature.
4. Replace 100% ethanol with xylene and incubate for 10 min. Since xylene is a toxic solvent, this step should be carried out in a fume hood.
5. Put the samples in histology cassettes and place in a vacuum molten paraffin chamber for 2 h.
6. Embed the samples in paraffin and allow the embedded blocks to solidify overnight at room temperature.
7. Cut the embedded samples at a thickness of 5 μm with a microtome.
8. Float the tissue sections on 45°C distilled water to remove wrinkles of the sections.
9. Mount the tissue sections on glass slides and bake on a 65°C hot plate for 2 h.
10. Deparaffinize the tissue sections on glass slides by immersing the slides in xylene for 15 min twice.
11. Rehydrate the sections by immersing in a series of 100, 95, and 70% (v/v) ethanol and distilled water for 15 min each time at room temperature.
12. Incubate the sections in 0.3% (v/v) H_2O_2 for 30 min at room temperature.
13. Wash in PBS for 5 min twice.
14. Incubate the sections for 30 min at room temperature with diluted blocking serum [1.5% (v/v) in PBS].

15. Remove excess blocking serum from the sections. The blocking serum remains in negative controls.

16. Incubate the sections for 90 min at room temperature or overnight at 4°C with diluted primary antibody solution (1:200 in the blocking serum solution for both S-100 Schwann cell antibody and NF-200 axon antibody). Negative controls are not incubated with the primary antibody solution.

17. Wash in PBS for 5 min twice.

18. Incubate the sections for 20 min at room temperature with diluted biotinylated secondary antibody solution (150 μl of blocking serum stock and 50 μl of biotinylated secondary antibody stock in PBS).

19. Wash in PBS for 5 min twice.

20. Incubate the sections for 30 min at room temperature with Vectastain ABC reagent.

21. Wash in distilled water for 5 min twice.

22. Incubate the sections in DAB solution until desired stain intensity develops.

23. Rinse the sections in tap water.

24. Counterstain with Gill's hematoxylin for 10 s and wash with tap water.

25. Dehydrate with a series of alcohol and xylene, and mount.

REFERENCES

1. Le Beau, J. M., Ellisman, M. H., and Powell, H. C. (1988). Ultrastructural and morphometric analysis of long-term peripheral nerve regeneration through silicone tubes. *J. Neurocytol.* **17**, 161–172.

2. Gibson, K. L., Daniloff, J. K., and Strain, G. M. (1989). Comparison of sciatic nerve regeneration through silicone tubes and nerve allografts. *Microsurgery* **10**, 126–129.

3. Fields, R. D., and Ellisman, M. H. (1986). Axons regenerated through silicone tube splices. *Exp. Neurol.* **92**, 48–60.

4. Jenq, C. B., and Coggeshall, R. E. (1986). Regeneration of transected rat sciatic nerves after using isolated nerve fragments as distal inserts in silicone tubes. *Exp. Neurol.* **91**, 154–162.

5. Bora, F. W., Bednar, J. M., Osterman, A. L., Brown, M. J., and Summer, A. J. (1987). Prosthetic nerve grafts: A resorbable tube as an alternative to autogenous nerve grafting. *J. Hand Surg.* **12A**, 685–692.

6. Harvey, A. R., Chen, M., Plant, G. W., and Dyson, S. E. (1994). Regrowth of axons within Schwann cell-filled polycarbonate tubes implanted into the damaged optic tract and cerebral cortex of rats. *Restorative Neurol. Neurosci.* **6**, 221–237.

7. Aebischer, P., Guenard, V., and Valentini, R. F. (1990). The morphology of regenerating peripheral nerves is modulated by the surface microgeometry of polymeric guidance channels. *Brain Res.* **531**, 211–218.

8. Evans, G. R., Brandt, K., Widmer, M. S., Lu, L., Meszlenyi, R. K., Gupta, P. K., Mikos, A. G., Hodges, J., Williams, J., Gurlek, A., Nabawi, A., Lohman, R., and Patrick, C. W., Jr. (1999). In vivo evaluation of poly(L-lactic acid) porous conduits for peripheral nerve regeneration. *Biomaterials* **20**, 1109–1115.

9. Hadlock, T., Elisseeff, J., Langer, R., Vacanti, J., and Cheney, M. (1998). A tissue-engineered conduit for peripheral nerve repair. *Arch. Otolaryngol. Head Neck Surg.* **124**, 1081–1086.

10. Widmer, M. S., Gupta, P. K., Lu, L., Meszlenyi, R. K., Evans, G. R., Brandt, K., Savel, T., Gurlek, A., Patrick, C. W., Jr., and Mikos, A. G. (1998). Manufacture of porous biodegradable polymer conduits by an extrusion process for guided tissue regeneration. *Biomaterials* **19**, 1945–1955.

11. Kiyotani, T., Nakamura, T., Shimizu, Y., and Endo, K. (1995). Experimental study of nerve regeneration in a biodegradable tube made from collagen and polyglycolic acid. *ASAIO J.* **41**, M657–M661.

12. Den Dunnen, W. F., Van der Lei, B., Schakenraad, J. M., Blaauw, E. H., Stokroos, I., Pennings, A. J., and Robinson, P. H. (1993). Long-term evaluation of nerve regeneration in a biodegradable nerve guide. *Microsurgery* **14**, 508–515.

13. den Dunnen, W. F., Stokroos, I., Blaauw, E. H., Holwerda, A., Pennings, A. J., Robinson, P. H., and Schakenraad, J. M. (1996). Light-microscopic and electron-microscopic evaluation of short-term nerve regeneration using a biodegradable poly(DL-lactide-ε-caprolacton) nerve guide. *J. Biomed. Mater. Res.* **31**, 105–115.

14. Nicoli Aldini, N., Perego, G., Cella, G. D., Maltarello, M. C., Fini, M., Rocca, M., and Giardino, R. (1996). Effectiveness of a bioabsorbable conduit in the repair of peripheral nerves. *Biomaterials* **17**, 959–962.

15. Hazari, A., Wiberg, M., Johansson-Ruden, G., Green, C., and Terenghi, G. (1999). A resorbable nerve conduit as an alternative to nerve autograft in nerve gap repair. *Br. J. Plast. Surg.* **52**, 653–657.

16. Cordeiro, P. G., Seckel, B. R., Lipton, S. A., D'Amore, P. A., Wagner, J., and Madison, R. (1989). Acidic fibroblast growth factor enhances peripheral nerve regeneration in vivo. *Plast. Reconstr. Surg.* **83**, 1013–1019.

17. Pu, L. L., Syed, S. A., Reid, M., Patwa, H., Goldstein, J. M., Forman, D. L., and Thomson, J. G. (1999). Effects of nerve growth factor on nerve regeneration through a vein graft across a gap. *Plast. Reconstr. Surg.* **104**, 1379–1385.

18. Bryan, D. J., Holway, A. H., Wang, K. K., Silva, A. E., Trantolo, D. J., Wise, D., and Summerhayes, I. C. (2000). Influence of glial growth factor and Schwann cells in a bioresorbable guidance channel on peripheral nerve regeneration. *Tissue Eng.* **6**, 129–138.

19. Fujimoto, E., Mizoguchi, A., Hanada, K., Yajima, M., and Ide, C. (1997). Basic fibroblast growth factor promotes extension of regenerating axons of peripheral nerve. In vivo experiments using a Schwann cell basal lamina tube model. *J. Neurocytol.* **26**, 511–528.

20. Ho, P. R., Coan, G. M., Cheng, E.T., Niell, C., Tarn, D. M., Zhou, H., Sierra, D., and Terris, D. J. (1998). Repair with collagen tubules linked with brain-derived neurotrophic factor and ciliary neurotrophic factor in a rat sciatic nerve injury model. *Arch. Otolaryngol. Head Neck Surg.* **124**, 761–766.

21. Welch, J. A., Kraus, K. H., Wells, M. R., Blunt, D. G., and Weremowitz, J. (1997). Effect of combined administration of insulin-like growth factor and platelet-derived growth factor on the regeneration of transected and anastomosed sciatic nerve in rats. *Am. J. Vet. Res.* **58**, 1033–1037.

22. Chamberlain, L. J., Yannas, I. V., Hsu, H. P., Strichartz, G. R., and Spector, M. (2000). Near-terminus axonal structure and function following rat sciatic nerve regeneration through a collagen–GAG matrix in a ten-millimeter gap. *J. Neurosci. Res.* **60**, 666–677.

23. Tong, X. J., Hirai, K., Shimada, H., Mizutani, Y., Izumi, T., Toda, N., and Yu, P. (1994). Sciatic nerve regeneration navigated by laminin–fibronectin double coated biodegradable collagen grafts in rats. *Brain Res.* **663**, 155–162.

24. Suzuki, K., Suzuki, Y., Tanihara, M., Ohnishi, K., Hashimoto, T., Endo, K., and Nishimura, Y. (2000). Reconstruction of rat peripheral nerve gap without sutures using freeze-dried alginate gel. *J. Biomed. Mater. Res.* **49**, 528–533.

25. Schense, J. C., Bloch, J., Aebischer, P., and Hubbell, J. A. (2000). Enzymatic incorporation of bioactive peptides into fibrin matrices enhances neurite extension. *Nat. Biotechnol.* **18**, 415–419.

26. Hadlock, T., Sundback, C., Hunter, D., Cheney, M., and Vacanti, J. P. (2000). A polymer foam conduit seeded with Schwann cells promotes guided peripheral nerve regeneration. *Tissue Eng.* **6**, 119–127.

27. Guenard, V., Kleitman, N., Morrissey, T. K., Bunge, R. P., and Aebischer, P. (1992). Syngeneic Schwann cells derived from adult nerves seeded in semipermeable guidance channels enhance peripheral nerve regeneration. *J. Neurosci.* **12**, 3310–3320.

28. Bryan, D. J., Wang, K. K., and Chakalis-Haley, D. P. (1996). Effect of Schwann cells in the enhancement of peripheral-nerve regeneration. *J. Reconstr. Microsurg.* **12**, 439–446.

29. Bunge, M. B., Bunge, R. P., Kleitman, N., and Dean, A. C. (1989). Role of peripheral nerve extracellular matrix in Schwann cell function and in neurite regeneration. *Dev. Neurosci.* **11**, 348–360.

30. Osawa, T., Tohyama, K., and Ide, C. (1990). Allogeneic nerve grafts in the rat, with special reference to the role of Schwann cell basal laminae in nerve regeneration. *J. Neurocytol.* **19**, 833–849.

31. Thalhammer, J. G., Vladimirova, M., Bershadsky, B., and Strichartz, G. R. (1995). Neurologic evaluation of the rat during sciatic nerve block with lidocaine. *Anesthesiology* **82**, 1013–1025.

32. Kohane, D. S., Yieh, J., Lu, N. T., Langer, R., Strichartz, G. R., and Berde, C. B. (1998). A re-examination of tetrodotoxin for prolonged duration local anesthesia. *Anesthesiology* **89**, 119–131.

33. Hadlock, T. A., Koka, R., Vacanti, J. P., and Cheney, M. L. (1999). A comparison of assessments of functional recovery in the rat. *J. Peripher. Nerv. Syst.* **4**, 258–264.

=CHAPTER 103

SPINAL CORD

Martin Oudega and Jacqueline Sagen

INTRODUCTION

Since the spinal cord provides a primary conduit for information exchange and inter-action between an organism and its environment, injuries to the spinal cord can have devastating consequences, severely limiting function and quality of life. The consequences of spinal cord injury are many-fold, including disruptions in sensory, motor, and autonomic realms. Traumatic injuries to the spinal cord can sever or damage long descending and ascending axons, segmental connections and local neurocircuitry, and intrinsic spinal neurons and glia. Thus, from a tissue engineering viewpoint, restoration of spinal cord function should include strategies for regenerating long axons, reestablishing connectivity between the severed ends of the spinal cord, and reestablishing local neurocircuitry and communication between the central and peripheral nervous systems. To accomplish this, it is likely that a combination of approaches will be necessary. This chapter describes several strategies for various aspects of spinal cord injury repair, particularly the use of bridges for reconnecting severed spinal axons, provision of neurotrophic factors or other soluble agents to enhance axonal regeneration or reduce axonal barriers, and provision of cells for replacement of lost or damaged local neural circuitries. The reader is also referred to other recent reviews [1–10].

AXON BRIDGES
SCHWANN CELL IMPLANTATION IN THE INJURED SPINAL CORD

In the mid-1970s, Richard Bunge recognized the possibility of culturing Schwann cells (SCs) from a patient's nerve for autologous implantation in injured spinal cord. SCs are known to secrete neurotrophic factors, express cell adhesion molecules, and generate several extracellular matrix molecules, which all can positively influence axonal regeneration. During the last three decades, the efficacy of SCs in promoting axonal growth in the injured adult spinal cord has been studied extensively [2,11–14]. To prepare SCs for this purpose (Fig. 103.1), 1-mm^2 pieces of sciatic nerves of adult Fischer rats are stripped of epineurium and connective tissue and kept in Dulbecco's Modified Eagle's Medium (DMEM, Gibco/BRL, Grand Island, NY) with 10% fetal calf serum (FCS, HyClone, Logan, UT), penicillin (50 U/ml), and streptomycin (50 μg/ml) for approximately 6 weeks. This allows fibroblasts to migrate away from the explants. The explants are then dissociated in the same medium and treated with Thy1.1 to eliminate remaining fibroblasts. Next, the SCs are stimulated to proliferate in D-10 medium; consisting of DMEM containing 10% fetal bovine serum, penicillin (50 U/ml), streptomycin (50 μg/ml) with the mitogens forskolin (2 μM; Sigma, St. Louis, MO) and bovine pituitary extract (20 μg/ml; Sigma). Throughout multiplication, the SCs are maintained on poly-(L-lysine)-coated 100-mm tissue culture dishes at 37°C and 6% CO_2, and passaged each time the cells reach confluency. At each passage, 2×10^6 cells per dish are replated for further expansion.

The day before implantation into the spinal cord, SCs of the third passage are mixed in the basal lamina matrix, Matrigel (30% in DMEM; 120×10^6 SC/ml; Fig. 103.1) and gently placed into a semipermeable polyacrylonitrile–poly(vinyl choride) (PAN-PVC) polymer tube

Methods of Tissue Engineering

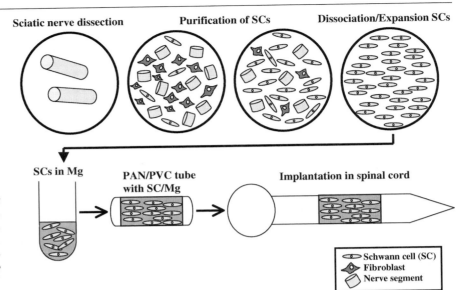

Fig. 103.1. Schematic drawing of the spinal cord implantation paradigm. Schwann cells (SCs) are purified from peripheral nerve segments, mixed with 30% Matrigel (Mg), seeded into a PAN-PVC tube, and implanted in the transected spinal cord. The cord stumps are inserted approximately 1 mm into the SC-filled PAN-PVC tube.

with an inner diameter of approximately 2.6 mm. The tube with the SC–Matrigel mixture is capped on both ends with PAN-PVC glue and kept overnight in DMEM at 37°C and 6% CO_2. During this period, a compact cable of longitudinally oriented SCs is formed within the Matrigel. The next day, the PAN-PVC tube containing the SC–Matrigel mixture is placed into a transection gap in the thoracic spinal cord and the cord stumps are eased approximately 1 mm into the tube (Fig. 103.1).

Within a few weeks, a tissue cable bridging the transection gap forms inside the polymer tube. This cable contains SCs, ensheathed unmyelinated and myelinated axons, blood vessels, and usually a few fibroblasts, meningeal cells, and macrophages [15–19]. Electron microscopy provided the estimate that approximately 20% of the total number of fibers are myelinated by the implanted SCs. Retrograde tracing experiments have revealed that the majority of the regenerated axons are derived from propriospinal neurons located as far rostral as the third cervical spinal level and as far caudal as the fourth sacral cord level [19]. Long spinal axon regeneration (e.g., from brain stem and higher centers), however, is more difficult to achieve using this approach. In addition, most findings indicate that once the regenerating axons have reached the distal host spinal cord, their further elongation is limited to the immediate vicinity of the graft–host border. A result important for future human application is that implantation of isolated human SCs in the nude (T-cell-deficient) rat was also found to elicit an axonal growth response [20].

IMPROVED SCHWANN CELL BRIDGE METHODS
Anti-inflammatory Agents

In most of the SC bridge studies, there is some loss of nervous tissue at the rostral and distal graft–host cord interface following implantation of the SC graft, which may restrict an optimal regenerative response. Such tissue loss can be limited by combining implantation of an SC graft with intravenous (jugular vein) administration of 30 mg/kg body weight of the anti-inflammatory agent, methylprednisolone at 5 min, 2 h, and 4 h after cord transection [15]. In addition, this neuroprotective treatment results in an improved axonal response [15]. The observed preservation of nervous tissue and enhanced axonal growth may be related to the methylprednisolone-induced decrease in the number of microglia/macrophages near the injury site [21]. Another promising anti-inflammatory agent is the cytokine interleukin 10 (IL-10), which has been shown to reduce lesion volume and improve functional recovery following spinal cord injury [22,23].

Neurotrophic Factors

While the studies just reviewed demonstrate that SC implantation may be a promising future surgical approach for repair of the injured human spinal cord, they also show that responding axons grow across the SC implant but not into the opposite cord stump, and that fibers from brain stem nuclei do not regenerate into thoracic SC grafts. These observations emphasize that SC grafting alone will not result in partial or complete recovery of injury-induced deficits in locomotor function.

Increasing the levels of neurotrophic factors in the graft environment can enhance the overall axonal growth response into an intraspinal graft. An SC graft similar to that already described, in combination with a 4-week infusion of brain-derived neurotrophic factor (BDNF) and neurotrophin 3 (NT-3), 12 μg/day, directly into the PAN-PVC tube, has been shown to increase the total number of fibers regenerating into the implant [18]. More importantly, this treatment also promotes fiber growth from neurons in 10 different brain stem nuclei, most of them deriving from the vestibulospinal nuclei [18]. Similar results have also been obtained with a 5-mm-long trail of SCs genetically modified to produce and secrete human BDNF implanted into the distal stump of the transected adult rat spinal cord [24]. This growth response appeared to be specific for the secreted neurotrophic factor, since many of the responding axons derived from neurons known to express *TrkB* [24]. Placement of SCs genetically modified to produce and secrete nerve growth factor (NGF) in a spinal cord hemisection was shown to also increase the axonal growth response [25,26].

Axonal growth from an intraspinal graft into the opposite cord stump can be achieved by increasing the levels of neurotrophic factors a short distance away from the graft. This has been clearly demonstrated in adult rats that received a complete transection of the right sciatic nerve 1 week before implantation of a predegenerated peripheral nerve bridge, 1–2 mm long, in the thoracic (T8) dorsal funiculus [27–29]. The nerve bridge was taken from the distal part of the transected right sciatic nerve [27]. To administer neurotrophic factors, a metal infusion device attached to an Alzet osmotic minipump is inserted approximately 1 mm into the dorsal spinal cord, 3–4 mm rostral to the graft–host cord interface immediately following bridge implantation [28]. The pump infuses neurotrophic factors such as NGF into the dorsal cord for 2 weeks. It was shown by means of the axonal tracer cholera toxin B-subunit that as a result of NGF infusion, approximately 50% of the sensory fibers deriving from the right sciatic nerve had regenerated into the dorsal funiculus white matter [28]. The number of axons decreases further away from the graft, and 10–20% reaches the infusion site but not farther, probably because of the chemoattractant effects of NGF. A largely similar response was seen after infusion of BDNF, NT-3, or a combination of all three neurotrophic factors [29]. Infusion of neurotrophins for more than 2 weeks may further enhance axonal growth into the spinal cord. Neurotrophic factor infusion directly into the dorsal spinal cord also promotes growth of descending fibers from an SC implant into the distal spinal cord [30]. In this particular study, the factors were infused over a 4-week period, and the reentering axons could be seen beyond the site of infusion.

Olfactory Ensheathing Glia

Olfactory ensheathing glia are unique cells in the olfactory system that have been found to support corticospinal tract regeneration in a very localized lesion [31,32]. In these studies, implantation of olfactory ensheathing glia between spinal cord stumps following a complete transection [33] or into an electrolytic lesion in the dorsal funiculus [31] promoted axonal growth across the injury site and into spinal tissue beyond. A combination of an intraspinal SC graft and olfactory ensheathing cells in the rostral and distal graft–host cord interface enables lengthy axon regeneration and growth into the distal cord [34]. The olfactory ensheathing cells were found to accompany the reentering axons, suggesting that as long as regenerating fibers are either not exposed or minimally exposed to interface tissue, they can grow across this growth barrier.

Neutralization of Growth Inhibitors

Clearly, the cellular and molecular characteristics of the graft–host cord interfaces are of crucial importance to axonal reentry into the spinal cord. It has been hypothesized that both a physical barrier (glial scarring) and chemical barrier (Nogo, chondroitin sulfate proteoglycans) exist at the injury site, impeding axonal regeneration [35,36]. Strategies such as the chemical neutralization or destruction of these growth inhibitors [37–41] may prove to be successful for reentry of fibers and, consequently enhancing the chances for functional recovery.

POLYMERS IN SPINAL CORD AXON BRIDGES

Natural and synthetic polymers have been used in spinal cord repair strategies as a matrix for cellular implants and/or a scaffold for regenerating axons. Fibrin [42], solid [43], or fluid collagen [44] were each found to promote axonal growth in the injured adult rat spinal cord. Other polymers that have been used in spinal cord repair approaches in the adult rat include carbon filaments [45], nitrocellulose membranes [45], polyacrylonitrile–poly(vinyl alcohol) tubes [46], polyacrylonitrile–poly(vinyl chloride) tubes [15–18,20], and polymethacrylate porous hydrogels [47].

Generally, a polymer structure used in spinal cord injury should fulfill several requirements. The device should be porous (to allow migration of cells and/or ingrowth of axons as well as exchange of nutrients), supportive of or even promoting axonal regeneration, three-dimensional, with preferably longitudinally oriented pores, biodegradable, and biocompatible (i.e., it should not induce an inflammatory or immune response above that normally seen in spinal cord injury).

Polymers that potentially fulfill all the foregoing requirements are those derived from lactic and glycolic acids, the aliphatic poly(α-hydroxy acids). The main advantage of this family of polymers is that their degradation and mechanical properties can be fully customized. These properties depend on the ratio of the L-lactic (L-LA), D-lactic (D-LA), and glycolic (GA) acid repeating units, their distribution along the macromolecular backbone, and their molecular weight distribution [48]. Moreover, following implantation, these polymers do not provoke immunological reactions [49] and are completely resorbed [50]. In the central nervous system, poly(α-hydroxy acid) implants have been used mainly as controlled delivery systems for bioactive (macro)molecules [51–53].

So far, aliphatic poly(α-hydroxy acids) have sporadically been used in the injured spinal cord. In an attempt to design biodegradable scaffolds for the implantation of SCs in the injured adult spinal cord, the biocampatibility of poly(α-hydroxy acids) with SCs and adult spinal tissue was investigated [54]. *In vitro*, it was demonstrated that PLA-GA and its breakdown products had no adverse effect on the morphology, survival, and proliferation of cultured rat Schwann cells [54]. *In vivo*, a 2-mm-long PLA-GA disk implanted in between the stumps of a transected adult rat thoracic spinal cord was shown to integrate well into spinal tissue at 2 weeks after implantation [54]. Up to 2 months postimplantation, the astrocytic and inflammatory responses near the lesion site were largely similar in experimental and controls animals. In another study, PLA needles with 100-μm-wide longitudinal grooves and embedded in fibrin glue were implanted in a similar complete adult spinal cord transection model [55]. Cell migration, angiogenesis, and axonal growth were observed within this implant. The longitudinal orientation of the needles and their macropores appeared to benefit blood vessel and axonal growth [55].

In the experiments in which SCs were implanted into the injured spinal cord, the SCs were usually enclosed in a nondegradable PAN-PVC tube. A tubular scaffold for cell implantation may improve axonal regeneration by limiting the formation of scar tissue and by allowing the accumulation of neurite growth promoting factors [12,56]. However, in some stages of the growth response, the presence of a tubular scaffold may actually restrict axonal growth or the maintenance of fibers that have grown into the implant as a result of constriction of the spinal cord, toxicity, or foreign body reaction. The use of a biodegradable tubular scaffold for SC implantation into the injured spinal cord may circumvent these detrimental effects. In 1998 it was shown that axonal regeneration and myelination do occur in an SC graft contained in a biodegradable PLA tube [57]. It was observed, however, that in time, damage by parts of the degrading scaffold caused a reduction in the number of axons within the tissue cable. Possibly the nonuniform breakdown of these tubes may result in the formation of large pieces of scaffold that can cause damage to the tissue cable inside. Future experiments need to focus on the development and use of biodegradable scaffolds that degrade uniformly.

CELL-BASED DELIVERY OF THERAPEUTIC MOLECULES

A second tissue approach to the restoration of function in the spinal cord following injury is the use of cellular implants to serve as a continual local source of therapeutic molecules. For this purpose, the cellular implants act as a biological minipump rather than necessarily integrating with the endogenous host neurocircuitry. Thus, the implants may be placed either directly into the host spinal parenchyma, or into the surrounding cerebrospinal fluid (CSF), which is potentially less invasive but may reduce host–graft interaction. Both approaches have been utilized in models of spinal cord pathology. Potential benefits of a cell-based delivery approach are several:

> Therapeutic agents can be delivered on a long-term basis, since living cells provide a continually renewable supply.
> Therapeutic agents with short biological half-lives or impeded passage into the CNS (e.g., neuropeptides, neurotrophic factors) can potentially be used.
> Cells can be genetically manipulated to deliver desired therapeutic agents.

This approach has been used in the delivery to the spinal cord of several agents for various indications, including trophic factors for spinal cord injury and amyotrophic lateral sclerosis (ALS), catecholamines for spinal cord injury and pain, and neuropeptides for pain.

CELL-BASED DELIVERY OF TROPHIC FACTORS

Tuszynski *et al.* [58] have pioneered the use of genetically modified neurotrophin-producing fibroblasts for improving regeneration in the cord. An enhanced regenerative response was found in grafts containing fibroblasts genetically modified to produce NGF [58–60], BDNF [60,61], or NT-3 [62]. In particular, intraspinally implanted NT-3-producing fibroblasts promote growth of corticospinal tract fibers in the intact gray matter around the grafted cells in a hemisected spinal cord [62]. Improved locomotor outcome was also noted. Fibroblasts secreting BDNF or NT-3 were also found to induce oligodendrocyte proliferation and myelination of regenerating axons when placed in a contused adult rat spinal cord [63]. Another group reported regeneration of rubrospinal fibers following transplantation of BDNF-producing fibroblast into a cervical hemisection cavity, with significant recovery in forelimb usage [61].

Immortalized cell lines have also been utilized for trophic factor delivery to the spinal cord. Baby hamster kidney (BHK) cells genetically engineered to produce CNTF were implanted in polymer capsules and implanted intrathecally in ALS patients [64]. Although patients did not improve, the findings revealed detectable levels of CNTF in the spinal CSF of these patients, supporting the feasibility of this approach. The goal of encapsulation is generally to allow free passage of small molecules such as nutrients into the device and release of the therapeutic factors from the device, while avoiding rejection by hindering the passage of larger molecules and cells from the host immune system. These immunoisolatory devices consist of three basic components, a permselective membrane, an internal matrix, and living cells. In this particular study, the encapsulation devices used consisted of a 5-cm hollow-fiber membrane fabricated from poly(ether sulfone) (PES, Akzo Nobel Faser AG) and connected with a silicon tether for retrieval. Approximately 100,000 cells in a collagen matrix were loaded in the devices. However, other encapsulation devices are possible and have been used in various studies, as discussed later and in other chapters in this volume.

Trophic factor secreting cell lines have also been implanted in the spinal subarachnoid space for the treatment of chronic pain [65]. To accomplish this, conditionally immortalized cell lines derived from fetal rat raphe were transfected with the temperature-sensitive mutant of the SV40 large-T antigen (tsTag) and BDNF. These cells undergo continual cell division between 32 and 34°C, but differentiate and become postmitotic when the temperature is raised to 37–39°C. The transplantation of these cells in the rat subarachnoid space reversed the behavioral hypersensitivity caused by a peripheral neuropathy.

CELL-BASED DELIVERY OF MONOAMINES
Monoaminergic Cells for Motor Function

Serotonergic or noradrenergic neurons from embryonic brain stem containing the embryonic raphe or locus coeruleus, respectively, have been utilized to support motor function in spinally injured animals [66–77]. For example, serotonergic fibers could be found surrounding motor neurons following medullary raphe grafts in the 5,7-dihydroxytryptamine-denervated spinal cord, and electrical stimulation of the grafts could excite motoneurons, suggesting that functional connections had been formed [68,69]. Similar survival and integration were reported with embryonic raphe cell suspensions implanted one week after complete spinal transection [73,74]. Noradrenergic neurons have also been utilized as intraspinal grafts in spinal-transected or 6-hydroxydopamine-denervated animals, with improvements in motor performance tasks observed [66,67,70–72,77]. Both raphe or locus coeruleus transplants were found to increase the excitability of the spinal stepping generator [78,79]. Intraspinally implanted adrenal chromaffin cells have also been used as a source of catecholamines to restore motor deficits [80].

Monoaminergic Cells for Pain

Adrenal medullary tissue and isolated cells have been used extensively to deliver monoamines and other antialgesic agents to the spinal cord via implantation into the spinal subarachnoid space [for review, see 81,82]. To accomplish this, host–graft integration is apparently not required, and the cells most likely produce their effects via release into the CSF or extracellular spaces. Thus, cells encapsulated in polymer membranes have also been used for this purpose, both in preclinical models and in early clinical trials [83–86]. An advantage of the latter approach is the potential ability to utilize cells obtained from xenogeneic sources, which may be more readily available in sufficient quantities than human donor cells. For example, large numbers of nearly pure populations of chromaffin cells can be isolated from adult or neonatal bovine adrenal glands. A cell preparation from 12 adult bovine adrenal glands yields approximately 3×10^8 chromaffin cells. Adrenal glands are perfused with collagenase P (concentration and timing determined empirically depending on gland age and size), and then the medullary tissue is peeled from the cortical tissue, minced and filtered through a nylon mesh, and purified by means of a self-generating Percoll density gradient [for details, see 87]. Cells removed from the middle band can be further purified by differential plating to remove rapidly adherent fibroblasts. The resultant cell population is approximately 95% chromaffin cells [87]. The cells can be maintained for several weeks in equal parts of Dulbecco's Modified Eagle's Medium (DMEM) and Ham's F-12 nutrient medium, with 5% fetal bovine serum (FBS), with occasional differential plating to limit fibroblast overgrowth. Alternatively, serum-free Chinese hamster ovary media will support chromaffin cell viability. For encapsulation and implantation in the rat spinal subarachnoid space, cells are resuspended in 2.0% alginate and loaded into prespun hollow fibers (PAN-PVC, 0.6 mm i.d., 0.75 mm o.d.). The ends are sealed by heat pinching, and the capsules are immersed in 1.0% $CaCl_2$ solution for 4–5 min to form a gel. Capsules are implanted via a slit in the dura mater following laminectomy at the L2 level, and threaded rostrally in the subarachnoid space until completely under the meninges. For implantation into the human spinal subarachnoid space, cells are similarly suspended in sterile alginate hydrogel and loaded into PAN-PVC hollow fibers (~1 mm in diameter and 7.0 cm long, 1–3 $\times 10^6$ cells/device). The ends of the tubes are sealed with a light-sensitive acrylic sealant. One end has a 20-cm silicone tether attached to the tube, with a titanium connector for retrieval. A 19-gauge Tonhy needle is needed to implant encapsulated chromaffin cell devices into the lumbar subarachnoid space of cancer pain patients under local anesthesia [see 86]. Microencapsulation of bovine chromaffin cells has also been recently reported for spinal subarachnoid implantation in cancer pain patients [88].

Another novel approach for implanting cells for delivery into the spinal subarachnoid space of agents such as monoamines is the use of supportive matrices. This approach has been utilized for implanting bovine chromaffin cells into both rat and nonhuman primate intrathecal space [89,90]. Approximately 3×10^6 purified bovine chromaffin cells were loaded into 0.5-mm-sized pieces of collagen foam matrix and injected via a needle into the macaque lumbar intrathecal space.

In addition to chromaffin cells, numerous other cell lines have been used to deliver monoamines, neuropeptides, trophic factors, and so on into the spinal subarachnoid space for treatment of pain. These include the AtT-20 cell derived from the mouse anterior pituitary, which releases high levels of opioid peptide, β-endorphin, and a genetically modified version of this cell line, AtT-20/hEnk, transfected with the human proenkephalin gene, hence able to secrete enkephalin as well as β-endorphin [91,92], catecholamine secreting cell line derived from the B16 melanoma [93], mouse neuroblastoma Neuro2A line, and the mouse pituitary AtT20 line, both transfected with the pro-opiomelanocortin gene [94,95], engineered embryonic carcinoma cell line P19 with dexamethasone-inducible expression of β-endorphin [96], and embryonic raphe-derived conditionally immortalized cell lines engineered to produce serotonin, brain-derived neurotrophic factor (BDNF), glutamate decarboxylase (to produce γ-aminobutyric acid), preproenkephalin, and preprogalanin [65,97,98].

These findings are particularly interesting in relation to a potential approach to the treatment of spinal cord injury pain, which can be extremely difficult to manage clinically with more traditional therapeutic strategies. Implantation into the rat spinal subarachnoid space of allogeneic adrenal medullary tissue or bovine chromaffin cells, either free floating or encapsulated, has been shown to reduce pain symptoms in three distinct spinal cord injury pain models [99–102]. Thus, tissue engineering approaches as applied in the spinal cord can also be important tools for improving quality of life after traumatic spinal cord injury.

Others

In addition to catecholamines, neuropeptides, and trophic factors, cell-based delivery of other bioactive molecules can also be imagined. For example, macrophages have been suggested as cells that may make the lesion milieu more favorable for neurite growth [103]. These cells secrete a variety of cytokines and growth factors. Macrophages have been activated *in vitro* by pieces of peripheral nerve and then injected into transected spinal cord [103]. This strategy leads to nerve fiber growth across the transected site, partial recovery of motor function, and evoked muscle responses. Alternatively, cells that are engineered to secrete specific cytokines could be designed. Cells may also be engineered to delivery anti-apoptotic agents, anti–growth inhibitor agents, angiogenic factors, and so on.

REPLACEMENT OF LOCAL NEURAL CIRCUITRIES
LOWER MOTOR NEURONS

In addition to destruction of long axons, damage to local spinal neurons and neurocircuitry is a consequence of injury to the spinal cord. Thus strategies to replace lost neuronal populations have also been explored. In particular, motor neuron replacement and integration in host circuitry is also potentially important in the treatment of neurological diseases such as ALS and poliomyelitis. Vrbová and colleagues have identified several factors, in both host and graft, that influence the survival of transplanted embryonic motoneurons [for review, see 104]. These include optimal donor gestational age (after the onset of motoneuron pool proliferation but before axonal extension into the periphery), provision of an appropriate target (i.e., denervated muscle), and provision of a conduit through which the implanted neurons can reach the target (e.g., peripheral nerve implanted in the vicinity of the graft). In addition, depletion of the host's own motoneuron pool may promote migration of embryonic motor neurons from the grafts to unoccupied sites. This approach has been used to transplant small fragments of embryonic rat ventral spinal cord (ED11-ED12) into the lumbar segment of motoneuron-depleted hosts, with host muscle and a length of attached nerve implanted as a target and conduit [105]. Results revealed surviving donor-derived cells, some with projections to the target muscle; however, most of these were smaller than adult motoneurons and had low expression of motoneuron markers such as choline acetyl-transferase (ChAT) and calcitonin gene-related peptides. In more recent studies, attempts to guide axons from grafted embryonic motoneurons to their target via a reimplanted ventral root revealed axons of graft origin in the ventral ramus and some reversal of the losses in muscle force 3–6 months post-transplantation [106].

Another approach that has been attempted is grafting cell suspensions from fetal spinal cord specifically enriched in motoneurons by density gradient purification or fractionation, using the low-affinity NGF receptor as a ligand [107–109]. When this approach was tried, the motoneurons survived, received host innervation, and in some cases migrated into host parenchyma, although ChAT expression was still generally low.

NEURAL STEM CELLS

An emergent breakthrough in recent years was the discovery in the CNS that stem cells possess the capacity to replace lost cellular populations [110–112]. Like stem cells identified in other biological systems, including the hematopoietic system and the gut, neural stem cells are capable of self-renewal, proliferation, and generation of a large number of progeny. In addition, when these cells are exposed to microenvironmental cues they have the capacity to give rise to progenitor neural cell lineages appropriate for their location and metabolic circumstances and can differentiate into the three major neural phenotypes *in vitro*: neurons, astrocytes, and oligodendrocytes (for details see Chapter 34 in this volume).

In spite of promising *in vitro* studies, both the survival and neuronal differentiation of neural stem cell or progenitor transplants to the adult CNS have proven challenging, possibly because of the lack of developmental cues in the embryonic environment normally encountered by these cells. Recent reports have shown that some neural progenitors can be detected 6–12 weeks following transplantation into the adult brain [113] or spinal cord [114], but the survival rate was low (1–5%) [114], and few cells appeared to differentiate into neurons. Even when these cells are placed in or near an injury site, which has been hypothesized to provide a more permissive environment by reason of reexpression of appropriate instructive cues [115], there may not be enough surviving cells to restore local circuitry at the injury site, and neuronal differentiation appears to be low. Evidence that sites of neurodegeneration may contain appropriate instructive factors is provided by studies showing robust engraftment of neural stem cell transplants into focal cytolytic lesions of the adult neocortex and approximately 15% differentiation into pyramidal neurons, in contrast to the intact neocortex, where only donor-derived glia and undifferentiated quiescent progenitors were detected [116]. However, in a more extensive hypoxic–ischemic brain injury, a majority of transplanted neural stem cells appeared to migrate away from the infarct site to the penumbra, where they differentiated into astrocytes expressing glial fibrillary acidic protein (GFAP) [117]. Similarly, proliferating populations of undifferentiated neural stem cells transplanted into the intact or contused adult spinal cord showed good survival, but the majority of cells differentiated into GFAP-positive astrocytes or remained as nestin-positive precursors, with no neurons or oligodendrocytes detected [118]. Thus, it is likely that further expansion of the transplanted stem cell population and the addition of appropriate differentiation factors will be necessary. Nevertheless, recent promising studies demonstrating neural stem cell survival and modest functional improvement in spinal cord injury models suggest that this avenue of research should be pursued [114,119].

An alternative source of neural precursors is embryonic stem (ES) cells, derived from the inner cell mass of blastocyte-stage embryos. These cells are totipotent and can differentiate into all tissues and cell types. The "4-/4+" protocol [120], can be used to induce undifferentiated ES cells propagated in the presence of leukemia inhibitory factor (LIF) to be differentiated toward neural precursors in the absence of LIF for 4 days followed by the addition of retinoic acid (500 nM) for 4 days. When this approach was used, neural differentiated mouse ES cells labeled with bromodeoxyuridine were partially trypsinized and transplanted into the syrinx formed in rat spinal cords 9 days after contusion injury [120]. Surviving ES cells were present 5 weeks after transplantation and had migrated up to 8 mm from the graft site. The cells were found to be primarily oligodendrocytes and astrocytes, but some Neu-N-labeled ES-derived cells were also identified. In addition, improvement in open-field locomotion was also observed in animals with ES transplants. In another recent report, fibroblast growth factor 2 (10 ng/ml) and platelet-derived glial factor (PDGF, 10 ng/ml) were used to differentiate ES cells toward glial precursors, for the purpose of generating a source of oligodendrocytes for myelinating transplants [121]. When these cells were transplanted into the spinal dorsal columns of myelin-deficient (*md*) rats, myelination was observed in the dorsal columns, gray matter, and ventral columns of most of the animals.

These studies indicate that provision of specific environmental cues may promote appropriate differentiation of transplanted precursor populations in disease or injury states. Thus, the use of either ES or neural stem cells as a graft source has the potential advantage over heterotypic grafts to fully differentiate and integrate within the host CNS. These transplants are likely to be most useful in combination with other approaches described earlier. For example, neural stem cells may be genetically modified to produce necessary neurotrophic factors or other therapeutic molecules. A clone of neural stem cells genetically modified to overproduce NT-3 was used in a 1999 investigation that revealed long distance migration and neuronal differentiation upon transplantation into the adult spinal cord [122]. As another example, a recent report described significant neurological recovery and histological evidence of tissue engineered spinal cord constructs when a combination of spinal cord neural stem cells suspended in a hydrogel scaffolding was implanted in a transected spinal cavity [123]. One may envision that a combination tissue engineering strategy taking advantage of recent advances in axon bridging, local neuronal and glial cell replacement, and provision of appropriate trophic and therapeutic molecules will ultimately prove most effective in repairing the injured spinal cord.

ACKNOWLEDGMENTS

Drs. M. B. Bunge, J.-M. Parel, and S. E. Gautier are thanked for their support in various parts of the studies described in this chapter. MO is a Werner Heumann Memorial International Scholar.

REFERENCES

1. Bregman, B. S. (1998). Regeneration in the spinal cord. *Curr. Opin. Neurobiol.* **8**, 800-807.
2. Bunge, M. B., and Kleitman, N. (1999). Neurotrophins and neuroprotection improve axonal regeneration into Schwann cell transplants placed in transected adult rat spinal cord. *In* "CNS Regeneration: Basic Science and Clinical Advances" (M. H. Tuszynski and J. Kordower, eds.), pp. 631–646. Academic Press, San Diego, CA.
3. Fawcett, J. W. (1998). Spinal cord repair: From experimental models to human application. *Spinal Cord* **36**, 811–817.
4. Giménez y Ribotta, M., and Privat, A. (1998). Biological interventions for spinal cord injury. *Curr. Opin. Neurol.* **11**, 647–654.
5. Plant, G. W., Bunge, M. B., and Ramon-Cueto, A. (2000). Transplantation of Schwann cells and ensheathing glia to improve regeneration in adult spinal cord. *In* "Nerve Regeneration" (N. A. Ingoglia and M. Murray, eds.), pp. 529–562. New York.
6. Sagen, J. (1998). Transplantation strategies for the treatment of pain. *In* "Cell Transplantation for Neurological Disorders: Toward Reconstruction of the Human Central Nervous System" (T. B. Freeman and H. Widner, eds.), pp. 231–251. Humana Press, Totowa , NJ.
7. Sagen, J., Bunge, M. B., and Kleitman, N. (2000). Transplantation strategies for treatment of spinal cord dysfunction and injury. *In* "Principles of Tissue Engineering" (R. P. Lanza, R. Langer and J. Vacanti, eds.), 2nd ed., pp. 799–820. Academic Press, San Diego, CA.
8. Schwab, M. E., and Bartholdi, D. (1996). Degeneration and regeneration of axons in the lesioned spinal cord. *Physiol. Rev.* **76**, 319–370.
9. Steeves, J. D. and Tetzlaff, W. (1998). Engines, accelerators, and brakes on functional spinal cord repair. *Ann. N.Y. Acad. Sci.* **860**, 412–424.
10. Zhang, S. C., Ge, B., and Duncan, I. D. (1999). Adult brain retains the potential to generate oligodendroglial progenitors with extensive myelination capacity. *Proc. Natl. Acad. Sci. U.S.A.* **96**, 4089–4094.
11. Bunge, R. P. (1993). Expanding roles for the Schwann cell: Ensheathment, myelination, trophism and regeneration. *Curr. Opin. Neurobiol.* **3**, 805–809.
12. Bunge, M. B. (1994). Transplantation of purified populations of Schwann cells into lesioned adult rat spinal cord. *J. Neurol.* **241**, 36–39.
13. Bunge, M. B., and Kleitman, N. (1997). Schwann cells as facilitators of axonal regeneration in CNS fiber tracts. *In* "Cell Biology and Pathology of Myelin" (Juurlink, B. H., *et al.*, eds.), pp. 319–333. Plenum, New York.
14. Guènard, V., Xu, X. M., and Bunge, M. B. (1993). The use of Schwann cell transplantation to foster central nervous system repair. *Semin. Neurosci.* **5**, 401–411.
15. Chen, A., Xu, X. M., Kleitman, N., and Bunge, M. B. (1996). Methylprednisolone administration improves axonal regeneration into Schwann cell grafts in transected adult rat thoracic spinal cord. *Exp. Neurol.* **138**, 261–276.

16. Oudega M., Xu, X. M., Guénard, V., Kleitman, N., and Bunge, M. B. (1997). A combination of insulin-like growth factor and platelet-derived growth factor enhances myelination but diminishes axonal regeneration into Schwann cell grafts in the adult rat spinal cord. *Glia* **19**, 247–258.

17. Xu, X. M., Guénard, V., Kleitman, N., and Bunge, M. B. (1995). Axonal regeneration into Schwann cell seeded guidance channels grafted into transected adult rat spinal cord. *J. Comp. Neurol.* **351**, 145–160.

18. Xu, X. M., Guénard, V., Kleitman, N., Aebischer, P., and Bunge, M. B. (1995). A combination of BDNF and NT-3 promotes supraspinal axonal regeneration into Schwann cell grafts in adult rat thoracic spinal cord. *Exp. Neurol.* **134**, 261–272.

19. Xu, X. M., Chen, A., Guénard, V., Kleitman, N., and Bunge, M. B. (1997). Bridging Schwann cell transplants promote axonal regeneration from both the proximal and distal stumps of transected adult rat spinal cord. *J. Neurocytol.* **26**, 1–16.

20. Guest, J. D., Rao, A., Olson, L., Bunge, M. B., and Bunge, R. P. (1997). The ability of human Schwann cell grafts to promote regeneration in the transected nude rat spinal cord. *Exp. Neurol.* **148**, 502–522.

21. Oudega, M. Vargas, C. G., Weber, A. B., Kleitman, N., and Bunge, M. B. (1999). Long-term effects of methylprednisolone following transection of adult rat spinal cord. *Eur. J. Neurosci.* **11**, 2453–2464.

22. Bethea, J. R., Nagashima, H., Acosta, M. C., Briceno, C., Gomez, F., Marcillo, A. E., Loor, K., Green, J., and Dietrich, W. D. (1999). Systemically administered interleukin-10 reduces tumor necrosis factor-alpha production and significantly improves functional recovery following traumatic brain injury in rats. *J. Neurotrauma* **16**, 851–853.

23. Brewer, K. L., Bethea, J. R., and Yezierski, R. P. (1999). Neuroprotective effects of interleukin-10 following excitotoxic spinal cord injury. *Exp. Neurol.* **159**, 484–493.

24. Menei, P., Montero-Menei, C., Whittemore, S. R., Bunge, R. P., and Bunge, M. B. (1998). Schwann cells genetically modified to secrete human BDNF promote enhanced axonal regrowth across transected adult rat spinal cord. *Eur. J. Neurosci.* **10**, 607–62l.

25. Tuszynski, M. H., Weidner, N., McCormack, M., Miller, I., Powell, H., and Conner, J. (1998). Grafts of genetically modified Schwann cells to the spinal cord: Survival, axon growth, and myelination. *Cell Transplant.* **7**, 187–196.

26. Weidner, N., Blesch, A., Grill, R. J., and Tuszynski, M. H. (1999). Nerve growth factor-hypersecreting Schwann cell grafts augment and guide spinal cord axonal growth and myelinate central nervous system axons in a phenotypically appropriate manner that correlates with expression of L1. *J. Comp. Neurol.* **413**, 495–506.

27. Oudega, M., Varon, S., and Hagg, T. (1994). Regeneration of adult rat sensory axons into intraspinal nerve grafts: Promoting effects of conditioning lesion and graft predegeneration. *Exp. Neurol.* **129**, 194–206.

28. Oudega, M., and Hagg, T. (1996). Nerve growth factor promotes regeneration of sensory axons into adult rat spinal cord. *Exp. Neurol.* **140**, 218–229.

29. Oudega, M., and Hagg, T. (1999). Neurotrophins promote regeneration of sensory axons in the adult rat spinal cord. *Brain Res.* **818**, 431–438.

30. Xu, X. M., Bamber, N. I., Li, H., Lu, X., Aebischer, P., and Oudega, M. (1998). Axonal regrowth and reentry into the distal spinal cord of adult rats following transplantation of Schwann cell seeded mini-channels and infusion of two neurotrophins, BDNF and NT-3, into the distal spinal cord. *Exp. Neurol.* **151**, 143–171.

31. Li, Y., Field, P. M., and Raisman, G. (1997). Repair of adult rat corticospinal tract by transplants of olfactory ensheathing cells. *Science* **277**, 2000–2002.

32. Li, Y., Field, P. M., and Raisman, G. (1998). Regeneration of adult rat corticospinal axons induced by transplanted olfactory ensheathing cells. *J. Neurosci.* **18**, 10514–10524.

33. Ramon-Cueto, A., Cordero, M. I., Santos-Benito, F. F., and Avila, J. (1999). Olfactory ensheathing glia transplants promote functional recovery and structural repair of transected adult rat spinal cords. *Soc. Neurosci. Abstr.* **25**, 295.5.

34. Ramon-Cueto, A., Plant, G. W., Avila, J., and Bunge, M. B. (1998). Long-distance axonal regeneration in the transected adult rat spinal cord is promoted by olfactory ensheathing glia transplants. *J. Neurosci.* **18**, 3803–3815.

35. Fawcett, J. W., and Asher, R. A. (1999). The glial scar and central nervous system repair. *Brain Res. Bull.* **49**, 377–391.

36. Schwab, M. E., Kapfhammer, J. P., and Bandlow, C. E. (1993). Inhibitors of neurite growth. *Annu. Rev. Neurosci.* **16**, 565–595.

37. Bregman, B. S., Kunkel-Bagden, E., Schnell, L., Dai, H. N., Gao, D., and Schwab, M. E. (1995). Recovery from spinal cord injury mediated by antibodies to neurite growth inhibitors. *Nature (London)* **378**, 498–501.

38. Schnell, L., Schneider, R., Kolbeck, R., Barde, Y. A., and Schwab, M. E. (1994). Neurotrophin-3 enhances sprouting of corticospinal tract during development and after adult spinal cord lesion. *Nature (London)* **367**, 170–173.

39. Schnell, L., and Schwab, M. E. (1990). Axonal regeneration in the rat spinal cord produced by an antibody against myelin-associated neurite growth inhibitors. *Nature (London)* **343**, 269–272.

40. Schnell, L., and Schwab, M. E. (1993). Sprouting and regeneration of lesioned corticospinal tract fibers in the adult rat spinal cord. *Eur. J. Neurosci.* **5**, 1156–1171.

41. Zuo, J., Neubauer, D., Dyess, K., Ferguson, T. A., and Muir, D. (1998). Degradation of chondroitin sulfate proteoglycan enhances the neurite-promoting potential of spinal cord tissue. *Exp. Neurol.* **154**, 654–662.

42. Cheng, H., Cao, Y., and Olson, L. (1996). Spinal cord repair in adult paraplegic rats: Partial restoration of hind limb function. *Science* **273**, 510–513.

43. Marchand, R., Woerly, S., Bertrand, L., and Valdes, N. (1993). Evaluation of two cross-linked collagen gels implanted in the transected spinal cord. *Brain Res. Bull.* **30**, 415–422.

44. Joosten, E. A. J., Bar, P. R., and Gispen, W. H. (1995). Collagen implant and corticospinal axonal growth after midthoracic spinal cord lesion in adult rat. *J. Neurosci. Res.* **41**. 481–490.

45. Houle, J. D. (1992). Regeneration of dorsal root axons is related to specific non-neuronal cells lining NGF-treated intraspinal nitrocellulose implants. *Exp. Neurol.* **118**, 133–142.

46. McCormack, M., Goddard, M., Guénard, V., and Aebischer, P. (1991). Comparison of dorsal and ventral spinal root regeneration through semipermeable guidance channels. *J. Comp. Neurol.* **313**, 449–456.

47. Woerly, S., Pinet, E., De Robertis, L., Bousmina, M., Laroche, G., Roitback, T., Vargova, L., and Sykova, E. (1998). Heterogeneous PHPMA hydrogels for tissue repair and axonal regeneration in the injured spinal cord. *J. Biomater. Sci. Polym.* **9**, 681–711.

48. Vert, M., Li, S. M., and Garreau, H. (1994). Attempts to map the structure and degradation characteristics of aliphatic polyesters derived from lactic and glycolic acids. *Biomater. Sci. Polym.* **6**, 639–649.

49. Menei, P., Daniel, V., Montero-Menei, C., Brouillard, M., Pouplart-Barthelaix, A., and Benoit, J.-P. (1993). Biodegradation and brain tissue reaction to poly(D,L-lactide-*co*-glycolide) microspheres. *Biomaterials* **14**, 470–478.

50. Athanasiou, K. A., Niederauer, G. G., and Agrawal, C. M. (1996). Sterilization, toxicity, biocompatibility and clinical applications of polylactic/polyglycolic acid copolymers *Biomaterials* **17**, 93–102.

51. Benoit, J.-P., Menei, P., Boisdron, M., Gamelin, E., and Guy, G. (1997). Radiosensitization of glioblastoma after intracranial implantaion of biodegradable 5-FU-loaded microspheres: Phase I/II clinical trial. *Proc. Int. Symp. Controlled Relat. Bioact. Mater.* **24**, 995.

52. Kou, J. H., Emmett, C., Shen, P., Aswani, S., Iwamoto, T., Vaghefi, F., Cain, G., and Sanders, L. (1997). Bioerosion and biocompatibility of poly(D,L-lactic-*co*-glycolic acid) implants in the brain. *J. Controlled Relat.* **43**, 123–130.

53. Menei, P., Benoit, J. P., Boisdron-Celle, M., Fournier, D., Mercier, P., and Guy, G. (1994). Drug targeting into the central nervous system by stereotactic implantation of biodegradable microspheres. *Neurosurgery* **34**, 1058–1064.

54. Gautier, S. E., Oudega, M., Fragoso,M., Chapon, P., Plant, G. W., Bunge, M. B., and Parel, J.-M. (1999). Poly(α-hydroxy acids) for application in the spinal cord: Resorbability and biocompatibility with adult rat Schwann cells and spinal cord. *J. Biomed. Mater. Res.* **42**, 642–654.

55. Maquet, V., Martin, D., Schultes, F., Franzen, R., Schoenen, J., Moonen, G., and Jérôme, R. (2001). Poly(D,L-lactide) foams modified by poly(ethylene oxide)-block-poly(D,L-lactide)copolymers and a-FGF: *In vitro* and *in vivo* evaluation for spinal cord regeneration. *Biomarker* **22**(10), 1137–1146.

56. Field, R. D., Le Beau, J. M., Longo, F. M., and Ellisman, M. H. (1989). Nerve regeneration through artificial tubular implants. *Prog. Neurobiol.* **33**, 87–134.

57. Oudega, M., Gautier, S. E., Fragoso, M., Plant, G. W., Bunge, M. B., and Parel, J.-M. (1998). In vitro and in vivo studies on bioresorbable poly (α-hydroxy acid) constructs for the application of Schwann cells in spinal cord regeneration. *Soc. Biomater.* p. 349.

58. Tuszynski, M. H., Peterson, D. A., Ray, J., Baird, A., Nakahara, Y., and Gage, F. H. (1994). Fibroblasts genetically modified to produce nerve growth factor induce robust neuritic ingrowth after grafting to the spinal cord. *Exp. Neurol.* **126**, 1–14.

59. Blesch, A., and Tuszynski, M. H. (1997). Robust growth of chronically injured spinal cord axons induced by grafts of genetically modified NGF-secreting cells. *Exp. Neurol.* **148**, 444–453.

60. Kim, D. H., Gutin, P. H., Noble, L. J., Nathan, D., Yu, J. S., and Nockels, R. P. (1996). Treatment with genetically engineered fibroblasts producing NGF or BDNF can accelerate recovery from traumatic spinal cord injury in the adult rat. *NeuroReport* **7**, 2221–2225.

61. Liu, Y., Kim, D., Himes, B. T., Chow, S. Y., Schallert, T., Murray, M., Tessler, A., and Fischer, I. (1999). Transplants of fibroblasts genetically modified to express BDNF promote regeneration of rat rubrospinal axons and recovery of forelimb function. *J. Neurosci.* **19**, 4370–4387.

62. Grill, R., Murai, K., Blesch, A., Gage, F. H., and Tuszynski, M. H. (1997). Cellular delivery of neurotrophin-3 promotes corticospinal axonal growth and partial functional recovery after spinal cord injury. *J. Neurosci.* **17**, 5560–5572.

63. McTigue, D. M., Horner, P. J., Stokes, B. T., and Gage, F. H. (1998). Neurotrophin-3 and brain-derived neurotrophic factor induce oligodendrocyte proliferation and myelination of regenerating axons in the contused adult rat spinal cord. *J. Neurosci.* **15**, 5354–5365.

64. Aebischer, P., Schluep, M., Déglon, N., Joseph, J. M., Hirt, L., Heyd, B., Goddard, M., Hammang, J. P., Zurn, A. D., Kato, A. C., Regli, F., and Baetge, E. E. (1996). Intrathecal delivery of CNTF using encapsulated genetically modified xenogeneic cells in amyotrophic lateral sclerosis patients. *Nat. Med.* **2**, 696–699.

65. Cejas, P., Martinez, M. A., Karmally, S., McKillop, M., McKillop, J., Plunkett, J. A., Oudega, M., and Eaton, M. J. (2000). Lumbar transplant of neurons genetically modified to secrete brain-derived neurotrophic factor attenuate allodynia and hyperalgesia after sciatic nerve constriction. *Pain* **86**, 195–210.

66. Buchanan, J. T., and Nornes, H. O. (1986). Transplants of embryonic brain stem containing the locus coeruleus into spinal cord enhance the hindlimb flexion reflex in adult rats. *Brain Res.* **381**, 225–236.

67. Commissiong, J. W., and Sauvé, Y. (1989). The physiological basis of transplantation of fetal catecholaminergic neurons in the transected spinal cord of the rat. *Comp. Biochem. Physiol.* **93A**, 301–307.

68. Foster, G. A., Schultzberg, M., Gage, F. H., Björklund, A., Hökfelt, T., Nornes, H., Cuello, A. C., Verhofstad, A. A., and Visser, T. J. (1985). Transmitter expression and morphological development of embryonic medullary and mesencephalic raphé neurones after transplantation to the adult rat central nervous system. *Exp. Brain Res.* **60**, 427–444.

69. Foster, G. A., Roberts, M. H. T., Wilkinson, L. S., Björklund, A., Gage, F. H., Hökfelt, T., Schultzberg, M., and Sharp, T. (1989). Structural and functional analysis of raphe neurone implants into denervated rat spinal cord. *Brain Res. Bull.* **22**, 131–137.

70. Moorman, S. J., Whalen, L. R., and Nornes, H. O. (1990). A neurotransmitter specific functional recovery mediated by fetal implants in the lesioned spinal cord of the rat. *Brain Res.* **508**, 194–198.

71. Nornes, H. O., Buchanan, J., and Björklund, A. (1988). Noradrenaline-containing transplants in the adult spinal cord of mammals. *Progr. Brain Res.* **78**, 181–186.

72. Nygren, L.-G., Olson, L., and Seiger, A. (1977). Monoaminergic reinnervation of transected spinal cord by homologous fetal brain grafts. *Brain Res.* **129**, 227–235.

73. Privat, A., Mansour, H., and Geffard, M. (1988). Transplantation of fetal serotonin neurons into the transected spinal cord of adult rats: Morphological development and functional influence. *Progr. Brain Res.* **78**, 155–166.

74. Privat, A., Mansour, H., Rajaofertra, N., and Geffard, M. (1989). Intraspinal transplants of serotonergic neurons in the adult rat. *Brain Res. Bull.* **22**, 123–129.

75. Reier, P. J., Stokes, B. T., Thompson, F. J., and Anderson, D. K. (1992). Fetal cell grafts into resection and contusion/compression injuries of the rat and cat spinal cord. *Exp. Neurol.* **115**, 177–188.

76. Reier, P. J., Anderson, D. K., Thompson, F. J., and Stokes, B. T. (1992). Neural tissue transplantation and CNS trauma: Anatomical and functional repair of the injured spinal cord. *J. Neurotrauma* **9**(1), S223–S248.

77. Yakovleff, A., Roby-Brami, A., Guézard, B., Mansour, H., Bussel, B., and Privat, A. (1989). Locomotion in rats transplanted with noradrenergic neurons. *Brain Res. Bull.* **22**, 112–121.

78. Feraboli-Lohnherr, D., Orsal, D., Yakovleff, A., Gimenez y Ribotta, M., and Privat, A. (1997). Recovery of locomotor activity in the adult chronic spinal rat after sublesional transplantation of embryonic nervous cells: Specific role of serotonergic neurons. *Exp. Brain Res.* **113**, 443–454.

79. Yakovleff, A., Cabelguen, J. M., Oral, D., Gimenez y Ribotta, M., Rajaofetra, N., Drian, M. J., Bussel, B., and Privat, A. (1995). Fictive motor activities in adult chronic spinal rats transplanted with embryonic brain stem neurons. *Exp. Brain Res.* **106**, 69–78.

80. Pulford, B. E., Mihajlov, A. R., Nornes, H. O., and Whalen, L. R. (1994). Effects of cultured adrenal chromaffin cell implants on hindlimb reflexes of the 6-OHDA lesioned rat. *J. Neural Transplant. Plast.* **5**, 89–102.

81. Hentall, I. D., and Sagen, J. (2000). The alleviation of pain by cell transplantation. *In* "Functional Neural Transplantation" (S. B. Dunnett and A. Björklund, eds.), 2nd ed. Elsevier, Amsterdam (in press).

82. Sagen, J. and Eaton, M. J. (2001). Cellular implantation for the treatment of chronic pain. *In* "Pain: Current Understanding, Emerging Therapies, and Novel Approaches to Drug Discovery" (C. Bountra, R. Munglani, and W. K. Schmidt, eds.). Dekker, New York (in press).

83. Buchser, E., Goddard, M., Heyd, B., Joseph, J. M., Favre, J., de Tribolet, N., Lysaght, M., and Aebischer, P. (1996). Immunoisolated xenogeneic chromaffin cell therapy for chronic pain: Initial experience. *Anesthesiology* **85**, 1005–1012.

84. Burgess, F. W., Goddard, M., Savarese, D., *et al.* (1996). Subarachnoid bovine adrenal chromaffin cell implants for cancer pain management. *Am. Pain Soc. Abstr.* **15**, A–33.

85. Décosterd, I., Buchser, E., Gilliard, N., Saydoff, J., Zurn, A. D., and Aebischer, P. (1998). Intrathecal implants of bovine chromaffin cells alleviate mechanical allodynia in a rat model of neuropathic pain. *Pain* **76**, 159–166.

86. Sagen, J., Lewis-Cullinan, C., Goddard, M., and Burgess, F. W. (2001). Encapsulated cell implants for pain surgery. *In* "Pain Surgery" (K. Burchiel, ed.). Thieme, Stuttgart (in press).

87. Czech, K. A., Pollak, R., Pappas, G. D., and Sagen, J. (1996). Bovine chromaffin cells for CNS transplants do not elicit xenogeneic T cell proliferative responses *in vitro*. *Cell Transplant.* **5**, 257–267.

88. He, L., Xue, Y., Wang, J., Wang, Z., Li, X., Zhangi, L., Dui, X., Zhu, J., Luo, Y., Zhong, D., and Li, Y. (2000). Subarachnoid microencapsulated bovine chromaffin cells xenograft for treatment of cancer pain: A report of preliminary studies. *Am. Soc. Neural Transplant. Repair Abstr.* **7**, 27.

89. Michalewicz, P., Laurito, C. E., Pappas, G. D., Lu, Y., and Yeomans, D. C. (1999). Purification of adrenal chromaffin cells increases antinociceptive efficacy of xenotransplants without immunosuppression. *Cell Transplant.* **8**, 103–109.

90. Yeomans, D. C., Lu, Y., Laurito, C. E., Votta-Belis, G., and Pappas, G. D. (2000). Analgesic effects of transplantation of bovine adrenal chromaffin cells in non-human primates. *Am. Soc. Neural Transplant. Repair Abstr.* **7**, 26.

91. Wu, H., Lester, B., Sun, Z., and Wilcox, G. L. (1994). Antinociception following implantation of mouse B16 melanoma cells in mouse and rat spinal cord. *Pain* **56**, 203–210.

92. Wu, H. H., McLoon, S. C., and Wilcox, G. L. (1993). Antinociception following implantation of AtT-20 and genetically modified AtT-20/hENK cells in rat spinal cord. *J. Neural Transplant. Plast.* **4**, 15–26.

93. Wu, H. H., Wilcox, G. L., and McLoon, S. C. (1994). Implantation of AtT-20 or genetically modified AtT-20-hENK cells in mouse spinal cord induced antinociception and opioid tolerance. *J. Neurosci.* **14**, 4806–4814.

94. Saitoh, Y., Eguchi, Y., Hagihara, Y., Arita, N., Watahiki, M., Tsujimoto, Y., and Hayakawa, T. (1998). Dose-dependent doxycycline-mediated adrenocorticotropic hormone secretion from encapsulated Tet-on proopiomelanocortin Neuro2A cells in the subarachnoid space. *Hum. Gene Ther.* **9**, 997–1002.

95. Saitoh, Y., Taki, T., Arita, N., Ohnishi, T., and Hayakawa, T. (1995). Analgesia induced by transplantation of encapsulated tumor cells secreting β-endorphin. *J. Neurosurg.* **82**, 630–634.

96. Ishii, K., Isono, M., Inoue, R., and Hori, S. (1999). Attempted gene therapy for intractable pain; dexamethasone-mediated exogenous control of β-endorphin secretion in genetically modified cells and their intrathecal transplantation. *Int. Meet. Neural Transplant. Repair* **7**, 27.

97. Eaton, M. J., Plunkett, J. A., Martinez, M. A., Karmally, S., Montañez, K., and Rabade, J. (1998). Changes in GAD and GABA immunoreactivity in the spinal dorsal horn after peripheral nerve injury and promotion of recovery by lumbar transplants of immortalized serotonergic neurons. *J. Chem. Neuroanat.* **16**, 57–72.

98. Eaton, M. J., Plunkett, J. A., Martinez, M. A., Lopez, T., Karmally, S., Cejas, P., and Whittemore, S. R. (1999). Transplants of neuronal cells bioengineered to synthesize GABA alleviate chronic neuropathic pain. *Cell Transplant.* **8**, 87–101.

99. Brewer, K. L., and Yezierski, R. P. (1998). Effects of adrenal medullary transplants on pain-related behaviors following excitotoxic spinal cord injury. *Brain Res.* **798**, 83–92.

100. Hains, B. C., Chastain, K. M., Everhart, A. W., and Hulsebosch, C. E. (1998). Reduction of chronic central pain following spinal cord injury by transplants of adrenal medullary chromaffin cells. *Soc. Neurosci. Abstr.* **24**, 1631.

101. Yu, W., Hao, J.-X., Xu, X.-J., Saydoff, J., Haegerstrand, A., and Wiesenfeld-Hallin, Z. (1998). Long-term alleviation of allodynia-like behaviors by intrathecal implantation of bovine chromaffin cells in rats with spinal cord injury. *Pain* **74**, 115–122.

102. Yu, W., Hao, J.-X., Xu, X.-J., Saydoff, J., Haegerstrand, A., and Wiesenfeld-Hallin, Z. (1998). Immunoisolating encapsulation of intrathecally implanted bovine chromaffin cells prolongs their survival and produces anti-allodynic effects in spinally injured rats. *Eur. J. Pain* **2**, 143–151.

103. Rapalino, O., Lazarov-Spiegler, O., Agranov, E., Velan, G. J., Yoles, E., Fraidakis, M., Solomon, A., Gepstein, R., Katz, A., Belkin, M., Hadani, M., and Schwartz, M. (1998). Implantation of stimulated homologous macrophages results in partial recovery of paraplegic rats. *Nat. Med.* **4**, 814–821.

104. Vrbová, G., Clowry, G., Nógrádi, A., and Sieradzan, K. (1994). "Transplantation of Neural Tissue into the Spinal Cord." R.G. Landes, Austin, TX.

105. Sieradzan, K., and Vrbová, G. (1993). Observations on the survival of grafted embryonic motoneurons in the spinal cord of developing rats. *Exp. Neurol.* **122**, 223–231.

106. Nógrádi, A., and Vrobová, G. (1996). Improved motor function of denervated rat hindlimb muscles induced by embryonic spinal cord grafts. *Eur. J. Neurosci.* **8**, 2198–2203.

107. Demierre, B., Ruiz-Flandes, P., Martinou, J.-C., and Kato, A. C. (1990). Grafting of embryonic motoneurones into adult spinal cord and brain. *Prog. Brain Res.* **82**, 233–237.

108. Peschanski, M. (1993). Spinal cord transplantation. *J. Neural Transplant. Plastic.* **4**, 109–111.

109. Peschanski, M., Nothias, F., and Cadusseau, J. (1992). Is there a therapeutic potential for intraspinal transplantation of fetal spinal neurons in motoneuronal disease? *Restorative Neurol. Neurosci.* **4**, 227.

110. Gage, F. H., Ray, J., and Fisher, L. J. (1995). Isolation, characterization, and use of stem cells from the CNS. *Annu. Rev. Neurosci.* **18**, 159–192.

111. McKay R. (1997). Stem cells in the central nervous system. *Science* **276**, 66–71.

112. Reynolds B., Tetzlaff W., and Weiss S. (1992). A multipotent EGF-responsive striatal embryonic progenitor cell produces neurons and astrocytes. *J. Neurosci.* **12**, 4565–4574.

113. Fricker, R. A., Carpenter, M. K., Winkler, C., Greco, C., Gates, M. A., and Björklund, A. (1999). Site-specific migration and neuronal differentiation of human neural progenitor cells after transplantation in the adult rat brain. *J. Neurosci.* **19**, 5990–6005.

114. Zompa, E. A., Cain, L. D., Everhart, A. W., Moyer, M. P., and Hulsebosch, C. E. (1997). Transplant therapy: Recovery of function after spinal cord injury. *J. Neurotrauma* **14**, 479–501.

115. Park, K. I., Liu, S., Flax, J. D., Nissim, S., Stieg, P. E., and Snyder, E. Y. (1999). Transplantation of neural progenitor and stem cells: Developmental insights may suggest new therapies for spinal cord and other CNS dysfunction. *J. Neurotrauma* **16**, 675–687.

116. Snyder, E. Y., Yoon, C., Flax, J. D., and Macklis, J. D. (1997). Multipotent neural precursors can differentiate toward replacement of neurons undergoing targeted apoptotic degeneration in adult mouse neocortex. *Proc. Natl. Acad. Sci. U.S.A.* **94**, 11663–11668.

117. Park, K. I., and Snyder, E. Y. (1999). Transplantation of human neural stem cells propagated by either genetic or epigenetic means, into hypoxic–ischemic brain injury. *Soc. Neurosci. Abstr.* **25**, 212.

118. Cao, Q. L., Loy, D. N., Zhang, Y. P., Howard, R. M., Magnuson, D. S.K., and Whittemore, S. R. (2000). Neural stem cell differentiation after engraftment into the normal and confused rat spinal cord. *Am. Soc. Neural Transplant. Abstr.* **7**, 33.

119. Marsala, M., Kakinohana, O., Cizkova, D., Yang, L. C., Webb, M., and Yaksh, T. L. (1999). Spinal ischemia-induced paraplegia: Differentiation of neuronal stem cells after intraspinal implantation. *Soc. Neurosci. Abstr.* **25**, 211.

120. McDonald, M. Liu, X.-Z., Qu, Y., Liu, S., Mickey, S. K., Turetsky, D., Gottlieb, D. I., and Choi, D. W. (1999). Transplanted embryonic stem cells survive, differentiate, and promote recovery in injured rat spinal cord. *Nat. Med.* **5**, 1410–1412.

121. Brüstle, O., Jones, K. N., Learish, R. D., Karram, K., Choudhary, K., Wiestler, O. D., Duncan, I. D., and McKay, R. D. G. (1999). Embryonic stem cell-derived glial precursors: A source of myelinating transplants. *Science* **285**, 754–756.

122. Liu, Y., Himes, B. T., Solowska, J., Moul, J., Chow, S. Y., Park, K. I, Tessler, A., Murray, M., Snyder, E. Y., and Fischer, I. (1999). Intraspinal delivery of neurotrophin-3 using neural stem cells genetically modified by recombinant retrovirus. *Exp. Neurol.* **158**, 9–26.

123. Vacanti, M. P. (2000). Neural stem cells. *In* "Principles of Tissue Engineering" (R. P. Lanza, R. Langer and J. Vacanti, eds.), 2nd ed., pp. 821–830. Academic Press, San Diego, CA.

CHAPTER 104

CRYOPRESERVED DERMAL IMPLANTS

Gail K. Naughton and Dawn R. Applegate

INTRODUCTION

Revolutionary advances in tissue engineering are redefining approaches to tissue repair and transplantation through creation of replacement tissues that remain biointeractive after implantation, imparting structure as well as physiological function to the damaged or diseased tissue. Maintenance of tissue integrity, functionality, and viability from product manufacture to end use has been accomplished through innovation in design and scale-up of both tissue growth and preservation processes. These first-of-a-kind systems have enabled the delivery of readily available, off-the-shelf, consistent, easy-to-use, safe, and effective replacement tissues, which afford clinical efficacy, feasibility of large-volume distribution, and cost-effectiveness. Skin replacement products are the most advanced, with several tissue-engineered wound care materials on the market in the United States and in several international communities [1–4]. Whereas traditional replacement devices provide structure to a tissue defect, tissue-engineered products are designed to replace structure as well as function, being biointeractive implants that can interact with the body's environment to restore function. For this reason, the maintenance of cell viability in these tissue-engineered constructs is crucial. Accordingly, the development of tissue-engineered products has required not only the design of closed, large-scale manufacturing systems but also the optimization of preservation processes to allow for the global distribution of products that will maintain their metabolic activity [5,6]. These fundamental discoveries from the development of pioneering fibroblast-based constructs are being leveraged to enable rapid advancement in the research and development of next-generation tissues, including cardiovascular, orthopedic, and organ replacement products. Learning curve advantages include synergies in mimicking physiological conditions *in vitro*, manufacturing design requirements, preservation processes, product attributes for clinical efficacy, clinical design, regulatory requirements, distribution methods, marketing and reimbursement approaches, and legal and patent strategies. This chapter offers an overview of these steps undertaken in bringing human fibroblast-based tissue-engineered products from concept through clinical trials and to the global marketplace. Emphasis will be placed upon specific manufacturing systems, cryopreservation processes, freezing effects on tissue function, and the unique mechanisms of action of these products *in vivo*.

DERMAL IMPLANTS

Skin, the body's largest organ, has been the first tissue-engineered organ to progress from lab bench to patient care. Skin consists of a continuously replicating epidermis and a nonreplicating dermis. The dermal (lower) layer of skin offers many potential advantages as a therapeutic implant. Whereas the epidermis regenerates after injury, the dermis after wounding is replaced with fibrotic scar tissue. In addition, epidermal keratinocytes and Langerhans cells carry surface antigens (HLA-DR) that can result in allograft rejection [7]. Implantation of allogeneic dermal tissue, however, does not stimulate an immune response [8–12]. Founded on these characteristics of native tissue repair and rejection processes, researchers at Advanced Tissue Sciences, Inc. designed Dermagraft®, an allogeneic, universal donor, human-based, metabolically active dermal implant for wound healing and other regenerative applications [6,13–15].

In addition to immunological and physiological advantages, the tissue-engineered dermis offers production advantages as well. The human diploid fibroblasts are readily harvested from neonatal foreskin tissue obtained from routine circumcisions. Moreover, fibroblasts have a proliferative potential of approximately 60 population doublings and have a history of use in therapeutics, biologics, and drug development [16–18]. Unlike orthopedic and cardiovascular tissues, dermal fibroblasts do not require stimulatory fluid forces to assemble into a human dermal equivalent. Nutrient diffusion is key in the manufacture of uniform, viable, tissue-engineered products. Dermal tissue equivalents are approximately several hundred micrometers in thickness and thereby amenable to diffusive nutrient and cryopreservative delivery, enabling the tissue to be grown in static systems and cryopreserved in large batch sizes. The diffusional characteristics of this dermal tissue also afford simple thawing and rinsing steps prior to implantation as well as rapid vascularization of the entire construct postimplantation.

DERMAGRAFT MANUFACTURING PROCESS
GROWTH PROCESS FOR DERMAGRAFT

Fibroblasts are enzymatically removed from neonatal foreskins and fully screened for pathogens and other infectious agents [14,19] according to the U.S. Food and Drug Administration (FDA) Points to Consider [20] as well as the Guidelines from the Committee for Proprietary Medicinal Products (CPMP) regulations [21]. After successful screening, fibroblasts are expanded and cryopreserved by standard methods [18,22]. Passage-8 cells are utilized for final product manufacture. In the closed manufacturing system, cells are seeded

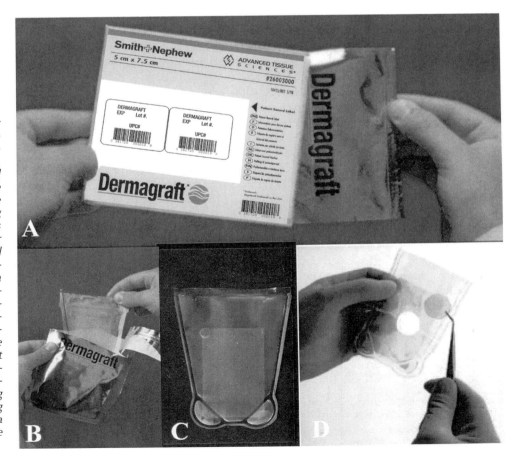

Fig. 104.1. Dermagraft bioreactor and final product packaging. Ethlene vinyl acetate (EVA) bioreactor into which one 2 in. × 3 in. piece of Vicryl scaffold is welded and cells are seeded to grow the dermal tissue is shown (C). The scaffold is held in place along the top with Z-welded spot welds and along the bottom with bar welds. For cryopreservation and delivery to the end user, the final grown tissue is placed in a trilaminate foil pouch for moisture and vapor barrier in long-term storage (B). The outer cardboard box (A) displays all necessary labeling, including removable lot number label for placement into the patient's record for traceability. The EVA bioreactor is designed to be translucent, enabling placement on the wound, tracing of the wound, and cutting of a piece of tissue to the desired size for implantation (D).

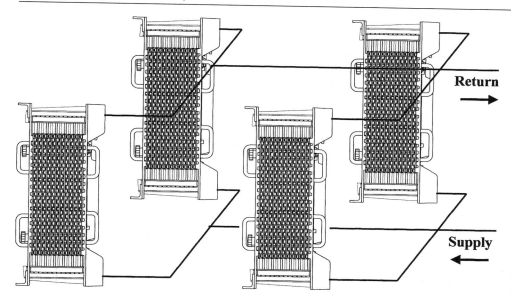

Fig. 104.2. Bioreactor manifold growth system for Dermagraft. Each manifold growth system contains multiple EVA processing units with multiple EVA bioreactors containing 2 in. × 3 in. mesh per processing unit. Processing units are attached vertically in each manifold for simplicity of feeding and draining. Manifold growth systems are connected in parallel (only four connections are shown for convenience) to form a manufacturing production lot.

onto 2 in. × 3 in. Vicryl® (Ethicon, Sommerville, NJ) scaffolds, which are laser-welded into EVA (ethylene vinyl acetate) bioreactors (Fig. 104.1C). Bar welds are used across the bottom of the scaffold, as are Z-welds across the top, to secure the scaffold in the center of the bioreactor to provide uniformity of environment and to prevent tissue contraction and rolling during growth. Manufacturing design has multiple bioreactors per process unit, with multiple process units per manifold growth system (Fig. 104.2). The corrugated design of the manifold growth system allows for uniform media distribution, removal of air bubbles, even heat conduction, and an unhindered supply of oxygen and carbon dioxide. Individual manifolds are rotated after initial cell seeding to result in uniform cell distribution across all scaffold surfaces. Several manifolds constitute one lot of product, thus allowing for the growth of over a thousand Dermagraft units per closed system (Fig. 104.2). During the manufacturing process, developing dermal tissues are fed with serum-containing cell culture medium [23,24], with several media changes over the growth period of approximately two weeks. This large-scale closed system offers major advantages, including the maintenance of asepsis during manufacture, the use of automated processes for tissue growth, and the ability to utilize in-line, in-process testing to assess cell growth and matrix deposition and to accurately predict optimal time for harvest.

The simulation of *in vivo* conditions through incorporation of the three-dimensional Vicryl scaffold in the physiological bioreactor system induces fibroblasts to not only divide, but also to secrete a variety of growth factors, and to deposit human collagens, glycosaminoglycans, and other matrix proteins to form a dermal tissue similar to native human dermis (Fig. 104.3). Characterization of cell growth and matrix deposition in this system has been reported elsewhere [14,15,19,23,24]. The progression of cell growth and deposition of matrix proteins during cultivation in the bioreactor system is shown in Fig. 104.3. At the conclusion of the growth period, the closed-production bioreactor system is separated to result in individual packaging of the 2 in.×3 in. final tissue products (Fig. 104.1C). This process allows for a uniform tissue product that is reproducible from lot to lot and allows for maintenance of asepsis, from introduction of cells to the bioreactor to delivery of a final tissue unit to the patient, while negating the need for aseptic repackaging of a delicate, living tissue.

CRYOPRESERVATION OF THE DERMAL IMPLANTS

To maintain tissue integrity and functionality of a transplant product until needed for use, the product must be preserved. Although maintenance of tissue integrity and viability for days to weeks for both engineered and native tissues has been reported with fresh preservation and protocols [25–28], this method often imposes impractical scheduling restraints for just-in-time manufacturing and end users. These fresh preservation procedures

Fig. 104.3. Progression of cell growth and matrix deposition to form an engineered human dermal tissue. Neonatal foreskin fibroblasts obtained from routine circumcisions are seeded onto a knitted Vicryl scaffold. By virtue of being on a three-dimensional scaffold, the cells not only grow and divide, but also secrete all the matrix proteins and growth factors characteristic of papillary human dermis, forming a completely human, engineered dermal tissue. Microscopic visualization of this cell growth and tissue formation process is also shown. Cells are visible as translucent dots that spread out between fiber bundles by day 1 post-seeding, growing to cover and close the mesh openings, and eventually enveloping the entire scaffold, at which point the cells move from the growth phase to the matrix deposition phase. A complete, human dermal replacement tissue is formed in approximately 2 weeks.

Growth of Engineered Human Dermal Tissue

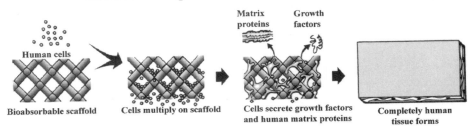

Microscopic Visualization of Cell Growth and Matrix Deposition

| 3 hr post cell seeding | 1 day post cell seeding | 2 days post cell seeding | 8 days post cell seeding |
| Cells attach to scaffold | Cells stretch out between fibers and divide | Cells fill in scaffold openings | Cells cover scaffold and begin secreting matrix |

usually result in unrealistic manufacturing and physician operating costs associated with unplanned losses, overproduction, changing demand, and management of patient scheduling. Moreover, nonpreserved or short-term-preserved products are shipped just-in-time and implanted at risk, since U.S. Pharmacopeia (USP) sterility tests [29], requiring a 14-day period, cannot be performed on these products. In addition to sterility, tests that ensure uniformity of tissue biochemical and biomechanical characteristics from lot to lot often cannot be performed owing to time constraints. Instead, it is necessary to use rapid testing methods, which are usually not quantitative or as reliable as standard testing procedures that normally require days or weeks to complete. Hence a benefit for products that undergo moderate- to long-term preservation is that their safety, characterization, and efficacy can be more rigorously proven and retained. Advancements in technologies such as lyophilization [30–32] and vitrification [33–36] have enabled long-term storage of acellular tissues. In addition, long-term unfrozen storage of cells and tissues has been shown to be feasible through the addition of natural protectants, such as the sugar trehalose [37,38]. Progress in the field continues at a steady pace. At present, however, only cryopreservation utilizing low concentrations of cryoprotectants has been proven for moderate- to long-term maintenance of cellular viability and functionality in native and engineered tissues.

In cryopreservation of tissue, a balance must be struck between minimizing exposure time to potentially toxic cryoprotectants and slow, stepwise addition of cryoprotectant to minimize osmotic forces. Practical experience with our tissue-engineered human dermal replacement, Dermagraft, revealed that tissue characteristics that meet product specifications for metabolic activity retention and matrix composition are maintained by use of a one-step cryoprotectant if the cryoprotectant, which is based on dimethyl sulfoxide (DMSO), is added at 4°C. At this lower temperature there is a slow diffusion of cryoprotectant into the tissue, which results in gentle exposure of the cells to a gradual increase in cryoprotectant and a concomitant increase in osmolarity and toxicity. However, postgrowth, prefreeze processing of large lots of Dermagraft at room temperature would enable cost reduction and simplification of equipment and process design. For this reason, a stepwise, room temperature addition method was developed that would afford equivalent viability to the one-step, 4°C process. The new process was designed based on estimates of nonosmotic cell volumes obtained from the published literature, along with calculations of stepwise addition protocols based on minimizing transient cell volume excursions according to the design of Arnaud and Pegg [39].

To provide a reproducible as well as robust process for cryopreserving over 1000 units at a time, a custom-designed, large-scale freezer was developed. This equipment utilizes high-accuracy, fast-feedback controllers combined with liquid nitrogen injection ports and a high-velocity air distribution system. Since multistep cooling and continuous changes in freeze rate have shown benefits over linear cooling rates in maintaining cell viability [40,41], the unit can be programmed for linear, stepwise, or nonlinear freezing profiles. The Dermagraft freezing unit is capable of freezing rates up to 100°C/min and temperature control within 0.5°C across the chamber. The equipment induces near-simultaneous nucleation of extracellular ice in all tissue products, obviating the need for manual nucleation methods (i.e., chilled rods), which are impractical in large-scale production. After cryopreservation, the human dermal implants are stored and shipped at −70°C. Custom-designed equipment was developed for transfer of product between freeze-down and storage freezers and between storage freezers and shipping containers. The design was based on quick, reliable, and reproducible transfers to prevent temperature gradients in the product as well as temperature increase of the product to detrimental levels, which occurs approximately 10 or more degrees above the storage temperature of −70°C. Specialized product racking systems were developed to facilitate quick freezer loading and unloading, rapid product transfers, and product identification and segregation in storage to simplify retrieval and traceability of frozen product and to satisfy inventory control regulatory requirements. Testing for final frozen product quality control includes USP sterility tests and endotoxin tests, as well as specific tissue assays for matrix proteins, growth factors, and cell metabolic activity.

PRODUCT PACKAGING

Final product packaging for large-scale engineered tissues must meet specific requirements for freezing, storage, product handling, global distribution, and physician use, including manufacturing lot identification for patient records and product thaw and preparation for implantation. The packaging needs to be symmetrical, uniform, and of low resistance, to enable even heat transfer during freeze-down while simultaneously providing some resistance to warming to allow for feasible handling times during product transfers. Moisture and vapor transfer during long-term storage, which can result in evaporation of freezing solutions, can be prevented by a variety of foil overwraps. Inks and removable labels that will persist for long periods of time at low temperature and high moisture are required so that the lot information can be peeled from the package and placed onto the patients' records for full product traceability. If the product is shipped internationally, the label must also contain information in multiple languages, or provision must be made to add specialized packaging such as an overwrap under ultracold conditions. The packaging should also be easy to open, to minimize handling time from freeze to thaw and to protect the product from handling damage. The final packaging for Dermagraft (Fig. 104.1A) was designed based on these requirements and has demonstrated its ability to protect cell viability, tissue integrity, and product asepsis for up to 5 months while providing end-user ease of product preparation and implantation.

FROZEN PRODUCT PREPARATION PROCEDURES FOR IMPLANTATION

For off-the-shelf tissue products such as Dermagraft that are delivered to the end user in a frozen state, product thawing and cryoprotectant removal steps, which are significant contributors to maintenance of product viability and tissue integrity, are in the hands of the end user. Because it is often not possible to visually determine whether product characteristics have been maintained post-thaw, and not practical or possible to have rapid assessment assays at the end-user site, preparation protocols must be designed to be reliable, reproducible, and foolproof. In addition, physicians demand that the procedures be quick, simple, and achievable with readily available supplies and minimum equipment requirements.

The preparation procedure developed for Dermagraft satisfies all the foregoing requirements (Fig. 104.4). Cell viability and tissue integrity from end-user storage to implantation are maintained by performing a simple, quick preparation procedure with readily available physicians' supplies and minimal specialty equipment. The Dermagraft bioreactor itself was specially designed to satisfy not only the requirements of tissue growth, but also

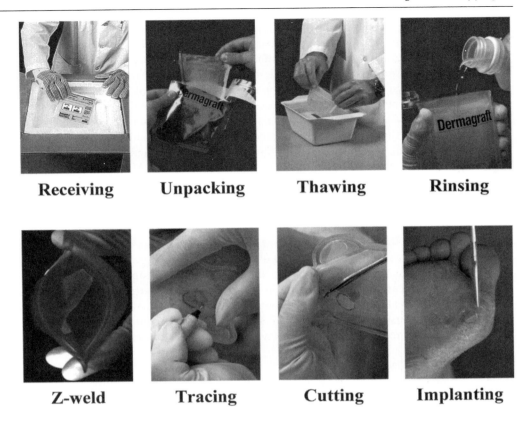

Receiving   **Unpacking**   **Thawing**   **Rinsing**

Z-weld   **Tracing**   **Cutting**   **Implanting**

Fig. 104.4. Product preparation steps for Dermagraft. The Dermagraft product arrives to the physician frozen on dry ice. The product is immediately removed from the shipping container, and the manufacturing lot label is peeled from the product box and placed in the patient record. The cardboard box is rapidly removed from the product, and the trilaminate foil bag is torn open to retrieve the bioreactor containing the frozen tissue. The bioreactor is quickly submerged in a warm water bath to thaw (~2 min). The product is placed in a protective stand and rinsed with saline to remove cryoprotectant. The tissue scaffold is Z-welded into the bioreactor bag to afford uniformity of fluid distribution during growth and rinsing and to maintain the tissue in the bioreactor. The bioreactor is translucent, enabling direct placement on the patient's wound and tracing of the wound shape onto the bag. The tissue is cut to the desired size, removed from the bioreactor, and placed into the wound bed. The implanted tissue is then covered with a moist wound dressing, which is changed daily to maintain the proper moisture level in the wound as well as to prevent infection.

those of product preparation. Dermagraft is retrieved from a −70°C storage freezer located at the clinical site or removed from a dry ice shipping container. Lot-specific information is peeled from the protective outer box and placed onto the patient's chart. The trilaminate foil pouch is removed from the outer box and torn open, and the bioreactor containing the dermal implant removed and immediately placed in a warm bath of water. Based on end-user preference, controlled temperature water baths or thermometers and water tubs that can be filled with water of the desired temperature from a faucet are supplied. Thawing is complete in approximately 2 min. Rapid thaw is critical to the maintenance of proper metabolic activity of the tissue postimplantation. The thawed product in the bioreactor is then placed in a holder, both to facilitate handling and to protect the tissue. The top of the bioreactor is cut open and the product is rinsed four times, using readily available normal saline to reduce the cryoprotectant concentration to below detectable levels. The holder is also used to keep the product submerged in fluid while awaiting implantation. The tissue scaffold is Z-welded into the EVA bioreactor to maintain the tissue in the center of the bioreactor, providing uniform delivery of nutrients during growth as well as efficient washing and anchoring of the product in the bioreactor during rinsing. The patient's wound is debrided, and the saline is drained from the bioreactor. By design, the bioreactor is translucent, enabling direct placement on and tracing of the wound. Dermagraft that is cut from the bioreactor, following the tracing to fit the wound, is implanted into the prepared wound bed. Dressings, designed to provide a wound environment that is either moist or absorbable depending on the type of wound being treated, are applied directly onto the dermal implant. In clinical trials, Dermagraft was applied to a freshly debrided wound once a week for up to 8 weeks and was maintained moist in the wound between applications.

MECHANISMS OF ACTION OF DERMAGRAFT

Any product designed to replace or restore tissue or organ function must provide critical therapeutic benefits, including restoration of normal physiology and utility. Stimulation of wound healing has been realized with topical application of growth factors in both animal and human models [42–46]. While application of a single growth factor or combination of several growth factors may prove effective in increasing cell proliferation, angiogenesis, or cell migration, it cannot substitute for the complex range of interactive events provided by a living, metabolically active, complete tissue. Dermagraft was designed to contain metabolically active fibroblasts to continuously deliver a cocktail of growth factors and structural and interactive human matrix components to the wound bed, thereby providing many of the elements critical to accelerating wound healing over an extended period of time. Five modes of action are believed to give Dermagraft its clinical efficacy. First, the matrix proteins comprising Dermagraft form a permissive substrate for cell migration. Second, metabolically active fibroblasts, distributed throughout the matrix, continuously secrete growth factors and matrix proteins into the wound bed, providing prolonged delivery of the therapy beyond the time of application. These fibroblasts, through their surface receptors, are believed to be able to respond to the wound environment and to be regulated as they would in native tissue, orchestrating the complex management of reactions required to achieve healing. Third, included in these secreted growth factors are angiogenic factors, necessary for promotion of rapid vascularization that is critical to wound healing. Fourth, the expression of inflammatory cytokines by the fibroblasts is important for arresting the inflammatory stage of wound healing and returning the chronic wound to an acute healing pathway. Fifth, the lack of immunological response to the allogeneic engineered tissue not only enables persistence of the tissue in the patient for months but also allows millions of patients to be treated from a single donor tissue source.

The components of this human dermal replacement tissue as they relate to four of the modes of action just outlined are given in Table 104.1. The primary component of skin is collagen, the major structural protein of the dermis. Dermagraft contains the matrix proteins collagen Types I and III, which provide structure to the tissue, as well as fibronectin, important for cell adhesion, spreading, migration, and mitogenesis, and tenascin, important in the initiation of wound healing and the control of cell adhesion. The matrix also contains the glycosaminoglycans veriscan, decorin, betaglycan, and syndecan, which are responsible for structuring collagen and for binding growth factors and hyaluronic acid, immobilizing them to the area in need. This matrix provides a substrate that, in concert with the expression of growth factors, promotes rapid reepithelialization of the wound. Once implanted, the living, metabolically active fibroblasts in Dermagraft are believed to continuously secrete these matrix proteins as well as a host of growth factors. Growth factors important to matrix synthesis include transforming growth factors β_1 and β_3 (TGF-β_1, -β_3). Mitogenic growth factors required to work in concert with matrix proteins to stimulate movement of fibroblasts and keratinocytes to close the wound include platelet-derived growth factor (PDGF) A chain, insulin-like growth factor 1 (IGF-1), keratinocyte growth factor (KGF), heparin-binding epidermal growth factor–like growth factor (HBEGF), and transforming growth factor α (TGF-α) [47,48]. These matrix proteins and growth factors work in concert to stimulate epithelial migration to close the wound as shown in Figs. 104.5 and 104.6. In Fig. 104.5, the epidermis surrounding a full-thickness wound bed of an athymic nude mouse treated with an explant from split-thickness skin shows that keratinocytes from the explant were unable to migrate away from the explant. Application of Dermagraft on top of the wound bed, under the explant, resulted in rapid stimulation of epithelial migration away from the explant over the adjacent tissue; the mobilized cells surrounding the rectangular dermal implant are visualized in white through use of a hematoxylin negative sur-

Table 104.1. Matrix Proteins, Glycosaminoglycans, and Growth Factors in Dermagraft, an Engineered Human Dermal Replacement Tissue

Material	Name	Function
Matrix proteins		
Collagens, Types I and III		Major structural protein of dermis
Fibronectin		Cell adhesion, spreading, migration, mitogenesis
Tenascin		Induced in wound healing; control of cell adhesion
Glysosaminoglycans		
Veriscan		Structural; binds hyaluronic acid and collagen
Decorin		Binds growth factors; influences collagen structure
Betaglycan		TGF-β Type III receptor
Syndecan		Binds growth factors; enhances activity
Growth Factors		
Matrix deposition factors		
Transforming growth factor β_1	TGF-β_1	Stimulates matrix deposition
Transforming growth factor β_3	TGF-β_3	Stimulates matrix deposition; antiscarring
Mitogenic factors		
Platelet-derived growth factor A chain	PDGF-A	Mitogen for fibroblasts, granulation tissue; chemotactic
Insulin-like growth factor 1	IGF-1	Mitogen for fibroblasts
Keratinocyte growth factor	KGF	Mitogen for keratinocytes
Heparin-binding epidermal growth factor–like growth factor	HBEGF	Mitogen for keratinocytes, fibroblasts
Transforming growth factor α	TGF-α	Mitogen for keratinocytes, fibroblasts
Angiogenic factors		
Vascular endothelial growth factor	VEGF	Angiogenesis
Hepatocyte growth factor	HGF	Angiogenesis
Basic fibroblast growth factor	bFGF	Angiogenesis
Secreted protein acidic and rich in cysteine	SPARC	Both anti- and proangiogenic
Interleukin 6	IL-6	Angiogenic; inflammatory cytokine; inhibitory effect on collagen synthesis
Interleukin 8	IL-8	Angiogenic; inflammatory cytokine; chemoattractive to neutrophils
Inflammatory cytokines		
Interleukin 6	IL-6	Angiogenic; inflammatory cytokine; inhibitory effect on collagen synthesis
Interleukin 8	IL-8	Angiogenic; inflammatory cytokine; chemoattractive to neutrophils
Granulocyte colony stimulating factor	G-CSF	Stimulates neutrophil production and maturation
Tumor necrosis factor α	TNF-α	Inflammatory cytokine

face stain [49]. This rapid migration of epidermis has been demonstrated clinically through weekly application of Dermagraft to diabetic foot ulcers (Fig. 104.6). Dermagraft is cut to fit the debrided diabetic ulcer wound bed and stimulates epithelial migration, with subsequent applications of Dermagraft being made into progressively smaller wound openings until the wound is completely reepithelialized and healed.

Wound Bed   DERMAGRAFT®

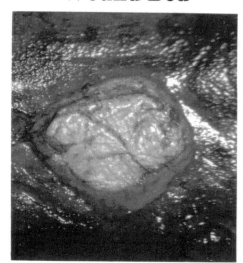
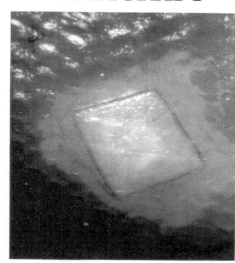

Fig. 104.5. Epidermal migration on wound bed and Dermagraft. A full-thickness wound bed of an athymic nude mouse treated with an explant from split-thickness skin (left) does not stimulate migration of keratinocytes from the explant to the surrounding tissue. Dermagraft applied onto the wound bed under the explant (right) stimulates rapid migration of keratinocytes from the explant across the surrounding tissue. The migration is visualized by a hematoxylin negative surface stain in which hematoxylin is used to stain the dermal substrate purple, whereupon the keratinocytes, which exclude the stain, appear white [49].

In addition to matrix proteins and growth factors affecting wound closure, Dermagraft aids in wound healing by delivery of angiogenic factors to stimulate new blood vessel formation in the wound. Included are vascular endothelial growth factor (VEGF), hepatocyte growth factor (HGF), basic fibroblast growth factor (bFGF), secreted protein acidic and rich in cysteine (SPARC), interleukin 6 (IL-6) and interleukin 8 (IL-8). To induce wound transition from a chronic to an acute healing pathway, Dermagraft secretes several inflammatory cytokines, including granulocyte colony stimulating factor (G-CSF) and tumor necrosis factor α (TNF-α) as well as IL-6 and IL-8, which are also angiogenic [47–51]. All these com-

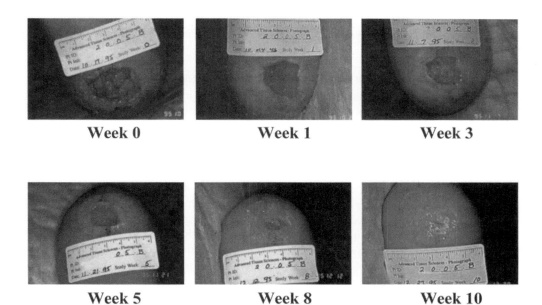

Week 0   **Week 1**   **Week 3**

Week 5   **Week 8**   **Week 10**

Fig. 104.6. Using Dermagraft as a dermal replacement in the healing of diabetic foot ulcers. In clinical studies, Dermagraft was applied to a sharply debrided diabetic foot ulcer once a week for up to 8 weeks. Dermagraft stimulated rapid reepithelialization from the wound edge, allowing the implantation of smaller pieces of the engineered tissue at each application until wound closure. While the once-a-week application was chosen for a variety of reasons at the onset of the clinical trial, results have shown that the product rapidly integrates into the wound bed, stimulating angiogenesis as well as epithelialization. Adding sequential product enables the wound to be gradually built up in thickness, with stimulation of angiogenesis through the tissue before application of another product, as opposed to application of multiple product simultaneously, which could result in starving the top layers of tissue of blood supply.

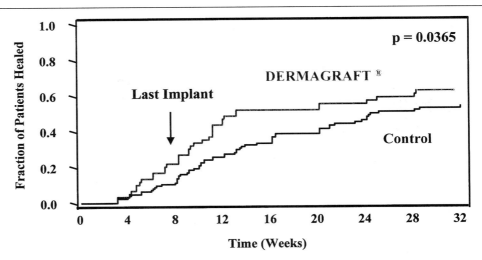

Fig. 104.7. Plot of fraction of patients healed versus treatment time for a Dermagraft diabetic foot ulcer clinical study. The progression of healing for Dermagraft, applied once a week for 8 weeks, versus control therapy for diabetic foot ulcers is shown by means of a Kaplan–Meiers survival curve. The healing rate for patients receiving Dermagraft continued to increase well beyond the time of the last implant (week 8), with the median time to complete wound closure being twice as fast in the Dermagraft group and the activity persisting for 6 months beyond the last application of engineered tissue. The data suggest that the fibroblast cells in Dermagraft are able to survive for long periods of time after implantation and to remain metabolically active, interacting with the wound bed and secreting growth factors and matrix proteins to stimulate wound healing.

ponents of Dermagraft are present in quantities similar to those found in native human papillary dermis. These components help to illustrate the unique biointeractive characteristics of a tissue-engineered implant, which are not offered by traditional wound coverings or other devices. These four explanations for the efficacy of Dermagraft are dependent on the presence of viable fibroblasts in the implant and their ability to persist in the wound. Various *in vitro*, preclinical, and clinical studies have been performed to further elucidate the role of these viable fibroblasts in the acceleration of healing induced by these living dermal implants.

PERSISTENCE OF DERMAGRAFT CELLS IN HOST WOUND SITES

The ability of live fibroblasts in implants to colonize wound beds, persist, and continue to secrete key growth factors is of tremendous importance in the induction of healing as well as in the continued maintenance of the healed tissue. Such colonization and persistence have been reported by several investigators [52–55] in various wound types. Since cells of the Dermagraft dermal implant are derived from foreskin fibroblasts obtained from routine circumcisions, the male-specific SRY Y-chromosome marker [56,57] was utilized to investigate the ability for Dermagraft-derived cells to remain in the wound bed of female recipients over several months after implantation. After a single Dermagraft implant in patients with venous ulcers, implant-derived fibroblasts were detected by means of a nested polymerase chain reaction assay [58,59] up to 6 months after implantation, the longest time point that was tested [48]. This fibroblast persistence was also indirectly observed in clinical trials on diabetic foot ulcers (Fig. 104.7). Dermagraft was applied to the diabetic foot ulcer once a week for 8 weeks, and the progress of wound healing was followed to 32 weeks, a full 6 months after the last dose [47,60,61]. Despite the long time between last implant and final follow-up, the healing rate for patients receiving Dermagraft continued to increase over that for the control patients. The ability of these cells to survive and to colonize the wound bed illustrates the lack of tissue rejection as well as the fibroblasts' ability to provide benefits well after initial wound closure has been achieved.

IMMUNOLOGICAL RESPONSE TO TISSUE-ENGINEERED DERMIS

Dermagraft implants have been clinically studied for over 10 years, and like other allogeneic dermal implants, have not demonstrated rejection. The persistence of viable cells

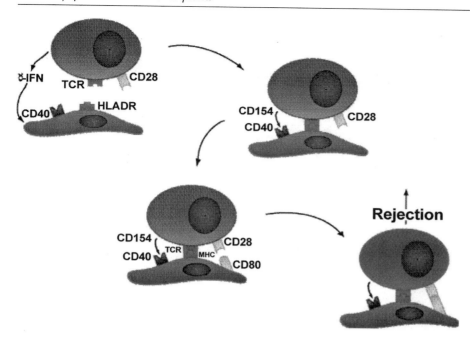

Fig. 104.8. Activation of graft rejection. The process of immunological rejection of an allogeneic engineered tissue exposed to γ-interferon is schematically presented. A γ-interferon challenge simulates the acute rejection pathway via binding of host T-cell receptors with the HLA class II antigens on the donor (fibroblast) cells. The interaction of CD40 on the allogeneic cells with host lymphocytic CD154 induces the presence of CD80 on the allogeneic cells and results in rejection. This γ-interferon challenge is used to test the mechanism by which allogeneic monolayer fibroblasts undergo rejection while allogeneic fibroblasts grown into three-dimensional tissues are not rejected.

from the Dermagraft implant 6 months after application further supports the lack of rejection. Since long-term persistence of cells from any transplant is key, studies were performed [62] to analyze the response of fibroblast cells in monolayer and in various three-dimensional constructs to γ-interferon, a substance known to induce HLA class II and CD40 antigens [7,63–69]. The acute rejection of allograft tissue is initiated by the binding of host T-cell receptors with the HLA class II antigen on the donor cells. The interaction of CD40 on the allogeneic cells with host lymphocytic CD154 induces the presence of CD80 or CD86 (co-stimulatory receptors) on the allogeneic cells and results in rejection (Fig. 104.8). Although fibroblasts normally do not express CD80, CD86, or HLA class II and have low levels of CD40 [64], studies have shown the induction of HLA class II and CD40 in the presence of γ-interferon [63–69]. An inflammatory environment, such as that present during wound healing, would be expected to affect these transplant antigens on the fibroblast surface.

The fibroblast cell source for Dermagraft has been used to demonstrate that both CD40 and HLA-DR are induced on monolayer cultures of neonatal foreskin fibroblasts in the presence of γ-interferon, whereas the same cells, when grown in three-dimensions to form Dermagraft, show little induction of either of these antigens [62]. When cells were removed from the Dermagraft constructs and cultured in monolayer, a response to γ-interferon was elicited. To assess the effect of three-dimensional culture on immunoprotection, fibroblasts were cultured in a bovine collagen gel [70] and exposed to γ-interferon. Fibroblasts in the bovine collagen gel system showed induction of CD40 and HLA-DR in the same manner as monolayer cultures. This data set supports the hypothesis that tissue-engineered implants produced by the seeding of fibroblasts on scaffolds with concomitant natural production of human extracellular matrix may elicit a less pronounced immune response when implanted as an allogeneic tissue than cells alone or implants based on bovine collagen. This lack of responsiveness of Dermagraft to γ-interferon challenge is supported by work on γ-interferon challenge of dermal fibroblasts *in vivo* [71]. In this work, γ-interferon injected intramuscularly induced HLA-DR expression in keratinocytes but did not induce a response in dermal fibroblasts. The lack of immune response to Dermagraft enables the product to persist in the wound site, to improve healing and contribute to long-term repair.

ANGIOGENIC ACTIVITY OF HUMAN-BASED DERMAL IMPLANTS

A variety of *in vitro* assays were utilized to test the correlation of angiogenic factor secretion with the induction of endothelial cell activity as well as new blood vessel formation.

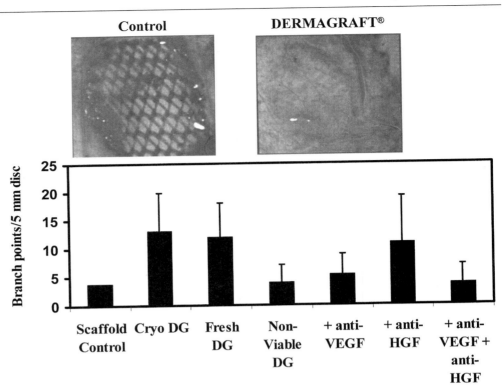

Fig. 104.9. Chick chorioallantoic membrane (CAM) model for angiogenesis. A disk (4–5 mm diameter) of Vicryl scaffold (left) or three-dimensional Dermagraft tissue (right) was placed on the membrane of a fertilized chick egg. After 72 h of incubation at 37°C, there was a significant induction of blood vessel formation in the Dermagraft-treated CAM as opposed to virtually no angiogenesis in the control Vicryl treated CAM. In addition, comparison of Dermagraft (DG) taken fresh from the manufacturing process (prepreservation) to cryopreserved, viable DG to nonviable DG shows that metabolically active fibroblasts are required to secrete growth factors that stimulate angiogenesis (quantitated as branch points per 5-mm disk of tissue). Addition of anti-VEGF nearly completely reverses the angiogenic effect, while anti-HGF results in a slight reduction of angiogenesis. In combination, the anti-VEGF and anti-HGF completely eliminate the angiogenic effect, providing evidence that these two growth factors, acting in concert, are primarily responsible for the angiogenic effect of Dermagraft.

In studies utilizing the chick chorioallantoic membrane model (Fig. 104.9), a statistically significant increase in new blood vessel branch points was seen at 24 h postapplication of fresh manufactured and cryopreserved, viable Dermagraft versus nonviable Dermagraft or scaffold alone [48,51]. The addition of anti-VEGF neutralizing antibody reduced the angiogenic effect by 90%, back to the level of the control, illustrating the relation between VEGF secreted by the fibroblasts and the new blood vessel formation [48,51]. The addition of anti-HGF neutralizing antibody reduced the angiogenesis slightly, but in combination with anti-VEGF completely eliminated the angiogenic effect of Dermagraft. This result corroborates the histological results shown in Fig. 104.9 and reported by others [72] in which observed blood vessel sizes are similar to those induced by addition of VEGF and smaller than those formed by stimulation by bFGF [73].

The Boyden chamber assay was used to further analyze the effect of the factors secreted by the human dermal implant on the migration of endothelial cells. Results showed a dose-dependent stimulation of endothelial cell activity, which was superior to that induced by exogenous VEGF or HGF alone [48], suggesting that a variety of angiogenic factors secreted by Dermagraft act in concert to stimulate endothelial cell activity. Several *ex vivo* models were developed to test the effect of growth factors secreted by Dermagraft on the expansion of wound tissues [74,75]. During an 11-day coculturing period, a significant increase was observed in both tissue expansion over the control (35.9 ± 6.8 vs 5.4 ± 1.5) and endothelial cell motility (62.8 for Dermagraft vs 3.5 ± 1.2) activities, which were correlated with levels of the angiogenic HGF present. Additional assessments of the angiogenesis-promoting characteristics of the growth factors secreted by Dermagraft involved use of a rat aortic ring assay, an endothelial tube formation assay, the Cytodex-2 bead assay, and the colloidal gold phagokinetic motility assay [74,75]. In the rat aortic ring assay, rings were embedded in a collagen matrix. Treatment of the rings with media conditioned by the tissue-engineered dermis resulted in a significant increase in new vessels growing into the collagen gel (5.9 ± 0.60 vessels, $P = 0.002$ vs control). The vascular endothelial tube formation assay showed that capillary tube formation was significantly increased when Dermagraft was added to the assay (relative tube length was 0.215 ± 0.04 mm for the negative control and 0.402 ± 0.055 mm with Dermagraft; $P = 0.012$). The living fibroblasts caused an effect that was also superior to results from the addition of the HGF positive control (0.347 ± 0.073 mm). Results from both bead motility assays showed a significant increase in human vascular endothelial cell motility after treatment with the dermal material. In the Cytodex-2 bead assay, mean absorbance at 340 nm was 0.105 ± 0.002 a.u. for the control versus 0.258 ± 0.013 a.u. for the treatment ($P = 0.005$). Similar results were noted with the colloidal gold assay ($P = 0.005$). These assays demonstrate that the fibroblasts in Dermagraft stimulate angiogenesis and regulate the motility of vascular endothelial cells, activities that are partly attributed to the secretion of HGF and VEGF by the fibroblasts. Based on the success seen with these bioassays, clinical feasibility studies were performed to assess whether Dermagraft would be effective at inducing angiogenesis *in vivo*.

Newton *et al.* [76] performed laser Doppler analysis on seven full-thickness diabetic foot ulcers of patients with peripheral neuropathy and with ulcer duration of 3 months to 12 years. Microvascular blood flow was assessed by means of laser Doppler imaging immediately before and after 2, 5, and 8 weeks of treatment with Dermagraft. An increase in mean blood flow was seen at the base of five out of seven ulcers over the 8-week treatment period ($P < 0.001$). Blood flow is present at the perimeter of the wound before treatment. After 5 weeks of Dermagraft treatment, blood flow begins penetrating the wound bed. Blood flow is supplied to the entire wound area by week 8 of treatment with Dermagraft. Five of the ulcers reached complete healing by 12 weeks, and the two remaining ulcers showed reduction in size by 25–30%. These preliminary results illustrate an increase in perfusion at the base of the diabetic foot ulcer following treatment with the tissue-engineered dermis and provide support for studying the potential to increase blood flow in other wound types.

ACTION OF INFLAMMATORY CYTOKINES EXPRESSED BY DERMAGRAFT

In addition to their role in promoting angiogenesis, inflammatory cytokines are crucial to arresting the inflammatory stage of wound healing and returning the chronic wound to an acute healing pathway. To further evaluate the mechanism of Dermagraft *in vivo*, Harding *et al.* [77] performed biopsies of the venous ulcer wound bed of control patients and patients implanted with Dermagraft. Histological cross-sections of biopsies from both sets of patients are shown in Fig. 104.10. The tissue of both patient groups at trial initiation demonstrated characteristics of a chronic nonhealing wound. A lack of blood vessels is seen, along with a large amount of inflammatory cells. Six weeks after weekly debridement, maintenance of a moist wound environment, and standard off-load-bearing measures, little improvement in wound bed appearance was noted in the control patients (Fig. 104.10D). By contrast, patients who received weekly implants of Dermagraft showed a significant decrease in inflammatory cells, an increase in blood vessels, and a healthy granulation tissue (Figure 104.10B). Both the laser Doppler study and the analysis of biopsy samples illustrate the unique benefits afforded by this tissue-engineered product. The ongoing secretion of a cocktail of growth factors in response to signals from the microenvironment of the wound bed induces angiogenesis, reduces inflammation, and enhances a healthy granulation tissue capable of supporting the migration and growth of the patient's own keratinocytes. Similar results have not been reported for traditional wound dressing treatments or for the topical application of single growth factors.

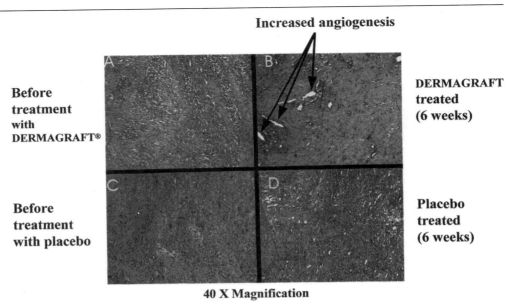

Fig. 104.10. Angiogenic and anti-inflammatory response of Dermagraft in the treatment of venous ulcers. Dermagraft secretes a variety of angiogenic and inflammatory cytokines important to wound healing. Shown are histological cross sections of biopsy samples from venous ulcer patients treated with Dermagraft. At the initiation of treatment, blood vessels are lacking and a large amount of inflammatory cells is present (A) and (C). Patients who received weekly implants of Dermagraft showed a significant decrease in inflammatory cells and an increase in blood vessels accompanied by the formation of a healthy granulation tissue (B). The control patient wound status (D) remained relatively unchanged. See also color insert. (By permission of Dr. Julian Adams, Smith & Nephew, Group Research Center, York, England.)

CRYOPRESERVATION OF AN ENGINEERED HUMAN DERMAL TISSUE

Cryopreserved human-based dermal implants have been studied in a wide variety of wounds. Initial clinical studies with Dermagraft for the treatment of chronic diabetic foot ulcers demonstrated the safety and effectiveness of the implant in wound healing [47,78] and correlated the extent of healing observed with the metabolic activity of the implant [15]. Dermagraft that possessed a low metabolic activity, as determined by the dimethylthiazol-diphenyltetrazolium bromide (MTT) reductase assay, after cryopreservation and thawing, showed low effectiveness in healing ulcers. By contrast, product in which approximately 60% of the fibroblast remained viable and metabolically active successfully accelerated wound closure. These results corroborated early *in vitro* and preclinical studies [24] of rate and normalcy of migration of keratinocytes from split-thickness meshed autografts. Concurrent experiments have helped to explain the need for a specific requirement for a defined percentage of viable cells in each implant. As mentioned earlier, metabolically active cells are necessary to colonize the wound bed, respond in an appropriate way to the changing wound bed environment, secrete angiogenic, anti-inflammatory, and other growth factors, and provide an optimal substrate for reepithelialization. Mansbridge *et al.* [50] reported on the specific relation between the metabolic activity of the cryopreserved dermal implant and its ability to supply key factors necessary for the healing of diabetic foot ulcers.

These various studies have led to further understanding of the mechanism by which cryopreserved tissue-engineered implants function and to the selection of key product specifications. Manufacturing and cryopreservation parameters have been adjusted to ensure the production and maintenance of these specifications through shipping and long-term storage at −70°C. While fibroblast cells in suspension are readily cryopreserved with nearly complete maintenance of cell viability post-thaw, fibroblasts frozen in monolayer or three-dimensional culture have heightened susceptibility to preservation damage, notably higher intracellular ice formation, likely due to changes in biophysical properties of the cells with surface attachment [79–86]. Cryopreservation of Dermagraft imparts damage only to the fibroblasts; matrix protein composition and structure are unaffected [5]. Retention of the majority of the cell viability in Dermagraft is achieved through alteration of several steps of the cryopreservation process, for suspensions of fibroblast cells including cryoprotectant addition and removal and freeze and thaw rates.

The studies on the mechanism of action of this cryopreserved dermal tissue have also resulted in research into the behavior of these implants in various wounds. Early clinical studies with metabolically active dermal implants have shown the product to be efficacious in the treatment of venous ulcers [77] and epidermolysis bullosa [87], and in generating healthy granulation tissues in wounds resulting from the removal of oral and facial tumors [88]. A clinical pilot study of Dermagraft as a substitute for autogenous grafts in the treatment of periodontal disease is providing early encouraging results [89] and raises the possibility of use of a cryopreserved tissue-engineered fibroblast-based product in nonskin wounds.

EFFECT OF CRYOPRESERVATION ON GROWTH FACTOR UP-REGULATION

Dermagraft is cryopreserved to afford off-the-shelf availability, reasonable manufacturing cost, and delivery to the patient of a rigorously tested, uniform, and reproducible tissue. The advantages are obvious, yet little has been done to correlate the effect of various preservation methods on subsequent graft take and survival. Research on cryopreserved Dermagraft indicates that preservation may also improve clinical efficacy.

Recently Liu *et al.* [90] studied the effects of cryopreservation on the stress response of fibroblasts in suspension as well as fibroblasts in living tissue-engineered implants. The results show that the fibroblast cells in both culture systems respond to cryopreservation in a manner similar to a reaction to elevated temperature or nutrient starvation in that the cells express heat shock proteins (hsp). Cryopreservation induced a more intense increase in both stress protein and mitogen activated protein kinase (MAP kinase) responses in the three-dimensional tissue-engineered material than in fibroblasts in suspension cultures. It has been hypothesized that the induction of a cellular stress response may enhance the ability of the fibroblasts to survive in the hostile environment of the wound bed [91]. For example, Liu *et al.* [90] showed that the stress proteins hsp 90, hsp 47, and GRP 78 are up-regulated in three-dimensional as opposed to monolayer culture and that all play a role in collagen, growth factor, and protein secretion, possibly assisting in the regeneration of cryopreservation-induced component denaturation in the tissue implant post-thaw. The induction of stress proteins has been demonstrated to enhance the survival of skin flaps and organs after transplantation [92,93], and preclinical studies have demonstrated that stress preconditioning confers a resistance to endotoxin-induced systemic shock [94] that has analogies in skin ulcers. In addition, diabetic wound healing has been shown to be enhanced by the induction of stress proteins [95]. Hence, post-thaw cells appear to focus on functions associated with wound repair.

Immediately upon thawing of Dermagraft, protein synthesis is inhibited by approximately 75% [50]. Total RNA, a measure of the amount of cell function committed to protein synthesis, decreases approximately 50% 24 h post-thaw, while the production of protein is decreased to a greater extent. This may be due to the differential recovery of various mRNAs. Cryopreservation of the tissue implants was shown to induce a number of growth factors, at the mRNA level, which are key to wound healing [90]. Determination of the cellular mRNA concentrations (TaqMan^{TM} procedure, ABI PRISM^{TM} 7700 Sequence Detector, Foster City, CA) showed that cryopreservation induced mRNA for KGF, PDGF A chain, and VEGF 5- to 20-fold at 48 h after thawing. At 5 days post-thaw, TGF-β was increased by 2.5-fold. Although it is not clear whether the mRNA expression is translated to protein secretion, the up-regulation of mRNA expression of these proteins may enable the cells to continue to secrete these proteins after cryopreservation, when protein synthesis as a whole is greatly reduced. The increase in the up-regulation of these growth

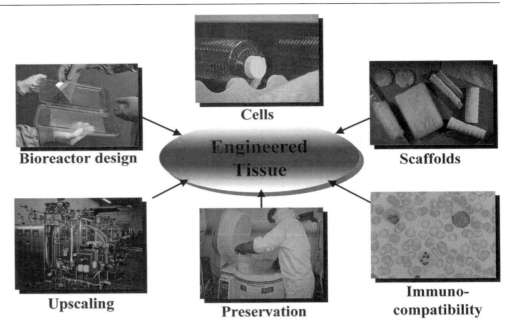

Fig. 104.11. Critical issues in tissue engineering. The successful large-scale production of a tissue-engineered implant that demonstrates clinical efficacy requires mastery of each of the fundamental requirements of engineered tissues. Included are the ability to procure and expand cells, to design biocompatible, bioreactive scaffolds to understand and solve immunological challenges, and to possess expertise in bioreactor design, and large-scale manufacturing and preservation methods to enable cost-effective, worldwide distribution.

factors may be key to the successful take of the cryopreserved tissue-engineered graft as well as to its efficacy postimplantation.

SUMMARY

Work on the development of human fibroblast-based tissue-engineered constructs has helped to establish critical issues that must be addressed in the creation of new engineered tissue and organs for use in repair and transplantation (Fig. 104.11). Cells must be derived from a source that will allow at least six passages prior to seeding onto a scaffold to enable cost-effective manufacture of an allogeneic tissue. Reproducible sources of tissues are also key. To date, neonatal foreskin has provided starting skin cells for dermal and full-thickness products in development and on the market. This starting material reduces variability by being of the same sex, age, and anatomical location. The neonatal fibroblasts have excellent proliferative potential, with one foreskin being capable of producing fibroblasts for at least 250,000 ft^2 of final tissue-engineered product. Scaffolds utilized are generally made of degradable suture material. The need to increase elasticity, porosity, and durability of future grafts may warrant new polymer/scaffold development. Individual closed-bioreactor systems that mimic physiological development conditions are a key component in functional engineering of tissue in which closed, large-scale manufacturing systems offer benefits that include more automated robotics, use of in-process metrics to assess tissue formation, and maintenance of sterility. An added benefit has been the ability to manufacture large numbers of products that are aseptic and highly reproducible from lot to lot. Cryopreservation of dermal product with a multiple-month shelf life is key to off-the-shelf availability, inventory control, and cost-effectiveness. Research has shown that added benefits of preservation include an up-regulation of specific growth factors and stress proteins that may provide enhanced function and longer construct survival post implantation. Natural, all-human constructs also have key nonimmunological characteristics. In addition, fibroblasts grown into three-dimensional engineered dermal tissues have never demonstrated rejection *in vivo*. As studies continue to elucidate the practicality as well as functional characteristics of human-based tissue-engineered equivalents, we will be able to update the design and to manufacture more sophisticated human-based products for tissue repair and transplantation.

REFERENCES

1. Arnst, C., and Carey, J. (1998). Biotech bodies. *Business Week*, **July 27**, 56–63.
2. Mooney, D. J., and Mikos, A. G. (1999). Growing new organs. *Sci. Am.* **280**(4), 60–65.

3. Naughton, G. K. (1999). Skin: The first tissue-engineered products, The Advanced Tissue Sciences Story. *Sci. Am.* **280**(4), 84–85.

4. Parenteau, N. (1999). Skin: The first tissue-engineered products, The Organogenesis Story. *Sci. Am.* **280**(4), 83–84.

5. Applegate, D. R., Liu, K., and Mansbridge, J. (1999). Practical considerations for large-scale cryopreservation of a tissue engineered human dermal replacement. *Adv. Heat Mass Transfer Biotechnol.* HTD-Vol. 363/BED-Vol. 44, 77–91.

6. Applegate, D. R. (1999). A tissue engineered, human dermal replacement for the treatment of diabetic foot ulcers. *Diabetes Today* **2**(2), 59–65.

7. Hansbrough, J. F. (1990). Current status of skin replacements for coverage of extensive burn wounds. *J. Trauma* **30**, S155–S162.

8. Mansbridge, J. (1998). Skin substitutes to enhance wound healing. *Exp. Opin. Invest. Drugs* **7**(5), 803–809.

9. Naughton, G. K., and Tolbert W. R. (1996–1997). Tissue engineering: Skin. *In* "Yearbook of Cell and Tissue Transplantation" (R. P. Lanza and W. L. Chick, eds.), pp. 265–274. Kluwer Academic Publishers, Norwell, MA.

10. Otto, W. R., Nanchahal, J., Lu, Q., Boddy, L., and Dover, N. (1995). Survival of allogeneic cells in cultured organotypic skin grafts. *Plast. Reconstr. Surg.* **96**, 166–176.

11. Cuono, C. B., Langdon, R., and Birchall, N, (1987). Composite autologous–allogeneic skin replacement: Development and clinical application. *Plast. Reconstr. Surg.* **80**, 626.

12. Cuono, C., Langdon, R., and McGuire, J. (1986). Use of cultured autografts and dermal allografts as skin replacement after burn injury. *Lancet* **1**, 1123–1124.

13. Applegate, D. R. (1999). A tissue engineered, human dermal replacement for the treatment of diabetic foot ulcers. Part 2. *Diabetes Today* **2**(3), 93–98.

14. Naughton, G. K. (1997). Skin and epithelia. *In* "Principles of Tissue Engineering" (R. P. Lanza, R. S. Langer, and W. L. Chick, eds.), pp. 769–782. Academic Press, San Diego, CA.

15. Naughton, G., Mansbridge, J., and Gentzkow, G. (1997). A metabolically active human dermal replacement for the treatment of diabetic foot ulcers. *Artif. Organs* **21**(11), 1203–1210.

16. Petricciani, J. C., and Hennessen, W. (1987). Cells, products, safety: Background papers for WHO study group on biologicals. *In* "Developments in Biological Standardization" (S. Karger, ed.), Vol. 68, pp. 19–25. Basel, Switzerland.

17. Petricciani, J. C., Hopps, H. E., and Chapple, P. J., eds. (1979). "Cell Substrates." Plenum, New York.

18. Kruse, P. F., Jr., and Patterson, M. K., Jr., eds. (1973). "Tissue Culture Methods and Applications." Academic Press, New York.

19. Naughton, G. K., Bartel, R., and Mansbridge, J. (1997). Synthetic biodegradable polymer scaffolds. *In* "Synthetic Biodegradable Polymer Scaffolds" (A. Atala and D. J. Mooney, eds.), pp. 121–147. Birkhaeuser, Boston.

20. U.S. Food and Drug Administration (1993). "Points to Consider in the Characterization of Cell Line Used to Produce Biologicals." U.S. Department of Health and Human Services, Bethesda, MD.

21. Committee for Proprietary Medicinal Products: Ad Hoc Working Party on Biotechnology/Pharmacy (1989). Notes to applicants for marketing authorizations on the production and quality control of monoclonal antibodies of murine origin intended for use in man. *J. Biol. Stand.* **17**, 213.

22. Jakoby, W. B., and Pastan, I. B., eds. (1979). Cell culture. "Methods in Enzymology," Vol. 58. Academic Press, New York.

23. Halberstdt, C. R., Hardin, R., Bezverkov, K., Snyder, D., Allen, L., and Landeen, L. (1994). The in vitro growth of a three-dimensional human dermal replacement using a single-pass perfusion system. *Biotechnol. Bioeng.* **43**, 740–746.

24. Landeen, L. K., Ziegler, F. C., Halberstadt, C., Cohen, R., and Slivka, S. R. (1992). Characterization of a human dermal replacement. *Wounds* **4**(5), 167–175.

25. Brockbank, K. G. M., Carpenter, J. F., and Dawson, P. E. (1992). Effects of storage temperature on viable bioprosthetic heart valves. *Cryobiology* **29**, 537–542.

26. Chang, P., Rosenquist, M. S., Lewis, R. W., II, and Kealey, G. P. (1998). A study of functional viability and metabolic degeneration of human skin stored at 4°C. *J. Burn Care Rehabil.* **19**(1), 25–28.

27. Cohen, I., Yemini, M., Wexler, M., and Brautbar, H. (1980). Prolonged survival of allogeneic mouse skin grafts following preservation, *Isr. Med. Sci.* **16**(9-10), 628–630.

28. May, S. R., and Wainwright, J. F. (1985). Integrated study of the structural and metabolic degeneration of skin during 4 degrees C storage in nutrient medium. *Cryobiology* **22**(1), 18–34.

29. U.S. Pharmacopeia (1995). "Sterility Testing," Vol. 23(71), 8th Suppl., United States Pharmacopeial Convention, Inc., Rockville, MD.

30. Dagalakis, N., Flink, J., Stasikelis, P., Burke, J. F., and Yannis, I. V. (1980). Design of an artificial skin. Part III. Control of pore structure. *J. Biomed. Mater. Res.* **14**(4), 511–528.

31. Roberts, M. (1976). The role of the skin bank. *Annu. Rev. Coll. Surg. Engl.* **58**(1), 70–74.

32. Sorensen, B., and Jemec, B. (1973), Freeze drying of skin. In vitro studies of vitality. *Scand. J. Plast. Reconstr. Surg.* **7**(1), 35–38.

33. Basil, A. R. (1982). A comparative study of glycerinized and lyophilized porcine skin in dressings for third-degree burns. *Plast. Reconstr. Surg.* **69**(6), 969–974.

34. Fahy, G. M., MacFarlane, D. R., Angell, C. A., and Meryman, H. T. (1984). Vitrification as an approach to cryopreservation. *Cryobiology* **21**, 407–426.

35. Fahy, G. M., Saur, J., and Williams, R. J. (1990). Physical problems with the vitrification of large biological systems. *Cryobiology* **27**, 492–510.

36. Kreis, R. W., Vloemans, A. F., Hoekstra, M. J., Mackie, D. P., and Hermans, R. P. (1989). The use of non-viable glycerol-preserved cadaver skin combined with widely expanded autografts in the treatment of extensive third-degree burns. *J. Trauma* **29**(1), 51–54.

37. Carbognani, P., Rusca, M., Spaggiari, L., Solli, P., Alfieri, R., Cattelani, L., Bobbio, A., and Bobbio, P. (1997). The effect of trehalose on human lung fibroblasts stored in Euro-Collins and low potassium dextran solutions. *J. Cardiovasc. Surg.* **38**(6), 669–671.

38. Kitahara, A. K, Suzuki, Y., Zhan, C. W., Wada, H., and Nishimura, Y. (1998). Evaluation of new improved solution containing trehalose in free skin flap storage. *Brit. J. Plast. Surg.* **51**(2), 118–121.

39. Arnaud, F. G., and Pegg, D. E. (1990). Permeation of glycerol and propane-1,2-diol in human platelets. *Cryobiology* **27**(2), 107–118.

40. McGann, L. E., and Farrant, J. (1976). Survival of tissue culture cells frozen by a two-step procedure to −196°C. I. Holding temperature and time. *Cryobiology* **13**, 261–268.

41. Pitt, R. E. (1992). Thermodynamics and intracellular ice formation. *In* "Advances in Low-Temperature Biology" (P. Steponkus, ed.), Vol. 1, pp. 63–99. JAI Press, London.

42. Steed, D. L., and the Diabetic Ulcer Study Group (1995). Clinical evaluation of recombinant human platelet derived growth factor for the treatment of lower extremity diabetic ulcers. *J. Vasc. Surg.* **21**, 71–81.

43. Brown, R. L., Breeden, M. P., and Greenhalgh, D. G. (1994). PDGF and TGF-α act synergistically to improve wound healing in the genetically diabetic mouse. *J. Surg. Res.* **56**, 562–570.

44. Albertson, S., Hummel, R. P., III, Breden, M. P., and Greenhalgh, D. G. (1993). PDGF and FGF reverse the healing impairment in protein-malnourished diabetic mice. *Surgery* **114**(2), 368–372.

45. Greenhalgh, D. G., Sprugel, K. H., Murray, M. J., and Ross R. (1990). PDGF and FGF stimulate healing in the genetically diabetic mouse, *Amer. J. Pathol.* **136**, 1235–1245.

46. Knighton, D. R., Ciresi, K., Fiegel, V. D., Schumerth, S., Butler, E., and Cerra, F. (1990). Stimulation of repair in chronic, nonhealing, cutaneous ulcers using platelet-derived wound healing formula. *Surg. Gynecol. Obstet.* **170**, 56–60.

47. Pollak, R. A., Edington, H., Jensen, J. L., Kroeker, R. O., and Gentzkow, G. D. (1997). A human dermal replacement for the treatment of diabetic foot ulcers. *Wounds* **9**(6), 175–183.

48. Mansbridge, J. N., Liu, K., Pinney, E,, Patch, R., Ratcliffe, A., and Naughton, G. K. (1999). Growth factors secreted by fibroblasts: Role in healing diabetic foot ulcers. *Diabetes, Obes. Metab.* **1**, 265–279.

49. Krejci-Papa, N. C., Hoang, A., and Hansbrough, J. F. (1999). Fibroblast sheets enable epithelialization of wounds that do not support keratinocyte migration. *Tissue Eng.* **5**(6), 555–561.

50. Mansbridge, J., Liu, K., Patch, R., Symons, K., and Pinney, E. (1998), Three-dimensional fibroblast culture implant for the treatment of diabetic foot ulcers: Metabolic activity and therapeutic range. *Tissue Eng.* **4**(4), 403–414.

51. Pinney, E., Liu, K., Sheeman, B., and Mansbridge, J. (2000). Human three-dimensional fibroblast cultures express angiogenic activity. *J. Cell. Physiol.* **183**, 74–82.

52. Hull, B., Sher, S., Friedman, L., Church, D., and Bell, E. (1981). Fibroblasts in a skin equivalent constructed in vitro persist after grafting. *J. Cell Biol.* **91**, 51a.

53. Sher, A. E., Hull, B. E., Rosen, S., Church, D., Friedman, L., and Bell, E. (1983). Acceptance of allogeneic fibroblasts in skin equivalent transplants. *Transplantation* **36**, 552–557.

54. Palmer, T. D., Rosman, G. J., Osborne, W. R. A., and Miller, A. D. (1991). Genetically modified skin fibroblasts persist long after transplantation but gradually inactivate introduced genes. *Proc. National Acad. Sci. U.S.A.* **88**, 1330–1334.

55. Lamme, E. N., van Leeuwen, R. T. J., Jonker, A., van Marle, J., and Middelkoop, E. (1998). Living skin substitutes: Survival and function of fibroblasts seeded in a dermal substitute in experimental wounds. *J. Invest. Dermatol.* **111**, 989–995.

56. Palmer, M. S., Berta, P., Sinclair, A. H., Pym, B., and Goodfellow, P. N. (1990). Comparison of human ZFY and ZFX transcripts. *Proc. National Acad. Sci. U.S.A.* **87**, 1681–1685.

57. Sasi, R., Fan, Y.-S., and Lin, C. C. (1991). Prenatal sexing and detection of ZFY gene sequences in sex chromosome disorders by polymerase chain reaction. *J. Clin. Lab. Anal.* **5**, 193–196.

58. Calzolari, E., Patracchini, P., Palazzi, P., *et al.* (1993). Characterization of a deleted Y-chromosome in a male with Turner stigmata. *Clin. Genet.* **43**, 16–22.

59. Petrovic, V., Nasioulas, S., Chow, C. W., Voullaire, L., Schmidt, M., and Dahl, H. (1992). Minute Y chromosome derived marker in a chiold with gonadoblastoma: Cytogenetic and DNA studies. *J. Med. Genet.* **29**, 542–546.

60. Gentzkow, G. D., Jensen, J. L., Pollak, R. A., Kroeker, R. O., Lerner, J. M., Lerner, M., Iwasaki, S. D., and the Dermagraft Diabetic Ulcer Study Group (1999). Improved healing of diabetic foot ulcers after grafting with a living human dermal replacement. *Wounds* **11**(3), 77–84.

61. Pham, H. T., Rosenblum, B. I., Lyons, T. E., Giurini, J. M., Chrzan, J. S., Habershaw, G. M., and Veves, A. (1999). Evaluation of a human skin equivalent for the treatment of diabetic foot ulcers in a prospective, randomized, clinical trial. *Wounds* **11**(4), 79–86.

62. Kern, A., Liu, K., and Mansbridge, J. (2001). Modification of fibroblast γ-interferon responses by extracellular matrix. *J. Invest. Dermatol.* **117**(1), 112–118.

63. Armendariz-Borunda, J., Endres, R. O., Ballou, L. R., and Postlethwaite, A. E. (1996). Transforming growth factor-beta inhibits interferon-gamma-induced HLA-DR expression by cultured human fibroblasts. *Int. J. Biochem. Cell Biol.* **28**, 1107–1116.

64. Fries, K. M., Sempowski, G. D., Gaspari, A. A., Blieden, T., Looney, R. J., and Phipps, R. P. (1995). CD40 expression by human fibroblasts, *Clin. Immunol. Immunopathol.* **77**, 42–51.

65. Gruschwitz, M. S., and Vieth, G. (1997). Up-regulation of class II major histocompatibility complex and intercellular adhesion molecule I expression on scleroderma fibroblasts and endothelial cells by interferon-gamma and tumor necrosis factor alpha in the early disease stage. *Arthritis Rheum.* **40**, 540–550.

66. Klett, Z. G., Elner, S. G., and Ehier, V. M. (1996). Differential expresssion of immunoreactive HLA-DR and ICAM-1 in human cultured orbital fibroblasts and orbital tissue. *Opththalmol. Plast. Reconstr. Surg.* **12**, 153–162.

67. Saunders, N. A., Smith, R. J., and Jetten, A. M. (1994). Differential responsiveness of human bronchial epithelial cells, lung carcinoma cells, and bronchial fibroblasts to interferon-gamma in vitro. *Am. J. Respir. Cell Mol. Biol.* **11**, 147–152.

68. Sempowski, G. D., Chess, P. R., and Phipps, R. P. (1997). CD40 is a functional activation antigen and B7-independent T cell costimulatory molecule on normal human lung fibroblasts. *J. Immunol.* **158**, 4670–4677.

69. Takahashi, K., Takigawa, M., Arai, J., Kurihara, H., and Murayama, Y. (1994). The inhibition of interferon-gamma-induced upregulation of HLA-DR expression on cultured human gingival fibroblasts by interleukin-1 beta or tumor necrosis factor-alpha. *J. Periodontol.* **65**, 336–341.

70. Bell, E., Sher, S., Hull, B. *et al.* (1983). The reconstitution of living skin. *J. Invest. Dermatol.* **81**, 2s–10s.

71. Morhenn, V. B., Pregerson-Rodan, K., Mullen, R. H., Woods, G. S., Nickoloff, B. J., Sherwin, S. A., and Farber, E. M. (1987). Use of recombinant interferon gamma administered intramuscularly for the treatment of psoriasis. *Arch. Dermatol. (Chicago)* **123**, 1633–1637.

72. Friedlander, M., Brooks, P. C., Shaffer, R. W., Kincaid, C. M., Varner, J. A., and Cheresh, D. A. (1995). Definition of two angiogenic pathways by distinct alpha v integrins. *Science* **270**, 1500–1502.

73. Brooks, P. C., Clark, R. A., and Cheresh, D. A. (1994). Requirement of vascular integrin alpha v beta 3 for angiogenesis. *Science* **264**, 569–571.

74. Martin, T. A., Harding, K. G., and Jiang, W. G. (1999). Regulation of angiogenesis and endothelial cell motility by matrix-bound fibroblasts. *Angiogenesis* **3**, 69–76.

75. Jiang, W. G., and Harding, K. G. (1998). Enhancement of wound tissue expansion and angiogenesis by matrix-embedded fibroblast (Dermagraft), a role of hepatocyte growth factor/scatter factor. *Int. J. Mol. Med.* **2**, 203–210.

76. Newton, D. J., Khan, F., Belch, J. J. F., Mitchell, M. R., and Leese, G. P. (2001). Improvement of blood flow in diabetic foot ulcers treated with Dermagraft dermal replacement therapy. *Am. Coll. Foot Ankle Surg. Annu. Sci. Semin.*, p. 37. New Orleans, LA.

77. Harding, K. G., Brassard, A., Dolynchuck, K. *et al.* (2000). Treatment of "hard to heal" venous leg ulcers with a human dermal replacement tissue (Dermagraft)—a clinical laboratory evaluation. *Handb. Abstr., 1st World Wound Heal. Congr.* Melbourne, p. 74.

78. Gentzkow, G. D., Iwasaki, S. D., Hershon, K. S., Mengel, M., Prendergast, J. J., Ricotta, J. J., Steed, D. P., and Lipkin, S. (1996). Use of Dermagraft, a cultured human dermis, to treat diabetic foot ulcers. *Diabetes Care* **19**(4), 350–354.

79. Hetzel, F. W., Kruuv, J., McGann, L. E., and Frey, H. E. (1973). Exposure of mammalian cells to physical damage: Effect of the state of adhesion on colony-forming potential. *Cryobiology* **10**, 206–211.

80. Homung, J., Muller, T., and Furhr, G. (1996). Cryopreservation of anchorage-dependent mammalian cells fixed to structured glass and silicone substrates. *Cryobiology* **33**(2), 260–270.

81. Hubel, A., Toner, M., Cravalho, E. G., Yarmush, M. L., and Tompkins, R. G. (1991). Intracellular ice formation during the freezing of hepatocytes cultured in a double collagen gel. *Biotechnol. Prog.* **7**, 554–559.

82. Porsche, A. M., Korber, C., and Rau, G. (1991). Freeze–thaw behavior of cultured (bovine corneal) endothelial cells: Suspensions vs. monolayer. *Cryobiology* **28**, 545.

83. Larese, A., Yang, H., Petrenko, A., and McGann, L. E. (1992). Intracellular ice formation is affected by cell-to-cell contact. *Cryobiology* **29**, 728.

84. Yarmush, M. L., Toner, M., Dunn, J. C. Y., Rotem, A., Hubel, A., and Tompkins, R. G. (1992). Hepatic tissue engineering. *Ann. N. Y. Acad. Sci.* **665**, 238–252.

85. Berger, W. K., and Uhrik, B. (1996). Freeze-induced shrinkage of individual cells and cell-to-cell propagation of intracellular ice in cell chains from salivary glands. *Experientia* **52**, 843–850.

86. Zieger, M. A., Tredget, E. E., and McGann, L. E. (1996). Mechanisms of cryo injury and cryoprotection in split-thickness skin. *Cryobiology* **33**(3), 376–389.

87. Williamson, D., and Sibbald, R. G. (2001). The role of a dermal skin substitute in the management of non-healing, chronic and unusual wounds. *Abstr., Symp. Adv. Wound Care*, p. C79. Las Vegas, NV.

88. Gath, H. J., Hell, B., Zarrinbal, R. *et al.* (2001). Regeneration of intraoral defects after tumor resection with a bio-engineered human dermal replacement (Dermagraft). *Plast. Reconstr. Surg.* (in press).

89. McGuire, M. (2001). Ongoing clinical studies, as practicing Periodontist of Private Practice and Associate Professor, Department of Periodontics, University of Texas Health Science Center, San Antonio and Houston, Houston.

90. Liu, K., Yang, Y., and Mansbridge, J. (2000). Comparison of the stress response to cryopreservation in monolayer and three-dimensional human fibroblast cultures: Stress proteins, MAP kinases, and growth factor gene expression. *Tissue Eng.* **6**(5), 539–554.

91. Morris, J. A., Dorner, A. J., Edwards, C. A., Hendershot, L. M., and Kaufman, R. J. (1997). Immunoglobulin binding protein (BiP) function is required to protect cells from endoplasmic reticulum stress but is not required for the secretion of selective proteins. *J. Biol. Chem.* **272**, 4327.

92. Perdrizet, G. A., Heffron, T. G., Buckingham, F. C., Salciunas, P. J., Gaber, A. O., Stuart, F. P., and Thistlethwaite, J. R. (1989). Stress conditioning: A novel approach to organ preservation. *Curr. Surg.* **46**, 23.

93. Koenig, W. J., Lohner, R. A., Perdrizet, G. A., Lohner, M. E., Schweitzer, R. T., and Lewis, V. L., Jr. (1992). Improving acute skin-flap survival through stress conditioning using heat shock and recover. *Plast. Reconstr. Surg.* **90**, 659.

94. Hotchkiss, R., Nunnally, I., Lindquist, S., Taulien, J., Perdrizet, G., and Karl, I. (1993). Hyperthermia protects mice against the lethal effects of endotoxin. *Am. J. Physiol.* **265**, R1447.

95. Farkas, B., Balogh, G., Ková, E., and Vigh, L. (1997). Roles of stress proteins in diabetic wound healing. *Australas. J. Dermatol.* **38**(Suppl. 2), 57.

CHAPTER 105

BILAYERED SKIN CONSTRUCTS

Janet Hardin-Young and Nancy L. Parenteau

A bioengineered skin replacement should ideally restore the functional properties of both dermis and epidermis. In 1981 Bell *et al.* [1] introduced the concept of a bilayered skin equivalent. Subsequent research, technology advances, and controlled clinical trials have now made the skin equivalent a clinical reality. The bilayered skin equivalent, Apligraf* (Graftskin, human skin construct) consists of a collagen lattice of bovine type I collagen containing human dermal fibroblasts and matrix proteins. The epidermis is stratified and differentiated, with a stratum corneum. The differentiated epidermis offers functional advantages. The construct can be meshed for clinical applications, is more resistant to topical agents, and is better able to persist as a graft in animals. If insury occurs, the construct is able to heal itself *in vitro*, exhibiting the cascade of growth factors found during normal wound healing. Wound healing is a dynamic process in which cells, factors, and extracellular matrix participate. The living construct is both active and interactive. This is particularly important in wounds that have lost their ability to heal. Apligraf has demonstrated safety and efficacy in the treatment of two types of chronic wound: venous leg ulcers and diabetic foot ulcers. It is approved by the U.S. Food and Drug Administration for these indications. Significant results in the difficult-to-treat wound foreshadow the impact tissue engineering will have in changing medical paradigms. In addition, the construct, made from allogeneic keratinocytes and fibroblasts, elicits no immune response. This is due to the lack of professional antigen-presenting cells, the presence of a functional co-stimulatory pathway in both the keratinocyte and fibroblast, and negative feedback by these cells.

INTRODUCTION

Advancements in cell culture technology and bioengineered materials have enabled researchers and biotechnology companies to develop new products to augment or replace traditional approaches of wound healing. There have been a variety of approaches involving the use of growth factors, matrix materials, dermal substitutes, and epidermal sheets [2], some of which are covered elsewhere in this book (Chapter 63). The goal in the development of a bilayered skin graft has been to re-create the beneficial characteristics of natural skin grafts while eliminating or minimizing their limitations. Ideally, a bioengineered skin replacement should restore the functional properties of the dermis and the epidermis. This chapter discusses current preclinical and clinical work in bilayered skin equivalents as it relates to the treatment of acute and chronic wounds and possibilities for the further development of skin substitutes.

One of the first attempts to replicate a full-thickness skin graft was by Bell *et al.* [1], who described a bilayered skin equivalent. Bell and colleagues recognized the potential benefit of providing both epidermis and dermis. The dermal component consisted of a lattice of type I collagen contracted by tractional forces of rat dermal fibroblasts trapped within the gelled collagen. The investigators demonstrated the ability of these early skin equivalents to take as a skin graft in rats. We have since developed methods of organotypic culture to provide a three-dimensional culture environment that controls collagen lattice contraction

---

*Apligraf is a registered trademark of Novartis.

and is permissive for proper epithelial differentiation. The resulting human skin equivalent (HSE) develops many of the structural, biochemical, and functional properties of human skin [3–6]. This construct is known commercially and clinically as Apligraf or Graftskin, human skin construct.

Boyce and Hansbrough [3], and separately, Eisenberg and Llewlyn [7], have modified the collagen–glycosaminoglycan sponge approach first proposed by Yannas *et al.* [8] as a dermal template to form bilayered composites seeded with fibroblasts and overlaid with epidermal keratinocytes; Boyce *et al.* used autologous cells harvested from the patient, and Eisenberg *et al.* used allogeneic cells from neonatal foreskin.

SCIENTIFIC RATIONALE

Autologous human skin is the gold standard for treatment of skin wounds. However, skin grafts are not always the perfect solution. They are limited with respect to the rigorous conditions needed for graft take, tissue availability, graft rejection, and conformability with the surrounding tissue with respect to thickness, pigmentation, and appendages.

The benefits of living skin tissue go beyond healing by graft take; indeed, small biopsy samples or pinch grafts of skin have been shown to be able to stimulate surrounding tissue to heal [9]. Tissue engineering gives us an opportunity to provide some of the benefits of a skin graft therapy without the clinical restrictions an autologous skin graft would impose. To better understand how this is possible, it is necessary to first review the mechanism of wound healing.

THE MECHANISM OF NORMAL WOUND HEALING

The immediate tissue response to wounding is clot formation. Simultaneously, there is a release of inflammatory cytokines, which regulate blood flow to the area, recruit lymphocytes and macrophages to fight infection, and later, stimulate angiogenesis and extracellular matrix deposition. These latter processes result in the formation of granulation tissue, a provisional matrix that serves as a scaffold into which cells migrate, repopulating the wounded area over time. Keratinocytes are stimulated to migrate over the wound bed, proliferate, and stratify, to restore the epidermal barrier. Fibroblasts are recruited into the provisional matrix through the action of inflammatory cytokines and growth factors such as platelet-derived growth factor (PDGF). They take on a myofibroblast phenotype, rich in smooth muscle actin [10]. The complex array of factors and their effects are summarized in Table 105.1.

Granulation tissue formation is important for normal healing in humans. It fills the wound with tissue that brings the needed inflammatory cells to supply protection against foreign invasion and provides a substrate for rapid reepithelialization, or wound closure. After the wound closes with epithelium, the granulation tissue remodels into scar tissue through the action of the myofibroblasts and factors such as transforming growth factor β (TGF-β, Table 105.1). Humans are unable to regenerate dermis, yet in the fetus, where TGF-β levels [11] and the inflammatory responses are relatively low [12,13] and the extracellular matrix is more permissive [14], healing occurs without scarring [15]. An engineered skin graft has the potential to cause a paradigm shift in how the wound heals by supplying the body with the needed living tissue for closure and repair.

THE CASE OF IMPAIRED WOUND HEALING

Wound healing requirements are different for acute and chronic wounds. In the chronic wound, the healing scenario is played out only with difficulty. There may no longer be a granulation tissue response, and the fibroblasts and keratinocytes at the wound perimeters (the indolent edge) may be senescent [16] or refractory to cytokine stimulation [17]. There are often attempts to debride the wound in an effort to stimulate the granulation tissue response anew. With proper dressings, such measures are sometimes effective in restimulating the healing response. However, for many(e.g., the long-standing venous leg ulcer patient) this is not enough. And for the diabetic ulcer patient, the time needed for healing may threaten the health of the limb.

Wound environments can be quite different. The poor healing response in chronic wounds is attributed to a number of factors, a key one being an imbalance of factors rather

Table 105.1. Factors Produced by Epidermal Keratinocytes and Fibroblasts and Their Autocrine and Paracrine Effects[a,b]

Factor	Origin	Biological effect
Inflammatory mediators		
Interleukin 1 (α, β) (IL-1)	Constitutively expressed by epidermal keratinocytes and dermal fibroblasts. Released upon inflammation or injury.	Primary inflammatory cytokine ↑ PDGF and KGF in fibroblasts ↕ fibroblast growth ↕ extracellular matrix synthesis in fibroblasts ↑ keratinocyte migration
Tumor necrosis factor α	Epidermal keratinocytes	Primary inflammatory cytokine ↑ fibroblast growth ↕ collagen biosynthesis by fibroblasts ↓ keratinocyte growth ▲ secondary cytokine release in both keratinocytes and fibroblasts
Interleukin 6	Epidermal keratinocytes and dermal fibroblasts	↑ tissue inhibitors of metalloproteases in fibroblasts ↑↑ comitogen for keratinocytes
Growth factors		

Epidermal growth factor–related molecules
  Transforming growth factor
  α (TGF-α)
  Amphiregulin
  Heparin binding epidermal growth factor
	Epidermal keratinocytes	All bind to same receptor (Erb-β ligands)
		↑ fibroblast proliferation
		↑ PDGF by fibroblasts
		↑collagen synthesis by fibroblasts
		↑ keratinocyte growth

Platelet-derived growth factor (PDGF)	Dermal fibroblasts	↑ fibroblast growth → fibroblasts ↑ collagenase by fibroblasts
Transforming growth factor β (1,2,3) (TGF-β)	Epidermal keratinocytes and dermal fibroblasts	↕ fibroblast growth ↑ extracellular matrix synthesis by fibroblasts ▲ TGF-β and PDGF in fibroblasts ↓ keratinocyte growth May modulate keratinocyte differentiation
Keratinocyte growth factor (KGF) (fibroblast growth factor 7)	Dermal fibroblasts	↑ keratinocyte growth ▲TGF-α in keratinocytes May modulate keratinocyte differentiation
Neu differentiation factor (NDF) (heregulin)	Epidermal keratinocytes and dermal fibroblasts	↑ keratinocyte growth
Nerve growth factor (NGF)	Epidermal keratinocytes and dermal fibroblasts	↑ keratinocyte growth † apoptosis in keratinocytes
Basic fibroblast growth factor (FGF-2)	Epidermal keratinocytes	↑↑ comitogen for keratinocytes Important in supporting the proliferative response without promoting progress toward differentiation
Insulin-like growth factor 1	Epidermal keratinocytes	↑↑ comitogen for keratinocytes ↑ may play a role in extracellular matrix synthesis by fibroblasts
Interferon α, β	Epidermal keratinocytes	↓ fibroblast growth ↓ collagen synthesis by fibroblasts ↓ keratinocyte growth ↑ MHC class II antigen expression in both cell types ▲ IL-1

[a]Much of what we have ascertained is from *in vitro* experiments, where factors are often studied singly and rarely in complex formulation. *In vivo*, the ratios of these factors and temporal sequence will influence their effect.

[b]*Key*: ↑, stimulates; ↓, inhibits; ↑↑, heavily stimulates; ↕, can stimulate or inhibit; ▲, induces; †, blocks; →, attracts.

Source: Adapted from Parenteau *et al.* [2].

than an insufficiency of any particular soluble factor [17]. A second element the presence in the wound of cells that may no longer be responsive to cytokines [16]. Our hypothesis was that a living HSE should have the potential to normalize this imbalance in the nonhealing wound through interaction and contribution of its own healing response, to effectively stimulate a recalcitrant wound to heal.

WHY BOTH LAYERS?

Epidermis plays a key role in the wound healing scenario and has demonstrated the potential to speed biological wound closure [18–21]. The only way to achieve biological wound closure, the key end point in clinical wound healing studies, is with epidermal coverage. Since the normal scenario does not favor regeneration of normal dermal tissue, even with epithelial coverage, a better outcome might logically be enabled by a living dermal element with fibroblasts of the appropriate phenotype and a noninflammatory, permissive, extracellular matrix. However, we must realize that since the action of the epidermis and dermis are intimately linked, their respective contributions to the restoration of skin tissue may be quite different if both are present. There is evidence [see, e.g., 22,23] that epidermal coverage plays a key role in the regulation of the underlying inflammatory response and the subsequent remodeling event. In animals, a dermal matrix containing fibroblasts has been shown to aid in the maintenance of a grafted epidermal cell population [24]. Therefore, the epidermal and dermal response in wound healing is dynamic and involves the interaction of many tissue factors that include extracellular matrix signals and inflammatory cytokines, as well as autocrine and paracrine factors produced by the dermal fibroblasts and epidermal keratinocytes [25–27]. The multiple factors and their *relationship to one another* serve to regulate growth and differentiation of keratinocytes, pro-inflammatory reactions, angiogenesis, remodeling, and deposition of extracellular matrix (Table 105.1).

Importantly, a living tissue has the potential to provide complex temporal control of factor delivery and effect. It can contribute the needed combination of chemical, structural, and last but not least, normal cellular elements. This dynamic relationship, not only between epidermis and dermis but also between graft and host, suggests that a construct containing both a living epidermis and dermis should behave differently, and arguably more broadly, from any single element alone. The dynamic nature of tissue interaction is what makes a tissue engineering answer to the problem of wound healing so powerful. Indeed, proof of our hypothesis regarding the potential clinical benefits of a differentiated HSE (Apligraf, Graftskin, human skin construct) has been clearly demonstrated in controlled clinical trials in two types of chronic wound [28–30], as well as in clinical studies for the treatment of burns [31], epidermolysis bullosa [32], and acute wounds [33,34]. Some of this work is summarized in this chapter.

DESIGN

To date, tissue engineering approaches have primarily focused on providing or imitating structural and biological characteristics of dermis and epidermis. The key features to be replicated in a bioengineered skin are as follows:

- A dermal or mesenchymal element capable of aiding appropriate dermal repair and epidermal support
- A living epidermis capable of easily achieving biological wound closure
- An epidermis capable of rapid reestablishment of barrier properties
- A permissive milieu for the components of the immune system, nervous system, and vasculature to repopulate
- A tissue capable of achieving normalization of structure and additional function such as reduction of long-term scarring and reestablishment of pigmentation

THE DESIGN AND MANUFACTURE OF THE HUMAN SKIN EQUIVALENT, APLIGRAF (GRAFTSKIN, HUMAN SKIN CONSTRUCT)
Cell Source

Historically, both autologous and allogeneic cells have been used to make skin equivalents [2]. Neonatal foreskin is the most commonly used tissue source for allogeneic keratinocytes and fibroblasts. Skin biopsy specimens from various locations on the body serve as the tissue source for autologous cells.

Isolation and Expansion of Cell Populations

Whether autologous or allogeneic, the expansion of keratinocyte and fibroblasts cell populations is critical. Our goal of commercializing a broadly available HSE is enabled by our use of allogeneic cells. A specific culture method was developed for neonatal foreskin keratinocytes that allows for the expansion of cells while minimizing terminal differentiation [35]. Sufficient quantities of cells allow for the development of frozen cell bank inventories, which are necessary for both commercialization and regulatory compliance. Frozen cell banks permit the necessary cell purity testing and safety testing for adventitious agents [36]. Purity of cell populations is critical. Interlaboratory differences in cell purity may have contributed to the conflicting results found in murine and human studies in the past. Consistent cell sources enable the reproducible production of relatively sophisticated organotypic tissues such as Apligraf (Fig. 105.1).

Immunological Properties of Allogeneic Keratinocytes and Fibroblasts

The use of allogeneic cells is necessary to have an off-the-shelf product. Fortunately, the immunological properties of keratinocytes and fibroblasts allow for the production of allogeneic skin equivalents that will not sensitize the patient. To understand this, some immunology background is necessary.

A hierarchy exists among cell types in regard to antigenicity (i.e., the ability to be recognized as foreign by the patient's immune system). Cells with the greatest potential to elicit an allogeneic response are classified as professional antigen-presenting cells (APC). Examples are dendritic cells, Langerhans cells, B cells, and endothelial cells. Cells classified as nonprofessional APC vary in their ability to induce graft rejection from moderate to minimal. Fibroblasts and keratinocytes are not professional APC and fail to stimulate the proliferation of allogeneic T cells [37–40]. Both these cell populations inherently do not express HLA class II molecules [37,41,42], or co-stimulatory molecules such as B7-1 [43]. The inability of keratinocytes and fibroblasts to induce proliferation of allogeneic T cells is primarily due to the lack of expression of co-stimulatory molecules, even though aberrant antigen processing and invariant chain expression may also contribute [44].

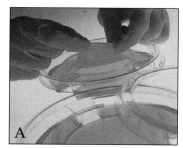 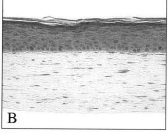

Fig. 105.1. The organotypic culture process results in an integrated tissue, similar in many ways to human skin. The physical integrity of the living construct gives it handling properties similar to those of a split-thickness skin graft. (A) Apligraf being removed from its shipping tray. Apligraf is shipped on nutrient agar at room temperature and has a packaged shelf life of 5 days. (B) Photomicrograph of the construct showing the bilayered structure of the skin equivalent with the in vivo–like organization of the stratified epidermal cells and the cellular dermis composed of collagen and fibroblasts. Hematoxylin and eosin stain; original magnification 200×.

Production of Human Skin Equivalent

The process used in the formation of the HSE has been covered in detail many times [27,45] and is outlined in Fig. 105.2. It involves preparation of cell banks of keratinocytes and fibroblasts from neonatal foreskin, production of a dermal component by combining the fibroblasts with purified type I collagen, and development of a differentiated epidermis through culture at the air–liquid interface. The culture of HSE proceeds best with minimal intervention. Normal cell populations appear to have an intrinsic ability to reexpress their differentiated program *in vitro* to an extent that is now only beginning to be appreciated [46]. A medium that supplies adequate amounts of nutrients, lipid precursors, vitamins, and minerals may be all that is required [45]. Another element is the environmental stimulus provided by culture at the air–liquid interface, which promotes differentiation and formation of the epidermal barrier [6]. The result is an integrated tissue containing two cell types that may be handled as you would a split-thickness skin graft (Fig. 105.1).

IMPORTANT FEATURES OF APLIGRAF

From the preceding discussion on wound healing and skin cell biology, it should be clear that the epidermal and dermal cells of a skin equivalent have the potential to contribute to clinical efficacy through their interaction and production of cytokines. However, there are less obvious features of skin that can also contribute to wound healing.

Barrier Properties

Barrier function is the single most important function of skin. It prevents desiccation and keeps out infection. Formation of the epidermal barrier also influences tissue physiology and turnover through control of water loss and its influence on epidermal physiology. During the airlift phase of organotypic culture *in vitro*, the differentiated skin equivalent forms a well-developed stratum corneum, providing significant barrier function [6]. Once grafted, the HSE achieves similar barrier function as normal human skin *in vivo*. Since barrier function requires a specific lipid composition and structure, these results suggest that lipid metabolism in the grafted epidermis normalizes during the first week after grafting.

Basement Membrane

Basement membrane (BM) is the structure that anchors the epidermis to the dermis. Without a functional basement membrane, the epidermis is fragile and subject to damage through sheer forces exerted by normal contact. Rapid formation of a basement membrane

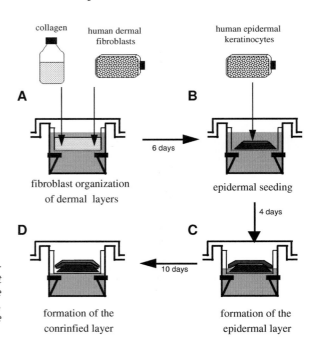

Fig. 105.2. Schematic representation of HSE production. (A) Human dermal fibroblasts are added to bovine type I collagen and allowed to contract over 6 days. (B) Human epidermal keratinocytes are seeded on top of the contracted collagen lattice. (C) An epidermal layer is grown submerged, to allow stratification of epidermal layers. (D) Cultures are raised to the air–liquid interface to allow maturation of the epidermis.

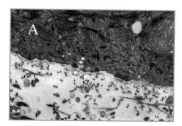
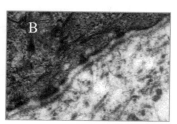

Fig. 105.3. HSE basement membrane formation in vitro and in vivo. Ultrastructural analysis of HSE basement membrane by transmission electron microscopy. (A) Mature HSE in vitro. The in vitro BM comprises a thin, discontinuous lamina densa with numerous gaps. A few hemidesmosomes are formed along the basal surface of keratinocytes, with some anchoring filaments visible between the hemidesmosomes and subjacent lamina densa. (B) HSE 14 days postgrafting onto an athymic mouse: a complete, continuous basement membrane can be visualized.

is therefore desirable for successful grafting, to provide epidermal strength and resistance to sheer. At maturity *in vitro*, HSE develops a significant but incomplete basement membrane structure [47]. The *in vitro* BM comprise a thin, discontinuous lamina densa with numerous gaps. A few hemidesmosomes were forming along the basal surface of keratinocytes, with some anchoring filaments visible between the hemidesmosomes and subjacent lamina densa. Final maturation of the BM occurs rapidly *in vivo*. Within 14 days of grafting, basement membrane structure was formed, and all the expected morphological features were apparent (Fig. 105.3). Thus, like the formation of barrier function, basement membrane begun *in vitro* was rapidly completed *in vitro*.

Effect of Construct Maturation on *In Vivo* Performance
In Vitro Maturation Adds Biological Robustness

Human skin is composed of a well-differentiated epidermis. The organotypic culture process is designed to induce maturation and differentiation of the epidermis *in vitro*. Does epidermal development *in vitro* affect clinical efficacy? In animal studies [48], when HSEs from different stages (4 days and 10 PAL) in the culture process were grafted onto mice, the less mature HSEs failed to engraft while the mature HSEs integrated and persisted on the animals. Analysis of cytokine gene expression, histology, and barrier function between the 4- and 10-day PAL HSE showed similar cytokine gene expression [unpublished data]. The histology differed in that the 4-day PAL HSE had not yet developed a stratum corneum. This lack of stratum corneum correlated with a decrease in barrier function. Preclinical studies of these types demonstrate the importance of *in vitro* development to the eventual clinical performance of the construct. Barrier function also allows the topical application of compounds such as antimicrobial agents without harm to the underlying living tissue.

Maturation Enhances Handling Properties

Another less obvious benefit of the stratum corneum is that it affords the bilayered construct with significant physical strength it would not otherwise have. The mechanical properties of the collagen matrix alone would not preserve tissue integrity upon meshing, for example. This physical property allows the doctor to treat Apligraf just as he would a split-thickness skin graft. Unlike epithelial sheets, which are very fragile and hard for the doctor to manipulate, the Apligraf can be cut, meshed, or perforated prior to placement on the patient. Also Apligraf can be sutured or stapled into place similar to split-thickness skin grafts.

Growth Factors

The correct balance of cytokines and their receptors is important to proper wound healing and homeostasis of skin. One advantage of using HSE to treat skin wounds as opposed to cytokine therapy is that the living cells are able to receive feedback from the wound environment to provide, stimulate, and reset the cytokine milieu. *In vitro*, conditioned culture medium taken from the HSE can stimulate keratinocyte migration and keratinocyte, fibroblast and endothelial cell proliferation [Ronfard, personal communication].

Table 105.2. Cytokine and Growth Factor Profile^a of HSE and HSE Cells

Cytokine	Keratinocytes^b	Fibroblasts^b	HSE	Normal skin
IL-1α	+	−	+	+
IL-6	−	+	+	+
IL-8	−	−	+	+
IL-11	−	+	+	+
IGF-I	−	−	+	+
IGF-II	−	+	+	+
PDGF	+	−	+	+
TGF-α	+	−	+	+
ECGF	−	+	+	+
VEGF	n.t.^{c}	n.t.^{c}	+	+
FGF-1	+	+	+	+
FGF-2	−	+	+	+
FGF-7	+	+	+	+
GM-CSF	n.t.^{c}	n.t.^{c}	+	+
TGF-β_1	−	+	+	+
TGF-β_3	−	+	+	+

^{a}mRNA detected by reverse transcriptase–polymerage chain reaction analysis.
^{b}Cells grown in monolayer cultures.
^{c}Not tested.

HSEs generate most, if not all, of the cytokines produced by fibroblasts and keratinocytes in normal skin. Importantly, both cell types are required to create the complete cytokine profile (Table 105.2).

Biocompatibility and Immunology

The biocompatibility of a skin construct is determined by the patient's imflammatory/remodeling response and the patient's specific immune response to components of the skin construct. Several of the materials that have been used to make the dermal portion of the skin constructs elicited an inflammatory response *in vivo* [2]. In contrast, the HSE integrates rapidly with host tissue *in vivo*, becoming vascularized with minimal inflammation [49].

Inconsistencies in the literature regarding the antigenicity of skin equivalents are due, in part, to the complexity of the biological and immunological factors that determine the ability of a graft to take and to persist over time. Among the properties that determine the antigenicity of an engineered allogeneic graft are purity of components (matrix and cellular) and the antigen-presenting capabilities of graft cells, as discussed earlier. Preclinical and clinical studies have assessed the immune response to intact skin constructs.

In animal studies, severe combined immunodeficient (SCID) mice, because they lack a functioning immune system, can successfully receive by transplantation a functioning human immune system without risk of rejection. SCID mice into which human leukocytes had been transplanted were used as an *in vivo* model to assess the immunogenicity of allogeneic HSEs [49]. In these studies, human skin was rejected, while allogeneic HSE survival was 100%.

The safety and immunological impact of the HSE (Apligraf, Graftskin) has also been assessed in clinical studies. There were no signs of rejection or bovine collagen–specific immune responses or response to alloantigens expressed on keratinocytes or fibroblasts.

CLINICAL PERFORMANCE

The commerically available HSE, Apligraf, has been studied clinically in a number of applications, including venous ulcers [29,50], dermatological excisions [33,34], burns [31], bullous disease [32], and diabetic ulcers [30]. True to our hypothesis, its effectiveness in the treatment of acute and chronic wounds has been attributed to its ability to interact with

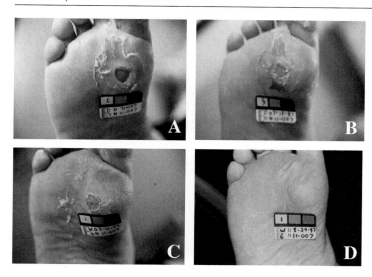

Fig. 105.4. (A) Pretreatment of Apligraf. (B) Apligraf applied. (C) Approximately 4 weeks after application. (D) Approximately 6 weeks after application.

the wound in multiple ways [28]. The integrated tissue architecture of Apligraf provides robustness needed for clinical manipulation.

The initial pivotal study by Falanga *et al.* [29], of Apligraf HSE for the treatment of venous leg ulcers, showed Apligraf to be safe and effective, healing more ulcers in shorter times than proved possible with conventional therapy. Data from this study served as the basis for approval by the U.S. Food and Drug Administration (FDA) for commercial use. Apligraf was later approved for use in the treatment of diabetic foot ulcers.

In the randomized, controlled, multicenter trial, diabetic ulcers were surgically debrided to viable tissue, treated with either Apligraf and saline-moistened dressings or saline-moistened dressings alone and were total-weight-off-loaded for the first 6 weeks of the study. Published results from a single center showed that Apligraf (n = 16) was significantly more effective than the control (n = 17) for frequency of closure (75% vs 41%, p = <0.05), and Apligraf significantly reduced the time to complete wound closure (median time, 38.5 days vs 91 days, p = <0.01) [30] (Fig. 105.4).

In addition, Apligraf has shown promise for the treatment of patients with the chronic hereditary blistering disease epidermolysis bullosa (EB) [32]. In an initial clinical study, 19 patients were treated with Apligraf. Results demonstrate that in comparison to routine dressing, Apligraf significantly improves time to healing and retards rate of reblistering.

In other laboratories, Hull *et al.* reported good results treating full-thickness burnwonnds with a bilayered construct consisting of a contracted collagen lattice and autologous instead of allogeneic keratinocytes [51]. In this study, the bilayered construct was grafted directly on wounds that had been excised and allowed to granulate; equivalent wounds treated with split-thickness autografts were evaluated for comparison. Using various modifications of the bilayered construct originally developed by Bell and colleagues [1], several groups have also achieved good clinical outcomes for the treatment of burn wounds [52,53]. Wassermann *et al.* [54], achieved disappointing results when they used a bilayered construct produced by a modification of Bell's method to treat eight patients with full-thickness burns. There was extensive lysis of the bilayered construct at 48 h postgraft, with no clinical benefit from treatment with the bioengineered skin substitute. The authors attributed the results to the presence of collagenases or proteases produced on the acute burn wound bed, which may have degraded the collagen matrix before take of the graft could be established.

In another study, a bilayered skin construct was used on three subjects with excised tattoos [55]. The construct consisted of allogeneic cells and a collagen lattice derived from an acid extract of rat tail collagen as the dermal matrix. According to the report, all three allografts were successful at the evaluation point of one month, with good cosmetic results and little contraction. Histological evaluation showed that the graft area had a fully differentiated and mature dermis without adnexae.

Another bilayered skin construct, Composite Cultured Skin (CCS), is composed of a porous bovine collagen sponge seeded with fibroblasts and a confluent layer of epidermal keratinocytes. The keratinocytes of the CCS are not exposed to air during culture and therefore do not develop a fully differentiated epidermis and stratum corneum comparable to human skin. CCS is being tested in the clinic for use in burns, chronic wounds, and skin blistering diseases. Eisenberg and Llewelyn have reported some success in the treatment of epidermolysis bullosa using this construct [7], despite the preclinical HSE studie described earlier, which showed that lack of barrier function can negatively influence the ability of tissues to survive *in vivo*.

Boyce and colleagues have developed a similar but more differentiated composite culture using a variation of the collagen–glycosaminoglycan matrix. A preliminary study on four patients with skin wounds of varied etiology reported stimulation of healing and graft persistence [56]. Composite cultures using autologous cells have also shown promise in the treatment of burns [57].

CONCLUSION

Bioengineered skin is now a reality and is becoming a routine therapy for the treatment of skin wounds. As one of the first living-tissue-engineered products to reach commercialization, its development has set the path for other cellular products, particularly those that will require regulation by the FDA to reach the marketplace. While the mechanism of action of bioengineered skin products has not been fully defined, research has demonstrated the importance of maintaining the synergy between cell populations found in the native tissue. Perhaps more importantly, the availability of a commercial bioengineered skin (Apligraf) demonstrates that a broadly available, effective medical device can be made through tissue engineering. It also illustrates the important immune tolerance of certain allogeneic cells, making the use of autologous cells unnecessary and for this application, undesirable, if one must utilize relatively sophisticated culture technology and deliver consistency, reasonable cost, ready availability, and reliable response.

THE FUTURE OF BIOENGINEERED SKIN GRAFTS

Tissue engineering has the potential to cause a paradigm shift in how we think of cutaneous wound healing. Our experience with the HSE also demonstrates that the response and interaction with the patient will vary with the human condition. While the construct takes readily as a skin graft in animal studies, it appears to respond in multiple and varying ways to a variety of human skin wounds, taking as a graft under noninflammatory conditions, acting to stimulate healing in others that do not favor graft take. Our experience also illustrates that important elements that have potential impact on the robustness of its clinical response and, therefore, its ultimate success, are not always obvious at first glance. By striving to recapitulate as many features of real skin as possible, we have been able to provide an effective therapy for a variety of human wound conditions. Our challenge will be to modify the skin equivalent to further enhance true graft take in a broader variety of human wound conditions to achieve even faster healing and repair. Our goal will also be to provide, as part of the skin equivalent, a dermal rudiment or precursor that is capable of directing true dermal regeneration. Our current research focuses on the control of fibroblast phenotype and matrix synthesis *in vitro*.

With future progress in stem cell research, it might also one day be possible to form bilayered skin constructs that naturally recapitulate appendageal structures under permissive organotypic culture conditions.

ACKNOWLEDGMENTS

The authors gratefully acknowledge the many scientists and clinicians who played a role in bringing Apligraf to the marketplace. In particular, the authors thank Cary Isaacs, Joseph Laning, Jeffrey Crews, Katrin Kriwet, Jeffrey Teumer, and Katherine Giovino for contributions highlighted in this chapter and Michelle Blinn for assistance in its preparation.

REFERENCES

1. Parenteau, N. L., Hardin-Young, J., and Ross, R. N. (2000). Skin. *In* "Principles of Tissue Engineering" (R. P. Lanza, R. S. Langer, and W. L. Chick, eds.), pp. 879–890. Academic Press, San Diego, CA.

2. Bell, E., Ehrlich, H. P., Buttle, D. J., and Nakatsuji, T. (1981). Living tissue formed in vitro and accepted as skin-equivalent tissue of full thickness. *Science* **211**, 1052–1054.

3. Boyce, S. T., and Hansbrough, J. F. (1988). Biologic attachment, growth, and differentiation of cultured human epidermal keratinocytes on a graftable collagen and chondroitin-6-sulfate substrate. *Surgery* **103**(4), 421–431.

4. Parenteau, N. L., Bilbo, P., Nolte, C. J., Mason, V. S., and Rosenberg, M. (1992). The organotypic culture of human skin keratinocytes and fibroblasts to achieve form and function. *Cytotechnology* **9**(1-3), 163–171.

5. Bilbo, P. R., Nolte, C. J. M., Oleson, M. A., Mason, V. S., and Parenteau, N. L. (1993). Skin in complex culture: The transition from "culture" phenotype to organotypic phenotype. *J. Toxicol.—Cutaneou Ocul. Toxicol.* **12**(2), 183–196.

6. Nolte, C. J., Oleson, M. A., Bilbo, P. R., and Parenteau, N. L. (1993). Development of a stratum corneum and barrier function in an organotypic skin culture. *Arch. Dermatol. Res.* **285**(8), 466–474.

7. Eisenberg, M., and Llewelyn, D. (1998). Surgical management of hands in children with recessive dystrophic epidermolysis bullosa: Use of allogeneic composite cultured skin grafts. *Br. J. Plast Surg.* **51**(8), 608–613.

8. Yannas, I. V., Burke, J. F., Orgill, D. P., and Skrabut, E. M. (1982). Wound tissue can utilize a polymeric template to synthesize a functional extension of skin. *Science* **215**, 174–176.

9. Poskitt, K. R., James, A. H., Lloyd-Davies, E. R., Walton, J., and McCollum, C. (1987). Pinch skin grafting or porcine dermis in venous ulcers: A randomized clinical trial. *Br. Med. J. (Clin. Res. Ed.)* **294**, 674–676.

10. Desmouliere, A., and Gabbiani, G. (1996). The role of the myofibroblast in wound healing and fibrocontractive diseases. *In* "The Molecular and Cellular Biology of Wound Repair." (R. A. F. Clark, ed.), pp. 391–414. Plenum, New York.

11. Whitby, D. J., and Ferguson, M. W. (1991). Immunohistochemical localization of growth factors in fetal wound healing. *Dev. Biol.* **147**(1), 207–215.

12. Armstrong, J. R., and Ferguson, M. W. (1995). Ontogeny of the skin and the transition from scar-free to scarring phenotype during wound healing in the pouch young of a marsupial, Monodelphis domestica. *Dev. Biol.* **169**(1), 242–260.

13. Cowin, A. J., Brosnan, M. P., Holmes, T. M., and Ferguson, M. W. (1998). Endogenous inflammatory response to dermal wound healing in the fetal and adult mouse. *Dev. Dyn.* **212**(3), 385–393.

14. Whitby, D. J., and Ferguson, M. W. (1991). The extracellular matrix of lip wounds in fetal, neonatal and adult mice. *Development (Cambridge, UK)* **112**(2), 651–668.

15. Longaker, M. T., Whitby, D. J., Ferguson, M. W., Harrison, M. R., Crombleholme, T. M., Langer, J. C., Cochrum, K. C., Verrier, E. D., and Stern, R. (1989). Studies in fetal wound healing. III. Early deposition of fibronectin distinguishes fetal from adult wound healing. *J. Pediatr. Surg.* **24**(8), 799–805.

16. Raffetto, J. D., Mendez, M. V., Phillips, T. J., Park, H. Y., and Menzoian, J. O. (1999). The effect of passage number on fibroblast cellular senescence in patients with chronic venous insufficiency with and without ulcer. *Am. J. Surg.* **178**(2), 107–112.

17. Hasan, A., Murata, H., Falabella, A., Ochoa, S., Zhou, L., Badiavas, E., and Falanga, V. (1997). Dermal fibroblasts from venous ulcers are unresponsive to the action of transforming growth factor-beta 1. *J. Dermatol. Sci.* **16**(1), 59–66.

18. Compton, C. C. (1993). Wound healing potential of cultured epithelium. *Wounds* **5**(2), 97–111.

19. Regauer, S., and Compton, C. C. (1990). Cultured keratinocyte sheets enhance spontaneous re-epithelialization in a dermal explant model of partial-thickness wound healing. *J. Invest. Dermatol.* **95**(3), 341–346.

20. Phillips, T. J., Kehinde, O., Green, H., and Gilchrest, B. A. (1989). Treatment of skin ulcers with cultured epidermal allografts. *J. Am. Acad. Dermatol.* **21**(2, pt. 1), 191–199.

21. De Luca, M., Albanese, E., Cancedda, R., Viacava, A., Faggioni, A., Zambruno, G., and Giannetti, A. (1992). Treatment of leg ulcers with cryopreserved allogeneic cultured epithelium. A multicenter study. *Arch. Dermatol. (Chicago)* **128**(5), 633–638.

22. Garner, W. L. (1998). Epidermal regulation of dermal fibroblast activity. *Plast. Reconstr. Surg.* **102**(1), 135–139.

23. Whitby, D. J., Longaker, M. T., Harrison, M. R., Adzick, N. S., and Ferguson, M. W. (1991). Rapid epithelialization of fetal wounds is associated with the early deposition of tenascin. *J. Cell Sci.* **99**(pt. 3), 583–586.

24. Coulomb, B., Lebreton, C., and Dubertret, L. (1989). Influence of human dermal fibroblasts on epidermalization. *J. Invest. Dermatol.* **92**(1), 122–125.

25. Slavin, J. (1996). The role of cytokines in wound healing. *J. Pathol.* **178**(1), 5–10.

26. Rothe, M., and Falanga, V. (1989). Growth factors. Their biology and promise in dermatologic diseases and tissue repair. *Arch. Dermatol.* **125**(10), 1390–1398.

27. Parenteau, N. (1994). Skin equivalents. *In* "Keratinocyte Methods" (I. Leigh and F. Watt, eds.), pp. 45–54. Cambridge University Press, Cambridge, UK.

28. Sabolinski, M. L., Alvarez, O., Auletta, M., Mulder, G., and Parenteau, N. L. (1996). Cultured skin as a "smart material" for healing wounds: Experience in venous ulcers. *Biomaterials* **17**, 311–320.

29. Falanga, V., Margolis, D., Alvarez, O., Auletta, M., Maggiacomo, F., Altman, M., Jensen, J., Sabolinski, M., and Hardin-Young, J. (1998). Rapid healing of venous ulcers and lack of clinical rejection with an allogeneic cultured human skin equivalent. Human Skin Equivalent Investigators Group. *Arch. Dermatol. (Chicago)* **134**(3), 293–300.

30. Pham, H. T., Rosenblum, B. I., Lyons, T. E., Giurini, J. M., Chrzan, J. S., Habershaw, G. M., and Veves, A. (1999). Evaluation of a human skin equivalent for the treatment of diabetic foot ulcers in a prospective randomized, clinical trial. *Wounds* **11**(4), 79–86.

31. Waymack, P., Duff, R. G., and Group, A. B. S. (2000). The effect of a tissue engineered bilayered living skin analog, over meshed split-thickness autografts on the healing of excised burn wounds. *Burns* (to be published).

32. Falabella, A. F., Schachner, L. A., Valencia, I. C., and Eaglstein, W. H. (1999). The use of tissue-engineered skin (Apligraf) to treat a newborn with epidermolysis bullosa. *Arch. Dermatol.* **135**(10), 1219–1222.

33. Muhart, M., McFalls, S., Kirsner, R., Elgart, G., Kerdel, F., Sabolinski, M., Hardin-Young, J., and Eaglstein, W. (1999). Behavior of tissue-engineered skin. A comparison of living skin equivalent, autograft, and occlusive dressing in human donor sites. *Arch. Dermatol. (Chicago)* **135**, 913–918.

34. Eaglstein, W. H., Alvarez, O. M., Auletta, M., Leffel, D., Rogers, G. S., Zitelli, J. A., *et al.* (1999). Acute excisional wounds treated with a tissue-engineered skin (Apligraf). *Dermatol. Surg.* **25**(3), 195–201.

35. Johnson, E. W., Meunier, S. F., Roy, C. J., and Parenteau, N. L. (1992). Serial cultivation of normal human keratinocytes: A defined system for studying the regulation of growth and differentiation. *In Vitro Cell. Dev. Biol.* **28A**(6), 429–435.

36. Wilkins, L. M., and Parenteau, N. L. (1999). Bioengineered skin: Manufacturing, safety, and quality control. *In* "Wound Healing and the Skin" (V. Falanga, ed.). Martin Dunitz Publishers, London (submitted for publication).

37. Laning, J. C., Isaacs, C. M., and Hardin-Young, J. (1997). Normal human keratinocytes inhibit the proliferation of unprimed T cells by TGF-β and PGE2, but not IL–10. *Cell. Immunol.* **175**(1), 16–24.

38. Nickoloff, B. J., Basham, T. Y., Merigan, T. C., Torseth, J. W., and Morhenn, V. B. (1986). Human keratinocyte-lymphocyte reactions in vitro. *J. Invest. Dermatol.* **87**(1), 11–18.

39. Niederwieser, D., Aubock, J., Troppmair, J., Herold, M., Schuler, G., Boeck, G., Lotz, J., Fritsch, P., and Huber, C. (1988). IFN-mediated induction of MHC antigen expression on human keratinocytes and its influence on in vitro alloimmune responses. *J. Immunol.* **140**(8), 2556–2564.

40. Theobald, V. A., Lauer, J. D., Kaplan, F. A., Baker, K. B., and Rosenberg, M. (1993). "Neutral allografts"—Lack of allogeneic stimulation by cultured human cells expressing MHC class I and class II antigens. *Transplantation* **55**(1), 128–133.

41. Gaspari, A. A., and Katz, S. I. (1988). Induction and functional characterization of class II MHC (Ia) antigens on murine keratinocytes. *J. Immunol.* **140**(9), 2956–2963.

42. Nickoloff, B. J., Basham, T. Y., Merigan, T. C., and Morhenn, V. B. (1985). Keratinocyte class II histocompatibility antigen expression. *Br. J. Dermatol.* **112**(3), 373–374.

43. Nickoloff, B. J., Mitra, R. S., Lee, K., Turka, L. A., Green, J., Thompson, C., and Shimizu, Y. (1993). Discordant expression of CD28 ligands, BB-1, and B7 on keratinocytes in vitro and psoriatic cells in vivo. *Am. J. Pathol.* **142**(4), 1029–1040.

44. Nickoloff, B. J., and Turka, L. A. (1994). Immunological functions of non-professional antigen-presenting cells: New insights from studies of T-cell interactions with keratinocytes. *Immunol. Today* **15**(10), 464–469.

45. Wilkins, L. M., Watson, S. R., Prosky, S. J., Meunier, S. F., and Parenteau, N. L. (1994). Development of a bilayered living skin construct for clinical applications. *Biotechnol. Bioeng.* **43**, 747–756.

46. Parenteau, N. L. (2000). Cell differentiation, animal. *In* "Encyclopedia of Cell Technology" (R. E. Spier, ed.), Vol. 1, pp. 365–377. Wiley, New York.

47. Nolte, C. J., Oleson, M. A., Hansbrough, J. F., Morgan, J., Greenleaf, G., and Wilkins, L. (1994). Ultrastructural features of composite skin cultures grafted onto athymic mice. *J. Anat.* **185**(pt. 2), 325–333.

48. Parenteau, N., Sabolinski, M., Prosky, S., Nolte, C., Oleson, M., Kriwet, K., and Bilbo, P. (1996). Biological and physical factors influencing the successful engraftment of a cultured human skin substitute. *Biotechnol. Bioeng.* **52**, 3–14.

49. Briscoe, D. M., Dharnidharka, V. R., Isaacs, C., Downing, G., Prosky, S., Shaw, P., Parenteau, N. L., and Hardin-Young, J. (1999). The allogeneic response to cultured human skin equivalent in the hu-PBL-SCID mouse model of skin rejection. *Transplant.* **67**(12), 1590–1599.

50. Sabolinski, M., Rovee, D., Parenteau, N. L., *et al.* (1995). The efficacy and safety of graftskin for the treatment of chronic venous ulcers. *Wound Repair Regeneration* **3**(1), 78.

51. Hull, B. E., Finley, R. K., and Miller, S. F. (1990). Coverage of full-thickness burns with bilayered skin equivalents: A preliminary clinical trial. *Surgery* **107**(5), 496–502.

52. Coulomb, B., Friteau, L., Baruch, J., Guilbaud, J., Chretien-Marquet, B., Glicenstein, J., Lebreton-Decoster, C., Bell, E. and Dubertret, L. (1998). Advantage of the presence of living dermal fibroblasts within in vitro reconstructed skin for grafting in humans. *Plast. Reconstr. Surg.* **101**(7), 1891–1903.

53. Kuroyanagi, Y., Kenmochi, M., Ishihara, S., Takeda, A., Shiraishi, A., Ootake, N., Uchinuma, E., Torikai, K., and Shioya, N. (1993). A cultured skin substitute composed of fibroblasts and keratinocytes with a collagen matrix: Preliminary results of clinical trials. *Ann. Plast. Surg.* **31**(4), 340–349; discussion: pp. 349–351.

54. Wassermann, D., Schlotterer, M., Toulon, A., Cazalet, C., Marien, M., Cherruau, B., and Jaffray, P. (1988). Preliminary clinical studies of a biological skin equivalent in burned patients. *Burns Incl. Therm. Inj.* **14**, 326–330.

55. Nanchahal, J., Otto, W. R., Dover, R., and Dhital, S. K. (1989). Cultured composite skin grafts: Biological skin equivalents permitting massive expansion. *Lancet* **2**, 191–193.

56. Boyce, S. T., and Williams, M. L. (1993). Lipid supplemented medium induces lamellar bodies and precursors of barrier lipids in cultured analogues of human skin. *J. Invest. Dermatol.* **101**(2), 180–184.

57. Boyce, S. T., Goretsky, M. J., Greenhalgh, D. G., Kagan, R. J., Rieman, M. T., and Warden, G. D. (1995). Comparative assessment of cultured skin substitutes and native skin autograft for treatment of full-thickness burns. *Ann. Surg.* **222**(6), 743–752.

CHAPTER 106

UTERUS

Carlos E. Baez and Anthony Atala

INTRODUCTION

Research on human endometrium in tissue culture started in the early 1900s. In 1926 an outgrowth of cells from endometrial explants was described [1]. Later studies dealt with the growth characteristics of these cells [2,3]. These studies found that explants obtained during the follicular phase of the menstrual cycle grew better than explants obtained in the luteal phase, and greater success was achieved with samples from women of childbearing age [2,3].

The development of uterine endometrial cell model systems that would exhibit structural and functional properties observed *in vivo*, and can be regulated *in vitro*, would be useful for tissue engineering applications.

ANATOMY AND HISTOLOGY

The uterus is a muscular organ in the female reproductive tract lined by glandular mucosa, which has a specialized vascularization. This hollow, pear-shaped organ is situated in the pelvic cavity interposed between the bladder and the rectum. The expanded upper portion is called the body or corpus. The area rostral to the point at which both oviducts join the uterus is often referred as the fundus. The constricted portion below the fundus is called the isthmus, below which there is a cylindrical portion called the cervix. The layers of this organ from internal to external are mucosa (endometrium), muscularis (myometrium), and serosa (perimetrium). Fluctuations in the levels of serum estradiol and progesterone cause all three layers to go through sequential structural cyclic changes.

The endometrium is approximately 5 mm thick but varies throughout the hormonal cycle. This layer is lined by a secretory simple columnar epithelium invaginated to form tubular uterine glands. Some ciliated columnar cells can also be found as part of the epithelium. The endometrium is composed of an upper stratum functionalis, which sheds during each menstrual cycle, and the deep stratum basalis, which is not sloughed off during menstruation. Coiled or spiral arteries that nurture a large capillary bed in the superficial endometrium supply vascularization of both strata. Although glandular and luminal epithelia are continuous with each other and appear to be morphologically similar by light and electron microscopy [4,5], they respond differently to hormonal stimulus.

The uterine epithelium is composed of quiescent and proliferating subpopulations, which show differential proliferative responses to estrogens and progesterones [6]. Administration of estrogen results in the recruitment of quiescent glandular cells into the cell cycle and decreases the rate of luminal cell loss. Progesterone induces acceleration in the rate of proliferation by shortening the cell cycle length in the glands and lumen [7].

The endometrial stroma resembles mesenchyme, containing stellate cells with large ovoid nuclei. Owing to decidual transformation, stromal cells are believed to play a role in implantation and in the maintenance of pregnancy through nutrition of the blastocyst, endocrine secretion (prolactin), and protection of the embryo [8].

The myometrium of the uterus is composed of four layers. The layers are not sharply demarcated because of complex interconnecting bundles, which are interspersed with considerable connective tissue. Four layer are easily recognizable:

> The stratum submucosum contains a thin layer beneath the submucosa with longitudinal fibers.
> The stratum vasculare, contains many large blood vessels that give it a spongy appearance, the fibers are circular and oblique.
> The stratum supravasculare has fibers that are mainly circular and longitudinal.
> The stratum subserosum consists of a thin longitudinal muscle layer.

The peritoneum consists of a single layer of flattened cells, which surround the oviduct and uterus. This thin layer also functions as a sheath over the nerves and vessels. The portion of the peritoneum, which surrounds the uterus and extends to the pelvic walls laterally, is called the perimetrium.

Several techniques can be explored for achieving the separation and culture of endometrial cells, including methods based on enzymatic digestion and mechanic dissociation [9–12].

PROTOCOLS FOR CULTURING ENDOMETRIAL CELLS
TRANSPORTATION OF ENDOMETRIAL TISSUE

Endometrial cells can be obtained from uterine biopsy or hysterectomy specimens. Biopsies should be transferred immediately to transport medium: DMEM/F-12 (Dulbecco's Modified Eogle's Medium with Ham's F-12 nutrient medium). Biopsies exceeding 2 cm in diameter will remain viable in this medium for up to 3 days at 4°C.

Method Using Explants

1. Remove the specimen from the transport medium with sterile forceps and place in a 10-cm petri dish.

2. Wash the specimen with three applications of 5 ml of sterile phosphate-buffered saline (PBS) containing gentamicin sulfate 50 (μg/ml) and amphotericin 5 (μg/ml).

3. Orient the specimen with the epithelial surface down and use curved iris scissors to dissect the submucosa from the muscle. A thin opaque white epithelial strip should be left after this procedure.

4. Mince the epithelial strip finely, into very small fragments.

5. Place the small fragments in a 10-cm petri dish and add equal parts of DMEM and F-12, as well as 10^{-8} M 17β-estradiol, 50 ng/ml progesterone, and 10 ng/ml epidermal growth factor (EGF).

6. Change the medium every 2–3 days.

Method Using Trypsin

1. Remove the specimen from the transport medium with sterile forceps and place in a 10-cm petri dish.

2. Wash with three applications of 5 ml of sterile PBS containing 50 μm/ml gentamicin sulfate and 5 μm/ml amphotericin.

3. Orient the specimen with the epithelial surface down on the culture dish. Use a scalpel to scrape off as much of the stroma and muscle as possible. After this procedure you should see a thin strip consisting of epithelial cells and submucosa.

4. Use curved iris scissors to mince the strip into small fragments.

5. Add 10 ml of trypsin–EDTA to the fragments and transfer to a beaker containing a sterile, plastio-coated, magnetic stir bar.

6. Place the beaker on a magnetic stirrer in an incubator or hot room at 37°C and stir slowly for 30–40 min.

7. Let the suspension sit at room temperature until all fragments have settled to the bottom of the beaker. Then use a 10-ml pipette to remove the supernatant containing single cells, and decant into a 50-ml tube through a 1-mm filter.

8. Add 10 ml of trypsin–EDTA to the fragments in the glass and repeat steps 6 and 7.

9. Add 15 ml of DMEM/F-12 medium and centrifuge at 80g for 5 min.

10. Remove supernatant and resuspend vigorously with a 10-ml pipette and medium.

11. Count the cells, using a counting chamber. Ignore large fragments and red blood cells. Cell viability can be assessed at this stage by trypan blue exclusion.

12. Plate cells at a density of 2×10^4 cells/cm^2.

Method Using a Collagen Extracellular Matrix

1. Use routine methods to culture fibroblasts at a density of 10^5/cm^2 [8]. It is important that the fibroblasts form a continuous even layer across the dish before this procedure in begun.

2. Prepare the conditioning medium, which consists of DMEM/F-12, 10% fetal bovin serum (FBS), 0.4 µg/ml hydrocortisone, 5 µg/ml insulin, 10 µg/ml EGF, and 1% penicillin–streptomycin.

3. When petri dishes are 90% confluent, change to a conditioning medium (DMEM/F-12, 75:25).

4. Incubate for 24 h.

5. Remove culture medium and add the same volume of 1% Triton X-100.

6. Wait 5–30 min or until fibroblasts detach.

7. Discard Triton X-100 and wash three times with sterile distilled water.

8. Dry remaining collagen extracellular matrix for immediate use or store at 4°C.

9. Added endometrial cells to this collagen layer.

10. Incubated cultures at 37°C in 5% CO$_2$ with DMEM/F-12 (50:50), adding 10^{-8} M 17β-estradiol, 50 ng/ml progesterone and 10 ng/ml EGF.

11. Change culture medium every 2–3 days.

Method Using a Fibroblast Feeder Cell Layer

1. The a feeder layer consists of lethally irradiated Swiss 3T3 cells at a density of 10^5 cells/cm^2 in complete medium. This layer can be prepared beforehand and cultured in Ham's F-12 with 10% FBS.

2. The medium consists of DMEM/F-12 (50:50) to which is added 17β-estradiol (10^{-8} M) and progesterone (50 ng/ml).

3. After the first medium change, add EGF (10 ng/ml).

4. After 14 days in culture, the endometrial cells colonies should be easily visible to the naked eye.

5. When the endometrial cells are confluent, separate from the feeder fibroblast layer by trypsinization.

Method Using Collagenase

1. Tissue is gently minced into small pieces (1–2 mm^3) and washed in fresh medium to remove mucus or red blood cells.

2. Add 10 ml of Ham's F-12 or DMEM/ F-12 culture medium containing 0.1% collagenase I A (470 U/mg, Sigma), 100 µg/ml streptomycin, 100 U/ml penicillin, and 0.25 µg/ml fungizone.

3. Shake for 2 h at 37°C.

4. Separate stromal from epithelial cells by filtration. The first filtration is performed through a 250-µm stainless steel sieve to remove mucous material or undigested tissue. The second filtration of the filtrate solution is performed through a 36-µm sieve.

5. The glands and epithelial cells that are trapped on the 36-µm sieve are backwashed from the sieve with medium and freed of collagenase by centrifugation (5 min, 75g).

6. The cell pellet is resuspended in medium, and the cell density is determined in a hemocytometer.

7. Stromal and epithelial cells are plated in different petri dishes at a density of 2–5 $\times$ 10^5 cells/cm^2.

8. The medium for the epithelial cells consists of DMEM/F-12 (50:50) with 10^{-8} M 17β-estradiol, 50 ng/ml progesterone, and 10 ng/ml EGF. The medium for the stromal cells consists of Ham's F-12 and 10% FBS.

9. Change media every 2–3 days.

PROTOCOLS FOR CULTIVATING MYOMETRIAL CELLS

1. Remove the specimen from the transport medium with sterile forceps and place in a 10-cm petri dish.

2. Orient the specimen with the epithelial surface down and use curved iris scissors to dissect the submucosa from the muscle.

3. Mince the muscle layer into very small fragments.

4. Seed the small fragments in a 10-cm petri dish and add DMEM and 10% FBS.

5. Change the medium every 3 days.

PROTOCOL FOR ENGINEERING UTERINE TISSUE

1. Materials are 1.2-cm-diameter poly(glycolic acid) (PGA) polymer and bladder submucosa of same size, sutured with Vicryl 5-0.

2. Place the polymers in 48 well plates with submucosal side facing up.

3. Seed endometrial cells, which were cultured beforehand in DMEM/F-12 (50:50) medium, 10^{-8} M 17β-estradiol, 50 ng/ml progesterone, and 10 ng/ml EGF in a concentration of 30 mM/ml.

4. After 8 h use forceps to transport each polymer to a 10-cm petri dish and add 15 ml of medium. Change medium daily for 5 days.

5. Place polymers into 48 well plates with PGA side facing up and seed muscle cells in a concentration of 50 mM/ml.

6. Implant scaffolds into the backs of mice.

IDENTIFICATION AND CHARACTERIZATION ENDOMETRIAL CELLS

Characterization of endometrial epithelia requires the use of specific differentiation markers. These markers include keratin intermediate filaments, intracytoplasmic glycogen, progesterone receptors, and estrogen receptors [15,18,20–45]. Cytokeratin intermediate filaments described in Refs. [5,6,13,14,16,17,19] are the most commonly used for characterization.

REFERENCES

1. Liszczak, T. M., *et al.* (1977). Ultrastructure of human endometrial epithelium in monolayer culture with and without steroid hormones. *In Vitro* **13**(6), 344–356.

2. Papanicolaou, G. a. M. B. (1958). Observations on the behaviour of human endometrial cells in tissue culture. *Am. J. Obstet. Gynecol.* **76**, 601–618.

3. Figge, D. (1960). Growth characteristics of human endometrium in tissue culture. *Obstet Gynecol.* **16**, 269–277.

4. Davies, J., and Hoffman, L. H. (1973). Studies on the progestational endometrium of the rabbit. I. Light microscopy, day 0 to day 13 of gonadotrophin-induced pseudopregnancy. *Am. J. Anat.* **137**(4), 423–445.

5. Davies, J., and Hoffman, L. H. (1975). Studies on the progestational endometrium of the rabbit. II. Electron microscopy, day 0 to day 13 of gonadotrophin-induced pseudopregnancy. *Am. J. Anat.* **142**(3), 335–365.

6. Conti, C. J., *et al.* (1984). Estrogen and progesterone regulation of proliferation, migration, and loss in different target cells of rabbit uterine epithelium. *Endocrinology (Baltimore)* **114**(2), 345–351.

7. Nawaz, S., *et al.* (1987). Hormonal regulation of cell death in rabbit uterine epithelium. *Am. J. Pathol.* **127**(1), 51–59.

8. Mizuno, K., *et al.* (1998). Establishment and characterization of in vitro decidualization in normal human endometrial stromal cells. *Osaka City Med. J.* **44**(1), 105–115.

9. Watson, H., Franks, S., and Bonney, R. C. (1994). Characterization of epidermal growth factor receptor in human endometrial cells in culture. *J. Reprod. Fertil.* **101**(2), 415–420.

10. Akoum, A., *et al.* (1996). Human endometrial cells cultured in a type I collagen gel. *J. Reprod. Med.* **41**(8), 555–561.

11. Barberini, F., Sartori, S., and Motta, P. (1978). Changes in the surface morphology of the rabbit endometrium related to the estrous and progestational stages of the reproductive cycle a scanning and transmission electron microscopic study. *Cell Tissue Res.* **190**(2), 207–222.

12. Bentin-Ley, U., *et al.* (1994). Isolation and culture of human endometrial cells in a three-dimensional culture system. *J. Reprod. Fertil.* **101**(2), 327–332.

13. Bongso, A., *et al.* (1988). Establishment of human endometrial cell cultures. *Hum. Reprod.* **3**(6), 705–713.

14. Bowen, J. A., *et al.* (1996). Characterization of a polarized porcine uterine epithelial model system. *Biol. Reprod.* **55**(3), 613–619.

15. Centola, G. M., Cisar, M., and Knab, D. R. (1984). Establishment and morphologic characterization of normal human endometrium in vitro. *In Vitro* **20**(6), 451–462.

16. Chatzaki, E., *et al.* (1994). Characterisation of the differential expression of marker antigens by normal and malignant endometrial epithelium. *Br. J. Cancer.* **69**(6), 1010–1014.

17. Classen-Linke, I., *et al.* (1997). Establishment of a human endometrial cell culture system and characterization of its polarized hormone responsive epithelial cells. *Cell Tissue Res.* **287**(1), 171–185.

18. Cooke, P. S., *et al.* (1986). Restoration of normal morphology and estrogen responsiveness in cultured vaginal and uterine epithelia transplanted with stroma. *Proc. Natl. Acad. Sci. U.S.A.* **83**(7), 2109–2113.

19. Gerschenson, L. E., and Fennell, R. H., Jr. (1982). A developmental view of endometrial hyperplasia and carcinoma based on experimental research. *Pathol. Res. Pract.* **174**(3), 285–296.

20. Glasser, S. R., and Mulholland, J. (1993). Receptivity is a polarity dependent special function of hormonally regulated uterine epithelial cells. *Microsc. Res. Tech.* **25**(2), 106–120.
   Hohn, H. P., *et al.* (1989). Rabbit endometrium in organ culture: Morphological evidence for progestational differentiation in vitro. *Cell Tissue Res.* **257**(3), 505–518.

21. Inaba, T., *et al.* (1988). Augmentation of the response of mouse uterine epithelial cells to estradiol by uterine stroma. *Endocrinology (Baltimore)* **123**(3), 1253–1258.

22. Irwin, J. C., *et al.* (1989). Hormonal regulation of human endometrial stromal cells in culture: An in vitro model for decidualization. *Fertil. Steril.* **52**(5), 761–768.

23. Johnson, G. A., *et al.* (1999). Development and characterization of immortalized ovine endometrial cell lines. *Biol. Reprod.* **61**(5), 1324–1330.

24. Kirk, D., *et al.* (1978). Normal human endometrium in cell culture. I. Separation and characterization of epithelial and stromal components in vitro. *In Vitro* **14**(8), 651–662.

25. Kleinman, D., *et al.* (1983). Human endometrium in cell culture: A new method for culturing human endometrium as separate epithelial and stromal components. *Arch. Gynecol.* **234**(2), 103–112.

26. Mangrulkar, R. S., *et al.* (1995). Isolation and characterization of heparin-binding growth factors in human leiomyomas and normal myometrium. *Biol. Reprod.* **53**(3), 636–646.

27. Marsh, M. M., *et al.* (1994). Production and characterization of endothelin released by human endometrial epithelial cells in culture. *J. Clin. Endocrinol. Metab.* **79**(6), 1625–1631.

28. Merviel, P., *et al.* (1995). Normal human endometrial cells in culture: Characterization and immortalization of epithelial and stromal cells by SV 40 large T antigen. *Biol. Cell* **84**(3), 187–193.

29. Merviel, P., Calvo, F., and Salat-Baroux, J. (1994). [Endometrial cell cultures: Their value in the understanding of the implantation mechanisms]. *Contraception Fertil. Sex.* **22**(1), 7–14.

30. Mulholland, J., Winterhager, E., and Beier, H. M. (1988). Changes in proteins synthesized by rabbit endometrial epithelial cells following primary culture. *Cell Tissue Res.* **252**(1), 123–132.

31. Osteen, K. G., *et al.* (1989). Development of a method to isolate and culture highly purified populations of stromal and epithelial cells from human endometrial biopsy specimens. *Fertil. Steril.* **52**(6), 965–972.

32. Pavlik, E. J., and Katzenellenbogen, B. S. (1978). Human endometrial cells in primary tissue culture: Estrogen interactions and modulation of cell proliferation. *J. Clin. Endocrinol. Metab.* **47**(2), 333–344.

33. Rajkumar, K., *et al.* (1983). Uteroglobin production by cultured rabbit uterine epithelial cells. *Endocrinology (Baltimore)* **112**(4), 1490–1498.

34. Rajkumar, K., *et al.* (1983). Effect of progesterone and 17 beta-estradiol on the production of uteroglobin by cultured rabbit uterine epithelial cells. *Endocrinology (Baltimore)* **112**(4), 1499–1505.

35. Ricketts, A. P., Hagensee, M., and Bullock, D. W. (1983). Characterization in primary monolayer culture of separated cell types from rabbit endometrium. *J. Reprod. Fertil.* **67**(1), 151–160.

36. Rotello, R. J., *et al.* (1992). Characterization of uterine epithelium apoptotic cell death kinetics and regulation by progesterone and RU 486. *Am. J. Pathol.* **140**(2), 449–456.

37. Ryan, I. P., Schriock, E. D., and Taylor, R. N. (1994). Isolation, characterization, and comparison of human endometrial and endometriosis cells in vitro. *J. Clin. Endocrinol. Metab.* **78**(3), 642–649.

38. Satyaswaroop, P. G., *et al.* (1979). Isolation and culture of human endometrial glands. *J. Clin. Endocrinol. Metab.* **48**(4), 639–641.

39. Schatz, F., *et al.* (2000). Human endometrial endothelial cells: Isolation, characterization, and inflammatory-mediated expression of tissue factor and type 1 plasminogen activator inhibitor. *Biol. Reprod.* **62**(3), 691–697.

40. Squier, C. A., and Kammeyer, G. A. (1983). The role of connective tissue in the maintenance of epithelial differentiation in the adult. *Cell Tissue Res.* **230**(3), 615–630.

41. Strowitzki, T., *et al.* (1996). Characterization of receptors for insulin-like growth factor type I on cultured human endometrial stromal cells: Downregulation by progesterone. *Gynecol. Endocrinol.* **10**(4), 229–240.

42. Tseng, L., Stolee, A., and Gurpide, E. (1972). Quantitative studies on the uptake and metabolism of estrogens and progesterone by human endometrium. *Endocrinology (Baltimore)* **90**(2), 390–404.
   Turyk, M. E., *et al.* (1989) Growth and characterization of epithelial cells from normal human uterine ectocervix and endocervix. *In Vitro Cell Dev. Biol.* **25**(6), 544–556.

43. Varma, V. A. (1982). *et al.,* Monolayer culture of human endometrium: Methods of culture and identification of cell types. *In Vitro* **18**(11), 911–918.

44. Vigano, P., *et al.* (1993). Culture of human endometrial cells: A new simple technique to completely separate epithelial glands. *Acta Obstet. Gynecol. Scand.* **72**(2), 87–92.

45. Zarmakoupis, P. N., *et al.* (1995). Inhibition of human endometrial stromal cell proliferation by interleukin 6. *Hum. Reprod.* **10**(9), 2395–2399.

CHAPTER 107

Jawbone

Yukihiko Kinoshita and Teruo Amagasa

The final target for the technology of jawbone reconstruction is to regenerate physiological bone that makes dental implants or accommodation of dentures possible. This chapter summarizes the strategies of bone tissue engineering and introduces the concept of jawbone reconstruction by means of bioabsorbable material and particulate cancellous bone and marrow (PCBM). Our studies indicate that a combination of poly(L-lactic acid) mesh tray or sheet and PCBM is useful for mandibular reconstruction.

To improve the efficiency of osteogenesis in the future, it is necessary to elucidate the molecular mechanisms involved in the formation of osteoblasts as well as their function manifestations, and subsequently, both to establish an optimal method of applying growth factors in the local site and to develop a suitable scaffold.

INTRODUCTION

The loss of the mandibular arch after tumor excisions or traumatic injuries leaves the patient not only with a remarkable deformity of the face but also with a mandible that does not function for chewing, swallowing, or speaking. In such a case, reconstruction has often been carried out by bone graft or implantation of artificial material. It has been considered that reconstruction by vascular or nonvascular autologous bone graft is a reliable procedure in these cases [1]. Nevertheless, as in other areas, there are some disadvantages, including donor site morbidity and inadequate supply. In the late 1990s distraction osteogenesis was applied in jawbone regeneration [2]. However, this procedure is time-consuming, and it is difficult to reproduce the complicated shape of the jawbone by itself. On the other hand, acknowledging rapid developments in molecular biology and material science, tissue engineering has been targeted for the regeneration of various organs and tissues [3]. Jawbone reconstruction is an attractive and natural target for tissue engineering, since there are numerous clinical indications, hence much necessity. This chapter briefly describes the strategies of bone tissue engineering and reports on a jawbone reconstruction engineering procedure that uses biodegradable polymer mesh and autogenic particulate cancellous bone and marrow.

STRATEGIES OF BONE TISSUE ENGINEERING

The strategies of bone tissue engineering are aimed at mimicking the natural process of bone repair in a bone fracture. In other words, bone tissue is regenerated in the desired site by making use of more than two or three individual parameters, namely, osteogenic progenitor cells, scaffolds for bone formation, and bioactive substances. Three general approaches have been applied to the art of tissue engineering of bone [4].

The first consists of therapies based on a scaffold (matrix). This approach, which introduces structural implants to replace the missing bone, is dependent on the recruitment of endogenous osteoprogenitor cells to regenerate bone. Titanium fiber metals and ceramics consisting of tricalcium phosphate and/or hydroxyapatite have been used for this purpose. These implants feature porosity, which facilitates the invasion of the bone. Also, the function of the obliteration membrane, which attempts to regenerate alveolar bone resorbed by periodontal disease, is being used to interrupt invasion of the gingival epithelium and connective tissue into the space of the bone defect. Specifically, this function is being used to secure space for osteoblasts and the remaining periodontal ligament to regenerate alveolar bone and cementum in a process called guided tissue regeneration (GTR) [5,6]. Poly(tetrafluoroethylene) and the bioabsorbable polymers poly(L-lactic acid) (PLLA) and poly(glycolic acid) (PGA) are used in this process. However, because these matrices lack biological activity such as osteoinduction, their application is limited. For example, it is difficult to apply matrices in the case of large bone defects or implanted sites where blood circulation is poor.

The second approach is a factor-based therapy. This procedure attempts to avoid the limitation inherent in the matrix-based therapy by directly providing osteoinductive stimuli to the bone defect site. Growth factors such as the bone morphogenetic proteins BMP-2, BMP-3, and BMP-7 osteogenic protein 1: (OP-1) are being studied well as osteoinductive proteins [7–15]. For clinical application it is necessary to establish a carrier system, or delivery system, in which the factor can be released effectively to the site in need of bone repair. When BMP, with a carrier system such as demineralized bone, porous hydroxyapatite, biodegradable polymer—poly(lactide-*co*-glycoside) and poly(lactide)—or collagen sponge is transplanted to the bone defect region of rodents, dogs, sheep, or monkeys, large bone defects in which regeneration is naturally impossible can be repaired [7–11]. However, the effectiveness of BMP varies according to age and species [10,11,14,15]. In particular, a large quantity of BMP is necessary to show a curative effect in primates [10,11,15]. Transplanting a profuse amount of BMP directly into a living body carries some risk and requires extreme care to avoid unexpected and harmful results.

The third approach is a cell-based therapy. In this procedure, osteoprogenitor cells are transplanted directly into the site at which an increase of bone is required. Bone marrow is known to contain many osteogenic precursors [16]. Therefore, its implantation was thought to have the potential to lead to effective bone formation. This was ascertained in various preclinical studies [17–19] and clinical research [20,21]. Because the cell-based approach does not depend on osteoprogenitor cells of the topographic to repair bone defects, this method is available for the patient in whom the host tissue bed is damaged—as for example, in the case of severe postoperative cicatricial tissue, after radiation therapy, and in the process of aging, diabetes, and the like. In a procedure, now being used in clinical applications, autogenous bone marrow is collected from the iliac crest, combined with a suitable biomaterial, and immediately transplanted to the site requiring bone regeneration. Marrow transfer, or grafting, of this type is comparatively easy and inexpensive. However, when healthy bone marrow decreases owing to aging and disease, osteogenic precursors decrease [22–24], and it may be difficult to obtain bone marrow rich in osteoprogenitor cells. Since 1990, many investigators have reported on techniques for the isolation of adult human and animal mesenchymal stem cells (MSCs) from bone marrow and for extensively expanding in number *in vitro* [19,25,26]. This technique has allowed surgeons to remove a small bone marrow biopsy samples from a patient, subsequent cell expansion and seeding onto a suitable carrier/scaffold. This procedure is minimally invasive and is very efficient, because the number of cells required is small. However, before reaching the stage of clinical application, it is necessary to wait for the development of a biodegradable scaffold that not only presents no harm to living body tissue but has sufficient mechanical strength, an appropriate rate of absorption, and minimal foreign body reaction.

USING POLY(L-LACTIC ACID) MESH AND AUTOGENIC PARTICULATE CANCELLOUS BONE AND MARROW TO RECONSTRUCT THE JAWBONE

Particulate cancellous bone and marrow (PCBM) has excellent properties as a bone graft: being rich in osteogenic progenitor cells and bone matrices, it has full bone formation ability, and the spontaneous regeneration of donor sites is also possible.

However, because PCBM does not, by itself, feature structural strength and the ability to hold its shape, it is necessary to provide a framework that will lead to bone formation to the desired shape and be able to support the newly formed bone while it acquires enough strength to withstand external force. Furthermore, it is desirable that this framework be biodegradable and disappear from the living body tissue after the process of bone repair is complete. Poly(L-lactic acid) (PLLA) has been found to be the best biodegradable macro-molecular material for imparting structural strength [6]. For the past decade or so, it has been used in osteosynthetic devices in orthopedic and oral surgery [27–30]. PLLA degrades to lactic acid within the body by nonenzymatic hydrolysis. The lactic acid in turn becomes incorporated in the tricarboxylic acid cycle and is excreted by the lungs and kidneys as carbon dioxide and water [31]. Therefore, PLLA is attractive for use as a framework in PCBM transplantation. However, it is important for clinical application to understand the state of tissue reaction and tumor genesis in the process of absorbing the biodegradable material. The authors developed a mesh manufactured from PLLA and examined its utility in jawbone reconstruction.

PRECLINICAL STUDY RESULTS
PLLA Mesh Reaction to Living Body Tissue

Monofilaments with diameters of 0.3 and 0.6 mm were fabricated from PLLA having a molecular weight of 205,000 Da by spinning at 245°C and drawing at 80°C (Gunze, Kyoto, Japan). These filaments were woven into PLLA mesh sheets (Table 107.1). The PLLA mesh can be cut with scissors and is easily molded by heating it up to about 70°C. PLLA mesh sheets were implanted subcutaneously into the backs of adult dogs [19] and rats, in addition to subperiosteally into the calvaria of rats. Histologically, one month after implantation, each of the monofilaments in the mesh sheets was completely surrounded by a thin fibrous capsule and the gaps between monofilaments were filled with capillary vessels, fibroblasts, and adipose tissue. Three months after implantation, macrophages were observed in direct contact with the monofilaments. After that, monofilaments degraded very slowly [19]. After 30 months post implantation, microscopical degradation and absorption of PLLA monofilaments continued in the mesh, with many macrophages in the circumferential tissue of hydrolyzed PLLA particles (Fig. 107.1). However, very little inflammatory cell infiltration was noted in the surrounding tissue of the particles, and no absorption was seen at the adjacent bone. Ultramicroscopically, macrophages were found to be highly packed with membrane-bound vacuoles containing PLLA particles; however their mitochondria and rough endoplasmic reticulum presented a normal structure (Fig. 107.2). No tumorgenesis was observed for any of the 50 samples implanted for 18–30 months in 10 rats. These results show that PLLA mesh is biodegraded and bioabsorbed at a very slow rate as a result of hydrolysis and phagocytosis by macrophages. This occurs without showing damage to living body tissue. Furthermore, the PLLA monofilaments in the living body still retained about 80% of the original strength for 3 months, which is sufficient for PCBM bone formation, although strength declines gradually after 3 months [19]. Thus it can be concluded that PLLA mesh is favorable as a biodegradable framework of particulate cancellous bone and marrow implantation.

Complete absorption of PLLA in living body tissue can 2–3 years [32,33], and in some cases, more than 5 years [34–36]. On the other hand, Bergsma *et al.* [34] found that 3.5–5.7 years postoperatively, the PLLA bone plates and screws used for zygomatic fractures provoked swelling due to nonspecific foreign body reaction to the degradation products of high molecular weight PLLA. The shape of the implant, the size, the molecular weight, the

Table 107.1. Specification of Poly(L-lactic acid) (PLLA) Mesh

Type	Monofilament diameter (mm)	Woven density (monofilaments/cm)	
		Longtudinal	Transverse
Thin	0.3	13.6	10.4
Thick	0.6	15.4	4.6

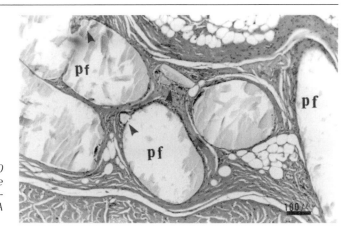

Fig. 107.1. Photomicrograph by light of PLLA mesh sheet, 30 months after subcutaneous implantation. PLLA particles (arrow) are surrounded by a large number of macrophages, very little inflammatory cell infiltration can be seen in the surrounding tissue of PLLA monofilaments (pf). Hematoxylin and eosin staining.

crystallinity, the degradation rate, and the blood circulation at the implanted site all influence tissue reaction of biodegradable materials [36–43]. Generally, the absorbent duration of PLLA with high molecular weight and high crystallinity is long, and its inflammatory reaction is slight. The orientation of the molecule and the crystal of the filament of the PLLA mesh developed during this time are comparatively high, with a low degradation rate. In addition, the contact surface with PLLA and the tissue is large, while degradation and absorption are well balanced. Therefore, it is suggested that rapid tissue reaction is minimal and not readily apparent. The late inflammatory reaction of PLLA, which was noted by Bergsma et al. [34], is probably caused by subcutaneous implantation of a large volume of high molecular weight (megadalton) PLLA that exceeds the cleaning capacity of the surrounding tissue.

Bone Tissue Engineering by PLLA Mesh and Autogenic Particulate Cancellous Bone and Marrow

A PLLA mesh cylinder filled with autogenic PCBM was implanted subcutaneously into the backs of adult dogs [19]. As a result, new bone formation was observed inside the cylinder heterotopically 1 week after implantation. Furthermore, bone formation reached a peak after 1 month, and fairly mature bone was almost completely formed in the same shape of the cylinders (Fig. 107.3). The interstitial tissue between new trabecular bones had abundant capillary vessel that infiltrated and grew through the mesh. The abundance of capillary vessel formation is important for the remolding of bone. These results show that PLLA mesh not only provides a framework for bone formation, but also assists in providing routes of nutrient supply and metabolite excretion. Furthermore, mandible continuity defects in adult dogs, where spontaneous recovery is impossible, were reconstructed using

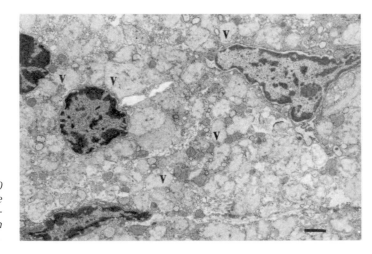

Fig. 107.2. Transmission electron micrograph, taken 30 months after a subcutaneous implantation of PLLA mesh. The cytoplasm of macrophages is highly packed with membrane-bound vacuoles (v) containing PLLA. Mitochondria and rough endoplasmic reticulum have normal appearance. Bar = 1 μm.

PLLA mesh and PCBM transplantation [41]. The continuity of the mandibular bone was established within 3 months postoperatively. Differences between the reconstructed part and the originally intact bone were almost in distinguishable after 6 months. These results show that a combination of PLLA and PCBM is available for reconstruction of the jawbone.

CLINICAL STUDY

Thirty-six patients underwent mandibular reconstruction by transplantation of PCBM with PLLA mesh tray/sheet in the seven hospitals between 1995 and 1998 (Figs. 107.4 and 107.5) [42]. Twenty-five patients were male and 11 female. The age of the patients ranged from 15 to 73 years (average age, 50.7 years). Seventeen patients had malignant tumors, 18 mandibular benign tumors, and 1 had a mandibular cyst. Seventeen patients had marginal resection of the mandible and 19 had segmental resection. Twenty-one patients had immediate reconstruction and 15 had secondary reconstruction ($\geqslant$1 year after the primary therapy).

Mandibular mesh trays were used in 28 cases and mesh sheets in 6 cases. The quantity of PCBM transplanted to individual patients was from 10 to 40 g. Treatment results of mandibular reconstruction were evaluated by X-ray findings (mainly by panoramic X-ray films) just after the operation and 6 months later. Effects were considered to be excellent if the area of osteogenesis, based on X-ray films 6 months after the surgery, was over two-thirds in comparison to right after operation. Effects were considered to be good when osteogenesis was less than two-thirds with no rereconstruction required. All other results were considered to be poor. As a result, 18 cases (50.0%) were judged as being excellent, 11 cases (31%) as being good and 7 cases as being poor (19%) (see Table 107.2). As for the relation between surgical method and treatment outcome, 14 (82.4%) of the 17 outcomes were good or excellent in the marginal resection group, whereas 12 (75.0%) of the 16 outcomes in the segmental resection group were in this category. Thus the incidence of an excellent effect was higher in the marginal resection group than in the segmental resection group. Seventeen (81%) of the 21 outcomes were excellent or good in the immediate reconstruction group, with 12 (80%) of the 15 in the secondary reconstruction group in that category. Among the outcomes cases considered to be poor, five patients developed infection (three had mandibular reconstruction in combination with reconstruction of the soft tissue with microsurgery), and two had loose fixation or inadequate blood circulation owing to repeated surgery. There were no cases of unacceptable reaction due to PLLA mesh tray/sheet during the follow-up from 1 to 5 years.

The indication for this mandibular reconstruction method by PCBM transplantation with PLLA mesh tray/sheet is the need for secondary reconstruction of a malignant tumor and reconstruction case of whole jawbone benign disease (Figs. 107.6 and 107.7). In the application, prevention of postoperative infection, security of the soft tissue in which blood circulation is good, and providing fixations are important for obtaining good results.

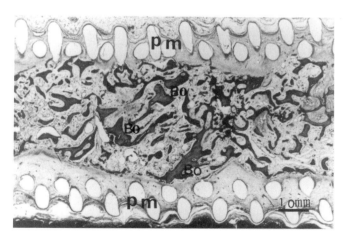

Fig. 107.3. Bone formation within the PLLA mesh cylinder (p_m) 1 month after implantation. Fairly mature bone (Bo) was almost completely formed in the same shape of the cylinders.

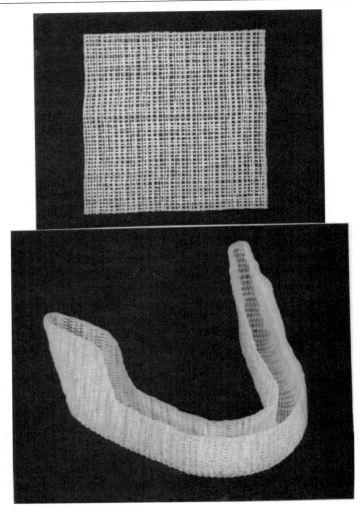

Fig. 107.4. PLLA mesh sheet and mandibular mesh tray.

This reconstruction method induces little surgical stress, and the operation is also comparatively easy. It is excellent for form recovery of the bone, especially for reconstruction of the chin, which is more difficult when a conventional iliac bone graft must be used. Moreover, it is also able to provide a good facial relationship between the upper and lower

Table 107.2. Results of Mandibular Reconstruction with PLLA Mesh and PCBM Transplantation

Method	Results			Total
	Excellent	Good	Poor	
Marginal resection	11 (64.8%)	3 (17.6%)	3 (17.6%)	17 (100%)
Segmental resection	6 (37.5%)	6 (37.5%)	4 (25.0%)	16 (100%)
Hemimandiblectomy	1 (33.3%)	2 (66.7%)		3 (100%)
Total	18 (50.0%)	11 (30.0%)	7 (19.4%)	36 (100%)

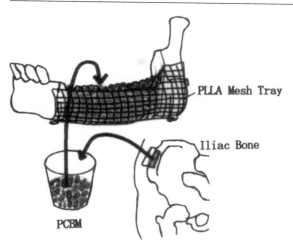

Fig. 107.5. Schematic drawing of mandibular reconstruction by means of PLLA mesh and particulate cancellous bone and marrow. PLLA mesh tray was adjusted to shape and size of the bone defect with cutting by scissors and warming at about 70°C. Next, the PLLA mesh tray was fixed to the residual bone with stainless steel wires and filled with PCBM taken from the iliac bone.

jaws, which is important for dental implantation or denture accommodation. By including the dental implant in the PLLA mesh system in advance, it might also be possible to give osseointegration of the dental implant simultaneously with the bone regeneration [43].

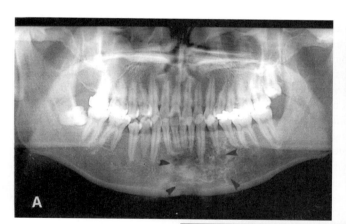

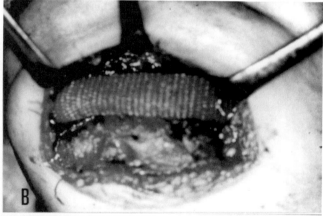

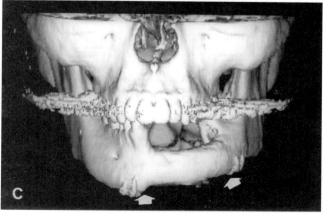

Fig. 107.6. This 27-year-old male had segmental resection of the mandible at the right second incisor to left first premolar to treat ameloblastoma (arrows) (A). Reconstruction of the mandible by PLLA mesh tray and PCBM transplantation followed immediately (B). The course after surgery was excellent. The reconstructed mandible, including chin, had good form (arrows in C) in this three-dimensional computed tomograph 3 years after operation, and the patient was wearing a partial denture.

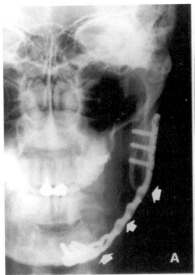
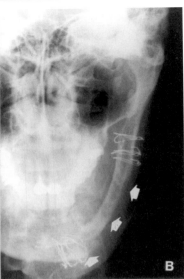

Fig. 107.7. (A) Reconstructed mandible of a 40-year-old male who held ameloblastoma of the left lower jaw (arrows). Fourteen months before this X-ray photo was taken, the patient underwent segmental resection of the mandible and reconstruction with titanium photos. (B) Discomfort in the area, of the surgery, however, led the patient to such secondary reconstruction of the lower jaw by PLLA mesh tray and PCBM; and the result was excellent. This X-ray photograph, taken 2 years after the surgery show mature bone formation with the cortical bone (arrows), and dental implantation had been scheduled.

IN THE FUTURE

It is anticipated that the effective use of bioactive factors that can promote vascularization and increase osteogenesis will result in the ossification of satisfactory amounts of bone while using only small quantities of bone marrow transplant. This process will promote bone regeneration in cases that involve poor blood circulation and will also have applications among patients of whose supply of osteogenic progenitor cells is reduced. In addition to improving angiogenicity, basic fibroblast growth factor (bFGF) enhances bone formation, by promoting the proliferation of mesenchymal precursor cells during osteogenesis [44–48]. Tabata *et al.* have reported that bFGF incorporated in gelatin hydrogels is released from the hydrogels together with degraded gelatin fragments in the body as a result of hydrogel degradation, and the retention period depends on the water content of the gelatin (extent of cross-linking) [47,48]. Yamada *et al.* used rabbit skull defects to demonstrate that the gelatin hydrogel is a promising matrix for effective induction of biological activity of bFGF for bone regeneration [46]. Since the use of bFGF incorporated in gelatin hydrogel in combination with PCBM is expected to further promote osteogenesis, this could also furnish a significant strategy for tissue engineering of the jawbone.

Research into artificial bone with a three-dimensional culture of marrow cells has also progressed [25,26,49]. This is one of the recent achievements of tissue engineering, and its application is expected to support the development of techniques for reconstruction of the jawbone. Technological development is also making further advances toward the universal production of osteogenic cells for use in bone tissue engineering. Eventually, genetic engineering of universal donor cells may provide growth factors for osteogenesis at the desired site [7,9,50].

ACKNOWLEDGMENTS

The authors express their appreciation to Drs. Kudo (Iwate Medical University), Okabe (Saitama Cancer Center), Nagayama (Tokushima University), Furuta (Tokushima Medical Pharmacological University), and Totsuka (Hokkaido University) for their cooperation in the clinical applications.

REFERENCES

1. Buckley, M. J., Agarwal, S., and Gasner, R. (1999). Tissue engineering and dentistry. *Clin. Plastic. Surg.* **6**, 657–662.
2. Yen, S. L. (1997). Distraction osteogenesis: Application to dentofacial orthopedics. *Semin. Orthodont.* **3**, 275–283.
3. Langer, R., and Vacanti, J. P. (1993). Tissue engineering. *Science* **14**, 920–926.
4. Bruder, S. P., and Fox, B. S. (1999). Tissue engineering bone. *Clin. Orthop. Relat. Res.* **367**, 68–83.
5. Aaboe, M., Pinholt, E. M., and Hjorting-Hansen, E. (1995). Healing of experimentally created defects: A review. *Br. J. Oral Maxillofac Surg.* **33**, 312–318.

6. Hutmacher, D., Hurzeler, M. B., and Schliephake, H. (1996). A review of material properties of biodegradable and bioresorbable polymers and devices for GTR and GBR applications. *Int. J. Oral Maxillofacial Implants* **11**, 667–678.

7. Boyan, B. D., Lohmann, C. H., Romero, J., and Schwarz, Z. (1999). Bone and cartilage tissue engineering. *Clin. Plast. Surg.* **26**, 629–645.

8. Ripamonti, U., Heever, V. D. B., Sampath, T. K., Tucker, M. M., Rueger, D. C., and Reddi, A. H. (1996). Complete regeneration of bone in the baboon by recombinant human osteogenic protein-1 (hOP-1, bone morphogenetic protein-7). *Growth Factors* **13**, 273–289.

9. Hollinger, J. O., and Winn, S. R. (1999). Tissue engineering of bone in the craniofacial complex. *Ann. N.Y. Acad. Sci.* **18**, 379–385.

10. Boyne, P. J., Marx, R. E., Nevins, M., Triplett, G., Lazaro, E., Lilly, L. C., Alder, M., and Nummikoski, P. (1997). A feasibility study evaluating rhBMP-2/absorbable collagen sponge for maxillary sinus floor augmentation. *Int. J. Periodont. Restorative Dent.* **17**, 11–25.

11. Howell, T. H., Fiorellini, J., Jones, A., Alder, M., Nummikoski, P., Lazaro, M., Lilly, L., and Cochran, D. (1997). A feasibility study evaluating rhBMP-2/absorbable collagen sponge device for local alveolar ridge preservation or augmentation. *Int. J.. Periodont. Restorative Dent.* **17**, 124–139.

12. Urist, M. R., and Dawson, E. (1981). Intertransverse process fusion with the aid of chemosterilized autolyzed antigen-extracted allogeneic (AAA) bone. *Clin. Orthop. Relat. Res.* **154**, 97–113.

13. Johnson, E. E., Urist, M. R., and Finerman, G. A. (1988). Repair of segmental defects of the tibia with cancellous bone grafts augmented with human bone morphogenetic protein. A preliminary report. *Clin. Orthop. Relat. Res.* **236**, 249–257.

14. Fleet, J. C., Cashman, K., Cox, K., and Rosen, V. (1996). The effects of aging on the bone inductive activity of recombinant human bone morphogenetic protein-2. *Endocrinology (Baltimore)* **137**, 4605–4610.

15. Cook, S. D., Wolfe, M. W., Salkeld, S. L., and Rueger, D. C. (1995). Effect of recombinant human osteogenic protein-1 on healing of segmental defects in non-human primates. *J. Bone. Jt. Surg. Am.* **77**, 734–750.

16. Beresford, J. N. (1989). Osteogenic stem cells and the stromal system of bone and marrow. *Clin. Orthop. Relat. Res.* **240**, 270–280.

17. Cummine, J., Armstrong, L., and Nade, S. (1983). Osteogenesis after bone and bone marrow transplantation. Studies of cellular behaviour using combined myelo-osseous grafts in the subscorbutic guinea pig. *Acta Orthop. Scand.* **54**, 235–241.

18. Ohgushi, H., Goldberg, V. M., and Caplan, A. I.(1989). Repair of bone defects with marrow cells and porous ceramic. Experiments in rats. *Acta Orthop. Scand.* **60**, 334–339.

19. Kinoshita,Y., Kirigakubo, M., Kobayashi, M., Tabata, T., Shimura, K., and Ikada, Y. (1993). Study on the efficacy of biodegradable poly(L-lactide) mesh for supporting transplanted particulate cancellous bone and marrow: Experimental involving subcutaneous implantation in dogs. *Biomaterials* **14**,729–736.

20. Schwarz, H. C. (1984). Mandibular reconstruction using the Dacron–urethane prosthesis and autogenic cancellous bone: Review of 32 cases. *Plast. Reconstr. Surg.***73**, 387–386.

21. Jackson, I. T., Scheker, L. R., Vandervord, J. G., and McLennan, J. G. (1981). Bone marrow grafting in the secondary closure of alveolar-palatal defects in children. *Br. J. Plast. Surg.* **34**, 422–425.

22. Inoue, K., Ohgushi, H., Yoshikawa, T., Okumura, M., Sempuku, T., Tamai, S., and Dohi, Y. (1997). The effect of aging on bone formation in porous hydroxyapatite: Biochemical and histological analysis. *J. Bone Miner. Res.* **12**, 989–994.

23. Quarto, R., Thomas, D., and Liang, C. T. (1995). Bone progenitor cell deficits and the age-associated decline in bone repair capacity. *Calcif. Tissue Int.* **56**, 123–129.

24. Tsuji, T., Hughes, F. J., McCulloch, C. A., and Melcher, A. H. (1990). Effects of donor age on osteogenic cells of rat bone marrow in vitro. *Mech. Ageing Dev.* **51**, 121–132.

25. Ishaug, S. L., Crane, G. M., Miller, M. J., Yasko, A. W., Yaszemski, M. J., and Mikos, A. G. (1997). Bone formation by three-dimensional stromal osteoblast culture in biodegradable polymer scaffolds. *J. Biomed. Mater. Res.* **36**, 17–28.

26. Ohgushi, H., and Caplan, A. I., (1999). Stem cell technology and bioceramics: From cell to gene engineering. *J. Biomed. Mater. Res.* **48**, 913–927.

27. Pihlajamaki, H., Bostman, O. M., Hirvensalo, E., Tormola, P., and Rokkanen, P. (1992). Absorbable pins of self-reinforced poly-L-lactide acid for fixation of fractures and osteotomies. *J. Bone Jt. Surg.* **74**, 853–857.

28. Cordewener, F. W., Bos, R. R. M., Rozema, F. R., and Houtman, W. A. (1996). Poly(L-lactide) implants for repair of human orbital floor defects. *J. Oral Maxillofacial Surg.* **54**, 9–13.

29. Bostman, O. M. (1998). Osteoarthritis of the ankle after foreign-body reaction to absorbable pins and screws. A three- to nine-year follow-up study. *J. Bone Jt. Surg. Br. Vol.* **80B**, 333–338.

30. Kallela, P., Laine, R., Surronen, P., Ranta, T., Iizuka, T., and Lindqvist, C. (1999). Osteotomy site healing following mandibular sagittal split osteotomy and rigid fixation with polylactide biodegradable screws. *Int. J. Oral Maxillofacial Surg.* **28**,166-170.

31. Hollinger, J. O., and Battistone, C. C. (1986). Biodegradable bone repair materials. Synthetic polymers and ceramics. *Clin. Orthop. Relat. Res.* **207**, 290–305.

32. Pihlajamaki, H., Bostman, O. M., Hirvensalo, E., Tormola, P., and Rokkanen, P. (1992). Absorbable pins of self-reinforced poly-L-lactide acid for fixation of fractures and osteotomies. *J. Bone Jt. Surg.* **74**, 853–857.

33. Suuronen, R., Pohjonen, T., Hietanen, J., and Lindqvist, C. (1998). A 5-year in vivo study of the biodegradation of polylactide plates. *J. Oral Maxillofacial Surg.* **56**, 604–614.

34. Bergsma, J. E., Bruijin, W. C. Rozema, F. R., Bos, R. R. M., and Boering, G. (1995). Late degradation tissue response to poly(L-lactide) bone and screws. *Biomaterials* **16**, 25–31.

35. Matsusue, T., Hanafusa, S., Yamamura, T., Shikinami, Y., and Ikada, Y. (1995). Tissue reaction of bioabsorbable ultra high strength poly(L-Lactide) rod. *Clin. Orthop. Relat. Res.* **317**, 246–253.

36. Sevastjanova, N. A., Mansurova, L. A., Dombrovska, L. E., and Slutsski, L. I. (1987). Biochemical of characterization of connective tissue reaction to synthetic polymer implants. *Biomaterials* **8**, 242–246.

37. Bergsma, J. E., Rozema, F. R., Bos, R. R. M., Boering, G., Bruijin, W. C., and Penning, A. J. (1995). In vivo degeneration and biocompatibility study of in vitro pre-degraded as-polymerized polylactide particles. *Biomaterials* **16**, 267–274.

38. Pinster, H., Hoppert, T., Gutwald, R., Mulling, J., and Reuther, J. (1994). Biodegradation von Polylactid-Osteosynthese-Materialien im Langseitversuch. *Dtsch. Zahn- Mund, Kiefergeschichts. Chir.* **18**, 50–53.

39. Dunnen, W. F. A., Robinson, P. H., Wessel, R., Pennings, A. J., Leeuwen, M. B. M., and Schakenraad, J. M. (1997). Long term evaluation of degradation and foreign-body reaction of subcutaneously implanted poly(DL-lactide-ε-caprolactone). *J. Biomed. Mater. Res.* **36**, 337–346.

40. Vert, M., Splenehauer, S.M., Li, G., and Guérin, P. (1992). Bioresorbability and biocompability of aliphatic polyesters. *J. Mater. Sci. Mater. Med.* **3**, 432–446.

41. Kinoshita, Y., Kobayashi, M., Hidaka, T., Shimura, K., and Ikada, Y. (1997). Reconstruction of mandibular continuity defects in dogs using poly(L-lactide) mesh and autogenic particulate cancellous bone and marrow: Preliminary report. *J. Oral Maxillofacial Surg.* **55**, 718–723.

42. Kinoshita, Y., Amagasa, T., Fujii, E., Ogura, I., Kudo, K., Okabe, S., Nagayama, M., Furuta, I., and Totsuka, Y. (1999). Reconstruction of the mandible with poly(L-lactide) mesh tray/sheet and transplantation of particulate cancellous bone and marrow. *Int. J. Oral Maxillofacial Surg.* **28**(Suppl. 1), 161.

43. Kinoshita, Y., Kobayashi, M., Fukuoka, S., Yokoya, S., and Ikada, Y. (1996). Functional reconstruction of the jawbones using poly(L-lactide) mesh and autogenic particulate cancellous bone and marrow. *Tissue Eng.* **2**, 327–341.

44. Wang, J. S., and Aspenberg, P. (1996). Basic fibroblast growth factor infused at different times during bone graft incorporation. Titanium chamber study in rats. *Acta Orthop Scand.* **67**, 229–361.

45. Kato, H., Matsuo, R., Komiyama, O., Tanaka, T., Inazu, M., Kitagawa, H., and Yoneda, T. (1995). Decreased mitogenic and osteogenic responsiveness of calvarial osteoblasts isolated from aged rats to basic fibroblast growth factor. *Gerontology Suppl.* **1**, 20–27.

46. Yamada, K., Tabata, Y., Yamamoto, K., Miyamoto, S., Nagata, I., Kikuchi, H., and Ikada, Y. (1997). Potential efficacy of basic fibroblast growth factor incorporated in biodegradable hydrogels for skull bone regeneration. *J. Neurosurg.* **86**, 871–875.

47. Tabata, Y., Yamada, K., Miyamoto, S., Nagata, I., Kikuchi, H., Aoyama, I., Tamura, M., and Ikada, Y. (1998). Bone regeneration by basic fibroblast growth factor complexed with biodegradable hydrogels. *Biomaterials* **19**, 807–815.

48. Tabata, Y., Nagano, A., and Ikada, Y. (1999). Biodegradation of hydrogel carrier incorporating fibroblast growth factor. *Tissue Eng.* **5**, 127–138.

49. Puelacher, W. C., Vacanti, J. P., Ferraro, N. F., Schloo, B., and Vacanti, C. A. (1996). Femoral shaft reconstruction using tissue-engineered growth of bone. *Int. J. Oral Maxillofacial Surg.* **25**, 223–228.

50. Lieberman, J. R., Le, L. Q., Wu, L., Finerman, G. A., Berk, A., Witte, O. N., and Stevenson, S. (1998). Regional gene therapy with a BMP-2-producing murine stromal cell line induces heterotopic and orthotopic bone formation in rodents. *J. Orthop. Res.* **16**, 330–339.

CHAPTER 108

PERIODONTAL APPLICATIONS

William V. Giannobile and Stephen J. Meraw

R egeneration of tooth-supporting structures destroyed by periodontitis is a major goal of periodontal therapy. Specifically, reconstitution of lost alveolar bone, tooth root cementum, and periodontal ligament is necessary to repair periodontal defects. A challenge in the regeneration of periodontal tissues is to develop predictable therapies to repair large "critical size" alveolar bone defects. Over the past 20 years, researchers have utilized advances made in materials science and molecular biology to apply to periodontology. Examples of current therapies include guiding tissue membranes, bone autografts/allografts, and polypeptide growth factors. These devices and regenerative molecules are in varying stages of development to treat advanced periodontal disease from preclinical studies to FDA-approved therapies. This chapter discusses emerging therapies in the areas of materials science, growth factor biology, and cell/gene therapy. Results from preclinical and clinical trials are reviewed. The chapter concludes with a future perspective on the use of novel tissue engineering approaches such as gene delivery of signaling molecules.

APPLICATION OF BIOMIMETICS AND TISSUE ENGINEERING TO PERIODONTOLOGY

The rapidly emerging field of tissue engineering has resulted in the development of a multitude of promising therapies in medicine [1]. Many of these approaches have reached dentistry in the promotion of tissue repair in the oral cavity, including soft and hard tissue defects. Oral tissue engineering has been applied to oral diseases and injuries such as salivary gland disorders and large craniofacial defects [2]. The rate-limiting step in the reengineering of periodontal structures is the targeting of signaling molecules and cells to the tooth root surface. The healing of a periodontal wound is complicated by several factors that limit predictable delivery of agents to the root surface; these include the perimucosal environment of the avascular mineralized tooth surface traversing the gingival soft tissue; a complex microbiota, which contaminates wounds at the soft–hard tissue interface and may affect release kinetics of delivered cells or molecules; occlusal forces on the tooth complex in transverse and axial planes, which disrupt the stability of the healing wound; the effects on the targeted delivery of devices (containing cells or factors) within these wounds; and the complexity of the tooth attachment apparatus (encompassing several stromal–cellular interactions) [3]. Therefore, the contributions of many factors, including signaling molecules, cells, scaffolds, and vasculature, dictate the degree of periodontal regeneration achieved (Fig. 108.1).

GRAFT MATERIALS AND DEVICES

The periodontium consists of those tissues involved in the support of the tooth. Most commonly, alteration of the periodontium results from a chronic infective process that leads to a slow, but progressive degeneration of the dental support. Advanced tissue destruction leading to tooth loss affects nearly 15% of the U.S. population [4]. Four different cellular phenotypes within the periodontium are capable of repopulation and direction of a regenerative response [5]. These four connective tissues are gingival connective tissue, periodontal ligament (PDL), cementum, and alveolar bone. The potential of regenerating lost periodontal support has led to numerous investigations, in addition to a paradigm shif from the

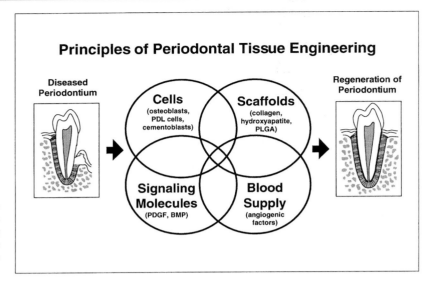

Fig. 108.1. Principles of periodontal tissue engineering. Critical components necessary for the reconstruction of lost periodontal support following decontamination of root surface microbial plaque include the combination of cells, scaffolds, signaling molecules, and blood supply.

formerly accepted end point of stability of compromised support to the more current view of regenerating lost periodontal structures (periodontal tissue engineering).

Procedures directed at regeneration of lost periodontium have included many techniques that focus on promoting a volumetric gain in alveolar bone, which serves as the foundational support for tooth anchorage. This has included the use of various graft materials, ranging from autogenous to heterogeneous, as well as alloplastic sources. In contrast to the transplantation or implantation of materials to the site of destroyed periodontia, other attempts at regeneration have examined healing following selective exclusion of proliferating epithelial cells from the potential space where bone formation is desired. The concept termed "guided tissue regeneration" originates from the orthopedic literature, which describes regeneration of bone defects following selective exclusion of the superficial soft tissues by placement of a barrier device to separate the supraperiosteal tissues from the underlying potential space in which osteogenesis is desired [6]. By placement of a barrier over the defect to inhibit ingrowth and dominance of the defect space by the rapidly migrating epithelial cells, the potential healing space for osteogenesis is maximized. Building upon these concepts, Melcher hypothesized the concept of directing a regenerative periodontal response by selective exclusion of the superficial soft tissues, epithelium, and gingival connective tissue, maximizing the potential defect space for regeneration of bone, PDL, and cementum [5] (Fig. 108.2).

BONE GRAFT BIOMATERIALS
Autografts

Autogenous bone grafts for periodontal regeneration have been widely used over the past 30 years, with donor tissue taken from a variety of intraoral and extraoral sites. The popularity of autogenous grafts is due to their osteoinductive, osteopromotive, and osteoconductive potential, which have resulted in demonstrable predictability of results in select clinical situations. Intraoral sources of autogenous bone have included the mandibular ramus, symphysis, healing tooth extraction sites, edentulous ridges, the maxillary tuberosity, and exostoses; extraoral sources have been restricted mainly to the iliac crest (reviewed in Brunsvold and Mellonig [7]). Studies in both animal and human models have demonstrated varying improvements in clinical parameters of tissue regeneration, including a percentage of defect bone fill that approaches complete or near-complete repair in some studies. Histological evidence of regenerated periodontal defects has shown that new PDL, cementum, and bone can be achieved with autogenous grafts [8]. Limitations in the use of autografts include a second surgical site and related potential increased morbidity, and volumetric limitations at the donor site.

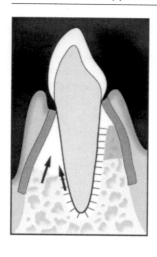

Fig. 108.2. Compartmentalized healing in periodontal wounds by Guided Tissue Regeneration. The barrier membrane separates superficial gingival connective tissue from underlying alveolar bone (left arrow) and periodontal ligament (right arrow) to allow their depopulation of the wound and repair the defect (right side of tooth).

Allografts

Attempts to address concerns of more extensive surgical procedures to procure autogenous grafts and limitations in the amount available for use have led to exploration of alternative sources of graft material. Bone allografts, which are commercially available, obviate concerns of volume restriction. The role of these materials in the healing of periodontal defects is controversial, however: although allografts are considered to be primarily osteoconductive biomaterials, some studies have supported the concept that they also possess limited osteoinduction [9,10]. The rationale that bone allografts may additionally be osteoinductive stems from studies showing that significant levels of bone morphogenetic proteins and growth factors are found in the graft material, possibly enhancing osteogenesis [11]. Histological evaluation of sites grafted with bone allografts in humans has demonstrated new PDL, root cementum, and bone [8]. However, other histological studies have shown the presence of nonresorbed residual graft material that failed to be replaced by new periodontal tissue [12].

The concern of disease transmissibility through use of foreign tissue as a graft source has also been addressed. Standard laboratory processing of these materials has been demonstrated to render HIV-spiked and AIDS donor bone completely free of detectable virus [13]. Variability also exists between allograft allotments secondary to the status of the individual donor; thus the question of consistent predictability of results may arise [9]. Several bone banks now test for osteoinductivity to ensure greater consistency of bioactivity between lots.

Alloplasts

The use of biocompatible synthetic graft materials is also advantageous with respect to volume availability and offers the added advantage of reduced potential for disease transmission. Several alloplasts are available for use in periodontal grafting as osteoconductive materials, most containing calcium phosphate ceramic compounds such as tricalcium phosphate (TCP), hydroxyapatite (HA), or biological "glasses" (e.g., Perioglas, Biogran). While several studies have shown that these materials hold promise with respect to improvement in clinical parameters [14], histological evidence from many groups has shown fibrous encapsulation of the graft particles, and slow resorption without regeneration of periodontal tissues [15]. Most clinicians consider alloplasts to be inert fillers that promote new clinical attachment without significant stimulation of osteogenesis. Further study is required to assess predictability of these materials in periodontal regeneration.

GUIDED TISSUE REGENERATION

The reconstitution of periodontal tissues through directed cellular exclusion has become an active area of investigation. Initial attempts looked at delaying the entry of the epithelium and gingival connective tissue into the periodontal defect by removing these tissues directly over the defect, although the tissues from the surrounding defect wound margins would eventually repopulate the area [16]. Current approaches aim to actually exclude these cells from the defect [17]. In this manner, a cell-occlusive barrier is employed to "compartmentalize" the periodontal defect, maximizing the potential volume for regeneration of PDL cells, cementoblasts, and osteoblasts, while eliminating or restricting competition into the defect from cellular phenotypes external to the barrier (epithelium and gingival connective tissue).

Cell-occlusive barriers used in GTR therapy are designed to remain in place long enough for the formation of new periodontal attachment, or for regeneration of PDL, bone, and cementum to occur. Clinical and histological evidence has demonstrated that periodontal new attachment may occur within 5–6 weeks after membrane placement [18]. Many initial studies utilized barriers, which were biocompatible, nonresorbable materials that included cellulose [19] and expanded poly(tetrafluorothylene) (ePTFE) membranes [20]. While the barriers proved to be useful in their ability to maintain a stable barrier function and to compartmentalize the defect, an additional surgical procedure is necessary to remove the membrane. Prolonged periods prior to membrane removal may result in an inflammatory response, potential bacterial contamination, and infection, followed by dehiscence of the soft tissues overlying the membrane, which commonly occurs with these materials [21].

A second generation of barrier membranes has been developed to address the concerns stemming from the additional surgical procedure. Several synthetic polymer and collagen membranes have been designed to undergo slow hydrolysis, hence to disappear gradually from the tissue site without requiring a second surgical procedure. Common materials include poly(lactic acid), poly(glycolic acid), polyglactin, and both soluble and insoluble collagen. Clinical trials have demonstrated significant resolution of periodontal defects with both resorbable and nonresorbable membranes, with no significant difference between type of membrane [17]. Results of periodontal defect resolution by means of GTR have varied and may be dependent on the remaining surface area of the vascularized recipient osseous site. The combination use of bone graft materials beneath the cell-occlusive membrane has shown modest improvement in clinical studies in comparison to monotherapies in interradicular defects associated with a multirooted tooth [22]. Data to support enhancement of regeneration within intraosseous periodontal defects are insufficient, however [23]. Benefits considered to occur from combination use of bone grafts underneath membranes include the reduction of barrier collapse into existing defect space and maximization of the total area for regeneration.

SIGNALING MOLECULES AND ATTACHMENT FACTORS

SIGNALING MOLECULES

Polypeptide growth factors are biological mediators that regulate crucial intra- and extracellular events involved in tissue repair, such as DNA synthesis, chemotaxis, differentiation, and matrix biosynthesis. Growth factors (GFs) exert their effects by binding to specific cell surface receptors that transduce signals to the nucleus by means of transduction pathways involving tyrosine kinase or serine/threonine kinase phosphorylation. The mechanisms of growth factor mediated signal transduction in periodontal cells are reviewed by Saygin, *et al.* [3]. Examples of GFs found locally in the tooth-supporting apparatus during healing and development include platelet-derived growth factor (PDGF), transforming growth factort-β (TGF-β), basic fibroblast growth factor (FGF-2), insulin-like growth factor I (IGF-1), and the bone morphogenetic proteins (BMPs) [24]. Comprehensive reviews on growth factors in periodontal regeneration can be found elsewhere [25,26].

The expression of various GFs, cytokines, and chemokines following acute and chronic tissue injury (such as that found in inflammatory periodontal disease) is significant in governing tissue homeostasis. Over the past 10 years, a plethora of preclinical and human pilot investigations have tested the ability of various signaling molecules to stimulate periodontal and peri-implant tissue regeneration. Basic and clinical research is in progress to better understand the role of GFs in oral wound repair. Several studies have examined the effects of GFs on periodontal wound healing such as PDGF [27], BMP-2 [28–30], and BMP-7 (ostogenic protein 1: OP-1) [31,32]. The local delivery of signaling molecules may augment the physiological induction of many of these molecules and their corresponding cognate receptors in oral wounds.

Platelet-Derived Growth Factor Based Therapies

Platelet-derived growth factor (PDGF) is an important molecule in the promotion of wound repair. PDGF is released from platelets after injury and is produced by numerous cell types during tissue regeneration. PDGF mediates its signal via two distinct high-affinity transmembrane receptors possessing intrinsic tyrosine kinase activity, termed PDGFR-α and -β [33]. PDGF is a potent mitogen and chemoattractant for many cell types, including those derived from the periodontium such as periodontal ligament (PDL) fibroblasts, gingival fibroblasts, and osteoblasts [34]. Some of the earliest growth factor studies on periodontal regeneration examined PDGF or PDGF combined with IGF-1. These investigations, reporting on natural disease in dogs and experimental periodontitis in nonhuman primates, found that this growth factor combination promoted new bone, cementum, and periodontal ligament [35–39]. Howell and colleagues reported the results of the first human clinical trial testing the safety and efficacy of administering two recombinant human growth factors: rhPDGF and rhIGF-1 [27]. This study examined patients diagnosed with advanced periodontal disease who were treated with either a combination of rhPDGF-BB and rhIGF-1 or surgery alone. The results revealed that subjects treated with rhPDGF/rhIGF-1 produced greater than 2.5 times more bone than those whose lesions were treated by conventional surgery. The GFs were shown to be safe and well tolerated by the subjects. Furthermore, furcation lesions (area located between tooth roots) improved most favorably to therapy, with nearly a fourfold increase in bone volume compared with paired controls. These results in humans were found to be highly consistent in comparison to preclinical studies performed in nonhuman primates [39]. In addition, the PDGF/IGF-1 combination stimulates bone regeneration in press-fit [40] and immediate extraction socket [41] dental implant sites.

Bone Morphogenetic Proteins (BMPs)

Bone morphogenetic proteins belong to the TGF-β superfamily of proteins. Most BMP members (BMPs 2–15) are osteoinductive and transduce signals through serine–threonine receptors and phosphorylation of intracellular SMAD (Sma and Mad) proteins [42]. The BMPs have been extensively evaluated in orthopedic models to induce osteogenesis [43]. In the first human study using a BMP to promote periodontal regeneration a single application of BMP-3 (osteogenin) was combined with demineralized bone allograft in a submerged tooth model [44]. The investigators found increased bone and cementum deposition in BMP-3 treated, periodontally involved, submerged teeth as assessed by human histology. The results demonstrated a trend for enhancement when osteogenin was used to stimulate new bone and attachment; however, the improvement was not significantly better than the use of a bone allograft material alone.

Bone morphogenetic protein 2 is the most thoroughly researched member of the TGF-β superfamily for the promotion of periodontal [28–30] and peri-implant bone regeneration [45–47]. Sigurdsson *et al.* reported the effects of rhBMP-2 on periodontal regeneration and found that BMP-2, delivered in synthetic bioabsorbable particles, dramatically stimulated osteogenesis and cementogenesis [28]. Near-complete regeneration of bone was noted in supra-alveolar (through and through furcations, or class III) submerged bone defects. More recently, our group demonstrated closure of class III furcations with the application of BMP-7/OP-1 to nonsubmerged periodontal lesions [32]. Furthermore, heightened stimulation of cellular "regenerative cementum" could be noted at 8 weeks after application of BMP-7/OP-1 (Fig. 108.3). Other investigators have demonstrated regeneration of deep class II furcation defects in baboons [48] and tooth fenestration defects in rats [49–52]. BMPs have also shown great promise in other intraoral orthotopic sites for bone regeneration, such as the maxillary sinus [53–55], extraction socket defects [56], and peri-implant sites [45–47]. In the craniofacial region, significant enhancement of bone regeneration has been reported in the cranium [57], body of the mandible [58], and ramus [59]. These re-

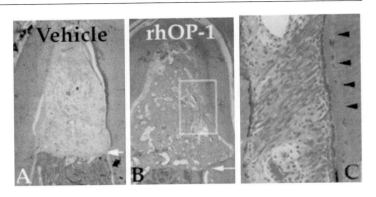

Fig. 108.3. Periodontal tissue regeneration stimulated by recombinant human osteogenic protein 1/bone manow protein 7 (rhOP-1/BMP-7) to treat critical sized supra-alveolar defects (0.5 cm high) created in the mondibles of dogs. Biopsy samples taken 2 months after surgery show results of application to osseons lesions of the Goldner's stained sagittal sections reveal the paucity of wound repair in vehicle, while rhOP-1/BMP lesions demonstrate profound regeneration of lost alveolar bone and periodontal ligament. (C) Higher magnification of (B), revealing formation of cellular "regenerative" cementum (arrows) along the root surface. Reprinted with permission from Giannobile et al. [32]

sults highlight the consistency of results obtained when BMPs are used in periodontal and craniofacial bone regeneration.

ATTACHMENT FACTORS
Enamel Matrix Proteins

Enamel matrix derivative (EMD) contains proteins belonging to the amelogenin family, which is the hydrophobic constituent of the enamel matrix proteins [60]. Early studies suggest that EMD is involved in the formation of acellular cementum during tooth development and that this matrix has the potential to induce regeneration of acellular cementum in periodontal disease [61]. EMD stimulates cellular proliferation, protein synthesis, and mineral nodule formation in several cell types, including PDL cells, osteoblasts, and cementoblasts [62,63]. In a tooth dehiscence model in nonhuman primates, Hammarström and colleagues tested the ability of EMD to affect periodontal wound healing [64]. When compared with carrier and EDTA root conditioning, the EMD-treated defects demonstrated significant enhancements of bone, cementum, and PDL. In human studies, EMD has demonstrated partial regeneration by human histology [65], safety in a multicenter trial of 10 test centers and 107 patients [66], and efficacy as shown by a placebo-controlled human trial of 33 subjects with paired intrabony defecte [67], The third study assessed EMD therapy coupled with flap surgery on periodontal wound healing and measured clinical attachment level change and subtraction radiography [67]. The EMD therapy promoted 66% defect fill that was maintained 3 years post-therapy, while paired control defects failed to show a change in radiographic bone level. Expanded studies in larger patient populations will be needed to further assess the effects of EMD treatment in intrabony periodontal defects.

P-15 Collagen Peptides

Work in both the preclinical and clinical arenas has suggested that the use of synthetic peptides relating to the biologically active domain of type I collagen promotes the attachment of mesenchymal cells to bone mineral substrata. The concept exploits the notion that collagen modulates cell proliferation, differentiation, and cell attachment. Collagen is frequently used as a substratum in three-dimensional culture systems to promote physiological behavior of cells in a situation similar to the normal wound healing environment. Qian and Bhatnagar first reported the use of a synthetic 15 amino acid sequence of type I collagen contained in residues 766–780 of the α_1 chain of type I collagen [68]. A series of reports demonstrated the potent ability of this molecule, termed P-15, to enhance cell attachment and proliferation to bone mineral. Furthermore, Bhatnagar *et al.* demonstrated the design of "biomimetic habitats" for tissue engineering with P-15 for periodontal ligament cells [69].

By constructing three-dimensional templates of P-15 with bovine anorganic bone mineral (ABM), they were able to develop a system for expanding PDL cells. This concept led to the development of a human clinical trial evaluating P-15 administration in periodontal subjects. Yukna *et al.* showed the stimulation of new bone and attachment by the single administration of P-15/ ABM in subjects with advanced periodontal disease [70]. P-15/ABM offers promise in the reconstruction of periodontal lesions.

CELL AND GENE THERAPY

CELL THERAPY

To date, cell therapy has not been employed to any great extent in periodontology. However, when one considers autologous bone grafting, for example, this is a form of cell therapy that also includes scaffolds for osteoinduction as well as signaling molecules. Applications for the development and characterization of tissue-engineered human oral mucosa equivalent were described by Izumi *et al.* in 2000 [71]. Thus, the premise is that expansion of cells in an *ex vivo* environment can be expanded *in vitro* for subsequent transplantation in areas of insufficient donor tissue.

Somerman and colleagues have also demonstrated the potential usage of cell therapy with cloned cementoblasts [72,73]. In a series of reports, the group has cloned and characterized a cementoblast cell line that possesses many of the phenotypic characteristics of tooth lining cells *in vivo*. By utilizing strategies to better understand the basic biological mechanisms involved in cementogenesis, strategies to improve periodontal wound healing may be developed [74].

GENE TRANSFER

The success of tissue engineering relies on the large-scale purification and production of signaling molecules, as well as methods to deliver these factors to their targets [75]. One problem with current growth factor delivery to periodontal wounds is the extremely short half-life of the factors. The factors remain in the periodontal defect only briefly, presumably owing to proteolytic breakdown, receptor-mediated endocytosis, and the solubility of the delivery vehicle. Therefore, the use of DNA delivery systems may serve as an alternative method of targeting proteins to an oral wound, since existing protein delivery systems provide such a short duration of action of the applied GF. Gene therapy has been applied to several diseases that result in significant tissue deficiencies. Therefore, gene therapy to promote repair and regeneration has become an active area of research. These somatic gene therapeutic approaches rely on two critical steps: first to deliver DNA to the appropriate target cells and second to elicit the therapeutic expression of transgene-encoded proteins [76]. In the context of wound repair, a transient expression of the transgene may be optimal to restore the tissue defect. Examples of methods of gene delivery for short-term expression of genes include adenovirus and DNA complexes [77]. Since the regulation of wound repair occurs in a controlled fashion over a short period of time, a goal of gene therapy in a compromised wound (i.e., periodontal disease) may benefit from an elevated and sustained production of GFs to promote tissue repair [78]. Under normal circumstances, periodontal lesions represent "critical size defects" (i.e., wounds that cannot completely repair), since the chronic disease process has prevented the tissues from predictably regenerating once significant disease has commenced [76]. The strategy seeks to utilize gene transfer to insert GFs or wound-healing cytokines into transiently expressing viral vectors or direct plasmid delivery by *in vivo* approaches. The use of this system is further substantiated on the basis of an attempt to more accurately deliver the growth factors for the length of time they are expressed following injury (up to 2 weeks as shown for PDGF and BMPs) [79–81]. Several groups have used plasmid DNA to deliver GF genes to healing skin [82,83], bone [84,85], and periodontal [78] wounds. The use of biodegradable polymers to deliver PDGF DNA offers promise in tissue repair [86]. Other groups have used *ex vivo* approaches to deliver BMP genes via adenovirus to orthotopic wounds, with impressive results [87–89]. Eriksson *et al.* have developed unique methods of transducing wounds by the *in vivo*, microseeding technique [90,91].

Much work is needed to optimize GF delivery by extending the duration of gene expression and increasing the number of cells transduced by the genes and, ultimately, to determine whether wound healing can be enhanced by gene transfer techniques. This possibility provides an exciting future for reconstructive periodontal therapy.

REFERENCES

1. Vacanti, J. P., and Langer, R. (1999). Tissue engineering: The design and fabrication of living replacement devices for surgical reconstruction and transplantation. *Lancet* 354(Suppl. 1), SI32–SI34.
2. Baum, B. J., and Mooney, D. J. (2000). The impact of tissue engineering on dentistry. *J. Am. Dent. Assoc.* 131, 309–318.
3. Saygin, N. E., Giannobile, W. V., and Somerman, M. J. (2000). Cell and molecular biology of cementum. *Periodontology 2000* 24, 73–98.
4. Williams, R. C. (1990). Periodontal disease. *N. Engl. J. Med.* 322, 373–382.
5. Melcher, A. H. (1976). On the repair potential of periodontal tissues. *J. Periodontol.* 47, 256–260.
6. Hurley, L. A., Stinchfield, F. E., Basset, A. L., and Lyon, W. (1959). The role of soft tissues in osteogenesis: An experimental study of canine spinal fusions. *J. Bone Jt. Surg. Am. Vol.* 41-A, 1243–1254.
7. Brunsvold, M. A., and Mellonig, J. T. (1999). Bone grafts and periodontal regeneration. *Periodontology 2000* 1, 80–91.
8. Bowers, G. M., Chadroff, B., Carnevale, R., Mellonig, J., Corio, R., Emerson, J., Stevens, M., and Romberg, E. (1989). Histologic evaluation of new attachment apparatus formation in humans. Part III. *J. Periodontol.* 60, 683–693.
9. Schwartz, Z., Somers, A., Mellonig, J. T., Carnes, D. L., Jr., Dean, D. D., Cochran, D. L., and Boyan, B. D. (1998). Ability of commercial demineralized freeze-dried bone allograft to induce new bone formation is dependent on donor age but not gender. *J. Periodontol.* 69, 470–478.
10. Schwartz, Z., Mellonig, J.T., Carnes, D. L., Jr., de la Fontaine, J., Cochran, D. L., Dean, D. D., and Boyan, B. D. (1996). Ability of commercial demineralized freeze-dried bone allograft to induce new bone formation. *J. Periodontol.* 67, 918–926.
11. Shigeyama, Y., D'Errico, J. A., Stone, R., and Somerman, M. J. (1995). Commercially-prepared allograft material has biological activity in vitro. *J. Periodontol.* 66, 478–487.
12. Becker, W., Becker, B. E., and Caffesse, R. (1994). A comparison of demineralized freeze-dried bone and autologous bone to induce bone formation in human extraction sockets. *J. Periodontol.* 65, 1128–1133, erratum: *Ibrd.* 66(4), 309 (1995).
13. Mellonig, J. T., Prewett, A. B., and Moyer, M. P. (1992). HIV inactivation in a bone allograft. *J. Periodontol.* 63, 979–983.
14. Lovelace, T. B., Mellonig, J. T., Meffert, R. M., Jones, A. A., Nummikoski, P. V., and Cochran, D. L. (1998). Clinical evaluation of bioactive glass in the treatment of periodontal osseous defects in humans. *J. Periodontol.* 69, 1027–1035.
15. Garrett, S. (1996). Periodontal regeneration around natural teeth. *Ann. Periodontol.* 1, 621–666.
16. Prichard, J. P. (1983). The diagnosis and management of vertical bony defects. *J. Periodontol.* 54, 29–35.
17. Wang, H. L., and MacNeil, R. L. (1998). Guided tissue regeneration. Absorbable barriers. *Dent. Clin. North Am.* 42, 505–522.
18. Stahl, S. S., Froum, S., and Tarnow, D. (1990). Human histologic responses to guided tissue regenerative techniques in intrabony lesions. Case reports on 9 sites. *J. Clin. Periodontol.* 17, 191–198.
19. Nyman, S., Lindhe, J., Karring, T., and Rylander, H. (1982). New attachment following surgical treatment of human periodontal disease. *J. Clin. Periodontol.* 9, 290–296.
20. Gottlow, J., Nyman, S., Karring, T., and Lindhe, J. (1984). New attachment formation as the result of controlled tissue regeneration. *J. Clin. Periodontol.* 11, 494–503.
21. Murphy, W. L., and Mooney, D. J. (1999). Controlled delivery of inductive proteins, plasmid DNA and cells from tissue engineering matrices. *J. Periodontal Res.* 34, 413–419.
22. Anderegg, C. R., Martin, S. J., Gray, J. L., Mellonig, J. T., and Gher, M. E. (1991). Clinical evaluation of the use of decalcified freeze-dried bone allograft with guided tissue regeneration in the treatment of molar furcation invasions. *J. Periodontol.* 62, 264–268.
23. Mellado, J. R., Salkin, L. M., Freedman, A. L., and Stein, M. D. (1995). A comparative study of ePTFE periodontal membranes with and without decalcified freeze-dried bone allografts for the regeneration of interproximal intraosseous defects. *J. Periodontol* 66, 751–755.
24. Thomadakis, G., Ramoshebi, L. N., Crooks, J., Rueger, D. C., and Ripamonti, U. (1999). Immunolocalization of bone morphogenetic protein-2 and -3 and osteogenic protein-1 during marine tooth root morphogenesis and in other craniofacial structures. *Eur. J. Oral. Sci.* 107, 368–377.
25. McCauley, L. K., and Somerman, M. J. (1998). Biologic modifiers in periodontal regeneration. *Dent. Clin. North. Am.* 42, 361–387.
26. Cochran, D. L., and Wozney, J. M. (1999). Biological mediators for periodontal regeneration. *Periodontology 2000* 19, 40–58.
27. Howell, T. H, Fiorellini, J. P., Paquette, D. W., Offenbacher, S., Giannobile, W. V., and Lynch, S. E. (1997). A phase I/II clinical trial to evaluate a combination of recombinant human platelet-derived growth factor-BB and recombinant human insulin-like growth factor-I in patients with periodontal disease. *J. Periodontol.* 68, 1186–1193.
28. Sigurdsson, T. J., Lee, M. B., Kubota, K., Turek, T. J., Wozney, J. M., and Wikesjo, U. M. (1995). Periodontal repair in dogs: Recombinant human bone morphogenetic protein-2 significantly enhances periodontal regeneration. *J. Periodontol.* 66, 131–138.
29. Sigurdsson, T. J., Nygaard, L., Tatakis, D. N., Fu, E., Turek, T. J., Jin, L., Wozney, J. M., and Wikesjo, U. M. (1996). Periodontal repair in dogs: Evaluation of rhBMP-2 carriers. *Int. J. Periodont. Restorative Dent.* 16, 524–537.

30. Kinoshita, A., Oda, S., Takahashi, K., Yokota, S., and Ishikawa, I. (1997). Periodontal regeneration by application of recombinant human bone morphogenetic protein-2 to horizontal circumferential defects created by experimental periodontitis in beagle dogs. *J. Periodontol.* **68**, 103–109.

31. Ripamonti, U., Heliotis, M., Rueger, D. C., and Sampath, T. K. (1996). Induction of cementogenesis by recombinant human osteogenic protein-1 (HOP-1/BMP-7) in the baboon (*Papio ursinus*). *Arch. Oral. Biol.* **41**, 1211–126.

32. Giannobile, W. V., Ryan, S., Shih, M. S., Su, D. L., Kaplan, P. L., and Chan, T. C. (1998). Recombinant human osteogenic protein-1 (OP-1) stimulates periodontal wound healing in class III furcation defects. *J. Periodontol.* **69**, 129–137.

33. Rosenkranz, S., and Kazlauskas, A. (1999). Evidence for distinct signaling properties and biological responses induced by the PDGF receptor alpha and beta subtypes. *Growth Factors* **16**, 201–216.

34. Piche, J. E., and Graves, D. T. (1989). Study of the growth factor requirements of human bone-derived cells: A comparison with human fibroblasts. *Bone* **10**, 131–138.

35. Lynch, S. E., de Castilla, G. R., Williams, R. C., Kiritsy, C. P., Howell,T. H., Reddy, M. S., and Antoniades, H. N. (1991). The effects of short-term application of a combination of platelet-derived and insulin-like growth factors on periodontal wound healing. *J. Periodontol.* **62**, 458–467.

36. Rutherford, R. B., Niekrash, C. E., Kennedy, J. E., and Charette, M. F. (1992). Platelet-derived and insulin-like growth factors stimulate regeneration of periodontal attachment in monkeys. *J. Periodontal. Res.* **27**, 285–290.

37. Giannobile, W. V., Finkelman, R. D., and Lynch, S. E. (1994). Comparison of canine and non-human primate animal models for periodontal regenerative therapy: Results following a single administration of PDGF/IGF-I. *J. Periodontol.* **65**, 1158–1168.

38. Park, J. B., Matsuura, M., Han, K. Y., Norderyd, O., Lin, W. L., Genco, R. J., and Cho, M. I. (1995). Periodontal regeneration in class III furcation defects of beagle dogs using guided tissue regenerative therapy with platelet-derived growth factor. *J. Periodontol.* **66**, 462–477.

39. Giannobile, W. V., Hernandez, R. A., Finkelman, R. D., Ryan, S., Kiritsy, C. P., D'Andrea, M., and Lynch, S. E. (1996). Comparative effects of platelet-derived growth factor-BB and insulin-like growth factor-I, individually and in combination, on periodontal regeneration in *Macaca fascicularis*. *J. Periodontal. Res.* **31**, 301–312.

40. Lynch, S. E., Buser, D., Hernandez, R. A., Weber, H. P., Stich, H., Fox, C. H., and Williams, R. C. (1991). Effects of the platelet-derived growth factor/insulin-like growth factor-I combination on bone regeneration around titanium dental implants. Results of a pilot study in beagle dogs. *J. Periodontol.* **62**, 710–716.

41. Becker, W., Lynch, S. E., Lekholm, U., Becker, B. E., Caffesse, R., Donath, K., and Sanchez, R. (1992). A comparison of ePTFE membranes alone or in combination with platelet-derived growth factors and insulin-like growth factor-I or demineralized freeze-dried bone in promoting bone formation around immediate extraction socket implants. *J. Periodontol.* **63**, 929—940.

42. Sakou, T. (1998). Bone morphogenetic proteins: From basic studies to clinical approaches. *Bone* **22**, 591–603.

43. Reddi, A. H. (1998). Initiation of fracture repair by bone morphogenetic proteins. *Clin. Orthop. Relat. Res., Suppl.*, S66–S72.

44. Bowers, G., Felton, F., Middleton, C., Glynn, D., Sharp, S., Mellonig, J., Corio, R., Emerson, J., Park, S., Suzuki, J., *et al.* (1991). Histologic comparison of regeneration of human intrabony defects when osteogenin is combined with demineralized freeze-dried bone allograft and with purified bovine collagen. *J. Periodontol.* **62**, 690–702.

45. Sigurdsson, T. J., Fu, E., Tatakis, D. N., Rohrer, M. D., and Wikesjo, U. M. (1997). Bone morphogenetic protein-2 for peri-implant bone regeneration and osseointegration. *Clin. Oral Implants Res.* **8**, 367–374.

46. Hanisch, O., Tatakis, D. N., Boskovic, M. M., Rohrer, M. D., and Wikesjo, U. M. (1997). Bone formation and re-osseointegration in peri-implantitis defects following surgical implantation of rhBMP-2. *Int. J. Oral Maxillofac. Implants* **12**, 604–610.

47. Cochran, D. L., Schenk, R., Buser, D., Wozney, J. M., and Jones, A. A. (1999). Recombinant human bone morphogenetic protein-2 stimulation of bone formation around endosseous dental implants. *J. Periodontol.* **70**, 139–150.

48. Ripamonti, U., Heliotis, M., van den Heever, B., and Reddi, A. H. (1994). Bone morphogenetic proteins induce periodontal regeneration in the baboon (*Papio ursinus*). *J. Periodontol. Res.* **29**, 439–445, erratum: *Ibid.* 30(2), 149–151 (1995).

49. King, G. N., King, N., Cruchley, A. T., Wozney, J. M., and Hughes, F. J. (1997). Recombinant human bone morphogenetic protein-2 promotes wound healing in rat periodontal fenestration defects. *J. Dent. Res.* **76**, 1460—1470.

50. King, G. N., King, N., and Hughes, F. J. (1998). The effect of root surface demineralization on bone morphogenetic protein-2-induced healing of rat periodontal fenestration defects. *J. Periodontol.* **69**, 561–570.

51. King, G. N., King, N., and Hughes, F. J. (1998). Effect of two delivery systems for recombinant human bone morphogenetic protein-2 on periodontal regeneration in vivo. *J. Periodontal. Res.* **33**, 226–236.

52. King, G. N., and Hughes, F. J. (1999). Effects of occlusal loading on ankylosis, bone, and cementum formation during bone morphogenetic protein-2-stimulated periodontal regeneration in vivo. *J. Periodontol.* **70**, 1125–1135.

53. Nevins, M., Kirker-Head, C., Wozney, J. A., Palmer, R., and Graham, D. (1996). Bone formation in the goat maxillary sinus induced by absorbable collagen sponge implants impregnated with recombinant human bone morphogenetic protein-2. *Int. J Periodont. Restorative Dent.* **16**, 8–19.

54. Hanisch, O., Tatakis, D. N., Rohrer, M. D., Wohrle, P. S., Wozney, J. M., and Wikesjo, U. M. (1997). Bone formation and osseointegration stimulated by rhBMP-2 following subantral augmentation procedures in nonhuman primates. *Int. J. Oral. Maxillofac. Implants* **12**, 785–792.

55. Boyne, P. J., Marx, R. E., Nevins, M., Triplett, G., Lazaro, E., Lilly, L. C., Alder, M., and Nummikoski, P. (1997). A feasibility study evaluating rhBMP-2/absorbable collagen sponge for maxillary sinus floor augmentation. *Int. J. Periodont. Restorative Dent.* **17**, 11–25.

56. Howell, T. H., Fiorellini, J., Jones, A., Alder, M., Nummikoski, P., Lazaro, M., Lilly, L., and Cochran, D. (1997). A feasibility study evaluating rhBMP- 2/absorbable collagen sponge device for local alveolar ridge preservation or augmentation. *Int. J. Periodont. Restorative Dent.* **17**, 124–139.

57. Ripamonti U., Van Den Heever, B., Sampath, T. K., Tucker, M. M., Rueger, D. C., and Reddi, A. H. (1996). Complete regeneration of bone in the baboon by recombinant human osteogenic protein-1 (hOP-1, bone morphogenetic protein-7). *Growth Factors* **13**, 273–289.

58. Toriumi, D. M., Kotler, H. S., Luxenberg, D. P., Holtrop, M. E., and Wang, E. A. (1991). Mandibular reconstruction with a recombinant bone-inducing factor. Functional, histologic, and biomechanical evaluation. *Arch. Otolaryngol. Head Neck Surg.* **117**, 1101–1112.

59. Higuchi, T., Kinoshita, A., Takahashi, K., Oda, S., and Ishikawa, I. (1999). Bone regeneration by recombinant human bone morphogenetic protein-2 in rat mandibular defects. An experimental model of defect filling. *J. Periodontol.* **70**, 1026–1031.

60. Fisher, L. W., and Termine, J. D. (1985). Noncollagenous proteins influencing the local mechanisms of calcification. *Clin. Orthop. Relat. Res.* **200**, 362–385.

61. Hammarström, L., Heijl, L., and Gestrelius, S. (1997). Periodontal regeneration in a buccal dehiscence model in monkeys after application of enamel matrix proteins. *J. Clin. Periodontol.* **24**, 669–677.

62. Gestrelius, S., Andersson, C., Lidstrom, D., Hammarström, L., and Somerman, M. (1997). In vitro studies on periodontal ligament cells and enamel matrix derivative. *J. Clin. Periodontol.* **24**, 685–692.

63. Tokiyasu, Y., Takata, T., Saygin, N. E., and Somerman, M. J. (2000). Enamel factors regulate expression of genes associated with cementoblasts. *J. Periodontol.* **71**, 1591–1600.

64. Hammarström, L. (1997). Enamel matrix, cementum development and regeneration. *J. Clin. Periodontol.* **24**, 658–668.

65. Heijl, L. (1997). Periodontal regeneration with enamel matrix derivative in one human experimental defect. A case report. *J. Clin. Periodontol.* **24**, 693–696.

66. Zetterstrom, O., Andersson, C., Eriksson, L., Fredriksson, A., Friskopp, J., Heden, G., Jansson, B., Lundgren, T., Nilveus, R., Olsson, A., Renvert, S., Salonen, L., Sjostrom, L., Winell, A., Ostgren, A., and Gestrelius, S. (1997). Clinical safety of enamel matrix derivative (EMDOGAIN) in the treatment of periodontal defects. *J. Clin. Periodontol.* **24**, 697–704.

67. Heijl, L., Heden, G., Svärdstrom, G., and Östgren, A. (1997). Enamel matrix derivative (EMDOGAIN®) in the treatment of intrabony periodontal defects. *J. Clin. Periodontol.* **24**, 705–714.

68. Qian, J. J., and Bhatnagar, R. S. (1996). Enhanced cell attachment to anorganic bone mineral in the presence of a synthetic peptide related to collagen. *J. Biomed. Mater. Res.* **31**, 545–554.

69. Bhatnagar, R. S., Qian, J. J., Wedrychowska, A., Sadeghi, M., Wu, Y. M., and Smith, N. (1999). Design of biomimetic habitats for tissue engineering with P-15, a synthetic peptide analogue of collagen. *Tissue Eng.* **5**, 53–65.

70. Yukna, R. A., Callan, D. P., Krauser, J. T., Evans, G. H., Aichelmann-Reidy, M. E., Moore, K., Cruz, R., and Scott, J. B. (1998). Multi-center clinical evaluation of combination anorganic bovine-derived hydroxyapatite matrix (ABM) cell binding peptide (P-15) as a bone replacement graft material in human periodontal osseous defects. 6 month results. *J. Periodontol.* **69**, 655–663.

71. Izumi, K., Terashi, H., Marcelo, C. L., and Feinberg, S. E. (2000). Development and characterization of a tissue-engineered human oral mucosa equivalent produced in a serum-free culture system. *J. Dent. Res.* **79**, 798–805.

72. D'Errico, J. A., MacNeil, R. L., Takata, T., Berry, J., Strayhorn, C., and Somerman, M. J. (1997). Expression of bone associated markers by tooth root lining cells, in situ and in vitro. *Bone* **20**, 117–126.

73. D'Errico, J. A., Ouyang, R., Berry, J. E., MacNeil, R. L., Strayhorn, C., Imperiale, M. J., Harris, N. L., Goldberg, H., and Somerman, M. J. (1999). Immortalized cementoblasts and periodontal ligament cells in culture. *Bone* **25**, 39–47.

74. Somerman, M. J., Ouyang, H. J., Berry, J. E., Saygin, N. E., Strayhorn, C. L., D'Errico, J. A., Hullinger, T., and Giannobile, W. V. (1999). Evolution of periodontal regeneration: From the roots point of view. *J. Periodontal Res.* **34**, 420–424.

75. Langer, R., and Vacanti, J. P. (1993). Tissue engineering. *Science* **260**, 920–926.

76. Bonadio, J., Goldstein, S. A., and Levy, R. J. (1998). Gene therapy for tissue repair and regeneration. *Adv. Drug. Del. Rev.* **33**, 53–69.

77. Schmid, S. I., and Hearing, P. (1999). Adenovirus DNA packaging: Construction and Analysis. *In* "Adenovirus Methods and Protocols" (W. S. M. Wold, ed.), pp. 47–60. Humana Press, Totowa, NJ.

78. Giannobile, W. V. (1999). Periodontal tissue regeneration by polypeptide growth factors and gene transfer. *In* "Tissue Engineering: Applications in Maxillofacial Surgery and Periodontics" (S. E. Lynch, R. J. Genco, and R. E. Marx, eds.), Vol. 1, pp. 231–243. Quintessence, Chicago.

79. Antoniades, H. N., Galanopolous, T., Neville-Golden, J., Kiritsy, C. P., and Lynch, S. E. (1991). Injury induces in vivo expression of platelet-derived growth factor (PDGF) and PDGF receptor mRNAs in skin epithelial cells and PDGF mRNA in connective tissue fibroblasts. *Proc. Natl. Acad. Sci. U.S.A.* **88**, 565–569.

80. Nakase, T., Nomura, S., Hashimoto, J., Yoshikawa, H., Takaoka, K. (1994). Transient and localized expression of bone morphogenetic protein-4 during fracture repair. *J. Bone Miner. Res.* **9**, 651–659.

81. Green, R. J., Usui, M. L., Hart, C. E., Ammons, W. F., and Narayanan, A. S. (1997). Immunolocalization of platelet-derived growth factor A and B chains and PDGF- alpha and beta receptors in human gingival wounds. *J. Periodontal Res.* **32**, 209–214.

82. Andree, C., Swain, W. F., Page, C. P., Macklin, M. D., Slama, J., Hatzis, D., and Eriksson, E. (1994). In vivo transfer and expression of a human epidermal growth factor gene accelerates wound repair. *Proc. Natl. Acad. U.S.A.* **91**, 12188–12192.

83. Eming, S. A., Whitsitt, J. S., He, L., Krieg, T., Morgan, J. R., and Davidson, J. M. (1999). Particle-mediated gene transfer of PDGF isoforms promotes wound repair. *J. Invest. Dermatol.* **112**, 297–302.

84. Fang, J., Zhu, Y. Y., Smiley, E., Bonadio, J., Rouleau, J. P., Goldstein, S. A., McCauley, L. K., Davidson, B. L., and Roessler, B. J. (1996). Stimulation of new bone formation by direct transfer of osteogenic plasmid genes. *Proc. Natl. Acad. Sci. U.S.A.* **93**, 5753–5758.

85. Bonadio, J., Smiley, E., Patil, P., and Goldstein, S. (1999). Localized, direct plasmid gene delivery in vivo: Prolonged therapy results in reproducible tissue regeneration. *Nat. Med.* **5**, 753–759.

86. Shea, L. D., Smiley, E., Bonadio, J., and Mooney, D. J. (1999). DNA delivery from polymer matrices for tissue engineering. *Nat. Biotechnol.* **17**, 551–554.

87. Lieberman, J. R., Le, L. Q., Wu, L., Finerman, G. A., Berk, A., Witte, O. N., and Stevenson, S. (1998). Regional gene therapy with a BMP-2-producing murine stromal cell line induces heterotopic and orthotopic bone formation in rodents. *J. Orthop. Res.* **16**, 330–339.

88. Lieberman, J. R., Daluiski, A., Stevenson, S., Wu, L., McAllister, P., Lee, Y. P., Kabo, J. M., Finerman, G. A., Berk, A. J., and Witte, O. N. (1999). The effect of regional gene therapy with bone morphogenetic protein-2-producing bone-marrow cells on the repair of segmental femoral defects in rats. *J. Bone Jt. Surg., Am. Vol.* **81**, 905–917.

89. Krebsbach, P. H., Gu, K., Franceschi, R. T., and Rutherford, R. B. (2000). Gene directed osteogenesis: BMP-transduced human fibroblasts form bone in vivo. *Hum. Gene Ther.* **11**, 1201–1210.

90. Slama, J., Andree, C., Svensjo, T., Swain, W. F., Macklin, M. U., and Eriksson, E. (1995). In vivo gene transfer with microseeding, *Surg. Forum* **46**, 702–705.

91. Eriksson, E., Yao, F., Svensjo, T., Winkler, T., Slamat J., Macklin, M. D., Andree, C., McGregor, M., Hinshaw, V., and Swain, W. F. (1998). In vivo gene transfer to skin and wound by microseeding, *J. Surg. Res.* **78**, 85–91.

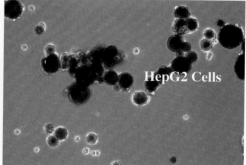

20μm 1000X

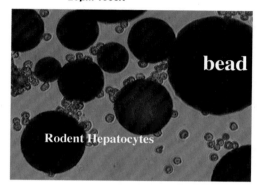

Chapter 11, Fig. 5. Cells on biodegradable microcarriers.

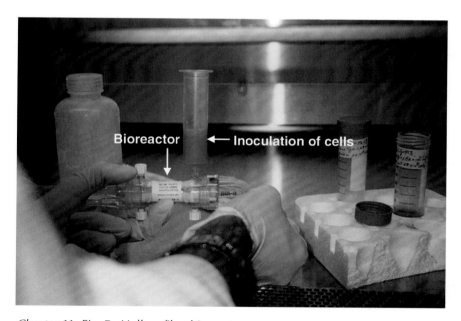

Chapter 11, Fig. 7. Hollow-fiber bioreactor.

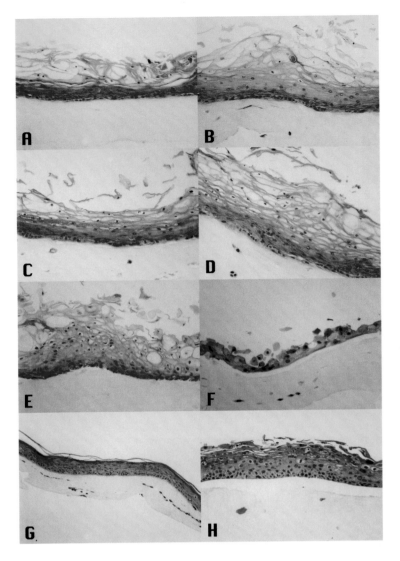

Chapter 18, Fig. 1. Hematoxylin and eosin staining pattern of organotypic (raft) culture tissues. (A)–(E) Ectocervical keratinocytes used were thawed into KGM medium and transferred to E medium with mitomycin C treated J2 3T3 feeder cells. Cultures were initiated on normal ectocervical keratinocyte rafts by using (A) 0.5 x 10⁶ keratinocytes, (B) 0.75 x 10⁶ keratinocytes, (C) 1.0 x 10⁶ keratinocytes, (D) 1.5 x 10⁶ keratinocytes, and (E) 1.75 x 10⁶ keratinocytes. (F) Normal ectocervical keratinocyte rafts, using the same ectocervical keratinocytes used for (A)–(E) except that these 1 x 10⁶ keratinocytes were thawed and continuously grown in KGM medium until used to initiate the raft culture. (G) Normal foreskin keratinocyte rafts, using 1 x 10⁶ keratinocytes to initiate the cultures. (H) Human papillomavirus type 18 productively infected ectocervical keratinocyte rafts [15].

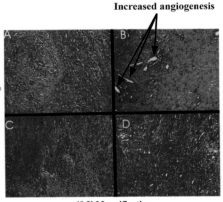

Before treatment with DERMAGRAFT®

DERMAGRAFT treated (6 weeks)

Before treatment with placebo

Placebo treated (6 weeks)

Increased angiogenesis

40 X Magnification

Chapter 104, Fig. 10. Angiogenic and anti-inflammatory response of Dermagraft in the treatment of venous ulcers. Dermagraft secretes a variety of angiogenic and inflammatory cytokines important to wound healing. Shown are histological cross sections of biopsy samples from venous ulcer patients treated with Dermagraft. At the initiation of treatment, blood vessels are lacking and a large amount of inflammatory cells is present (A) and (C). Patients who received weekly implants of Dermagraft showed a significant decrease in inflammatory cells and an increase in blood vessels accompanied by the formation of a healthy granulation tissue (B). The control patient wound status (D) remained relatively unchanged. (By permission of Dr. Julian Adams, Smith & Nephew, Group Research Center, York, England.)

Rhim, S., 133, 136, 137, 139
Rhoades, H. M., 1130, 1134
Rhodes, N. P., 255, 260
Rhodes, S. J., 439, 441, 449
Rialland, L., 155, 197
Ribaudo, R. K., 488, 499
Ricci, E., 1093, 1098
Rice, H., 466, 469
Rice, R., 450, 1089, 1096
Rich, A., 138, 140, 348, 355
Rich, I. N., 477, 484
Richard, F. M., 508, 513
Richard, J. S., 405, 409
Richards, C. L., 445, 450, 1089, 1096, 1101, 1105
Richards, J., 405, 406, 409
Richards, J. M., 86, 94
Richards, J. S., 405, 406, 409
Richardson, A., 391, 395
Richardson, J. L., 445, 550
Richardson, P. M., 386, 395
Richardson, R. L., 885, 889
Richardson, R. R., Jr., 1114, 1122
Richardson, S. C. W., 565, 573
Richelson, E., 80, 84
Richler, C., 1086, 1094
Richon, V. M., 49, 52
Richter, H., 1113, 1122
Ricketts, A. P., 1193
Rickinson, A., 492, 501
Rickwood, D., 23, 26, 32, 33
Ricordi, C., 22, 31, 86, 94, 203, 206, 209, 214–217, 980, 985
Ricotta, J., 235, 240, 1170, 1175
Riddell, S., 492, 500
Riddell, S. R., 489, 492, 499
Ried, T., 45, 47, 51
Rieger, J., 229, 232, 989, 995
Riegler, M., 237, 242
Rieman, M. T., 1186, 1188
Riesbeck, K., 991, 996, 997
Riesle, J., 97, 101, 108, 109, 110
Riester, D., 632, 646
Riethmacher, D., 441, 451
Rietze, R. L., 462, 468
Rifas, L., 461, 468
Rifkind, R. A., 49, 52
Rigamonti, W., 1005, 1010
Rigby, G. W., 857, 869
Riggs, A. D., 399, 402
Riley, J. L., 490, 500
Riley, J. M., 1067, 1072
Rilo, H. R., 86, 94
Rimarachin, J., 458, 460
Rinaudo, M., 655, 662
Ring, C., 793, 801
Ringden, O., 155, 196
Ringel, I., 616, 617
Rink, R. C., 511, 514, 1007, 1010
Ripamonti, U., 1208, 1209, 1212–1214

Rippy, M. K., 511, 514, 1007, 1010
Risau, W., 455–457, 459, 909, 913, 991, 997
Risberg, B., 350, 356
Risbridger, G. P., 398, 400, 402
Risch, R. M., 81, 84
Riss, I. L., 789, 799
Rissel, M., 152, 153, 197
Ritchie, K. A., 477, 481, 484
Ritchie, S. C., 893, 902
Ritsila, V. A., 1066, 1072
Ritson, A., 20, 30
Rittgers, S. E., 350, 356
Rivard, C.-H., 577, 583
Rivarola, M. A., 398, 401, 402
River–Bastide, M., 300, 305
Riviere, D., 141, 149
Rivoltini, L., 489, 499
Rizzo, A.-M. T., 649, 652
Rizzo, J., 1112, 1122
Rizzo, L., 1118, 1120, 1124
Rizzo, R. J., 892, 902
Rizzuto, R., 1090, 1097
Ro, M. S., 190, 200
Robb, G. L., 141, 148, 285, 286, 881, 888
Robb, R. J., 487, 499
Robbins, P., 439, 448–450, 453, 1068, 1069, 1073, 1085–1086, 1090–1091, 1094, 1097
Robbins, P. F., 489, 499
Robblee, L. S., 1111, 1122
Roberson, K. M., 255, 260
Robert, E., 1027, 1037
Roberts, C., 759, 762, 770
Roberts, C. J., 760, 761, 762, 770
Roberts, D. L., 21, 31
Roberts, H. R., 842, 855
Roberts, K. M., 507, 512
Roberts, M., 1160, 1173
Roberts, M. H. T., 1148, 1153
Roberts, P. E., 4, 16
Roberts, R. A., 980, 985
Roberts, S., 989, 995
Roberts, T., 825, 828, 831, 833, 1115, 1123
Roberts, T. S., 1115, 1123
Robertson, B., 693, 696
Robertson, C. N., 255, 260
Robertson, D., 444, 451
Robertson, D. M., 401, 403
Robertson, E., 416, 420
Robertson, G. M. S., 205, 215, 216
Robertson, G. S. M., 205, 215, 216
Robertson, N. E., 789, 799
Robertson, S., 166, 197
Robertson, T. A., 1100, 1104
Robin, J. B., 1113, 1122
Robiner, E., 490, 500

Robinson, B., 620, 621, 625, 627
Robinson, B. P., 697, 703
Robinson, D., 542, 551, 673, 679
Robinson, E. J., 249, 253, 259, 260
Robinson, H. T., 237, 242
Robinson, P. H., 1135, 1141, 1197, 1198, 1204
Robinson, S. W., 1102, 1105
Robitaille, R., 809, 810, 813
Roblero, L., 789, 800
Rocca, M., 1135, 1141
Rocha, V., 875, 878
Roche, E., 412, 420
Roche, P., 334, 342
Rochels, R., 927, 939
Rochford, R., 492, 501
Rochira, M., 541, 550
Rodeheaver, G. T., 556, 557, 562
Rodemann, H. P., 999, 1002
Roder, J. C., 411, 420
Rodet, C., 493, 501
Rodgers, B. J., 322, 329, 475
Rodgers, R. J., 405, 409
Rodkay, W. G., 1059, 1069
Rodrigo, J. J., 1059, 1069
Rodrigues, M. M., 927, 939
Rodrigue–Way, A., 1128, 1133
Rodriguez, A., 663, 677, 1031, 1033, 1034, 1038
Rodriguez, B. A., 754, 756, 766
Rodriguez, E. C., 764, 770
Rodriguez–Sanchez, D., 858, 869
Roeder, R., 900, 903
Roehrich, J. M., 967, 975
Roessler, B. J., 1067, 1072, 1211, 1215
Rogan, E. M., 44, 51
Rogatsch, H., 255, 260
Rogers, C., 351, 352, 357
Rogers, C. E., 859, 869
Rogers, G. S., 1180, 1184, 1188
Rogers, J. A., 1118, 1124
Rohrer, M. D., 1209, 1213
Rohrer, M. J., 349, 356
Rohrig, K., 285, 286
Rohwedel, J., 153, 154, 155, 198, 412, 420
Roig–Lopez, J. L., 373, 382
Roitback, T., 1146, 1153
Rojansky, N., 875, 878
Rojkind, M., 174, 199
Rokkanen, P., 610, 616, 1197, 1203
Roland, W. D., 325, 326, 330, 340, 344
Role, L. W., 373, 382
Rolle, J., 1007, 1010
Rollins, M. D., 952, 962

Rollins, T. E., 231, 233
Roman, M., 1092, 1097, 1101, 1104
Romani, N., 493, 501
Romaniuk, P., 990, 996
Romano, A. R., 347, 354
Romberg, E., 1206, 1212
Romero, J., 1196, 1202, 1203
Romijn, J. C., 254, 256, 260, 288, 292
Rompre, P., 367, 370
Ron, D., 50, 53
Ron, E., 611, 613, 616
Roncari, D. A., 141, 149, 285, 286, 885, 889
Ronco, C., 987, 989, 995
Rondeau, E., 229, 233
Ronel, S. H., 689, 695
Ronnett, G. V., 80, 83, 83
Ronteltap, C., 492, 501
Rooney, C., 492, 501
Rooney, P., 541, 550
Roontga, V., 754, 759, 766
Roos, E., 1102, 1106
Roque, M. A., 161, 198
Rorabeck, C., 109, 111
Rorke, E. A., 268, 271
Rosati, L., 587, 588, 590
Rosdy, M., 4, 16
Rose, R., 548, 553
Rose, T. L., 1111, 1122
Rose–Caprara, V., 347, 355
Roseman, S., 565, 566, 573
Rosen, A. L., 966, 974
Rosen, E. M., 50, 53
Rosen, H., 609, 616
Rosen, J., 688, 695, 828, 833
Rosen, M. A., 253, 260
Rosen, S., 233, 264, 270, 999, 1002, 1166, 1174
Rosen, V., 444, 451, 1196, 1203
Rosenberg, A., 490, 500
Rosenberg, E., 169, 171, 199
Rosenberg, L., 169, 171, 172, 200, 208, 209, 217
Rosenberg, M., 4, 16, 979, 984, 1178, 1181, 1187, 1188
Rosenberg, N., 366, 369
Rosenberg, R. D., 457, 460, 911, 913
Rosenberg, S. A., 487, 488, 489, 490, 499, 500
Rosenberg, W. S., 842, 855
Rosenblatt, D. J., 446, 452
Rosenblatt, J., 440, 449, 605, 607, 877, 879, 1086, 1087, 1095
Rosenbloom, I., 113, 120
Rosenbloom, J., 243, 246
Rosenblum, B. I., 1166, 1174, 1180, 1184, 1185, 1187
Rosener, V. M., 525, 533

Rosenheck, K., 372, 373, 381, 382
Rosenkranz, S., 1209, 1213
Rosenquist, M. S., 1159, 1173
Rosenthal, A. M., 349, 355
Rosenthal, E. L., 507, 512
Rosenthal, N. S., 335, 336, 343, 462, 468
Rosenthal, P., 190, 198
Roses, A. D., 1085, 1086, 1090, 1094
Rosman, G. J., 1166, 1174
Ross, C. J. D., 825, 832
Ross, R., 362, 369, 905, 912, 1163, 1174
Ross, R. D., 1115, 1123
Ross, R. N., 1177, 1185, 1187
Rossant, J., 411, 420, 456, 458, 459, 460
Rosser, A. E., 424, 428
Rosser, M. P., 249, 259
Rossetti, F., 1031, 1036, 1038, 1039, 1068, 1072
Rossi, A., 561, 563
Rossi, M. A., 507, 512
Rostein, L. E., 842, 854
Rotello, R. J., 1193
Rotem, A., 127, 130, 1170, 1175
Rotem, C., 294, 297, 297, 919, 922
Roth, G. J., 477, 481, 484
Roth, J. A., 487, 499
Roth, S., 924, 925
Roth, T. P., 653, 661
Rothe, G., 496, 501
Rothe, M., 1180, 1187
Rothman, A., 291, 292
Rothmund, M., 825, 831
Rothschild, M. A., 525, 533, 956, 962
Rothwell, J. C., 457, 460
Rotter, N., 323, 329, 1068, 1073
Rouabhia, M., 362, 369
Rouais, F., 50, 53, 367, 370
Rouanet, P., 440, 450
Roubin, G. S., 345, 346, 351, 353, 354, 357
Rouche, A., 1089, 1096, 1101, 1105
Roudiere, J. L., 350, 356
Rouiller–Fabre, V., 400, 401, 402
Rouleau, J. P., 1211, 1215
Roulet, A., 1090, 1096
Rousche, P. J., 1116, 1123
Rouslahti, E., 754, 755, 757, 760, 765, 766, 769
Rout, P. G. J., 338, 343
Roux, F., 50, 52
Roux, R. C., 649, 651
Rovee, D., 1184, 1188
Rovira, I. I., 22, 31
Rovner, A. S., 290, 292

Sugawara, T., 752, 759, 760, 761, 765, 770
Sugawara, Y., 349, 355, 886, 889
Suggs, L. J., 649, 651, 652
Sugihara, H., 247, 258, 282, 286, 936, 940
Sugihara, K., 755, 761, 767
Sugimura, K., 763, 770
Sugimura, T., 199
Sugino, I. K., 1117, 1124
Sugino, N., 993, 998
Sugishita, K., 294, 297
Sugishita, Y., 294, 297
Sugita, H., 445, 451, 1090, 1096
Sugitachi, A., 561, 563
Sugiura, K., 155, 198
Suh, J. K., 1068, 1069, 1073
Suh, N. P., 715, 724, 735, 737, 739
Suhara, Y., 445, 451
Sukenik, C. N., 754, 760, 761, 765
Sukoyan, M. A., 411, 419
Sul, H. S., 141, 149
Sullivan, A., 155, 199, 429, 436, 456, 457, 458, 459, 991, 997
Sullivan, K., 1103, 1106
Sullivan, S., 511, 514, 689, 695, 841, 842, 854, 900, 903, 906, 912
Sulston, J., 421, 427
Summer, A. J., 1135, 1141
Summerhayes, I. C., 1135, 1141
Sun, A. M., 689, 695, 781, 786–787, 789, 795, 796, 799–801, 803, 807–810, 813, 825, 827, 831–832, 842, 854–855, 858–861, 863, 868–869, 870, 1021, 1024–1025
Sun, D., 858, 869, 1034, 1039
Sun, F. X., 258, 261
Sun, J., 916, 921
Sun, T.-T., 137, 140
Sun, W., 347, 355, 759, 761, 769, 991, 997
Sun, X. R., 398, 400, 402
Sun, Y., 803, 807, 809–810, 813, 842, 854
Sun, Z., 171, 197, 1149, 1154
Sunada, Y., 1085, 1094
Sundback, C., 186, 197, 980, 985, 1135, 1142
Sundelin, J., 754, 760, 766
Sundstrom, E., 155, 196
Sung, H. W., 509, 513
Sung, J. L., 153, 154, 200
Sunness, J. S., 1118, 1120, 1124
Sunny, M. C., 838, 839
Suprasanna, P., 867, 871

Surronen, P., 1197, 1203
Sussman, M., 462, 468
Sussman, N. L., 190, 200
Sutherland, D. E., 208, 217, 977, 984
Sutherland, D. R., 25, 33, 444, 451
Sutherland, H. J., 80, 83, 83
Sutherland, I. W., 540, 549, 653, 662
Sutkowski, D. M., 247, 253, 258, 260
Sutton, R., 203, 204, 214, 215
Suuronen, R., 1197, 1203
Suzuki, A., 983, 985, 986
Suzuki, H., 199
Suzuki, J., 1209, 1213
Suzuki, K., 757, 759, 761, 768, 858, 869, 1135, 1142
Suzuki, M., 858, 869
Suzuki, S., 754, 766, 838, 839, 858, 869, 1114, 1115, 1122, 1123
Suzuki, T., 229, 233, 757, 759, 761, 768, 919, 922
Suzuki, Y., 1135, 1142, 1160, 1173
Svärdstrom, G., 1210, 1214
Svendsen, C. N., 424, 428
Svensjo, T., 1211, 1215
Swaffield, M. N., 430, 437, 982, 985
Swain, W. F., 1211, 1214, 1215
Swales, N. J., 155, 200
Swann, D. A., 542, 551
Swanson, C., 157, 199, 204, 205, 208, 215, 216, 688, 695
Swanson, L. W., 391, 395
Swartz, R., 490, 491, 500
Swasdison, S., 1103, 1106
Sweeney, D. F., 934, 940
Sweetenham, J. W., 991, 997
Swensson, B., 927, 939
Swensson, O., 927, 939
Swiergiel, J. J., 153, 155, 200, 201, 411, 420
Swynghedauw, B., 919, 922
Sy, M. S., 251, 260
Sydorak, G. R., 562
Syed, S. A., 1135, 1141
Syftestad, G. T., 697, 703
Sykova, E., 1146, 1153
Symons, K., 97, 110, 1165, 1170, 1171, 1174
Sys, S. U., 919, 922
Szachowicz, E., 620, 621, 625, 627, 697, 703
Szczebny, T. M., 168, 200
Szebani, J., 972, 975
Szeles, C., 507, 512
Szigeti, I., 1069, 1073
Szilvassay, S. J., 479, 484

Szurman, P., 1113, 1114, 1122

T

Taavitsainen, P., 958, 963
Tabata, T., 1196, 1197, 1198, 1203
Tabata, Y., 511, 514, 621, 622, 627, 1202, 1204
Tabatabaei, S., 253, 260
Tabatabay, C., 619, 622, 627
Tabbal, S., 875, 878, 878
Tabone, E., 401, 403
Tachibana, M., 1000, 1003
Taddeucci, P., 542, 547, 550
Tagawa, M., 1060, 1070
Taguchi, I., 756, 767
Tague, L., 50, 52
Tahara, H., 447, 453
Tai, I. T., 789, 800, 825, 832, 1021, 1025
Tai, J., 688, 695, 1020, 1024
Tajbakhsh, S., 446, 452
Tajima, K., 321, 328
Tajima, S., 654, 662
Tajima, Y., 399, 402
Tajiri, D. T., 190, 193, 198
Takada, M., 1034, 1039
Takada, Y., 983, 985, 986
Takagi, A., 1090, 1096
Takagi, T., 781, 786, 803, 807, 842, 854
Takahara, H., 654, 662
Takahashi, J., 1118, 1124
Takahashi, K., 1167, 1174, 1208, 1209, 1213, 1214
Takahashi, M., 1118, 1124
Takahashi, N., 444, 451
Takahashi, R., 705, 709, 714
Takahashi, S., 982, 985
Takahashi, T., 294, 297, 456–457, 459, 756, 767, 816, 822, 991, 997
Takai, M., 644, 647
Takakuda, K., 705, 707, 713
Takakura, Y., 334, 342, 640, 647
Takano, H., 781, 786, 919, 922
Takaoka, K., 1211, 1214
Takashina, M., 50, 53, 175, 200
Takasuga, H., 249, 253, 259
Takata, K., 229, 233
Takata, T., 1210–1211, 1214
Takaue, Y., 27, 34
Takayama, K., 816, 822
Takayama, R., 990, 996
Takayuki, A., 667–668, 678
Takebe, K., 50, 53
Takebe, Y., 489, 500
Takeda, A., 1185, 1188
Takeda, K., 22, 31
Takeda, T., 792, 800, 979, 980–982, 984–985
Takeda, Y., 792, 800

Takei, Y., 548, 553
Takeichi, M., 50, 53
Takemura, S., 640, 647
Takenaka, I., 249, 253, 259
Takenami, T., 187–188, 197
Taketa, S., 249, 253, 259
Taki, T., 1149, 1154
Takigawa, M., 326, 327, 330, 1167, 1174
Takiguchi, M., 412, 420
Takimoto, N., 944, 949
Takimoto, Y., 944, 949
Takiuchi, H., 990, 996
Takka, S., 789, 799, 800
Talbot, J., 445, 452
Talke, A. M., 190, 198
Talman, E. A., 506, 512
Talsma, H., 760–761, 770
Tam, J. P., 636, 646
Tam, N. N. C., 257, 261
Tam, S. K., 293, 297
Tam, V. K., 992, 998
Tam, Y. K., 80, 84
Tamada, J. A., 609, 610, 616, 1033, 1038
Tamai, M., 1121, 1124–1125
Tamai, S., 338, 339, 343, 1196, 1203
Tamargo, R., 611, 616
Tamooka, Y., 257, 261
Tamura, H., 568, 574
Tamura, K., 836, 839, 943, 949
Tamura, M., 1202, 1204
Tan, H. S., 838, 839
Tan, J., 672, 678
Tan, S., 103, 111, 751, 757, 764
Tan, S. A., 1092, 1097
Tanagho, E. A., 511, 514
Tanaka, H., 788, 789, 799, 799
Tanaka, J., 705, 707, 713
Tanaka, K., 175, 200, 948, 949
Tanaka, T., 322, 329, 665, 667, 668, 670, 674, 678, 679, 1202, 1204
Tanaka, Y., 154, 201, 1086, 1095
Tanel, R. E., 663, 677, 907, 912, 923, 924, 925
Tang, L., 752, 756, 757, 765
Tang, R., 587, 589, 590
Taniguchi, H., 983, 985, 986
Taniguchi, T., 489, 499
Tanihara, M., 757, 759, 761, 768, 1135, 1142
Taniuchi, M., 99, 110
Tannenbaum, G. S., 789, 800
Tano, Y., 136, 137, 140
Tanzawa, H., 990, 996
Tanzer, M. L., 322, 329
Taparelli, F., 540, 549
Tarcha, P., 610, 614, 615, 616

Tardif, F., 1086, 1090, 1091, 1095, 1101, 1105
Tark, K. C., 884, 888
Tarella, C., 24, 32, 49, 52
Tarn, D. M., 1135, 1142
Tarnow, D., 1208, 1212
Tarnowski, C., 334, 342
Tarrant, S. F., 338, 343
Tartarkiewicz, K., 810, 813
Tashiro, K., 991, 997
Tatakis, D. N., 1208, 1209, 1212, 1213
Tatarkiewicz, K., 809, 810, 811, 812, 813
Tate, A., 992, 998
Tateishi, T., 334, 342
Tatekawa, I., 543, 551
Tateno, C., 154–155, 164, 200
Tatoud, R., 256, 261
Taub, M., 168, 200, 231, 233
Taubman, M. B., 291, 292, 345, 353
Tauc, M., 28, 29, 34, 35
Taulien, J., 1171, 1175
Tavassoli, M., 461, 466, 467, 469
Tawes, R. L., Jr., 562
Tay, B. K. B., 705–707, 713
Tay, K. P., 1075, 1081
Taylor, B. M., 548, 553
Taylor, D., 857, 868, 1102, 1105–1106
Taylor, D. D., 446, 453
Taylor, G., 80, 84, 577, 582, 583
Taylor, J. E., 919, 922
Taylor, L. D., 664, 665, 677
Taylor, M. S., 619, 620, 622, 627
Taylor, R. M., 424, 428
Taylor, R. N., 1193
Taylor, W. G., 50, 53
Taylor–Papadimitriou, J., 47, 52
Taylor–Robinson, D., 90, 95
Tchoukalova, T. D., 29, 35
Teahan, C., 1079, 1081
Teboul, L., 445, 451
Teerds, K., 398, 402
Tegeler, J., 542, 550, 755, 761, 767
Teich, N. M., 757, 768
Teichert-Kuliszewska, K., 141, 149
Teirstein, P. S., 345, 346, 353
Teixeira, M. D. R., 507, 512
Tellier, M., 557, 558, 559, 562
Temple, S., 155, 200, 421, 427
Templeton, D., 967, 975
Templin, S., 490, 500
Tempst, P., 447, 453
Tendler, M., 28, 34

SUBJECT INDEX

excretory function, 1000–1001
extracorporeal bioartificial kidney, 993–995
goals, 987, 994
hemofiltration
artificial membranes, 989–990
bioartificial hemofilter, 990–991
implanted cell–polymer scaffolds, 1000
kidney function
effector synthesis, 999
glomerular filtration, 988
nephron parts, 987
volume homeostasis, 988–989
requirements, 989

L

L-Lactide
p-dioxanone-based copolymers, 592
trimethylene carbonate block copolymer
synthesis
L-2, 587
TL-1, 587
TMC-LG, 588
LAK, see Lymphocyte-activated killer cell
LAL assay, see Limulus amoebocyte lysate assay
Lateral ganglionic eminence (LGE) cells
γ-aminobutyric acid neurotransmission, 1128,
1133
fetal pig cell isolation, 1128
rat transplant model
behavioral assessment, 1129–1130
histology, 1130
lesioning, 1129
pain studies, 1132–1133
preparation, 1128–1129
stroke studies, 1130–1131
transplantation surgery, 1129
LCST, see Lower critical solution temperature
Leukemia inhibitory factor (LIF), preparation, 417
Leydig cell
function, 397
isolation
animal cells, 397–398
human cells, 399
microencapsulation for testosterone therapy
capsule materials, 1021–1022
encapsulation in alginate–polylysine,
1023–1024
implantation sites, 1022–1023
rationale, 1021
testosterone synthesis, 1022–1023
plating and culture, 400–401
steroid synthesis assay, 399–400
viability assay, 399
LGE cells, see Lateral ganglionic eminence cells
Lidocaine, bioartifical liver metabolism assay,
957–958
LIF, see Leukemia inhibitory factor
Limulus amoebocyte lysate (LAL) assay, endotoxin,
92
Liposome, hyaluronic acid composites, 547
Liver cell, see also Bioartifical liver; Hepatocyte
antigenic profiles, 165
centrifugal elutriation, 164
engineering potential, 151
immunoselection, 164–165
isolation from perfused liver

combing, 159
culture, 160
human cells, 162
media, 158
Nytex filtration, 159–160
maturation
diploid cells, 153–154, 156
lineage model, 152–154
polyploid cells, 153–154, 156
properties of cells at different stages, 155–156
stem cell, see Hepatic stem cell
messenger RNA synthesis requirements, 185
minimally deviant cell line isolation and
establishment, 168
monolayer culture
classical culture conditions, 179
completely defined conditions
mature cells, 180
soluble factor and matrix component
requirements, 182
stem cells, 180
tumor cells, 180–181
defined conditions
mature cells, 181, 185
soluble factor and matrix component
requirements, 182
stem cells, 181
tumor cells, 185
differentiation, soluble factor and matrix
component requirements, 183–184
perfusion of liver
collagenase preparations, 155, 157
overview, 155
perfusion technique
animal liver, 159
human liver, 161–162
setup, 158–159
solutions, 157–158
supplies, 157
surgery
animal liver, 159
human liver, 160
ploidy analysis
fixed cells, 163
fluorescent dyes, 162–163
nuclei isolation and analysis, 163
viable cells, 163
size fractionation, 164
three-dimensional culture
collagen-coated poly(lactic-co-glycolic) acid
beads
adult rodent liver cells, 186–187
hepatoma cell line culture, 186
hollow-fiber bioreactors
cartridges and vendors, 192
conditioning, 190, 191
functional analysis, 193
inoculation, 192
setup, 190
termination of experiment, 193
types, 191
spheroids
applications, 187
formation techniques, 187–190
materials, 189
medium, 188
time requirements for formation, 188

viability assay, 188–189
Liver tissue engineering
culture system, 982–983
goals, 978, 984
hepatocyte
coculture, 979–980
culture, 979
stimulating factors, 981–982
implantation of constructs, 981
rationale, 977
scaffolds
cell seeding, 980
materials, 978–979
stem cell utilization, 983
steps, 978
vascularization, 981, 984
LNL, see Lymph node lymphocyte
Lower critical solution temperature (LCST)
injectable scaffold development principles, 665
poly(N-isopropylacrylamide), 664–666, 669–670
polyphosphazene behavior, 601–602
Lymph node lymphocyte (LNL)
culture, see T cell
features, 488
immunotherapy, 488
Lymphocyte-activated killer cell (LAK)
culture, see T cell
features, 487–488
immunotherapy, 488

M

Macrophage
spinal cord regeneration therapies, 1149
targeted adhesion, 757
MACS, see Magnetic activated cell sorting
Magnetic activated cell sorting (MACS)
bead design, 25–26
cellular effects, 26
hematopoietic cells, 26
negative selection, 25
principles, 25
quantitative limitations, 26
MAH cell, features, 375
Matrigel, preparation, 173
Mechanical tissue disruption, cell isolation, 20–21
Media
Food and Drug Administration regulation of
components, 4
nutrient requirement optimization, 28–29
osmolality in cell selection, 28
purity requirements, 91–92
replacement indications, 38–39
selection for primary cell culture, 40
selective media, 47
Melt molding, principles, 734–735
Membrane lamination
advantages over microencapsulation, 688
encapsulated cell therapy applications, 687–688,
694
polymers, structure–property relationships,
688–689
processing
diffusion-induced precipitation, 690
phase inversion, 689
post-treatments of dense films, 690
thermal gelation phase inversion, 690